ORGANIC COMPOUNDS

Family						
Ether	Amine	Aldehyde	Ketone	Carboxylic Acid	Ester	Amide
CH_3OCH_3	CH_3NH_2	$CH_3\overset{\displaystyle O}{\overset{\|}{C}}H$	$CH_3\overset{\displaystyle O}{\overset{\|}{C}}CH_3$	$CH_3\overset{\displaystyle O}{\overset{\|}{C}}OH$	$CH_3\overset{\displaystyle O}{\overset{\|}{C}}OCH_3$	$CH_3\overset{\displaystyle O}{\overset{\|}{C}}NH_2$
Methoxy-methane	Methan-amine	Ethanal	Propanone	Ethanoic Acid	Methyl ethanoate	Ethanamide
Dimethyl ether	Methyl-amine	Acetal-dehyde	Acetone	Acetic acid	Methyl acetate	Acetamide
ROR	RNH_2 R_2NH R_3N	$R\overset{\displaystyle O}{\overset{\|}{C}}H$	$R\overset{\displaystyle O}{\overset{\|}{C}}R$	$R\overset{\displaystyle O}{\overset{\|}{C}}OH$	$R\overset{\displaystyle O}{\overset{\|}{C}}OR$	$R\overset{\displaystyle O}{\overset{\|}{C}}NH_2$ $R\overset{\displaystyle O}{\overset{\|}{C}}NHR$ $R\overset{\displaystyle O}{\overset{\|}{C}}NR_2$
$-\overset{\|}{\underset{\|}{C}}-O-\overset{\|}{\underset{\|}{C}}-$	$-\overset{\|}{\underset{\|}{C}}-N-$	$\overset{\displaystyle O}{\overset{\|}{\underset{\|}{C}}}-H$	$-\overset{\|}{\underset{\|}{C}}-\overset{\displaystyle O}{\overset{\|}{C}}-\overset{\|}{\underset{\|}{C}}-$	$-\overset{\displaystyle O}{\overset{\|}{C}}-OH$	$-\overset{\displaystyle O}{\overset{\|}{C}}-O-\overset{\|}{\underset{\|}{C}}-$	$-\overset{\displaystyle O}{\overset{\|}{C}}-N-$

ORGANIC CHEMISTRY

FOURTH EDITION
ORGANIC CHEMISTRY

T. W. GRAHAM SOLOMONS

University of South Florida

WILEY

JOHN WILEY & SONS

New York   Chichester   Brisbane   Toronto   Singapore

COVER PHOTO: Lithium diisopropylamide molecular graphic produced by SYBYL/MENDYL Molecular Modeling Software, courtesy of Tripos Associates, Inc., St. Louis, MO.

CHAPTER OPENING PHOTO CREDITS

Chapter 1   Dennis Kunkel/Phototake.
Chapter 2   Michael Siegel/Phototake.
Chapter 3   Dr. E. R. Degginger.
Chapters 4–10 Manfred Kage/
   Peter Arnold, Inc.
Chapters 11 & 12   Dr. E. R. Degginger.
Chapter 13   Manfred Kage/Peter
   Arnold, Inc.
Chapter 14   Dr. E. R. Degginger.
Chapters 15 & 16   Manfred Kage/
   Peter Arnold, Inc.

Chapter 17   Martin Rotker/Phototake.
Chapter 18   Dennis Kunkel/Phototake.
Chapters 19 & 20   Dr. E. R. Degginger.
Chapter 21   Dennis Kunkel/Phototake.
Chapter 22   Phillip A. Harrington/
   Peter Arnold, Inc.
Chapter 23   W. C. Still, Columbia
   University
Chapter 24   David Gnizak/Phototake.

Production supervisor: Elizabeth A. Austin
Cover & interior designer: Dawn L. Stanley
Illustrations: John Balbalis with the assistance of the Wiley Illustration Department
Photo editor: Safra Nimrod
Manuscript editor: Jeannette Stiefel under the supervision of Deborah Herbert

Library of Congress Cataloging in Publication Data:

Solomons, T. W. Graham.
   Organic chemistry.

   Bibliography: p. B-1
   Includes index.
   1. Chemistry, Organic.   I. Title.
QD251.2.S66   1988   547   87-29443
ISBN 0-471-83659-1

Printed in the United States of America
10 9 8 7 6 5 4 3

About the Author

T. W. GRAHAM SOLOMONS   received his doctorate in organic chemistry from Duke University in 1959 after which he became a Sloan Foundation Postdoctoral Fellow at the University of Rochester. In 1960 he became a charter member of the faculty at the University of South Florida where he teaches today. In 1973 he became Professor of Chemistry. For several years he was director of an NSF-sponsored Undergraduate Research Participation Program at USF. His research interests are in the areas of heterocyclic chemistry and unusual aromatic compounds. He has received several awards for distinguished teaching.

He has also spent several years in England as a visiting member of the faculty at the University of Sussex. He and his wife Judith have a 500-year-old farmhouse in Sussex and this has been where he has done most of the writing for his textbooks. They have two young sons and an older daughter who is a geophysicist.

TO THE STUDENT: A Study Guide for the textbook is available through your college bookstore under the title Study Guide to accompany *ORGANIC CHEMISTRY,* Fourth Edition by T. W. Graham Solomons and Jack E. Fernandez. The Study Guide can help you with course material by acting as a tutorial, review, and study aid. If the Study Guide is not in stock, ask the bookstore manager to order a copy for you.

For Judith

Acknowledgments

I am grateful to the many people who have provided the reviews that have guided me in preparing this and the earlier editions of my textbooks:

Winfield M. Baldwin
University of Georgia

David Ball
California State University, Chico

Paul A. Barks
North Hennepin State Junior College

Harold Bell
Virginia Polytechnic Institute
and State University

Newell S. Bowman
The University of Tennessee

Edward M. Burgess
Georgia Institute of Technology

Robert Carlson
University of Minnesota

William D. Closson
State University of New York at Albany

Phillip Crews
University of California, Santa Cruz

James Damewood
University of Delaware

O. C. Dermer
Oklahoma State University

Robert C. Duty
Illinois State University

Stuart Fenton
University of Minnesota

Jeremiah P. Freeman
Notre Dame University

M. K. Gleicher
Oregon State University

Wayne Guida
Eckerd College

Philip L. Hall
Virginia Polytechnic Institute
and State University

Lee Harris
University of Arizona

William H. Hersh
University of California, Los Angeles

Jerry A. Hirsch
Seton Hall University

John Holum
Augsburg College

Stanley N. Johnson
Orange Coast College

John F. Keana
University of Oregon

David H. Kenny
Michigan Technological University

Robert C. Kerber
State University of New York
at Stony Brook

Karl R. Kopecky
The University of Alberta

Paul J. Kropp
University of North Carolina
at Chapel Hill

John A. Landgrebe
University of Kansas

Allan K. Lazarus
Trenton State College

Philip W. LeQuesne
Northeastern University

Robert Levine
University of Pittsburgh

Samuel G. Levine
North Carolina State University

John Mangravite
West Chester University

Jerry March
Adelphi University

John L. Meisenheimer
Eastern Kentucky University

Gerado Molina
Universidad de Puerto Rico

Everett Nienhouse
Ferris State College

John Otto Olson
Camrose Lutheran College

Allen Pinhas
University of Cincinnati

William A. Pryor
Louisiana State University

Thomas R. Riggs
University of Michigan

Stephen Rodemeyer
California State University, Fresno

Yousry Sayed
University of North Carolina
at Wilmington

Ronald Starkey
University of Wisconsin — Green Bay

James G. Traynham
Louisiana State University

Daniel Trifan
Fairleigh Dickinson University

Desmond M. S. Wheeler
University of Nebraska

James K. Whitesell
The University of Texas at Austin

Joseph Wolinski
Purdue University

Darrell J. Woodman
University of Washington

I thank my colleagues at the University of South Florida for the many helpful suggestions that they have offered. In this regard, I think especially of: Raymond N. Castle, Jack E. Fernandez, George R. Jurch, Leon Mandell, Terence C. Owen, Douglas J. Raber, Stewart W. Schneller, George R. Wenzinger, and Robert D. Whitaker.

I am much indebted to Jeannette Stiefel for copyediting and proofreading.

I would also like to thank Mary Woodward and Ann Richards of Tripos Associates, Inc., for their help in creating the computer graphics used on the cover.

I am grateful to many people at Wiley for their help, especially Dennis Sawicki, Chemistry Editor, Elizabeth Austin, Production Supervisor, Linda Indig, Production Manager, and John Balbalis, Illustration Designer. I also thank Dawn Stanley for her cover design.

And, finally, I thank my wife, Judith Taylor Solomons, for her encouragement and for her editing, proofreading, and typing.

T. W. GRAHAM SOLOMONS

Preface

As I continue to teach, I continue to learn. I learn from my students as, each year, I guide them through an introduction to the fascinating science of organic chemistry. I also learn from my colleagues, here at U.S.F. and elsewhere, who have kindly offered many helpful suggestions about revisions, additions, reorganizations, and so on. In preparing this new edition I have drawn upon these lessons in the hope that by doing so I will make this new edition even more useful to my students, to my colleagues, and to their students.

This fourth edition of *Organic Chemistry* differs in several significant ways from its predecessor. There is much new material: there are new problems; there are many new pedagogical features including a full-color format; and in the first half of the book there is a major new approach to the introduction of reaction mechanisms. In the paragraphs that follow I shall point out these differences, give the reasons for them, and describe the main features of this new edition.

AN EARLIER TREATMENT OF STEREOCHEMISTRY

In a major departure from earlier editions of this text, I now introduce all of the main concepts of stereochemistry much earlier. The beginning ideas about the shapes of molecules are described in Chapters 1 and 2. Chapter 3 then focuses on stereochemistry from the point of view of conformational analysis with special emphasis on the conformations of cyclohexane derivatives. Chapter 4 now introduces the concepts of molecular chirality, enantiomerism, optical activity, $(R)-(S)$ designations, diastereomerism, and so on. At this point, as most of the important principles of stereochemistry have been developed, these principles can be applied and reapplied to topics in the chapters that follow. The first application comes in the very next chapter when the mechanisms of S_N1 and S_N2 reactions are introduced, and where now the stereochemistry of these important reactions can be described completely and effectively. In doing this I can now use chiral molecules to illustrate how S_N2 reactions lead to the inversion of configuration and how S_N1 reactions can occur with racemization.

MECHANISMS ARE NOW INTRODUCED WITH IONIC REACTIONS

Because I believe that ionic reactions are fundamentally simpler than free radical reactions, in this new edition I have used a simple S_N2 reaction to introduce the important ideas associated with reaction mechanisms. This introduction is given in Chapter 5 and afterward, in the same chapter, the students also encounter simple examples of S_N1 reactions and of E1 and E2 reactions. I have tested this new approach to mechanisms in my shorter textbook *Fundamentals of Organic Chemistry,* Second Edition, and users of that text have found that it works very well.

The development of ionic chemistry is continued in Chapters 6 to 8 where elimination reactions are studied further and where addition reactions of carbon–carbon double and triple bonds become one main theme. Then in Chapter 9 free radical reactions, both substitutions and additions, are introduced. From this point in the text onward, because the basic principles have been developed, ionic and free radical reactions can be described and compared again and again in the chapters that follow.

AN EARLIER TREATMENT OF ALCOHOLS AND ETHERS

In this new edition the chemistry of alcohols and ethers is considered in Chapter 8 instead of Chapter 15. This change has the benefit of increasing the variety of reaction types and functional groups that can be explored in the first term of the course. Covering alcohols and ethers earlier also permits a greater scope for the introduction of ionic mechanisms and for reinforcing the principles of stereochemistry that were developed in Chapter 4.

COVERAGE OF ALKENES AND ALKYNES ARE COMBINED

Many users of earlier editions of this text have suggested that the chemistry of alkynes does not warrant a chapter by itself, and that this chemistry might be covered more efficiently in the chapters devoted to alkenes. It has been argued, too, that a combined treatment of alkenes and alkynes is more effective because a comparison of the chemistry of carbon–carbon double and triple bonds becomes more obvious. Responding to these suggestions, I have taken an approach in this edition that combines discussions of the chemistry of alkenes and alkynes in Chapters 6 and 7. Chapter 6 is concerned mainly with the syntheses and the structures of molecules containing double and triple bonds, while Chapter 7 focuses mainly on the addition reactions that these molecules undergo.

AN EXPANDED TREATMENT OF ORGANIC SYNTHESIS

The principles involved in planning organic syntheses are now set out in a new section (Section 7.18). Here all of the important aspects of synthesis are discussed — construction of the carbon skeleton, functional group interconversions, control of regiochemistry, and control of stereochemistry. These aspects are all related through retrosynthetic analysis, and the students see several simple syntheses that illustrate these ideas. These principles are then reiterated many times in other sections that are devoted to synthesis and in many of the sample problems where the specific focus is a multistep synthesis. The students are also given numerous opportunities to develop their skills in planning syntheses through problems given within and at the ends of chapters.

A NEW SPECIAL TOPIC ON LITHIUM ENOLATES

The usefulness of lithium enolates in organic synthesis has been so amply demonstrated in recent years that I have written a new special topic showing how lithium enolates are synthesized and how they can be used in directed aldol syntheses and in α-selenenation reactions. This special topic is just one of many that can be included in the course at the discretion of the instructor. Even if the special topics are not included, they offer those students, whose interests have been stimulated an opportunity to go beyond the material given in most textbooks, and thereby to enhance their understanding of organic chemistry.

A FULL COLOR FORMAT

A full-color format is now used throughout the text. One great advantage of this format is the way it can be used to draw the readers' attention to those portions of molecules that are undergoing change in a chemical reaction or a synthesis. This kind of use makes mechanisms and syntheses much clearer. Full color also greatly enhances the illustrations in the book. My goal in using color has been to make this book more effective as a teaching device, not just to make it more attractive (although color does this as well).

Examples of the way the full-color format has been employed are the following:

In illustrations of molecular models carbon atoms are black, hydrogen atoms are red, oxygen atoms are light blue, nitrogen atoms are dark blue, and chlorine atoms are green.

In treatments of molecular orbital theory, the phase signs of orbitals are differentiated by using red if the sign is (+) and blue if it is (−).

In the discussion of S_N1 and S_N2 reactions in Chapter 5, nucleophiles are red and leaving groups are blue.

In the discussions of electrophilic aromatic substitution in Chapter 12, electron-releasing groups are shown in green, electron-withdrawing groups are red, and the attacking electrophile is blue.

In illustrating spectra, different colors have been used to coordinate with groups giving rise to that spectral feature. In other illustrations involving spectra, the following color scheme applies: proton nmr spectra are printed blue with the integral curve black, infrared spectra are red, visible–ultraviolet spectra are violet, carbon-13 nmr spectra are green, and mass spectra are brown orange.

The photographs selected for the chapter opening pages in this book were made using several microscopic techniques, such as viewing crystals with polarized light or looking at microorganisms through a scanning electron microscope. The stunning colors are all created from the basic three colors of white light, as seen in the Chapter 13 opener.

The beauty of these substances, from vitamins to pigments, is revealed here in a way which only an artist can bring to light. The students, however, will discover the beauty of the science of organic chemistry as they advance through the chapters of this book.

I hope that these new features will, for the students who read this book, achieve their intended goal: to make organic chemistry as interesting and exciting for them as it continues to be for me.

T. W. GRAHAM SOLOMONS

Contents

CHAPTER THREE
ALKANES AND CYCLOALKANES. CONFORMATIONAL ANALYSIS

CHAPTER FOUR
STEREOCHEMISTRY. CHIRAL MOLECULES

CHAPTER FIVE
IONIC REACTIONS — NUCLEOPHILIC SUBSTITUTION AND ELIMINATION REACTIONS OF ALKYL HALIDES

CHAPTER SIX
ALKENES AND ALKYNES I. PROPERTIES AND SYNTHESIS

CHAPTER SEVEN
ALKENES AND ALKYNES II. ADDITION REACTIONS OF CARBON–CARBON MULTIPLE BONDS

CHAPTER EIGHT
ALCOHOLS AND ETHERS

CHAPTER NINE
FREE RADICAL REACTIONS

CHAPTER TEN
CONJUGATED UNSATURATED SYSTEMS

CHAPTER ELEVEN
AROMATIC COMPOUNDS I: THE PHENOMENON OF AROMATICITY

CHAPTER TWELVE
AROMATIC COMPOUNDS II: ELECTROPHILIC AROMATIC SUBSTITUTION

CHAPTER THIRTEEN
SPECTROSCOPIC METHODS OF STRUCTURE DETERMINATION

CHAPTER FOURTEEN
PHENOLS AND ARYL HALIDES. NUCLEOPHILIC AROMATIC SUBSTITUTION

CHAPTER FIFTEEN
ORGANIC OXIDATION AND REDUCTION REACTIONS.
ORGANOMETALLIC COMPOUNDS

CHAPTER SIXTEEN
ALDEHYDES AND KETONES I. NUCLEOPHILIC
ADDITIONS TO THE CARBONYL GROUP

CHAPTER SEVENTEEN
ALDEHYDES AND KETONES II. REACTIONS AT THE α CARBON. ALDOL REACTIONS

CHAPTER EIGHTEEN
CARBOXYLIC ACIDS AND THEIR DERIVATIVES: NUCLEOPHILIC SUBSTITUTION AT ACYL CARBON

CHAPTER NINETEEN
AMINES

CHAPTER TWENTY
SYNTHESIS AND REACTIONS OF β-DICARBONYL COMPOUNDS: MORE CHEMISTRY OF ENOLATE IONS

CHAPTER TWENTY-ONE
CARBOHYDRATES

CHAPTER TWENTY-TWO
LIPIDS

CHAPTER TWENTY-THREE
AMINO ACIDS AND PROTEINS

CHAPTER TWENTY-FOUR
NUCLEIC ACIDS AND PROTEIN SYNTHESIS

ORGANIC
CHEMISTRY

Urea Crystals 40x. This photomicrograph is of a urea crystal. The importance of urea in the history of organic chemistry is described in Section 1.2A.

Carbon Compounds and Chemical Bonds

1.1 INTRODUCTION

Organic chemistry is the study of *the compounds of carbon.* The compounds of carbon are the central substances of which all living things on this planet are made. Carbon compounds include DNA, the giant molecules that contain all the genetic information for a given species. Carbon compounds also make up the proteins of our muscle and skin. They make up the enzymes that catalyze the reactions that occur in our bodies. Together with oxygen in the air we breathe, carbon compounds in our diets furnish the energy that sustains life.

Not only are we composed largely of organic compounds, not only are we derived from and nourished by them, *we also live in an Age of Organic Chemistry.* The clothing we wear, whether a natural substance such as wool or cotton or a synthetic such as nylon or a polyester, is made up of carbon compounds. Many of the materials that go into the houses that shelter us are organic. The gasoline that propels our automobiles, the rubber of their tires, and the plastic of their interiors are all organic. Most of the medicines that help us cure diseases and relieve suffering are organic. Organic pesticides help us eliminate many of the agents that spread diseases in both plants and animals.

Organic chemicals are also factors in some of our most serious problems. Many of the organic chemicals introduced into the environment have had consequences far beyond those originally intended. A number of insecticides, widely used for many years, have now been banned because they harm many species other than insects and they pose a danger to humans. Organic compounds called polychlorobiphenyls (PCBs) are responsible for pollution of the Hudson River that may take years and enormous amounts of money to reverse. Organic compounds used as propellants for aerosols have been banned because they threatened to destroy the ozone layer of the outer atmosphere, a layer that protects us from extremely harmful radiation.

Thus for good or bad, organic chemistry is associated with nearly every aspect of our lives. We would be wise to understand it as best we can.

1.2 THE DEVELOPMENT OF ORGANIC CHEMISTRY AS A SCIENCE

Humans have used organic compounds and their reactions for thousands of years. Their first deliberate experience with an organic reaction probably dates from their discovery of fire. The ancient Egyptians used organic compounds (indigo and alizarin) to dye cloth. The famous "royal purple" used by the Phoenicians was also an organic substance, obtained from mollusks. The fermentation of grapes to produce ethyl alcohol and the acidic qualities of "soured wine" are both described in the Bible and were probably known earlier.

As a science, organic chemistry is less than 200 years old. Most historians of science date its origin to the early part of the nineteenth century, a time in which an erroneous belief was dispelled.

1.2A Vitalism

During the 1780s scientists began to distinguish between **organic compounds** and **inorganic compounds.** Organic compounds were defined as compounds that could be obtained from *living organisms.* Inorganic compounds were those that came from *nonliving sources.* Along with this distinction, a belief called "vitalism" grew. According to this idea, the intervention of a "vital force" was necessary for the synthesis of an organic compound. Such synthesis, chemists held then, could take place only in living organisms. It could not take place in the test tubes and flasks of a chemistry laboratory.

Between 1828 and 1850 a number of compounds that were clearly "organic" were synthesized from sources that were clearly "inorganic." The first of these syntheses was accomplished by Friedrich Wöhler in 1828. Wöhler found that the organic compound urea (a constituent of urine) could be made by evaporating an aqueous solution containing the inorganic compound ammonium cyanate.

$$\underset{\textbf{Ammonium cyanate}}{NH_4{}^+NCO^-} \xrightarrow{\text{heat}} \underset{\textbf{Urea}}{H_2N-\overset{\overset{\textstyle O}{\|}}{C}-NH_2}$$

Although "vitalism" died slowly after Wöhler's synthesis, its passing made possible the flowering of the science of organic chemistry that has occurred since 1850.

1.2B Empirical and Molecular Formulas

Even while vitalism persisted* extremely important advances were made in the development of qualitative and quantitative methods for analyzing organic substances. In 1784 Antoine Lavoisier first showed that organic compounds were composed primarily of carbon, hydrogen, and oxygen. Between 1811 and 1831, *quantitative* methods for determining the composition of organic compounds were developed by Justus Liebig, J. J. Berzelius, and J. B. A. Dumas. A great confusion was dispelled in 1860 when Stanislao Cannizzaro showed that the earlier hypothesis of Amedeo Avogadro (1811) could be used to distinguish between **empirical** and **molecular formulas.** As a result, many molecules that had appeared earlier to have the same formula were seen to be composed of different numbers of atoms. For example, ethylene, cyclopentane, and cyclohexane all have the same empirical formula: CH_2. However, they have molecular formulas of C_2H_4, C_5H_{10}, and C_6H_{12}, respectively. Appendix A of the Study Guide that accompanies this book contains a review of how empirical and molecular formulas are determined and calculated.

---

*It is a belief still held today by some groups. While there are sound arguments made against foods contaminated with pesticides, it is impossible to argue that "natural" vitamin C, for example, is healthier than the "synthetic" vitamin, since they are structurally identical.

1.3 THE STRUCTURAL THEORY OF ORGANIC CHEMISTRY

Between 1858 and 1861, August Kekulé, Archibald Scott Couper, and Alexander M. Butlerov, working independently, laid the basis for one of the most fundamental theories in organic chemistry: **the structural theory.** Two central ideas make up this theory:

1. The atoms of the elements in organic compounds can form a fixed number of bonds. The measure of this ability is called **valence.** Carbon is *tetravalent;* that is, carbon atoms form four bonds. Oxygen is *divalent;* oxygen atoms form two bonds. Hydrogen and usually the halogens are *monovalent;* their atoms form only one bond.

$$-\overset{|}{\underset{|}{C}}- \qquad\qquad -O- \qquad\qquad H- \qquad Cl-$$

Carbon atoms   Oxygen atoms   Hydrogen and halogen
are tetravalent   are divalent   atoms are monovalent

2. A carbon atom can use one or more of its valences to form bonds to other carbon atoms.

Carbon–carbon bonds

$$-\overset{|}{\underset{|}{C}}-\overset{|}{\underset{|}{C}}- \qquad\qquad \overset{\diagdown}{\diagup}C=C\overset{\diagup}{\diagdown} \qquad\qquad -C\equiv C-$$

Single bond   Double bond   Triple bond

In his original publication Couper represented these bonds by lines much in the same way that most of the formulas in this book are drawn. In his textbook (published in 1861), Kekulé gave the science of organic chemistry its modern definition: *A study of the compounds of carbon.*

We can appreciate the importance of the structural theory if we consider now one simple example. These are two compounds that have the *same* molecular formula, C_2H_6O, but these compounds have strikingly different properties. (See Table 1.1.) One compound, called *dimethyl ether,* is a gas at room temperature; the other compound, called *ethyl alcohol,* is a liquid. Dimethyl ether does not react with sodium; ethyl alcohol does, and the reaction produces hydrogen gas.

Because the molecular formula for these two compounds is the same, it gives us no basis for understanding the differences between them. The structural theory

TABLE 1.1 Properties of ethyl alcohol and dimethyl ether

	ETHYL ALCOHOL C_2H_6O	DIMETHYL ETHER C_2H_6O
Boiling point, °C[a]	78.5	−24.9
Melting point, °C	−117.3	−138
Reaction with sodium	Displaces hydrogen	No reaction

[a]Unless otherwise stated all temperatures in this text are given in degrees Celsius.

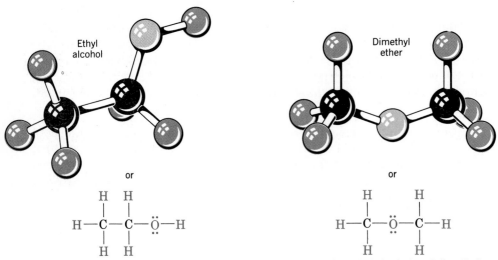

remedies this situation, however. It does so by giving us **structural formulas** for the two compounds and these structural formulas (Fig. 1.1) are different.

One glance at the structural formulas for these two compounds reveals their difference. The two compounds differ in their **connectivity:** The atoms of ethyl alcohol are connected in a way that is different from those of dimethyl ether. In ethyl alcohol there is a C—C—O linkage; in dimethyl ether the linkage is C—O—C. Ethyl alcohol has a hydrogen atom attached to oxygen; in dimethyl ether all of the hydrogen atoms are attached to carbon. It is the hydrogen atom covalently bonded to oxygen in ethyl alcohol that is displaced when this alcohol reacts with sodium:

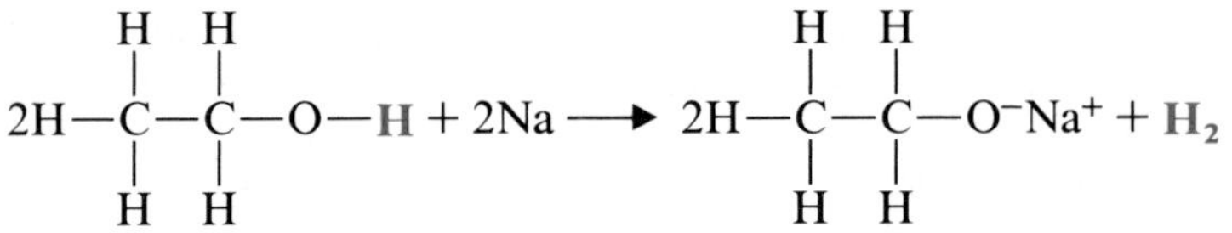

$$2H-\overset{\overset{\displaystyle H}{|}}{\underset{\underset{\displaystyle H}{|}}{C}}-\overset{\overset{\displaystyle H}{|}}{\underset{\underset{\displaystyle H}{|}}{C}}-O-H + 2Na \longrightarrow 2H-\overset{\overset{\displaystyle H}{|}}{\underset{\underset{\displaystyle H}{|}}{C}}-\overset{\overset{\displaystyle H}{|}}{\underset{\underset{\displaystyle H}{|}}{C}}-O^-Na^+ + H_2$$

This is just the way water reacts with sodium:

$$2H-O-H + 2Na \longrightarrow 2H-O^-Na^+ + H_2$$

Hydrogen atoms that are covalently bonded to carbon are normally unreactive toward sodium. As a result, none of the hydrogen atoms in dimethyl ether is displaced by sodium.

> The hydrogen atom attached to oxygen also accounts for the fact that ethyl alcohol is a liquid at room temperature. As we shall see in Section 2.17, this hydrogen atom allows molecules of ethyl alcohol to form hydrogen bonds to each other and gives ethyl alcohol a boiling point much higher than that of dimethyl ether.

1.4 ISOMERISM. CONSTITUTIONAL ISOMERS

More than 4 million organic compounds have now been isolated in a pure state and have been characterized on the basis of their physical and chemical properties.

Additional compounds are added to this list by the tens of thousands each year. A look into *Chemical Abstracts* or Beilstein's *Handbuch der Organischen Chemie,* where known organic compounds are catalogued, shows that there are dozens and sometimes hundreds of *different compounds that have the same molecular formula.* Such compounds are called **isomers.** Different compounds with the same molecular formula are said to be **isomeric,** and this phenomenon is called **isomerism.**

Ethyl alcohol and dimethyl ether are examples of what are now called **constitutional isomers.*** *Constitutional isomers are different compounds that have the same molecular formula, but differ in their connectivity, that is, in the way in which their atoms are bonded together.* Constitutional isomers usually have different physical properties (e.g., melting point, boiling point, density) and different chemical properties. The differences, however, may not always be as large as those between ethyl alcohol and dimethyl ether.

1.5 THE TETRAHEDRAL SHAPE OF METHANE

In 1874, the structural formulas originated by Kekulé, Couper, and Butlerov were expanded into three dimensions by the independent work of J. H. van't Hoff and J. A. Le Bel. Van't Hoff and Le Bel proposed that the four bonds of the carbon atom in methane, for example, are arranged in such a way that they would point toward the corners of a regular tetrahedron, the carbon atom being placed at its center (Fig. 1.2). The necessity for knowing the arrangement of the atoms in space, taken together with an understanding of the order in which they are connected, is central to an understanding of organic chemistry, and we shall have much more to say about this later, in Chapters 3 and 4.

1.6 CHEMICAL BONDS. THE OCTET RULE

The first explanations of the nature of chemical bonds were advanced by G. N. Lewis (of the University of California, Berkeley) and W. Kössel (of the University of Munich) in 1916. Two major types of chemical bonds were proposed.

1. The **ionic** (or **electrovalent**) bond, formed by the transfer of one or more electrons from one atom to another to create ions.
2. The **covalent** bond, a bond that results when atoms share electrons.

The central idea in their work on bonds is that atoms without the electronic configuration of a noble gas generally react in a way that produces such a configuration.

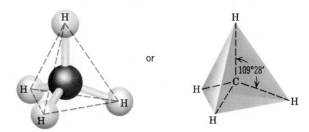

FIGURE 1.2 The terahedral structure of methane.

___

*An older term for isomers of this type was **structural isomers.** The International Union of Pure and Applied Chemistry (IUPAC) now recommends that use of the term "structural" when applied to isomers of this type be abandoned.

The concepts and explanations that arise from the original propositions of Lewis and Kössel are satisfactory for explanations of many of the problems we deal with in organic chemistry today. For this reason we shall review these two types of bonds in more modern terms.

1.6A Ionic Bonds

Atoms may gain or lose electrons and form charged particles called *ions.* An ionic bond is a force of attraction between oppositely charged ions. One source of such ions is the interaction of atoms of widely differing electronegativities (Table 1.2). *Electronegativity measures the ability of an atom to attract electrons.* Notice in Table 1.2 that electronegativity increases as we go across a horizontal row of the Periodic Table from left to right:

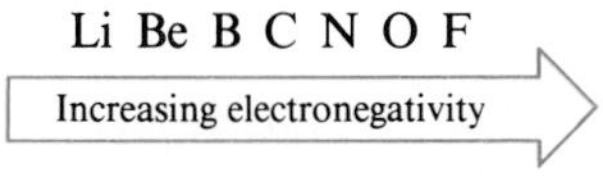

Li Be B C N O F

⟩ Increasing electronegativity ⟩

and that it decreases as we go down a vertical column:

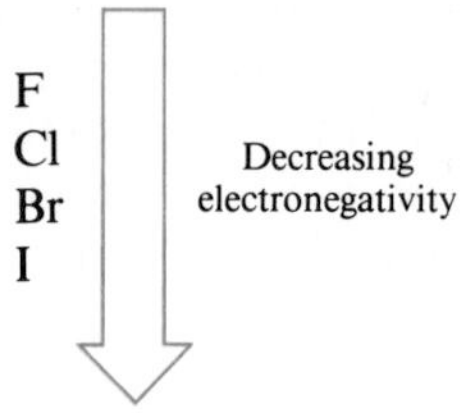

F
Cl   Decreasing
Br   electronegativity
I

An example of the formation of an ionic bond is the reaction of lithium atoms and fluorine atoms.

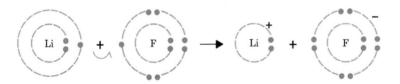

Lithium, a typical metal, has a very low electronegativity; fluorine, a nonmetal, is the most electronegative element of all. The loss of an electron (a negatively charged species) by the lithium atom leaves a lithium cation (Li^+); the gain of an electron by

TABLE 1.2   Electronegativities of some of the elements

						H
						2.1
Li	Be	B	C	N	O	F
1.0	1.5	2.0	2.5	3.0	3.5	4.0
Na	Mg	Al	Si	P	S	Cl
0.9	1.2	1.5	1.8	2.1	2.5	3.0
K						Br
0.8						2.8

the fluorine atom gives a fluoride anion (F^-). Why do these ions form? In terms of the Lewis–Kössel theory both atoms achieve the electronic structure of a noble gas by becoming ions. The lithium cation with two electrons in its valence shell is like an atom of the noble gas helium, and the fluoride anion with eight electrons in its valence shell is like an atom of the noble gas neon. Moreover, crystalline lithium fluoride forms from the individual lithium and fluoride ions. In this process negative fluoride ions become surrounded by positive lithium ions, and positive lithium ions by negative fluoride ions. In this crystalline state, the ions have substantially lower energies than the atoms from which they were formed. Lithium and fluorine are thus "stabilized" when they react to form crystalline lithium fluoride.

Ionic substances, because of their strong internal electrostatic forces, are usually very high melting solids, often having melting points above 1000 °C. In polar solvents, such as water, the ions are solvated, and such solutions usually conduct an electric current.

1.6B Covalent Bonds

When two or more atoms of the same or similar electronegativities react, a complete transfer of electrons does not occur. In these instances the atoms achieve noble gas structures by *sharing electrons.* *Covalent* bonds form between the atoms and the products are called *molecules.* Molecules may be represented by electron-dot formulas but, more conveniently, by dash formulas where each dash represents a pair of electrons shared by two atoms. Some examples are shown here. These formulas

$$\textbf{H}_2 \qquad \text{H:H} \quad \text{or} \quad \text{H}-\text{H} \qquad \textbf{Cl}_2 \qquad \text{:}\overset{..}{\underset{..}{\text{Cl}}}\text{:}\overset{..}{\underset{..}{\text{Cl}}}\text{:} \quad \text{or} \quad \text{:}\overset{..}{\underset{..}{\text{Cl}}}-\overset{..}{\underset{..}{\text{Cl}}}\text{:}$$

$$\textbf{HCl} \qquad \text{H:}\overset{..}{\underset{..}{\text{Cl}}}\text{:} \quad \text{or} \quad \text{H}-\overset{..}{\underset{..}{\text{Cl}}}\text{:} \qquad \textbf{CH}_4 \qquad \text{H:}\overset{\text{H}}{\underset{\text{H}}{\overset{..}{\text{C}}}}\text{:H} \quad \text{or} \quad \text{H}-\overset{\text{H}}{\underset{\text{H}}{\text{C}}}-\text{H}$$

are often called **Lewis structures;** in writing them we show only the outer level electrons.

In certain cases, multiple covalent bonds are formed; for example,

$$\textbf{N}_2 \qquad \text{:N::N:} \quad \text{or} \quad \text{:N}\equiv\text{N:}$$

and ions themselves may contain covalent bonds.

$$\overset{+}{\textbf{NH}}_4 \qquad \text{H:}\overset{\text{H}}{\underset{\text{H}}{\overset{..}{\text{N}}}}\overset{+}{\text{:H}} \quad \text{or} \quad \text{H}-\overset{\text{H}}{\underset{\text{H}}{\text{N}^+}}-\text{H}$$

1.6C Writing Lewis Structures

When we write Lewis structures (electron-dot formulas) we assemble the molecule or ion from the constituent atoms showing only the valence electrons (i.e., the electrons of the outermost shell). By having the atoms share or transfer electrons, we try to give each atom the electronic structure of a noble gas. For example, we give hydrogen atoms two electrons because by doing so we give them the structure of helium. We give carbon, nitrogen, oxygen, and fluorine atoms eight electrons because by doing this we give them the electronic structure of neon. The number of valence electrons of an atom can be obtained from the Periodic Table because it is equal to the group number of the atom. Carbon, for example, is in Group **IVA** and it has four valence electrons; fluorine, in Group **VIIA** has seven; hydrogen in Group **IA**, has one.

SAMPLE PROBLEM

Write the Lewis structure of CH_3F.

Answer:

1. We find the total number of valence electrons of all the atoms:

$$4 + 3(1) + 7 = 14$$
$$\uparrow \qquad \uparrow \qquad \uparrow$$
$$C \quad H_3 \quad F$$

2. We use pairs of electrons to form bonds between all atoms that are bonded to each other. We represent these bonding pairs with lines. In our example this requires four pairs of electrons (8 of our 14 valence electrons).

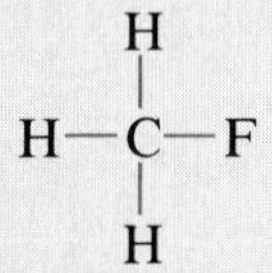

$$\begin{array}{c} H \\ | \\ H-C-F \\ | \\ H \end{array}$$

3. We then add the remaining electrons in pairs so as to give each hydrogen 2 electrons (a duet) and every other atom 8 electrons (an octet). In our example, we assign the remaining 6 valence electrons to the fluorine atom in three nonbonding pairs.

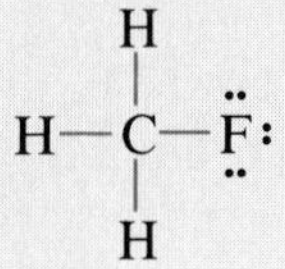

$$\begin{array}{c} H \\ | \\ H-C-\ddot{\underset{\cdot\cdot}{F}}: \\ | \\ H \end{array}$$

If the structure is an ion, we add or subtract electrons to give it the proper charge.

SAMPLE PROBLEM

Write the Lewis structure for the chlorate ion ($ClO_3{}^-$).

Answer:

1. We find the total number of valence electrons of all the atoms including the extra electron needed to give the ion a negative charge:

$$7 + 3(6) + 1 = 26$$

$$\overset{\uparrow}{\text{Cl}} \quad \overset{\uparrow}{\text{O}_3} \quad \overset{\uparrow}{e^-}$$

2. We use three pairs of electrons to form bonds between the chlorine atom and the three oxygen atoms:

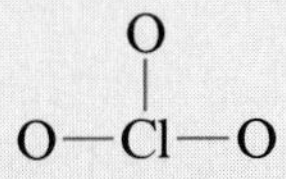

$$
\begin{array}{c}
\text{O} \\
| \\
\text{O}-\text{Cl}-\text{O}
\end{array}
$$

3. We then add the remaining 20 electrons in pairs so as to give each atom an octet.

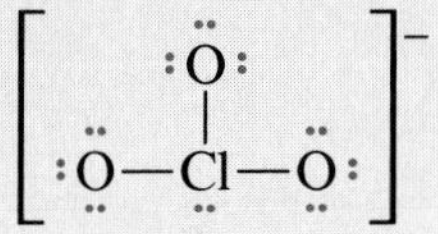

$$
\left[
\begin{array}{c}
:\ddot{\text{O}}: \\
| \\
:\ddot{\text{O}}-\ddot{\text{Cl}}-\ddot{\text{O}}:
\end{array}
\right]^{-}
$$

If necessary, we use multiple bonds to give atoms the noble gas structure. The carbonate ion ($CO_3{}^{2-}$) illustrates this.

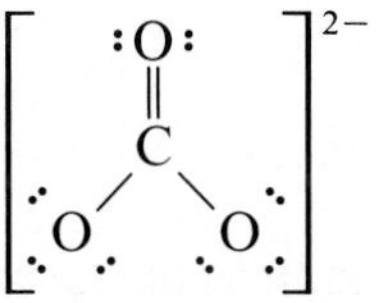

$$
\left[
\begin{array}{c}
:\ddot{\text{O}}: \\
\| \\
\text{C} \\
\diagup \quad \diagdown \\
\ddot{\text{O}} \qquad \ddot{\text{O}}:
\end{array}
\right]^{2-}
$$

The organic molecules ethene (C_2H_4) and ethyne (C_2H_2) have a double and triple bond, respectively.

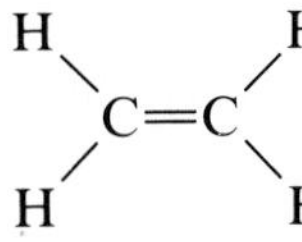

$$
\begin{array}{cc}
\text{H} & \text{H} \\
\diagdown \quad \diagup \\
\text{C}=\text{C} \\
\diagup \quad \diagdown \\
\text{H} & \text{H}
\end{array}
\quad \text{and} \quad \text{H}-\text{C}\equiv\text{C}-\text{H}
$$

1.6D Exceptions to the Octet Rule

Atoms of elements beyond the second period of the Periodic Table can accommodate more than eight electrons in their outer shell. Examples are the compounds PCl_5 and SF_6.

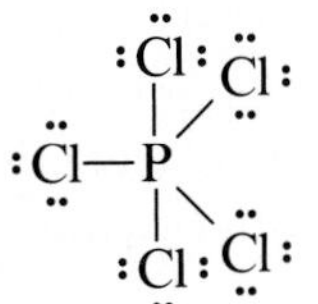

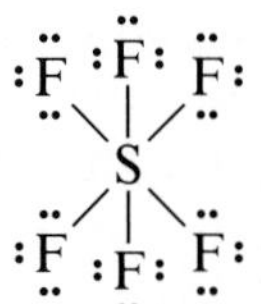

$$
\begin{array}{cc}
:\ddot{\text{Cl}}:\; \ddot{\text{Cl}}: & :\ddot{\text{F}}\; :\ddot{\text{F}}:\; \ddot{\text{F}}: \\
| \diagup & \diagdown | \diagup \\
:\ddot{\text{Cl}}-\text{P} & \text{S} \\
| \diagdown & \diagup | \diagdown \\
:\ddot{\text{Cl}}:\; \ddot{\text{Cl}}: & :\ddot{\text{F}}\; :\ddot{\text{F}}:\; \ddot{\text{F}}:
\end{array}
$$

Some highly reactive molecules or ions have atoms with fewer than eight electrons in their outer shell. An example is boron trifluoride (BF_3). In the BF_3 molecule the central boron atom has only six electrons around it.

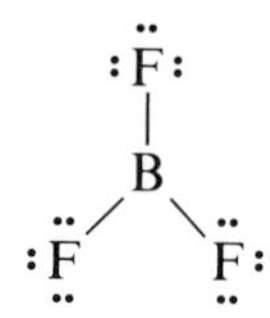

$$\overset{\displaystyle :\overset{\displaystyle ..}{F}:}{\underset{\displaystyle :\overset{..}{F}\diagup \quad \diagdown \overset{..}{F}:}{\overset{|}{B}}}$$

Finally, one point needs to be stressed: Before we can write some Lewis structures, *we must know how the atoms are connected to each other.* Consider nitric acid, for example. Even though the formula for nitric acid is often written HNO_3, the hydrogen is actually connected to an oxygen, not to the nitrogen. The structure is $HONO_2$ and not HNO_3. Thus the correct Lewis structure is

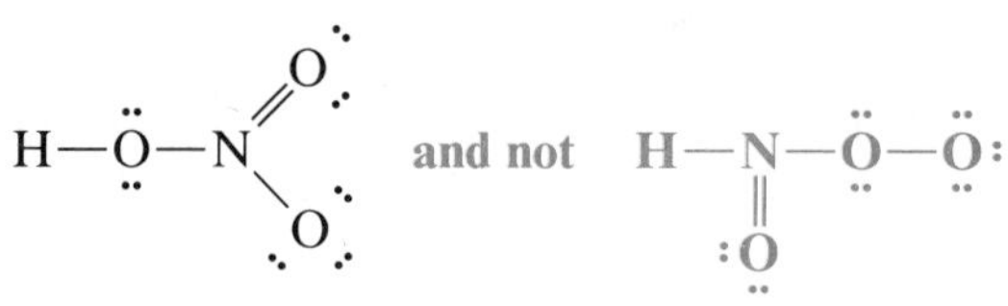

$$H-\overset{..}{\underset{..}{O}}-N\diagdiagup\overset{\displaystyle \overset{..}{O}\colon}{} \qquad \text{and not} \qquad H-\overset{..}{\underset{\displaystyle \overset{\|}{\underset{..}{O}}}{N}}-\overset{..}{\underset{..}{O}}-\overset{..}{\underset{..}{O}}\colon$$

This knowledge comes ultimately from experiments. If you have forgotten the structures of some of the common inorganic molecules and ions (such as those listed in Problem 1.1), this may be a good time for a review of the relevant portions of your general chemistry text.

PROBLEM 1.1

Write electron-dot and dash formulas for each of the following molecules or ions. In each case show how the atoms achieve the noble gas structure.

(a) HBr   (e) H_2O_2 (i) NF_3 (m) NH_4Cl ($NH_4^+Cl^-$)
(b) Br_2 (f) SiH_4 (j) CH_3Cl (n) NaOH (Na^+OH^-)
(c) CO_2 (g) NH_3 (k) H_2O
(d) CH_4 (h) PCl_3 (l) OH^-

1.7 IONS CONTAINING COVALENT BONDS: FORMAL CHARGE

Some positive or negative ions consist of atoms held together by covalent bonds. In Problem 1.1, for example, you were asked to write electron-dot formulas for the ammonium ion (NH_4^+) and the hydroxide ion (OH^-). It will be helpful at this point, as a review, to show how the charges on ions can be calculated. Let us begin with a simple example: The ammonium ion.

There are several ways to calculate the charge on the ammonium ion. The most fundamental is to calculate the arithmetic sum of all of the nuclear protons (positive charges) and of all of the extranuclear electrons (negative charges). This approach is shown in the following figure:

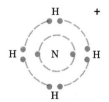

One nitrogen atom = 7 protons
Four hydrogen atoms = 4 protons
Total protons = 11
Number of outer level electrons = 8
Number of inner level electrons = 2
Total electrons = 10
Net charge = $(+11) + (-10) = +1$

1.7A Formal Charge

Another way of calculating the charge on a polyatomic ion is based on the concept of **formal charge.** This procedure is really a method for electron bookkeeping. In this approach, the formal charge on each atom is calculated first. Then the arithmetic sum of all of the formal charges gives the charge on the ion as a whole. The formal charge of each atom is calculated by taking the group number of that atom (from the Periodic Table) and subtracting the number of electrons associated with it using the following formula:

$$\begin{matrix} \text{Formal} \\ \text{charge} \end{matrix} = \begin{matrix} \text{group} \\ \text{number} \\ \text{of atom} \end{matrix} - \left[\frac{1}{2}\left(\begin{matrix} \text{number of} \\ \text{shared} \\ \text{electrons} \end{matrix}\right) + \left(\begin{matrix} \text{number of} \\ \text{unshared} \\ \text{electrons} \end{matrix}\right) \right]$$

Let us consider the ammonium ion again. First we calculate the formal charge on each atom:

Formal charge on the hydrogen atom = $+1 - [\frac{1}{2}(2) + (0)] = 0$

Formal charge on the nitrogen atom = $+5 - [\frac{1}{2}(8) + (0)] = +1$

Then we calculate the formal charge on the ion as a whole:

$$\begin{matrix} \overset{..}{\text{H}} \\ \text{H:}\overset{..}{\text{N}}\text{:}^{+}\text{H} \\ \overset{..}{\text{H}} \end{matrix}$$

Formal charge on each hydrogen atom = $0 \times 4 = 0$
Formal charge on nitrogen atom   = $+1 \times 1 = +1$
Total charge on the ion   = $+1$

A moment's reflection will reveal why this formula works: The group number of an element is nothing more than the kernel charge of that element. The kernel charge is defined as the charge on that portion of the atom that includes the nuclear protons and the inner level electrons. For a nitrogen atom (Fig. 1.3) the kernel charge is $+5$.

Taking the sum $\frac{1}{2}$ (number of shared electrons) + (number of unshared electrons) is nothing more than a way of apportioning the valence electrons. We divide shared electrons equally between the atoms that share them; we assign unshared pairs

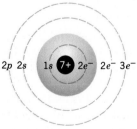

Kernel charge = $+7 - 2 = +5$

FIGURE 1.3 The kernel charge of the nitrogen atom is the algebraic sum of the electrical charges of the protons and the inner level (1s) electron.

directly to the atom that possesses them. The ammonium ion has no unshared pairs; all of the valence electrons are divided between the sharing atoms:

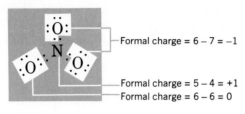

For hydrogen:	kernel charge	$= +1$
	electron	$= \underline{-1}$
	Formal charge	$= 0$
For nitrogen:	kernel charge	$= +5$
	4 electrons	$= \underline{-4}$
	Formal charge	$= +1$

Let us consider an ion that has unshared pairs. The nitrate ion (NO_3^-) may be written in the following way:

Formal charge $= 6 - 7 = -1$

Formal charge $= 5 - 4 = +1$
Formal charge $= 6 - 6 = 0$

$$\text{Charge on ion} = 2(-1) + 1 + 0 = -1$$

Notice then, for *ions,* the net charge on the ion as a whole is equal to the sum of the formal charges on each individual atom.

Molecules, of course, have no net electrical charge. Molecules, by definition, are neutral. Therefore, the sum of the formal charges on each atom making up a molecule must be zero. Consider the following examples:

Ammonia

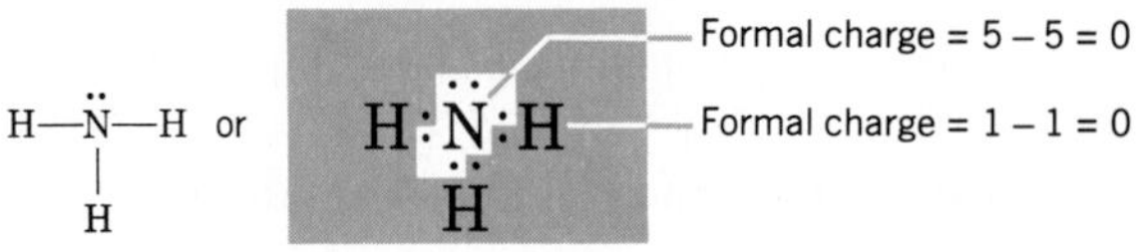

H—N̈—H or H:N:H

Formal charge $= 5 - 5 = 0$
Formal charge $= 1 - 1 = 0$

$$\text{Charge on molecule} = 0 + 3(0) = 0$$

Water

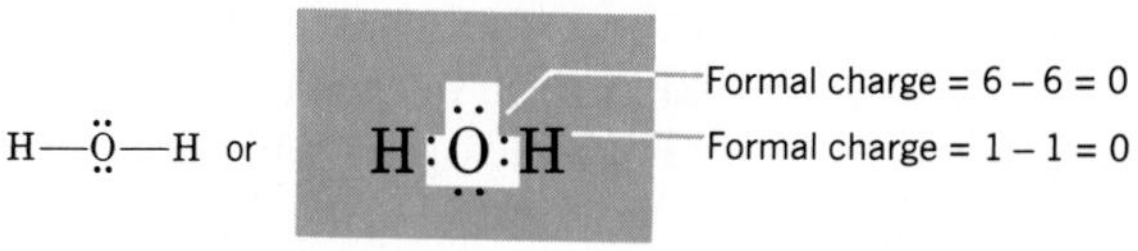

H—Ö—H or H:O:H

Formal charge $= 6 - 6 = 0$
Formal charge $= 1 - 1 = 0$

$$\text{Charge on molecule} = 0 + 2(0) = 0$$

PROBLEM 1.2

Calculate the formal charge on each atom, and verify the total charge on the molecule or ion, for each of the following species:

(a) BH_4^- (e) CO_3^{2-} (i) $:CH_2$ (a carbene)

(b) OH^- (f) $:CH_3^-$ (a carbanion) (j) $:\ddot{N}H_2^-$

(c) BF_4^- (g) CH_3^+ (a carbocation)

(d) H_3O^+ (h) $\cdot CH_3$ (a free radical)

1.7B Summary of Formal Charges

With this background it should now be clear that each time an oxygen atom of the type $-\overset{\cdot\cdot}{\underset{\cdot\cdot}{O}}:$ appears in a molecule or ion it will have a formal charge of -1, and that each time an oxygen atom of the type $=\overset{\cdot\cdot}{O}:$ or $-\overset{\cdot\cdot}{\underset{\cdot\cdot}{O}}-$ appears it will have a formal charge of zero. Similarly; $-\overset{|}{\underset{|}{N}}-$ will be $+1$, and $-\overset{\cdot\cdot}{\underset{|}{N}}-$ will be zero. It is much easier to memorize these common structures than to calculate their formal charges each time they are encountered. These common structures are summarized in Table 1.3.

PROBLEM 1.3

Using the chart given in Table 1.3, determine the formal charge on each colored atom of the following molecules and ions. (*Remember:* With respect to formal charge, $-\overset{\cdot\cdot}{\underset{\cdot\cdot}{O}}-$ is equal to $=\overset{\cdot\cdot}{O}:$, $-\overset{|}{N}-$ is equal to $\equiv N:$, and so on.)

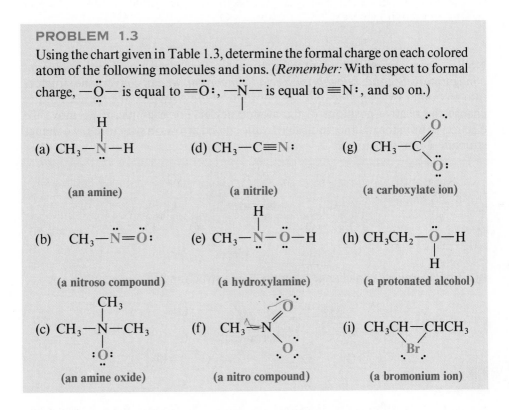

(a) $CH_3-\overset{H}{\underset{\cdot\cdot}{N}}-H$

(an amine)

(b) $CH_3-\overset{\cdot\cdot}{N}=\overset{\cdot\cdot}{O}:$

(a nitroso compound)

(c) $CH_3-\overset{CH_3}{\underset{:\overset{\cdot\cdot}{O}:}{N}}-CH_3$

(an amine oxide)

(d) $CH_3-C\equiv N:$

(a nitrile)

(e) $CH_3-\overset{H}{\underset{\cdot\cdot}{N}}-\overset{\cdot\cdot}{\underset{\cdot\cdot}{O}}-H$

(a hydroxylamine)

(f) $CH_3=N\overset{\overset{\cdot\cdot}{O}}{\underset{\overset{\cdot\cdot}{O}:}{}}$

(a nitro compound)

(g) $CH_3-C\overset{\overset{\cdot\cdot}{O}}{\underset{\overset{\cdot\cdot}{O}:}{}}$

(a carboxylate ion)

(h) $CH_3CH_2-\overset{\cdot\cdot}{\underset{H}{O}}-H$

(a protonated alcohol)

(i) $CH_3CH-CHCH_3$ over $\overset{}{Br}$

(a bromonium ion)

TABLE 1.3 A summary of formal charges

GROUP	FORMAL CHARGE OF +1	FORMAL CHARGE OF 0	FORMAL CHARGE OF −1
3		$-B-$	$-\overset{-}{B}-$
4	$-\overset{+}{C}-$	$-C-$	$-\overset{-}{C}:-$
5	$-\overset{+}{N}-$ $=\overset{+}{N}-$ $\equiv\overset{+}{N}-$	$-\overset{\cdot\cdot}{N}-$ $=\overset{\cdot\cdot}{N}-$ $\equiv N:$	$-\overset{\cdot\cdot}{N}^{-}-$ $=\overset{\cdot\cdot}{N}:^{-}$
6	$-\overset{\cdot\cdot}{O}^{+}-$ $=\overset{+}{O}-$	$-\overset{\cdot\cdot}{\underset{\cdot\cdot}{O}}-$ $=\overset{\cdot\cdot}{O}:$	$-\overset{\cdot\cdot}{\underset{\cdot\cdot}{O}}:^{-}$
7	$-\overset{\cdot\cdot}{X}^{+}-$	$-\overset{\cdot\cdot}{X}:$ (X = F, Cl, Br, or I)	$:\overset{\cdot\cdot}{X}:^{-}$

1.8 RESONANCE

One problem with Lewis structures is that they impose an artificial **location** on the electrons. As a result, more than one *equivalent* Lewis structure can be written for many molecules and ions. Consider, for example, the carbonate ion (CO_3^{2-}). We can write three *different* but *equivalent* structures, **1–3**.

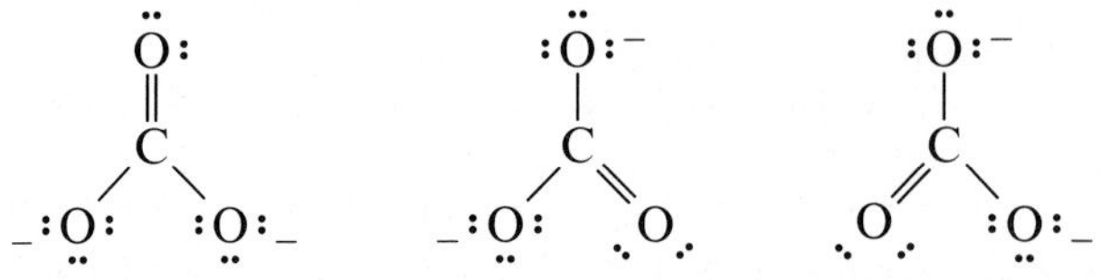

Notice two important features of these structures. First, each atom has the noble gas configuration. Second, *and this is especially important,* we can convert one structure into any other by *changing only the positions of the electrons.* We do not need to change the relative positions of the atomic nuclei. For example, if we move the electron pairs in the manner indicated by the curved arrows in structure **1**, we change structure **1** into structure **2**:

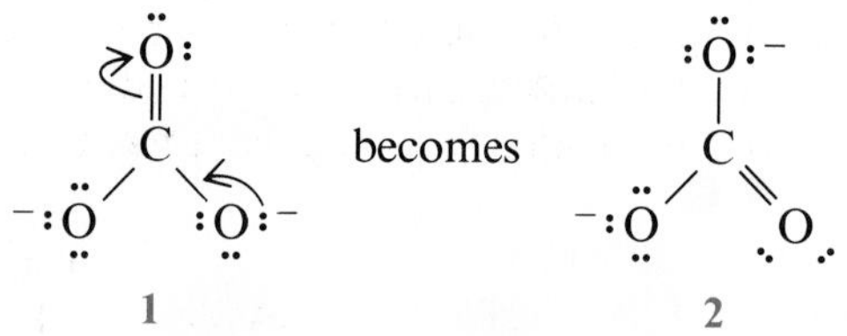

In a similar way we can change structure **2** into structure **3**:

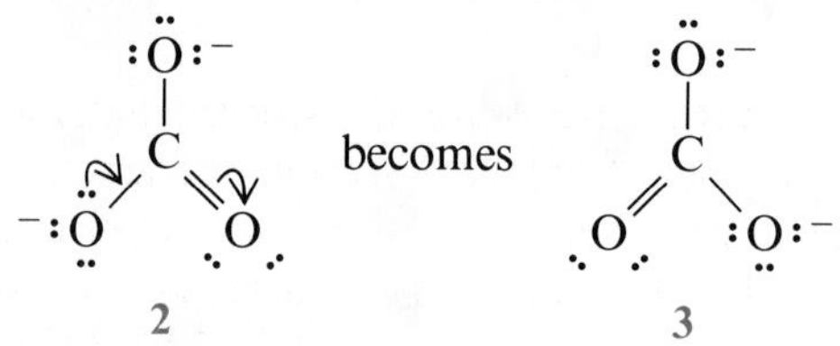

Structures **1–3**, although not identical, *are equivalent.* None of them, however, fits important data about the carbonate ion.

 X-ray studies have shown that carbon–oxygen double bonds are shorter than single bonds. The same kind of study of the carbonate ion shows, however, that all of its carbon–oxygen bonds are of equal length. One is not shorter than the others. All are equivalent. Clearly none of the three structures agrees with this evidence. In each structure, **1–3**, one carbon–oxygen bond is a double bond and the other two are single bonds. None of the structures is correct. How, then, should we represent the carbonate ion?

 A situation like this can be handled by a theory called **resonance theory.** Resonance theory says that whenever a molecule or ion can be represented by two or more Lewis structures *that differ only in the positions of the electrons,* two things will be true:

1. None of these structures, which we call **resonance structures** or **resonance contributors,** will be a correct representation for the molecule. None will be in complete accord with the physical or chemical properties of the substance.

2. The actual molecule or ion will be better represented by a *hybrid of these structures.*

Resonance structures, then, are not structures for the actual molecule or ion; they exist only in theory. As such they can never be isolated. No single contributor adequately represents the molecule or ion. In resonance theory we view the carbonate ion, which is, of course, a real entity, as having a structure that is a **hybrid** of these three **hypothetical** resonance structures.

What would a hybrid of structures **1** to **3** be like? Look at the structures and look especially at a particular carbon–oxygen bond, say, the one at the top. This carbon–oxygen bond is a double bond in one structure (**1**) and a single bond in the other two (**2** and **3**). The actual carbon–oxygen bond, since it is a hybrid, must be something in between a double bond and a single bond. Because the carbon–oxygen bond is a single bond in two of the structures and a double bond in only one it must be more like a single bond than a double bond. It must be like a one- and one-third bond. We could call it a partial double bond. And, of course, what we have just said about any one carbon–oxygen bond will be equally true of the other two. Thus all of the carbon–oxygen bonds of the carbonate ion are partial double bonds, and *all are equivalent.* All of them *should be* the same length, and this is exactly what experiments tell us. They are all 1.31 Å long.

One other important point: By convention, when we draw resonance structures, we connect them by double-headed arrows to indicate clearly that they are hypothetical, not real. For the carbonate ion we write them this way:

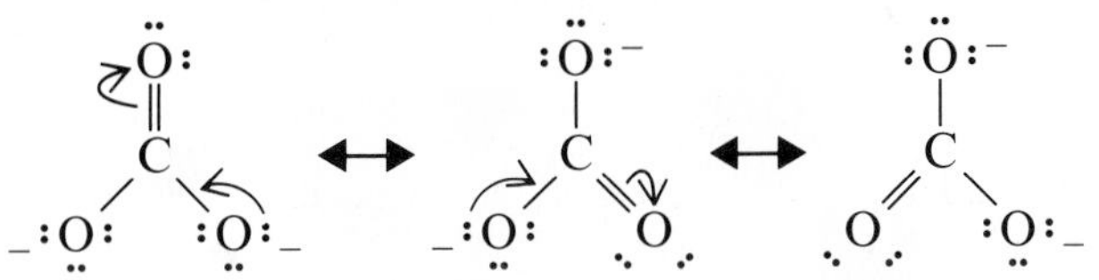

We should not let these arrows, or the word "resonance," mislead us into thinking that the carbonate ion fluctuates between one structure and another. These structures exist only on paper; therefore, the carbonate ion cannot fluctuate among them. It is also important to distinguish between resonance and **an equilibrium.** In an equilibrium between two, or more, species, it is quite correct to think of different structures and moving (or fluctuating) atoms, *but not in the case of resonance* (as in the carbonate ion). Here the atoms do not move, and the "structures" exist only on paper. An equilibrium is indicated by ⇌ and resonance by ↔.

How can we write the structure of the carbonate ion in a way that will indicate its actual structure? We may do two things: we may write all of the resonance structures as we have just done and let the reader mentally fashion the hybrid or we may write a non-Lewis structure that attempts to represent the hybrid. For the carbonate ion we might do the following:

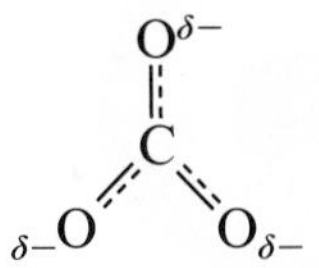

The bonds are indicated by a combination of a solid line and a dashed line. This is to indicate that the bonds are something in between a single bond and a double bond. As a rule, we use a solid line whenever a bond appears in all structures, and a dashed line when a bond exists in one or more but not all. We also place a $\delta-$ (read partial minus) beside each oxygen to indicate that something less than a full negative charge resides on each oxygen atom. (In this instance each oxygen atom has two thirds of a full negative charge.)

SAMPLE PROBLEM

The following is one way of writing the structure of the nitrate ion.

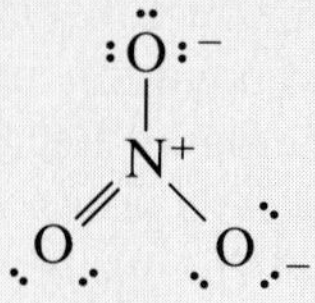

However, considerable physical evidence indicates that all three nitrogen-oxygen bonds are equivalent and that they have a bond distance between that expected for a nitrogen-oxygen single bond and a nitrogen-oxygen double bond. Explain this in terms of resonance theory.

Answer:

We recognize that if we move the electron pairs in the following way, we can write three *different* but *equivalent* structures for the nitrate ion:

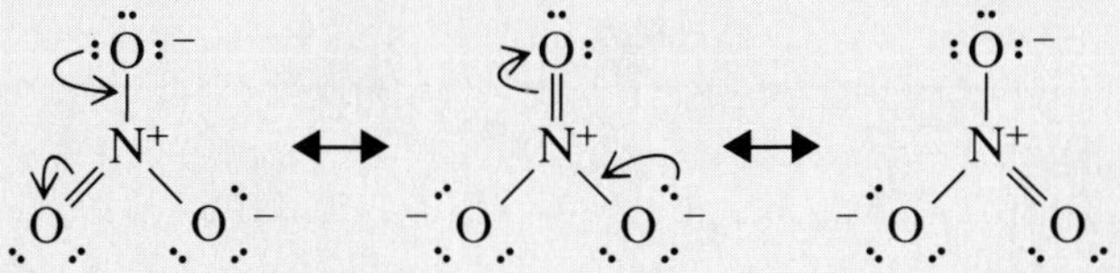

Since these structures differ from one another *only in the positions of their electrons,* they are *resonance structures* or *resonance contributors.* As such, no single structure taken alone will adequately represent the nitrate ion. The actual molecule will be best represented by a *hybrid of these three structures.* We might write this hybrid in the following way to indicate that all of the bonds are equivalent and that they are more than single bonds and less than double bonds. We also indicate that each oxygen atom bears an equal partial negative charge. This charge distribution corresponds to what we find experimentally.

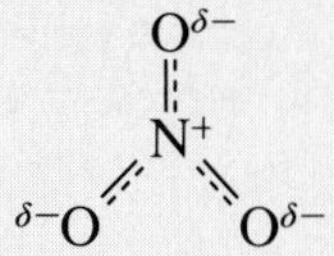

Hybrid structure for the nitrate ion

PROBLEM 1.4

(a) Write two resonance structures for sulfur dioxide (SO_2). (Sulfur dioxide is a major air pollutant.)   (b) Would you expect the sulfur-oxygen bonds of SO_2 to be equivalent and thus of equal length? Explain.

1.9 POLAR COVALENT BONDS

When two atoms of different electronegativities form a covalent bond, the electrons are not shared equally between them. The atom with greater electronegativity draws the electron pair closer to it, and a **polar covalent bond** results. (One definition of *electronegativity* is *the ability of an element to attract electrons that it is sharing in a covalent bond.*) An example of such a polar covalent bond is the one in hydrogen chloride. The chlorine atom, with its greater electronegativity, pulls the bonding electrons closer to it. This makes the hydrogen atom somewhat electron deficient and gives it a *partial* positive charge ($\delta+$). The chlorine atom becomes somewhat electron rich and bears a *partial* negative charge ($\delta-$).

$$\overset{\delta+}{H} \quad \overset{\delta-}{:\ddot{Cl}:}$$

Because the hydrogen chloride molecule has a partially positive end and a partially negative end, it is a dipole, and it has a **dipole moment.** The dipole moment (μ) is a physical property that can be measured experimentally. The unit is the *Debye* and is

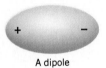

A dipole

abbreviated D. (The unit is named after Peter Debye, a Dutch physical chemist, who made many contributions to our understanding of polar molecules.)

The direction of polarity of a polar bond is sometimes symbolized by $\longmapsto$. In HCl this is expressed in the following way:

$$\text{(positive end)} \longmapsto \text{(negative end)}$$
$$\text{H}-\text{Cl}$$
$$\mu = 1.08 \text{ D}$$

Dipole moments, as we shall see in Section 1.19, are very useful quantities in accounting for physical properties of compounds.

PROBLEM 1.5

Predict the direction of the dipole (if any) in the following molecules:
(a) HBr, (b) ICl, (c) H_2, (d) Cl_2.

1.10 QUANTUM MECHANICS

In 1926 a new theory of atomic and molecular structure was advanced independently and almost simultaneously by three men: Erwin Schrödinger, Werner Heisenberg, and Paul Dirac. This theory, called **wave mechanics** by Schrödinger or **quantum**

mechanics by Heisenberg, has become the basis from which we derive our modern understanding of bonding in molecules.

The formulation of quantum mechanics that Schrödinger advanced is the form that is most often used by chemists. In Schrödinger's publication the motion of the electrons was described in terms that took into account the wave nature of the electron.* Schrödinger developed a way to convert the mathematical expression for the total energy of the system consisting of one proton and one electron — the hydrogen atom — into another expression called a **wave equation.** This equation was then solved to yield not one but a series of solutions called **wave functions.**

Wave functions are most often denoted by the Greek letter psi (ψ), and each wave function (ψ function) corresponds to a different state for the electron. Corresponding to each state, and calculable from the wave equation for the state, is a particular energy.

Each state is a sublevel where one or two electrons can reside. *The solutions to the wave equation for a hydrogen atom can also be used* (with appropriate modifications) *to give sublevels for the electrons of higher elements.*

A wave equation is simply a tool for calculating two important properties: These are the energy associated with the state and the relative probability of an electron residing at particular places in the sublevel (Section 1.11). When the value of a wave equation is calculated for a particular point in space relative to the nucleus, the result may be a positive number or a negative number (or zero). These signs are sometimes called **phase signs.** They are characteristic of all equations that describe waves. We do not need to go into the mathematics of waves here, but a simple analogy will help us understand the nature of these phase signs.

Imagine a wave moving across a lake. As it moves along, the wave has crests and troughs; that is, it has regions where the wave rises above the average level of the lake or falls below it (Fig. 1.4). Now, if an equation were to be written for this wave, the wave function (ψ) would be plus (+) in regions where the wave is above the average level of the lake (i.e., in crests) and it would be minus (−) in regions where the wave is below the average level (i.e., in troughs). The relative magnitude of ψ (called the amplitude) will be related to the distance the wave rises above or falls below the average level of the lake. At the places where the wave is exactly at the average level of the lake, the wave function will be zero. Such a place is called a **node.**

One other characteristic of waves is their ability to reinforce each other or to interfere with one another. Imagine two waves approaching each other as they move across a lake. If the waves meet so that a crest meets a crest, that is, so that *waves of the same phase sign meet each other,* the waves **reinforce** each other, they add together, and the resulting wave is larger than either individual wave. On the other hand, if a crest meets a trough, that is, if waves of opposite sign meet, the waves **interfere** with each other, they subtract from each other, and the resulting wave is smaller than either individual wave. (If the two waves of opposite sign meet in precisely the right way, complete cancellation can occur.)

The wave functions that describe the motion of an electron in an atom or molecule are, of course, different from the equations that describe waves moving across lakes. And when dealing with the electron we should be careful not to take analogies like this too far. Electron wave functions, however, are like the equations that describe water waves in that they have phase signs and nodes, and *they undergo reinforcement and interference.*

---

*The idea that the electron has the properties of a wave as well as those of a particle was proposed by Louis de Broglie in 1923.

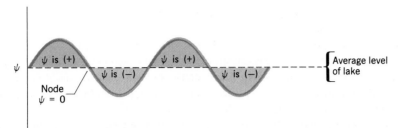

FIGURE 1.4 A wave moving across a lake is viewed along a slice through the lake. For this wave the wave function, ψ, is plus (+) in crests and minus (−) in troughs. At the average level of the lake it is zero; these places are called nodes.

1.11 ATOMIC ORBITALS

For a short time after Schrödinger's proposal in 1926, a precise physical interpretation for the electron wave function eluded early practitioners of quantum mechanics. It remained for Max Born, a few months later, to point out that the square of ψ *could* be given a precise physical meaning. According to Born, ψ^2 for a particular location (x,y,z) expresses the **probability** of finding an electron at that particular location in space. If ψ^2 is large in a unit volume of space, the probability of finding an electron in that volume is great — we say that the **electron probability density** is large. Conversely if ψ^2 for some other unit volume of space is small, the probability of finding an electron there is low. Plots of ψ^2 in three dimensions generate the shapes of the familiar s, p, and d atomic orbitals.

The f orbitals are practically never used in organic chemistry, and we shall not concern ourselves with them in this book. The d orbitals will be discussed briefly later when we discuss compounds in which d orbital interactions are important. The s and p orbitals are, by far, the most important in the formation of organic molecules and, at this point, we shall limit our discussion to them.

An orbital is a region of space where the probability of finding an electron is large. The shapes of s and p orbitals are shown in Fig. 1.5. There is a finite, but very

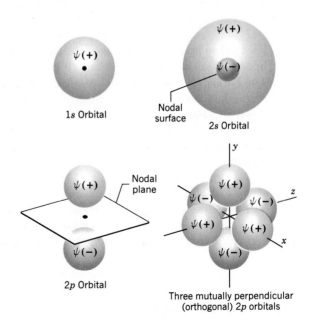

FIGURE 1.5 The shapes of some s and p orbitals.

small, probability of finding an electron at greater distances from the nucleus. The volumes that we typically use to illustrate an orbital are those volumes that would contain the electron 90 to 95% of the time.

Both the $1s$ and $2s$ orbitals are spheres (as are all higher s orbitals). The sign of the wave function, ψ_{1s}, is positive (+) over the entire $1s$ orbital (Fig. 1.5). The $2s$ orbital contains a nodal surface, that is, an area where $\psi = 0$. In the inner portion of the $2s$ orbital, ψ_{2s} is negative.

The $2p$ orbitals have the shape of two almost-touching spheres. The phase sign of the wave function, ψ_{2p}, is positive in one lobe (or sphere) and negative in the other. A nodal plane separates the two lobes of a p orbital, and the three p orbitals are arranged in space so that their axes are mutually perpendicular.

You should not associate the sign of the wave function with anything having to do with electrical charge. As we said earlier the (+) and (−) signs associated with ψ are simply the arithmetic signs of the wave function in that region of space. The (+) and (−) signs do not imply a greater or lesser probability of finding an electron either. The probability of finding an electron is ψ^2, and ψ^2 is always positive. (Squaring a negative number always makes it positive.) Thus the probability of finding the electron in the (−) lobe of a p orbital is the same as that of the (+) lobe. The significance of the (+) and (−) signs will become clear later when we see how atomic orbitals combine to form molecular orbitals and when we see how covalent bonds are formed.

> There is a relationship between the number of nodes of an orbital and its energy: ***The greater the number of nodes, the greater the energy.*** We can see an example here; the $2s$ and $2p$ orbitals have one node each and they have greater energy than a $1s$ orbital, which has no nodes.

The relative energies of the lower energy orbitals are as follows. Electrons in $1s$ orbitals have the lowest energy because they are closest to the positive nucleus. Electrons in $2s$ orbitals are next lowest in energy. Electrons of $2p$ orbitals have equal but still higher energy. (Orbitals of equal energy are said to be **degenerate orbitals**.)

We can use these relative energies to arrive at the electronic configuration of any atom in the first two rows of the Periodic Table. We need only follow a few simple rules.

1. **The aufbau principle:** Orbitals are filled so that those of lowest energy are filled first. (*Aufbau* is German for "building up.")
2. **The Pauli exclusion principle:** A maximum of two electrons may be placed in each orbital *but only when the spins of the electrons are paired.* An electron spins about its own axis. For reasons that we cannot develop here, an electron is permitted only one or another of only two possible spin orientations. We usually show these orientations by arrows, either ↑ or ↓. Thus two spin-paired electrons would be designated ↑↓. Unpaired electrons, which are not permitted in the same orbital, are designated ↑↑ (or ↓↓).
3. **Hund's rule:** When we come to orbitals of equal energy (degenerate orbitals) such as the three p orbitals, we add one electron to each *with their spins unpaired* until each of the degenerate orbitals contains one electron. Then we begin adding a second electron to each degenerate orbital so that the spins are paired.

If we apply these rules to some of the second-row elements of the Periodic Table, we get the results shown in Fig. 1.6.

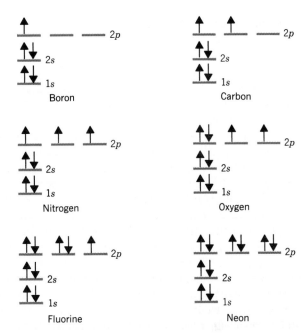

FIGURE 1.6 The electron configuration of some second-row elements.

PROBLEM 1.6

Taking into account the fact that electrons repel each other and thus prefer to stay as far apart as possible, provide an explanation for Hund's rule.

1.12 MOLECULAR ORBITALS

For the organic chemist the greatest utility of atomic orbitals is in using them to understand how atoms combine to form molecules. We shall have much more to say about this subject in subsequent chapters for, as we have already said, covalent bonds are central to the study of organic chemistry. First, however, we shall concern ourselves with a very simple case: The covalent bond that is formed when two hydrogen atoms combine to form a hydrogen molecule. We shall see that the description of the formation of the H—H bond is the same as, or at least very similar to, the description of bonds in more complex molecules.

Let us begin by examining what happens to the total energy of two hydrogen atoms with electrons of opposite spins when they are brought closer and closer together. This can best be shown with the curve shown in Fig. 1.7.

When the atoms of hydrogen are relatively far apart (**I**) their total energy is simply that of two isolated hydrogen atoms. As the hydrogen atoms move closer together (**II**), each nucleus increasingly attracts the other's electron. This attraction more than compensates for the repulsive force between the two nuclei (or the two electrons), and the result of this attraction *is to lower the energy of the total system.* When the two nuclei are 0.74 Å apart (**III**), the most stable (lowest energy) state is obtained. This distance, 0.74 Å, corresponds to the *bond length* for the hydrogen molecule. If the nuclei are moved closer together (**IV**) the repulsion of the two positively charged nuclei predominates, and the energy of the system rises.

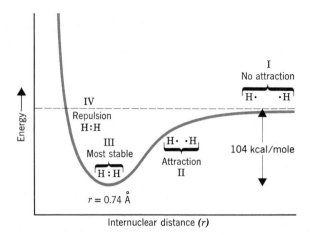

FIGURE 1.7 The potential energy of the hydrogen molecule as a function of internuclear distance.

There is one serious problem with this explanation of bond formation. We have assumed that the electrons are essentially motionless and that as the nuclei come together they will be stationary in the region between the two nuclei. Electrons do not behave that way. Electrons move about, and according to the **Heisenberg uncertainty principle,** we cannot know simultaneously the position and momentum of an electron. That is, we cannot pin the electrons down as precisely as our explanation suggests.

We avoid this problem when we use quantum mechanics, because we describe the electron in terms of wave functions (ψ) and in terms of probabilities (ψ^2) of finding it at particular places. By treating the electron in this way we do not violate the uncertainty principle, because we do not talk about where the electron is precisely. We talk instead about where the *electron probability density* is large or small.

Thus a better explanation for what happens when two hydrogen atoms combine to form a hydrogen molecule is the following: As the hydrogen atoms approach each other, their $1s$ orbitals (ψ_{1s}) begin to overlap. As the atoms move closer together, orbital overlap increases until the **atomic orbitals** combine to become **molecular orbitals (MO's).** The molecular orbitals that are formed encompass both nuclei and, in them, the electrons can move about both nuclei. They are not restricted to the vicinity of one nucleus or the other as they were in the separate atomic orbitals. Molecular orbitals, like atomic orbitals, *may contain a maximum of two spin-paired electrons.*

When atomic orbitals combine to form molecular orbitals, *the number of molecular orbitals that result always equals the number of atomic orbitals that combine.* Thus in the formation of a hydrogen molecule the *two* atomic orbitals combine to produce *two* molecular orbitals. Two orbitals result because the mathematical properties of wave functions permit them to be combined by either *addition* or *subtraction.* That is, they can combine either *in* or *out of* phase. What are the natures of these new molecular orbitals?

One molecular orbital, called the **bonding molecular orbital** (ψ_{molec}), is formed when the atomic orbitals combine in the way shown in Fig. 1.8. Here atomic orbitals combine by *addition,* and this means that atomic *orbitals of the same phase sign overlap.* Such overlap leads to *reinforcement* of the wave function in the region between the two nuclei. Reinforcement of the wave function not only means that the value of ψ is larger between the two nuclei, it means that ψ^2 is larger as well. Moreover, since ψ^2 expresses the probability of finding an electron in this region of space,

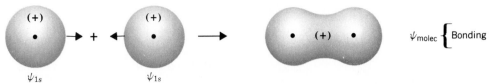

FIGURE 1.8 The overlapping of two hydrogen $1s$ atomic orbitals to form a bonding molecular orbital.

we can now understand how orbital overlap of this kind leads to bonding. It does so by increasing the electron probability density in exactly the right place—in the region of space between the nuclei. When the electron density is large here, the attractive force of the nuclei for the electrons more than offsets the repulsive force acting between the two nuclei (and between the two electrons). This extra attractive force is, of course, the "glue" that holds the atoms together.

The second molecular orbital, called the **antibonding molecular orbital** (ψ^*_{molec}), is formed by subtraction in the way shown in Fig. 1.9. [Subtraction means that the phase sign of one orbital has been changed from (+) to (−).] Here, because *orbitals of opposite signs overlap,* the wave functions *interfere* with each other in the region between the two nuclei and a node is produced. At the node $\psi = 0$, and on either side of the node ψ is small. This means that in the region between the nuclei ψ^2 is also small. Thus if electrons were to occupy the antibonding orbital, the electrons would avoid the region between the nuclei. There would be only a small attractive force of the nuclei for the electrons. Repulsive forces (between the two nuclei and between the two electrons) would be greater than the attractive forces. Having electrons in the antibonding orbital would not tend to hold the atoms together; it would tend to make them fly apart.

What we have just described has its counterpart in a mathematical treatment called the LCAO (linear combination of atomic orbitals) method. In the LCAO treatment, wave functions for the atomic orbitals are combined in a linear fashion (by addition or subtraction) in order to obtain new wave functions for the molecular orbitals.

Molecular orbitals, like atomic orbitals, correspond to particular energy states for an electron. Calculations show that the relative energy of an electron in the bonding molecular orbital of the hydrogen molecule is substantially less than its energy in a ψ_{1s} atomic orbital. These calculations also show that the energy of an electron in the antibonding molecular orbital is substantially greater than its energy in a ψ_{1s} atomic orbital.

An energy diagram for the molecular orbitals of the hydrogen molecule is shown in Fig. 1.10. Notice that electrons are placed in molecular orbitals in the same way that they were in atomic orbitals. Two electrons (with their spins opposed)

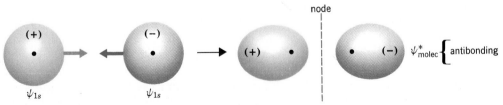

FIGURE 1.9 The overlapping of two hydrogen $1s$ atomic orbitals to form an antibonding molecular orbital.

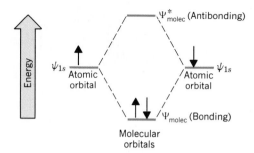

FIGURE 1.10 Energy diagram for the hydrogen molecule. Combination of two atomic orbitals. ψ_{1s} gives two molecular orbitals, ψ_{molec} and ψ_{molec}^*. The energy of ψ_{molec} is lower than that of the separate atomic orbitals, and in the lowest electronic energy state of molecular hydrogen it contains both electrons.

occupy the bonding molecular orbital, where their total energy is less than in the separate atomic orbitals. This is the *lowest electronic energy state* or **ground state** of the hydrogen molecule. (An electron may occupy the antibonding orbital in what is called an **excited state** for the molecule. This state forms when the molecule in the ground state absorbs a photon of light of proper energy.)

1.13 ENERGY CHANGES

Since we will be talking frequently about the energies of chemical systems, perhaps we should pause here for a brief review. *Energy* is defined as the capacity to do work. The two fundamental types of energy are **kinetic energy** and **potential energy.**

Kinetic energy is the energy an object has because of its motion; it equals one half the object's mass multiplied by the square of its velocity (i.e., $\frac{1}{2} mv^2$).

Potential energy is stored energy. It exists only when an attractive or repulsive force exists between objects. Two balls attached to each other by a spring can have their potential energy increased when the spring is stretched or compressed (Fig. 1.11). If the spring is stretched, an attractive force will exist between the balls. If it is compressed, a repulsive force will exist. In either instance releasing the balls will cause the potential energy (stored energy) of the balls to be converted into kinetic energy (energy of motion).

Chemical energy is a form of potential energy. It exists because attractive and repulsive electrical forces exist between different pieces of the molecules. Nuclei attract electrons, nuclei repel each other, and electrons repel each other.

It is usually impractical (and often impossible) to describe the *absolute* amount of potential energy contained by a substance. Thus we usually think in terms of their *relative potential energies.* We say that one system has *more* or *less* potential energy than another.

Another term that chemists frequently use in this context is the term **stability** or **relative stability. *The relative stability of a system is inversely related to its relative***

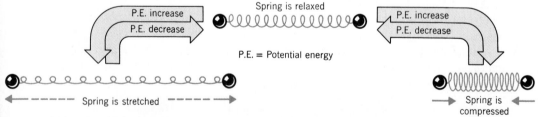

FIGURE 1.11 Potential energy (P.E.) exists between objects that either attract or repel each other. When the spring is either stretched or compressed, the P.E. of the two balls increases. (Adapted with permission from J. E. Brady and G. E. Humiston, *General Chemistry: Principles and Structure*, 1st ed., Wiley, New York, p. 18.)

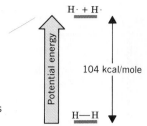

FIGURE 1.12 The relative potential energies of hydrogen atoms and hydrogen molecules.

potential energy. The *more* potential energy an object has, the *less stable* it is. Consider, as an example, the relative potential energy and the relative stability of snow when it lies high on a mountainside and when it lies serenely in the valley below. Because of the attractive force of gravity, the snow high on the mountain *has greater potential energy and is much less stable* than the snow in the valley. This greater potential energy of the snow on the mountainside can become converted to the enormous kinetic energy of an avalanche. By contrast, the snow in the valley with its lower potential energy and with its greater stability is incapable of releasing such energy.

1.13A Potential Energy and Covalent Bonds

Atoms and molecules possess potential energy — often called chemical energy — that can be released as heat when they react. Because heat is associated with molecular motion, this release of heat results from a change from potential energy to kinetic energy.

From the standpoint of covalent bonds, the state of greatest potential energy is the state of free atoms, the state in which the atoms are not bonded to each other at all. This is true because the formation of a chemical bond is always accompanied by the lowering of the potential energy of the atoms (cf. Fig. 1.7). Consider as an example the formation of hydrogen molecules from hydrogen atoms:

$$H \cdot + H \cdot \longrightarrow H—H + 104 \text{ kcal/mole} \quad (435 \text{ kJ/mole})*$$

The potential energy of the atoms decreases by 104 kcal/mole as the covalent bonds form. This potential energy change is illustrated graphically in Fig. 1.12.

A convenient way to represent the relative potential energies of molecules is in terms of their relative **enthalpies** or **heat contents,** *H.* (*Enthalpy* comes from the German word *enthalten* meaning to contain.) The difference in relative enthalpies of reactants and products in a chemical change is called the enthalpy change and is symbolized by $\Delta H°$. [The Δ (delta) in front of a quantity usually means the difference, or change, in the quantity. The superscript ° indicates that the measurement is made under standard conditions.]

By convention, the sign of $\Delta H°$ for **exothermic** reactions (those evolving heat) is negative. **Endothermic** reactions (those which absorb heat) have a positive $\Delta H°$. The heat of reaction, $\Delta H°$, measures the change in enthalpy of the atoms of the reactants as they are converted to products. For an exothermic reaction the atoms have a smaller enthalpy as products than they do as reactants. For endothermic reactions, the reverse is true.

---

*A kilocalorie of energy (1000 cal) is the amount of energy in the form of heat required to raise by 1 °C the temperature of 1 kg (1000 g) of water at 15 °C. The unit of energy in SI units is the joule, J, and 1 cal = 4.184 J. (Thus 1 kcal = 4.184 kJ.)

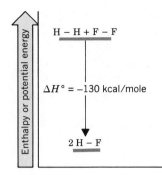

FIGURE 1.13 The energy change that accompanies the reaction $H_2 + F_2 \longrightarrow 2HF$.

We express the exothermic reaction of hydrogen atoms to form hydrogen molecules this way:

$$H\cdot + H\cdot \longrightarrow H{-}H \quad \Delta H° = -104 \text{ kcal/mole} \quad (-435 \text{ kJ/mole})$$

The reverse reaction is endothermic. Energy has to be supplied to break the covalent bonds:

$$H{-}H \longrightarrow H\cdot + H\cdot \quad \Delta H° = +104 \text{ kcal/mole} \quad (+435 \text{ kJ/mole})$$

The covalent bond of a fluorine molecule is weaker than that of a hydrogen molecule. Less energy is released when it forms and, consequently, less energy is required to break it:

$$F\cdot + F\cdot \longrightarrow F{-}F \quad \Delta H° = -38 \text{ kcal/mole} \quad (-159 \text{ kJ/mole})$$

$$F{-}F \longrightarrow F\cdot + F\cdot \quad \Delta H° = +38 \text{ kcal/mole} \quad (+159 \text{ kJ/mole})$$

The bond of hydrogen fluoride is very strong:

$$H\cdot + F\cdot \longrightarrow H{-}F \quad \Delta H° = -136 \text{ kcal/mole} \quad (-569 \text{ kJ/mole})$$

$$H{-}F \longrightarrow H\cdot + F\cdot \quad \Delta H° = +136 \text{ kcal/mole} \quad (+569 \text{ kJ/mole})$$

When molecules react with each other, the reaction can be either exothermic or endothermic. Generally speaking, if the bonds of the products are collectively stronger than those of the reactants, the reaction will be exothermic. An example is the reaction of hydrogen with fluorine:

$$H{-}H + F{-}F \longrightarrow 2H{-}F \quad \Delta H° = -130 \text{ kcal/mole} \quad (-544 \text{ kJ/mole})$$

Because the bonds of the H—F molecules are collectively stronger than those of hydrogen and fluorine, the reaction evolves heat. The heat of reaction or enthalpy change is negative. The product molecules with their stronger bonds have lower potential energy than those of the reactants (Fig. 1.13).

1.14 ORBITAL HYBRIDIZATION AND THE STRUCTURE OF METHANE: sp^3 HYBRIDIZATION

The s and p orbitals used in the quantum mechanical description of the carbon atom, given in Section 1.11, were based on calculations for hydrogen atoms. These simple s and p orbitals do not, when taken alone, provide a satisfactory explanation for the *tetravalent-tetrahedral* carbon of methane. However, a satisfactory description of methane's structure that is based on quantum mechanics *can* be obtained through an approach called **orbital hybridization.** Orbital hybridization, in its simplest terms, is nothing more than a mathematical approach that involves the combining of individual wave functions for *s* and *p* orbitals to obtain wave functions for new orbitals. The new orbitals have, *in varying proportions,* the properties of the original orbitals taken separately. These new orbitals are called **hybrid orbitals.**

According to quantum mechanics the electronic configuration of a carbon atom in its lowest energy state—called the *ground state*—is that given here.

$$C \quad \underline{\uparrow\downarrow} \quad \underline{\uparrow\downarrow} \quad \underline{\uparrow} \quad \underline{\uparrow} \quad \underline{}$$
$$\quad 1s \quad 2s \quad 2p \quad 2p \quad 2p$$

Ground state of a carbon atom

The valence electrons of a carbon atom (those used in bonding) are those of the *outer level,* that is, the 2*s* and 2*p* electrons.

Hybrid orbitals that account for methane's structure can be obtained by combining the wave functions of the 2*s* orbital of carbon with those of the three 2*p* orbitals. The mathematical process of hybridization can be approximated by the illustration that is shown in Fig. 1.14.

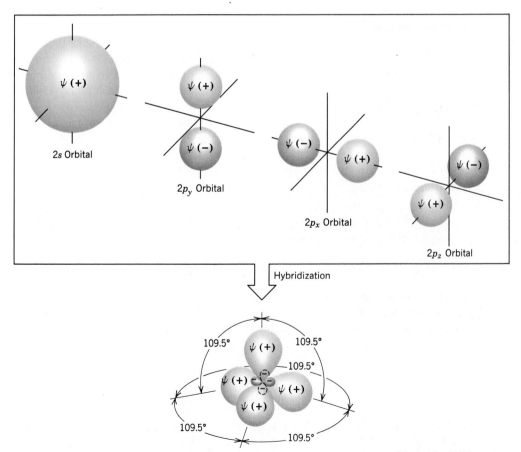

FIGURE 1.14 Hybridization of atomic orbitals of a carbon atom to produce *sp*³-hybrid orbitals.

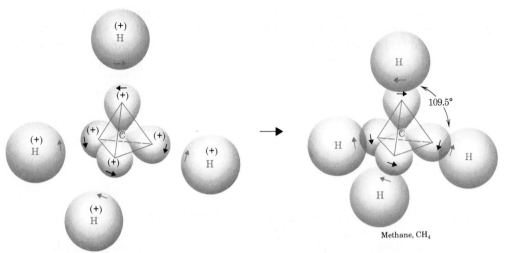

FIGURE 1.15 The formation of methane from an sp^3-hybridized carbon atom. In orbital hybridization we combine orbitals, *not* electrons. The electrons can then be replaced in the hybrid orbitals as necessary for bond formation, but always in accordance with the Pauli principle of no more than two electrons (with opposite spin) in each orbital. In this illustration we have placed one electron in each of the hybrid carbon orbitals. In this illustration, too, we have shown only the bonding molecular orbital of each C—H bond because these are the orbitals that contain the electrons in the lowest energy state of the molecule.

In this process, four orbitals are mixed—or hybridized—and four new hybrid orbitals are obtained. The hybrid orbitals are called sp^3 orbitals to indicate that they have one part the character of an *s* orbital and three parts the character of a *p* orbital. The mathematical treatment of orbital hybridization also shows that *the four sp^3 orbitals should be oriented at angles of 109.5° with respect to each other.* This is precisely the spatial orientation of the four hydrogen atoms of methane.

If, in our imagination, we visualize the formation of methane from an sp^3-hybridized carbon atom and four hydrogen atoms, the process might be like that shown in Fig. 1.15. For simplicity we show only the formation of the *bonding molecular orbital* for each carbon–hydrogen bond. We see that an sp^3-hybridized carbon gives a *tetrahedral structure for methane, and one with four equivalent C—H bonds.*

PROBLEM 1.7

(a) Consider a carbon atom in its ground state. Would such an atom offer a satisfactory model for the carbon of methane? If not, why not? (*Hint:* Consider whether or not a ground state carbon atom could be tetravalent, and consider the bond angles that would result if it were to combine with hydrogen atoms.)
(b) What about a carbon atom in the excited state:

C ⥮ ↑   ↑   ↑   ↑

1s   2s $2p_x$ $2p_y$ $2p_z$

Excited state of a carbon atom

Would such an atom offer a satisfactory model for the carbon of methane? If not, why not?

In addition to accounting properly for the shape of methane, the theory of orbital hybridization also explains the very strong bonds that are formed between carbon and hydrogen. To see how this is so, consider the shape of the individual sp^3 orbital shown in the following figure:

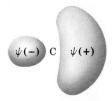

Because the sp^3 orbital has the character of a p orbital, the positive lobe of the sp^3 orbital is large and is extended quite far into space.

It is the positive lobe of the sp^3 orbital that overlaps with the positive $1s$ orbital of hydrogen to form the bonding molecular orbital of a carbon–hydrogen bond. Because the positive lobe of the sp^3 orbital is large and is extended into space, the

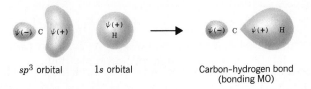

overlap between it and the $1s$ orbital of hydrogen is also large, and the resulting carbon–hydrogen bond is quite strong.

The bond formed from the overlap of an sp^3 orbital and a $1s$ orbital is an example of a **sigma bond** (Fig. 1.16). The term *sigma bond* is a general term applied to those bonds in which orbital overlap gives a bond that is *circularly symmetrical in cross section when viewed along the bond axis*. ***All purely single bonds are sigma bonds.***

> From this point on we shall often show only the bonding molecular orbitals because they are the ones that contain the electrons when the molecule is in its lowest energy state. Consideration of antibonding orbitals is important when a molecule absorbs light and in explaining certain reactions. We shall point these instances out later.

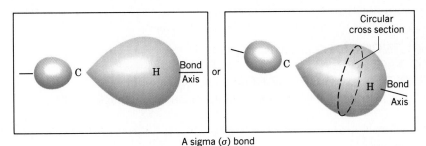

A sigma (σ) bond

FIGURE 1.16 A sigma (σ) bond.

1.15 ORBITAL HYBRIDIZATION AND THE STRUCTURE OF BORON TRIFLUORIDE: sp^2 HYBRIDIZATION

Boron trifluoride (in the following figure) has a triangular (trigonal planar) shape with three equivalent boron–fluorine bonds. In its ground state the boron atom has the following electronic configuration:

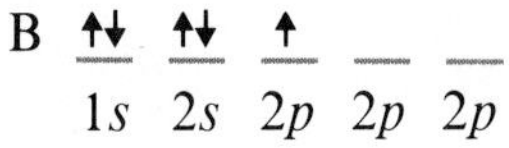

B ↿⇂ ↿⇂ ↿ __ __
1s 2s 2p 2p 2p

Boron atom ground state

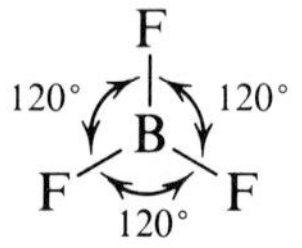

F
120° / | ↘ 120°
B
F ↘ ↙ F
120°

Triangular structure of BF₃

Clearly, the s and p orbitals of the ground state will not account for the trivalent and triangularly bonded boron of BF_3.

PROBLEM 1.8

(a) What valence would you expect a boron atom in its ground state to have?
(b) Consider an excited state of boron in which one $2s$ electron is promoted to a vacant $2p$ orbital. Show how this state of boron also fails to account for the structure of BF_3.

Once again we use orbital hybridization. Here, however, we combine the $2s$ orbital with only two of the $2p$ orbitals. Mixing three orbitals as shown in Fig. 1.17 gives three equivalent hybrid orbitals and these orbitals are sp^2 orbitals. They have one part the character of an s orbital and two parts the character of a p orbital. Calculations show that these orbitals are pointed toward the corners of an equilateral triangle with angles of 120° between their axes. These orbitals, then, are just what we need to account for the trivalent, trigonal planar boron atom of boron trifluoride.

By placing one of the valence electrons in each of the three sp^2 orbitals and allowing these orbitals to overlap with an orbital containing one electron from each of three fluorine atoms, we obtain the structure shown in Fig. 1.18. Notice that the boron atom still has a vacant p orbital, the one that we did not hybridize.

We shall see in Section 2.4 that sp^2 hybridization occurs with carbon atoms that form double bonds.

1.16 ORBITAL HYBRIDIZATION AND THE STRUCTURE OF BERYLLIUM HYDRIDE: sp HYBRIDIZATION

Beryllium hydride (BeH_2) is a linear molecule; the bond angle is 180°.

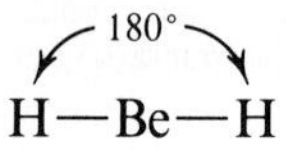

180°
H—Be—H

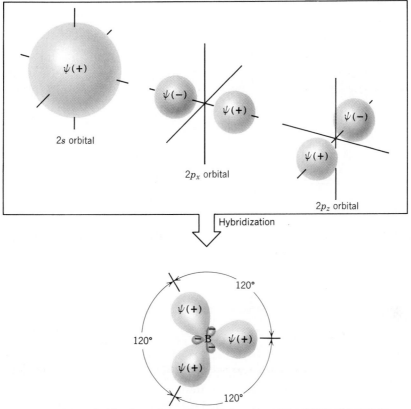

FIGURE 1.17 Hybridization of one $2s$ orbital and two $2p$ orbitals of boron to produce three sp^2-hybrid orbitals.

In its ground state the beryllium atom has the following electronic configuration:

$$\text{Be} \quad \underline{\uparrow\downarrow} \quad \underline{\uparrow\downarrow} \quad \underline{\quad} \ \underline{\quad} \ \underline{\quad}$$
$$\qquad\ \ 1s \quad 2s \quad 2p \ \ 2p \ \ 2p$$

In order to account for the structure of BeH_2 we again need orbital hybridization. Here (Fig. 1.19) we hybridize one s orbital with one p orbital and obtain two sp orbitals. Calculations show that these sp orbitals are oriented at an angle of $180°$. The

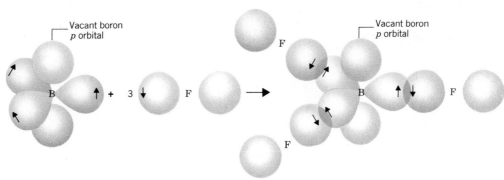

FIGURE 1.18 The formation of the bonding MO's of boron trifluoride from an sp^2-hybridized boron atom and three fluorine atoms.

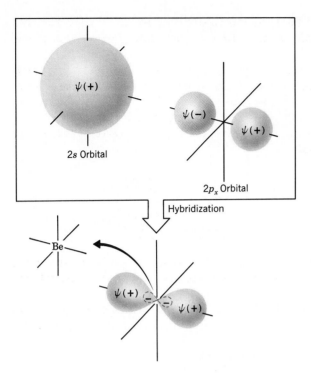

FIGURE 1.19 Hybridization of one 2s orbital and one 2p orbital of beryllium to produce two *sp*-hybrid orbitals.

two *p* orbitals that were not mixed are vacant. Beryllium can use these hybrid orbitals to form bonds to two hydrogen atoms in the way shown in Fig. 1.20.

We shall see in Section 6.3 that *sp* hybridization occurs with carbon atoms that form triple bonds.

1.17 A SUMMARY OF IMPORTANT CONCEPTS THAT COME FROM QUANTUM MECHANICS

1. An **atomic orbital (AO)** corresponds to a region of space about the nucleus of a single atom where there is a high probability of finding an electron. Atomic orbitals called *s* orbitals are spherical; those called *p* orbitals are like two almost-touching spheres. Orbitals can hold a maximum of two electrons when

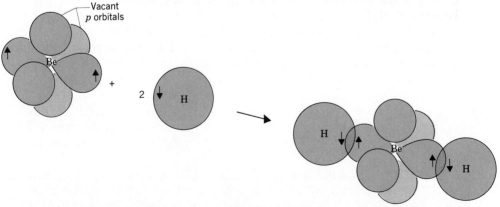

FIGURE 1.20 The formation of the bonding MO's of BeH_2 from an *sp*-hybridized beryllium atom and two hydrogen atoms.

their spins are paired. Orbitals are described by a wave function, ψ, and each orbital has a characteristic energy. The phase signs associated with an orbital may be $(+)$ or $(-)$.

2. When atomic orbitals overlap, they combine to form **molecular orbitals (MO's).** Molecular orbitals correspond to regions of space encompassing two (or more) nuclei where electrons are to be found. Like atomic orbitals, molecular orbitals can hold up to two electrons if their spins are paired.

3. When atomic orbitals with the same phase sign interact they combine to form a **bonding molecular orbital:**

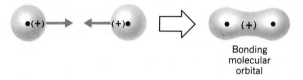

Bonding
molecular
orbital

The electron probability density of a bonding molecular orbital is large in the region of space between the two nuclei where the negative electrons hold the positive nuclei together.

4. An **antibonding molecular orbital** forms when orbitals of opposite phase sign overlap:

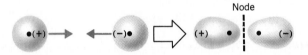

An antibonding orbital has higher energy than a bonding orbital. The electron probability density of the region between the nuclei is small and it contains a **node**—a region where $\psi = 0$. Thus, having electrons in an antibonding orbital does not help hold the nuclei together. The internuclear repulsions tend to make them fly apart.

5. The **energy of electrons** in a bonding molecular orbital is less than the energy of the electrons in their separate atomic orbitals. The energy of electrons in an antibonding orbital is greater than that of electrons in their separate atomic orbitals.

6. The **number of molecular orbitals** always equals the number of atomic orbitals from which they are formed. Combining two atomic orbitals will always yield two molecular orbitals—one bonding and one antibonding.

7. **Hybrid orbitals** form by mixing (hybridizing) the wave functions for orbitals of a different type (i.e., s and p orbitals) but from the same atom.

8. Hybridizing three p orbitals with one s orbital yields four sp^3 **orbitals.** Atoms that are sp^3 hybridized direct the axes of their four sp^3 orbitals toward the corners of a tetrahedron. The carbon of methane is sp^3 hybridized and **tetrahedral.**

9. Hybridizing two p orbitals with one s orbital yields three sp^2 **orbitals.** Atoms that are sp^2 hybridized point the axes of three sp^2 orbitals toward the corners of an equilateral triangle. The boron atom in BF_3 is sp^2 hybridized and **trigonal planar.**

10. Hybridizing one p orbital with one s orbital yields two sp **orbitals.** Atoms that are sp hybridized orient the axes of their two sp orbitals in opposite directions (at an angle of $180°$). The beryllium atom of BeH_2 is sp hybridized and BeH_2 is a **linear** molecule.

11. **A sigma bond** (a type of single bond) is one in which the electron density has circular symmetry when viewed along the bond axis. In general, the skeletons of organic molecules are constructed of atoms linked by sigma bonds.

1.18 MOLECULAR GEOMETRY: THE VALENCE SHELL ELECTRON-PAIR REPULSION (VSEPR) MODEL

We have been discussing the geometry of molecules on the basis of theories that arise from quantum mechanics. It is possible, however, to predict the arrangement of atoms in molecules and ions on the basis of a theory called the **valence shell electron-pair repulsion (VSEPR) theory.** Consider the following examples found in Sections 1.18A to 1.18F.

We apply VSEPR theory in the following way:

1. We consider molecules (or ions) in which the central atom is covalently bonded to two or more atoms or groups.
2. We consider all of the electron pairs of the central atom — both those that are shared in covalent bonds, called **bonding pairs,** and those that are unshared, called **nonbonding pairs** or **unshared pairs.**
3. Because electron pairs repel each other, the electron pairs of the valence shell tend to stay as far apart as possible.
4. We arrive at the geometry of the molecule by considering all of the electron pairs, bonding and nonbonding, but we describe the shape of the molecule or ion by referring to the positions of the nuclei (or atoms) and not by the positions of the electron pairs.

Consider the following examples.

1.18A Methane

The valence shell of methane contains four pairs of bonding electrons. Only a tetrahedral orientation will allow four pairs of electrons to have the maximum possible separation (Fig. 1.21). Any other orientation, for example, a square planar arrangement, places the electron pairs closer together.

Thus, in the case of methane, the valence shell electron-pair repulsion model accommodates what we have known since the discovery of van't Hoff and Le Bel (Section 1.5): The molecule of methane has a tetrahedral shape.

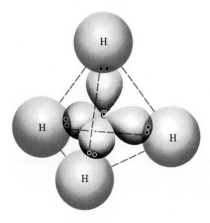

FIGURE 1.21 A tetrahedral shape for methane allows the maximum separation of the four bonding electron pairs.

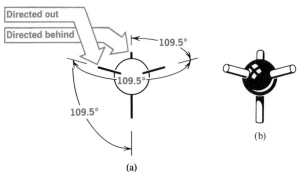

FIGURE 1.22 (*a*) Line-and-circle representation of a tetrahedral atom. (*b*) Ball-and-stick model.

PROBLEM 1.9

Part of the reasoning that led van't Hoff and Le Bel to propose a tetrahedral shape for molecules of methane was based on the number of compounds that are theoretically possible for substituted methanes, that is, for compounds in which one or more hydrogen atoms of methane have been replaced by some other group. For example, only one compound of the type CH_2X_2 has ever been found. (a) Is this consistent with a tetrahedral shape? (b) With a square planar shape? Explain.

The bond angles for any atom that has a regular tetrahedral structure are 109.5°. One way of representing a tetrahedral atom is shown in Fig. 1.22. This representation is derived from the familiar ball-and-stick model. The lines that intersect the circle are directed out of the plane of the paper.

1.18B Ammonia

The geometry of a molecule of ammonia is a **trigonal pyramid.** The bond angles in a molecule of ammonia are 107°, a value very close to the tetrahedral angle (109.5°). We can write a general tetrahedral structure for the electron pairs of ammonia by placing the nonbonding pair at one corner (Fig. 1.23). A *tetrahedral arrangement* of the electron pairs explains the *trigonal pyramidal* arrangement of the four atoms.

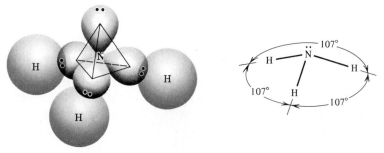

FIGURE 1.23 The tetrahedral arrangement of the electron pairs of an ammonia molecule that results when the nonbonding electron pair is considered to occupy one corner. This arrangement of electron pairs explains the trigonal pyramidal shape of the NH_3 molecule.

1.18C Water

A molecule of water has an **angular** or **bent geometry.** The H—O—H bond angle in a molecule of water is 105°, an angle that is also quite close to the 109.5° bond angles of methane.

We can write a general tetrahedral structure for the electron pairs of a molecule of water *if we place the two nonbonding electron pairs at corners of the tetrahedron.* Such a structure is shown in Fig. 1.24. A *tetrahedral arrangement* of the electron pairs accounts for the *angular arrangement* of the three atoms.

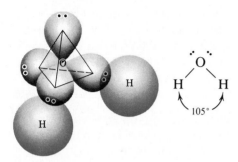

FIGURE 1.24 A tetrahedral arrangement of the electron pairs of a molecule of water that results when the pairs of nonbonding electrons are considered to occupy corners. This arrangement accounts for the angular shape of the H_2O molecule.

1.18D Boron Trifluoride

Boron, a Group **IIIA** element, has only three outer level electrons. In the compound boron trifluoride (BF_3) these three electrons are shared with three fluorine atoms. As a result, the boron atom in BF_3 has only six electrons (three bonding pairs) around it. Maximum separation of three bonding pairs occurs when they occupy the corners of an equilateral triangle. Consequently, in the boron trifluoride molecule the three fluorine atoms lie in a plane at the corners of an equilateral triangle (Fig. 1.25). Boron trifluoride is said to have a *trigonal planar structure.* The bond angles are 120°.

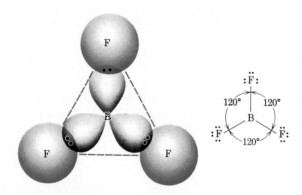

FIGURE 1.25 The triangular (trigonal planar) shape of boron fluoride maximally separates the three bonding pairs.

1.18E Beryllium Hydride

The central beryllium atom of BeH_2 has only two electron pairs around it; both electron pairs are bonding pairs. These two pairs are maximally separated when they are on opposite sides of the central atom as shown in the following structures. This arrangement of the electron pairs accounts for the linear geometry of the BeH_2 molecule and its bond angle of 180°.

$$H\!:\!Be\!:\!H \quad \text{or} \quad \overset{\frown{\ 180°\ }}{H\!-\!Be\!-\!H}$$

Linear geometry of BeH$_2$

PROBLEM 1.10

Predict the general shapes of the following molecules and ions:

(a) SiH_4 (c) CH_3^+ (e) CCl_4 (g) BH_4^- (i) BH_3

(b) $:CH_3^-$ (d) BF_4^- (f) $BeCl_2$ (h) BeH_2

1.18F Carbon Dioxide

The valence shell electron-pair repulsion method can also be used to predict the shapes of molecules containing multiple bonds if we assume that *all of the electrons of a multiple bond act as though they were a single unit,* and, therefore, are located in the region of space between the two atoms joined by a multiple bond.

    This principle can be illustrated with the structure of a molecule of carbon dioxide (CO_2). The central carbon atom of carbon dioxide is bonded to each oxygen atom by a double bond. Carbon dioxide is known to have a linear shape; the bond angle is 180°.

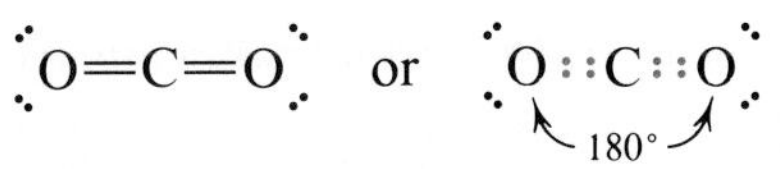

$$\ddot{O}=C=\ddot{O} \quad or \quad \ddot{O}::C::\ddot{O}$$

The four electrons of each double bond act as a single unit and are maximally separated from each other

Such a structure is consistent with a maximum separation of the two groups of four bonding electrons. (The nonbonding pairs associated with the oxygen atoms have no effect on the shape.)

PROBLEM 1.11

Predict the bond angles of   (a) $CH_2=CH_2$,   (b) $HC\equiv CH$,   (c) $HC\equiv N$, (d) SO_2, and   (e) SO_3.

    The shapes of several simple molecules and ions as predicted by VSEPR theory are shown in Table 1.4. In this table we have also included the hybridization state of the central atom.

TABLE 1.4 Shapes of molecules and ions from VSEPR theory

NUMBER OF ELECTRON PAIRS			HYBRIDIZATION STATE OF	SHAPE OF MOLECULE	
Bonding	Nonbonding	Total	CENTRAL ATOM	OR ION*	EXAMPLES
2	0	2	sp	Linear	BeH_2
3	0	3	sp^2	Trigonal planar	BF_3, CH_3^+
4	0	4	sp^3	Tetrahedral	CH_4, NH_4^+
3	1	4	$\sim sp^3$	Trigonal pyramidal	NH_3, CH_3^-
2	2	4	$\sim sp^3$	Angular	H_2O

*Excluding nonbonding pairs.

TABLE 1.5 Dipole moments of some simple molecules

FORMULA	μ (D)	FORMULA	μ (D)
H_2	0	CH_4	0
Cl_2	0	CH_3Cl	1.87
HF	1.91	CH_2Cl_2	1.55
HCl	1.08	$CHCl_3$	1.02
HBr	0.80	CCl_4	0
HI	0.42	NH_3	1.47
BF_3	0	NF_3	0.24
CO_2	0	H_2O	1.85

1.19 POLAR AND NONPOLAR MOLECULES

In our discussion of dipole moments in Section 1.9, we restricted our attention to simple diatomic molecules. Any *diatomic* molecule in which the two atoms are *different* (and thus have different electronegativities) will, of necessity, have a dipole moment. If we examine Table 1.5, however, we find that a number of molecules (e.g., CCl_4, CO_2) consist of more than two atoms, have *polar* bonds, *but have no dipole moment.* Now that we have an understanding of the shapes of molecules we can understand how this can occur.

Consider a molecule of carbon tetrachloride (CCl_4). Because the electronegativity of chlorine is greater than that of carbon, each of the carbon–chlorine bonds in CCl_4 is polar. Each chlorine atom has a partial negative charge, and the carbon atom is considerably positive. Because a molecule of carbon tetrachloride is tetrahedral (Fig. 1.26), however, *the center of positive charge and the center of negative charge coincide, and the molecule has no net dipole moment.*

This result can be illustrated in a slightly different way: If we use arrows ($\longmapsto$) to represent the direction of polarity of each bond, we get the arrangement of bond moments shown in Fig. 1.27. Since the bond moments are vectors of equal magnitude arranged tetrahedrally, their effects cancel. Their vector sum is zero. The molecule has *no net dipole moment.*

The chloromethane molecule (CH_3Cl) has a net dipole moment of 1.87 D. Since carbon and hydrogen have electronegativities (Table 1.2) that are nearly the same, the contribution of three C—H bonds to the net dipole is negligible. The electronegativity difference between carbon and chlorine is large, however, and this highly polar C—Cl bond accounts for most of the dipole moment of CH_3Cl (Fig. 1.28).

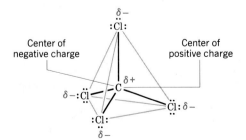

FIGURE 1.26 Charge distribution in carbon tetrachloride.

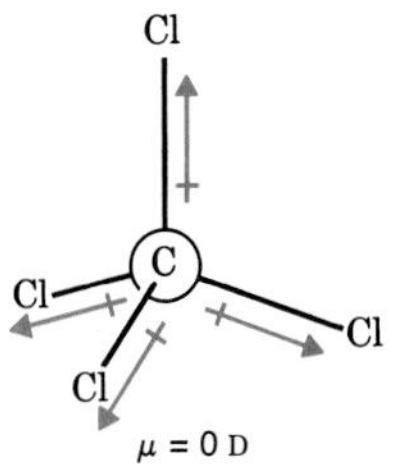

Cl
Cl—C—Cl
Cl
$\mu = 0$ D

FIGURE 1.27 A tetrahedral orientation of equal bond moments causes their effects to cancel.

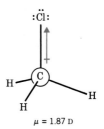

:Cl:
H—C—H
H
$\mu = 1.87$ D

FIGURE 1.28 The dipole moment of chloromethane arises mainly from the highly polar carbon–chlorine bond.

PROBLEM 1.12

A molecule of carbon dioxide (CO_2) is linear (Section 1.18F). Show how this accounts for the fact that CO_2 has no dipole moment.

PROBLEM 1.13

Sulfur dioxide (cf. Problem 1.4) has a dipole moment. (The dipole moment of SO_2 is 1.63 D.) What does this fact confirm about the shape of an SO_2 molecule that you predicted in Problem 1.11?

Unshared pairs of electrons make large contributions to the dipole moments of water and ammonia. Because an unshared pair has no atom attached to it to partially neutralize its negative charge, an unshared electron pair contributes a large moment directed away from the central atom (Fig. 1.29). (The O—H and N—H moments are also appreciable.)

PROBLEM 1.14

Nitrogen trifluoride ($:NF_3$) has a shape very much like that of ammonia. It has, however, a very low dipole moment ($\mu = 0.24$ D). How can you explain this?

PROBLEM 1.15

Boron trifluoride (BF_3) has no dipole moment. How can this be explained?

FIGURE 1.29 Bond moments and the resulting dipole moment of water and ammonia.

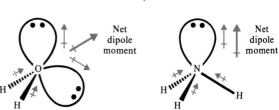

1.20 THE REPRESENTATION OF STRUCTURAL FORMULAS

Organic chemists use a variety of ways to write structural formulas. The most common types of formulas are shown in Fig. 1.30. The **dot structure** shows all of the valence electrons, but writing it is tedious and time consuming. The other formulas are more convenient and are, therefore, more often used.

In fact, we often omit unshared pairs when we write formulas unless there is a reason to include them. For example,

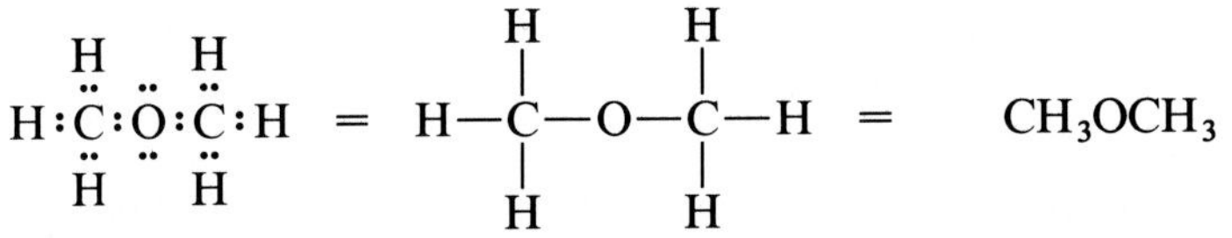

$$
\begin{array}{ccc}
\text{H} \quad \text{H} & \text{H} \quad\quad \text{H} & \\
\text{H:}\overset{..}{\text{C}}\text{:}\overset{..}{\text{O}}\text{:}\overset{..}{\text{C}}\text{:H} = & \text{H}-\overset{|}{\underset{|}{\text{C}}}-\text{O}-\overset{|}{\underset{|}{\text{C}}}-\text{H} = & CH_3OCH_3 \\
\text{H} \quad \text{H} & \text{H} \quad\quad \text{H} & \\
\text{Dot structure} & \text{Dash formula} & \text{Condensed formula}
\end{array}
$$

1.20A Dash Structural Formulas

If we look at the ball-and-stick model for propyl alcohol given in Fig. 1.30 and compare it with the formulas given there, we find that the chain of atoms is straight in all the formulas. In the model, which corresponds more accurately to the actual shape of the molecule, the chain of atoms is not at all straight. Also of importance is this: *Atoms joined by single bonds can rotate relatively freely with respect to one another.* (We discuss this point further in Section 2.2B.) This relatively free rotation means that the chain of atoms in propyl alcohol can assume a variety of arrangements like those that follow:

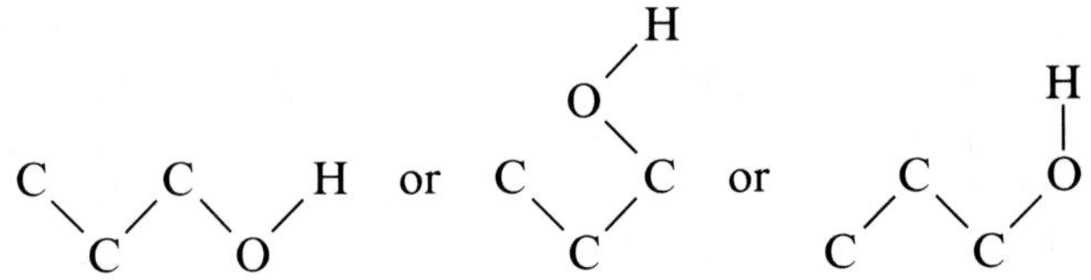

It also means that all of the dash structures that follow are *equivalent* and all represent propyl alcohol.

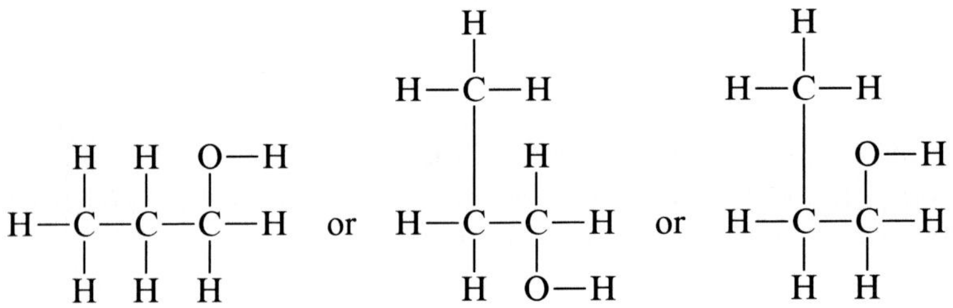

Equivalent dash formulas for propyl alcohol

Structural formulas show the way in which the atoms are attached to each other. They show what is called the **connectivity** of the atoms. *Constitutional isomers (Section 1.4) have different connectivity, and, therefore, must have different structural formulas.*

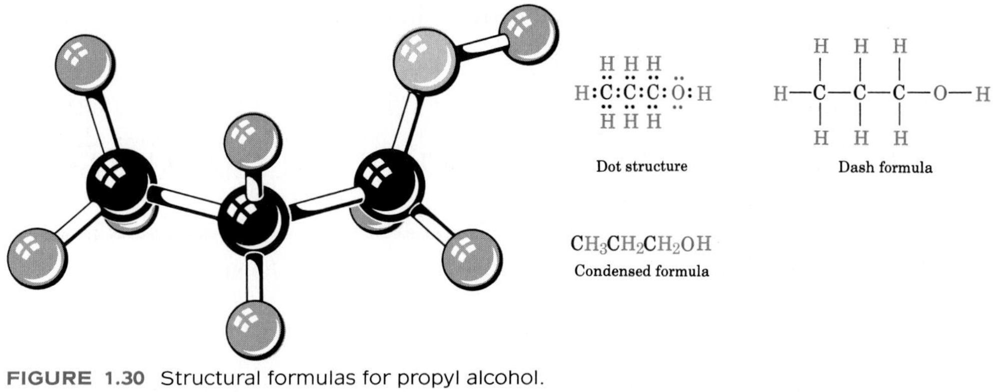

$$H : \overset{\overset{\displaystyle H}{|}}{\underset{\underset{\displaystyle H}{|}}{C}} : \overset{\overset{\displaystyle H}{|}}{\underset{\underset{\displaystyle H}{|}}{C}} : \overset{\overset{\displaystyle H}{|}}{\underset{\underset{\displaystyle H}{|}}{C}} : \ddot{\overset{..}{O}} : H$$

Dot structure

$$H - \overset{\overset{\displaystyle H}{|}}{\underset{\underset{\displaystyle H}{|}}{C}} - \overset{\overset{\displaystyle H}{|}}{\underset{\underset{\displaystyle H}{|}}{C}} - \overset{\overset{\displaystyle H}{|}}{\underset{\underset{\displaystyle H}{|}}{C}} - O - H$$

Dash formula

$CH_3CH_2CH_2OH$
Condensed formula

FIGURE 1.30 Structural formulas for propyl alcohol.

Consider the compound called isopropyl alcohol, whose formula we might write in a variety of ways:

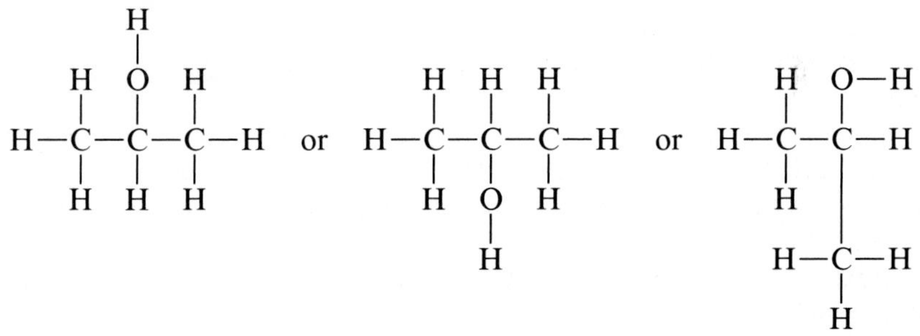

Equivalent dash formulas for isopropyl alcohol

Isopropyl alcohol is a constitutional isomer (Section 1.4) of propyl alcohol because its atoms are connected in a different order and both compounds have the same molecular formula, C_3H_8O. In isopropyl alcohol the O—H group is attached to the central carbon; in propyl alcohol it is attached to an end carbon.

One other point: In problems you will often be asked to write structural formulas for all the isomers with a given molecular formula. Do not make the error of writing several equivalent formulas, like those that we have just shown, mistaking them for different constitutional isomers.

PROBLEM 1.16

There are actually three constitutional isomers with the molecular formula C_3H_8O. We have seen two of them in propyl alcohol and isopropyl alcohol. Write a dash formula for the third isomer.

1.20B Condensed Structural Formulas

Condensed structural formulas are easier to write than dash formulas, and when we become familiar with them, they will impart all the information that is contained in the dash structure. In condensed formulas all of the hydrogen atoms that are attached to a particular carbon are written immediately after that carbon. In fully condensed formulas, all of the atoms that are attached to the carbon are usually written immediately after that carbon. For example,

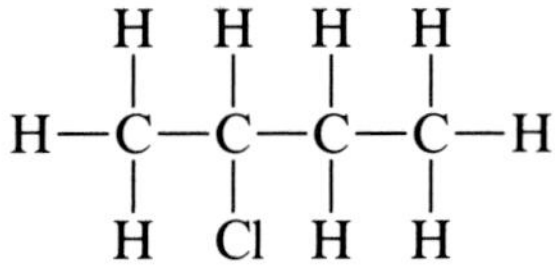

$$
\begin{array}{cccc}
H & H & H & H \\
| & | & | & | \\
H-C-C-C-C-H \\
| & | & | & | \\
H & Cl & H & H
\end{array}
\qquad
CH_3CHCH_2CH_3 \quad \text{or} \quad CH_3CHClCH_2CH_3
$$

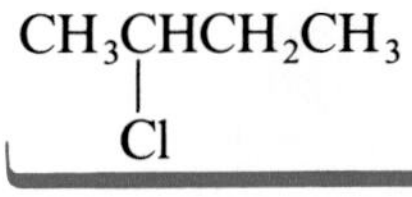

Dash formula   Condensed formulas

The condensed formula for isopropyl alcohol can be written in four different ways:

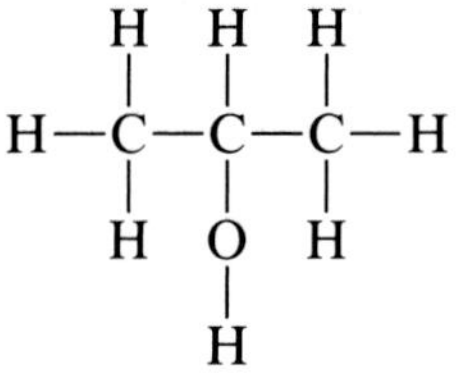

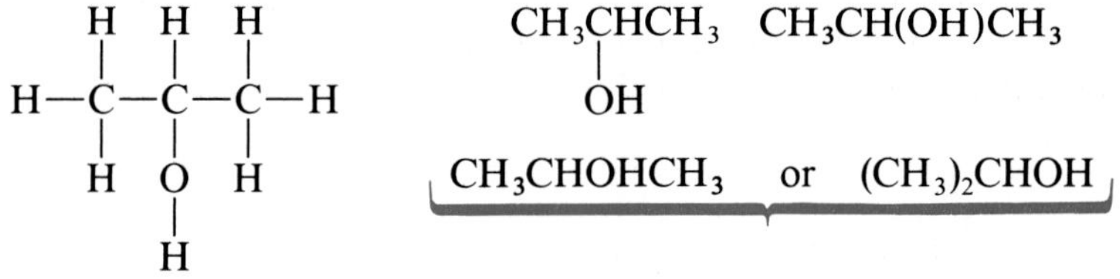

$$
\begin{array}{ccc}
H & H & H \\
| & | & | \\
H-C-C-C-H \\
| & | & | \\
H & O & H \\
 & | & \\
 & H &
\end{array}
\qquad
\begin{array}{l}
CH_3CHCH_3 \quad CH_3CH(OH)CH_3 \\
| \\
OH \\
\\
CH_3CHOHCH_3 \quad \text{or} \quad (CH_3)_2CHOH
\end{array}
$$

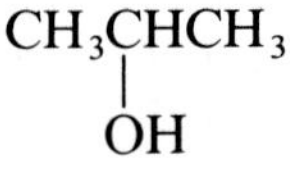

Dash formula   Condensed formulas

SAMPLE PROBLEM

Write a condensed structural formula for the compound that follows:

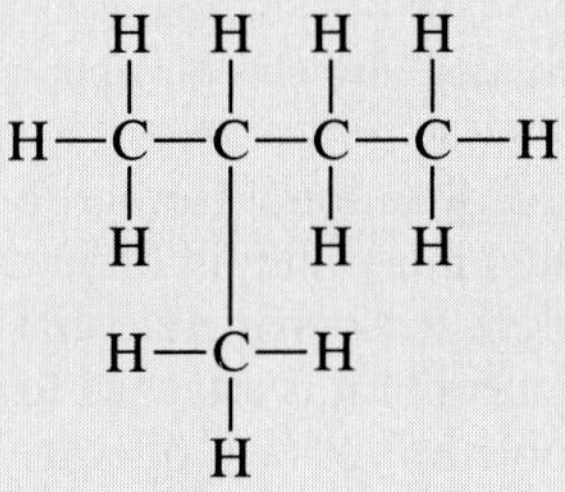

$$
\begin{array}{cccc}
H & H & H & H \\
| & | & | & | \\
H-C-C-C-C-H \\
| & | & | & | \\
H & | & H & H \\
 & H-C-H & & \\
 & | & & \\
 & H & &
\end{array}
$$

Answer:

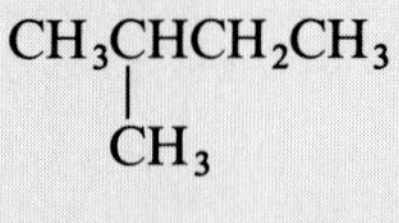

$$
\begin{array}{l}
CH_3CHCH_2CH_3 \quad \text{or} \quad CH_3CH(CH_3)CH_2CH_3 \quad \text{or} \quad (CH_3)_2CHCH_2CH_3 \\
| \\
CH_3
\end{array}
$$

$$
\text{or} \quad CH_3CH_2CH(CH_3)_2 \quad \text{or} \quad CH_3CH_2CHCH_3
$$
$$
\phantom{\text{or} \quad CH_3CH_2CH(CH_3)_2 \quad \text{or} \quad CH_3CH_2C}|
$$
$$
\phantom{\text{or} \quad CH_3CH_2CH(CH_3)_2 \quad \text{or} \quad CH_3CH_2C}CH_3
$$

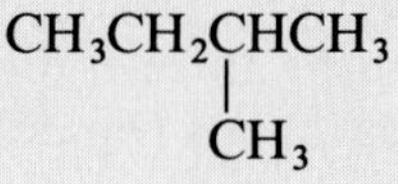

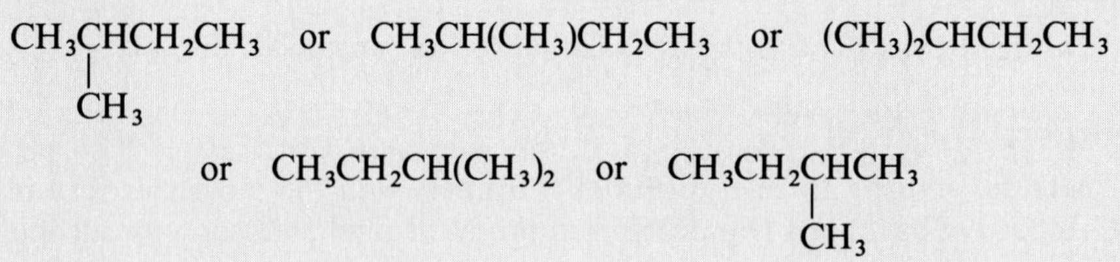

1.20C Cyclic Molecules

Organic compounds not only have their carbon atoms arranged in chains, they can also have them arranged in rings. The compound called cyclopropane has its carbon atoms arranged in a three-membered ring. (Cyclopropane has been used as an anesthetic.)

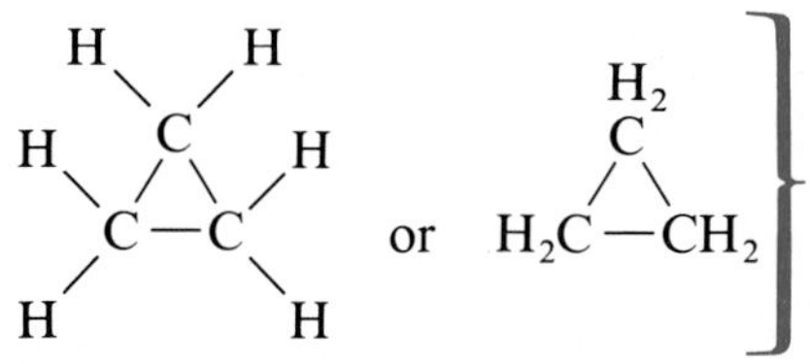

Formulas for cyclopropane

1.20D Bond-Line Formulas

More and more organic chemists are using a very simplified formula called a **bond-line formula** to represent structural formulas. The bond-line representation is the quickest of all to write because it shows only the carbon skeleton. The number of hydrogen atoms necessary to fulfill the carbon atoms' valences are assumed to be present, but we do not write them in. Other atoms (e.g., O, Cl, N) *are* written in. Each intersection of two or more lines and the end of a line represents a carbon atom unless some other atom is written in.

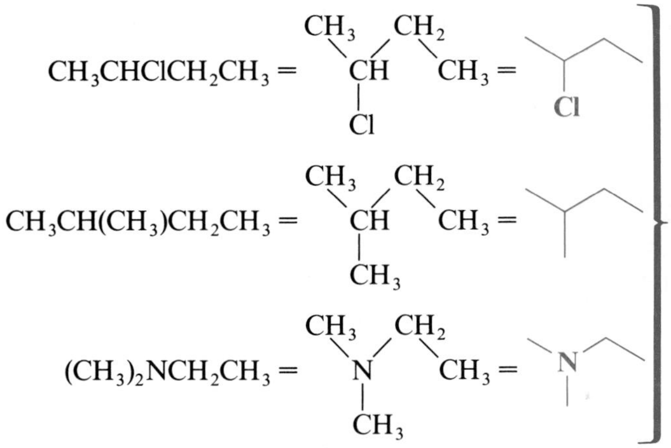

Bond-line formulas

Bond-line formulas are often used for cyclic compounds:

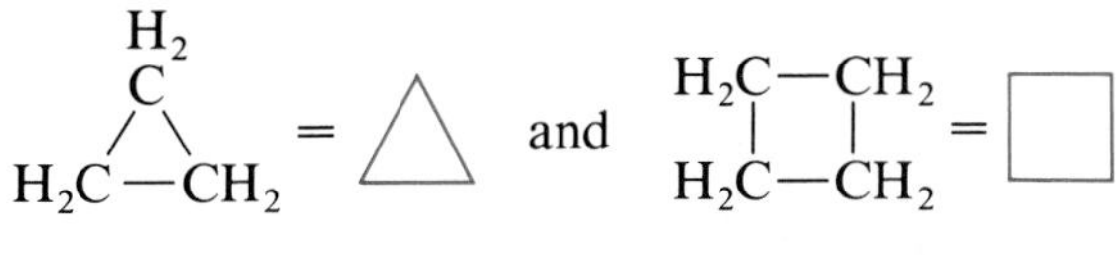

Multiple bonds are also indicated in bond-line formulas. For example,

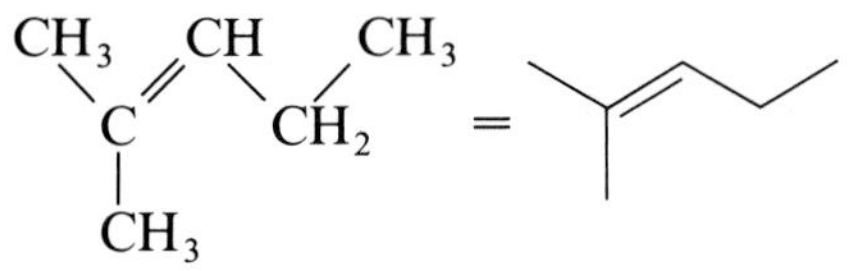

SAMPLE PROBLEM

Write the bond-line formula for $CH_3CHCH_2CH_2CH_2OH$.
$\qquad\qquad\qquad\quad |$
$\qquad\qquad\qquad CH_3$

Answer:
First, we outline the carbon skeleton, including the OH group as follows:

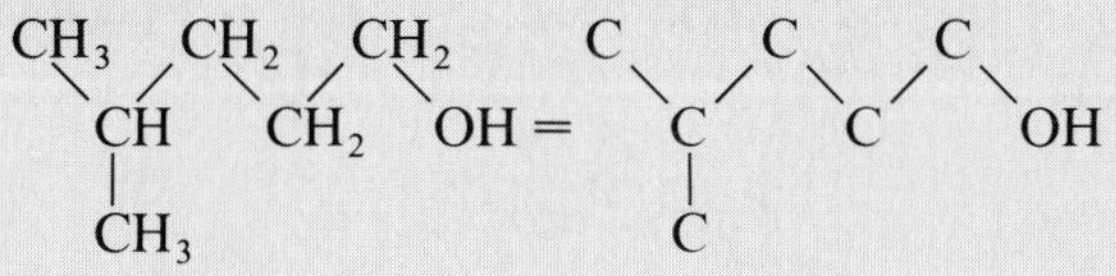

$$CH_3 \quad CH_2 \quad CH_2 \qquad C \qquad C \qquad C$$
$$CH \qquad CH_2 \quad OH = \quad C \qquad C \qquad OH$$
$$CH_3 \qquad\qquad C$$

Thus, the bond-line formula is

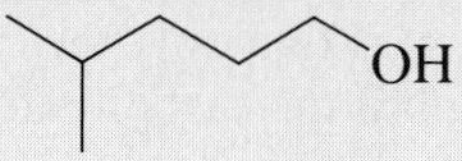

OH

PROBLEM 1.17

Rewrite each of the following condensed structural formulas, as *dash formulas* and as *bond-line formulas:*

(a) $CH_3CCl_2CH_2CH_3$ (f) $CH_3CH_2CH_2CH_2OH$

(b) $CH_3CH(CH_2Cl)CH_2CH_3$

(c) $(CH_3)_3CCH_2CH_3$

(d) $CH_3CHClCHClCH_3$

$$\quad\quad\quad\quad\quad\quad\quad\quad\quad\quad\quad O$$
$$\quad\quad\quad\quad\quad\quad\quad\quad\quad\quad\quad \|$$
(e) $CH_3CH(OH)CH_2CH_3$ (g) $CH_3CCH_2CH(CH_3)_2$

(h) $CH_3CH_2CH(OH)CH(CH_3)_2$

PROBLEM 1.18

Are any of the compounds listed in Problem 1.17 constitutional (structural) isomers of each other? If so, which ones are?

PROBLEM 1.19

Write dash formulas for each of the following bond-line formulas:

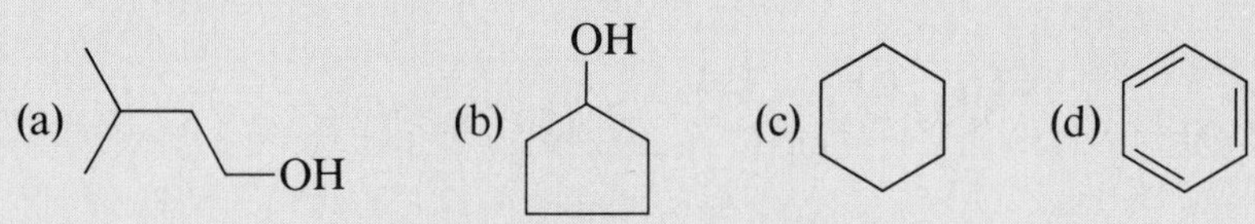

(a) OH   (b) OH   (c)   (d)

1.20E Three-Dimensional Formulas

None of the formulas that we have described so far tells us how the atoms are arranged in space. There are, however, two frequently used formulations that do impart the three-dimensional structure of a molecule. They are the Alexander (or circle-and-line formula) and the wedge-line–dashed wedge formula (Fig. 1.31). In wedge-line–dashed wedge formula, atoms that project out of the plane of the paper are connected

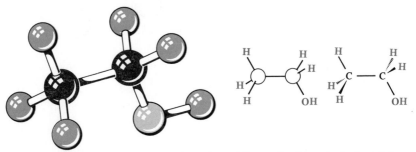

FIGURE 1.31 The circle-and-line formula and the wedge-line–dashed wedge formula for ethanol.

by a wedge (—►), those that lie behind the plane are connected with a dashed wedge (⸳⸳⸳), and those atoms in the plane of the paper are connected by a line.

PROBLEM 1.20

Write three-dimensional (wedge-line–dashed wedge) formulas for each of the following: (a) CH_3Cl, (b) CH_2Cl_2, (c) CH_2BrCl, (d) CH_3CH_2Cl.

ADDITIONAL PROBLEMS

1.21

Show an electron-dot formula, including any formal charge, for each of the following compounds:

(a) CH_3NCS (c) CH_3ONO_2 (e) CH_2CO (g) KNH_2 (i) CH_2O

(b) CH_3CNO (d) CH_3NCO (f) CH_2N_2 (h) NaN_3 (j) HCO_2H

1.22

(a) Write out the ground state electron configuration for each of the following atoms.
(b) Make a sketch of the atom showing the orbital arrangement, shape, and the disposition of the electrons in s and p orbitals.
(1) Be,   (2) B,   (3) C,   (4) N,   (5) O.

1.23

Give the formal charge (if one exists) on each atom of the following:

(a) CH_3—$\ddot{O}$—$\overset{\overset{\displaystyle \ddot{O}:}{\|}}{\underset{\underset{\displaystyle \ddot{O}:}{\|}}{S}}$—$\ddot{O}:$ (c) $:\ddot{O}$—$\overset{\overset{\displaystyle \ddot{O}:}{\|}}{\underset{\underset{\displaystyle \ddot{O}:}{\|}}{S}}$—$\ddot{O}:$

(b) CH_3—$\overset{\overset{\displaystyle :\ddot{O}:}{|}}{\underset{\underset{\displaystyle ..}{}}{S}}$—$CH_3$ (d) CH_3—$\overset{\overset{\displaystyle \ddot{O}:}{\|}}{\underset{\underset{\displaystyle \ddot{O}:}{\|}}{S}}$—$\ddot{O}:$

1.24

Write a condensed structural formula for each compound given here.

(a) 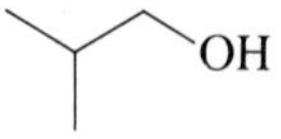 [structure: isopropyl-CH₂-OH branched]   (c) 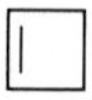 [cyclobutane square with double bond]

(b) 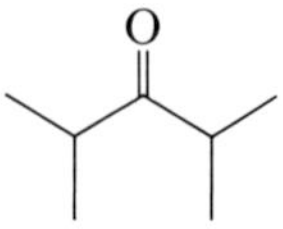 [structure with C=O ketone branched]   (d) 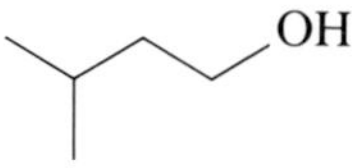 [structure with OH]

1.25

What is the molecular formula for each of the compounds given in Problem 1.24?

1.26

Consider each pair of structural formulas that follow and state whether the two formulas represent the same compound, whether they represent different compounds that are constitutional isomers of each other, or whether they represent different compounds that are not isomeric.

(a) 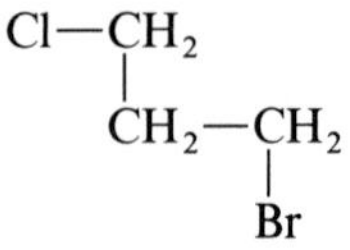 Cl—CH₂   and $\overset{\displaystyle H\quad H\quad H}{\underset{\displaystyle H\quad H\quad H}{H-C-C-C-Br}}$
$\quad\quad\ \ $ CH₂—CH₂
$\quad\quad\quad\quad\ $ Br

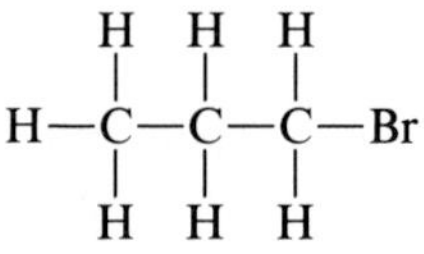

(b) 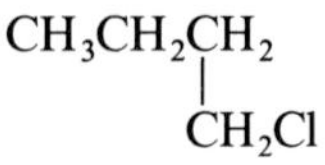 CH₃CH₂CH₂   and   ClCH₂CH(CH₃)₂
$\quad\quad\quad$ CH₂Cl

(c) $\overset{\displaystyle H}{\underset{\displaystyle Cl}{H-C-Cl}}$ and $\overset{\displaystyle H}{\underset{\displaystyle H}{Cl-C-Cl}}$

(d) $\overset{\displaystyle H\ \ H\ \ H}{\underset{\displaystyle H\ \ H}{F-C-C-C-H}}$ and   CH₂FCH₂CH₂CH₂F
$\quad\quad\quad\quad\quad\overset{\displaystyle H-C-F}{\underset{\displaystyle H}{\ }}$

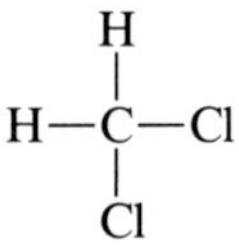

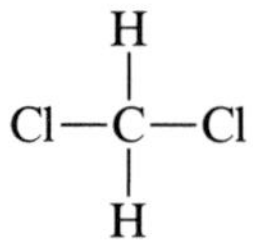

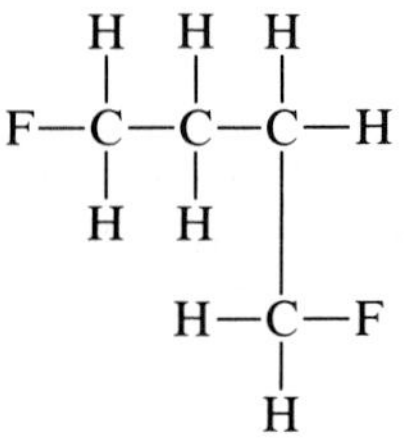

(e) $\overset{\displaystyle CH_3}{\underset{\displaystyle CH_3}{CH_3-C-CH_3}}$ and   (CH₃)₃C—CH₃

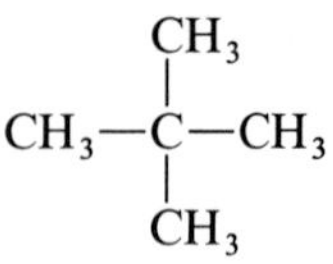

(f) CH₂=CHCH₂CH₃   and $\overset{\displaystyle CH_3}{\underset{\displaystyle H_2C-CH_2}{CH}}$

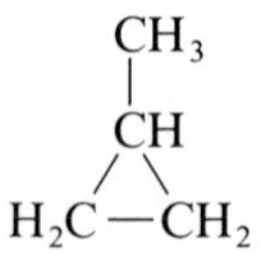

(g) CH₃OCH₂CH₃   and $\overset{\displaystyle O}{CH_3-C-CH_3}$

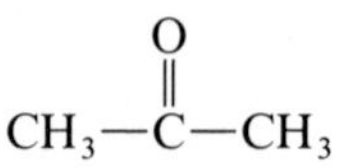

(h) 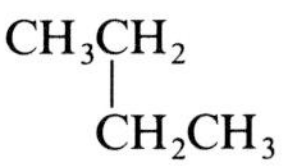 CH$_3$CH$_2$ and   CH$_3$CH$_2$CH$_2$CH$_3$

   |
   CH$_2$CH$_3$

(i) CH$_3$OCH$_2$CH$_3$ and

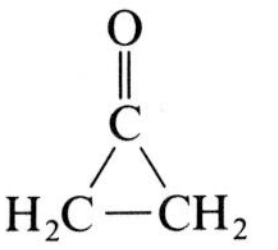

$$\underset{H_2C-CH_2}{\overset{\overset{\textstyle O}{\|}}{C}}$$

(j) CH$_2$ClCHClCH$_3$ and   CH$_3$CHClCH$_2$Cl

(k) CH$_3$CH$_2$CHClCH$_2$Cl   and   CH$_3$CHCH$_2$Cl

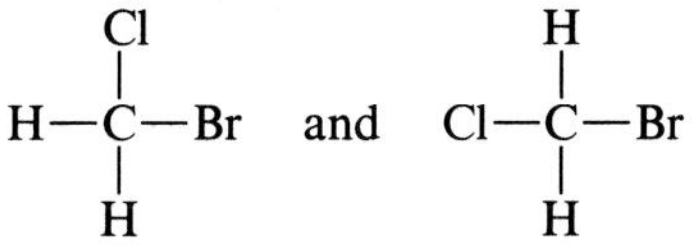

                                      |
                                      CH$_2$Cl

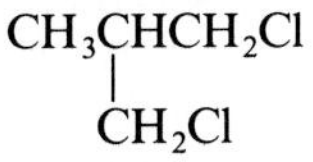

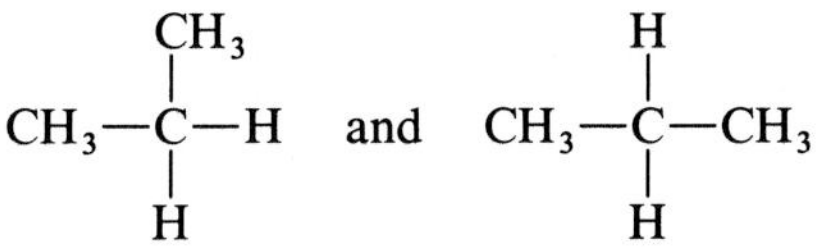

(l) 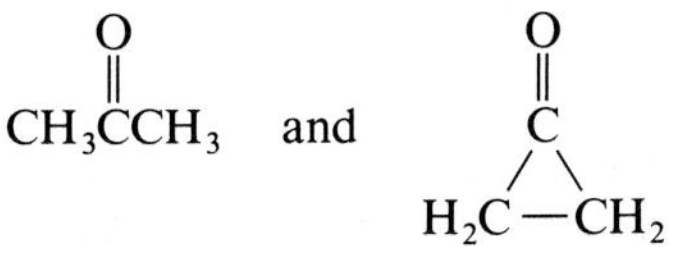 CH$_3\overset{\overset{\textstyle O}{\|}}{C}CH_3$ and $\underset{H_2C-CH_2}{\overset{\overset{\textstyle O}{\|}}{C}}$

(m) $\underset{H}{\overset{Cl}{H-\underset{|}{\overset{|}{C}}-Br}}$ and $\underset{H}{\overset{H}{Cl-\underset{|}{\overset{|}{C}}-Br}}$

(n) $\underset{H}{\overset{CH_3}{CH_3-\underset{|}{\overset{|}{C}}-H}}$ and $\underset{H}{\overset{H}{CH_3-\underset{|}{\overset{|}{C}}-CH_3}}$

(o)   and

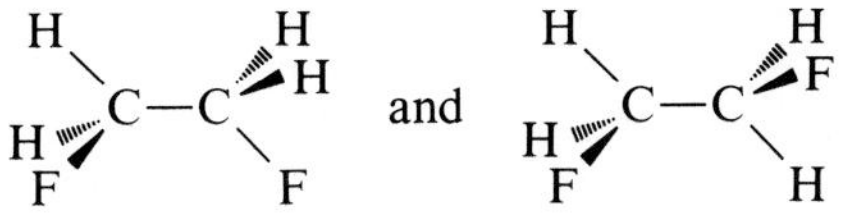

(p) 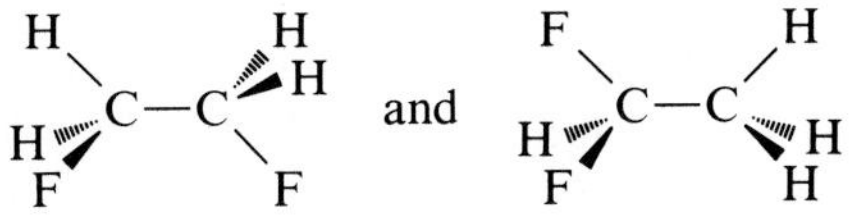   and

1.27

Write a three-dimensional formula for each of the following molecules. If the molecule has a net dipole moment, indicate its direction with an arrow, $\longmapsto$. If the molecule has no net dipole moment, you should so state. (You may ignore the small polarity of C—H bonds in working this and similar problems.)

(a) CH$_3$F   (c) CHF$_3$ (e) CH$_2$FCl   (g) BeF$_2$ (i) CH$_3$OH

(b) CH$_2$F$_2$ (d) CF$_4$ (f) BCl$_3$ (h) CH$_3$OCH$_3$ (j) CH$_2$O

1.28

Rewrite each of the following using bond-line formulas:

(a) 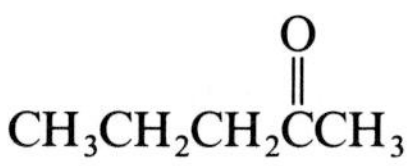 CH$_3$CH$_2$CH$_2\overset{\overset{\textstyle O}{\|}}{C}CH_3$ (c) (CH$_3$)$_3$CCH$_2$CH$_2$CH$_2$OH

(b) 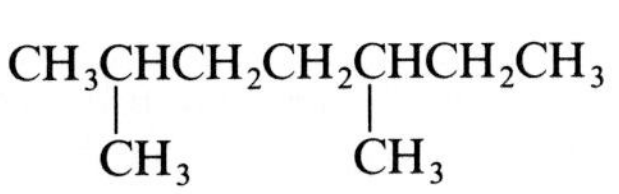 CH$_3\underset{CH_3}{\overset{|}{C}}HCH_2CH_2\underset{CH_3}{\overset{|}{C}}HCH_2CH_3$ (d) 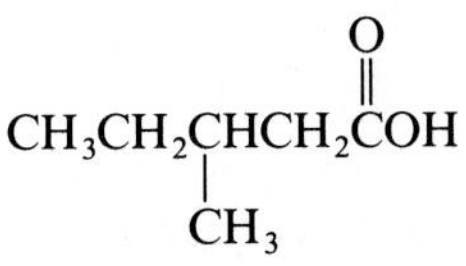 CH$_3$CH$_2\underset{CH_3}{\overset{|}{C}}HCH_2\overset{\overset{\textstyle O}{\|}}{C}$OH

(e) $CH_2=CHCH_2CH_2CH=CHCH_3$ (f)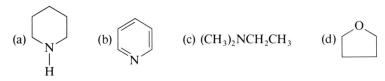

1.29
Write a dash formula for each of the following showing any unshared electron pairs:

(a) (b) (c) $(CH_3)_2NCH_2CH_3$ (d)

1.30
Write structural formulas of your choice for all of the constitutional isomers with the molecular formula C_4H_8.

1.31
Write two resonance structures for the nitrite ion (NO_2^-), and show the formal charge on each atom in each structure. Do these structures account for the fact that the nitrogen bonds are of equal length?

1.32
(a) Taking into account the shape of an ammonia molecule (Section 1.18B), in what kind of orbital would you expect the unshared electron pair to be found? (b) In what kind(s) of orbitals would you expect the electron pairs of a water molecule to be found? Explain.

1.33
Chloromethane (CH_3Cl) has a larger dipole moment ($\mu = 1.87$ D) than fluoromethane (CH_3F) ($\mu = 1.81$ D), even though fluorine is more electronegative than chlorine. Explain.

1.34
Cyanic acid ($H-O-C\equiv N$) and isocyanic acid ($H-N=C=O$) differ in the positions of their electrons but their structures do not represent resonance structures. (a) Explain. (b) Loss of a proton from cyanic acid yields the same anion as that obtained by loss of a proton from isocyanic acid. Explain.

1.35
Boron trifluoride reacts readily with ammonia to form a compound, BF_3NH_3. (a) What factors account for this reaction taking place so readily? (b) What formal charge is present on boron in the product? (c) On nitrogen? (d) What hybridization state would you expect for boron in the product? (e) For nitrogen?

1.36
Consider the substances, H_3O^+, H_2O, and OH^-; or NH_3 and NH_4^+; or H_2S and SH^-, and describe the relationship between formal charge and acid strength.

1.37
Ozone (O_3) is found in the upper atmosphere where it absorbs highly energetic ultraviolet (UV) light and thus provides the surface of the earth with a protective screen (cf. Section

9.11). (a) Given that ozone molecules are not cyclic, and that all of the electrons are paired, write resonance structures for ozone. (b) Would you expect the two oxygen – oxygen bonds of ozone to be equivalent? (c) Given that ozone has a dipole moment ($\mu = 0.52$ D) what does this indicate about the shape of an ozone molecule? (d) Is this the shape you would predict on the basis of VSEPR theory? Explain.

1.38

In Problem 1.2 you wrote the Lewis structure for the methyl cation, (CH_3^+). In Problem 1.10 you were able to predict its shape on the basis of VSEPR theory. Now describe the methyl cation in terms of orbital hybridization. (Pay special attention to the hybridization state of the carbon atom and be sure to include any vacant orbitals.)

CHAPTER TWO

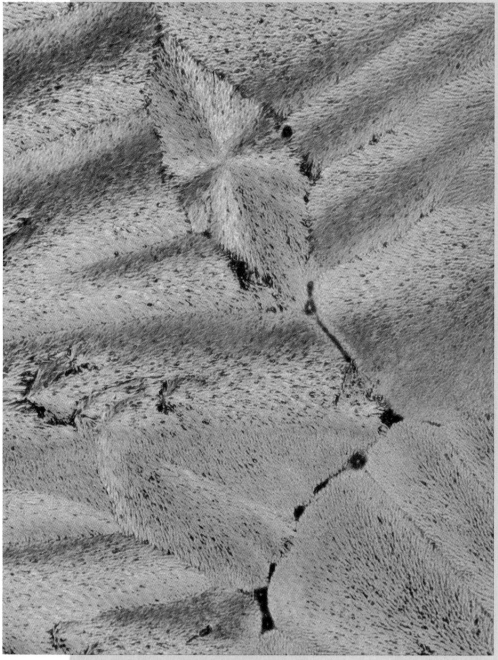

Penicillin 130x. Penicillins, the familiar antibiotics, are produced by moulds of the *Penicillium* species. The chemical structures of penicillins are described in Section 18.8.

Representative
Carbon Compounds

2.1 CARBON–CARBON COVALENT BONDS

Carbon's ability to form strong covalent bonds to other carbon atoms is the single property of the carbon atom that—more than any other—accounts for the very existence of a field of study called organic chemistry. It is this property too that accounts in part for carbon being the element around which most of the molecules of living organisms are constructed. Carbon's ability to form strong bonds to other carbon atoms and to form strong bonds to hydrogen, oxygen, sulfur, and nitrogen atoms as well, provides the necessary versatility of structure that makes possible the vast number of different molecules required for complex living organisms.

2.2 METHANE AND ETHANE.
REPRESENTATIVE ALKANES

Methane (CH_4) and ethane (C_2H_6) are two members of a broad family of organic compounds called **hydrocarbons.** Hydrocarbons, as the name implies, are compounds whose molecules contain only carbon and hydrogen atoms. Methane and ethane also belong to a subgroup of hydrocarbons known as **alkanes** whose members do not have multiple bonds between carbon atoms. *Hydrocarbons whose molecules have a carbon–carbon double bond* are called *alkenes,* and *those with a carbon–carbon triple bond* are called *alkynes.* Hydrocarbons that contain a special ring that we shall introduce in Section 2.7 and study in Chapter 11 are called aromatic hydrocarbons.

Generally speaking, compounds, such as the alkanes, whose molecules contain only single bonds are often referred to as **saturated** compounds and those whose molecules contain multiple bonds are called **unsaturated** compounds. We shall see why in Section 6.5.

2.2A Sources of Methane

Methane was one major component of the early atmosphere of this planet. Methane is still found in the atmosphere of Earth, but no longer in appreciable amounts. It is, however, a major component of the atmosphere of Jupiter, Saturn, Uranus, and Neptune. Recently, methane has also been detected in interstellar space—far from the earth (10^{16} km) in a celestial body that emits radio waves in the constellation Orion.

On Earth, methane is the major component of natural gas, along with ethane and other low molecular weight alkanes. The United States is currently using its large reserves of natural gas at a very high rate. Because the components of natural gas are important in industry, efforts are being made to develop coal-gasification processes to provide alternative sources.

Some living organisms produce methane from carbon dioxide and hydrogen. These very primitive creatures called *methanogens* may be the Earth's oldest organisms, and they may represent a separate form of evolutionary development. Methanogens can survive only in an anaerobic (i.e., oxygen-free) environment. They have been found in ocean trenches, in mud, in sewage, and in cows' stomachs.

2.2B The Structure of Ethane

The bond angles at the carbon atoms of ethane, and of all alkanes, are also tetrahedral like those in methane. In the case of ethane (Fig. 2.1), each carbon atom is at one corner of the other carbon atom's tetrahedron; hydrogen atoms are situated at the other three corners.

A satisfactory model for ethane (and for other alkanes as well) can be provided by sp^3-hybridized carbon atoms (Section 1.14). Figure 2.2 shows how we might imagine the bonding molecular orbitals of an ethane molecule being constructed from two sp^3-hybridized carbon atoms and six hydrogen atoms.

The carbon–carbon bond of ethane is a *sigma bond* (Section 1.14), formed by two overlapping sp^3 orbitals. (The carbon–hydrogen bonds are also sigma bonds. They are formed from overlapping carbon sp^3 orbitals and hydrogen s orbitals.)

Because a sigma bond (i.e., any nonmultiple bond) has circular symmetry along the bond axis, *rotation of groups joined by a single bond does not usually require a*

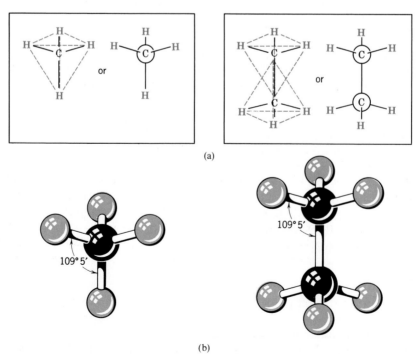

(a)

(b)

FIGURE 2.1 (a) Two ways of representing the structures of methane and ethane that show the tetrahedral arrangements of the atoms around carbon. (b) Ball-and-stick models of methane and ethane.

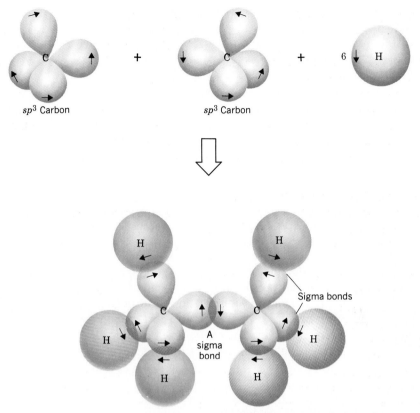

FIGURE 2.2 The formation of the bonding molecular orbitals of ethane from two sp^3-hybridized carbon atoms and six hydrogen atoms. All of the bonds are sigma bonds. (Antibonding sigma molecular orbitals—called σ^* orbitals—are formed in each instance as well, but for simplicity these are not shown.)

large amount of energy. Consequently, groups joined by single bonds rotate relatively freely with respect to one another. (We discuss this point further in Section 3.6.)

2.3 ALKENES: COMPOUNDS CONTAINING THE CARBON–CARBON DOUBLE BOND; ETHENE AND PROPENE

The carbon atoms of virtually all of the molecules that we have considered so far have used their four valence electrons to form four single covalent bonds to four other atoms. We find, however, that many important organic compounds exist in which carbon atoms share more than two electrons with another atom. In molecules of these compounds some bonds that are formed are multiple covalent bonds. When two carbon atoms share two pairs of electrons, for example, the result is a carbon–carbon double bond.

$$\overset{..}{\underset{..}{C}} :: \overset{..}{\underset{..}{C}} \quad \text{or} \quad \overset{\diagdown}{\diagup} C = C \overset{\diagup}{\diagdown}$$

Hydrocarbons whose molecules contain a carbon–carbon double bond are called **alkenes.** Ethene (C_2H_4) and propene (C_3H_6) are both alkenes. (Ethene is also called ethylene, and propene is sometimes called propylene.)

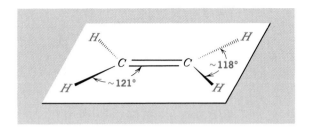

FIGURE 2.3 The structure and bond angles of ethene.

Ethene   **Propene**

In ethene the only carbon–carbon bond is a double bond. Propene has one carbon–carbon single bond and one carbon–carbon double bond.

The spatial arrangement of the atoms of alkenes is different from that of alkanes. The six atoms of ethene are coplanar, and the arrangement of atoms around each carbon atom is triangular (Fig. 2.3). In the next section we shall see how the structure of ethene can be explained on the basis of the same kind of orbital hybridization, sp^2, that we learned about for boron trifluoride (Section 1.15).

2.4 ORBITAL HYBRIDIZATION AND THE STRUCTURE OF ALKENES

We can account for the structure of the carbon–carbon double bond in terms of orbital hybridization. The basis for our model is sp^2-*hybridized carbon* atoms.*

Formation of the sp^2-hybridized carbon atoms for our model can be visualized in the following way (Fig. 2.4). One electron from a carbon atom in its ground state is promoted from a $2s$ orbital to a $2p$ orbital. Then the $2s$ orbital is hybridized with two of the $2p$ orbitals. One $2p$ orbital is left unhybridized. One electron is then placed in each of the sp^2-hybrid orbitals and one electron remains in the $2p$ orbital.

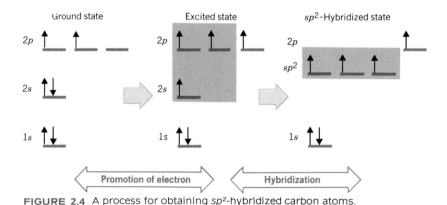

FIGURE 2.4 A process for obtaining sp^2-hybridized carbon atoms.

*An alternative model for the carbon–carbon double bond is discussed in a recent report by W. E. Palke, *J. Am. Chem. Soc.*, **108**, 6543–6544 (1986).

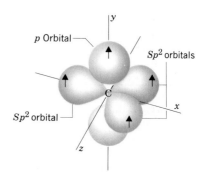

FIGURE 2.5 An *sp²*-hybridized carbon atom.

The three sp^2 orbitals that result from hybridization are directed towards the corners of a regular triangle (with angles of 120° between them). The carbon p orbital that is not hybridized is perpendicular to the plane of the triangle formed by the hybrid sp^2 orbitals (Fig. 2.5).

In our model for ethene (Fig. 2.6) we see that two sp^2-hybridized carbon atoms form a sigma (σ) bond between them by the overlap of one sp^2 orbital from each. The remaining sp^2 orbitals of each carbon atom form σ bonds to four hydrogen atoms through overlap with the 1s orbitals of the hydrogen atoms. These five bonds account for 10 of the 12 bonding electrons of ethene, and they are called the **σ-bond framework.** The bond angles that we would predict on the basis of sp^2-hybridized carbon atoms (120° all around) are quite close to the bond angles that are actually found (Fig. 2.3).

The remaining two bonding electrons in our model are located in the p orbitals of each carbon atom. We can better visualize how these p orbitals interact with each other if we replace the σ bonds by lines. This is shown in Fig. 2.7. We see that the parallel p orbitals *overlap above and below the plane of the σ framework.* This sideways overlap of the p orbitals results in a new type of covalent bond, known as a **pi (π) bond.** Note the difference in shape of the bonding molecular orbital of a π bond as contrasted to that of a σ bond. A σ bond has cylindrical symmetry about a line connecting the two bonded nuclei. A π bond has a nodal plane passing through the two bonded nuclei.

According to molecular orbital theory, both bonding and antibonding π molecular orbitals are formed when p orbitals interact in this way to form a π bond. The

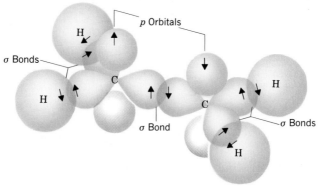

FIGURE 2.6 A model for the bonding molecular orbitals of ethene formed from two *sp²*-hybridized carbon atoms and four hydrogen atoms.

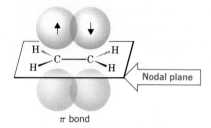

π bond

FIGURE 2.7 The overlapping *p* orbitals of ethene to make a π bond.

bonding π orbital (Fig. 2.8) results when *p*-orbital lobes of like signs overlap; the antibonding π orbital is formed when *p*-orbital lobes of opposite signs overlap.

The bonding π orbital is the lower-energy orbital and contains both π electrons (with opposite spins) in the ground state of the molecule. The region of greatest probability of finding the electrons in the bonding π orbital is a region generally

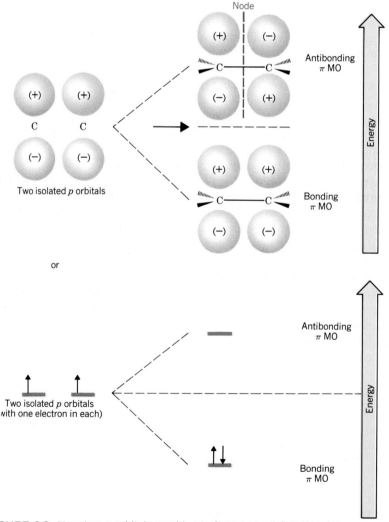

FIGURE 2.8 How two *p* orbitals combine to form two π (pi) molecular orbitals. There is no node in the bonding MO, and it is of lower energy. The higher-energy antibonding MO contains a node.

situated above and below the plane of the σ-bond framework between the two carbon atoms. The antibonding π^* orbital is of higher energy, and it is not occupied by electrons when the molecule is in the ground state. It can become occupied, however, if the molecule absorbs light of the right frequency, and an electron is promoted from the lower energy level to the higher one. The antibonding π^* orbital has a nodal plane between the two carbon atoms.

To summarize: In our model based on orbital hybridization, the carbon–carbon double bond consists of two different kinds of bonds, *a σ bond and a π bond.* The σ bond is formed by two overlapping sp^2 orbitals end-to-end and is symmetrical about an axis linking the two carbon atoms. The π bond is formed by a sideways overlap of two p orbitals; it has a nodal plane like a p orbital. In the ground state the electrons of the π bond are located between the two carbon atoms but generally above and below the plane of the σ-bond framework.

Electrons of the π bond have greater energy than electrons of the σ bond. The relative energies of the σ and π molecular orbitals (with the electrons in the ground state) are shown in the following figure. (The σ^* orbital is the antibonding sigma orbital.)

2.4A Restricted Rotation and the Double Bond

The $\sigma-\pi$ model for the carbon–carbon double bond also accounts for an important property of the double bond: *There is a large barrier to free rotation associated with groups joined by a double bond.* Maximum overlap between the p orbitals of a π bond occurs when the axes of the p orbitals are exactly parallel. Rotating one carbon of the double bond 90° (Fig. 2.9) breaks the π bond, for then the axes of the p orbitals are perpendicular and there is no net overlap between them. Estimates based on thermochemical calculations indicate that the strength of the π bond is 63 kcal/mole. This, then, is the barrier to rotation of the double bond. It is markedly higher than the rotational barrier of groups joined by carbon–carbon single bonds (3–6 kcal/mole). While groups joined by single bonds rotate relatively freely at room temperature, those joined by double bonds do not.

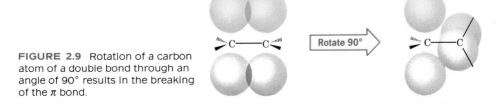

FIGURE 2.9 Rotation of a carbon atom of a double bond through an angle of 90° results in the breaking of the π bond.

2.4B Cis–Trans Isomerism

Restricted rotation of groups joined by a double bond causes a new type of isomerism that we illustrate with the two dichloroethenes written in the following structures.

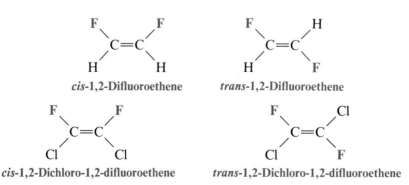

cis-1,2-Dichloroethene   *trans*-1,2-Dichloroethene

These two compounds are isomers; they are different compounds that have the same molecular formula. We can tell that they are different compounds by trying to superpose a model of one on a model of the other. We find that it cannot be done. By superpose we mean that we attempt to place one model on the other *so that all parts of each coincide.*

We indicate that they are different isomers by attaching the prefixes *cis* or *trans* to their names (*cis*, Latin: on this side; *trans*, Latin: across). *cis*-1,2-Dichloroethene and *trans*-1,2-dichloroethene are not constitutional isomers because the connectivity of the atoms is the same in each. The two compounds **differ only in the arrangement of their atoms in space.** Isomers of this kind are classified formally as **stereoisomers,** but often they are called simply cis–trans isomers. (We shall study stereoisomerism in detail in Chapters 3 and 4.)

The structural requirements for cis–trans isomerism will become clear if we consider a few additional examples. 1,1-Dichloroethene and 1,1,2-trichloroethene do not show this type of isomerism.

1,1-Dichloroethene
(no cis–trans isomerism)

1,1,2-Trichloroethene
(no cis–trans isomerism)

1,2-Difluoroethene and 1,2-dichloro-1,2-difluoroethene do exist as cis–trans isomers. Notice that we designate the isomer with two identical groups on the same side as being cis.

cis-1,2-Difluoroethene   *trans*-1,2-Difluoroethene

cis-1,2-Dichloro-1,2-difluoroethene   *trans*-1,2-Dichloro-1,2-difluoroethene

Clearly, then, *cis – trans isomerism of this type is not possible if one carbon atom of the double bond bears two identical groups.*

PROBLEM 2.1

Which of the following alkenes can exist as cis–trans isomers? Write their structures.

(a) $CH_2{=}CHCH_2CH_3$ (c) $CH_2{=}C(CH_3)_2$

(b) $CH_3CH{=}CHCH_3$ (d) $CH_3CH_2CH{=}CHCl$

Cis–trans isomers have different physical properties. They have different melting points and boiling points, and often cis–trans isomers differ markedly in the magnitude of their dipole moments. Table 2.1 summarizes some of the physical properties of two pairs of cis–trans isomers.

TABLE 2.1 Physical properties of cis–trans isomers

COMPOUND	MELTING POINT (°C)	BOILING POINT (°C)	DIPOLE MOMENT (D)
cis-1,2-Dichloroethene	−80	60	1.90
trans-1,2-Dichloroethene	−50	48	0
cis-1,2-Dibromoethene	−53	112.5	1.35
trans-1,2-Dibromoethene	− 6	108	0

PROBLEM 2.2

(a) How do you explain the fact that *trans*-1,2-dichloroethene and *trans*-1,2-dibromoethene have no dipole moments ($\mu = 0$), whereas the corresponding cis isomers have rather large dipole moments (for *cis*-1,2-dichloroethene, $\mu = 1.90$ D, and for *cis*-1,2-dibromoethene, $\mu = 1.35$ D)? (b) Account for the fact that *cis*-1,2-dichloroethene has a larger dipole moment than *cis*-1,2-dibromoethene.

PROBLEM 2.3

Write structural formulas for (a) all of the compounds that could be obtained by replacing one hydrogen of propene with chlorine; (b) all of the compounds that could be obtained by replacing two hydrogen atoms of propene with chlorine; (c) three hydrogens; (d) four hydrogens; (e) five hydrogens. (f) In each instance [(a)–(e)] designate pairs of cis–trans isomers.

2.5 ALKYNES: COMPOUNDS CONTAINING THE CARBON–CARBON TRIPLE BOND; ETHYNE (ACETYLENE) AND PROPYNE

Hydrocarbons in which two carbon atoms share three pairs of electrons between them, and are thus bonded by a triple bond, are called **alkynes.** The two simplest alkynes are ethyne and propyne.

$$H—C{\equiv}C—H \qquad CH_3—C{\equiv}C—H$$

Ethyne
(acetylene)
(C_2H_2)

Propyne
(C_3H_4)

Ethyne, a compound that is also called acetylene, consists of linear molecules. The H—C≡C bond angles of ethyne molecules are 180°.

$$H{\frown}C{\equiv}C{\frown}H$$
$$180° \quad 180°$$

2.6 ORBITAL HYBRIDIZATION AND THE STRUCTURE OF ALKYNES

We can account for the structure of ethyne on the basis of orbital hybridization as we did for ethane and ethene. In our model for ethane (Section 2.2B) we saw that the carbon orbitals are sp^3 hybridized, and in our model for ethene (Section 2.4) we saw that they are sp^2 hybridized. In our model for ethyne we shall see that the carbon atoms are *sp hybridized* and resemble the hybrid orbitals of BeH_2 (Section 1.16).

The mathematical process for obtaining the sp-hybrid orbitals of ethyne can be visualized in the following way (Fig. 2.10). After promotion of an electron, the $2s$ orbital and one $2p$ orbital of carbon are hybridized to form two sp orbitals. The remaining two $2p$ orbitals are not hybridized. Calculations show that the sp-hybrid orbitals have their large positive lobes oriented at an angle of 180° with respect to each other. The $2p$ orbitals that were not hybridized are perpendicular to the axis that passes through the center of the two sp orbitals (Fig. 2.11).

We envision the bonding molecular orbitals of ethyne being formed in the following way (Fig. 2.12). Two carbon atoms overlap sp orbitals to form a sigma bond

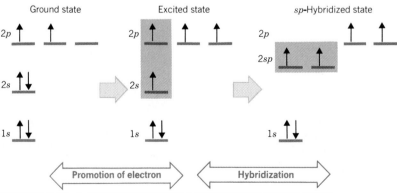

FIGURE 2.10 A process for obtaining sp-hybridized carbon atoms.

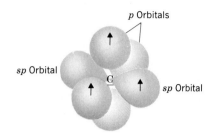

FIGURE 2.11 An *sp*-hybridized carbon atom.

between them (this is one bond of the triple bond). The remaining two *sp* orbitals at each carbon atom overlap with *s* orbitals from hydrogen atoms to produce two sigma C—H bonds. The two *p* orbitals on each carbon atom also overlap side to side to form two π bonds. These are the other two bonds of the triple bond. If we replace the σ bonds of this illustration with lines, it is easier to see how the *p* orbitals overlap. Thus we see that the carbon–carbon triple bond consists of two π bonds and one σ bond.

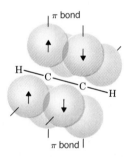

2.6A Bond Lengths of Ethyne, Ethene, and Ethane

The carbon–carbon triple bond is shorter than the carbon–carbon double bond, and the carbon–carbon double bond is shorter than the carbon–carbon single bond. The carbon–hydrogen bonds of ethyne are also shorter than those of ethene, and the carbon–hydrogen bonds of ethene are shorter than those of ethane. This illustrates a general principle: *The shortest C—H bonds are associated with those atoms having atomic orbitals with the greatest s character.* The *sp* orbitals of ethyne—50% *s* (and 50% *p*) in character—form the shortest C—H bonds. The *sp*³ orbitals of ethane— 25% *s* (and 75% *p*) in character form the longest C—H bonds. The differences in bond lengths and bond angles of ethyne, ethene, and ethane are summarized in Fig. 2.13.

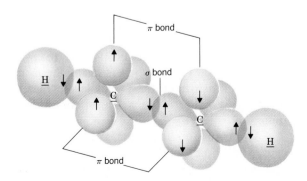

FIGURE 2.12 Formation of the bonding molecular orbitals of ethyne from two *sp*-hybridized carbon atoms and two hydrogen atoms. (Antibonding orbitals are formed as well but these have been omitted for simplicity.)

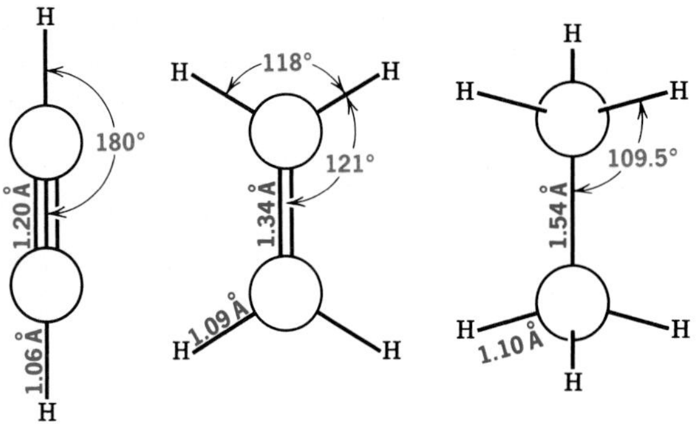

FIGURE 2.13 Bond angles and bond lengths of ethyne, ethene, and ethane.

2.7 BENZENE: A REPRESENTATIVE AROMATIC HYDROCARBON

In Chapter 11 we shall study a group of cyclic hydrocarbons known as **aromatic hydrocarbons.** Benzene is a typical example. Its structure is usually written in one of the following ways.

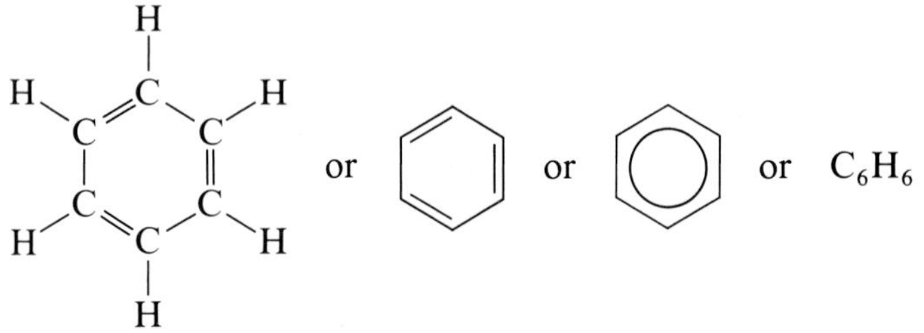

When the benzene ring is found attached to some other group of atoms in a molecule, it is called a **phenyl group** and it is represented as follows:

Because the properties of benzene and other aromatic hydrocarbons are so special (and so interesting), we shall defer any further discussions until Chapter 11, when we can take them up in detail.

2.8 FUNCTIONAL GROUPS

One great advantage of the structural theory is that it enables us to classify the vast number of organic compounds into a relatively small number of families based on their structures. (The end papers inside the front cover of this text give the most important of these families.) The molecules of compounds in a particular family are characterized by the presence of a certain arrangement of atoms called a **functional group.**

A functional group is the part of a molecule where most of its chemical reactions occur. It is the part that effectively determines the compound's chemical properties (and many of its physical properties as well). The functional group of an alkene, for example, is its carbon–carbon double bond. When we study the reactions of alkenes in greater detail in Chapter 7, we shall find that most of the chemical reactions of alkenes are the chemical reactions of the carbon–carbon double bond.

The functional group of an alkyne is its carbon–carbon triple bond. Alkanes do not have a functional group. Their molecules have carbon–carbon single bonds and carbon–hydrogen bonds, but these bonds are present in molecules of almost all organic molecules, and C—C and C—H bonds are, in general, much less reactive than common functional groups.

2.8A Alkyl Groups and the Symbol R

Alkyl groups are the groups that we identify for purposes of naming compounds. They are made (on paper) by removing a hydrogen atom from an alkane:

$$CH_4 \xrightarrow{-H} CH_3-$$

Methane   Methyl group

$$CH_3CH_3 \xrightarrow{-H} CH_3CH_2- \quad (or \quad C_2H_5-)$$

Ethane   Ethyl group

$$CH_3CH_2CH_3 \xrightarrow[\text{carbon}]{-H \text{ at end}} CH_3CH_2CH_2-$$

Propane   Propyl group

$$CH_3CH_2CH_3 \xrightarrow[\text{carbon}]{-H \text{ at middle}} CH_3CHCH_3 \left(or \quad CH_3\overset{\displaystyle CH_3}{\underset{|}{CH}}- \quad or \quad (CH_3)_2CH- \right)$$

Isopropyl group

The methyl group, the ethyl group, the propyl group, and the isopropyl group are all alkyl groups. Their names and structures must be learned.

We can simplify much of our future discussion if, at this point, we introduce a symbol that is widely used in designating general structures of organic molecules: The symbol R. *R is used as a general symbol to represent any alkyl group.* For example, R might be a methyl group, an ethyl group, a propyl group, or an isopropyl group.

CH_3-	Methyl	All of
CH_3CH_2-	Ethyl	these
$CH_3CH_2CH_2-$	Propyl	can be by R
$CH_3\underset{|}{C}HCH_3$	Isopropyl	designated

Thus, the general formula for an alkane is R—H.

Using R, we can write also a general formula for any monosubstituted alkene (i.e., one having only one alkyl group attached to a doubly bonded carbon) such as propene. We write the formula in the following way:

$$R-CH=CH_2$$

Similarly, we can write a general formula for any monosubstituted alkyne (i.e., one with only one alkyl group attached to the triply bonded carbon atom) such as propyne:

$$R—C≡CH$$

2.9 ALKYL HALIDES OR HALOALKANES

Alkyl halides are compounds in which a halogen atom (fluorine, chlorine, bromine, or iodine) replaces a hydrogen atom of an alkane. For example, CH_3Cl and CH_3CH_2Br are alkyl halides. Alkyl halides are also called **haloalkanes.**

Alkyl halides are classified as being primary (1°), secondary (2°), or tertiary (3°).* *This classification is based on the carbon to which the halogen is directly attached.* If the carbon that bears the halogen is attached to only one other carbon, the carbon is said to be a **primary carbon** and the alkyl halide is classified as a **primary alkyl halide.** If the carbon that bears the halogen is itself attached to two other carbons, then the carbon is a **secondary carbon** and the alkyl halide is a **secondary alkyl** halide, and so on. Examples of primary, secondary, and tertiary alkyl halides are the following:

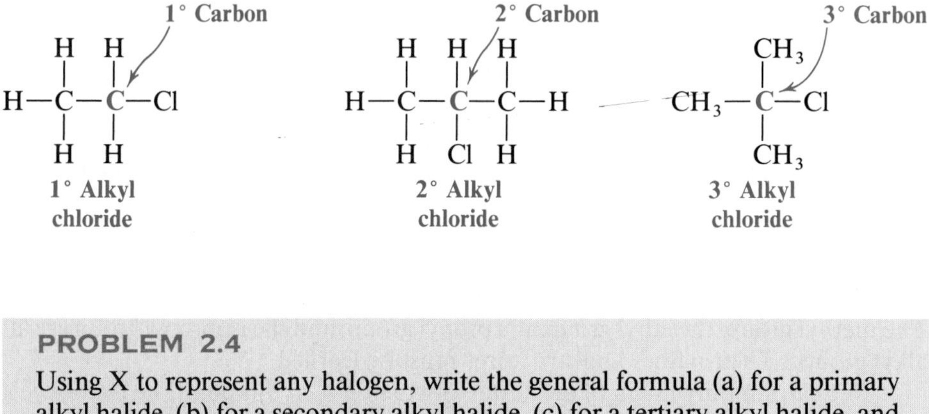

1° Carbon   2° Carbon   3° Carbon

| 1° Alkyl chloride | 2° Alkyl chloride | 3° Alkyl chloride |

PROBLEM 2.4

Using X to represent any halogen, write the general formula (a) for a primary alkyl halide, (b) for a secondary alkyl halide, (c) for a tertiary alkyl halide, and (d) for *any* alkyl halide regardless of its classification?

PROBLEM 2.5

Although we shall discuss the naming of organic compounds later when we consider the individual families in detail, one method of naming alkyl halides is so straightforward that it is worth describing here. We simply give the name of the alkyl group attached to the halogen and add the word *bromide, chloride,* and so forth. Write formulas for (a) propyl chloride and (b) isopropyl bromide. What are names for (c) CH_3CH_2F, (d) CH_3CHICH_3, and (e) CH_3I?

*Although we use the symbols 1°, 2°, 3°, we do not *say* first degree, second degree, and third degree; we say *primary, secondary,* and *tertiary.*

2.10 ALCOHOLS

Methyl alcohol (more systematically called methanol) has the structural formula CH_3OH and is the simplest member of a family of organic compounds known as **alcohols.** The characteristic functional group of this family is the hydroxyl group (OH) attached to a tetrahedral carbon atom. Another example of an alcohol is ethyl alcohol, CH_3CH_2OH (also called ethanol).

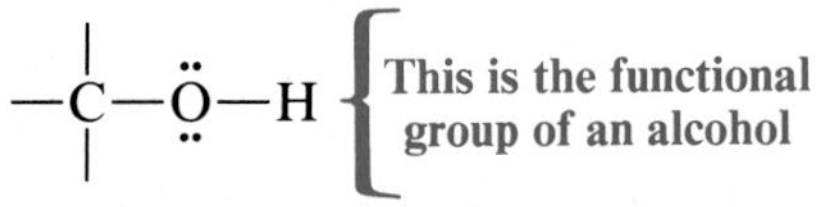

$$-\overset{|}{\underset{|}{C}}-\ddot{\underset{\cdot\cdot}{O}}-H \quad \left\{ \begin{array}{l} \text{This is the functional} \\ \text{group of an alcohol} \end{array} \right.$$

Alcohols may be viewed in two ways structurally: (1) as hydroxy derivatives of alkanes, and (2) as alkyl derivatives of water. Ethyl alcohol, for example, can be seen as an ethane molecule in which one hydrogen has been replaced by a hydroxyl group, or as a water molecule in which one hydrogen has been replaced by an ethyl group. That the latter way of regarding ethyl alcohol is valid is shown by the fact that the C—O—H bond angle of ethyl alcohol is similar in size to the H—O—H bond angle of water.

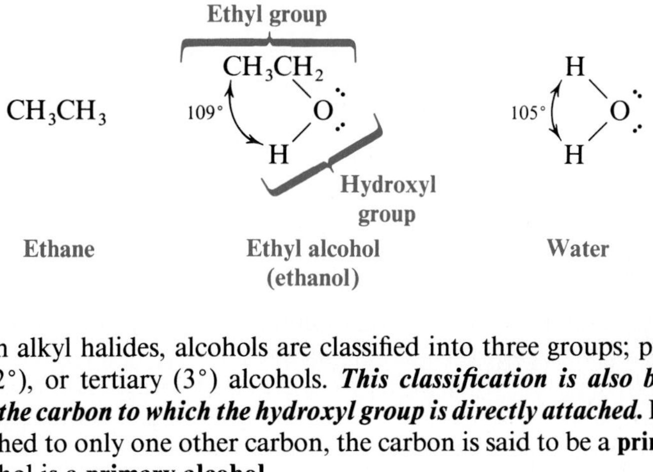

Ethyl group

CH_3CH_3 CH_3CH_2 H

Ethane   Ethyl alcohol (ethanol)   Water

Hydroxyl group

As with alkyl halides, alcohols are classified into three groups; primary (1°), secondary (2°), or tertiary (3°) alcohols. ***This classification is also based on the condition of the carbon to which the hydroxyl group is directly attached.*** If the carbon is itself attached to only one other carbon, the carbon is said to be a **primary carbon** and the alcohol is a **primary alcohol.**

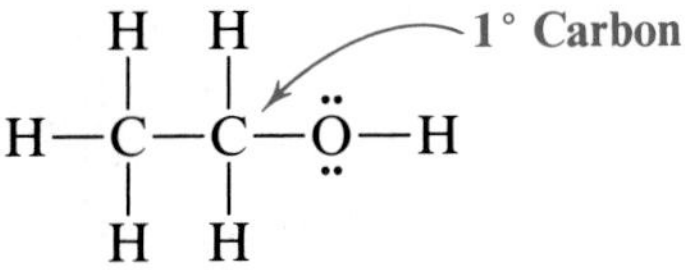

1° Carbon

$$H-\overset{\overset{\displaystyle H}{|}}{\underset{\underset{\displaystyle H}{|}}{C}}-\overset{\overset{\displaystyle H}{|}}{\underset{\underset{\displaystyle H}{|}}{C}}-\ddot{\underset{\cdot\cdot}{O}}-H$$

A primary alcohol

CH_2OH

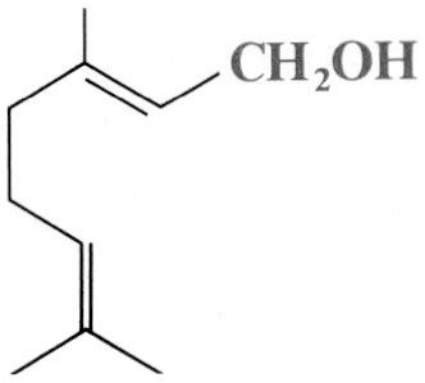

Geraniol
(a 1° alcohol with the odor of roses)

If the carbon atom that bears the hydroxyl group also has two other carbon atoms attached to it, this carbon is called a secondary carbon, and the alcohol is a secondary alcohol, and so on.

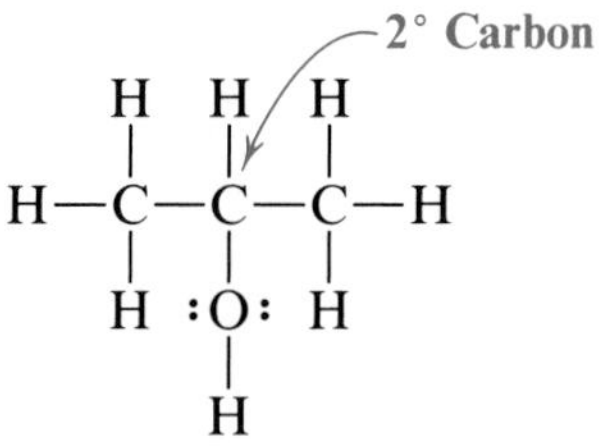

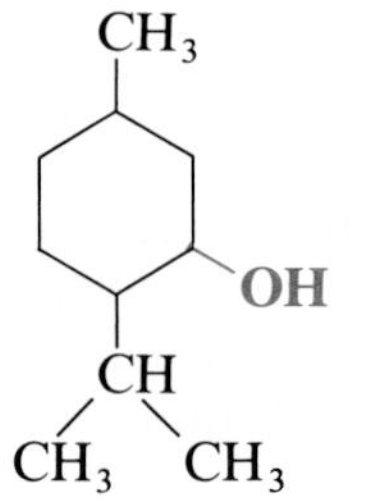

Menthol
(a 2° alcohol found
in peppermint oil)

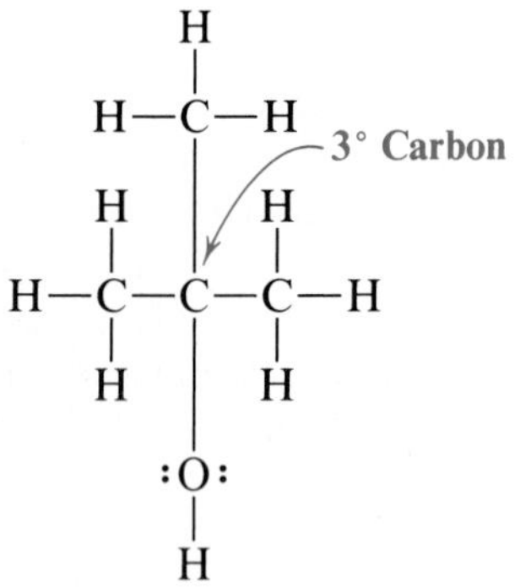

A tertiary (3°) alcohol

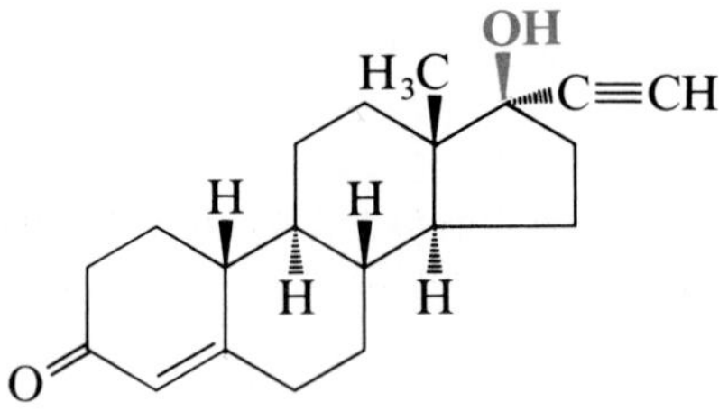

Norethindrone
(an oral contraceptive that contains a 3° alcohol
group, as well as a ketone group and carbon–
carbon double and triple bonds)

PROBLEM 2.6

Using the symbol R, write a general formula for (a) a primary alcohol, (b) a secondary alcohol, and (c) a tertiary alcohol.

PROBLEM 2.7

One way of naming alcohols is to name the alkyl group that is attached to the —OH and add the word *alcohol*. Write the structures of (a) propyl alcohol and (b) isopropyl alcohol.

2.11 ETHERS

Ethers have the general formula R—O—R or R—O—R' where R' may be an alkyl group different from R. They can be thought of as derivatives of water in which both hydrogen atoms have been replaced by alkyl groups. The bond angle at the oxygen atom of an ether is only slightly larger than that of water.

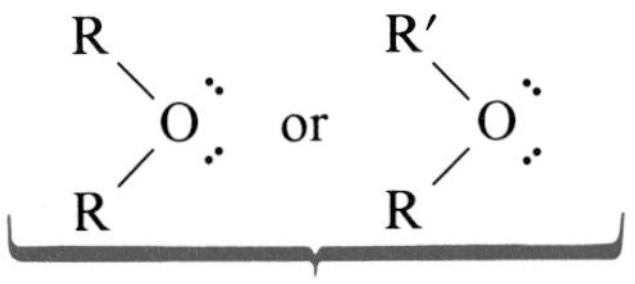

General formula for an ether

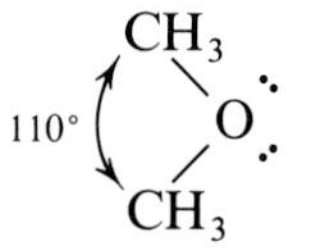

Dimethyl ether
(a typical ether)

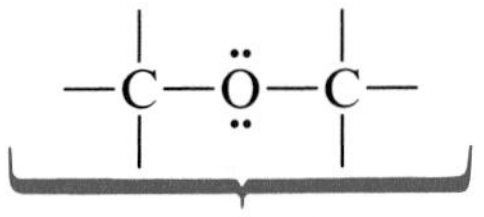

The functional group
of an ether

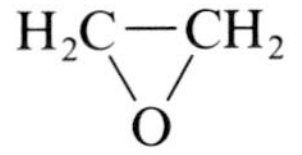

H$_2$C—CH$_2$ with O below

Ethylene
oxide

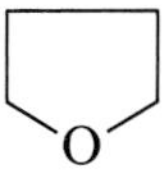

Tetrahydro-
furan

Two cyclic ethers

PROBLEM 2.8

One way of naming ethers is to name the two alkyl groups attached to the oxygen atom and add the word *ether*. If the two alkyl groups are the same, we use the prefix *di-*, for example, as in *dimethyl ether*. Write structural formulas for (a) ethyl methyl ether, (b) dipropyl ether, (c) isopropyl methyl ether. (d) What name would you give to $CH_3CH_2OCH_2CH_2CH_3$? (e) To $(CH_3)_2CHOCH_2CH_2CH_3$?

2.12 AMINES

Just as alcohols and ethers may be considered as organic derivatives of water, amines may be considered as organic derivatives of ammonia.

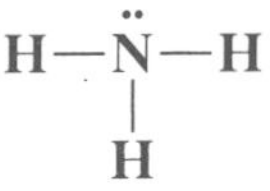

$$H-\overset{\cdot\cdot}{N}-H \qquad R-\overset{\cdot\cdot}{N}-H \qquad C_6H_5CH_2CHCH_3 \qquad H_2NCH_2CH_2CH_2CH_2NH_2$$

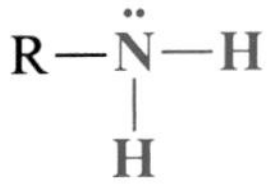

Ammonia   An amine   Amphetamine   Putrescine
(a dangerous stimulant)   (found in decaying meat)

Amines are classified as primary, secondary, or tertiary amines. This classification is based on *the number of organic groups that are attached to the nitrogen atom:*

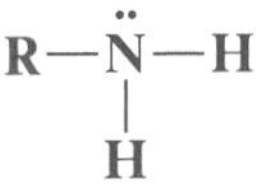

$$R-\overset{\cdot\cdot}{N}-H \qquad R-\overset{\cdot\cdot}{N}-H \qquad R-\overset{\cdot\cdot}{N}-R''$$

A primary (1°)   A secondary (2°)   A tertiary (3°)
amine   amine   amine

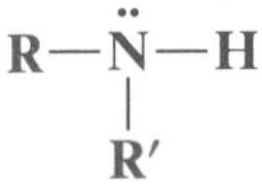

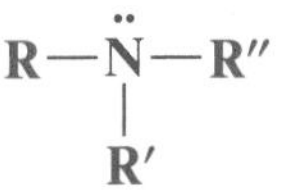

Notice that this is quite different from the way alcohols and alkyl halides are classified. Isopropylamine, for example, is a primary amine even though its —NH$_2$ group is attached to a secondary carbon atom. It is a primary amine because only one organic group is attached to the nitrogen atom.

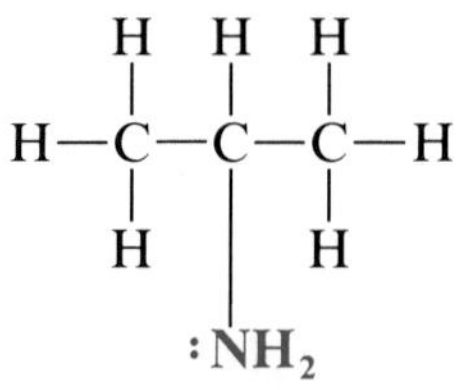

$$H-\underset{\underset{\displaystyle :NH_2}{\overset{\displaystyle |}{\underset{\displaystyle |}{C}}}}{\overset{\displaystyle H}{\overset{\displaystyle |}{\underset{\displaystyle |}{C}}}}-\underset{\overset{\displaystyle H}{|}}{\overset{\displaystyle H}{\overset{\displaystyle |}{C}}}-\underset{\overset{\displaystyle H}{|}}{\overset{\displaystyle H}{\overset{\displaystyle |}{C}}}-H$$

Isopropylamine
(a 1° amine)

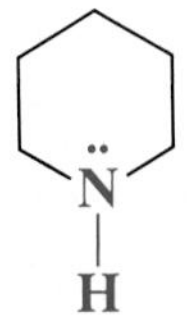

A cyclic 2° amine

PROBLEM 2.9

One way of naming amines is to name the alkyl groups attached to the nitrogen atom, using the prefixes *di-* and *tri-* if the groups are the same. Then *-amine* is added as a suffix (not as a separate word). An example is isopropylamine given previously. Write formulas for (a) dimethylamine, (b) triethylamine, and (c) ethylmethylpropylamine. What are names for (d) $(CH_3)_2CHNHCH_3$, (e) $(CH_3CH_2CH_2)_2NCH_3$, and (f) $(CH_3)_2CHNH_2$?

PROBLEM 2.10

Which amines in Problem 2.9 are (a) primary amines, (b) secondary amines, and (c) tertiary amines?

If we consider the unshared electron pair of an amine as being at one corner, the general shape of an amine (see following figure) is like that of ammonia; it is tetrahedral. The C—N—C bond angles of trimethylamine are 108.7°, a value very close to the H—C—H bond angles of methane. Thus, for all practical purposes, the nitrogen atom of an amine can be considered to be sp^3 hybridized. This means that the unshared electron pair occupies an sp^3 orbital, and thus it is considerably extended into space. This is important because, as we shall see, the unshared electron pair is involved in almost all of the reactions of amines.

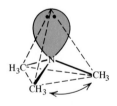

Bond angle = 108.7°

PROBLEM 2.11

(a) What general hybridization state would you expect for the oxygen atom of an alcohol or an ether (cf. Sections 2.10 and 2.11)? (b) What kind of orbitals would you expect the unshared electron pairs to occupy?

2.13 ALDEHYDES AND KETONES

Aldehydes and ketones both contain the **carbonyl group** — a group in which a carbon atom has a double bond to oxygen.

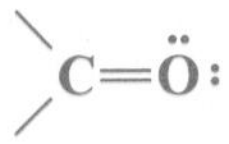

$$\begin{array}{c} \diagdown \\ \diagup \end{array} C = \ddot{O}:$$

The carbonyl group

The carbonyl group in aldehydes is bonded to at least one *hydrogen atom,* and in ketones it is bonded to *two carbon atoms.* Using R, we can designate the general formula for an aldehyde as

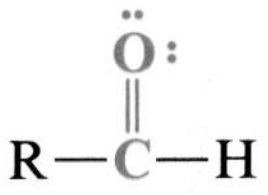

$$\begin{array}{c} \ddot{O}: \\ \parallel \\ R-C-H \end{array} \qquad \text{R may also be H}$$

and the general formula for a ketone as

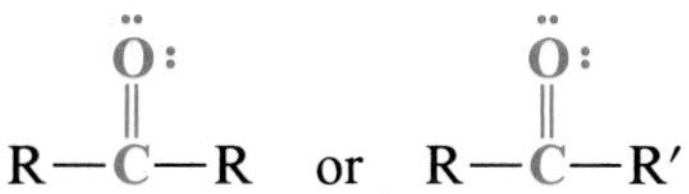

$$\begin{array}{ccc} \ddot{O}: & & \ddot{O}: \\ \parallel & & \parallel \\ R-C-R & \text{or} & R-C-R' \end{array}$$

(where R′ may be an alkyl group different from R).

Some examples of aldehydes and ketones are

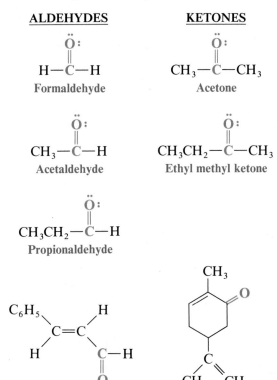

ALDEHYDES

$$\begin{array}{c} \ddot{O}: \\ \parallel \\ H-C-H \end{array}$$
Formaldehyde

$$\begin{array}{c} \ddot{O}: \\ \parallel \\ CH_3-C-H \end{array}$$
Acetaldehyde

$$\begin{array}{c} \ddot{O}: \\ \parallel \\ CH_3CH_2-C-H \end{array}$$
Propionaldehyde

KETONES

$$\begin{array}{c} \ddot{O}: \\ \parallel \\ CH_3-C-CH_3 \end{array}$$
Acetone

$$\begin{array}{c} \ddot{O}: \\ \parallel \\ CH_3CH_2-C-CH_3 \end{array}$$
Ethyl methyl ketone

trans-Cinnamaldehyde
(present in cinnamon)

Carvone
(from spearmint)

Aldehydes and ketones have a trigonal planar arrangement of groups around the carbonyl carbon atom. The carbon atom is sp^2 hybridized. In formaldehyde, for example, the bond angles are as follows:

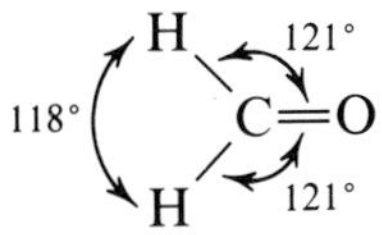

2.14 CARBOXYLIC ACIDS, AMIDES, AND ESTERS

2.14A Carboxylic Acids

Carboxylic acids have the general formula $R\overset{\overset{\displaystyle O}{\|}}{-C}-O-H$. The functional group, $\overset{\overset{\displaystyle O}{\|}}{-C}-O-H$, is called the **carboxyl group** (**carbo**nyl + hydro**xyl**). (Colloquially, carboxylic acids are often just called "organic acids.")

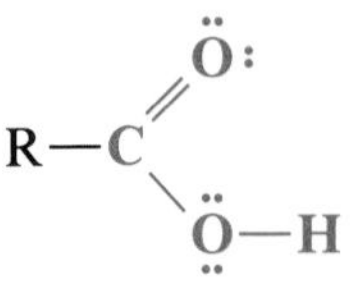

RCO$_2$H   or   RCOOH

A carboxylic
acid

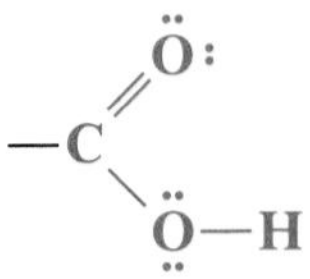

−CO$_2$H   or   −COOH

The carboxyl
group

Examples of carboxylic acids are formic acid and acetic acid.

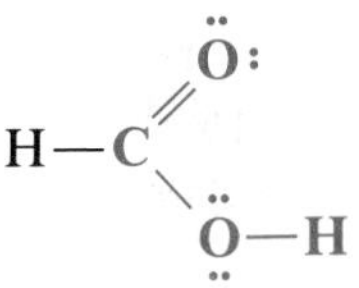

HCO$_2$H   or   HCOOH

Formic acid

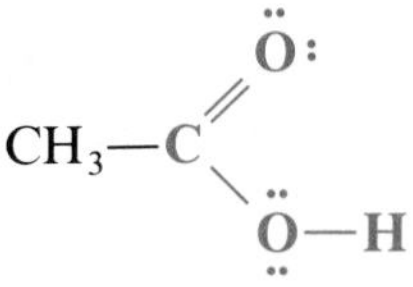

CH$_3$CO$_2$H   or   CH$_3$COOH

Acetic acid

Formic acid is an irritating liquid produced by ants. (The sting of the ant is caused, in part, by formic acid being injected under the skin.) Acetic acid, the substance responsible for the sour taste of vinegar, is produced when certain bacteria act on the ethyl alcohol of wine and cause the ethyl alcohol to be oxidized by air.

2.14B Amides
Amides have the formulas RCONH$_2$, RCONHR', or RCONR'R". Specific examples are the following:

$$CH_3C \overset{\overset{\displaystyle \ddot{O}:}{\parallel}}{\underset{\displaystyle \ddot{N}H_2}{\diagdown}} \qquad CH_3C \overset{\overset{\displaystyle \ddot{O}:}{\parallel}}{\underset{\displaystyle \ddot{N}HCH_3}{\diagdown}} \qquad CH_3C \overset{\overset{\displaystyle \ddot{O}}{\parallel}}{\underset{\displaystyle \underset{\displaystyle CH_3}{|}}{\ddot{N}-CH_3}}$$

Acetamide   N-Methylacetamide   N,N-Dimethylacetamide

The N- and N,N- indicate that the substituents are attached to the nitrogen atom.

2.14C Esters

Esters have the general formula RCO_2R' (or $RCOOR'$).

$$R-C \overset{\overset{\displaystyle \ddot{O}:}{\parallel}}{\underset{\displaystyle \ddot{O}-R'}{\diagdown}} \qquad \text{or} \quad RCO_2R' \quad \text{or} \quad RCOOR'$$

General formula for an ester

$$CH_3-C \overset{\overset{\displaystyle O}{\parallel}}{\underset{\displaystyle \ddot{O}CH_2CH_3}{\diagdown}} \qquad \text{or} \quad CH_3CO_2CH_2CH_3 \quad \text{or} \quad CH_3COOCH_2CH_3$$

A specific ester called ethyl acetate

2.15 ACID–BASE REACTIONS: BRØNSTED–LOWRY ACIDS AND BASES

Involved at some point in many of the reactions that occur with organic compounds are **acid–base reactions.** For this reason, we need to review some of the essential principles of acid–base chemistry.

2.15A Strong Acids and Bases

According to the Brønsted–Lowry theory, an acid is a substance that can donate a proton, and a base is a substance that can accept a proton. Let us consider, as an example of this concept, the reaction that occurs when gaseous hydrogen chloride dissolves in water.

$$H-\underset{\displaystyle |}{\overset{\displaystyle \ddot{O}:}{\underset{\displaystyle H}{}}} + H-\ddot{C}l: \longrightarrow H-\underset{\displaystyle |}{\overset{\displaystyle \ddot{O}^+}{\underset{\displaystyle H}{}}}-H + :\ddot{C}l:^-$$

Base   Acid   Conjugate   Conjugate
proton   *proton*   acid   base
acceptor   *donor*   of water   of HCl

Hydrogen chloride, a very strong acid, transfers its proton to water. Water acts as a base and accepts the proton. The products that result from this reaction are the hydronium ion (H_3O^+) and the chloride ion (Cl^-). The reaction, for all practical purposes, goes to completion.

The molecule or ion that forms when an acid loses its proton is called **the conjugate base** of that acid. The chloride ion, therefore, is the conjugate base of HCl.

The molecule or ion that forms when a base accepts a proton is called **the conjugate acid** of that base. The hydronium ion, therefore, is the conjugate acid of water.

Other strong acids that show essentially complete proton transfer when dissolved in water are hydrogen bromide, hydrogen iodide, nitric acid, perchloric acid, and sulfuric acid.*

$$HBr + H_2O \longrightarrow H_3O^+ + Br^-$$

$$HI + H_2O \longrightarrow H_3O^+ + I^-$$

$$HNO_3 + H_2O \longrightarrow H_3O^+ + NO_3^-$$

$$HClO_4 + H_2O \longrightarrow H_3O^+ + ClO_4^-$$

$$H_2SO_4 + H_2O \longrightarrow H_3O^+ + HSO_4^-$$

$$HSO_4^- + H_2O \longrightarrow H_3O^+ + SO_4^{2-}$$

Because sulfuric acid has two protons that it can transfer to a base, it is called a diprotic (or dibasic) acid. The proton transfer is stepwise; the first proton transfer occurs completely and the second nearly so.

Hydronium ions (H_3O^+) and hydroxide ions (OH^-) are the strongest acids and bases that are capable of existence in aqueous solutions in significant quantities. When sodium hydroxide (a crystalline compound consisting of sodium ions and hydroxide ions) dissolves in water, the result is a solution consisting of solvated sodium ions and hydroxide ions.

$$NaOH_{(solid)} \xrightarrow{\text{water}} Na_{(aq)}^+ + OH_{(aq)}^-$$

When an aqueous solution of sodium hydroxide is mixed with an aqueous solution of hydrogen chloride (hydrochloric acid), the reaction that occurs is between hydroxide ions and hydronium ions. The sodium ions and chloride ions are called spectator ions because they play no part in the acid–base reaction.

Net Reaction
$$OH^- + H-O^+-H \longrightarrow 2HOH$$
$$\overset{|}{H}$$

Spectator Ions $Na^+ +$ Cl^- $Na^+ + Cl^-$

What we have just said about hydrochloric acid and aqueous sodium hydroxide is true when aqueous solutions of all strong acids and strong bases are mixed. The net ionic reaction that occurs is simply

$$H_3O^+ + OH^- \longrightarrow 2H_2O$$

2.15B Weak Acids and Bases

In contrast to the strong acids, such as HCl and H_2SO_4, acetic acid is a much weaker acid. When acetic acid dissolves in water, the following reaction does not proceed to completion.

---

*The extent to which an acid transfers the protons to a base like water is a measure of its "strength" as an acid. Acid strength is thus a measure of the percentage of ionization and not of concentration.

$$\overset{\displaystyle :\overset{..}{O}}{\underset{}{\underset{\displaystyle \|}{}}}\quad\quad\quad\overset{\displaystyle \overset{..}{O}:}{\underset{\displaystyle \|}{}}$$

$$CH_3\overset{}{C}-\overset{..}{\underset{..}{O}}H + H_2O \rightleftharpoons CH_3\overset{}{C}-\overset{..}{\underset{..}{O}}:^- + H_3O^+$$

Experiments show that in a 0.1 M solution of acetic acid at 25 °C only 1% of the acetic acid molecules transfer their protons to water.

Because the reaction that occurs in an aqueous solution of acetic acid is an equilibrium, we can describe it with an expression for the equilibrium constant.

$$K_{eq} = \frac{[H_3O^+][CH_3CO_2^-]}{[CH_3CO_2H][H_2O]}$$

For dilute aqueous solutions the concentration of water is essentially constant (~ 55 mole/L), so we can rewrite the expression for the equilibrium constant in terms of a new constant, K_a, called **the acidity constant.**

$$K_a = K_{eq}[H_2O] = \frac{[H_3O^+][CH_3CO_2^-]}{[CH_3CO_2H]}$$

At 25 °C, the acidity constant for acetic acid is 1.8×10^{-5}.

We can write similar expressions for any weak acid dissolved in water. Using a generalized hypothetical acid (HA) the reaction in water is

$$HA + H_2\overset{..}{O}: \rightleftharpoons H_3O:^+ + A^-$$

and the expression for the acidity constant is

$$K_a = \frac{[H_3O^+][A^-]}{[HA]}$$

In this standard form the molar concentrations of the products of the proton transfer reaction are written in the numerator of the expression and the molar concentration of the undissociated acid is written in the denominator. For this reason, a large value of K_a means the acid is a strong acid, and a small value of K_a means the acid is a weak acid. If the acidity constant is greater than 10, the acid will be, for all practical purposes, completely dissociated in water.

PROBLEM 2.12

Trifluoroacetic acid (CF_3CO_2H) has a $K_a = 1$ at 25 °C. (a) What are the molar concentrations of hydronium ion and trifluoroacetate ion ($CF_3CO_2^-$) in a 0.1 M aqueous solution of trifluoroacetic acid? (b) What percentage of the trifluoroacetic acid is ionized?

Chemists often express the acidity constant, K_a, as its negative logarithm, pK_a.

$$pK_a = -\log K_a$$

This is analogous to expressing the hydronium ion concentration as pH.

$$pH = -\log[H_3O^+]$$

Note that the *larger* the value of the pK_a, the *weaker* is the acid.

Although acetic acid and other carboxylic acids containing fewer than five carbon atoms are soluble in water, many other carboxylic acids of higher molecular weight are not appreciably soluble in water. Because of their acidity, however, *water-insoluble acids dissolve in aqueous sodium hydroxide;* they do so by reacting to form *water-soluble* sodium salts.

$$C_7H_{15}CO_2H + Na^+OH^- \longrightarrow C_7H_{15}CO_2{}^-Na^+ + H_2O$$

Water insoluble   Water soluble

Water, itself, is a very weak acid and undergoes self-ionization even in the absence of added acids or bases.

$$H_2\ddot{O}: + H_2\ddot{O}: \rightleftharpoons H_3O:^+ + :\ddot{O}H^-$$

The self-ionization of pure water produces concentrations of hydronium and hydroxide ions equal to 10^{-7} mole/L at 25 °C. Since the concentration of water in pure water is 55.5 mole/L we can calculate the K_a for water.

$$K_a = \frac{[H_3O^+][OH^-]}{[H_2O]} \qquad K_a = \frac{(10^{-7})(10^{-7})}{55.5} = 1.8 \times 10^{-16}$$

In our discussion so far we have dealt only with the strengths of acids. Arising as a natural corollary to this is a principle that allows us to estimate the strengths of bases as well. Simply stated, *the conjugate base of a strong acid will be a weak base, and the conjugate base of a weak acid will be a strong base.* Moreover, *the stronger the acid, the weaker will be its conjugate base* and vice versa.

As examples of this principle consider the following:

1. The chloride ion (Cl^-), bromide ion (Br^-), iodide ion (I^-), nitrate ion ($NO_3{}^-$), and perchlorate ion ($ClO_4{}^-$) are all conjugate bases of very strong acids ($K_a > 10$) and, thus, all are very weak bases. In aqueous solution these anions have virtually no affinity for protons and exist as simple solvated ions.

2. The hydroxide ion (OH^-) is the conjugate base of water; water is a very weak acid ($K_a \approx 10^{-16}$). The hydroxide ion is therefore a strong base. It is, as we have said, the strongest base that can exist in aqueous solutions.

3. The acetate ion ($CH_3CO_2^-$) is the conjugate base of the weak acid, acetic acid ($K_a \approx 10^{-5}$). As a result, the acetate ion is a moderately strong base.

Amines, like ammonia, are weak bases. The reaction of an aqueous solution of ethylamine with a strong aqueous acid, for example, is very similar to that of aqueous ammonia.

$$:NH_3 + H_3O:^+ \rightleftharpoons NH_4^+ + H_2\ddot{O}:$$
Ammonia   Ammonium
ion

$$CH_3CH_2\ddot{N}H_2 + H_3O:^+ \rightleftharpoons CH_3CH_2\overset{+}{N}H_3 + H_2\ddot{O}:$$
Ethylamine   Ethylaminium
ion

While ethylamine and most amines of low molecular weight are very soluble in water, high-molecular-weight amines, such as hexylamine ($C_6H_{13}NH_2$), have limited water solubility. However, such water-soluble amines dissolve readily in hydrochloric acid because the acid–base reaction produces a soluble salt.

$$C_6H_{13}\ddot{N}H_2 + H{-}\overset{\displaystyle \ddot{O}^+}{\underset{\displaystyle H}{\vert}}{-}H + Cl^- \longrightarrow C_6H_{13}\overset{+}{N}H_3Cl^- + H_2O$$

Hexylamine   Hexylaminium chloride
slight water   water-soluble
solubility   salt

Much of what we have discussed in this section has been a review of concepts that you studied in your general chemistry course. If these ideas are still unclear, this may be a good time to refer to the appropriate sections of your general chemistry text.

SAMPLE PROBLEM

Write net ionic equations for the reactions that take place   (a) when benzoic acid ($C_6H_5CO_2H$) dissolves in aqueous sodium hydroxide and   (b) when aniline ($C_6H_5NH_2$) dissolves in aqueous sulfuric acid.

Answer:

(a) $C_6H_5CO_2H + OH^- \longrightarrow C_6H_5CO_2^- + H_2O$

(b) $C_6H_5NH_2 + H_3O^+ \longrightarrow C_6H_5NH_3^+ + H_2O$

2.15C Acids and Bases in Nonaqueous Solutions

Many of the organic acid–base reactions that we describe in this text occur in solutions other than aqueous solution. Many of the acids that we encounter are much weaker acids than water. The conjugate bases of these very weak acids are exceedingly powerful bases. Reactions of organic compounds, for example, are sometimes carried out in liquid ammonia (the liquefied gas, not the aqueous solution commonly used in your general chemistry laboratory). The acidity constant, K_a, of ammonia is 10^{-33} (Table 2.2); thus liquid ammonia is a weaker acid than water by a factor of 10^{-17}. The

TABLE 2.2 Scale of acidities and basicities

	ACID	APPROXIMATE		CONJUGATE BASE	
		K_a	pK_a		
	CH_3CH_3	10^{-50}	50	$CH_3CH_2^-$	
	$CH_2{=}CH_2$	10^{-44}	44	$CH_2{=}CH^-$	
	H_2	10^{-35}	35	H^-	
INCREASING	NH_3	10^{-33}	33	NH_2^-	INCREASING
STRENGTH	$HC{\equiv}CH$	10^{-25}	25	$HC{\equiv}C^-$	STRENGTH
OF ACID	CH_3CH_2OH	10^{-18}	18	$CH_3CH_2O^-$	OF CONJUGATE BASE
	H_2O	10^{-16}	16	HO^-	
	CH_3CO_2H	10^{-5}	5	$CH_3CO_2^-$	
	CF_3CO_2H	1	0	$CF_3CO_2^-$	
	HNO_3	20	-1.3	NO_3^-	
	H_3O^+	50	-1.7	H_2O	
	HCl	10^7	-7	Cl^-	
	H_2SO_4	10^9	-9	HSO_4^-	
	HI	10^{10}	-10	I^-	
	$HClO_4$	10^{10}	-10	ClO_4^-	
	$SbF_5{\cdot}FSO_3H$	$>10^{12}$	-12	$SbF_5{\cdot}FSO_3^-$	

most powerful base that can be employed in liquid ammonia is the amide ion (NH_2^-), the conjugate base of ammonia. The amide ion, because it is the conjugate base of an extremely weak acid, is an extremely powerful base. In liquid ammonia, and in the presence of $NaNH_2$, even the hydrocarbon ethyne will react as an acid.

$$H-C{\equiv}C-H + \quad :NH_2^- \quad \xrightarrow{\text{liq.}}_{NH_3} \quad H-C{\equiv}C:^- + \quad :NH_3$$

Stronger acid	Stronger	Weaker	Weaker
$K_a = 10^{-25}$	base	base	acid
	(from $NaNH_2$)		$K_a = 10^{-33}$

This example also allows us to illustrate a general principle of acid–base reactions: *The equilibrium will always favor the formation of the weaker acid and weaker base.*

All alkynes with the general formula $RC{\equiv}CH$ have acidity constants of about 10^{-25}; thus, all react with sodium or potassium amide ($NaNH_2$ or KNH_2) in liquid ammonia in the same way that ethyne does. The general reaction is

$$RC{\equiv}CH + :NH_2^- \xrightarrow{\text{liq. } NH_3} RC{\equiv}C:^- + \quad :NH_3$$

Stronger acid	Stronger base	Weaker base	Weaker acid
$K_a = 10^{-25}$			$K_a = 10^{-33}$

Ethane, and all other alkanes, are much weaker acids than ammonia. The acidity constant for ethane is approximately 10^{-50}. This means that the conjugate base of ethane, the ethanide ion ($CH_3CH_2:^-$), is a much stronger base than the amide ion. If ammonia were added to a solution containing ethanide ions (from CH_3CH_2Li in hexane, for example), the following reaction would take place:

$$CH_3CH_2{:}^- + \quad {:}NH_3 \quad \xrightarrow{\text{hexane}} \quad CH_3CH_3 + {:}\overset{..}{N}H_2{}^-$$

Stronger	Stronger	Weaker	Weaker
base	acid	acid	base
	$K_a = 10^{-33}$	$K_a = 10^{-50}$	

Generally speaking, the order of acidity of some of the important weak acids that we shall encounter is as follows:

$$RH < RCH{=}CH_2 < H_2 < NH_3 < RC{\equiv}CH < ROH < H_2O < RCO_2H$$

Increasing acidity

This means that the respective conjugate bases have the following order of basicity:

$$R{:}^- > RCH{=}CH{:}^- > H{:}^- > {:}NH_2{}^- > RC{\equiv}C{:}^- > RO{:}^- > {:}OH^- > RCO_2{}^-$$

Increasing basicity

The acidity constants of some of the molecules we have considered so far are listed in Table 2.2. All of the proton acids that we consider in this book have strengths in between that of ethane and that of $SbF_5 \cdot FSO_3H$ (called a "superacid"). On the other hand, all of the bases that we shall consider will be weaker in strength than the ethanide ion. As you examine Table 2.2, however, take care not to lose sight of the vast spectrum of acidities and basicities that it represents.

SAMPLE PROBLEM

Write the net ionic reaction that takes place when sodium hydride (NaH) is dissolved in ethyl alcohol.

Answer:

Sodium hydride consists of a sodium ion (Na^+) and a hydride ion (${:}H^-$). The sodium ion will be a spectator ion, but the hydride ion, because it is the conjugate base of a very weak acid, H_2 (cf. Table 2.2) will be a very powerful base. The hydride ion will readily react with a molecule of ethyl alcohol to remove the proton bonded to oxygen, thus producing an ethoxide ion ($CH_3CH_2O^-$) and H_2.

$$CH_3CH_2OH + \quad {:}H^- \quad \longrightarrow \quad CH_3CH_2O^- + \quad H_2{\uparrow}$$

Stronger	Stronger	Weaker	Weaker
acid	base	base	acid
$K_a = 10^{-18}$			$K_a = 10^{-35}$

2.15D Acid–Base Reactions and the Preparation of Deuterium- and Tritium-Labeled Compounds

Chemists often use compounds in which deuterium or tritium atoms have replaced one or more hydrogen atoms of the compound as a method of "labeling" or identifying particular hydrogen atoms. Deuterium (^2H) and tritium (^3H) are isotopes of hydrogen with masses of 2 and 3 atomic mass units, respectively.

For most chemical purposes, deuterium and tritium atoms in a molecule behave in the same way that ordinary hydrogen atoms behave. The extra mass associated with a deuterium or tritium atom often makes its position in a molecule easy to locate by certain spectroscopic methods that we shall study later. (Their extra mass may also cause compounds with deuterium or tritium atoms to react more slowly than compounds with ordinary hydrogen atoms. This effect, called an "isotope effect," can be useful in studying the details of many reactions.)

One way to introduce a deuterium or tritium atom into a specific location in a molecule is through the acid–base reaction that takes place when a very strong base is treated with D_2O or T_2O. For example, treating a solution containing $(CH_3)_2CHLi$ (isopropyllithium) with D_2O results in the formation of propane labeled with deuterium at the central carbon atom:

$$
\begin{array}{ccc}
\underset{\substack{\text{Isopropyl-}\\\text{lithium}\\\text{(stronger}\\\text{base)}}}{CH_3\overset{\displaystyle CH_3}{\overset{|}{CH}}\text{:}^-Li^+} + \underset{\substack{\text{(stronger}\\\text{acid)}}}{D_2O} & \xrightarrow{\text{hexane}} & \underset{\substack{\text{2-Deuterio-}\\\text{propane}\\\text{(weaker}\\\text{acid)}}}{CH_3\overset{\displaystyle CH_3}{\overset{|}{CH}}\!-\!D} + \underset{\substack{\text{(weaker}\\\text{base)}}}{OD^-}
\end{array}
$$

D_2O and T_2O have approximately the same K_a as H_2O. Hexane, the solvent for this reaction, is an alkane (C_6H_{14}) with an acidity approximately the same as that of methane or ethane ($K_a \simeq 10^{-50}$). The hexane does not participate in the acid–base reaction.

SAMPLE PROBLEM

Assuming you have available propyne, a solution of sodium amide in liquid ammonia, and T_2O, show how you would prepare the tritium-labeled compound ($CH_3C{\equiv}CT$).

Answer:
First add the propyne to the sodium amide in liquid ammonia. The following acid–base reaction will take place:

$$
\underset{\substack{\text{Stronger}\\\text{acid}}}{CH_3C{\equiv}CH} + \underset{\substack{\text{Stronger}\\\text{base}}}{NH_2{}^-} \xrightarrow{\text{liq. ammonia}} \underset{\substack{\text{Weaker}\\\text{base}}}{CH_3C{\equiv}C\text{:}^-} + \underset{\substack{\text{Weaker}\\\text{acid}}}{NH_3}
$$

Then adding T_2O (a much stronger acid than NH_3) to the solution will produce $CH_3C{\equiv}CT$.

$$
\underset{\substack{\text{Stronger}\\\text{base}}}{CH_3C{\equiv}C\text{:}^-} + \underset{\substack{\text{Stronger}\\\text{acid}}}{T_2O} \xrightarrow{\text{liq. ammonia}} \underset{\substack{\text{Weaker}\\\text{acid}}}{CH_3C{\equiv}CT} + \underset{\substack{\text{Weaker}\\\text{base}}}{OT^-}
$$

PROBLEM 2.13

Complete the following acid–base reactions:

(a) $HC \equiv CH + NaH \xrightarrow{\text{hexane}}$

(b) The solution obtained in (a) $+ D_2O \longrightarrow$

(c) $CH_3CH_2Li + D_2O \xrightarrow{\text{hexane}}$

(d) $CH_3CH_2OH + NaH \xrightarrow{\text{hexane}}$

(e) The solution obtained in (d) $+ T_2O \longrightarrow$

(f) $CH_3CH_2CH_2Li + D_2O \xrightarrow{\text{hexane}}$

2.16 LEWIS ACID–BASE THEORY

Acid–base theory was broadened immensely by G. N. Lewis in 1923. Striking at what he called "the cult of the proton," Lewis proposed that acids be defined as **electron-pair receptors** and bases as **electron-pair donors.** In the Lewis theory, not only is the proton an acid, but many other species are as well. Aluminum chloride and boron trifluoride, for example, react with amines in much the same way a proton does. Using curved arrows to show the donation of an electron pair we can give the following examples:

$$H^+ \quad + \quad :NH_3 \quad \longrightarrow \quad H-NH_3^+$$

Lewis acid   Lewis base
(electron-pair   (electron-pair
acceptor)   donor)

$$Cl-\overset{\overset{\textstyle Cl}{|}}{\underset{\underset{\textstyle Cl}{|}}{Al}} \quad + \quad :NH_3 \quad \longrightarrow \quad Cl-\overset{\overset{\textstyle Cl}{|}}{\underset{\underset{\textstyle Cl}{|}}{Al^-}}-NH_3^+$$

Lewis acid   Lewis base
(electron-pair   (electron-pair
acceptor)   donor)

$$F-\overset{\overset{\textstyle F}{|}}{\underset{\underset{\textstyle F}{|}}{B}} \quad + \quad :NH_3 \quad \longrightarrow \quad F-\overset{\overset{\textstyle F}{|}}{\underset{\underset{\textstyle F}{|}}{B^-}}-NH_3^+$$

Lewis acid   Lewis base
(electron-pair   (electron-pair
acceptor)   donor)

In these examples, aluminum chloride and boron trifluoride accept the electron pair of ammonia just as the proton does, by using it to form a covalent bond to the nitrogen atom. They do this because the central aluminum and boron atoms have only a sextet of electrons and are thus electron deficient. When they accept an electron pair, aluminum chloride and boron trifluoride are, in the Lewis definition, *acting as acids.*

Bases are much the same in both the Lewis theory and the Brønsted–Lowry theory, because in the Brønsted–Lowry theory a base must possess a pair of electrons in order to accept a proton.

The Lewis theory, by virtue of its broader definition of acids, allows acid–base theory to include all of the Brønsted–Lowry reactions and, as we shall see, a great many others.

Any *electron-deficient* atom can act as a Lewis acid. Many compounds containing Group **IIIA** elements such as boron and aluminum are Lewis acids because these compounds have an incomplete octet of electrons in their outer shell. Many other compounds that have vacant orbitals act as Lewis acids. Zinc and ferric halides are frequently used as Lewis acids in organic reactions. Two examples that we shall study later are the following:

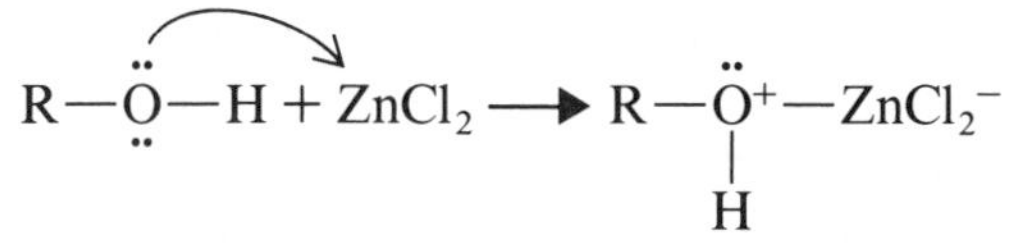

$$R-\overset{..}{\underset{..}{O}}-H + ZnCl_2 \longrightarrow R-\overset{..}{\underset{\underset{H}{|}}{O^+}}-ZnCl_2{}^-$$

Lewis base   Lewis acid

$$:\overset{..}{\underset{..}{Br}}-\overset{..}{\underset{..}{Br}}: + FeBr_3 \longrightarrow :\overset{..}{\underset{..}{Br}}-\overset{..}{\underset{..}{Br^+}}-FeBr_3{}^-$$

Lewis base   Lewis acid

PROBLEM 2.14

Write equations showing the Lewis acid–base reaction that takes place when:

(a) BF_3 reacts with an alcohol.

(b) $AlCl_3$ reacts with a tertiary amine.

(c) BF_3 reacts with a ketone.

In each instance you should use a curved arrow to indicate the donation of an electron pair.

PROBLEM 2.15

Identify the Lewis acid and Lewis base in each of the following reactions:

(a) $CH_3Cl + AlCl_3 \longrightarrow CH_3Cl^+{-}AlCl_3{}^-$

(b) $ROH + H^+ \longrightarrow ROH_2{}^+$

(c) $(CH_3)_3C^+ + Cl^- \longrightarrow (CH_3)_3C{-}Cl$

(d) 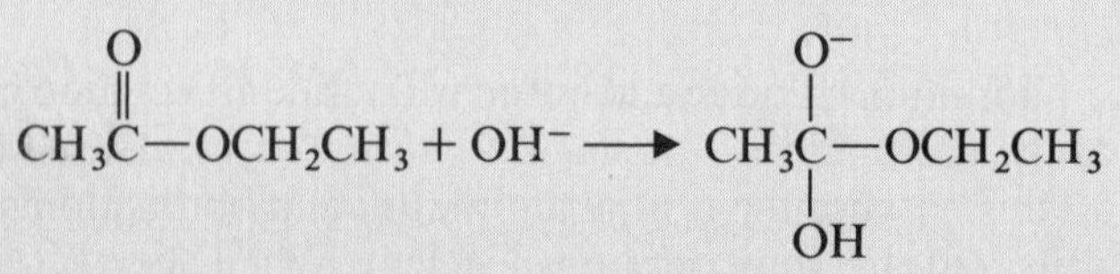 $CH_3\overset{\overset{\displaystyle O}{\|}}{C}-OCH_2CH_3 + OH^- \longrightarrow CH_3\overset{\overset{\displaystyle O^-}{|}}{\underset{\underset{\displaystyle OH}{|}}{C}}-OCH_2CH_3$

(e) $CH_2{=}CH_2 + H^+ \longrightarrow CH_3{-}CH_2{}^+$

$$\text{(f)} \ CH_3CH_2:^- + CH_3{-}\overset{\overset{\textstyle O}{\|}}{C}{-}H \longrightarrow CH_3{-}\underset{\underset{\textstyle CH_2CH_3}{|}}{\overset{\overset{\textstyle O^-}{|}}{C}}{-}H$$

2.16A The Use of Curved Arrows in Illustrating Reactions

In the previous subsection we showed the donation of an electron pair by the Lewis bases with a curved arrow. This type of notation is commonly used by organic chemists in writing reactions to show the direction of *electron flow*. Organic chemists are interested in what is called the **mechanisms of reactions.** That is, they are interested in step-by-step descriptions of the way chemical reactions occur. Inevitably in chemical reactions of organic compounds, certain covalent bonds are formed and others are broken. The curved-arrow notation is very useful in indicating the manner in which bonds are made and cleaved.

We shall not begin a study of reaction mechanisms in detail until we reach Chapter 5. At this point, however, we can illustrate the basic ideas of the curved-arrow notation with simple Lewis acid–base reactions. We begin by writing the reaction of water with hydrogen chloride (Section 2.15A) with curved arrows.

$$H{-}\overset{\cdot\cdot}{\underset{\underset{\textstyle H}{|}}{O}}{:} \ + \quad \overset{\delta+}{H}{-}\overset{\cdot\cdot}{\underset{\cdot\cdot}{Cl}}{:}^{\delta-} \longrightarrow H{-}\overset{\cdot\cdot}{\underset{\underset{\textstyle H}{|}}{O}}{}^+{-}H + :\overset{\cdot\cdot}{\underset{\cdot\cdot}{Cl}}{:}^-$$

This bond breaks   This bond is formed

Here we see that the water molecule uses one of its unshared electron pairs (shown in color) to form a bond to the hydrogen of HCl. It does this because the negatively charged electrons are attracted to the positively charged hydrogen. As this happens, the hydrogen–chlorine bond of HCl breaks, and the chlorine of HCl departs with the electron pair that formerly bonded it to the hydrogen. We use a curved arrow to indicate this bond cleavage as well. The products of the reaction, therefore, are a hydronium ion and a chloride ion, and the curved arrows help us to see how they are formed.

Written here are several other acid–base reactions illustrated with the curved-arrow notation.

$$CH_3\overset{\overset{\textstyle :\overset{\cdot\cdot}{O}}{\|}}{C}{-}\overset{\cdot\cdot}{\underset{\cdot\cdot}{O}}{-}H \ + \ :\overset{\cdot\cdot}{\underset{\underset{\textstyle H}{|}}{O}}{-}H \rightleftharpoons CH_3\overset{\overset{\textstyle :\overset{\cdot\cdot}{O}}{\|}}{C}{-}\overset{\cdot\cdot}{\underset{\cdot\cdot}{O}}{:}^- + H{-}\overset{\cdot\cdot}{\underset{\underset{\textstyle H}{|}}{O}}{}^+{-}H$$

Acid   Base

$$H{-}C{\equiv}C{-}H \ + \ ^-:\overset{}{\underset{\underset{\textstyle H}{|}}{N}}{-}H \longrightarrow H{-}C{\equiv}C:^- + H{-}\overset{\cdot\cdot}{\underset{\underset{\textstyle H}{|}}{N}}{-}H$$

Acid   Base

$$\underset{\text{Base}}{\overset{\displaystyle CH_3}{\underset{|}{CH_3CH:^-Li^+}}} + \underset{\text{Acid}}{D\overset{\curvearrowright}{-}\overset{..}{\underset{..}{O}}-D} \longrightarrow \overset{\displaystyle CH_3}{\underset{|}{CH_3CH}}-D + {}^-:\overset{..}{\underset{..}{O}}-D$$

You might now want to rewrite the equations in Problem 2.15 using the curved-arrow notation.

2.17 PHYSICAL PROPERTIES AND MOLECULAR STRUCTURE

So far, we have said little about one of the most obvious characteristics of organic compounds, that is, *their physical state.* Whether a particular substance is a solid, or a liquid, or a gas would certainly be one of the first observations that we would note in any experimental work. The temperatures at which transitions occur between physical states, that is, melting points and boiling points, are also among the more easily measured physical properties. Melting points and boiling points are also useful in identifying and isolating organic compounds.

Suppose, for example, we have just carried out the synthesis of an organic compound that is known to be a liquid at room temperature and 1-atm pressure. If we know the boiling point of our desired product, and the boiling points of other by-products and solvents that may be present in the reaction mixture, we can decide whether or not simple distillation will be a feasible method for isolating our product.

In another instance our product might be a solid. In this case, in order to isolate the substance by crystallization, we need to know its melting point and its solubility in different solvents.

The physical constants of known organic substances are easily found in handbooks and journals.* Table 2.3 lists the melting and boiling points of some of the compounds that we have discussed in this chapter.

Often in the course of research, however, the product of a synthesis is a new compound—one that has never been described before. In these instances, success in isolating the new compound depends on making reasonably accurate estimates of its melting point, boiling point, and solubilities. Estimations of these macroscopic physical properties are based on the most likely structure of the substance and on the forces that act between molecules and ions. What are these forces? How do they affect the melting point, boiling point, and solubilities of a compound?

2.17A Ion–Ion Forces

The **melting point** of a substance is the temperature at which an equilibrium exists between the well-ordered crystalline state and the more random liquid state. If the substance is an ionic compound, such as sodium acetate (Table 2.3), the forces that hold the ions together in the crystalline state are the strong electrostatic lattice forces that act between the positive and negative ions in the orderly crystalline structure. In Fig. 2.14 each sodium ion is surrounded by negatively charged acetate ions, and each acetate ion is surrounded by positive sodium ions. A large amount of thermal energy is required to break up the orderly structure of the crystal into the disorderly open

---

*Two useful handbooks are *Handbook of Chemistry,* N. A. Lange, Ed., McGraw-Hill, New York and *CRC Handbook of Chemistry and Physics,* CRC, Boca Raton, FL.

TABLE 2.3

COMPOUND	STRUCTURE	mp (°C)	bp (°C) (1 atm)
Methane	CH_4	-182	-162
Ethane	CH_3CH_3	-183	-88.2
Ethene	$CH_2{=}CH_2$	-169	-102
Ethyne	$HC{\equiv}CH$	-82	84 subl.[a]
Chloromethane	CH_3Cl	-97	-23.7
Chloroethane	CH_3CH_2Cl	-138.7	13.1
Ethyl alcohol	CH_3CH_2OH	-115	78.5
Acetaldehyde	CH_3CHO	-121	20
Acetic acid	CH_3CO_2H	16.6	118
Sodium acetate	CH_3CO_2Na	324	dec.[a]
Ethylamine	$CH_3CH_2NH_2$	-80	17
Diethyl ether	$(CH_3CH_2)_2O$	-116	34.6
Ethyl acetate	$CH_3CO_2CH_2CH_3$	-84	77

[a]In this table dec. = decompose and subl. = sublimes.

structure of a liquid. As a result, the temperature at which sodium acetate melts is quite high, 324 °C. The *boiling points* of ionic compounds are higher still, so high that most ionic organic compounds decompose before they boil. Sodium acetate shows this behavior.

2.17B Dipole–Dipole Forces

Most organic molecules are not fully ionic, but rather have a *permanent dipole moment* resulting from a nonuniform distribution of the bonding electrons (Section 1.19). Chloromethane and chloroethane are examples of molecules with permanent dipoles. In these compounds, the attractive forces between molecules are much easier to visualize. In the liquid or solid state, dipole–dipole attractions cause the molecules to orient themselves so that the positive end of one molecule is directed toward the negative end of another (Fig. 2.15).

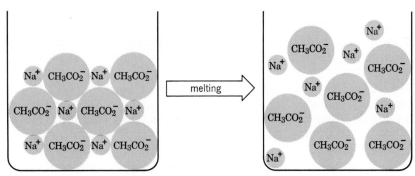

FIGURE 2.14 The melting of sodium acetate.

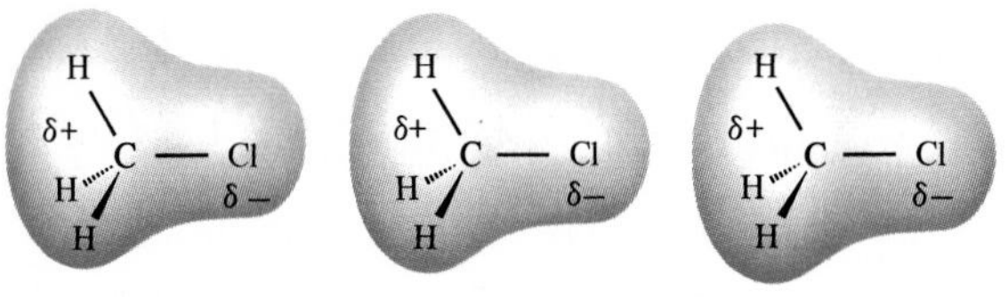

FIGURE 2.15 Dipole–dipole interactions between chloromethane molecules.

2.17C Hydrogen Bonds

Very strong dipole–dipole attractions occur between hydrogen atoms bonded to small, strongly electronegative atoms (O, N, F, or Cl) and nonbonding electron pairs on other such electronegative atoms (Fig. 2.16). This type of intermolecular force is called a **hydrogen bond.** The hydrogen bond (bond dissociation energy about 1–9 kcal/mole) is weaker than an ordinary covalent bond, but is much stronger than the dipole–dipole interactions that occur in chloromethane.

FIGURE 2.16 The hydrogen bond. Z is a strongly electronegative element, usually oxygen, nitrogen, fluorine, or chlorine.

Hydrogen bonding accounts for the fact that ethyl alcohol has a much higher boiling point ($+78.5$ °C) than dimethyl ether (-24.9 °C) even though the two compounds have the same molecular weight. Molecules of ethyl alcohol, because they have a hydrogen atom covalently bonded to an oxygen atom, can form strong hydrogen bonds to each other.

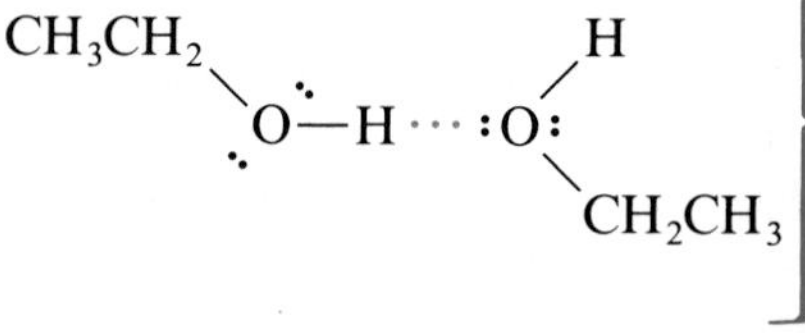

The dotted bond is a hydrogen bond. Strong hydrogen bonding is limited to molecules having a hydrogen atom attached to O, N, F, or Cl atoms

Molecules of dimethyl ether, because they lack a hydrogen atom attached to a strongly electronegative atom, cannot form strong hydrogen bonds to each other. In dimethyl ether the intermolecular forces are weaker dipole–dipole interactions.

PROBLEM 2.16

Explain why $(CH_3)_3N$ (trimethylamine) has a considerably lower boiling point (3 °C) than $CH_3CH_2CH_2NH_2$ (propylamine) (49 °C), even though these two compounds have the same molecular weight.

A factor (in addition to polarity and hydrogen bonding) that affects the *melting point* of many organic compounds is the compactness and rigidity of their individual molecules. Molecules that are symmetrical and rigid fit easily into a crystal lattice and thus their compounds have higher melting points. *tert*-Butyl alcohol, for example, has a much higher melting point than the other isomeric alcohols shown here.

$$CH_3-\overset{\overset{\displaystyle CH_3}{|}}{\underset{\underset{\displaystyle CH_3}{|}}{C}}-OH$$

tert-Butyl alcohol
mp, 25 °C

$$CH_3CH_2CH_2CH_2OH$$

Butyl alcohol
mp, −90 °C

$$CH_3\overset{\overset{\displaystyle CH_3}{|}}{CH}CH_2OH$$

Isobutyl alcohol
mp, −108 °C

$$CH_3CH_2\overset{\overset{\displaystyle CH_3}{|}}{CH}OH$$

sec-Butyl alcohol
mp, −114 °C

PROBLEM 2.17

Which compound would you expect to have the higher melting point, propane or cyclopropane? Explain your answer.

2.17D van der Waals Forces

If we consider a substance like methane where the particles are nonpolar molecules, we find that the melting point and boiling point are very low: −183 °C and −162 °C, respectively. Rather than ask, "Why does methane melt and boil at low temperatures?" a more appropriate question might be "Why does methane, a nonionic, nonpolar substance, become a liquid or a solid at all?" The answer to this question can be given in terms of attractive intermolecular forces called **van der Waals forces** (or **London forces**).

An accurate account of the nature of van der Waals forces requires the use of quantum mechanics. We can, however, visualize the origin of these forces in the following way. The average distribution of charge in a nonpolar molecule (like methane) over a period of time is uniform. At any given instant, however, *because electrons move,* the electrons and thus the charge may not be uniformly distributed. Electrons may, in one instant, be slightly accumulated on one part of the molecule and, as a consequence, *a small temporary dipole will occur* (Fig. 2.17). This temporary dipole in one molecule can induce opposite (attractive) dipoles in surrounding molecules. It does this because the negative (or positive) charge in a portion of one molecule will distort the electron cloud of an adjacent portion of another molecule causing an opposite charge to develop there. These temporary dipoles change constantly, but the net result of their existence is to produce attractive forces between nonpolar molecules, and thus make possible the existence of their liquid and solid states.

One important factor that determines the magnitude of van der Waals forces is the relative **polarizability** of the electrons of the atoms involved. By polarizability we mean *the ability of the electrons to respond to a changing electric field.* Relative polarizability depends on how loosely or tightly the electrons are held. In the halogen family, for example, polarizability increases in the order F < Cl < Br < I. Fluorine atoms show a very low polarizability because their electrons are very tightly held; they are close to the nucleus. Iodine atoms are easily polarized. Their electrons are far from the nucleus. Atoms with unshared pairs are generally more polarizable than those

FIGURE 2.17 Temporary dipoles and induced dipoles in nonpolar molecules resulting from a nonuniform distribution of electrons at a given instant.

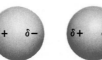

TABLE 2.4 Attractive energies in simple molecular solids

| MOLECULE | DIPOLE MOMENT (D) | ATTRACTIVE ENERGIES (kcal/mole) | | MELTING POINT (°C) | BOILING POINT (°C) |
		DIPOLE– DIPOLE	VAN DER WAALS		
H_2O	1.85	8.7[a]	2.1	0	100
NH_3	1.47	3.3[a]	3.5	−78	−33
HCl	1.08	0.8[a]	4.0	−115	−85
HBr	0.80	0.2	5.2	−88	−67
HI	0.42	0.006	6.7	−51	−35

[a]These dipole–dipole attractions are called hydrogen bonds.

with bonding pairs. Thus a halogen substituent is more polarizable than an alkyl group of comparable size. Table 2.4 gives the relative magnitude of van der Waals forces and dipole–dipole interactions for several simple compounds.

The *boiling point* of a liquid is the temperature at which the vapor pressure of the liquid equals the pressure of the atmosphere above it. For this reason, the boiling points of liquids are *pressure dependent,* and boiling points are always reported as occurring at a particular pressure, as 1 atm (or at 760 Torr), for example. A substance that boils at 150 °C at 1-atm pressure will boil at a substantially lower temperature if the pressure is reduced to, for example, 0.01 Torr (a pressure easily obtained with a vacuum pump). The normal boiling point given for a liquid is its boiling point at 1 atm.

In passing from a liquid to a gaseous state the individual molecules (or ions) of the substance must separate considerably. Because of this, we can understand why ionic organic compounds often decompose before they boil. The thermal energy required to completely separate (volatilize) the ions is so great that chemical reactions (decompositions) occur first.

Nonpolar compounds, where the intermolecular forces are very weak, usually boil at low temperatures even at 1-atm pressure. This is not always true, however, because of other factors that we have not yet mentioned: the effects of molecular weight and molecular size. Heavier molecules require greater thermal energy in order to acquire velocities sufficiently great to escape the liquid surface, and because their surface areas are usually much greater, intermolecular van der Waals attractions are also much larger. These factors explain why nonpolar ethane (bp, −88.2 °C) boils higher than methane (bp, −162 °C) at a pressure of 1 atm. It also explains why, at 1 atm, the even heavier and larger nonpolar molecule decane ($C_{10}H_{22}$) boils at +174 °C.

2.17E Solubilities

Intermolecular forces are of primary importance in explaining the **solubilities** of substances. Dissolution of a solid in a liquid is, in many respects, like the melting of a solid. The orderly crystal structure of the solid is destroyed, and the result is the formation of the more disorderly arrangement of the molecules (or ions) in solution. In the process of dissolving, too, the molecules or ions must be separated from each other, and energy must be supplied for both changes. The energy required to overcome lattice energies and intermolecular or interionic attractions comes from the formation of new attractive forces between solute and solvent.

Consider the dissolution of an ionic substance as an example. Here both the lattice energy and interionic attractions are large. We find that water and only a few other very polar solvents are capable of dissolving ionic compounds. These solvents dissolve ionic compounds by **hydrating** or **solvating** the ions (Fig. 2.18).

Water molecules, by virtue of their great polarity, as well as their very small compact shape, can very effectively surround the individual ions as they are freed from the crystal surface. Positive ions are surrounded by water molecules with the negative end of the water dipole pointed toward the positive ion; negative ions are solvated in exactly the opposite way. Because water is highly polar, and because water is capable of forming strong hydrogen bonds, the *dipole–ion* attractive forces are also large. The energy supplied by the formation of these forces is great enough to overcome both the lattice energy and interionic attractions of the crystal.

A rule of thumb for predicting solubilities is that "like dissolves like." Polar and ionic compounds tend to dissolve in polar solvents. Polar liquids are generally miscible with each other. Nonpolar solids are usually soluble in nonpolar solvents. On the other hand, nonpolar solids are insoluble in polar solvents. Nonpolar liquids are usually mutually miscible, but nonpolar liquids and polar liquids "like oil and water" do not mix.

We can understand why this is true if we understand that when substances of similar polarities are mixed, the "new" intermolecular forces that form in the solution are very much like those that existed in the separate substances. The miscibility of nonpolar carbon tetrachloride with a nonpolar alkane would be an example. Very polar water molecules are probably capable of inducing polarities in alkane molecules that are sufficiently large to form attractive forces between them. Water and alkanes are not soluble in each other, however, because dissolution of the alkane in water requires the separation of strongly attractive water molecules from each other.

Ethanol and water, by contrast, are miscible in all proportions. In this example, both molecules are highly polar and the new attractive forces are as strong as those they replace and, in this instance, both compounds are capable of forming strong hydrogen bonds.

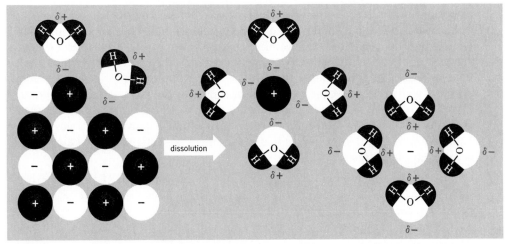

FIGURE 2.18 The dissolution of an ionic solid in water showing the hydration of positive and negative ions by the very polar water molecules.

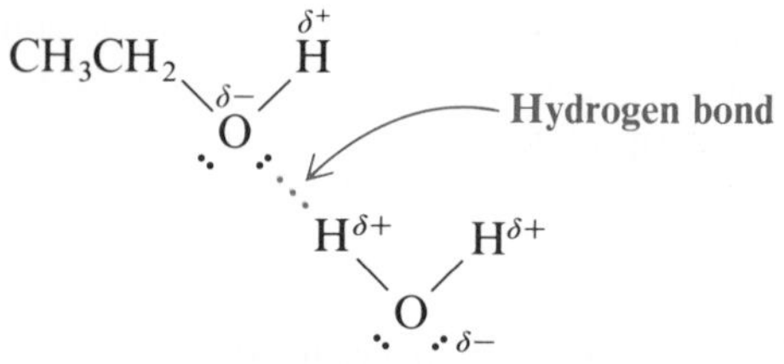

If the carbon chain of an alcohol is long, however, we find that the alcohol is much less soluble in water. Decyl alcohol (see following structure) with a chain of 10 carbon atoms is only very slightly soluble in water. Decyl alcohol resembles an alkane more than it does water. The long carbon chain of decyl alcohol is said to be **hydrophobic** (*hydro,* water; *phobic,* fearing or avoiding—"water avoiding"). Only the OH group, a rather small part of the molecule, is **hydrophilic** (*philic,* loving or seeking—"water seeking"). On the other hand, decyl alcohol is quite soluble in nonpolar solvents.

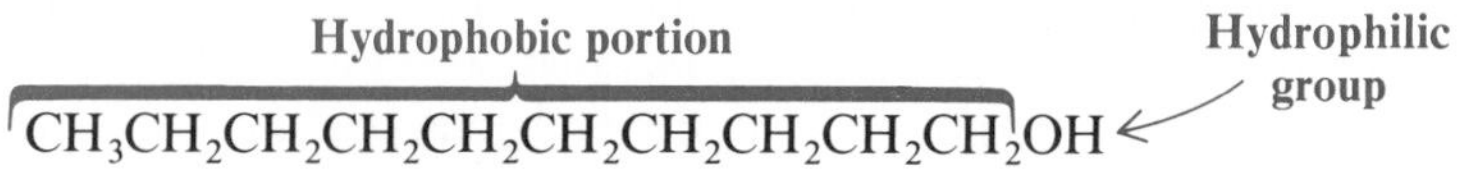

2.17F Guidelines for Water Solubility

Organic chemists usually define a compound as water soluble if at least 3 g of the organic compound dissolve in 100 mL of water. We find that for compounds containing nitrogen or oxygen atoms—and thus capable of forming strong hydrogen bonds—the following approximate guidelines hold: Compounds with one to three carbon atoms are water soluble, compounds with four or five carbon atoms are borderline, and compounds with six carbon atoms or more are insoluble.

2.17G Intermolecular Forces in Biochemistry

Later, after we have had a chance to examine in detail the properties of the molecules that make up living organisms, we shall see how intermolecular forces are extremely important in the functioning of cells. Hydrogen bond formation, the hydration of polar groups, and the tendency of nonpolar groups to avoid a polar environment all cause complex protein molecules to fold in precise ways—ways that allow them to function as biological catalysts of incredible efficiency. The same factors allow molecules of hemoglobin to assume the shape needed to transport oxygen. They allow proteins and molecules called glycosphingolipids to function as cell membranes. Hydrogen bonding alone gives molecules of certain carbohydrates a globular shape that makes them highly efficient food reserves in animals. It gives molecules of other carbohydrates a rigid linear shape that makes them perfectly suited to be structural components in plants.

2.18 SUMMARY OF ATTRACTIVE ELECTRIC FORCES

The attractive forces occurring between molecules and ions that we have studied so far are summarized in Table 2.5.

TABLE 2.5 Attractive electric forces

ELECTRIC FORCE	RELATIVE STRENGTH	TYPE	EXAMPLE
Cation–anion (in a crystal)	Very strong	$\oplus \ominus$	Lithium fluoride crystal lattice
Covalent bonds	Strong (36–125 kcal/mole)	Shared electron pairs	H—H (104 kcal/mole) CH_3—CH_3 (88 kcal/mole) I—I (36 kcal/mole)
Ion–dipole	Moderate	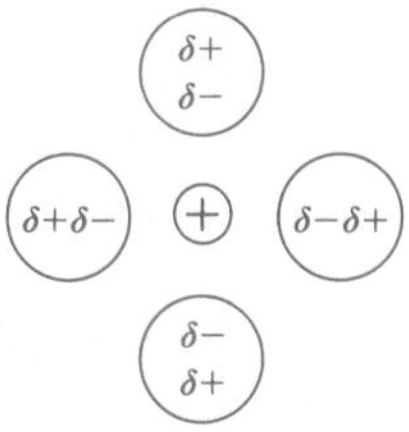	Na^+ in water (see Fig. 2.18)
Dipole–dipole (including hydrogen bonds)	Moderate-weak (1–9 kcal/mole)	$\overset{\delta-}{-Z}\!:\cdots\overset{\delta+}{H}-$ and	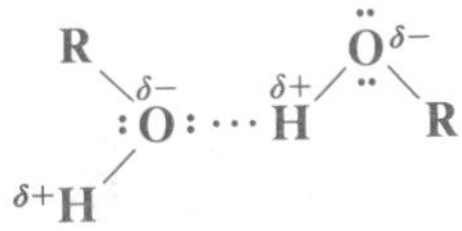
van der Waals	Variable	Fluctuating dipole	Interactions between methane molecules

2.19 SUMMARY OF IMPORTANT FAMILIES OF ORGANIC COMPOUNDS

A summary of the important families of organic compounds including their functional groups is given on the inside of the front cover of this book.

ADDITIONAL PROBLEMS

2.18

Classify each of the following compounds as an alkane, alkene, alkyne, alcohol, or aldehyde, and so forth.

(a) $CH_3C\equiv CCH_3$ (b) $CH_3\underset{\overset{|}{CH_3}}{CH}CH_2\overset{\overset{O}{\parallel}}{C}OH$ (c) 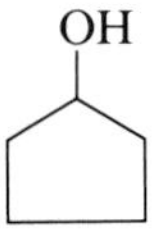 [cyclopentane with OH] (d) 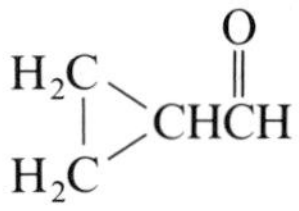 $\underset{H_2C}{\overset{H_2C}{\diagdown}}{\diagup}CHCH\overset{O}{\parallel}$

(e) $CH_3\underset{\overset{|}{CH_3}}{CH}CH_2CH_2CH_2CH_2CH_2CH_2CH_2CH_2CH_2CH_2CH_2CH_2CH_3$

(a sex attractant of the female tiger moth)

(f) 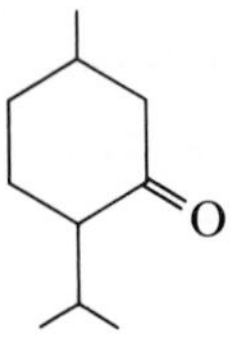 (obtained from peppermint oil)

2.19
Identify all of the functional groups in each of the following compounds:

(a) 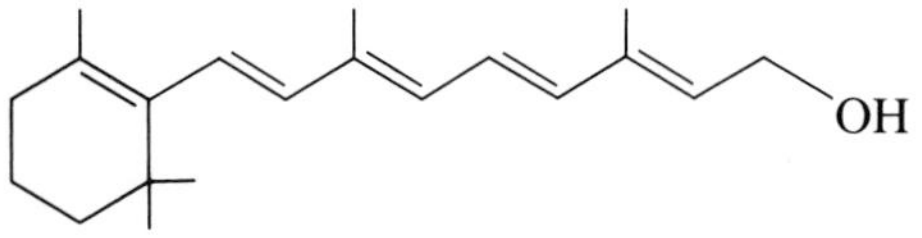 OH

(vitamin A$_1$)

(b) 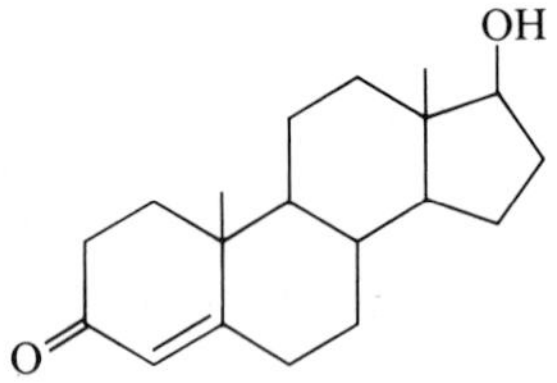 (testosterone, a male sex hormone)

(c) 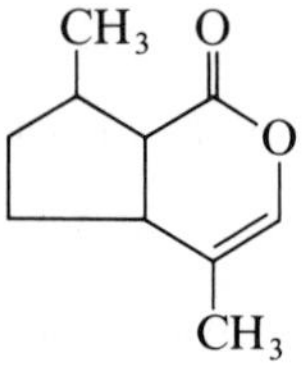 (nepetalactone, one constituent of catnip)

(d) $\left[NH-(CH_2)_6-NH\overset{\overset{\text{O}}{\|}}{C}-(CH_2)_4-\overset{\overset{\text{O}}{\|}}{C} \right]_n$

(a nylon)

(e)

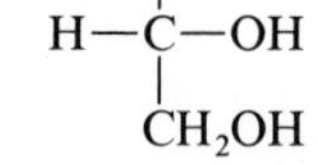

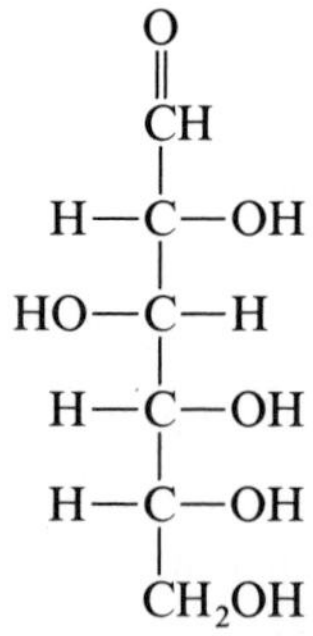

$$
\begin{array}{c}
\overset{\overset{\displaystyle\text{O}}{\|}}{\text{CH}} \\
| \\
\text{H}-\text{C}-\text{OH} \\
| \\
\text{HO}-\text{C}-\text{H} \\
| \\
\text{H}-\text{C}-\text{OH} \\
| \\
\text{H}-\text{C}-\text{OH} \\
| \\
\text{CH}_2\text{OH}
\end{array}
$$

(glucose, a sugar)

(f) $CH_2{=}CH{-}O{-}CH{=}CH_2$

(an anesthetic)

(g) 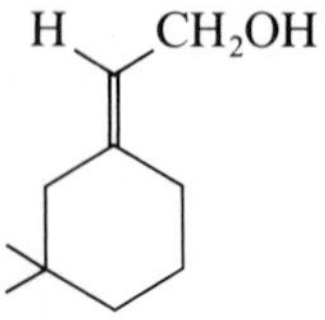 (male boll weevil sex attractant)

(h)

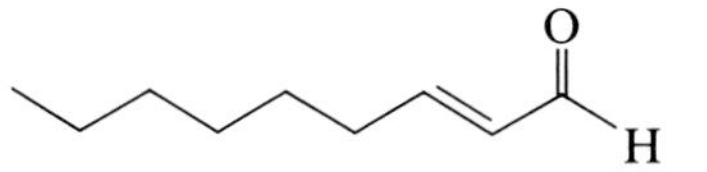

(a cockroach repellent found
in sliced cucumbers)

(i)

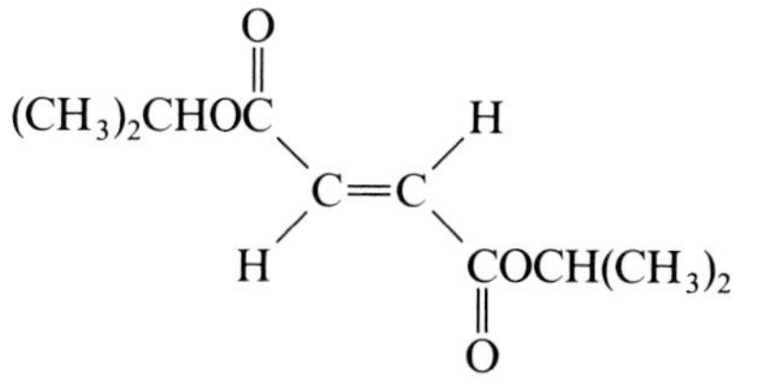

(a synthetic cockroach repellent)

2.20

There are four alkyl bromides with the formula C_4H_9Br. Write their structural formulas and classify each as to whether it is a primary, secondary, or tertiary alkyl bromide.

2.21

There are seven isomeric compounds with the formula $C_4H_{10}O$. Write their structures and classify each compound according to its functional group.

2.22

Write structural formulas for four compounds with the formula C_3H_6O and classify each according to its functional group.

2.23

Classify the following alcohols as primary, secondary, or tertiary.

(a) $CH_3CH_2CH_2CH_2OH$

(b) $CH_3CH_2CH_2CH(OH)CH_3$

(c) $CH_3CH_2C(OH)(CH_3)_2$

(d) $CH_3CH(CH_3)CH(OH)CH_3$

(e)

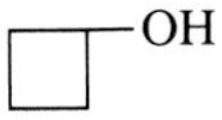

(f)

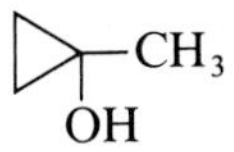

2.24

Classify the following amines as primary, secondary, or tertiary.

(a) $CH_3CH_2NHCH_2CH_3$

(b) $CH_3CH(NH_2)CH_2CH_3$

(c) 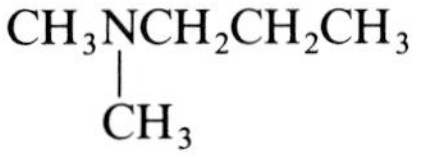 $CH_3NCH_2CH_2CH_3$
     |
     CH_3

(d)

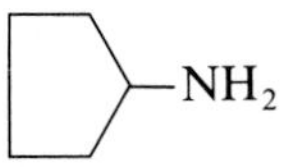

(e)

2.25

Write structural formulas for each of the following:

(a) An ether with the formula C_3H_8O

(b) A primary alcohol with the formula C_3H_8O

(c) A secondary alcohol with the formula C_3H_8O

(d) Two esters with the formula $C_4H_8O_2$

(e) A primary alkyl halide with the formula C_4H_9X

(f) A secondary alkyl halide with the formula C_4H_9X

(g) A tertiary alkyl halide with the formula C_4H_9X

(h) An aldehyde with the formula C_4H_8O

(i) A ketone with the formula C_4H_8O

(j) A primary amine with the formula $C_4H_{11}N$

(k) A secondary amine with the formula $C_4H_{11}N$

(l) A tertiary amine with the formula $C_4H_{11}N$

(m) An amide of ammonia with the formula C_4H_9NO

(n) An N-substituted amide with the formula C_4H_9NO

(o) A tertiary alcohol with the formula C_4H_8O containing no multiple bonds

2.26

Write net ionic equations for the following acid–base reactions:

(a) $HCl_{(aq)} + Na_2CO_{3(aq)} \longrightarrow H_2O + CO_2 + 2NaCl_{(aq)}$

(b) $HBr_{(aq)} + CH_3\overset{\overset{O}{\|}}{C}ONa_{(aq)} \longrightarrow CH_3\overset{\overset{O}{\|}}{C}OH_{(aq)} + NaBr_{(aq)}$

(c) $Na_2CO_{3(aq)} + H_2O \longrightarrow NaHCO_{3(aq)} + NaOH_{(aq)}$

(d) $NaH + H_2O \longrightarrow H_2 + NaOH_{(aq)}$

(e) $CH_3Li + H_2O \longrightarrow CH_4 + LiOH_{(aq)}$

(f) $CH_3Li + HC{\equiv}CH \longrightarrow LiC{\equiv}CH + CH_4$

(g) $HCl_{(aq)} + NH_{3(aq)} \longrightarrow NH_4Cl_{(aq)}$

(h) $NH_4Cl + NaNH_2 \xrightarrow{NH_3} 2NH_3 + NaCl$

(i) $CH_3CH_2ONa + H_2O \longrightarrow CH_3CH_2OH_{(aq)} + NaOH_{(aq)}$

2.27

Explain why almost all oxygen-containing organic compounds dissolve in concentrated sulfuric acid.

2.28

Which compound in each of the following pairs would have the higher boiling point?

(a) Ethyl alcohol (CH_3CH_2OH) or methyl ether (CH_3OCH_3)

(b) Ethylene glycol ($HOCH_2CH_2OH$) or ethyl alcohol (CH_3CH_2OH)

(c) Pentane (C_5H_{12}) or heptane (C_7H_{16})

(d) Acetone $\left(CH_3\overset{\overset{O}{\|}}{C}CH_3\right)$ or propyl alcohol ($CH_3CH_2CH_2OH$)

(e) cis-1,2-Dichloroethene, $\overset{Cl}{\underset{H}{}}C{=}C\overset{Cl}{\underset{H}{}}$ or trans-1,2-dichloroethene, $\overset{Cl}{\underset{H}{}}C{=}C\overset{H}{\underset{Cl}{}}$

(f) Propionic acid $\left(CH_3CH_2\overset{\overset{O}{\|}}{C}OH\right)$ or methyl acetate $\left(CH_3\overset{\overset{O}{\|}}{C}{-}OCH_3\right)$

2.29

There are four amides with the formula C_3H_7NO. (a) Write their structures. (b) One of these amides has a melting and boiling point that is substantially lower than that of the other three. Which amide is this? Explain your answer.

2.30

What reaction will take place if ethyl alcohol is added to a solution of $HC \equiv C:^-Na^+$ in liquid ammonia?

2.31

(a) The K_a of acetic acid is 1.8×10^{-5}. What is its pK_a? (b) What is the K_a of an acid whose pK_a is 13?

2.32

Acid HA has a pK_a equal to 20; acid HB has a pK_a equal to 10. (a) Which is the stronger acid? (b) Will an acid–base reaction with an equilibrium lying to the right take place if $Na^+A:^-$ is added to HB? Explain your answer.

2.33

$(CF_3)_3N$ is a weaker base than $(CH_3)_3N$. Explain.

2.34

Cyclic compounds of the general type shown here are called lactones. What functional group do they contain?

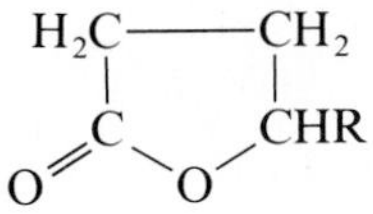

2.35

Hydrogen fluoride has a dipole moment of 1.82 D; its boiling point is 19.34 °C. Ethyl fluoride (CH_3CH_2F) has an almost identical dipole moment and has a larger molecular weight, yet its boiling point is -37.7 °C. Explain.

CHAPTER THREE

Triphenylmethane. Triphenylmethane is $(C_6H_5)_3CH$. The C_6H_5— group, called the phenyl group, is discussed in Section 2.7 and in Chapter 11.

Alkanes and Cycloalkanes. Conformational Analysis

3.1 INTRODUCTION TO ALKANES AND CYCLOALKANES

We noted earlier that the family of organic compounds called hydrocarbons can be divided into several groups based on the type of bond that exists between the individual carbon atoms. Those hydrocarbons in which all of the carbon–carbon bonds are single bonds are called *alkanes;* those hydrocarbons that contain a carbon–carbon double bond are called *alkenes;* and those with a carbon–carbon triple bond are called *alkynes.*

Cycloalkanes are alkanes in which the carbon atoms are arranged in a ring. Alkanes have the general formula C_nH_{2n+2}; cycloalkanes have two fewer hydrogen atoms and thus have the general formula C_nH_{2n}.

Alkanes and cycloalkanes are so similar that many of their basic properties can be considered side by side. Some differences remain, however, and certain structural features arise from the rings of cycloalkanes that are more conveniently studied separately. We shall point out the chemical and physical similarities of alkanes and cycloalkanes as we go along.

3.2 SHAPES OF ALKANES

A general tetrahedral orientation of groups—and thus sp^3 hybridization—is the rule for the carbon atoms of all alkanes and cycloalkanes. Using circle-and-line formulas, we can represent the shapes of alkanes as shown in Fig. 3.1.

Butane and pentane are examples of alkanes that are sometimes called "straight-chain" alkanes. One glance at their three-dimensional structures shows that because of the tetrahedral carbon atoms their chains are zigzagged and not at all straight. Indeed, the structures that we have depicted in Fig. 3.1 are the straightest possible arrangements of the chains, for rotations about the carbon–carbon single bonds produce arrangements that are even less straight. The better description is **unbranched.** This means that each carbon atom within the chain is bonded to no more than two other carbon atoms and that unbranched alkanes contain only primary and secondary carbon atoms. Unbranched alkanes used to be called "normal" alkanes or *n*-alkanes, but the prefix *n*- is archaic and should not be used now.

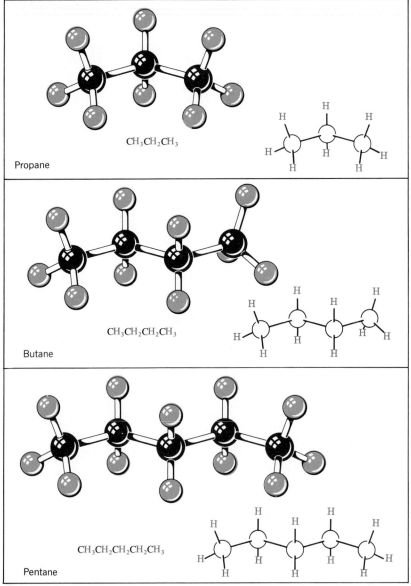

FIGURE 3.1 Ball-and-stick models and line-and-circle formulas for three simple alkanes.

Isobutane and isopentane (Fig. 3.2) are examples of branched-chain alkanes. In neopentane (Section 3.3C) the central carbon atom is bonded to four carbon atoms.

Butane and isobutane have the same molecular formula: C_4H_{10}. The two compounds have their atoms connected in a different order and are, therefore, *constitutional isomers.* Pentane, isopentane, and neopentane are also constitutional isomers. They, too, have the same molecular formula (C_5H_{12}) but have different structures.

PROBLEM 3.1

Write condensed structural formulas for all of the constitutional isomers of C_6H_{14}. Compare your answers with the condensed structural formulas given in Table 3.1.

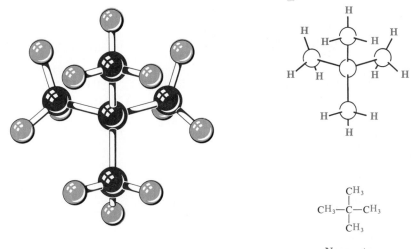

$$CH_3-\underset{\underset{\textstyle CH_3}{|}}{\overset{\overset{\textstyle CH_3}{|}}{C}}-CH_3$$

Neopentane

Constitutional isomers, as stated earlier, have different physical properties. The differences may not always be large, but constitutional isomers will always be found to have different melting points, boiling points, densities, indexes of refraction, and so forth. Table 3.1 gives some of the physical properties of the C_4H_{10}, C_5H_{12}, and C_6H_{14} isomers.

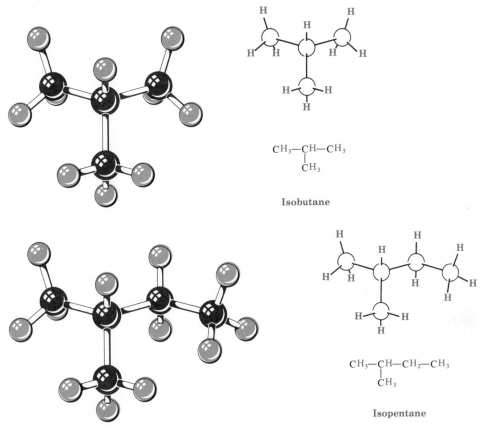

$$CH_3-\underset{\underset{\textstyle CH_3}{|}}{CH}-CH_3$$

Isobutane

$$CH_3-\underset{\underset{\textstyle CH_3}{|}}{CH}-CH_2-CH_3$$

Isopentane

FIGURE 3.2 Ball-and-stick models and line-and-circle formulas for two branched-chain alkanes. In each of the compounds one carbon atom is attached to three other carbon atoms.

TABLE 3.1 Physical constants of the butane, pentane, and hexane isomers

MOLECULAR FORMULA	STRUCTURAL FORMULA	mp (°C)	bp (°C)[a] (1 atm)	DENSITY[b] (g/mL)	INDEX OF REFRACTION[c] (n_D 20 °C)
C_4H_{10}	$CH_3CH_2CH_2CH_3$	−138.3	−0.5	0.6012^0	1.3543
C_4H_{10}	CH_3CHCH_3 | CH_3	−159	−12	0.603^0	—
C_5H_{12}	$CH_3CH_2CH_2CH_2CH_3$	−129.72	36	0.6262^{20}	1.3579
C_5H_{12}	$CH_3CHCH_2CH_3$ | CH_3	−160	27.9	0.6197^{20}	1.3537
C_5H_{12}	CH_3 | $CH_3{-}C{-}CH_3$ | CH_3	−20	9.45	0.61350^{20}	1.3476
C_6H_{14}	$CH_3CH_2CH_2CH_2CH_2CH_3$	−95	68	0.65937^{20}	1.3748
C_6H_{14}	$CH_3CHCH_2CH_2CH_3$ | CH_3	−153.67	60.3	0.6532^{20}	1.3714
C_6H_{14}	$CH_3CH_2CHCH_2CH_3$ | CH_3	−118	63.265	0.6643^{20}	1.3765
C_6H_{14}	$CH_3CH{-}CHCH_3$ | | CH_3 CH_3	−128.8	58	0.6616^{20}	1.3750
C_6H_{14}	CH_3 | $CH_3{-}C{-}CH_2CH_3$ | CH_3	−98	49.7	0.6492^{20}	1.3688

[a]Unless otherwise indicated, all boiling points are at 1 atm or 760 Torr.

[b]The superscript indicates the temperature at which the density was measured.

[c]The index of refraction is a measure of the ability of the alkane to bend (refract) light rays. The values reported are for light of the D line of the sodium spectrum (n_D).

3.3 IUPAC NOMENCLATURE OF ALKANES, ALKYL HALIDES, AND ALCOHOLS

The development of a formal system for naming organic compounds did not come about until near the end of the nineteenth century. Prior to that time many organic compounds had already been discovered. The names given these compounds sometimes reflected a source of the compound. Acetic acid, for example, can be obtained from vinegar; it got its name from the Latin word for vinegar, *acetum*. Formic acid can be obtained from some ants; it got its name from the Latin word for ants, *formicae.* Ethanol (or ethyl alcohol) was at one time called grain alcohol because it was obtained by the fermentation of grains.

These older names for organic compounds are now called "common" or "trivial" names. Many of these names are still widely used by chemists, biochemists, and the companies that sell chemicals. For this reason it is still necessary to learn the common names for some of the common compounds. We shall point out these common names as we go along, and we shall use them occasionally. Most of the time, however, the names that we shall use will be those called IUPAC names.

The formal system of nomenclature used today is one proposed by the International Union of Pure and Applied Chemistry (IUPAC). This system was first developed in 1892 and has been revised at irregular intervals to keep it up to date. Underlying the IUPAC system of nomenclature for organic compounds is a fundamental principle: *Each different compound should have a different name.* Thus, through a systematic set of rules, the IUPAC system provides different names for the more than 6 million known organic compounds, and names can be devised for any one of millions of other compounds yet to be synthesized. In addition, the IUPAC system is simple enough to allow any chemist familiar with the rules (or with the rules at hand) to write the name for any compound that might be encountered. In the same way, one is also able to derive the structure of a given compound from its IUPAC name.

The IUPAC system for naming alkanes is not difficult to learn, and the principles involved are used in naming compounds in other families as well. For these reasons we begin our study of the IUPAC system with the rules for naming alkanes, alkyl halides, and alcohols.

The names for several of the unbranched alkanes are listed in Table 3.2. The ending for all of the names of alkanes is *-ane.* The prefixes of the names of most of the alkanes (above C_4) are of Greek and Latin origin. Learning the prefixes is like learning to count in organic chemistry. Thus, one, two, three, four, five, becomes meth-, eth-, prop-, but-, pent-.

3.3A Nomenclature of Alkyl Groups

If on paper we remove one hydrogen atom from an alkane, we obtain what is called **an alkyl group.** These alkyl groups have names that end in **-yl.** When the alkane is **unbranched,** and the hydrogen atom that is removed is a **terminal** hydrogen atom, the names are straightforward:

ALKANE		ALKYL GROUP	ABBREVIATION
CH_3—H Methane	becomes	CH_3— Methyl	Me
CH_3CH_2—H Ethane	becomes	CH_3CH_2— Ethyl	Et
$CH_3CH_2CH_2$—H Propane	becomes	$CH_3CH_2CH_2$— Propyl	Pr
$CH_3CH_2CH_2CH_2$—H Butane	becomes	$CH_3CH_2CH_2CH_2$— Butyl	Bu

For alkanes with more than two carbon atoms, however, more than one derived group is possible. Two groups can be derived from propane; the propyl group (above)

TABLE 3.2 The unbranched alkanes

NAME	NUMBER OF CARBON ATOMS	STRUCTURE
Methane	1	CH_4
Ethane	2	CH_3CH_3
Propane	3	$CH_3CH_2CH_3$
Butane	4	$CH_3(CH_2)_2CH_3$
Pentane	5	$CH_3(CH_2)_3CH_3$
Hexane	6	$CH_3(CH_2)_4CH_3$
Heptane	7	$CH_3(CH_2)_5CH_3$
Octane	8	$CH_3(CH_2)_6CH_3$
Nonane	9	$CH_3(CH_2)_7CH_3$
Decane	10	$CH_3(CH_2)_8CH_3$
Undecane	11	$CH_3(CH_2)_9CH_3$
Dodecane	12	$CH_3(CH_2)_{10}CH_3$
Tridecane	13	$CH_3(CH_2)_{11}CH_3$
Tetradecane	14	$CH_3(CH_2)_{12}CH_3$
Pentadecane	15	$CH_3(CH_2)_{13}CH_3$
Hexadecane	16	$CH_3(CH_2)_{14}CH_3$
Heptadecane	17	$CH_3(CH_2)_{15}CH_3$
Octadecane	18	$CH_3(CH_2)_{16}CH_3$
Nonadecane	19	$CH_3(CH_2)_{17}CH_3$
Eicosane	20	$CH_3(CH_2)_{18}CH_3$
Heneicosane	21	$CH_3(CH_2)_{19}CH_3$
Docosane	22	$CH_3(CH_2)_{20}CH_3$
Tricosane	23	$CH_3(CH_2)_{21}CH_3$
Triacontane	30	$CH_3(CH_2)_{28}CH_3$
Hentriacontane	31	$CH_3(CH_2)_{29}CH_3$
Tetracontane	40	$CH_3(CH_2)_{38}CH_3$
Pentacontane	50	$CH_3(CH_2)_{48}CH_3$
Hexacontane	60	$CH_3(CH_2)_{58}CH_3$
Heptacontane	70	$CH_3(CH_2)_{68}CH_3$
Octacontane	80	$CH_3(CH_2)_{78}CH_3$
Nonacontane	90	$CH_3(CH_2)_{88}CH_3$
Hectane	100	$CH_3(CH_2)_{98}CH_3$

is derived by removal of a terminal hydrogen atom, and the isopropyl group (below) is derived by removal of an inner hydrogen atom.

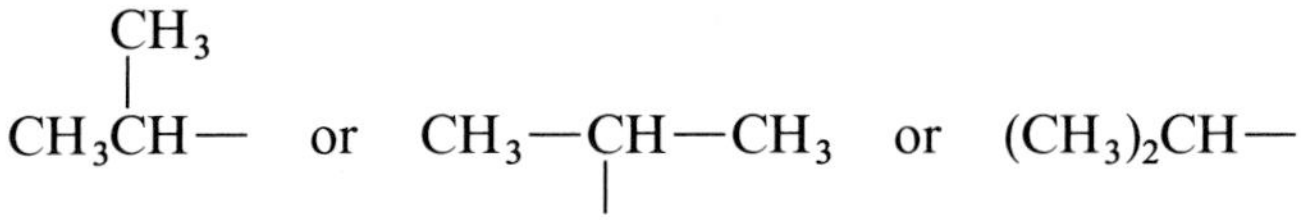

$$CH_3CH- \quad \text{or} \quad CH_3-CH-CH_3 \quad \text{or} \quad (CH_3)_2CH-$$

Isopropyl group

There are four alkyl groups that contain four carbon atoms each:

$$CH_3CH_2CH_2CH_2- \; [\text{or } CH_3(CH_2)_2CH_2-]$$
Butyl

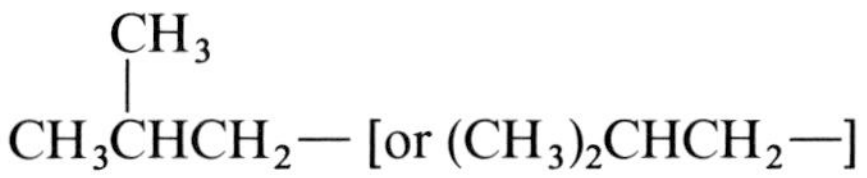

$$CH_3CHCH_2- \; [\text{or } (CH_3)_2CHCH_2-]$$
Isobutyl

$$CH_3CH_2CH- \; [\text{or } CH_3CH_2CH-] \qquad CH_3-C- \; [\text{or } (CH_3)_3C-]$$

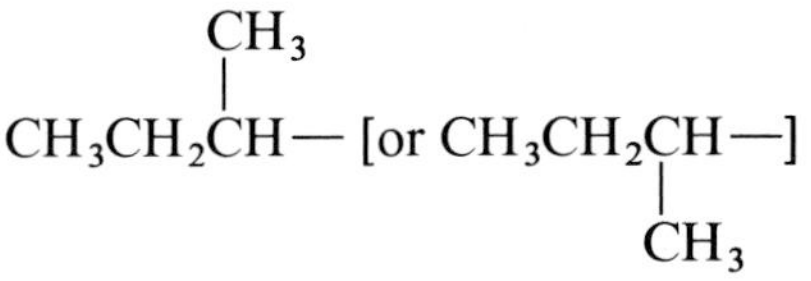

sec-**Butyl** 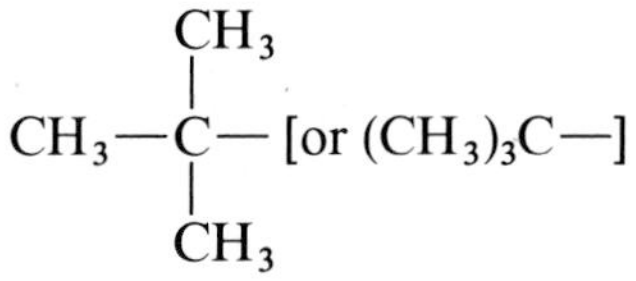   *tert*-**Butyl**

All of these groups should be memorized so well that you are able to recognize them when they are written backward or upside down. As an aid in learning these groups it may be helpful to point out the following characteristics.

1. The base name of any group relates to the total number of carbon atoms in the group. (The propyl and isopropyl groups have *three* carbon atoms and all of the "butyl" groups have *four* carbon atoms.)

2. The isopropyl and isobutyl groups both have an iso structure* shown here.

$$CH_3-CH- \qquad\qquad CH_3CH-CH_2-$$

Iso structures

*So, too, do the isopentyl ($CH_3CHCH_2CH_2-$) and isohexyl ($CH_3CHCH_2CH_2CH_2-$) groups, respectively. These names are also provided in the IUPAC system, and the IUPAC system allows the names isobutane (for CH_3CHCH_3), isopentane (for $CH_3CHCH_2CH_3$), and isohexane (for $CH_3CHCH_2CH_2CH_3$). The iso system stops after isohexyl (for groups) and isohexane (for alkanes), however.

3. The incomplete valence of the *sec*-butyl group is directed away from a *secondary* carbon atom.

$$CH_3CH_2\overset{\overset{\displaystyle CH_3 \;\text{— Secondary carbon atom}}{|}}{CH}—$$

sec-**Butyl**

4. The incomplete valence of the *tert*-butyl group is directed away from a *tertiary* carbon atom.

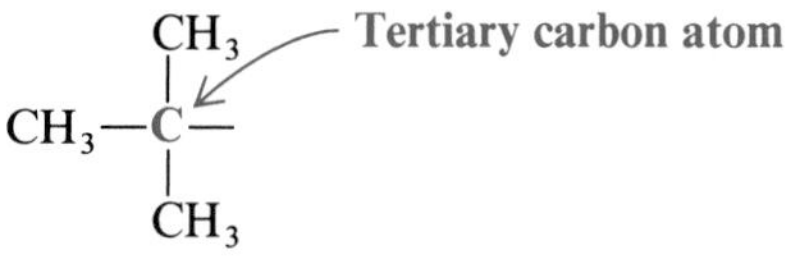

$$CH_3—\overset{\overset{\displaystyle CH_3 \;\text{— Tertiary carbon atom}}{|}}{\underset{\underset{\displaystyle CH_3}{|}}{C}}{—}$$

tert-**Butyl**

3.3B Nomenclature of Branched-Chain Alkanes

Branched-chain alkanes are named according to the following rules:

1. **Locate the longest continuous chain of carbon atoms; this chain determines the base name for the alkane.**

   We designate the following compound, for example, as a *hexane* because the longest continuous chain contains six carbon atoms.

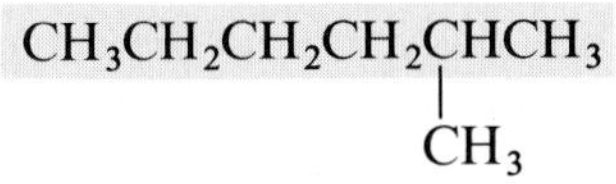

   $$CH_3CH_2CH_2CH_2\underset{\underset{\displaystyle CH_3}{|}}{CH}CH_3$$

   The longest continuous chain may not always be obvious from the way the formula is written. Notice, for example, that the following alkane is designated as a *heptane* because the longest chain contains seven carbon atoms.

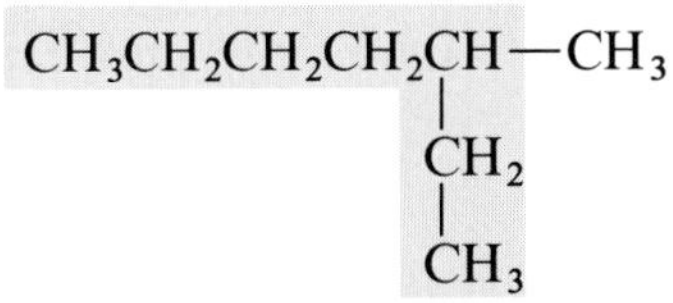

   $$CH_3CH_2CH_2CH_2\underset{\underset{\underset{\displaystyle CH_3}{|}}{\underset{\displaystyle CH_2}{|}}}{CH}—CH_3$$

2. **Number the longest chain beginning with the end of the chain nearer the substituent.**

   Applying this rule, we number the two alkanes that we illustrated previously in the following way.

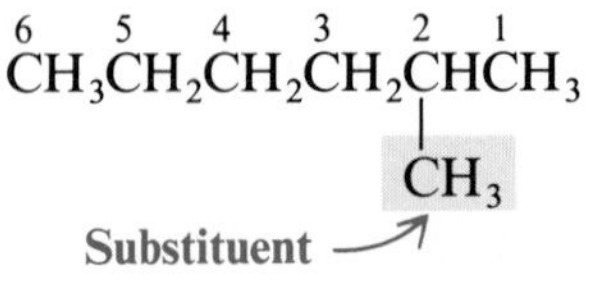

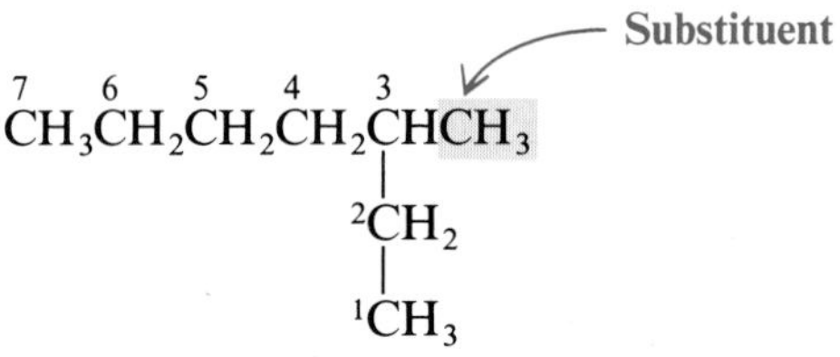

$$\overset{6}{C}H_3\overset{5}{C}H_2\overset{4}{C}H_2\overset{3}{C}H_2\overset{2}{\underset{\underset{\displaystyle CH_3}{|}}{C}}H\overset{1}{C}H_3$$

Substituent —↗

— Substituent ↙

$$\overset{7}{C}H_3\overset{6}{C}H_2\overset{5}{C}H_2\overset{4}{C}H_2\overset{3}{\underset{\underset{\underset{\displaystyle {}^1CH_3}{|}}{{}^2CH_2}}{C}}HCH_3$$

3. **Use the numbers obtained by application of rule 2 to designate the location of the substituent group.** The base name is placed last, and the substituent group, preceded by the number designating its location on the chain, is placed first. Numbers are separated from words by a hyphen. Our two examples are 2-methylhexane and 3-methylheptane, respectively.

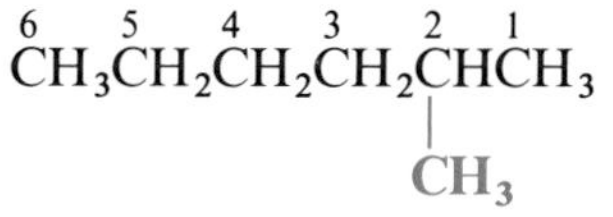

$$\overset{6}{C}H_3\overset{5}{C}H_2\overset{4}{C}H_2\overset{3}{C}H_2\overset{2}{C}HCH_3$$
$$|$$
$$CH_3$$

$$\overset{7}{C}H_3\overset{6}{C}H_2\overset{5}{C}H_2\overset{4}{C}H_2\overset{3}{C}HCH_3$$
$$|$$
$$\overset{2}{C}H_2$$
$$|$$
$$\overset{1}{C}H_3$$

2-Methylhexane 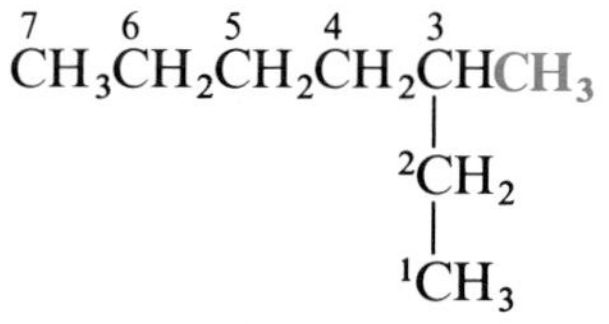   3-Methylheptane

4. **When two or more substituents are present, give each substituent a number corresponding to its location on the longest chain.** For example, we designate the following compound as 4-ethyl-2-methylhexane.

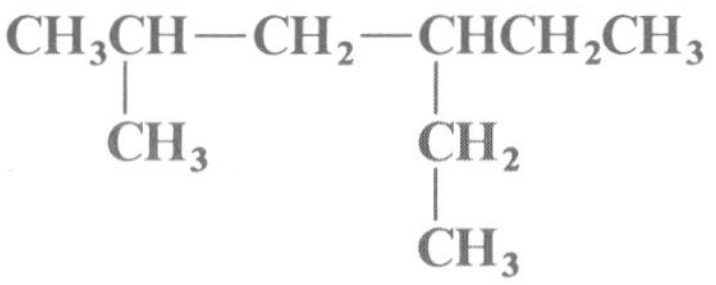

$$CH_3CH-CH_2-CHCH_2CH_3$$
$$|\qquad\qquad\quad|$$
$$CH_3\qquad\quad CH_2$$
$$|$$
$$CH_3$$

4-Ethyl-2-methylhexane

The groups should be listed *alphabetically* (i.e., ethyl before methyl).* In deciding on alphabetical order disregard multiplying prefixes such as "di" and "tri" and disregard structure-defining prefixes that are written in italics and are separated from the name by a hyphen such as *sec-* and *tert-* and consider the initial letter(s) of the substituent name. Thus ethyl precedes dimethyl, and *tert*-butyl precedes ethyl, but ethyl precedes isobutyl.

5. **When two or more substituents are present on the same carbon atom, use that number twice.**

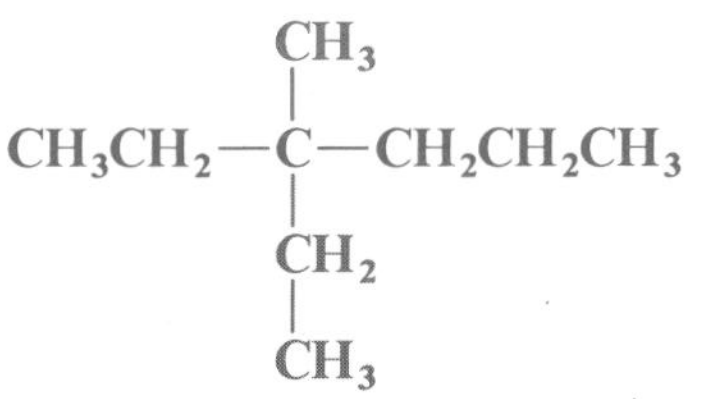

$$CH_3$$
$$|$$
$$CH_3CH_2-C-CH_2CH_2CH_3$$
$$|$$
$$CH_2$$
$$|$$
$$CH_3$$

3-Ethyl-3-methylhexane

6. **When two or more substituents are identical, indicate this by the use of the prefixes di-, tri-, tetra-,** and so on. Then make certain that each and every substituent has a number. Commas are used to separate numbers from each other.

*Some handbooks also list the groups in order of increasing size or complexity (i.e., methyl before ethyl). Alphabetical listing, however, is now by far the most widely used system.

$$CH_3CH \underline{\quad\quad} CHCH_3$$
$$\qquad | \qquad |$$
$$\qquad CH_3 \quad CH_3$$
2,3-Dimethylbutane

$$\qquad\quad CH_3$$
$$\qquad\quad |$$
$$CH_3CHCHCHCH_3$$
$$\quad\quad | \quad\quad |$$
$$\quad\quad CH_3 \quad CH_3$$
2,3,4-Trimethylpentane

$$\qquad\quad CH_3 \quad CH_3$$
$$\qquad\quad | \qquad |$$
$$CH_3CCH_2CCH_3$$
$$\qquad\quad | \qquad |$$
$$\qquad\quad CH_3 \quad CH_3$$
2,2,4,4-Tetramethylpentane

Application of these six rules allows us to name most of the alkanes that we shall encounter. Two other rules, however, may be required occasionally.

7. **When two chains of equal length compete for selection as the base chain, choose the chain with the greater number of substituents.**

$$\overset{7}{CH_3}\overset{6}{CH_2} - \overset{5}{CH} - \overset{4}{CH} - \overset{3}{CH} - \overset{2}{CH} - \overset{1}{CH_3}$$
$$\qquad\quad | \quad\quad | \quad\quad | \quad\quad |$$
$$\qquad\quad CH_3 \quad CH_2 \quad CH_3 \quad CH_3$$
$$\qquad\qquad\quad\quad |$$
$$\qquad\qquad\quad\quad CH_2$$
$$\qquad\qquad\quad\quad |$$
$$\qquad\qquad\quad\quad CH_3$$

2,3,5-Trimethyl-4-propylheptane
(four substituents)
(*not* 4-*sec*-butyl-2,3-dimethylheptane)
(three substituents)

8. **When branching first occurs at an equal distance from either end of the longest chain, choose the name that gives the lower number at the first point of difference.**

$$\overset{6}{CH_3} - \overset{5}{CH} - \overset{4}{CH_2} - \overset{3}{CH} \underline{\quad\quad} \overset{2}{CH} - \overset{1}{CH_3}$$
$$\qquad\quad | \qquad\qquad\quad | \quad\quad |$$
$$\qquad\quad CH_3 \qquad\quad CH_3 \quad CH_3$$

2,3,5-Trimethylhexane
(*not* 2,4,5-trimethylhexane)

PROBLEM 3.2

(a) Give correct IUPAC names for all the C_6H_{14} isomers in Table 3.1.
(b) Write structural formulas for the nine isomers of C_7H_{16} and give IUPAC names for each. (*Hint:* You may find it helpful to name each compound as you write its structure. This will help you to decide whether or not two structures are really different. If their IUPAC names are different then so are the structures.)

3.3C Classification of Hydrogen Atoms

The hydrogen atoms of an alkane are classified on the basis of the carbon atom to which they are attached. A hydrogen atom attached to a primary carbon atom is a primary hydrogen atom, and so forth. The following compound, 2-methylbutane, has primary, secondary, and tertiary hydrogen atoms.

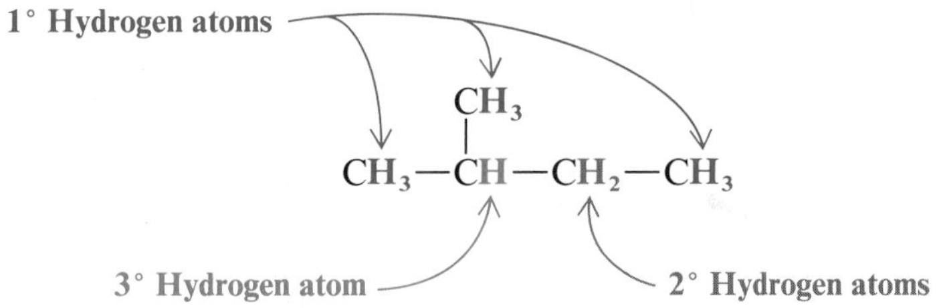

1° Hydrogen atoms

$$CH_3$$
$$|$$
$$CH_3-CH-CH_2-CH_3$$

3° Hydrogen atom

2° Hydrogen atoms

On the other hand, 2,2-dimethylpropane, a compound that is often called **neopentane,** has only primary hydrogen atoms.

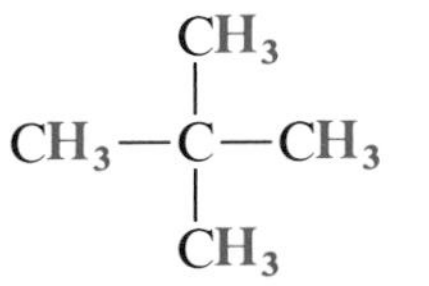

$$CH_3$$
$$|$$
$$CH_3-C-CH_3$$
$$|$$
$$CH_3$$

2,2-Dimethylpropane
(neopentane)

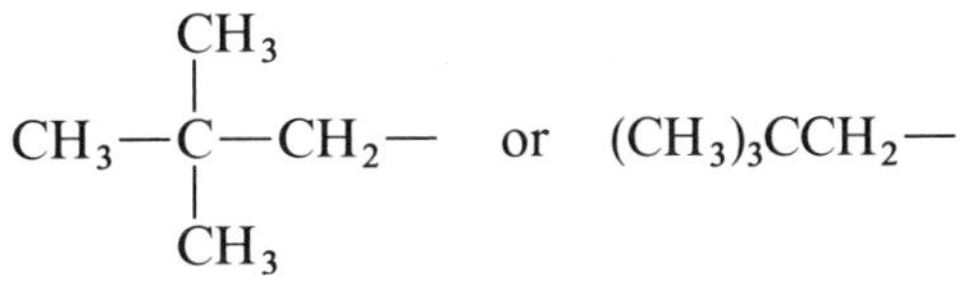

$$CH_3$$
$$|$$
$$CH_3-C-CH_2- \quad or \quad (CH_3)_3CCH_2-$$
$$|$$
$$CH_3$$

The neopentyl
group

3.3D Nomenclature of Alkyl Halides

Alkanes bearing halogen substituents are named, in the IUPAC **substitutive** system, as **haloalkanes;** for example,

$$CH_3CH_2Cl$$

Chloroethane

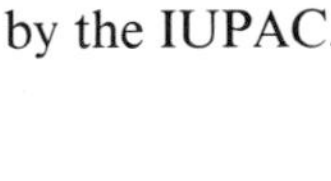

$$CH_3CHCH_3$$
$$|$$
$$Br$$

2-Bromopropane

$$CH_3$$
$$|$$
$$CH_3CHCHCH_2CH_3$$
$$|$$
$$Cl$$

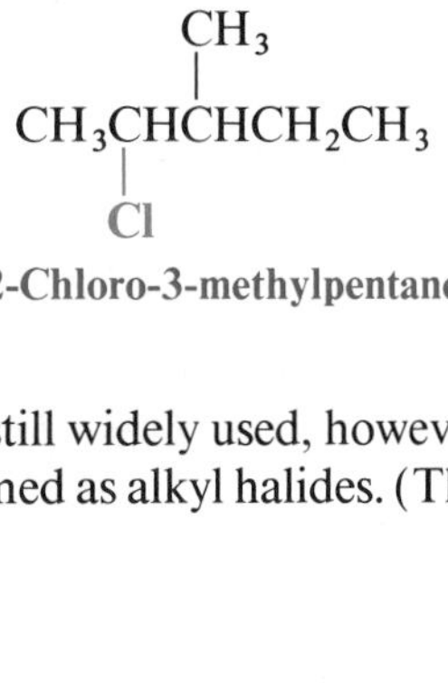

2-Chloro-3-methylpentane

Common names for many simple haloalkanes are still widely used, however. In this common nomenclature system haloalkanes are named as alkyl halides. (The following names are also accepted by the IUPAC.)

$$CH_3CH_2Cl$$

Ethyl
chloride

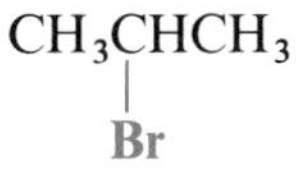

$$CH_3CHCH_3$$
$$|$$
$$Br$$

Isopropyl
bromide

$$(CH_3)_3CBr$$

tert-Butyl
bromide

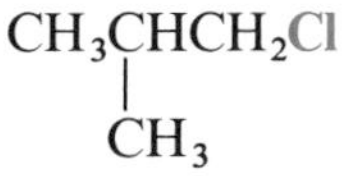

$$CH_3CHCH_2Cl$$
$$|$$
$$CH_3$$

Isobutyl
chloride

$$CH_3$$
$$|$$
$$CH_3CCH_2Br$$
$$|$$
$$CH_3$$

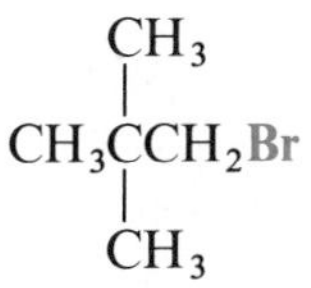

Neopentyl
bromide

PROBLEM 3.3

Give IUPAC substitutive names for all of the isomers of (a) C_4H_9Cl and (b) $C_5H_{11}Br$.

3.3E Nomenclature of Alcohols

The IUPAC rules for naming alcohols are as follows. These are the rules for what are called *substitutive* names. (The names such as methyl alcohol and ethyl alcohol are also approved by the IUPAC.)

1. Select the longest continuous carbon chain *to which the hydroxyl is directly attached.* Change the name of the alkane corresponding to this chain by dropping the final *e* and adding *ol.* This gives the base name of the alcohol.
2. Number the longest continuous carbon chain so as to give the carbon atom bearing the hydroxyl group the lower number. Indicate the position of the hydroxyl group by using this number; indicate the positions of other substituents by using the numbers corresponding to their positions along the carbon chain.

The following examples show how these rules are applied.

$$\overset{3}{C}H_3\overset{2}{C}H_2\overset{1}{C}H_2OH \qquad \overset{1}{C}H_3\overset{2}{C}HCH_2\overset{3}{C}H_2\overset{4}{C}H_3 \qquad \overset{5}{C}H_3\overset{4}{C}HCH_2\overset{3}{C}H_2\overset{2}{C}H_2\overset{1}{C}H_2OH$$

$$\qquad\qquad\qquad\qquad\qquad OH \qquad\qquad\qquad\qquad CH_3$$

1-Propanol   2-Butanol   4-Methyl-1-pentanol
(*not* 2-methyl-5-pentanol)

$$CH_3$$

$$\overset{3}{C}lCH_2\overset{2}{C}H_2\overset{1}{C}H_2OH \qquad \overset{1}{C}H_3\overset{2}{C}HCH_2\overset{3}{|}\overset{4}{C}\overset{5}{C}H_3$$

$$\qquad\qquad\qquad\qquad\qquad OH \qquad CH_3$$

3-Chloro-1-propanol   4,4-Dimethyl-2-pentanol

PROBLEM 3.4

Give IUPAC substitutive names for all of the isomeric alcohols with the formulas (a) $C_4H_{10}O$ and (b) $C_5H_{12}O$.

Simple alcohols are often called by *common* names that are also approved by the IUPAC. We have seen several examples already (Section 2.10). In addition to *methyl alcohol, ethyl alcohol,* and *isopropyl alcohol,* there are several others including the following:

$$CH_3CH_2CH_2OH \qquad CH_3CH_2CH_2CH_2OH \qquad CH_3CH_2CHCH_3$$

$$\qquad\qquad\qquad\qquad\qquad\qquad\qquad\qquad\qquad\qquad OH$$

Propyl alcohol   Butyl alcohol   *sec*-Butyl alcohol

$$CH_3 \qquad\qquad CH_3 \qquad\qquad CH_3$$

$$CH_3-C-OH \qquad CH_3CHCH_2OH \qquad CH_3CCH_2OH$$

$$CH_3 \qquad\qquad\qquad\qquad\qquad\qquad CH_3$$

tert-Butyl alcohol   Isobutyl alcohol   Neopentyl alcohol

Alcohols containing two hydroxyl groups are commonly called glycols. In the IUPAC substitutive system they are named as **diols.**

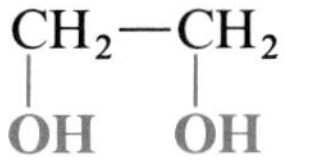

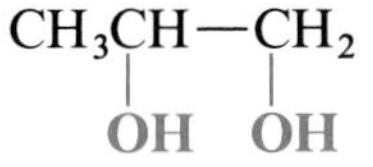

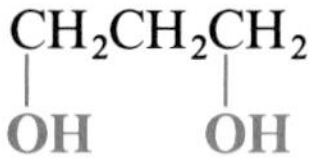

	CH_2-CH_2	CH_3CH-CH_2	$CH_2CH_2CH_2$
	OH OH	OH OH	OH OH
Common	Ethylene glycol	Propylene glycol	Trimethylene glycol
Substitutive	1,2-Ethanediol	1,2-Propanediol	1,3-Propanediol

3.4 NOMENCLATURE OF CYCLOALKANES

3.4A Monocyclic Compounds

Cycloalkanes with only one ring are named by attaching the prefix cyclo to the names of the alkanes possessing the same number of carbon atoms. For example,

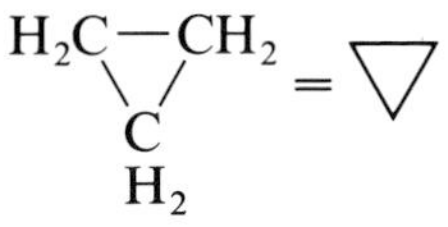

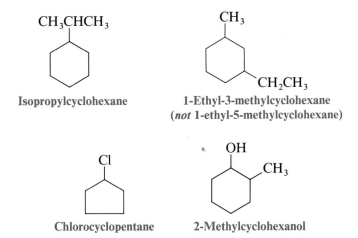

Cyclopropane 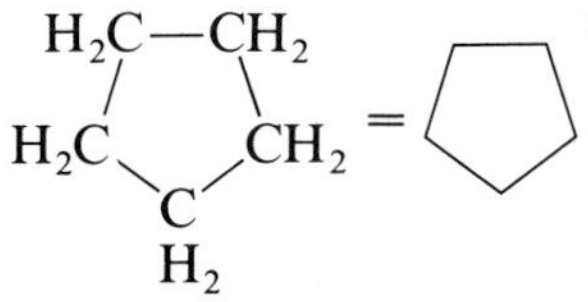   Cyclopentane

Naming substituted cycloalkanes is straightforward: we name them as *alkylcycloalkanes, halocycloalkanes, alkylcycloalkanols,* and so on. If only one substituent is present, it is not necessary to designate its position. When more than one substituent is present, we number the ring *beginning with one substituent* in the way that gives the next substituent the lowest number possible.

CH_3CHCH_3

Isopropylcyclohexane

CH_3

CH_2CH_3

1-Ethyl-3-methylcyclohexane
(*not* 1-ethyl-5-methylcyclohexane)

Cl

Chlorocyclopentane

OH

CH_3

2-Methylcyclohexanol

Occasionally, it is convenient and appropriate to name *cycloalkyl groups* as substituents on alkane chains or on other rings. Cycloalkyl groups are derived, on paper, by removing a hydrogen atom from a cycloalkane.

PROBLEM 3.5

Give names for the following substituted cycloalkanes:

(a) 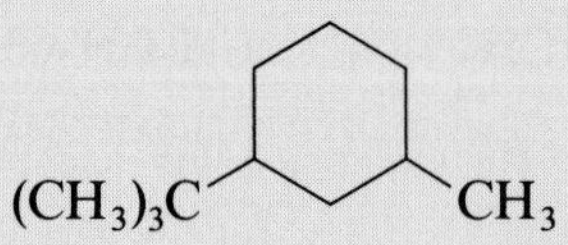 $(CH_3)_3C$ — (cyclohexane with) CH_3

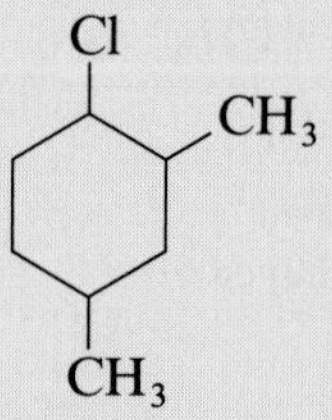

(b) 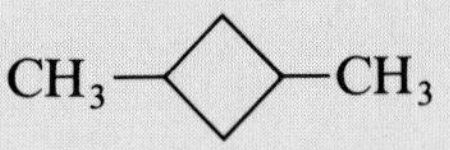 CH_3—◇—CH_3

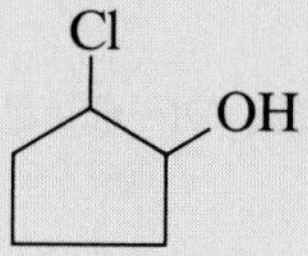

(c) 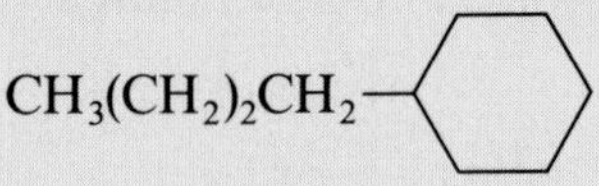 $CH_3(CH_2)_2CH_2$—

(d) Cl, CH_3, CH_3

(e) Cl, OH

(f) 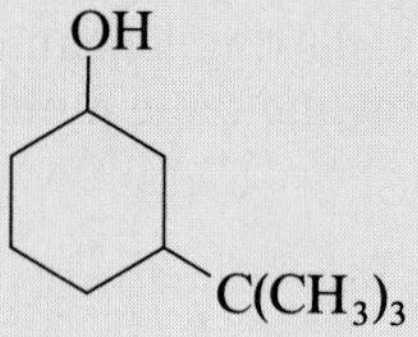 OH, $C(CH_3)_3$

3.4B Bicyclic Compounds

We name compounds containing two fused or bridged rings as **bicycloalkanes** and we use the name of the alkane corresponding to the total number of carbon atoms in the rings as the base name. The following compound, for example, contains seven carbon atoms and is, therefore, a bicycloheptane. The carbon atoms common to both rings are called bridgeheads, and each bond, or chain of atoms connecting the bridgehead atoms, is called a bridge.

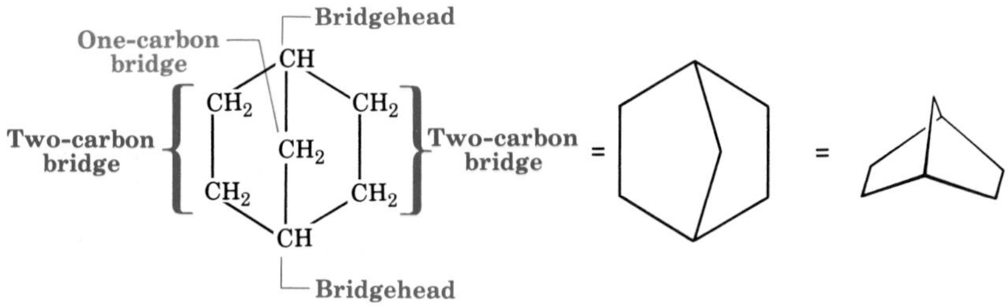

One-carbon bridge — CH — Bridgehead

Two-carbon bridge { CH_2 CH_2 CH_2 CH_2 CH_2 } Two-carbon bridge = ◇ = △

CH — Bridgehead

A bicycloheptane

Then we interpose in the name an expression in brackets that denotes the number of carbon atoms in each bridge (in descending order). For example,

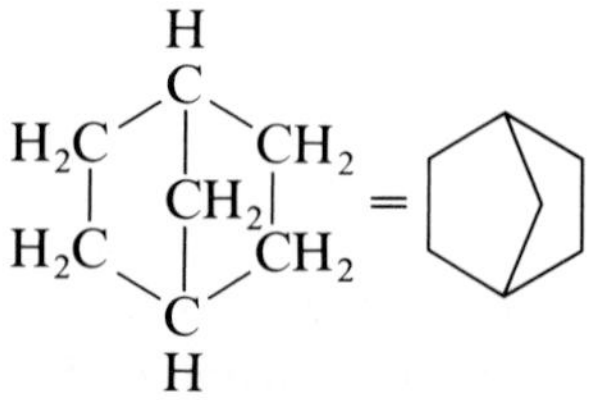

$$H \quad C$$
$$H_2C \quad CH_2$$
$$CH_2 \quad = $$
$$H_2C \quad CH_2$$
$$C$$
$$H$$

Bicyclo[2.2.1]heptane
(also called *norbornane*)

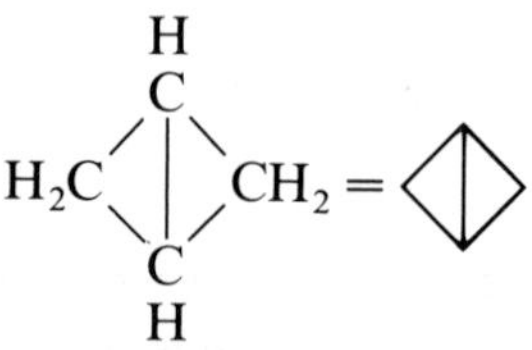

$$H \quad C$$
$$H_2C \quad CH_2 = ◇$$
$$C$$
$$H$$

Bicyclo[1.1.0]butane

If substituents are present, we number the bridged ring system beginning at one bridgehead, proceeding first along the longest bridge to the other bridgehead, then along the next longest bridge back to the first bridgehead: The shortest bridge is numbered last:

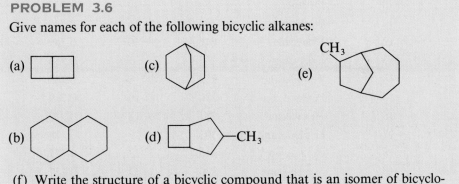

8-Methylbicyclo[3.2.1]octane   8-Methylbicyclo[4.3.0]nonane

PROBLEM 3.6

Give names for each of the following bicyclic alkanes:

(a)   (c)   (e)

(b)   (d) —CH₃

(f) Write the structure of a bicyclic compound that is an isomer of bicyclo-[2.2.1]heptane and give its name.

3.5 PHYSICAL PROPERTIES OF ALKANES AND CYCLOALKANES

If we examine the unbranched alkanes in Table 3.2 we notice that each alkane differs from the preceding one by one $-CH_2-$ group. Butane, for example, is $CH_3(CH_2)_2CH_3$ and pentane is $CH_3(CH_2)_3CH_3$. A series of compounds like this, where each member differs from the next member by a constant unit, is called a **homologous series.** Members of a homologous series are called **homologs.**

At room temperature (25 °C) and 1-atm pressure the first four members of the homologous series of unbranched alkanes (Table 3.3) are gases; the C_5 to C_{17} unbranched alkanes (pentane to heptadecane) are liquids; and the unbranched alkanes with 18 and more carbon atoms are solids.

Boiling Points. The boiling points of the unbranched alkanes show a regular increase with increasing molecular weight (Fig. 3.3). Branching of the alkane chain, however, dramatically lowers the boiling point. (Examples of the effect of chain branching can be seen in Table 3.1.) Cycloalkanes, however, have higher boiling points than unbranched alkanes with the same number of carbon atoms (Table 3.4).

Melting Points. The unbranched alkanes do not show the same smooth increase in melting points with increasing molecular weight (black line Fig. 3.4) that they show in their boiling points. There is an alternation as one progresses from an

TABLE 3.3 Physical constants of unbranched alkanes

NUMBER OF CARBON ATOMS	NAME	bp (°C) (1 atm)	mp (°C)	DENSITY d^{20} (g/mL)
1	Methane	−161.5	−182	
2	Ethane	−88.6	−183	
3	Propane	−42.1	−188	
4	Butane	−0.5	−138	
5	Pentane	36.1	−130	0.626
6	Hexane	68.7	−95	0.659
7	Heptane	98.4	−91	0.684
8	Octane	125.7	−57	0.703
9	Nonane	150.8	−54	0.718
10	Decane	174.1	−30	0.730
11	Undecane	195.9	−26	0.740
12	Dodecane	216.3	−10	0.749
13	Tridecane	235.4	−5.5	0.756
14	Tetradecane	253.5	6	0.763
15	Pentadecane	270.5	10	0.769
16	Hexadecane	287	18	0.773
17	Heptadecane	303	22	0.778
18	Octadecane	316.7	28	0.777
19	Nonadecane	330	32	0.777
20	Eicosane	343	36.8	0.789

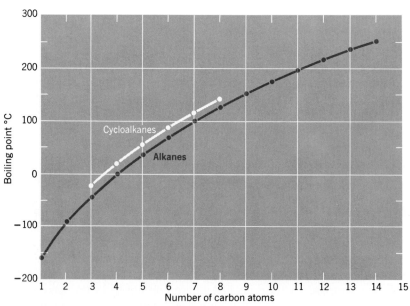

FIGURE 3.3 Boiling points of unbranched alkanes (in red) and cycloalkanes (in white).

TABLE 3.4 Physical constants of cycloalkanes

NUMBER OF CARBON ATOMS	NAME	bp (°C) (1 atm)	mp (°C)	DENSITY d^{20}(g/mL)	REFRACTIVE INDEX (n_D^{20})
3	Cyclopropane	−33	−126.6		
4	Cyclobutane	13	−90		1.4260
5	Cyclopentane	49	−94	0.751	1.4064
6	Cyclohexane	81	6.5	0.779	1.4266
7	Cycloheptane	118.5	−12	0.811	1.4449
8	Cyclooctane	149	13.5	0.834	

unbranched alkane with an even number of carbon atoms to the next one with an odd number of carbon atoms. For example, propane (mp, − 188 °C) melts lower than ethane (mp, − 183 °C) and also lower than methane (mp, − 182 °C). Butane, (mp, − 138 °C) melts 53 °C higher than propane and only 5 °C lower than pentane (mp, − 130 °C). If, however, the even- and odd-numbered alkanes are plotted on *separate* curves (white and red lines in Fig. 3.4), there *is* a smooth increase in melting point with increasing molecular weight.

X-ray-diffraction studies have revealed the reason for this apparent anomaly. Alkane chains with an even number of carbon atoms pack more closely in the crystalline state. As a result, attractive forces between individual chains are greater and melting points are higher.

The effect of chain branching on the melting points of alkanes is more difficult to predict. Generally, however, branching that produces highly symmetrical structures results in abnormally high melting points. The compound 2,2,3,3-tetramethylbutane, for example, melts at 100.7 °C. Its boiling point is only six degrees higher, 106.3 °C.

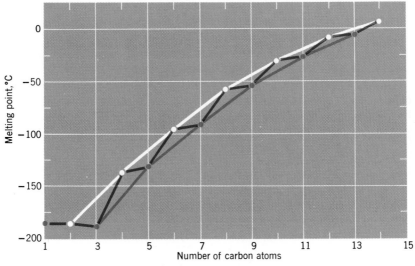

FIGURE 3.4 Melting points of unbranched alkanes.

$$CH_3 \quad CH_3$$
$$| \qquad |$$
$$CH_3-\overset{\displaystyle |}{\underset{\displaystyle |}{C}}-\overset{\displaystyle |}{\underset{\displaystyle |}{C}}-CH_3$$
$$| \qquad |$$
$$CH_3 \quad CH_3$$

2,2,3,3-Tetramethylbutane

Cycloalkanes also have much higher melting points than their open-chain counterparts (Table 3.4). Because of their greater symmetry they pack more tightly into a crystal lattice.

Density. As a class, the alkanes and cycloalkanes are the least dense of all groups of organic compounds. All alkanes and cycloalkanes have densities considerably less than 1.00 g/mL (the density of water at 4 °C). As a result, petroleum (a mixture of hydrocarbons rich in alkanes) floats on water.

Solubility. Alkanes and cycloalkanes are almost totally insoluble in water because of their very low polarity and their inability to form hydrogen bonds. Liquid alkanes and cycloalkanes are soluble in one another, and they generally dissolve in solvents of low polarity. Good solvents for them are benzene, carbon tetrachloride, chloroform, and other hydrocarbons.

3.6 SIGMA BONDS AND ROTATION ABOUT BONDS

Groups bonded only by a sigma bond (i.e., by a single bond) can undergo rotation about that bond with respect to each other and this rotation does not affect the overlap of the constituent orbitals. The different arrangements of the atoms in space that result from rotations of groups about single bonds are called **conformations** of the molecule. An analysis of the energy changes that a molecule undergoes as groups rotate about single bonds is called a **conformational analysis.***

Let us consider the ethane molecule as an example. Obviously an infinite number of different conformations could result from rotations of the CH_3 groups about the carbon–carbon bond. These different conformations, however, are not all of equal stability. The conformation (Fig. 3.5) in which the hydrogen atoms attached to each carbon atom are perfectly staggered when viewed from one end of the molecule along the carbon–carbon bond axis is the *most stable* conformation (i.e., it is the conformation of *lowest potential energy*). This is easily explained in terms of repulsive interactions between bonding pairs of electrons. The staggered conformation allows the maximum possible separation of the electron pairs of the six carbon–hydrogen bonds and therefore it has the lowest energy.

In Fig. 3.5*b* we have drawn what is called a **Newman projection formula†** for ethane. In writing a Newman projection we imagine ourselves viewing the molecule from one end directly along the carbon–carbon bond axis. The bonds of the front carbon atom are represented as $\curlyvee$ and those of the back atom as $\curlyveedownarrow$.

---

*Conformational analysis owes its modern origins largely to the work of O. Hassel of Norway and D. H. R. Barton of Great Britain. Hassel and Barton won the Nobel Prize in 1969, mainly for their contributions in this area. The idea that certain conformations of molecules will be favored, however, originated from the work of van't Hoff.

†These formulas are named after their inventor Melvin S. Newman of the Ohio State University.

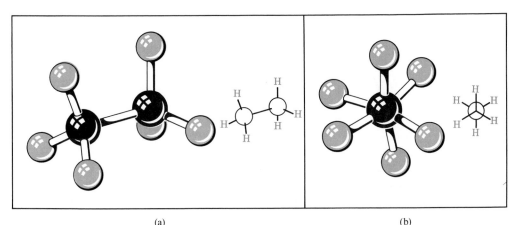

(a)   (b)

FIGURE 3.5 (*a*) The staggered conformation of ethane. (*b*) The Newman projection formula for the staggered conformation.

The least stable conformation of ethane is the **eclipsed conformation** (Fig. 3.6). When viewed from one end along the carbon–carbon bond axis, the hydrogen atoms attached to each carbon atom in the eclipsed conformation are in direct opposition to each other. This conformation requires the maximum repulsive interaction between the electrons of the six carbon–hydrogen bonds. It is, therefore, of highest energy and has the least stability.

We represent this situation graphically by plotting the energy of an ethane molecule as a function of rotation about the carbon–carbon bond. The energy changes that occur are illustrated in Fig. 3.7.

In ethane the difference in energy between the staggered and eclipsed conformations is 2.8 kcal/mole (12 kJ/mole). This small barrier to rotation is called the **torsional barrier** of the single bond. Unless the temperature is extremely low (−250 °C) many ethane molecules (at any given moment) will have enough energy to surmount this barrier. Some molecules will wag back and forth with their atoms in staggered or nearly staggered conformations. The more energetic ones, however, will rotate through eclipsed conformations to other staggered conformations.

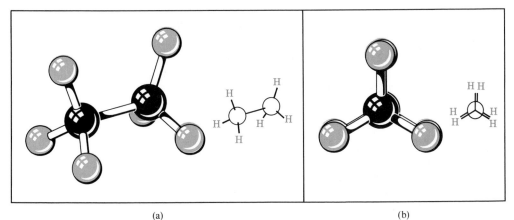

(a)   (b)

FIGURE 3.6 (*a*) The eclipsed conformation of ethane. (*b*) The Newman projection formula for the eclipsed conformation.

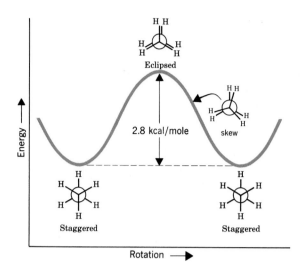

FIGURE 3.7 Potential energy changes that accompany rotation of groups about the carbon–carbon bond of ethane.

What does all this mean about ethane? We can answer this question in two different ways. If we consider a single molecule of ethane, we can say, for example, that it will spend most of its time in the lowest energy, staggered conformation or in a conformation very close to being staggered. Many times every second, however, it will acquire enough energy through collisions with other molecules to surmount the tortional barrier and it will rotate through an eclipsed conformation. If we speak in terms of a large number of ethane molecules (a more realistic situation), we can say that at any given moment most of the molecules will be in staggered or nearly staggered conformations.

If we consider substituted ethanes (G is a group or atom other than hydrogen) such as GCH_2CH_2G, the barriers to rotation are somewhat larger but they are still far too small to allow isolation of the different staggered conformations or **conformers** (see following figure), even at temperatures considerably below room temperature.

These conformers cannot be
isolated except at extremely
low temperatures

3.7 STEREOCHEMISTRY AND A CONFORMATIONAL ANALYSIS OF BUTANE

When we concern ourselves with *the three-dimensional aspects of molecular structure,* we are involved in the field of study called **stereochemistry.** We have already had some experience with stereochemistry because we began considering the shapes of molecules as early as Chapter 1. In this chapter, however, we begin our study in earnest as we take a detailed look at the *conformations* of alkanes and cycloalkanes. In many of the chapters that follow we shall see some of the consequences of this stereochemistry in the reactions that these molecules undergo. In Chapter 4, we shall see further basic principles of stereochemistry when we examine the properties of molecules that, because of their shape, are said to possess "handedness" or **chirality.** Let us begin, however, with a relatively simple molecule, butane, and study its conformations and their relative energies.

3.7A A Conformational Analysis of Butane

The study of the energy changes that occur in a molecule when groups rotate about single bonds is called *conformational analysis.* We saw the results of such a study for ethane in Section 3.6. Ethane has a slight barrier (2.8 kcal/mole) to free rotation about the carbon–carbon single bond. This barrier causes the energy of the ethane molecule to rise to a maximum when rotation brings the hydrogen atoms into an eclipsed conformation. This barrier to free rotation in ethane is called the **torsional strain** of an eclipsed conformation of the molecule.

If we consider rotation about the C-2—C-3 bond of butane, torsional strain plays a part, too. There are, however, additional factors. To see what these are, we should look at the important conformations of butane **I – VI.**

I	**II**	**III**
Anti conformation	An eclipsed conformation	A *gauche* conformation
IV	**V**	**VI**
An eclipsed conformation	A *gauche* conformation	An eclipsed conformation

The **anti conformation (I)** does not have torsional strain because the groups are staggered and the methyl groups are far apart. Therefore, the *anti* conformation is the most stable. The methyl groups in the **gauche conformations** are close enough to each other that the van der Waals forces between them are *repulsive;* the electron clouds of the two groups are so close that they repel each other. This repulsion causes the *gauche* conformations to have approximately 0.8 kcal/mole (3.3 kJ/mole) more energy than the *anti* conformation.

The eclipsed conformations (**II, IV,** and **VI**) represent energy maxima in the potential energy diagram (Fig. 3.8). Eclipsed conformations **II** and **VI** not only have torsional strain, they have van der Waals repulsions arising from the eclipsed methyl groups and hydrogen atoms. Eclipsed conformation **IV** has the greatest energy of all

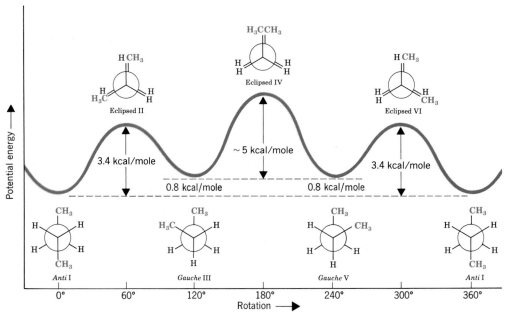

FIGURE 3.8 Energy changes that arise from rotation of the C-2—C-3 bond of butane.

because, in addition to torsional strain, there is the large van der Waals repulsive force between the eclipsed methyl groups.

While the barriers to rotation in a butane molecule are larger than those of an ethane molecule (Section 3.6), they are still far too small to permit isolation of the *gauche* and *anti* conformations at normal temperatures. Only at extremely low temperatures would the molecules have insufficient energies to surmount these barriers.

> We saw earlier that van der Waals forces can be *attractive*. Here, however, we find that they can also be *repulsive*. Whether or not van der Waals interactions lead to attraction or repulsion depends on the distance that separates the two groups. As two nonpolar groups are brought closer and closer together, the first effect is one in which a momentarily unsymmetrical distribution of electrons in one group induces an opposite polarity in the other. The opposite charges induced in those portions of the two groups that are in closest proximity lead to attraction between them. This attraction increases to a maximum as the internuclear distance of the two groups decreases. The internuclear distance at which the attractive force is at a maximum is equal to the sum of what are called the *van der Waals radii* of the two groups. The van der Waals radius of a group is, in effect, a measure of its size. If the two groups are brought still closer — closer than the sum of their van der Waals radii — the interaction between them becomes repulsive: Their electron clouds begin to penetrate each other, and strong electron–electron interactions begin to occur.

PROBLEM 3.7

Sketch a curve similar to that in Fig. 3.8 showing in general terms the energy changes that occur when the groups rotate about one carbon–carbon bond of propane. You need not concern yourself with the actual numerical values of the energy changes, but you should label all maxima and minima with the appropriate conformations.

3.8 THE RELATIVE STABILITIES OF CYCLOALKANES: RING STRAIN

Cycloalkanes do not all have the same relative stability. Data from heats of combustion (Section 3.8A) show that cyclohexane is the most stable cycloalkane and cyclopropane and cyclobutane are much less stable. The relative instability of cyclopropane and cyclobutane is a direct consequence of their cyclic structures and for this reason their molecules are said to possess **ring strain.** To see how this can be demonstrated experimentally, we need to examine the relative heats of combustion of cycloalkanes.

3.8A Heats of Combustion

The **heat of combustion** of a compound is the enthalpy change for the complete oxidation of the compound.

For a hydrocarbon complete oxidation means converting it to carbon dioxide and water. This can be accomplished experimentally, and the amount of heat evolved can be accurately measured in a device called a calorimeter. For methane, for example, the heat of combustion is -192 kcal/mole (-803 kJ/mole).

$$CH_4 + 2\,O_2 \longrightarrow CO_2 + 2H_2O \qquad \Delta H° = -192 \text{ kcal/mole} \quad (-803 \text{ kJ/mole})$$

For isomeric hydrocarbons, complete combustion of 1 mole of each will require the same amount of oxygen and will yield the same number of moles of carbon dioxide and water. We can, therefore, use heats of combustion to measure the relative stabilities of the isomers.

Consider, as an example, the combustion of butane and isobutane:

$$CH_3CH_2CH_2CH_3 + 6\tfrac{1}{2}\,O_2 \longrightarrow 4CO_2 + 5H_2O \qquad \Delta H° = -687.5 \text{ kcal/mole}$$
$$(C_4H_{10}) \hspace{6cm} (-2877 \text{ kJ/mole})$$

$$CH_3CHCH_3 + 6\tfrac{1}{2}\,O_2 \longrightarrow 4CO_2 + 5H_2O \qquad \Delta H° = -685.5 \text{ kcal/mole}$$
$$\overset{|}{CH_3} \hspace{6cm} (-2868 \text{ kJ/mole})$$
$$(C_4H_{10})$$

Since butane liberates more heat on combustion than isobutane, it must contain relatively more potential energy. Isobutane, therefore, must be *more stable.* Figure 3.9 illustrates this comparison.

3.8B Heats of Combustion of Cycloalkanes

The cycloalkanes constitute a *homologous series;* each member of the series differs from the one immediately preceding it by the constant amount of one $-CH_2-$

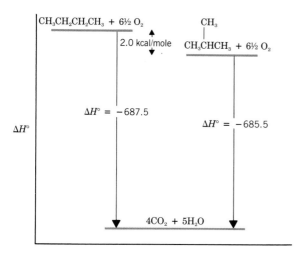

ΔH°

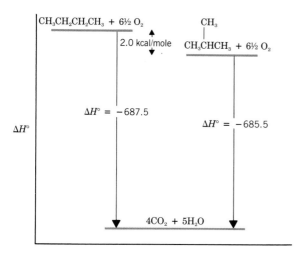

FIGURE 3.9 Heats of combustion show that isobutane is more stable than butane by 2.0 kcal/mole (8.4 kJ/mole).

group. Thus, the general equation for combustion of a cycloalkane can be formulated as follows:

$$(CH_2)_n + \tfrac{3}{2}\, n\, O_2 \longrightarrow nCO_2 + nH_2O + \text{heat}$$

Because the cycloalkanes are not isomeric, their heats of combustion cannot be compared directly. However, we can calculate the amount of heat evolved *per CH_2 group*. On this basis, the stabilities of the cycloalkanes become directly comparable. The results of such an investigation are given in Table 3.5.

Several observations emerge from a consideration of these results.

1. Cyclohexane has the lowest heat of combustion per CH_2 group (157.4 kcal/mole). This amount does not differ from that of unbranched alkanes, which, having no ring, can have no ring strain. We can assume, therefore, that cyclohexane has no ring strain and that it can serve as our standard for comparison with other cycloalkanes. We can calculate ring strain for the other cycloalkanes

TABLE 3.5 Heats of combustion of cycloalkanes

CYCLOALKANE $(CH_2)_n$	n	HEAT OF COMBUSTION		HEAT OF COMBUSTION/CH_2 GROUP	
		(kcal/mole)	(kJ/mole)	(kcal/mole)	(kJ/mole)
Cyclopropane	3	499.8	2091	166.6	697.5
Cyclobutane	4	655.9	2744	164.0	686.2
Cyclopentane	5	793.5	3220	158.7	664.0
Cyclohexane	6	944.5	3952	157.4	658.6
Cycloheptane	7	1108.2	4636.7	158.3	662.3
Cyclooctane	8	1269.2	5310.3	158.6	663.6
Cyclononane	9	1429.5	5981.0	158.8	664.4
Cyclodecane	10	1586.0	6635.8	158.6	663.6
Cyclopentadecane	15	2362.5	9984.7	157.5	659.0
Unbranched alkane				157.4	658.6

TABLE 3.6 Ring strain of cycloalkanes

| | RING STRAIN | |
CYCLOALKANE	(kcal/mole)	(kJ/mole)
Cyclopropane	27.6	115
Cyclobutane	26.4	110
Cyclopentane	6.5	27
Cyclohexane	0	0
Cycloheptane	6.4	27
Cyclooctane	10.0	42
Cyclononane	12.9	54
Cyclodecane	12.0	50
Cyclopentadecane	1.5	6

(Table 3.6) by multiplying 157.4 kcal/mole by n and then subtracting the result from the heat of combustion of the cycloalkane.

2. The combustion of cyclopropane evolves the greatest amount of heat per CH_2 group. Therefore, molecules of cyclopropane must have the greatest ring strain (27.6 kcal/mole, cf. Table 3.6). Since cyclopropane molecules evolve the greatest amount of heat energy per CH_2 group on combustion, they must contain the greatest amount of potential energy per CH_2 group. Thus what we call ring strain is a form of potential energy that the cyclic molecule contains. The more ring strain a molecule possesses, the more potential energy it has and the less stable it is compared to its ring homologs.

3. The combustion of cyclobutane evolves the second largest amount of heat per CH_2 group and, therefore, cyclobutane has the second largest amount of ring strain (26.3 kcal/mole).

4. While other cycloalkanes possess ring strain to varying degrees, the relative amounts are not large. Cyclopentane and cycloheptane have about the same modest amount of ring strain. Rings of 8, 9, and 10 members have slightly larger amounts of ring strain and then the amount falls off. A 15-membered ring has only a very slight amount of ring strain.

3.9 THE ORIGIN OF RING STRAIN IN CYCLOPROPANE AND CYCLOBUTANE: ANGLE STRAIN AND TORSIONAL STRAIN

The carbon atoms of alkanes are sp^3 hybridized. The normal tetrahedral bond angle of an sp^3-hybridized atom is 109.5°. In cyclopropane (a molecule with the shape of a regular triangle) the internal angles must be 60° and therefore they must depart from this ideal value by a very large amount—by 49.5°.

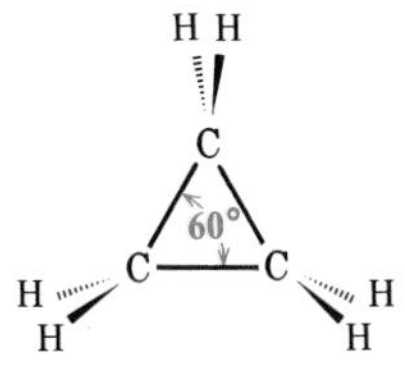

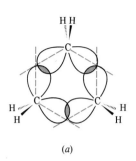

(a)

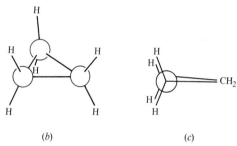

(b)   (c)

FIGURE 3.10 (a) Orbital overlap in the carbon–carbon bonds of cyclopropane cannot occur perfectly end-on. This leads to weaker "bent" bonds and to angle strain. (b) A line-and-circle drawing of the cyclopropane ring. (c) A Newman projection formula as viewed along one carbon–carbon bond shows the eclipsed hydrogens. (Viewing along either of the other two bonds would show the same picture.)

This compression of the internal bond angle causes what chemists call **angle strain.** Angle strain exists in a cyclopropane ring because the sp^3 orbitals of the carbon atoms cannot overlap as effectively (Fig. 3.10*a*) as they do in alkanes (where perfect end-on overlap is possible). The bonds of cyclopropane are often described as being "bent." Orbital overlap is less effective even though the carbon atoms adopt a hybridization state that is not purely sp^3 (it contains more *p* character). The carbon–carbon bonds of cyclopropane are weaker, and as a result, the molecule has greater potential energy.

While angle strain accounts for most of the ring strain in cyclopropane, it does not account for it all. Because the ring is (of necessity) planar, the hydrogen atoms of the ring are all *eclipsed* (Fig. 3.10*b* and *c*), and the molecule has torsional strain as well.

Cyclobutane also has considerable angle strain. The internal angles are 88° — a departure of more than 21° from the normal tetrahedral bond angle. The cyclobutane ring is not planar but is slightly "folded" (Fig. 3.11*a*). If the cyclobutane ring were planar, the angle strain would be somewhat less (the internal angles would be 90° instead of 88°), but torsional strain would be considerably larger because all eight hydrogens would be eclipsed. By folding or bending slightly the cyclobutane ring

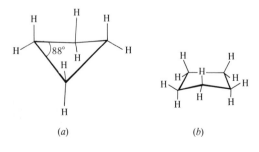

(a)   (b)

FIGURE 3.11 (a) The "folded" or "bent" conformation of cyclobutane.
(b) The "bent" or "envelope" form of cyclopentane. In this structue the front carbon atom is bent upwards. In actuality, the molecule is flexible and shifts conformations constantly.

relieves more of its torsional strain than it gains in the slight increase in its angle strain.

3.9A Cyclopentane

The internal angles of a regular pentagon are 108°, a value very close to the normal tetrahedral bond angles of 109.5°. Therefore, if cyclopentane molecules were planar, they would have very little angle strain. Planarity, however, would introduce considerable torsional strain because all 10 hydrogen atoms would be eclipsed. Consequently, like cyclobutane, cyclopentane assumes a slightly bent conformation in which one or two of the atoms of the ring are out of the plane of the others (Fig. 3.11b). This relieves some of the torsional strain. Slight twisting of carbon–carbon bonds can occur with little change in energy, and causes the out-of-plane atoms to move into plane and causes others to move out. Therefore, the molecule is flexible and shifts rapidly, from one conformation to another. With little torsional strain and angle strain, cyclopentane is almost as stable as cyclohexane.

3.10 CONFORMATIONS OF CYCLOHEXANE

There is considerable evidence that the most stable conformation of the cyclohexane ring is the "chair" conformation illustrated in Fig. 3.12.* In this nonplanar structure

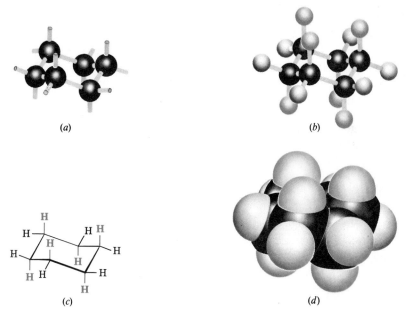

(a)

(b)

(c)

(d)

FIGURE 3.12 Representations of the chair conformation of cyclohexane: (a) carbon skeleton only; (b) carbon and hydrogen atoms; (c) line drawing; (d) space-filling model of cyclohexane. Notice that there are two types of hydrogen substituents—those that project up or down (shown in red) and those that lie generally in the plane of the ring (shown black or gray). We shall discuss this further in Section 3.12.

*An understanding of this and subsequent discussions of conformational analysis can be aided immeasurably through the use of a molecular model such as the Theta Molecular Model Set for Organic Chemistry, developed by Ronald Starkey and available from John Wiley & Sons, Inc.

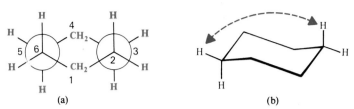

(a)   (b)

FIGURE 3.13 (a) A Newman projection of the chair conformation of cyclohexane. (Comparison with an actual molecular model will make this formulation clearer and will show that similar staggered arrangements are seen when other carbon–carbon bonds are chosen for sighting.) (b) Illustration of large separation between hydrogen atoms at opposite corners of the ring (designated C-1 and C-4) when the ring is in the chair conformation.

the carbon–carbon bond angles are all 109.5° and are thereby free of angle strain. The chair conformation is free of torsional strain as well. When viewed along the carbon–carbon–carbon bonds of any side (viewing the structure from an end, Fig. 3.13), the atoms are seen to be perfectly staggered. Moreover, the hydrogen atoms at opposite corners of the cyclohexane ring are maximally separated.

By simple rotations about the carbon–carbon single bonds of the ring, the chair conformation can assume another shape called the "boat" conformation (Fig. 3.14). The boat conformation is like the chair conformation in that it is also free of angle strain.

The boat conformation, however, is not free of torsional strain. When a model of the boat conformation is viewed down carbon-carbon bond axes along either side (Fig. 3.15a), the hydrogen substituents at those carbon atoms, are found to be

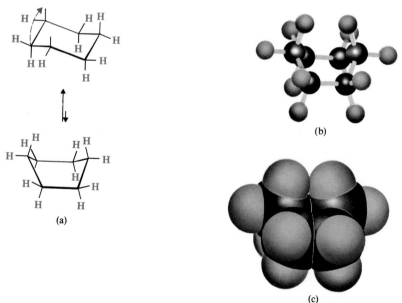

(a)

(b)

(c)

FIGURE 3.14 (a) The boat conformation of cyclohexane is formed by "flipping" one end of the chair form up (or down). This flip requires only rotations about carbon–carbon single bonds. (b) Ball-and-stick model of the boat conformation. (c) A space-filling model.

(a)   (b)

FIGURE 3.15 (a) Illustration of the eclipsed conformation of the boat conformation of cyclohexane. (b) Flagpole interaction of the C-1 and C-4 hydrogen atoms of the boat conformation.

eclipsed. Additionally, two of the hydrogen atoms on C-1 and C-4 are close enough to each other to cause van der Waals repulsion (Fig. 3.15b). This latter effect has been called the "flagpole" interaction of the boat conformation. Torsional strain and flagpole interactions cause the boat conformation to have considerably higher energy than the chair conformation.

Although it is more stable, the chair conformation is much more rigid than the boat conformation. The boat conformation is quite flexible. By flexing to a new form — the twist conformation (Fig. 3.16) — the boat conformation can relieve some of its torsional strain and, at the same time, reduce the flagpole interactions. Thus, the twist conformation has a lower energy than the boat conformation. *The stability gained by flexing is insufficient, however, to cause the twist conformation of cyclohexane to be more stable than the chair conformation.* The chair conformation is estimated to be lower in energy than the twist conformation by approximately 5 kcal/mole (21 kJ/mole).

The energy barriers between the chair, boat, and twist conformations of cyclohexane are low enough (Fig. 3.17) to make their separation impossible at room temperature. At room temperature the thermal energies of the molecules are great enough to cause approximately 1 million interconversions to occur each second and, *because of its greater stability, more than 99% of the molecules are estimated to be in a chair conformation at any given moment.*

3.11 CONFORMATIONS OF HIGHER CYCLOALKANES

Cycloheptane, cyclooctane, and cyclononane and other higher cycloalkanes also exist in nonplanar conformations. The small instabilities of these higher cycloalkanes (Table 3.6) appear to be caused primarily by torsional strain and van der Waals repulsions between hydrogen atoms across rings. The nonplanar conformations of these rings, however, are essentially free of angle strain.

X-ray crystallographic studies of cyclodecane reveal that the most stable conformation has carbon–carbon–carbon bond angles of 117°. This indicates some

FIGURE 3.16 (a) Carbon skeleton and (b) line drawing of the twist conformation of cyclohexane.

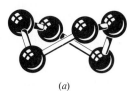

(a)

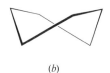

(b)

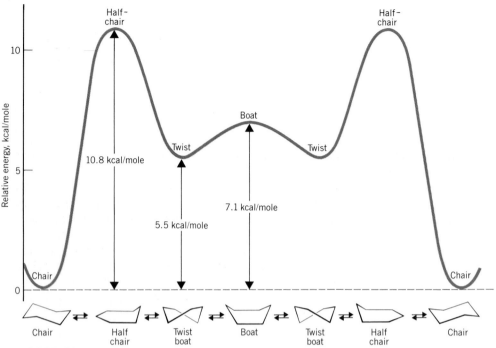

FIGURE 3.17 The relative energies of the various conformations of cyclohexane. The positions of maximum energy are conformations called half-chair conformations, in which the carbon atoms of one end of the ring have become coplanar.

angle strain. The wide bond angles apparently allow the molecule to expand and thereby minimize unfavorable repulsions between hydrogen atoms across the ring.

There is very little free space in the center of a cycloalkane unless the ring is quite large. Calculations indicate that cyclooctadecane, for example, is the smallest ring through which a $-CH_2CH_2CH_2-$ chain can be threaded. Molecules have been synthesized, however, which have large rings threaded on chains and which have large rings that are interlocked like links in a chain. These latter molecules are called **catenanes**.

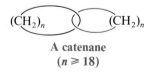

A catenane
($n \geqslant 18$)

3.12 SUBSTITUTED CYCLOHEXANES: AXIAL AND EQUATORIAL HYDROGEN ATOMS

The six-membered ring is the most common ring found among nature's organic molecules. For this reason, we shall give it special attention. We have already seen that the chair conformation of cyclohexane is the most stable one and that it is the predominant conformation of the molecules in a sample of cyclohexane. With this

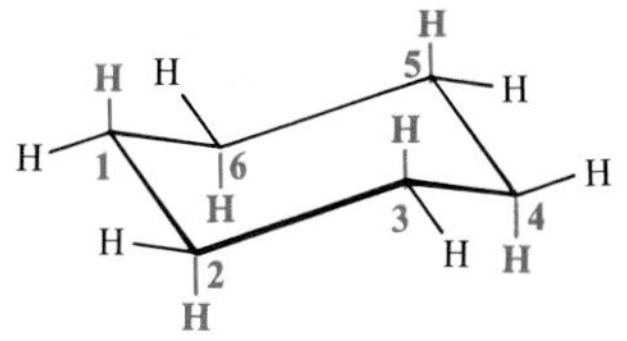

FIGURE 3.18 The chair conformation of cyclohexane. The axial hydrogen atoms are shown in color.

fact in mind, we are in a position to undertake a limited analysis of the conformations of substituted cyclohexanes.

If we look carefully at the chair conformation of cyclohexane (Fig. 3.18), we can see that there are only two different kinds of hydrogen atoms. One hydrogen atom attached to each of the six carbon atoms lies in a plane generally defined by the ring of carbon atoms. These hydrogen atoms, by analogy with the equator of the earth, are called **equatorial** hydrogen atoms. Six other hydrogen atoms, one on each carbon, are oriented in a direction that is generally perpendicular to the average plane of the ring. These hydrogen atoms, again by analogy with the earth, are called **axial** hydrogen atoms. There are three axial hydrogen atoms on each face of the cyclohexane ring and their orientation (up or down) alternates from one carbon atom to the next.

STUDY AID

You should now learn how to draw the important chair conformation. Notice (Fig. 3.19) the sets of parallel lines that constitute the bonds of the ring and the equatorial hydrogen atoms. Notice, too, that when drawn this way, the axial bonds are all vertical, and when the vertex of the ring points up, the axial bond is up; when the vertex is down, the axial bond is down.

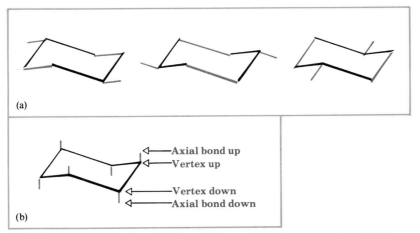

(a)

(b)

Axial bond up
Vertex up

Vertex down
Axial bond down

FIGURE 3.19 (a) Sets of parallel lines that constitute the ring and equatorial C—H bonds of the chair conformation. (b) The axial bonds are all vertical. When the vertex of the ring points up, the axial bond is up and vice versa.

We saw in Section 3.10 (and Fig. 3.17) that at room temperature, the cyclohexane ring rapidly flips back and forth between two *equivalent* chain conformations. An important thing to notice now is that **when the ring flips all of the bonds that were axial become equatorial and vice versa:**

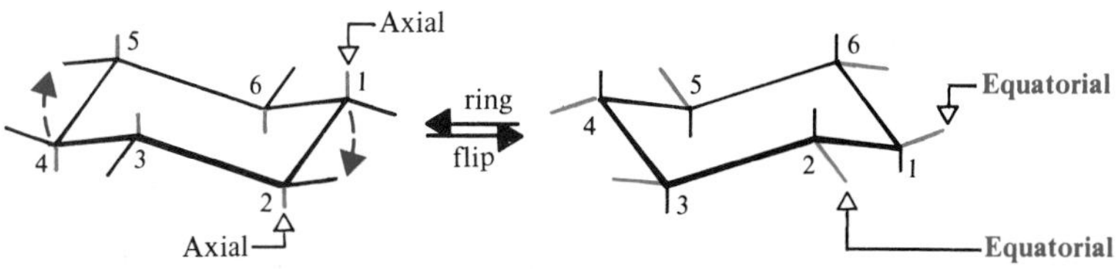

The question one might next ask is what is the most stable conformation of a cyclohexane derivative *in which one hydrogen atom has been replaced by a substituent?* That is, what is the most stable conformation of a monosubstituted cyclohexane? We can answer this question by considering methylcyclohexane as an example.

Methylcyclohexane has two possible chair conformations (Fig. 3.20), and these are interconvertible through partial rotations about the single bonds of the ring. In one conformation (Fig. 3.20*a*) the methyl group occupies an *axial* position, and in the other the methyl group occupies an *equatorial* position. Studies indicate that the conformation with the methyl group equatorial is more stable than the conformation with the methyl group axial by about 1.8 kcal/mole. Thus, in the equilibrium mixture, the conformation with the methyl group in the equatorial position is the predominant one. Calculations (see Special Topic A) show that it constitutes about 93% of the equilibrium mixture.

The greater stability of methylcyclohexane with an equatorial methyl group can be understood through an inspection of the two forms as they are shown in Fig. 3.20*b* and *c*.

Studies done with scale models of the two conformations show that when the methyl group is axial, it is so close to the two axial hydrogen atoms on the same side of

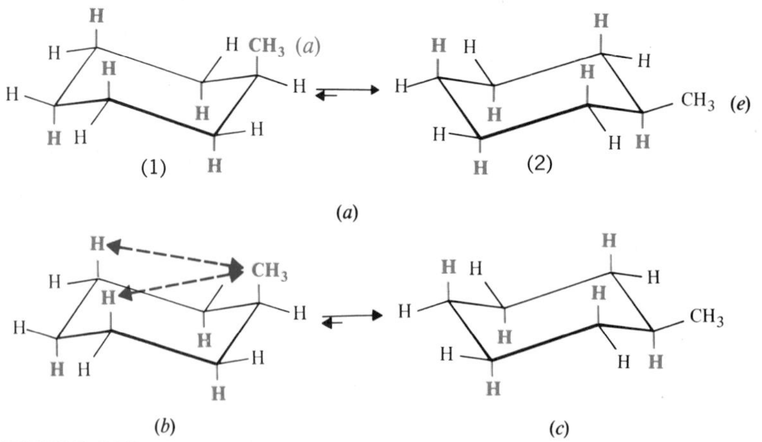

FIGURE 3.20 (*a*) The conformations of methylcyclohexane with the methyl group axial (1) and equatorial (2). (*b*) 1,3-Diaxial interactions between the two axial hydrogen atoms and the axial methyl group in the axial conformation of methylcyclohexane. (*c*) Less crowding occurs in the equatorial conformation.

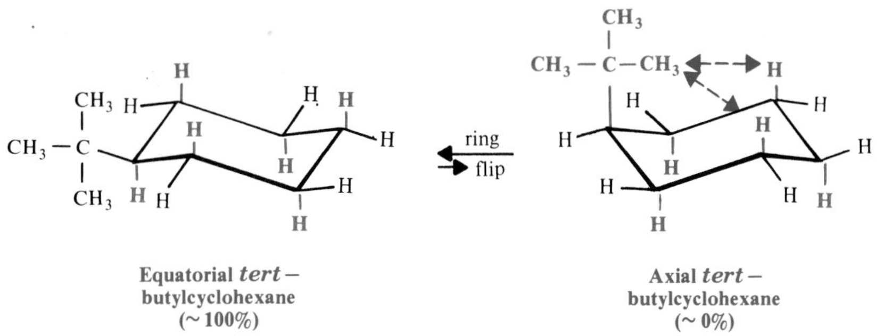

Equatorial *tert*-
butylcyclohexane
(~100%)

Axial *tert*-
butylcyclohexane
(~0%)

FIGURE 3.21 Diaxial interactions with the large *tert*-butyl group axial cause the conformation with the *tert*-butyl group equatorial to be present almost exclusively.

the molecule that the van der Waals forces between them are repulsive. Similar studies with other substituents indicates that *there will generally be less repulsive interaction when the groups are equatorial rather than axial.*

In cyclohexane derivatives with larger alkyl substituents, this effect is even more pronounced. The conformation of *tert*-butylcyclohexane with the *tert*-butyl group equatorial is estimated to be more than 5 kcal/mole more stable than the axial form (Fig. 3.21). This large energy difference between the two conformations means that, at room temperature, virtually 100% of the molecules of *tert*-butylcyclohexane have the *tert*-butyl group in the equatorial position.

3.13 DISUBSTITUTED CYCLOALKANES: CIS–TRANS ISOMERISM

The presence of two substituents on the ring of a molecule of any cycloalkane allows for the possibility of cis–trans isomerism. We can see this most easily if we begin by examining cyclopentane derivatives because the cyclopentane ring is essentially planar. (At any given moment the ring of cyclopentane is, of course, slightly bent, but we know that the various bent conformations are rapidly interconverted. Over a period of time, the average conformation of the cyclopentane ring is planar.) Since the planar representation is much more convenient for an initial presentation of cis–trans isomerism in cycloalkanes, we shall use it here.

Let us consider 1,2-dimethylcyclopentane as an example. We can write the structures shown in Fig. 3.22. In the first structure the methyl groups are on the same side of the ring, that is, they are cis. In the second structure the methyl groups are on opposite sides of the ring; they are trans.

The *cis*- and *trans*-1,2-dimethylcyclopentanes are stereoisomers: They differ from each other only in the arrangement of the atoms in space. The two forms cannot be interconverted without breaking carbon–carbon bonds. Bond breaking requires a

FIGURE 3.22 *Cis*- and *trans*-1,2-dimethylcyclopentanes.

cis-1, 2-Dimethylcyclopentane
bp 99.5 °C

trans-1, 2-Dimethylcyclopentane
bp 91.9 °C

great deal of energy, and does not occur at temperatures even considerably above room temperature. As a result, the cis and trans forms can be separated, placed in separate bottles, and kept indefinitely.

1,3-Dimethylcyclopentanes show cis–trans isomerism as well:

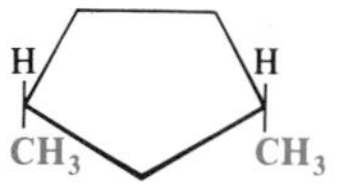

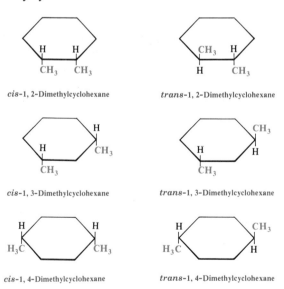

cis-1, 3-Dimethylcyclopentane 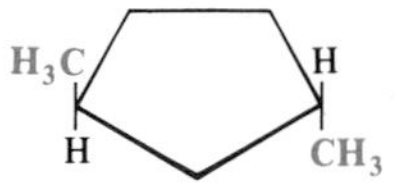   trans-1, 3-Dimethylcyclopentane

The physical properties of cis–trans isomers are different; they have different melting points, boiling points, and so on. Table 3.7 lists these physical constants of the dimethylcyclohexanes.

PROBLEM 3.8

Write structures for the cis and trans isomers of (a) 1,2-dimethylcyclopropane and (b) 1,2-dibromocyclobutane.

The cyclohexane ring is, of course, not planar. A "time average" of the various interconverting chair conformations would, however, be planar and, as with cyclopentane, this planar representation is convenient for introducing the topic of cis–trans isomerism of cyclohexane derivatives. The planar representations of the 1,2-, 1,3-, and 1,4-dimethylcyclohexane isomers follow:

cis-1, 2-Dimethylcyclohexane   trans-1, 2-Dimethylcyclohexane

cis-1, 3-Dimethylcyclohexane   trans-1, 3-Dimethylcyclohexane

cis-1, 4-Dimethylcyclohexane   trans-1, 4-Dimethylcyclohexane

3.13A Cis–Trans Isomerism and Conformational Structures

If we consider the *actual* conformations of these isomers, the structures are somewhat more complex. Beginning with *trans*-1,4-dimethylcyclohexane, because it is easiest

TABLE 3.7 Physical constants of *cis*- and *trans*-disubstituted cyclohexane derivatives

SUBSTITUENTS	ISOMER	mp (°C)	bp (°C)^a
1,2-Dimethyl-	cis	-50.1	130.04^{760}
1,2-Dimethyl-	trans	-89.4	123.7^{760}
1,3-Dimethyl-	cis	75.6	120.1^{760}
1,3-Dimethyl-	trans	-90.1	123.5^{760}
1,2-Dichloro-	cis	-6	93.5^{22}
1,2-Dichloro-	trans	-7	74.7^{16}

^{a}The pressures (in units of Torr) at which the boiling points were measured are given as superscripts.

to visualize, we find there are two possible chair conformations (Fig. 3.23). In one conformation both methyl groups are axial; in the other both are equatorial. The diequatorial conformation is, as we would expect it to be, the more stable conformation, and it represents the structure of at least 99% of the molecules at equilibrium.

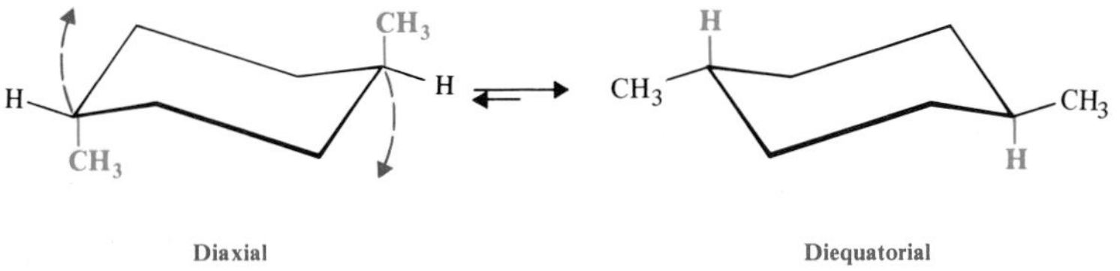

Diaxial   Diequatorial

FIGURE 3.23 The two chair conformations of *trans*-1,4-dimethylcyclohexane. (*Note:* all other C—H bonds have been omitted for clarity.)

That the diaxial form of *trans*-1,4-dimethylcyclohexane is a trans isomer is easy to see; the two methyl groups are clearly on opposite sides of the ring. The trans relationship of the methyl groups in the diequatorial form is not as obvious, however. The trans relationship of the methyl groups becomes more apparent if we imagine ourselves "flattening" the molecule by turning one end up and the other down.

A second *and general* way to recognize a *trans*-disubstituted cyclohexane is to notice that one group is attached by the *upper* bond (of the two to its carbon) and one by the *lower* bond.

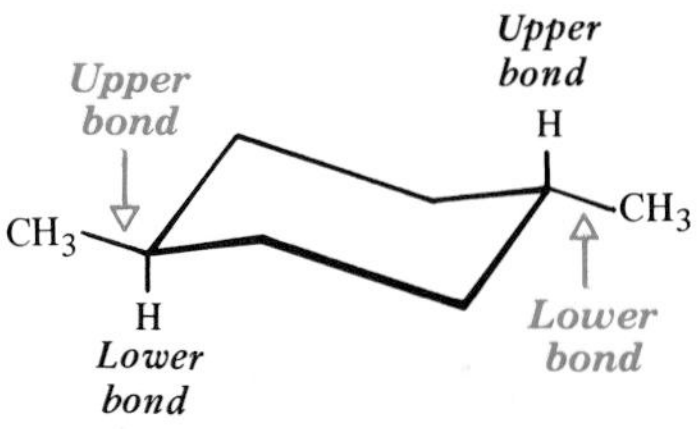

trans-1, 4-Dimethylcyclohexane

In a *cis*-disubstituted cyclohexane both groups will be attached by an upper bond or both by a lower bond. For example,

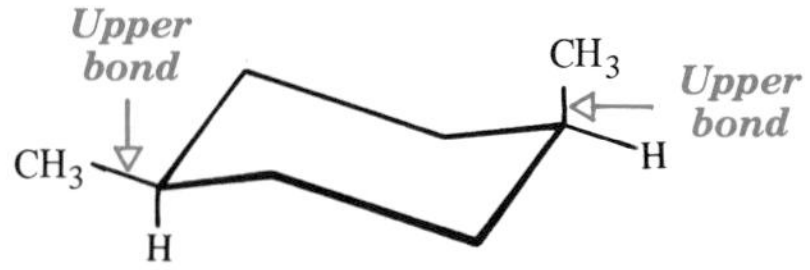

cis-1, 4-Dimethylcyclohexane

cis-1,4-Dimethylcyclohexane actually exists in two *equivalent* chair conformations (Fig. 3.24). The cis relationship of the methyl groups, however, precludes the possibility of a structure with both groups in an equatorial position. One group is axial in either conformation.

Axial–equatorial

FIGURE 3.24   Equivalent conformations of *cis*-1,4-dimethylcyclohexane.

SAMPLE PROBLEM

Consider each of the following conformational structures and tell whether each is cis or trans.

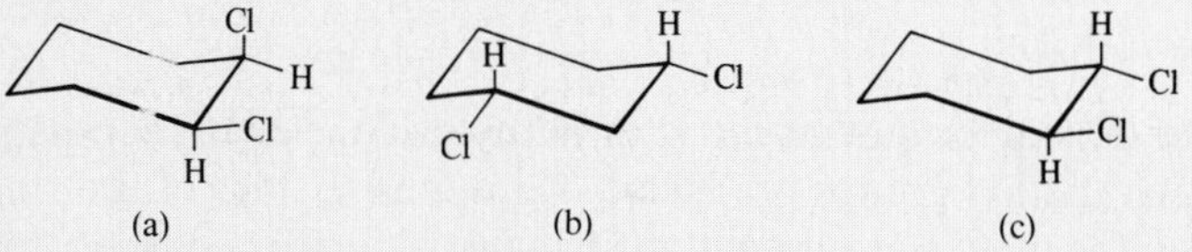

Answer:
(a) Each chlorine atom is attached by the upper bond at its carbon; therefore, both chlorine atoms are on the same side of the molecule and this is a cis isomer. This is *cis*-1,2-dichlorocyclohexane.   (b) Here both chlorine atoms are attached by a lower bond; therefore, in this example, too, both chlorine atoms are on the same side of the molecule and this, too, is a cis isomer. It is *cis*-1,3-dichlorocyclohexane.   (c) Here one chlorine atom is attached by a lower bond and one by an upper bond. The two chlorine atoms, therefore, are on opposite sides of the molecule, and this is a trans isomer. It is *trans*-1,2-dichlorocyclohexane.

PROBLEM 3.9

(a) Write structural formulas for the two chair conformations of *cis*-1-*tert*-butyl-4-methylcyclohexane.   (b) Are these two conformations equivalent? (c) If not, which would be more stable?   (d) Which would be the preferred conformation at equilibrium?

trans-1,3-Dimethylcyclohexane is like the *cis*-1,4-compound in that no conformation is possible with both methyl groups in the favored equatorial position. The following two conformations are of equal energy and are equally populated at equilibrium.

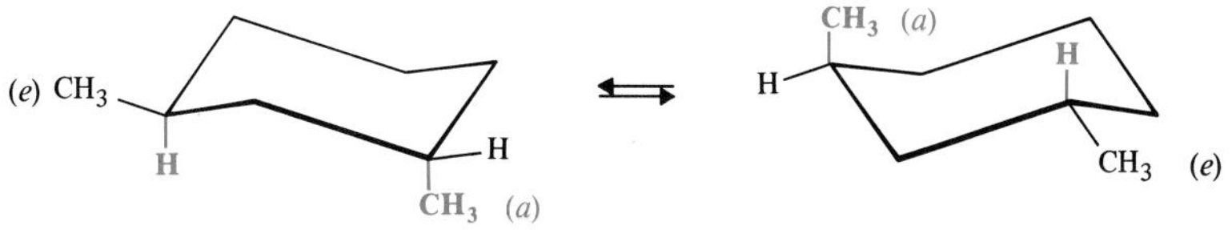

trans-1, 3-Dimethylcyclohexane

If, however, we consider some other *trans*-1,3-disubstituted cyclohexane in which one group is larger than the other, the conformation of lower energy is the one having the larger group in the equatorial position. For example, the more stable conformation of *trans*-1-*tert*-butyl-3-methylcyclohexane, shown here, has the large *tert*-butyl group occupying the equatorial position.

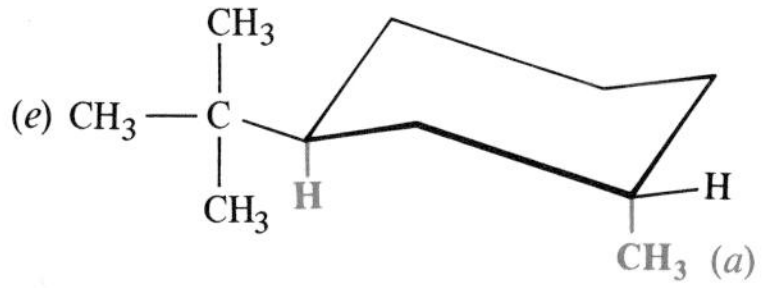

PROBLEM 3.10

(a) Write chair conformations for *cis*- and *trans*-1,2-dimethylcyclohexane.
(b) For which isomer (cis or trans) are the two conformations equivalent?
(c) For the isomer where the two conformations are not equivalent, which conformation is more stable? (d) Which conformation would be more highly populated at equilibrium? (Check your answer with Table 3.8.)

The different conformations of the dimethylcyclohexanes are summarized in Table 3.8. The more stable conformation, where one exists, is set in heavy type.

3.14 BICYCLIC AND POLYCYCLIC ALKANES

Many of the molecules that we encounter in our study of organic chemistry contain more than one ring (Section 3.4B). One of the most important bicyclic systems

TABLE 3.8 Conformations
of dimethylcyclohexanes

COMPOUND	cis ISOMER			trans ISOMER		
1,2-Dimethyl-	*a,e*	or	*e,a*	*e,e*	or	*a,a*
1,3-Dimethyl-	*e,e*	or	*a,a*	*a,e*	or	*e,a*
1,4-Dimethyl-	*a,e*	or	*e,a*	*e,e*	or	*a,a*

is bicyclo[4.4.0]decane, a compound that is usually called by its common name, *decalin*.

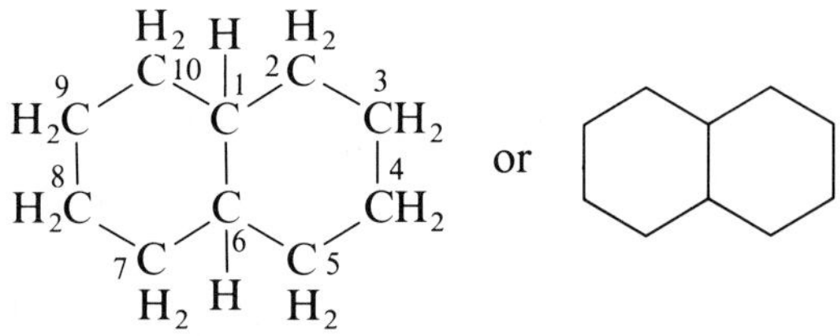

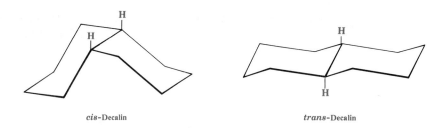

Decalin (or bicyclo[4.4.0]decane)
(carbon atoms 1 and 6 are bridgehead carbon atoms)

Decalin shows cis–trans isomerism:

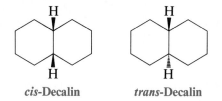

cis–Decalin   *trans*–Decalin

In *cis*-decalin the two hydrogen atoms attached to the bridgehead atoms lie on the same side of the ring; in *trans*-decalin they are on opposite sides. We often indicate this by writing their structures in the following way:

cis-Decalin   *trans*-Decalin

Simple rotations of groups about carbon–carbon bonds do not interconvert *cis*- and *trans*-decalins. In this respect they resemble the isomeric *cis*- and *trans*-disubstituted cyclohexanes. (We can, in fact, regard them as being *cis*- or *trans*-1,2-disubstituted cyclohexanes in which the 1,2-substituents are the two ends of a four-carbon bridge, that is, $-CH_2CH_2CH_2CH_2-$.)

The *cis*- and *trans*-decalins can be separated. *cis*-Decalin boils at 195 °C (at 760 Torr) and *trans*-decalin boils at 185.5 °C (at 760 Torr).

Adamantane (see the following figure) is a tricyclic system that contains a three-dimensional array of cyclohexane rings, all of which are in the chair form. Extending the structure of adamantane in three dimensions gives the structure of diamond. The great hardness of diamond results from the fact that the entire diamond crystal is actually one very large molecule — a molecule that is held together by millions of strong covalent bonds.

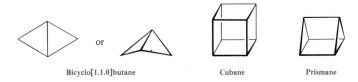

Adamantane   A portion of the diamond structure

One goal of research in recent years has been the synthesis of unusual, and sometimes highly strained, cyclic hydrocarbons. Among those that have been prepared are the compounds that follow:

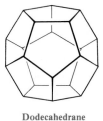

Bicyclo[1.1.0]butane   Cubane   Prismane

In 1982, Leo A. Paquette and his co-workers at the Ohio State University announced the successful synthesis of the "most complex, symmetric, and aesthetically appealing" molecule called dodecahedrane. Apparently because of its great symmetry, dodecahedrane melts above 450 °C.

Dodecahedrane

3.15 SOURCES OF ALKANES: PETROLEUM

The primary source of alkanes is petroleum. Petroleum is a complex mixture of organic compounds, most of which are alkanes and aromatic hydrocarbons (cf. Chapter 12). It also contains small amounts of oxygen-, nitrogen-, and sulfur-containing compounds.

TABLE 3.9 Typical fractions obtained by distillation of petroleum

BOILING RANGE OF FRACTION (°C)	NUMBER OF CARBON ATOMS PER MOLECULE	USE
Below 20	C_1—C_4	Natural gas, bottled gas, petrochemicals
20–60	C_5—C_6	Petroleum ether, solvents
60–100	C_6—C_7	Ligroin, solvents
40–200	C_5—C_{10}	Gasoline (straight-run gasoline)
175–325	C_{12}—C_{18}	Kerosene and jet fuel
250–400	C_{12} and higher	Gas oil, fuel oil, and diesel oil
Nonvolatile liquids	C_{20} and higher	Refined mineral oil, lubricating oil, grease
Nonvolatile solids	C_{20} and higher	Paraffin wax, asphalt, and tar

Adapted with permission from John R. Holum, *Organic Chemistry: A Brief Course,* Wiley, New York, 1975.

3.15A Petroleum Refining

The first step in refining petroleum is distillation; the object here is to separate the petroleum into fractions based on the volatility of its components. Complete separation into fractions containing individual compounds is economically impractical and virtually impossible technically. More than 500 different compounds are contained in the petroleum distillates boiling below 200 °C and many have almost the same boiling points. Thus the fractions taken contain mixtures of alkanes of similar boiling points (cf. Table 3.9). Mixtures of alkanes, fortunately, are perfectly suitable for uses as fuels, solvents, and lubricants, the primary uses of petroleum.

3.15B Cracking

The demand for gasoline is much greater than that supplied by the gasoline fraction of petroleum. Important processes in the petroleum industry, therefore, are concerned with converting hydrocarbons from other fractions into gasoline. When a mixture of alkanes from the gas oil (C_{12} and higher) fraction are heated at very high temperatures (~ 500 °C) in the presence of a variety of catalysts, the molecules break apart and rearrange to smaller, more highly branched alkanes containing 5 to 10 carbon atoms (see Table 3.9). This process is called **catalytic cracking.** Cracking can also be done in the absence of a catalyst—called **thermal cracking**—but in this process the products tend to have unbranched chains, and alkanes with unbranched chains have a very low "octane rating" (cf. Section 7.11). Gasoline components can also be made from smaller molecules (alkenes) by a process called alkylation. We shall see how this is done later.

3.16 CHEMICAL REACTIONS OF ALKANES

Alkanes, as a class, are characterized by a general inertness to many chemical reagents. Carbon–carbon and carbon–hydrogen bonds are quite strong; they do not break unless alkanes are heated to very high temperatures (cf. Section 3.15B). Because carbon and hydrogen atoms have nearly the same electronegativity, the carbon–hydrogen bonds of alkanes are only slightly polarized. As a consequence, they are generally unaffected by most bases. Molecules of alkanes have no unshared electrons to offer sites for attack by acids. This low reactivity of alkanes toward many reagents accounts for the fact that alkanes were originally called *paraffins* (Latin: *parum affinis,* little affinity).

The term paraffin, however, is probably not an appropriate one. We all know that alkanes react vigorously with oxygen when an appropriate mixture is ignited. This combustion occurs in the cylinders of automobiles, and in oil furnaces, for example. When heated, alkanes also react with chlorine, and they react explosively with fluorine.

3.17 THE REACTIONS OF ALKANES WITH HALOGENS: SUBSTITUTION REACTIONS

Methane, ethane, and other alkanes react with the first three members of the halogen family: fluorine, chlorine, and bromine. Alkanes do not react appreciably with iodine. With methane the reaction produces a mixture of halomethanes and a hydrogen halide.

$$
\underset{\text{Methane}}{H-\overset{\overset{\displaystyle H}{|}}{\underset{\underset{\displaystyle H}{|}}{C}}-H} + \underset{\text{Halogen}}{X_2} \xrightarrow[\substack{\text{or} \\ \text{light}}]{\text{heat}} \underset{\text{Halomethane}}{H-\overset{\overset{\displaystyle H}{|}}{\underset{\underset{\displaystyle H}{|}}{C}}-X} + \underset{\text{Dihalomethane}}{H-\overset{\overset{\displaystyle X}{|}}{\underset{\underset{\displaystyle H}{|}}{C}}-X}
$$

$$
+ \underset{\text{Trihalomethane}}{H-\overset{\overset{\displaystyle X}{|}}{\underset{\underset{\displaystyle X}{|}}{C}}-X} + \underset{\text{Tetrahalomethane}}{X-\overset{\overset{\displaystyle X}{|}}{\underset{\underset{\displaystyle X}{|}}{C}}-X} + \underset{\text{Hydrogen halide}}{H-X}
$$

X = F, Cl, or Br

The reaction of an alkane with a halogen is called **halogenation.** The general reaction to produce a monohaloalkane can be written as follows:

$$
R-H + X_2 \longrightarrow R-X + HX
$$

In these reactions a halogen atom replaces one or more of the hydrogen atoms of the alkane. Reactions of this type, *in which one atom or group replaces another,* are called **substitution reactions.**

$$
-\overset{|}{\underset{|}{C}}-H + Cl_2 \longrightarrow -\overset{|}{\underset{|}{C}}-Cl + H-Cl \quad \text{A substitution reaction}
$$

3.17A Halogenation Reactions

One complicating characteristic of alkane halogenations is that multiple substitution reactions almost always occur. As we saw at the beginning of this section, the halogenation of methane produces a mixture of monohalomethane, dihalomethane, trihalomethane, and tetrahalomethane.

This happens because all hydrogen atoms attached to carbon are capable of reacting with fluorine, chlorine, or bromine.

Let us consider the reaction that takes place between chlorine and methane as an example. If we mix methane and chlorine (both substances are gases at room temperature) and then either heat the mixture or irradiate it with light, a reaction begins to occur vigorously. At the outset, the only compounds that are present in the mixture are chlorine and methane, and the only reaction that can take place is one that produces chloromethane and hydrogen chloride.

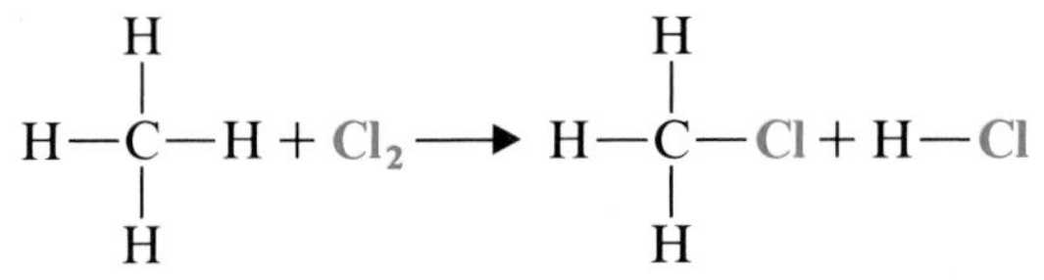

$$\underset{\underset{H}{|}}{\overset{\overset{H}{|}}{H-C-H}} + Cl_2 \longrightarrow \underset{\underset{H}{|}}{\overset{\overset{H}{|}}{H-C-Cl}} + H-Cl$$

As the reaction progresses, however, the concentration of chloromethane in the mixture increases, and a second substitution reaction begins to occur. Chloromethane reacts with chlorine to produce dichloromethane.

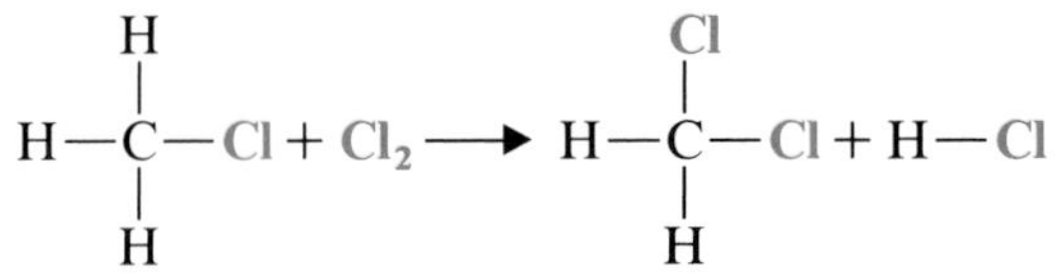

$$\underset{\underset{H}{|}}{\overset{\overset{H}{|}}{H-C-Cl}} + Cl_2 \longrightarrow \underset{\underset{H}{|}}{\overset{\overset{Cl}{|}}{H-C-Cl}} + H-Cl$$

Dichloromethane can then produce trichloromethane,

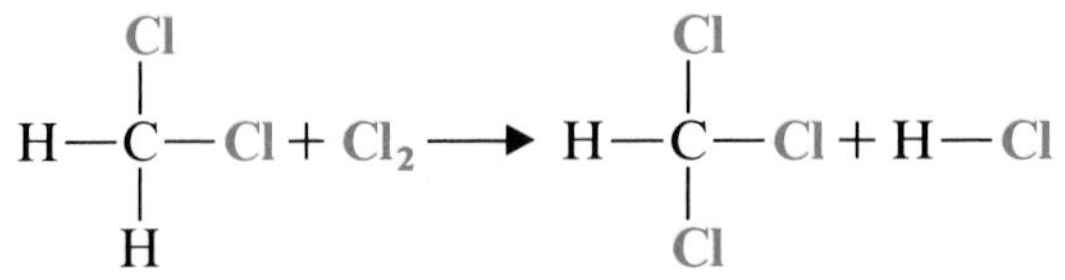

$$\underset{\underset{H}{|}}{\overset{\overset{Cl}{|}}{H-C-Cl}} + Cl_2 \longrightarrow \underset{\underset{Cl}{|}}{\overset{\overset{Cl}{|}}{H-C-Cl}} + H-Cl$$

and trichloromethane, as it accumulates in the mixture, can react with chlorine to produce tetrachloromethane.

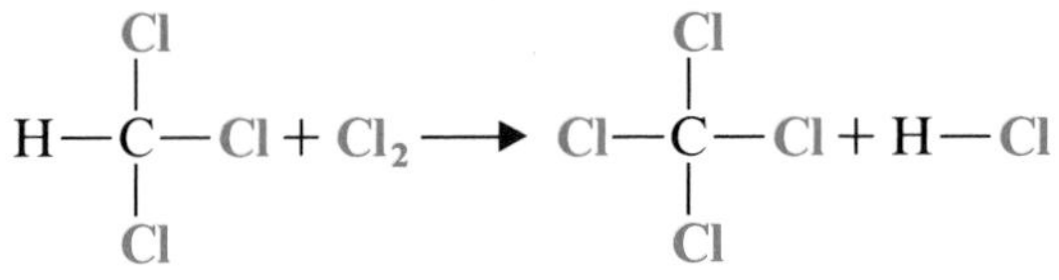

$$\underset{\underset{Cl}{|}}{\overset{\overset{Cl}{|}}{H-C-Cl}} + Cl_2 \longrightarrow \underset{\underset{Cl}{|}}{\overset{\overset{Cl}{|}}{Cl-C-Cl}} + H-Cl$$

Each time a substitution of —Cl for —H takes place a molecule of H—Cl is produced.

SAMPLE PROBLEM

If the goal of a synthesis is to prepare chloromethane (CH_3Cl), its formation can be maximized and the formation of CH_2Cl_2, $CHCl_3$, and CCl_4 minimized by using a large excess of methane in the reaction mixture. Explain why this is possible.

Answer:
The use of a large excess of methane maximizes the probability that chlorine will attack methane molecules because the concentration of methane in the

mixture will always be relatively large. It also minimizes the probability that chlorine will attack molecules of CH_3Cl, CH_2Cl_2, and so on, because their concentrations will always be relatively small. After the reaction is over, the unreacted excess methane can be recovered and recycled.

When ethane and chlorine react, similar substitution reactions occur. Ultimately all six hydrogen atoms of ethane may be replaced. We notice in the following diagram that the second substitution reaction of ethane results in the formation of two different molecules: 1,1-dichloroethane and 1,2-dichloroethane. These two molecules have the same molecular formula ($C_2H_4Cl_2$) but these atoms are connected in a different way. They are **constitutional isomers.**

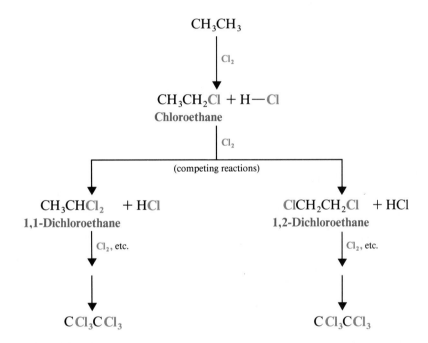

Chlorination of most alkanes whose molecules contain three carbon atoms or more gives a mixture of isomeric monochloro products as well as more highly halogenated compounds. Chlorine is relatively **unselective;** it does not discriminate among the different types of hydrogen atoms (primary, secondary, and tertiary) in an alkane. An example is the light-promoted chlorination of isobutane.

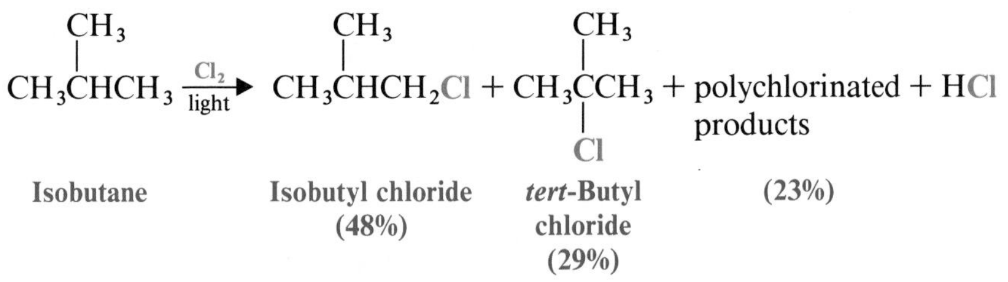

Because alkane chlorinations usually yield a complex mixture of products, they are not generally useful as synthetic methods when our goal is the preparation of a specific alkyl chloride. An exception is the halogenation of an alkane (or cycloalkane) whose hydrogen atoms *are all equivalent*.* Neopentane, for example, can form only one monohalogenation product, and the use of a large excess of neopentane minimizes polychlorination.

$$
\underset{\substack{\text{Neopentane} \\ \text{(excess)}}}{CH_3-\overset{\displaystyle CH_3}{\underset{\displaystyle CH_3}{C}}-CH_3} + Cl_2 \xrightarrow[\substack{\text{or} \\ \text{light}}]{\text{heat}} \underset{\text{Neopentyl chloride}}{CH_3-\overset{\displaystyle CH_3}{\underset{\displaystyle CH_3}{C}}-CH_2Cl} + HCl
$$

In a similar way cyclobutane yields cyclobutyl chloride.

$$
\square + Cl_2 \xrightarrow[\substack{\text{or} \\ \text{light}}]{\text{heat}} \square^{Cl} + HCl
$$

(excess)

3.17B Selectivity of Bromine

Bromine is generally less reactive toward alkanes than chlorine, and bromine is *more selective* in the site of attack when it does react. Bromine shows a much greater ability to discriminate among the different types of hydrogen atoms and preferentially replaces tertiary hydrogen atoms. The reaction of isobutane and bromine, for example, gives almost exclusive replacement of the tertiary hydrogen atom.

$$
CH_3-\overset{\displaystyle CH_3}{\underset{\displaystyle H}{C}}-CH_3 \xrightarrow[\text{light, 127 °C}]{Br_2} \underset{(>99\%)}{CH_3-\overset{\displaystyle CH_3}{\underset{\displaystyle Br}{C}}-CH_3} + \underset{(\text{trace})}{CH_3-\overset{\displaystyle CH_3}{\underset{\displaystyle H}{C}}-CH_2Br} + HBr
$$

A similar result is obtained with 2-methylbutane.

$$
CH_3\overset{\displaystyle CH_3}{\underset{}{CH}}CH_2CH_3 \xrightarrow[\text{light}]{Br_2} \underset{(\sim 93\%)}{CH_3\overset{\displaystyle CH_3}{\underset{\displaystyle Br}{C}}CH_2CH_3} + \underset{(\sim 6\%)}{CH_3\overset{\displaystyle CH_3}{\underset{\displaystyle Br}{CH}}CHCH_3} + HBr
$$

*Equivalent hydrogen atoms are defined as those which on replacement by some other group (e.g., chlorine) yield the same compound.

PROBLEM 3.11

When propane reacts with chlorine the reaction yields (in addition to more highly chlorinated compounds) *two compounds with the molecular formula C_3H_7Cl.* One compound has a boiling point of 46.6 °C, the other a boiling point of 36.5 °C. (a) Write possible structural formulas for these compounds. (b) Can you tell which one is which? (c) When the compound with a boiling point of 46.6 °C is allowed to react with chlorine, the reaction yields three compounds with the formula $C_3H_6Cl_2$ (again in addition to more highly chlorinated compounds). When the compound with boiling point 36.5 °C reacts with chlorine, only two isomers with the formula $C_3H_6Cl_2$ are produced. Show how this information allows you to assign structural formulas to the two C_3H_7Cl compounds. (d) Write all of the reactions involved.

PROBLEM 3.12

Three isomeric alkanes **A**, **B**, and **C** have the formula C_5H_{12}. **A** reacts with chlorine to yield only one compound with the formula $C_5H_{11}Cl$. **B** reacts with chlorine to yield three isomers with the formula $C_5H_{11}Cl$, and **C** reacts to yield four isomers with the formula $C_5H_{11}Cl$. Assign structural formulas to **A**, **B**, and **C**, and show the isomers of $C_5H_{11}Cl$ formed when each reacts with chlorine. (Other more highly chlorinated compounds are also formed but these are not important for this problem.)

We shall discuss alkane halogenations in more detail in Chapter 9.

3.17C Halogenation of Cycloalkanes

The reactions of cycloalkanes are similar to those of alkanes. Cyclopentane and cyclohexane, for example,

Cl_2
heat, light
Cl
+ HCl
Br_2
heat, light
Br
+ HBr

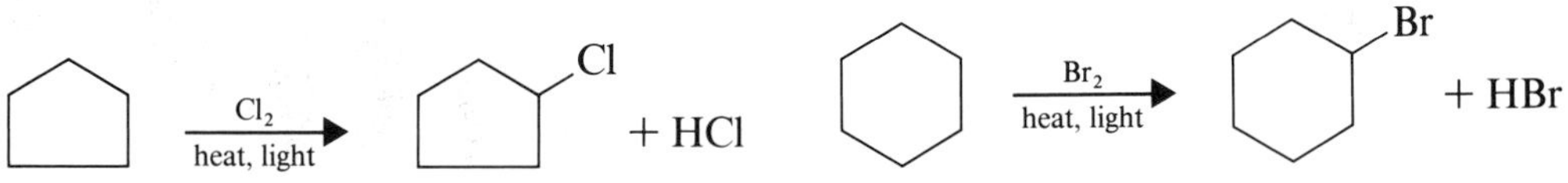

Cyclopentane (excess) — Chlorocyclopentane — Cyclohexane (excess) — Bromocyclohexane

undergo halogenation by substitution.

These monohalocycloalkanes can be prepared in good yield when an excess of the cycloalkane is used (to minimize polyhalogenation). This happens because a simple cycloalkane can yield only one monohalo product. With a substituted cycloalkane such as methylcyclopentane, however, chlorination produces a complex mixture of products:

CH_3
Cl_2
heat, light
CH_2Cl
CH_3
Cl
CH_3
Cl
CH_3
Cl
+ HCl

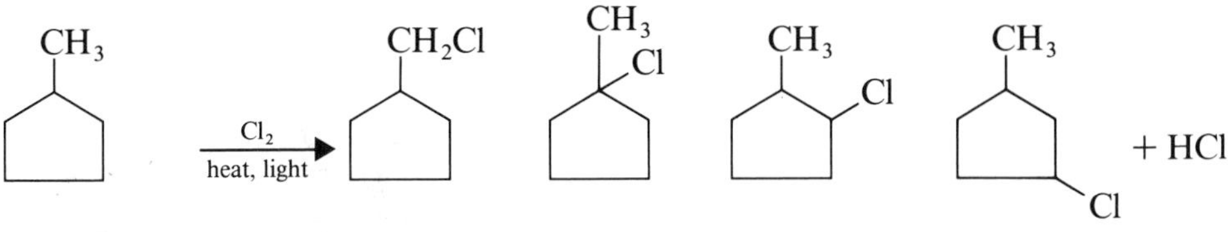

Methylcyclopentane (excess) — (cis and trans) — (cis and trans)

PROBLEM 3.13

Write conformational structures for all of the dichlorocyclohexanes that you would expect to obtain when cyclohexane is chlorinated with excess chlorine.

PROBLEM 3.14

Although chlorination of methylcyclopentane (just cited) yields a complex mixture, bromination of methylcyclopentane yields one predominant product. What is this product?

3.18 SYNTHESIS OF ALKANES AND CYCLOALKANES

Mixtures of alkanes as they are obtained from petroleum are suitable as fuels. However, in our laboratory work we often have the need for a pure sample of a particular alkane. For these purposes, the chemical preparation — or synthesis — of that particular alkane is often the most reliable way of obtaining it. The preparative method that we choose should be one that will lead to the desired product alone or, at least, to products that can be easily and effectively separated.

Several such methods are available, and two are outlined here. In subsequent chapters we shall encounter others.

1. **Hydrogenation of alkenes.** Alkenes react with hydrogen in the presence of metal catalysts such as nickel and platinum to produce alkanes.

The general reaction is one in which the atoms of the hydrogen molecule add to each atom of the carbon–carbon double bond of the alkene. This converts the alkene to an alkane.

General Reaction

$$
\underset{\text{Alkene}}{\overset{\displaystyle \diagup \text{C} \diagdown}{\underset{\diagup \text{C} \diagdown}{\|}}}
+
\overset{\text{H}}{\underset{\text{H}}{|}}
\xrightarrow[\text{solvent}]{\text{Pt or Ni}}
\underset{\text{Alkane}}{
\begin{array}{c} | \\ -\text{C}-\text{H} \\ | \\ -\text{C}-\text{H} \\ | \end{array}}
$$

The reaction is usually carried out by dissolving the alkene in a solvent such as ethyl alcohol (C_2H_5OH), adding the metal catalyst, and then exposing the mixture to hydrogen gas under pressure in a special apparatus. (We shall have much more to say about this reaction — called hydrogenation — in Chapter 6.)

Specific Examples

$$
\underset{\text{Propene}}{CH_3CH{=}CH_2} + H{-}H \xrightarrow[\substack{C_2H_5OH \\ (25\,°C,\,50\,atm)}]{Ni} \underset{\text{Propane}}{CH_3\underset{\overset{|}{H}}{CH}{-}\underset{\overset{|}{H}}{CH_2}}
$$

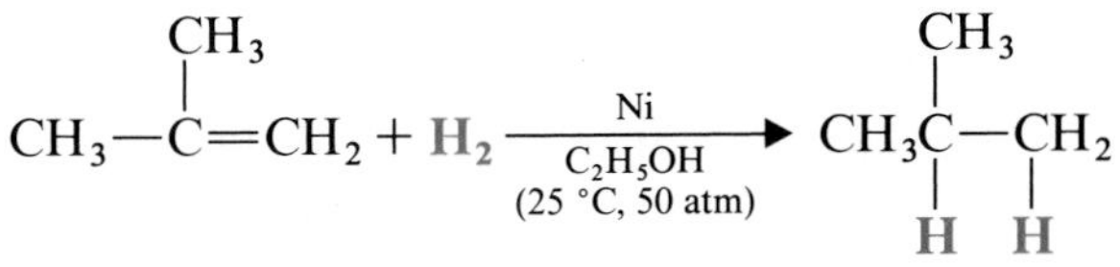

$$CH_3-\overset{\overset{\displaystyle CH_3}{|}}{C}=CH_2 + H_2 \xrightarrow[\substack{C_2H_5OH \\ (25\ °C,\ 50\ atm)}]{Ni} CH_3\overset{\overset{\displaystyle CH_3}{|}}{\underset{\underset{\displaystyle H}{|}}{C}}-\overset{\overset{}{}}{\underset{\underset{\displaystyle H}{|}}{C}}H_2$$

2-Methylpropene   Isobutane

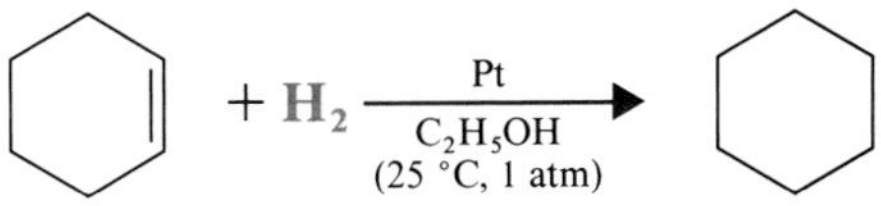

$$\bigcirc + H_2 \xrightarrow[\substack{C_2H_5OH \\ (25\ °C,\ 1\ atm)}]{Pt} \bigcirc$$

Cyclohexene   Cyclohexane

PROBLEM 3.15

Three different alkenes will react with hydrogen in the presence of a platinum or nickel catalyst to yield butane. What are their structures? Show the reactions.

---

2. **Reduction of alkyl halides.** Most alkyl halides react with zinc and aqueous acid to produce an alkane. In this reaction zinc acts as a reducing agent and causes the halogen of the alkyl halide to be replaced by hydrogen. (Since hydrogen is less electronegative than a halogen, the alkyl halide is said to be **reduced.**) The general reaction is as follows:

General Reaction

$$2R-X + Zn + 2H^+ \longrightarrow 2R-H + ZnX_2$$

or* $\quad R-X \xrightarrow[(-ZnX_2)]{Zn,\ H^+} R-H$

Specific Examples

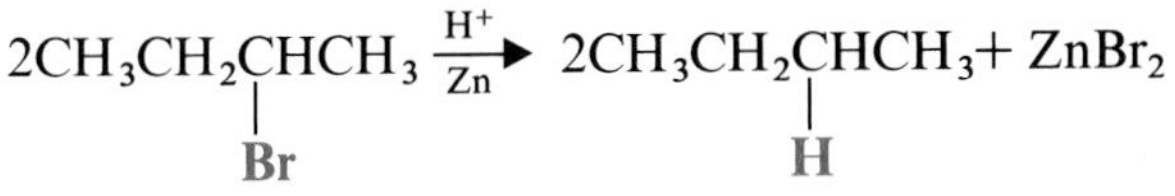

$$2CH_3CH_2\overset{}{\underset{\underset{\displaystyle Br}{|}}{C}}HCH_3 \xrightarrow[Zn]{H^+} 2CH_3CH_2\overset{}{\underset{\underset{\displaystyle H}{|}}{C}}HCH_3 + ZnBr_2$$

sec-Butyl bromide   Butane
(2-bromobutane)

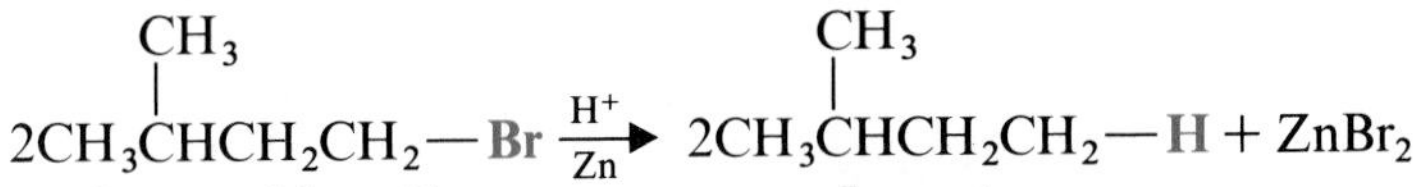

$$2CH_3\overset{\overset{\displaystyle CH_3}{|}}{C}HCH_2CH_2-Br \xrightarrow[Zn]{H^+} 2CH_3\overset{\overset{\displaystyle CH_3}{|}}{C}HCH_2CH_2-H + ZnBr_2$$

Isopentyl bromide   Isopentane
(1-bromo-3-methylbutane)   (2-methylbutane)

---

*This illustrates the way organic chemists often write abbreviated equations for chemical reactions. The organic reactant is shown on the left and the organic product on the right. The reagents necessary to bring about the transformation are written over (or under) the arrow. The equations are often left unbalanced and sometimes by-products (in this case, ZnX_2) are either omitted or are placed under the arrow in parentheses with a minus sign, for example, $(-ZnX_2)$.

PROBLEM 3.16

In addition to isopentyl bromide (just cited), three other alkyl halides will yield isopentane when they are treated with zinc and aqueous acid. What are their structures? Show the reactions.

3.19 SOME IMPORTANT TERMS AND CONCEPTS

Alkanes are hydrocarbons with the general formula C_nH_{2n+2}. Molecules of alkanes have no rings (i.e., they are **acyclic**) and they have only single bonds between carbon atoms. Their carbon atoms are sp^3 hybridized.

Cycloalkanes are hydrocarbons with the general formula C_nH_{2n} whose molecules have their carbon atoms arranged into a ring. They have only single bonds between carbon atoms, and their carbon atoms are sp^3 hybridized.

Conformational analysis is a study of the energy changes that occur in a molecule when groups rotate about single bonds.

Torsional strain refers to a small barrier to free rotation about the carbon–carbon single bond. For ethane this barrier is 2.8 kcal/mole (11.7 kJ/mole).

van der Waals forces are weak forces that act between nonpolar molecules or between parts of the same molecule. Bringing two groups together first results in an *attractive* van der Waals force between them because a temporary unsymmetrical distribution of electrons in one group induces an opposite polarity in the other. When the groups are brought closer than their *van der Waals radii,* the force between them becomes repulsive because their electron clouds begin to interpenetrate each other. The methyl groups of the *gauche* form of butane, for example, are close enough for the van der Waals forces to be repulsive.

Ring strain is a strain that gives certain cycloalkanes greater potential energy than others. The principal sources of ring strain are *torsional strain* and *angle strain.*

Angle strain is strain introduced into a molecule because some factor (e.g., ring size) causes the bond angles of its atoms to deviate from the normal bond angle. The normal bond angles of an sp^3 carbon are 109.5°, but in cyclopropane, for example, one pair of bonds at each carbon atom is constrained to a much smaller angle. This introduces considerable angle strain into the molecule causing molecules of cyclopropane to have greater potential energy per CH_2 group than cycloalkanes with less (or no) angle strain.

Cis–trans isomerism is a type of stereoisomerism possible with certain alkenes (Section 2.4B) and with disubstituted cycloalkanes. Cis–trans isomers of 1,2-dimethylcyclopropane are shown here.

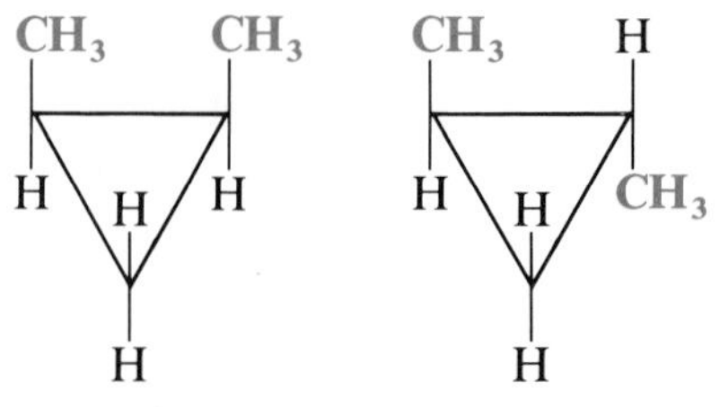

cis   trans

These two isomers can be separated and they have different physical properties. The two forms cannot be interconverted without breaking carbon–carbon bonds.

Conformations of molecules of cyclohexane. The most stable conformation is a chair conformation. Twist conformations and boat conformations have greater potential energy. These conformations (chair, boat, and twist) can be interconverted by rotations of single bonds. In a sample of cyclohexane more than 99% of the molecules are in a chair conformation at any given moment. A group attached to a carbon atom of a molecule of cyclohexane in a chair conformation can assume either of two positions: *axial* or *equatorial,* and these are interconverted when the ring flips from one chair conformation to another. A group has more room when it is equatorial; thus most of the molecules of substituted cyclohexanes at any given moment will be in the chair conformation that has the largest group (or groups) equatorial.

ADDITIONAL PROBLEMS

3.17
Write a structural formula for each of the following compounds:

(a) 2,3-Dichloropentane

(b) *tert*-Butyl iodide

(c) 3-Ethylpentane

(d) 2,3,4-Trimethyldecane

(e) 4-Isopropylnonane

(f) 1,1-Dimethylcyclopropane

(g) *cis*-1,2-Dimethylcyclobutane

(h) *trans*-1,3-Dimethylcyclobutane

(i) Isopropylcyclohexane (most stable conformation)

(j) *trans*-1-Isopropyl-3-methylcyclohexane (most stable conformation)

(k) Isohexyl chloride

(l) 2,2,4,4-Tetramethyloctane

(m) Neopentyl chloride

(n) Isopentane

3.18
Name each of the following compounds by the IUPAC system:

(a) $CH_3CH(C_2H_5)CH(CH_3)CH_2CH_3$

(b) $CH_3CH(CH_3)CH_2CH_3$

(c)

(d) $CH_3CH_2CH(C_2H_5)CH_3$

(e)

(f)

(g)

3.19
Write the structure and give the IUPAC name of an alkane or cycloalkane with the formula: (a) C_5H_{12} that has only primary hydrogen atoms (i.e., hydrogen atoms attached to primary carbon atoms), (b) C_5H_{12} that has only one tertiary hydrogen atom, (c) C_5H_{12} that has only primary and secondary hydrogen atoms, (d) C_5H_{10} that has only secondary hydrogen atoms, and (e) C_6H_{14} that has only primary and tertiary hydrogen atoms.

3.20
Three different alkenes yield 2-methylbutane when they are hydrogenated in the presence of a metal catalyst. Give their structures and write equations for the reactions involved.

3.21

An alkane with the formula C_6H_{14} can be synthesized by treating (in separate reactions) five different alkyl chlorides $(C_6H_{13}Cl)$ with zinc and aqueous acid. Give the structure of the alkane and the structures of the alkyl chlorides.

3.22

An alkane with the formula C_6H_{14} can be prepared by reduction (with Zn and H^+) of only two alkyl chlorides $(C_6H_{13}Cl)$ and by the hydrogenation of only two alkenes (C_6H_{12}). Write the structure of this alkane, give its IUPAC name, and show the reactions.

3.23

Four different cycloalkenes will all yield methylcyclopentane when subjected to catalytic hydrogenation. What are their structures? Show the reactions.

3.24

The heats of combustion of three pentane (C_5H_{12}) isomers are $CH_3(CH_2)_3CH_3$, 845.2 kcal/mole; $CH_3CH(CH_3)CH_2CH_3$, 843.4 kcal/mole; and $(CH_3)_3CCH_3$, 840.0 kcal/mole. Which isomer is most stable? Construct a diagram such as that in Fig. 3.9 showing the relative potential energies of the three compounds.

3.25

Tell what is meant by a homologous series and illustrate your answer by writing a homologous series of alkyl halides.

3.26

Write the structures of two chair conformations of 1-*tert*-butyl-1-methylcyclohexane. Which conformation is more stable? Explain your answer.

3.27

Ignoring compounds with double bonds, write structural formulas and give names for all of the isomers with the formula C_5H_{10}.

3.28

Write structures for the following bicyclic alkanes.
(a) Bicyclo[1.1.0]butane   (c) 2-Chlorobicyclo[3.2.0]heptane
(b) Bicyclo[2.1.0]pentane   (d) 7-Methylbicyclo[2.2.1]heptane

3.29

The carbon–carbon bond angles of isobutane are ∼ 111.5°. These angles are larger than those expected from regular (or undistorted) tetrahedral carbon (i.e., 109.5°). Explain.

3.30

Sketch approximate potential energy diagrams for rotations about   (a) the C-2—C-3 bond of 2,3-dimethylbutane,   (b) the C-2—C-3 bond of 2,2,3,3-tetramethylbutane, and   (c) the C-2—C-3 bond of 2-methylbutane.

3.31

Without referring to tables, decide which member of each of the following pairs has the higher boiling point. Explain your answers.
(a) Hexane or isohexane   (d) Ethane or chloroethane
(b) Hexane or pentane   (e) Propane or ethyl alcohol
(c) Pentane or neopentane

3.32

cis-1,2-Dimethylcyclopropane has a larger heat of combustion than *trans*-1,2-dimethylcyclopropane. (a) Which compound is more stable? (b) Give a reason that would explain your answer to part (a).

3.33

Write structural formulas for   (a) the two chair conformations of *cis*-1-isopropyl-3-methylcyclohexane,   (b) the two chair conformations of *trans*-1-isopropyl-3-methylcyclohexane, and   (c) designate which conformation in part (a) and (b) is more stable.

3.34

Which member of each of the following pairs of compounds would be more stable?   (a) *cis*- or *trans*-1,2-Dimethylcyclohexane,   (b) *cis*- or *trans*-1,3-dimethylcyclohexane, and   (c) *cis*- or *trans*-1,4-dimethylcyclohexane.

3.35

Norman L. Allinger of the University of Georgia has obtained evidence indicating that, while *cis*-1,3-di-*tert*-butylcyclohexane exists predominantly in a chair conformation, *trans*-1,3-di-*tert*-butylcyclohexane adopts a twist-boat conformation. Explain.

3.36

The important sugar glucose exists in the following cyclic form:

$$CH_2OH$$
$$|$$
$$CH$$
$$HOCH \qquad O$$
$$|\qquad\qquad |$$
$$HOCH \qquad CHOH$$
$$CH$$
$$|$$
$$OH$$

The six-membered ring of glucose has the chair conformation. In one isomer, β-glucose, all of the secondary hydroxyl groups and the —CH_2OH group are equatorial. Write a structure for β-glucose.

*3.37

When 1,2-dimethylcyclohexene (below) is allowed to react with hydrogen in the presence of a platinum catalyst, the product of the reaction is a cycloalkane that has a melting point of −50 °C and a boiling point of 130 °C (at 760 Torr).   (a) What is the structure of the product of this reaction?   (b) Consult an appropriate table and tell which stereoisomer it is.   (c) What does this experiment suggest about the mode of addition of hydrogen to the double bond?

1,2-Dimethylcyclohexene

*3.38

When cyclohexene is dissolved in an appropriate solvent and allowed to react with chlorine, the product of the reaction, $C_6H_{10}Cl_2$, has a melting point of −7 °C and a boiling point (at 16 Torr) of 74 °C.   (a) Which stereoisomer is this?   (b) What does this experiment suggest about the mode of addition of chlorine to the double bond?

*An asterisk beside a problem indicates that it is somewhat more challenging. Your instructor may tell you that these problems are optional.

CHAPTER FOUR

Tartaric Acid 40x. Tartaric acid, an important compound in the
history of stereochemistry, was the object of a study by Louis
Pasteur in 1848. This study, as described in Sections 4.13 and
4.14, led to the discovery of enantiomerism.

Stereochemistry. Chiral Molecules

4.1 INTRODUCTION

Almost every reaction in a living cell requires a particular protein to act as a catalyst. These catalysts are called enzymes, and each consists of an extremely large molecule that has a very definite shape. The compound whose reaction the enzyme catalyzes is called the substrate, and its molecules also have a particular shape. For catalysis to happen, the substrate molecule has to fit momentarily onto the surface of the enzyme molecule and form a complex. The reaction takes place while this complex exists. If anything is wrong with the shapes of either the substrate or the enzyme, there can be no fit and, therefore, no reaction.

Pictured here are two highly schematic ways of representing these ideas about shape and fit. The substrate is shown by a tetrahedral carbon atom holding four different atoms or groups with each group given a unique shape. The only difference in the representations of the two substrate molecules is that two of the groups on the tetrahedral carbon atom are switched; otherwise, their structures are identical. They are just as similar as the left hand is to the right hand. In fact, we can say that the two substrate molecules shown here have **chirality** or "handedness."

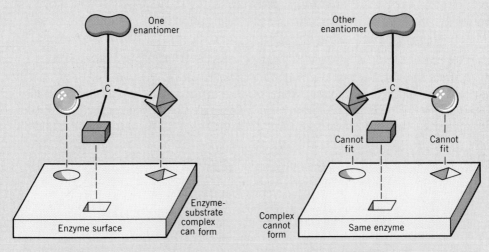

We shall explore the significance of molecular chirality or handedness in this chapter. Nearly all of the organic molecules of living organisms have "handedness." So do the molecules of almost everything that we eat. What difference does this all make? Quite literally, the difference between life and death! To see why, examine the figure again. Beneath each substrate is a representation of the *same* enzyme. Each enzyme in the body comes in only one shape as far as its catalytic powers are concerned. Notice that only one form of the substrate can fit onto this enzyme. This illustrates one of the central facts about all enzyme-catalyzed reactions—the substrate molecules can be of only one handedness if they have this geometrical property at all. If by great effort we managed to prepare a diet made up of molecules all having a handedness opposite to that which naturally occurs, we'd starve to death on it!

Molecular chirality, however, is not just associated with the molecules of nature. Many simpler molecules, many of those that we encounter commonly in the laboratory, have this property as well. After a brief review of isomerism in the next section, we shall begin our study of chiral molecules.

4.2 ISOMERISM: CONSTITUTIONAL ISOMERS AND STEREOISOMERS

Isomers are different compounds that have the same molecular formula. In our study of carbon compounds, thus far, most of our attention has been directed toward those isomers that we have called constitutional isomers.

Constitutional isomers are isomers that differ because their atoms are connected in a different order. They are said to have a different **connectivity.** Several examples of constitutional isomers are the following:

MOLECULAR FORMULA	CONSTITUTIONAL ISOMERS		
C_4H_{10}	$CH_3CH_2CH_2CH_3$	and	$CH_3\overset{\overset{\displaystyle CH_3}{\mid}}{C}HCH_3$
	Butane		Isobutane
C_3H_7Cl	$CH_3CH_2CH_2Cl$	and	$CH_3\underset{\underset{\displaystyle Cl}{\mid}}{C}HCH_3$
	1-Chloropropane		2-Chloropropane
C_2H_6O	CH_3CH_2OH	and	CH_3OCH_3
	Ethanol		Dimethyl ether

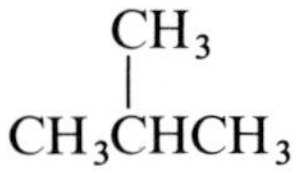

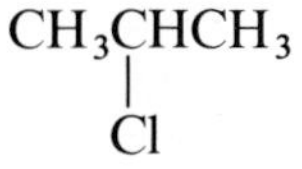

Stereoisomers are not constitutional isomers—they have their constituent atoms connected in the same way. *Stereoisomers differ only in arrangement of their atoms in space.* The cis and trans isomers of alkenes are stereoisomers (Section 2.4B); we can see that this is true if we examine the *cis-* and *trans-*1,2-dichloroethenes shown here.

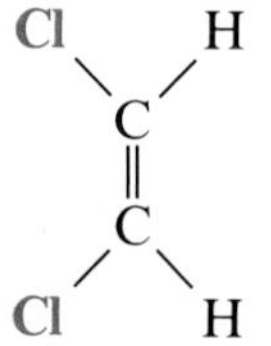

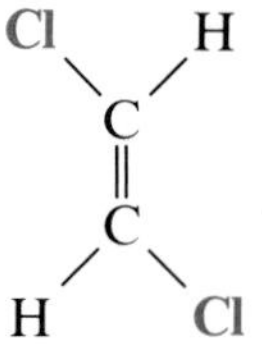

cis-1,2-Dichloroethene and *trans*-1,2-dichloroethene are isomers because both compounds have the same molecular formula ($C_2H_2Cl_2$) but they are different. They are *not* constitutional isomers, because the order of connections of the atoms in both compounds is the same. Both compounds have two central carbon atoms joined by a double bond, and both compounds have one chlorine atom and one hydrogen atom attached to the two central atoms. The *cis*-1,2-dichloroethene and *trans*-1,2-dichloroethene isomers differ only in the arrangement of their atoms in space. In *cis*-1,2-dichloroethene the hydrogen atoms are on the same side of the molecule, and in *trans*-1,2-dichloroethene the hydrogen atoms are on opposite sides. Thus, *cis*-1,2-dichloroethene and *trans*-1,2-dichloroethene are stereoisomers (see Section 2.4B).

Stereoisomers can be subdivided into two general categories: **enantiomers** and **diastereomers.** Enantiomers are stereoisomers whose molecules *are mirror reflections of each other.* Diastereomers are stereoisomers whose molecules *are not mirror reflections of each other.*

Molecules of *cis*-1,2-dichloroethene and *trans*-1,2-dichloroethene *are not* mirror reflections of each other. If one holds a model of *cis*-1,2-dichloroethene up to a mirror, the model that one sees in the mirror is not *trans*-1,2-dichloroethene. But *cis*-1,2-dichloroethene and *trans*-1,2-dichloroethene *are* stereoisomers and, since they are not related to each other as an object and its mirror reflection, they are diastereomers.

Cis and trans isomers of cycloalkanes furnish us with another example of stereoisomers that are diastereomers of each other. Consider the following two compounds.

cis-1, 2-Dimethylcyclopentane   *trans*-1, 2-Dimethylcyclopentane
(C_7H_{14})   (C_7H_{14})

These two compounds are isomers of each other *because they are different compounds* and *because they have the same molecular formula* (C_7H_{14}). They are not constitutional isomers because their atoms are joined in the same way. They are, therefore, *stereoisomers; they differ only in the arrangement of their atoms in space.* They are not enantiomers because their molecules are not mirror reflections of each other. They are, therefore *diastereomers.* (In Section 4.12 we shall find that the *trans*-1,2-dimethylcyclopentane also has an enantiomer.)

SUBDIVISION OF ISOMERS

ISOMERS
(Different compounds with
same molecular formula)

Constitutional isomers	**Stereoisomers**
formerly *Structural isomers* (Isomers whose atoms have a different connectivity)	(Isomers that have the same connectivity but that differ in the arrangement of their atoms in space)

Enantiomers	**Diastereomers**
(Stereoisomers that are mirror reflections of each other)	(Stereoisomers that are not mirror reflections of each other)

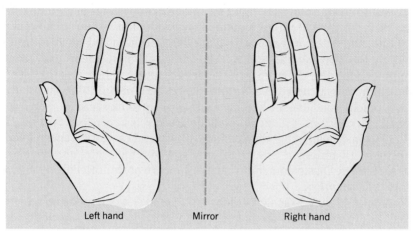

FIGURE 4.1 The mirror reflection of a left hand is a right hand.

4.3 ENANTIOMERS AND CHIRAL MOLECULES

Enantiomers occur only with those compounds whose molecules are **chiral. A chiral molecule is defined as one that is not superposable * on its mirror reflection.**

The word chiral comes from the Greek word *cheir,* meaning "hand." Chiral objects (including molecules) are said to possess "handedness." The term chiral is used to describe molecules of enantiomers because they are related to each other in the same way that a left hand is related to a right hand. When you view your left hand in a mirror, the mirror reflection of your left hand is a right hand (Fig. 4.1). Your left and right hands, moreover, are not superposable (Fig. 4.2). (This fact becomes obvious when one attempts to put a "left-handed" glove on a right hand or vice versa.)

Many familiar objects are chiral and the chirality of some of these objects is clear because we normally speak of them as having "handedness." We speak, for

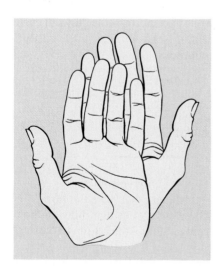

FIGURE 4.2 Left and right hands are not superposable.

Remember: To be *superposable* means that we can place one thing on top of the other *so that all parts of each coincide* (cf. Section 2.4B).

example, of nuts and bolts as having right- or left-handed threads or of a propeller as having a right- or left-handed pitch. The chirality of many other objects is not obvious in this sense, but becomes obvious when we apply the test of nonsuperposability of the object and its mirror reflection.

Objects (and molecules) that *are* superposable on their mirror images are **achiral.** Most socks, for example, are achiral whereas gloves are chiral.

PROBLEM 4.1

Classify the following objects as to whether they are chiral or achiral.

(a) Screw   (c) Fork   (e) Foot   (g) Shoe

(b) Plain spoon   (d) Cup   (f) Ear   (h) Spiral staircase

The chirality of molecules can be demonstrated with relatively simple compounds. Consider, for example, 2-butanol.

$$CH_3CHCH_2CH_3$$
$$|$$
$$OH$$

2-Butanol

Until now, we have presented the formula just written as though it represented only one compound and we have not mentioned that molecules of 2-butanol are chiral. Because they are, there are actually two different 2-butanols and these two 2-butanols are enantiomers. We can understand this if we examine the drawings and models in Fig. 4.3.

If model **I** is held before a mirror, model **II** is seen in the mirror and vice versa. Models **I** and **II** are not superposable on each other; therefore they represent different, but isomeric, molecules. ***Because models I and II are nonsuperposable mirror reflections of each other, the molecules that they represent are enantiomers.***

PROBLEM 4.2

(a) If models are available, construct the 2-butanols represented in Fig. 4.3 and demonstrate for yourself that they are not mutually superposable.   (b) Make similar models of 2-propanol ($CH_3CHOHCH_3$). Are they superposable?   (c) Is 2-propanol chiral?   (d) Would you expect to find enantiomeric forms of 2-propanol?

How do we know when to expect the possibility of enantiomers? A pair of enantiomers is possible for all molecules that contain *a single tetrahedral atom that has four different groups attached to it.* In 2-butanol (Fig. 4.4) this atom is carbon-2. The four different groups that are attached to carbon-2 of 2-butanol are a hydroxyl group, a hydrogen atom, a methyl group, and an ethyl group.

An important property of enantiomers such as these is that *interchanging any two groups at the tetrahedral atom that bears four different groups converts one*

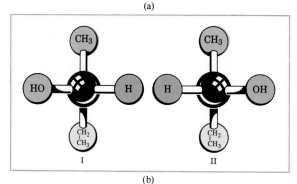

(a)

(b)

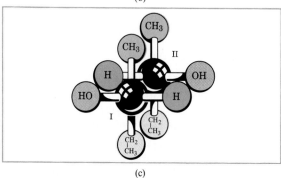

(c)

FIGURE 4.3 (a) Three-dimensional drawings of the 2-butanol enantiomers I and II. (b) Models of the 2-butanol enantiomers. (c) An unsuccessful attempt to superpose models of I and II.

enantiomer into the other. In Fig. 4.3b it is easy to see that interchanging the hydroxyl group and the hydrogen atom converts one enantiomer into the other. You should now convince yourself with models that interchanging any other two groups has the same result.

Because interchanging two groups at carbon-2 converts one stereoisomer into another, carbon-2 is an example of what is called a **stereocenter.** A **stereocenter** is defined as **an atom bearing groups of such nature that an interchange of any two groups will produce a stereoisomer.** Carbon-2 of 2-butanol is an example of a **tetrahedral stereocenter.** Not all stereocenters are tetrahedral, however. The carbon atoms of *cis-* and *trans-*1,2-dichloroethene (Section 4.2) are examples of *trigonal planar*

(hydrogen)

(methyl) $\overset{1}{C}H_3 - \overset{2}{\underset{|}{\overset{|}{C}}} - \overset{3}{C}H_2\overset{4}{C}H_3$ (ethyl)

            $\overset{H}{\underset{OH}{}}$

(hydroxyl)

FIGURE 4.4 The tetrahedral carbon atom of 2-butanol that bears four different groups. (By convention such atoms are often designated with an asterisk.)

stereocenters because an interchange of groups at either atom also produces a stereoisomer (a diastereomer). In this chapter, however, we shall concern ourselves exclusively with tetrahedral stereocenters.

When we discuss interchanging groups like this, we must take care to notice that what we are describing is *something we do to a molecular model* or *something we do on paper*. An interchange of groups in a real molecule, if it can be done, requires breaking covalent bonds, and this is something that requires a large input of energy. This means that enantiomers such as the 2-butanol enantiomers ***do not interconvert*** spontaneously.

> Prior to 1984, *tetrahedral atoms* with four different groups were called *chiral atoms* or *asymmetric atoms*. Then, in an important publication, K. Mislow and J. Siegel (of Princeton University) pointed out that the use of terms like this has represented a source of conceptual confusion in stereochemistry that has existed from the time of van't Hoff (Section 4.3A). Chirality is a geometric property that pervades and affects all parts of a chiral molecule. All of the atoms of 2-butanol, for example, are in a chiral environment and, therefore, all are said to be *chirotopic*. When we consider an atom such as carbon-2 of 2-butanol in the way that we describe here, however, we are considering it as a *stereocenter* and, therefore, we should designate it as such, and not as a "chiral atom." Further consideration of these issues is beyond our scope here, but those interested may wish to read the original paper; cf. K. Mislow and J. Siegel, *J. Am. Chem. Soc.,* **106,** 3319–3328 (1984).

Figure 4.5 demonstrates the validity of the generalization that enantiomeric compounds are possible whenever a molecule contains a single tetrahedral stereocenter.

PROBLEM 4.3

Demonstrate the validity of what we have represented in Fig. 4.5 by constructing models. Arrange four different colored atoms at each corner of a tetrahedral carbon atom. Demonstrate for yourself that **III** and **IV** are related as an object and its mirror reflection *and that they are not superposable* (i.e., that **III** and **IV** are chiral molecules and are enantiomers). (a) Replace one atom on each model so that each model has two atoms of the same color arranged around the central carbon atom. Are the molecules that these models represent mirror reflections of each other? (b) Are they superposable? (c) Are they chiral? (d) Are they enantiomers?

If all of the tetrahedral atoms in a molecule have two or more groups attached that *are the same* the molecule does not have a stereocenter. The molecule is superposable on its mirror image and is **achiral**. An example of a molecule of this type is 2-propanol; carbon atoms 1 and 3 bear three identical hydrogen atoms and the central atom bears two identical methyl groups. If we write three-dimensional formulas for 2-propanol, we find (Fig. 4.6) that one structure can be superposed on its mirror reflection.

Thus, we would not predict the existence of enantiomeric forms of 2-propanol, and experimentally only one form of 2-propanol has ever been found.

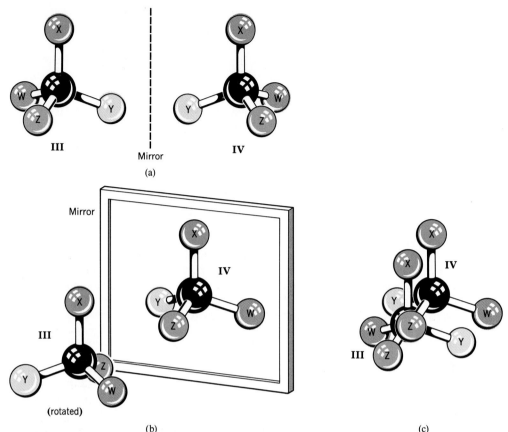

FIGURE 4.5 A demonstration of chirality of a generalized molecule containing one tetrahedral stereocenter. (*a*) The four different groups around the carbon atom in **III** and **IV** are arbitrary. (*b*) **III** is rotated and placed in front of a mirror. **III** and **IV** are found to be related as an object and its mirror reflection. (*c*) **III** and **IV** are not superposable; therefore, the molecules that they represent are chiral and are enantiomers.

PROBLEM 4.4

Some of the molecules listed here have stereocenters; some do not. Write three-dimensional formulas for the enantiomers of those molecules that do have stereocenters.

(a) 1-Chloropropane   (e) 2-Bromobutane

(b) Bromochloroiodomethane   (f) 1-Chloropentane

(c) 1-Chloro-2-methylpropane   (g) 2-Chloropentane

(d) 2-Chloro-2-methylpropane   (h) 3-Chloropentane

4.3A Historical Origin of Stereochemistry

In 1877, Hermann Kolbe (of the University of Leipzig), one of the most eminent organic chemists of the time, wrote the following:

> Not long ago, I expressed the view that the lack of general education and of thorough training in chemistry was one of the causes of the deterioration of chemical research

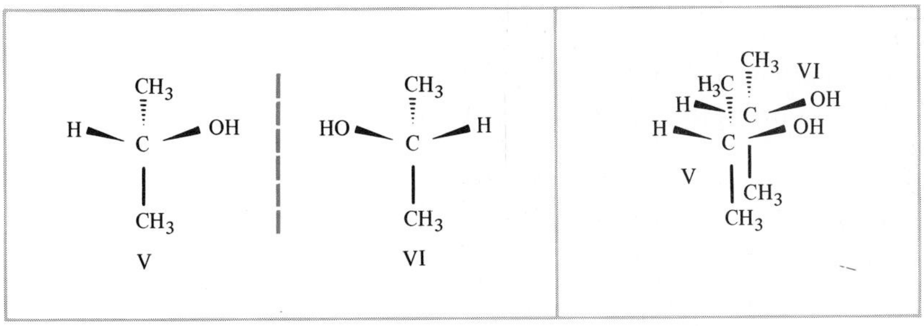

FIGURE 4.6   (a) 2-Propanol **(V)** and its mirror reflection **(VI)**. (b) When one is rotated, the two structures are superposable and thus do not represent enantiomers. They represent two molecules of the same compound. 2-Propanol does not have a stereocenter.

in Germany. . . . Will anyone to whom my worries seem exaggerated please read, if he can, a recent memoir by a Herr van't Hoff on 'The Arrangements of Atoms in Space,' a document crammed to the hilt with the outpourings of a childish fantasy. . . . This Dr. J. H. van't Hoff, employed by the Veterinary College at Utrecht, has, so it seems, no taste for accurate chemical research. He finds it more convenient to mount his Pegasus (evidently taken from the stables of the Veterinary College) and to announce how, on his bold flight to Mount Parnassus, he saw the atoms arranged in space.

Kolbe, nearing the end of his career, was reacting to a publication of a 22-year-old Dutch scientist. This publication had appeared two years earlier in September 1874, and in it, van't Hoff had argued that the spatial arrangement of four groups around a central carbon atom is tetrahedral. A young French scientist, J. A. Le Bel, had independently advanced the same idea in a publication in November 1874. Within 10 years after Kolbe's comments, however, abundant evidence had accumulated that substantiated the "childish fantasy" of van't Hoff. Later in his career (in 1901), and for other work, van't Hoff was named the first recipient of the Nobel Prize for chemistry.

Together, the publications of van't Hoff and Le Bel marked an important turn in a field of study that is concerned with the structures of molecules in three dimensions: *stereochemistry.* Stereochemistry, as we shall see in Section 4.14, had been founded earlier by Louis Pasteur.

It was reasoning based on many observations such as those we presented earlier in this section that led van't Hoff and Le Bel to the conclusion that the spatial orientation of groups around carbon atoms is tetrahedral when a carbon atom is bonded to four other atoms. The following information was available to van't Hoff and Le Bel.

1. Only one compound with the general formula CH_3X is ever found.
2. Only one compound with the formula CH_2X_2 or CH_2XY is ever found.
3. Two enantiomeric compounds with the formula CHXYZ are found.

PROBLEM 4.5

(a) Prove to yourself the correctness of the reasoning of van't Hoff and Le Bel by writing tetrahedral representations for carbon compounds of the three types given previously. (b) How many isomers would be possible in each instance if the carbon atom were at the center of a square? (c) At the center of a rectangle? (d) At one corner of a regular pyramid?

4.4 SYMMETRY ELEMENTS: PLANES OF SYMMETRY

The ultimate way to test for molecular chirality is to construct models of the molecule and its mirror reflection and then determine whether they are superposable. If the two models are superposable, the molecule that they represent is achiral. If the models are not superposable, then the molecules that they represent are chiral. We can apply this test with actual models, as we have just described, or we can apply it by drawing three-dimensional structures and attempting to superpose them in our minds.

There are other aids, however, that will assist us in recognizing chiral molecules. We have mentioned one already: the presence of a *single* tetrahedral stereocenter. The other aids are based on the absence in the molecule of certain symmetry elements. A molecule **will not be chiral,** for example, if it possesses **a plane of symmetry.**

A **plane of symmetry** (also called a **mirror plane**) is defined as ***an imaginary plane that bisects a molecule in such a way that the two halves of the molecule are mirror reflections of each other.*** For example, 2-chloropropane has a plane of symmetry (Fig. 4.7a), while 2-chlorobutane does not (Fig. 4.7b). **All molecules with a plane of symmetry are achiral.**

PROBLEM 4.6

Which of the objects listed in Problem 4.1 possess a plane of symmetry and are, therefore, achiral?

PROBLEM 4.7

Write three-dimensional formulas and designate a plane of symmetry for all of the achiral molecules in Problem 4.4. (In order to be able to designate a plane of symmetry you may have to write the molecule in an appropriate conformation.

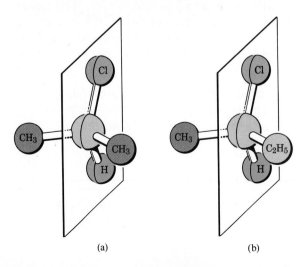

FIGURE 4.7 (*a*) 2-Chloropropane has a plane of symmetry and is achiral. (*b*) 2-Chlorobutane does not possess a plane of symmetry and is chiral.

(a)   (b)

This is permissible with all of these molecules because they have only single bonds and groups joined by single bonds are capable of essentially free rotation at room temperature.)

4.5 NOMENCLATURE OF ENANTIOMERS: THE (*R*)–(*S*) SYSTEM

The two enantiomers of 2-butanol are the following:

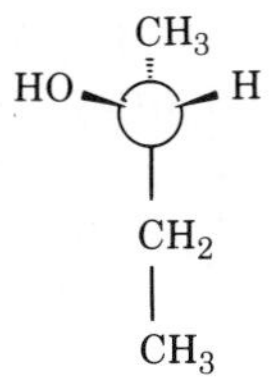

I 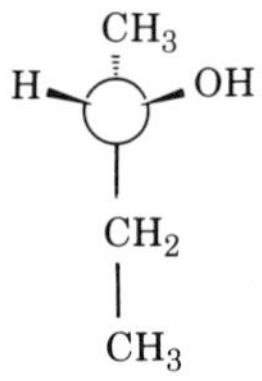   II

If we name these two enantiomers using only the IUPAC system of nomenclature that we have learned so far, both enantiomers will have the same name—2-butanol (or *sec*-butyl alcohol) (Section 3.3E). This is undesirable because *each compound must have its own distinct name.* Moreover, the name that is given a compound should allow a chemist who is familiar with the rules of nomenclature to write the structure of the compound from its name alone. Given the name 2-butanol, a chemist could write either structure **I** or structure **II**.

Three chemists, R. S. Cahn (England), C. K. Ingold (England), and V. Prelog (Switzerland), devised a system of nomenclature that, when added to the IUPAC system, solves both of these problems. This system, called the (*R*)–(*S*) system, or the Cahn–Ingold–Prelog system, is now widely used and is part of the IUPAC rules.

According to the system one enantiomer of 2-butanol should be designated (*R*)-2-butanol and the other enantiomer should be designated (*S*)-2-butanol. [(*R*) and (*S*) are from the Latin words *rectus* and *sinister,* meaning right and left, respectively.]

(*R*) and (*S*) designations are assigned as follows:

1. Each of the four groups attached to the stereocenter is assigned a **priority** or **preference** *a, b, c,* or *d*. Priority is first assigned on the basis of the **atomic number** of the atom that is directly attached to the stereocenter. The group with the lowest atomic number is given the lowest priority, *d*; the group with next highest atomic number is given the next higher priority, *c*; and so on. (In the case of isotopes, the isotope of greatest atomic mass has highest priority.)

We can illustrate the application of this rule with the 2-butanol enantiomer, **I**.

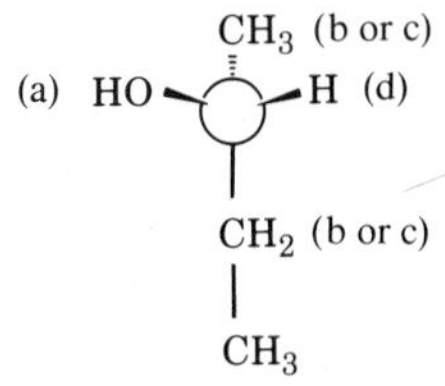

I

Oxygen has the highest atomic number of the four atoms attached to the stereocenter and is assigned the highest priority, *a*. Hydrogen has the lowest atomic number and is assigned the lowest priority, *d*. A priority cannot be assigned for the methyl group and the ethyl group because the atom that is directly attached to the stereocenter is a carbon atom in both groups.

2. When a priority cannot be assigned on the basis of the atomic number of the atoms that are directly attached to the stereocenter, then the next sets of atoms in the unassigned groups are examined. This process is continued until a decision can be made. *We assign a priority at the first point of difference.*

When we examine the methyl group of enantiomer **I**, we find that the next set of atoms consists of three hydrogen atoms (**H, H, H**). In the ethyl group of **I** the next set of atoms consists of one carbon atom and two hydrogen atoms (**C, H, H**). Carbon has a higher atomic number than hydrogen so we assign the ethyl group the higher priority, *b*, and the methyl group the lower priority, *c*.

$$
\begin{array}{c}
\text{H} \\
| \quad\quad \text{(c)} \quad (\text{H, H, H}) \\
\text{H—C—H} \\
\text{(a)} \quad \text{HO} \diagdown \!\!\!\!\!\!\!\diagup \text{H} \quad \text{(d)} \quad (\text{C, H, H}) > (\text{H, H, H}) \\
| \\
\text{H—C—H} \\
| \quad\quad \text{(b)} \quad (\text{C, H, H}) \\
\text{H—C—H} \\
| \\
\text{H} \\
\textbf{I}
\end{array}
$$

3. We now rotate the formula (or model) so that the group with lowest priority (*d*) is directed away from us.

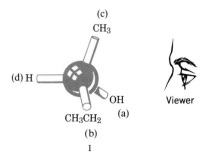

Then we trace a path from *a* to *b* to *c*. If, as we do this, the direction of our finger (or pencil) is *clockwise,* the enantiomer is designated (*R*). If the direction is *counterclockwise,* the enantiomer is designated (*S*). On this basis the 2-butanol enantiomer **I** is (*R*)-2-butanol.

Arrows are clockwise

PROBLEM 4.8

Apply the procedure just given to the 2-butanol enantiomer **II** and show that it is (S)-2-butanol.

PROBLEM 4.9

Give (R) and (S) designations for each pair of enantiomers given as answers to Problem 4.4.

The first three rules of the Cahn–Ingold–Prelog system allow us to make an (R) or (S) designation for most compounds containing single bonds. For compounds containing multiple bonds one other rule is necessary.

4. Groups containing double or triple bonds are assigned priorities as if both atoms were duplicated or triplicated, that is,

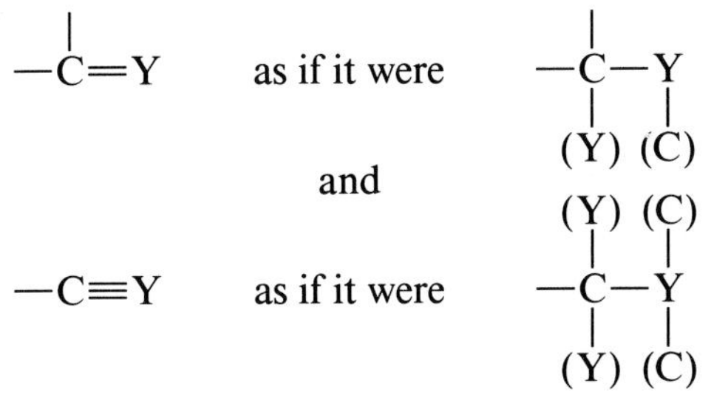

$$-\overset{|}{\underset{}{C}}=Y \quad \text{as if it were} \quad -\overset{|}{\underset{\underset{(Y)\ (C)}{|\ \ |}}{C}}-Y$$

and

$$-C\equiv Y \quad \text{as if it were} \quad -\overset{\overset{(Y)\ (C)}{|\ \ |}}{\underset{\underset{(Y)\ (C)}{|\ \ |}}{C}}-Y$$

where the symbols in parentheses are duplicate or triplicate representations of the atoms at the other end of the double bond.

Thus, the vinyl group, $-CH=CH_2$, is of higher priority than the isopropyl group, $-CH(CH_3)_2$.

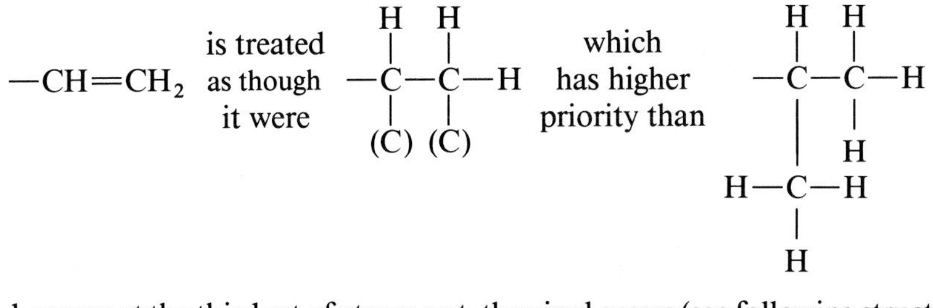

$$-CH=CH_2 \quad \begin{matrix}\text{is treated}\\ \text{as though}\\ \text{it were}\end{matrix} \quad \overset{\overset{H\ \ H}{|\ \ |}}{\underset{\underset{(C)\ (C)}{|\ \ |}}{C}}-\overset{}{\underset{}{C}}-H \quad \begin{matrix}\text{which}\\ \text{has higher}\\ \text{priority than}\end{matrix} \quad \begin{matrix}\overset{H\ \ H}{|\ \ |}\\ -C-C-H\\ |\ \ \ \ |\\ \ \ \ \ \ H\\ \ \ \ \ |\\ H-C-H\\ |\\ H\end{matrix}$$

because at the third set of atoms out, the vinyl group (see following structure) is C, H, H, whereas the isopropyl group along either branch is H, H, H. (At the first and second set of atoms both groups are the same: C, then C, C, H.)

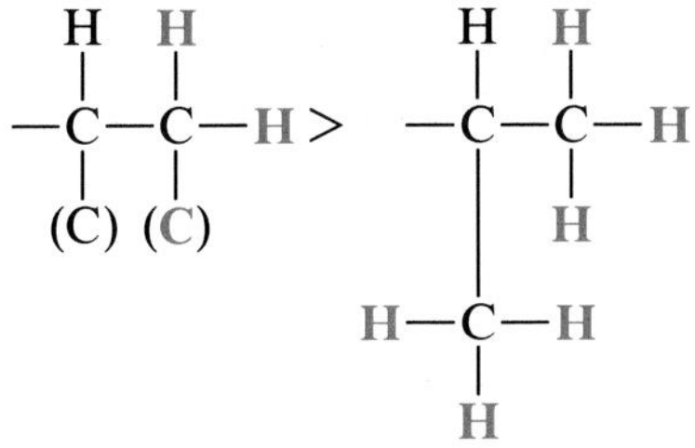

$$\overset{\overset{H\ \ H}{|\ \ |}}{\underset{\underset{(C)\ (C)}{|\ \ |}}{C}}-C-H > \begin{matrix}\overset{H\ \ H}{|\ \ |}\\ -C-C-H\\ \ \ \ \ |\\ \ \ \ \ H\\ \ \ \ |\\ H-C-H\\ |\\ H\end{matrix}$$

$$\begin{matrix}\textbf{C, H, H} & > & \textbf{H, H, H}\\ \textbf{Vinyl group} & & \textbf{Isopropyl group}\end{matrix}$$

Other rules exist for more complicated structures, but we shall not study them here.

PROBLEM 4.10

An important compound in stereochemistry and biology is the compound glyceraldehyde (see the following structure). Write three-dimensional formulas for the glyceraldehyde enantiomers and give each its proper (R)–(S) designation.

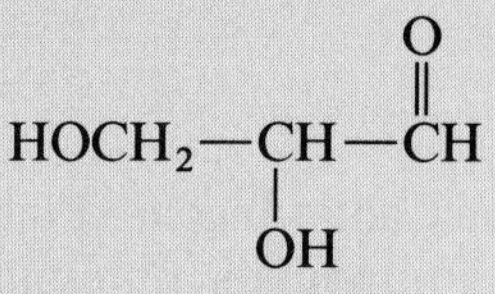

$$HOCH_2-\overset{\displaystyle |}{\underset{\displaystyle OH}{CH}}-\overset{\displaystyle O}{\overset{\displaystyle \|}{CH}}$$

Glyceraldehyde

PROBLEM 4.11

Assign (R) or (S) designations to each of the following compounds:

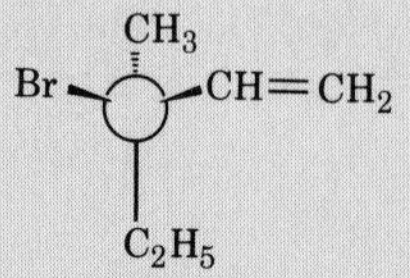

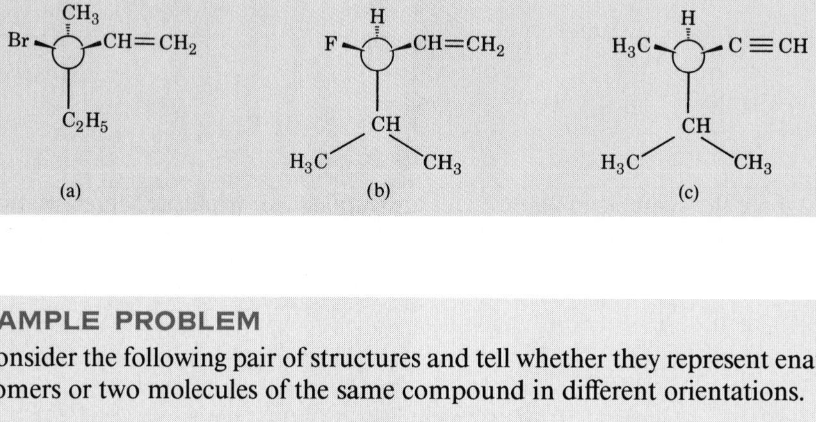

(a)

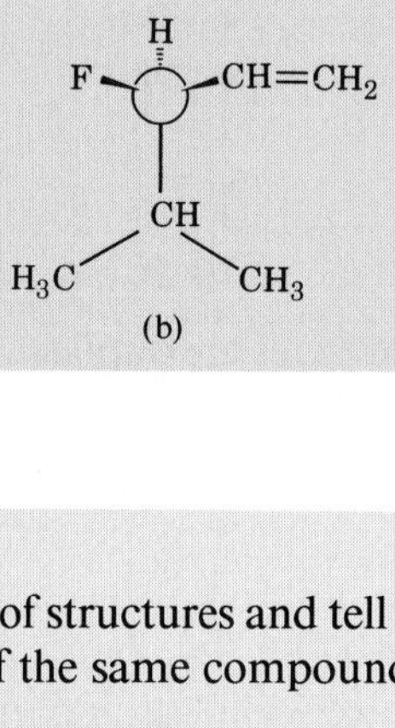

(b)

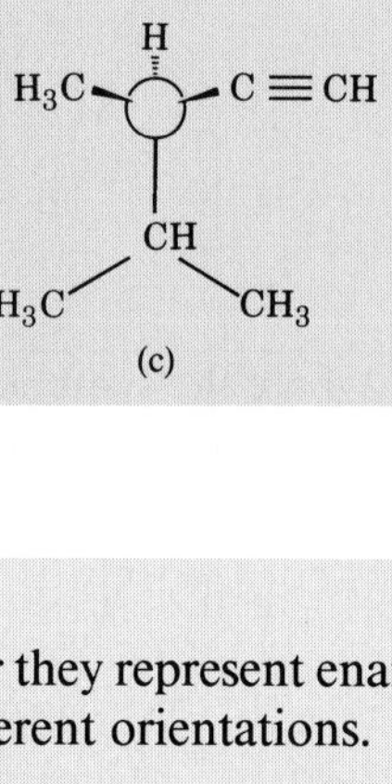

(c)

SAMPLE PROBLEM

Consider the following pair of structures and tell whether they represent enantiomers or two molecules of the same compound in different orientations.

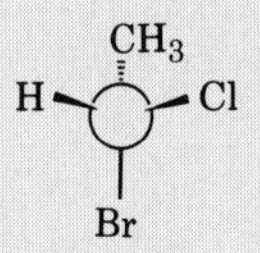

A 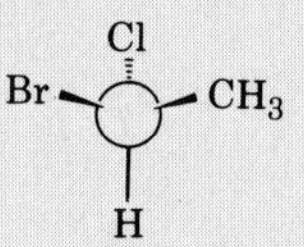   B

Answer:

One way to approach this kind of problem is to take one structure and turn it (in your mind) until it either becomes identical with the other or until it becomes the mirror reflection of the other. We might begin by rotating **B** about the C*—Cl axis until the bromine is at the bottom. Then rotation about the vertical axis through the C*—Br bond makes **B** identical with **A**.

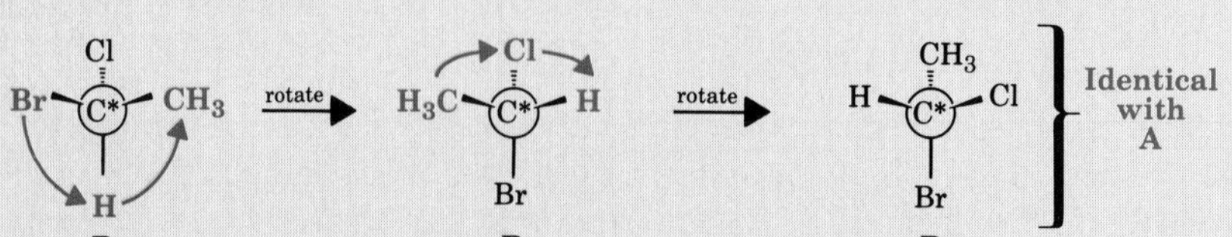

Identical with A

Another approach is to recognize that exchanging two groups at the stereocenter *inverts the configuration of* that carbon atom and converts a structure *with only one stereocenter* into its enantiomer; a second exchange recreates the original molecule. So we proceed this way, keeping track of how many exchanges are required to convert **B** into **A**. In this instance we find that two exchanges are required, and, again, we conclude that **A** and **B** are the same.

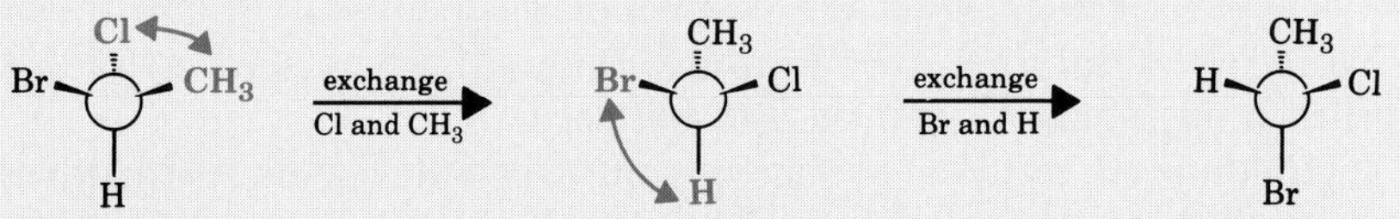

A useful check is to name each compound including its (R)–(S) designation. If the names are the same, then the structures are the same. In this instance both structures are (R)-1-bromo-1-chloroethane.

PROBLEM 4.12

Tell whether the two structures in each pair represent enantiomers or two molecules of the same compound in different orientations.

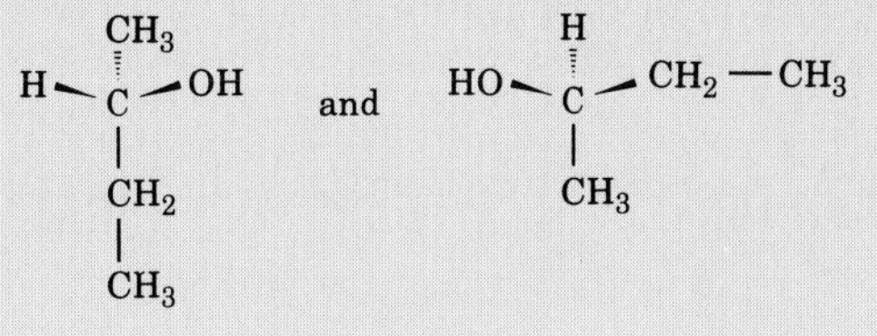

4.6 PROPERTIES OF ENANTIOMERS: OPTICAL ACTIVITY

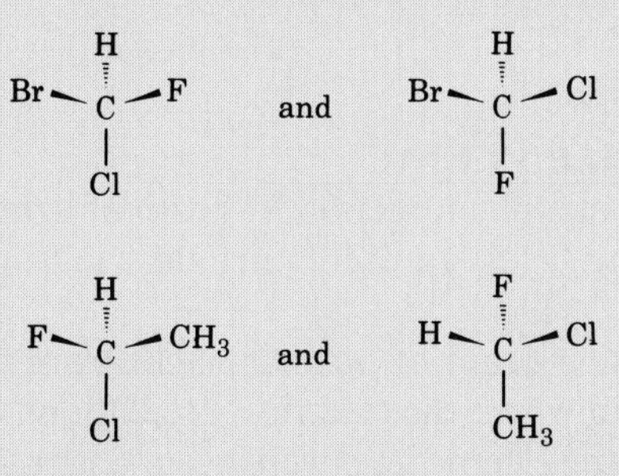

The molecules of enantiomers are not superposable one on the other, and on this basis alone, we have concluded that enantiomers are different compounds. How are they different? Do enantiomers resemble constitutional isomers and diastereomers in having different melting and boiling points? The answer is *no*. Enantiomers have *identical* melting and boiling points. Do enantiomers have different indexes of refraction, different solubilities in common solvents, different infrared spectra, and different rates of reaction with ordinary reagents? The answer to each of these questions is also no.

TABLE 4.1 Physical properties of (*R*)- and (*S*)-2-butanol

PHYSICAL PROPERTY	(*R*)-2-BUTANOL	(*S*)-2-BUTANOL
Boiling point (1 atm)	99.5 °C	99.5 °C
Density (g/mL at 20 °C)	0.808	0.808
Index of refraction (20 °C)	1.397	1.397

We can see examples if we examine Table 4.1 where some of the physical properties of the 2-butanol enantiomers are listed.

Enantiomers show different behavior only when they interact with other chiral substances. Enantiomers show different rates of reaction toward other chiral molecules — that is, toward reagents that consist of a single enantiomer or an excess of a single enantiomer. Enantiomers also show different solubilities in solvents that consist of a single enantiomer or an excess of a single enantiomer.

One easily observable way in which enantiomers differ is in *their behavior toward plane-polarized light.* When a beam of plane-polarized light passes through an enantiomer, the plane of polarization **rotates.** Moreover, separate enantiomers rotate the plane of plane-polarized light equal amounts *but in opposite directions.* Because of their effect on plane-polarized light, separate enantiomers are said to be **optically active compounds.**

In order to understand this behavior of enantiomers we need to understand the nature of plane-polarized light. We also need to understand how an instrument called a **polarimeter** operates.

4.6A Plane-Polarized Light

Light is an electromagnetic phenomenon. A beam of light consists of two mutually perpendicular oscillating fields: an oscillating electric field and an oscillating magnetic field (Fig. 4.8).

If we were to view a beam of ordinary light from one end, and if we could actually see the planes in which the electrical oscillations were occurring, we would find that oscillations of the electric field were occurring in all possible planes perpendicular to the direction of propagation (Fig. 4.9). (The same would be true of the magnetic field.)

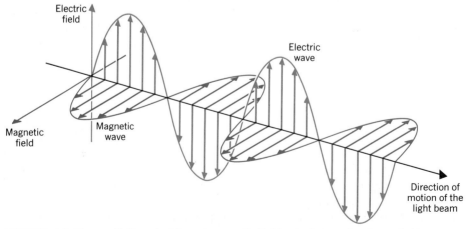

Electric field

Electric wave

Magnetic field

Magnetic wave

Direction of motion of the light beam

FIGURE 4.8 The oscillating electric and magnetic fields of a beam of ordinary light.

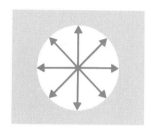

FIGURE 4.9 Oscillation of the electrical field of ordinary light occurs in all possible planes perpendicular to the direction of propagation.

When ordinary light is passed through a polarizer, the polarizer interacts with the electrical field so that the electrical field of the light that emerges from the polarizer (and the magnetic field perpendicular to it) is oscillating only in one plane. Such light is called plane-polarized light (Fig. 4.10).

> The lenses of Polaroid sunglasses have this effect. You can demonstrate for yourself that this is true with two pairs of Polaroid sunglasses. If two lenses are placed one on top of the other so that the axes of polarization coincide, then light passes through both normally. Then if one lens is rotated 90° with respect to the other, no light passes through.

4.6B The Polarimeter

The device that is used for measuring the effect of plane-polarized light on optically active compounds is a polarimeter. A sketch of a polarimeter is shown in Fig. 4.11. The principal working parts of a polarimeter are (1) a light source (usually a sodium lamp), (2) a polarizer, (3) a tube for holding the optically active substance (or solution) in the light beam, (4) an analyzer, and (5) a scale for measuring the number of degrees that the plane of polarized light has been rotated.

The analyzer of a polarimeter (Fig. 4.11) is nothing more than another polarizer. If the tube of the polarimeter is empty, or if an optically *inactive* substance is present, the axes of the plane-polarized light and the analyzer will be exactly parallel when the instrument reads 0°, and the observer will detect the maximum amount of light passing through. If, by contrast, the tube contains an optically active substance, a solution of one enantiomer, for example, the plane of polarization of the light will be rotated as it passes through the tube. In order to detect the maximum brightness of light the observer will have to rotate the axis of the analyzer in either a clockwise or counterclockwise direction. If the analyzer is rotated in a clockwise direction, the rotation, α (measured in degrees), is said to be positive (+). If the rotation is counterclockwise, the rotation is said to be negative (−). A substance that rotates plane-polarized light in the clockwise direction is also said to be **dextrorotatory,** and one that rotates plane-polarized light in a counterclockwise direction is said to be **levorotatory** (from the Latin: *dexter,* right and *laevus,* left).

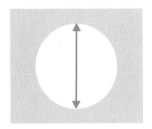

FIGURE 4.10 The plane of oscillation of the electrical field of plane-polarized light. In this example the plane of polarization is vertical.

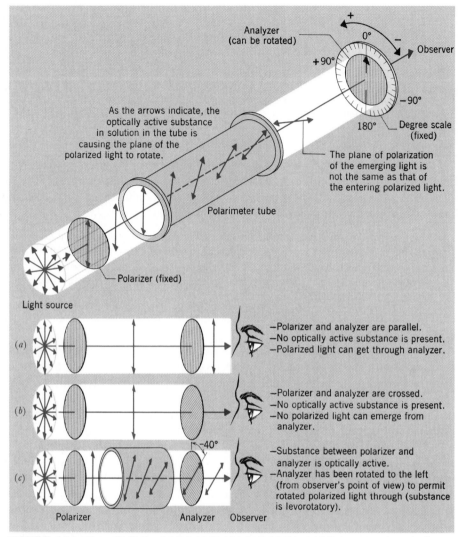

FIGURE 4.11 The principal working parts of a polarimeter and the measurement of optical rotation. (From John R. Holum, *Organic Chemistry: A Brief Course*, Wiley, New York, 1975, p. 316.)

4.6C Specific Rotation

The number of degrees that the plane of polarization is rotated as the light passes through a solution of an enantiomer depends on the number of chiral molecules that it encounters. This, of course, depends on the length of the tube and the concentration of the enantiomer. In order to place measured rotations on a standard basis, chemists calculate a quantity called the **specific rotation, $[\alpha]$**, by the following equation:

$$[\alpha] = \frac{\alpha}{c \cdot l}$$

where $[\alpha]$ = the specific rotation

α = the observed rotation

c = the concentration of the solution in grams per milliliter of solution (or density in g/mL for neat liquids)

l = the length of the tube in decimeters (1 dm = 10 cm)

The specific rotation also depends on the temperature and the wavelength of light that is employed. Specific rotations are reported so as to incorporate these quantities as well. A specific rotation might be given as follows:

$$[\alpha]_D^{25} = +3.12°$$

This means that, the D line* of a sodium lamp was used for the light, a temperature of 25 °C was maintained, and that a sample containing 1.00 g/mL of the optically active substance, in a 1-dm tube, produced a rotation of 3.12° in a clockwise direction.

The specific rotations of (R)-2-butanol and (S)-2-butanol are given here.

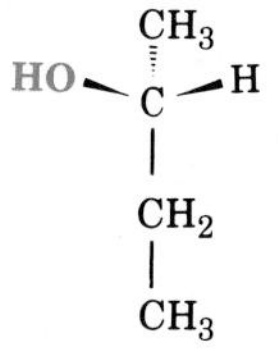

(R)-2-Butanol
$[\alpha]_D^{25} = -13.52°$

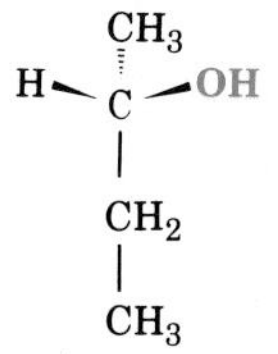

(S)-2-Butanol
$[\alpha]_D^{25} = +13.52°$

The direction of rotation of plane-polarized light is often incorporated into the names of optically active compounds. The following two sets of enantiomers show how this is done.

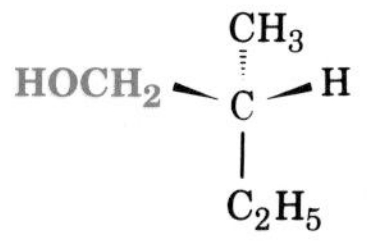

(R)-(+)-2-Methyl-1-butanol
$[\alpha]_D^{25} = +5.756°$

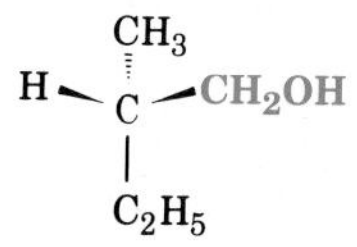

(S)-(−)-2-Methyl-butanol
$[\alpha]_D^{25} = -5.756°$

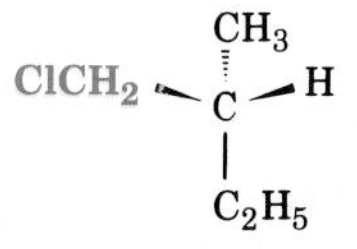

(R)-(−)-1-Chloro-2-methylbutane
$[\alpha]_D^{25} = -1.64°$

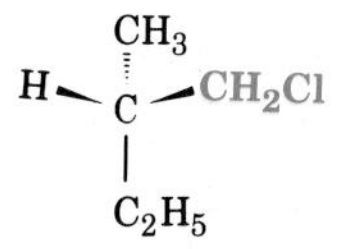

(S)-(+)-1-Chloro-2-methylbutane
$[\alpha]_D^{25} = +1.64°$

The previous compounds also illustrate an important principle: *No obvious correlation exists between the configurations of enantiomers and the direction [(+) or (−)] in which they rotate plane-polarized light.*

*Wavelength = 589.6 nm.

(R)-$(+)$-2-Methyl-1-butanol and (R)-$(-)$-1-chloro-2-methylbutane have the same *configuration,* that is, they have the same general arrangement of their atoms in space. They have, however, an opposite effect on the direction of rotation of the plane of plane-polarized light.

$$
\underset{\text{(R)-(+)-2-Methyl-1-butanol}}{\text{HOCH}_2 \diagdown \underset{\overset{|}{\text{C}_2\text{H}_5}}{\overset{\overset{\text{CH}_3}{\vdots}}{\text{C}}} \diagup \text{H}}
\qquad
\begin{matrix}\text{Same}\\\text{configuration}\end{matrix}
\qquad
\underset{\text{(R)-(-)-1-Chloro-2-methylbutane}}{\text{ClCH}_2 \diagdown \underset{\overset{|}{\text{C}_2\text{H}_5}}{\overset{\overset{\text{CH}_3}{\vdots}}{\text{C}}} \diagup \text{H}}
$$

These same compounds also illustrate a second important principle: *No necessary correlation exists between the (R) and (S) designation and the direction of rotation of plane-polarized light.* (R)-2-Methyl-1-butanol is dextrorotatory, $(+)$, and (R)-1-chloro-2-methylbutane is levorotatory, $(-)$.

A method based on the measurement of optical rotation measured at many different wavelengths, called optical rotatory dispersion, has been used to correlate configurations of chiral molecules. A discussion of the technique of optical rotatory dispersion, however, is beyond the scope of this text.

4.7 THE ORIGIN OF OPTICAL ACTIVITY

It is not possible to give a complete, condensed account of the origin of the optical activity observed for separate enantiomers. An insight into the source of this phenomenon can be obtained, however, by comparing what occurs when a beam of plane-polarized light passes through a solution of *achiral* molecules with what occurs when a beam of plane-polarized light passes through a solution of *chiral* molecules.

Almost all *individual* molecules, whether chiral or achiral, are theoretically capable of producing a slight rotation of the plane of plane-polarized light. The direction and magnitude of the rotation produced by an individual molecule depends, in part, on its orientation at the precise moment that it encounters the beam. In a solution, of course, billions of molecules are in the path of the light beam and at any given moment these molecules will be present in all possible orientations. If the beam of plane-polarized light passes through a solution of the achiral compound 2-propanol, for example, it should encounter at least two molecules in the exact orientations shown in Fig. 4.12. The effect of the first encounter might be to produce

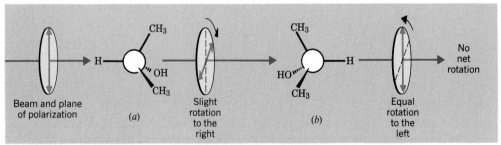

FIGURE 4.12 A beam of plane-polarized light encountering a molecule of 2-propanol (an achiral molecule) in orientation (*a*) and then a second molecule in the mirror-reflection orientation (*b*). The beam emerges from these two encounters with no net rotation of its plane of polarization.

a very slight rotation of the plane of polarization to the right. Before the beam emerges from the solution, however, it should encounter at least one molecule of 2-propanol that is in exactly the mirror reflection orientation of the first. The effect of this second encounter will be to produce an equal and opposite rotation of the plane: A rotation that exactly cancels the first rotation. The beam, therefore, emerges with no net rotation.

What we have just described for the two encounters shown in Fig. 4.12 can be said of all possible encounters of the beam with molecules of 2-propanol. Because so many molecules are present, it is statistically certain that *for each encounter with a particular orientation there will be an encounter with a molecule that is in a mirror-reflection orientation.* The result of all of these encounters will be such that all of the rotations produced by individual molecules will be canceled and 2-propanol will be found to be **optically inactive.**

What, then, is the situation when a beam of plane-polarized light passes through a solution of one enantiomer of a chiral compound? We can answer this question by considering what might occur when plane-polarized light passes through a solution of pure (R)-2-butanol. Figure 4.13 illustrates one possible encounter of a beam of plane-polarized light with a molecule of (R)-2-butanol.

When a beam of plane-polarized light passes through a solution of (R)-2-butanol, *no molecule is present that can ever be exactly oriented as a mirror reflection of any given orientation of an (R)-2-butanol molecule.* The only molecules that could do this would be molecules of (S)-2-butanol, and they are not present. Exact cancellation of the rotations produced by all of the encounters of the beam with random orientations of (R)-2-butanol does not happen and, as a result, a net rotation of the plane of polarization is observed. (R)-2-Butanol is found to be *optically active.*

4.7A Racemic Forms

The net rotation of the plane of polarization that we observe for a solution consisting of molecules of (R)-2-butanol alone would not be observed if we passed the beam through a solution that contained equimolar amounts of (R)-2-butanol and (S)-2-butanol. In the latter instance, molecules of (S)-2-butanol would be present in a quantity equal to those of (R)-2-butanol and for every possible orientation of one enantiomer, a molecule of the other enantiomer would be in a mirror-reflection

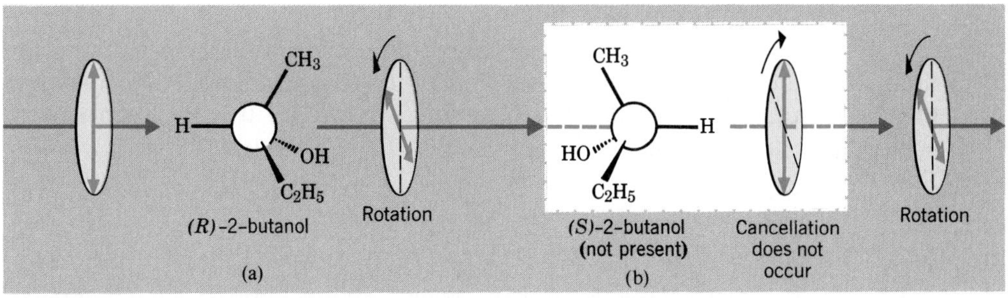

FIGURE 4.13 (a) A beam of plane-polarized light encounters a molecule of (R)-2-butanol (a chiral molecule) in a particular orientation. This encounter produces a slight rotation of the plane of polarization. (b) Exact cancellation of this rotation requires that a second molecule be oriented as an exact mirror reflection. This cancellation does not occur because the only molecule that could ever be oriented as an exact mirror reflection at the first encounter is a molecule of (S)-2-butanol, which is not present. As a result, a net rotation of the plane of polarization occurs.

orientation. Exact cancellations of all rotations would occur, and the solution of the equimolar mixture of enantiomers would be *optically inactive.*

An equimolar mixture of two enantiomers is called a **racemic form** (either a **racemate** or a **racemic mixture**). A racemic form shows no rotation of plane-polarized light; as such, it is often designated as being ($\pm$). A racemic form of (R)-($-$)-2-butanol and (S)-($+$)-2-butanol might be indicated as

$$(\pm)\text{-2-Butanol} \quad \text{or as} \quad (\pm)\text{-CH}_3\text{CH}_2\text{CHOHCH}_3$$

4.7B Enantiomeric Purity, Optical Purity, and Enantiomeric Excess

A sample of an optically active substance that consists of a single enantiomer is said to be **enantiomerically pure.** An enantiomerically pure sample of (S)-($+$)-2-butanol shows a specific rotation of $+13.52°$ ($[\alpha]_D^{25} = +13.52°$). On the other hand, a sample of (S)-($+$)-2-butanol that contains less than an equimolar amount of (R)-($-$)-2-butanol will show a specific rotation that is less than $+13.52°$ but greater than $0°$. Such a sample is said to have an *enantiomeric purity* less than 100%. The **percent enantiomeric purity** is defined as follows:

Percent enantiomeric purity

$$= \frac{\text{moles of one enantiomer} - \text{moles of other enantiomer}}{\text{moles of both enantiomers}} \times 100$$

The percent enantiomeric purity is also often called the **enantiomeric excess** and it is equal to *the percent optical purity.* The percent optical purity is defined in terms of specific rotations:

$$\text{Percent optical purity*} = \frac{\text{observed specific rotation}}{\text{specific rotation of the pure enantiomer}} \times 100$$

Let us suppose, for example, that the sample showed a specific rotation of $+6.76°$. We would then say that the optical purity of the (S)-($+$)-2-butanol is 50%.

$$\text{Optical purity} = \frac{+6.76°}{+13.52°} \times 100 = 50\%$$

When we say the optical purity of this mixture is 50%, we also mean that 50% of the mixture consists of the racemic form, ($\pm$)-2-butanol, and the other 50% consists of the enantiomer, (S)-($+$)-2-butanol. Therefore, we can say that the enantiomeric excess of the mixture is 50%.

PROBLEM 4.13

What relative molar proportions of (S)-($+$)-2-butanol and (R)-($-$)-2-butanol would give a specific rotation, $[\alpha]_D^{25}$, equal to $+6.76°$?

---

*The term *optical purity* is applied to a single enantiomer or to mixtures of enantiomers only. It should not be applied to mixtures in which some other compound is present.

4.8 THE SYNTHESIS OF ENANTIOMERS

Many times in the course of working in the organic laboratory a reaction carried out with reactants whose molecules are achiral results in the formation of products whose molecules are chiral. In the absence of any chiral influence (from the solvent or a catalyst), the outcome of such a reaction is the formation of a racemic form. The reason: The chiral molecules of the product are obtained as a 50:50 mixture of enantiomers.

An example is the synthesis of 2-butanol by the nickel-catalyzed hydrogenation of 2-butanone. In this reaction the hydrogen molecule adds across the carbon–oxygen double bond in much the same way that it adds to a carbon–carbon double bond (Section 3.18).

$$CH_3CH_2\overset{\|}{\underset{O}{C}}CH_3 + H-H \xrightarrow{Ni} (\pm)\ CH_3CH_2\overset{*}{\underset{OH}{C}HCH_3}$$

2-Butanone (achiral molecules)	Hydrogen (achiral molecules)	(±)-2-Butanol [chiral molecules but 50:50 mixture (R) and (S)]

Molecules of neither reactant (2-butanone nor hydrogen) are chiral. The molecules of the product (2-butanol) are chiral. The product, however, is obtained as a racemic form because the two enantiomers, (R)-(−)-2-butanol and (S)-(+)-2-butanol, are obtained in equal amounts.

> This is not the result if reactions like this are carried out in the presence of a chiral influence such as an optically active solvent or, as we shall see later, an enzyme. The nickel catalyst used in this reaction does not exert a chiral influence.

Figure 4.14 shows why a racemic form of 2-butanol is obtained. Hydrogen, adsorbed on the surface of the nickel catalyst, adds with equal facility at either face of

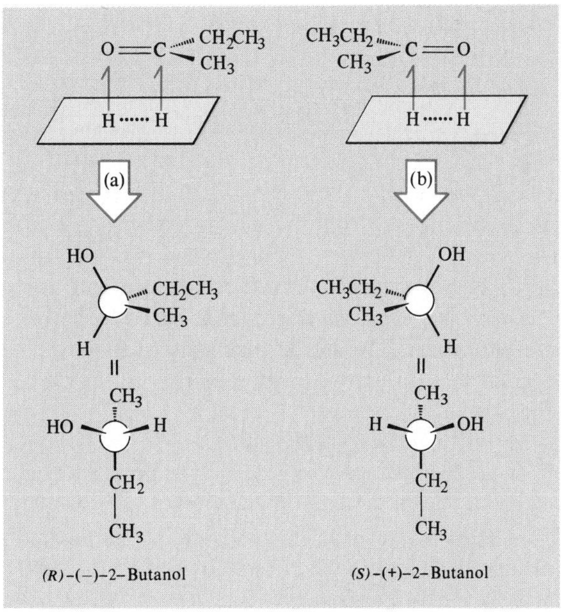

FIGURE 4.14 The reaction of 2-butanone with hydrogen in the presence of a nickel catalyst. The reaction rate by path (a) is equal to that by path (b). (R)-(−)-2-butanol and (S)-(+)-2-butanol are produced in equal amounts, as a racemic form.

2-butanone. Reaction at one face produces one enantiomer; reaction at the other face produces the other enantiomer, and the two reactions occur at the same rate.

Another example of the same type is the following catalytic hydrogenation of pyruvic acid to produce a racemic form of lactic acid.

$$CH_3\underset{\underset{O}{\|}}{C}CO_2H \quad + \quad H_2 \quad \xrightarrow{Ni} \quad (\pm)\ CH_3\underset{\underset{OH}{|}}{C}HCO_2H$$

Pyruvic acid	Hydrogen	$(\pm)$-Lactic acid
(achiral molecules)	(achiral molecules)	[chiral molecules but 50:50 mixture (R) and (S) forms]

PROBLEM 4.14

What products would you expect from the following reaction? What relative amounts of each product would be formed?

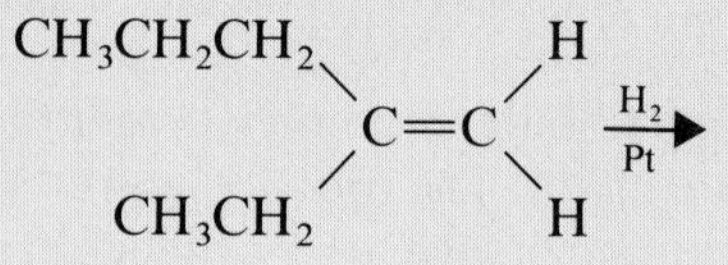

$$\begin{array}{c} CH_3CH_2CH_2 \\ \diagdown \\ CH_3CH_2 \diagup \end{array} C=C \begin{array}{c} H \\ \diagup \\ \diagdown H \end{array} \xrightarrow[Pt]{H_2}$$

4.8A Stereoselective Reactions

The stereochemical course of reactions that occur in living cells is often dramatically different from those that are carried out in the glassware of the laboratory. The laboratory hydrogenation of pyruvic acid in the presence of a nickel catalyst, as we have just seen, produces a racemic form of lactic acid. In muscle cells, however, the conversion of pyruvic acid to lactic acid is catalyzed by an enzyme called *lactic acid dehydrogenase.** The enzyme-catalyzed reaction results in the formation of a single enantiomer, (S)-$(+)$-lactic acid:

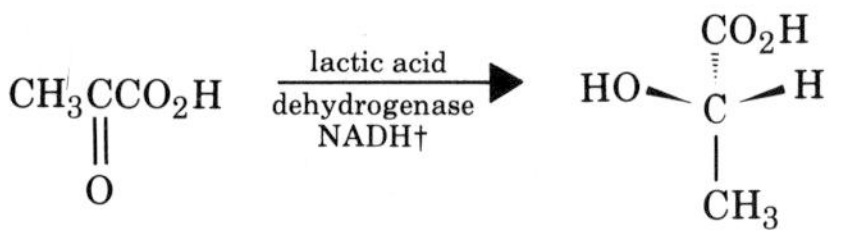

$$CH_3\underset{\underset{O}{\|}}{C}CO_2H \quad \xrightarrow[\substack{dehydrogenase \\ NADH\dagger}]{lactic\ acid} \quad HO\diagdown \underset{\underset{CH_3}{|}}{\overset{CO_2H}{C}} \diagup H$$

Pyruvic acid	(S)-$(+)$-Lactic acid $[\alpha] = +3.82$

The enzyme-catalyzed reaction is said to be **stereoselective.** *A stereoselective reaction is a reaction that yields exclusively (or predominantly) only one of a set of stereoisomers.* The stereoselectivity of the enzyme-catalyzed reaction is a result of a special relation between the reacting molecule and the enzyme surface as shown in Fig. 4.15.

*The name of the enzyme is derived from the fact that this enzyme also catalyzes the reverse reaction, that is, the dehydrogenation of (S)-$(+)$-lactic acid to produce pyruvic acid.

†The reducing agent in this reaction is a compound (NADH) called the reduced form of nicotinamide adenine dinucleotide (NAD⁺) (Section 11.11). NADH is a cofactor of the enzyme.

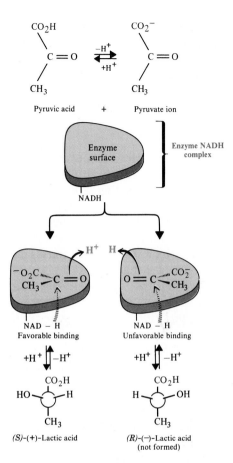

FIGURE 4.15 A mechanism that accounts for stereoselectivity in the enzyme-catalyzed reduction of pyruvic acid to (S)-(+)-lactic acid. Pyruvate ion can bind itself to the surface of the enzyme NADH complex in only one favorable way. As a result, NADH can transfer hydride ions (H:⁻) to only one surface of the pyruvate ion. This leads to the formation only of (S)-(+)-lactic acid. (R)-(−)-Lactic acid is not produced because its formation would require an unfavorable binding between the NADH enzyme complex.

4.9 MOLECULES WITH MORE THAN ONE STEREOCENTER

Thus far all of the chiral molecules that we have considered have contained only one stereocenter. Many organic molecules, especially those important in biology, contain more than one stereocenter. Cholesterol (Section 22.4B), for example, contains eight stereocenters. (Can you locate them?) We can begin, however, with simpler molecules. Let us consider 2,3-dibromopentane shown here—a structure that has two stereocenters.

$$CH_3\overset{*}{C}H\overset{*}{C}HCH_2CH_3$$
$$\underset{Br\ \ Br}{|\quad |}$$

2,3-Dibromopentane

There is a useful rule that helps us to know how many stereoisomers to expect from structures like this one. *The total number of stereoisomers will not exceed 2^n where n is equal to the number of stereocenters.* For this structural formula we should not expect more than four stereoisomers ($2^2 = 4$).

Our next task is to write three-dimensional formulas for the stereoisomers of the compound. We begin by writing a three-dimensional formula for one stereoisomer and then by writing the formula for *its* mirror reflection.

$$
\begin{array}{cc}
\text{CH}_3 & \text{CH}_3 \\
\text{H} \diagdown \overset{|}{\text{C}} \diagup \text{Br} & \text{Br} \diagdown \overset{|}{\text{C}} \diagup \text{H} \\
| & | \\
\text{H} \diagup \overset{|}{\text{C}} \diagdown \text{Br} & \text{Br} \diagup \overset{|}{\text{C}} \diagdown \text{H} \\
\text{C}_2\text{H}_5 & \text{C}_2\text{H}_5 \\
\mathbf{1} & \mathbf{2}
\end{array}
$$

It is helpful to follow certain conventions when we write these three-dimensional formulas. For example, we usually write our structures in eclipsed conformations. When we do this we do not mean to imply that eclipsed conformations are the most stable ones—they most certainly are not. We write eclipsed conformations because, as we shall see later, they make it easy for us to recognize planes of symmetry when they are present. We also write the longest carbon chain in a generally vertical orientation on the page; this makes the structures that we write directly comparable. As we do these things, however, *we must remember that molecules can rotate in their entirety* and that *at normal temperatures rotations about all single bonds are also possible.* If rotations of the structure itself or rotations of groups joined by single bonds make one structure superposable with another, then *the structures do not represent different compounds;* instead, they represent different orientations or different conformations of two molecules of the same compound.

Since structures **1** and **2** are not superposable, they represent different compounds. Since structures **1** and **2** differ *only* in the arrangement of their atoms in space, they represent stereoisomers. Structures **1** and **2** are also mirror reflections of each other, thus **1** and **2** represent enantiomers.

Structures **1** and **2** are not the only possible structures, however. We find that we can write a structure **3** that is different from either **1** or **2**, and we can write a structure **4** that is a nonsuperposable mirror reflection of structure **3**.

$$
\begin{array}{cc}
\text{CH}_3 & \text{CH}_3 \\
\text{Br} \diagdown \overset{|}{\text{C}} \diagup \text{H} & \text{H} \diagdown \overset{|}{\text{C}} \diagup \text{Br} \\
| & | \\
\text{H} \diagup \overset{|}{\text{C}} \diagdown \text{Br} & \text{Br} \diagup \overset{|}{\text{C}} \diagdown \text{H} \\
\text{C}_2\text{H}_5 & \text{C}_2\text{H}_5 \\
\mathbf{3} & \mathbf{4}
\end{array}
$$

Structures **3** and **4** correspond to another pair of enantiomers. Structures **1** to **4** are all different, so there are a total of four stereoisomers of 2,3-dibromopentane. At this point you should convince yourself that there are no other stereoisomers by writing other structural formulas. You will find that rotation of the single bonds (or of the entire structure) of any other arrangement of the atoms will cause the structure to become superposable with one of the structures that we have written here. Better yet, using different-colored balls, make molecular models as you work this out.

The compounds represented by structures **1** to **4** are all optically active compounds. Any one of them, if placed separately in a polarimeter, would show optical activity.

The compounds represented by structures **1** and **2** are enantiomers. The compounds represented by structures **3** and **4** are also enantiomers. But what is the isomeric relation between the compounds represented by **1** and **3**?

We can answer this question by observing that **1** and **3** *are stereoisomers* and that they *are not mirror reflections of each other.* They are, therefore, *diastereomers.* **Diastereomers have different physical properties**—different melting points and boiling points, different solubilities, and so forth. In this respect these diastereomers are just like diastereomeric alkenes such as *cis-* and *trans-*2-butene.

PROBLEM 4.15

(a) What is the stereoisomeric relation between compounds **2** and **3**? (b) Between **1** and **4**? (c) Between **2** and **4**? (d) Make a table showing all of the stereoisomeric relations between all possible pairs of compounds **1** to **4**. (e) Would compounds **1** and **2** have the same boiling point? (f) Would compounds **1** and **3**?

4.9A Meso Compounds

A structure with two stereocenters will not always have four possible stereoisomers. Sometimes there are only *three.* This happens because some molecules with stereocenters are, overall, *achiral.*

To understand this, let us write stereochemical formulas for 2,3-dibromobutane shown here.

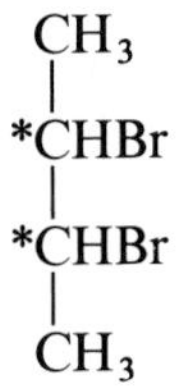

$$CH_3$$
$$|$$
$$*CHBr$$
$$|$$
$$*CHBr$$
$$|$$
$$CH_3$$

2,3-Dibromobutane

We begin in the same way as we did before. We write the formula for one stereoisomer and for its mirror reflection.

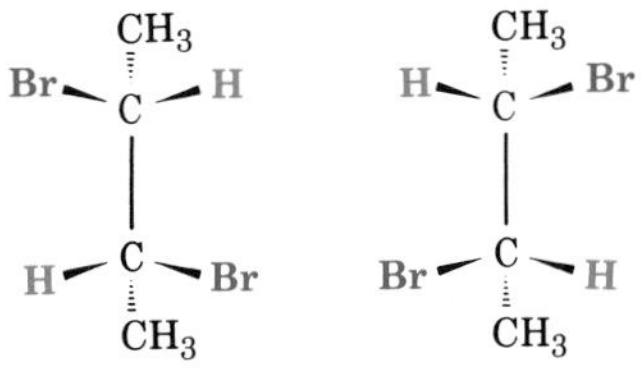

A   B

Structures **A** and **B** are nonsuperposable and represent a pair of enantiomers.

When we write structure **C** (see the following structure) and its mirror reflection **D**, however, the situation is different. *The two structures are superposable.* This means that **C** and **D** do not represent a pair of enantiomers. Formulas **C** and **D** represent two different orientations of the same compound.

$$
\begin{array}{cc}
\underset{\mathrm{CH_3}}{H-C-Br} & \underset{\mathrm{CH_3}}{Br-C-H}
\end{array}
$$

This structure when turned in the plane end for end can be superposed on **C**

C   **D**

The molecule represented by structure **C** (or **D**) is not chiral even though it contains tetrahedral atoms with four different attached groups. Such molecules are called **_meso compounds._** Meso compounds, *because they are achiral,* are optically inactive.

The ultimate test for molecular chirality is to construct a model (or write the structure) of the molecule and then test whether or not the model (or structure) is superposable on its mirror reflection. If it is, the molecule is achiral: If it *is not,* the molecule is chiral.

We have already carried out this test with structure **C** and found that it is achiral. We can also demonstrate that **C** is achiral in another way. Figure 4.16 shows that structure **C** *has a plane of symmetry* (Section 4.4).

PROBLEM 4.16

Which of the following would be optically active?

(a) **A** alone   (c) **C** alone

(b) **B** alone   (d) An equimolar mixture of **A** and **B**

PROBLEM 4.17

Shown here are formulas for compounds **A, B,** or **C** written in noneclipsed conformations. In each instance tell which compound (**A, B,** or **C**) each formula represents.

(1)   **(2)**   **(3)**

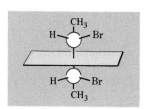

FIGURE 4.16 The plane of symmetry of *meso*-2,3-dibromobutane. This plane divides the molecule into halves that are mirror reflections of each other.

PROBLEM 4.18

Write three-dimensional formulas for all of the stereoisomers of each of the following compounds.

(a) $CH_3CHClCHClCH_3$

(b) $CH_3CHBrCHClCH_3$

(c) $CH_3CHBrCHBrCH_2Br$

(d) $CH_2BrCHBrCHBrCH_2Br$

(e) $CH_3CHClCHClCHClCH_3$

(f) In answers to parts (a)–(e) label pairs of enantiomers and meso compounds.

4.10 NAMING COMPOUNDS WITH MORE THAN ONE STEREOCENTER

If a compound has more than one tetrahedral stereocenter, we analyze each center separately and decide whether it is (R) or (S). Then, using numbers, we tell which designation refers to which carbon atom.

Consider the stereoisomer **A** of 2,3-dibromobutane.

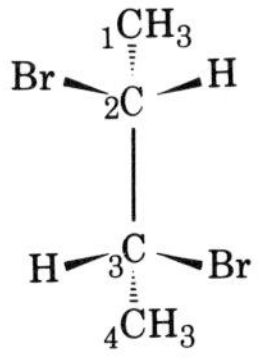

$$\underset{4CH_3}{\overset{1CH_3}{\underset{H-3C-Br}{\overset{Br-2C-H}{|}}}}\qquad \begin{array}{l}\textbf{A}\\ \textbf{2,3-Dibromobutane}\end{array}$$

When this formula is rotated so that the group of lowest priority attached to carbon-2 is directed away from the viewer it resembles the following.

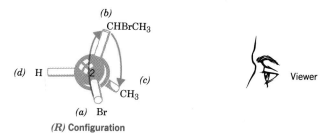

(R) Configuration

The order of progression from the group of highest priority to that of next highest priority (from — Br, to — $CHBrCH_3$, to — CH_3) is clockwise. So carbon-2 has the (R) configuration.

When we repeat this procedure with carbon-3 we find that carbon-3 also has the (R) configuration.

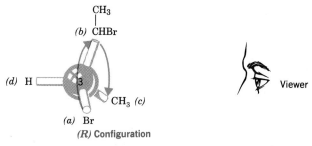

(R) Configuration

Compound **A**, therefore, is $(2R, 3R)$-2,3-dibromobutane.

PROBLEM 4.19

Give names that include (*R*) and (*S*) designations for compounds **B** and **C** in Section 4.9A.

PROBLEM 4.20

Give names that include (*R*) and (*S*) designations for your answers to Problem 4.18 (a) and (b).

4.11 FISCHER PROJECTION FORMULAS

In writing structures for chiral molecules thus far, we have used only three-dimensional formulas, and we shall continue to do so until we study carbohydrates in Chapter 21. The reason: Three-dimensional formulas are unambiguous and can be manipulated on paper in any way that we wish, as long as we do not break bonds. Their use, moreover, teaches us to see molecules (in our mind's eye) in three dimensions, and this ability will serve us well.

Chemists sometimes represent structures for chiral molecules with *two-dimensional formulas* called **Fischer projection formulas.** These two-dimensional formulas are especially useful for compounds with several stereocenters because they save space and are easy to write. Their use, however, requires a rigid adherence to certain conventions. ***Used carelessly, these projection formulas can easily lead to incorrect conclusions.***

The Fischer projection formula for (2*R*, 3*R*)-2,3-dibromobutane is written as follows:

Three-dimensional formula

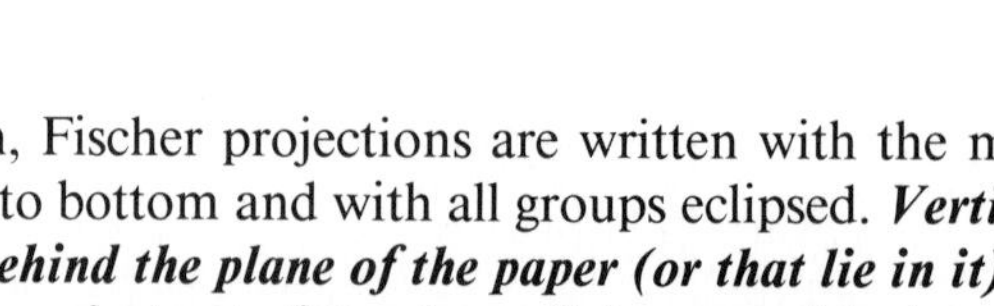

Fischer projection formula

A A

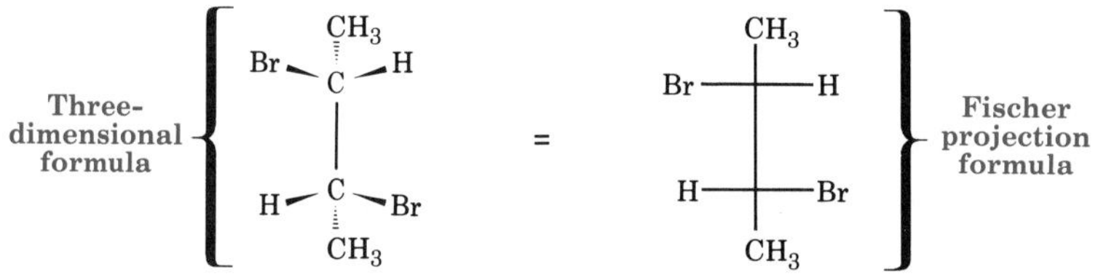

By convention, Fischer projections are written with the main carbon chain extending from top to bottom and with all groups eclipsed. ***Vertical lines represent bonds that project behind the plane of the paper (or that lie in it). Horizontal lines represent bonds that project out of the plane of the paper.*** The intersection of vertical and horizontal lines represents a carbon atom, usually one that is a stereocenter.

In using Fischer projections to test the superposability for two structures, we are permitted to rotate them in the plane of the paper by 180° *but by no other angle.* We must always keep them in the plane of the paper, and ***we are not allowed to flip them over.***

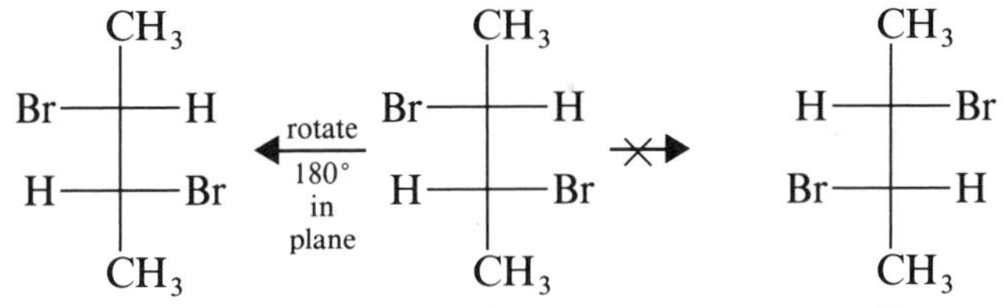

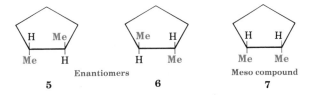

A	A	B
Same structure		*Not the same*

Not the same
(Flipping the projection
formula over creates the
projection formula for
the enantiomer of A)

Your instructor will advise you about the use you are to make of Fischer projections.

4.12 STEREOISOMERISM OF CYCLIC COMPOUNDS

Because the cyclopentane ring is essentially planar, cyclopentane derivatives offer a convenient starting point for a discussion of the stereoisomerism of cyclic compounds. For example, 1,2-dimethylcyclopentane has two stereocenters and exists in three stereoisomeric forms **5**, **6**, and **7**.

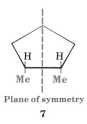

5 **6** **7**

Enantiomers   Meso compound

The trans compound exists as a pair of enantiomers **5** and **6**. *cis*-1,2-Dimethylcyclopentane is a meso compound. It has a plane of symmetry that is perpendicular to the plane of the ring.

Plane of symmetry

7

PROBLEM 4.21

(a) Is the *trans*-1,2-dimethylcyclopentane (**5**) superposable on its mirror image (i.e., on compound **6**)? (b) Is the *cis*-1,2-dimethylcyclopentane (**7**) superposable on its mirror image? (c) Is the *cis*-1,2-dimethylcyclopentane a chiral molecule? (d) Would *cis*-1,2-dimethylcyclopentane show optical activity? (e) What is the stereoisomeric relation between **5** and **7**? (f) Between **6** and **7**?

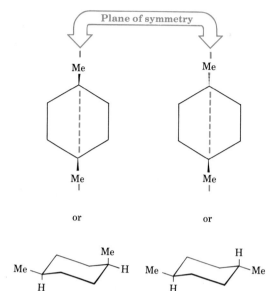

cis-1, 4-Dimethylcyclohexane   trans-1, 4-Dimethylcyclohexane

FIGURE 4.17 The cis and trans forms of 1,4-dimethylcyclohexane are diastereomers of each other. Both compounds are achiral.

PROBLEM 4.22

Write structural formulas for all of the stereoisomers of 1,3-dimethylcyclopentane. Label pairs of enantiomers and meso compounds if they exist.

4.12A Cyclohexane Derivatives

1,4-Dimethylcyclohexanes. If we examine a formula of 1,4-dimethylcyclohexane we find that it does not contain any tetrahedral atoms with four different groups. However, we learned in Section 3.13A that 1,4-dimethylcyclohexane exists as cis–trans isomers. The cis and trans forms (Fig. 4.17) are *diastereomers.* Neither compound is chiral and, therefore, neither is optically active. Notice that both the cis and trans forms of 1,4-dimethylcyclohexane have a plane of symmetry.

1,3-Dimethylcyclohexanes. A 1,3-dimethylcyclohexane has two stereocenters; we can, therefore, expect as many as four stereoisomers ($2^2 = 4$). In reality there are only three. *cis*-1,3-Dimethylcyclohexane has a plane of symmetry (Fig. 4.18) and is achiral. *trans*-1,3-Dimethylcyclohexane does not have a plane of sym-

FIGURE 4.18 *cis*-1,3-Dimethylcyclohexane has a plane of symmetry and is, therefore, achiral.

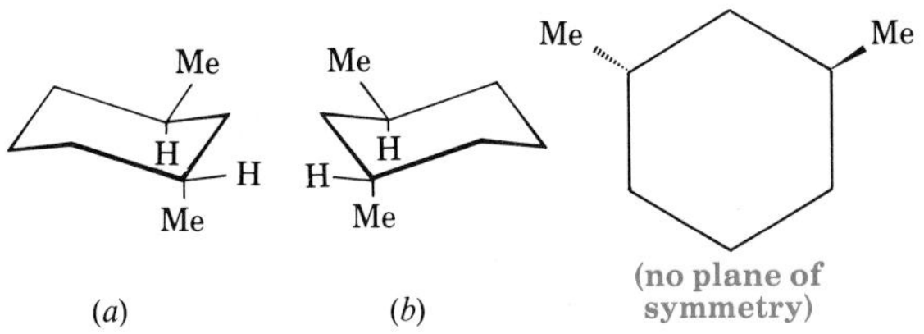

FIGURE 4.19 *trans*-1,3-Dimethyl-cyclohexane does not have a plane of symmetry and exists as a pair of enantiomers. The two structures (*a*) and (*b*) shown here are not superposable as they stand, and flipping the ring of either structure does not make it superposable on the other.

(*a*)   (*b*)   (no plane of symmetry)

metry and exists as a pair of enantiomers (Fig. 4.19). You may want to make models of the *trans*-1,3-dimethylcyclohexane enantiomers. Having done so, convince yourself that they cannot be superposed as they stand, and that they cannot be superposed after one enantiomer has undergone a ring flip.

1,2-Dimethylcyclohexanes. A 1,2-dimethylcyclohexane also has two stereocenters and again we might expect as many as four stereoisomers. However, again we find that there are only three. *trans*-1,2-Dimethylcyclohexane (Fig. 4.20) exists as a pair of enantiomers. Its molecules do not have a plane of symmetry.

With *cis*-1,2-dimethylcyclohexane, the situation is somewhat more complex. If we consider the two conformational structures (*a*) and (*b*) shown in Fig. 4.21 we find that these two mirror-image structures are not superposable one on the other, *but they are interconvertible by a ring flip*. (You should prove this to yourself with models.) Therefore, while the two structures represent enantiomers *they cannot be separated*. They simply represent *different conformations of the same compound*. If we consider the structure for *cis*-1,2-dimethylcyclohexane with a planar ring shown in Fig. 4.21 (and a time average of the ring of the two chair conformations is planar) we find that the structure has a plane of symmetry. On this basis we would not expect to find a pair of separable enantiomers.

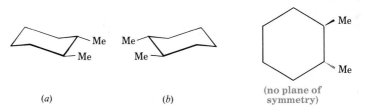

(*a*)   (*b*)   (no plane of symmetry)

FIGURE 4.20 *trans*-1,2-Dimethylcyclohexane has no plane of symmetry and exists as a pair of enantiomers (*a*) and (*b*). (Notice that we have written the most stable conformations for (*a*) or (*b*). A ring flip of either (*a*) or (*b*) would cause both methyl groups to become axial.)

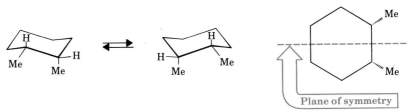

Plane of symmetry

FIGURE 4.21 *cis*-1,2-Dimethylcyclohexane exists as two interconvertible chair conformations (*a*) and (*b*). A planar representation of the ring has a plane of symmetry.

PROBLEM 4.23

Write formulas for all of the isomers of each of the following. Designate pairs of enantiomers and achiral compounds where they exist.

(a) 1-Bromo-2-chlorocyclohexane (c) 1-Bromo-4-chlorocyclohexane

(b) 1-Bromo-3-chlorocyclohexane

PROBLEM 4.24

Give the $(R)-(S)$ designation for each compound given as an answer to Problem 4.23.

4.13 RELATING CONFIGURATIONS THROUGH REACTIONS IN WHICH NO BONDS TO THE STEREOCENTER ARE BROKEN

Reactions in which **no bonds to the stereocenter are broken** must proceed with **retention of configuration.** We can illustrate this by the reaction of (*R*)-(−)-2-butanol with acetyl chloride shown in Fig. 4.22.

The reaction of acetyl chloride with an alcohol (Section 18.7) is an easy way to synthesize an ester. The reaction of acetyl chloride with (*R*)-(−)-2-butanol produces the ester, (*R*)-(−)-*sec*-butyl acetate. It is known (we shall see how later), that the reaction does not involve cleavage of any of the bonds to the stereocenter. We can be certain, therefore, that the general spatial arrangement of the groups in the product will be the same as that of the reactant. The configuration of the reactant is said to be *retained* in the product; the reaction is said to proceed with *retention of configuration.*

In the example that we just cited, the optically active reactant and product both rotate plane-polarized light in the same direction. This is not always the case, as the following example illustrates.

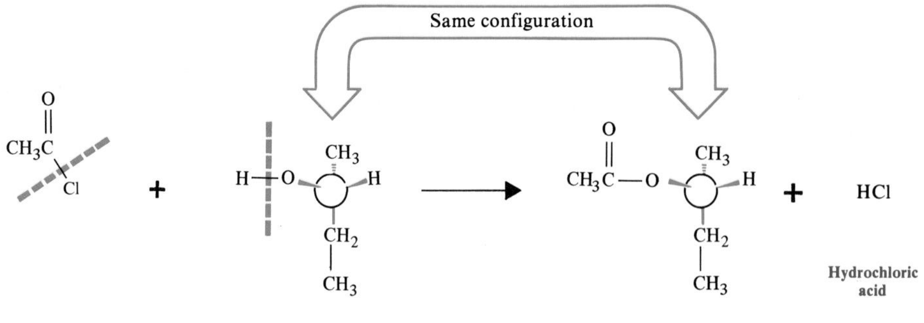

Acetyl chloride

(*R*)-(−)-2-Butanol
$[\alpha]_D^{25} = -13.52°$

(*R*)-(−)-*sec*-Butyl acetate
$[\alpha]_D^{25} = -25.43°$

FIGURE 4.22 The reaction of (*R*)-(−)-2-butanol with acetyl chloride. The bonds that are broken are shown with dashed lines. A bond between one of the carbonyl groups and the chlorine atom of acetyl chloride is broken. The only bond that is broken in (*R*)-(−)-2-butanol is the one between the hydrogen atom and oxygen atom of the hydroxyl group. No bonds to the stereocenter (shown in color) are broken.

$$
\begin{array}{ccc}
\text{H}-\overset{\text{CH}_3}{\underset{\text{C}_2\text{H}_5}{\bigcirc}}-\text{CH}_2\text{—OH} + \text{H—Cl} & \xrightarrow[\substack{\text{(retention of}\\\text{configuration)}}]{\text{ZnCl}_2} & \text{H}-\overset{\text{CH}_3}{\underset{\text{C}_2\text{H}_5}{\bigcirc}}-\text{CH}_2\text{—Cl}
\end{array}
$$

(S)-(−)-2-Methyl-1-butanol   (S)-(+)-1-Chloro-2-methylbutane
$[\alpha] = -5.756°$ $[\alpha] = +1.64°$

And in some instances the (R) and (S) designation may change even though the reaction proceeds with retention of configuration.

$$
\begin{array}{ccc}
\text{H}-\overset{\text{CH}_2\text{—Br}}{\underset{\underset{\text{CH}_3}{\overset{|}{\text{CH}_2}}}{\bigcirc}}\text{—OH} & + \text{Zn} \xrightarrow[\substack{\text{(retention of}\\\text{configuration)}}]{\text{H}^+} & \text{H}-\overset{\text{CH}_3}{\underset{\underset{\text{CH}_3}{\overset{|}{\text{CH}_2}}}{\bigcirc}}\text{—OH} + \text{ZnBr}_2
\end{array}
$$

(R)-1-Bromo-2-butanol   (S)-2-Butanol

In this example the $(R)-(S)$ designation changes because the $—\text{CH}_2\text{Br}$ group of the reactant ($—\text{CH}_2\text{Br}$ has a higher priority than $—\text{CH}_2\text{CH}_3$) changes to a $—\text{CH}_3$ group in the product ($—\text{CH}_3$ has a lower priority than $—\text{CH}_2\text{CH}_3$).

4.13A Relative and Absolute Configurations

Reactions in which no bonds to the stereocenter are broken are useful in relating configurations of chiral molecules. That is, they allow us to demonstrate that certain compounds have the same **relative configuration.** In each of the three examples that we have just cited, the products of the reactions have the same *relative configurations* as the reactants.

Before 1951 only relative configurations of chiral molecules were known. No one prior to that time had been able to demonstrate with certainty what the actual spatial arrangement of groups was in any chiral molecule. To say this another way, no one had been able to determine the **absolute configuration** of an optically active compound.

Configurations of chiral molecules were related to each other *through reactions of known stereochemistry.* Attempts were also made to relate all configurations back to a single compound that had been chosen arbitrarily to be the standard. This standard compound was glyceraldehyde.

$$
\begin{array}{c}
\overset{\text{O}}{\overset{\|}{\text{CH}}} \\
| \\
\text{*CHOH} \\
| \\
\text{CH}_2\text{OH}
\end{array}
$$
Glyceraldehyde

Glyceraldehyde molecules have one tetrahedral stereocenter; therefore, glyceraldehyde exists as a pair of enantiomers.

$$
\begin{array}{ccc}
\overset{\displaystyle O}{\overset{\|}{C}}-H & & \overset{\displaystyle O}{\overset{\|}{C}}-H \\
H-\overset{|}{\underset{|}{C}}-OH & \text{and} & HO-\overset{|}{\underset{|}{C}}-H \\
CH_2OH & & CH_2OH
\end{array}
$$

(R)-Glyceraldehyde   (S)-Glyceraldehyde

In the older system for designating configurations (R)-glyceraldehyde was called D-glyceraldehyde and (S)-glyceraldehyde was called L-glyceraldehyde.

One glyceraldehyde enantiomer is dextrorotatory (+) and the other, of course, is levorotatory (−). Before 1951 no one could be sure, however, which configuration belonged to which enantiomer. Chemists decided arbitrarily to assign the (R) configuration to the (+)-enantiomer. Then configurations of other molecules were related to one glyceraldehyde enantiomer or the other through reactions of known stereochemistry.

For example, the configuration of (−)-lactic acid can be related to (+)-glyceraldehyde through the following sequence of reactions.

This bond is broken

$$
\underset{\text{(+)-Glyceraldehyde}}{\overset{\displaystyle O}{\overset{\|}{C}}+H \atop \displaystyle H-\overset{|}{\underset{|}{C}}-OH \atop CH_2OH}
\xrightarrow[\text{(oxidation)}]{\text{HgO}}
\underset{\text{(−)-Glyceric acid}}{\overset{\displaystyle O}{\overset{\|}{C}}-OH \atop \displaystyle H-\overset{|}{\underset{|}{C}}-OH \atop CH_2OH}
\xleftarrow[H_2O]{\text{HNO}_2}
\underset{\text{(+)-Isoserine}}{\overset{\displaystyle O}{\overset{\|}{C}}-OH \atop \displaystyle H-\overset{|}{\underset{|}{C}}-OH \atop CH_2+NH_2}
\xrightarrow[\text{HBr}]{\text{HNO}_2}
$$

This bond is broken

This bond is broken

$$
\underset{\substack{\text{(−)-3-Bromo-2-hydroxy-}\\ \text{propanoic acid}}}{\overset{\displaystyle O}{\overset{\|}{C}}-OH \atop \displaystyle H-\overset{|}{\underset{|}{C}}-OH \atop CH_2+Br}
\xrightarrow{\text{Zn,H}^+}
\underset{\text{(−)-Lactic acid}}{\overset{\displaystyle O}{\overset{\|}{C}}-OH \atop \displaystyle H-\overset{|}{\underset{|}{C}}-OH \atop CH_3}
$$

The stereochemistry of all of these reactions is known. Because bonds to the stereocenter (shown in red) are not broken in any of them, they all proceed with retention of configuration. If the assumption is made that the configuration of (+)-glyceraldehyde is as follows:

$$
\overset{\displaystyle O}{\overset{\|}{CH}} \atop H-\overset{|}{\underset{|}{C}}-OH \atop CH_2OH
$$

(R)-(+)-Glyceraldehyde

then the configuration of (−)-lactic acid is

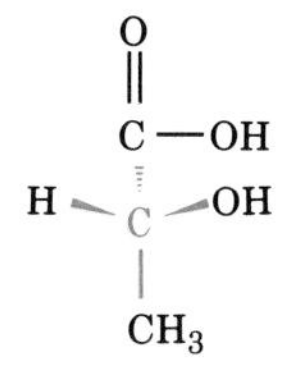

$$
\begin{array}{c}
O \\
\parallel \\
C-OH \\
H \diagdown \underset{C}{} \diagup OH \\
| \\
CH_3
\end{array}
$$

(R)-(−)-Lactic acid

PROBLEM 4.25

(a) Write three-dimensional structures for the relative configurations of (−)-glyceric acid and (−)-3-bromo-2-hydroxypropanoic acid. (b) What is the (R)–(S) designation of (−)-glyceric acid? (c) Of (+)-isoserine? (d) Of (−)-3-bromo-2-hydroxypropanoic acid?

The configuration of (−)-glyceraldehyde was also related through reactions of known stereochemistry to (+)-tartaric acid.

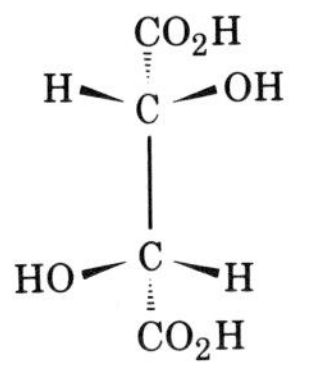

$$
\begin{array}{c}
CO_2H \\
H \diagdown \underset{C}{} \diagup OH \\
| \\
HO \diagup \overset{C}{} \diagdown H \\
CO_2H
\end{array}
$$

(+)-Tartaric acid

In 1951 J. M. Bijvoet, the director of the van't Hoff Laboratory of the University of Utrecht in Holland, using a special technique of X-ray diffraction, was able to show conclusively that (+)-tartaric acid had the absolute configuration shown above. This meant that the original arbitrary assignment of the configurations of (+)- and (−)-glyceraldehyde was also correct. It also meant that the configurations of all of the compounds that had been related to one glyceraldehyde enantiomer or the other were now absolute configurations.

4.14 SEPARATION OF ENANTIOMERS: RESOLUTION

So far we have left unanswered an important question about optically active compounds and racemic forms: How are enantiomers separated? Enantiomers have identical solubilities in ordinary solvents, and they have identical boiling poir ts. Consequently, the conventional methods for separating organic compounds such as crystallization and distillation fail when applied to a racemic form.

It was, in fact, Louis Pasteur's separation of a racemic form of a salt of tartaric acid in 1848 that led to the discovery of the phenomenon called enantiomerism. Pasteur, consequently, is often considered to be the founder of the field of stereochemistry.

Tartaric acid is one of the by-products of wine making, and Pasteur had obtained a sample of racemic tartaric acid from the owner of a chemical plant. In the course of his investigation Pasteur began examining the crystal structure of the sodium ammonium salt of racemic tartaric acid. He noticed that two types of crystals were present. One was identical with crystals of the sodium ammonium salt of (+)-tartaric acid that had been discovered earlier and had been shown to be dextrorotatory. Crystals of the other type were *non*superposable mirror reflections of the first kind. The two types of crystals were actually chiral. Using tweezers and a magnifying glass, Pasteur separated the two kinds of crystals, dissolved them in water, and placed the solutions in a polarimeter. The solution of crystals of the first type was dextrorotatory, and the crystals themselves proved to be identical with the sodium ammonium salt of (+)-tartaric acid that was already known. The solution of crystals of the second type was levorotatory: it rotated plane-polarized light in the opposite direction and by an equal amount. The crystals of the second type were the sodium ammonium salt of (−)-tartaric acid. The chirality of the crystals themselves disappeared, of course, as the crystals dissolved into their solutions *but the optical activity* remained. Pasteur reasoned, therefore, that the molecules themselves must be chiral.

Pasteur's discovery of enantiomerism and his demonstration that the optical activity of the two forms of tartaric acid was a property of the molecules themselves led, in 1874, to the proposal of the tetrahedral structure of carbon by van't Hoff and Le Bel.

Unfortunately, few organic compounds give chiral crystals as do the (+)- and (−)-tartaric acid salts. Few organic compounds crystallize into separate crystals (containing separate enantiomers) that are visibly chiral like the crystals of the sodium ammonium salt of tartaric acid. Pasteur's method, therefore, is not one that is generally applicable.

The most useful procedure for separating enantiomers is based on allowing a racemic form to react with a single enantiomer of some other compound. This changes a *racemic form into a mixture of diastereomers;* and **diastereomers, because they have different melting and boiling points, can be separated by conventional means.** We shall see how this is done in Chapter 19. The separation of the enantiomers of a racemic form is called **resolution.**

4.15 COMPOUNDS WITH STEREOCENTERS OTHER THAN CARBON

Any tetrahedral atom with four different groups attached to it is a stereocenter. Listed here are general formulas of compounds whose molecules contain stereocenters

$$R_4\!\!-\!\!\underset{\underset{R_3}{|}}{\overset{\overset{R_1}{|}}{Si}}\!\!-\!\!R_2 \qquad R_4\!\!-\!\!\underset{\underset{R_3}{|}}{\overset{\overset{R_1}{|}}{Ge}}\!\!-\!\!R_2 \qquad R_4\!\!-\!\!\underset{\underset{R_3}{|}}{\overset{\overset{R_1}{|}}{N}}\!\!\overset{+}{-}\!\!R_2 \quad X^-$$

other than carbon. Silicon and germanium are in the same group of the Periodic Table as carbon. They form tetrahedral compounds as carbon does. When four different groups are situated around the central atom in silicon, germanium, and nitrogen compounds, the molecules are chiral and the enantiomers can be separated.

4.16 CHIRAL MOLECULES THAT DO NOT POSSESS A TETRAHEDRAL ATOM WITH FOUR DIFFERENT GROUPS

A molecule is chiral if it is not superposable on its mirror reflection. The presence of a tetrahedral atom with four different groups is only one focus that will confer chirality on a molecule. Most of the molecules that we shall encounter do have such stereocenters. Many chiral molecules are known, however, that do not. An example is 1,3-dichloroallene.

Allenes are compounds whose molecules contain the following double bond sequence.

$$\diagdown C = C = C \diagup$$

The planes of the π bonds of allenes are perpendicular to each other.

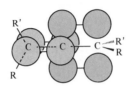

This geometry of the π bonds causes the groups attached to the end carbon atoms to lie in perpendicular planes and, because of this, allenes with different substituents on the end carbon atoms are chiral (Fig. 4.23). (Allenes do not show cis–trans isomerism.)

FIGURE 4.23 Enantiomeric forms of 1,3-dichloroallene. These two molecules are nonsuperposable mirror reflections of each other and are, therefore, chiral. They do not possess a tetrahedral atom with four different groups, however.

4.17 SOME IMPORTANT TERMS AND CONCEPTS

Stereochemistry. Chemical studies that take into account the spatial aspects of molecules.

Isomers are different compounds that have the same molecular formula. All isomers fall into either of two groups: *constitutional* isomers or *stereoisomers.*

Constitutional isomers (formerly called **structural isomers**) are isomers that have their atoms connected in a different order.

Stereoisomers have their atoms joined in the same order but differ in the way their atoms are arranged in space. Stereoisomers can be subdivided into two categories: *enantiomers* and *diastereomers.*

Enantiomers are stereoisomers that are related as an object and its mirror reflection. Enantiomers occur only with compounds whose molecules are chiral, that is, with molecules that are *not* superposable on their mirror reflections. Separate enantiomers rotate the plane of plane polarized light and are said to be *optically active.* They have equal but opposite specific rotation.

Diastereomers are stereoisomers that are not enantiomers, that is, they are stereoisomers that are not related as an object and its mirror reflection.

Chirality is equivalent to "handedness." A chiral molecule is one that is not superposable on its mirror reflection. An *achiral* molecule is one that can be superposed on its mirror reflection. Any tetrahedral atom that has four different attached groups is a **stereocenter.** A pair of enantiomers is possible for all molecules that contain a single tetrahedral stereocenter. For molecules with more than one stereocenter, the number of stereoisomers will not exceed 2^n where n is the number of stereocenters.

Plane of symmetry. An imaginary plane that bisects a molecule in such a way that the two halves of the molecule are mirror reflections of each other. Any molecule that has a plane of symmetry will be achiral.

Configuration. The particular arrangement of atoms (or groups) in space that is characteristic of a given stereoisomer. The configuration at each stereocenter can be designated as (R) or (S) using the rules given in Section 4.5.

Racemic form (racemate, or racemic mixture). An equimolar mixture of enantiomers.

Meso compound. An optically inactive compound whose molecules are achiral even though they contain tetrahedral atoms with four different attached groups.

Stereoselective reaction. One that yields exclusively (or predominantly) one of a set of stereoisomers.

Resolution. The separation of the enantiomers of a racemic form.

ADDITIONAL PROBLEMS
4.26
Give definitions of each of the following terms and examples that illustrate their meaning:

(a) Isomers	(f) Meso compound	(k) Achiral molecule
(b) Constitutional isomers	(g) Racemic form	(l) Optical activity
(c) Stereoisomers	(h) Plane of symmetry	(m) Dextrorotatory
(d) Diastereomers	(i) Stereocenter	(n) Retention of configuration
(e) Enantiomers	(j) Chiral molecule	

4.27
Consider the following pairs of structures. Identify the relation between them by describing them as representing enantiomers, diastereomers, constitutional isomers, or two molecules of the same compound.

(a)

$$CH_3$$
$$H \quad C \quad Br$$
$$Cl$$

and

$$CH_3$$
$$H \quad C \quad Cl$$
$$Br$$

(b)

$$CH_3$$
$$H \quad C \quad Br$$
$$Cl$$

and

$$CH_3$$
$$Cl \quad C \quad H$$
$$Br$$

(*c*) 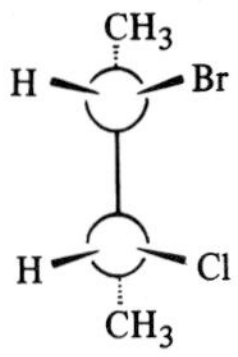 and 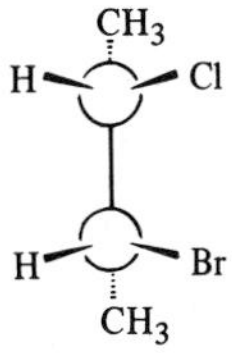

(*d*) 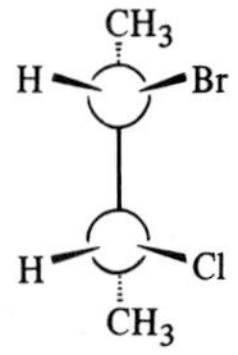 and 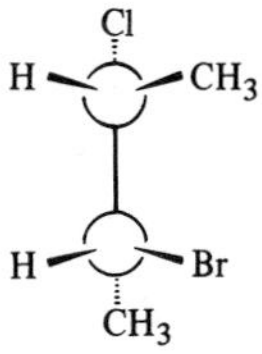

(*e*) 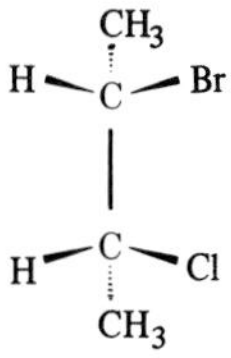 and 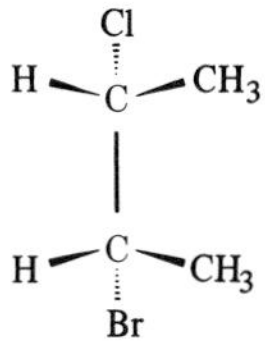

(*f*) 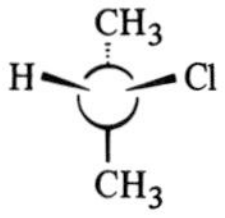 and 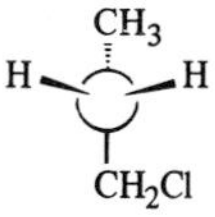

(*g*) 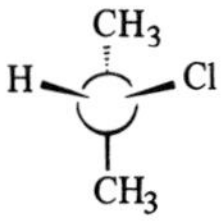 and 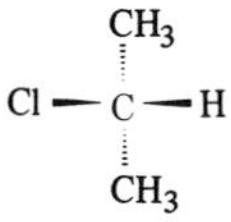

(*h*) 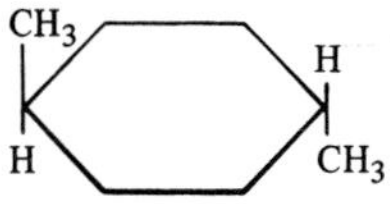 and 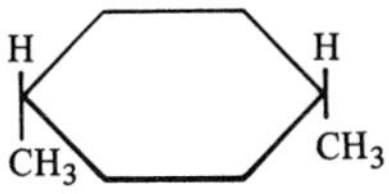

(*i*) 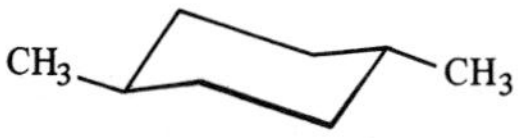 and 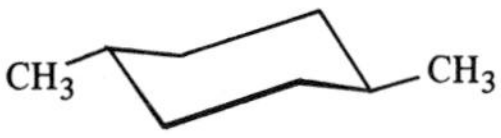

(*j*) 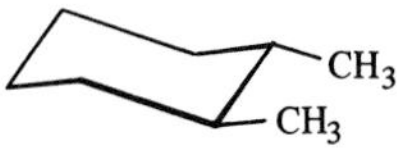 and 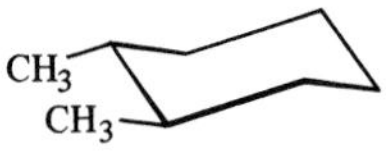

(*k*) 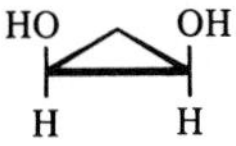 and 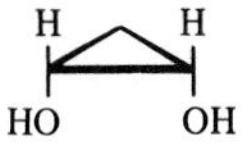

(*l*) 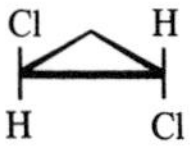 and 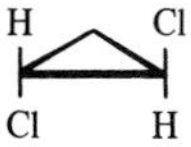

(m)

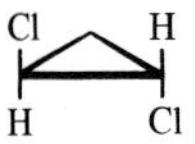

and

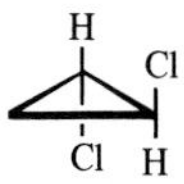

(n)

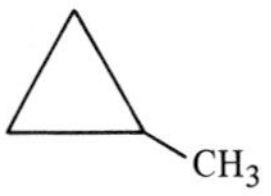

and

(o)

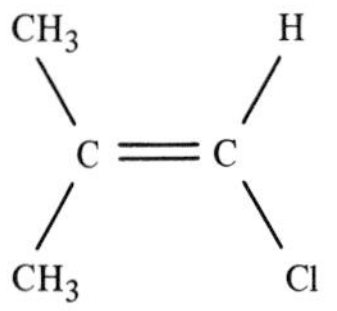

and

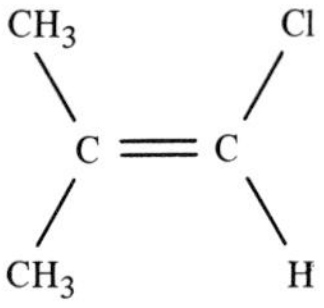

(p)

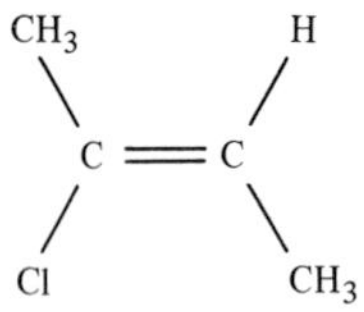

and

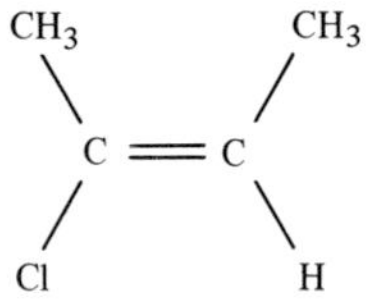

(q)

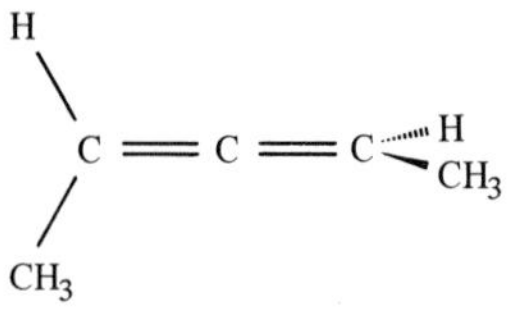

and

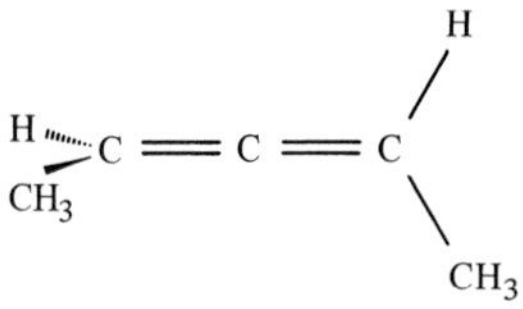

4.28

There are four dimethylcyclopropane isomers. (a) Write three-dimensional formulas for them. (b) Which dimethylcyclopropane isomers would, if taken separately, show optical activity? (c) If a mixture consisting of 1 mole of each of the four dimethylcyclopropane isomers were subjected to fractional distillation, how many fractions would be obtained? (d) How many of these fractions would show optical activity?

4.29

(Use models to solve this problem.) (a) Write a conformational structure for the most stable conformation of *trans*-1,2-cyclohexanediol and write its mirror reflection. (b) Are these two molecules superposable? (c) Are they interconvertible through a ring "flip"? (d) Repeat the process in (a) with *cis*-1,2-cyclohexanediol. (e) Are these structures superposable? (f) Are they interconvertible?

4.30

(Use models to solve this problem.) (a) Write a conformational structure for the most stable conformation of *trans*-1,4-cyclohexanediol and for its mirror reflection. (b) Are these structures superposable? (c) Do they represent enantiomers? (d) Does *trans*-1,4-cyclohexanediol have a stereoisomer, and if so, what is it? (e) Is this stereoisomer chiral?

4.31

(Use models to solve this problem.) Write conformational structures for all of the stereo-isomers of 1,3-cyclohexanediol. Label pairs of enantiomers and meso compounds if they exist.

4.32

Tartaric acid [$HO_2CCH(OH)CH(OH)CO_2H$] was an important compound in the history of stereochemistry. Two naturally occurring forms of tartaric acid are optically inactive. One form has a melting point of 206 °C, the other a melting point of 140 °C. The inactive tartaric acid with a melting point of 206 °C can be separated into two optically active forms of tartaric acid with the same melting point (170 °C). One optically active tartaric acid has $[\alpha]_D^{25} = +12°$, the other $[\alpha]_D^{25} = -12°$. All attempts to separate the other inactive tartaric acid (melting point 140 °C) into optically active compounds fail. (a) Write the three-dimensional structure of the tartaric acid with melting point 140 °C. (b) What are possible structures for the optically active tartaric acids with melting points of 170 °C? (c) Can you be sure which tartaric acid in (b) has a positive rotation and which has a negative rotation? (d) What is the nature of the form of tartaric acid with a melting point of 206 °C?

CHAPTER FIVE

Vitamin B$_6$. Vitamin B$_6$ is a nitrogen-containing a heterocyclic compound called pyridoxine. The chemical structure of pyridoxine is given in Section 19.4.

Ionic Reactions — Nucleophilic Substitution and Elimination Reactions of Alkyl Halides

5.1 INTRODUCTION: HOMOLYSIS AND HETEROLYSIS OF COVALENT BONDS

In this chapter we shall begin to look at some of the important reactions that organic compounds undergo. As we examine these reactions we shall not only want to know what the products are, we shall also be interested in *how the reaction takes place*. We shall be interested in what chemists call the ***mechanism of the reaction — the events that take place at the molecular level as reactants become products.*** If the reaction takes place in more than one step, then what are these steps, and what kinds of **intermediates** intervene between reactants and products?

5.1A Homolysis and Heterolysis of Covalent Bonds

Reactions of organic compounds almost inevitably involve the making and breaking of covalent bonds. If we consider a hypothetical molecule $A:B$, its covalent bond may break in three possible ways:

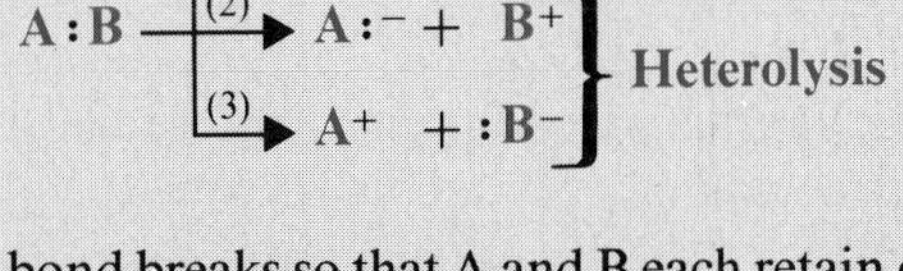

$$A:B \begin{cases} \xrightarrow{(1)} A\cdot \ + \ B\cdot & \text{Homolysis} \\ \xrightarrow{(2)} A:^- + \ B^+ \\ \xrightarrow{(3)} A^+ \ + :B^- \end{cases} \text{Heterolysis}$$

In (1) above the bond breaks so that A and B each retain one of the electrons of the bond, and cleavage leads to the neutral fragments $A\cdot$ and $B\cdot$. This type of bond breaking is called **homolysis** (Gr: *homo-*, the same, *+lysis,* loosening or cleavage); the bond is said to have broken *homolytically.* The neutral fragments $A\cdot$ and $B\cdot$ are called **free radicals,** or often simply radicals. Radicals always contain unpaired electrons.

In (2) and (3) bond cleavage leads to charged fragments or **ions** ($A:^-$ and B^+ or A^+ and $:B^-$). This kind of bond cleavage is called **heterolysis** (Gr: *hetero-*, different, *+ lysis*); the bond is said to have broken *heterolytically.*

5.1B Reactive Intermediates in Organic Chemistry

Organic reactions that take place in more than one step involve the formation of an *intermediate* — one that results from either homolysis or heterolysis of a bond. Homolysis of a bond to carbon leads to an intermediate known as a carbon *radical* (or free radical).

$$-\overset{|}{\underset{|}{C}}\!:\!Z \xrightarrow{\text{homolysis}} \quad -\overset{|}{\underset{|}{C}}\!\cdot \quad + \, Z\cdot$$

A carbon radical
(or *free* radical)

Heterolysis of a bond to carbon can lead either to a trivalent carbon cation or carbon anion.

$$-\overset{|}{\underset{|}{C}}\!:\!Z \xrightarrow{\text{heterolysis}} \left\{ \begin{array}{l} -\overset{|}{\underset{|}{C}}{}^{+} \quad + \, :\!Z^{-} \\[2ex] \text{Carbocation} \\ \text{(or } carbonium ion\text{)} \\[3ex] -\overset{|}{\underset{|}{C}}\!:^{-} \, + \, Z^{+} \\[2ex] \text{Carbanion} \end{array} \right.$$

Trivalent carbon cations are called either **carbocations** or **carbenium ions.*** The term *carbocation* has a clear and distinct meaning. The newer term *carbenium ion* has not yet found wide usage. Because of this, we shall always refer to trivalent, positively charged species such as $-\overset{|}{\underset{|}{C}}{}^{+}$ as carbocations.

Carbon anions are called **carbanions.**

Carbon radicals and carbocations are electron-deficient species. A carbon radical has seven electrons in its valence shell; a carbocation has only six and is positively charged. As a consequence, both species are **electron-seeking reagents** called **electrophiles.** *In their reactions they seek the extra electron or electrons that will give them a stable octet.*

Carbanions are usually strong **bases** and strong **nucleophiles.** *Nucleophiles are Lewis bases—they are electron-pair donors.* *Carbanions, therefore seek either a proton or some other positively charged center to neutralize their negative charge.*

Carbon radicals, carbocations, and carbanions are usually highly reactive species. In most instances they exist only as short-lived intermediates in an organic reaction. Under certain conditions, however, these species may exist long enough for chemists to study them using special techniques.

A few carbon radicals, carbocations, and carbanions are stable enough to be isolated. This only happens, however, when special groups are attached to the central carbon atom that allow the charge or the odd electron to be stabilized.

5.1C Ionic Reactions and Free Radical Reactions

Most reactions of organic compounds can be placed into either of two broad categories: **ionic reactions** or **free radical reactions.** As these names suggest, these categories arise from the kinds of reagents that are used to bring about the reactions and from the

---

*An older term, *carbonium ion,* is no longer used because it has taken on a different meaning.

kinds of reactive intermediates that form in them. In ionic reactions the bonds of the reacting molecules undergo **heterolysis;** in free radical reactions, they undergo *homolysis.*

We shall discuss free radical reactions in detail in Chapter 9. In this chapter and in the ones that immediately follow, we concern ourselves only with ionic reactions.

5.2 NUCLEOPHILIC SUBSTITUTION REACTIONS

There are many reactions of the general type shown here.

$$Nu:^- \ + \ R-\ddot{X}: \longrightarrow R-Nu+ \quad :\ddot{X}:^-$$

| Nucleophile | Alkyl halide (substrate) | Product | Halide ion |

Following are some examples:

$$H\ddot{O}:^- + CH_3-\ddot{C}l: \longrightarrow CH_3-\ddot{O}H + :\ddot{C}l:^-$$

$$CH_3\ddot{O}:^- + CH_3CH_2-\ddot{B}r: \longrightarrow CH_3CH_2-\ddot{O}CH_3 + :\ddot{B}r:^-$$

$$:\ddot{I}:^- + CH_3CH_2CH_2-\ddot{C}l: \longrightarrow CH_3CH_2CH_2-\ddot{I}: + :\ddot{C}l:^-$$

In this type of reaction a ***nucleophile, a species with an unshared electron pair,*** reacts with an alkyl halide (called the **substrate**) by replacing the halogen substituent. A *substitution reaction* takes place and the halogen substituent, called the leaving group, departs as a halide ion. Because the substitution reaction is initiated by a nucleophile, it is called a **nucleophilic substitution reaction.**

In nucleophilic substitution reactions the carbon–halogen bond of the substrate undergoes *heterolysis,* and the unshared pair of the nucleophile is used to form a new bond to the carbon atom:

$$Nu:^- + \ R \vert :\ddot{X}: \ \longrightarrow \ Nu:R + :\ddot{X}:^-$$

Leaving Group

Nucleophile

Heterolysis occurs here

One of the questions we shall want to address later in this chapter is, when does the carbon–halogen bond break? Does it break at the same time that the new bond between the nucleophile and the carbon forms?

$$Nu:^- + R:\ddot{X}: \longrightarrow \overset{\delta-}{Nu}\text{---}R\text{---}\overset{\delta-}{\ddot{X}}: \longrightarrow Nu:R + :\ddot{X}:^-$$

Or does the carbon–halogen bond break first?

$$R:X \longrightarrow R^+ + :\ddot{X}:^-$$

then

$$Nu:^- + R^+ \longrightarrow Nu:R$$

We shall find that the answer depends primarily on the structure of the alkyl halide.

5.3 NUCLEOPHILES

The word nucleophile comes from *nucleus* (the positive part of an atom) plus *phile* from the Greek word *philein* meaning to love. Thus a nucleophile is a reagent that "loves" or "seeks" a positive center in an organic molecule. In an alkyl halide the positive center is the carbon atom to which the halogen is attached. This carbon atom bears a partial positive charge because the electronegative halogen pulls the electrons of the carbon–halogen bond in its direction.

$$—\overset{|}{\underset{|}{C}}^{\delta+}\!\!\longrightarrow\!\!\ddot{\underset{\cdot\cdot}{X}}\!:^{\delta-}$$

This is the positive center that the nucleophile seeks　　　*The electronegative halogen polarizes the C—X bond*

Any molecule or negative ion that has an unshared pair of electrons is a potential nucleophile. The hydroxide ion, for example, is a nucleophile and it can react with an alkyl halide to form an alcohol.*

$$HO:^- + R—X \longrightarrow R—OH + :X^-$$

Specific Example

$$HO:^- + CH_3CH_2Br \longrightarrow CH_3CH_2OH + :Br^-$$

Although nucleophiles are often negatively charged ions, they do not have to be. Neutral molecules with unshared electron pairs act as nucleophiles as well. Water molecules can act as nucleophiles, for example, and water can react with certain alkyl halides to yield alcohols.

$$H—\underset{|}{\overset{}{O}}: + R—X \longrightarrow R—\overset{+}{\underset{|}{O}}—H + :X^-$$

Nucleophile　　Alkyl halide　　Alkyloxonium ion

$$\underset{H_2O}{\updownarrow}$$

$$R—OH + H_3O^+ + :X^-$$

Specific Example

$$H—\underset{|}{\overset{}{O}}: + (CH_3)_3C—Cl \longrightarrow (CH_3)_3C—\overset{+}{\underset{|}{O}}—H + :Cl^-$$

$$\underset{H_2O}{\updownarrow}$$

$$(CH_3)_3C—OH + H_3O^+ + :Cl^-$$

*A hydroxide ion actually has three unshared electron pairs, not just one, as we have shown in this reaction. A halide ion actually has four. They are $H\ddot{O}:^-$ and $:\ddot{\underset{\cdot\cdot}{X}}:^-$, respectively. But since only one electron pair is important in this reaction, it is convenient to write the reaction the way we have. In the future we shall often neglect to write in all of the unshared electron pairs. When we do this, it does not mean that they are not there; it just means that we have left them out because they were not important in the reaction being illustrated.

In this reaction the first product is an alkyloxonium ion, $R\!-\!\overset{\underset{|}{H}}{O}{}^{+}\!-\!H$, which then loses a proton to a water molecule to form an alcohol. Alkyloxonium ions are like hydronium ions, $H\!-\!\overset{\underset{|}{H}}{O}{}^{+}\!-\!H$, and just as hydronium ions can donate protons to bases, so too can alkyloxonium ions. Just as hydronium ions are "protonated" water molecules, alkyloxonium ions of the type $R\!-\!\overset{\underset{|}{H}}{O}{}^{+}\!-\!H$ are often called "protonated alcohols." Both hydronium ions and alkyloxonium ions are strong Brønsted acids.

PROBLEM 5.1

Write an electron dot structure for each of the following molecules and ions that shows that each is a potential nucleophile:

(a) Ethyl alcohol, C_2H_5OH

(b) Ethoxide ion, $C_2H_5O^-$

(c) Ammonia, NH_3

(d) Methylamine, CH_3NH_2

(e) Cyanide ion, CN^-

(f) Acetic acid, CH_3CO_2H

(g) Acetate ion, $CH_3CO_2^-$

(h) Formic acid, HCO_2H

(i) Formate ion, HCO_2^-

(j) Ethanethiol, C_2H_5SH

(k) Ethanethiolate ion, $C_2H_5S^-$

(l) Azide ion, N_3^-

5.3A Leaving Groups

Alkyl halides are not the only substances that can act as substrates in nucleophilic substitution reactions. We shall see later that other compounds can also react in the same way. To be reactive — that is, to be able to act as the substrate in a nucleophilic substitution reaction — a molecule must have a good **leaving group.** In alkyl halides the leaving group is the halogen substituent — it leaves as a halide ion. *To be a good leaving group the substituent must be able to leave as a relatively stable, weakly basic molecule or ion.* Because halide ions are relatively stable and are very weak bases, they are good leaving groups. Other groups can function as good leaving groups as well. We can write more general equations for nucleophilic substitution reactions using L to represent a leaving group.

$$Nu\!:^- + R\!-\!L \longrightarrow R\!-\!Nu + :L^-$$

or

$$Nu\!: + R\!-\!L \longrightarrow R\!-\!Nu^+ + :L^-$$

Specific Examples

$$HO\!:^- + CH_3\!-\!Cl \longrightarrow CH_3\!-\!OH + :Cl^-$$

$$H_3N\!: + CH_3\!-\!Br \longrightarrow CH_3\!-\!NH_3{}^+ + :Br^-$$

Later we shall also see reactions where the substrate bears a positive charge and a reaction like the following takes place:

$$Nu\!: + R\!-\!L^+ \longrightarrow R\!-\!Nu^+ + :L$$

Specific Example

$$\underset{\underset{H}{|}}{CH_3-\overset{..}{O}:} + \underset{\underset{H}{|}}{CH_3-\overset{+}{O}-H} \longrightarrow \underset{\underset{H}{|}}{CH_3-\overset{+}{O}-CH_3} + \underset{\underset{H}{|}}{:\overset{..}{O}-H}$$

Nucleophilic substitution reactions will be more understandable and useful if we know something about their mechanisms. How does the nucleophile replace the leaving group? Does the reaction take place in one step, or is more than one step involved? If more than one step is involved, what kinds of intermediates are formed? Which steps are fast and which are slow? In order to answer these questions we need to know something about the rates of chemical reactions.

5.4 REACTION RATES. KINETICS

With a chemical reaction we are usually concerned with two features: *the extent* to which it takes place and *the rate* of the reaction. By the extent of a reaction, we mean how much of the reactants will be converted to products if equilibrium is established between them. By the rate of reaction, we mean how rapidly will the reactants be converted to products. In simpler terms then, with chemical reactions we are concerned with "how far" and "how fast."

Reaction rates or kinetics are important in guiding our predictions about the eventual outcome of chemical reactions. Many reactions that have favorable energy changes occur so slowly as to be imperceptible. In other instances several reaction pathways may compete with each other, and the actual distribution of products of such a reaction may be governed not by a position of equilibrium — or by the magnitude of $\Delta H°$ — *but by the rate of the reaction that occurs most rapidly.*

In order to understand the factors that affect the rate of a chemical reaction, let us consider as a hypothetical example the following simple nucleophilic substitution reaction:

$$A:^- + B-C \longrightarrow A-B + :C^-$$

Let us also assume that $A:^-$ and $B-C$ react by colliding with each other.

The rate of the reaction can be determined experimentally by measuring the rate at which $A:^-$ or $B-C$ disappears from the reaction mixture, or by measuring the rate at which $A-B$ or $C:^-$ is formed in the mixture. We can make any of these measurements in the laboratory by simply withdrawing small samples of the mixture at measured intervals of time and analyzing the samples for the concentrations of $A:^-$, $B-C$, $A-B$, or $C:^-$.

We can then analyze our data to find certain relationships. In this instance, we would find that the overall rate of the reaction is proportional to the concentrations of $A:^-$ *and* $B-C$ present in the mixture at any given moment, that is,

$$Rate \propto [A:^-][B-C]$$

This proportionality can be expressed as an equation by the introduction of a proportionality constant, k, called the rate constant:

$$Rate = k[A:^-][B-C]$$

TABLE 5.1 Hypothetical second-order reaction

$$A:^- + B\!\!-\!\!C \longrightarrow A\!\!-\!\!B + :C^-$$

EXPERIMENT NUMBER	INITIAL CONCENTRATION		INITIAL RATE OF FORMATION OF A—B or :C⁻ (mole L⁻¹ s⁻¹)
	A:⁻ (mole L⁻¹)	B—C (mole L⁻¹)	
(1)	1	1	0.01
(2)	2	1	0.02
(3)	1	2	0.02
(4)	2	2	0.04

This reaction rate equation is said to be **second order.*** The kinds of numbers that we might get from a kinetic study of this reaction are shown in Table 5.1.

5.4A Collision Theory of Reaction Rates

Much of what we observe about the rates of reactions is explained by the collision theory of reaction rates. As its name suggests, the basic premise of this theory is that for a chemical reaction to take place the reacting particles must collide. *Not all collisions between molecules are effective, however.* By "effective" we mean that a collision produces products. The rate at which molecules collide is called the **collision frequency.** For a gas at room temperature and 1-atm pressure, the collision frequency is about 10^{31} collisions L⁻¹ s⁻¹. This is an enormous number—so large, in fact, that if all collisions did lead to reaction then all chemical reactions would take place exceedingly rapidly.

Collisions between particles may not be effective for two reasons. First, the colliding molecules may not be properly oriented when the collision takes place. Second, the colliding molecules may not have enough energy to bring about a reaction between them. On the outside, molecules have electron clouds, and molecules, therefore, repel each other. If slowly moving molecules collide with each other, they simply bounce apart because of the repulsive forces between their electron clouds. Only when faster molecules collide will their electron clouds penetrate each other enough to allow new covalent bonds to form between them and to allow existing bonds to break. The total collision frequency is *the number of collisions between* A:⁻ *and* B—C *that occur in each unit volume of the reaction mixture in each second.* To determine the frequency of successful collisions we must multiply the total collision frequency by two other factors: by a factor *that measures the fraction of collisions in*

---

*In general the overall order of a reaction is equal to the sum of the exponents *a* and *b* in the rate equation

$$\text{Rate} = k[A]^a[B]^b$$

If in some other reaction, for example, we found that the

$$\text{Rate} = k[A]^2[B]$$

then we would say that the reaction rate is second order with respect to [A], first order with respect to [B], and third order overall.

which the orientation of the colliding molecules with respect to each other allows a reaction to take place; and a factor that measures the fraction of collisions in which the collision energy is greater than a certain minimum amount, called the energy of activation. Thus the rate of reaction can be expressed in the following way:

$$
\text{Reaction rate} =
\begin{bmatrix}
\text{total number} \\
\text{of collisions} \\
\text{between A}:^- \\
\text{and B—C} \\
\text{per liter per} \\
\text{second}
\end{bmatrix}
\times
\begin{bmatrix}
\text{fraction of} \\
\text{collisions} \\
\text{with the} \\
\text{correct ori-} \\
\text{entation}
\end{bmatrix}
\times
\begin{bmatrix}
\text{fraction of} \\
\text{collisions} \\
\text{with suffi-} \\
\text{cient energy} \\
\text{to allow a} \\
\text{reaction}
\end{bmatrix}
$$

5.4B The Collision Frequency

Many experiments show that the rates of chemical reactions are directly proportional to the collision frequency—the greater the collision frequency, the faster the reaction. Two major factors that determine the magnitude of the collision frequency are concentration and temperature.

The more concentrated the reacting molecules are the greater will be the number of collisions occurring in each unit volume of the mixture each second. A simple analogy will help make this clear. Consider, for example, a room in which blindfolded people walk about randomly. It is easy to see that if the room is crowded (a high concentration of people), people will bump into each other more often (have a higher collision frequency) than they will if only a few are present.

The higher the temperature, the faster the molecules move and, as a result, the greater are the number of collisions in a unit volume per unit of time. In our analogy, this would correspond to having our blindfolded people run rather than walk.

5.4C The Orientation or Probability Factor

A great deal of experimental evidence indicates that collisions between molecules must occur in a particular way in order to be effective. While the orientation factor for a particular reaction is difficult to estimate, it is easy to see why it should exist. Consider again the reaction of $A:^-$ with a molecule of B—C to produce A—B and $C:^-$.

There are, naturally, an infinite number of ways that collisions between $A:^-$ and molecules of B—C could occur. Some of these possibilities are shown in Fig. 5.1. Because the reaction that we are considering is one in which $A:^-$ forms a bond to B, it is reasonable to assume that in order for the collision to be effective, $A:^-$ must collide with the B end of the B—C molecule. Collisions in which the $A:^-$ collides with the C end of the B—C molecule are not likely to be effective.

5.4D The Energy Factor
and the Energy of Activation

Not all collisions produce a chemical reaction even when the colliding particles have the proper orientation. A collision will lead to a product only when the colliding particles bring to the collision a certain minimum amount of energy called **the energy of activation** (and abbreviated E_{act}). We know that molecules possess energy called kinetic energy because of their motion through space.

In a properly aligned collision this kinetic energy, if it is large enough, can be used to provide the energy of activation. If the sum of the kinetic energies of the colliding particles is too small, no reaction occurs.

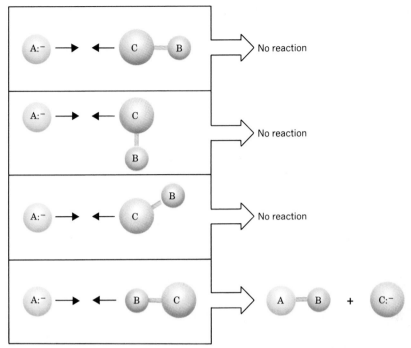

FIGURE 5.1 Some of the possible orientations of collisions between A:$^-$ and molecules of B—C.

5.5 KINETICS OF A NUCLEOPHILIC SUBSTITUTION REACTION. AN S$_N$2 REACTION

Let us now examine an actual nucleophilic substitution reaction: The reaction of methyl chloride with hydroxide ion.

$$CH_3—Cl + OH^- \xrightarrow{H_2O} CH_3—OH + Cl^-$$

When methyl chloride reacts with aqueous sodium hydroxide, experiments have shown that the rate depends on the concentrations of both methyl chloride and hydroxide ion. Doubling the concentration of methyl chloride while keeping the concentration of the hydroxide ion constant, causes the rate of the reaction to *double.* Doubling the concentration of hydroxide ion while keeping the concentration of methyl chloride constant, causes the rate to *double.* And, finally, when *both concentrations* (hydroxide ion and methyl chloride) are doubled, the rate increases *four times.* (In all of these experiments the temperature is the same.) Thus the rate equation for the reaction must be first order with respect to methyl chloride, first order with respect to hydroxide ion, and second order overall.

$$CH_3—Cl + OH^- \xrightarrow{H_2O} CH_3—OH + Cl^-$$

$$Rate \propto [CH_3Cl][OH^-]$$

$$Rate = k[CH_3Cl][OH^-]$$

We can conclude, therefore, that the reaction involves a collision of a methyl chloride molecule and a hydroxide ion, and that the reaction is **bimolecular.** We call this kind of reaction an S_N2 reaction, meaning Substitution, Nucleophilic, bimolecular.

5.6 A MECHANISM FOR THE S_N2 REACTION

A mechanism for the S_N2 reaction — one that is based on ideas proposed in the 1930s by Sir Christopher Ingold* of the University College, London — is outlined here.

Transition state

According to this mechanism the nucleophile approaches the carbon bearing the leaving group from the **backside,** directly opposite the leaving group. The orbital that contains the electron pair of the nucleophile begins to overlap with an empty (antibonding) orbital of the carbon atom bearing the leaving group. As the reaction progresses the bond between the nucleophile and carbon atom grows, and the bond between the carbon atom and the leaving group weakens. As this happens the leaving group is pushed away. The formation of the bond between the nucleophile and the carbon atom provides most of the energy necessary to break the bond between the carbon atom and the leaving group. We can represent this mechanism with methyl chloride and hydroxide ion in the following way:

Transition
state

The Ingold mechanism for the S_N2 reaction involves only one step. There are no intermediates. The reaction proceeds through the formation of an unstable arrangement of the reacting atoms called the **transition state.**

Transition
state
(not an intermediate)

The transition state is one in which both the nucleophile and the leaving group are partially bonded to the carbon atom undergoing attack. Because this transition state involves both the nucleophile (e.g., a hydroxide ion) and the substrate (e.g., a molecule of methyl chloride), this mechanism accounts for the second-order reaction kinetics that we observe.

The transition state of a chemical reaction has an extremely brief existence. It lasts only as long as the time required for one vibration of a molecule, about 10^{-12} s.

---

*Ingold and his co-workers were the pioneers in this field. Their work provided the foundation on which our understanding of nucleophilic substitution and elimination is built.

The energy and structure of the transition state are highly important aspects of any chemical reaction. We shall, therefore, examine this subject further in the next section.

5.7 TRANSITION STATE THEORY. POTENTIAL ENERGY DIAGRAMS

The reaction between methyl chloride and hydroxide ion is *exothermic;* when 1 mole of each reactant in aqueous solution is converted to the products, about 18 kcal of heat is evolved.

$$CH_3—Cl_{(H_2O)} + OH^-_{(H_2O)} \longrightarrow CH_3—OH_{(H_2O)} + Cl^-_{(H_2O)} \quad \Delta H° = -18 \text{ kcal/mole}$$

This means that the enthalpy of the products in their solvated state is **lower** than that of the reactants in their solvated state. In energy terms, we can say, therefore, that the reaction has gone **downhill.**

Considerable experimental evidence exists that shows that all reactions *in which covalent bonds are broken* **must go up an energy hill first,** before they can go downhill. This will be true even though the reaction is exothermic overall.

We can represent this graphically by plotting the potential energy of reacting particles against what is called the **reaction coordinate.** Such a graph is given in Fig. 5.2. We have chosen as our example a generalized S_N2 reaction.

The reaction coordinate is a quantity that measures the **progress of the reaction.** In this instance the B—C distance could be used as the reaction coordinate because as the reaction progresses the B—C distance becomes longer.

In our illustration (Fig. 5.2), we can see that **an energy barrier** exists between reactants and products. It is, in a sense, an energy hill that the reacting species must traverse in order to become products. The height of this barrier (in kilocalories per mole) above the level of reactants is called **the energy of activation.**

The top of the energy hill corresponds to the energy of the **transition state.**

The difference in potential energy between the reactants and the transition state is the energy of activation, E_{act}. The difference in potential energy between the reactants and products is the heat of reaction, $\Delta H°$. For our example the energy level of the products is lower than that of the reactants. In terms of our analogy, we can say that the reactants on one energy plateau must traverse an energy hill (the energy of activation) in order to arrive at the lower energy plateau of products.

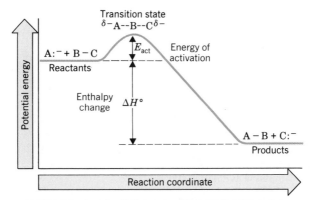

FIGURE 5.2 A potential energy diagram for the exothermic S_N2 reaction of A:$^-$ with B—C.

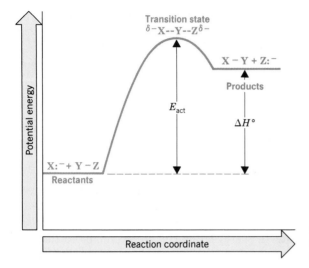

FIGURE 5.3 A potential energy diagram for the endothermic S_N2 reaction of $X:^-$ with $Y-Z$.

If a reaction in which covalent bonds are broken is endothermic overall (Fig. 5.3), there will still be an energy of activation. That is, if the products have greater potential energy than reactants, the transition state will have a potential energy even higher. In other words, in the **uphill** (endothermic) reaction an even larger energy hill intervenes between the reactants on the lower plateau and the products on the higher one.

When a three-dimensional plot of potential energy versus the reaction coordinate is made, the transition state is found to resemble a mountain pass or *col* (Fig. 5.4) rather than the top of an energy hill as we have shown previously in our two-di-

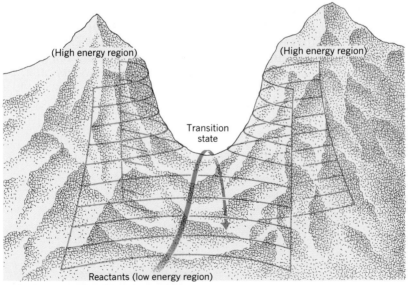

FIGURE 5.4 Mountain pass or col analogy for the transition state. (Adapted with permission from J. E. Leffler and E. Grunwald, *Rates and Equilibria of Organic Reactions*, Wiley, New York, 1963, p. 6.)

mensional plot. That is, the reactants and products appear to be separated by an energy barrier resembling a mountain range. While an infinite number of possible routes lead from reactants to products, the transition state lies at the top of the route that requires the lowest (energy) climb. Whether or not the pass is a wide or narrow one depends on the orientation or probability factor. A wide pass means that there is a relatively large probability that collisions will occur with an orientation that allows a reaction to take place. A narrow pass means just the opposite.

5.7A Reaction Rates: An Explanation of the Effect of Temperature Changes

The existence of an energy of activation explains why most chemical reactions occur much more rapidly at higher temperatures. The increase in reaction rate with temperature is usually much larger than can be explained by a simple increase in collision frequency. *For many reactions a 10° increase in the temperature will cause the reaction rate to double.*

This dramatic increase in reaction rate results from a large increase in the number of collisions between molecules that have enough energy to surmount the energy barrier (E_{act}) at the higher temperature. The kinetic energies of molecules at a given temperature are not all the same (Fig. 5.5). A few molecules will have low energies, a few will have very large energies, and most will have kinetic energies near the average. Increasing the temperature by only a small amount will cause a large increase in the number of molecules with large kinetic energies. Consequently, a modest temperature increase will produce a large increase in the number of collisions with energy sufficient to lead to reaction.

Let us designate a particular *minimal* collision energy (E_{act}) that is required to bring about a reaction between the colliding molecules. Then the number of collisions having sufficient energy to cause a reaction is proportional to the area under that portion of the curve that represents collision energies greater than or equal to E_{act}. If, at the same time, we also examine the distribution of collision energies at two different temperatures, T_1 and T_2, where T_2 is a higher temperature than T_1, we get the result shown in Fig. 5.6.

Because of the shapes of the curves, and because the curve for the higher temperature is shifted to higher kinetic energies, the number of collisions with energies great enough to cause a reaction, that is, with energies greater than E_{act}, *is very much larger at the higher temperature.* The area under curve T_1 with energy greater than E_{act} is quite small, but that under curve T_2 with energy greater than E_{act} is quite large. It is this difference that accounts, primarily, for the very dramatic increase in reaction rate that can occur with a very modest increase in temperature.

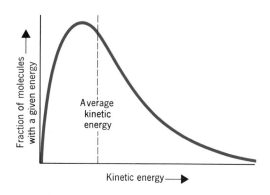

FIGURE 5.5 A plot showing the distribution of kinetic energies among molecules in the gas phase.

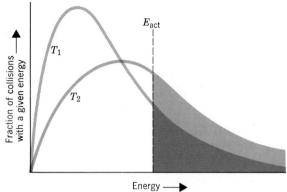

FIGURE 5.6 The distribution of collision energies at two different temperatures, T_1 and T_2 ($T_2 > T_1$). The number of collisions with energies greater than the activation energy is indicated by the appropriately shaded area under each curve.

5.7B Reaction Rates: The Effect of the Energy of Activation

The energy factor accounts for another important observation: *There is a relation between the reaction rate and the magnitude of the energy of activation for different reactions occurring at the same temperature.* ***A reaction with a lower energy of activation will occur very much faster than one with a higher energy of activation.*** Because of the way the energies of molecules are distributed, there will be many more collisions with enough energy to surmount a lower energy hill than a higher one.

Generally speaking, if a reaction has an energy of activation less than 20 kcal/mole, it will take place at room temperature or below. If the energy of activation is greater than 20 kcal/mole, heating will be required to cause the reaction to take place at a reasonable rate.

A potential energy diagram for the reaction of methyl chloride with hydroxide ion is shown in Fig. 5.7. The energy of activation is about 26 kcal/mole, which means that the reaction will be essentially complete in a matter of several hours if the reaction is carried out at 50 °C.

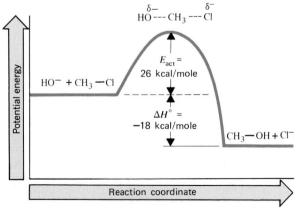

FIGURE 5.7 A potential energy diagram for the reaction of methyl chloride with hydroxide ion.

5.8 THE STEREOCHEMISTRY OF S$_N$2 REACTIONS

As we learned earlier (Section 5.6), in an S$_N$2 reaction *the nucleophile must attack from the backside, that is, from the side directly opposite the leaving group.* This mode of attack (see following figure) causes **a change in the configuration** of the carbon atom that is the object of nucleophilic attack. (The configuration of an atom *is the particular arrangement of groups around that atom in space,* Section 4.6C) As the displacement takes place, the configuration of the carbon atom under attack **inverts** — it is turned inside out in much the same way that an umbrella is turned inside out, or inverts, when caught in a strong wind.

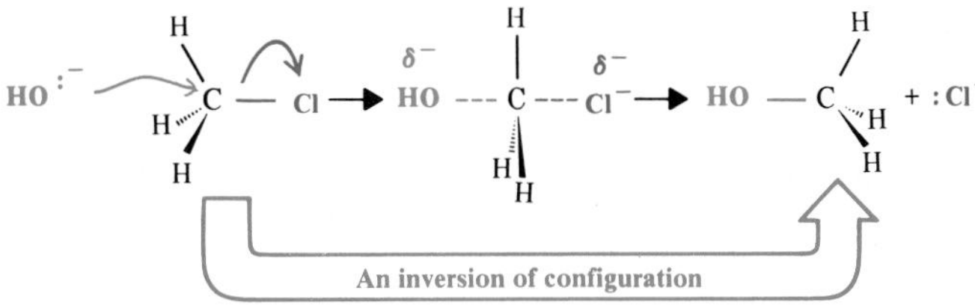

An inversion of configuration

With a molecule like methyl chloride, however, there is no way to prove that attack by the nucleophile inverts the configuration of the carbon atom because one form of methyl chloride is identical to its inverted form. With a cyclic molecule like *cis*-1-chloro-3-methylcyclopentane, however, we can observe the results of a *configuration inversion.* When *cis*-1-chloro-3-methylcyclopentane reacts with hydroxide ion in an S$_N$2 reaction the product is *trans*-3-methylcyclopentanol. *The hydroxide ion ends up being bonded on the opposite side of the ring from the chloride it replaces:*

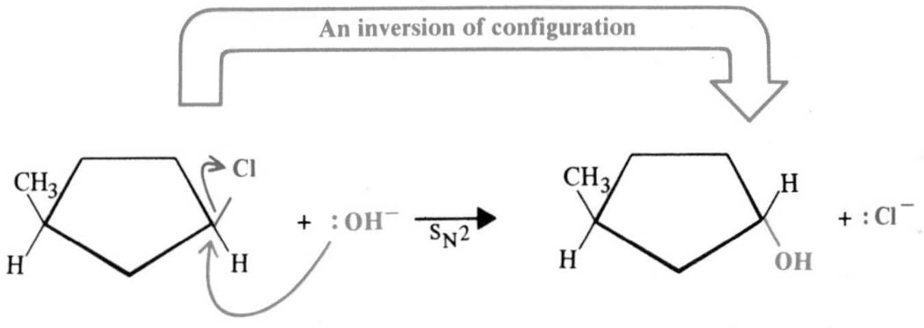

An inversion of configuration

cis — 1 — Chloro — 3 —
methylcyclopentane

trans — 3 — Methylcyclopentanol

Presumably, the transition state for this reaction is like that shown here.

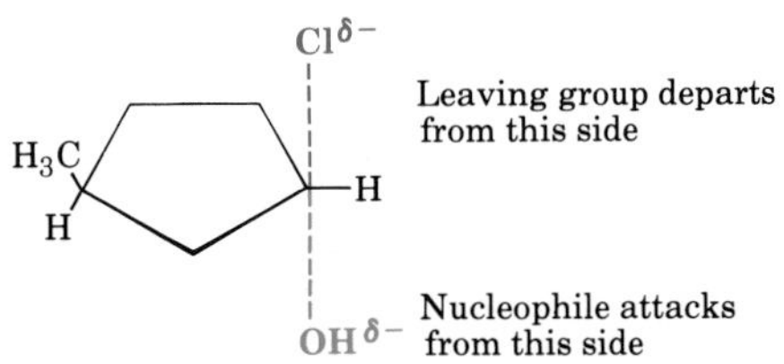

Leaving group departs
from this side

Nucleophile attacks
from this side

PROBLEM 5.2

What product would result from the reaction just given, if attack by the hydroxide ion had occurred from the same side as the leaving group—that is, what product would have been formed if retention of configuration had taken place?

We can also observe an inversion of configuration with an acyclic molecule *when the S_N2 reaction takes place at a stereocenter.* Here, too, we find that S_N2 *reactions always lead to inversion of configuration.*

A compound that contains one stereocenter and, therefore, exists as a pair of enantiomers is 2-bromooctane. These enantiomers have been obtained separately and are known to have the configurations and rotations shown here.

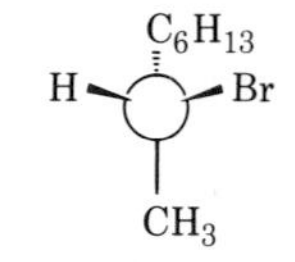

C_6H_{13}

H ⟋⟍ Br

CH_3

(R)-(−)-2-Bromooctane
$[\alpha] = -34.25°$

C_6H_{13}

Br ⟍⟋ H

CH_3

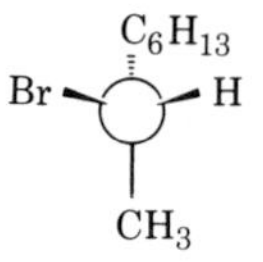

(S)-(+)-2-Bromooctane
$[\alpha] = +34.25°$

The alcohol 2-octanol is also chiral. The configurations and rotations of the 2-octanol enantiomers have also been determined:

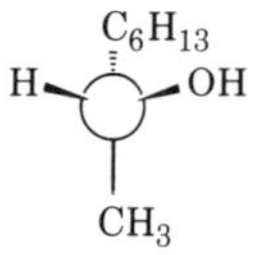

C_6H_{13}

H ⟋⟍ OH

CH_3

(R)-(−)-2-Octanol
$[\alpha] = -9.90°$

C_6H_{13}

HO ⟍⟋ H

CH_3

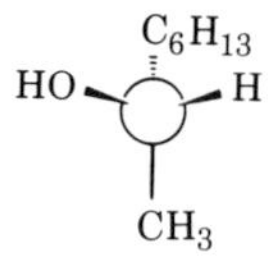

(S)-(+)-2-Octanol
$[\alpha] = +9.90°$

When (*R*)-(−)-2-bromooctane reacts with sodium hydroxide, the substitution product that is obtained from the reaction is only (*S*)-(+)-2-octanol. The following reaction is S_N2 and takes place with *complete inversion of configuration.*

An inversion of configuration

C_6H_{13}

H ⟋⟍ Br

CH_3

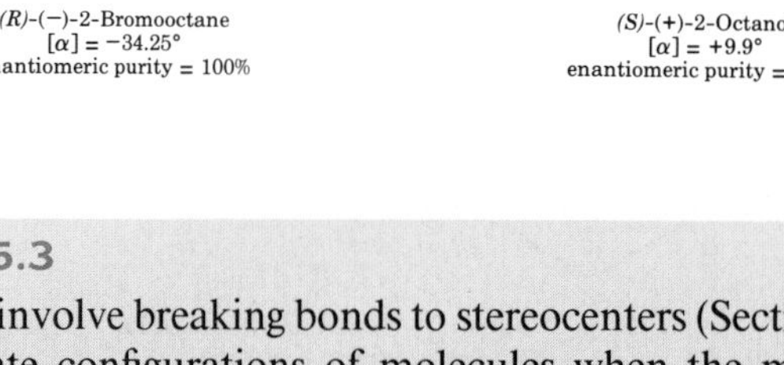

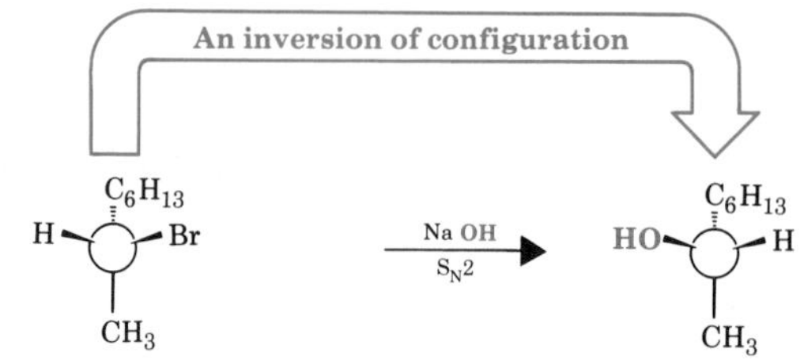

(R)-(−)-2-Bromooctane
$[\alpha] = -34.25°$
enantiomeric purity = 100%

$\xrightarrow[\text{$S_N2$}]{\text{Na OH}}$

C_6H_{13}

HO ⟍⟋ H

CH_3

(S)-(+)-2-Octanol
$[\alpha] = +9.9°$
enantiomeric purity = 100%

PROBLEM 5.3

Reactions that involve breaking bonds to stereocenters (Section 4.13) can also be used to relate configurations of molecules when the mechanism of the reaction *and its stereochemistry* are known. (a) Illustrate how this is true by assigning configurations to the 2-chlorobutane enantiomers based on the following data. [The configuration of (−)-2-butanol is given in Section 4.6C.]

$$\text{(+)-2-Chlorobutane} \xrightarrow[\text{S_N2}]{\text{OH}^-} \text{(−)-2-Butanol}$$

$[\alpha]_D^{25} = +36.00°$ $[\alpha]_D^{25} = -13.52°$

enantiomerically pure   enantiomerically pure

(b) When optically pure (+)-2-chlorobutane is allowed to react with potassium iodide in acetone in an S_N2 reaction, the 2-iodobutane that is produced has a minus rotation. What is the configuration of (−)-2-iodobutane? Of (+)-2-iodo-butane?

5.9 THE REACTION OF *TERT*-BUTYL CHLORIDE WITH HYDROXIDE ION: AN S_N1 REACTION

When *tert*-butyl chloride reacts with sodium hydroxide in a mixture of water and acetone, the kinetic results are quite different. The rate of formation of *tert*-butyl alcohol is dependent on the concentration of *tert*-butyl chloride, but it is *independent of the concentration of hydroxide ion.* Doubling the *tert*-butyl chloride concentration *doubles* the rate of the reaction, but changing the hydroxide ion concentration (within limits) has no appreciable effect. *tert*-Butyl chloride reacts by substitution at virtually the same rate in pure water (where the hydroxide ion is $10^{-7}\ M$) as it does in 0.05 M aqueous sodium hydroxide (where the hydroxide ion concentration is 500,000 times larger). (We shall see in Section 5.10 that the important nucleophile in this reaction is a molecule of water.)

Thus the rate equation for this substitution reaction is first order with respect to *tert*-butyl chloride and *first order overall.*

$$(CH_3)_3C\!-\!Cl + OH^- \xrightarrow[\text{H}_2\text{O}]{\text{acetone}} (CH_3)_3C\!-\!OH + Cl^-$$

$$\text{Rate} \propto [(CH_3)_3CCl]$$

$$\text{Rate} = k[(CH_3)_3CCl]$$

We can conclude, therefore, that hydroxide ions do not participate in the transition state of the step that controls the rate of the reaction, and that only molecules of *tert*-butyl chloride are involved. This reaction is **unimolecular.** We call this type of reaction an S_N1 reaction (Substitution, Nucleophilic, unimolecular).

How can we explain an S_N1 reaction in terms of a mechanism? To do so we shall need to consider the possibility that the mechanism involves more than one step. But what kind of kinetic results should we expect from a multistep reaction? Let us consider this point further.

5.9A Multistep Reactions and the Rate-Determining Step

If a reaction takes place in a series of steps, and if the first step is intrinsically slower than all the others, then the rate of the overall reaction will be essentially the same as the rate of this slow step. This slow step, consequently, is called the **rate-limiting step** or the **rate-determining step.**

Consider a multistep reaction such as the following:

Step 1   Reactant $\xrightarrow{\text{slow}}$ intermediate-1

Step 2   Intermediate-1 $\xrightarrow{\text{fast}}$ intermediate-2

Step 3   Intermediate-2 $\xrightarrow{\text{fast}}$ product

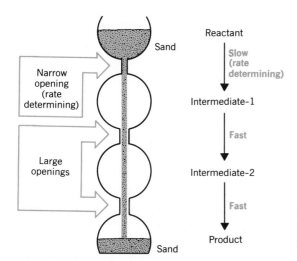

FIGURE 5.8 A modified hourglass that serves as an analogy for a multistep reaction. The overall rate is limited by the rate of the slow step.

When we say that the first step is intrinsically slow, we mean that the rate constant for step 1 is very much smaller than the rate constant for step 2 or for step 3:

Step 1	Rate $= k_1$ [Reactant]
Step 2	Rate $= k_2$ [Intermediate-1]
Step 3	Rate $= k_3$ [Intermediate-2]

$$k_1 \ll k_2 \quad \text{or} \quad k_3$$

When we say that steps 2 and 3 are *fast,* we mean that because their rate constants are larger, they could (in theory) take place rapidly if the concentrations of the two intermediates ever became high. In actuality, the concentrations of the intermediates are always very small because of the slowness of step 1, and steps 2 and 3 actually occur at the same rate as step 1.

An analogy may help clarify this. Imagine an hourglass modified in the way shown in Fig. 5.8. The opening between the top chamber and the one just below is considerably smaller than the other two. The overall rate at which sand falls from the top to the bottom of the hourglass is limited by the rate at which sand passes through this small orifice. This step, in the passage of sand, is analogous to the rate-determining step of the multistep reaction.

5.10 A MECHANISM FOR THE S$_N$1 REACTION

The mechanism for the reaction of *tert*-butyl chloride (Section 5.9) apparently involves three steps. Two distinct **intermediates** are formed. The first step is the slow step—it is the rate-determining step. In it a molecule of *tert*-butyl chloride ionizes and becomes a *tert*-butyl cation and a chloride ion. Carbocation formation in general takes place slowly because it is usually a highly endothermic (uphill) process.

$$\text{Step 1} \qquad \underset{\underset{\text{CH}_3}{|}}{\overset{\overset{\text{CH}_3}{|}}{\text{CH}_3-\text{C}-\text{Cl}}} \xrightarrow[\text{(rate-determining step)}]{\text{slow}} \underset{\underset{\text{CH}_3}{|}}{\overset{\overset{\text{CH}_3}{|}}{\text{CH}_3-\text{C}^+}} + \ :\text{Cl}^-$$

The next two steps are the following:

Step 2 $CH_3-\overset{\overset{\displaystyle CH_3}{|}}{\underset{\underset{\displaystyle CH_3}{|}}{C}}{}^+ + :OH_2 \xrightarrow{\text{fast}} CH_3-\overset{\overset{\displaystyle CH_3}{|}}{\underset{\underset{\displaystyle CH_3}{|}}{C}}-\overset{+}{O}H_2$

Step 3 $CH_3-\overset{\overset{\displaystyle CH_3}{|}}{\underset{\underset{\displaystyle CH_3}{|}}{C}}-OH_2{}^+ + H_2\overset{\cdot\cdot}{O} \xrightarrow{\text{fast}} CH_3-\overset{\overset{\displaystyle CH_3}{|}}{\underset{\underset{\displaystyle CH_3}{|}}{C}}-OH + H_3O^+$

In the second step the intermediate *tert*-butyl cation reacts rapidly with water to produce a *tert*-butyloxonium ion (another intermediate) which, in the third step, rapidly transfers a proton to a molecule of water producing *tert*-butyl alcohol.

The first step requires heterolytic cleavage of the carbon–chlorine bond. Because no other bonds are formed in this step, it should be highly endothermic and it should have a high energy of activation. That it takes place at all is largely because of the ionizing ability of the solvent, water. Experiments indicate that in the gas phase (i.e., in the absence of a solvent), the energy of activation is 151 kcal/mole! In aqueous solution, however, the energy of activation is much lower—about 20 kcal/mole. Water molecules surround and stabilize the cation and anion that are produced (cf. Section 2.17E).

Even though the *tert*-butyl cation produced in step 1 is stabilized by solvation, it is still a highly reactive species. Almost immediately after it is formed, it reacts with one of the surrounding water molecules to form the *tert*-butyloxonium ion, $(CH_3)_3COH_2{}^+$. (It may also occasionally react with a hydroxide ion, but water molecules are far more plentiful.)

A potential energy diagram for the S$_N$1 reaction of *tert*-butyl chloride and water is given in Fig. 5.9.

The important transition state for the S$_N$1 reaction is the transition state of the rate-determining step [TS(1)]. In it the carbon–chlorine bond of *tert*-butyl chloride is

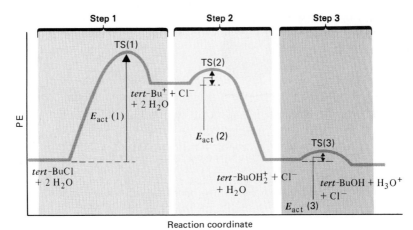

Reaction coordinate

FIGURE 5.9 A potential energy diagram for the S$_N$1 reaction of $(CH_3)_3C$—Cl (*tert*-Bu—Cl). The energy of activation of the first step, $E_{act(1)}$, is much larger than $E_{act(2)}$ or $E_{act(3)}$. TS(1) stands for transition state (1), and so on.

largely broken and ions are beginning to develop:

$$CH_3 - \overset{\displaystyle CH_3}{\underset{\displaystyle CH_3}{\overset{|}{\underset{|}{C}}}}{}^{\delta+} \text{---} Cl^{\delta-}$$

The solvent (water) stabilizes these developing ions by solvation.

5.11 CARBOCATIONS

Beginning in the 1920s much evidence began to accumulate implicating simple alkyl cations as intermediates in a variety of ionic reactions. However, because alkyl cations are highly unstable and highly reactive, they were in all instances studied prior to 1962 very short lived, transient species that could not be observed directly.* However, in 1962 George A. Olah (now at the University of Southern California) and his co-workers published the first of a series of papers describing experiments in which alkyl cations were prepared in an environment in which they were reasonably stable and in which they could be observed by a number of spectroscopic techniques. We shall see the results of some of these spectroscopic studies in Chapter 13.

5.11A The Structure of Carbocations

Considerable experimental evidence indicates that the structure of carbocations is **trigonal planar** like that of BF_3 (Section 1.15). Just as the trigonal planar structure of BF_3 can be accounted for on the basis of sp^2 hybridization so, too (Fig. 5.10), can the trigonal planar structure of carbocations.

The central carbon atom in a carbocation is electron deficient; it has only six electrons in its outside energy level. In our model (Fig. 5.10) these six electrons are used to form sigma covalent bonds to hydrogen atoms (or to alkyl groups). The p orbital contains no electrons.

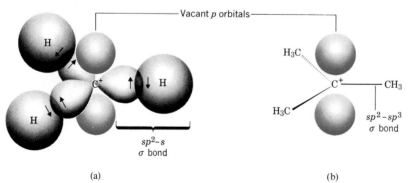

(a)   (b)

FIGURE 5.10 (a) Orbital structure of the methyl cation. The bonds are sigma bonds (σ) formed by overlap of the carbon atom's three sp^2 orbitals with the $1s$ orbitals of the hydrogen atoms. The p orbital is vacant. (b) A dash-line-wedge representation of the *tert*-butyl cation. The bonds between carbon atoms are formed by overlap of sp^3 orbitals of the methyl groups with sp^2 orbitals of the central carbon atom.

---

*As we shall learn later, carbocations bearing aromatic groups can be much more stable; one of these had been studied as early as 1901.

5.11B The Relative Stabilities of Carbocations

A large body of experimental evidence indicates that the relative stabilities of carbocations are related to the number of alkyl groups attached to the positively charged trivalent carbon atom. Tertiary carbocations are the most stable, and the methyl cation is the least stable. The overall order of stability is as follows:

$$R\overset{R}{\underset{R}{-\overset{|}{\underset{|}{C^+}}}} > R\overset{R}{\underset{H}{-\overset{|}{\underset{|}{C^+}}}} > R\overset{H}{\underset{H}{-\overset{|}{\underset{|}{C^+}}}} > H\overset{H}{\underset{H}{-\overset{|}{\underset{|}{C^+}}}}$$

Most stable			Least stable
3°	> 2°	> 1°	> methyl

    This order of stability of carbocations can be explained on the basis of a law of physics that states that *a charged system is stabilized when the charge is dispersed or delocalized.* Alkyl groups, when compared to hydrogen atoms, are **electron releasing.** This means that alkyl groups will shift electron density toward a positive charge. Through electron release, *alkyl groups* attached to the positive carbon atom of a carbocation **delocalize** the positive charge. In doing so, the attached alkyl groups assume part of the positive charge themselves and thus *stabilize* the carbocation. We can see how this occurs by inspecting Fig. 5.11.

    In the *tert*-butyl cation (see the following structure) three electron-releasing methyl groups surround the central carbon atom and assist in delocalizing the positive charge. In the isopropyl cation there are only two attached methyl groups that can serve to delocalize the charge. In the ethyl cation there is only one attached methyl group, and in the methyl cation there is none at all. As a result, *the delocalization of charge and the order of stability of the carbocations parallel the number of attached methyl groups.*

$$\delta^+CH_3 \rightarrow \overset{\overset{\delta^+}{CH_3}}{\underset{\underset{\delta^+}{CH_3}}{C^{\delta^+}}} \quad \text{is more stable than} \quad \delta^+CH_3 \rightarrow \overset{\overset{\delta^+}{CH_3}}{\underset{H}{C^{\delta^+}}} \quad \text{is more stable than} \quad \delta^+CH_3 \rightarrow \overset{H}{\underset{H}{C^{\delta^+}}} \quad \text{is more stable than} \quad H-\overset{H}{\underset{H}{C^+}}$$

tert-Butyl cation (3°)	Isopropyl cation (2°)	Ethyl cation (1°)	Methyl cation
most stable			least stable

The relative stability of carbocations is 3° > 2° > 1° > methyl

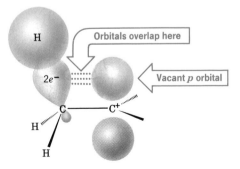

FIGURE 5.11 *How a methyl group helps stabilize the positive charge of a carbocation.* Electron density from one of the carbon–hydrogen sigma bonds of the methyl group flows into the vacant *p* orbital of the carbocation because the orbitals can partly overlap. Shifting electron density in this way makes the *sp²*-hybridized carbon of the carbocation somewhat less positive and the hydrogens of the methyl group assumes some of the positive charge. Delocalization (dispersal) of the charge in this way leads to greater stability.

5.12 THE STEREOCHEMISTRY OF S$_N$1 REACTIONS

Because the carbocation formed in the first step of an S$_N$1 reaction has a trigonal planar structure (Section 5.11A) when it reacts with a nucleophile, it may do so from either the frontside or backside (see following figure). With the *tert*-butyl cation this makes no difference because the same product is formed by either mode of attack.

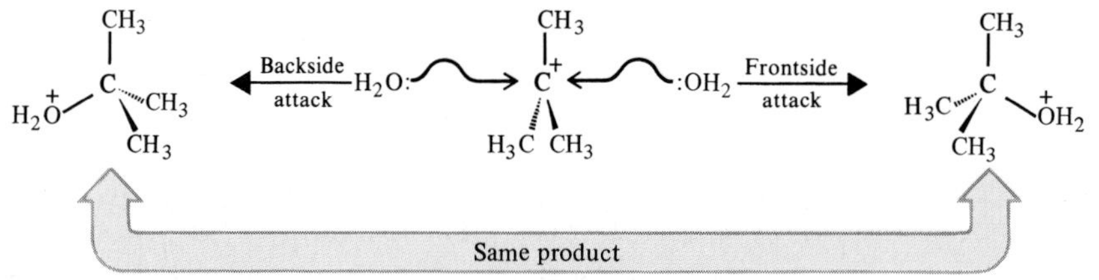

With some cations, however, different products arise from the two reaction possibilities. We shall study this point further next.

5.12A Reactions that Involve Racemization

A reaction that transforms an optically active compound into a racemic form is said to proceed with **racemization.** If the original compound loses all of its optical activity in the course of the reaction, chemists describe the reaction as having taken place with *complete* racemization. If the original compound loses only part of its optical activity, as would be the case if an enantiomer were only partially converted to a racemic form, then chemists describe this as proceeding with *partial* racemization.

Racemization will take place *whenever the reaction causes chiral molecules to be converted to an achiral intermediate.*

Examples of this type of reaction are S$_N$1 reactions in which the leaving group departs from a stereocenter. These reactions almost always result in extensive

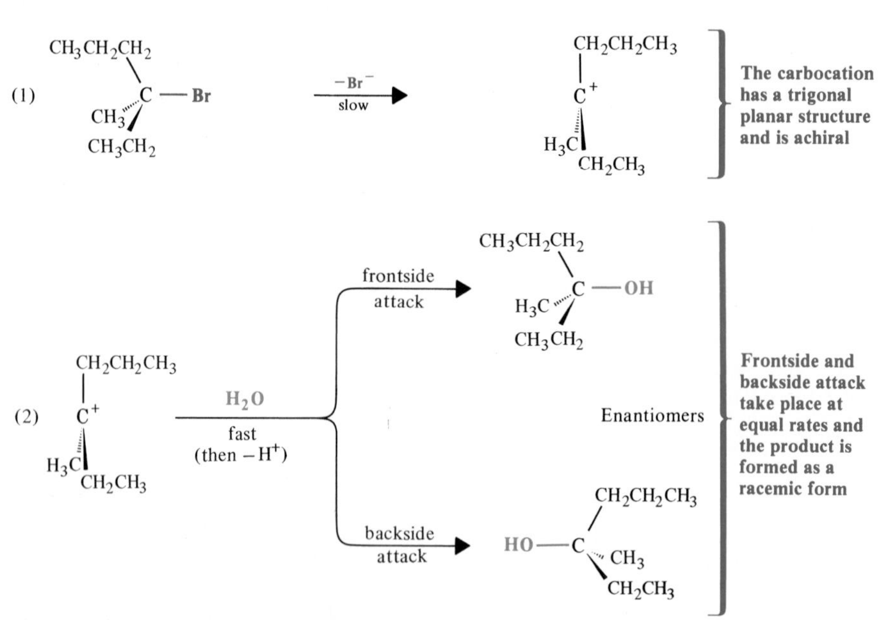

FIGURE 5.12 The S$_N$1 reaction of 3-bromo-3-methylhexane proceeds with racemization because the intermediate carbocation is achiral.

and sometimes complete racemization. For example, heating optically active (S)-3-bromo-3-methylhexane with aqueous acetone results in the formation of 3-methyl-3-hexanol as a racemic form.

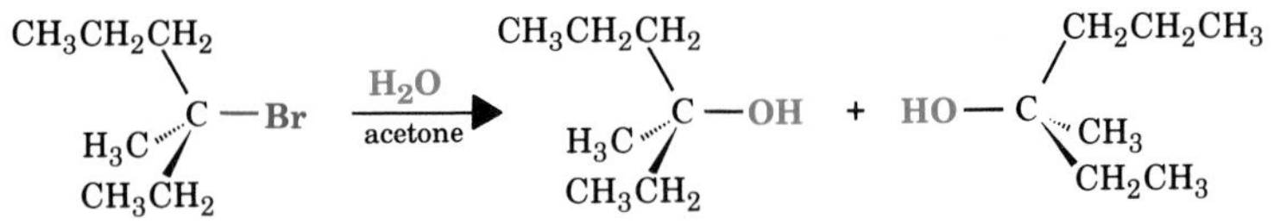

(S)-3-Bromo-3-methylhexane (optically active)	(S)-3-Methyl-3-hexanol	(R)-3-Methyl-3-hexanol
	(optically inactive, a racemic form)	

The reason: The S$_N$1 reaction proceeds through the formation of an intermediate carbocation (Fig. 5.12) and the carbocation, because of its trigonal planar configuration, *is achiral.* It reacts with water with equal ease from either side to form the enantiomers of 3-methyl-3-hexanol in equal amounts.

PROBLEM 5.4

Keeping in mind that carbocations have a trigonal planar structure, (a) write a structure for the carbocation intermediate and (b) write structures for the alcohol (or alcohols) you would expect from the following reaction:

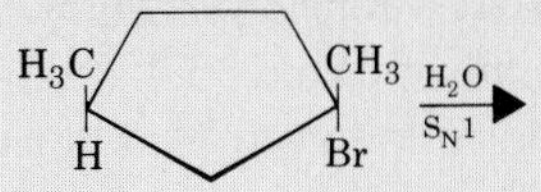

5.12B Solvolysis

The S$_N$1 reaction of an alkyl halide with water is an example of **solvolysis.** A solvolysis is a nucleophilic substitution in which *the nucleophile is a molecule of the solvent* (*solvent + lysis:* cleavage by the solvent). Since the solvent in this instance is water, we could also call the reaction a **hydrolysis.** If the reaction had taken place in ethanol we would call the reaction an **ethanolysis.**

EXAMPLES OF SOLVOLYSIS

$$(CH_3)_3C—Br + H_2O \longrightarrow (CH_3)_3C—OH + HBr$$

$$(CH_3)_3C—Cl + CH_3OH \longrightarrow (CH_3)_3C—OCH_3 + HCl$$

$$(CH_3)_3C—Cl + H\overset{\text{O}}{\overset{\|}{C}}OH \longrightarrow (CH_3)_3C—O\overset{\text{O}}{\overset{\|}{C}}H + HCl$$

These reactions all involve the initial formation of a carbocation and the subsequent reaction of that cation with a molecule of the solvent. In the last example the solvent is formic acid (HCO_2H) and the following steps take place:

Step 1 $$(CH_3)_3C—Cl \xrightarrow{\text{slow}} (CH_3)_3C^+ + Cl^-$$

Step 2 $(CH_3)_3C^+ + \overset{..}{H}\overset{..}{O}-\overset{\overset{\displaystyle O}{\|}}{C}H \xrightarrow{\text{fast}} (CH_3)_3C-\overset{+}{\underset{\underset{\displaystyle H}{|}}{O}}-\overset{\overset{\displaystyle O}{\|}}{C}H$

Step 3 $(CH_3)_3C-\overset{+}{\underset{\underset{\displaystyle H}{|}}{O}}-\overset{\overset{\displaystyle O}{\|}}{C}H \xrightarrow{\text{fast}} (CH_3)_3C-O-\overset{\overset{\displaystyle O}{\|}}{C}H + H^+$ ■

PROBLEM 5.5

(a) What product would be obtained from the ethanolysis of *tert*-butyl chloride? (b) Outline the steps of this S_N1 reaction.

5.13 FACTORS AFFECTING THE RATES OF S_N1 AND S_N2 REACTIONS

Now that we have an understanding of the mechanisms of S_N2 and S_N1 reactions, our next task is to explain why methyl chloride reacts by an S_N2 mechanism and *tert*-butyl chloride by an S_N1 mechanism. We would also like to be able to predict which pathway—S_N1 or S_N2—would be followed by the reaction of any alkyl halide with any nucleophile under varying conditions.

The answer to this kind of question is to be found in the *relative rates of the reactions that occur.* If a given alkyl halide and nucleophile react *rapidly* by an S_N2 mechanism but *slowly* by an S_N1 mechanism under a given set of conditions, then an S_N2 pathway will be followed by most of the molecules. On the other hand another alkyl halide and another nucleophile may react very slowly (or not at all) by an S_N2 pathway. If they react rapidly by an S_N1 mechanism, then the reactants will follow an S_N1 pathway.

Experiments have shown that a number of factors affect the relative rates of S_N1 and S_N2 reactions. The most important factors are

1. The structure of the substrate.
2. The concentration and reactivity of the nucleophile (for bimolecular reactions only).
3. The effect of the solvent.
4. The nature of the leaving group.

5.13A The Effect of the Structure of the Substrate

S_N2 Reactions. Simple alkyl halides show the following general order of reactivity in S_N2 reactions:

$$\text{methyl} > \text{primary} > \text{secondary} > (\text{tertiary})$$

Methyl halides react most rapidly and tertiary halides react so slowly as to be unreactive by the S_N2 mechanism. Table 5.2 gives the relative rates of typical S_N2 reactions.

TABLE 5.2 Relative rates of reactions
of alkyl halides in S$_N$2 reactions

SUBSTITUENT	COMPOUND	RELATIVE RATE
Methyl	CH_3—X	30
1°	CH_3CH_2—X	1
2°	$(CH_3)_2CHX$	0.02
Neopentyl	$(CH_3)_3CCH_2X$	0.00001
3°	$(CH_3)_3CX$	~0

Neopentyl halides, even though they are primary halides, are very unreactive.

$$CH_3-\underset{\underset{\displaystyle CH_3}{|}}{\overset{\overset{\displaystyle CH_3}{|}}{C}}-CH_2-X$$

A neopentyl halide

The important factor behind this order of reactivity is a **steric effect.** A steric effect is an effect on relative rates caused by the space-filling properties of those parts of a molecule attached at or near the reacting site. One kind of steric effect — the kind that is important here — is called **steric hindrance.** By this we mean that the spatial arrangement of the atoms or groups at or near the reacting site of a molecule hinders or retards a reaction.

For particles (molecules and ions) to react, their reactive centers must be able to come within bonding distance of each other. Although most molecules are reasonably flexible, very large and bulky groups can often hinder the formation of the required transition state. In some cases they can prevent its formation altogether.

An S$_N$2 reaction requires an approach by the nucleophile to a distance within bonding range of the carbon atom bearing the leaving group. Because of this, bulky substituents on *or near* that carbon atom have a dramatic inhibiting effect (Fig. 5.13). Of the simple alkyl halides, methyl halides react most rapidly in S$_N$2 reactions because only three small hydrogen atoms interfere with the approaching nucleophile. Neopentyl and tertiary halides are the least reactive because bulky groups present a

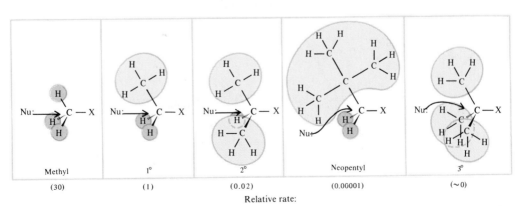

FIGURE 5.13 Steric effects in the S$_N$2 reaction.

strong hindrance to the approaching nucleophile. (Tertiary substrates, for all practical purposes, do not react by an S_N2 mechanism.)

S_N1 Reactions. *The primary factor that determines the reactivity of organic substrates in an S_N1 reaction is the relative stability of the carbocation that is formed.*

Except for those reactions that take place in strong acids (which we shall study later), the only organic compounds that undergo reaction by an S_N1 path at a reasonable rate are *those that are capable of forming relatively stable carbocations.* Of the simple alkyl halides that we have studied so far, this means (for all practical purposes) that only tertiary halides react by an S_N1 mechanism. (Later we shall see that certain organic halides, called *allylic halides* and *benzylic halides,* can also react by an S_N1 mechanism because they can form relatively stable carbocations, cf. Section 10.7.)

Tertiary carbocations are stabilized because three alkyl groups release electrons to the positive carbon atom and thereby disperse its charge (see Section 5.11B).

Formation of a relatively stable carbocation is important in an S_N1 reaction because it means that the energy of activation for the slow step (i.e., $R-X \longrightarrow R^+ + X^-$) will be low enough for the overall reaction to take place at a reasonable rate. This step is **endothermic** (see Fig. 5.9 again) and according to a postulate made by G. S. Hammond (then at the California Institute of Technology) and J. E. Leffler (Florida State University) *the transition state of an endothermic step should bear a strong resemblance to the product of that step.* Since the product of this step (actually an intermediate) is a carbocation, any factor that stabilizes it—such as dispersal of the positive charge by electron-releasing groups—will also stabilize the transition state in which the charge is developing. *Stabilization of the transition state lowers its potential energy and, therefore, lowers the energy of activation.*

The Hammond–Leffler Postulate can be better understood through consideration of the potential energy versus reaction coordinate diagrams given in Fig. 5.14.

Highly exothermic and highly endothermic steps are shown in Fig. 5.14 because they illustrate the Hammond–Leffler postulate most dramatically. In the highly exothermic step the energy levels of the reactants and the transition state are close to each other. The transition state also lies close to the reactants along the reaction coordinate. This means that in the highly exothermic step, bond breaking has not proceeded very far when the transition state is reached.

The transition states of highly endothermic steps, on the other hand, lie close to the products on the potential energy coordinate and *along the reaction coordinate.* In

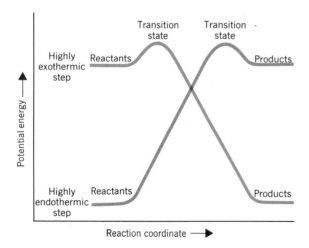

FIGURE 5.14 Energy diagrams for highly exothermic and highly endothermic steps of reactions. (From William A. Pryor, *Introduction to Free Radical Chemistry* © 1966. Reprinted by permission of Prentice-Hall, Inc., Englewood Cliffs, NJ, p. 53.)

highly endothermic steps the bond has broken to a considerable extent by the time the transition state is reached. The step that leads to the formation of a carbocation in an S$_N$1 reaction is highly endothermic, and, therefore, the transition state of this step strongly resembles the carbocation that is formed in it. The R—X bond is almost broken when the transition state is reached, and the carbon atom has acquired a substantial positive charge. Three electron-releasing alkyl groups stabilize this transition state, lowering its energy enough to make the S$_N$1 reaction of a tertiary substrate possible.

$$^{\delta+}CH_3$$
$$\downarrow$$
$$^{\delta+}CH_3 \rightarrow C^{\delta+} \text{------} X^{\delta-}$$
$$\uparrow$$
$$^{\delta+}CH_3$$

Transition state of an S$_N$1 reaction

5.13B The Effect of the Concentration and Strength of the Nucleophile

Since the nucleophile does not participate in the rate-determining step of an S$_N$1 reaction, the rates of S$_N$1 reactions are unaffected by either the concentration or the identity of the nucleophile. The rates of S$_N$2 reactions, however, depend on *both* the concentration *and* the identity of the attacking nucleophile. We saw in Section 5.5 how increasing the concentration of the nucleophile increases the rate of an S$_N$2 reaction. We can now examine how the rate of an S$_N$2 reaction depends on the identity of the nucleophile.

We describe nucleophiles as being *strong* or *weak*. When we do this we are really describing their relative reactivities in S$_N$2 reactions. A strong nucleophile is one that reacts rapidly with a given substrate. A weak nucleophile is one that reacts slowly with the same substrate under the same reaction conditions.

The methoxide ion, for example, is a strong nucleophile. It reacts relatively rapidly with methyl iodide to produce dimethyl ether.

$$CH_3O^- + CH_3I \xrightarrow{\text{rapid}} CH_3OCH_3 + I^-$$

Methanol, on the other hand, is a weak nucleophile. Under the same conditions it reacts very slowly with methyl iodide.

$$CH_3OH + CH_3I \xrightarrow{\text{very slow}} CH_3\overset{+}{\underset{\underset{H}{|}}{O}}CH_3 + I^-$$

The relative strengths of nucleophiles can be correlated with two structural features:

1. **A negatively charged nucleophile is always a stronger nucleophile than its conjugate acid.** Thus OH$^-$ is a stronger nucleophile than H$_2$O and RO$^-$ is stronger than ROH.

2. **In a group of nucleophiles in which the nucleophilic atom is the same, nucleophilicities parallel basicities.** Oxygen compounds, for example, show the following order of reactivity:

$$RO^- > HO^- \gg RCO_2^- > ROH > H_2O$$

This is also their order of basicity. An alkoxide ion (RO^-) is a slightly stronger base than a hydroxide ion (HO^-), a hydroxide ion is a much stronger base than a carboxylate ion (RCO_2^-), and so on.

5.13C Solvent Effects on S_N2 Reactions. Polar Protic and Aprotic Solvents

The relative strengths of nucleophiles do not always parallel their basicities *when the nucleophilic atoms are not the same.* When we examine the relative nucleophilicity of compounds within the same group of the Periodic Table, we find that *in solvents such as alcohols and water* the nucleophile with the larger nucleophilic atom is stronger. Thiols ($R-SH$) are stronger nucleophiles than alcohols (ROH); RS^- ions are stronger than RO^- ions; and the halide ions show the following order:

$$I^- > Br^- > Cl^- > F^-$$

This effect is related to the strength of the interactions between the nucleophile and its surrounding layer of solvent molecules. A molecule of a solvent such as water or an alcohol—called a **protic solvent**—*has a hydrogen atom attached to an atom of a strongly electronegative element (oxygen).* Molecules of protic solvents can, therefore, form hydrogen bonds to nucleophiles in the following way:

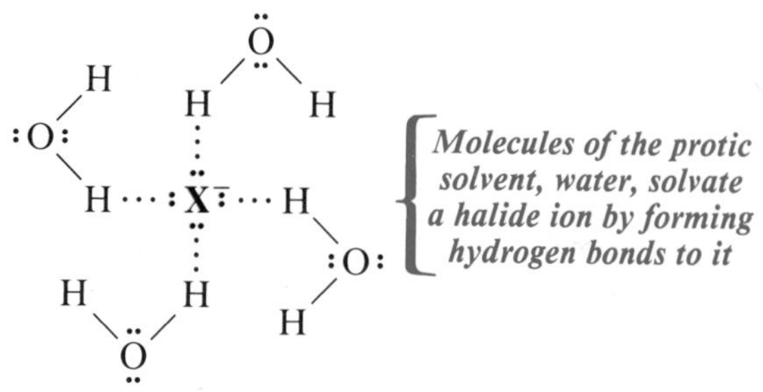

Molecules of the protic solvent, water, solvate a halide ion by forming hydrogen bonds to it

A small nucleophile, such as a fluoride ion, because its charge is more concentrated, is more strongly solvated than a larger one. Hydrogen bonds to a small atom are stronger than those to a large atom. For a nucleophile to react it must shed some of its solvent molecules because it must closely approach the carbon bearing the leaving group. A large ion, because the hydrogen bonds between it and the solvent are weaker, can shed some of its solvent molecules more easily and thus it will be more nucleophilic.

Polar Aprotic Solvents

Aprotic solvents are those solvents whose molecules do not have a hydrogen atom that is attached to an atom of a strongly electronegative element. Most aprotic solvents (benzene, the alkanes, and so on) are relatively nonpolar, and they do not dissolve most ionic compounds. (In Section 8.19 we shall see how they can be induced to do so, however.) In recent years a number of **polar aprotic solvents** have come into wide use by chemists; *they are especially useful in S_N2 reactions.* Several examples *are shown here.*

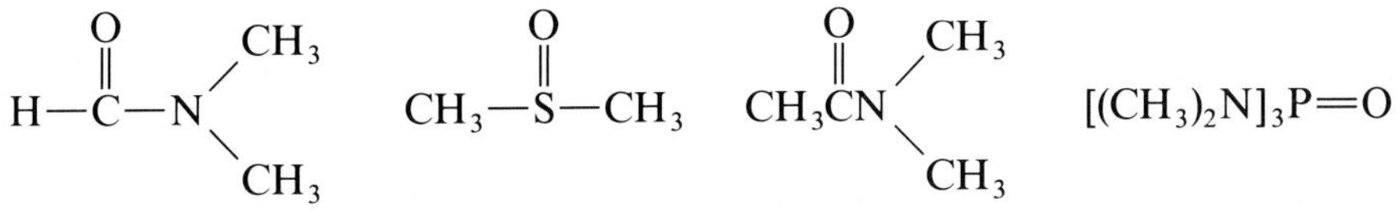

$$
\begin{array}{cccc}
\underset{\substack{\displaystyle \text{O} \\ \parallel}}{\text{H}-\text{C}-\text{N}} \Big\langle \begin{array}{l} \text{CH}_3 \\ \text{CH}_3 \end{array} & \text{CH}_3-\overset{\displaystyle \text{O}}{\overset{\parallel}{\text{S}}}-\text{CH}_3 & \text{CH}_3\overset{\displaystyle \text{O}}{\overset{\parallel}{\text{C}}}\text{N} \Big\langle \begin{array}{l} \text{CH}_3 \\ \text{CH}_3 \end{array} & [(\text{CH}_3)_2\text{N}]_3\text{P}{=}\text{O}
\end{array}
$$

N,N-Dimethylformamide Dimethyl sulfoxide Dimethylacetamide Hexamethylphosphoric
(DMF)   (DMSO)   (DMA)   triamide
                                                                   (HMPT)

All of these solvents (DMF, DMSO, DMA, HMPT) dissolve ionic compounds, and they solvate cations very well. They do so in the same way that protic solvents solvate cations: by orienting their negative ends around the cation and by donating unshared electron pairs to vacant orbitals of the cation:

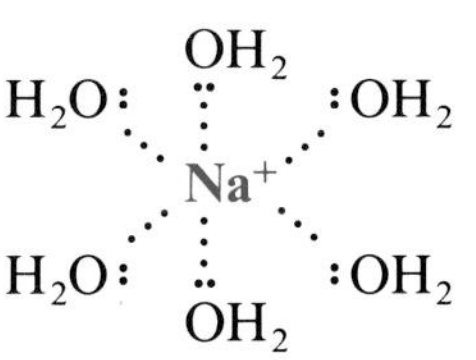

A sodium ion solvated   A sodium ion solvated by
by molecules of the   molecules of the aprotic
protic solvent water   solvent dimethyl sulfoxide

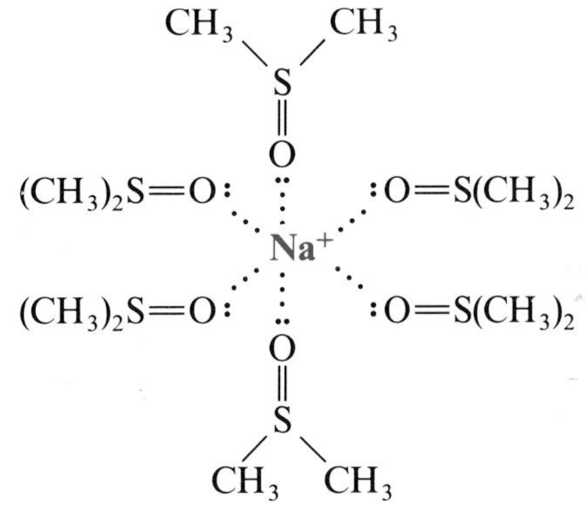

However, because they cannot form hydrogen bonds, *aprotic solvents do not solvate anions to any appreciable extent.* In these solvents anions are unencumbered by a layer of solvent molecules and, therefore, they are poorly stabilized by solvation. These "naked" anions are highly reactive both *as bases and nucleophiles.* In dimethyl sulfoxide, for example, the relative order of reactivity of halide ions is

$$\text{F}^- > \text{Cl}^- > \text{Br}^- > \text{I}^-$$

This is the opposite of their strength as nucleophiles in alcohol or water solutions:

$$\text{I}^- > \text{Br}^- > \text{Cl}^- > \text{F}^-$$

The rates of S_N2 reactions generally are vastly increased when they are carried out in polar aprotic solvents. The increase in rate can be as large as a millionfold.

PROBLEM 5.6

Classify the following solvents as being protic or aprotic: Formic acid, $\underset{\substack{\displaystyle \text{O} \\ \parallel}}{\text{HCOH}}$;

acetone, $CH_3\overset{\displaystyle O}{\overset{\|}{C}}CH_3$; acetonitrile, $CH_3C\equiv N$; formamide, $HC\overset{\displaystyle O}{\overset{\|}{N}}H_2$; sulfur dioxide, SO_2; ammonia, NH_3; trimethylamine, $N(CH_3)_3$; ethylene glycol, $HOCH_2CH_2OH$.

PROBLEM 5.7

Would you expect the reaction of propyl bromide with sodium cyanide (NaCN), that is,

$$CH_3CH_2CH_2Br + NaCN \longrightarrow CH_3CH_2CH_2CN + NaBr$$

to occur faster in dimethylformamide (DMF) or in ethanol? Explain your answer.

PROBLEM 5.8

Which would you expect to be the stronger nucleophile in a protic solvent: (a) the amide ion (NH_2^-) or ammonia? (b) RS^- or RSH? (c) PH_3 or NH_3?

5.13D Solvent Effects on S_N1 Reactions.
The Ionizing Ability of the Solvent

Because of its ability to solvate cations *and* anions so effectively, the use of a **polar protic solvent** will greatly increase the rate of ionization of an alkyl halide *in any S_N1 reaction.* It does this because solvation stabilizes the transition state leading to the intermediate carbocation and halide ion more than it does the reactants; thus the energy of activation is lower. The transition state for this endothermic step is one in which separated charges are developing and thus it resembles the ions that are ultimately produced.

$$(CH_3)_3C\!-\!Cl \qquad (CH_3)_3\overset{\delta+}{C}\cdots\overset{\delta-}{Cl} \longrightarrow (CH_3)_3C^+ + Cl^-$$

Reactant   Transition state   Products

Separated charges are developing

A rough indication of a solvent's polarity is a quantity called the **dielectric constant.** The dielectric constant is a measure of the solvent's ability to insulate opposite charges from each other. The greater the dielectric constant, the greater is the solvent's ability to insulate both ions and developing ions. Table 5.3 gives the dielectric constants of some common solvents.

Water is the most effective solvent for promoting ionization, but most organic compounds do not dissolve appreciably in water. They usually dissolve, however, in alcohols, and quite often mixed solvents are used. Methanol–water and ethanol–water are common mixed solvents for nucleophilic substitution reactions.

TABLE 5.3 Dielectric constants of common solvents

	SOLVENT	FORMULA	DIELECTRIC CONSTANT
	Water	H_2O	80
	Formic acid	$\overset{\displaystyle O}{\overset{\|}{H\text{C}OH}}$	59
	Dimethyl sulfoxide	$\overset{\displaystyle O}{\overset{\|}{CH_3\text{S}CH_3}}$	49
	Dimethylformamide	$\overset{\displaystyle O}{\overset{\|}{H\text{C}N(CH_3)_2}}$	37
	Acetonitrile	$CH_3C{\equiv}N$	36
	Methanol	CH_3OH	33
	Hexamethylphosphoric triamide (HMPT)	$[(CH_3)_2N]_3P{=}O$	30
	Ethanol	CH_3CH_2OH	24
	Acetone	$\overset{\displaystyle O}{\overset{\|}{CH_3\text{C}CH_3}}$	21
	Acetic acid	$\overset{\displaystyle O}{\overset{\|}{CH_3\text{C}OH}}$	6

Increasing solvent polarity (↑)

PROBLEM 5.9

When *tert*-butyl bromide undergoes solvolysis in a mixture of methanol and water, the rate of solvolysis (measured by the rate at which bromide ions form in the mixture) *increases* when the percentage of water in the mixture is increased. (a) Explain this occurrence. (b) Provide an explanation for the observation that the rate of the S$_N$2 reaction of ethyl chloride with potassium iodide in methanol and water *decreases* when the percentage of water in the mixture is increased.

5.13E The Nature of the Leaving Group

The best leaving groups are those that become the most stable molecules or ions after they depart. This means, in general, that the best leaving groups are the ions or molecules that are the weakest bases. Of the halogens, an iodide ion is the best leaving group and a fluoride ion is the poorest:

$$I^- > Br^- > Cl^- \gg F^-$$

The order is the opposite of the basicity:

$$F^- \gg Cl^- > Br^- > I^-$$

Other weak bases that are good leaving groups are alkanesulfonate ions and alkyl sulfate ions:

$$\underset{\text{An alkanesulfonate ion}}{\overset{\displaystyle \overset{O}{\underset{O}{\overset{\|}{\underset{\|}{S}}}}}{^{-}O-S-R}} \qquad \underset{\text{An alkyl sulfate ion}}{\overset{\displaystyle \overset{O}{\underset{O}{\overset{\|}{\underset{\|}{S}}}}}{^{-}O-S-O-R}}$$

These anions are all the conjugate bases of very strong acids.

The trifluoromethanesulfonate ion ($CF_3SO_3^-$, commonly called the **triflate ion**) is one of the best leaving groups known to chemists. It is the anion of CF_3SO_3H, an exceedingly strong acid—one that is much stronger than sulfuric acid.

$$CF_3SO_3^-$$

Triflate ion
(a "super" leaving group)

Strongly basic ions rarely act as leaving groups. The hydroxide ion, for example, is a strong base and thus reactions like the following do not take place:

$$X:^- \longrightarrow R\overset{\frown}{-}OH \;\not\!\!\!\longrightarrow\; R-X + OH^-$$

(This reaction does not take place because the leaving group is a strongly basic hydroxide ion)

However, when an alcohol is dissolved in a strong acid it can react with a halide ion. Because the acid protonates the —OH group of the alcohol, the leaving group no longer needs to be a hydroxide ion; it is now a molecule of water—a much weaker base than a hydroxide ion.

$$X:^- \longrightarrow R\overset{\frown}{-}\underset{\underset{H}{|}}{\overset{+}{O}H} \longrightarrow R-X + H_2O$$

(This reaction takes place because the leaving group is a weak base)

PROBLEM 5.10

The reaction of methyl chloride with aqueous sodium hydroxide to produce methyl alcohol and sodium chloride is essentially irreversible—the reaction proceeds virtually to completion. What factors account for this?

Very powerful bases such as hydride ions ($H:^-$) and alkanide ions ($R:^-$) never act as leaving groups. Therefore, reactions such as the following never take place:

$$Nu:^- + CH_3CH_2\overset{\frown}{-}H \;\not\!\!\!\longrightarrow\; CH_3CH_2-Nu + H:^- \left.\begin{array}{c} \\ \\ \end{array}\right\} \begin{array}{l}\text{These are}\\ \text{not leaving}\\ \text{groups}\end{array}$$

or $\quad Nu:^- + CH_3\overset{\frown}{-}CH_3 \;\not\!\!\!\longrightarrow\; CH_3-Nu + CH_3:^-$

5.13F Summary. S$_N$1 versus S$_N$2

Reactions of alkyl halides by an S$_N$1 mechanism are favored by the use of substrates that can form relatively stable carbocations, by the use of weak nucleophiles, and by the use of highly ionizing solvents. S$_N$1 mechanisms, therefore, are important in solvolysis reactions of tertiary halides, especially when the solvent is highly polar. In a solvolysis the nucleophile is weak because it is a neutral molecule (of the solvent) rather than an anion.

If we want to favor the reaction of an alkyl halide by an S$_N$2 mechanism, we should use a relatively unhindered alkyl halide, a strong nucleophile, a polar aprotic solvent, and a high concentration of the nucleophile. For substrates, the order of reactivity in S$_N$2 reactions is

$$\underset{\text{Methyl}}{CH_3-X} > \underset{1°}{C-CH_2-X} > \underset{2°}{\overset{\overset{\displaystyle C}{\displaystyle |}}{C-CH-X}}$$

Tertiary halides do not react by an S$_N$2 mechanism.

The effect of the leaving group is the same in both S$_N$1 and S$_N$2 reactions: alkyl iodides react fastest; fluorides react slowest. (Because alkyl fluorides react so slowly, they are seldom used in nucleophilic substitution reactions.)

$$R-I > R-Br > R-Cl \qquad S_N1 \quad \text{or} \quad S_N2$$

5.14 ORGANIC SYNTHESIS. FUNCTIONAL GROUP TRANSFORMATIONS USING S$_N$2 REACTIONS

The process of making one compound from another is called **synthesis.** When, for one reason or another, we find ourselves in need of an organic compound that is not available in the stockroom, or perhaps even of one that has never been made before, the task of synthesis starts. We shall have the job of making the compound we need from other compounds that are available.

S$_N$2 reactions are highly useful in organic synthesis because they enable us to convert one functional group into another—a process that is called a **functional group transformation** or a **functional group interconversion.** With the following S$_N$2 reactions the functional group of a methyl, primary, or secondary alkyl halide can be transformed into that of an alcohol, ether, thiol, thioether, nitrile, ester, and so on. (*Note:* The use of the prefix *thio* in a name means that a sulfur atom has replaced an oxygen atom in the compound.)

	$\xrightarrow{\text{OH}^-}$ R—OH	Alcohol
	$\xrightarrow{\text{R'O}^-}$ R—OR'	Ether
	$\xrightarrow{\text{SH}^-}$ R—SH	Thiol
	$\xrightarrow{\text{R'S}^-}$ R—SR'	Thioether
R—X	$\xrightarrow{\text{CN}^-}$ R—C≡N	Nitrile
(R = Me, 1°, or 2°) (X = Cl, Br, or I)	$\xrightarrow{\overset{\overset{\displaystyle O}{\displaystyle \parallel}}{\text{R'CO}^-}}$ R—OCR' (with $\overset{\displaystyle O}{\overset{\displaystyle \parallel}{}}$)	Ester
	$\xrightarrow{\text{R}_3'\text{N}}$ R—NR$_3'^+$ X$^-$	Quaternary ammonium halide
	$\xrightarrow{\text{N}_3^-}$ R—N$_3$	Alkyl azide

Alkyl chlorides and bromides are also easily converted to alkyl iodides by nucleophilic substitution reactions.

$$R-Cl \atop R-Br \Bigg\} \xrightarrow{\ I^-\ } R-I \quad (+ Cl^- \ \text{or} \ Br^-)$$

One other aspect of the S_N2 reaction that is of great importance in synthesis is its **stereochemistry** (Section 5.8). S_N2 reactions always occur **with inversion of configuration** at the atom that bears the leaving group. This means that when we use S_N2 reactions in syntheses we can be sure of the configuration of our product if we know the configuration of our reactant. For example, suppose we need a sample of the nitrile shown here with the (S) configuration.

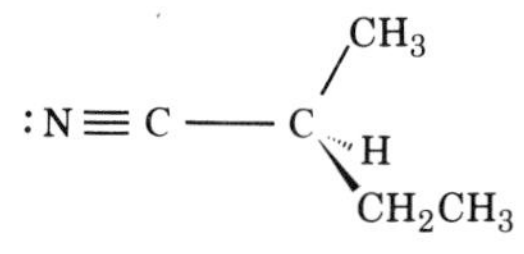

$$: N \equiv C \longrightarrow C \overset{\displaystyle CH_3}{\underset{\displaystyle CH_2CH_3}{\overset{|}{\underset{|}{\rule{0pt}{0pt}}}}} H$$

(S)-2-Methylbutanoic acid

If we have available (R)-2-bromobutane, we can carry out the following synthesis:

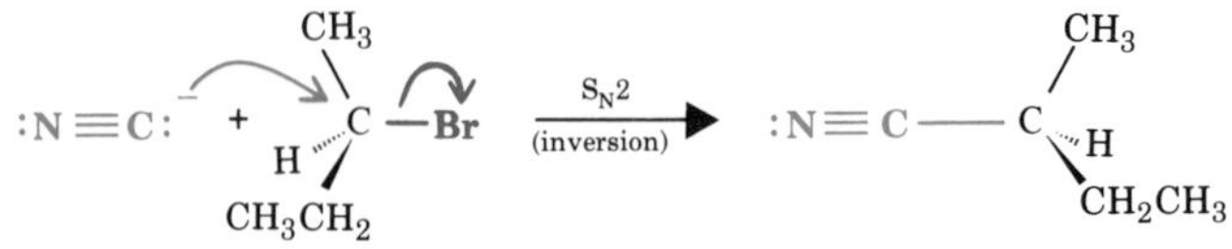

$$: N \equiv C : \ + \ \overset{CH_3}{\underset{CH_3CH_2}{\overset{|}{C}}} \overset{\displaystyle H}{} - Br \xrightarrow[\text{(inversion)}]{S_N2} \ : N \equiv C \longrightarrow C \overset{\displaystyle CH_3}{\underset{\displaystyle CH_2CH_3}{\overset{|}{\underset{|}{\rule{0pt}{0pt}}}}} H$$

(R)-2-Bromobutane   *(S)*-2-Methylbutanenitrile

PROBLEM 5.11

Starting with (S)-2-bromobutane, outline syntheses of each of the following compounds:

(a) (R)-CH$_3$CHCH$_2$CH$_3$
      |
     OCH$_2$CH$_3$

(b) (R)-CH$_3$CHCH$_2$CH$_3$
      |
     OCCH$_3$
      ‖
      O

(c) (R)-CH$_3$CHCH$_2$CH$_3$
      |
     SH

(d) (R)-CH$_3$CHCH$_2$CH$_3$
      |
     SCH$_3$

5.14A The Unreactivity of Vinylic and Aryl Halides

Compounds that have a halogen atom attached to one carbon atom of a double bond are called **vinylic halides.** Those that have a halogen atom attached to a benzene ring (Section 2.7) are called **phenyl halides** and they belong to a larger class of compounds that we shall study in Chapter 14 called **aryl halides.**

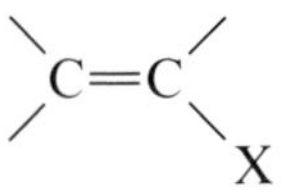

$$\overset{\diagdown}{\diagup} C = C \overset{\diagup}{\underset{\diagdown}{}} X \qquad\qquad \langle\!\bigcirc\!\rangle - X \quad \text{or} \quad C_6H_5X \quad \text{or} \quad PhX^*$$

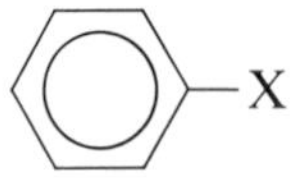

A vinylic halide   Phenyl halide
               (an aryl halide)

*The symbol Ph— is often used for a benzene ring.

Vinylic halides and aryl halides are generally unreactive in S_N1 or S_N2 reactions. Vinylic and aryl cations are relatively unstable and do not form readily. This explains the unreactivity of vinylic and aryl halides in S_N1 reactions. The carbon–halogen bond of a vinylic or aryl halide is stronger than that of an alkyl halide (we shall see why later) and the electrons of the double bond or benzene ring repel the approach of a nucleophile from the back side. These factors explain the unreactivity of a vinylic or aryl halide in an S_N2 reaction.

5.15 ELIMINATION REACTIONS OF ALKYL HALIDES

In an elimination reaction the fragments of some molecule (YZ) are removed (eliminated) from adjacent atoms of the reactant. This elimination leads to the introduction of a multiple bond:

$$-\overset{\displaystyle |}{\underset{\displaystyle Y}{C}}-\overset{\displaystyle |}{\underset{\displaystyle Z}{C}}- \xrightarrow[(-YZ)]{\text{elimination}} \overset{\displaystyle \diagdown}{\diagup}C=C\overset{\displaystyle \diagup}{\diagdown}$$

5.15A Dehydrohalogenation

A widely used method for synthesizing alkenes is the elimination of HX from adjacent atoms of an alkyl halide. Heating the alkyl halide with a strong base causes the reaction to take place. The following are two examples.

$$\underset{\underset{\displaystyle Br}{|}}{CH_3CHCH_3} \xrightarrow[C_2H_5OH, \ 55\ °C]{C_2H_5ONa} CH_2{=}CH{-}CH_3 + NaBr + C_2H_5OH$$
$$(79\%)$$

$$CH_3{-}\overset{\displaystyle \underset{\displaystyle |}{CH_3}}{\underset{\displaystyle \underset{\displaystyle |}{CH_3}}{C}}{-}Br \xrightarrow[C_2H_5OH, \ 25\ °C]{C_2H_5ONa} CH_3{-}\overset{\displaystyle \underset{\displaystyle |}{CH_3}}{C}{=}CH_2 + NaBr + C_2H_5OH$$
$$(91\%)$$

Reactions like these are not limited to the elimination of hydrogen bromide. Chloroalkanes also undergo the elimination of hydrogen chloride, iodoalkanes undergo the elimination of hydrogen iodide and, in all cases, alkenes are produced. When the elements of a hydrogen halide are eliminated from a haloalkane in this way, the reaction is often called **dehydrohalogenation.**

$$-\overset{\displaystyle \overset{\displaystyle H}{|}}{\underset{\displaystyle |}{C^\beta}}{-}\overset{\displaystyle |}{\underset{\displaystyle \underset{\displaystyle X}{||}}{C^\alpha}}{-} + \ :B^- \longrightarrow \overset{\displaystyle \diagdown}{\diagup}C{=}C\overset{\displaystyle \diagup}{\diagdown} + H\!:\!B + \ :X^-$$

(a base)

Dehydrohalogenation

In these eliminations, as in S_N1 and S_N2 reactions, there is a leaving group and an attacking particle (the base) that possesses an electron pair.

Chemists often call the carbon atom that bears the halogen atom (see previous reaction) the **alpha (α) carbon atom** and any carbon atom adjacent to it a **beta (β) carbon atom**. A hydrogen atom attached to the β carbon atom is called a **β hydrogen atom**. Since the hydrogen atom that is eliminated in dehydrohalogenation is from the β carbon atom, these reactions are often called **β eliminations**. They are also often referred to as **1,2 eliminations**.

We shall have more to say about dehydrohalogenation in the next chapter, but we can examine some basic features here.

5.15B Bases Used in Dehydrohalogenation

Various strong bases have been used for dehydrohalogenations. Potassium hydroxide dissolved in ethyl alcohol is a reagent sometimes used, but the sodium salts of alcohols often offer distinct advantages.

The sodium salt of an alcohol (a sodium alkoxide) can be prepared by treating an alcohol with sodium metal:

$$2R\!-\!\overset{..}{\underset{..}{O}}H + 2Na \longrightarrow 2R\!-\!\overset{..}{\underset{..}{O}}:^-Na^+ + H_2$$

Alcohol   Sodium
                                alkoxide

This reaction involves the displacement of hydrogen from the alcohol and is, thus, an **oxidation–reduction reaction.** Sodium, an alkali metal, is a very powerful reducing agent and always displaces hydrogen atoms that are bonded to oxygen atoms. The vigorous (at times explosive) reaction of sodium with water is of the same type.

$$2H\overset{..}{\underset{..}{O}}H + 2Na \longrightarrow 2H\overset{..}{\underset{..}{O}}:^-Na^+ + H_2$$

Sodium
hydroxide

Sodium alkoxides can also be prepared by allowing an alcohol to react with sodium hydride (NaH). The hydride ion ($H:^-$) is a very strong base.

$$R\!-\!\overset{..}{\underset{..}{O}}H + Na^+:H^- \longrightarrow R\!-\!\overset{..}{\underset{..}{O}}:^-Na^+ + H_2$$

Sodium (and potassium) alkoxides are usually prepared by using an excess of the alcohol, and the excess alcohol becomes the solvent for the reaction. Sodium ethoxide is frequently employed in this way.

$$2CH_3CH_2OH + 2Na \longrightarrow 2CH_3CH_2O^-Na^+ + H_2$$

Ethyl alcohol   Sodium ethoxide
(excess)

Potassium *tert*-butoxide is another highly effective dehydrohalogenating reagent.

$$2CH_3\overset{\displaystyle CH_3}{\underset{\displaystyle CH_3}{\overset{|}{\underset{|}{C}}}}\!-\!\overset{..}{\underset{..}{O}}H + 2K \longrightarrow 2CH_3\overset{\displaystyle CH_3}{\underset{\displaystyle CH_3}{\overset{|}{\underset{|}{C}}}}\!-\!\overset{..}{\underset{..}{O}}:^-K^+ + H_2$$

tert-Butyl alcohol   Potassium *tert*-butoxide
(excess)

5.15C Mechanisms of Dehydrohalogenations

Elimination reactions occur by a variety of mechanisms. With alkyl halides, two mechanisms are especially important because they are closely related to the S_N2 and S_N1 reactions that we have just studied. One mechanism is a bimolecular mechanism called the E2 reaction; the other is a unimolecular mechanism called the E1 reaction.

5.16 THE E2 REACTION

When isopropyl bromide is heated with sodium ethoxide in ethanol to form propene, the reaction rate depends on the concentration of isopropyl bromide and on the concentration of ethoxide ion. The rate equation is first order in each reactant and second order overall.

$$\text{Rate} \propto [CH_3CHBrCH_3][C_2H_5O^-]$$

$$\text{Rate} = k[CH_3CHBrCH_3][C_2H_5O^-]$$

From this we infer that the transition state for the rate-determining step must involve both the alkyl halide and the alkoxide ion. The reaction must be bimolecular. Considerable experimental evidence indicates that the reaction takes place in the following way:

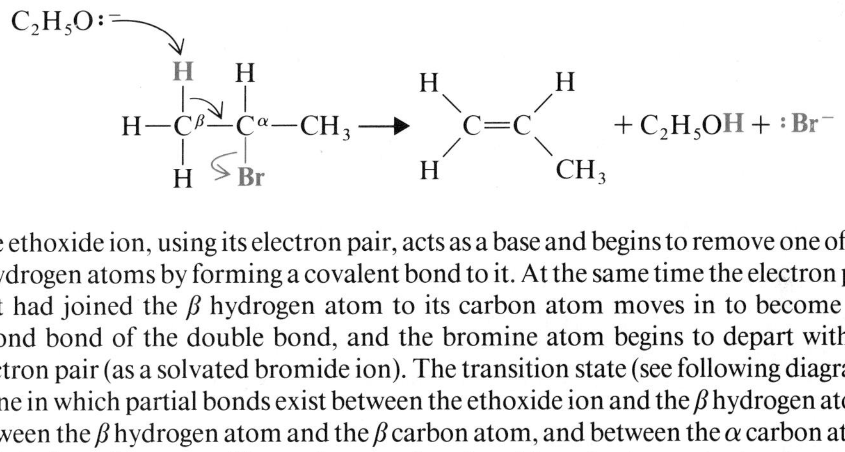

The ethoxide ion, using its electron pair, acts as a base and begins to remove one of the β hydrogen atoms by forming a covalent bond to it. At the same time the electron pair that had joined the β hydrogen atom to its carbon atom moves in to become the second bond of the double bond, and the bromine atom begins to depart with its electron pair (as a solvated bromide ion). The transition state (see following diagram) is one in which partial bonds exist between the ethoxide ion and the β hydrogen atom, between the β hydrogen atom and the β carbon atom, and between the α carbon atom and the bromine atom. The carbon–carbon bond has also begun to develop some double-bond character.

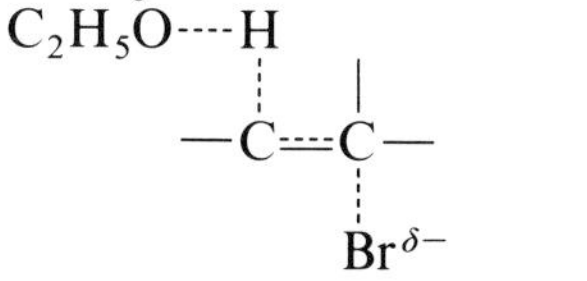

Transition state for an E2 reaction

5.17 THE E1 REACTION

Eliminations may take a different pathway from that given in the previous section. Heating *tert*-butyl chloride with 80% aqueous ethanol at 25 °C, for example, gives *substitution products* in 83% yield and an elimination product (2-methylpropene) in 17% yield.

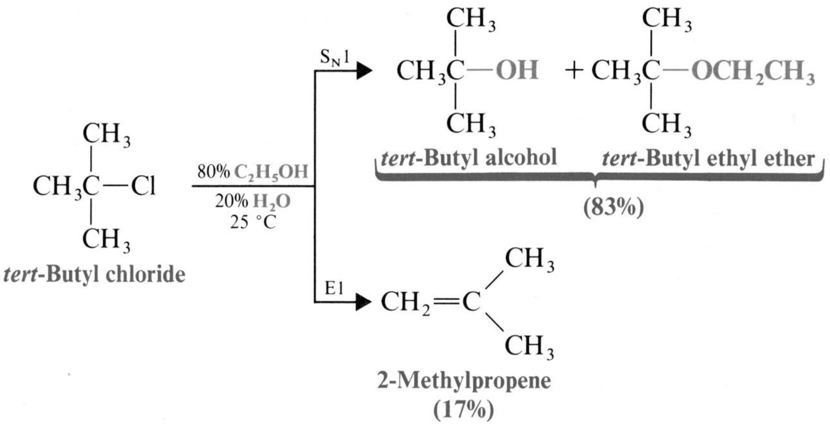

CH₃C—Cl (tert-Butyl chloride) $\xrightarrow[\text{25 °C}]{\substack{80\% \text{ C}_2\text{H}_5\text{OH} \\ 20\% \text{ H}_2\text{O}}}$

$\xrightarrow{\text{S}_\text{N}1}$ CH₃C—OH + CH₃C—OCH₂CH₃

tert-Butyl alcohol *tert*-Butyl ethyl ether

(83%)

$\xrightarrow{\text{E1}}$ CH₂=C(CH₃)(CH₃)

2-Methylpropene
(17%)

The initial step for both reactions is the formation of a *tert*-butyl cation. This is also the rate-determining step for both reactions; thus both reactions are unimolecular.

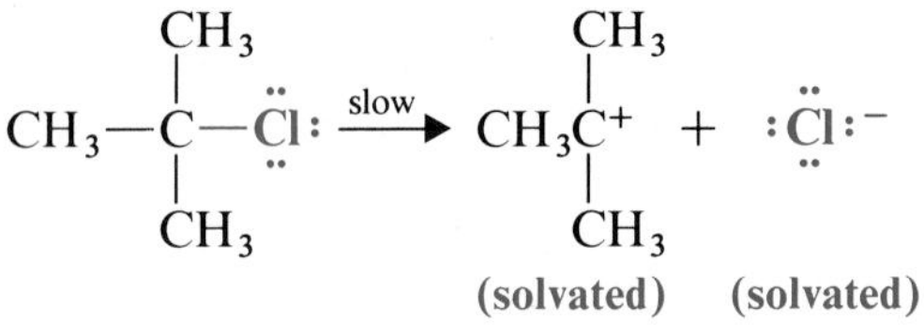

CH₃—C—Cl: $\xrightarrow{\text{slow}}$ CH₃C⁺ + :Cl:⁻

(solvated)   (solvated)

Whether substitution or elimination takes place depends on the next step (the fast step). If a solvent molecule reacts as a nucleophile at the positive carbon atom of the *tert*-butyl cation, the product is *tert*-butyl alcohol or *tert*-butyl ethyl ether and the reaction is S$_\text{N}$1.

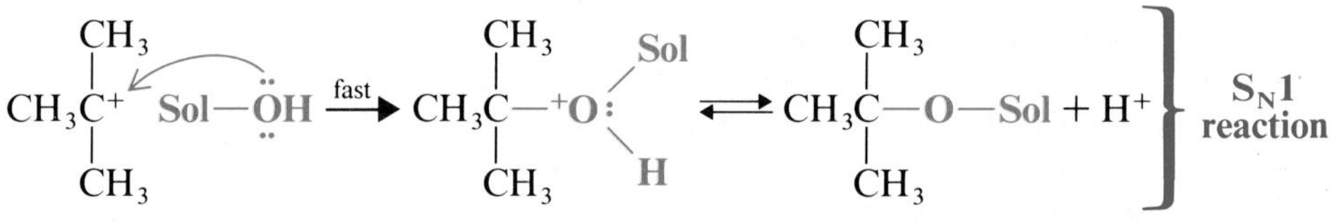

CH₃C⁺ Sol—ÖH $\xrightarrow{\text{fast}}$ CH₃C—⁺Ö: ⇌ CH₃C—O—Sol + H⁺ } S$_\text{N}$1 reaction

(Sol = H— or CH₃CH₂—)

If, however, a solvent molecule acts as a base and abstracts one of the β hydrogen atoms as a proton, the product is 2-methylpropene and the reaction is E1.

E1 reactions always accompany S$_\text{N}$1 reactions.

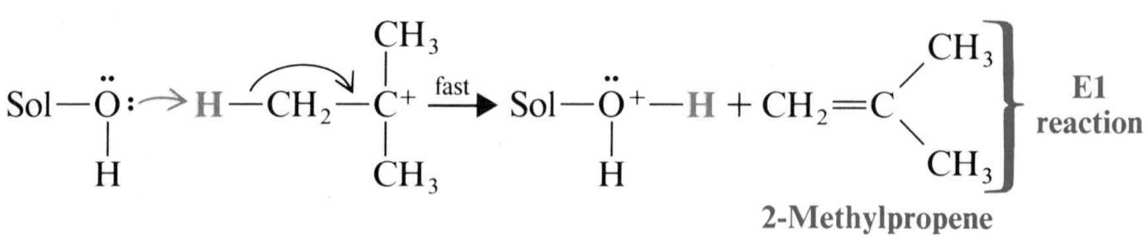

Sol—Ö: → H—CH₂—C⁺ $\xrightarrow{\text{fast}}$ Sol—Ö⁺—H + CH₂=C(CH₃)(CH₃) } E1 reaction

2-Methylpropene

5.18 SUBSTITUTION VERSUS ELIMINATION

Because the reactive part of a nucleophile or a base is an unshared electron pair, all nucleophiles are potential bases and all bases are potential nucleophiles. It should not be surprising, then, that nucleophilic substitution reactions and elimination reactions often compete with each other.

5.18A S$_N$2 versus E2

Since eliminations occur best by an E2 path when carried out with a high concentration of a strong base (and thus a high concentration of a strong nucleophile), substitution reactions by an S$_N$2 path often compete with the elimination reaction. When the nucleophile (base) attacks a β hydrogen atom, elimination occurs. When the nucleophile attacks the carbon atom bearing the leaving group, substitution results.

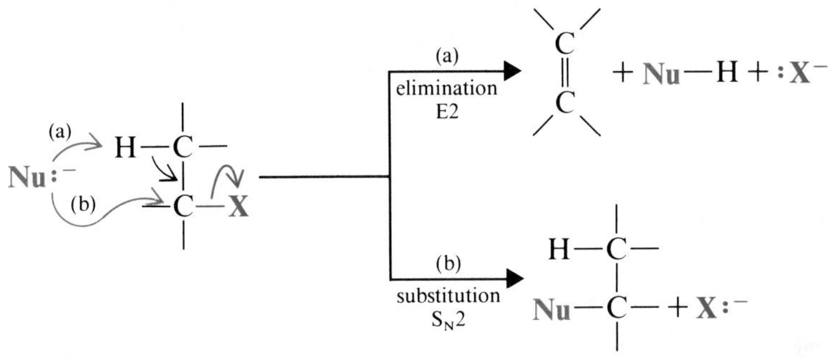

When the substrate is a primary halide and the base is ethoxide ion, substitution is highly favored.

$$CH_3CH_2O^-Na^+ + CH_3CH_2Br \xrightarrow[(-NaBr)]{\underset{55\ °C}{C_2H_5OH}} CH_3CH_2OCH_2CH_3 + CH_2{=}CH_2$$

(90%)   (10%)
S$_N$2   E2

With secondary halides, however, the elimination reaction is favored.

$$C_2H_5O^-Na^+ + CH_3CHCH_3 \xrightarrow[(-NaBr)]{\underset{55\ °C}{C_2H_5OH}} CH_3CHCH_3 + CH_2{=}CHCH_3$$

with Br below first, and O–C$_2$H$_5$ below product

(21%)   (79%)
S$_N$2   E2

With tertiary halides an S$_N$2 reaction cannot take place and thus the elimination reaction is highly favored, especially when the reaction is carried out at higher temperatures. Any substitution that occurs probably takes place through an S$_N$1 mechanism.

$$C_2H_5O^-Na^+ + CH_3\overset{\overset{\displaystyle CH_3}{|}}{\underset{\underset{\displaystyle Br}{|}}{C}}CH_3 \xrightarrow[(-NaBr)]{\underset{25\ °C}{C_2H_5OH}} CH_3\overset{\overset{\displaystyle CH_3}{|}}{\underset{\underset{\displaystyle C_2H_5}{\underset{|}{O}}}{C}}CH_3 + CH_2{=}\overset{\overset{\displaystyle CH_3}{|}}{C}CH_3$$

(9%)   (91%)
S$_N$1   E2 + E1

$$C_2H_5O^-Na^+ + CH_3\underset{\underset{Br}{|}}{\overset{\overset{CH_3}{|}}{C}}CH_3 \xrightarrow[\substack{55\ °C \\ (-NaBr)}]{C_2H_5OH} CH_2{=}\underset{\underset{(100\%)}{}}{\overset{\overset{CH_3}{|}}{C}}CH_3 + C_2H_5OH$$

(100%)
E2 + E1

Increasing the reaction temperature is one way of favorably influencing an elimination reaction of an alkyl halide. Another way is to use a strong sterically hindered base such as the *tert*-butoxide ion. The bulky methyl groups of the *tert*-butoxide ion appear to inhibit its reacting by substitution, so elimination reactions take precedence. We can see an example of this effect in the following two reactions. The relatively unhindered methoxide ion reacts with octadecyl bromide primarily by *substitution;* the bulky *tert*-butoxide ion gives mainly *elimination.*

$$CH_3O^- + CH_3(CH_2)_{15}CH_2CH_2{-}Br \xrightarrow[65\ °C]{CH_3OH}$$

$$CH_3(CH_2)_{15}CH{=}CH_2 + CH_3(CH_2)_{15}CH_2CH_2OCH_3$$

(1%)   (99%)
E2 S_N2

$$CH_3{-}\underset{\underset{CH_3}{|}}{\overset{\overset{CH_3}{|}}{C}}{-}O^- + CH_3(CH_2)_{15}CH_2CH_2{-}Br \xrightarrow[40\ °C]{(CH_3)_3COH}$$

$$CH_3(CH_2)_{15}CH{=}CH_2 + CH_3(CH_2)_{15}CH_2CH_2{-}O{-}\underset{\underset{CH_3}{|}}{\overset{\overset{CH_3}{|}}{C}}{-}CH_3$$

(85%)   (15%)
E2 S_N2

Another factor that affects the relative rates of E2 and S_N2 reactions is the relative basicity and polarizability of the base/nucleophile. Use of a strong, slightly polarizable base such as amide ion (NH_2^-) or alkoxide ion (especially a hindered one) tends to increase the likelihood of elimination (E2). Use of a weakly basic ion such as a chloride ion (Cl^-) or an acetate ion ($CH_3CO_2^-$) or a weakly basic and highly polarizable one such as Br^-, I^-, or RS^- increases the likelihood of substitution (S_N2). Acetate ion, for example, reacts with isopropyl bromide almost exclusively by the S_N2 path:

$$CH_3\overset{\overset{O}{\|}}{C}{-}O^- + CH_3\overset{\overset{CH_3}{|}}{C}H{-}Br \longrightarrow CH_3\overset{\overset{O}{\|}}{C}{-}O{-}\overset{\overset{CH_3}{|}}{C}HCH_3 + Br^-$$

(~100%)
S_N2

The more strongly basic ethoxide ion (Table 2.2 and Section 5.15B) reacts with the same compound mainly by an E2 mechanism.

5.18B Tertiary Halides. S_N1 versus E1

Because the E1 reaction and the S_N1 reaction proceed through the formation of a common intermediate, the two types respond in similar ways to factors affecting reactivities. E1 reactions are favored with substrates that can form stable carbocations (i.e., tertiary halides); they are also favored by the use of weak nucleophiles (bases) and they are generally favored by the use of polar solvents.

It is usually difficult to influence the relative partition between S_N1 and E1 products because the energy of activation for either reaction of the carbocation (loss of a proton or combination with a molecule of the solvent) is very small.

In most unimolecular reactions the S_N1 reaction is favored over the E1 reaction, especially at lower temperatures. *In general, however, substitution reactions of tertiary halides are not very useful as synthetic methods. Such halides undergo eliminations much too easily.*

Increasing the temperature of the reaction favors reaction by the E1 mechanism at the expense of the S_N1 mechanism. *If the elimination product is desired, however, it is more convenient to add a strong base and force an E2 reaction to take place instead.*

5.18C Overall Summary

We can summarize the most important reaction pathways according to the type of substrate in the following way:

CH_3X	RCH_2X	R $\|$ $RCHX$	R $\|$ $R-C-X$ $\|$ R
Methyl	1°	2°	3°
← ——— Bimolecular reactions only ———→			← S_N1/E1 or E2 →
Gives S_N2 reactions.	Gives mainly S_N2 except with a hindered strong base [e.g., $(CH_3)_3CO^-$] and then gives mainly E2.	Gives mainly S_N2 with weak bases (e.g., I^-, CN^-, RCO_2^-) and mainly E2 with strong bases (e.g., RO^-).	No S_N2 reaction. In solvolysis gives S_N1/E1, and at lower temperatures S_N1 is favored. When a strong base (e.g., RO^-) is used, E2 predominates.

SAMPLE PROBLEM

Illustrating a Multistep Synthesis

Starting with alkanes of four carbon atoms or fewer and any needed solvents or inorganic reagents, outline a synthesis of *tert*-butyl methyl ether $(CH_3)_3COCH_3$.

Answer:

Since we know of no methods for making an ether directly from an alkane, we conclude that a multistep synthesis will be required. We begin our thinking by working backward. We try to think of a way of making the final product from other compounds. We recall (Section 5.18A) that we can make an ether by the nucleophilic substitution reaction between an alkoxide ion (RO^-) and an alkyl halide ($R'—X$). In theory, at least, there are two ways that we could synthesize *tert*-butyl methyl ether:

Method 1 $(CH_3)_3CO^- + CH_3—Br \longrightarrow (CH_3)_3COCH_3 + Br^-$

Method 2 $CH_3O^- + (CH_3)_3C—Br \overset{}{\nrightarrow} (CH_3)_3COCH_3 + Br^-$

The second method will fail, however. Because the substrate is a tertiary halide, treating it with the strong base (CH_3O^-) will cause an E2 reaction to take place. The reaction will yield an alkene [$CH_2{=}C(CH_3)_2$] instead of methyl *tert*-butyl ether.

So we choose method 1 and begin thinking of ways to make the needed reagents, $(CH_3)_3CO^-$ and CH_3Br. Synthesizing methyl bromide is easy (Section 3.17). We simply allow methane to react with bromine (using an excess of methane to minimize polybromination):

$$CH_3—H \xrightarrow[\substack{\text{heat or light} \\ (-HBr)}]{Br_2} CH_3—Br$$

We know that we can make a *tert*-butoxide ion (as sodium or potassium *tert*-butoxide) by treating *tert*-butyl alcohol with sodium hydride or potassium metal (Section 5.15B):

$$(CH_3)_3COH \xrightarrow[(-H_2)]{NaH} (CH_3)_3CO^-Na^+$$

or

$$(CH_3)_3COH \xrightarrow[(-H_2)]{K} (CH_3)_3CO^-K^+$$

We also know that we can make *tert*-butyl alcohol by an S_N1 hydrolysis of *tert*-butyl bromide, and that we can obtain *tert*-butyl bromide by brominating isobutane:

$$(CH_3)_3C—H \xrightarrow[\substack{\text{heat or light} \\ (-HBr)}]{Br_2} (CH_3)_3C—Br \xrightarrow[(-HBr)]{H_2O} (CH_3)_3C—OH$$

The first step of this sequence works well because of the great selectivity of bromine for tertiary hydrogen atoms in alkane brominations (Section 3.17B).

So now our synthesis is complete. (In Section 7.5 we shall study a better method for making *tert*-butyl alcohol.)

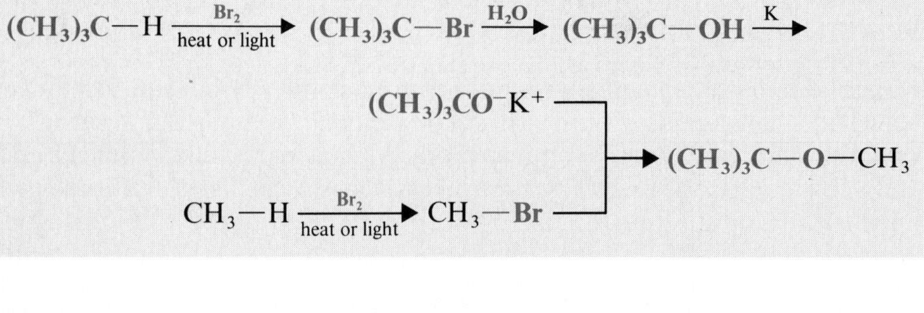

$$(CH_3)_3C—H \xrightarrow[\text{heat or light}]{Br_2} (CH_3)_3C—Br \xrightarrow{H_2O} (CH_3)_3C—OH \xrightarrow{K}$$

$$(CH_3)_3CO^-K^+ \rule[0.5ex]{1em}{0.4pt}$$

$$\longrightarrow (CH_3)_3C—O—CH_3$$

$$CH_3—H \xrightarrow[\text{heat or light}]{Br_2} CH_3—Br \rule[0.5ex]{1em}{0.4pt}$$

5.19 PHYSICAL PROPERTIES
OF ORGANIC HALIDES

Alkyl and aryl halides have very low solubilities in water, but as we might expect, they are miscible with each other and with other relatively nonpolar solvents. Methylene chloride (CH_2Cl_2), chloroform ($CHCl_3$), and carbon tetrachloride (CCl_4) are often used as solvents for nonpolar and moderately polar organic compounds. Many chloroalkanes including $CHCl_3$ and CCl_4 have a cumulative toxicity, however, and should, therefore, be used only in fume hoods and with great care.

Methyl iodide (bp, 42 °C) is the only monohalomethane that is a liquid at room temperature and atmospheric pressure. Ethyl bromide (bp, 38 °C) and ethyl iodide (bp, 72 °C) are both liquids, but ethyl chloride (bp, 13 °C) is a gas. The propyl chlorides, bromides, and iodides are all liquids. In general alkyl chlorides, bromides, and iodides tend to have boiling points near those of alkanes of similar molecular weights.

Polyfluoroalkanes, however, tend to have unusually low boiling points. Hexafluoroethane boils at −79 °C, even though its molecular weight (138) is near that of decane (bp, 174 °C).

Table 5.4 lists some of the physical properties of organic halides.

The electrons of polyfluoroalkanes are tightly held because of fluorine's great electronegativity. Consequently, molecules of polyfluoroalkanes are not easily polarized, and therefore, van der Waals attractions between them are small.

5.20 SOME IMPORTANT TERMS AND CONCEPTS

Carbocation. A positive ion formed by heterolysis of a bond to a carbon atom as follows:

$$-\overset{|}{\underset{|}{C}}-X \longrightarrow -\overset{|}{\underset{|}{C}}{}^{+} + :X^{-}$$

Carbocations show the relative stabilities:

$$3° > 2° > 1° > \text{methyl}$$

Nucleophile. A negative ion or a molecule that has an unshared pair of electrons. In a chemical reaction a nucleophile attacks a positive center of some other molecule or positive ion.

Nucleophilic substitution reaction (abbreviated as S_N reaction). A substitution reaction brought about when a nucleophile reacts with a *substrate* that bears a *leaving group.*

S_N2 reaction. A nucleophilic substitution reaction for which the rate-determining step is *bimolecular* (i.e., the transition state involves two species). The reaction of methyl chloride with hydroxide ion is an S_N2 reaction. According to the Ingold mechanism it takes place in a *single step* as follows:

$$\text{HO}:^- + \text{CH}_3-\text{Cl} \longrightarrow \overset{\delta-}{\text{HO}}\text{---CH}_3\text{---}\overset{\delta-}{\text{Cl}} \longrightarrow \text{HO}-\text{CH}_3 + :\text{Cl}^-$$

Transition
state

TABLE 5.4 Organic halides

GROUP	FLUORIDE		CHLORIDE		BROMIDE		IODIDE	
	bp (°C)	DENSITY (g/mL)	bp (°C)	DENSITY (g/mL)	bp (°C)	DENSITY (g/mL)	bp (°C)	DENSITY (g/mL)
Methyl	−78.4	0.84^{-60}	−23.8	0.92^{20}	3.6	1.73^{0}	42.5	2.28^{20}
Ethyl	−37.7	0.72^{20}	13.1	0.91^{15}	38.4	1.46^{20}	72	1.95^{20}
Propyl	+2.5	0.78^{-3}	46.6	0.89^{20}	70.8	1.35^{20}	102	1.74^{20}
Isopropyl	−9.4	0.72^{20}	34	0.86^{20}	59.4	1.31^{20}	89.4	1.70^{20}
Butyl	32	0.78^{20}	78.4	0.89^{20}	101	1.27^{20}	130	1.61^{20}
sec-Butyl			68	0.87^{20}	91.2	1.26^{20}	120	1.60^{20}
Isobutyl			69	0.87^{20}	91	1.26^{20}	119	1.60^{20}
tert-Butyl	12	0.75^{12}	51	0.84^{20}	73.3	1.22^{20}	100 dec.[a]	1.57^{0}
Pentyl	62	0.79^{20}	108.2	0.88^{20}	129.6	1.22^{20}	155^{740}	1.52^{20}
Neopentyl			84.4	0.87^{20}	105	1.20^{20}	127 dec.[a]	1.53^{13}
$CH_2=CH-$	−72	0.68^{26}	−13.9	0.91^{20}	16	1.52^{14}	56	2.04^{20}
$CH_2=CHCH_2-$	−3		45	0.94^{20}	70	1.40^{20}	102–103	1.84^{22}
C_6H_5-	85	1.02^{20}	132	1.10^{20}	155	1.52^{20}	189	1.82^{20}
$C_6H_5CH_2-$	140	1.02^{25}	179	1.10^{25}	201	1.44^{22}	93^{10}	1.73^{25}

[a]Decompose is abbreviated as dec.

The order of reactivity of alkyl halides in S_N2 reactions is

$$CH_3-X > RCH_2X > R_2CHX$$

Methyl   1°   2°

S_N1 reaction. A nucleophilic substitution reaction for which the rate-determining step is *unimolecular*. The hydrolysis of *tert*-butyl chloride is an S_N1 reaction that takes place in three steps as follows. The rate-determining step is step 1.

Step 1 $(CH_3)_3CCl \xrightarrow{\text{slow}} (CH_3)_3C^+ + Cl^-$

Step 2 $(CH_3)_3C^+ + H_2O: \xrightarrow{\text{fast}} (CH_3)_3C\overset{+}{O}H_2$

Step 3 $(CH_3)_3C\overset{+}{O}H_2 + H_2O \xrightarrow{\text{fast}} (CH_3)_3COH + H_3O^+$

S_N1 reactions are important with tertiary halides and with other substrates that can form relatively stable carbocations.

Solvolysis. A nucleophilic substitution reaction in which the nucleophile is a molecule of the solvent.

Steric effect. An effect on relative reaction rates caused by the space-filling properties of those parts of a molecule attached at or near the reacting site. *Steric hindrance* is an important effect in S_N2 reactions. It explains why methyl halides are most reactive and tertiary halides are least reactive.

Elimination reaction. A reaction in which the fragments of some molecule are eliminated from adjacent atoms of the reactant to give a multiple bond. Dehydrohalogenation is an elimination reaction in which HX is eliminated from an alkyl halide, leading to the formation of an alkene.

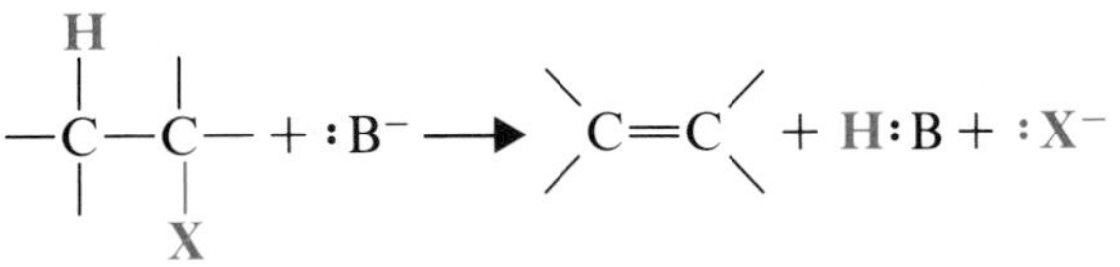

$$-\overset{\overset{\displaystyle H}{|}}{\underset{|}{C}}-\overset{|}{\underset{\underset{\displaystyle X}{|}}{C}}- \; + :B^- \longrightarrow \; \overset{\diagdown}{\diagup}C=C\overset{\diagup}{\diagdown} \; + H:B + :X^-$$

E1 reaction. A unimolecular elimination. The first step of an E1 reaction, formation of a carbocation, is the same as that of an S_N1 reaction, consequently E1 and S_N1 reactions compete with each other. E1 reactions are important when tertiary halides are subjected to solvolysis in polar solvents especially at higher temperatures. The steps in the E1 reaction of *tert*-butyl chloride are the following:

Step 1 $CH_3-\overset{\overset{\displaystyle CH_3}{|}}{\underset{\underset{\displaystyle CH_3}{|}}{C}}-Cl \xrightarrow{\text{slow}} CH_3-\overset{\overset{\displaystyle CH_3}{|}}{\underset{\underset{\displaystyle CH_3}{|}}{C}}{}^+ + :Cl^-$

Step 2 $Sol-\overset{..}{O}H + H-CH_2-\overset{\overset{\displaystyle CH_3}{|}}{\underset{\underset{\displaystyle CH_3}{|}}{C}}{}^+ \longrightarrow CH_2=C\overset{\diagup CH_3}{\diagdown CH_3} + Sol-\overset{+}{O}H_2$

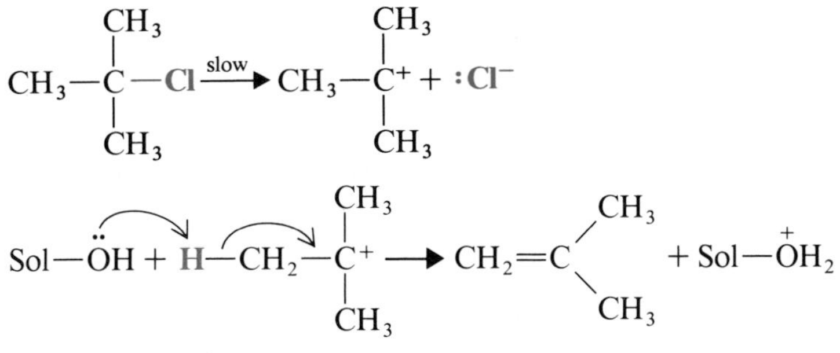

E2 reaction. A bimolecular elimination that often competes with S_N2 reactions. E2 reactions are favored by the use of a high concentration of a strong, bulky, and slightly polarizable base. The order of reactivity of alkyl halides toward E2 reactions is $3° \gg 2° > 1°$. The mechanism of the E2 reaction involves a single step:

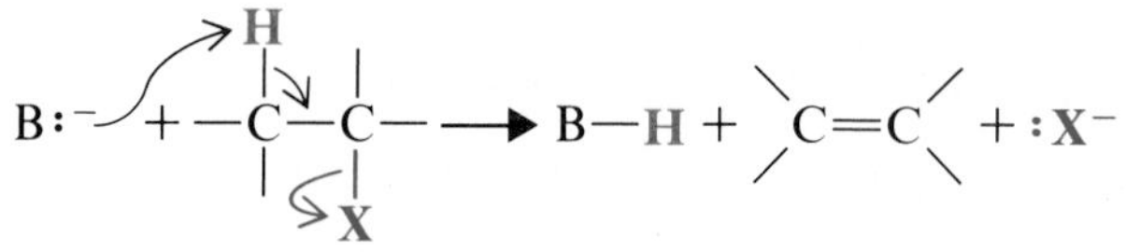

$$B: \overset{\curvearrowright}{} + -\overset{H}{\underset{X}{\overset{|}{\underset{|}{C}}}}-\overset{|}{\underset{|}{C}}- \longrightarrow B-H + \,\overset{\textstyle\diagdown}{}C=C\overset{\textstyle\diagup}{} + :X^-$$

ADDITIONAL PROBLEMS

5.12
Show how you might use a nucleophilic substitution reaction of propyl bromide to synthesize each of the following compounds. (You may use any other compounds that are necessary.)

(a) $CH_3CH_2CH_2OH$

(b) $CH_3CH_2CH_2I$

(c) $CH_3CH_2OCH_2CH_2CH_3$

(d) $CH_3CH_2CH_2-S-CH_3$

(e) $CH_3\overset{\overset{\textstyle O}{\|}}{C}OCH_2CH_2CH_3$

(f) $CH_3CH_2CH_2N_3$

(g) $CH_3-\overset{\overset{\textstyle CH_3}{|}}{\underset{\underset{\textstyle CH_3}{|}}{N^+}}-CH_2CH_2CH_3 \ Br^-$

(h) $CH_3CH_2CH_2CN$

(i) $CH_3CH_2CH_2SH$

5.13
Which alkyl halide would you expect to react more rapidly by an S_N2 mechanism? Explain your answer.

(a) $CH_3CH_2CH_2CH_2Br$ or $CH_3CH_2\underset{\underset{\textstyle Br}{|}}{C}HCH_3$

(b) $CH_3CH_2\underset{\underset{\textstyle Br}{|}}{C}HCH_3$ or $CH_3\underset{\underset{\textstyle Br}{|}}{\overset{\overset{\textstyle CH_3}{|}}{C}}CH_3$

(c) $CH_3CH_2CH_2Cl$ or $CH_3CH_2CH_2Br$

(d) $CH_3\underset{\underset{\textstyle CH_3}{|}}{C}HCH_2CH_2Br$ or $CH_3CH_2\underset{\underset{\textstyle CH_3}{|}}{C}HCH_2Br$

(e) CH_3CH_2Cl or $CH_2{=}CHCl$

5.14
Which S_N2 reaction of each pair would you expect to take place more rapidly in a protic solvent? Explain your answer.

(a) $CH_3CH_2CH_2Br + CH_3OH \longrightarrow CH_3CH_2CH_2OCH_3 + HBr$

     or

     $CH_3CH_2CH_2Br + CH_3O^- \longrightarrow CH_3CH_2CH_2OCH_3 + Br^-$

(b) $CH_3CH_2I + OH^- \longrightarrow CH_3CH_2OH + I^-$
or
$CH_3CH_2I + SH^- \longrightarrow CH_3CH_2SH + I^-$
(c) $CH_3Br + CH_3OH \longrightarrow CH_3OCH_3 + HBr$
or
$CH_3Br + CH_3SH \longrightarrow CH_3SCH_3 + HBr$
(d) $CH_3CH_2I + CH_3S^-(1.0\ M) \longrightarrow CH_3CH_2SCH_3 + I^-$
or
$CH_3CH_2I + CH_3S^-(2.0\ M) \longrightarrow CH_3CH_2SCH_3 + I^-$

5.15

Which S_N1 reaction would you expect to take place more rapidly? Explain your answer.

(a) $(CH_3)_3CI + CH_3OH \longrightarrow (CH_3)_3COCH_3 + HI$
or
$(CH_3)_3CCl + CH_3OH \longrightarrow (CH_3)_3COCH_3 + HCl$
(b) $(CH_3)_3CBr + H_2O \longrightarrow (CH_3)_3COH + HBr$
or
$(CH_3)_3CBr + CH_3OH \longrightarrow (CH_3)_3COCH_3 + HBr$
(c) $(CH_3)_3CCl + CH_3O^-(0.01\ M) \xrightarrow[CH_3OH]{} (CH_3)_3COCH_3 + Cl^-$
or
$(CH_3)_3CCl + CH_3O^-(0.001\ M) \xrightarrow[CH_3OH]{} (CH_3)_3COCH_3 + Cl^-$
(d) $(CH_3)_3CCl + H_2O \longrightarrow (CH_3)_3COH + HCl$
or
$(CH_3)_2C\!\!=\!\!CHCl + H_2O \longrightarrow (CH_3)_2C\!\!=\!\!CHOH + HCl$

5.16

With methane, ethane, and/or cyclopentane as your organic starting materials and using any needed solvents or inorganic reagents, outline syntheses of each of the following. More than one step may be necessary and you need not repeat steps carried out in earlier parts of this problem.

(a) CH_3I	(d) CH_3CH_2OH	(g) CH_3CN	(j) $CH_3OCH_2CH_3$
(b) CH_3CH_2I	(e) CH_3SH	(h) CH_3CH_2CN	(k) Cyclopentene
(c) CH_3OH	(f) CH_3CH_2SH	(i) CH_3OCH_3	

5.17

Listed here are several hypothetical nucleophilic substitution reactions. None is synthetically useful because the product indicated is *not* formed at an appreciable rate. In each case account for the failure of the reaction to take place.

(a) $HO^- + CH_3CH_3 \xrightarrow{\times} CH_3CH_2OH + H:^-$
(b) $HO^- + CH_3CH_2CH_3 \xrightarrow{\times} CH_3CH_2OH + CH_3:^-$

(c)
$$
\begin{array}{c}
\overset{H_2}{\underset{}{C}} \\
H_2C \diagup \quad \diagdown CH_2 \\
| \qquad\qquad | \\
H_2C \!\!-\!\!-\!\!-\!\! CH_2
\end{array}
+ H_2O \xrightarrow{\times} CH_3CH_2CH_2CH_2CH_2OH
$$

(d) $CN^- + (CH_3)_3CBr \xrightarrow{\times} (CH_3)_3C\!-\!CN + Br^-$
(e) $CH_3CH\!\!=\!\!CHBr + CH_3S^- \xrightarrow{\times} CH_3CH\!\!=\!\!CHSCH_3 + Br^-$
(f) $Cl^- + CH_3OCH_3 \xrightarrow{\times} CH_3Cl + CH_3O^-$
(g) $NH_3 + CH_3CH_2\overset{+}{O}H_2 \xrightarrow{\times} CH_3CH_2NH_3^+ + H_2O$
(h) $CH_3:^- + CH_3CH_2OH \xrightarrow{\times} CH_3CH_2CH_3 + OH^-$

5.18

You are given the task of preparing propene by dehydrohalogenating one of the halopropanes (i.e., $CH_3CH_2CH_2Br$ or $CH_3CHBrCH_3$). Which halide would you choose to give the alkene in maximum yield? Why?

5.19

Your task is to prepare isopropyl methyl ether $[CH_3OCH(CH_3)_2]$ by one of the following reactions. Which reaction would give the better yield? Explain your choice.

(1) $CH_3ONa + (CH_3)_2CHI \longrightarrow CH_3OCH(CH_3)_2$
(2) $(CH_3)_2CHONa + CH_3I \longrightarrow CH_3OCH(CH_3)_2$

5.20

Which product (or products) would you expect to obtain from each of the following reactions? In each case give the mechanism (S_N1, S_N2, E1, or E2) by which each product is formed and predict the relative amount of each (i.e., would the product be the only product, the major product, a minor product, etc.?).

(a) $CH_3CH_2CH_2CH_2Br + CH_3O^- \xrightarrow[CH_3OH]{50\ °C}$

(b) $CH_3CH_2CH_2CH_2Br + (CH_3)_3CO^- \xrightarrow[(CH_3)_3COH]{50\ °C}$

(c) $(CH_3)_3CO^- + CH_3I \xrightarrow[(CH_3)_3COH]{50\ °C}$

(d) $(CH_3)_3CI + CH_3O^- \xrightarrow[CH_3OH]{50\ °C}$

(e)

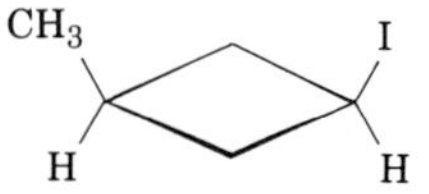

$+ CH_3O^- \xrightarrow[CH_3OH]{50\ °C}$

(f)

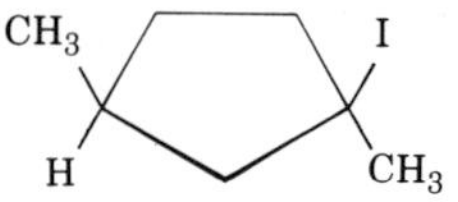

$\xrightarrow[CH_3OH]{25\ °C}$

(g) $CH_3CH_2CHCH_2CH_3 + C_2H_5O^- \xrightarrow[C_2H_5OH]{50\ °C}$
         |
         Br

(h) $(CH_3)_3CO^- + CH_3CHCH_3 \xrightarrow[(CH_3)_3COH]{50\ °C}$
                        |
                        Br

(i) $HO^- + (R)$-2-bromobutane $\xrightarrow{25\ °C}$

(j) (S)-3-bromo-3-methylhexane $\xrightarrow[CH_3OH]{25\ °C}$

(k) (S)-2-bromooctane $+ I^- \xrightarrow[CH_3OH]{50\ °C}$

5.21

Write conformational structures for the substitution products of the following deuterium-labeled compounds:

(a)

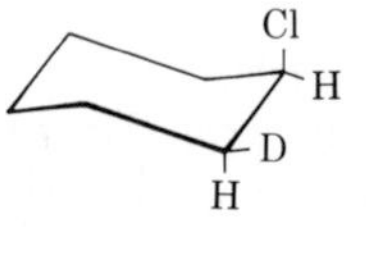

$\xrightarrow[CH_3OH]{I^-}$?

(b)

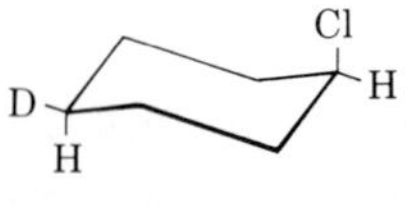

$\xrightarrow[CH_3OH]{I^-}$?

(c) 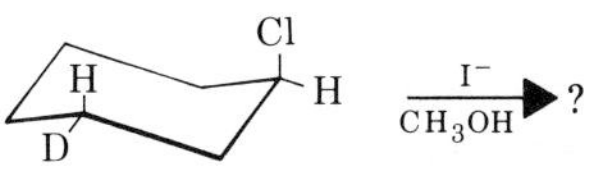 $\xrightarrow[\text{CH}_3\text{OH}]{\text{I}^-}$?   (d) 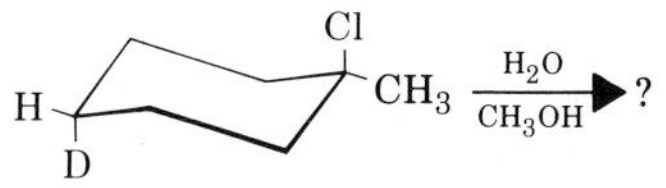 $\xrightarrow[\text{CH}_3\text{OH}]{\text{H}_2\text{O}}$?

5.22

Although ethyl bromide and isobutyl bromide are both primary halides, ethyl bromide undergoes S_N2 reactions more than 10 times faster than isobutyl bromide. When each compound is treated with a strong base/nucleophile ($CH_3CH_2O^-$), isobutyl bromide gives a greater yield of elimination products than substitution products, whereas with ethyl bromide this behavior is reversed. What factor accounts for these results?

5.23

Consider the reaction of I^- with CH_3CH_2Cl. (a) Would you expect the reaction to be S_N1 or S_N2? The rate constant for the reaction at 60 °C is 5×10^{-5} L mole^{-1} s^{-1}. (b) What is the reaction rate if $[I^-] = 0.1$ mole/L^{-1} and $[CH_3CH_2Cl] = 0.1$ mole/L^{-1}? (c) If $[I^-] = 0.1$ mole/L^{-1} and $[CH_3CH_2Cl] = 0.2$ mole/L^{-1}? (d) If $[I^-] = 0.2$ mole/L^{-1} and $[CH_3CH_2Cl] = 0.1$ mole/L^{-1}? (e) If $[I^-] = 0.2$ mole/L^{-1} and $[CH_3CH_2Cl] = 0.2$ mole/L^{-1}?

5.24

Which reagent in each pair listed here would be the stronger nucleophile in a protic solvent?

(a) CH_3NH^- or CH_3NH_2

(b) CH_3O^- or $CH_3\overset{\displaystyle O}{\overset{\|}{C}}O^-$

(c) CH_3SH or CH_3OH

(d) $(C_6H_5)_3N$ or $(C_6H_5)_3P$

(e) H_2O or H_3O^+

(f) NH_3 or NH_4^+

(g) H_2S or HS^-

(h) $CH_3\overset{\displaystyle O}{\overset{\|}{C}}O^-$ or OH^-

5.25

Write mechanisms that account for the products of the following reactions:

(a) $HOCH_2CH_2Br \xrightarrow[\text{H}_2\text{O}]{\text{OH}^-} H_2C\overset{\displaystyle O}{\underset{}{\diagdown\!\!\diagup}}CH_2$

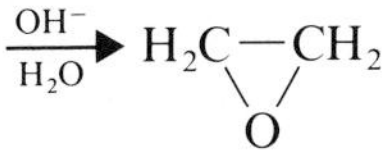

(b) $H_2NCH_2CH_2CH_2CH_2Br \xrightarrow[\text{H}_2\text{O}]{\text{OH}^-}$ pyrrolidine ring

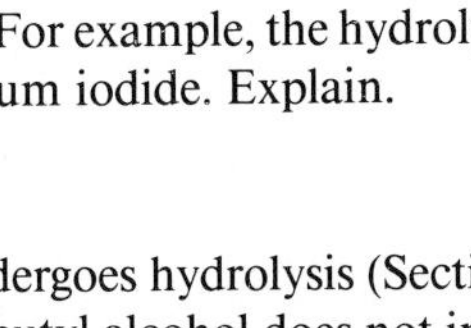

5.26

Many S_N2 reactions of alkyl chlorides and alkyl bromides are catalyzed by the addition of sodium or potassium iodide. For example, the hydrolysis of methyl bromide takes place much faster in the presence of sodium iodide. Explain.

5.27

When *tert*-butyl chloride undergoes hydrolysis (Section 5.10) in aqueous sodium hydroxide, the rate of formation of *tert*-butyl alcohol does not increase appreciably as the hydroxide ion concentration is increased. Increasing hydroxide ion concentration, however, causes a marked increase in the rate of disappearance of *tert*-butyl chloride. Explain.

5.28

(a) Consider the general problem of converting a tertiary alkyl halide to an alkene, for example, the conversion of *tert*-butyl chloride to 2-methylpropene. What experimental conditions would you choose to insure that elimination is favored over substitution? (b) Consider the opposite problem, that of carrying out a substitution reaction on a tertiary alkyl halide. Use as your example the conversion of *tert*-butyl chloride to *tert*-butyl ethyl ether. What experimental conditions would you employ to insure the highest possible yield of the ether?

5.29

Bridged cyclic compounds like those shown here are extremely *unreactive* in S_N2 reactions.

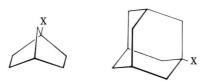

(a) Give a reason that will explain this. (b) How can you explain the fact that compounds of this type are also less reactive in S_N1 reactions than similar noncyclic compounds? (Consider the fact that carbocations are generally sp^2 hybridized.)

5.30

When CH_3Br reacts with CN^- the major product is CH_3CN, but some CH_3NC is formed as well. Write the Lewis structure for both products and explain.

*5.31

The relative rates of ethanolysis of several primary alkyl halides are as follows: CH_3CH_2Br, 1.0; $CH_3CH_2CH_2Br$, 0.28; $(CH_3)_2CHCH_2Br$, 0.030; $(CH_3)_3CCH_2Br$, 0.0000042. (a) Are these reactions S_N1 or S_N2? (b) What factor will account for these relative reactivities?

*5.32

In contrast to S_N2 reactions, S_N1 reactions show relatively little nucleophile selectivity. That is, when more than one nucleophile is present in the reaction medium, S_N1 reactions show only a slight tendency to discriminate between weak nucleophiles and strong nucleophiles, whereas S_N2 reactions show a marked tendency to discriminate. (a) Provide an explanation for this behavior. (b) Show how your answer accounts for the fact that $CH_3CH_2CH_2CH_2Cl$ reacts with 0.01 M NaCN in ethanol to yield primarily $CH_3CH_2CH_2CH_2CN$, whereas under the same conditions, $(CH_3)_3CCl$ reacts to give primarily $(CH_3)_3COCH_2CH_3$.

*5.33

When *tert*-butyl bromide undergoes S_N1 hydrolysis, adding a "common ion" (e.g., NaBr) to the aqueous solution has no effect on the rate. On the other hand when $(C_6H_5)_2CHBr$ undergoes S_N1 hydrolysis, adding NaBr retards the reaction. Given that the $(C_6H_5)_2CH^+$ cation is known to be much more stable than the $(CH_3)_3C^+$ cation (and we shall see why in Section 12.10B), provide an explanation for the different behavior of the two compounds.

*5.34

When the alkyl bromides (listed here) were subjected to hydrolysis in a mixture of ethanol and water (80% C_2H_5OH/20% H_2O) at 55 °C, the rates of the reaction showed the following order:

$$(CH_3)_3CBr > CH_3Br > CH_3CH_2Br > (CH_3)_2CHBr$$

Provide an explanation for this order of reactivity.

A BIOLOGICAL NUCLEOPHILIC SUBSTITUTION REACTION: BIOLOGICAL METHYLATION

The cells of living organisms synthesize many of the compounds they need from smaller molecules. Often these biosyntheses resemble the syntheses organic chemists carry out in their laboratories. Let us examine one example now.

Many reactions take place in the cells of plants and animals that involve the transfer of a methyl group from an amino acid called methionine to some other compound. That this transfer takes place can be demonstrated experimentally by feeding a plant or animal methionine containing a radioactive carbon atom (^{14}C) in its methyl group. Later, other compounds containing the "labeled" methyl group can be isolated from the organism. Some of the compounds that get their methyl groups from methionine are the following. The radioactively labeled carbon atom is shown in color.

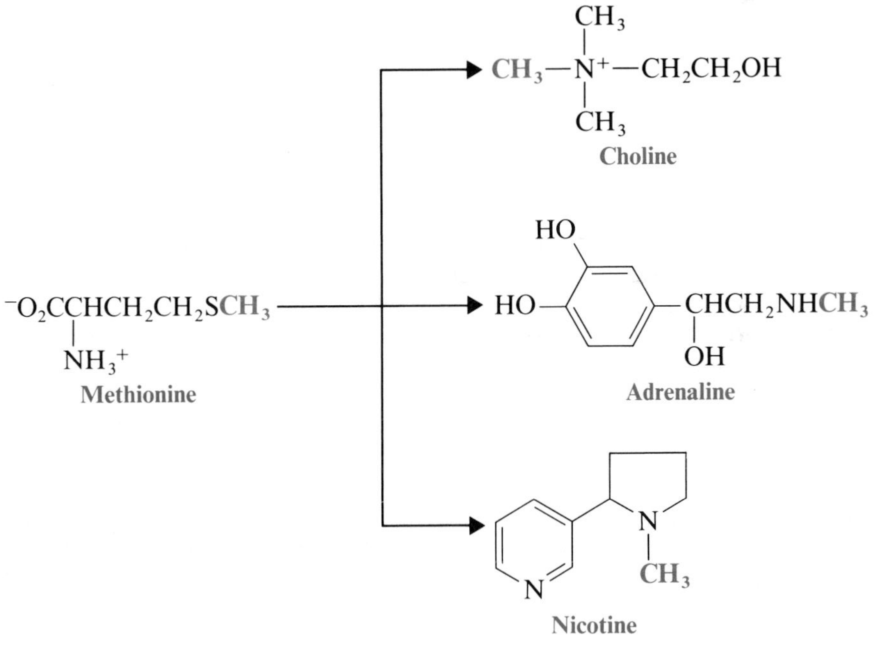

$$CH_3-N^+-CH_2CH_2OH$$

with CH_3 groups above and below the nitrogen.

Choline

$$^-O_2CCHCH_2CH_2SCH_3$$
with NH_3^+ below.

Methionine

Adrenaline structure: HO and HO on benzene ring, with $-CHCH_2NHCH_3$ and OH.

Adrenaline

Nicotine structure with pyridine ring and N—CH_3.

Nicotine

Choline is important in the transmission of nerve impulses, adrenaline causes blood pressure to increase, and nicotine is the compound contained in tobacco that makes smoking tobacco addictive. (In larger doses nicotine is poisonous.)

The transfer of the methyl group from methionine to these other compounds does not take place directly. The actual methylating agent is not methionine; it is S-adenosylmethionine,* a compound that results when methionine reacts with

---

*The prefix S is a locant meaning "on the sulfur atom" and should not be confused with the (S) used to distinguish absolute configuration. Another example of this kind of locant is N meaning "on the nitrogen atom" (Section 5.13C and Special Topic A).

adenosine triphosphate (ATP):

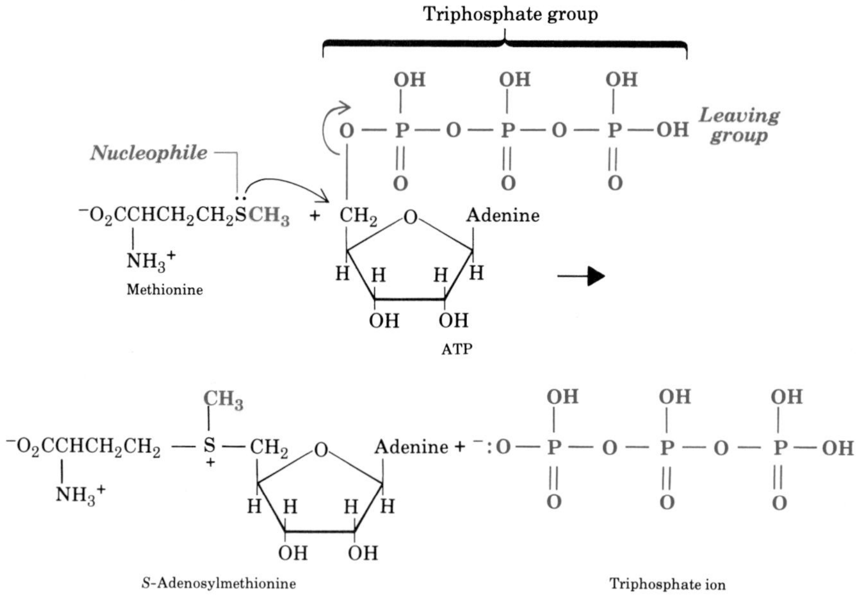

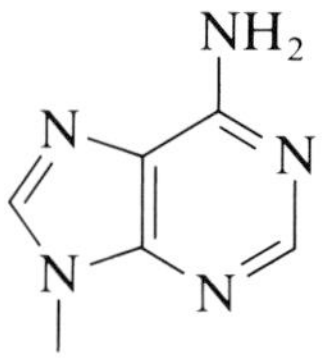

This reaction is a nucleophilic substitution reaction. The nucleophilic atom is the sulfur atom of methionine. The leaving group is the weakly basic triphosphate group of adenosine triphosphate. The product, *S*-adenosylmethionine, contains a methyl-sulfonium group, $CH_3-\overset{|}{S^+}-$.

S-Adenosylmethionine then acts as the substrate for other nucleophilic substitution reactions. In the biosynthesis of choline, for example, it transfers its methyl group to a nucleophilic nitrogen atom of 2-(*N*, *N*-dimethylamino)ethanol:

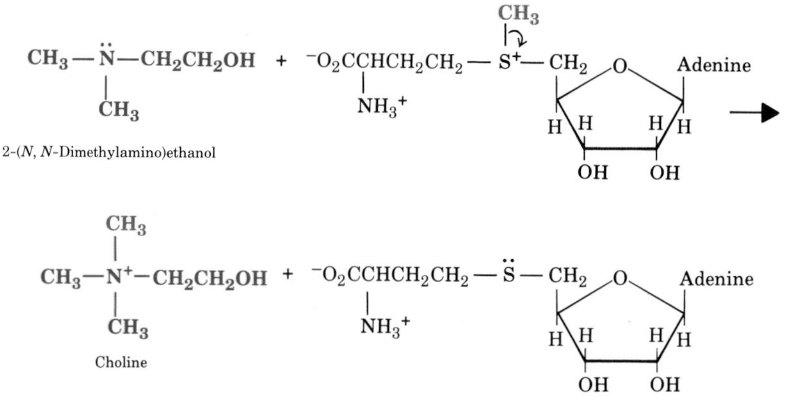

These reactions appear complicated only because the structures of the nucleophiles and substrates are complex. Yet conceptually they are simple and they illustrate many of the principles we have encountered in Chapter 5. In them we see how nature makes use of the high nucleophilicity of sulfur atoms. We also see how a weakly basic group (e.g., the triphosphate group of ATP) functions as a leaving group. In the reaction of 2-(N,N-dimethylamino)ethanol we see that the more basic $(CH_3)_2N$ group acts as the nucleophile rather than the less basic —OH group. And when a nucleophile attacks S-adenosylmethionine, we see that the attack takes place at the less hindered CH_3—group rather than at one of the more hindered —CH_2—groups.

PROBLEM A.1

(a) What is the leaving group when 2-(N,N-dimethylamino)ethanol reacts with S-adenosylmethionine? (b) What would the leaving group have to be if methionine itself were to react with 2-(N,N-dimethylamino)ethanol? (c) Of what special significance is this difference?

SPECIAL TOPIC B

ELEMENTARY THERMODYNAMICS: $\Delta H°$, $\Delta S°$, AND $\Delta G°$

In Chapter 5 we devoted a great deal of attention to the question of how fast reactions occur. We saw that most reactions have an energy barrier between reactants and products and that the most important factor determining how rapidly a reaction takes place is the height of that barrier, the energy of activation, E_{act}.

When we consider the question of how far a reaction goes, one important factor is the relative energy (stability) of the reactants and products. We learned that the spontaneous tendency for reactions to proceed downhill (in energy terms) favors the formation of those products that are downhill. The energy change with which we have been mainly concerned is the enthalpy change, $\Delta H°$. The change in enthalpy is a function largely of the potential energy in chemical bonds. Reactions in which strong bonds are formed at the expense of weak bonds are exothermic. Those in which weak bonds are formed from strong bonds are endothermic. Consider the reaction of methane with chlorine as an example. ($DH°$ is the energy required in kcal/mole to break the bond homolytically.)

$$CH_3{-}H \ + \ Cl{-}Cl \ \longrightarrow \ CH_3{-}Cl \ + \ H{-}Cl$$
$$(DH° = 104) \quad (DH° = 58) \qquad (DH° = 83.5) \quad (DH° = 103)$$

$$\Delta H° = (104 + 58) - (83.5 + 103) = -24.5 \text{ kcal/mole}$$

Because the bonds of CH_3Cl and HCl are collectively stronger than those of CH_4 and Cl_2, the reaction is exothermic. In this case the reaction is highly exothermic; $\Delta H° = -24.5$ kcal/mole.

Many studies of a large number of chemical reactions have shown that reactions that have a negative enthalpy change as large as this almost always go to completion at equilibrium. A *small* negative value for $\Delta H°$, however, is not a guarantee that a reaction will produce substantial proportions of products at equilibrium. To account for this we need to consider two other thermodynamic quantities; $\Delta G°$, the *free-energy change,* and $\Delta S°$, *the entropy change.* To see how these thermodynamic quantities determine the position of equilibrium of a chemical reaction let us consider the general reaction written here:

$$A + B \rightleftharpoons C$$

An expression can be written for the equilibrium constant, K_{eq},

$$K_{eq} = \frac{[C]}{[A][B]}$$

where [A] and [B] are the equilibrium concentrations of the reactants A and B in moles per liter, and [C] is the equilibrium concentration of the product C in moles per liter.

By convention, the equilibrium concentration of the products of a reaction are written in the numerator of this expression and those of the reactants are written in

TABLE B.1 Relation between $\Delta G°$ and K_{eq}
at 25°C for reactions of the type X $\rightleftharpoons$ Y

$\Delta G°$, (kcal/mole)	−0.41	−0.65	−0.82	−0.95	−1.4	−1.8	−2.7	−4.1	−5.5
K_{eq}	2	3	4	5	10	20	100	1000	10,000
Percentage of Y at equilibrium	67	75	80	83	91	95	99	99.9	99.99

Adapted from E. Eliel, *Stereochemistry of Carbon Compounds,* Wiley, New York, 1962, p. 207.

the denominator. This means that a large value of K_{eq} is associated with a reaction that goes essentially to completion, and a small value of K_{eq} is associated with a reaction that produces only small amounts of products at equilibrium.

PROBLEM B.1

Consider the simple reaction:

$$X \rightleftharpoons Y$$

Assume an initial concentration of X equal to 1.0 mole/L, and calculate the equilibrium concentration of Y that is formed if the equilibrium constant is (a) 10, (b) 1, (c) 10^{-3}. In each case, what percentage of the reactant is converted to product?

The equilibrium constant for a reaction is directly related to the thermodynamic quantity called the free-energy change for the reaction. The free-energy change is symbolized as $\Delta G°$ and the relationship between $\Delta G°$ and the equilibrium constant, K_{eq}, is

$$\log K_{eq} = \frac{\Delta G°}{-2.3RT}$$

where R is the gas constant (1.986 cal/deg-mole) and T is the absolute temperature.
It is easy to show with this equation that *the more negative $\Delta G°$ is, the larger will be the value of the equilibrium constant, K_{eq}. And the larger the value for K_{eq}, of course, the more the reaction will favor the formation of products.* Table B.1 gives the results of several calculations for the simple equilibrium X $\rightleftharpoons$ Y.
What Table B.1 shows us is that a relatively modest negative $\Delta G°$ (e.g., −4.1 kcal/mole) means that the product will be the major component (99.9%) present at equilibrium. For a reaction X $\rightleftharpoons$ Y, it means for all practical purposes that the reaction goes to completion.

PROBLEM B.2

Table B.1 is especially useful when applied to equilibria involving two forms of a cyclic molecule such as those we studied in Chapter 3. (a) The free-energy change $\Delta G°$ is −5 to −6 kcal/mole for the boat $\rightleftharpoons$ chair form of cyclohexane. What does this mean about the relative population of the two forms at equilibrium? (b) For the axial $\rightleftharpoons$ equatorial form of ethylcyclohexane, $\Delta G° \approx -1.8$ kcal/mole, what percentage of the axial form is present at equilibrium?

The free-energy change ($\Delta G°$) and the enthalpy change ($\Delta H°$) are related to one another through another equation that involves a third thermodynamic quantity, the entropy change ($\Delta S°$), and the absolute temperature, T. This important equation is

$$\Delta G° = \Delta H° - T\Delta S°$$

The entropy of any system is a measure of the relative disorder of that system. The more random (less ordered) a system is, the greater is its entropy. Thus, a *positive* entropy change ($+\Delta S°$) is always associated with a change from a more ordered arrangement to a more disordered one. A negative entropy change ($-\Delta S°$) always accompanies the reverse process.

For a chemical system the relative disorder (or randomness) of the molecules can be related to the number of *degrees of freedom* available to the molecules and their constituent atoms. Degrees of freedom are associated with ways in which *movement* or *changes in relative position* can occur. Molecules have three sorts of degrees of freedom: translational degrees of freedom associated with movements of the whole molecule through space, rotational degrees of freedom associated with the tumbling motions of the molecule, and vibrational degrees of freedom associated with the stretching and bending motion of atoms about the bonds that connect them. If the

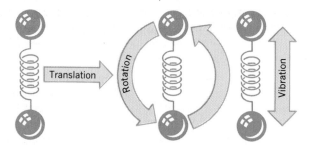

atoms of the products of a reaction have more degrees of freedom available than they did as reactants, the entropy change for the reaction will be positive. If, on the other hand, the atoms of the products are more constrained (have fewer degrees of freedom) than they were as reactants, a negative $\Delta S°$ will result.

With these ideas and relationships in mind, we are now in a position to understand how these three thermodynamic quantities, $\Delta H°$, $\Delta G°$, and $\Delta S°$, account for the course of a chemical reaction. For reactions occurring near room temperature,* if $\Delta H°$ has a large negative value ($\Delta H° > -15$ kcal/mole), a small negative entropy

*At very high temperatures the entropy term ($T\Delta S°$) in the equation $\Delta G° = \Delta H° - T\Delta S°$ becomes much more important because T becomes very large. Most of the reactions that we shall consider, however, are carried out between 250 and 450 K. Under these conditions $\Delta H°$ is the important term if it is more negative than -15 kcal/mole.

change (associated with the atoms being more constrained in the products) will not generally result in a positive free-energy change ($+\Delta G°$). For these reactions $\Delta G°$ will be negative and the formation of products at equilibrium will be favored:

$\Delta G° = \qquad\qquad \Delta H° \qquad\qquad - \qquad\qquad [T \qquad\qquad \times \qquad \Delta S°]$
$\Delta G° = \text{(large negative quantity)} - [\text{(positive quantity)} \qquad \times \text{(small negative quantity)}]$

$\Delta G° = \text{(large negative quantity)} - \text{(smaller negative quantity)}$
$\Delta G° = \text{negative quantity}$

We can now see how, in some reactions that are endothermic (have a $+\Delta H$), the formation of products at equilibrium will still be favored. These reactions have an accompanying positive entropy change large enough to offset the unfavorable enthalpy change. This can occur when the atoms of the products are much less constrained than those of the reactants. In such cases the large positive entropy term $(T) \times (+\Delta S°)$ more than compensates for the unfavorable $+\Delta H°$.

$\Delta G° = \qquad\qquad \Delta H° \qquad\qquad - \qquad\qquad [T \qquad\qquad \times \qquad \Delta S°]$
$\Delta G° = \text{(small positive quantity)} - [\text{(positive quantity)} \qquad \times \text{(large positive quantity)}]$

$\Delta G° = \text{(small positive quantity)} - \text{(larger positive quantity)}$
$\Delta G° = \text{negative quantity}$

For many reactions the entropy change is so small that the term $T\Delta S°$ is almost zero and $\Delta H° \simeq \Delta G°$. This happens when the reaction is one in which the number of degrees of freedom available to molecules of reactants and to molecules of products is essentially the same. Consider the reaction of methane and chlorine again:

$$CH_4 + Cl_2 \longrightarrow CH_3Cl + HCl$$

Here, 2 moles of product molecules are formed from the same number of moles of reactant molecules. Thus the number of translational degrees of freedom available to products and reactants will be approximately the same. Moreover, CH_3Cl is a tetrahedral molecule like CH_4 and HCl is a diatomic molecule like Cl_2. This means that vibrational and rotational degrees of freedom available to products and reactants should also be approximately the same. The actual entropy change for this reaction is quite small, $\Delta S° = +0.67$ cal K^{-1} mole^{-1}. Thus at room temperature (298 K) the $T\Delta S$ term is only $+0.2$ kcal/mole. The enthalpy change for the reaction and the free-energy change are almost equal, $\Delta H° = -24.5$ kcal/mole and $\Delta G° = -24.7$ kcal/mole.

Since the free-energy change for the chlorination of methane is such a large negative number, the equilibrium constant is enormous, $K_{eq} \simeq 10^{18}$.

PROBLEM B.3

The reaction of ethyne with hydrogen to produce ethene,

$$H-C\equiv C-H_{(g)} + H_{2(g)} \longrightarrow CH_2=CH_{2(g)}$$

has the following values of $\Delta H°$ and $\Delta S°$ at 27 °C (300 K): $\Delta H° = -41.7$ kcal/mole, $\Delta S° = -26.6$ cal K^{-1} mole^{-1}. (a) Calculate the value of $\Delta G°$ for this reaction. (*Remember:* 1000 cal = 1.0 kcal.) (b) Would you expect the position of equilibrium to favor the formation of ethene? (c) Does the entropy change for this reaction, taken alone, favor the formation of ethene? (d) How can you account for the negative entropy change?

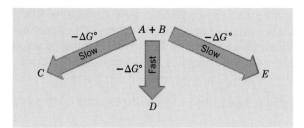

FIGURE B.1 The hypothetical reaction of A and B to form different products, C, D, and E. All of the reactions have favorable free-energy changes. D is the product that is actually obtained, however, because it is produced by the reaction that proceeds most rapidly. E and C are not isolated in appreciable amounts because the reactions by which they are formed are slow.

The reaction in Problem B.3 takes place in the gas phase and is therefore, relatively simple. For reactions taking place in solution the situation is much more complex because an analysis must also take into account enthalpy and entropy changes of the solvent.

Remember: A negative free-energy change for a reaction only ensures that a substantial proportion of products will be formed *when the reaction reaches equilibrium.* The overall free-energy change for a reaction tells us nothing, however, about how long that particular reaction will take to reach equilibrium. The reaction of hydrogen and oxygen to produce water, for example,

$$2\ H_2 + O_2 \longrightarrow 2H_2O$$

has associated with it a very large negative free-energy change. However, it also has a large energy of activation. In the absence of a catalyst (or applied heat), however, mixtures of hydrogen and oxygen react so slowly that the formation of water is essentially imperceptible.

For many other reactions (Fig. B.1) pathways to several different products may have favorable free-energy changes. In such cases, the product that we actually obtain in greatest amount will be the one that comes from the pathway with lowest energy of activation. It will form the most rapidly. We can see from this that a knowledge of the factors that determine reaction rates is of considerable importance as well.

CHAPTER SIX

An Organic Pigment.

Alkenes and Alkynes I. Properties and Synthesis

6.1 INTRODUCTION

Alkenes are hydrocarbons whose molecules contain the carbon–carbon double bond. An old name for this family of compounds that is still often used is the name *olefins*. Ethene, the simplest olefin (alkene), was called olefiant gas (Latin: *oleum,* oil + *facere,* to make) because gaseous ethene (C_2H_4) reacts with chlorine to form $C_2H_4Cl_2$, a liquid (oil).

Hydrocarbons whose molecules contain the carbon–carbon triple bond are called alkynes. The common name for this family is *acetylenes,* after the first member, $HC\equiv CH$.

6.2 NOMENCLATURE OF ALKENES AND CYCLOALKENES

Many older names for alkenes are still in common use. Propene is often called propylene, and 2-methylpropene frequently bears the name isobutylene.

$$CH_2{=}CH_2 \qquad CH_3CH{=}CH_2 \qquad CH_3{-}\overset{\overset{\displaystyle CH_3}{|}}{C}{=}CH_2$$

IUPAC: Ethene   *IUPAC:* Propene   *IUPAC:* 2-Methylpropene
or ethylene*   or propylene*   *Common:* Isobutylene

The IUPAC rules for naming alkenes are similar in many respects to those for naming alkanes:

1. **Determine the base name by selecting the longest chain that contains the double bond and change the ending of the name of the alkane of identical length from ane to ene.** Thus, if this longest chain contains five carbon atoms, the base name for the alkene is *pentene;* if it contains six carbon atoms, the base name is *hexene,* and so on.

2. **Number the chain so as to include both carbon atoms of the double bond, and begin numbering at the end of the chain nearer the double bond. Designate the**

---

*The IUPAC system also retains the name ethylene and propylene when no substituents are present.

location of the double bond by using the number of the first atom of the double bond as a prefix:

$$\overset{1}{C}H_2=\overset{2}{C}H\overset{3}{C}H_2\overset{4}{C}H_3 \qquad CH_3CH=CHCH_2CH_2CH_3$$

1-Butene
(*not* 3-butene)

2-Hexene
(*not* 4-hexene)

3. **Indicate the locations of the substituent groups by the numbers of the carbon atoms to which they are attached.**

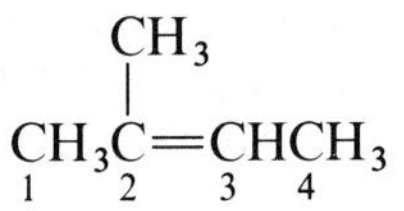

$$\begin{array}{c} CH_3 \\ | \\ \overset{1}{C}H_3\overset{2}{C}=\overset{3}{C}H\overset{4}{C}H_3 \end{array} \qquad \begin{array}{c} CH_3 \qquad\quad CH_3 \\ | \qquad\qquad | \\ \overset{1}{C}H_3\overset{2}{C}=\overset{3}{C}H\overset{4}{C}H_2\overset{5}{C}H\overset{6}{C}H_3 \end{array}$$

2-Methyl-2-butene
(*not* 3-methyl-2-butene)

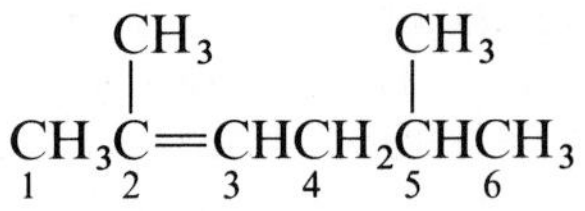

2,5-Dimethyl-2-hexene
(*not* 2,5-dimethyl-4-hexene)

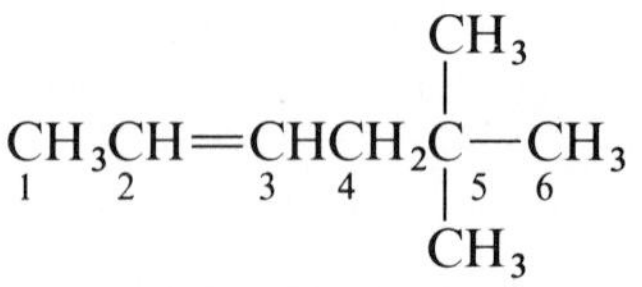

$$\begin{array}{c} CH_3 \\ | \\ \overset{1}{C}H_3\overset{2}{C}H=\overset{3}{C}H\overset{4}{C}H_2\overset{5}{C}-\overset{6}{C}H_3 \\ | \\ CH_3 \end{array} \qquad \overset{4}{C}H_3\overset{3}{C}H=\overset{2}{C}H\overset{1}{C}H_2Cl$$

5,5-Dimethyl-2-hexene

1-Chloro-2-butene

4. **Number substituted cycloalkenes in the way that gives the carbon atoms of the double bond the 1- and 2- positions and that also gives the substituent groups the lower numbers at the first point of difference.** With substituted cycloalkenes it is not necessary to specify the position of the double bond since it will always be on C-1. The two examples listed here illustrate the application of these rules.

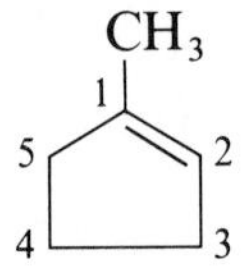

1-Methylcyclopentene
(*not* 2-methylcyclopentene)

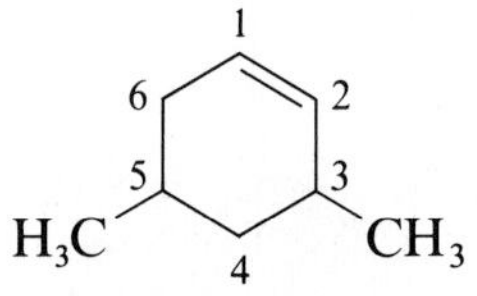

3,5-Dimethylcyclohexene
(*not* 4,6-dimethylcyclohexene)

5. Two frequently encountered alkenyl groups are the *vinyl group* and the *allyl group.*

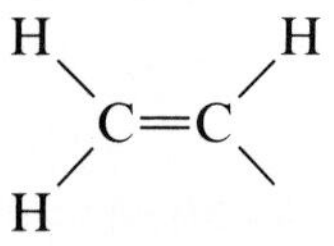

or   or

$$CH_2=CH- \qquad\qquad CH_2=CHCH_2-$$

The vinyl group   **The allyl group**

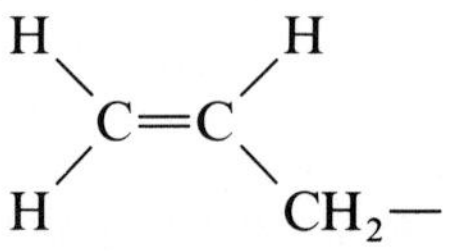

The following examples illustrate how these names are employed:

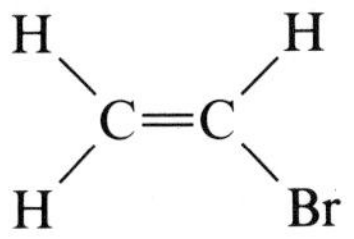

$$
\begin{array}{cc}
\underset{H}{\overset{H}{\diagdown}}C=C\underset{Br}{\overset{H}{\diagup}} & \underset{H}{\overset{H}{\diagdown}}C=C\underset{CH_2Cl}{\overset{H}{\diagup}}
\end{array}
$$

Bromoethene
or
vinyl bromide
(common)

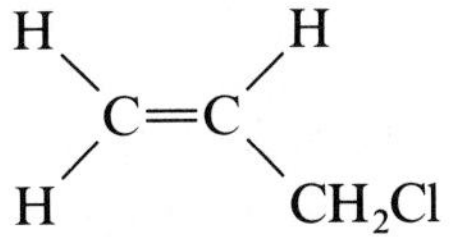

3-Chloropropene
or
allyl chloride
(common)

6. Designate the geometry of a double bond with two identical groups with the prefixes *cis*- and *trans*-. If two identical groups (usually hydrogen atoms) are on the same side of the double bond, it is **cis;** if they are on opposite sides, it is **trans.**

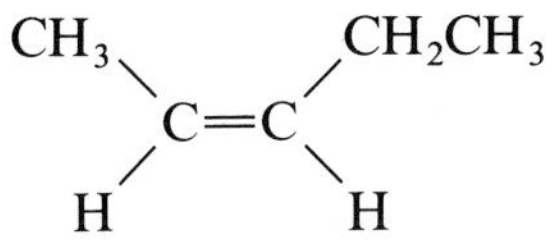

$$
\begin{array}{cc}
\underset{H}{\overset{CH_3}{\diagdown}}C=C\underset{H}{\overset{CH_2CH_3}{\diagup}} & \underset{H}{\overset{CH_3}{\diagdown}}C=C\underset{CH_2CH_3}{\overset{H}{\diagup}}
\end{array}
$$

cis-2-Pentene

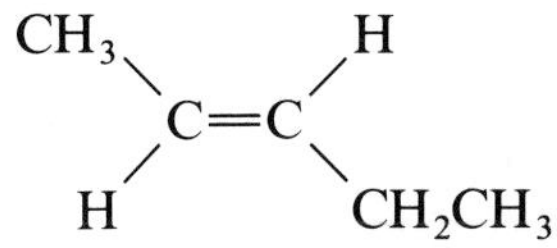

trans-2-Pentene

In the next section we shall see another method for designating the geometry of the double bond.

PROBLEM 6.1

Give IUPAC names for the following alkenes:

(a) $\underset{CH_3}{\overset{CH_3}{\diagdown}}C=C\underset{CH_3}{\overset{H}{\diagup}}$

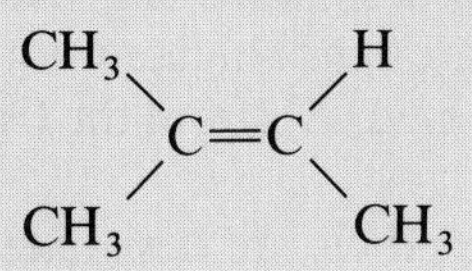

(c) $\underset{CH_3}{\overset{CH_3}{\diagdown}}C=C\underset{Br}{\overset{H}{\diagup}}$

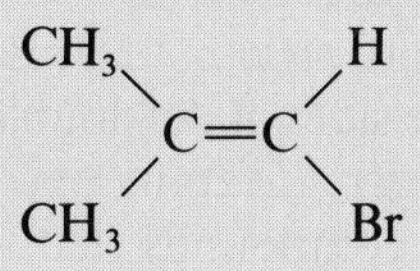

(b) $\underset{H}{\overset{CH_3CH_2CH_2}{\diagdown}}C=C\underset{H}{\overset{CH_2CH_2CH_3}{\diagup}}$

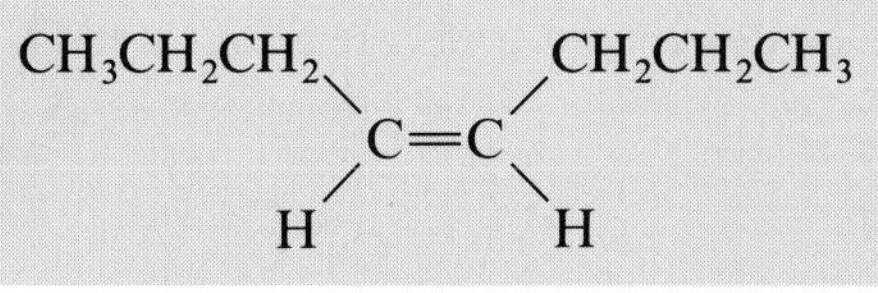

(d) [cyclohexene ring with CH₃ substituent]

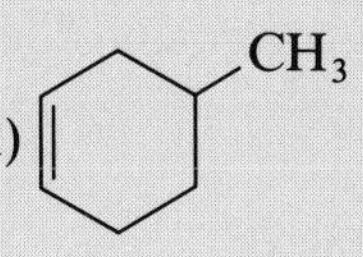

PROBLEM 6.2

Write structural formulas for

(a) *cis*-3-Hexene

(b) *trans*-2-Pentene

(c) 3-Ethylcyclohexene

(d) Vinylcyclohexane

(e) 4,4-Dimethyl-1-hexene

(f) 3-Methylcyclopentene

(g) 3-Chloro-1-octene

(h) 1,2-Dimethylcyclohexene

(i) 1,3-Dimethylcyclopentene

(j) 1,5-Dibromocyclohexene

6.2A The (E)–(Z) System
for Designating Alkene Diastereomers

The terms cis and trans, when used to designate the stereochemistry of alkene diastereomers, are unambiguous only when applied to disubstituted alkenes. If the
alkene is trisubstituted or tetrasubstituted, the terms cis and trans are either ambiguous or do not apply at all. Consider the following alkene as an example.

$$Br\diagdown C=C\diagup Cl$$
$$H\diagup\diagdown F$$

A

It is impossible to decide whether **A** is cis or trans since no two groups are the same.

A newer system is based on the priorities of groups in the Cahn–Ingold–Prelog
convention (Section 4.5). This system, called the (E)–(Z) system, applies to alkene
diastereomers of all types. In the (E)–(Z) system, we examine the two groups attached to one carbon atom of the double bond and decide which has higher priority.
Then we repeat that operation at the other carbon atom.

Higher → Cl   F   F   Cl ← Higher   Cl > F
priority   priority

Higher → Br   H   Higher → Br   H   Br > H
priority   (Z)   priority   (E)

We take the group of higher priority on one carbon atom and compare it with the
group of higher priority on the other carbon atom. If the two groups of higher priority
are on the same side of the double bond, the alkene is designated (Z) (from the
German word *zusammen,* meaning together). If the two groups of higher priority are
on opposite sides of the double bond, the alkene is designated (E) (from the German
word *entgegen,* meaning opposite).

Most compounds that would normally be designated cis are designated (Z).
Similarly most compounds that would be designated trans are designated (E).

CH₃ > H

$$CH_3\diagdown C=C\diagup CH_3$$
$$H\diagup\diagdown H$$
(Z)-2-Butene
(*cis*-2-butene)

$$CH_3\diagdown C=C\diagup H$$
$$H\diagup\diagdown CH_3$$
(E)-2-Butene
(*trans*-2-butene)

There are exceptions, however.

$$Cl\diagdown C=C\diagup Cl$$
$$H\diagup\diagdown Br$$
(E)-1-Bromo-1,2-dichloroethene
(1-bromo-*cis*-1,2-dichloroethene)

Cl > H
Br > Cl

$$Cl\diagdown C=C\diagup Br$$
$$H\diagup\diagdown Cl$$
(Z)-1-Bromo-1,2-dichloroethene
(1-bromo-*trans*-1,2-dichloroethene)

(a)
$$\begin{array}{c}Cl \\ \end{array}\overset{}{C}=\overset{}{C}\begin{array}{c}H \\ CH_2CH_3\end{array}$$
    Br

(c)
$$CH_3 \qquad CH_2CH_3$$
$$C=C$$
$$H \qquad CH(CH_3)_2$$

(b)
$$I \qquad Br$$
$$C=C$$
$$Cl \qquad CH_3$$

(d)
$$Cl \qquad CH_3$$
$$C=C$$
$$F \qquad CH_2CH_3$$

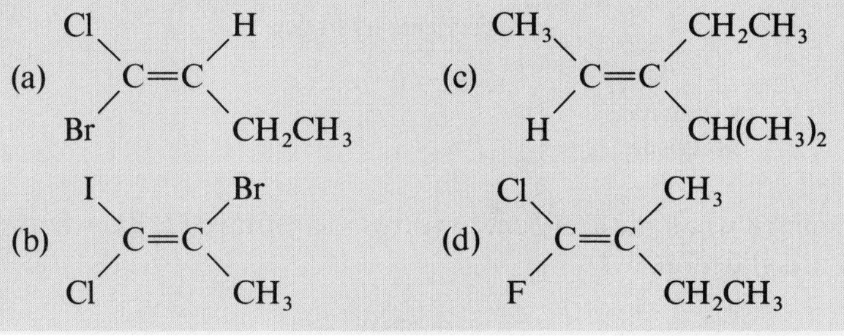

6.3 NOMENCLATURE OF ALKYNES

6.3A IUPAC Nomenclature

Alkynes are named in much the same way as alkenes. Unbranched alkynes, for example, are named by replacing the **-ane** of the name of the corresponding alkane with the ending **-yne.** The chain is numbered in order to give the carbon atoms of the triple bond the lowest possible numbers. The lower number of the two carbon atoms of the triple bond is used to designate the location of the triple bond. The IUPAC names of three unbranched alkynes are shown here.

$$H{-}C{\equiv}C{-}H \qquad CH_3CH_2C{\equiv}CCH_3 \qquad H{-}C{\equiv}CCH_2CH{=}CH_2$$
Ethyne or   2-Pentyne   1-Penten-4-yne†
acetylene*

The locations of substituent groups of branched alkynes and substituted alkynes are also indicated with numbers.

$$\overset{3}{C}l{-}\overset{2}{C}H_2\overset{1}{C}{\equiv}\overset{}{C}H \qquad \overset{4}{C}H_3\overset{3}{C}{\equiv}\overset{2}{C}\overset{1}{C}H_2Cl$$
3-Chloropropyne   1-Chloro-2-butyne

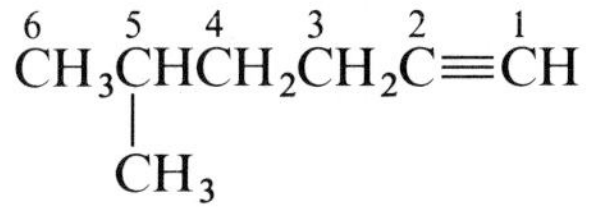

$$\overset{6}{C}H_3\overset{5}{C}H\overset{4}{C}H_2\overset{3}{C}H_2\overset{2}{C}{\equiv}\overset{1}{C}H \qquad \begin{array}{c}CH_3 \\ | \\ CH_3CCH_2C{\equiv}CH \\ | \\ CH_3\end{array}$$
    |
   CH_3
5-Methyl-1-hexyne 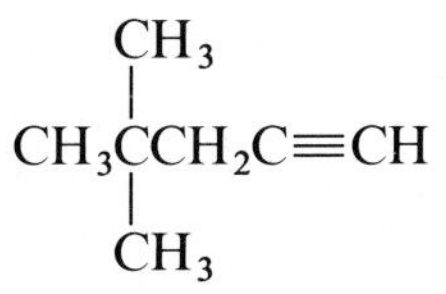   4,4-Dimethyl-1-pentyne

*The name acetylene is retained by the IUPAC system for the compound HC≡CH and is used much more frequently than the name ethyne.

†Where there is a choice the double bond is given the lower number.

Monosubstituted acetylenes or 1-alkynes are called **terminal alkynes,** and the hydrogen attached to the carbon of the triple bond is called the acetylenic hydrogen.

Acetylenic hydrogen

$$R—C≡C—H$$

A terminal
acetylene

The anion obtained when the acetylenic hydrogen is removed is known as *an alkynide ion* or an acetylide ion.

$$R—C≡C:^-$$ $$CH_3C≡C:^-$$

An alkynide ion   The propynide ion
(an acetylide ion)

6.4 PHYSICAL PROPERTIES OF ALKENES AND ALKYNES

Alkenes and alkynes have physical properties similar to those of corresponding alkanes. The lower-molecular-weight alkenes and alkynes (Tables 6.1 and 6.2) are gases at room temperature. Being relatively nonpolar themselves, alkenes and alkynes dissolve in nonpolar solvents or in solvents of low polarity. Alkenes and alkynes are only *very slightly soluble* in water (with alkynes being slightly more soluble than alkenes). The densities of alkenes and alkynes are less than that of water.

6.5 HYDROGENATION OF ALKENES

Alkenes react with hydrogen in the presence of a variety of finely divided metal catalysts (cf. Section 3.18). The reaction that takes place is an **addition reaction;** one atom of hydrogen *adds* to each carbon atom of the double bond. The catalysts that are most commonly used are finely divided metals. Without a catalyst the reaction does not take place at an appreciable rate. (We shall see how the catalyst functions in Section 6.6.)

TABLE 6.1 Physical constants of alkenes

NAME	FORMULA	mp (°C)	bp (°C)	DENSITY d_4^{20} (g/mL)
Ethene	$CH_2=CH_2$	-169	-104	0.384^a
Propene	$CH_3CH=CH_2$	-185	-47	0.514
1-Butene	$CH_3CH_2CH=CH_2$	-185	-6.3	0.595
(Z)-2-Butene	$CH_3CH=CHCH_3$ (cis)	-139	3.7	0.621
(E)-2-Butene	$CH_3CH=CHCH_3$ (trans)	-106	0.9	0.604
1-Pentene	$CH_3(CH_2)_2CH=CH_2$	-165	30	0.641
2-Methyl-1-butene	$CH_2=C(CH_3)CH_2CH_3$	-138	31	0.650
1-Hexene	$CH_3(CH_2)_3CH=CH_2$	-140	63	0.673
1-Heptene	$CH_3(CH_2)_4CH=CH_2$	-119	94	0.697

^{a}Density at -10 °C.

TABLE 6.2 Physical constants of alkynes

NAME	FORMULA	mp (°C)	bp (°C)	DENSITY d_4^{20} (g/mL)
Acetylene	$HC\equiv CH$	-81.8	-83.6	—
Propyne	$CH_3C\equiv CH$	-101.51	-23.2	—
1-Butyne	$CH_3CH_2C\equiv CH$	-125.7	8.1	—
2-Butyne	$CH_3C\equiv CCH_3$	-32.3	27	0.691
1-Pentyne	$CH_3(CH_2)_2C\equiv CH$	-90	39.3	0.695
2-Pentyne	$CH_3CH_2C\equiv CCH_3$	-101	55.5	0.714
1-Hexyne	$CH_3(CH_2)_3C\equiv CH$	-132	71	0.715
2-Hexyne	$CH_3(CH_2)_2C\equiv CCH_3$	-88	84	0.730
3-Hexyne	$CH_3CH_2C\equiv CCH_2CH_3$	-101	81.8	0.724

$$CH_2{=}CH_2 + H_2 \xrightarrow[\substack{\text{or Pt} \\ 25\,°C}]{\text{Ni, Pd}} CH_3{-}CH_3$$

$$CH_3CH{=}CH_2 + H_2 \xrightarrow[\substack{\text{or Pt} \\ 25\,°C}]{\text{Ni, Pd}} CH_3CH_2{-}CH_3$$

The product that results from the addition of hydrogen to an alkene is an alkane. Alkanes have only single bonds and contain the maximum number of hydrogen atoms that a hydrocarbon can possess. For this reason, alkanes are said to be **saturated compounds.** Alkenes, because they contain a double bond and possess fewer than the maximum number of hydrogens, are capable of adding hydrogen and are said to be **unsaturated.** The process of adding hydrogen to an alkene is sometimes described as being one of **reduction.** Most often, however, the term used to describe the addition of hydrogen is **catalytic hydrogenation.**

6.6 HYDROGENATION: THE FUNCTION OF THE CATALYST

Hydrogenation of an alkene is an exothermic reaction ($\Delta H° \cong -30$ kcal/mole).

$$R{-}CH{=}CH{-}R + H_2 \xrightarrow{\text{hydrogenation}} R{-}CH_2{-}CH_2{-}R + \text{heat}$$

Hydrogenation reactions usually have high energies of activation. The reaction of an alkene with molecular hydrogen does not take place at room temperature in the absence of a catalyst, but often *does* take place at room temperature when a metal catalyst is added. The catalyst provides a new pathway for the reaction with a *lower energy of activation* (Fig. 6.1).

The most commonly used catalysts for hydrogenation (finely divided platinum, nickel, palladium, rhodium, and ruthenium) apparently serve to adsorb hydrogen molecules on their surfaces. This adsorption of hydrogen is essentially a chemical reaction; unpaired electrons on the surface of the metal *pair* with the electrons of hydrogen (Fig. 6.2a) and bind the hydrogen to the surface. The collision of an alkene with the surface bearing adsorbed hydrogen causes adsorption of the alkene as well

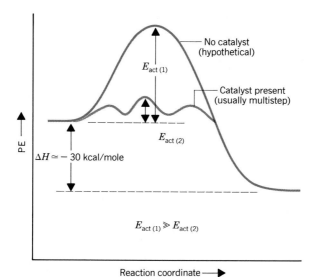

FIGURE 6.1 Potential energy diagram for the hydrogen-ation of an alkene in the presence of a catalyst and the hypothetical reaction in the absence of a catalyst. The energy of activation for the uncatalyzed reaction [$E_{act(1)}$] is very much larger than the largest energy of activation for the catalyzed reaction [$E_{act(2)}$].

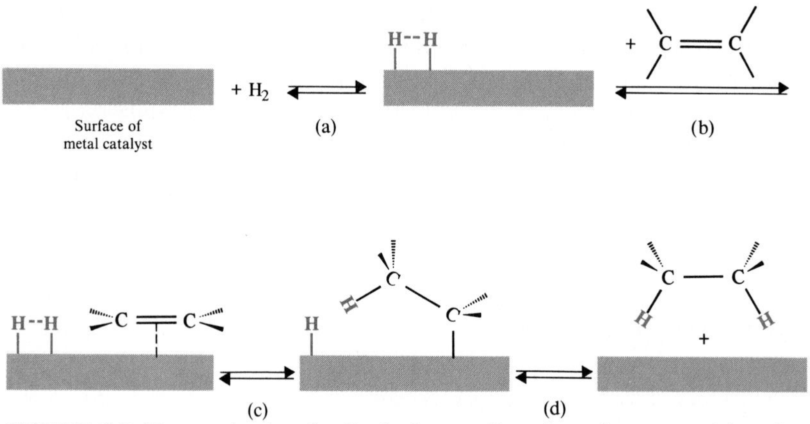

FIGURE 6.2 The mechanism for the hydrogenation of an alkene as catalyzed by finely divided platinum metal. Notice that both hydrogen atoms add from the same side of the double bond.

(Fig. 6.2*b*). A stepwise transfer of hydrogen atoms takes place, and this produces an alkane before the organic molecule leaves the catalyst surface (Fig. 6.2*c* and *d*). As a consequence, *both hydrogen atoms usually add from the same side of the molecule.* This mode of addition is called a **syn** addition (Section 6.6A).

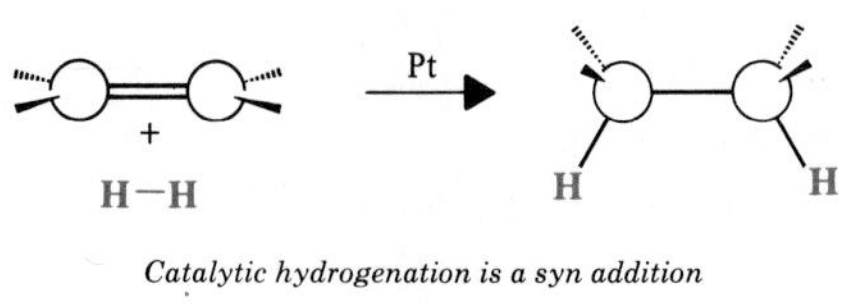

Catalytic hydrogenation is a syn addition

6.6A Syn and Anti Additions

An addition that places the parts of the adding reagent on the same side (or face) of the reactant is called a **syn addition.** We have just seen that the platinum-catalyzed addition of hydrogen ($X = Y = H$) is a syn addition.

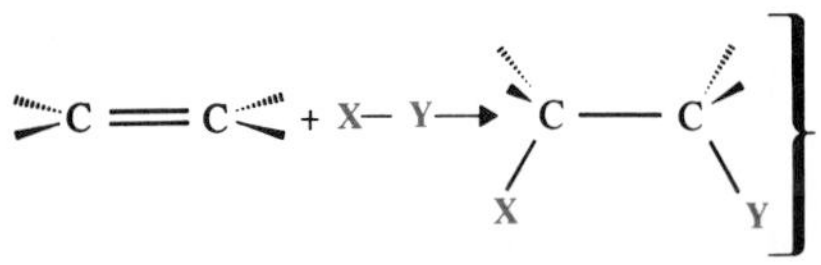

$$\sim C = C \sim + \; X-Y \longrightarrow \; \begin{array}{c} C - C \\ X \qquad Y \end{array} \left.\begin{array}{c} \\ \\ \\ \end{array}\right] \begin{array}{l} A \\ syn \\ addition \end{array}$$

The opposite of a syn addition is an **anti addition.** An anti addition places the parts of the adding reagent on opposite faces of the reactant.

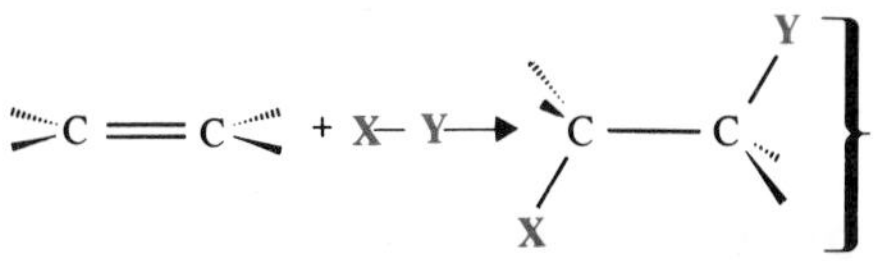

$$\sim C = C \sim + \; X-Y \longrightarrow \; \begin{array}{c} \qquad Y \\ C - C \\ X \end{array} \left.\begin{array}{c} \\ \\ \\ \end{array}\right\} \begin{array}{l} An \\ anti \\ addition \end{array}$$

In the next chapter we shall study a number of important syn and anti additions.

6.7 HYDROGENATION OF ALKYNES

Depending on the conditions and the catalyst employed, one or two molar equivalents of hydrogen will add to a carbon–carbon triple bond. When a platinum catalyst is used, the alkyne generally reacts with two molar equivalents of hydrogen to give an alkane.

$$CH_3C\equiv CCH_3 \xrightarrow[H_2]{Pt} [CH_3CH=CHCH_3] \xrightarrow[H_2]{Pt} CH_3CH_2CH_2CH_3$$

However, hydrogenation of an alkyne to an alkene can be accomplished through the use of special catalysts or reagents. Moreover, these special methods allow the preparation of either (E) or (Z) alkenes from disubstituted acetylenes.

6.7A Syn Addition of Hydrogen.
Synthesis of *cis*-Alkenes

A catalyst that permits hydrogenation of an alkyne to an alkene is the nickel boride compound called P-2 catalyst. This catalyst can be prepared by the reduction of nickel acetate with sodium borohydride.

$$\overset{\overset{\displaystyle O}{\displaystyle \|}}{Ni(OCCH_3)_2} \xrightarrow[C_2H_5OH]{NaBH_4} Ni_2B \; (P\text{-}2)$$

Hydrogenation of alkynes in the presence of P-2 catalyst causes **syn addition of hydrogen** to take place and the alkene that is formed from an alkyne with an internal triple bond has the (Z) or cis configuration. The hydrogenation of 3-hexyne (see the following reaction) illustrates this method. The reaction takes place on the surface of the catalyst (Section 6.6) accounting for the syn addition.

$$CH_3CH_2C{\equiv}CCH_2CH_3 \xrightarrow[\text{(syn addition)}]{H_2/Ni_2B \text{ (P-2)}}$$

$$\underset{H}{\overset{CH_3CH_2}{\diagdown}}C{=}C\underset{H}{\overset{CH_2CH_3}{\diagup}}$$

3-Hexyne

(97%)
(Z)-3-Hexene
(cis-3-hexene)

Other specially conditioned catalysts can be used to prepare *cis*-alkenes from disubstituted acetylenes. Metallic palladium deposited on calcium carbonate can be used in this way after it has been conditioned with lead acetate and quinoline (Section 19.1B). This special catalyst is known as Lindlar's catalyst.

$$R{-}C{\equiv}C{-}R \xrightarrow[\substack{\text{quinoline} \\ \text{(syn addition)}}]{\substack{H_2, \text{ Pd/CaCO}_3 \\ \text{(Lindlar's catalyst)}}} \underset{H}{\overset{R}{\diagdown}}C{=}C\underset{H}{\overset{R}{\diagup}}$$

6.7B Anti Addition of Hydrogen. Synthesis of *trans*-Alkenes

An **anti addition** of hydrogen atoms to the triple bond occurs when alkynes are reduced with lithium or sodium metal in ammonia or ethylamine at low temperatures. This reaction called a **dissolving metal reduction** produces an (E) or *trans*-alkene.

$$CH_3(CH_2)_2{-}C{\equiv}C{-}(CH_2)_2CH_3 \xrightarrow[\text{(2) NH}_4\text{Cl}]{\text{(1) Li, C}_2\text{H}_5\text{NH}_2, -78\ °C}$$

$$\underset{H}{\overset{CH_3(CH_2)_2}{\diagdown}}C{=}C\underset{(CH_2)_2CH_3}{\overset{H}{\diagup}}$$

4-Octyne

(52%)
(E)-4-Octene
(trans-4-octene)

The mechanism for this reduction is shown in the following outline. It involves successive electron transfers from the lithium (or sodium) atom and proton transfers from the amine (or ammonia). In the first step, the lithium atom transfers an electron to the alkyne to produce an intermediate that bears a negative charge and has an unpaired electron, called a **radical anion**. In the second step, the amine transfers a proton to produce a **vinylic radical**. Then, transfer of another electron gives a **vinylic anion**. It is this step that determines the stereochemistry of the reaction. The *trans*-vinylic anion is formed preferentially because it is more stable; the bulky alkyl groups are farther apart. Protonation of the *trans*-vinylic anion leads to the *trans*-alkene.

$$R{-}C{\equiv}C{-}R \xrightarrow{Li\cdot} \underset{R}{\overset{R}{\diagdown}}C{=}\overset{\cdot}{C} \xrightarrow{H{-}NHR} \underset{H}{\overset{R}{\diagdown}}C{=}\overset{\cdot}{C}\overset{R}{\diagup} \xrightarrow{Li\cdot}$$

Radical anion   **Vinylic radical**

$$\underset{H}{\overset{R}{\diagdown}}C{=}C\overset{R}{\diagup} \xrightarrow{H{-}NHR} \underset{H}{\overset{R}{\diagdown}}C{=}C\underset{R}{\overset{H}{\diagup}}$$

***trans*-Vinylic anion**   ***trans*-Alkene**

6.8 MOLECULAR FORMULAS OF HYDROCARBONS: THE INDEX OF HYDROGEN DEFICIENCY

Alkenes whose molecules contain only one double bond have the general formula C_nH_{2n}. They are isomeric with cycloalkanes. For example, 1-hexene and cyclohexane have the same molecular formula (C_6H_{12}):

$$CH_2=CHCH_2CH_2CH_2CH_3$$

1-Hexene
(C_6H_{12})

Cyclohexane
(C_6H_{12})

Cyclohexane and 1-hexene are constitutional isomers.

Alkynes and alkenes with two double bonds (alkadienes) have the general formula C_nH_{2n-2}. Hydrocarbons with one triple bond and one double bond (alkenynes) and alkenes with three double bonds (alkatrienes) have the general formula C_nH_{2n-4}, and so forth.

$$CH_2=CH-CH=CH_2 \qquad CH_2=CH-CH=CH-CH=CH_2$$

1,3-Butadiene
(C_4H_6)

1,3,5-Hexatriene
(C_6H_8)

A chemist working with an unknown hydrocarbon can obtain considerable information about its structure from its molecular formula and its **index of hydrogen deficiency.** The index of hydrogen deficiency is defined as the number of *pairs* of hydrogen atoms that must be subtracted from the molecular formula of the corresponding alkane to give the molecular formula of the compound under consideration.

For example, both cyclohexane and 1-hexene have an index of hydrogen deficiency equal to one (meaning one *pair* of hydrogen atoms). The corresponding alkane (i.e., the alkane with the same number of carbon atoms) is hexane.

C_6H_{14} = formula of corresponding alkane (hexane)
$\underline{C_6H_{12}}$ = formula of compound (1-hexene or cyclohexane)
H_2 = difference = 1 pair of hydrogen atoms

Index of hydrogen deficiency = 1

The index of hydrogen deficiency of acetylene or of 1,3-butadiene equals 2; the index of hydrogen deficiency of 1,3,5-hexatriene equals 3. (Do the calculations.)

Determining the number of rings present in a given compound is easily done experimentally. Molecules with double bonds and triple bonds add hydrogen readily at room temperature in the presence of a platinum catalyst. **Each double bond consumes one molar equivalent of hydrogen; each triple bond consumes two. Rings are not affected by hydrogenation at room temperature.** Hydrogenation, therefore, allows us to distinguish between rings on the one hand and double or triple bonds on the other. Consider as an example two compounds with the molecular formula C_6H_{12}: 1-hexene and cyclohexane. 1-Hexene reacts with one molar equivalent of hydrogen to yield hexane; under the same conditions cyclohexane does not react.

$$CH_2\!=\!CH(CH_2)_3CH_3 + H_2 \xrightarrow[25\,°C]{Pt} CH_3(CH_2)_4CH_3$$

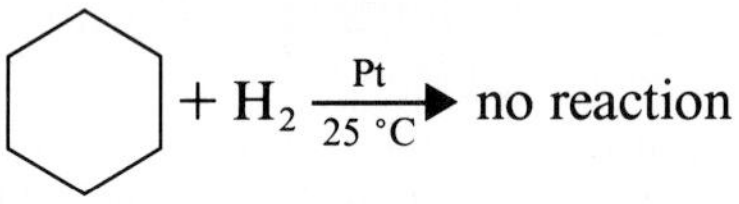

$+ H_2 \xrightarrow[25\,°C]{Pt}$ no reaction

Or consider another example. Cyclohexene and 1,3-hexadiene have the same molecular formula (C_6H_{10}). Both compounds react with hydrogen in the presence of a catalyst, but cyclohexene, because it has a ring and only one double bond, reacts with only one molar equivalent. 1,3-Hexadiene adds two molar equivalents.

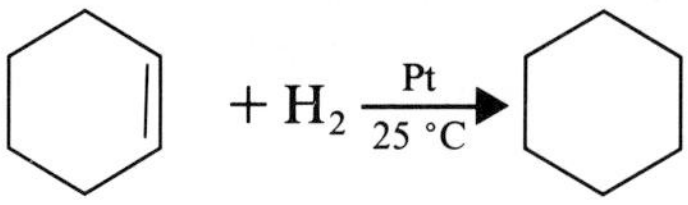

$+ H_2 \xrightarrow[25\,°C]{Pt}$

Cyclohexene

$$CH_2\!=\!CHCH\!=\!CHCH_2CH_3 + 2H_2 \xrightarrow[25\,°C]{Pt} CH_3(CH_2)_4CH_3$$
1,3-Hexadiene

PROBLEM 6.5

(a) What is the index of hydrogen deficiency of 2-hexene? (b) Of methylcyclopentane? (c) Does the index of hydrogen deficiency reveal anything about the location of the double bond in the chain? (d) About the size of the ring? (e) What is the index of hydrogen deficiency of 2-hexyne? (f) In general terms, what structural possibilities exist for a compound with the molecular formula $C_{10}H_{16}$?

PROBLEM 6.6

Zingiberene, a fragrant compound isolated from ginger, has the molecular formula $C_{15}H_{24}$ and is known not to contain any triple bonds. (a) What is the index of hydrogen deficiency of zingiberene? (b) When zingiberene is subjected to catalytic hydrogenation using an excess of hydrogen, 1 mole of zingiberene absorbs 3 moles of hydrogen and produces a compound with the formula $C_{15}H_{30}$. How many double bonds does a molecule of zingiberene have? (c) How many rings?

6.9 RELATIVE STABILITIES OF ALKENES

6.9A Heats of Hydrogenation

Hydrogenation also provides a way to measure the relative stabilities of certain alkenes. The reaction of an alkene with hydrogen is an exothermic reaction; the enthalpy change involved is called **the heat of hydrogenation.** Most alkenes have heats of hydrogenation near -30 kcal/mole. Individual alkenes, however, have heats of hydrogenation that may differ from this value by more than 2 kcal/mole.

$$\text{C}=\text{C} + \text{H}-\text{H} \xrightarrow{\text{Pt}} -\overset{|}{\underset{\text{H}}{\text{C}}}-\overset{|}{\underset{\text{H}}{\text{C}}}- \qquad \Delta H° \simeq -30 \text{ kcal/mole}$$

These differences permit the measurement of the relative stabilities of alkene isomers *when hydrogenation converts them to the same product.*

Consider, as examples, the three butene isomers that follow:

$$\text{CH}_3\text{CH}_2\text{CH}=\text{CH}_2 + \text{H}_2 \xrightarrow{\text{Pt}} \text{CH}_3\text{CH}_2\text{CH}_2\text{CH}_3 \qquad \Delta H° = -30.3 \text{ kcal/mole}$$

1-Butene   **Butane**
(C₄H₈)

$$\text{CH}_3\text{C}=\text{C}\text{CH}_3 \,(\text{H, H}) + \text{H}_2 \xrightarrow{\text{Pt}} \text{CH}_3\text{CH}_2\text{CH}_2\text{CH}_3 \qquad \Delta H° = -28.6 \text{ kcal/mole}$$

cis-**2-Butene**   **Butane**
(C₄H₈)

$$\text{CH}_3\text{C}=\text{C}\,(\text{H, CH}_3) + \text{H}_2 \xrightarrow{\text{Pt}} \text{CH}_3\text{CH}_2\text{CH}_2\text{CH}_3 \qquad \Delta H° = -27.6 \text{ kcal/mole}$$

trans-**2-Butene**   **Butane**
(C₄H₈)

In each reaction the product (butane) is the same. In each case, too, one of the reactants (hydrogen) is the same. Different amounts of *heat* are evolved in each reaction, however, and these differences must be related to different relative stabilities (different heat contents) of the individual butenes. 1-Butene evolves the greatest amount of heat when hydrogenated, and *trans*-2-butene evolves the least. Therefore 1-butene must have the greatest potential energy and be the least stable isomer. *trans*-2-Butene must have the lowest potential energy and be the most stable isomer. The potential energy (and stability) of *cis*-2-butene falls in between. The order of stabilities of the butenes is easier to see if we examine the potential energy diagram in Fig. 6.3.

The greater stability of the *trans*-2-butene when compared to *cis*-2-butene illustrates a general pattern found in cis–trans alkene pairs. The 2-pentenes, for example, show the same stability relationship: **trans isomer > cis isomer.**

$$\text{CH}_3\text{CH}_2\text{C}=\text{C}\,(\text{CH}_3, \text{H, H}) + \text{H}_2 \xrightarrow{\text{Pt}} \text{CH}_3\text{CH}_2\text{CH}_2\text{CH}_2\text{CH}_3 \qquad \Delta H° = -28.6 \text{ kcal/mole}$$

cis-**2-Pentene**   **Pentane**

$$\text{CH}_3\text{CH}_2\text{C}=\text{C}\,(\text{H, H, CH}_3) + \text{H}_2 \xrightarrow{\text{Pt}} \text{CH}_3\text{CH}_2\text{CH}_2\text{CH}_2\text{CH}_3 \qquad \Delta H° = -27.6 \text{ kcal/mole}$$

trans-**2-Pentene**   **Pentane**

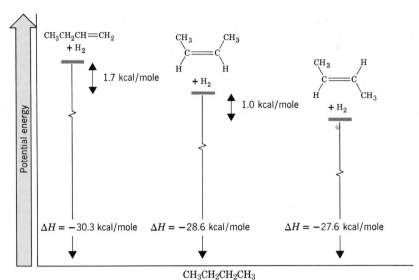

FIGURE 6.3 A potential energy diagram for the three butene isomers. The order of stability is *trans*-2-butene > *cis*-2-butene > 1-butene.

The greater potential energy of cis isomers can be attributed to strain caused by the crowding of two alkyl groups on the same side of the double bond (Fig. 6.4).

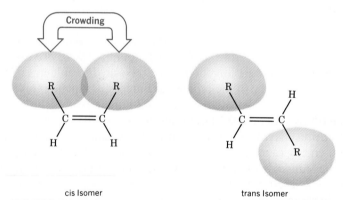

cis Isomer   trans Isomer

FIGURE 6.4 *cis*- and *trans*-Alkene isomers. The less stable cis isomer has greater strain.

6.9B Relative Stabilities from Heats of Combustion

When hydrogenation of isomeric alkenes does not yield the same alkane, *heats of combustion can be used to measure their relative stabilities.* For example, 2-methylpropene cannot be compared directly with the other butene isomers (1-butene, *cis*-, and *trans*-2-butene) because on hydrogenation 2-methylpropene yields isobutane, *not butane:*

$$CH_3\overset{\overset{\displaystyle CH_3}{|}}{C}=CH_2 \;+\; H_2 \xrightarrow{\text{Pt}} CH_3\overset{\overset{\displaystyle CH_3}{|}}{C}HCH_3$$

2-Methylpropene   Isobutane

Isobutane and butane do not have the same potential energy so a direct comparison of heats of hydrogenation is not possible.

However, when 2-methylpropene is subjected to complete combustion, the products are the same as those produced by the other butene isomers. Each isomer consumes 6 molar equivalents of oxygen and *produces 4 molar equivalents of CO_2 and 4 molar equivalents of H_2O.* Comparison of the heats of combustion shows that 2-methylpropene is the most stable of the four isomers because it evolves the least heat.

$$CH_3CH_2CH{=}CH_2 + 6\ O_2 \longrightarrow 4CO_2 + 4H_2O \qquad \Delta H° = -649.8\ \text{kcal/mole}$$

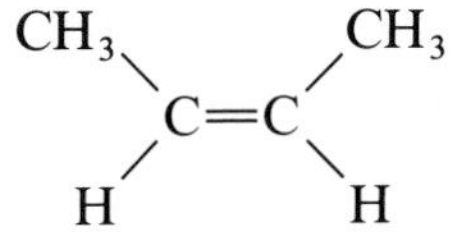

$$\underset{H}{\overset{CH_3}{\diagdown}}C{=}C\underset{H}{\overset{CH_3}{\diagup}} + 6\ O_2 \longrightarrow 4CO_2 + 4H_2O \qquad \Delta H° = -648.1\ \text{kcal/mole}$$

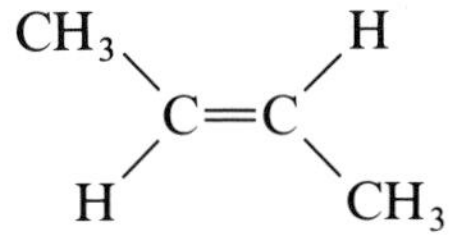

$$\underset{H}{\overset{CH_3}{\diagdown}}C{=}C\underset{CH_3}{\overset{H}{\diagup}} + 6\ O_2 \longrightarrow 4CO_2 + 4H_2O \qquad \Delta H° = -647.1\ \text{kcal/mole}$$

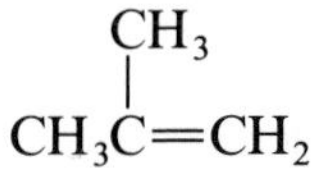

$$\overset{\overset{\textstyle CH_3}{|}}{CH_3C}{=}CH_2 + 6\ O_2 \longrightarrow 4CO_2 + 4H_2O \qquad \Delta H° = -646.1\ \text{kcal/mole}$$

The heat evolved by each of the other three isomers, moreover, confirms the order of stability measured by heats of hydrogenation. Therefore, the stability of the butene isomers overall is

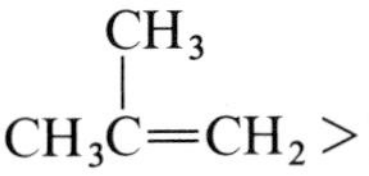

$$\overset{\overset{\textstyle CH_3}{|}}{CH_3C}{=}CH_2 > \textit{trans-}CH_3CH{=}CHCH_3 > \textit{cis-}CH_3CH{=}CHCH_3 > CH_3CH_2CH{=}CH_2$$

6.9C Overall Relative Stabilities of Alkenes

Studies of numerous alkenes reveal a pattern of stabilities that is related to the number of alkyl groups attached to the carbon atoms of the double bond. **The greater the number of attached alkyl groups, (i.e., the more highly substituted the carbon atoms of the double bond) the greater is the alkene's stability.** This order of stabilities can be given in general terms as follows:*

Relative Stabilities of Alkenes

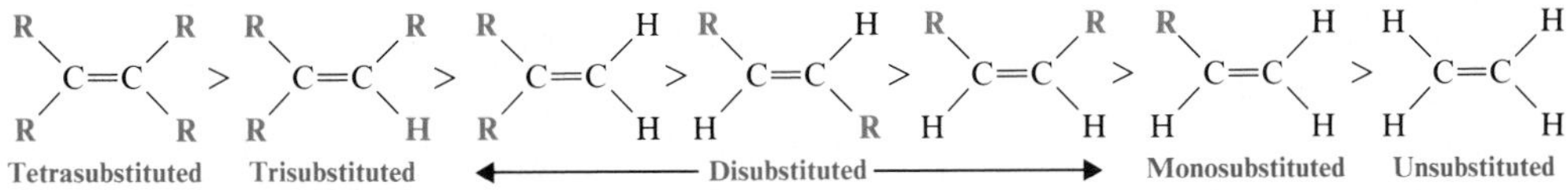

| Tetrasubstituted | Trisubstituted | ← | Disubstituted | → | Monosubstituted | Unsubstituted |

*This order of stabilities may seem contradictory when compared with the explanation given for the relative stabilities of cis and trans isomers. Although a detailed explanation of the trend given here is beyond our scope, the relative stabilities of substituted alkenes can be rationalized. Part of the explanation can be given in terms of the electron-releasing effect of alkyl groups, an effect that satisfies the electron-withdrawing properties of the sp^2-hybridized carbon atoms of the double bond.

PROBLEM 6.7

Heats of hydrogenation of three alkenes are as follows:

2-methyl-1-butene (−28.5 kcal/mole)

3-methyl-1-butene (−30.3 kcal/mole)

2-methyl-2-butene (−26.9 kcal/mole)

(a) Write the structure of each alkene and classify it as to whether its doubly bonded atoms are monosubstituted, disubstituted, trisubstituted, and so on. (b) Write the product formed when each alkene is hydrogenated. (c) Can heats of hydrogenation be used to relate the relative stabilities of these three alkenes? (d) If so, what is the predicted order of stability? If not, why not? (e) What other alkene isomers are possible for these alkenes? Write their structures. (f) What data would be necessary to relate the stabilities of all of these isomers?

PROBLEM 6.8

Predict the more stable alkene of each pair. (a) 1-Heptene or *cis*-2-heptene, (b) *cis*-2-heptene or *trans*-2-heptene, (c) *trans*-2-heptene or 2-methyl-2-hexene, and (d) 2-methyl-2-hexene or 2,3-dimethyl-2-pentene.

PROBLEM 6.9

Reconsider the pairs of alkenes given in Problem 6.8. For which pairs could you use heats of hydrogenation to determine their relative stabilities? For which pairs would you be required to use heats of combustion?

6.10 CYCLOALKENES

The rings of cycloalkenes containing five carbon atoms or fewer exist only in the cis form (Fig. 6.5). The introduction of a trans double bond into rings this small would, if it were possible, introduce greater strain than the bonds of the ring atoms could accommodate. *trans*-Cyclohexene might resemble the structure shown in Fig. 6.6. There is evidence that it can be formed as a very reactive short-lived intermediate in some chemical reactions.

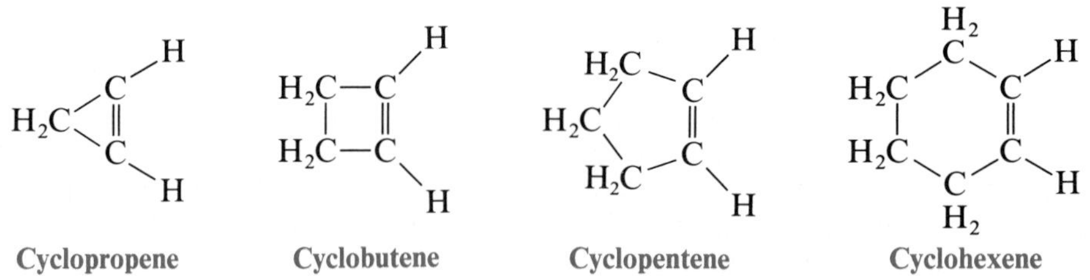

Cyclopropene   Cyclobutene   Cyclopentene   Cyclohexene

FIGURE 6.5 *cis*-Cycloalkanes.

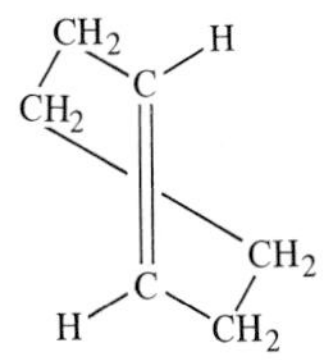

FIGURE 6.6 Hypothetical *trans*-cyclohexene. This molecule is apparently too highly strained to exist at room temperature.

trans-Cycloheptene has been observed with instruments called spectrometers, but it is a substance with a very short lifetime and has not been isolated.

trans-Cyclooctene (Fig. 6.7) has been isolated, however. Here the ring is large enough to accommodate the geometry required by a trans double bond and still be stable at room temperature.

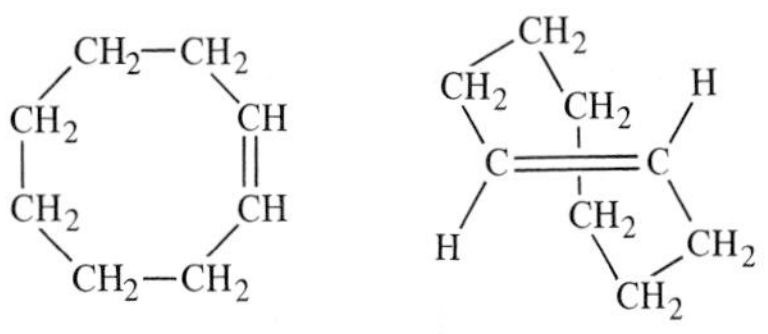

FIGURE 6.7 The *cis* and *trans* forms of cyclooctene.   *cis*-Cyclooctene   *trans*-Cyclooctene

6.11 SYNTHESIS OF ALKENES THROUGH ELIMINATION REACTIONS

Eliminations are the most widely used reactions for synthesizing alkenes. In this chapter we shall examine three methods: dehydrohalogenation of alkyl halides $(-HX)$, dehydration of alcohols $(-H_2O)$, and debromination of dibromoalkanes $(-Br_2)$.

6.12 SYNTHESIS OF ALKENES BY DEHYDROHALOGENATION OF ALKYL HALIDES

Synthesis of an alkene by dehydrohalogenation is almost always better achieved by an E2 reaction:

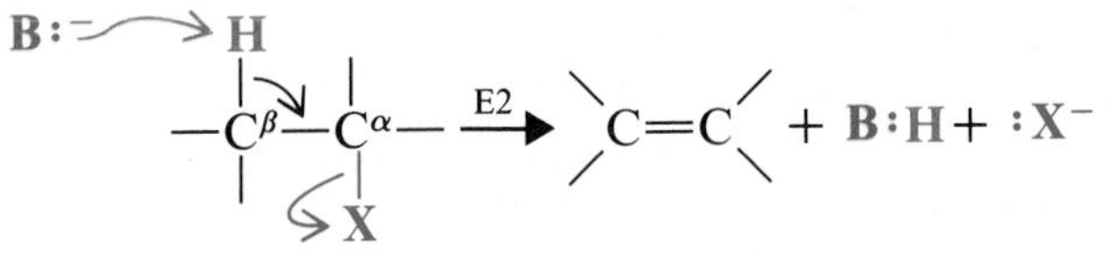

The reason for this choice is that dehydrohalogenation by an E1 mechanism is too variable. Too many competing events are possible, one being rearrangement of the carbon skeleton (Section 6.15). In order to bring about an E2 reaction, use a secondary or tertiary alkyl halide if possible. (If the synthesis must begin with a primary halide, then use a bulky base.) To try to avoid E1 conditions use a high concentration of a strong, relatively nonpolarizable base such as an alkoxide ion, and use a relatively nonpolar solvent such as an alcohol. To favor elimination generally, use a relatively high temperature. The typical reagents for dehydrohalogenation are sodium ethoxide in ethanol and potassium *tert*-butoxide in *tert*-butyl alcohol. Potassium hydroxide in ethanol is also used sometimes; in this reagent the reactive bases probably include the ethoxide ion formed by the following equilibrium.

$$OH^- + C_2H_5OH \rightleftharpoons H_2O + C_2H_5O^-$$

6.12A E2 Reactions: The Orientation of the Double Bond in the Product. Zaitsev's Rule

In earlier examples of dehydrohalogenations (Sections 5.15–5.16) only a single elimination product was possible. For example:

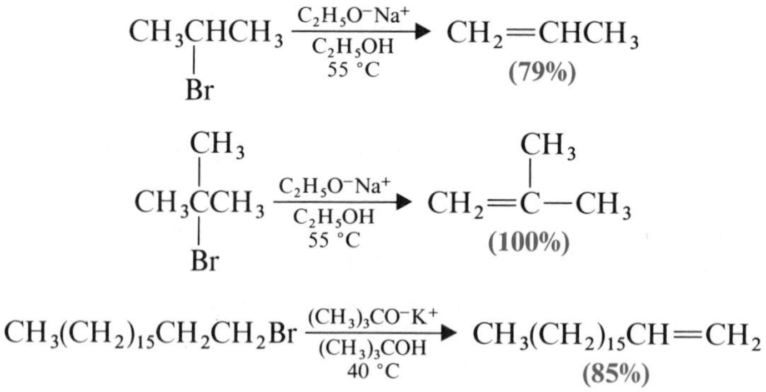

$$CH_3\underset{\overset{|}{Br}}{CH}CH_3 \xrightarrow[\substack{C_2H_5OH \\ 55\ °C}]{C_2H_5O^-Na^+} CH_2{=}CHCH_3$$
$$(79\%)$$

$$CH_3\underset{\overset{|}{Br}}{\overset{\overset{\displaystyle CH_3}{|}}{C}}CH_3 \xrightarrow[\substack{C_2H_5OH \\ 55\ °C}]{C_2H_5O^-Na^+} CH_2{=}\overset{\overset{\displaystyle CH_3}{|}}{C}{-}CH_3$$
$$(100\%)$$

$$CH_3(CH_2)_{15}CH_2CH_2Br \xrightarrow[\substack{(CH_3)_3COH \\ 40\ °C}]{(CH_3)_3CO^-K^+} CH_3(CH_2)_{15}CH{=}CH_2$$
$$(85\%)$$

Dehydrohalogenation of most alkyl halides, however, yields more than one product. For example, dehydrohalogenation of 2-bromo-2-methylbutane can yield two products: 2-methyl-2-butene or 2-methyl-1-butene.

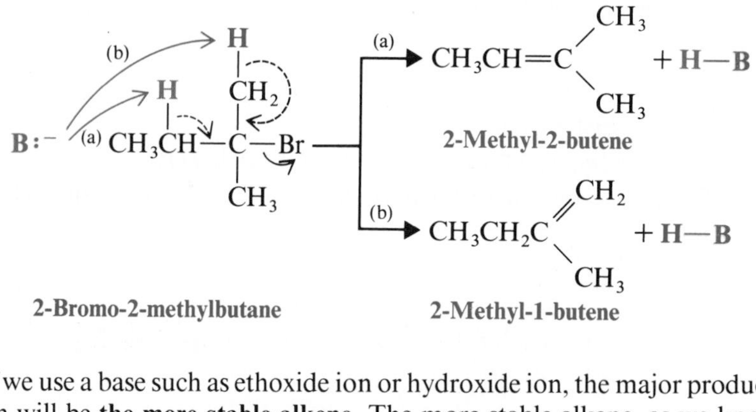

2-Bromo-2-methylbutane   2-Methyl-2-butene   2-Methyl-1-butene

If we use a base such as ethoxide ion or hydroxide ion, the major product of the reaction will be **the more stable alkene.** The more stable alkene, as we know from Section 6.9, has the more highly substituted double bond.

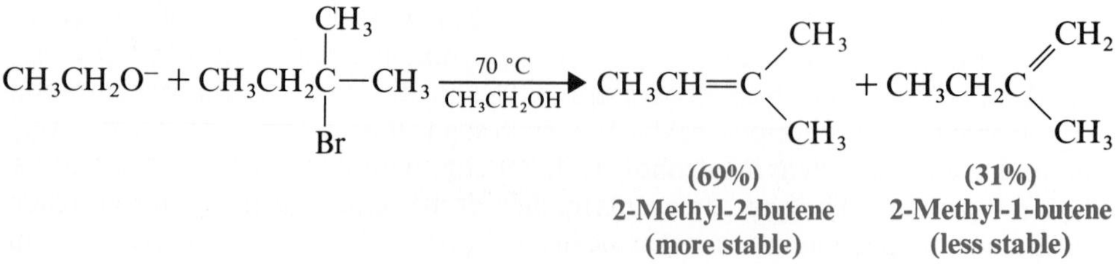

$$CH_3CH_2O^- + CH_3CH_2\underset{\overset{|}{Br}}{\overset{\overset{\displaystyle CH_3}{|}}{C}}{-}CH_3 \xrightarrow[CH_3CH_2OH]{70\ °C} CH_3CH{=}C\overset{\diagup CH_3}{\diagdown CH_3} + CH_3CH_2C\overset{\diagup\!\!\diagup CH_2}{\diagdown CH_3}$$

$$\qquad\qquad\qquad\qquad\qquad\qquad\qquad (69\%) \qquad\qquad (31\%)$$
$$\qquad\qquad\qquad\qquad\qquad\qquad 2\text{-Methyl-2-butene} \quad 2\text{-Methyl-1-butene}$$
$$\qquad\qquad\qquad\qquad\qquad\qquad (\text{more stable}) \qquad (\text{less stable})$$

2-Methyl-2-butene is a trisubstituted alkene (three methyl groups are attached to carbon atoms of the double bond), whereas 2-methyl-1-butene is only disubstituted. 2-Methyl-2-butene is the major product.

The reason for this behavior appears to be related to double-bond character that develops in the transition state (cf. Section 5.16) for each reaction:

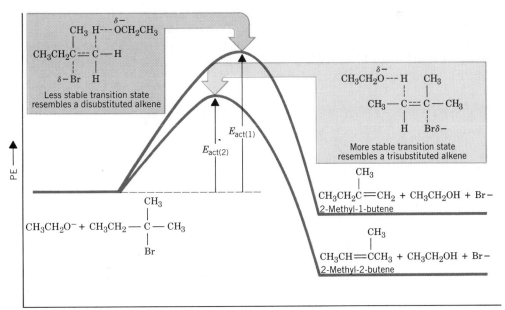

Transition state for an E2 reaction
The carbon–carbon bond has some of the character of a double bond.

The transition state for the reaction leading to 2-methyl-2-butene (Fig. 6.8) resembles the product of the reaction: A trisubstituted alkene. The transition state for the reaction leading to 2-methyl-1-butene resembles its product: A disubstituted alkene. Because the transition state leading to 2-methyl-2-butene resembles a more stable alkene, this transition state is more stable. Because this transition state is more stable (occurs at lower potential energy), the energy of activation for this reaction is lower and 2-methyl-2-butene is formed faster. This explains why 2-methyl-2-butene is the major product.

Whenever an elimination occurs to give the most stable, more highly substituted alkene, chemists say that the elimination follows the **Zaitsev rule,** named for the nineteenth-century Russian chemist A. N. Zaitsev (1841–1910) who formulated it. (Zaitsev's name is also transliterated as Zaitzev, Saytzeff, Saytseff, or Saytzev.)

FIGURE 6.8 Reaction (2) leading to the more stable alkene occurs faster than reaction (1) leading to the less stable alkene; $E_{act(2)}$ is less than $E_{act(1)}$.

PROBLEM 6.10

Dehydrohalogenation of 2-bromobutane with potassium hydroxide in ethyl alcohol yields a mixture of 2-butene and 1-butene. (a) Which butene would you expect to predominate? (b) The 2-butene formed in the reaction is a mixture of *cis*-2-butene and *trans*-2-butene. Which 2-butene would you expect to predominate?

6.12B An Exception to Zaitsev's Rule

Carrying out dehydrohalogenations with a base such as potassium *tert*-butoxide in *tert*-butyl alcohol favors the formation of **the less substituted alkene:**

$$CH_3-\overset{\overset{\displaystyle CH_3}{|}}{\underset{\underset{\displaystyle CH_3}{|}}{C}}-O^- + CH_3CH_2-\overset{\overset{\displaystyle CH_3}{|}}{\underset{\underset{\displaystyle CH_3}{|}}{C}}-Br \xrightarrow[\text{(CH}_3)_3\text{COH}]{75\ ^\circ C} CH_3CH=C\overset{\displaystyle CH_3}{\underset{\displaystyle CH_3}{\big\langle}} + CH_3CH_2C\overset{\displaystyle CH_2}{\underset{\displaystyle CH_3}{\big\langle}}$$

(27.5%)   (72.5%)

2-Methyl-2-butene   2-Methyl-1-butene

(more substituted)   (less substituted)

The reasons for this behavior are complicated but seem to be related in part to the steric bulk of the base and to the fact that in *tert*-butyl alcohol the base is associated with solvent molecules and thus made even larger. The large *tert*-butoxide ion appears to have difficulty removing one of the internal (2°) hydrogen atoms. It removes one of the more exposed (1°) hydrogen atoms of the methyl group instead.

6.12C The Stereochemistry of E2 Reactions: The Orientation of Groups in the Transition State

Considerable experimental evidence indicates that the five atoms involved in the transition state of an E2 reaction (including the base) must lie in the same plane. There are two ways that this can happen:

anti periplanar
transition state
(preferred)

syn periplanar
transition state
(only with certain
rigid molecules)

Evidence also indicates that of these two arrangements for the transition state, the arrangement called the **anti periplanar** conformation is the preferred one. (The **syn periplanar** transition state occurs only with rigid molecules that are unable to assume the anti arrangement.)

The requirement for coplanarity of the H—C—C—L unit arises from a need for proper overlap of orbitals in the developing π bond of the alkene that is being formed.

Part of the evidence for the preferred anti periplanar arrangement of groups comes from experiments done with cyclic molecules. As examples, let us consider the different behavior shown in E2 reactions by two compounds containing cyclohexane rings that have the common names *neomenthyl chloride* and *menthyl chloride.*

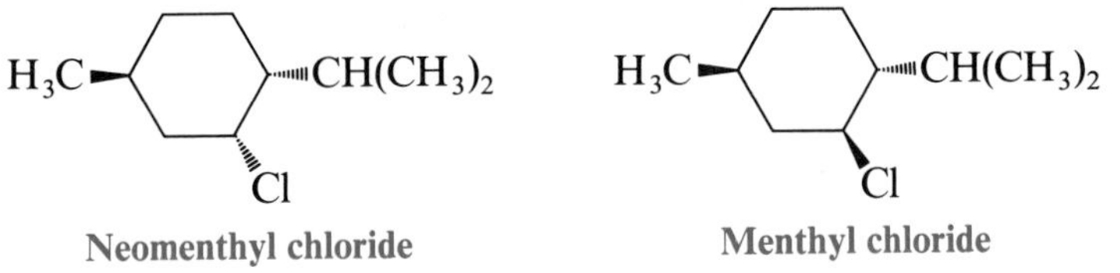

Neomenthyl chloride

Menthyl chloride

The β hydrogen and the leaving group on a cyclohexane ring can assume an anti periplanar conformation **only when they are both axial:**

Here the β hydrogen
and the chlorine are
both axial. This
allows an anti periplanar
transition state.

A Newman projection
formula showing that
the β hydrogen and
the chlorine are anti
periplanar when they
are both axial.

Neither an axial–equatorial nor an equatorial–equatorial orientation of the groups allows the formation of an anti periplanar transition state.

In the more stable conformation of neomenthyl chloride (see following figure), the alkyl groups are both equatorial and the chlorine is axial. There are also axial hydrogen atoms on both C-2 and C-4. The base can attack either of these hydrogen atoms and achieve an anti periplanar transition state for an E2 reaction. Products corresponding to each of these transition states (2-menthene and 3-menthene) are formed rapidly. In accordance with Zaitsev's rule, 3-menthene (with the more highly substituted double bond) is the major product.

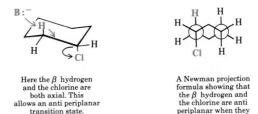

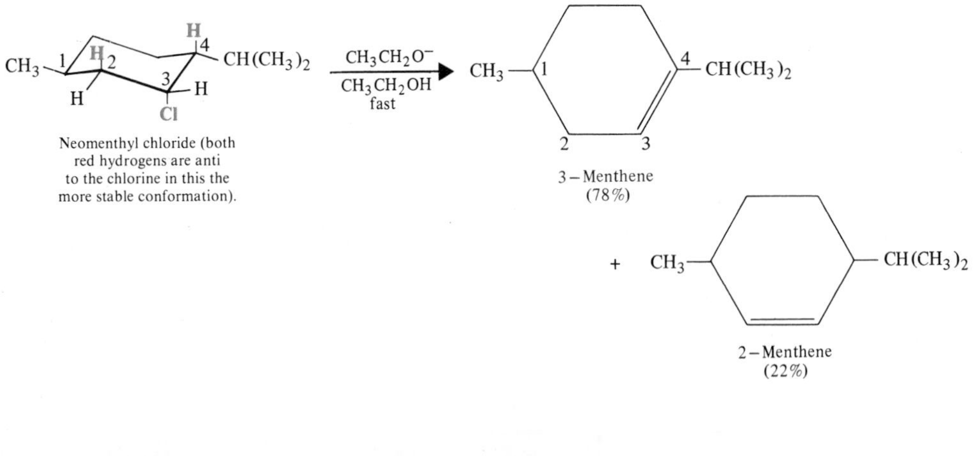

Neomenthyl chloride (both
red hydrogens are anti
to the chlorine in this the
more stable conformation).

3 – Menthene
(78%)

+   CH₃

2 – Menthene
(22%)

On the other hand, the more stable conformation of menthyl chloride has all three groups (including the chlorine) equatorial. For the chlorine to become axial, menthyl chloride has to assume a conformation in which the large isopropyl group and the methyl group are also axial. This conformation is of much higher energy, and the activation energy for the reaction is large because it includes the energy necessary for the conformational change. Consequently, menthyl chloride undergoes an E2 reaction very slowly, and the product is entirely 2-menthene (contrary to Zaitsev's Rule).

Menthyl chloride
more stable
conformation (anti
elimination is
not possible.)

Menthyl chloride
less stable
conformation
(red H and Cl
are axial and
can undergo
elimination.)

2-Menthene
(100%)

PROBLEM 6.12

When *cis*-1-bromo-4-*tert*-butylcyclohexane is treated with sodium ethoxide in ethanol, it reacts rapidly; the product is 4-*tert*-butylcyclohexene. Under the same conditions, *trans*-1-bromo-4-*tert*-butylcyclohexane reacts very slowly. Write conformational structures and explain the difference in reactivity of these cis–trans isomers.

PROBLEM 6.13

(a) When *cis*-1-bromo-2-methylcyclohexane undergoes an E2 reaction, two products (cycloalkenes) are formed. What are these two cycloalkenes, and which would you expect to be the major product? Write conformational structures showing how each is formed. (b) When *trans*-1-bromo-2-methylcyclohexane reacts in an E2 reaction, only one cycloalkene is formed. What is this product? Write conformational structures showing why it is the only product.

6.13 SYNTHESIS OF ALKENES BY DEHYDRATION OF ALCOHOLS

Heating most alcohols with a strong acid causes them to lose a molecule of water (to **dehydrate**) and form an alkene:

$$-\overset{|}{\underset{H}{C}}-\overset{|}{\underset{OH}{C}}- \xrightarrow[\text{heat}]{H^+} \overset{\diagdown}{}C=C\overset{\diagup}{} + H_2O$$

The reaction is an **elimination** and is favored at higher temperatures (Section 5.18). The most commonly used acids in the laboratory are Brønsted acids—proton donors such as sulfuric acid and phosphoric acid. Lewis acids such as alumina (Al_2O_3) are often used in industrial, gas-phase dehydrations.

Dehydration reactions of alcohols show several important characteristics, which we shall soon explain.

1. **The experimental conditions — temperature and acid concentration — that are required to bring about dehydration are closely related to the structure of the individual alcohol.** Alcohols in which the hydroxyl group is attached to a primary carbon (primary alcohols) are the most difficult to dehydrate. Dehydration of ethyl alcohol, for example, requires concentrated sulfuric acid and a temperature of 180 °C.

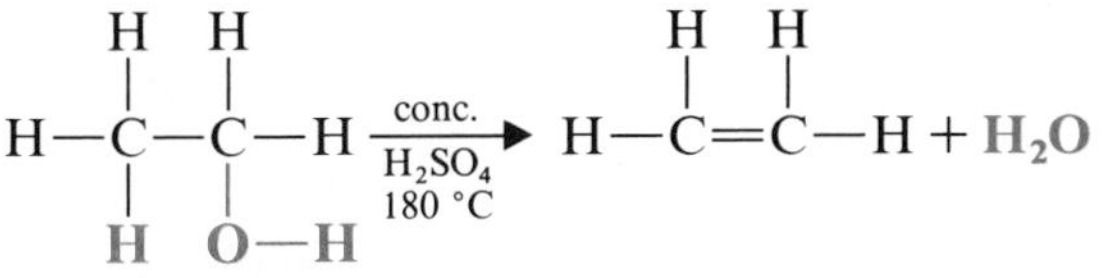

$$H-\overset{\overset{\displaystyle H}{|}}{\underset{\underset{\displaystyle H}{|}}{C}}-\overset{\overset{\displaystyle H}{|}}{\underset{\underset{\displaystyle O-H}{|}}{C}}-H \xrightarrow[\substack{H_2SO_4 \\ 180\ °C}]{\text{conc.}} H-\overset{\overset{\displaystyle H}{|}}{C}=\overset{\overset{\displaystyle H}{|}}{C}-H + H_2O$$

Ethyl alcohol
(a 1° alcohol)

Secondary alcohols usually dehydrate under milder conditions. Cyclohexanol, for example, dehydrates in 85% phosphoric acid at 165 to 170 °C.

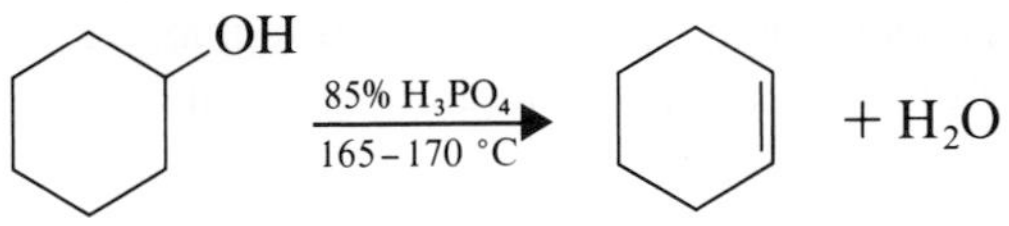

Cyclohexanol $\xrightarrow[165-170\ °C]{85\%\ H_3PO_4}$ **(80%) Cyclohexene** $+ H_2O$

Tertiary alcohols are usually so easily dehydrated that extremely mild conditions can be used. *tert*-Butyl alcohol, for example, dehydrates in 20% aqueous sulfuric acid at a temperature of 85 °C.

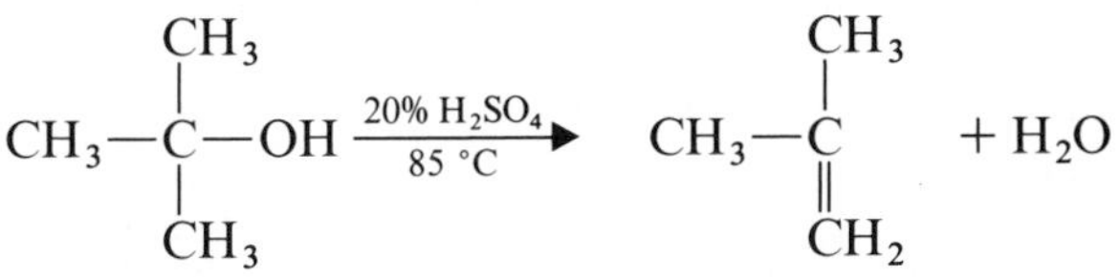

$$CH_3-\overset{\overset{\displaystyle CH_3}{|}}{\underset{\underset{\displaystyle CH_3}{|}}{C}}-OH \xrightarrow[85\ °C]{20\%\ H_2SO_4} CH_3-\overset{\overset{\displaystyle CH_3}{|}}{\underset{\underset{\displaystyle CH_2}{||}}{C}} + H_2O$$

tert-**Butyl alcohol**   **(84%) 2-Methylpropene**

Thus, overall, the relative ease with which alcohols undergo dehydration is in the following order:

Ease of Dehydration

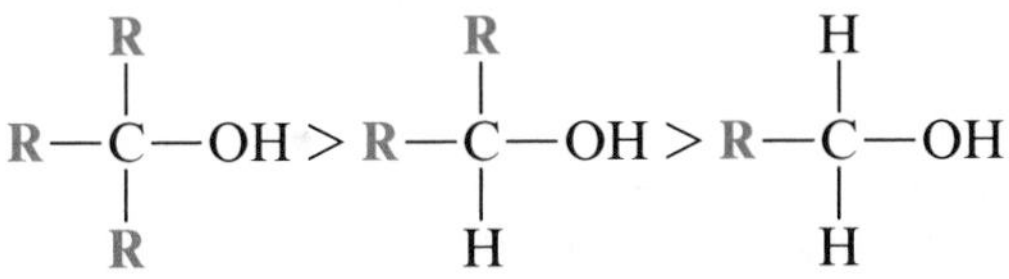

$$R-\overset{\overset{\displaystyle R}{|}}{\underset{\underset{\displaystyle R}{|}}{C}}-OH > R-\overset{\overset{\displaystyle R}{|}}{\underset{\underset{\displaystyle H}{|}}{C}}-OH > R-\overset{\overset{\displaystyle H}{|}}{\underset{\underset{\displaystyle H}{|}}{C}}-OH$$

3° Alcohol   **2° Alcohol**   **1° Alcohol**

This behavior, as we shall see in Section 6.14, is related to the stability of the carbocation formed in each reaction.

2. **Certain alcohols dehydrate to give more than one product.** When 1-butanol dehydrates, for example, three products form: *trans*-2-butene, *cis*-2-butene, and 1-butene. (As we shall see in Section 6.15, this reaction involves the migration of a hydrogen atom from one carbon atom to the next.)

$$CH_3CH_2CH_2CH_2OH \xrightarrow[\substack{170\ °C}]{\substack{conc. \\ H_2SO_4}}$$

$$\underset{\substack{\textit{trans-}\textbf{2-Butene} \\ \textbf{(major product)}}}{\overset{CH_3}{\underset{H}{}} C=C \overset{H}{\underset{CH_3}{}}} + \underset{\substack{\textit{cis-}\textbf{2-Butene} \\ \textbf{(minor product)}}}{\overset{CH_3}{\underset{H}{}} C=C \overset{CH_3}{\underset{H}{}}}$$

$$+ \underset{\substack{\textbf{1-Butene} \\ \textbf{(minor product)}}}{\overset{CH_3CH_2}{\underset{H}{}} C=C \overset{H}{\underset{H}{}}} + H_2O$$

Notice that this reaction illustrates Zaitsev's rule and yields the most stable alkene as the major product. *The applicability of Zaitsev's rule is characteristic of alcohol dehydrations.*

3. **Some primary and secondary alcohols also undergo rearrangements of their carbon skeleton during dehydration.** Such a rearrangement occurs in the dehydration of 3,3-dimethyl-2-butanol.

$$\underset{\substack{\textbf{3,3-Dimethyl-2-butanol}}}{CH_3 - \underset{\underset{CH_3}{|}}{\overset{\overset{CH_3}{|}}{C}} - \underset{\underset{OH}{|}}{CH} - CH_3} \xrightarrow[\substack{80\ °C}]{\substack{85\%\ H_3PO_4}} \underset{\substack{(80\%) \\ \textbf{2,3-Dimethyl-2-butene}}}{CH_3 - \overset{\overset{CH_3}{|}}{C} = \overset{\overset{CH_3}{|}}{C} - CH_3} + \underset{\substack{(20\%) \\ \textbf{2,3-Dimethyl-1-butene}}}{CH_2 = \overset{\overset{CH_3}{|}}{C} - \overset{\overset{CH_3}{|}}{C}HCH_3}$$

Notice that the carbon skeleton of the reactant is

$$\underset{\underset{C}{|}}{C - \overset{\overset{C}{|}}{C} - C - C} \qquad \text{while that of the products is} \qquad C - \overset{\overset{C}{|}}{C} - \overset{\overset{C}{|}}{C} - C$$

We shall see in Section 6.15 that this reaction involves the migration of a methyl group from one carbon to the next.

6.13A Mechanism of Alcohol Dehydration. An E1 Reaction

Explanations for all of these observations can be based on a step-wise mechanism originally proposed by F. Whitmore (of the Pennsylvania State University). The mechanism is *an E1 reaction in which the substrate is a protonated alcohol (or an alkyloxonium ion,* see Section 5.3A). Consider the dehydration of *tert*-butyl alcohol as an example.

Step 1

$$CH_3-\underset{\underset{CH_3}{|}}{\overset{\overset{CH_3}{|}}{C}}-\overset{..}{\underset{..}{O}}-H + H-\overset{\overset{H}{|}}{\underset{H}{O}}:^+ \rightleftharpoons CH_3-\underset{\underset{CH_3}{|}}{\overset{\overset{CH_3\ \ H}{|}}{C}}-\overset{+}{\underset{..}{O}}-H + H:\overset{..}{\underset{..}{O}}-H$$

Protonated alcohol
or alkyloxonium ion

In this step, an acid–base reaction, a proton is rapidly transferred from the acid to one of the unshared electron pairs of the alcohol. In dilute sulfuric acid the acid is a hydronium ion; in concentrated sulfuric acid the proton donor is sulfuric acid itself. This step is characteristic of all reactions of an alcohol with a strong acid.

The presence of the positive charge on the oxygen of the protonated alcohol weakens all bonds from oxygen including the carbon–oxygen bond, and in step 2 the carbon–oxygen bond breaks. The leaving group is a molecule of water:

Step 2

$$CH_3-\underset{\underset{CH_3}{|}}{\overset{\overset{CH_3\ \ H}{|}}{C}}-\overset{+}{\underset{..}{O}}-H \rightleftharpoons CH_3-\underset{\underset{CH_3}{|}}{\overset{\overset{CH_3\ \ H}{|}}{C^+}} + :\overset{..}{O}-H$$

A carbocation

The carbon–oxygen bond breaks **heterolytically.** The bonding electrons depart with the water molecule and leave behind a carbocation. The carbocation is, of course, highly reactive because the central carbon atom has only six electrons in its valence level, not eight.

Finally, in step 3, the carbocation transfers a proton to a molecule of water. The result is the formation of a hydronium ion and an alkene.

Step 3

$$CH_3-\underset{\underset{CH_3}{|}}{\overset{\overset{H-\overset{\overset{H}{|}}{C}-H}{|}}{C^+}} + :\overset{..}{O}-H \rightleftharpoons CH_3-\underset{\underset{CH_3}{|}}{\overset{\overset{CH_2}{||}}{C}} + H-\overset{+}{\underset{..}{O}}-H$$

2-Methylpropene

In step 3, also an acid–base reaction, any one of the nine protons available at the three methyl groups can be transferred to a molecule of water. The electron pair that bonded the hydrogen atom to the carbon atom in the carbocation becomes the second bond of the double bond of the alkene. Notice that this step restores an octet of electrons to the central carbon atom.

PROBLEM 6.14

(a) What would the leaving group have to be for the alcohol itself (rather than the protonated alcohol) to undergo dehydration? (b) How does your answer explain the requirement for an acid catalyst in alcohol dehydrations?

By itself, the Whitmore mechanism does not explain the observed order of reactivity of alcohols: **tertiary** > **secondary** > **primary.** Taken alone, it does not explain the formation of more than one product in the dehydration of certain alcohols nor the occurrence of a rearranged carbon skeleton in the dehydration of others. But when coupled with what is known about *the stability of carbocations,* the Whitmore mechanism *does* eventually account for all of these observations.

6.14 CARBOCATION STABILITY AND THE TRANSITION STATE

We saw in Section 5.11 that the order of stability of carbocations is tertiary > secondary > primary > methyl:

$$
\underset{3°}{\overset{R}{\underset{R}{R-\overset{|}{\underset{|}{C}}{}^{+}}}} > \underset{2°}{\overset{H}{\underset{R}{R-\overset{|}{\underset{|}{C}}{}^{+}}}} > \underset{1°}{\overset{H}{\underset{H}{R-\overset{|}{\underset{|}{C}}{}^{+}}}} > \underset{Methyl}{\overset{H}{\underset{H}{H-\overset{|}{\underset{|}{C}}{}^{+}}}}
$$

In the dehydration of alcohols (i.e., following steps 1–3 in the forward direction) the slowest step is step 2 because as we shall see, it is *a highly endothermic step:* the formation of the carbocation from the protonated alcohol. The first and third steps are simple acid–base reactions. Proton-transfer reactions of this type occur very rapidly.

General Mechanism for the Acid-Catalyzed Dehydration of an Alcohol

Step 1

$$-\overset{|}{\underset{H}{C}}-\overset{|}{\underset{|}{C}}-\overset{..}{\underset{..}{O}}H + H_3O:^+ \rightleftharpoons -\overset{|}{\underset{H}{C}}-\overset{|}{\underset{|}{C}}-\overset{H}{\overset{|}{\underset{..}{O}{}^{+}}}-H + H_2\overset{..}{O}: \qquad \text{Fast}$$

Step 2

$$-\overset{|}{\underset{H}{C}}-\overset{|}{\underset{|}{C}}\overset{H}{\overset{|}{\underset{..}{O}{}^{+}}}-H \rightleftharpoons -\overset{|}{\underset{H}{C}}-\overset{|}{\underset{+}{C}}- + H_2\overset{..}{O}: \qquad \begin{array}{c}\text{Slow}\\ \text{(rate determining)}\end{array}$$

Step 3

$$-\overset{|}{\underset{\underset{H}{|}}{C}}\overset{|}{\underset{+}{C}}- + H_2\overset{..}{O}: \rightleftharpoons -C=C- + H_3O:^+ \qquad \text{Fast}$$

Because step 2 is, then, the rate-determining step, it is the step that determines the reactivity of alcohols toward dehydration. With this in mind, we can now understand why tertiary alcohols are the most easily dehydrated. The formation of a tertiary carbocation is easiest because the energy of activation for step 2 of a reaction leading to a tertiary carbocation is lowest (see Fig. 6.9).

The reactions by which carbocations are formed from protonated alcohols are all highly *endothermic.* According to the Hammond-Leffler postulate (Section

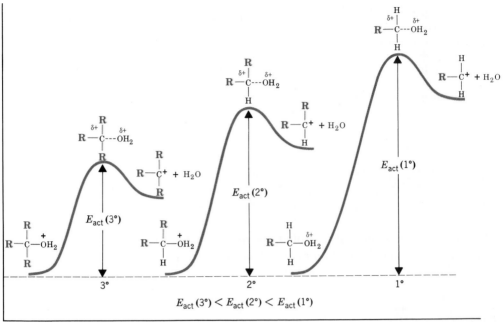

FIGURE 6.9 Potential energy diagrams for the formation of carbocations from protonated tertiary, secondary, and primary alcohols. The relative energies of activation are tertiary < secondary < primary.

5.13A), there should be a strong resemblance between the transition state and the product in each case. Of the three, *the transition state that leads to the tertiary carbocation is lowest in potential energy because it resembles the most stable product.* By contrast, the transition state that leads to the primary carbocation occurs at highest potential energy because it resembles the least stable product. In each instance, moreover, the transition state is stabilized by the same factor that stabilizes the carbocation itself: **By delocalization of the charge.** We can understand this if we examine the process by which the transition state is formed.

$$
\underset{\substack{\text{Protonated}\\\text{alcohol}}}{-\overset{|}{\underset{|}{C}}-\overset{H}{\underset{\cdot\cdot}{\overset{|}{O}}}{}^{+}-H} \rightleftarrows \underset{\substack{\text{Transition}\\\text{state}}}{-\overset{|}{\underset{|}{C}}{}^{\delta+}---\overset{H}{\underset{\cdot\cdot}{\overset{|}{O}}}{}^{\delta+}-H} \rightleftarrows \underset{\text{Carbocation}}{-\overset{|}{\underset{|}{C}}{}^{+} \quad +:\overset{H}{\underset{\cdot\cdot}{O}}-H}
$$

The oxygen atom of the protonated alcohol bears a full positive charge. As the transition state develops this oxygen atom begins to separate from the carbon atom to which it is attached. The carbon atom, because it is losing the electrons that bonded it to the oxygen atom, begins to develop a partial positive charge. This developing positive charge *is most effectively delocalized in the transition state leading to a tertiary carbocation because of the presence of three electron-releasing alkyl groups.* The positive charge is less effectively delocalized in the transition state leading to a secondary carbocation (*two* electron-releasing groups) and is least effectively delocalized in the transition state leading to a primary carbocation (*one* electron-releasing group).

$$\overset{\delta+}{R}\downarrow$$
$$\overset{\delta+}{R}\rightarrow C^{\delta+}\text{---}\overset{..}{\underset{\delta+}{O}}\text{---}H$$
$$\overset{\uparrow}{\underset{\delta+}{R}}$$

**Transition state leading
to 3° carbocation
(most stable)**

$$\overset{\delta+}{R}\downarrow$$
$$\overset{\delta+}{R}\rightarrow C^{\delta+}\text{---}\overset{..}{\underset{\delta+}{O}}\text{---}H$$
$$\overset{}{\underset{H}{}}$$

**Transition state leading
to 2° carbocation**

$$\overset{H}{\downarrow}\quad H$$
$$\overset{\delta+}{R}\rightarrow C^{\delta+}\text{---}\overset{..}{\underset{\delta+}{O}}\text{---}H$$
$$\overset{}{\underset{H}{}}$$

**Transition state leading
to 1° carbocation
(least stable)**

6.15 CARBOCATION STABILITY AND THE OCCURRENCE OF MOLECULAR REARRANGEMENTS

With an understanding of carbocation stability and its effect on transition states behind us, we now proceed to explain the rearrangements of carbon skeletons that occur in some alcohol dehydrations. For example, let us consider again the rearrangement that occurs when 3,3-dimethyl-2-butanol is dehydrated.

$$CH_3\text{---}\underset{\underset{CH_3}{|}}{\overset{\overset{CH_3}{|}}{C}}\text{---}\underset{\underset{OH}{|}}{CH}\text{---}CH_3 \xrightarrow[\text{heat}]{85\%\ H_3PO_4} CH_3\text{---}\underset{\overset{CH_3}{|}}{C}=\underset{\overset{CH_3}{|}}{C}\text{---}CH_3 + CH_2=\underset{\overset{CH_3}{|}}{C}\text{---}CHCH_3$$

3,3-Dimethyl-2-butanol   **2,3-Dimethyl-2-butene**   **2,3-Dimethyl-1-butene**
                               (major product)   **(minor product)**

The first step of this dehydration is the formation of the protonated alcohol in the usual way:

Step 1 $$CH_3\text{---}\underset{\underset{:\overset{..}{O}\text{---}H}{|}}{\overset{\overset{CH_3}{|}}{C}}\text{---}CH\text{---}CH_3 + H\text{---}\underset{\underset{H}{|}}{\overset{\overset{H}{|}}{O}}:^+ \rightleftharpoons CH_3\text{---}\underset{\underset{:\overset{}{O}H_2}{|}}{\overset{\overset{CH_3}{|}}{C}}\text{---}CHCH_3 + H_2\overset{..}{O}:$$

Protonated alcohol

In the second step the protonated alcohol loses water and a secondary carbocation forms:

Step 2 $$CH_3\text{---}\underset{\underset{CH_3\ :OH_2}{|}}{\overset{\overset{CH_3}{|}}{C}}\text{---}CH\text{---}CH_3 \rightleftharpoons CH_3\text{---}\underset{\underset{CH_3}{|}}{\overset{\overset{CH_3}{|}}{C}}\text{---}\overset{+}{C}HCH_3 + H_2\overset{..}{O}:$$

A 2° carbocation

Now the rearrangement occurs. Before anything else can happen to it, *a less stable, secondary carbocation rearranges to a more stable tertiary carbocation.*

Step 3 $$CH_3\text{---}\underset{\underset{CH_3}{|}}{\overset{\overset{CH_3}{|}}{C}}\text{---}\overset{+}{C}HCH_3 \rightarrow CH_3\text{---}\underset{\underset{CH_3}{|}}{\overset{\overset{CH_3}{|}}{C}}^{\delta+}\cdots\overset{\delta+}{C}HCH_3 \rightarrow CH_3\text{---}\overset{+}{\underset{\underset{CH_3}{|}}{C}}\text{---}CH\text{---}CH_3$$

**2° Carbocation
(less stable)**   **Transition state**   **3° Carbocation
(more stable)**

The rearrangement occurs through the migration of an alkyl group (methyl) from the carbon atom adjacent to the one with the positive charge. The methyl group migrates **with its pairs of electrons,** that is, as a methyl anion, $^-\!:CH_3$. After the migration is complete, the carbon atom that the methyl anion left has become a carbocation, and the positive charge on the carbon atom to which it migrated has been neutralized. Because a group migrates from one carbon to the next, this kind of rearrangement is often called **a 1,2-shift.**

In the transition state the shifting methyl is partly bonded to both carbon atoms by the pair of electrons with which it migrates. It never leaves the carbon skeleton.

The final step of the reaction is the loss of proton from the new carbocation and the formation of an alkene. This step, however, can occur in two ways.

Step 4

$$\text{H}-\overset{(a)}{\overset{\frown}{\text{CH}_2}}-\overset{+}{\underset{\underset{\text{CH}_3}{|}}{\text{C}}}-\overset{\overset{\text{H}\ (b)}{\frown}}{\underset{\underset{\text{CH}_3}{|}}{\text{C}}}-\text{CH}_3$$

$$\xrightarrow{(a)} \quad \underset{\underset{\text{CH}_3}{|}\ \underset{\text{CH}_3}{|}}{\text{CH}_2=\text{C}-\text{CHCH}_3}$$

(minor product) — **Less stable alkene**

$$\xrightarrow{(b)} \quad \underset{\underset{\text{CH}_3}{|}\ \underset{\text{CH}_3}{|}}{\text{CH}_3-\text{C}=\text{C}-\text{CH}_3}$$

(major product) — **More stable alkene**

The more favored route is dictated by the type of alkene being formed. Path (b) leads to the highly stable tetrasubstituted alkene, and this is the path followed by most of the carbocations. Path (a), on the other hand, leads to a less stable, disubstituted alkene and produces the minor product of the reaction. *The formation of the more stable alkene is the general rule (Zaitsev's rule) in the acid-catalyzed dehydration reactions of alcohols.*

When we examine the rearrangement that occurs in the dehydration of butyl alcohol (Section 6.13), we find that the mechanism is similar to that for 3,3-dimethyl-2-butanol. In this instance, however, **the migrating group is a hydride ion, $H\!:^-$** — *a group that also migrates with a pair of electrons.*

Step 1 $\quad CH_3CH_2CH_2CH_2\overset{..}{\underset{..}{O}}H + H^+ \;\rightleftharpoons\; CH_3CH_2CH_2CH_2\overset{..}{\underset{..}{O}}H_2{}^+$

Step 2 $\quad CH_3CH_2CH_2CH_2-\overset{..}{O}H_2{}^+ \;\rightleftharpoons\; CH_3CH_2CH_2CH_2{}^+ + H_2O$

A 1° carbocation

Step 3 $\quad CH_3CH_2\overset{+}{\text{CH}}-\overset{}{\text{CH}_2} \longrightarrow CH_3CH_2\overset{\delta+}{\text{CH}}\cdots\overset{\delta+}{\text{CH}_2} \longrightarrow CH_3CH_2\overset{+}{\text{CH}}-CH_2$

$\qquad\qquad\qquad\ \ |\qquad\qquad\qquad\qquad\ |\qquad\qquad\qquad\qquad\quad |$
$\qquad\qquad\qquad\ \ \text{H}\qquad\qquad\qquad\qquad\ \text{H}\qquad\qquad\qquad\qquad\quad \text{H}$

1° Carbocation $\qquad\qquad$ Transition state $\qquad\qquad$ 2° Carbocation

Step 4

$$CH_3\overset{+}{\text{CH}}-\overset{}{\text{CH}}-\overset{}{\text{CH}_2}$$
$$\ \ \ |\qquad\quad |$$
$$\ \ \ \text{H}\qquad\ \ \text{H}$$
$$\underset{(a)}{\overset{\frown}{}}\ \overset{..}{\underset{..}{\text{O}}}\ \underset{(b)}{\overset{\frown}{}}$$
$$\ \ \text{H}\qquad\ \ \text{H}$$

$\xrightarrow{(a)} \quad CH_3CH=CHCH_3 + CH_3CH=CHCH_3 + H_3O^+$

$\qquad\qquad$ trans $\qquad\qquad\qquad\quad$ cis
$\qquad$ (major product) $\qquad\quad$ (minor product)
$\qquad$ *Most stable* $\qquad\qquad$ *Less stable*
$\qquad$ *alkene* $\qquad\qquad\qquad$ *alkene*

$\xrightarrow{(b)} \quad CH_3CH_2CH=CH_2 + H_3O^+$

$\qquad\qquad$ (minor product)
$\qquad\qquad$ *Least stable alkene*

Steps 2 and 3 may actually occur as a single step. In step 4 three different alkenes are possible products: *trans*-2-butene, *cis*-2-butene, and 1-butene. Of these, *trans*-2-butene is the most stable and it is, therefore, the major product.

Studies of thousands of reactions involving carbocations show that rearrangements like those just described are general phenomena. *They occur almost invariably when the migration of an alkyl group or hydride ion can lead to a more stable carbocation.* The following are examples:

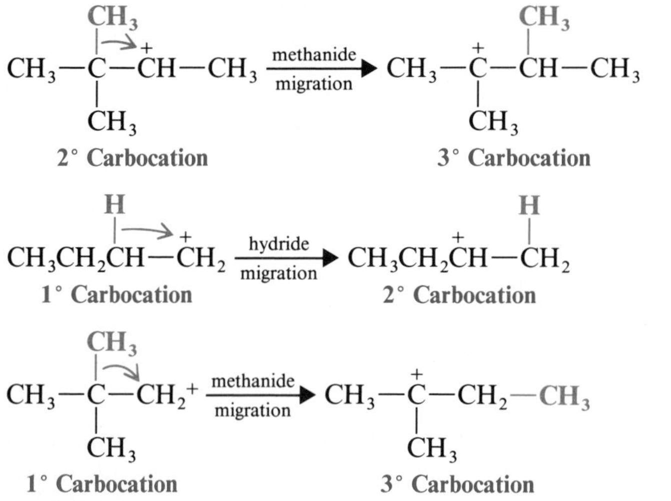

Rearrangements of carbocations can also lead to a change in ring size, as the following example shows:

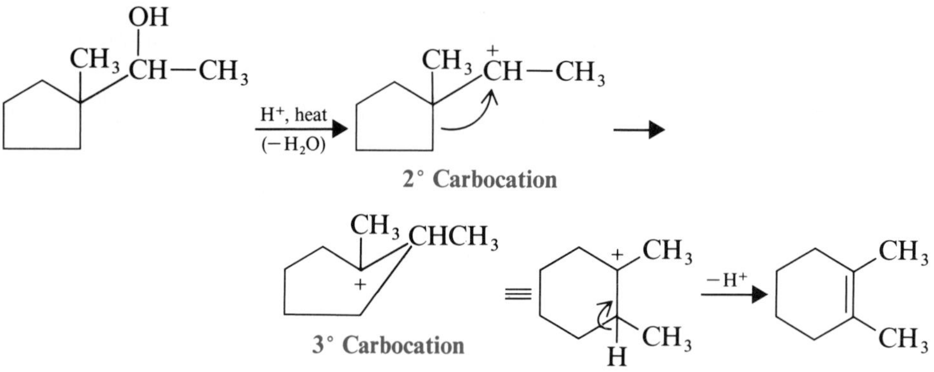

PROBLEM 6.15

Acid-catalyzed dehydration of neopentyl alcohol, $(CH_3)_3CCH_2OH$, yields 2-methyl-2-butene. Outline a mechanism showing all steps in its formation.

PROBLEM 6.16

Heating neopentyl iodide, $(CH_3)_3CCH_2I$, in formic acid (a solvent of very high ionizing ability) slowly leads to the formation of 2-methyl-2-butene. Propose a mechanism for this reaction.

PROBLEM 6.17

When the compound called *isoborneol* is heated with 50% sulfuric acid, the product of the reaction is the compound called camphene and not bornylene as one might expect. Using models to assist you, write a step-by-step mechanism showing how camphene is formed.

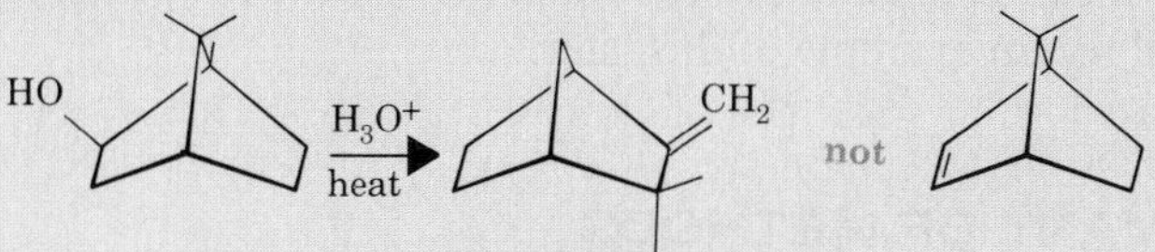

Isoborneol   Camphene   Bornylene

6.16 THE FORMATION OF ALKENES
BY DEBROMINATION OF VICINAL DIBROMIDES

Vicinal (or **vic**) dihalides are dihalo compounds in which the halogens are situated on adjacent carbon atoms. The name **geminal** (or **gem**) dihalide is used for those dihalides where both halogen atoms are attached to the same carbon atom.

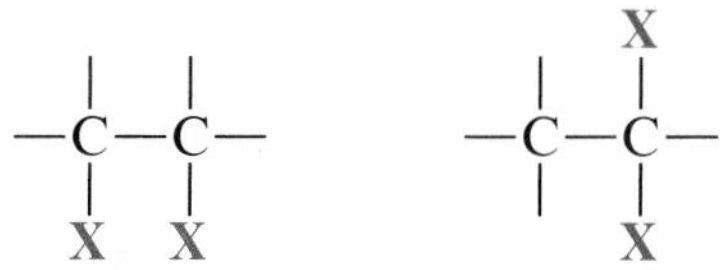

A *vic*-dihalide   A *gem*-dihalide

vic-Dibromides undergo **debromination** when they are treated with a solution of sodium iodide in acetone or with a mixture of zinc dust in acetic acid (or ethanol).

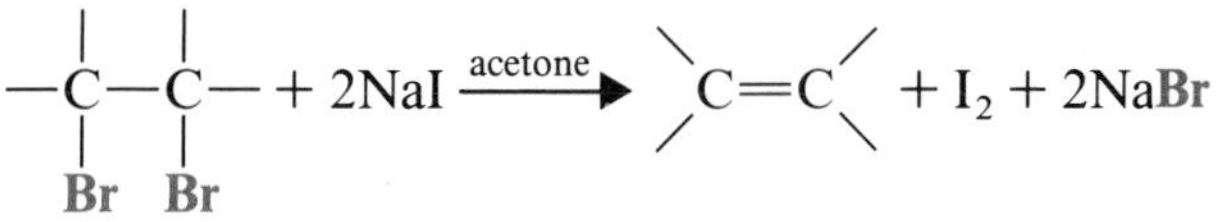

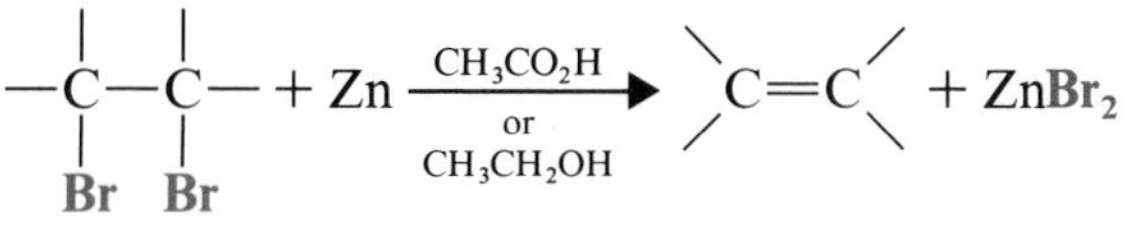

Debromination by sodium iodide takes place by an E2 mechanism similar to that for dehydrohalogenation.

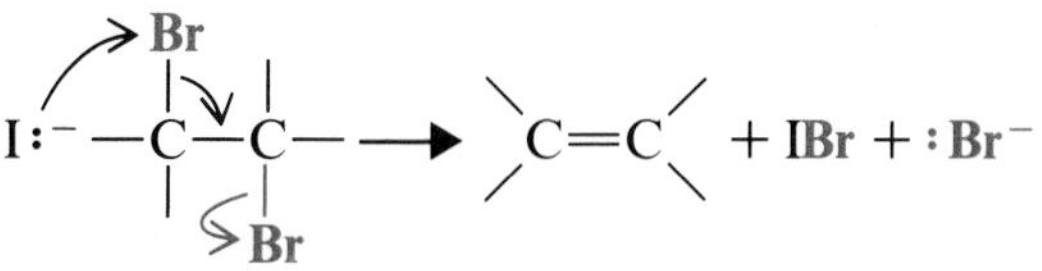

then

$$I^- + IBr \longrightarrow I_2 + Br^-$$

Debromination by zinc takes place on the surface of the metal and the mechanism is uncertain. Other electropositive metals (Na, Ca, Mg, for example) also cause debromination of *vic*-dibromides.

vic-Dibromides are usually prepared by the addition of bromine to an alkene (cf. Section 7.6). Consequently, dehalogenation of a *vic*-dibromide is of little use as a general preparative reaction. Bromination followed by debromination is useful, however, in the purification of alkenes (see Problem 6.41) and in "protecting" the double bond. We shall see an example of this later.

6.17 SUMMARY OF METHODS FOR THE PREPARATION OF ALKENES

In this chapter we described four general methods for the preparation of alkenes.

1. **Dehydrohalogenation of alkyl halides (Section 6.12)**

   General Example

$$-\underset{\underset{H}{|}}{C}-\underset{\underset{X}{|}}{C}- \xrightarrow[\substack{\text{heat} \\ (-\text{HX})}]{\text{base}} \quad \underset{/}{\overset{\backslash}{C}}=\underset{\backslash}{\overset{/}{C}}$$

   Specific Examples

$$CH_3CH_2\underset{\underset{Br}{|}}{C}HCH_3 \xrightarrow[C_2H_5OH]{C_2H_5ONa} CH_3CH=CHCH_3 + CH_3CH_2CH=CH_2$$

   (cis and trans 81%)   (19%)

$$CH_3CH_2\underset{\underset{Br}{|}}{C}HCH_3 \xrightarrow[70\,°C]{(CH_3)_3COK} CH_3CH=CHCH_3 + CH_3CH_2CH=CH_2$$

   (cis and trans, 47%)   (53%)
   Disubstituted alkenes   Monosubstituted alkene

2. **Dehydration of alcohols (Sections 6.13–6.15)**

   General Example

$$-\underset{\underset{H}{|}}{C}-\underset{\underset{OH}{|}}{C}- \xrightarrow[\text{heat}]{\text{acid}} \quad \underset{/}{\overset{\backslash}{C}}=\underset{\backslash}{\overset{/}{C}} + H_2O$$

   Specific Examples

$$CH_3CH_2OH \xrightarrow[180\,°C]{\text{conc. } H_2SO_4} CH_2=CH_2 + H_2O$$

$$CH_3\underset{\underset{CH_3}{|}}{\overset{\overset{CH_3}{|}}{C}}-OH \xrightarrow[85\,°C]{20\%\ H_2SO_4} CH_3\overset{\overset{CH_3}{|}}{C}=CH_2 + H_2O$$

   (83%)

3. **Dehalogenation of *vic*-dibromides (Section 6.16)**

   General Example

$$-\underset{\underset{Br}{|}}{C}-\underset{\underset{Br}{|}}{C}- \xrightarrow[CH_3CO_2H]{Zn} \quad \underset{/}{\overset{\backslash}{C}}=\underset{\backslash}{\overset{/}{C}} + ZnBr_2$$

4. Hydrogenation of alkynes (Section 6.7B)

General Example

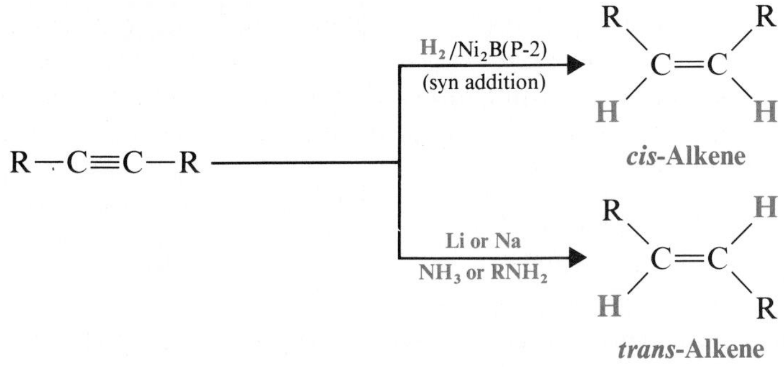

$$R-C\equiv C-R \xrightarrow[\text{(syn addition)}]{H_2/Ni_2B(P-2)}$$

R, R $C=C$ H, H

cis-Alkene

$$\xrightarrow[\text{NH}_3 \text{ or RNH}_2]{\text{Li or Na}}$$

R, H $C=C$ H, R

trans-Alkene

In subsequent chapters we shall see a number of other methods for alkene synthesis.

6.18 THE ACIDITY OF ACETYLENE AND TERMINAL ALKYNES

The hydrogen atoms of acetylene are considerably more acidic than those of ethene or ethane (Section 2.15C).

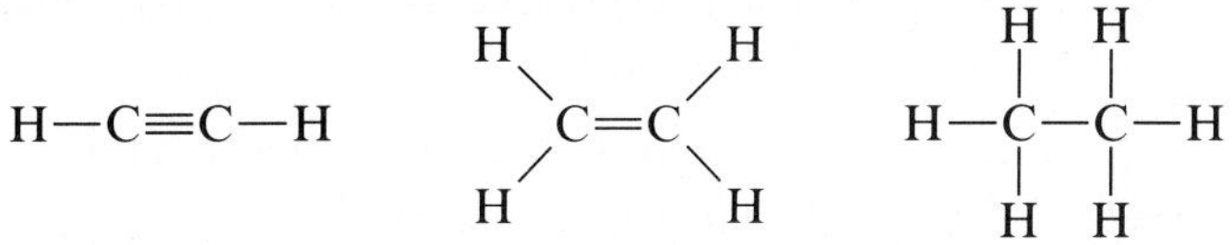

$$H-C\equiv C-H$$

H, H $C=C$ H, H

$H-\overset{\displaystyle H}{\underset{\displaystyle H}{C}}-\overset{\displaystyle H}{\underset{\displaystyle H}{C}}-H$

$K_a \simeq 10^{-25}$
$(pK_a = 25)$

$K_a \simeq 10^{-44}$
$(pK_a = 44)$

$K_a \simeq 10^{-50}$
$(pK_a = 50)$

We can account for this order of acidities on the basis of the hybridization state of carbon in each compound. Electrons of $2s$ orbitals have lower energy than those in $2p$ orbitals because electrons in $2s$ orbitals tend on the average to be much closer to the nucleus than electrons in $2p$ orbitals. With hybrid orbitals, therefore, *having more s character means that the electrons will, on the average, be lower in energy because they will be closer to the nucleus.* The sp orbitals of the sigma bonds of acetylene have 50% s character (because they arise from one s orbital and one p orbital), those of sp^2 orbitals have 33.3% s character, while those of sp^3 orbitals have only 25% s character. This means, in effect, that the sp carbon atoms of acetylene act as if they were the most electronegative when compared to sp^2 and sp^3 carbon atoms. (Remember that electronegativity measures an atom's ability to hold bonding electrons close to its nucleus, and having electrons closer to the nucleus *makes them more stable.*) The order of electronegativity of carbon in each hybridization state is $sp > sp^2 > sp^3$.

Now we can see how the order of relative acidities of acetylene, ethene, and ethane parallels the effective electronegativity of carbon:

Relative Acidity

$$HC\equiv CH > CH_2=CH_2 > CH_3CH_3$$

Being most electronegative, the carbon atom of acetylene is best able to accommodate the electron pair in the anion left after the proton is lost.

$$HC\equiv CH + B:^- \rightleftharpoons B-H + HC\equiv C:^-$$

> This anion is the most stable because the *sp* carbon atom is most electronegative. Therefore, acetylene is most acidic

$$CH_2=CH_2 + B:^- \rightleftharpoons B-H + CH_2=CH:^-$$

$$CH_3CH_3 + B:^- \rightleftharpoons B-H + CH_3CH_2:^-$$

> This anion is the least stable because the *sp³* carbon atom is the least electronegative. Therefore, ethane is least acidic

Notice that the explanation given here is the same as that given to account for the relative acidities of HF, H_2O, and NH_3.

Relative Acidity

$$HF > H_2O > NH_3$$

Since the atoms being compared are in the same horizontal row of the Periodic Table, the acidity of these hydrides parallels the electronegativity of their atoms:

$$F > O > N$$

Because fluorine is most electronegative, fluoride ions are the most stable (more stable than OH^- or NH_2^-) and consequently HF is the strongest acid.

The order of basicities of the anions is opposite their relative stability. The ethanide ion is least stable and therefore it is the most basic.

Relative Basicity

$$CH_3CH_2:^- > CH_2=CH:^- > HC\equiv C:^-$$

What we have said about acetylene and acetylide ions is true of any terminal alkyne ($RC\equiv CH$) and any alkynide ion ($RC\equiv C:^-$). If we include other hydrogen compounds of the first-row elements of the Periodic Table, we can write the following orders of relative acidities and basicities:

Relative Acidity

$$H-\overset{..}{\underset{..}{O}}H > H-\overset{..}{\underset{..}{O}}R > H-C\equiv CR > H-\overset{..}{N}H_2 > H-CH=CH_2 > H-CH_2CH_3$$

Relative Basicity

$$^-:\overset{..}{\underset{..}{O}}H < ^-:\overset{..}{\underset{..}{O}}R < ^-:C\equiv CR < ^-:\overset{..}{N}H_2 < ^-:CH=CH_2 < ^-:CH_2CH_3$$

We see from the order just given that while terminal alkynes are more acidic than ammonia, they are less acidic than alcohols and are considerably less acidic than water.

> The arguments just made apply only to acid–base reactions that take place in solution. In the gas phase, acidities and basicities are very much different. For example, in the gas phase the hydroxide ion is a stronger base than the acetylide ion. The explanation for this shows us again the important roles solvents play in reactions that involve ions (cf. Section 5.13). In solution, smaller ions (e.g., hydroxide ions) are more effectively solvated than larger ones (e.g., acetylide ions).

Because they are more effectively solvated, smaller ions are more stable and are therefore less basic. In the gas phase, large ions are stabilized by polarization of their bonding electrons and the bigger a group is the more polarizable it will be. Thus in the gas phase larger ions are less basic.

PROBLEM 6.18

Predict the products of the following acid–base reactions. If the equilibrium would not result in the formation of appreciable amounts of products, you should so indicate. In each case label the stronger acid, the stronger base, the weaker acid, and the weaker base.

(a) $H—C{\equiv}C—H + NaNH_2 \longrightarrow$

(b) $CH_2{=}CH_2 + NaNH_2 \longrightarrow$

(c) $CH_3CH_3 + NaNH_2 \longrightarrow$

(d) $H—C{\equiv}C\!:^-Na^+ + CH_3CH_2OH \longrightarrow$

(e) $H—C{\equiv}C\!:^-Na^+ + H_2O \longrightarrow$

6.19 REPLACEMENT OF THE ACETYLENIC HYDROGEN ATOM OF TERMINAL ALKYNES

Sodium acetylide and other sodium alkynides can be prepared by treating terminal alkynes with sodium amide in liquid ammonia.

$$H—C{\equiv}C—H + NaNH_2 \xrightarrow{\text{liq. NH}_3} H—C{\equiv}C\!:^-Na^+ + NH_3$$

$$CH_3C{\equiv}C—H + NaNH_2 \xrightarrow{\text{liq. NH}_3} CH_3C{\equiv}C\!:^-Na^+ + NH_3$$

These are acid–base reactions. The amide ion, by virtue of its being the anion of the very weak acid, ammonia ($K_a \approx 10^{-33}$), is able to remove the acetylenic protons of terminal alkynes ($K_a \approx 10^{-25}$). These reactions, for all practical purposes, go to completion.

Sodium alkynides are useful intermediates for the synthesis of other alkynes. These syntheses can be accomplished by treating the sodium alkynide with a primary alkyl halide.

$$R—C{\equiv}C\!:^-Na^+ + R'CH_2{-}Br \longrightarrow R—C{\equiv}C—CH_2R' + NaBr$$

| Sodium | Primary | Mono- or disubstituted |
| alkynide | alkyl halide | acetylene |

(R or R′ or both may be hydrogen)

The following examples illustrate this synthesis of alkynes:

$$HC{\equiv}C\!:^-Na^+ + CH_3{-}Br \xrightarrow[\text{5 h}]{\text{liq. NH}_3} H—C{\equiv}C—CH_3 + NaBr$$

(84%)
Propyne

$$CH_3CH_2C\equiv C:^- Na^+ + CH_3CH_2{-}Br \xrightarrow[\text{6 h}]{\text{liq. NH}_3} CH_3CH_2C\equiv CCH_2CH_3 + NaBr$$

(75%)

3-Hexyne

In all of these examples the alkynide ion acts as a nucleophile and displaces a halide ion from the primary alkyl halide. The result is **an S$_N$2 reaction** (Section 5.5).

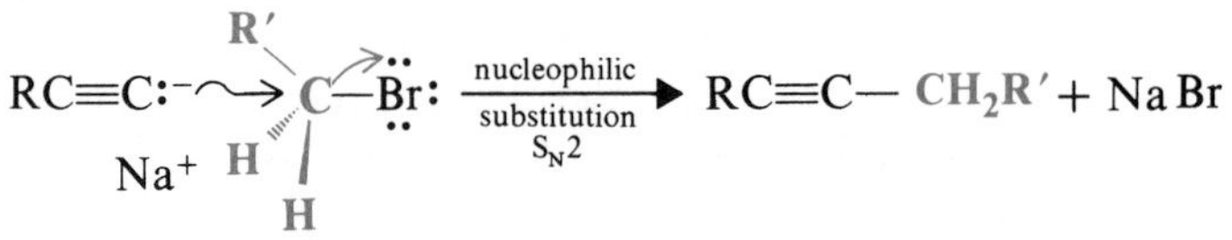

$$RC\equiv C:^- \rightarrow \underset{\underset{\text{H}}{\overset{\text{R'}}{|}}}{C}{-}\ddot{\overset{..}{B}}r: \xrightarrow[\substack{\text{substitution}\\ S_N2}]{\text{nucleophilic}} RC\equiv C{-}CH_2R' + Na\,Br$$

Na$^+$ H

Sodium   1° Alkyl
alkynide   halide

The unshared electron pair of the alkynide ion attacks the backside of the carbon atom that bears the halogen atom and forms a bond to it. The halogen atom departs as a halide ion.

   When secondary or tertiary halides are used, the alkynide ion acts as a base rather than as a nucleophile, and the major result is an **E2 elimination** (Section 5.16). The products of the elimination are an alkene and the alkyne from which the sodium alkynide was originally formed.

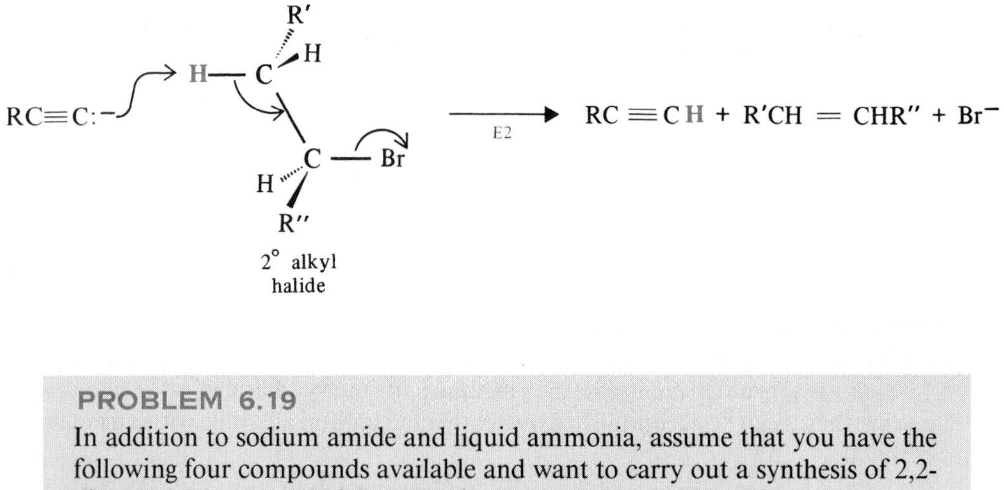

$$RC\equiv C:^- \quad H{-}\underset{\underset{R''}{\overset{}{\underset{|}{C}{-}Br}}}{\overset{R'}{\overset{|}{C}{-}H}} \xrightarrow{\text{E2}} RC\equiv CH + R'CH = CHR'' + Br^-$$

2° alkyl
halide

PROBLEM 6.19

In addition to sodium amide and liquid ammonia, assume that you have the following four compounds available and want to carry out a synthesis of 2,2-dimethyl-3-hexyne. Which synthetic route would you choose?

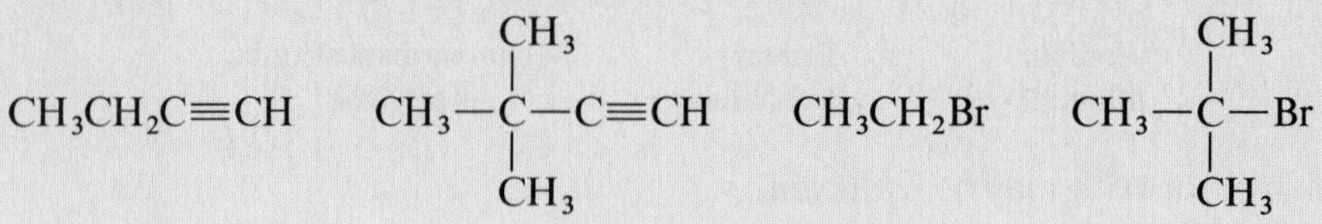

$$CH_3CH_2C\equiv CH \quad\quad CH_3{-}\underset{\underset{CH_3}{|}}{\overset{\overset{CH_3}{|}}{C}}{-}C\equiv CH \quad\quad CH_3CH_2Br \quad\quad CH_3{-}\underset{\underset{CH_3}{|}}{\overset{\overset{CH_3}{|}}{C}}{-}Br$$

6.20 OTHER METAL ACETYLIDES

Acetylene and terminal alkynes also form metal derivatives with silver and copper(I) ions.

$$RC\equiv CH + Cu(NH_3)_2^+ + OH^- \xrightarrow{H_2O} R-C\equiv CCu + H_2O + 2NH_3$$

$$RC\equiv CH + Ag(NH_3)_2^+ + OH^- \xrightarrow{H_2O} RC\equiv CAg + H_2O + 2NH_3$$

Silver and copper alkynides differ from sodium alkynides in several ways. The metal–carbon bond in silver and copper alkynides is largely covalent. As a result, silver and copper alkynides are poor bases and poor nucleophiles. Silver and copper alkynides can be prepared in water, whereas sodium alkynides react vigorously with water.

$$R-C\equiv CNa + H_2O \longrightarrow R-C\equiv CH + NaOH$$

Silver and copper alkynides are also quite insoluble in water and precipitate when they are prepared. This is the basis for an old and still convenient test for terminal alkynes as well as a method for separating terminal alkynes from alkynes that have an internal triple bond.

$$R-C\equiv C-R + Ag(NH_3)_2^+OH^- \longrightarrow \text{no precipitate}$$

$$R-C\equiv CH + Ag(NH_3)_2^+OH^- \longrightarrow RC\equiv CAg\downarrow$$

Once a separation has been carried out, the terminal alkyne can be regenerated by treating the alkynide with sodium cyanide (or with a strong acid).

$$R-C\equiv CAg + 2CN^- + H_2O \longrightarrow R-C\equiv CH + Ag(CN)_2^- + OH^-$$

Silver and copper alkynides must be handled cautiously; when dry they are likely to explode.

6.21 SYNTHESIS OF ALKYNES

6.21A From Other Alkynes

Alkynes can be prepared from other terminal alkynes through the nucleophilic substitution reaction that we saw in Section 6.19.

$$R-C\equiv CH \xrightarrow[\text{liq. NH}_3]{\text{NaNH}_2} R-C\equiv C:^-Na^+ \xrightarrow[\substack{\text{primary}\\\text{halide}}]{R'-X} R-C\equiv C-R'$$

6.21B By Elimination Reactions

Alkynes can also be synthesized from alkenes. In this method an alkene is first treated with bromine to form a *vic*-dibromo compound.

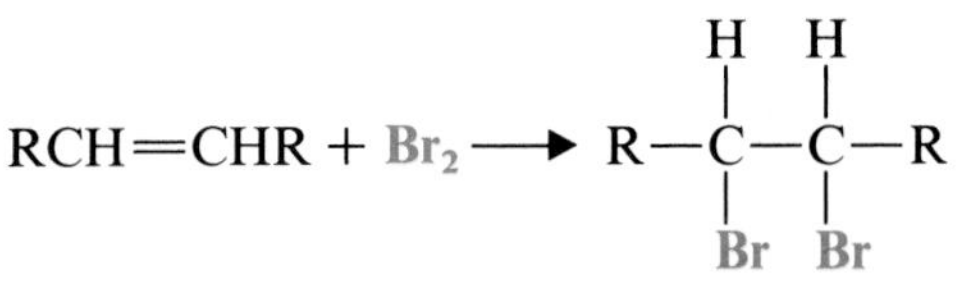

$$RCH=CHR + Br_2 \longrightarrow R-\underset{\underset{Br}{|}}{\overset{\overset{H}{|}}{C}}-\underset{\underset{Br}{|}}{\overset{\overset{H}{|}}{C}}-R$$

vic-Dibromide

Then the *vic* dibromide is dehydrohalogenated through its reaction with a strong base. The dehydrohalogenation occurs in two steps. The first step yields a bromoalkene.

Step 1

$$R-\overset{\underset{|}{H}}{\underset{Br}{C}}-\overset{\underset{|}{H}}{\underset{Br}{C}}-R \quad :NH_2^- \longrightarrow R-\overset{H}{\underset{Br}{C}}=\overset{}{\underset{}{C}}-R + NH_3 + Br^-$$

vic-Dibromide   Strong   Bromoalkene
               base

The second step is more difficult; it yields an alkyne.

Step 2 $H_2\ddot{N}:^-$ $R-\overset{H}{\underset{\underset{Br}{|}}{C}}=\overset{}{\underset{}{C}}-R \longrightarrow R-C\equiv C-R + NH_3 + Br^-$

Strong   Alkyne
base

Depending on the conditions, these two dehydrohalogenations may be carried out as separate reactions or they may be carried out consecutively in a single mixture. The strong base, sodium amide, is capable of effecting both dehydrohalogenations in a single reaction mixture. (At least two molar equivalents of sodium amide per mole of the dihalide must be used, and if the product is a terminal alkyne, three molar equivalents must be used because the terminal alkyne will react with sodium amide as it is formed in the mixture.) Dehydrohalogenations with sodium amide are usually carried out in liquid ammonia or in an inert medium such as mineral oil.

The following example illustrates this method.

$$CH_3CH_2CH{=}CH_2 \xrightarrow[CCl_4]{Br_2} CH_3CH_2\underset{\underset{Br}{|}}{C}HCH_2Br \xrightarrow[\substack{mineral\ oil\\110-160\ °C}]{NaNH_2}$$

$$\begin{bmatrix} CH_3CH_2CH{=}CHBr \\ + \\ CH_3CH_2\underset{\underset{Br}{|}}{C}{=}CH_2 \end{bmatrix} \xrightarrow[\substack{mineral\ oil\\110-160\ °C}]{NaNH_2} [CH_3CH_2C{\equiv}CH] \xrightarrow{NaNH_2}$$

$$CH_3CH_2C{\equiv}C:^-Na^+ \xrightarrow{NH_4Cl} CH_3CH_2C{\equiv}CH + NH_3 + NaCl$$

Ketones can be converted to *gem*-dichlorides through their reaction with phosphorus pentachloride, and these can also be used to synthesize alkynes.

Methyl cyclohexyl   (70–80%)   (46%)
ketone   A *gem*-dichloride   Cyclohexylacetylene

PROBLEM 6.20

Outline all steps in a synthesis of propyne from each of the following:

(a) CH_3COCH_3 (c) $CH_3CHBrCH_2Br$

(b) $CH_3CH_2CHBr_2$ (d) $CH_3CH{=}CH_2$

ADDITIONAL PROBLEMS

6.21

Each of the following names is incorrect. Tell how and give the correct name.

(a) *cis*-3-Pentene

(b) 1,1,2,2-Tetramethylethene

(c) 2-Methylcycloheptene

(d) 1-Methyl-1-heptene

(e) 3-Methyl-2-butene

(f) 4,5-Dichlorocyclopentene

6.22

Write a structural formula for each of the following:

(a) 1-Methylcyclobutene

(b) 3-Methylcyclopentene

(c) 2,3-Dimethyl-2-pentene

(d) *trans*-2-Hexene

(e) *cis*-3-Heptene

(f) 3,3,3-Trichloropropene

(g) Isobutylene

(h) Propylene

(i) 4-Cyclopentyl-1-pentene

(j) Cyclopropylethene

6.23

Write structural formulas and give IUPAC names for all alkene isomers of   (a) C_5H_{10} and (b) C_6H_{12}.   (c) What other isomers are possible for C_5H_{10} and C_6H_{12}? Write their structures.

6.24

Give the IUPAC names for each of the following:

(a)

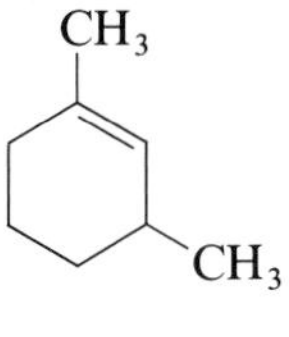

(c) CH_3CH_2
        $\diagdown$
          $C{=}CH_2$
        $\diagup$
   $CH_3CH_2CH_2$

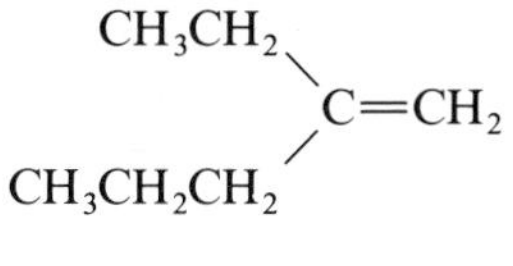

(b) $CH_2{=}CCH_2CH_3$
          |
       $CH_2CH_2CH_3$

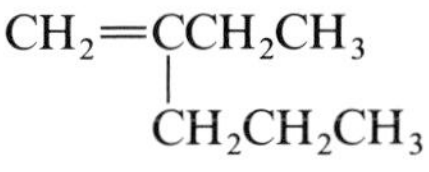

(d)

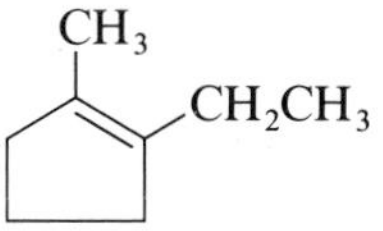

6.25

Outline a synthesis of propene from each of the following:

(a) Propyl chloride

(b) Isopropyl chloride

(c) Propyl alcohol

(d) Isopropyl alcohol

(e) 1,2-Dibromopropane

6.26

Outline a synthesis of cyclopentene from each of the following:

(a) Bromocyclopentane

(b) 1,2-Dichlorocyclopentane

(c) Cyclopentanol [i.e., $(CH_2)_4CHOH$]

6.27

Starting with 1-methylcyclohexene and using any other needed reagents, outline a synthesis of the following deuterium-labeled compound.

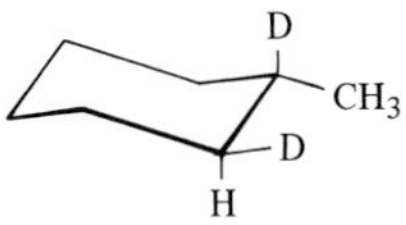

6.28

When *trans*-2-methylcyclohexanol (see following reaction) is subjected to acid-catalyzed dehydration, the major product is 1-methylcyclohexene:

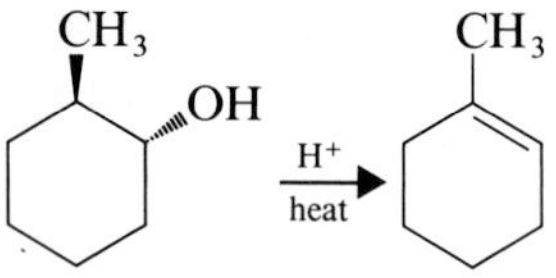

However, when *trans*-1-bromo-2-methylcyclohexane is subjected to dehydrohalogenation, the major product is 3-methylcyclohexene:

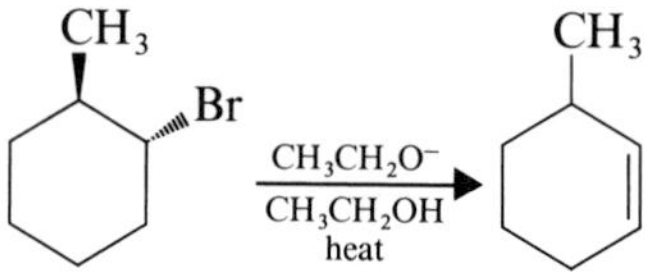

Account for the different products of these two reactions.

6.29

Write structural formulas for the products of the following reactions. If more than one product is possible, tell which one would be the major product.

(a) 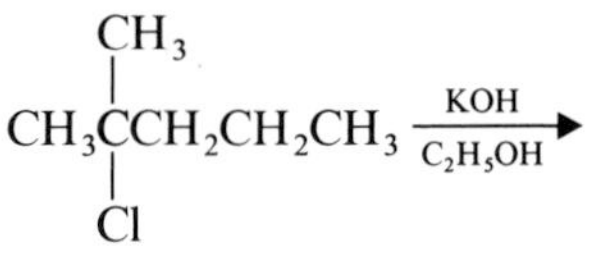 $\underset{\underset{Cl}{|}}{CH_3\overset{\overset{CH_3}{|}}{C}CH_2CH_2CH_3} \xrightarrow[C_2H_5OH]{KOH}$

(b) $\underset{\underset{Cl}{|}}{CH_3CHCH_2CH_2CH_3} \xrightarrow[C_2H_5OH]{C_2H_5ONa}$

(c) $\underset{\underset{OH}{|}}{CH_3\overset{\overset{CH_3}{|}}{C}CH_2CH_2CH_3} \xrightarrow[heat]{H_3PO_4}$

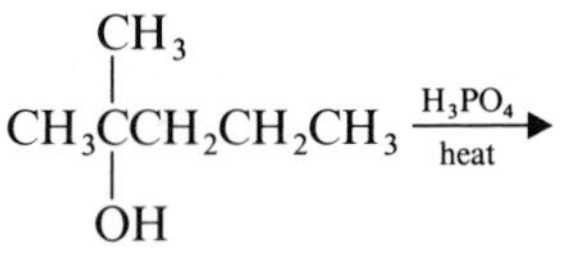

(d) 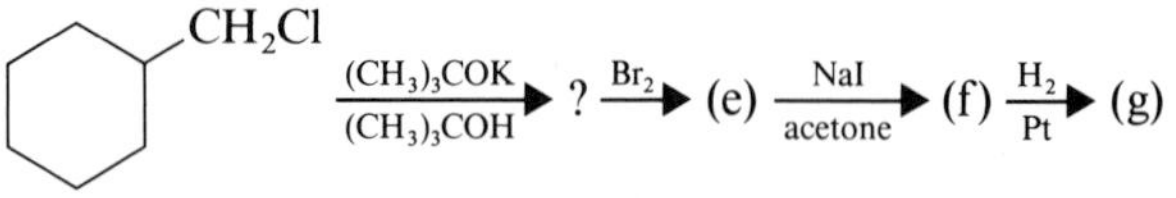 $\xrightarrow[(CH_3)_3COH]{(CH_3)_3COK}$? $\xrightarrow{Br_2}$ (e) $\xrightarrow[acetone]{NaI}$ (f) $\xrightarrow{H_2}{Pt}$ (g)

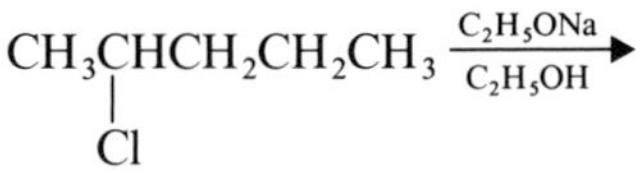

6.30

Cyclohexane and 1-hexene have the same molecular formula. Suggest a simple chemical reaction that will distinguish one from the other.

6.31

For which of the following compounds is cis–trans isomerism possible? Where cis–trans isomerism is possible, write structural formulas for the isomeric compounds.

(a) 1-Butene
(b) 2-Methylpropene
(c) 2-Heptene
(d) 2-Methyl-2-heptene
(e) 1-Chloro-1-butene
(f) 1,1-Dichloro-1-butene

(g)

(i)

(h)

6.32

(a) Arrange the following alkenes in order of their relative stabilities:

trans-3-hexene;   1-hexene;   2-methyl-2-pentene;   cis-2-hexene;   2,3-dimethyl-2-butene

(b) For which of the alkenes listed in part (a) could comparative heats of hydrogenation be used to measure their relative stabilities?

6.33

Which compound would you expect to have the larger heat of hydrogenation (in kilocalories per mole): cis-cyclooctene or trans-cyclooctene? Explain.

6.34

When cis-2-butene is heated to a temperature greater than 300 °C, a mixture of two isomeric 2-butenes results. (a) What chemical change takes place? (b) Which butene isomer would you expect to predominate in the mixture when equilibrium is established between them?

6.35

Write the structural formulas for the alkenes that could be formed when each of the following alkyl halides is subjected to dehydrohalogenation by the action of ethoxide ion in ethanol. When more than one product results, designate the major product. (Neglect cis–trans isomerism in this problem.)

(a) $CH_3CHBrCH(CH_3)CH_3$
(b) $CH_3CH_2CH(CH_3)CH_2Br$
(c) $CH_3CH_2CHBrCH_2CH_3$
(d) $CH_3CHBrCH_2CH_2CH_3$

(e)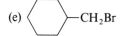

(f)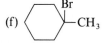

6.36

Starting with an appropriate alkyl halide and base, outline syntheses that would yield each of the following alkenes as the major (or only) product.

(a) 1-Pentene
(b) 3-Methyl-1-butene
(c) 2,3-Dimethyl-1-butene
(d) 4-Methylcyclohexene
(e) 1-Methylcyclopentene

6.37

Give structural formulas for the alcohol or alcohols that, on dehydration, would yield each of the following alkenes as the major product.

(a) 2-Methylpropene
(b) 2,3-Dimethyl-2-butene
(c) Cyclopentene
(d) Cyclohexene
(e) trans-2-Butene
(f) 1-Methylcyclopentene

6.38

Arrange the following alcohols in order of their reactivity toward acid-catalyzed dehydration and explain your reasoning:

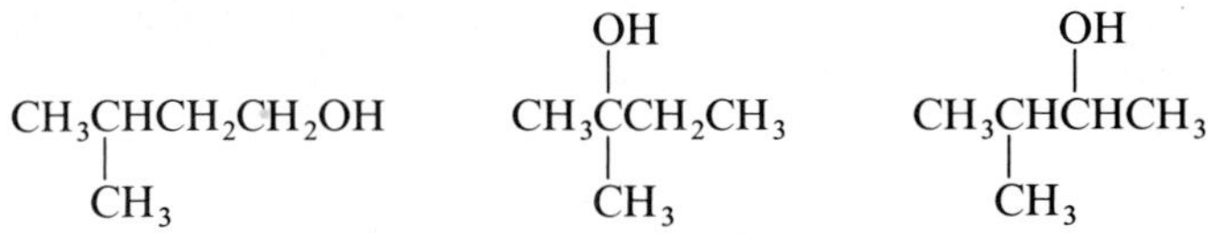

$$CH_3CHCH_2CH_2OH \qquad CH_3CCH_2CH_3 \qquad CH_3CHCHCH_3$$
$$\underset{CH_3}{|} \qquad\qquad \underset{CH_3}{\overset{OH}{|}} \qquad\qquad \underset{CH_3}{\overset{OH}{|}}$$

6.39

(a) When 1,2-dimethylcyclopentene reacts with hydrogen in the presence of finely divided platinum, only one of the isomeric 1,2-dimethylcyclopentanes forms in appreciable amounts. Which is it? (Assume a mechanism similar to that in Fig. 6.2.) (b) What predominant product would you expect from a similar hydrogenation of 1,2-dimethylcyclohexene? Write a conformational formula for this product.

6.40

Write step-by-step mechanisms that account for each product of the following reactions and explain the relative proportions of the isomers obtained in each instance.

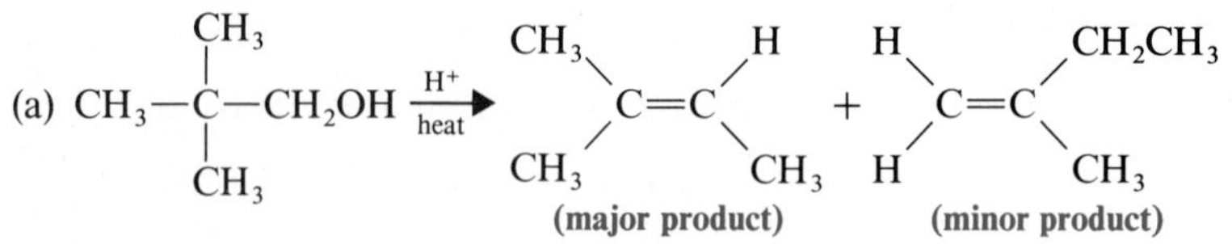

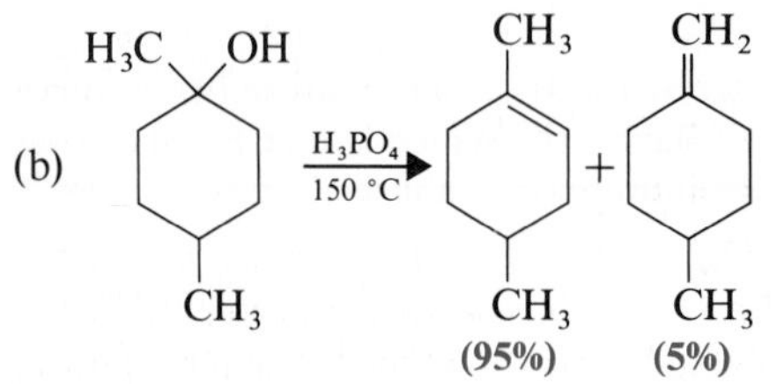

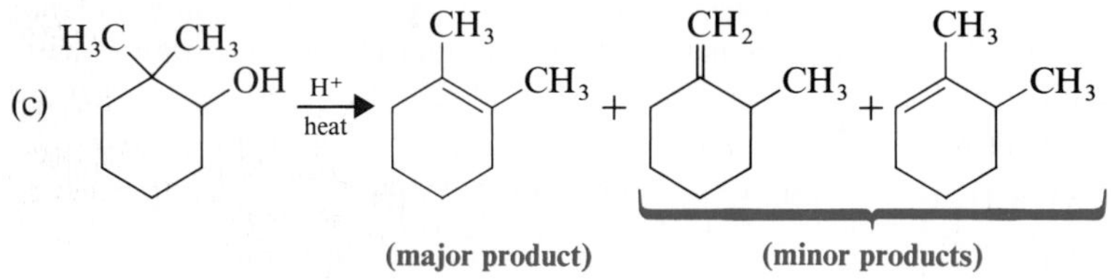

(a)

(major product) (minor product)

(b)

(95%)   (5%)

(c)

(major product)   (minor products)

6.41

Cholesterol is an important steroid found in nearly all body tissues; it is also the major component of gallstones. Impure cholesterol can be obtained from gallstones by extracting them with an organic solvent. The crude cholesterol thus obtained can be purified by (a) treatment with Br_2 in $CHCl_3$, (b) careful crystallization of the product, and (c) treatment of the latter with zinc in ethyl alcohol. What reactions are involved in this procedure?

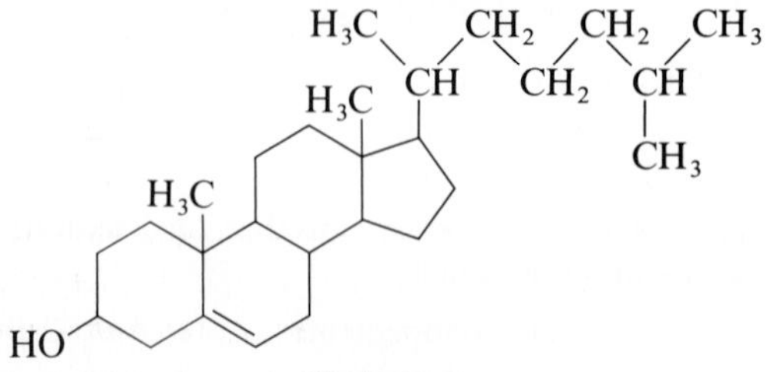

Cholesterol

6.42

Caryophyllene, a compound found in oil of cloves, has the molecular formula $C_{15}H_{24}$ and has no triple bonds. Reaction of caryophyllene with an excess of hydrogen in the presence of a platinum catalyst produces a compound with the formula $C_{15}H_{28}$. How many (a) double bonds, and (b) rings does a molecule of caryophyllene have?

6.43

Squalene, an important intermediate in the biosynthesis of steroids, has the molecular formula $C_{30}H_{50}$ and has no triple bonds. (a) What is the index of hydrogen deficiency of squalene? (b) Squalene undergoes catalytic hydrogenation to yield a compound with the molecular formula $C_{30}H_{62}$. How many double bonds does a molecule of squalene have? (c) How many rings?

6.44

Reconsider the interconversion of *cis*-2-butene and *trans*-2-butene given in Problem 6.34. (a) What is the value of $\Delta H°$ for the reaction, *cis*-2-butene $\longrightarrow$ *trans*-2-butene? (b) What minimum value of E_{act} would you expect for this reaction? (c) Sketch a potential energy diagram for the reaction and label $\Delta H°$ and E_{act}.

6.45

Propose structures for compounds **E–H**. (a) Compound **E** has the molecular formula C_5H_8 and is optically active. On catalytic hydrogenation **E** yields **F**. **F** has the molecular formula C_5H_{10}, is optically inactive, and cannot be resolved into separate enantiomers. (b) Compound **G** has the molecular formula C_6H_{10} and is optically active. **G** contains no triple bonds. On catalytic hydrogenation **G** yields **H**. **H** has the molecular formula C_6H_{14}, is optically inactive, and cannot be resolved into separate enantiomers.

6.46

Compounds **I** and **J** both have the molecular formula C_7H_{14}. **I** and **J** are both optically active and both rotate plane-polarized light in the same direction. On catalytic hydrogenation **I** and **J** yield the same compound **K**(C_7H_{16}). **K** is optically active. Propose possible structures for **I, J**, and **K**.

6.47

Compounds **L** and **M** have the molecular formula C_7H_{14}. **L** and **M** are optically inactive, are nonresolvable, and are diastereomers of each other. Catalytic hydrogenation of either **L** or **M** yields **N**. **N** is optically inactive but can be resolved into separate enantiomers. Propose possible structures for **L, M**, and **N**.

CHAPTER SEVEN

Vitamin B_1 40x. Vitamin B_1, also called Thiamine, is found in most whole grains; a deficiency of Vitamin B_1 causes the disease beriberi.

Alkenes and Alkynes II. Addition Reactions of Carbon–Carbon Multiple Bonds

7.1 INTRODUCTION. ADDITIONS TO ALKENES

The most commonly encountered reaction of compounds containing a carbon–carbon double bond is **addition.** Addition may involve a symmetrical reagent:

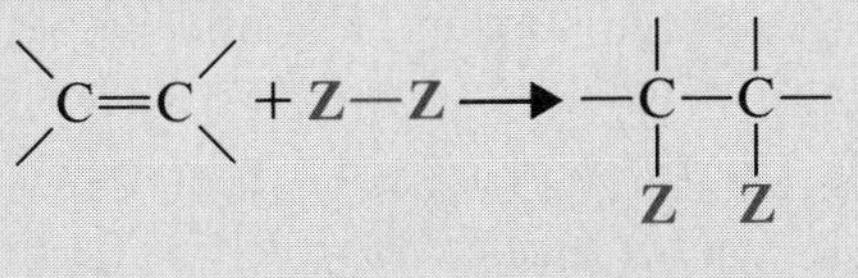

$$\text{C=C} + \text{Z-Z} \longrightarrow -\underset{\underset{Z}{|}}{C}-\underset{\underset{Z}{|}}{C}-$$

or an unsymmetrical reagent:

$$\text{C=C} + \text{Z-Y} \longrightarrow -\underset{\underset{Z}{|}}{C}-\underset{\underset{Y}{|}}{C}-$$

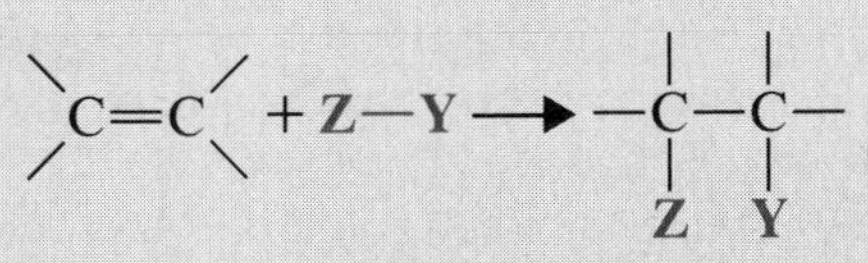

Examples of addition of symmetrical reagents are the additions of hydrogen (Section 6.5) or bromine (Section 7.6):

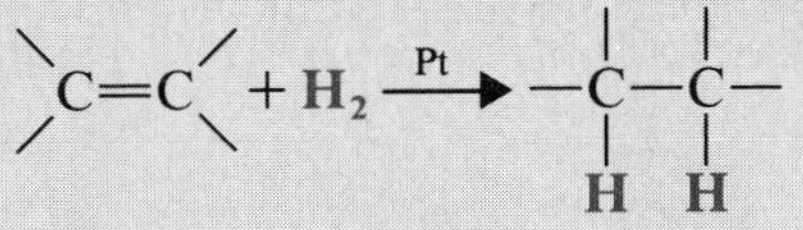

$$\underset{\text{Alkene}}{\text{C=C}} + \text{H}_2 \xrightarrow{\text{Pt}} \underset{\text{Alkane}}{-\underset{\underset{H}{|}}{C}-\underset{\underset{H}{|}}{C}-}$$

$$\underset{\text{Alkene}}{\text{C=C}} + \text{Br}_2 \xrightarrow{\text{CCl}_4} \underset{\textit{vic}\text{-Dibromide}}{-\underset{\underset{Br}{|}}{C}-\underset{\underset{Br}{|}}{C}-}$$

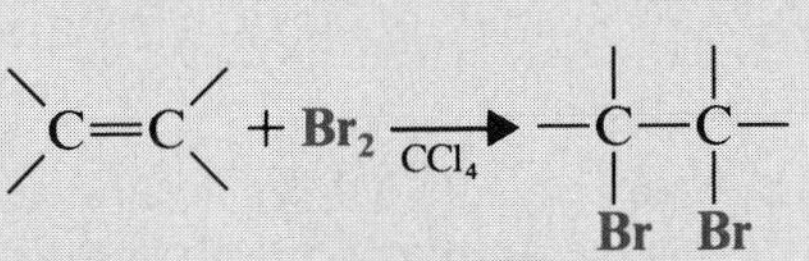

Examples of additions of unsymmetrical reagents are the additions of hydrogen halides (Section 7.2) or the acid-catalyzed addition of water (Section 7.5):

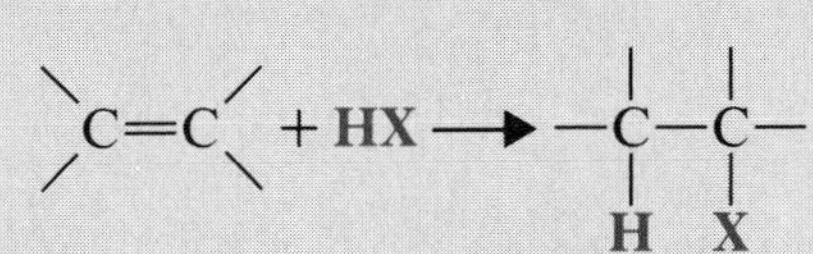

$$\underset{\text{Alkene}}{\text{C=C}} + \text{HX} \longrightarrow \underset{\text{Alkyl halide}}{-\underset{\underset{H}{|}}{C}-\underset{\underset{X}{|}}{C}-}$$

297

$$\underset{\text{Alkene}}{\diagdown C = C \diagup} + H\text{--}OH \xrightarrow{H^+} \underset{\substack{|\quad| \\ H\quad OH \\ \text{Alcohol}}}{\text{--}\overset{|}{C}\text{--}\overset{|}{C}\text{--}}$$

The large number of reactions of alkenes that fall into the category of addition reactions can be accounted for on the basis of two characteristics of the carbon–carbon double bond:

1. An addition reaction results in the conversion of one π bond (Section 2.4) and one σ bond into two σ bonds. The result of this change is usually energetically

$$\diagdown C = C \diagup + X\text{--}Y \longrightarrow \text{--}\overset{|}{\underset{|}{C}}\text{--}\overset{|}{\underset{|}{C}}\text{--}$$

| π bond | σ bond | $\longrightarrow$ | 2 σ bonds |

Bonds broken   **Bonds formed**

favorable. The heat evolved in making 2 σ bonds exceeds that needed to break one σ and one π bond, and addition reactions are usually exothermic.

2. The electrons of the π bond are exposed because the π orbital has considerable p character. Thus, a π bond is particularly susceptible to electron-seeking reagents. Such electron-seeking reagents are said to be **electrophilic** (electron loving) and are called **electrophiles.** Electrophiles include positive reagents such as the proton (H^+) or neutral reagents such as bromine (because it can be polarized), and the Lewis acids BF_3 and $AlCl_3$. Metal ions that contain vacant orbitals—the silver ion (Ag^+), the mercuric ion (Hg^{2+}), and the platinum ion (Pt^{2+}), for example—also act as electrophiles.

Strong acids (proton donors) react with alkenes to yield *carbocations.* In this reaction the proton acts as an electrophile and attacks the π bond. It uses the π electrons to form a σ bond to one carbon of the alkene. We represent this reaction in the following way:

$$H^+ \quad \diagdown C = C \diagup \longrightarrow \underset{\text{Carbocation}}{H\text{--}\overset{|}{\underset{|}{C}}\text{--}\overset{|}{\underset{|}{C}}{}^+}$$

The carbocation thus formed goes on to react further. For example, if the proton donor is HCl, the carbocation can combine with a chloride ion to form an alkyl chloride:

$$H\text{--}\overset{|}{\underset{|}{C}}\text{--}\overset{|}{\underset{|}{C}}{}^+ + :Cl^- \longrightarrow \underset{\text{Alkyl halide}}{H\text{--}\overset{|}{\underset{|}{C}}\text{--}\overset{|}{\underset{|}{C}}\text{--}Cl}$$

Electrophiles are Lewis acids: They are molecules or ions that can accept an electron pair. Nucleophiles are molecules or ions that can furnish an electron pair (i.e., Lewis bases). Any reaction of an electrophile also involves a nucleophile. In the protonation of an alkene the electrophile is the proton donated by an acid; the nucleophile is the alkene.

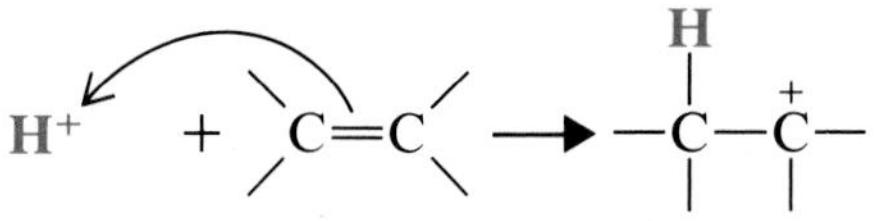

$$H^+ \quad + \quad \begin{array}{c} \diagdown \\ \diagup \end{array} C{=}C \begin{array}{c} \diagup \\ \diagdown \end{array} \longrightarrow \quad \overset{\overset{\displaystyle H}{|}}{\underset{|}{-C}}\overset{+}{\underset{|}{-C-}}$$

Electrophile   Nucleophile

In the next step, the reaction of the carbocation with a chloride ion, the carbocation is the electrophile and the chloride ion is the nucleophile.

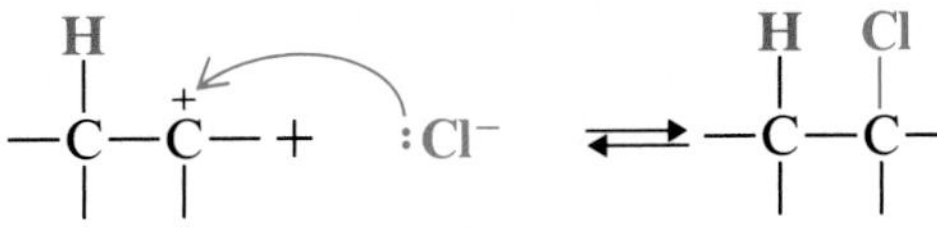

$$\overset{\overset{\displaystyle H}{|}}{\underset{|}{-C}}\overset{+}{\underset{|}{-C-}} + \quad :Cl^- \quad \rightleftharpoons \overset{\overset{\displaystyle H}{|}\ \overset{\displaystyle Cl}{|}}{\underset{|\ \ \ |}{-C-C-}}$$

Electrophile   Nucleophile

7.2 ADDITION OF HYDROGEN HALIDES TO ALKENES: MARKOVNIKOV'S RULE

Hydrogen halides (HF, HCl, HBr, and HI) add readily to the double bond of alkenes:

$$\begin{array}{c} \diagdown \\ \diagup \end{array} C{=}C \begin{array}{c} \diagup \\ \diagdown \end{array} + HX \qquad \overset{\overset{\displaystyle |}{|}\ \overset{\displaystyle |}{|}}{\underset{\underset{\displaystyle H}{|}\ \underset{\displaystyle X}{|}}{-C-C-}}$$

Two examples are shown here.

$$CH_3CH{=}CHCH_3 + HCl \longrightarrow CH_3CH_2\underset{\underset{\displaystyle Cl}{|}}{CHCH_3}$$

2-Butene   2-Chlorobutane
(cis or trans)

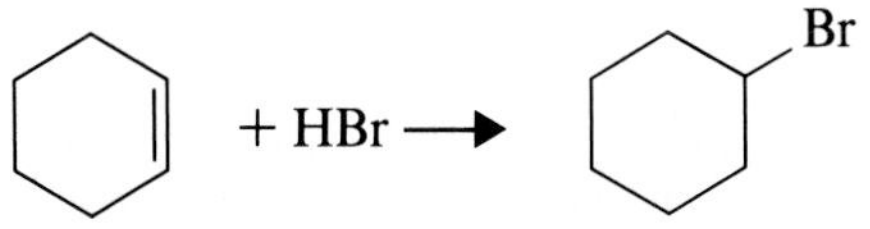

Cyclohexene   Cyclohexyl bromide

In carrying out these reactions, the hydrogen halide may be dissolved in acetic acid and mixed with the alkene, or gaseous hydrogen halide may be bubbled directly into the alkene with the alkene, itself, being used as the solvent.

The addition of HX to an unsymmetrical alkene could conceivably occur in two ways. In practice, however, one product usually predominates. The addition of HCl to propene, for example, could conceivably lead to either 1-chloropropane or 2-chloropropane. The actual product, however, is 2-chloropropane.

$$CH_2{=}CHCH_3 + HCl \longrightarrow CH_3\underset{\underset{\displaystyle Cl}{|}}{CHCH_3} \qquad (not\ ClCH_2CH_2CH_3)$$

2-Chloropropane   1-Chloropropane

When 2-methylpropene reacts with HCl, the product is *tert*-butyl chloride, not isobutyl chloride.

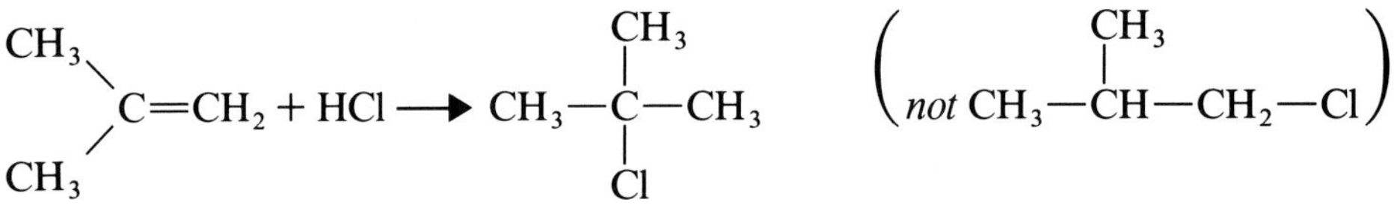

$$CH_3 \backslash C=CH_2 + HCl \longrightarrow CH_3-\underset{\underset{Cl}{|}}{\overset{\overset{CH_3}{|}}{C}}-CH_3 \quad \left(not \; CH_3-\underset{\overset{|}{CH_3}}{CH}-CH_2-Cl \right)$$

2-Methylpropene   *tert*-**Butyl chloride**   **Isobutyl chloride**
(isobutylene)

Consideration of many examples like this led the Russian chemist Vladimir Markovnikov (in 1869) to formulate what is now known as **Markovnikov's rule.** As originally stated, this rule said that in the addition of HX to an alkene, *the hydrogen atom adds to the carbon atom of the double bond with the greater number of hydrogen atoms.* We can illustrate this original statement of the rule with propene:

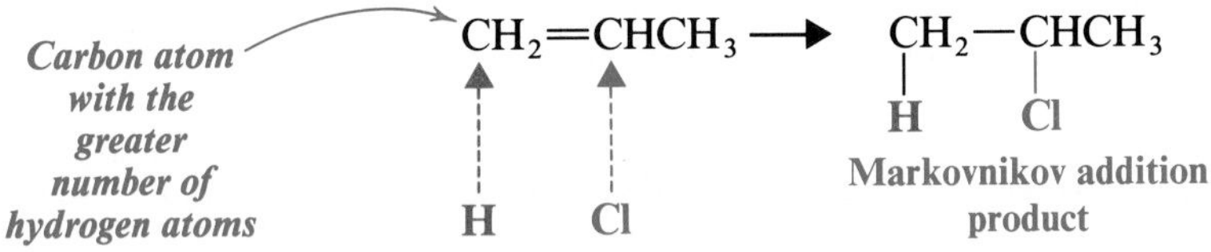

Carbon atom with the greater number of hydrogen atoms

$$CH_2{=}CHCH_3 \longrightarrow CH_2-CHCH_3$$
$$\overset{|}{H} \quad \overset{|}{Cl}$$

H   **Cl**   **Markovnikov addition product**

Reactions that illustrate Markovnikov's rule are said to be *Markovnikov additions.*

A mechanism for addition of a hydrogen halide to an alkene involves the following two steps:

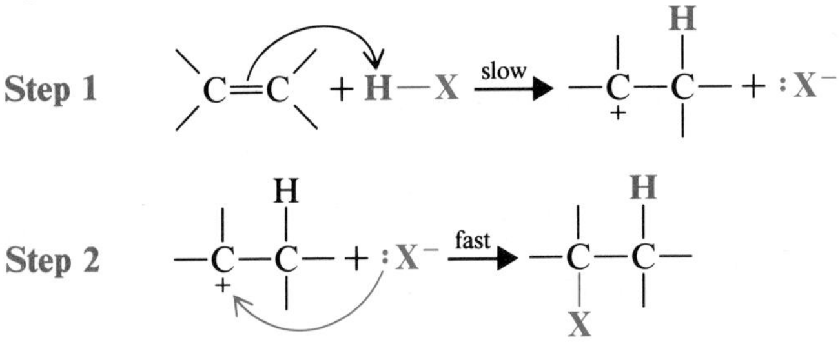

Step 1 $\overset{\backslash}{/}C{=}C\overset{/}{\backslash} + H-X \xrightarrow{\text{slow}} -\overset{|}{\underset{+}{C}}-\overset{\overset{H}{|}}{\underset{|}{C}}- + :X^-$

Step 2 $-\overset{\overset{H}{|}}{\underset{+}{C}}-\overset{|}{\underset{|}{C}}- + :X^- \xrightarrow{\text{fast}} -\overset{\overset{H}{|}}{\underset{|}{C}}-\overset{|}{\underset{X}{C}}-$

The important step—because it is the **rate-determining step**—is step 1. In step 1 the alkene accepts a proton from the hydrogen halide and forms a carbocation. This step (Fig. 7.1) is highly endothermic and has a high energy of activation. Consequently, it takes place slowly. In step 2 the highly reactive carbocation stabilizes itself by combining with a halide ion. This exothermic step has a very low energy of activation and takes place very rapidly.

7.2B Theoretical Explanation of Markovnikov's Rule

If the alkene that undergoes addition of a hydrogen halide is an unsymmetrical alkene such as propene, then step 1 could conceivably lead to two different carbocations:

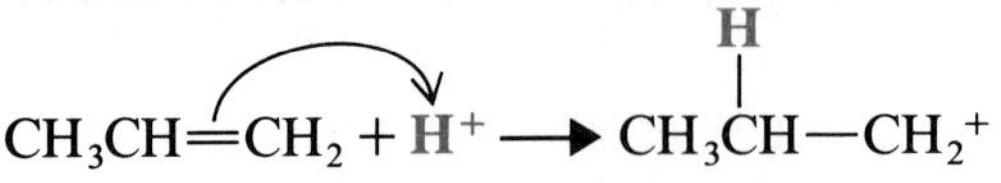

$$CH_3CH{=}CH_2 + H^+ \longrightarrow CH_3CH-CH_2^+ \qquad \text{1° Carbocation}$$
$$\overset{\overset{H}{|}}{} \qquad \qquad \text{(less stable)}$$

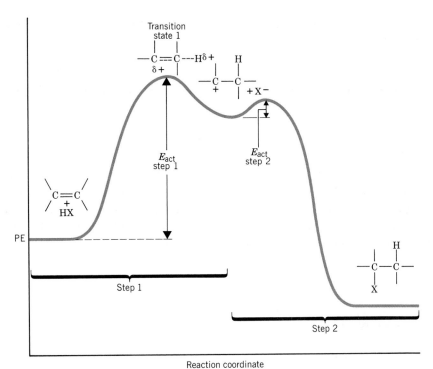

FIGURE 7.1 Potential energy diagram for the addition of HX to an alkene. The energy of activation for step 1 is much larger than for step 2.

$$CH_3CH\!=\!CH_2 + H^+ \longrightarrow CH_3\overset{+}{C}H\!-\!CH_2\!-\!H \qquad 2° \text{ Carbocation (more stable)}$$

These two carbocations are not of equal stability, however. The secondary carbocation is *more stable*, and it is the greater stability of the secondary carbocation that accounts for the correct prediction of the overall addition by Markovnikov's rule. In the addition of HCl to propene, for example, the reaction takes the following course:

$$CH_3CH\!=\!CH_2 \xrightarrow[\text{slow}]{H^+}$$

$$\xcancel{\longrightarrow}\ CH_3CH_2CH_2{}^+ \xrightarrow{\text{Cl}^-} CH_3CH_2CH_2Cl$$
$$1° \qquad\qquad \textbf{1-Chloropropane}\ (\textit{not}\ \textbf{formed})$$

$$\longrightarrow CH_3\overset{+}{C}HCH_3 \xrightarrow[\text{fast}]{\text{Cl}^-} CH_3CHCH_3$$
$$\underset{\text{Cl}}{|}$$
$$2° \qquad\qquad \textbf{2-Chloropropane}\ (\textbf{actual product})$$

|— Step 1 ————————+———————— Step 2 —|

The ultimate product of the reaction is 2-chloropropane because the more stable secondary carbocation is formed preferentially in the first step.

The more stable carbocation predominates because it is formed faster. We can understand why this is true if we examine the potential energy diagrams in Fig. 7.2.

The reaction (Fig. 7.2) leading to the secondary carbocation (and ultimately to 2-chloropropane) has the lower energy of activation. That is reasonable because its transition state resembles the more stable carbocation. The reaction leading to the primary carbocation (and ultimately to 1-chloropropane) has a higher energy of activation because its transition state resembles a less stable primary carbocation. This second reaction is much slower and does not compete with the first reaction.

The reaction of HCl with 2-methylpropene produces only *tert*-butyl chloride, and for the same reason. Here, in the first step (i.e., the attachment of the proton) the choice is even more pronounced—between a tertiary carbocation and a primary carbocation.

Thus, isobutyl chloride is *not* obtained as a product of the reaction because its formation would require the formation of a primary carbocation. Such a reaction would have a much higher energy of activation than that leading to a tertiary carbocation.

$$
\begin{array}{c}
CH_3 \\
\diagdown \\
C = C \\
\diagup \diagdown \\
CH_3 H
\end{array}
\begin{array}{c}
H \\
\diagup \\

\end{array}
$$

2-Methylpropene

HCl

$$
\begin{array}{cc}
CH_3 & CH_3 \\
| & | \\
CH_3\!-\!\overset{+}{C}\!-\!CH_2\!-\!H & CH_3\!-\!C\!-\!CH_2{}^{+} \\
 & | \\
 & H \\
\textbf{3° Carbocation} & \textbf{1° Carbocation}
\end{array}
$$

Cl⁻   Cl⁻

$$
\begin{array}{cc}
CH_3 & CH_3 \\
| & | \\
CH_3\!-\!C\!-\!CH_2\!-\!H & CH_3\!-\!C\!-\!CH_2\!-\!Cl \\
| & | \\
Cl & H
\end{array}
$$

tert-**Butyl chloride**   **Isobutyl chloride**
(actual product)   (not formed)

Because carbocations are formed in the addition of HX to an alkene, rearrangements sometimes occur (see Problem 7.3).

7.2C Modern Statement of Markovnikov's Rule

With this understanding of the mechanism for the ionic addition of hydrogen halides to alkenes behind us, we are now in a position to give the following modern statement of Markovnikov's rule: ***In the ionic addition of an unsymmetrical reagent to a double***

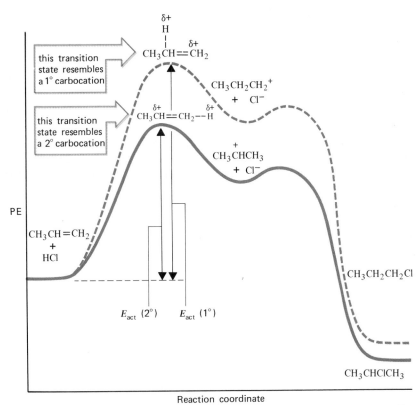

FIGURE 7.2 Potential energy diagram for the addition of HCl to propene E_{act} (2°) is less than E_{act} (1°).

bond, the positive portion of the adding reagent attaches itself to a carbon atom of the double bond so as to yield the more stable carbocation as an intermediate. Because this is the step that occurs first (prior to the addition of the negative portion of the adding reagent), it is the step that determines the overall orientation of the reaction.

Notice that this formulation of Markovnikov's rule allows us to predict the outcome of the addition of a reagent such as ICl. Because of the greater electronegativity of chlorine, the positive portion of this molecule is iodine. The addition of ICl to 2-methylpropene takes place in the following way and produces 2-chloro-1-iodo-2-methylpropane.

CH₃ ... 2-Methylpropene ... 2-Chloro-1-iodo-2-methylpropane

PROBLEM 7.1

Give the structure and name of the product that would be obtained from the ionic addition of ICl to propene.

PROBLEM 7.2

Outline mechanisms for the ionic additions   (a) of HI to 1-butene,   (b) of IBr to 2-methyl-2-butene, and   (c) of HCl to 1-methylcyclohexene.

PROBLEM 7.3

The addition of hydrogen chloride to 3,3-dimethyl-1-butene (see following reaction) yields two products: 3-chloro-2,2-dimethylbutane and 2-chloro-2,3-dimethylbutane. Write mechanisms that account for the formation of each product.

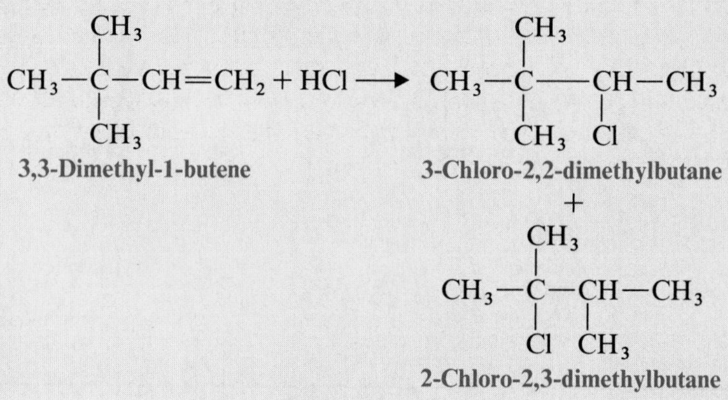

$$CH_3-\overset{\overset{\displaystyle CH_3}{|}}{\underset{\underset{\displaystyle CH_3}{|}}{C}}-CH=CH_2 + HCl \longrightarrow CH_3-\overset{\overset{\displaystyle CH_3}{|}}{\underset{\underset{\displaystyle CH_3}{|}}{C}}-\overset{}{\underset{\underset{\displaystyle Cl}{|}}{C}H}-CH_3$$

3,3-Dimethyl-1-butene   3-Chloro-2,2-dimethylbutane

+

$$CH_3-\overset{\overset{\displaystyle CH_3}{|}}{\underset{\underset{\displaystyle Cl}{|}}{C}}-\overset{}{\underset{\underset{\displaystyle CH_3}{|}}{C}H}-CH_3$$

2-Chloro-2,3-dimethylbutane

7.2D Regioselective Reactions

Chemists describe reactions like the Markovnikov additions of hydrogen halides to alkenes as being *regioselective. Regio* comes from the Latin word *regionem* meaning direction. When a reaction *that can potentially yield two or more constitutional isomers actually produces only one* (or a predominance of one), the reaction is said to be *regioselective.* The addition of HX to an unsymmetrical alkene such as propene could conceivably yield two constitutional isomers, for example. However, as we have seen, the reaction yields only one, and therefore it is regioselective.

7.2E An Exception to Markovnikov's Rule

In Chapter 9 we shall study an exception to Markovnikov's rule. This exception concerns the addition of HBr to alkenes **when the addition is carried out in the presence of peroxides** (i.e., compounds with the general formula ROOR). When alkenes are treated with HBr in the presence of peroxides the addition occurs in an anti-Markovnikov manner in the sense that the hydrogen atom becomes attached to the carbon atom with the fewer hydrogen atoms. With propene, for example, the addition takes place as follows:

$$CH_3CH=CH_2 + HBr \xrightarrow{\text{ROOR}} CH_3CH_2CH_2Br$$

In Section 9.9 we shall find that this addition occurs by *a free radical mechanism,* and not by the ionic mechanism given in Section 7.2. This anti-Markovnikov addition occurs **only when HBr is used in the presence of peroxides** and does not occur significantly with HF, HCl, and HI even when peroxides are present.

7.3 STEREOCHEMISTRY OF THE IONIC ADDITION TO AN ALKENE

Consider the following addition of HCl to 1-butene and notice that the reaction leads to the formation of a product, 2-chlorobutane, that contains a stereocenter.

$$CH_3CH_2CH{=}CH_2 + HCl \longrightarrow CH_3CH_2\overset{*}{C}HCH_3$$
$$\underset{Cl}{|}$$

The product, therefore, can exist as a pair of enantiomers. The question now arises as to how will these enantiomers be formed. Will one enantiomer be formed in greater amounts than the other? The answer is *no*; the carbocation that is formed in the first step of the addition (see following figure) is trigonal planar and *is achiral* (a model will show that it has a plane of symmetry). When the chloride ion reacts with this achiral carbocation in the second step, **reaction is equally likely at either face.** The reactions leading to the two enantiomers occur at the same rate, and the enantiomers, therefore, are produced in equal amounts **as a racemic form.**

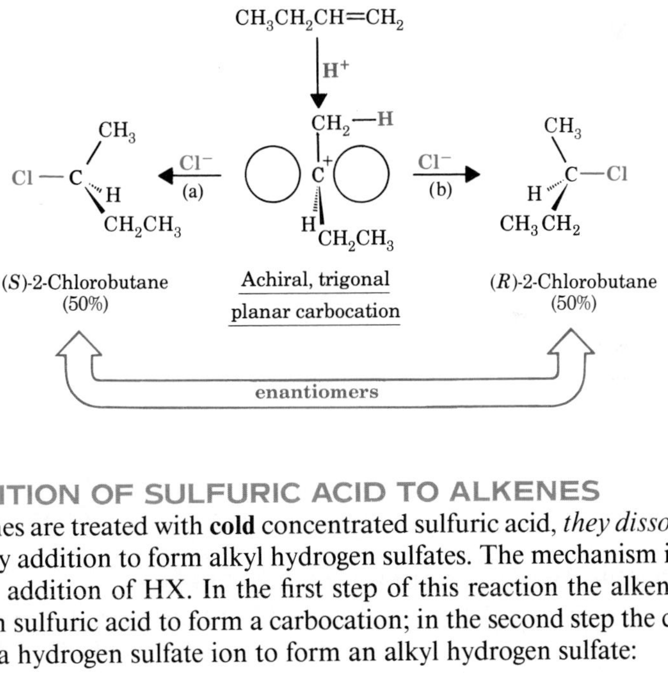

7.4 ADDITION OF SULFURIC ACID TO ALKENES

When alkenes are treated with **cold** concentrated sulfuric acid, *they dissolve* because they react by addition to form alkyl hydrogen sulfates. The mechanism is similar to that for the addition of HX. In the first step of this reaction the alkene accepts a proton from sulfuric acid to form a carbocation; in the second step the carbocation reacts with a hydrogen sulfate ion to form an alkyl hydrogen sulfate:

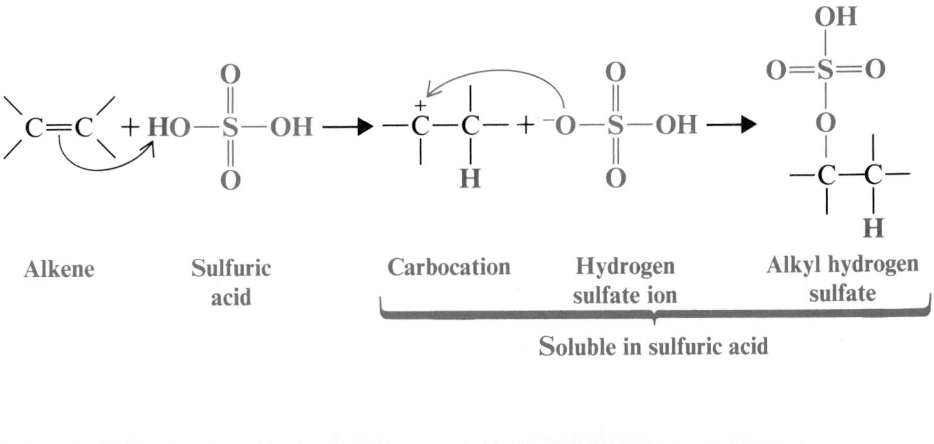

The addition of sulfuric acid is also regioselective, and it follows Markovnikov's rule. Propene, for example, reacts to yield isopropyl hydrogen sulfate rather than propyl hydrogen sulfate.

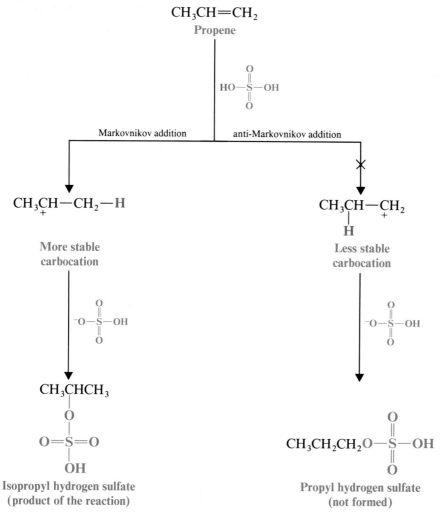

7.4A Alcohols from Alkyl Hydrogen Sulfates

Alkyl hydrogen sulfates can be easily hydrolyzed to alcohols by **heating** them with water. The overall result of the addition of sulfuric acid to an alkene followed by hydrolysis is the Markovnikov addition of H— and —OH.

$$CH_3CH=CH_2 \xrightarrow[\text{H}_2\text{SO}_4]{\text{cold}} CH_3CHCH_3 \xrightarrow{\text{H}_2\text{O, heat}} CH_3CHCH_3 + H_2SO_4$$
$$\qquad\qquad\quad OSO_3H \qquad\qquad\qquad OH$$

PROBLEM 7.4

In one industrial synthesis of ethyl alcohol, ethene is first dissolved in 95% sulfuric acid. In a second step water is added and the mixture is heated. Outline the reactions involved.

7.5 ADDITION OF WATER TO ALKENES: ACID-CATALYZED HYDRATION

The acid-catalyzed addition of water to the double bond of an alkene (hydration of an alkene) is a method for the preparation of low-molecular-weight alcohols that has its greatest utility in large-scale industrial processes. The acids most commonly used to catalyze the hydration of alkenes are sulfuric acid and phosphoric acid. These reactions, too, are usually regioselective, and the addition of water to the double bond follows Markovnikov's rule. In general the reaction takes the form that follows:

$$\diagdown C{=}C \diagup + HOH \xrightarrow{H^+} -\underset{H}{\overset{|}{C}}-\underset{OH}{\overset{|}{C}}-$$

An example is the hydration of 2-methylpropene.

$$CH_3-\underset{\overset{|}{CH_3}}{\overset{\overset{\textstyle CH_3}{|}}{C}}{=}CH_2 + HOH \xrightarrow[25\,°C]{H^+} CH_3-\underset{\overset{|}{OH}}{\overset{\overset{\textstyle CH_3}{|}}{C}}-CH_2-H$$

2-Methylpropene   *tert*-Butyl alcohol
(isobutylene)

Because the reactions follow Markovnikov's rule, acid-catalyzed hydrations of alkenes do not yield primary alcohols except in the special case of the hydration of ethene.

$$CH_2{=}CH_2 + HOH \xrightarrow[300\,°C]{H_3PO_4} CH_3CH_2OH$$

The mechanism for the hydration of an alkene is simply the reverse of the mechanism for the dehydration of an alcohol. We can illustrate this by giving the mechanism for the **hydration** of 2-methylpropene and by comparing it with the mechanism for the **dehydration** of *tert*-butyl alcohol given in Section 6.13.

Step 1

$$CH_3-\underset{\overset{|}{CH_3}}{\overset{\overset{\textstyle CH_2}{\|}}{C}} + H{-}\overset{..}{\underset{+}{O}}{-}H \underset{}{\overset{slow}{\rightleftharpoons}} CH_3-\underset{\overset{|}{CH_3}}{\overset{\overset{\textstyle CH_2-H}{|}}{C^+}} + :\overset{..}{O}{-}H$$

Step 2

$$CH_3-\underset{\overset{|}{CH_3}}{\overset{\overset{\textstyle CH_3}{|}}{C^+}} + :\overset{\overset{\textstyle H}{|}}{O}{-}H \overset{fast}{\rightleftharpoons} CH_3-\underset{\overset{|}{CH_3}}{\overset{\overset{\textstyle CH_3}{|}}{C}}-\underset{+}{\overset{\overset{\textstyle H}{|}}{O}}{-}H$$

Step 3

$$CH_3-\underset{\overset{|}{CH_3}}{\overset{\overset{\textstyle CH_3}{|}}{C}}-\overset{\overset{\textstyle H}{}}{\underset{..}{O^+}}{-}H + :\overset{..}{O}{-}H \overset{fast}{\rightleftharpoons} CH_3-\underset{\overset{|}{CH_3}}{\overset{\overset{\textstyle CH_3}{|}}{C}}-\overset{..}{\underset{..}{O}}{-}H + H{-}\underset{+}{\overset{..}{O}}{-}H$$

The rate-determining step in the *hydration* mechanism is step 1: the formation of the carbocation. It is this step, too, that accounts for the Markovnikov addition of water to the double bond. The reaction produces *tert*-butyl alcohol because step 1 leads to the formation of the more stable *tert*-butyl cation rather than the much less stable isobutyl cation:

$$
CH_3-\underset{\underset{CH_3}{|}}{\overset{\overset{CH_2}{||}}{C}} \;+\; H-\underset{\overset{+}{\cdot\cdot}}{O}-H \underset{slow}{\overset{very}{\rightleftharpoons}} CH_3\underset{\underset{CH_3}{|}}{\overset{\overset{CH_2^+}{|}}{C}}-H \;+\; :\underset{\cdot\cdot}{O}-H
$$

For all practical purposes this reaction does not take place because it produces a 1° carbocation

The reactions whereby *alkenes are hydrated or alcohols are dehydrated* are reactions in which the ultimate product is governed by the position of an equilibrium. Therefore, in the *dehydration of an alcohol* it is best to use a concentrated acid so that the concentration of water is low. (The water can be removed as it is formed, and it helps to use a high temperature.) In the *hydration of an alkene* it is best to use dilute acid so that the concentration of water is high. (It also usually helps to use a lower temperature.)

PROBLEM 7.5

(a) Show all steps in the acid-catalyzed hydration of propene. (b) Account for the fact that the product of the reaction is isopropyl alcohol (in accordance with Markovnikov's rule) and not propyl alcohol, that is,

$$
H_2O + CH_3-CH{=}CH_2 \xrightarrow{H^+} CH_3\underset{\overset{|}{OH}}{C}HCH_3 \qquad (not\; CH_3CH_2CH_2OH)
$$

Isopropyl
alcohol

Propyl
alcohol

One complication associated with alkene hydrations is the occurrence of **rearrangements.** Because the reaction involves the formation of a carbocation in the first step, the carbocation formed initially invariably rearranges to a more stable one if such a rearrangement is possible. An illustration is the formation of 2,3-dimethyl-2-butanol as the major product when 3,3-dimethyl-1-butene is hydrated:

$$
CH_3-\underset{\underset{CH_3}{|}}{\overset{\overset{CH_3}{|}}{C}}-CH{=}CH_2 \xrightarrow[H_2O]{H_2SO_4} CH_3-\underset{\underset{CH_3}{|}}{\overset{\overset{OH}{|}}{C}}-\underset{\underset{CH_3}{|}}{CH}-CH_3
$$

3,3-Dimethyl-1-butene

2,3-Dimethyl-2-butanol
(major product)

PROBLEM 7.6

Outline all steps in a mechanism showing how 2,3-dimethyl-2-butanol is formed in the acid-catalyzed hydration of 3,3-dimethyl-1-butene.

PROBLEM 7.7

The following order of reactivity is observed when the following alkenes are subjected to acid-catalyzed hydration.

$$(CH_3)_2C{=}CH_2 > CH_3CH{=}CH_2 > CH_2{=}CH_2$$

Explain this order of reactivity.

PROBLEM 7.8

When 2-methylpropene (isobutylene) is dissolved in methyl alcohol containing a strong acid, a reaction takes place to produce *tert*-butyl methyl ether, $CH_3OC(CH_3)_3$. Write a mechanism that accounts for this.

The occurrence of carbocation rearrangements limits the utility of alkene hydrations as a laboratory method for preparing alcohols. In the next chapter we shall study two very useful laboratory syntheses. One, called **oxymercuration–demercuration,** allows the Markovnikov addition of H— and —OH *without rearrangements.* Another, called **hydroboration–oxidation,** permits the *anti-Markovnikov* and *syn addition* of H— and —OH.

7.6 ADDITION OF BROMINE AND CHLORINE TO ALKENES

In the absence of light, *alkanes* do not react appreciably with chlorine or bromine at room temperature. If we add an alkane to a solution of bromine in carbon tetrachloride, the red-brown color of the bromine will persist in the solution as long as we keep the mixture away from sunlight and as long as the solution is not heated.

$$\underset{\substack{\text{Alkane}\\\text{(colorless)}}}{R{-}H} + \underset{\substack{\text{Bromine}\\\text{(red brown)}}}{Br_2} \xrightarrow[\text{in the dark, CCl}_4]{\text{room temperature}} \text{no appreciable reaction}$$

On the other hand, if we expose the reactants to sunlight, the bromine color will fade slowly. If we now place a small piece of moist blue litmus paper in the region above the liquid, the litmus paper will turn red because of the hydrogen bromide that evolves as the alkane and bromine react. (Hydrogen bromide is not very soluble in carbon tetrachloride.)

$$\underset{\substack{\text{Alkane}\\\text{(colorless)}}}{R{-}H} + \underset{\substack{\text{Bromine}\\\text{(red brown)}}}{Br_2} \xrightarrow[\text{sunlight, CCl}_4]{\text{room temperature}} \underset{\substack{\text{Alkyl halide}\\\text{(colorless)}}}{R{-}Br} + \underset{\substack{\text{(detected by}\\\text{moist blue litmus)}}}{HBr}$$

The behavior of **alkenes** toward bromine in carbon tetrachloride contrasts markedly with that of alkanes *and is a useful test for carbon–carbon multiple bonds.* Alkenes react rapidly with bromine at room temperature and in the *absence of light.*

If we add bromine to an alkene, the red-brown color of the bromine disappears almost instantly as long as the alkene is present in excess. If we test the atmosphere above the solution with moist blue litmus paper, we shall find that no hydrogen bromide is present. The reaction is one of addition. (Alkynes, as we shall see in Section 7.13, also undergo addition of bromine.)

$$\begin{array}{c}\diagdown \\ \diagup \end{array}C=C\begin{array}{c}\diagup \\ \diagdown \end{array} + Br_2 \xrightarrow[\text{in the dark, CCl}_4]{\text{room temperature}} \left.\begin{array}{c} | \quad | \\ -C-C- \\ | \quad | \\ Br \quad Br \end{array}\right\}$$

An alkene   *vic*-Dibromide
(colorless)   (colorless)

Rapid decolorization of Br_2/CCl_4 is a test for alkenes and alkynes

The addition reaction between alkenes and chlorine or bromine is a general one. The products are vicinal dihalides.

$$CH_3CH{=}CHCH_3 + Cl_2 \xrightarrow{-9\,°C} CH_3\underset{Cl}{\underset{|}{CH}}\underset{Cl}{\underset{|}{CH}}CH_3 \qquad (100\%)$$

$$CH_3CH_2CH{=}CH_2 + Cl_2 \xrightarrow{-9\,°C} CH_3CH_2\underset{Cl}{\underset{|}{CH}}{-}\underset{Cl}{\underset{|}{CH}}_2 \qquad (97\%)$$

$$\bigcirc\!\!\!| + Br_2 \xrightarrow[CCl_4/C_2H_5OH]{-5\,°C}$$

+ enantiomer   (95%)

trans-1,2-Dibromocyclohexane
(as a racemic form)

7.6A Mechanism of Halogen Addition

One mechanism that has been proposed for halogen addition is **an ionic mechanism.***
In the first step the exposed electrons of the π bond of the alkene attack the halogen in the following way:

$$\begin{array}{c} \diagdown \\ C \\ \| \\ C \\ \diagup \end{array} \quad \overset{\delta+}{:\!\overset{..}{Br}} \!-\! \overset{\delta-}{\overset{..}{Br}:} \quad \rightleftharpoons \quad \begin{array}{c} C \\ | \;\; \rangle \overset{.\,+}{Br} \\ C \end{array} + :\overset{..}{\underset{..}{Br}}:^-$$

Bromonium   Bromide
ion   ion

As the π electrons of the alkene approach the bromine molecule, the electrons of the bromine–bromine bond drift in the direction of the bromine atom more distant from the approaching alkene. The bromine molecule becomes *polarized* as a result. The more distant bromine develops a partial negative charge; the nearer bromine becomes partially positive. Polarization weakens the bromine–bromine bond, caus-

---

*There is evidence that in the absence of oxygen some reactions between alkenes and chlorine proceed through a free radical mechanism. We shall not discuss this mechanism here, however.

ing it to *break heterolytically.* A bromide ion departs, and a *bromonium ion* forms. In the bromonium ion a positively charged bromine atom is bonded to two carbon atoms by *two pairs of electrons:* one pair from the π bond of the alkene, the other pair from the bromine atom (one of its unshared pairs).

In the second step, one of the bromide ions produced in step 1 attacks one of the carbon atoms of the bromonium ion. The nucleophilic attack results in the formation of a *vic*-dibromide by opening the three-membered ring.

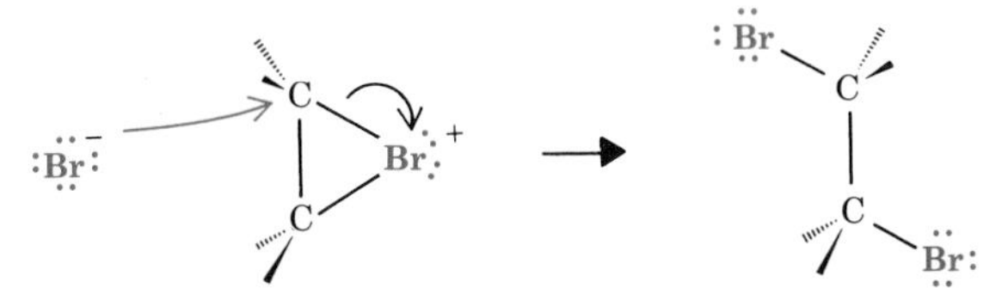

vic-Dibromide

This ring opening (see preceding figure) is an S_N2 reaction. The bromide ion, acting as a nucleophile, uses a pair of electrons to form a bond to one carbon atom of the bromonium ion while the positive bromine of the bromonium ion acts as a leaving group.

7.7 ANTI ADDITION OF HALOGENS

The addition of bromine to cyclopentene provides additional evidence for bromonium ion intermediates in bromine additions. When cyclopentene reacts with bromine in carbon tetrachloride, **anti addition** occurs, and the product of the reaction is *trans*-1,2-dibromocyclopentane.

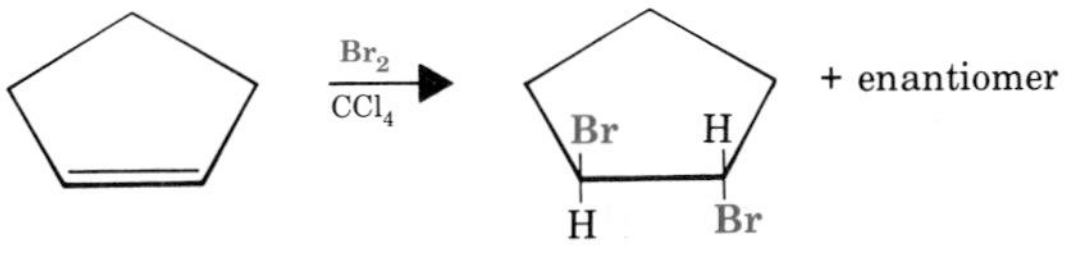

trans-1, 2-Dibromo-
cyclopentane

This anti addition of bromine to cyclopentene can be explained by a mechanism that involves the formation of a bromonium ion in the first step. In the second step, a bromide ion attacks a carbon atom of the ring from the side opposite that of the bromonium ion. The reaction is an S_N2 reaction. Nucleophilic attack by the bromide ion causes **inversion of the configuration of the carbon being attacked** (Section 5.8). This inversion of configuration at one carbon atom of the ring leads to the formation of *trans*-1,2-dibromocyclopentane.

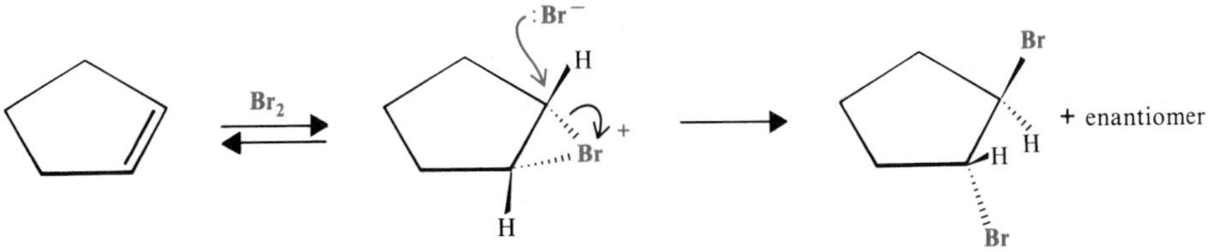

Bromonium
ion

trans-1, 2-Dibromo-
cyclopentane

When cyclohexene undergoes addition of bromine, the product is *trans*-1,2-dibromocyclohexane (Section 7.6). In this instance, too, *anti* addition results from the intermediate formation of a bromonium ion followed by S_N2 attack by a bromide ion. The reaction is shown here. Notice in cyclohexene that C-1, C-2, C-3, and C-6 all lie in the same plane. One of the other carbon atoms is above this plane and the other is below it. (Two such conformations are possible.)

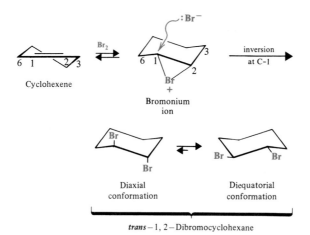

Cyclohexene

Bromonium
ion

Diaxial
conformation

Diequatorial
conformation

trans — 1, 2 — Dibromocyclohexane

Notice, too, that the initial product of the reaction is the *diaxial conformer.* This rapidly converts into the diequatorial form, and when equilibrium is reached the diequatorial form predominates. We saw earlier (Section 6.12C) that when cyclohexane derivatives undergo elimination, the required conformation is the diaxial one. Here we find that when cyclohexene undergoes addition (the opposite of elimination), the initial product is also diaxial.

7.7A Stereospecific Reactions

The anti addition of a halogen to an alkene provides us with an example of what is called a **stereospecific reaction.**

A reaction is *stereospecific* when *a particular stereoisomeric form of the starting material reacts in such a way that it gives a specific stereoisomeric form of the product.* It does this because the reaction mechanism requires that the configurations of the atoms involved change in a characteristic way.

Consider the reactions of *cis*- and *trans*-2-butene with bromine shown at the top of the next page. When *trans*-2-butene adds bromine, the product is the meso compound, (2R,3S)-2,3-dibromobutane. When *cis*-2-butene adds bromine, the product is a *racemic form* of (2R,3R)-2,3-dibromobutane and (2S,3S)-2,3-dibromobutane.

The reactants *cis*-2-butene and *trans*-2-butene are stereoisomers; they are *diastereomers.* The product of reaction (1) (2R,3S)-2,3-dibromobutane is a meso compound, and it is a stereoisomer of either of the products of reaction (2) (the enantiomeric 2,3-dibromobutanes). Thus, by definition, both reactions are stereospecific. One stereoisomeric form of the reactant (e.g., *trans*-2-butene) gives one product (the meso compound) while the other stereoisomeric form of the reactant (*cis*-2-butene) gives a stereoisomerically different product (the enantiomers).

We can better understand the results of these two reactions if we examine their mechanisms.

Figure 7.3 shows how *cis*-2-butene adds bromine to yield intermediate bromonium ions that are achiral. (The bromonium ion has a plane of symmetry. Can

(1) trans-2-Butene $\xrightarrow[\text{CCl}_4]{\text{Br}_2}$ (2R, 3S)-2, 3-Dibromobutane (a meso compound)

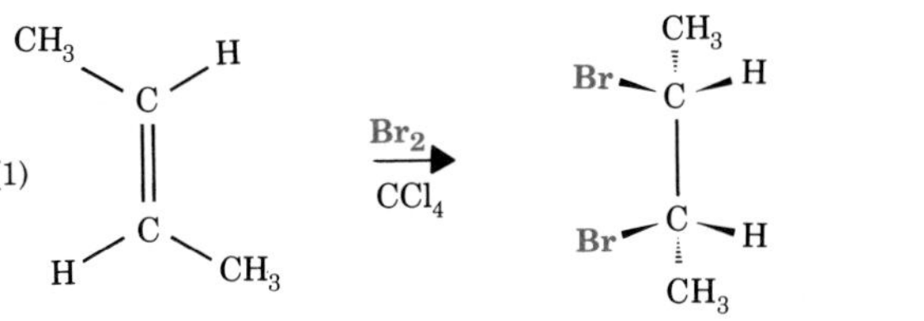

(2) cis-2-Butene $\xrightarrow[\text{CCl}_4]{\text{Br}_2}$ (2R, 3R) + (2S, 3S)

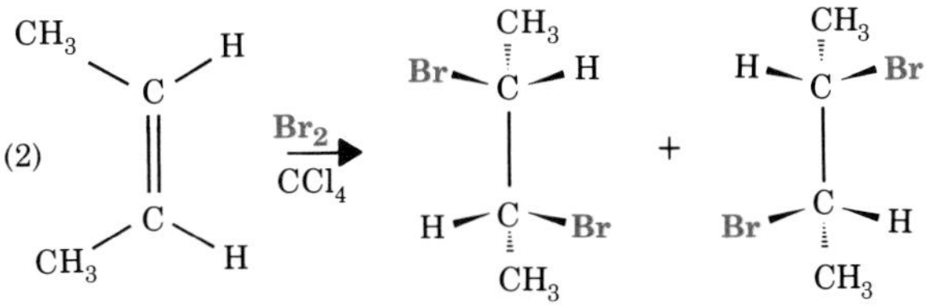

Enantiomeric 2,3-Dibromobutanes
(a racemic form)

you find it?) These bromonium ions can then react with bromide ions by either path (a) or by path (b). Reaction by path (a) yields one 2,3-dibromobutane enantiomer; reaction by path (b) yields the other enantiomer. Reaction occurs at the same rate by either path; therefore the two enantiomers are produced in equal amounts (as a racemic form).

Figure 7.4 shows how *trans*-2-butene reacts at the bottom face to yield an intermediate bromonium ion that is chiral. (Reaction at the other face would produce the enantiomeric bromonium ion.) Reaction of this chiral bromonium ion (or its enantiomer) with a bromide ion either by path (a) or by path (b) yields the same compound, the *meso*-2,3-dibromobutane.

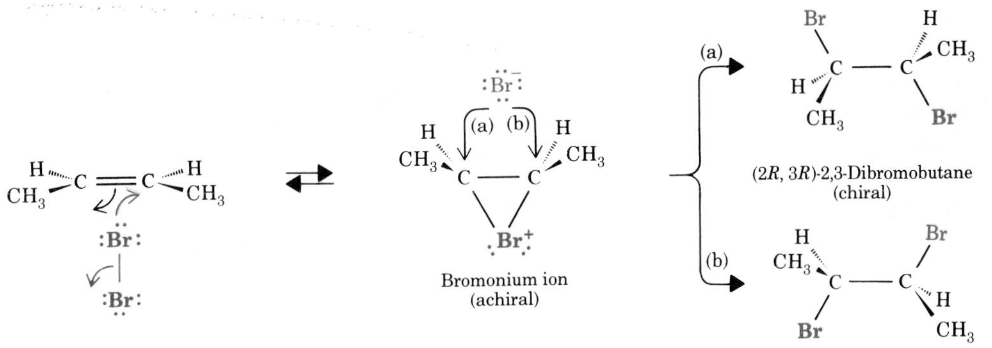

(2R, 3R)-2,3-Dibromobutane
(chiral)

(2S, 3S)-2,3-Dibromobutane
(chiral)

Bromonium ion
(achiral)

cis-2-Butene reacts with bromine to yield achiral bromonium ions and bromide ions. [Reaction at the other face of the alkene (top) would yield the same bromonium ions.]

The bromonium ions react with the bromide ions at equal rates by paths (a) and (b) to yield the two enantiomers in equal amounts (*i.e.*, as the racemic form).

FIGURE 7.3 A mechanism showing how *cis*-2-butene reacts with bromine to yield the enantiomeric 2,3-dibromobutanes.

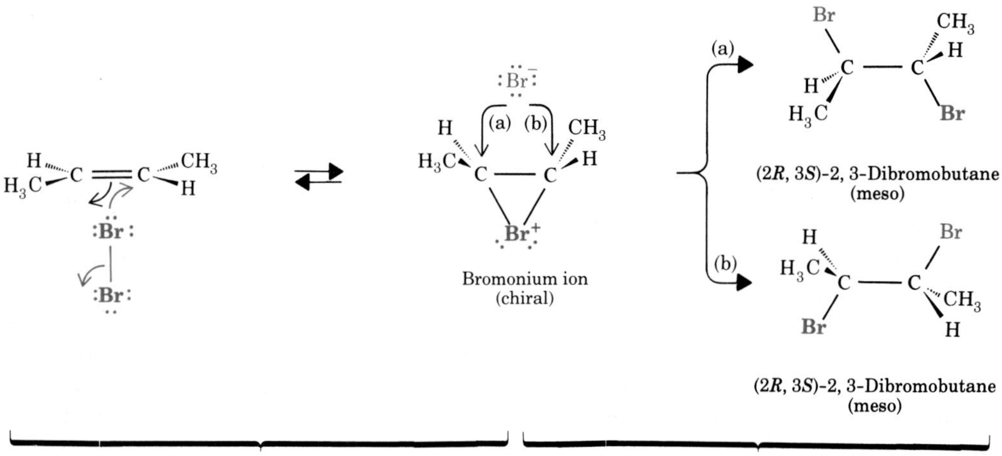

trans-2-Butene reacts with bromine to yield chiral bromonium ions and bromide ions. [Reaction at the other face (top) would yield the enantiomer of the bromonium ion as shown here.]

When the bromonium ions react by either path (a) or (b), they yield the same achiral meso compound. [Reaction of the enantiomer of the intermediate bromonium ion would produce the same result.]

FIGURE 7.4 A mechanism showing how *trans*-2-butene reacts with bromine to yield the *meso*-2-dibromobutane.

PROBLEM 7.9

In Section 7.7 you studied a mechanism for the formation of one enantiomer of *trans*-1,2-dibromocyclopentane when bromine adds to cyclopentene. You should now write a mechanism showing how the other enantiomer forms.

7.8 HALOHYDRIN FORMATION

If the halogenation of an alkene is carried out in aqueous solution (rather than in carbon tetrachloride), the major product of the overall reaction is not a *vic*-dihalide, but rather it is a **halo alcohol** called a **halohydrin.** In this case, molecules of the solvent become reactants, too.

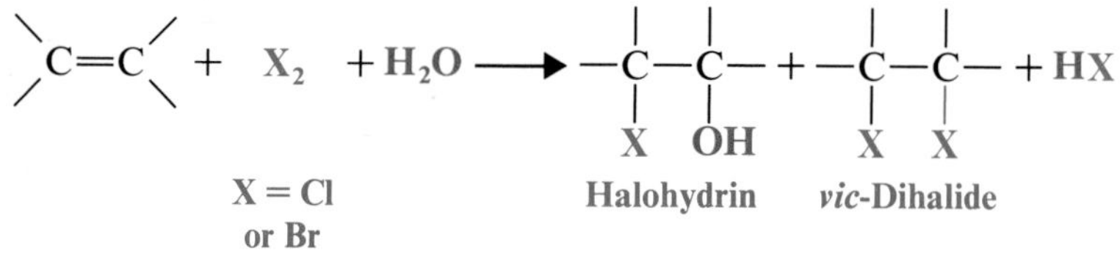

Halohydrin formation can be explained by the following mechanism:

Step 1

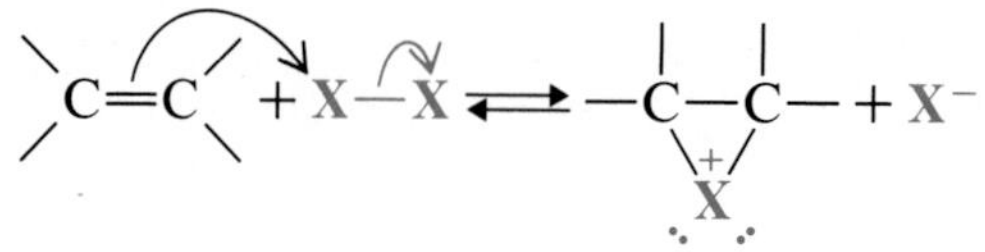

Step 2

$$-\underset{X}{\underset{|}{\overset{|}{C}}}-\overset{+}{\underset{|}{C}}\underset{X}{\overset{|}{\Big(}} + H_2O: \longrightarrow -\underset{X}{\overset{|}{\underset{|}{C}}}-\underset{|}{\overset{\overset{+}{O}H_2}{\underset{|}{C}}}- \xrightarrow{-H^+} -\underset{X}{\overset{|}{\underset{|}{C}}}-\underset{|}{\overset{OH}{\underset{|}{C}}}-$$

The first step is the same as that for halogen addition. In the second step, however, the two mechanisms differ. In halohydrin formation, water acts as the nucleophile and attacks one carbon atom of the halonium ion. The three-membered ring opens, and a protonated halohydrin is produced. Loss of a proton then leads to the formation of the halohydrin itself.

Water, because of its unshared electron pairs, acts as a nucleophile in this and in many other reactions. In this instance water molecules far outnumber halide ions because water is the solvent for the reactants. This accounts for the halohydrin being the major product.

PROBLEM 7.10

Outline a mechanism that accounts for the formation of *trans*-2-chlorocyclo-pentanol from cyclopentene and chlorine in aqueous solution.

+ enantiomer

trans – 2 – Chlorocyclopentanol

If the alkene is unsymmetrical, the halogen ends up on the carbon atom with the greater number of hydrogen atoms. Bonding in the intermediate bromonium ion is apparently *unsymmetrical*. The more highly substituted carbon atom bears the greater positive charge because it resembles the more stable carbocation. Consequently, water attacks this carbon atom preferentially. The greater positive charge on the tertiary carbon permits a pathway with a lower energy of activation even though attack at the primary carbon atom is less hindered.

$$\underset{CH_3}{\overset{CH_3}{\diagup}}C=CH_2 \xrightarrow{Br_2} CH_3-\underset{\underset{\delta+}{Br}}{\overset{CH_3}{\underset{|}{\overset{\delta+|}{C}}}}\overset{:OH_2}{\diagdown}CH_2 \longrightarrow CH_3-\underset{\underset{|}{CH_3}}{\overset{\overset{+}{O}H_2}{\underset{|}{C}}}-CH_2Br \xrightarrow{-H^+} CH_3-\underset{\underset{|}{CH_3}}{\overset{OH}{\underset{|}{C}}}-CH_2Br$$

(73%)

PROBLEM 7.11

When ethene gas is passed into an aqueous solution containing bromine and sodium chloride, the products of the reaction are $BrCH_2CH_2Br$, $BrCH_2CH_2OH$, *and* $BrCH_2CH_2Cl$. Write mechanisms showing how each product is formed.

7.9 EPOXIDES: EPOXIDATION OF ALKENES

Epoxides are cyclic ethers with three-membered rings. In IUPAC nomenclature epoxides are called **oxiranes.** The simplest epoxide has the common name ethylene oxide.

$$-\overset{|}{C}-\overset{|}{C}- \quad\quad H_2\overset{2}{C}-\overset{3}{C}H_2$$

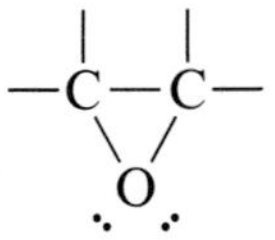

An epoxide

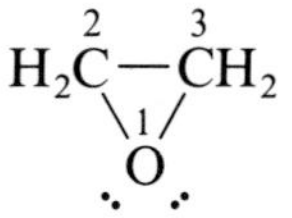

IUPAC name: Oxirane
Common name: Ethylene oxide

The most widely used method for synthesizing epoxides is the reaction of an alkene with an organic **peroxy acid** (sometimes called simply a **peracid**), a process that is called **epoxidation.**

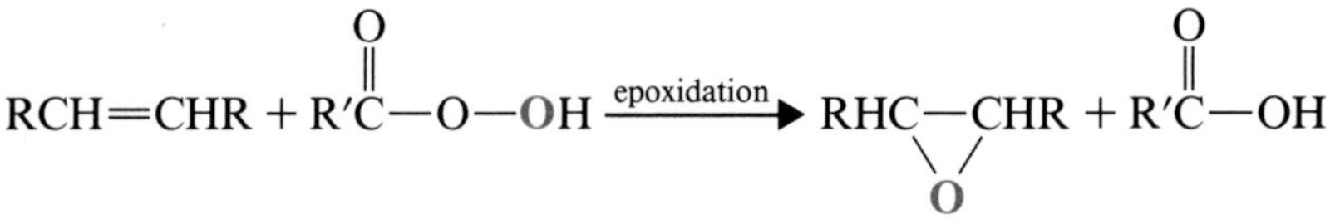

$$RCH{=}CHR + R'\overset{O}{\overset{\|}{C}}{-}O{-}OH \xrightarrow{\text{epoxidation}} RHC\overset{}{-}CHR + R'\overset{O}{\overset{\|}{C}}{-}OH$$

An alkene   A peroxy acid   An epoxide
(or oxirane)

In this reaction the peroxy acid transfers an oxygen atom to the alkene. The following mechanism has been proposed.

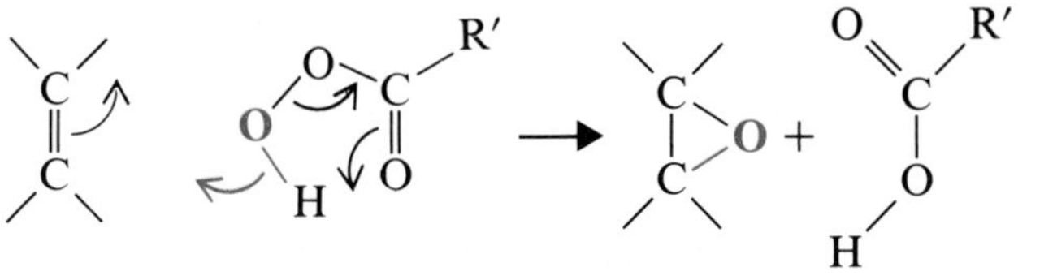

The addition of oxygen to the double bond in an epoxidation reaction is, of necessity, a **syn** addition. In order to form a three-membered ring, the oxygen atom must add to both carbon atoms of the double bond at the same face.

The peroxy acids most commonly used are peroxyacetic acid ($CH_3\overset{O}{\overset{\|}{C}}OOH$) and peroxybenzoic acid ($C_6H_5\overset{O}{\overset{\|}{C}}OOH$). Cyclohexene, for example, reacts with peroxybenzoic acid to give cyclohexene oxide in a quantitative yield.

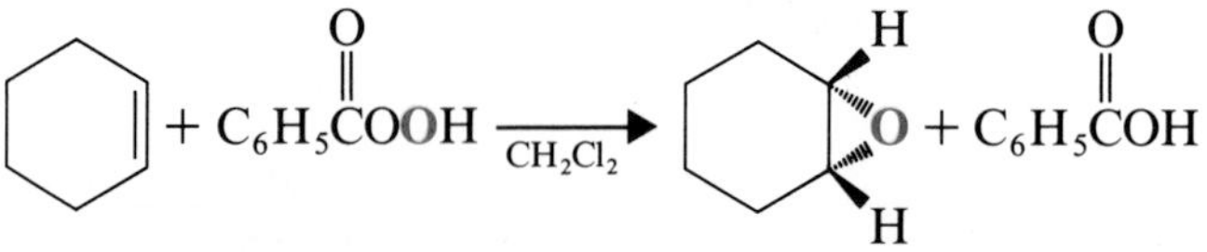

Peroxybenzoic
acid

(100%)
Cyclohexene
oxide

The reaction of alkenes with peroxy acids takes place in a stereospecific way: *cis*-2-butene, for example, yields only *cis*-2,3-dimethyloxirane, and *trans*-2-butene yields only the *trans*-2,3-dimethyloxiranes.

Step 1

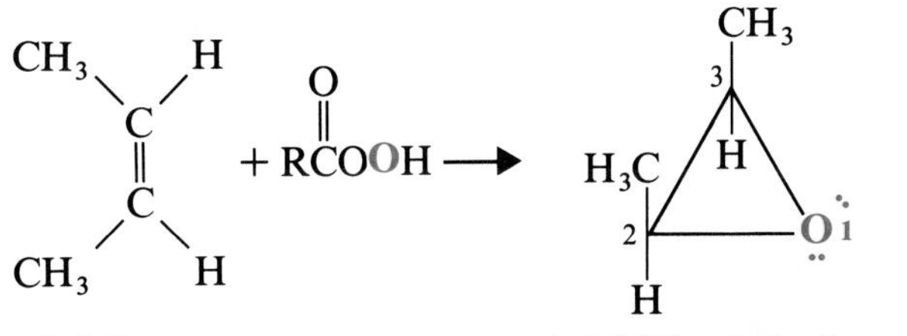

cis-2-Butene

cis-2,3-Dimethyloxirane
(a meso compound)

Step 2

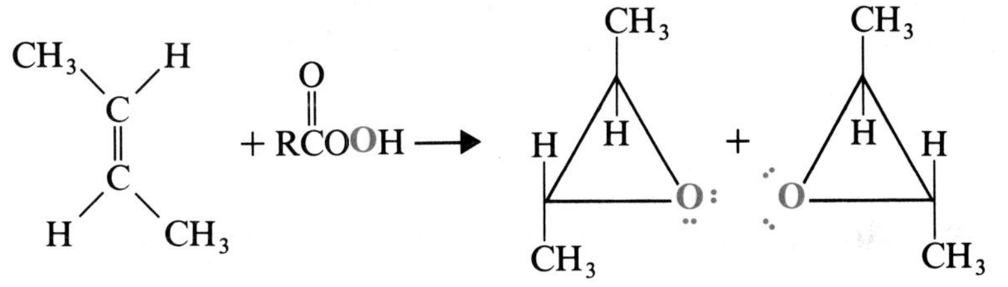

trans-2-Butene

Enantiomeric *trans*-2,3-dimethyloxiranes

7.9A Acid-Catalyzed Hydrolysis of Epoxides

Although most ethers react with few reagents, the strained three-membered ring of epoxides makes them highly susceptible to ring-opening reactions. Ring opening takes place through cleavage of one of the carbon–oxygen bonds. It can be initiated by either electrophiles or nucleophiles, or catalyzed by either acids or bases. The acid-catalyzed hydrolysis of an epoxide, for example, is a useful procedure for preparing vicinal-dihydroxy compounds called 1,2-diols or glycols.

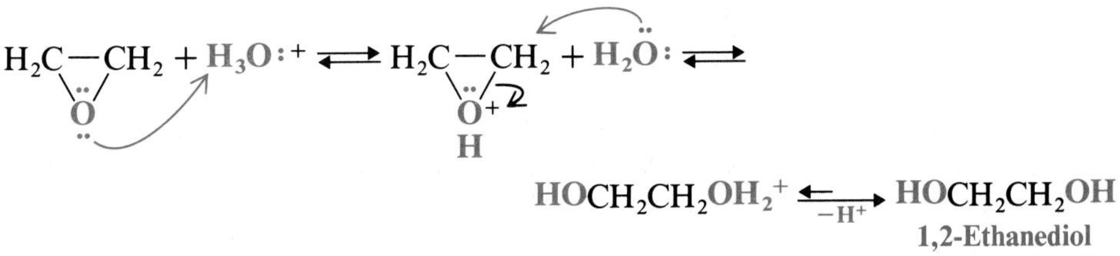

$$HOCH_2CH_2OH_2^+ \underset{-H^+}{\overset{}{\rightleftharpoons}} HOCH_2CH_2OH$$

1,2-Ethanediol
(ethylene glycol)

Notice the similarity between this mechanism and that given for the ring opening of the bromonium ion given in Section 7.6A.

7.9B Anti Hydroxylation of Alkenes

Epoxidation of cyclopentene produces cyclopentene oxide:

(1)

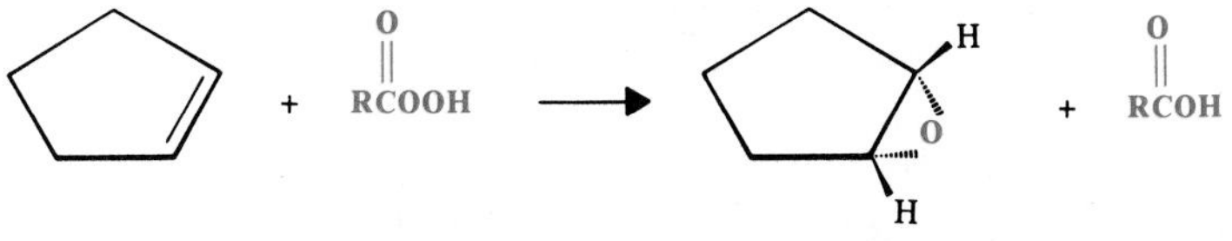

Cyclopentene

Cyclopentene oxide

Acid-catalyzed hydrolysis of cyclopentene oxide yields a *trans*-diol, *trans*-1,2-cyclopentanediol. Water acting as a nucleophile attacks the protonated epoxide from the side opposite the epoxide group. The carbon atom being attacked undergoes an inversion of configuration. We show (here) only one carbon atom being attacked. Attack at the other carbon atom is equally likely and produces the enantiomeric form of *trans*-1,2-cyclopentanediol.

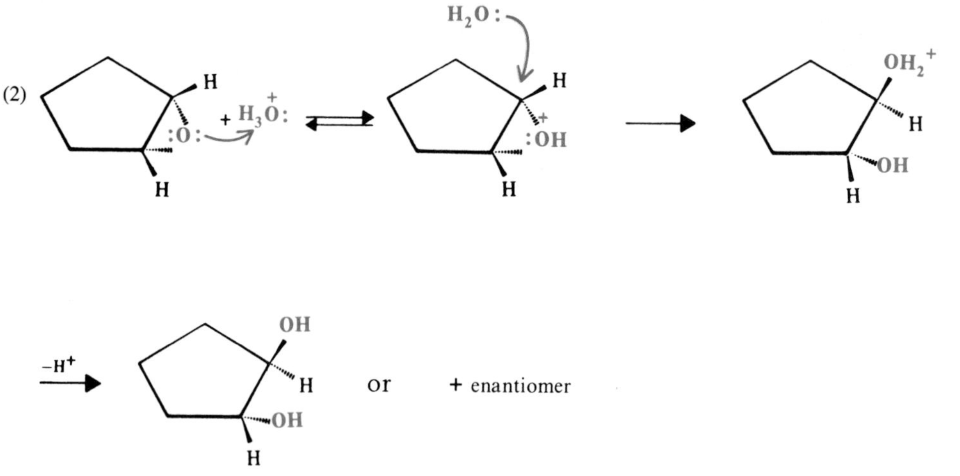

trans-1, 2-Cyclopentanediol

Epoxidation followed by acid-catalyzed hydrolysis gives us, therefore, a method of **hydroxylating** a double bond (i.e., a method for adding a hydroxyl group to each carbon atom). The stereochemistry of this technique parallels closely the stereochemistry of the bromination of cyclopentene given earlier (Section 7.7).

PROBLEM 7.12

Outline a mechanism similar to the one just given that shows how the enantiomeric form of *trans*-1,2-cyclopentanediol is produced.

SAMPLE PROBLEM

Earlier in this section we showed the epoxidation of *cis*-2-butene to yield *cis*-2,3-dimethyloxirane and epoxidation of *trans*-2-butene to yield *trans*-2,3-dimethyloxirane. (a) Now consider acid-catalyzed hydrolysis of these two epoxides and show what product or products would result from each. (b) Are these reactions stereospecific?

Answer:

The meso compound, *cis*-2,3-dimethyloxirane (Fig. 7.5), yields on hydrolysis (2R,3R)-2,3-butanediol and (2S,3S)-2,3-butanediol. These products are enantiomers. Since the attack by water at either carbon [path (a) or path (b)] occurs at the same rate, the product is obtained in a racemic form.

When either of the *trans*-2,3-dimethyloxirane enantiomers undergoes acid-catalyzed hydrolysis, the only product that is obtained is the meso compound, (2R,3S)-2,3-butanediol. The hydrolysis of one enantiomer is shown in Fig. 7.6. (You might construct a similar diagram showing the hydrolysis of the other enantiomer to convince yourself that it, too, yields the same product.)

Since both steps in this method for the conversion of an alkene to a diol (glycol) are stereospecific (i.e., both the epoxidation step and the acid-catalyzed hydrolysis), the net result is a stereospecific anti hydroxylation of the double bond (Fig. 7.7).

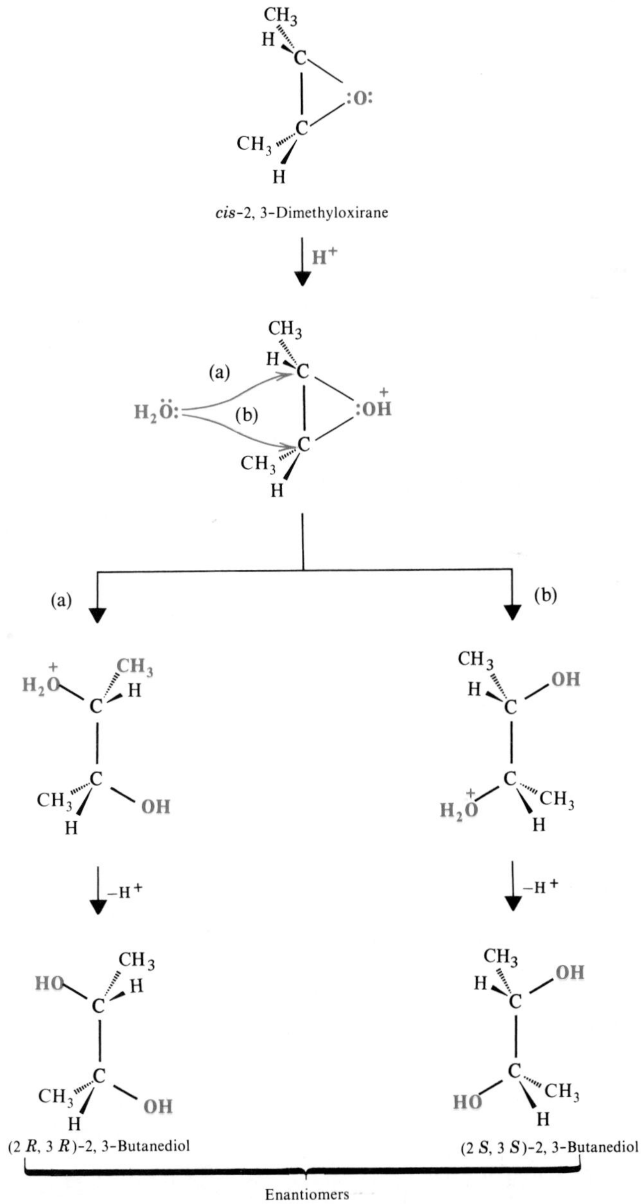

FIGURE 7.5 Acid-catalyzed hydrolysis of *cis*-2,3-dimethyloxirane yields (2R,3R)-2,3-butanediol by path (a) and (2S,3S)-2,3-butanediol by path (b). (Use models to convince yourself.)

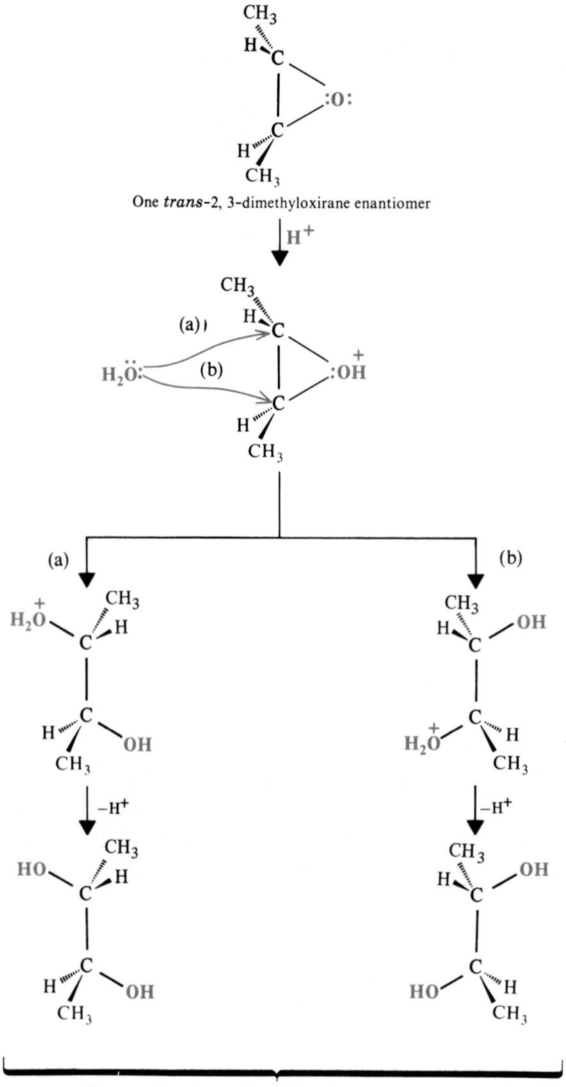

One *trans*-2, 3-dimethyloxirane enantiomer

FIGURE 7.6 The acid-catalyzed hydrolysis of one *trans*-2,3-di-methyloxirane enantiomer produces the meso compound, (2*R*,3*S*)-2,3-butanediol, by path (a) or (b). Hydrolysis of the other enantiomer (or the racemic modification) would yield the same product. (You should use models to convince yourself that the two structures given for the products above do represent the same compound.)

These molecules are identical: they both represent the meso compound (2*R*, 3*S*)-2, 3-butanediol

7.10 OXIDATIONS OF OTHER ALKENES

Alkenes undergo a number of other reactions in which the carbon–carbon double bond is oxidized. Potassium permanganate or osmium tetroxide, for example, can also be used to oxidize alkenes to **1,2-diols** called **glycols.**

$$CH_2{=}CH_2 + KMnO_4 \xrightarrow[\text{OH}^-]{\text{cold}} \begin{array}{c} H_2C{-}CH_2 \\ | \quad | \\ OH \; OH \end{array}$$

Ethene

1,2-Ethanediol
(ethylene glycol)

$$CH_3CH{=}CH_2 \xrightarrow[\text{(2) } Na_2SO_4/H_2O]{\text{(1) } OsO_4} \begin{array}{c} CH_3CH{-}CH_2 \\ | \quad | \\ OH \quad OH \end{array}$$

Na_2SO_3

Propene

1,2-Propanediol
(propylene glycol)

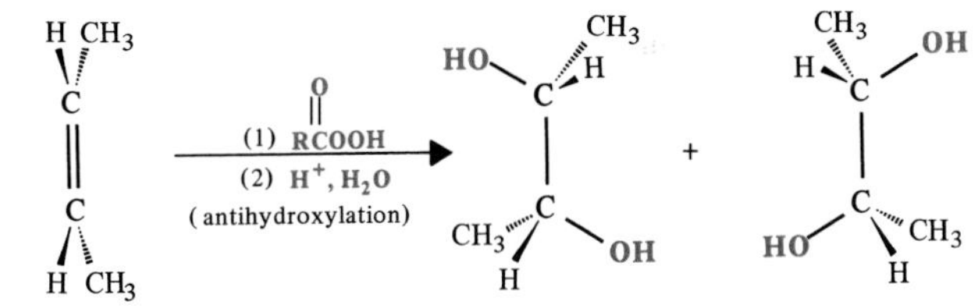

cis-2-Butene   Enantiomeric 2, 3-butanediols

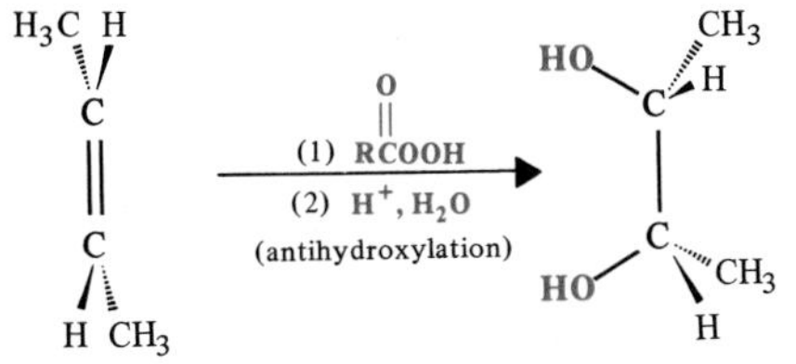

trans-2-Butene   meso-2, 3-Butanediol

FIGURE 7.7 The overall result of epoxidation followed by acid-catalyzed hydrolysis is a stereospecific anti hydroxylation of the double bond. *cis*-2-Butene yields the enantiomeric 2,3-butanediols; *trans*-2-butene yields the meso compound.

7.10A Syn Hydroxylation of Alkenes

The mechanisms for the formation of glycols by permanganate ion and osmium tetroxide oxidations first involve the formation of cyclic intermediates. Then in several steps cleavage at the oxygen–metal bond takes place (at the dashed lines in the following reactions) ultimately producing the glycol and MnO_2 or osmium metal.

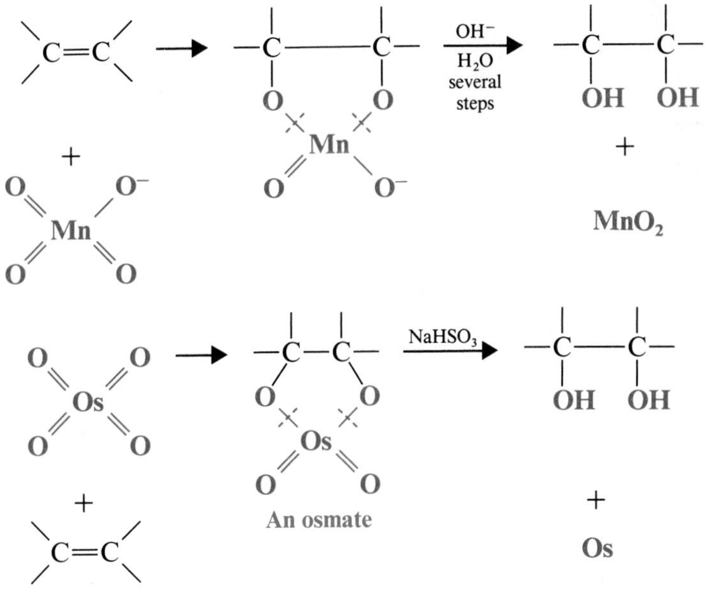

The course of these reactions is **syn hydroxylation.** This can be seen, readily, when cyclopentene reacts with cold dilute potassium permanganate (in base) or with osmium tetroxide (followed by treatment with $NaHSO_3$ or Na_2SO_3). The product in either case is *cis*-1,2-cyclopentanediol. (*cis*-1,2-Cyclopentanediol is a meso compound.)

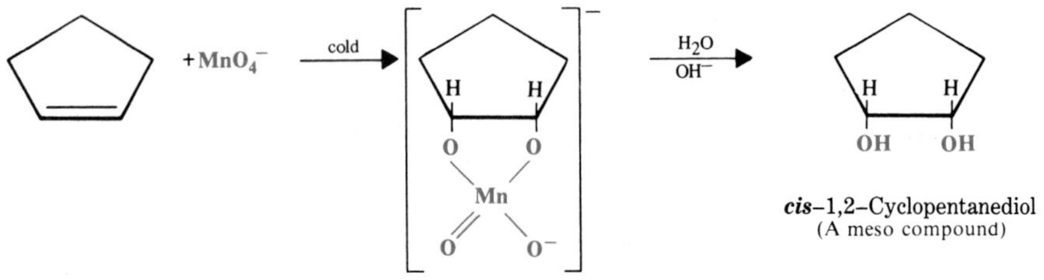

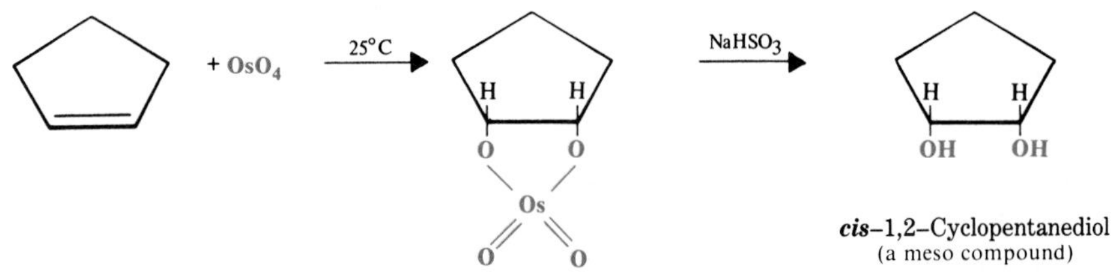

cis–1,2–Cyclopentanediol
(A meso compound)

cis–1,2–Cyclopentanediol
(a meso compound)

Of the two reagents used for syn hydroxylation, osmium tetroxide gives the higher yields. Unfortunately, however, osmium tetroxide is highly toxic and very expensive. Potassium permanganate is a very powerful oxidizing agent and, as we shall see in Section 7.10B, *it is easily capable of causing further oxidation of the glycol.* Limiting the reaction to hydroxylation alone is often difficult, but is usually attempted by using cold, dilute, and basic solutions of potassium permanganate. Even so, yields are sometimes very low.

PROBLEM 7.13

(a) What product(s) would you expect from syn hydroxylation of *cis*-2-butene? (b) Of *trans*-2-butene? (c) Are these reactions stereospecific? Explain your answer.

7.10B Oxidative Cleavage of Alkenes

Alkenes are oxidatively cleaved to salts of carboxylic acids *by hot permanganate solutions.* We can illustrate this reaction with the oxidative cleavage of either *cis*- or *trans*-2-butene to two molar equivalents of acetate ion. The intermediate in this reaction may be a glycol that is oxidized further with cleavage at the carbon–carbon bond.

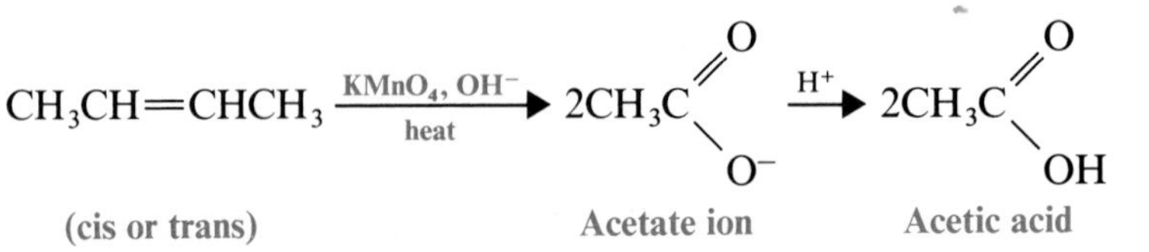

Acidification of the mixture, after the oxidation is complete, produces 2 moles of acetic acid for each mole of 2-butene.

The terminal CH_2 group of a 1-alkene is completely oxidized to carbon dioxide and water by hot permanganate. A disubstituted carbon atom of a double bond becomes the $\diagup\diagdown C=O$ group of a ketone (Section 2.13).

$$\underset{\substack{CH_3CH_2C=CH_2}}{\overset{\substack{CH_3}}{}} \xrightarrow[\text{(2) H}^+]{\substack{\text{(1) KMnO}_4,\ \text{OH}^- \\ \text{heat}}} \underset{}{\overset{\substack{CH_3}}{}} CH_3CH_2C=O + CO_2 + H_2O$$

The oxidative cleavage of alkenes has frequently been used to prove the location of the double bond in an alkene chain or ring. We can see how this might be done with the following example.

EXAMPLE
An unknown alkene with the formula C_8H_{16} was found, on oxidation with hot permanganate, to yield a five-carbon carboxylic acid (pentanoic acid) and a three-carbon carboxylic acid (propanoic acid).

$$C_8H_{16} \xrightarrow[\text{(2) H}^+]{\substack{\text{(1) KMnO}_4,\ \text{H}_2\text{O}, \\ \text{OH}^-,\ \text{heat}}} \underset{\text{Pentanoic acid}}{CH_3CH_2CH_2CH_2\overset{\overset{\displaystyle O}{\|}}{C}-OH} + \underset{\text{Propanoic acid}}{CH_3CH_2\overset{\overset{\displaystyle O}{\|}}{C}-OH}$$

We can see, then, from what we know about oxidative cleavage by permanganate, that the original alkene must have been either *cis-* or *trans*-3-octene:

$$CH_3CH_2CH=CHCH_2CH_2CH_2CH_3$$
3-Octene
(cis or trans)   ■

7.10C Ozonization (Ozonolysis) of Alkenes
A more widely used method for locating the double bond of an alkene involves the use of ozone (O_3). Ozone reacts vigorously with alkenes to form unstable compounds called *initial ozonides,* which rearrange spontaneously to form compounds known as **ozonides.**

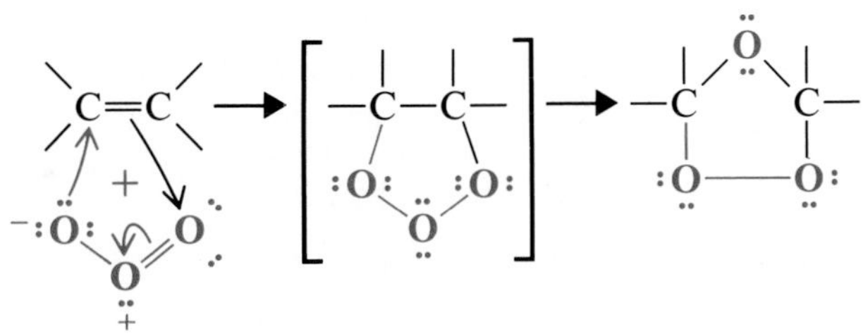

Initial ozonide   Ozonide

This rearrangement is thought to occur through dissociation of the initial ozonide into reactive fragments that recombine to yield the ozonide.

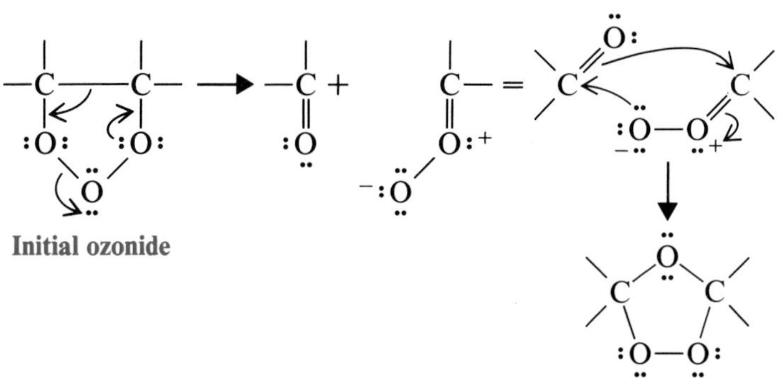

Initial ozonide

Ozonide

Ozonides, themselves, are very unstable compounds and low-molecular-weight ozonides often explode violently. Because of this property they are not usually isolated, but are reduced directly by treatment with zinc and water. The reduction produces carbonyl compounds (either aldehydes or ketones, see Section 2.13) that can be safely isolated and identified.

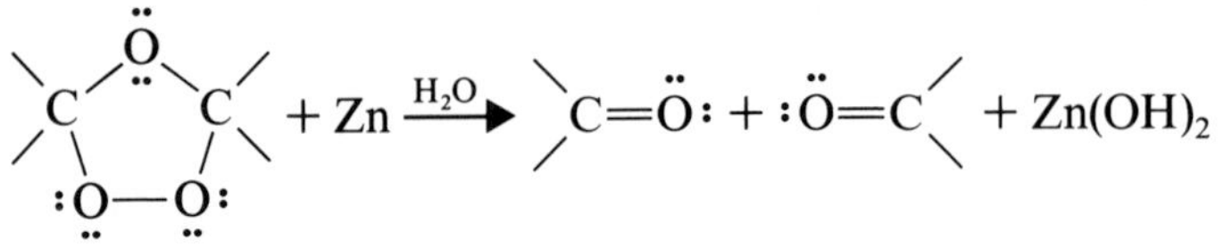

Ozonide

Aldehydes and/or ketones

The identities of the aldehydes or ketones disclose the location of the double bond in the original alkene. The following examples will illustrate the kinds of products that result from ozonization and subsequent treatment with zinc and water.

$$\underset{\text{2-Methyl-2-butene}}{CH_3\overset{\overset{\displaystyle CH_3}{|}}{C}{=}CHCH_3} \xrightarrow[\text{(2) Zn/}H_2O]{\text{(1) }O_3} \underset{\text{Acetone}}{CH_3\overset{\overset{\displaystyle CH_3}{|}}{C}{=}O} + \underset{\text{Acetaldehyde}}{CH_3\overset{\overset{\displaystyle O}{\|}}{C}H}$$

$$\underset{\text{3-Methyl-1-butene}}{CH_3\overset{\overset{\displaystyle CH_3}{|}}{C}H{-}CH{=}CH_2} \xrightarrow[\text{(2) Zn/}H_2O]{\text{(1) }O_3} \underset{\text{Isobutyraldehyde}}{CH_3\overset{\overset{\displaystyle CH_3}{|}}{C}H{-}\overset{\overset{\displaystyle O}{\|}}{C}H} + \underset{\text{Formaldehyde}}{H\overset{\overset{\displaystyle O}{\|}}{C}H}$$

PROBLEM 7.14

Write the general structures of the alkenes that would produce the following products when treated with ozone and then with zinc and water.

(a) $CH_3CH_2\overset{\overset{\displaystyle O}{\|}}{C}H + CH_3\overset{\overset{\displaystyle O}{\|}}{C}H$

(b) $CH_3{-}\underset{\underset{\displaystyle CH_3}{|}}{C}{=}O$ only (2 moles are produced from 1 mole of alkene)

(c) $CH_3CH_2\overset{O}{\underset{\underset{CH_3}{|}}{\overset{||}{C}}H-\overset{O}{\overset{||}{C}}H + H\overset{O}{\overset{||}{C}}H$

(d) $H-\overset{O}{\overset{||}{C}}CH_2CH_2CH_2CH_2\overset{O}{\overset{||}{C}}-H$ only

7.11 REACTION OF ALKENES WITH CARBOCATIONS

Heating isobutylene (2-methylpropene) in 60% sulfuric acid at 70 °C causes the formation of two main products—two isomeric compounds called *diisobutylenes.* The diisobutylenes are examples of **dimers** (*di-* = two + Gr., *meros* = part); isobutylene is said to have been dimerized by the reaction. (The dashed lines through the structures divide them into their original units.)

$$2CH_3-\underset{\underset{CH_3}{|}}{\overset{CH_3}{\overset{|}{C}}}=CH_2 \xrightarrow[70\ °C]{60\%\ H_2SO_4} CH_3-\underset{\underset{CH_3}{|}}{\overset{CH_3}{\overset{|}{C}}}\!\!+\!\!CH_2-C\!\!\begin{smallmatrix}CH_2\\ \\CH_3\end{smallmatrix} + CH_3\!\!\begin{smallmatrix}H\\ \diagdown\\ \diagup\\C\\ \diagup \diagdown\\ CH_3\ CH_3\end{smallmatrix}\!\!C=C\!\!\begin{smallmatrix}CH_3\\ \\CH_3\end{smallmatrix}$$

(80%)   (20%)

Diisobutylenes

In this reaction the less highly substituted alkene is the major product because it is apparently the more stable. In molecules of the more highly substituted alkene considerable internal repulsion exists between the large *tert*-butyl group and a *cis*-methyl group. This is an exception to the general rule concerning alkene stability (Section 6.9).

At one time the dimerization of isobutylene was important in the petroleum industry because hydrogenation of the mixture of dimers produces the single product called *isooctane.**

$$CH_3-\underset{\underset{CH_3}{|}}{\overset{CH_3}{\overset{|}{C}}}-CH_2-C\!\!\begin{smallmatrix}CH_2\\ \\CH_3\end{smallmatrix} + CH_3-\underset{\underset{CH_3}{|}}{\overset{CH_3}{\overset{|}{C}}}-CH=C\!\!\begin{smallmatrix}CH_3\\ \\CH_3\end{smallmatrix} \xrightarrow[Ni]{H_2} CH_3-\underset{\underset{CH_3}{|}}{\overset{CH_3}{\overset{|}{C}}}-CH_2-\overset{CH_3}{\overset{|}{C}}H-CH_3$$

"Isooctane"
(2,2,4-trimethylpentane)

2,2,4-Trimethylpentane ("isooctane") burns very smoothly (without knocking) in internal combustion engines and is used as one of the standards by which the octane rating of gasolines is established. According to this scale 2,2,4-

*Although calling this compound isooctane is incorrect, the name has been used for many years in the petroleum industry. The correct IUPAC name is 2,2,4-trimethylpentane.

trimethylpentane has an octane rating of 100. Heptane, a compound that produces much knocking when it is burned in an internal combustion engine, is given an octane rating of zero. Mixtures of 2,2,4-trimethylpentane and heptane are used as standards for octane ratings between 0 and 100. A gasoline, for example, that has the same characteristics in an engine as a mixture of 87% 2,2,4-trimethylpentane–13% heptane would be rated as 87-octane gasoline.

The mechanism by which isobutylene dimerizes (see following reactions), is straightforward. The second step shows us another property of carbocations: *Their ability to react with alkenes.*

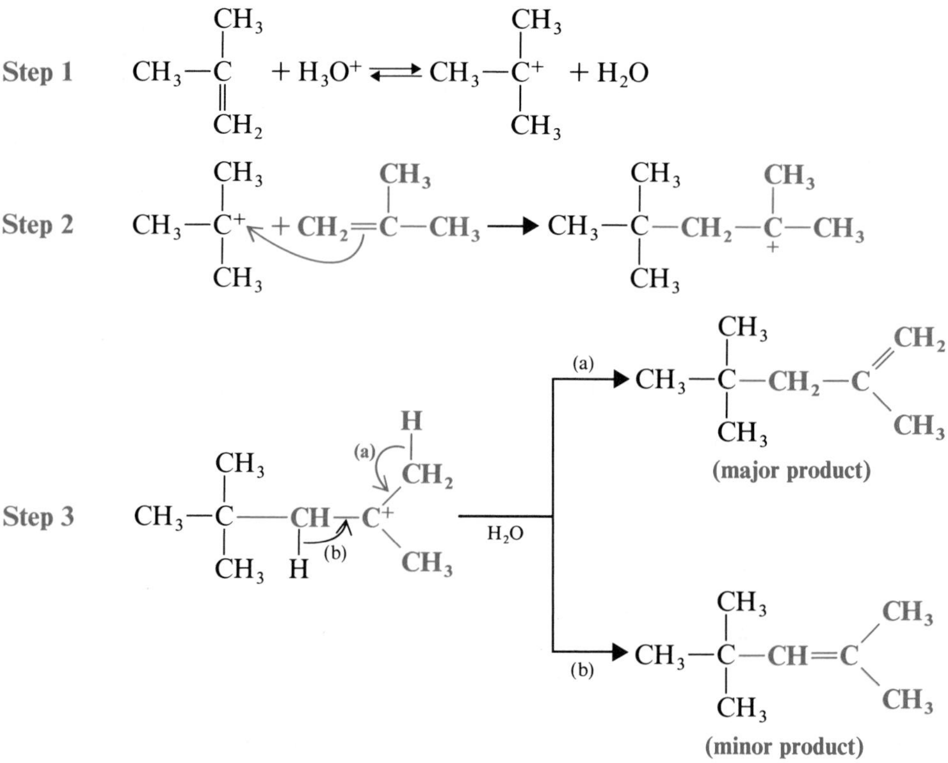

Step 1

$$CH_3-\underset{\underset{CH_2}{|}}{\overset{\overset{CH_3}{|}}{C}} \ + H_3O^+ \rightleftharpoons CH_3-\underset{\underset{CH_3}{|}}{\overset{\overset{CH_3}{|}}{C^+}} \ + H_2O$$

Step 2

$$CH_3-\underset{\underset{CH_3}{|}}{\overset{\overset{CH_3}{|}}{C^+}} + CH_2{=}\underset{}{\overset{\overset{CH_3}{|}}{C}}-CH_3 \longrightarrow CH_3-\underset{\underset{CH_3}{|}}{\overset{\overset{CH_3}{|}}{C}}-CH_2-\underset{+}{\overset{\overset{CH_3}{|}}{C}}-CH_3$$

Step 3

$$CH_3-\underset{\underset{CH_3}{|}}{\overset{\overset{CH_3}{|}}{C}}\underset{\underset{H}{|}}{\overset{}{\longrightarrow}}\overset{(a)}{\underset{(b)}{CH}}-\overset{\overset{(a)\ \ \overset{H}{|}\ CH_2}{}}{C^+}{\overset{}{\underset{CH_3}{}}}$$

(a) → $CH_3-\underset{\underset{CH_3}{|}}{\overset{\overset{CH_3}{|}}{C}}-CH_2-\underset{\diagdown CH_3}{\overset{\diagup CH_2}{C}}$

(major product)

$\xrightarrow{H_2O}$

(b) → $CH_3-\underset{\underset{CH_3}{|}}{\overset{\overset{CH_3}{|}}{C}}-CH{=}\underset{\diagdown CH_3}{\overset{\diagup CH_3}{C}}$

(minor product)

In step 1 the acid donates a proton to isobutylene to give a *tert*-butyl cation. We have seen similar reactions several times before. In step 2 the carbocation — acting as an electrophile — attacks another molecule of isobutylene. Isobutylene acts as a nucleophile; its π electrons form a σ bond to the *tert*-butyl cation. The larger carbocation that is formed as a result may lose a proton in two different ways (step 3) to form the mixture of diisobutylenes.

The reaction of a carbocation with an alkene can also lead to **cyclization** (the formation of a ring). In the example that follows, treating the open-chain compound called pseudoionone with acid yields the two cyclic compounds α-ionone and β-ionone.

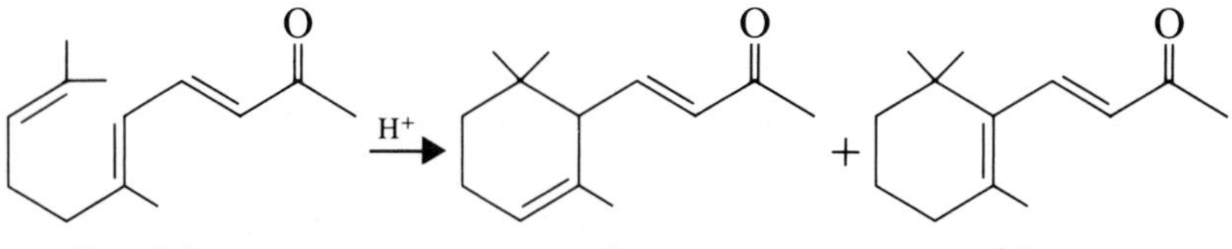

| Pseudoionone | α-Ionone | β-Ionone |

The mechanism involves several steps and is similar to the dimerization of isobutylene. In the first step, one of the double bonds of pseudoionone accepts a proton to

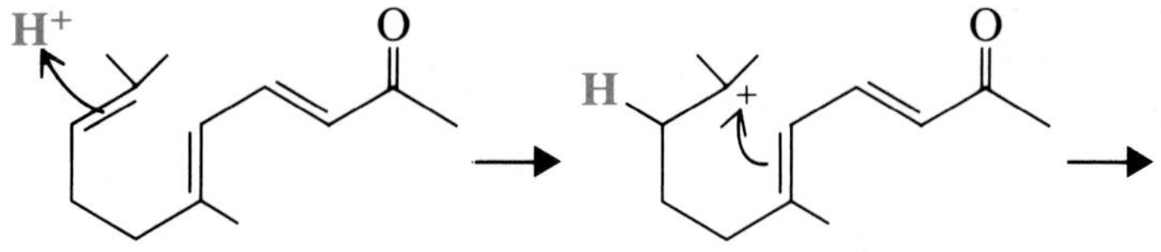
H+
O
H
+
O

Pseudoionone

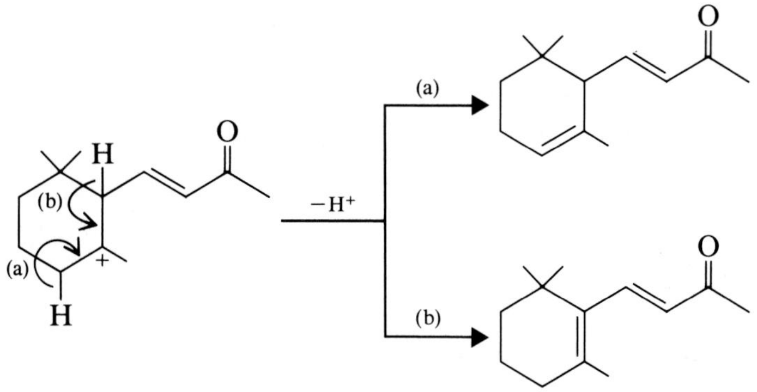
O
H
(b)
(a)
+
H
−H+
(a)
(b)
O
O

α-Ionone

β-Ionone

form a tertiary carbocation. In the next step this carbocation reacts with one of the remaining double bonds to form a six-membered ring containing a new tertiary carbocation. Finally, this carbocation loses a proton by either of two paths (a) or (b) generating α-ionone and β-ionone.

PROBLEM 7.15

Write a mechanism that accounts for the following cyclization reaction:

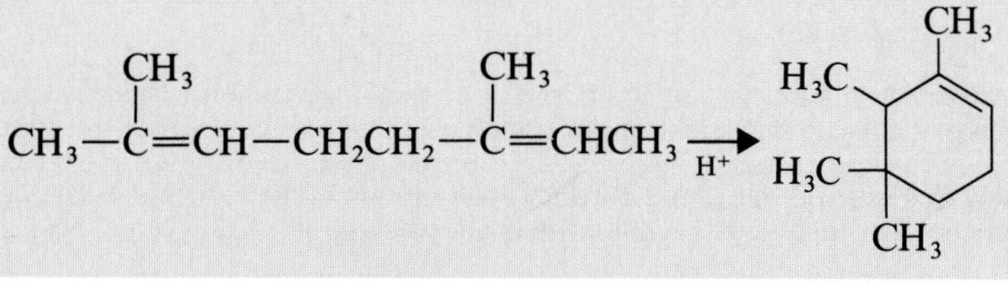
CH3
CH3
CH3—C=CH—CH2CH2—C=CHCH3
H+
CH3
H3C
H3C
CH3

7.12 SUMMARY OF ADDITION REACTIONS OF ALKENES

The stereochemistry and regiospecificity (where appropriate) of the addition reactions of alkenes that we have studied thus far are summarized in Figure 7.8. We have used 1-methylcyclopentene as the starting alkene.

7.13 ADDITION OF HALOGENS TO ALKYNES

Alkynes show the same kind of reactions toward chlorine and bromine that alkenes do: ***They react by addition.*** However, with alkynes the addition may occur once or twice, depending on the number of molar equivalents of halogen we employ.

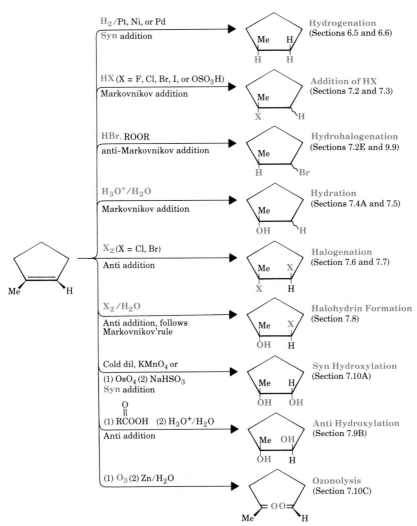

FIGURE 7.8 A summary of addition reactions of alkenes with 1-methylcyclo-pentene as the organic substrate. A bond designated ∿ means that the stereo-chemistry of the group is unspecified. For brevity we have shown the structure of only one enantiomer of the product even though racemic forms would be produced in all instances in which the product is chiral.

It is usually possible to prepare a dihaloalkene by simply adding one molar equivalent of the halogen.

$$CH_3CH_2CH_2CH_2C\equiv CCH_2OH \xrightarrow[\substack{CCl_4 \\ 0\,°C}]{Br_2\ (1\ mole)} CH_3CH_2CH_2CH_2CBr=CBrCH_2OH$$
$$(80\%)$$

Most additions of chlorine and bromine to alkynes are anti additions and yield *trans*-dihaloalkenes. Addition of bromine to acetylenedicarboxylic acid, for example, gives the trans isomer in 70% yield.

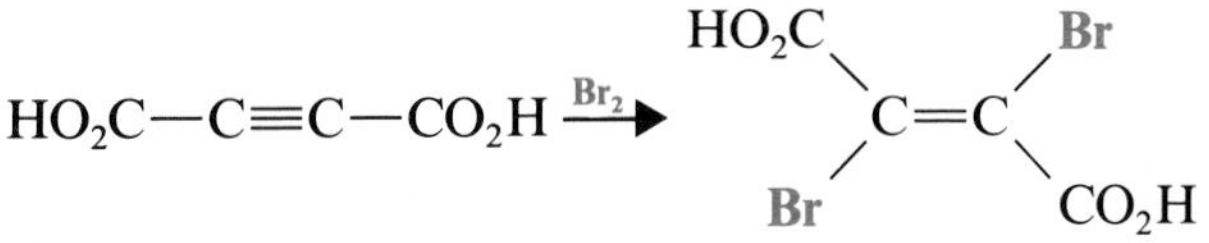

$$HO_2C-C\equiv C-CO_2H \xrightarrow{Br_2}$$

HO₂C, Br
  C=C
Br, CO₂H

Acetylenedicarboxylic
acid

(70%)

PROBLEM 7.16

Alkenes are more reactive than alkynes toward addition of electrophilic reagents (i.e., Br_2, Cl_2, HCl, etc.). Yet when alkynes are treated with one molar equivalent of these same electrophilic reagents, it is easy to stop the addition at the "alkene stage." This appears to be a paradox and yet it is not. Explain.

7.14 ADDITION OF HYDROGEN HALIDES TO ALKYNES

Alkynes react with hydrogen chloride and hydrogen bromide to form haloalkenes or geminal dihalides depending, once again, on whether one or two molar equivalents of the hydrogen halide are used. **Both additions are regioselective and follow Markovnikov's rule:**

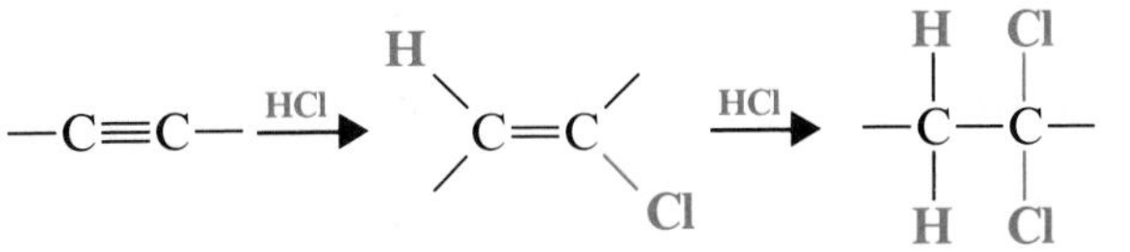

$$-C\equiv C- \xrightarrow{HCl} \quad \substack{H \\ C=C \\ Cl} \quad \xrightarrow{HCl} \quad \substack{H\ Cl \\ -C-C- \\ H\ Cl}$$

Chloroalkene   *gem*-Dichloride

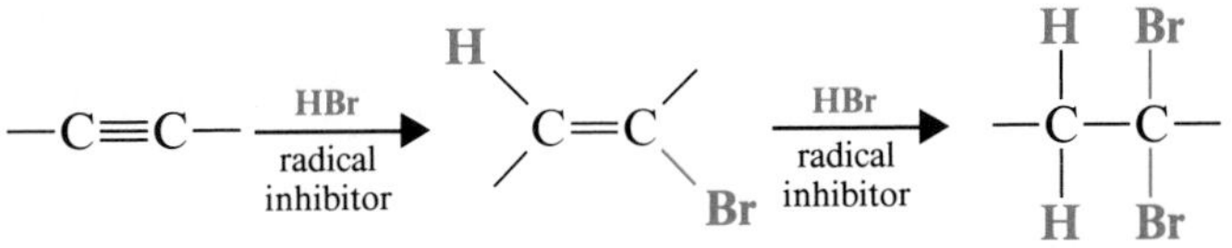

$$-C\equiv C- \xrightarrow[\substack{radical \\ inhibitor}]{HBr} \quad \substack{H \\ C=C \\ Br} \quad \xrightarrow[\substack{radical \\ inhibitor}]{HBr} \quad \substack{H\ Br \\ -C-C- \\ H\ Br}$$

Bromoalkene   *gem*-Dibromide

The hydrogen atom of the hydrogen halide becomes attached to the carbon atom that has the greater number of hydrogen atoms. Propyne, for example, reacts with one molar equivalent of hydrogen chloride to yield 2-chloropropene and with two molar equivalents to yield 2,2-dichloropropane.

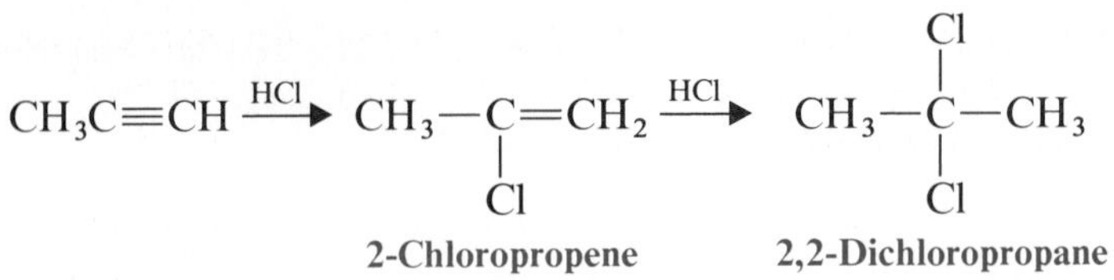

$$CH_3C\equiv CH \xrightarrow{HCl} CH_3-\underset{\underset{Cl}{|}}{C}=CH_2 \xrightarrow{HCl} CH_3-\underset{\underset{Cl}{|}}{\overset{\overset{Cl}{|}}{C}}-CH_3$$

2-Chloropropene   2,2-Dichloropropane

The initial addition of a hydrogen halide to an alkyne usually occurs in an **anti** manner. This is especially likely if an ionic halide corresponding to the halogen of the hydrogen halide is present in the mixture.

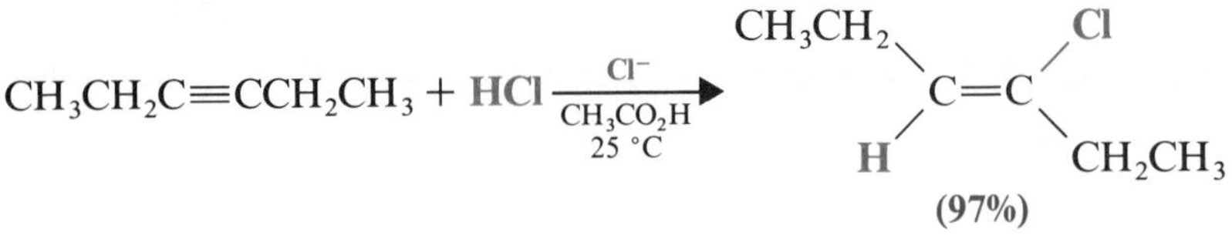

$$CH_3CH_2C\equiv CCH_2CH_3 + HCl \xrightarrow[\substack{CH_3CO_2H \\ 25\,°C}]{Cl^-} \overset{CH_3CH_2}{\underset{H}{}} \diagdown \underset{}{C}=C \diagup \overset{Cl}{\underset{CH_2CH_3}{}}$$

(97%)

The mechanism of the addition of HX to an alkyne involves the formation of an intermediate **vinylic cation,** which subsequently reacts with the halide ion to produce the haloalkene. Vinylic cations are much less stable than corresponding alkyl cations.

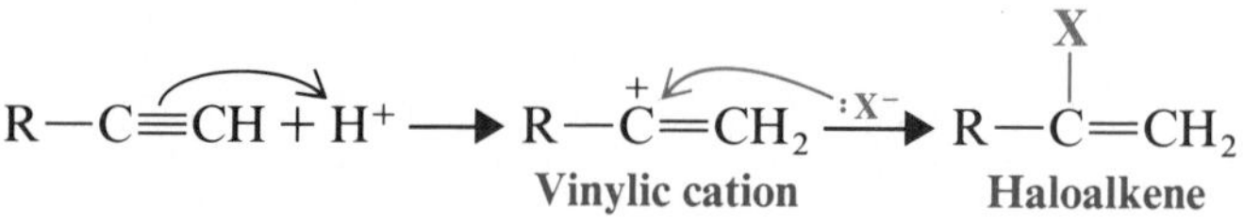

$$R-C\equiv CH + H^+ \longrightarrow R-\overset{+}{C}=CH_2 \xrightarrow{:X^-} R-\underset{\underset{X}{|}}{C}=CH_2$$

Vinylic cation   **Haloalkene**

The haloalkene can react further to yield the *gem*-dihalide:

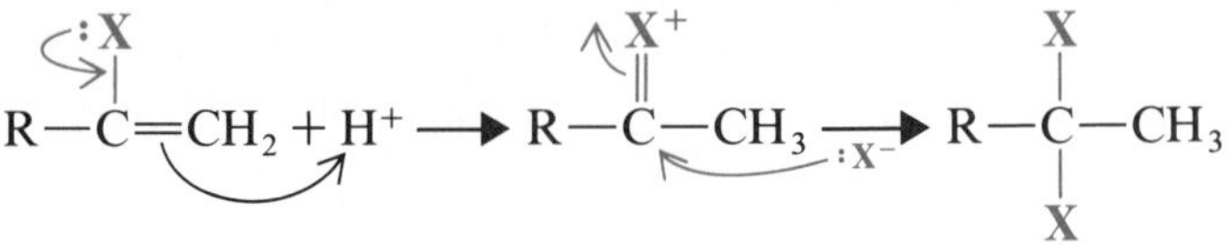

$$R-\overset{:X}{\underset{}{C}}=CH_2 + H^+ \longrightarrow R-\overset{X^+}{\underset{}{C}}-CH_3 \underset{:X^-}{\longrightarrow} R-\underset{\underset{X}{|}}{\overset{\overset{X}{|}}{C}}-CH_3$$

In this step the electron pair of the halogen helps stabilize the intermediate cation.

Anti-Markovnikov addition of hydrogen bromide to alkynes occurs when peroxides are present in the reaction mixture. These reactions take place through a free radical mechanism (Section 9.9).

$$CH_3CH_2CH_2CH_2C\equiv CH \xrightarrow[\text{peroxides}]{HBr} CH_3CH_2CH_2CH_2CH=CHBr$$

(74%)

7.15 ADDITION OF WATER TO ALKYNES

Alkynes add water readily when the reaction is catalyzed by strong acids and mercuric (Hg^{2+}) ions. Aqueous solutions of sulfuric acid and mercuric sulfate are often used for this purpose. The vinylic alcohol that is initially produced is usually unstable, and it rearranges rapidly to an aldehyde or a ketone. The rearrangement involves the loss of a proton from the hydroxyl group, the addition of a proton to the adjacent carbon atom, and the relocation of the double bond.

$$
-C\equiv C- + H-OH \xrightarrow[\text{H}_2\text{SO}_4]{\text{HgSO}_4}
\left[
\begin{array}{c}
\overset{\displaystyle H}{\underset{\displaystyle OH}{-C=C-}}
\end{array}
\right]
\longrightarrow
\overset{\displaystyle H}{\underset{\displaystyle H \quad O}{-C-C-}}
$$

A vinylic   Aldehyde
alcohol   or
(unstable)   ketone

This kind of rearrangement, known as a **tautomerization,** is acid catalyzed and occurs in the following way.

$$
H^+ + -C=C- \longrightarrow -C-C- \longrightarrow -C-C- + H^+
$$

Vinylic   Aldehyde
alcohol   or
                ketone

The vinylic alcohol accepts a proton at one carbon atom of the double bond to yield a cationic intermediate that then loses a proton from the oxygen atom to produce an aldehyde or ketone.

Vinylic alcohols are often called **enols** (-*en*, the ending for alkenes, plus -*ol*, the ending for alcohols). The product of the rearrangement is often a ketone, and these rearrangements are known as **keto-enol tautomerizations.**

$$
-\overset{|}{C}=\overset{|}{C}- \underset{\overset{\text{H}^+}{\longleftarrow}}{} -\overset{|}{C}-\overset{|}{C}-
$$

Enol form   Keto form

We examine this phenomenon in greater detail in Section 17.2.

When acetylene itself undergoes addition of water, the product is an aldehyde.

$$
H-C\equiv C-H + H_2O \xrightarrow[\text{H}_2\text{SO}_4]{\text{HgSO}_4}
\left[
\begin{array}{c}
\overset{H}{\underset{H}{}}C=C\overset{H}{\underset{OH}{}}
\end{array}
\right]
\longrightarrow
H-\overset{\displaystyle H}{\underset{\displaystyle H \quad O}{C-C}}-H
$$

Acetaldehyde

This method has been important in the commercial production of acetaldehyde.

The addition of water to alkynes also follows Markovnikov's rule—the hydrogen atom becomes attached to the carbon atom with the greater number of hydrogen atoms. Therefore, when higher terminal alkynes are hydrated, ketones, rather than aldehydes, are the products

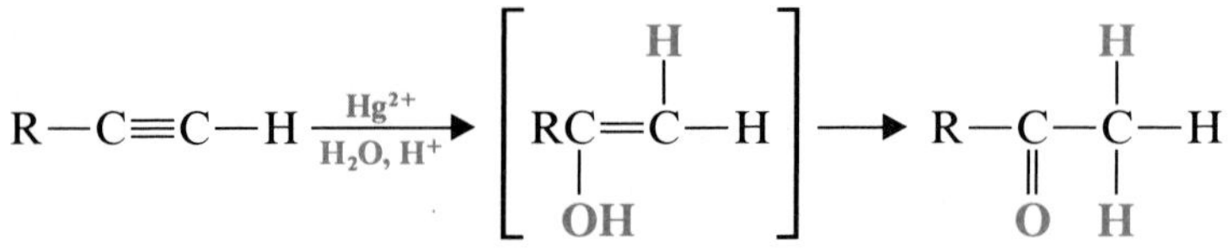

$$R-C \equiv C-H \xrightarrow[H_2O,\ H^+]{Hg^{2+}} \left[\begin{array}{c} \overset{\displaystyle H}{\underset{\displaystyle OH}{RC=C-H}} \end{array} \right] \longrightarrow R-\overset{\displaystyle H}{\underset{\displaystyle O\ H}{C-C-H}}$$

A ketone

Two examples of this ketone synthesis are listed here.

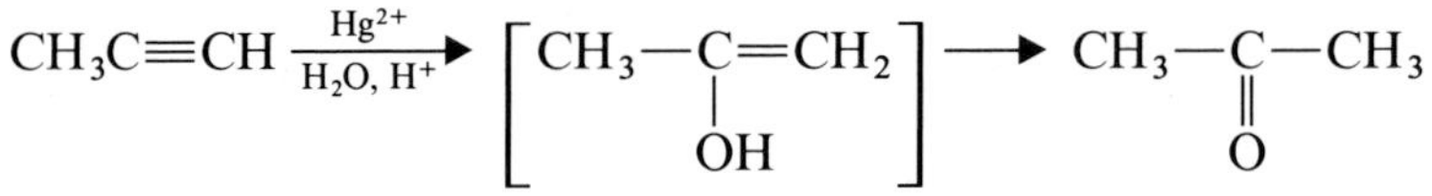

$$CH_3C \equiv CH \xrightarrow[H_2O,\ H^+]{Hg^{2+}} \left[\begin{array}{c} \underset{\displaystyle OH}{CH_3-C=CH_2} \end{array} \right] \longrightarrow CH_3-\underset{\displaystyle O}{C}-CH_3$$

Acetone

$$CH_3CH_2CH_2CH_2C \equiv CH \xrightarrow[\substack{H_2SO_4 \\ H_2O}]{HgSO_4} CH_3CH_2CH_2CH_2\underset{\displaystyle O}{C}CH_3$$

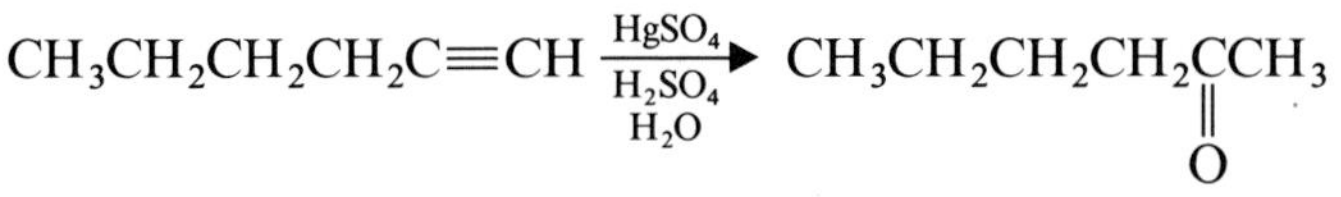

(80%)

7.16 OXIDATIVE CLEAVAGE OF ALKYNES

Treating alkynes with ozone or with basic potassium permanganate leads to cleavage at the carbon–carbon triple bond. The products are carboxylic acids.

$$R-C \equiv C-R' \xrightarrow[(2)\ H_2O]{(1)\ O_3} RCO_2H + R'CO_2H$$

or

$$R-C \equiv C-R' \xrightarrow[(2)\ H^+]{(1)\ KMnO_4,\ OH^-} RCO_2H + R'CO_2H$$

PROBLEM 7.17

Give the name and structure of each of the following alkynes used in the following reactions.

(a) 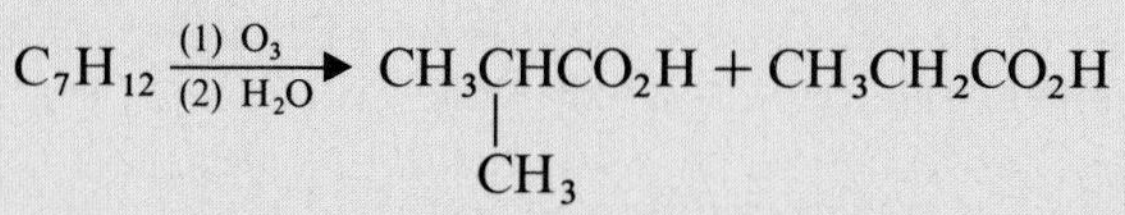 $C_7H_{12} \xrightarrow[(2)\ H_2O]{(1)\ O_3} CH_3\underset{\displaystyle CH_3}{CH}CO_2H + CH_3CH_2CO_2H$

(b) $C_8H_{12} \xrightarrow[(2)\ H_2O]{(1)\ O_3} HO_2C-(CH_2)_6-CO_2H$ only

(c) 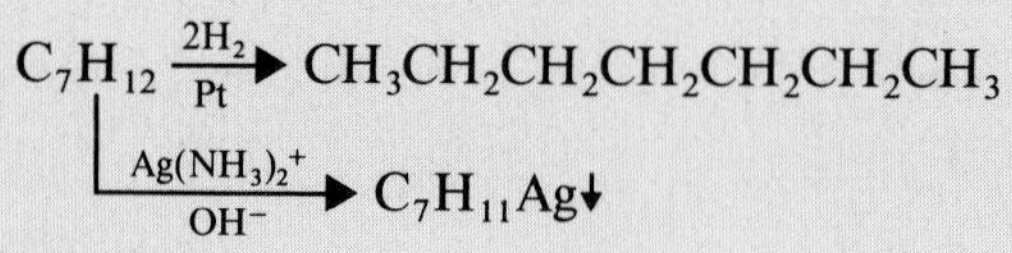 $C_7H_{12} \xrightarrow{2H_2}{Pt} CH_3CH_2CH_2CH_2CH_2CH_2CH_3$

$\xrightarrow[OH^-]{Ag(NH_3)_2^+} C_7H_{11}Ag \downarrow$

7.17 SUMMARY OF ADDITION REACTIONS OF ALKYNES

Figure 7.9 summarizes the addition reactions of alkynes.

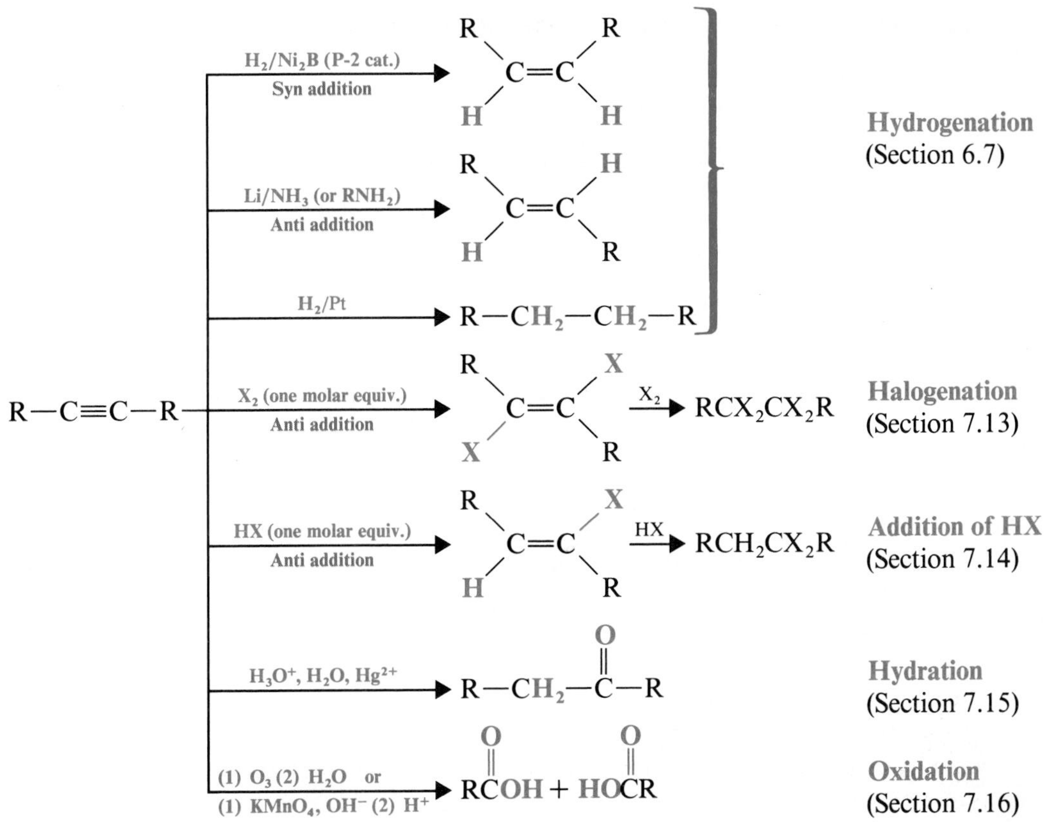

FIGURE 7.9 A summary of the addition reactions of alkynes.

7.18 PLANNING ORGANIC SYNTHESES

Most of the more than 6 million organic compounds that are now known have come about because organic chemists have synthesized them. Only a small fraction of these compounds have been isolated from natural sources. Even then, in the isolation of naturally occurring compounds, the final step in the proof of their structures is the synthesis of the compound by an unambiguous route from simpler molecules.

Synthesis is carried out for many other reasons. We may need a particular compound to test some hypothesis about a reaction mechanism. In this case we may need a particular compound with a "labeled" atom (e.g., a deuterium or tritium atom) at a certain position. Or, we may synthesize a compound because we think it will have some beneficial use. Most of the compounds now used to cure diseases, for example, have come about in this way.

Designing a synthesis is intellectually challenging, and when done successfully, it is intellectually rewarding. The syntheses that you are asked to design in this book have a teaching purpose. At first, you will be asked to design syntheses of relatively simple compounds that are, in most instances, commercially available. Nonetheless, successfully planning these syntheses can offer an intellectual challenge and reward.

In planning syntheses we are required to think backward, to work our way backwards from relatively complex molecules to simpler ones. We carry out what is called a **retrosynthetic analysis.** Many of the syntheses that you will be asked to plan will specify that you begin with only certain small molecules. This restriction should help you learn the retrosynthetic approach.

Thus far you have learned only a few of the basic reactions of organic chemistry. As you progress through this course you will learn many others, and your skill in planning complex syntheses will increase accordingly.

In planning syntheses we must consider at least four interrelated aspects:

1. Construction of the carbon skeleton.
2. Functional group interconversions.
3. Control of regiochemistry.
4. Control of stereochemistry.

Let us suppose, for example, that we are required to synthesize 1-bromobutane from compounds of two carbon atoms or fewer. Reasoning backward we might proceed in the following way. 1-Bromobutane can be synthesized from 1-butene in a regioselective way by the addition of HBr in the presence of peroxides (Section 7.2E):

$$CH_3CH_2CH_2CH_2Br \xleftarrow{\text{HBr, ROOR}} CH_3CH_2CH=CH_2$$
1-Bromobutane   **1-Butene**

1-Butene can be synthesized from 1-butyne by hydrogenation using the Ni_2B (P-2) catalyst (Section 6.7A). This step is a functional group interconversion:

$$CH_3CH_2CH=CH_2 \xleftarrow{\text{H}_2/\text{Ni}_2\text{B (P-2)}} CH_3CH_2C\equiv CH$$
1-Butene   **1-Butyne**

Finally, 1-butyne can be synthesized from two-carbon compounds by alkylating sodium acetylide with ethyl bromide (Section 6.19). This step constructs the carbon skeleton and places a functional group (the triple bond) at the end of the chain, where we need it.

$$CH_3CH_2C\equiv CH \xleftarrow{\text{CH}_3\text{CH}_2\text{Br}} Na^+:C\equiv CH \xleftarrow{\text{NaNH}_2} HC\equiv CH$$
1-Butyne

Consider another example: the synthesis of 2-pentanone, $CH_3CH_2CH_2\overset{\overset{\textstyle O}{\|}}{C}CH_3$, from compounds of three carbon atoms or fewer.

Thus far we have studied only one method for the synthesis of ketones: the hydration of alkynes (Section 7.15). Taking into account the regiochemistry of this reaction, that the addition of water follows Markovnikov's rule, we conclude that the last step in our synthesis should be

$$CH_3CH_2CH_2\overset{\overset{\textstyle O}{\|}}{C}CH_3 \longleftarrow \left[CH_3CH_2CH_2\overset{\overset{\textstyle OH}{|}}{C}=CH_2\right]$$
2-Pentanone   **Enol**

$$\xleftarrow{\text{H}_3\text{O}^+,\ \text{Hg}^{2+},\ \text{H}_2\text{O}} CH_3CH_2CH_2C\equiv CH$$
1-Pentyne

1-Pentyne can be synthesized by alkylating sodium acetylide with 1-bromopropane.

$$CH_3CH_2CH_2C\equiv CH \longleftarrow CH_3CH_2CH_2Br + NaC\equiv CH$$

Consider one more example: A synthesis that requires stereochemical control, the synthesis of the enantiomeric 2,3-butanediols from compounds of three carbon atoms or fewer.

Here we can see two ways to the final compound: the anti hydroxylation of *cis*-2-butene (Section 7.9B);

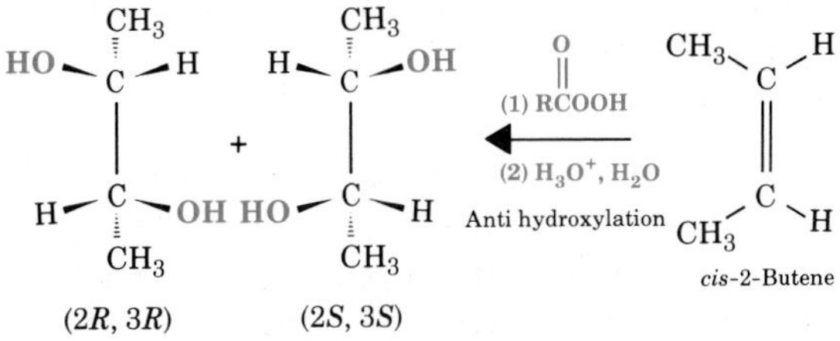

(2R, 3R)   (2S, 3S)

Enantiomeric 2, 3-butanediols

or, the syn hydroxylation of *trans*-2-butene (Section 7.10A):

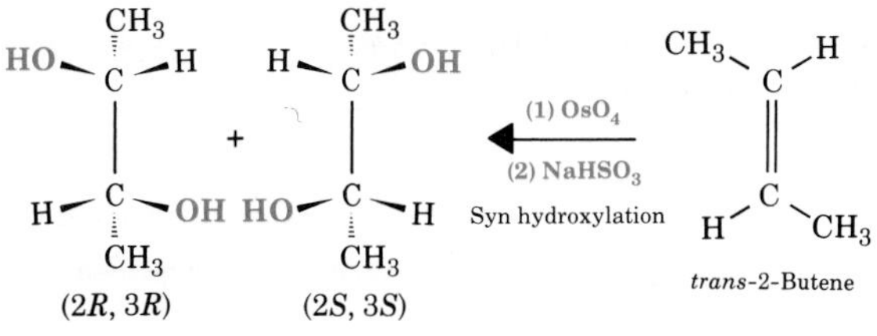

(2R, 3R)   (2S, 3S)

Enantiomeric 2, 3-butanediols

Both of these reactions are stereospecific and produce the enantiomeric 2,3-butanediols that we require as a racemic form.

We can synthesize *cis*-2-butene by hydrogenating 2-butyne using Ni_2B (P-2) catalyst:

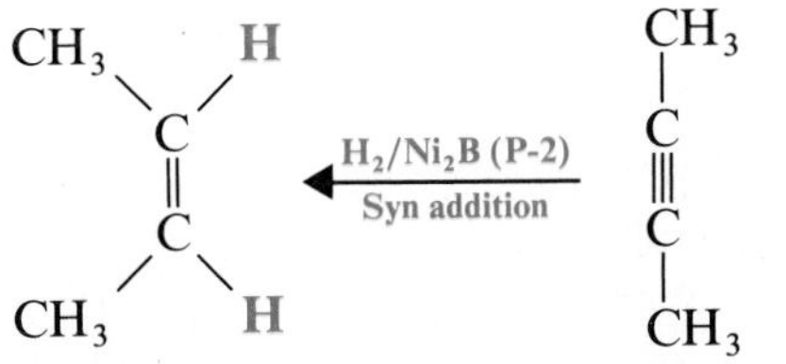

cis-2-Butene   2-Butyne

Or, we could synthesize *trans*-2-butene by treating 2-butyne with lithium and ammonia.

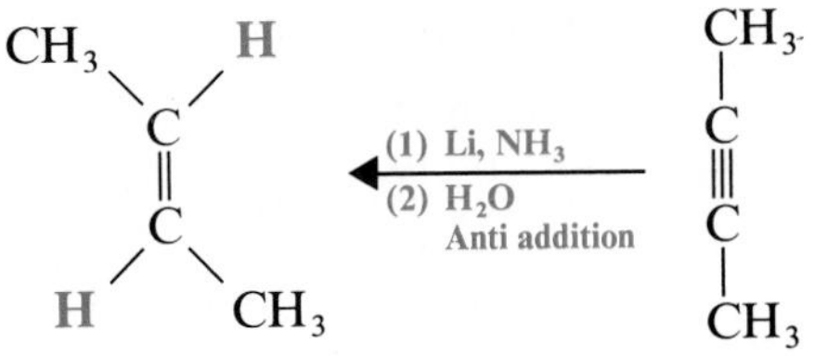

trans-2-Butene   2-Butyne

And, we can synthesize 2-butyne by akylating sodium propynide with methyl iodide:

$$CH_3C{\equiv}CCH_3 \xleftarrow[\text{(2) CH}_3\text{I}]{\text{(1) NaNH}_2} CH_3C{\equiv}CH$$

SAMPLE PROBLEM

Illustrating a Stereospecific Multistep Synthesis

Starting with compounds of two carbon atoms or fewer, outline a stereospecific synthesis of *meso*-3,4-dibromohexane.

Answer:

We begin by working backward from the product. The addition of bromine to an alkene is stereospecifically anti. Therefore, adding bromine to *trans*-3-hexene will give *meso*-3,4-dibromohexane:

trans-3-Hexene   meso-3, 4-Dibromohexane

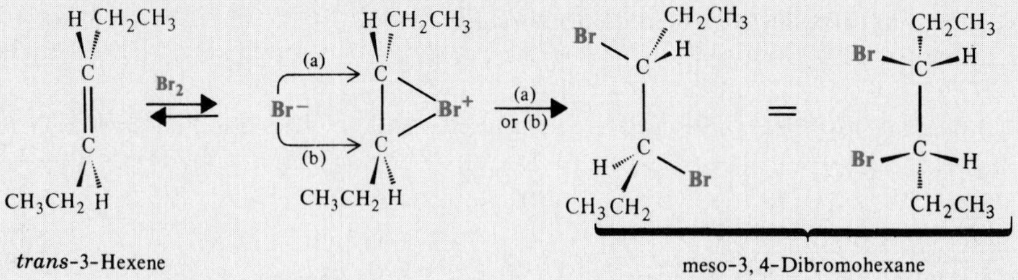

We can make *trans*-3-hexene in a stereospecific way from 3-hexyne by reducing it with lithium in ammonia (Section 6.7). Again the addition is anti.

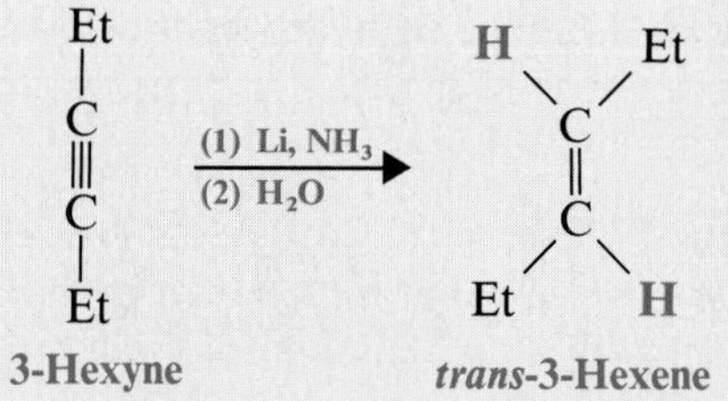

3-Hexyne   *trans*-3-Hexene

3-Hexyne can be made from acetylene and ethyl bromide by successive alkylations using sodium amide as a base:

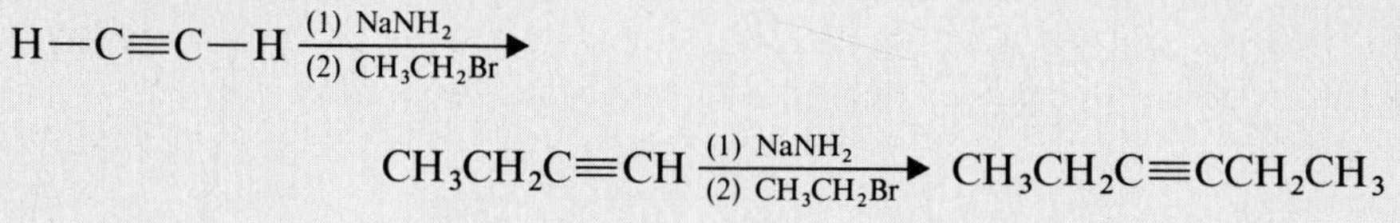

$$H-C\equiv C-H \xrightarrow[\text{(2) } CH_3CH_2Br]{\text{(1) } NaNH_2}$$

$$CH_3CH_2C\equiv CH \xrightarrow[\text{(2) } CH_3CH_2Br]{\text{(1) } NaNH_2} CH_3CH_2C\equiv CCH_2CH_3$$

PROBLEM 7.18

How would you modify the procedure given in the sample problem so as to synthesize a racemic form of (3*R*,4*R*)- and (3*S*,4*S*)-3,4-dibromohexane?

7.19 SIMPLE CHEMICAL TESTS FOR ALKANES, ALKENES, ALKYNES, ALKYL HALIDES, AND ALCOHOLS

Very often in the course of laboratory work we need to decide what functional groups are present in a compound that we have isolated. We may have isolated a compound from a synthesis, for example, and the presence of a particular functional group may tell us whether our synthesis has succeeded or failed. Or we may have isolated a compound from some natural material. Before we subject it to elaborate procedures for structure determination, it is often desirable to know something about the kind of compound we have.

Spectroscopic methods are available that will do all of these things for us, and we shall study these procedures in Chapter 13. Spectrometers are expensive instruments, however. It is helpful, therefore, to have simpler means to identify a particular functional group.

Very often this can be done by a simple chemical test. Such a test will often consist of a single reagent that, when mixed with the compound in question, will indicate the presence of a particular functional group. Not all reactions of a functional group serve as chemical tests, however. To be useful the reaction must proceed with a clear signal; a color change, the evolution of a gas, or the appearance of a precipitate.

7.19A Chemical Tests

A number of reagents that are used as tests for some of the functional groups that we have studied so far are summarized in the following sections. We are restricting our attention at this point to alkanes, alkenes, alkynes, alkyl halides, and alcohols.

7.19B Concentrated Sulfuric Acid (Section 7.4)

Alkenes, alkynes, and alcohols are protonated and, therefore, dissolve when they are added to cold concentrated sulfuric acid. Alkanes and alkyl halides are insoluble in cold concentrated sulfuric acid.

7.19C Bromine in Carbon Tetrachloride (Section 7.6)

Alkenes and alkynes both add bromine at room temperature and in the absence of light. Alkanes, alkyl halides, and alcohols do not react with bromine unless the reaction mixture is heated or exposed to strong irradiation. Thus, rapid decolorization of bromine in carbon tetrachloride at room temperature and in the absence of strong irradiation by light indicates the presence of a carbon–carbon double bond or a carbon–carbon triple bond.

7.19D Cold Dilute Potassium Permanganate (Section 7.10)

Alkenes and alkynes are oxidized by cold dilute solutions of potassium permanganate. If the alkene or alkyne is present in excess, the deep-purple color of the permanganate solution disappears and is replaced by the brown color of precipitated manganese dioxide.

Alkanes, alkyl halides, and pure alcohols do not react with cold dilute potassium permanganate. When these compounds are tested, the purple color is not discharged and a precipitate of manganese dioxide does not appear. (Impure alcohols often contain aldehydes and aldehydes give a positive test with cold dilute potassium permanganate.)

Cold dilute potassium permanganate is often called Baeyer's reagent.

7.19E Alcoholic Silver Nitrate

Alkyl and allylic halides (Section 10.2) react with silver ion to form a precipitate of silver halide. Ethyl alcohol is a convenient solvent because it dissolves silver nitrate and the alkyl halide. It does not dissolve the silver halide.

$$R—X + AgNO_3 \xrightarrow{\text{alcohol}} AgX\downarrow + R^+ \longrightarrow \text{other products}$$

Alkyl
halide

$$R—CH=CHCH_2X + AgNO_3 \xrightarrow{\text{alcohol}} AgX\downarrow + RCH=CH\overset{+}{C}H_2 \longrightarrow \text{other products}$$

An allylic halide

Vinylic halides (Section 5.14A) and phenyl halides (Chapter 11) do not give a silver halide precipitate when treated with silver nitrate in alcohol because vinylic cations and phenyl cations are very unstable and, therefore, do not form readily.

7.19F Silver Nitrate in Ammonia

Silver nitrate reacts with aqueous ammonia to give a solution containing $Ag(NH_3)_2OH$. This reacts with terminal alkynes to form a precipitate of the silver alkynide (Section 6.20).

$$R—C\equiv CH + Ag(NH_3)_2^+ + OH^- \longrightarrow R—C\equiv CAg\downarrow + HOH + 2NH_3$$

Nonterminal alkynes do not give a precipitate. *Silver alkynides can be distinguished from silver halides on the basis of their solubility in nitric acid; silver alkynides dissolve, whereas silver halides do not.*

ADDITIONAL PROBLEMS
7.19

Write structural formulas for the products that form when 1-pentene reacts with each of the following reagents:

(a) HCl

(b) Br_2 in CCl_4, room temperature

(c) H_3O^+, H_2O, heat

(d) Cold concentrated H_2SO_4

(e) Cold concentrated H_2SO_4, then H_2O and heat

(f) HBr

(g) HI

(h) H_2, Pt

(i) Br_2 in CCl_4, then KI in acetone

(j) Dilute $KMnO_4$, OH^-, cold

(k) OsO_4, then $NaHSO_3/H_2O$

(l) $KMnO_4$, OH^-, heat, then H^+

(m) O_3, then Zn, H_2O

(n) Br_2 in H_2O

(o) HBr, peroxides

7.20

Repeat Problem 7.19 using cyclopentene instead of 1-pentene.

7.21

Give the structure of the products that you would expect from the reaction of 1-pentyne with:

(a) One molar equivalent of Br_2

(b) One molar equivalent of HCl

(c) Two molar equivalents of HCl

(d) One molar equivalent of HBr and peroxides

(e) H_2O, H^+, Hg^{2+}

(f) H_2, $Ni_2B(P-2)$

(g) $NaNH_2$ in liq. NH_3

(h) $NaNH_2$ in liq. NH_3, then CH_3I

(i) $Ag(NH_3)_2OH$

(j) $Cu(NH_3)_2OH$

7.22

Give the structure of the products you would expect from the reaction (if any) of 3-hexyne with:

(a) One molar equivalent of HCl

(b) Two molar equivalents of HCl

(c) One molar equivalent of Br_2

(d) Two molar equivalents of Br_2

(e) $Ni_2B(P-2)$, H_2

(f) One molar equivalent of HBr

(g) Li/NH_3

(h) H_2O, H^+, Hg^{2+}

(i) $Ag(NH_3)_2OH$

(j) Two molar equivalents of H_2, Pt

(k) $KMnO_4$, OH^-, then H^+

(l) O_3, H_2O

(m) $NaNH_2$, liq. NH_3

7.23

Show how each of the following compounds might be transformed into 1-pentyne:

(a) 1-Pentene

(b) 1-Chloropentane

(c) 1-Chloro-1-pentene

(d) 1,1-Dichloropentane

(e) 1-Bromopropane and acetylene

7.24

Starting with 2-methylpropene (isobutylene) and using any other needed reagents, outline a synthesis of each of the following:

(a) $(CH_3)_3COH$

(b) $(CH_3)_3CCl$

(c) $(CH_3)_3CBr$

(d) $(CH_3)_2CHCH_2Br$

(e) $(CH_3)_2CHCH_2I$

(f) $(CH_3)_2CHCH_2CN$

(g) $(CH_3)_3CF$

(h) $(CH_3)_2C(OH)CH_2Cl$

7.25

Myrcene, a fragrant compound found in bayberry wax, has the formula $C_{10}H_{16}$ and is known not to contain any triple bonds. (a) What is the index of hydrogen deficiency of myrcene? When treated with excess hydrogen and a platinum catalyst, myrcene is converted to a compound **(A)** with the formula $C_{10}H_{22}$. (b) How many rings does myrcene contain? (c) How many double bonds? Compound **A** can be identified as 2,6-dimethyloctane. Ozonolysis of myrcene followed by treatment with zinc and water yields 2 moles of formaldehyde (HCHO), 1 mole of acetone (CH_3COCH_3), and a third compound **(B)** with the formula $C_5H_6O_3$. (d) What is the structure of myrcene? (e) Of compound **B**?

7.26

When either *cis*- or *trans*-2-butene is treated with hydrogen chloride in ethyl alcohol, one of the products of the reaction is *sec*-butyl ethyl ether. Write a mechanism that accounts for the formation of this product.

7.27

When alkenes add HX, the relative rates of reaction are $R_2C=CH_2 > RCH=CH_2 > CH_2=CH_2$. What factor accounts for this?

7.28

Write the products and show how many molar equivalents of each would be formed when squalene is subjected to ozonolysis and the ozonide is subsequently treated with zinc and water.

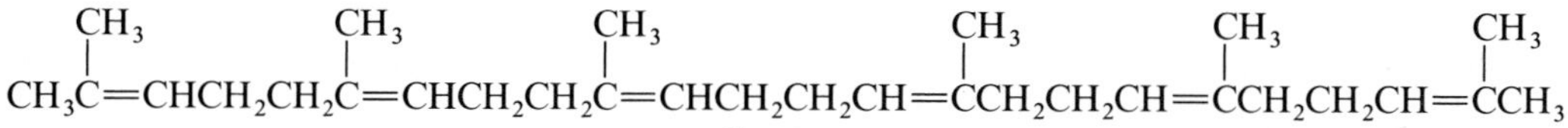

$$\underset{\underset{Squalene}{}}{CH_3C=CHCH_2CH_2\overset{CH_3}{\overset{|}{C}}=CHCH_2CH_2\overset{CH_3}{\overset{|}{C}}=CHCH_2CH_2CH=\overset{CH_3}{\overset{|}{C}}CH_2CH_2CH=\overset{CH_3}{\overset{|}{C}}CH_2CH_2CH=\overset{CH_3}{\overset{|}{C}}CH_3}$$

7.29

Arrange the following alkenes in order of their reactivity toward acid-catalyzed hydration and explain your reasoning:

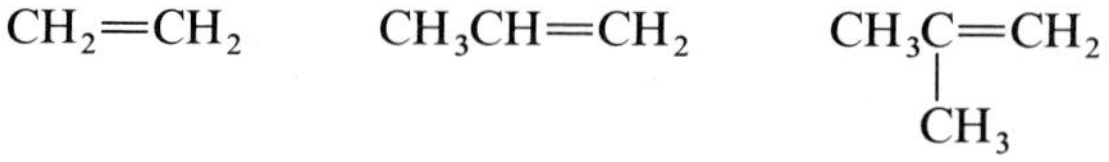

$$CH_2=CH_2 \qquad CH_3CH=CH_2 \qquad CH_3\overset{}{C}=CH_2$$
$$\qquad\qquad\qquad\qquad\qquad\qquad\qquad\qquad \overset{|}{CH_3}$$

7.30

(a) When treated with strong acid at 25 °C, either *cis*-2-butene or *trans*-2-butene is converted to a mixture of *trans*-2-butene, *cis*-2-butene, and 1-butene. *trans*-2-Butene predominates in the mixture. (The mixture contains 74% *trans*-2-butene, 23% *cis*-2-butene, and 3% 1-butene.) Write mechanisms for the reactions that occur, and account for the relative amounts of the alkene isomers that are formed. (b) When treated with strong acid, 1-butene is converted to the same mixture of alkenes referred to in part (a). How can you explain this? (c) Can you also explain why 2-methylpropene is *not* formed in either of the reactions referred to in parts (a) or (b) even through 2-methylpropene is more stable than 1-butene?

7.31

Write a mechanism that explains the course of the following reaction:

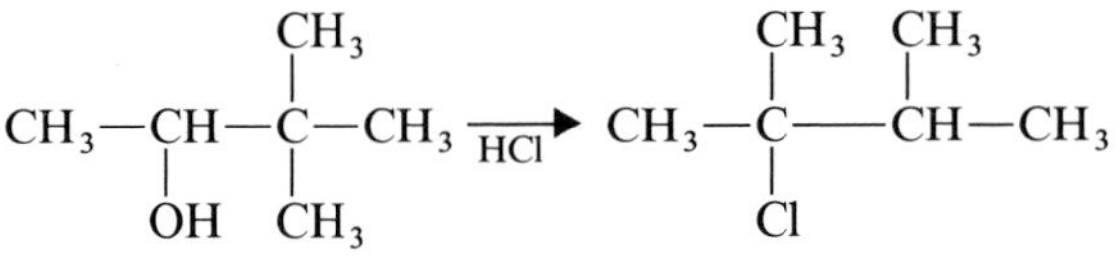

$$CH_3-\underset{\underset{OH}{|}}{\overset{\overset{CH_3}{|}}{CH}}-\underset{\underset{CH_3}{|}}{\overset{}{C}}-CH_3 \xrightarrow{HCl} CH_3-\underset{\underset{Cl}{|}}{\overset{\overset{CH_3}{|}}{C}}\overset{\overset{CH_3}{|}}{CH}-CH_3$$

7.32

A cycloalkene reacts with hydrogen and a catalyst to yield methylcyclohexane. On vigorous oxidation with potassium permanganate the cycloalkene yields only

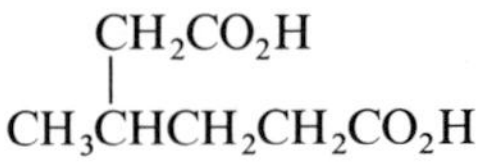

$$CH_3\overset{}{C}HCH_2CH_2CO_2H$$
$$\overset{\overset{CH_2CO_2H}{|}}{}$$

What is the structure of the cycloalkene?

7.33

Outline all steps in a laboratory synthesis of each of the following compounds. You should begin with the organic compound indicated, and you may use any needed solvents or inorganic compounds. These syntheses may require more than one step and should be designed to give reasonably good yields of reasonably pure products.

(a) Propene from propane

(d) 2-Methylpropene from 2-methylpropane

(b) 2-Bromopropane from propane

(e) *tert*-Butyl alcohol from 2-methylpropane

(c) 1-Bromopropane from propane

(f) 1,2-Dichlorobutane from 1-chlorobutane

(g) 2-Bromoethanol from ethyl bromide

(h) + enantiomer from cyclopentane

trans

(i) 2-Bromobutane from 1-bromobutane

7.34

Vicinal halo alcohols (halohydrins) can be synthesized by treating epoxides with HX. (a) Show how you would use this method to synthesize 2-chlorocyclopentanol from cyclopentene. (b) Would you expect the product to be *cis*-2-chlorocyclopentanol or *trans*-2-chlorocyclopentanol, that is, would you expect a net syn addition or a net anti addition of —Cl and —OH? Explain.

7.35

Pheromones are substances secreted by animals (especially insects) that produce a specific behavioral reaction in other members of the same species. Pheromones are effective at very low concentrations and include sex attractants, warning substances, and "aggregation" compounds. After many years of research, the sex attractant of the gypsy moth has been identified and synthesized in the laboratory. This sex pheromone is unusual in that it appears to be equally attractive to male and female gypsy moths. (It promises to be useful in their control even though this may seem somewhat unfair.) The final step in the synthesis of the pheromone involves treatment of *cis*-2-methyl-7-octadecene with a peroxy acid. What is the structure of the gypsy moth sex pheromone?

7.36

The green peach aphid is repelled by its own defensive pheromone. (It is also repelled by other squashed aphids.) This alarm pheromone has been isolated and has been shown to have the molecular formula $C_{15}H_{24}$. On catalytic hydrogenation it absorbs 4 moles of hydrogen and yields 2,6,10-trimethyldodecane, that is,

$$\underset{CH_3}{\underset{|}{CH_3CHCH_2CH_2CH_2}}\underset{CH_3}{\underset{|}{CHCH_2CH_2CH_2}}\underset{CH_3}{\underset{|}{CHCH_2CH_3}}$$

When subjected to ozonolysis followed by treatment with zinc and water, 1 mole of the alarm pheromone produces: 2 moles of formaldehyde, $H{-}\overset{O}{\overset{\|}{C}}{-}H$; 1 mole of acetone, $CH_3\overset{O}{\overset{\|}{C}}CH_3$; 1 mole of $CH_3\overset{O}{\overset{\|}{C}}CH_2CH_2\overset{O}{\overset{\|}{C}}H$; and 1 mole of $H\overset{O}{\overset{\|}{C}}CH_2CH_2\overset{O}{\overset{\|}{C}}{-}\overset{O}{\overset{\|}{C}}H$.

Neglecting cis–trans isomerism, propose a structure for the green peach aphid alarm pheromone.

7.37

(a) What product would you expect to form when isobutyl bromide is heated with $(CH_3)_3COK/(CH_3)_3COH$? (b) Can you suggest a method for the conversion of isobutyl bromide into *tert*-butyl bromide?

7.38

When cyclopentene is allowed to react with bromine in an aqueous solution of sodium chloride, the products of the reaction are *trans*-1,2-dibromocyclopentane, the trans-bromohydrin of cyclopentene, *and trans-1-bromo-2-chlorocyclopentane.* Write a mechanism that explains the formation of this last product.

7.39

The structures of the diisobutylene dimers (Section 7.11) were determined by F. C. Whitmore and his students on the basis of the products formed when each dimer was subjected to ozonolysis. Show how this might have been done.

7.40

In an industrial process, propene is heated with phosphoric acid at 205 °C under a pressure of 1000 atm. The major products of the reaction are two isomers with the molecular formula $C_{12}H_{24}$. Propose structures for the isomers and write a mechanism that explains their formation. (*Note:* At one time these isomers were used extensively in the synthesis of a nonbiodegradable detergent.)

7.41

Write stereochemical formulas for all of the products that you would expect from each of the following reactions. (You may find models helpful.)

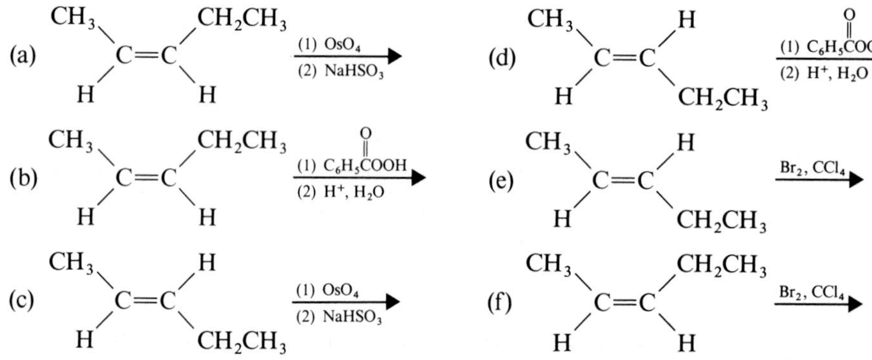

(a)
$$\underset{H}{\overset{CH_3}{>}}C=C\underset{H}{\overset{CH_2CH_3}{<}} \xrightarrow[\text{(2) NaHSO}_3]{\text{(1) OsO}_4}$$

(b)
$$\underset{H}{\overset{CH_3}{>}}C=C\underset{H}{\overset{CH_2CH_3}{<}} \xrightarrow[\text{(2) H}^+,\ \text{H}_2\text{O}]{\text{(1) C}_6\text{H}_5\text{COOH}}$$

(c)
$$\underset{H}{\overset{CH_3}{>}}C=C\underset{CH_2CH_3}{\overset{H}{<}} \xrightarrow[\text{(2) NaHSO}_3]{\text{(1) OsO}_4}$$

(d)
$$\underset{H}{\overset{CH_3}{>}}C=C\underset{CH_2CH_3}{\overset{H}{<}} \xrightarrow[\text{(2) H}^+,\ \text{H}_2\text{O}]{\text{(1) C}_6\text{H}_5\text{COOH}}$$

(e)
$$\underset{H}{\overset{CH_3}{>}}C=C\underset{CH_2CH_3}{\overset{H}{<}} \xrightarrow{\text{Br}_2,\ \text{CCl}_4}$$

(f)
$$\underset{H}{\overset{CH_3}{>}}C=C\underset{H}{\overset{CH_2CH_3}{<}} \xrightarrow{\text{Br}_2,\ \text{CCl}_4}$$

7.42

Give (R)–(S) designations for each different compound given as an answer to Problem 7.41.

7.43

Describe with equations a simple test that would distinguish between each of the following pairs of compounds. (In each case tell what you would see.)
(a) Propane and propyne
(b) Propene and propyne
(c) 1-Bromopropene and 2-bromopropane
(d) 2-Bromo-2-butene and 1-butyne
(e) 1-Butyne and 2-butyne
(f) 2-Butyne and butyl alcohol
(g) 2-Butyne and 2-bromobutane
(h) $CH_3C\equiv CCH_2OH$ and $CH_3CH_2CH_2CH_2OH$
(i) $CH_3CH=CHCH_2OH$ and $CH_3CH_2CH_2CH_2OH$

7.44

Three compounds **A**, **B**, and **C** all have the formula C_5H_8. All three compounds rapidly decolorize bromine in carbon tetrachloride, all three give a positive test with dilute $KMnO_4$, and all three are soluble in cold concentrated sulfuric acid. Compound **A** gives a precipitate when treated with ammoniacal silver nitrate, but compounds **B** and **C** do not. Compounds **A** and **B** both yield pentane (C_5H_{12}) when they are treated with excess hydrogen in the presence of a platinum catalyst. Under these same conditions, compound **C** absorbs only 1 mole of hydrogen and gives a product with the formula C_5H_{10}. (a) Suggest possible structures for **A**, **B**, and **C**. (b) Are other structures possible for **B** and **C**? (c) Oxidative cleavage of **B** with hot, basic $KMnO_4$ gives, after acidification, acetic acid and $CH_3CH_2CO_2H$. What is the structure of **B**? (d) Oxidative cleavage of **C** with ozone gives $HO_2CCH_2CH_2CH_2CO_2H$. What is the structure of **C**?

7.45
Starting with 3-methyl-1-butyne and any inorganic reagents, show how the following compounds could be synthesized:

$$
\begin{array}{lll}
& \overset{\displaystyle CH_3}{\underset{\displaystyle |}{}} & \overset{\displaystyle CH_3}{\underset{\displaystyle |}{}} & \overset{\displaystyle CH_3}{\underset{\displaystyle |}{}} \\
\end{array}
$$

(a) $CH_3\overset{|}{\underset{\underset{\textstyle Cl}{|}}{C}H}C=CH_2$ (c) $CH_3\overset{|}{\underset{\underset{\textstyle Cl}{|}}{C}H}CHCH_3$ (e) $CH_3\overset{|}{C}HCClBrCH_3$

(f) $CH_3\overset{|}{\underset{\underset{\textstyle}{}}{C}}HCO_2H$

(b) $CH_3\overset{|}{\underset{\underset{\textstyle}{CH_3}}{C}}HCH_2CH_2Br$ (d) $CH_3\overset{|}{\underset{\underset{\textstyle}{CH_3}}{C}}HCCl_2CH_2Cl$

7.46
Ricinoleic acid, a compound that can be isolated from castor oil, has the structure $CH_3(CH_2)_5CHOHCH_2CH=CH(CH_2)_7CO_2H$. (a) How many stereoisomers of this structure are possible? (b) Write these structures.

7.47
There are two dicarboxylic acids with the general formula $HO_2CCH=CHCO_2H$. One dicarboxylic acid is called maleic acid; the other is called fumaric acid. In 1880, Kekulé found that on treatment with cold dilute $KMnO_4$, maleic acid yields *meso*-tartaric acid and that fumaric acid yields ($\pm$)-tartaric acid. Show how this information allows one to write stereochemical formulas for maleic acid and fumaric acid.

7.48
Use your answers to the preceding problem to predict the stereochemical outcome of the addition of bromine to maleic acid and to fumaric acid. (a) Which dicarboxylic acid would add bromine to yield a meso compound? (b) Which would yield a racemic form?

7.49
An optically active compound **A** (assume that it is dextrorotatory) has the molecular formula $C_7H_{11}Br$. **A** reacts with hydrogen bromide, in the absence of peroxides, to yield isomeric products, **B** and **C**, with molecular formula $C_7H_{12}Br_2$. **B** is optically active; **C** is not. Treating **B** with 1 mole of potassium *tert*-butoxide yields (+)**A**. Treating **C** with 1 mole of potassium *tert*-butoxide yields ($\pm$)**A**. Treating **A** with potassium *tert*-butoxide yields **D** (C_7H_{10}). Subjecting 1 mole of **D** to ozonolysis followed by treatment with zinc and water yields 2 moles of formaldehyde and 1 mole of 1,3-cyclopentanedione.

1,3-Cyclopentanedione

Propose stereochemical formulas for **A**, **B**, **C**, and **D** and outline the reactions involved in these transformations.

7.50

A naturally occurring antibiotic called mycomycin has the structure shown here. Mycomycin is optically active. Explain this by writing structures for the enantiomeric forms of mycomycin.

$$HC\equiv C-C\equiv C-CH=C=CH-(CH=CH)_2CH_2CO_2H$$

Mycomycin

7.51

An optically active compound **D** has the molecular formula C_6H_{10}. The compound gives a precipitate when treated with a solution containing $Ag(NH_3)_2OH$. On catalytic hydrogenation **D** yields $E(C_6H_{14})$. **E** is optically inactive and cannot be resolved. Propose structures for **D** and **E**.

DIVALENT CARBON COMPOUNDS. CARBENES

In recent years considerable research has been devoted to investigating the structures and reactions of a group of compounds in which carbon forms only *two bonds.* These neutral divalent carbon compounds are called *carbenes.*

Most carbenes are highly unstable compounds that are capable of only fleeting existence. Soon after carbenes are formed they usually react with another molecule. The reactions of carbenes are especially interesting because, in many instances, the reactions show a remarkable degree of stereospecificity. The reactions of carbenes are also of great synthetic use in the preparation of compounds that have three-membered rings.

C.1 Structure of Methylene

The simplest carbene is the compound called methylene (CH_2). Methylene can be prepared by the decomposition of diazomethane* (CH_2N_2). This decomposition can be accomplished by heating diazomethane (thermolysis) or by irradiating it with light of a wavelength that it can absorb (photolysis).

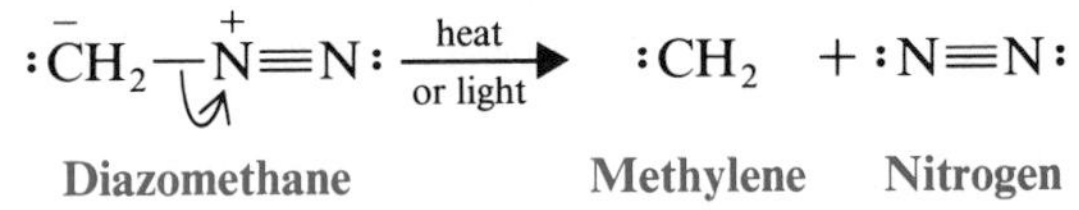

$$:\overset{-}{C}H_2\!-\!\overset{+}{N}\!\equiv\!N: \xrightarrow[\text{or light}]{\text{heat}} \quad :CH_2 \;\; + :N\!\equiv\!N:$$

Diazomethane   Methylene   Nitrogen

C.2 Reactions of Methylene

Methylene reacts with alkenes by adding to the double bond to form cyclopropanes.

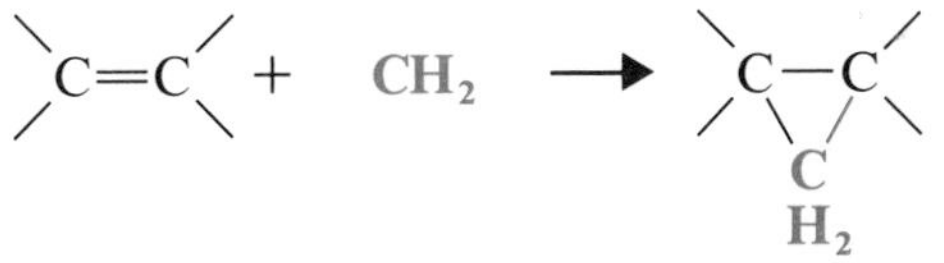

Alkene   Methylene   Cyclopropane

C.3 Reactions of Other Carbenes: Dihalocarbenes

Dihalocarbenes are also frequently employed in the synthesis of cyclopropane derivatives from alkenes. Most reactions of dihalocarbenes are stereospecific.

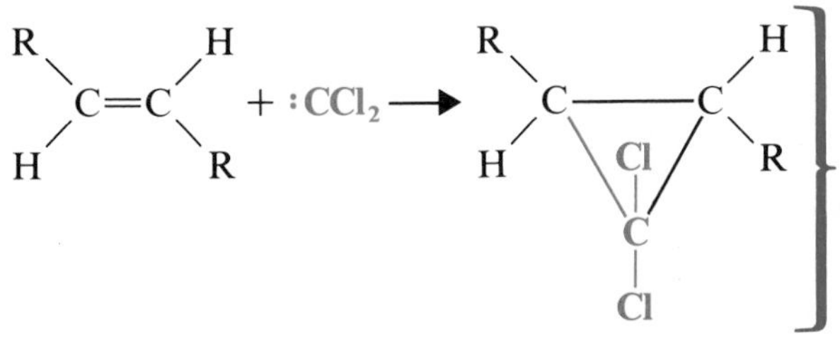

The addition of $:CX_2$ is stereospecific. If the R groups of the alkene are trans, they will be trans in the product

*Diazomethane is a resonance hybrid of the three structures, **I**, **II**, and **III**, shown below.

$$:\overset{-}{C}H_2\!-\!\overset{+}{N}\!\equiv\!N: \;\longleftrightarrow\; CH_2\!=\!\overset{+}{N}\!=\!\overset{..}{\underset{..}{N}}: \;\longleftrightarrow\; \overset{..}{\overset{-}{C}}H_2\!-\!\overset{+}{N}\!=\!\overset{..}{N}:$$

I   II   III

We have chosen resonance structure **I** to illustrate the decomposition of diazomethane because with **I** it is readily apparent that heterolytic cleavage of the carbon–nitrogen bond results in the formation of methylene and molecular nitrogen.

Dichlorocarbene can be synthesized by the **α *elimination*** of the elements of hydrogen chloride from chloroform. This reaction resembles the β elimination reactions by which alkenes are synthesized from alkyl halides (Section 5.15).

$$R\overset{..}{\underset{..}{O}}:^-K^+ + H:CCl_3 \rightleftharpoons R\overset{..}{\underset{..}{O}}:H + \,^-:CCl_3 + K^+ \quad \,^-:CCl_3$$

$$\xrightarrow[slow]{} \quad :CCl_2 \quad + :\overset{..}{\underset{..}{Cl}}:^-$$
Dichlorocarbene

Compounds *with a β hydrogen* react by β elimination preferentially. Compounds with no β hydrogen (such as chloroform) react by α elimination.

A variety of cyclopropane derivatives have been prepared by generating dichlorocarbene in the presence of alkenes. Cyclohexene, for example, reacts with dichlorocarbene generated by treating chloroform with potassium *tert*-butoxide to give a bicyclic product.

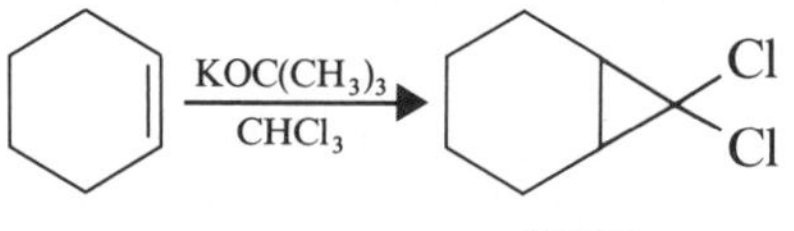

(59%)

C.4 Carbenoids: The Simmons–Smith Cyclopropane Synthesis

A useful cyclopropane synthesis has been developed by H. E. Simmons and R. D. Smith of the du Pont Company. In this synthesis diiodomethane and a zinc–copper couple are stirred together with an alkene. The diiodomethane and zinc react to produce a carbenelike species called a *carbenoid*.

$$CH_2I_2 + Zn(Cu) \longrightarrow ICH_2ZnI$$
A carbenoid

The carbenoid then brings about the stereospecific addition of a CH_2 group directly to the double bond.

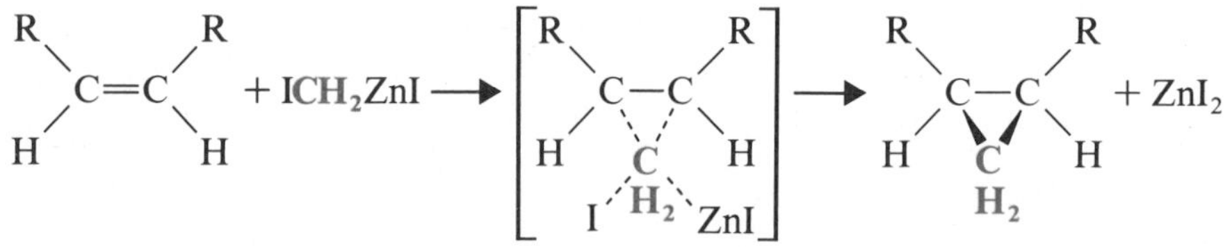

This synthesis has been used widely. One example is the synthesis of methyl dihydrosterculate from methyl oleate.

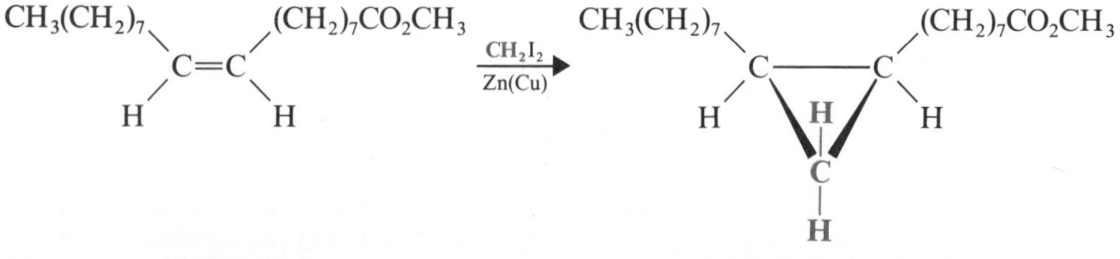

Methyl oleate   **Methyl dihydrosterculate**

Methyl dihydrosterculate is related to sterculic acid, an interesting compound that has been isolated from the kernel oil of the tropical tree *Sterculia foetida*. Sterculic acid was the first naturally occurring compound found to have the highly strained cyclopropene ring.

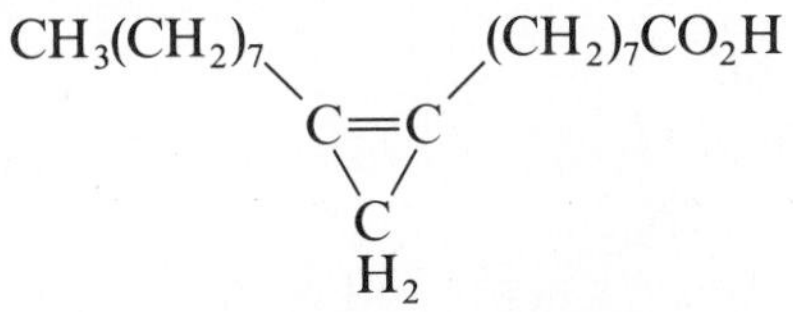

Sterculic acid

Sterculic acid, itself, has been synthesized using the Simmons–Smith method.

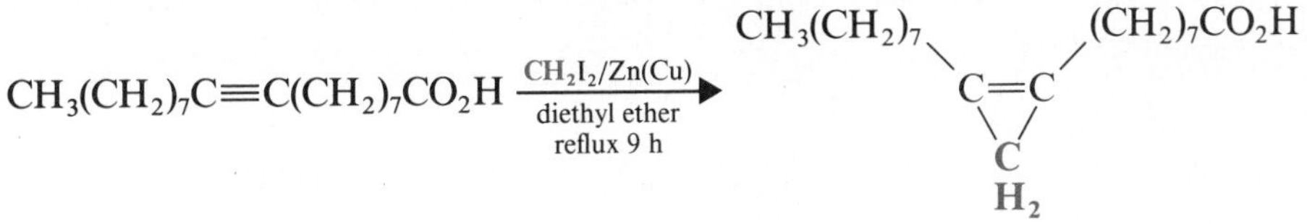

Stearolic acid   (4%)

This reaction illustrates the addition of a carbenoid to a carbon–carbon triple bond.

The zinc–copper couple used in the Simmons–Smith synthesis can also be prepared *in situ* (in the reaction mixture), as the following example illustrates.

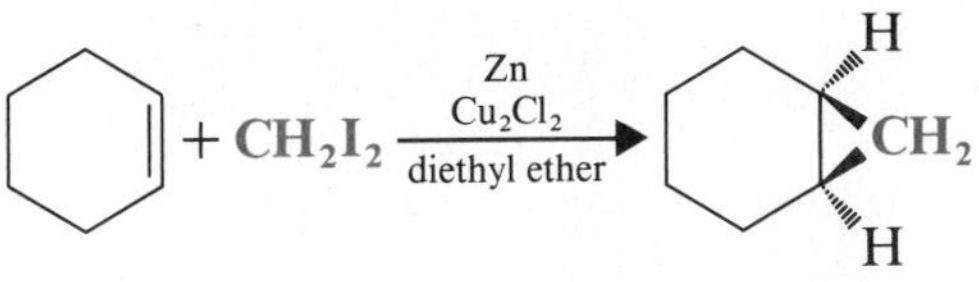

(92%)

PROBLEM C.1

How might the following compounds be synthesized?

(a)

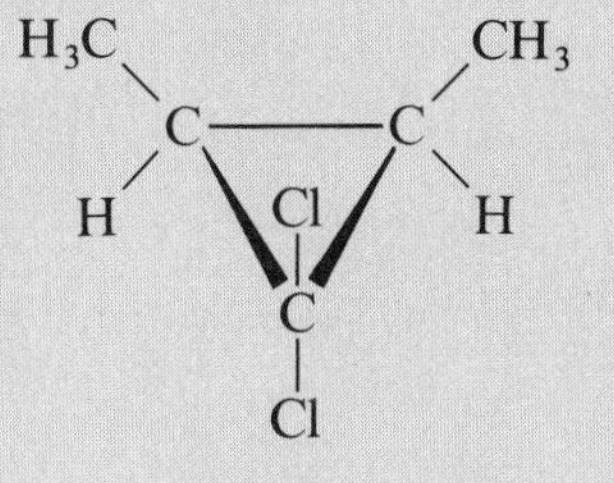

(c)

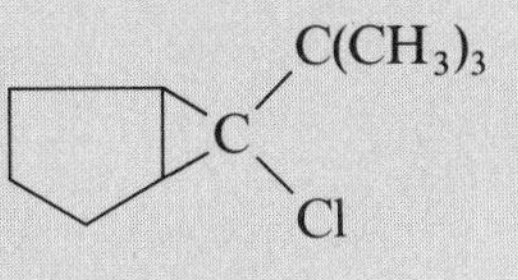

(b)

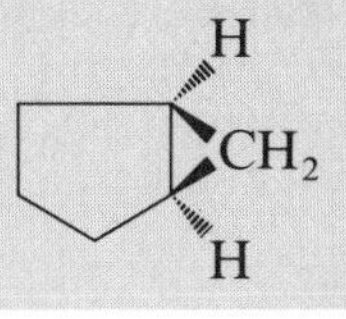

(d)

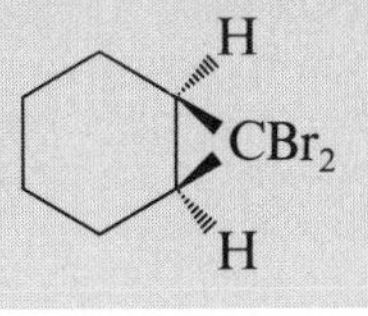

CHAPTER EIGHT

Ion exchange spheres. Ion exchange resins, which consist of tiny spheres, are highly useful in effecting separations of organic and inorganic compounds (Section 23.5) and in organic synthesis (Section 23.8D).

Alcohols and Ethers

8.1 STRUCTURE AND NOMENCLATURE

Alcohols are compounds whose molecules have a hydroxyl group attached to a *saturated* carbon atom.* The saturated carbon atom may be that of a simple alkyl group:

$$CH_3OH \qquad CH_3CH_2OH \qquad CH_3\underset{\underset{\displaystyle OH}{|}}{C}HCH_3 \qquad CH_3\underset{\underset{\displaystyle OH}{|}}{\overset{\overset{\displaystyle CH_3}{|}}{C}}CH_3$$

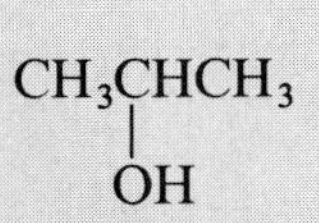

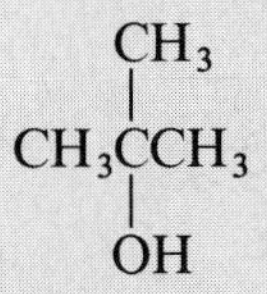

Methanol	Ethanol	2-Propanol	2-Methyl-2-propanol
(methyl alcohol)	(ethyl alcohol)	(isopropyl alcohol)	(*tert*-butyl alcohol)
	a 1° alcohol	*a 2° alcohol*	*a 3° alcohol*

The carbon atom may be a saturated carbon atom of an alkenyl or alkynyl group, or the carbon atom may be a saturated carbon atom that is attached to a benzene ring.

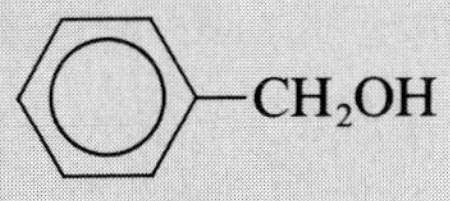

$$\text{C}_6\text{H}_5-CH_2OH \qquad CH_2{=}CHCH_2OH \qquad H-C{\equiv}CCH_2OH$$

Benzyl alcohol	2-Propenol	2-Propynol
a benzylic alcohol	(allyl alcohol)	(propargyl alcohol)
	an allylic alcohol	

Compounds that have a hydroxyl group attached directly to a benzene ring are called *phenols*. (Phenols will be discussed in detail in Chapter 14.)

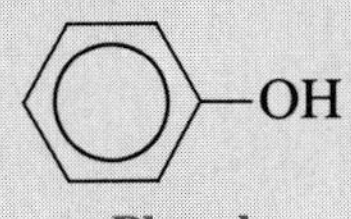

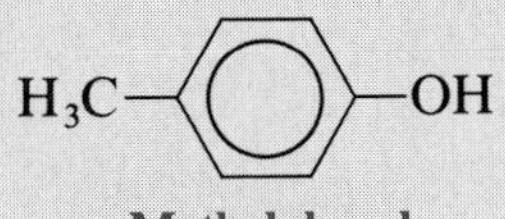

$$\text{C}_6\text{H}_5-OH \qquad H_3C-\text{C}_6\text{H}_4-OH \qquad Ar-OH$$

Phenol	p-Methylphenol	*General formula*
	a substituted phenol	*for a phenol*

---

* Compounds in which a hydroxyl group is attached to an unsaturated carbon atom of a double bond (i.e., C=C) are called enols, cf. Section 7.15.

349

Ethers differ from alcohols in that the oxygen atom of an ether is bonded to two carbon atoms. The hydrocarbon groups may be alkyl, alkenyl, vinyl, alkynyl, or aryl. Several examples are shown here.

CH_3CH_2—O—CH_2CH_3 CH_2=$CHCH_2$—O—CH_3
Diethyl ether   **Allyl methyl ether**

CH_2=CH—O—CH=CH_2 —OCH_3

Divinyl ether   **Anisole**

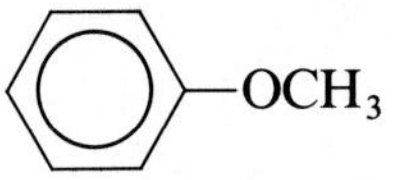

8.1A Nomenclature of Alcohols

We studied the IUPAC system of nomenclature for alcohols in Section 3.3E. As a review consider the following example.

SAMPLE PROBLEM

Give IUPAC substitutive names for the following alcohols:

(a) $CH_3CHCH_2CHCH_2OH$
        $|$ $|$
       CH_3 CH_3

(c) CH_3CHCH_2CH=CH_2
        $|$
       OH

(b) $CH_3CHCH_2CHCH_3$
        $|$ $|$
       OH C_6H_5

Answer:

The longest chain *to which the hydroxyl group is attached* gives us the *base name.* The ending is -ol. We then number *the longest chain from the end that gives the carbon bearing the hydroxyl group the lower number.* Thus, the names are

(a) $\overset{5}{C}H_3\overset{4}{C}HCH_2\overset{2}{C}H\overset{1}{C}H_2OH$
          $|$ $|$
         CH_3 CH_3
     2,4-Dimethyl-1-pentanol

(c) $\overset{1}{C}H_3\overset{2}{C}HCH_2\overset{4}{C}H$=$\overset{5}{C}H_2$
          $|$
         OH
     4-Penten-2-ol

(b) $\overset{1}{C}H_3\overset{2}{C}HCH_2\overset{4}{C}HCH_3$
          $|$ $|$
         OH C_6H_5
     4-Phenyl-2-pentanol

In common functional class nomenclature (Section 2.10) alcohols are called **alkyl alcohols** such as methyl alcohol, ethyl alcohol, and so on.

8.1B Nomenclature of Ethers

Ethers are frequently given common names and to do so is quite easy. One simply names in alphabetical order both groups that are attached to the oxygen atom.

$$CH_3OCH_2CH_3 \qquad CH_3CH_2OCH_2CH_3 \qquad C_6H_5O\overset{\displaystyle CH_3}{\underset{\displaystyle CH_3}{C}}-CH_3$$

Ethyl methyl ether   **Diethyl ether**   *tert*-**Butyl phenyl ether**

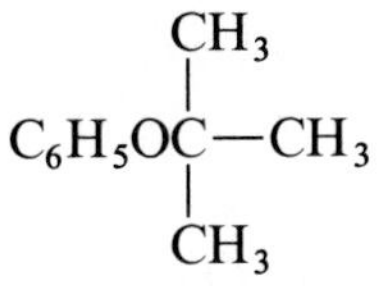

IUPAC names are seldom used for simple ethers. IUPAC names are used for complicated ethers, however, and for compounds with more than one ether linkage. In the IUPAC system ethers are named as alkoxyalkanes, alkoxyalkenes, and alkoxyarenes. The RO— group is an **alkoxy** group.

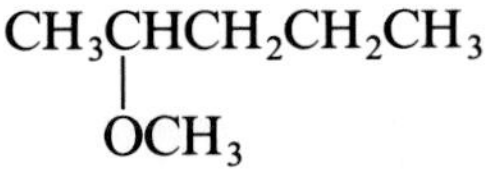

$$CH_3\overset{\displaystyle}{\underset{\displaystyle OCH_3}{CH}}CH_2CH_2CH_3 \qquad CH_3CH_2O-\!\!\left\langle\!\!\bigcirc\!\!\right\rangle\!\!-CH_3 \qquad CH_3OCH_2CH_2OCH_3$$

2-Methoxypentane 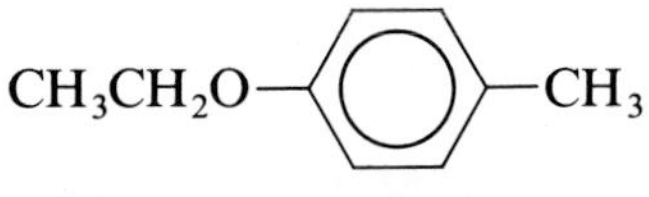   **1-Ethoxy-4-methylbenzene**   **1,2-Dimethoxyethane**

Two cyclic ethers that are frequently used as solvents have the common names tetrahydrofuran (THF) and 1,4-dioxane.

Tetrahydrofuran   **1,4-Dioxane**
(oxacyclopentane)   **(1,4-dioxacyclohexane)**

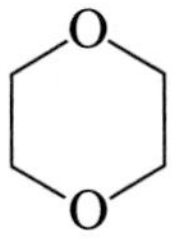

PROBLEM 8.1

Give appropriate names for all of the alcohols and ethers with the formulas
(a) C_3H_6O, (b) C_4H_8O, and (c) $C_5H_{12}O$.

8.2 PHYSICAL PROPERTIES OF ALCOHOLS AND ETHERS

The physical properties of a number of alcohols and ethers are given in Tables 8.1 and 8.2.

Ethers have boiling points that are roughly comparable with those of hydrocarbons of the same molecular weight. For example, the boiling point of diethyl ether (MW = 74) is 34.6 °C; that of pentane (MW = 72) is 36 °C. Alcohols, on the other hand, have much higher boiling points than comparable ethers or hydrocarbons. The boiling point of butyl alcohol (MW = 74) is 117.7 °C. We learned the reason for this

TABLE 8.1 Physical properties of ethers

NAME	FORMULA	mp (°C)	bp (°C)	DENSITY d_4^{20} (g/mL)
Dimethyl ether	CH_3OCH_3	−138	−24.9	0.661
Ethyl methyl ether	$CH_3OCH_2CH_3$		10.8	0.697
Diethyl ether	$CH_3CH_2OCH_2CH_3$	−116	34.6	0.714
Dipropyl ether	$(CH_3CH_2CH_2)_2O$	−122	90.5	0.736
Diisopropyl ether	$(CH_3)_2CHOCH(CH_3)_2$	−86	68	0.725
Dibutyl ether	$(CH_3CH_2CH_2CH_2)_2O$	−97.9	141	0.769
1,2-Dimethoxyethane	$CH_3OCH_2CH_2OCH_3$	−68	83	0.863
Tetrahydrofuran	$(CH_2)_4O$	−108	65.4	0.888
1,4-Dioxane	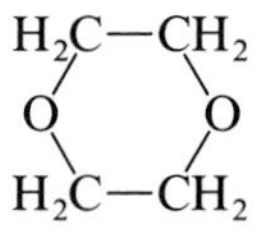 H₂C—CH₂ O O H₂C—CH₂	11	101	1.033
Anisole (methoxybenzene)	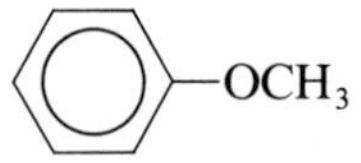 ⬡—OCH₃	−37.3	158.3	0.994

behavior in Section 2.17C; the molecules of alcohols can associated with each other through **hydrogen bonding** while those of ethers and hydrocarbons cannot.

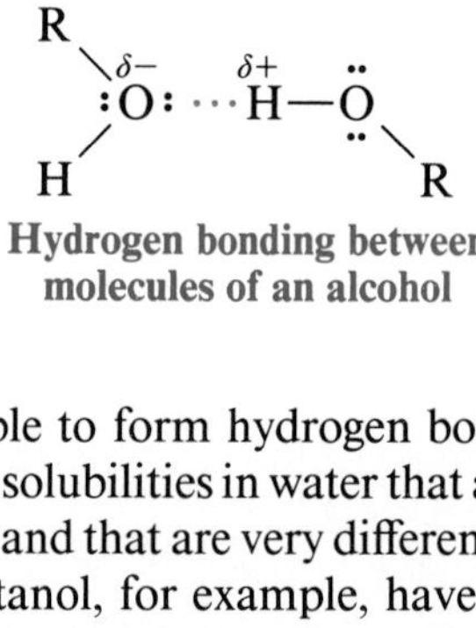

Hydrogen bonding between
molecules of an alcohol

Ethers, however, *are* able to form hydrogen bonds with compounds such as water. Ethers, therefore, have solubilities in water that are similar to those of alcohols of the same molecular weight and that are very different from those of hydrocarbons.

Diethyl ether and 1-butanol, for example, have the same solubility in water, approximately 8.0 g/100 mL at room temperature. Pentane, by contrast, is virtually insoluble in water.

Methyl alcohol, ethyl alcohol, both propyl alcohols, and *tert*-butyl alcohol are completely miscible with water (Table 8.2). The remaining butyl alcohols have solubilities in water between 7.9 and 12.5 g/100 mL. The solubility of alcohols in water gradually decreases as the hydrocarbon portion of the molecule lengthens; long-chain alcohols are more "alkanelike" and are, therefore, less like water.

One of the reasons we are interested in solubilities in water is that we need to learn the structural features that help or hinder a substance's solubility in this important solvent. This solvent is the medium, for example, of the cells of all living organisms.

TABLE 8.2 Physical properties of alcohols

COMPOUND	NAME	mp (°C)	bp (°C) (1 atm)	DENSITY d_4^{20} (g/mL)	WATER SOLUBILITY (g/100 mL H_2O)
Monohydroxy Alcohols					
CH_3OH	Methyl alcohol	−97	64.7	0.792	∞
CH_3CH_2OH	Ethyl alcohol	−117	78.3	0.789	∞
$CH_3CH_2CH_2OH$	Propyl alcohol	−126	97.2	0.804	∞
$CH_3CH(OH)CH_3$	Isopropyl alcohol	−88	82.3	0.786	∞
$CH_3CH_2CH_2CH_2OH$	Butyl alcohol	−90	117.7	0.810	7.9
$CH_3CH(CH_3)CH_2OH$	Isobutyl alcohol	−108	108.0	0.802	10.0
$CH_3CH_2CH(OH)CH_3$	*sec*-Butyl alcohol	−114	99.5	0.808	12.5
$(CH_3)_3COH$	*tert*-Butyl alcohol	25	82.5	0.789	∞
$CH_3(CH_2)_3CH_2OH$	Pentyl alcohol	−78.5	138.0	0.817	2.4
$CH_3(CH_2)_4CH_2OH$	Hexyl alcohol	−52	156.5	0.819	0.6
$CH_3(CH_2)_5CH_2OH$	Heptyl alcohol	−34	176	0.822	0.2
$CH_3(CH_2)_6CH_2OH$	Octyl alcohol	−15	195	0.825	0.05
$CH_3(CH_2)_7CH_2OH$	Nonyl alcohol	−5.5	212	0.827	—
$CH_3(CH_2)_8CH_2OH$	Decyl alcohol	6	228	0.829	—
$CH_2{=}CHCH_2OH$	Allyl alcohol	−129	97	0.855	∞
$(CH_2)_4CHOH$	Cyclopentanol	−19	140	0.949	—
$(CH_2)_5CHOH$	Cyclohexanol	24	161.5	0.962	3.6
$C_6H_5CH_2OH$	Benzyl alcohol	−15	205	1.046	4
Diols and Triols					
CH_2OHCH_2OH	Ethylene glycol	−12.6	197	1.113	∞
$CH_3CHOHCH_2OH$	Propylene glycol	−59	187	1.040	∞
$CH_2OHCH_2CH_2OH$	Trimethylene glycol	−30	215	1.060	∞
$CH_2OHCHOHCH_2OH$	Glycerol	18	290	1.261	∞

PROBLEM 8.2

How can you account for the fact that the boiling point of ethylene glycol is much higher than that of either propyl alcohol or isopropyl alcohol even though all three compounds have roughly the same molecular weight?

8.3 IMPORTANT ALCOHOLS AND ETHERS

8.3A Methanol

At one time, most methanol was produced by the destructive distillation of wood (i.e., heating wood to a high temperature in the absence of air). It was because of this method of preparation that methanol came to be called "wood alcohol." Today, most methanol is prepared by the catalytic hydrogenation of carbon monoxide. This reaction takes place under high pressure and at a temperature of 300 to 400 °C.

$$CO + 2H_2 \xrightarrow[\substack{200-300 \text{ atm} \\ ZnO-Cr_2O_3}]{300-400 \text{ °C}} CH_3OH$$

Methanol is highly toxic. Ingestion of even small quantities of methanol can cause blindness; large quantities cause death. Methanol poisoning can also occur by inhalation of the vapors or by prolonged exposure of the skin.

8.3B Ethanol

Ethanol can be made by the fermentation of sugars, and it is the alcohol of all alcoholic beverages. The synthesis of ethanol in the form of wine by the fermentation of the sugars of fruit juices was probably man's first accomplishment in the field of organic synthesis. Sugars from a wide variety of sources can be used in the preparation of alcoholic beverages. Often, these sugars are from grains, and it is this derivation that accounts for ethanol having the synonym "grain alcohol."

Fermentation is usually carried out by adding yeast to a mixture of sugars and water. Yeast contains enzymes that promote a long series of reactions that ultimately convert a simple sugar ($C_6H_{12}O_6$) to ethanol and carbon dioxide.

$$C_6H_{12}O_6 \xrightarrow{\text{yeast}} 2CH_3CH_2OH + 2CO_2$$
$$(\sim 95\% \text{ yield})$$

Fermentation alone does not produce beverages with an ethanol content greater than 12–15% because the yeast is destroyed at higher concentrations. To produce beverages of higher alcohol content the aqueous solution must be distilled. Brandy, whiskey, and vodka are produced in this way. The "proof" of an alcoholic beverage is simply twice the percentage of ethanol (by volume). One hundred proof whiskey is 50% ethanol. The flavors of the various distilled liquors result from other organic compounds that distill with the alcohol and water.

Distillation of a solution of ethanol and water will not yield ethanol more concentrated than 95%. A mixture of 95% ethanol and 5% water boils at a lower temperature (78.15 °C) than either pure ethanol (bp, 78.3 °C) or pure water (bp, 100 °C). Such a mixture is an example of an **azeotrope.*** Pure ethanol can be prepared by adding benzene to the mixture of 95% ethanol and water and then distilling this solution. Benzene forms a different azeotrope with ethanol and water that is 7.5% water. This azeotrope boils at 64.9 °C and allows removal of the water (along with some ethanol). Eventually pure ethanol distills over. Pure ethanol is called **absolute alcohol**.

Ethanol is quite cheap, but when it is used for beverages it is highly taxed. (The tax is greater than $20 per gallon in most states.) Federal law requires that some ethanol used for scientific and industrial purposes be adulterated or "denatured" to make it undrinkable. Various denaturants are used including methanol.

Ethanol is an important industrial chemical. Most ethanol for industrial purposes is produced by the acid-catalyzed hydration of ethene.

---

*Azeotropes can also have boiling points that are higher than that of either of the pure components.

$$CH_2=CH_2 + H_2O \xrightarrow{acid} CH_3CH_2OH$$

Ethanol is a *hypnotic* (sleep producer). It depresses activity in the upper brain even though it gives the illusion of being a stimulant. Ethanol is also toxic, but it is much less toxic than methyl alcohol. In rats the lethal dose of ethanol is 13.7 g/kg of body weight. Abuse of the use of ethanol is a major drug problem in most countries.

8.3C Ethylene Glycol

Ethylene glycol has a low molecular weight and a high boiling point and is miscible with water. These properties make ethylene glycol an ideal automobile antifreeze. Much ethylene glycol is sold for this purpose under a variety of trade names.

8.3D Diethyl Ether

Diethyl ether is a very low-boiling, highly flammable liquid. Care should always be taken when diethyl ether is used in the laboratory, because open flames or sparks from light switches can cause explosive combustion of mixtures of diethyl ether and air.

Most ethers react slowly with oxygen in the air to form organic peroxides. These peroxides, which may accumulate in ethers that have been left standing for long periods in contact with air, are dangerously explosive and may detonate without warning when ether solutions are distilled to near dryness. Since ethers are used frequently in extractions, one should take care to test for and decompose any peroxides present in the ether before a distillation is carried out. (Consult a laboratory manual for instructions.)

Diethyl ether was first employed as a surgical anesthetic by C. W. Long of Jefferson, Georgia, in 1842. Long's use of diethyl ether was not published, but shortly thereafter, diethyl ether was introduced into surgical use at the Massachusetts General Hospital in Boston by J. C. Warren.

The most popular modern anesthetic is halothane ($CF_3CHBrCl$). Unlike diethyl ether, halothane is not flammable.

8.4 SYNTHESIS OF ALCOHOLS FROM ALKENES

We have already studied one method for the synthesis of alcohols from alkenes: **acid-catalyzed hydration** (or by the addition of sulfuric acid followed by hydrolysis, which amounts to the same thing).

8.4A Hydration of Alkenes (discussed in Section 7.5)

Alkenes add water in the presence of an acid catalyst. The addition follows Markovnikov's rule; thus, except for the hydration of ethylene, the reaction produces secondary and tertiary alcohols. The reaction is reversible and the mechanism for the hydration of an alkene is simply the reverse of that for the dehydration of an alcohol (Section 6.13).

Alkene   Alcohol

PROBLEM 8.3

Show how you would prepare each of the following alcohols by acid-catalyzed hydration of the appropriate alkene.

(a) *tert*-Butyl alcohol   (c) Cyclopentanol

(b) 2-Hexanol   (d) 1-Methylcyclohexanol

PROBLEM 8.4

When 3,3-dimethyl-1-butene is subjected to acid-catalyzed hydration the major product is 2,3-dimethyl-2-butanol. How can you explain this result?

In the sections that follow we shall study two new methods for synthesizing alcohols from alkenes. One of these methods, **oxymercuration–demercuration** (Section 8.5), complements acid-catalyzed hydration in that it gives us an additional method for Markovnikov addition of H— and —OH, and one that in many instances gives better yields than acid-catalyzed hydration. The other method, **hydroboration–oxidation** (Section 8.7), *gives us a method for the net anti-Markovnikov addition of H— and —OH.*

8.5 ALCOHOLS FROM ALKENES THROUGH OXYMERCURATION–DEMERCURATION

A highly useful laboratory procedure for synthesizing alcohols from alkenes is a two-step method called **oxymercuration–demercuration.**

Alkenes react with mercuric acetate in a mixture of tetrahydrofuran (THF) and water to produce (hydroxyalkyl)mercury compounds. These (hydroxyalkyl)mercury compounds can be reduced to alcohols with sodium borohydride:

$$-\underset{|}{\overset{|}{C}}=\underset{|}{\overset{|}{C}}-\ +\ H_2O\ +\ Hg\left(O\overset{\overset{\displaystyle O}{\|}}{C}CH_3\right)_2 \xrightarrow[\text{oxymercuration}]{\text{THF}} -\underset{\underset{\displaystyle HO}{|}}{\overset{|}{C}}-\underset{\underset{\displaystyle Hg-O\overset{\overset{\displaystyle O}{\|}}{C}CH_3}{|}}{\overset{|}{C}}-\ \ +\ CH_3\overset{\overset{\displaystyle O}{\|}}{C}OH$$

$$-\underset{\underset{\displaystyle HO}{|}}{\overset{|}{C}}-\underset{\underset{\displaystyle Hg-O\overset{\overset{\displaystyle O}{\|}}{C}CH_3}{|}}{\overset{|}{C}}-\ \ +\ OH^-\ +\ NaBH_4 \xrightarrow[\text{demercuration}]{} -\underset{\underset{\displaystyle HO}{|}}{\overset{|}{C}}-\underset{\underset{\displaystyle H}{|}}{\overset{|}{C}}-\ +\ Hg\ +\ CH_3\overset{\overset{\displaystyle O}{\|}}{C}O^-$$

In the first step, **oxymercuration,** water and mercuric acetate add to the double bond; in the second step, **demercuration,** sodium borohydride reduces the acetoxymercuri group and replaces it with hydrogen. (The acetate group is often abbreviated —OAc.)

Both steps can be carried out in the same vessel, and both reactions take place very rapidly at room temperature or below. The first step—oxymercuration— usually goes to completion within a period of 20 s to 10 min. The second step— demercuration—normally requires less than an hour. The overall reaction gives alcohols in very high yields, usually greater than 90%.

Oxymercuration–demercuration is also highly regioselective. The net orientation of the addition of the elements of water, H— and —OH, *is in accordance with Markovnikov's rule.* The H— becomes attached to the carbon atom of the double bond with the greater number of hydrogen atoms:

$$
R-\overset{\overset{\displaystyle H}{|}}{C}=\overset{\overset{\displaystyle H}{|}}{C}-H \quad \xrightarrow[\text{(2) NaBH}_4,\text{ OH}^-]{\text{(1) Hg(OAc)}_2/\text{THF–H}_2\text{O*}} \quad R-\overset{\overset{\displaystyle H}{|}}{\underset{\underset{\displaystyle HO}{|}}{C}}-\overset{\overset{\displaystyle H}{|}}{\underset{\underset{\displaystyle H}{|}}{C}}-H
$$
$$
+
$$
$$
HO-H
$$

The following specific examples are illustrations:

$$
CH_3(CH_2)_2CH=CH_2 \xrightarrow[\substack{\text{THF–H}_2\text{O} \\ \text{(15 s)}}]{\text{Hg(OAc)}_2} CH_3(CH_2)_2\underset{\underset{\displaystyle OH}{|}}{CH}-\underset{\underset{\displaystyle HgOAc}{|}}{CH_2} \xrightarrow[\substack{\text{OH}^- \\ \text{(1 h)}}]{\text{NaBH}_4}
$$

1-Pentene

$$
CH_3(CH_2)_2\underset{\underset{\displaystyle OH}{|}}{CH}CH_3 + Hg
$$
$$
\textbf{(93\%)}
$$
$$
\textbf{2-Pentanol}
$$

1-Methylcyclo-
pentene

1-Methylcyclo-
pentanol

Rearrangements of the carbon skeleton seldom occur in oxymercuration–demercuration. The following oxymercuration–demercuration of 3,3-dimethyl-1-butene is a striking example illustrating this feature.

$$
\overset{\overset{\displaystyle CH_3}{|}}{\underset{\underset{\displaystyle CH_3}{|}}{CH_3C}}-CH=CH_2 \xrightarrow[\text{(2) NaBH}_4,\text{ OH}^-]{\text{(1) Hg(OAc)}_2/\text{THF–H}_2\text{O}} \overset{\overset{\displaystyle CH_3}{|}}{\underset{\underset{\displaystyle CH_3}{|}}{CH_3C}}-\underset{\underset{\displaystyle OH}{|}}{CH}CH_3
$$

**3,3-Dimethyl-1-
butene**

**(94%)
3,3-Dimethyl-2-
butanol**

Analysis of the mixture of products by gas-liquid chromatography failed to reveal the presence of any 2,3-dimethyl-2-butanol. The acid-catalyzed hydration of 3,3-dimethyl-1-butene, by contrast, gives 2,3-dimethyl-2-butanol as the major product (Section 7.5).

A mechanism that accounts for the orientation of addition in the oxymercuration stage, and one that also explains the general lack of accompanying rearrange-

---

*Writing reagents above and below the arrow like this $\xrightarrow[\text{(2) NaBH}_4,\text{ OH}^-]{\text{(1) Hg(OAc)}_2/\text{THF–H}_2\text{O}}$ means that two steps are involved.

ments, is shown here. According to this mechanism, the first step of the oxymercuration reaction is an electrophilic attack by the mercury species, $\overset{+}{H}gOAc$, at the less substituted carbon of the double bond (i.e., at the carbon atom that bears the greater number of hydrogen atoms). We can illustrate this step using 1-pentene in the following example:

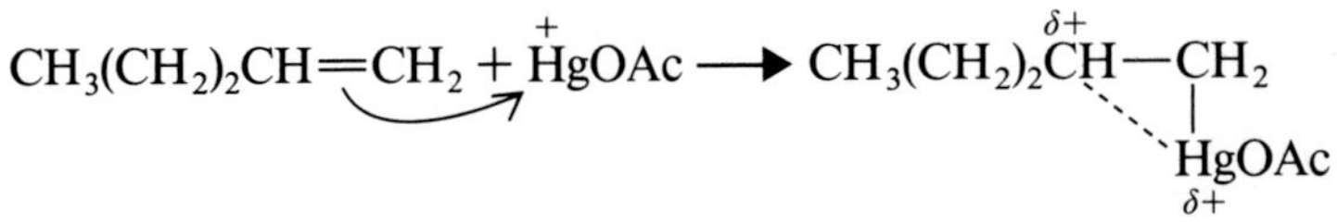

$$CH_3(CH_2)_2CH\!=\!CH_2 + \overset{+}{H}gOAc \longrightarrow CH_3(CH_2)_2\overset{\delta+}{CH}\!-\!CH_2$$
$$\underset{\delta+}{\overset{|}{HgOAc}}$$

1-Pentene	Mercury-bridged carbocation

The mercury-bridged carbocation produced in this way then reacts very rapidly with water to produce a (hydroxyalkyl)mercury compound.

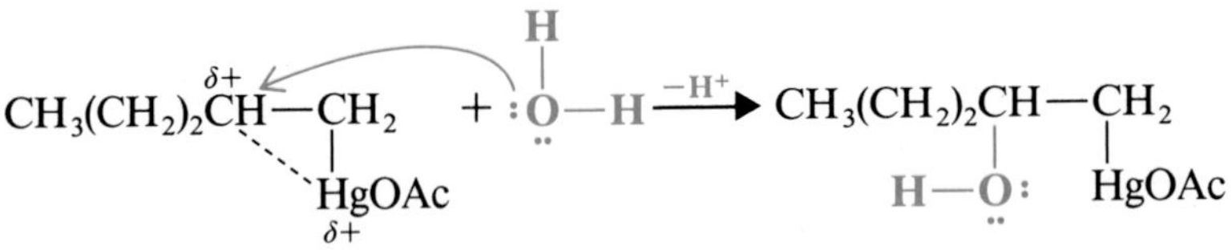

$$CH_3(CH_2)_2\overset{\delta+}{CH}\!-\!CH_2 \;+\; :\!\overset{H}{\underset{\cdot\cdot}{O}}\!-\!H \xrightarrow{-H^+} CH_3(CH_2)_2CH\!-\!CH_2$$
$$\overset{|}{\underset{\delta+}{HgOAc}} \qquad\qquad\qquad H\!-\!\underset{\cdot\cdot}{O}: \quad HgOAc$$

Mercury-bridged carbocation	(Hydroxyalkyl)mercury compound

Calculations indicate that mercury-bridged carbocations like those formed in this reaction retain much of the positive charge on the mercury moiety. Only a small portion of the positive charge resides on the more substituted carbon atom. The charge is large enough to account for the observed Markovnikov addition, but it is too small to allow the usual rapid carbon-skeleton rearrangements that take place with more fully developed carbocations.

The mechanism for the replacement of mercury by hydrogen is not well understood. Free radicals are thought to be involved.

PROBLEM 8.5

Starting with an appropriate alkene, show all steps in the synthesis of each of the following alcohols by oxymercuration–demercuration.

(a) *tert*-Butyl alcohol (b) Isopropyl alcohol (c) 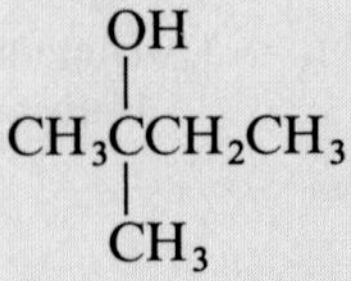 $CH_3\overset{\overset{\displaystyle OH}{|}}{\underset{\underset{\displaystyle CH_3}{|}}{C}}CH_2CH_3$

PROBLEM 8.6

When an alkene is treated with mercuric trifluoroacetate $[Hg(O_2CCF_3)_2]$ in THF containing an alcohol (ROH), the product is an (alkoxyalkyl) mercury compound. Treating this product with $NaBH_4/OH^-$ results in the formation of an ether. The overall process is called *solvomercuration–demercuration*.

$$\ce{>C=C<} \xrightarrow[\text{solvomercuration}]{\ce{Hg(O_2CCF_3)_2/THF - ROH}} \underset{\underset{\ce{Hg(O_2CCF_3)}}{|}}{\overset{\overset{\ce{RO}}{|}}{\ce{-C-C-}}} \xrightarrow[\text{demercuration}]{\ce{NaBH_4, OH^-}} \underset{\underset{\ce{H}}{|}}{\overset{\overset{\ce{RO}}{|}}{\ce{-C-C-}}}$$

| Alkene | (Alkoxyalkyl)mercuric trifluoroacetate | Ether |

(a) Outline a likely mechanism for the solvomercuration step of this ether synthesis. (b) Show how you would use solvomercuration–demercuration to prepare *tert*-butyl methyl ether.

8.6 HYDROBORATION: SYNTHESIS OF ORGANOBORANES

The addition of a compound containing a hydrogen–boron bond, $\ce{H-B<}$ (called a

boron hydride), to an alkene is the starting point for a number of highly useful synthetic procedures. This addition, called **hydroboration,** was discovered by Herbert C. Brown* (of Purdue University). In its simplest terms, hydroboration can be represented as follows:

$$\ce{>C=C<} + \ce{H-B<} \xrightarrow{\text{hydroboration}} \underset{\underset{\ce{B-}}{|}}{\underset{\ce{H}}{|}}{\ce{-C-C-}}$$

| Alkene | Boron hydride | Organo- borane |

Hydroboration can be carried out by using the boron hydride ($\ce{B_2H_6}$) called **diborane.** It is much more convenient, however, to use a solution of diborane in tetrahydrofuran (THF). When diborane dissolves in THF the $\ce{B_2H_6}$ dissociates to produce two molecules of a complex between $\ce{BH_3}$ (called **borane**) and tetrahydrofuran:

$$\ce{B_2H_6} + \ce{:O} \longrightarrow \ce{O:BH_3}$$

| Diborane | THF | THF:BH$_3$ |

BH_3 is a Lewis acid (because the boron has only six electrons in its valence shell). It accepts an electron pair from the oxygen atom of THF

Solutions containing the $\ce{THF:BH_3}$ complex can be obtained commercially. Hydroboration reactions are usually carried out in ethers; either in diethyl ether, $\ce{(C_2H_5)_2O}$, or in some higher-molecular-weight ether such as "diglyme," $\ce{(CH_3OCH_2CH_2)_2O}$, *di*ethylene *gly*col di*me*thyl ether.

---

*Brown's discovery of hydroboration led to his being a co-winner of the Nobel Prize for Chemistry in 1979.

Great care must be used in handling diborane and alkylboranes because they ignite spontaneously in air. The solution of THF∶BH₃ is considerably less prone to spontaneous ignition but still must be used in an inert atmosphere and with care.

8.6A Mechanism of Hydroboration

When a 1-alkene such as propene is treated with a solution containing the $THF \colon BH_3$ complex, the boron hydride adds successively to the double bonds of three molecules of the alkene to form a trialkylborane:

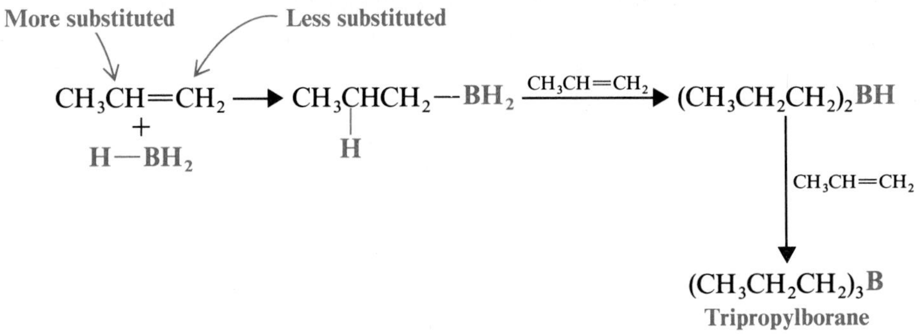

More substituted ⟶ Less substituted

$$CH_3CH{=}CH_2 \longrightarrow CH_3\underset{\underset{H}{|}}{C}HCH_2{-}BH_2 \xrightarrow{CH_3CH=CH_2} (CH_3CH_2CH_2)_2BH$$

$$+$$

$$H{-}BH_2$$

$\xrightarrow{CH_3CH=CH_2}$

$$(CH_3CH_2CH_2)_3B$$
Tripropylborane

In each addition step *the boron atom becomes attached to the less substituted carbon atom of the double bond,* and a hydrogen atom is transferred from the boron atom to the other carbon atom of the double bond. Thus, hydroboration is regioselective and it is **anti-Markovnikov** (the hydrogen atom becomes attached to the carbon atom with **fewer** hydrogen atoms).

Other examples that illustrate this tendency for the boron atom to become attached to the less substituted carbon atom are shown here. The percentages designate where the boron atom becomes attached.

Less substituted   Less substituted

$$CH_3CH_2\underset{\overset{|}{CH_3}}{C}{=}CH_2 \qquad CH_3\underset{\overset{|}{CH_3}}{C}{=}CHCH_3$$

1%   99%   2%   98%

This observed attachment of boron to the less substituted carbon atom of the double bond seems to result in part from **steric factors** — the bulky boron-containing group can approach the less substituted carbon atom more easily.

The actual addition of BH_3 to the double bond begins with a donation of π electrons from the double bond to the vacant p orbital of BH_3 (Fig. 8.1). In the next step this complex changes to one in which the boron atom is partially bonded to the

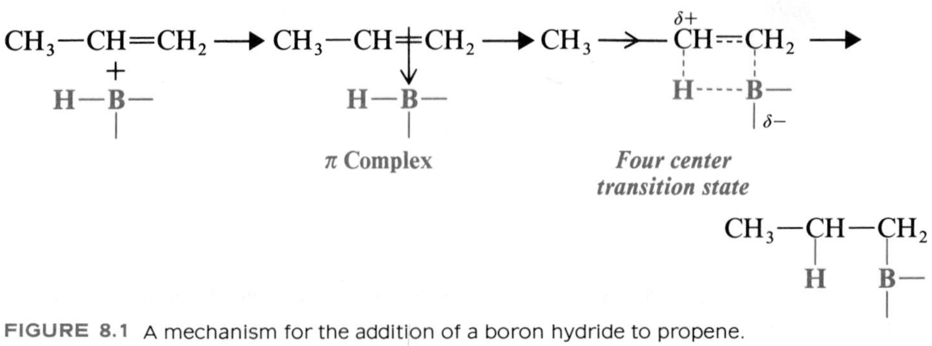

$$CH_3{-}CH{=}CH_2 \longrightarrow CH_3{-}CH\overset{\vert}{=}CH_2 \longrightarrow CH_3 \longrightarrow \overset{\delta+}{C}H{=\!=}CH_2 \longrightarrow$$

$$+$$

$$H{-}B{-}$$

$$H{-}B{-}$$

$$H{-}\!-\!-\!-B{-}$$

$$\mid \delta-$$

π Complex   *Four center transition state*

$$CH_3{-}\underset{\overset{|}{H}}{C}H{-}\underset{\overset{|}{B{-}}}{C}H_2$$

FIGURE 8.1 A mechanism for the addition of a boron hydride to propene.

less substituted carbon atom of the double bond and one hydrogen atom is partially bonded to the other carbon atom. Calculations indicate that as the transition state is approached, electrons shift in the direction of the boron atom and away from the more substituted carbon atom of the double bond. This makes the more substituted carbon atom develop a partial positive charge *and because it bears an electron-releasing alkyl group, it is better able to accommodate this positive charge.* Thus **electronic factors** augment the steric factors in determining the orientation of the addition.

8.6B The Stereochemistry of Hydroboration

The transition state for hydroboration requires that *the boron atom and the hydrogen atom add to the same face of the double bond.* (Look at Fig. 8.1 again.) The addition, therefore, is a **syn** addition:

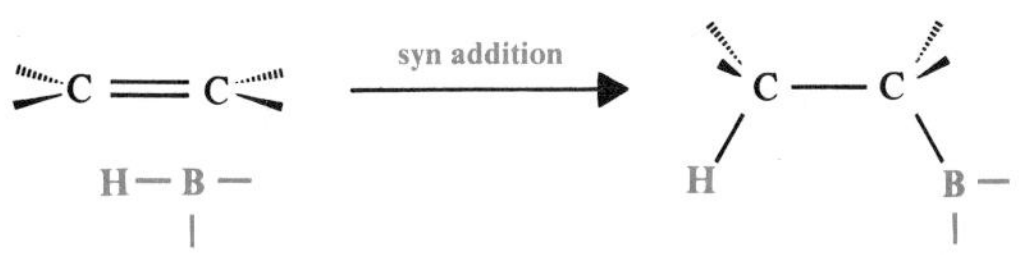

We can see the results of a syn addition in the hydroboration of 1-methylcyclopentene. (We also see again that the addition is anti-Markovnikov.)

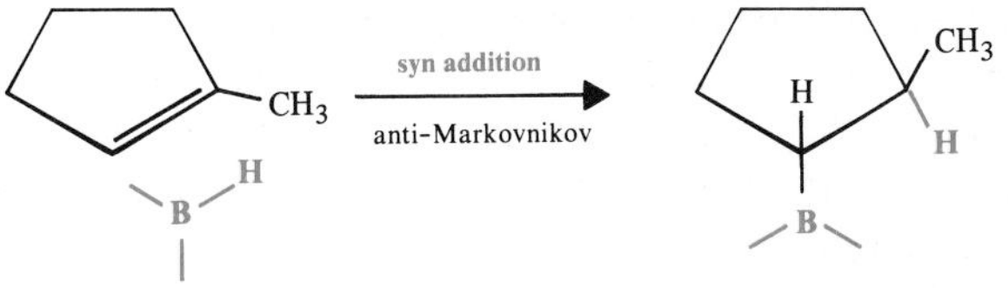

PROBLEM 8.7

Starting with an appropriate alkene, show the synthesis of (a) tributylborane, (b) triisobutylborane, and (c) tri-*sec*-butylborane. (d) Show the stereochemistry involved in the hydroboration of 1-methylcyclohexene.

PROBLEM 8.8

Treating a hindered alkene such as 2-methyl-2-butene with THF : BH_3 leads to the formation of a dialkylborane instead of a trialkylborane. When 2 moles of 2-methyl-2-butene add to 1 mole of BH_3, the product formed has the nickname "disiamylborane." Write its structure. (The name "disiamyl" comes from "*di-secondary-isoamyl*" a completely unsystematic and unacceptable name. The name "amyl" is an old common name for a five-carbon alkyl group.) As we shall see later (Section 16.4), disiamylborane is a useful reagent in certain syntheses.

8.6C Isomerization of Organoboranes

Heating an organoborane in which the boron atom is bonded to an internal carbon atom causes a remarkable isomerization to take place. The boron atom migrates along the alkyl group until it reaches the least hindered position at the end. For example, heating the organoborane obtained from 3-hexene yields the organoborane that one would expect to obtain from 1-hexene:

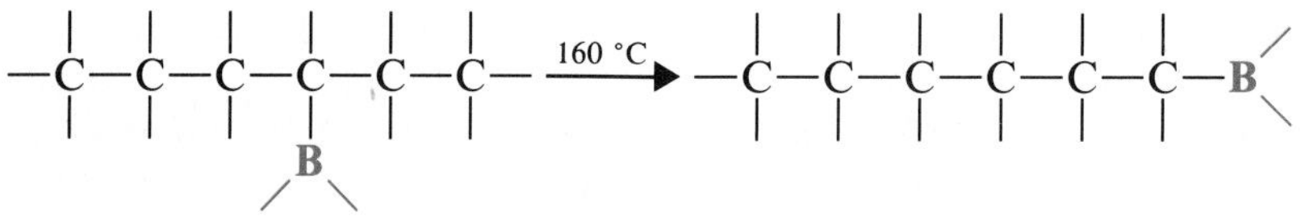

Even more remarkable is the following isomerization.

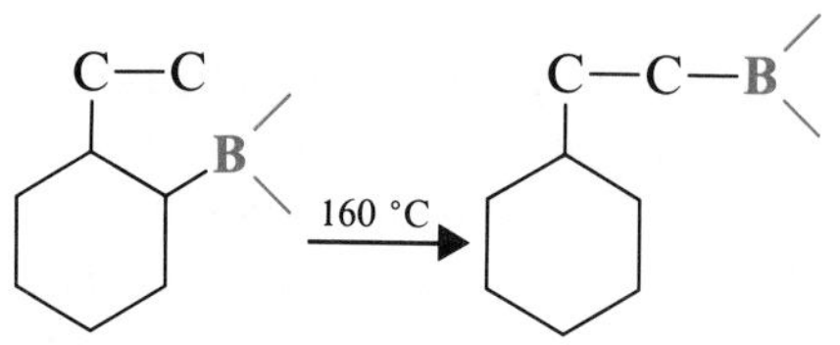

These isomerizations occur because hydroboration is reversible. The organoborane partially dissociates to yield an alkene and a boron hydride. Addition then occurs again and the process repeats itself until the boron atom ends up at the least hindered atom in the molecule.

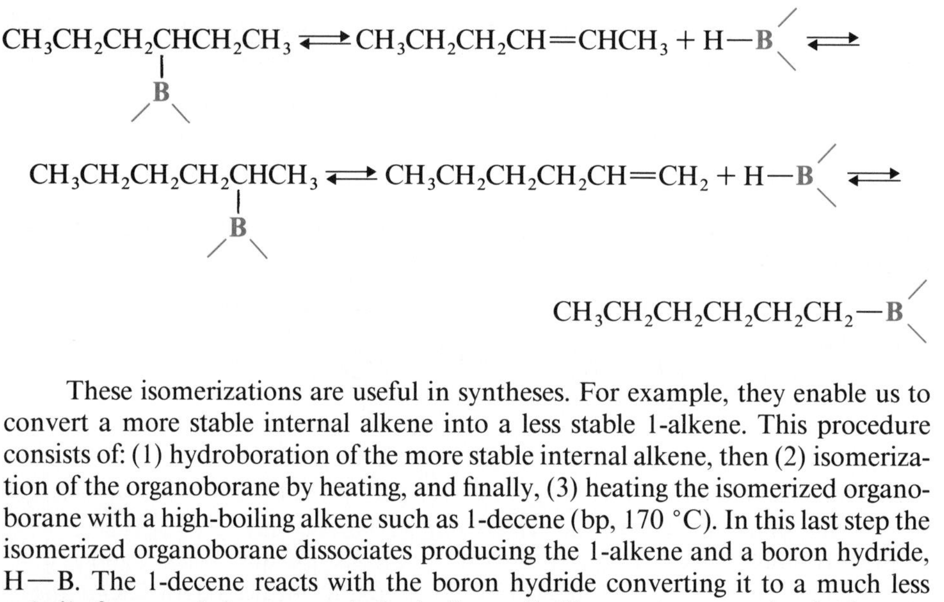

These isomerizations are useful in syntheses. For example, they enable us to convert a more stable internal alkene into a less stable 1-alkene. This procedure consists of: (1) hydroboration of the more stable internal alkene, then (2) isomerization of the organoborane by heating, and finally, (3) heating the isomerized organoborane with a high-boiling alkene such as 1-decene (bp, 170 °C). In this last step the isomerized organoborane dissociates producing the 1-alkene and a boron hydride, H—B. The 1-decene reacts with the boron hydride converting it to a much less volatile form, and the more volatile 1-alkene distills out. An example of this technique is the conversion of 2-pentene into 1-pentene:

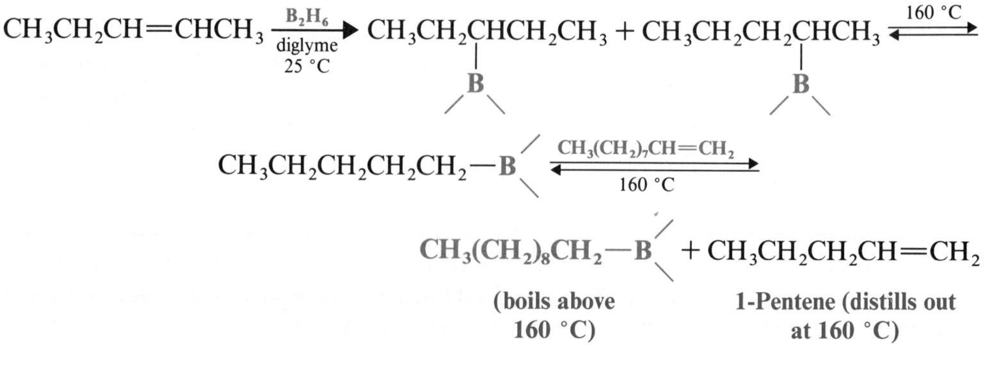

PROBLEM 8.9

Outline a procedure for the conversion of 4,4-dimethyl-2-pentene into 4,4-dimethyl-1-pentene.

8.7 ALCOHOLS FROM ALKENES THROUGH HYDROBORATION–OXIDATION

Addition of the elements of water to a double bond can also be achieved in the laboratory through the use of diborane or $THF\!:\!BH_3$. The addition of water is indirect and two reactions are involved. The first is the addition of boron hydride to the double bond, **hydroboration**; the second is the **oxidation** and hydrolysis of the organoboron intermediate to an alcohol and boric acid. We can illustrate these steps with the hydroboration–oxidation of propene.

$$3CH_3CH\!=\!CH_2 \xrightarrow[\text{hydroboration}]{\text{THF}:BH_3} (CH_3CH_2CH_2)_3B \xrightarrow[\text{oxidation}]{H_2O_2/OH^-} 3CH_3CH_2CH_2OH$$

$$\text{Propene} \qquad\qquad \text{Tripropylborane} \qquad\qquad \text{Propyl alcohol}$$

The alkylboranes produced in the hydroboration step usually are not isolated. They are oxidized and hydrolyzed to alcohols in the same reaction vessel by the addition of hydrogen peroxide in an aqueous base.

$$R_3B \xrightarrow[\substack{\text{NaOH, 25 °C} \\ \text{oxidation}}]{H_2O_2} 3R\!-\!OH + Na_3BO_3$$

The mechanism for the oxidation step begins with the addition of a hydroperoxide ion (HOO^-) to the electron-deficient boron atom.

$$
\begin{array}{c}
\text{R} \\
| \\
R\!-\!B\!+\!{}^-\!O\!-\!OH \\
| \\
\text{R}
\end{array}
\longrightarrow
\left[
\begin{array}{c}
\text{R} \\
| \\
R\!-\!B\!-\!O\!-\!OH \\
| \\
\text{R}
\end{array}
\right]^-
$$

The resulting compound is unstable, and loses a hydroxide ion. At the same time that this happens *an alkyl group migrates from the boron atom to the oxygen atom.*

$$
\left[
\begin{array}{c}
\text{R} \\
| \\
R\!-\!B\!-\!O\!-\!OH \\
| \\
\text{R}
\end{array}
\right]^-
\longrightarrow
\begin{array}{c}
\text{R} \\
| \\
R\!-\!B\!-\!O\!-\!R + OH^-
\end{array}
$$

This step, as we shall examine later, *takes place with retention of configuration of the alkyl group.* Repetitions of these two steps occur until all of the alkyl groups have become attached to oxygen atoms. The result is the formation of a trialkyl borate, an ester, $B(OR)_3$. This ester then undergoes basic hydrolysis to produce three molecules of the alcohol and a borate ion.

$$B(OR)_3 + 3\ OH^- \xrightarrow{H_2O} 3ROH + BO_3{}^{3-}$$

Because hydroboration reactions are regioselective, the net result of hydroboration–oxidation is an apparent **anti-Markovnikov addition of water.** As a consequence, *hydroboration–oxidation gives us a method for the preparation of alcohols* *that cannot normally be obtained through the acid-catalyzed hydration of alkenes or by oxymercuration–demercuration.* For example, acid-catalyzed hydration (or oxymercuration–demercuration) of 1-hexene yields 2-hexanol:

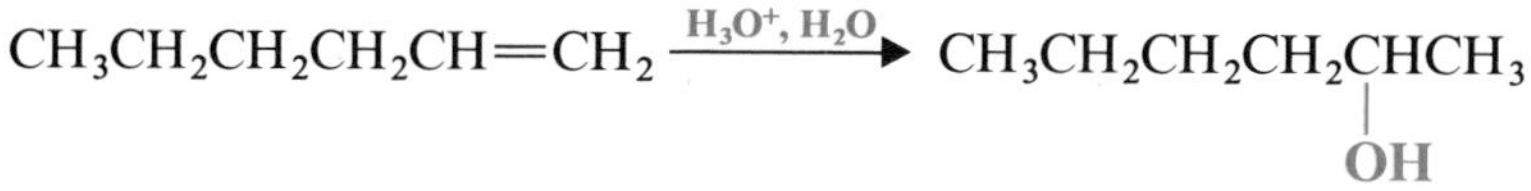

$$CH_3CH_2CH_2CH_2CH{=}CH_2 \xrightarrow{H_3O^+,\ H_2O} CH_3CH_2CH_2CH_2\underset{\underset{OH}{|}}{C}HCH_3$$

1-Hexene   2-Hexanol

Hydroboration–oxidation, by contrast, yields 1-hexanol:

$$CH_3CH_2CH_2CH_2CH{=}CH_2 \xrightarrow[(2)\ H_2O_2,\ OH^-]{(1)\ THF\colon BH_3} CH_3CH_2CH_2CH_2CH_2CH_2OH$$

1-Hexene   1-Hexanol (90%)

Other examples of hydroboration–oxidation are the following:

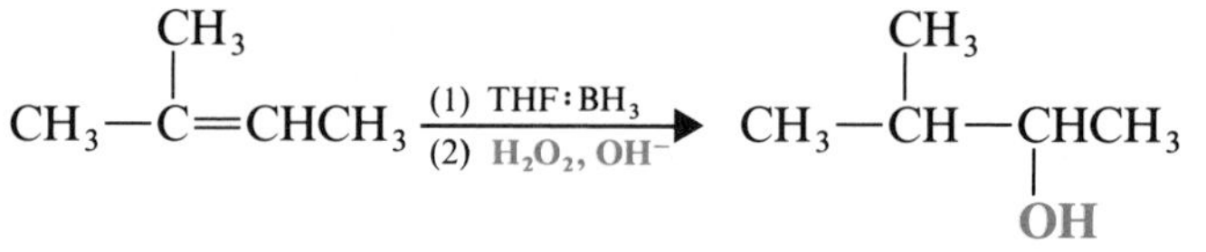

$$\underset{\text{2-Methyl-2-butene}}{CH_3{-}\underset{\underset{CH_3}{|}}{C}{=}CHCH_3} \xrightarrow[(2)\ H_2O_2,\ OH^-]{(1)\ THF\colon BH_3} \underset{\text{3-Methyl-2-butanol (59\%)}}{CH_3{-}\underset{\underset{CH_3}{|}}{C}H{-}\underset{\underset{OH}{|}}{C}HCH_3}$$

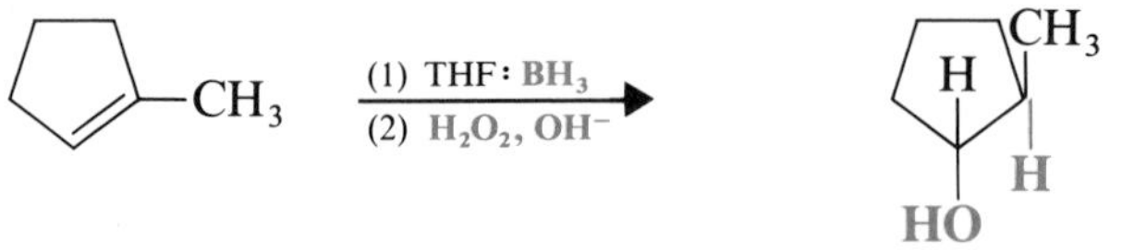

1-Methylcyclopentene   *trans*-2-Methylcyclopentanol (86%)

8.7A The Stereochemistry of the Oxidation of Organoboranes

Because the oxidation step in the hydroboration–oxidation synthesis of alcohols takes place with retention of configuration, ***the hydroxyl group replaces the boron atom where it stands in the organoboron compound.*** The net result of the two steps (hydroboration and oxidation) is the ***syn addition*** of —H and —OH. We can see this if we examine the hydroboration–oxidation of 1-methylcyclopentene (Fig. 8.2).

PROBLEM 8.10

Show how you might employ hydroboration–oxidation reactions to carry out the following syntheses.

(a) 1-Butene $\longrightarrow CH_3CH_2CH_2CH_2OH$

(b) 2-Methyl-2-butene $\longrightarrow CH_3\underset{\underset{OH}{|}}{\overset{\overset{CH_3}{|}}{C}}HCHCH_3$

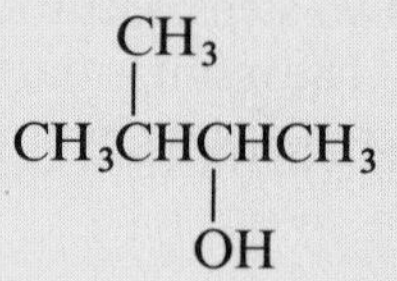

(c) 1-Methylcyclohexene $\longrightarrow$

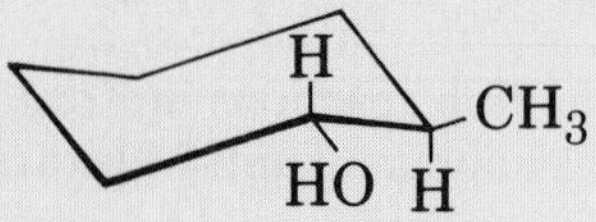

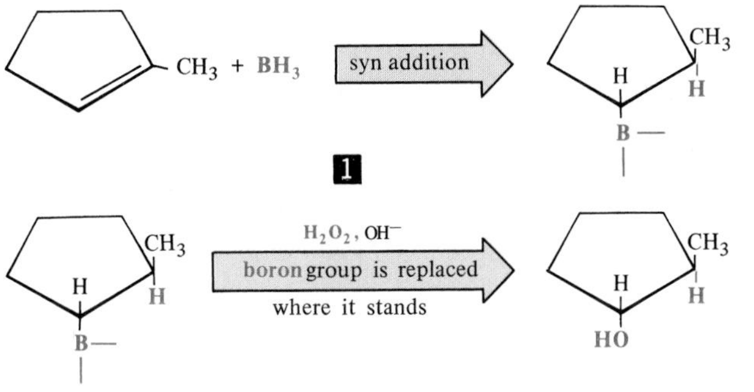

FIGURE 8.2 The hydroboration–oxidation of 1-methylcyclopentene. The first reaction is a syn addition of boron hydride. (In this illustration we have shown the boron and hydrogen both entering from the bottom side of 1-methylcyclopentene. The reaction also takes place from the top side at an equal rate.) In the second reaction the boron atom is replaced by a hydroxyl group on the same side of the molecule. The product is a trans compound (*trans*-2-methylcyclopentanol) and the overall result is the syn addition of —H and —OH.

8.7B Protonolysis of Organoboranes

Heating an organoborane with acetic acid causes cleavage of the carbon–boron bond in the following way:

$$R—B— \xrightarrow[\text{heat}]{CH_3CO_2H} R—H + CH_3\overset{\|}{\underset{O}{C}}—O—B—$$

Organoborane   Alkane

In this reaction, hydrogen replaces boron *where it stands* in the organoborane. The stereochemistry of this reaction, therefore, is like that of the oxidation of organoboranes, and it can be very useful in introducing deuterium or tritium in a specific way.

PROBLEM 8.11

Starting with any alkene (or cycloalkene) you choose, and assuming you have deuterioacetic acid (CH_3CO_2D) available, outline syntheses of the following deuterium-labeled compounds.

(a) CH$_3$CHCH$_2$D
   with CH$_3$ substituent

(b) 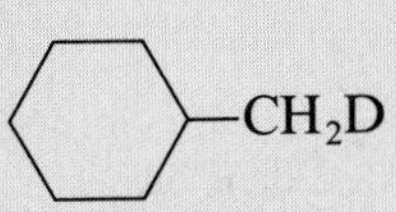 [cyclohexyl]—CH$_2$D

(c) 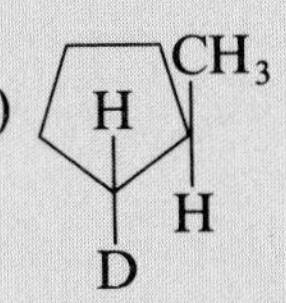 [cyclopentane with CH$_3$, H, H, D substituents]

(d) Assume you also have available $(BD_3)_2$ and CH_3CO_2T. Can you suggest a synthesis of the following?

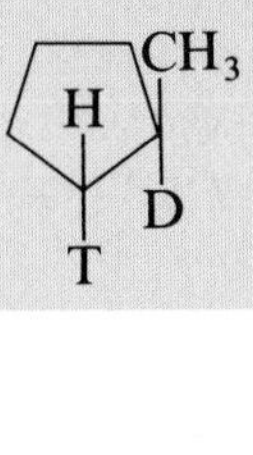

[cyclopentane ring with CH$_3$, H, D, T substituents]

SAMPLE PROBLEM

Illustrating a Multistep Synthesis

Starting with methylcyclopentane and using any other required reagents, outline a synthesis of the following alcohol.

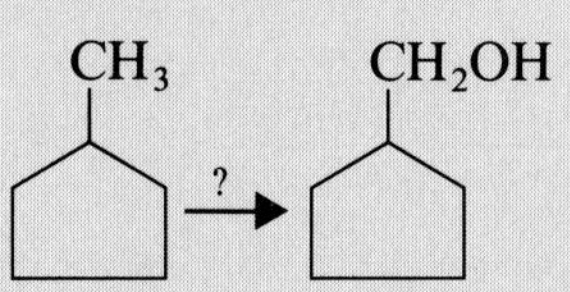

Answer:

In planning a multistep synthesis, it is often useful to work backward, always keeping in mind the starting compound. In this instance we need to prepare a primary alcohol. By now we know of several methods for the synthesis of alcohols: (1) *From primary or secondary alkyl halides by nucleophilic substitution with hydroxide ion* (Section 5.6). To use this method we would need the following alkyl halide.

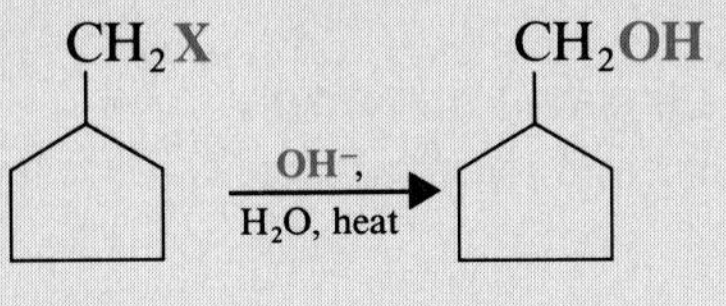

Unfortunately, this alkyl halide is not easy to prepare from methylcyclopentane. Chlorination of methylcyclopentane (Section 3.17A) would lead to a complex mixture of monochloro isomers:

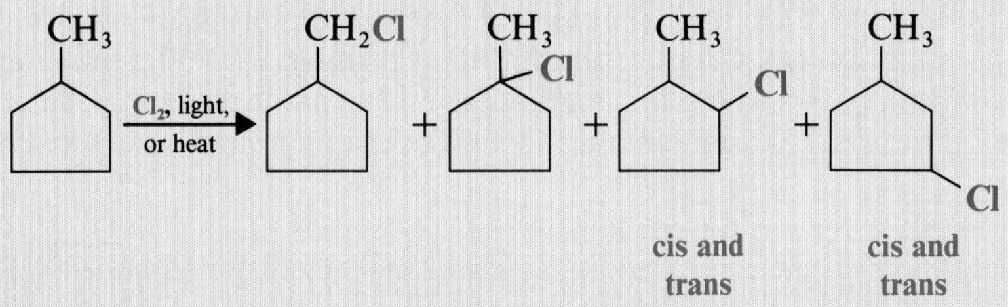

while bromination (Section 3.17B) should result in a relatively selective replacement of the tertiary hydrogen:

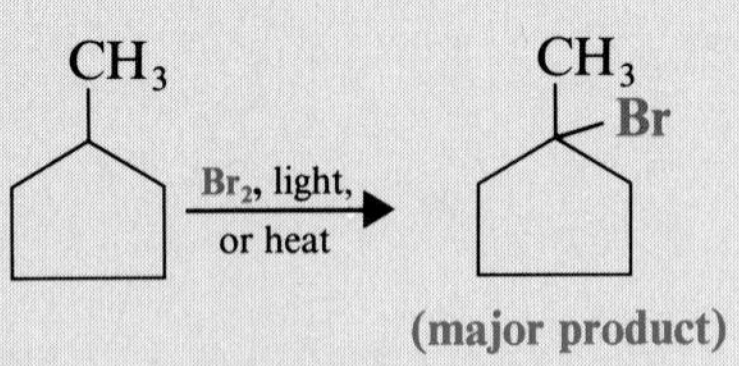

(2) and (3) *From alkenes by acid-catalyzed hydration or by oxymercuration–demercuration.* Both of these methods cause addition of water in accordance with Markovnikov's rule and (except with ethene) do not yield primary alcohols. Thus even if we were able to synthesize the following alkene, these two methods would lead to the wrong alcohol.

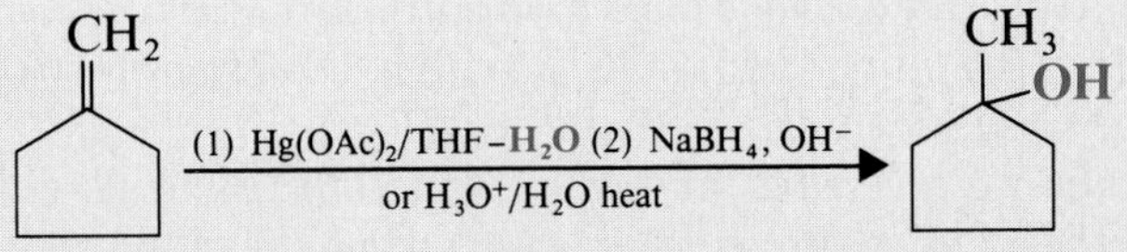

(4) *From organoboranes by oxidation* (Section 8.7A). To do this we need the following organoborane.

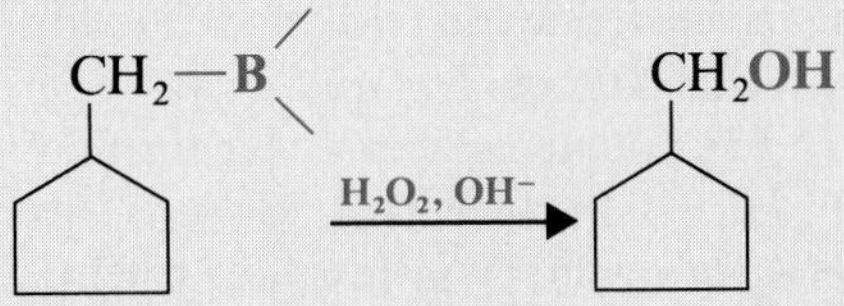

This organoborane can be prepared from 1-methylcyclopentene by hydroboration followed by isomerization.

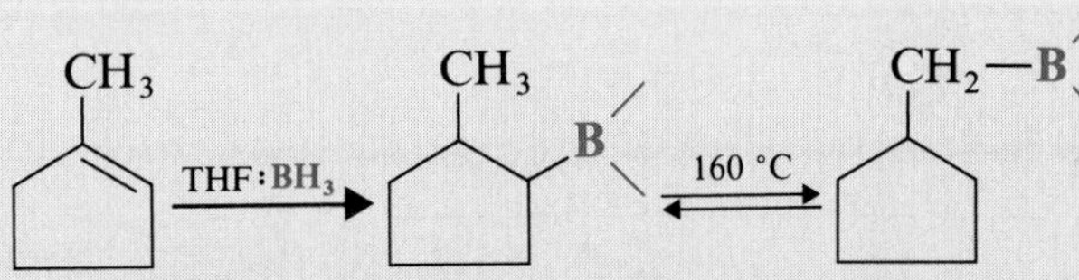

And 1-methylcyclopentene can be prepared from 1-methylcyclopentane by bromination followed by dehydrohalogenation.

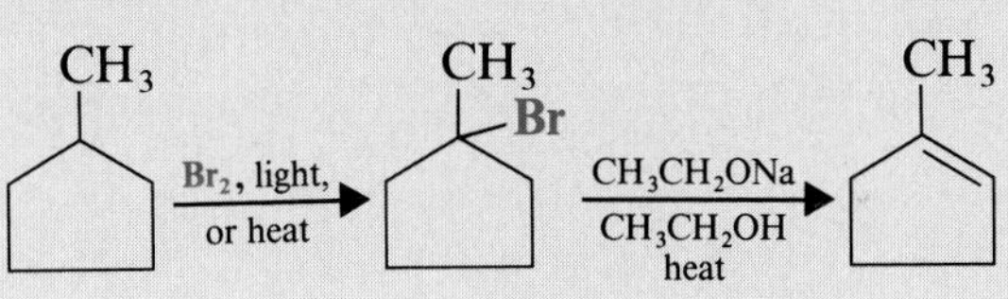

PROBLEM 8.12

Starting with 2-bromo-2-methylpentane, outline a synthesis of the following alcohol.

$$CH_3CHCH_2CH_2CH_2OH$$
$$\overset{\displaystyle CH_3}{|}$$

8.8 ALKENES IN SYNTHESIS

We have studied reactions in this chapter that can be extremely useful in designing syntheses. For example, if we want to **hydrate a double bond in a Markovnikov orientation,** we have three methods for doing so: (1) *oxymercuration–demercuration* (Section 8.5), (2) *acid-catalyzed hydration* (Section 7.5), and (3) *addition of sulfuric acid followed by hydrolysis* (Section 7.4). Of these methods oxymercuration–demercuration is the most useful in the laboratory because it is easy to carry out and because it *is not accompanied by rearrangements.*

If we want to **hydrate a double bond in an anti-Markovnikov orientation,** we can use *hydroboration–oxidation* (Section 8.7). With hydroboration–oxidation we can also achieve a *syn addition of the H— and —OH groups.*

Hydroboration is reversible, and when heated to 160 °C, *organoboranes isomerize to move the boron atom to the end of the chain* (Section 8.6C). This gives us a method for synthesizing primary alcohols from alkenes with internal double bonds (see Problem 8.12) and for isomerizing alkenes with internal double bonds to 1-alkenes (Section 8.6C). Remember, too, **the boron group of an organoborane can be replaced by hydrogen, deuterium, or tritium** (Section 8.7B), and that hydroboration, itself, involves a *syn addition of H— and —B.*

If we want to **add HX to a double bond in a Markovnikov sense** (Section 7.2), we treat the alkene with HF, HCl, HBr, or HI.

If we want to **add HBr in an anti-Markovnikov orientation** (Sections 7.2D and 9.9), we treat the alkene with HBr *and a peroxide.* (The other hydrogen halides do not undergo anti-Markovnikov addition when peroxides are present.)

We can **add bromine or chlorine to a double bond** (Section 7.6), and the addition is an *anti addition* (Section 7.7). We can also **add X— and —OH to a double bond**

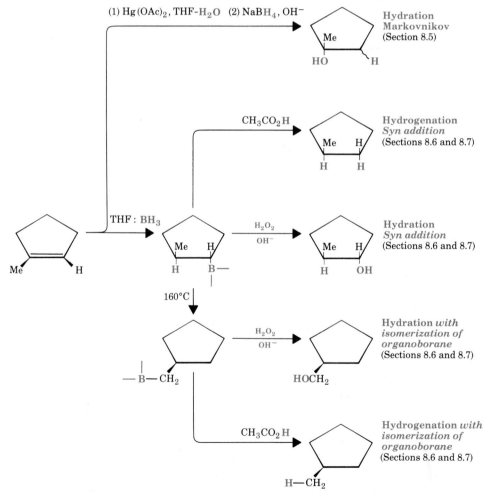

FIGURE 8.3 Oxymercuration–demercuration and hydroboration reactions of 1-methyl-cyclopentene. This figure supplements Fig. 7.8.

(i.e., synthesize a halohydrin) by carrying out the bromination or chlorination in water (Section 7.8). This addition, too, is an *anti addition.*

If we want to carry out a **syn hydroxylation of a double bond,** we can use either $KMnO_4$ in a cold, dilute, and basic solution or use OsO_4 followed by $NaHSO_3$ (Section 7.10A). Of these two methods, the latter is preferable because of the tendency of $KMnO_4$ to overoxidize the alkene and cause cleavage at the double bond.

Anti hydroxylation of a double bond can be achieved by converting the alkene to an *epoxide* and then carrying out an acid-catalyzed hydrolysis (Section 7.9B). Equations for these reactions are given in Figs. 7.8 and 8.3.

8.9 REACTIONS OF ALCOHOLS

We can classify the reactions of alcohols into two general groups: (1) those reactions that take place with cleavage at the O—H bond, and (2) those that result in cleavage at the C—O bond.

$$-\overset{|}{\underset{|}{C}}-O\!\!\mid\!\!H \qquad\qquad -\overset{|}{\underset{|}{C}}\!\!\mid\!\!O-H$$

O—H bond cleavage   C—O bond cleavage

We begin our study of reactions of alcohols with reactions in which the O—H bond is broken.

8.10 O—H BOND CLEAVAGE.
ALCOHOLS AS ACIDS

Alcohols are weak acids (Section 2.15C). For reactions taking place in solution, the acidity constants of most alcohols are of the order of 10^{-18}. This means that alcohols are slightly weaker acids than water ($K_a \sim 10^{-16}$), but they are much stronger acids than terminal alkynes ($K_a \sim 10^{-25}$) or ammonia ($K_a \sim 10^{-34}$). Alcohols are, of course, very much stronger acids than alkanes ($K_a \sim 10^{-50}$).

Relative Acidity

$$H_2O > ROH > RC{\equiv}CH > NH_3 > H_2 > RH$$

The conjugate base of an alcohol is an **alkoxide ion.** Sodium and potassium alkoxides can be prepared by treating alcohols with sodium or potassium metal or with sodium hydride (Section 5.15B). Because an alcohol is a weaker acid than water, the alkoxide ion is a stronger base than the hydroxide ion.

Relative Basicity

$$R^- > H^- > NH_2^- > RC{\equiv}C^- > RO^- > OH^-$$

PROBLEM 8.13

Write equations for the acid–base reactions that would occur if ethanol were added to each of the following compounds. In each equation label the stronger acid, the stronger base, and so on.

(a) $CH_3C{\equiv}CNa$ (b) $CH_3CH_2CH_2CH_2Li$ (c) NaH

Sodium and potassium alkoxides are often used as bases in organic syntheses (Section 5.15B). We use alkoxides when we carry out reactions that require stronger bases than hydroxide ion, but do not require exceptionally powerful bases such as the amide ion or the anion of an alkane. We also use alkoxide ions when (for reasons of solubility) we need to carry out a reaction in an alcohol solvent rather than in water.

8.11 O—H BOND CLEAVAGE.
FORMATION OF INORGANIC ESTERS

Shown here is an example of a reaction in which an alcohol reacts with an acyl chloride to form an ester of a carboxylic acid. We shall study the properties of carboxylic acid esters in detail in Chapter 18.

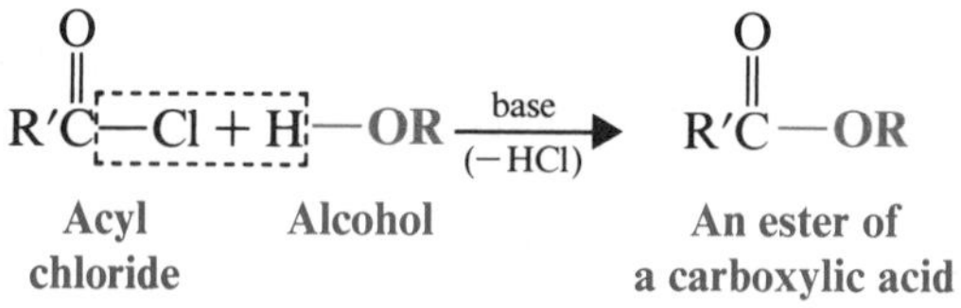

$$
\underset{\substack{\text{Acyl}\\\text{chloride}}}{R'\overset{O}{\overset{\|}{C}}-Cl} + \underset{\text{Alcohol}}{H-OR} \xrightarrow[(-\text{HCl})]{\text{base}} \underset{\substack{\text{An ester of}\\\text{a carboxylic acid}}}{R'\overset{O}{\overset{\|}{C}}-OR}
$$

8.11A Alkyl Sulfonates. Tosylates and Mesylates

Alcohols also form esters when they react with certain derivatives of inorganic acids. Among the more important esters of this type are **sulfonates.** Ethyl alcohol, for example, reacts with methanesulfonyl chloride to form *ethyl methanesulfonate* and with p-toluenesulfonyl chloride to form *ethyl p-toluenesulfonate.* These reactions also involve cleavage of the O—H bond of the alcohol *and not the C—O bond.*

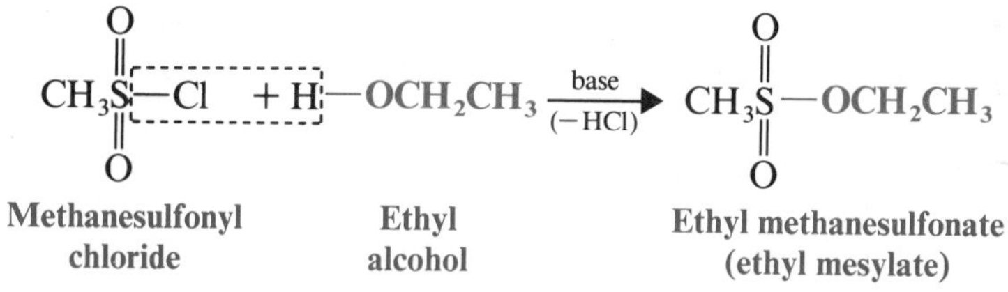

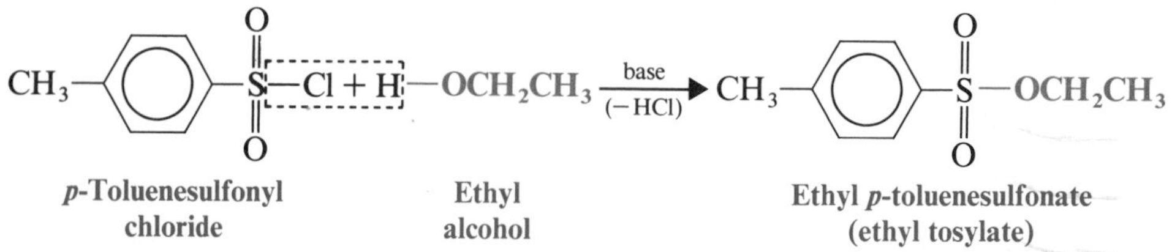

$$
\underset{\substack{\text{Methanesulfonyl}\\\text{chloride}}}{CH_3\overset{O}{\underset{O}{\overset{\|}{\underset{\|}{S}}}}-Cl} + \underset{\substack{\text{Ethyl}\\\text{alcohol}}}{H-OCH_2CH_3} \xrightarrow[(-\text{HCl})]{\text{base}} \underset{\substack{\text{Ethyl methanesulfonate}\\\text{(ethyl mesylate)}}}{CH_3\overset{O}{\underset{O}{\overset{\|}{\underset{\|}{S}}}}-OCH_2CH_3}
$$

$$
\underset{\substack{\text{p-Toluenesulfonyl}\\\text{chloride}}}{CH_3-\!\!\bigcirc\!\!-\overset{O}{\underset{O}{\overset{\|}{\underset{\|}{S}}}}-Cl} + \underset{\substack{\text{Ethyl}\\\text{alcohol}}}{H-OCH_2CH_3} \xrightarrow[(-\text{HCl})]{\text{base}} \underset{\substack{\text{Ethyl p-toluenesulfonate}\\\text{(ethyl tosylate)}}}{CH_3-\!\!\bigcirc\!\!-\overset{O}{\underset{O}{\overset{\|}{\underset{\|}{S}}}}-OCH_2CH_3}
$$

The mechanism that follows (using methanesulfonyl chloride as the example) accounts for the fact that the C—O bond of the alcohol does not break.

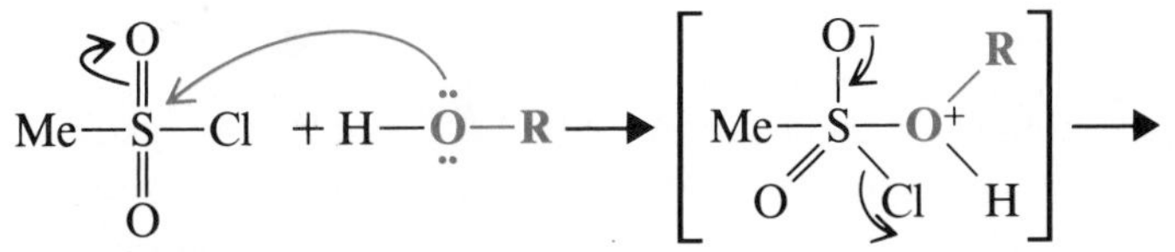

Methanesulfonyl   Alcohol
chloride

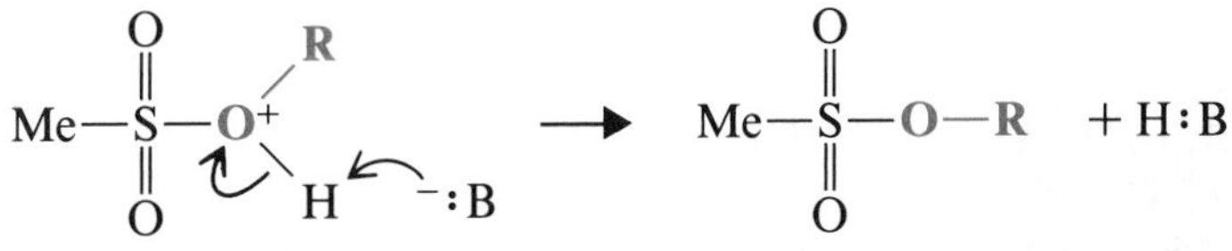

(a base)   Alkyl methanesulfonate

PROBLEM 8.14

Suggest an experiment using an isotopically labeled alcohol that would prove that the formation of an alkyl sulfonate does not cause cleavage at the C—O bond of the alcohol.

Sulfonyl chlorides are usually prepared by treating sulfonic acids with phosphorus pentachloride. (We shall study syntheses of sulfonic acids in Chapter 12.)

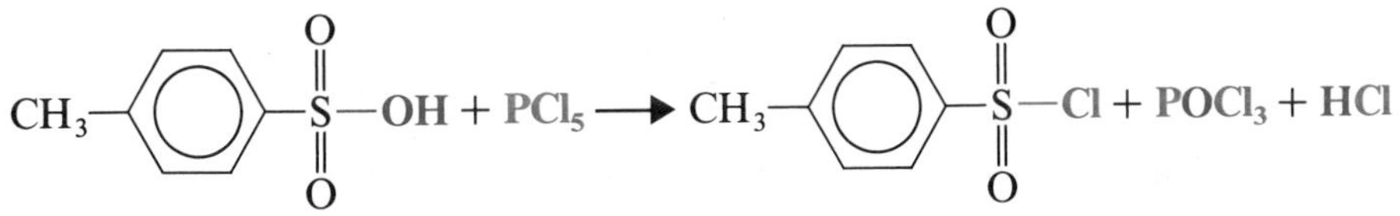

p-Toluenesulfonic   *p*-Toluenesulfonyl chloride
acid   (tosyl chloride)

Methanesulfonyl chloride and *p*-toluenesulfonyl chloride are used so often that organic chemists have shortened their rather long names to "mesyl chloride" and "tosyl chloride," respectively. The methanesulfonyl group is often called a "mesyl" group and the *p*-toluenesulfonyl group is called a "tosyl" group. Methanesulfonates are known as "mesylates" and *p*-toluenesulfonates are known as "tosylates."

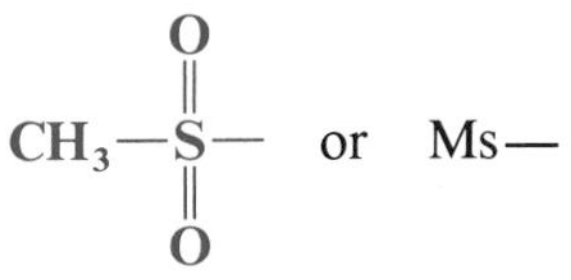

The mesyl group 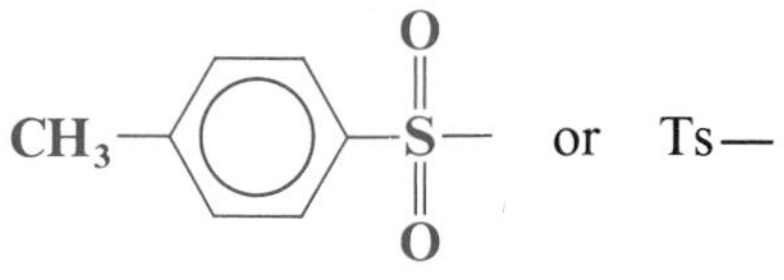   The tosyl group

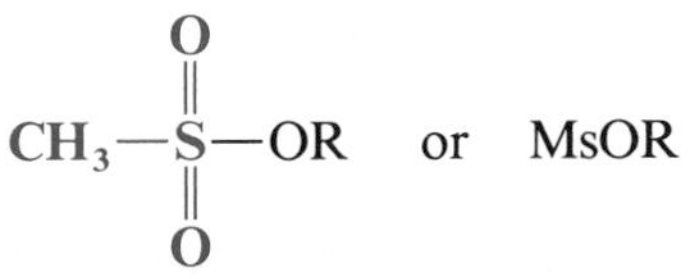

An alkyl mesylate 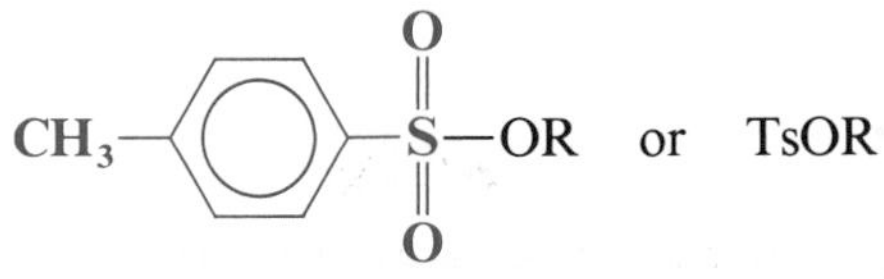   An alkyl tosylate

---

PROBLEM 8.15

Starting with toluenesulfonic acid or methanesulfonic acid and any necessary alcohols or inorganic reagents, show how you would prepare each of the following sulfonates: (a) Methyl *p*-toluenesulfonate, (b) isopropyl *p*-toluenesulfonate, and (c) *tert*-butyl methanesulfonate.

8.11B Alkyl Sulfonates in S_N2 Reactions

Alkyl sulfonates are frequently used as substrates for nucleophilic substitution reactions because sulfonate ions are excellent leaving groups.

$$Nu:^- + RCH_2-O-\overset{\overset{O}{\|}}{\underset{\underset{O}{\|}}{S}}-R' \longrightarrow Nu-CH_2R + \ ^-O-\overset{\overset{O}{\|}}{\underset{\underset{O}{\|}}{S}}-R'$$

Alkyl sulfonate
(tosylate, mesylate, etc.)

Sulfonate ion
(very weak base —
a good leaving group)

The trifluoromethanesulfonate ion ($CF_3SO_2O^-$) is one of the best of all known leaving groups. Alkyl trifluoromethanesulfonates—called *alkyl triflates*—react extremely rapidly in nucleophilic substitution reactions. The triflate ion is such a good leaving group that even vinylic triflates undergo S_N1 reactions and yield vinylic cations.

$$\underset{\text{Vinylic triflate}}{\overset{OSO_2CF_3}{\underset{/}{\overset{\backslash}{C}}=\underset{\backslash}{\overset{/}{C}}}} \xrightarrow{\text{solvolysis}} \underset{\text{Vinylic cation}}{\overset{\backslash}{\underset{/}{C}}=C^+-} + \underset{\text{Triflate ion}}{^-OSO_2CF_3}$$

Alkyl sulfonates give us an indirect method for carrying out nucleophilic substitution reactions on alcohols. We first convert the alcohol to an alkyl sulfonate and then we allow the sulfonate to react with a nucleophile. When the carbon atom bearing the —OH is a stereocenter, the first step—sulfonate formation—proceeds with **retention of configuration** because no bonds to the stereocenter are broken. Only the O—H bond breaks. The second step—if the reaction is S_N2—proceeds with *inversion of configuration.*

Step 1 $\quad \underset{H}{\overset{R}{\underset{\diagup}{\overset{\diagdown}{C}}}}-O+H + \boxed{Cl+Ts} \xrightarrow[-\ HCl]{\text{retention}} \underset{H}{\overset{R}{\underset{\diagup}{\overset{\diagdown}{C}}}}-O-Ts$

Step 2 Nu:$^-$ $+$ $\underset{H}{\overset{R}{\underset{\diagup}{\overset{\diagdown}{C}}}}-O-Ts \xrightarrow[S_N2]{\text{inversion}}$ Nu $-$ $\underset{R'}{\overset{R}{C}}\diagdown H$ $+$ ^-O-Ts

Alkyl sulfonates (tosylates, etc.) undergo all the nucleophilic substitution reactions that alkyl halides do.

PROBLEM 8.16

Show the configurations of products formed when (a) (*R*)-2-butanol is converted to a tosylate, and (b) when this tosylate reacts with hydroxide ion by an S_N2 reaction. (c) Converting *cis*-4-methylcyclohexanol to a tosylate and then allowing the tosylate to react with LiCl (in an appropriate solvent) yields *trans*-1-chloro-4-methylcyclohexane. Outline the stereochemistry of these steps.

8.11C Alkyl Phosphates

Alcohols react with phosphoric acid to yield alkyl phosphates:

$$\text{ROH} + \underset{\substack{\text{Phosphoric} \\ \text{acid}}}{\text{HO}-\overset{\displaystyle O}{\underset{\displaystyle OH}{\overset{\|}{P}}}-\text{OH}} \xrightarrow[(-\text{H}_2\text{O})]{} \underset{\substack{\text{Alkyl dihydrogen} \\ \text{phosphate}}}{\text{RO}-\overset{\displaystyle O}{\underset{\displaystyle OH}{\overset{\|}{P}}}-\text{OH}} \xrightarrow[(-\text{H}_2\text{O})]{\text{ROH}}$$

$$\underset{\substack{\text{Dialkyl hydrogen} \\ \text{phosphate}}}{\text{RO}-\overset{\displaystyle O}{\underset{\displaystyle OR}{\overset{\|}{P}}-\text{OH}}} \xrightarrow[(-\text{H}_2\text{O})]{\text{ROH}} \underset{\substack{\text{Trialkyl} \\ \text{phosphate}}}{\text{RO}-\overset{\displaystyle O}{\underset{\displaystyle OR}{\overset{\|}{P}}-\text{OR}}}$$

When phosphoric acid is heated, it forms phosphoric *anhydrides* called diphosphoric acid and triphosphoric acid.

$$2\,\text{HO}-\overset{\displaystyle O}{\underset{\displaystyle OH}{\overset{\|}{P}}}-\text{OH} \xrightarrow[-\text{H}_2\text{O}]{} \underset{\substack{\text{Diphosphoric acid} \\ \text{(pyrophosphoric acid)}}}{\text{HO}-\overset{\displaystyle O}{\underset{\displaystyle OH}{\overset{\|}{P}}}-\text{O}-\overset{\displaystyle O}{\underset{\displaystyle OH}{\overset{\|}{P}}-\text{OH}}} \qquad \textit{Anhydride linkage}$$

$$3\,\text{HO}-\overset{\displaystyle O}{\underset{\displaystyle OH}{\overset{\|}{P}}}-\text{OH} \xrightarrow[(-2\text{H}_2\text{O})]{} \underset{\text{Triphosphoric acid}}{\text{HO}-\overset{\displaystyle O}{\underset{\displaystyle OH}{\overset{\|}{P}}}-\text{O}-\overset{\displaystyle O}{\underset{\displaystyle OH}{\overset{\|}{P}}}-\text{O}-\overset{\displaystyle O}{\underset{\displaystyle OH}{\overset{\|}{P}}-\text{OH}}} \qquad \textit{Anhydride linkages}$$

These phosphoric acid anhydrides can also react with alcohols to form esters such as the following:

$$\underset{\substack{\text{An alkyl trihydrogen} \\ \text{diphosphate}}}{\text{RO}-\overset{\displaystyle O}{\underset{\displaystyle OH}{\overset{\|}{P}}}-\text{O}-\overset{\displaystyle O}{\underset{\displaystyle OH}{\overset{\|}{P}}-\text{OH}}} \qquad \underset{\substack{\text{An alkyl tetrahydrogen} \\ \text{triphosphate}}}{\text{RO}-\overset{\displaystyle O}{\underset{\displaystyle OH}{\overset{\|}{P}}}-\text{O}-\overset{\displaystyle O}{\underset{\displaystyle OH}{\overset{\|}{P}}}-\text{O}-\overset{\displaystyle O}{\underset{\displaystyle OH}{\overset{\|}{P}}-\text{OH}}}$$

Esters of phosphoric acids are extremely important in biochemical reactions. Especially important are triphosphate esters. Although hydrolysis of the ester group or of one of the anhydride linkages of an alkyl triphosphate is exothermic, these reactions occur very slowly in aqueous solutions. Near pH 7, these triphosphates exist as negatively charged ions and hence are much less susceptible to nucleophilic attack. Alkyl triphosphates are, consequently, relatively stable compounds in the aqueous medium of a living cell.

$$ROH + HO-\overset{\overset{\displaystyle O}{\|}}{\underset{\underset{\displaystyle OH}{|}}{P}}-O-\overset{\overset{\displaystyle O}{\|}}{\underset{\underset{\displaystyle OH}{|}}{P}}-O-\overset{\overset{\displaystyle O}{\|}}{\underset{\underset{\displaystyle OH}{|}}{P}}-OH$$

Ester linkage

$$RO-\overset{\overset{\displaystyle O}{\|}}{\underset{\underset{\displaystyle OH}{|}}{P}}-O-\overset{\overset{\displaystyle O}{\|}}{\underset{\underset{\displaystyle OH}{|}}{P}}-O-\overset{\overset{\displaystyle O}{\|}}{\underset{\underset{\displaystyle OH}{|}}{P}}-OH \xrightarrow[\text{slow}]{H_2O}$$

Anhydride linkages

$$RO-\overset{\overset{\displaystyle O}{\|}}{\underset{\underset{\displaystyle OH}{|}}{P}}-OH + HO-\overset{\overset{\displaystyle O}{\|}}{\underset{\underset{\displaystyle OH}{|}}{P}}-O-\overset{\overset{\displaystyle O}{\|}}{\underset{\underset{\displaystyle OH}{|}}{P}}-OH$$

$$RO-\overset{\overset{\displaystyle O}{\|}}{\underset{\underset{\displaystyle OH}{|}}{P}}-O-\overset{\overset{\displaystyle O}{\|}}{\underset{\underset{\displaystyle OH}{|}}{P}}-OH + HO-\overset{\overset{\displaystyle O}{\|}}{\underset{\underset{\displaystyle OH}{|}}{P}}-OH$$

Enzymes, on the other hand, are able to catalyze reactions of these triphosphates in which the energy made available when their anhydride linkages break helps the cell make other chemical bonds. We have more to say about this in Chapter 21 when we discuss the important triphosphate called adenosine triphosphate (or ATP).

8.12 REACTIONS OF ALCOHOLS INVOLVING C—O BOND CLEAVAGE

We have already seen one reaction that involves cleavage at the C—O bond of an alcohol.

8.12A Dehydration of Alcohols

When alcohols are heated with strong acids, they undergo elimination of water (dehydration) and form alkenes (Sections 6.13–6.15). The mechanism (discussed earlier) is given here.

$$-\overset{|}{\underset{\underset{\displaystyle H}{|}}{C}}-\overset{|}{\underset{\underset{\displaystyle OH}{|}}{C}}- \xrightleftharpoons[-H^+]{+H^-} -\overset{|}{\underset{\underset{\displaystyle H}{|}}{C}}-\overset{|}{\underset{\underset{\displaystyle {}^+OH_2}{|}}{C}}- \xrightleftharpoons[+H_2O]{-H_2O} -\overset{|}{\underset{\underset{\displaystyle H}{|}}{C}}-\overset{|}{\underset{\displaystyle +}{C}}- \xrightleftharpoons[+H^+]{-H^+} \overset{\diagdown}{\diagup}C=C\overset{\diagup}{\diagdown}$$

PROBLEM 8.18

When 3,3-dimethyl-2-butanol is treated with 85% phosphoric acid, the following products are obtained: 3,3-dimethyl-1-butene (0.4%), 2,3-dimethyl-3-butene (20%), and 2,3-dimethyl-2-butene (80%). (a) Write a mechanism that accounts for the formation of each product. (b) Why is 2,3-dimethyl-2-butene the major product?

8.13 C—O BOND CLEAVAGE.
ALKYL HALIDES FROM ALCOHOLS

Alcohols react with a variety of reagents to yield alkyl halides. The most commonly used reagents are hydrogen halides (HCl, HBr, or HI), phosphorus tribromide (PBr_3), and thionyl chloride ($SOCl_2$). Examples of the use of these reagents are shown here. All of these reactions result in cleavage at the C—O bond of the alcohol.

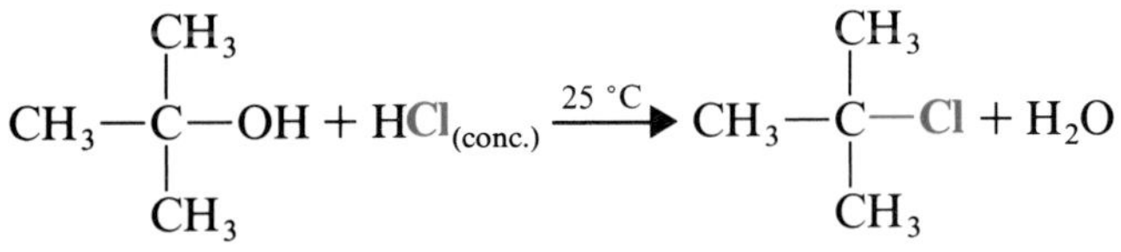

$$CH_3-\underset{\underset{CH_3}{|}}{\overset{\overset{CH_3}{|}}{C}}-OH + HCl_{(conc.)} \xrightarrow{25\ °C} CH_3-\underset{\underset{CH_3}{|}}{\overset{\overset{CH_3}{|}}{C}}-Cl + H_2O$$
$$(94\%)$$

$$CH_3CH_2CH_2CH_2OH + HBr_{(conc.)} \xrightarrow[\text{reflux}]{H_2SO_4} CH_3CH_2CH_2CH_2Br$$
$$(95\%)$$

$$3(CH_3)_2CHCH_2OH + PBr_3 \xrightarrow[4\ h]{-10\ to\ 0\ °C} 3(CH_3)_2CHCH_2Br + H_3PO_3$$
$$(55-60\%)$$

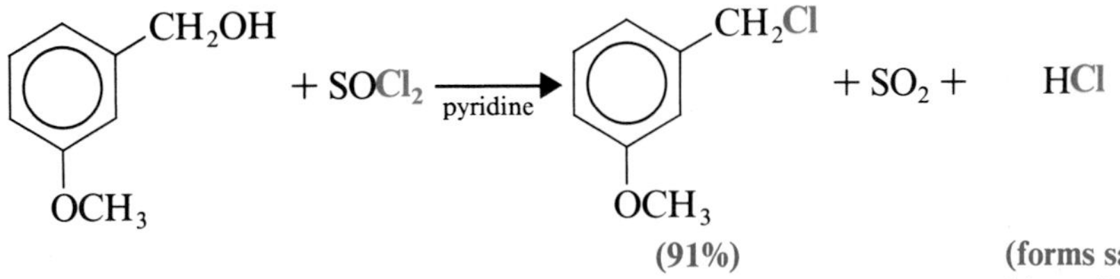

$$+ SOCl_2 \xrightarrow{\text{pyridine}}$$

$$(91\%) \qquad\qquad (\text{forms salt with pyridine})$$

8.14 ALKYL HALIDES FROM THE REACTIONS
OF ALCOHOLS WITH HYDROGEN HALIDES

When alcohols react with a hydrogen halide, a substitution takes place producing an alkyl halide and water:

$$R \overset{|}{\underset{|}{+}} OH + HX \longrightarrow R-X + H_2O$$

The order of reactivity of the hydrogen halides is HI > HBr > HCl (HF is generally unreactive), and the order of reactivity of alcohols is $3° > 2° > 1° <$ methyl.

The reaction is *acid catalyzed.* Alcohols react with the strongly acidic hydrogen halides, HCl, HBr, and HI, but they do not react with nonacidic NaCl, NaBr, or NaI. Primary and secondary alcohols can be converted to alkyl iodides and bromides by allowing them to react with a mixture of a sodium halide and sulfuric acid. This mixture generates the hydrogen halide in the mixture (*in situ*).

$$ROH + NaX \xrightarrow{H_2SO_4} RX + NaHSO_4 + H_2O$$

8.14A Mechanisms of the Reactions of Alcohols with HX

Secondary, tertiary, allylic, and benzylic alcohols appear to react by a mechanism that involves the formation of a carbocation—one that is recognizable *as an S_N1-type reaction with the protonated alcohol acting as the substrate.* We illustrate this mechanism with the reaction of *tert*-butyl alcohol and hydrochloric acid.

The first two steps are the same as in the mechanism for the dehydration of an alcohol (Section 6.13). The alcohol accepts a proton and then the protonated alcohol dissociates to form a carbocation and water.

Step 1

Step 2

In step 3 the mechanisms for the dehydration of an alcohol and the formation of an alkyl halide differ. In dehydration reactions the carbocation loses a proton in an E1-type reaction to form an alkene. In the formation of an alkyl halide, the carbocation reacts with a nucleophile (a halide ion) in an S_N1-type reaction.

Step 3

How can we account for the different course of these two reactions?

When we dehydrate alcohols we usually carry out the reaction in concentrated sulfuric acid. The only nucleophiles present in this reaction mixture are water and hydrogen sulfate (HSO_4^-) ions. Both are poor nucleophiles and both are usually present in low concentrations. Under these conditions, the highly reactive carbocation stabilizes itself by losing a proton and becoming an alkene. The net result is *an E1 reaction.*

> In the reverse reaction, that is, the hydration of an alkene (Section 7.4), the carbocation *does* react with a nucleophile. It reacts with water. Alkene hydrations are carried out in dilute sulfuric acid where the water concentration is high. In some instances, too, carbocations may react with HSO_4^- ions or with sulfuric acid, itself, when they do they form alkyl hydrogen sulfates ($R-OSO_2OH$).

When we convert an alcohol to an alkyl halide, we carry out the reaction in the presence of acid and *in the presence of halide ions.* Halide ions are good nucleophiles (much stronger nucleophiles than water), and since they are present in high concentration, most of the carbocations stabilize themselves by accepting the electron pair of a halide ion. The overall result is an S_N1 reaction.

These two reactions, dehydration and the formation of an alkyl halide, also furnish us another example of the competition between nucleophilic substitution and elimination (cf. Section 5.18). Very often, in conversions of alcohols to alkyl halides, we find that the reaction is accompanied by the formation of some alkene (i.e., by elimination). The activation energies for these two reactions of carbocations are not very different from one another. Thus, not all of the carbocations react with nucleophiles; some stabilize themselves by losing protons.

Not all acid-catalyzed conversions of alcohols to alkyl halides proceed through the formation of carbocations. Primary alcohols and methyl alcohol apparently react through a mechanism that we recognize as *an S_N2 type.* In these reactions the function of the acid is to produce *a protonated alcohol.* The halide ion then displaces a molecule of water (a good leaving group) from carbon; this produces an alkyl halide.

$$:\overset{..}{\underset{..}{X}}:^- + -\overset{|}{\underset{|}{C}}-\overset{H}{\underset{..}{\overset{|}{O^+}}}-H \longrightarrow :\overset{..}{\underset{..}{X}}-\overset{|}{\underset{|}{C}}- + :\overset{H}{\underset{..}{\overset{|}{O}}}-H$$

(protonated 1°   (a good
or methyl   leaving
alcohol)   group)

Although halide ions (particularly iodide and bromide ions) are strong nucleophiles, they are not strong enough to carry out substitution reactions with alcohols themselves. That is, reactions of the following type do not occur to any appreciable extent.

$$:\overset{..}{\underset{..}{Br}}:^- + -\overset{|}{\underset{|}{C}}-\overset{..}{\underset{..}{O}}H \;\overset{}{\nrightarrow}\; :\overset{..}{\underset{..}{Br}}-\overset{|}{\underset{|}{C}}- + :\overset{..}{\underset{..}{O}}H^-$$

They do not occur because the leaving group would have to be a strongly basic hydroxide ion.

> The reverse reaction, that is, the reaction of an alkyl halide with hydroxide ion, does occur and is a method for the synthesis of alcohols. We saw this reaction in Chapter 5.

We can see now why the reactions of alcohols with hydrogen halides are acid catalyzed. With tertiary and secondary alcohols the function of the acid is to help produce a carbocation. With methyl alcohol and primary alcohols, the function of the acid is to produce a substrate in which the leaving group is a weakly basic water molecule rather than a strongly basic hydroxide ion.

As we might expect, many reactions of alcohols with hydrogen halides, particularly those in which carbocations are formed, *are accompanied by rearrangements.*

> Because the chloride ion is a weaker nucleophile than bromide or iodide ions, hydrogen chloride does not react with primary or secondary alcohols unless zinc chloride or some similar Lewis acid is added to the reaction mixture as well. Zinc chloride, a good Lewis acid, forms a complex with the alcohol through association with an unshared pair of electrons on the oxygen atom. This provides a better leaving group for the reaction than H_2O.

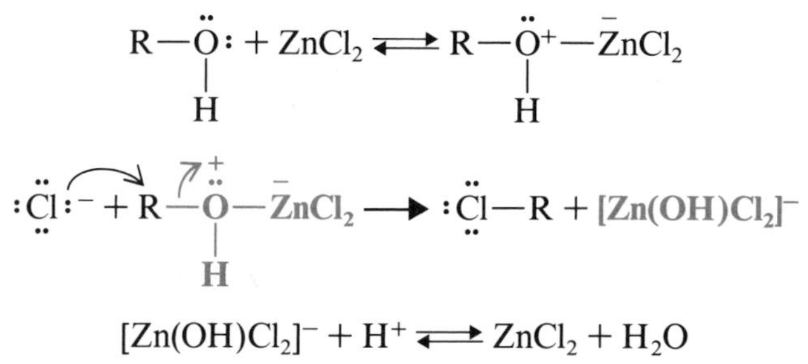

$$R-\overset{..}{\underset{\underset{H}{|}}{O}}: + ZnCl_2 \rightleftharpoons R-\overset{..}{\underset{\underset{H}{|}}{O}}{}^+-\overset{-}{Z}nCl_2$$

$$:\overset{..}{\underset{..}{Cl}}:^- + R-\overset{+\,..}{\underset{\underset{H}{|}}{O}}-\overset{-}{Z}nCl_2 \longrightarrow :\overset{..}{\underset{..}{Cl}}-R + [Zn(OH)Cl_2]^-$$

$$[Zn(OH)Cl_2]^- + H^+ \rightleftharpoons ZnCl_2 + H_2O$$

PROBLEM 8.19

(a) What factor explains the observation that tertiary alcohols react with HX faster than secondary alcohols? (b) What factor explains the observation that methyl alcohol reacts with HX faster than a primary alcohol?

PROBLEM 8.20

Treating 3-methyl-2-butanol (see following reaction) with HBr yields 2-bromo-2-methylbutane as the sole product. Outline a mechanism for the reaction.

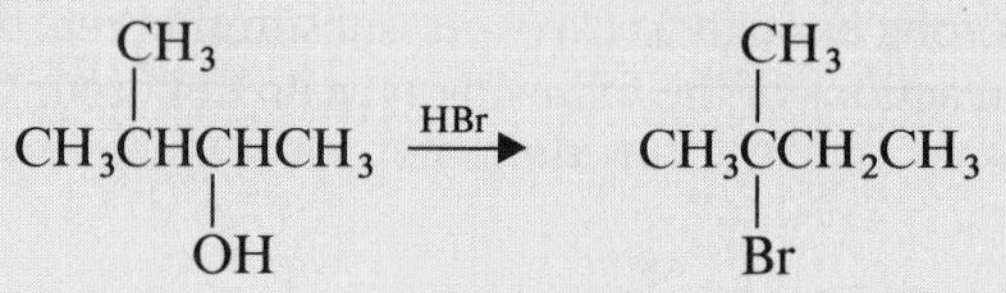

$$\underset{\text{3-Methyl-2-butanol}}{\overset{\overset{\displaystyle CH_3}{|}}{CH_3CHCHCH_3}} \overset{HBr}{\longrightarrow} \underset{\text{2-Bromo-2-methylbutane}}{\overset{\overset{\displaystyle CH_3}{|}}{CH_3CCH_2CH_3}}$$

(with OH below the left structure and Br below the right structure)

8.15 ALKYL HALIDES FROM THE REACTIONS OF ALCOHOLS WITH PBr₃ OR SOCl₂

Primary and secondary alcohols react with phosphorus tribromide to yield alkyl halides.

$$3R\overset{|}{}OH + PBr_3 \longrightarrow 3R-Br + H_3PO_3$$
$$(1° \text{ or } 2°)$$

Unlike the reaction of an alcohol with HBr, the reaction of an alcohol with PBr₃ does not involve the formation of a carbocation and *usually occurs without rearrangement* of the carbon skeleton (especially if the temperature is kept below 0 °C). For this reason phosphorus tribromide is often preferred as a reagent for the transformation of an alcohol to the corresponding alkyl bromide.

The mechanism for the reaction involves the initial formation of a protonated alkyl dibromophosphite (see following reaction) by a nucleophilic displacement on phosphorus; the alcohol acts as the nucleophile:

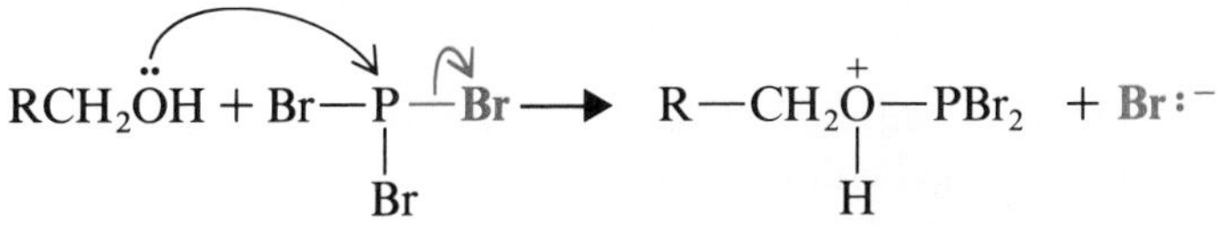

$$RCH_2\overset{..}{O}H + Br-\overset{|}{\underset{Br}{P}}-Br \longrightarrow R-CH_2\overset{+}{\underset{H}{O}}-PBr_2 + Br:^-$$

**Protonated
alkyl dibromophosphite**

Then a bromide ion acts as a nucleophile and displaces HOPBr₂.

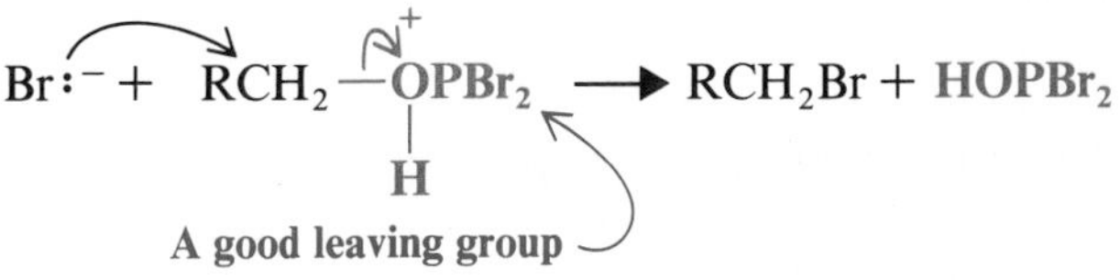

$$Br:^- + RCH_2-\overset{+}{\underset{H}{O}}PBr_2 \longrightarrow RCH_2Br + HOPBr_2$$

A good leaving group

The HOPBr₂ can react with more alcohol so the net result is the conversion of 3 moles of alcohol to alkyl bromide by 1 mole of phosphorus tribromide.

Thionyl chloride (SOCl₂) converts primary and secondary alcohols to alkyl chlorides (usually without rearrangement).

$$R-OH + SOCl_2 \xrightarrow{\text{reflux}} R-Cl + SO_2 + HCl$$
$$(1° \text{ or } 2°)$$

Often a tertiary amine is added to the mixture to promote the reaction by reacting with the HCl (cf. Section 8.13).

$$R_3N: + HCl \longrightarrow R_3NH^+ + Cl^-$$

The reaction mechanism involves initial formation of the alkyl chlorosulfite:

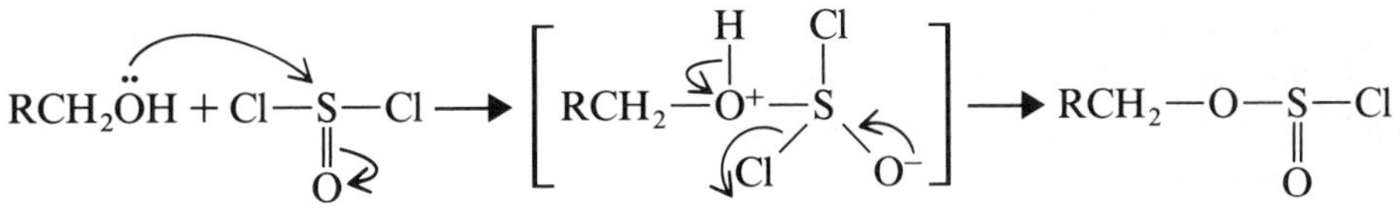

$$RCH_2\overset{..}{O}H + Cl-\overset{\|}{\underset{O}{S}}-Cl \longrightarrow \left[RCH_2-\overset{H}{\underset{Cl}{\overset{+}{O}}}-\overset{Cl}{\underset{O^-}{S}} \right] \longrightarrow RCH_2-O-\overset{\|}{\underset{O}{S}}-Cl$$

**Alkyl
chlorosulfite**

Then a chloride ion (from $R_3N + HCl \longrightarrow R_3NH^+ + Cl^-$) can bring about an S$_N$2 displacement of a very good leaving group, ClSO₂⁻, which, by decomposing (to the gas, SO₂, and Cl⁻ ion), helps drive the reaction to completion.

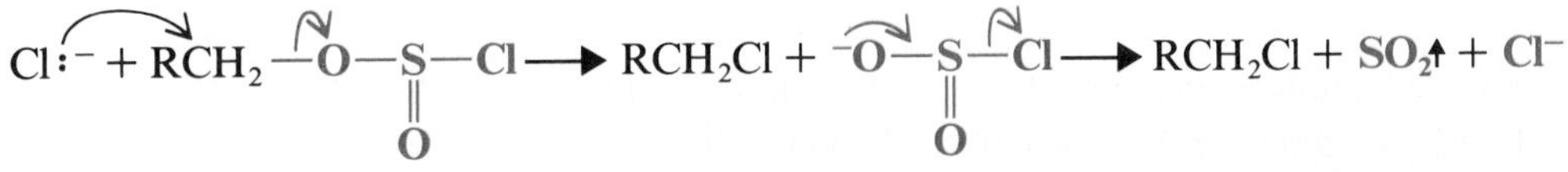

$$Cl:^- + RCH_2-\overset{\|}{\underset{O}{O}}-S-Cl \longrightarrow RCH_2Cl + ^-O-\overset{\|}{\underset{O}{S}}-Cl \longrightarrow RCH_2Cl + SO_2\uparrow + Cl^-$$

SAMPLE PROBLEM

Starting with alcohols, outline a synthesis of each of the following. (a) Benzyl bromide, (b) cyclohexyl chloride, and (c) butyl bromide.

Possible Answers:

(a) $C_6H_5CH_2OH \xrightarrow{PBr_3} C_6H_5CH_2Br$

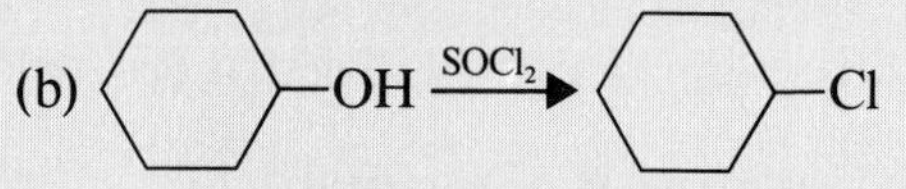

(b) ⬡—OH $\xrightarrow{SOCl_2}$ ⬡—Cl

(c) $CH_3CH_2CH_2CH_2OH \xrightarrow{PBr_3} CH_3CH_2CH_2CH_2Br$

8.16 POLYHYDROXY ALCOHOLS

Compounds containing two hydroxyl groups are called **"glycols"** or **diols.** The simplest possible diol is the unstable compound methylene glycol or methanediol.

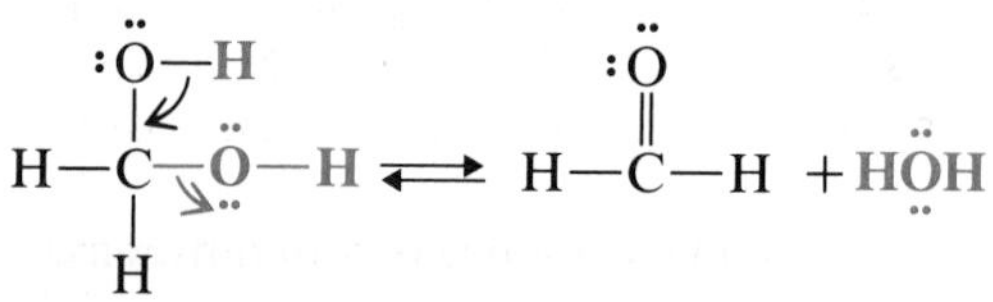

$$H-\overset{\overset{\displaystyle :\ddot{O}-H}{|}}{\underset{\underset{\displaystyle H}{|}}{C}}\!\!-\!\ddot{O}-H \rightleftarrows H-\overset{\overset{\displaystyle :\ddot{O}}{\|}}{C}-H \;+\; H\ddot{O}H$$

Methanediol   **Formaldehyde**
(a *gem*-diol)

Methanediol is a *gem*-diol (a diol that has both hydroxyl groups attached to the same carbon atom. The prefix *gem* comes from *geminal,* which comes from the Latin, *gemini,* meaning twins). Most *gem*-diols are unstable except in an aqueous solution (Section 16.7). When the water is removed, the *gem*-diol dehydrates and forms an aldehyde or ketone. When methanediol dehydrates, it produces formaldehyde.

    gem-Diols with strong electron-withdrawing groups can usually be isolated. One example is chloral hydrate (2,2,2-trichloro-1,1-ethanediol). Chloral hydrate is occasionally used as a sleep-inducing drug.

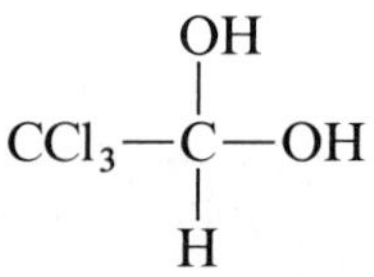

$$CCl_3-\overset{\overset{\displaystyle OH}{|}}{\underset{\underset{\displaystyle H}{|}}{C}}-OH$$

Chloral hydrate
(mp, 57 °C)

    vic-Diols (*vicinal* diols) do not dehydrate readily as *gem*-diols do. *vic*-Diols can be prepared by the hydroxylation of alkenes (cf. Sections 7.9 and 7.10).

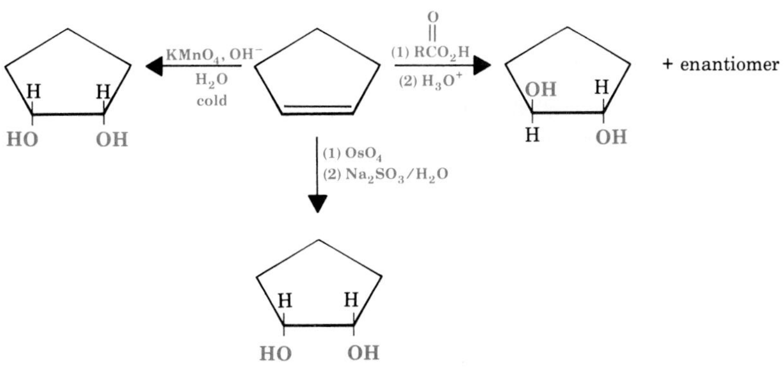

An important triol is the compound glycerol (1,2,3-propanetriol).

$$CH_2OH$$
$$|$$
$$CHOH$$
$$|$$
$$CH_2OH$$
Glycerol

Glycerol esters are, as we shall see, very important compounds in biochemistry. Glycerol itself is a viscous hygroscopic liquid with a high boiling point. In moderate amounts, glycerol is nontoxic. It is often used as a moistening agent in food, tobacco, and cosmetics.

The ester formed when glycerol reacts with nitric acid is the well-known explosive nitroglycerin.

$$CH_2ONO_2$$
$$|$$
$$CHONO_2$$
$$|$$
$$CH_2ONO_2$$
Glyceryl trinitrate or "nitroglycerin"

Nitroglycerin is very sensitive to shock but becomes much more stable and, therefore, safer when it is absorbed by sawdust or diatomaceous earth. In this form nitroglycerin is called "dynamite." Dynamite was invented by the Swedish industrial chemist Alfred Nobel.

In 1895 Nobel established a trust fund for the purpose of awarding annual prizes for exceptional contributions to the fields of chemistry, physics, medicine, and literature and to the cause of world peace. Prizes have been awarded since 1900. The first recipient of the Nobel Prize for chemistry was J. H. van't Hoff (Section 4.3A). Marie Sklodowska Curie, a Polish chemist who worked in France, won two Nobel Prizes for science: one for her work in physics (1903) and one for her work in chemistry (1911). Professor Linus Pauling (then at the California Institute of Technology) is the only person to have won two Nobel Prizes in distinctly separate areas. Pauling won the Nobel Prize for his contributions to chemistry in 1954, and for his contributions toward world peace in 1962.

8.17 SYNTHESIS OF ETHERS

8.17A Ethers by Intermolecular Dehydration of Alcohols

Alcohols can dehydrate to form alkenes. We studied this in Sections 6.13–6.15. Primary alcohols can also dehydrate to form ethers.

$$R-OH + HO-R \xrightarrow[(-H_2O)]{H^+} R-O-R$$

Dehydration to an ether usually takes place at a lower temperature than dehydration to the alkene, and dehydration to the ether can be aided by distilling the ether as it is formed. Diethyl ether is made commercially by dehydration of ethyl alcohol. Diethyl ether is the predominant product at 140 °C; ethene is the major product at 180 °C:

$$CH_3CH_2OH \begin{cases} \xrightarrow[180\ °C]{H_2SO_4} CH_2{=}CH_2 \quad \text{Ethene} \\[2ex] \xrightarrow[140\ °C]{H_2SO_4} CH_3CH_2OCH_2CH_3 \quad \text{Diethyl ether} \end{cases}$$

The formation of the ether occurs by an S_N2 mechanism with one molecule of the alcohol acting as the nucleophile and with another protonated molecule of the alcohol acting as the substrate.

$$CH_3CH_2\overset{..}{O}H + CH_3CH_2-\overset{+}{O}H_2 \rightleftharpoons CH_3CH_2-\underset{\underset{H}{|}}{\overset{+}{O}}-CH_2CH_3 + H_2O \rightleftharpoons$$

$$CH_3CH_2OCH_2CH_3 + H_3O^+$$

This method of preparing ethers is of limited usefulness, however. Attempts to synthesize ethers with secondary alkyl groups by intermolecular dehydration of secondary alcohols are usually unsuccessful because alkenes form too easily. Attempts to make ethers with tertiary alkyl groups lead exclusively to the alkenes. And, finally, this method is not useful for the preparation of unsymmetrical ethers from primary alcohols because the reaction leads to a mixture of products.

$$\underbrace{ROH + R'OH}_{1°\ alcohols} \xrightleftharpoons[H_2SO_4]{} \begin{matrix} ROR \\ + \\ ROR' \\ + \\ R'OR' \end{matrix} + H_2O$$

PROBLEM 8.21

An exception to what we have just said has to do with syntheses of unsymmetrical ethers in which one alkyl group is a *tert*-butyl group and the other group is primary. This synthesis can be accomplished by adding *tert*-butyl alcohol to a mixture of the primary alcohol and H_2SO_4 at room temperature. Give a likely mechanism for this reaction and explain why it is successful.

8.17B The Williamson Synthesis of Ethers

An important route to unsymmetrical ethers is a nucleophilic substitution reaction known as the Williamson synthesis. This synthesis consists of an S_N2 reaction of a sodium alkoxide with alkyl halide, alkyl sulfonate, or alkyl sulfate:

$$R{-}O^-Na^+ + R'{-}L \longrightarrow R{-}O{-}R' + Na^+L^-$$
$$(L = Br, I, OSO_2R'', \text{ or } OSO_2OR')$$

The following reaction is a specific example of the Williamson synthesis.

$$CH_3CH_2CH_2OH + Na \longrightarrow CH_3CH_2CH_2\ddot{O}{:}^-Na^+ + \tfrac{1}{2}H_2$$

Propyl alcohol   Sodium propoxide

$$\downarrow CH_3CH_2I$$

$$CH_3CH_2OCH_2CH_2CH_3 + Na^+ I^-$$
(70%)
Ethyl propyl ether

The usual limitations of S_N2 reactions apply here. Best results are obtained when the alkyl halide, sulfonate, or sulfate is primary (or methyl). If the substrate is tertiary, elimination is the exclusive result. Substitution is also favored over elimination at lower temperatures.

PROBLEM 8.22

(a) Outline two methods for preparing isopropyl methyl ether by a Williamson synthesis. (b) One method gives a much better yield of the ether than the other. Explain which is the better method and why.

PROBLEM 8.23

The two syntheses of 2-ethoxy-1-phenylpropane shown here give products with opposite optical rotations.

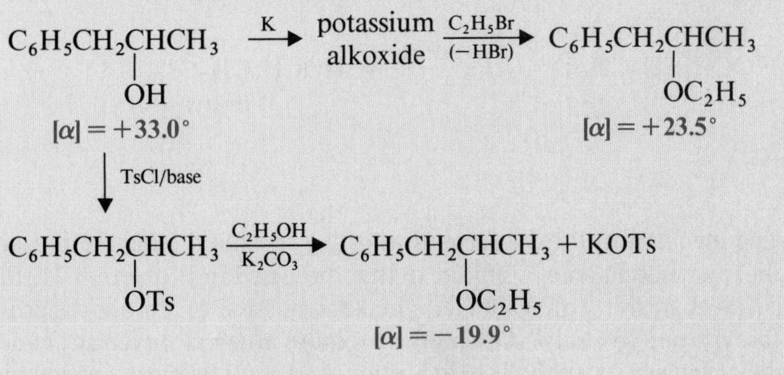

$$C_6H_5CH_2\underset{\underset{\textstyle [\alpha] = +33.0^\circ}{OH}}{CHCH_3} \xrightarrow{\text{K}} \text{potassium}\atop\text{alkoxide} \xrightarrow[(-HBr)]{C_2H_5Br} C_6H_5CH_2\underset{\underset{\textstyle [\alpha] = +23.5^\circ}{OC_2H_5}}{CHCH_3}$$

$$\downarrow TsCl/base$$

$$C_6H_5CH_2\underset{OTs}{CHCH_3} \xrightarrow[K_2CO_3]{C_2H_5OH} C_6H_5CH_2\underset{\underset{\textstyle [\alpha] = -19.9^\circ}{OC_2H_5}}{CHCH_3} + KOTs$$

How can you explain this result?

Write a mechanism that explains the formation of tetrahydrofuran from the reaction of 4-chloro-1-butanol and aqueous sodium hydroxide.

Epoxides can by synthesized by treating halohydrins with aqueous base. For example, treating $ClCH_2CH_2OH$ with aqueous sodium hydroxide yields ethylene oxide. (a) Propose a mechanism for this reaction. (b) *trans*-2-Chlorocyclohexanol reacts readily with sodium hydroxide to yield cyclohexene oxide. *cis*-2-Chlorocyclohexanol does not undergo this reaction, however. How can you account for this difference?

8.17C *tert*-Butyl Ethers by Alkylation of Alcohols

Primary alcohols can be converted to *tert*-butyl ethers by dissolving them in a strong acid such as sulfuric acid and then adding isobutylene to the mixture. (This procedure minimizes dimerization and polymerization of the isobutylene.)

$$
RCH_2OH + CH_2{=}CCH_3 \xrightarrow{H_2SO_4} RCH_2O-CCH_3
$$

$$
\begin{array}{c}
CH_3 \\
|
\end{array}
\qquad
\begin{array}{c}
CH_3 \\
|
\end{array}
$$

tert-Butyl protecting group

This method is often used to "protect" the hydroxyl group of a primary alcohol while another reaction is carried out on some other part of the molecule. The protecting *tert*-butyl group can be removed easily by treating the ether with dilute aqueous acid.

Suppose, for example, we wanted to prepare 4-pentyn-1-ol from 3-bromo-1-propanol and sodium acetylide. If we allow them to react directly, the strongly basic sodium acetylide will react first with the hydroxyl group.

$$HOCH_2CH_2CH_2Br + NaC{\equiv}CH \longrightarrow BrCH_2CH_2CH_2ONa + HC{\equiv}CH$$
3-Bromo-1-propanol

However, if we protect the —OH group first, the synthesis becomes feasible.

$$HOCH_2CH_2CH_2Br \xrightarrow[\text{(2) } CH_2=C(CH_3)_2]{\text{(1) } H_2SO_4} (CH_3)_3COCH_2CH_2CH_2Br \xrightarrow{NaC{\equiv}CH}$$

$$(CH_3)_3COCH_2CH_2CH_2C{\equiv}CH \xrightarrow{H_3O^+/H_2O} HOCH_2CH_2CH_2C{\equiv}CH + (CH_3)_3COH$$
4-Pentyn-1-ol

(a) The mechanism for the formation of the *tert*-butyl ether from a primary alcohol and isobutylene is similar to that discussed in Problem 8.21. Propose such a mechanism. (b) What factor makes it possible to remove the protecting *tert*-butyl group so easily? (Other ethers require much more forcing conditions for their cleavage, as we shall see in Section 8.18.) (c) Propose a mechanism for the removal of the protecting *tert*-butyl group.

8.17D Trimethylsilyl Ethers. Silylation

A hydroxyl group can also be protected in neutral or basic solutions by converting it to a trimethylsilyl ether group, $-OSi(CH_3)_3$. This reaction, called **silylation,** is done by allowing the alcohol to react with trimethylchlorosilane in the presence of a tertiary amine:

$$R-OH + \quad (CH_3)_3SiCl \quad \xrightarrow{(CH_3CH_2)_3N} \quad R-O-Si(CH_3)_3$$

Trimethylchlorosilane

This protecting group can also be removed with aqueous acid.

$$R-O-Si(CH_3)_3 \xrightarrow{H_3O^+/H_2O} R-OH + (CH_3)_3SiOH$$

Converting an alcohol to a trimethylsilyl ether also makes it much more volatile. (Why?) This increased volatility makes the alcohol (as a trimethylsilyl ether) much more amenable to analysis by gas-liquid chromatography.

8.17E Ethers by Epoxidation

Cyclic ethers with three-membered rings (epoxides) can be synthesized by treating an alkene with a peroxy acid (Section 7.9).

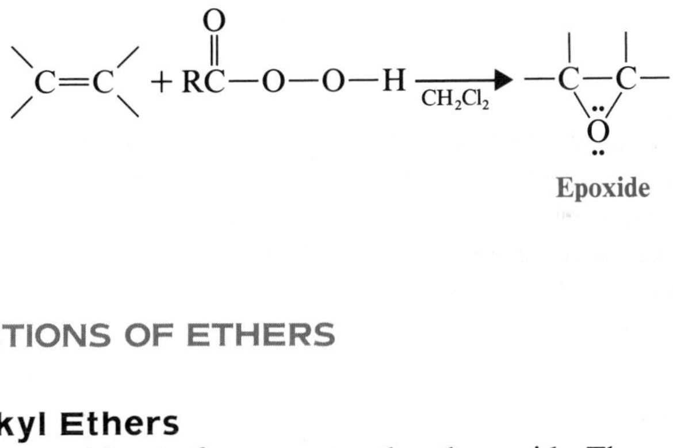

Epoxide

8.18 REACTIONS OF ETHERS

8.18A Dialkyl Ethers

Dialkyl ethers react with very few reagents other than acids. The only reactive sites that molecules of a dialkyl ether present to another reactive substance are the $C-H$ bonds of the alkyl groups and the $-\overset{..}{\underset{..}{O}}-$ group of the ether linkage. Ethers resist attack by nucleophiles (why?) and by bases. This lack of reactivity, coupled with the ability of ethers to solvate cations (by donating an electron pair from their oxygen atom) makes ethers especially useful as solvents for many reactions.

Ethers are like alkanes in that they undergo halogenation reactions (Chapter 9), but these are of little synthetic importance.

The oxygen of the ether linkage makes ethers basic. Ethers can react with proton donors to form **oxonium salts.**

$$CH_3CH_2\overset{..}{\underset{..}{O}}CH_2CH_3 + HBr \rightleftharpoons CH_3CH_2-\overset{+}{\underset{\underset{H}{|}}{\overset{..}{O}}}-CH_2CH_3 \; Br^-$$

An oxonium salt

Heating dialkyl ethers with very strong acids (HI, HBr, and H_2SO_4) causes them to undergo reactions in which the carbon–oxygen bond breaks. Diethyl ether, for example, reacts with hot concentrated hydrobromic acid to give two molar equivalents of ethyl bromide.

$$CH_3CH_2OCH_2CH_3 + 2HBr \longrightarrow 2CH_3CH_2Br + H_2O \qquad \text{Cleavage of an ether}$$

The mechanism for this reaction begins with formation of an oxonium ion. Then an S_N2 reaction with a bromide ion acting as the nucleophile produces ethyl alcohol and ethyl bromide.

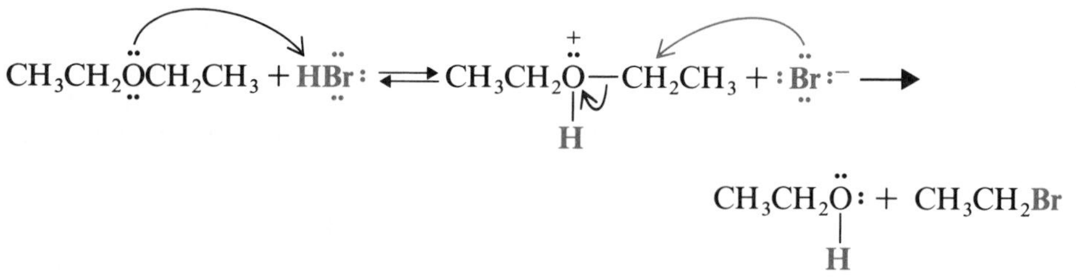

$$CH_3CH_2\overset{..}{\underset{..}{O}}CH_2CH_3 + H\overset{..}{\underset{..}{Br}} \rightleftharpoons CH_3CH_2\overset{+}{\underset{|\,H}{O}}{-}CH_2CH_3 + :\overset{..}{\underset{..}{Br}}:^- \longrightarrow$$

$$CH_3CH_2\overset{..}{\underset{|\,H}{O}}: + CH_3CH_2Br$$

Ethyl alcohol   Ethyl bromide

In the next step the ethyl alcohol (just formed) reacts with HBr to form a second molar equivalent of ethyl bromide.

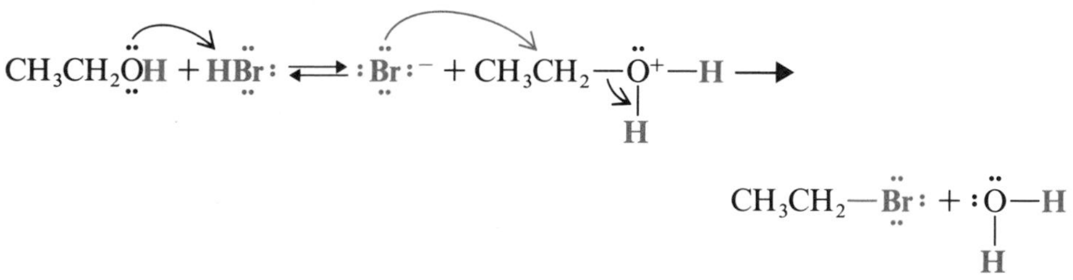

$$CH_3CH_2\overset{..}{\underset{..}{O}}H + H\overset{..}{\underset{..}{Br}} \rightleftharpoons :\overset{..}{\underset{..}{Br}}:^- + CH_3CH_2{-}\overset{+}{\underset{|\,H}{O}}{-}H \longrightarrow$$

$$CH_3CH_2{-}\overset{..}{\underset{..}{Br}}: + :\overset{..}{\underset{|\,H}{O}}{-}H$$

PROBLEM 8.27

When an ether is treated with *cold* concentrated HI, cleavage occurs as follows:

$$R{-}O{-}R + HI \longrightarrow ROH + RI$$

When mixed ethers are used, the alcohol and alkyl iodide that form depend on the nature of the alkyl groups. Explain the following observations. (a) When (R)-2-methoxybutane reacts, the products are methyl iodide and (R)-2-butanol. (b) When *tert*-butyl methyl ether reacts, the products are methanol and *tert*-butyl iodide.

8.18B Epoxides

The highly strained three-membered ring in molecules of epoxides makes them much more reactive toward nucleophilic substitution than other ethers. We have seen an example: Acid-catalyzed hydration of epoxides (Section 7.9A) is a method for preparing *vic*-diols (glycols).

Acid catalysis assists epoxide ring opening by providing a better leaving group (an alcohol) at the carbon atom undergoing nucleophilic attack. This catalysis is especially important if the nucleophile is a weak nucleophile such as water or an alcohol:

Acid-Catalyzed Ring Opening

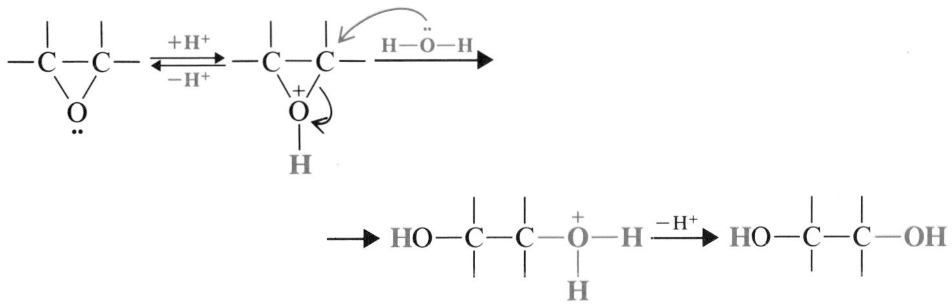

In the absence of an acid catalyst the leaving group must be a strongly basic alkoxide ion. Such reactions do not occur with other ethers, but they are possible with epoxides (because of ring strain), provided that the attacking nucleophile is also a strong base such as an alkoxide ion.

Base-Catalyzed Ring Opening

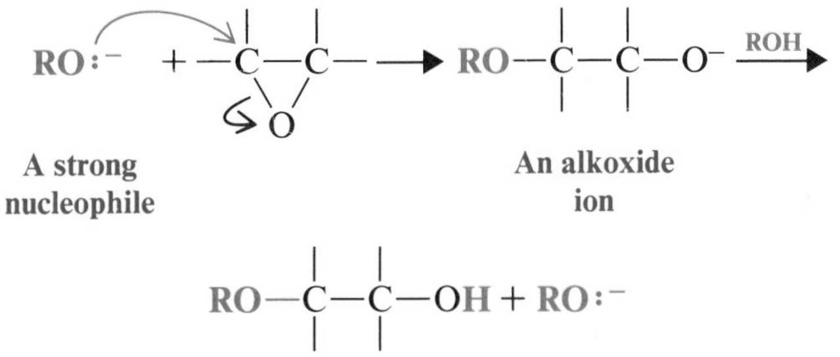

If the epoxide is unsymmetrical, in **base-catalyzed ring-opening** attack by the alkoxide ion occurs primarily *at the less substituted carbon atom.* For example, propylene oxide reacts with an alkoxide ion primarily at its primary carbon atom:

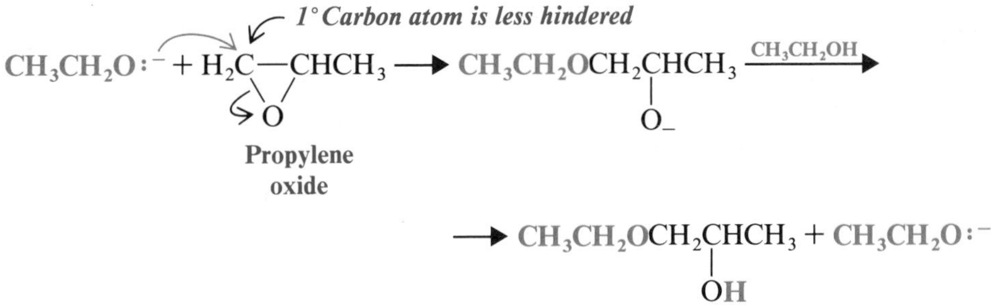

This is just what we should expect: The reaction is, after all, an S_N2 reaction, and as we learned earlier (Section 5.13A), primary substrates react more rapidly in S_N2 reactions because they are less sterically hindered.

In the **acid-catalyzed ring opening** of an unsymmetrical epoxide the nucleophile attacks primarily *at the more substituted carbon atom.* For example,

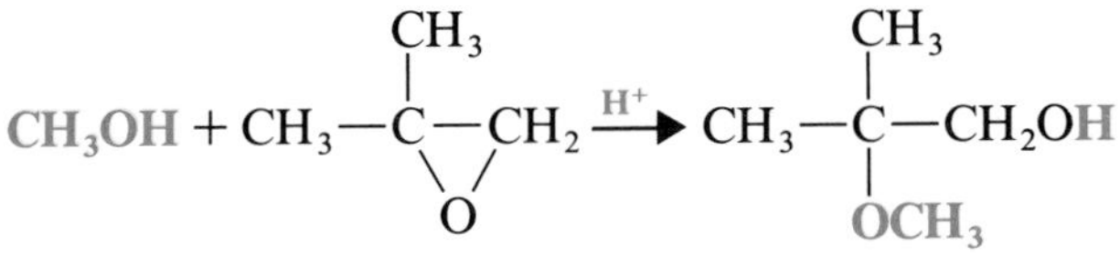

$$CH_3OH + CH_3-\underset{\underset{O}{|}}{\overset{\overset{CH_3}{|}}{C}}-CH_2 \xrightarrow{H^+} CH_3-\underset{\underset{OCH_3}{|}}{\overset{\overset{CH_3}{|}}{C}}-CH_2OH$$

The reason: Bonding in the protonated epoxide (see following reaction) is unsymmetrical with the more highly substituted carbon atom bearing a considerable positive charge. The nucleophile, therefore, attacks this carbon atom even though it is more highly substituted.

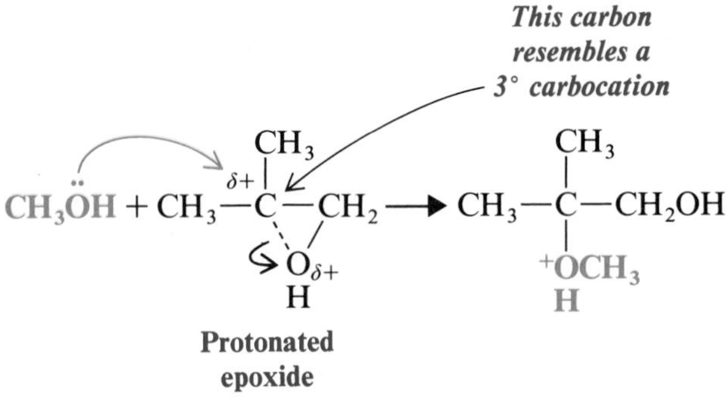

This carbon resembles a 3° carbocation

$$CH_3\ddot{O}H + CH_3-\underset{\underset{H}{\overset{\delta+}{O}}}{\overset{\overset{CH_3}{|}}{\underset{\delta+}{C}}}-CH_2 \longrightarrow CH_3-\underset{\underset{H}{\overset{+}{O}CH_3}}{\overset{\overset{CH_3}{|}}{C}}-CH_2OH$$

Protonated epoxide

The more highly substituted carbon atom bears a greater positive charge because it resembles a more stable tertiary carbocation. [Notice how this reaction (and its explanation) resembles that given for halohydrin formation from unsymmetrical alkenes in Section 7.8.]

PROBLEM 8.28

Propose structures for each of the following products:

(a) Ethylene oxide $\xrightarrow[CH_3OH]{H^+}$ $C_3H_8O_2$ (an industrial solvent called Methyl Cellosolve)

(b) Ethylene oxide $\xrightarrow[CH_3CH_2OH]{H^+}$ $C_4H_{10}O_2$ (Ethyl Cellosolve)

(c) Ethylene oxide $\xrightarrow[H_2O]{KI}$ C_2H_5IO

(d) Ethylene oxide $\xrightarrow{NH_3}$ C_2H_7NO

(e) Ethylene oxide $\xrightarrow[CH_3OH]{CH_3ONa}$ $C_3H_8O_2$

PROBLEM 8.29

Treating isobutylene oxide $\left[\begin{matrix} H_2C-C(CH_3)_2 \\ \diagdown \diagup \\ O \end{matrix}\right]$ with sodium methoxide in methanol gives primarily 1-methoxy-2-methyl-2-propanol. What factor accounts for this result?

PROBLEM 8.30

When sodium ethoxide reacts with epichlorohydrin, labeled with ^{14}C as shown by the asterisk in **I**, the major product is an epoxide bearing the label as in **II**. Provide an explanation for this reaction.

$$Cl-CH_2-\underset{\underset{O}{\diagdown \diagup}}{CH}-\overset{*}{C}H_2 \xrightarrow{\text{NaOC}_2\text{H}_5} C_2H_5O\overset{*}{C}H_2-\underset{\underset{O}{\diagdown \diagup}}{CH}-CH_2$$

$$\text{I} \qquad\qquad\qquad \text{II}$$

Epichlorohydrin

8.19 CROWN ETHERS. NUCLEOPHILIC SUBSTITUTION REACTIONS IN NONPOLAR APROTIC SOLVENTS BY PHASE-TRANSFER CATALYSIS

When we studied the effect of the solvent on nucleophilic substitution reactions in Section 5.13A, we found that S_N2 reactions take place much more rapidly in polar aprotic solvents such as dimethyl sulfoxide and dimethylformamide. The reason: *In these polar aprotic solvents the nucleophile is only very slightly solvated and is, consequently, highly reactive.*

This increased reactivity of nucleophiles is a distinct advantage. Reactions that might have taken many hours or days are often over in a matter of minutes. There are, unfortunately, certain disadvantages that accompany the use of solvents such as dimethyl sulfoxide and dimethylformamide. These solvents have very high boiling points, and as a result, they are often difficult to remove after the reaction is over. Purification of these solvents is also time consuming, and they are expensive. At high temperatures certain of these polar aprotic solvents decompose.

In some ways the ideal solvent for an S_N2 reaction would be a *nonpolar* aprotic solvent such as a hydrocarbon or chlorinated hydrocarbon. They have low boiling points, they are cheap, and they are relatively stable.

Until recently, nonpolar aprotic solvents such as a hydrocarbon or chlorinated hydrocarbon were seldom used for nucleophilic substitution reactions because of their inability to dissolve ionic compounds. This situation has changed with the development of a procedure called **phase-transfer catalysis.**

With phase-transfer catalysis, we usually use two immiscible phases that are in contact — often an aqueous phase containing an ionic reactant and an organic phase (benzene, $CHCl_3$, etc.) containing the organic substrate. Normally the reaction of two substances in separate phases like this is inhibited because of the inability of the

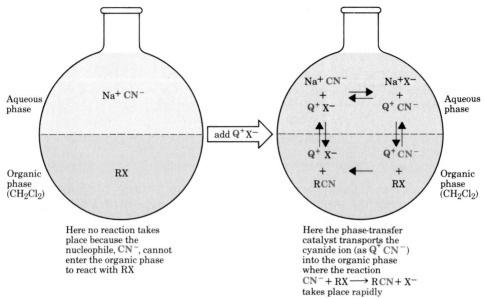

Here no reaction takes place because the nucleophile, CN^-, cannot enter the organic phase to react with RX

Here the phase-transfer catalyst transports the cyanide ion (as $Q^+ CN^-$) into the organic phase where the reaction
$$CN^- + RX \longrightarrow RCN + X^-$$
takes place rapidly

FIGURE 8.4 Phase-transfer catalysis of the S_N2 reaction between sodium cyanide and an alkyl halide.

reagents to come together. Adding a phase-transfer catalyst solves this problem by transferring the ionic reactant into the organic phase. And again, because the reaction medium is aprotic, an S_N2 reaction occurs rapidly.

An example of phase-transfer catalysis is outlined in Fig. 8.4. The phase-transfer catalyst (Q^+X^-) is usually a quaternary ammonium halide ($R_4N^+X^-$) such as tetrabutylammonium halide ($CH_3CH_2CH_2CH_2)_4N^+X^-$. The phase-transfer catalyst causes the transfer of the nucleophile (for example, CN^-) as an ion pair [Q^+CN^-] into the organic phase. This transfer apparently takes place because the cation (Q^+) of the ion pair, with its four alkyl groups, resembles a hydrocarbon in spite of its positive charge. It is said to be **lipophilic** — it prefers a nonpolar environment to an aqueous one. In the organic phase the nucleophile of the ion pair (CN^-) reacts with the organic substrate RX. The cation (Q^+) then migrates back into the aqueous phase to complete the cycle. This process continues until all of the nucleophile or the organic substrate has reacted.

An example of a nucleophilic substitution reaction carried out with phase-transfer catalysis is the reaction of 1-chlorooctane (in decane) and sodium cyanide (in water). The reaction (at 105 °C) is complete in less than 2 h and gives a 95% yield of the substitution product.

$$CH_3(CH_2)_7Cl \text{ (in decane)} \xrightarrow[\text{aqueous NaCN, 105 °C}]{R_4N^+Br^-} CH_3(CH_2)_7CN$$
$$(95\%)$$

Many other nucleophilic substitution reactions have been carried out in a similar way.

Phase-transfer catalysis, however, is not limited to nucleophilic substitutions. Many other types of reactions are also amenable to phase-transfer catalysis. Oxidations of alkenes dissolved in benzene can be accomplished in excellent yield using potassium permanganate (in water) when a quaternary ammonium salt is present:

$$CH_3(CH_2)_5CH{=}CH_2 \text{ (benzene)} \xrightarrow[\text{aqueous KMnO}_4,\ 35\ °C]{R_4N^+X^-} CH_3(CH_2)_5CO_2H$$
$$(99\%)$$

Potassium permanganate can also be transferred to benzene by quaternary ammonium salts for the purpose of chemical tests. The resulting "purple benzene" can be used as a test reagent for unsaturated compounds. As an unsaturated compound is added to the benzene solution of $KMnO_4$, the purple color disappears and the solution becomes brown (because of the presence of MnO_2), indicating a positive test (see Section 7.19D).

PROBLEM 8.31

Outline a scheme such as the one shown in Fig. 8.4 showing how the reaction of $CH_3(CH_2)_7Cl$ with cyanide ion (just shown) takes place by phase-transfer catalysis. Be sure to indicate which ions are present in the organic phase, which are in the aqueous phase, and which pass from one phase to the other.

8.19A Crown Ethers

Compounds called **crown ethers** are also phase-transfer catalysts and are able to transport ionic compounds into an organic phase. Crown ethers are cyclic polymers of ethylene glycol such as the 18-crown-6 that follows:

18-Crown-6

Crown ethers are named as x-crown-y where x is the total number of atoms in the ring and y is the number of oxygen atoms. The relationship between the crown ether and the ion that it transports is called a **host–guest** relationship. The crown ether acts as the **host,** and the coordinated cation is the **guest.**

When crown ethers coordinate with a metal cation, they thereby convert the metal ion into a species with a hydrocarbonlike exterior. The crown ether 18-crown-6, for example, coordinates very effectively with potassium ions because the cavity size is correct and because the six oxygen atoms are ideally situated to donate their electron pairs to the central ion.

Crown ethers render many salts soluble in nonpolar solvents. Salts such as KF, KCN, and CH_3CO_2K, for example, can be transferred into aprotic solvents by using catalytic amounts of 18-crown-6. In the organic phase the relatively unsolvated anions of these salts can carry out a nucleophilic substitution reaction on an organic substrate.

$$K^+CN^- + RCH_2X \xrightarrow[\text{benzene}]{\text{18-crown-6}} RCH_2CN + K^+X^-$$

$$C_6H_5CH_2Cl + KF \xrightarrow[\text{acetonitrile}]{\text{18-crown-6}} C_6H_5CH_2F + K^+Cl^-$$
$$(100\%)$$

Crown ethers can also be used as phase-transfer catalysts for many other types of reactions. The following reaction is one example of the use of a crown ether in an oxidation.

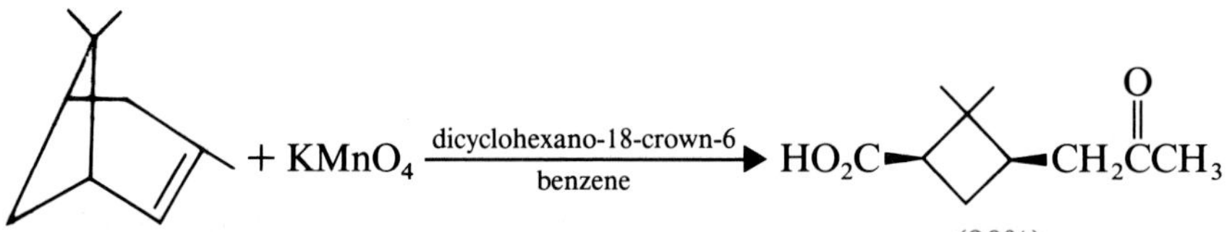

$$+ KMnO_4 \xrightarrow[\text{benzene}]{\text{dicyclohexano-18-crown-6}} HO_2C\text{—}\underset{}{\text{—}}CH_2CCH_3$$
$$(90\%)$$

Dicyclohexano-18-crown-6 has the following structure:

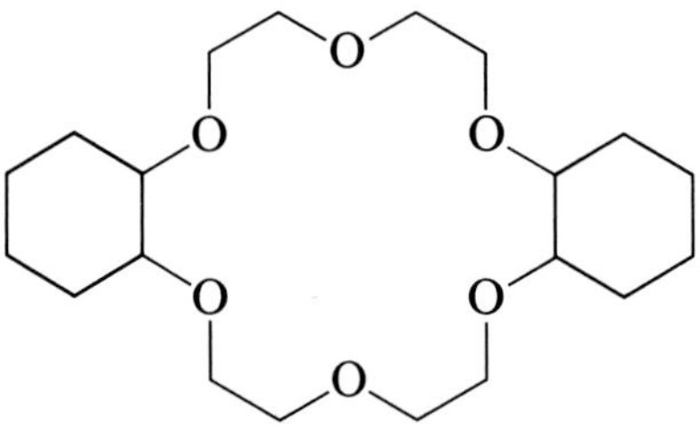

Dicyclohexano-18-crown-6

PROBLEM 8.32

Write structures for (a) 15-crown-5 and (b) 12-crown-4.

8.19B Transport Antibiotics and Crown Ethers

There are several antibiotics, most notably *nonactin* and *valinomycin,* that coordinate with metal cations in a manner similar to that of crown ethers. Normally, cells must maintain a gradient between the concentrations of sodium and potassium ions inside and outside the cell wall. Potassium ions are "pumped" in; sodium ions are pumped out. The cell membrane, in its interior, is like a hydrocarbon, because it consists in this region primarily of the hydrocarbon portions of lipids (Chapter 22). The transport of hydrated sodium and potassium ions through the cell membrane is slow, and this transport requires an expenditure of energy by the cell. Nonactin, for example, upsets the concentration gradient by coordinating more strongly with potassium ions than with sodium ions. Because the potassium ions are bound in the interior of the nonactin, this host–guest complex becomes hydrocarbonlike on its surface and passes readily through the interior of the membrane. The cell membrane thereby becomes permeable to potassium ions, and the essential concentration gradient is destroyed.

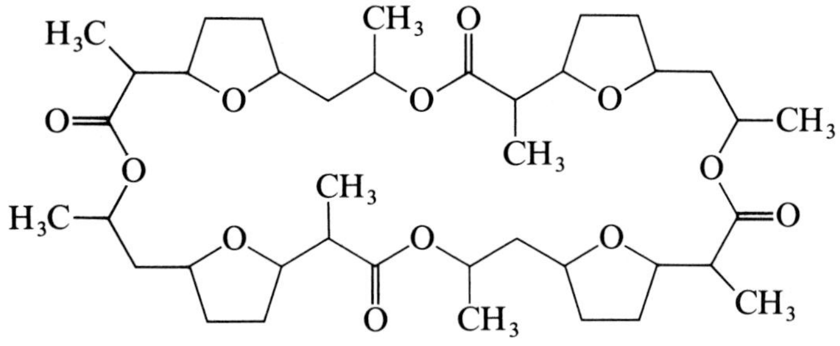

Nonactin

8.20 SUMMARY OF REACTIONS OF ALCOHOLS AND ETHERS

Most of the important reactions of alcohols and ethers that we have studied thus far are summarized in Fig. 8.5 on the next page.

ADDITIONAL PROBLEMS

8.33

Give an IUPAC substitutive name for each of the following alcohols:

(a) $(CH_3)_3CCH_2CH_2OH$

(b) $CH_2=CHCH_2\overset{\displaystyle CH_3}{\underset{\displaystyle |}{C}}HOH$

(c) $HOCH_2\overset{\displaystyle |}{\underset{\displaystyle CH_3}{C}}HCH_2CH_2OH$

(d) $C_6H_5CH_2CH_2OH$

(e) 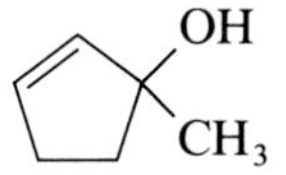 [structure of cyclopentene with OH and CH₃]

(f) 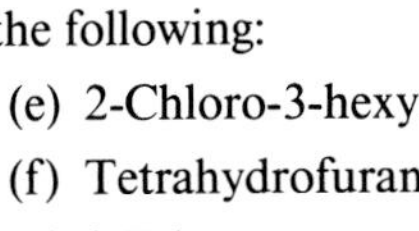 [cyclohexane structure with H, OH, CH₃]

8.34

Write structural formulas for each of the following:

(a) (Z)-2-Buten-1-ol

(b) (R)-1,2,4-Butanetriol

(c) (1R,2R)-1,2-Cyclopentanediol

(d) 1-Ethylcyclobutanol

(e) 2-Chloro-3-hexyn-1-ol

(f) Tetrahydrofuran

(g) 2-Ethoxypentane

(h) Ethyl phenyl ether

(i) Diisopropyl ether

(j) 2-Ethoxyethanol

8.35

Starting with each of the following, outline a practical synthesis of 1-butanol.

(a) 1-Butene   (b) 2-Butene   (c) 1-Chlorobutane   (d) 2-Chlorobutane   (e) 1-Butyne

8.36

Show how you might prepare 2-bromobutane from

(a) 2-Butanol, $CH_3CH_2CHOHCH_3$

(b) 1-Butanol, $CH_3CH_2CH_2CH_2OH$

(c) 1-Butene

(d) 1-Butyne

8.37

Show how you might prepare 1-bromobutane from each of the compounds listed in Problem 8.36.

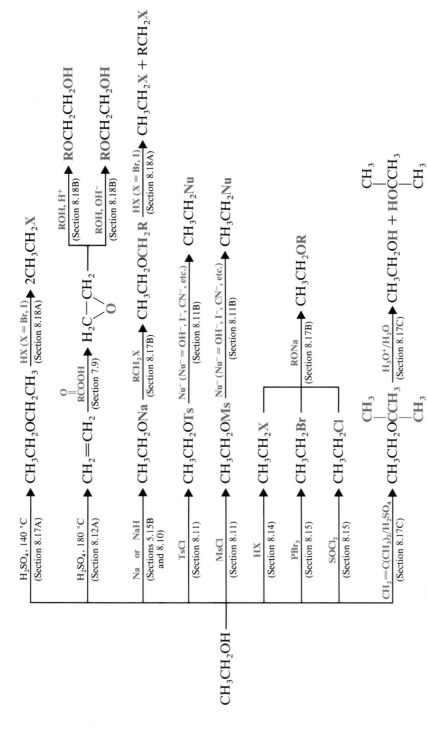

FIGURE 8.5 A summary of important reactions of alcohols and ethers starting with ethanol.

8.38

Show how you might carry out the following transformations:

(a) Cyclohexanol ⟶ chlorocyclohexane

(b) Cyclohexene ⟶ chlorocyclohexane

(c) 1-Methylcyclohexene ⟶ 1-bromo-1-methylcyclohexane

(d) 1-Methylcyclohexene ⟶ *trans*-2-methylcyclohexanol

(e) 1-Bromo-1-methylcyclohexane ⟶ cyclohexylmethanol

8.39

Give structures and names for the compounds that would be formed when 1-propanol is treated with each of the following reagents:

(a) Sodium metal

(b) Sodium metal, then 1-bromobutane

(c) Methanesulfonyl chloride

(d) *p*-Toluenesulfonyl chloride

(e) Product of (c), then CH_3ONa

(f) Product of (d), then KI

(g) Phosphorus trichloride

(h) Thionyl chloride

(i) Sulfuric acid at 140 °C

(j) Refluxing concentrated hydrobromic acid

8.40

Give structures and names for the compounds that would be formed when 2-propanol is treated with each of the reagents given in Problem 8.39.

8.41

What compounds would you expect to be formed when each of the following ethers is refluxed with excess concentrated hydrobromic acid?

(a) Ethyl methyl ether   (b) Ethyl *tert*-butyl ether   (c) Tetrahydrofuran   (d) 1,4-Dioxane

8.42

Write a mechanism that accounts for the following reaction:

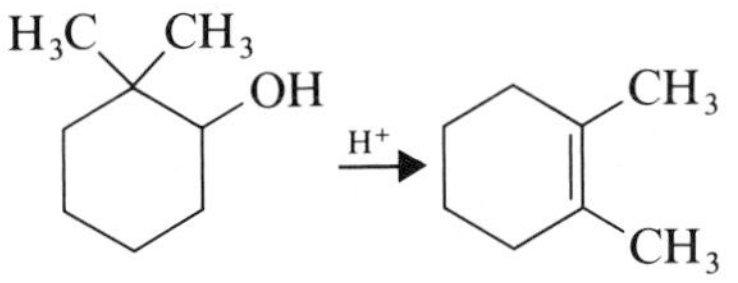

8.43

Show how you would utilize the hydroboration–oxidation procedure to prepare each of the following alcohols:

(a) 3,3-Dimethyl-1-butanol

(b) 1-Hexanol

(c) 2-Phenylethanol

(d) *trans*-2-Methylcyclopentanol

8.44

Write three-dimensional formulas for the product formed when 1-methylcyclohexene is treated with each of the following reagents. In each case, designate the location of deuterium or tritium atoms.

(a) (1) THF:BH_3 (2) CH_3CO_2T

(b) (1) THF:BD_3 (2) CH_3CO_2D

(c) (1) THF:BD_3 (2) NaOH, H_2O_2, H_2O

8.45

Starting with 2-methylpropene (isobutylene) and using any other needed reagents, outline a synthesis of each of the following:

(a) $(CH_3)_2CHCH_2OH$ (c) $(CH_3)_2CDCH_2T$

(b) $(CH_3)_2CHCH_2T$ (d) $(CH_3)_2CHCH_2OCH_2CH_3$

8.46

Show how you would use oxymercuration–demercuration to prepare each of the following alcohols from the appropriate alkene:

(a) 2-Pentanol   (c) 3-Methyl-3-pentanol

(b) 1-Cyclopentylethanol   (d) 1-Ethylcyclopentanol

8.47

Give stereochemical formulas for each product **A–L** and answer the questions given in parts (b) and (g).

(a) 1-Methylcyclobutene $\xrightarrow[\text{(2) H}_2\text{O}_2,\ \text{OH}^-]{\text{(1) THF: BH}_3}$ **A** $(C_5H_{10}O)$ $\xrightarrow[\text{OH}^-]{\text{TsCl}}$

B $(C_{12}H_{16}SO_3)$ $\xrightarrow{\text{OH}^-}$ **C** $(C_5H_{10}O)$

(b) What is the stereoisomeric relationship between **A** and **C**?

(c) **B** $(C_{12}H_{16}SO_3)$ $\xrightarrow{\text{I}^-}$ **D** (C_5H_9I)

(d) *trans*-4-Methylcyclohexanol $\xrightarrow[\text{OH}^-]{\text{MsCl}}$ **E** $(C_8H_{16}SO_3)$ $\xrightarrow{\text{HC} \equiv \text{CNa}}$

F (C_9H_{14}) $\xrightarrow{\text{H}_3\text{O}^+,\ \text{Hg}^{2+},\ \text{H}_2\text{O}}$ **G** $(C_9H_{16}O)$

(e) (*R*)-2-Butanol $\xrightarrow{\text{NaH}}$ **[H** $(C_4H_9ONa)]$ $\xrightarrow{\text{CH}_3\text{I}}$ **J** $(C_5H_{12}O)$

(f) (*R*)-2-Butanol $\xrightarrow{\text{MsCl}}$ **K** $(C_5H_{12}SO_3)$ $\xrightarrow{\text{CH}_3\text{ONa}}$ **L** $(C_5H_{12}O)$

(g) What is the stereoisomeric relationship between **J** and **L**?

8.48

When the 3-bromo-2-butanol with the stereochemical structure **A** is treated with concentrated HBr it yields *meso*-2,3-dibromobutane; a similar reaction of the 3-bromo-2-butanol **B** yields (±)-2,3-dibromobutane. This classic experiment performed in 1939 by S. Winstein and H. J. Lucas was the starting point for a series of investigations of what are called *neighboring group effects* (cf. Special Topic O). Propose mechanisms that will account for the stereochemistry of these reactions.

A   **B**

CHAPTER NINE

Vitamin B$_6$. The chemical structure of Vitamin B$_6$ is described in Section 19.4.

Free Radical Reactions

9.1 INTRODUCTION

Thus far all of the reactions whose mechanisms we have studied have been **ionic reactions.** Ionic reactions are those in which covalent bonds break **heterolytically** and in which ions are involved as reactants, intermediates, and products. As pointed out earlier (Section 5.1A), another broad category of reactions involves **homolysis** of covalent bonds with the production of molecular fragments possessing unpaired electrons, called **radicals** or **free radicals.**

$$-\overset{|}{\underset{|}{C}} : Z \xrightarrow{\text{homolysis}} \quad -\overset{|}{\underset{|}{C}} \cdot \quad + Z \cdot$$

A carbon radical
(or free radical)

We shall now study the mechanisms of a number of important reactions of this type. Before doing so, however, it will be useful to examine the energy changes that occur when covalent bonds break homolytically.

9.2 HOMOLYTIC BOND DISSOCIATION ENERGIES

When atoms combine to form molecules, energy is released as covalent bonds form. The molecules of the products have lower enthalpy than the separate atoms. When hydrogen atoms combine to form hydrogen molecules, for example, the reaction is *exothermic;* it evolves 104 kcal of heat for every mole of hydrogen that is produced. Similarly, when chlorine atoms combine to form chlorine molecules, the reaction evolves 58 kcal/mole of chlorine produced.

$$
\begin{aligned}
\text{H} \cdot + \text{H} \cdot &\longrightarrow \text{H}-\text{H} & \Delta H^\circ &= -104 \text{ kcal/mole} \\
\text{Cl} \cdot + \text{Cl} \cdot &\longrightarrow \text{Cl}-\text{Cl} & \Delta H^\circ &= -\ 58 \text{ kcal/mole}
\end{aligned}
\Bigg\}
\begin{array}{l}
\textit{Bond formation} \\
\textit{is an exothermic process}
\end{array}
$$

To break covalent bonds, energy must be supplied. Reactions in which only bond breaking occurs are always endothermic. The energy required to break the covalent bonds of hydrogen or chlorine homolytically is exactly equal to that evolved when the separate atoms combine to form molecules. In the bond cleavage reaction, however, ΔH° is positive.

$$H—H \longrightarrow H\cdot + H\cdot \qquad \Delta H° = +104 \text{ kcal/mole}$$

$$Cl—Cl \longrightarrow Cl\cdot + Cl\cdot \qquad \Delta H° = +58 \text{ kcal/mole}$$

The energies required to break covalent bonds homolytically have been determined experimentally for many types of covalent bonds. These energies are called **homolytic bond dissociation energies,** and they are usually abbreviated by the symbol

TABLE 9.1 Single-bond homolytic dissociation energies $DH°$ at 25 °C

$$A:B \longrightarrow A\cdot + B\cdot$$

BOND BROKEN (shown in red)	kcal/ mole	kJ/ mole	BOND BROKEN (shown in red)	kcal/ mole	kJ/ mole
H—H	104	435	$(CH_3)_2CH—H$	94.5	395
D—D	106	444	$(CH_3)_2CH—F$	105	439
F—F	38	159	$(CH_3)_2CH—Cl$	81	339
Cl—Cl	58	243	$(CH_3)_2CH—Br$	68	285
Br—Br	46	192	$(CH_3)_2CH—I$	53	222
I—I	36	151	$(CH_3)_2CH—OH$	92	385
H—F	136	569	$(CH_3)_2CH—OCH_3$	80.5	337
H—Cl	103	431	$(CH_3)_2CHCH_2—H$	98	410
H—Br	87.5	366	$(CH_3)_3C—H$	91	381
H—I	71	297	$(CH_3)_3C—Cl$	78.5	328
$CH_3—H$	104	435	$(CH_3)_3C—Br$	63	264
$CH_3—F$	108	452	$(CH_3)_3C—I$	49.5	207
$CH_3—Cl$	83.5	349	$(CH_3)_3C—OH$	90.5	379
$CH_3—Br$	70	293	$(CH_3)_3C—OCH_3$	78	326
$CH_3—I$	56	234	$C_6H_5CH_2—H$	85	356
$CH_3—OH$	91.5	383	$CH_2=CHCH_2—H$	85	356
$CH_3—OCH_3$	80	335	$CH_2=CH—H$	108	452
$CH_3CH_2—H$	98	410	$C_6H_5—H$	110	460
$CH_3CH_2—F$	106	444	$HC\equiv C—H$	125	523
$CH_3CH_2—Cl$	81.5	341	$CH_3—CH_3$	88	368
$CH_3CH_2—Br$	69	289	$CH_3CH_2—CH_3$	85	356
$CH_3CH_2—I$	53.5	224	$CH_3CH_2CH_2—CH_3$	85	356
$CH_3CH_2—OH$	91.5	383	$CH_3CH_2—CH_2CH_3$	82	343
$CH_3CH_2—OCH_3$	80	335	$(CH_3)_2CH—CH_3$	84	351
			$(CH_3)_3C—CH_3$	80	335
$CH_3CH_2CH_2—H$	98	410	HO—H	119	498
$CH_3CH_2CH_2—F$	106	444	HOO—H	90	377
$CH_3CH_2CH_2—Cl$	81.5	341	HO—OH	51	213
$CH_3CH_2CH_2—Br$	69	289	$CH_3CH_2O—OCH_3$	44	184
$CH_3CH_2CH_2—I$	53.5	224	$CH_3CH_2O—H$	103	431
$CH_3CH_2CH_2—OH$	91.5	383			
$CH_3CH_2CH_2—OCH_3$	80	335	$CH_3\overset{\displaystyle O}{\overset{\|}{C}}—H$	87	364

$DH°$. The homolytic bond dissociation energies of hydrogen and chlorine, for example, might be written in the following way.

$$H\!-\!H \qquad\qquad Cl\!-\!Cl$$
$$(DH° = 104\ kcal/mole) \qquad (DH° = 58\ kcal/mole)$$

The homolytic bond dissociation energies of a variety of covalent bonds are listed in Table 9.1.

9.2A Homolytic Bond Dissociation Energies and Heats of Reaction

Bond dissociation energies have, as we shall see, a variety of uses. They can be used, for example, to calculate the enthalpy change $(\Delta H°)$ for a reaction. To make such a calculation (see following reaction) we must remember that for bond breaking $\Delta H°$ is positive and for bond formation $\Delta H°$ is negative. Let us consider, for example, the reaction of hydrogen and chlorine to produce 2 moles of hydrogen chloride. From Table 9.1 we get the following values of $DH°$.

$$H\!-\!H \quad + \quad Cl\!-\!Cl \quad \longrightarrow \quad 2H\!-\!Cl$$
$$(DH° = 104) \quad (DH° = 58) \qquad\qquad (DH° = 103) \times 2$$
$$+162\ kcal/mole\ is\ required \qquad -206\ kcal/mole\ is\ evolved$$
$$for\ bond\ cleavage \qquad\qquad in\ bond\ formation$$

Overall, the reaction is exothermic:

$$\Delta H° = (-206\ kcal/mole + 162\ kcal/mole) = -44\ kcal/mole$$

For the purpose of our calculation, we have assumed a particular pathway, that amounts to:

$$H\!-\!H \longrightarrow 2H\cdot$$

and

$$Cl\!-\!Cl \longrightarrow 2Cl\cdot$$

then

$$2H\cdot + 2Cl\cdot \longrightarrow 2H\!-\!Cl$$

This is not the way the reaction actually occurs. Nonetheless, the heat of reaction, $\Delta H°$, is a thermodynamic quantity that is dependent *only* on the initial and final states of the reacting molecules. $\Delta H°$ is independent of the path followed and, for this reason, our calculation is valid.

PROBLEM 9.1

Calculate the heat of reaction, $\Delta H°$, for the following reactions:

(a) $H_2 + Br_2 \longrightarrow 2HBr$

(b) $CH_3CH_3 + F_2 \longrightarrow CH_3CH_2F + HF$

(c) $CH_3CH_3 + I_2 \longrightarrow CH_3CH_2I + HI$

(d) $CH_4 + Cl_2 \longrightarrow CH_3Cl + HCl$

(e) $(CH_3)_3CH + Cl_2 \longrightarrow (CH_3)_3CCl + HCl$

(f) $(CH_3)_3CH + Br_2 \longrightarrow (CH_3)_3CBr + HBr$

(g) $CH_3CH_2CH_3 \longrightarrow CH_3CH_2\cdot + CH_3\cdot$

(h) $2CH_3CH_2\cdot \longrightarrow CH_3CH_2CH_2CH_3$

9.2B Homolytic Bond Dissociation Energies and the Relative Stabilities of Free Radicals

Homolytic bond dissociation energies also provide us with a convenient way to estimate the relative stabilities of free radicals. If we examine the data given in Table 9.1, we find the following values of $DH°$ for the primary and secondary C—H bonds of propane:

$$CH_3CH_2CH_2—H \qquad\qquad (CH_3)_2CH—H$$
$$(DH° = 98 \text{ kcal/mole}) \qquad (DH° = 94.5 \text{ kcal/mole})$$

This means that for the reaction in which the designated C—H bonds are broken homolytically, the values of $\Delta H°$ are those given here.

$$CH_3CH_2CH_2—H \longrightarrow CH_3CH_2CH_2\cdot + H\cdot \qquad \Delta H° = +98 \text{ kcal/mole}$$
Propyl radical
(a 1° radical)

$$CH_3\underset{\underset{H}{|}}{C}HCH_3 \longrightarrow CH_3\overset{\cdot}{C}HCH_3 + H\cdot \qquad \Delta H° = +94.5 \text{ kcal/mole}$$
Isopropyl radical
(a 2° radical)

These reactions resemble each other in two respects: They both begin with the same alkane (propane), and they both produce an alkyl radical and a hydrogen atom. They differ, however, in the amount of energy required and in the type of carbon radical being produced.* These two differences are related to each other.

More energy must be supplied to produce a primary carbon radical (the propyl radical) from propane than is required to produce a secondary carbon radical (the isopropyl radical) from the same compound. This must mean that the primary radical has absorbed more energy and thus has greater *potential energy.* Because the relative stability of a chemical species is inversely related to its potential energy, the secondary radical must be the *more stable* radical (Fig. 9.1*a*). In fact, the secondary isopropyl radical is more stable than the primary propyl radical by 3.5 kcal/mole.

We can use the data in Table 9.1 to make a similar comparison of the *tert*-butyl radical (a 3° radical) and the isobutyl radical (a 1° radical) relative to isobutane.

$$CH_3—\underset{\underset{H}{|}}{\overset{\overset{CH_3}{|}}{C}}—CH_2—H \longrightarrow CH_3\overset{\overset{CH_3}{|}}{\underset{}{\overset{\cdot}{C}}}CH_3 + H\cdot \qquad \Delta H° = +91 \text{ kcal/mole}$$
tert-Butyl
radical
(a 3° radical)

$$CH_3—\underset{\underset{H}{|}}{\overset{\overset{CH_3}{|}}{C}}—CH_2—H \longrightarrow CH_3\overset{\overset{CH_3}{|}}{\underset{}{C}}H\overset{\cdot}{C}H_2 + H\cdot \qquad \Delta H° = +98 \text{ kcal/mole}$$
Isobutyl radical
(a 1° radical)

*Carbon radicals are classified as being 1°, 2°, or 3° on the basis of the carbon atom that has the unpaired electron.

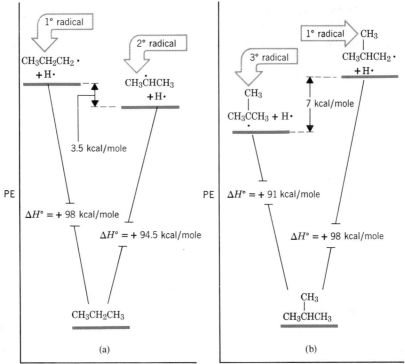

FIGURE 9.1 (a) A comparison of the potential energies of the propyl
radical (+H ·) and the isopropyl radical (+H ·) relative to propane. The
isopropyl radical—a 2° radical—is more stable than the 1° radical by
3.5 kcal/mole. (b) A comparison of the potential energies of the *tert*-
butyl radical (+H ·) and the isobutyl radical (+H ·) relative to isobutane.
The 3° radical is more stable than the 1° radical by 7 kcal/mole.

Here we find (Fig. 9.1b) that the difference in stability of the two radicals is even
larger. The tertiary radical is more stable than the primary radical by 7 kcal/mole.

The kind of pattern that we find in these examples is found with alkyl radicals
generally; overall their relative stabilities are the following:

Tertiary > Secondary > Primary > Methyl

$$\underset{\overset{|}{C}}{\overset{\overset{C}{|}}{C-C\cdot}} > \underset{\overset{|}{H}}{\overset{\overset{C}{|}}{C-C\cdot}} > \underset{\overset{|}{H}}{\overset{\overset{H}{|}}{C-C\cdot}} > \underset{\overset{|}{H}}{\overset{\overset{H}{|}}{H-C\cdot}}$$

PROBLEM 9.2

(a) Sketch diagrams similar to those in Fig. 9.1 showing the potential energy of
$(CH_3)_2CH\cdot + H\cdot$ relative to propane and the potential energy of $CH_3CH_2\cdot +$
$H\cdot$ relative to ethane. Align the two diagrams so that the potential energy of the
alkane is the same in each. What does this indicate about the stability of an ethyl
radical and an isopropyl radical relative to the alkane from which each is
derived? (b) Repeat this process by drawing potential energy diagrams show-

ing the energy of $CH_3CH_2 \cdot + H \cdot$ relative to ethane and of $CH_3 \cdot + H \cdot$ relative to methane. What do these graphs indicate about the relative stabilities of an ethyl radical and a methyl radical? (c) Make similar sketches that compare an ethyl radical with a propyl radical. (d) Account for the similarity of the potential energy diagrams in part (c).

PROBLEM 9.3

One can also estimate the relative stabilities of free radicals by comparing the homolytic bond dissociation energies of the C—X bonds of haloalkanes. Show how this can be done with CH_3—Cl, CH_3CH_2—Cl, $(CH_3)_2CH$—Cl, and $(CH_3)_3C$—Cl.

9.3 CHLORINATION OF METHANE: THE REACTION MECHANISM

When we studied alkanes (Section 3.17), we found that they react with the halogens (fluorine, chlorine, and bromine) to form haloalkanes. These *halogenation* reactions are substitution reactions that take place by a free radical mechanism. Let us begin our study of free radical chemistry by examining a simple example of an alkane halogenation: the reaction of methane with chlorine that takes place in the gas phase.

Several important experimental observations can be made about this reaction:

$$CH_4 + Cl_2 \longrightarrow CH_3Cl + HCl \ (+ \ CH_2Cl_2, \ CHCl_3, \ \text{and} \ CCl_4)$$

1. **The reaction is promoted by heat or light.** At room temperature methane and chlorine do not react at a perceptible rate as long as the mixture is kept away from light. Methane and chlorine do react, however, at room temperature if the gaseous reaction mixture is irradiated with ultraviolet light, and methane and chlorine do react in the dark, if the gaseous mixture is heated to temperatures greater than 100 °C.
2. **The light-promoted reaction is highly efficient.** A relatively small number of light photons permits the formation of relatively large amounts of chlorinated product.

A mechanism that is consistent with these observations has several steps. The first step involves the fragmentation of a chlorine molecule, by heat or light, into two chlorine atoms.

$$\text{Step 1} \qquad :\!\overset{..}{\underset{..}{Cl}}\!:\!\overset{..}{\underset{..}{Cl}}\!: \xrightarrow[\substack{\text{or} \\ \text{light}}]{\text{heat}} 2:\!\overset{..}{\underset{..}{Cl}}\cdot$$

Chlorine is known, from other evidence, to undergo such reactions. It can be shown, moreover, that the frequency of light that promotes the chlorination of methane is a frequency that is absorbed by chlorine molecules and not by methane molecules.

The second and third steps of the mechanism are as follows:*

$$\text{Step 2}$$

$$\ddot{\text{Cl}} \cdot + \text{H} : \underset{\underset{\text{H}}{|}}{\overset{\overset{\text{H}}{|}}{\text{C}}} - \text{H} \longrightarrow \text{H} : \ddot{\text{Cl}} : + \cdot \underset{\underset{\text{H}}{|}}{\overset{\overset{\text{H}}{|}}{\text{C}}} - \text{H}$$

$$\text{Step 3}$$

$$\text{H} - \underset{\underset{\text{H}}{|}}{\overset{\overset{\text{H}}{|}}{\text{C}}} \cdot + : \ddot{\text{Cl}} : \ddot{\text{Cl}} : \longrightarrow \text{H} - \underset{\underset{\text{H}}{|}}{\overset{\overset{\text{H}}{|}}{\text{C}}} : \ddot{\text{Cl}} : + \cdot \ddot{\text{Cl}} :$$

Step 2 is the abstraction of a hydrogen atom from the methane molecule by a chlorine atom. This step results in the formation of a molecule of hydrogen chloride and a methyl radical.

In step 3 the highly reactive methyl radical reacts with a chlorine molecule by abstracting a chlorine atom. This results in the formation of a molecule of chloromethane (one of the ultimate products of the reaction) and a *chlorine atom.* This latter product is particularly significant, for the chlorine atom formed in step 3 can attack another methane molecule and cause a repetition of step 2. Then, step 3 is repeated, and so forth for hundreds or thousands of times. (With each repetition of step 3 a molecule of chloromethane is produced.) This type of sequential, stepwise mechanism, in which each step generates the reactive intermediate that causes the next step to occur, is called a *chain reaction.*

Step 1 is called the **chain-initiating step.** In the chain-initiating step *radicals are created.* Steps 2 and 3 are called **chain-propagating steps.** In chain-propagating steps *one radical generates another.*

Chain Initiation

$$\text{Step 1} \qquad \text{Cl}_2 \xrightarrow[\text{light}]{\text{heat or}} 2\text{Cl} \cdot$$

Chain Propagation

$$\text{Step 2} \qquad \text{CH}_4 + \text{Cl} \cdot \longrightarrow \text{H} - \text{Cl} + \text{CH}_3 \cdot$$

$$\text{Step 3} \qquad \text{CH}_3 \cdot + \text{Cl}_2 \longrightarrow \text{CH}_3\text{Cl} + \text{Cl} \cdot$$

The chain nature of the reaction accounts for the observation that the light-promoted reaction is highly efficient. The presence of a relatively few atoms of chlorine at any given moment is all that is needed to cause the formation of many thousands of molecules of chloromethane.

What causes the chains to terminate? Why does one photon of light not promote the chlorination of all of the methane molecules present? We know that this does not happen because we find that at low temperatures continuous irradiation is required or the reaction slows and stops. The answer to these questions is the existence of *chain-terminating steps:* steps that occur infrequently, but occur often enough to *use up one or both of the reactive intermediates.* The continuous replace-

---

*These conventions are used in illustrating reaction mechanisms in this text.
1. Arrows ⌒ or ⌒⇀ always show the direction of movement of electrons.
2. Single-barbed arrows ⌒ show the attack (or movement) of an unpaired electron.
3. Double-barbed arrows ⌒⇀ show the attack (or movement) of an electron pair.

ment of intermediates used up by chain-terminating steps requires continuous irradiation. Plausible chain-terminating steps are

Chain Termination

$$H-\overset{\overset{\displaystyle H}{|}}{\underset{\underset{\displaystyle H}{|}}{C}}\cdot \; + \; \cdot\ddot{C}l \colon \longrightarrow H-\overset{\overset{\displaystyle H}{|}}{\underset{\underset{\displaystyle H}{|}}{C}}\colon\ddot{C}l \colon$$

$$H-\overset{\overset{\displaystyle H}{|}}{\underset{\underset{\displaystyle H}{|}}{C}}\cdot \; + \; \cdot\overset{\overset{\displaystyle H}{|}}{\underset{\underset{\displaystyle H}{|}}{C}}-H \longrightarrow H-\overset{\overset{\displaystyle H}{|}}{\underset{\underset{\displaystyle H}{|}}{C}}\colon\overset{\overset{\displaystyle H}{|}}{\underset{\underset{\displaystyle H}{|}}{C}}-H$$

and
$$\colon\ddot{C}l\cdot \; + \; \cdot\ddot{C}l\colon \longrightarrow \colon\ddot{C}l\colon\ddot{C}l\colon$$

This last step probably occurs least frequently. The two chlorine atoms are highly energetic; as a result, the simple diatomic chlorine molecule that is formed has to dissipate its excess energy rapidly by colliding with some other molecule or the walls of the container. Otherwise it simply flies apart again. By contrast, chloromethane and ethane, formed in the other two chain-terminating steps, can dissipate their excess energy through vibrations of their C—H bonds.

Our free radical mechanism also explains how the reaction of methane with chlorine produces the more highly halogenated products, CH_2Cl_2, $CHCl_3$, and CCl_4 (as well as additional HCl). As the reaction progresses, chloromethane (CH_3Cl) accumulates in the mixture and its hydrogen atoms, too, are susceptible to abstraction by chlorine. Thus chloromethyl radicals are produced that lead to dichloromethane (CH_2Cl_2)

Step 2a
$$Cl\cdot \; + \; H\colon\overset{\overset{\displaystyle Cl}{|}}{\underset{\underset{\displaystyle H}{|}}{C}}-H \longrightarrow H\colon Cl \; + \; \cdot\overset{\overset{\displaystyle Cl}{|}}{\underset{\underset{\displaystyle H}{|}}{C}}-H$$

Step 3a
$$H-\overset{\overset{\displaystyle Cl}{|}}{\underset{\underset{\displaystyle H}{|}}{C}}\cdot \; + \; Cl\colon Cl \longrightarrow H-\overset{\overset{\displaystyle Cl}{|}}{\underset{\underset{\displaystyle H}{|}}{C}}\colon Cl \; + \; Cl\cdot$$

Then step 2a is repeated, then step 3a is repeated, and so on. Each repetition of step 2a yields a molecule of HCl, and each repetition of step 3a yields a molecule of CH_2Cl_2.

PROBLEM 9.4
Write mechanisms showing how $CHCl_3$ and CCl_4 might be formed in the reaction mixture when methane is chlorinated.

PROBLEM 9.5

When methane is chlorinated, among the products are found traces of chloro-ethane. How is it formed? Of what significance is its formation?

PROBLEM 9.6

If our goal is to synthesize CCl_4 in maximum yield, this can be accomplished by using a large excess of chlorine. Explain.

9.4 CHLORINATION OF METHANE: ENERGY CHANGES

We saw in Section 9.2A that we can calculate the overall heat of reaction from bond dissociation energies. We can also calculate the heat of reaction for each individual step of a mechanism.

Chain Initiation

Step 1 $Cl-Cl \longrightarrow 2Cl\cdot$ $\Delta H° = +58$ kcal/mole
             $(DH° = 58)$

Chain Propagation

Step 2 $CH_3-H \longrightarrow H-Cl$ $\Delta H° = +1$ kcal/mole
            $(DH° = 104)$ $(DH° = 103)$

Step 3 $Cl-Cl \longrightarrow CH_3-Cl$ $\Delta H° = -25.2$ kcal/mole
            $(DH° = 58)$ $(DH° = 83.5)$

Chain Termination

         $CH_3\cdot + Cl\cdot \longrightarrow CH_3-Cl$ $\Delta H° = -83.5$ kcal/mole
                   $(DH° = 83.5)$

         $CH_3\cdot + \cdot CH_3 \longrightarrow CH_3-CH_3$ $\Delta H° = -88$ kcal/mole
                   $(DH° = 88)$

         $Cl\cdot + Cl\cdot \longrightarrow Cl-Cl$ $\Delta H° = -58$ kcal/mole
                   $(DH° = 58)$

In the chain-initiating step only one bond is broken—the bond between two chlorine atoms—and no bonds are formed. The heat of reaction for this step is simply the bond dissociation energy for a chlorine molecule, and it is highly endothermic.

In the chain-terminating steps bonds are formed, but no bonds are broken. As a result, all of the chain-terminating steps are highly exothermic.

Each of the chain-propagating steps, on the other hand, requires the breaking of one bond and the formation of another. The value of $\Delta H°$ for each of these steps is the difference between the bond dissociation energy of the bond that is broken and the bond dissociation energy for the bond that is formed. The first chain-propagating step

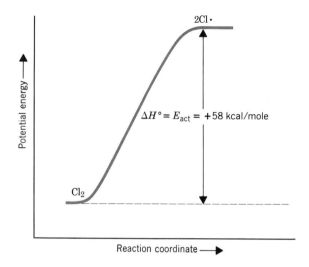

FIGURE 9.3 The potential energy diagram for the dissociation of a chlorine molecule into chlorine atoms.

4. The energy of activation for a **gas-phase** reaction in which **radicals combine to form molecules is usually zero.*** In reactions of this type the problem of non-simultaneous bond formation and bond rupture does not exist; only one process occurs: that of bond formation. All of the chain-terminating steps in the chlorination of methane fall into this category. An example is the combination of two methyl radicals to form a molecule of ethane.

$$2CH_3 \cdot \longrightarrow CH_3 - CH_3 \qquad \Delta H° = -88 \text{ kcal/mole}$$
$$(DH° = 88) \qquad E_{act} = 0$$

Figure 9.4 illustrates the potential energy changes that occur in this reaction.

In Section 9.6 we shall see how we can estimate energies of activation by taking advantage of the fact that the transition states of reactions resemble the reactants and products.

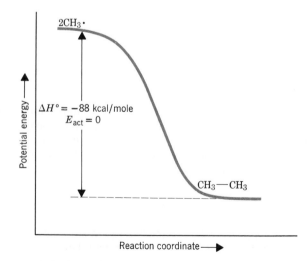

FIGURE 9.4 The potential energy diagram for the combination of two methyl radicals to form a molecule of ethane.

*This rule also applies only to radical reactions taking place in the gas phase. It does not apply to reactions taking place in solution, especially where ions are involved.

PROBLEM 9.9

When gaseous ethane is heated to a very high temperature, free radical reactions take place that produce (among other products) methane and butane. This type of change is called thermal cracking. Among the reactions that take place when ethane undergoes thermal cracking are the following:

(1) $CH_3CH_3 \longrightarrow 2CH_3 \cdot$

(2) $CH_3 \cdot + CH_3CH_3 \longrightarrow CH_4 + CH_3CH_2 \cdot$

(3) $2CH_3CH_2 \cdot \longrightarrow CH_3CH_2CH_2CH_3$

(4) $CH_3CH_2 \cdot \longrightarrow CH_2{=}CH_2 + H \cdot$

(5) $CH_3 \cdot + H \cdot \longrightarrow CH_4$

(a) For which reaction(s) would you expect E_{act} to equal zero? (b) For which would you expect E_{act} to be greater than zero? (c) For which would you expect E_{act} to equal $\Delta H°$?

PROBLEM 9.10

Sketch potential energy diagrams for the following reactions. Label the heat of reaction ($\Delta H°$) and the energy of activation (E_{act}) in each case. [Notice that the reactions in (1) and (2) are the reverse of those shown in Fig. 9.2.]

(1) $CH_3 \cdot + HCl \longrightarrow CH_3{-}H + Cl \cdot$

(2) $CH_3 \cdot + HBr \longrightarrow CH_3{-}H + Br \cdot$

(3) $CH_3{-}CH_3 \longrightarrow 2CH_3 \cdot$

(4) $Br{-}Br \longrightarrow 2Br \cdot$

(5) $2Cl \cdot \longrightarrow Cl{-}Cl$

9.5 REACTION OF METHANE WITH OTHER HALOGENS

The *reactivity* of one substance toward another is measured by the *rate* at which the two substances react. A reagent that reacts very rapidly with a particular substance is said to be highly reactive toward that substance. One that reacts slowly or not at all under the same experimental conditions (e.g., concentration, pressure, and temperature) is said to have a low relative reactivity or to be unreactive. The reactions of the halogens (fluorine, chlorine, bromine, and iodine) with methane show a wide spread of relative reactivities. Fluorine is most reactive—so reactive, in fact, that without special precautions mixtures of fluorine and methane explode. Chlorine is the next most reactive. However, the chlorination of methane is easily controlled by the judicious control of heat and light. Bromine is much less reactive toward methane than chlorine, and iodine is so unreactive that for all practical purposes we can say that no reaction takes place at all.

If the mechanisms for fluorination, bromination, and iodination of methane are the same as for its chlorination, we can explain the wide variation in reactivity of the halogens by a careful examination of $\Delta H°$ and E_{act} for each step.

Fluorination

	$\Delta H°$(kcal/mole)	E_{act}(kcal/mole)

Chain Initiation

$$F_2 \longrightarrow 2F\cdot$$

| | $+38$ | $+38$ |

Chain Propagation

$$F\cdot + CH_4 \longrightarrow HF + CH_3\cdot \qquad -32 \qquad +1.2$$
$$CH_3\cdot + F_2 \longrightarrow CH_3F + F\cdot \qquad \underline{-70} \qquad \text{small}$$
$$\text{Overall } \Delta H° = -102$$

The chain-initiating step in fluorination is highly endothermic and thus has a high energy of activation.

If we did not know otherwise, we might carelessly conclude from the energy of activation of the chain-initiating step alone that fluorine would be quite unreactive toward methane. (If we then proceeded to try the reaction, as a result of this careless assessment, the results would be literally disastrous.) We know, however, that the chain-initiating step occurs only infrequently relative to the chain-propagating steps. One initiating step is able to produce thousands of fluorination reactions. As a result, the high activation energy for this step is not an impediment to the reaction.

Chain-propagating steps, by contrast, cannot afford to have high energies of activation. If they do, the highly reactive intermediates are consumed by chain-terminating steps before the chains progress very far. Both of the chain-propagating steps in fluorination have very small energies of activation. This allows a relatively large fraction of energetically favorable collisions even at room temperature. Moreover, the overall heat of reaction, $\Delta H°$, is very large. This means that as the reaction occurs, a large quantity of heat is evolved. This heat may accumulate in the mixture faster than it dissipates to the surroundings, causing the temperature to rise and with it a rapid increase in the frequency of additional chain-initiating steps that would generate additional chains. These two factors, the low energy of activation for the chain-propagating steps, and the large overall heat of reaction, account for the high reactivity of fluorine toward methane.*

Chlorination

	$\Delta H°$(kcal/mole)	E_{act}(kcal/mole)

Chain Initiation

$$Cl_2 \longrightarrow 2Cl\cdot$$

| | $+58$ | $+58$ |

Chain Propagation

$$Cl\cdot + CH_4 \longrightarrow HCl + CH_3\cdot \qquad +1 \qquad +3.8$$
$$CH_3\cdot + Cl_2 \longrightarrow CH_3Cl + Cl\cdot \qquad \underline{-25.5} \qquad \text{small}$$
$$\text{Overall } \Delta H° = -24.5$$

*Fluorination reactions can be controlled. This is usually accomplished by diluting both the hydrocarbon and the fluorine with an inert gas such as helium before bringing them together. The reaction is also carried out in a reactor packed with copper shot. The copper, by absorbing the heat produced, moderates the reaction.

The higher energy of activation of the first chain-propagating step (the hydrogen abstraction step) in chlorination of methane ($+3.8$ kcal/mole), versus the lower energy of activation ($+1.2$ kcal/mole) in fluorination, partly explains the lower reactivity of chlorine. The greater energy required to break the chlorine–chlorine bond in the initiating step ($+58$ kcal/mole for Cl_2 versus $+38$ kcal/mole for F_2) has some effect, too. However, the much greater overall heat of reaction in fluorination probably plays the greatest role in accounting for the much greater reactivity of fluorine.

Bromination

	$\Delta H°$(kcal/mole)	E_{act}(kcal/mole)
Chain Initiation		
$Br_2 \longrightarrow 2Br\cdot$	$+46$	$+46$
Chain Propagation		
$Br\cdot + CH_4 \longrightarrow HBr + CH_3\cdot$	$+16.5$	$+18.6$
$CH_3\cdot + Br_2 \longrightarrow CH_3Br + Br\cdot$	-24	small
Overall $\Delta H° = -7.5$		

In contrast to chlorination, the hydrogen-atom abstraction step in bromination has a very high energy of activation ($E_{act} = 18.6$ kcal/mole). This means that only a very tiny fraction of all of the collisions between bromine atoms and methane molecules will be energetically effective even at a temperature of 300 °C. Bromine, as a result, is much less reactive toward methane than chlorine even though the net reaction is slightly exothermic.

Iodination

	$\Delta H°$(kcal/mole)	E_{act}(kcal/mole)
Chain Initiation		
$I_2 \longrightarrow 2I\cdot$	$+36$	$+36$
Chain Propagation		
$I\cdot + CH_4 \longrightarrow HI + CH_3\cdot$	$+31$	$+33.5$
$CH_3\cdot + I_2 \longrightarrow CH_3I + I\cdot$	-20	small
Overall $\Delta H° = -11$		

The thermodynamic quantities for iodination of methane make it clear that the chain-initiating step is not responsible for the observed order of reactivities: $F_2 > Cl_2 > Br_2 > I_2$. The iodine–iodine bond is even weaker than the fluorine–fluorine bond. On this basis alone, one would predict that iodine would be the most reactive of the halogens. This clearly is not the case. Once again, it is the hydrogen-atom abstraction step that correlates with the experimentally determined order of reactivities. The energy of activation of this step in the iodine reaction (33.5 kcal/mole) is so large that only two collisions out of every 10^{12} have sufficient energy to produce reactions at 300 °C. As a result, iodination is not a feasible reaction experimentally.

Before we leave this topic, one further point needs to be made. We have given explanations of the relative reactivities of the halogens toward methane that have been based on energy considerations alone. This has been possible *only because the reactions are quite similar and thus have similar orientation factors* (Section 5.4C). Had the reactions been of different types, this kind of analysis would not have been proper and might have given incorrect explanations.

9.6 HALOGENATION OF HIGHER ALKANES

Higher alkanes react with halogens by the same kind of chain mechanisms as those that we have just seen. Ethane, for example, reacts with chlorine to produce chloroethane (ethyl chloride). The mechanism is as follows:

Chain Initiation

Step 1 $Cl_2 \xrightarrow[\text{or heat}]{\text{light}} 2Cl\cdot$

Chain Propagation

Step 2 $CH_3CH_2\!:\!H + \cdot Cl \longrightarrow CH_3CH_2\cdot + H\!:\!Cl$

Step 3 $CH_3CH_2\cdot + Cl\!:\!Cl \longrightarrow CH_3CH_2\!:\!Cl + Cl\cdot$

Then steps 2, 3, 2, 3, and so on.

Chain Termination

$$CH_3CH_2\cdot + \cdot Cl \longrightarrow CH_3CH_2\!:\!Cl$$

$$CH_3CH_2\cdot + \cdot CH_2CH_3 \longrightarrow CH_3CH_2\!:\!CH_2CH_3$$

$$Cl\cdot + \cdot Cl \longrightarrow Cl\!:\!Cl$$

PROBLEM 9.11

The energy of activation for the hydrogen-atom abstraction step in the chlorination of ethane is 1.0 kcal/mole. (a) Use the homolytic bond dissociation energies in Table 9.1 to calculate $\Delta H°$ for this step. (b) Sketch a potential energy diagram for the hydrogen-atom abstraction step in the chlorination of ethane similar to that for the chlorination of methane shown in Fig. 9.2a. (c) When an equimolar mixture of methane and ethane are chlorinated, the reaction yields far more ethyl chloride than methyl chloride ($\sim$ 400 molecules of ethyl chloride for every molecule of methyl chloride). Explain this greater yield of ethyl chloride.

PROBLEM 9.12

When ethane is chlorinated 1,1-dichloroethane and 1,2-dichloroethane as well as more highly chlorinated ethanes are formed in the mixture (cf. Section 3.17A). Write chain mechanisms accounting for the formation of 1,1-dichloroethane and 1,2-dichloroethane.

Chlorination of most alkanes whose molecules contain more than two carbon atoms gives a mixture of isomeric monochloro products (as well as more highly chlorinated compounds). Several examples follow. The percentages given are based on the total amount of monochloro products formed in each reaction.

$$CH_3CH_2CH_3 \xrightarrow[\text{light, 25 °C}]{Cl_2} CH_3CH_2CH_2Cl + CH_3\overset{\displaystyle |}{\underset{\displaystyle Cl}{C}}HCH_3$$

Propane	(45%)	(55%)
	Propyl chloride	Isopropyl chloride

$$\underset{\text{Isobutane}}{CH_3\overset{\displaystyle \overset{CH_3}{|}}{C}HCH_3} \xrightarrow[\substack{\text{light}\\25\text{ °C}}]{Cl_2} \underset{\substack{(63\%)\\\text{Isobutyl chloride}}}{CH_3\overset{\displaystyle \overset{CH_3}{|}}{C}HCH_2Cl} + \underset{\substack{(37\%)\\\textit{tert}\text{-Butyl}\\\text{chloride}}}{CH_3\overset{\displaystyle \overset{CH_3}{|}}{\underset{\displaystyle Cl}{C}}CH_3}$$

$$\underset{\text{2-Methylbutane}}{CH_3\overset{\displaystyle \overset{CH_3}{|}}{\underset{\displaystyle H}{C}}CH_2CH_3} \xrightarrow[300\text{ °C}]{Cl_2} \underset{\substack{(30\%)\\\text{1-Chloro-2-methyl-}\\\text{butane}}}{ClCH_2\overset{\displaystyle \overset{CH_3}{|}}{C}HCH_2CH_3} + \underset{\substack{(22\%)\\\text{2-Chloro-2-methyl-}\\\text{butane}}}{CH_3\overset{\displaystyle \overset{CH_3}{|}}{\underset{\displaystyle Cl}{C}}CH_2CH_3}$$

$$+ \underset{\substack{(33\%)\\\text{2-Chloro-3-methyl-}\\\text{butane}}}{CH_3\overset{\displaystyle \overset{CH_3}{|}}{C}H\overset{\displaystyle |}{\underset{\displaystyle Cl}{C}}HCH_3} + \underset{\substack{(15\%)\\\text{1-Chloro-3-methyl-}\\\text{butane}}}{CH_3\overset{\displaystyle \overset{CH_3}{|}}{C}HCH_2CH_2Cl}$$

The ratios of products that we obtain from chlorination reactions of higher alkanes are not identical with what we would expect if all the hydrogen atoms of the alkane were equally reactive. We find that there is a correlation between reactivity of different hydrogen atoms and the type of hydrogen atom (1°, 2°, or 3°) being replaced. The tertiary hydrogen atoms of an alkane are most reactive, secondary hydrogen atoms are next most reactive, and primary hydrogen atoms are the least reactive (see Problem 9.13).

PROBLEM 9.13

If we examine just the monochloro products of the reaction of isobutane with chlorine just given, we find that isobutyl chloride represents 63% of the mono-chlorinated product, while *tert*-butyl chloride represents 37%. Explain how this demonstrates that the tertiary hydrogen atom is more reactive. (*Hint:* Consider what percentages of the butyl chlorides would be obtained if the nine primary hydrogen atoms and the single tertiary hydrogen atom were all equally reactive.)

We can account for the relative reactivities of the primary, secondary, and tertiary hydrogen atoms in a chlorination reaction on the basis of the homolytic bond dissociation energies we saw earlier (Table 9.1). Of the three types, breaking a tertiary C—H bond requires the least energy, and breaking a primary C—H bond requires the most. Since the step in which the C—H bond is broken (i.e., the hydrogen-atom abstraction step) determines the location or orientation of the chlorination, we would expect the E_{act} for abstracting a tertiary hydrogen atom to be least and E_{act} for abstracting a primary hydrogen atom to be greatest. Thus tertiary hydrogen atoms should be most reactive, secondary hydrogen atoms should be the next most reactive, and primary hydrogen atoms should be the least reactive.

The differences in the rates with which primary, secondary, and tertiary hydrogen atoms are replaced by chlorine are not large, however. Chlorine, as a result, does not discriminate between the different types of hydrogen atoms in a way that makes chlorination of higher alkanes a generally useful laboratory procedure. (Alkane chlorinations do find use in some industrial processes, especially in those instances where mixtures of alkyl chlorides can be used.)

PROBLEM 9.14

Chlorination reactions of certain higher alkanes can be used for laboratory preparations. Examples are the preparation of neopentyl chloride from neopentane and cyclopentyl chloride from cyclopentane. What structural feature of these molecules makes this possible?

PROBLEM 9.15

The hydrogen-atom abstraction steps for most alkane chlorinations are exothermic. Show that this is true by calculating $\Delta H°$ for the reaction where Cl· abstracts (a) a primary hydrogen of ethane, (b) a secondary hydrogen of propane, and (c) a primary hydrogen of propane.

9.6A Selectivity of Bromine

Bromine is less reactive toward alkanes in general than chlorine, but bromine is more *selective* in the site of attack when it does react. As we saw in Section 3.17B, bromine shows a much greater ability to discriminate among the different types of hydrogen atoms. The reaction of isobutane and bromine, for example, gives almost exclusive replacement of the tertiary hydrogen atom.

$$CH_3-\underset{\underset{H}{|}}{\overset{\overset{CH_3}{|}}{C}}-CH_3 \xrightarrow[\text{light, 127 °C}]{Br_2} CH_3-\underset{\underset{Br}{|}}{\overset{\overset{CH_3}{|}}{C}}-CH_3 + CH_3-\underset{\underset{H}{|}}{\overset{\overset{CH_3}{|}}{C}}-CH_2Br$$

$$(>99\%) \qquad\qquad (trace)$$

Compare this with the reaction of isobutane with chlorine.

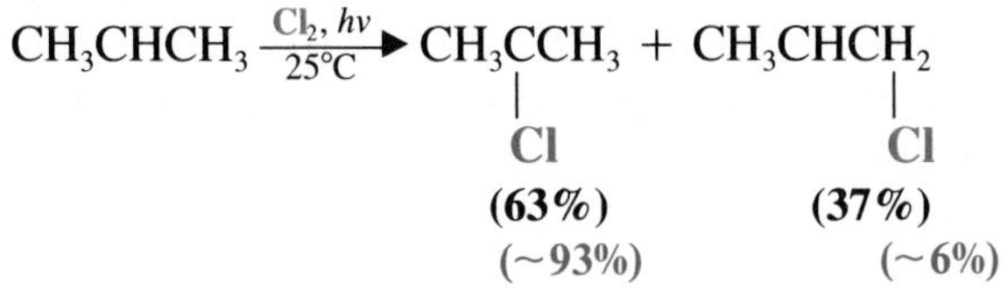

$$CH_3CHCH_3 \xrightarrow[25°C]{Cl_2,\ h\nu} CH_3CCH_3 + CH_3CHCH_2$$

$$\begin{array}{cc} | & | \\ Cl & Cl \\ (63\%) & (37\%) \\ (\sim 93\%) & (\sim 6\%) \end{array}$$

The greater selectivity of bromine can be explained in terms of transition state theory, and bromine's greater selectivity is directly related to its lower reactivity.* According to the Hammond-Leffler postulate (Section 5.13A) *the structure of the transition state of an endothermic step of a reaction resembles the products of that step more than it does the reactants. For an exothermic step the structure of the transition state is more like the reactants than the products.*

In a highly exothermic step the energy levels of the reactants and the transition state are close to each other (see Fig. 5.14 again). The transition state also lies close to the reactants along the *reaction coordinate.* This means that in the highly exothermic step, bond breaking has not proceeded very far when the transition state is reached. In the exothermic reactions of isobutane molecules with chlorine atoms, for example, relatively little carbon–hydrogen bond breaking has developed in the transition states. The transition states for the two hydrogen-atom abstraction steps might resemble those shown here.

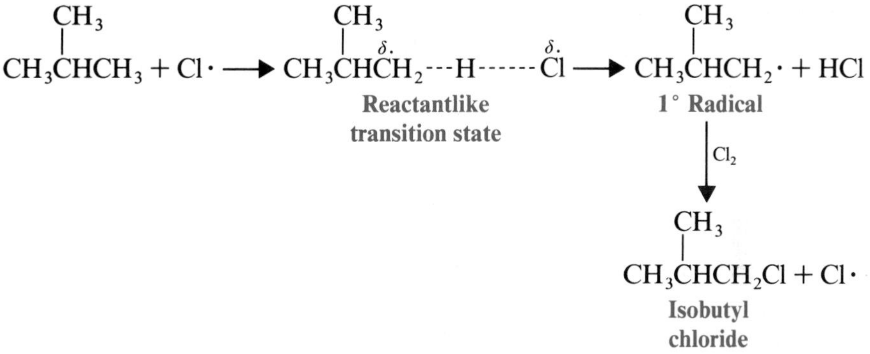

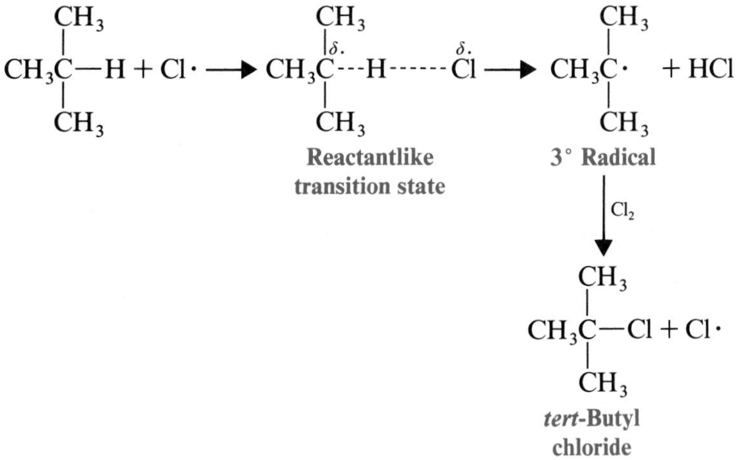

$$CH_3CHCH_3 + Cl\cdot \longrightarrow CH_3\overset{CH_3}{\overset{|}{C}}HCH_2\overset{\delta\cdot}{\cdots}H\text{------}\overset{\delta\cdot}{Cl} \longrightarrow CH_3CHCH_2\cdot + HCl$$

Reactantlike
transition state   1° Radical

$$\Big| Cl_2$$

$$\begin{array}{c} CH_3 \\ | \\ CH_3CHCH_2Cl + Cl\cdot \end{array}$$

Isobutyl
chloride

$$CH_3\overset{CH_3}{\underset{CH_3}{\overset{|}{C}}}-H + Cl\cdot \longrightarrow CH_3\overset{CH_3}{\underset{CH_3}{\overset{|\delta\cdot}{C}}}\cdots H\text{------}\overset{\delta\cdot}{Cl} \longrightarrow CH_3\overset{CH_3}{\underset{CH_3}{\overset{|}{C}}}\cdot + HCl$$

Reactantlike
transition state   3° Radical

$$\Big| Cl_2$$

$$CH_3\overset{CH_3}{\underset{CH_3}{\overset{|}{C}}}-Cl + Cl\cdot$$

tert-Butyl
chloride

*While reactivity and selectivity are related in these alkane halogenations, they are not necessarily related in other reactions.

Because the transition states in both cases are reactantlike in both structure and energy, they show relatively little resemblance to the products of the hydrogen-atom abstraction step, a 1° radical and a 3° radical. And because the reactants in both cases are the same, the exact type of C—H bond being broken (primary or tertiary) has a relatively small influence on the relative rates of the reactions. The two reactions proceed with similar (but not identical) rates because their respective activation energies are quite similar (Fig. 9.5).

The transition states of highly endothermic steps, on the other hand, lie close to the products on the potential energy coordinate and *along the reaction coordinate.* In highly endothermic steps the bond has broken to a considerable extent by the time the transition state is reached. The two hydrogen-atom abstraction steps in the reaction of isobutane with bromine are both highly endothermic. In these reactions considerable carbon–hydrogen bond breaking has occurred when the transition state is reached. These transition states might be depicted in the following way:

$$CH_3CHCH_3 + Br\cdot \longrightarrow CH_3\overset{CH_3}{\underset{}{C}}HCH_2\text{------}H\text{---}Br \longrightarrow CH_3\overset{CH_3}{\underset{}{C}}HCH_2\cdot + HBr$$

Productlike transition state   **1° Radical**

$$\downarrow Br_2$$

$$CH_3\overset{CH_3}{\underset{}{C}}HCH_2Br + Br\cdot$$

Isobutyl bromide

$$CH_3\overset{CH_3}{\underset{CH_3}{C}}-H + Br\cdot \longrightarrow CH_3\overset{CH_3}{\underset{CH_3}{C}}\text{------}H\text{---}Br \longrightarrow CH_3\overset{CH_3}{\underset{CH_3}{C}}\cdot + HBr$$

Productlike transition state   **3° Radical**

$$\downarrow Br_2$$

$$CH_3\overset{CH_3}{\underset{CH_3}{C}}-Br + Br\cdot$$

tert-**Butyl bromide**

Because the transition states for both steps in bromination are productlike in structure and energy, and because the products of each hydrogen-atom abstraction step are, in fact, quite different (a 1° radical versus a 3° radical), the type of C—H bond being broken will have a marked influence on the relative rates of the reactions. In fact, they proceed with very different rates. Abstraction of the 3° hydrogen takes place much faster. Bromine, as a result, discriminates more effectively between the primary and tertiary hydrogen atoms. A comparison of potential energy diagrams for the abstraction of the primary and tertiary hydrogen atoms by bromine is given in Fig. 9.6.

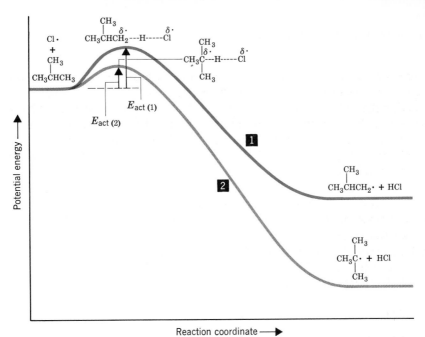

FIGURE 9.5 Potential energy diagrams for the two hydrogen-atom abstraction steps in the reaction of isobutane with Cl · . Both steps are exothermic and both transition states resemble the reactants. The activation energies are similar, but because 3° C—H bonds are broken more easily than 1° C—H bonds, reaction (2) has a lower activation energy and proceeds at a somewhat faster rate.

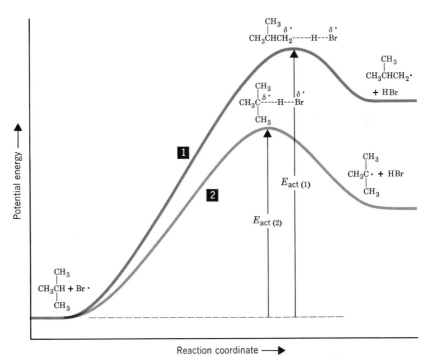

FIGURE 9.6 Potential energy diagrams for the two hydrogen-atom abstraction steps in the reaction of isobutane with Br · . Both steps are highly endothermic and, in both, the transition states resemble the products. Since the products—a 3° radical and a 1° radical—have quite different potential energies (stabilities), the transition states for the two steps are also quite different. The transition state for reaction (1) resembles a 1° radical. It occurs at a much higher potential energy than the transition state for reaction (2) because the transition state for reaction (2) resembles a much more stable 3° radical. The activation energy for reaction (2) is much lower than that for reaction (1). Reaction (2), consequently, proceeds at a much faster rate. The ultimate product that arises from reaction (2) is *tert*-butyl bromide, and this is the predominant product of the reaction.

PROBLEM 9.16

Fluorine is far less selective than bromine and is even less selective than chlorine. The products that one obtains from alkane fluorinations are, in fact, almost those that one would expect if the different types of hydrogen were equally reactive. Explain.

9.7 THE GEOMETRY OF FREE RADICALS

Experimental evidence indicates that the geometrical structure of most alkyl radicals is trigonal planar at the carbon having the unpaired electron. This structure can be accommodated by an sp^2-hybridized central carbon. In an alkyl radical, the p orbital contains the unpaired electron (Fig. 9.7).

9.8 A FREE RADICAL REACTION THAT GENERATES A TETRAHEDRAL STEREOCENTER

We saw earlier (in Section 4.8) that when achiral molecules react to produce a compound with a single tetrahedral stereocenter, the product will be obtained as a racemic form. This will always be true in the absence of any chiral influence on the reaction such as an enzyme or the use of a chiral solvent.

We can now examine another reaction that illustrates this principle, the free radical chlorination of pentane.

$$CH_3CH_2CH_2CH_2CH_3 \xrightarrow[\text{(achiral)}]{Cl_2} CH_3CH_2CH_2CH_2CH_2Cl$$

Pentane   1-Chloropentane
(achiral)   (achiral)

$$+ CH_3CH_2CH_2\overset{*}{C}HClCH_3 + CH_3CH_2CHClCH_2CH_3$$

(±)-2-Chloropentane   3-Chloropentane
(a racemic form)   (achiral)

The reaction will produce the products shown here, as well as more highly chlorinated products. (We can use an excess of pentane to minimize multiple chlorinations.) Neither 1-chloropentane nor 3-chloropentane contains a stereocenter, but 2-chloropentane does, and it is *obtained as a racemic form*. If we examine the mechanism in Fig. 9.8 we shall see why.

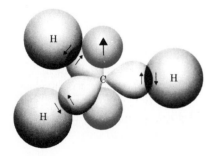

FIGURE 9.7 sp^2-Hybridized carbon atom at the center of a methyl radical showing the odd electron in one lobe of the half-filled p orbital. It could be pictured in the other lobe.

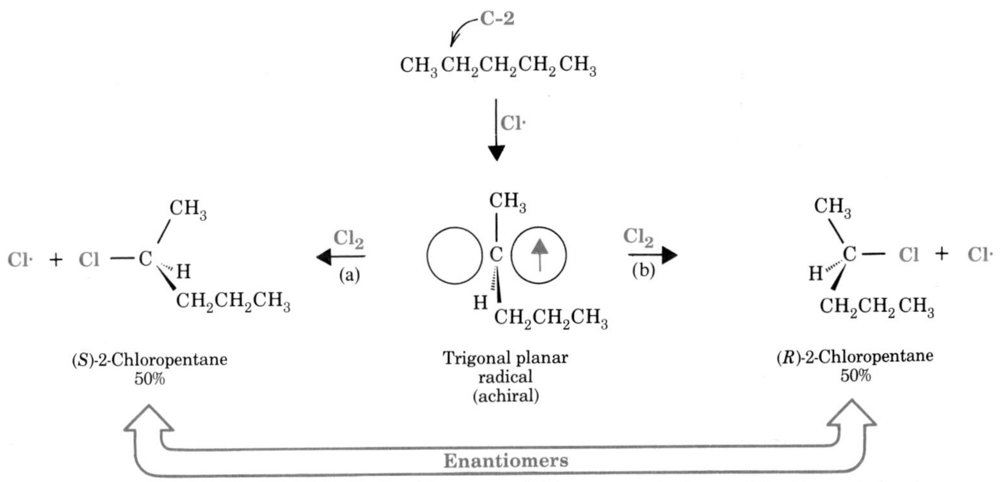

FIGURE 9.8 How chlorination of C-2 of pentane yields a racemic form of 2-chloropentane. Abstraction of a hydrogen atom from C-2 produces a trigonal planar free radical that is achiral. This radical then reacts with chlorine at either face [by path (a) or (b)]. Because the radical is achiral the probability of reaction by either path is the same, therefore, the two enantiomers are produced in equal amounts.

9.8A Generation of a Second Stereocenter in a Free Radical Halogenation

Let us now examine what happens when a chiral molecule (containing one stereo-center) reacts so as to yield a product with a second stereocenter. As an example consider what happens when (S)-2-chloropentane undergoes chlorination at C-3 (other products are formed, of course by chlorination at other carbon atoms). The results of chlorination at C-3 are shown in Fig. 9.9.

The products of the reactions are (2S,3S)-2,3-dichloropentane and (2S,3R)-2,3-dichloropentane. These two compounds are **diastereomers.** (They are stereo-isomers but they are not mirror reflections of each other.) The two diastereomers are *not* produced in equal amounts. Because the intermediate free radical itself is chiral, reactions at the two faces are not equally likely. The radical will react with chlorine to a greater extent at one face than the other (although we cannot easily predict which). That is, the presence of a stereocenter in the radical (at C-2) influences the reaction that introduces the new stereocenter (at C-3).

Both of the 2,3-dichloropentane diastereomers are chiral, and therefore, each would exhibit optical activity. Moreover, because the two compounds are *diastereomers,* they will have different physical properties (e.g., different melting points and boiling points) and will be separable by conventional means (by gas-liquid chroma-tography or by careful fractional distillation).

PROBLEM 9.17

Consider the chlorination of (S)-2-chloropentane at C-4. (a) Write stereo-chemical structures for the products that would be obtained and give each its proper (R)–(S) designation. (b) What is the stereoisomeric relationship be-tween these products? (c) Are both products chiral? (d) Are both optically active? (e) Could the products be separated by conventional means? (f) What other dichloropentanes would be obtained by chlorination of (S)-2-chloropen-tane? (g) Which of these are optically active?

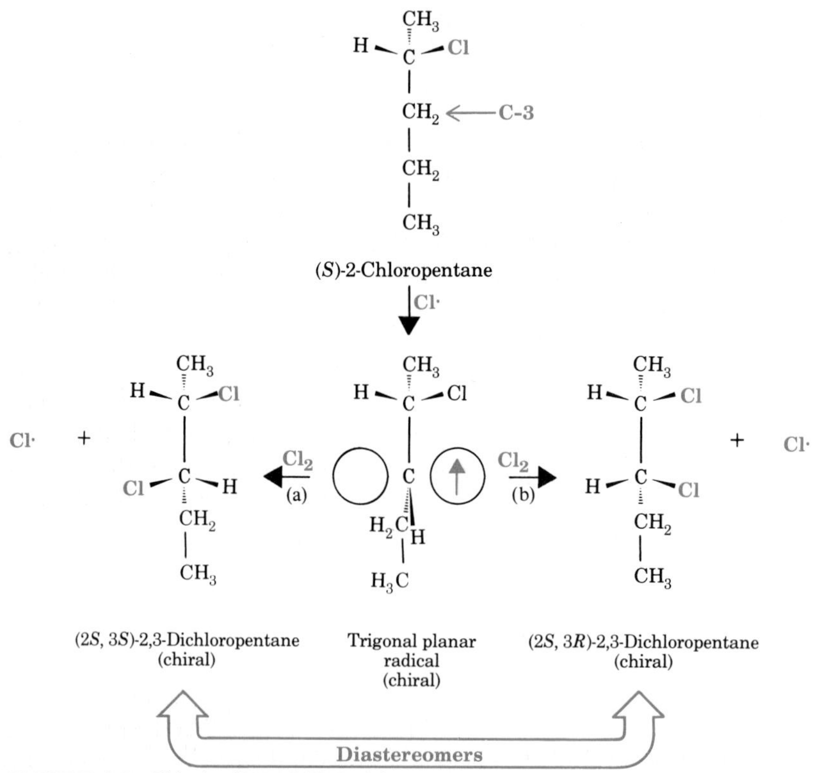

FIGURE 9.9 Chlorination of (S)-2-chloropentane at C-3. Abstraction of a hydrogen atom from C-3 produces a free radical that is chiral (it contains a stereocenter at C-2). This chiral radical can then react with chlorine at one face [path (a)] to produce (2S, 3S)-2,3-dichloropentane and at the other face [path (b)] to yield (2S,3R)-2,3-dichloropentane. These two compounds are diastereomers and they are not produced in equal amounts. Each product is chiral and each alone would be optically active.

PROBLEM 9.18

Consider the chlorination of pentane itself using sufficient chlorine to cause dichlorination. After the reaction is over you isolate all of the isomers with the molecular formula $C_5H_{10}Cl_2$ and subject this mixture to careful fractional distillation. (a) Tell how many fractions you would obtain, and what each fraction would contain. (b) Which (if any) of these fractions would show optical activity?

PROBLEM 9.19

We show the chlorination of 2-methylbutane yielding 1-chloro-2-methyl-butane, 2-chloro-2-methylbutane, 2-chloro-3-methylbutane, and 1-chloro-3-methylbutane on page 415. (a) Assuming that these compounds were separated after the reaction by fractional distillation, tell whether any fractions would show optical activity. (b) Would any of these fractions be resolvable into enantiomers?

9.9 FREE RADICAL ADDITION TO ALKENES: THE ANTI-MARKOVNIKOV ADDITION OF HYDROGEN BROMIDE

Before 1933, the orientation of the addition of hydrogen bromide to alkenes was the subject of much confusion. At times addition occurred in accordance with Markovnikov's rule; at other times it occurred in just the opposite manner. Many instances were reported where, under what seemed to be the same experimental conditions, Markovnikov additions were obtained in one laboratory and anti-Markovnikov additions in another. At times even the same chemist would obtain different results using the same conditions but on different occasions.

The mystery was solved in 1933 by the research of M. S. Kharasch and F. R. Mayo (of the University of Chicago). The explanatory factor turned out to be organic peroxides present in the alkenes—peroxides that were formed by the action of atmospheric oxygen on the alkenes. Kharasch and Mayo found that when alkenes that contained peroxides reacted with hydrogen bromide, anti-Markovnikov addition of hydrogen bromide occurred.

$$R-\overset{\cdot\cdot}{\underset{\cdot\cdot}{O}}-\overset{\cdot\cdot}{\underset{\cdot\cdot}{O}}-R$$

An organic peroxide

Under these conditions, for example, propene yields 1-bromopropane. In the absence of peroxides, or in the presence of compounds that would "trap" free radicals, normal Markovnikov addition occurs.

$$CH_3CH{=}CH_2 + HBr \xrightarrow{\text{ROOR}} CH_3CH_2CH_2Br \qquad \text{Anti-Markovnikov addition}$$

$$CH_3CH{=}CH_2 + HBr \xrightarrow[\text{peroxides}]{\text{no}} CH_3\overset{|}{\underset{|}{C}}HCH_3 \qquad \text{Markovnikov addition}$$
$$\underset{\text{2-Bromopropane}}{Br}$$

Hydrogen fluoride, hydrogen chloride, and hydrogen iodide *do not* give anti-Markovnikov addition even when peroxides are present.

According to Kharasch and Mayo, the mechanism for anti-Markovnikov addition of hydrogen bromide is a *free radical chain reaction* initiated by peroxides:

Chain Initiation

Step 1 $\qquad R-\overset{\cdot\cdot}{\underset{\cdot\cdot}{O}}{:}\overset{\cdot\cdot}{\underset{\cdot\cdot}{O}}-R \xrightarrow{\text{heat}} 2R-\overset{\cdot\cdot}{\underset{\cdot\cdot}{O}}\cdot$

Step 2 $\qquad R-\overset{\cdot\cdot}{\underset{\cdot\cdot}{O}}\cdot + H{:}\overset{\cdot\cdot}{\underset{\cdot\cdot}{Br}}{:} \longrightarrow R-\overset{\cdot\cdot}{\underset{\cdot\cdot}{O}}{:}H + {:}\overset{\cdot\cdot}{\underset{\cdot\cdot}{Br}}\cdot$

Chain Propagation

Step 3 $\qquad {:}\overset{\cdot\cdot}{\underset{\cdot\cdot}{Br}}\cdot + CH_2{=}CHCH_3 \longrightarrow {:}\overset{\cdot\cdot}{\underset{\cdot\cdot}{Br}}{:}CH_2\overset{\cdot}{C}HCH_3$
$$\qquad\qquad\qquad\qquad\qquad\qquad\qquad \text{2° Free radical}$$

Step 4 $\qquad {:}\overset{\cdot\cdot}{\underset{\cdot\cdot}{Br}}-CH_2\overset{\cdot}{C}HCH_3 + H{:}\overset{\cdot\cdot}{\underset{\cdot\cdot}{Br}}{:} \longrightarrow BrCH_2\overset{|}{\underset{|}{C}}HCH_3 + \cdot\overset{\cdot\cdot}{\underset{\cdot\cdot}{Br}}{:}$
$$\qquad\qquad\qquad\qquad\qquad\qquad\qquad\qquad\qquad\qquad\qquad H$$
$$\qquad\qquad\qquad\qquad\qquad\qquad \text{1-Bromopropane}$$

Step 1 is the simple homolytic cleavage of the peroxide molecule to produce two alkoxyl radicals. The oxygen–oxygen bond of peroxides is weak and such reactions are known to occur readily.

$$R—O{:}O—R \longrightarrow \quad 2R—O{\cdot} \qquad \Delta H° \cong +35 \text{ kcal/mole}$$

Peroxide   Alkoxyl radical

Step 2 of the mechanism, abstraction of a hydrogen atom by the radical, is exothermic and has a low energy of activation.

$$R—\ddot{O}{\cdot} + H{:}\ddot{B}r{:} \longrightarrow R—\ddot{O}{:}H + {:}\ddot{B}r{\cdot} \qquad \Delta H° \cong -23 \text{ kcal/mole}$$
$$E_{\text{act}} \text{ is low}$$

Step 3 of the mechanism determines the final orientation of bromine in the product. It occurs as it does because a *more stable secondary radical* is produced and because *attack at the primary carbon atom is less hindered.* Had the bromine attacked propene at the secondary carbon atom, a less stable, primary radical would have been the result,

$$Br{\cdot} + CH_2{=}CHCH_3 \xrightarrow{\quad\times\quad} {\cdot}CH_2CHCH_3$$
$$\underset{Br}{|}$$

1° **Free radical**

and attack at the secondary carbon atom would have been more hindered.

Step 4 of the mechanism is simply the abstraction of a hydrogen atom from hydrogen bromide by the radical produced in step 3. This hydrogen-atom abstraction produces a bromine atom that can bring about step 3 again, then step 4 occurs again—a chain reaction.

We can now see the contrast between the two ways that HBr can add to an alkene. In the absence of peroxides, the reagent that attacks the double bond is a proton. Because a proton is small, steric effects are unimportant. It attaches itself to a carbon atom in the way that yields the more stable carbocation. The result is Markovnikov addition.

Ionic Addition

$$CH_3CH{=}CH_2 \xrightarrow{H^+} CH_3\overset{+}{C}HCH_3 \xrightarrow{Br^-} CH_3CHCH_3$$
$$\underset{Br}{|}$$

More stable   Markovnikov
carbocation   product

In the presence of peroxides, the reagent that attacks the double bond is a larger bromine atom. It attaches itself to the less-hindered carbon atom in the way that yields the more stable free radical. The result is anti-Markovnikov addition.

Free Radical Addition

$$CH_3CH{=}CH_2 \xrightarrow{Br{\cdot}} CH_3\overset{\cdot}{C}HCH_2Br \xrightarrow{HBr} CH_3CH_2CH_2Br + Br{\cdot}$$

More stable   Anti-Markovnikov
free radical   product

9.9A Other Free Radical Additions to Alkenes

Many other molecules, other than hydrogen bromide, add to alkenes under the influence of a peroxide initiator. Tetrabromomethane, for example, reacts with 1-octene to yield 1,1,1,3-tetrabromononane.

$$CH_3(CH_2)_5CH=CH_2 + CBr_4 \xrightarrow{ROOR} CH_3(CH_2)_5\overset{|}{\underset{Br}{C}}HCH_2CBr_3$$

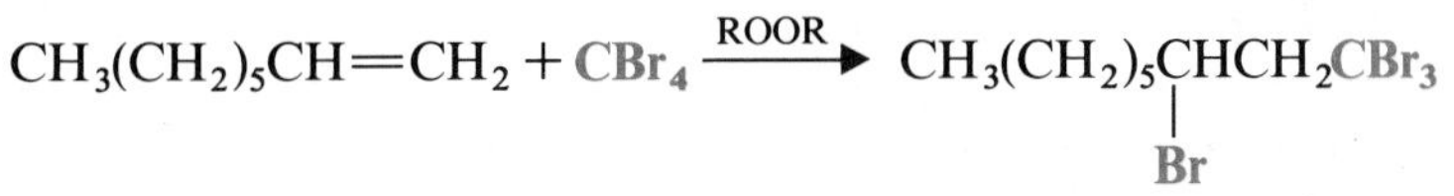

1-Octene   1,1,1,3-Tetrabromononane

The mechanism for this reaction is as follows:

Chain Initiation

Step 1 $RO—OR \longrightarrow 2RO\cdot$

Step 2 $RO\cdot + Br—CBr_3 \longrightarrow ROBr + \cdot CBr_3$

Chain Propagation

Step 3 $CH_3(CH_2)_5CH=CH_2 + \cdot CBr_3 \longrightarrow CH_3(CH_2)_5\overset{\cdot}{C}HCH_2—CBr_3$

Step 4 $CH_3(CH_2)_5\overset{\cdot}{C}HCH_2—CBr_3 + Br—CBr_3 \longrightarrow$

$$CH_3(CH_2)_5\overset{|}{\underset{Br}{C}}HCH_2—CBr_3 + CBr_3\cdot$$

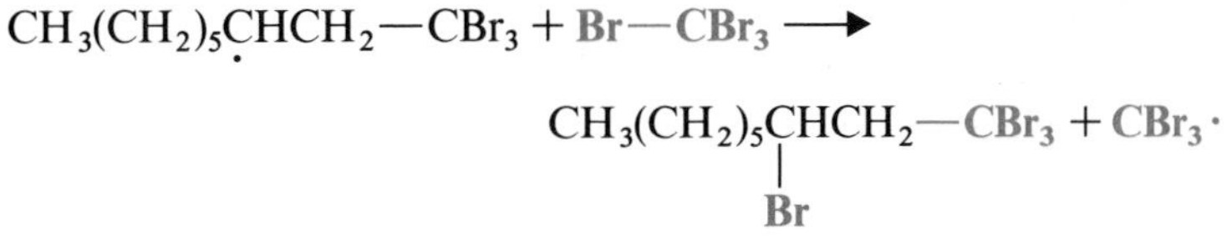

Other examples of free radical additions to alkenes are the following:

$$CH_3CH_2CH_2CH=CH_2 + HCCl_3 \xrightarrow{peroxides} CH_3CH_2CH_2CH_2CH_2—CCl_3$$
1,1,1-Trichlorohexane

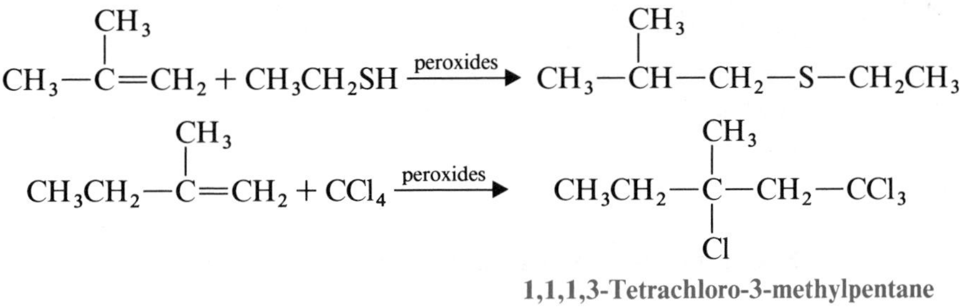

$$CH_3—\overset{\overset{CH_3}{|}}{C}=CH_2 + CH_3CH_2SH \xrightarrow{peroxides} CH_3—\overset{\overset{CH_3}{|}}{C}H—CH_2—S—CH_2CH_3$$

$$CH_3CH_2—\overset{\overset{CH_3}{|}}{C}=CH_2 + CCl_4 \xrightarrow{peroxides} CH_3CH_2—\overset{\overset{CH_3}{|}}{\underset{\underset{Cl}{|}}{C}}—CH_2—CCl_3$$
1,1,1,3-Tetrachloro-3-methylpentane

PROBLEM 9.20

Write free radical, chain-reaction mechanisms that account for the products formed in each of the reactions listed here.

Free radicals also cause alkenes to add to each other to form large molecules called addition polymers. These reactions are described in the next section.

9.10 FREE RADICAL POLYMERIZATION OF ALKENES. ADDITION POLYMERS

Polymers are substances that consist of very large molecules called **macromolecules** that are made up of many repeating subunits. The molecular subunits that are used to synthesize polymers are called **monomers,** and the reactions by which momomers are joined together are called **polymerizations.** Many polymerizations can be initiated by free radicals.

Ethylene, for example, is the monomer that is used to synthesize the familiar polymer called *polyethylene.*

$$m\text{CH}_2{=}\text{CH}_2 \xrightarrow{\text{polymerization}} -\text{CH}_2\text{CH}_2\text{+}(\text{CH}_2\text{CH}_2)_n\text{CH}_2\text{CH}_2-$$

Ethylene — Monomeric units —
monomer
Polyethylene
polymer

(*m* and *n* are large numbers)

Because polymers such as polyethylene are made by addition reactions, they are often called **addition polymers.** Let us now examine in some detail how polyethylene is made.

Ethylene polymerizes by a free radical mechanism when it is heated at a pressure of 1000 atm with a small amount of an organic peroxide. The peroxide dissociates to produce free radicals, which in turn initiate chains.

Chain Initiation

Step 1
$$\text{R}-\overset{\overset{\text{O}}{\|}}{\text{C}}-\text{O}\!:\!\text{O}-\overset{\overset{\text{O}}{\|}}{\text{C}}-\text{R} \longrightarrow 2\text{R}\!:\!\overset{\overset{\text{O}}{\|}}{\text{C}}-\text{O}\cdot \longrightarrow 2\text{CO}_2 + 2\text{R}\cdot$$

Step 2
$$\text{R}\cdot + \text{CH}_2{=}\text{CH}_2 \longrightarrow \text{R}\!:\!\text{CH}_2-\text{CH}_2\cdot$$

Chain Propagation

Step 3
$$\text{R}-\text{CH}_2\text{CH}_2\cdot + n\text{CH}_2{=}\text{CH}_2 \longrightarrow \text{R}\text{+}(\text{CH}_2\text{CH}_2)_n\text{CH}_2\text{CH}_2\cdot$$

Chains propagate by adding successive ethylene units, until their growth is stopped by combination or disproportionation.

Chain Termination

Step 4
$$2\text{R}-(\text{CH}_2\text{CH}_2)_n\text{CH}_2\text{CH}_2\cdot$$

combination $\longrightarrow [\text{R}-(\text{CH}_2\text{CH}_2)_n\text{CH}_2\text{CH}_2]_2$

disproportionation $\longrightarrow \text{R}-(\text{CH}_2\text{CH}_2)_n\text{CH}{=}\text{CH}_2 +$
$$\text{R}-(\text{CH}_2\text{CH}_2)_n\text{CH}_2\text{CH}_3$$

The free radical at the end of the growing polymer chain can also abstract a hydrogen atom from itself by what is called "back biting." This leads to chain branching.

Chain Branching

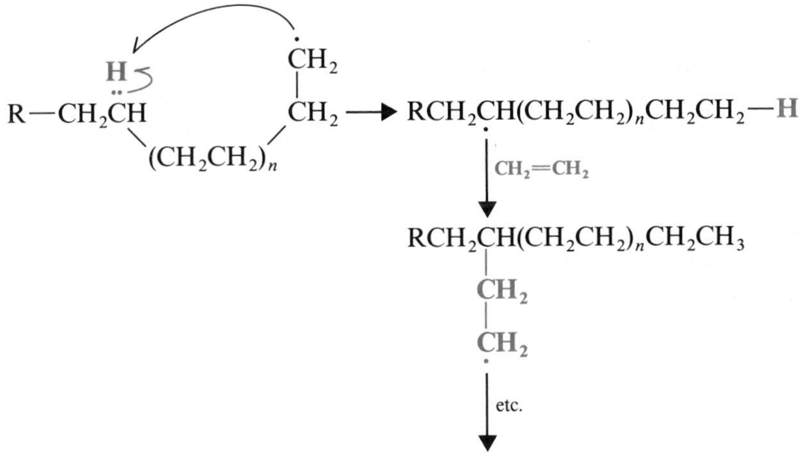

The polyethylene produced by free radical polymerization is not generally useful unless it has a molecular weight of nearly 1,000,000. Very high-molecular-weight polyethylene can be obtained by using a low concentration of the initiator. This initiates the growth of only a few chains and ensures that each chain will have a large excess of the monomer available. More initiator may be added as chains terminate during the polymerization and, in this way, new chains are begun.

Polyethylene has been produced commercially since 1943. It is used in manufacturing flexible bottles, films, sheets, and insulation for electric wires. Polyethylene produced by free radical polymerization has a softening point of about 110 °C.

Polyethylene can be produced in a different way using catalysts that are organometallic complexes of transition metals called **Ziegler–Natta catalysts.** In this process no free radicals are produced, no back biting occurs, and, consequently, there is no chain branching. The polyethylene that is produced is of higher density, has a higher melting point, and has greater strength. (Ziegler–Natta catalysts are discussed in greater detail in Special Topic D.)

Another familiar polymer is *polystyrene.* The monomer used in making polystyrene is phenylethene, a compound commonly known as *styrene.*

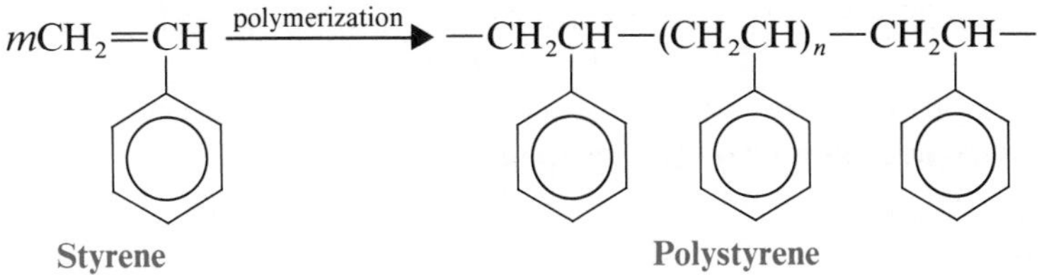

Styrene   Polystyrene

Addition polymers are discussed in much more detail in Special Topic D. Table 9.2 lists several other common addition polymers.

9.11 OTHER IMPORTANT FREE RADICAL CHAIN REACTIONS

Free radical chain mechanisms are important in understanding many other organic reactions. We shall see other examples in later chapters, but let us examine two here: the combustion of alkanes and some reactions of chlorofluoromethanes that have threatened the protective layer of ozone in the stratosphere.

TABLE 9.2 Other common addition polymers

MONOMER	POLYMER	NAMES
$CH_2\!\!=\!\!CHCH_3$	$\mathrm{+CH_2\!\!-\!\!CH\,\!)_n}$ $\quad\quad\;\; CH_3$	Polypropylene
$CH_2\!\!=\!\!CHCl$	$\mathrm{+CH_2\!\!-\!\!CH\,\!)_n}$ $\quad\quad\;\; Cl$	Poly(vinyl chloride), PVC
$CH_2\!\!=\!\!CHCN$	$\mathrm{+CH_2\!\!-\!\!CH\,\!)_n}$ $\quad\quad\;\; CN$	Polyacrylonitrile, *Orlon*
$CF_2\!\!=\!\!CF_2$	$\mathrm{+CF_2\!\!-\!\!CF_2\,\!)_n}$	Polytetrafluoroethene, *Teflon*
$\quad\;\; CH_3$ $\quad\;\;\; \mid$ $CH_2\!\!=\!\!CCO_2CH_3$	$\quad\;\; CH_3$ $\quad\;\;\; \mid$ $\mathrm{+CH_2\!\!-\!\!C\,\!)_n}$ $\quad\;\; CO_2CH_3$	Poly(methyl methacrylate), *Lucite, Plexiglas, Perspex*

9.11A Combustion of Alkanes

When alkanes react with oxygen (e.g., in oil furnaces and in internal combustion engines) a complex series of reactions takes place ultimately converting the alkane to carbon dioxide and water (Section 3.8A). Although our understanding of the detailed mechanism of combustion is incomplete, we do know that the important reactions occur by free radical chain mechanisms with chain-propagating steps like the following reactions.

$$R\cdot + O_2 \longrightarrow R\!\!-\!\!OO\cdot$$

$$R\!\!-\!\!OO\cdot + R\!\!-\!\!H \longrightarrow R\!\!-\!\!OOH + R\cdot$$

One product of the second step is $R\!\!-\!\!OOH$, called an alkyl hydroperoxide. The oxygen–oxygen bond of an alkyl hydroperoxide is quite weak, and it can break and produce free radicals that can initiate other chains:

$$RO\!\!-\!\!OH \longrightarrow RO\cdot + HO\cdot$$

9.11B Freons and Ozone Depletion

In the stratosphere at altitudes of about 25 km, very high-energy (very short wavelength) ultraviolet light converts diatomic oxygen (O_2) into ozone (O_3). The reactions that take place may be represented as follows:

Step 1 $O_2 + h\nu \longrightarrow O + O$

Step 2 $O + O_2 + M \longrightarrow O_3 + M + \text{heat}$

where M is some other particle that can absorb some of the energy released in the second step.

The ozone produced in step 2 can also interact with high-energy ultraviolet light in the following way.

Step 3 $O_3 + h\nu \longrightarrow O_2 + O + \text{heat}$

The oxygen atom formed in step 3 can cause a repetition of step 2 and so forth. The net result of these steps is to convert highly energetic ultraviolet light into heat. This is important because the existence of this cycle shields the earth from radiation that is destructive to living organisms. This shield makes life possible on the earth's surface. Even a relatively small increase in high-energy ultraviolet radiation at the earth's surface would cause a large increase in the incidence of skin cancers.

Production of chlorofluoromethanes (and of chlorofluoroethanes) called *Freons* began in 1930. These compounds have been used as refrigerants, solvents, and propellants in aerosol cans. Typical Freons are trichlorofluoromethane, $CFCl_3$ (called Freon-11), and dichlorodifluoromethane, CF_2Cl_2 (called Freon-12).

By 1974 world Freon production was about 2 billion pounds annually. Most Freon, even that used in refrigeration, eventually makes its way into the atmosphere where it diffuses unchanged into the stratosphere. In June 1974 F. S. Rowland and M. J. Molina published an article indicating, for the first time, that in the stratosphere Freon is able to initiate free radical chain reactions that can upset the natural ozone balance. The reactions that take place are the following. (Freon-12 is used as an example.)

Chain Initiation

$$\text{Step 1} \qquad CF_2Cl_2 + h\nu \longrightarrow CF_2Cl\cdot + Cl\cdot$$

Chain Propagation

$$\text{Step 2} \qquad Cl\cdot + O_3 \longrightarrow ClO\cdot + O_2$$

$$\text{Step 3} \qquad ClO\cdot + O \longrightarrow O_2 + Cl\cdot$$

In the chain-initiating step, ultraviolet light causes homolytic cleavage of one C—Cl bond of the Freon. The chlorine atom, thus produced, is the real villain; it can set off a chain reaction that destroys thousands of molecules of ozone before it diffuses out of the stratosphere or reacts with some other substance.

In 1975 a study by the National Academy of Science supported the predictions of Rowland and Molina and since January 1978 the use of Freons in aerosol cans in the United States has been banned. Many other countries still allow this use of Freons, however.

In 1985 a hole was discovered in the ozone layer above Antarctica. Studies done in 1986 and 1987 suggest that chlorine atom destruction of the ozone may be a factor in the formation of the hole.

9.12 SOME IMPORTANT TERMS AND CONCEPTS

Free radicals (or radicals) are reactive intermediates that have an unpaired electron. The relative stability of carbon free radicals is as follows:

$$\underset{3^\circ}{\overset{\displaystyle C}{\underset{\displaystyle C}{C-\overset{\displaystyle |}{\underset{\displaystyle |}{C}}\cdot}}} > \underset{2^\circ}{\overset{\displaystyle C}{\underset{\displaystyle H}{C-\overset{\displaystyle |}{\underset{\displaystyle |}{C}}\cdot}}} > \underset{1^\circ}{\overset{\displaystyle H}{\underset{\displaystyle H}{C-\overset{\displaystyle |}{\underset{\displaystyle |}{C}}\cdot}}} > \underset{\text{Methyl}}{\overset{\displaystyle H}{\underset{\displaystyle H}{H-\overset{\displaystyle |}{\underset{\displaystyle |}{C}}\cdot}}}$$

Bond dissociation energy (abbreviated DH°) is the amount of energy required for homolysis of a covalent bond.

Halogenations of alkanes are substitution reactions in which a halogen replaces one (or more) of the alkane's hydrogen atoms.

$$RH + X_2 \longrightarrow RX + HX$$

The reactions occur by a free radical mechanism.

Chain Initiation

Step 1 $X_2 \longrightarrow 2X \cdot$

Chain Propagation

Step 2 $RH + X \cdot \longrightarrow R \cdot + HX$

Step 3 $R \cdot + X_2 \longrightarrow RX + X \cdot$

Chain reactions are reactions whose mechanisms involve a series of steps with each step producing a reactive intermediate that causes the next step to occur. The halogenation of an alkane is a chain reaction.

Anti-Markovnikov addition of HBr to alkenes occurs when the reaction is initiated by peroxides. The mechanism is a free radical chain reaction:

Step 1 $RO-OR \longrightarrow 2RO \cdot$

Step 2 $RO \cdot + HBr \longrightarrow ROH + Br \cdot$

Step 3 $Br \cdot + R-CH=CH_2 \longrightarrow R\underset{\cdot}{C}H-CH_2Br$

More stable
radical

Step 4 $R\underset{\cdot}{C}H-CH_2Br + HBr \longrightarrow RCH_2CH_2Br + Br \cdot$

then steps 3, 4, 3, 4, and so on.

A *polymer* is a large molecule made up of many repeating subunits.

An *addition polymer* is a polymer synthesized by an addition reaction. Polyethylene and polystyrene are examples of addition polymers. They are made by **polymerizing** the alkene **monomers** ethylene and styrene, respectively.

ADDITIONAL PROBLEMS

9.21

The free radical reaction of propane with chlorine yields (in addition to more highly halogenated compounds) 1-chloropropane and 2-chloropropane. Write chain-initiating and chain-propagating steps showing how each compound is formed.

9.22

Starting with isobutane show how each of the following could be synthesized. (You need not repeat the synthesis of a compound prepared in an earlier part of this problem.)

(a) *tert*-Butyl bromide   (d) Isobutyl iodide

(b) 2-Methylpropene   (e) Isobutyl alcohol   (two ways)

(c) Isobutyl bromide   (f) *tert*-Butyl alcohol

(g) Isobutyl methyl ether

$$\overset{\overset{\displaystyle CH_3}{|}}{(h)\ \ CH_3CHCH_2}\overset{\overset{\displaystyle O}{\|}}{OCCH_3}$$

(i) $\overset{\overset{\displaystyle CH_3}{|}}{CH_3CHCH_2CN}$

(j) $\overset{\overset{\displaystyle CH_3}{|}}{CH_3CHCH_2SCH_3}$ (two ways)

(k) $\overset{\overset{\displaystyle CH_3}{|}}{CH_3\underset{\underset{\displaystyle Br}{|}}{C}CH_2CBr_3}$

(l) Polyisobutylene

9.23

In the presence of peroxides, 1-octene reacts with each of the following compounds to yield the product indicated. Write mechanisms for each reaction.

(a) 1-Octene + $CBrCl_3 \xrightarrow{ROOR} CH_3(CH_2)_5\underset{\underset{\displaystyle Br}{|}}{C}HCH_2CCl_3$

(b) 1-Octene + $CHCl_3 \xrightarrow{ROOR} CH_3(CH_2)_7CCl_3$

(c) 1-Octene + $CCl_4 \xrightarrow{ROOR} CH_3(CH_2)_5\underset{\underset{\displaystyle Cl}{|}}{C}HCH_2CCl_3$

9.24

Bromination of 2-methylbutane yields a predominance of one product with the formula $C_5H_{11}Br$. What is its structure? Show how this product could be used to synthesize each of the following. (You need not repeat steps carried out in earlier parts.)

(a) 2-Methyl-1-butene
(b) 2-Methyl-2-butene
(c) 1-Bromo-2-methylbutane
(d) 2-Bromo-3-methylbutane
(e) 3-Methyl-1-butene
(f) 1-Bromo-3-methylbutane
(g) 1-Iodo-2-methylbutane
(h) 2-Iodo-2-methylbutane
(i) 1-Iodo-3-methylbutane
(j) 2-Chloro-1-iodo-2-methylbutane

(k) $\overset{\overset{\displaystyle CH_3}{|}}{CH_3CH_2CHCH_2}OH$

(l) $\overset{\overset{\displaystyle CH_3}{|}}{CH_3CHCH_2CH_2}OH$

(m) $\overset{\overset{\displaystyle CH_3}{|}}{CH_3CHCHCH_3}$
$\quad\quad\quad\ \underset{\underset{\displaystyle OH}{|}}{}$

(n) $\overset{\overset{\displaystyle CH_3}{|}}{CH_3CH\underset{\underset{\displaystyle Br}{|}}{C}HCH_2CBr_3}$

(o) $\overset{\overset{\displaystyle O}{\|}}{CH_3CCH_2CH_3}$

(p) $\overset{\overset{\displaystyle CH_3}{|}}{CH_3CH}-\overset{\overset{\displaystyle O}{\|}}{CH}$

(q) $HOCH_2\underset{\underset{\displaystyle OH}{|}}{\overset{\overset{\displaystyle CH_3}{|}}{C}}CH_2CH_3$

(r) $\overset{\overset{\displaystyle CH_3}{|}}{CH_3C}\underset{\underset{\displaystyle O}{\diagdown\diagup}}{-}CHCH_3$

9.25

In addition to more highly fluorinated products, fluorination of 2-methylbutane yields a mixture of compounds with the formula $C_5H_{11}F$. (a) Taking stereochemistry into account, how many different isomers with the formula $C_5H_{11}F$ would you expect to be produced? (b) If the mixture of $C_5H_{11}F$ isomers were subjected to fractional distillation how many fractions would you expect to obtain? (c) Which fractions would be optically inactive? (d) Which would you be able to resolve into enantiomers?

9.26

Fluorination of (R)-2-fluorobutane yields a mixture of isomers with the formula $C_4H_8F_2$. (a) How many different isomers would you expect to be produced? Write their structures. (b) If the mixture of $C_4H_8F_2$ isomers were subjected to fractional distillation how many fractions would you expect to obtain? (c) Which of these fractions would be optically active?

9.27

Starting with acetylene and any other needed reagents show how you might synthesize each of the following. (You need not repeat steps carried out earlier.)

(a) 1-Bromopropane   (c) 1-Hexyne   (e) 3-Heptyne

(b) 1-Bromobutane   (d) 2-Heptyne   (f) 3-Octyne

9.28

Peroxides are often used to initiate free radical chain reactions such as alkane halogenations. (a) Examine the bond energies in Table 9.1 and give reasons that will explain why peroxides are especially effective as free radical initiators. (b) Illustrate your answer by outlining how di-*tert*-butyl peroxide, $(CH_3)_3CO-OC(CH_3)_3$, might initiate an alkane halogenation.

9.29

When an alkene $(RCH=CH_2)$ and hydrogen sulfide are irradiated with light (of wavelength that can be absorbed by H_2S), a chain reaction takes place producing a thiol, RCH_2CH_2SH. (a) Outline a possible mechanism for this reaction. (b) A side product of the reaction is a thioether, $(RCH_2CH_2)_2S$. Suggest how it is formed.

9.30

Free radical fluorination of methane occurs in the absence of light. A mechanism that has been proposed for the dark reaction is

$$CH_4 + F_2 \xrightarrow{slow} CH_3\cdot + HF + F\cdot$$

$$CH_3\cdot + F\cdot \xrightarrow{fast} CH_3F$$

(a) Basing your answer on bond dissociation energies, assess the likelihood of the reaction occurring by this mechanism. (b) What is the likelihood of a similar mechanism occurring when a mixture of methane and chlorine is heated in the dark?

9.31

Use bond dissociation energies in Table 9.1 to account for the following: (a) Thermal cracking of a C—H bond of methane requires a higher temperature (~ 1200 °C) than does a similar breaking of a C—H bond of ethane (500–600 °C). (b) When ethane undergoes homolysis at high temperatures, the C—C bond breaks more readily than the C—H bonds. (c) When butane "cracks" the reaction $CH_3CH_2CH_2CH_3 \longrightarrow 2CH_3CH_2\cdot$ occurs more readily than the reaction $CH_3CH_2CH_2CH_3 \longrightarrow CH_3CH_2CH_2\cdot + CH_3\cdot$.

9.32

When propane is heated to a very high temperature, it undergoes thermal cracking through homolysis of C—C and C—H bonds. The major products of the reaction are methane and ethene. A chain mechanism has been proposed for this reaction. (a) Which of the following reactions is most likely to be the major chain-initiating step? Explain your answer by estimating activation energies for each reaction.

$$CH_3CH_2CH_3 \longrightarrow CH_3CH_2 \cdot + CH_3 \cdot$$

$$CH_3CH_2CH_3 \longrightarrow CH_3CH_2CH_2 \cdot + H \cdot$$

$$CH_3CH_2CH_3 \longrightarrow CH_3\overset{\cdot}{C}HCH_3 + H \cdot$$

Possible chain-propagating steps are

Step 1 $CH_3 \cdot + CH_3CH_2CH_3 \longrightarrow CH_4 + \cdot CH_2CH_2CH_3$

Step 2 $\cdot CH_2 \!\!-\!\! CH_2 \!\!:\!\! CH_3 \longrightarrow CH_2 \!\!=\!\! CH_2 + \cdot CH_3$

(b) Both reactions have reasonably low activation energies (low enough to occur at very high temperatures). Show that this is likely for step 1 by calculating $\Delta H°$ for step 1. (c) An alternative to step 1 is

$$CH_3 \cdot + CH_3CH_2CH_3 \longrightarrow CH_4 + CH_3\overset{\cdot}{C}HCH_3$$

Comment on the likelihood of this reaction occurring in terms of energy and probability factors.

*9.33

The following reactions show comparisons between two sets of similar reactions. In each set we compare reactions in which a hydrogen atom is abstracted from methane and from ethane. In the first set **(A)** the abstracting agent is a methyl radical; in the second set **(B)** it is a bromine atom. (a) Sketch energy diagrams for each set of reactions taking the Hammond-Leffler postulate into account. Take care to locate each transition state properly not only along the energy axis but along the reaction coordinate as well. For convenience in making comparisons, you should align the curves so that the potential energies of the reactants are the same. (b) For which reaction will bond breaking have occurred to the *least* extent when the transition state is reached? (c) To the *greatest* extent? (d) To what approximate extent will bond breaking have occurred in reaction **A** step 1? (e) For which set of reactions will the transition states more resemble products? (f) Notice that the difference in $\Delta H°$ for the two sets of reactions is the same (6 kcal/mole). Why is this so? (g) The difference in E_{act} for the first set of reactions is relatively small (2.8 kcal/mole). For the second set of reactions, however, the difference in E_{act} is large (5.0 kcal/mole); it is nearly as large as the difference in $\Delta H°$. Explain.

		$\Delta H°$ (kcal/mole)	E_{act} (kcal/mole)
(A) Step 1	$CH_3 \cdot + H{-}CH_3 \longrightarrow CH_3{-}H + \cdot CH_3$	0	14.5
Step 2	$CH_3 \cdot + H{-}CH_2CH_3 \longrightarrow CH_3{-}H + \cdot CH_2CH_3$	-6.0	11.7
	Difference	6.0	2.8
(B) Step 1	$Br \cdot + H{-}CH_3 \longrightarrow Br{-}H + \cdot CH_3$	16.5	18.6
Step 2	$Br \cdot + H{-}CH_2CH_3 \longrightarrow Br{-}H + \cdot CH_2CH_3$	10.5	13.6
	Difference	6.0	5.0

SPECIAL TOPIC D

POLYMERIZATION OF ALKENES: ADDITION POLYMERS

The names *Orlon, Plexiglas, Lucite, polyethylene,* and *Teflon* are now familiar names to most of us. These "plastics" or polymers are used in the construction of many objects around us — from the clothing we wear to portions of the houses we live in. Yet all of these compounds were unknown 50 years ago. The development of the processes by which synthetic polymers are made, more than any other single factor, has been responsible for the remarkable growth of the chemical industry in this century.

At the same time, some scientists are now expressing concern about the reliance we have placed on these synthetic materials. Because they are the products of laboratory and industrial processes rather than processes that occur in nature, nature often has no way of disposing of many of them. Although progress has been made in the development of "biodegradable plastics" in recent years, many materials are still used that are not biodegradable. Although most of these objects are combustible, incineration is not always a feasible method of disposal because of attendant air pollution.

Not all polymers are synthetic. Many naturally occurring compounds are polymers as well. Silk and wool are polymers that we call proteins. The starches of our diet are polymers and so is the cellulose of cotton and wood.

Polymers are compounds that consist of very large molecules made up of many repeating subunits. The molecular subunits that are used to synthesize polymers are called *monomers,* and the reactions by which monomers are joined together are called polymerization reactions.

Propylene, for example, can be polymerized to form *polypropylene.* This polymerization occurs by an addition reaction and, as a consequence, polymers such as polypropylene are called *addition polymers.*

Propylene → Polypropylene

As we might expect, alkenes are convenient starting materials for the preparation of addition polymers. The addition reactions occur through free radical, cationic, or anionic mechanisms depending on how they are initiated. The following examples illustrate these mechanisms. All of these reactions are chain reactions.

Free Radical Polymerization

Cationic Polymerization

Anionic Polymerization

$$Z:^- + \; \underset{\diagup}{\overset{\diagdown}{C}}=\underset{\diagdown}{\overset{\diagup}{C}} \; \longrightarrow \; Z-\overset{|}{\underset{|}{C}}-\overset{|}{\underset{|}{C}}:^- \; \text{-------} \; \overset{\overset{\diagdown}{C}=\overset{\diagup}{C}}{\longrightarrow} \; Z-\overset{|}{\underset{|}{C}}-\overset{|}{\underset{|}{C}}-\overset{|}{\underset{|}{C}}-\overset{|}{\underset{|}{C}}:^- \; \text{-------} \; \longrightarrow \; \text{etc.}$$

Free radical polymerization of chloroethene (vinyl chloride) produces a polymer called poly(vinyl chloride).

$$\underset{\text{Vinyl chloride}}{CH_2{=}\underset{\underset{Cl}{|}}{CH}} \; \longrightarrow \; \underset{\text{Poly(vinyl chloride)}}{\left(CH_2{-}\underset{\underset{Cl}{|}}{CH}\right)_n}$$

This reaction produces a polymer that has a molecular weight of about 1,500,000 and that is a hard, brittle, and rigid material. In this form it is often used to make pipes, rods, and phonograph records. Poly(vinyl chloride) can be softened by mixing it with esters (called plasticizers). The softer material is used for making "vinyl leather," plastic raincoats, shower curtains, and garden hoses.

Exposure to vinyl chloride has been linked to the development of a rare cancer of the liver called angiocarcinoma. This link was first noted in 1974 and 1975 among workers in vinyl chloride factories. Since that time, standards have been set to limit workers' exposure to less than one part per million average over an 8-h day. The Food and Drug Administration has banned the use of poly(vinyl chloride) in packages for food. [There is evidence that poly(vinyl chloride) contains traces of vinyl chloride.]

Acrylonitrile ($CH_2{=}CHCN$) polymerizes to form polyacrylonitrile or *Orlon*. The initiator for the polymerization is a mixture of ferrous sulfate and hydrogen peroxide. These two compounds react to produce hydroxyl radicals ($\cdot OH$), which act as chain initiators.

$$\underset{\text{Acrylonitrile}}{CH_2{=}\underset{\underset{CN}{|}}{CH}} \; \xrightarrow[\text{H-O-O-H}]{\text{FeSO}_4} \; \underset{\substack{\text{Polyacrylonitrile} \\ \text{(Orlon)}}}{\left(CH_2{-}\underset{\underset{CN}{|}}{CH}\right)_n}$$

Polyacrylonitrile decomposes before it melts, thus melt spinning cannot be used for the production of fibers. Polyacrylonitrile, however, is soluble in *N,N*-dimethylformamide, and these solutions can be used to spin fibers. Fibers produced in this way are used in making carpets and clothing.

Teflon is made by polymerizing tetrafluoroethylene in aqueous suspension.

$$nCF_2{=}CF_2 \; \xrightarrow[\substack{H_2O_2 \\ H_2O}]{Fe^{2+}} \; (CF_2{-}CF_2)_n$$

The reaction is highly exothermic and water helps dissipate the heat that is produced. Teflon has a melting point (327 °C) that is unusually high for an addition polymer. It is also highly resistant to chemical attack and has a low coefficient of friction. Because of these properties Teflon is used in greaseless bearings, in liners for pots and pans, and in many special situations that require a substance that is highly resistant to corrosive chemicals.

Vinyl alcohol is an unstable compound that rearranges spontaneously to acetaldehyde (cf. Section 7.15). Consequently, the water-soluble polymer, poly(vinyl alcohol), cannot be made directly. It can be made, however, by an indirect method that

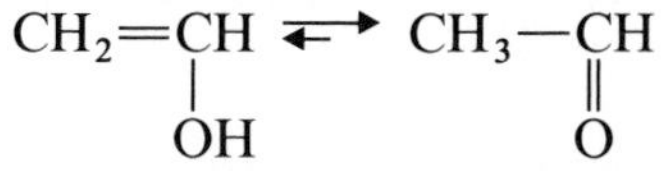

$$CH_2=CH \rightleftharpoons CH_3-CH$$
$$\underset{OH}{|} \qquad \underset{O}{\|}$$

Vinyl alcohol   **Acetaldehyde**

begins with the polymerization of vinyl acetate to poly(vinyl acetate). This is then hydrolyzed to poly(vinyl alcohol). Hydrolysis is rarely carried to completion, however, because the presence of a few ester groups helps confer water solubility on the product. The ester groups apparently help keep the polymer chains apart and this

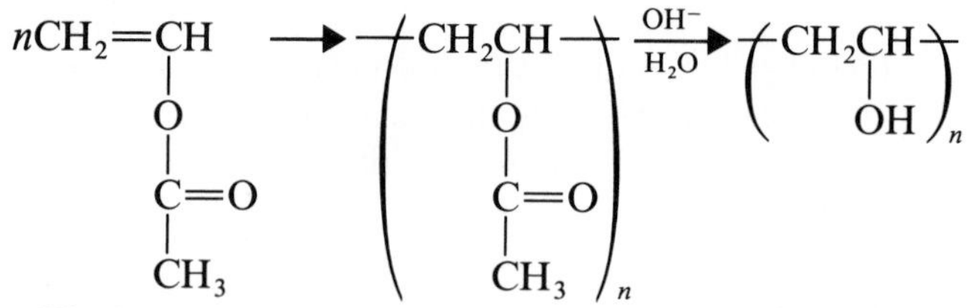

Vinyl acetate   **Poly(vinyl acetate)**   **Poly(vinyl alcohol)**

permits hydration of the hydroxyl groups. Poly(vinyl alcohol) in which 10% of the ester groups remain dissolves readily in water. Poly(vinyl alcohol) is used to manufacture water-soluble films and adhesives. Poly(vinyl acetate) is used as an emulsion in water-base paints.

A polymer with excellent optical properties can be made by the free radical polymerization of methyl methacrylate. Poly(methyl methacrylate) is marketed under the names *Lucite, Plexiglas,* and *Perspex.*

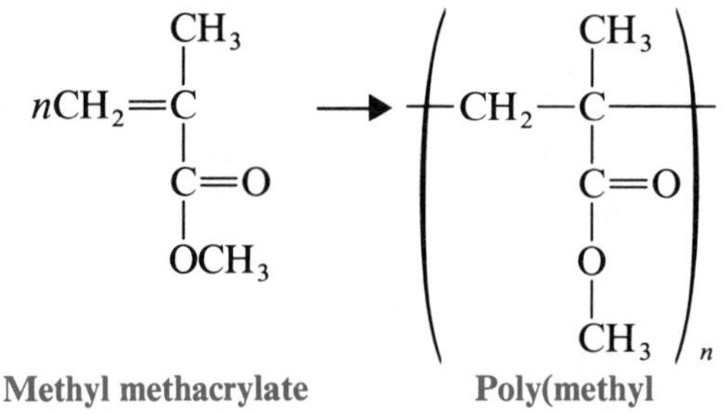

Methyl methacrylate   **Poly(methyl methacrylate)**

A mixture of vinyl chloride and vinylidene chloride polymerizes to form what is known as a *copolymer.* The familiar *Saran Wrap* used in food packaging is made by polymerizing a mixture in which the vinylidene chloride predominates.

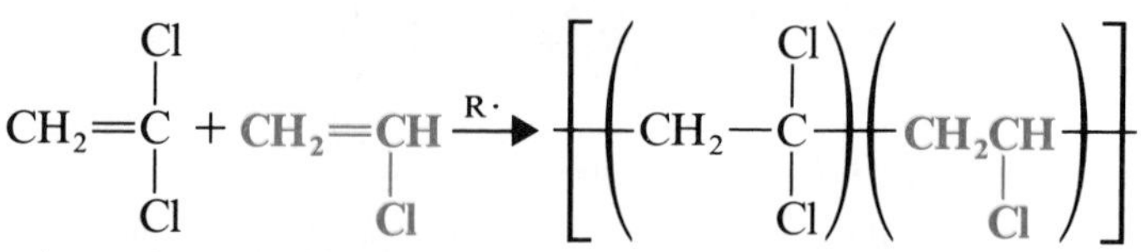

(excess)   **Vinyl**   **Saran Wrap**
Vinylidene   **chloride**
chloride

The subunits do not necessarily alternate regularly along the polymer chain.

PROBLEM D.1

Can you suggest an explanation that accounts for the fact that the radical polymerization of propylene occurs in a head-to-tail fashion

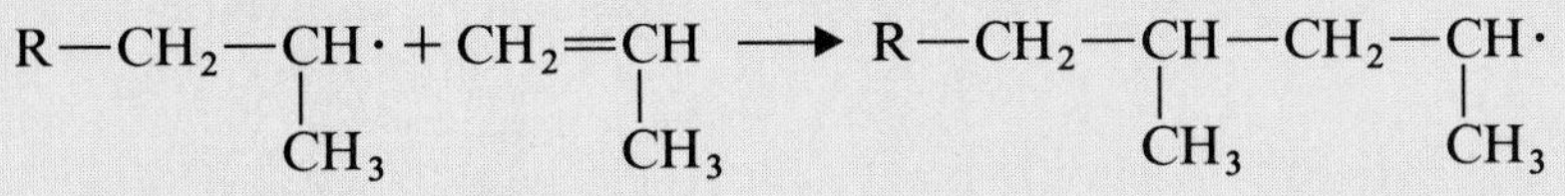

$$R-CH_2-CH\cdot + CH_2=CH \longrightarrow R-CH_2-CH-CH_2-CH\cdot$$
$$\underset{\text{"Head"}}{\overset{|}{CH_3}} \quad \underset{\text{"Tail"}}{\overset{|}{CH_3}} \qquad\qquad \overset{|}{CH_3} \qquad \overset{|}{CH_3}$$

rather than the head-to-head manner, shown here?

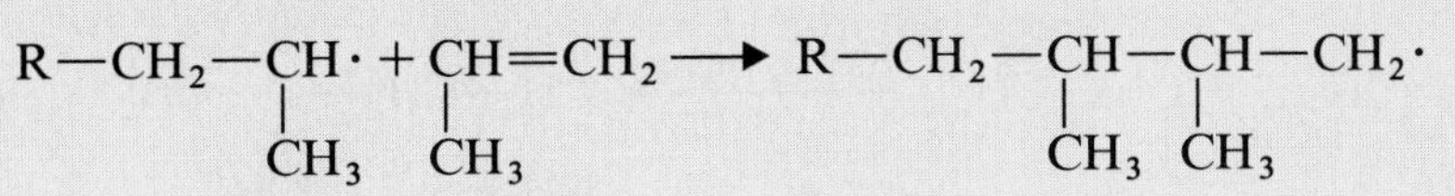

$$R-CH_2-CH\cdot + CH=CH_2 \longrightarrow R-CH_2-CH-CH-CH_2\cdot$$
$$\underset{\text{"Head"}}{\overset{|}{CH_3}} \quad \underset{\text{"Head"}}{\overset{|}{CH_3}} \qquad\qquad \overset{|}{CH_3} \ \overset{|}{CH_3}$$

PROBLEM D.2

Outline general methods for the synthesis of each of the following polymers by free radical polymerization. Assume that the appropriate monomers are available.

(a) Poly(vinyl fluoride) (Tedlar), $+CH_2CHF\frac{}{}_n$

(b) Poly(chlorotrifluoroethylene) (Kel—F), $+CF_2-CFCl\frac{}{}_n$

(c) *Viton,* a copolymer of hexafluoropropene, $CF_2=CFCF_3$, and vinylidene fluoride ($CH_2=CF_2$).

Alkenes also polymerize when they are treated with strong acids. The growing chains in acid-catalyzed polymerizations are *cations* rather than free radicals. The following reactions illustrate the cationic polymerization of isobutylene.

Step 1 $\quad H_2O + BF_3 \rightleftharpoons H^+ + BF_3(OH)^-$

Step 2 $\quad H^+ + CH_2=\underset{\underset{CH_3}{|}}{\overset{\overset{CH_3}{|}}{C}} \longrightarrow CH_3-\underset{\underset{CH_3}{|}}{\overset{\overset{CH_3}{|}}{C^+}}$

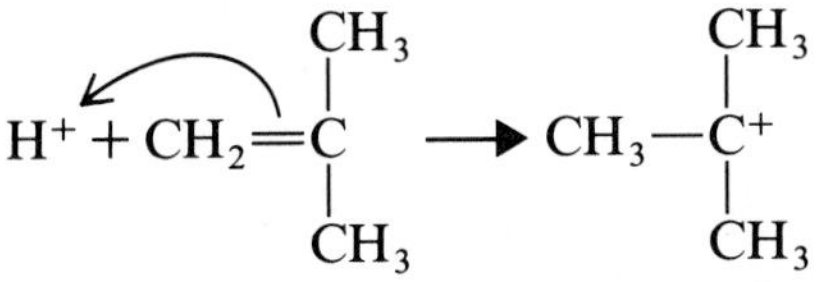

Step 3 $\quad CH_3-\underset{\underset{CH_3}{|}}{\overset{\overset{CH_3}{|}}{C^+}} + CH_2=\underset{\underset{CH_3}{|}}{\overset{\overset{CH_3}{|}}{C}} \longrightarrow CH_3-\underset{\underset{CH_3}{|}}{\overset{\overset{CH_3}{|}}{C}}-CH_2-\underset{\underset{CH_3}{|}}{\overset{\overset{CH_3}{|}}{C^+}}$

Step 4 $\quad CH_3-\underset{\underset{CH_3}{|}}{\overset{\overset{CH_3}{|}}{C}}-CH_2-\underset{\underset{CH_3}{|}}{\overset{\overset{CH_3}{|}}{C^+}} \xrightarrow{\underset{CH_3}{\overset{CH_3}{CH_2=C}}} CH_3-\underset{\underset{CH_3}{|}}{\overset{\overset{CH_3}{|}}{C}}-CH_2-\underset{\underset{CH_3}{|}}{\overset{\overset{CH_3}{|}}{C}}-CH_2-\underset{\underset{CH_3}{|}}{\overset{\overset{CH_3}{|}}{C^+}} \xrightarrow{etc.}$

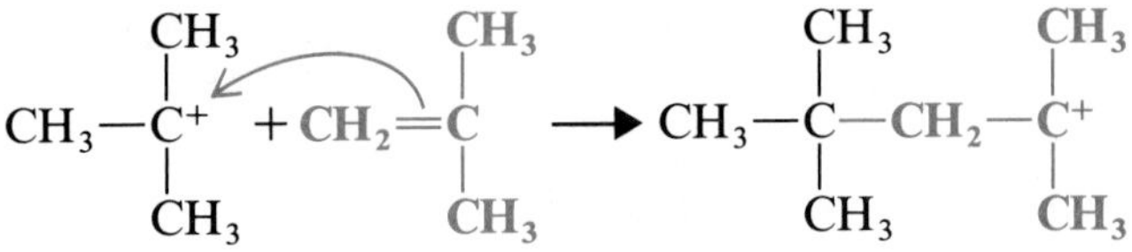

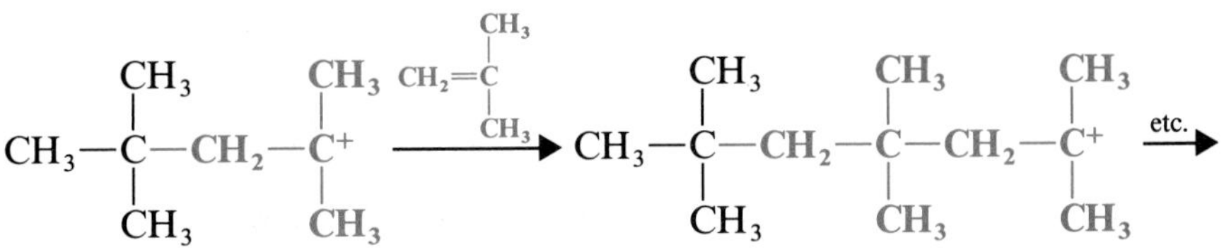

The catalysts used for cationic polymerizations are usually Lewis acids that contain a small amount of water. The polymerization of isobutylene illustrates how the catalyst (BF_3 and H_2O) functions to produce growing cationic chains.

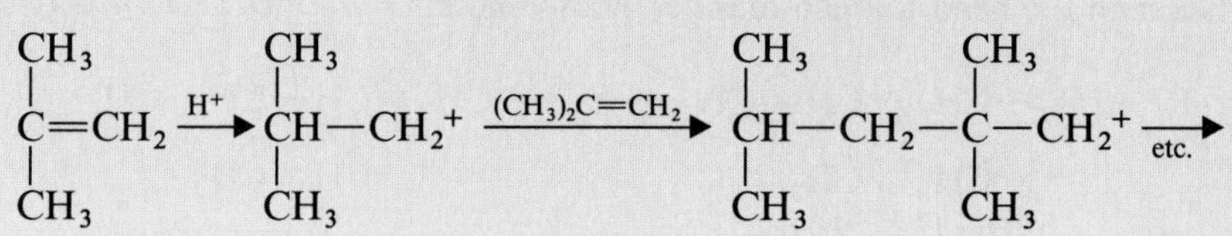

Alkenes containing electron-withdrawing groups polymerize in the presence of strong bases. Acrylonitrile, for example, polymerizes when it is treated with sodium amide ($NaNH_2$) in liquid ammonia. The growing chains in this polymerization are anions.

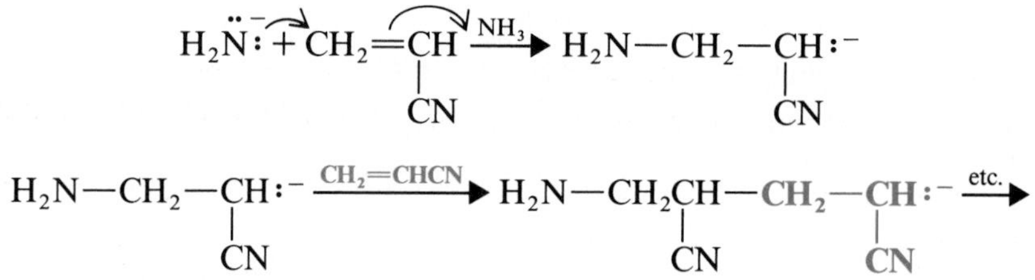

$$H_2\ddot{N}:^- + CH_2{=}CH \xrightarrow{NH_3} H_2N-CH_2-\underset{CN}{CH}:^-$$
$$\underset{CN}{}$$

$$H_2N-CH_2-\underset{CN}{CH}:^- \xrightarrow{CH_2=CHCN} H_2N-CH_2\underset{CN}{CH}-CH_2-\underset{CN}{CH}:^- \xrightarrow{etc.}$$

Anionic polymerization of acrylonitrile is less important in commercial production than the free radical process we illustrated earlier.

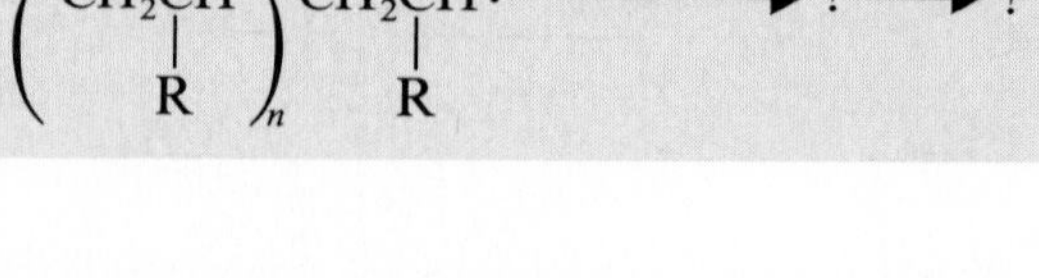

HINT: Ethylene oxide, $H_2C\overset{\diagup\diagdown}{\underset{O}{}}CH_2$, can also be polymerized by anions. The

reaction involves ring opening of the highly strained three-membered ring (cf. Section 9.18B).

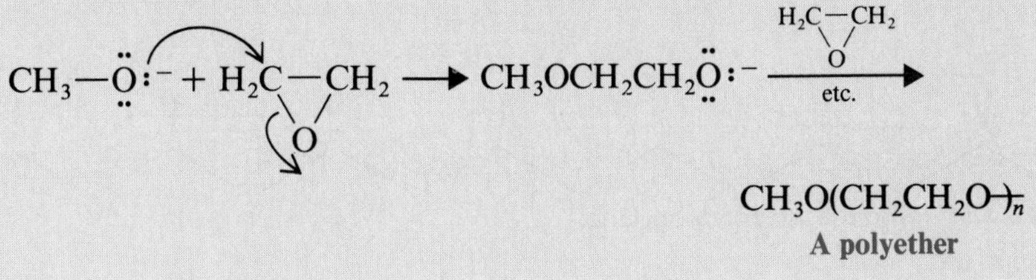

$$CH_3-\overset{..}{\underset{..}{O}}{:}^- + H_2C\overset{\frown}{\underset{\underset{\displaystyle O}{\diagup\diagdown}}{}}CH_2 \longrightarrow CH_3OCH_2CH_2\overset{..}{\underset{..}{O}}{:}^-\ \underset{\text{etc.}}{\overset{\displaystyle H_2C\overset{\diagup\diagdown}{\underset{O}{}}CH_2}{\longrightarrow}}$$

$$CH_3O(CH_2CH_2O)_{\overline{n}}$$
A polyether

D.1 Stereochemistry of Addition Polymers

Head-to-tail polymerization of propylene produces a polymer in which every other atom is a stereocenter. Many of the physical properties of the polypropylene produced in this way depend on the stereochemistry of these stereocenters.

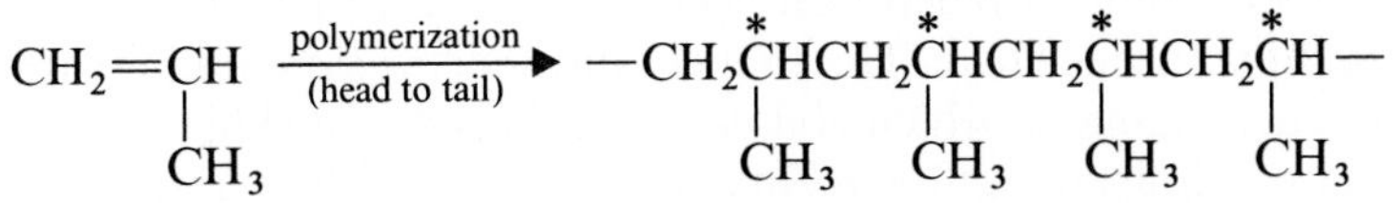

$$CH_2{=}\underset{\displaystyle CH_3}{\underset{|}{CH}} \xrightarrow[\text{(head to tail)}]{\text{polymerization}} -CH_2\overset{*}{\underset{\displaystyle CH_3}{\underset{|}{CH}}}CH_2\overset{*}{\underset{\displaystyle CH_3}{\underset{|}{CH}}}CH_2\overset{*}{\underset{\displaystyle CH_3}{\underset{|}{CH}}}CH_2\overset{*}{\underset{\displaystyle CH_3}{\underset{|}{CH}}}-$$

There are three general arrangements of the methyl groups and hydrogen atoms along the chain. These arrangements are described as being *atactic, syndiotactic,* and *isotactic.*

If the stereochemistry at the stereocenters is random (Fig. D.1), the polymer is said to be atactic (*a*, without + Greek: *taktikos,* order).

In atactic polypropylene the methyl groups are randomly disposed on either side of the stretched carbon chain. [If we were to arbitrarily designate one end of the chain as having higher preference than the other, we could give (*R*)–(*S*) designations (Section 4.5) to the stereocenters. In atactic polypropylene the sequence of (*R*)–(*S*) designations along the chain is random.

Polypropylene produced by free radical polymerization at high pressures is atactic. Because the polymer is atactic, it is noncrystalline, has a low softening point, and has poor mechanical properties.

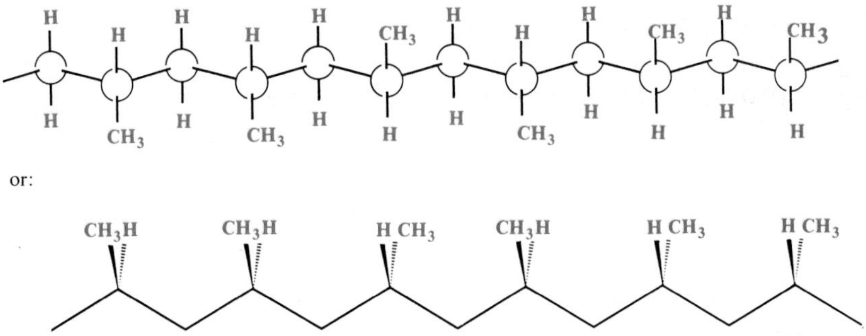

FIGURE D.1 Atactic polypropylene. (In this illustration a "stretched" carbon chain is used for clarity.)

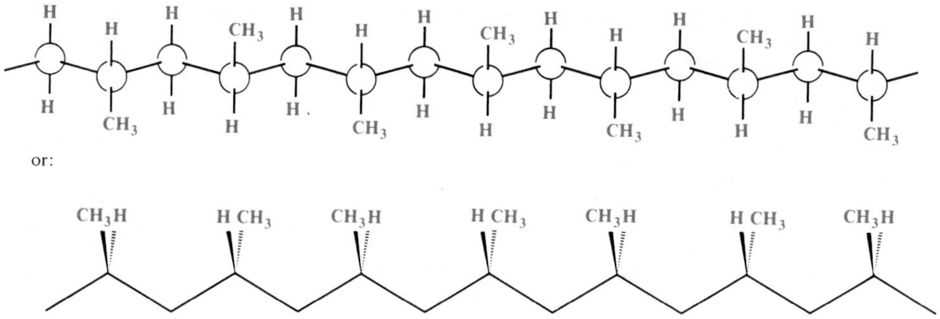

or:

FIGURE D.2   Syndiotactic polypropylene.

A second possible arrangement of the groups along the carbon chain is that of *syndiotactic* polypropylene. In syndiotactic polypropylene the methyl groups alternate regularly from one side of the stretched chain to the other (Fig. D.2). If we were to arbitrarily designate one end of the chain of syndiotactic polypropylene as having higher preference, the configuration of the stereocenters would alternate, (R), (S), (R), (S), (R), (S), (R), (S), and so on.

The third possible arrangement of stereocenters is the *isotactic* arrangement shown in Fig. D.3. In the isotactic arrangement all of the methyl groups are on the same side of the stretched chain. The configurations of the stereocenters are either all (R) or all (S) depending on which end of the chain is assigned higher preference.

The names isotactic and syndiotactic come from the Greek term *taktikos* (order) plus *iso* (same) and *syndyo* (two together).

Before 1953 isotactic and syndiotactic addition polymers were unknown. In that year, however, a German chemist, Karl Ziegler, and an Italian chemist, Giulio Natta, announced independently the discovery of catalysts that permit stereochemical control of polymerization reactions.* The Ziegler–Natta catalysts, as they are now called, are prepared from transition metal halides and a reducing agent. The catalysts most commonly used are prepared from titanium tetrachloride ($TiCl_4$) and a trialkylaluminum (R_3Al).

Ziegler–Natta catalysts are generally employed as suspended solids, and polymerization probably occurs at metal atoms on the surfaces of the particles. The mechanism for the polymerization is an ionic mechanism, but its details are not fully understood. There is evidence that polymerization occurs through an insertion of the alkene monomer between the metal and the growing polymer chain.

*Ziegler and Natta were awarded the Nobel Prize for their discoveries in 1963.

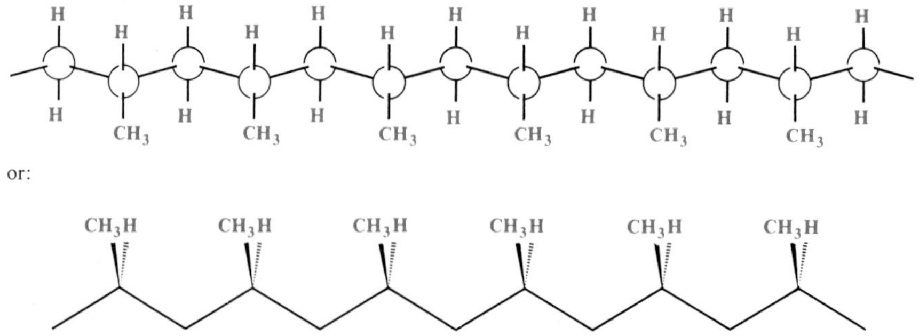

or:

FIGURE D.3   Isotactic polypropylene.

   Both syndiotactic and isotactic polypropylene have been made using Ziegler–Natta catalysts. The polymerizations occur at much lower pressures and the polymers that are produced are much higher melting than atactic polypropylene. Isotactic polypropylene, for example, melts at 175 °C. Isotactic and syndiotactic polymers are also much more crystalline than atactic polymers. The regular arrangement of groups along the chains allows them to fit together better in a crystal structure.

   Atactic, syndiotactic, and isotactic forms of poly(methyl methacrylate) (p. 436) are known. The atactic form is a noncrystalline glass. The crystalline syndiotactic and isotactic forms melt at 160 and 200 °C, respectively.

PROBLEM D.5

Write structural formulas for portions of the chain of: (a) Atactic poly(methyl methacrylate),   (b) syndiotactic poly(methyl methacrylate), and   (c) isotactic poly(methyl methacrylate).

CHAPTER TEN

Vitamin A in Polarized Light. Vitamin A, a conjugated unsaturated compound, is an important compound in vision. The chemical structure of Vitamin A is described in Section 22.3 and the role of Vitamin A in the photochemistry of vision is discussed in Special Topic N.

Conjugated Unsaturated Systems

10.1 INTRODUCTION

In our study of the reactions of alkenes in Chapter 7 we saw how important the π bond is in understanding the chemistry of unsaturated compounds. In this chapter we shall study a special group of unsaturated compounds and again we shall find that the π bond is the important part of the molecule. Here we shall examine ***species that have a p orbital on an atom adjacent to a double bond.*** The p orbital may be one that contains a single electron as in the allyl radical ($CH_2{=}CHCH_2\cdot$) (Section 10.2); it may be a vacant p orbital as in the allyl cation ($CH_2{=}CHCH_2^+$) (Section 10.4); or it may be the p orbital of another double bond as in 1,3-butadiene ($CH_2{=}CH{-}CH{=}CH_2$) (Section 10.9). We shall see that having a p orbital on an atom adjacent to a double bond allows the formation of an extended π bond—one that encompasses more than two nuclei.

Systems that have a p orbital on an atom adjacent to a double bond—molecules with delocalized π bonds—are called **conjugated unsaturated systems.** This general phenomenon is called **conjugation.** As we shall see, conjugation gives these systems special properties. We shall find, for example, that conjugated radicals, ions, or molecules are more stable than nonconjugated ones. We shall demonstrate this with the allyl radical, and allyl cation, and 1,3-butadiene. Conjugation also allows molecules to undergo unusual reactions, and we shall study these, too, including an important reaction for forming rings called the Diels–Alder reaction (Section 10.12).

10.2 ALLYLIC SUBSTITUTION AND THE ALLYL RADICAL

When propene reacts with bromine or chlorine at low temperatures, the reaction that takes place is the usual addition of halogen to the double bond.

$$CH_2{=}CH{-}CH_3 + X_2 \xrightarrow[\substack{CCl_4 \\ \text{(addition reaction)}}]{\text{low temperature}} \underset{\substack{| \quad\; | \\ X \quad X}}{CH_2{-}CH{-}CH_3}$$

However, when propene reacts with chlorine or bromine at very high temperatures or under conditions in which the concentration of the halogen is very small, the reaction that occurs is a **substitution.** These two examples illustrate how we can often change the course of an organic reaction simply by changing the conditions. (They also illustrate the need for specifying the conditions of a reaction carefully when we report experimental results.)

$$CH_2=CH-CH_3 + X_2 \xrightarrow[\substack{\text{or} \\ \text{low conc. of } X_2 \\ \text{(substitution reaction)}}]{\text{high temperature}} CH_2=CH-CH_2X + HX$$

Propene

In this substitution a halogen atom replaces one of the hydrogen atoms of the methyl group of propene. These hydrogen atoms are called the **allylic hydrogen atoms** and the substitution reaction is known as an **allylic substitution.***

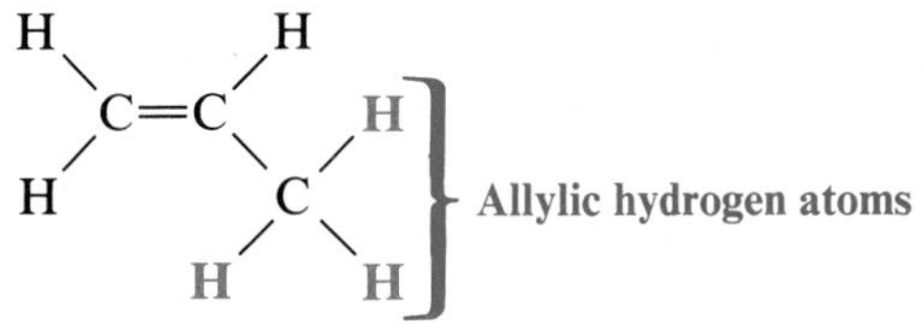

Propene undergoes allylic chlorination when propene and chlorine react in the gas phase at 400 °C.

$$CH_2=CH-CH_3 + Cl_2 \xrightarrow[\text{gas phase}]{400\ °C} CH_2=CH-CH_2Cl + HCl$$

Allyl chloride

Propene undergoes allylic bromination when it is treated with *N*-bromosuccinimide in carbon tetrachloride at room temperature. The reaction is initiated by light or peroxides.

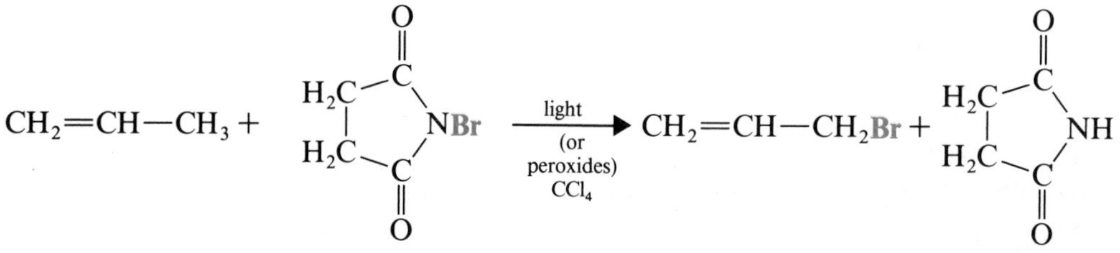

N-Bromosuccinimide   Allyl bromide   Succinimide
(NBS)

N-Bromosuccinimide is nearly insoluble in CCl_4 and provides a constant but very low concentration of bromine in the reaction mixture. It does this by reacting very

---

*These are general terms as well. The hydrogen atoms of any saturated carbon atom adjacent to a double bond, that is,

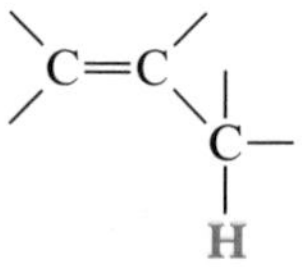

are called *allylic* hydrogen atoms and any reaction in which an allylic hydrogen atom is replaced is called an *allylic substitution.*

rapidly with the hydrogen bromide formed in the substitution reaction. Each molecule of HBr that is formed is replaced by one molecule of Br_2.

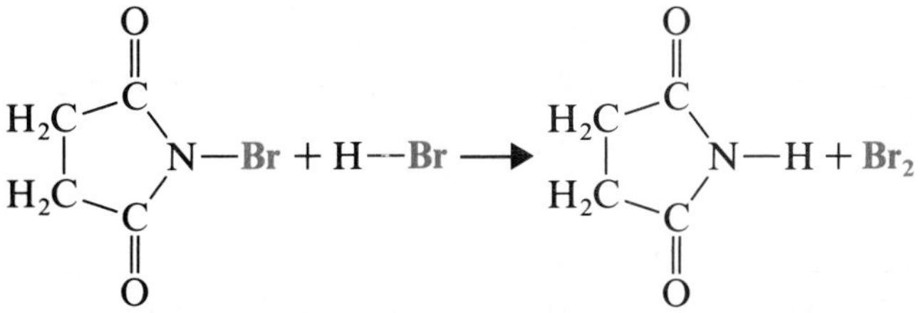

Under these conditions, that is, *in a nonpolar solvent and with a very low
concentration of bromine,* very little bromine adds to the double bond; it reacts by
substitution and replaces an allylic hydrogen atom instead.

The mechanism for these substitution reactions is the same as the chain mechanism for alkane halogenations that we saw in Chapter 9. In the chain-initiating step,
the halogen (chlorine or bromine) molecule dissociates into halogen atoms.

Chain-Initiating Step

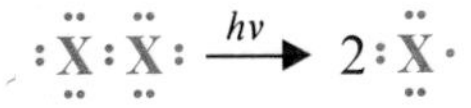

$$:\overset{..}{\underset{..}{X}}:\overset{..}{\underset{..}{X}}: \xrightarrow{h\nu} 2:\overset{..}{\underset{..}{X}}\cdot$$

In the first chain-propagating step the halogen atom abstracts one of the allylic
hydrogen atoms.

First Chain-Propagating Step

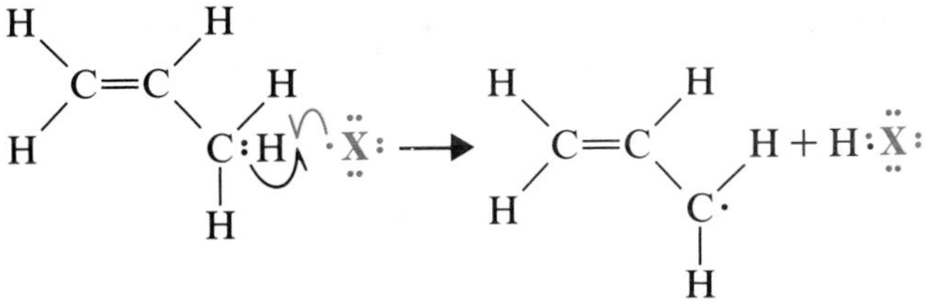

Allyl radical

The radical that is produced in this step is called an *allyl radical.**
In the second chain-propagating step the allyl radical reacts with a molecule of
the halogen.

Second Chain-Propagating Step

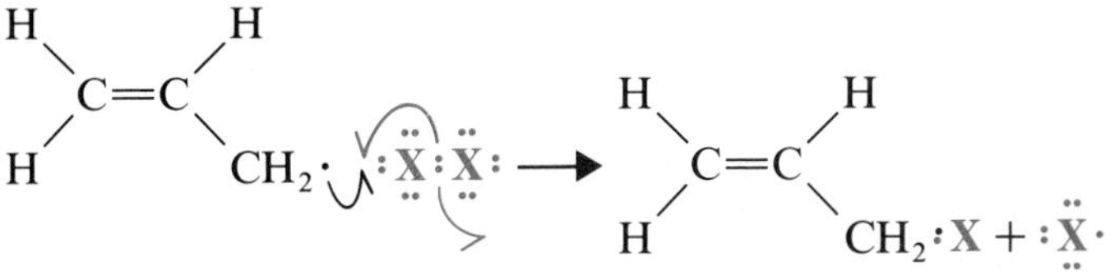

Allyl halide

This step results in the formation of a molecule of allyl halide and a halogen atom.
The halogen atom then brings about a repetition of the first chain-propagating step.

---

*A radical of the general type $-\overset{|}{C}=\overset{|}{C}-\overset{|}{\underset{\cdot}{C}}-$ is called an *allylic* radical.

The chain reaction continues until the usual chain-terminating steps consume the radicals.

The reason for substitution at the allylic hydrogen atoms of propene will be more understandable if we examine the bond dissociation energy of an allylic carbon–hydrogen bond and compare it with the bond dissociation energies of other carbon–hydrogen bonds (cf. Table 9.1.)

$$CH_2=CHCH_2{-}H \longrightarrow CH_2=CHCH_2\cdot + H\cdot \qquad DH° = 85 \text{ kcal/mole}$$

Propene   Allyl radical

$$(CH_3)_3C{-}H \longrightarrow (CH_3)_3C\cdot + H\cdot \qquad DH° = 91 \text{ kcal/mole}$$

Isobutane   3° Radical

$$(CH_3)_2CH{-}H \longrightarrow (CH_3)_2CH\cdot + H\cdot \qquad DH° = 94.5 \text{ kcal/mole}$$

Propane   2° Radical

$$CH_3CH_2CH_2{-}H \longrightarrow CH_3CH_2CH_2\cdot + H\cdot \qquad DH° = 98 \text{ kcal/mole}$$

Propane   1° Radical

$$CH_2=CH{-}H \longrightarrow CH_2=CH\cdot + H\cdot \qquad DH° = 108 \text{ kcal/mole}$$

Ethene   Vinyl radical

We see that an allylic carbon–hydrogen bond of propene is broken with greater ease than even the tertiary carbon–hydrogen bond of isobutane and with far greater ease than a vinylic carbon–hydrogen bond.

$$CH_2=CH{-}CH_2{-}H + \cdot X \longrightarrow CH_2=CH{-}CH_2\cdot + HX \qquad E_{act} \text{ is low}$$

Allyl radical

$$X\cdot + H{-}CH=CH{-}CH_3 \longrightarrow \cdot CH=CH{-}CH_3 + HX \qquad E_{act} \text{ is high}$$

Vinylic radical

The ease with which an allylic carbon–hydrogen bond is broken means that relative to primary, secondary, tertiary, and vinylic free radicals the allyl radical is the *most stable* (Fig. 10.1).

Relative stability   allylic or allyl $> 3° > 2° > 1° >$ vinyl

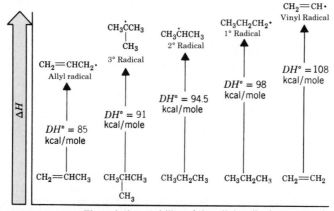

FIGURE 10.1 The relative stability of the allyl radical compared to 1°, 2°, 3°, and vinyl radicals. (The stabilities of the radicals are relative to the hydrocarbon from which each was formed, and the overall order of stability is allyl $> 3° > 2° > 1° >$ vinyl.)

OPTIONAL MATERIAL

Why, we might ask, does a low concentration of bromine favor allylic substitution over addition? To understand this we must recall the mechanism for addition and notice that in the first step only one atom of the bromine molecule becomes attached to the alkene *in a reversible step.*

$$Br-Br + \begin{array}{c} \diagdown C \diagup \\ \parallel \\ \diagup C \diagdown \end{array} \rightleftharpoons \overset{+}{Br}\begin{array}{c} \diagup C- \\ \mid \\ \diagdown C- \end{array} + Br^- \longrightarrow \begin{array}{c} -\overset{\mid}{C}-Br \\ \mid \\ Br-\overset{\mid}{C}- \end{array}$$

The other atom (from the bromide ion) becomes attached in the second step. Now, if the concentration of bromine is low, the equilibrium for the first step will lie far to the left. Moreover, even when the bromonium ion forms, the probability of its finding a bromide ion in its vicinity will also be low. These two factors slow the addition so that allylic substitution competes successfully.

The use of a nonpolar solvent also slows addition. Since there are no polar molecules to solvate (and thus stabilize) the bromide ion formed in the first step, the bromide ion uses a bromine molecule as a substitute:

$$2Br_2 + \begin{array}{c} \diagdown C \diagup \\ \parallel \\ \diagup C \diagdown \end{array} \underset{\text{solvent}}{\overset{\text{nonpolar}}{\rightleftharpoons}} \overset{+}{Br}\begin{array}{c} \diagup C- \\ \mid \\ \diagdown C- \end{array} + Br_3^-$$

This means that in a nonpolar solvent the rate equation is second order with respect to bromine,

$$\text{rate} = k\left[\begin{array}{c} \diagdown \\ \diagup \end{array} C{=}C \begin{array}{c} \diagup \\ \diagdown \end{array} \right][Br_2]^2$$

and that the low bromine concentration has an even more pronounced effect in slowing the rate of addition.

To understand why a high temperature favors allylic substitution over addition requires a study of Special Topic B concerning the effect of entropy changes on equilibria. The addition reaction, because it combines two molecules into one, has a substantial negative entropy change. At low temperatures, the $T\,\Delta S°$ term in $\Delta G° = \Delta H° - T\,\Delta S°$, is not large enough to offset the favorable $\Delta H°$ term. But as the temperature is increased, the $T\,\Delta S°$ term becomes more significant, $\Delta G°$ becomes more positive, and the equilibrium becomes more unfavorable.

10.3 THE STABILITY OF THE ALLYL RADICAL

An explanation of the stability of the allyl radical can be approached in two ways: in terms of molecular orbital theory and in terms of resonance theory (Section 1.8). As we shall see soon, both approaches give us equivalent descriptions of the allyl radical. The molecular orbital approach is easier to visualize, so we shall begin with it. (As preparation for this section, it would be a good idea to review the molecular orbital theory given in Sections 1.12 and 2.4.)

10.3A Molecular Orbital Description of the Allyl Radical

As an allylic hydrogen atom is abstracted from propene, the sp^3-hybridized carbon atom of the methyl group changes its hybridization state to sp^2 (cf. Section 9.7). The p orbital of this new sp^2-hybridized carbon atom overlaps with the p orbital of the central carbon atom. Thus, in the allyl radical three p orbitals overlap to form a set of π molecular orbitals that encompass all three carbon atoms. The new p orbital of the allyl radical is said to be *conjugated* with those of the double bond and the allyl radical is said to be a *conjugated unsaturated system*.

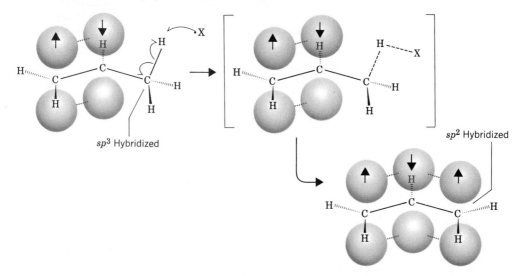

sp³ Hybridized

sp² Hybridized

The unpaired electron of the allyl radical and the two electrons of the π bond are **delocalized** over all three carbon atoms. This delocalization of the unpaired electron accounts for the greater stability of the allyl radical when compared to primary, secondary, and tertiary radicals. Although some delocalization occurs in primary, secondary, and tertiary radicals, delocalization is not as effective because it occurs through σ bonds.

The diagram in Figure 10.2 illustrates how the three p orbitals of the allyl radical combine to form three π molecular orbitals. (*Remember:* The number of molecular orbitals that results always equals the number of atomic orbitals that combine, cf. Section 1.11.) The bonding π molecular orbital is of lowest energy; it encompasses all three carbon atoms and is occupied by two spin-paired electrons. This bonding π orbital is the result of having p orbitals with lobes of the same sign overlap between adjacent carbon atoms. This type of overlap, as we recall, increases the π electron density in the regions between the atoms where it is needed for bonding. The non-bonding π orbital is occupied by one unpaired electron and it has a node at the central carbon atom. This node means that the unpaired electron is located in the vicinity of carbon atoms **1** and **3** only. The antibonding π molecular orbital results when orbital lobes of opposite sign overlap between adjacent carbon atoms: Such overlap means that in the antibonding π orbital there is a node between each pair of carbon atoms. This antibonding orbital of the allyl radical is of highest energy and is empty in the ground state of the radical.

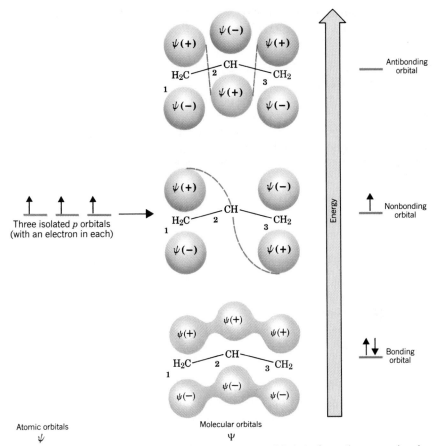

FIGURE 10.2 The combination of three atomic p orbitals to form three π molecular orbitals in the allyl radical. The bonding π molecular orbital is formed by the combination of the three p orbitals with lobes of the same sign overlapping above and below the plane of the atoms. The nonbonding π molecular orbital has a node at carbon **2**. The antibonding π molecular orbital has two nodes: between carbon atoms **1** and **2**, and between carbon atoms **2** and **3**.

We can illustrate the picture of the allyl radical given by molecular orbital theory in simpler terms with the following structure.

$$\tfrac{1}{2}\cdot H_2 \underset{1}{C} \quad \overset{\overset{\displaystyle H}{|}}{\underset{2}{C}} \quad \underset{3}{CH_2}\cdot\tfrac{1}{2}$$

We indicate with dotted lines that both carbon–carbon bonds are partial double bonds. This accommodates one of the things that molecular orbital theory tells us: *That there is a π bond encompassing all three atoms.* We also place the symbol $\tfrac{1}{2}\cdot$ beside carbon atoms **1** and **3**. This denotes a second thing molecular orbital theory tells us: *that the unpaired electron spends its time in the vicinity of carbon atoms 1 and 3.* Finally, implicit in the molecular orbital picture of the allyl radical is this: The two ends of the allyl radical are *equivalent.* This aspect of the molecular orbital description is also implicit in the formula just given.

10.3B Resonance Description of the Allyl Radical

Earlier in this section we wrote the structure of the allyl radical as **A**.

$$
\begin{array}{c}
\text{H} \\
\text{C} \\
\text{H}_2\text{C} \diagup \quad \diagdown \text{CH}_2 \cdot \\
\text{A}
\end{array}
$$

However, we might just as well have written the equivalent structure, **B**.

$$
\begin{array}{c}
\text{H} \\
\text{C} \\
\cdot \text{H}_2\text{C} \diagup \quad \diagdown \text{CH}_2 \\
\text{B}
\end{array}
$$

In writing structure **B** we do not mean to imply that we have simply taken structure **A** and turned it over. What we have done is moved the electrons in the following way:

$$
\begin{array}{c}
\text{H} \\
\text{C} \\
\text{H}_2\text{C} \diagup \quad \diagdown \text{CH}_2 \cdot
\end{array}
$$

We have not moved the atomic nuclei themselves.

Resonance theory (Section 1.8) tells us that whenever we can write two structures for a chemical entity ***that differ only in the positions of the electrons,*** the entity cannot be represented by either structure alone but is a *hybrid* of both. We can represent the hybrid in two ways: We can write both structures **A** and **B**, and connect them with a double-headed arrow, a special sign in resonance theory, that indicates they are resonance structures.

$$
\begin{array}{ccc}
\text{H} & & \text{H} \\
\overset{2}{\text{C}} & & \overset{2}{\text{C}} \\
\text{H}_2\underset{1}{\text{C}} \diagup \quad \diagdown \underset{3}{\text{CH}_2} \cdot & \longleftrightarrow & \cdot \text{H}_2\underset{1}{\text{C}} \diagup \quad \diagdown \underset{3}{\text{CH}_2} \\
\text{A} & & \text{B}
\end{array}
$$

Or we can write a single structure, **C**, that blends the features of both resonance structures.

$$
\begin{array}{c}
\text{H} \\
\overset{2}{\text{C}} \\
\tfrac{1}{2} \cdot \text{H}_2\underset{1}{\text{C}} \diagup \quad \diagdown \underset{3}{\text{CH}_2} \cdot \tfrac{1}{2} \\
\text{C}
\end{array}
$$

We see, then, that resonance theory gives us exactly the same picture of the allyl radical as we got from molecular orbital theory. Structure **C** describes the carbon–carbon bonds of the allyl radical as partial double bonds. The resonance structures **A** and **B** also tell us that the unpaired electron is associated only with carbon atoms **1** and **3**. We indicate this in structure **C** by placing a $\frac{1}{2}\cdot$ beside carbon atoms **1** and **3**.* Because resonance structures **A** and **B** are equivalent, carbon atoms *1 and 3 are also equivalent.*

Another rule in resonance theory is that *whenever equivalent resonance structures* can be written for a chemical species, *the chemical species is much more stable than either resonance structure (when taken alone) would indicate.* If we were to examine either **A** or **B** alone, we might decide that they resemble primary radicals. Thus, we might estimate the stability of the allyl radical as approximately that of a primary radical. In doing so, we would greatly underestimate the stability of the allyl radical. Resonance theory tells us, however, that since **A** and **B** are *equivalent resonance structures,* the allyl radical should be much more stable than either, that is, much more stable than a primary radical. This correlates with what experiments have shown to be true: The allyl radical is even more stable than a tertiary radical.

PROBLEM 10.1

(a) What product(s) would you expect to obtain if propene labeled with ^{14}C at carbon atom **1** were subjected to allylic chlorination or bromination? (b) Explain your answer.

$$^{14}CH_2{=}CHCH_3 + X_2 \xrightarrow[\substack{\text{or} \\ \text{low conc. of } X_2}]{\text{high temperature}} \ ?$$

(c) If more than one product would be obtained, what relative proportions would you expect?

10.4 THE ALLYL CATION

Although we cannot go into the experimental evidence here, the allyl cation $(CH_2{=}CHCH_2^+)$ is an unusually stable carbocation. It is even more stable than a secondary carbocation and is almost as stable as a tertiary carbocation. In general terms, the relative order of stabilities of carbocations is that given here.

Relative Order of Carbocation Stability

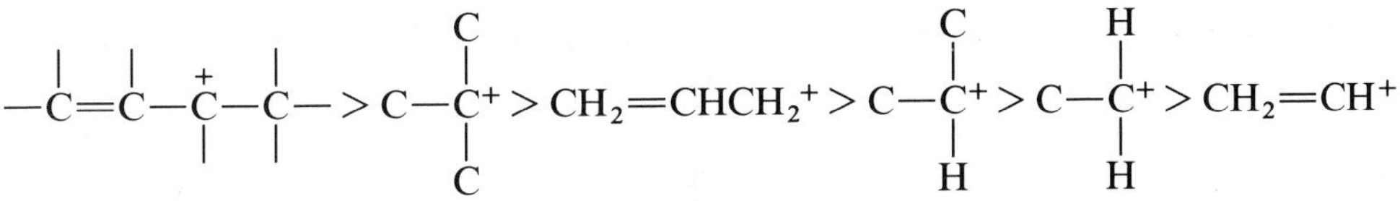

$$-\overset{|}{\underset{|}{C}}{=}\overset{|}{\underset{|}{C}}-\overset{+}{\underset{|}{C}}-\overset{|}{\underset{|}{C}}- > \ \overset{\overset{\displaystyle C}{|}}{C-\underset{\underset{\displaystyle C}{|}}{C}^+} > \ CH_2{=}CHCH_2^+ > \ \overset{\overset{\displaystyle C}{|}}{C-\underset{\underset{\displaystyle H}{|}}{C}^+} > \ \overset{\overset{\displaystyle H}{|}}{C-\underset{\underset{\displaystyle H}{|}}{C}^+} > \ CH_2{=}CH^+$$

Allylic   >   3°   >   Allyl   >   2°   >   1°   >   Vinyl

---

*A resonance structure such as the one shown below would indicate that an unpaired electron is associated with carbon atom **2**. This structure is not a proper resonance structure because resonance theory dictates that *all resonance structures must have the same number of unpaired electrons* (cf. Section 10.5).

$$\cdot CH_2{-}\overset{\displaystyle \cdot}{C}H{-}CH_2\cdot$$

(an incorrect resonance structure)

As we might expect, the unusual stability of the allyl cation and other allylic cations can also be accounted for in terms of molecular orbital or resonance theory.

The molecular orbital description of the allyl cation is shown in Fig. 10.3.

The bonding π molecular orbital of the allyl cation, like that of the allyl radical (Fig. 10.2), contains two spin-paired electrons. The nonbonding π molecular orbital of the allyl cation, however, is empty. Since an allyl cation is what we would get if we removed an electron from an allyl radical, we can say, in effect, that we remove the electron from the nonbonding molecular orbital.

$$CH_2{=}CHCH_2 \cdot \xrightarrow{\ -e^-\ } CH_2{=}CHCH_2{}^+$$

Removal of an electron from a nonbonding orbital (cf. Fig. 10.2) is known to require less energy than removal of an electron from a bonding orbital. In addition, the positive charge that forms on the allyl cation is *effectively delocalized* between carbon atoms **1** and **3**. Thus, in molecular orbital theory these two factors, the ease of removal of a nonbonding electron, and the delocalization of charge, account for the unusual stability of the allyl cation.

Resonance theory depicts the allyl cation as a hybrid of structures **D** and **E** represented here.

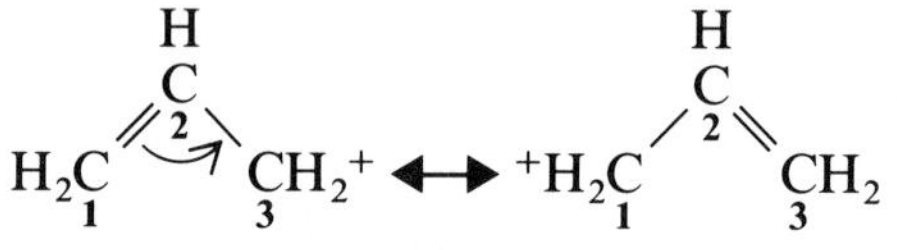

Because **D** and **E** are *equivalent* resonance structures, resonance theory predicts that the allyl cation should be unusually stable. Since the positive charge is located on carbon atom **3** in **D** and on carbon atom **1** in **E**, resonance theory also tells us that the positive charge should be delocalized over both carbon atoms. Carbon atom **2** carries none of the positive charge. The hybrid structure **F** (see following structure) includes charge and bond features of both **D** and **E**.

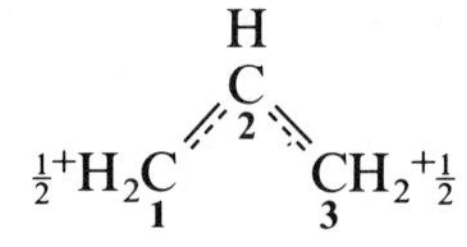

PROBLEM 10.2

(a) Write structures corresponding to **D**, **E**, and **F** for the carbocation shown.

$$CH_3{-}\overset{+}{C}H{-}CH{=}CH_2$$

(b) This carbocation appears to be even more stable than a tertiary carbocation; how can you explain this?   (c) What product(s) would you expect to be formed if this carbocation reacted with a chloride ion?

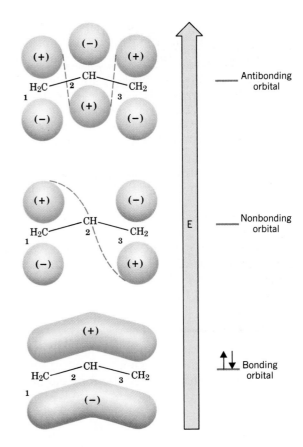

FIGURE 10.3 The π molecular orbitals of the allyl cation. The allyl cation, like the allyl radical (Fig. 10.2), is a conjugated unsaturated system.

10.5 SUMMARY OF RULES FOR RESONANCE

We have used resonance theory extensively in earlier sections of this chapter because we have been describing radicals and ions with delocalized electrons (and charges) in π bonds. Resonance theory is especially useful with systems like this, and we shall use it again and again in the chapters that follow. We had an introduction to resonance theory in Section 1.8 and it should be helpful now to summarize the rules for writing resonance structures and for estimating the relative contribution a given structure will make to the overall hybrid.

1. **Resonance structures exist only on paper.** They have no real existence of their own. Resonance structures are useful because they allow us to describe molecules, radicals, and ions for which a single Lewis structure is inadequate. We write two or more Lewis structures, calling them resonance structures or resonance contributors. We connect these structures by double-headed arrows (⟷), and we say that the real molecule, radical, or ion is like a hybrid of all of them.

2. **In writing resonance structures we are only allowed to move electrons.** The positions of the nuclei of the atoms must remain the same in all of the structures. Structure **3** is not a resonance structure for the allylic cation, for example, because in order to form it we would have to move a hydrogen atom and this is not permitted.

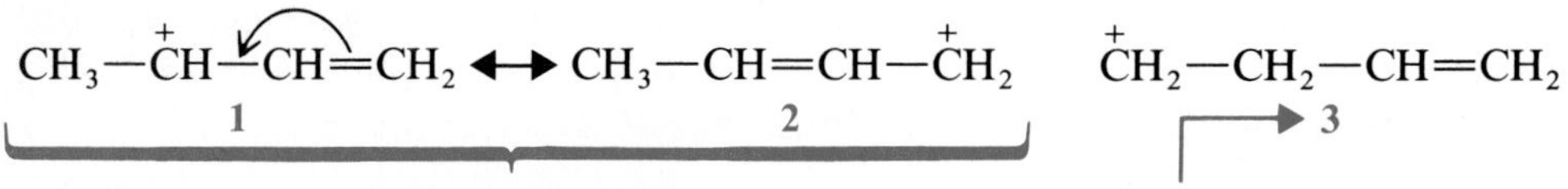

$$CH_3-\overset{+}{C}H-CH{=}CH_2 \longleftrightarrow CH_3-CH{=}CH-\overset{+}{C}H_2 \qquad \overset{+}{C}H_2-CH_2-CH{=}CH_2$$

1	**2**	**3**

These are resonance structures for the allylic cation formed when 1,3-butadiene accepts a proton

This is not a proper resonance structure for the allylic cation because a hydrogen atom has been moved

3. **All of the structures must be proper Lewis structures.** We should not write structures in which carbon has five bonds, for example.

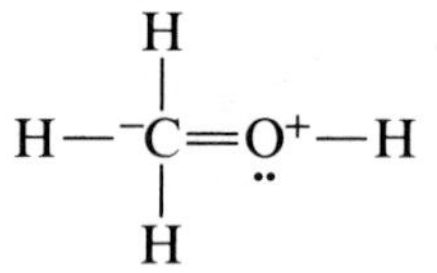

$$H-\overset{\underset{\displaystyle H}{|}}{\underset{\underset{\displaystyle H}{|}}{C}}{=}\overset{+}{\underset{\displaystyle \cdot\cdot}{O}}-H$$

This is not a proper resonance structure for methanol because carbon has five bonds. Elements of the first major row of the Periodic Table cannot have more than eight electrons in their valence shell

4. **All resonance structures must have the same number of unpaired electrons.** The following structure is not a resonance structure for the allyl radical because it contains three unpaired electrons and the allyl radical contains only one.

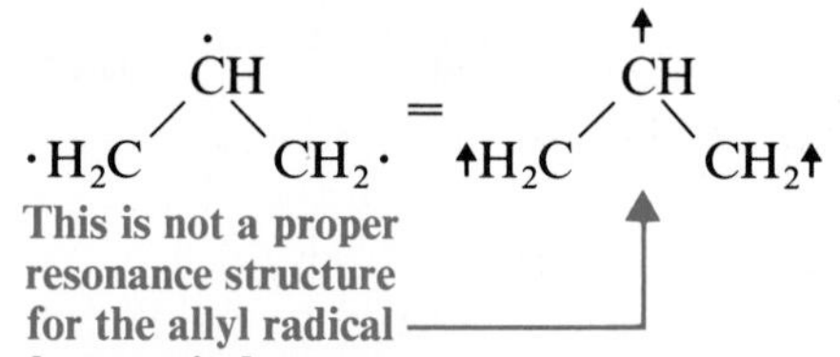

$$\overset{\displaystyle \cdot CH}{\underset{\displaystyle \cdot H_2C \qquad CH_2 \cdot}{\diagup\diagdown}} \quad = \quad \overset{\displaystyle \uparrow CH}{\underset{\displaystyle \uparrow H_2C \qquad CH_2\uparrow}{\diagup\diagdown}}$$

This is not a proper resonance structure for the allyl radical because it does not contain the same number of unpaired electrons as $CH_2{=}CHCH_2\cdot$

5. **All atoms that are a part of the delocalized system must lie in a plane or be nearly planar.** For example, 2,3-di-*tert*-butyl-1,3-butadiene behaves like a *nonconjugated* diene because the large *tert*-butyl groups twist the structure and prevent the double bonds from lying in the same plane. Because they are not in the same plane, the *p* orbitals at C-2 and C-3 do not overlap and delocalization (and therefore resonance) is prevented.

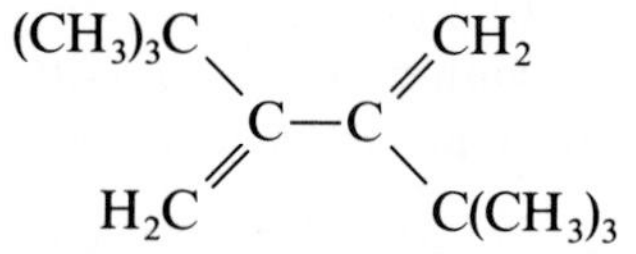

$$\underset{H_2C}{\overset{(CH_3)_3C}{\diagdown}}\overset{\diagup CH_2}{\underset{\diagdown C(CH_3)_3}{C{-}C}}$$

2,3-Di-*tert*-butyl-1,3-butadiene

6. **The energy of the actual molecule is lower than the energy that might be estimated for any contributing structure.** The actual allyl cation, for example, is more stable than either resonance structure **4** or **5** taken separately would

indicate. Structures **4** and **5** resemble primary carbocations and yet the allyl cation is more stable (has lower energy) than a secondary carbocation. Chemists often call this kind of stabilization *resonance stabilization.*

$$CH_2\!=\!CH\!-\!CH_2^+ \longleftrightarrow {}^+CH_2\!-\!CH\!=\!CH_2$$
$$\qquad\quad\ \ 4 \qquad\qquad\qquad\qquad 5$$

In the next chapter we shall find that benzene is highly resonance stabilized because it is a hybrid of the two equivalent forms that follow:

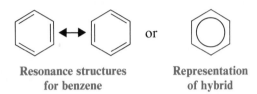

Resonance structures	Representation
for benzene	of hybrid

7. **Equivalent resonance structures make equal contributions to the hybrid, and a system described by them has a large resonance stabilization.** Structures **4** and **5** make equal contributions to the allylic cation because they are equivalent. They also make a large stabilizing contribution and account for allylic cations being unusually stable. The same can be said about the contributions made by the equivalent structures **A** and **B** (Section 10.3B) for the allyl radical and by the equivalent structures for benzene.

8. Structures that are not equivalent do not make equal contributions. Generally speaking, **the more stable a structure is (when taken by itself), the greater is its contribution to the hybrid.** For example, the following cation is a hybrid of structures **6** and **7**. Structure **6** makes a greater contribution than **7** because structure **6** resembles a tertiary carbocation while structure **7** resembles a primary cation and tertiary cations are more stable.

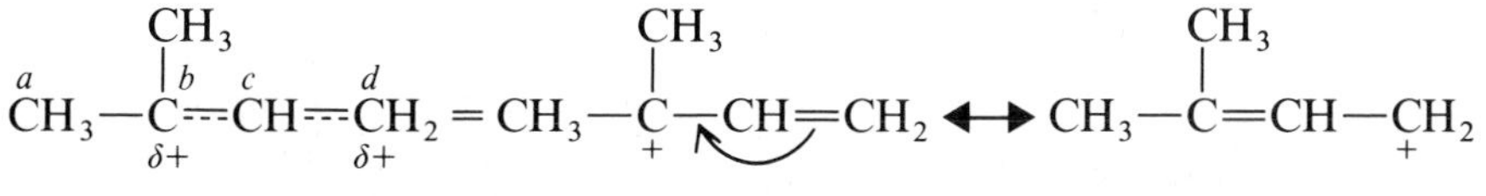

$$\overset{\displaystyle CH_3}{\underset{a}{CH_3}\!-\!\overset{b}{\underset{\delta+}{C}}\!\overset{c}{=\!\!=\!}CH\!\overset{d}{=\!\!=\!}\underset{\delta+}{CH_2}} = CH_3\!-\!\overset{CH_3}{\underset{+}{C}}\!\curvearrowright\!CH\!=\!CH_2 \longleftrightarrow CH_3\!-\!\overset{CH_3}{C}\!=\!CH\!-\!\underset{+}{CH_2}$$
$$\qquad\qquad\qquad 6 \qquad\qquad\qquad\qquad\qquad 7$$

That **6** makes a larger contribution means that the partial positive charge on carbon b of the hybrid will be larger than the partial positive charge on carbon d. It also means that the bond between carbons c and d will be more like a double bond than the bond between carbons b and c.

The following rules will help us in making decisions about the relative stabilities of resonance structures.

a. **The more covalent bonds a structure has, the more stable it is.** This is exactly what we would expect because we know that forming a covalent bond lowers the energy of atoms. This means that of the following structures for 1,3-butadiene, **8**, is by far the most stable and makes by far the largest contribution because it contains one more bond. (It is also most stable for the reason given under rule **c.**)

$$CH_2\!=\!CH\!-\!CH\!=\!CH_2 \longleftrightarrow \overset{+}{C}H_2\!-\!CH\!=\!CH\!-\!\overset{..}{C}H_2 \longleftrightarrow \overset{..}{C}H_2\!-\!CH\!=\!CH\!-\!\overset{+}{C}H_2$$

8   9   10

This structure is the most stable because it contains more covalent bonds

b. **Structures in which all of the atoms have a complete valence shell of electrons** (i.e., the noble gas structure) **are especially stable and make large contributions to the hybrid.** Again, this is what we would expect from what we know about bonding. This means, for example, that **12** makes a larger stabilizing contribution to the cation below than **11** because all of its atoms have a complete valence shell. (Notice too that **12** has more covalent bonds than **11**, cf. rule **a.**)

$$\overset{+}{C}H_2\!-\!\overset{..}{\underset{..}{O}}\!-\!CH_3 \longleftrightarrow CH_2\!=\!\overset{+}{\underset{..}{O}}\!-\!CH_3$$

11   12

Here this carbon atom has only six electrons   Here the carbon atom has eight electrons

c. **Charge separation decreases stability.** Separating opposite charges requires energy. Therefore, structures in which opposite charges are separated have greater energy (lower stability) than those that have no charge separation. This means that of the following two structures for vinyl chloride, structure **13** makes a larger contribution because it does not have separated charges. (This does not mean that structure **14** does not contribute to the hybrid, it just means that the contribution made by **14** is smaller.)

$$CH_2\!=\!CH\!-\!\overset{..}{\underset{..}{C}l}: \longleftrightarrow :\overset{-}{C}H_2\!-\!CH\!=\!\overset{..}{\underset{..}{C}l}:^+$$

13   14

PROBLEM 10.3

Give the important resonance structures for each of the following:

(a) $CH_2\!=\!\underset{\underset{CH_3}{|}}{C}\!-\!CH_2\cdot$

(b) $CH_2\!=\!CH\!-\!\underset{+}{C}H\!-\!CH\!=\!CH_2$

(c) [benzene ring with radical]

(d) [cyclohexadienyl cation]

(e) $CH_3CH\!=\!CH\!-\!CH\!=\!\overset{+}{\underset{..}{O}}H$

(f) $CH_2\!=\!CH\!-\!Cl$

(g) [benzyl cation CH_2^+]

(h) $^-\!:CH_2\!-\!\overset{\overset{O}{\|}}{C}\!-\!CH_3$

(i) $CH_3\!-\!S\!-\!CH_2^+$

(j) $CH_3\!-\!NO_2$

PROBLEM 10.4

From each set of resonance structures that follow, designate the one that would contribute most to the hybrid and explain your choice.

(a)
$$\underset{\underset{CH_3CH_2\overset{\displaystyle CH_3}{|}}{}}{CH_3CH_2C=CH-CH_2{}^+} \longleftrightarrow \underset{\underset{CH_3CH_2\overset{\displaystyle CH_3}{|}}{+}}{CH_3CH_2C-CH=CH_2}$$

(b)

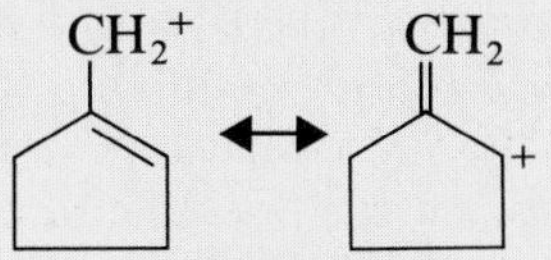

(c) $\overset{+}{C}H_2-\overset{..}{N}(CH_3)_2 \longleftrightarrow CH_2=\overset{+}{N}(CH_3)_2$

(d) $CH_3-\overset{\overset{\displaystyle O}{\|}}{C}-O-H \longleftrightarrow CH_3-\overset{\overset{\displaystyle O^-}{|}}{C}=O^+-H$

(e) $\underset{\cdot}{C}H_2CH=CHCH=CH_2 \longleftrightarrow CH_2=CH\underset{\cdot}{C}HCH=CH_2 \longleftrightarrow$

$$CH_2=CHCH=CH\underset{\cdot}{C}H_2$$

(f) $:NH_2-C\equiv N: \longleftrightarrow {}^+NH_2=C=\overset{..}{N}:{}^-$

PROBLEM 10.5

The following keto and enol forms differ in the positions for their electrons but they are not resonance structures. Explain why they are not.

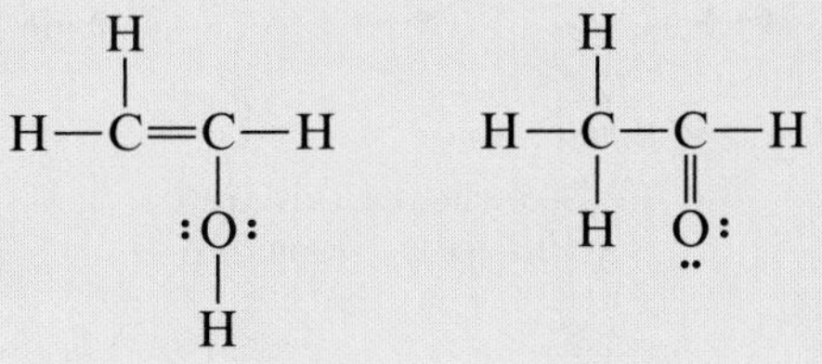

$$\underset{\underset{H}{|}}{H-\overset{\overset{\displaystyle H}{|}}{C}=\underset{\underset{:O:}{\|}}{C}-H} \qquad \underset{\underset{H \quad O:}{|}}{H-\overset{\overset{\displaystyle H}{|}}{C}-\overset{\|}{C}-H}$$

Enol form   Keto form

10.6 BENZYLIC RADICALS AND CATIONS

Removal of a hydrogen atom from the methyl group of toluene (methylbenzene) produces a radical called the **benzyl radical:**

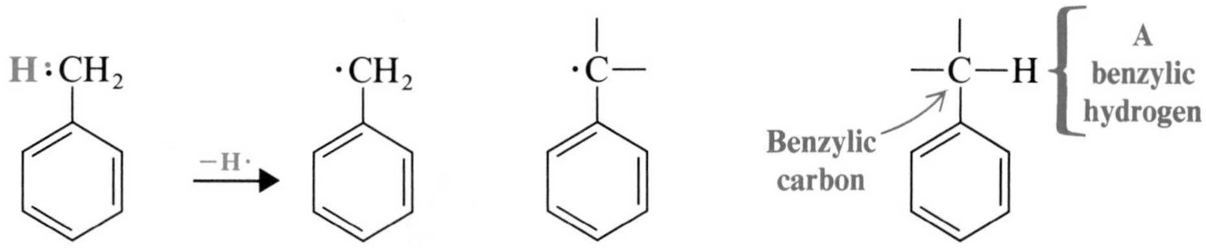

Toluene   The benzyl   A benzylic
(methylbenzene)   radical   radical

The name benzyl radical is used as a specific name for the radical produced in this reaction. The general name **benzylic radical** applies to all radicals that have an unpaired electron on the side chain carbon atom that is directly attached to the benzene ring. The hydrogen atoms of the carbon atom directly attached to the benzene ring are called **benzylic hydrogen atoms.**

Removal of an electron from a benzylic radical produces a **benzylic cation:**

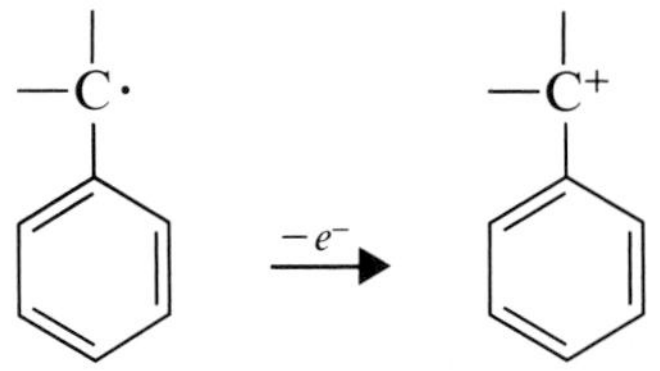

Benzylic radical   Benzylic cation

Benzylic radicals and benzylic cations are *conjugated unsaturated systems* and *both are unusually stable.* They have approximately the same stabilities as allylic radicals and allylic cations. This exceptional stability of benzylic radicals and cations is easily explained by resonance theory. (It can also be explained by molecular orbital theory, but we shall not go into this here.) In the case of each of the following entities several resonance structures can be written. Each entity, therefore, is highly stabilized by resonance.

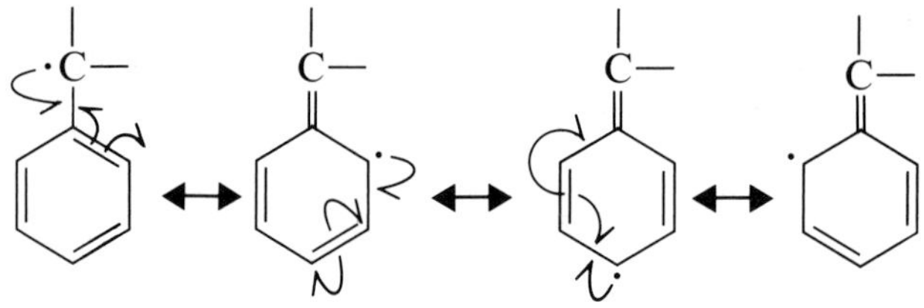

Benzylic radicals are
stabilized by resonance

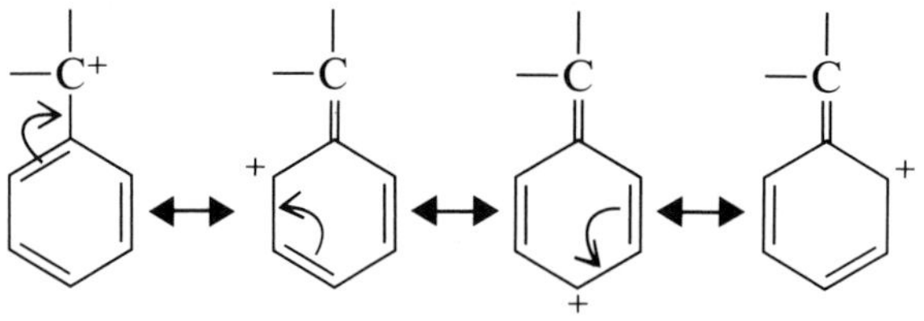

Benzylic cations are
stabilized by resonance

10.7 ALLYLIC AND BENZYLIC HALIDES IN NUCLEOPHILIC SUBSTITUTION REACTIONS

Allylic and benzylic halides can be classified in the same way that we have classified other organic halides:

$$-\overset{|}{\underset{}{C}}=\overset{|}{\underset{}{C}}-CH_2X \qquad -\overset{|}{\underset{}{C}}=\overset{|}{\underset{}{C}}-\overset{\overset{R}{|}}{\underset{}{C}}HX \qquad -\overset{|}{\underset{}{C}}=\overset{|}{\underset{}{C}}-\overset{\overset{R}{|}}{\underset{\underset{R'}{|}}{C}}X$$

1° Allylic   2° Allylic   3° Allylic

$$ArCH_2X \qquad \overset{\overset{R}{|}}{Ar}CHX \qquad \overset{\overset{R}{|}}{Ar}\underset{\underset{R'}{|}}{C}X$$

1° Benzylic   2° Benzylic   3° Benzylic

All of these compounds undergo nucleophilic substitution reactions. As with other tertiary halides (Section 5.13B), the steric hindrance associated with having three bulky groups on the carbon bearing the halogen prevents tertiary allylic and tertiary benzylic halides from reacting by an S_N2 mechanism. They react with nucleophiles only by an S_N1 mechanism.

Primary and secondary allylic and benzylic halides can react either by an S_N2 mechanism or by an S_N1 mechanism in ordinary nonacidic solvents. We would expect these halides to react by an S_N2 mechanism because they are structurally similar to primary and secondary alkyl halides. (Having only one or two groups attached to the carbon bearing the halogen does not prevent S_N2 attack.) But primary and secondary allylic and benzylic halides can also react by an S_N1 mechanism because they can form relatively stable carbocations and in this regard they differ from primary and secondary alkyl halides.*

Overall we can summarize the effect of structure on reactivity of alkyl, allylic, and benzylic halides in the following way:

These halides give mainly S_N2 reactions.	These halides give mainly S_N1 reactions.
$CH_3-X \qquad R-CH_2-X \qquad R-\overset{}{\underset{\underset{R'}{\|}}{C}H-X}$	$R'-\overset{\overset{R}{\|}}{\underset{\underset{R''}{\|}}{C}}-X$
These halides may give either S_N1 or S_N2 reactions.	
$Ar-CH_2-X \qquad Ar-\underset{\underset{R}{\|}}{C}H-X$	$Ar-\overset{\overset{R}{\|}}{\underset{\underset{R'}{\|}}{C}}-X$
$-\overset{\|}{\underset{}{C}}=\overset{\|}{\underset{}{C}}CH_2-X \qquad -\overset{\|}{\underset{}{C}}=\overset{\|}{\underset{}{C}}\underset{\underset{R}{\|}}{C}H-X$	$-\overset{\|}{\underset{}{C}}=\overset{\|}{\underset{}{C}}-\overset{\overset{R}{\|}}{\underset{\underset{R'}{\|}}{C}}-X$

*There is some dispute as to whether 2° alkyl halides react by an S_N1 mechanism to any appreciable extent in ordinary nonacidic solvents such as mixtures of water and alcohol or acetone, but it is clear that reaction by an S_N2 mechanism is, for all practical purposes, the more important pathway.

PROBLEM 10.6

Account for the following observations: (a) When 1-chloro-2-butene is allowed to react with a relatively concentrated solution of sodium ethoxide in ethyl alcohol, the reaction rate depends on the concentration of the allylic halide and on the concentration of ethoxide ion. The product of the reaction is almost exclusively $CH_3CH=CHCH_2OCH_2CH_3$. (b) When 1-chloro-2-butene is allowed to react with very dilute solutions of sodium ethoxide in ethyl alcohol (or with ethyl alcohol alone), the reaction rate is independent of the concentration of ethoxide ion; it depends only on the concentration of the allylic halide. Under these conditions the reaction produces a mixture of $CH_3CH=CHCH_2OCH_2CH_3$ and $CH_3CHCH=CH_2$. (c) In the presence

$$OCH_2CH_3$$

of traces of water 1-chloro-2-butene is slowly converted to a mixture of 1-chloro-2-butene and 3-chloro-1-butene.

PROBLEM 10.7

1-Chloro-3-methyl-2-butene undergoes hydrolysis in a mixture of water and dioxane at a rate that is more than a thousand times that of 1-chloro-2-butene. (a) What factor accounts for the difference in reactivity? (b) What products would you expect to obtain? [Dioxane is a cyclic ether (below) that is miscible with water in all proportions and is a convenient co-solvent for conducting reactions like these.]

Dioxane

PROBLEM 10.8

Primary halides of the type $ROCH_2X$ apparently undergo S_N1 type reactions, whereas most primary halides do not. Can you propose a resonance explanation for the ability of halides of the type $ROCH_2X$ to undergo S_N1 reactions?

PROBLEM 10.9

The following chlorides undergo solvolysis in ethanol at the relative rates given in parentheses. How can you explain these results?

$C_6H_5CH_2Cl$	$C_6H_5CHCH_3$	$(C_6H_5)_2CHCl$	$(C_6H_5)_3CCl$
	Cl		
(0.08)	(1)	(300)	(3×10^6)

10.8 ALKADIENES AND POLYUNSATURATED HYDROCARBONS

Many hydrocarbons are known whose molecules contain more than one double or triple bond. A hydrocarbon whose molecules contain two double bonds is called an **alkadiene;** one whose molecules contain three double bonds is called an **alkatriene,** and so on. Colloquially, these compounds are often referred to as simply as "dienes" or "trienes." A hydrocarbon with two triple bonds is called an **alkadiyne,** and a hydrocarbon with a double and triple bond is called an **alkenyne.**

The following examples of polyunsaturated hydrocarbons illustrate how specific compounds are named.

$$\overset{1}{C}H_2=\overset{2}{C}=\overset{3}{C}H_2 \qquad \overset{1}{C}H_2=\overset{2}{C}H-\overset{3}{C}H=\overset{4}{C}H_2$$

1,2-Propadiene   **1,3-Butadiene**
(allene)

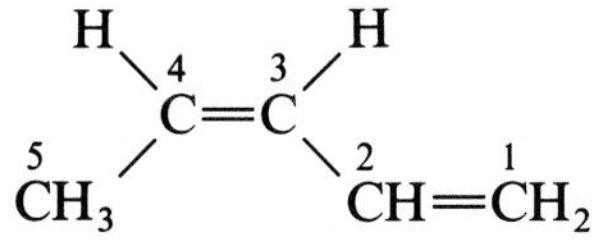

(3Z)-1,3-Pentadiene   **(2E,4E)-2,4-Hexadiene**
(*cis*-1,3-pentadiene)   (*trans,trans*-2,4-hexadiene)

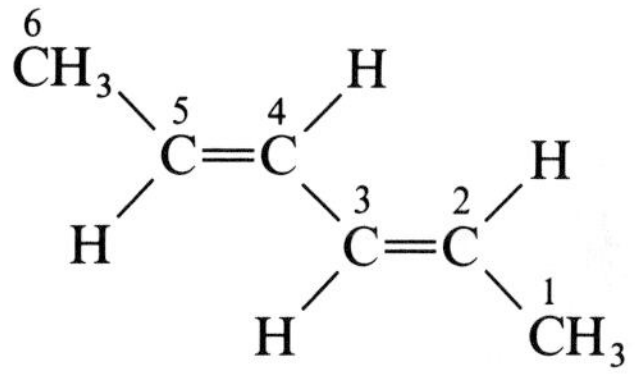

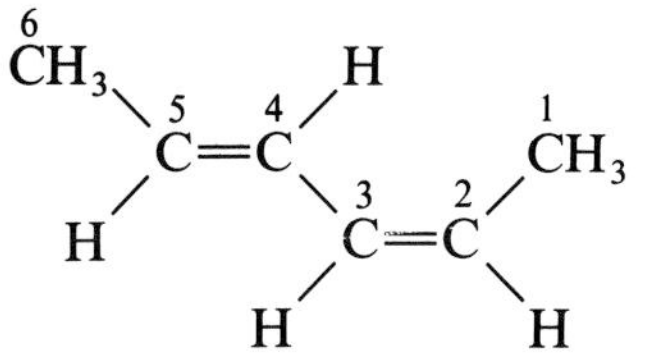

(2Z,4E)-2,4-Hexadiene $H\overset{5}{C}\equiv\overset{4}{C}-\overset{3}{C}H_2\overset{2}{C}H=\overset{1}{C}H_2$
(*cis,trans*-2,4-hexadiene)

                    1-Penten-4-yne

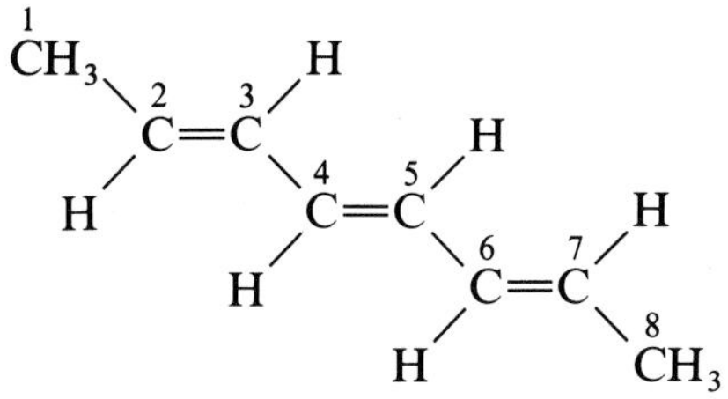

(2E,4E,6E)-2,4,6-Octatriene
(*trans,trans,trans*-2,4,6-octatriene)

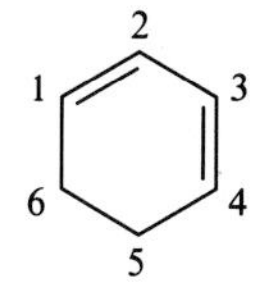

1,3-Cyclohexadiene 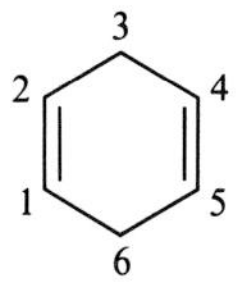   **1,4-Cyclohexadiene**

The multiple bonds of polyunsaturated compounds are classified as being **cumulated, conjugated,** or **isolated.** The double bonds of allene (1,2-propadiene) are said to be cumulated because one carbon (the central carbon) participates in two double bonds. Hydrocarbons whose molecules have cumulated double bonds are called **cumulenes.** The name **allene** (Section 4.16) is also used as a class name for molecules with two cumulated double bonds.

$$CH_2{=}C{=}CH_2 \qquad \overset{\diagdown}{\diagup}C{=}C{=}C\overset{\diagup}{\diagdown}$$

Allene   A cumulated
                   diene

An example of a conjugated diene is 1,3-butadiene. In conjugated polyenes the double and single bonds *alternate* along the chain.

$$CH_2{=}CH{-}CH{=}CH_2 \qquad \overset{\diagdown}{\diagup}C{=}C\overset{\diagup}{\diagdown}$$

1,3-Butadiene   A conjugated diene

(2*E*,4*E*,6*E*)-2,4,6-Octatriene is an example of a conjugated alkatriene.

If one or more saturated carbon atoms intervene between the double bonds of an alkadiene, the double bonds are said to be *isolated.* An example of an isolated diene is 1,4-pentadiene.

$$\overset{\diagdown}{\diagup}C{=}C\overset{\diagup}{\diagdown}\underset{(CH_2)_n}{}C{=}C\overset{\diagup}{\diagdown} \qquad CH_2{=}CH{-}CH_2{-}CH{=}CH_2$$

An isolated diene   1,4-Pentadiene
(*n* ≠ 0)

PROBLEM 10.10

(a) Which other compounds in Section 10.8 are conjugated dienes? (b) Which other compound is an isolated diene? (c) Which compound is an isolated enyne?

In Chapter 4 we saw that appropriately substituted cumulated dienes (allenes) give rise to chiral molecules. Cumulated dienes have had some commercial importance and cumulated double bonds are occasionally found in naturally occurring molecules. In general, cumulated dienes are less stable than isolated dienes.

The double bonds of isolated dienes behave just as their name suggests—as isolated "enes." They undergo all of the reactions of alkenes; and except for the fact that they are capable of reacting twice, their behavior is not unusual. Conjugated dienes are far more interesting because we find that their double bonds interact with each other. This interaction leads to unexpected properties and reactions. We shall, therefore, consider the chemistry of conjugated dienes in detail.

10.9 1,3-BUTADIENE: ELECTRON DELOCALIZATION

10.9A Bond Lengths of 1,3-Butadiene

The carbon–carbon bond lengths of 1,3-butadiene have been determined and are shown here.

$$\overset{1}{CH_2}=\!\!=\!\!\overset{2}{CH}\underset{1.47\ \text{Å}}{\text{———}}\overset{3}{CH}=\!\!=\!\!\overset{4}{CH_2}$$
$$\underset{1.34\ \text{Å}}{} \qquad \underset{1.34\ \text{Å}}{}$$

The C-1—C-2 bond and the C-3—C-4 bond are (within experimental error) the same length as the carbon–carbon double bond of ethene. The central bond of 1,3-butadiene (1.47 Å), however, is considerably shorter than the single bond of ethane (1.54 Å).

This should not be surprising. All of the carbon atoms of 1,3-butadiene are sp^2 hybridized and, as a result, the central bond of butadiene results from overlapping sp^2 orbitals. And, as we know, a sigma bond that is sp^3-sp^3 is *longer.* There is, in fact, a steady decrease in bond length of carbon–carbon single bonds as the hybridization state of the bonded atoms changes from sp^3 to sp (Table 10.1).

10.9B Conformations of 1,3-Butadiene

There are two possible planar conformations of 1,3-butadiene: the *s*-cis and the *s*-trans conformations.

s-cis Conformation of 1,3-butadiene   *s*-trans Conformation of 1,3-butadiene

These are not true cis and trans forms since the *s*-cis and *s*-trans conformations of 1,3-butadiene can be interconverted through rotation about the single bond (hence the prefix *s*). The *s*-trans conformation is the predominant one at room temperature.

10.9C Molecular Orbitals of 1,3-Butadiene

The central carbon atoms of 1,3-butadiene (Fig. 10.4) are close enough for overlap to occur between the *p* orbitals of carbon-2 and carbon-3. This overlap is not as great as

TABLE 10.1 Carbon–carbon single bond lengths and hybridization state

COMPOUND	HYBRIDIZATION STATE	BOND LENGTH (Å)
H_3C—CH_3	sp^3-sp^3	1.54
CH_2=CH—CH_3	sp^2-sp^3	1.50
CH_2=CH—CH=CH_2	sp^2-sp^2	1.47
HC≡C—CH_3	sp-sp^3	1.46
HC≡C—CH=CH_2	sp-sp^2	1.43
HC≡C—C≡CH	sp-sp	1.37

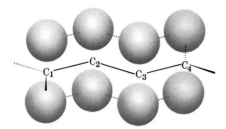

FIGURE 10.4 The *p* orbitals of 1,3-butadiene.

that between the orbitals of C-1 and C-2 (or those of C-3 and C-4). The C-2–C-3 orbital overlap, however, gives the central bond partial double bond character and allows the four π electrons of 1,3-butadiene to be delocalized over all four atoms.

Figure 10.5 shows how the four *p* orbitals of 1,3-butadiene combine to form a set of four π molecular orbitals.

Two of the π molecular orbitals of 1,3-butadiene are bonding molecular orbitals. In the ground state these orbitals hold the four π electrons with two spin-paired

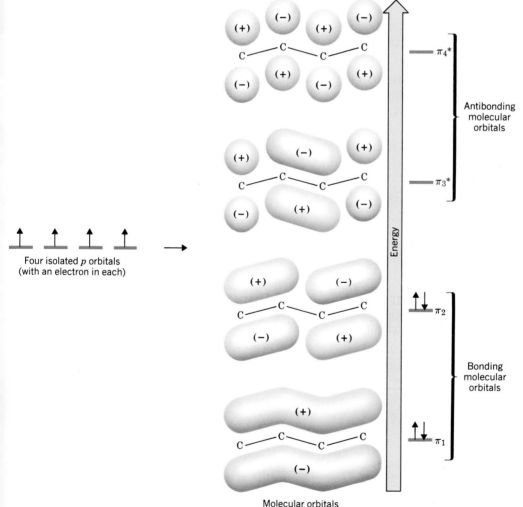

FIGURE 10.5 Formation of the π molecular orbitals of 1,3-butadiene from four isolated *p* orbitals.

electrons in each. The other two π molecular orbitals are antibonding molecular orbitals. In the ground state these orbitals are unoccupied. An electron can be excited from the highest occupied molecular orbital to the lowest empty molecular orbital when 1,3-butadiene absorbs light with a wavelength of 217 nm. (We shall study the absorption of light by unsaturated molecules in Chapter 13.)

The delocalized bonding that we have just described for 1,3-butadiene is characteristic of all conjugated polyenes.

10.10 THE STABILITY OF CONJUGATED DIENES

Conjugated alkadienes are thermodynamically more stable than isomeric isolated alkadienes. Two examples of this extra stability of conjugated dienes can be seen in an analysis of the heats of hydrogenation given in Table 10.2.

In itself, 1,3-butadiene cannot be compared directly with an isolated diene of the same chain length. However, a comparison can be made between the heat of hydrogenation of 1,3-butadiene and that obtained when 2 moles of 1-butene are hydrogenated.

$$\Delta H° \text{ (kcal/mole)}$$

$$2CH_2\!=\!CHCH_2CH_3 + 2H_2 \longrightarrow 2CH_3CH_2CH_2CH_3 \ (2 \times -30.3) = -60.6$$
$$\text{1-Butene}$$

$$CH_2\!=\!CHCH\!=\!CH_2 + 2H_2 \longrightarrow CH_3CH_2CH_2CH_3 \qquad\qquad = \underline{-57.1}$$
$$\text{1,3-Butadiene} \hspace{3cm} \text{Difference} \quad 3.5 \text{ kcal/mole}$$

Because 1-butene has the same kind of monosubstituted double bond as either of those in 1,3-butadiene, we might expect that hydrogenation of 1,3-butadiene would liberate the same amount of heat (-60.6 kcal/mole) as 2 moles of 1-butene. We find, however, that 1,3-butadiene liberates only 57.1 kcal/mole, 3.5 kcal/mole *less* than expected. We conclude, therefore, that conjugation imparts some extra stability to the conjugated system (Fig. 10.6).

An assessment of the stabilization that conjugation provides *trans*-1,3-pentadiene can be made by comparing the heat of hydrogenation of *trans*-1,3-pentadiene to the sum of the heats of hydrogenation of 1-pentene and *trans*-2-pentene. This way we are comparing double bonds of comparable types.

TABLE 10.2 Heats of hydrogenation of alkenes and alkadienes

COMPOUND	H_2 (moles)	$\Delta H°$ (kcal/mole)	(kJ/mole)
1-Butene	1	-30.3	-126.8
1-Pentene	1	-30.1	-125.9
trans-2-Pentene	1	-27.6	-115.5
1,3-Butadiene	2	-57.1	-238.9
trans-1,3-Pentadiene	2	-54.1	-226.4
1,4-Pentadiene	2	-60.8	-254.4
1,5-Hexadiene	2	-60.5	-253.1

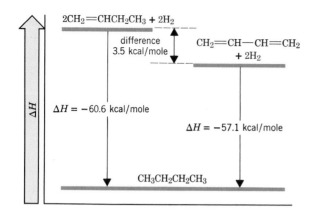

FIGURE 10.6 Heats of hydrogenation of 2 moles of 1-butene and 1 mole of 1,3-butadiene.

$$CH_2\!\!=\!\!CHCH_2CH_2CH_3 \qquad \Delta H° = -30.1 \text{ kcal/mole}$$
1-Pentene

$$\Delta H° = -27.6 \text{ kcal/mole}$$
$$\overline{\text{Sum} = -57.7 \text{ kcal/mole}}$$

trans-2-Pentene

$$\Delta H° = -54.1 \text{ kcal/mole}$$
$$\overline{\text{Difference} = 3.6 \text{ kcal/mole}}$$

trans-1,3-Pentadiene

We see from these calculations that conjugation affords *trans*-1,3-pentadiene an extra stability of 3.6 kcal/mole, a value that is very close to the one we obtained for 1,3-butadiene (3.5 kcal/mole).

When calculations like these are carried out for other conjugated dienes, similar results are obtained; *conjugated dienes are found to be more stable than isolated dienes.* The question, then, is this: What is the source of the extra stability associated with conjugated dienes? There are two factors that contribute. The extra stability of conjugated dienes arises in part from the stronger central bond that they contain, and in part from the additional delocalization of the π electrons that occurs in conjugated dienes.

10.11 ELECTROPHILIC ATTACK AND CONJUGATED DIENES: 1,4 ADDITION

Not only are conjugated dienes somewhat more stable than nonconjugated dienes, they also display special behavior when they react with electrophilic reagents. For example, 1,3-butadiene reacts with 1 mole of hydrogen chloride to produce two products: 3-chloro-1-butene and 1-chloro-2-butene.

$$CH_2\!\!=\!\!CH\!\!-\!\!CH\!\!=\!\!CH_2 \xrightarrow[25\,°C]{HCl} CH_3\!\!-\!\!\underset{\underset{Cl}{|}}{CH}\!\!-\!\!CH\!\!=\!\!CH_2 + CH_3\!\!-\!\!CH\!\!=\!\!CH\!\!-\!\!CH_2Cl$$
1,3-Butadiene

3-Chloro-1-butene (78%)   1-Chloro-2-butene (22%)

If only the first product (3-chloro-1-butene) were formed, we would not be particularly surprised. We would conclude that hydrogen chloride had added to one double bond of 1,3-butadiene in the usual way.

$$\overset{1}{CH_2}=\overset{2}{CH}-\overset{3}{CH}=\overset{4}{CH_2} \xrightarrow{\text{1,2 addition}} CH_2-CH-CH=CH_2$$

$$+$$
$$H-Cl$$

$$\qquad\qquad\qquad\qquad\qquad\qquad\qquad\qquad\quad \underset{H}{|}\quad \underset{Cl}{|}$$

3-Chloro-1-butene

It is the second product, 1-chloro-2-butene, that is unusual. Its double bond is between the central atoms, and the elements of hydrogen chloride have added to carbon atoms 1 and 4.

$$\overset{1}{CH_2}=\overset{2}{CH}-\overset{3}{CH}=\overset{4}{CH_2} \xrightarrow{\text{1,4 addition}} CH_2-CH=CH-CH_2$$

$$+$$
$$H-Cl$$

$$\qquad\qquad\qquad\qquad\qquad\qquad\qquad\qquad\quad \underset{H}{|}\qquad\qquad \underset{Cl}{|}$$

1-Chloro-2-butene

This unusual behavior of 1,3-butadiene can be attributed directly to the stability and the delocalized nature of an allylic cation (Section 10.4). In order to see this, consider a mechanism for the addition of hydrogen chloride.

Step 1

$$H^+ + CH_2=CH-CH=CH_2 \longrightarrow CH_3-\underset{+}{CH}-CH=CH_2 \longleftrightarrow CH_3-CH=CH-\underset{+}{CH_2}$$

An allylic cation
equivalent to

$$CH_3-\underset{\delta+}{CH}=\!\!=\!\!CH=\!\!=\underset{\delta+}{CH_2}$$

Step 2

$$CH_3\underset{\delta+}{CH}=\!\!=\!\!CH=\!\!=\underset{\delta+}{CH_2} + \overset{..}{\underset{..}{Cl}}:^-$$

(a) $\longrightarrow$ $CH_3\underset{Cl}{\overset{|}{CH}}-CH=CH_2$ 1,2 Addition

(b) $\longrightarrow$ $CH_3CH=CHCH_2Cl$ 1,4 Addition

In step 1 a proton adds to one of the terminal carbon atoms of 1,3-butadiene to form, as usual, the more stable carbocation, in this case a resonance-stabilized allylic cation. Addition to one of the inner carbon atoms would have produced a much less stable primary cation, one that could not be stabilized by resonance.

$$CH_2=CH-CH=CH_2 \xrightarrow{} {}^+CH_2-CH_2-CH=CH_2$$

$$\qquad\qquad\quad \searrow_{H^+}$$

A 1° carbocation

In step 2 a chloride ion forms a bond to one of the carbon atoms of the allylic cation that bears a partial positive charge. Reaction at one carbon atom results in the 1,2-addition product; reaction at the other gives the 1,4-addition product.

PROBLEM 10.11

(a) What products would you expect to obtain if hydrogen chloride were allowed to react with a 2,4-hexadiene, $CH_3CH=CHCH=CHCH_3$? (b) With 1,3-pentadiene, $CH_2=CHCH=CHCH_3$? (Neglect cis-trans isomerism.)

1,3-Butadiene shows 1,4-addition reactions with electrophilic reagents other than hydrogen chloride. Two examples are shown here, the addition of hydrogen bromide and the addition of bromine.

$$CH_2=CHCH=CH_2 \xrightarrow[40\ °C]{HBr} CH_3CHBrCH=CH_2 + CH_3CH=CHCH_2Br$$
$$\qquad\qquad\qquad\qquad\qquad (20\%) \qquad\qquad\qquad (80\%)$$

$$CH_2=CHCH=CH_2 \xrightarrow[-15\ °C]{Br_2} CH_2BrCHBrCH=CH_2 + CH_2BrCH=CHCH_2Br$$
$$\qquad\qquad\qquad\qquad\qquad (54\%) \qquad\qquad\qquad (46\%)$$

Reactions of this type are quite general with other conjugated dienes. Conjugated trienes often show 1,6 addition. An example is the 1,6 addition of bromine to 1,3,5-cyclooctatriene:

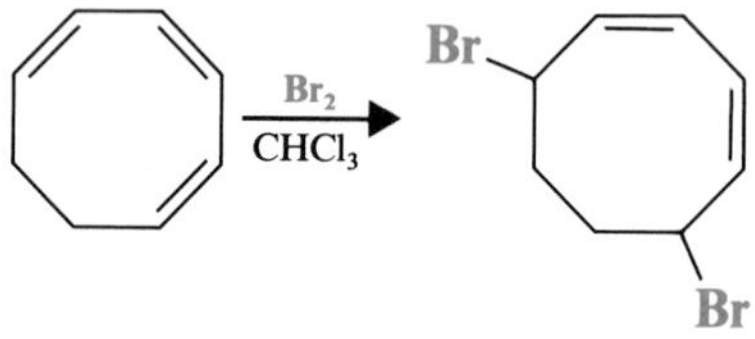

$(>68\%)$

10.11A Rate Control Versus Equilibrium Control of a Chemical Reaction

The addition of hydrogen bromide to 1,3-butadiene is interesting in another respect. The relative amounts of 1,2- and 1,4-addition products that we obtain are dependent on the temperature at which we carry out the reaction.

When 1,3-butadiene and hydrogen bromide react at a low temperature ($-80\ °C$), the major reaction is 1,2 addition; we obtain about 80% of the 1,2 product and only about 20% of the 1,4 product. At a higher temperature (40 °C) the result is reversed. The major reaction is 1,4 addition; we obtain about 80% of the 1,4 product and only about 20% of the 1,2 product.

When the mixture formed at the lower temperature is brought to the higher temperature, moreover, the relative amounts of the two products change. This new reaction mixture eventually contains the same proportion of products given by the reaction carried out at the higher temperature.

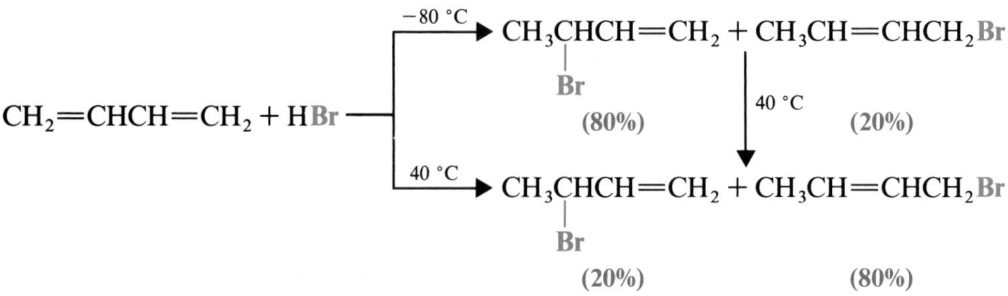

It can also be shown that at the higher temperature and in the presence of hydrogen bromide, the 1,2-addition product rearranges to the 1,4 product and that an equilibrium exists between them.

$$CH_3CHCH=CH_2 \underset{}{\overset{80\ °C,\ HBr}{\rightleftharpoons}} CH_3CH=CHCH_2Br$$

Br

1,2-Addition   **1,4-Addition**

product   product

Because this equilibrium favors the 1,4-addition product, *it must be more stable.*

The reactions of hydrogen bromide with 1,3-butadiene serve as a striking illustration of the way that the outcome of a chemical reaction can be determined, in one instance, by relative rates of competing reactions and, in another, by the relative stabilities of the final products. At the lower temperature, the relative amounts of the products of the addition are determined by the relative rates at which the two additions occur: 1,2 addition occurs faster so the 1,2-addition product is the major product. At the higher temperature, the relative amounts of the products are determined by the position of an equilibrium: The 1,4-addition product is the more stable, so it is the major product.

This behavior of 1,3-butadiene and hydrogen bromide can be more fully understood if we examine the diagram shown in Fig. 10.7.

The step that determines the overall outcome of the reaction is the step in which the hybrid allylic cation combines with a bromide ion, that is,

$$CH_2=CH-CH=CH_2 \xrightarrow{H^+}$$

$$CH_3-\overset{\delta+}{CH}\cdots CH\cdots\overset{\delta+}{CH_2} \longrightarrow$$

$$\xrightarrow{Br^-} CH_3-\underset{\underset{\text{1,2 Product}}{Br}}{CH}-CH=CH_2$$

$$\xrightarrow{Br^-} CH_3-CH=CH-CH_2Br$$

1,4 Product

This step determines the regioselectivity of the reaction

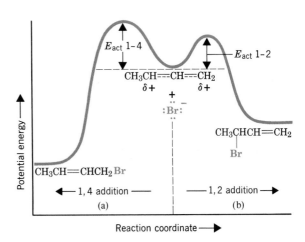

FIGURE 10.7 Potential energy versus reaction coordinate diagram for the reactions of an allylic cation with a bromide ion. One reaction pathway (*a*) leads to the 1,4-addition product and the other (*b*) leads to the 1,2-addition product.

We see in Fig. 10.7 that, for this step, the energy of activation leading to the 1,2-addition product is less than the energy of activation leading to the 1,4-addition product, even though the 1,4 product is more stable. At low temperatures, a larger fraction of collisions between the intermediate ions will have enough energy to cross the lower barrier (leading to the 1,2 product), and only a very small fraction of collisions will have enough energy to cross the higher barrier (leading to the 1,4 product). In either case (and this is the *key point*), whichever barrier is crossed, product formation is *irreversible* because there is not enough energy available to lift either product out of its deep potential energy valley. Since 1,2 addition occurs faster, the 1,2 product predominates and the reaction is said to be under **rate control** or **kinetic control.**

At higher temperatures, the intermediate ions have sufficient energy to cross both barriers with relative ease. More importantly, however, *both reactions are reversible.* Sufficient energy is also available to take the products back over their energy barriers to the intermediate level of allylic cations and bromide ions. The 1,2 product is still formed faster, but being less stable than the 1,4 product, it also reverts to the allylic cation faster. Under these conditions, that is, at higher temperatures, the relative proportions of the products *do not reflect* the relative heights of the energy barriers leading from allylic cation to products. Instead, *they reflect the relative stabilities of the products themselves.* Since the 1,4 product is more stable, it is formed at the expense of the 1,2 product because the overall change from 1,2 product to 1,4 product is energetically favored. Such a reaction is said to be under **equilibrium control** or **thermodynamic control.**

Before we leave this subject one final point should be made. This example clearly demonstrates that predictions of relative reaction rates made on the basis of product stabilities alone can be wrong. This is not always the case, however. For many reactions in which a common intermediate leads to two or more products, the most stable product is formed fastest.

PROBLEM 10.12

(a) Can you suggest a possible explanation for the fact that the 1,2-addition reaction of 1,3-butadiene and hydrogen bromide occurs faster than 1,4 addition? (*Hint:* Consider the relative contributions that the two forms $CH_3\overset{+}{C}HCH=CH_2$ and $CH_3CH=CHCH_2{}^+$ make to the resonance hybrid of the allylic cation.) (b) How can you account for the fact that the 1,4-addition product is more stable?

10.12 THE DIELS–ALDER REACTION: A 1,4-CYCLOADDITION REACTION OF DIENES

In 1928 two German chemists, Otto Diels and Kurt Alder, developed a 1,4-cyclo-addition reaction of dienes that has since come to bear their names. The reaction proved to be one of such great versatility and synthetic utility that Diels and Alder were awarded the Nobel Prize for Chemistry in 1950.

An example of the Diels–Alder reaction is the reaction that takes place when 1,3-butadiene and maleic anhydride are heated together at 100 °C. The product is obtained in quantitative yield.

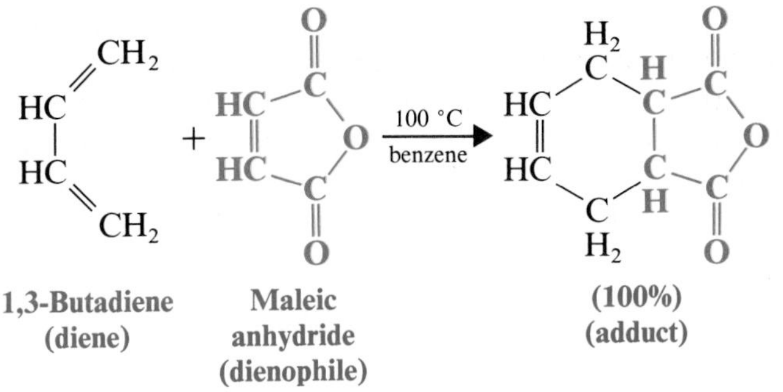

1,3-Butadiene   Maleic   (100%)
  (diene)   anhydride   (adduct)
              (dienophile)

or

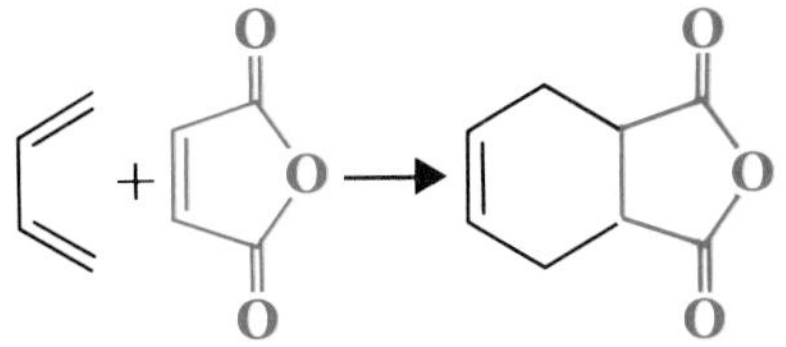

In general terms, the reaction is one between a conjugated **diene** (a 4π-electron system) and a compound containing a double bond (a 2π-electron system) called a **dienophile** (diene + Greek: *philein,* to love). The product of a Diels–Alder reaction is often called an **adduct.** In the Diels–Alder reaction, two new σ bonds are formed at the expense of two π bonds of the diene and dienophile. Since σ bonds are usually stronger than π bonds, formation of the adduct is usually favored energetically, *but most Diels–Alder reactions are reversible.*

We can account for all of the bond changes in a Diels–Alder reaction by using curved arrows in the following way:

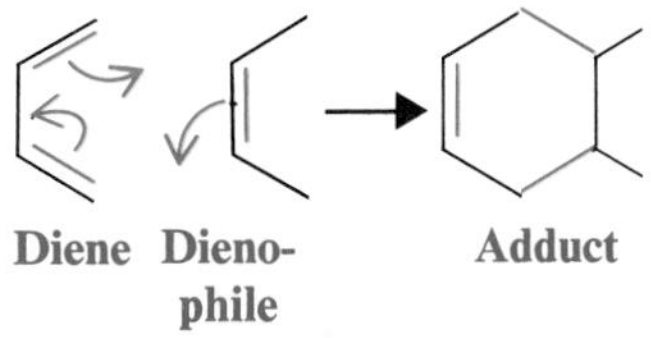

Diene Dieno-   Adduct
        phile

We do not intend to imply a mechanism by the use of these curved arrows; they are used only to keep account of the electrons. (The mechanism of the Diels–Alder reaction is discussed in Special Topic O.)

The simplest example of a Diels–Alder reaction is the one that takes place between 1,3-butadiene and ethylene. This reaction, however, takes place much more slowly than the reaction of butadiene with maleic anhydride and must also be carried out under pressure.

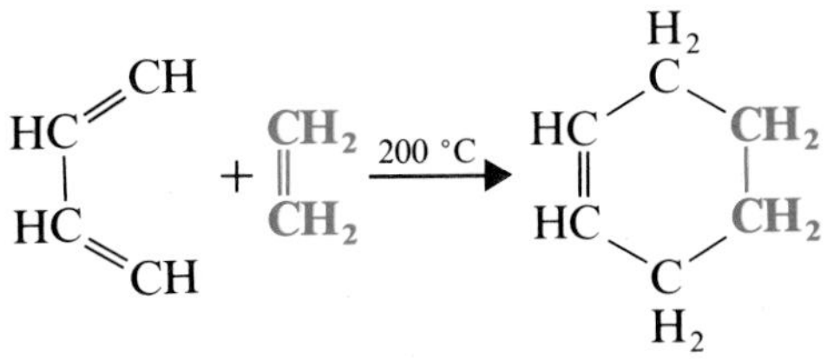

(20%)

Alder originally stated that the Diels–Alder reaction is favored by the presence of electron-withdrawing groups in the dienophile and by electron-releasing groups in the diene. Maleic anhydride, a very potent dienophile, has two carbonyl groups on carbons adjacent to the double bond. Carbonyl groups are electron withdrawing because of the electronegativity of their oxygen atoms and because resonance structures such as the following contribute to the hybrid. The comparative yields of the two examples that we have given (1,3-butadiene + maleic anhydride and 1,3-butadiene + ethylene) illustrate the help that electron-withdrawing groups in the dienophile gives the Diels–Alder reaction.

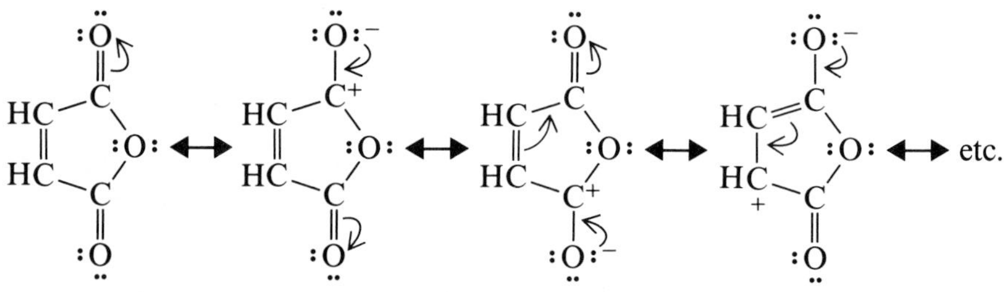

The helpful effect of electron-releasing groups in the diene can also be demonstrated; 2,3-dimethyl-1,3-butadiene, for example, is nearly five times as reactive in Diels–Alder reactions as is 1,3-butadiene. When 2,3-dimethyl-1,3-butadiene reacts with propenal (acrolein) at only 30 °C, the adduct is obtained in quantitative yield.

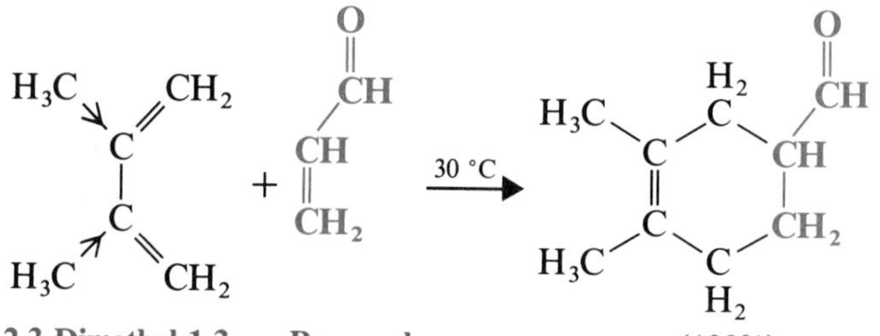

| 2,3-Dimethyl-1,3-butadiene | Propenal (acrolein) | (100%) |

Research (by C. K. Bradsher of Duke University) has shown that the locations of electron-withdrawing and electron-releasing groups in the dienophile and diene can be reversed without reducing the yields of the adducts. Dienes with electron-withdrawing groups have been found to react readily with dienophiles containing electron-releasing groups. Additional facts about the reaction are these.

The Diels–Alder reaction is highly stereospecific:

1. **The reaction is a *syn* addition and the configuration of the dienophile is *retained* in the product.** Two examples that illustrate this aspect of the reaction are shown here.

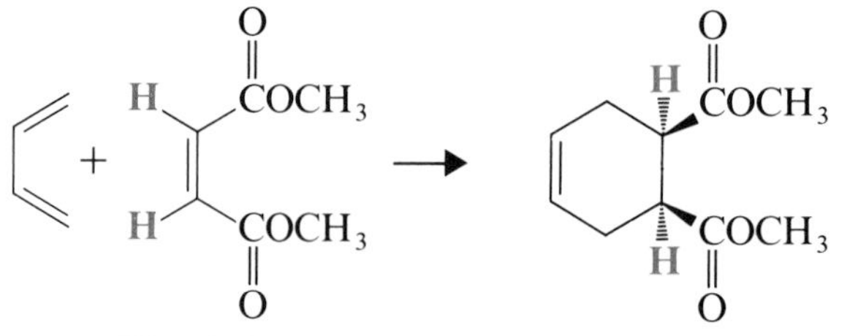

| Dimethyl maleate (a *cis*-dienophile) | Dimethyl *cis*-4-cyclohexene-1,2-dicarboxylate |

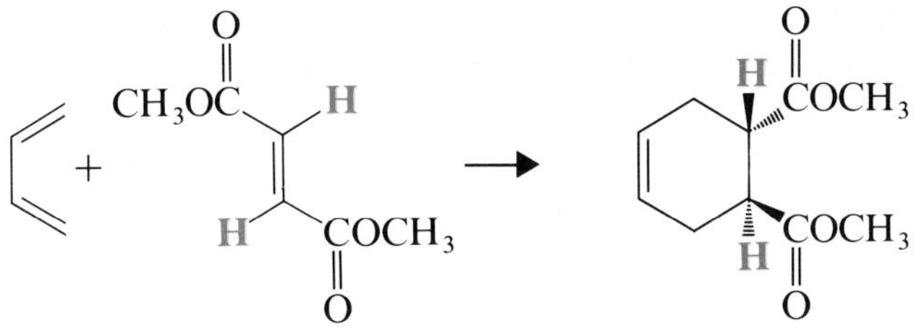

Dimethyl fumarate
(a *trans*-dienophile)

Dimethyl *trans*-4-cyclohexene-
1,2-dicarboxylate

In the first example, a dienophile with cis ester groups reacts with 1,3-butadiene to give an adduct with cis ester groups. In the second example just the reverse is true. A *trans* dienophile gives a trans adduct.

2. **The diene, of necessity, must react in the *s*-cis conformation rather than the *s*-trans.**

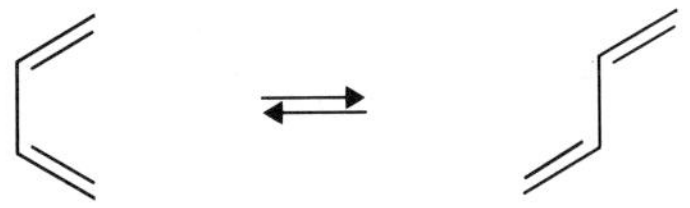

s-cis Conformation   *s*-trans Conformation

Reaction in the *s*-trans conformation would, if it occurred, produce a six-membered ring with a highly strained trans double bond. This course of the Diels–Alder reaction has never been observed.

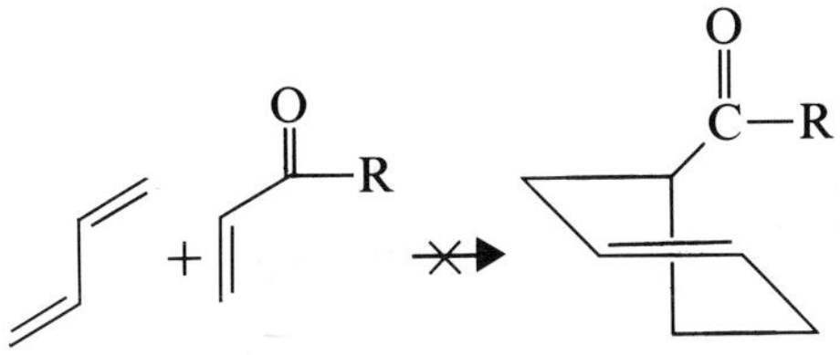

Highly strained

Cyclic dienes in which the double bonds are held in the *s*-cis configuration are usually highly reactive in the Diels–Alder reaction. Cyclopentadiene, for example, reacts with maleic anhydride at room temperature to give the following adduct in quantitative yield.

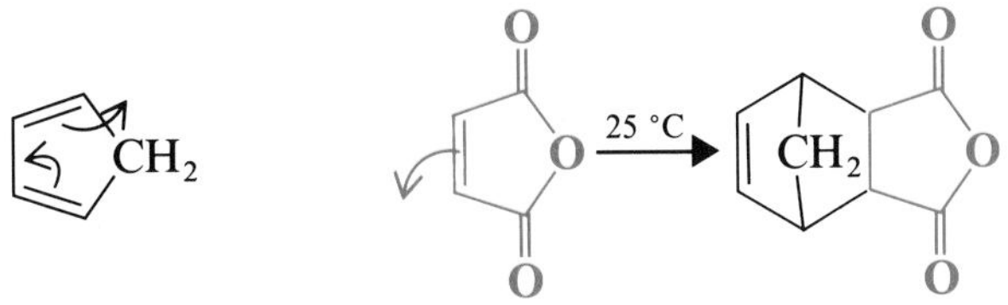

Cyclopentadiene   Maleic
anhydride

(100%)

Cyclopentadiene is so reactive that on standing at room temperature it slowly undergoes a Diels–Alder reaction with itself.

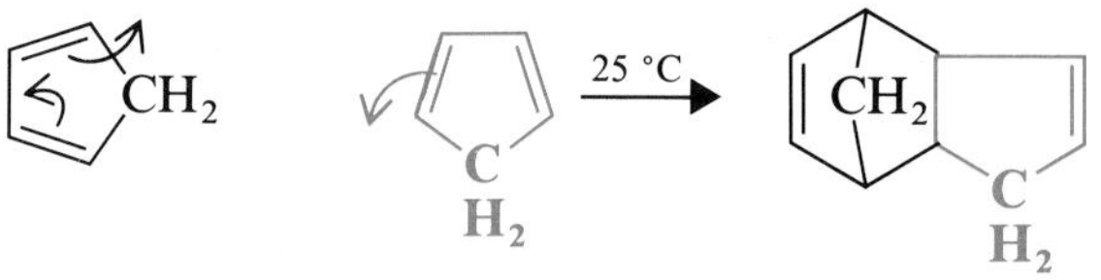

"Dicyclopentadiene"

The reaction is reversible, however. When "dicyclopentadiene" is distilled, it dissociates into 2 moles of cyclopentadiene.

The reactions of cyclopentadiene illustrate a third stereochemical characteristic of the Diels–Alder reaction.

3. **The Diels–Alder reaction occurs primarily in an *endo* rather than an *exo* fashion when the reaction is kinetically controlled** (cf. Problem 10.29). *Endo* and *exo* are terms used to designate the stereochemistry of bridged rings such as norbornane. The point of reference is the "shortest" bridge. A group that is on the same side of the six-membered ring as the one-carbon bridge is said to be *exo;* if it is on the opposite side, it is *endo.**

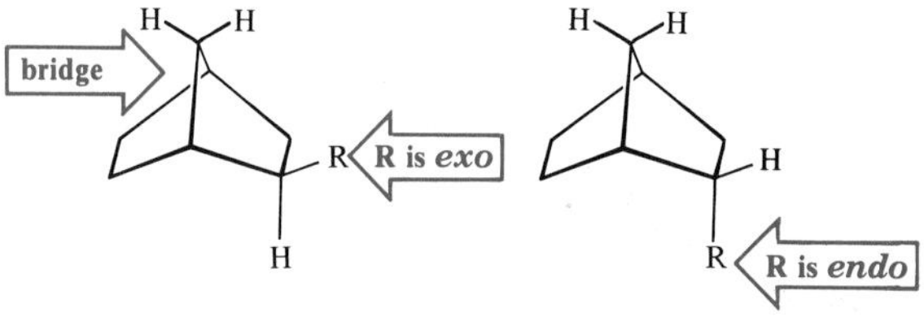

In the Diels–Alder reaction of cyclopentadiene with maleic anhydride the major product is the one in which the anhydride linkage, $-\overset{\|}{\underset{O}{C}}-O-\overset{\|}{\underset{O}{C}}-$, has

assumed the *endo* configuration. See the following illustration. This favored *endo* stereochemistry seems to arise from favorable interactions between the π electrons of the developing double bond in the diene and the π electrons of unsaturated groups of the dienophile. In this example, the π electrons of the $-\overset{\|}{\underset{O}{C}}-O-\overset{\|}{\underset{O}{C}}-$ linkage of the

anhydride interact with the π electrons of the developing double bond in cyclopentadiene.

*In general, the *exo* substituent is always on the side opposite the *longer* bridge of a bicyclic structure (*exo,* outside; *endo,* inside). For example,

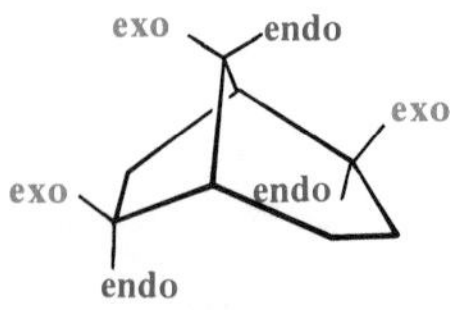

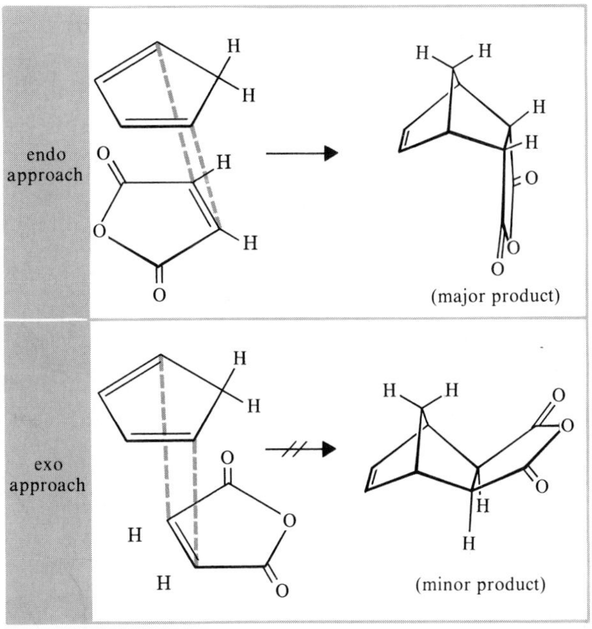

(major product)

(minor product)

PROBLEM 10.13

The dimerization of cyclopentadiene also occurs in an *endo* way. (a) Show how this happens. (b) Which π electrons interact? (c) What is the three-dimensional structure of the product?

PROBLEM 10.14

What products would you expect from the following reactions?

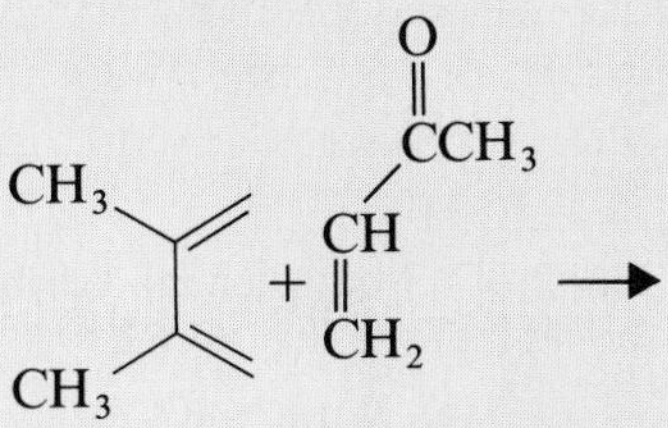

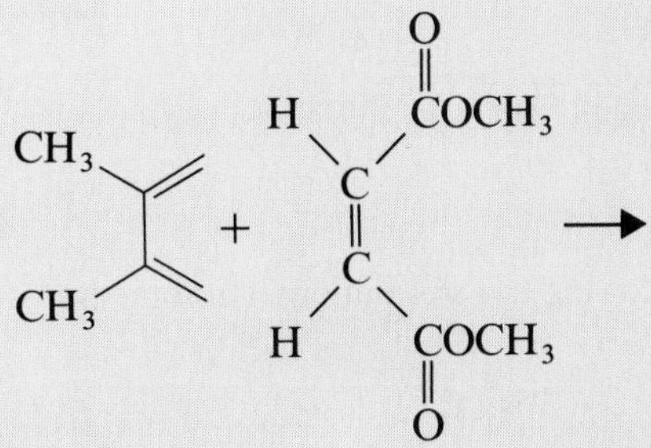

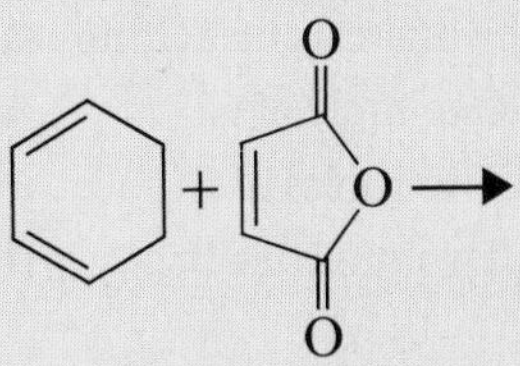

PROBLEM 10.15

Which diene and dienophile would you employ to synthesize the following compound?

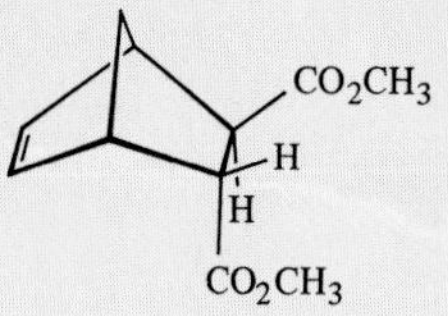

PROBLEM 10.16

Diels–Alder reactions also take place with triple-bonded (acetylenic) dienophiles. Which diene and which dienophile would you use to prepare:

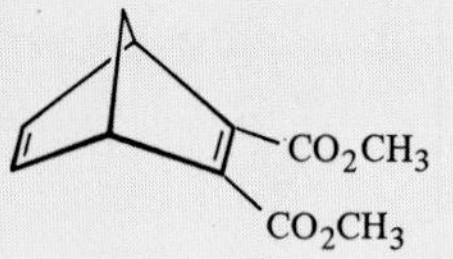

ADDITIONAL PROBLEMS

10.17

Outline a synthesis of 1,3-butadiene starting from

(a) 1,4-Dibromobutane

(b) $HOCH_2(CH_2)_2CH_2OH$

(c) $CH_2{=}CHCH_2CH_2OH$

(d) $CH_2{=}CHCH_2CH_2Cl$

(e) $CH_2{=}CHCHClCH_3$

(f) $CH_2{=}CHCH(OH)CH_3$

(g) $HC{\equiv}CCH{=}CH_2$

10.18

What product would you expect from the following reaction?

$$(CH_3)_2C{-}C(CH_3)_2 + 2KOH \xrightarrow[\text{heat}]{\text{ethanol}}$$
$$\underset{Cl}{|} \quad \underset{Cl}{|}$$

10.19

What products would you expect from the reaction of 1 mole of 1,3-butadiene and each of the following reagents? (If no reaction would occur, you should indicate that as well.)

(a) One mole of Cl_2

(b) Two moles of Cl_2

(c) Two moles of Br_2

(d) Two moles of H_2, Ni

(e) $Ag(NH_3)_2{}^+OH^-$

(f) One mole of Cl_2 in H_2O

(g) Hot $KMnO_4$

(h) H^+, H_2O

10.20

Show how you might carry out each of the following transformations. (In some transformations several steps may be necessary.)

(a) 1-Butene $\longrightarrow$ 1,3-butadiene

(b) 1-Pentene $\longrightarrow$ 1,3-pentadiene

(c) $CH_3CH_2CH_2CH_2OH \longrightarrow CH_2BrCH=CHCH_2Br$

(d) $CH_3CH=CHCH_3 \longrightarrow CH_3CH=CHCH_2Br$

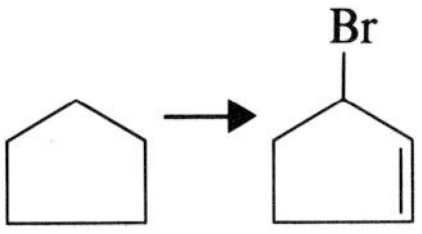

(f)

(e)

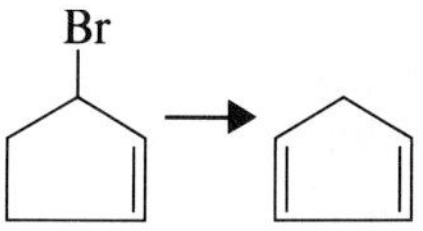

10.21

Conjugated dienes react with free radicals by both 1,2 and 1,4 addition. Account for this fact by using the peroxide-catalyzed addition of 1 mole of HBr to 1,3-butadiene as an illustration.

10.22

Outline a simple chemical test that would distinguish between the members of each of the following pairs of compounds.

(a) 1,3-Butadiene and 1-butyne

(b) 1,3-Butadiene and butane

(c) Butane and $CH_2=CHCH_2CH_2OH$

(d) 1,3-Butadiene and $CH_2=CHCH_2CH_2Br$

(e) $CH_2BrCH=CHCH_2Br$ and $CH_3CBr=CBrCH_3$

10.23

(a) The hydrogen atoms attached to C-3 of 1,4-pentadiene are unusually susceptible to abstraction radicals. How can you account for this? (b) Can you also provide an explanation for the fact that the protons attached to carbon-3 of 1,4-pentadiene are more acidic than the methyl hydrogen atoms of propene?

10.24

When 2-methyl-1,3-butadiene (isoprene) undergoes a 1,4 addition of hydrogen chloride, the major product that is formed is 1-chloro-3-methyl-2-butene. Little or no 1-chloro-2-methyl-2-butene is formed. How can you explain this?

10.25

Which diene and dienophile would you employ in a synthesis of each of the following?

(a)

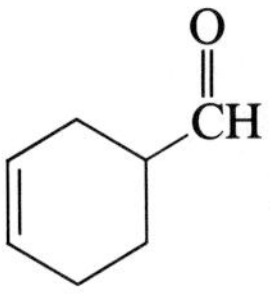

(b)

(c)

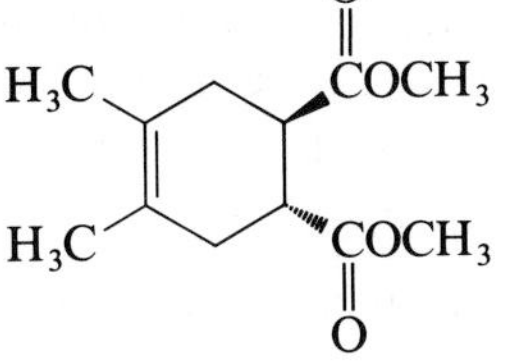

(d)

(e)

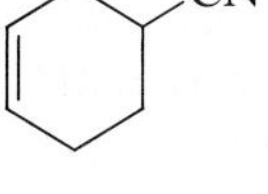

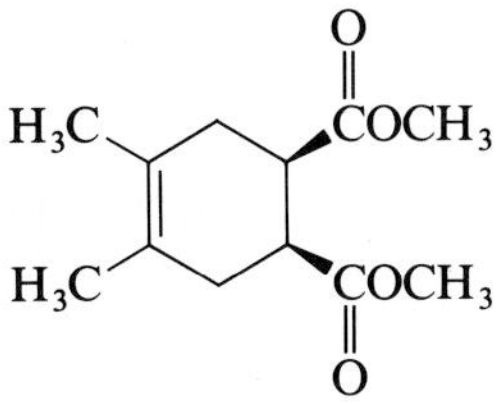

(f)

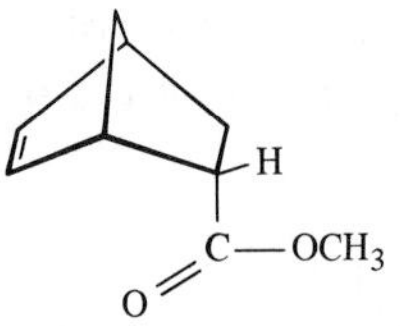

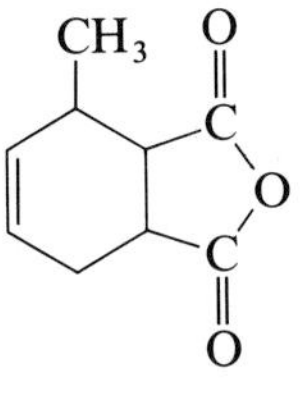

10.26

Account for the fact that neither of the following compounds undergoes a Diels–Alder reaction with maleic anhydride.

$$HC{\equiv}C{-}C{\equiv}CH \quad \text{or} \quad$$ =CH$_2$

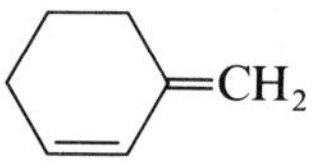

10.27

Acetylenic compounds may be used as dienophiles in the Diels–Alder reaction (cf. Problem 10.16). Write structures for the adducts that you expect from the reaction of 1,3-butadiene with:

(a) $CH_3OCC{\equiv}CCOCH_3$ (dimethyl acetylenedicarboxylate)

(b) $CF_3C{\equiv}CCF_3$ (hexafluoro-2-butyne)

10.28

Cyclopentadiene undergoes a Diels–Alder reaction with ethene at 160–180 °C. Write the structure of the product of this reaction.

10.29

When furan and maleimide undergo a Diels–Alder reaction at 25 °C, the major product is the *endo* adduct **G**. When the reaction is carried out at 90 °C, however, the major product is the *exo* isomer **H**. The *endo* adduct isomerizes to the *exo* adduct when it is heated to 90 °C. Propose an explanation that will account for these results.

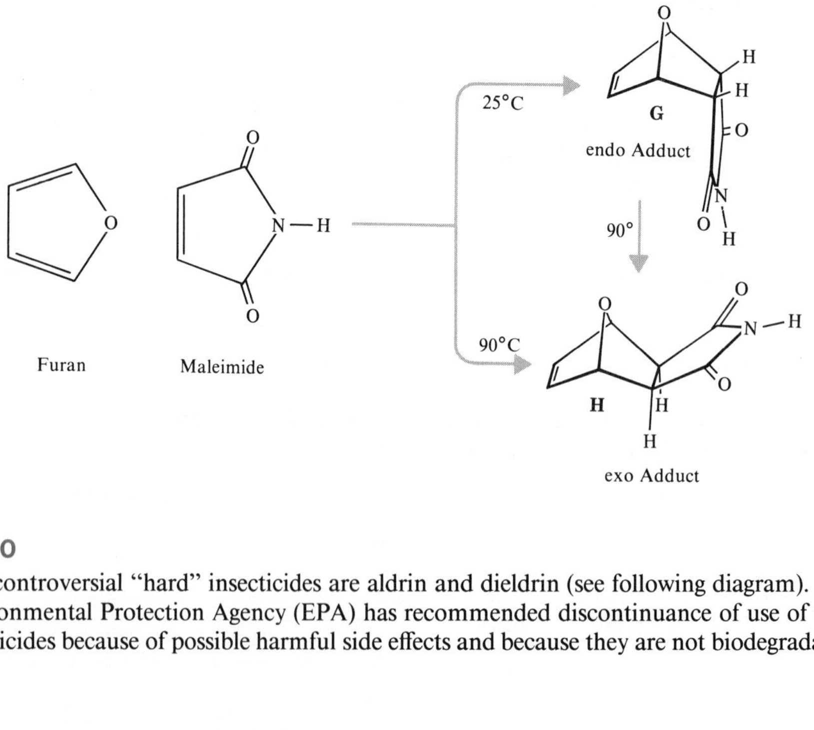

Furan   Maleimide   endo Adduct   exo Adduct

10.30

Two controversial "hard" insecticides are aldrin and dieldrin (see following diagram). [The Environmental Protection Agency (EPA) has recommended discontinuance of use of these insecticides because of possible harmful side effects and because they are not biodegradable.]

The commercial synthesis of aldrin begins with hexachlorocyclopentadiene and norbornadiene. Dieldrin is synthesized from aldrin. Show how these syntheses might be carried out.

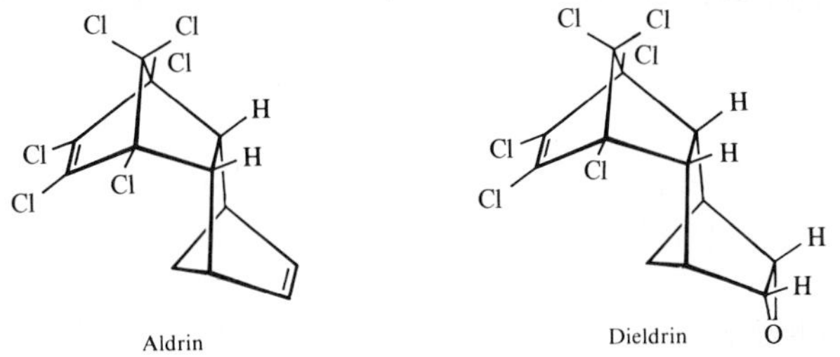

Aldrin

Dieldrin

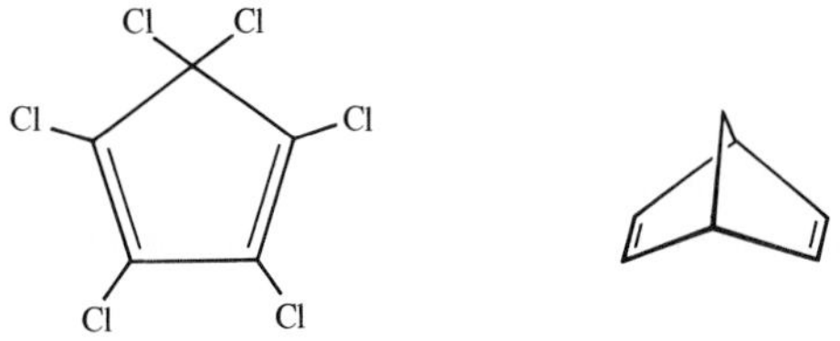

Hexachlorocyclopentadiene

Norbornadiene

10.31

(a) Norbornadiene for the aldrin synthesis (Problem 10.30) can be prepared from cyclopentadiene and acetylene. Show the reaction involved. (b) It can also be prepared by allowing cyclopentadiene to react with vinyl chloride and treating the product with base. Outline this synthesis.

10.32

Two other hard insecticides (cf. Problem 10.30) are chlordan and heptachlor. Their commercial syntheses begin with cyclopentadiene and hexachlorocyclopentadiene. Show how these syntheses might be carried out.

Chlordan

Heptachlor

10.33

Isodrin, an isomer of aldrin, is obtained when cyclopentadiene reacts with the hexachloronorbornadiene, shown here. Propose a structure for isodrin.

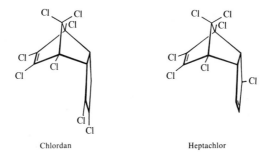

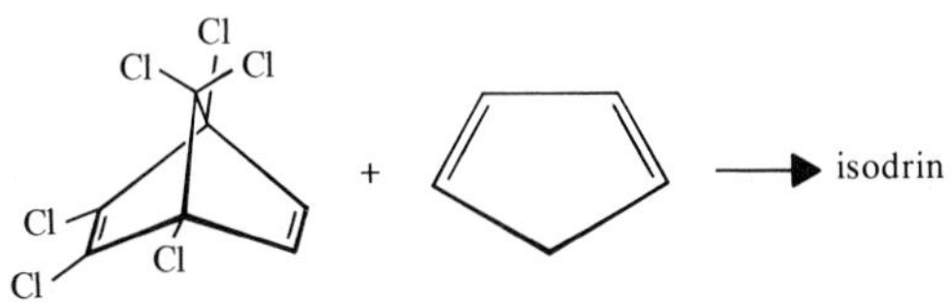

+ ⟶ isodrin

10.34

When $CH_3CH=CHCH_2OH$ is treated with concentrated HCl, two products are produced, $CH_3CH=CHCH_2Cl$ and $CH_3CHClCH=CH_2$. Outline a mechanism that will explain this.

10.35

When a solution of 1,3-butadiene in CH_3OH is treated with chlorine, the products are $ClCH_2CH=CHCH_2OCH_3$ (30%) and $ClCH_2\overset{\displaystyle |}{\underset{\displaystyle OCH_3}{C}}HCH=CH_2$ (70%). Write a mechanism that accounts for their formation.

10.36

Dehydrohalogenation of *vic*-dihalides (with the elimination of 2 moles of HX) normally leads to an alkyne rather than to a conjugated diene. However, when 1,2-dibromocyclohexane is dehydrohalogenated, 1,3-cyclohexadiene is produced in good yield. What factor accounts for this?

10.37

When 1-pentene reacts with *N*-bromosuccinimide, two products with the formula C_5H_9Br are obtained. What are these products and how are they formed?

10.38

Treating either 1-chloro-3-methyl-2-butene or 3-chloro-3-methyl-1-butene with Ag_2O in water gives (in addition to AgCl) the same mixture of alcohols: $(CH_3)_2C=CHCH_2OH$ (15%) and $(CH_3)_2\overset{\displaystyle |}{\underset{\displaystyle OH}{C}}CH=CH_2$ (85%). (a) Write a mechanism that accounts for the formation of these products. (b) What might explain the relative proportions of the two alkenes that are formed?

10.39

The heat of hydrogenation of allene is 71.3 kcal/mole while that of propyne is 69.3 kcal/mole. (a) Which compound is more stable? (b) Treating allene with a strong base causes it to isomerize to propyne. Explain.

*10.40

Mixing furan (Problem 10.29) with maleic anhydride in ether yields a crystalline solid with a melting point of 125 °C. When melting of this compound takes place, however, one can notice that the melt evolves a gas. If the melt is allowed to resolidify, one finds that it no longer melts at 125 °C but instead it melts at 56 °C. Consult an appropriate chemistry handbook and provide an explanation for what is taking place.

FIRST REVIEW PROBLEM SET

1. Provide a reasonable mechanism for the following reactions:

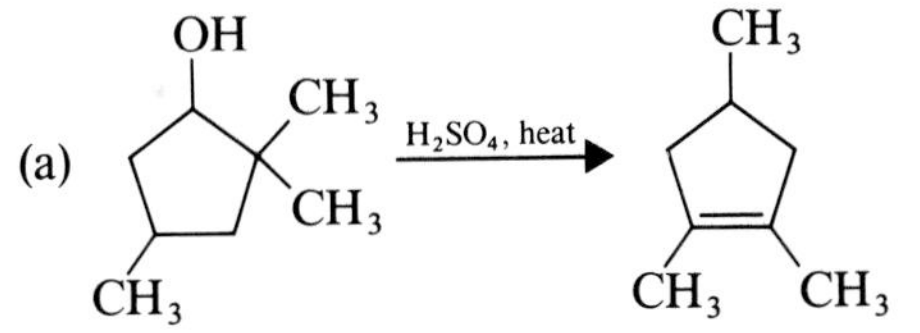

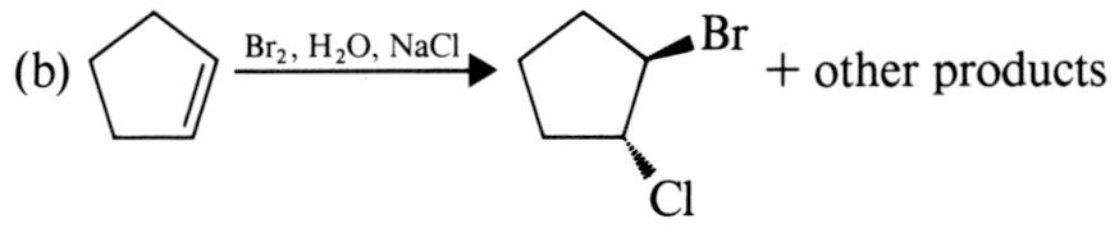

(a) [structure of cyclopentanol with OH, CH$_3$, CH$_3$, CH$_3$ groups] $\xrightarrow{\text{H}_2\text{SO}_4,\ \text{heat}}$ [cyclopentene product with CH$_3$ groups]

(b) [cyclopentene] $\xrightarrow{\text{Br}_2,\ \text{H}_2\text{O},\ \text{NaCl}}$ [cyclopentane with Br and Cl] + other products

(c) What other products would you expect from the reaction given in (b)?

2. What are compounds **A–C**?

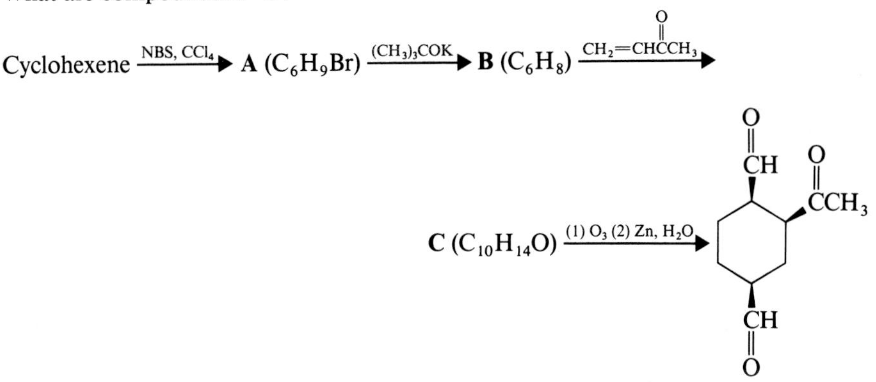

Cyclohexene $\xrightarrow{\text{NBS, CCl}_4}$ **A** (C$_6$H$_9$Br) $\xrightarrow{(\text{CH}_3)_3\text{COK}}$ **B** (C$_6$H$_8$) $\xrightarrow{\text{CH}_2=\text{CHCCH}_3}$

C (C$_{10}$H$_{14}$O) $\xrightarrow{(1)\ \text{O}_3\ (2)\ \text{Zn, H}_2\text{O}}$ [cyclohexane ring substituted with CH=O, CCH$_3$ (C=O), and CH=O groups]

3. Given the following data:

	CH$_2$=CH$_2$	CH$_3$CH$_2$Cl	CH$_2$=CHCl
C—Cl bond length	—	1.76 Å	1.69 Å
C=C bond length	1.34 Å	—	1.38 Å
C—C bond length	—	1.54 Å	—
Dipole moment	0	2.05 D	1.44 D

Use resonance theory to explain each of the following: (a) The shorter C—Cl bond length of CH$_2$=CHCl when compared to CH$_3$CH$_2$Cl. (b) The longer C=C bond of CH$_2$=CHCl when compared to CH$_2$=CH$_2$. (c) The greater dipole moment of CH$_3$CH$_2$Cl when compared to CH$_2$=CHCl.

4. The following is a synthesis of "muscalure," the sex-attractant pheromone of the common house fly. Give the structure of each intermediate and of muscalure itself.

CH$_3$(CH$_2$)$_{11}$CH$_2$Br $\xrightarrow{\text{HC}\equiv\text{CNa}}$ **A** (C$_{15}$H$_{28}$) $\xrightarrow{\text{NaNH}_2}$ **B** (C$_{15}$H$_{27}$Na) $\xrightarrow{\text{1-bromooctane}}$

C (C$_{23}$H$_{44}$) $\xrightarrow{\text{H}_2,\ \text{Ni}_2\text{B (P-2)}}$ muscalure (C$_{23}$H$_{46}$)

5. Write structures for the diastereomers of 2,3-diphenyl-2-butene and assign each diastereomer its (E) or (Z) designation. Hydrogenation of one of these diastereomers using a palladium catalyst produces a racemic form; similar treatment of the other produces a meso compound. On the basis of these experiments, tell which diastereomer is (E) and which is (Z).

6. A hydrocarbon (**A**) has the formula C_7H_{10}. On catalytic hydrogenation, **A** is converted to **B** (C_7H_{12}). On treatment with cold, dilute, and basic $KMnO_4$, **A** is converted to **C** ($C_7H_{12}O_2$). When heated with $KMnO_4$ in basic solution, followed by acidification, either **A** or **C** produces the meso form of 1,3-cyclopentanedicarboxylic acid (see the following structure). Give structural formulas for **A**–**C**.

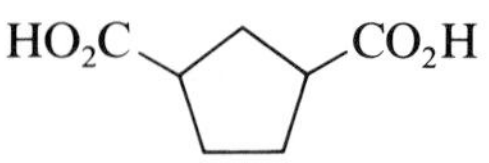

HO_2C CO_2H

1,3-Cyclopentanedicarboxylic acid

7. Starting with propyne, and using any other required reagents, show how you would synthesize each of the following compounds. You need not repeat steps carried out in earlier parts of this problem.
   (a) 2-Butyne
   (b) *cis*-2-Butene
   (c) *trans*-2-Butene
   (d) 1-Butene
   (e) 1,3-Butadiene
   (f) 1-Bromobutane
   (g) 2-Bromobutane
   (h) $(2R,3S)$-2,3-Dibromobutane
   (i) $(2R,3R)$- and $(2S,3S)$-2,3-Dibromobutane (as a racemic form)
   (j) *meso*-2,3-Butanediol
   (k) (Z)-2-Bromo-2-butene

8. Bromination of 2-methylbutane yields predominantly one product with the formula $C_5H_{11}Br$. What is this product? Show how you could use this compound to synthesize each of the following. (You need not repeat steps carried out in earlier parts.)
   (a) 2-Methyl-2-butene
   (b) 2-Methyl-2-butanol
   (c) 3-Methyl-2-butanol
   (d) 3-Methyl-1-butanol
   (e) 3-Methyl-1-butene
   (f) 3-Methyl-1-butyne
   (g) 1-Bromo-3-methylbutane
   (h) 2-Chloro-3-methylbutane
   (i) 2-Chloro-2-methylbutane
   (j) 1-Iodo-3-methylbutane

   (k) $CH_3\overset{\displaystyle O}{\overset{\|}{C}}CH_3$ and $CH_3\overset{\displaystyle O}{\overset{\|}{C}}H$

   (l) $(CH_3)_2CH\overset{\displaystyle O}{\overset{\|}{C}}H$

   (m) $(CH_3)_2CH\overset{\displaystyle O}{\overset{\|}{C}}CH_3$

9. An alkane (**A**) with the formula C_6H_{14} reacts with chlorine to yield three compounds with the formula $C_6H_{13}Cl$, **B**, **C**, and **D**. Of these only **C** and **D** undergo dehydrohalogenation with sodium ethoxide in ethanol to produce an alkene. Moreover, **C** and **D** yield the same alkene **E** (C_6H_{12}). Hydrogenation of **E** produces **A**. Treating **E** with HCl produces a compound (**F**) that is an isomer of **B**, **C**, and **D**. Treating **F** with Zn and acetic acid gives a compound (**G**) that is isomeric with **A**. Propose structures for **A**–**G**.

10. Compound **A** (C_4H_6) reacts with hydrogen and a platinum catalyst to yield butane. Compound **A** decolorizes Br_2 in CCl_4 and aqueous $KMnO_4$, but it does not react with $Ag(NH_3)_2^+$. On treatment with hydrogen and Ni_2B (P-2 catalyst), **A** is converted to **B** (C_4H_8). When **B** is treated with OsO_4 followed by treatment with $NaHSO_3$, **B** is converted to **C** ($C_4H_{10}O_2$). Compound **C** cannot be resolved. Provide structures for **A–C**.

11. Dehalogenation of *meso*-2,3-dibromobutane occurs when it is treated with potassium iodide in ethanol. The product is *trans*-2-butene. Similar dehalogenation of either of the enantiomeric forms of 2,3-dibromobutane produces *cis*-2-butene. Give a mechanistic explanation of these results.

12. Dehydrohalogenation of *meso*-1,2-dibromo-1,2-diphenylethane by the action of sodium ethoxide in ethanol yields (*E*)-1-bromo-1,2-diphenylethene. Similar dehydrohalogenation of either of the enantiomeric forms of 1,2-dibromo-1,2-diphenylethane yields (*Z*)-1-bromo-1,2-diphenylethene. Provide an explanation for the results.

13. Give conformational structures for the major product formed when 1-*tert*-butylcyclohexene reacts with each of the following reagents. If the product would be obtained as a racemic form you should so indicate.
    (a) Br_2, CCl_4
    (b) OsO_4, then aqueous $NaHSO_3$
    (c) $C_6H_5CO_3H$, then H_3O^+, H_2O
    (d) $THF{:}BH_3$, then H_2O_2, OH^-
    (e) $Hg(OAc)_2$ in $THF–H_2O$, then $NaBH_4$, OH^-
    (f) Br_2, H_2O
    (g) ICl
    (h) O_3, then Zn, H_2O (conformational structure not required)
    (i) D_2, Pt
    (j) $THF{:}BD_3$, then CH_3CO_2T

14. Give structures for **A–D**.

$$\underset{\underset{\overset{|}{Br}}{\overset{|}{CH_3}}}{CH_3CCH_2CH_2CH_3} \xrightarrow{\text{EtO}^-/\text{EtOH}} \textbf{A} \ (C_6H_{12}) \text{ major product} \xrightarrow{\text{THF:BH}_3}$$

$$\textbf{B} \ (C_6H_{13})_2BH \xrightarrow{160\ °C} \textbf{C} \ (C_6H_{13})_2BH \xrightarrow{H_2O_2,\ OH^-} \textbf{D} \ (C_6H_{14}O)$$

15. (*R*)-3-Methyl-1-pentene is treated separately with the following reagents, and the products in each case are separated by fractional distillation. Write appropriate formulas for all of the components of each fraction, and tell whether each fraction would be optically active.
    (a) Br_2, CCl_4 (d) $THF{:}BH_3$, then H_2O_2, OH^-
    (b) H_2, Pt (e) $Hg(OAc)_2$, $THF–H_2O$, then $NaBH_4$, OH^-
    (c) OsO_4, then $NaHSO_3$ (f) Perbenzoic acid, then H_3O^+, H_2O

16. Compound **A** ($C_8H_{15}Cl$) exists as a racemic form. Compound **A** does not decolorize either Br_2/CCl_4 or dilute aqueous $KMnO_4$. When **A** is treated with zinc and acetic acid, and the mixture is separated by gas-liquid chromatography, two fractions **B** and **C** are obtained. The components of both fractions have the formula C_8H_{16}. Fraction **B** consists of a racemic form and can be resolved. Fraction **C** cannot be resolved. Treating **A** with sodium ethoxide in ethanol converts **A** into **D** (C_8H_{14}). Hydrogenation of **D** using a platinum catalyst yields **C**. Ozonolysis of **D** followed by treatment with zinc and water yields.

$$\underset{\text{O}}{\overset{\text{O}}{\underset{\|}{\overset{\|}{CH_3CCH_2CH_2CH_2CH_2CCH_3}}}}$$

Propose structures for **A, B, C,** and **D** including, where appropriate, their stereochemistry.

17. There are nine stereoisomers of 1,2,3,4,5,6-hexachlorocyclohexane. Seven of these isomers are meso compounds, and two are a pair of enantiomers. (a) Write structures for all of these stereoisomers, labeling meso forms and the pair of enantiomers. (b) One of these stereoisomers undergoes E2 reactions much more slowly than any of the others. Which isomer is this and why does it react so slowly in an E2 reaction?

18. In addition to more highly fluorinated products, fluorination of 2-methylbutane yields a mixture of compounds with the formula $C_5H_{11}F$. (a) How many different isomers with the formula $C_5H_{11}F$ would you expect to be produced, taking stereochemistry into account? (b) If the mixture of $C_5H_{11}F$ isomers were subjected to fractional distillation, how many fractions would you expect to obtain? (c) Which fractions would be optically inactive? (d) Which would you be able to resolve into enantiomers?

19. Fluorination of (R)-2-fluorobutane yields a mixture of isomers with the formula $C_4H_8F_2$. (a) How many different isomers would you expect to be produced? Write their structures. (b) If the mixture of $C_4H_8F_2$ isomers were subjected to fractional distillation, how many fractions would you expect to obtain? (c) Which of these fractions would be optically active?

20. There are two optically inactive (and nonresolvable) forms of 1,3-di-*sec*-butylcyclohexane. Write their structures.

21. When the following deuterium-labeled isomer undergoes elimination, the reaction yields *trans*-2-butene and *cis*-2-butene-*2-d* (as well as some 1-butene-*3-d*). The reaction does not yield *cis*-2-butene or *trans*-2-butene-*2-d*.

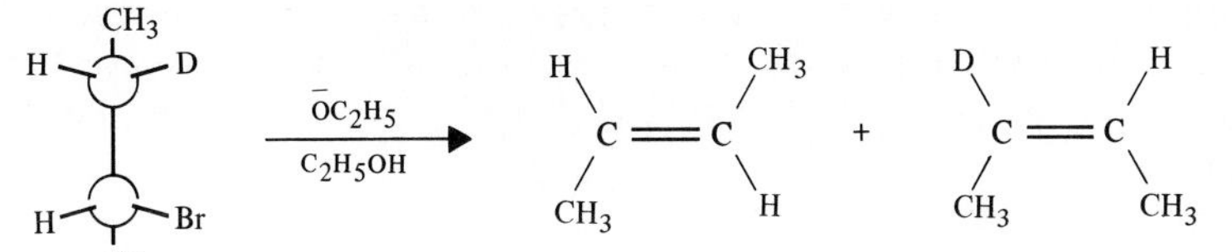

trans-2-Butene   *cis*-2-Butene-2-*d*

(+ $CH_3CHDCH{=}CH_2$)

but no

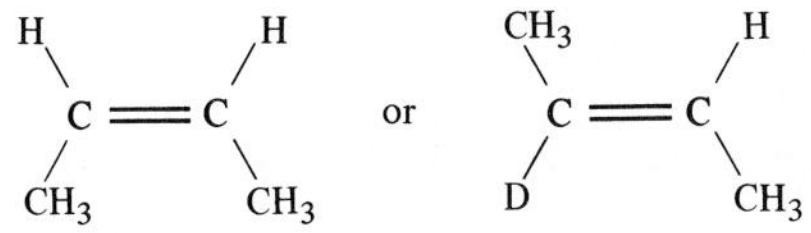

cis-2-Butene   *trans*-2-Butene-2-*d*

How can you explain these results?

CHAPTER ELEVEN

p-Bromonitrobenzene Crystals. *p*-Bromonitrobenzene is a typical derivative of benzene. The nomenclature of benzene derivatives is described in Section 11.8. When you study Chapter 12 you will learn how to synthesize *p*-nitrobenzene and many other compounds similar to it.

Aromatic Compounds I: The Phenomenon of Aromaticity

11.1 INTRODUCTION

In 1825 Michael Faraday, a scientist most often remembered for his work with electricity, isolated a new substance from a gas used at that time for illumination. This compound, now called **benzene,** was an example of a new class of organic substances called **aromatic compounds.**

In 1834 benzene was shown to have the empirical formula CH. Later its molecular formula was shown to be C_6H_6. This in itself was a surprising discovery. Benzene has *only* as many hydrogen atoms as it has carbon atoms! Most compounds that were then known had a far greater proportion of hydrogen atoms, usually twice as many. Benzene, with a formula of C_6H_6, or C_nH_{2n-6}, must consist of *highly unsaturated* molecules. It has an index of hydrogen deficiency equal to four.

During the latter part of the nineteenth century the Kekulé–Couper–Butlerov theory of valence was systematically applied to all known organic compounds. One result of this effort was the placing of organic compounds in either of two broad categories; compounds were classified as being either **aliphatic** or **aromatic.** To be classified as aliphatic meant then that the chemical behavior of a compound was "fatlike." (Now it means that the compound reacts like an alkane, an alkene, an alkyne, or one of their derivatives.) To be classified as aromatic meant then that the compound had a low hydrogen-to-carbon ratio and that it was "fragrant." Most of the early aromatic compounds were obtained from balsams, resins, or essential oils. Included among these were benzaldehyde (from oil of bitter almonds), benzoic acid and benzyl alcohol (from gum benzoin), and toluene (from tolu balsam).

Kekulé was the first to recognize that these early aromatic compounds all contain a six-carbon unit and that they retain this six-carbon unit through most chemical transformations and degradations. Benzene was eventually recognized as being the parent compound of this new series.

Since this new group of compounds proved to be distinctive in ways that are far more important than their odors, the term *aromatic* began to take on a purely chemical connotation. We shall see in this chapter that the meaning of aromatic has evolved as chemists have learned more about the reactions and properties of aromatic compounds.

11.2 REACTIONS OF BENZENE

The benzene molecule is highly unsaturated, and because of this we might expect that it would react accordingly. We might expect, for example, that benzene would decolorize bromine in carbon tetrachloride by *adding* bromine; that it would decolorize aqueous potassium permanganate by being *oxidized;* that it would *add* hydrogen easily in the presence of a catalyst; and that it would *add* water in the presence of acids.

Benzene does none of these. When benzene is treated with bromine in carbon tetrachloride in the dark or with aqueous potassium permanganate or with dilute acids, none of the expected reactions occurs. Benzene does add hydrogen in the presence of finely divided nickel, but only at high temperatures and under high pressures.

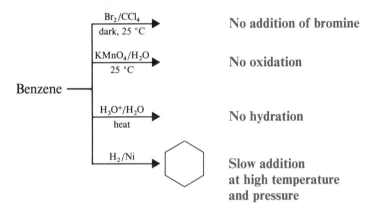

Benzene *does* react with bromine but only in the presence of a Lewis-acid catalyst such as ferric bromide. Most surprisingly, however, it reacts not by addition but by *substitution.*

Substitution

$$C_6H_6 + Br_2 \xrightarrow{FeBr_3} C_6H_5Br + HBr \qquad \text{Observed}$$

Addition

$$C_6H_6 + Br_2 \xrightarrow{\quad} C_6H_6Br_2 + C_6H_6Br_4 + C_6H_6Br_6 \qquad \text{Not observed}$$

When benzene reacts with bromine *only one monobromobenzene* is formed. That is, only one compound with the formula C_6H_5Br is found among the products. Similarly, when benzene is chlorinated *only one monochlorobenzene* results.

Two possible explanations can be given for these observations. The first is that only one of the six hydrogen atoms in benzene is reactive toward these reagents. The second is that all six hydrogen atoms in benzene are equivalent, and replacing any one of them with a substituent results in the same product. As we shall see, the second explanation is correct.

PROBLEM 11.1

The following are several compounds that have the formula C_6H_6. (a) For which of these compounds, if any, would a substitution of bromine for hydrogen yield only one *mono*bromo product? (b) Which of these compounds would you expect to react with bromine by substitution alone?

$$H-C\equiv C-CH_2CH_2-C\equiv C-H$$
(a)

$$H-C\equiv C-CH_2-C\equiv C-CH_3$$
(d)

$$CH_2=CH-CH=CH-C\equiv CH$$
(b)

$$CH_2=CH-C\equiv C-CH=CH_2$$
(e)

$$CH_3CH_2-C\equiv C-C\equiv C-H$$
(c)

11.3 THE KEKULÉ STRUCTURE FOR BENZENE

In 1865 August Kekulé proposed a structure for benzene. Kekulé suggested that the carbon atoms of benzene are in a ring, that they are bonded to each other by alternating single and double bonds, and that one hydrogen atom is attached to each carbon atom. This structure satisfied the requirements that carbon atoms form four bonds, and that all of the hydrogen atoms of benzene are equivalent.

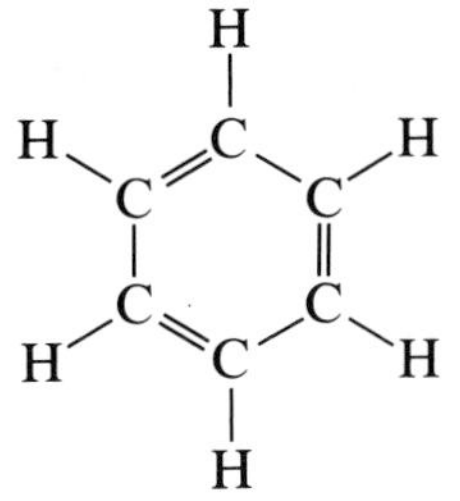

The Kekulé formula for benzene

A problem soon arose with the Kekulé structure, however. The Kekulé structure predicts that there should be two different 1,2-dibromobenzenes. In one of these hypothetical compounds (below), the carbon atoms that bear the bromines are separated by a single bond, and in the other they are separated by a double bond. *Only one 1,2-dibromobenzene, however, has ever been found.*

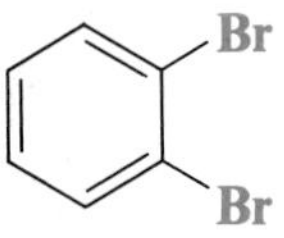

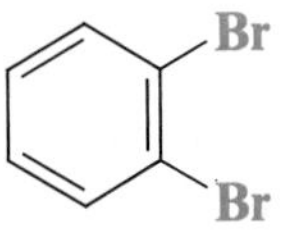

In order to accommodate this objection, Kekulé proposed that the two forms of benzene (and of benzene derivatives) are in a state of equilibrium, and that this equilibrium is so rapidly established that it prevents isolation of the separate compounds.

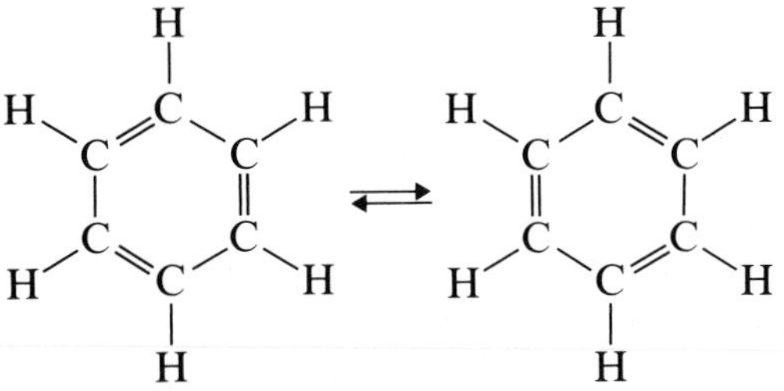

Thus, the two 1,2-dibromobenzenes would also be rapidly equilibrated, and this would explain why chemists had not been able to isolate the two forms.

We now know that this proposal was wrong and that *no such equilibrium exists.* Nonetheless, the Kekulé formulation of benzene's structure was an important step forward and, for very practical reasons, it is still used today. We understand its meaning differently, however.

The tendency of benzene to react by substitution rather than addition gave rise to another concept of aromaticity. For a compound to be called aromatic meant, experimentally, that it gave substitution reactions rather than addition reactions even though it was highly unsaturated.

Before 1900, chemists assumed that the ring of alternating single and double bonds was the structural feature that gave rise to the aromatic properties. Since benzene and benzene derivatives (i.e., compounds with six-membered rings) were the only aromatic compounds known, chemists naturally sought other examples. The compound cyclooctatetraene seemed to be a likely candidate.

Cyclooctatetraene

In 1911, Richard Willstätter succeeded in synthesizing cyclooctatetraene. Willstätter found, however, that it is not at all like benzene. Cyclooctatetraene reacts with bromine by addition, it adds hydrogen readily, it decolorizes solutions of potassium permanganate, and thus it is clearly *not aromatic.* While these findings must have been a keen disappointment to Willstätter, they were very significant for what they did not prove. Chemists, as a result, had to look deeper to discover the origin of benzene's aromaticity.

11.4 THE STABILITY OF BENZENE

We have seen that benzene shows unusual behavior by undergoing substitution reactions when, on the basis of its Kekulé structure, we should expect it to undergo addition. Benzene is unusual in another sense: It is *more stable* than the Kekulé structure suggests. To see how, consider the following thermochemical results.

Cyclohexene, a six-membered ring containing one double bond, can be hydrogenated easily to cyclohexane. When the $\Delta H°$ for this reaction is measured it is found to be -28.6 kcal/mole, very much like that of any similarly substituted alkene.

$$+ H_2 \xrightarrow{\text{Pt}} \qquad \qquad \Delta H° = -28.6 \text{ kcal/mole}$$

Cyclohexene   Cyclohexane

We would expect that hydrogenation of 1,3-cyclohexadiene would liberate roughly twice as much heat and thus have a $\Delta H°$ equal to about -57.2 kcal/mole. When this experiment is done, the result is a $\Delta H° = -55.4$ kcal/mole. This result is quite close to what we calculated, and the difference can be explained by taking into account the fact that compounds containing conjugated double bonds are usually somewhat more stable than those that contain isolated double bonds (Section 10.10).

<table>
<tr><td></td><td></td><td></td><td>Calculated</td></tr>
<tr><td></td><td>$+ 2H_2 \xrightarrow{\text{Pt}}$</td><td></td><td>$\Delta H° = (2 \times -28.6) = -57.2$ kcal/mole</td></tr>
<tr><td></td><td></td><td></td><td>Observed</td></tr>
<tr><td></td><td></td><td></td><td>$\Delta H° = -55.4$ kcal/mole</td></tr>
<tr><td>1,3-Cyclohexadiene</td><td></td><td>Cyclohexane</td><td></td></tr>
</table>

If we extend this kind of thinking, and if benzene is simply 1,3,5-cyclohexatriene, we would predict that benzene would liberate approximately 85.8 kcal/mole (3×-28.6) when it is hydrogenated. When the experiment is actually done the result is surprisingly different. The reaction is exothermic, but only by 49.8 kcal/mole.

<table>
<tr><td></td><td></td><td></td><td>Calculated</td></tr>
<tr><td></td><td>$+ 3H_2 \xrightarrow{\text{Pt}}$</td><td></td><td>$\Delta H° = (3 \times -28.6) = -85.8$ kcal/mole</td></tr>
<tr><td></td><td></td><td></td><td>Observed $\Delta H° = -49.8$ kcal/mole</td></tr>
<tr><td>Benzene</td><td></td><td>Cyclohexane</td><td>Difference $= \quad 36.0$ kcal/mole</td></tr>
</table>

When these results are represented as in Fig. 11.1, it becomes clear that benzene is much more stable than we calculated it to be. Indeed, it is more stable than the hypothetical 1,3,5-cyclohexatriene by 36 kcal/mole. This difference between the

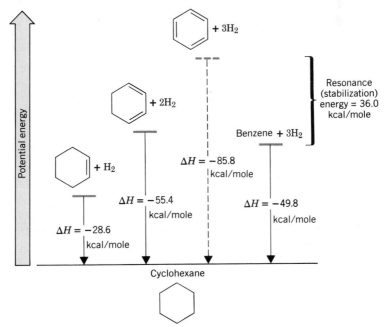

FIGURE 11.1 Relative stabilities of cyclohexene, 1,3-cyclohexadiene, 1,3,5-cyclohexatriene (hypothetically), and benzene.

amount of heat actually released and that calculated on the basis of the Kekulé structure is now called the **resonance energy** of the compound.

11.5 MODERN THEORIES
OF THE STRUCTURE OF BENZENE

It was not until the development of quantum mechanics in the 1920s that the unusual behavior and stability of benzene began to be understood. Quantum mechanics, as we have seen, produced two ways of viewing bonds in molecules: resonance theory and molecular orbital theory. We now look at both of these as they apply to benzene.

11.5A The Resonance Explanation
of the Structure of Benzene

A basic postulate of resonance theory (Sections 1.8 and 10.5) is that whenever two or more structures can be written for a molecule *differing only in the positions of the electrons,* none of the structures will be in complete accord with the compound's chemical and physical properties. If we recognize this, we can now understand the true nature of the two Kekulé structures (**I** and **II**) for benzene. The two Kekulé structures differ only in the positions of the electrons. Structures **I** and **II**, then, do not represent two separate molecules in equilibrium as Kekulé had proposed. Instead, they are the closest we can get to a structure for benzene within the limitations of its molecular formula, the classical rules of valence, and the fact that the six hydrogen atoms are chemically equivalent. Resonance theory, fortunately, does not stop with telling us when to expect this kind of trouble; it also gives us a way out. Resonance theory tells us to use structures **I** and **II** as resonance contributors to a picture of the real molecule of benzene. As such **I** and **II** should be connected with a double-headed arrow and not with two separate ones (because we must reserve the symbol of two separate arrows for chemical equilibria). Resonance contributors, we emphasize again, are not in equilibrium. They are not structures of real molecules. They are the closest we can get if we are bound by simple rules of valence, and they are very useful in helping us visualize the actual molecule as a hybrid.

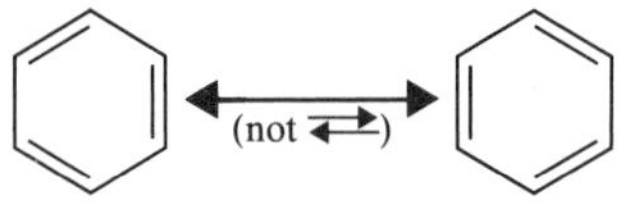

$$\text{I} \qquad\qquad \text{II}$$

Look at the structures carefully. All of the single bonds in structure **I** are double bonds in structure **II**. If we blend **I** and **II**, that is, if we fashion a hybrid of them, then the carbon–carbon bonds in benzene are neither single bonds nor double bonds. Rather, they have a bond order between that of a single bond and that of a double bond. This is exactly what we find experimentally. Spectroscopic measurements show that molecules of benzene are planar and that all of its carbon–carbon bonds

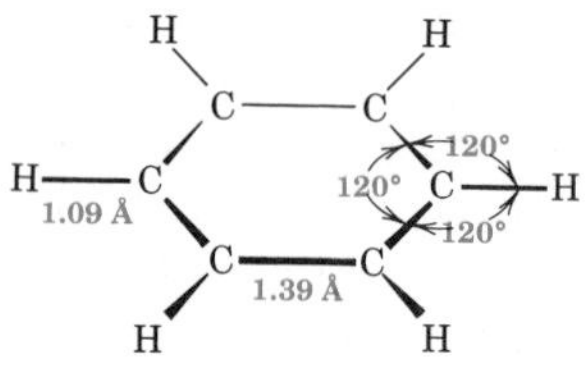

FIGURE 11.2 Bond lengths and angles in benzene.

are of equal length. Moreover, the carbon–carbon bond lengths in benzene (Fig. 11.2) are 1.39 Å, a value in between that for a carbon–carbon single bond between sp^2-hybridized atoms (1.47 Å) (cf. Table 10.1) and that for a carbon–carbon double bond (1.33 Å).

The hybrid structure is represented by inscribing a circle in the hexagon, and it is this new formula **(III)** that is most often used for benzene today. There are times, however, when an accounting of the electrons must be made, and for these purposes we may use one or the other of the Kekulé structures. We do this simply because the electron count in a Kekulé structure is obvious, while the number of electrons represented by a circle or portion of a circle is ambiguous.

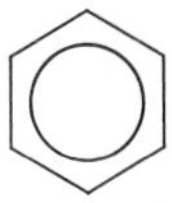

III

PROBLEM 11.2

If benzene were 1,3,5-cyclohexatriene, the carbon–carbon bonds would be alternately long and short as indicated in the following structures. However, to consider the structures here as resonance contributors (or to connect them by a double-headed arrow) violates a basic principle of resonance theory. Explain.

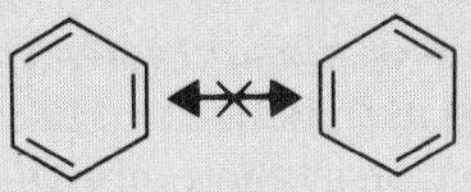

Resonance theory (Section 10.5) also tells us that whenever equivalent resonance structures can be drawn for a molecule, the molecule (or hybrid) is much more stable than any of the resonance structures could be individually if they could exist. In this way resonance theory accounts for the much greater stability of benzene when compared to the hypothetical 1,3,5-cyclohexatriene. For this reason the extra stability associated with benzene is called its *resonance energy.*

11.5B The Molecular Orbital Explanation of the Structure of Benzene

The fact that the bond angles of the carbon atoms in the benzene ring are 120° strongly suggests that the carbon atoms are sp^2 *hybridized.* If we accept this suggestion and construct a planar six-membered ring from sp^2 carbon atoms as shown in Fig. 11.3, another picture of benzene begins to emerge. Because the carbon–carbon bond lengths are all 1.39 Å, the p orbitals are close enough to overlap effectively. The p orbitals overlap equally all around the ring.

According to molecular orbital theory, the six overlapping p orbitals combine to form a set of six π molecular orbitals. Molecular orbital theory also allows us to calculate the relative energies of the π molecular orbitals. These calculations are beyond the scope of our discussion, but the energy levels are shown in Fig. 11.4.

A molecular orbital, as we have seen, can accommodate two electrons if their spins are opposed. Thus, the electronic structure of the ground state of benzene is

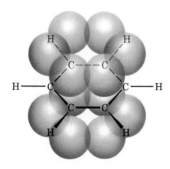

FIGURE 11.3 Overlapping *p* orbitals in benzene.

obtained by adding the six electrons to the π molecular orbitals starting with those of lowest energy, as shown in Fig. 11.4. Notice that in benzene, all of the bonding orbitals are filled, all of the electrons have their spins paired, and there are no electrons in antibonding orbitals. Benzene is, thus, said to have a *closed bonding shell* of delocalized π electrons. This closed bonding shell accounts, in part, for the stability of benzene. (The shapes of the molecular orbitals are given in Fig. 11.5.)

11.6 HÜCKEL'S RULE.
THE (4*n* + 2) π ELECTRON RULE

In 1931 the German physicist E. Hückel carried out a series of mathematical calculations based on the kind of theory that we have just described. Hückel concerned himself with the general situation of *planar monocyclic* rings in which each atom of the ring has an available *p* orbital as in benzene. His calculations indicate that planar rings containing (4*n* + 2) π electrons, where $n = 0, 1, 2, 3, \ldots$, and so on, have closed shells of delocalized electrons like benzene, and should have substantial resonance energies. In other words, *planar monocyclic rings with 2, 6, 10, 14, . . . , etc., delocalized π electrons should be aromatic.*

The fact that cyclooctatetraene lacks aromatic properties correlates with Hückel's rule. Cyclooctatetraene has a total of 8 π electrons. On the basis of Hückel's calculations, monocyclic rings with 8 π electrons are not predicted to have a closed shell of π electrons or substantial resonance energies and they are not expected to be

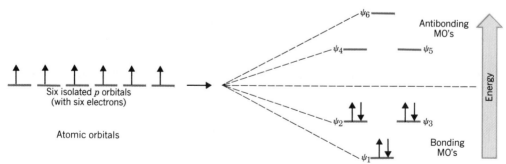

FIGURE 11.4 How six *p* atomic orbitals (one from each carbon of the benzene ring) combine to form six π molecular orbitals. Three of the molecular orbitals have energies lower than that of an isolated *p* orbital; these are the bonding molecular orbitals. Three of the molecular orbitals have energies higher than that of an isolated *p* orbital; these are the antibonding molecular orbitals. Orbitals ψ_2 and ψ_3 have the same energy and are said to be degenerate; the same is true of orbitals ψ_4 and ψ_5.

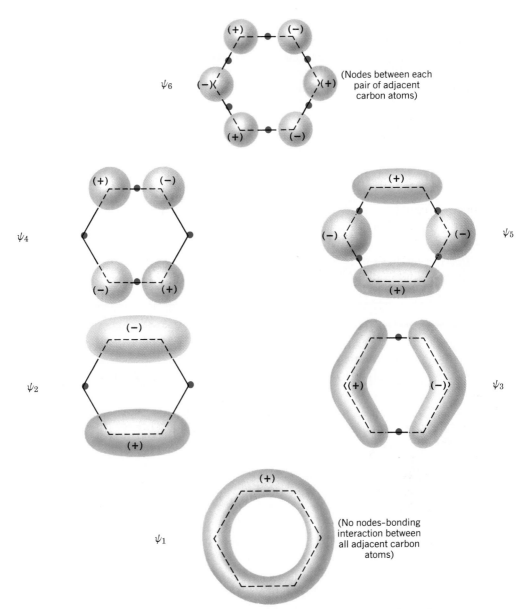

FIGURE 11.5 Shapes of the π molecular orbitals of benzene as viewed from above.

aromatic. A molecule of cyclooctatetraene, we now know, is not planar but is shaped like a tub (see the following figure).

The bonds of cyclooctatetraene are alternately long and short; electron diffraction studies indicate that they are 1.50 and 1.35 Å, respectively. These values are those of single bonds and double bonds, and are not at all like the hybrid bonds of benzene. Cyclooctatetraene is then, clearly, a simple cyclic polyene.

11.6A The Annulenes

The name annulene has been proposed as a general name for monocyclic compounds that can be represented by structures having alternating single and double bonds. The ring size of an annulene is indicated by a number in brackets. Thus, benzene is [6]annulene and cyclooctatetraene is [8]annulene.* Hückel's rule predicts that an-

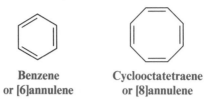

Benzene
or [6]annulene

Cyclooctatetraene
or [8]annulene

nulenes will be aromatic, provided their molecules have $(4n + 2)$ π electrons and have a planar carbon skeleton.

Before 1960 the only annulenes that were available to test Hückel's predictions were benzene and cyclooctatetraene. During the 1960s, and largely as a result of research by the late Franz Sondheimer, a number of large-ring annulenes were synthesized, and the predictions of Hückel's rule were verified.

Consider the [14], [16], [20], [22], and [24]annulenes as examples. Of these, *as Hückel's rule predicts,* the [14], [18], and [22]annulenes ($4n + 2$ when $n = 3, 4, 5$, respectively) have been found to be aromatic. The [16]annulene and the [24]annulene are not aromatic. They are $4n$ compounds, not $4n + 2$ compounds.

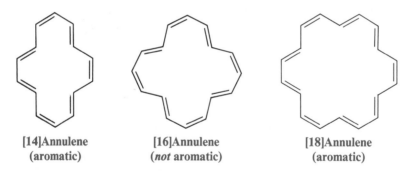

[14]Annulene
(aromatic)

[16]Annulene
(*not* aromatic)

[18]Annulene
(aromatic)

Examples of [10] and [12]annulenes have also been synthesized and none are aromatic. We would not expect [12]annulenes to be aromatic since they have 12 π electrons and, thus, do not obey Hückel's rule. The following [10]annulenes would be expected to be aromatic on the basis of electron count, but their rings are not planar.

4

5

6

[10]Annulenes
None are aromatic because none are planar

---

*These names are seldom used for benzene and cyclooctatetraene. They are often used, however, for conjugated rings of 10 or more carbon atoms.

The [10]annulene (**4**) has two trans double bonds. The carbon atoms of its ring are prevented from becoming coplanar because the two hydrogen atoms in the center of the ring interfere with each other. The [10]annulene with all cis double bonds (**5**) and the [10]annulene with one trans double bond (**6**) are not planar because of ring strain.

After many unsuccessful attempts, [4]annulene (or cyclobutadiene) has also been synthesized. As we would expect, it is a highly reactive compound and *it is not aromatic.*

Cyclobutadiene
or [4]annulene
(*not* aromatic)

11.6B Aromatic Ions

In addition to the neutral molecules that we have discussed so far, there are a number of monocyclic species that bear either a positive or a negative charge. Some of these ions show unexpected stabilities that suggest that they, too, are aromatic. Hückel's rule is helpful in accounting for the properties of these ions as well. We shall consider two examples: the cyclopentadienyl anion and the cycloheptatrienyl cation.

Cyclopentadiene is not aromatic; however, it is unusually acidic for a hydrocarbon. (The K_a for cyclopentadiene is 10^{-15} and, by contrast, the K_a for cycloheptatriene is 10^{-36}.) Because of its acidity, cyclopentadiene can be converted to its anion by treatment with moderately strong bases. The cyclopentadienyl anion, moreover, is unusually stable and nuclear magnetic resonance (nmr) spectroscopy (Chapter 13) shows that all five hydrogen atoms in the cyclopentadienyl anion are equivalent.

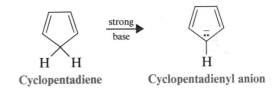

Cyclopentadiene   Cyclopentadienyl anion

The orbital structure of cyclopentadiene (Fig. 11.6) shows why cyclopentadiene, itself, is not aromatic. Not only does it not have the proper number of π electrons, but the π electrons cannot be delocalized about the entire ring because of the intervening sp^3-hybridized —CH_2— group with no available *p* orbital.

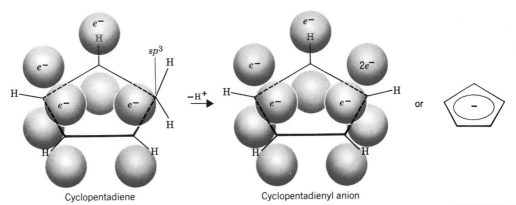

Cyclopentadiene   Cyclopentadienyl anion

FIGURE 11.6 The *p* orbitals of cyclopentadiene and of the cyclopentadienyl anion.

On the other hand, if the $-CH_2-$ carbon atom becomes sp^2 hybridized after it loses a proton (Fig. 11.6), the two electrons left behind can occupy the new p orbital that is produced. Moreover, this new p orbital can overlap with the p orbitals on either side of it and give rise to a ring with *six* delocalized π electrons. Because the electrons are delocalized, all of the hydrogen atoms are equivalent and this agrees with what nmr spectroscopy tells us.

Six is, of course, a Hückel number ($4n + 2$, where $n = 1$), and the cyclopentadienyl anion is, in fact, an **aromatic anion.** The unusual acidity of cyclopentadiene is a result of the unusual stability of its anion.

PROBLEM 11.3

(a) Write all of the resonance structures for the cyclopentadienyl anion.
(b) Can the equivalence of the five hydrogen atoms be explained by resonance?

Cycloheptatriene (Fig. 11.7) has six π electrons. However, the six π electrons of cycloheptatriene cannot be fully delocalized because of the presence of the $-CH_2-$ group—a group that does not have an available p orbital (Fig. 11.7).

When cycloheptatriene is treated with a reagent that can abstract a hydride ion, it is converted to the cycloheptatrienyl (or tropylium) cation. The loss of a hydride ion from cycloheptatriene occurs with unexpected ease, and the cycloheptatrienyl cation is found to be unusually stable. The nmr spectrum of the cycloheptatrienyl cation

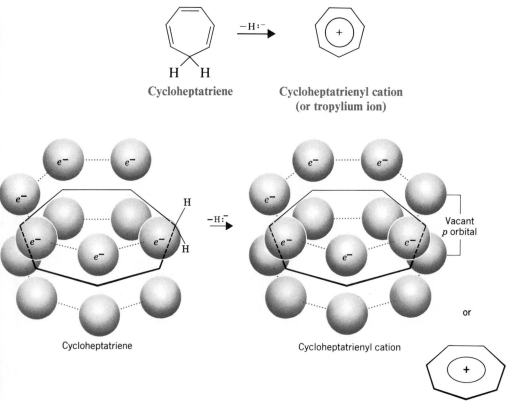

Cycloheptatriene   **Cycloheptatrienyl cation (or tropylium ion)**

FIGURE 11.7 The p orbitals of cycloheptatriene and of the cycloheptatrienyl (tropylium) cation.

indicates that all seven hydrogen atoms are equivalent. If we look closely at Fig. 11.7, we see how we can account for these observations.

As a hydride ion is removed from the —CH$_2$— group of cycloheptatriene, a vacant p orbital is created, and the carbon atom becomes sp^2 hybridized. The cation that results has seven overlapping p orbitals containing *six* delocalized π electrons. The cycloheptatrienyl cation is, therefore, an aromatic cation, and all of its hydrogen atoms should be equivalent; again this is exactly what we find experimentally.

PROBLEM 11.4

The conversion of cycloheptatriene to the cycloheptatrienyl cation can be accomplished by treating cycloheptatriene with triphenylcarbenium perchlorate [(C$_6$H$_5$)$_3$C$^+$ClO$_4^-$]. (a) Write the structure of triphenylcarbenium perchlorate and show how it abstracts a hydride ion from cycloheptatriene. (b) What other product is formed in this reaction? (c) What anion is associated with the cycloheptatrienyl cation that is produced?

PROBLEM 11.5

Tropylium bromide (C$_7$H$_7$Br) is insoluble in nonpolar solvents but dissolves readily in water. When an aqueous solution of tropylium bromide is treated with silver nitrate, a precipitate of AgBr forms immediately. The melting point of tropylium bromide is above 200 °C, quite high for an organic compound. How can you account for these facts?

11.6C Aromatic, Antiaromatic, and Nonaromatic Compounds

What do we mean when we say that a compound is aromatic? We mean that its π electrons are *delocalized* over the entire ring and that it is *stabilized* by the π-electron delocalization.

One of the best ways to determine whether or not the π electrons of a cyclic system are delocalized is through the use of nuclear magnetic resonance spectroscopy. It provides direct physical evidence of whether or not the π electrons are delocalized. We shall have more to say about how this is done in Chapter 13.

But what do we mean by saying that a compound is stabilized by π-electron delocalization? We have an idea of what this means from our comparison of the heat of hydrogenation of benzene and that calculated for the hypothetical 1,3,5-cyclohexatriene. We saw that benzene—in which the π electrons are delocalized—is much more stable than 1,3,5-cyclohexatriene (a model in which the π electrons are not delocalized). We call the energy difference between them the resonance energy (delocalization energy) or stabilization energy.

In order to make similar comparisons for other aromatic compounds we need to choose proper models. But what should these models be?

One proposal is that we should compare the π-electron energy of the cyclic system with that of the corresponding open-chain compound. This approach is particularly useful because it furnishes us with models not only for annulenes but for aromatic cations and anions as well. (Corrections need to be made, of course, when the cyclic system is strained.)

When we use this approach we take as our model a linear chain of sp^2-hybridized atoms that carries the same number of π electrons as our cyclic compound. Then we imagine ourselves removing two hydrogen atoms from the end of this chain and joining the ends to form a ring. If the ring has *lower* π-electron energy than the open chain, then the ring is *aromatic*. If the ring and chain have *the same* π-electron energy, then the ring is *nonaromatic*. If the ring has *greater* π-electron energy than the open chain, then the ring is *antiaromatic*.

The actual calculations and experiments used in determining π-electron energies are beyond our scope, but we can study four examples that illustrate how this approach has been used.

Cyclobutadiene. For cyclobutadiene we consider the change in π-electron energy for the following *hypothetical* transformation.

1,3-Butadiene
4 π electrons

π-electron energy increases

Cyclobutadiene
4 π electrons (antiaromatic)

$+ H_2$

Calculations indicate and experiments appear to confirm that the π-electron energy of cyclobutadiene is higher than that of its open-chain counterpart. Thus cyclobutadiene is classified as being antiaromatic.

Benzene. Here our comparison is based on the following hypothetical transformation.

1,3,5-Hexatriene
6 π electrons

π-electron energy decreases

Benzene
6 π electrons (aromatic)

$+ H_2$

Calculations indicate and experiments confirm that benzene has a much lower π-electron energy than 1,3,5-hexatriene. Benzene is classified as being aromatic on the basis of this comparison as well.

Cyclopentadienyl Anion. Here we use a linear anion for our hypothetical transformation:

$HC:^-$

6 π electrons

π-electron energy decreases

Cyclopentadienyl anion
6 π electrons (aromatic)

$+ H_2$

Both calculations and experiments confirm that the cyclic anion has a lower π-electron energy than its open-chain counterpart. Therefore the cyclopentadienyl anion is classified as being aromatic.

PROBLEM 11.6

(a) What open-chain compound would you use for comparison in assessing the π-electron energy of the cycloheptatrienyl cation? (b) Both theory and experiments indicate that the cycloheptatrienyl cation has a lower π-electron energy than its open-chain counterpart. What conclusion does this justify?

PROBLEM 11.7

(a) The cyclopentadienyl cation is apparently *antiaromatic*. Show what this means through a comparison of π-electron energies of the cyclic and open-chain compounds. (b) Make a similar comparison for the cyclopropenyl

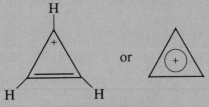

Cyclopropenyl cation

cation—a system that is known from experiments to be *aromatic*. (c) What does Hückel's rule predict for the cyclopropenyl cation? (d) Calculations and experiments indicate that the cyclopropenyl *anion* is *antiaromatic*. What does this mean?

11.7 OTHER AROMATIC COMPOUNDS

11.7A Benzenoid Aromatic Compounds

In addition to those that we have seen so far, there are many other examples of aromatic compounds. Representatives of one broad class of aromatic compounds, called **polycyclic benzenoid aromatic hydrocarbons,** are illustrated in Fig. 11.8.

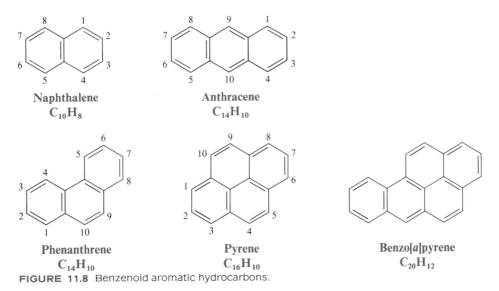

Naphthalene
$C_{10}H_8$

Anthracene
$C_{14}H_{10}$

Phenanthrene
$C_{14}H_{10}$

Pyrene
$C_{16}H_{10}$

Benzo[*a*]pyrene
$C_{20}H_{12}$

FIGURE 11.8 Benzenoid aromatic hydrocarbons.

All of these consist of molecules having two or more benzene rings *fused* together. A close look at one, naphthalene, will illustrate what we mean by this.

According to resonance theory, a molecule of naphthalene can be considered to be a hybrid of three Kekulé structures. One of these Kekulé structures is shown in Fig. 11.9. There are two carbon atoms in naphthalene (C-9 and C-10) that are common to both rings. These two atoms are said to be at the points of *ring fusion.* They direct all of their bonds toward other carbon atoms and do not bear hydrogen atoms.

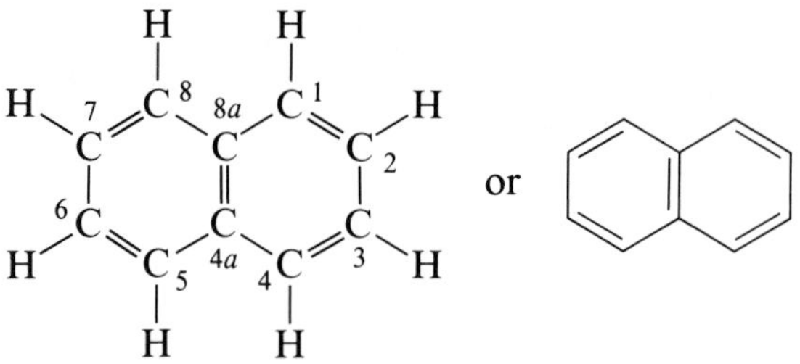

FIGURE 11.9 One Kekulé structure for naphthalene.

PROBLEM 11.8

(a) Write the three resonance structures for naphthalene. (b) The C-1—C-2 bond of naphthalene is shorter than the C-2—C-3 bond. Do the resonance structures you have written account for this? Explain.

Molecular orbital calculations for naphthalene begin with the model shown in Fig. 11.10. The *p* orbitals overlap around the periphery of both rings and across the points of ring fusion.

When molecular orbital calculations are carried out for naphthalene using the model shown in Fig. 11.10, the results of the calculations correlate well with our experimental knowledge of naphthalene. The calculations indicate that delocalization of the 10 π electrons over the two rings produces a structure with considerably

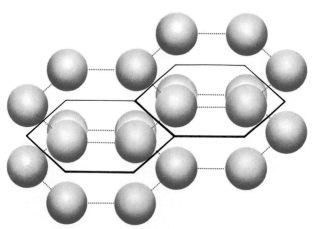

FIGURE 11.10 The *p* orbitals of naphthalene.

lower energy than that calculated for any individual Kekulé structure. Naphthalene, consequently, has a substantial resonance energy. Based on what we know about benzene, moreover, naphthalene's tendency to react by substitution rather than addition and to show other properties associated with aromatic compounds is understandable.

Anthracene and phenanthrene are isomers. In anthracene the three rings are fused in a linear way, and in phenanthrene they are fused so as to produce an angular molecule. Both of these molecules also show large resonance energies and chemical properties typical of aromatic compounds.

Pyrene is also aromatic. Pyrene itself has been known for a long time; a pyrene derivative, however, has been the object of research that shows another interesting application of Hückel's rule.

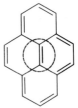

FIGURE 11.11 One Kekulé structure for pyrene. The internal double bond is enclosed in a dotted circle for emphasis.

In order to understand this particular research we need to pay special attention to the Kekulé structure for pyrene (Fig. 11.11). The total number of π electrons in pyrene is 16 (8 double bonds = 16 π electrons). Sixteen is a non-Hückel number, but Hückel's rule is intended to be applied only to monocyclic compounds and pyrene is clearly tetracyclic. If we disregard the internal double bond of pyrene, however, and look only at the periphery, we see that the periphery resembles a monocyclic ring with 14 π electrons. The periphery is, in fact, very much like that of the following [14]annulene. Fourteen *is* a Hückel number ($4n + 2$, where $n = 3$) and one might then predict that the periphery of pyrene would be aromatic by itself, in the absence of the internal double bond.

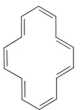

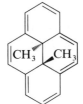

[14]Annulene   *trans*-15,16-Dimethyldihydropyrene

This prediction was confirmed when V. Boekelheide (University of Oregon) synthesized the very unusual *trans*-15,16-dimethyldihydropyrene and showed that it is aromatic.

11.7B Nonbenzenoid Aromatic Compounds

Naphthalene, phenanthrene, and anthracene are examples of *benzenoid* aromatic compounds. On the other hand, the cyclopentadienyl anion, the cycloheptatrienyl cation, *trans*-15,16-dimethyldihydropyrene, and the aromatic annulenes (except for [6]annulene) are classified as **nonbenzenoid aromatic compounds.**

Another example of a *nonbenzenoid* aromatic hydrocarbon is the compound azulene. Azulene has a resonance energy of 49 kcal/mole.

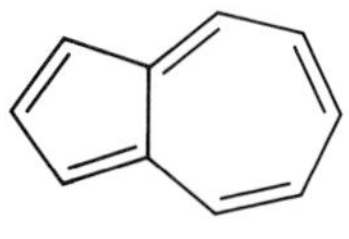

Azulene

This deep-blue hydrocarbon (its name is derived from the word *azure*) is an isomer of naphthalene. It has the same number of π electrons as naphthalene and, for this reason, azulene is also said to be *isoelectronic* with naphthalene. In addition to its deep-blue color (naphthalene by contrast is colorless), azulene differs from naphthalene in another respect that seems, at first, to be peculiar. Azulene is found to have a substantial dipole moment. The dipole moment of azulene is 1.0 D, whereas the dipole moment of naphthalene is zero.

That azulene has a dipole moment at all indicates that charge separation exists in the molecule. If we recognize this and begin writing resonance structures for azulene that involve charge separation, we find that we can write a number of structures like the one shown in Fig. 11.12.

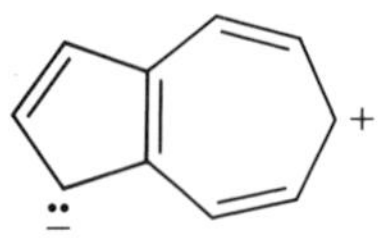

FIGURE 11.12 One resonance structure for azulene that has separated charges.

When we inspect this resonance structure, we see that the five-membered ring is very much like the *aromatic* cyclopentadienyl anion and that the seven-membered ring resembles the *aromatic* cycloheptatrienyl cation. If resonance structures of this type contribute to the overall hybrid for azulene, then we not only understand why azulene has a dipole moment, but we also have some insight into why it is aromatic. Such speculation is strengthened by the results of studies done with substituted azulenes that show quite conclusively that the five-membered ring is negatively charged and the seven-membered ring is positive.

PROBLEM 11.9

Diphenylcyclopropenone (**I**) has a much larger dipole moment than benzophenone (**II**). Can you think of an explanation that would account for this?

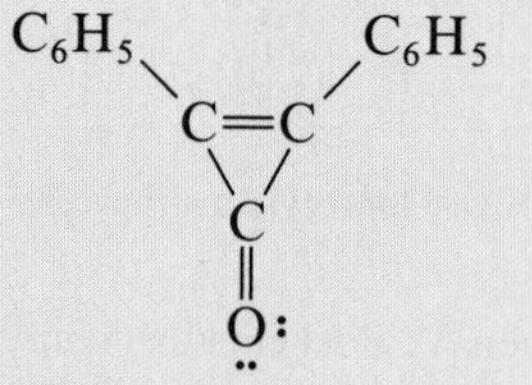

$$C_6H_5 \diagdown \quad C_6H_5$$

I 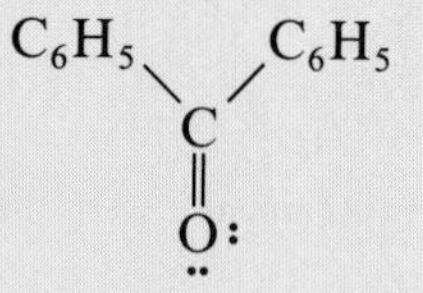   **II**

11.8 NOMENCLATURE OF BENZENE DERIVATIVES

Two systems are used in naming monosubstituted benzenes. In certain compounds, *benzene* is the parent name and the substituent is simply indicated by a prefix. We have, for example,

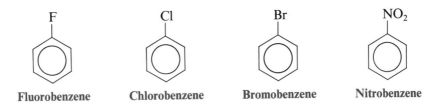

Fluorobenzene   Chlorobenzene   Bromobenzene   Nitrobenzene

For other compounds, the substituent and the benzene ring taken together may form a new parent name. Methylbenzene is usually called *toluene,* hydroxybenzene is almost always called *phenol,* and aminobenzene is almost always called *aniline.* These and other examples are indicated here.

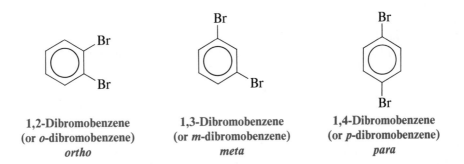

Toluene   Phenol   Aniline

Benzenesulfonic acid   Benzoic acid   Acetophenone   Anisole

When two substituents are present, their relative positions are indicated by the prefixes **ortho, meta,** and **para** (abbreviated **o-**, **m-**, and **p-**) or by the use of numbers.* For the dibromobenzenes we have

1,2-Dibromobenzene   1,3-Dibromobenzene   1,4-Dibromobenzene
(or *o*-dibromobenzene)   (or *m*-dibromobenzene)   (or *p*-dibromobenzene)
       ortho   *meta*   *para*

*Numbers can be used for two or more substituents, but ortho, meta, and para must never be used for more than two.

and for the nitrobenzoic acids:

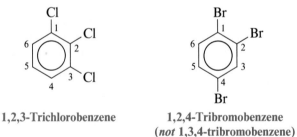

2-Nitrobenzoic acid	**3-Nitrobenzoic acid**	**4-Nitrobenzoic acid**
(or *o*-nitrobenzoic acid)	(or *m*-nitrobenzoic acid)	(or *p*-nitrobenzoic acid)

The dimethylbenzenes are called *xylenes.*

1,2-Dimethylbenzene	**1,3-Dimethylbenzene**	**1,4-Dimethylbenzene**
(or *o*-xylene)	(or *m*-xylene)	(or *p*-xylene)

If more than two groups are present on the benzene ring, their positions must be indicated by the use of *numbers.* As examples, consider the following two compounds.

1,2,3-Trichlorobenzene	**1,2,4-Tribromobenzene**
	(*not* 1,3,4-tribromobenzene)

We notice, too, that the benzene ring is numbered so as to give **the lowest possible numbers to the substituents.**

When more than two substituents are present and the substituents are different, they are listed in alphabetical order (or in order of increasing complexity or size; see Section 3.3B).

When a substituent is one that when taken together with the benzene ring gives a new base name, that substituent is assumed to be in position 1 and the new parent name is used:

3,5-Dinitrobenzoic acid	**2,4-Difluorobenzenesulfonic acid**

When the benzene ring is named as a substituent, it is called a **phenyl group.** For example,

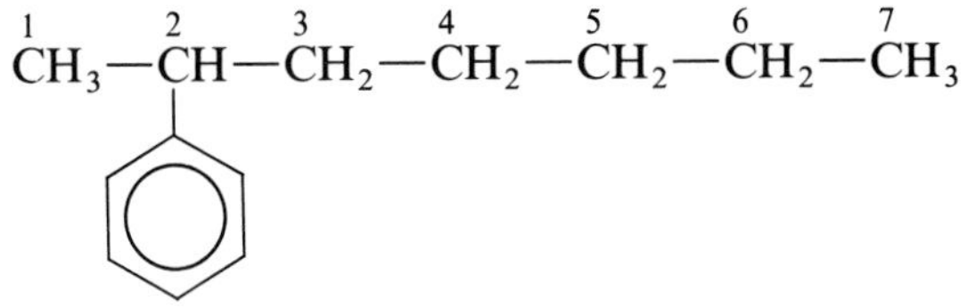

$$\overset{1}{C}H_3-\overset{2}{C}H-\overset{3}{C}H_2-\overset{4}{C}H_2-\overset{5}{C}H_2-\overset{6}{C}H_2-\overset{7}{C}H_3 \qquad CH_3C=CHCH_3$$

2-Phenylheptane
(*not* 2-benzylheptane)

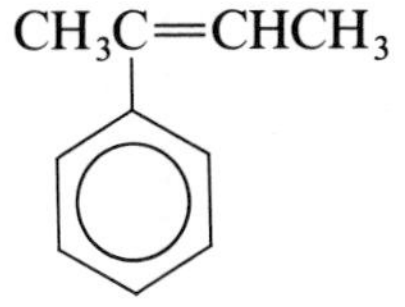

2-Phenyl-2-butene

The phenyl group is often abbreviated as C_6H_5-, $Ph-$, or $\phi-$. The name **benzyl** is reserved for the group derived from toluene by the removal of one hydrogen atom of the methyl group.

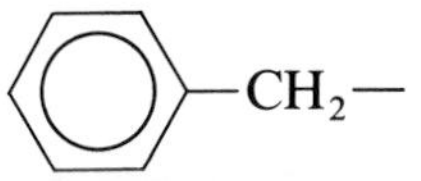

$$\text{—CH}_2\text{—} \qquad \text{—CH}_2\text{Cl}$$

The benzyl group
(or the phenylmethyl group)

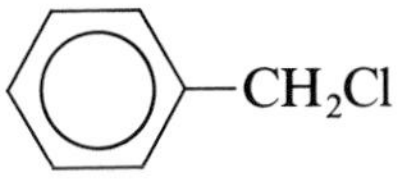

Benzyl chloride
(or phenylmethyl chloride)

11.9 REDUCTION OF AROMATIC COMPOUNDS. THE BIRCH REDUCTION

Hydrogenation of benzene under pressure using a metal catalyst such as nickel results in the addition of three molar equivalents of hydrogen and the formation of cyclohexane (Section 11.2). The intermediate cyclohexadienes and cyclohexene cannot be isolated because these undergo catalytic hydrogenation faster than benzene does.

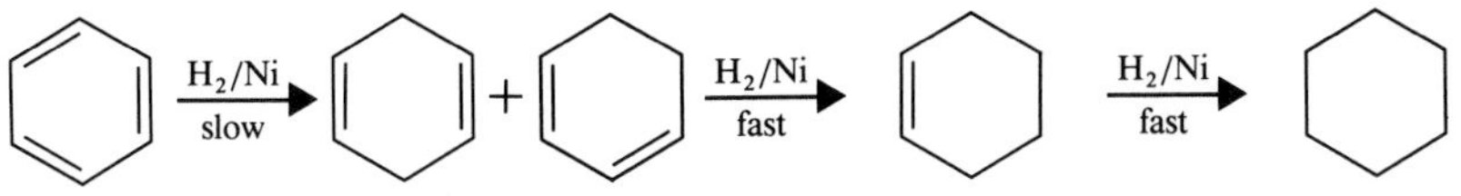

$$\xrightarrow[\text{slow}]{H_2/Ni} \quad + \quad \xrightarrow[\text{fast}]{H_2/Ni} \quad \xrightarrow[\text{fast}]{H_2/Ni}$$

Benzene   **Cyclohexadienes**   **Cyclohexene**   **Cyclohexane**

11.9A The Birch Reduction

Benzene can be reduced to 1,4-cyclohexadiene by treating it with an alkali metal (sodium, lithium, or potassium) in a mixture of liquid ammonia and an alcohol.

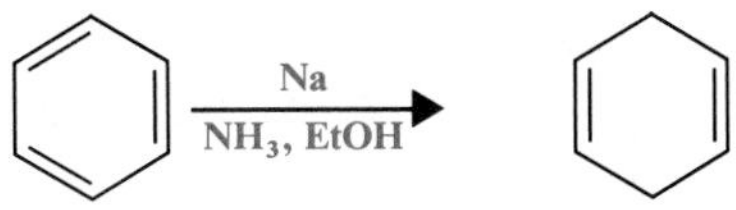

$$\xrightarrow[NH_3, \text{ EtOH}]{Na}$$

Benzene   **1,4-Cyclohexadiene**

This is another dissolving metal reduction and the mechanism for it resembles the mechanism for the reduction of alkynes that we studied in Section 6.7B. A sequence of electron transfers from the alkali metal and proton transfers from the alcohol takes place. (See the following reaction sequence.) The first electron transfer produces a delocalized benzene radical anion. Protonation produces a cyclohexadienyl radical (also a delocalized species). Transfer of another electron leads to the formation of a delocalized cyclohexadienyl anion and protonation of this produces the 1,4-cyclohexadiene.

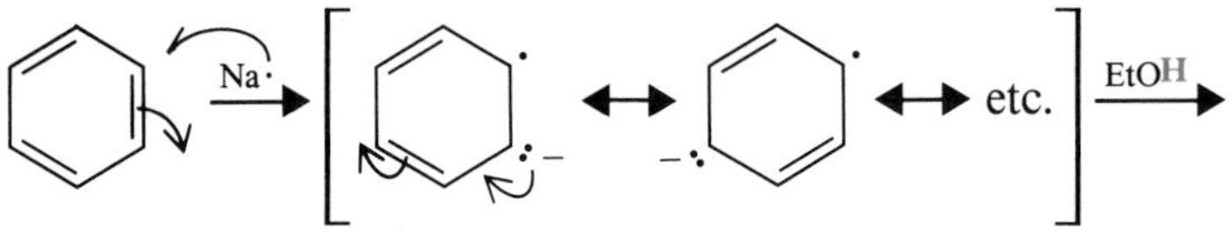

Benzene   **Benzene anion radical**

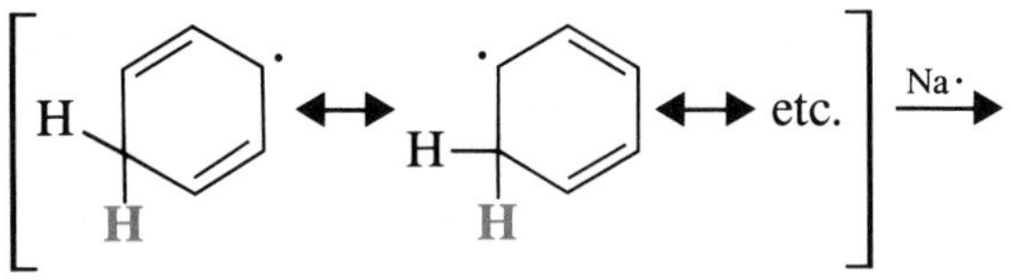

Cyclohexadienyl radical

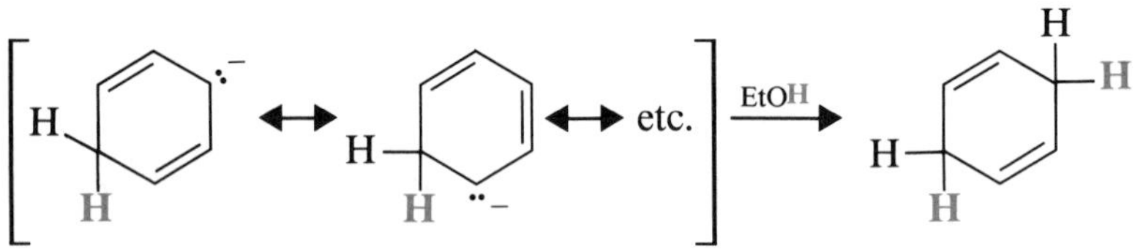

Cyclohexadienyl anion   **1,4-Cyclohexadiene**

Formation of a 1,4-cyclohexadiene in a reaction of this type is quite general, but the reason for its formation in preference to the more stable conjugated 1,3-cyclohexadiene is not understood.

Dissolving metal reductions of this type were developed by the Australian chemist A. J. Birch and have come to be known as **Birch Reductions.**

Substituent groups on the benzene ring influence the course of the reaction. Birch reduction of methoxybenzene (anisole) produces the following result:

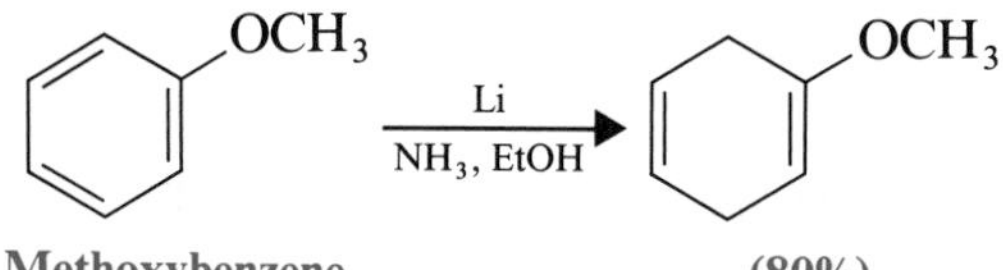

Methoxybenzene   **(80%)**
(anisole)   **1-Methoxy-1,4-**
          cyclohexadiene

Reduction of 1,2-dimethylbenzene (*o*-xylene) gives 1,2-dimethyl-1,4-cyclohexadiene:

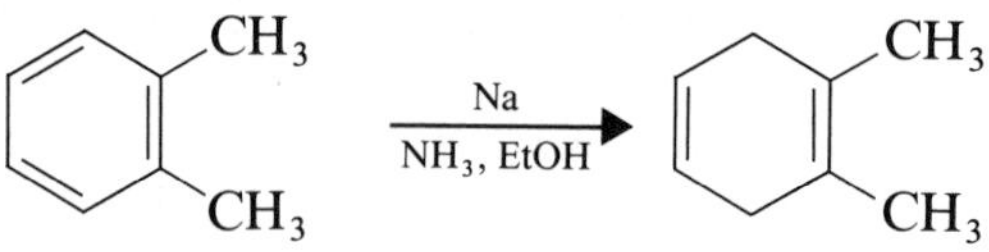

1,2-Dimethylbenzene   **(77–92%)**
          1,2-Dimethyl-1,4-
          cyclohexadiene

Birch reduction of sodium benzoate, however, yields a product with the substituent on the saturated carbon atom.

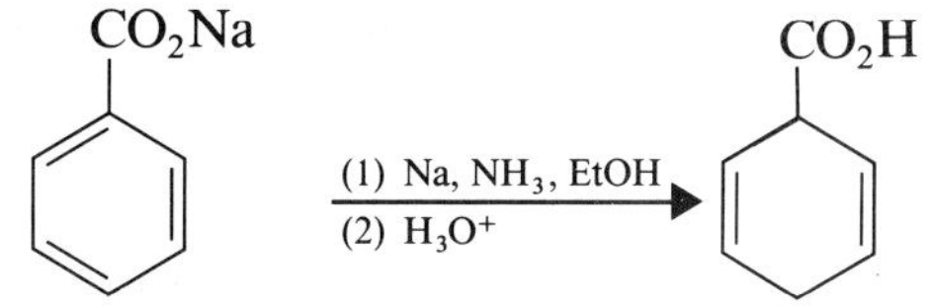

Sodium benzoate → (1) Na, NH₃, EtOH (2) H₃O⁺ → (89–95%)

PROBLEM 11.10

That the product of the Birch reduction of benzene is 1,4-cyclohexadiene and not 1,3-cyclohexadiene can be demonstrated by ozonolysis. (a) Explain how this can be done by showing the products that would be obtained in each instance. (b) What would ozonolysis of the product obtained from the Birch reduction of 1,2-dimethylbenzene yield?

PROBLEM 11.11

Birch reduction of toluene yields a product **X** with the molecular formula C_7H_{10}. On ozonolysis **X** is transformed into CH_3COCH_2CHO and $OHCCH_2CHO$. What is the structure of **X**?

PROBLEM 11.12

Acidic hydrolysis of the Birch reduction product obtained from methoxybenzene (i.e., 1-methoxy-1,4-cyclohexadiene) yields 2-cyclohexenone. Propose a mechanism for this reaction. (*Hint:* Refer to Section 7.15, and also recall that molecules with conjugated double bonds are more stable that those with isolated double bonds.)

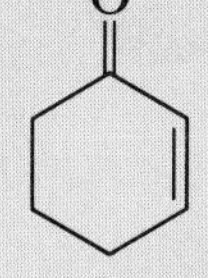

2-Cyclohexenone

PROBLEM 11.13

Syn hydroxylation of 1,4-cyclohexadiene with OsO_4 yields two products. (a) Write the structures of these products and (b) tell whether either product could be resolved into separate enantiomers.

11.10 HETEROCYCLIC AROMATIC COMPOUNDS

Almost all of the cyclic molecules that we have discussed so far have had rings composed solely of carbon atoms. However, in molecules of many cyclic compounds an element other than carbon is present in the ring. These compounds are called

heterocyclic compounds. Heterocyclic molecules are quite commonly encountered in nature. For this reason, and because the structures of some of these molecules are closely related to the compounds that we discussed earlier, we shall now describe a few examples.

Heterocyclic compounds containing nitrogen, oxygen, and sulfur are by far the most common. Four important examples are given here in their Kekulé forms. *These four compounds are all aromatic.*

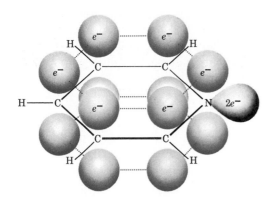

Pyridine   Pyrrole   Furan   Thiophene

If we examine these structures, we shall see that pyridine is electronically related to benzene, and that pyrrole, furan, and thiophene are related to the cyclopentadienyl anion.

The nitrogen atoms in molecules of both pyridine and pyrrole are sp^2 hybridized. In pyridine (Fig. 11.13) the sp^2-hybridized nitrogen donates one electron to the π system. This electron, together with one from each of the five carbon atoms, gives pyridine a sextet of electrons like benzene. The two unshared electrons of the nitrogen of pyridine are in an sp^2 orbital and are not a part of the aromatic sextet. These electrons confer on pyridine the properties of a weak base.

FIGURE 11.13 The orbital structure of pyridine.

SAMPLE PROBLEM

Write resonance structures for pyridine.

Answer:

We can write the following structures, which are analogous to the Kekulé structures for benzene.

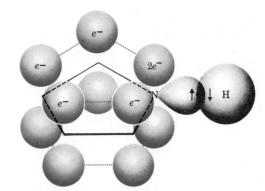

FIGURE 11.14 The orbital structure of pyrrole. (Compare with the orbital structure of the cyclopentadienyl anion in Fig. 11.6.)

In pyrrole (Fig. 11.14) the electrons are arranged differently. Because only four π electrons are contributed by the carbon atoms of the pyrrole ring, the sp^2-hybridized nitrogen must contribute two electrons to give an aromatic sextet. Because these electrons are a part of the aromatic sextet, they are not available for donation to a proton. Thus, in aqueous solution, pyrrole is not appreciably basic.

Furan and thiophene are structurally quite similar to pyrrole. The oxygen atom in furan and the sulfur atom in thiophene are sp^2 hybridized. In both compounds the p orbital of the heteroatom donates two electrons to the π system. The oxygen and sulfur atoms of furan and thiophene carry an unshared pair of electrons in an sp^2 orbital (Fig. 11.15).

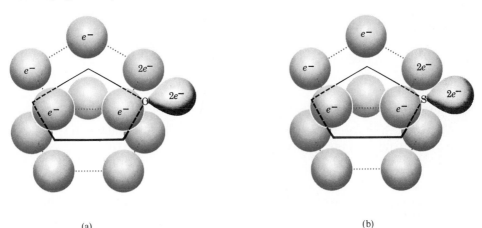

(a)   (b)

FIGURE 11.15 The orbital structures of furan (a) and thiophene (b).

11.11 AROMATIC COMPOUNDS IN BIOCHEMISTRY

Compounds with aromatic rings occupy numerous and important positions in reactions that occur in living systems. It would be impossible to describe them all in this chapter. We shall, however, point out a few examples now and we shall see others later.

Two amino acids necessary for protein synthesis contain the benzene ring:

$$\text{⬡—CH}_2\text{CHCO}_2^- \qquad\qquad \text{HO—⬡—CH}_2\text{CHCO}_2^-$$
$$\overset{|}{\text{NH}_3^+} \qquad\qquad\qquad\qquad\qquad \overset{|}{\text{NH}_3^+}$$

Phenylalanine   **Tyrosine**

A third aromatic amino acid, tryptophan, contains a benzene ring fused to a pyrrole ring. (This aromatic ring system is called an indole system, cf. Section 19.1B.)

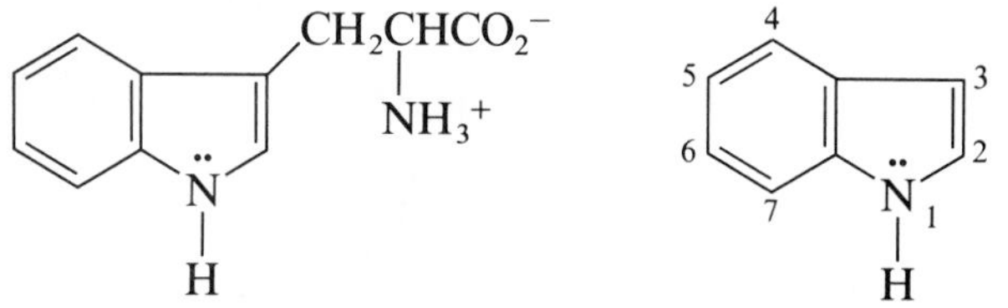

Tryptophan   Indole

It appears that humans, because of the course of evolution, do not have the biochemical ability to synthesize the benzene ring. As a result, phenylalanine and tryptophan derivatives are essential in the human diet. Because tyrosine can be synthesized from phenylalanine by an enzyme known as *phenylalanine hydroxylase,* it is not essential in the diet as long as phenylalanine is present.

Heterocyclic aromatic compounds are also present in many biochemical systems. Derivatives of purine and pyrimidine are essential parts of DNA and RNA.

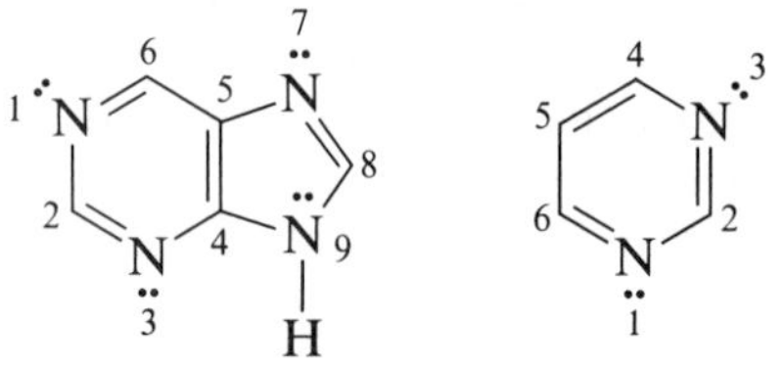

Purine   Pyrimidine

DNA is the molecule responsible for the storage of genetic information and RNA is prominently involved in the synthesis of enzymes (Chapter 23).

PROBLEM 11.14

Classify each nitrogen atom in the purine molecule as to whether it is of the pyridine type or of the pyrrole type.

Both a pyridine derivative (nicotinamide) and a purine derivative (adenine) are present in one of the most important coenzymes in biological oxidations. This molecule, **nicotinamide adenine dinucleotide (NAD$^+$),** is shown in Fig. 11.16.

NAD$^+$, together with another compound in the liver (an apoenzyme), is capable of oxidizing alcohols to aldehydes. While the overall change is quite complex, a look at one aspect of it will illustrate a *biological use* of the extra stability (resonance or delocalization energy) associated with an aromatic ring.

A simplified version of the oxidation of an alcohol to an aldehyde is illustrated here:

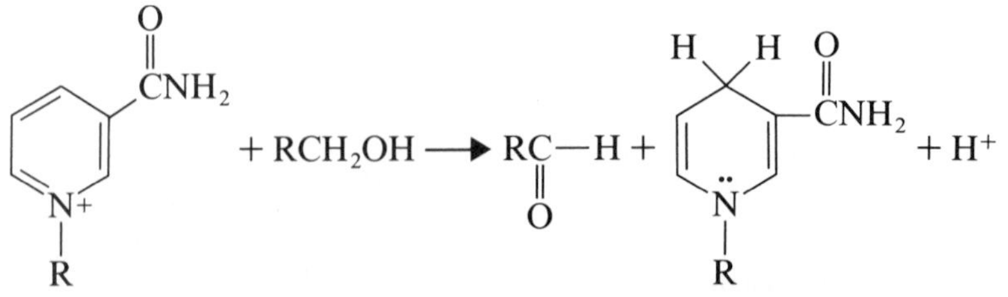

NAD$^+$ 1° Alcohol   Aldehyde   NADH

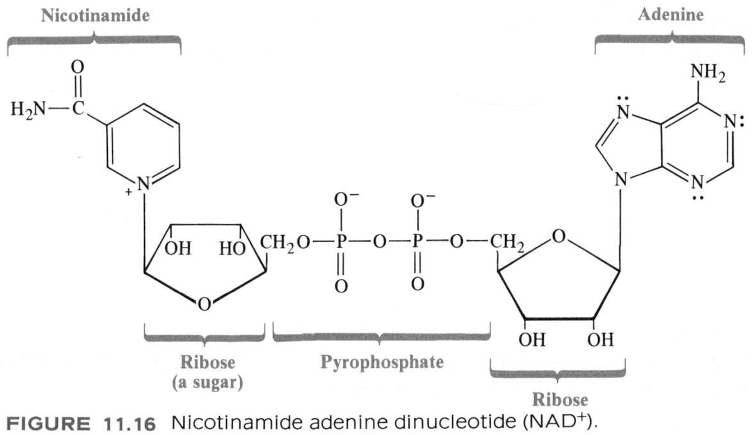

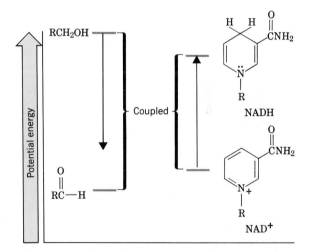

FIGURE 11.16 Nicotinamide adenine dinucleotide (NAD$^+$).

The *aromatic* pyridine ring (actually a *pyridinium* ring, because it is positively charged) in NAD$^+$ is converted to a *nonaromatic* ring in NADH. The extra stability of the pyridine ring is lost in this change; and, as a result, the potential energy of NADH is greater than that of NAD$^+$. The conversion of the alcohol to the aldehyde, however, occurs with a decrease in potential energy. Because these reactions are coupled in biological systems (Fig. 11.17), a portion of the potential energy contained in the alcohol becomes chemically contained in NADH. This stored energy in NADH is used to bring about other biochemical reactions that require energy and that are necessary to life.

FIGURE 11.17 Potential energy diagram for the biologically coupled oxidation of an alcohol and reduction of nicotinamide adenine dinucleotide.

Although many aromatic compounds are essential to life, others are hazardous. Many are quite toxic and several benzenoid compounds, including benzene itself, are **carcinogenic** (i.e., cancer causing). Two other examples are benzo[*a*]pyrene and 7-methylbenz[*a*]anthracene.

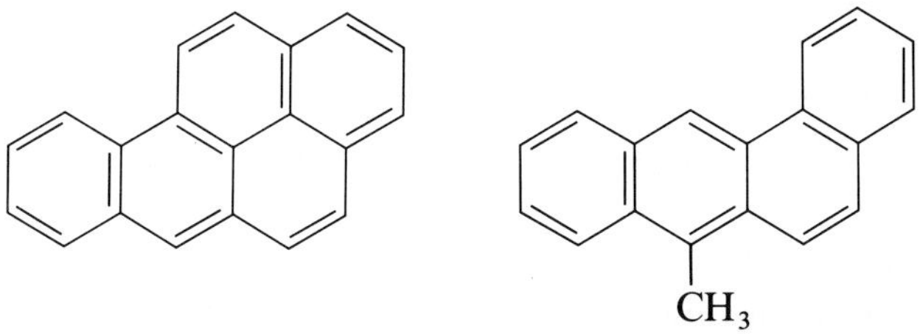

Benzo[*a*]pyrene   **7-Methylbenz[*a*]anthracene**

The hydrocarbon benzo[*a*]pyrene has been found in cigarette smoke and in the exhaust from automobiles. It is also formed in the incomplete combustion of any fossil fuel. It is found on charcoal-broiled steaks and exudes from asphalt streets on a hot summer day. Benzo[*a*]pyrene is so carcinogenic that one can induce skin cancers in mice with almost total certainty by simply shaving an area of the body of the mouse and applying a coating of benzo[*a*]pyrene.

11.12 A SUMMARY OF IMPORTANT TERMS AND CONCEPTS

An *aliphatic compound* is a compound such as an alkane, alkene, alkyne, cyclo-alkane, cycloalkene, or any of their derivatives.

An *aromatic compound* traditionally means one "having the chemistry typified by benzene." The molecules of aromatic compounds are cyclic and conjugated. They have a stability significantly greater than that of a hypothetical resonance structure (e.g., a Kekulé structure). Many aromatic compounds react with electrophilic re-agents (Br_2, HNO_3, H_2SO_4) by substitution rather than addition even though they are unsaturated (Chapter 12). A modern definition of any aromatic compound compares the energy of the π electrons of the cyclic conjugated molecule or ion with that of its open-chain counterpart. If on ring closure the π-electron energy *decreases,* the molecule is classified as being **aromatic,** if it *increases* the molecule is classified as being **antiaromatic,** and if it remains the same the molecule is classified as being **nonaromatic.**

The *resonance energy* of an aromatic compound is the difference in energy between the actual aromatic compound and that calculated for one of the hypothetical reso-nance structures (e.g., a Kekulé structure). The resonance energy is also referred to as *stabilization energy* or *delocalization energy.*

Hückel's rule states that planar monocyclic conjugated rings with $(4n + 2)$ π elec-trons (i.e., with 2, 6, 10, 14, 18, or 22 π electrons) should be aromatic. Hückel's rule has an upper limit.

An *annulene* is a monocyclic compound that can be represented by a structure having alternating single and double bonds. For example, cyclobutadiene is [4]annulene and benzene is [6]annulene.

A *benzenoid* aromatic compound is one whose molecules contain benzene rings or fused benzene rings. Examples are benzene, naphthalene, anthracene, and phen-anthrene.

A *Nonbenzenoid* aromatic compound is one whose molecules contain a ring that is not six membered. Examples are [14]annulene, azulene, the cyclopentadienyl anion, and the cycloheptatrienyl cation.

A *heterocyclic* compound is one whose molecules have a ring containing an element other than carbon. Some heterocyclic compounds (e.g., pyridine, pyrrole, thiophene) are aromatic.

ADDITIONAL PROBLEMS
11.15
Draw structural formulas for the following:

(a) 4-Nitrobenzenesulfonic acid   (b) *o*-Chlorotoluene

(c) *m*-Dichlorobenzene

(d) *p*-Dinitrobenzene

(e) 4-Bromo-1-methoxybenzene

(f) *m*-Nitrobenzoic acid

(g) *p*-Iodophenol

(h) 2-Chlorobenzoic acid

(i) 2-Bromonaphthalene

(j) 9-Chloroanthracene

(k) 3-Nitrophenanthrene

(l) 4-Nitropyridine

(m) 2-Methylpyrrole

(n) 2,4-Dichloro-1-nitrobenzene

(o) *p*-Nitrobenzyl bromide

(p) *o*-Chloroaniline

(q) 2,5-Dibromo-3-nitrobenzoic acid

(r) 1,3,5-Trimethylbenzene (mesitylene)

(s) *p*-Hydroxybenzoic acid

(t) Vinylbenzene (styrene)

(u) Benzo[*a*]pyrene

(v) 2-Phenylcyclohexanol

(w) 2,4,6-Trinitrotoluene (TNT)

(x) A [12]annulene

(y) A [14]annulene

(z) An [18]annulene

11.16

Write structural formulas and give names for all of the following:

(a) Trichlorobenzenes

(b) Dibromonitrobenzenes

(c) Dichlorotoluenes

(d) Monochloronaphthalenes

(e) Nitropyridines

(f) Methylfurans

(g) Chlorodinitrobenzenes

(h) Chlorodimethylbenzenes

(i) Cresols (methylphenols)

11.17

(a) Write the five principal resonance structures for phenanthrene. (b) On the basis of these can you speculate about the length of the C-9–C-10 bond? (c) About its double-bond character? (d) Phenanthrene, in contrast to most aromatic molecules, tends to *add* 1 mole of bromine to form a molecule with the formula $C_{14}H_{10}Br_2$. How can you account for this behavior?

11.18

2-Methoxy-1,2,3-trimethylcyclopropane **(III)** has been found to react with fluoroboric acid to yield methanol and a compound with the formula $C_6H_9^+ BF_4^-$ **(IV)**.

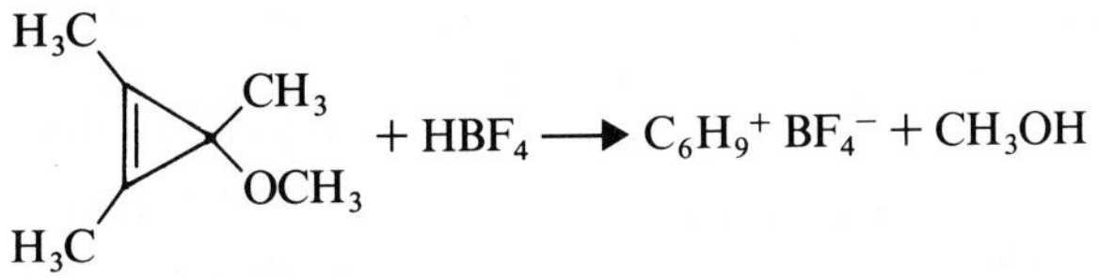

$$\text{(III)} + HBF_4 \longrightarrow C_6H_9^+ BF_4^- + CH_3OH$$

III   IV

(a) What is the structure of $C_6H_9^+ BF_4^-$? (b) How can you account for its formation?

11.19

Diphenylcyclopropenone (cf. Problem 11.9) reacts with hydrogen bromide to form a stable crystalline hydrobromide. What is its structure?

11.20

Cyclooctatetraene has been shown by Thomas Katz of Columbia University to react with two molar equivalents of potassium to yield an unusually stable compound with the formula

$2K^+C_8H_8^{2-}$ (**V**). The nmr spectrum of **V** indicates that all of its hydrogen atoms are equivalent. (a) What is the structure of **V**? (b) How can you account for the formation of **V**?

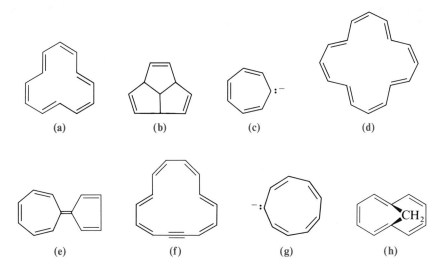

$$2K + \quad \longrightarrow \quad K_2C_8H_8$$

Cyclooctatetraene   V

11.21

Indicate whether each of the following molecules or ions would or would not be aromatic. Explain your answer in each instance.

(a)   (b)   (c)   (d)

(e)   (f)   (g)   (h)

*11.22

The relative energies of the π molecular orbitals of conjugated monocyclic systems can be derived in a relatively simple way. One inscribes a regular polygon (corresponding to the ring size) in a circle so that *one corner of the polygon is at the bottom.* The points where the corners of the polygon touch the circle correspond to the energy levels of the π molecular orbitals of the system. With benzene, for example, the following method furnishes the energy levels given in Fig. 11.4.

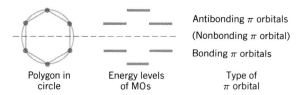

Antibonding π orbitals

(Nonbonding π orbital)

Bonding π orbitals

Polygon in   Energy levels   Type of
circle   of MOs π orbital

A horizontal line halfway up the circle divides the bonding orbitals from the antibonding orbitals. (If an orbital falls on this line, it is a nonbonding orbital.) Use this method to derive the relative energies of the π molecular orbitals of the following molecules, radicals, and ions. In each case you should also give the π electron distribution of the lowest energy state. For the purpose of this exercise, you should assume that all of the structures are planar.

(a) The cyclopropenyl cation   (d) Cyclobutadiene

(b) The cyclopropenyl radical   (e) The cyclobutadienyl dication

(c) The cyclopropenyl anion   (f) The cyclopentadienyl cation

(g) The cyclopentadienyl anion

(h) The cycloheptatrienyl cation

(i) The cycloheptatrienyl anion

(j) Cyclooctatetraene

(k) The cyclooctatetraenyl dication

(l) The cyclooctatetraenyl dianion

(m) Which of the systems (a–l) have unpaired electrons?

(n) Which have closed bonding shells of π electrons?

(o) Which systems would you expect to be aromatic?

(p) Does your answer to (o) correspond to the predictions of Hückel's rule?

*11.23

Cycloheptatrienone (**I**) is very stable. Cyclopentadienone (**II**) by contrast is quite unstable and rapidly undergoes a Diels–Alder reaction with itself. (a) Propose an explanation for the different stabilities of these two compounds. (b) Write the structure of the Diels–Alder adduct of cyclopentadienone.

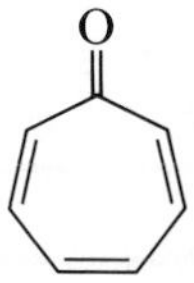

I 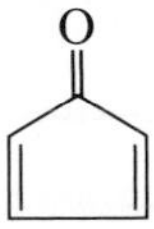   II

CHAPTER TWELVE

Polarized Crystal of p-Nitrophenylacetic Acid. When crystallized from water, p-nitrophenylacetic acid, p—NO_2—$C_6H_4CH_2CO_2H$, is obtained as long pale yellow needles, which melt at 153 °C.

Aromatic Compounds II: Electrophilic Aromatic Substitution

12.1 ELECTROPHILIC AROMATIC SUBSTITUTION REACTIONS

Aromatic hydrocarbons are known generally as **arenes**. An **aryl group** is one derived from an arene by removal of a hydrogen atom and its symbol is Ar—. Thus arenes are designated ArH just as alkanes are designated RH.

The most characteristic reactions of benzenoid arenes are the substitution reactions that occur when they react with electrophilic reagents. These reactions are of the general type shown below.

$$ArH + E^+ \longrightarrow Ar—E + H^+$$

or

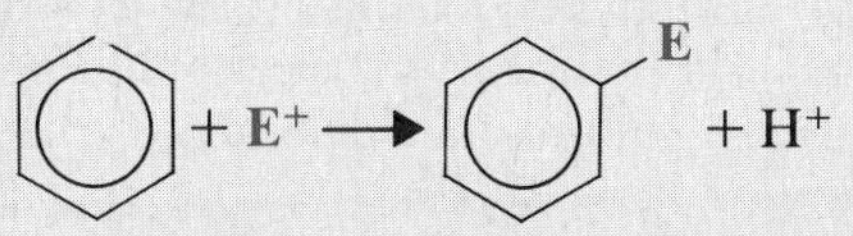

Electrophilic aromatic substitution reactions allow us to introduce a wide variety of groups into aromatic rings and because of this, they give us synthetic access to a vast number of aromatic compounds that would not be available otherwise.

In this chapter we shall see how we can use electrophilic aromatic substitution reactions to introduce:

1. A halo substituent, —X (Section 12.3)
2. A nitro substituent, —NO$_2$ (Section 12.4)
3. A sulfonic acid group, —SO$_3$H (Section 12.5)
4. An alkyl group, —R (Section 12.6)
5. An acyl group, —COR (Section 12.7)

All of these reactions involve an attack on the benzene ring by an electron-deficient species—by an electrophile.

12.2 A GENERAL MECHANISM FOR ELECTROPHILIC AROMATIC SUBSTITUTION: ARENIUM IONS

Benzene is susceptible to electrophilic attack primarily because of its exposed π electrons. In this respect benzene resembles an alkene, for in the reaction of an alkene with an electrophile the site of attack is the exposed π bond.

We saw in Chapter 11, however, that benzene differs from an alkene in a very significant way. Benzene's closed shell of six π electrons gives it a special stability. So while benzene is susceptible to electrophilic attack, it undergoes *substitution reactions* rather than *addition reactions*. Substitution reactions allow the aromatic sextet of π electrons to be regenerated after attack by the electrophile has occurred. We can see how this happens if we examine a general mechanism for electrophilic aromatic substitution.

A considerable body of experimental evidence indicates that electrophiles attack the π system of benzene to form a *delocalized nonaromatic carbocation* known as an **arenium ion** (or sometimes as a σ *complex*).

Step 1

sp^3 Hybridized

Benzene   Electrophile   Arenium ion
(or σ complex)

In step 1 the electrophile takes two electrons of the π system to form a σ bond between it and one carbon atom of the benzene ring. This interrupts the cyclic system of π electrons, because in the formation of the arenium ion one carbon becomes sp^3 hybridized and, thus, no longer has an available p orbital. The four remaining π electrons of the arenium ion are delocalized over the five remaining sp^2-carbon atoms.

In step 2 the arenium ion loses a proton from the carbon atom that bears the electrophile. The two electrons that bonded this proton to carbon become a part of the π system. The carbon atom that bears the electrophile become sp^2 hybridized again and a benzene derivative with six fully delocalized π electrons is formed.

Step 2

We can represent both steps of this mechanism with one of the Kekulé formulas for benzene. When we do this it becomes much easier to account for the π electrons. The first step can be shown in the following way:

Arenium ion
(a carbocation)

By using a Kekulé structure we can also see that the arenium ion is a hybrid of three *allylic-type* resonance structures, each of which has a positive charge *on a carbon that is ortho or para to the site of electrophilic attack.*

We can represent the second step, the loss of a proton, with any one of the resonance structures for the arenium ion.

When we do this we can see clearly that two electrons from the carbon–hydrogen bond serve to regenerate the ring of alternating single and double bonds.

PROBLEM 12.1

Show the second step of this mechanism using each of the other two resonance structures for the arenium ion.

There is firm experimental evidence that the arenium ion is a true *intermediate* in electrophilic substitution reactions. It is not a transition state. This means that in a potential energy diagram (Fig. 12.1) the arenium ion lies in an energy valley between two transition states.

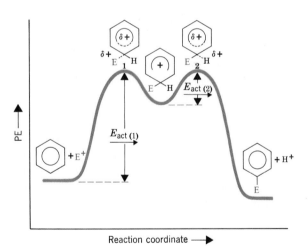

FIGURE 12.1 The potential energy diagram for an electrophilic aromatic substitution reaction. The arenium ion is a true intermediate lying between transition states **1** and **2**. In transition state **1** the bond between the electrophile and one carbon atom of the benzene ring is only partially formed. In transition state **2** the bond between the same benzene carbon atom and its hydrogen atom is partially broken.

The energy of activation for the reaction leading from benzene and the electrophile (E^+) to the arenium ion [$E_{act(1)}$] has been shown to be much greater than the energy of activation [$E_{act(2)}$] leading from the arenium ion to the final product. This is consistent with what we would expect. The reaction leading from benzene and an electrophile to the arenium ion is highly endothermic, because the benzene ring loses its resonance energy. The reaction leading from the arenium ion to the substituted benzene, by contrast, is highly exothermic because in it the benzene ring regains its resonance energy.

Of the following two steps, step 1 — the formation of the arenium ion — is the rate-determining step in electrophilic aromatic substitution.

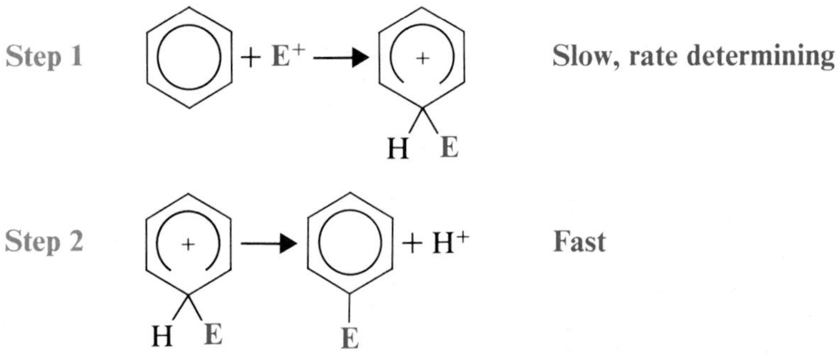

Step 1

Slow, rate determining

Step 2

Fast

Step 2, the loss of a proton, occurs rapidly relative to step 1 and has no effect on the overall rate of reaction.

12.3 HALOGENATION OF BENZENE

Benzene does not react with bromine or chlorine unless a Lewis acid is present in the mixture. (As a consequence, benzene does not decolorize a solution of bromine in carbon tetrachloride.) When Lewis acids are present, however, benzene reacts readily with bromine or chlorine, and the reactions give bromobenzene and chlorobenzene in good yields.

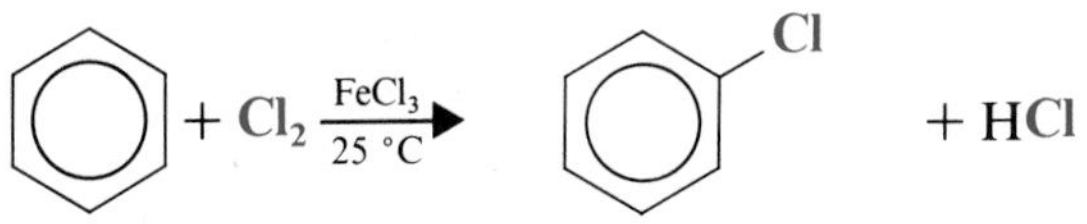

Chlorobenzene (90%)

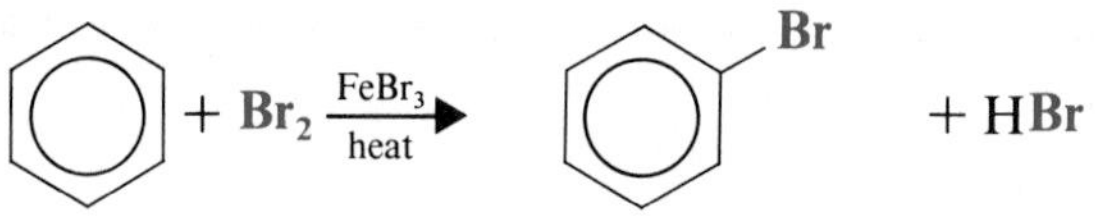

Bromobenzene (75%)

The Lewis acids most commonly used to effect chlorination and bromination reactions are $FeCl_3$, $FeBr_3$, and $AlCl_3$. Ferric chloride and ferric bromide are usually generated in the reaction mixture by adding iron to it. The iron then reacts with halogen to produce the ferric halide:

$$2Fe + 3X_2 \longrightarrow 2FeX_3$$

The mechanism for aromatic bromination is as follows:

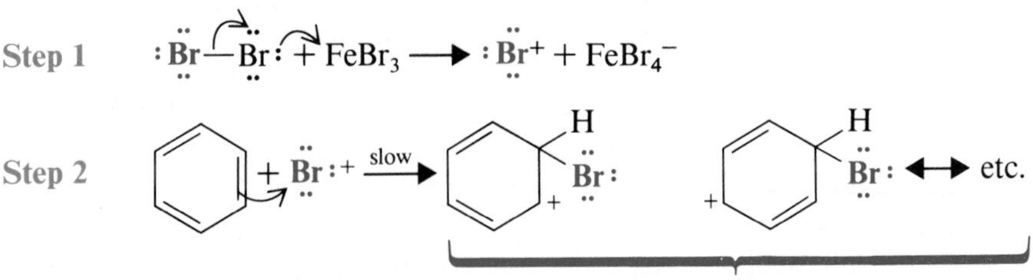

Step 1

$$:\!\overset{..}{\underset{..}{Br}}\!\!-\!\!\overset{..}{\underset{..}{Br}}\!: + FeBr_3 \longrightarrow :\!\overset{..}{\underset{..}{Br}}{}^+ + FeBr_4{}^-$$

Step 2

Arenium ion

Step 3

The function of the Lewis acid can be seen in step 1. The ferric bromide reacts with bromine to produce a positive bromine ion, Br^+ (and $FeBr_4^-$). In step 2 this Br^+ ion attacks the benzene ring to produce an arenium ion. Then finally in step 3 the arenium ion transfers a proton to $FeBr_4^-$. This results in the formation of bromobenzene and hydrogen bromide — the products of the reaction. At the same time this step regenerates the catalyst — $FeBr_3$.

The mechanism of the chlorination of benzene in the presence of ferric chloride is analogous to the one for bromination. Ferric chloride serves the same purpose in aromatic chlorinations as ferric bromide does in aromatic brominations. It assists in the generation and transfer of a positive halogen ion.

Fluorine reacts so rapidly with benzene that aromatic fluorination requires special conditions and special types of apparatus. Even then, it is difficult to limit the reaction to monofluorination. Fluorobenzene can be made, however, by an indirect method that we shall see in Chapter 18.

Iodine, on the other hand, is so unreactive that a special technique has to be used to effect direct iodination; the reaction has to be carried out in the presence of an oxidizing agent such as nitric acid:

86%

The purpose of the nitric acid is to oxidize the I_2 to I^+ ions; these then attack the benzene ring in the same way as Br^+ ions and Cl^+ ions.

12.4 NITRATION OF BENZENE

Benzene reacts slowly with hot concentrated nitric acid to yield nitrobenzene. The reaction is much faster if it is carried out by heating benzene with a mixture of concentrated nitric acid and concentrated sulfuric acid.

(85%)

Concentrated sulfuric acid increases the rate of the reaction by increasing the concentration of the electrophile — the nitronium ion (NO_2^+).

Step 1 $H-\overset{..}{\underset{..}{O}}-NO_2 + HOSO_3H \rightleftharpoons H-\overset{..}{O^+}-NO_2 + HSO_4^-$
            (H_2SO_4) $\quad\quad\quad\quad | $
                                                $\quad\quad\quad\quad H$

Step 2 $H-\overset{..}{O^+}-NO_2 + H_2SO_4 \rightleftharpoons \overset{+}{N}O_2 + H_3O^+ + HSO_4^-$
            $\quad\;\; |$ Nitronium
            $\quad\;\; H$ ion

In step 1 nitric acid acts as a base and accepts a proton from the stronger acid, sulfuric acid. In step 2 the protonated nitric acid dissociates and produces a nitronium ion.

The nitronium ion reacts with benzene by attacking the π cloud and forming an arenium ion.

Step 3

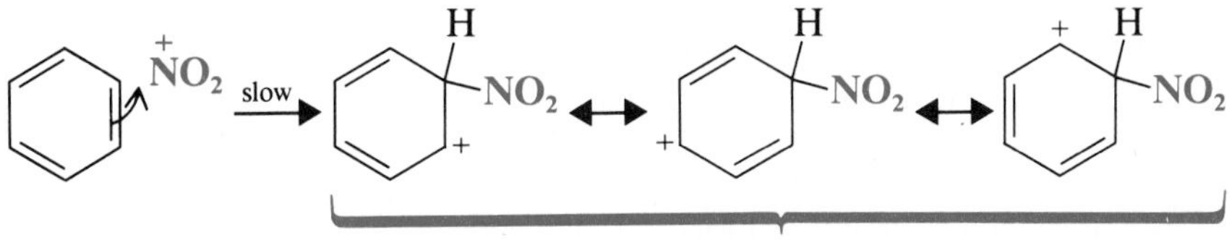

Arenium ion

The arenium ion then transfers a proton to some base in the mixture such as HSO_4^- and becomes nitrobenzene.

Step 4

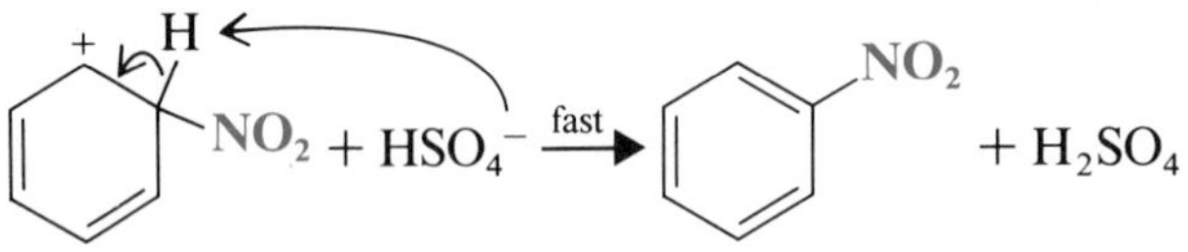

PROBLEM 12.2

Write equations that show how nitronium ions might be formed in nitration reactions in which concentrated nitric acid is used by itself.

12.5 SULFONATION OF BENZENE

Benzene reacts with fuming sulfuric acid at room temperature to produce benzenesulfonic acid. Fuming sulfuric acid is sulfuric acid that contains added sulfur trioxide (SO_3). Sulfonation also takes place in concentrated sulfuric acid alone, but more slowly.

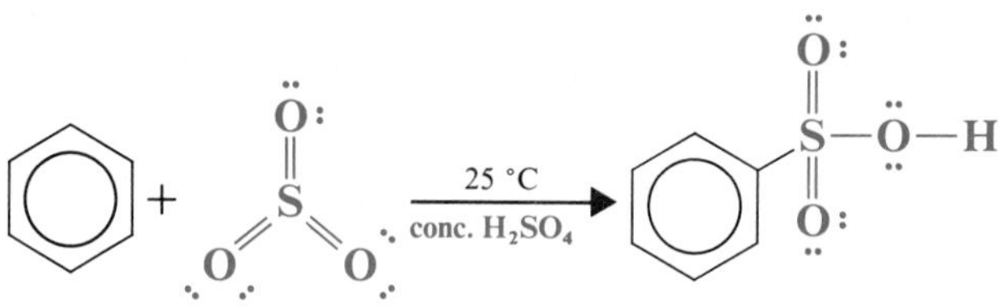

Sulfur
trioxide

(56%)
Benzenesulfonic acid

In either reaction the electrophile appears to be sulfur trioxide. In concentrated sulfuric acid, sulfur trioxide is produced in the following equilibrium in which H_2SO_4 acts as both an acid and a base.

Step 1 $2H_2SO_4 \rightleftharpoons SO_3 + H_3O^+ + HSO_4^-$

When sulfur trioxide reacts with benzene the following steps occur.

Step 2

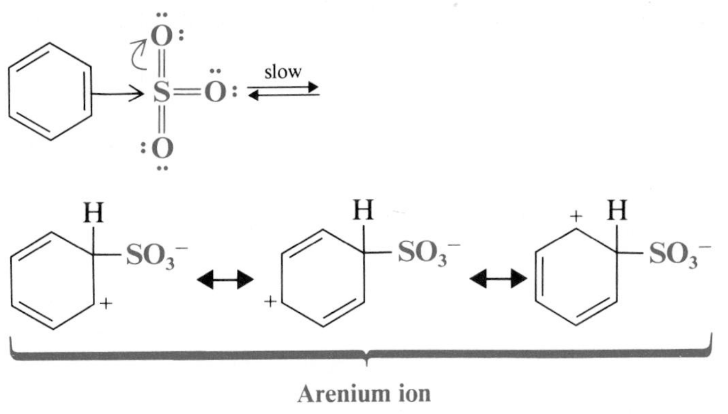

Arenium ion

Step 3

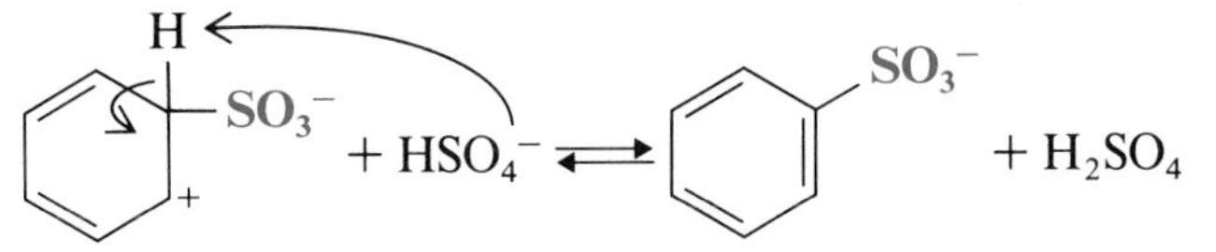

+ HSO$_4^-$ ⇌ + H$_2$SO$_4$ Fast

Step 4

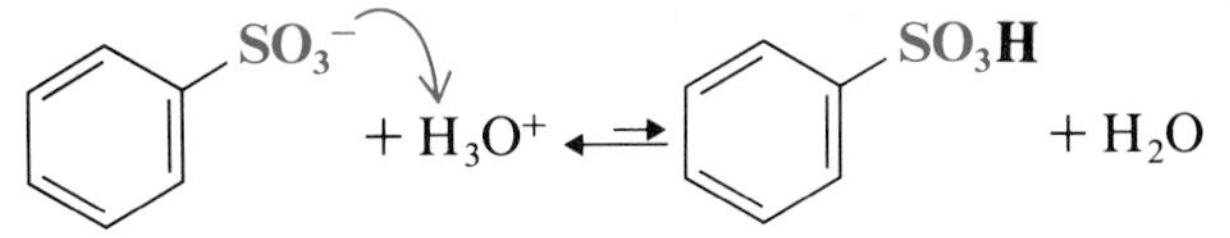

+ H$_3$O$^+$ ⇌ + H$_2$O   Fast

All of the steps are equilibria, including step 1 in which sulfur trioxide is formed from sulfuric acid. This means that the overall reaction is an equilibrium as well. In concentrated sulfuric acid, the overall equilibrium is the sum of steps 1 to 4.

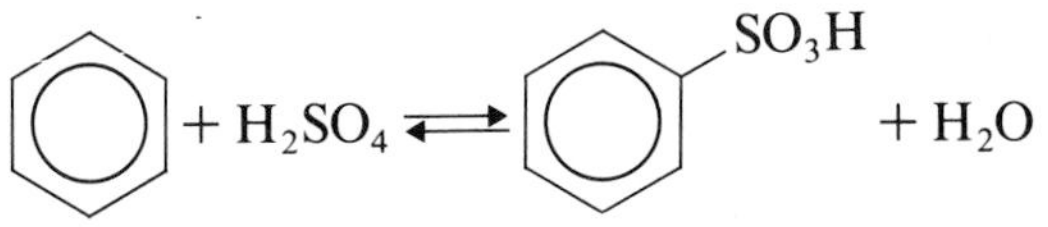

+ H$_2$SO$_4$ ⇌ + H$_2$O

In fuming sulfuric acid, step 1 is unimportant because the dissolved sulfur trioxide reacts directly.

Because all of the steps are equilibria, the position of equilibrium can be influenced by the conditions we employ. If we want to sulfonate benzene we use concentrated sulfuric acid or — better yet — fuming sulfuric acid. Under these conditions the position of equilibrium lies appreciably to the right and we obtain benzenesulfonic acid in good yield.

On the other hand, we may want to remove a sulfonic acid group from a benzene ring. To do this we employ dilute sulfuric acid and usually pass steam through the mixture. Under these conditions — with a high concentration of water — the equilibrium lies appreciably to the left and desulfonation occurs. The equilibrium is shifted even further to the left with volatile aromatic compounds because the aromatic compound distills with the steam.

We shall see later that sulfonation and desulfonation reactions are often used in synthetic work. We may, for example, introduce a sulfonic acid group into a benzene ring to influence the course of some further reaction. Later, we may remove the sulfonic acid group by desulfonation. [Of the five groups that we shall learn to put on a benzene ring in this chapter (—X, —NO$_2$, —SO$_3$H, —R, —COR) the sulfonic acid group is the only one that can be replaced by a hydrogen atom in one step.]

PROBLEM 12.3

In most desulfonation reactions the electrophile is a proton. Other electrophiles may be used, however. (a) Show all steps of the desulfonation reaction that would occur when benzenesulfonic acid is desulfonated with deuterium sulfate (D_2SO_4) dissolved in D_2O. (b) When benzenesulfonic acid reacts with bromine in the presence of ferric bromide, bromobenzene is obtained from the reaction mixture. What is the electrophile in this reaction? Show all steps in the mechanism for this desulfonation.

12.6 FRIEDEL–CRAFTS ALKYLATION

In 1877 a French chemist, Charles Friedel, and his American collaborator, James M. Crafts, discovered new methods for the preparation of alkylbenzenes (ArR) and acylbenzenes (ArCOR). These reactions are now called the Friedel–Crafts alkylation and acylation reactions. We shall study the Friedel–Crafts alkylation reaction here and take up the Friedel–Crafts acylation reaction in Section 12.7.

A general equation for a Friedel–Crafts alkylation reaction is the following:

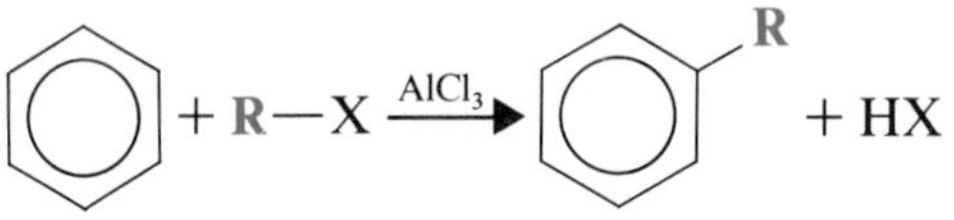

$$\text{C}_6\text{H}_6 + R\!-\!X \xrightarrow{\text{AlCl}_3} \text{C}_6\text{H}_5R + HX$$

The mechanism for the reaction (shown in the following steps—with isopropyl chloride as R—X) starts with the formation of a carbocation (step 1). The carbocation then acts as an electrophile (step 2) and attacks the benzene ring to form an arenium ion. The arenium ion (step 3) then loses a proton to generate isopropylbenzene.

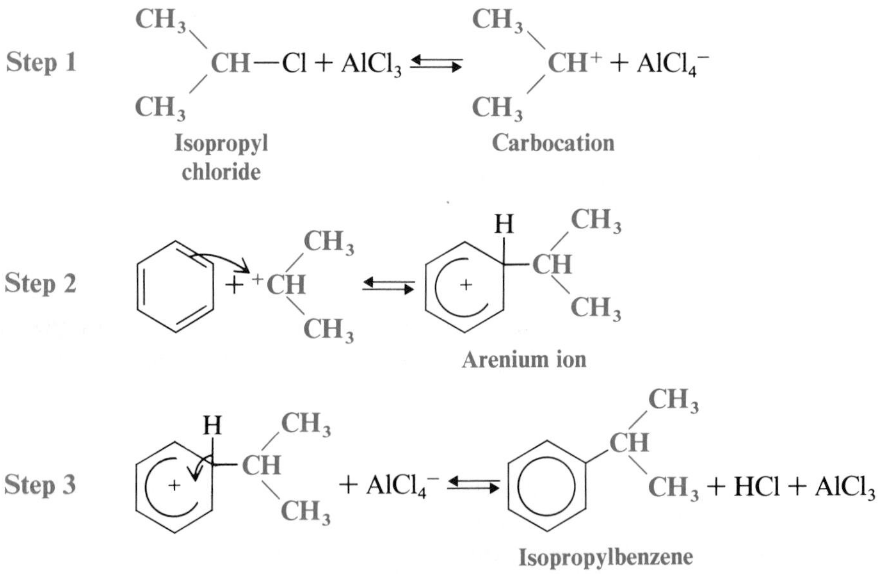

Step 1

Isopropyl chloride → Carbocation

Step 2

Arenium ion

Step 3

Isopropylbenzene

When R—X is a primary halide, a simple carbocation probably does not form. Rather, the aluminum chloride forms a complex with the alkyl halide and this complex acts as the electrophile. The complex is one in which the carbon–halogen bond is nearly broken—and one in which the carbon atom has a considerable positive charge:

$$\overset{\delta+}{R\overset{}{C}H_2}\cdots Cl\!:\!\overset{\delta-}{AlCl_3}$$

Even though this complex is not a simple carbocation, it acts as if it were and it transfers a positive alkyl group to the aromatic ring. As we shall see in Section 12.12B, these complexes are so carbocationlike that they also undergo typical carbocation rearrangements.

Friedel–Crafts alkylations are not restricted to the use of alkyl halides and aluminum chloride. Many other pairs of reagents that form carbocations (or carbocationlike species) may be used as well. These possibilities include the use of a mixture of an alkene and an acid.

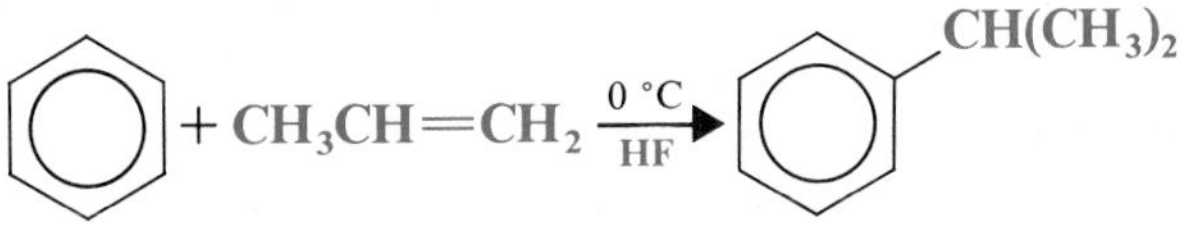

Propene   (84%)
Isopropylbenzene

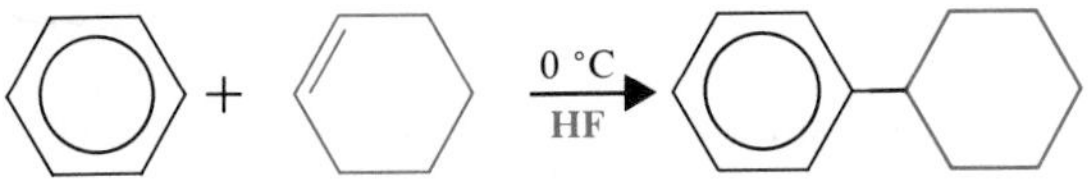

Cyclohexene   (62%)
Cyclohexylbenzene

A mixture of an alcohol and an acid may also be used

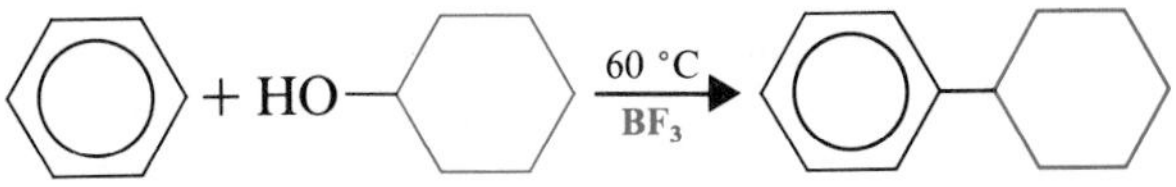

Cyclohexanol   (56%)
Cyclohexylbenzene

There are several important limitations of the Friedel–Crafts alkylation reaction. These are discussed in Section 12.12B.

PROBLEM 12.4

Assume that carbocations are involved and propose step-by-step mechanisms for both of the syntheses of cyclohexylbenzene given previously.

12.7 FRIEDEL–CRAFTS ACYLATION

The $\overset{O}{\overset{\|}{R C}}\!-$ group is called an **acyl group,** and a reaction whereby an acyl group is introduced into a compound is called an **acylation** reaction. Two common acyl groups are the acetyl group and the benzoyl group.

$$
\underset{\substack{\text{Acetyl} \\ \text{group} \\ \text{(ethanoyl group)}}}{CH_3\overset{\displaystyle O}{\overset{\|}{C}}-}
\qquad\qquad
\underset{\substack{\text{Benzoyl} \\ \text{group}}}{\text{⬡}-\overset{\displaystyle O}{\overset{\|}{C}}-}
$$

The Friedel–Crafts acylation reaction is an effective means of introducing an acyl group into an aromatic ring. The reaction is often carried out by treating the aromatic compound with an acyl halide. Unless the aromatic compound is one that is highly reactive, the reaction requires the addition of at least one equivalent of a Lewis acid (such as $AlCl_3$) as well. The product of the reaction is an aryl ketone.

$$
\text{⬡} + CH_3\overset{\displaystyle O}{\overset{\|}{C}}-Cl \xrightarrow[\substack{\text{excess} \\ \text{benzene} \\ 80\ °C}]{AlCl_3} \underset{\substack{(97\%) \\ \text{Acetophenone} \\ \text{(methyl phenyl ketone)}}}{\text{⬡}-\overset{\displaystyle O}{\overset{\|}{C}}CH_3} + HCl
$$

$$
\underset{\substack{\text{Acetyl} \\ \text{chloride}}}{}
$$

Acyl chlorides, also called **acid chlorides,** are easily prepared by treating carboxylic acids with thionyl chloride ($SOCl_2$) or phosphorus pentachloride (PCl_5).

$$
\underset{\substack{\text{Acetic} \\ \text{acid}}}{CH_3\overset{\displaystyle O}{\overset{\|}{C}}OH} + \underset{\substack{\text{Thionyl} \\ \text{chloride}}}{SOCl_2} \xrightarrow{80\ °C} \underset{\substack{(80-90\%) \\ \text{Acetyl} \\ \text{chloride}}}{CH_3\overset{\displaystyle O}{\overset{\|}{C}}Cl} + SO_2 + HCl
$$

$$
\underset{\substack{\text{Benzoic} \\ \text{acid}}}{\text{⬡}-\overset{\displaystyle O}{\overset{\|}{C}}OH} + \underset{\substack{\text{Phosphorus} \\ \text{pentachloride}}}{PCl_5} \longrightarrow \underset{\substack{(90\%) \\ \text{Benzoyl} \\ \text{chloride}}}{\text{⬡}-\overset{\displaystyle O}{\overset{\|}{C}}Cl} + POCl_3 + HCl
$$

Friedel–Crafts acylations can also be carried out using carboxylic acid anhydrides. For example:

$$
\text{⬡} + \underset{\substack{\text{Acetic anhydride} \\ \text{(a carboxylic acid} \\ \text{anhydride)}}}{\begin{array}{c} CH_3\overset{\displaystyle O}{\overset{\|}{C}} \\ {}^{\diagdown} \\ {}_{\diagup}O \\ CH_3\underset{\displaystyle O}{\underset{\|}{C}} \end{array}} \xrightarrow[\substack{\text{excess benzene} \\ 80\ °C}]{AlCl_3} \underset{\substack{(82-85\%) \\ \text{Acetophenone}}}{\text{⬡}-\overset{\displaystyle O}{\overset{\|}{C}}CH_3} + CH_3\overset{\displaystyle O}{\overset{\|}{C}}OH
$$

In most Friedel–Crafts acylations the electrophile appears to be an **acylium ion** formed from an acyl halide in the following way:

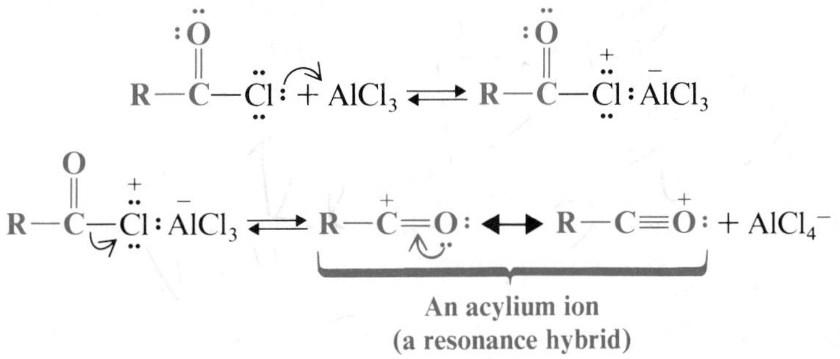

$$R-\overset{\overset{\displaystyle :\ddot{O}}{\|}}{C}-\overset{..}{\underset{..}{Cl}}: + AlCl_3 \rightleftharpoons R-\overset{\overset{\displaystyle :\ddot{O}}{\|}}{C}-\overset{+}{\underset{..}{\overset{..}{Cl}}}: \overset{-}{AlCl_3}$$

$$R-\overset{\overset{\displaystyle O}{\|}}{C}-\overset{+}{\underset{..}{\overset{..}{Cl}}}: \overset{-}{AlCl_3} \rightleftharpoons R-\overset{+}{C}=\ddot{O}: \longleftrightarrow R-C\equiv\overset{+}{O}: + AlCl_4^-$$

An acylium ion
(a resonance hybrid)

PROBLEM 12.5

Show how an acylium ion could be formed from an acid anhydride.

The remaining steps in the Friedel–Crafts acylation of benzene are the following:

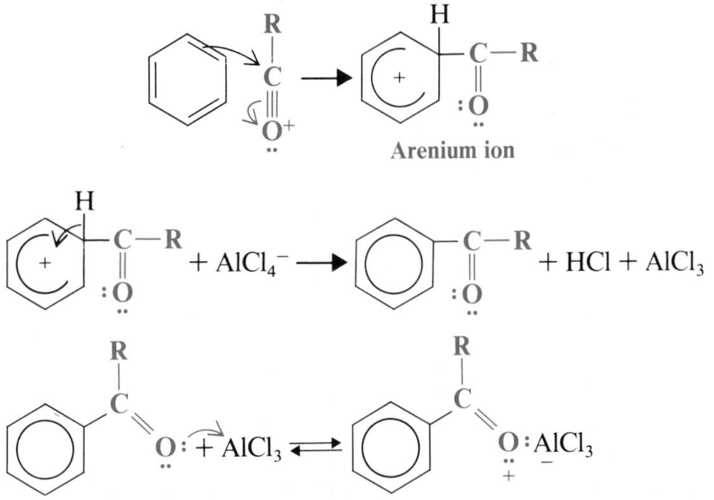

Arenium ion

In the last step aluminum chloride (a Lewis acid) forms a complex with the ketone (a Lewis base). After the reaction is over, treating the complex with water liberates the ketone.

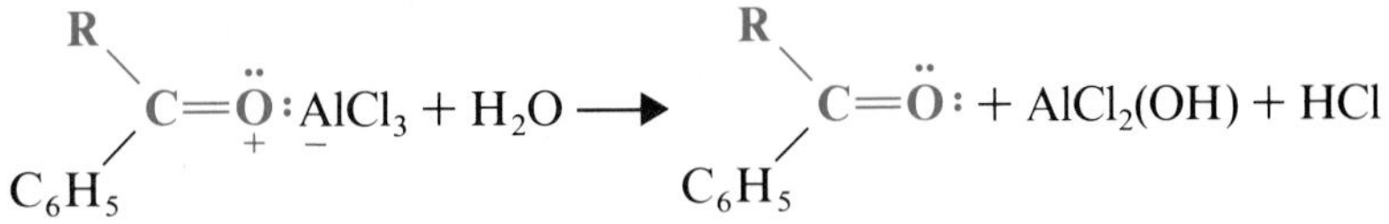

$$\underset{C_6H_5}{\overset{R}{\diagdown}}C=\ddot{O}:\underset{+}{AlCl_3} + H_2O \longrightarrow \underset{C_6H_5}{\overset{R}{\diagdown}}C=\ddot{O}: + AlCl_2(OH) + HCl$$

Several important synthetic applications of the Friedel–Crafts acylation reaction are given in Section 12.12C.

12.8 EFFECT OF SUBSTITUENTS: REACTIVITY AND ORIENTATION

When substituted benzenes undergo electrophilic attack, groups already on the ring affect both the rate of the reaction and the site of attack. We say, therefore, that substituent groups affect both **reactivity** and **orientation** in electrophilic aromatic substitutions.

We can divide substituent groups into two classes according to their influence on the reactivity of the ring. Those that cause the ring to be more reactive than benzene itself we call **activating groups.** Those that cause the ring to be less reactive than benzene we call **deactivating groups.**

We also find that we can divide substituent groups into two classes according to the way they influence the orientation of attack by the incoming electrophile. Substituents in one class tend to bring about electrophilic substitution primarily at the positions *ortho* and *para* to themselves. We call these groups *ortho - para directors* because they tend to *direct* the incoming group into the ortho and para positions. Substituents in the second category tend to direct the incoming electrophile to the *meta* position. We call these groups *meta directors.*

Several examples will illustrate more clearly what we mean by these terms.

12.8A Activating Groups — Ortho–Para Directors

The methyl group is an **activating** group and **an ortho - para director.** Toluene reacts considerably faster than benzene in all electrophilic substitutions.

An activating group

**More reactive than benzene
toward electrophilic substitution**

We observe the greater reactivity of toluene in several ways. We find, for example, that with toluene, milder conditions — lower temperatures and lower concentrations of the electrophile — can be used in electrophilic substitutions than with benzene. We also find that under the same conditions, toluene reacts faster than benzene. In nitration, for example, toluene reacts 25 times as fast as benzene.

We find, moreover, that when toluene undergoes electrophilic substitution, most of the substitution takes place at its ortho and para positions. When we nitrate toluene with nitric and sulfuric acids, we get mononitrotoluenes in the following relative proportions.

Ortho	Para	Meta
(59%)	(37%)	(4%)
o-Nitrotoluene	*p*-Nitrotoluene	*m*-Nitrotoluene

Of the mononitrotoluene obtained from the reaction, 96% (59% + 37%) has the nitro group in an ortho or para position. Only 4% has the nitro group in a meta position.

PROBLEM 12.6

What percentage of each nitrotoluene would you expect if substitution were to take place on a purely *statistical* basis?

Predominant substitution at the ortho and para positions of toluene is not restricted to nitration reactions. The same behavior is observed in halogenation, sulfonation, and so forth.

All alkyl groups are activating groups, and they are all also ortho–para directors. The methoxyl group, CH_3O—, and the acetamido group, CH_3CONH—, are strong activating groups and both are ortho–para directors.

The hydroxyl group and the amino group are very powerful activating groups and are also powerful ortho–para directors. Phenol and aniline react with bromine in water (no catalyst is required) to produce products in which both of the ortho positions and the para position are substituted. These tribromo products are obtained in nearly quantitative yield.

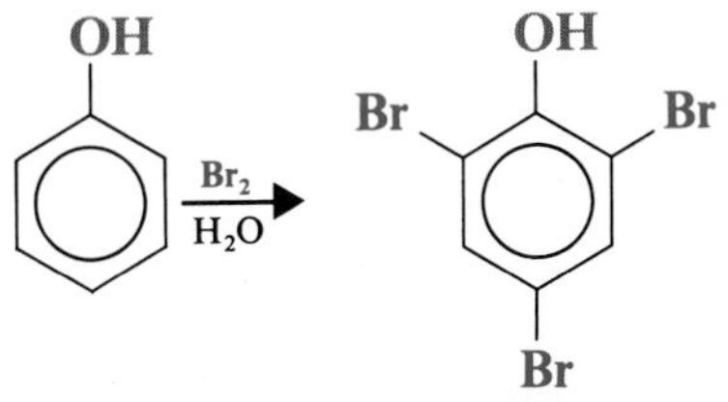

2,4,6-Tribromophenol
(~100%)

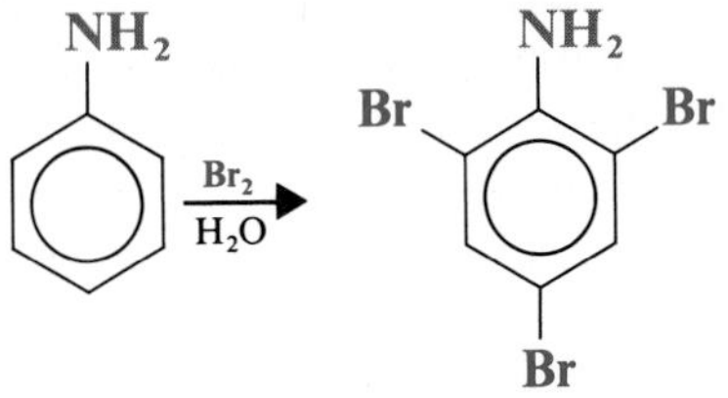

2,4,6-Tribromoaniline
(~100%)

12.8B Deactivating Groups — Meta Directors

The nitro group is a very strong **deactivating group.** Nitrobenzene undergoes nitration at a rate only 10^{-4} times that of benzene. The nitro group is a meta director. When nitrobenzene is nitrated with nitric and sulfuric acids, 93% of the substitution occurs at the meta position.

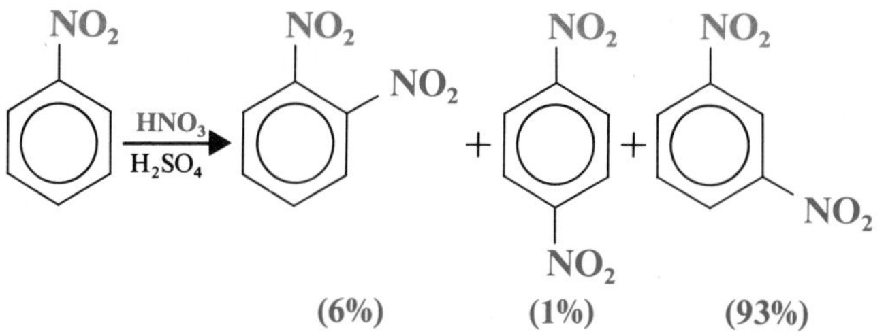

(6%)   (1%)   (93%)

The carboxyl group ($-CO_2H$), the sulfo group ($-SO_3H$), and the trifluoro-methyl group ($-CF_3$) are also deactivating groups; they are also meta directors.

12.8C Halo Substituents: Deactivating Ortho–Para Directors

The chloro and bromo groups are weak deactivating groups. Chlorobenzene and bromobenzene undergo nitration at rates that are, respectively, 33 and 30 times slower than for benzene. The chloro and bromo groups are ortho–para directors, however. The relative percentages of monosubstituted products that are obtained when chlorobenzene is chlorinated, brominated, nitrated, and sulfonated are shown in Table 12.1.

TABLE 12.1 Electrophilic substitutions of chlorobenzene

REACTION	ORTHO PRODUCT (%)	PARA PRODUCT (%)	TOTAL ORTHO AND PARA (%)	META PRODUCT (%)
Chlorination	39	55	94	6
Bromination	11	87	98	2
Nitration	30	70	100	—
Sulfonation	—	100	100	—

Similar results are obtained from electrophilic substitutions of bromobenzene.

12.8D Classification of Substituents

Studies like the ones that we have presented in this section have been done for a number of other substituted benzenes. The effects of these substituents on reactivity and orientation are included in Table 12.2.

PROBLEM 12.7

What would be the major monochloro product (or products) formed when each of the following compounds reacts with chlorine in the presence of ferric chloride?

(a) Ethylbenzene, $C_6H_5CH_2CH_3$

(b) (Trifluoromethyl)benzene, $C_6H_5CF_3$

(c) Trimethylphenylammonium chloride, $C_6H_5\overset{+}{N}(CH_3)_3Cl^-$

(d) Methyl benzoate, $C_6H_5CO_2CH_3$

TABLE 12.2 Effect of substituents on electrophilic aromatic substitution

ORTHO–PARA DIRECTORS	META DIRECTORS
Strongly Activating	**Moderately Deactivating**
$—\ddot{N}H_2, —\ddot{N}HR, —\ddot{N}R_2$	$—C{\equiv}N$
$—\ddot{O}H, —\ddot{O}{:}^-$	$—SO_3H$
	$—CO_2H, —CO_2R$
Moderately Activating	$—CHO, —COR$
$—\ddot{N}HCOCH_3, —\ddot{N}HCOR$	**Strongly Deactivating**
$—\ddot{O}CH_3, —\ddot{O}R$	$—NO_2$
	$—NR_3^+$
Weakly Activating	$—CF_3, —CCl_3$
$—CH_3, —C_2H_5, —R$	
$—C_6H_5$	
Weakly Deactivating	
$—\ddot{\underset{..}{F}}{:}, —\ddot{\underset{..}{C}l}{:}, —\ddot{\underset{..}{B}r}{:}, —\ddot{\underset{..}{I}}{:}$	

12.9 THEORY OF ELECTROPHILIC AROMATIC SUBSTITUTION

12.9A Reactivity: The Effect of Electron-Releasing and Electron-Withdrawing Groups

We have now seen that certain groups *activate* the benzene ring toward electrophilic substitution, while other groups *deactivate* the ring. When we say that a group activates the ring, what we mean, of course, is that the group increases the relative rate of the reaction. We mean that an aromatic compound with an activating group reacts faster in electrophilic substitutions than benzene. When we say that a group deactivates the ring, we mean that an aromatic compound with a deactivating group reacts slower than benzene.

We have also seen that we can account for relative reaction rates by examining the transition state for the rate-determining steps. We know that any factor that increases the energy of the transition state relative to that of the reactants decreases the relative rate of the reaction. It does this because it increases the energy of activation of the reaction. In the same way, any factor that decreases the energy of the transition state relative to that of the reactants lowers the energy of activation and increases the relative rate of the reaction.

The rate-determining step in electrophilic substitutions of substituted benzenes is the step that results in the formation of the arenium ion. We can write the formula for a substituted benzene in a generalized way if we use the letter **S** to represent any ring substituent including hydrogen. (If **S** is hydrogen the compound is benzene itself.) We can also write the structure for the arenium ion in the way shown here. By this formula we mean that **S** can be in any position — ortho, meta, or para — relative to the electrophile, **E**. Using these conventions, then, we are able to write the rate-determining step for electrophilic aromatic substitution in the following general way.

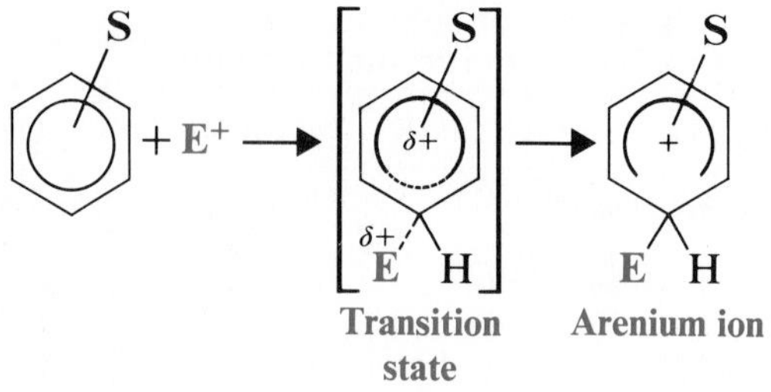

When we examine this step for a large number of reactions, we find that the relative rates of the reactions depend on whether **S withdraws** or **releases** electrons. If **S** is an electron-releasing group (relative to hydrogen), the reaction occurs faster than the corresponding reaction of benzene. If **S** is an electron-withdrawing group, the reaction is slower than that of benzene.

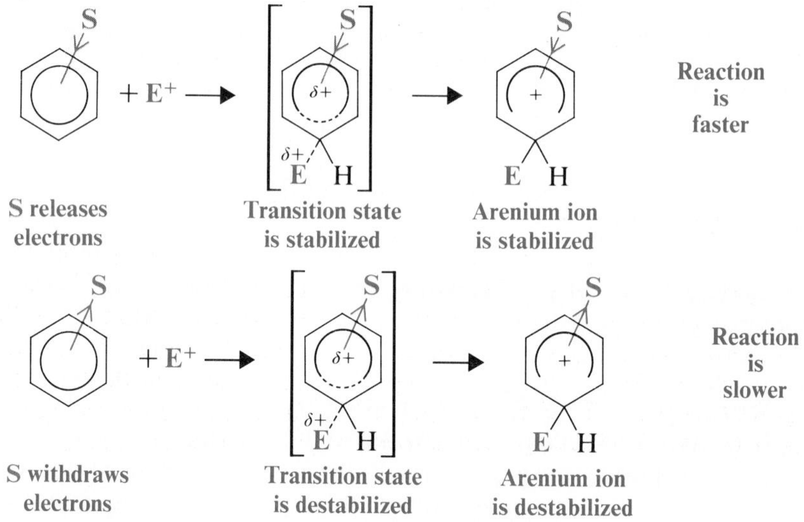

It appears, then, that the substituent, **S**, must affect the stability of the transition state relative to that of the reactants. Electron-releasing groups apparently make the transition state more stable, while electron-withdrawing groups make it less stable. That this is so is entirely reasonable, because the transition state resembles the arenium ion, and the arenium ion is a delocalized *carbocation.*

This effect illustrates another application of the Hammond-Leffler Postulate (Section 5.13A). The arenium ion is a high-energy intermediate, and the step that leads to it is a *highly endothermic step.* Thus, according to the Hammond-Leffler Postulate there should be a strong resemblance between the arenium ion itself and the transition state leading to it.

Since the arenium ion is positively charged, we would expect an electron-releasing group to stabilize it *and the transition state leading to the arenium ion,* for the transition state is a developing delocalized carbocation. We can make the same kind of arguments about the effect of electron-withdrawing groups. An electron-withdrawing group should make the arenium ion *less stable* and in a corresponding way it should make the transition state leading to the arenium ion *less stable.*

Figure 12.2 shows how the electron-withdrawing and electron-releasing abilities of substituents affect the relative energies of activation of electrophilic aromatic substitution reactions.

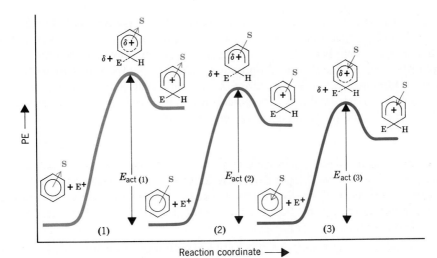

FIGURE 12.2 Energy profiles for the formation of the arenium ion in three electrophilic aromatic substitution reactions. In (1), **S** is an electron-withdrawing group. In (2) **S** = H. In (3) **S** is an electron-releasing group. $E_{act(1)} > E_{act(2)} > E_{act(3)}$.

12.9B Inductive and Resonance Effects: Theory of Orientation

We can account for the electron-withdrawing and electron-releasing properties of groups on the basis of two factors: *inductive effects and resonance effects.* We shall also see that these two factors determine orientation in aromatic substitution reactions.

The **inductive effect** of a substituent **S** arises from the electrostatic interaction of the polarized S-ring bond with the developing positive charge in the ring as it is attacked by an electrophile. If, for example, **S** is a more electronegative atom (or group) than carbon, then the ring will be at the positive end of the dipole:

$$\overset{\delta-}{S} \overset{\delta+}{\longleftarrow} \bigcirc \qquad \text{(e.g., S = F, Cl, Br)}$$

Attack by an electrophile will be retarded because this will lead to an additional full positive charge on the ring. The halogens are all more electronegative than carbon and exert an electron-withdrawing inductive effect. Other groups have an electron-withdrawing inductive effect because the atom directly attached to the ring bears a full or partial positive charge. Examples are the following:

$$\overset{+}{\rightarrow}NR_3 \text{ (R = alkyl or H)} \qquad \rightarrow \overset{X^{\delta-}}{\underset{X^{\delta-}}{\overset{\uparrow}{C}}} \overset{\delta+}{\rightarrow} X^{\delta-} \qquad \rightarrow \overset{O}{\underset{O_-}{N^+}} \qquad \rightarrow \overset{O^-}{\underset{O}{\overset{|}{S^+}}} - OH$$

$$\rightarrow \overset{\overset{..}{O}:}{\overset{\|}{C}} - G \longleftrightarrow \rightarrow \overset{:\overset{..}{O}:^-}{\overset{|}{C^+}} - G \qquad \text{(G = H, R, OH, or OR)}$$

Electron-withdrawing groups with a full or partial charge on the atom attached to the ring

The **resonance effect** of a substituent S refers to the possibility that the presence of S may increase or decrease the resonance stabilization of the intermediate arenium ion. The S substituent may, for example, cause one of the three contributors to the resonance hybrid for the arenium ion to be better or worse than the case when S is hydrogen. Moreover, when S is an atom bearing one or more nonbonding electron pairs, it may lend extra stability to the arenium ion by providing a *fourth* resonance contributor in which the positive charge resides on S.

This electron-donating resonance effect applies with increasing strength in the following order:

| Most electron donating | $-\ddot{N}H_2$, $-\ddot{N}R_2$ > $-\ddot{O}H$, $-\ddot{O}R$ > $-\ddot{X}$: | Least electron donating |

When X = F, this order is related to the electronegativity of the atoms with the nonbonding pair. The more electronegative the atom is (fluorine is most electronegative), the less able it is to accept the positive charge. When X = Cl, Br, or I, the relatively poor electron-donating ability of the halogens by resonance is understandable on a different basis: These atoms (Cl, Br, and I) are all larger than carbon and, therefore, the orbitals that contain the nonbonding pairs are further from the nucleus and do not overlap well with the $2p$ orbital of carbon. (This is a general phenomenon; Resonance effects are not transmitted well between atoms of different rows on the Periodic Table.)

12.9C Meta-Directing Groups

All meta-directing groups have either a partial positive charge or a full positive charge on the atom directly attached to the ring. As a typical example let us consider the trifluoromethyl group.

The trifluoromethyl group, because of the three highly electronegative fluorine atoms, is strongly electron withdrawing. It is a strong deactivating group and a powerful meta director in electrophilic aromatic substitution reactions. We can account for both of these characteristics of the trifluoromethyl group in the following way.

The trifluoromethyl group affects reactivity by causing the transition state leading to the arenium ion to be highly unstable. It does this by withdrawing electrons from the developing carbocation thus increasing the positive charge in the ring.

(Trifluoromethyl)benzene   Transition state   Arenium ion

We can understand how the trifluoromethyl group affects *orientation* in electrophilic aromatic substitution if we examine the resonance structures for the arenium ion that would be formed when an electrophile attacks the ortho, meta, and para positions of (trifluoromethyl)benzene.

Ortho attack

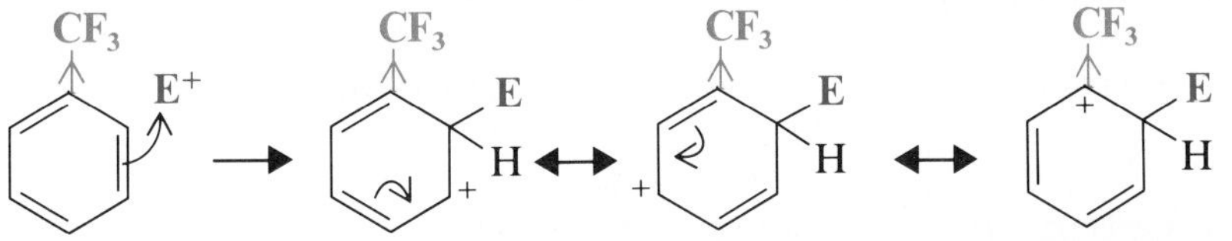

Highly unstable
contributor

Meta attack

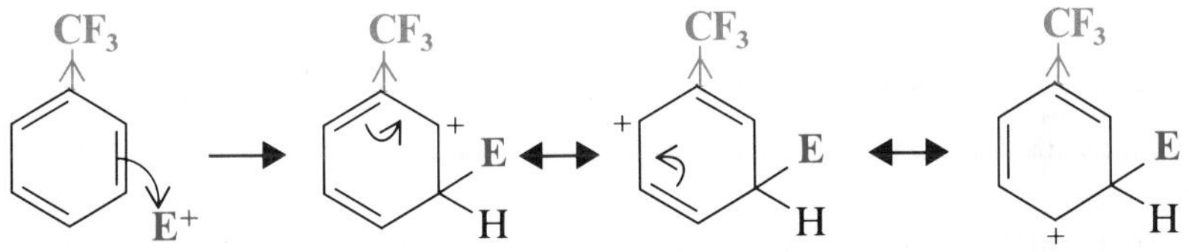

Para attack

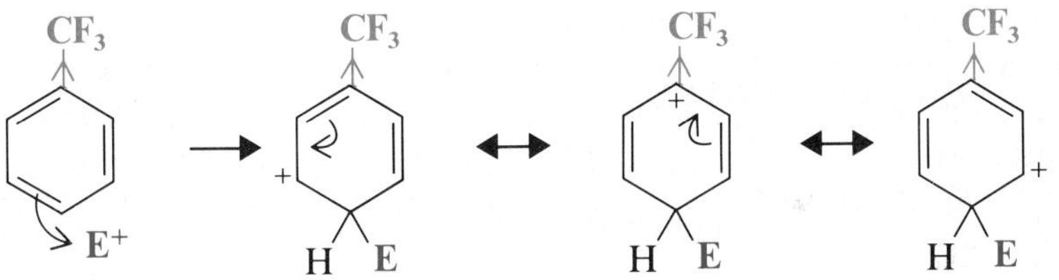

Highly unstable
contributor

We see in the resonance structures for the arenium ion arising from ortho and para attack that *one contributing structure is highly unstable relative to all the others because the positive charge is located on the ring carbon that bears the electron-withdrawing group.* We see *no* such highly unstable resonance structure in the arenium ion arising from meta attack. This means that the arenium ion formed by meta attack should be the most stable of the three. By the usual reasoning we would also expect the transition state leading to the meta-substituted arenium ion to be the most stable and, therefore, that meta attack would be favored. This is exactly what we find experimentally. The trifluoromethyl group is a powerful meta director.

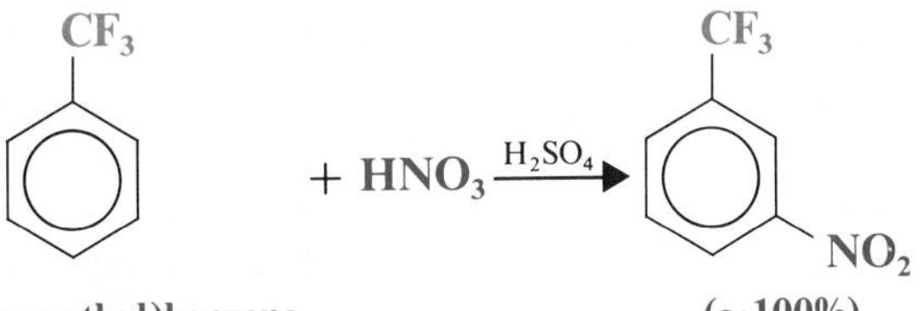

(Trifluoromethyl)benzene
(benzotrifluoride)

$(\sim 100\%)$

Bear in mind, however, that meta substitution is favored only in the sense that *it is the least unfavorable of three unfavorable pathways.* The energy of activation for substitution at the meta position of (trifluoromethyl)benzene is less than that for attack at an ortho or para position, but it is still far greater than that for an attack on benzene. Substitution occurs at the meta position of (trifluoromethyl)benzene faster than substitution takes place at the ortho and para positions, but it occurs much more slowly than it does with benzene.

The nitro group, the carboxyl group, and other meta-directing groups are all powerful electron-withdrawing groups and all act in a similar way.

12.9D Ortho–Para-Directing Groups

Except for the alkyl and phenyl substituents, all of the ortho–para-directing groups in Table 12.2 are of the following general type:

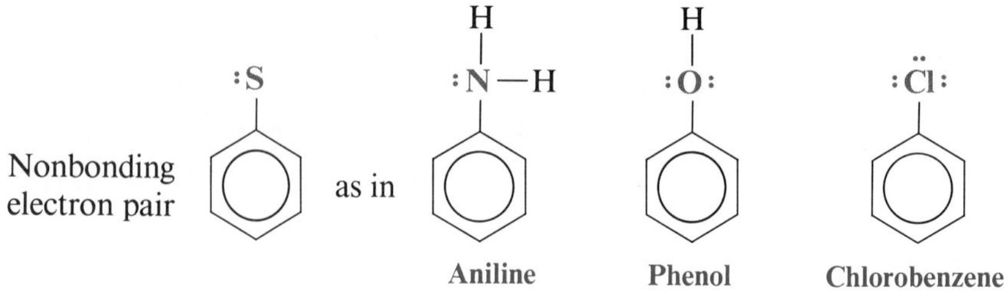

All of these ortho–para directors have at least one pair of nonbonding electrons on the atom adjacent to the benzene ring.

This structural feature — an unshared electron pair on the atom adjacent to the ring — determines the orientation and influences reactivity in electrophilic substitution reactions.

The *directive effect* of these groups with an unshared pair is predominantly caused by an electron-releasing resonance effect. The resonance effect, moreover, operates primarily in the arenium ion and, consequently, in the transition state leading to it.

Except for the halogens, the primary effect on reactivity of these groups is also caused by an electron-releasing resonance effect. And, again, this effect operates primarily in the transition state leading to the arenium ion.

In order to understand these resonance effects let us begin by recalling the effect of the amino group on electrophilic aromatic substitution reactions. The amino group is not only a powerful activating group, it is also a powerful ortho–para director. We saw earlier (Section 12.8A) that aniline reacts with bromine at room temperature and in the absence of a catalyst to yield a product in which both ortho positions and the para position are substituted.

The inductive effect of the amino group makes it slightly electron withdrawing. Nitrogen, as we know, is more electronegative than carbon. The difference between the electronegativities of nitrogen and carbon in aniline is not large, however, because the carbon of the benzene ring is sp^2 hybridized and thus is somewhat more electronegative than it would be if it were sp^3 hybridized.

The resonance effect of the amino group is far more important than its inductive effect in electrophilic aromatic substitution, and this resonance effect makes the

amino group electron releasing. We can understand this effect if we write the resonance structures for the arenium ions that would arise from ortho, meta, and para
attack on aniline.

Ortho attack

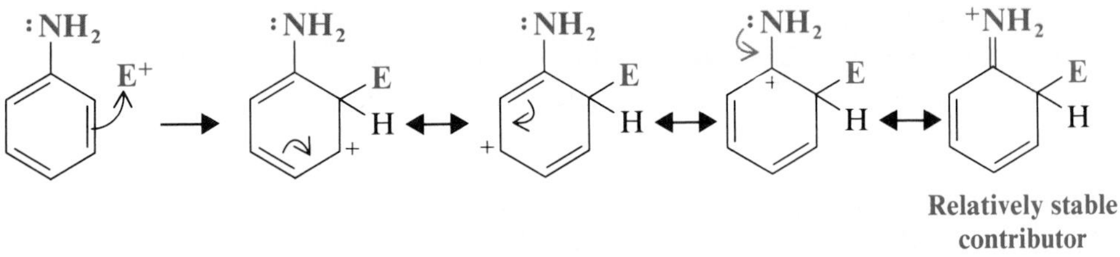

Relatively stable
contributor

Meta attack

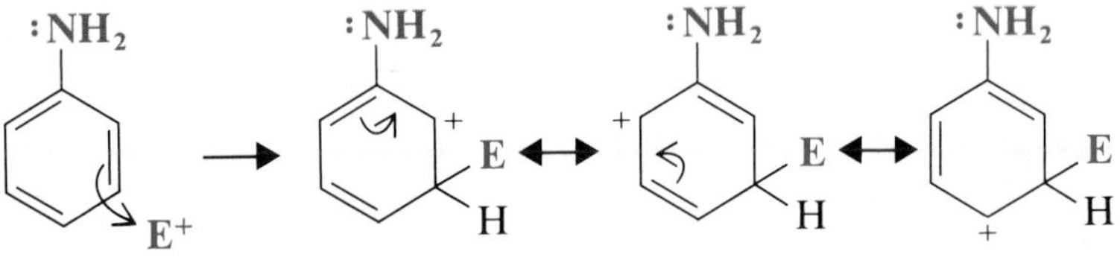

Para attack

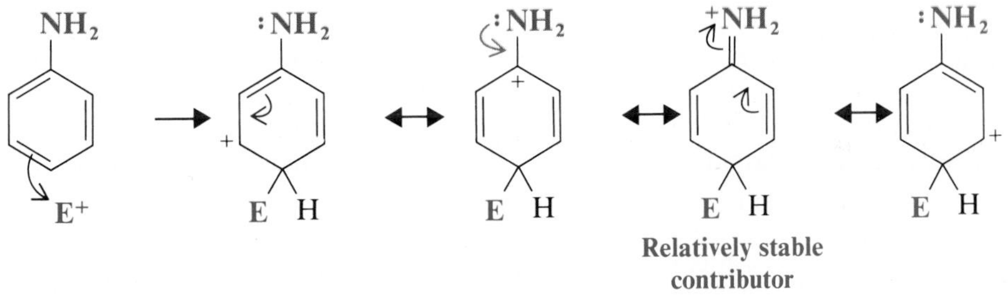

Relatively stable
contributor

We see that four reasonable resonance structures can be written for the arenium
ions resulting from ortho and para attack, whereas only three can be written for the
arenium ion that results from meta attack. This, in itself, suggests that the ortho- and
para-substituted arenium ions should be more stable. Of greater importance, however, are the relatively stable structures that contribute to the hybrid for the ortho-
and para-substituted arenium ions. In these structures, nonbonding pairs of electrons
from nitrogen form an extra bond to the carbon of the ring. This extra bond—and
the fact that every atom in each of these structures has a complete outer octet of
electrons—makes these structures the most stable of all of the contributors. Because
these structures are unusually stable, they make a large—*and stabilizing*—
contribution to the hybrid. This means, of course, that the ortho- and para-substituted arenium ions themselves are considerably more stable than the arenium ion
that results from the meta attack. The transition states leading to the ortho- and
para-substituted arenium ions occur at unusually low potential energies. As a result,
electrophiles react at the ortho and para positions very rapidly.

PROBLEM 12.8

(a) Write resonance structures for the arenium ions that would result from electrophilic attack on the ortho, meta, and para positions of phenol. (b) Can you account for the fact that phenol is highly susceptible to electrophilic attack? (c) Can you account for the fact that the hydroxyl group is an ortho and para director? (d) Would you expect the phenoxide ion, $C_6H_5O^-$, to be more or less reactive than phenol in electrophilic substitution? (e) Explain.

PROBLEM 12.9

(a) Ignore resonance structures involving electrons of the ring and write *one* other resonance structure for acetanilide. (Your structure will contain $+$ and $-$

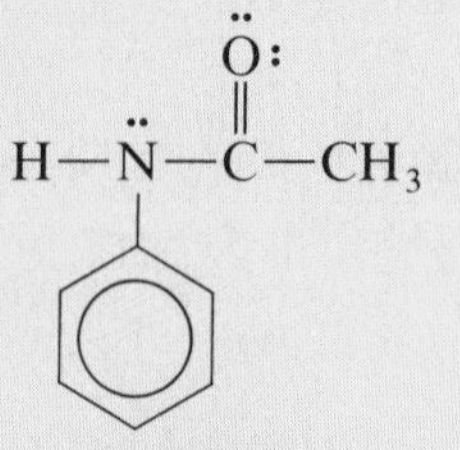

Acetanilide 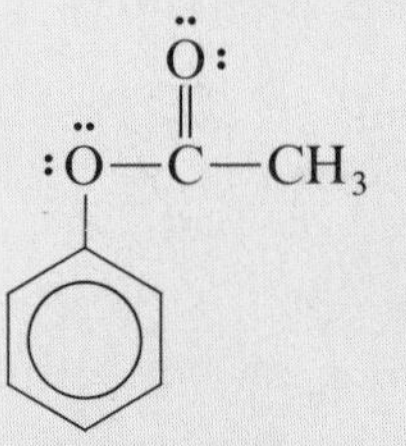   Phenyl acetate

charges.) (b) Acetanilide is less reactive toward electrophilic substitution than aniline. How can you explain this on the basis of the resonance structure you have just written? (c) Acetanilide, however, is much more reactive than benzene and the acetamido group, CH_3CONH—, is an ortho–para director. Can you account for these facts in terms of resonance structures that involve the ring? (d) Would you expect phenyl acetate to be *more* or *less* reactive than phenol? Explain. (e) What kind of directional influence would you expect the acetoxy group, CH_3C—O—, to show? (f) Would you expect phenyl acetate to be *more* or *less* reactive in electrophilic substitution than benzene? Explain.

The directive and reactivity effects of halo substituents may, at first, seem to be contradictory. *The halo groups are the only ortho–para directors that are deactivating groups.* All other deactivating groups (Table 12.2) are meta directors. We can readily account for the behavior of halo substituents, however, if we assume that their electron-withdrawing inductive effect influences reactivity and their electron-donating resonance effect governs orientation.

Let us apply these assumptions specifically to chlorobenzene. The chloro group is highly electronegative. Thus, we would expect a chloro group to withdraw electrons from the benzene ring and thereby deactivate it.

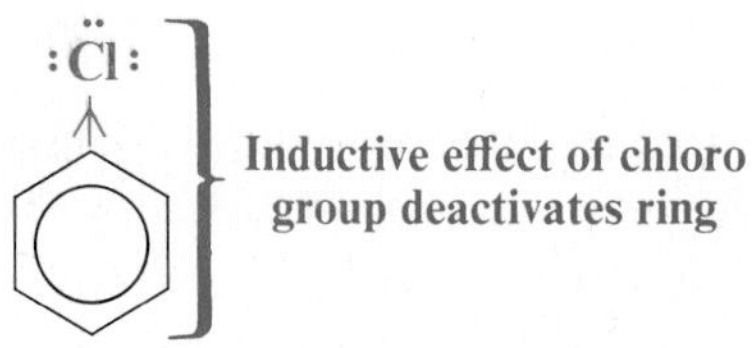

Inductive effect of chloro group deactivates ring

On the other hand, when electrophilic attack does take place, the chloro group stabilizes the arenium ions resulting from ortho and para attack. The chloro group does this in the same way as amino groups and hydroxyl groups do — *by donating an unshared pair of electrons.* These electrons give rise to relatively stable resonance structures contributing to the hybrids for the ortho- and para-substituted arenium ions (Section 12.9E).

Ortho attack

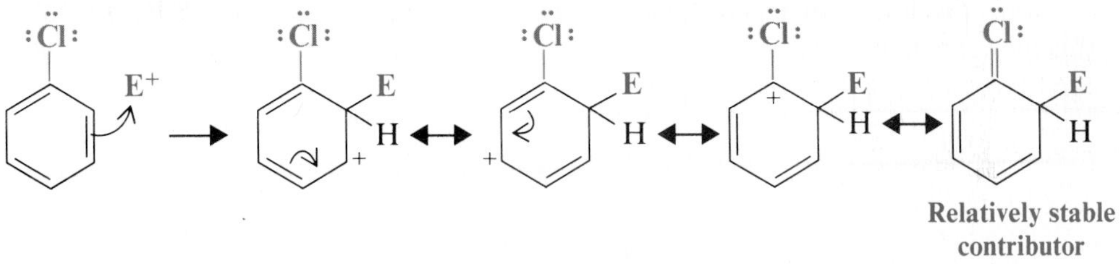

Relatively stable contributor

Meta attack

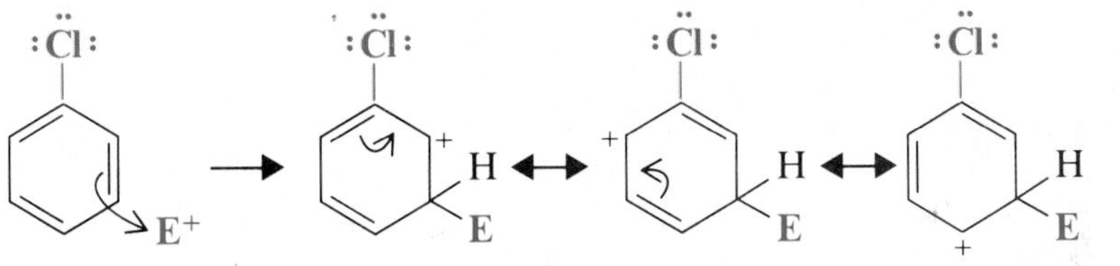

Para attack

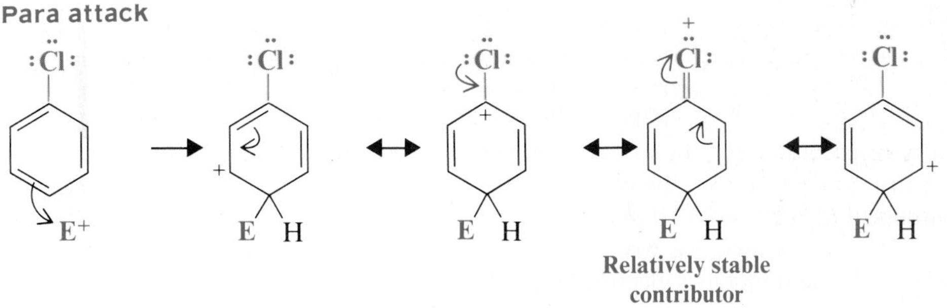

Relatively stable contributor

What we have said about chlorobenzene is, of course, true of bromobenzene.

We can summarize the inductive and resonance effects of halo substituents in the following way. Through their electron-withdrawing inductive effect halo groups make the ring more positive than that of benzene. This causes the energy of activation for any electrophilic aromatic substitution reaction to be greater than that for benzene, and, therefore, halo groups are deactivating. Through their electron-donating resonance effect, however, halo substituents cause the energies of activation leading to ortho and para substitution to be lower than the energy of activation leading to meta substitution. This makes halo substituents ortho–para directors.

PROBLEM 12.10

Chloroethene adds hydrogen chloride more slowly than ethene and the product is 1,1-dichloroethane. How can you explain this using resonance and inductive effects?

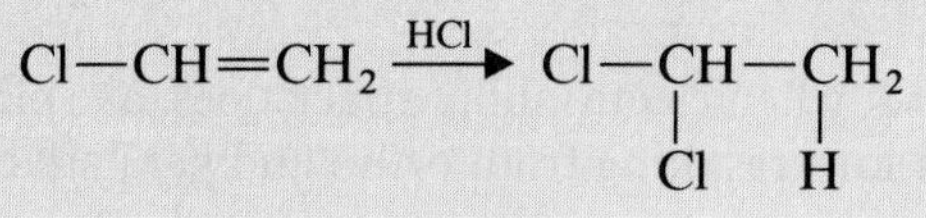

$$Cl-CH=CH_2 \xrightarrow{\text{HCl}} Cl-\underset{\underset{Cl}{|}}{CH}-\underset{\underset{H}{|}}{CH_2}$$

12.9E Ortho–Para Direction and Reactivity of Alkylbenzenes

Alkyl groups can be much better electron-releasing groups than hydrogen. Because of this they can activate a benzene ring toward electrophilic substitution by stabilizing the transition state leading to the arenium ion:

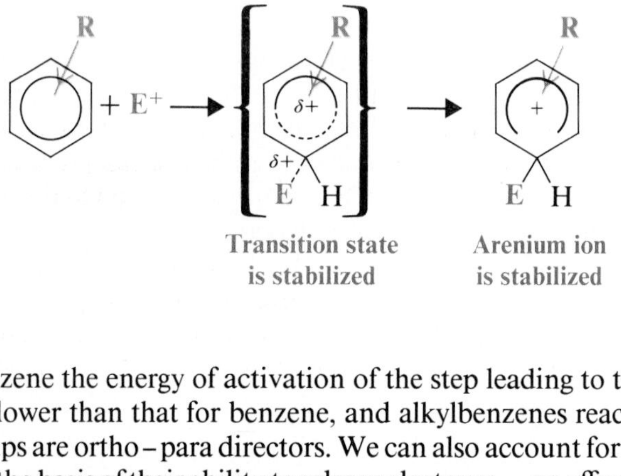

Transition state   Arenium ion
is stabilized   is stabilized

For an alkylbenzene the energy of activation of the step leading to the arenium ion (just shown) is lower than that for benzene, and alkylbenzenes react faster.

Alkyl groups are ortho–para directors. We can also account for this property of alkyl groups on the basis of their ability to release electrons—an effect that is particularly important when the alkyl group is attached directly to a carbon that bears a positive charge. (Recall the ability of alkyl groups to stabilize carbocations that we discussed in Section 5.11B and in Fig. 5.11.)

If, for example, we write resonance structures for the arenium ions formed when toluene undergoes electrophilic substitution, we get the following results:

Ortho attack

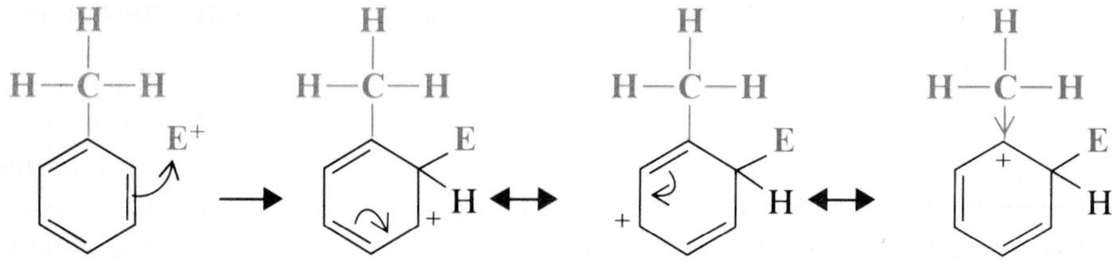

Relatively
stable contributor

Meta attack

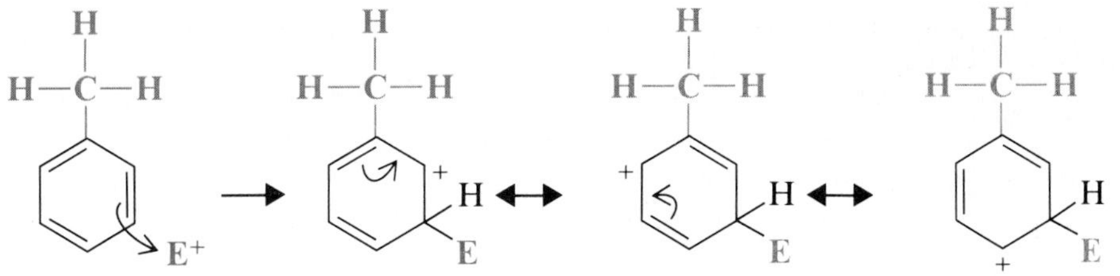

Para attack

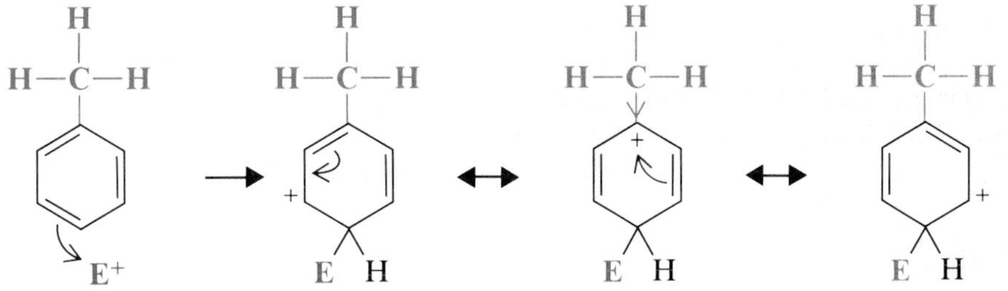

Relatively
stable contributor

In ortho attack and para attack we find that we can write resonance structures in which the methyl group is directly attached to a positively charged carbon of the ring. These structures are more *stable* relative to any of the others because in them the stabilizing influence of the methyl group (by electron release) is most effective. These structures, therefore, make a large (stabilizing) contribution to the overall hybrid for ortho- and para-substituted arenium ions. No such relatively stable structure contributes to the hybrid for the meta-substituted arenium ion, and as a result it is less stable than the ortho- or para-substituted arenium ion. Since the ortho- and para-substituted arenium ions are more stable, the transition states leading to them occur at lower energy and ortho and para substitution take place most rapidly.

PROBLEM 12.11

Write resonance structures for the arenium ions formed when ethylbenzene undergoes electrophilic attack.

PROBLEM 12.12

Resonance structures can also be used to account for the fact that a phenyl group is an ortho–para director and that it is an activating group. Show how this is possible.

12.9F Summary of Effects on Orientation and Reactivity

We can summarize the effects that groups have on orientation and reactivity in the following way.

Full or partial (+) charge on directly attached atom	At least one nonbonding pair on directly attached atom		Alkyl or aryl
	Halogen	$-\overset{\cdot\cdot}{\underset{}{N}}H_2$, $-\overset{\cdot\cdot}{\underset{\cdot\cdot}{O}}H$, etc.	

←meta directing →←————— ortho–para directing ————————→
←————— deactivating ————————→←———— activating ————————→

12.10 REACTIONS OF THE SIDE CHAIN OF ALKYLBENZENES

Hydrocarbons that consist of both aliphatic and aromatic groups are also known as **arenes.** Toluene, ethylbenzene, and isopropylbenzene are **alkylbenzenes.**

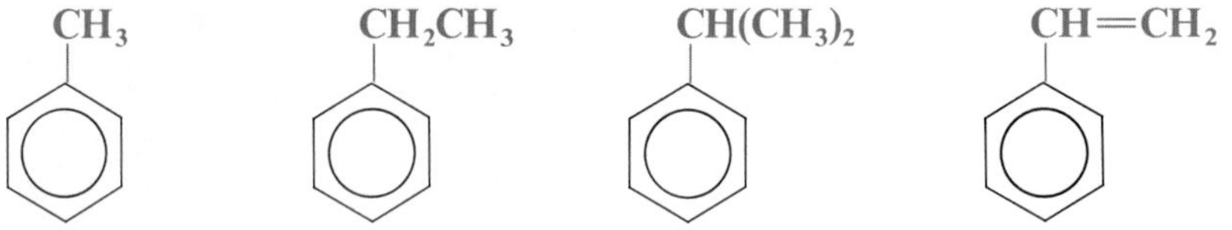

Toluene
(methylbenzene)

Ethylbenzene

Isopropyl-
benzene

Styrene
(phenylethene or
vinylbenzene)

Styrene is an example of an **alkenylbenzene.** The aliphatic portion of these compounds is commonly called the **side chain.**

Styrene is one of the most important industrial chemicals — more than 6 billion lb are produced each year. The starting material for the commercial synthesis of styrene is ethylbenzene, produced by Friedel–Crafts alkylation of benzene:

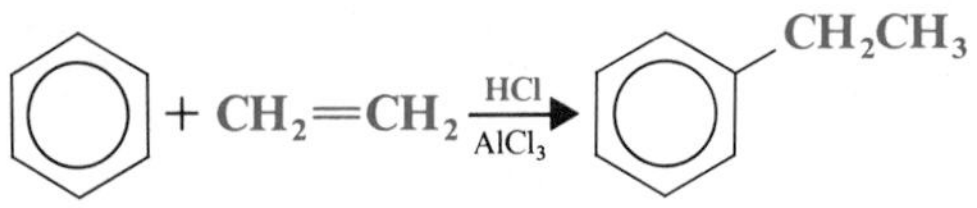

Ethylbenzene

Ethylbenzene is then dehydrogenated in the presence of a catalyst (zinc oxide or chromium oxide) to produce styrene.

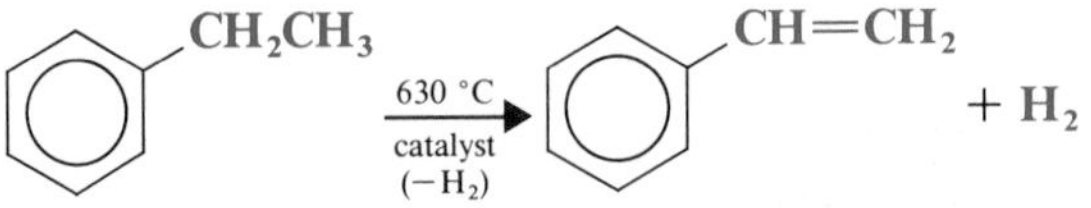

(90–92% yield)
Styrene

Most styrene is polymerized (Special Topic D) to the familiar plastic, polystyrene.

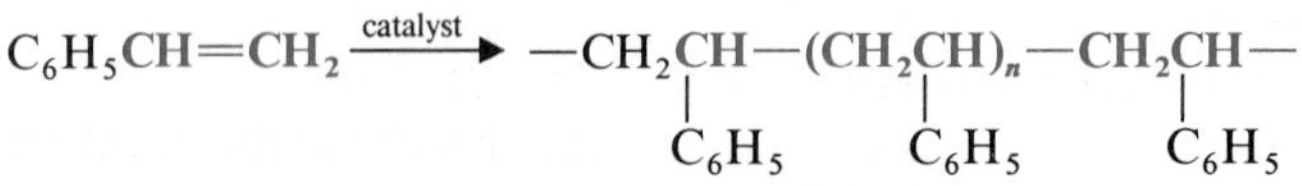

$$C_6H_5CH{=}CH_2 \xrightarrow{\text{catalyst}} -CH_2CH-(CH_2CH)_n-CH_2CH-$$
$$\underset{C_6H_5}{|}\quad\underset{C_6H_5}{|}\quad\underset{C_6H_5}{|}$$

Polystyrene

12.10A Halogenation of the Side Chain. Benzylic Radicals

We have seen that bromine and chlorine replace hydrogen atoms on the ring of toluene when the reaction takes place in the presence of a Lewis acid. In ring halogenations the electrophiles are *positive* chlorine or bromine ions or they are Lewis-acid complexes that have positive halogens. These positive electrophiles attack the π electrons of the benzene ring and aromatic substitution takes place.

Chlorine and bromine can also be made to replace hydrogens of the methyl group of toluene. Side-chain halogenation takes place when the reaction is carried out *in the absence of Lewis acids* and under conditions that favor the formation of free radicals. When toluene reacts with *N*-bromosuccinimide (NBS) in the presence of peroxides, for example, the major product is benzyl bromide. [*N*-Bromosuccinimide furnishes a low concentration of Br_2 (cf. Section 10.2).]

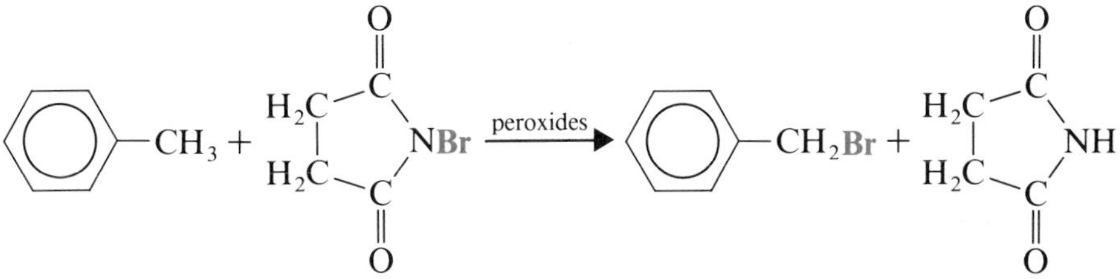

NBS

(64%)
Benzyl bromide
(α-bromotoluene)

Side-chain chlorination of toluene also takes place in the gas phase at 400 to 600 °C or in the presence of ultraviolet light. When an excess of chlorine is used, multiple chlorinations of the side chain occur.

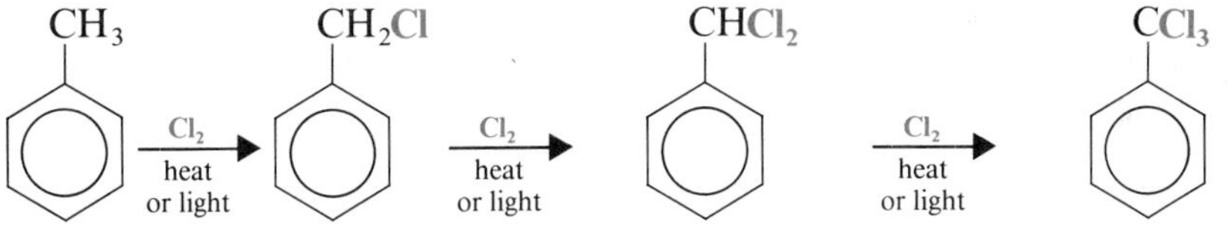

CH$_3$	CH$_2$Cl	CHCl$_2$	CCl$_3$
	Benzyl chloride	(Dichloromethyl)-benzene	(Trichloromethyl)-benzene

These halogenations take place through the same free radical mechanisms we saw for alkanes in Section 9.3. The halogens dissociate to produce halogen atoms and then the halogen atoms initiate chains by abstracting hydrogens of the methyl group.

Chain Initiation

Step 1 $X_2 \xrightarrow[\text{or light}]{\substack{\text{peroxides,}\\ \text{heat,}}} 2X\cdot$

Chain Propagation

Step 2 $C_6H_5CH_3 + X\cdot \longrightarrow C_6H_5CH_2\cdot + HX$
                                Benzyl
                                radical

Step 3 $C_6H_5CH_2\cdot + X_2 \longrightarrow C_6H_5CH_2X + X\cdot$
                Benzyl   Benzyl
                radical   halide

Abstraction of a hydrogen from the methyl group of toluene produces *a benzyl radical.* The benzyl radical then reacts with a halogen molecule to produce a benzyl halide and a halogen atom. The halogen atom then brings about a repetition of step 2, then step 3 occurs again, and so on.

Benzylic halogenations are similar to allylic halogenations (Section 10.2) in that they involve the formation of *unusually stable free radicals* (Section 10.6). Benzylic and allylic radicals are even more stable than tertiary radicals.

The greater stability of benzylic radicals accounts for the fact that when ethylbenzene is halogenated the major product is the 1-halo-1-phenylethane. The benzylic radical is formed faster than the 1° radical:

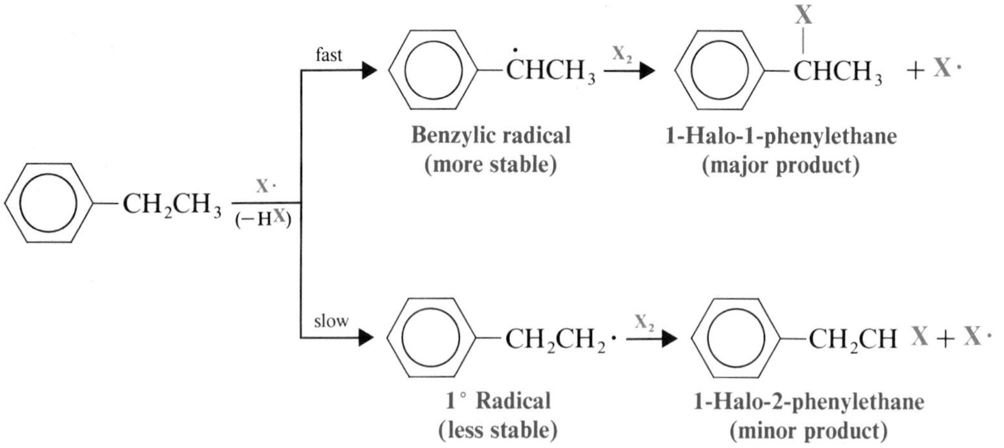

Benzylic radical
(more stable)

1-Halo-1-phenylethane
(major product)

1° Radical
(less stable)

1-Halo-2-phenylethane
(minor product)

PROBLEM 12.13

When propylbenzene reacts with chlorine in the presence of ultraviolet radiation, the major product is 1-chloro-1-phenylpropane. Both 2-chloro-1-phenylpropane and 1-chloro-3-phenylpropane are minor products. Write the structure of the radical leading to each product and account for the fact that 1-chloro-1-phenylpropane is the major product.

SAMPLE PROBLEM

Illustrating a Multistep Synthesis

Starting with ethylbenzene, outline a synthesis of phenylacetylene $(C_6H_5C{\equiv}CH)$.

Answer:

Working backward, that is, using *retrosynthetic analysis,* we realize that we could make phenylacetylene by dehydrohalogenating either of the following compounds using sodium amide in mineral oil (Section 6.21B).

$$C_6H_5CBr_2CH_3 \xrightarrow[\text{(2) H}^+]{\text{(1) NaNH}_2, \text{ mineral oil, heat}} C_6H_5C{\equiv}CH$$

$$C_6H_5CHBrCH_2Br \xrightarrow[\text{(2) H}^+]{\text{(1) NaNH}_2, \text{ mineral oil, heat}} C_6H_5C{\equiv}CH$$

We could make the first compound from ethylbenzene by allowing it to react with two moles of *N*-bromosuccinimide (NBS).

$$C_6H_5CH_2CH_3 \xrightarrow{\text{NBS, peroxides}} C_6H_5CBr_2CH_3$$

We could make the second compound by adding bromine to styrene, and we could make styrene from ethylbenzene as follows:

$$C_6H_5CH_2CH_3 \xrightarrow{\text{NBS, peroxides}} C_6H_5CHBrCH_3 \xrightarrow{\text{KOH, heat}}$$

$$C_6H_5CH{=}CH_2 \xrightarrow{\text{Br}_2,\text{ CCl}_4} C_6H_5CHBrCH_2Br$$

PROBLEM 12.14

Starting with phenylacetylene ($C_6H_5C{\equiv}CH$), outline a synthesis of (a) 1-phenylpropyne,   (b) 1-phenyl-1-butyne,   (c) (*Z*)-1-phenylpropene,   and (d) (*E*)-1-phenylpropene.

12.10B Benzyl and Benzylic Cations

Recall that benzyl and benzylic cations are unusually stable carbocations; they are approximately as stable as tertiary cations (Section 10.6).

PROBLEM 12.15

Write resonance structures for the *benzyl cation* ($C_6H_5CH_2{}^+$) that account for its unusual stability.

12.11 ALKENYLBENZENES

12.11A Stability of Conjugated Alkenylbenzenes

Alkenylbenzenes that have their double bond conjugated with the benzene ring are more stable than those that do not.

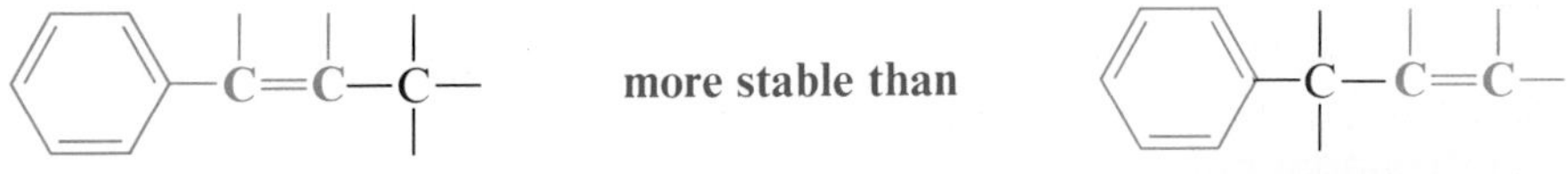

Conjugated system   more stable than   Nonconjugated system

Part of the evidence for this comes from acid-catalyzed alcohol dehydrations, which are known to yield the most stable alkene (Section 6.15). For example, dehydration of the following alcohol yields exclusively the conjugated system.

$$\text{C}_6\text{H}_5-\underset{\underset{\text{OH}}{|}}{\overset{\overset{\text{H}}{|}}{\text{C}}}-\underset{\overset{\text{H}}{|}}{\overset{\overset{\text{H}}{|}}{\text{C}}}-\overset{}{\text{C}}- \xrightarrow[\text{(}-\text{H}_2\text{O)}]{\text{H}^+,\text{ heat}} \text{C}_6\text{H}_5-\text{C}=\text{C}-\underset{\overset{\text{H}}{|}}{\text{C}}-$$

Because conjugation always lowers the energy of an unsaturated system by allowing the π electrons to be delocalized, this behavior is just what we would expect.

12.11B Additions to the Double Bond of Alkenylbenzenes

In the presence of peroxides, hydrogen bromide adds to the double bond of 1-phenyl-propene to give 2-bromo-1-phenylpropane as the major product.

$$\text{C}_6\text{H}_5-\text{CH}=\text{CHCH}_3 \xrightarrow[\text{peroxides}]{\text{HBr}} \text{C}_6\text{H}_5-\text{CH}_2\underset{\overset{|}{\text{Br}}}{\text{CHCH}_3}$$

1-Phenylpropene	2-Bromo-1-phenylpropane

In the absence of peroxides, HBr adds in just the opposite way.

$$\text{C}_6\text{H}_5-\text{CH}=\text{CHCH}_3 \xrightarrow[\text{(no peroxides)}]{\text{HBr}} \text{C}_6\text{H}_5-\underset{\overset{|}{\text{Br}}}{\text{CHCH}_2\text{CH}_3}$$

1-Phenylpropene	1-Bromo-1-phenylpropane

The addition of hydrogen bromide to 1-phenylpropene proceeds through a benzylic radical in the presence of peroxides, and through a benzylic cation in their absence (cf. 1 and 2 as follows).

    1. **Hydrogen bromide addition in the presence of peroxides.**

Chain Initiation

Step 1 $\text{R}-\text{O}-\text{O}-\text{R} \longrightarrow 2\text{R}-\text{O}\cdot$

Step 2 $\text{RO}\cdot + \text{H}-\text{Br} \longrightarrow \text{R}-\text{O}-\text{H} + \text{Br}\cdot$

Step 3 $\text{Br}\cdot + \text{C}_6\text{H}_5\text{CH}=\text{CHCH}_3 \longrightarrow \text{C}_6\text{H}_5\overset{\cdot}{\text{C}}\text{H}-\underset{\overset{|}{\text{Br}}}{\text{CHCH}_3}$

                                                 A benzylic radical

Chain Propagation

Step 4 $\text{C}_6\text{H}_5\overset{\cdot}{\text{C}}\underset{\overset{|}{\text{Br}}}{\text{HCHCH}_3} + \text{H}-\text{Br} \longrightarrow \text{C}_6\text{H}_5\text{CH}_2\underset{\overset{|}{\text{Br}}}{\text{CHCH}_3} + \text{Br}\cdot$

                                                         2-Bromo-1-phenylpropane

The mechanism for the addition of hydrogen bromide to 1-phenylpropene in the presence of peroxides is a chain mechanism analogous to the one we discussed when we described anti-Markovnikov addition in Section 9.9. The step that determines the orientation of the reaction is the first chain-propagating step. Bromine attacks the second carbon atom of the chain because by doing so the reaction produces a more stable benzylic radical. Had the bromine atom attacked the double bond in the opposite way, a less stable secondary radical would have been formed.

$$C_6H_5CH{=}CHCH_3 + Br\cdot \ \xrightarrow{\ \times\ } \ C_6H_5CH{-}\overset{\displaystyle\cdot}{C}HCH_3$$
$$\underset{\displaystyle Br}{|}$$

A secondary radical

2. **Hydrogen bromide addition in the absence of peroxides.**

$$C_6H_5CH{=}CHCH_3 + HBr \longrightarrow C_6H_5\overset{+}{C}HCH_2CH_3 + Br^-$$

A benzylic cation

$$\downarrow$$

$$C_6H_5CHCH_2CH_3$$
$$\underset{\displaystyle Br}{|}$$

In the absence of peroxides hydrogen bromide adds through an ionic mechanism. The step that determines the orientation in the ionic mechanism is the first, where the proton attacks the double bond to give the more stable benzylic cation. Had the proton attacked the double bond in the opposite way, a less stable secondary cation would have been formed.

$$C_6H_5CH{=}CHCH_3 + HBr \ \xrightarrow{\ \times\ } \ C_6H_5CH{-}\overset{+}{C}HCH_3 + Br^-$$
$$\underset{\displaystyle H}{|}$$

A secondary cation

PROBLEM 12.16

(a) What would you expect to be the major product when 1-phenylpropene reacts with HCl? (b) When it is subjected to oxymercuration–demercuration?

12.11C Oxidation of the Side Chain

Strong oxidizing agents oxidize toluene to benzoic acid. The oxidation can be carried out by the action of hot alkaline potassium permanganate. This method gives benzoic acid in almost quantitative yield.

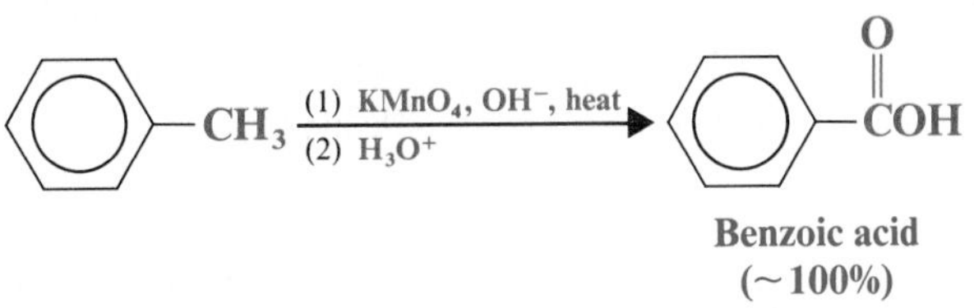

Benzoic acid
($\sim 100\%$)

An important characteristic of side-chain oxidations is that oxidation takes place initially at the benzylic carbon; **alkylbenzenes with alkyl groups longer than methyl are ultimately degraded to benzoic acids.**

$$\text{(An alkylbenzene)} \quad C_6H_5-CH_2CH_2CH_2R \xrightarrow[\text{(2) } H_3O^+]{\substack{\text{(1) } KMnO_4, OH^- \\ \text{heat}}} C_6H_5-\overset{\displaystyle O}{\overset{\|}{C}}-OH \quad \text{(Benzoic acid)}$$

Side-chain oxidations are similar to benzylic halogenations, because in the first step the oxidizing agent abstracts a benzylic hydrogen. Once oxidation is begun at the benzylic carbon, it continues at that site. Ultimately, the oxidizing agent oxidizes the benzylic carbon to a carboxyl group and, in the process, it cleaves off the remaining carbon atoms of the side chain. (*tert*-Butylbenzene is resistant to side-chain oxidation. Why?)

Side-chain oxidation is not restricted to alkyl groups. **Alkenyl, alkynyl, and acyl groups are oxidized by hot alkaline potassium permanganate in the same way.**

$$\left.\begin{array}{c} C_6H_5CH=CHCH_3 \\[1ex] \text{or} \\[1ex] C_6H_5C\equiv CCH_3 \\[1ex] \text{or} \\[1ex] C_6H_5\overset{\displaystyle O}{\overset{\|}{C}}CH_2CH_3 \end{array}\right\} \xrightarrow[\text{(2) } H_3O^+]{\text{(1) } KMnO_4, OH^-, \text{heat}} C_6H_5\overset{\displaystyle O}{\overset{\|}{C}}OH$$

12.11D Oxidation of the Benzene Ring

The benzene ring of an alkylbenzene can be converted to a carboxyl group by ozonolysis, followed by treatment with hydrogen peroxide:

$$R-C_6H_5 \xrightarrow[\text{(2) } H_2O_2]{\text{(1) } O_3, CH_3CO_2H} R-\overset{\displaystyle O}{\overset{\|}{C}}OH$$

12.12 SYNTHETIC APPLICATIONS

The substitution reactions of aromatic rings and the reactions of the side chains of alkyl- and alkenylbenzenes, when taken together, offer us a powerful set of reactions for organic synthesis. By using these reactions skillfully, we shall be able to synthesize a large number of benzene derivatives.

Part of the skill in planning a synthesis is in deciding the order in which reactions should be carried out. Let us suppose, for example, that we want to synthesize *o*-bromonitrobenzene. We can see very quickly that we should introduce the bromine into the ring first because it is an ortho–para director.

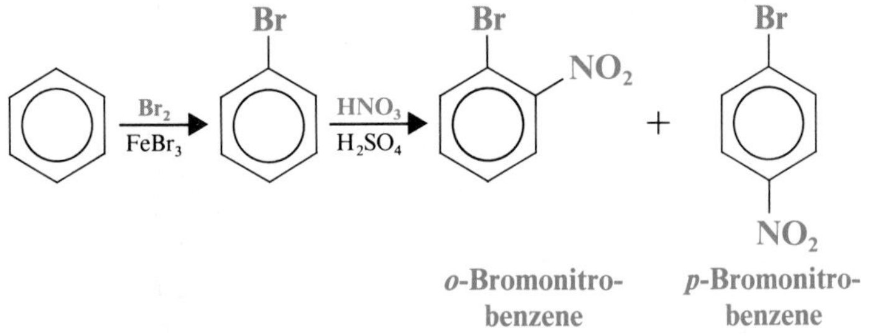

o-Bromonitro-
benzene

p-Bromonitro-
benzene

The ortho and para compounds that we get as products can be separated by various methods. However, had we introduced the nitro group first, we would have obtained *m*-bromonitrobenzene as the major product.

Other examples in which choosing the proper order for the reactions are important are the syntheses of the *ortho-, meta-,* and *para*-nitrobenzoic acids. We can synthesize the *ortho-* and *para*-nitrobenzoic acids from toluene by nitrating it, separating the *ortho-* and *para*-nitrotoluenes, and then oxidizing the methyl groups to carboxyl groups.

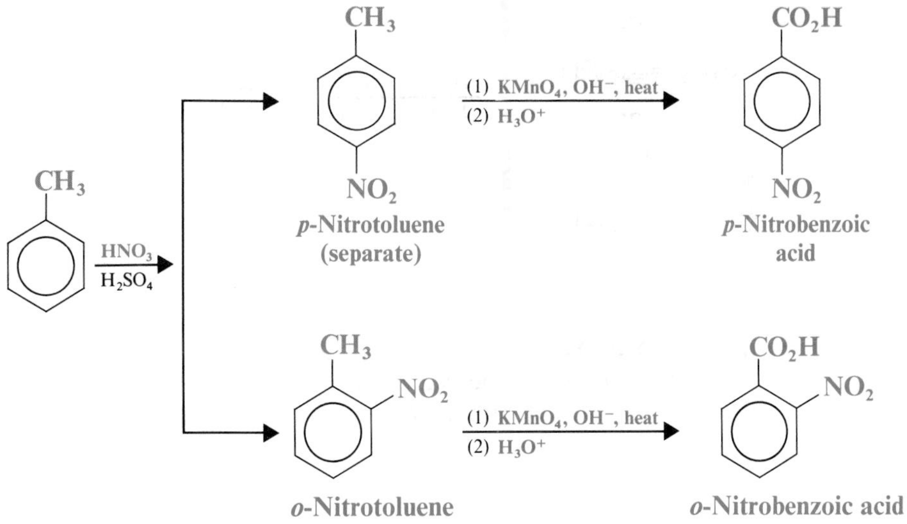

p-Nitrotoluene
(separate)

p-Nitrobenzoic
acid

o-Nitrotoluene

o-Nitrobenzoic acid

We can synthesize *m*-nitrobenzoic acid by reversing the order of the reactions.

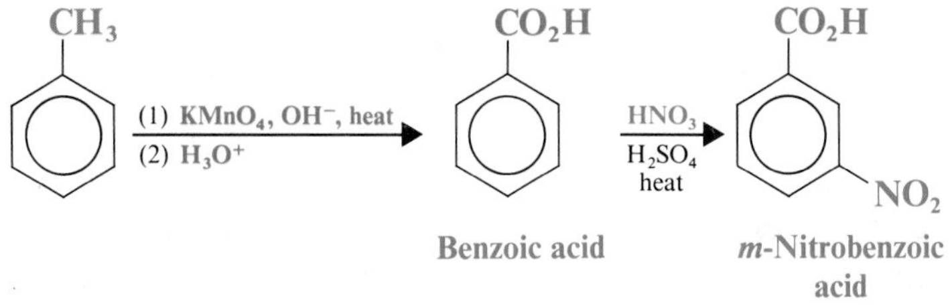

Benzoic acid

m-Nitrobenzoic
acid

SAMPLE PROBLEM

Starting with toluene, outline a synthesis of (a) 1-bromo-2-(trichloromethyl)-benzene, (b) 1-bromo-3-(trichloromethyl)benzene, and (c) 1-bromo-4-(trichloromethyl)benzene.

Answer:

Compounds (a) and (c) can be obtained by ring bromination of toluene followed by chlorination of the side chain using three molar equivalents of chlorine:

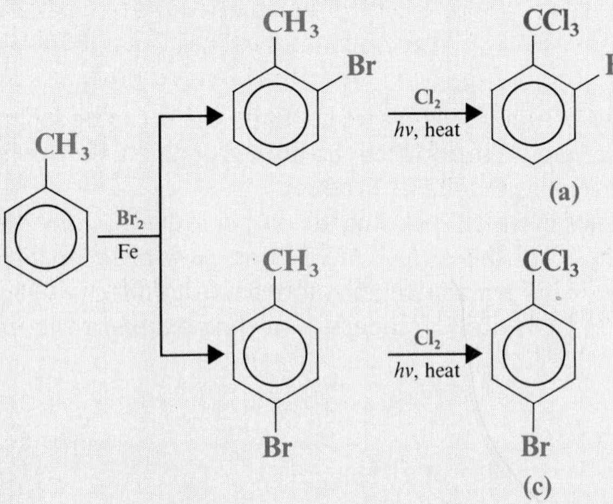

To make compound (b) we reverse the order of the reactions. By converting the side chain to a —CCl_3 group first, we create a meta director, which causes the bromine to enter the correct position.

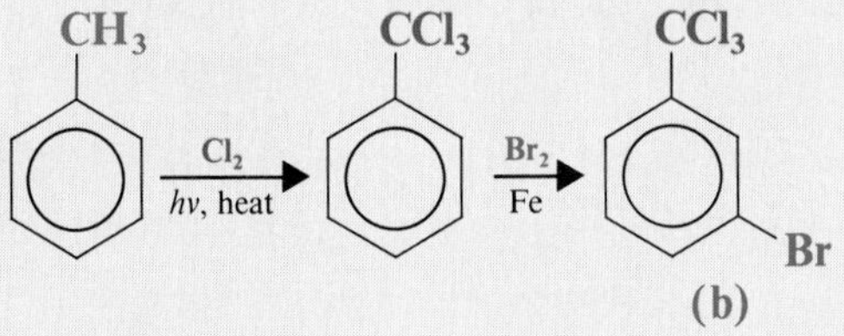

PROBLEM 12.17

Suppose you needed to synthesize 1-(*p*-chlorophenyl)propene from propylbenzene.

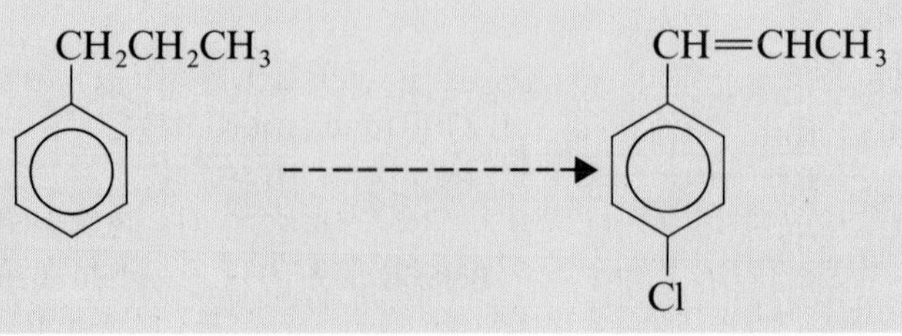

You could introduce the double bond into the side chain through a benzylic halogenation and subsequent dehydrohalogenation. You could introduce the chlorine into the benzene ring through a Lewis-acid catalyzed chlorination. Which reaction would you carry out first? Why?

Very powerful activating groups such as amino groups and hydroxyl groups cause the benzene ring to be so reactive that undesirable reactions take place. Some reagents used for electrophilic substitution reactions, such as nitric acid, are also strong *oxidizing agents.* (Both electrophiles and oxidizing agents seek electrons.) Thus, amino groups and hydroxyl groups not only activate the ring toward electrophilic substitution, they also activate it toward oxidation. Nitration of aniline, for example, results in considerable destruction of the benzene ring because it is oxidized by the nitric acid. Direct nitration of aniline, consequently, is not a satisfactory method for the preparation of *o-* and *p-*nitroaniline.

Treating aniline with acetyl chloride, CH_3COCl, or acetic anhydride, $(CH_3CO)_2O$, converts aniline to acetanilide. The amino group is converted to an acetamido group ($-NHCOCH_3$), a group that is only moderately activating and one that does not make the ring highly susceptible to oxidation. With acetanilide direct nitration becomes possible.

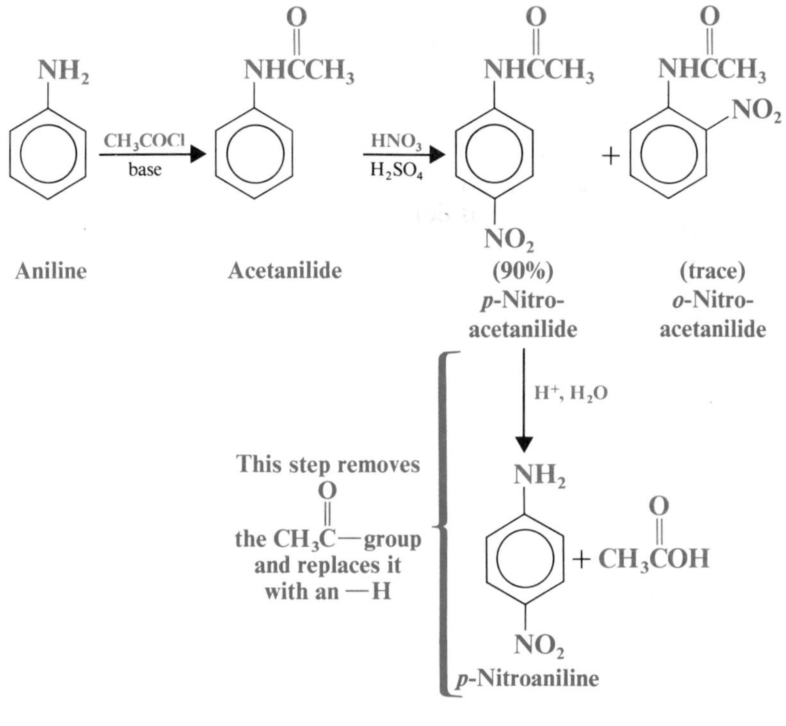

Nitration of acetanilide gives *p-*nitroacetanilide in excellent yield with only a trace of the ortho isomer. Acidic hydrolysis of *p-*nitroacetanilide (Section 18.8F) removes the acetyl group and gives *p-*nitroaniline, also in good yield.

Suppose, however, that we need *o-*nitroaniline. The synthesis that we just outlined would obviously not be a satisfactory method, for only a trace of *o-*nitroacetanilide is obtained in the nitration reaction. (The acetamido group is purely a para director in many reactions. Bromination of acetanilide, for example, gives *p*-bromoacetanilide almost exclusively.)

We can synthesize *o*-nitroacetanilide, however, through the reactions that follow:

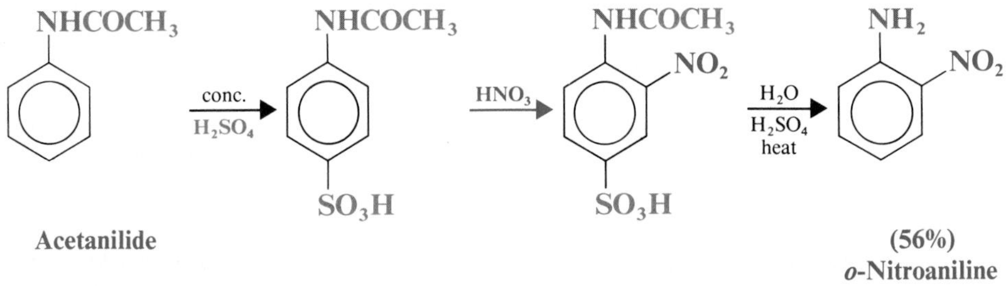

Acetanilide

(56%)
o-Nitroaniline

Here we see how a sulfonic acid group can be used as a "blocking group." We can remove the sulfonic acid group by desulfonation of a later stage. In this example, the reagent used for desulfonation (dilute H_2SO_4) also conveniently removes the acetyl group that we employed to "protect" the benzene ring from oxidation by nitric acid.

12.12A Orientation in Disubstituted Benzenes

The problem of orientation is somewhat more complicated when two substituents are present on a benzene ring. We find, however, that in many instances we can make very good estimates of the outcomes of the reaction by relatively simple analyses.

If two groups are located so that their directive effects reinforce each other, then predicting the outcome is easy. Consider the examples shown here. In each case the entering substituent is directed by both groups into the position indicated by the arrows.

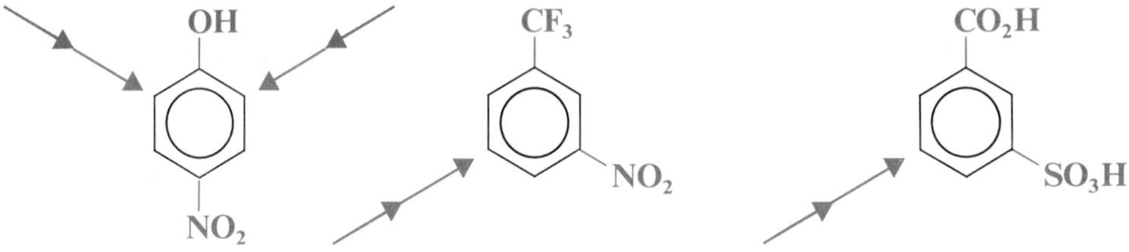

When the directive effects of two groups oppose each other, **the more powerful activating group** (Table 12.2) **generally determines the outcome of the reaction.** Let us consider, as an example, the orientation of electrophilic substitution of *p*-methylacetanilide. The acetamido group is a much stronger activating group than the methyl group. The following example shows that the acetamido group determines the outcome of the reaction. Substitution occurs primarily at the position ortho to the acetamido group.

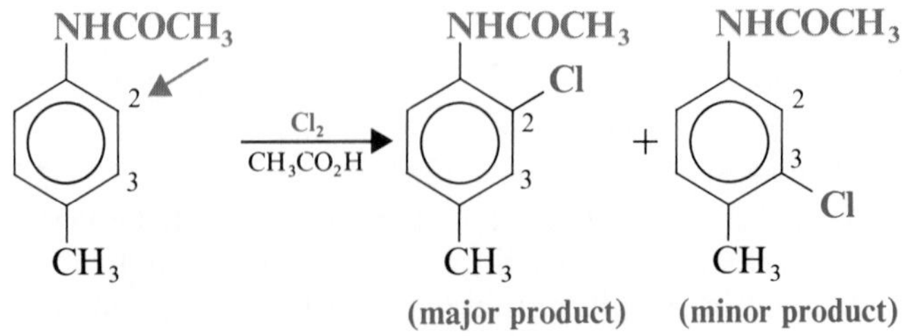

(major product)   (minor product)

Because all ortho–para- directing groups are more activating than meta direc-tors, **the ortho–para director determines the orientation of the incoming group.** For example,

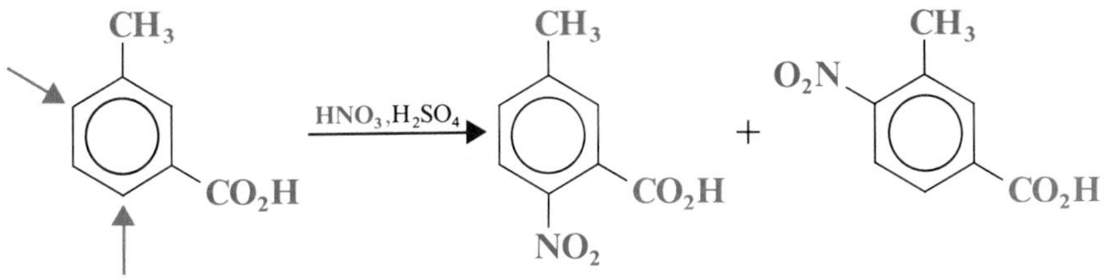

When two opposing groups have approximately the same directive effect, the results are not nearly so clear-cut. The following reaction is a typical example.

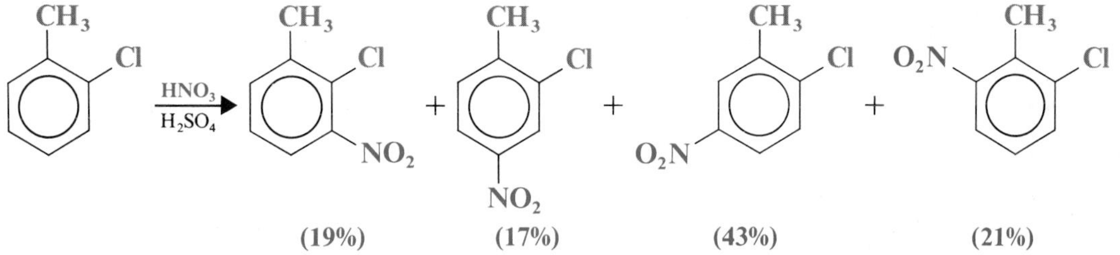

(19%)   (17%)   (43%)   (21%)

Steric effects are also important in aromatic substitutions. **Substitution does not occur to an appreciable extent between meta substituents if another position is open.** A good example of this effect can be seen in the nitration of m-bromochlorobenzene.

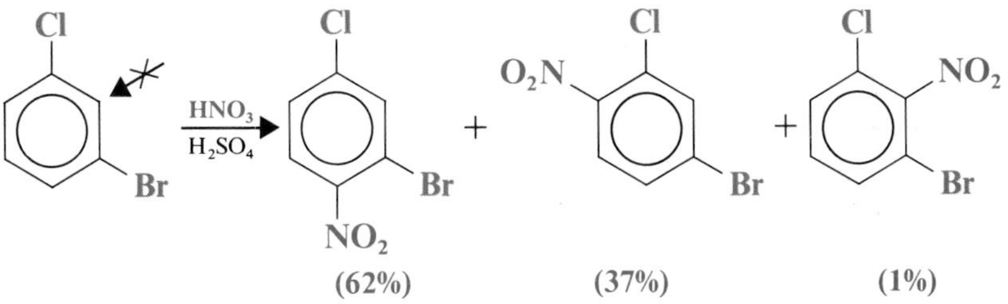

(62%)   (37%)   (1%)

Only 1% of the mononitro product has the nitro group between the bromine and chlorine.

PROBLEM 12.18

Predict the major product (or products) that would be obtained when each of the following compounds is nitrated.

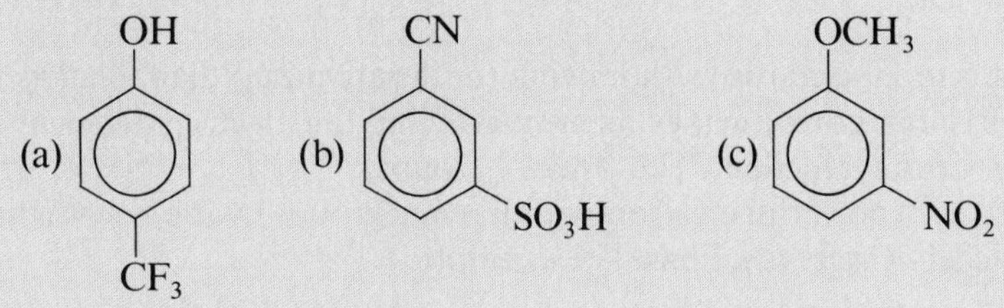

12.12B Limitations of Friedel–Crafts Alkylation Reactions

Several restrictions limit the usefulness of Friedel–Crafts alkylations.

1. **When the carbocation formed from an alkyl halide, alkene, or alcohol can rearrange to a more stable carbocation, it usually does so and the major product obtained from the reaction is usually the one from the more stable carbocation.**

When benzene is alkylated with butyl bromide, for example, some of the developing butyl cations rearrange by a hydride shift — some developing 1° carbocations (see following reactions) become more stable 2° carbocations. Then benzene reacts with both kinds of carbocations to form both butylbenzene and *sec*-butylbenzene:

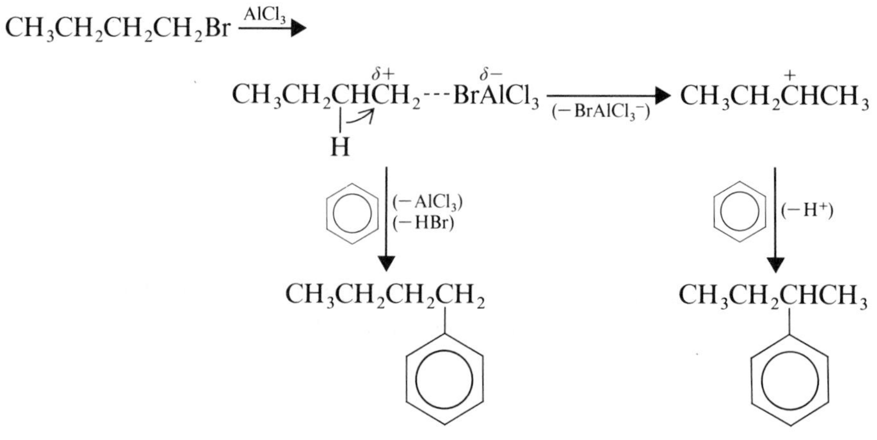

$$CH_3CH_2CH_2CH_2Br \xrightarrow{AlCl_3}$$

$$CH_3CH_2\overset{\delta+}{C}HCH_2\cdots\overset{\delta-}{Br}AlCl_3 \xrightarrow[(-BrAlCl_3^-)]{} CH_3CH_2\overset{+}{C}HCH_3$$

with H shift shown below the carbon

$(-AlCl_3)$ $(-HBr)$ → $CH_3CH_2CH_2CH_2$—

$(-H^+)$ → $CH_3CH_2CHCH_3$—

(32–36%)
Butylbenzene

(64–68%)
sec-Butylbenzene

2. **Friedel–Crafts reactions do not occur when powerful electron-withdrawing groups** (Section 12.9) **are present on the aromatic ring or when the ring bears an** $—NH_2$, $—NHR$, **or** $—NR_2$ **group.** This applies to alkylations and *acylations.*

NO_2 $\quad$ $^+N(CH_3)_3$ $\quad$ NH_2

These do not undergo
Friedel–Crafts reactions

Any substituent more electron withdrawing (or deactivating) than a halogen, that is, **any meta-directing group, makes an aromatic ring too electron deficient to undergo a Friedel–Crafts reaction.** The amino groups, $—NH_2$, $—NHR$, and $—NR_2$, are changed into powerful electron-withdrawing groups by the Lewis acids used to catalyze Friedel–Crafts reactions. For example:

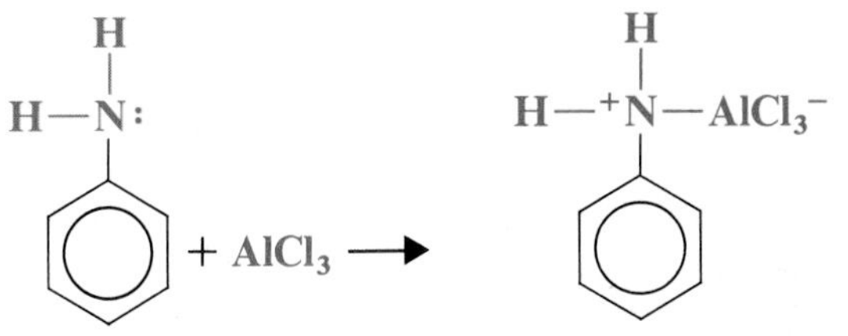

Does not undergo
a Friedel–Crafts reaction

3. **Aryl and vinylic halides cannot be used as the halide component because they do not form carbocations readily** (cf. Section 14.11).

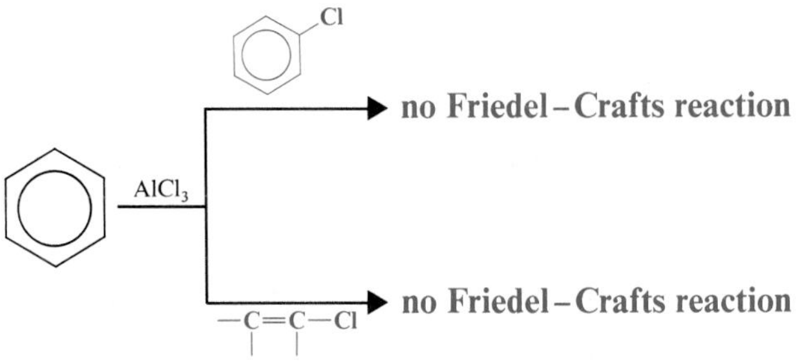

→ no Friedel–Crafts reaction

→ no Friedel–Crafts reaction

4. **Polyalkylations often occur.** Alkyl groups are electron-releasing groups, and once one is introduced into the benzene ring it activates the ring toward further substitution (cf. Section 12.8).

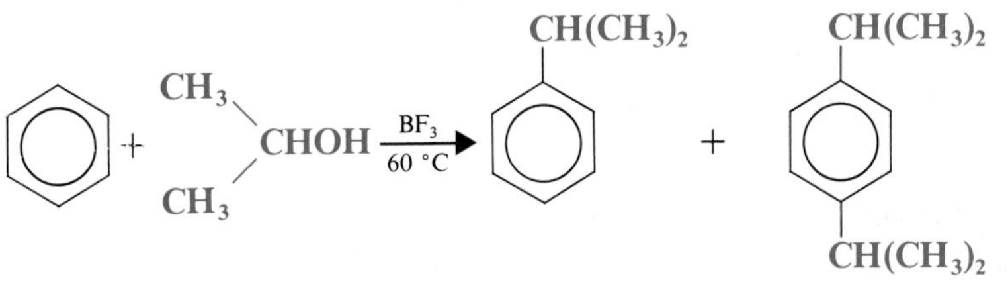

(24%)
Isopropyl-
benzene

(14%)
p-Diisopropylbenzene

Polyacylations are not a problem in Friedel–Crafts acylations, however. The acyl group (RCO—) by itself is a deactivating group, and when it forms a complex with $AlCl_3$ in the last step of the reaction (Section 12.7), it is made even more deactivating. This strongly inhibits further substitution and makes mono-acylation easy.

PROBLEM 12.19

When benzene reacts with neopentyl chloride, $(CH_3)_3CCH_2Cl$, in the presence of aluminum chloride, the major product is 2-methyl-2-phenylbutane, not neopentylbenzene. Explain this result.

PROBLEM 12.20

When benzene reacts with propyl alcohol in the presence of boron trifluoride, both propylbenzene and isopropylbenzene are obtained as products. Write a mechanism that accounts for this.

12.12C Synthetic Applications of Friedel–Crafts Acylations

Rearrangements of the carbon chain do not occur in Friedel–Crafts acylations. The acylium ion, because it is stabilized by resonance, is more stable than most other carbocations. Thus, there is no driving force for a rearrangement. Because rearrangements do not occur, Friedel–Crafts acylations often give us much better routes to unbranched alkylbenzenes than do Friedel–Crafts alkylations.

As an example, let us consider the problem of synthesizing propylbenzene. If we attempt this synthesis through a Friedel–Crafts alkylation, a rearrangement occurs and the major product is isopropylbenzene (see also Problem 12.20).

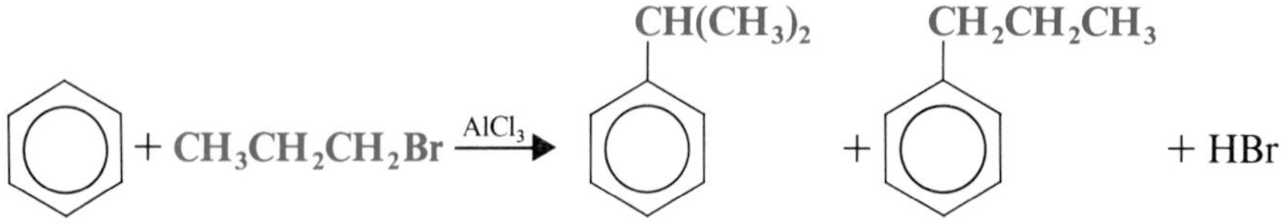

Isopropylbenzene (major product)   Propylbenzene (minor product)

By contrast, the Friedel–Crafts acylation of benzene with propanoyl chloride produces a ketone with an unrearranged carbon chain in excellent yield.

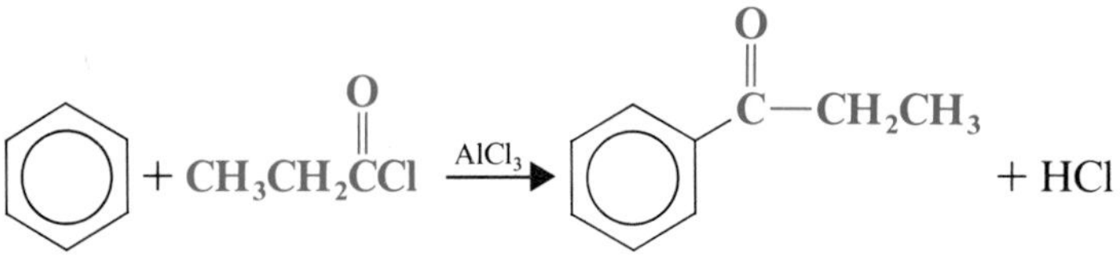

Propanoyl chloride   (90%) Ethyl phenyl ketone

This ketone can then be reduced to propylbenzene by several methods. One general method—called **the Clemmensen reduction**—consists of refluxing the ketone with hydrochloric acid containing amalgamated zinc. [*Caution:* As we shall discuss later (Section 19.5), zinc and hydrochloric acid will also reduce nitro groups to amino groups.]

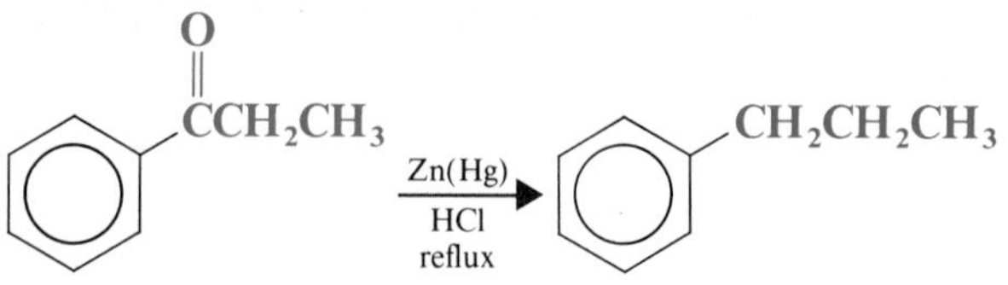

Ethyl phenyl ketone   (80%) Propylbenzene

or

$$\text{ArCR} \xrightarrow[\text{HCl, reflux}]{\text{Zn(Hg)}} \text{ArCH}_2\text{R}$$

When cyclic anhydrides are used as one component, the Friedel–Crafts acylation provides a means of adding a new ring to an aromatic compound. One illustration is shown here.

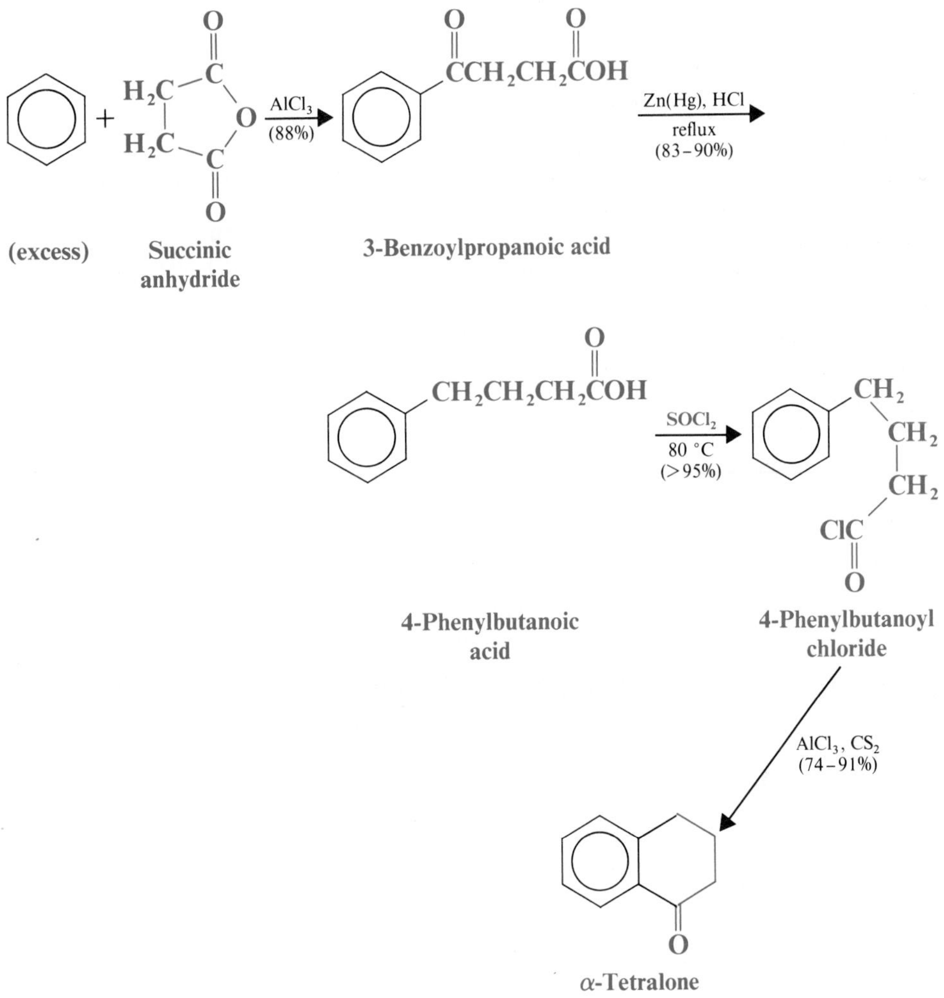
(excess)
Succinic anhydride
$AlCl_3$ (88%)
3-Benzoylpropanoic acid
Zn(Hg), HCl reflux (83–90%)
4-Phenylbutanoic acid
$SOCl_2$ 80 °C (>95%)
4-Phenylbutanoyl chloride
$AlCl_3$, CS_2 (74–91%)
α-Tetralone

PROBLEM 12.21

Starting with benzene and the appropriate acid chloride or anhydride, outline a synthesis of each of the following:

(a) Hexylbenzene
(b) Isobutylbenzene
(c) Diphenylmethane
(d) Anthrone

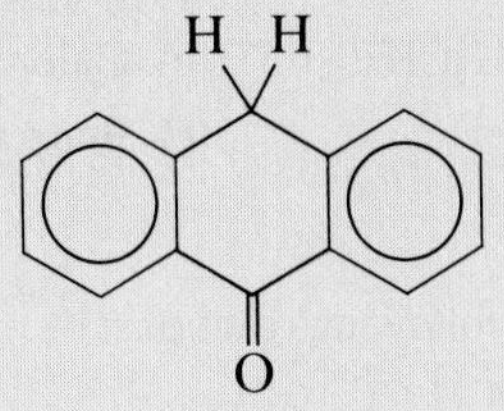

Anthrone

12.13 SUMMARY OF REACTIONS

Figures 12.3 and 12.4 summarize many of the important reactions described in this chapter.

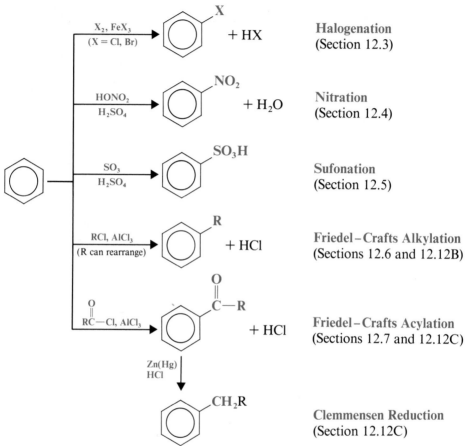

FIGURE 12.3 A summary of reactions of benzene and its derivatives.

ADDITIONAL PROBLEMS

12.22

Outline ring bromination, nitration, and sulfonation reactions of the following compounds. In each case give the structure of the major reaction product or products. Also indicate whether the reaction would occur faster or slower than the corresponding reaction of benzene.

(a) Anisole, $C_6H_5OCH_3$

(b) (Difluoromethyl)benzene, $C_6H_5CHF_2$

(c) Ethylbenzene

(d) Nitrobenzene

(e) Chlorobenzene

(f) Benzenesulfonic acid

12.23

Predict the major products of the following reactions:

(a) Sulfonation of p-$CH_3C_6H_4COCH_3$

(b) Nitration of m-dichlorobenzene

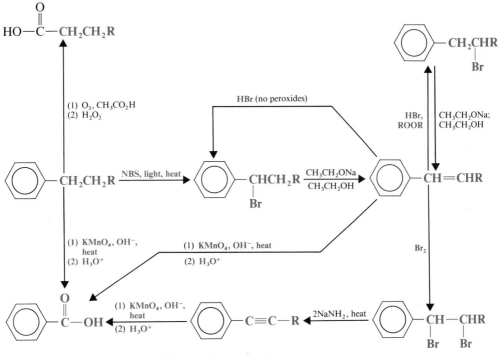

FIGURE 12.4 A summary of the reactions of arenes.

(c) Nitration of 1,3-dimethoxybenzene
(d) Monobromination of p-CH$_3$CONHC$_6$H$_4$NH$_2$
(e) Nitration of p-HO$_3$SC$_6$H$_4$OH

(f) Nitration of (benzene)—CH$_2$—(benzene)—CO$_2$H

(g) Chlorination of C$_6$H$_5$CCl$_3$

12.24
Give the structures of the major products of the following reactions:
(a) Styrene + HCl ⟶
(b) 2-Bromo-1-phenylpropane + C$_2$H$_5$ONa ⟶
(c) C$_6$H$_5$CH$_2$CHOHCH$_2$CH$_3$ $\xrightarrow{H^+, \text{ heat}}$
(d) Product of (c) + HBr $\xrightarrow{\text{peroxides}}$
(e) Product of (c) + H$_2$O $\xrightarrow[\text{heat}]{H^+}$
(f) Product of (c) + H$_2$ (1 molar equivalent) $\xrightarrow[25\,°C]{Pt}$
(g) Product of (f) $\xrightarrow[(2)\ H_3O^+]{(1)\ KMnO_4,\ OH^-,\ heat}$

12.25
Starting with benzene, outline a synthesis of each of the following:

(a) Isopropylbenzene
(b) *tert*-Butylbenzene
(c) Propylbenzene
(d) Butylbenzene
(e) 1-*tert*-Butyl-4-chlorobenzene
(f) 1-Phenylcyclopentene
(g) *trans*-2-Phenylcyclopentanol
(h) *m*-Dinitrobenzene

(i) *m*-Bromonitrobenzene
(j) *p*-Bromonitrobenzene
(k) *p*-Chlorobenzenesulfonic acid
(l) *o*-Chloronitrobenzene
(m) *m*-Nitrobenzenesulfonic acid
(n) $CH_3CH_2CH_2CH_2CO_2H$
(o) $CH_3CH_2CHDCHDCO_2H$
(p) $DCH_2CH_2CH_2CH_2CO_2H$

12.26
Starting with styrene, outline a synthesis of each of the following:

(a) $C_6H_5CHClCH_2Cl$
(b) $C_6H_5CH_2CH_3$
(c) $C_6H_5CHOHCH_2OH$
(d) $C_6H_5CO_2H$
(e) $C_6H_5CHOHCH_3$

(f) $C_6H_5CHBrCH_3$
(g) $C_6H_5CH_2CH_2OH$
(h) $C_6H_5CH_2CH_2D$
(i) $C_6H_5CH_2CH_2Br$
(j) $C_6H_5CH_2CH_2I$

(k) $C_6H_5CH_2CH_2CN$
(l) $C_6H_5CHDCH_2D$
(m) Cyclohexylbenzene
(n) $C_6H_5CH_2CH_2OCH_3$

12.27
Starting with toluene, outline a synthesis of each of the following:

(a) *m*-Chlorobenzoic acid
(b) *p*-Acetyltoluene
(c) 2-Bromo-4-nitrotoluene
(d) *p*-Bromobenzoic acid
(e) 1-Chloro-3-(trichloromethyl)benzene

(f) *p*-Isopropyltoluene
(g) 1-Cyclohexyl-4-methylbenzene
(h) 2,4,6-Trinitrotoluene (TNT)
(i) 4-Chloro-2-nitrobenzoic acid
(j) 1-Butyl-4-methylbenzene

12.28
Starting with aniline, outline a synthesis of each of the following:

(a) *p*-Bromoaniline
(b) *o*-Bromoaniline

(c) 2-Bromo-4-nitroaniline
(d) 4-Bromo-2-nitroaniline

(e) 2,4,6-Tribromoaniline

12.29
(a) Which ring of benzanilide would you expect to undergo electrophilic substitution more readily? (b) Write resonance structures that explain your choice.

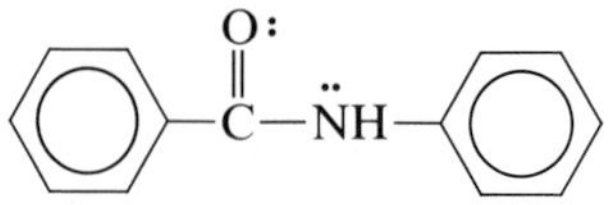

Benzanilide

12.30
What products would you expect from the nitration of phenyl benzoate?

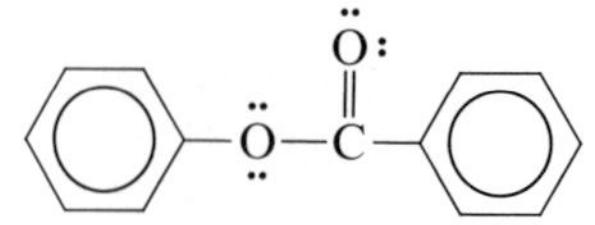

Phenyl benzoate

12.31

Naphthalene can be synthesized from benzene through the following sequence of reactions. Write the structure of each intermediate.

Benzene + succinic anhydride $\xrightarrow{\text{AlCl}_3}$ **A** $\xrightarrow[\text{HCl}]{\text{Zn(Hg)}}$ **B** $\xrightarrow{\text{SOCl}_2}$
$(C_{10}H_{10}O_3)$ $(C_{10}H_{12}O_2)$

C $\xrightarrow{\text{. AlCl}_3}$ **D** $\xrightarrow[\text{HCl}]{\text{Zn(Hg)}}$ **E** $\xrightarrow[\text{peroxides}]{\text{NBS}}$
$(C_{10}H_{11}ClO)$ $(C_{10}H_{10}O)$ $(C_{10}H_{12})$

F $\xrightarrow[\text{heat}]{\text{KOH, ethanol}}$ **G** $\xrightarrow[\text{heat}]{\text{Pt}}$ naphthalene + H$_2$
$(C_{10}H_{11}Br)$ $(C_{10}H_{10})$

12.32

Anthracene and many other polycyclic aromatic compounds have been synthesized by a cyclization reaction known as the *Bradsher reaction* or *aromatic cyclodehydration.* This method, developed by C. K. Bradsher of Duke University, can be illustrated by the conversion of an *o*-benzylphenyl ketone to a substituted anthracene.

An *o*-benzylphenyl   Substituted anthracene
ketone

An arenium ion is an intermediate in this reaction and the last step involves the dehydration of an alcohol. Propose a mechanism for the Bradsher reaction.

12.33

Propose structures for compounds **G–I**.

G $\xrightarrow[\text{conc. H}_2\text{SO}_4]{\text{conc. HNO}_3}$ **H** $\xrightarrow[\text{heat}]{\text{H}_3\text{O}^+, \text{H}_2\text{O}}$ **I**

$(C_6H_6S_2O_8)$ $(C_6H_5NS_2O_{10})$ $(C_6H_5NO_4)$

12.34

2,6-Dichlorophenol has been isolated from the females of two species of ticks (*Amblyomma americanum* and *A. maculatum*), where it apparently serves as a sex attractant. Each female tick yields about 5 ng of 2,6-dichlorophenol. Assume that you need larger quantities than this, and outline a synthesis of 2,6-dichlorophenol from phenol. (*Hint:* When phenol is sulfonated at 100 °C, the product is chiefly *p*-hydroxybenzenesulfonic acid.)

*12.35

The addition of a hydrogen halide (hydrogen bromide or hydrogen chloride) to 1-phenyl-1,3-butadiene produces (only) 1-phenyl-3-halo-1-butene. (a) Write a mechanism that accounts for the formation of this product. (b) Is this 1,4 addition or 1,2 addition to the butadiene system? (c) Is the product of the reaction consistent with the formation of the most stable intermediate carbocation?   (d) Does the reaction appear to be under kinetic control or equilibrium control? Explain.

12.36

We have seen that benzene undergoes ring substitution when it reacts with chlorine in the presence of a Lewis acid. However, benzene can be made to undergo *addition* of chlorine by irradiating a mixture of benzene and chlorine with ultraviolet light. The addition reaction produces a mixture of 1,2,3,4,5,6-hexachlorocyclohexanes. One of these hexachlorocyclohexanes is *lindane,* a very effective (but potentially hazardous) insecticide. The chloro groups of lindane at carbon atoms 1, 2, and 3 are equatorial; those at 4, 5, and 6 are axial.   (a) Write the structure of lindane.   (b) Would you expect lindane to exist in enantiomeric forms?   (c) If not, why not?   (d) One isomeric form of 1,2,3,4,5,6-hexachlorocyclohexane isomer does exist in enantiomeric forms. Write its structure.

12.37

Naphthalene undergoes electrophilic attack at the 1 position much more rapidly than it does at the 2 position.

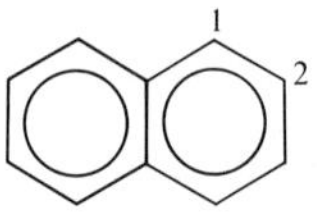

Naphthalene

The greater reactivity at the 1 position can be accounted for by writing resonance structures for the ring that undergoes electrophilic attack. Show how this is possible.

*12.38

Write mechanisms that account for the products of the following reactions:

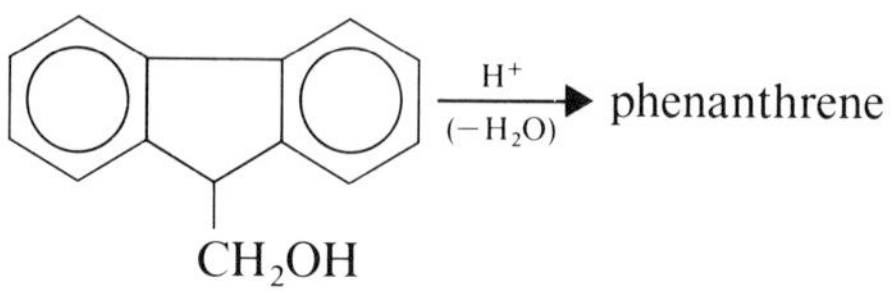

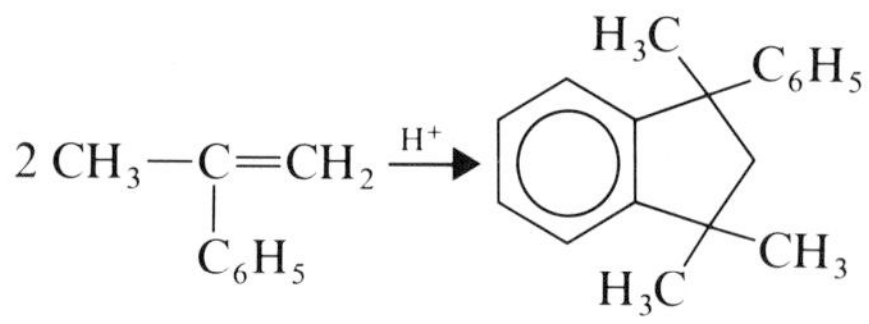

*12.39

At one time extensive use was made of detergents manufactured from the propene tetramers whose synthesis we saw in Problem 7.40. In the industrial process, heating benzene and a mixture of the propene tetramers with aluminum chloride at 35–45 °C gave a mixture of isomers with the molecular formula $C_{18}H_{30}$. This mixture was then heated with sulfuric acid and the products of this reaction (isomers with the molecular formula $C_{18}H_{30}SO_3$) were treated with aqueous sodium hydroxide to give the detergent (also a mixture of isomers). Propose structures for the detergent isomers and write the reactions involved in their formation. (*Note:* Use of these detergents was discontinued a number of years ago because they proved not to be biodegradable.)

*12.40

The compound phenylbenzene (C_6H_5—C_6H_5) is called *biphenyl* and the rings are numbered in the following manner.

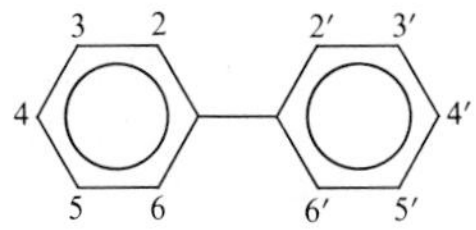

Use models to answer the following questions about substituted biphenyls. (a) When certain large groups occupy three or four of the *ortho* positions (i.e., 2,6,2', and 6'), the substituted biphenyl may exist in enantiomeric forms. An example of a biphenyl that exists in enantiomeric forms is the compound in which the following substituents are present. 2-NO_2, 6-CO_2H, 2'-NO_2, 6'-CO_2H. What factors account for this? (b) Would you expect a biphenyl with 2-Br, 6-CO_2H, 2'-CO_2H, 6'-H to exist in enantiomeric forms? (c) The biphenyl with 2-NO_2, 6-NO_2, 2'-CO_2H, 6'-Br cannot be resolved into enantiomeric forms. Explain.

*12.41

Give structures (including stereochemistry where appropriate) for compounds **A** to **G**.

(a) Benzene + $CH_3CH_2\overset{\displaystyle O}{\overset{\|}{C}}Cl \longrightarrow$ **A** $\xrightarrow[0\,°C]{PCl_5}$

$\quad$ **B** ($C_9H_{10}Cl_2$) $\xrightarrow[\substack{\text{mineral oil,}\\\text{heat}}]{\text{2NaNH}_2}$ **C** (C_9H_8) $\xrightarrow{H_2,\ Ni_2B\ (P\text{-}2)}$ **D** (C_9H_{10})

(b) **C** $\xrightarrow[\text{(2) H}^2]{\text{(1) Li, liq. NH}_3}$ **E** (C_9H_{10})

(c) **D** $\xrightarrow[2-5\,°C]{Br_2,\ CCl_4}$ **F** + enantiomer (major products)

(d) **E** $\xrightarrow[2-5\,°C]{Br_2,\ CCl_4}$ **G** + enantiomer (major products)

*12.42

Friedel–Crafts acylation of azulene gives mainly one isomer:

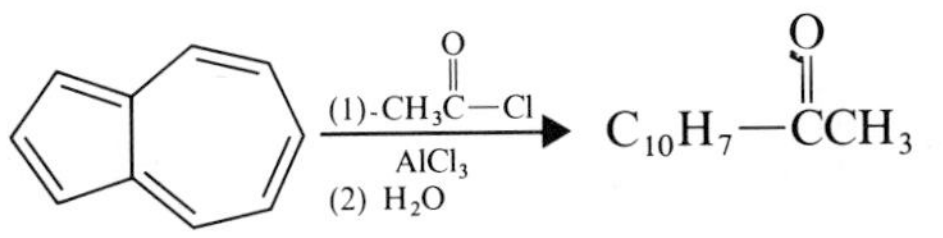

$\qquad\qquad$ Azulene $\qquad\qquad\qquad\qquad$ (mainly one isomer)

One ring of azulene is attacked by $CH_3C{\equiv}O^+$ preferentially because an especially stable arenium ion forms. (a) What is the structure of this arenium ion and (b) why is it especially stable? (c) What is the structure of the acetylazulene that forms as the major product?

CHAPTER THIRTEEN

Interference Colors. This is a portion of the electromagnetic spectrum (Section 13.1) in what is called the visible region.

Spectroscopic Methods
of Structure Determination

The names of most forms of electromagnetic energy have become familiar terms. The *X rays* used in medicine, the *light* that we see, the *ultraviolet* rays that produce sunburns, and the *radio* and *radar* waves used in communication are all different forms of the same phenomenon: electromagnetic radiation.

According to quantum mechanics, electromagnetic radiation has a dual and seemingly contradictory nature. Electromagnetic radiation has the properties of both a wave and a particle. Electromagnetic radiation can be described as a wave occurring simultaneously in electrical and magnetic fields. It can also be described as if it consisted of particles called quanta or photons. Different experiments disclose these two different aspects of electromagnetic radiation. They are not seen together in the same experiment.

A wave is usually described in terms of its **wavelength** (λ) or its **frequency** (v). A simple wave is shown in Fig. 13.1. The distance between consecutive crests (or troughs) is the wavelength. The number of full cycles of the wave that pass a given point each second, as the wave moves through space, is called the *frequency*.

The frequencies of electromagnetic waves are usually reported in cycles per second or **hertz.*** The wavelengths of electromagnetic radiation are expressed in either meters (m), millimeters (1 mm = 10^{-3} m), micrometers (1 μm = 10^{-6} m), or nanometers (1 nm = 10^{-9} m). [An older term for micrometer is *micron* (abbreviated μ) and an older term for nanometer is *millimicron.*]

The energy of a quantum of electromagnetic energy is directly related to its frequency.

$$E = hv$$

where h = Planck's constant, 6.63×10^{-34} J $\cdot$ s,

and $$v = \text{the frequency (Hz)}$$

---

*The term hertz (after the German physicist H. R. Hertz), abbreviated Hz, is now often used in place of *cycles per second.* Frequency of electromagnetic radiation is also sometimes expressed in *wavenumbers,* that is, the number of waves per centimeter.

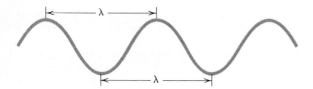

FIGURE 13.1 A simple wave and the wavelength, λ.

This means that the higher the frequency of radiation the greater is its energy. X rays, for example, are much more energetic than rays of visible light. The frequencies of X rays are of the order of 10^{19} Hz, while those of visible light are of the order of 10^{15} Hz.

Since $v = c/\lambda$, the energy of electromagnetic radiation is inversely proportional to its wavelength.

$$E = \frac{hc}{\lambda} \qquad (c = \text{the velocity of light})$$

Thus, per quantum, electromagnetic radiation of long wavelength has low energy, while that of short wavelength has high energy. X rays have wavelengths of the order of 0.1 nm, while visible light has wavelengths between 400 and 750 nm.*

It may be helpful to point out, too, that for visible light, wavelengths (and, thus, frequencies) are related to what we perceive as colors. The light that we call red light has a wavelength of approximately 750 nm. The light we call violet light has a wavelength of approximately 400 nm. All of the other colors of the visible spectrum (the rainbow) lie in between these wavelengths.

The different regions of the electromagnetic spectrum are shown in Fig. 13.2. Nearly every portion of the electromagnetic spectrum from the region of X rays to those of microwave and radio wave has been used in elucidating structures of atoms and molecules. Later in this chaper we discuss the use that can be made of the infrared and radio regions when we take up infrared spectroscopy and nuclear magnetic resonance spectroscopy. At this point we direct our attention to electromagnetic radiation in the near ultraviolet and visible regions and see how it interacts with conjugated polyenes.

13.2 VISIBLE AND ULTRAVIOLET SPECTROSCOPY

When electromagnetic radiation in the ultraviolet and visible regions passes through a compound containing multiple bonds, a portion of the radiation is usually absorbed by the compound. Just how much of the radiation is absorbed depends on the wavelength of the radiation and the structure of the compound. The absorption of radiation is caused by the subtraction of energy from the radiation beam when electrons in orbitals of lower energy are excited into orbitals of higher energy.

Instruments called visible–ultraviolet (UV) spectrometers are used to measure the amount of light absorbed at each wavelength of the visible and ultraviolet region. In these instruments a beam of light is split; half the beam (the sample beam) is directed through a transparent cell containing a solution of the compound being analyzed, and half (the reference beam) is directed through an identical cell that does

*A convenient formula that relates wavelength (in nm) to the energy of electromagnetic radiation is the following:

$$E \text{ (in kcal/mole)} = \frac{2.86 \times 10^4 \text{ kcal} \cdot \text{nm/mole}}{\text{wavelength in nanometers}}$$

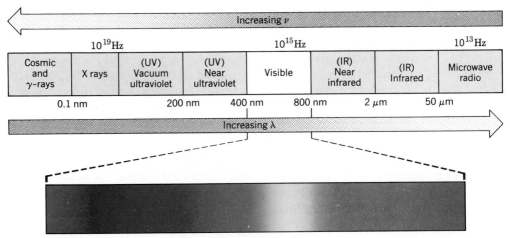

FIGURE 13.2 The electromagnetic spectrum.

not contain the compound but contains the solvent. Solvents are chosen to be transparent in the region being analyzed. The instrument is designed so that it can make a comparison of the intensities of the two beams at each wavelength of the region. If the compound absorbs light at a particular wavelength, the intensity of the sample beam (I_S) will be less than that of the reference beam (I_R). The instrument indicates this by producing a graph—a plot of the wavelength of the entire region *versus* the absorbance (A) of light at each wavelength. [The absorbance at a particular wavelength is defined by the equation: $A_\lambda = \log(I_R/I_S)$.] Such a graph is called an **absorption spectrum.**

A typical ultraviolet absorption spectrum, that of 2,5-dimethyl-2,4-hexadiene, is shown in Fig. 13.3. It shows a broad absorption band in the region between 210 and 260 nm. The absorption is at a maximum at 242.5 nm. It is this wavelength that is usually reported in the chemical literature.

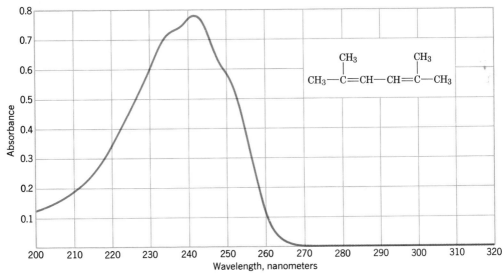

FIGURE 13.3 The ultraviolet absorption spectrum of 2,5-dimethyl-2,4-hexadiene in methanol. (Spectrum courtesy of Sadtler Research Laboratories, Philadelphia.)

In addition to reporting the wavelength of maximum absorption (λ_{max}), chemists often report another quantity that indicates the strength of the absorption, called the **molar absorptivity, ε.*

The molar absorptivity is simply the proportionality constant that relates the observed absorbance (A) at a particular wavelength (λ) to the molar concentration (C) of the sample and the length (l) (in centimeters), of the path of the light beam through the sample cell.

$$A = \varepsilon \times C \times l \quad \text{or} \quad \varepsilon = \frac{A}{C \times l}$$

For 2,5-dimethyl-2,4-hexadiene dissolved in methanol the molar absorptivity at the wavelength of maximum absorbance (242.5 nm) is 13,100 M^{-1} cm^{-1}. In the chemical literature this would be reported as

2,5-dimethyl-2,4-hexadiene, $\lambda_{max}^{methanol}$ 242.5 nm $\quad\quad$ ($\varepsilon = 13,100$)

As we noted earlier, when compounds absorb light in the ultraviolet and visible regions, electrons are excited from lower electronic energy levels to higher ones. For this reason, visible and ultraviolet spectra are often called **electronic spectra.** The absorption spectrum of 2,5-dimethyl-2,4-hexadiene is a typical electronic spectrum because the absorption band (or peak) is very broad. Most absorption bands in the visible and ultraviolet region are broad because each electronic energy level has associated with it vibrational and rotational levels. Thus, electron transitions may occur from any of several vibrational and rotational states of one electronic level to any of several vibrational and rotational states of a higher level.

Alkenes and nonconjugated dienes usually have absorption maxima below 200 nm. Ethene, for example, gives an absorption maximum at 171 nm; 1,4-pentadiene gives an absorption maximum at 178 nm. These absorptions occur at wavelengths that are out of the range of operation of most visible–ultraviolet spectrometers because they occur where the oxygen in air also absorbs. Special air-free techniques must be employed in measuring them.

Compounds whose molecules contain *conjugated* multiple bonds have maxima at wavelengths longer than 200 nm. For example, 1,3-butadiene absorbs at 217 nm. This longer-wavelength absorption by conjugated dienes is a direct consequence of conjugation.

We can understand how conjugation of multiple bonds brings about absorption of light at longer wavelengths if we examine Fig. 13.4.

When a molecule absorbs light at its longest wavelength, an electron is excited from its highest occupied molecular orbital (HOMO) to the lowest unoccupied molecular orbital (LUMO). For most alkenes and alkadienes the highest occupied molecular orbital is a bonding π orbital and the lowest unoccupied molecular orbital is an antibonding π^* orbital. The wavelength of the absorption maximum is determined by the difference in energy between these two levels. The energy gap between the highest occupied molecular orbital and lowest unoccupied molecular orbital of ethene is greater than that between the corresponding orbitals of 1,3-butadiene. Thus, the $\pi \longrightarrow \pi^*$ electron excitation of ethene requires absorption of light of greater energy (shorter wavelength) than the corresponding $\pi_2 \longrightarrow \pi_3^*$ excitation in 1,3-butadiene. The energy difference between the highest occupied molecular orbi-

*In older literature, the molar absorptivity (ε) is often referred to as the molar extinction coefficient.

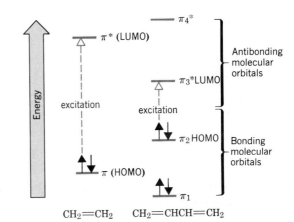

FIGURE 13.4 The relative energies of the π molecular orbitals of ethene and 1,3-butadiene (Section 10.9).

tals and the lowest unoccupied molecular orbitals of the two compounds is reflected in their absorption spectra. Ethene has its λ_{max} at 171 nm; 1,3-butadiene has a λ_{max} at 217 nm.

The narrower gap between the highest occupied molecular orbital and the lowest unoccupied molecular orbital in 1,3-butadiene results from the conjugation of the double bonds. Molecular orbital calculations indicate that a much larger gap should occur in isolated alkadienes. This is borne out experimentally. Isolated alkadienes give absorption spectra similar to those of alkenes. Their λ_{max} are at shorter wavelengths, usually below 200 nm. As we mentioned, 1,4-pentadiene has its λ_{max} at 178 nm.

Conjugated alkatrienes absorb at longer wavelengths than conjugated alkadienes, and this too can be accounted for in molecular orbital calculations. The energy gap between the highest occupied molecular orbital and the lowest unoccupied molecular orbital of an alkatriene is even smaller than that of an alkadiene. In fact, there is a general rule that states that *the greater the number of conjugated multiple bonds a compound contains, the longer will be the wavelength at which the compound absorbs light.*

Polyenes with eight or more conjugated double bonds absorb light in the visible region of the spectrum. For example, β-carotene, a precursor of Vitamin A and a compound that imparts its orange color to carrots, has 11 conjugated double bonds; β-carotene has an absorption maximum at 497 nm, well into the visible region. Light of 497 nm has a blue-green color; this is the light that is absorbed by β-carotene. We perceive the complementary color of blue green, which is red orange.

β-Carotene

Lycopene, a compound partly responsible for the red color of tomatoes, also has 11 conjugated double bonds. Lycopene has an absorption maximum at 505 nm, and it absorbs there intensely. (Approximately 0.02 g of lycopene can be isolated from 1 kg of fresh, ripe tomatoes.)

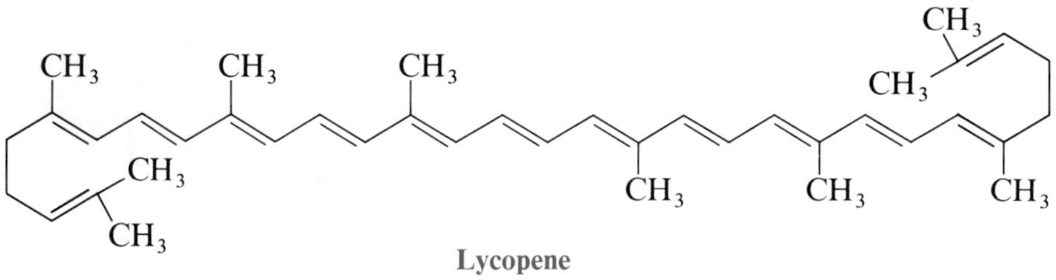

Lycopene

Table 13.1 gives the values of λ_{max} for a number of unsaturated compounds.

TABLE 13.1 Long-wavelength absorption maxima of unsaturated hydrocarbons

COMPOUND	STRUCTURE	λ_{max} (nm)	ε_{max}
Ethene	$CH_2{=}CH_2$	171	15,530
trans-3-Hexene	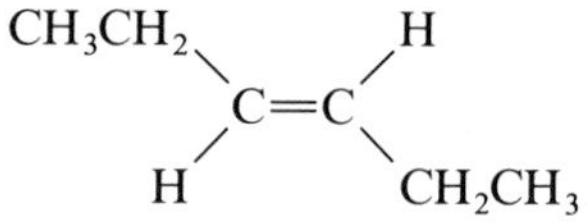	184	10,000
Cyclohexene	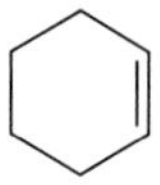	182	7,600
1-Octene	$CH_3(CH_2)_5CH{=}CH_2$	177	12,600
1-Octyne	$CH_3(CH_2)_5C{\equiv}CH$	185	2,000
1,3-Butadiene	$CH_2{=}CHCH{=}CH_2$	217	21,000
cis-1,3-Pentadiene	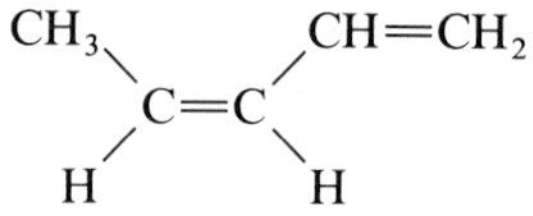	223	22,600
trans-1,3-Pentadiene	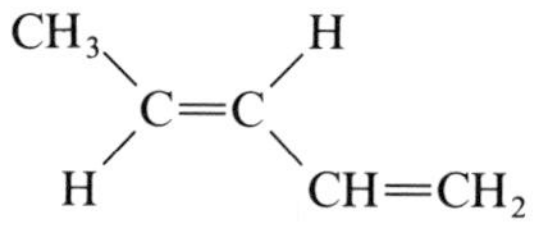	223.5	23,000
1-Buten-3-yne	$CH_2{=}CHC{\equiv}CH$	228	7,800
1,4-Pentadiene	$CH_2{=}CHCH_2CH{=}CH_2$	178	17,000
1,3-Cyclopentadiene		239	3,400
1,3-Cyclohexadiene		256	8,000
trans-1,3,5-Hexatriene	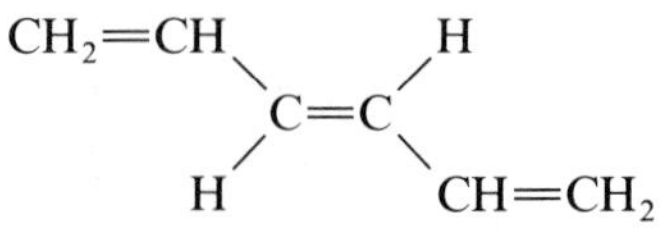	274	50,000

Compounds with carbon–oxygen double bonds also absorb light in the ultraviolet region. Acetone, for example, has a broad absorption peak at 280 nm that corresponds to the excitation of an electron from one of the unshared pairs (a nonbonding or "n" electron) to the π^* orbital of the carbon–oxygen double bond:

$$CH_3 \!\!\diagdown \atop CH_3 \!\!\diagup \!\!\! C{=}O\!: \quad \xrightarrow{\;n \longrightarrow \pi^*\;} \quad CH_3 \!\!\diagdown \atop CH_3 \!\!\diagup \!\!\! \overset{\cdot}{C}{=}\overset{\cdot}{O}\cdot$$

Acetone

$\lambda_{max} = 280$

$\varepsilon_{max} = 15$

Compounds in which the carbon–oxygen double bond is conjugated with a carbon–carbon double bond have absorption maxima corresponding to $n \longrightarrow \pi^*$ excitations and $\pi \longrightarrow \pi^*$ excitations. The $n \longrightarrow \pi^*$ absorption maximum occurs at longer wavelengths but is much weaker.

$$CH_2{=}CH{-}\underset{\underset{CH_3}{|}}{C}{=}O$$

$n \longrightarrow \pi^*\ \lambda_{max} = 324$ nm, $\varepsilon_{max} = 24$

$\pi \longrightarrow \pi^*\ \lambda_{max} = 219$ nm, $\varepsilon_{max} = 3600$

PROBLEM 13.1

Two compounds, **A** and **B**, have the same molecular formula C_6H_8. Both **A** and **B** decolorize bromine in carbon tetrachloride and both give positive tests with cold dilute potassium permanganate. Both **A** and **B** react with 2 moles of hydrogen in the presence of platinum to yield cyclohexane. **A** shows an absorption maximum at 256 nm, while **B** shows no absorption maximum beyond 200 nm. What are the structures of **A** and **B**?

PROBLEM 13.2

Three compounds, **D**, **E**, and **F**, have the same molecular formula C_5H_6. In the presence of a platinum catalyst, all three compounds absorb 3 moles of hydrogen and yield pentane. Compounds **E** and **F** give a precipitate when treated with ammoniacal silver nitrate; compound **D** gives no reaction. Compounds **D** and **E** show an absorption maximum near 230 nm. Compound **F** shows no absorption maximum beyond 200 nm. Propose structures for **D**, **E**, and **F**.

13.3 PROTON NUCLEAR MAGNETIC RESONANCE SPECTROSCOPY

The hydrogen nucleus, or proton, has magnetic properties. When one places a compound containing hydrogen in a very strong magnetic field and simultaneously irradiates it with electromagnetic energy, the hydrogen nuclei of the compound may

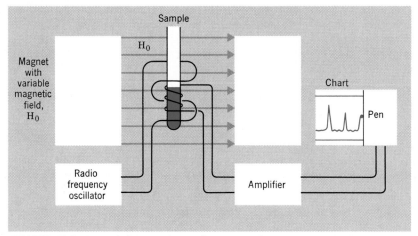

FIGURE 13.5 Essential parts of a nuclear magnetic resonance spectrometer.

absorb energy through a process known as **magnetic resonance.†** This absorption of energy, like all processes that occur on the atomic and molecular scale, is *quantized.* Absorption of energy does not occur until the strength of the magnetic field and the frequency of electromagnetic radiation are at specific values.

Instruments known as nuclear magnetic resonance (nmr) spectrometers (Fig. 13.5) allow chemists to measure the absorption of energy by hydrogen nuclei. These instruments use very powerful magnets and irradiate the sample with electromagnetic radiation in the radio frequency region.

Nuclear magnetic resonance spectrometers are usually designed so that they irradiate the compound with electromagnetic energy of a constant frequency while the magnetic field strength is varied. When the magnetic field reaches the correct strength, the nuclei absorb energy and resonance occurs. This causes a tiny electrical current to flow in an antenna coil surrounding the sample. The instrument then amplifies this current and displays it as a signal (a peak or series of peaks) on a strip of calibrated chart paper.

If hydrogen nuclei were stripped of their electrons and isolated from other nuclei, all hydrogen nuclei (protons) would absorb energy at the same magnetic field strength for a given frequency of electromagnetic radiation. If this were the case, nuclear magnetic resonance spectrometers would only be very expensive instruments for analysis for hydrogen.

Fortunately, the nuclei of hydrogen atoms of compounds of interest to the organic chemist are not stripped of their electrons, and they are not isolated from each other. Some hydrogen nuclei are in regions of greater electron density than others. Because of this, the protons of these compounds absorb energy at *slightly different* magnetic field strengths. The actual field strength at which absorption occurs is highly dependent on the magnetic environment of each proton. This magnetic environment depends on two factors: magnetic fields generated by circulating electrons and magnetic fields that result from other nearby protons (or other magnetic nuclei).

We shall discuss the theory of nuclear magnetic resonance in more detail later.

---

†Magnetic resonance is an entirely different phenomenon from the resonance theory that we have discussed in earlier chapters.

Before we do, however, it will be helpful to examine the proton nuclear magnetic resonance spectra of some simple compounds.

Figure 13.6 shows the proton nuclear magnetic resonance spectrum of *p*-xylene. The spectrum is the blue line. The black line is called the "integral curve" and we shall explain this later.

Magnetic field strength is measured along the bottom of the spectrum on a delta (δ) scale in units of parts per million (ppm) and along the top in hertz (cycles per second). We shall have more to say about these units later; for the moment, we need only point out that the externally applied magnetic field strength increases from left to right. A signal that occurs at $\delta = 7$ ppm occurs at a lower external magnetic field strength than one that occurs at $\delta = 2$ ppm. Signals on the left of the spectrum are also said to occur **downfield** and those on the right are said to be **upfield.**

The spectrum in Fig. 13.6 shows a small signal at $\delta = 0$ ppm. This is caused by a compound that has been added to the sample to allow calibration of the instrument.

The first feature we want to notice is the relation between the number of signals in the spectrum and the number of different types of hydrogen atoms in the compound.

p-Xylene has only *two* different types of hydrogen atoms, and it gives only *two* signals in its nuclear magnetic resonance spectrum.

The two different types of hydrogen atoms of *p*-xylene are the hydrogen atoms of the methyl groups and the hydrogen atoms of the benzene ring. The six methyl hydrogen atoms of *p*-xylene are all *equivalent* and they are in a different environment from the four hydrogen atoms of the ring. The six methyl hydrogen atoms give rise to the signal that occurs at $\delta = 2.30$ ppm. The four hydrogen atoms of the benzene ring are also equivalent; they give rise to the signal at $\delta = 7.05$ ppm.

Next, we want to examine the relative magnitude of the peaks (or signals), for these are often helpful in assigning peaks to particular groups of hydrogen atoms. What is important here is not necessarily the height of each peak, but *the area underneath it.* These areas, when accurately measured (the spectrometers do this automatically), are in the same ratio as the number of hydrogen atoms causing each

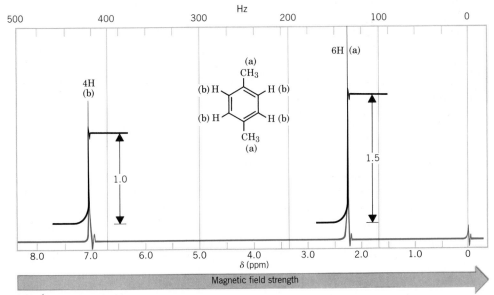

FIGURE 13.6 The proton nmr spectrum of *p*-xylene. (Spectrum courtesy of Varian Associates, Palo Alto, CA.)

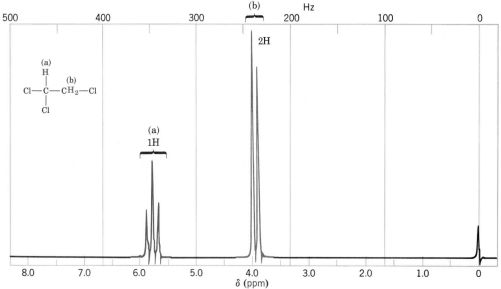

FIGURE 13.7 The proton nmr spectrum of 1,1,2-trichloroethane. (Spectrum courtesy of Varian Associates, Palo Alto, CA.)

signal. We can see, however, without measuring, that the area under the signal for the methyl hydrogen atoms of *p*-xylene (6H) is larger than that for the phenyl hydrogen atoms (4H). When these areas are measured accurately they are found to be in a ratio of 1.5:1 or 3:2 or 6:4.

The black line superimposed on each peak in the spectrum of *p*-xylene shows one way that nuclear magnetic spectrometers display the relative area under each peak. This line, called **the integral curve,** rises by an amount that is proportional to the area under each peak. Figure 13.6 shows how these heights are measured. In this case the ratio of heights on the integral curve is 1.5:1 or 3:2 or 6:4.

A third feature of proton nmr spectra that provides us with information about the structure of a compound can be illustrated if we examine the spectrum for 1,1,2-trichloroethane (Fig. 13.7).

In Fig. 13.7 we have an example of signal splitting. Signal splitting is a phenomenon that arises from magnetic influences of hydrogens on atoms adjacent to those bearing the hydrogen atoms causing the general signal. The signal (*b*) from the two equivalent hydrogen atoms of the —CH_2Cl group is split into two peaks (a doublet) by the magnetic influence of the hydrogen of the —$CHCl_2$ group. Conversely, the signal (*a*) from the hydrogen of the —$CHCl_2$ group is split into three peaks (a triplet) by the magnetic influences of the two equivalent hydrogens of the —CH_2Cl group.

At this point signal splitting may seem like an unnecessary complication. As we gain experience in interpreting proton nmr spectra we shall find that because signal splitting occurs in a predictable way, it often provides us with important information about the structure of the compound.

Now that we have had an introduction to the important features of proton nmr spectra, we are in a position to consider them in greater detail.

13.4 NUCLEAR SPIN: THE ORIGIN OF THE SIGNAL

We are already familiar with the concept of electron spin and with the fact that the spins of electrons confer on them the spin quantum states of $+\frac{1}{2}$ or $-\frac{1}{2}$. Electron spin

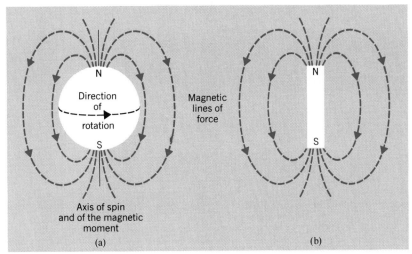

FIGURE 13.8 (a) The magnetic field associated with a spinning proton. (b) The spinning proton resembles a tiny bar magnet.

is the basis for the Pauli exclusion principle (Section 1.11); it allows us to understand how two electrons with paired spins may occupy the same atomic or molecular orbital.

The nuclei of certain isotopes also spin and therefore these nuclei possess spin quantum numbers, I. The nucleus of ordinary hydrogen, ^1H (i.e., a proton), is like the electron; its spin quantum number I is $\frac{1}{2}$ and it can assume either of two spin states: $+\frac{1}{2}$ or $-\frac{1}{2}$. These correspond to the magnetic moments allowed for $I = \frac{1}{2}$, $m = +\frac{1}{2}$ or $-\frac{1}{2}$. Other nuclei with spin quantum numbers $I = \frac{1}{2}$ are ^{13}C, ^{19}F, and ^{31}P. Some nuclei, such as ^{12}C, ^{16}O, and ^{32}S, have no spin ($I = 0$) and these nuclei do not give an nmr spectrum. Other nuclei have spin quantum numbers greater than $\frac{1}{2}$. In our treatment here, however, we shall be primarily concerned with the spectra that arise from protons and from ^{13}C, both of which have $I = \frac{1}{2}$. We shall begin with proton spectra.

Since the proton is electrically charged, the spinning proton generates a tiny magnetic moment—one that coincides with the axis of spin (Fig. 13.8). This tiny magnetic moment confers on the spinning proton the properties of a tiny bar magnet.

In the absence of a magnetic field (Fig. 13.9a), the magnetic moments of the protons of a given sample are randomly oriented. When a compound containing hydrogen (and thus protons) is placed in an applied external magnetic field, however, the protons may assume one of two possible orientations with respect to the external

FIGURE 13.9 (a) In the absence of a magnetic field the magnetic moments of protons (represented by arrows) are randomly oriented. (b) When an external magnetic field (H_0) is applied the protons orient themselves. Some are aligned with the applied field (α spin state) and some against it (β spin state).

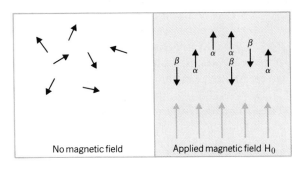

No magnetic field

Applied magnetic field H_0

(a)

(b)

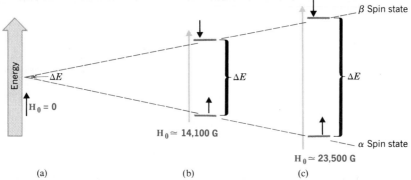

FIGURE 13.10 The energy difference between the two spin states of a proton depends on the strength of the applied external magnetic field, H_0. (*a*) If there is no applied field ($H_0 = 0$), there is no energy difference between the two states. (*b*) If $H_0 \simeq 14,100$ G, the energy difference corresponds to that of electromagnetic radiation of 60×10^6 Hz (60 MHz). (*c*) In a magnetic field of approximately 23,500 G, the energy difference corresponds to electromagnetic radiation of 100×10^6 Hz (100 MHz). Instruments are available that operate at these and even higher frequencies (as high as 500 MHz).

magnetic field. The magnetic moment of the proton may be aligned "with" the external field or "against" it (Fig. 13.9*b*). These alignments correspond to the two spin states mentioned earlier.

As we might expect, the two alignments of the proton in an external field are not of equal energy. When the proton is aligned with the magnetic field, its energy is lower than when it is aligned against the magnetic field.

Energy is required to "flip" the proton from its lower-energy state (with the field) to its higher-energy state (against the field). In a nuclear magnetic resonance spectrometer this energy is supplied by electromagnetic radiation in the radio frequency region. When this energy absorption occurs, the nuclei are said to be *in resonance* with the electromagnetic radiation. The energy required is proportional to the strength of the magnetic field (Fig. 13.10). One can show by relatively simple calculations, that in a magnetic field of approximately 14,100 G, electromagnetic radiation of 60×10^6 cycles per second (cps) (60 MHz) supplies the correct amount of energy for protons.*

As we mentioned earlier, nmr spectrometers are designed so that the radio frequency is kept constant (at 60 MHz, for example) and the magnetic field is varied. When the magnetic field is tuned to precisely the right strength, the protons of a particular set in the molecule flip from one state to the other. In doing so they absorb radio frequency energy. This flipping of the protons generates a small electric current in a coil of wire surrounding the sample. After being amplified, the current is displayed as a signal in the spectrum.

13.5 SHIELDING AND DESHIELDING OF PROTONS

All protons do not absorb energy at the same external magnetic field strength. The two spectra that we examined earlier demonstrate this for us. The aromatic protons of

*The relationship between the frequency of the radiation, v, and the strength of the magnetic field, H_0, is,

$$v = \frac{\mu H_0}{2\pi}$$

where μ is the magnetogyric ratio. For a proton, $\mu = 26,753$ radians. $\text{s}^{-1} \cdot \text{G}^{-1}$.

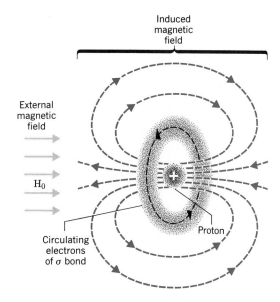

FIGURE 13.11 The circulations of the electrons of a C—H bond under the influence of an external magnetic field. The electron circulations generate a small magnetic field (an induced *field*) that shields the proton from the external field.

p-xylene absorb at lower field strength ($\delta = 7.05$ ppm); the various alkyl protons of *p*-xylene and 1,1,2-trichloroethane all absorb at higher magnetic fields strengths.

The general position of a signal in a nuclear magnetic resonance spectrum — that is, the strength of the magnetic field required to bring about absorption of energy — can be related to electron densities and electron circulations in the compounds. Under the influence of an external magnetic field the electrons move in certain preferred paths. Because they do, and because electrons are charged particles, they generate tiny magnetic fields.

We can see how this happens if we consider the electrons around the proton in a σ bond of a C—H group. In doing so, we shall oversimplify the situation by assuming that σ electrons move in generally circular paths. The magnetic field generated by these σ electrons is shown in Fig. 13.11.

The small magnetic field generated by the electrons is called **an induced field.** *At the proton, the induced magnetic field opposes the external magnetic field.* This means that the actual magnetic field sensed by the proton is slightly less than the external field. The electrons are said *to shield* the proton.

A proton shielded by electrons will not, of course, absorb at the same external field strength as a proton that has no electrons. A shielded proton will absorb *at higher external field strengths;* the external field must be made larger by the spectrometer in order to compensate for the small induced field (Fig. 13.12).

The extent to which a proton is shielded by the circulation of σ electrons depends on the relative electron density around the proton. This electron density

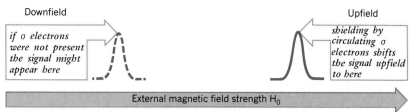

FIGURE 13.12 Shielding by σ electrons causes proton nmr absorption to be shifted to higher external magnetic field strengths.

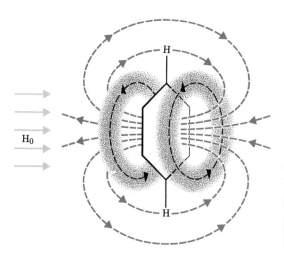

H_0

FIGURE 13.13 The induced magnetic field of the π electrons of benzene deshields the benzene protons. Deshielding occurs because at the location of the protons the induced field is in the same direction as the applied field.

depends largely on the presence or absence of electronegative groups. Electronegative groups withdraw electron density from the C—H bond, particularly if they are attached to the same carbon. We can see an example of this effect in the spectrum of 1,1,2-trichloroethane (Fig. 13.7). The proton of carbon-1 absorbs at a lower magnetic field strength ($\delta = 5.77$ ppm) than the protons of carbon-2 ($\delta = 3.95$ ppm). Carbon-1 bears two highly electronegative chloro groups whereas carbon-2 bears only one. The protons of carbon-2, consequently, are more effectively shielded because the σ electron density around them is greater.

The circulations of delocalized π electrons generate magnetic fields that can either **shield or deshield** nearby protons. Whether shielding or deshielding occurs depends on the location of the proton in the *induced* field. The aromatic protons of benzene derivatives (Fig. 13.13) are *deshielded* because their locations are such that the induced magnetic field reinforces the applied magnetic field.

Because of this deshielding effect the absorption of energy by phenyl protons occurs downfield at relatively low magnetic field strength. The protons of benzene itself absorb at $\delta = 7.27$ ppm. The aromatic protons of *p*-xylene (Fig. 13.6) absorb at $\delta = 7.05$ ppm.

The deshielding of external aromatic protons that results from the circulating π electrons is one of the best pieces of physical evidence that we have for π electron delocalization in aromatic rings. In fact, low field strength proton absorption is often used as a criterion for aromaticity in newly synthesized conjugated cyclic compounds.

Not all aromatic protons absorb at low magnetic field strengths, however. Large-ring aromatic compounds have been synthesized that have hydrogens *in the*

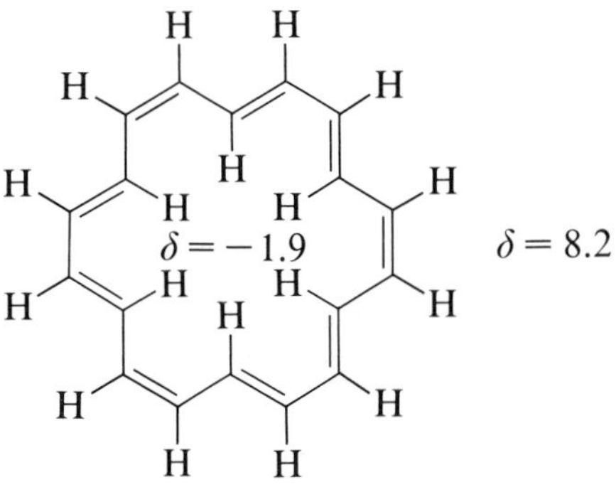

$\delta = -1.9$ $\delta = 8.2$

FIGURE 13.14 [18]Annulene. The internal protons are highly shielded and absorb at $\delta = -1.9$ ppm. The external protons are highly deshielded and absorb at $\delta = 8.2$ ppm.

center of the ring (in the π electron cavity). The protons of these internal hydrogen atoms absorb at unusually high magnetic field strengths because they are highly shielded by the opposing induced field in the center of the ring (cf. Fig. 13.13). These internal protons often absorb at field strengths greater than that used for the reference point, $\delta = 0$. The internal protons of [18]annulene (Fig. 13.14) absorb at $\delta = -1.9$ ppm.

PROBLEM 13.3

The methyl protons of *trans*-15,16-dimethyldihydropyrene (Section 11.7B) absorb at very high magnetic field strengths, $\delta = -4.2$. Can you account for this?

Pi-electron circulations also *shield* the protons of acetylene causing them to absorb at higher magnetic field strengths than we might otherwise expect. If we were to consider *only* the relative electronegativities of carbon in its three hybridization states, we might expect the following order of protons attached to each type of carbon:

(low field strength) $sp < sp^2 < sp^3$ (high field strength)

In fact, acetylenic protons absorb between $\delta = 2.0$ and $\delta = 3.0$ and the order is

(low field strength) $sp^2 < sp < sp^3$ (high field strength)

This upfield shift of the absorption of acetylenic protons is a result of shielding produced by the circulating π electrons of the triple bond. The origin of this shielding is illustrated in Fig. 13.15.

13.6 THE CHEMICAL SHIFT

We see now that shielding and deshielding effects cause the absorptions of protons to be shifted from the position at which a bare proton would absorb (i.e., a proton stripped of its electrons). Since these shifts result from the circulations of electrons in *chemical* bonds, they are called **chemical shifts.**

Chemical shifts are measured with reference to the absorption of protons of reference compounds. A reference is used because it is impractical to measure the actual value of the magnetic field at which absorptions occur. The reference compound most often used is tetramethylsilane (TMS). A small amount of tetramethylsilane is usually added to the sample whose spectrum is being measured, and the signal from the 12 equivalent protons of tetramethylsilane is used to establish the zero point on the delta scale.

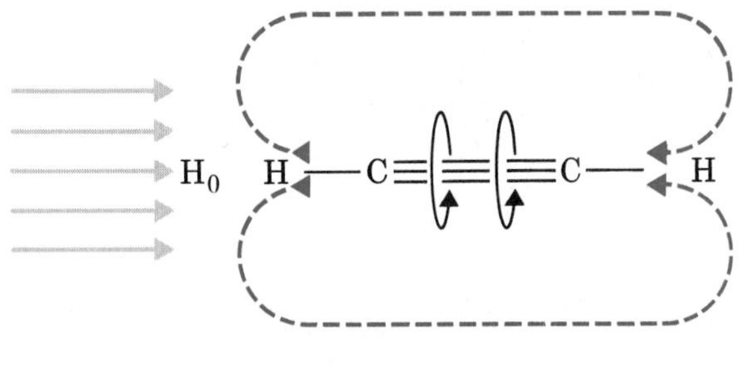

FIGURE 13.15 The shielding of acetylenic protons by π electron circulations. Shielding causes acetylenic protons to absorb further upfield than vinylic protons.

$$Si(CH_3)_4$$

Tetramethylsilane
(TMS)

Tetramethylsilane was chosen as a reference compound for several reasons. It has 12 hydrogen atoms and, therefore, a very small amount of tetramethylsilane gives a relatively large signal. Because the hydrogen atoms are all equivalent, they give a *single signal*. Since silicon is less electronegative than carbon the protons of tetramethylsilane are in regions of high electron density. They are, as a result, highly shielded, and the signal from tetramethylsilane occurs in a region of the spectrum where few other hydrogen atoms absorb. Thus, their signal seldom interferes with the signals from other hydrogen atoms. Tetramethylsilane, like an alkane, is relatively inert. Finally, it is volatile; its boiling point is 27 °C. After the spectrum has been determined, the tetramethylsilane can be removed easily by evaporation.

Chemical shifts are measured in hertz (cycles per second), as if the frequency of the electromagnetic radiation were being varied. In actuality it is the magnetic field

TABLE 13.2 Approximate proton chemical shifts

TYPE OF PROTON	CHEMICAL SHIFT (δ, ppm)
1° Alkyl, RCH_3	0.8–1.0
2° Alkyl, RCH_2R	1.2–1.4
3° Alkyl, R_3CH	1.4–1.7
Allylic, $R_2C{=}C{-}CH_3$ $\quad\quad\quad\quad\mid$ $\quad\quad\quad\quad R$	1.6–1.9
Benzylic, $ArCH_3$	2.2–2.5
Alkyl chloride, RCH_2Cl	3.6–3.8
Alkyl bromide, RCH_2Br	3.4–3.6
Alkyl iodide, RCH_2I	3.1–3.3
Ether, $ROCH_2R$	3.3–3.9
Alcohol, $HOCH_2R$	3.3–4.0
Ketone, $RCCH_3$ $\quad\quad\quad\;\;\|$ $\quad\quad\quad\;\;O$	2.1–2.6
Aldehyde, RCH $\quad\quad\quad\;\;\|$ $\quad\quad\quad\;\;O$	9.5–9.6
Vinylic, $R_2C{=}CH_2$	4.6–5.0
Vinylic, $R_2C{=}CH$ $\quad\quad\quad\quad\mid$ $\quad\quad\quad\quad R$	5.2–5.7
Aromatic, ArH	6.0–9.5
Acetylenic, $RC{\equiv}CH$	2.5–3.1
Alcohol hydroxyl, ROH	0.5–6.0[a]
Carboxylic, $RCOH$ $\quad\quad\quad\quad\|$ $\quad\quad\quad\quad O$	10–13[a]
Phenolic, $ArOH$	4.5–7.7[a]
Amino, $R{-}NH_2$	1.0–5.0[a]

[a]The chemical shifts of these protons vary in different solvents and with temperature and concentration.

that is changed. But since the values of frequency and the strength of the magnetic field are mathematically proportional, frequency units (hertz) are appropriate ones.

The chemical shift of a proton, when expressed in hertz, is proportional to the strength of the external magnetic field. Since spectrometers with different magnetic field strengths are commonly used, it is desirable to express chemical shifts in a form that is independent of the strength of the external field. This can be done easily by dividing the chemical shift by frequency of the spectrometer, with both numerator and denominator of the fraction expressed in frequency units (hertz). Since chemical shifts are always very small (typically less than 500 Hz) compared with the total field strength (commonly the equivalent of 30, 60, and 100 *million* Hz), it is convenient to express these fractions in units of *parts per million* (ppm). This is the origin of the delta scale for the expression of chemical shifts relative to TMS.

$$\delta = \frac{(\text{observed shift from TMS in hertz}) \times 10^6}{(\text{operating frequency of the instrument in hertz})}$$

Table 13.2 gives the *approximate* values of proton chemical shifts for some common hydrogen-containing groups.

13.7 CHEMICAL SHIFT EQUIVALENT AND NONEQUIVALENT PROTONS

Two or more protons that are in identical environments have the same chemical shift and, therefore, give only one proton nmr signal. How do we know when protons are in the same environment? For most compounds, protons that are in the same environment are also equivalent in chemical reactions. That is, **chemically equivalent** protons are **chemical shift equivalent** in proton nmr spectra.

We saw, for example, that the six methyl protons of *p*-xylene give a single proton nmr signal. We probably recognize, intuitively, that these six hydrogen atoms are chemically equivalent. We can demonstrate their equivalence, however, by replacing each hydrogen in turn with some other group. If in making these substitutions we get the same compound from each replacement, then the protons are chemically equivalent and are chemical shift equivalent. The replacements can be replacements that occur in an actual chemical reaction or they can be purely imaginary. For the methyl hydrogen atoms of *p*-xylene we can think of an actual chemical reaction that demonstrates their equivalence, *benzylic bromination.* Benzylic bromination produces the same monobromo product regardless of which of the six hydrogen atoms is replaced.

etc.

We can also think of a chemical reaction that demonstrates the equivalence of the four aromatic hydrogen atoms of *p*-xylene, *ring bromination.* Once again, we get the same compound regardless of which of the four hydrogen atoms is replaced.

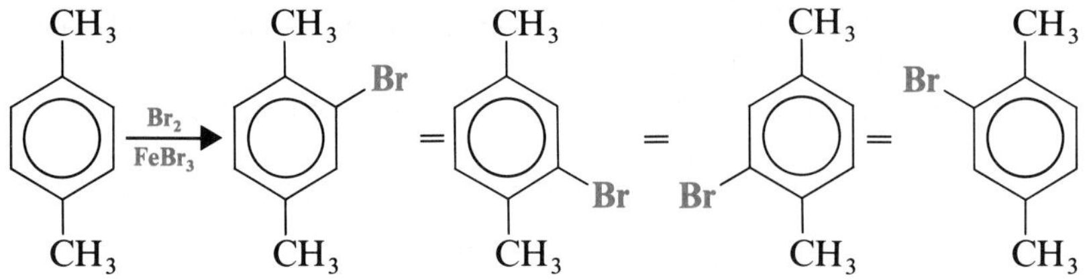

PROBLEM 13.4

How many different sets of equivalent protons do each of the following compounds have? How many signals would each compound give in its proton nmr spectrum?

(a) CH_3CH_3

(b) $CH_3CH_2CH_3$

(c) CH_3OCH_3

(d) 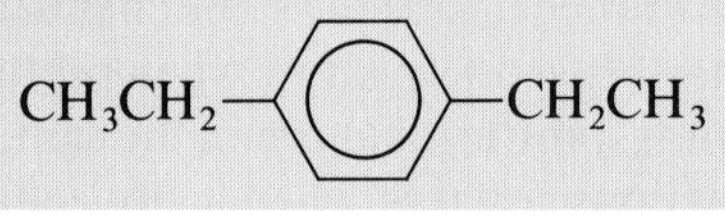 CH_3CH_2—⟨○⟩—CH_2CH_3

(e) $CH_3\overset{\overset{\displaystyle O}{\|}}{C}$—$OCH_3$

(f) $CH_3\overset{\overset{\displaystyle O}{\|}}{C}$—$OCH(CH_3)_2$

13.7A Enantiotopic and Diastereotopic Hydrogen Atoms

If replacement of each of two hydrogen atoms by the same group yields compounds that are enantiomers, the two hydrogen atoms are said to be **enantiotopic**. *The protons of enantiotopic hydrogen atoms have the same chemical shift and give only one proton nmr signal.**

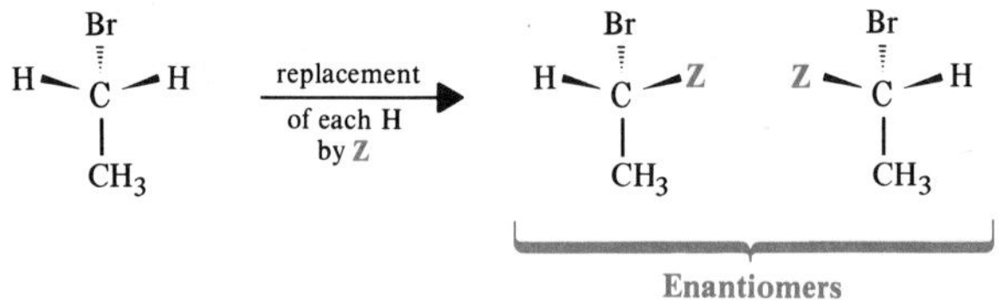

The two hydrogen atoms of the —CH_2Br group of ethyl bromide are enantiotopic. Ethyl bromide, then, gives two signals in its pmr spectrum. The three equivalent protons of the CH_3— group give one signal; the two enantiotopic protons of the —CH_2Br group give the other signal. (The proton nmr spectrum of ethyl bromide as

---

*Enantiotopic hydrogen atoms may not have the same chemical shift if the compound is dissolved in an optically active solvent. However, most proton nmr spectra are determined using optically inactive solvents and in this situation enantiotopic protons have the same chemical shift.

we shall see, actually consists of seven peaks. This is a result of signal splitting, which will be explained in Section 13.8.)

If replacement of each of two hydrogen atoms by a group, **Z**, gives compounds that are diastereomers, the two hydrogens are said to be **diastereotopic.** Except for accidental coincidence, ***diastereotopic protons do not have the same chemical shift and give rise to different proton nmr signals.***

The two protons of the $=CH_2$ group of chloroethene are diastereotopic.

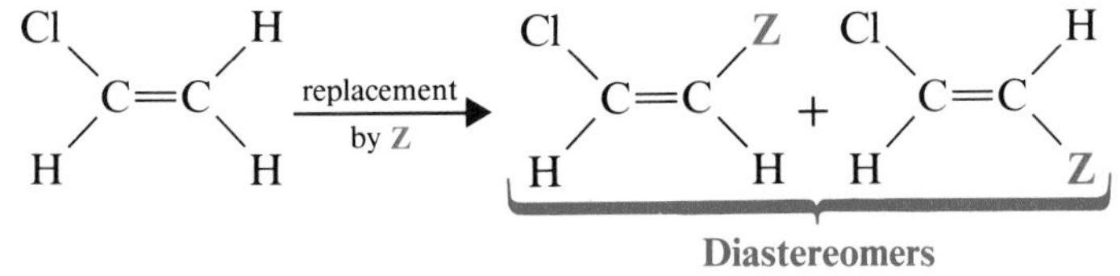

Diastereomers

Chloroethene, then, should give signals from three nonequivalent protons; one for the proton of the $ClCH=$ group, and one for each of the diastereotopic protons of the $=CH_2$ group.

The two methylene ($-CH_2-$) protons of the following *sec*-butyl alcohol are also diastereotopic. We can illustrate this with one enantiomer of *sec*-butyl alcohol in the following way:

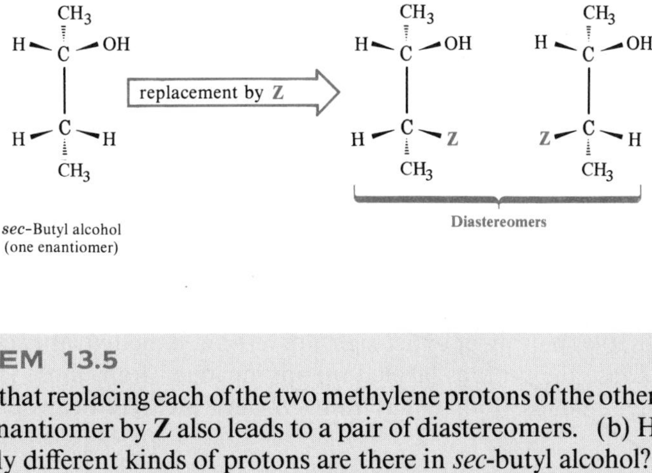

sec-Butyl alcohol
(one enantiomer)

Diastereomers

PROBLEM 13.5

(a) Show that replacing each of the two methylene protons of the other *sec*-butyl alcohol enantiomer by **Z** also leads to a pair of diastereomers. (b) How many chemically different kinds of protons are there in *sec*-butyl alcohol? (c) How many proton nmr signals would you expect to find in the spectrum of *sec*-butyl alcohol?

PROBLEM 13.6

How many proton nmr signals would you expect from each of the following compounds? (Neglect signal splitting.)

(a) $CH_3CH_2CH_2CH_3$

(b) CH_3CH_2OH

(c) $CH_3CH=CH_2$

(d) *trans*-2-Butene

(e) 1,2-Dibromopropane

(f) 1,1-Dimethylcyclopropane

(g) *trans*-1,2-Dimethylcyclopropane

(h) *cis*-1,2-Dimethylcyclopropane

(i) 1-Pentene

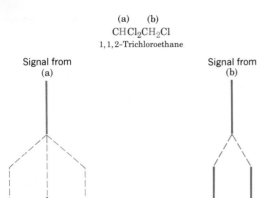

$$\underset{1,1,2\text{-Trichloroethane}}{\overset{\text{(a)} \quad \text{(b)}}{CHCl_2CH_2Cl}}$$

Signal from (a) — Split into a triplet by the two (b) protons

Signal from (b) — Split into a doublet by the (a) proton

FIGURE 13.16 Signal splitting in 1,1,2-trichloroethane.

13.8 SIGNAL SPLITTING: SPIN–SPIN COUPLING

Signal splitting is caused by magnetic fields of protons on nearby atoms. We have seen an example of signal splitting in the spectrum of 1,1,2-trichloroethane (Fig. 13.7). The signal from the two equivalent protons of the $-CH_2Cl$ group of 1,1,2-trichloroethane is split into two peaks by the single proton of the $CHCl_2-$ group. The signal from the proton of the $CHCl_2-$ group is split into three peaks by the two protons of the $-CH_2Cl$ group. This is further illustrated in Fig. 13.16.

Signal splitting arises from a phenomenon known as **spin–spin coupling,** which we shall soon examine. Spin–spin coupling effects are transferred primarily through the bonding electrons and *are not usually observed if the coupled protons are separated by more than three σ bonds.* Thus, we observe signal splitting from the protons of *adjacent σ-bonded* atoms as in 1,1,2-trichloroethane (Fig. 13.7). However, we would not observe splitting of either signal of *tert*-butyl methyl ether (see following structure) because the protons labeled (*b*) are separated from those labeled (*a*) by more than three σ bonds. Both signals from *tert*-butyl methyl ether would be singlets.

$$\overset{(a)}{CH_3}$$
$$\overset{(a)}{CH_3}-\overset{|}{\underset{|}{C}}-O-\overset{(b)}{CH_3}$$
$$\underset{(a)}{CH_3}$$

tert-**Butyl methyl ether**
(no signal splitting)

Signal splitting is not observed for protons that are chemically equivalent or enantiotopic. That is, signal splittings do not occur between protons that have *exactly the same chemical shift.* Thus, we would not expect, and do not find, signal splitting in the signal from the six equivalent hydrogen atoms of ethane.

$$CH_3CH_3 \qquad \text{(no signal splitting)}$$

Nor do we find signal splitting occurring from enantiotopic protons of methoxy-acetonitrile (Fig. 13.17).

> There is a subtle distinction between *spin–spin coupling* and signal splitting. Spin–spin coupling often occurs between sets of protons that have the same chemical shift (and this coupling can be detected by methods that we shall not go into here). However, spin–spin coupling *leads to signal splitting only when the sets of protons have different chemical shifts.*

Let us now explain how signal splitting arises from coupled sets of protons that are not chemical shift equivalent.

We have seen that protons can be aligned in only two ways in an external magnetic field: with the field or against it. Therefore, the magnetic moment of a proton on an adjacent atom may affect the magnetic field at the proton whose signal we are observing in only one of two ways. The occurrence of these two slightly different effects causes the appearance of a smaller peak somewhat upfield (from where the signal might have occurred) and another peak somewhat downfield.

Figure 13.18 shows how two possible orientations of a neighboring proton, H_b, split the signal of the proton H_a. (H_b and H_a are not equivalent.)

The separation of these peaks in frequency units is called the **coupling constant** and is abbreviated J_{ab}. Coupling constants are generally reported in hertz (cycles per second). Because coupling is caused entirely by internal forces, the magnitudes of coupling constants *are not* dependent on the magnitude of the applied field. Coupling constants measured on an instrument operating at 60 MHz will be the same as those measured on an instrument operating at 100 MHz.

When we determine proton nmr spectra we are, of course, observing effects produced by billions of molecules. Since the difference in energy between the two possible orientations of the proton of H_b is very small, the two orientations will be present in roughly (but not exactly) equal numbers. The signal that we observe from H_a is, therefore, split into two peaks of roughly equal intensity, *a 1:1 doublet.*

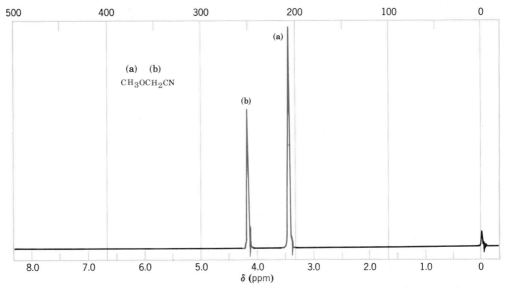

FIGURE 13.17 The proton nmr spectrum of methoxyacetonitrile. The signal of the enantiotopic protons (*b*) is not split. (Spectrum courtesy of Varian Associates, Palo Alto, CA.)

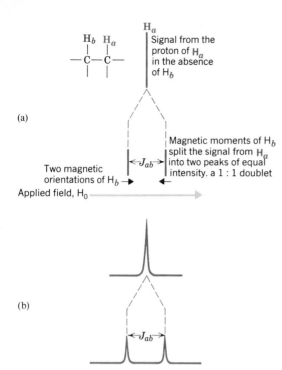

(a)

Two magnetic orientations of H_b →
Applied field, H_0 ————→

(b)

FIGURE 13.18 Signal splitting arising from spin–spin coupling with one nonequivalent proton of a neighboring hydrogen atom. A theoretical analysis is shown in (a) and the actual appearance of the spectrum in (b). The distance between the centers of the peaks of the doublet is called the coupling constant, J_{ab}. J_{ab} is measured in hertz (cycles per second). The magnitudes of coupling constants are *not* dependent upon the magnitude of the applied field and their values, in hertz, are the same regardless of the operating frequency of the spectrometer.

PROBLEM 13.7

Sketch the proton nmr spectrum of $CHBr_2CHCl_2$. Which signal would you expect to occur at lower magnetic field strength; that of the proton of the $CHBr_2—$ group or of the $—CHCl_2$ group? Why?

Two equivalent protons on an adjacent carbon (or carbon atoms) split the signal from an absorbing proton into a 1 : 2 : 1 *triplet*. Figure 13.19 illustrates how this pattern occurs.

In compounds of either type (Fig. 13.19), both protons may be aligned with the applied field. This orientation causes a peak to appear at a lower applied field strength

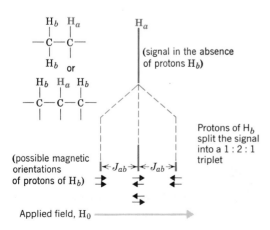

(possible magnetic orientations of protons of H_b)

Applied field, H_0 ————→

FIGURE 13.19 Two equivalent protons (H_b) on an adjacent carbon atom split the signal from H_a into a 1 : 2 : 1 triplet.

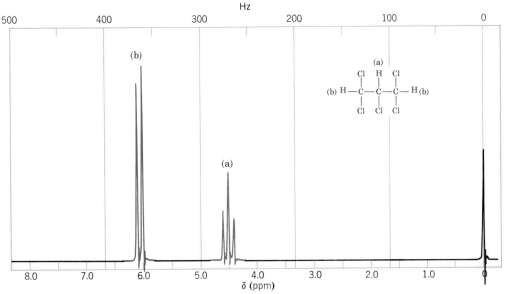

FIGURE 13.20 The proton nmr spectrum of 1,1,2,3,3-pentachloropropane. (Spectrum courtesy of Varian Associates, Palo Alto, CA.)

than would occur in the absence of the two hydrogen atoms H_b. Conversely, both protons may be aligned against the applied field. This orientation of the protons of H_b causes a peak to appear at higher applied field strengths than would occur in their absence. Finally, there are two ways in which the two protons may be aligned in which one opposes the applied field and one reinforces it. These arrangements do not displace the signal. Since the probability of this last arrangement is twice that of either of the other two, the center peak of the triplet is twice as intense.

The proton of the —$CHCl_2$ group of 1,1,2-trichloroethane is an example of a proton of the type having two equivalent protons on an adjacent carbon. The signal from the —$CHCl_2$ group (Fig. 13.7) appears as a $1:2:1$ triplet and, as we would expect, the signal of the —CH_2Cl group of 1,1,2-trichloroethane is split into a $1:1$ doublet by the proton of the —$CHCl_2$ group.

The spectrum of 1,1,2,3,3-pentachloropropane (Fig. 13.20) is similar to that of 1,1,2-trichloroethane in that it also consists of a $1:2:1$ triplet and a $1:1$ doublet. The two hydrogen atoms H_b of 1,1,2,3,3-pentachloropropane are equivalent even though they are on separate carbon atoms.

PROBLEM 13.8

The relative positions of the doublet and triplet of 1,1,2-trichloroethane (Fig. 13.7) and 1,1,2,3,3-pentachloropropane (Fig. 13.20) are reversed. Explain this.

Three equivalent protons (H_b) on a neighboring carbon split the signal from the H_a into a $1:3:3:1$ quartet. This pattern is shown in Fig. 13.21.

The signal from two equivalent protons of the —CH_2Br group of ethyl bromide (Fig. 13.22) appears as a $1:3:3:1$ quartet because of this type of signal splitting. The

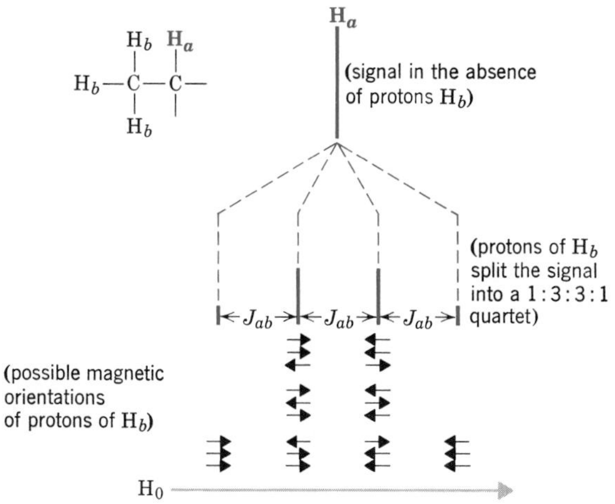

FIGURE 13.21 Three equivalent protons (H_b) on an adjacent carbon split the signal from H_a into a 1:3:3:1 quartet.

three equivalent protons of the CH_3— group are split into a 1:2:1 triplet by the two protons of the —CH_2Br group.

The kind of analysis that we have just given can be extended to compounds with even larger numbers of equivalent protons on adjacent atoms. These analyses show that *if there are n equivalent protons on adjacent atoms these will split a signal into n + 1 peaks.* (We may not always see all of these peaks in actual spectra, however, because some of them may be very small.)

PROBLEM 13.9

What kind of proton nmr spectrum would you expect the following compound to give?

$$(Cl_2CH)_3CH$$

Sketch the spectrum showing the splitting patterns and relative position of each signal.

PROBLEM 13.10

Propose structures for each of the compounds whose spectra are shown in Fig. 13.23, and account for the splitting pattern of each signal.

The splitting patterns shown in Fig. 13.23 are fairly easy to recognize because in each compound there are only two sets of nonequivalent hydrogen atoms. One feature present in all spectra, however, will help us recognize splitting patterns in more complicated spectra: the **reciprocity of coupling constants.**

The separation of the peaks in hertz gives us the value of the coupling constants. Therefore, if we look for doublets, triplets, quartets, and so on, that have *the same*

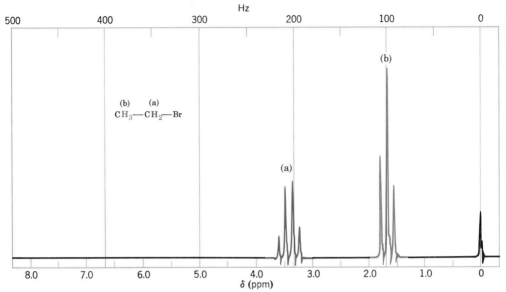

FIGURE 13.22 The proton nmr spectrum of ethyl bromide.

coupling constants, the chances are good that these multiplets are related to each other because they arise from reciprocal spin-spin couplings.

The two sets of protons of an ethyl group, for example, appear as a triplet and a quartet as long as the ethyl group is attached to an atom that does not bear any hydrogen atoms. The spacings of the peaks of the triplet and the quartet of an ethyl group will be the same because the coupling constants, J_{ab}, are the same (Fig. 13.24).

Proton nmr spectra have other features, however, that are not at all helpful when we try to determine the structure of a compound.

1. Signals may overlap. This happens when the chemical shifts of the signals are very nearly the same. In the spectrum of ethyl chloroacetate (Fig. 13.25) we see that the singlet of the —CH_2Cl group falls directly on top of one of the outer-most peaks of the ethyl quartet.
2. Spin–spin couplings between the protons of nonadjacent atoms may occur. This long-range coupling happens frequently when π bonded atoms intervene between the atoms bearing the coupled protons.
3. The splitting patterns of aromatic groups are difficult to analyze. A monosub-stituted benzene ring (a phenyl group) has three different kinds of protons.

The chemical shifts of these protons may be so similar that the phenyl group gives a signal that resembles a singlet. Or the chemical shifts may be different and because of long-range couplings, the phenyl group appears as a very com-plicated multiplet.

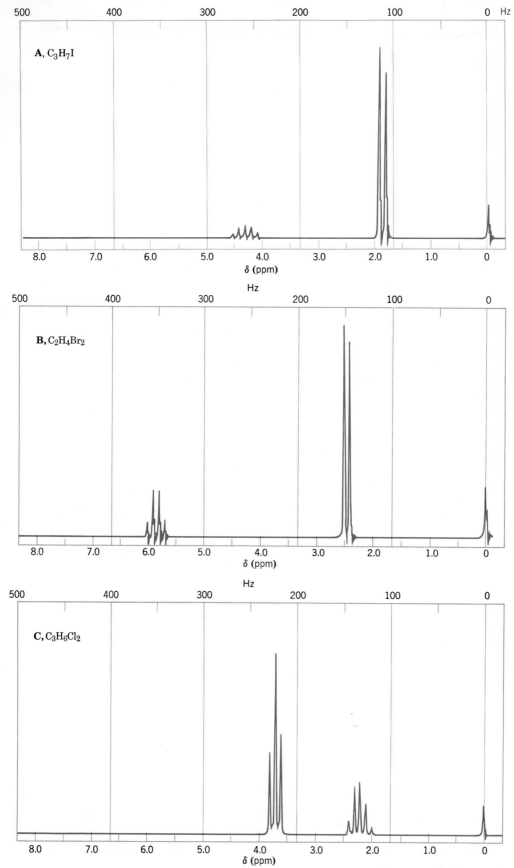

FIGURE 13.23 Proton nmr spectra for Problem 13.10. (Spectra courtesy of Varian Associates, Palo Alto, CA.)

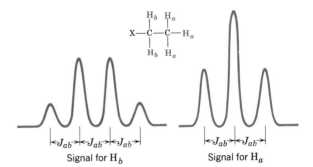

FIGURE 13.24 A theoretical splitting pattern for an ethyl group. For an example, see the spectrum of ethyl bromide (Fig. 13.22).

Signal for H_b Signal for H_a

Disubstituted benzenes show a range of complicated splitting patterns. In many instances these patterns can be analyzed by using techniques that are beyond the scope of our discussion here. We shall see later in this chapter that infrared spectroscopy gives us a relatively easy method for deciding whether the substituents of disubstituted benzenes are ortho, meta, or para to each other.

In all of the proton nmr spectra that we have considered so far, we have restricted our attention to signal splittings arising from interactions of only two sets of equivalent protons on adjacent atoms. What kind of patterns should we expect from compounds in which more than two sets of equivalent protons are interacting? We cannot answer this question completely because of limitations of space, but we can give an example that illustrates the kind of analysis that is involved. Let us consider a 1-substituted propane.

$$\overset{(a)}{CH_3}-\overset{(b)}{CH_2}-\overset{(c)}{CH_2}-Z$$

Here, there are three sets of equivalent protons. We have no problem in deciding what kind of signal splitting to expect from the protons of the CH_3— group or the

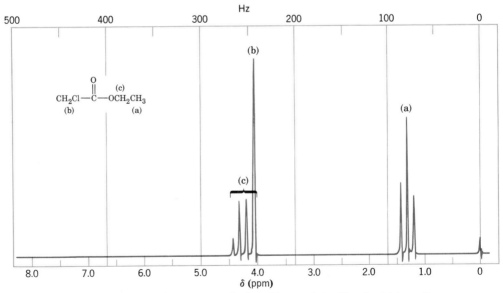

FIGURE 13.25 The proton nmr spectrum of ethyl chloroacetate. The singlet from the protons of (b) falls on one of the outermost peaks of the quartet from (c). (Spectrum courtesy of Varian Associates, Palo Alto, CA.)

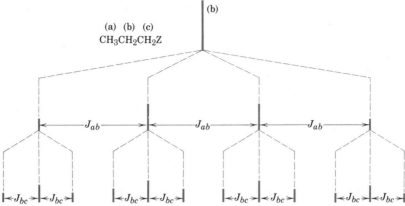

(a) (b) (c)
$CH_3CH_2CH_2Z$

FIGURE 13.26 The splitting pattern that would occur for the (b) protons of $CH_3CH_2CH_2Z$ if J_{ab} is much larger than J_{bc}. Here $J_{ab} = 3J_{bc}$.

—CH_2Z group. The methyl group is spin–spin coupled only to the two protons of the central —CH_2— group. Therefore, the methyl group should appear as a triplet. The protons of the —CH_2Z group are similarly coupled only to the two protons of the central —CH_2— group. Thus, the protons of the —CH_2Z group should also appear as a triplet.

But what about the protons of the central —CH_2— group (b)? They are spin–spin coupled with the three protons at (a) and with two protons at (c). The protons at (a) and (c), moreover, are not equivalent. If the coupling constants J_{ab} and J_{bc} have quite different values, then the protons at (b) could be split into a quartet by the three protons (a) and each line of the quartet could be split into a triplet by the two protons (c) (Fig. 13.26).

It is unlikely, however, that we would observe as many as 12 peaks in an actual spectrum because the coupling constants are such that peaks usually fall on top of peaks. The proton nmr spectrum of 1-nitropropane (Fig. 13.27) is typical of 1-substi-

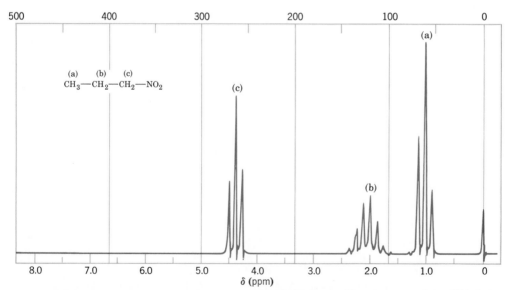

(a)   (b)   (c)
CH_3—CH_2—CH_2—NO_2

δ (ppm)

FIGURE 13.27 The proton nmr spectrum of 1-nitropropane. (Spectrum courtesy of Varian Associates, Palo Alto, CA.)

tuted propane compounds. We see that the (*b*) protons are split into six major peaks, each of which shows a slight sign of further splitting.

PROBLEM 13.11

Carry out an analysis like that shown in Fig. 13.26 and show how many peaks the signal from (*b*) would be split into if $J_{ab} = 2J_{bc}$ and if $J_{ab} = J_{bc}$. (*Hint:* In both cases peaks will fall on top of peaks so that the total number of peaks in the signal is fewer than 12.)

The presentation we have given here applies only to what are called *first-order spectra*. In first-order spectra, the distance in hertz, Δv, that separates the coupled signals is very much larger than the coupling constant, *J*. That is, $\Delta v \gg J$. In *second-order spectra* (which we have not discussed) Δv approaches *J* in magnitude and the situation becomes much more complex. The number of peaks increases and the intensities are not those that might be expected from first-order considerations.

13.9 PROTON NUCLEAR MAGNETIC RESONANCE SPECTRA OF COMPOUNDS CONTAINING FLUORINE AND DEUTERIUM

The fluorine (^{19}F) nucleus has spin quantum numbers of $+\frac{1}{2}$ and $-\frac{1}{2}$. In this respect ^{19}F nuclei resemble protons and fluorine magnetic spectra can be observed. When measured at the same radio frequency, the signals from ^{19}F absorptions occur at considerably different magnetic field strengths than those of protons, so we do not see peaks due to ^{19}F absorption in proton magnetic resonance spectra. We do, however, see splitting of proton signals caused by spin–spin couplings between protons and ^{19}F nuclei. The signal from the two protons of 1,2-dichloro-1,1-difluoroethane, for example, is split into a triplet by the fluorine atoms on the adjacent carbon (Fig. 13.28).

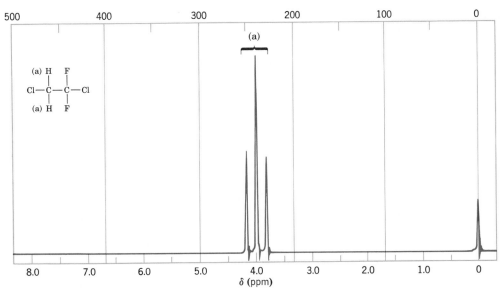

FIGURE 13.28 The proton nmr spectrum of 1,2-dichloro-1,1-difluoroethane. The J_{HF} coupling constant is ~ 12 Hz. (Spectrum courtesy of Varian Associates, Palo Alto, CA.)

The nucleus of a deuterium atom (a deuteron) has a much smaller magnetic moment than a proton, and signals from deuteron absorption do not occur in proton magnetic resonance spectra.

Spin–spin couplings between deuterons and protons are small but the presence of deuterium on an adjacent atom can cause splitting of the proton signal.

PROBLEM 13.12

Sketch the proton nmr spectra that you would expect from each of the following compounds: (a) $CH_3CF_2CH_3$, (b) CH_3CF_2Cl, (c) CH_3CFCl_2, and (d) CH_3CF_3.

13.10 PROTON NUCLEAR MAGNETIC RESONANCE SPECTRA AND RATE PROCESSES

J. D. Roberts (of the California Institute of Technology), a pioneer in the application of nmr spectroscopy to problems of organic chemistry, has compared the nuclear magnetic resonance spectrometer to a camera with a relatively slow shutter speed. Just as a camera with a slow shutter speed blurs photographs of objects that are moving rapidly, the nuclear magnetic resonance spectrometer blurs its picture of molecular processes that are occurring rapidly.

What are some of the rapid processes that occur in organic molecules?

At temperatures near room temperature, groups connected by carbon–carbon single bonds rotate very rapidly. Because of this, when we determine spectra of compounds with single bonds, the spectra that we obtain often reflect the individual hydrogen atoms in their average environment — that is, in an environment that is an average of all the environments that the protons have as a result of the group rotations.

To see an example of this effect, let us consider the spectrum of ethyl bromide again. The most stable conformation of ethyl bromide is the one in which the groups are perfectly staggered. In this staggered conformation one hydrogen of the methyl group (in red in following figure) is in a different environment from that of the other

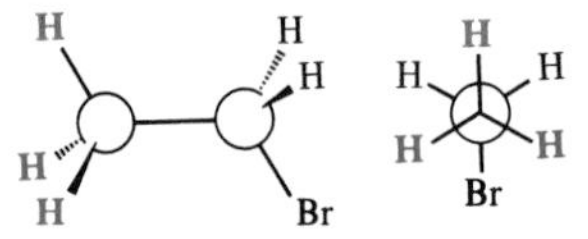

two methyl hydrogen atoms. If the nmr spectrometer were to detect this particular conformation of ethyl bromide, it would show the proton of this hydrogen of the methyl group at *a different chemical shift*. We know, however, that in the spectrum of ethyl bromide (Fig. 13.22), the three protons of the methyl group give *a single signal* (a signal that is split into a triplet by spin–spin coupling with the two protons of the adjacent carbon).

The methyl protons of ethyl bromide give a single signal because at room temperature the groups connected by the carbon–carbon single bond rotate approximately 1 million times each second. The "shutter speed" of the nmr spectrometer is too slow to "photograph" this rapid rotation; instead, it photographs the methyl hydrogen atoms in their average environments, and in this sense, it gives us a blurred picture of the methyl group.

Rotations about single bonds slow down as the temperature of the compound is lowered. Sometimes, this slowing of rotations allows us to "see" the different conformations of a molecule when we determine the spectrum at a sufficiently low temperature.

An example of this phenomenon, and one that also shows the usefulness of deuterium labeling, can be seen in the low temperature proton nmr spectra of cyclohexane and of undecadeuteriocyclohexane. (These experiments originated with F. A. L. Anet of the University of California, Los Angeles, another pioneer in the applications of nmr spectroscopy to organic chemistry, especially to conformational analysis.)

Undecadeuteriocyclohexane

At room temperature, ordinary cyclohexane gives one signal because interconversions between the various chair forms occur very rapidly. At low temperatures, however, ordinary cyclohexane gives a very complex proton nmr spectrum. At low temperatures interconversions are slow; the axial and equatorial protons have different chemical shifts; and complex spin–spin couplings occur.

At -100 °C, however, undecadeuteriocyclohexane gives only two signals of equal intensity. These signals correspond to the axial and equatorial hydrogen atoms of the following two chair conformations. Interconversions between these conformations occur at this low temperature, but they happen slowly enough for the nmr spectrometer to detect the individual conformations.

PROBLEM 13.13

What kind of proton nmr spectrum would you expect to obtain from undecadeuteriocyclohexane at room temperature?

Another example of a rapidly occurring process can be seen in proton nmr spectra of ethanol. The proton nmr spectrum of ordinary ethanol shows the hydroxyl proton as a singlet and the protons of the —CH_2— group as a quartet (Fig. 13.29). In ordinary ethanol we observe *no signal splitting arising from coupling between the hydroxyl proton and the protons of the —CH_2— group even though they are on adjacent atoms.*

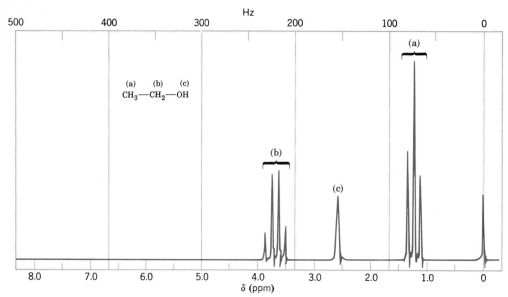

FIGURE 13.29 The proton nmr spectrum of ordinary ethanol. (Spectrum courtesy of Varian Associates, Palo Alto, CA.)

If we examine a proton nmr spectrum of *very pure* ethanol, however, we find that the signal from the hydroxyl proton is split into a triplet, and that the signal from the protons of —CH_2— group is split into a multiplet of eight peaks. Clearly, in very pure ethanol the spin of the proton of the hydroxyl group couples with the spins of the protons of the —CH_2— groups.

Whether or not coupling occurs between the hydroxyl protons and the methylene protons depends on the length of time the proton spends on a particular ethanol molecule. Protons attached to electronegative atoms with lone pairs such as oxygen can undergo rapid **chemical exchange.** That is, they can be transferred rapidly from one molecule to another. The chemical exchange in very pure ethanol is slow and, as a consequence, we see the signal splitting of and by the hydroxyl proton in the spectrum. In ordinary ethanol, acidic and basic impurities catalyze the chemical exchange; the exchange occurs so rapidly that the hydroxyl proton gives an unsplit signal and those of the methylene protons are split only by coupling with the protons of the methyl group. We say, then, that rapid exchange causes **spin decoupling.**

Spin decoupling is often found in the proton nmr spectra of alcohols, amines, and carboxylic acids.

PROBLEM 13.14

Apply principles that you have learned in this and in earlier sections, and assign structures to each of the compounds in Fig. 13.30.

13.11 PROTON NMR SPECTRA OF CARBOCATIONS

Olah (Section 5.11) has developed methods for preparing carbocations under conditions where they are stable enough to be studied by nmr spectroscopy. Olah has

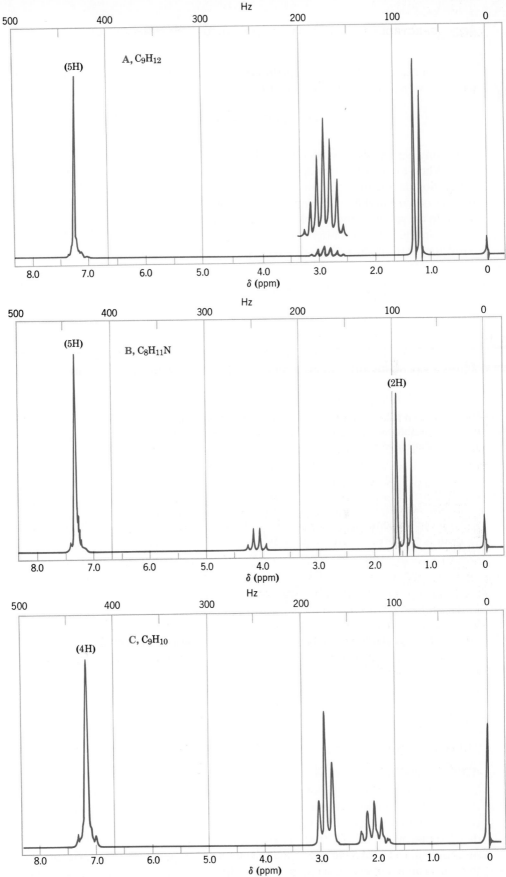

FIGURE 13.30 Proton nmr spectra for Problem 13.12. (Spectra courtesy of Varian Associates, Palo Alto, CA.)

found, for example, that in liquid sulfur dioxide, alkyl fluorides react with antimony pentafluoride to yield solutions of carbocations.

$$R\!-\!F + SbF_5 \xrightarrow[\text{liq. } SO_2]{} R^+ SbF_6^-$$

Antimony pentafluoride is a powerful Lewis acid.

When the proton nmr spectrum of *tert*-butyl fluoride is measured in liquid sulfur dioxide, the nine protons appear as a doublet centered at δ 1.3. (Why do the protons of *tert*-butyl fluoride appear as a doublet?) When antimony pentafluoride is added to the solution, the doublet at δ 1.35 is replaced by a singlet at δ 4.35. Both the change in the splitting pattern of the methyl protons and the downfield shift are consistent with the formation of a *tert*-butyl cation. (Why?)

$$\underset{\substack{\delta\ 1.35 \\ \text{(doublet)}}}{(CH_3)_3C\!-\!F} + SbF_5 \xrightarrow[SO_2]{\text{liq.}} \underset{\substack{\delta\ 4.35 \\ \text{(singlet)}}}{(CH_3)_3C^+}\ SbF_6^-$$

When a solution of isopropyl fluoride in liquid sulfur dioxide is treated with antimony pentafluoride, an even more remarkable downfield shift occurs.

$$\underset{\substack{\delta\ 1.23 \qquad\ \delta\ 4.64 \\ \text{(doublet of \ (multiplet)} \\ \text{doublets)}}}{(CH_3)_2CHF} + SbF_5 \longrightarrow \underset{\substack{\delta\ 5.03 \quad \delta\ 13.50 \\ \text{(doublet) (septet)}}}{(CH_3)_2CH^+} \qquad SbF_6^-$$

PROBLEM 13.15

When 2-methyl-1,1-diphenyl-2-propanol in liquid sulfur dioxide was treated with the "superacid" $FSO_3H\!-\!SbF_5$, the spectrum of the solution showed the proton nmr absorptions that follow:

$$\underset{\substack{\text{2-Methyl-1,1-diphenyl-} \\ \text{2-propanol}}}{\overset{\substack{H \qquad\ OH \\ | \qquad\ \ | }}{C_6H_5\!-\!\underset{\substack{| \\ C_6H_5}}{C}\!-\!\!-\!\underset{\substack{| \\ CH_3}}{C}\!-\!CH_3}} \xrightarrow[\text{liq. } SO_2]{FSO_3H-SbF_5} \qquad ?$$

Doublet, δ 1.48 (6H)
Multiplet, δ 4.45 (1H)
Multiplet, δ 8.0 (10H)

What carbocation is formed in this reaction?

13.12 CARBON-13 NMR SPECTROSCOPY

The most abundant isotope of the element carbon is carbon-12 (^{12}C) (natural abundance ~99%). Nuclei of carbon-12 have no net magnetic spin and therefore they cannot give nmr signals. This is not true, however, of nuclei of the much less abundant isotope of carbon, ^{13}C (natural abundance ~1%). Carbon-13 nuclei have a net magnetic spin and can give nmr signals. The nuclei of ^{13}C are like the nuclei of ^1H in that they can assume spin states of $+\frac{1}{2}$ or $-\frac{1}{2}$.

The low natural abundance of ^{13}C means that highly sensitive spectrometers employing special techniques must be used to measure ^{13}C spectra. These spectrometers have now become widely available, and ^{13}C spectroscopy is rapidly becoming another powerful method for determining the structures of organic molecules.

With ^{1}H spectroscopy (proton nmr) we obtain indirect information about the carbon skeleton of an organic molecule because *most* (but not all) of the carbon atoms have at least one attached hydrogen. In ^{13}C spectroscopy, we observe the carbon skeleton directly and, therefore, we see peaks arising from all of the carbon atoms, whether they bear hydrogen atoms or not.

One great advantage of ^{13}C spectroscopy is the wide range of chemical shifts over which ^{13}C nuclei absorb. In ^{13}C spectroscopy signals from organic compounds are spread over a chemical shift range of 200 ppm, compared with a range less than 20 ppm in proton spectra. Carbon-13 spectra are generally simpler because signals are less likely to overlap.

The very low natural abundance of ^{13}C has an important effect that further simplifies ^{13}C spectra. Because of its low natural abundance, there is a very low probability that two adjacent carbon atoms will both have ^{13}C nuclei. Therefore, in ^{13}C spectra we do not observe spin–spin couplings between the carbon nuclei.

Electronic techniques are also available to allow *decoupling* of spin–spin interactions between ^{13}C nuclei and ^{1}H nuclei. Thus, it is possible to obtain ^{13}C spectra in which all carbon resonances appear as singlets. Spectra obtained in this mode of operation of the spectrometer are called **proton-decoupled** spectra.

Carbon-13 spectrometers can also be operated in another mode, one that allows one-bond couplings between ^{13}C and ^{1}H nuclei to occur. This mode of operation is called **proton off-resonance decoupling**. It produces spectra in which —CH_3 groups appear as quartets, —CH_2— groups appear as triplets ⟍CH⟋ groups as doublets, and carbon atoms with no attached hydrogen atoms as singlets.

An excellent illustration of the application of ^{13}C spectroscopy is shown in spectra of 4-(N,N-diethylamino)benzaldehyde (Fig. 13.31). Bear in mind as you examine Fig. 13.31 that *in ^{13}C spectra the areas under signals are not always proportional to the number of atoms causing the signal.*

4-(*N,N*-Diethylamino)benzaldehyde

The bottom spectrum in Fig. 13.31 is the *proton-decoupled* spectrum in which all the signals from 4-(N,N-diethylamino)benzaldehyde appear as singlets. The triplet centered at δ 79 (ppm) is caused by the solvent, $CDCl_3$. (The ^{13}C signal of $CDCl_3$ is split into a triplet by coupling with the deuterium atom, spin quantum number = 1, spin states $+1, 0, -1$). The signal at δ 0 arises from $(CH_3)_4Si$.

The top spectrum in Fig. 13.31 is the *proton off-resonance decoupled* spectrum. It shows us immediately which signals belong to the ^{13}C nuclei of the ethyl groups. The triplet at δ 47 is caused by the equivalent —CH_2— groups and the quartet at δ 13 arises from the equivalent —CH_3 groups.

The two singlets in the top spectrum at δ 126 and δ 154 correspond to the carbon atoms of the benzene ring that do not bear hydrogen atoms, (*b*) and (*e*). The

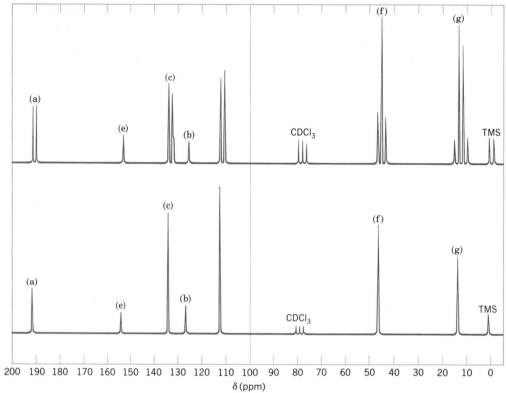

FIGURE 13.31 The ^{13}C nmr spectra of 4-(N,N-diethylamino)benzaldehyde,

HC—⬡—$N(CH_2CH_3)_2$. (Spectra courtesy of Philip L. Fuchs, Purdue University.)

The top spectrum is the proton off-resonance decoupled spectrum. In it, the multiplicity (i.e., the number of peaks in an individual signal) of each signal helps to match the signal with the carbon atom responsible for it. The multiplicity of the signal is one greater than the number of hydrogen atoms bonded to the carbon atom giving the signal. The bottom spectrum is the proton-decoupled spectrum. In it, all the signals from the compound being analyzed are singlets.

greater electronegativity of nitrogen (when compared to carbon) causes the signal from (e) to be further downfield (at δ 154). The doublet at δ 193 arises from the carbon of the aldehyde group. Its chemical shift is the most downfield of all the peaks because of the great electronegativity of its attached oxygen and because of resonance contribution of the second structure that follows. Both factors cause the electron density at this carbon to be very low and, therefore, the carbon is not well shielded.

$$H—\overset{\overset{\displaystyle :\ddot{O}:}{\|}}{C}— \longleftrightarrow H—\overset{\overset{\displaystyle :\ddot{O}:^{-}}{|}}{\underset{+}{C}}—$$

Resonance contributors for an aldehyde group

This leaves the signals at δ 112 and δ 135 and the two sets of carbon atoms of the benzene ring labeled (c) and (d) to be accounted for. Both signals appear as doublets in the proton off-resonance decoupled spectrum because both types of carbon have one attached hydrogen. But which signal belongs to which set of carbon atoms? Here we find another interesting application of resonance theory.

If we write resonance structures **A** to **D** involving the unshared electron pair of the amino group, we see that contributions made by **B** and **D** increase the electron

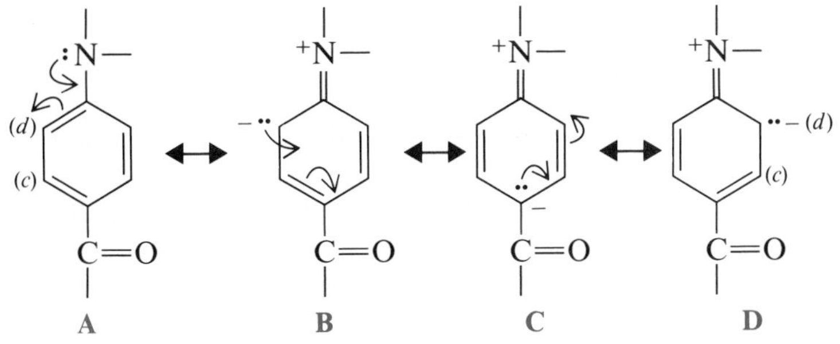

density at the set of carbon atoms labeled (*d*). On the other hand, writing structures **E** to **H** involving the aldehyde group shows us that contributions made by **F** and **H**

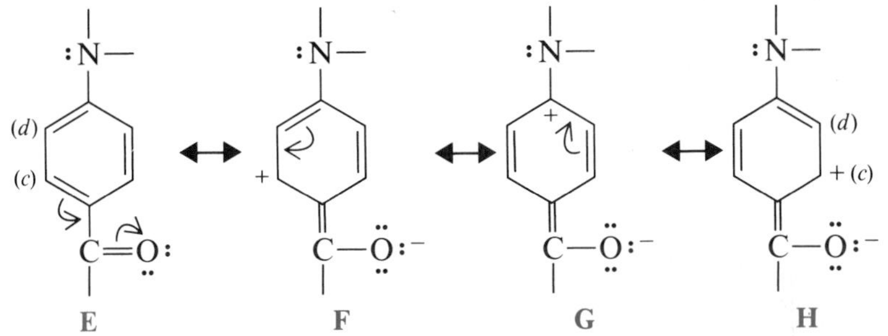

decrease the electron density at the set of carbon atoms labeled (*c*). (Other resonance structures are possible but are not pertinent to the argument here.)

Increasing the electron density at a carbon should increase its shielding and should shift its signal upfield. Therefore, we assign the signal at δ 112 to the set of carbon atoms labeled (*d*). Conversely, decreasing the electron density at a carbon should shift its signal downfield, so we assign the signal at δ 135 to the set labeled (*c*).

Table 13.3 gives *approximate* carbon-13 chemical shifts for a variety of carbon-containing groups.

Carbon-13 spectroscopy can be especially useful in recognizing a compound with a high degree of symmetry. The following sample problem illustrates one such application.

SAMPLE PROBLEM

The proton-decoupled ^{13}C spectrum given in Fig. 13.32 is of a tribromobenzene $(C_6H_3Br_3)$. Which tribromobenzene is it?

Answer:

There are three possible tribromobenzenes:

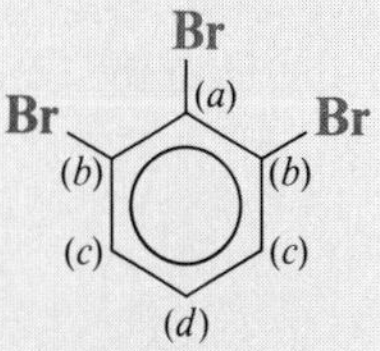

1,2,3-Tribromobenzene 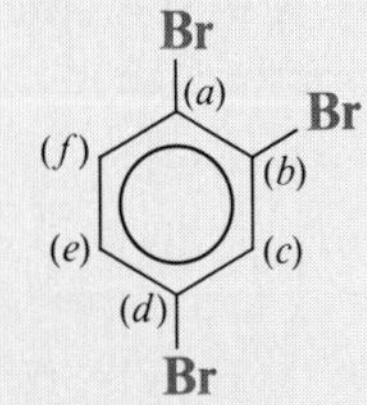   1,2,4-Tribromobenzene 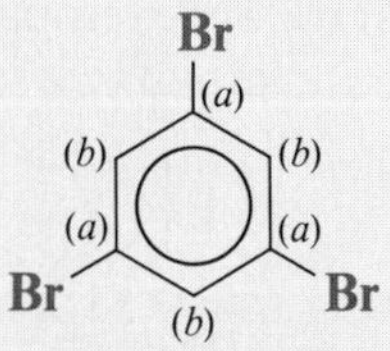   1,3,5-Tribromobenzene

Our spectrum (Fig. 13.32) consists of only two signals, indicating that only two different types of carbon atoms are present in the compound. Only 1,3,5-tribromobenzene has a degree of symmetry such that it would give only two signals, and, therefore, it is the correct answer. 1,2,3-Tribromobenzene would give four ^{13}C signals and 1,2,4-tribromobenzene would give six.

TABLE 13.3 Approximate carbon-13 chemical shifts

TYPE OF CARBON ATOM	CHEMICAL SHIFT (δ, ppm)
1° Alkyl, RCH$_3$	0–40
2° Alkyl, RCH$_2$R	10–50
3° Alkyl, RCHR$_2$	15–50
Alkyl halide or amine, $-\overset{\vert}{\underset{\vert}{C}}-X$ $\left(X = Cl, Br, or \overset{\vert}{N}-\right)$	10–65
Alcohol or ether, $-\overset{\vert}{\underset{\vert}{C}}-O$	50–90
Alkyne, $-C\equiv$	60–90
Alkene, $\overset{\diagdown}{\underset{\diagup}{C}}=$	100–170
Aryl,  $\langle\bigcirc\rangle C-$	100–170
Nitriles, $-C\equiv N$	120–130
Amides, $-\overset{O}{\overset{\parallel}{C}}-\overset{\vert}{N}-$	150–180
Carboxylic acids, esters, $-\overset{O}{\overset{\parallel}{C}}-O$	160–185
Aldehydes, ketones, $-\overset{O}{\overset{\parallel}{C}}-$	182–215

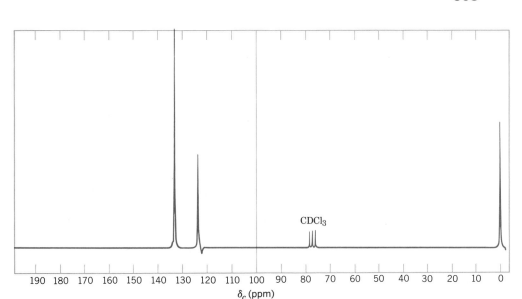

FIGURE 13.32 The ¹³C spectrum of a tribromobenzene. (Spectrum adapted with permission from L. F. Johnson and W. C. Jankowski, *Carbon-13 NMR Spectra: A Collection of Assigned, Coded, and Indexed Spectra*, Wiley–Interscience, New York, 1972.)

PROBLEM 13.16

Explain how ¹³C spectroscopy could be used to distinguish the *ortho-*, *meta-*, and *para-*dibromobenzene isomers one from another.

PROBLEM 13.17

Compounds **A**, **B**, and **C** are isomers with the formula $C_5H_{11}Cl$. Their proton-decoupled ¹³C spectra are given in Fig. 13.33. The letters *s*, *d*, *t*, and *q* give the multiplicities of each peak in the proton off-resonance decoupled ¹³C spectrum. Give structures for **A**, **B**, and **C**.

13.13 INFRARED SPECTROSCOPY

We saw in Section 13.2 that many organic compounds absorb radiation in the visible and ultraviolet regions of the electromagnetic spectrum. We also saw that when compounds absorb radiation of the visible and ultraviolet regions, electrons are excited from lower-energy molecular orbitals to higher ones.

Organic compounds also absorb electromagnetic energy in the infrared region of the spectrum. Infrared radiation does not have sufficient energy to cause the excitation of electrons, but it does cause atoms and groups of atoms of organic compounds to vibrate faster about the covalent bonds that connect them. These vibrations are *quantized*, and as they occur, the compounds absorb infrared energy in particular regions of the spectrum.

Infrared spectrometers operate in a manner similar to that of visible–ultraviolet spectrometers. A beam of infrared radiation is passed through the sample

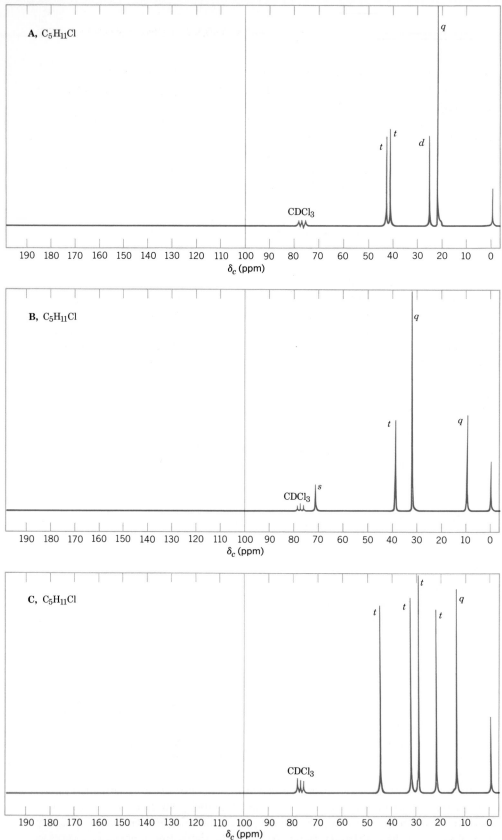

FIGURE 13.33 The ^{13}C nmr spectra compounds **A**, **B**, and **C**, Problem 13.17. The letters indicate the multiplicities of the signals in the proton off-resonance decoupled spectra (s = singlet, d = doublet, t = triplet, q = quartet). (Adapted with permission from L. F. Johnson and W. C. Jankowski, *Carbon-13 NMR Spectra: A Collection of Assigned, Coded, and Indexed Spectra*, Wiley–Interscience, New York, 1972.)

and is constantly compared with a reference beam as the frequency of the incident radiation is varied. The spectrometer plots the results as a graph showing absorption versus frequency or wavelength.

The location of an infrared absorption band (or peak) can be specified in **frequency units** by its wavenumber $\bar{v}$, measured in reciprocal centimeters (cm^{-1}), or by its **wavelength,** λ, measured in micrometers (μm; old name micron, μ). The wavenumber is the number of cycles of the wave in each centimeter along the light beam, and the wavelength is the length of the wave, crest to crest.

$$\bar{v} = \frac{1}{\lambda} \text{ (with } \lambda \text{ in cm)} \quad \text{or} \quad \bar{v} = \frac{10,000}{\lambda} \text{ (with } \lambda \text{ in } \mu m)$$

In their vibrations covalent bonds behave as if they were tiny springs connecting the atoms. When the atoms vibrate they can do so only at certain frequencies, as if the bonds were "tuned." Because of this, covalently bonded atoms have only particular vibrational energy levels. The excitation of a molecule from one vibrational energy level to another occurs only when the compound absorbs infrared radiation of a particular energy, meaning a particular wavelength or frequency (since $\Delta E = hv$).

Molecules can vibrate in a variety of ways. Two atoms joined by a covalent bond can undergo a stretching vibration where the atoms move back and forth as if joined by a spring.

A stretching vibration

Three atoms can also undergo in a variety of stretching and bending vibrations:

Symmetric stretching

Asymmetric stretching

An in-plane bending vibration

An out-of-plane bending vibration

The *frequency* of a given stretching vibration and thus *its location in an infrared spectrum* can be related to two factors. These are *the masses of the bonded atoms—* light atoms vibrate at higher frequencies than heavier ones — *and the relative stiffness of the bond.* Triple bonds are stiffer (and vibrate at higher frequencies) than double bonds and double bonds are stiffer (and vibrate at higher frequencies) than single bonds. We can see some of these effects in Table 13.4. Notice that stretching frequen-

TABLE 13.4 Characteristic infrared absorptions of groups

GROUP		FREQUENCY RANGE (cm^{-1})	INTENSITY[a]
A. Alkyl			
C—H (stretching)		2853–2962	(m–s)
Isopropyl, $—CH(CH_3)_2$		1380–1385	(s)
	and	1365–1370	(s)
tert-Butyl, $—C(CH_3)_3$		1385–1395	(m)
	and	~1365	(s)
B. Alkenyl			
C—H (stretching)		3010–3095	(m)
C=C (stretching)		1620–1680	(v)
$R—CH=CH_2$ (out-of-plane C—H bendings)		985–1000	(s)
	and	905–920	(s)
$R_2C=CH_2$ (out-of-plane C—H bendings)		880–900	(s)
cis-RCH=CHR (out-of-plane C—H bendings)		675–730	(s)
trans-RCH=CHR (out-of-plane C—H bendings)		960–975	(s)
C. Alkynyl			
≡C—H (stretching)		~3300	(s)
C≡C (stretching)		2100–2260	(v)
D. Aromatic			
Ar—H (stretching)		~3030	(v)
Aromatic substitution type (C—H out-of-plane bendings)			
Monosubstituted		690–710	(very s)
	and	730–770	(very s)
o-Disubstituted		735–770	(s)
m-Disubstituted		680–725	(s)
	and	750–810	(very s)
p-Disubstituted		800–840	(very s)
E. Alcohols, Phenols, and Carboxylic Acids			
O—H (stretching)			
Alcohols, phenols (dilute solutions)		3590–3650	(sharp, v)
Alcohols, phenols (hydrogen bonded)		3200–3550	(broad, s)
Carboxylic acids (hydrogen bonded)		2500–3000	(broad, v)
F. Aldehydes, Ketones, Esters, and Carboxylic Acids			
C=O (stretching)		1630–1780	(s)
Aldehydes		1690–1740	(s)
Ketones		1680–1750	(s)
Esters		1735–1750	(s)
Carboxylic acids		1710–1780	(s)
Amides		1630–1690	(s)

TABLE 13.4 (continued)

GROUP	FREQUENCY RANGE (cm⁻¹)	INTENSITY[a]
G. Amines		
N—H	3300–3500	(m)
H. Nitriles		
C≡N	2220–2260	(m)

[a]Abbreviations: s = strong, m = medium, w = weak, v = variable, ~ = approximately.

cies of groups involving hydrogen (a light atom) such as C—H, N—H, and O—H all occur at relatively high frequencies:

GROUP	BOND	FREQUENCY RANGE (cm⁻¹)
Alkyl	C—H	2853–2962
Alcohol	O—H	3590–3650
Amine	N—H	3300–3500

Notice too, that triple bonds vibrate at higher frequencies than double bonds:

BOND	FREQUENCY RANGE (cm⁻¹)
C≡C	2100–2260
C≡N	2220–2260
C=C	1620–1680
C=O	1630–1780

The infrared spectra of even relatively simple compounds contain many absorption peaks. It can be shown that a nonlinear molecule of n atoms has $3n - 6$ possible *fundamental* vibrational modes that can be responsible for the absorption of infrared radiation. This means that, theoretically, methane has 9 possible fundamental absorption peaks and benzene has 30.

Not all molecular vibrations result in the absorption of infrared energy. ***In order for a vibration to occur with the absorption of infrared energy, the dipole moment of the molecule must change as the vibration occurs.*** Thus when the four hydrogen atoms of methane vibrate symmetrically, methane does not absorb infrared energy. Symmetrical vibrations of the carbon–carbon double and triple bonds of ethene and ethyne do not result in the absorption of infrared radiation, either.

Vibrational absorption may occur outside the region measured by a particular infrared spectrophotometer and vibrational absorptions may occur so closely together that peaks fall on top of peaks. These factors, together with the absence of absorptions because of vibrations that have no dipole moment change, cause infrared spectra to contain fewer peaks than the formula $3n - 6$ would predict.

However, other factors bring about even more absorption peaks. Overtones (harmonics) of fundamental absorption bands may be seen in infrared spectra even though these overtones occur with greatly reduced intensity. Bands called combination bands and difference bands also appear in infrared spectra.

Because infrared spectra contain so many peaks, the possibility that two compounds will have the same infrared spectrum is exceedingly small. It is because of this that an infrared spectrum has been called the "fingerprint" of a molecule. Thus, with organic compounds, if two pure samples give different infrared spectra, one can be certain that they are different compounds. If they give the same infrared spectrum then they are the same compound.

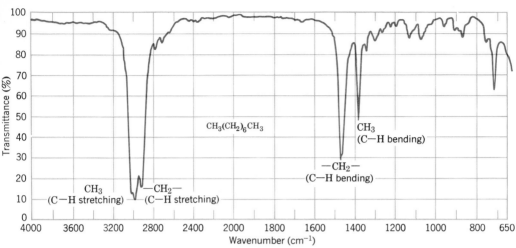

FIGURE 13.34 The infrared spectrum of octane. (Notice that, in infrared spectra, the peaks are "upside down." This is simply a result of the way infrared spectrophotometers operate.)

In the hands of one skilled in their interpretation, infrared spectra contain a wealth of information about the structures of compounds. We show some of the information that can be gathered from the spectra of octane and toluene in Figs. 13.34 and 13.35. We have neither the time nor the space here to develop the skill that would lead to complete interpretations of infrared spectra, but we can learn how to recognize the presence of absorption peaks in the infrared spectrum that result from vibrations of characteristic functional groups in the compound. By doing only this, however, we shall be able to use the information we gather from infrared spectra in a powerful way, particularly when we couple it with the information we gather from nuclear magnetic resonance spectra.

Let us now see how we can apply the data given in Table 13.4 to the interpretation of infrared spectra.

13.13A Hydrocarbons

All hydrocarbons give absorption peaks in the 2800–3300-cm^{-1} region that are associated with carbon–hydrogen stretching vibrations. We can use these peaks in

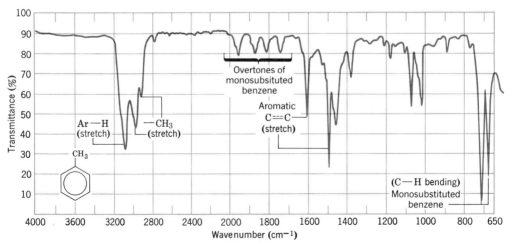

FIGURE 13.35 The infrared spectrum of toluene.

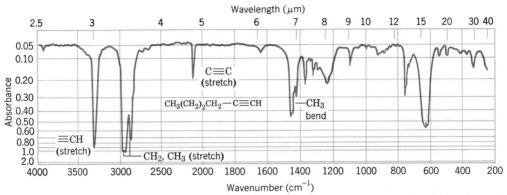

FIGURE 13.36 The infrared spectrum of 1-hexyne. (Spectrum courtesy of Sadtler Research Laboratories, Inc., Philadelphia.)

interpreting infrared spectra because the exact location of the peak depends on the strength (and stiffness) of the C—H bond, which in turn depend on the hybridization state of the carbon that bears the hydrogen. We have already seen that C—H bonds involving sp-hybridized carbon are strongest and those involving sp^3-hybridized carbon are weakest. The order of bond strength is

$$sp > sp^2 > sp^3$$

This too is the order of the bond stiffness.

The carbon–hydrogen stretching peaks of hydrogen atoms attached to sp-hybridized carbon atoms occur at highest frequencies, ~ 3300 cm^{-1}. Thus, $\equiv$C—H groups of terminal alkynes give peaks in this region. We can see the absorption of the acetylenic hydrogen of 1-hexyne at 3320 cm^{-1} in Fig. 13.36.

The carbon–hydrogen stretching peaks of hydrogen atoms attached to sp^2-hybridized carbon atoms occur in the 3000–3100-cm^{-1} region. Thus, alkenyl hydrogen atoms and the C—H groups of aromatic rings give absorption peaks in this region. We can see the alkenyl C—H absorption peak of 3080 cm^{-1} in the spectrum of 1-hexene (Fig. 13.37) and we can see the C—H absorption of the aromatic hydrogen atoms at 3090 cm^{-1} in the spectrum of toluene (Fig. 13.35).

The carbon–hydrogen stretching bands of hydrogen atoms attached to sp^3-hybridized carbon atoms occur at lowest frequencies, in the 2800–3000-cm^{-1} region.

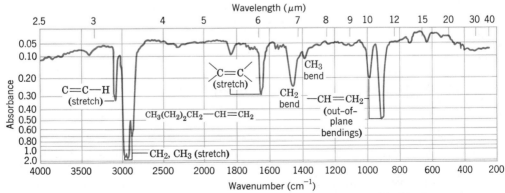

FIGURE 13.37 The infrared spectrum of 1-hexene. (Spectrum courtesy of Sadtler Research Laboratories, Inc., Philadelphia.)

We can see methyl and methylene absorption peaks in the spectra of octane (Fig. 13.34), toluene (Fig. 13.35), 1-hexyne (Fig. 13.36), and 1-hexene (Fig. 13.37).

Hydrocarbons also give absorption peaks in their infrared spectra that result from carbon–carbon bond stretchings. Carbon–carbon single bonds normally give rise to very weak peaks that are usually of little use in assigning structures. More useful peaks arise from multiple carbon–carbon bonds, however. Carbon–carbon double bonds give absorption peaks in the 1620–1680-cm^{-1} region and carbon–carbon triple bonds give absorption peaks between 2100 and 2260 cm^{-1}. These absorptions are not usually strong ones and they will not be present at all if the double or triple bond is symmetrically substituted. (No dipole moment change will be associated with the vibration.) The stretchings of the carbon–carbon bonds of benzene rings usually give a set of characteristic sharp peaks in the 1450–1600-cm^{-1} region.

Absorptions arising from carbon–hydrogen bending vibrations of alkenes occur in the 600–1000-cm^{-1} region. The exact location of these peaks can often be used to determine the *nature of the double bond and its configuration.*

Monosubstituted alkenes give two strong peaks in the 905–920- and the 985–1000-cm^{-1} regions. Disubstituted alkenes of the type R$_2$C=CH$_2$ give a strong peak in the 880–900-cm^{-1} range. *cis*-Alkenes give an absorption peak in the 675–730-cm^{-1} region and *trans*-alkenes give a peak between 960 and 975 cm^{-1}. These ranges for the carbon–hydrogen bending vibrations can be used with fair reliability for alkenes that do not have an electron-releasing or electron-withdrawing substituent (other than an alkyl group) on one of the carbon atoms of the double bond. When electron-releasing or electron-withdrawing substituents are present on a double-bond carbon, the bending absorption peaks may be shifted out of the regions we have given.

13.13B Substituted Benzenes

Ortho-, meta-, and para-substituted benzenes give absorption peaks in the 680–840-cm^{-1} region that characterize their substitution patterns. **Ortho-substituted benzenes** show a strong absorption peak arising from bending motions of the aromatic hydrogen atoms between 735 and 770 cm^{-1}. **Meta-substituted benzenes** show two peaks; one strong peak between 680 and 725 cm^{-1} and one very strong peak between 750 and 810 cm^{-1}. **Para-substituted benzenes** give a single very strong absorption between 800 and 840 cm^{-1}

Monosubstituted benzenes give two very strong peaks, between 690 and 710 cm^{-1} and between 730 and 770 cm^{-1} (see Fig. 13.35).

PROBLEM 13.18

Four benzenoid compounds, all with the formula C$_7$H$_7$Br, gave the following infrared peaks in the 680–840-cm^{-1} region.

A, 740 cm^{-1} (s)   **C**, 680 cm^{-1} (s) and 760 cm^{-1} (very s)
B, 800 cm^{-1} (very s)   **D**, 693 cm^{-1} (very s) and 765 cm^{-1} (very s)

Propose structures for **A**, **B**, **C**, and **D**.

13.13C Other Functional Groups

Infrared spectroscopy gives us an invaluable method for recognizing quickly and simply the presence of certain functional groups in a molecule. One important functional group that gives a prominent absorption peak in infrared spectra is the **carbonyl group** $\diagdown$C$=$O. This group is present in aldehydes, ketones, esters, carboxylic acids, amides, and so forth. The carbon–oxygen double bond stretching frequency of all these groups gives a strong peak between 1630 and 1780 cm^{-1}. The exact location of the peak depends on whether it arises from an aldehyde, ketone, ester, and so forth. These locations are the following and we shall have more to say about carbonyl absorption peaks when we discuss these compounds in later chapters.

$$
\begin{array}{ccc}
\underset{\substack{\text{Aldehyde}\\1690-1740\ \text{cm}^{-1}}}{\overset{\displaystyle O \atop \displaystyle \|}{\text{R—C—H}}} &
\underset{\substack{\text{Ketone}\\1680-1750\ \text{cm}^{-1}}}{\overset{\displaystyle O \atop \displaystyle \|}{\text{R—C—R}}} &
\underset{\substack{\text{Ester}\\1735-1750\ \text{cm}^{-1}}}{\overset{\displaystyle O \atop \displaystyle \|}{\text{R—C—OR}}}
\end{array}
$$

$$
\begin{array}{cc}
\underset{\substack{1710-1780\ \text{cm}^{-1}\\\text{Carboxylic acid}}}{\overset{\displaystyle O \atop \displaystyle \|}{\text{R—C—OH}}} &
\underset{\substack{1630-1690\ \text{cm}^{-1}\\\text{Amide}}}{\overset{\displaystyle O \atop \displaystyle \|}{\text{R—C—NH}_2}}
\end{array}
$$

The **hydroxyl groups** of alcohols and phenols are also easy to recognize in infrared spectra by their O—H stretching absorptions. These bonds also give us direct evidence for hydrogen bonding. If an alcohol or phenol is present as a very dilute solution in CCl$_4$, O—H absorption occurs as a very sharp peak in the 3590–3650-cm^{-1} region. In very dilute solution, formation of intermolecular hydrogen bonds does not take place because the molecules are too widely separated. The sharp peak in the 3590–3650-cm^{-1} region, therefore, is attributed to "free" (unassociated) hydroxyl groups. Increasing the concentration of the alcohol or phenol causes the sharp peak to be replaced by a broad band in the 3200–3550-cm^{-1} region. This absorption is attributed to OH groups that are associated through intermolecular hydrogen bonding.

Very dilute solutions of **amines** also give sharp peaks in the 3300–3500-cm^{-1} region arising from free N—H stretching vibrations. Primary amines give two sharp peaks; secondary amines give only one.

$$
\begin{array}{cc}
\text{RNH}_2 & \text{R}_2\text{NH}\\
1°\ \text{Amine} & 2°\ \text{Amine}\\
\text{Two peaks in} & \text{One peak in}\\
3300-3500\text{-cm}^{-1} & 3300-3500\text{-cm}^{-1}\\
\text{region} & \text{region}
\end{array}
$$

Hydrogen bonding causes these peaks to broaden. The NH groups of **amides** also give similar absorption peaks.

ADDITIONAL PROBLEMS
13.19

Listed here are proton nmr absorption peaks for several compounds. Propose a structure that is consistent with each set of data. (In some cases characteristic infrared absorptions are given as well.)

(a) $C_4H_{10}O$ proton nmr spectrum
singlet, δ 1.28 (9H)
singlet, δ 1.35 (1H)

(b) C_3H_7Br proton nmr spectrum
doublet, δ 1.71 (6H)
septet, δ 4.32 (1H)

(c) C_4H_8O proton nmr spectrum   IR spectrum
triplet, δ 1.05 (3H)   1720 cm^{-1}
singlet, δ 2.13 (3H)
quartet, δ 2.47 (2H)

(d) C_7H_8O proton nmr spectrum   IR spectrum
singlet, δ 2.43 (1H)   broad peak in
singlet, δ 4.58 (2H)   3200–3550-cm^{-1}
multiplet, δ 7.28 (5H)   region

(e) C_4H_9Cl proton nmr spectrum
doublet, δ 1.04 (6H)
multiplet, δ 1.95 (1H)
doublet, δ 3.35 (2H)

(f) $C_{15}H_{14}O$ proton nmr spectrum   IR spectrum
singlet, δ 2.20 (3H)   strong peak
singlet, δ 5.08 (1H)   near 1720 cm^{-1}
multiplet, δ 7.25 (10H)

(g) $C_4H_7BrO_2$ proton nmr spectrum   IR spectrum
triplet, δ 1.08 (3H)   broad peak in
multiplet, δ 2.07 (2H)   2500–3000-cm^{-1}
triplet, δ 4.23 (1H)   region and a peak
singlet, δ 10.97 (1H)   at 1715 cm^{-1}

(h) C_8H_{10} proton nmr spectrum
triplet, δ 1.25 (3H)
quartet, δ 2.68 (2H)
multiplet, δ 7.23 (5H)

(i) $C_4H_8O_3$ proton nmr spectrum   IR spectrum
triplet, δ 1.27 (3H)   broad peak in
quartet, δ 3.66 (2H)   2500–3000-cm^{-1}
singlet, δ 4.13 (2H)   region and a peak
singlet, δ 10.95 (1H)   at 1715 cm^{-1}

(j) $C_3H_7NO_2$ proton nmr spectrum
doublet, δ 1.55 (6H)
septet, δ 4.67 (1H)

(k) $C_4H_{10}O_2$ proton nmr spectrum
singlet, δ 3.25 (6H)
singlet, δ 3.45 (4H)

(l) $C_5H_{10}O$ proton nmr spectrum   IR spectrum
doublet, δ 1.10 (6H)   strong peak
singlet, δ 2.10 (3H)   near 1720 cm^{-1}
septet, δ 2.50 (1H)

(m) C_8H_9Br proton nmr spectrum
                      doublet, δ 2.0 (3H)
                      quartet, δ 5.15 (1H)
                      multiplet, δ 7.35 (5H)

13.20

The infrared spectrum of compound **E** (C_8H_6) is shown in Fig. 13.38. **E** decolorizes bromine in carbon tetrachloride and gives a precipitate when treated with ammoniacal silver nitrate. What is the structure of **E**?

13.21

The proton nmr spectrum of cyclooctatetraene consists of a single line located at δ 5.78. What does this suggest about electron delocalization in cyclooctatetraene?

13.22

Give a structure for compound **F** that is consistent with the proton nmr and IR spectra in Fig. 13.39.

13.23

Propose structures for the compounds **G** and **H** whose proton nmr spectra are shown in Figs. 13.40 and 13.41.

13.24

Propose a structure for compound **I** whose proton nmr and infrared spectra are given in Figs. 13.42 and 13.43.

13.25

A two-carbon compound (**J**) contains only carbon, hydrogen, and chlorine. Its infrared spectrum is relatively simple and shows the following absorbance peaks: 3125 cm^{-1}(m), 1625 cm^{-1}(m), 1280 cm^{-1}(m), 820 cm^{-1}(s), 695 cm^{-1}(s). The proton nmr spectrum of **J** consists of a singlet at δ 6.3. Using Table 13.4, make as many infrared assignments as you can and propose a structure for compound **J**.

13.26

When dissolved in CDCl$_3$, a compound (**K**) with the molecular formula $C_4H_8O_2$ gives a proton nmr spectrum that consists of a doublet at δ 1.35, a singlet at δ 2.15, a broad singlet at δ 3.75 (1H), and a quartet at δ 4.25 (1H). When dissolved in D$_2$O, the compound gives a

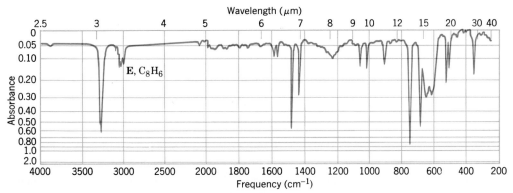

FIGURE 13.38 The infrared spectrum of compound **E**, Problem 13.20. (Spectrum courtesy of Sadtler Research Laboratories, Inc., Philadelphia.)

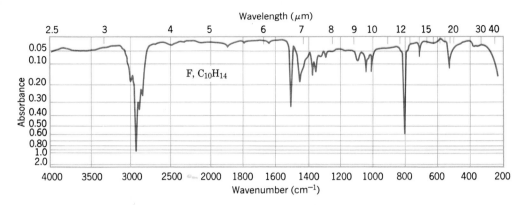

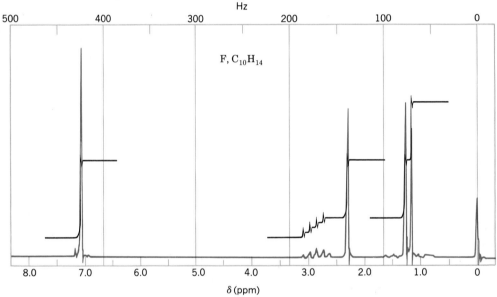

FIGURE 13.39 The proton nmr and IR spectra of compound **F**, Problem 13.22. (Proton nmr spectrum adapted from Varian Associates, Palo Alto, CA. IR spectrum adapted from Sadtler Research Laboratories, Inc., Philadelphia.)

similar proton nmr spectrum with the exception that the signal at δ 3.75 has disappeared. The infrared spectrum of the compound shows a strong absorption peak near 1720 cm^{-1}. (a) Propose a structure for compound **K** and (b) explain why the nmr signal at δ 3.75 disappears when D_2O is used as the solvent.

13.27

A compound (**L**) with the molecular formula C_9H_{10} decolorizes bromine in carbon tetrachloride and gives an infrared absorption spectrum that includes the following absorption peaks: 3035 cm^{-1}(m), 3020 cm^{-1}(m), 2925 cm^{-1}(m), 2853 cm^{-1}(w), 1640 cm^{-1}(m), 990 cm^{-1}(s), 915 cm^{-1}(s), 740 cm^{-1}(s), 695 cm^{-1}(s). The proton nmr spectrum of **L** consists of:

Doublet δ 3.1 (2H)   Multiplet δ 5.1   Multiplet δ 7.1 (5H)
Multiplet δ 4.8   Multiplet δ 5.8

The ultraviolet spectrum shows a maximum at 255 nm. Propose a structure for compound **L** and make assignments for each of the infrared peaks.

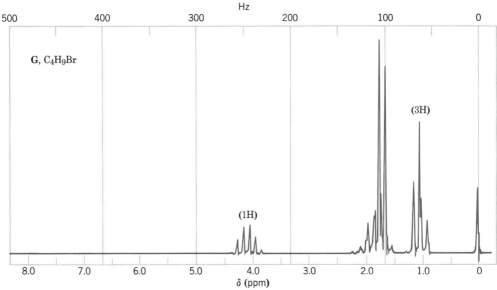

FIGURE 13.40 The proton nmr spectrum of compound **G**, Problem 13.23. (Spectrum courtesy of Varian Associates, Palo Alto, CA.)

13.28

Assume that in a certain proton nmr spectrum, you find two peaks of roughly equal intensity. You are not certain whether these two peaks are *singlets* arising from uncoupled protons at different chemical shifts, or whether they are two peaks of a *doublet* that arises from protons coupling with a single adjacent proton. What simple experiment would you perform to distinguish between these two possibilities?

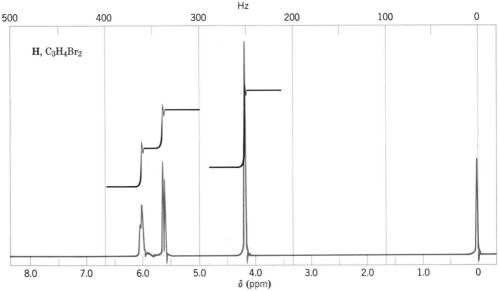

FIGURE 13.41 The proton nmr spectrum of compound **H**, Problem 13.23. (Spectrum courtesy of Varian Association, Palo Alto, CA.)

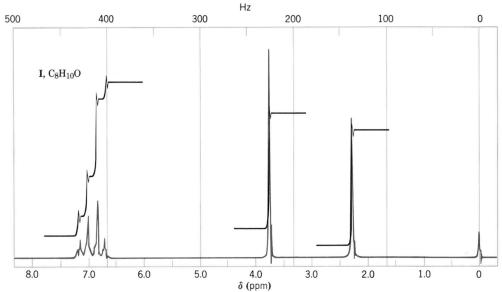

FIGURE 13.42 The proton nmr spectrum of compound **I**, Problem 13.24. (Spectrum courtesy of Varian Associates, Palo Alto, CA.)

13.29

Compound **M** has the molecular formula C_9H_{12}. The proton nmr spectrum of **M** is given in Fig. 13.44 and the infrared spectrum in Fig. 13.45. Propose a structure for **M**.

13.30

A compound (**N**) with the molecular formula $C_9H_{10}O$ gives a positive test with cold dilute aqueous potassium permanganate. The proton nmr spectrum of **N** is shown in Fig. 13.46 and the infrared spectrum of **N** is shown in Fig. 13.47. Propose a structure for **N**.

*13.31

When 2,3-dibromo-2,3-dimethylbutane is treated with SbF_5 in liquid SO_2 at $-60\ °C$, the proton nmr spectrum does not show the two signals that would be expected of a carbocation

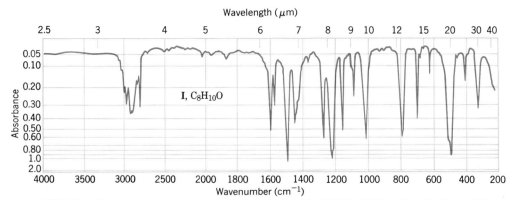

FIGURE 13.43 The infrared spectrum of compound **I**, Problem 13.24. (Spectrum courtesy of Sadtler Research Laboratories, Inc., Philadelphia.)

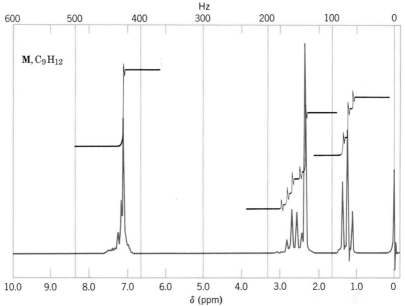

FIGURE 13.44 The proton nmr spectrum of compound **M**, Problem 13.29. (Spectrum courtesy of Aldrich Chemical Co., Milwaukee, WI.)

like $CH_3\overset{+}{C}Br—\overset{+}{C}CH_3$. Instead only one signal (at δ 2.9) is observed. What carbocation is
$\underset{CH_3}{|} \quad \underset{CH_3}{|}$
formed in this reaction and of what special significance is this experiment?

13.32
Compound **O** (C_6H_8) reacts with two molar equivalents of hydrogen in the presence of a catalyst to produce **P** (C_6H_{12}). The proton-decoupled ^{13}C spectrum of **O** consists of two singlets, one at 26.0 and one at 124.5. In the proton off-resonance ^{13}C spectrum of **O** the signal at 26.0 appears as a triplet and the one at 124.5 appears as a doublet. Propose structures for **O** and **P**.

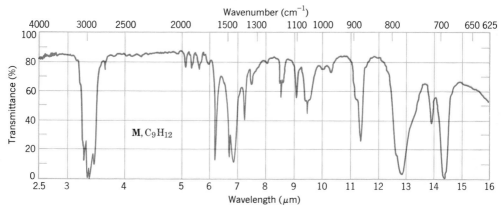

FIGURE 13.45 The infrared spectrum of compound **M**, Problem 13.29. (Spectrum courtesy of Aldrich Chemical Co., Milwaukee, WI.)

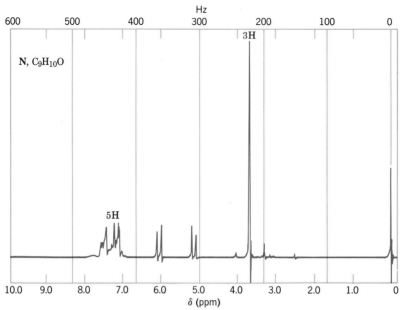

FIGURE 13.46 The proton nmr spectrum of compound **N**, Problem 13.30. (Spectrum courtesy of Aldrich Chemical Co., Milwaukee, WI.)

13.33

Compound **Q**, has the molecular formula C_7H_8. On catalytic hydrogenation **Q** is converted to **R** (C_7H_{12}). The proton-decoupled ^{13}C spectrum of **Q** is given in Fig. 13.48. Propose structures for **Q** and **R**.

13.34

Compound **S** (C_8H_{16}) decolorizes a solution of bromine in carbon tetrachloride. The proton-decoupled ^{13}C spectrum of **S** is given in Fig. 13.49. Propose a structure for **S**.

13.35

Compound **T** (C_5H_8O) has a strong infrared absorption band at 1745 cm^{-1}. The proton-decoupled ^{13}C spectrum of **T** is given in Fig. 13.50. Propose a structure for **T**.

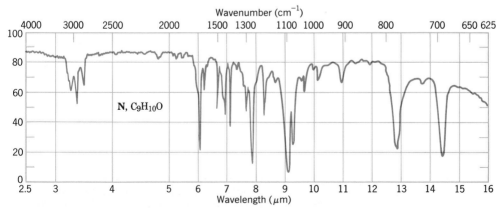

FIGURE 13.47 The infrared spectrum of compound **N**, Problem 13.30. (Spectrum courtesy of Aldrich Chemical Co., Milwaukee, WI.)

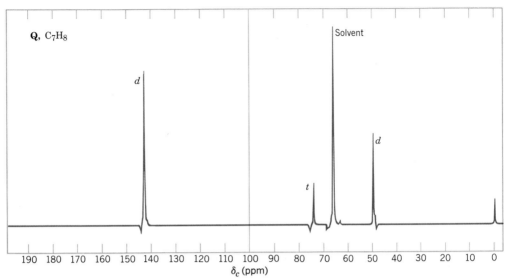

FIGURE 13.48 The proton-decoupled ^{13}C nmr spectrum of compound **Q**, Problem 13.33. The letters s, d, and q refer to the signal splitting (singlet, doublet, quartet) in the proton off-resonance decoupled spectrum. (Adapted from L. F. Johnson and W. C. Jankowski, *Carbon-13 NMR Spectra: A Collection of Assigned, Coded, and Indexed Spectra*, Wiley–Interscience, New York, 1972.)

13.36

The infrared and proton nmr spectra for compound **X** (C_8H_{10}) are given in Fig. 13.51. Propose a structure for compound **X**.

13.37

The infrared and proton nmr spectra of compound **Y** ($C_9H_{12}O$) are given in Fig. 13.52. Propose a structure for **Y**.

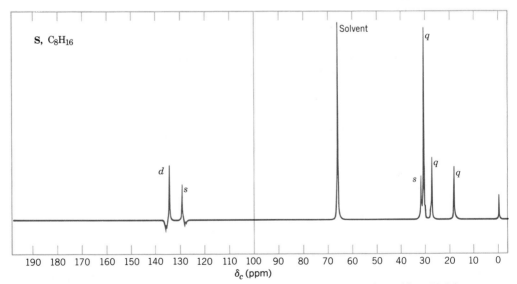

FIGURE 13.49 The proton-decoupled ^{13}C nmr spectrum of compound **S**, Problem 13.34. The letters s, d, t, and q refer to signal splitting (singlet, doublet, triplet, and quartet) in the proton off-resonance decoupled spectrum. (Adapted from L. F. Johnson and W. C. Jankowski, *Carbon-13 NMR Spectra: A Collection of Assigned, Coded, and Indexed Spectra*, Wiley–Interscience, New York, 1972.)

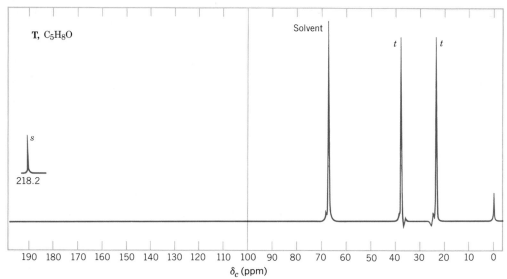

FIGURE 13.50 The proton-decoupled ^{13}C nmr spectrum of compound **T**, Problem 13.35. The letters s and t refer to the signal splitting (singlet and triplet) in the proton off-resonance decoupled spectrum. (Adapted from L. F. Johnson and W. C. Jankowski, *Carbon-13 NMR Spectra: A Collection of Assigned, Coded, and Indexed Spectra*, Wiley–Interscience, New York, 1972.)

13.38

The following [14]annulene obeys Hückel's rule. Its proton nmr spectrum shows signals

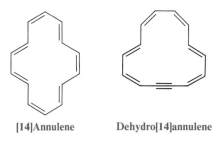

[14]Annulene   Dehydro[14]annulene

at $\delta 7.78$ (10H) and $\delta - 0.61$ (4H). Dehydro[14]annulene gives proton nmr signals at $\delta 8.0$ (10H) and $\delta 0.0$ (2H). How can you account for the relative intensities of the signals given by the two compounds?

13.39

(a) When butyl fluoride and *sec*-butyl fluoride are treated separately with excess SbF_5, their solutions give identical proton nmr spectra. These spectra, moreover, are the same as that obtained when *tert*-butyl fluoride is treated with SbF_5. Explain these results. (b) Treating the eight isomeric fluoropentanes with excess SbF_5 furnishes solutions that give identical proton nmr spectra. What species is formed in these reactions? Sketch the spectrum that you would expect to obtain.

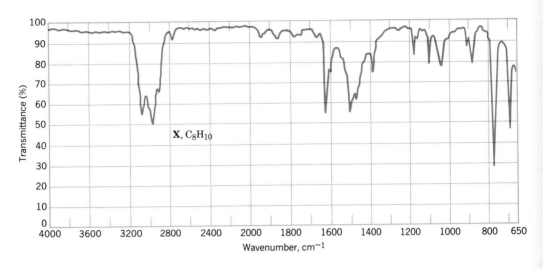

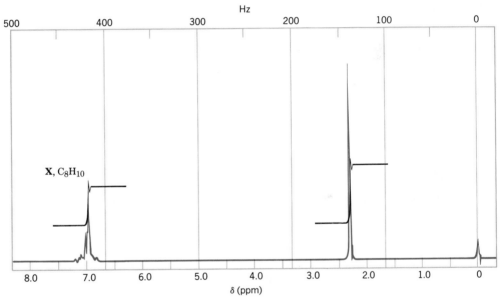

FIGURE 13.51 The infrared and proton nmr spectra of compound **X**, Problem 13.36. (Spectra courtesy of Varian Associates, Palo Alto, CA.)

13.40

(a) How many peaks would you expect to find in the proton nmr spectrum of caffeine?

Caffeine

(b) What characteristic peaks would you expect to find at the infrared spectrum of caffeine?

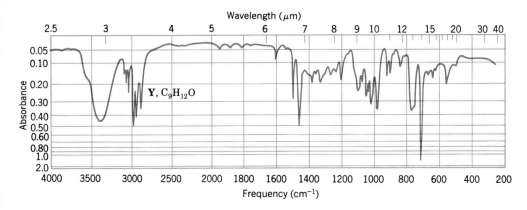

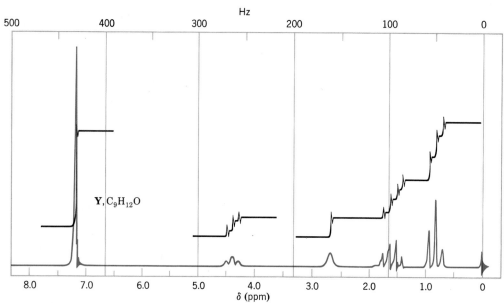

FIGURE 13.52 The infrared and proton nmr spectra of compound **Y**, Problem 13.37. (Infrared spectrum courtesy of Sadtler Research Laboratories, Philadelphia. The proton nmr spectrum courtesy of Varian Associates, Palo Alto, CA.)

13.41

3,4-Dichloro-1,2,3,4-tetramethylcyclobutene, **I** (see following structure), gives a proton nmr signal at $\delta 1.15$ corresponding to the protons labeled **A** and a signal at δ 1.26 corresponding to

the protons labeled **B**. When **I** is added to SbF_5—SO_2 at -78 °C, a pale-yellow solution is formed whose proton nmr spectrum shows the following singlets: $\delta 2.05$ (3H), $\delta 2.20$ (3H), $\delta 2.65$ (6H). After several minutes, these peaks begin to be replaced by a sharp singlet at $\delta 3.68$. Recall that SbF_5 is a powerful Lewis acid and explain what is taking place.

*13.42

Given the following information predict the appearance of the proton nmr spectrum given by the vinyl hydrogen atoms of p-chlorostyrene.

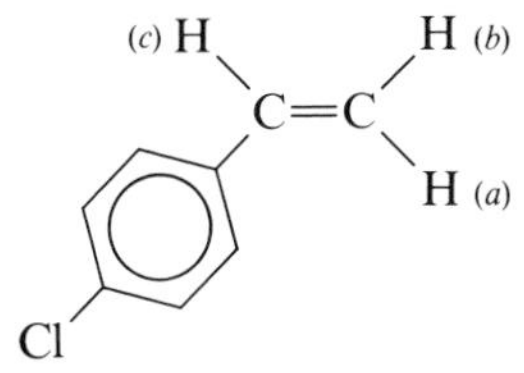

Deshielding by the induced magnetic field of the ring is greatest at proton (c) ($\delta6.7$) and is least at proton (b) ($\delta5.3$). The chemical shift of (a) is $\sim\delta5.7$. The coupling constants have the following approximate magnitudes: $J_{ac} \approx 18$ Hz, $J_{bc} \approx 11$ Hz, and $J_{ab} \approx 2$ Hz. (These coupling constants are typical of those given by vinylic systems: coupling constants for trans hydrogen atoms are larger than those for cis hydrogen atoms and coupling constants for geminal hydrogen atoms are very small.)

SPECIAL TOPIC E

MASS SPECTROSCOPY

E.1 The Mass Spectrometer

In a mass spectrometer (Fig. E.1) molecules in the gaseous state under low pressure are bombarded with a beam of high-energy electrons. The energy of the beam of electrons is usually 70 eV (electron volts) and one of the things this bombardment can do is dislodge one of the electrons of the molecule and produce a positively charged ion called *the molecular ion.*

$$M \quad + \quad e^- \quad \longrightarrow \quad M^{\ddagger} \quad + 2e^-$$

Molecule   *High-energy*   *Molecular*
                electron   *ion*

The molecular ion is not only a cation, but because it contains an odd number of electrons, it also is a free radical. Thus it belongs to a general group of ions called *radical cations.* If, for example, the molecule under bombardment is a molecule of ammonia, the following reaction will take place.

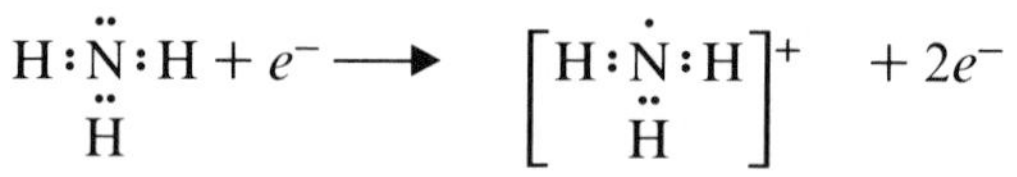

$$H\!:\!\overset{\cdot\cdot}{\underset{\overset{\cdot\cdot}{H}}{N}}\!:\!H + e^- \longrightarrow \left[H\!:\!\overset{\cdot}{\underset{\overset{\cdot\cdot}{H}}{N}}\!:\!H\right]^+ \quad + 2e^-$$

Molecular ion, $M^{\ddagger}$
(a radical cation)

An electron beam with an energy of 70 eV (~ 1600 kcal/mole) not only dislodges electrons from molecules, producing molecular ions, it also imparts to the molecular ions considerable surplus energy. Not all molecular ions will have the same amount of surplus energy, but for most, the surplus will be far in excess of that required to break covalent bonds (50–100 kcal/mole). Thus, soon after they are formed, most molecular ions literally fly apart — they undergo *fragmentation.* Fragmentation can take place in a variety of ways depending on the nature of the particular molecular ion, and as we shall see later, the way a molecular ion fragments can give us highly useful information about the structure of a complex molecule. Even with a relatively simple molecule like ammonia, however, fragmentation can produce several new cations. The molecular ion can eject a hydrogen atom, for example, and produce the cation NH_2^+.

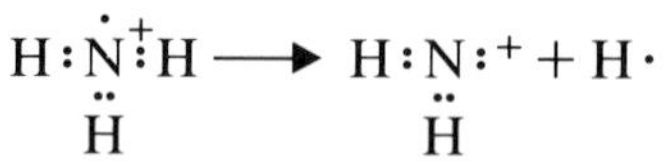

$$H\!:\!\overset{\cdot\,+}{\underset{\overset{\cdot\cdot}{H}}{N}}\!:\!H \longrightarrow H\!:\!\overset{\cdot\cdot}{\underset{\overset{\cdot\cdot}{H}}{N}}\!:^+ + H\cdot$$

This NH_2^+ cation can then lose a hydrogen atom to produce $NH^{\ddagger}$, which can lead, in turn, to N^+.

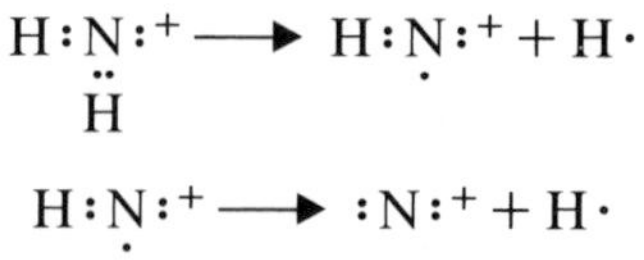

$$H\!:\!\overset{\cdot\cdot}{\underset{\overset{\cdot\cdot}{H}}{N}}\!:^+ \longrightarrow H\!:\!\overset{\cdot\cdot}{\underset{\cdot}{N}}\!:^+ + H\cdot$$

$$H\!:\!\overset{\cdot\cdot}{\underset{\cdot}{N}}\!:^+ \longrightarrow :\!\overset{\cdot\cdot}{\underset{\cdot}{N}}\!:^+ + H\cdot$$

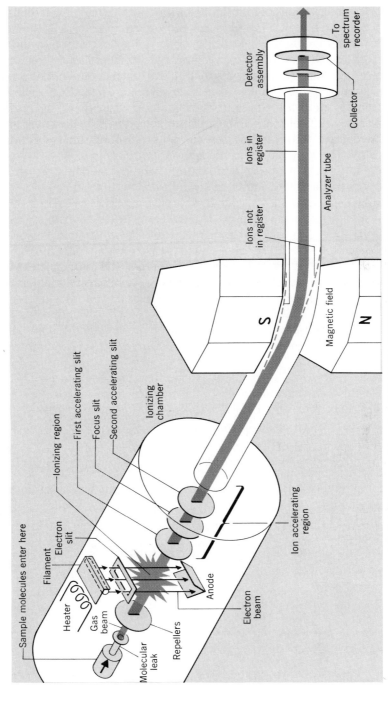

FIGURE E.1 Mass spectrometer. Schematic diagram of CEG model 21-103. The magnetic field that brings ions of varying mass-to-charge ratios into register is perpendicular to the page. (From John R. Holum, *Organic Chemistry: A Brief Course*, Wiley, New York, 1975. Used with permission.)

Labels in figure: Sample molecules enter here; Filament; Electron slit; Heater; Gas beam; Molecular leak; Repellers; Anode; Electron beam; Ion accelerating region; Ionizing region; First accelerating slit; Focus slit; Second accelerating slit; Ionizing chamber; S; N; Magnetic field; Ions not in register; Ions in register; Analyzer tube; Detector assembly; Collector; To spectrum recorder

The mass spectrometer then *sorts* these cations on the basis of their mass/charge or m/e ratio. Since for all practical purposes the charge on all of the ions is $+1$, this amounts to sorting them on the basis of their mass. The conventional mass spectrometer does this by accelerating the ions through a series of slits and then it sends the ion beam into a curved tube (see Fig. E.1 again). This curved tube passes through a variable magnetic field and the magnetic field exerts an influence on the moving ions. Depending on its strength at a given moment, the magnetic field will cause ions with a particular m/e ratio to follow a curved path that exactly matches the curvature of the tube. These ions are said to be "in register." Because they are in register, these ions pass through another slit and impinge on an ion collector where the intensity of the ion beam is measured electronically. The intensity of the beam is simply a measure of the relative abundance of the ions with a particular m/e ratio. Some mass spectrometers are so sensitive that they can detect the arrival of a *single ion.*

The actual sorting of ions takes place in the magnetic field, and this sorting takes place because laws of physics govern the paths followed by charged particles when they move through magnetic fields. Generally speaking, a magnetic field such as this will cause ions moving through it to move in a path that represents part of a circle. The radius of curvature of this circular path is related to the m/e ratio of the ions, to the strength of the magnetic field (H, gauss) and to the accelerating voltage. If we keep the accelerating voltage constant and progressively increase the magnetic field, ions whose m/e ratios are progressively larger will travel in a circular path that exactly matches that of the curved tube. Hence by steadily increasing H, ions with progressively increasing m/e ratios will be brought into register and thus will be detected at the ion collector. Since, as we said earlier, the charge on nearly all of the ions is unity, this means that *ions of progressively increasing mass arrive at the collector and are detected.*

What we have described is called "magnetic focusing" (or "magnetic scanning"), and all of this is done automatically by the mass spectrometer. The spectrometer displays the results by plotting a series of peaks of varying intensity in which each peak corresponds to ions of a particular m/e ratio. This display (Fig. E.2) is one form of a *mass spectrum.*

Ion sorting can also be done with "electrical focusing." In this technique, the magnetic field is held constant and the accelerating voltage is varied. Both methods, of course, accomplish the same thing, and some high-resolution mass spectrometers employ both techniques.

To summarize: A mass spectrometer bombards organic molecules with a beam of high-energy electrons causing them to ionize and fragment. It then separates the resulting mixture of ions on the basis of their mass/charge ratio and records the relative abundance of each ionic fragment. It displays this result as a plot of ion abundance versus m/e.

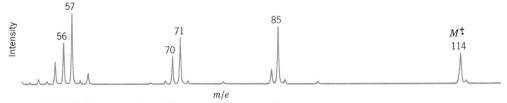

FIGURE E.2 A portion of the mass spectrum of octane.

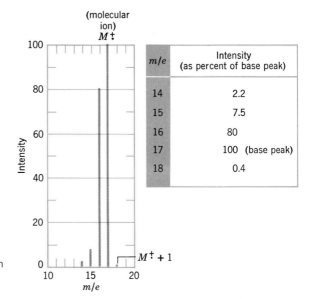

m/e	Intensity (as percent of base peak)
14	2.2
15	7.5
16	80
17	100 (base peak)
18	0.4

FIGURE E.3 The mass spectrum of NH_3 presented as a bar graph and in tabular form.

E.2 The Mass Spectrum

Mass spectra are usually published as bar graphs or in tabular form, as illustrated in Fig. E.3 for the mass spectrum of ammonia. In either presentation, the most intense peak — called the *base peak* — is arbitrarily assigned an intensity of 100%. The intensities of all other peaks are given proportionate values, as percentages of the base peak.

The masses of the ions given in a mass spectrum are those that we would calculate for the ion by assigning the constituent atoms' *masses rounded off to the nearest whole number*. For the commonly encountered atoms the nearest whole-number masses are

$$H = 1 \qquad O = 16$$
$$C = 12 \qquad F = 19$$
$$N = 14$$

In the mass spectrum of ammonia we see peaks at $m/e = 14, 15, 16,$ and 17. These correspond to the molecular ion and to the fragments we saw earlier.

$$NH_3 \xrightarrow{-e^-} [NH_3]^{\ddagger} \xrightarrow{-H \cdot} [NH_2]^+ \xrightarrow{-H \cdot} [NH]^{\ddagger} \xrightarrow{-H \cdot} [N]^+$$
$$m/e \quad = \qquad 17 \qquad\qquad 16 \qquad\qquad 15 \qquad\qquad 14$$
$$\qquad\qquad (molecular\ ion)$$

By convention we express,

$$H \!:\! \overset{\cdot}{\underset{H}{N}} \!:\! H^+ \quad \text{as} \quad [NH_3]^{\ddagger}$$

$$H \!:\! \underset{H}{\overset{\cdot\cdot}{N}} \!:\!^+ \quad \text{as} \quad [NH_2]^+$$

$$H \!:\! \overset{\cdot}{N} \!:\!^+ \quad \text{as} \quad [NH]^{\ddagger}$$

and

$$: N \!:\!^+ \quad \text{as} \quad [N]^+$$

TABLE E.1 Principal stable isotopes of common elements[a]

ELEMENT	MOST COMMON ISOTOPE		NATURAL ABUNDANCE OF OTHER ISOTOPES (BASED ON 100 ATOMS OF MOST COMMON ISOTOPE)			
Carbon	^{12}C	100	^{13}C	1.08		
Hydrogen	^{1}H	100	^{2}H	0.016		
Nitrogen	^{14}N	100	^{15}N	0.38		
Oxygen	^{16}O	100	^{17}O	0.04	^{18}O	0.20
Fluorine	^{19}F	100				
Sulfur	^{32}S	100	^{33}S	0.78	^{34}S	4.40
Chlorine	^{35}Cl	100	^{37}Cl	32.5		
Bromine	^{79}Br	100	^{81}Br	98.0		
Iodine	^{127}I	100				

[a]Data obtained from R. M. Silverstein, G. C. Bassler, and T. C. Morrill, *Spectrometric Identification of Organic Compounds,* 3rd ed., Wiley, New York, 1974, p. 13.

In the case of ammonia, the base peak is the peak arising from the molecular ion. This is not always the case, however; in many of the spectra that we shall see later the base peak (the most intense peak) will be at an m/e value different from that of the molecular ion. This happens because in many instances the molecular ion fragments so rapidly that some other ion at a smaller m/e value produces the most intense peak. In a few cases the molecular ion peak is extremely small, and sometimes it is absent altogether.

One other feature in the spectrum of ammonia requires explanation: The small peak that occurs at m/e 18. In the bar graph we have labeled this peak $M^{\ddagger} + 1$ to indicate that it is one mass unit greater than the molecular ion. The $M^{\ddagger} + 1$ peak appears in the spectrum because most elements (e.g., nitrogen and hydrogen) have more than one naturally occurring isotope (Table E.1). Although most of the NH_3

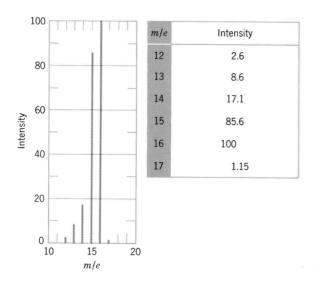

m/e	Intensity
12	2.6
13	8.6
14	17.1
15	85.6
16	100
17	1.15

FIGURE E.4 Mass spectrum for Problem E.1.

molecules in a sample of ammonia are composed of $^{14}N^1H_3$, a small but detectable fraction of molecules are composed of $^{15}N^1H_3$. (A very tiny fraction of molecules are also composed of $^{14}N^1H_2{}^2H$.) These molecules — $^{15}N^1H_3$ or $^{14}N^1H_2{}^2H$ — produce molecular ions at m/e 18, that is at $M^{\ddagger} + 1$.

The spectrum of ammonia begins to show us with a simple example how the masses (or m/e's) of individual ions can give us information about the composition of the ions and how this information can allow us to arrive at possible structures for a compound. Problems E.1 to E.3 will allow us further practice with this technique.

PROBLEM E.1

Propose a structure for the compound whose mass spectrum is given in Fig. E.4 and make reasonable assignments for each peak.

PROBLEM E.2

Propose a structure for the compound whose mass spectrum is given in Fig. E.5 and make reasonable assignments for each peak.

PROBLEM E.3

The compound whose mass spectrum is given in Fig. E.6 contains three elements, one of which is fluorine. Propose a structure for the compound and make reasonable assignments for each peak.

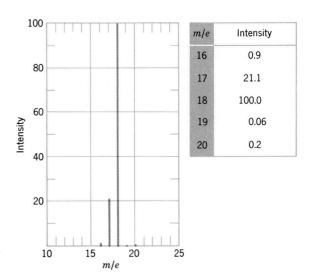

m/e	Intensity
16	0.9
17	21.1
18	100.0
19	0.06
20	0.2

FIGURE E.5 Mass spectrum for Problem E.2.

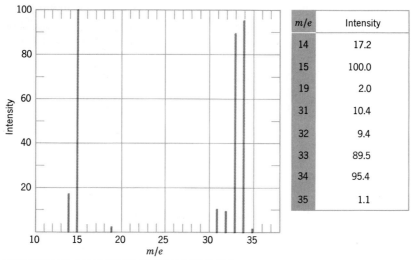

m/e	Intensity
14	17.2
15	100.0
19	2.0
31	10.4
32	9.4
33	89.5
34	95.4
35	1.1

FIGURE E.6 Mass spectrum for Problem E.3.

E.3 Determination of Molecular Formulas and Molecular Weights

E.3A The Molecular Ion and Isotopic Peaks

Look at Table E.1 for a moment. Notice that most of the common elements found in organic compounds have naturally occurring *heavier* isotopes. For three of the elements—carbon, hydrogen, and nitrogen—the principal heavier isotope is one mass unit greater than the most common isotope. The presence of these elements in a compound will give rise to a small isotopic peak one unit greater than the molecular ion—at $M^+ + 1$. For four of the elements—oxygen, sulfur, chlorine, and bromine—the principal heavier isotope is two mass units greater than the most common isotope. The presence of these elements in a compound gives rise to an isotopic peak at $M^+ + 2$.

$$M^+ + 1 \text{ Element:} \quad C, H, N$$

$$M^+ + 2 \text{ Elements:} \quad O, S, Br, Cl$$

Isotopic peaks give us one method for determining molecular formulas. To understand how this can be done, let us begin by noticing that the isotope abundances in Table E.1 are based on 100 atoms of the normal isotope. Now let us suppose, as an example, that we have 100 molecules of methane (CH_4). On the average there will be 1.08 molecules that contain ^{13}C and 4×0.016 molecules that contain ^2H. Altogether then, these heavier isotopes should contribute an $M^+ + 1$ peak whose intensity is ~1.14% of the intensity of the peak for the molecular ion.

$$1.08 + 4(0.016) \simeq 1.14\%$$

This correlates well with the observed intensity of the $M^+ + 1$ peak in the actual spectrum of methane given in Fig. E.4.

For molecules with a modest number of atoms we can determine molecular formulas in the following way. If the M^+ peak is not the base peak the first thing we do

m/e	Intensity (as percent of base peak)	m/e	Intensity (as percent of $M^{\ddagger}$)
27	59.0	72	$M^{\ddagger}$ — 100.0
28	15.0	73	$M^{\ddagger}+1$ — 4.5
29	54.0	74	$M^{\ddagger}+2$ — 0.3
39	23.0		Recalculated to base on $M^{\ddagger}$
41	60.0		
42	12.0		
43	79.0		
44	100.0 (base)		
72	73.0 $M^{\ddagger}$		
73	3.3		
74	0.2		

FIGURE E.7 Mass spectrum of an unknown compound.

with the mass spectrum of an unknown compound is to recalculate the intensities of the $M^{\ddagger}+1$ and $M^{\ddagger}+2$ to express them as percentages of the intensity of the $M^{\ddagger}$ peak. Consider, for example, the mass spectrum given in Fig. E.7. The $M^{\ddagger}$ peak at $m/e=72$ is not the base peak. Therefore, we need to recalculate the intensities of the peaks in our spectrum at m/e 72, 73, and 74 as percents of the peak at m/e 72. We do this by dividing each intensity by the intensity of the $M^{\ddagger}$ peak (which is 73%) and multiply by 100. These results are shown here and in the second column of Fig. E.7.

m/e	Intensity % of $M^{\ddagger}$
72	$73.0/73 \times 100 = 100$
73	$3.3/73 \times 100 = 4.5$
74	$0.2/73 \times 100 = 0.3$

Then we use the following guides to determine the molecular formula.

1. **Is $M^{\ddagger}$ odd or even? According to the nitrogen rule, if it is even, then the compound must contain an even number of nitrogen atoms (zero is an even number).**

   For our unknown, $M^{\ddagger}$ is even. The compound must have an even number of nitrogen atoms.

2. **The relative abundance of the $M^{\ddagger}+1$ peak indicates the number of carbon atoms. Number of C atoms = relative abundance of $(M^{\ddagger}+1)/1.1$**

   For our unknown (Fig. E.7), Number of C atoms $= \dfrac{4.5}{1.1} = 4$

   (This formula works because ^{13}C is the most important contributor to the $M^{\ddagger}+1$ peak and the approximate natural abundance of ^{13}C is 1.1%.)

3. **The relative abundance of the $M^{\ddagger} + 2$ peak indicates the presence (or absence) of S, (4.4%); Cl, (33%); or Br (98%)** (see Table E.1).

For our known, $M^{\ddagger} + 2 = 0.2\%$; thus we can assume that S, Cl, and Br are absent.

4. **The molecular formula can now be established by determining the number of hydrogen atoms and adding the appropriate number of oxygen atoms, if necessary.**

For our unknown the $M^{\ddagger}$ peak at m/e 72 gives us the molecular weight. It also tells us (since it is even) that nitrogen is absent because C_4N_2 has a molecular weight (76) greater than that of our compound.

For a molecule composed of C and H only:

$H = 72 - (4 \times 12) = 24$, but C_4H_{24} is impossible.

For a molecule composed of C, H, and one O:

$H = 72 - (4 \times 12) - 16 = 8$ and thus our unknown has the molecular formula C_4H_8O.

PROBLEM E.4

(a) Write structural formulas for at least 14 stable compounds that have the formula C_4H_8O. (b) The infrared spectrum of the unknown compound shows a strong peak near 1730 cm^{-1}. Which structures now remain as possible formulas for the compound? (We continue with this compound in Problem E.14.)

PROBLEM E.5

Determine the molecular formula of the following compound. (The complete mass spectrum of this compound is given in Fig. E.18; cf. Problem E.19.)

m/e	INTENSITY (AS % OF BASE PEAK)
86 $M^{\ddagger}$	10.00
87	0.56
88	0.04

PROBLEM E.6

(a) What approximate intensities would you expect for the $M^{\ddagger}$ and $M^{\ddagger} + 2$ peaks of CH_3Cl? (b) For the $M^{\ddagger}$ and $M^{\ddagger} + 2$ peaks of CH_3Br? (c) An organic compound gives an $M^{\ddagger}$ peak at m/e 122 and a peak of nearly equal intensity at m/e 124. What is a likely molecular formula for the compound?

PROBLEM E.7

Use the mass spectral data given in Fig. E.8 to determine the molecular formula for the compound.

FIGURE E.8 Mass spectrum for Problem E.7.

m/e	INTENSITY (AS % OF BASE PEAK)
14	8.0
15	38.6
18	16.3
28	39.7
29	23.4
42	46.6
43	10.7
44	100.0 (base)
73	86.1 $M^{\ddagger}$
74	3.2
75	0.2

PROBLEM E.8

(a) Determine the molecular formula of the compound whose mass spectrum is given here. (b) The proton nmr spectrum of this compound consists only of a large doublet and a small septet. What is the structure of the compound?

m/e	INTENSITY (AS % OF BASE PEAK)
27	34
39	11
41	22
43	100 (base)
63	26
65	8
78	24 $M^{\ddagger}$
79	0.8
80	8

As the number of atoms in a molecule increases, calculations like this become more and more complex and time consuming. Fortunately, however, these calculations can be done readily with computers, and tables are now available that give relative values for the $M^{\ddagger} + 1$ and $M^{\ddagger} + 2$ peaks for all combinations of common

TABLE E.2 Relative intensities of $M^+ + 1$ and $M^+ + 2$ peaks for various combinations of C, H, N, and O for masses 72 and 73

M^+	FORMULAS	PERCENTAGE OF M^+ INTENSITY		M^+	FORMULAS	PERCENTAGE OF M^+ INTENSITY	
		$M^+ + 1$	$M^+ + 2$			$M^+ + 1$	$M^+ + 2$
72	CH_2N_3O	2.30	0.22	73	CHN_2O_2	1.94	0.41
	CH_4N_4	2.67	0.03		CH_3N_3O	2.31	0.22
	$C_2H_2NO_2$	2.65	0.42		CH_5N_4	2.69	0.03
	$C_2H_4N_2O$	3.03	0.23		C_2HO_3	2.30	0.62
	$C_2H_6N_3$	3.40	0.04		$C_2H_3NO_2$	2.67	0.42
	$C_3H_4O_2$	3.38	0.44		$C_2H_5N_2O$	3.04	0.23
	C_3H_6NO	3.76	0.25		$C_2H_7N_3$	3.42	0.04
	$C_3H_8N_2$	4.13	0.07		$C_3H_5O_2$	3.40	0.44
	C_4H_8O	4.49	0.28		C_3H_7NO	3.77	0.25
	$C_4H_{10}N$	4.86	0.09		$C_3H_9N_2$	4.15	0.07
	C_5H_{12}	5.60	0.13		C_4H_9O	4.51	0.28
					$C_4H_{11}N$	4.88	0.10
					C_6H	6.50	0.18

Data from J. H. Beynon, *Mass Spectrometry and Its Application to Organic Chemistry,* Elsevier, Amsterdam, 1960.

elements with molecular formulas up to mass 500. Part of the data obtained from one of these tables is given in Table E.2. Use Table E.2 to check the results of our example (Fig. E.7) and your answer to Problem E.7.

E.3B High-Resolution Mass Spectroscopy

All of the spectra that we have described so far were determined on what are called "low-resolution" mass spectrometers. These spectrometers, as we noted earlier, measure m/e values to the nearest whole-number mass unit. Most laboratories are equipped with this type of mass spectrometer.

Many laboratories, however, are equipped with the more expensive "high-resolution" mass spectrometers. These spectrometers can measure m/e values to three or four decimal places and thus they provide an extremely accurate method for determining molecular weights. And because molecular weights can be measured so accurately, these spectrometers also allow us to determine molecular formulas.

The determination of a molecular formula by an accurate measurement of a molecular weight is possible because the actual masses of atomic particles (nuclides) are not integers (see Table E.3). Consider, as examples, the three molecules, O_2, N_2H_4, and CH_3OH. The actual atomic masses of the molecules are all different.

$$O_2 = 2(15.9949) = 31.9898$$

$$N_2H_4 = 2(14.0031) + 4(1.00783) = 32.0375$$

$$CH_4O = 12.0000 + 4(1.00783) + 15.9949 = 32.0262$$

TABLE E.3 Exact masses of nuclides

ISOTOPE	MASS
^1H	1.00783
^2H	2.01410
^{12}C	12.00000 (std)
^{13}C	13.00336
^{14}N	14.0031
^{15}N	15.0001
^{16}O	15.9949
^{17}O	16.9991
^{18}O	17.9992
^{19}F	18.9984
^{32}S	31.9721
^{33}S	32.9715
^{34}S	33.9679
^{35}Cl	34.9689
^{37}Cl	36.9659
^{79}Br	78.9183
^{81}Br	80.9163
^{127}I	126.9045

High-resolution mass spectrometers are available that are capable of measuring mass with an accuracy of 1 part in 40,000. Thus, such a spectrometer can easily distinguish among these three molecules and, in effect, tell us the molecular formula.

E.4 Fragmentation

In most instances the molecular ion is a highly energetic species, and in the case of a complex molecule a great many things can happen to it. The molecular ion can break apart in a variety of ways and the fragments that are produced can then undergo further fragmentation and so on. In a certain sense mass spectroscopy is a "brute force" technique. Striking an organic molecule with 70-eV electrons is a little like firing a howitzer at a house made of matchsticks. That fragmentation takes place in any sort of predictable way is truly remarkable—and yet it does. Many of the same factors that govern ordinary chemical reactions seem to apply to fragmentation processes, and many of the principles that we have learned about the relative stabilities of carbocations, free radicals, and molecules will help us to make some sense out of what takes place. And as we learn something about what kind of fragmentations to expect, we shall be much better able to use mass spectra as aids in determining the structures of organic molecules.

We cannot, of course, in the limited space that we have here, look at these processes in great detail, but we can examine some of the more important ones.

As we begin, keep two important principles in mind. (1) The reactions that take place in a mass spectrometer are usually *unimolecular*—that is, they involve only a *single* molecular fragment. This is true because the pressure in a mass spectrometer is kept so low (~ 10^{-6} Torr) that reactions requiring bimolecular collisions usually do not occur. (2) The relative ion abundances, as measured by peak intensities, are extremely important. We shall see that the appearance of certain prominent peaks in the spectrum gives us important information about the structures of the fragments produced and about their original locations in the molecule.

E.4A Fragmentation by Cleavage at a Single Bond

One important type of fragmentation is the simple cleavage at a single bond. With a radical cation this cleavage can take place in at least two ways; each way produces a *cation* and a *free radical.* Only the cations are detected in a mass spectrometer. (The free radicals, because they are not charged, do not move and, therefore, are not detected.) With the molecular ion obtained from propane, for example, two possible modes of cleavage are

$$CH_3CH_2\overset{+}{\cdot}CH_3 \longrightarrow CH_3CH_2^+ + \cdot CH_3$$
$$m/e\ 29$$

$$CH_3CH_2\overset{+}{\cdot}CH_3 \longrightarrow CH_3CH_2\cdot + {}^+CH_3$$
$$m/e\ 15$$

These two modes of cleavage do not take place at equal rates, however. While the relative abundance of cations produced by such a cleavage is influenced both by the stability of the carbocation and by the stability of the free radical, *the carbocation's stability is more important.** In the spectrum of propane the peak at *m/e* 29 ($CH_3CH_2^+$) is the most intense peak; the peak at *m/e* 15 (CH_3^+) has an intensity of only 5.6%. This reflects the greater stability of $CH_3CH_2^+$ when compared to CH_3^+.

E.4B Fragmentation Equations

Before we go further, we need to examine some of the conventions that are used in writing equations for fragmentation reactions. In the two equations for cleavage at the single bond of propane that we have just written, we have localized the odd electron and the charge on one of the carbon–carbon sigma bonds of the molecular ion. When we write structures this way, the choice of just where to localize the odd electron and the charge is sometimes arbitrary. When possible, however, we write the structure showing the molecular ion that would result from the removal of one of the most loosely held electrons of the original molecule. Just which electrons these are can usually be estimated from ionization potentials (Table E.4). [The ionization potential of a molecule is the amount of energy (in eV) required to remove an electron from the molecule.] As we might expect, ionization potentials indicate that the nonbonding electrons of nitrogen and oxygen and the pi electrons of alkenes and aromatic molecules are held more loosely than the electrons of carbon–carbon and carbon–hydrogen sigma bonds. Thus the convention of localizing the odd electron and charge is especially applicable when the molecule contains an oxygen, nitrogen, double bond, or aromatic ring. If the molecule contains only carbon–carbon and carbon–hydrogen sigma bonds, and if it contains a great many of these, then the choice of where to localize the odd electron and the charge is so arbitrary as to be

---

*This can be demonstrated through thermochemical calculations that we cannot go into here. The interested student is referred to F. W. McLafferty, *Interpretation of Mass Spectra,* 2nd ed., Benjamin, Reading, MA, 1973, p. 41 and pp. 210–211.

TABLE E.4 Ionization potentials of selected molecules

COMPOUND	IONIZATION POTENTIAL (ELECTRON VOLTS)
$CH_3(CH_2)_3NH_2$	8.7
C_6H_6	9.2
C_2H_4	10.5
CH_3OH	10.8
C_2H_6	11.5
CH_4	12.7

impractical. In these instances we usually resort to another convention: We write the formula for the radical cation in brackets and place the odd electron and charge outside. Using this convention we would write the two fragmentation reactions of propane in the following way:

$$[CH_3CH_2CH_3]^{\ddagger} \longrightarrow CH_3CH_2^+ + \cdot CH_3$$

$$[CH_3CH_2CH_3]^{\ddagger} \longrightarrow CH_3CH_2\cdot + CH_3^+$$

PROBLEM E.9

The most intense peak in the mass spectrum of 2,2-dimethylbutane occurs at m/e 57.   (a) What carbocation does this peak represent?   (b) Using the convention that we have just described, write an equation that shows how this carbocation arises from the molecular ion.

Figure E.9 shows us the kind of fragmentation a longer-chain alkane can undergo. The example here is hexane and we see a reasonably abundant molecular ion at m/e 86 accompanied by a small $M^{\ddagger} + 1$ peak. There is also a smaller peak at m/e 71 ($M^{\ddagger} - 15$) corresponding to the loss of $\cdot CH_3$, and the base peak is at m/e 57 ($M^{\ddagger} - 29$) corresponding to the loss of $\cdot CH_2CH_3$. The other prominent peaks are at m/e 43 ($M^{\ddagger} - 43$) and m/e 29 ($M^{\ddagger} - 57$) corresponding to the loss of $\cdot CH_2CH_2CH_3$ and $\cdot CH_2CH_2CH_2CH_3$, respectively. The important fragmentations are just the ones we would expect:

$$[CH_3CH_2CH_2CH_2CH_2CH_3]^{\ddagger} \longrightarrow$$

$$\longrightarrow CH_3CH_2CH_2CH_2CH_2^+ + \cdot CH_3$$
$$m/e\ 71$$

$$\longrightarrow CH_3CH_2CH_2CH_2^+ + \cdot CH_2CH_3$$
$$m/e\ 57$$

$$\longrightarrow CH_3CH_2CH_2^+ + \cdot CH_2CH_2CH_3$$
$$m/e\ 43$$

$$\longrightarrow CH_3CH_2^+ + \cdot CH_2CH_2CH_2CH_3$$
$$m/e\ 29$$

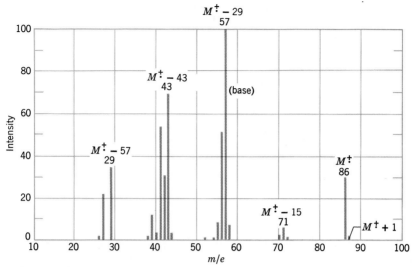

FIGURE E.9 Mass spectrum of hexane.

Chain branching increases the likelihood of cleavage at a branch point because a more stable carbocation can result. When we compare the mass spectrum of 2-methylbutane (Fig. E.10) with the spectrum of hexane, we see a much more intense peak at $M^{\ddagger} - 15$. Loss of a methyl radical from the molecular ion of 2-methylbutane can give a secondary carbocation:

$$\left[\begin{array}{c} CH_3 \\ | \\ CH_3CHCH_2CH_3 \end{array} \right]^{\ddagger} \longrightarrow CH_3\overset{+}{C}HCH_2CH_3 + \cdot CH_3$$

$$\begin{array}{cc} m/e\ 72 & m/e\ 57 \\ M^{\ddagger} & M^{\ddagger} - 15 \end{array}$$

whereas with hexane, loss of a methyl radical can yield only a primary carbocation.

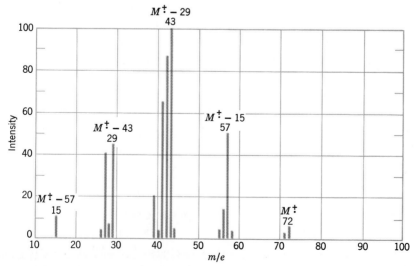

FIGURE E.10 The mass spectrum of 2-methylbutane.

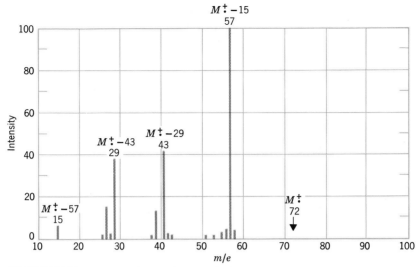

FIGURE E.11 Mass spectrum of neopentane.

With neopentane (Fig. E.11), this effect is even more dramatic. Loss of a methyl radical by the molecular ion produces a *tertiary* carbocation, and this reaction takes place so readily that virtually none of the molecular ions survive long enough to be detected.

$$
\left[
\begin{array}{c}
CH_3 \\
| \\
CH_3-C-CH_3 \\
| \\
CH_3
\end{array}
\right]^{+\cdot}
\longrightarrow
\begin{array}{c}
CH_3 \\
| \\
CH_3-C^+ \quad + \cdot CH_3 \\
| \\
CH_3
\end{array}
$$

$$
\begin{array}{cc}
m/e\ 72 & \qquad m/e\ 57 \\
M^{+\cdot} & \qquad M^{+\cdot}-15
\end{array}
$$

PROBLEM E.10

In contrast to 2-methylbutane and neopentane, the mass spectrum of 3-methylpentane (not given) has a peak of very low intensity at $M^{+\cdot} - 15$. It has a peak of very high intensity at $M^{+\cdot} - 29$, however. Explain.

Carbocations stabilized by resonance are usually also prominent in mass spectra. Several ways that resonance-stabilized carbocations can be produced are outlined in the following list.

1. Alkenes frequently undergo fragmentations that yield allylic cations.

$$
CH_2\overset{+\cdot}{=}CH-CH_2:R \longrightarrow \overset{+}{C}H_2-CH=CH_2 + \cdot R
$$
$$
m/e\ 41
$$

2. Carbon–carbon bonds next to an atom with an unshared electron pair usually break readily because the resulting carbocation is resonance stabilized.

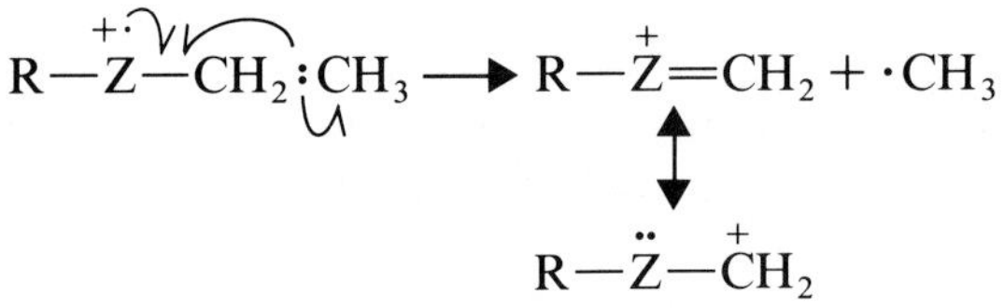

$$R-\overset{+}{\overset{\cdot\cdot}{Z}}-CH_2 \overset{\cap}{:} CH_3 \longrightarrow R-\overset{+}{Z}=CH_2 + \cdot CH_3$$

$$R-\overset{\cdot\cdot}{Z}-\overset{+}{CH_2}$$

$Z = N$, O, or S; R may also be H

3. Carbon–carbon bonds next to the carbonyl group of an aldehyde or ketone break readily because resonance-stabilized ions called acylium ions are produced.

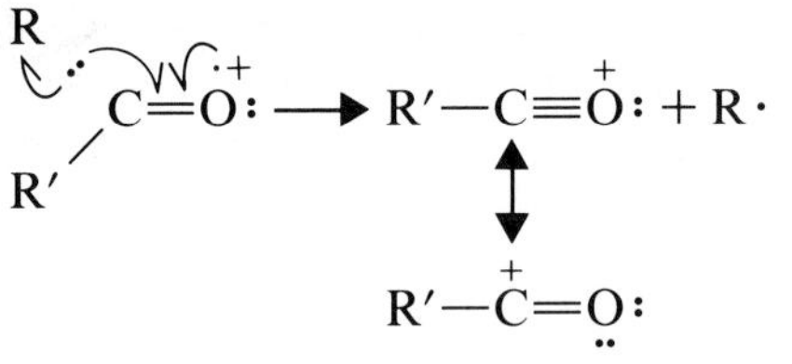

$$\overset{R}{\underset{R'}{>}}C\overset{\cdot+}{=}\overset{\cdot\cdot}{O}: \longrightarrow R'-C\equiv\overset{+}{O}: + R\cdot$$

$$R'-\overset{+}{C}=\overset{\cdot\cdot}{O}:$$

Acylium ion

or

$$\overset{R}{\underset{R'}{>}}C\overset{\cdot+}{=}\overset{\cdot\cdot}{O}: \longrightarrow R-C\equiv\overset{+}{O}: + R'\cdot$$

$$R-\overset{+}{C}=\overset{\cdot\cdot}{O}:$$

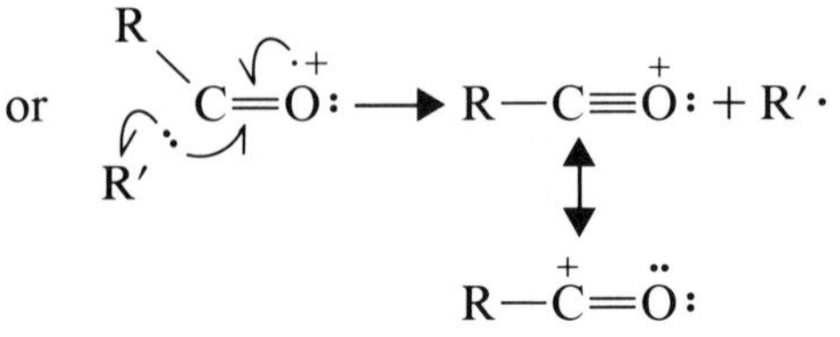

Acylium ion

4. Alkyl-substituted benzenes undergo loss of a hydrogen atom or methyl group to yield the relatively stable tropylium ion (cf. Section 11.6). This fragmentation gives a prominent peak (sometimes the base peak) at m/e 91.

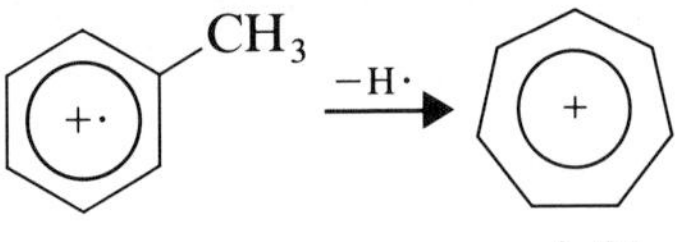

m/e 91

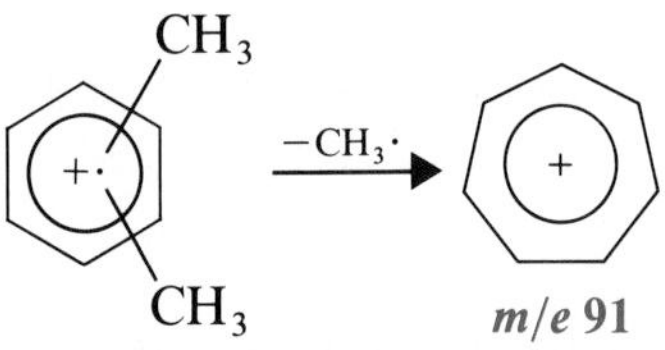

m/e 91

5. Substituted benzenes also lose their substituent and yield a phenyl cation at m/e 77.

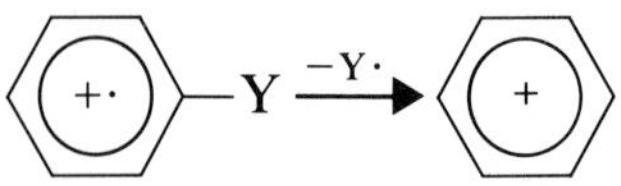

m/e 77

$$Y = \text{halogen}, -NO_2, -\overset{\overset{\textstyle O}{\|}}{C}R, -R, \text{ and so on}$$

PROBLEM E.11

The mass spectrum of 4-methyl-1-hexene (not given) shows intense peaks at m/e 57 and m/e 41. What fragmentation reactions account for these peaks?

PROBLEM E.12

Explain the following observations that can be made about the mass spectra of alcohols: (a) The molecular ion peak of a primary or secondary alcohol is very small; with a tertiary alcohol it is usually undetectable. (b) Primary alcohols show a prominent peak at m/e 31. (c) Secondary alcohols usually give prominent peaks at m/e 45, 59, 73, and so on. (d) Tertiary alcohols have prominent peaks at m/e 59, 73, 87, and so on.

PROBLEM E.13

The mass spectra of isopropyl butyl ether and propyl butyl ether are given in Figs. E.12 and E.13. (a) Which spectrum represents which ether? (b) Explain your choice.

E.4C Fragmentation by Cleavage of Two Bonds

Many peaks in mass spectra can be explained by fragmentation reactions that involve the breaking of two covalent bonds. When a radical–cation undergoes this type of fragmentation the products are *a new radical–cation* and *a neutral molecule.* Some important examples are the following:

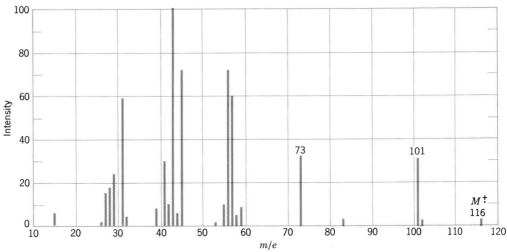

FIGURE E.12 Mass spectrum for Problem E.13.

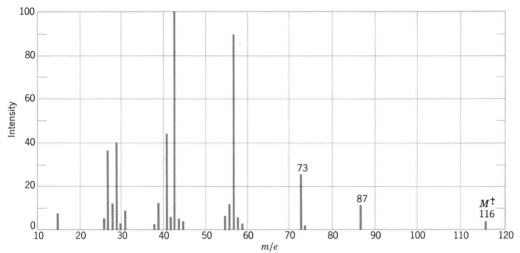

FIGURE E.13 Mass spectrum for Problem E.13.

1. Alcohols frequently show a prominent peak at $M^{+} - 18$. This corresponds to the loss of a molecule of water.

$$R—\overset{H}{\underset{M^{+}}{CH}}—\overset{\overset{\cdot\cdot}{+}\overset{\cdot\cdot}{O}H}{CH_2} \longrightarrow R—CH\overset{\cdot+}{=}CH_2 + H—\overset{\cdot\cdot}{\underset{\cdot\cdot}{O}}—H$$
$$\qquad\qquad\qquad\quad M^{+} - 18$$

or $[R—CH_2—CH_2—OH]^{+} \longrightarrow [R—CH=CH_2]^{+} + H_2O$
$\qquad\qquad M^{+}\qquad\qquad\qquad\qquad\quad M^{+} - 18$

2. Cycloalkenes can undergo a retro-Diels–Alder reaction that produces an alkene and an alkadienyl radical cation.

$$\left[\vcenter{\hbox{⬡}}\right]^{+} \longrightarrow \left[\vcenter{\hbox{▱}}\right]^{+} + \overset{CH_2}{\underset{CH_2}{\|}}$$

3. Carbonyl compounds with a hydrogen on their γ carbon undergo a fragmentation called the *McLafferty arrangement*.

$$\left[\begin{array}{c} O\diagdown^H \\ Y—C \diagup CHR \\ H_2C^{\diagup}CH_2 \end{array}\right]^{+} \longrightarrow \left[\begin{array}{c} H \\ O\diagup \\ Y—C\diagdown_{CH_2} \end{array}\right]^{+} + RCH=CH_2$$

Y may be R, H, OR, OH, and so on

In addition to these reactions, we frequently find peaks in mass spectra that result from the elimination of other small stable neutral molecules, for example, H_2, NH_3, CO, HCN, H_2S, alcohols and alkenes.

ADDITIONAL PROBLEMS
E.14
Reconsider Problem E.4 and the spectrum given in Fig. E.7. Important clues to the structure of this compound are the peaks at m/e 44 (the base peak) and m/e 29. Propose a structure for the compound and write fragmentation equations showing how these peaks arise.

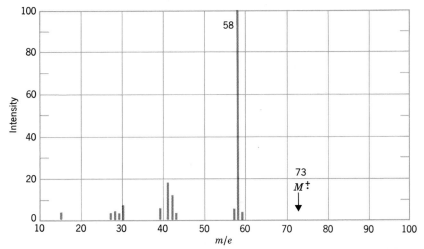

FIGURE E.14 Mass spectrum of compound **A** (Problem E.16).

E.15

The homologous series of primary amines, $CH_3(CH_2)_nNH_2$, from CH_3NH_2 to $CH_3(CH_2)_{13}NH_2$ all have their base (largest) peak at m/e 30. What ion does this peak represent and how is it formed?

E.16

The mass spectrum of compound **A** is given in Fig. E.14. The proton nmr spectrum of **A** consists of two singlets with area ratios of $9:2$. The larger singlet is at $\delta 1.2$, the smaller one at δ 1.3. Propose a structure for compound **A**.

E.17

The mass spectrum of compound **B** is given in Fig. E.15. The infrared spectrum of **B** shows a broad peak between 3200 and 3550 cm^{-1}. The proton nmr spectrum of **B** shows the following peaks: a triplet at $\delta 0.9$, a singlet at $\delta 1.1$, and a quartet at $\delta 1.6$. The area ratios of these peaks is $3:7:2$, respectively. Propose a structure for **B**.

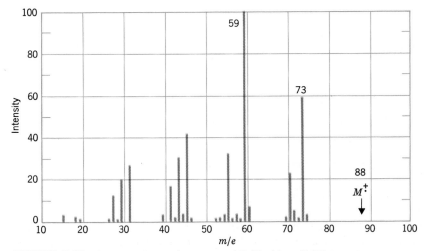

FIGURE E.15 Mass spectrum of compound **B** (Problem E.17).

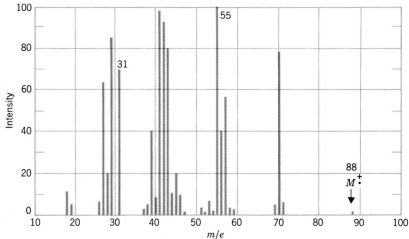

FIGURE E.16 Mass spectrum of compound **C** (Problem E.18).

E.18

The mass spectrum of compound **C** is given in Fig. E.16. Compound **C** is an isomer of **B** and the infrared spectrum of **C** also shows a broad peak in the 3200–3550-cm^{-1} region. The proton nmr spectrum of **C** is given in Fig. E.17. Propose a structure for **C**.

E.19

The mass spectrum of compound **D** is given in Fig. E.18. (**D** is also the subject of Problem E.5.) **D** shows a strong infrared peak at 1710 cm^{-1}. The proton nmr spectrum of **D** is given in Fig. E.19. Propose a structure for **D**.

E.20

Propose a structure for compound **E** whose mass spectrum is given in Fig. E.20.

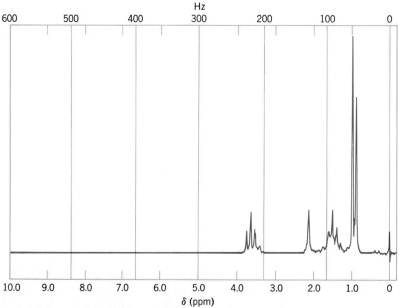

FIGURE E.17 The proton nmr spectrum of compound **C** (Problem E.18). (Courtesy of Aldrich Chemical Co., Milwaukee, WI.)

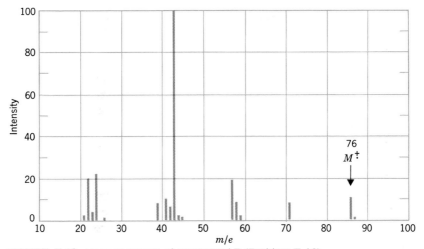

FIGURE E.18   Mass spectrum of compound **D** (Problem E.19).

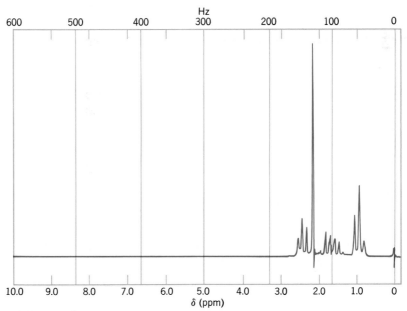

FIGURE E.19   The proton nmr spectrum of compound **D** (Problem E.19).
(Courtesy of Aldrich Chemical Co., Milwaukee, WI.)

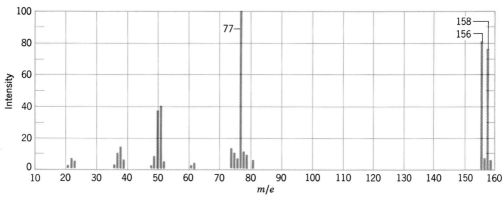

FIGURE E.20   The mass spectrum of compound **E** (Problem E.20).

CHAPTER FOURTEEN

Thymol Crystals. Thymol is a substituted phenol that can be obtained from ordinary garden thyme, *Thymus vulgaris*. The structure of thymol is given in Section 14.2.

Phenols and Aryl Halides. Nucleophilic Aromatic Substitution

14.1 STRUCTURE AND NOMENCLATURE OF PHENOLS

Compounds that have a hydroxyl group directly attached to a benzene ring are called **phenols.** Thus, **phenol** is the specific name for hydroxybenzene and it is the general name for the family of compounds derived from hydroxybenzene:

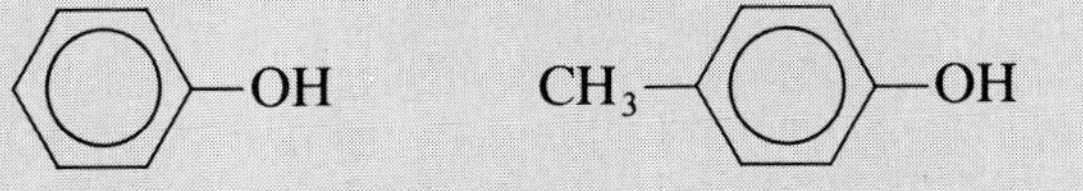

Phenol

p-Cresol (or 4-methylphenol)
(a phenol)

Compounds that have a hydroxyl group attached to a polycyclic benzenoid ring are chemically similar to phenols, but they are called **naphthols** and **phenanthrols,** for example.

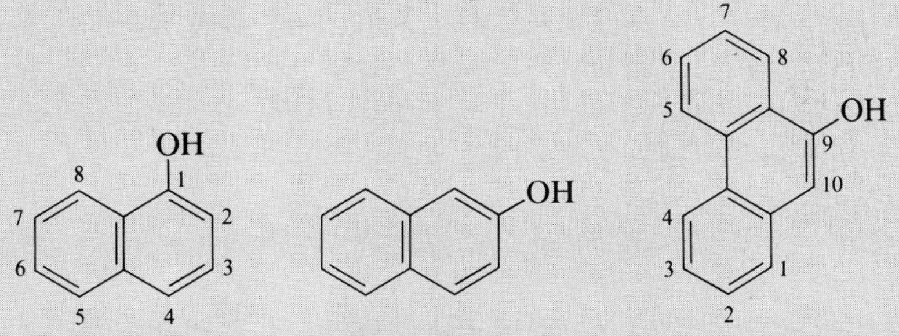

1-Naphthol
(α-naphthol)

2-Naphthol
(β-naphthol)

9-Phenanthrol

14.1A Nomenclature of Phenols

We studied the nomenclature of some of the phenols in Chapter 11. In many compounds *phenol* is the base name:

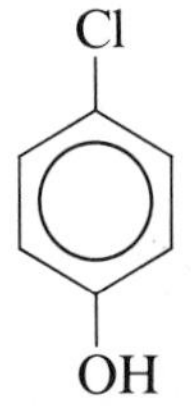

4-Chlorophenol
(*p*-chlorophenol)

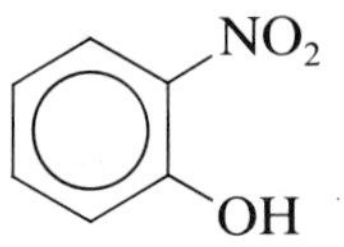

2-Nitrophenol
(*o*-nitrophenol)

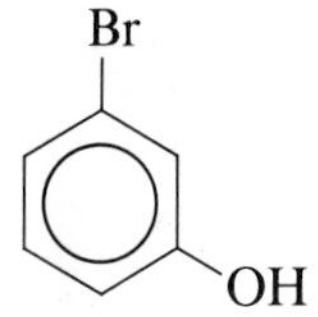

3-Bromophenol
(*m*-bromophenol)

The methylphenols are commonly called *cresols:*

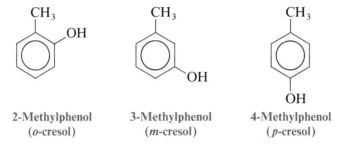

2-Methylphenol
(*o*-cresol)

3-Methylphenol
(*m*-cresol)

4-Methylphenol
(*p*-cresol)

The benzenediols also have common names:

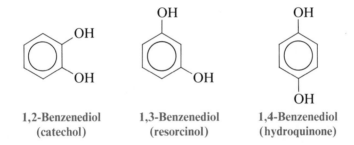

1,2-Benzenediol
(catechol)

1,3-Benzenediol
(resorcinol)

1,4-Benzenediol
(hydroquinone)

14.2 NATURALLY OCCURRING PHENOLS

Phenols and related compounds occur widely in nature. Tyrosine is an amino acid that occurs in proteins. Methyl salicylate is found in oil of wintergreen, eugenol is found in oil of cloves, and thymol is found in thyme.

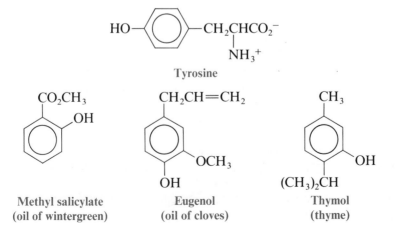

Tyrosine

Methyl salicylate
(oil of wintergreen)

Eugenol
(oil of cloves)

Thymol
(thyme)

The urushiols are blistering agents (vesicants) found in poison ivy.

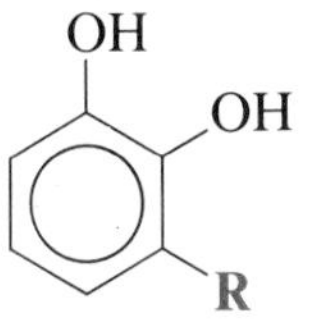

$$R = -(CH_2)_{14}CH_3,$$
$$-(CH_2)_7CH=CH(CH_2)_5CH_3, \quad \text{or}$$
$$-(CH_2)_7CH=CHCH_2CH=CH(CH_2)_2CH_3$$

Urushiols

Estradiol is a female sex hormone and the tetracyclines are important antibiotics.

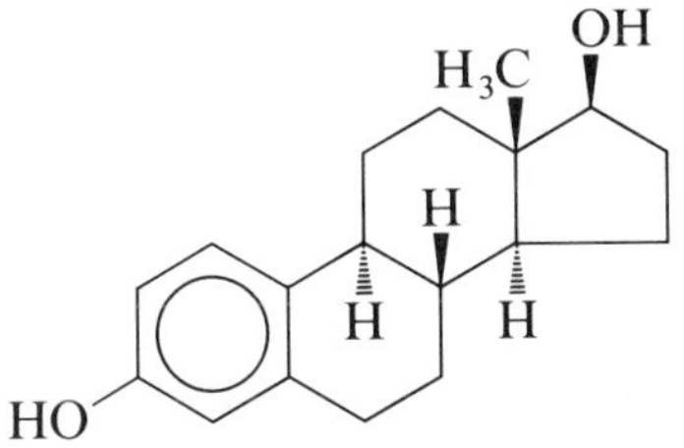

Estradiol

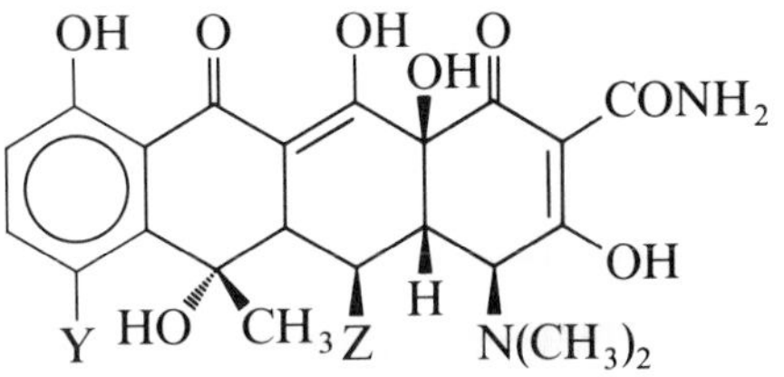

Tetracyclines
(Y = Cl, Z = H; Aureomycin)
(Y = H, Z = OH; Terramycin)

14.3 PHYSICAL PROPERTIES OF PHENOLS

The presence of hydroxyl groups in the molecules of phenols means that phenols are like alcohols (Section 8.2) in being able to form strong intermolecular hydrogen bonds. This hydrogen bonding causes phenols to be associated and, therefore, to have higher boiling points than hydrocarbons of the same molecular weight. For example, phenol (bp, 182 °C) has a boiling point more than 70 °C higher than toluene (bp, 110.6 °C), even though the two compounds have almost the same molecular weight.

The ability to form strong hydrogen bonds to molecules of water confers on phenols a modest solubility in water. Table 14.1 lists the physical properties of a number of common phenols.

14.4 SYNTHESIS OF PHENOLS

Phenol is a highly important industrial chemical; it serves as the raw material for a large number of commercial products ranging from aspirin to a variety of plastics. Worldwide production of phenol is more than 3 million tons per year! Several methods have been used to synthesize phenol commercially.

1. **Hydrolysis of Chlorobenzene (Dow Process).** In this process chlorobenzene is heated at 350 °C (under high pressure) with aqueous sodium hydroxide. The reaction produces sodium phenoxide, which on acidification yields phenol. The mechanism for the reaction probably involves the formation of benzyne (Section 14.11B).

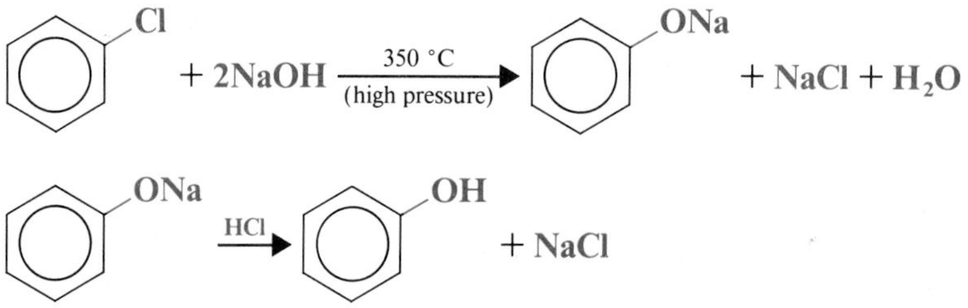

2. **Alkali Fusion of Sodium Benzenesulfonate.** This, the first commercial process for synthesizing phenol, was developed in Germany in 1890. Sodium benzenesulfonate is melted (fused) with sodium hydroxide (at 350 °C) to produce sodium phenoxide. Acidification then yields phenol.

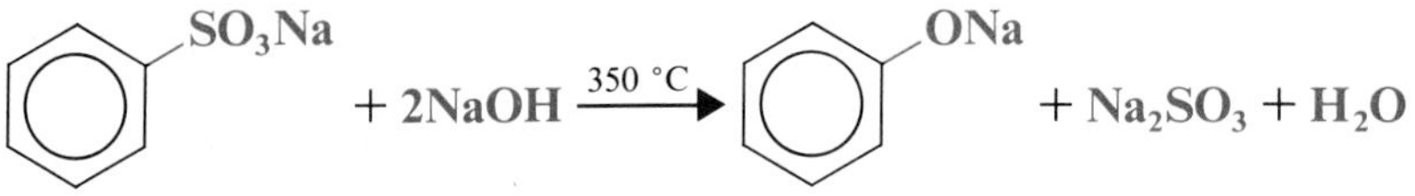

This procedure can also be used in the laboratory and works quite well for the preparation of *p*-cresol, as the following example shows. However, the condi-

TABLE 14.1 Physical properties of phenols

NAME	FORMULA	mp (°C)	bp (°C)	WATER SOLUBILITY (g/100 mL/H_2O)
Phenol	C_6H_5OH	43	182	9.3
o-Cresol	o-$CH_3C_6H_4OH$	30	191	2.5
m-Cresol	m-$CH_3C_6H_4OH$	11	201	2.6
p-Cresol	p-$CH_3C_6H_4OH$	35.5	201	2.3
o-Chlorophenol	o-ClC_6H_4OH	8	176	2.8
m-Chlorophenol	m-ClC_6H_4OH	33	214	2.6
p-Chlorophenol	p-ClC_6H_4OH	43	220	2.7
o-Nitrophenol	o-$O_2NC_6H_4OH$	45	217	0.2
m-Nitrophenol	m-$O_2NC_6H_4OH$	96		1.4
p-Nitrophenol	p-$O_2NC_6H_4OH$	114		1.7
2,4-Dinitrophenol	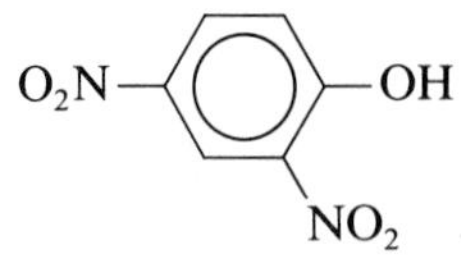	113		0.6
2,4,6-Trinitrophenol (picric acid)	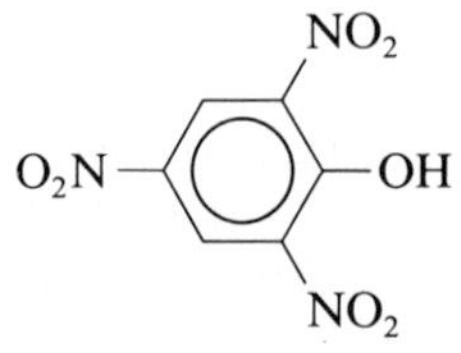	122		1.4

tions required to bring about the reaction are so vigorous that its use in the preparation of many phenols is limited.

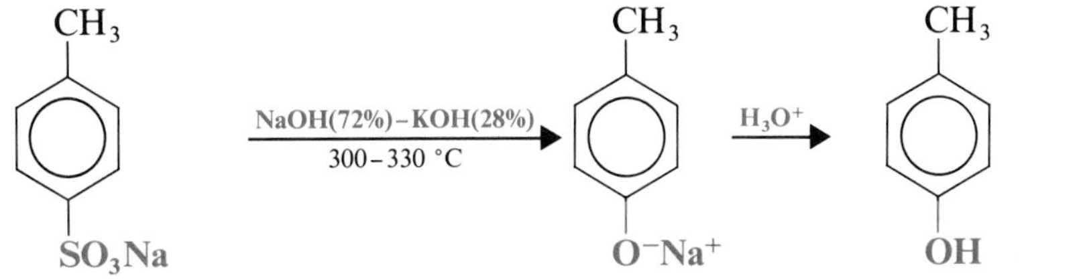

Sodium
p-toluenesulfonate

(63–70% overall)
p-Cresol

3. **From Cumene Hydroperoxide.** This process illustrates industrial chemistry at its best. Overall it is a method for converting two relatively inexpensive organic compounds — benzene and propene — into two more valuable ones — phenol and acetone. The only other substance consumed in the process is oxygen from air. Most of the worldwide production of phenol is now based on this method.

The synthesis begins with the Friedel–Crafts alkylation of benzene with propene to produce cumene (isopropylbenzene).

Then cumene is oxidized to cumene hydroperoxide:

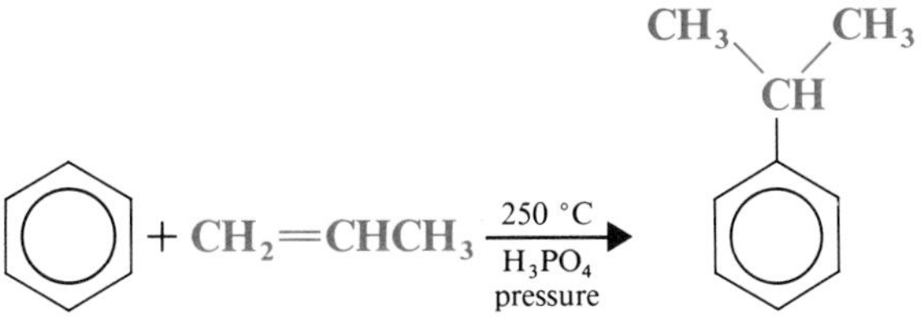

$$\text{Reaction 2} \quad C_6H_5-\underset{\underset{CH_3}{|}}{\overset{\overset{CH_3}{|}}{C}}H + O_2 \xrightarrow{95-135 \text{ °C}} C_6H_5-\underset{\underset{CH_3}{|}}{\overset{\overset{CH_3}{|}}{C}}-O-O-H$$

Cumene hydroperoxide

Finally, when treated with 10% sulfuric acid, cumene hydroperoxide undergoes a hydrolytic rearrangement that yields phenol and acetone:

$$\text{Reaction 3} \quad C_6H_5-\underset{\underset{CH_3}{|}}{\overset{\overset{CH_3}{|}}{C}}-O-OH \xrightarrow[50-90 \text{ °C}]{H^+, H_2O} C_6H_5OH + \underset{\underset{CH_3}{|}}{\overset{\overset{CH_3}{|}}{C}}=O$$

Phenol   Acetone

The mechanism of each of these reactions require some comment. The first is a familiar one. The isopropyl carbocation generated by the reaction of propene with the acid (H_3PO_4) alkylates benzene in a typical electrophilic aromatic substitution:

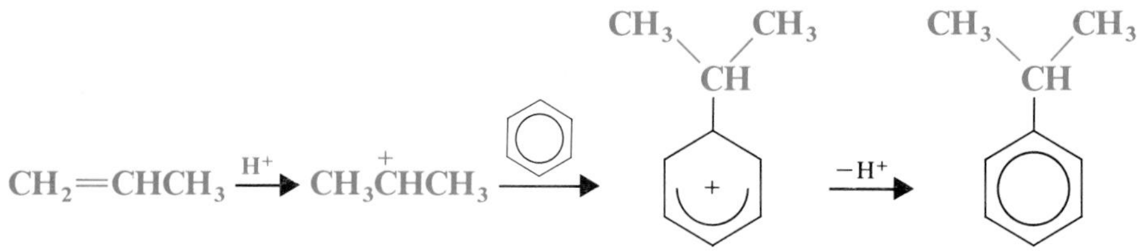

$$CH_2{=}CHCH_3 \xrightarrow{H^+} CH_3\overset{+}{C}HCH_3 \xrightarrow{\text{(benzene)}} \left[\underset{+}{\text{(ring)}}\right] \xrightarrow{-H^+} \text{(isopropylbenzene)}$$

The second reaction is a free radical chain reaction. A radical initiator abstracts the benzylic hydrogen atom of cumene producing a 3° benzylic free radical. Then a chain reaction with oxygen produces cumene hydroperoxide:

Chain Initiation

$$\text{Step 1}\qquad C_6H_5-\underset{CH_3}{\overset{CH_3}{\underset{|}{\overset{|}{C}}}}-H + R{\cdot} \longrightarrow C_6H_5-\underset{CH_3}{\overset{CH_3}{\underset{|}{\overset{|}{C}}}}{\cdot}\;\; + R-H$$

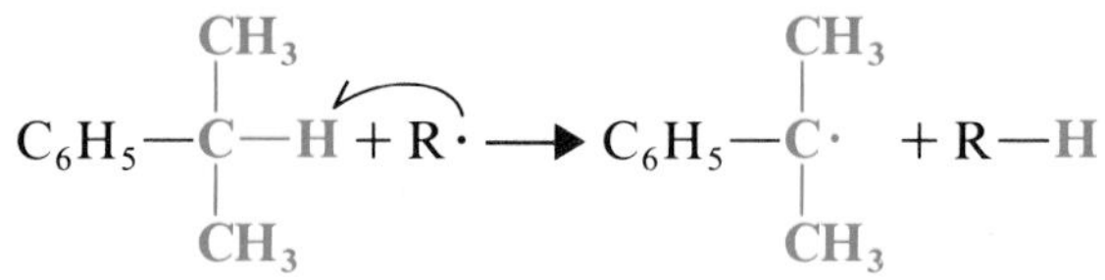

Chain Propagation

$$\text{Step 2}\qquad C_6H_5-\underset{CH_3}{\overset{CH_3}{\underset{|}{\overset{|}{C}}}}{\cdot}\;\; + O_2 \longrightarrow C_6H_5-\underset{CH_3}{\overset{CH_3}{\underset{|}{\overset{|}{C}}}}-O-O{\cdot}$$

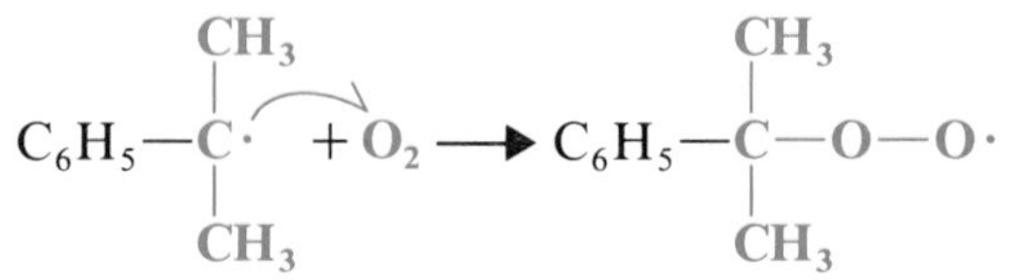

$$\text{Step 3}\qquad C_6H_5-\underset{CH_3}{\overset{CH_3}{\underset{|}{\overset{|}{C}}}}-O-O{\cdot} + H-\underset{CH_3}{\overset{CH_3}{\underset{|}{\overset{|}{C}}}}-C_6H_5 \longrightarrow$$

$$C_6H_5-\underset{CH_3}{\overset{CH_3}{\underset{|}{\overset{|}{C}}}}-O-O-H + C_6H_5-\underset{CH_3}{\overset{CH_3}{\underset{|}{\overset{|}{C}}}}{\cdot}$$

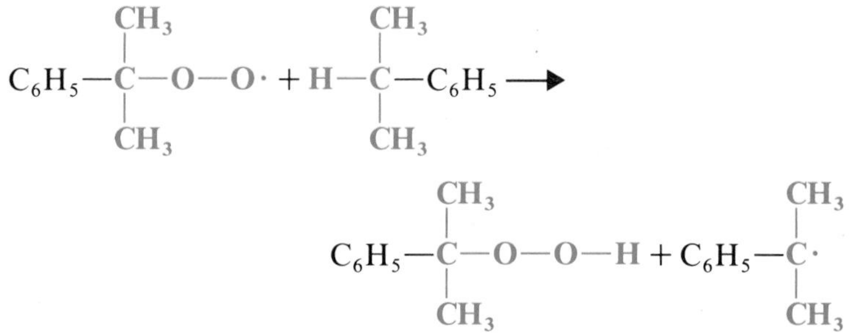

Then, Step 2, Step 3, Step 2, Step 3, etc.

The third reaction—the hydrolytic rearrangement—resembles the carbocation rearrangements that we have studied before. In this instance, however, the rearrangement involves the migration of a phenyl group to *a cationic oxygen atom.* Phenyl groups have a much greater tendency to migrate to a cationic center than do methyl groups (see Section 16.12A). The following equations show all the steps of the mechanism.

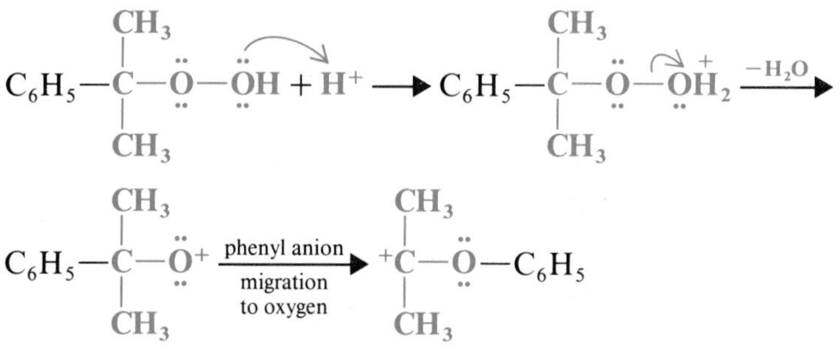

$$C_6H_5-\underset{CH_3}{\overset{CH_3}{\underset{|}{\overset{|}{C}}}}-\ddot{\underset{\cdot\cdot}{O}}-\ddot{\underset{\cdot\cdot}{O}}H + H^+ \longrightarrow C_6H_5-\underset{CH_3}{\overset{CH_3}{\underset{|}{\overset{|}{C}}}}-\ddot{\underset{\cdot\cdot}{O}}-\overset{+}{\underset{\cdot\cdot}{O}}H_2 \xrightarrow{-H_2O}$$

$$C_6H_5-\underset{CH_3}{\overset{CH_3}{\underset{|}{\overset{|}{C}}}}-\overset{+}{\underset{\cdot\cdot}{O}}{\cdot} \xrightarrow[\substack{\text{migration} \\ \text{to oxygen}}]{\text{phenyl anion}} \overset{+}{\underset{CH_3}{\overset{CH_3}{\underset{|}{\overset{|}{C}}}}}-\ddot{\underset{\cdot\cdot}{O}}-C_6H_5$$

$$\xrightarrow{\text{H}_2\text{O}} \text{H}-\overset{+}{\underset{\cdot\cdot}{\text{O}}}-\overset{\overset{\displaystyle\text{H}}{|}}{\underset{\underset{\displaystyle\text{CH}_3}{|}}{\text{C}}}-\overset{\overset{\displaystyle\text{CH}_3}{|}}{\underset{\cdot\cdot}{\text{O}}}-\text{C}_6\text{H}_5 \rightleftharpoons \text{H}-\overset{\cdot\cdot}{\underset{\cdot\cdot}{\text{O}}}-\overset{\overset{\displaystyle\text{CH}_3}{|}}{\underset{\underset{\displaystyle\text{CH}_3}{|}}{\text{C}}}-\overset{\overset{\displaystyle\text{H}}{|}}{\overset{+}{\underset{\cdot\cdot}{\text{O}}}}-\text{C}_6\text{H}_5$$

$$\xrightarrow{-\text{H}^+} \overset{\cdot\cdot}{\underset{\cdot\cdot}{\text{O}}}=\overset{\overset{\displaystyle\text{CH}_3}{|}}{\underset{\underset{\displaystyle\text{CH}_3}{|}}{\text{C}}} + \text{H}\overset{\cdot\cdot}{\underset{\cdot\cdot}{\text{O}}}\text{C}_6\text{H}_5$$

Acetone   Phenol

The second and third steps of the mechanism may actually take place at the same time, that is; the loss of H_2O and the migration of C_6H_5- may be concerted.

4. **Other Methods.** In Chapter 18 we shall discuss a useful laboratory procedure for synthesizing phenols from arylamines.

14.5 REACTIONS OF PHENOLS AS ACIDS

14.5A Strength of Phenols as Acids

Although phenols are structurally similar to alcohols, they are much stronger acids. The acidity constants of most alcohols are of the order of 10^{-18}. However, as we see in Table 14.2, the acidity constants of phenols are of the order of 10^{-11} or greater.

We can explain the greater acidity of phenols relative to alcohols through the application of resonance theory. Before we do this, however, it will be helpful if we review acid–base theory in general terms.

TABLE 14.2 The acidity constants of phenols

NAME	ACIDITY CONSTANT K_a (in H_2O at 25 °C)
Phenol	1.3×10^{-10}
o-Cresol	6.3×10^{-11}
m-Cresol	9.8×10^{-11}
p-Cresol	6.7×10^{-11}
o-Chlorophenol	7.7×10^{-9}
m-Chlorophenol	1.6×10^{-9}
p-Chlorophenol	6.3×10^{-10}
o-Nitrophenol	6.8×10^{-8}
m-Nitrophenol	5.3×10^{-9}
p-Nitrophenol	7×10^{-8}
2,4-Dinitrophenol	1.1×10^{-4}
2,4,6-Trinitrophenol (picric acid)	4.2×10^{-1}
1-Naphthol	4.9×10^{-10}
2-Naphthol	2.8×10^{-10}

Acid–base reactions are under **equilibrium control.** That is, they are reversible reactions, and the relative concentrations of conjugate acid and base formed are governed by the position of an equilibrium. We have seen other reactions under equilibrium control: for example, 1,4 additions to conjugated dienes (Section 10.11) and aromatic sulfonations (Section 12.5). We have also seen that we can account for reactions that are under equilibrium control by comparing the relative energies of the products and the reactants, rather than the relative magnitudes of energies of activation. The reason for this comparison is simple; in reactions under equilibrium control both the products and the reactants have sufficient energy to surmount the energy barrier between them.

Thus, with acid–base reactions, we can assess the effects of variations in structure on K_a by estimating the energy difference between the acid and its conjugate base. Structural factors that stabilize the conjugate base, A$:^-$, more than they stabilize the acid, HA, cause the value of K_a to be larger. (HA will be, therefore, a stronger acid.) Conversely, structural factors that stabilize the acid more than they do the conjugate base cause the value of K_a to be smaller. (In this case HA will be a weaker acid.)

Resonance effects are often an important factor in acid–base reactions, and resonance effects usually stabilize the conjugate base more than they stabilize the acid. We can see a vivid example of this if we return now to one of our original tasks and account for the greater acidity of a phenol relative to that of an alcohol.

Let us compare two *superficially* similar compounds, cyclohexanol and phenol.

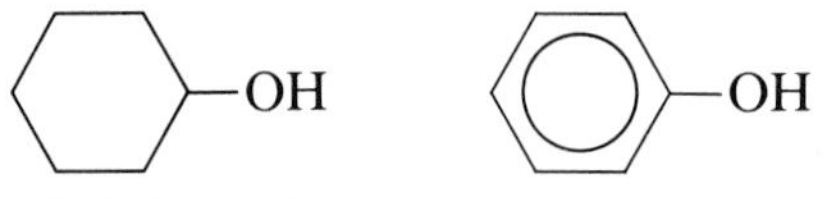

Cyclohexanol
$K_a \simeq 10^{-18}$

Phenol
$K_a = 1.3 \times 10^{-10}$

Although phenol is a weak acid when compared with a carboxylic acid such as acetic acid ($K_a = 10^{-5}$), phenol is a much stronger acid than cyclohexanol (by a factor of almost 10^8).

We can write resonance structures for both the phenol and phenoxide ion such as those shown here. (No analogous resonance structures are possible for cyclohexanol, of course.)

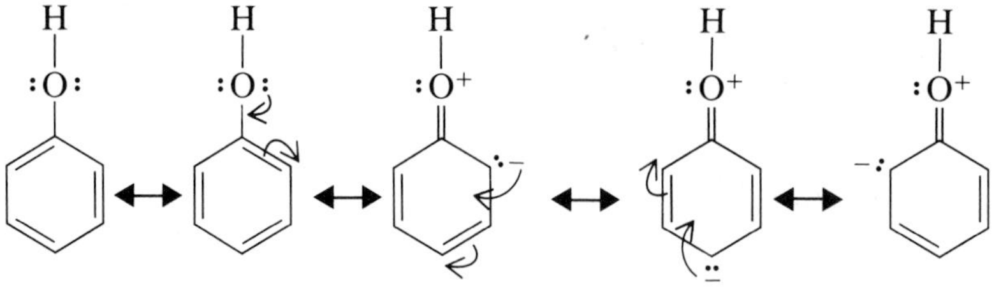

Resonance structures for phenol

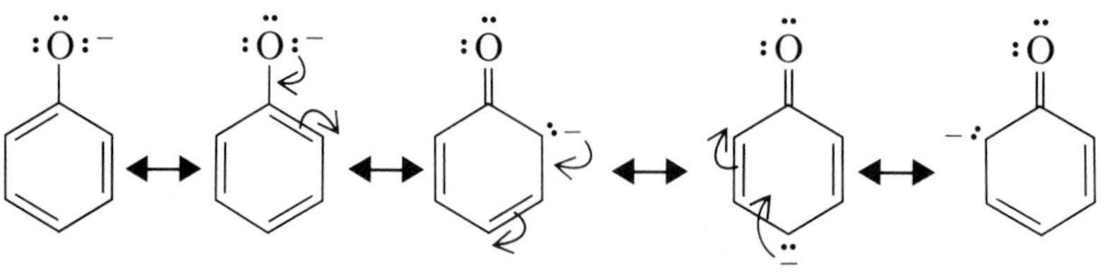

1a 1b 2 3 4

These structures have separated charges

Resonance structures for the phenoxide ion

5a 5b 6 7 8

With the exception of the Kekulé structures (**1a** and **1b**), we see that the resonance structures for phenol (**2–4**) require separation of opposite charges while the corresponding resonance structures (**6–8**) for the phenoxide ion do not. *Energy is required to separate opposite charges,* and therefore we can conclude that this type of resonance makes a *smaller* stabilizing contribution to phenol than it does to the phenoxide ion. That is, *resonance stabilizes the conjugate base* (the phenoxide ion) *more than it does the acid* (phenol).

Resonance stabilization of the phenoxide ion is particularly important because the negative charge is *delocalized* over the benzene ring. It is not localized on the oxygen atom as it would be in the anion of cyclohexanol:

Phenol
(moderate resonance
stabilization)

Phenoxide ion
(large resonance
stabilization —
charge is delocalized)

Anion is
stabilized
more than
the acid —
K_a is larger

Cyclohexanol
(no resonance stabilization)

Cyclohexyloxide ion
(no resonance stabilization —
charge is localized)

Neither acid
nor anion
is stabilized —
K_a is smaller

Since resonance stabilizes the phenoxide ion more than it does phenol itself, ionization of phenol is a less endothermic reaction than the corresponding ionization of cyclohexanol (Fig. 14.1). The equilibrium between phenol and its conjugate base will, as a consequence, favor the formation of the conjugate base and the hydronium ion to a greater extent than the corresponding equilibrium involving cyclohexanol (where neither the alcohol nor the alkoxide ion is resonance stabilized). Both compounds are weak acids, but, of the two, phenol is the stronger.

In the analysis given here, we have made our comparisons solely on the basis of enthalpy changes ($\Delta H°$'s) for the two reactions. Such an analysis contains a simplification. In actuality, the equilibrium constant for a reaction is directly related to the standard free-energy change for the reaction, $\Delta G°$ (cf. Special Topic B). The equation that relates these two quantities is

$$\log K_{eq} = -\frac{\Delta G°}{2.303\ RT}$$

where R is the gas constant and T is the absolute temperature.

Since $\Delta G°$ is related both to the enthalpy change $\Delta H°$ and to the entropy change $\Delta S°$, that is,

$$\Delta G° = \Delta H° - T\Delta S°$$

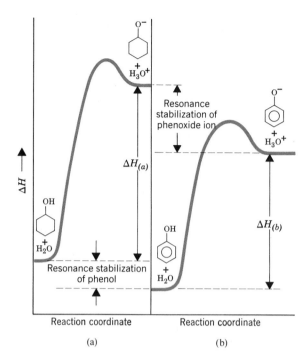

FIGURE 14.1 Potential energy diagrams for the ionization of cyclohexanol (a) and phenol (b). Resonance stabilizes the phenoxide ion to a greater extent than it does phenol itself; thus, the ionization of phenol is a less endothermic reaction than the ionization of cyclohexanol, that is, $\Delta H_{(a)} > \Delta H_{(b)}$. Phenol, therefore, is the stronger acid. (The curves are aligned for comparison only.)

entropy effects can be important. In the two reactions considered here, the entropy changes have been of the same order of magnitude and have, therefore, canceled out. In other reactions, however, this may not be the case and, in fact, entropy changes may be the predominant influence.

PROBLEM 14.1

The carbon–oxygen bond of phenol is much stronger than that of an alcohol. Phenol, for example, is not converted to bromobenzene when it is refluxed with concentrated hydrobromic acid. Similar treatment of cyclohexanol, however, does give bromocyclohexane.

$$\text{C}_6\text{H}_5\text{—OH} + \text{HBr} \xrightarrow{\text{reflux}} \text{no reaction}$$

$$\text{C}_6\text{H}_{11}\text{—OH} + \text{HBr} \xrightarrow{\text{reflux}} \text{C}_6\text{H}_{11}\text{—Br} + \text{H}_2\text{O}$$

Although resonance structures **2** and **4** do not make as great a contribution to the phenol hybrid as the corresponding structures (**6–8**) do to the phenoxide ion, structures **2** to **4** do help us understand why the carbon–oxygen bond of phenols is very strong. Explain.

PROBLEM 14.2

If we examine Table 14.2 we see that phenols having electron-withdrawing groups (Cl— or O_2N—) attached to the benzene ring are more acidic than phenol itself. On the other hand, those phenols bearing electron-releasing groups (e.g., CH_3—) are less acidic than phenol. Account for this trend on the basis of resonance and inductive effects. [Notice that 2,4,6-trinitrophenol (called *picric acid*) is exceptionally acidic ($K_a = 4.2 \times 10^{-1}$)—over 20,000 times as acidic as acetic acid ($K_a = 1.8 \times 10^{-5}$).]

14.5B Distinguishing and Separating Phenols from Alcohols and Carboxylic Acids

Because phenols are more acidic than water, the following reaction goes essentially to completion and produces water-soluble sodium phenoxide.

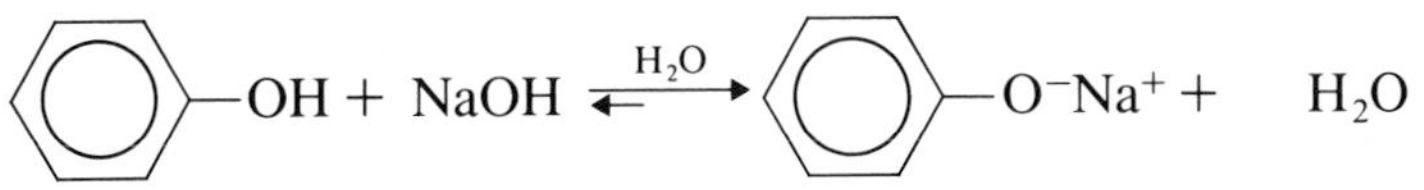

(slightly soluble) (soluble)
Stronger acid Stronger Weaker Weaker acid
$K_a \simeq 10^{-10}$ base base $K_a \simeq 10^{-16}$

The corresponding reaction of cyclohexanol with aqueous sodium hydroxide does not occur to any appreciable extent because cyclohexanol is a weaker acid than water.

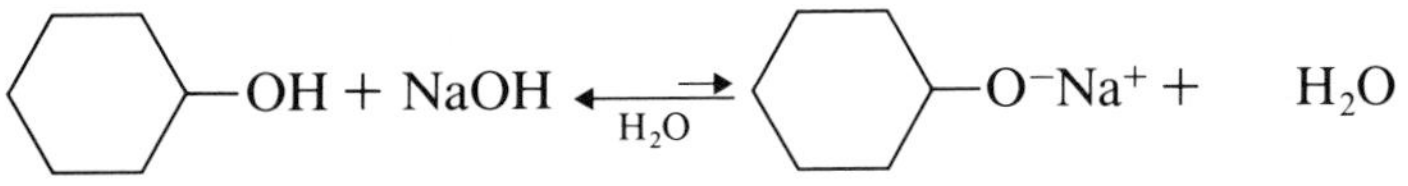

(very slightly soluble) (soluble)
Weaker acid Weaker Stronger Stronger acid
$K_a \simeq 10^{-18}$ base base $K_a \simeq 10^{-16}$

The fact that phenols dissolve in aqueous sodium hydroxide, whereas most alcohols with six carbon atoms or more do not, gives us a convenient means for distinguishing and separating phenols from most alcohols. (Alcohols with five carbon atoms or fewer are quite soluble in water—some are infinitely so—and thus they dissolve in aqueous sodium hydroxide even though they are not converted to sodium alkoxides in appreciable amounts.)

Most phenols, however, are not soluble in aqueous sodium bicarbonate ($NaHCO_3$), but carboxylic acids are soluble. Thus, aqueous $NaHCO_3$ provides a method for distinguishing and separating most phenols from carboxylic acids.

PROBLEM 14.3

The apparent acidity constant for the first ionization of carbonic acid ($H_2CO_3 + H_2O \rightleftharpoons HCO_3^- + H_3O^+$) is 4.3×10^{-7}. Which of the following compounds would you expect to dissolve in aqueous sodium bicarbonate (aq. $NaHCO_3$)? Explain your answers.

(a) Phenol   (c) *o*-Chlorophenol   (e) 2,4,6-Trinitrophenol

(b) *p*-Cresol   (d) 2,4-Dinitrophenol   (f) Benzoic acid ($K_a = 6.4 \times 10^{-5}$)

14.6 OTHER REACTIONS OF THE O—H GROUP OF PHENOLS

Phenols react with carboxylic acid anhydrides and acid chlorides to form esters. These reactions are quite similar to those of alcohols (Section 8.11).

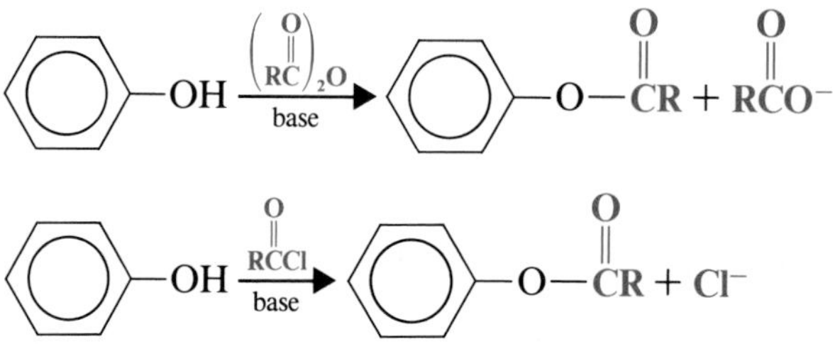

14.6A Phenols in the Williamson Synthesis

Phenols can be converted to ethers through the Williamson synthesis (Section 8.17B). Because phenols are more acidic than alcohols, they can be converted to sodium phenoxides through the use of sodium hydroxide (rather than metallic sodium, the reagent used to convert alcohols to alkoxide ions).

General Reaction

$$ArOH \xrightarrow{NaOH} ArO^-Na^+ \xrightarrow[(X = Cl, Br, I, OSO_2OR \text{ or } OSO_2R')]{R-X} ArOR + NaX$$

Specific Examples

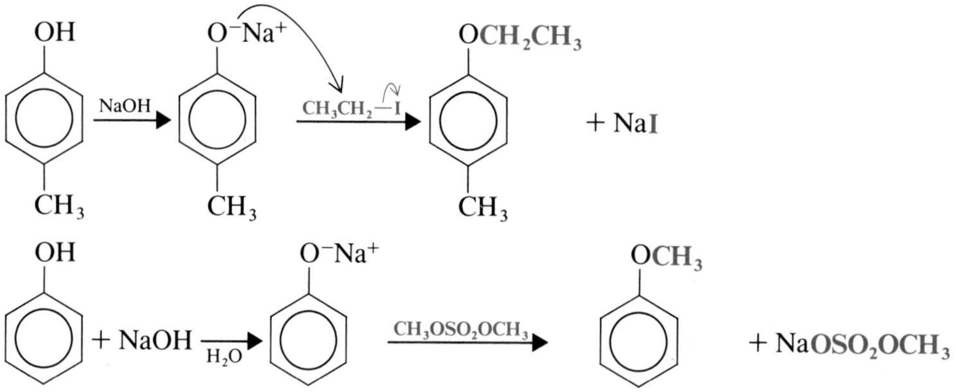

Anisole
(methoxybenzene)

14.7 CLEAVAGE OF ALKYL ARYL ETHERS

We learned in Section 8.17A that when dialkyl ethers are heated with excess concentrated HBr or HI, the ethers are cleaved and alkyl halides are produced from both alkyl groups.

$$R-O-R' \xrightarrow[\text{heat}]{\text{conc. HX}} R-X + R'-X + H_2O$$

When alkyl aryl ethers react with strong acids such as HI and HBr, the reaction produces an alkyl halide and a phenol. The phenol does not react further to produce an aryl halide because the carbon–oxygen bond is very strong (cf. Problem 14.1) and because phenyl cations do not form readily.

General Reaction

$$Ar\!-\!O\!-\!R \xrightarrow[\text{heat}]{\text{conc. HX}} Ar\!-\!OH + R\!-\!X$$

Specific Example

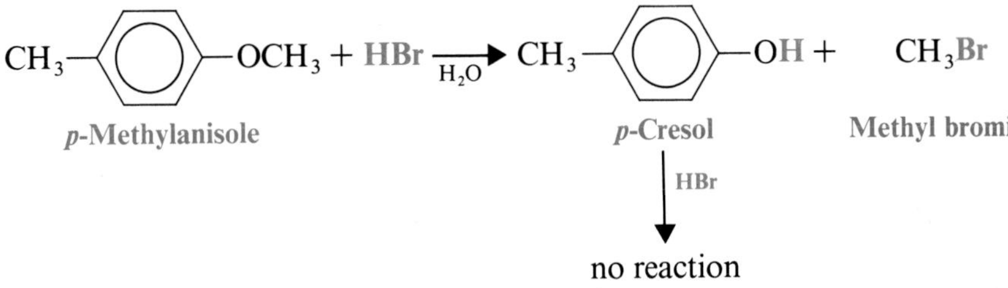

CH$_3$—⟨◯⟩—OCH$_3$ + HBr $\xrightarrow{H_2O}$ CH$_3$—⟨◯⟩—OH + CH$_3$Br

p-Methylanisole   *p*-Cresol   Methyl bromide

$\downarrow$ HBr

no reaction

14.8 REACTIONS OF THE BENZENE RING OF PHENOLS

Bromination. The hydroxyl group is a powerful activating group—and an ortho–para director—in electrophilic substitutions. Phenol itself reacts with bromine in aqueous solution to yield 2,4,6-tribromophenol in nearly quantitative yield. Note that a Lewis acid is not required for the bromination of this highly activated ring.

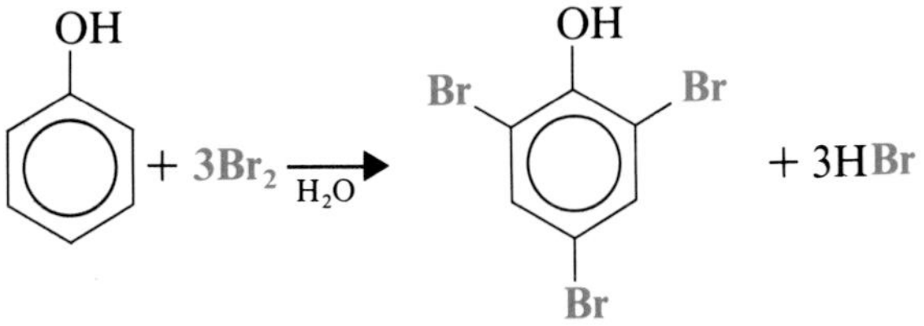

$(\sim 100\%)$
2,4,6-Tribromophenol

Monobromination of phenol can be achieved by carrying out the reaction in carbon disulfide at a low temperature, conditions that reduce the electrophilic reactivity of bromine. The major product is the para isomer.

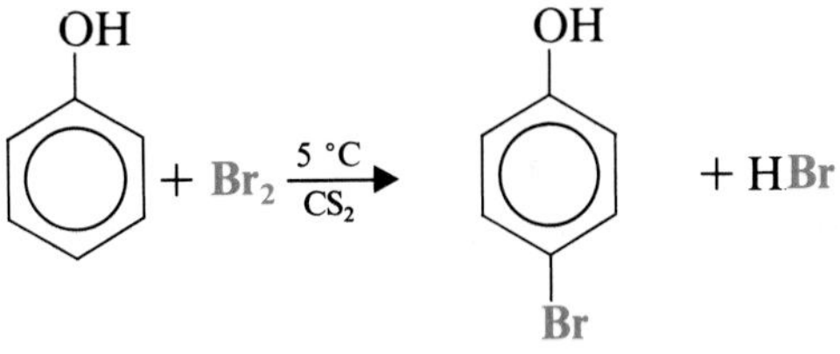

$(80–84\%)$
p-Bromophenol

Nitration. Phenol reacts with dilute nitric acid to yield a mixture of *o*- and *p*-nitrophenol. Although the yield is relatively low (because of oxidation of the ring), the ortho and para isomers can be separated by steam distillation. *o*-Nitrophenol is the more volatile isomer because its hydrogen bonding (see the following structures) is *intramolecular*. *p*-Nitrophenol is less volatile because intermolecular hydrogen bonding causes association among its molecules. Thus, *o*-nitrophenol passes over with the steam, and *p*-nitrophenol remains in the distillation flask.

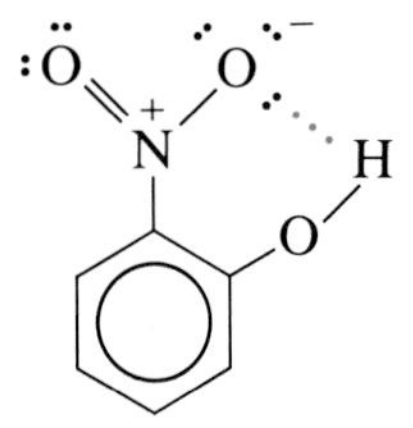

o-**Nitrophenol**
(more volatile because of
intramolecular hydrogen bonding)

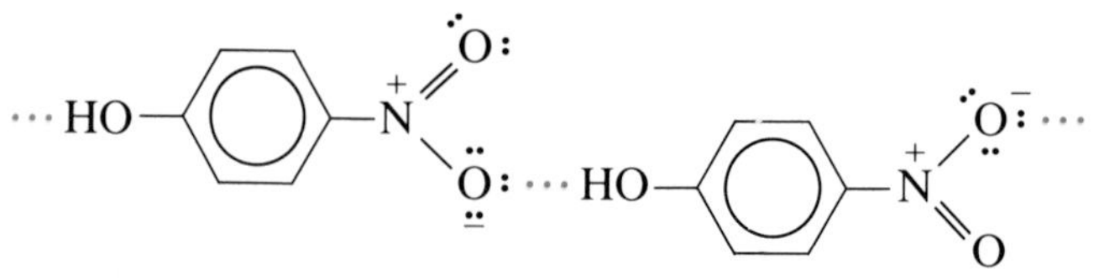

p-**Nitrophenol**
(less volatile because of
intermolecular hydrogen bonding)

Sulfonation. Phenol reacts with concentrated sulfuric acid to yield mainly the ortho-sulfonated product if the reaction is carried out at 25 °C and mainly the para-sulfonated product at 100 °C. This is another example of equilibrium versus rate control of a reaction.

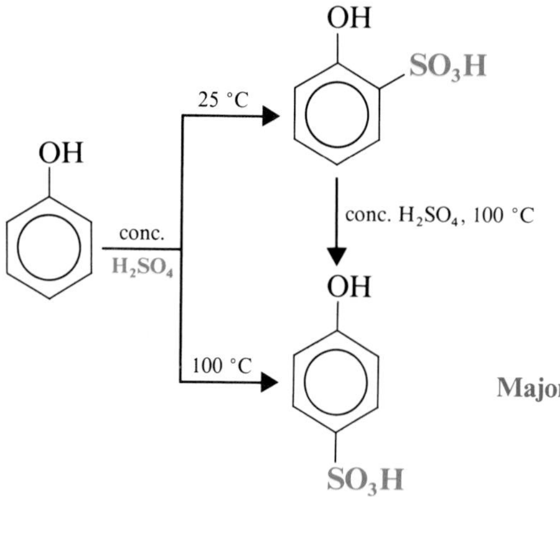

Major product, rate control

Major product, equilibrium control

PROBLEM 14.4

(a) Which sulfonic acid (see previous reactions) is more stable? (b) For which sulfonation (ortho or para) is the energy of activation lower?

Kolbe Reaction. The phenoxide ion is even more susceptible to electrophilic aromatic substitution than phenol itself. (Why?) Use is made of the high reactivity of the phenoxide ring in a reaction called the *Kolbe reaction.* In the Kolbe reaction carbon dioxide acts as the electrophile.

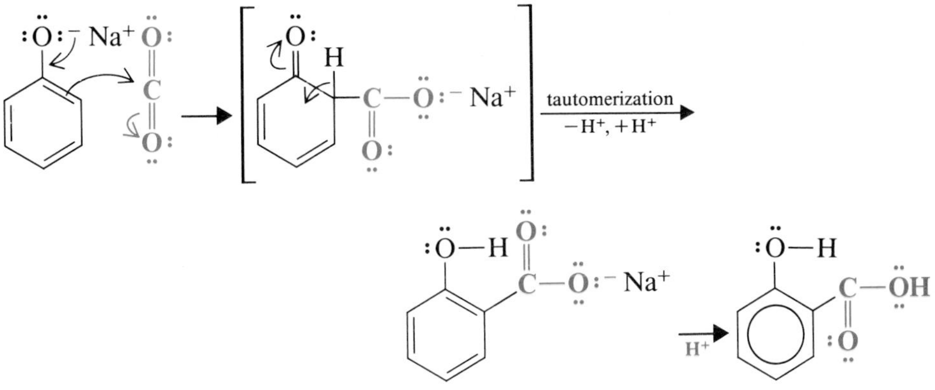

Sodium salicylate   Salicylic acid

The reaction is usually carried out by allowing sodium phenoxide to absorb carbon dioxide and then heating the product to 125 °C under a pressure of several atmospheres of carbon dioxide. The unstable intermediate undergoes a proton shift (a keto-enol tautomerization; see Sections 7.15 and 17.2) that leads to sodium salicylate. Subsequent acidification of the mixture produces *salicylic acid.*

Reaction of salicylic acid with acetic anhydride yields the widely used pain reliever — *aspirin.*

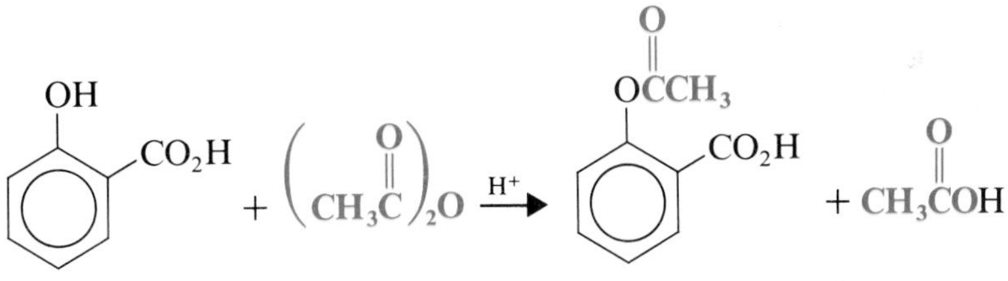

Salicylic   Acetic   Acetylsalicylic acid
acid   anhydride   (aspirin)

14.9 THE CLAISEN REARRANGEMENT

Heating allyl phenyl ether to 200 °C effects an intramolecular reaction called a **Claisen rearrangement.** The product of the rearrangement is *o*-allylphenol:

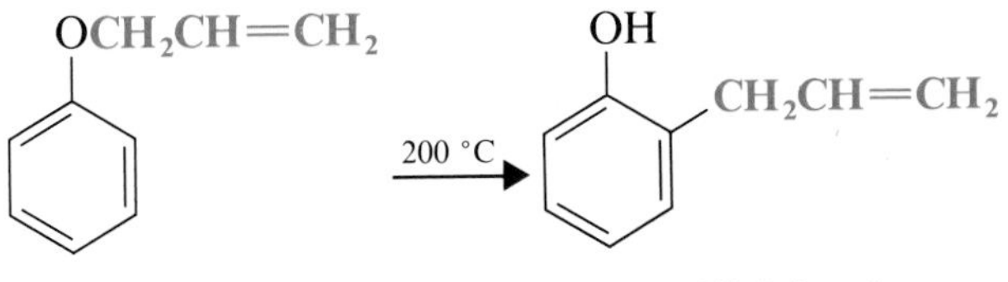

Allyl phenyl ether   *o*-Allylphenol

The reaction takes place through a **concerted rearrangement** in which the bond between carbon-3 of the allyl group and the ortho position of the benzene ring forms at the same time that the carbon–oxygen bond of the allyl phenyl ether breaks. The product of this rearrangement is an unstable intermediate that, like the unstable intermediate in the Kolbe Reaction (Section 14.8) undergoes a proton shift (a keto–enol tautomerization, see Sections 7.15 and 17.2) that leads to the *o*-allylphenol.

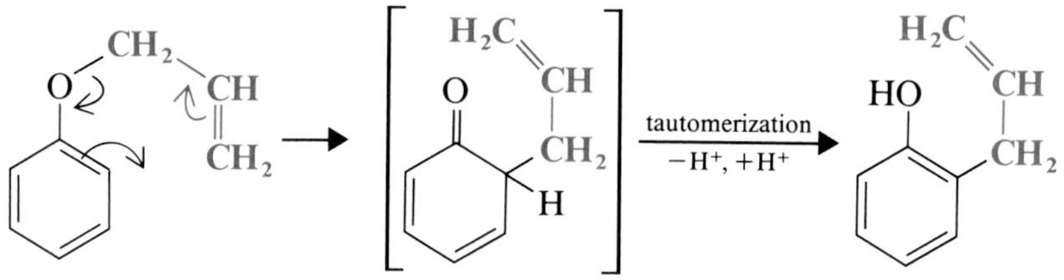

Unstable intermediate

That only carbon-3 of the allyl group becomes bonded to the benzene ring was demonstrated by carrying out the rearrangement with allyl phenyl ether containing ^{14}C at carbon-3. All of the product of this reaction had the labeled carbon atom bonded to the ring.

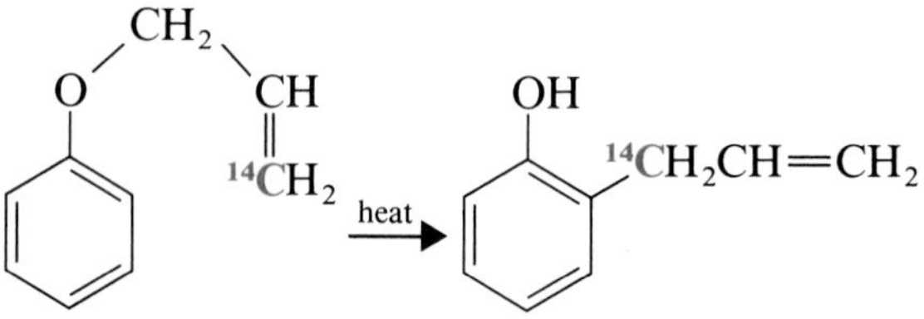

Only product

PROBLEM 14.5

The labeling experiment just described eliminates from consideration a mechanism in which the allyl phenyl ether dissociates to produce an allyl cation (Section 10.4) and a phenoxide ion, which then subsequently undergo a Friedel–Crafts alkylation (Section 12.6) to produce the *o*-allylphenol. Explain how this alternative mechanism can be discounted by showing the product (or products) that would result from it.

PROBLEM 14.6

Show how you would synthesize allyl phenyl ether through a Williamson synthesis (Section 14.6A) starting with phenol and allyl bromide.

A Claisen rearrangement also takes place when allyl vinyl ethers are heated. For example:

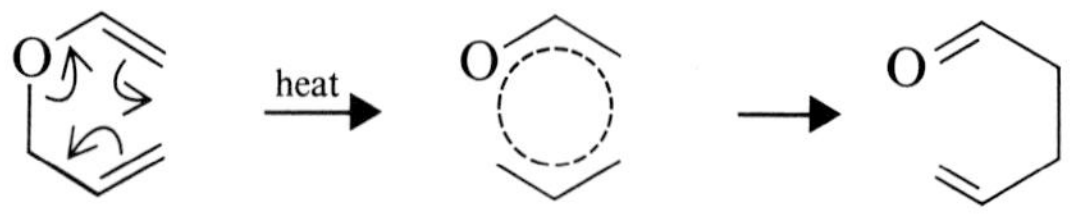

| Allyl vinyl ether | Aromatic transition state | 4-Pentenal |

The transition state for the Claisen rearrangement involves a cycle of six orbitals and six electrons. Having six electrons suggests that the transition state has aromatic character (Section 11.6). Other reactions of this general type are known and they are called **pericyclic reactions.**

Another similar pericyclic reaction is the **Cope rearrangement** shown here.

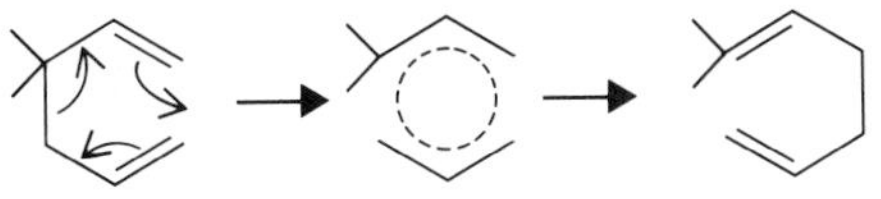

3,3-Dimethyl-
1,5-hexadiene

Aromatic
transition
state

2-Methyl-2,6-
heptadiene

The Diels–Alder reaction (Section 10.12) is also a pericyclic reaction. The transition state for the Diels–Alder reaction also involves six orbitals and six electrons.

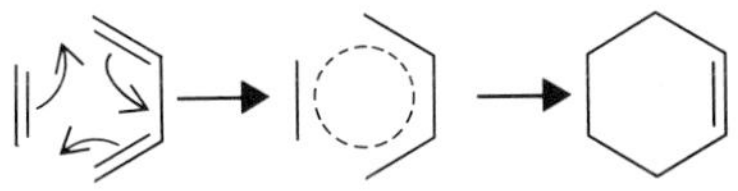

Aromatic
transition
state

We shall discuss the mechanism of the Diels–Alder reaction further in Special Topic N.

14.10 QUINONES

Oxidation of hydroquinone (1,4-benzenediol) produces a compound known as *p*-benzoquinone. The oxidation can be brought about by mild oxidizing agents, and overall the oxidation amounts to the removal of a pair of electrons ($2e^-$) and two protons from hydroquinone. (Another way of visualizing the oxidation is as the loss of a hydrogen molecule, $H:H$, making it a dehydrogenation.)

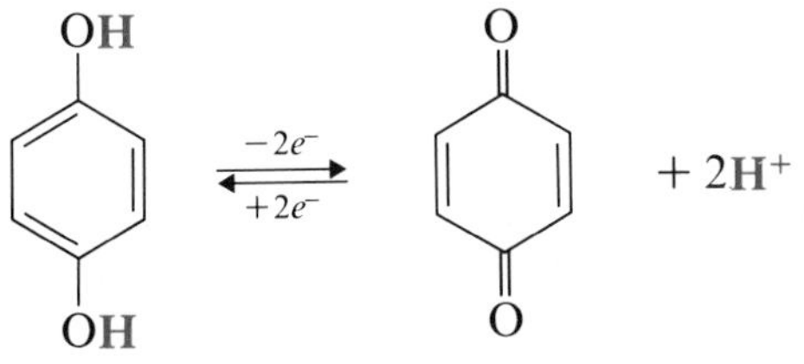

Hydroquinone $\qquad$ *p*-Benzoquinone

This reaction is reversible; *p*-benzoquinone is easily reduced by mild reducing agents to hydroquinone.

Nature makes much use of this type of reversible oxidation–reduction to transport a pair of electrons from one substance to another in enzyme-catalyzed reactions. Important compounds in this respect are the compounds called **ubiquinones** (from *ubiquitous* + quinone—these quinones are found everywhere in biological systems). Ubiquinones are also called coenzymes Q.

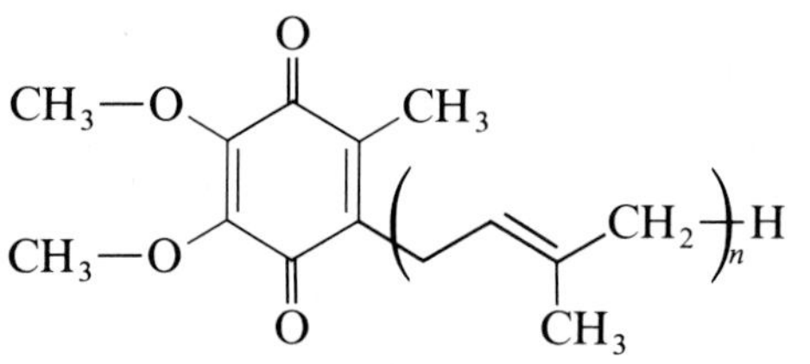

Ubiquinones ($n = 6-10$)
(coenzymes Q)

Vitamin K_1, the important dietary factor that is instrumental in maintaining the coagulant properties of blood, contains a 1,4-naphthoquinone structure.

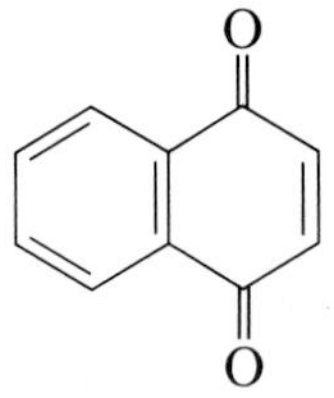

1,4-Naphthoquinone

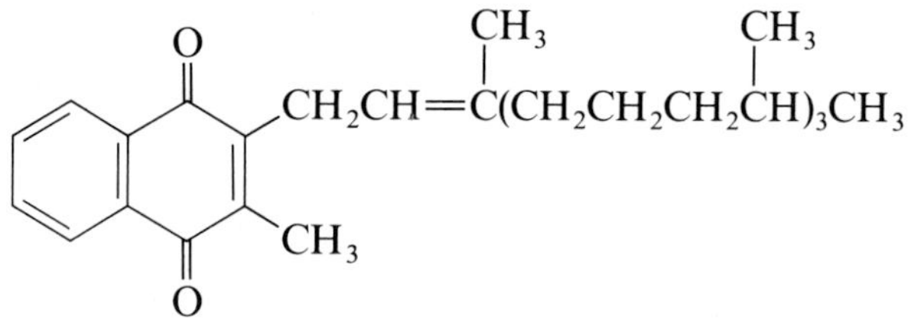

Vitamin K_1

PROBLEM 14.7

p-Benzoquinone and 1,4-naphthoquinone act as dienophiles in Diels–Alder reactions. Give the structures of the products of the following reactions: (a) *p*-Benzoquinone + 1,3-butadiene, (b) 1,4-Naphthoquinone + butadiene, and (c) *p*-Benzoquinone + 1,3-cyclopentadiene.

PROBLEM 14.8

Outline a possible synthesis of the following compound.

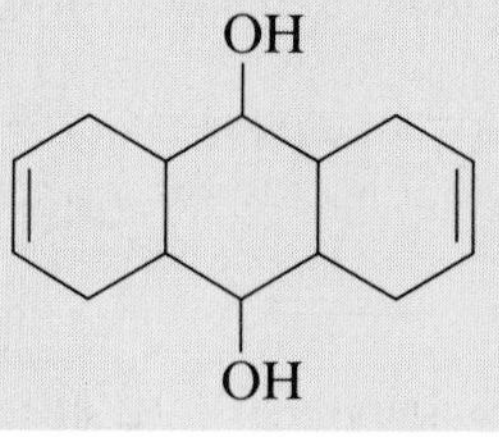

14.11 ARYL HALIDES AND NUCLEOPHILIC AROMATIC SUBSTITUTION

Simple aryl halides are like vinylic halides (Section 5.14A) in that they are relatively unreactive toward nucleophilic substitution under conditions that give facile nucleophilic substitution with alkyl halides. Chlorobenzene, for example, can be boiled with sodium hydroxide for days without producing a detectable amount of phenol (or sodium phenoxide). Similarly, when vinyl chloride is heated with sodium hydroxide, no substitution occurs:

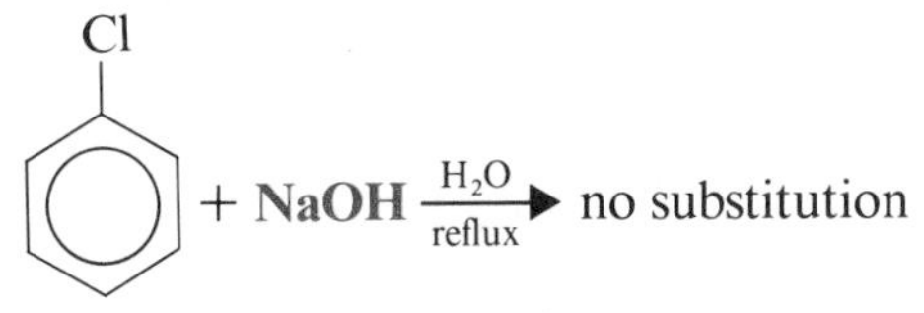

$$CH_2{=}CHCl + NaOH \xrightarrow[\text{reflux}]{\text{H}_2\text{O}} \text{no substitution}$$

Aryl halides and vinylic halides do not give a positive test (a silver halide precipitate) when treated with alcoholic silver nitrate (Section 7.19E).

We can understand this lack of reactivity on the basis of several factors. The benzene ring of an aryl halide prevents backside attack in an S_N2 reaction:

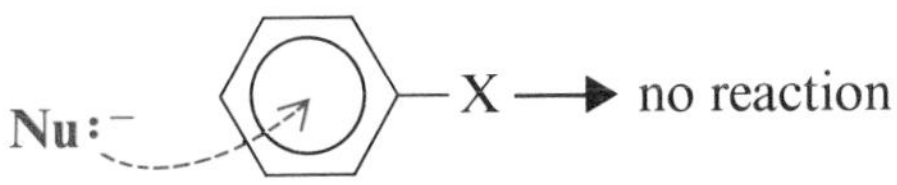

Phenyl cations are very unstable; thus S_N1 reactions do not occur. The carbon–halogen bonds of aryl (and vinylic) halides are shorter and stronger than those of alkyl, allylic, and benzylic halides. Stronger carbon–halogen bonds mean that bond breaking by either an S_N1 or S_N2 mechanism will require more energy.

Two effects make the carbon–halogen bonds of aryl and vinylic halides shorter and stronger. (1) The carbon of either type of halide is sp^2 hybridized and thus the electrons of the carbon orbital are closer to the nucleus than those of an sp^3-hybridized carbon. (2) Resonance of the type shown here strengthens the carbon–halogen bond by giving it *double-bond character*.

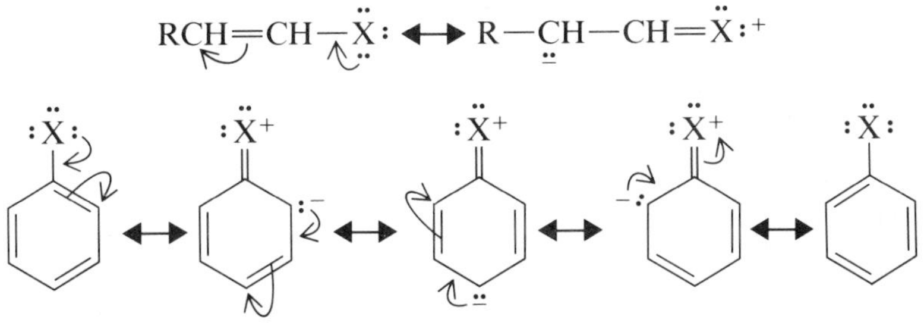

Having said all this, we shall find in the next two subsections that *aryl halides can be remarkably reactive toward nucleophiles* if they bear certain substituents or when we allow them to react under the proper conditions.

14.11A Nucleophilic Aromatic Substitution by Addition–Elimination: The S_NAr Mechanism

Nucleophilic substitution reactions of aryl halides *do* occur readily when an electronic factor makes the aryl carbon susceptible to nucleophilic attack. *Nucleophilic substitution can occur when strong electron-withdrawing groups are ortho or para to the halogen atom:*

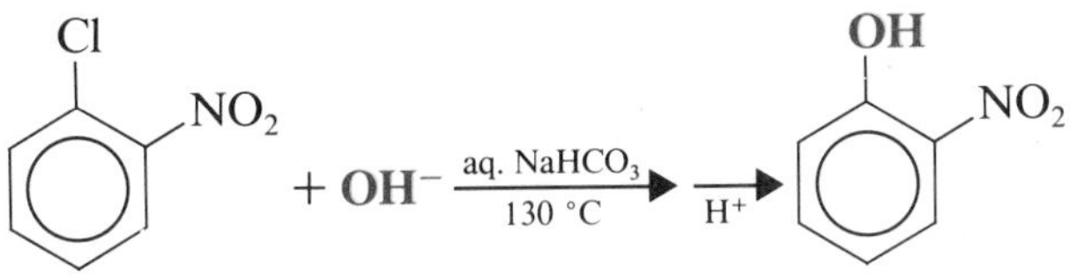

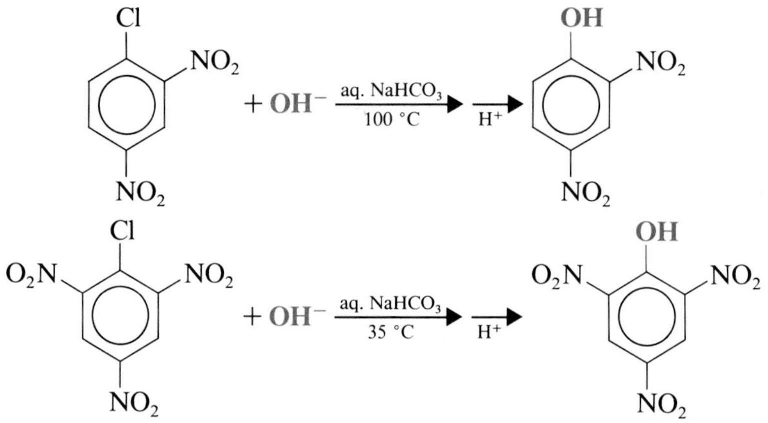

We also see in these examples that the temperature required to bring about the reaction is related to the number of ortho or para nitro groups. Of the three compounds, *o*-nitrochlorobenzene requires the highest temperature (*p*-nitrochlorobenzene reacts at 130 °C as well) and 2,4,6-trinitrochlorobenzene requires the lowest temperature.

A meta-nitro group does not produce a similar activating effect. For example, *m*-nitrochlorobenzene gives no corresponding reaction.

The mechanism that operates in these reactions is an *addition–elimination* mechanism involving the formation of a delocalized *carbanion* called a **Meisenheimer complex** after the German chemist, Jacob Meisenheimer, who proposed its correct structure. In the following first step addition of a hydroxide ion to *p*-nitrochlorobenzene, for example, produces the delocalized carbanion; then elimination of a chloride ion yields the substitution product as the aromaticity of the ring is recovered. This mechanism is called the $S_N Ar$ mechanism.

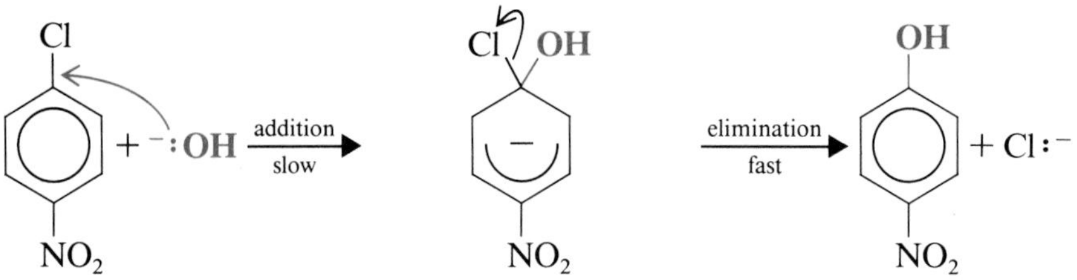

Delocalized
carbanion
(Meisenheimer complex)

The delocalized carbanion is stabilized by *electron-withdrawing groups* in the positions ortho and para to the halogen atom. If we examine the following resonance structures, we can see how.

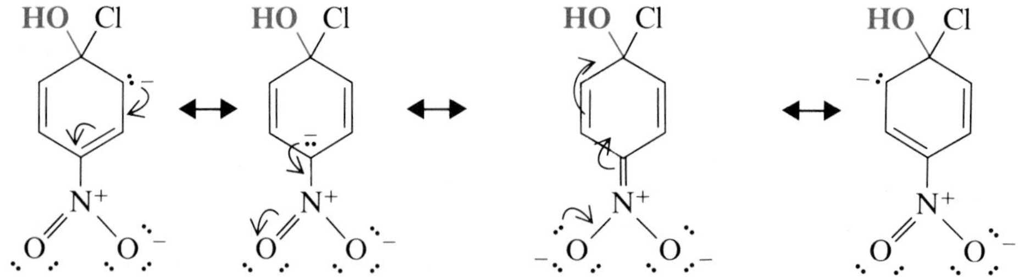

Especially stable
(negative charges
are both on oxygen atoms)

PROBLEM 14.9

What products would you expect from each of the following nucleophilic substitution reactions?

(a) p-Nitrochlorobenzene + CH_3ONa $\xrightarrow[100\ °C]{CH_3OH}$

(b) o-Nitrochlorobenzene + CH_3NH_2 $\xrightarrow[160\ °C]{C_2H_5OH}$

(c) 2,4-Dinitrochlorobenzene + $C_6H_5NH_2$ $\xrightarrow[95\ °C]{C_2H_5OH}$

14.11B Nucleophilic Aromatic Substitution Through an Elimination–Addition Mechanism: Benzyne

Although aryl halides such as chlorobenzene and bromobenzene do not react with most nucleophiles under ordinary circumstances, they do react under highly forcing conditions. Chlorobenzene can be converted to phenol by heating it with aqueous sodium hydroxide in a pressurized reactor at 350 °C (Section 14.4).

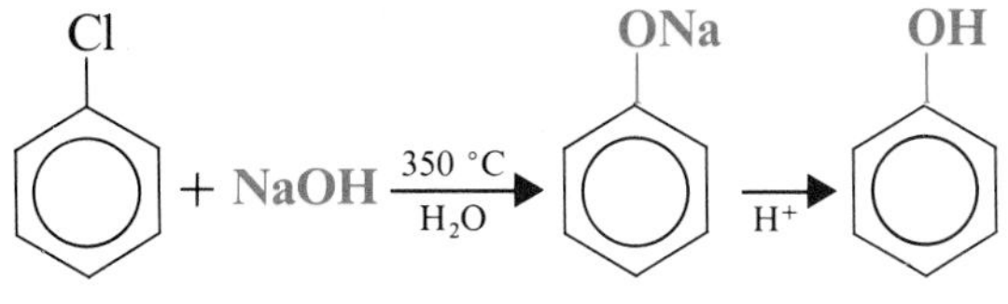

Phenol

Bromobenzene reacts with the very powerful base, NH_2^-, in liquid ammonia:

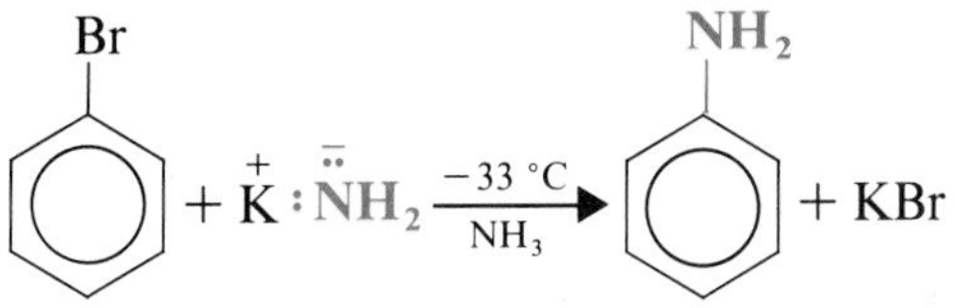

Aniline

These reactions take place through an **elimination–addition mechanism** that involves the formation of an interesting intermediate called *benzyne*. We can illustrate this mechanism with the reaction of bromobenzene and amide ion.

In the first step (see the following mechanism) the amide ion initiates an elimination by abstracting one of the ortho protons because they are the most acidic. The negative charge that develops on the ortho carbon is stabilized by the inductive effect of the bromine. The anion then loses a bromide ion. This elimination produces the highly unstable, and thus highly reactive, **benzyne.** Benzyne then reacts with any available nucleophile (in this case, an amide ion) by a two-step addition reaction to produce aniline.

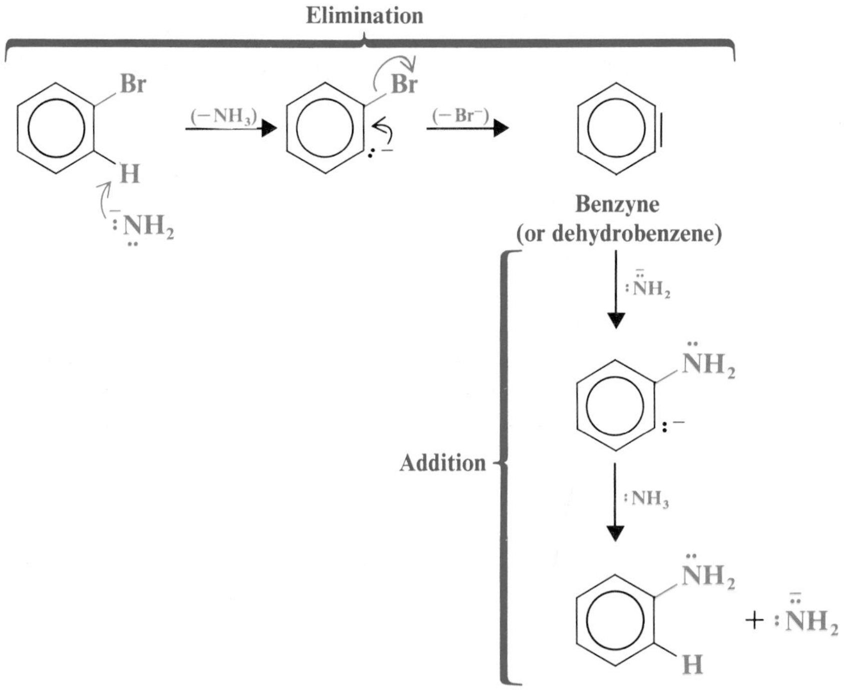

Elimination

Benzyne
(or dehydrobenzene)

Addition

The nature of benzyne itself will become clearer if we examine the following orbital diagram.

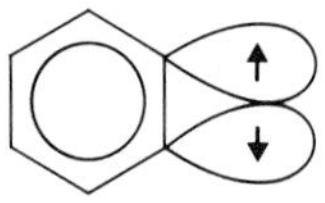

Benzyne

The extra bond in benzyne results from the overlap of sp^2 orbitals on adjacent carbon atoms of the ring. The axes of these sp^2 orbitals lie in the same plane as that of the ring, and consequently they do not overlap with the π orbitals of the aromatic system. They do not appreciably disturb the aromatic system and they do not make an appreciable resonance contribution to it. The extra bond is weak. Even though the ring hexagon is probably somewhat distorted in order to bring the sp^2 orbitals closer together, overlap between them is not large. Benzyne, as a result, is highly unstable and highly reactive. It has never been isolated.

What, then, is some of the evidence for the existence of benzyne and for an elimination–addition mechanism in some nucleophilic aromatic substitutions?

The first piece of clear-cut evidence was an experiment done by J. D. Roberts (Section 13.10) in 1953—one that marked the beginning of benzyne chemistry. Roberts showed that when ^{14}C-labeled bromobenzene is treated with amide ion in liquid ammonia, the aniline that is produced has the label equally divided between the 1 and 2 positions. This result is consistent with the elimination–addition mechanism on page 668 but is, of course, not at all consistent with a direct displacement or with an addition–elimination mechanism. (Why?)

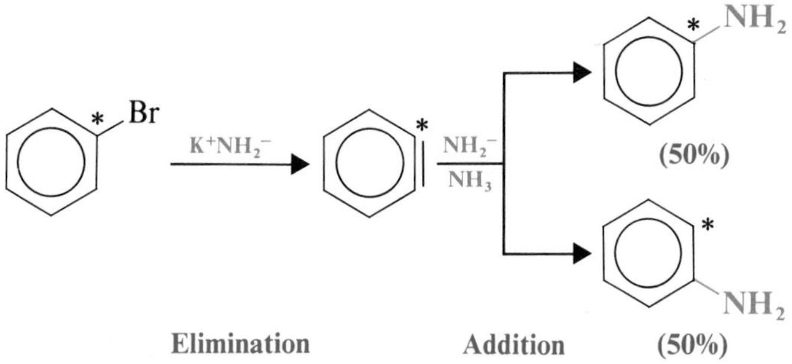

Elimination   Addition   (50%)

An even more striking illustration can be seen in the following reaction. When the ortho derivative **1** is treated with sodium amide, the only organic product obtained is *m*-(trifluoromethyl)aniline.

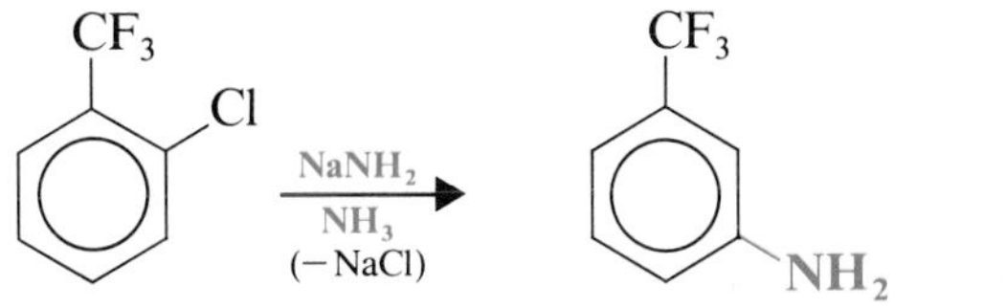

1   *m*-(Trifluoromethyl)aniline

This result can also be explained by an elimination-addition mechanism. The first step produces the benzyne **2**:

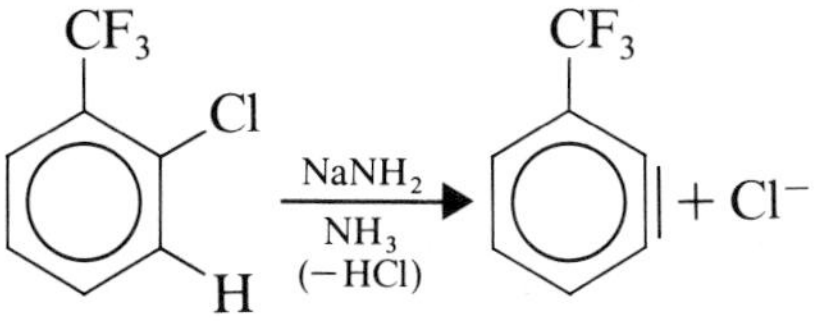

1   **2**

This benzyne then adds an amide ion in the way that produces the more stable carbanion **3** rather than the less stable carbanion **4**.

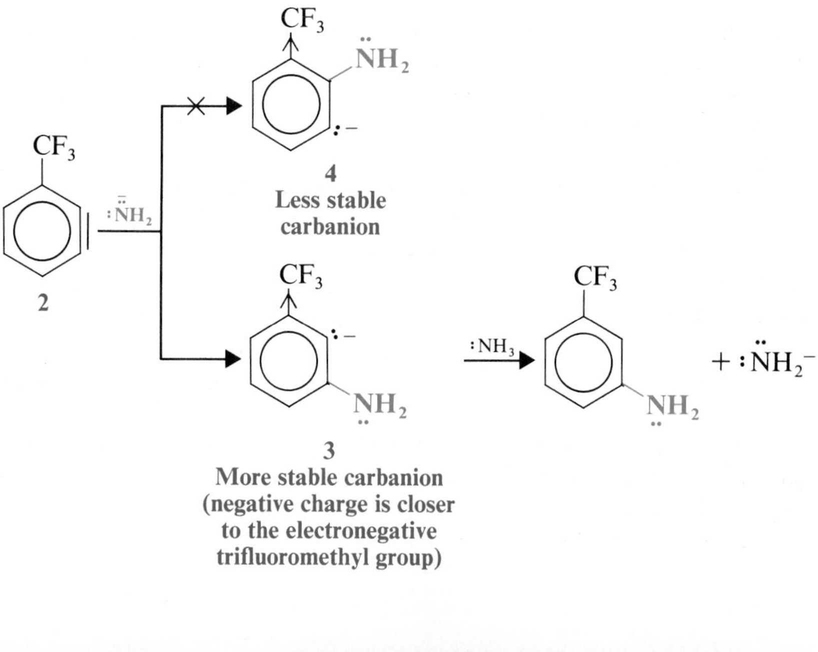

4
Less stable carbanion

3
More stable carbanion (negative charge is closer to the electronegative trifluoromethyl group)

Carbanion **3** then accepts a proton from ammonia to form *m*-(trifluoromethyl)-aniline.

Carbanion **3** is more stable than **4** because the carbon atom bearing the negative charge is closer to the highly electronegative trifluoromethyl group. The trifluoromethyl group stabilizes the negative charge through its inductive effect. (Resonance effects are not important here because the sp^2 orbital that contains the electron pair does not overlap with the π orbitals of the aromatic system.)

Benzyne intermediates have been "trapped" through the use of Diels–Alder reactions. When benzyne is generated in the presence of the diene *furan,* the product is a Diels–Alder adduct.

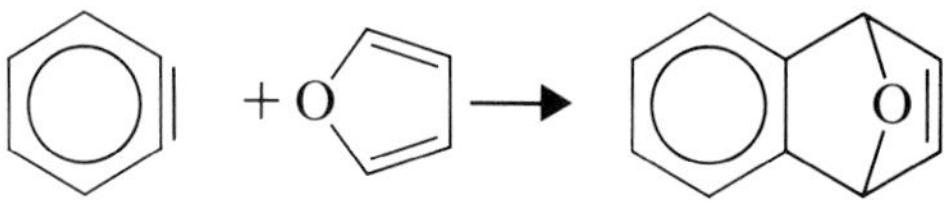

Benzyne
(generated by
an elimination
reaction)

Furan

Diels–Alder adduct

PROBLEM 14.10

(a) When *p*-chlorotoluene is heated with aqueous sodium hydroxide at 340 °C, *p*-cresol and *m*-cresol are obtained in equal amounts. Write a mechanism that would account for this result. (b) What does this suggest about the mechanism of the Dow Process (Section 14.4) for the synthesis of phenols?

PROBLEM 14.11

When 2-bromo-3-methylanisole is treated with amide ion in liquid ammonia, no substitution takes place. This has been interpreted as providing evidence for the elimination–addition mechanism. Explain.

ADDITIONAL PROBLEMS

14.12

What products would be obtained from each of the following acid–base reactions?

(a) Sodium ethoxide in ethanol + phenol $\longrightarrow$

(b) Phenol + aqueous sodium hydroxide $\longrightarrow$

(c) Sodium phenoxide + aqueous hydrochloric acid $\longrightarrow$

(d) Sodium phenoxide + H_2O + CO_2 $\longrightarrow$

14.13

Complete the following equations:

(a) Phenol + Br_2 $\xrightarrow{5\ °C,\ CS_2}$

(b) Phenol + conc. H_2SO_4 $\xrightarrow{25\ °C}$

(c) Phenol + conc. H_2SO_4 $\xrightarrow{100\ °C}$

(d)  CH_3—⟨⟩—OH + p-toluenesulfonyl chloride $\xrightarrow{OH^-}$

(e) Phenol + Br_2 $\xrightarrow{H_2O}$

(f) Phenol +

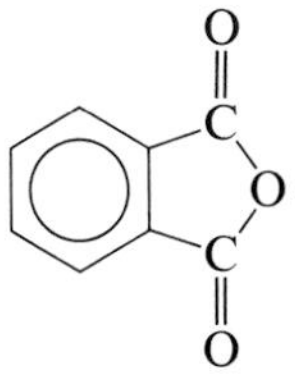

$\longrightarrow$

(g) p-Cresol + Br_2 $\xrightarrow{H_2O}$

(h) Phenol + $C_6H_5\overset{\overset{\textstyle O}{\|}}{C}Cl$ $\xrightarrow{base}$

(i) Phenol + $(C_6H_5\overset{\overset{\textstyle O}{\|}}{C})_2O$ $\xrightarrow{base}$

(j) Phenol + NaOH $\longrightarrow$

(k) Product of (j) + $CH_3OSO_2OCH_3$ $\longrightarrow$

(l) Product of (j) + CH_3I $\longrightarrow$

(m) Product of (j) + $C_6H_5CH_2Cl$ $\longrightarrow$

14.14

Describe a simple chemical test that could be used to distinguish between the members of each of the following pairs of compounds:

(a) p-Cresol and benzyl alcohol

(b) Phenol and cyclohexane

(c) Cyclohexanol and cyclohexene

(d) Allyl phenyl ether and phenyl propyl ether

(e) Methoxybenzene (anisole) and p-cresol

(f) 2,4,6-Trinitrophenol (picric acid) and 2,4,6-trimethylphenol

14.15

Thymol (see following structure) can be obtained from thyme oil. Thymol is an effective disinfectant and is used in many antiseptic preparations. (a) Suggest a synthesis of thymol from m-cresol and propylene. (b) Suggest a method for transforming thymol into menthol.

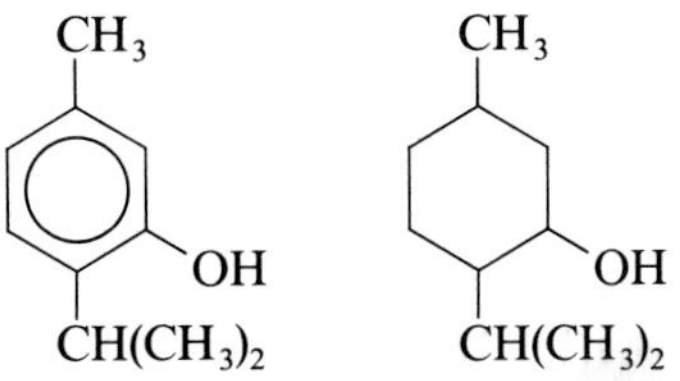

Thymol   Menthol

14.16

Carvacrol is another naturally occurring phenol, and it is an isomer of thymol (Problem 14.15). Carvacrol can be synthesized from *p*-cymene (*p*-isopropyltoluene) by ring sulfonation and treating the sulfonic acid with fused alkali. Explain why this synthesis yields mainly carvacrol and very little thymol.

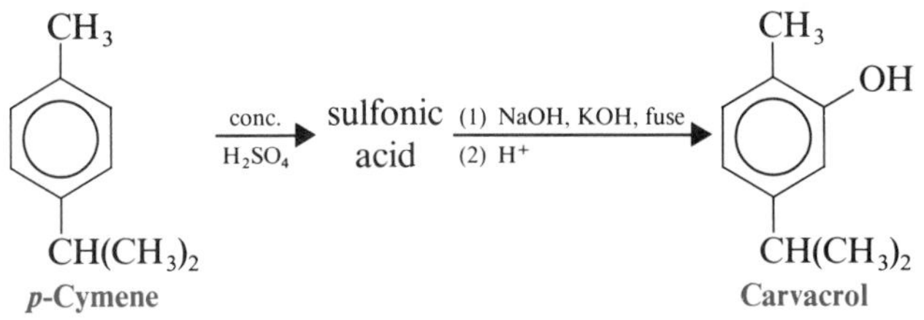

14.17

A widely used synthetic antiseptic is 4-*hexylresorcinol.* Suggest a synthesis of 4-hexylresorcinol from resorcinol and hexanoic acid.

14.18

Anethole (see following structure) is the chief component of anise oil. Suggest a synthesis of anethole from anisole and propanoic acid.

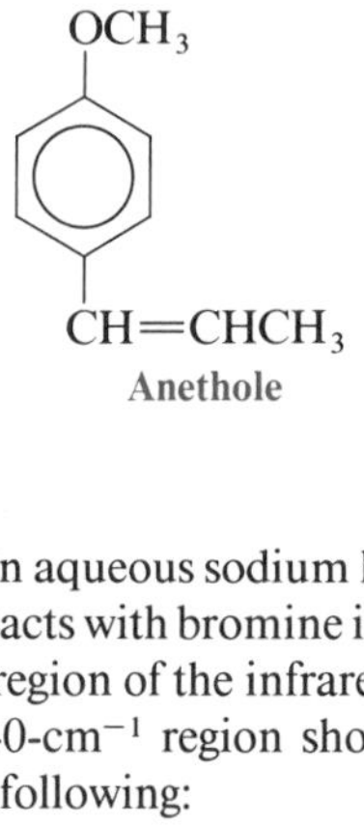

Anethole

14.19

A compound X $(C_{10}H_{14}O)$ dissolves in aqueous sodium hydroxide but is insoluble in aqueous sodium bicarbonate. Compound X reacts with bromine in water to yield a dibromo derivative, $C_{10}H_{12}Br_2O$. The 3000–4000-cm^{-1} region of the infrared spectrum of X shows a broad peak centered at 3250 cm^{-1}; the 680–840-cm^{-1} region shows a strong peak at 830 cm^{-1}. The proton nmr spectrum of X gives the following:

Singlet	δ 1.3 (9H)
Singlet	δ 4.9 (1H)
Multiplet	δ 7.0 (4H)

What is the structure of X?

14.20

The widely used antioxidant and food preservative called **BHA** (Butylated HydroxyAnisole) is actually a mixture of 2-*tert*-butyl-4-methoxyphenol and 3-*tert*-butyl-4-methoxyphenol. **BHA** is synthesized from *p*-methoxyphenol and 2-methylpropene. (a) Suggest how this is done. (b) Another widely used antioxidant is **BHT** (Butylated Hydroxy Toluene). **BHT** is actually 2,6-di-*tert*-butyl-4-methylphenol, and the raw materials used in its production are *p*-cresol and 2-methylpropene. What reaction is used here?

14.21

The herbicide **2,4-D** (cf. Special Topic H) can be synthesized from phenol and chloroacetic acid. Outline the steps involved.

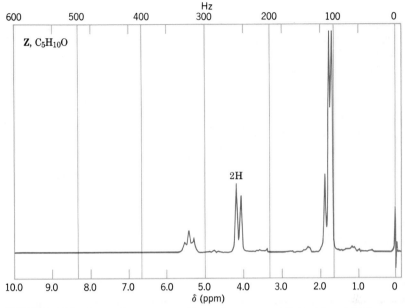

FIGURE 14.2 The proton nmr spectrum of compound **Z**, Problem 14.22. (Spectrum courtesy of Aldrich Chemical Co., Milwaukee, WI.)

2,4-D
(2,4-dichlorophenoxyacetic acid)

Chloroacetic acid

*14.22

Compound **Z** ($C_5H_{10}O$) decolorizes bromine in carbon tetrachloride. The infrared spectrum of **Z** shows a broad peak in the 3200–3600-cm^{-1} region. The proton nmr spectrum of **Z** is given in Fig. 14.2. Propose a structure for **Z**.

SPECIAL TOPIC F

THIOLS, THIOETHERS, AND THIOPHENOLS

Sulfur is directly below oxygen in Group **VI** of the Periodic Table and, as we might expect, there are sulfur counterparts of the oxygen compounds that we studied in Chapters 8 and 14 such as thiols, thioethers, and thiophenols.

These and other important examples of organosulfur compounds are the following:

$$R-SH \qquad R-S-R' \qquad ArSH \qquad R-S-S-R' \qquad R-\overset{\displaystyle R'}{\underset{|}{S^+}}-R''$$

| Thiols | Thioethers | Thiophenols | Disulfides | Trialkylsulfonium ions |

$$R-\overset{\displaystyle O}{\underset{\|}{S}}-R' \qquad R-\overset{\displaystyle O}{\underset{\underset{\displaystyle O}{\|}}{\overset{\|}{S}}}-R' \qquad R-\overset{\displaystyle S}{\underset{\|}{C}}-R' \qquad R-\overset{\displaystyle O}{\underset{\|}{S}}-OH \qquad R-\overset{\displaystyle O}{\underset{\underset{\displaystyle O}{\|}}{\overset{\|}{S}}}-OH$$

| Sulfoxides | Sulfones | Thioketones | Sulfinic acids | Sulfonic acids |

The sulfur counterpart of an alcohol is called a *thiol* or a *mercaptan.* The name mercaptan comes from the Latin, *mercurium captans,* meaning "capturing mercury." Mercaptans react with mercuric ions and the ions of other heavy metals to form precipitates. The compound CH_2CHCH_2OH, known as British Anti-Lewisite
   $\quad\quad\quad\quad\quad\quad\quad\quad\quad\quad\quad\quad\quad\quad\quad\;\;$ | |
   $\quad\quad\quad\quad\quad\quad\quad\quad\quad\quad\quad\quad\quad\quad\quad\;$ SH SH
(BAL), was developed as an antidote for poisonous arsenic compounds used as war gases. BAL is also an effective antidote for mercury poisoning.

Several simple thiols are shown below.

$$CH_3CH_2SH \qquad CH_3CH_2CH_2SH \qquad CH_3\overset{\displaystyle CH_3}{\underset{|}{CH}}CH_2CH_2SH \qquad CH_2{=}CHCH_2SH$$

| Ethanethiol | 1-Propanethiol | 3-Methyl-1-butanethiol | 2-Propene-1-thiol |
| (ethyl mercaptan) | (propyl mercaptan) | (isopentyl mercaptan) | (allyl mercaptan) |

Compounds of sulfur, in general, and the low-molecular-weight thiols, in particular, are noted for their disagreeable odors. Anyone who has passed anywhere near a general chemistry laboratory when hydrogen sulfide (H_2S) was being used has noticed the strong odor of that substance—the odor of rotten eggs. Another sulfur compound, 3-methyl-1-butanethiol, is one unpleasant constituent of the liquid that skunks use as a defensive weapon. 1-Propanethiol evolves from freshly chopped onions, and allyl mercaptan is one of the compounds responsible for the odor and flavor of garlic.

Aside from their odors, analogous sulfur and oxygen compounds show other chemical differences. These arise largely from the following features of sulfur compounds.

1. The sulfur atom is larger and more polarizable than the oxygen atom. As a result, sulfur compounds are more powerful nucleophiles and compounds containing —SH groups are stronger acids than their oxygen analogs. The ethanethiolate ($CH_3CH_2\ddot{S}{:}^-$), for example, is a much stronger nucleophile when it reacts at carbon atoms than is the ethoxide ion ($CH_3CH_2O^-$). On the other hand, since ethanol is a weaker acid than ethanethiol, the ethoxide ion is the stronger of the two conjugate bases.

2. The bond dissociation energy of the S—H bond of thiols (~ 80 kcal/mole) is much less than that of the O—H bond of alcohols (~ 100 kcal/mole). The weakness of the S—H bond allows thiols to undergo an oxidative coupling reaction when they react with mild oxidizing agents; the product is a disulfide:

$$2RS—H + H_2O_2 \longrightarrow RS—SR + 2H_2O$$
   A thiol   A disulfide

   Alcohols do not undergo an analogous reaction. When alcohols are treated with oxidizing agents, oxidation takes place at the weaker C—H (~ 85 kcal/mole) bond rather than at the strong O—H bond.

3. Because sulfur atoms are easily polarized they can stabilize a negative charge on an adjacent atom. This means that hydrogen atoms on carbon atoms that are adjacent to an alkylthio group are more acidic than those adjacent to an alkyloxy group. Thioanisole, for example, reacts with butyllithium in the following way:

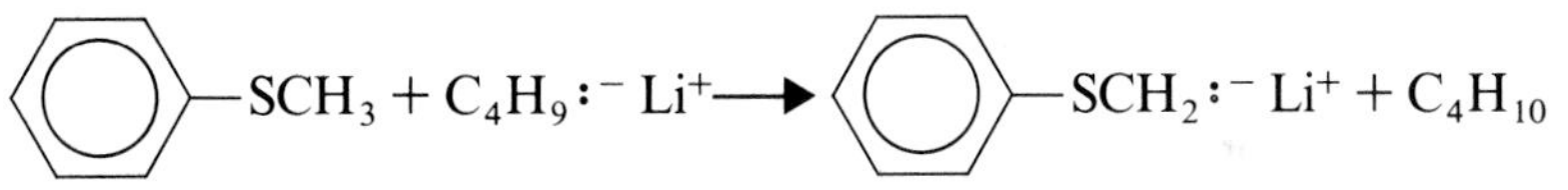

   Thioanisole

   Anisole ($CH_3OC_6H_5$) does not undergo an analogous reaction.

   The $\diagdown S{=}O$ group of sulfoxides and the positive sulfur of sulfonium ions are even more effective in delocalizing negative charge on an adjacent atom:

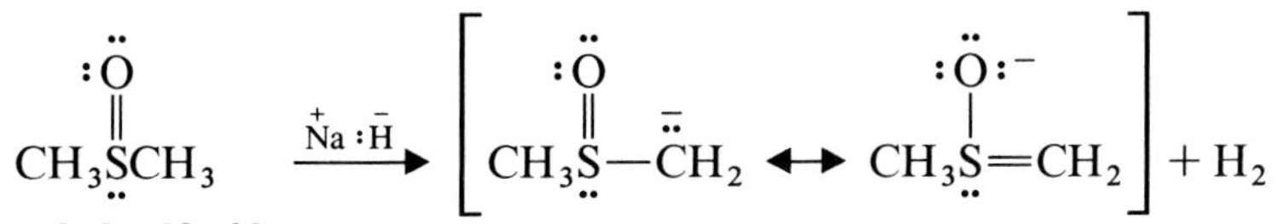

Dimethyl sulfoxide

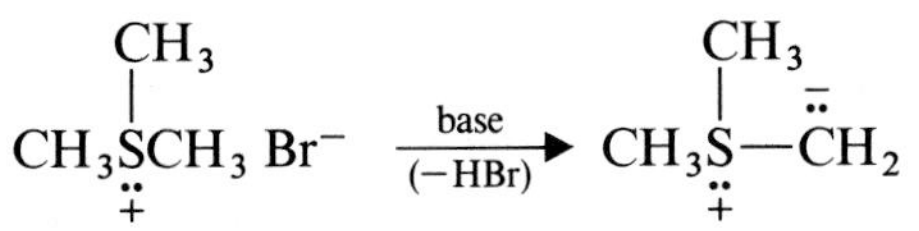

Trimethylsulfonium   An ylide*
bromide

---

*An ylide is a neutral molecule that can be represented as a resonance hybrid, one structure of which has a negative carbon atom directly attached to a positive heteroatom.

The anions formed in the reactions just given are of synthetic use. They can be used to synthesize epoxides, for example (cf. Section 16.10B).

F.1 Preparation of Thiols

Alkyl bromides and iodides react with potassium hydrogen sulfide to form thiols. (Potassium hydrogen sulfide can be generated by passing gaseous H_2S into an alcoholic solution of potassium hydroxide.)

$$R—Br + KOH + \underset{\text{(excess)}}{H_2S} \xrightarrow[\text{heat}]{C_2H_5OH} R—SH + KBr + H_2O$$

The thiol that forms is sufficiently acidic to form a thiolate ion in the presence of potassium hydroxide. Thus, if excess H_2S is not employed in the reaction, the major product of the reaction will be a thioether. The thioether results from the following reactions:

$$R—SH + KOH \longrightarrow R—\ddot{S}{:}^- \; K^+ + H_2O$$

$$R—\ddot{S}{:}^- \; K^+ + R—\ddot{Br}{:} \longrightarrow R—\ddot{S}—R + KBr$$
$$\text{Thioether}$$

Alkyl halides also react with thiourea to form (stable) S-alkylisothiouronium salts. These can be used to prepare thiols.

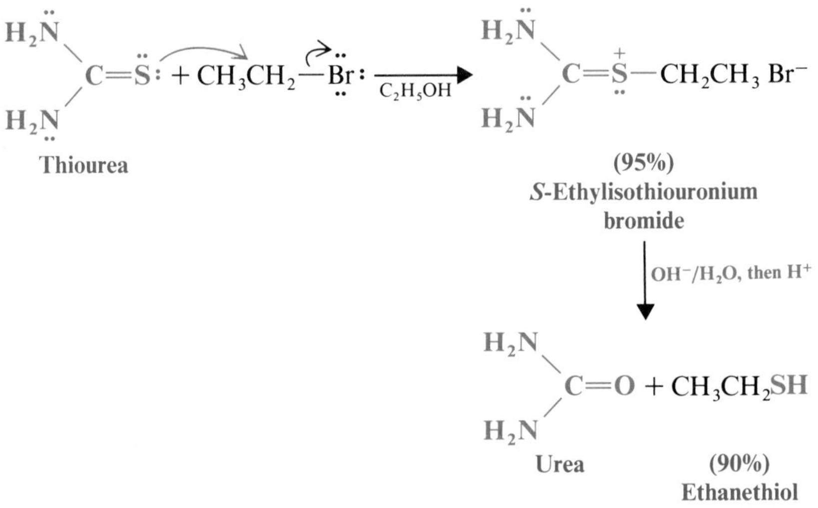

$$\begin{array}{c} H_2\ddot{N} \\ \diagdown \\ \diagup \\ H_2\ddot{N} \end{array} C{=}\ddot{S}{:} + CH_3CH_2{-}\ddot{Br}{:} \xrightarrow{C_2H_5OH} \begin{array}{c} H_2\ddot{N} \\ \diagdown \\ \diagup \\ H_2\ddot{N} \end{array} C{=}\overset{+}{\ddot{S}}{-}CH_2CH_3 \; Br^-$$

Thiourea

(95%)
S-Ethylisothiouronium
bromide

$$\downarrow \text{OH}^-/\text{H}_2\text{O, then H}^+$$

$$\begin{array}{c} H_2N \\ \diagdown \\ \diagup \\ H_2N \end{array} C{=}O + CH_3CH_2SH$$

Urea   (90%)
Ethanethiol

F.2 Physical Properties of Thiols

Thiols form very weak hydrogen bonds; their hydrogen bonds are not nearly as strong as those of alcohols. Because of this, low-molecular-weight thiols have lower boiling points than corresponding alcohols. Ethanethiol, for example, boils more than 40 °C lower than ethanol (37 versus 78 °C). The relative weakness of hydrogen bonds between molecules of thiols is also evident when we compare the boiling points of ethanethiol and its isomer dimethyl sulfide:

$$\begin{array}{cc} CH_3CH_2SH & CH_3SCH_3 \\ \text{bp, 37 °C} & \text{bp, 38 °C} \end{array}$$

TABLE F.1 Physical properties of thiols

COMPOUND	STRUCTURE	mp (°C)	bp (°C)
Methanethiol	CH_3SH	−123	6
Ethanethiol	CH_3CH_2SH	−144	37
1-Propanethiol	$CH_3CH_2CH_2SH$	−113	67
2-Propanethiol	$(CH_3)_2CHSH$	−131	58
1-Butanethiol	$CH_3(CH_2)_2CH_2SH$	−116	98

Physical properties of several thiols are given in Table F.1.

F.3 Thiols and Disulfides in Biochemistry

Thiols and disulfides are important compounds in living cells, and in many biochemical oxidation–reduction reactions they are interconverted.

$$2RSH \underset{[H]}{\overset{[O]}{\rightleftharpoons}} R-S-S-R$$

Lipoic acid, for example, an important cofactor in biological oxidations, undergoes this oxidation-reduction reaction.

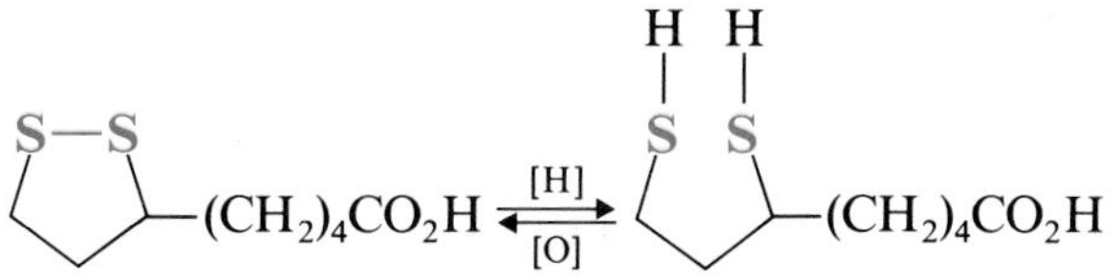

Lipoic acid   Dihydrolipoic acid

The amino acids *cysteine* and *cystine* are interconverted in a similar way.

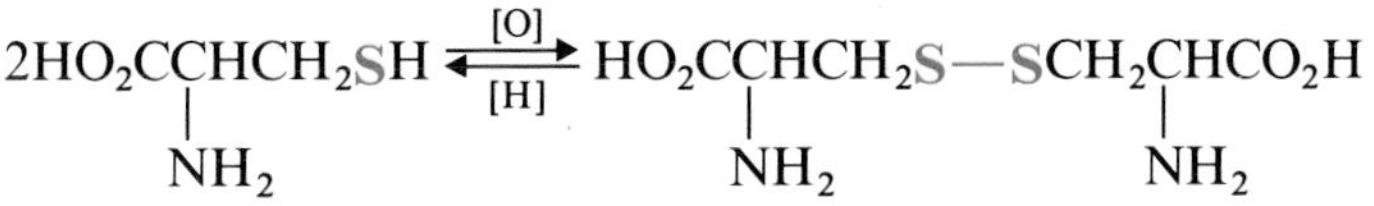

Cysteine   Cystine

As we shall see later, the disulfide linkages of cystine units are important in determining the overall shapes of protein molecules

PROBLEM F.1

Give structures for the products of the following reactions:

(a) Benzyl bromide + thiourea ⟶

(b) Product of (a) + OH^-/H_2O then H^+ ⟶

(c) Product of (b) + H_2O_2 ⟶

(d) Product of (b) + NaOH ⟶

(e) Product of (d) + benzyl bromide ⟶

PROBLEM F.2

Allyl disulfide, $CH_2{=}CHCH_2S{-}SCH_2CH{=}CH_2$, is another important component of oil of garlic. Suggest a synthesis of allyl disulfide starting with allyl bromide.

PROBLEM F.3

Starting with allyl alcohol, outline a synthesis of British Anti-Lewisite, $CH_2SHCHSHCH_2OH$.

PROBLEM F.4

A synthesis of lipoic acid (see structure just given) is outlined here. Supply the missing reagents and intermediates.

$$Cl{-}\overset{\overset{\textstyle O}{\|}}{C}(CH_2)_4CO_2C_2H_5 \xrightarrow[\text{AlCl}_3]{\text{CH}_2=\text{CH}_2} \text{(a) } C_{10}H_{17}ClO_3 \xrightarrow{\text{NaBH}_4}$$

$$\underset{\underset{\textstyle OH}{|}}{ClCH_2CH_2CH}(CH_2)_4CO_2C_2H_5 \xrightarrow{\text{(b)}} \underset{\underset{\textstyle Cl}{|}}{ClCH_2CH_2CH}(CH_2)_4CO_2C_2H_5 \xrightarrow[\text{(d)}]{\text{(c)}}$$

$$\underset{\underset{\textstyle SCH_2C_6H_5}{|}}{C_6H_5CH_2SCH_2CH_2CH}(CH_2)_4CO_2H \xrightarrow[\text{(2) H}^+]{\text{(1) Na, NH}_3}$$

$$\text{(e) } C_8H_{16}S_2O_2 \xrightarrow{O_2} \text{lipoic acid}$$

PROBLEM F.5

One chemical-warfare agent used in World War I is a powerful vesicant called "mustard gas." (The name comes from its mustardlike odor; mustard gas, however, is not a gas but a high-boiling liquid that was dispersed as a mist of tiny droplets.) Mustard gas can be synthesized from ethylene oxide in the following manner. Outline the reactions involved.

$$2H_2C\overset{}{\underset{O}{-}}CH_2 + H_2S \longrightarrow C_4H_{10}SO_2 \xrightarrow[\text{ZnCl}_2]{\text{HCl}} C_4H_8SCl_2$$

$$\text{"Mustard gas"}$$

CHAPTER FIFTEEN

Leucomycin. Leucomycin is an antibiotic isolated from *Streptomyces kitasaoensis*.

Organic Oxidation and Reduction Reactions. Organometallic Compounds

15.1 OXIDATION–REDUCTION REACTIONS IN ORGANIC CHEMISTRY

Reduction of an organic molecule usually corresponds to increasing its hydrogen content or to decreasing its oxygen content. For example, converting a carboxylic acid to an aldehyde is a reduction because the oxygen content is decreased.

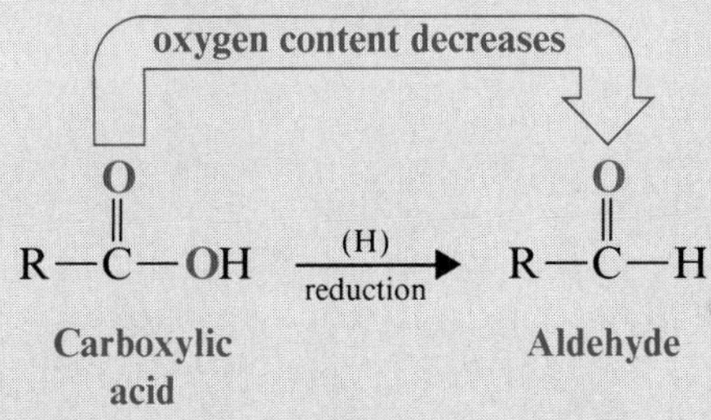

$$
\underset{\substack{\text{Carboxylic}\\\text{acid}}}{R-\overset{\overset{\textstyle O}{\|}}{C}-OH} \quad \xrightarrow[\text{reduction}]{\text{(H)}} \quad \underset{\text{Aldehyde}}{R-\overset{\overset{\textstyle O}{\|}}{C}-H}
$$

oxygen content decreases

Converting an aldehyde to an alcohol is also a reduction.

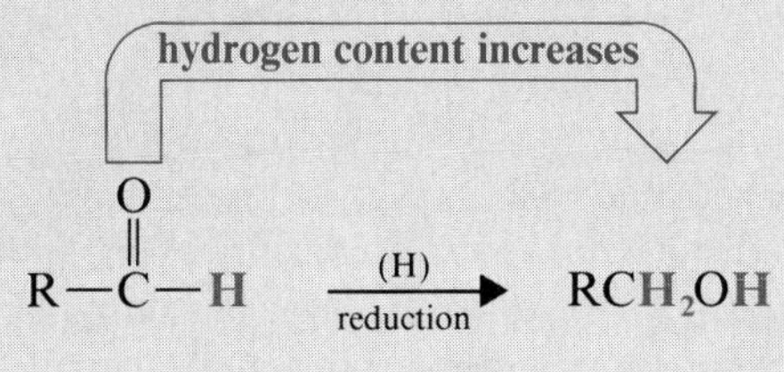

$$
R-\overset{\overset{\textstyle O}{\|}}{C}-H \quad \xrightarrow[\text{reduction}]{\text{(H)}} \quad RCH_2OH
$$

hydrogen content increases

Converting an alcohol to an alkane is also a reduction.

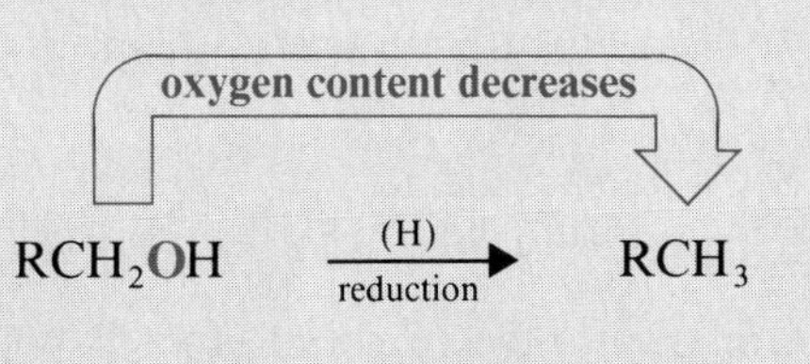

$$
RCH_2OH \quad \xrightarrow[\text{reduction}]{\text{(H)}} \quad RCH_3
$$

oxygen content decreases

In these examples we have used the symbol (H) to indicate that a reduction of the organic compound has taken place. We do this when we want to write a general equation without specifying what the reducing agent is.

The opposite of reduction is **oxidation.** *Thus, increasing the oxygen content of an organic molecule or decreasing its hydrogen content is an oxidation* of the organic molecule. The reverse of each reaction that we have given is an oxidation of the organic molecule, and we can summarize these oxidation–reduction reactions as follows below. We use the symbol (O) to indicate in a general way that the organic molecule has been oxidized.

$$\underset{\substack{\text{Lowest}\\\text{oxidation}\\\text{state}}}{RCH_3} \underset{(H)}{\overset{(O)}{\rightleftarrows}} RCH_2OH \underset{(H)}{\overset{(O)}{\rightleftarrows}} R\overset{\overset{\textstyle O}{\|}}{C}H \underset{(H)}{\overset{(O)}{\rightleftarrows}} \underset{\substack{\text{Highest}\\\text{oxidation}\\\text{state}}}{R\overset{\overset{\textstyle O}{\|}}{C}OH}$$

Oxidation of an organic compound may be more broadly defined as a reaction that increases its content of any element more electronegative than carbon. For example, replacing hydrogen atoms by chlorine atoms is an oxidation:

$$Ar-CH_3 \underset{(H)}{\overset{(O)}{\rightleftarrows}} Ar-CH_2Cl \underset{(H)}{\overset{(O)}{\rightleftarrows}} Ar-CHCl_2 \underset{(H)}{\overset{(O)}{\rightleftarrows}} Ar-CCl_3$$

Of course, when an organic compound is reduced, something else — **the reducing agent** — must be oxidized. And when an organic compound is oxidized, something else — **the oxidizing agent** — is reduced. These oxidizing and reducing agents are often inorganic compounds, and in the next two sections we shall see what some of them are.

PROBLEM 15.1

One method for assigning an oxidation state to a carbon atom of an organic compound is to base that assignment on the groups attached to the carbon; a bond to hydrogen (or anything less electronegative than carbon) makes it -1, a bond to oxygen, nitrogen, or halogen (or to anything more electronegative than carbon) makes it $+1$, and a bond to another carbon 0. Thus the carbon of methane is assigned an oxidation state of -4, and that of carbon dioxide $+4$. (a) Use this method to assign oxidation states to the carbon atoms of methyl alcohol (CH_3OH), formic acid ($H\overset{\overset{\textstyle O}{\|}}{C}OH$), and formaldehyde ($H\overset{\overset{\textstyle O}{\|}}{C}H$). (b) Arrange the compounds methane, carbon dioxide, methyl alcohol, formic acid, and formaldehyde in order of increasing oxidation state. (c) What change in oxidation state accompanies the reaction, methyl alcohol $\longrightarrow$ formaldehyde? (d) Is this an oxidation or a reduction? (e) When H_2CrO_4 acts as an oxidizing agent in this reaction, the chromium of H_2CrO_4 becomes Cr^{3+}. What change in oxidation state does chromium undergo?

ONE

PROBLEM 15.2

(a) Use the method described in the preceding problem to assign oxidation states to each carbon of ethyl alcohol and to each carbon of acetaldehyde. (b) What do these numbers reveal about the site of oxidation when ethyl alcohol is oxidized to acetaldehyde? (c) Repeat this procedure for the oxidation of acetaldehyde to acetic acid.

PROBLEM 15.3

(a) Although we have described the hydrogenation of an alkene as an addition reaction, organic chemists often refer to it as a "reduction." Refer to the method described in Problem 15.1 and explain. (b) Make similar comments about the reversible reaction:

$$CH_3-\overset{\overset{\displaystyle O}{\|}}{C}-H + H_2 \underset{}{\overset{Ni}{\rightleftharpoons}} CH_3CH_2OH$$

15.1A Balancing Oxidation–Reduction Equations

A method for balancing organic oxidation–reduction reactions is described in the Study Guide that accompanies this text.

15.2 ALCOHOLS BY REDUCTION OF CARBONYL COMPOUNDS

Primary and secondary alcohols can be synthesized by the reduction of a variety of compounds that contain the carbonyl $\left(\underset{/}{\overset{\backslash}{}}C{=}O \right)$ group. Several general examples are shown here.

$$R-\overset{\overset{\displaystyle O}{\|}}{C}-OH \xrightarrow{\text{(H)}} R-CH_2OH$$

Carboxylic acid 1° Alcohol

$$R-\overset{\overset{\displaystyle O}{\|}}{C}-OR' \xrightarrow{\text{(H)}} R-CH_2OH \; (+ R'OH)$$

Ester 1° Alcohol

$$R-\overset{\overset{\displaystyle O}{\|}}{C}-H \xrightarrow{\text{(H)}} R-CH_2OH$$

Aldehyde 1° Alcohol

$$R-\overset{\overset{\displaystyle O}{\|}}{C}-R' \xrightarrow{\text{(H)}} R-\underset{\underset{\displaystyle OH}{|}}{CH}-R'$$

Ketone 2° Alcohol

Reductions of carboxylic acids are the most difficult, and prior to 1946 direct reduction of acids was not possible. However, the discovery in 1946 of the powerful reducing agent **lithium aluminum hydride** gave organic chemists a method for reducing acids to primary alcohols in excellent yields.

$$4RCO_2H + 3LiAlH_4 \xrightarrow{Et_2O} [(RCH_2O)_4Al]Li + 4H_2 + 2LiAlO_2$$

Lithium aluminum hydride

$$\xrightarrow[H_2O]{} 4RCH_2OH + Al(OH)_3 + LiOH$$

Two examples are the lithium aluminum hydride reductions of acetic acid and 2,2-dimethylpropanoic acid.

$$\underset{\substack{\text{Acetic acid}}}{CH_3\overset{\displaystyle O}{\overset{\|}{C}}-OH} \xrightarrow[\text{(2) H}_2\text{O}]{\text{(1) LiAlH}_4/\text{Et}_2\text{O}} \underset{\substack{(100\%)\\ \text{Ethyl alcohol}}}{CH_3CH_2OH}$$

$$\underset{\substack{\text{2,2-Dimethylpropanoic}\\ \text{acid}}}{CH_3-\overset{\displaystyle CH_3}{\underset{\displaystyle CH_3}{\overset{\|}{\underset{\|}{C}}}}-CO_2H} \xrightarrow[\text{(2) H}_2\text{O}]{\text{(1) LiAlH}_4/\text{Et}_2\text{O}} \underset{\substack{(92\%)\\ \text{Neopentyl alcohol}}}{CH_3\overset{\displaystyle CH_3}{\underset{\displaystyle CH_3}{\overset{\|}{\underset{\|}{C}}}}-CH_2OH}$$

Esters can be reduced by high-pressure hydrogenation (a reaction preferred for industrial processes and often referred to as "hydrogenolysis" because a carbon–oxygen bond is cleaved in the process), or through the use of lithium aluminum hydride.

$$R-\overset{\displaystyle O}{\overset{\|}{C}}OR' + H_2 \xrightarrow[\substack{175\ ^\circ C\\ 5000\ psi}]{CuO\cdot CuCr_2O_4} RCH_2OH + R'OH$$

$$R-\overset{\displaystyle O}{\overset{\|}{C}}OR' \xrightarrow[\text{(2) H}_2\text{O}]{\text{(1) LiAlH}_4/\text{Et}_2\text{O}} RCH_2OH + R'OH$$

The last method is the one most commonly used now in small-scale laboratory synthesis.

Aldehydes and ketones can also be reduced to alcohols by hydrogen and a metal catalyst, by sodium in alcohol, and by lithium aluminum hydride. The reducing agent most often used, however, is sodium borohydride ($NaBH_4$).

$$4R\overset{\displaystyle O}{\overset{\|}{C}}H + NaBH_4 + 3H_2O \longrightarrow 4RCH_2OH + NaH_2BO_3$$

$$\underset{\substack{\text{Butanal}}}{CH_3CH_2CH_2\overset{\displaystyle O}{\overset{\|}{C}}H} \xrightarrow[\text{H}_2\text{O}]{NaBH_4} \underset{\substack{(85\%)\\ \text{1-Butanol}}}{CH_3CH_2CH_2CH_2OH}$$

$$\underset{\substack{\text{2-Butanone}}}{CH_3CH_2\underset{\displaystyle O}{\overset{\|}{C}}CH_3} \xrightarrow[\text{H}_2\text{O}]{NaBH_4} \underset{\substack{(87\%)\\ \text{2-Butanol}}}{CH_3CH_2\underset{\displaystyle OH}{C}HCH_3}$$

The key step in the reduction of a carbonyl compound by either lithium aluminum hydride or sodium borohydride is the transfer of a **hydride ion** from the metal to the carbonyl carbon. In this transfer the hydride ion acts as a *nucleophile.* Since compounds of trivalent boron and aluminum are Lewis acids, they react as electrophiles at the carbonyl oxygen and facilitate the hydride transfer. The mechanism for the reduction of a ketone by sodium borohydride is illustrated here.

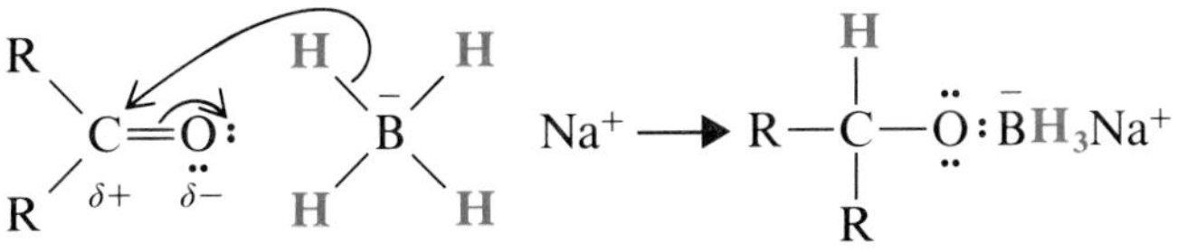

This step is then repeated until all hydrogen atoms attached to boron have been transferred. The boron complex decomposes in water to form the secondary alcohol.

$$(R_2CHO)_4B^- \ Na^+ + 3H_2O \longrightarrow 4RCHR + NaH_2BO_3$$
$$\underset{\displaystyle OH}{|}$$

Sodium borohydride is a milder reducing agent than lithium aluminum hydride. Lithium aluminum hydride will reduce acids, esters, aldehydes, and ketones; but sodium borohydride will reduce only aldehydes, ketones, and acyl chlorides. *Lithium aluminum hydride reacts violently with water* and therefore reductions with lithium aluminum hydride must be carried out in anhydrous solutions, usually in anhydrous ether. (Water is added cautiously after the reaction is over to decompose the aluminum complex.)* Sodium borohydride reductions, by contrast, can be carried out in water or alcohol solutions.

PROBLEM 15.4

Which reducing agent ($NaBH_4$ or $LiAlH_4$) would you use to carry out each of the following transformations?

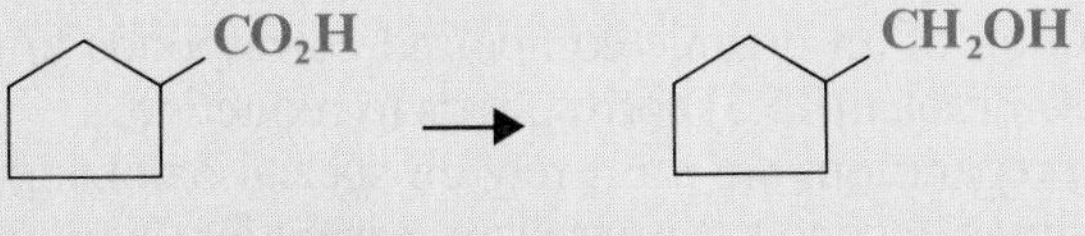

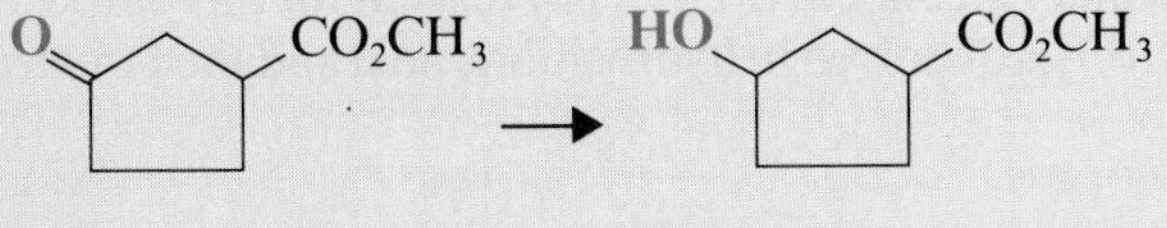

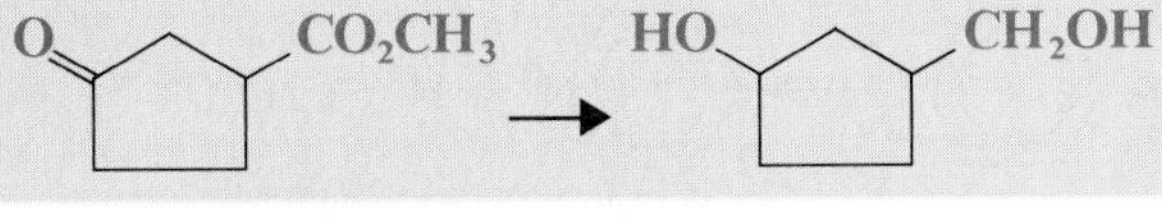

*Unless special precautions are taken, lithium aluminum hydride reductions can be very dangerous. You should consult an appropriate laboratory manual before attempting such a reduction, and the reaction should be carried out on a small scale.

15.3 OXIDATION OF ALCOHOLS

15.3A Oxidation of 1° Alcohols:
$RCH_2OH \longrightarrow RCHO$

Primary alcohols can be oxidized to aldehydes and carboxylic acids.

$$R—CH_2OH \xrightarrow{(O)} \underset{\substack{\\ \text{Aldehyde}}}{R—\overset{\displaystyle O}{\overset{\|}{C}}—H} \xrightarrow{(O)} \underset{\substack{\\ \text{Carboxylic acid}}}{R—\overset{\displaystyle O}{\overset{\|}{C}}—OH}$$
$$\underset{\text{1° Alcohol}}{}$$

The oxidation of aldehydes to carboxylic acids usually takes place with milder oxidizing agents than those required to oxidize primary alcohols to aldehydes; thus it is difficult to stop the oxidation at the aldehyde stage. One way of avoiding this problem is to remove the aldehyde as soon as it is formed. This can often be done because aldehydes have lower boiling points than alcohols (why?), and therefore aldehydes can be distilled from the reaction mixture as they are formed. An example is the synthesis of butanal from 1-butanol using a mixture of $K_2Cr_2O_7$ and sulfuric acid:

$$\underset{\substack{\text{1-Butanol}\\ \text{bp, 117.5 °C}}}{CH_3CH_2CH_2CH_2OH} \xrightarrow[H_2SO_4]{K_2Cr_2O_7} \underset{\substack{\text{Butanal}\\ \text{bp, 75.7 °C}}}{CH_3CH_2CH_2\overset{\displaystyle O}{\overset{\|}{C}H}} \qquad \text{(50\% yield)}$$

This procedure does not give good yields with aldehydes that boil above 100 °C, however.

An industrial process for preparing low-molecular-weight aldehydes is **dehydrogenation** of a primary alcohol:

$$CH_3CH_2OH \xrightarrow[300\ °C]{Cu} CH_3\overset{\displaystyle O}{\overset{\|}{C}H} + H_2$$

[Notice that dehydrogenation of an organic compound corresponds to oxidation whereas hydrogenation (cf. Problem 15.3) corresponds to reduction.]

In most laboratory preparations we must rely on special oxidizing agents to prepare aldehydes from primary alcohols. An excellent reagent for this purpose is the compound formed when CrO_3 is dissolved in hydrochloric acid and then treated with pyridine.

$$CrO_3 + HCl + \underset{\substack{\text{Pyridine}\\ (C_5H_5N)}}{\left\langle\bigcirc\right\rangle N\colon} \longrightarrow \underset{\substack{\text{Pyridinium chlorochromate}\\ \text{(PCC)}}}{\left\langle\bigcirc\right\rangle N^+—H \quad CrO_3Cl^-}$$

This compound, called **pyridinium chlorochromate** [abbreviated (PCC)], when dissolved in CH_2Cl_2, will oxidize a primary alcohol to an aldehyde and stop at that stage.

$$\underset{\substack{\text{2-Ethyl-2-methyl-1-}\\\text{butanol}}}{(C_2H_5)_2\overset{\overset{\displaystyle CH_3}{\vert}}{C}\!-\!CH_2OH} + PCC \xrightarrow[25\ °C]{CH_2Cl_2} \underset{\text{2-Ethyl-2-methylbutanal}}{(C_2H_5)_2\overset{\overset{\displaystyle CH_3}{\vert}}{C}\!-\!\overset{\overset{\displaystyle O}{\|}}{C}H}$$

Pyridinium chlorochromate (PCC) also does not attack double bonds.

One reason for the success of oxidation with pyridinium chlorochromate is that the oxidation can be carried out in a solvent such as CH_2Cl_2, in which PCC is soluble. Aldehydes themselves are not nearly as easily oxidized as are the *aldehyde hydrates, RCH(OH)$_2$*, that form (Section 16.7A) when aldehydes are dissolved in water, the usual medium for oxidation by chromium compounds.

$$RCHO + H_2O \rightleftharpoons RCH(OH)_2$$

15.3B Oxidation of 1° Alcohols: RCH$_2$OH $\longrightarrow$ RCO$_2$H

Primary alcohols can be oxidized to **carboxylic acids** by potassium permanganate. The reaction is usually carried out in basic aqueous solution from which MnO_2 precipitates as the oxidation takes place. After the oxidation is complete, filtration allows removal of the MnO_2 and acidification of the filtrate gives the carboxylic acid.

$$R\!-\!CH_2OH + KMnO_4 \xrightarrow[\substack{H_2O\\\text{heat}}]{OH^-} RCO_2^-K^+ + MnO_2$$
$$\downarrow H^+$$
$$RCO_2H$$

15.3C Oxidation of 2° Alcohols:
$$\underset{\textbf{RCHR}'}{\overset{\textbf{OH}}{\vert}} \longrightarrow \underset{}{\textbf{R}\!-\!\overset{\overset{\textbf{O}}{\|}}{\textbf{C}}\!-\!\textbf{R}'}$$

Secondary alcohols can be oxidized to ketones. The reaction usually stops at the ketone stage because further oxidation requires the breaking of a carbon–carbon bond.

$$\underset{\text{2° Alcohol}}{R\!-\!\overset{\overset{\displaystyle OH}{\vert}}{C}H\!-\!R'} \xrightarrow{(O)} \underset{\text{Ketone}}{R\!-\!\overset{\overset{\displaystyle O}{\|}}{C}\!-\!R'}$$

Various oxidizing agents based on chromium(VI) have been used to oxidize secondary alcohols to ketones. The most commonly used reagent is chromic acid (H_2CrO_4). Chromic acid is usually prepared by adding chromium (VI) oxide (CrO_3) or sodium dichromate ($Na_2Cr_2O_7$) to aqueous sulfuric acid. Oxidations of secondary alcohols are generally carried out in acetone or acetic acid solutions. The balanced equation is shown here.

$$3\ \overset{R}{\underset{R}{>}}CHOH + 2H_2CrO_4 + 6H^+ \longrightarrow 3\ \overset{R}{\underset{R}{>}}C\!=\!O + 2Cr^{3+} + 8H_2O$$

As chromic acid oxidizes the alcohol to the ketone, chromium is reduced from the $+6$ oxidation state (H_2CrO_4) to the $+3$ oxidation state (Cr^{3+}).* Chromic acid oxidations of secondary alcohols generally give ketones in excellent yields if the temperature is controlled. A specific example is the oxidation of cyclooctanol to cyclooctanone.

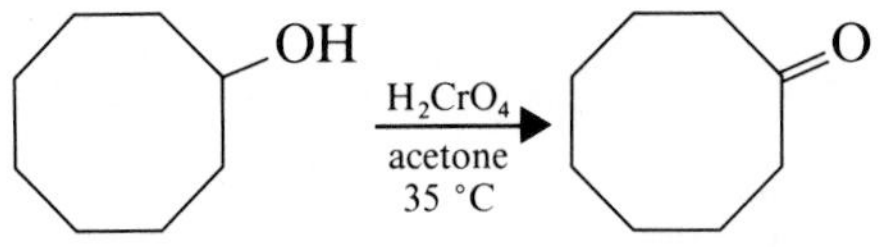

Cyclooctanol $\xrightarrow[\substack{\text{acetone} \\ 35\,°C}]{H_2CrO_4}$ (92–96%) Cyclooctanone

The following mechanism for chromic acid oxidations involves initial formation of a chromate ester. Then loss of a proton and an $HCrO_3^-$ ion produces the ketone.

$$R_2CHOH + {}^-OCrO_3H + H^+ \longrightarrow R_2CHOCrO_3H + H_2O$$

Chromate ester

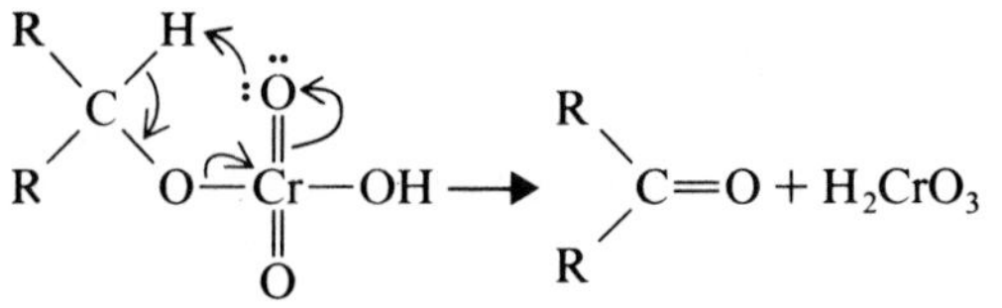

Chromate ester   Ketone

Subsequent reactions with chromium species ultimately reduce the chromium to Cr^{3+}.

The use of CrO_3 in aqueous acetone is usually called the **Jones oxidation** (or oxidation by the Jones reagent). This procedure rarely affects double bonds present in the molecule.

15.3D Oxidation of 3° Alcohols

Tertiary alcohols can be oxidized, but only under very forcing conditions. Tertiary alcohols are not easily oxidized because they do not have a hydrogen atom attached to the carbon atom that bears the —OH group. Oxidations of tertiary alcohols, when they do occur, involve cleavage of the carbon–carbon bond and are of little synthetic utility.

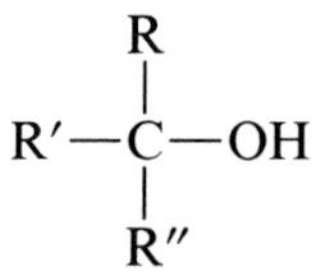

Difficult to oxidize

3° Alcohol

15.3E A Chemical Test for Primary and Secondary Alcohols

The relative ease of oxidation of primary and secondary alcohols compared with the difficulty of oxidizing tertiary alcohols forms the basis for a convenient chemical test.

*It is the color change that accompanies this change in oxidation state that allows chromic acid to be used as a test for primary and secondary alcohols (Section 15.3E).

Primary and secondary alcohols are rapidly oxidized by a solution of CrO_3 in aqueous sulfuric acid. Chromic oxide (CrO_3) dissolves in aqueous sulfuric acid to give a clear orange solution containing $Cr_2O_7^{2-}$ ions. A positive test is indicated when this clear orange solution becomes opaque and takes on a greenish cast within 2 seconds.

RCH$_2$OH
 or + CrO$_3$/aqueous H$_2$SO$_4$ $\longrightarrow$ greenish opaque solution containing Cr^{3+} and oxidation products
RCHOH
|
R

$\underbrace{\hspace{6cm}}$ $\underbrace{\hspace{6cm}}$
Clear orange solution   Greenish-opaque solution

Not only will this test distinguish primary and secondary alcohols from tertiary alcohols, it will distinguish primary and secondary alcohols from most other compounds except aldehydes.

PROBLEM 15.5

Show how each of the following transformations could be accomplished.

(a)

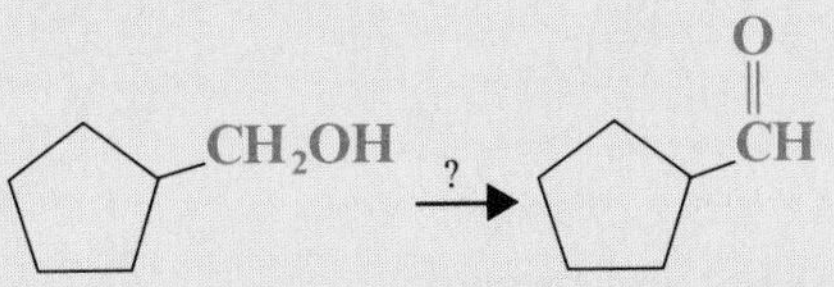

(b)

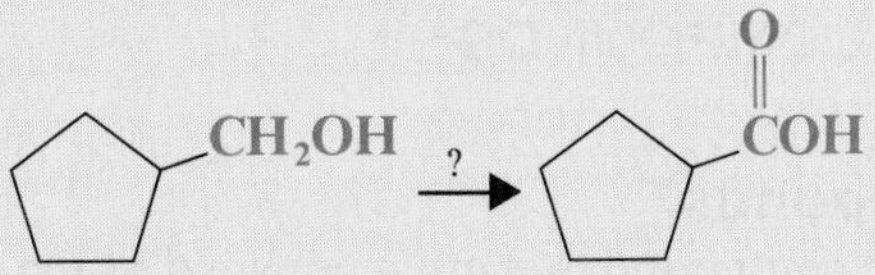

(c)

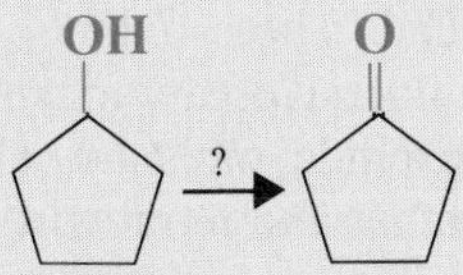

(d)

15.4 ORGANOMETALLIC COMPOUNDS

Compounds that contain carbon–metal bonds are called **organometallic compounds.** The natures of the carbon–metal bonds vary widely, ranging from bonds that are essentially ionic to those that are primarily covalent. While the structure of the organic portion of the organometallic compound has some effect on the nature of the carbon–metal bond, the identity of the metal itself is of far greater importance. Carbon–sodium and carbon–potassium bonds are largely ionic in character; carbon–lead, carbon–tin, carbon–thallium, and carbon–mercury bonds are essentially covalent. Carbon–lithium and carbon–magnesium bonds lie between these extremes.

$$
\begin{array}{ccc}
\overset{|}{-}\!\underset{|}{C}\!:^{-}\overset{+}{M} & \overset{|}{-}\!\underset{|}{C}\!:\!M \\
\text{Primarily ionic} & (M = Mg \text{ or } Li) & \overset{|}{-}\!\underset{|}{C}\!-\!M \\
(M = Na^{+} \text{ or } K^{+}) & & \text{Primarily covalent} \\
& & (M = Pb, Sn, Hg, \text{ or } Tl)
\end{array}
$$

The reactivity of organometallic compounds increases with the percent ionic character of the carbon–metal bond. Alkylsodium and alkylpotassium compounds are highly reactive and are among the most powerful of bases. They react explosively with water and burst into flame when exposed to air. Organomercury and -lead compounds are much less reactive; they are often volatile and are stable in air. They are all poisonous. They are generally soluble in nonpolar solvents. Tetraethyllead, for example, is used as an "antiknock" compound in gasoline.

Organometallic compounds of lithium and magnesium are of great importance in organic synthesis. They are relatively stable in ether solutions, but their carbon–metal bonds have considerable ionic character. Because of this ionic nature, the carbon atom of an organolithium or organomagnesium compound that is bonded to the metal atom is a strong base and powerful nucleophile. We shall soon see reactions that illustrate both of these properties.

15.5 PREPARATION OF ORGANOLITHIUM AND ORGANOMAGNESIUM COMPOUNDS

15.5A Organolithium Compounds

Organolithium and organomagnesium compounds are often prepared by the reduction of organic halides with lithium or magnesium metal. These reductions are usually carried out in ether solvents, and since organolithium and organomagnesium compounds are strong bases, care must be taken to exclude moisture. (Why?) The ethers most commonly used as solvents are diethyl ether and tetrahydrofuran. (Tetrahydrofuran is a cyclic ether.)

$$CH_3CH_2\overset{..}{\underset{..}{O}}CH_2CH_3$$

Diethyl ether
(Et$_2$O)

$$
\begin{array}{c}
H_2C\!-\!\!-\!\!CH_2 \\
|\qquad\quad| \\
H_2C\qquad CH_2 \\
\diagdown\underset{..}{\overset{..}{O}}\diagup
\end{array}
$$

Tetrahydrofuran
(THF)

For example, butyl bromide reacts with lithium metal in diethyl ether to give a solution of butyllithium.

$$CH_3CH_2CH_2CH_2Br + 2Li \xrightarrow[\text{Et}_2\text{O}]{-10\ °C} CH_3CH_2CH_2CH_2Li + LiBr$$

Butyl bromide
(80–90%)
Butyllithium

Other organolithium compounds, such as methyllithium, ethyllithium, and phenyllithium, can be prepared in the same general way.

$$\underset{\text{(or Ar—X)}}{\text{R—X}} + 2\text{Li} \xrightarrow{\text{ether}} \underset{\text{(or ArLi)}}{\text{RLi}} + \text{LiX}$$

The order of reactivity of halides is $RI > RBr > RCl$. (Alkyl and aryl fluorides are seldom used in the preparation of organolithium compounds.)

> Most organolithium compounds slowly attack ethers by bringing about an elimination reaction.

$$\overset{\delta-}{R} \overset{\delta+}{:Li} + H{-}CH_2{-}CH_2{-}OCH_2CH_3 \longrightarrow RH + CH_2{=}CH_2 + \overset{+}{Li}\overset{-}{O}CH_2CH_3$$

> For this reason, ether solutions of organolithium reagents are not usually stored but are used immediately after preparation. Organolithium compounds are much more stable in hydrocarbon solvents. Several alkyl- and aryllithium reagents are commercially available in hexane, paraffin wax, or mineral oil.

15.5B Grignard Reagents

Organomagnesium halides were discovered by the French chemist Victor Grignard in 1900. Grignard received the Nobel Prize for his discovery in 1912, and organomagnesium halides are now called **Grignard reagents** in his honor. Grignard reagents have great use in organic synthesis.

Grignard reagents are usually prepared by the reaction of an organic halide and magnesium metal (turnings) in an ether solvent.

$$\left. \begin{array}{l} \text{RX} + \text{Mg} \xrightarrow{\text{ether}} \text{RMgX} \\[2mm] \text{ArX} + \text{Mg} \xrightarrow{\text{ether}} \text{ArMgX} \end{array} \right\} \begin{array}{l} \text{Grignard} \\ \text{reagents} \end{array}$$

The order of reactivity of halides with magnesium is also $RI > RBr > RCl$. Very few organomagnesium fluorides have been prepared. Aryl Grignard reagents are more easily prepared from aryl bromides and aryl iodides than from aryl chlorides, which react very sluggishly.

Grignard reagents are seldom isolated but are used for further reactions in ether solution. The ether solutions can be analyzed for the content of the Grignard reagent, however, and the yields of Grignard reagents are almost always very high (85–95%). Two examples are shown here.

$$\text{CH}_3\text{I} + \text{Mg} \xrightarrow[\text{35 °C}]{\text{Et}_2\text{O}} \underset{\substack{(95\%) \\ \text{Methylmagnesium} \\ \text{iodide}}}{\text{CH}_3\text{MgI}}$$

$$\text{C}_6\text{H}_5\text{Br} + \text{Mg} \xrightarrow[\text{35 °C}]{\text{Et}_2\text{O}} \underset{\substack{(95\%) \\ \text{Phenylmagnesium} \\ \text{bromide}}}{\text{C}_6\text{H}_5\text{MgBr}}$$

The actual structures of Grignard reagents are more complex than the general formula RMgX indicates. Experiments done with radioactive magnesium have established that, for most Grignard reagents, there is an equilibrium between an alkylmagnesium halide and a dialkylmagnesium.

$$2RMgX \rightleftharpoons R_2Mg + MgX_2$$

$$\underset{\text{halide}}{\underset{\text{Alkylmagnesium}}{}} \qquad \underset{\text{Dialkylmagnesium}}{}$$

For convenience in this text, however, we shall write the formula for the Grignard reagent as though it were simply RMgX.

A Grignard reagent forms a complex with its ether solvent; the structure of the complex can be represented as follows:

$$
\begin{array}{ccc}
R & & R \\
 \diagdown \ddot{} \diagup & \\
 & \underset{\ddot{}}{O} & \\
R-&Mg&-X \\
 & \underset{\ddot{}}{O} & \\
 \diagup \ddot{} \diagdown & \\
R & & R
\end{array}
$$

Complex formation with molecules of ether is an important factor in the formation and stability of Grignard reagents. Organomagnesium compounds can be prepared in nonethereal solvents, but the preparations are more difficult.

The mechanism by which Grignard reagents are formed is still not fully understood. The most likely path, however, appears to be the following two-step radical mechanism.

$$R-X + :Mg \longrightarrow R\cdot + \cdot MgX$$

$$R\cdot + \cdot MgX \longrightarrow RMgX$$

15.6 REACTIONS OF ORGANOLITHIUM AND ORGANOMAGNESIUM COMPOUNDS

15.6A Reactions with Compounds Containing Acidic Hydrogen Atoms

Grignard reagents and organolithium compounds are very strong bases. They react with any compound that has a hydrogen more acidic than the hydrogen atoms of the hydrocarbon from which the Grignard reagent or organolithium is derived. We can understand how these reactions occur if we represent the Grignard reagent and organolithium compounds in the following ways:

$$\overset{\delta-}{R} \overset{\delta+}{:MgX} \quad \text{and} \quad \overset{\delta-}{R} \overset{\delta+}{:Li}$$

When we do this, we can see that the reactions of Grignard reagents with water and alcohols are nothing more than acid–base reactions; they lead to the formation of the weaker conjugate acid and weaker conjugate base. The Grignard reagent behaves as if it contained the anion of an alkane, *as if it contained a carbanion.*

$$\overset{\delta-}{R}:\overset{\delta+}{MgX} + \overset{..}{H}:\overset{..}{O}H \longrightarrow R:H + \overset{..}{H}\overset{..}{O}: \quad + Mg^{2+} + X^-$$

Grignard	Water	Alkane	Hydroxide ion
reagent	(stronger	(weaker	(weaker
(stronger	acid)	acid)	base)
base)			

$$\overset{\delta-}{R}:\overset{\delta+}{MgX} + \overset{..}{H}:\overset{..}{O}R \longrightarrow R:H + \overset{..}{R}\overset{..}{O}: \quad + Mg^{2+} + X^-$$

Grignard	Alcohol	Alkane	Alkoxide ion
reagent	(stronger	(weaker	(weaker
(stronger	acid)	acid)	base)
base)			

PROBLEM 15.6

Write similar equations for the reactions that take place when butyllithium is treated with (a) water, (b) ethanol. Designate the stronger and weaker acids and the stronger and weaker bases.

PROBLEM 15.7

Assuming you have *tert*-butyl bromide, magnesium, dry ether, and deuterium oxide (D_2O) available, show how you might synthesize the following deuterium-labeled alkane.

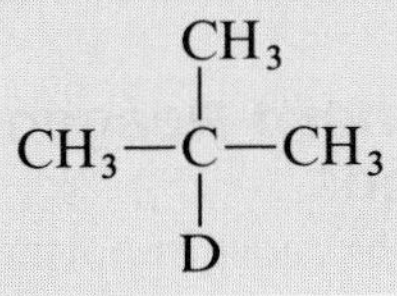

$$\begin{array}{c} CH_3 \\ | \\ CH_3-C-CH_3 \\ | \\ D \end{array}$$

Grignard reagents and organolithium compounds abstract protons that are much less acidic than those of water and alcohols. They react with the terminal hydrogen atoms of 1-alkynes, for example, and this is a useful method for the preparation of alkynylmagnesium halides and alkynyllithiums. These reactions are also acid–base reactions.

$$RC{\equiv}CH + \overset{\delta-}{R'}:\overset{\delta+}{MgX} \longrightarrow RC{\equiv}\overset{\delta-}{C}:\overset{\delta+}{MgX} + R':H$$

Terminal	Grignard	Alkynylmagnesium	Alkane
alkyne	reagent	halide	(weaker
(stronger	(stronger	(weaker	acid)
acid)	base)	base)	

$$R{-}C{\equiv}CH + \overset{\delta-}{R'}:\overset{\delta+}{Li} \longrightarrow R{-}C{\equiv}\overset{\delta-}{C}:\overset{\delta+}{Li} + R':H$$

Terminal	Alkyl-	Alkynyllithium	Alkane
alkyne	lithium	(weaker	(weaker
(stronger	(stronger	base)	acid)
acid)	base)		

The fact that these reactions go to completion is not surprising when we recall that the acidity constants of alkanes are of the order of $10^{-45} - 10^{-50}$, while those of terminal alkynes are 10^{-25} (Table 2.2).

Grignard reagents are not only strong bases, they are also *powerful nucleophiles.* Reactions in which Grignard reagents act as nucleophiles are by far the most important. At this point, let us consider general examples that illustrate the ability of a Grignard reagent to act as a nucleophile by attacking saturated and unsaturated carbon atoms.

15.6B Reactions of Grignard Reagents with Ethylene Oxide

Grignard reagents carry out nucleophilic attack at a saturated carbon when they react with ethylene oxide. These reactions take the general form shown here and give us a convenient synthesis of primary alcohols.

The nucleophilic alkyl group of the Grignard reagent attacks the partially positive carbon of the ethylene oxide ring. Because it is highly strained, the ring opens, and the reaction leads to the salt of a primary alcohol. Subsequent acidification produces the alcohol.

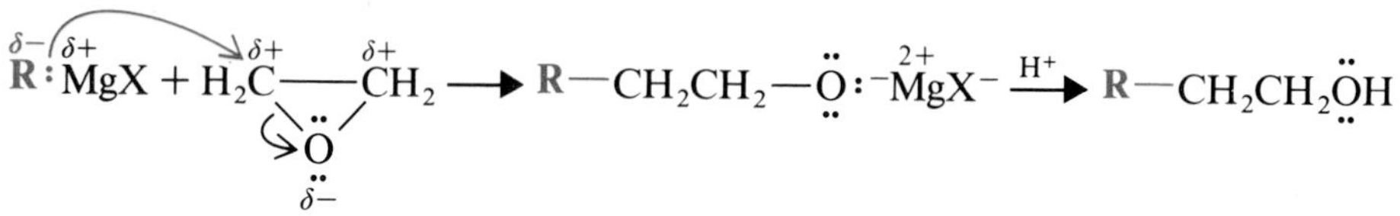

$$R:MgX + H_2C\!-\!\!-\!\!-\!CH_2 \longrightarrow R\!-\!CH_2CH_2\!-\!\ddot{O}:^-\overset{2+}{M}gX^- \overset{H^+}{\longrightarrow} R\!-\!CH_2CH_2\ddot{O}H$$

Ethylene oxide   A primary alcohol

15.6C Reactions of Grignard Reagents with Carbonyl Compounds

From a synthetic point of view, the most important reactions of Grignard reagents and organolithium compounds are those in which these reagents act as nucleophiles and attack an unsaturated carbon — *especially the carbon of a carbonyl group.*

The carbon atom of a carbonyl group is positively charged because the electronegativity of the oxygen atom causes the second structure shown here to make a large contribution to the resonance hybrid.

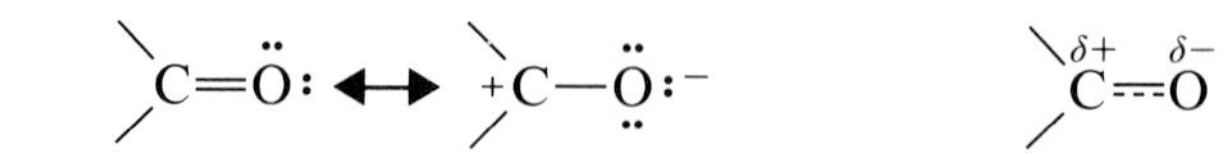

$$\diagdown\!\!\!\diagup C\!=\!\ddot{O}: \longleftrightarrow ^+\!C\!-\!\ddot{O}:^- \qquad \overset{\delta+\quad\delta-}{\diagdown\!\!\!\diagup C\!=\!\!=\!O}$$

Resonance structures of the carbonyl group   Hybrid

Compounds that contain carbonyl groups are, therefore, highly susceptible to nucleophilic attack. Grignard reagents react with carbonyl compounds (aldehydes and ketones) in the following way:

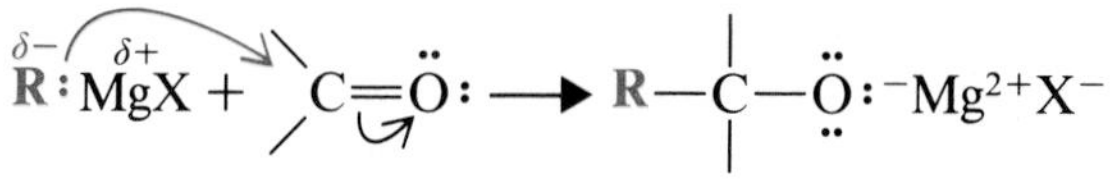

$$R:MgX + \diagdown\!\!\!\diagup C\!=\!\ddot{O}: \longrightarrow R\!-\!\overset{|}{\underset{|}{C}}\!-\!\ddot{O}:^-Mg^{2+}X^-$$

This reaction is a nucleophilic addition to the carbon–oxygen double bond. The nucleophilic carbon of the Grignard reagent uses its electron pair to form a bond to the carbonyl carbon. The carbonyl carbon can accept this electron pair because one pair of electrons of the carbon–oxygen double bond can shift out to the oxygen.

The product formed when a Grignard reagent adds to a carbonyl group is an alkoxide ion $R-\overset{|}{\underset{|}{C}}-\overset{..}{\underset{..}{O}}:^-$ that is associated with $Mg^{2+}X^-$. When water or dilute acid is added to the reaction mixture after the Grignard addition is over, an acid–base reaction takes place to produce an alcohol.

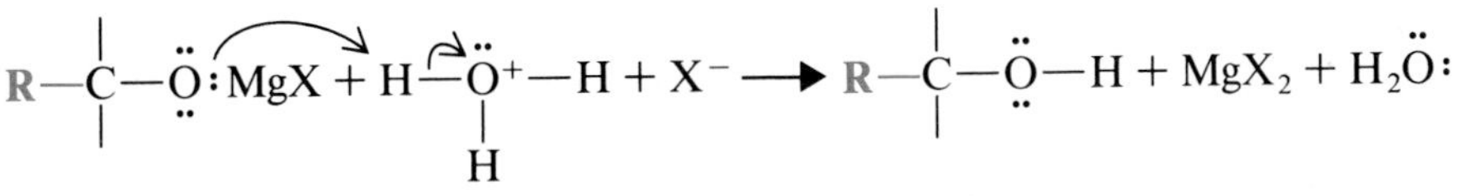

$$R-\overset{|}{\underset{|}{C}}-\overset{..}{\underset{..}{O}}:MgX + H-\overset{+}{\overset{..}{\underset{|}{O}}}-H + X^- \longrightarrow R-\overset{|}{\underset{|}{C}}-\overset{..}{\underset{..}{O}}-H + MgX_2 + H_2\overset{..}{\underset{..}{O}}:$$
$$\quad\quad\quad\quad\quad\quad\quad\quad\quad H$$

Magnesium halide
alkoxide

Alcohol

15.7 ALCOHOLS FROM GRIGNARD REAGENTS

Grignard additions to carbonyl compounds are especially useful because they can be used to prepare primary, secondary, or tertiary alcohols.

1. A Grignard reagent reacts with formaldehyde, for example, to give a **primary alcohol.**

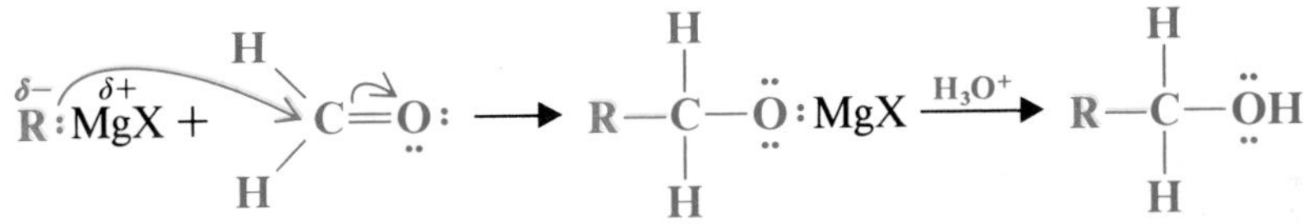

Formaldehyde   1° Alcohol

2. Grignard reagents react with higher aldehydes to give **secondary alcohols.**

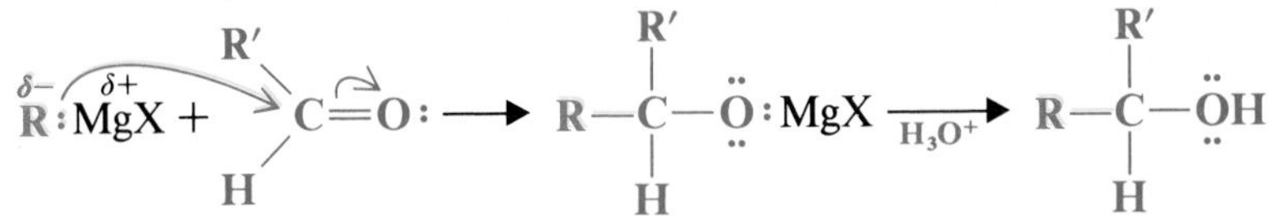

Higher
aldehyde   2° Alcohol

3. And Grignard reagents react with ketones to give **tertiary alcohols.**

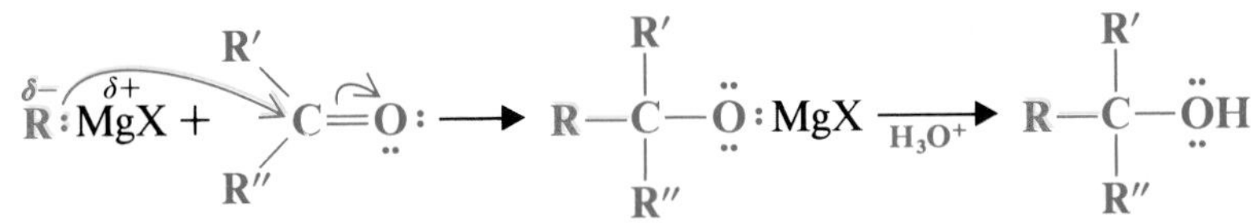

Ketone   3° Alcohol

Specific examples of these reactions are shown here.

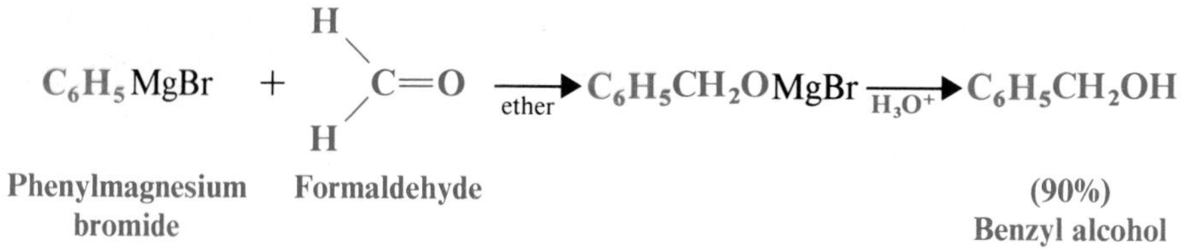

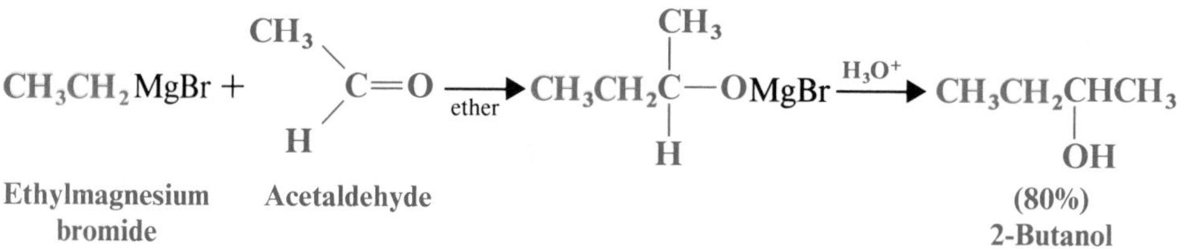

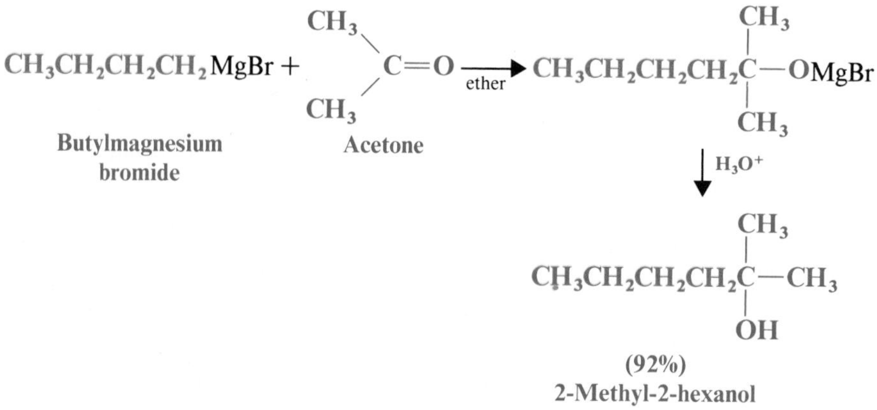

$$C_6H_5MgBr \ + \ \underset{H}{\overset{H}{>}}C=O \xrightarrow[\text{ether}]{} C_6H_5CH_2OMgBr \xrightarrow[H_3O^+]{} C_6H_5CH_2OH$$

Phenylmagnesium   Formaldehyde   (90%)
    bromide   Benzyl alcohol

$$CH_3CH_2MgBr + \underset{H}{\overset{CH_3}{>}}C=O \xrightarrow[\text{ether}]{} CH_3CH_2\underset{H}{\overset{CH_3}{\underset{|}{\overset{|}{C}}}}-OMgBr \xrightarrow{H_3O^+} CH_3CH_2\underset{OH}{\overset{|}{CHCH_3}}$$

Ethylmagnesium   Acetaldehyde   (80%)
    bromide   2-Butanol

$$CH_3CH_2CH_2CH_2MgBr + \underset{CH_3}{\overset{CH_3}{>}}C=O \xrightarrow[\text{ether}]{} CH_3CH_2CH_2CH_2\underset{CH_3}{\overset{CH_3}{\underset{|}{\overset{|}{C}}}}-OMgBr$$

Butylmagnesium   Acetone
    bromide

$$\Big\downarrow H_3O^+$$

$$CH_3CH_2CH_2CH_2\underset{OH}{\overset{CH_3}{\underset{|}{\overset{|}{C}}}}-CH_3$$

(92%)
2-Methyl-2-hexanol

4. A Grignard reagent also adds to the carbonyl group of an ester. The initial product is unstable and it loses a magnesium alkoxide to form a ketone. Ketones are more reactive toward Grignard reagents than esters. Therefore as soon as a molecule of the ketone is formed in the mixture, it reacts with a second molecule of the Grignard reagent. After hydrolysis, **the product is a tertiary alcohol with two identical alkyl groups,** groups that correspond to the alkyl portion of the Grignard reagent.

$$R\!:\!MgX + \underset{R''\ddot{O}}{\overset{R'}{>}}C=\ddot{O}: \longrightarrow \left[R-\underset{:O-R''}{\overset{R'}{\underset{|}{\overset{|}{C}}}}-\ddot{O}-MgX \right] \xrightarrow[\text{spontaneously}]{-R''OMgX}$$

                 Ester   Initial product
                                       (unstable)

$$\left[\underset{R}{\overset{R'}{>}}C=\ddot{O}: \right] \xrightarrow{RMgX} R-\underset{R}{\overset{R'}{\underset{|}{\overset{|}{C}}}}-\ddot{O}MgX \xrightarrow{H_3O^+} R-\underset{R}{\overset{R'}{\underset{|}{\overset{|}{C}}}}-OH$$

      A ketone   Salt of an   3° Alcohol
                             alcohol
                          (not isolated)

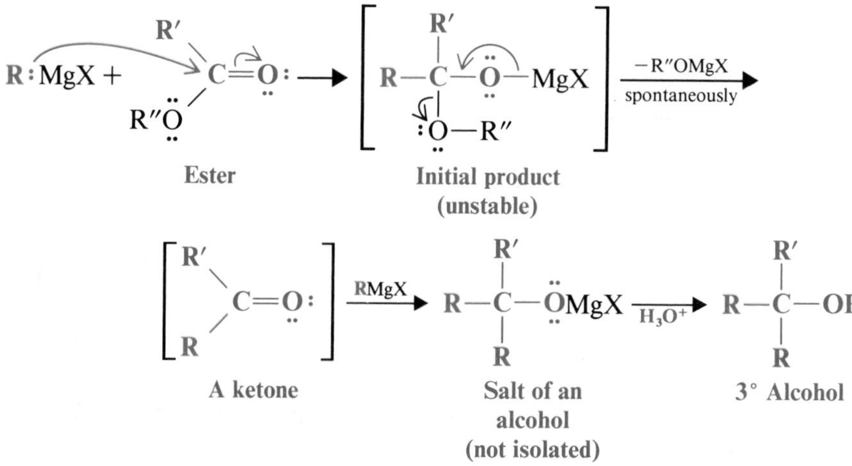

A specific example of this reaction is shown here.

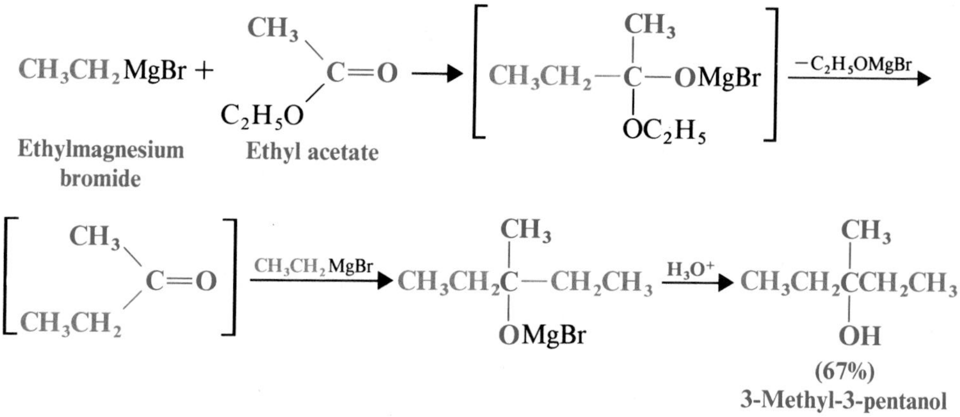

$$CH_3CH_2MgBr + \underset{\underset{\text{Ethyl acetate}}{C_2H_5O}}{\overset{CH_3}{\diagdown}}C{=}O \longrightarrow \left[CH_3CH_2{-}\underset{OC_2H_5}{\overset{CH_3}{\underset{|}{\overset{|}{C}}}}{-}OMgBr \right] \xrightarrow{-C_2H_5OMgBr}$$

Ethylmagnesium
bromide

$$\left[\underset{CH_3CH_2}{\overset{CH_3}{\diagdown}}C{=}O \right] \xrightarrow{CH_3CH_2MgBr} CH_3CH_2\underset{OMgBr}{\overset{CH_3}{\underset{|}{\overset{|}{C}}}}{-}CH_2CH_3 \xrightarrow{H_3O^+} CH_3CH_2\overset{CH_3}{\underset{OH}{\overset{|}{\underset{|}{C}}}}CH_2CH_3$$

(67%)
3-Methyl-3-pentanol

PROBLEM 15.8

Phenylmagnesium bromide reacts with benzoyl chloride, $C_6H_5\overset{\overset{O}{\parallel}}{C}Cl$, to form triphenylmethanol, $(C_6H_5)_3COH$. This reaction is typical of the reaction of Grignard reagents with acyl chlorides, and the mechanism is similar to that for the reaction of a Grignard reagent with an ester just shown. Show the steps that lead to the formation of triphenylmethanol.

15.7A Planning a Grignard Synthesis

By using Grignard synthesis skillfully we can synthesize almost any alcohol we wish. In planning a Grignard synthesis we must simply choose the correct Grignard reagent and the correct aldehyde, ketone, ester, or epoxide. We do this by examining the alcohol we wish to prepare and by paying special attention to the groups attached to the carbon atom bearing the —OH group. Many times there may be more than one way of carrying out the synthesis. In these cases our final choice will probably be dictated by the availability of starting compounds. Let us consider an example.

EXAMPLE

Suppose we want to prepare 3-phenyl-3-pentanol. We examine its structure and we see that the groups attached to the carbon atom bearing the —OH are a *phenyl group*

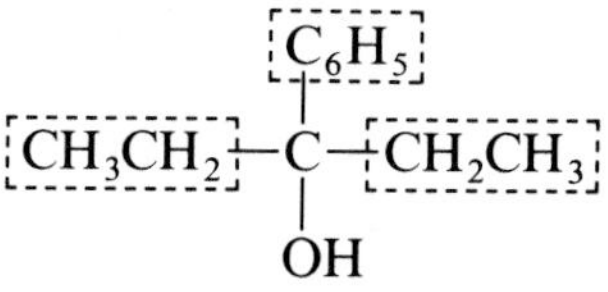

$$CH_3CH_2{-}\underset{OH}{\overset{C_6H_5}{\underset{|}{\overset{|}{C}}}}{-}CH_2CH_3$$

3-Phenyl-3-pentanol

and *two ethyl groups.* This means that we can synthesize this compound in different ways.

1. We can use a ketone with two ethyl groups (3-pentanone) and allow it to react with phenylmagnesium bromide:

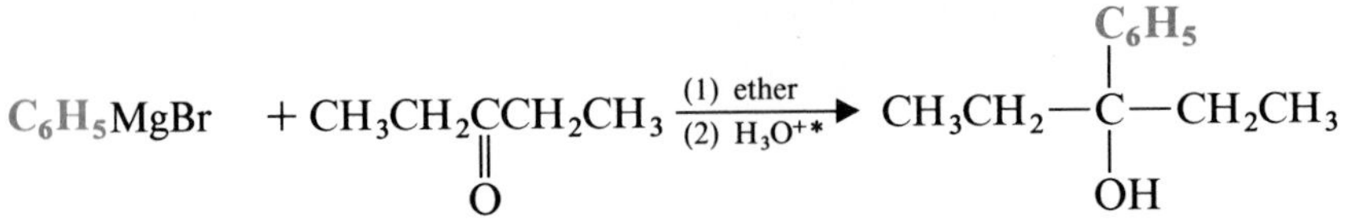

$$C_6H_5MgBr \quad + CH_3CH_2CCH_2CH_3 \xrightarrow[\text{(2) } H_3O^{+*}]{\text{(1) ether}} \quad CH_3CH_2-\overset{\overset{\displaystyle C_6H_5}{|}}{\underset{\underset{\displaystyle OH}{|}}{C}}-CH_2CH_3$$

Phenylmagnesium   3-Pentanone   3-Phenyl-3-pentanol
bromide

2. We can use a ketone containing an ethyl group and a phenyl group (ethyl phenyl ketone) and allow it to react with ethylmagnesium bromide:

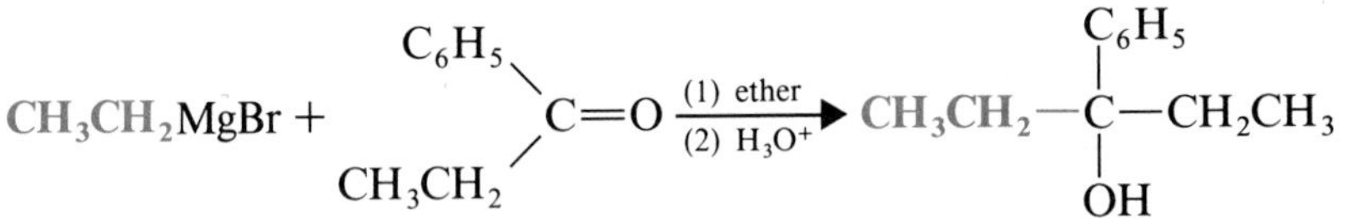

$$CH_3CH_2MgBr + \quad \underset{CH_3CH_2}{\overset{C_6H_5}{>}}C=O \xrightarrow[\text{(2) } H_3O^+]{\text{(1) ether}} CH_3CH_2-\overset{\overset{\displaystyle C_6H_5}{|}}{\underset{\underset{\displaystyle OH}{|}}{C}}-CH_2CH_3$$

Ethylmagnesium   Ethyl phenyl   3-Phenyl-3-pentanol
bromide   ketone

3. Or we can use an ester of benzoic acid and allow it to react with 2 molar equivalents of ethylmagnesium bromide:

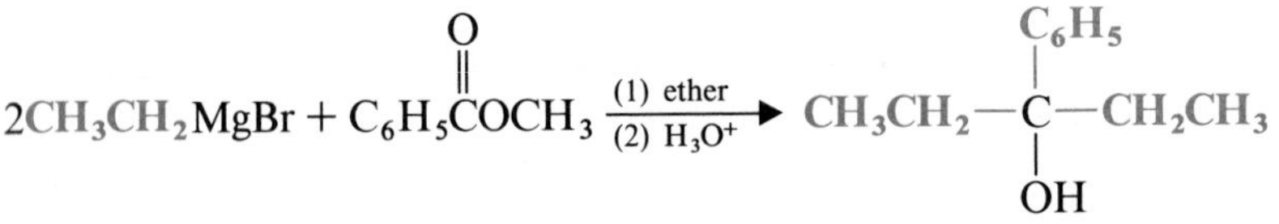

$$2CH_3CH_2MgBr + C_6H_5\overset{\overset{\displaystyle O}{||}}{C}OCH_3 \xrightarrow[\text{(2) } H_3O^+]{\text{(1) ether}} CH_3CH_2-\overset{\overset{\displaystyle C_6H_5}{|}}{\underset{\underset{\displaystyle OH}{|}}{C}}-CH_2CH_3$$

Ethylmagnesium   Methyl   3-Phenyl-3-pentanol
bromide   benzoate

All of these methods will be likely to give us our desired compound in yields greater than 80%. ■

SAMPLE PROBLEM
Illustrating a Multistep Synthesis
Using an alcohol of no more than four carbon atoms as your only organic starting material, outline a synthesis of **A**.

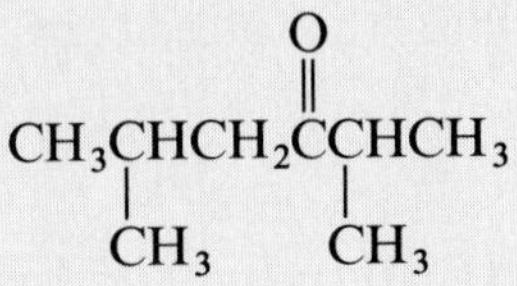

$$\underset{\underset{\displaystyle CH_3}{|}}{CH_3CHCH_2}\overset{\overset{\displaystyle O}{||}}{C}\underset{\underset{\displaystyle CH_3}{|}}{CHCH_3}$$

A

*By writing (2) H_3O^+ under the arrow in equations like these, we mean that in a second step, after the Grignard reagent has reacted with the ketone, we add dilute aqueous acid to convert the salt of the alcohol (ROMgX) to the alcohol itself.

Answer:

We can construct the carbon skeleton from two four-carbon atom compounds using a Grignard reaction. Then oxidation of the alcohol produced will yield the desired ketone.

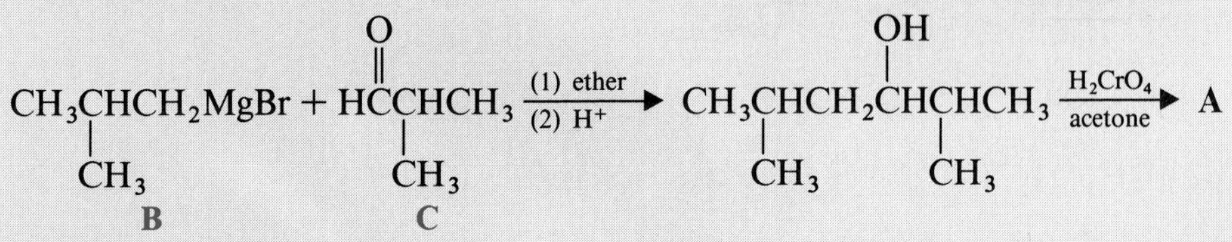

$$\underset{\underset{B}{\overset{CH_3}{|}}}{CH_3CHCH_2MgBr} + \underset{\underset{C}{\overset{CH_3}{|}}}{\overset{\overset{O}{\parallel}}{HCCHCH_3}} \xrightarrow[\text{(2) } H^+]{\text{(1) ether}} \underset{\overset{|}{CH_3} \quad \overset{|}{CH_3}}{CH_3CHCH_2CHCHCH_3} \xrightarrow[\text{acetone}]{H_2CrO_4} A$$

We can synthesize the Grignard reagent (**B**) and the aldehyde (**C**) from isobutyl alcohol.

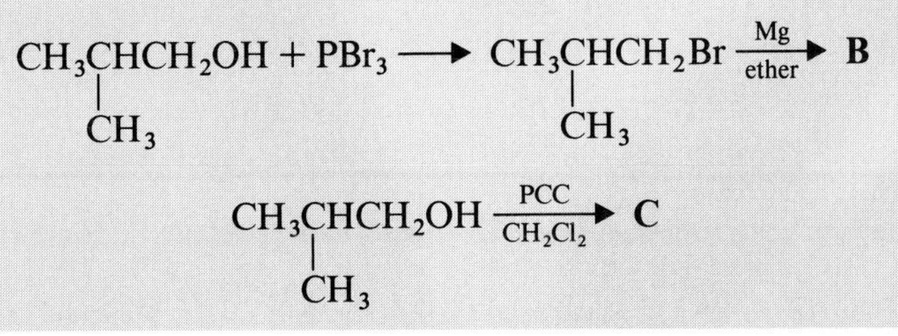

$$\underset{\overset{|}{CH_3}}{CH_3CHCH_2OH} + PBr_3 \longrightarrow \underset{\overset{|}{CH_3}}{CH_3CHCH_2Br} \xrightarrow[\text{ether}]{Mg} B$$

$$\underset{\overset{|}{CH_3}}{CH_3CHCH_2OH} \xrightarrow[\text{CH}_2\text{Cl}_2]{PCC} C$$

SAMPLE PROBLEM

Illustrating a Multistep Synthesis

Starting with benzene and using any other needed reagents, outline a synthesis of 2-phenylethanol ($C_6H_5CH_2CH_2OH$).

Answer:

Working backwards, we remember (Section 15.6B) that we can synthesize 2-phenylethanol by treating phenylmagnesium bromide with ethylene oxide. (Adding ethylene oxide to a Grignard reagent is a very useful method for adding a —CH_2CH_2OH unit to an organic group.)

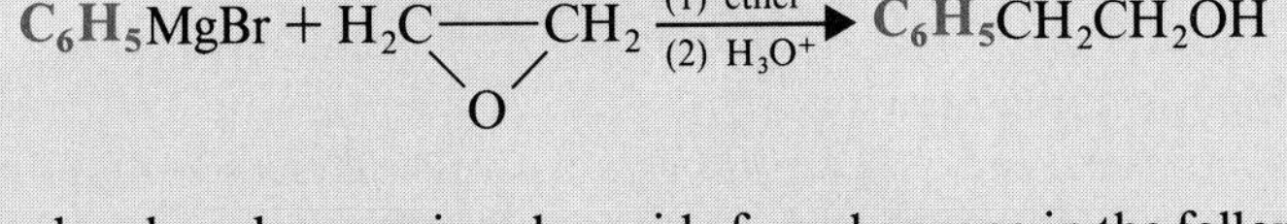

$$C_6H_5MgBr + H_2C\underset{\overset{\diagdown}{O}\diagup}{----}CH_2 \xrightarrow[\text{(2) } H_3O^+]{\text{(1) ether}} C_6H_5CH_2CH_2OH$$

We can make phenylmagnesium bromide from benzene in the following way:

$$C_6H_5-H \xrightarrow{Br_2, FeBr_3} C_6H_5Br \xrightarrow[\text{ether}]{Mg} C_6H_5MgBr$$

PROBLEM 15.9

Show how Grignard reactions could be used to synthesize each of the following compounds. (You must start with a Grignard reagent and you may use any other compounds needed.)

(a) *tert*-Butyl alcohol   (two ways)

(b) $CH_3CH_2CH_2CHOHCH_3$ (two ways)

$$\overset{\displaystyle CH_3}{\underset{\displaystyle OH}{\text{(c) } C_6H_5\overset{|}{\underset{|}{C}}CH_2CH_3}}$$ (three ways)   .

(d) $CH_3CH_2CH_2CH_2CH_2CH_2OH$ (two ways)

PROBLEM 15.10

Outline synthesis of each of the following. Permitted starting materials are benzene, ethylene oxide, and alcohols or esters of four carbon atoms or fewer. You may use any inorganic reagents and oxidizing agents such as pyridinium chlorochromate (PCC).

$$\text{(a) } C_6H_5\overset{O}{\overset{\|}{\underset{\underset{\displaystyle CH_3}{|}}{C}}H}CH \qquad \text{(c) } C_6H_5\overset{O}{\overset{\|}{C}}H \qquad \text{(e) } C_6H_5\overset{OH}{\overset{|}{\underset{\underset{\displaystyle CH_3}{|}}{C}}H}CHCH_3$$

$$\text{(b) } C_6H_5\overset{H}{\overset{|}{\underset{\underset{\displaystyle OH}{|}}{C}}}CH_2CH_3 \qquad \text{(d) } C_6H_5\overset{OH}{\overset{|}{\underset{\underset{\displaystyle C_6H_5}{|}}{C}}}CH_2CH_3$$

15.7B Restrictions on the Use of Grignard Reagents

While the Grignard synthesis is one of the most versatile of all general synthetic procedures, it is not without its limitations. Most of these limitations arise from the very feature of the Grignard reagent that makes it so useful, its *extraordinary reactivity as a nucleophile and a base.*

The Grignard reagent is a very powerful base; in effect it contains a carbanion. Thus, it is not possible to prepare a Grignard reagent from an organic group that contains an *acidic hydrogen;* and by an acidic hydrogen we mean any hydrogen more acidic than the hydrogen atoms of an alkane or alkene. We cannot, for example, prepare a Grignard reagent from a compound containing an —OH group, an —NH— group, an —SH group, a —CO_2H group, or an —SO_3H group. If we were to attempt to prepare a Grignard reagent from an organic halide containing any of these groups, the formation of the Grignard reagent would simply fail to take place. (Even if a Grignard reagent were to form, it would immediately react with the acidic group.)

Since Grignard reagents are powerful nucleophiles we cannot prepare a Grignard reagent from any organic halide that contains a carbonyl, epoxy, nitro, or cyano (—CN) group. If we were to attempt to carry out this kind of reaction, any Grignard reagent that formed would only react with the unreacted starting material.

$$-OH, -NH_2, -NHR, -CO_2H, -SO_3H, -SH, -C\equiv C-H$$

$$\underset{\substack{\parallel \\ O}}{-CH}, \underset{\substack{\parallel \\ O}}{-CR}, \underset{\substack{\parallel \\ O}}{-COR}, \underset{\substack{\parallel \\ O}}{-CNH_2}, -NO_2, -C\equiv N, -\underset{\substack{\diagup \\ O}}{\overset{\mid \quad \mid}{C-C}}-$$

> Grignard reagents containing these groups cannot be prepared

This means that when we prepare Grignard reagents, we are effectively limited to alkyl halides or to analogous organic halides containing carbon–carbon double bonds, internal triple bonds, ether linkages, and $-NR_2$ *groups.*

Grignard reactions are so sensitive to acidic compounds that when we prepare a Grignard reagent we must take special care to exclude moisture from our apparatus, and we must use an anhydrous ether as our solvent.

As we saw earlier, acetylenic hydrogens are acidic enough to react with Grignard reagents. This is a limitation that we can use, however. We can make acetylenic Grignard reagents by allowing terminal alkynes to react with alkyl Grignard reagents (cf. Section 15.6A). We can then use these acetylenic Grignard reagents to carry out other syntheses. For example,

$$C_6H_5C\equiv CH + C_2H_5MgBr \longrightarrow C_6H_5C\equiv CMgBr + C_2H_6\uparrow$$

$$C_6H_5C\equiv CMgBr + C_2H_5\overset{\overset{\textstyle O}{\parallel}}{C}H \xrightarrow[(2)\ H^+]{} C_6H_5C\equiv C-\underset{\underset{\textstyle OH}{\mid}}{C}HC_2H_5$$

(52%)

When we plan Grignard syntheses we must also take care not to plan a reaction in which a Grignard reagent is treated with an aldehyde, ketone, or ester that contains an acidic group (other than when we deliberately let it react with a terminal alkyne). If we were to do this, the Grignard reagent would simply react as a base with the acidic hydrogen rather than react at the carbonyl or epoxide carbon as a nucleophile. If we were to treat 4-hydroxy-2-butanone with methylmagnesium bromide, for example, the following reaction would take place first,

$$CH_3MgBr + HOCH_2CH_2\overset{\overset{\textstyle O}{\parallel}}{C}CH_3 \longrightarrow CH_4\uparrow + BrMgOCH_2CH_2\overset{\overset{\textstyle O}{\parallel}}{C}CH_3$$

4-Hydroxy-2-butanone

rather than

$$CH_3MgBr + HOCH_2CH_2\overset{\overset{\textstyle O}{\parallel}}{C}CH_3 \xmapsto{\quad\times\quad} HOCH_2CH_2\overset{\overset{\textstyle CH_3}{\mid}}{\underset{\underset{\textstyle OMgBr}{\mid}}{C}}-CH_3$$

If we are prepared to waste one molar equivalent of the Grignard reagent, we can treat 4-hydroxy-2-butanone with two molar equivalents of the Grignard reagent and thereby get addition to the carbonyl group.

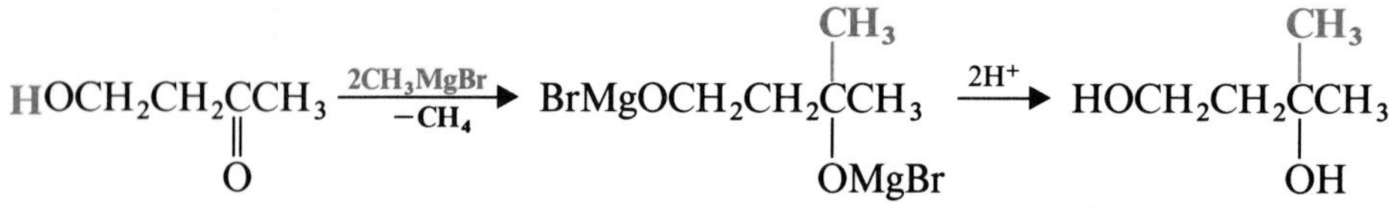

$$HOCH_2CH_2CCH_3 \xrightarrow[-CH_4]{2CH_3MgBr} BrMgOCH_2CH_2CCH_3 \xrightarrow{2H^+} HOCH_2CH_2CCH_3$$

This technique is sometimes employed in small-scale reactions when the Grignard reagent is inexpensive and the other reagent is expensive.

15.7C The Use of Lithium Reagents

Organolithium reagents (RLi) react with carbonyl compounds in the same way as Grignard reagents and thus provide an alternative method for preparing alcohols.

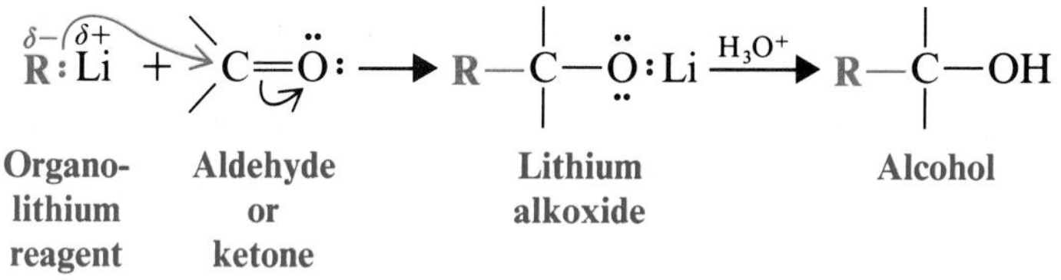

$$\overset{\delta-}{R}\!:\!\overset{\delta+}{Li} + \overset{}{\underset{}{>}}C=\ddot{O}\!: \longrightarrow R-\overset{|}{\underset{|}{C}}-\ddot{O}\!:\!Li \xrightarrow{H_3O^+} R-\overset{|}{\underset{|}{C}}-OH$$

| Organo-lithium reagent | Aldehyde or ketone | Lithium alkoxide | Alcohol |

Organolithium reagents have the advantage of being somewhat more reactive than Grignard reagents.

15.7D The Use of Sodium Alkynides

Sodium alkynides also react with aldehydes and ketones to yield alcohols. An example is the following:

$$CH_3C\equiv CH \xrightarrow[-NH_3]{NaNH_2} CH_3C\equiv CNa$$

Then,

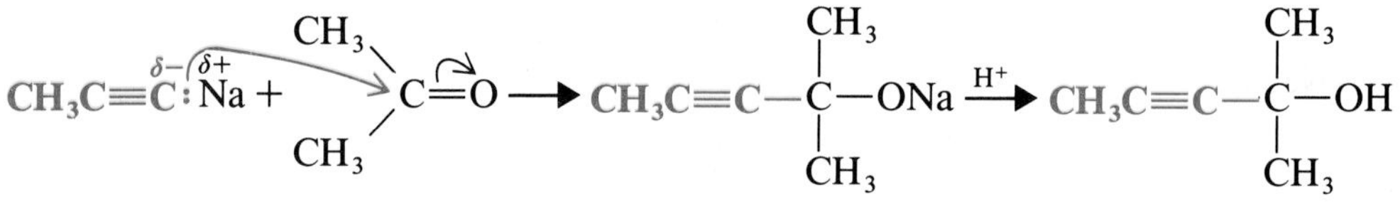

$$CH_3C\equiv\overset{\delta-}{C}\!:\!\overset{\delta+}{Na} + \underset{CH_3}{\overset{CH_3}{>}}C=O \longrightarrow CH_3C\equiv C-\underset{CH_3}{\overset{CH_3}{\overset{|}{C}}}-ONa \xrightarrow{H^+} CH_3C\equiv C-\underset{CH_3}{\overset{CH_3}{\overset{|}{C}}}-OH$$

SAMPLE PROBLEM
Illustrating a Multistep Syntheses

Starting with hydrocarbons, alcohols, aldehydes, ketones, or esters containing six carbon atoms or fewer and using any other needed reagents, outline a synthesis of each of the following:

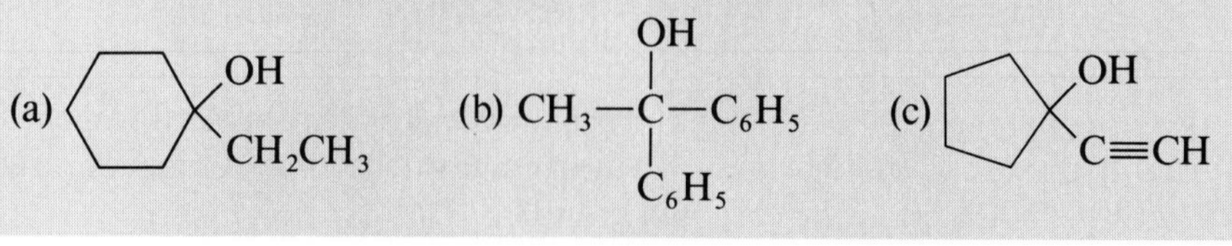

(a) cyclohexane with OH and CH$_2$CH$_3$ substituents

(b) $CH_3-\underset{C_6H_5}{\overset{OH}{\overset{|}{\underset{|}{C}}}}-C_6H_5$

(c) cyclopentane with OH and C≡CH substituents

Answers:

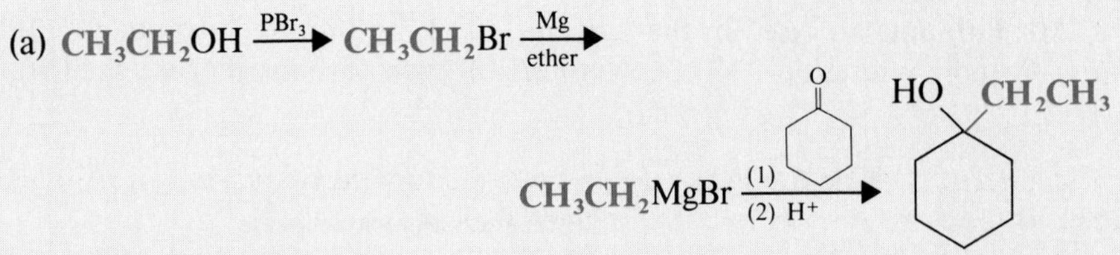

(a) $CH_3CH_2OH \xrightarrow{PBr_3} CH_3CH_2Br \xrightarrow[ether]{Mg}$

$CH_3CH_2MgBr \xrightarrow[(2) \ H^+]{(1)}$ [image of reaction with cyclohexanone giving 1-ethylcyclohexanol: HO, CH₂CH₃]

(b) $C_6H_6 \xrightarrow{Br_2 \ / Fe} C_6H_5Br \xrightarrow[ether]{Mg} C_6H_5MgBr \xrightarrow[(2) \ H^+]{(1) \ CH_3\overset{O}{\overset{\|}{C}}CH_3} CH_3-\underset{\underset{C_6H_5}{|}}{\overset{\overset{OH}{|}}{C}}-C_6H_5$

(c) $HC{\equiv}CH \xrightarrow{NaNH_2} HC{\equiv}CNa \xrightarrow[(2) \ H^+]{(1)}$ [image of reaction with cyclopentanone: HO, C≡CH]

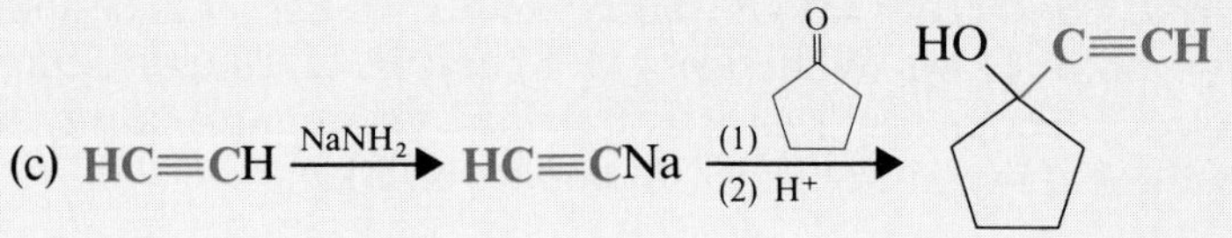

15.8 REACTION OF GRIGNARD AND ORGANOLITHIUM REAGENTS WITH LESS ELECTROPOSITIVE METAL HALIDES

Grignard reagents and alkyllithium reagents react with a number of halides of less electropositive elements to produce new organometallic compounds. Since Grignard reagents and alkyllithiums are easily prepared, these reactions furnish us with useful syntheses of alkyl derivatives of mercury, zinc, cadmium, copper, silicon, and phosphorus, for example. In general terms the reactions using Grignard reagents take the form shown here.

$$n RMgX + MX_n \longrightarrow R_nM + nMgX_2$$
(M is less electropositive than Mg)

Several specific examples are shown here.

$$2CH_3MgCl + HgCl_2 \longrightarrow \underset{\text{Dimethylmercury}}{(CH_3)_2Hg} + 2MgCl_2$$

$$2CH_3CH_2MgBr + ZnCl_2 \longrightarrow \underset{\substack{(100\%) \\ \text{Diethylzinc}}}{(CH_3CH_2)_2Zn} + 2MgClBr$$

$$2C_6H_5MgBr + CdCl_2 \longrightarrow \underset{\substack{(>83\%) \\ \text{Diphenylcadmium)}}}{(C_6H_5)_2Cd} + 2MgClBr$$

$$3CH_3CH_2CH_2CH_2MgBr + PCl_3 \longrightarrow \underset{\substack{(57\%) \\ \text{Tributylphosphine}}}{(CH_3CH_2CH_2CH_2)_3P} + 3MgClBr$$

$$4CH_3CH_2MgCl + SiCl_4 \longrightarrow \underset{\substack{(80\%) \\ \text{Tetraethylsilane}}}{(CH_3CH_2)_4Si} + 4MgCl_2$$

We shall see in Chapter 16 that organocadmium compounds are useful reagents for preparing ketones.

Alkyllithiums are used in the preparation of lithium dialkylcuprates for the Corey–Posner, Whitesides–House synthesis of hydrocarbons that we shall study next.

$$2CH_3Li + CuI \xrightarrow[Et_2O]{0\ °C} (CH_3)_2CuLi$$
$$\text{Lithium dimethylcuprate}$$

15.9 LITHIUM DIALKYLCUPRATES: THE COREY–POSNER, WHITESIDES–HOUSE SYNTHESIS

A highly versatile method for the synthesis of hydrocarbons from organic halides has been developed by E. J. Corey (Harvard University) and G. H. Posner (The Johns Hopkins University) and by G. M. Whitesides (MIT) and H. O. House (Georgia Institute of Technology). The overall synthesis provides, for example, a way for coupling the alkyl groups of two alkyl halides to produce an alkane:

$$R{-}X + R'{-}X \xrightarrow[\substack{\text{steps} \\ (-2X)}]{\text{several}} R{-}R'$$

In order to accomplish this coupling, we must transform one alkyl halide into a lithium dialkylcuprate (R_2CuLi). This transformation requires two steps. First, the alkyl halide is treated with lithium metal in an ether solvent to convert the alkyl halide into an alkyllithium (RLi).

$$R{-}X + 2Li \xrightarrow{\text{ether}} RLi + LiX$$
$$\text{Alkyllithium}$$

Then the alkyllithium is treated with cuprous iodide (CuI). This converts it to the lithium dialkylcuprate.

$$2RLi + CuI \longrightarrow R_2CuLi + LiI$$
$$\text{Alkyllithium} \qquad\qquad \text{Lithium} \\ \text{dialkylcuprate}$$

When the lithium dialkylcuprate is treated with the second alkyl halide ($R'{-}X$), coupling takes place between one alkyl group of the lithium dialkylcuprate and the alkyl group of the alkyl halide, $R'{-}X$.

$$R_2CuLi + R'{-}X \longrightarrow R{-}R' + RCu + LiX$$
$$\text{Lithium} \qquad \text{Alkyl halide} \qquad \text{Alkane} \\ \text{dialkylcuprate}$$

For the last step to give a good yield of the alkane, the alkyl halide $R'{-}X$ must be either a methyl halide, a primary alkyl halide, or a secondary cycloalkyl halide. The alkyl groups of the lithium dialkylcuprate may be methyl, 1°, 2°, or 3°.* Moreover, the two alkyl groups being coupled need not be different.

---

*Special techniques, which we shall not discuss here, are required when R is tertiary. For an excellent review of these reactions see Gary H. Posner, "Substitution Reactions Using Organocopper Reagents," *Organic Reactions,* Vol. 22, Wiley, New York, 1975.

The overall scheme for this alkane synthesis is as follows:

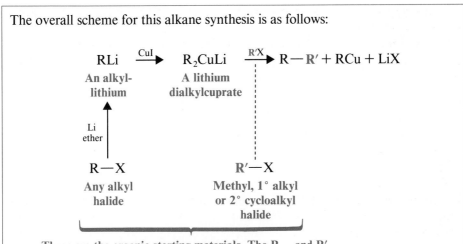

Two specific examples are outlined here.

$(CH_3)_2CuLi + CH_3CH_2CH_2CH_2CH_2{-}I \xrightarrow[\text{3.5 h, 25 °C}]{\text{diethyl ether}}$

$$CH_3{-}CH_2CH_2CH_2CH_2CH_3$$
(98%)

$(CH_3CH_2CH_2CH_2)_2CuLi + CH_3CH_2CH_2CH_2CH_2CH_2CH_2{-}Cl \xrightarrow[\text{5 days, 0 °C}]{\text{diethyl ether}}$

$$CH_3CH_2CH_2CH_2{-}CH_2CH_2CH_2CH_2CH_2CH_2CH_3$$
(75%)

SAMPLE PROBLEM
Outline the synthesis of 2-methylpropane (isobutane) from a propyl halide.

Answer:
We are required to start with a three-carbon (propyl) halide. Thus, we must combine it with a one-carbon compound. The Corey–Posner, Whitesides–House synthesis offers us a method for doing this. In theory there are two ways we might carry out the synthesis. We could (1) prepare lithium dimethylcuprate and allow it to react with an isopropyl halide, or (2) we could prepare a lithium diisopropylcuprate and allow it to react with a methyl halide.

Method 1

$$CH_3I \xrightarrow{\text{Li}} CH_3Li \xrightarrow{\text{CuI}} (CH_3)_2CuLi \xrightarrow{(CH_3)_2CHBr} CH_3CHCH_3$$
$$\underset{CH_3}{|}$$
(poor yield)

Method 2

$(CH_3)_2CHBr \xrightarrow{\text{Li}} (CH_3)_2CHLi \xrightarrow{\text{CuI}} [(CH_3)_2CH]_2CuLi \xrightarrow{CH_3I} CH_3CHCH_3$
$\qquad\qquad\qquad\qquad\qquad\qquad\qquad\qquad\qquad\qquad\qquad\qquad\qquad\qquad\qquad\ \ \mid$
$\qquad\qquad\qquad\qquad\qquad\qquad\qquad\qquad\qquad\qquad\qquad\qquad\qquad\qquad\ CH_3$
$\qquad\qquad\qquad\qquad\qquad\qquad\qquad\qquad\qquad\qquad\qquad\qquad\qquad\ \text{(good yield)}$

The second method is far better because this synthesis gives poor yields if a secondary (or tertiary) halide is used in the last step.

PROBLEM 15.11

Outline a synthesis of each of the following alkanes from appropriate alkyl halides using the Corey–Posner, Whitesides–House method:
(a) Propane   (b) Butane   (c) 2-Methylbutane   (d) 2,7-Dimethyloctane

---

PROBLEM 15.12

Outline methods showing how hexane could be prepared starting with:
(a) A bromopropane   (c) A bromopentane   (e) A hexene
(b) A bromobutane   (d) A bromohexane

Lithium dialkylcuprates couple with other organic groups. Coupling reactions of lithium dimethylcuprate with a cycloalkyl and a cycloallylic halide are shown here.

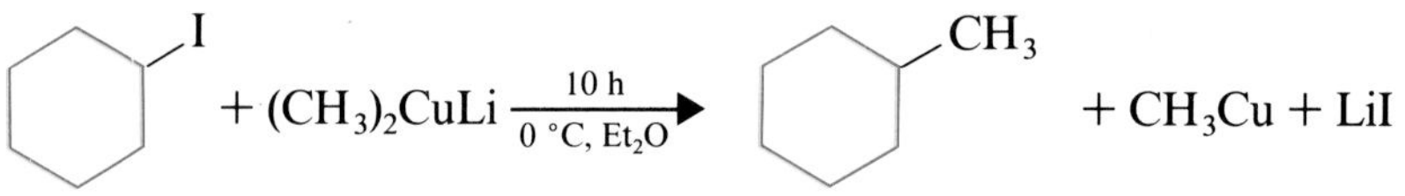

$$\text{cyclohexyl-I} + (CH_3)_2CuLi \xrightarrow[\text{0 °C, Et}_2\text{O}]{\text{10 h}} \text{methylcyclohexane} + CH_3Cu + LiI$$

(75%)
Methylcyclohexane

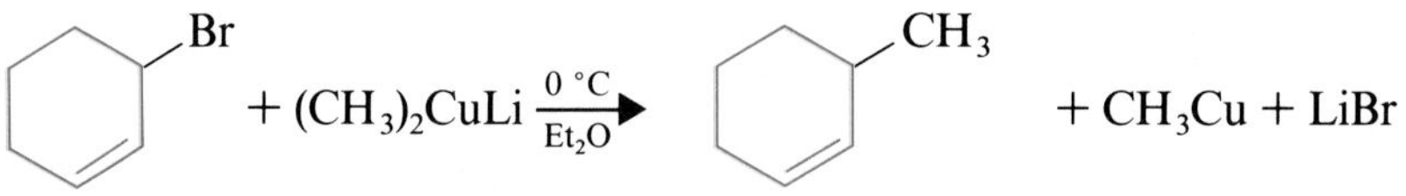

$$\text{cyclohexenyl-Br} + (CH_3)_2CuLi \xrightarrow[\text{Et}_2\text{O}]{\text{0 °C}} \text{3-methylcyclohexene} + CH_3Cu + LiBr$$

(75%)
3-Methylcyclohexene

Vinylic halides couple with lithium dialkylcuprates in a stereospecific way as the following example illustrates.

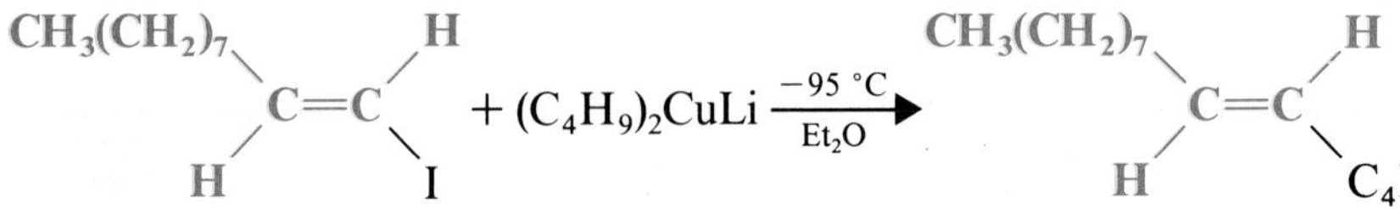

$$\begin{array}{c} CH_3(CH_2)_7 \\ \diagdown \\ C=C \\ \diagup \diagdown \\ H I \end{array} \begin{array}{c} H \end{array} + (C_4H_9)_2CuLi \xrightarrow[\text{Et}_2\text{O}]{-95\ °C} \begin{array}{c} CH_3(CH_2)_7 \\ \diagdown \\ C=C \\ \diagup \diagdown \\ H C_4H_9 \end{array} \begin{array}{c} H \end{array}$$

(E)-1-Iodo-1-decene   (74%)
(E)-5-Tetradecene

Lithium dialkylcuprates also couple with aryl halides:

$$(CH_3CH_2CH_2CH_2)_2CuLi + C_6H_5I \longrightarrow$$

$$CH_3CH_2CH_2CH_2-C_6H_5 + CH_3CH_2CH_2CH_2Cu + LiI$$
(75%)

We can summarize the coupling reactions of lithium dialkylcuprates in the following way:

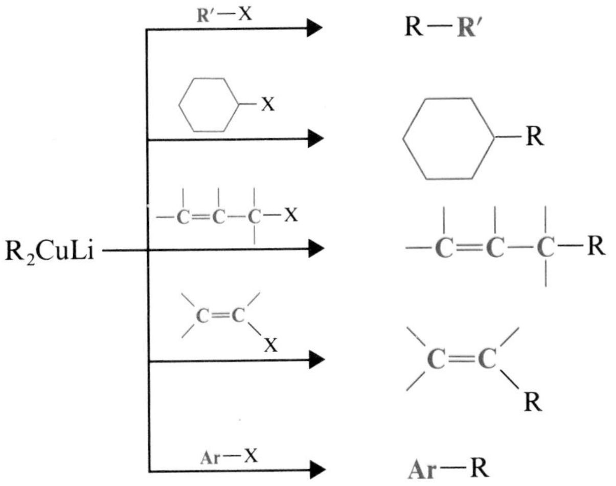

The chemistry of transition metal organic compounds is discussed further in Special Topic G.

ADDITIONAL PROBLEMS

15.13
What product would be formed from the reaction of isobutyl bromide, $(CH_3)_2CHCH_2Br$, with each of the following reagents?

(a) OH^-, H_2O

(b) CN^-, alcohol

(c) $(CH_3)_3CO^-$, $(CH_3)_3COH$

(d) CH_3O^-, CH_3OH

(e) Li, ether, then $CH_3\overset{\overset{\displaystyle O}{\|}}{C}CH_3$, then H_3O^+

(f) Mg, ether, then $CH_3\overset{\overset{\displaystyle O}{\|}}{C}H$, then H_3O^+

(g) Mg, ether, then $CH_3\overset{\overset{\displaystyle O}{\|}}{C}OCH_3$, then H_3O^+

(h) Mg, ether, then $H_2\overset{O}{\overset{\diagdown}{C}}{-}\overset{}{CH_2}$, then H_3O^+

(i) Mg, ether, then $H-\overset{\overset{\displaystyle O}{\|}}{C}-H$, then H_3O^+

(j) Li, ether, then CH_3OH

(k) Li, ether, then $CH_3C\equiv CH$

15.14
What products would you expect from the reaction of ethylmagnesium bromide (CH_3CH_2MgBr) with each of the following reagents?

(a) H_2O

(b) D_2O

(c) $C_6H_5\overset{\overset{\displaystyle O}{\|}}{C}H$, then H_3O^+

(d) $C_6H_5\overset{O}{\overset{\|}{C}}C_6H_5$, then H_3O^+

(e) $C_6H_5\overset{O}{\overset{\|}{C}}OCH_3$, then H_3O^+

(f) $C_6H_5\overset{O}{\overset{\|}{C}}CH_3$, then H_3O^+

(g) $CH_3CH_2C\equiv CH$, then $CH_3\overset{O}{\overset{\|}{C}}H$, then H_3O^+

(h) Cyclopentadiene

(i) $HgCl_2$

(j) $CdCl_2$

(k) PCl_3

15.15

What products would you expect from the reaction of propyllithium ($CH_3CH_2CH_2Li$) with each of the following reagents?

(a) $(CH_3)_2CH\overset{O}{\overset{\|}{C}}H$, then H_3O^+

(b) $(CH_3)_2CH\overset{O}{\overset{\|}{C}}CH_3$, then H_3O^+

(c) 1-Pentyne, then $CH_3\overset{O}{\overset{\|}{C}}CH_3$, then H_3O^+

(d) Ethyl alcohol

(e) CuI, then $CH_2=CHCH_2Br$

(f) CuI, then cyclopentyl bromide

(g) CuI, then (Z)-1-iodopropene

(h) CuI, then CH_3I

(i) CH_3CO_2D

(j) $SiCl_4$

(k) $ZnCl_2$

15.16

Show how you might prepare each of the following alcohols through a Grignard synthesis. (Assume that you have available any necessary organic halides, aldehydes, ketones, esters, and epoxides as well as any necessary inorganic reagents.)

(a) $CH_3CH_2\underset{\underset{CH_3}{|}}{\overset{\overset{CH_3}{|}}{C}}OH$ (three ways)

(b) 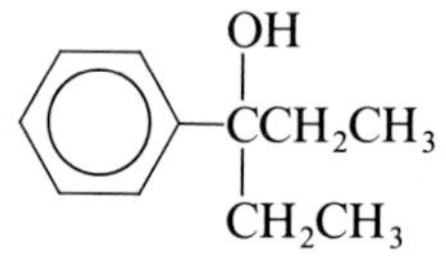 phenyl group $-\underset{\underset{CH_2CH_3}{|}}{\overset{\overset{OH}{|}}{C}}CH_2CH_3$ (three ways)

(c) 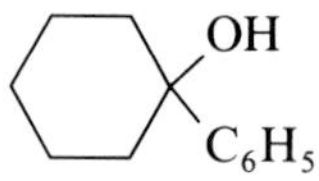 cyclohexyl with $\overset{OH}{\underset{C_6H_5}{}}$

(d) 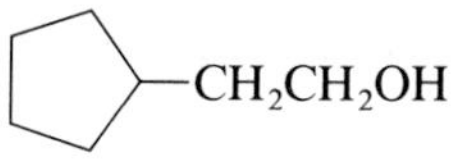 cyclopentyl $-CH_2CH_2OH$

(e) 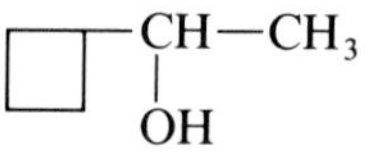 cyclobutyl $-\underset{\underset{OH}{|}}{CH}-CH_3$ (two ways)

15.17

Outline all steps in a synthesis that would transform isopropyl alcohol [$CH_3CH(OH)CH_3$] into each of the following:

(a) $(CH_3)_2CHCH(OH)CH_3$

(b) $(CH_3)_2CHCH_2OH$

(c) $(CH_3)_2CHCH_2CH_2Cl$

(d) $(CH_3)_2CHCH(OH)CH(CH_3)_2$

(e) CH_3CHDCH_3

(f) 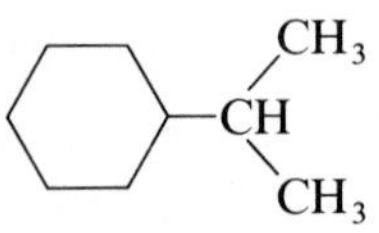 cyclohexyl $-\overset{\overset{CH_3}{\diagup}}{\underset{\diagdown CH_3}{CH}}$

15.18

Although $(C_6H_5)_2CHCl$ is a secondary halide, it undergoes S_N1 solvolysis in 80% ethanol at a rate that is 1300 times that for the tertiary halide, $(CH_3)_3CCl$. Explain.

15.19

When Grignard reagents are prepared from allyl halides, that is,

$$RCH{=}CHCH_2X + Mg \xrightarrow{\text{ether}} RCH{=}CHCH_2MgX$$

unavoidable by-products of the reactions are compounds with the formula $RCH{=}CHCH_2CH_2CH{=}CHR$. By-product formation is especially prevalent when concentrated solutions are used. Explain.

15.20

How might you carry out the following transformations?

(a) Bromocyclopentane $\longrightarrow$ methylcyclopentane

(b) 3-Bromocyclopentene $\longrightarrow$ 3-methylcyclopentene

(c) Allyl bromide $\longrightarrow$ 1-pentene

(d) (E)-2-Iodo-2-butene $\longrightarrow$ (E)-3-methyl-2-heptene

15.21

What products would you expect from the following reactions?

(a) Phenyllithium + acetic acid $\longrightarrow$

(b) Phenyllithium + methyl alcohol $\longrightarrow$

(c) Methylmagnesium bromide + ammonia $\longrightarrow$

(d) (Four molar equivalents) methylmagnesium bromide + $SiCl_4 \longrightarrow$

(e) (Three molar equivalents) phenylmagnesium bromide + $PCl_3 \longrightarrow$

(f) (Two molar equivalents) ethylmagnesium bromide + $CdCl_2 \longrightarrow$

(g) Phenylmagnesium bromide + $H{-}\overset{\overset{\textstyle O}{\|}}{C}{-}H \xrightarrow{\text{(2) } H_3O^+}$

15.22

Propose simple chemical tests that could be used to distinguish between the following pairs of compounds.

(a) Allyl bromide and propyl bromide

(b) Benzyl bromide and p-bromotoluene

(c) Vinyl chloride and benzyl chloride

(d) Phenyllithium and diphenylmercury

(e) Bromobenzene and bromocyclohexane

15.23

Show how each of the following transformations could be carried out.

(a) Styrene $\longrightarrow$ 1-phenylethanol   (two ways)

(b) Styrene $\longrightarrow$ 2-phenylethanol

(c) Styrene $\longrightarrow$ 1-methoxy-2-phenylethane

(d) 1-Phenylethanol $\longrightarrow$ ethyl 1-phenylethyl ether

(e) Phenylacetic acid $(C_6H_5CH_2CO_2H) \longrightarrow$ 2-phenylethanol

(f) Methyl phenyl ketone $(C_6H_5COCH_3) \longrightarrow$ 1-phenylethanol

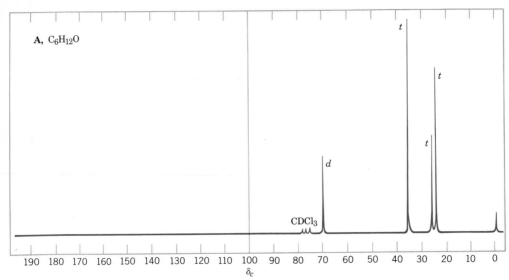

A, $C_6H_{12}O$

190 180 170 160 150 140 130 120 110 100 90 80 70 60 50 40 30 20 10 0

δ_c

FIGURE 15.1 The proton-decoupled ^{13}C spectrum of compound **A**, Problem 15.25. The letters *d* and *t* refer to the splitting of the signals (doublet and triplet) in the proton off-reso-nance decoupled spectrum. (Adapted from L. F. Johnson and W. C. Jankowski, *Carbon-13 NMR Spectra: A Collection of Assigned, Coded, and Indexed Spectra*. Wiley–Interscience, New York, 1972.)

(g) Toluene $\longrightarrow$ 2-phenylethanol

(h) Benzene $\longrightarrow$ 2-phenylethanol

(i) Methyl phenylacetate $(C_6H_5CH_2CO_2CH_3) \longrightarrow$ 2-phenylethanol

15.24

Show how 1-butanol could be transformed into each of the following compounds. (You may use any necessary inorganic reagents and you need not show the synthesis of a particular compound more than once.)

(a) 1-Butene

(b) 2-Butanol

(c) 2-Butanone $(CH_3COCH_2CH_3)$

(d) 1-Bromobutane

(e) 2-Bromobutane

(f) 1-Pentanol

(g) 1-Hexene

(h) 3-Methyl-3-heptanol

(i) Butanal $(CH_3CH_2CH_2CHO)$

(j) 4-Octanol

(k) 3-Methyl-4-heptanol

(l) Pentanoic acid $(CH_3CH_2CH_2CH_2CO_2H)$

(m) Butyl *sec*-butyl ether

(n) Dibutyl ether   (two ways)

(o) Butyllithium

(p) Octane

15.25

Compound **A** gives a positive test when treated with chromic oxide in aqueous sulfuric acid. The molecular formula of **A** is $C_6H_{12}O$, and its proton-decoupled ^{13}C spectrum is given in Fig. 15.1. Propose a structure for **A**.

TRANSITION METAL ORGANIC COMPOUNDS

G.1 Introduction

One of the most active areas of chemical research in recent years has involved studying compounds in which a bond exists between the carbon atom of an organic group and a transition metal. This field, which combines aspects of organic chemistry and inorganic chemistry, has led to many important applications in organic synthesis. Many of these transition metal organic compounds act as catalysts of extraordinary selectivity.

The transition metals are defined as those elements that have partly filled *d* (or *f*) shells, either in the elemental state or in their important compounds. The transition metals that are of most concern to organic chemists are those shown in the shaded portion of the Periodic Table given in Fig. G.1. Transition metals react with a variety of molecules or groups, called *ligands* to form *transition metal complexes.* In forming a complex, the ligands donate electrons to vacant orbitals of the metal. The bonds between the ligand and the metal range from bonds that are very weak to those that are very strong. The bonds are covalent but often have considerable polar character.

Transition metal complexes can assume a variety of geometries depending on the metal and on the number of ligands around it. Rhodium can form complexes

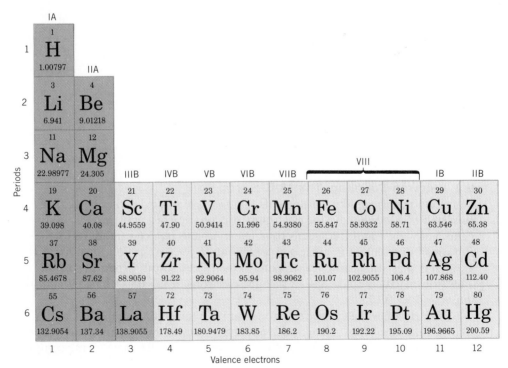

FIGURE G.1 Important transition elements are shown in the green and yellow portion of the Periodic Table. Given across the bottom are the total number of valence electrons (*s* and *d*) of each element.

TABLE G.1 Common ligands in transition metal complexes[a]

LIGAND	COUNT AS	NUMBER OF ELECTRONS DONATED
Negatively charged ligands		
H	H:⁻	2
R	R:⁻	2
X	X:⁻	2
Allyl		4
Cyclopentadienyl, Cp	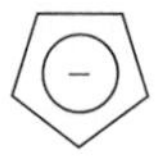	6
Electrically neutral ligands		
Carbonyl (carbon monoxide)	CO	2
Phosphine	R_3P or Ph_3P	2
Alkene	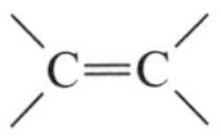	2
Diene		4
Benzene		6

[a]Adapted from J. Schwartz and J. A. Labinger, *J. Chem. Educ.*, **57**, 170 (1980).

with four ligands, for example, that are *square planar.* On the other hand, rhodium can form complexes with five or six ligands that are trigonal bipyramidal or octahedral. These typical shapes are shown below with the letter L used to indicate a ligand.

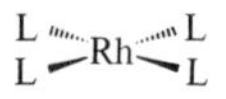

| Square planar rhodium complex | Trigonal bipyramidal rhodium complex | Octahedral rhodium complex |

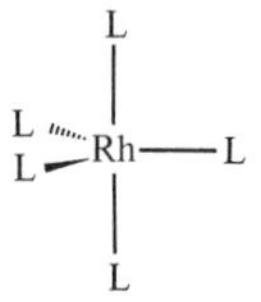

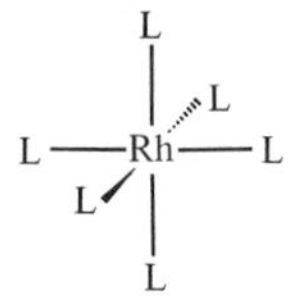

G.2 Electron Counting. The 18-Electron Rule

Transition metals are like the elements that we have studied earlier in that they are most stable when they have the electronic configuration of a noble gas. In addition to *s* and *p* orbitals, transition metals have five *d* orbitals (which can hold a total of 10 electrons). Therefore, the noble gas configuration for a transition metal is *18 electrons,* not 8 as with carbon, nitrogen, oxygen, and so on. When the metal of a transition metal complex has 18 valence electrons, it is said to be *coordinatively saturated.**

To determine the valence electron count of a transition metal in a complex, we take the total number of valence electrons of the metal in the elemental state (see Fig. G.1) and subtract from this number the oxidation state of the metal in the complex.

*We do not usually show the unshared electron pairs of a metal complex in our structures, because to do so would make the structure unnecessarily complicated.

This gives us what is called the d electron count, d^n. The oxidation state of the metal is the charge that would be left on the metal if all the ligands (Table G.1) were removed.

$$d^n = \frac{\text{total number of valence electrons}}{\text{of the elemental metal}} - \frac{\text{oxidation state of}}{\text{the metal in the complex}}$$

Then to get the total valence electron count of the metal *in the complex,* we add to d^n the number of electrons donated by all of the ligands. Table G.1 gives the number of electrons donated by several of the most common ligands.

$$\frac{\text{total number of valence electrons}}{\text{of the metal in the complex}} = d^n + \frac{\text{electrons donated}}{\text{by ligands}}$$

Let us now work out the valence electron count of two examples.

EXAMPLE A

Consider iron pentacarbonyl, $Fe(CO)_5$, a toxic liquid that forms when finely divided iron reacts with carbon monoxide.

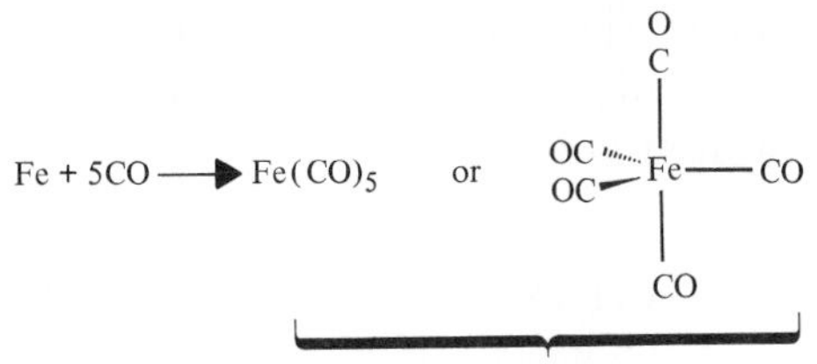

$$Fe + 5CO \longrightarrow Fe(CO)_5 \quad \text{or}$$

Iron pentacarbonyl

From Fig. G.1 we find that an iron atom in the elemental state has 8 valence electrons. We arrive at the oxidation state of iron in iron pentacarbonyl by noting that the charge on the complex as a whole is zero (it is not an ion), and that the charge on each CO ligand is also zero. Therefore, the iron is in the zero oxidation state.

Using these numbers, we can now calculate d^n and, from it, the total number of valence electrons of the iron in the complex.

$$d^n = 8 - 0 = 8$$

$$\frac{\text{total number of}}{\text{valence electrons}} = d^n + 5(CO) = 8 + 5(2) = 18$$

We find that the iron of $Fe(CO)_5$ has 18 valence electrons and is, therefore, coordinatively saturated. ■

EXAMPLE B

Consider the rhodium complex $Rh[(C_6H_5)_3P]_3H_2Cl$, a complex that, as we shall see later, is an intermediate in certain alkene hydrogenations.

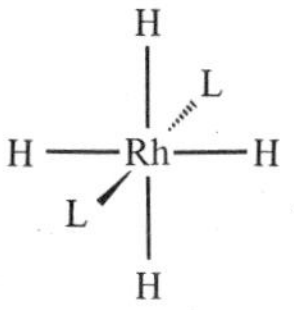

$L = Ph_3P$ [i. e., $(C_6H_5)_3P$]

The oxidation state of rhodium in the complex is $+3$. (The two hydrogen atoms and the chlorine are each counted as -1, and the charge on each of the triphenyl-phosphine ligands is zero. Removing all the ligands would leave a Rh^{3+} ion.) From Fig. G.1 we find that in the elemental state, rhodium has nine valence electrons. We can now calculate d^n for the rhodium of the complex.

$$d^n = 9 - 3 = 6$$

Each of the six ligands of the complex donates two electrons to the rhodium in the complex, and therefore, the total number of valence electrons of the rhodium is 18. The rhodium of $Rh[(C_6H_5)_3P]_3H_2Cl$ is coordinatively saturated.

$$\text{total number of valence} \atop \text{electrons of rhodium} = d^n + 6(2) = 6 + 12 = 18 \quad ∎$$

G.3 Metallocenes: Organometallic Sandwich Compounds

Cyclopentadiene reacts with phenylmagnesium bromide to give the Grignard re-agent of cyclopentadiene. This reaction is not unusual for it is simply another acid–base reaction like those we saw earlier. The methylene hydrogen atoms of cyclopen-tadiene are much more acidic than the hydrogen atoms of benzene and, therefore, the reaction goes to completion. (The methylene hydrogen atoms of cyclopentadiene are acidic relative to ordinary methylene hydrogen atoms because the cyclopentadienyl anion is aromatic; cf. Section 11.6B.)

Cyclopentadiene	Phenylmagnesium bromide		Cyclopenta-dienylmagnesium bromide	Benzene

$$\text{Cyclopentadiene} + C_6H_5MgBr \xrightarrow{\text{ether}} \overset{2+}{\text{MgBr}^-} + C_6H_6$$

When the Grignard reagent of cyclopentadiene is treated with ferrous chloride, a reaction takes place that produces a product called *ferrocene.*

$$2 \; \overset{2+ \; -}{\text{MgBr}} + FeCl_2 \longrightarrow (C_5H_5)_2Fe + 2MgBrCl$$

Ferrocene
(71% overall yield
from cyclopentadiene)

Ferrocene is an orange solid with a melting point of 174 °C. It is a highly stable compound; ferrocene can be sublimed at 100 °C and is not damaged when heated to 400 °C.

Many studies, including X-ray analysis, show that ferrocene is a compound in which the iron(II) ion is located between two cyclopentadienyl rings.

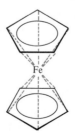

The carbon–carbon bond distances are all 1.40 Å and the carbon–iron bond distances are all 2.04 Å. Because of their structures, molecules such as ferrocene have been called "sandwich" compounds.

The carbon–iron bonding in ferrocene results from overlap between the inner lobes of the p orbitals of the cyclopentadienyl anions and $3d$ orbitals of the iron atom. Studies have shown, moreover, that this bonding is such that the rings of ferrocene are capable of essentially free rotation about an axis that passes through the iron atom and that is perpendicular to the rings.

The iron of ferrocene has 18 valence electrons and is, therefore, coordinatively saturated. We calculate this number as follows:

Iron has eight valence electrons in the elemental state and its oxidation state in ferrocene is $+2$. Therefore, $d^n = 6$.

$$d^n = 8 - 2 = 6$$

Each cyclopentadienyl ligand of ferrocene donates 6 electrons to the iron. Therefore, for the iron, the valence electron count is 18.

$$\text{total number of valence electrons} = d^n + 2(\text{Cp}) = 6 + 2(6) = 18$$

Ferrocene is an *aromatic compound.* It undergoes a number of electrophilic aromatic substitutions, including sulfonation and Friedel–Crafts acylation.

The discovery of ferrocene (in 1951) was followed by the preparation of a number of similar aromatic compounds. These compounds, as a class, are called *metallocenes.*[*] Metallocenes with five-, six-, seven-, and even eight-membered rings have been synthesized from metals as diverse as zirconium, manganese, cobalt, nickel, chromium, and uranium.

"Half-sandwich" compounds have been prepared through the use of metal carbonyls. Several are shown here.

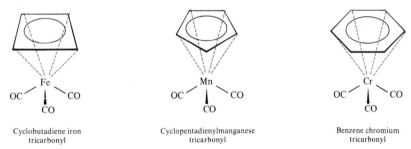

| Cyclobutadiene iron
tricarbonyl | Cyclopentadienylmanganese
tricarbonyl | Benzene chromium
tricarbonyl |

Although cyclobutadiene itself is *not* stable, the cyclobutadiene iron tricarbonyl is.

PROBLEM G.1

The metal of each of the previously given half-sandwich compounds is coordinatively saturated. Show that this is true by working out the valence electron count for the metal in each complex.

---

*Ernst O. Fischer (of the Technical University, Munich) and Geoffrey Wilkinson (of Imperial College, London) received the Nobel Prize in 1973 for their pioneering work (performed independently) on the chemistry of organometallic sandwich compounds—or metallocenes.

G.4 Reactions of Transition Metal Complexes

Much of the chemistry of organic transition metal compounds will be more under-standable if we are able to follow the mechanisms of the reactions that occur. These mechanisms will, in most cases, amount to nothing more than a sequence of reactions, each of which represents *a fundamental reaction type that is characteristic of a transition metal complex.* Let us examine three of the fundamental reaction types now. In each instance we shall use steps that occur when an alkene is hydrogenated using a catalyst called Wilkinson's catalyst. Later (in Section G.5) we shall examine the entire hydrogenation mechanism.

1. **Ligand Dissociation–Association (Ligand Exchange).** A transition metal complex can lose a ligand (by dissociation) and combine with another ligand (by association). In the process it undergoes *ligand exchange.* For example, the rhodium complex that we encountered in Example B can react with an alkene (in this example with ethene) as follows:

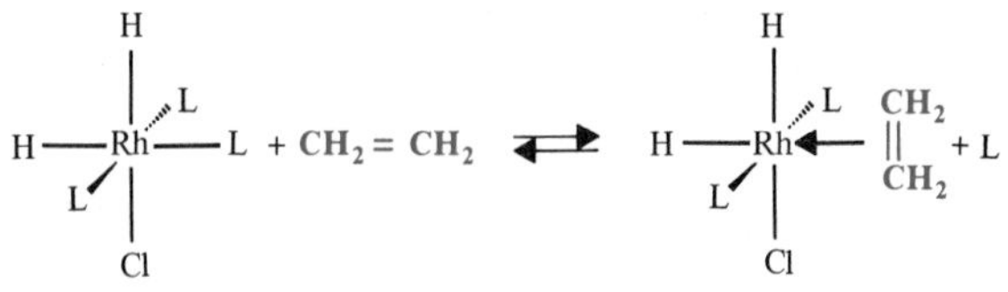

$$L = Ph_3P \, [\, i. \, e., \, (C_6H_5)_3P\,]$$

Two steps are actually involved. In the first step, one of the triphenylphosphine ligands dissociates. This leads to a complex in which the rhodium has only 16 electrons and is, therefore, coordinatively *unsaturated.*

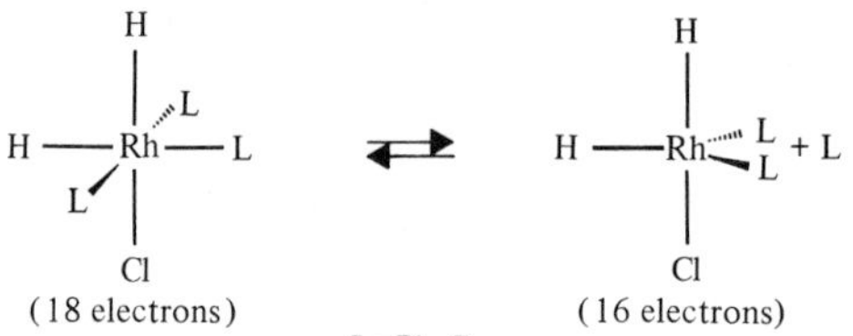

$$L = Ph_3P$$

In the second step, the rhodium associates with the alkene to become coordinatively saturated again.

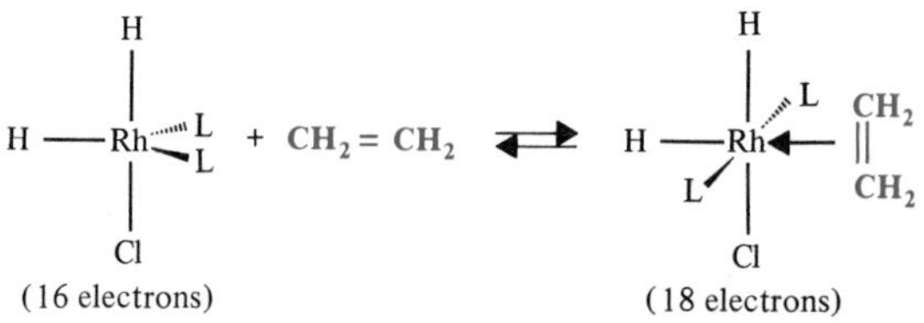

The complex between the rhodium and the alkene is called a *π-complex.* In it, two electrons are donated by the alkene to the rhodium. Alkenes are often called π-donors to distinguish them from σ-donors such as $Ph_3P:$, $Cl:$, and so on.

In a π-complex such as the one just given, there is also a donation of electrons from a populated *d* orbital of the metal back to the vacant π* orbital of the alkene. This kind of donation is called "back-bonding."

2. **Insertion–Deinsertion.** An unsaturated ligand such as an alkene can undergo *insertion* into a bond between the metal of a complex and a hydrogen or a carbon. These reactions are reversible, and the reverse reaction is called *deinsertion.*

The following is an example of insertion–deinsertion.

(18 electrons)   (16 electrons)

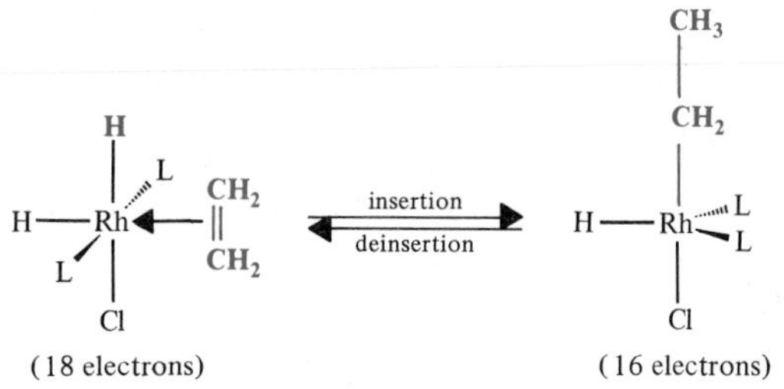

In this process a π bond (between the rhodium and the alkene) and a σ bond (between the rhodium and the hydrogen) are exchanged for a new σ bond (between rhodium and carbon). The valence electron count of the rhodium decreases from 18 to 16.

This insertion–deinsertion occurs in a stereospecific way, as a *syn addition* of the M–H unit to the alkene.

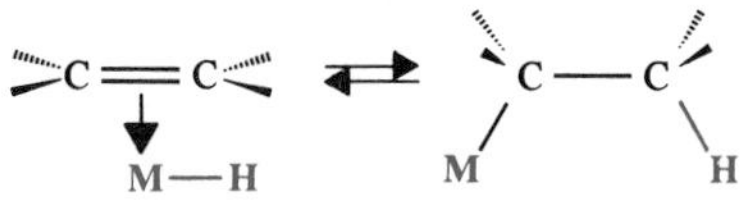

3. **Oxidative Addition–Reductive Elimination.** Coordinatively unsaturated metal complexes can undergo oxidative addition of a variety of substrates in the following way.†

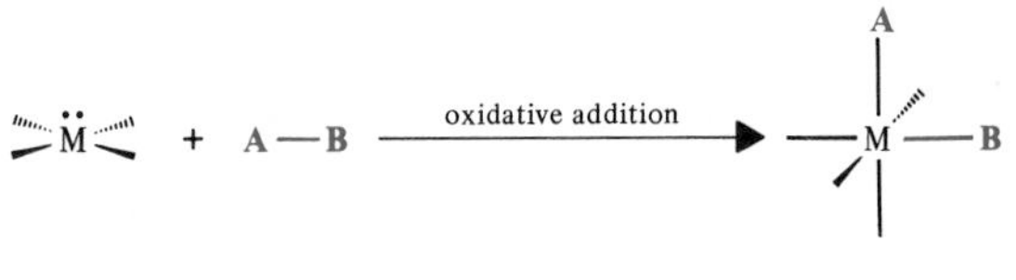

The substrates, A—B, can be H—H, H—X, R—X, RCO—H, RCO—X, and a number of other compounds.

In this type of oxidative addition, the metal of the complex undergoes an increase in the number of its valence electrons *and in its oxidation state.* Consider, as an example, the oxidative addition of hydrogen to the rhodium complex that follows (L = Ph_3P).

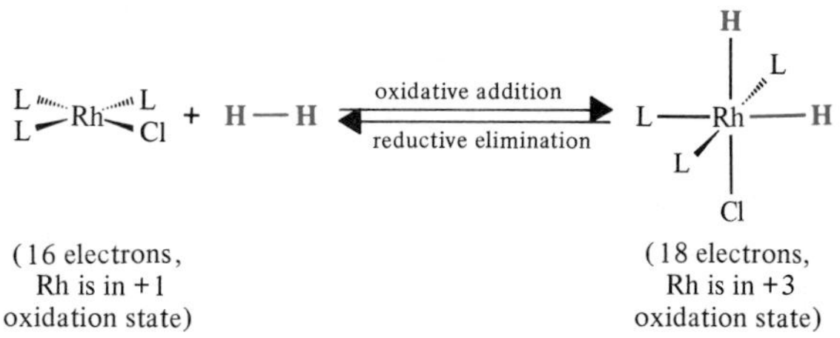

(16 electrons,   (18 electrons,
Rh is in +1   Rh is in +3
oxidation state)   oxidation state)

---

†Coordinatively saturated complexes also undergo oxidative addition. We shall see an example in Section G.8.

Reductive elimination is the reverse of oxidative addition. With this background, we are now in a position to examine a few interesting applications of transition metal complexes in organic synthesis.

G.5 Homogeneous Hydrogenation

Until now, all of the hydrogenations that we have examined have been heterogeneous processes. Two phases have been involved in the reaction: the solid phase of the catalyst (Pt, Pd, Ni, etc.) containing the adsorbed hydrogen, and the liquid phase of the solution containing the unsaturated compound. In homogeneous hydrogenation using a transition metal complex such as $Rh[(C_6H_5)_3P]_3Cl$ (called Wilkinson's catalyst), hydrogenation takes place *in a single phase*—in solution.

When Wilkinson's catalyst is used to carry out the hydrogenation of an alkene, the following steps take place ($L = Ph_3P$).

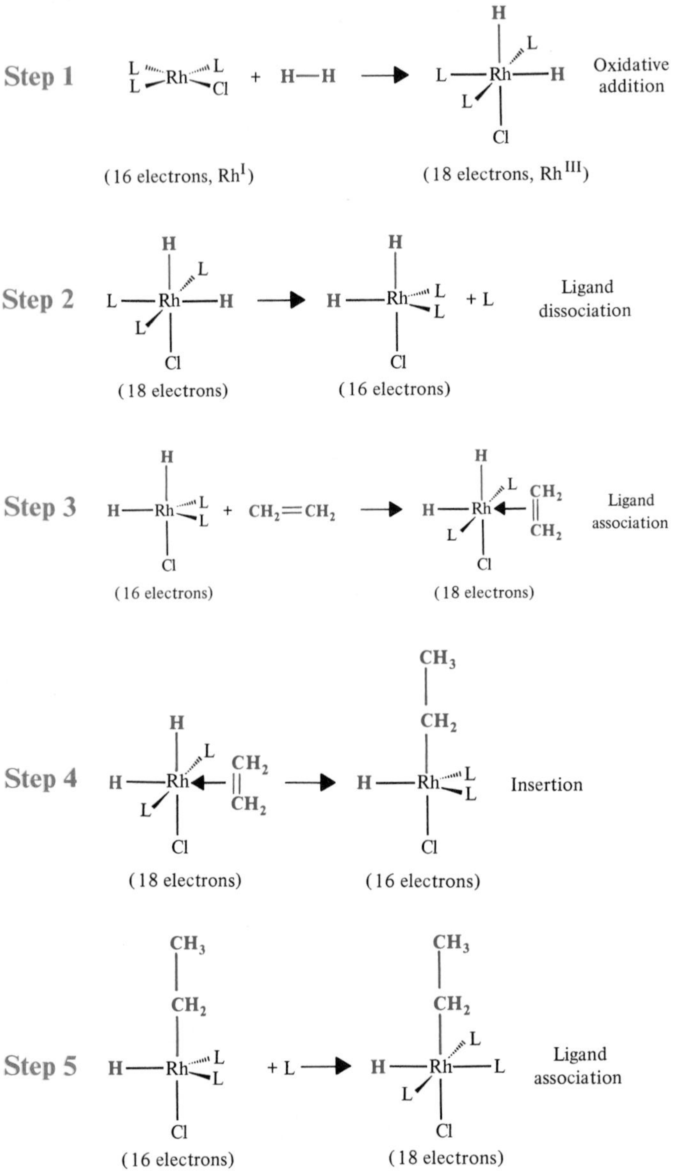

Step 6

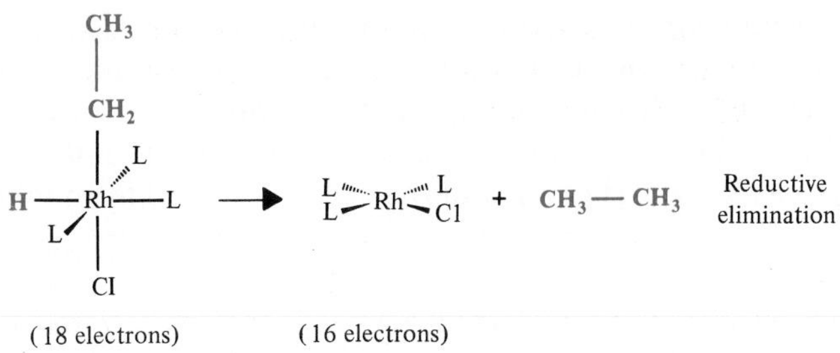

$$\underset{\text{(18 electrons)}}{\text{H}-\overset{\overset{\displaystyle CH_3}{\overset{|}{\underset{|}{CH_2}}}}{\underset{\underset{\displaystyle Cl}{|}}{\overset{\displaystyle L}{\underset{\displaystyle L}{Rh}}}}-L} \longrightarrow \underset{\text{(16 electrons)}}{\overset{L}{\underset{L}{Rh}}\overset{L}{\underset{Cl}{}}} + CH_3-CH_3 \qquad \text{Reductive elimination}$$

Then steps 1, 2, 3, 4, 5, 6, and so on.

Step 6 regenerates the catalyst, which can then cause hydrogenation of another molecule of the alkene.

Because the insertion step 4 and the reductive elimination step 6 are stereospecific, the net result of hydrogenation using Wilkinson's catalyst is a *syn addition* of hydrogen to the alkene. The following example (with D_2 in place of H_2) illustrates this aspect.

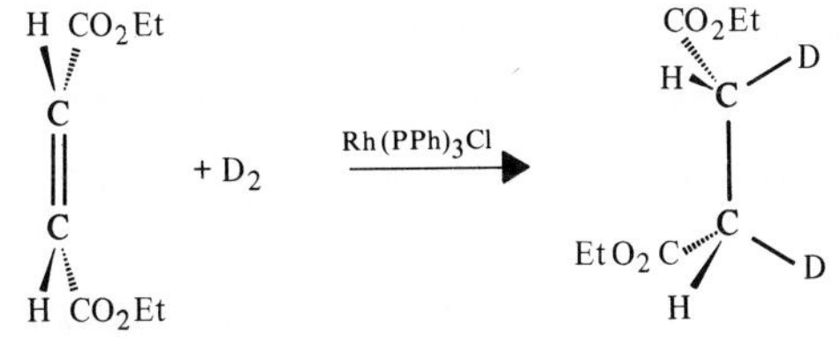

A *cis* alkene
(diethyl maleate)

A meso compound

PROBLEM G.2

What product (or products) would be formed if the *trans* alkene corresponding to the *cis* alkene (see previous reaction) had been hydrogenated with D_2 and Wilkinson's catalyst?

G.6 Carbon–Carbon-Bond-Forming Reactions
Using Rhodium Complexes

Rhodium complexes have also been used to synthesize compounds in which the formation of a carbon–carbon bond is required. An example is the synthesis that follows:

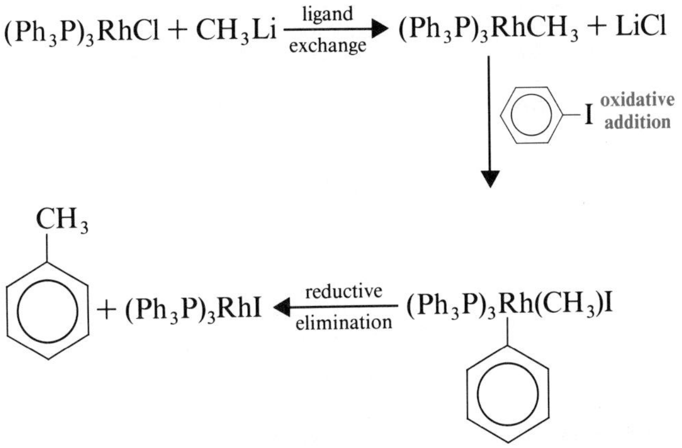

$$(Ph_3P)_3RhCl + CH_3Li \xrightarrow[\text{exchange}]{\text{ligand}} (Ph_3P)_3RhCH_3 + LiCl$$

The first step, *a ligand exchange,* occurs by a combination of ligand associa-tion–dissociation steps and incorporates the methyl group into the coordination sphere of the rhodium. The next step, *an oxidative addition,* incorporates the phenyl group into the rhodium coordination sphere. Then, in the last step, *a reductive elimination* joins the methyl group and the benzene ring to form toluene.

PROBLEM G.3

Give the total valence electron count for rhodium in each complex in the synthesis outlined previously.

Another example is the following ketone synthesis.

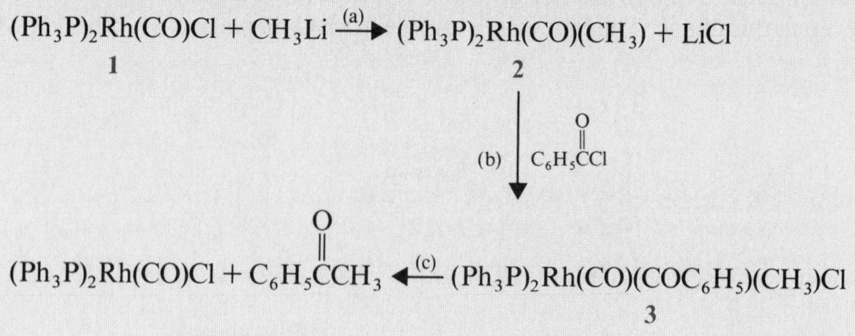

$$(Ph_3P)_2Rh(CO)Cl + CH_3Li \xrightarrow{\text{(a)}} (Ph_3P)_2Rh(CO)(CH_3) + LiCl$$
$$\text{1} \qquad\qquad\qquad\qquad\qquad\qquad \text{2}$$

$$(b) \downarrow \; C_6H_5\overset{\overset{\displaystyle O}{\|}}{C}Cl$$

$$(Ph_3P)_2Rh(CO)Cl + C_6H_5\overset{\overset{\displaystyle O}{\|}}{C}CH_3 \xleftarrow{\text{(c)}} (Ph_3P)_2Rh(CO)(COC_6H_5)(CH_3)Cl$$
$$\text{3}$$

PROBLEM G.4

Give the valence electron count and the oxidation state of rhodium in the complexes labeled **1, 2, 3**; then describe each step (a), (b), and (c) as to its fundamental type (oxidative addition, ligand exchange, etc.).

Still another carbon–carbon bond-forming reaction (below) illustrates the stereospecificity of these reactions.

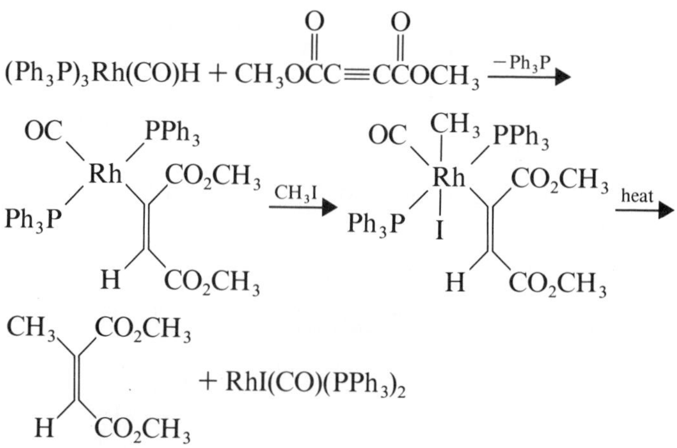

$$(Ph_3P)_3Rh(CO)H + CH_3O\overset{\overset{\displaystyle O}{\|}}{C}C{\equiv}C\overset{\overset{\displaystyle O}{\|}}{C}OCH_3 \xrightarrow{-Ph_3P}$$

$$+ RhI(CO)(PPh_3)_2$$

PROBLEM G.5

Give, in detail, a possible mechanism for the synthesis just outlined, describing each step according to its fundamental type.

PROBLEM G.6

The actual mechanism of the Corey–Posner, Whitesides–House synthesis is not known with certainty. One possible mechanism involves the oxidative addition of $R'-X$ or $Ar-X$ to R_2CuLi followed by a reductive elimination to generate $R-R'$ or $R-Ar$. Outline the steps in this mechanism using $(CH_3)_2CuLi$ and C_6H_5I.

G.7 Hydroformylation. The Oxo Reaction

An industrial process for the synthesis of aldehydes that we shall study in Section 16.4B is called *hydroformylation* or the *oxo reaction*. In this reaction an alkene reacts with carbon monoxide and hydrogen in the presence of the cobalt catalyst, $HCo(CO)_4$ as follows:

$$RCH=CH_2 + CO + H_2 \xrightarrow[\text{2000 psi, 110–150 °C}]{HCo(CO)_4} RCH_2CH_2CHO + \text{isomers}$$

The mechanism for hydroformylation has been studied extensively by Milton Orchin (University of Cincinnati) and is quite complex, if we take into account all of the reactions that occur. Essential to the mechanism, however, are the following fundamental steps.

Step 1 $HCo(CO)_4 \rightleftharpoons HCo(CO)_3 + CO$ Ligand dissociation

Step 2 $RCH=CH_2 + HCo(CO)_3 \rightleftharpoons \underset{\underset{HCo(CO)_3}{\downarrow}}{RCH=CH_2}$ Ligand association

Step 3 $\underset{\underset{HCo(CO)_3}{\downarrow}}{RCH=CH_2} \rightleftharpoons RCH_2CH_2Co(CO)_3$ Insertion

Step 4 $RCH_2CH_2Co(CO)_3 + CO \rightleftharpoons RCH_2CH_2Co(CO)_4$ Ligand association

Step 5 $RCH_2CH_2Co(CO)_4 \rightleftharpoons RCH_2CH_2COCo(CO)_3$ Insertion

Step 6 $RCH_2CH_2COCo(CO)_3 + H_2 \rightleftharpoons RCH_2CH_2CO\overset{\overset{H}{|}}{\underset{\underset{H}{|}}{Co}}(CO)_3$ Oxidative addition

$$\text{Step 7} \quad \underset{\overset{\displaystyle |}{H}}{\overset{\overset{\displaystyle H}{|}}{RCH_2CH_2COCo(CO)_3}} \rightleftharpoons RCH_2CH_2CHO + HCo(CO)_3 \quad \begin{array}{l}\text{Reductive}\\\text{elimination}\end{array}$$

$$\text{Step 8} \quad HCo(CO)_3 + CO \rightleftharpoons HCo(CO)_4 \quad \begin{array}{l}\text{Ligand}\\\text{association}\end{array}$$

An important step in the sequence just given is step 5. In this step a carbonyl group is inserted between the metal and the coordinated alkyl group. (One can also consider this step as a migration of the alkyl group from the metal to the carbon of a coordinated CO ligand.)

Carbonyl insertion reactions are reversible and can be exploited synthetically as "decarbonylation" reactions. In the following example decarbonylation of benzaldehyde leads to the formation of benzene.

$$C_6H_5\overset{\overset{\displaystyle O}{\|}}{C}H + (Ph_3P)_3RhCl \xrightarrow{-Ph_3P}$$

$$C_6H_5RhH(CO)(Ph_3P)_2Cl \longrightarrow C_6H_6 + Rh(CO)(Ph_3P)_2Cl$$

PROBLEM G.7

Outline a detailed mechanism for the decarbonylation of benzaldehyde and describe each step according to its fundamental type.

G.8 Disodium Tetracarbonylferrate. Collman's Reagent

Iron pentacarbonyl reacts with sodium to produce disodium tetracarbonylferrate, $Na_2Fe(CO)_4$, a highly versatile compound for organic synthesis.

$$Fe(CO)_5 + 2Na \xrightarrow{THF} \underset{\substack{\textbf{1}\\\text{Collman's}\\\text{reagent}}}{Na_2Fe(CO)_4}$$

This reagent, discovered by James P. Collman (Stanford University), reacts with alkyl halides in the following way:

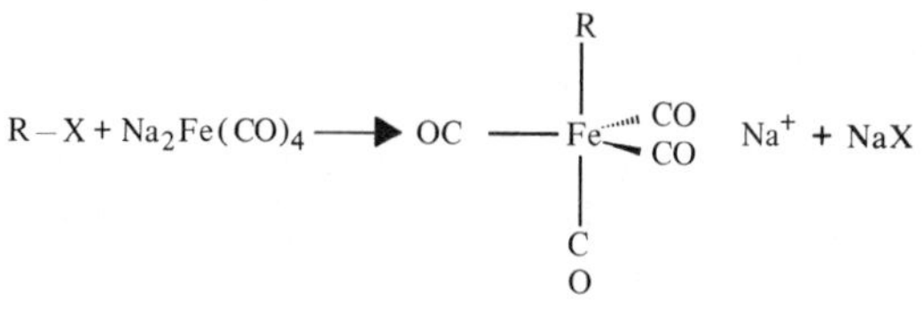

$$R-X + Na_2Fe(CO)_4 \longrightarrow OC-\underset{\underset{\underset{\textbf{2}}{\overset{\displaystyle O}{C}}}{\overset{\displaystyle |}{|}}}{\overset{\overset{\displaystyle R}{|}}{Fe}}\overset{CO}{\underset{CO}{\diagdown}} \quad Na^+ + NaX$$

This reaction can be considered an oxidative addition to a *coordinatively saturated* metal complex. Equivalently, however, it can be viewed as an S_N2 attack on the carbon of the alkyl halide with an anion of $Na_2Fe(CO)_4$ acting as the nucleophile. Evidence supporting this view is the order of reactivities of alkyl halides:

$$CH_3X > RCH_2X > R_2CHX \quad \text{and} \quad RI > RBr > RCl$$

Furthermore, the reaction has been shown to take place with *inversion of configuration of the alkyl group.*

The alkyltetracarbonylferrate anion **(2)** undergoes insertion of carbon monoxide to yield **3**, and **3** can be converted to aldehydes, ketones, esters, and carboxylic acids.

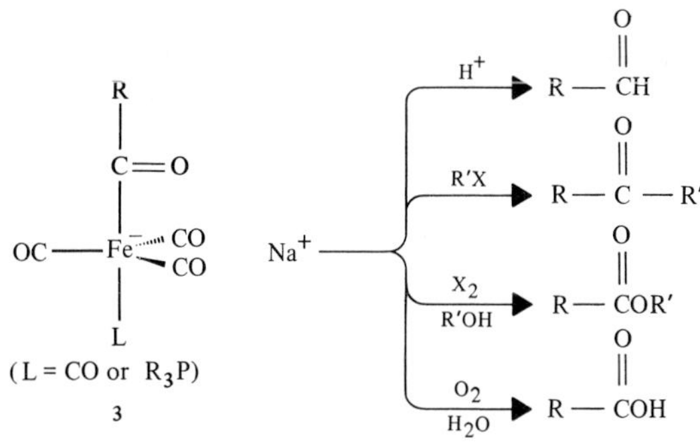

Compound **3** can also be synthesized from the reaction of **1** with an acid chloride.

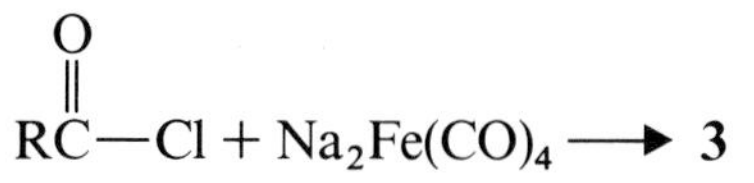

The alkyltetracarbonylferrate anion **2** also undergoes useful reactions.

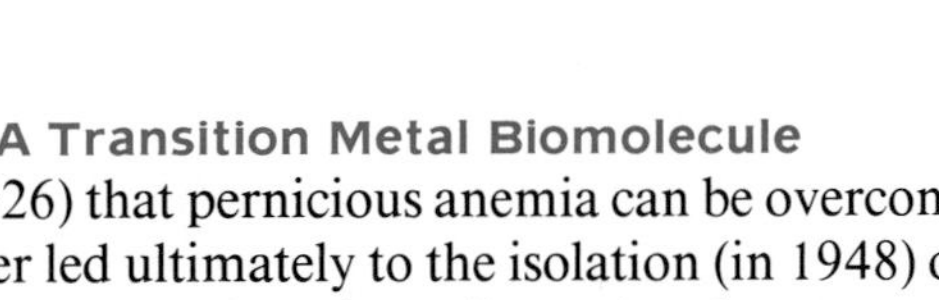

G.9 Vitamin B₁₂. A Transition Metal Biomolecule

The discovery (in 1926) that pernicious anemia can be overcome by the ingestion of large amounts of liver led ultimately to the isolation (in 1948) of the curative factor, called vitamin B_{12}. The complete three-dimensional structure of vitamin B_{12} (Fig. G.2) was elucidated in 1956 through the X-ray studies of Dorothy Hodgkin (Nobel Prize, 1964), and in 1972 the synthesis of this complicated molecule was announced by R. B. Woodward (Harvard University) and A. Eschenmoser (Swiss Federal Institute of Technology). The synthesis took 11 years and involved more than 90 separate reactions. One hundred co-workers took part in the project.

Vitamin B_{12} is the only known biomolecule that possesses a carbon–metal bond. In the stable commercial form of the vitamin, a cyano group is bonded to the

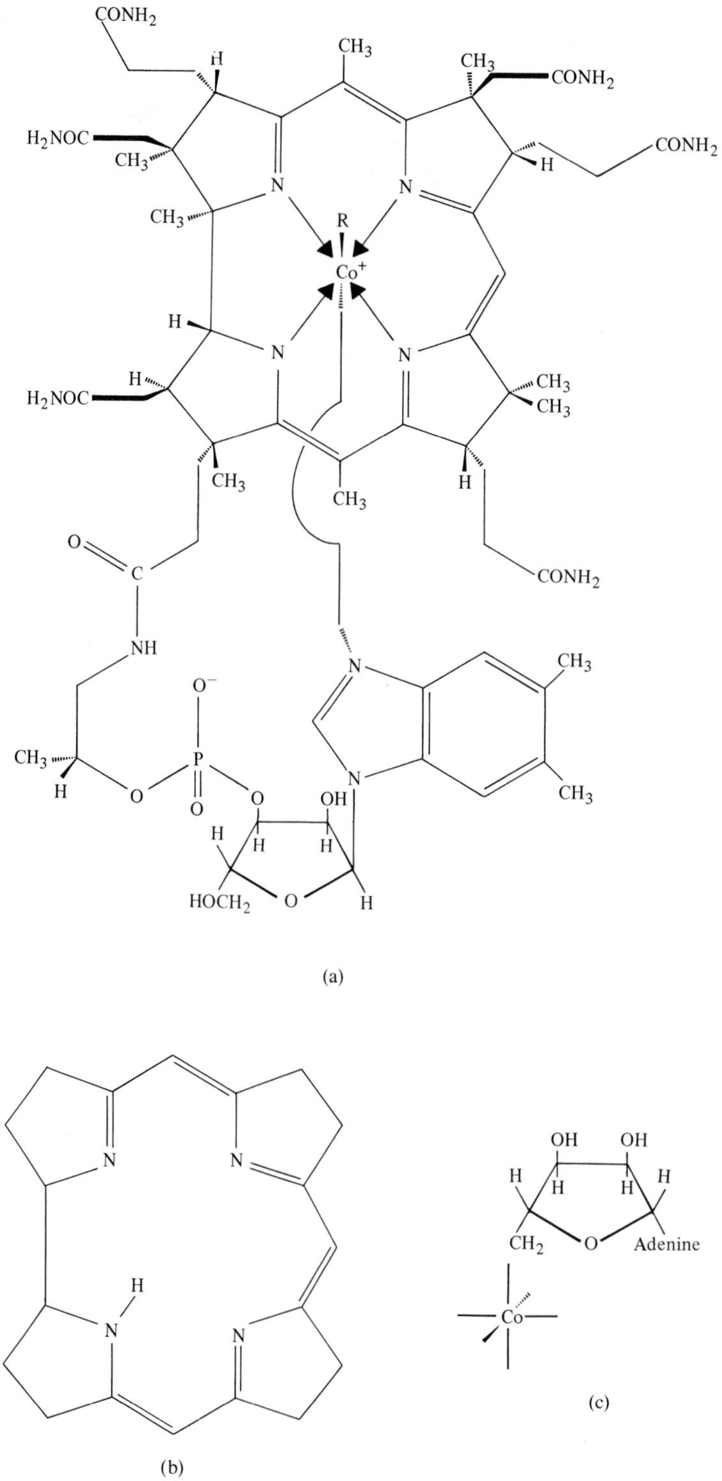

(a)

(b)

(c)

FIGURE G.2 (*a*) The structure of vitamin B_{12}. In the commercial form of the vitamin (cyano-cobalamin), R = CN. (*b*) The corrin ring system. (*c*) In the biologically active form of the vitamin (5′-deoxy-adenosyl-cobalamin) the 5′-carbon atom of 5′-deoxyadenosine is coordinated to the cobalt atom. For the structure of adenine see Section 24.2.

cobalt, and the cobalt is in the $+3$ oxidation state. The core of the vitamin B_{12} molecule is a *corrin ring* with various attached side groups. The corrin ring consists of four pyrrole subunits, the nitrogen of each of which is coordinated to the central cobalt. The sixth ligand (below the corrin ring in Fig. G.2) is a nitrogen of a heterocyclic molecule called 5,6-dimethylbenzimidazole.

The cobalt of vitamin B_{12} can be reduced to a $+2$ or a $+1$ oxidation state. When the cobalt is in the $+1$ oxidation state, vitamin B_{12} (called B_{12s}) becomes one of the most powerful nucleophiles known, being more nucleophilic than methanol by a factor of 10^{14}.

Acting as a nucleophile, vitamin B_{12s} reacts with adenosine triphosphate (Fig. 21.2) to yield the biologically active form of the vitamin (Fig. G.2c).

SPECIAL TOPIC H

ORGANIC HALIDES AND ORGANOMETALLIC COMPOUNDS IN THE ENVIRONMENT

H.1 Organic Halides as Insecticides

Since the discovery of the insecticidal properties of DDT in 1942, vast quantities of chlorinated hydrocarbons have been sprayed over the surface of the earth in an effort to destroy insects. These efforts initially met with incredible success in ridding large areas of the earth of disease-carrying insects, particularly those of typhus and malaria. As time has passed, however, we have begun to understand that this prodigious use of chlorinated hydrocarbons has not been without harmful — indeed tragic — side effects. Chlorinated hydrocarbons are usually highly stable compounds and are only slowly destroyed by natural processes in the environment. As a result, many chloro-organic insecticides will remain in the environment for years. These persistent pesticides are called "hard" pesticides.

Chlorohydrocarbons are also fat soluble and tend to accumulate in the fatty tissues of most animals. The food chain that runs from plankton to small fish to larger fish to birds and to larger animals including man tends to magnify the concentrations of chloroorganic compounds at each step.

The chlorohydrocarbon DDT is prepared from inexpensive starting materials, chlorobenzene and trichloroacetaldehyde. The reaction is catalyzed by acid.

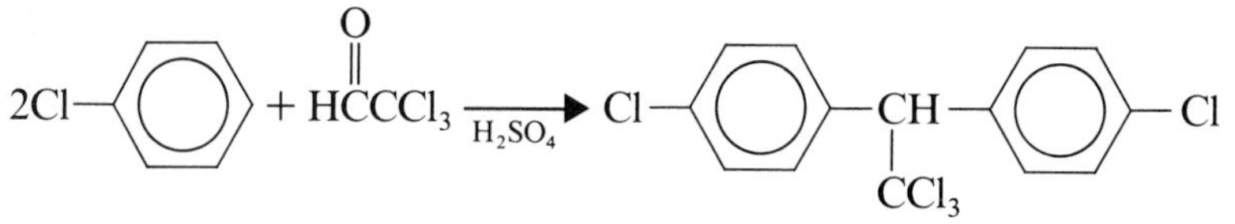

DDT
[1,1,1-Trichloro-2,2-
bis(*p*-chlorophenyl)ethane]

In nature the principal decomposition product of DDT is DDE.

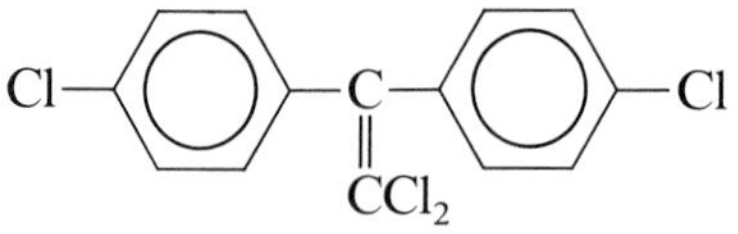

DDE
[1,1-Dichloro-2,2-
bis(*p*-chlorophenyl)ethene]

Estimates indicate that nearly 1 billion lb of DDE are now spread throughout the world ecosystem. One pronounced environmental effect of DDE has been in its action on egg-shell formation of many birds. DDE inhibits the enzyme *carbonic anhydrase* that controls the calcium supply for shell formation. As a consequence, the shells are often very fragile and do not survive to the time of hatching. During the late 1940s the populations of eagles, falcons, and hawks dropped dramatically. There can be little doubt that DDE was primarily responsible.

DDE also accumulates in the fatty tissues of man. Although man appears to have a short-range tolerance to moderate DDE levels, the long-range effects are far from certain.

Other hard insecticides are aldrin, dieldrin, and chlordan. Aldrin can be manufactured through the Diels–Alder reaction of hexachlorocyclopentadiene and norbornadiene.

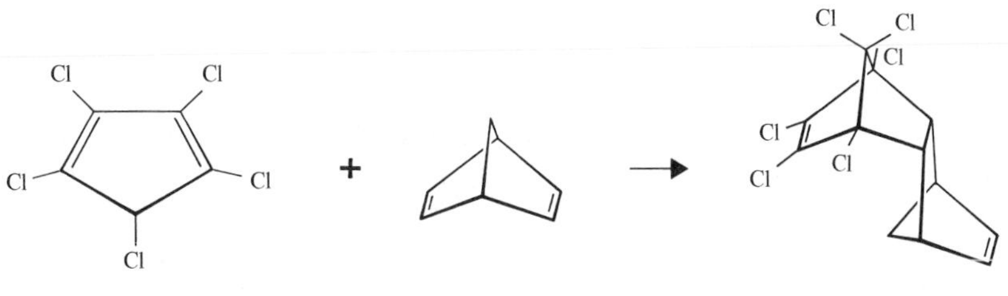

| Hexachloro-cyclopentadiene | Norbornadiene | Aldrin |

Chlordan can be made by adding chlorine to the unsubstituted double bond of the Diels–Alder adduct obtained from hexachlorocyclopentadiene and cyclopentadiene. Dieldrin can be made by converting an aldrin double bond to an epoxide. (This reaction also takes place in nature.)

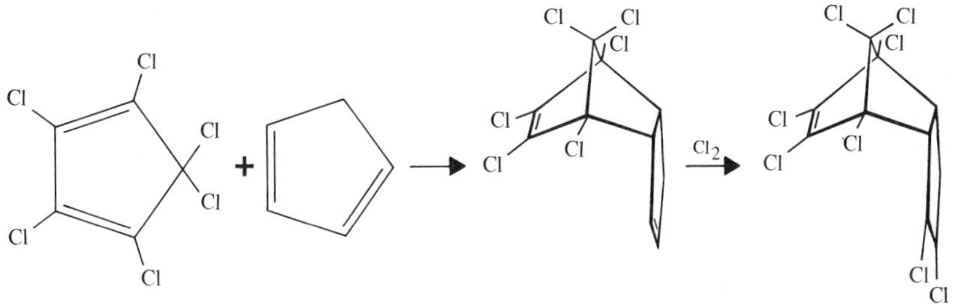

Chlordan

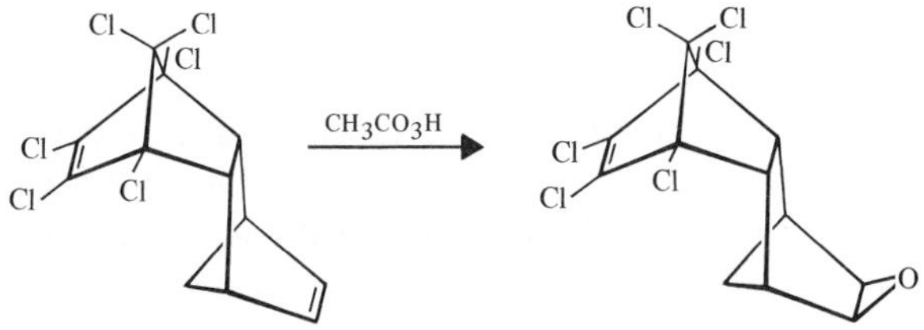

Aldrin   Dieldrin

During the 1970s the Environmental Protection Agency (EPA) banned the use of DDT, aldrin, dieldrin, and chlordan because of known or suspected hazards to human life. All of the compounds are suspected of causing cancers.

PROBLEM H.1

The mechanism for the formation of DDT from chlorobenzene and trichloroacetaldehyde in sulfuric acid involves two electrophilic aromatic substitution reactions. In the first electrophilic substitution reaction the electrophile is protonated trichloroacetaldehyde. In the second the electrophile is a carbocation. Propose a mechanism for the formation of DDT.

PROBLEM H.2

What kind of reaction is involved in the conversion of DDT to DDE?

Mirex, Kepone, and lindane are also hard insecticides whose use has been banned.

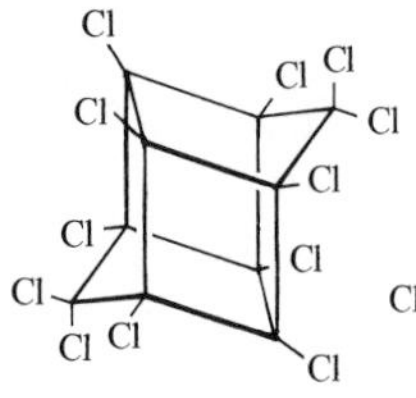

Mirex 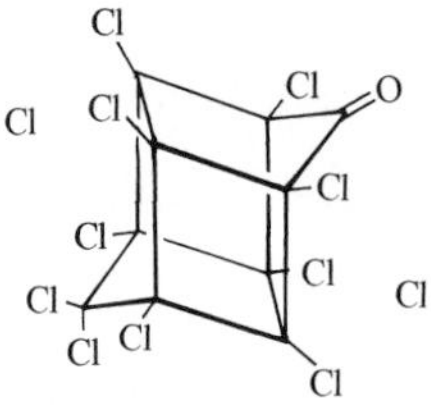   Kepone 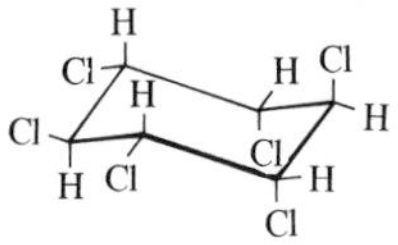   Lindane

H.2 Organic Halides as Herbicides

Other chlorinated organic compounds have been used extensively as herbicides. The following two examples are 2,4-D and 2,4,5-T.

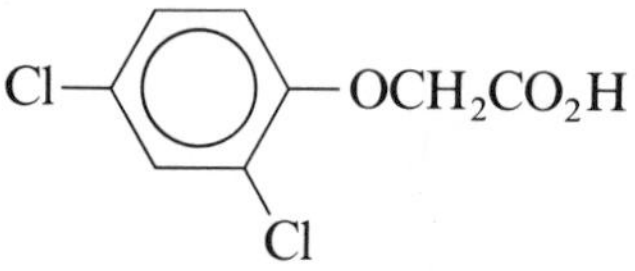

2,4-D
(2,4-Dichlorophenoxy-
acetic acid)

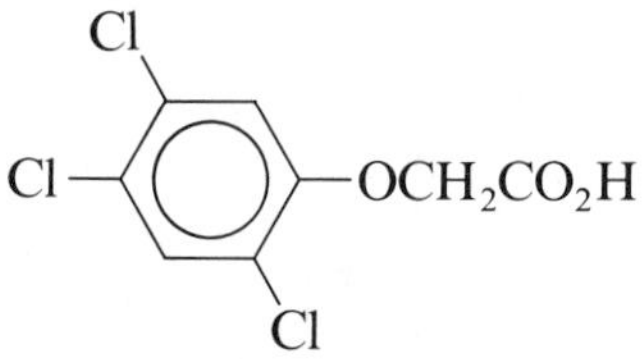

2,4,5-T
(2,4,5-Trichlorophenoxy-
acetic acid)

Enormous quantities of these two compounds were used as defoliants in the jungles of Indochina during the Vietnam war. Some samples of 2,4,5-T have been shown to be a teratogen (a fetus-deforming agent). This teratogenic effect was the result of an impurity present in commercial 2,4,5-T, the compound 2,3,7,8-tetra-chlorodibenzodioxin. 2,3,7,8-Tetrachlorodibenzodioxin is also highly toxic; it is more toxic, for example, than cyanide ion, strychnine, and the nerve gases.

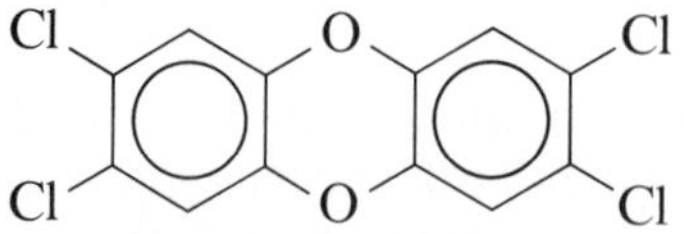

2,3,7,8-Tetrachlorodibenzodioxin
(also called TCDD)

This dioxin is also highly stable; it persists in the environment and because of its fat solubility can be passed up the food chain. In sublethal amounts it can cause a disfiguring skin disease called chloracne.

In July 1976 an explosion at a chemical plant in Seveso, Italy, caused the release of between 22 and 132 lb of this dioxin into the atmosphere. The plant was engaged in the manufacture of 2,4,5-trichlorophenol (used in making 2,4,5-T) using the follow-ing method:

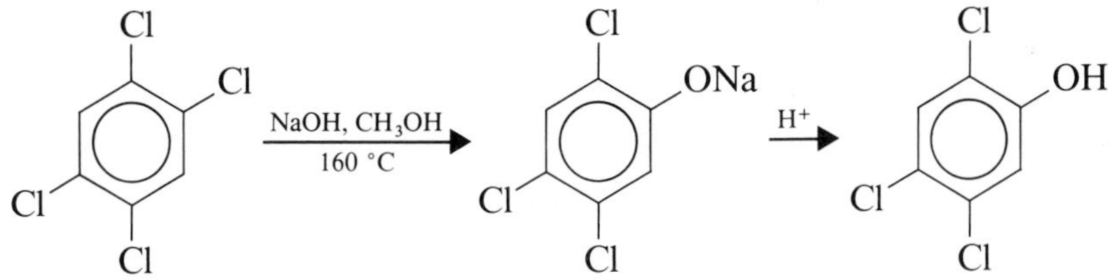

1,2,4,5-Tetra-
chlorobenzene

Sodium 2,4,5-
trichlorophenoxide

2,4,5-Trichloro-
phenol

The temperature of the first reaction must be very carefully controlled; if it is not, this dioxin forms in the reaction mixture:

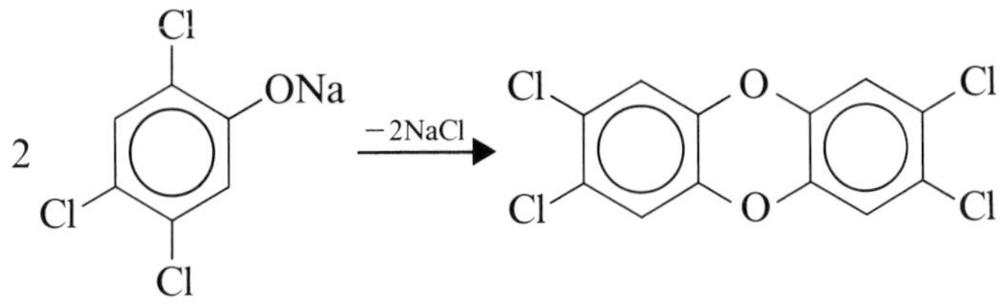

Apparently at the Italian factory the temperature got out of control causing the pressure to build up. Eventually a valve opened and released a cloud of trichlorophenols and the dioxin into the atmosphere. Many wild and domestic animals were killed and many people, especially children, were afflicted with severe skin rashes.

PROBLEM H.3

(a) Assume that the ortho and para chlorine atoms provide enough activation by electron withdrawal for nucleophilic substitution to occur by an addition–elimination pathway and outline a possible mechanism for the conversion of 1,2,4,5-tetrachlorobenzene to sodium 2,4,5-trichlorophenoxide. (b) Do the same for the conversion of 2,4,5-trichlorophenoxide to the dioxin of Section H.2.

PROBLEM H.4

2,4,5-T is made by allowing sodium 2,4,5-trichlorophenoxide to react with sodium chloroacetate ($ClCH_2COONa$). (This produces the sodium salt of 2,4,5-T, which, on acidification, gives 2,4,5-T itself.) What kind of mechanism accounts for the reaction of sodium 2,4,5-trichlorophenoxide with $ClCH_2COONa$? Write the equation.

H.3 Germicides

2,4,5-Trichlorophenol is also used in the manufacture of hexachlorophene, a germicide once widely used in soaps, shampoos, deodorants, mouthwashes, aftershave lotions, and other over-the-counter products.

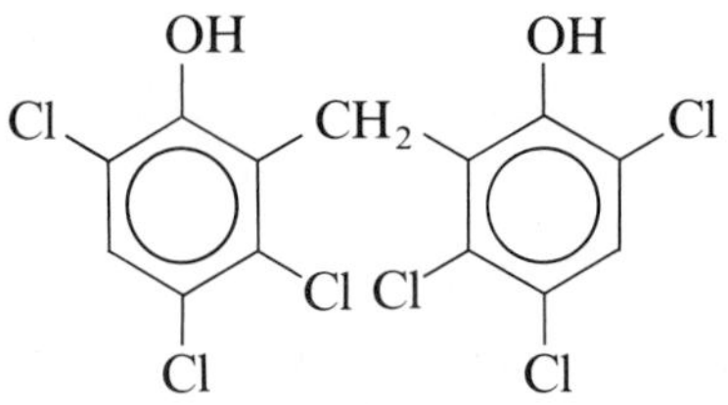

Hexachlorophene

Hexachlorophene is absorbed intact through the skin and tests with experimental animals have shown that it causes brain damage. Since 1972, the use of hexachlorophene in cleansers and cosmetics sold over the counter has been banned by the Food and Drug Administration.

H.4 Polychlorinated Biphenyls (PCBs)

Mixtures of polychlorinated biphenyls have been produced and used commercially since 1929. In these mixtures, biphenyls with chlorine atoms at any of the numbered positions (see following structure) may be present. In all, there are 210 possible compounds. A typical commercial mixture may contain as many as 50 different PCBs. Mixtures are usually classified on the basis of their chlorine content, and most industrial mixtures contain from 40 to 60% chlorine.

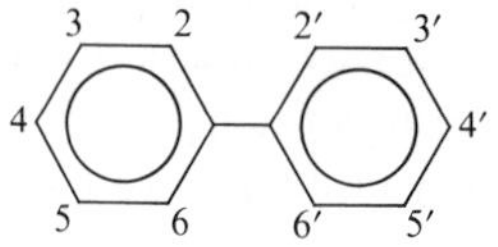

Biphenyl

PCBs have had a multitude of uses: as heat exchange agents in transformers; in capacitors, thermostats, and hydraulic systems; as plasticizers in polystyrene coffee cups, frozen food bags, bread wrappers, and plastic liners for baby bottles. They have been used in printing inks, in carbonless carbon paper, and as waxes for making molds for metal castings. Between 1929 and 1972 about 500,000 metric tons of PCBs were manufactured.

Although they were never intended for release into the environment, PCBs have become, perhaps more than any other chemical, the most widespread pollutant. They have been found in rain water, in many species of fish, birds, and others animals (including polar bears) all over the globe, and in human tissue.

PCBs are highly persistent and being fat soluble tend to accumulate in the food chain. Fish that feed in PCB-contaminated waters, for example, have PCB levels 1000 to 100,000 times the level of the surrounding water, and this amount is further magnified in birds that feed on the fish. The toxicity of PCBs depends on the composition of the individual mixture. The largest incident of human poisoning by PCBs occurred in Japan in 1968 when about 1000 people ingested a cooking oil accidentally contaminated with PCBs.

As late as 1975 industrial concerns were legally discharging PCBs into the Hudson River. In 1977 the Environmental Protection Agency banned the direct discharge into waterways, and since 1979 their manufacture, processing, and distribution have been prohibited.

H.5 Polybromobiphenyls (PBBs)

Polybromobiphenyls are bromine analogs of PCBs that have been used as flame retardants. In 1973, in Michigan, a mistake at a chemical company led to PBBs being mixed into animal feeds that were sold to farmers. Before the mistake was recognized, PBBs had affected thousands of dairy cattle, hogs, chickens, and sheep, necessitating their destruction.

H.6 Organometallic Compounds

With few exceptions, organometallic compounds are toxic. This toxicity varies greatly depending on the nature of the organometallic compound and the identity of the metal. Organic compounds of arsenic, antimony, lead, thallium, and mercury are toxic because the metal ions, themselves, are toxic. Certain organic derivatives of silicon are toxic even though silicon and most of its inorganic compounds are nontoxic.

Early in this century the recognition of the biocidal effects of organoarsenic compounds led Paul Ehrlich to his pioneering work in chemotherapy. Ehrlich sought compounds (which he called "magic bullets") that would show greater toxicity toward disease-causing microorganisms than they would toward their hosts. Ehrlich's research led to the development of Salvarsan and Neosalvarsan, two organoarsenic compounds that were used successfully in the treatment of diseases caused by spirochetes (e.g., syphilis) and trypanosomes (e.g., sleeping sickness). Salvarsan and Neosalvarsan are no longer used in the treatment of these diseases; they have been displaced by safer and more effective antibiotics. Ehrlich's research, however, initiated the field of chemotherapy (cf. Section 19.11).

Many microorganisms actually synthesize organometallic compounds, and this discovery has an alarming ecological aspect. Mercury metal is toxic, but mercury metal is also unreactive. In the past untold tons of mercury metal present in industrial wastes have been disposed of by simply dumping such wastes into lakes and streams. Since mercury is toxic, many bacteria protect themselves from its effect by converting mercury metal to methylmercury ions (CH_3Hg^+) and to gaseous dimethylmercury $(CH_3)_2Hg$. These organic mercury compounds are passed up the food chain (with modification) through fish to humans where methylmercury ions act as a deadly nerve poison. Between 1953 and 1964, 116 people in Minamata, Japan, were poisoned by eating fish containing methylmercury compounds. Arsenic is also methylated by organisms to the poisonous dimethylarsine, $(CH_3)_2AsH$.

Ironically, chlorinated hydrocarbons appear to inhibit the biological reactions that bring about mercury methylation. Lakes polluted with organochlorine pesticides show significantly lower mercury methylation. While this particular interaction of two pollutants may, in a certain sense, be beneficial, it is also instructive of the complexity of the environmental problems that we may face.

Tetraethyllead and other alkyllead compounds have been used as antiknock agents in gasoline since 1923. Although this use is now being phased out, more than 1 trillion lb of lead have been introduced into the atmosphere. In the northern hemisphere, gasoline burning alone has spread about 10 mg of lead on each square meter of the earth's surface. In highly industrialized areas the amount of lead per square meter is probably several hundred times higher. Because of the well-known toxicity of lead, these facts are of great concern.

CHAPTER SIXTEEN

Citric Acid Crystals. Citric acid, $HO_2CCH_2C(OH)(CO_2H)CH_2CO_2H$, got its name because it occurs in citrus fruits. Citric acid is also widely distributed in plant and animal tissues and fluids.

Aldehydes and Ketones I. Nucleophilic Additions to the Carbonyl Group

16.1 INTRODUCTION

Except for formaldehyde, the simplest aldehyde, all aldehydes have a carbonyl group,

$$\overset{O}{\underset{\parallel}{-C-}},$$ bonded on one side to a carbon, and on the other side to a hydrogen. In ketones, the carbonyl group is situated between two carbon atoms.

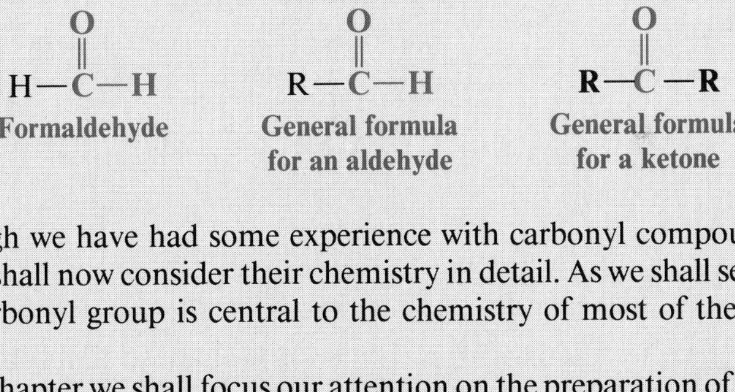

$$
\begin{array}{ccc}
\overset{O}{\underset{\parallel}{H-C-H}} & \overset{O}{\underset{\parallel}{R-C-H}} & \overset{O}{\underset{\parallel}{R-C-R}} \\
\text{Formaldehyde} & \text{General formula} & \text{General formula} \\
 & \text{for an aldehyde} & \text{for a ketone}
\end{array}
$$

Although we have had some experience with carbonyl compounds in earlier chapters, we shall now consider their chemistry in detail. As we shall see, the chemistry of the carbonyl group is central to the chemistry of most of the chapters that follow.

In this chapter we shall focus our attention on the preparation of aldehydes and ketones, on their physical properties, and especially on *the nucleophilic addition reactions that take place at their carbonyl groups*. In the next chapter we shall study the chemistry of aldehydes and ketones *that results from the acidity of the hydrogen atoms on the carbon atoms adjacent to their carbonyl groups*.

16.2 NOMENCLATURE OF ALDEHYDES AND KETONES

In the IUPAC system we name aliphatic aldehydes substitutively by replacing the final **e** of the name of the corresponding alkane with **al**. Since the aldehyde group must be at the end of the chain of carbon atoms, there is no need to indicate its position. When other substituents are present, however, we must designate the carbonyl group carbon as position 1. Many aldehydes also have common names; these are given here in parentheses. These common names are derived from the common names for the corresponding carboxylic acids (Section 18.2) and some of them are retained by the IUPAC as acceptable names.

$$\underset{\text{Methanal}}{\underset{\text{(formaldehyde)}}{H—\overset{\overset{O}{\|}}{C}—H}} \qquad \underset{\text{Ethanal}}{\underset{\text{(acetaldehyde)}}{CH_3\overset{\overset{O}{\|}}{C}—H}} \qquad \underset{\text{Propanal}}{\underset{\text{(propionaldehyde)}}{CH_3CH_2\overset{\overset{O}{\|}}{C}—H}}$$

$$\underset{\text{5-Chloropentanal}}{ClCH_2CH_2CH_2CH_2\overset{\overset{O}{\|}}{C}—H} \qquad \underset{\text{Phenylethanal}}{\underset{\text{(phenylacetaldehyde)}}{C_6H_5CH_2\overset{\overset{O}{\|}}{C}—H}}$$

When the carbonyl group is attached to an aromatic ring, we name the compound as a benzaldehyde, toluadehyde, naphthaldehyde, and so on.

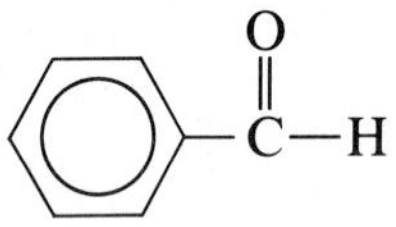

Benzaldehyde 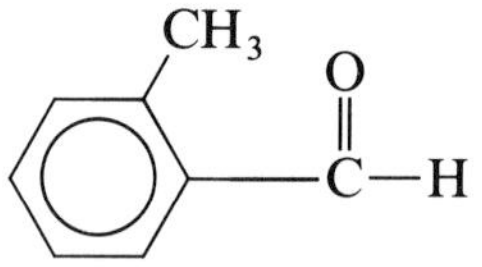 2-Methylbenzaldehyde 2-Naphthaldehyde
(*o*-tolualdehyde)

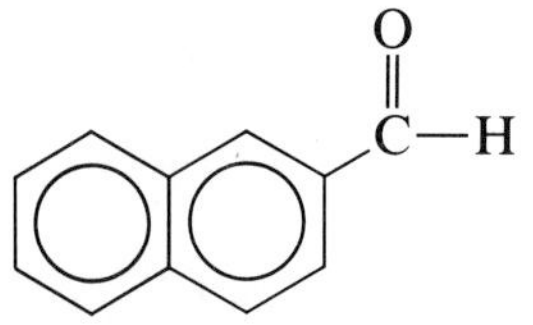

We name aliphatic ketones substitutively by replacing the final **e** of the name of the corresponding alkane with **one**. We then number the chain in the way that gives the carbonyl carbon atom the lower possible number, and we use this number to designate its position.

$$\underset{\underset{\text{(acetone)}}{\text{2-Propanone}}}{CH_3\underset{\underset{O}{\|}}{C}CH_3} \qquad \underset{\underset{\text{(ethyl methyl ketone)}}{\text{2-Butanone}}}{CH_3CH_2\underset{\underset{O}{\|}}{C}CH_3} \qquad \underset{\underset{\text{(methyl propyl ketone)}}{\text{2-Pentanone}}}{CH_3\overset{\overset{O}{\|}}{C}CH_2CH_2CH_3} \qquad \underset{\underset{\text{(\textit{not} 1-penten-4-one)}}{\text{4-Penten-2-one}}}{CH_3\overset{\overset{O}{\|}}{C}CH_2CH=CH_2}$$

Common names for ketones (in parentheses) are obtained simply by separately naming the two groups attached to the carbonyl group and adding the word **ketone** as a separate word.

Some aryl ketones have special names.

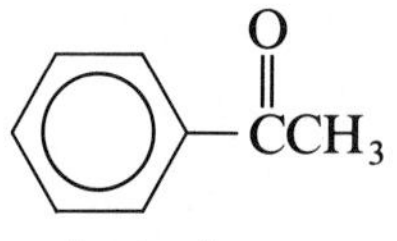

Acetophenone   Benzophenone
(methyl phenyl ketone) (diphenyl ketone)

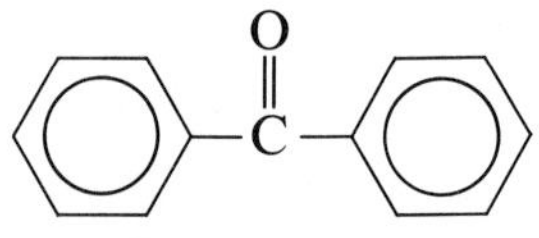

When we find it necessary to name the $—\overset{\overset{O}{\|}}{C}H$ group as a substituent, we call it the **methanoyl** or **formyl group.** It is often written —CHO. When $R\overset{\overset{O}{\|}}{C}—$ groups are named as substituents, they are called **alkanoyl** or **acyl groups.**

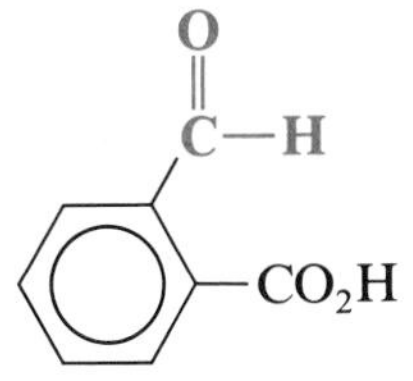

$$\underset{\substack{\text{2-Methanoylbenzoic acid}\\ (\textit{o}\text{-formylbenzoic acid})}}{\overset{\displaystyle \overset{\text{O}}{\underset{\|}{\text{C}}}-\text{H}}{\bigcirc\!\!-\text{CO}_2\text{H}}}$$

2-Methanoylbenzoic acid
(*o*-formylbenzoic acid)

$$\text{CH}_3\overset{\text{O}}{\underset{\|}{\text{C}}}-\bigcirc\!\!-\text{SO}_3\text{H}$$

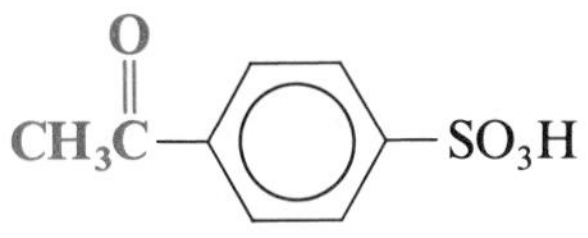

4-Ethanoylbenzenesulfonic acid
(*p*-acetylbenzenesulfonic acid)

PROBLEM 16.1

(a) Give IUPAC names for the seven isomeric aldehydes and ketones with the formula $C_5H_{10}O$. (b) How many aldehydes and ketones contain a benzene ring and have the formula C_8H_8O? (c) What are their names and structures?

16.3 PHYSICAL PROPERTIES

The carbonyl group is a polar group; therefore aldehydes and ketones have higher boiling points than hydrocarbons of the same molecular weight. However, since aldehydes and ketones cannot have strong hydrogen bonds between their molecules, they have lower boiling points than corresponding alcohols.

$CH_3CH_2CH_2CH_3$	$CH_3CH_2\overset{\text{O}}{\underset{\|}{\text{C}}}H$	$CH_3\overset{\text{O}}{\underset{\|}{\text{C}}}CH_3$	$CH_3CH_2CH_2OH$
Butane	Propanal	Acetone	1-Propanol
bp, $-0.5\ °C$	bp, $49\ °C$	bp, $56.1\ °C$	bp, $97.2\ °C$

PROBLEM 16.2

Which compound in each of the following pairs listed has the higher boiling point? (Answer this problem without consulting tables.)

(a) Pentanal or 1-pentanol   (d) Acetophenone or 2-phenylethanol

(b) 2-Pentanone or 2-pentanol   (e) Benzaldehyde or benzyl alcohol

(c) Pentane or pentanal

The carbonyl oxygen atom allows molecules of aldehydes and ketones to form strong hydrogen bonds to molecules of water. As a result, low-molecular-weight aldehydes and ketones show appreciable solubilities in water. Acetone and acetaldehyde are soluble in water in all proportions.

Table 16.1 lists the physical properties of a number of common aldehydes and ketones.

TABLE 16.1 Physical properties of aldehydes and ketones

FORMULA	NAME	mp (°C)	bp (°C)	SOLUBILITY IN WATER
HCHO	Formaldehyde	−92	−21	Very soluble
CH_3CHO	Acetaldehyde	−125	21	∞
CH_3CH_2CHO	Propanal	−81	49	Very soluble
$CH_3(CH_2)_2CHO$	Butanal	−99	76	Soluble
$CH_3(CH_2)_3CHO$	Pentanal	−91.5	102	Sl. soluble
$CH_3(CH_2)_4CHO$	Hexanal	−51	131	Sl. soluble
C_6H_5CHO	Benzaldehyde	−26	178	Sl. soluble
$C_6H_5CH_2CHO$	Phenylacetaldehyde	33	193	Sl. soluble
CH_3COCH_3	Acetone	−95	56.1	∞
$CH_3COCH_2CH_3$	2-Butanone	−86	79.6	Very soluble
$CH_3COCH_2CH_2CH_3$	2-Pentanone	−78	102	Soluble
$CH_3CH_2COCH_2CH_3$	3-Pentanone	−39	102	Soluble
$C_6H_5COCH_3$	Acetophenone	21	202	Insoluble
$C_6H_5COC_6H_5$	Benzophenone	48	306	Insoluble

Some aromatic aldehydes obtained from natural sources have very pleasant fragrances. Some of these are the following:

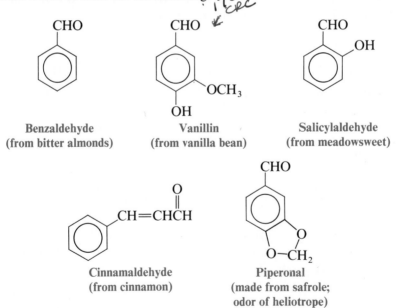

Benzaldehyde
(from bitter almonds)

Vanillin
(from vanilla bean)

Salicylaldehyde
(from meadowsweet)

Cinnamaldehyde
(from cinnamon)

Piperonal
(made from safrole;
odor of heliotrope)

16.4 SYNTHESIS OF ALDEHYDES

16.4A Laboratory Syntheses

An aldehyde can be prepared by oxidizing a primary alcohol, or by reducing an acyl chloride, ester, or nitrile. However, because aldehydes, themselves, are easily oxidized and reduced, we must use special reagents for these procedures to avoid over oxidation or over reduction. An aldehyde can also be synthesized by the anti-Markovnikov addition of water to a terminal alkyne. These methods are summarized in the following chart and discussed in detail in the paragraphs that follow:

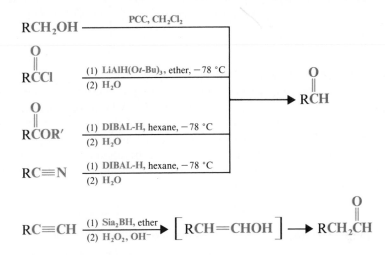

1. **Aldehydes from Primary Alcohols:** $RCH_2OH \longrightarrow RCHO$ (discussed in Section 15.3). Primary alcohols can be oxidized to aldehydes using pyridinium chlorochromate (PCC) in CH_2Cl_2 as a solvent. The following is an example:

$$CH_3(CH_2)_5CH_2OH + C_5H_5NH^+CrO_3Cl^- \xrightarrow{CH_2Cl_2} CH_3(CH_2)_5CHO$$

1-Heptanol	PCC	(93%)
		Heptanal

2. **Aldehydes from Acyl Chlorides:** $RCOCl \longrightarrow RCHO$. Acyl chlorides can be reduced to aldehydes by treating them with lithium tri-*tert*-butoxyaluminum hydride, $LiAlH[OC(CH_3)_3]_3$, at -78 °C. (Carboxylic acids can be converted to acyl chlorides by using $SOCl_2$, Section 12.7.)

$$\underset{RCOH}{\overset{O}{\|}} \xrightarrow{SOCl_2} \underset{RCCl}{\overset{O}{\|}} \xrightarrow[\text{(2) H}_2\text{O}]{\text{(1) LiAlH(O}t\text{-Bu)}_3\text{, ether, }-78\text{ °C}} \underset{RCH}{\overset{O}{\|}}$$

The following is a specific example:

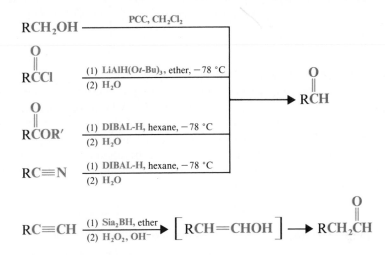

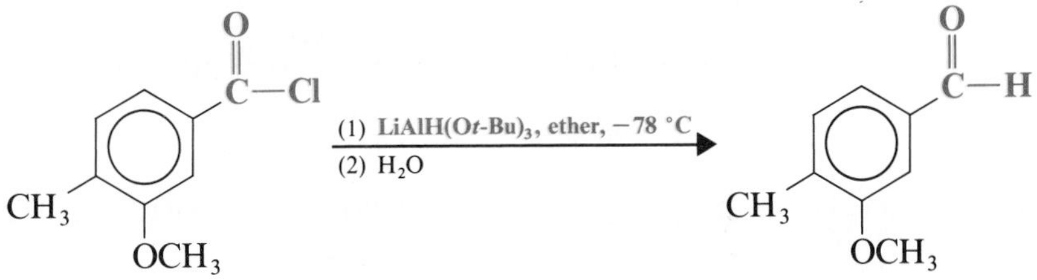

Lithium tri-*tert*-butoxyaluminum hydride is much less reactive than $LiAlH_4$; consequently, it does not reduce the aldehyde that is formed to a primary alcohol. Mechanistically, the reduction is brought about by the transfer of a hydride ion from the aluminum atom to the carbonyl carbon of the acyl chloride (cf. Section 15.2). Subsequent hydrolysis frees the aldehyde.

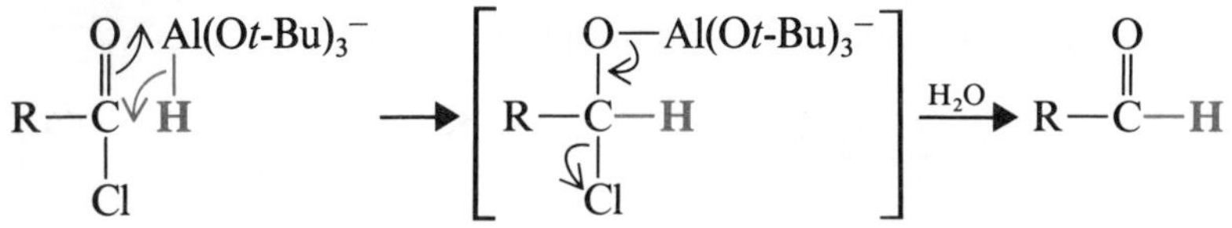

3. **Aldehydes from Esters and Nitriles:** $RCO_2R' \longrightarrow$ RCHO and RC≡N $\longrightarrow$ RCHO. Esters and nitriles can be reduced to aldehydes by use of the reagent diisobutylaluminum hydride, $[(CH_3)_2CHCH_2]_2AlH$ [abbreviated $(i\text{-}Bu_2AlH)$ or **DIBAL-H**]. Carefully controlled amounts of the reagent must be used to avoid over reduction and the reactions must be carried out at low temperatures. Both reductions result in the formation of a relatively stable intermediate by the addition of a hydride ion to the carbonyl carbon of the ester or to the carbon of the —C≡N group of the nitrile. Hydrolysis of the intermediate liberates the aldehyde. Schematically the reactions can be viewed this way:

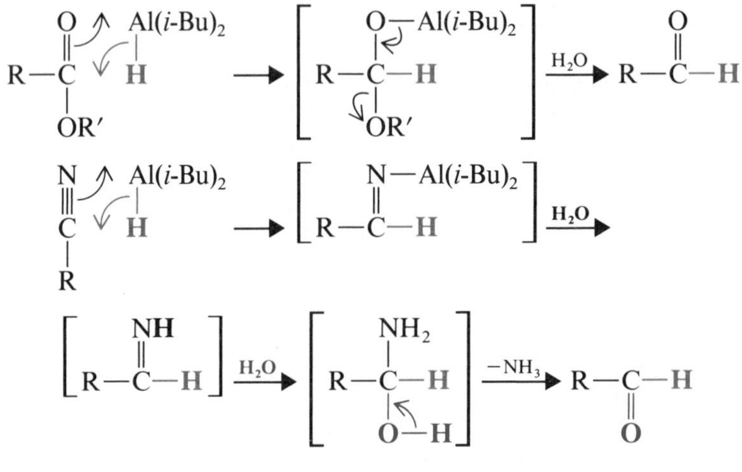

The following specific examples illustrate these syntheses.

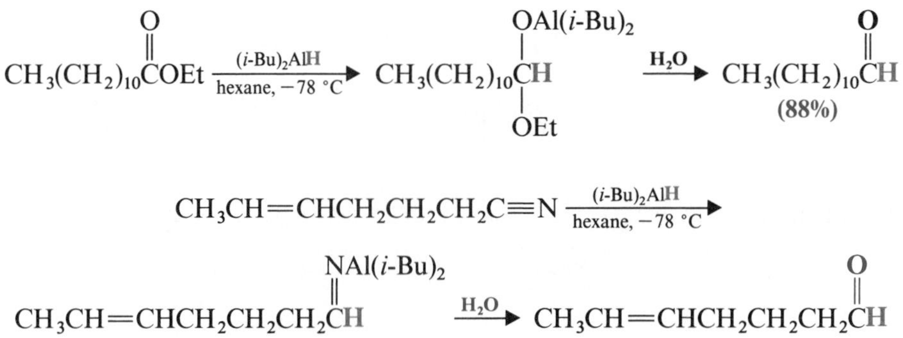

4. **Aldehydes from Terminal Alkynes:** RC≡CH $\longrightarrow$ RCH_2CHO. Aldehydes can also be prepared by hydroboration (Section 8.6) of terminal alkynes using a hindered borane followed by oxidation with hydrogen peroxide (H_2O_2) in basic solution.

A commonly used hindered borane for this purpose is the compound called **disiamylborane.** (Siamyl is a common name for the 2,3-dimethylpropyl group.)

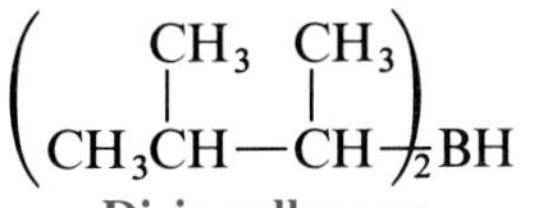

$$\left(\begin{array}{cc} CH_3 & CH_3 \\ | & | \\ CH_3CH-CH \end{array} \right)_2 BH \qquad \text{abbreviated as } Sia_2BH$$

Disiamylborane

Disiamylborane adds only once to an alkyne to yield a vinylic borane:

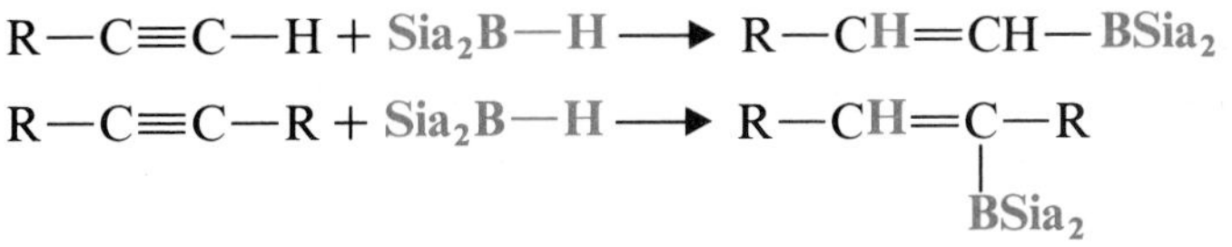

$$R-C\equiv C-H + Sia_2B-H \longrightarrow R-CH=CH-BSia_2$$

$$R-C\equiv C-R + Sia_2B-H \longrightarrow R-CH=\underset{\underset{BSia_2}{|}}{C}-R$$

These vinylic boranes can then be oxidized with H_2O_2 in basic solution to produce aldehydes or ketones (via the enol):

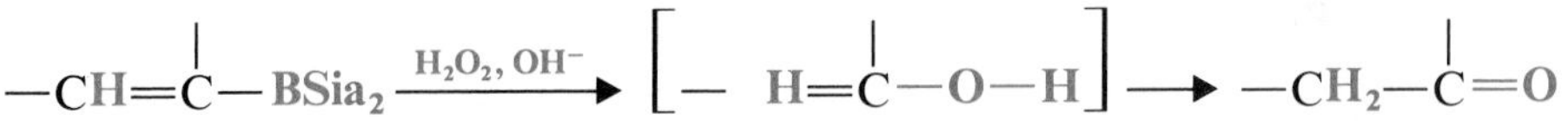

$$-CH=\underset{\underset{|}{|}}{C}-BSia_2 \xrightarrow{H_2O_2,\ OH^-} \left[-\ H=\underset{|}{\overset{|}{C}}-O-H \right] \longrightarrow -CH_2-\underset{|}{\overset{|}{C}}=O$$

Vinylic borane   **Enol**   **Aldehyde or ketone**

An example is the conversion of 1-hexyne into the aldehyde, hexanal. In the first step hydroboration with Sia_2BH occurs so as to yield the least-hindered vinylic borane:

$$CH_3(CH_2)_3C\equiv CH + Sia_2BH \longrightarrow CH_3(CH_2)_3CH=CH-BSia_2$$

1-Hexyne   **(100%)**

Oxidation of this product with alkaline hydrogen peroxide produces the aldehyde via the enol.

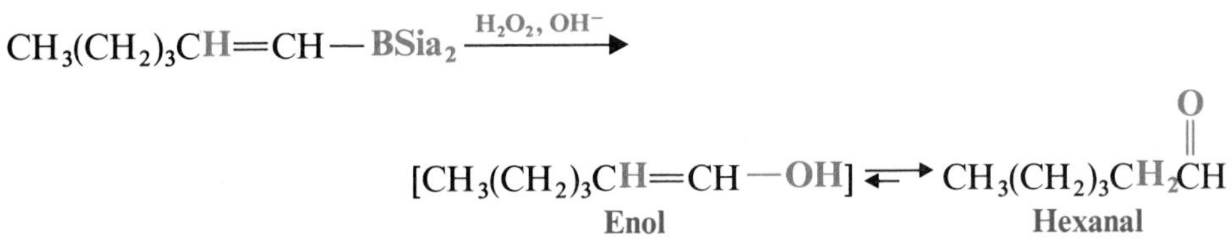

$$CH_3(CH_2)_3CH=CH-BSia_2 \xrightarrow{H_2O_2,\ OH^-}$$

$$[CH_3(CH_2)_3CH=CH-OH] \rightleftharpoons CH_3(CH_2)_3CH_2\overset{\overset{O}{||}}{C}H$$

Enol   **Hexanal**

Overall, this procedure provides an anti-Markovnikov addition of H— and —OH to an alkyne, and thus it is a convenient synthesis of aldehydes.

PROBLEM 16.3

What other organic product would you expect to obtain from the oxidation of $CH_3(CH_2)_3CH=CH-BSia_2$ with hydrogen peroxide in base?

PROBLEM 16.4

Hydroboration of an alkyne with an internal triple bond, followed by proton-olysis of the vinylic borane with acetic acid (Section 8.7B), provides another method (Section 6.7A) for synthesizing a *cis*-alkene. Show how you would use this method to prepare *cis*-3-hexene from 3-hexyne.

PROBLEM 16.5

Show how you would synthesize propanal from each of the following: (a) 1-Propanol, (b) propyne, and (c) propanoic acid ($CH_3CH_2CO_2H$).

16.4B An Industrial Process.
Hydroformylation—The Oxo Reaction

An important process used in the chemical industry to synthesize aldehydes from alkenes is one called *hydroformylation* or the *Oxo reaction.* In this process a terminal alkene reacts with carbon monoxide and hydrogen in the presence of a cobalt cata-lyst, $HCo(CO)_4$.

$$RCH{=}CH_2 + CO + H_2 \xrightarrow[\text{2000 psi, 110--150 °C}]{HCo(CO)_4} RCH_2CH_2CHO + \text{isomers}$$

EXAMPLE

$$CH_3CH{=}CH_2 + CO + H_2 \xrightarrow[\text{2000 psi, 110--150 °C}]{HCo(CO)_4} CH_3CH_2CH_2\overset{\overset{\displaystyle O}{\|}}{C}H + CH_3\overset{\overset{\displaystyle O}{\|}}{C}HCH$$
$$\underset{\substack{| \\ CH_3 \\ \text{(minor} \\ \text{product)}}}{}$$

This reaction is called *hydroformylation* because the effect of it is to add a hydrogen atom (—H) and a formyl group (—CHO) across the double bond.

If the hydroformylation is carried out at a higher temperature (185 °C), the products consist principally of alcohols. Under these conditions the $HCo(CO)_4$ cata-lyst acts as a hydrogenation catalyst and causes the aldehydes to be reduced to primary alcohols. The mechanism of hydroformylation is discussed in Section G.7.

$$RCH_2CH_2CHO + \underset{\substack{| \\ CH_3}}{RCHCHO} \xrightarrow[HCo(CO)_4]{H_2} RCH_2CH_2CH_2OH + \underset{\substack{| \\ CH_3}}{RCHCH_2OH} \quad \blacksquare$$

16.5 SYNTHESIS OF KETONES

We have seen four laboratory methods for the preparation of ketones in earlier chapters.

 1. **Ketones (and Aldehydes) by Ozonolysis of Alkenes** (discussed in Section 7.10).

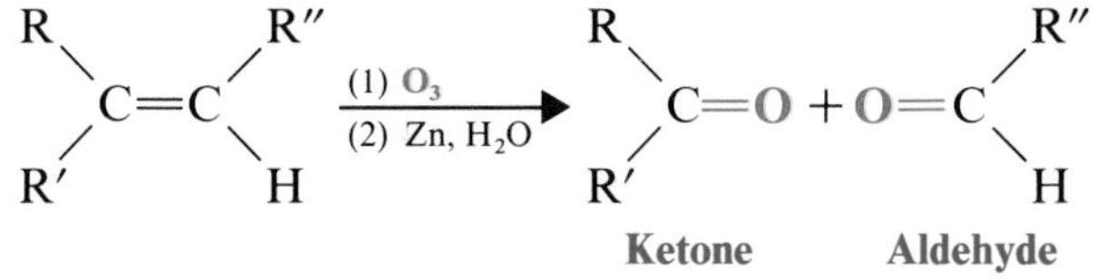

$$\underset{R'}{\overset{R}{>}}C=C\underset{H}{\overset{R''}{<}} \xrightarrow[\text{(2) Zn, H}_2\text{O}]{\text{(1) O}_3} \underset{R'}{\overset{R}{>}}C=O + O=C\underset{H}{\overset{R''}{<}}$$

Ketone   Aldehyde

2. **Ketones from Friedel–Crafts Acylations** (discussed in Section 12.7).

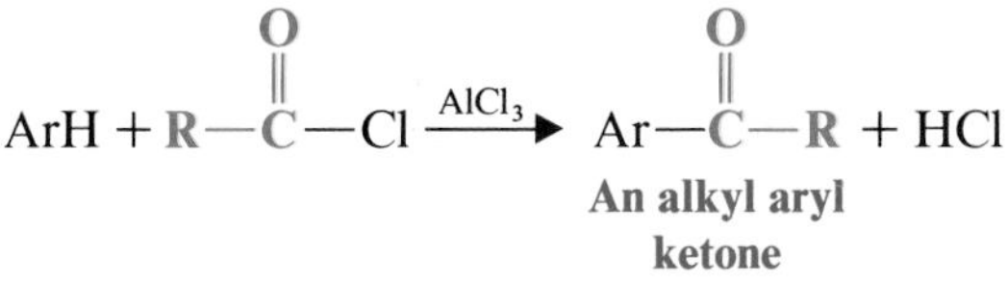

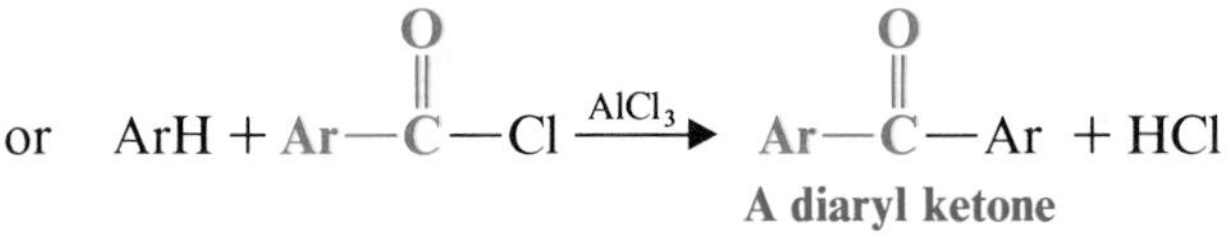

$$ArH + R-\overset{\overset{\displaystyle O}{\|}}{C}-Cl \xrightarrow{\text{AlCl}_3} Ar-\overset{\overset{\displaystyle O}{\|}}{C}-R + HCl$$

An alkyl aryl
ketone

or $$ArH + Ar-\overset{\overset{\displaystyle O}{\|}}{C}-Cl \xrightarrow{\text{AlCl}_3} Ar-\overset{\overset{\displaystyle O}{\|}}{C}-Ar + HCl$$

A diaryl ketone

3. **Ketones from Oxidations of Secondary Alcohols** (discussed in Section 15.3).

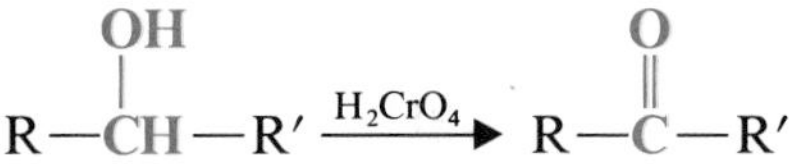

$$R-\overset{\overset{\displaystyle OH}{|}}{C}H-R' \xrightarrow{\text{H}_2\text{CrO}_4} R-\overset{\overset{\displaystyle O}{\|}}{C}-R'$$

4. **Ketones from Alkynes** (discussed in Section 7.15). Alkynes can be converted to ketones by hydration (with H_3O^+, Hg^{2+}, and H_2O).

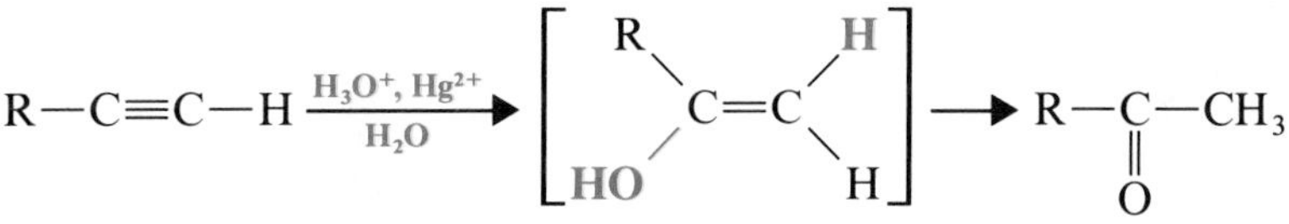

$$R-C\equiv C-H \xrightarrow[\text{H}_2\text{O}]{\text{H}_3\text{O}^+, \text{Hg}^{2+}} \left[\underset{HO}{\overset{R}{>}}C=C\underset{H}{\overset{H}{<}} \right] \longrightarrow R-\overset{\overset{\displaystyle }{}}{\underset{\underset{\displaystyle O}{\|}}{C}}-CH_3$$

Three other laboratory methods for the preparation of ketones are based on the use of organometallic compounds.

5. **Ketones from Lithium Dialkylcuprates.** When an ether solution of a lithium dialkylcuprate is treated with an acyl chloride at $-78\ °C$, the product is a ketone. This ketone synthesis is a variation of the Corey–Posner, Whitesides–House alkane synthesis (Section 15.9).

General Reaction

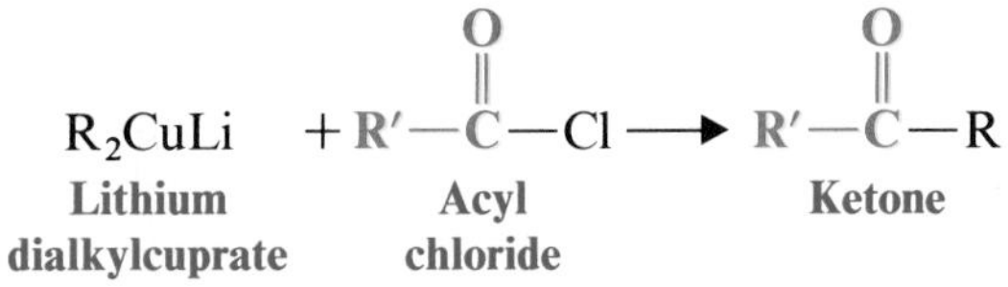

$$R_2CuLi + R'-\overset{\overset{\displaystyle O}{\|}}{C}-Cl \longrightarrow R'-\overset{\overset{\displaystyle O}{\|}}{C}-R$$

Lithium   Acyl   Ketone
dialkylcuprate   chloride

Specific Example

$$\text{C}_6\text{H}_{11}-\overset{\overset{\displaystyle O}{\|}}{C}-Cl + (CH_3)_2CuLi \xrightarrow[\text{ether}]{-78\ °C} \text{C}_6\text{H}_{11}-\overset{\overset{\displaystyle O}{\|}}{C}-CH_3$$

(81%)

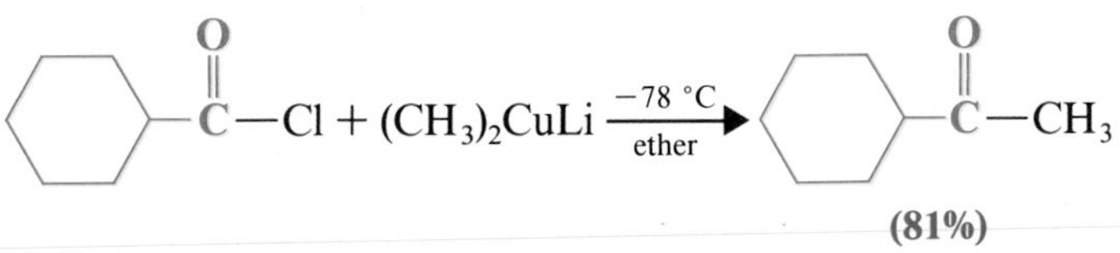

6. **Ketones from the Reaction of Nitriles with RMgX or RLi.** Treating a nitrile $(R—C\equiv N)$ with either a Grignard reagent or an organolithium reagent followed by hydrolysis yields a ketone.

General Reactions

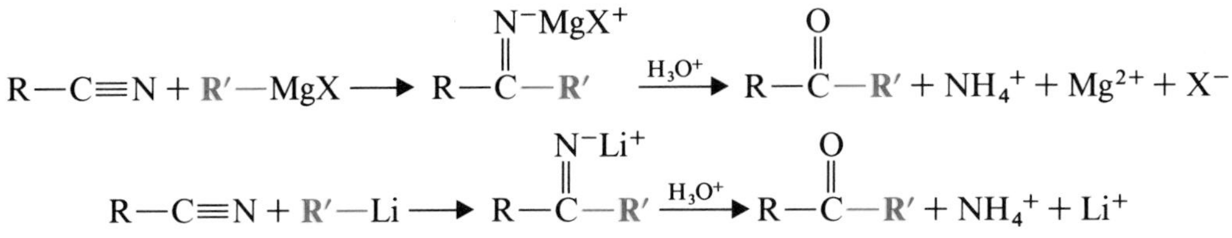

$$R—C\equiv N + R'—MgX \longrightarrow R—\overset{\overset{\displaystyle N^-MgX^+}{\|}}{C}—R' \xrightarrow{H_3O^+} R—\overset{\overset{\displaystyle O}{\|}}{C}—R' + NH_4^+ + Mg^{2+} + X^-$$

$$R—C\equiv N + R'—Li \longrightarrow R—\overset{\overset{\displaystyle N^-Li^+}{\|}}{C}—R' \xrightarrow{H_3O^+} R—\overset{\overset{\displaystyle O}{\|}}{C}—R' + NH_4^+ + Li^+$$

Specific Examples

$$C_6H_5—C\equiv N + CH_3CH_2CH_2CH_2Li \xrightarrow[(2) \; H_3O^+]{(1) \; ether} C_6H_5—\overset{\overset{\displaystyle O}{\|}}{C}—CH_2CH_2CH_2CH_3$$

$$\underset{\underset{\displaystyle CH_3}{|}}{CH_3CH}—C\equiv N + C_6H_5MgBr \xrightarrow[(2) \; H_3O^+]{(1) \; ether} \underset{\underset{\displaystyle CH_3}{|}}{CH_3CH}—\overset{\overset{\displaystyle O}{\|}}{C}—C_6H_5$$

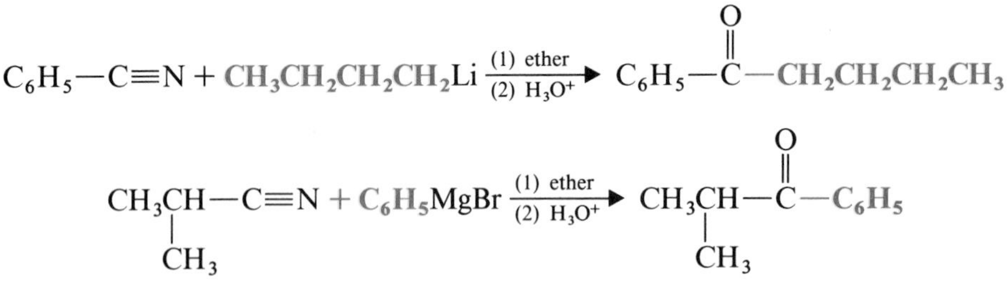

Even though a nitrile has a triple bond, addition of the Grignard or lithium reagent takes place only once. The reason: If addition took place twice, this would place a double negative charge on the nitrogen:

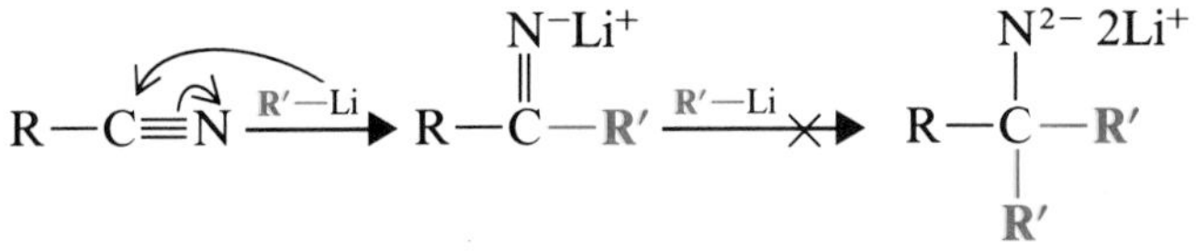

$$R—C\equiv N \xrightarrow{R'—Li} R—\overset{\overset{\displaystyle N^-Li^+}{\|}}{C}—R' \overset{R'—Li}{\nrightarrow} R—\underset{\underset{\displaystyle R'}{|}}{\overset{\overset{\displaystyle N^{2-} \; 2Li^+}{|}}{C}}—R'$$

(does not form)

7. **Ketones from the Reaction of a Carboxylic Acid with RLi.** Treating a carboxylic acid with two molar equivalents of an alkyl- or aryllithium (but not a Grignard reagent) yields a ketone.

General Reaction

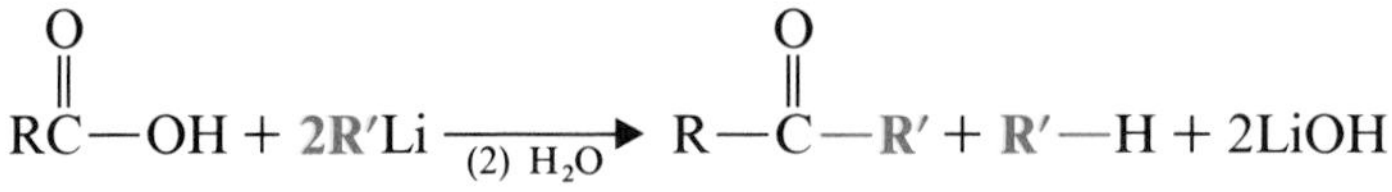

$$R\overset{\overset{\displaystyle O}{\|}}{C}—OH + 2R'Li \xrightarrow{(2) \; H_2O} R—\overset{\overset{\displaystyle O}{\|}}{C}—R' + R'—H + 2LiOH$$

Specific Examples

$$CH_3CH_2CH_2CH_2CH_2\overset{\overset{\displaystyle O}{\|}}{C}OH + 2CH_3Li \xrightarrow{(2) \; H_2O}$$

$$CH_3CH_2CH_2CH_2CH_2\overset{\overset{\displaystyle O}{\|}}{C}CH_3 + CH_3—H + 2LiOH$$

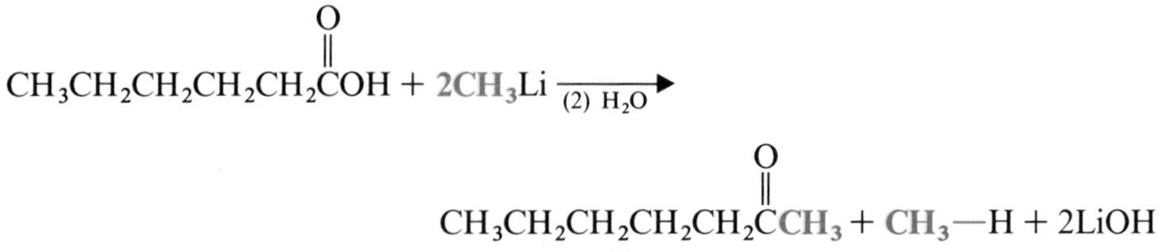

(83%)

The first step of this synthesis is an acid–base reaction between the carboxylic acid and the organolithium reagent. Then the second molar equivalent of the organolithium adds to the lithium carboxylate formed in the first step. (The reaction can be started with a lithium carboxylate itself, and this avoids wasting one molar equivalent of the organolithium.) Finally, addition of water leads to a *gem*-diol, which spontaneously loses water to produce a ketone (Section 8.16).

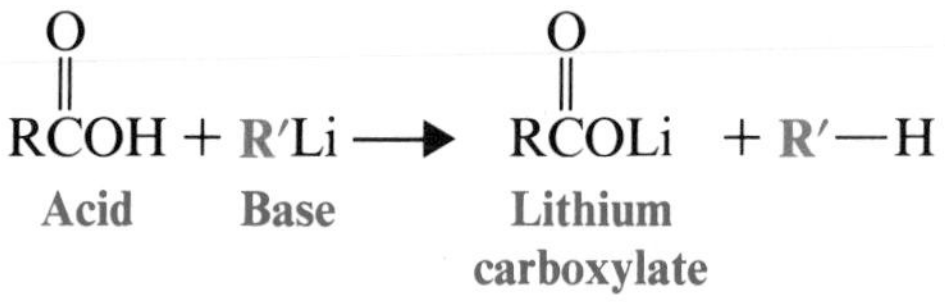

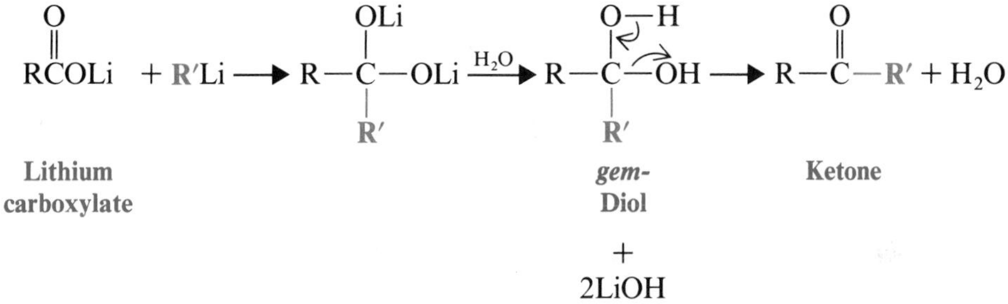

$$
\underset{\substack{\text{Acid}}}{\overset{\substack{O\\\parallel}}{RCOH}} + \underset{\substack{\text{Base}}}{R'Li} \longrightarrow \underset{\substack{\text{Lithium}\\\text{carboxylate}}}{\overset{\substack{O\\\parallel}}{RCOLi}} + R'{-}H
$$

$$
\underset{\substack{\text{Lithium}\\\text{carboxylate}}}{\overset{\substack{O\\\parallel}}{RCOLi}} + R'Li \longrightarrow \underset{\substack{|\\R'}}{R-\overset{\substack{OLi\\|}}{C}-OLi} \xrightarrow{H_2O} \underset{\substack{|\\R'\\ \textit{gem-}\\\text{Diol}}}{R-\overset{\substack{O-H\\|}}{C}-OH} \longrightarrow \underset{\substack{\text{Ketone}}}{\overset{\substack{O\\\parallel}}{R-C-R'}} + H_2O
$$

$$
+
$$
$$
2LiOH
$$

SAMPLE PROBLEM

Illustrating a Multistep Synthesis

With 1-butanol as your only organic starting compound, outline a synthesis of 5-nonanone.

Answer:

5-Nonanone can be synthesized by adding butylmagnesium bromide to the following nitrile.

$$
CH_3CH_2CH_2CH_2C\equiv N + CH_3CH_2CH_2CH_2MgBr \xrightarrow[\text{(2) } H_3O^+]{}
$$

$$
\underset{\substack{\text{5-Nonanone}}}{CH_3(CH_2)_3\overset{\substack{O\\\parallel}}{C}(CH_2)_3CH_3}
$$

The nitrile can be synthesized from butyl bromide and sodium cyanide in an S_N2 reaction.

$$
CH_3CH_2CH_2CH_2Br + NaCN \longrightarrow CH_3CH_2CH_2CH_2C\equiv N + NaBr
$$

Butyl bromide also can be used to prepare the Grignard reagent.

$$
CH_3CH_2CH_2CH_2Br + Mg \xrightarrow{\text{ether}} CH_3CH_2CH_2CH_2MgBr
$$

And, finally, butyl bromide can be prepared from 1-butanol.

$$
CH_3CH_2CH_2CH_2OH \xrightarrow{PBr_3} CH_3CH_2CH_2CH_2Br
$$

PROBLEM 16.6

Which reagents would you use to carry out each of the following reactions?

(a) Benzene ⟶ bromobenzene ⟶ phenylmagnesium bromide ⟶ benzyl alcohol ⟶ benzaldehyde

(b) Toluene ⟶ benzoic acid ⟶ benzoyl chloride ⟶ benzaldehyde

(c) Ethyl bromide ⟶ 1-butyne ⟶ butanal

(d) 2-Butyne ⟶ butanone

(e) 1-Phenylethanol ⟶ acetophenone

(f) Benzene ⟶ acetophenone

(g) Benzoyl chloride ⟶ acetophenone

(h) Benzoic acid ⟶ acetophenone

(i) Benzyl bromide ⟶ $C_6H_5CH_2CN$ ⟶ 1-phenyl-2-butanone

(j) $C_6H_5CH_2CN$ ⟶ 2-phenylethanal

(k) $CH_3(CH_2)_4CO_2CH_3$ ⟶ hexanal

16.6 GENERAL CONSIDERATION OF THE REACTIONS OF CARBONYL COMPOUNDS

Before we continue our detailed study of carbonyl compounds, it will be helpful if we examine the structures of carbonyl compounds in a general way. As we do this we shall find that certain structural features of carbonyl compounds underlie—*and thus unify*—most of the reactions that we shall discuss in this chapter and the next.

16.6A Structure of the Carbonyl Group

The carbonyl carbon atom is sp^2 hybridized; thus it, and the three atoms attached to it, lie in the same plane. The bond angles between the three attached atoms are what we would expect of a trigonal planar structure: They are approximately 120°.

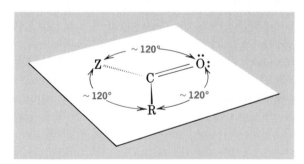

The carbonyl carbon atom bears a substantial partial *positive* charge; the carbonyl oxygen atom bears a substantial partial *negative* charge. This charge distribution arises because the electronegative oxygen atom makes the resonance contribution of the second structure shown here a significant one.

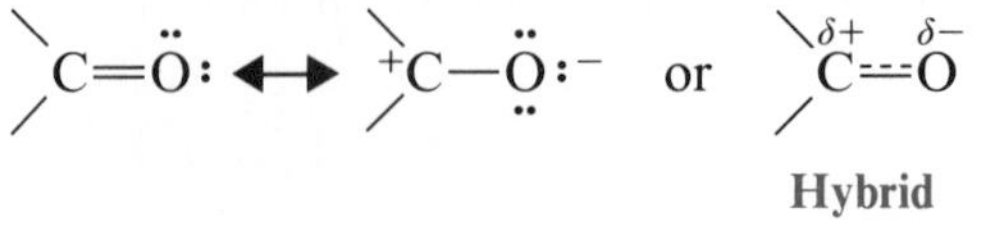

$$\mathrm{\overset{\diagdown}{\underset{\diagup}{C}}{=}\ddot{O}\!: \;\longleftrightarrow\; {}^{+}\overset{\diagdown}{\underset{\diagup}{C}}{-}\underset{\cdot\cdot}{\ddot{O}}\!:^{-}} \quad \text{or} \quad \mathrm{\overset{\diagdown}{\underset{\diagup}{C}}\!\!{=}\!\!\overset{\delta+\ \ \ \delta-}{O}}$$

Hybrid

Resonance structures for the carbonyl group

Evidence for this polarity of the carbon–oxygen bond can be found in the rather large dipole moments associated with carbonyl compounds. Acetone, for example, has a dipole moment of 2.88 D.

16.6B Nucleophilic Addition to the Carbon–Oxygen Double Bond

One highly characteristic reaction of aldehydes and ketones is a *nucleophilic addition* to the carbon–oxygen double bond.

$$\begin{array}{c} R \\ \diagdown \\ \diagup \\ H \end{array} C=O + H-Nu \rightleftarrows R-\overset{\overset{\displaystyle Nu}{|}}{\underset{\underset{\displaystyle H}{|}}{C}}-OH$$

Specific Examples

$$\begin{array}{c} CH_3 \\ \diagdown \\ \diagup \\ H \end{array} C=O + H-OCH_2CH_3 \rightleftarrows CH_3-\overset{\overset{\displaystyle OCH_2CH_3}{|}}{\underset{\underset{\displaystyle H}{|}}{C}}-OH$$

(a hemiacetal, see
Section 16.7)

$$\begin{array}{c} CH_3 \\ \diagdown \\ \diagup \\ CH_3 \end{array} C=O + H-CN \rightleftarrows CH_3-\overset{\overset{\displaystyle CN}{|}}{\underset{\underset{\displaystyle CH_3}{|}}{C}}-OH$$

(a cyanohydrin, see
Section 16.9)

Aldehydes and ketones are especially susceptible to nucleophilic addition because of the structural features that we have just mentioned. The trigonal planar arrangement of groups around the carbonyl carbon atom means that the carbonyl carbon atom is relatively open to attack from above or below. The positive charge on the carbonyl carbon atom means that it is especially susceptible to attack by a nucleophile. The negative charge on the carbonyl oxygen atom means that nucleophilic addition is susceptible to acid catalysis. We can visualize nucleophilic addition to the carbon–oxygen double bond occurring in either of two general ways:

1. When the reagent is a particularly strong nucleophile, addition will usually take place in the following way:

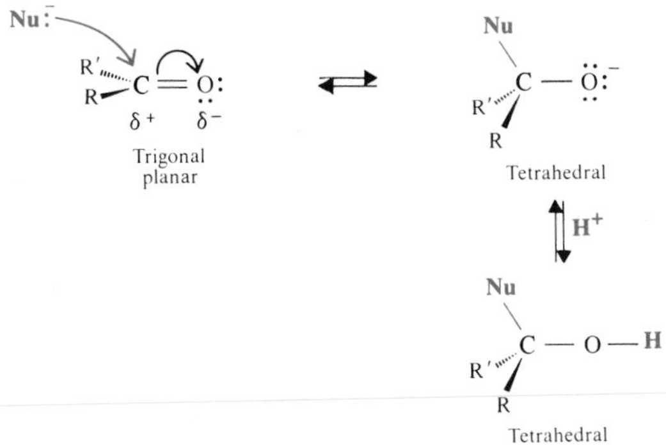

In this type of addition the nucleophile uses its electron pair to form a bond to the carbonyl carbon atom. As this happens an electron pair of the carbon–oxygen π bond shifts out to the carbonyl oxygen atom and the hybridization state of the carbon changes from sp^2 to sp^3. *The important aspect of this step is the ability of the carbonyl oxygen atom to accommodate the electron pair of the carbon–oxygen double bond.*

In the second step the oxygen atom associates with an electrophile (usually a proton). This happens because the oxygen atom is now much more basic; it carries a full negative charge, and it resembles the oxygen atom of an alkoxide ion. (In some reactions the oxygen atom of the carbonyl group actually becomes the oxygen atom of an alkoxide ion.)

2. A second general mechanism that operates in nucleophilic additions to carbon–oxygen double bonds is an acid-catalyzed mechanism:

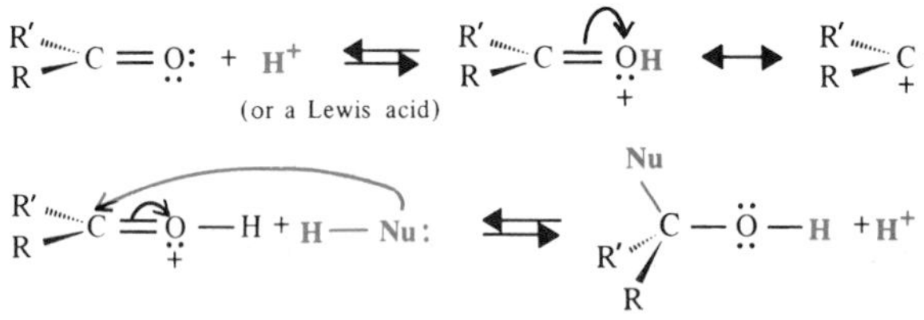

This mechanism operates when carbonyl compounds are treated with *strong acids* in the presence of *weak nucleophiles.* In the first step the acid attacks an electron pair of the carbonyl oxygen atom: The resulting protonated carbonyl compound, an **oxonium ion,** is highly reactive toward nucleophilic attack at the carbonyl carbon atom (in the second step) because the carbonyl carbon atom carries more positive charge than it does in the unprotonated compound.

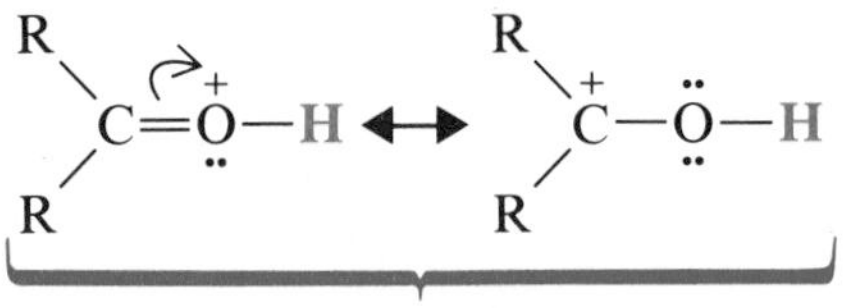

An oxonium ion

16.6C Reversibility of Nucleophilic Additions to the Carbon–Oxygen Double Bond

Many nucleophilic additions to carbon–oxygen double bonds are reversible; the overall results of these reactions will depend, therefore, on the position of an equilibrium. This behavior contrasts markedly with most electrophilic additions to carbon–carbon double bonds and with nucleophilic substitutions at saturated carbon atoms. The latter reactions are essentially irreversible, and overall results are a function of relative reaction rates.

16.6D Subsequent Reactions of Addition Products

Nucleophilic addition to a carbon-oxygen double bond may lead to a product that is stable under the reaction conditions that we employ. If this is the case we are then able to isolate products with the following general structure:

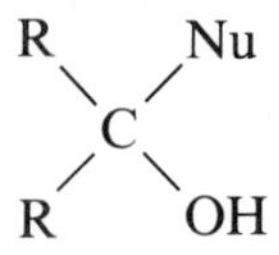

In other reactions the product formed initially may be unstable and may spontaneously undergo subsequent reactions. Even if the initial addition product is stable, however, we may deliberately bring about a subsequent reaction by changing the reaction conditions. When we begin our study of specific reactions, we shall see that one common subsequent reaction is an *elimination reaction,* especially *dehydration.*

PROBLEM 16.7

The reaction of an aldehyde or ketone with a Grignard reagent (Section 15.7) is a nucleophilic addition to the carbon–oxygen double bond.   (a) What is the nucleophile?   (b) The magnesium portion of the Grignard reagent plays an important part in this reaction. What is its function?   (c) What product is formed initially?   (d) What product forms when water is added?

PROBLEM 16.8

The reactions of aldehydes and ketones with $LiAlH_4$ and $NaBH_4$ (Section 15.2) are nucleophilic additions to the carbonyl group. What is the nucleophile in these reactions?

16.7 THE ADDITION OF WATER AND ALCOHOLS: HYDRATES, ACETALS, AND KETALS

16.7A Aldehyde Hydrates. *Gem*-Diols

Dissolving an aldehyde such as acetaldehyde in water causes the establishment of an equilibrium between the aldehyde and its **hydrate**. This hydrate is in actuality a *gem*-diol (Section 8.16).

$$\underset{\text{Acetaldehyde}}{\overset{CH_3}{\underset{H}{\diagdown}}C=O + H_2O} \rightleftarrows \underset{\substack{\text{Hydrate} \\ \text{(a }gem\text{-diol)}}}{\overset{CH_3}{\underset{H}{\diagdown}}C\overset{O-H}{\underset{O-H}{\diagup}}}$$

The *gem*-diol results from a nucleophilic addition of water to the carbonyl group of the aldehyde:

$$\underset{H}{\overset{CH_3}{\diagdown}}C=\ddot{O}: + :\ddot{O}H_2 \rightleftarrows \underset{H}{\overset{CH_3}{\diagdown}}C\overset{\overset{..}{O}H_2{}^+}{\underset{\underset{..}{\ddot{O}}{}^-}{\diagup}} \rightleftarrows \underset{H}{\overset{CH_3}{\diagdown}}C\overset{\overset{..}{O}H}{\underset{\ddot{O}H}{\diagup}}$$

In this step water attacks the carbonyl carbon atom

In this step a proton is lost from the positive oxygen atom and a proton is gained at the negative oxygen atom

PROBLEM 16.9

Dissolving formaldehyde in water leads to a solution containing primarily the *gem*-diol, $CH_2(OH)_2$. Show the steps in its formation from formaldehyde.

The addition of water is subject to catalysis by both acids and bases. That is, addition takes place much more rapidly in the presence of small amounts of acids or bases than it does in pure water.

The mechanism for the **base-catalyzed reaction** is as follows:

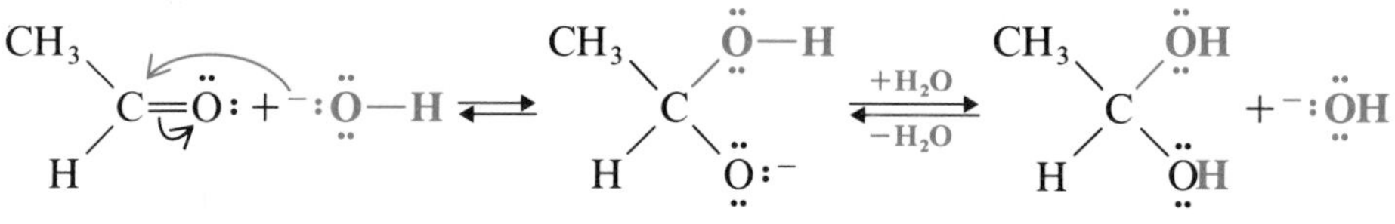

The important factor here in increasing the rate is the greater nucleophilicity of the hydroxide ion when compared to water.

The **acid-catalyzed mechanism** involves an initial rapid protonation of the carbonyl oxygen atom:

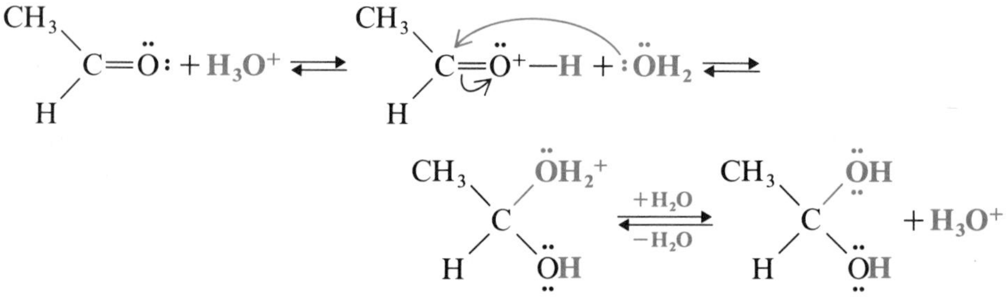

Protonation makes the carbonyl carbon atom more susceptible to attack by water, and here this factor is the key to the rate acceleration.

PROBLEM 16.10

When acetone is dissolved in water containing ^{18}O instead of ordinary ^{16}O (i.e., $H_2{}^{18}O$ instead of $H_2{}^{16}O$), the acetone soon begins to acquire ^{18}O and becomes $CH_3\overset{^{18}O}{\overset{\|}{C}}CH_3$. The formation of this oxygen-labeled acetone is catalyzed by traces of strong acids and by strong bases (e.g., OH^-). Show the steps that explain both the acid-catalyzed reaction and the base-catalyzed reaction.

As we discussed earlier (Section 8.16), it is not possible to isolate most *gem*-diols from the aqueous solutions in which they are formed. Evaporation of the water, for example, simply displaces the overall equilibrium to the left and the *gem*-diol (or hydrate) reverts to the carbonyl compound.

$$\underset{\substack{H \qquad OH}}{\overset{\substack{R \qquad OH}}{C}} \xrightarrow{\text{distillation}} \underset{H}{\overset{R}{C}}=O + H_2O$$

[An exception to this general instability of carbonyl hydrates is furnished by the compound called chloral hydrate (Section 8.16).]

16.7B Hemiacetals and Hemiketals

Dissolving an aldehyde in an alcohol causes the establishment of an equilibrium between these two compounds and a new compound called a **hemiacetal:**

$$\underset{\substack{\text{Aldehyde}}}{\overset{R}{\underset{H}{C}}=\overset{..}{\overset{..}{O}}:} + \underset{\text{Alcohol}}{H\overset{..}{\overset{..}{O}}-R'} \rightleftharpoons \underset{\substack{H \qquad \overset{..}{\overset{..}{O}}:^-}}{\overset{\substack{R \qquad \overset{\overset{H}{|}}{O^+}-R'}}{C}} \rightleftharpoons \underset{\substack{H \qquad \overset{..}{\underset{..}{O}} \quad H}}{\overset{\substack{R \qquad \overset{..}{\overset{..}{O}}-R'}}{C}}$$

Hemiacetal
(usually too unstable to isolate)

The essential structural features of a hemiacetal are an —OH and an —OR group attached to the same carbon atom (and since this carbon atom came from an aldehyde, the carbon also has one hydrogen atom attached to it).

Most open-chain hemiacetals are not sufficiently stable to allow their isolation. Cyclic hemiacetals with five- or six-membered rings, however, are usually much more stable:

$$\underset{..}{H\overset{..}{O}CH_2CH_2CH_2CH} \rightleftharpoons \cdots \rightleftharpoons \cdots$$

Most simple sugars (Chapter 21) exist primarily in a cyclic hemiacetal form.

Ketones undergo similar reactions when they are dissolved in an alcohol. The products (also unstable in open-chain compounds) are called **hemiketals.**

$$\underset{\substack{\text{Ketone}}}{\overset{R}{\underset{R}{C}}=\overset{..}{\overset{..}{O}}:} + H\overset{..}{\overset{..}{O}}-R' \rightleftharpoons \underset{\substack{R \qquad \overset{..}{\underset{..}{O}}-H}}{\overset{\substack{R \qquad \overset{\overset{H}{|}}{O^+}-R'}}{C}} \rightleftharpoons \underset{\substack{R \qquad \overset{..}{\underset{..}{O}}-H}}{\overset{\substack{R \qquad \overset{..}{\overset{..}{O}}-R'}}{C}}$$

Hemiketal

The formation of hemiacetals and hemiketals is also catalyzed by acids and bases:

Acid-Catalyzed Hemiacetal Formation

$$
\begin{array}{c}
\underset{H}{\overset{CH_3}{\diagdown}}C=\ddot{O}: \; \underset{-H^+}{\overset{+H^+}{\rightleftarrows}} \; \underset{H}{\overset{CH_3}{\diagdown}}\overset{+}{C}=\ddot{O}H \; \underset{-CH_3\ddot{O}H}{\overset{+CH_3\ddot{O}H}{\rightleftarrows}} \; \underset{H}{\overset{CH_3}{\diagdown}}\underset{\ddot{O}-H}{\overset{\overset{\overset{H}{\mid}}{\overset{+}{O}-CH_3}}{C}} \; \underset{+H^+}{\overset{-H^+}{\rightleftarrows}}
\end{array}
$$

$$
\underset{H}{\overset{CH_3}{\diagdown}}\underset{\ddot{O}-H}{\overset{\ddot{O}-CH_3}{C}}
$$

Base-Catalyzed Hemiacetal Formation

$$
\underset{H}{\overset{CH_3}{\diagdown}}C=\ddot{O}: + {}^{-}:\ddot{O}CH_3 \rightleftarrows \underset{H}{\overset{CH_3}{\diagdown}}\underset{\ddot{O}:^{-}}{\overset{\ddot{O}CH_3}{C}} \; \underset{-CH_3\ddot{O}H}{\overset{+CH_3\ddot{O}H}{\rightleftarrows}}
$$

$$
\underset{H}{\overset{CH_3}{\diagdown}}\underset{\ddot{O}-H}{\overset{\ddot{O}-CH_3}{C}} \; + CH_3\ddot{O}:^{-}
$$

16.7C Acetals and Cyclic Ketals

If we take an alcohol solution of an aldehyde and pass into it a small amount of gaseous HCl the hemiacetal forms, and then a second reaction takes place. The hemiacetal reacts with a second molar equivalent of the alcohol to produce an **acetal.** An acetal has two —OR groups attached to the same —CH group.

$$
\underset{\underset{\text{Hemiacetal}}{}}{R-\underset{\overset{|}{H}}{\overset{\overset{OH}{|}}{C}}-OR'} \; \xrightarrow[R'-OH]{HCl_{(g)}} \; \underset{\underset{\text{An acetal}}{}}{R-\underset{\overset{|}{H}}{\overset{\overset{OR'}{|}}{C}}-OR' + H_2O}
$$

The mechanism for acetal formation involves acid-catalyzed formation of the hemiacetal, then an acid-catalyzed elimination of water, followed by a second *addition* of the alcohol and loss of a proton.

$$
\underset{H}{\overset{R}{\diagdown}}C=\ddot{O}: \; \underset{-H^+}{\overset{+H^+}{\rightleftarrows}} \; \underset{H}{\overset{R}{\diagdown}}C=\overset{+}{\ddot{O}}-H \; \underset{-R'-\ddot{O}H}{\overset{+R'-\ddot{O}H}{\rightleftarrows}} \; \underset{H}{\overset{R}{\diagdown}}\underset{\underset{\overset{|}{H}}{\overset{+}{O}-R'}}{\overset{\ddot{O}-H}{C}} \; \underset{+H^+}{\overset{-H^+}{\rightleftarrows}} \; \underset{H}{\overset{R}{\diagdown}}\underset{\ddot{O}-R'}{\overset{\ddot{O}-H}{C}}
$$

Acid-catalyzed formation of a hemiacetal

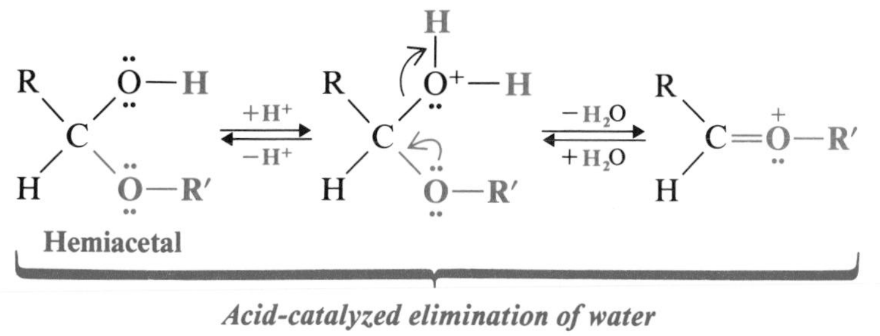

Acid-catalyzed elimination of water

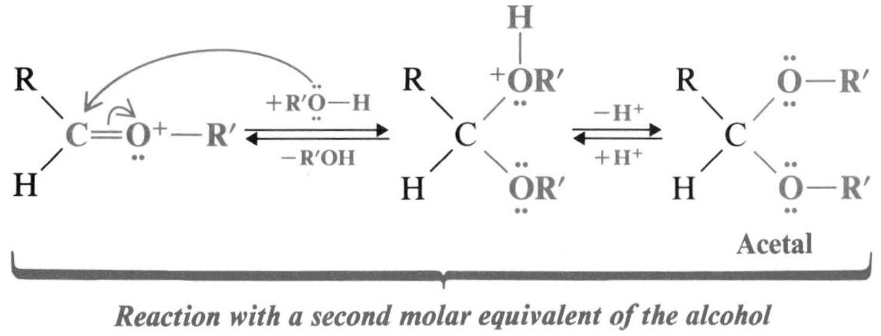

Reaction with a second molar equivalent of the alcohol

PROBLEM 16.11

Write out all the steps (as just shown) for the formation of an acetal from benzaldehyde and methanol in the presence of an acid catalyst.

All steps in the formation of an acetal from an aldehyde are reversible. If we dissolve an aldehyde in a large excess of an anhydrous alcohol and add a small amount of an anhydrous acid (e.g., gaseous HCl or conc. H_2SO_4), the equilibrium will strongly favor the formation of an acetal. After the equilibrium is established, we can isolate the acetal by neutralizing the acid and evaporating the excess alcohol.

If we then place the acetal in water and add a small amount of acid, all of the steps reverse. Under these conditions (an excess of water), the equilibrium favors the formation of the aldehyde. The acetal undergoes *hydrolysis.*

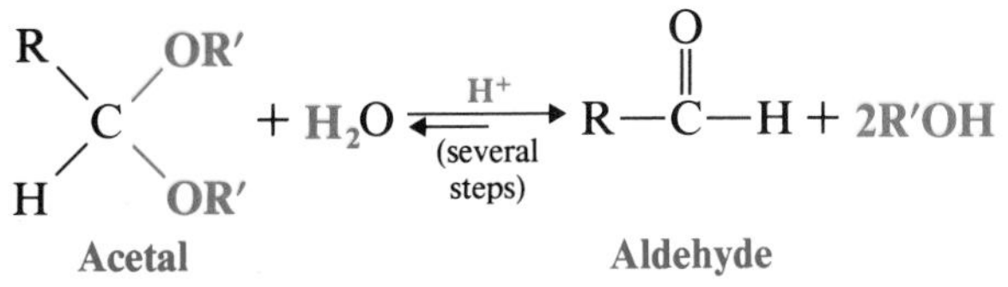

Ketal formation is not favored when ketones are treated with simple alcohols and gaseous HCl. Cyclic ketal formation *is* favored, however, when a ketone is treated with an excess of a 1,2-diol and a trace of acid.

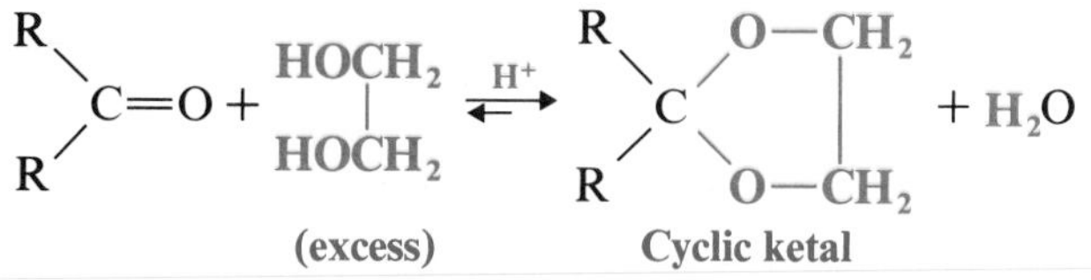

This reaction, too, can be reversed by treating the ketal with aqueous acid.

$$\underset{R}{\overset{R}{>}}C\overset{O-CH_2}{\underset{O-CH_2}{<}} \quad H_2O \underset{H^+}{\overset{}{\rightleftarrows}} \quad \underset{R}{\overset{R}{>}}C{=}O + \begin{matrix} CH_2OH \\ | \\ CH_2OH \end{matrix}$$

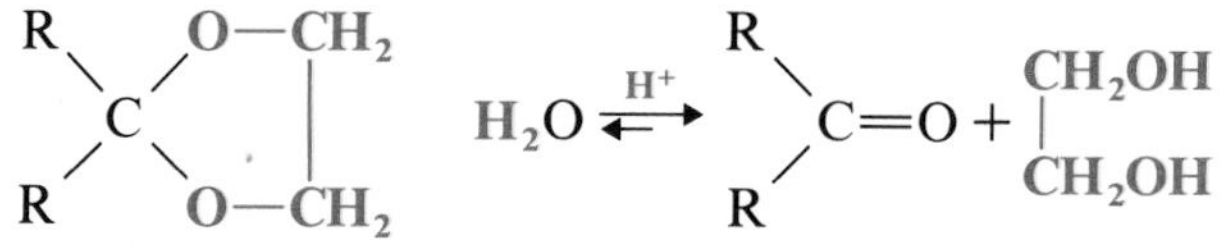

PROBLEM 16.12

Outline all steps in the mechanism for the formation of a cyclic ketal from acetone and ethylene glycol in the presence of gaseous HCl.

16.7D Acetals and Cyclic Ketals as Protecting Groups

Although acetals and cyclic ketals are hydrolyzed to aldehydes and ketones in aqueous acid, *they are stable in basic solutions.*

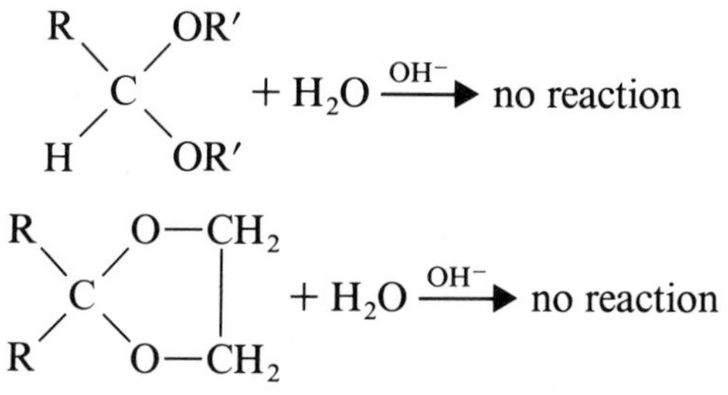

$$\underset{H}{\overset{R}{>}}C\overset{OR'}{\underset{OR'}{<}} + H_2O \overset{OH^-}{\longrightarrow} \text{no reaction}$$

$$\underset{R}{\overset{R}{>}}C\overset{O-CH_2}{\underset{O-CH_2}{<}} + H_2O \overset{OH^-}{\longrightarrow} \text{no reaction}$$

Because of this property, acetals and ketals give us a convenient method for protecting aldehyde and ketone groups from undesired reactions in basic solutions. (Acetals and ketals are really *gem*-diethers and, like ethers, they are relatively unreactive toward bases.) We can convert an aldehyde or ketone to an acetal or cyclic ketal, carry out a reaction on some other part of the molecule, and then hydrolyze the acetal or ketal with aqueous acid.

As an example, let us consider the problem of converting

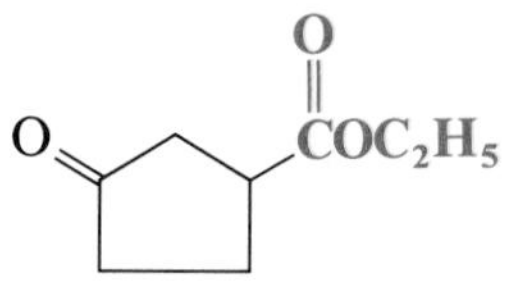

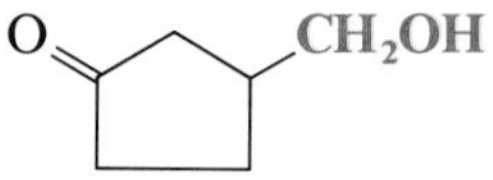

Keto groups are more easily reduced than ester groups. Any reducing agent (e.g., $LiAlH_4$ or H_2/Ni) that will reduce the ester group of **A** will reduce the keto group as well. But if we "protect" the keto group by converting it to a cyclic ketal (the ester group does not react), we can reduce the ester group in basic solution without affecting the cyclic ketal. After we finish the ester reduction, we can hydrolyze the cyclic ketal and obtain our desired product, **B.**

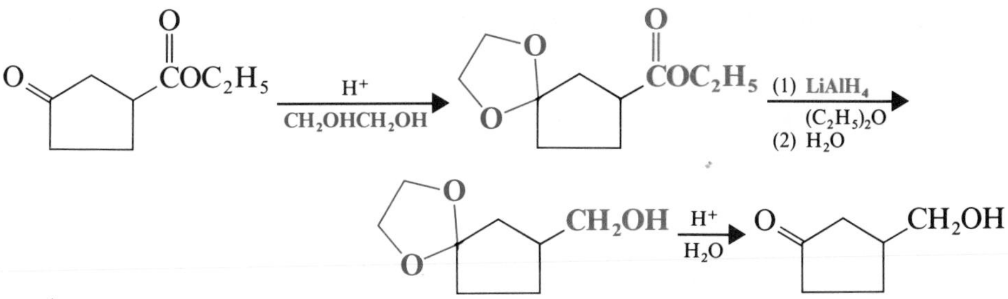

PROBLEM 16.13

What product would be obtained if **A** were treated with lithium aluminum hydride without first converting it to a cyclic ketal?

PROBLEM 16.14

(a) Show how you might use a cyclic ketal in carrying out the following transformation:

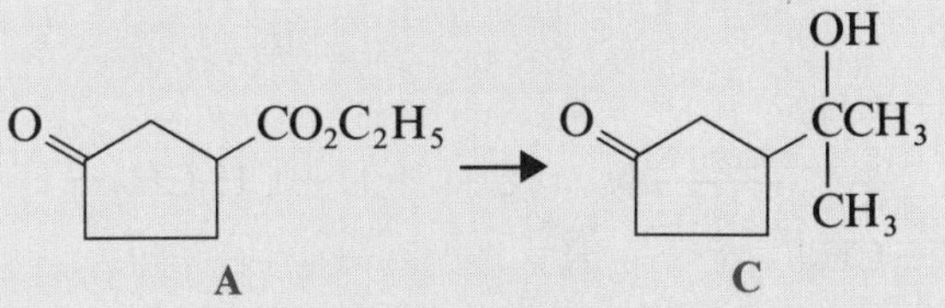

(b) Why would a direct addition of methylmagnesium bromide to **A** fail to give **C**?

PROBLEM 16.15

Dihydropyran reacts readily with an alcohol in the presence of a trace of anhydrous HCl or H_2SO_4, to form a tetrahydropyranyl ether.

Dihydropyran   **Tetrahydropyranyl ether**

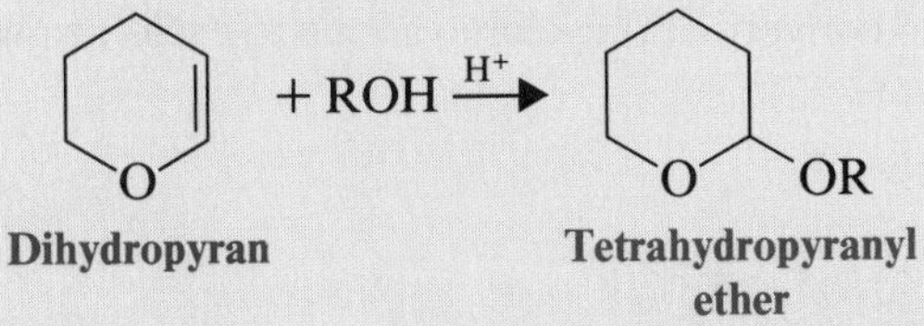

(a) Write a plausible mechanism for this reaction.   (b) Tetrahydropyranyl ethers are stable in aqueous base but hydrolyze rapidly in aqueous acid to yield the original alcohol and another compound. Explain. (What is the other compound?)   (c) The tetrahydropyranyl group can be used as a protecting group for alcohols and phenols. Show how you might use it in a synthesis of 5-methyl-1,5-hexanediol starting with 4-chloro-1-butanol.

16.7E Thioacetals and Thioketals:

$$RCR' \longrightarrow RCH_2R$$

(with O double-bonded above RCR')

Aldehydes and ketones react with thiols to form *thioacetals* and *thioketals*.

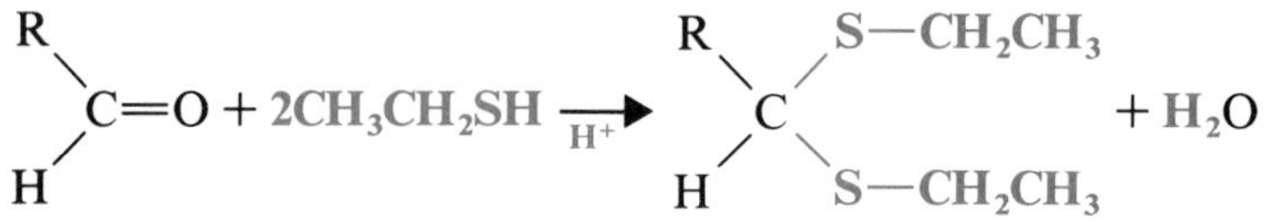

$$\underset{H}{\overset{R}{\diagdown}}C=O + 2CH_3CH_2SH \xrightarrow{H^+} \underset{H}{\overset{R}{\diagdown}}C\underset{S-CH_2CH_3}{\overset{S-CH_2CH_3}{\diagup}} + H_2O$$

Thioacetal

$$\underset{R}{\overset{R}{\diagdown}}C=O + HSCH_2CH_2SH \xrightarrow{BF_3} \underset{R}{\overset{R}{\diagdown}}C\underset{S-CH_2}{\overset{S-CH_2}{\diagup}}\Big| + H_2O$$

Cyclic thioketal

Thioacetals and thioketals are important in organic synthesis because they react with Raney nickel to yield hydrocarbons.* These reactions (i.e., thioacetal or thioketal formation and subsequent "desulfurization") give us an additional method for

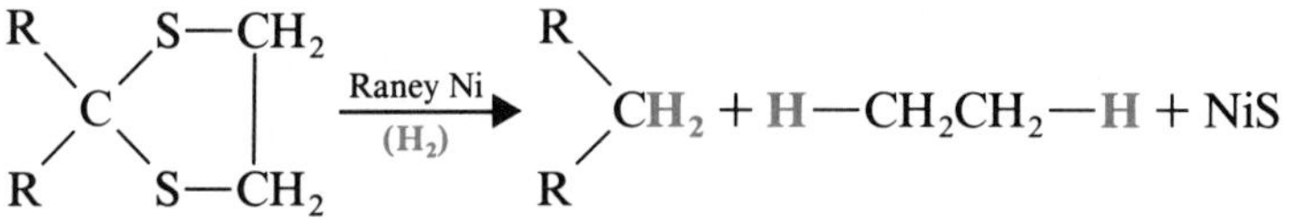

$$\underset{R}{\overset{R}{\diagdown}}C\underset{S-CH_2}{\overset{S-CH_2}{\diagup}}\Big| \xrightarrow[\text{(H}_2)]{\text{Raney Ni}} \underset{R}{\overset{R}{\diagdown}}CH_2 + H-CH_2CH_2-H + NiS$$

converting carbonyl groups of aldehydes and ketones to —CH₂— groups. The other method we have studied is the **Clemmensen reduction** (Section 12.12C). In the next section (16.8C), we shall see how this can also be accomplished with the **Wolff–Kishner reduction.**

PROBLEM 16.16

Show how you might use thioacetal and thioketal formation and Raney nickel desulfurization to convert: (a) cyclohexanone to cyclohexane; (b) benzaldehyde to toluene.

16.8 THE ADDITION OF DERIVATIVES OF AMMONIA

Aldehydes and ketones react with a number of derivatives of ammonia in the general way shown in the following sequence:

---

*Raney nickel is a special nickel catalyst that contains adsorbed hydrogen.

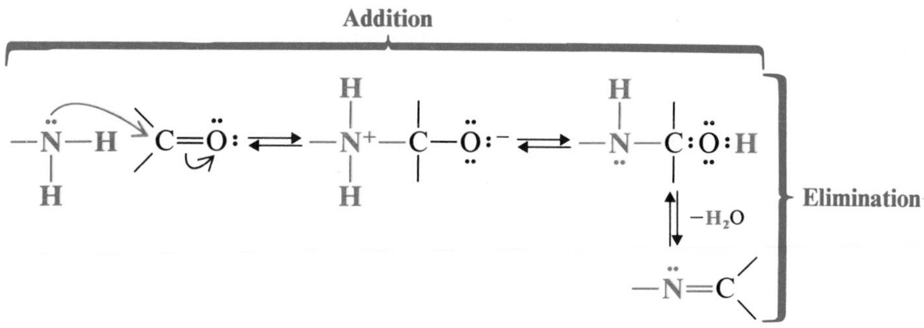

Table 16.2 lists several important examples of this general reaction.

16.8A 2,4-Dinitrophenylhydrazones, Semicarbazones, and Oximes

The products of the reactions of aldehydes and ketones with 2,4-dinitrophenylhydrazine, semicarbazide, and hydroxylamine are often used to identify unknown aldehydes and ketones. These compounds, that is, 2,4-dinitrophenylhydrazones, semicarbazones, and oximes, are usually relatively insoluble solids that have sharp characteristic melting points. Table 16.3 gives representative examples from the very extensive tables of these derivatives.

16.8B Imines and Enamines

Aldehydes and ketones react with primary amines to form **imines.** (Such *N*-substituted imines are also called Schiff bases.)

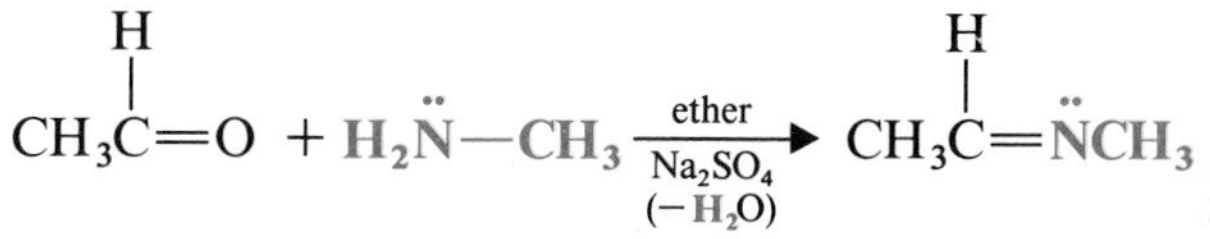

$$
\underset{\substack{\text{Acetaldehyde}}}{\overset{\overset{\displaystyle H}{|}}{CH_3C}=O} + \underset{\substack{\text{Methylamine}}}{H_2\ddot{N}-CH_3} \xrightarrow[\substack{Na_2SO_4 \\ (-H_2O)}]{\text{ether}} \underset{\substack{(40\%) \\ \text{Acetaldimine} \\ \text{(an imine)}}}{\overset{\overset{\displaystyle H}{|}}{CH_3C}=\ddot{N}CH_3}
$$

Imines are important in many biochemical reactions because many enzymes use an $-NH_2$ group of an amino acid to react with an aldehyde or ketone to form an imine linkage.

Imines are also formed as intermediates in a useful synthesis of amines (Section 19.5).

Aldehydes and ketones react with secondary amines to form **enamines** (*ene* + *amine*):

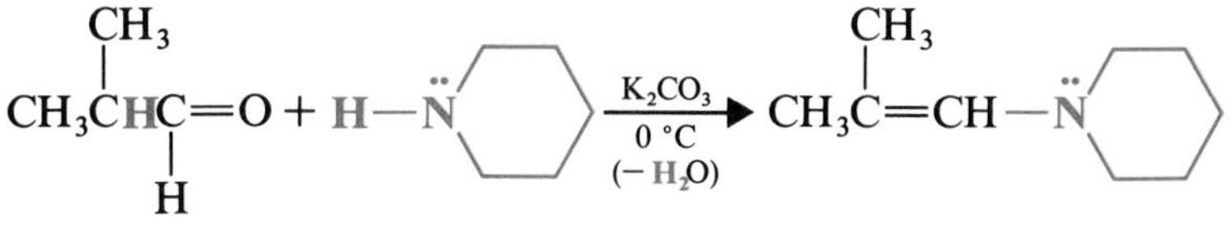

$$
\underset{\substack{}}{\overset{\overset{\displaystyle CH_3}{|}}{CH_3CHC}=O} + H-\ddot{N}\hspace{-0.3em}\bigcirc \xrightarrow[\substack{0\,°C \\ (-H_2O)}]{K_2CO_3} \underset{\substack{(75\%) \\ \text{An enamine}}}{\overset{\overset{\displaystyle CH_3}{|}}{CH_3C}=CH-\ddot{N}\hspace{-0.3em}\bigcirc}
$$

Piperidine

Enamines are useful synthetic intermediates. We shall discuss their chemistry in Section 20.10.

TABLE 16.2 Reactions of aldehydes and ketones with derivatives of ammonia

1. Reaction with Hydroxylamine

General Reaction

$$\ce{\overset{\diagdown}{\underset{\diagup}{C}}=O} + \ce{H2N-OH} \longrightarrow \ce{\overset{\diagdown}{\underset{\diagup}{C}}=N-OH} + \ce{H2O}$$

Aldehyde or   Hydroxylamine   An oxime
ketone

Specific Example

$$\ce{\overset{CH3}{\underset{H}{\diagdown}}C=O} + \ce{H2NOH} \longrightarrow \ce{\overset{CH3}{\underset{H}{\diagdown}}C=NOH} + \ce{H2O}$$

Acetaldehyde   Acetaldoxime

2. Reactions with Hydrazine, Phenylhydrazine, and 2,4-Dinitrophenylhydrazine

General Reactions
Aldehyde or
ketone

$$\ce{\overset{\diagdown}{\underset{\diagup}{C}}=O} + \ce{H2NNH2} \longrightarrow \ce{\overset{\diagdown}{\underset{\diagup}{C}}=NNH2} + \ce{H2O}$$

Hydrazine   A hydrazone

$$\ce{\overset{\diagdown}{\underset{\diagup}{C}}=O} + \ce{H2NNHC6H5} \longrightarrow \ce{\overset{\diagdown}{\underset{\diagup}{C}}=NNHC6H5} + \ce{H2O}$$

Phenylhydrazine   A phenylhydrazone

$$\ce{\overset{\diagdown}{\underset{\diagup}{C}}=O} + \ce{H2NNH}-\underset{NO_2}{\overset{}{\bigcirc}}-NO_2 \longrightarrow \ce{\overset{\diagdown}{\underset{\diagup}{C}}=NNH}-\underset{NO_2}{\overset{}{\bigcirc}}-NO_2 + \ce{H2O}$$

2,4-Dinitrophenylhydrazine   A 2,4-dinitrophenylhydrazone

Specific Examples

$$\ce{\overset{C6H5}{\underset{CH3CH2}{\diagdown}}C=O} + \ce{H2NNH2} \xrightarrow{heat} \ce{\overset{C6H5}{\underset{CH3CH2}{\diagdown}}C=NNH2} + \ce{H2O}$$

Propiophenone   Propiophenone hydrazone

$$\ce{\overset{C6H5}{\underset{CH3}{\diagdown}}C=O} + \ce{H2NNHC6H5} \xrightarrow[\ce{CH3CO2H}]{\ce{H3O+}} \ce{\overset{C6H5}{\underset{CH3}{\diagdown}}C=NNHC6H5} + \ce{H2O}$$

Acetophenone   Acetophenone phenylhydrazone

$$\ce{\overset{C6H5}{\underset{H}{\diagdown}}C=O} + \ce{H2NNH}-\underset{NO_2}{\overset{}{\bigcirc}}-NO_2 \xrightarrow[\substack{C_2H_5OH \\ H_2O}]{HCl} \ce{\overset{C6H5}{\underset{H}{\diagdown}}C=NNH}-\underset{NO_2}{\overset{}{\bigcirc}}-NO_2 + \ce{H2O}$$

Benzaldehyde   Benzaldehyde
                2,4-dinitrophenylhydrazone

TABLE 16.2 (continued)

3. Reaction with Semicarbazide

General Reaction

$$\text{\Large$\diagdown$}C=O + H_2NNH\overset{\overset{\displaystyle O}{\|}}{C}NH_2 \longrightarrow \text{\Large$\diagdown$}C=NNH\overset{\overset{\displaystyle O}{\|}}{C}NH_2 + H_2O$$

Aldehyde   Semicarbazide   A semicarbazone
or ketone

Specific Example

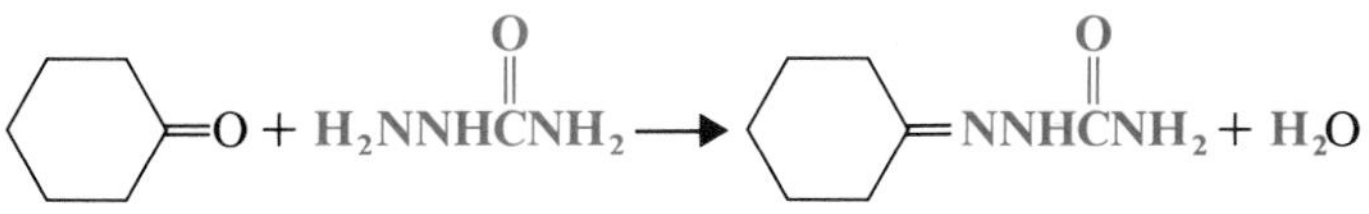

Cyclohexanone
semicarbazone

TABLE 16.3 Derivatives of aldehydes and ketone

ALDEHYDE OR KETONE	mp (°C) OF 2,4-DINITRO-PHENYLHYDRAZONE	mp (°C) OF SEMICARBAZONE	mp (°C) OF OXIME
Acetaldehyde	168.5	162	46.5
Acetone	128	187 dec.	61
Benzaldehyde	237	222	35
o-Tolualdehyde	195	208	49
m-Tolualdehyde	211	204	60
p-Tolualdehyde	233	234	79
Phenylacetaldehyde	121	156	103

16.8C Hydrazones: The Wolff–Kishner Reduction

Hydrazones are occasionally used to identify aldehydes and ketones. But unlike 2,4-dinitrophenylhydrazones, simple hydrazones often have low melting points. Hydrazones, however, are the basis for a useful method to reduce carbonyl groups of aldehydes and ketones to —CH$_2$— groups, called the **Wolff–Kishner reduction:**

$$\text{\Large$\diagdown$}C=\ddot{O}: + H_2N-NH_2 \xrightarrow[\text{heat}]{\text{base}} \left[\text{\Large$\diagdown$}C=N-NH_2 \right] + H_2O$$

Aldehyde   Hydrazone
or ketone   (not isolated)

$$\text{\Large$\diagdown$}CH_2 + N_2$$

Specific Example

$$\text{C}_6\text{H}_5-\overset{\overset{\text{O}}{\|}}{\text{C}}\text{CH}_2\text{CH}_3 + \text{H}_2\text{NNH}_2 \xrightarrow[\substack{\text{triethylene glycol} \\ 200 \text{ °C}}]{\text{NaOH}} \text{C}_6\text{H}_5-\text{CH}_2\text{CH}_2\text{CH}_3$$

$$(82\%)$$

The Wolff–Kishner reduction can be accomplished at much lower temperatures if dimethyl sulfoxide is used as the solvent.

The Wolff–Kishner reduction complements the Clemmensen reduction (Section 12.12C) and the reduction of thioacetals, (Section 16.7E), because all three reactions convert $\diagdown$C=O groups into $-\text{CH}_2-$ groups. The Clemmensen reduction takes place in strongly acidic media and can be used for those compounds that are sensitive to base. The Wolff–Kishner reduction takes place in strongly basic solutions and can be used for those compounds that are sensitive to acid. The reduction of thioacetals takes place in neutral solution and can be used for compounds that are sensitive to both acids and bases.

> The mechanism of the Wolff–Kishner reduction is as follows: The first step is the formation of the hydrazone. Then the strong base brings about an isomerization of the hydrazone to a derivative with the structure $\diagdown$CH—N=NH. This derivative then undergoes the base-catalyzed elimination of a molecule of nitrogen. The loss of the especially stable molecule of nitrogen provides the driving force for the reaction.

1. $-\overset{|}{\underset{|}{\text{C}}}=\text{O} + \text{H}_2\text{N}-\text{NH}_2 \rightleftharpoons -\overset{|}{\underset{|}{\text{C}}}=\text{N}-\text{NH}_2 + \text{H}_2\text{O}$

2. $-\overset{|}{\underset{|}{\text{C}}}=\text{N}-\text{NH}_2 \underset{\text{H}_2\text{O}}{\overset{\text{OH}^-}{\rightleftharpoons}} \left[-\overset{|}{\underset{|}{\text{C}}}=\text{N}-\overset{..}{\text{N}}\text{H} \longleftrightarrow -\overset{|}{\underset{|}{\overset{..}{\text{C}}}}-\text{N}=\text{NH} \right] \underset{\text{OH}^-}{\overset{\text{H}_2\text{O}}{\rightleftharpoons}}$

$-\overset{\overset{\text{H}}{|}}{\underset{|}{\text{C}}}-\text{N}=\text{N}-\text{H} \underset{\text{H}_2\text{O}}{\overset{\text{OH}^-}{\rightleftharpoons}} -\overset{\overset{\text{H}}{|}}{\underset{|}{\text{C}}}-\text{N}=\text{N}:^- \xrightarrow{-\text{N}_2} -\overset{\overset{\text{H}}{|}}{\underset{|}{\text{C}}}:^- \xrightarrow{\text{H}_2\text{O}} -\overset{\overset{\text{H}}{|}}{\underset{|}{\text{C}}}-\text{H}$

16.9 THE ADDITION OF HYDROGEN CYANIDE AND OF SODIUM BISULFITE

16.9A Hydrogen Cyanide Addition

Hydrogen cyanide adds to the carbonyl groups of aldehydes and most ketones to form compounds called **cyanohydrins.** Ketones in which the carbonyl group is highly hindered do not undergo this reaction.

$$
\left.
\begin{array}{l}
\underset{\displaystyle \overset{O}{\|}}{R C H} + HCN \rightleftharpoons \underset{\displaystyle H}{\overset{\displaystyle R}{\diagup}} C \overset{\displaystyle OH}{\underset{\displaystyle CN}{\diagdown}} \\[20pt]
\underset{\displaystyle \overset{O}{\|}}{R-C-R'} + HCN \rightleftharpoons \underset{\displaystyle R'}{\overset{\displaystyle R}{\diagup}} C \overset{\displaystyle OH}{\underset{\displaystyle CN}{\diagdown}}
\end{array}
\right\} \quad \text{Cyanohydrins}
$$

The addition of hydrogen cyanide itself takes place very slowly because HCN is a poor nucleophile. The addition of potassium cyanide, or any base that can generate cyanide ions from the weak acid HCN, causes a dramatic increase in the rate of reaction. This effect was discovered in 1903 by the British chemist Arthur Lapworth, and in his studies of the addition of HCN, Lapworth became one of the originators of the mechanistic view of organic chemistry. Lapworth assumed that the addition was ionic in nature (a remarkable insight considering that Lewis and Kössel's theories of bonding were some 13 years in the future).

He proposed "that the formation of cyanohydrins is to be represented as a comparatively slow union of the negative cyanide ion with carbonyl, followed by almost instantaneous combination of the complex with hydrogen."*

$$
\begin{array}{ccc}
\diagup C = \ddot{O}: + \ ^- :C \equiv N: & \xrightarrow{\text{slow}} & \diagup C \overset{\ddot{O}:^-}{\diagdown} \\
& & C \equiv N:
\end{array}
\quad
\begin{array}{c}
\xrightarrow[-H-C\equiv N:]{+H-C\equiv N:}
\end{array}
\quad
\begin{array}{c}
\diagup C \overset{\ddot{O}-H}{\diagdown} \\
C \equiv N:
\end{array}
+ \ ^- :C \equiv N:
$$

The cyanide ion, being a stronger nucleophile, is able to attack the carbonyl carbon atom much more rapidly than HCN itself and this is the source of its catalytic effect. Once the addition of cyanide ion has taken place, the strongly basic alkoxide oxygen atom of the intermediate removes a proton from any available acid. If this acid is HCN, this step regenerates the cyanide ion.

Bases stronger than cyanide ion catalyze the reaction by converting HCN ($K_a \approx 10^{-9}$) to cyanide ion in an acid–base reaction. The cyanide ions, thus formed, can go on to attack the carbonyl group.

$$
B: \ ^- + H - C \equiv N: \rightleftharpoons B:H + \ ^- :C \equiv N:
$$

Liquid hydrogen cyanide can be used for this reaction, but since HCN is very toxic and volatile, it is safer to generate it in the reaction mixture. This can be done by mixing the aldehyde or ketone with aqueous sodium cyanide and then slowly adding sulfuric acid to the mixture. *Even with this procedure, however, great care must be taken and the reaction must be carried out in a very efficient fume hood.*

Cyanohydrins are useful intermediates in organic synthesis. Depending on the conditions used, acidic hydrolysis converts cyanohydrins to α-hydroxy acids or to α,β-unsaturated acids. (The mechanism for this hydrolysis is discussed in Section 18.8H.)

---

*A. Lapworth, *J. Chem. Soc.*, **83**, 995 (1903). For a fine review of Lapworth's work see M. Saltzman, *J. Chem. Educ.*, **49**, 750 (1972).

$$CH_3CH_2-\overset{\overset{O}{\|}}{C}-CH_3 \xrightarrow{HCN} CH_3CH_2-\overset{\overset{HO}{|}}{\underset{\underset{CH_3}{|}}{C}}-CN \xrightarrow[\underset{heat}{H_2O}]{HCl} CH_3CH_2-\overset{\overset{HO}{|}}{\underset{\underset{CH_3}{|}}{C}}-\overset{\overset{O}{\|}}{C}OH$$

α-Hydroxy acid

$$\downarrow \overset{95\% \ H_2SO_4}{\underset{heat}{}}$$

$$CH_3CH=\overset{}{\underset{\underset{CH_3}{|}}{C}}-\overset{\overset{O}{\|}}{C}OH$$

α,β-Unsaturated acid

Treating acetone cyanohydrin with acid and methanol converts it to methyl methacrylate. Methyl methacrylate is the starting material for the synthesis of the polymer known as Plexiglas or Lucite (Special Topic D).

$$CH_3-\overset{\overset{O}{\|}}{C}-CH_3 \xrightarrow{HCN} CH_3-\overset{\overset{OH}{|}}{\underset{\underset{CH_3}{|}}{C}}-CN \xrightarrow[CH_3OH]{H_2SO_4} CH_2=\overset{}{\underset{\underset{CH_3}{|}}{C}}-\overset{\overset{O}{\|}}{C}OCH_3$$

Acetone   (78%)   (90%)

          Acetone cyanohydrin   Methyl methacrylate

Reducing a cyanohydrin with lithium aluminum hydride gives a β-amino alcohol:

PROBLEM 16.17

(a) Show how you might prepare lactic acid ($CH_3CHOHCO_2H$) from acetaldehyde through a cyanohydrin intermediate. (b) What stereoisomeric form of lactic acid would you expect to obtain?

16.9B Sodium Bisulfite Addition

Sodium bisulfite ($NaHSO_3$) adds to a carbonyl group in a way that is very similar to the addition of HCN.

$$\overset{\overset{O}{\|}}{\underset{/ \ \ \backslash}{C}} \quad \underset{Na^+}{:\overset{-}{S}O_3H} \rightleftharpoons -\overset{\overset{:\overset{..}{O}:^- Na^+}{|}}{\underset{|}{C}}-SO_3H \rightleftharpoons -\overset{\overset{:\overset{..}{O}-H}{|}}{\underset{|}{C}}-SO_3^- \ Na^+$$

Bisulfite addition
product

    This reaction takes place with aldehydes and some ketones. Most aldehydes react with one molar equivalent of sodium bisulfite to give the addition product in 70–90% yield. Under the same conditions, methyl ketones give yields varying from 12 to 56%. Most higher ketones do not give bisulfite addition products in appreciable amounts because the addition is very sensitive to steric hindrance. However, since the reaction involves an equilibrium, yields from aldehydes and methyl ketones can be improved by using an excess of sodium bisulfite.

Because bisulfite addition compounds are crystalline salts, a bisulfite addition reaction is often used in separating aldehydes and methyl ketones from other substances. Since bisulfite addition is reversible, the aldehyde or methyl ketone can be regenerated, after a separation has been made, by adding either an acid or a base. These additions displace the equilibrium to the left by converting the HSO_3^- ion to SO_2 (in acid) or to SO_3^{2-} (in base).

16.10 THE ADDITION OF YLIDES

16.10A The Wittig Reaction

Aldehydes and ketones react with phosphorus ylides to yield *alkenes* and triphenylphosphine oxide. (An ylide is a neutral molecule having a negative carbon adjacent to a positive heteroatom.)

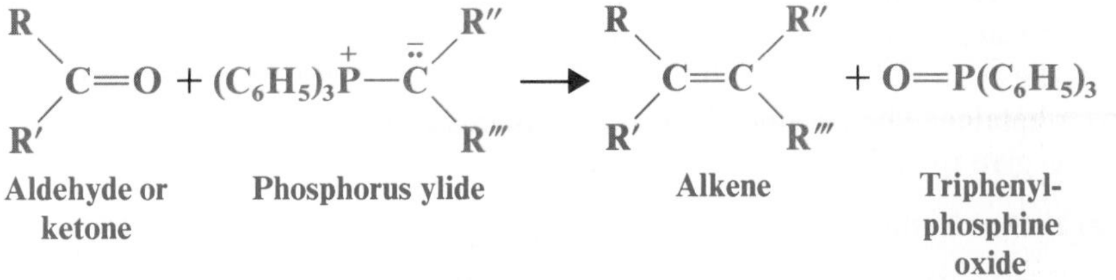

$$\underset{\substack{\text{Aldehyde or}\\\text{ketone}}}{\overset{R}{\underset{R'}{>}}C{=}O} + \underset{\text{Phosphorus ylide}}{(C_6H_5)_3\overset{+}{P}{-}\overset{\cdot\cdot}{\underset{R'''}{\overset{R''}{C}}}} \longrightarrow \underset{\text{Alkene}}{\overset{R}{\underset{R'}{>}}C{=}C\overset{R''}{\underset{R'''}{<}}} + \underset{\substack{\text{Triphenyl-}\\\text{phosphine}\\\text{oxide}}}{O{=}P(C_6H_5)_3}$$

This reaction, known as the **Wittig reaction,*** has proved to be a valuable method for synthesizing alkenes. The Wittig reaction is applicable to a wide variety of compounds, and it gives a great advantage over most other alkene syntheses in that no ambiguity exists as to the location of the double bond in the product.

Phosphorus ylides are easily prepared from triphenylphosphine and alkyl halides. Their preparation involves two reactions:

General Reaction

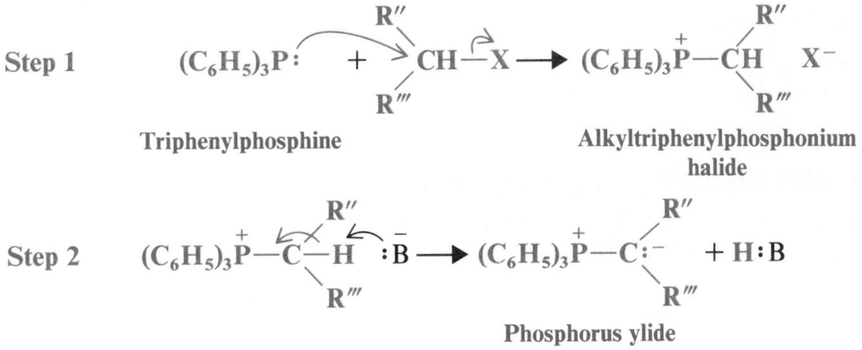

Step 1 $\quad$ $(C_6H_5)_3P\colon$ $\quad + \underset{R'''}{\overset{R''}{>}}CH{-}X \longrightarrow (C_6H_5)_3\overset{+}{P}{-}\underset{R'''}{\overset{R''}{C}}H \quad X^-$

$\qquad\qquad$ Triphenylphosphine $\qquad\qquad\qquad\qquad$ Alkyltriphenylphosphonium halide

Step 2 $\quad$ $(C_6H_5)_3\overset{+}{P}{-}\underset{R'''}{\overset{R''}{C}}{-}H \quad \colon\!\ddot{B} \longrightarrow (C_6H_5)_3\overset{+}{P}{-}\underset{R'''}{\overset{R''}{C}}\colon^- \quad + H\colon B$

$\qquad\qquad\qquad\qquad\qquad\qquad\qquad\qquad\qquad$ Phosphorus ylide

Specific Example

Step 1 $\quad$ $(C_6H_5)_3P\colon + CH_3Br \xrightarrow{C_6H_6} (C_6H_5)_3\overset{+}{P}{-}CH_3 \ Br^-$

$\qquad\qquad\qquad\qquad\qquad\qquad\qquad\qquad\qquad$ (89%)

$\qquad\qquad\qquad\qquad\qquad\qquad\qquad$ Methyltriphenylphosphonium bromide

---

*Discovered in 1954 by George Wittig, then at the University of Tübingen. Wittig was a co-winner of the Nobel prize for chemistry in 1979.

Step 2 $(C_6H_5)_3\overset{+}{P}-CH_3 + C_6H_5Li \longrightarrow (C_6H_5)_3\overset{+}{P}-CH_2:^- + C_6H_6$
                    Br^- $+ LiBr$

The first reaction is a nucleophilic substitution reaction. Triphenylphosphine acts as a nucleophile and displaces a halide ion from the alkyl halide to give an alkyltriphenylphosphonium salt. The second reaction is an acid–base reaction. A strong base (usually an alkyllithium, or phenyllithium) removes a proton from the carbon that is attached to phosphorus to give the ylide.

Phosphorus ylides can be represented as a hybrid of the two resonance structures shown here. Quantum mechanical calculations indicate that the contribution

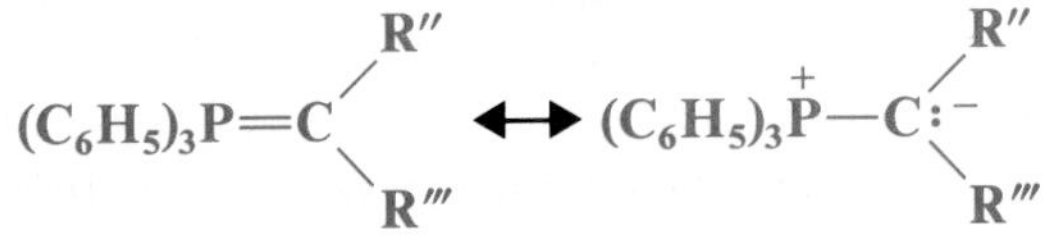

$$(C_6H_5)_3P{=}C\overset{\displaystyle R''}{\underset{\displaystyle R'''}{\Big\langle}} \longleftrightarrow (C_6H_5)_3\overset{+}{P}{-}C{:}^-\overset{\displaystyle R''}{\underset{\displaystyle R'''}{\Big\langle}}$$

made by the first structure is relatively unimportant. The contribution made to the hybrid by the second structure explains the reaction of the ylide with an aldehyde or ketone because the ylide acts as a nucleophile—in effect as a carbanion—and attacks the carbonyl carbon. This step yields an intermediate with separated charges called a **betaine**. The betaine then loses triphenylphosphine oxide (often spontaneously) to give the alkene.

General Mechanism

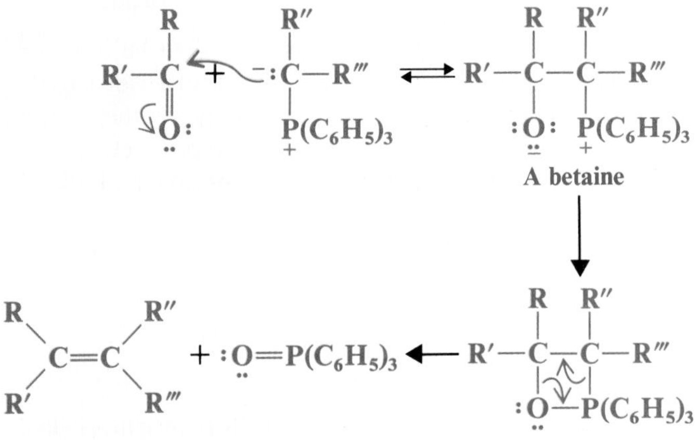

A betaine

Specific Example

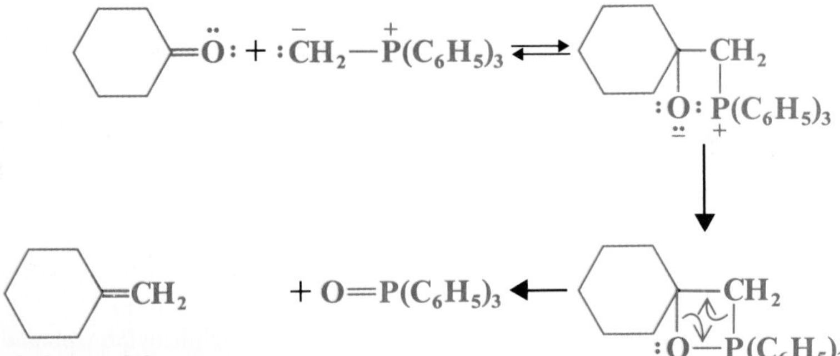

Methylenecyclohexane
(86% from cyclohexanone
and methyltriphenylphosphonium
bromide)

The elimination of triphenylphosphine oxide from the betaine may occur in two separate steps as we have shown previously, or both steps may occur simultaneously.

While Wittig syntheses may appear to be complicated, in actual practice they are easy to carry out. Most of the steps can be carried out in the same reaction vessel, and the entire synthesis can be accomplished in a matter of hours.

The overall result of a Wittig synthesis is

$$\begin{array}{c} R \\ \diagdown \\ R' \diagup \end{array} C{=}O \;+\; \begin{array}{c} X \quad R'' \\ \diagdown \diagup \\ C \\ \diagup \diagdown \\ H \quad R''' \end{array} \xrightarrow[\text{steps}]{\text{several}} \begin{array}{c} R \qquad R'' \\ \diagdown \diagup \\ C{=}C \\ \diagup \diagdown \\ R' \qquad R''' \end{array}$$

Planning a Wittig synthesis begins with recognizing in the desired alkene what can be the aldehyde or ketone component and what can be the halide component. Any or all of the R groups may be hydrogen. The halide component must be a primary, secondary, or methyl halide.

SAMPLE PROBLEM

Outline a Wittig synthesis of 2-methyl-1-phenyl-1-propene.

Answer:

We examine the structure of the compound, paying attention to the groups on each side of the double bond.

$$\begin{array}{c} \qquad\quad CH_3 \\ \qquad\quad | \\ C_6H_5CH{=}CCH_3 \end{array}$$

2-Methyl-1-phenyl-1-propene

Two general approaches to the synthesis are possible:

(a)
$$\begin{array}{c} C_6H_5 \\ \diagdown \\ H \diagup \end{array} C{=}O \;+\; \begin{array}{c} X \quad CH_3 \\ \diagdown \diagup \\ C \\ \diagup \diagdown \\ H \quad CH_3 \end{array} \longrightarrow \begin{array}{c} C_6H_5 \qquad CH_3 \\ \diagdown \diagup \\ C{=}C \\ \diagup \diagdown \\ H \qquad CH_3 \end{array}$$

or

(b)
$$\begin{array}{c} C_6H_5 \quad H \\ \diagdown \diagup \\ C \\ \diagup \diagdown \\ H \quad X \end{array} \;+\; O{=}C\begin{array}{c} CH_3 \\ \diagdown \\ \diagup \\ CH_3 \end{array} \longrightarrow \begin{array}{c} C_6H_5 \qquad CH_3 \\ \diagdown \diagup \\ C{=}C \\ \diagup \diagdown \\ H \qquad CH_3 \end{array}$$

In (a) we first make the ylide from a 2-halopropane and then allow it to react with benzaldehyde.

(a) $(CH_3)_2CHBr + (C_6H_5)_3P \longrightarrow (CH_3)_2CH{-}P(C_6H_5)_3{}^+Br^- \xrightarrow{RLi}$

$(CH_3)_2\overset{\cdot\cdot}{\underset{}{C}}{-}\overset{+}{P}(C_6H_5)_3 \xrightarrow{C_6H_5CHO} (CH_3)_2C{=}CHC_6H_5 + (C_6H_5)_3P{=}O$

In (b) we would make the ylide from a benzyl halide and allow it to react with acetone.

(b) $C_6H_5CH_2Br + (C_6H_5)_3P \longrightarrow C_6H_5CH_2{-}P(C_6H_5)_3{}^+Br^- \xrightarrow{RLi}$

$C_6H_5\overset{\cdot\cdot}{\underset{}{C}}H{-}\overset{+}{P}(C_6H_5)_3 \xrightarrow{(CH_3)_2C{=}O} C_6H_5CH{=}C(CH_3)_2 + (C_6H_5)_3P{=}O$

PROBLEM 16.18

In addition to triphenylphosphine assume that you have available as starting materials any necessary aldehydes, ketones, and organic halides. Show how you might synthesize each of the following alkenes using the Wittig reaction:

(a) $C_6H_5\underset{\underset{CH_3}{|}}{C}=CH_2$

(b) $C_6H_5\underset{\underset{CH_3}{|}}{C}=CHCH_3$

(c) $\underset{CH_3}{\overset{CH_3}{\diagdown}}C=CH_2$

(d)

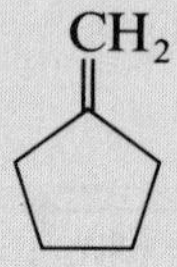

(e) $CH_3CH_2CH=\underset{\underset{|}{\overset{CH_3}{|}}}{C}CH_2CH_3$

(f) $C_6H_5CH=CHCH=CH_2$

(g) $C_6H_5CH=CHC_6H_5$

PROBLEM 16.19

Triphenylphosphine can be used to convert epoxides to alkenes, for example,

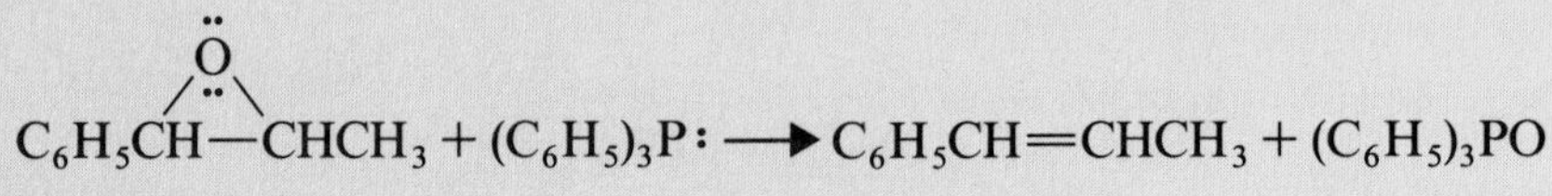

$$C_6H_5\overset{\overset{\overset{..}{O}}{\diagup\diagdown}}{CH}-CHCH_3 + (C_6H_5)_3P\colon \longrightarrow C_6H_5CH=CHCH_3 + (C_6H_5)_3PO$$

Propose a likely mechanism for this reaction.

PROBLEM 16.20

The Wittig reaction can be used in the synthesis of aldehydes, for example,

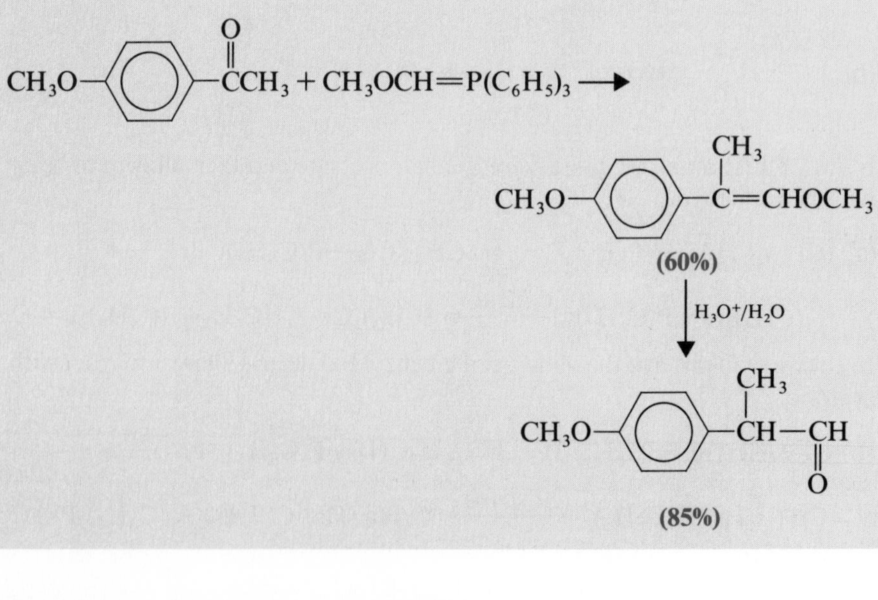

$$CH_3O-\!\!\left\langle\bigcirc\right\rangle\!\!-\overset{\overset{O}{\parallel}}{C}CH_3 + CH_3OCH=P(C_6H_5)_3 \longrightarrow$$

$$CH_3O-\!\!\left\langle\bigcirc\right\rangle\!\!-\underset{\underset{CH_3}{\overset{\overset{CH_3}{|}}{}}}{C}=CHOCH_3$$

(60%)

$$\Big\downarrow H_3O^+/H_2O$$

$$CH_3O-\!\!\left\langle\bigcirc\right\rangle\!\!-\underset{\underset{CH_3}{\overset{CH_3}{|}}}{CH}-\overset{\overset{}{}}{CH}\!\!\underset{\underset{O}{\parallel}}{}$$

(85%)

(a) How would you prepare $CH_3OCH\!=\!P(C_6H_5)_3$?   (b) Show with a mechanism how the second reaction produces an aldehyde.   (c) How would you use this method to prepare ⬡CHO from cyclohexanone?

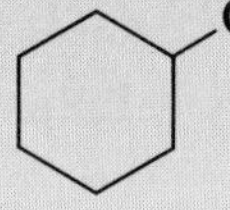

16.10B The Addition of Sulfur Ylides   *skip*

Sulfur ylides also react as nucleophiles at the carbonyl carbon of aldehydes and ketones. The betaine that forms usually decomposes to an *epoxide* rather than to an alkene.

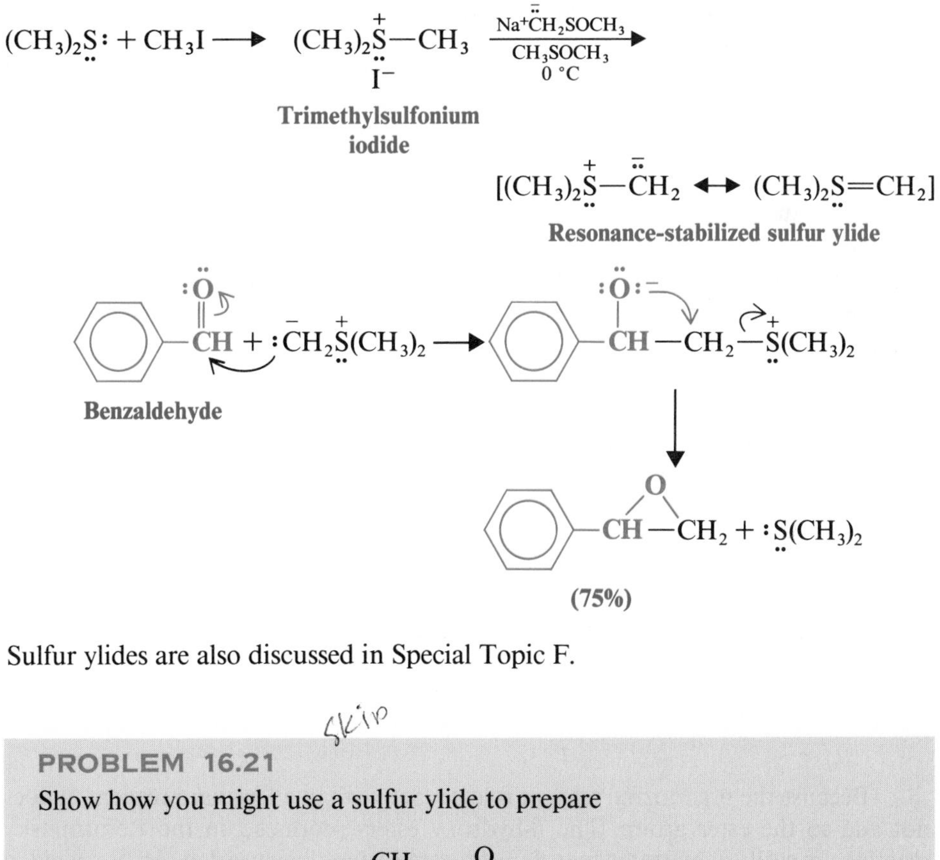

Trimethylsulfonium iodide

$$[(CH_3)_2\overset{+}{S}\!-\!\overset{..}{\overset{-}{C}H_2} \longleftrightarrow (CH_3)_2\overset{..}{S}\!=\!CH_2]$$

Resonance-stabilized sulfur ylide

Benzaldehyde

(75%)

Sulfur ylides are also discussed in Special Topic F.

skip

PROBLEM 16.21

Show how you might use a sulfur ylide to prepare

(a) 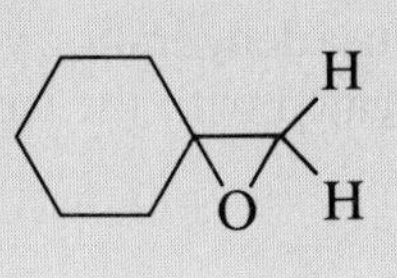   (b)

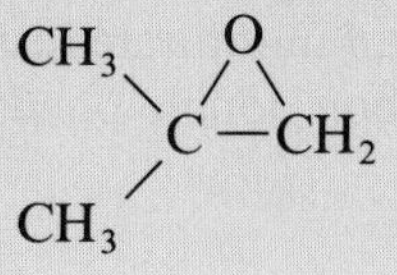

16.11 THE ADDITION OF ORGANOMETALLIC REAGENTS: THE REFORMATSKY REACTION

In Section 15.7 we studied the addition of Grignard reagents, organolithium compounds, and sodium alkynides to aldehydes and ketones. These reactions, as we saw then, can be used to produce a wide variety of alcohols:

$$\overset{\delta-}{R} \overset{\delta+}{:MgX} + \overset{\frown}{\underset{/}{\overset{\backslash}{C}}{=}O} \longrightarrow R{-}\overset{|}{\underset{|}{C}}{-}OMgX \xrightarrow{H_2O} R{-}\overset{|}{\underset{|}{C}}{-}OH$$

$$\overset{\delta-}{R} \overset{\delta+}{:Li} + \overset{\frown}{\underset{/}{\overset{\backslash}{C}}{=}O} \longrightarrow R{-}\overset{|}{\underset{|}{C}}{-}OLi \xrightarrow{H_2O} R{-}\overset{|}{\underset{|}{C}}{-}OH$$

$$RC{\equiv}\overset{\delta-}{C}\overset{\delta+}{:Na} + \overset{\frown}{\underset{/}{\overset{\backslash}{C}}{=}O} \longrightarrow RC{\equiv}C{-}\overset{|}{\underset{|}{C}}{-}ONa \xrightarrow{H_2O} RC{\equiv}C{-}\overset{|}{\underset{|}{C}}{-}OH$$

We shall now examine a similar reaction that involves the addition of an organozinc reagent to the carbonyl group of an aldehyde or ketone. This reaction, called the *Reformatsky reaction,* extends the carbon skeleton of an aldehyde or ketone and yields β-hydroxy esters. It involves treating an aldehyde or ketone with an α-bromo ester in the presence of zinc metal; the solvent most often used is benzene. The initial product is a zinc alkoxide, which must be hydrolyzed to yield the β-hydroxy ester.

$$\underset{\substack{\text{Aldehyde} \\ \text{or} \\ \text{ketone}}}{\overset{\backslash}{\underset{/}{C}}{=}O} + \underset{\substack{\alpha\text{-Bromo} \\ \text{ester}}}{Br{-}\overset{|}{\underset{|}{C}}{-}CO_2R} \xrightarrow[\text{benzene}]{Zn} \overset{BrZnO}{{-}\overset{|}{\underset{|}{C}}{-}\overset{|}{\underset{|}{C}}{-}CO_2R} \xrightarrow{H_3O^+} \underset{\substack{\beta\text{-Hydroxy} \\ \text{ester}}}{\overset{HO}{{-}\overset{|}{\underset{|}{C}}{}^\beta{-}\overset{|}{\underset{|}{C}}{}^\alpha{-}CO_2R}}$$

The intermediate in the reaction appears to be an organozinc reagent that adds to the carbonyl group in a manner analogous to that of a Grignard reagent.

$$Br{-}\overset{|}{\underset{|}{C}}{-}CO_2R \xrightarrow[\text{benzene}]{Zn} \overset{\delta+}{BrZn}{:}\overset{\delta-}{\overset{|}{\underset{|}{C}}}{-}CO_2R \longrightarrow$$

$$\overset{BrZnO}{{-}\overset{|}{\underset{|}{C}}{-}\overset{|}{\underset{|}{C}}{-}CO_2R} \xrightarrow{H_3O^+} \overset{HO}{{-}\overset{|}{\underset{|}{C}}{-}\overset{|}{\underset{|}{C}}{-}CO_2R}$$

Because the organozinc reagent is less reactive than a Grignard reagent, it does not add to the ester group. The β-hydroxy esters produced in the Reformatsky reaction are easily dehydrated to α,β-unsaturated esters, because dehydration yields a system in which the carbon–carbon double bond is conjugated with the carbon–oxygen double bond of the ester.

$$\underset{\substack{\beta\text{-Hydroxy} \\ \text{ester}}}{\overset{HO}{{-}\overset{|}{\underset{|}{C}}{-}\overset{|}{\underset{\underset{H}{|}}{C}}{-}\overset{O}{\overset{\|}{C}OR}} \xrightarrow[\substack{\text{heat} \\ (-H_2O)}]{H_3O^+} \underset{\substack{\alpha,\beta\text{-Unsaturated} \\ \text{ester}}}{{-}\overset{|}{C}{=}\overset{|}{C}{-}\overset{O}{\overset{\|}{C}OR}}$$

Examples of the Reformatsky reaction are the following:

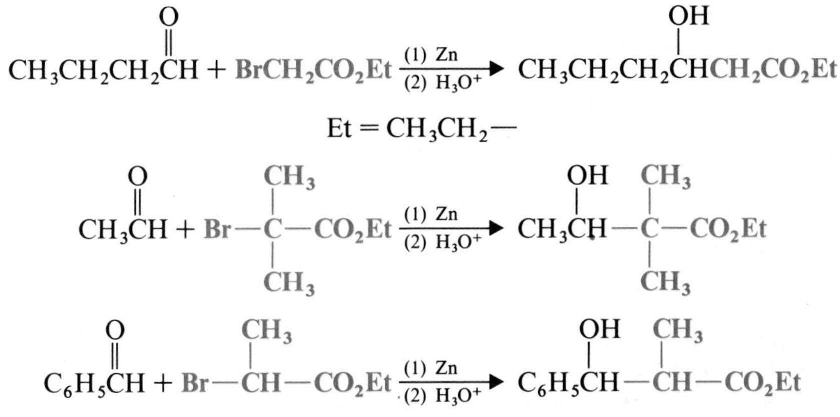

$$\underset{\substack{\| \\ O}}{CH_3CH_2CH_2CH} + BrCH_2CO_2Et \xrightarrow[\text{(2) H}_3O^+]{\text{(1) Zn}} \underset{\substack{| \\ OH}}{CH_3CH_2CH_2CHCH_2CO_2Et}$$

$$Et = CH_3CH_2-$$

$$\underset{\substack{\| \\ O}}{CH_3CH} + \underset{\substack{| \\ CH_3}}{Br-\underset{\substack{| \\ CH_3}}{C}-CO_2Et} \xrightarrow[\text{(2) H}_3O^+]{\text{(1) Zn}} \underset{\substack{| \\ OH}}{CH_3CH}-\underset{\substack{| \\ CH_3}}{C}-CO_2Et$$

$$\underset{\substack{\| \\ O}}{C_6H_5CH} + \underset{\substack{| \\ CH_3}}{Br-CH-CO_2Et} \xrightarrow[\text{(2) H}_3O^+]{\text{(1) Zn}} \underset{\substack{| \\ OH}}{C_6H_5CH}-\underset{\substack{| \\ CH_3}}{CH}-CO_2Et$$

skip

PROBLEM 16.22

Show how you would use a Reformatsky reaction in the synthesis of each of the following compounds:

(a) 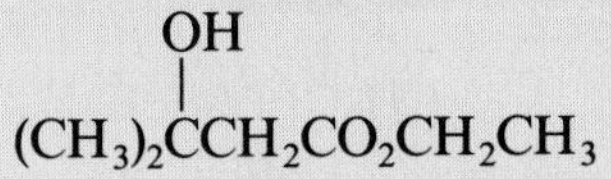 $\underset{\substack{| \\ OH}}{(CH_3)_2CCH_2CO_2CH_2CH_3}$

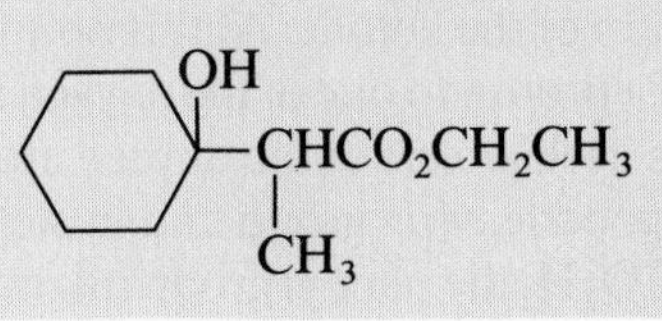

(c) $CH_3CH_2CH_2CH_2CO_2CH_2CH_3$

(b) cyclohexyl—$\underset{\substack{| \\ CH_3}}{\overset{OH}{CH}CO_2CH_2CH_3}$

16.12 OXIDATION OF ALDEHYDES AND KETONES

Aldehydes are much more easily oxidized than ketones. Aldehydes are readily oxidized by strong oxidizing agents such as potassium permanganate, and they are also oxidized by such mild oxidizing agents as silver oxide.

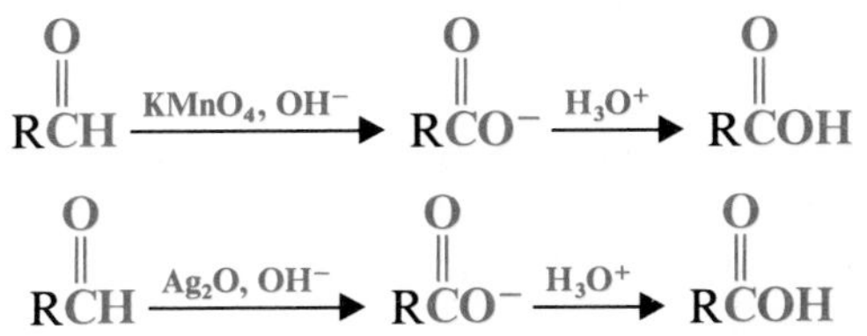

$$\underset{\substack{\| \\ O}}{RCH} \xrightarrow{KMnO_4,\ OH^-} \underset{\substack{\| \\ O}}{RCO^-} \xrightarrow{H_3O^+} \underset{\substack{\| \\ O}}{RCOH}$$

$$\underset{\substack{\| \\ O}}{RCH} \xrightarrow{Ag_2O,\ OH^-} \underset{\substack{\| \\ O}}{RCO^-} \xrightarrow{H_3O^+} \underset{\substack{\| \\ O}}{RCOH}$$

Notice that in these oxidations aldehydes lose the hydrogen that is attached to the carbonyl carbon atom. Because ketones lack this hydrogen, they are more resistant to oxidation.

16.12A The Baeyer–Villiger Oxidation of Aldehydes and Ketones

Both aldehydes and ketones are oxidized by peroxy acids. This reaction, called the *Baeyer–Villiger oxidation,* is especially useful with ketones, because it converts them to carboxylic esters. For example, treating acetophenone with a peroxy acid converts it to the ester, phenyl acetate.

$$C_6H_5-\overset{\overset{\displaystyle O}{\|}}{C}-CH_3 \xrightarrow{RCOOH} C_6H_5-O-\overset{\overset{\displaystyle O}{\|}}{C}-CH_3$$

Acetophenone   Phenyl acetate

The mechanism for this reaction involves the following steps:

$$CH_3-\overset{\overset{\displaystyle O}{\|}}{\underset{C_6H_5}{C}} + :\overset{\overset{\displaystyle H}{|}}{O}-O-\overset{\overset{\displaystyle O}{\|}}{C}-R \overset{(1)}{\rightleftharpoons} CH_3-\overset{\overset{\displaystyle O-H}{|}}{\underset{C_6H_5}{C}}-O-O-\overset{\overset{\displaystyle O}{\|}}{C}-R \overset{(2)}{\underset{H^+}{\rightarrow}}$$

$$CH_3-\overset{\overset{\displaystyle O-H}{|}}{\underset{C_6H_5}{C}}-O-O-\overset{\overset{\displaystyle O-H^+}{\|}}{C}-R \xrightarrow[(3a)]{-R\overset{\overset{\displaystyle O}{\|}}{C}-OH} CH_3-\overset{\overset{\displaystyle O-H}{|}}{\underset{C_6H_5}{C}}-\overset{..}{\overset{..}{O}}:^+ \xrightarrow[\substack{(3b)}]{\substack{phenyl \\ migration}}$$

$$CH_3\overset{\overset{\displaystyle O}{\|}}{C}-O-C_6H_5 + H^+$$

In step 1 the peroxy acid adds to the carbonyl group of the ketone. At this point there are several equilibria involving the attachment of a proton to one of the oxygen atoms of this addition product. When a proton attaches itself to one of the oxygen atoms of the carboxylic acid portion, it makes this part a good leaving group and in step 3a it departs. Simultaneously with the departure of RCO_2H, the phenyl group migrates (as an anion) to the electron-deficient oxygen that is being created (step 3b). After that, the loss of a proton produces the ester. The mechanism for this reaction is similar to the one that occurs in the cumene hydroperoxide synthesis of phenol (Section 14.4).

Step 3b shows that a phenyl group has a greater tendency to migrate than a methyl group; otherwise the product would have been $C_6H_5COOCH_3$ and not $CH_3COOC_6H_5$. This tendency of a group to migrate is called its **migratory aptitude.** Studies of the Baeyer–Villiger oxidation and other reactions have shown that the migratory aptitude of groups is H > phenyl > 3 ° alkyl > 2 ° alkyl > 1 ° alkyl > methyl. In all cases, this order is for groups migrating with their electron pairs, that is, as anions.

PROBLEM 16.23

When benzaldehyde reacts with a peroxy acid, the product is benzoic acid. The mechanism for this reaction is analogous to the one just given for the oxidation of acetophenone, and the outcome illustrates the greater migratory aptitude of a hydrogen atom compared to phenyl. Outline all the steps involved.

PROBLEM 16.24

Give the structure of the product that would result from a Baeyer–Villiger oxidation of cyclopentanone.

PROBLEM 16.25

What would be the major product formed in the Baeyer–Villiger oxidation of 3-methyl-2-butanone?

16.13 CHEMICAL AND SPECTROSCOPIC ANALYSIS FOR ALDEHYDES AND KETONES

Aldehydes and ketones can be differentiated from noncarbonyl compounds through their reactions with derivatives of ammonia (Section 16.8). Semicarbazide, 2,4-dinitrophenylhydrazine, and hydroxylamine react with aldehydes and ketones to form precipitates. Semicarbazones and oximes are usually colorless, while 2,4-dinitrophenylhydrazones are usually orange. The melting points of these derivatives can also be used in identifying specific aldehydes and ketones.

The ease with which aldehydes undergo oxidation provides a useful test that differentiates aldehydes from most ketones.

16.13A Tollens' Test (Silver Mirror Test)

Mixing aqueous silver nitrate with aqueous ammonia produces a solution known as Tollens' reagent. The reagent contains the diamminosilver(I) ion, $Ag(NH_3)_2^+$. Although this ion is a very weak oxidizing agent, it will oxidize aldehydes to carboxylate ions. As it does this, silver is reduced from the $+1$ oxidation state [of $Ag(NH_3)_2^+$] to metallic silver. If the rate of reaction is slow and the walls of the vessel are clean, metallic silver deposits on the walls of the test tube as a mirror; if not, it deposits as a gray to black precipitate. Tollens' reagent gives a negative result with all ketones except α-hydroxy ketones.

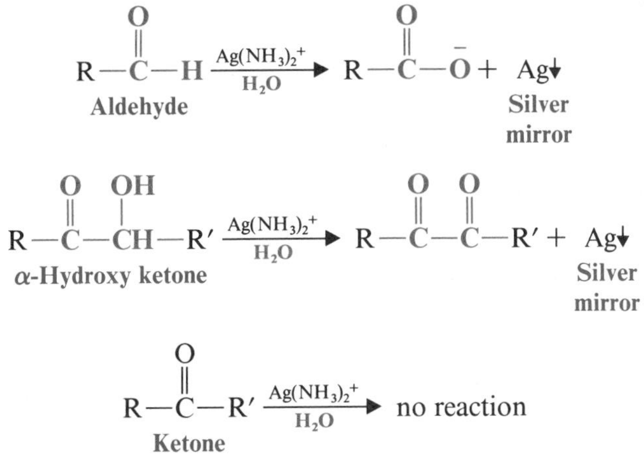

16.13B Spectroscopic Properties of Aldehydes and Ketones

Carbonyl groups of aldehydes and ketones give rise to very strong C=O stretching bands in the 1665–1780-cm^{-1} region of the **infrared (IR) spectrum.** The exact location of the peak (Table 16.4) depends on the structure of the aldehyde or ketone.

The CHO group of aldehydes also gives two weak bands in the 2700–2775 and 2820–2900-cm^{-1} region of the infrared spectrum.

TABLE 16.4 Carbonyl stretching bands of aldehydes and ketones

C=O STRETCHING FREQUENCIES (cm^{-1})			
R—CHO	1720–1740	RCOR	1705–1720
Ar—CHO	1695–1715	ArCOR	1680–1700
$-\overset{\textstyle\mid}{C}=\overset{\textstyle\mid}{C}-CHO$	1680–1690	$-\overset{\textstyle\mid}{C}=\overset{\textstyle\mid}{C}-COR$	1665–1680
		Cyclohexanone	1715
		Cyclopentanone	1751

The aldehydic proton gives a signal far downfield ($\delta = 9-10$) in **proton nmr spectra.** An example is given in the proton nmr spectrum of acetaldehyde (Fig. 16.1).

The carbonyl groups of saturated aldehydes and ketones give a weak absorption band in the **ultraviolet (UV) region** between 270 and 300 nm. This band is shifted to longer wavelengths (300–350 nm) when the carbonyl group is conjugated with a double bond.

16.14 SUMMARY OF THE ADDITION REACTIONS OF ALDEHYDES AND KETONES

Table 16.5 summarizes the nucleophilic addition reactions of aldehydes and ketones that occur at the carbonyl carbon atom that we have studied so far. In the next chapter we shall see other examples.

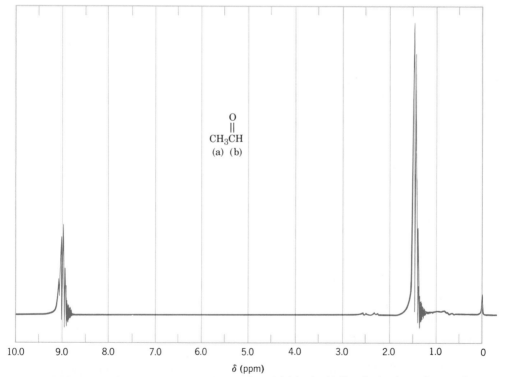

$$\overset{\textstyle O}{\overset{\textstyle \|}{CH_3CH}}$$
(a) (b)

FIGURE 16.1 The proton nmr spectrum of acetaldehyde. Notice that spin–spin coupling between the aldehydic proton and the protons of the methyl group cause both signals to be split. The aldehydic proton occurs as a quartet and the methyl protons occur as a doublet. The coupling constant is about 3 Hz.

TABLE 16.5 Nucleophilic addition reactions of aldehydes and ketones

1. Addition of Organometallic Compounds

   General Reaction

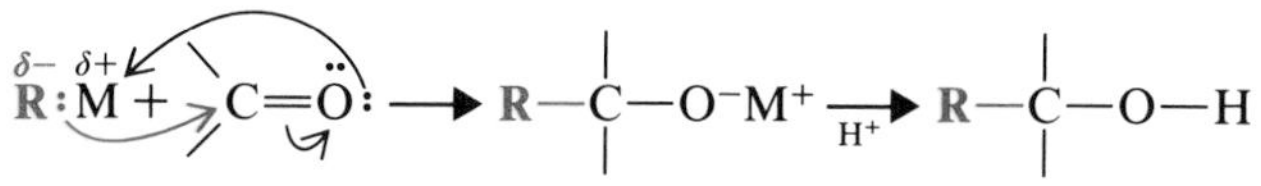

   $$\overset{\delta-\ \delta+}{R:M} + \ >C=\ddot{O}: \longrightarrow R-\overset{|}{\underset{|}{C}}-O^-M^+ \xrightarrow[H^+]{} R-\overset{|}{\underset{|}{C}}-O-H$$

   Specific Example Using a Grignard Reagent (Section 15.6)

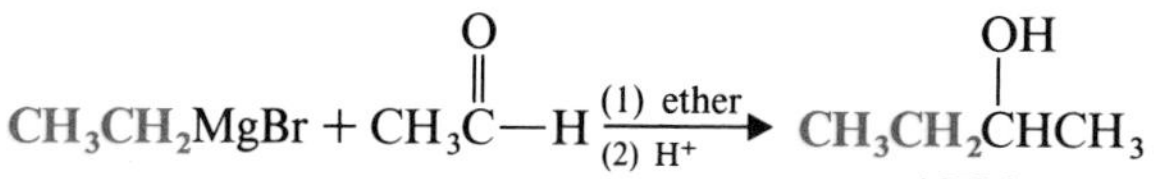

   $$CH_3CH_2MgBr + CH_3\overset{O}{\overset{\|}{C}}-H \xrightarrow[(2)\ H^+]{(1)\ ether} CH_3CH_2\overset{OH}{\overset{|}{C}}HCH_3$$
   (67%)

   Specific Example Using the Reformatsky Reaction (Section 16.11)

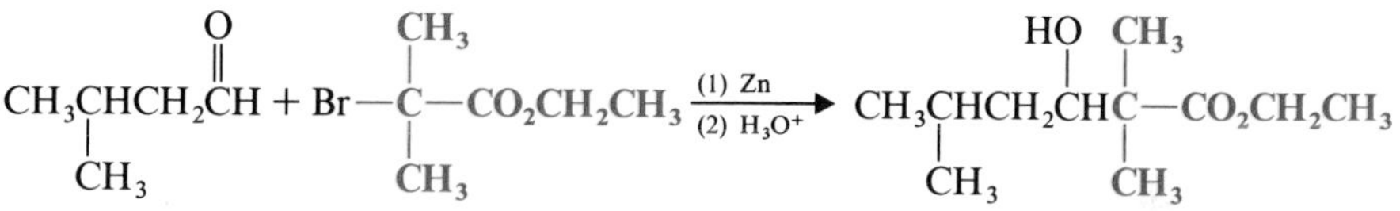

   $$CH_3\underset{\underset{CH_3}{|}}{C}HCH_2\overset{O}{\overset{\|}{C}}H + Br-\underset{\underset{CH_3}{|}}{\overset{\overset{CH_3}{|}}{C}}-CO_2CH_2CH_3 \xrightarrow[(2)\ H_3O^+]{(1)\ Zn} CH_3\underset{\underset{CH_3}{|}}{C}HCH_2\overset{HO}{\overset{|}{C}}H\overset{CH_3}{\overset{|}{C}}-CO_2CH_2CH_3$$
   (65%)

2. Addition of Hydride Ion

   General Reaction

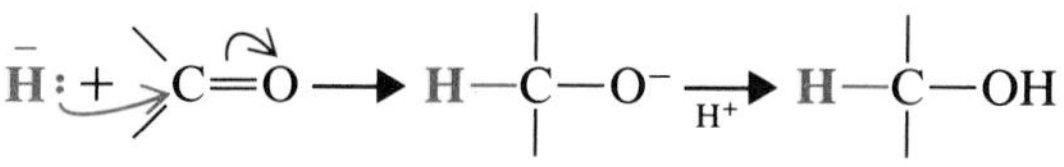

   $$\bar{H}: + \ >C=O \longrightarrow H-\overset{|}{\underset{|}{C}}-O^- \xrightarrow[H^+]{} H-\overset{|}{\underset{|}{C}}-OH$$

   Specific Examples Using Metal Hydrides (Section 15.2)

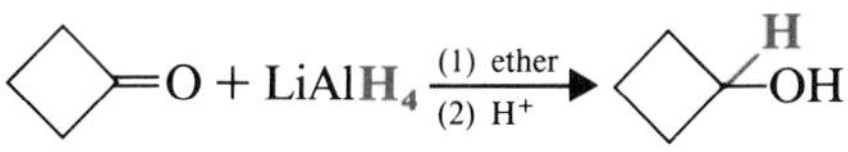

   $$\square{=}O + LiAlH_4 \xrightarrow[(2)\ H^+]{(1)\ ether} \square\overset{H}{\underset{}{-}}OH$$
   (90%)

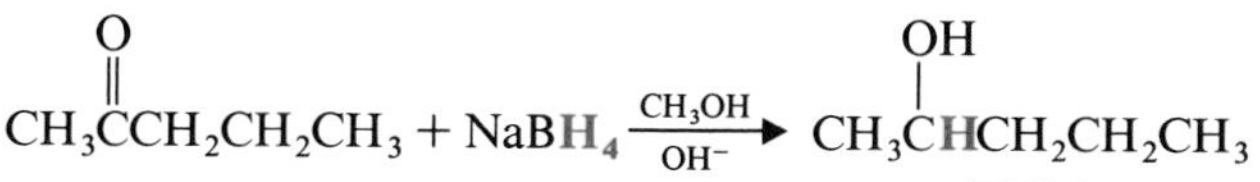

   $$CH_3\overset{O}{\overset{\|}{C}}CH_2CH_2CH_3 + NaBH_4 \xrightarrow[OH^-]{CH_3OH} CH_3\overset{OH}{\overset{|}{C}}HCH_2CH_2CH_3$$
   (100%)

3. Addition of Hydrogen Cyanide and Sodium Bisulfite (Section 16.9)

   General Reaction

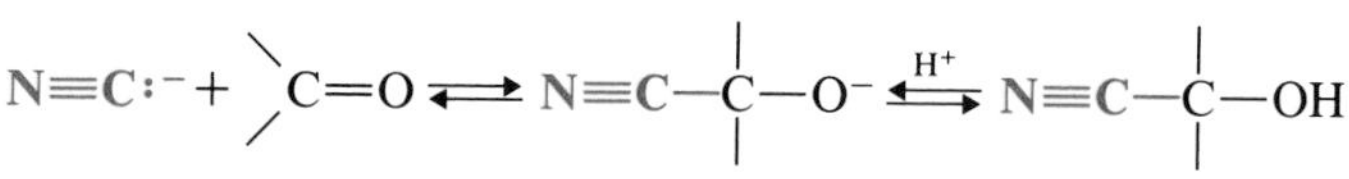

   $$N{\equiv}C:^- + \ >C=O \rightleftharpoons N{\equiv}C-\overset{|}{\underset{|}{C}}-O^- \overset{H^+}{\rightleftharpoons} N{\equiv}C-\overset{|}{\underset{|}{C}}-OH$$

   Specific Example

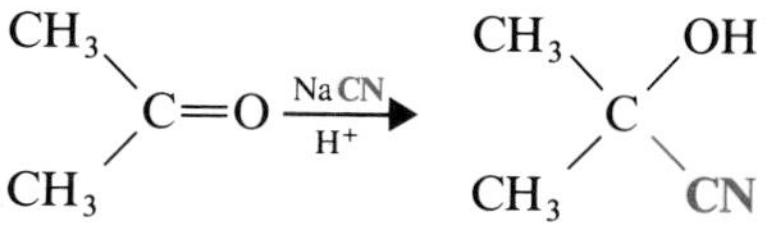

   $$\overset{CH_3}{\underset{CH_3}{>}}C=O \xrightarrow[H^+]{NaCN} \overset{CH_3}{\underset{CH_3}{>}}C\overset{OH}{\underset{CN}{<}}$$
   (78%)
   Acetone cyanohydrin

   (continued)

TABLE 16.5 (continued)

General Reaction

$$NaHSO_3 + \overset{\diagdown}{\underset{\diagup}{}}C=O \rightleftharpoons \overset{\diagdown}{\underset{\diagup}{}}\overset{\overset{\textstyle SO_3^- \, Na^+}{|}}{\underset{\underset{\textstyle OH}{|}}{C}}$$

Specific Example

$$\underset{\textstyle H}{\overset{\textstyle CH_3}{\diagdown}}C=O + NaHSO_3 \rightleftharpoons \underset{\underset{\textstyle H \quad OH}{}}{\overset{\overset{\textstyle CH_3 \quad SO_3^- \, Na^+}{}}{C}}$$
$$(88\%)$$

4. Addition of Ylides (Section 16.10)

The Wittig Reaction

$$Ar_3\overset{+}{P}-\overset{..}{\underset{|}{C}}- + \overset{\diagdown}{\underset{\diagup}{}}C=O \rightleftharpoons -\overset{|}{\underset{\underset{\textstyle Ar_3P^+}{|}}{C}}-\overset{|}{\underset{\underset{\textstyle O^-}{|}}{C}}- \xrightarrow{-Ar_3PO} \overset{\diagdown}{\underset{\diagup}{}}C=C\overset{\diagup}{\diagdown}$$

5. Addition of Alcohols (Section 16.7)

General Reaction

$$R-\overset{..}{\underset{..}{O}}-H + \overset{\diagdown}{\underset{\diagup}{}}C=O \rightleftharpoons R-O-\overset{|}{\underset{|}{C}}-OH \xrightarrow[H^+]{ROH} R-O-\overset{|}{\underset{|}{C}}-O-R$$

Hemiacetal   Acetal or
or hemiketal   ketal

Specific Example

$$C_2H_5OH + CH_3\overset{\overset{\textstyle O}{\|}}{C}H \rightleftharpoons C_2H_5O-\overset{\overset{\textstyle CH_3}{|}}{\underset{\underset{\textstyle H}{|}}{C}}-OH \xrightarrow[H^+]{C_2H_5OH} C_2H_5O-\overset{\overset{\textstyle CH_3}{|}}{\underset{\underset{\textstyle H}{|}}{C}}-OC_2H_5$$

6. Addition of Derivatives of Ammonia (Section 16.8)

General Reaction

$$-\overset{..}{\underset{\underset{\textstyle H}{|}}{N}}-H + \overset{\diagdown}{\underset{\diagup}{}}C=O \rightleftharpoons -\overset{..}{\underset{\underset{\textstyle H}{|}}{N}}-\overset{|}{\underset{|}{C}}-OH \xrightarrow{-H_2O} -\overset{..}{N}=C\overset{\diagup}{\diagdown}$$

Specific Examples

$$CH_3\overset{\overset{\textstyle O}{\|}}{C}H + NH_2OH \longrightarrow CH_3CH=NOH$$
Acetaldoxime

$$C_6H_5\overset{\overset{\textstyle O}{\|}}{C}H + H_2NNHC_6H_5 \longrightarrow C_6H_5CH=NNHC_6H_5$$
Benzaldehyde phenylhydrazone

ADDITIONAL PROBLEMS

16.26

Give a structural formula and another acceptable name for each of the following compounds:

(a) Formaldehyde
(b) Acetaldehyde
(c) Phenylacetaldehyde
(d) Acetone
(e) Ethyl methyl ketone

(f) Acetophenone
(g) Benzophenone
(h) Salicylaldehyde
(i) Vanillin
(j) Diethyl ketone

(k) Ethyl isopropyl ketone
(l) Diisopropyl ketone
(m) Dibutyl ketone
(n) Dipropyl ketone
(o) Cinnamaldehyde

16.27

Write structural formulas for the products formed when propanal reacts with each of the following reagents:

(a) $NaBH_4$ in aqueous NaOH
(b) C_6H_5MgBr, then H_2O
(c) $LiAlH_4$, then H_2O
(d) Ag_2O, OH^-
(e) $(C_6H_5)_3P$=CH_2
(f) H_2 and Pt
(g) $HOCH_2CH_2OH$ and H^+
(h) CH_3CH=$P(C_6H_5)_3$

(i) (1) $BrCH_2CO_2C_2H_5$, Zn; (2) H_3O^+
(j) $Ag(NH_3)_2^+$
(k) Hydroxylamine
(l) Semicarbazide
(m) Phenylhydrazine
(n) Cold dilute $KMnO_4$
(o) $HSCH_2CH_2SH$, H^+
(p) $HSCH_2CH_2SH$, H^+, then Raney nickel

16.28

Give structural formulas for the products formed (if any) from the reaction of acetone with each reagent in Problem 16.27.

16.29

What products would be obtained from each of the following reactions of acetophenone?

(a) Acetophenone + HNO_3 $\xrightarrow[H_2SO_4]{}$

(b) Acetophenone + $C_6H_5NHNH_2$ $\xrightarrow[OH^-]{}$

(c) Acetophenone + CH_2=$P(C_6H_5)_3$ $\longrightarrow$

(d) Acetophenone + $NaBH_4$ $\xrightarrow[OH^-]{H_2O}$

(e) Acetophenone + C_6H_5MgBr $\xrightarrow[(2) H_2O]{}$

16.30

(a) Give three methods for synthesizing phenyl propyl ketone from benzene and any other needed reagents. (b) Give three methods for transforming phenyl propyl ketone into butylbenzene.

16.31

Show how you would convert benzaldehyde into each of the following. You may use any other needed reagents, and more than one step may be required.

(a) Benzyl alcohol
(b) Benzoic acid
(c) Benzoyl chloride
(d) Benzophenone
(e) Acetophenone
(f) 1-Phenylethanol

(g) 3-Methyl-1-phenyl-1-butanol
(h) Benzyl bromide
(i) Toluene
(j) $C_6H_5CH(OCH_3)_2$
(k) $C_6H_5CH^{18}O$
(l) C_6H_5CHDOH

(m) $C_6H_5CH(OH)CN$
(n) C_6H_5CH=NOH
(o) C_6H_5CH=$NNHC_6H_5$
(p) C_6H_5CH=$NNHCONH_2$
(q) C_6H_5CH=$CHCH$=CH_2
(r) $C_6H_5CH(OH)SO_3Na$

16.32

Show how ethyl phenyl ketone ($C_6H_5COCH_2CH_3$) could be synthesized from each of the following:

(a) Benzene   (c) Benzonitrile, C_6H_5CN (e) Benzoic acid

(b) Benzoyl chloride   (d) Benzaldehyde

16.33

Show how benzaldehyde could be synthesized from each of the following:

(a) Benzyl alcohol   (c) Phenylethyne   (e) $C_6H_5CO_2CH_3$

(b) Benzoic acid   (d) Phenylethene (styrene)   (f) $C_6H_5C{\equiv}N$

16.34

When acetaldehyde is converted to a cyanohydrin by treating it with NaCN in H_2O and then adding sulfuric acid, the rate of the reaction is found to depend on the concentrations of acetaldehyde and cyanide ion. Write a mechanism that is consistent with this finding, and tell which step of the mechanism is the slow step.

16.35

Give structures for compounds **A**–**E**.

Cyclohexanol $\xrightarrow[\text{acetone}]{H_2CrO_4}$ **A** $(C_6H_{10}O)$ $\xrightarrow[\text{(2) } H_3O^+]{\text{(1) } CH_3MgI}$ **B** $(C_7H_{14}O)$ $\xrightarrow[\text{heat}]{H^+}$

C (C_7H_{12}) $\xrightarrow[\text{(2) Zn, } H_2O]{\text{(1) } O_3}$ **D** $(C_7H_{12}O_2)$ $\xrightarrow[\text{(2) } H^+]{\text{(1) } Ag_2O, OH^-}$ **E** $(C_7H_{12}O_3)$

16.36

Give structures for **F**–**J**.

Cyclopentanone $\xrightarrow[\text{(2) } H_3O^+]{\text{(1) } NaC{\equiv}CH}$ **F** $(C_7H_{10}O)$ $\xrightarrow[\text{(2) } H_2O_2, OH^-]{\text{(1) } Sia_2BH}$ **G** $(C_7H_{12}O_2)$ $\xrightarrow{CH_3OH, H^+}$

H $(C_9H_{18}O_3)$ $\xrightarrow[\text{(2) } CH_3I]{\text{(1) } NaH}$ **I** $(C_{10}H_{20}O_3)$ $\xrightarrow[H_2O]{H^+}$ **J** $(C_8H_{14}O_2)$

16.37

The following reaction sequence shows how the carbon chain of an aldehyde may be lengthened by two carbon atoms. What are the intermediates **K**–**N**?

Propanal $\xrightarrow[\text{(2) } H^+]{\text{(1) } BrCH_2CO_2Et, Zn}$ **K** $(C_7H_{14}O_3)$ $\xrightarrow{H^+, \text{heat}}$ **L** $(C_7H_{12}O_2)$ $\xrightarrow{H_2, Pt}$

M $(C_7H_{14}O_2)$ $\xrightarrow[\text{(2) } H_2O]{\text{(1) } LiAlH_4}$ **N** $(C_5H_{12}O)$ $\xrightarrow[CH_2Cl_2]{PCC}$ pentanal

16.38

Warming piperonal (Section 16.3) with dilute aqueous HCl converts it to a compound with the formula $C_7H_6O_3$. What is this compound and what type of reaction is involved?

16.39

Starting with benzyl bromide, show how you would synthesize each of the following:

(a) $C_6H_5CH_2CHOHCH_3$ (c) $C_6H_5CH{=}CH{-}CH{=}CHC_6H_5$

(b) $C_6H_5CH_2CH_2CHO$ (d) $C_6H_5CH_2COCH_2CH_3$

16.40

Compounds **A** and **D** do not give positive Tollens' tests; however, compound **C** does. Give structures for **A–D**.

4-Bromobutanal $\xrightarrow{\text{HOCH}_2\text{CH}_2\text{OH, H}^+}$ **A** $(C_6H_{11}O_2Br)$ $\xrightarrow{\text{Mg, ether}}$

[**B** $(C_6H_{11}MgO_2Br)$] $\xrightarrow[\text{(2) H}_3\text{O}^+\text{, H}_2\text{O}]{\text{(1) CH}_3\text{CHO}}$ **C** $(C_6H_{12}O_2)$ $\xrightarrow[\text{H}^+]{\text{CH}_3\text{OH}}$ **D** $(C_7H_{14}O_2)$

16.41

Provide the missing steps in the following synthesis:

HO—⟨◯⟩—CH$_2$OH $\xrightarrow{\text{(a)}}$ CH$_3$O—⟨◯⟩—CH$_2$OH $\xrightarrow{\text{(b)}}$? $\xrightarrow{\text{(c)}}$

CH$_3$O—⟨◯⟩—$\underset{\text{CH}_3}{\overset{\text{OH}}{\text{CHCHCO}_2\text{Et}}}$ $\xrightarrow{\text{(d)}}$ CH$_3$O—⟨◯⟩—$\underset{\text{CH}_3}{\overset{\text{OH}}{\text{CHCHCH}_2\text{OH}}}$

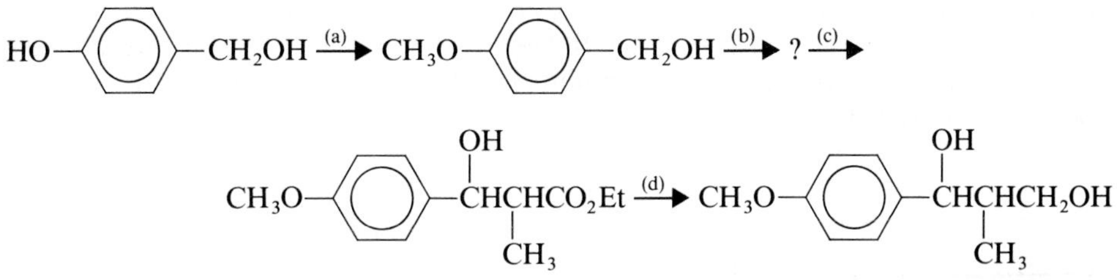

16.42

Outlined here is a synthesis of glyceraldehyde (Section 4.13A). What are the intermediates **A–C**, and what stereoisomeric form of glyceraldehyde would you expect to obtain?

CH$_2$=CHCH$_2$OH $\xrightarrow[\text{CH}_2\text{Cl}_2]{\text{PCC}}$ **A** (C_3H_4O) $\xrightarrow{\text{CH}_3\text{OH, H}^+}$

B $(C_5H_{10}O_2)$ $\xrightarrow[\text{cold, dilute}]{\text{KMnO}_4\text{, OH}^-}$ **C** $(C_5H_{12}O_4)$ $\xrightarrow[\text{H}_2\text{O}]{\text{H}_3\text{O}^+}$ glyceraldehyde

16.43

Consider the reduction of (*R*)-3-phenyl-2-pentanone by sodium borohydride. After the reduction is complete, the mixture is separated by gas-liquid chromatography into two fractions. These fractions contain isomeric compounds, and each isomer is optically active. What are these two isomers, and what is the stereoisomeric relationship between them?

16.44

The structure of the sex pheromone (attractant) of the female tsetse fly has been confirmed by the following synthesis. Compound **C** appears to be identical to the natural pheromone in all respects (including the response of the male tsetse fly). Provide structures for **A, B,** and **C**.

BrCH$_2$(CH$_2$)$_7$CH$_2$Br $\xrightarrow[\text{(2) 2RLi}]{\text{(1) 2(C}_6\text{H}_5)_3\text{P}}$ **A** $(C_{45}H_{46}P_2)$ $\xrightarrow{2\text{CH}_3(\text{CH}_2)_{11}\overset{\overset{\text{O}}{\|}}{\text{C}}\text{CH}_3}$

B $(C_{37}H_{72})$ $\xrightarrow{\text{H}_2\text{, Pt}}$ **C** $(C_{37}H_{76})$

16.45

Outline simple chemical tests that would distinguish between each of the following:

(a) Benzaldehyde and benzyl alcohol

(b) Hexanal and 2-hexanone

(c) 2-Hexanone and hexane

(d) 2-Hexanol and 2-hexanone

(e) $C_6H_5CH=CHCOC_6H_5$ and $C_6H_5COC_6H_5$

(f) Pentanal and diethyl ether

(g) 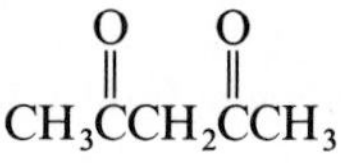 and 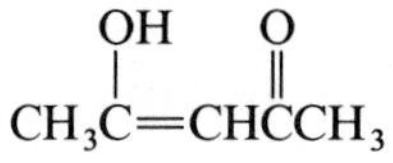

(h) 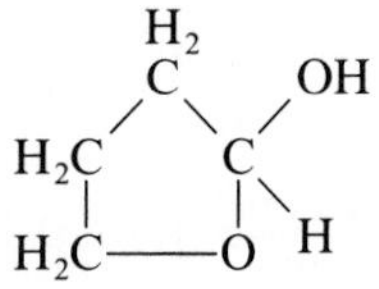 and 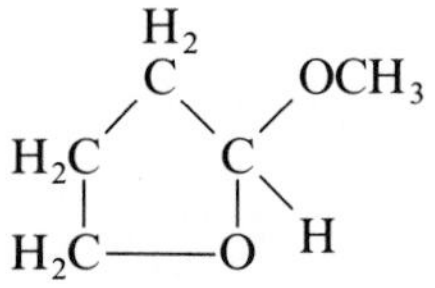

16.46

Compounds **W** and **X** are isomers; they have the molecular formula C_9H_8O. The infrared spectrum of each compound shows a strong absorption band near 1715 cm^{-1}. Oxidation of either compound with hot, basic potassium permanganate followed by acidification yields phthalic acid. The proton nmr spectrum of **W** shows a multiplet at δ 7.3 and a singlet at δ 3.4. The proton nmr spectrum of **X** shows a multiplet at δ 7.5, a triplet at δ 3.1, and a triplet at δ 2.5. Propose structures for **W** and **X**.

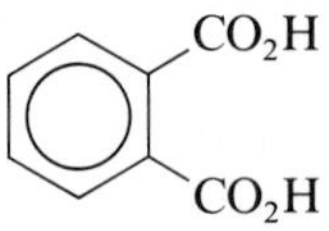

Phthalic acid

16.47

Compounds **Y** and **Z** are isomers with the molecular formula $C_{10}H_{12}O$. The infrared spectrum of each compound shows a strong absorption band near 1710 cm^{-1}. The proton nmr spectra of **Y** and **Z** are given in Figs. 16.2 and 16.3. Propose structures for **Y** and **Z**.

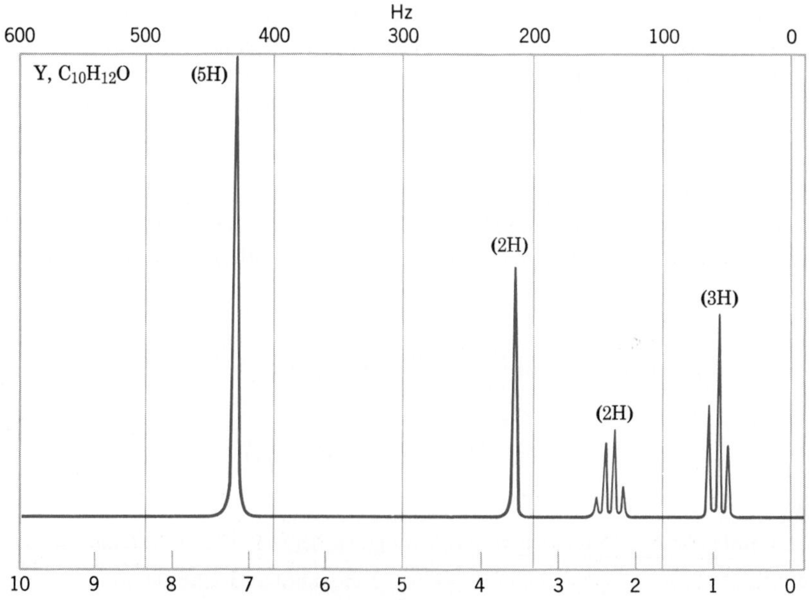

FIGURE 16.2 The proton nmr spectrum of compound **Y**, Problem 16.47. (Courtesy Aldrich Chemical Co., Milwaukee, WI.)

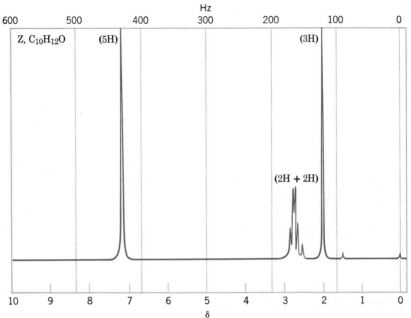

FIGURE 16.3 The proton nmr spectrum of compound **Z**, Problem 16.47. (Courtesy Aldrich Chemical Co., Milwaukee, WI.)

16.48

Compound **A** ($C_9H_{18}O$) forms a phenylhydrazone, but gives a negative Tollens' test. The infrared spectrum of **A** has a strong band near 1710 cm^{-1}. The proton-decoupled ^{13}C nmr spectrum of **A** is given in Fig. 16.4. Propose a structure for **A**.

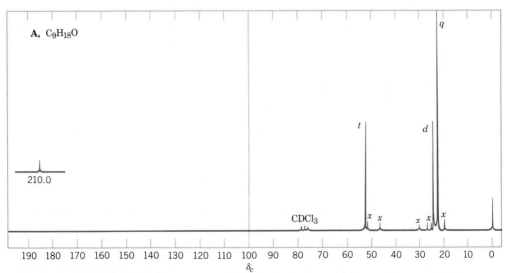

FIGURE 16.4 The proton-decoupled ^{13}C nmr spectrum of compound **A**, Problem 16.48. The letters *d*, *t*, and *q* refer to the splitting of the signal (doublet, triplet, and quartet) in the proton off-resonance decoupled spectrum. A signal marked with an *x* arises from an impurity and should be ignored. (Adapted from L. F. Johnson and W. C. Jankowski, *Carbon-13 NMR Spectra: A Collection of Assigned, Coded, and Indexed Spectra*, Wiley–Interscience, New York, 1972.)

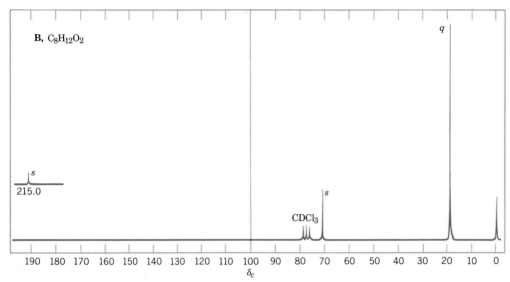

FIGURE 16.5 The proton-decoupled ¹³C nmr spectrum of compound **B**, Problem 16.49. The letters *s* and *q* refer to the splitting of the signal (singlet and quartet) in the proton off-reso-nance decoupled spectrum. (Adapted from L. F. Johnson and W. C. Jankowski, *Carbon-13 NMR Spectra: Assigned, Coded, and Indexed Spectra*, Wiley–Interscience, New York, 1972.)

16.49

Compound **B** ($C_8H_{12}O_2$) shows a strong carbonyl absorption in its infrared spectrum. The proton-decoupled ¹³C nmr spectrum of **B** is given in Fig. 16.5. Propose a structure for **B**.

*16.50

When semicarbazide ($H_2NNHCONH_2$) reacts with a ketone (or an aldehyde) to form a semicarbazone (Section 16.8) only one nitrogen atom of semicarbazide acts as a nucleophile and attacks the carbonyl carbon atom of the ketone. The product of the reaction, conse-quently, is $R_2C=NNHCONH_2$ rather than $R_2C=NCONHNH_2$. What factor accounts for the fact that two nitrogen atoms of semicarbazide are relatively nonnucleophilic?

*16.51

Dimethyl sulfoxide reacts with methyl iodide to give trimethylsulfoxonium iodide. Trimethyl-sulfoxonium iodide forms an ylide when it is treated with sodium hydride.

$$CH_3\ddot{S}CH_3 + CH_3-I \longrightarrow \underset{\underset{O}{\|}}{CH_3\overset{CH_3}{\underset{|}{\overset{+}{S}}}CH_3I^-} \xrightarrow{\text{NaH}} \underset{\underset{O}{\|}}{CH_2=\overset{CH_3}{\underset{|}{S}}CH_3} + NaI + H_2$$

Trimethylsulfoxonium iodide   Ylide

E. J. Corey of Harvard University has shown that this ylide reacts with ketones to give epoxides, in the manner shown in the following example:

(71%)

Propose a mechanism for this epoxide synthesis.

CHAPTER SEVENTEEN

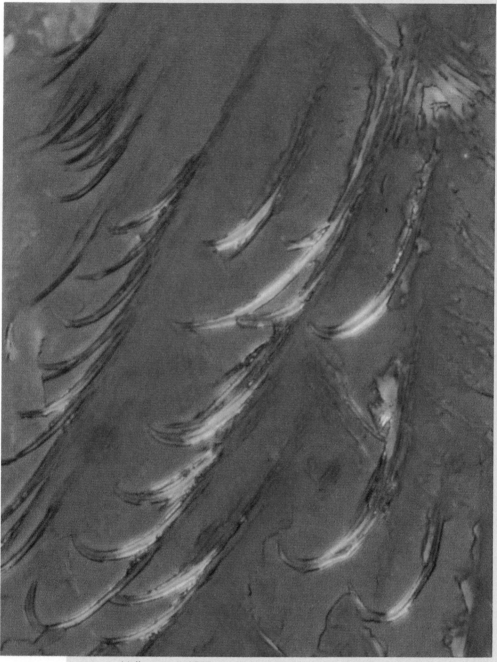

Valium crystal 250x. Valium, also called diazepam, is a powerful tranquilizer with a chemical structure similar to that of chlordiazepoxide given in Section 19.4.

Aldehydes and Ketones II. Reactions at the α Carbon. Aldol Reactions

In the chapter that we have just finished, we found that one important characteristic of aldehydes and ketones is their ability to undergo nucleophilic addition at their carbonyl groups.

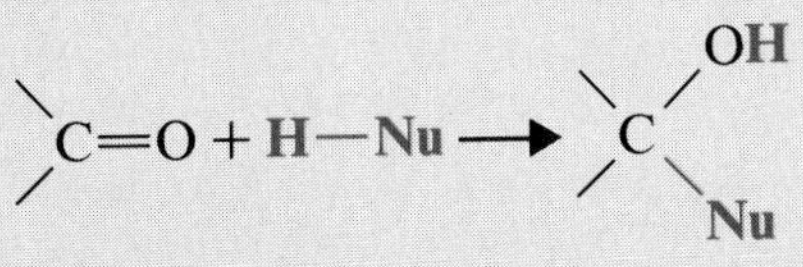

$$\begin{array}{c} \diagdown \\ \diagup \end{array} C{=}O + H{-}Nu \longrightarrow \begin{array}{c} \diagdown \\ \diagup \end{array} C \begin{array}{c} {}^{OH} \\ {}_{Nu} \end{array} \qquad \textit{Nucleophilic addition}$$

A second important characteristic of carbonyl compounds is an unusual acidity of hydrogen atoms on carbon atoms adjacent to the carbonyl group. (These hydrogen atoms are usually called the α **hydrogens,** and the carbon to which they are attached is called the α **carbon.**)

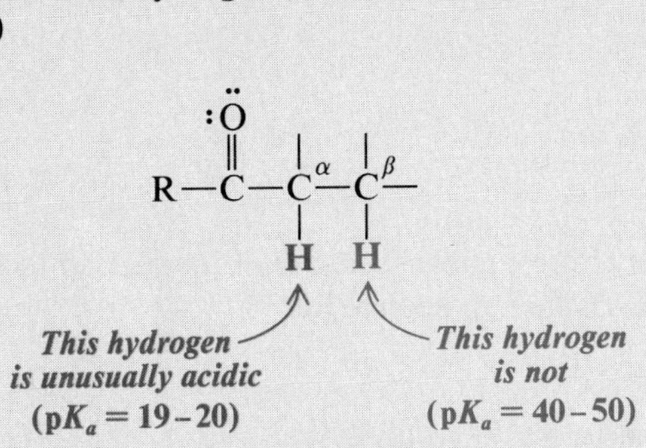

$$R{-}\overset{\overset{\displaystyle :\ddot{O}}{\|}}{C}{-}\underset{\underset{\displaystyle H}{|}}{\overset{\overset{\displaystyle |}{}}{C}}{}^{\alpha}{-}\underset{\underset{\displaystyle H}{|}}{\overset{\overset{\displaystyle |}{}}{C}}{}^{\beta}{-}$$

This hydrogen is unusually acidic ($pK_a = 19{-}20$) *This hydrogen is not* ($pK_a = 40{-}50$)

When we say that the α hydrogens are acidic, *we mean that they are unusually acidic for hydrogen atoms attached to carbon.* The pK_a's for the α hydrogens of most simple aldehydes or ketones are of the order of $19{-}20$ ($K_a = 10^{-19}{-}10^{-20}$). This means that they are more acidic than hydrogen atoms of acetylene, $pK_a = 25$ ($K_a = 10^{-25}$) and are far more acidic than the hydrogens of ethene ($pK_a = 44$) or of ethane ($pK_a = 50$).

The reason for the unusual acidity of the α hydrogens of carbonyl compounds is straightforward: When a carbonyl compound loses an α proton, the anion that is produced is *stabilized by resonance. The negative charge of the anion is delocalized.*

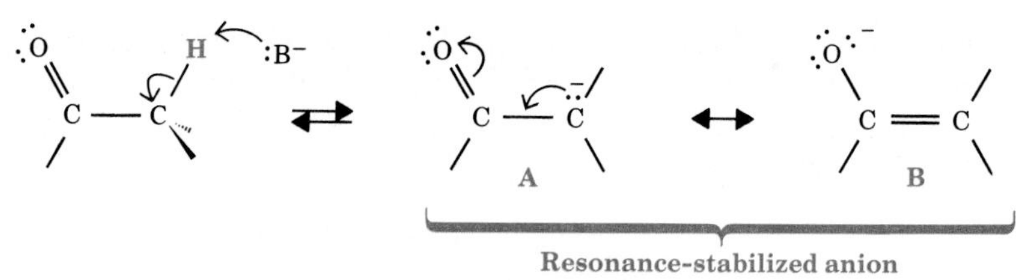

Resonance-stabilized anion

We see from this reaction that two resonance structures, **A** and **B**, can be written for the anion. In structure **A** the negative charge is on carbon and in structure **B** the negative charge is on oxygen. Both structures contribute to the hybrid. Structure **A** is favored by the greater strength of its carbon–oxygen π bond relative to the weaker carbon–carbon π bond of **B**. However, we would expect structure **B** to make a substantial contribution to the hybrid because oxygen, being highly electronegative, is better able to accommodate the negative charge. We can depict the hybrid in the following way:

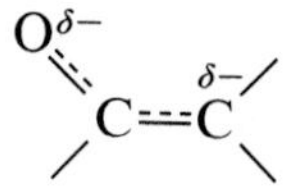

When this resonance-stabilized anion accepts a proton, it can do so in either of two ways: it can accept the proton at carbon to form the original carbonyl compound in what is called the **keto form,** or it may accept the proton at oxygen to form an **enol.**

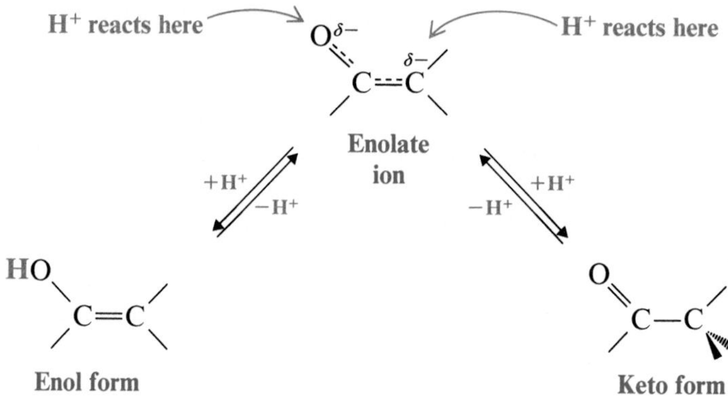

Enol form   **Keto form**

Both of these reactions are reversible. Because of its relation to the enol, the resonance-stabilized anion is called an **enolate ion.**

17.2 KETO AND ENOL TAUTOMERS

The keto and enol forms of carbonyl compounds are constitutional isomers, but of a special type. Because they are easily interconverted in the presence of traces of acids and bases, chemists use a special term to describe this type of constitutional isomerism. Interconvertible keto and enol forms are said to be **tautomers,** and their interconversion is called **tautomerization.**

Under most circumstances, we encounter keto–enol tautomers in a state of equilibrium. (The surfaces of ordinary laboratory glassware are able to catalyze the interconversion and establish the equilibrium.) For simple monocarbonyl compounds such as acetone and acetaldehyde, the amount of the enol form present at equilibrium is *very small.* In acetone it is much less than 1%; in acetaldehyde the enol concentration is too small to be detected. The greater stability of the following keto forms of monocarbonyl compounds can be related to the greater strength of the carbon–oxygen π bond compared to the carbon–carbon π bond (~ 87 kcal/mole versus ~ 60 kcal/mole).

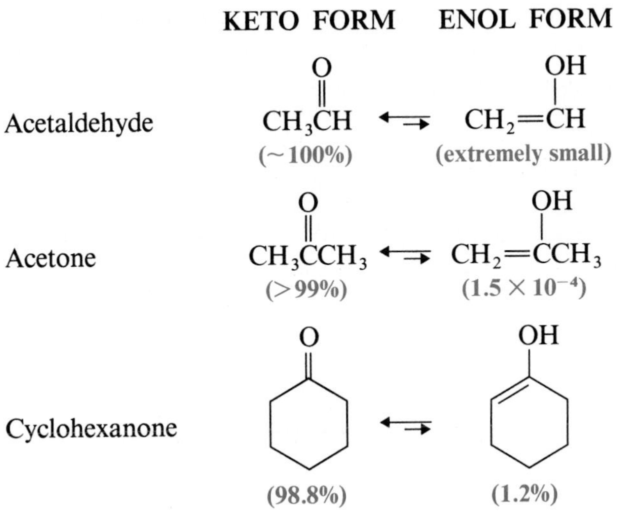

KETO FORM **ENOL FORM**

Acetaldehyde

$$\underset{(\sim 100\%)}{CH_3\overset{\displaystyle O}{\overset{\|}{C}}H} \quad \longleftrightarrow \quad \underset{\text{(extremely small)}}{CH_2{=}\overset{\displaystyle OH}{\overset{|}{C}}H}$$

Acetone

$$\underset{(>99\%)}{CH_3\overset{\displaystyle O}{\overset{\|}{C}}CH_3} \quad \longleftrightarrow \quad \underset{(1.5\times 10^{-4})}{CH_2{=}\overset{\displaystyle OH}{\overset{|}{C}}CH_3}$$

Cyclohexanone

(98.8%) (1.2%)

In compounds whose molecules have two carbonyl groups separated by one saturated carbon (called β-dicarbonyl compounds), the amount of enol present at equilibrium is far higher. For example, 2,4-pentanedione exists in the enol form to an extent of 76%.

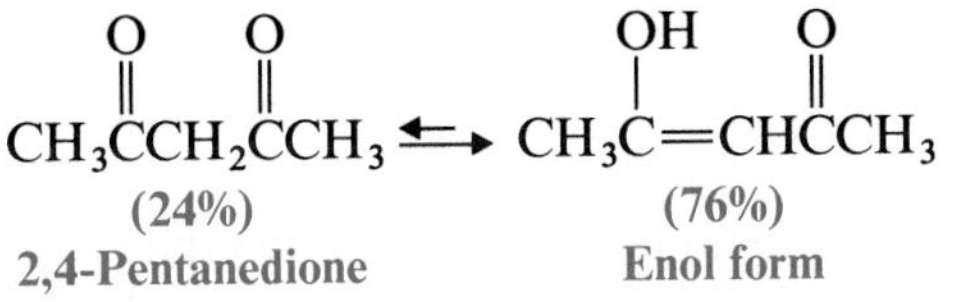

$$\underset{\substack{(24\%) \\ \textbf{2,4-Pentanedione}}}{CH_3\overset{\displaystyle O}{\overset{\|}{C}}CH_2\overset{\displaystyle O}{\overset{\|}{C}}CH_3} \quad \longleftrightarrow \quad \underset{\substack{(76\%) \\ \textbf{Enol form}}}{CH_3\overset{\displaystyle OH}{\overset{|}{C}}{=}CH\overset{\displaystyle O}{\overset{\|}{C}}CH_3}$$

The greater stability of the enol form of β-dicarbonyl compounds can be attributed to stability gained through resonance stabilization of the conjugated double bonds and (in a cyclic form) through hydrogen bonding.

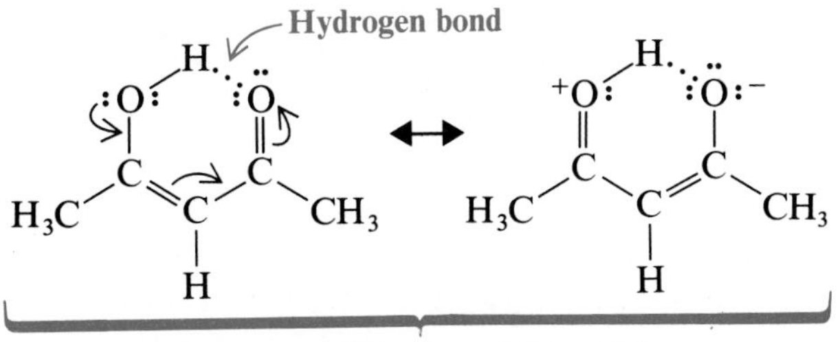

Resonance stabilization of the enol form

PROBLEM 17.1

For all practical purposes, the compound 2,4-cyclohexadien-1-one exists totally in its enol form. Write the structure of 2,4-cyclohexadien-1-one and of its enol form. What special factor accounts for the stability of the enol form?

17.3 REACTIONS VIA ENOLS AND ENOLATE IONS

17.3A Racemization

When a solution of (+)-*sec*-butyl phenyl ketone (see following reaction) in aqueous ethanol is treated with either acids or bases, the solution gradually loses its optical activity. After a time, isolation of the ketone shows that it has been racemized.

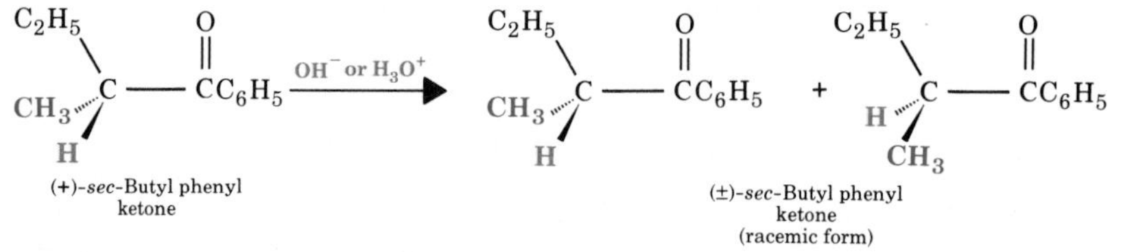

Racemization takes place in the presence of acids or bases because the ketone slowly but reversibly changes to its enol *and the enol is achiral.* When the enol reverts to the keto form, it produces equal amounts of the two enantiomers.

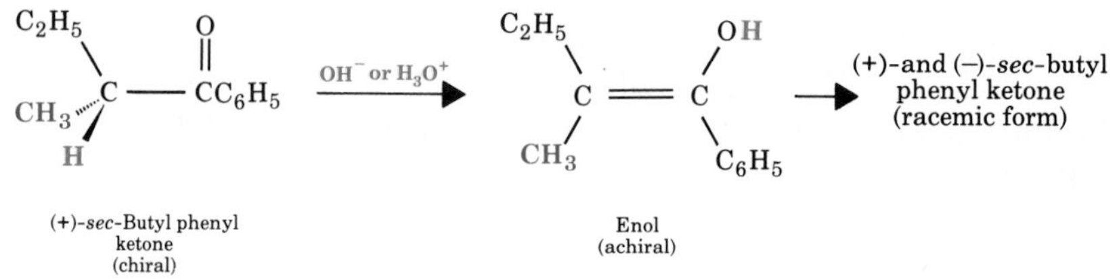

Base catalyzes the formation of an enol through the intermediate formation of an enolate ion:

Base-Catalyzed Enolization

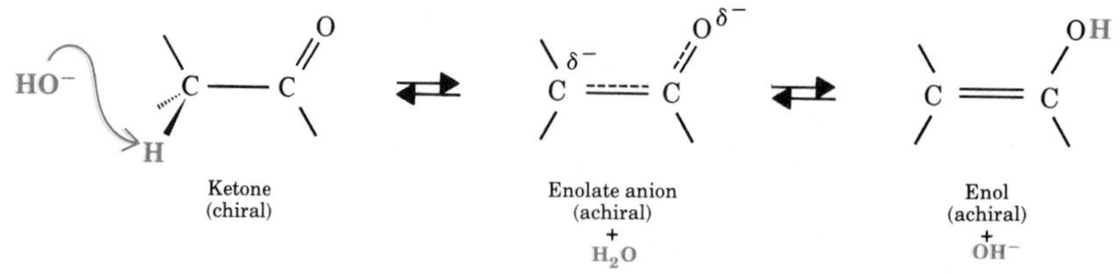

Acid can catalyze enolization in the following way:

Acid-Catalyzed Enolization

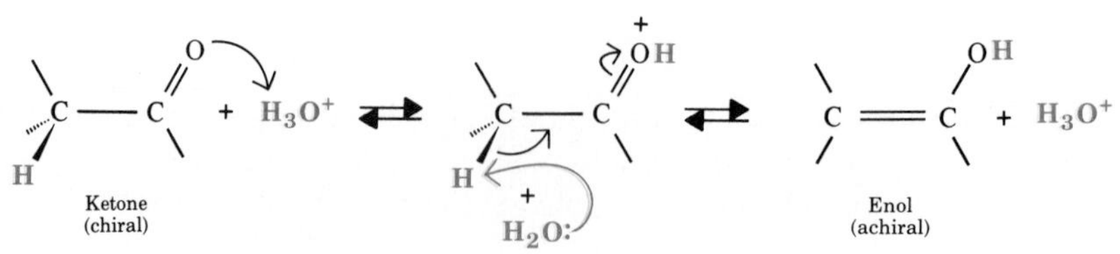

PROBLEM 17.2

Would you expect optically active ketones such as the following to undergo acid- or base-catalyzed racemization? Explain your answer.

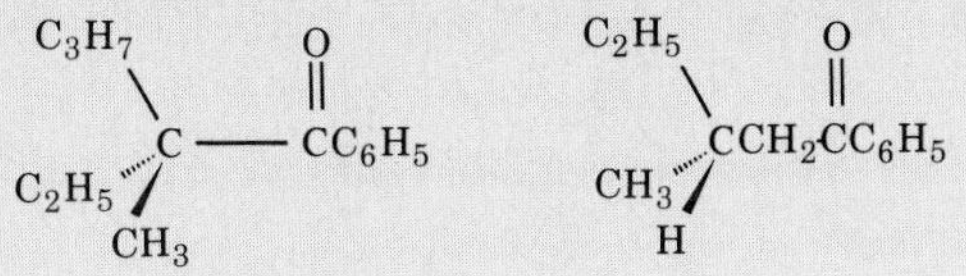

PROBLEM 17.3

When *sec*-butyl phenyl ketone is treated with either OD^- or D_3O^+ in the presence of D_2O, the ketone undergoes hydrogen–deuterium exchange and produces:

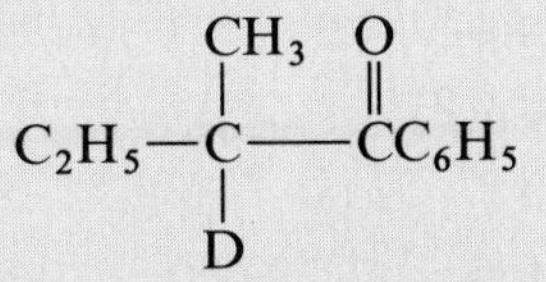

$$C_2H_5-\underset{\underset{D}{|}}{\overset{\overset{CH_3}{|}}{C}}-\overset{\overset{O}{||}}{C}C_6H_5$$

Write mechanisms that account for this behavior.

17.3B Halogenation of Ketones

Ketones that have an α hydrogen react readily with halogen atoms by substitution. The rates of these halogenation reactions *increase when acids or bases are added and substitution takes place almost exclusively at the α carbon:*

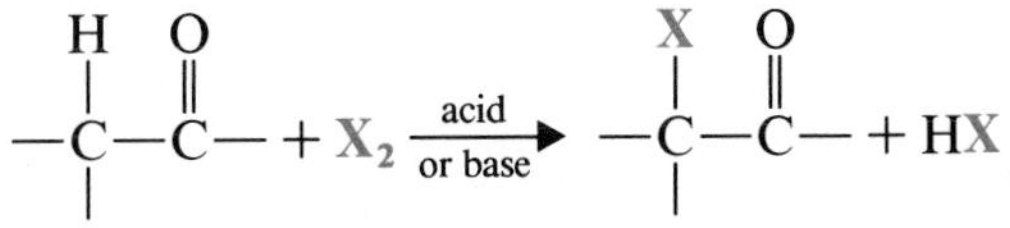

$$-\overset{\overset{H}{|}}{\underset{|}{C}}-\overset{\overset{O}{||}}{C}- + X_2 \xrightarrow[\text{or base}]{\text{acid}} -\overset{\overset{X}{|}}{\underset{|}{C}}-\overset{\overset{O}{||}}{C}- + HX$$

This behavior of ketones can be accounted for in terms of two related properties that we have already encountered: the acidity of the α hydrogens of ketones and the tendency of ketones to form enols.

Base-Promoted Halogenation

In the presence of bases, halogenation takes place through the slow formation of an enolate ion or an enol, followed by a rapid reaction of the enolate ion or enol with halogen.

1. 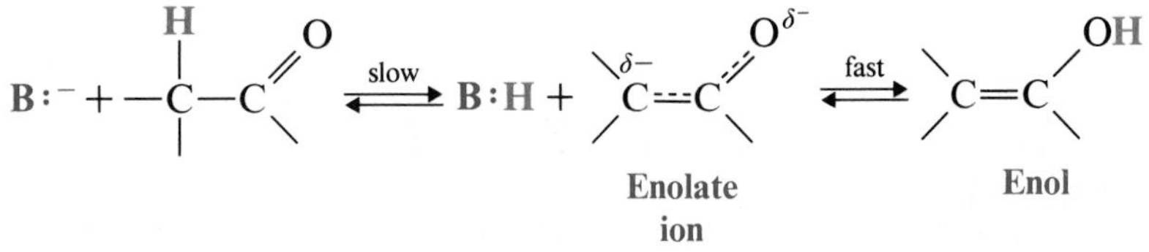 $B:^- + -\overset{\overset{H}{|}}{\underset{|}{C}}-C\overset{\diagup O}{\diagdown} \underset{\text{slow}}{\rightleftarrows} B:H + \overset{\delta-}{C} \cdots C \overset{O^{\delta-}}{\diagdown} \underset{\text{fast}}{\rightleftarrows} C=C\overset{OH}{\diagdown}$

     Enolate   Enol
     ion

2a.

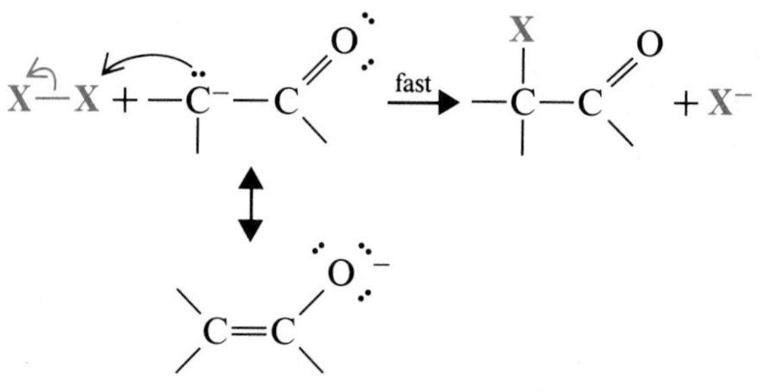

Enolate ion

or

2b.

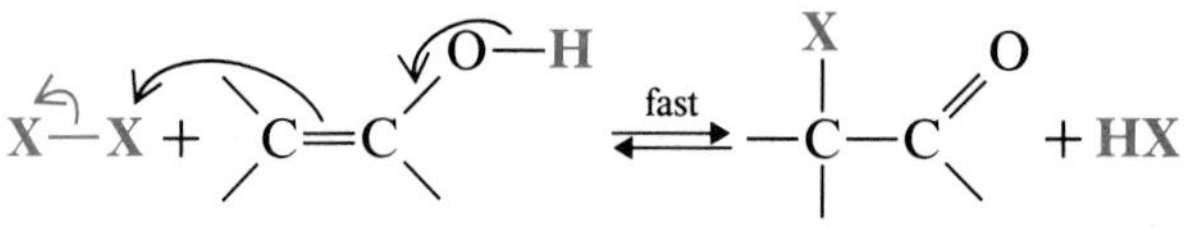

Enol

Acid-Catalyzed Halogenation

In the presence of acids, halogenation takes place through slow formation of an enol followed by rapid reaction of the enol with the halogen.

1.

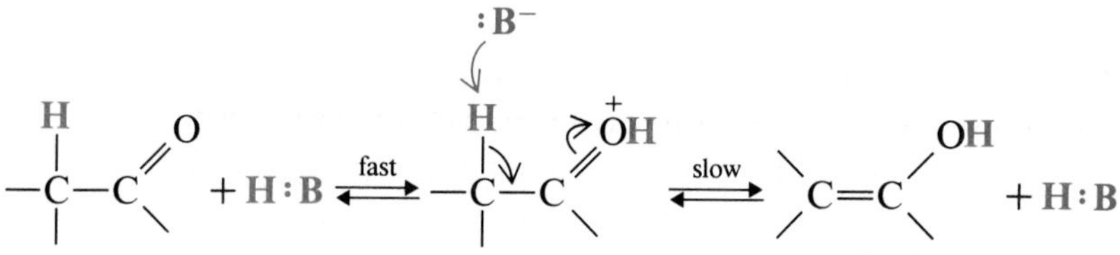

Enol

2.

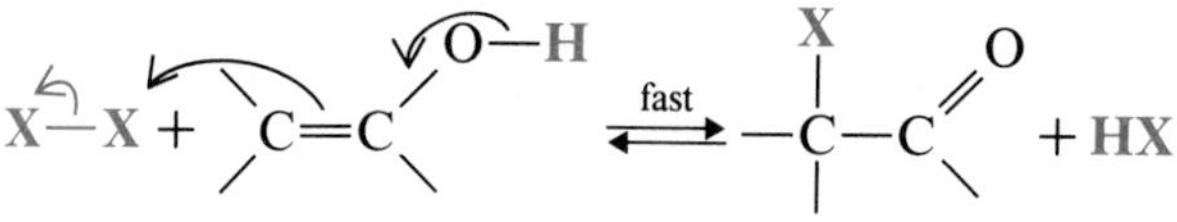

Part of the evidence that supports these mechanisms comes from studies of reaction kinetics. Both base-promoted and acid-catalyzed halogenations of ketones *show initial rates that are independent of the halogen concentration.* The mechanisms that we have written are in accord with this observation: In both instances the slow step of the mechanism occurs prior to the intervention of the halogen.

PROBLEM 17.4

Why do we say that the first reaction is "base promoted" rather than "base catalyzed?"

PROBLEM 17.5

Additional evidence for the halogenation mechanisms that we just presented comes from the following facts: (a) Optically active *sec*-butyl phenyl ketone undergoes acid-catalyzed racemization at a rate exactly equivalent to the rate at which it undergoes acid-catalyzed halogenation. (b) *sec*-Butyl phenyl ketone undergoes acid-catalyzed iodination at the same rate that it undergoes acid-catalyzed bromination. (c) *sec*-Butyl phenyl ketone undergoes base-catalyzed hydrogen–deuterium exchange at the same rate that it undergoes base-promoted halogenation. Explain how each of these observations supports the mechanisms that we have presented.

17.4 THE HALOFORM REACTION

When methyl ketones react with halogens in the presence of base (cf. Section 17.3), multiple halogenations always occur at the carbon of the methyl group. Multiple

$$
C_6H_5-\overset{\overset{\displaystyle O}{\|}}{C}-\overset{\overset{\displaystyle H}{|}}{\underset{\underset{\displaystyle H}{|}}{C}}-H + 3X_2 + 3\ OH^- \xrightarrow[\text{base}]{} C_6H_5-\overset{\overset{\displaystyle O}{\|}}{C}-\overset{\overset{\displaystyle X}{|}}{\underset{\underset{\displaystyle X}{|}}{C}}-X + 3X^- + 3H_2O
$$

halogenations occur because introduction of the first halogen (owing to its electronegativity) makes the remaining α hydrogens on the methyl carbon more acidic.

$$
C_6H_5-\overset{\overset{\displaystyle O}{\|}}{C}-\overset{\overset{\displaystyle H}{}}{\underset{\underset{\displaystyle H}{|}}{C}}-H + :\bar{B} \rightleftharpoons C_6H_5-\overset{\overset{\displaystyle O}{\|}}{C}-\overset{\overset{\displaystyle \cdot\cdot}{}}{\underset{\underset{\displaystyle H}{|}}{C}}-H \xrightarrow{X-X} C_6H_5-\overset{\overset{\displaystyle O}{\|}}{C}-\overset{\overset{\displaystyle X}{|}}{\underset{\underset{\displaystyle H}{|}}{C}}-H + X^-
$$

Enolate ion

$$
C_6H_5-\overset{\overset{\displaystyle O}{\|}}{C}-\overset{\overset{\displaystyle X}{|}}{\underset{\underset{\displaystyle H}{|}}{C}}-H + :\bar{B} \rightleftharpoons C_6H_5-\overset{\overset{\displaystyle O}{\|}}{C}-\overset{\overset{\displaystyle X}{|}}{\underset{\underset{\displaystyle H}{|}}{C}}:^- \xrightarrow{X-X} C_6H_5-\overset{\overset{\displaystyle O}{\|}}{C}-\overset{\overset{\displaystyle X}{|}}{\underset{\underset{\displaystyle H}{|}}{C}}-X
$$

Acidity is increased by electron-withdrawing halogen

$:\bar{B}$, then X_2

$$
C_6H_5-\overset{\overset{\displaystyle O}{\|}}{C}-\overset{\overset{\displaystyle X}{|}}{\underset{\underset{\displaystyle X}{|}}{C}}-X
$$

When methyl ketones react with halogens in aqueous sodium hydroxide (i.e., in *hypohalite solutions**), an additional reaction takes place. Hydroxide ion attacks the

---

*Dissolving a halogen in aqueous sodium hydroxide produces a solution containing sodium hypohalite (NaOX) because of the following equilibrium:

$$
X_2 + 2NaOH \rightleftharpoons NaOX + NaX + H_2O
$$

carbonyl carbon atom of the trihalo ketone and causes a cleavage at the carbon–carbon bond between the carbonyl group and the trihalomethyl group, a moderately good leaving group. This cleavage ultimately produces a carboxylate ion and a *haloform* (i.e., either $CHCl_3$, $CHBr_3$, or CHI_3). The initial step is a nucleophilic attack by hydroxide ion on the carbonyl carbon atom. In the next step carbon–

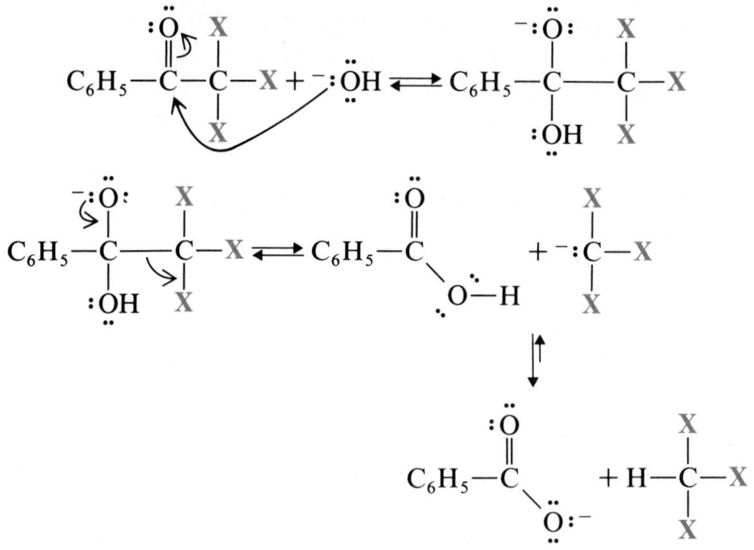

Carboxylate   Haloform
ion

carbon bond cleavage occurs and the haloform anion ($:CX_3^-$) departs. This step can occur because the haloform anion is unusually stable; its negative charge is dispersed by the three electronegative halogen atoms. In the last step, a proton transfer takes place between the carboxylic acid and the haloform anion.

The haloform reaction is of synthetic utility as a means of converting methyl ketones to carboxylic acids. When the haloform reaction is used in synthesis, chlorine and bromine are most commonly used as the halogen component. Chloroform ($CHCl_3$) and bromoform ($CHBr_3$) are both liquids and are easily separated from the acid.

17.4A The Iodoform Test

The haloform reaction using iodine and aqueous sodium hydroxide is called the *iodoform test.* The iodoform test was once frequently used in structure determinations (before the advent of nmr spectral analysis) because it allows identification of the following two groups:

$$-\underset{\underset{O}{\|}}{C}-CH_3 \quad \text{and} \quad -\underset{\underset{OH}{|}}{CH}-CH_3$$

Compounds containing either of these groups react with iodine in sodium hydroxide to give bright yellow precipitates of *iodoform* (CHI_3, mp, 119 °C). Compounds containing the $-CHOHCH_3$ group give a positive iodoform test because they are first oxidized to methyl ketones:

$$-\underset{\underset{OH}{|}}{CH}CH_3 + I_2 + 2\ OH^- \longrightarrow -\underset{\underset{O}{\|}}{C}CH_3 + 2\ I^- + 2H_2O$$

Methyl ketones then react with iodine and hydroxide ion to produce iodoform:

$$-\underset{\underset{O}{\|}}{C}-CH_3 + 3\ I_2 + 3\ OH^- \longrightarrow -\underset{\underset{O}{\|}}{C}-CI_3 + 3\ I^- + 3H_2O$$

$$-\underset{\underset{O}{\|}}{C}-CI_3 + OH^- \longrightarrow -\underset{\underset{O}{\|}}{C}-O^- + \ CHI_3\!\downarrow$$

Yellow
precipitate

The group to which the $-COCH_3$ or $-CHOHCH_3$ function is attached can be aryl, alkyl, or hydrogen. Thus, even ethyl alcohol and acetaldehyde give positive iodoform tests.

PROBLEM 17.6

Which of the following compounds would give a positive iodoform test?

(a) Acetone   (e) 3-Pentanone   (i) Methyl 2-naphthyl ketone

(b) Acetophenone   (f) 1-Phenylethanol   (j) 3-Pentanol

(c) Pentanal   (g) 2-Phenylethanol

(d) 2-Pentanone   (h) 2-Butanol

17.5 THE ALDOL REACTION: THE ADDITION OF ENOLATE IONS TO ALDEHYDES AND KETONES

When acetaldehyde reacts with dilute sodium hydroxide at room temperature (or below), a dimerization takes place producing 3-hydroxybutanal. Since 3-hydroxy-

$$2CH_3\underset{\underset{O}{\|}}{C}H \xrightarrow[\text{5 °C}]{\text{10\% NaOH, H}_2\text{O}} CH_3\underset{\underset{OH}{|}}{C}HCH_2\underset{\underset{O}{\|}}{C}H$$

(50%)
3-Hydroxybutanal
"aldol"

butanal is both an **alde**hyde and an alco**hol**, it has been given the common name "**aldol**," and reactions of this general type have come to be known as **aldol additions** (or **aldol reactions**).

The mechanism for the aldol addition illustrates two important characteristics of carbonyl compounds: the acidity of their α hydrogens and the tendency of their carbonyl groups to undergo nucleophilic addition.

In the first step, the base (hydroxide ion) abstracts a proton from the α carbon of one molecule of acetaldehyde to give a resonance-stabilized enolate ion.

Step 1 $HO\colon^- + H-CH_2\underset{\underset{O}{\|}}{C}H \rightleftharpoons HOH + \left[\ \colon\!\overset{-}{C}H_2-\underset{\underset{O}{\|}}{C}H \longleftrightarrow CH_2\!=\!\underset{\underset{\colon\overset{..}{O}\colon^-}{|}}{C}H\ \right]$

Enolate ion

In the second step the enolate ion acts as a nucleophile—*as a carbanion*—and attacks the carbonyl carbon atom of a second molecule of acetaldehyde. This step gives an alkoxide ion.

Step 2 $CH_3\overset{\overset{\displaystyle :\ddot{O}}{\|}}{C}H + :\bar{C}H_2\overset{\overset{\displaystyle \ddot{O}:}{\|}}{-}CH \rightleftharpoons CH_3\overset{\overset{\displaystyle :\ddot{O}:^-}{|}}{C}H\overset{\overset{\displaystyle O}{\|}}{CH_2C}H$

An alkoxide ion

$$CH_2\overset{\overset{\displaystyle :\ddot{O}:^-}{|}}{=}CH$$

In the third step, the alkoxide ion abstracts a proton from water to form aldol. This step takes place because the alkoxide ion is a stronger base than a hydroxide ion.

Step 3 $CH_3\overset{\overset{\displaystyle :\ddot{O}:^-}{|}}{C}H\overset{\overset{\displaystyle \ddot{O}:}{\|}}{CH_2C}H + H\ddot{O}H \rightleftharpoons CH_3\overset{\overset{\displaystyle :\ddot{O}H}{|}}{C}H\overset{\overset{\displaystyle \ddot{O}:}{\|}}{CH_2C}H + :\ddot{O}H$

Stronger base Aldol Weaker base

17.5A Dehydration of Addition Product

If the basic mixture containing the aldol (in the previous example) is heated, dehydration takes place and crotonaldehyde (2-butenal) is formed. Dehydration occurs readily because of the acidity of the remaining α hydrogens (even though the leaving group is a hydroxide ion) *and because the product is stabilized by having conjugated double bonds.*

$$\ddot{H}\ddot{O}:^- + CH_3\overset{\overset{\displaystyle OH}{|}}{C}H\overset{\overset{}{}}{-}\underset{\underset{\displaystyle H}{|}}{C}H\overset{\overset{\displaystyle O}{\|}}{-}CH \xrightarrow{-H_2O} CH_3\overset{\overset{\displaystyle OH}{|}}{C}H\overset{}{-}CH\overset{\overset{\displaystyle O}{\|}}{-}CH \xrightarrow{-OH^-} CH_3CH=CH\overset{\overset{\displaystyle O}{\|}}{-}CH$$

Crotonaldehyde (2-butenal)

In some aldol reactions, dehydration occurs so readily that we cannot isolate the product in the aldol form; we obtain the derived *enal* instead. An **aldol condensation** occurs rather than an aldol *addition.* A condensation reaction is one in which molecules are joined through the intermolecular elimination of a small molecule such as water or an alcohol.

Addition product   *Condensation product*

$$2RCH_2\overset{\overset{\displaystyle O}{\|}}{C}H \xrightarrow{\text{base}} \left[RCH_2\overset{\overset{\displaystyle OH}{|}}{C}H\underset{\underset{\displaystyle R}{|}}{C}H\overset{\overset{\displaystyle O}{\|}}{C}H \right] \xrightarrow{-H_2O} RCH_2CH=\underset{\underset{\displaystyle R}{|}}{C}\overset{\overset{\displaystyle O}{\|}}{-}CH$$

Not isolated An enal

17.5B Synthetic Applications

The aldol reaction is a general reaction of aldehydes that possess an α hydrogen. Propanal, for example, reacts with aqueous sodium hydroxide to give 3-hydroxy-2-methylpentanal.

$$2CH_3CH_2\overset{\overset{\displaystyle O}{\|}}{C}H \xrightarrow[0-10\ °C]{OH^-} CH_3CH_2\overset{\overset{\displaystyle OH}{|}}{C}H\overset{}{C}H\overset{\overset{\displaystyle O}{\|}}{C}H$$
$$\underset{\displaystyle CH_3}{|}$$

Propanal	(55–60%)
	3-Hydroxy-2-methylpentanal

PROBLEM 17.7

(a) Show all steps in the aldol addition that occur when propanal

$(CH_3CH_2\overset{\overset{\displaystyle O}{\|}}{C}H)$ is treated with base. (b) How can you account for the fact that

the product of the aldol addition is $CH_3CH_2\overset{\overset{\displaystyle OH}{|}}{C}H\overset{}{C}H\overset{\overset{\displaystyle O}{\|}}{C}H$ and not
$\underset{\displaystyle CH_3}{}$

$CH_3CH_2\overset{\overset{\displaystyle OH}{|}}{C}HCH_2CH_2\overset{\overset{\displaystyle O}{\|}}{C}H$? (c) What product would be formed if the reaction
mixture were heated?

The aldol reaction is important in organic synthesis because it gives us a method for linking two smaller molecules by introducing a carbon–carbon bond between them. Because aldol products contain two functional groups, —OH and —CHO, we can use them to carry out a number of subsequent reactions. Examples are the following:

$$2RCH_2\overset{\overset{\displaystyle O}{\|}}{C}H \xrightarrow[H_2O]{OH^-} RCH_2\overset{\overset{\displaystyle OH}{|}}{C}H\overset{}{\underset{\displaystyle R}{C}H}\overset{\overset{\displaystyle O}{\|}}{C}H \xrightarrow{NaBH_4} RCH_2\overset{\overset{\displaystyle OH}{|}}{C}H\overset{}{\underset{\displaystyle R}{C}H}CH_2OH$$

Aldehyde	An aldol	A 1,3-diol

$$H^+ \downarrow -H_2O$$

$$RCH_2CH_2\overset{\overset{\displaystyle }{}}{\underset{\displaystyle R}{C}H}CH_2OH \xleftarrow[\substack{high\\pressure}]{H_2/Ni} RCH_2CH=\overset{\overset{\displaystyle O}{\|}}{\underset{\displaystyle R}{C}}CH \xrightarrow{LiAlH_4{}^*} RCH_2CH=\overset{}{\underset{\displaystyle R}{C}}CH_2OH$$

A saturated alcohol	An α, β-unsaturated aldehyde	An allylic alcohol

$$\downarrow H_2,\ Pd—C$$

$$RCH_2CH_2\overset{\overset{\displaystyle O}{\|}}{\underset{\displaystyle R}{C}H}CH$$

An aldehyde

*LiAlH$_4$ reduces the carbonyl group of α, β-unsaturated aldehydes and ketones cleanly. NaBH$_4$ often reduces the double bond as well.

PROBLEM 17.8

One industrial process for the synthesis of 1-butanol begins with acetaldehyde. Show how this synthesis might be carried out.

PROBLEM 17.9

Show how each of the following products could be synthesized from butanal:
(a) 3-Hydroxy-2-ethylhexanal
(b) 2-Ethyl-2-hexen-1-ol
(c) 2-Ethyl-1-hexanol
(d) 2-Ethyl-1,3-hexanediol (the insect repellent "6–12")

Ketones also undergo base-catalyzed aldol additions, but for them the equilibrium is unfavorable. This complication can be overcome, however, by carrying out the reaction in a special apparatus that allows the product to be removed from contact with the base as it is formed. This removal of product displaces the equilibrium to the right and permits successful aldol additions with many ketones. Acetone, for example, reacts as follows:

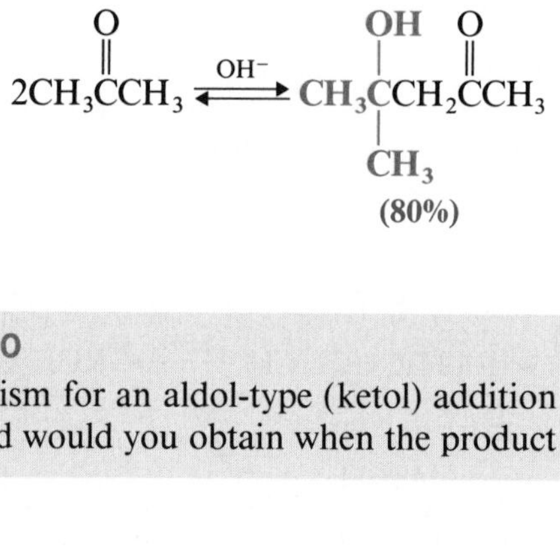

$$2CH_3\overset{O}{\overset{\|}{C}}CH_3 \underset{}{\overset{OH^-}{\rightleftharpoons}} CH_3\overset{OH}{\underset{\underset{CH_3}{|}}{\overset{|}{C}}}CH_2\overset{O}{\overset{\|}{C}}CH_3$$
$$(80\%)$$

PROBLEM 17.10

(a) Write a mechanism for an aldol-type (ketol) addition of acetone in base.
(b) What compound would you obtain when the product is dehydrated?

17.5C The Reversibility of Aldol Additions

The aldol addition is reversible. If, for example, the aldol addition product obtained from acetone (see Problem 17.10) is heated with a strong base, it reverts to an equilibrium mixture that consists largely (~95%) of acetone. This type of reaction is called a *retro-aldol* reaction.

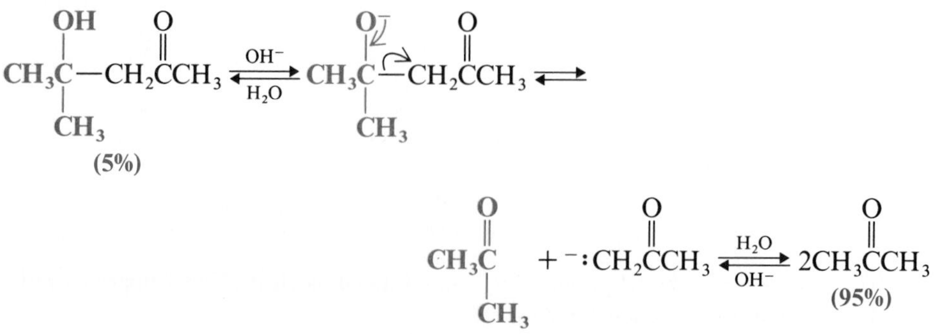

$$CH_3\overset{OH}{\underset{\underset{CH_3}{|}}{\overset{|}{C}}}-CH_2\overset{O}{\overset{\|}{C}}CH_3 \underset{H_2O}{\overset{OH^-}{\rightleftharpoons}} CH_3\overset{O^-}{\underset{\underset{CH_3}{|}}{\overset{|}{C}}}CH_2\overset{O}{\overset{\|}{C}}CH_3 \rightleftharpoons$$
$$(5\%)$$

$$CH_3\overset{O}{\underset{\underset{CH_3}{|}}{\overset{\|}{C}}} + {}^-:CH_2\overset{O}{\overset{\|}{C}}CH_3 \underset{OH^-}{\overset{H_2O}{\rightleftharpoons}} 2CH_3\overset{O}{\overset{\|}{C}}CH_3$$
$$(95\%)$$

17.6 CROSSED ALDOL REACTIONS

An aldol reaction that starts with two different carbonyl compounds is called a **crossed aldol reaction.** Crossed aldol reactions using aqueous sodium hydroxide solutions are of little synthetic importance if both reactants have α hydrogens, because these reactions give a complex mixture of products. If, for example, we were to carry out a crossed aldol addition using acetaldehyde and propanal, we would obtain at least four products.

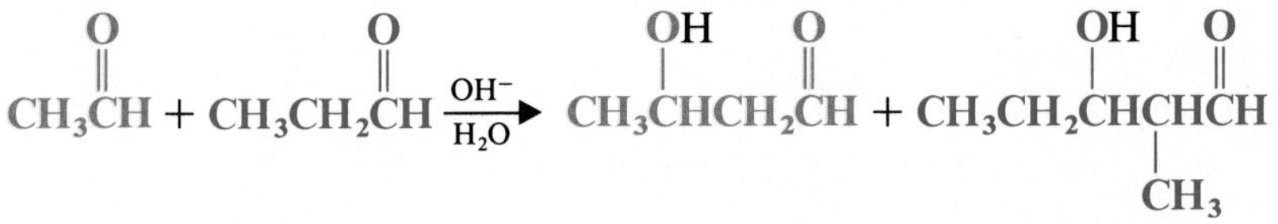

$$CH_3CH + CH_3CH_2CH \xrightarrow[H_2O]{OH^-} CH_3CHCH_2CH + CH_3CH_2CHCHCH$$

3-Hydroxybutanal
(from 2 molecules
of acetaldehyde)

3-Hydroxy-2-
methylpentanal
(from 2 molecules
of propanal)

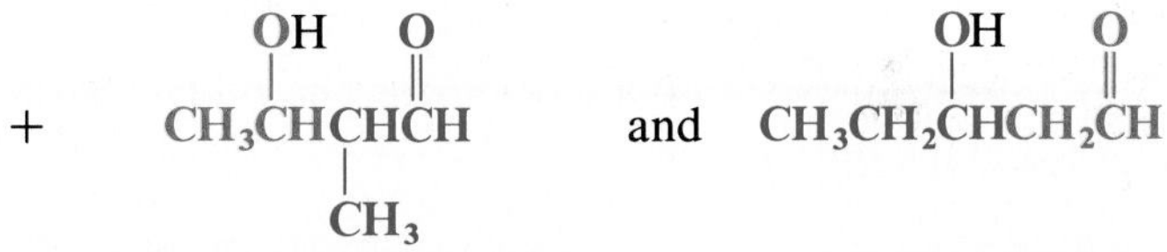

$$+ \quad CH_3CHCHCH \qquad \text{and} \quad CH_3CH_2CHCH_2CH$$

3-Hydroxy-2-methyl-butanal
3-Hydroxypentanal
(from 1 molecule of acetaldehyde and 1 molecule of propanal)

SAMPLE PROBLEM

Show how each of the four products just given is formed in the crossed aldol addition between acetaldehyde and propanal.

Answer:

In the basic aqueous solution, four organic entities will be present: molecules of acetaldehyde, molecules of propanal, enolate ions derived from acetaldehyde, and enolate ions derived from propanal.

We have already seen (Section 17.5) how a molecule of acetaldehyde can react with its enolate ion to form 3-hydroxybutanal (aldol).

Reaction 1 $$CH_3CH \; + \; ^-:CH_2CH \longrightarrow CH_3CHCH_2CH \xrightarrow{H_2O}$$

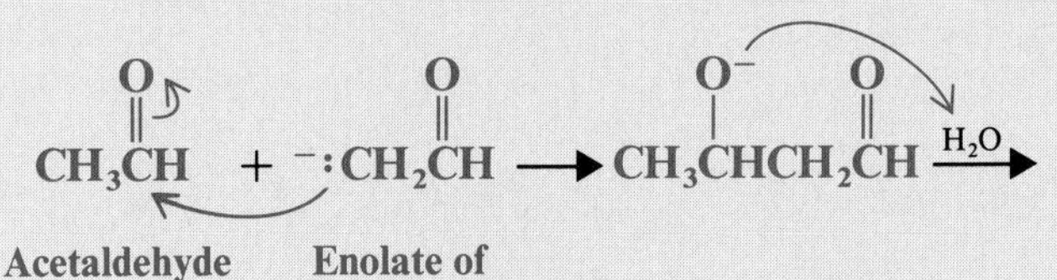

Acetaldehyde   Enolate of
acetaldehyde

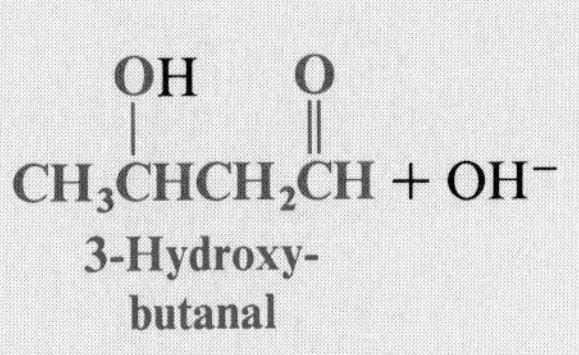

$$CH_3CHCH_2CH + OH^-$$

3-Hydroxy-
butanal

We have also seen (Problem 17.7) how propanal can react with its enolate ion to form 3-hydroxy-2-methylpentanal.

Reaction 2

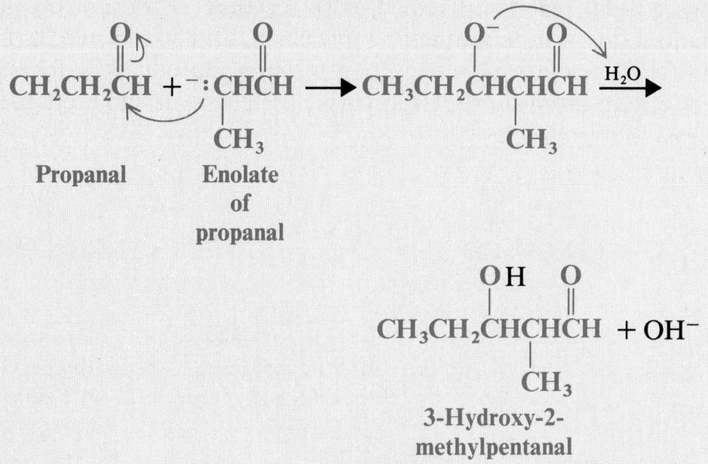

$$\underset{\substack{\text{Propanal}}}{CH_2CH_2CH} + \underset{\substack{\text{Enolate} \\ \text{of} \\ \text{propanal}}}{^-\!:CHCH} \longrightarrow CH_3CH_2CHCHCH \xrightarrow{H_2O}$$

with CH_3 groups on the enolate carbon.

$$\underset{\substack{CH_3 \\ \text{3-Hydroxy-2-} \\ \text{methylpentanal}}}{CH_3CH_2CHCHCH} + OH^-$$

with OH and O, and CH_3 substituent.

Acetaldehyde can also react with the enolate of propanal. This reaction leads to the third product, 3-hydroxy-2-methylbutanal.

Reaction 3

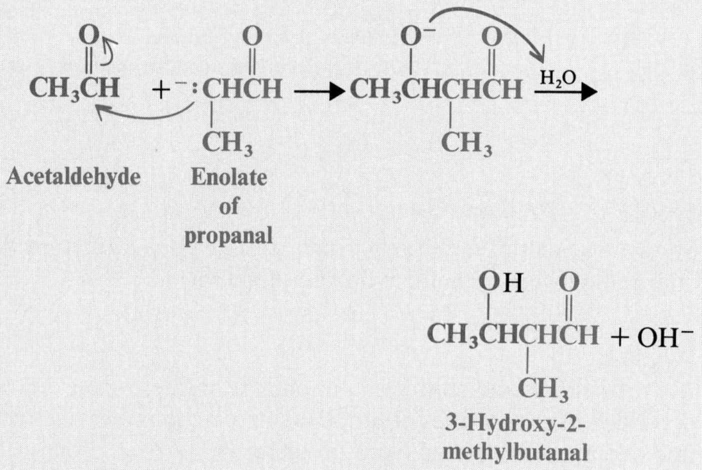

$$\underset{\substack{\text{Acetaldehyde}}}{CH_3CH} + \underset{\substack{\text{Enolate} \\ \text{of} \\ \text{propanal}}}{^-\!:CHCH} \longrightarrow CH_3CHCHCH \xrightarrow{H_2O}$$

$$\underset{\substack{CH_3 \\ \text{3-Hydroxy-2-} \\ \text{methylbutanal}}}{CH_3CHCHCH} + OH^-$$

And finally, propanal can react with the enolate of acetaldehyde. This reaction accounts for the fourth product.

Reaction 4

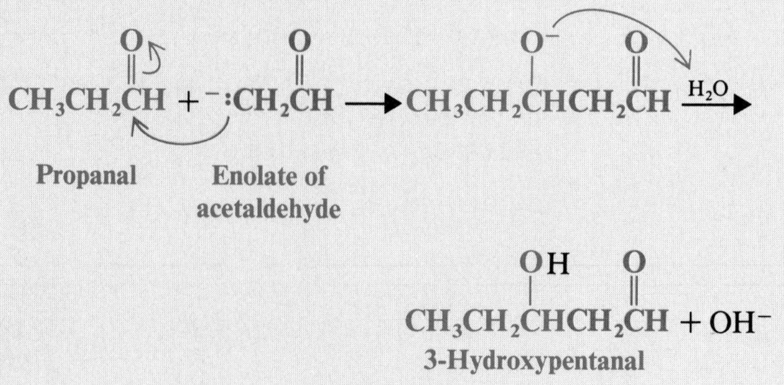

$$\underset{\substack{\text{Propanal}}}{CH_3CH_2CH} + \underset{\substack{\text{Enolate of} \\ \text{acetaldehyde}}}{^-\!:CH_2CH} \longrightarrow CH_3CH_2CHCH_2CH \xrightarrow{H_2O}$$

$$\underset{\substack{\text{3-Hydroxypentanal}}}{CH_3CH_2CHCH_2CH} + OH^-$$

17.6A Practical Crossed Aldol Reactions

Crossed aldol reactions are practical, with bases such as NaOH, when one reactant does not have an α hydrogen and, thus, cannot undergo self-condensation. We can avoid other side reactions by placing this component in base and then slowly adding the reactant with an α hydrogen to the mixture. Under these conditions the concentration of the reactant with an α hydrogen will always be low and most of it will be present as an enolate ion. The main reaction that will take place is one between this enolate ion and the component that has no α hydrogen. The examples listed in Table 17.1 illustrate this technique.

TABLE 17.1 Crossed aldol reactions

THIS REACTANT WITH NO α HYDROGEN IS PLACED IN BASE	THIS REACTANT WITH AN α HYDROGEN IS ADDED SLOWLY		PRODUCT
O‖ C₆H₅CH Benzaldehyde	+	O‖ CH₃CH₂CH Propionaldehyde $\xrightarrow[10\,°C]{OH^-}$	CH₃ O‖ C₆H₅CH=C—CH (68%) 2-Methyl-3-phenyl-2-propenal (α-methylcinnamaldehyde)
O‖ C₆H₅CH Benzaldehyde	+	O‖ C₆H₅CH₂CH Phenylacetaldehyde $\xrightarrow[20\,°C]{OH^-}$	O‖ C₆H₅CH=CCH C₆H₅ 2,3-Diphenyl-2-propenal
O‖ HCH Formaldehyde	+	CH₃ O‖ CH₃CH—CH 2-Methylpropanal $\xrightarrow[40\,°C]{dil.\ Na_2CO_3}$	CH₃ O‖ CH₃—C—CH CH₂OH (>64%) 3-Hydroxy-2,2-dimethylpropanal

As the examples in Table 17.1 also show, the crossed aldol reaction is often accompanied by dehydration. Whether or not dehydration occurs can, at times, be determined by our choice of reaction conditions, but *dehydration is especially easy when it leads to an extended conjugated system.*

PROBLEM 17.11

Show how you could use a crossed aldol reaction to synthesize cinnamaldeh
(C₆H₅CH=CHCHO).

PROBLEM 17.12

When excess formaldehyde in basic solution is treated with acetaldehyde, the following reaction takes place:

$$3HCH + CH_3CH \xrightarrow[40\ °C]{dil.\ Na_2CO_3} HOCH_2-\overset{CH_2OH}{\underset{CH_2OH}{C}}-CHO$$

(82%)

Write a mechanism that accounts for the formation of the product.

17.6B Claisen–Schmidt Reactions

When ketones are used as one component, the crossed aldol reactions are called **Claisen–Schmidt reactions.** These reactions are practical when bases such as sodium hydroxide are used because, under these conditions ketones do not self-condense appreciably. (The equilibrium is unfavorable; cf. Section 17.5C.)

Two examples of Claisen–Schmidt reactions are the following:

$$C_6H_5CH + CH_3CCH_3 \xrightarrow[100\ °C]{OH^-} C_6H_5CH=CHCCH_3$$

(70%)
4-Phenyl-3-buten-2-one
(benzalacetone)

$$C_6H_5CH + CH_3CC_6H_5 \xrightarrow[20\ °C]{OH^-} C_6H_5CH=CHCC_6H_5$$

(85%)
1,3-Diphenyl-2-propen-1-one
(benzalacetophenone)

In both of these reactions dehydration occurs readily because the double bond that forms is conjugated both with the carbonyl group and with the benzene ring. The conjugated system is thereby extended.

An important step in a commercial synthesis of vitamin A makes use of a Claisen–Schmidt reaction between geranial and acetone:

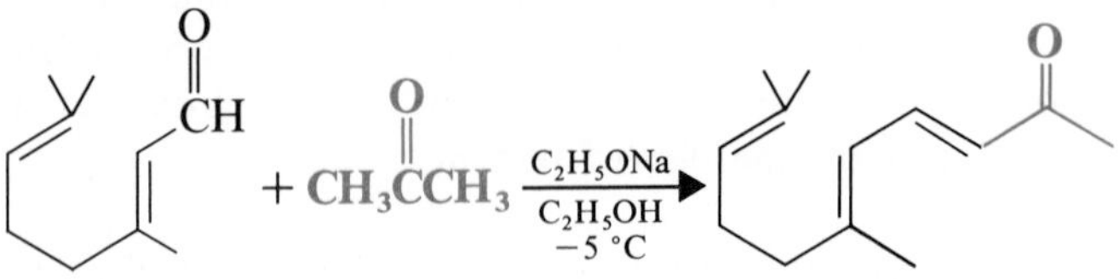

Geranial $+ CH_3CCH_3 \xrightarrow[\substack{C_2H_5OH \\ -5\ °C}]{C_2H_5ONa}$ (49%)
Pseudoionone

Geranial is a naturally occurring aldehyde that can be obtained from lemongrass oil. Notice, in this reaction, too, dehydration occurs readily because dehydration extends the conjugated system.

In Special Topic I we shall study another method of carrying out crossed aldol reactions based on the use of lithium enolates.

PROBLEM 17.13

When pseudoionone is treated with BF_3 in acetic acid, ring closure takes place and α- and β-ionone are produced. This is the next step in the vitamin A synthesis.

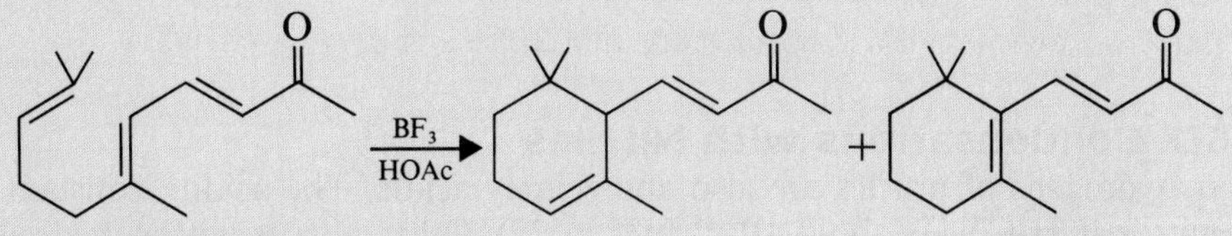

Pseudoionone   α-Ionone   β-Ionone

(a) Write mechanisms that explain the formation of α- and β-ionone. (b) β-Ionone is the major product. How can you explain this? (c) Which ionone would you expect to absorb at longer wavelengths in the visible–ultraviolet region? Why?

17.6C Condensations with Nitroalkanes

The α hydrogens of nitroalkanes are appreciably acidic ($K_a \approx 10^{-10}$). The acidity of these hydrogen atoms, like the α hydrogens of aldehydes and ketones, can be explained by resonance stabilization of the anion that is produced.

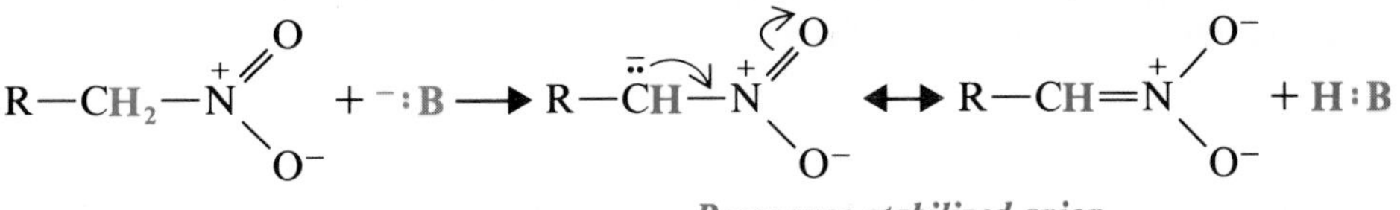

Resonance-stabilized anion

Nitroalkanes that have α hydrogens undergo base-catalyzed condensations with aldehydes and ketones that resemble aldol condensations. An example is the condensation of benzaldehyde with nitromethane.

$$C_6H_5\overset{\overset{\textstyle O}{\|}}{C}H + CH_3NO_2 \xrightarrow{OH^-} C_6H_5CH=CHNO_2$$

This condensation is especially useful because the nitro group of the product can be easily reduced to an amino group. One technique that will bring about this transformation is to use hydrogen and a nickel catalyst. This combination not only reduces the nitro group but also reduces the double bond:

$$C_6H_5CH=CHNO_2 \xrightarrow{H_2,\ Ni} C_6H_5CH_2CH_2NH_2$$

PROBLEM 17.14

Assuming that you have available the required aldehydes, ketones, and nitro-alkanes, show how you would synthesize each of the following:

(a) $C_6H_5CH{=}\underset{\underset{\textstyle CH_3}{|}}{C}NO_2$ (b) $HOCH_2CH_2NO_2$ (c)

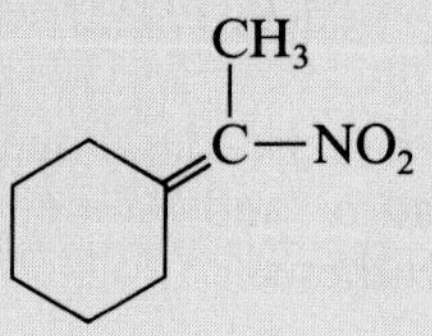

17.6D Condensations with Nitriles

The α hydrogens of nitriles are also appreciably acidic. The acidity constant for acetonitrile (CH_3CN) is about 10^{-25} ($pK_a \sim 25$). Other nitriles with α hydrogens show comparable acidities, and consequently these nitriles undergo condensations of the aldol type. An example is the condensation of benzaldehyde with phenylaceto-nitrile.

$$C_6H_5\overset{\overset{\textstyle O}{\|}}{C}H + C_6H_5CH_2CN \xrightarrow{\text{EtO}^-/\text{EtOH}} C_6H_5CH{=}\underset{\underset{\textstyle C_6H_5}{|}}{C}{-}CN$$

PROBLEM 17.15

(a) Write resonance structures for the anion of acetonitrile that account for its being much more acidic than ethane.   (b) Give a step-by-step mechanism for the condensation of benzaldehyde with acetonitrile.

17.7 CYCLIZATIONS VIA ALDOL CONDENSATIONS

The aldol condensation also offers a convenient way to synthesize molecules with five- and six-membered rings (and sometimes even larger rings). This can be done by an intramolecular aldol condensation using a dialdehyde, a keto aldehyde, or a diketone as the substrate. For example, the following keto aldehyde cyclizes to yield 1-cyclopentenyl methyl ketone.

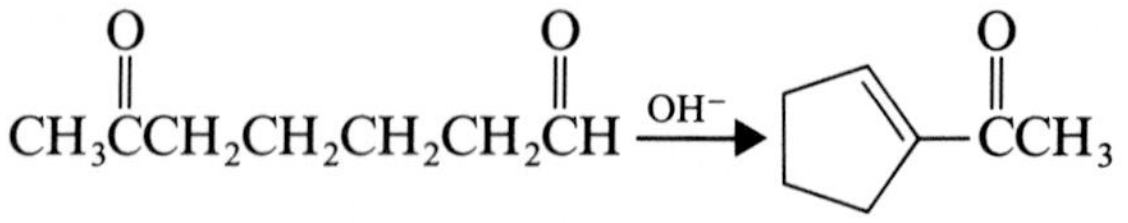

$$CH_3\overset{\overset{\textstyle O}{\|}}{C}CH_2CH_2CH_2CH_2\overset{\overset{\textstyle O}{\|}}{C}H \xrightarrow{\text{OH}^-}$$

(73%)

This reaction almost certainly involves the formation of at least three different enolates. However, it is the following enolate from the ketone side of the molecule that adds to the aldehyde group that leads to the product.

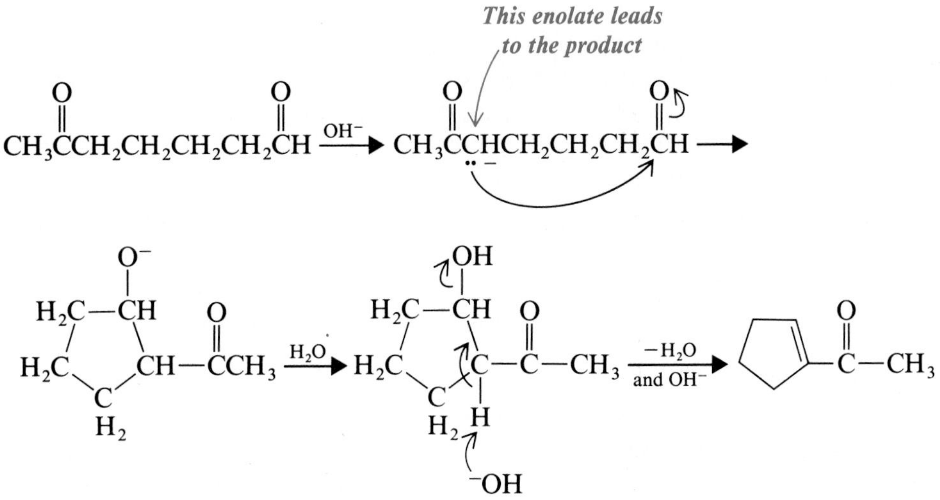

*This enolate leads
to the product*

$$CH_3\overset{O}{\underset{\|}{C}}CH_2CH_2CH_2CH_2CH \xrightarrow{OH^-} CH_3\overset{O}{\underset{\|}{C}}\overset{..}{\underset{-}{C}}HCH_2CH_2CH_2CH \longrightarrow$$

The reason the aldehyde group undergoes addition preferentially may arise from the greater activity of aldehydes toward nucleophilic addition generally. The carbonyl carbon atom of a ketone is less positive (and therefore less reactive toward a nucleophile) because it bears two electron-releasing alkyl groups; it is also more sterically hindered.

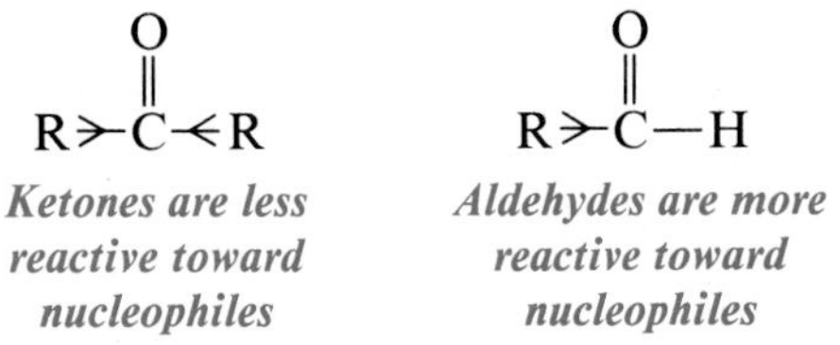

$$R \blacktriangleright \overset{O}{\underset{\|}{C}} \blacktriangleleft R \qquad\qquad R \blacktriangleright \overset{O}{\underset{\|}{C}} - H$$

*Ketones are less
reactive toward
nucleophiles*

*Aldehydes are more
reactive toward
nucleophiles*

In reactions of this type, five-membered rings form far more readily than seven-membered rings.

PROBLEM 17.16

Assuming that dehydration occurs in all instances, write the structures of the two other products that might have resulted from the previous aldol cyclization. (One of these products will have a five-membered ring and the other will have a seven-membered ring.)

PROBLEM 17.17

What starting compound would you use in an aldol cyclization to prepare each of the following?

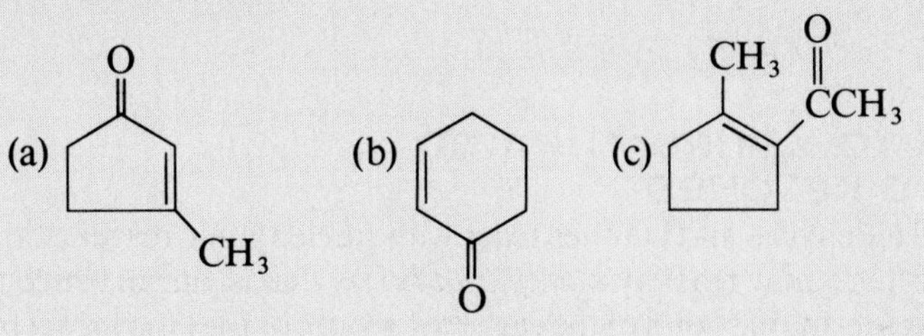

(a)

(b)

(c)

17.8 ACID-CATALYZED ALDOL CONDENSATIONS

Aldol condensations can also be brought about with acid catalysis. Treating acetone with hydrogen chloride, for example, leads to the formation of 4-methyl-3-penten-2-one, the aldol condensation product. In general, acid-catalyzed aldol reactions lead to dehydration of the initially formed aldol addition product.

The mechanism begins with acid-catalyzed formation of the enol:

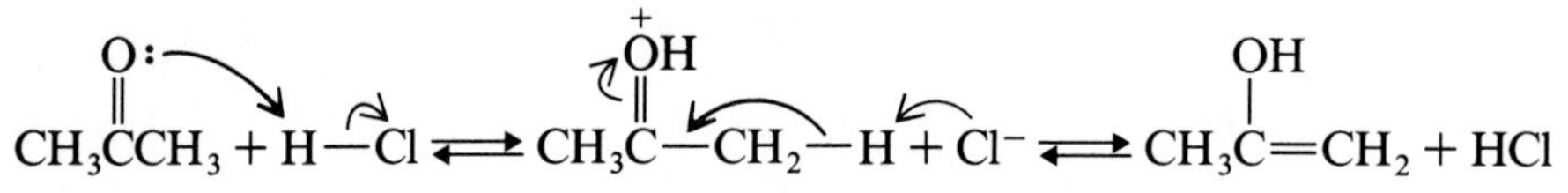

$$CH_3CCH_3 + H-Cl \rightleftharpoons CH_3C-CH_2-H + Cl^- \rightleftharpoons CH_3C=CH_2 + HCl$$

Then the enol adds to the protonated carbonyl group of another molecule of acetone:

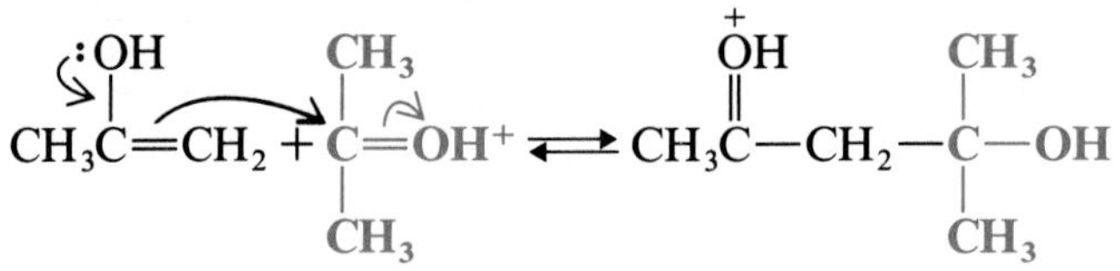

$$CH_3C=CH_2 + C=OH^+ \rightleftharpoons CH_3C-CH_2-C-OH$$

Then dehydration occurs leading to the product.

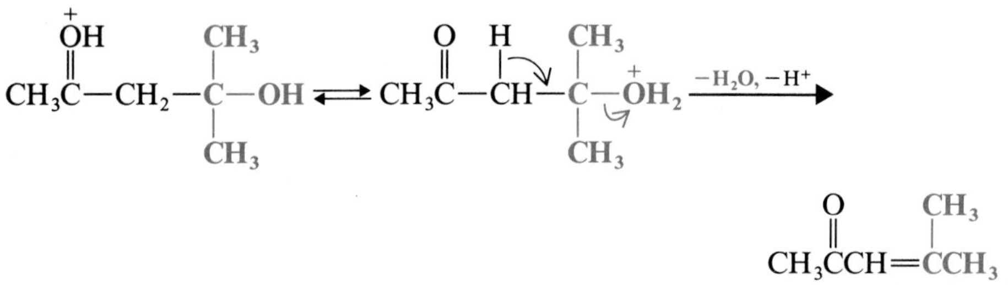

$$CH_3C-CH_2-C-OH \rightleftharpoons CH_3C-CH-C-OH_2 \xrightarrow{-H_2O, -H^+}$$

$$CH_3CCH=CCH_3$$
4-Methyl-3-penten-2-one

PROBLEM 17.18

The acid-catalyzed aldol condensation of acetone (just shown) also produces some 2,6-dimethyl-2,5-heptadien-4-one. Give a mechanism that explains the formation of this product.

PROBLEM 17.19

Heating acetone with sulfuric acid leads to the formation of mesitylene (1,3,5-trimethylbenzene). Propose a mechanism for this reaction.

17.9 ADDITIONS TO α,β-UNSATURATED ALDEHYDES AND KETONES

When α,β-unsaturated aldehydes and ketones react with nucleophilic reagents, they may do so in two ways: They may react by a *simple addition,* that is, one in which the nucleophile adds across the double bond of the carbonyl group; or they may react by a *conjugate addition.* These two processes resemble the 1,2- and the 1,4-addition reactions of conjugated dienes (Section 10.11).

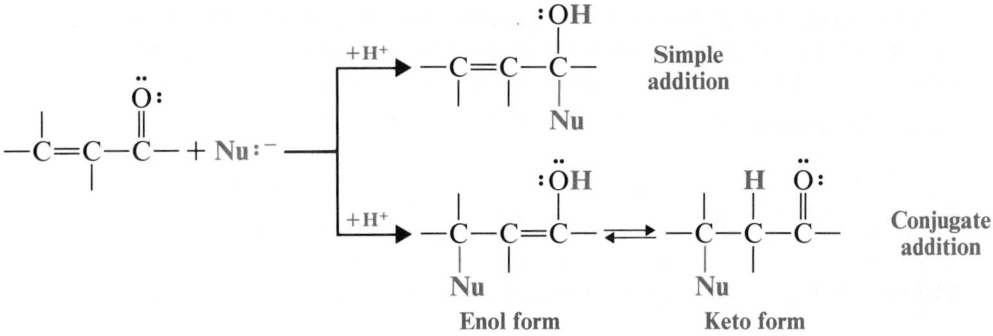

$$-\overset{|}{C}=\overset{|}{C}-\overset{\overset{\displaystyle\ddot{O}:}{\|}}{C}-\ +\ Nu\!:^-$$

Simple addition:

$$-\overset{|}{C}=\overset{|}{C}-\overset{\overset{\displaystyle:\ddot{O}H}{|}}{\underset{\underset{\displaystyle Nu}{|}}{C}}-$$

Conjugate addition:

$$-\overset{\overset{\displaystyle:\ddot{O}H}{|}}{C}-\overset{|}{C}=\overset{|}{\underset{\underset{\displaystyle Nu}{|}}{C}}- \ \rightleftharpoons \ -\overset{\overset{\displaystyle H}{|}}{C}-\overset{\overset{\displaystyle\ddot{O}:}{\|}}{\underset{\underset{\displaystyle Nu}{|}}{C}}-\overset{}{C}-$$

Enol form   Keto form

In many instances both modes of addition occur in the same mixture. As an example, let us consider the Grignard reaction shown here.

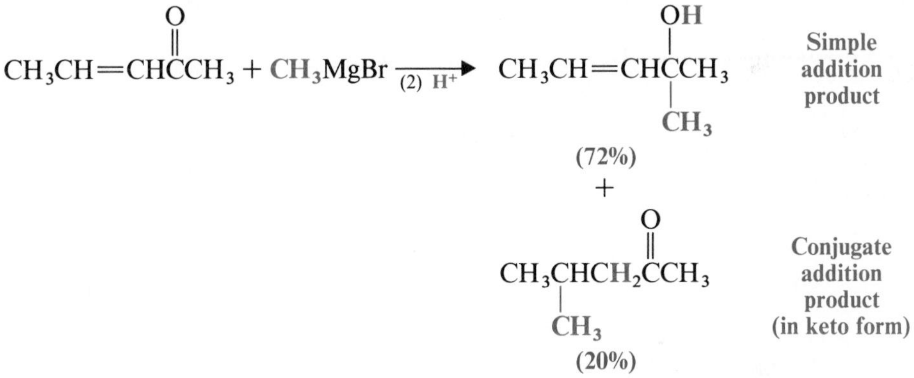

$$CH_3CH=CH\overset{\overset{\displaystyle O}{\|}}{C}CH_3 + CH_3MgBr \xrightarrow[(2)\ H^+]{} \ CH_3CH=CH\overset{\overset{\displaystyle OH}{|}}{\underset{\underset{\displaystyle CH_3}{|}}{C}}CH_3$$

Simple addition product

(72%)

+

$$CH_3\underset{\underset{\displaystyle CH_3}{|}}{CH}CH_2\overset{\overset{\displaystyle O}{\|}}{C}CH_3$$

Conjugate addition product (in keto form)

(20%)

In this example we see that simple addition is favored, but that conjugate addition accounts for a substantial amount of the product.

If we examine the resonance structures that contribute to the overall hybrid for an α,β-unsaturated aldehyde or ketone (see structures **A**–**C**), we shall be in a better position to understand these reactions.

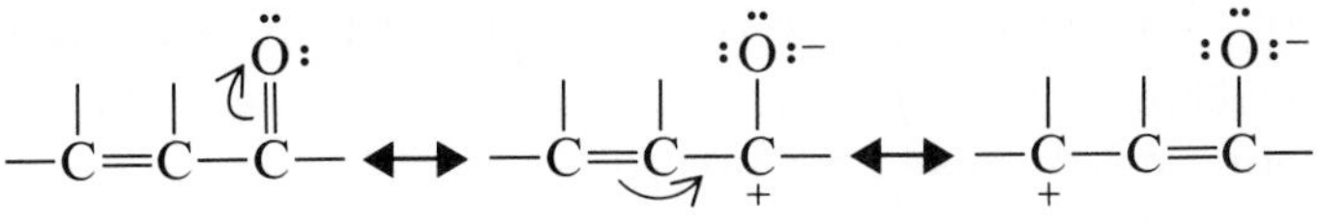

$$-\overset{|}{C}=\overset{|}{C}-\overset{\overset{\displaystyle\ddot{O}:}{\|}}{C}- \ \longleftrightarrow \ -\overset{|}{C}=\overset{|}{\underset{+}{C}}-\overset{\overset{\displaystyle:\ddot{O}:^-}{|}}{C}- \ \longleftrightarrow \ -\overset{|}{\underset{+}{C}}-\overset{|}{C}=\overset{\overset{\displaystyle:\ddot{O}:^-}{|}}{C}-$$

A   **B**   **C**

Although structures **B** and **C** involve separated charges, they make a significant contribution to the hybrid because, in each, the negative charge is carried by electronegative oxygen. Structures **B** and **C** not only indicate that the oxygen of the hybrid should bear a partial negative charge, but they also indicate that *both the carbonyl carbon and the β carbon should bear a partial positive charge.* They indicate that we should represent the hybrid in the following way:

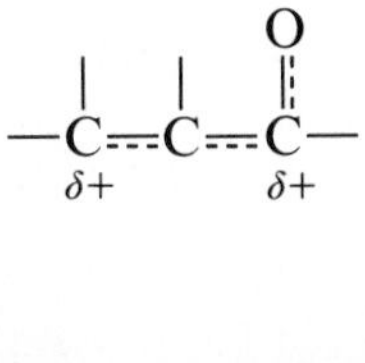

$$-\underset{\delta+}{\overset{|}{C}}\cdots\overset{|}{C}\cdots\underset{\delta+}{\overset{\overset{\displaystyle\overset{\delta-}{O}}{\|}}{C}}-$$

This structure tells us that we should expect an electrophilic reagent to attack the carbonyl oxygen and a nucleophilic reagent to attack either the carbonyl carbon or the β carbon.

This expectation is exactly what happens in the Grignard reactions that we saw earlier. The electrophilic magnesium attacks the carbonyl oxygen; the nucleophilic carbon of the Grignard reagent attacks either the carbonyl carbon or the β carbon.

Simple Addition

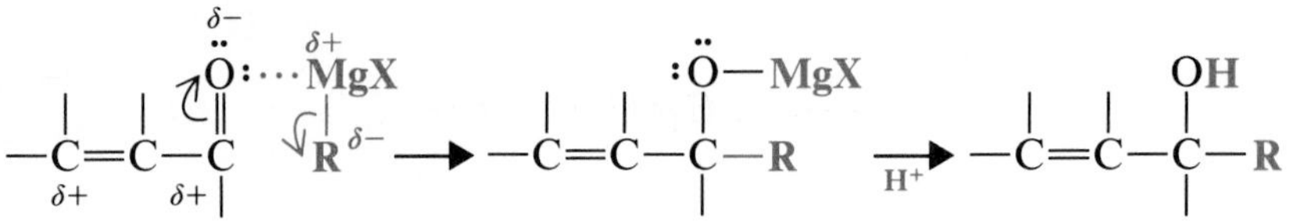

Conjugate Addition

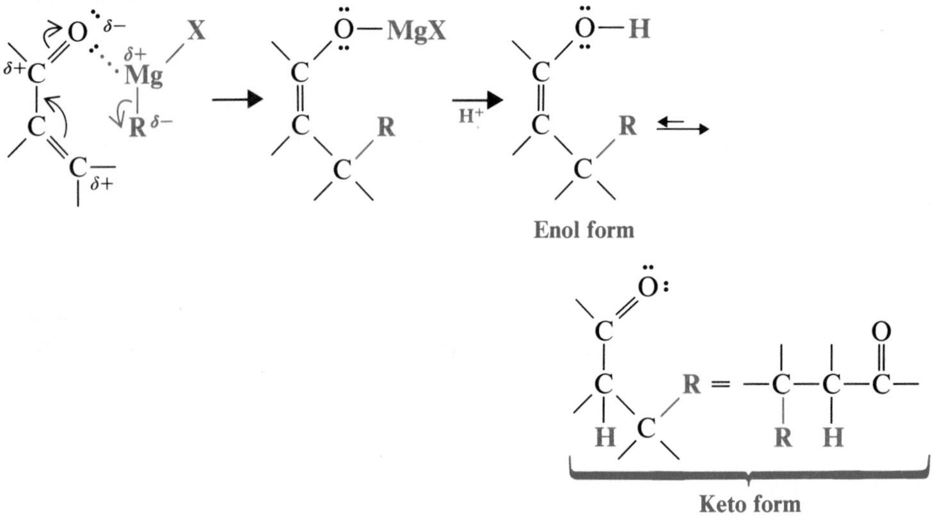

Enol form

Keto form

17.9A Addition of Other Nucleophiles

Grignard reagents are not the only nucleophilic reagents that add in a conjugate manner to α,β-unsaturated aldehydes and ketones. Almost every nucleophilic reagent that adds at the carbonyl carbon of a simple aldehyde or ketone is capable of adding at the β carbon of an α,β-unsaturated carbonyl compound. In many instances conjugate addition is the major reaction path:

$$C_6H_5CH{=}CHCC_6H_5 + CN^- \xrightarrow[CH_3CO_2H]{C_2H_5OH} C_6H_5CH{-}CH_2CC_6H_5$$

$$(95\%)$$

Mechanism

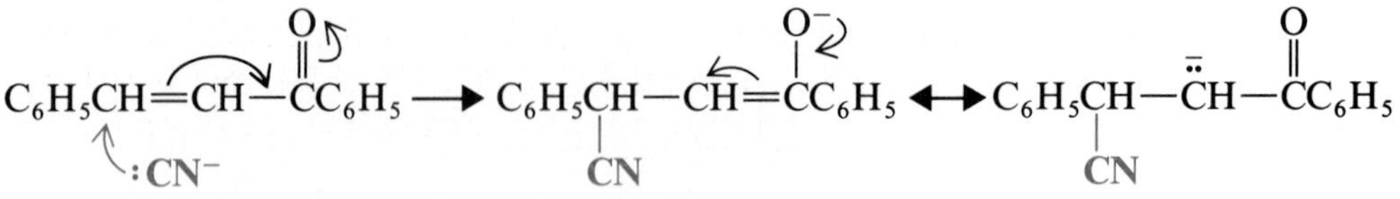

then,

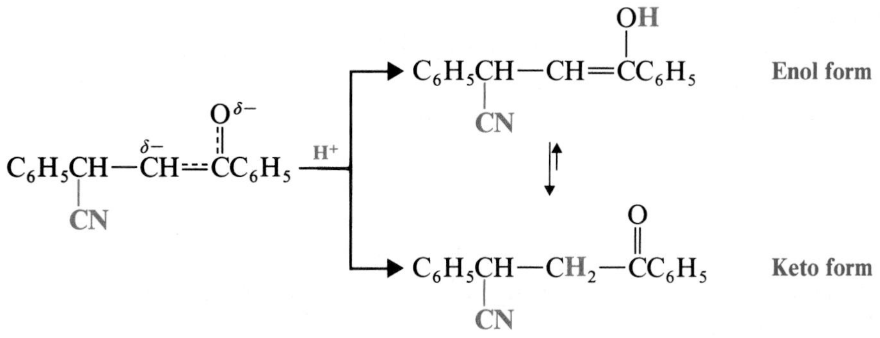

$$C_6H_5CH-CH=CC_6H_5 \quad \text{Enol form}$$
with OH and CN substituents

$$C_6H_5CH-CH \doublebond CC_6H_5 \xrightarrow{\;H^+\;}$$ (with δ− labels, O^{δ−}, CN)

$$C_6H_5CH-CH_2-CC_6H_5 \quad \text{Keto form}$$
with O and CN substituents

Another example is the following:

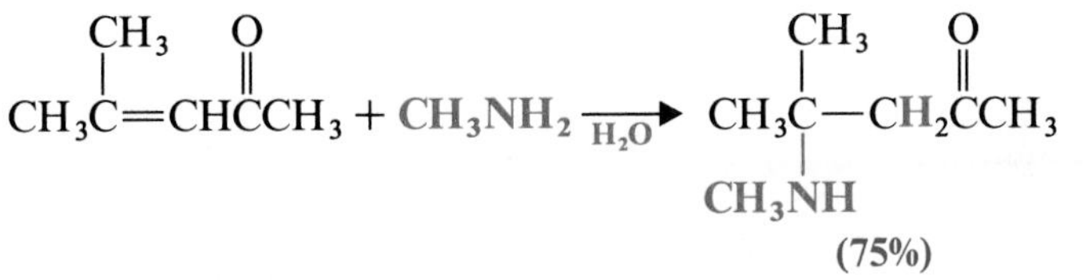

$$CH_3C=CHCCH_3 + CH_3NH_2 \xrightarrow{H_2O} CH_3C-CH_2CCH_3$$
(with CH_3 and O groups; product has CH_3NH)

(75%)

17.9B Michael Additions

Conjugate additions of enolate ions to α,β-unsaturated carbonyl compounds are known generally as Michael additions. An example is the addition of cyclohexanone to $C_6H_5CH=CHCOC_6H_5$ shown here.

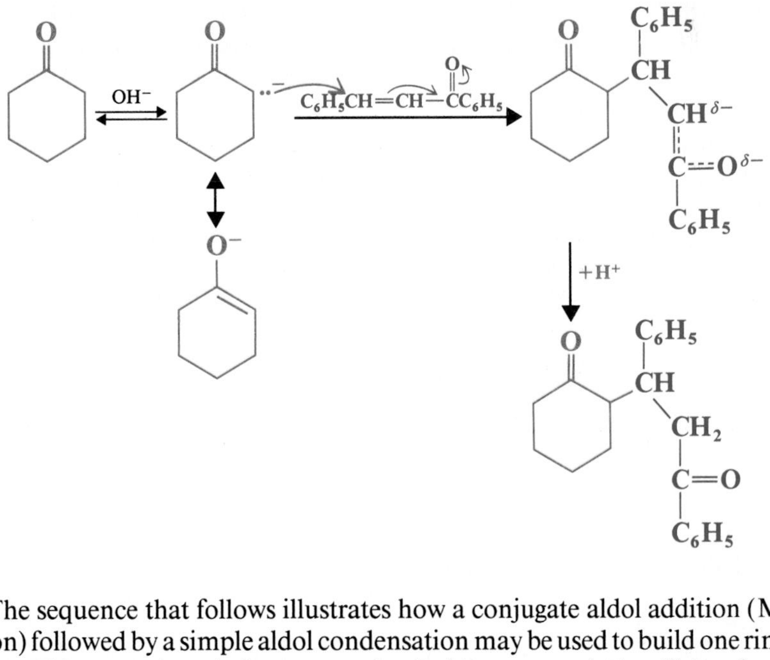

The sequence that follows illustrates how a conjugate aldol addition (Michael addition) followed by a simple aldol condensation may be used to build one ring on to another. This procedure is known as the *Robinson annulation* ("ring forming") reaction.

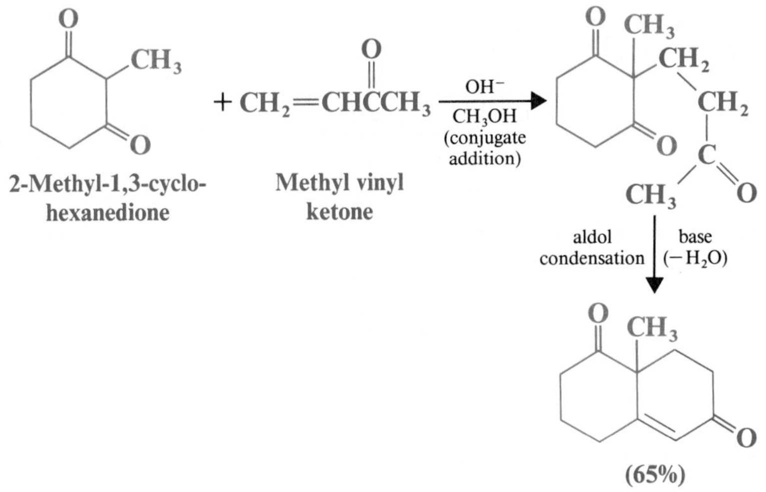

2-Methyl-1,3-cyclo-
hexanedione

Methyl vinyl
ketone

OH⁻
CH₃OH
(conjugate
addition)

aldol
condensation

base
(−H₂O)

(65%)

PROBLEM 17.20

(a) Propose step-by-step mechanisms for both transformations of the Robinson annulation sequence just shown.　(b) Would you expect 2-methyl-1,3-cyclo-hexanedione to be more or less acidic than cyclohexanone? Explain your answer.

PROBLEM 17.21

What product would you expect to obtain from the base-catalyzed Michael reaction (a) of 1,3-diphenyl-2-propen-1-one (Section 17.6B) and acetophenone?　(b) of 1,3-diphenyl-2-propen-1-one and cyclopentadiene? Show all steps in each mechanism.

PROBLEM 17.22

When acrolein reacts with hydrazine, the product is a dihydropyrazole:

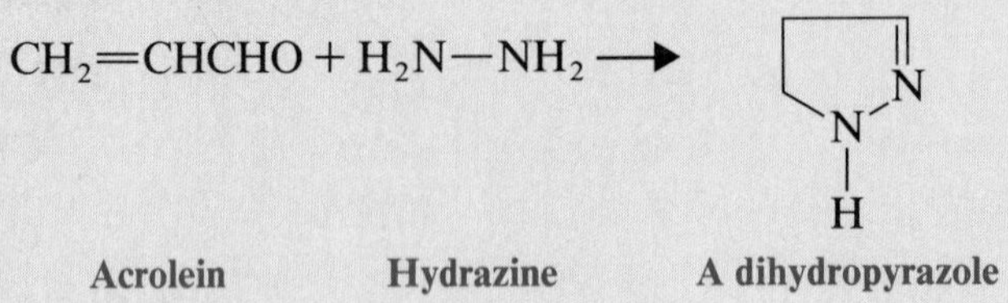

$$CH_2{=}CHCHO + H_2N{-}NH_2 \longrightarrow$$

Acrolein　　　Hydrazine　　　A dihydropyrazole

Suggest a mechanism that explains this reaction.

We shall study further examples of the Michael addition in Chapter 20.

17.9C Conjugate Addition of Organocopper Reagents

Organocopper reagents, either RCu or R_2CuLi, add to α,β-unsaturated carbonyl compounds and, unlike Grignard reagents, **organocopper reagents add almost exclusively in the conjugate manner.**

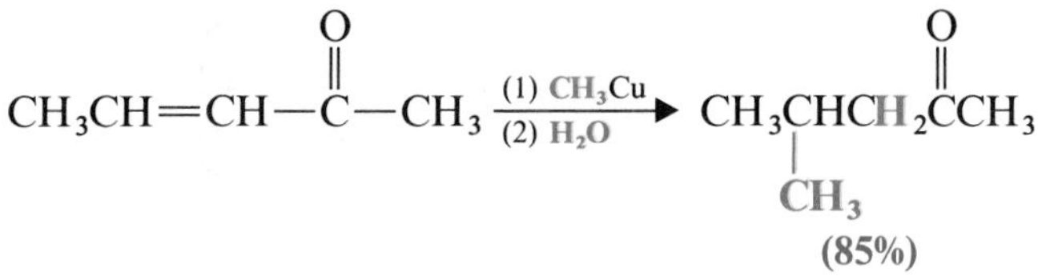

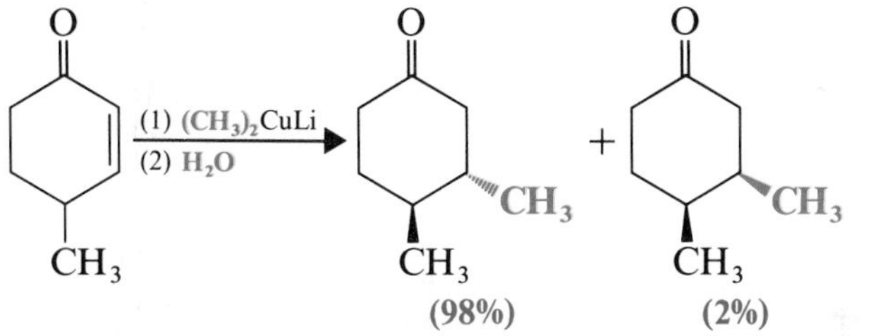

$$CH_3CH{=}CH{-}\overset{\overset{\displaystyle O}{\|}}{C}{-}CH_3 \xrightarrow[\text{(2) } H_2O]{\text{(1) } CH_3Cu} CH_3\underset{\underset{\displaystyle CH_3}{|}}{CH}CH_2\overset{\overset{\displaystyle O}{\|}}{C}CH_3$$

(85%)

(98%)   (2%)

With an alkyl-substituted cyclic α,β-unsaturated ketone, as the example just cited shows, lithium dialkylcuprates add predominantly in the less-hindered way to give the product with the alkyl groups trans to each other.

ADDITIONAL PROBLEMS

17.23

Give structural formulas for the products of the reaction (if one occurs) when propanal is treated with each of the following reagents:

(a) OH^-, H_2O

(b) C_6H_5CHO, OH^-

(c) HCN

(d) $NaBH_4$

(e) $HOCH_2CH_2OH$, H^+

(f) Ag_2O, OH^-, then H^+

(g) CH_3MgI, then H^+

(h) $Ag(NH_3)_2{}^+OH^-$, then H^+

(i) NH_2OH

(j) $C_6H_5\overset{-}{C}H{-}\overset{+}{P}(C_6H_5)_3$

(k) C_6H_5Li, then H^+

(l) $HC{\equiv}CNa$, then H^+

(m) $HSCH_2CH_2SH$, H^+, then Raney Ni, H_2

(n) $CH_3CH_2CHBrCO_2Et$ and Zn, then H^+

17.24

Give structural formulas for the products of the reaction (if one occurs) when acetone is treated with each reagent of the preceding problem.

17.25

What products would form when 4-methylbenzaldehyde reacts with each of the following?

(a) CH_3CHO, OH^-

(b) $CH_3C{\equiv}CNa$, then H^+

(c) CH_3CH_2MgBr, then H^+

(d) Cold dilute $KMnO_4$, OH^-, then H^+

(e) Hot $KMnO_4$, OH^-, then H^+

(f) $^-{:}CH_2{-}\overset{+}{P}(C_6H_5)_3$

(g) $CH_3COC_6H_5$, OH^-

(h) $BrCH_2CO_2Et$ and Zn, then H^+

17.26

Show how each of the following transformations could be accomplished. You may use any other required reagents.

(a) $CH_3COC(CH_3)_3 \longrightarrow C_6H_5CH=CHCOC(CH_3)_3$

(b) 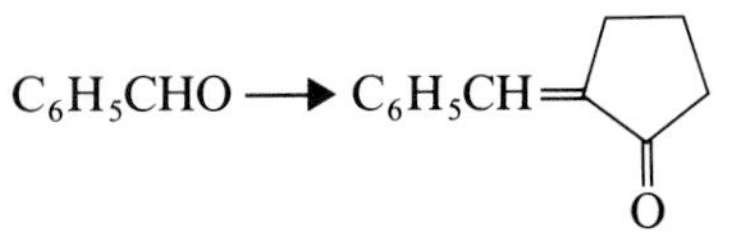 $C_6H_5CHO \longrightarrow C_6H_5CH=$

(c) $C_6H_5CHO \longrightarrow C_6H_5CH_2\underset{\underset{CH_3}{|}}{C}HNH_2$

(d) 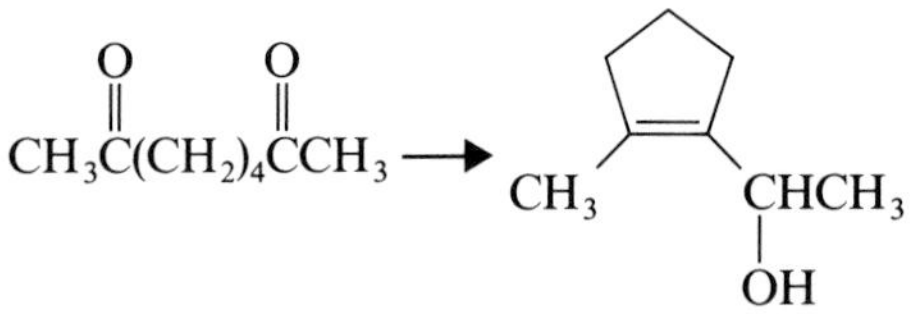 $CH_3\overset{\overset{O}{\|}}{C}(CH_2)_4\overset{\overset{O}{\|}}{C}CH_3 \longrightarrow$

(e) 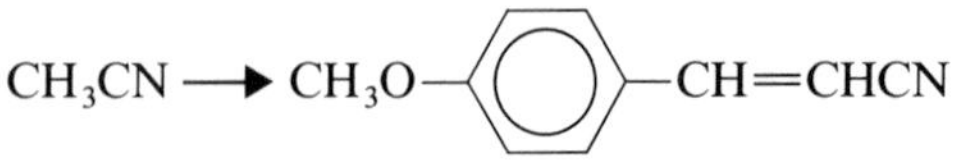 $CH_3CN \longrightarrow CH_3O-$ $-CH=CHCN$

(f) $CH_3CH_2CH_2CH_2\overset{\overset{O}{\|}}{C}H \longrightarrow CH_3(CH_2)_3CH=\overset{\overset{CH_2OH}{|}}{C}(CH_2)_2CH_3$

(g)

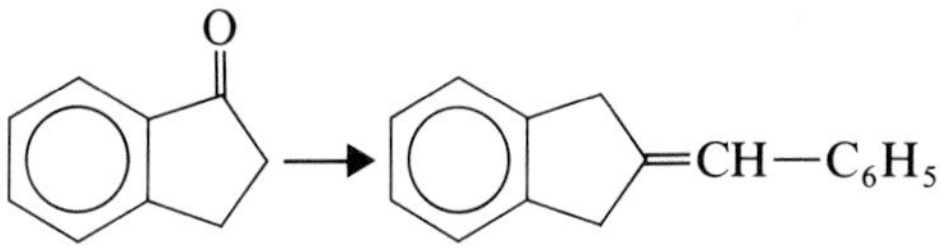

17.27

The following reaction illustrates the Robinson annulation reaction (Section 17.9B). Give mechanisms for the steps that occur.

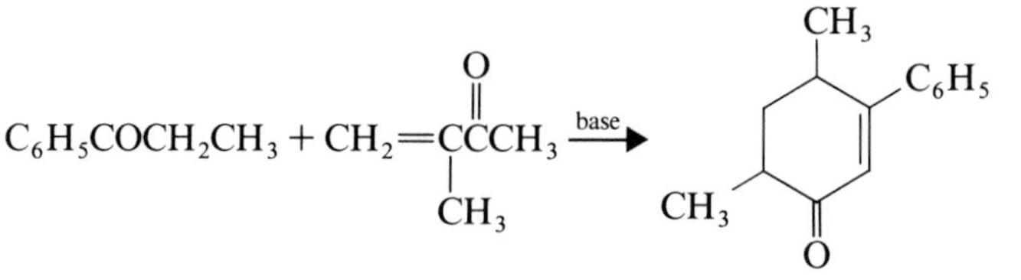

$$C_6H_5COCH_2CH_3 + CH_2=\overset{\overset{O}{\|}}{\underset{\underset{CH_3}{|}}{C}}CCH_3 \xrightarrow{\text{base}}$$

17.28

Write structural formulas for **A**, **B**, and **C**.

$$HC\equiv CH \xrightarrow[\substack{(2)\ CH_3COCH_3 \\ (3)\ H^+}]{(1)\ NaNH_2} A\ (C_5H_8O) \xrightarrow[H_2O]{Hg^{2+},\ H_3O^+} B\ (C_5H_{10}O_2) \xrightarrow{C_6H_5CHO,\ OH^-} C\ (C_{12}H_{14}O_2)$$

17.29

The hydrogen atoms of the γ carbon of crotonaldehyde are appreciably acidic ($K_a \sim 10^{-20}$).

(a) Write resonance structures that will explain this fact.

$$\overset{\gamma}{C}H_3\overset{\beta}{C}H=\overset{\alpha}{C}HCHO$$

Crotonaldehyde

(b) Write a mechanism that accounts for the following reaction:

$$C_6H_5CH=CHCHO + CH_3CH=CHCHO \xrightarrow[C_2H_5OH]{\text{base}} C_6H_5(CH=CH)_3CHO$$

(87%)

17.30
What reagents would you use to bring about each step of the following syntheses?

(a)

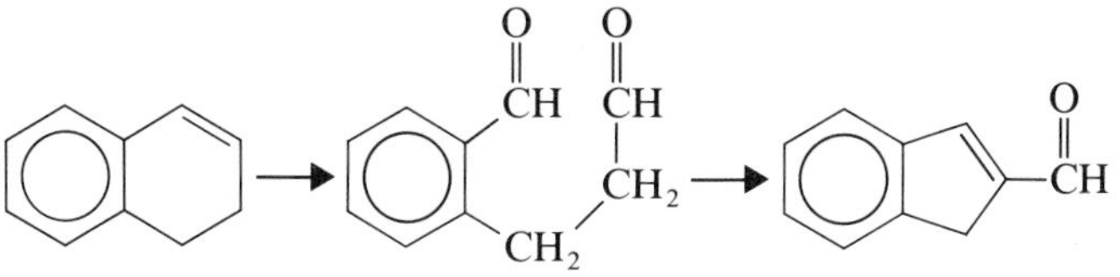

(b)

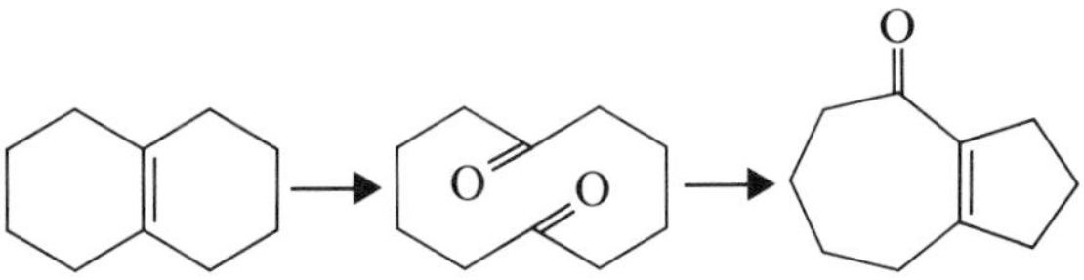

(c)

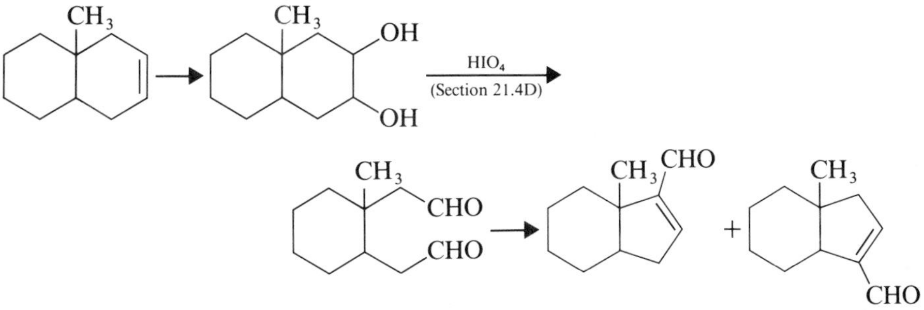

$$\xrightarrow[\text{(Section 21.4D)}]{\text{HIO}_4}$$

(d)

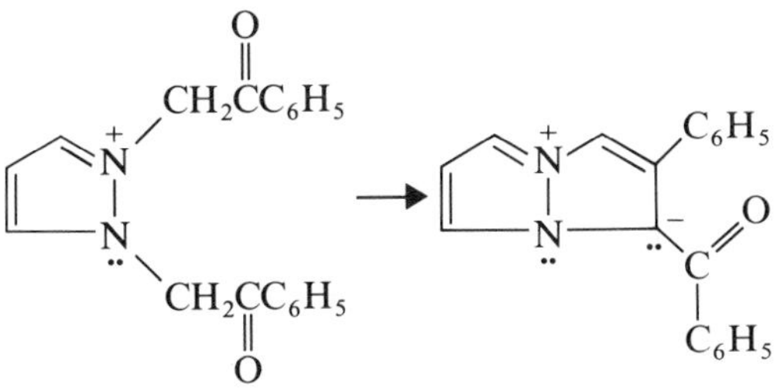

17.31
(a) Infrared spectroscopy gives an easy method for deciding whether the product obtained from the addition of a Grignard reagent to an α,β-unsaturated ketone is the simple addition product or the conjugate addition product. Explain. (What peak or peaks would you look for?) (b) How might you follow the rate of the following reaction using ultraviolet spectroscopy?

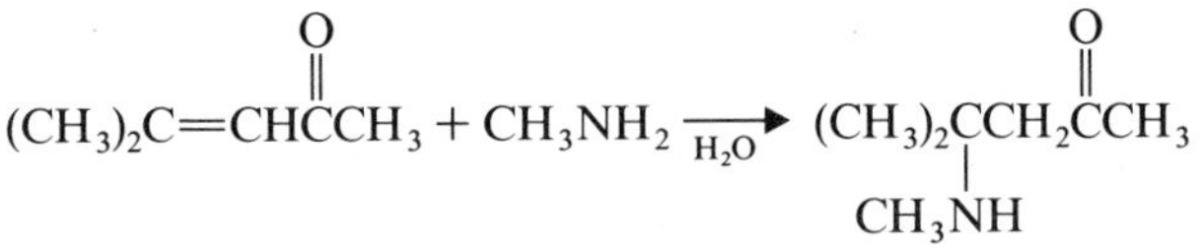

$$(CH_3)_2C=CHCCH_3 + CH_3NH_2 \xrightarrow{H_2O} (CH_3)_2CCH_2CCH_3$$
$$\overset{|}{CH_3NH}$$

17.32

(a) A compound **U** ($C_9H_{10}O$) gives a negative iodoform test. The infrared spectrum of **U** shows a strong absorption peak at 1690 cm^{-1}. The proton nmr spectrum of **U** gives the following:

Triplet	$\delta 1.2$ (3H)
Quartet	$\delta 3.0$ (2H)
Multiplet	$\delta 7.7$ (5H)

What is the structure of **U**?

(b) A compound **V** is an isomer of **U**. Compound **V** gives a positive iodoform test; its infrared spectrum shows a strong peak at 1705 cm^{-1}. The proton nmr spectrum of **V** gives the following:

Singlet	$\delta 2.0$ (3H)
Singlet	$\delta 3.5$ (2H)
Multiplet	$\delta 7.1$ (5H)

What is the structure of **V**?

17.33

Compound **A** has the molecular formula $C_6H_{12}O_3$ and shows a strong infrared absorption peak at 1710 cm^{-1}. When treated with iodine in aqueous sodium hydroxide, **A** gives a yellow precipitate. When **A** is treated with Tollens' reagent, no reaction occurs; however, if **A** is treated first with water containing a drop of sulfuric acid and then treated with Tollens' reagent, a silver mirror forms in the test tube. Compound **A** shows the following proton nmr spectrum.

Singlet	$\delta 2.1$
Doublet	$\delta 2.6$
Singlet	$\delta 3.2$ (6H)
Triplet	$\delta 4.7$

Write a structure for **A**.

*17.34

Treating a solution of *cis*-1-decalone with base causes an isomerization to take place. When the system reaches equilibrium, the solution is found to contain about 95% *trans*-1-decalone and about 5% *cis*-1-decalone. Explain this isomerization.

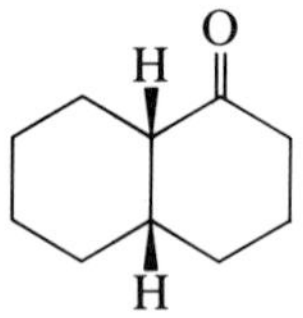

cis-1-Decalone

LITHIUM ENOLATES IN ORGANIC SYNTHESIS

I.1 Enolate Ions

Enolate ions are formed when a carbonyl compound with an α hydrogen is treated with a base (Section 17.1). The extent to which the enolate ion forms will depend on the strength of the base used. If the base employed is a weaker base than the enolate ion, then the equilibrium will lie to the left. This will be the case, for example, when a ketone is treated with aqueous solution containing sodium hydroxide.

$$\underset{\substack{\text{Weaker acid} \\ (pK_a = 20)}}{CH_3-\overset{\overset{\displaystyle O}{\|}}{C}-CH_3} + \underset{\text{Weaker base}}{Na^+OH^-} \;\rightleftharpoons\; \underset{\text{Stronger base}}{CH_3-\overset{\overset{\displaystyle O^{\delta-}\,Na^+}{\|}}{C}=CH_2{}^{\delta-}} + \underset{\substack{\text{Stronger acid} \\ (pK_a = 16)}}{H_2O}$$

On the other hand, if a very strong base is employed, the equilibrium will lie far to the right. One very useful strong base, for converting ketones to enolates is lithium diisopropylamide, $(i\text{-}C_3H_7)_2N^-Li^+$.

$$\underset{\substack{\text{Stronger acid} \\ (pK_a = 20)}}{CH_3-\overset{\overset{\displaystyle O}{\|}}{C}-CH_3} + \underset{\text{Stronger base}}{(i\text{-}C_3H_7)_2N^-Li^+} \longrightarrow \underset{\text{Weaker base}}{CH_3-\overset{\overset{\displaystyle O^{\delta-}\,Li^+}{\|}}{C}=CH_2{}^{\delta-}} + \underset{\substack{\text{Weaker acid} \\ (pK_a = 38)}}{(i\text{-}C_3H_7)_2NH}$$

Lithium diisopropylamide (abbreviated **LDA**) can be prepared by dissolving diisopropylamine in a solvent such as diethyl ether or THF, and treating it with an alkyllithium.

$$\underset{\substack{\text{Stronger acid} \\ (pK_a = 38)}}{(i\text{-}C_3H_7)_2NH} + \underset{\substack{\text{Stronger base}}}{CH_3{}^-Li^+} \xrightarrow{\text{ether}} \underset{\text{Weaker base}}{(i\text{-}C_3H_7)_2N^-Li^+} + \underset{\substack{\text{Weaker acid} \\ (pK_a = 50)}}{CH_4\uparrow}$$

I.1A Enolate Ions as Ambident Nucleophiles

Because enolate ions have a partial negative charge on an oxygen atom they can react in nucleophilic substitution reactions as if they were **alkoxide ions.** Because they have a partial negative charge on a carbon atom they can also react as **carbanions.** Nucleophiles like this, *those that are capable of reacting at two sites,* are called **ambident nucleophiles.**

This site reacts as an alkoxide ion

$$CH_3-\overset{\overset{\displaystyle O^{\delta-}}{\|}}{C}=\overset{\delta-}{CH_2}$$

This site reacts as a carbanion

Just how an enolate ion reacts depends, in part, on the substrate with which it reacts. *One substrate that tends to react almost exclusively at the oxygen atom of an enolate is chlorotrimethylsilane, $(CH_3)_3SiCl$.*

$$CH_3\overset{\overset{\displaystyle O^{\delta-}}{\|}}{C}\text{---}\overset{\delta-}{C}H_2 + (CH_3)_3Si\text{---}Cl \xrightarrow{\text{THF}} CH_3\overset{\overset{\displaystyle OSi(CH_3)_3}{|}}{C}=CH_2 + Cl^-$$

(85%)
Enol trimethylsilyl
ether

This reaction, called **silylation** (cf. Section 8.17D), is a nucleophilic substitution at the silicon atom by the oxygen atom of the enolate, and it takes place as it does because the oxygen–silicon bond that forms in the enol trimethylsilyl ether is very strong (much stronger than a carbon–silicon bond). This factor makes formation of the enol trimethylsilyl ether highly exothermic, and, consequently, the energy of activation for reaction at the oxygen atom is lower than that for reaction of the α carbon.

Enolate ions display their carbanionic character when they react with alkyl halides. In these reactions the major product is usually the one in which alkylation occurs at the carbon atom.

$$R\text{---}\overset{\overset{\displaystyle O^{\delta-}}{\|}}{C}\text{---}\overset{\delta-}{C}HR' + R''CH_2\text{---}X \longrightarrow R\text{---}\overset{\overset{\displaystyle O}{\|}}{C}\text{---}\overset{\overset{\displaystyle CH_2R''}{|}}{C}HR'$$

C-alkylated
product
(major product)

Alkylation reactions like these have an important limitation: Because the reactions are S_N2 reactions and because enolate ions are strong bases, *successful alkylations occur only when primary alkyl, primary benzylic, and primary allylic halides are used.* With secondary and tertiary halides, elimination becomes the main course of the reaction.

PROBLEM I.1

Write structures for the *C*-alkylated and *O*-alkylated products that form when the enolate derived from cyclohexanone reacts with benzyl bromide.

I.1B Regioselective Formation of Enolate Ions

An unsymmetrical ketone such as 2-methylcyclohexanone can form two possible enolates. Just which enolate will be formed predominantly depends on the base used and on the conditions employed. The enolate *with the more highly substituted double bond is the **thermodynamically more stable enolate*** in the same way that an alkene with the more highly substituted double bond is the more stable alkene (Section 6.9). This enolate, called the **thermodynamic enolate,** will be formed predominantly under conditions that permit the establishment of an equilibrium. This will generally be the case if the enolate is produced using a relatively weak base in a protic solvent.

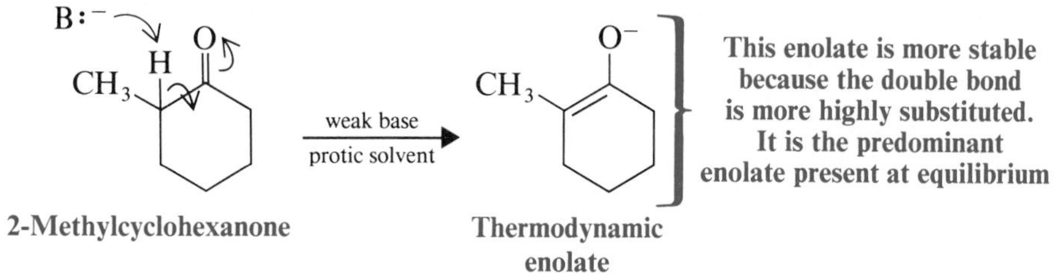

2-Methylcyclohexanone   Thermodynamic
                              enolate

This enolate is more stable because the double bond is more highly substituted. It is the predominant enolate present at equilibrium

On the other hand, *the enolate with the less substituted double bond is usually formed faster,* because removal of the hydrogen necessary to produce this enolate is less sterically hindered. This enolate, called the **kinetic enolate,** is formed predominantly when the reaction is kinetically controlled (or rate controlled).

The kinetically favored enolate can be formed cleanly through the use of lithium diisopropylamide (LDA). This strong, sterically hindered base rapidly removes the proton from the less substituted α carbon of the ketone. The following sample, using 2-methylcyclohexanone, is an illustration. The solvent for the reaction is 1,2-dimethoxyethane ($CH_3OCH_2CH_2OCH_3$) abbreviated **DME.**

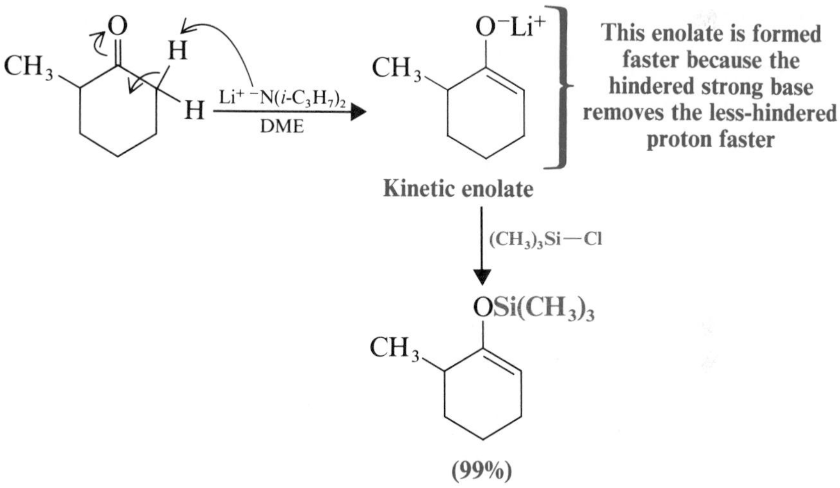

Kinetic enolate

This enolate is formed faster because the hindered strong base removes the less-hindered proton faster

(99%)

The example just given also shows how the enolate ion can be "trapped" by converting it to the enol trimethylsilyl ether. This procedure is especially useful because the enol trimethylsilyl ether can be purified, if necessary, and then converted back to an enolate. One way of achieving this conversion is by treating the enol trimethylsilyl ether with a solution containing fluoride ions.

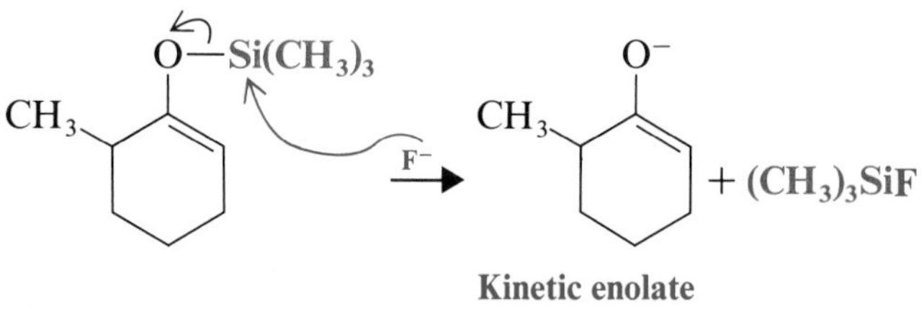

Kinetic enolate

This reaction is a nucleophilic substitution at the silicon atom brought about by a fluoride ion. Fluoride ions have an extremely high affinity for silicon atoms because Si—F bonds are very strong.

Another way to convert an enol trimethylsilyl ether back to an enolate is to treat it with methyllithium.

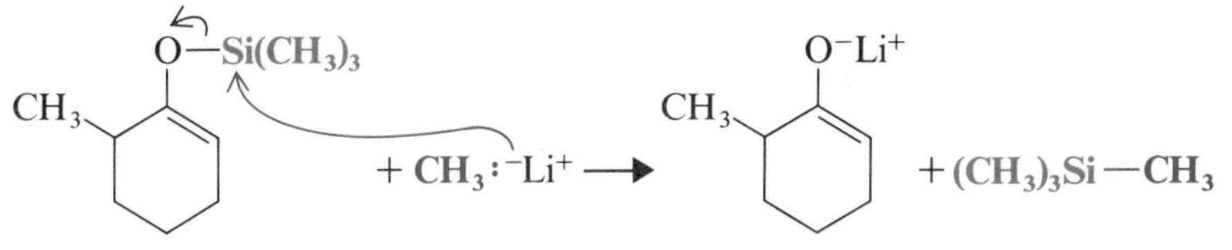

I.2 Lithium Enolates in Directed Aldol Reactions

One of the most effective and versatile ways to bring about a crossed aldol reaction is to use a lithium enolate obtained from a ketone as one component and an aldehyde or ketone as the other. An example of what is called a **directed aldol reaction** is shown in Fig. I.1.

Regioselectivity can be achieved when unsymmetrical ketones are used in directed aldol reactions by generating the kinetic enolate using lithium diisopropylamide. This will ensure production of the enolate in which the proton has been removed from the less-substituted α carbon. Two examples are the following:

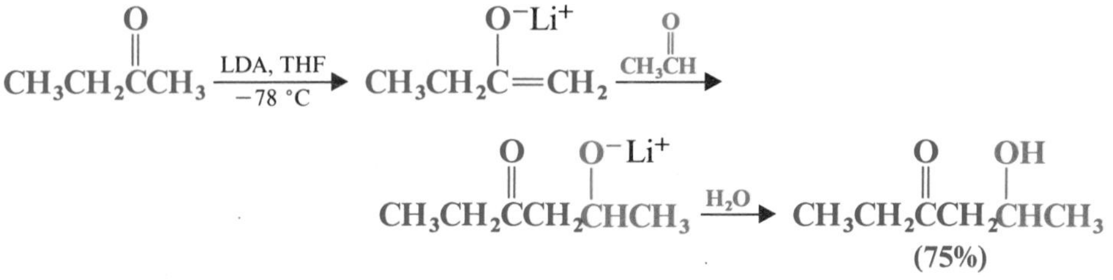

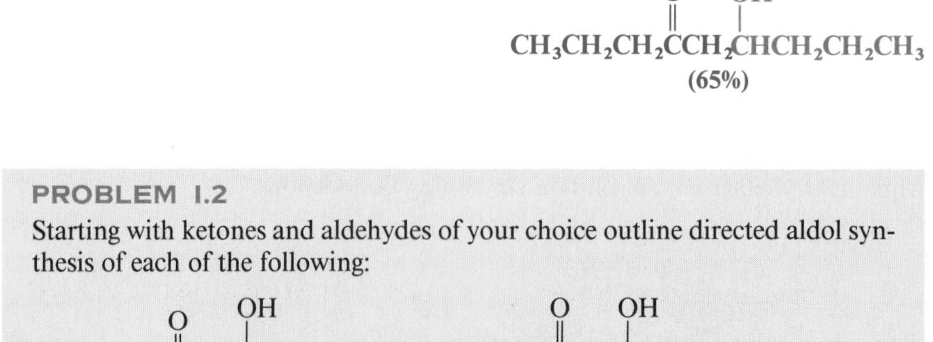

PROBLEM I.2

Starting with ketones and aldehydes of your choice outline directed aldol synthesis of each of the following:

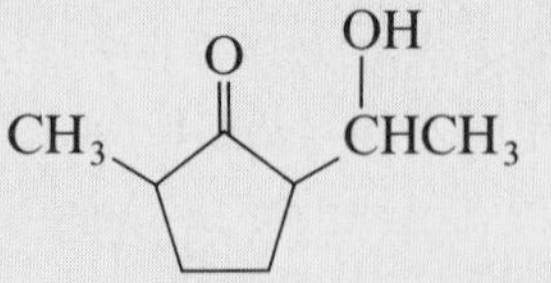

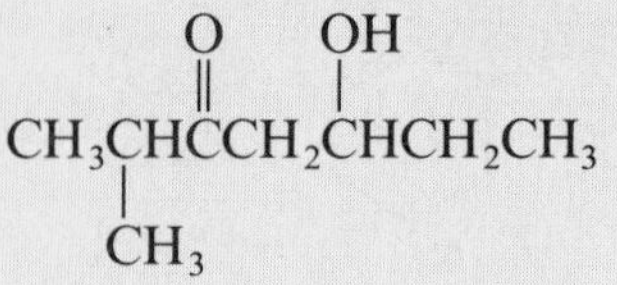

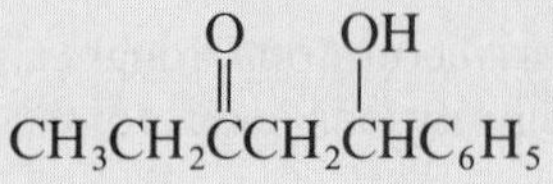

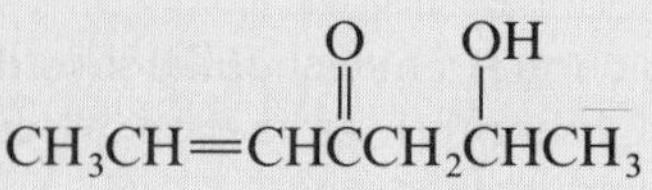

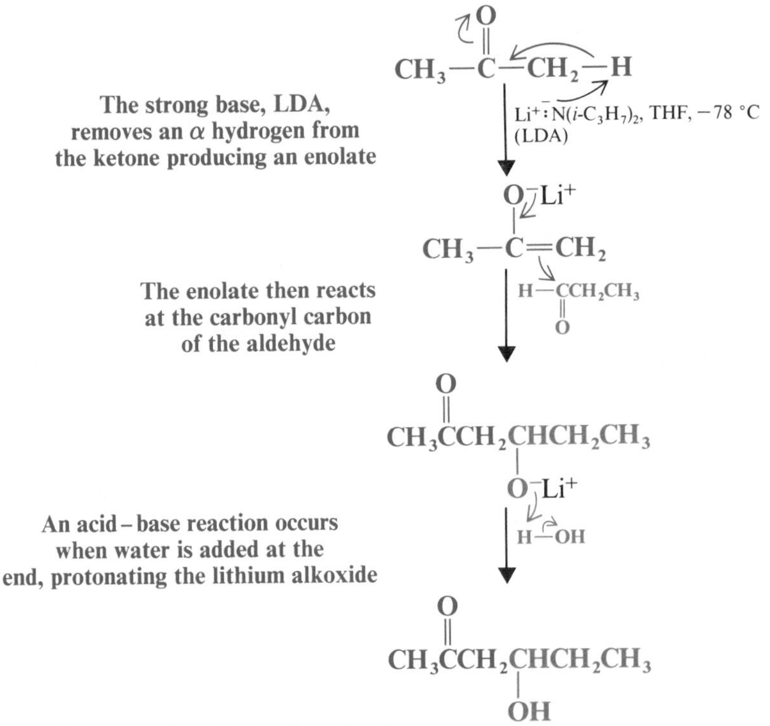

$$CH_3-\overset{\overset{\displaystyle O}{\|}}{C}-CH_2-H$$

The strong base, LDA, removes an α hydrogen from the ketone producing an enolate

$Li^+ : \bar{N}(i\text{-}C_3H_7)_2$, THF, $-78\ ^\circ C$
(LDA)

$$CH_3-\overset{\overset{\displaystyle O^-Li^+}{\|}}{C}=CH_2$$

$H-CCH_2CH_3$
$\overset{\|}{O}$

The enolate then reacts at the carbonyl carbon of the aldehyde

$$CH_3\overset{\overset{\displaystyle O}{\|}}{C}CH_2CHCH_2CH_3$$

$\overset{|}{O^-Li^+}$

$H-OH$

An acid–base reaction occurs when water is added at the end, protonating the lithium alkoxide

$$CH_3\overset{\overset{\displaystyle O}{\|}}{C}CH_2CHCH_2CH_3$$
$\overset{|}{OH}$

FIGURE I.1 A directed aldol synthesis.

PROBLEM I.3

The compounds called α-bisabolanone and ocimenone have both been synthesized by directed aldol syntheses. In both syntheses one starting compound was $(CH_3)_2C=CHCOCH_3$. Choose other appropriate starting compounds and outline syntheses of (a) α-bisabolanone and (b) ocimenone.

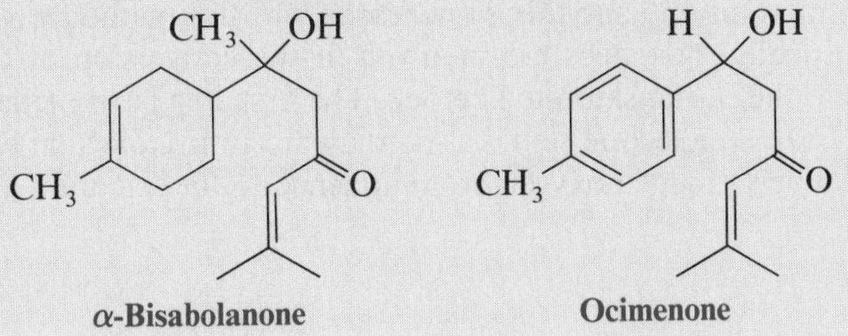

α-**Bisabolanone**

Ocimenone

PROBLEM I.4

Treating the enol trimethylsilyl ether derived from cyclohexanone with benzaldehyde and tetrabutylammonium fluoride, $(C_4H_9)_4N^+F^-$ (abbreviated TBAF), gave the following product. Outline the steps that occur in this reaction.

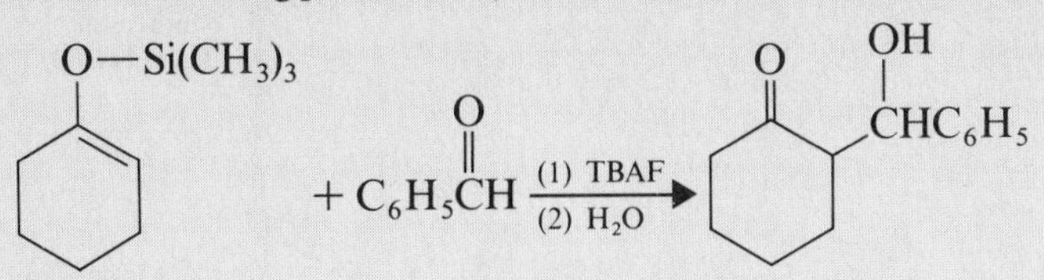

$$\text{[enol TMS ether of cyclohexanone]} + C_6H_5\overset{\overset{\displaystyle O}{\|}}{C}H \xrightarrow[\text{(2) } H_2O]{\text{(1) TBAF}} \text{[product]}$$

I.3 α-Selenation: A Synthesis
of α,β-Unsaturated Carbonyl Compounds

Lithium enolates react with benzeneselenenyl bromide (C_6H_5SeBr) (or with C_6H_5SeCl) to yield products containing a C_6H_5Se— group at the α position.

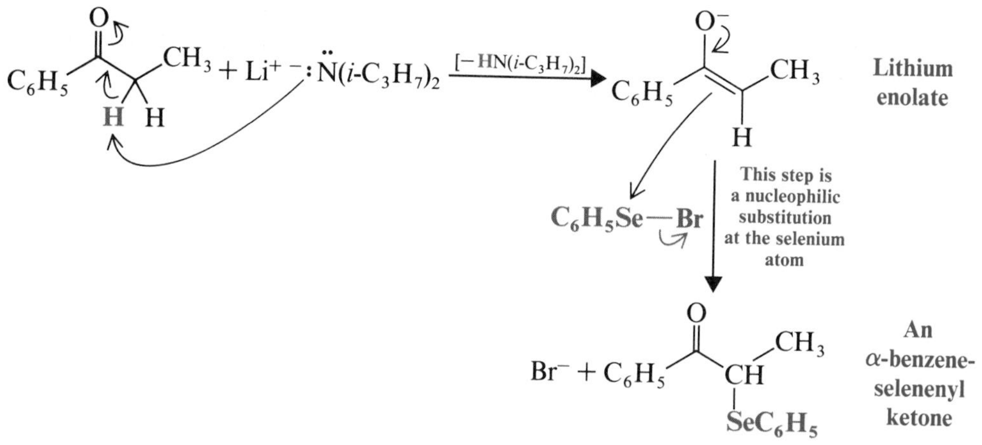

Treating the α-phenylseleno ketone with hydrogen peroxide at room temperature converts it to an α,β-unsaturated ketone.

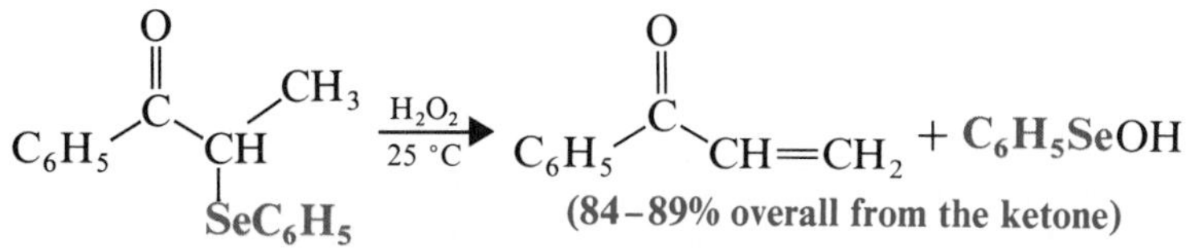

$(84–89\% \text{ overall from the ketone})$

These are very mild conditions for the introduction of a double bond (room temperature and a neutral solution), and this is one reason why this method is a valuable one.

Mechanistically, two steps are involved in the conversion of the α-phenylseleno ketone to the α,β-unsaturated ketone. The first step is an oxidation brought about by the H_2O_2. The second step is a spontaneous intramolecular elimination in which the negatively charged oxygen atom attached to the selenium atom acts as a base.

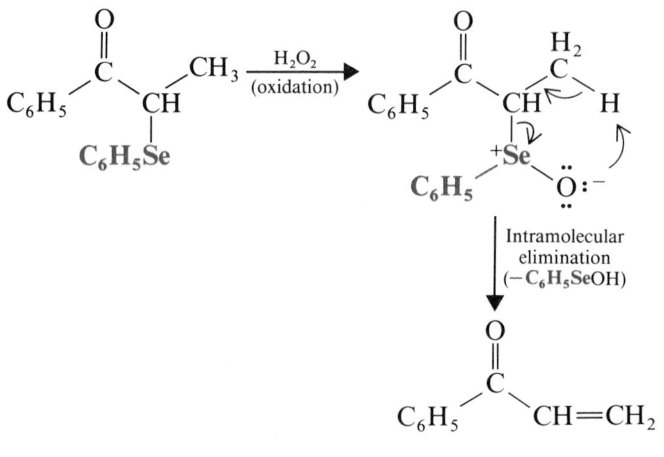

When we study the Cope elimination in Section 19.13B, we shall find another example of this kind of intramolecular elimination.

PROBLEM I.5

Starting with 2-methylcyclohexanone, show how you would use α-selenation in a synthesis of the following compound:

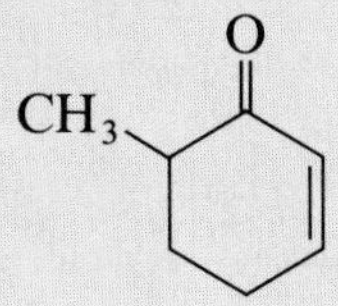

CHAPTER EIGHTEEN

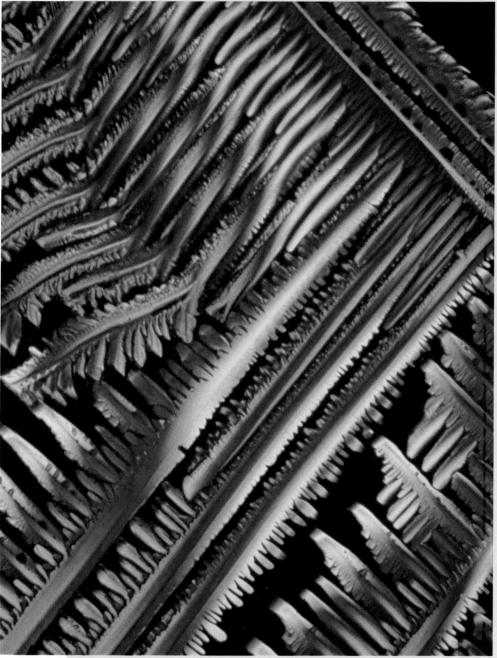

4-Aminobutyric acid and pyrogallic acid crystals. Pyrogallic acid is 1,2,3-trihydroxybenzoic acid.

Carboxylic Acids and Their Derivatives: Nucleophilic Substitution at Acyl Carbon

18.1 INTRODUCTION

The carboxyl group $-\overset{\overset{\displaystyle O}{\|}}{C}OH$ (abbreviated $-CO_2H$ or $-COOH$) is one of the most widely occurring functional groups in chemistry and biochemistry. Not only are carboxylic acids themselves important, but the carboxyl group is the parent group of a large family of related compounds.

All of these carboxylic acid derivatives contain the acyl group, $R\overset{\overset{\displaystyle O}{\|}}{C}-$. As a result, they are often called *acyl compounds*. They are called *carboxylic acid derivatives* because they are derived from a carboxylic acid by replacing the $-OH$ of $R\overset{\overset{\displaystyle O}{\|}}{C}OH$ by some other group (Table 18.1).

18.2 NOMENCLATURE AND PHYSICAL PROPERTIES

18.2A Carboxylic Acids

IUPAC systematic or substitutive names for carboxylic acids are obtained by dropping the final *e* of the name of the alkane corresponding to the longest chain in the

TABLE 18.1 Carboxylic acid derivatives

STRUCTURE	NAME	STRUCTURE	NAME
$R-\overset{\overset{\displaystyle O}{\|}}{C}-Cl$	Acyl (or acid) chloride	$R-\overset{\overset{\displaystyle O}{\|}}{C}-NH_2$	
$R-\overset{\overset{\displaystyle O}{\|}}{C}-O-\overset{\overset{\displaystyle O}{\|}}{C}-R$	Acid anhydride	$R-\overset{\overset{\displaystyle O}{\|}}{C}-NHR'$	Amide
$R-\overset{\overset{\displaystyle O}{\|}}{C}-O-R'$	Ester	$R-\overset{\overset{\displaystyle O}{\|}}{C}-NR_2$	

acid and by adding -*oic acid.* The carboxyl carbon atom is assigned number 1. The examples listed here illustrate how this is done.

$$\overset{6}{\text{CH}_3}\overset{5}{\text{CH}_2}\overset{4}{\text{CHCH}_2}\overset{3}{\text{CH}_2}\overset{2}{\text{CH}_2}\overset{1}{\overset{\displaystyle O}{\overset{\displaystyle \|}{\text{C}}}}\text{OH}$$
$$\underset{\displaystyle \text{CH}_3}{|}$$

4-Methylhexanoic acid

$$\overset{6}{\text{CH}_3}\overset{5}{\text{CH}}=\overset{4}{\text{CHCH}_2}\overset{3}{\text{CH}_2}\overset{2}{\overset{\displaystyle O}{\overset{\displaystyle \|}{\text{C}}}}\text{OH}$$

4-Hexenoic acid

Many carboxylic acids have common names that are derived from Latin or Greek words that indicate one of their natural sources (Table 18.2). Methanoic acid is called formic acid (from the Latin, *formica,* or ant). Ethanoic acid is called acetic acid (from the Latin, *acetum,* or vinegar). Butanoic acid is one compound responsible for the odor of rancid butter, thus its common name is butyric acid (from the Latin, *butyrum,* or butter). Hexanoic acid is one compound associated with the odor of goats, hence its common name, caproic acid (from the Latin *caper,* or goat). Pentanoic acid, as a result of its occurrence in valerian, a perennial herb, is named valeric acid. Octadecanoic acid takes its common name, stearic acid, from the Greek word *stear,* for tallow.

Most of these common names have been with us for a long time and some are likely to remain in common usage for even longer, so it is helpful to be familiar with them. In this text we shall always refer to methanoic acid and ethanoic acid as formic acid and acetic acid. However, in almost all other instances we shall use IUPAC systematic or substitutive names.

Carboxylic acids are polar substances. Their molecules can form strong hydrogen bonds with each other and with water. As a result, carboxylic acids generally have high boiling points, and low-molecular-weight carboxylic acids show appreciable solubility in water. The first four carboxylic acids (Table 18.2) are miscible with water in all proportions. As the length of the carbon chain increases, water solubility declines.

18.2B Carboxylic Salts

Salts of carboxylic acids are named as -*ates;* in both common and systematic names, -*ate* replaces -*ic acid.* Thus, CH_3CO_2Na is sodium acetate or sodium ethanoate.

Sodium and potassium salts of most carboxylic acids are readily soluble in water. This is true even of the long-chain carboxylic acids. Sodium or potassium salts of long-chain carboxylic acids are the major ingredients of soap (cf. Section 22.2B).

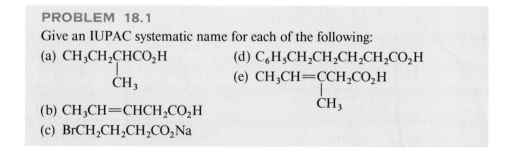

PROBLEM 18.1

Give an IUPAC systematic name for each of the following:

(a) $CH_3CH_2CHCO_2H$
    $\quad\quad\quad\quad |$
    $\quad\quad\quad CH_3$

(b) $CH_3CH=CHCH_2CO_2H$

(c) $BrCH_2CH_2CH_2CO_2Na$

(d) $C_6H_5CH_2CH_2CH_2CH_2CO_2H$

(e) $CH_3CH=CCH_2CO_2H$
    $\quad\quad\quad\quad\quad |$
    $\quad\quad\quad\quad CH_3$

18.2C Acidity of Carboxylic Acids

Most unsubstituted carboxylic acids have K_a's in the range of $10^{-4}-10^{-5}$ ($pK_a =$ 4–5) as seen in Table 18.2. The K_a of water is $\sim 10^{-16}$ and the apparent K_a of H_2CO_3 is $\sim 10^{-7}$. These relative acidities mean that carboxylic acids react readily with aqueous solutions of sodium hydroxide and sodium bicarbonate to form soluble sodium salts. We can use solubility tests, therefore, to distinguish water-insoluble carboxylic acids from water-insoluble phenols and alcohols. Water-insoluble carboxylic acids will dissolve in either aqueous sodium hydroxide or aqueous sodium bicarbonate:

Benzoic acid (water insoluble) Stronger acid + NaOH → Sodium benzoate (water soluble) Weaker base + H₂O

(water insoluble) Stronger acid + NaHCO₃ → (water soluble) Weaker base + CO₂↑ + H₂O

Water-insoluble phenols (Section 14.5) dissolve in aqueous sodium hydroxide but (except for some nitrophenols) do not dissolve in aqueous sodium bicarbonate. Water-insoluble alcohols do not dissolve in either aqueous sodium hydroxide or sodium bicarbonate.

PROBLEM 18.2

(a) When excess carbon dioxide is passed into a solution of sodium benzoate and sodium p-methylphenoxide in aqueous sodium hydroxide, p-methylphenol separates from the solution (as an oil) but sodium benzoate remains in solution. Write equations that will provide an explanation for this. (b) Given the following reagents—aqueous sodium hydroxide, aqueous hydrochloric acid, ether, and carbon dioxide—explain how you would separate a mixture of benzoic acid, p-methylphenol, and 1-methylcyclohexanol.

When we studied phenols in Section 14.5, we saw there that we can account for relative acidities of molecules on the basis of the relative stabilities of the acids and of their conjugate bases. Any factor that stabilizes the conjugate base of an acid more than it stabilizes the acid itself will increase the strength of the acid. The greater acidity of carboxylic acids, when compared with alcohols, for example, arises from an extra resonance stability associated with the carboxylate ion. To understand this

TABLE 18.2 Carboxylic acids

STRUCTURE	SYSTEMATIC NAME	COMMON NAME	mp (°C)	bp (°C)	WATER SOLUBILITY (g/100 mL H_2O) 25 °C	K_a (at 25 °C)
HCO_2H	Methanoic acid	Formic acid	8	100.5	∞	1.77×10^{-4}
CH_3CO_2H	Ethanoic acid	Acetic acid	16.6	118	∞	1.76×10^{-5}
$CH_3CH_2CO_2H$	Propanoic acid	Propionic acid	−21	141	∞	1.34×10^{-5}
$CH_3(CH_2)_2CO_2H$	Butanoic acid	Butyric acid	−6	164	∞	1.54×10^{-5}
$CH_3(CH_2)_3CO_2H$	Pentanoic acid	Valeric acid	−34	187	4.97	1.52×10^{-5}
$CH_3(CH_2)_4CO_2H$	Hexanoic acid	Caproic acid	−3	205	1.08	1.43×10^{-5}
$CH_3(CH_2)_6CO_2H$	Octanoic acid	Caprylic acid	16	239	0.07	1.28×10^{-5}
$CH_3(CH_2)_8CO_2H$	Decanoic acid	Capric acid	31	269	0.015	1.43×10^{-5}
$CH_3(CH_2)_{10}CO_2H$	Dodecanoic acid	Lauric acid	44	179^{18}	0.006	5.01×10^{-6}
$CH_3(CH_2)_{12}CO_2H$	Tetradecanoic acid	Myristic acid	59	200^{20}	0.002	
$CH_3(CH_2)_{14}CO_2H$	Hexadecanoic acid	Palmitic acid	63	219^{17}	0.0007	3.46×10^{-7}
$CH_3(CH_2)_{16}CO_2H$	Octadecanoic acid	Stearic acid	70	383	0.0003	
CH_2ClCO_2H	Chloroethanoic acid	Chloroacetic acid	63	189	Very soluble	1.40×10^{-3}
$CHCl_2CO_2H$	Dichloroethanoic acid	Dichloroacetic acid	10.8	192	Very soluble	3.32×10^{-2}
CCl_3CO_2H	Trichloroethanoic acid	Trichloroacetic acid	56.3	198	Very soluble	2.00×10^{-1}
$CH_3CHClCO_2H$	2-Chloropropanoic acid	α-Chloropropionic acid		186	Soluble	1.47×10^{-3}
$CH_2ClCH_2CO_2H$	3-Chloropropanoic acid	β-Chloropropionic acid	61	204	Soluble	1.04×10^{-4}
$C_6H_5CO_2H$	Benzoic acid	Benzoic acid	122	250	0.34	6.46×10^{-5}
$p\text{-}CH_3C_6H_4CO_2H$	4-Methylbenzoic acid	p-Toluic acid	180	275	0.03	4.33×10^{-5}
$p\text{-}Cl_6H_4CO_2H$	4-Chlorobenzoic acid	p-Chlorobenzoic acid	242		0.009	1.04×10^{-4}
$p\text{-}NO_2C_6H_4CO_2H$	4-Nitrobenzoic acid	p-Nitrobenzoic acid	242		0.03	3.93×10^{-4}
	1-Naphthoic acid	α-Naphthoic acid	160	300	Insoluble	2.00×10^{-4}
	2-Naphthoic acid	β-Naphthoic acid	185	> 300	Insoluble	6.80×10^{-5}

effect let us examine the principal resonance structures for a carboxylic acid (**1** and **2**) and for a carboxylate ion (**3** and **4**).

This structure has separated charges

These structures are equivalent

A carboxylic acid can be represented by two *nonequivalent* structures, one of which involves charge separation. The carboxylate ion, on the other hand, is a hybrid

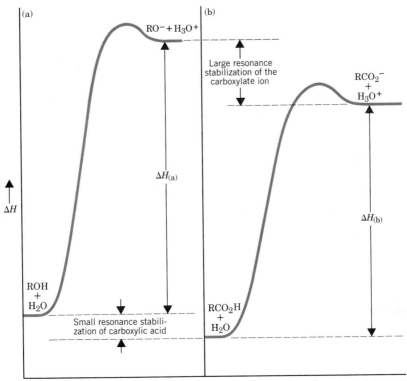

FIGURE 18.1 Potential energy diagrams for the ionization of an alcohol *(a)* and the corresponding carboxylic acid *(b)*. Resonance stabilizes the carboxylate ion to a greater extent than it stabilizes the acid itself; therefore, the ionization of the acid is less endothermic than the ionization of the corresponding alcohol that is, $\Delta H_{(a)} > \Delta H_{(b)}$. The carboxylic acid is, therefore, the stronger acid. (The curves are aligned for comparison only.)

of two *equivalent* structures neither of which involves separated charges. Resonance theory (Section 10.5) tells us that because structures without separated charges are more stable, we should expect much greater resonance stabilization from the carboxylate ion than for the acid. Since resonance stabilizes the carboxylate ion more than it does the acid, the ionization of a carboxylic acid is energetically more favorable than the corresponding ionization of an alcohol (Fig. 18.1), where neither the alcohol nor the alkoxide ion is resonance stabilized.

$$R-C \overset{O^{\delta-}}{\underset{O^{\delta+}-H}{\big\langle}} + H_2O \rightleftharpoons R-C \overset{O^{\delta-}}{\underset{O^{\delta-}}{\big\langle}} + H_3O^+$$

Small resonance stabilization   *Large resonance stabilization*

Carboxylate ion is stabilized more than the acid — K_a is larger

$$R-OH + H_2O \rightleftharpoons R-O^- + H_3O^+$$

No resonance stabilization   *No resonance stabilization*

Neither the alcohol nor the alkoxide ion is stabilized — K_a is smaller

CARBOXYLIC ACIDS AND THEIR DERIVATIVES

PROBLEM 18.3

We see in Table 18.2 that carboxylic acids having electron-withdrawing groups are stronger than unsubstituted acids. The chloroacetic acids, for example, show the following order of acidities:

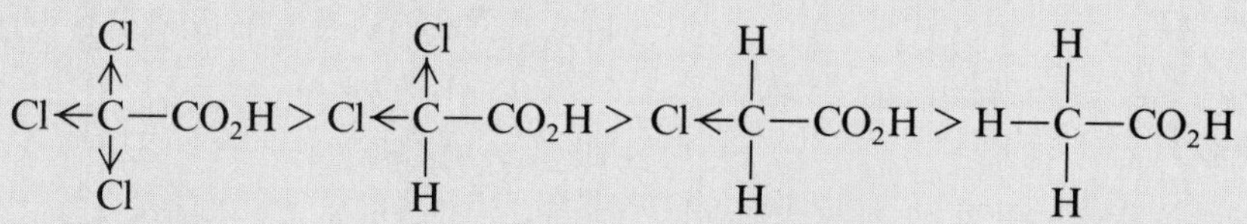

$$\underset{\underset{\displaystyle Cl}{\big|}}{\overset{\overset{\displaystyle Cl}{\big|}}{Cl{\leftarrow}C}}{-}CO_2H > \underset{\underset{\displaystyle H}{\big|}}{\overset{\overset{\displaystyle Cl}{\big|}}{Cl{\leftarrow}C}}{-}CO_2H > \underset{\underset{\displaystyle H}{\big|}}{\overset{\overset{\displaystyle H}{\big|}}{Cl{\leftarrow}C}}{-}CO_2H > \underset{\underset{\displaystyle H}{\big|}}{\overset{\overset{\displaystyle H}{\big|}}{H{-}C}}{-}CO_2H$$

K_a	2.0×10^{-1}	3.32×10^{-2}	1.40×10^{-3}	1.76×10^{-5}
pK_a	0.7	1.48	2.85	4.75

Explain this acid-strengthening effect of electron-withdrawing groups.

PROBLEM 18.4

Which acid of each pair shown here would you expect to be stronger?

(a) CH_3CO_2H or CH_2FCO_2H

(b) CH_2FCO_2H or CH_2ClCO_2H

(c) CH_2ClCO_2H or CH_2BrCO_2H

(d) $CH_2ClCH_2CH_2CO_2H$ or $CH_3CHClCH_2CO_2H$

(e) $CH_3CH_2CHClCO_2H$ or $CH_3CHClCH_2CO_2H$

(f) 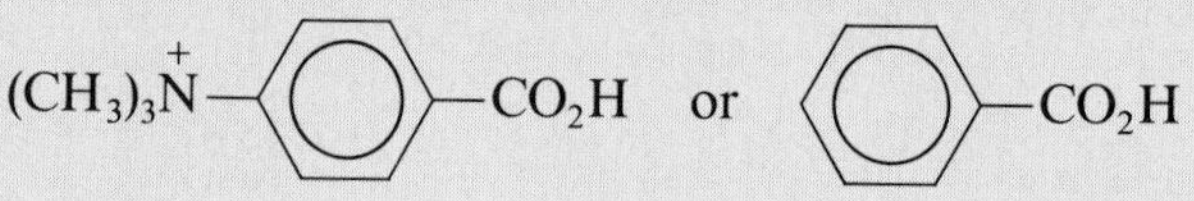 $(CH_3)_3\overset{+}{N}{-}\langle\bigcirc\rangle{-}CO_2H$ or $\langle\bigcirc\rangle{-}CO_2H$

(g) 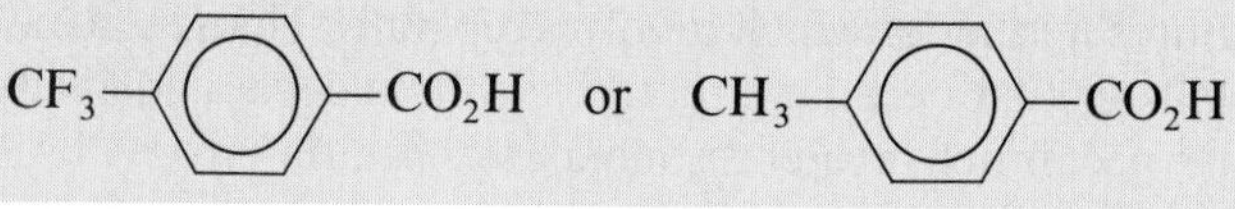 $CF_3{-}\langle\bigcirc\rangle{-}CO_2H$ or $CH_3{-}\langle\bigcirc\rangle{-}CO_2H$

18.2D Dicarboxylic Acids

Dicarboxylic acids are named as **alkanedioic acids** in the IUPAC systematic or substitutive system. Most simple dicarboxylic acids have common names (Table 18.3), and these are the names that we shall use.

PROBLEM 18.5

Suggest explanations for the following facts: (a) K_1 for all of the dicarboxylic acids in Table 18.3 are higher than the K_a for monocarboxylic acids with the same number of carbon atoms. (b) The difference between K_1 and K_2 for dicarboxylic acids of type $HO_2C(CH_2)_nCO_2H$ decreases as n increases.

TABLE 18.3 Dicarboxylic acids

STRUCTURE	COMMON NAME	mp (°C)	K_a (at 25 °C) K_1	K_2
$HO_2C—CO_2H$	Oxalic acid	189 dec.	5.9×10^{-2}	6.4×10^{-5}
$HO_2CCH_2CO_2H$	Malonic acid	136	1.4×10^{-3}	2.0×10^{-6}
$HO_2C(CH_2)_2CO_2H$	Succinic acid	187	6.9×10^{-5}	2.5×10^{-6}
$HO_2C(CH_2)_3CO_2H$	Glutaric acid	98	4.6×10^{-5}	3.9×10^{-6}
$HO_2C(CH_2)_4CO_2H$	Adipic acid	153	3.7×10^{-5}	2.4×10^{-6}
cis-$HO_2C—CH{=}CH—CO_2H$	Maleic acid	131	1.4×10^{-2}	8.6×10^{-7}
trans-$HO_2C—CH{=}CH—CO_2H$	Fumaric acid	287	9.3×10^{-4}	3.6×10^{-5}
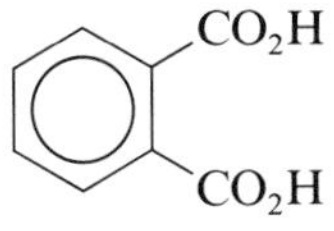	Phthalic acid	206–208 dec.	1.3×10^{-3}	3.9×10^{-6}
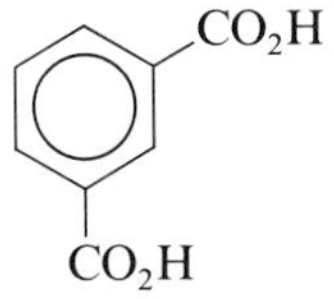	Isophthalic acid	345–348	2.9×10^{-4}	2.5×10^{-5}
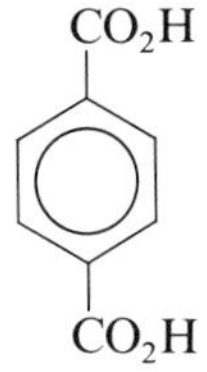	Terephthalic acid	Sublimes	3.1×10^{-4}	1.5×10^{-5}

18.2E Esters

The names of esters are derived from the names of the alcohol (with the ending **-yl**) and the acid (with the ending **-ate** or **-oate**). The portion of the name derived from the alcohol comes first.

$CH_3C(=O)—OCH_2CH_3$

Ethyl acetate or ethyl ethanoate

$CH_3CH_2C(=O)—OC(CH_3)_2—CH_3$

***tert*-Butyl propanoate**

$CH_3C(=O)OCH{=}CH_2$

Vinyl acetate or ethenyl ethanoate

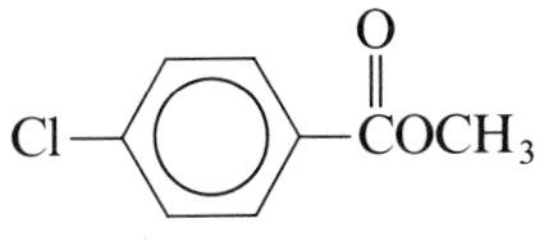

Methyl *p*-chlorobenzoate

$CH_3CH_2OC(=O)CH_2C(=O)OCH_2CH_3$

Diethyl malonate

Esters are polar compounds but lacking a hydrogen attached to oxygen, their molecules cannot form strong hydrogen bonds to each other. As a result, esters have boiling points that are lower than those of acids and alcohols of comparable molecu-

TABLE 18.4 Carboxylic esters

NAME	STRUCTURE	mp (°C)	bp (°C)	SOLUBILITY IN WATER (g/100 mL at 20 °C)
Methyl formate	HCO_2CH_3	−99	31.5	Very soluble
Ethyl formate	$HCO_2CH_2CH_3$	−79	54	soluble
Methyl acetate	$CH_3CO_2CH_3$	−99	57	24.4
Ethyl acetate	$CH_3CO_2CH_2CH_3$	−82	77	7.39 (25 °C)
Propyl acetate	$CH_3CO_2CH_2CH_2CH_3$	−93	102	1.89
Butyl acetate	$CH_3CO_2CH_2(CH_2)_2CH_3$	−74	125	1.0 (22 °C)
Ethyl propanoate	$CH_3CH_2CO_2CH_2CH_3$	−73	99	1.75
Ethyl butanoate	$CH_3(CH_2)_2CO_2CH_2CH_3$	−93	120	0.51
Ethyl pentanoate	$CH_3(CH_2)_3CO_2CH_2CH_3$	−91	145	0.22
Ethyl hexanoate	$CH_3(CH_2)_4CO_2CH_2CH_3$	−68	168	0.063
Methyl benzoate	$C_6H_5CO_2CH_3$	−12	199	0.15
Ethyl benzoate	$C_6H_5CO_2CH_2CH_3$	−35	213	0.08
Phenyl acetate	$CH_3CO_2C_6H_5$		196	V. sl. soluble
Methyl salicylate	$o\text{-}HOC_6H_4CO_2CH_3$	−9	223	0.74 (30°C)

lar weight. The boiling points (Table 18.4) of esters are about the same as those of comparable aldehydes and ketones.

Unlike the low-molecular-weight acids, esters usually have pleasant odors, some resembling those of fruits, and these are used in the manufacture of synthetic flavors:

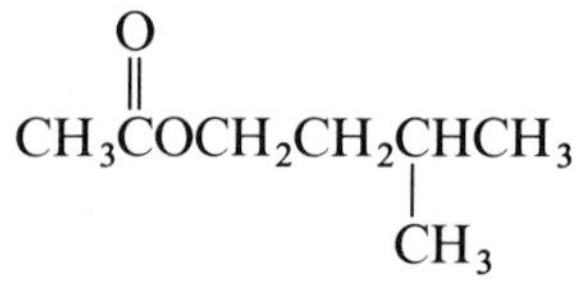

$$CH_3\overset{O}{\overset{\|}{C}}OCH_2CH_2\overset{|}{\underset{CH_3}{CH}}CH_3$$

Isopentyl acetate
(used in synthetic banana flavor)

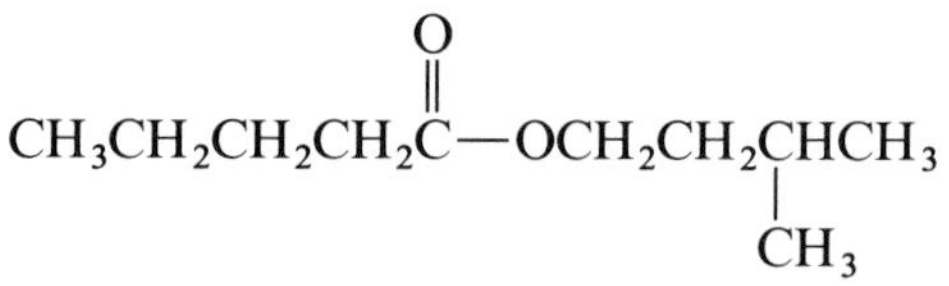

$$CH_3CH_2CH_2CH_2\overset{O}{\overset{\|}{C}}-OCH_2CH_2\overset{|}{\underset{CH_3}{CH}}CH_3$$

Isopentyl pentanoate
(used in synthetic apple flavor)

18.2F Carboxylic Anhydrides

Most anhydrides are named by dropping the word **acid** from the name of the carboxylic acid and then adding the word **anhydride.**

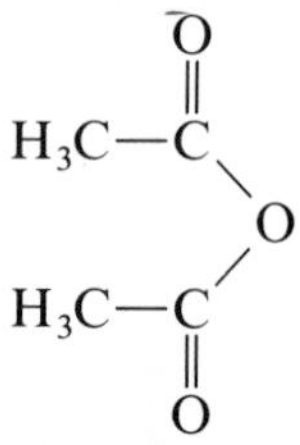

Acetic anhydride
(ethanoic anhydride)
mp, −73 °C

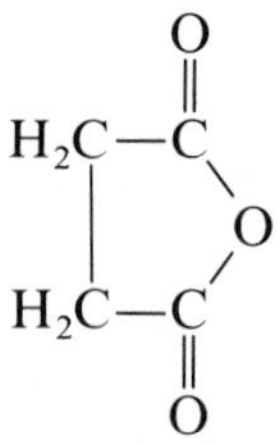

Succinic anhydride
mp, 121 °C

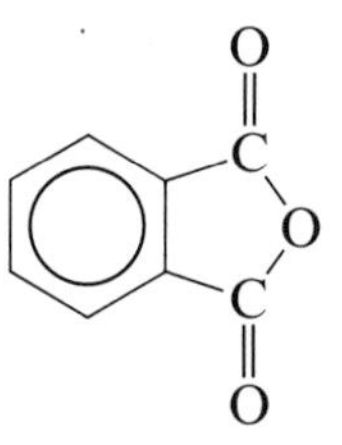

Phthalic anhydride
mp, 131 °C

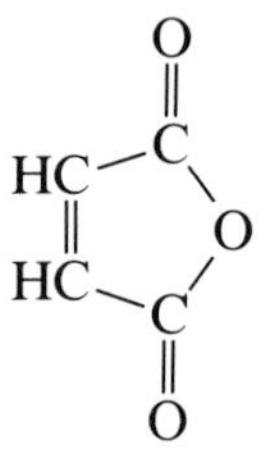

Maleic anhydride
mp, 53 °C

18.2G Acyl Chlorides

Acyl chlorides are also called **acid chlorides.** They are named by dropping **-ic acid** from the name of the acid and then adding **-yl chloride.** Examples are

$$
\underset{\substack{\text{Acetyl chloride} \\ \text{(ethanoyl chloride)} \\ \text{mp, } -112\ ^\circ\text{C; bp, } 51\ ^\circ\text{C}}}{\text{CH}_3\overset{\displaystyle O}{\overset{\|}{\text{C}}}\!\!-\!\text{Cl}}
\qquad
\underset{\substack{\text{Propanoyl chloride} \\ \text{mp, } -94\ ^\circ\text{C; bp, } 80\ ^\circ\text{C}}}{\text{CH}_3\text{CH}_2\overset{\displaystyle O}{\overset{\|}{\text{C}}}\!\!-\!\text{Cl}}
\qquad
\underset{\substack{\text{Benzoyl chloride} \\ \text{mp, } -1\ ^\circ\text{C; bp, } 197\ ^\circ\text{C}}}{\text{C}_6\text{H}_5\overset{\displaystyle O}{\overset{\|}{\text{C}}}\!\!-\!\text{Cl}}
$$

Acyl chlorides and carboxylic anhydrides have boiling points in the same range as esters of comparable molecular weight.

18.2H Amides

Amides that have no substituent on nitrogen are named by dropping **-ic acid** from the common name of the acid (or *-oic acid* from the substitutive name) and then adding **-amide.** Alkyl groups on the nitrogen atom of amides are named as substituents and the named substituent is prefaced by *N-*, or *N,N-*. Examples are

$$
\underset{\substack{\text{Acetamide or} \\ \text{ethanamide} \\ \text{mp, } 82\ ^\circ\text{C; bp, } 221\ ^\circ\text{C}}}{\text{CH}_3\overset{\displaystyle O}{\overset{\|}{\text{C}}}\!\!-\!\text{NH}_2}
\qquad
\underset{\substack{N,N\text{-Dimethylacetamide} \\ \text{mp, } -20\ ^\circ\text{C; bp, } 166\ ^\circ\text{C}}}{\text{CH}_3\overset{\displaystyle O}{\overset{\|}{\text{C}}}\!\!-\!\text{N}\!\!\begin{smallmatrix}\text{CH}_3\\ \\ \text{CH}_3\end{smallmatrix}}
\qquad
\underset{\substack{N\text{-Ethylacetamide} \\ \text{bp, } 205\ ^\circ\text{C}}}{\text{CH}_3\overset{\displaystyle O}{\overset{\|}{\text{C}}}\!\!-\!\text{NHC}_2\text{H}_5}
$$

$$
\underset{\substack{\text{Benzamide} \\ \text{mp, } 130\ ^\circ\text{C; bp, } 290\ ^\circ\text{C}}}{\text{C}_6\text{H}_5\overset{\displaystyle O}{\overset{\|}{\text{C}}}\!\!-\!\text{NH}_2}
$$

Molecules of amides with one (or no) substituent on nitrogen are able to form strong hydrogen bonds to each other and, consequently, such amides have high melting points and boiling points. Molecules of *N,N*-disubstituted amides cannot form strong hydrogen bonds to each other; they have lower melting points and boiling points.

$$
\underset{\substack{\text{R}'}}{\text{R}-\text{C}}\overset{\displaystyle \overset{\cdot\cdot\delta-}{\text{O}}\cdots\overset{\delta+}{\text{H}}-\overset{\displaystyle \text{R}'}{\underset{\displaystyle}{\text{N}}}-\overset{\displaystyle O}{\overset{\|}{\text{C}}}-\text{R}}{\underset{\text{NH}}{\big\backslash}}
$$

Hydrogen bonding
between molecules of an amide

PROBLEM 18.6

Give IUPAC names for the following:

(a) $HO_2C(CH_2)_5CO_2H$ (e) $CH_3CH_2CH_2COCl$

(b) $CH_3CH_2CH_2CO_2CH_3$ (f) $CH_3CH_2CONH_2$

(c) $CH_3CH_2CHClCO_2H$ (g) $CH_3CH_2CH_2CONHCH_3$

(d) $(CH_3CH_2CO)_2O$ (h) $C_6H_5CH_2CH_2CH_2COCl$

PROBLEM 18.7

Write structural formulas for the following:

(a) Methyl propanoate   (e) Dimethyl phthalate

(b) Ethyl *p*-nitrobenzoate   (f) Dipropyl maleate

(c) Dimethyl malonate   (g) *N,N*-Dimethylformamide

(d) *N,N*-Dimethylbenzamide   (h) 2-Bromopropanoyl bromide

18.2I Spectroscopic Properties of Acyl Compounds

Infrared spectroscopy is of considerable importance in identifying carboxylic acids and their derivatives. The $C{=}O$ stretching band is one of the most prominent in their infrared spectra since it is always a strong band. The $C{=}O$ stretching band occurs at different frequencies for acids, esters, and amides, and its precise location is often helpful in structure determination. Table 18.5 gives the location of this band for most acyl compounds.

The hydroxyl groups of carboxylic acids also give rise to a broad peak in the $2500{-}2700\text{-cm}^{-1}$ region arising from O—H stretching vibrations. The N—H stretching vibrations of amides absorb between 3140 and 3500 cm^{-1}.

The acidic protons of carboxylic acids usually absorb very far downfield (δ 10–12) in their proton nmr spectra.

18.3 PREPARATION OF CARBOXYLIC ACIDS

Most of the methods for the preparation of carboxylic acids are familiar ones.

1. By oxidation of alkenes (discussed in Section 7.10).

$$RCH{=}CHR' \xrightarrow[\text{heat}]{\text{KMnO}_4} RCO_2H + R'CO_2H$$

or

$$RCH{=}CHR' \xrightarrow[\text{(2) H}_2\text{O}_2]{\text{(1) O}_3} RCO_2H + R'CO_2H$$

TABLE 18.5 Carbonyl stretching absorptions of acyl compounds

TYPE OF COMPOUND	FREQUENCY RANGE (cm^{-1})
Carboxylic Acids	
$R-CO_2H$	1700–1725
$-\overset{\displaystyle \vert}{C}=C-CO_2H$	1690–1715
$ArCO_2H$	1680–1700
Acid Anhydrides	
$R-\overset{\overset{\displaystyle O}{\|\|}}{C}-O-\overset{\overset{\displaystyle O}{\|\|}}{C}-R$	1800–1850 and 1740–1790
$Ar-\overset{\overset{\displaystyle O}{\|\|}}{C}-O-\overset{\overset{\displaystyle O}{\|\|}}{C}-Ar$	1780–1860 and 1730–1780
Acyl Chlorides	
$R-\overset{\overset{\displaystyle O}{\|\|}}{C}Cl$ and $Ar-\overset{\overset{\displaystyle O}{\|\|}}{C}Cl$	1780–1850
Esters	
$R-\overset{\overset{\displaystyle O}{\|\|}}{C}-OR'$	1735–1750
$Ar-\overset{\overset{\displaystyle O}{\|\|}}{C}-OR'$	1715–1730
Amides	
$R\overset{\overset{\displaystyle O}{\|\|}}{C}-NH_2$, $R\overset{\overset{\displaystyle O}{\|\|}}{C}NHR$, and $R\overset{\overset{\displaystyle O}{\|\|}}{C}NR_2$	1630–1690
Carboxylate Ions	
RCO_2^-	1550–1630

2. **By oxidation of aldehydes and primary alcohols** (discussed in Sections 15.3 and 16.12).

$$R-CHO \xrightarrow[\text{(2) } H^+]{\text{(1) } Ag_2O \text{ or } Ag(NH_3)_2^+OH^-} RCO_2H$$

$$RCH_2OH \xrightarrow[\text{(2) } H^+]{\overset{\text{(1) } KMnO_4,\ OH^-}{\text{heat}}} RCO_2H$$

3. **By oxidation of alkylbenzenes** (discussed in Section 12.10).

$$\text{C}_6\text{H}_5-\text{CH}_3 \xrightarrow[\text{(2) H}^+]{\substack{\text{(1) KMnO}_4,\ \text{OH}^- \\ \text{heat}}} \text{C}_6\text{H}_5-\text{CO}_2\text{H}$$

4. **By oxidation of methyl ketones** (discussed in Section 17.4).

$$\overset{\text{O}}{\overset{\|}{\text{Ar}-\text{C}-\text{CH}_3}} \xrightarrow[\text{(2) H}^+]{\text{(1) X}_2/\text{NaOH}} \overset{\text{O}}{\overset{\|}{\text{Ar}-\text{COH}}} + \text{CHX}_3$$

5. **By hydrolysis of cyanohydrins and other nitriles.** We saw, in Section 16.9A, that aldehydes and ketones can be converted to cyanohydrins, and that these can be hydrolyzed to α-hydroxy acids. In the hydrolysis the $-\text{CN}$ group is converted to a $-\text{CO}_2\text{H}$ group. The mechanism of hydrolysis is discussed in Section 18.8H.

$$\underset{\text{R}}{\overset{\text{R}}{>}}\text{C}=\text{O} + \text{HCN} \rightleftharpoons \underset{\text{R}}{\overset{\text{R}}{>}}\text{C}\underset{\text{CN}}{\overset{\text{OH}}{<}} \xrightarrow[\text{H}_2\text{O}]{\text{H}^+} \text{R}-\overset{\text{OH}}{\underset{\text{R}}{\overset{|}{\underset{|}{\text{C}}}}}-\text{CO}_2\text{H}$$

Nitriles can also be prepared by nucleophilic substitution reactions of alkyl halides with sodium cyanide. Hydrolysis of the nitrile yields a carboxylic acid *with a chain one carbon atom longer* than the original alkyl halide.

General Reaction

$$\text{R}-\text{CH}_2\text{X} + \text{CN}^- \longrightarrow \text{RCH}_2\text{CN} \xrightarrow[\substack{\text{H}_2\text{O} \\ \text{heat}}]{\text{H}^+} \text{RCH}_2\text{CO}_2\text{H} + \text{NH}_4^+$$

$$\xrightarrow[\substack{\text{H}_2\text{O} \\ \text{heat}}]{\text{OH}^-} \text{RCH}_2\text{CO}_2^- + \text{NH}_3$$

Specific Examples

$$\text{HOCH}_2\text{CH}_2\text{Cl} \xrightarrow[(80\%)]{\text{NaCN}} \underset{\substack{\text{3-Hydroxy-} \\ \text{propanenitrile}}}{\text{HOCH}_2\text{CH}_2\text{CN}} \xrightarrow[\substack{\text{(2) H}_3\text{O}^+ \\ (75-80\%)}]{\text{(1) OH}^-,\ \text{H}_2\text{O}} \underset{\substack{\text{3-Hydroxypropanoic} \\ \text{acid}}}{\text{HOCH}_2\text{CH}_2\text{CO}_2\text{H}}$$

$$\text{BrCH}_2\text{CH}_2\text{CH}_2\text{Br} \xrightarrow[(77-86\%)]{\text{NaCN}} \underset{\text{Pentanedinitrile}}{\text{NCCH}_2\text{CH}_2\text{CH}_2\text{CN}} \xrightarrow[(83-85\%)]{\text{H}_3\text{O}^+} \underset{\text{Glutaric acid}}{\text{HO}_2\text{CCH}_2\text{CH}_2\text{CH}_2\text{CO}_2\text{H}}$$

This synthetic method is generally limited to the use of *primary alkyl* halides. The cyanide ion is a relatively strong base, and the use of a secondary or tertiary alkyl halide leads primarily to an alkene (through elimination) rather than to a nitrile (through substitution). Aryl halides (except for those with ortho and para nitro groups) do not react with sodium cyanide.

6. **By carbonation of Grignard reagents.** Grignard reagents react with carbon dioxide to yield magnesium carboxylates. Acidification produces carboxylic acids.

$$R\text{—}X + Mg \xrightarrow[\text{ether}]{} RMgX \xrightarrow{CO_2} RCO_2MgX \xrightarrow{H^+} RCO_2H$$

or

$$Ar\text{—}Br + Mg \xrightarrow[\text{ether}]{} ArMgBr \xrightarrow{CO_2} ArCO_2MgBr \xrightarrow{H^+} ArCO_2H$$

This synthesis of carboxylic acids is applicable to primary, secondary, tertiary, allyl, benzyl, and aryl halides, provided they have no groups incompatible with a Grignard reaction (cf. Section 15.7B).

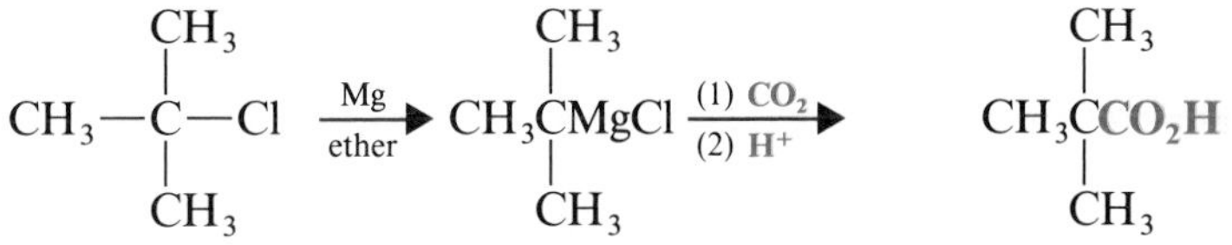

tert-Butyl chloride

(79–80% overall)
2,2-Dimethylpropanoic acid

$$CH_3CH_2CH_2CH_2Cl \xrightarrow[\text{ether}]{Mg} CH_3CH_2CH_2CH_2MgCl \xrightarrow[\text{(2) } H^+]{\text{(1) } CO_2} CH_3CH_2CH_2CH_2CO_2H$$

Butyl chloride

(80% overall)
Pentanoic acid

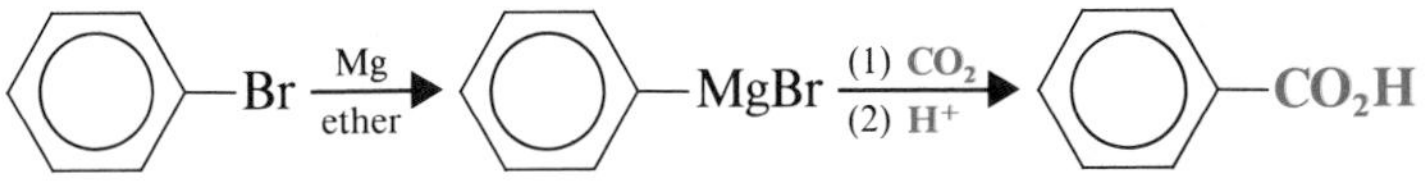

(85%)
Benzoic acid

PROBLEM 18.8

Show how you would prepare each of the following carboxylic acids through a Grignard synthesis.

(a) Phenylacetic acid   (c) 3-Butenoic acid   (e) Hexanoic acid

(b) 2,2-Dimethylpentanoic acid   (d) *p*-Toluic acid

PROBLEM 18.9

(a) Which of the carboxylic acids in Problem 18.8 could be prepared by a nitrile synthesis as well? (b) Which synthesis, Grignard or nitrile, would you choose to prepare $HOCH_2CH_2CH_2CH_2CO_2H$ from $HOCH_2CH_2CH_2CH_2Br$? Why?

18.4 NUCLEOPHILIC SUBSTITUTIONS AT ACYL CARBON

In our study of carbonyl compounds in the last chapter, we saw that a characteristic reaction of aldehydes and ketones is one of *nucleophilic addition* to the carbon–oxygen double bond.

Nucleophilic addition

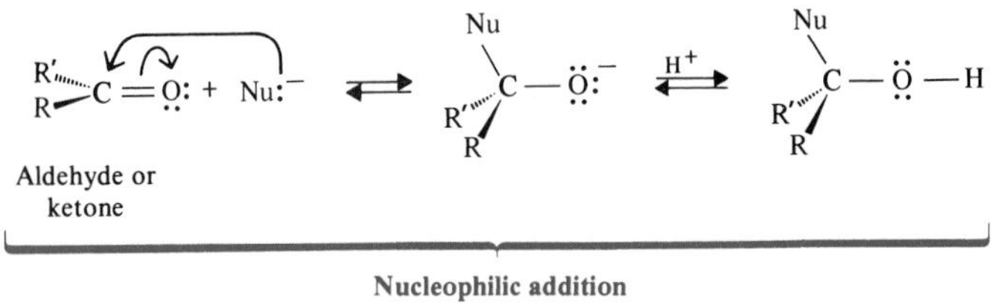

As we study carboxylic acids and their derivatives in this chapter we shall find that their reactions are characterized by **nucleophilic substitutions** that take place at their acyl (carbonyl) carbon atoms. We shall encounter many reactions of the following general type.

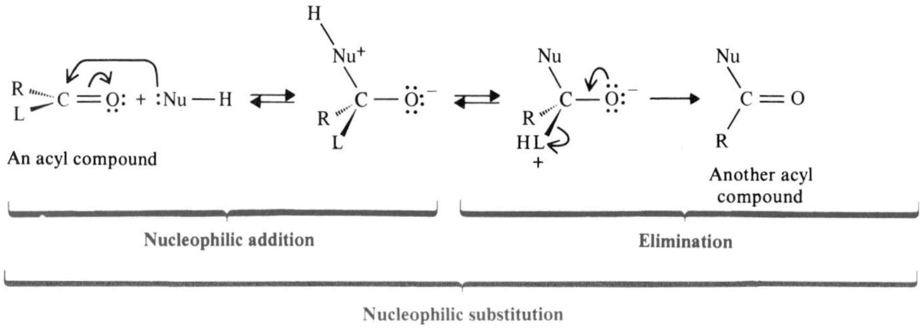

Although the final results obtained from the reactions of acyl compounds with nucleophiles (substitutions) differ from those obtained from aldehydes and ketones (additions), the two reactions have one characteristic in common. *The initial step in both reactions involves nucleophilic addition at the carbonyl carbon atom.* With both groups of compounds this initial attack is facilitated by the same factors: The relative steric openness of the carbonyl carbon atom and the ability of the carbonyl oxygen atom to accommodate an electron pair of the carbon–oxygen double bond.

It is after the initial nucleophilic attack has taken place that the two reactions differ. The tetrahedral intermediate formed from an aldehyde or ketone usually accepts a proton to form a stable addition product. By contrast, the intermediate formed from an acyl compound usually *eliminates* a leaving group; this **elimination** leads to regeneration of the carbon–oxygen double bond and a *substitution product.* The overall process in the case of **acyl substitution** occurs, therefore by a **nucleophilic addition–elimination** mechanism.

Acyl compounds react as they do because they all have good leaving groups attached to the carbonyl carbon atom: An acyl chloride, for example, generally reacts by losing *a chloride ion*—a very weak base, and thus, a very good leaving group.

EXAMPLE

The reaction of an acyl chloride with water.

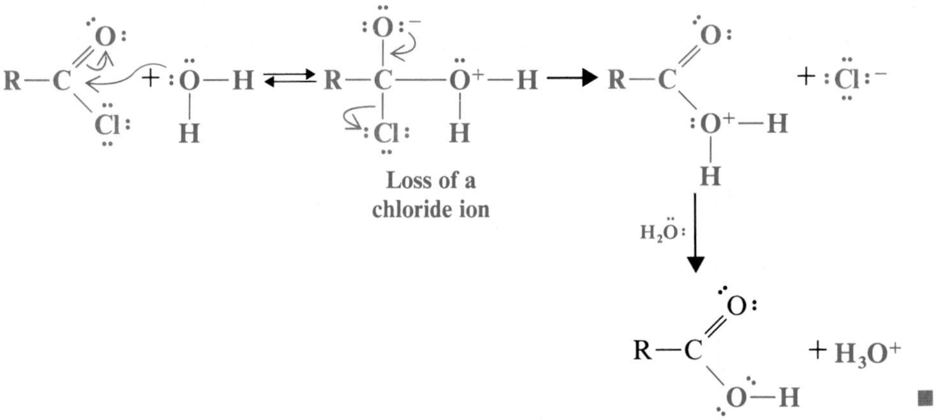

Loss of a
chloride ion

An acid anhydride generally reacts by losing *a carboxylate ion* or a molecule of a *carboxylic acid*—both are weak bases and good leaving groups.

EXAMPLE

The reaction of a carboxylic acid anhydride with an alcohol.

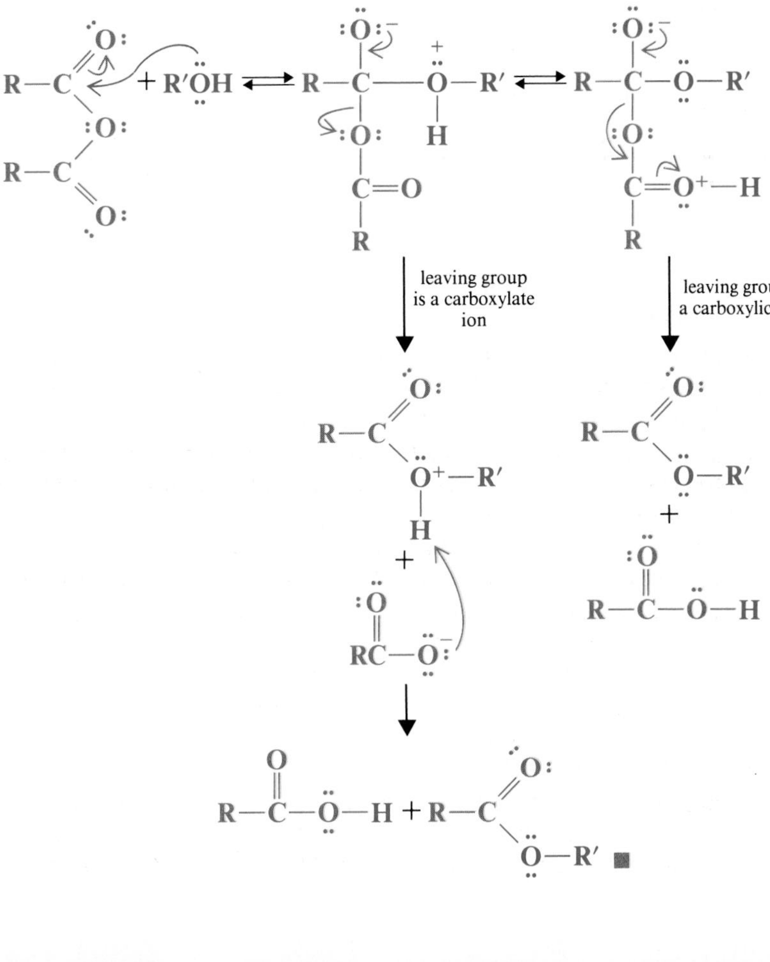

leaving group
is a carboxylate
ion

leaving group is
a carboxylic acid

As we shall see later, esters generally undergo nucleophilic substitution by losing a molecule of an *alcohol,* acids react by losing a molecule of *water,* and amides react by losing a molecule of *ammonia* or of an *amine.* All of the molecules lost in these reactions are weak bases and are reasonably good leaving groups.

For an aldehyde or ketone to react by substitution, the tetrahedral intermediate would have to eject a hydride ion ($H:^-$) or an alkanide ion ($R:^-$). Both are *very powerful bases* and both are, therefore, *very poor leaving groups.*

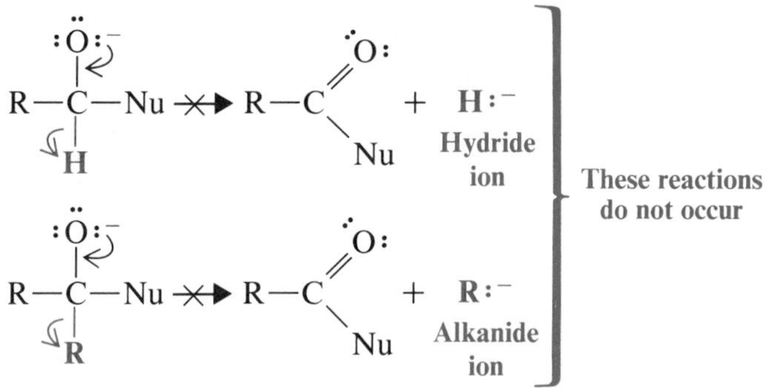

18.4A Relative Reactivity of Acyl Compounds

Of the acid derivatives that we shall study in this chapter, acyl chlorides are the most reactive toward nucleophilic substitution and amides are the least reactive. In general, the overall order of reactivity is

| Acyl chloride | Acid anhydride | Ester | Amide |

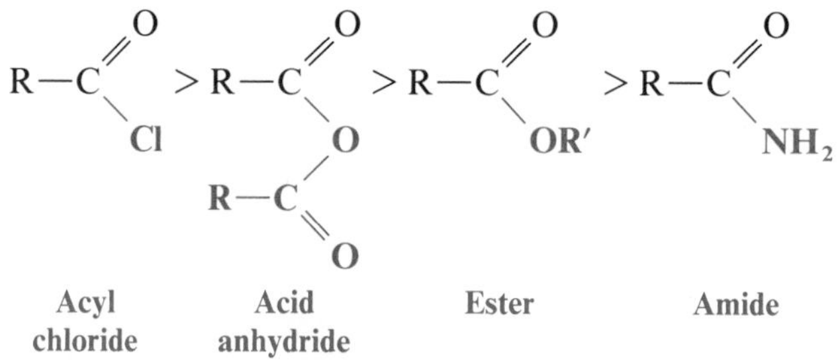

One way that we can account for this overall order of reactivity is by taking into account the basicity of the leaving group in each instance.

The general order of reactivity of acid derivatives roughly parallels the basicities of the leaving groups. When acyl chlorides react, the leaving group is a *chloride ion.* When acid anhydrides react, the leaving group is a carboxylic acid or a carboxylate ion. When esters react, the leaving group is an alcohol, and when amides react, the leaving group is an amine (or ammonia). Of all of these bases, chloride ions are the weakest bases and acyl chlorides are the most reactive acyl compounds. Amines (or ammonia) are the strongest bases and amides are the least reactive acyl compounds.

18.4B Synthesis of Acid Derivatives

As we begin now to explore the syntheses of carboxylic acid derivatives we shall find that in many instances one acid derivative can be synthesized through a nucleophilic substitution reaction of another. The order of reactivities that we have presented gives us a clue as to which syntheses are practical and which are not. In general, ***less***

reactive acyl compounds can be synthesized from more reactive ones, but the reverse is usually difficult and, when possible, requires special conditions.

18.5 SYNTHESIS OF ACYL CHLORIDES

Since acyl chlorides are the most reactive of the acid derivatives, we must use special reagents to prepare them. We use other acid chlorides, *the acid chlorides of inorganic acids:* we use $\mathbf{PCl_5}$ (an acid chloride of phosphoric acid), $\mathbf{PCl_3}$ (an acid chloride of phosphorous acid), and $\mathbf{SOCl_2}$ (an acid chloride of sulfurous acid).

All of these reagents react with carboxylic acids to give acyl chlorides in good yield.

General Reactions

$$\mathrm{R\overset{O}{\overset{\|}{C}}OH} + \underset{\text{Thionyl chloride}}{\mathrm{SOCl_2}} \longrightarrow \mathrm{R{-}\overset{O}{\overset{\|}{C}}{-}Cl} + \mathrm{SO_2} + \mathrm{HCl}$$

$$\mathrm{3R\overset{O}{\overset{\|}{C}}OH} + \underset{\text{Phosphorus trichloride}}{\mathrm{PCl_3}} \longrightarrow \mathrm{3R\overset{O}{\overset{\|}{C}}Cl} + \mathrm{H_3PO_3}$$

$$\mathrm{R\overset{O}{\overset{\|}{C}}OH} + \underset{\text{Phosphorus pentachloride}}{\mathrm{PCl_5}} \longrightarrow \mathrm{R\overset{O}{\overset{\|}{C}}Cl} + \mathrm{POCl_3} + \mathrm{HCl}$$

These reactions all involve nucleophilic substitutions by chloride ion on a highly reactive intermediate: an acyl chlorosulfite, an acyl chlorophosphite, or an acyl chlorophosphate. Thionyl chloride, for example, reacts with a carboxylic acid in the following way:

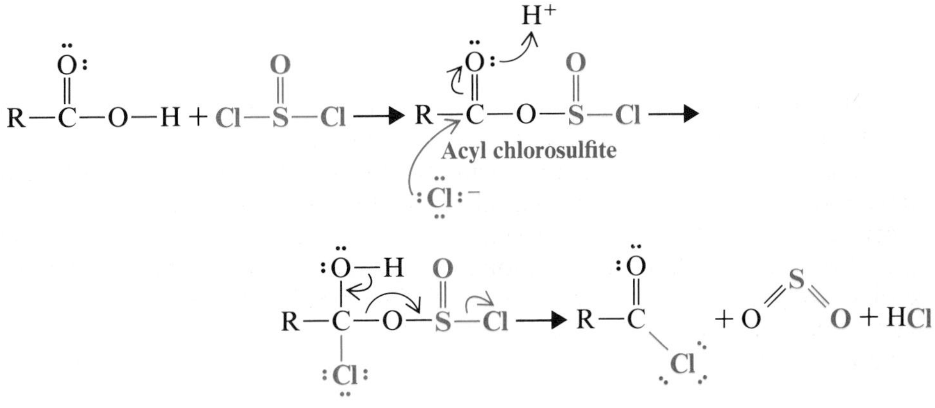

18.6 SYNTHESIS OF CARBOXYLIC ACID ANHYDRIDES

Carboxylic acids react with acyl chlorides in the presence of pyridine to give carboxylic acid anhydrides.

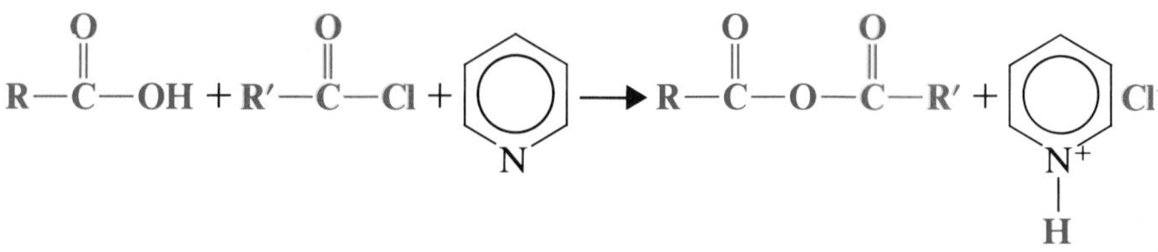

This method is frequently used in the laboratory for the preparation of anhydrides. The method is quite general and can be used to prepare mixed anhydrides (R ≠ R′) or simple anhydrides (R = R′).

Sodium salts of carboxylic acids also react with acyl chlorides to give anhydrides:

In this reaction a carboxylate ion acts as a nucleophile and brings about a nucleophilic substitution reaction at the acyl carbon of the acyl chloride.

Cyclic anhydrides can sometimes be prepared by simply heating the appropriate dicarboxylic acid. This method succeeds, however, only when anhydride formation leads to a five- or six-membered ring.

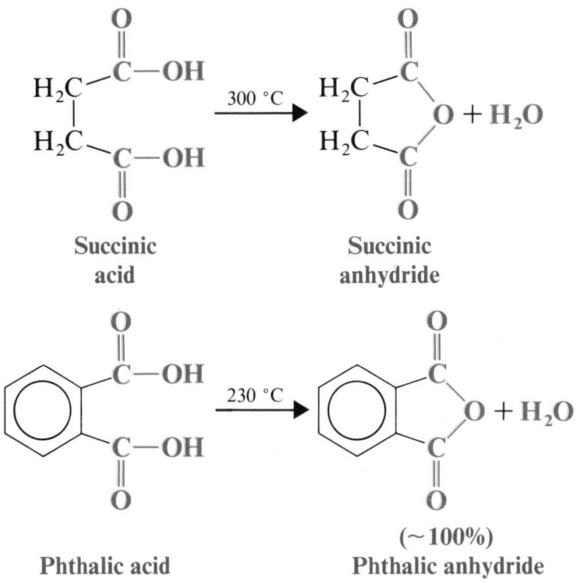

Succinic acid → Succinic anhydride

Phthalic acid → Phthalic anhydride (~100%)

PROBLEM 18.10

When maleic acid is heated to 200 °C, it loses water and becomes maleic anhydride. Fumaric acid, a diastereomer of maleic acid, requires a much higher temperature before it dehydrates; when it does it also yields maleic anhydride. Provide an explanation for these observations.

18.7 ESTERS

18.7A Synthesis of Esters: Esterification

Carboxylic acids react with alcohols to form esters through a condensation reaction known as **esterification:**

General Reaction

$$\underset{R-C-OH}{\overset{O}{\parallel}} + R'-OH \underset{\longrightarrow}{\overset{H^+}{\rightleftarrows}} \underset{R-C-OR'}{\overset{O}{\parallel}} + H_2O$$

Specific Examples

$$\underset{CH_3COH}{\overset{O}{\parallel}} + CH_3CH_2OH \overset{H^+}{\rightleftarrows} \underset{CH_3COCH_2CH_3}{\overset{O}{\parallel}} + H_2O$$
Acetic acid   Ethyl alcohol   Ethyl acetate

$$\underset{C_6H_5COH}{\overset{O}{\parallel}} + \quad CH_3OH \quad \overset{H^+}{\rightleftarrows} \underset{C_6H_5COCH_3}{\overset{O}{\parallel}} + H_2O$$
Benzoic acid   Methyl alcohol   Methyl benzoate

Esterification reactions are acid catalyzed. They proceed very slowly in the absence of strong acids, but reach equilibrium within a matter of a few hours when an acid and an alcohol are refluxed with a small amount of concentrated sulfuric acid or hydrogen chloride. Since the position of equilibrium controls the amount of the ester formed, the use of an excess of either the carboxylic acid or the alcohol increases the yield based on the limiting reagent. Just which component we choose to use in excess will depend on its availability and cost. The yield of an esterification reaction can also be increased by removing water from the reaction mixture as it is formed.

When benzoic acid reacts with methanol that has been labeled with ^{18}O, the labeled oxygen appears in the ester: This result reveals just which bonds break in the esterification.

$$\underset{C_6H_5C}{\overset{O}{\parallel}}{\vdash}OH + CH_3-^{18}O{\vdash}H \overset{H^+}{\rightleftarrows} \underset{C_6H_5C}{\overset{O}{\parallel}}-^{18}OCH_3 + H_2O$$

The results of the labeling experiment and the fact that esterifications are acid catalyzed are both consistent with the mechanism that follows. This mechanism is typical of acid-catalyzed nucleophilic substitution reactions at acyl carbon atoms.

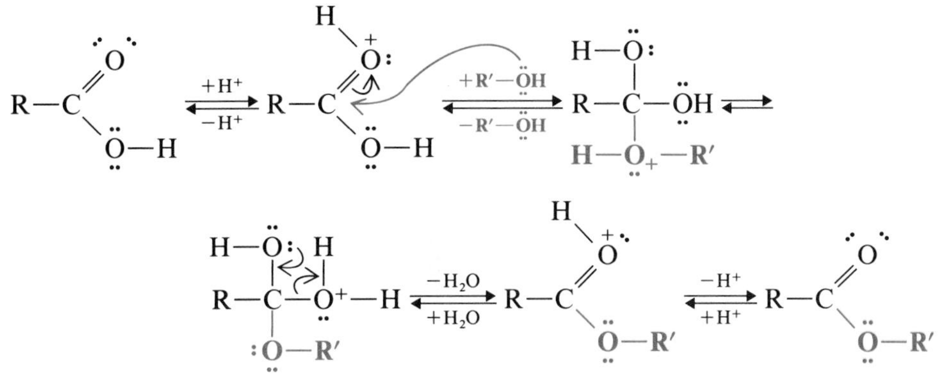

If we follow the forward reactions in this mechanism, we have the mechanism for the *acid-catalyzed esterification of an acid.* If, however, we follow the reverse reactions, we have the mechanism for the *acid-catalyzed hydrolysis of an ester:*

$$
\underset{\substack{\| \\ O}}{R-C-OR'} + H_2O \overset{H_3O^+}{\rightleftharpoons} \underset{\substack{\| \\ O}}{R-C-OH} + R'-OH
$$

Which result we obtain will depend on the conditions we choose. If we want to esterify an acid, we use an excess of the alcohol and if possible remove the water as it is formed. If we want to hydrolyze an ester, we use a large excess of water; that is, we reflux the ester with dilute aqueous HCl or dilute aqueous H_2SO_4.

PROBLEM 18.11

Where would you expect to find the labeled oxygen if you carried out an acid-catalyzed hydrolysis of methyl benzoate in ^{18}O-labeled water?

Steric factors strongly affect the rates of acid-catalyzed hydrolyses of esters. Large groups near the reaction site, whether in the alcohol component or the acid component, slow both reactions markedly. Tertiary alcohols, for example, react so slowly in esterifications that they usually undergo elimination instead.

Esters from Acyl Chlorides. Esters can also be synthesized by the reaction of acyl chlorides with alcohols. Since acyl chlorides are much more reactive toward nucleophilic substitution than carboxylic acids, the reaction of an acyl chloride and an alcohol occurs rapidly and does not require an acid catalyst. Pyridine is usually added to the reaction mixture to react with the HCl that forms.

General Reaction

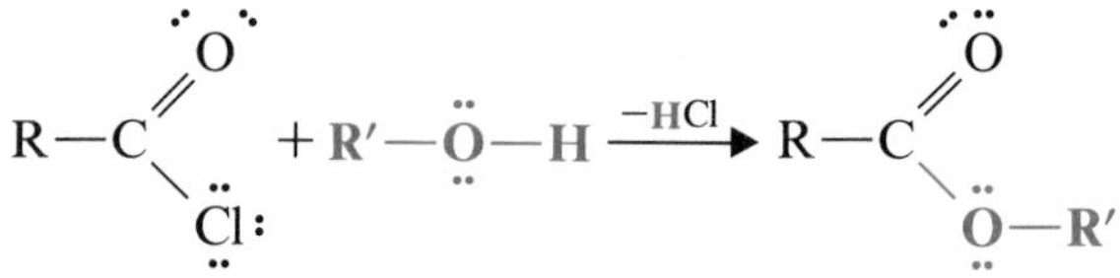

Specific Example

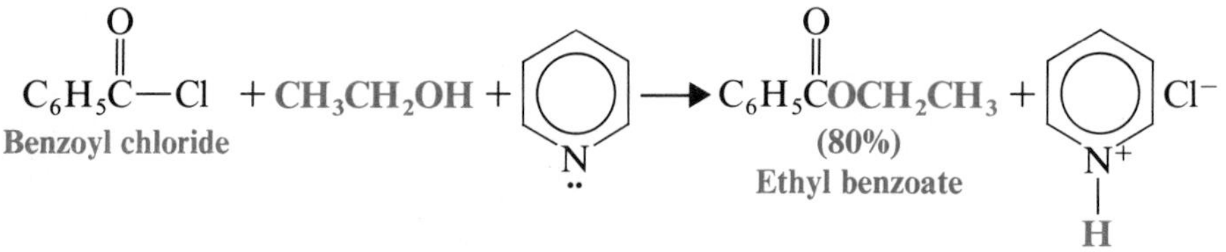

Benzoyl chloride   Ethyl benzoate   (80%)

Esters from Carboxylic Acid Anhydrides. Carboxylic acid anhydrides also react with alcohols to form esters in the absence of an acid catalyst.

General Reaction

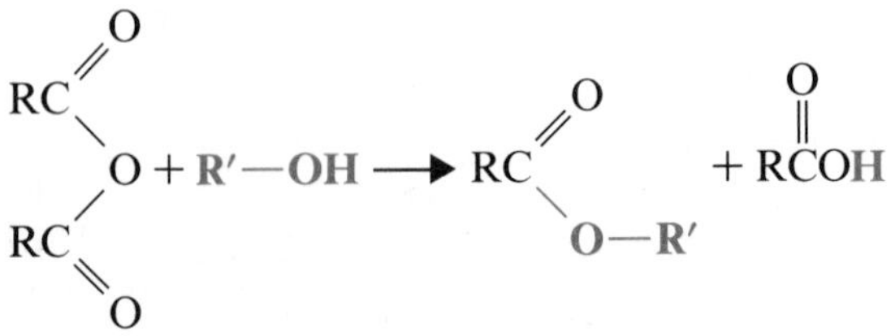

Specific Example

$$(CH_3\overset{O}{\underset{||}{C}})_2O + C_6H_5CH_2OH \longrightarrow CH_3\overset{O}{\underset{||}{C}}OCH_2C_6H_5 + CH_3CO_2H$$

Acetic   Benzyl   Benzyl acetate
anhydride   alcohol

The reaction of an alcohol with an anhydride or an acyl chloride is often the best method for preparing an ester.

Cyclic anhydrides react with one molar equivalent of an alcohol to form compounds that are both *esters and acids.*

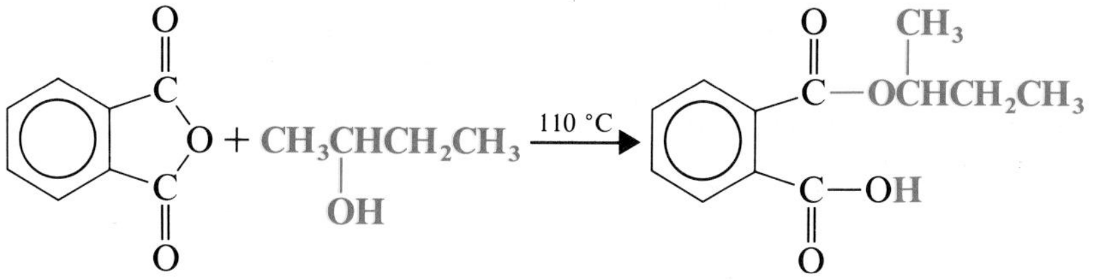

Phthalic   *sec*-Butyl alcohol   (97%)
anhydride   *sec*-Butyl hydrogen phthalate

PROBLEM 18.12

Esters can also be synthesized by *transesterification:*

$$R-\overset{O}{\underset{||}{C}}-OR' + R''-OH \underset{}{\overset{H^+, \text{ heat}}{\rightleftharpoons}} R\overset{O}{\underset{||}{C}}-OR'' + R'-OH$$

High-boiling   High-boiling   Higher-boiling   Low-boiling
ester   alcohol   ester   alcohol

In this procedure we shift the equilibrium to the right by allowing the low-boiling alcohol to distill from the reaction mixture. The mechanism for transesterification is similar to that for an acid-catalyzed esterification (or an acid-catalyzed ester hydrolysis). Write a mechanism for the following transesterification.

$$CH_2{=}CH\overset{O}{\underset{||}{C}}OCH_3 + CH_3CH_2CH_2CH_2OH \overset{H^+}{\rightleftharpoons}$$

Methyl acrylate   Butyl alcohol

$$CH_2{=}CH\overset{O}{\underset{||}{C}}OCH_2CH_2CH_2CH_3 + CH_3OH$$

(94%)   Methyl
Butyl acrylate   alcohol

18.7B Base-Promoted Hydrolysis of Esters: Saponification

Esters not only undergo acid hydrolysis, they also undergo *base-promoted hydrolysis.* Base-promoted hydrolysis is sometimes called *saponification.* Refluxing an ester with aqueous sodium hydroxide, for example, produces an alcohol and the sodium salt of the acid:

$$RC\overset{O}{\underset{\parallel}{}}-OR' + NaOH \xrightarrow{H_2O} R-\overset{O}{\underset{\parallel}{C}}-O^-Na^+ + R'OH$$

Ester   Sodium carboxylate   Alcohol

The carboxylate ion is very unreactive toward nucleophilic substitution because it is negatively charged. Base-promoted hydrolysis of an ester, as a result, is an essentially irreversible reaction.

The mechanism for the base-promoted hydrolysis of an ester also involves a nucleophilic substitution at the acyl carbon:

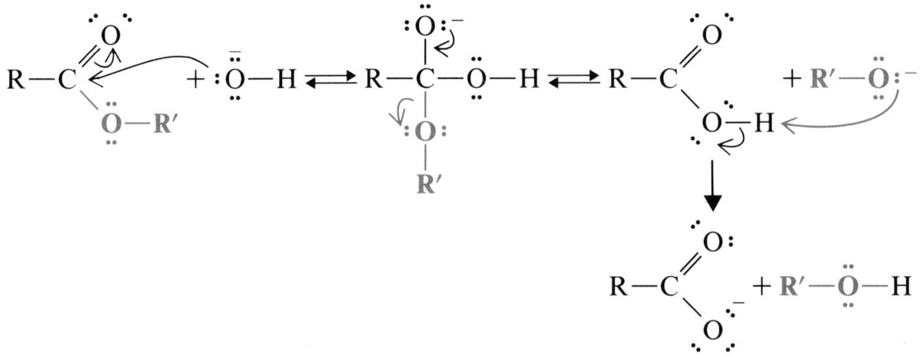

Part of the evidence that nucleophilic attack occurs at the acyl carbon comes from studies in which esters of chiral alcohols were subjected to base-promoted hydrolysis. The reaction of an ester with hydroxide ion could conceivably occur in two ways: reaction could take place through a nucleophilic substitution at the acyl carbon of the acid component (path A) or through a nucleophilic substitution at the alkyl carbon of the alcohol portion (path B). Reaction by path A should lead to retention of configuration in the alcohol. Reaction by path B should lead to an inversion of configuration of the alcohol. *Inversion of configuration is almost never observed.* In almost every instance basic hydrolysis of a carboxylic ester of a chiral alcohol proceeds with *retention of configuration.*

Path A: Nucleophilic substitution at the acyl carbon

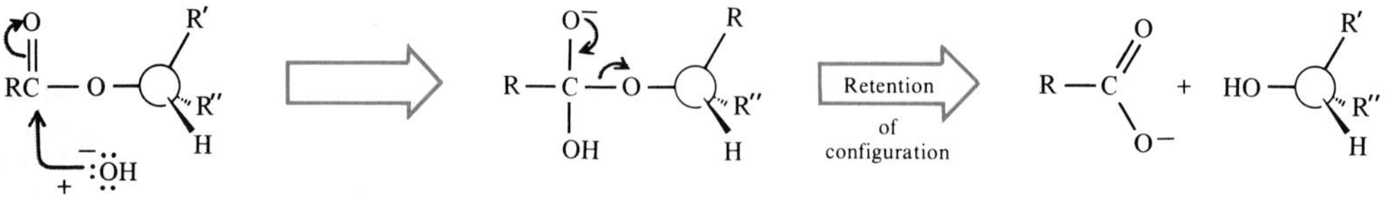

Path B: Nucleophilic substitution at the alkyl carbon

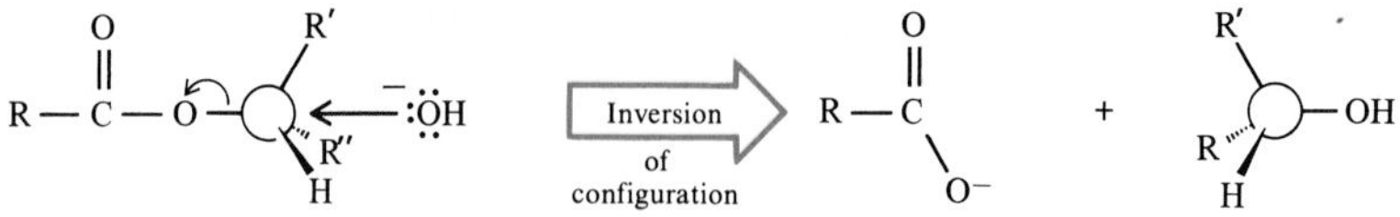

(This reaction seldom occurs with esters of carboxylic acids)

Although nucleophilic attack at the alkyl carbon seldom occurs with esters of carboxylic acids, it is the preferred mode of attack with esters of sulfonic acids (Section 8.11).

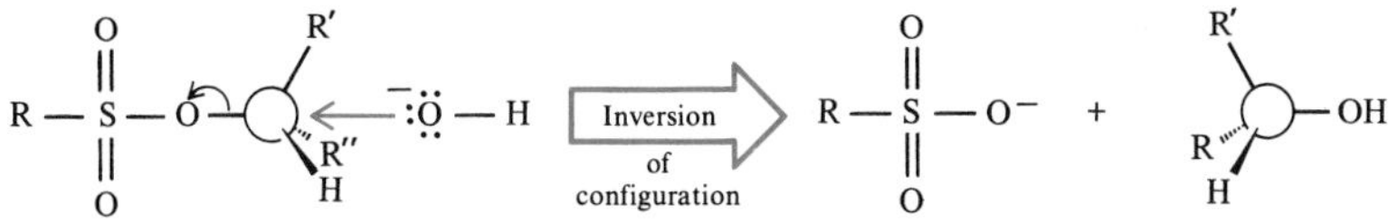

An alkyl sulfonate

PROBLEM 18.13

(a) Write stereochemical formulas for compounds **A** to **F**.

1. *cis*-3-Methylcyclopentanol + $C_6H_5SO_2Cl \longrightarrow$

$$\textbf{A} \xrightarrow[\text{heat}]{OH^-} \textbf{B} + C_6H_5SO_3^-$$

2. *cis*-3-Methylcyclopentanol + $C_6H_5\overset{\displaystyle O}{\overset{\displaystyle \|}{C}}-Cl \longrightarrow$

$$\textbf{C} \xrightarrow[\text{reflux}]{OH^-} \textbf{D} + C_6H_5CO_2^-$$

3. (R)-2-Bromooctane + $CH_3CO_2^-Na^+ \longrightarrow \textbf{E} + NaBr$

$$\downarrow \begin{array}{c} OH^-, H_2O \\ (\text{reflux}) \end{array}$$

$$\textbf{F}$$

4. (R)-2-Bromooctane + $OH^- \xrightarrow{\text{alcohol}} \textbf{F} + Br^-$

(b) Which of the last two methods, **(3)** or **(4)**, would you expect to give a higher yield of **F**? Why?

PROBLEM 18.14

Base-promoted hydrolysis of methyl mesitoate occurs through an attack on the alcohol carbon instead of the acyl carbon.

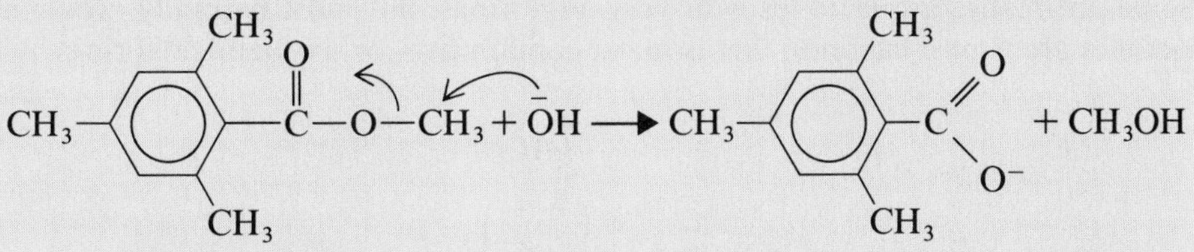

Methyl mesitoate

(a) Can you suggest a reason that will account for this unusual behavior?
(b) Suggest an experiment with labeled compounds that would confirm this mode of attack.

18.7C Lactones

Carboxylic acids whose molecules have a hydroxyl group on a γ- or δ-carbon undergo an intramolecular esterification to give cyclic esters known as γ- or *δ-lactones.*

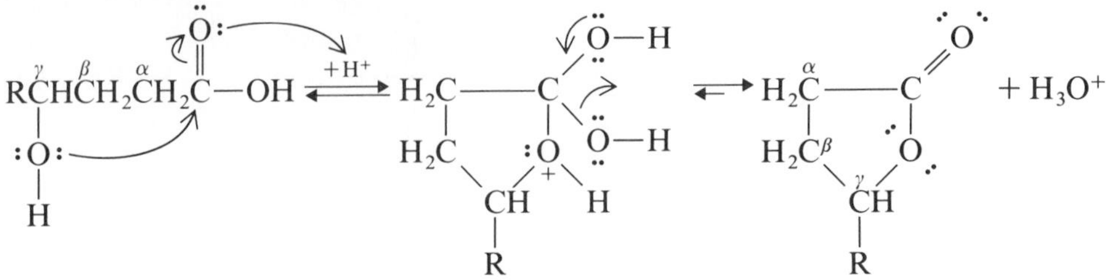

A γ-hydroxy acid   A γ-lactone

A δ-hydroxy acid   A δ-lactone

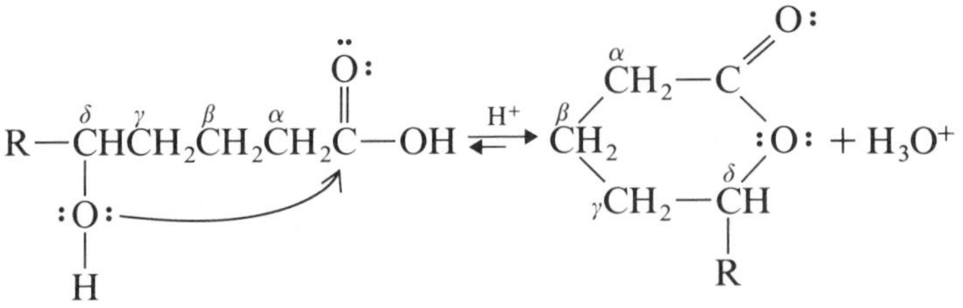

Lactones are hydrolyzed by aqueous base just as other esters are. Acidification of the sodium salt, however, may lead spontaneously back to the γ- or δ-lactone, particularly if excess acid is used.

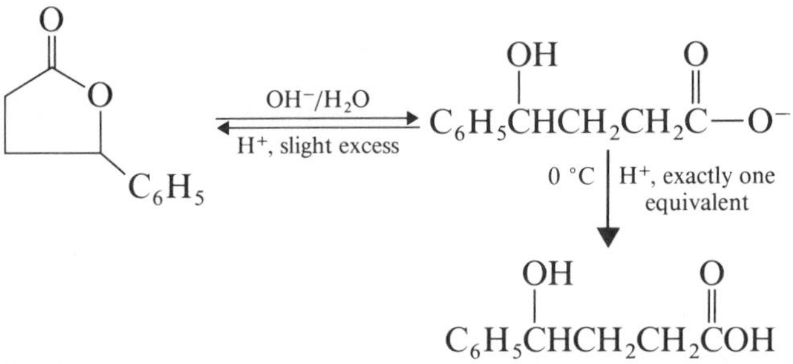

$$\xrightarrow[\text{H}^+,\text{ slight excess}]{\text{OH}^-/\text{H}_2\text{O}} \quad C_6H_5\overset{\text{OH}}{\underset{}{\text{CH}}}CH_2CH_2\overset{\text{O}}{\underset{}{\text{C}}}\!\!-\!\text{O}^-$$

$$0\ ^\circ\text{C} \;\Big|\; \text{H}^+,\text{ exactly one equivalent}$$

$$C_6H_5\overset{\text{OH}}{\underset{}{\text{CH}}}CH_2CH_2\overset{\text{O}}{\underset{}{\text{C}}}\text{OH}$$

Many lactones occur in nature. Vitamin C (below), for example, is a γ-lactone. Some antibiotics are lactones with very large rings, but most naturally occurring lactones are γ- or δ-lactones; that is, most contain five- or six-membered rings.

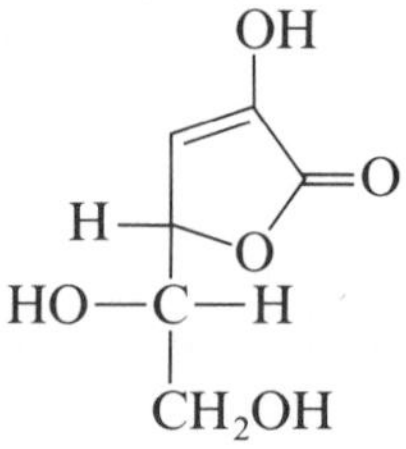

Vitamin C

β-Lactones (lactones with four-membered rings) have been detected as intermediates in some reactions and several have been isolated. They are highly reactive, however. If one attempts to prepare a β-lactone from a β-hydroxy acid, β elimination usually occurs instead:

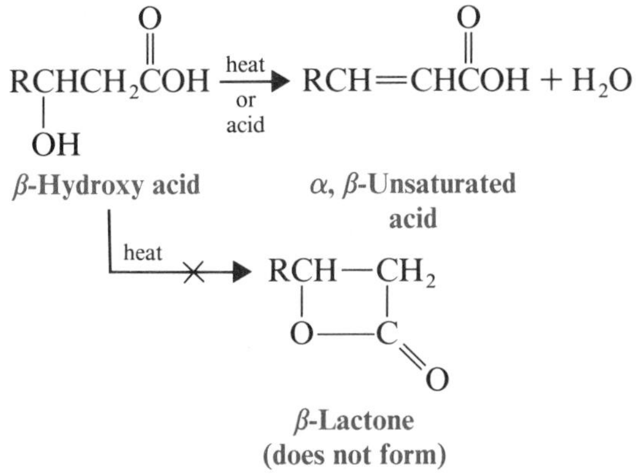

$$\underset{\substack{| \\ \text{OH}}}{\text{RCHCH}_2\overset{\displaystyle O}{\overset{\|}{\text{C}}}\text{OH}} \xrightarrow[\substack{\text{or} \\ \text{acid}}]{\text{heat}} \text{RCH}{=}\text{CH}\overset{\displaystyle O}{\overset{\|}{\text{C}}}\text{OH} + \text{H}_2\text{O}$$

β-Hydroxy acid   α, β-Unsaturated
                                         acid

heat ──✗──▶ RCH──CH₂
                          | |
                          O──C
                                  ‖
                                  O

β-Lactone
(does not form)

When α-hydroxy acids are heated, they form cyclic diesters called *lactides.*

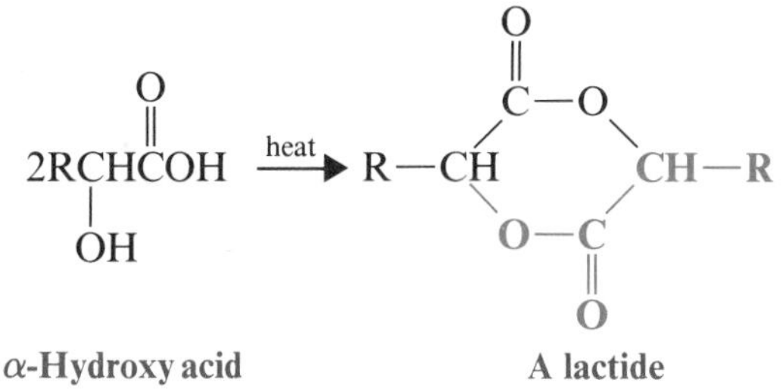

$$2\underset{\substack{| \\ \text{OH}}}{\text{RCH}\overset{\displaystyle O}{\overset{\|}{\text{C}}}\text{OH}} \xrightarrow{\text{heat}} \text{R}{-}\text{CH}\underset{\substack{\\ \text{O}{-}\text{C} \\ \|\\ \text{O}}}{\overset{\substack{\text{O} \\ \|\\ \text{C}{-}\text{O}}}{}} \text{CH}{-}\text{R}$$

α-Hydroxy acid   A lactide

α-Lactones occur as intermediates in some reactions (cf. Special Topic L).

18.8 AMIDES

18.8A Synthesis of Amides

Amides can be prepared in a variety of ways starting with acyl chlorides, acid anhydrides, esters, carboxylic acids, and carboxylic salts. All of these methods involve nucleophilic substitution reactions by ammonia or an amine at an acyl carbon. As we might expect, acid chlorides are the most reactive and carboxylate ions are the least.

18.8B Amides from Acyl Chlorides

Primary amines, secondary amines, and ammonia all react rapidly with acid chlorides to form amides:

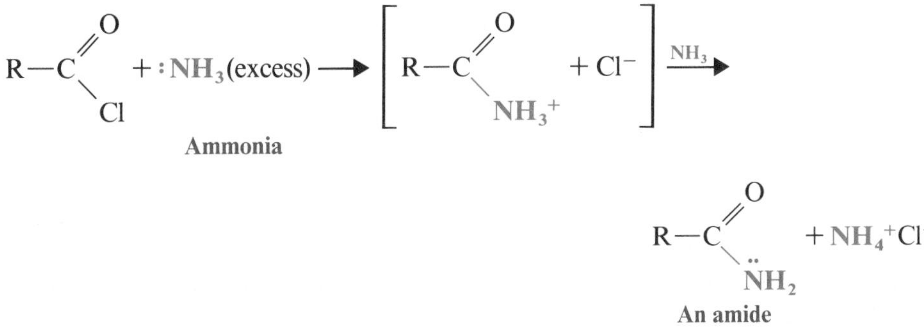

$$\text{R}{-}\overset{\displaystyle O}{\overset{\|}{\underset{\displaystyle \text{Cl}}{\text{C}}}} + :\text{NH}_3(\text{excess}) \longrightarrow \left[\text{R}{-}\overset{\displaystyle O}{\overset{\|}{\underset{\displaystyle \text{NH}_3{}^+}{\text{C}}}} + \text{Cl}^- \right] \xrightarrow{\text{NH}_3}$$

Ammonia

$$\text{R}{-}\overset{\displaystyle O}{\overset{\|}{\underset{\displaystyle \ddot{\text{N}}\text{H}_2}{\text{C}}}} + \text{NH}_4{}^+\text{Cl}^-$$

An amide

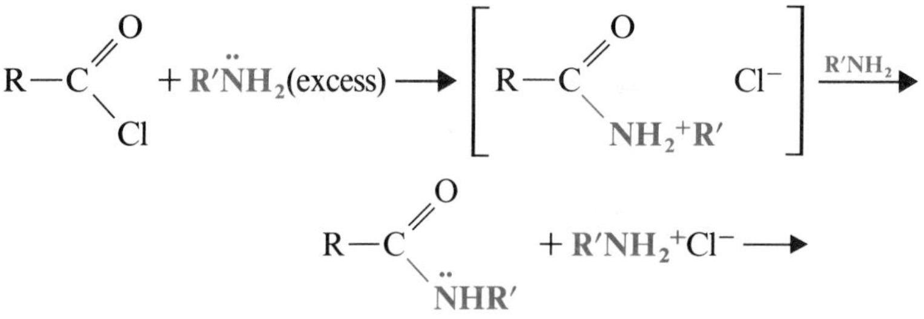

$$R-C\underset{Cl}{\overset{O}{\Vert}} + R'\ddot{N}H_2(\text{excess}) \longrightarrow \left[R-C\underset{NH_2^+R'}{\overset{O}{\Vert}} \quad Cl^- \right] \xrightarrow{R'NH_2}$$

$$R-C\underset{\ddot{N}HR'}{\overset{O}{\Vert}} + R'NH_2^+Cl^- \longrightarrow$$

An N-substituted
amide

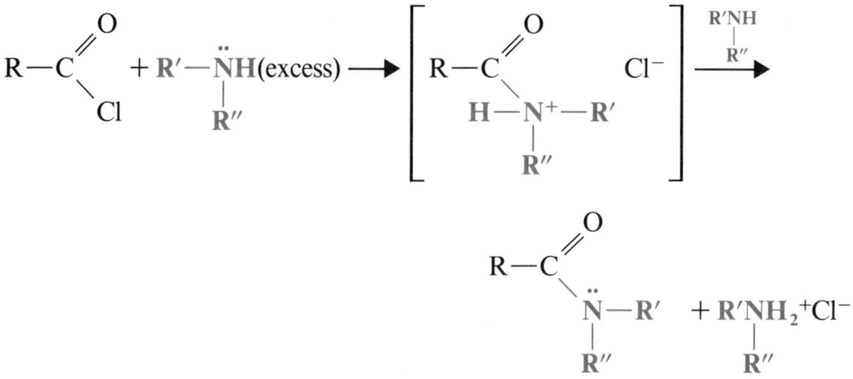

$$R-C\underset{Cl}{\overset{O}{\Vert}} + R'-\underset{R''}{\ddot{N}H}(\text{excess}) \longrightarrow \left[R-C\underset{H-N^+-R'}{\overset{O}{\Vert}} \quad Cl^- \underset{R''}{} \right] \xrightarrow[R'']{R'NH}$$

$$R-C\underset{\underset{R''}{\ddot{N}-R'}}{\overset{O}{\Vert}} + R'NH_2^+Cl^-$$

An N,N-disubstituted
amide

Since acyl chlorides are easily prepared from carboxylic acids this is one of the most widely used laboratory methods for the synthesis of amides. The reaction between the acyl chloride and the amine (or ammonia) usually takes place at room temperature (or below) and produces the amide in high yield.

Acyl chlorides also react with tertiary amines by a nucleophilic substitution reaction. The acylammonium ion that forms, however, is not stable in the presence of water or any hydroxylic solvent.

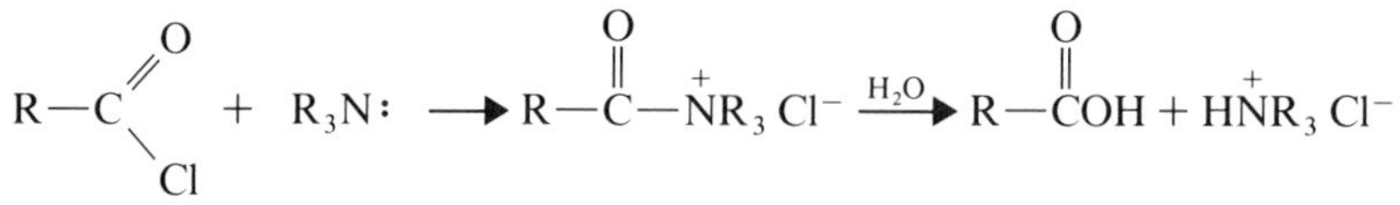

$$R-C\underset{Cl}{\overset{O}{\Vert}} + R_3N\colon \longrightarrow R-\overset{O}{\overset{\Vert}{C}}-\overset{+}{N}R_3\ Cl^- \xrightarrow{H_2O} R-\overset{O}{\overset{\Vert}{C}}OH + H\overset{+}{N}R_3\ Cl^-$$

Acyl chloride   3° Amine   Acylammonium ion

Acylpyridinium ions are probably involved as intermediates in those reactions of acyl chlorides that are carried out in the presence of pyridine.

18.8C Amides from Carboxylic Anhydrides

Acid anhydrides react with ammonia and with primary and secondary amines and form amides through reactions that are analogous to those of acyl chlorides.

$$\left(\overset{O}{\underset{RC}{\overset{\Vert}{}}} \right)_2 O + 2\ddot{N}H_3 \longrightarrow R\overset{O}{\overset{\Vert}{C}}-\ddot{N}H_2 + RCO_2^-NH_4^+$$

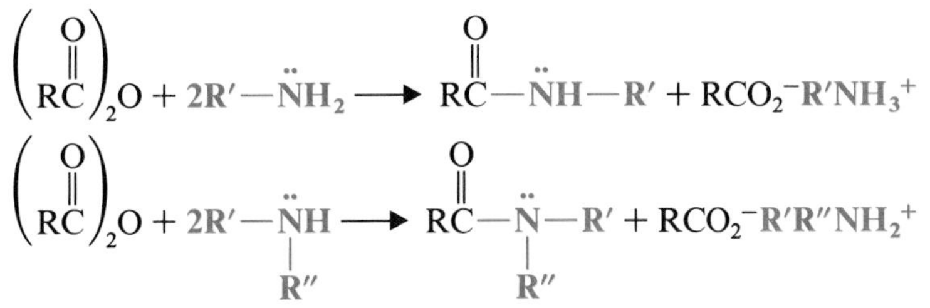

$$\left(\overset{\overset{\displaystyle O}{\parallel}}{RC}\right)_2 O + 2R'{-}\ddot{N}H_2 \longrightarrow \overset{\overset{\displaystyle O}{\parallel}}{RC}{-}\ddot{N}H{-}R' + RCO_2^-R'NH_3^+$$

$$\left(\overset{\overset{\displaystyle O}{\parallel}}{RC}\right)_2 O + 2R'{-}\underset{\underset{\displaystyle R''}{|}}{\ddot{N}H} \longrightarrow \overset{\overset{\displaystyle O}{\parallel}}{RC}{-}\underset{\underset{\displaystyle R''}{|}}{\ddot{N}}{-}R' + RCO_2^-R'R''NH_2^+$$

Cyclic anhydrides react with ammonia or an amine in the same general way as acyclic anhydrides; however, the reaction produces a product that is both an amide and an ammonium salt. Acidifying the ammonium salt gives a compound that is both an amide and an acid:

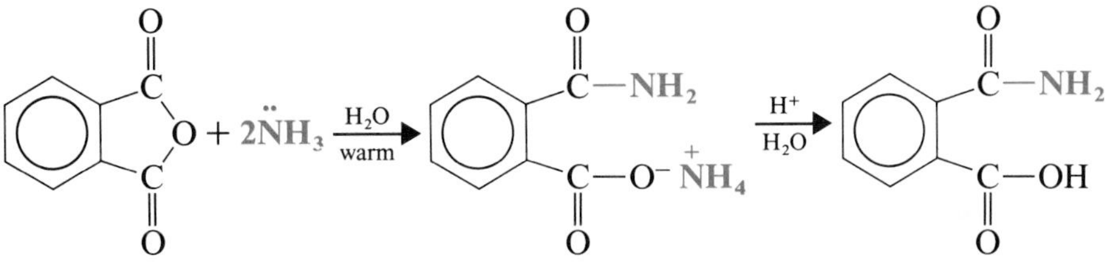

| Phthalic anhydride | (94%) Ammonium phthalamate | (81%) Phthalamic acid |

Heating the amide acid causes dehydration to occur and gives an *imide*. Imides contain the linkage $-\overset{\overset{\displaystyle O}{\parallel}}{C}{-}NH{-}\overset{\overset{\displaystyle O}{\parallel}}{C}{-}$.

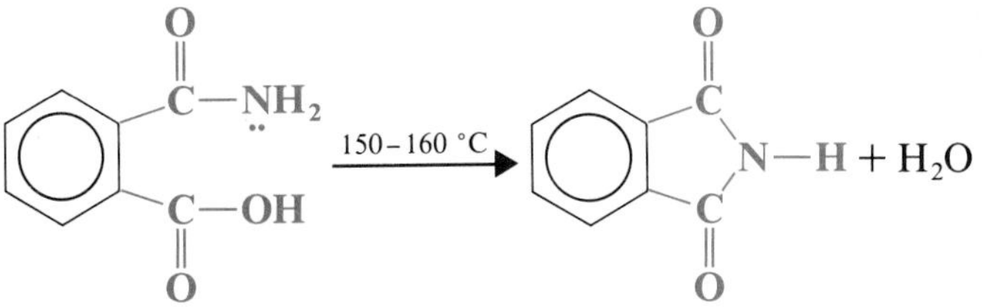

Phthalamic acid   (~100%) Phthalimide

18.8D Amides from Esters

Esters undergo nucleophilic substitution at their acyl carbons when they are treated with ammonia or with primary and secondary amines. These reactions take place more slowly than those of acyl chlorides and anhydrides, but they are synthetically useful.

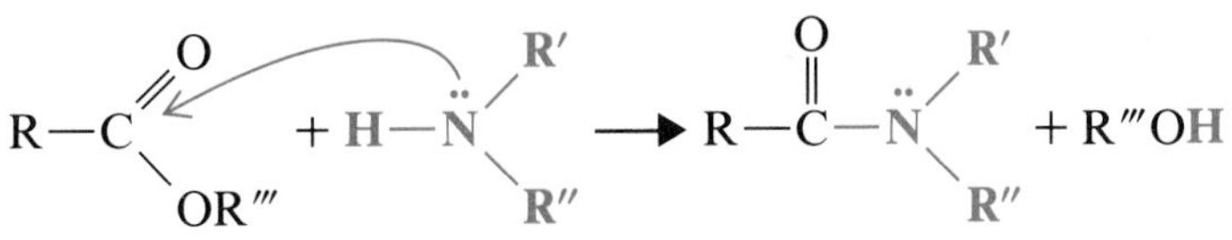

R' and/or R''
may be H

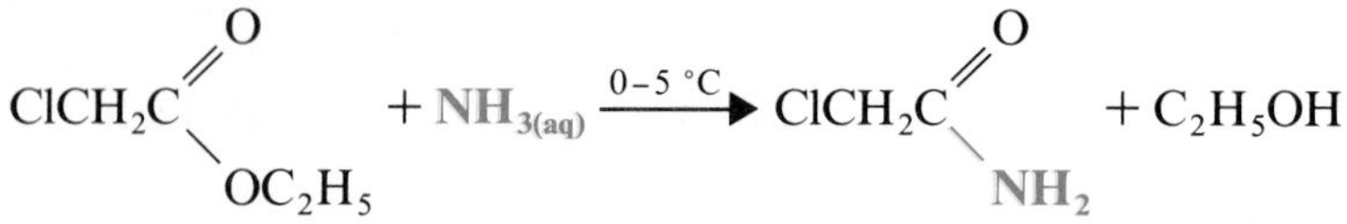

$$\text{ClCH}_2\text{C} \overset{\displaystyle O}{\underset{\displaystyle OC_2H_5}{\big\langle}} + \text{NH}_{3(aq)} \xrightarrow{0-5\ ^\circ C} \text{ClCH}_2\text{C} \overset{\displaystyle O}{\underset{\displaystyle NH_2}{\big\langle}} + \text{C}_2\text{H}_5\text{OH}$$

Ethyl chloroacetate   (62–87%)
Chloroacetamide

18.8E Amides from Carboxylic Acids and Ammonium Carboxylates

Carboxylic acids react with aqueous ammonia to form ammonium salts.

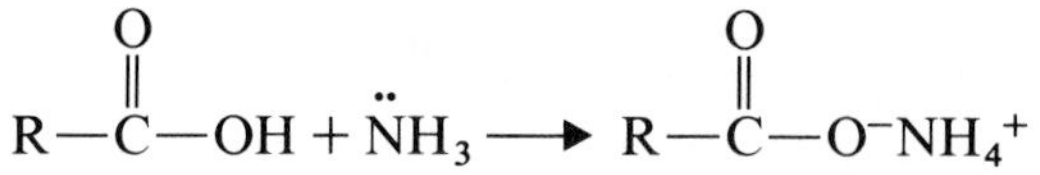

$$R-\overset{\displaystyle O}{\overset{\|}{C}}-OH + \ddot{N}H_3 \longrightarrow R-\overset{\displaystyle O}{\overset{\|}{C}}-O^-NH_4^+$$

An ammonium
carboxylate

Because of the low reactivity of the carboxylate ion toward nucleophilic substitution, further reaction does not usually take place in aqueous solution. However, if we evaporate the water and subsequently heat the dry salt, dehydration produces an amide.

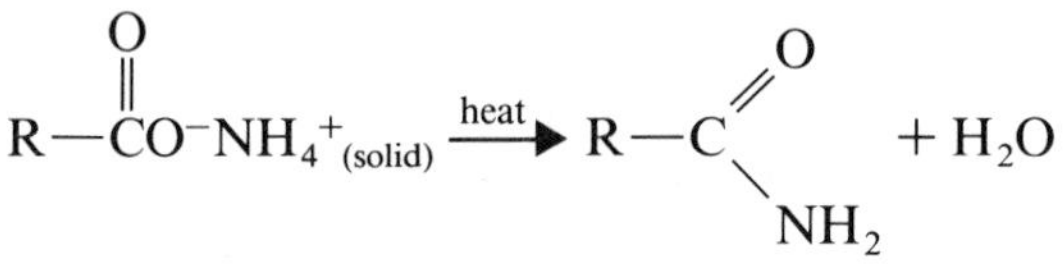

$$R-\overset{\displaystyle O}{\overset{\|}{C}}O^-NH_4{}^+{}_{(solid)} \xrightarrow{heat} R-C \overset{\displaystyle O}{\underset{\displaystyle NH_2}{\big\langle}} + H_2O$$

This is generally a poor method for preparing amides. A much better method is to convert the acid to an acyl chloride and then treat the acyl chloride with ammonia or an amine (Section 18.8B).

Amides are of great importance in biochemistry. The linkages that join individual amino acids together to form proteins are primarily amide linkages. As a consequence, much research has been done to find new and mild ways for amide synthesis. One especially useful reagent is the compound dicyclohexylcarbodiimide, $C_6H_{11}-N=C=N-C_6H_{11}$. Dicyclohexylcarbodiimide promotes amide formation by reacting with the carboxyl group of an acid and activating it toward nucleophilic substitution.

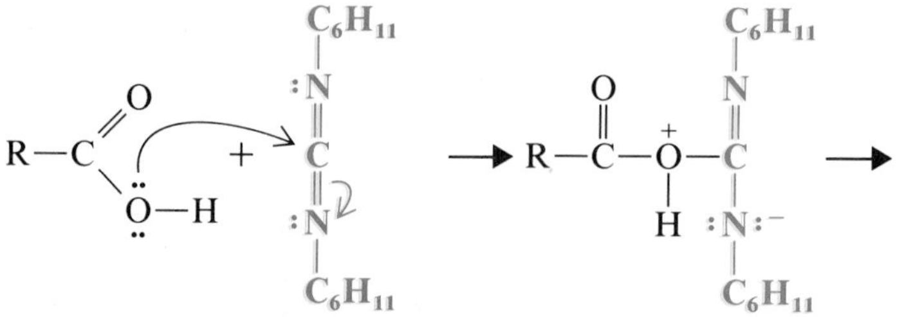

Dicyclohexyl-
carbodiimide
(DCC)

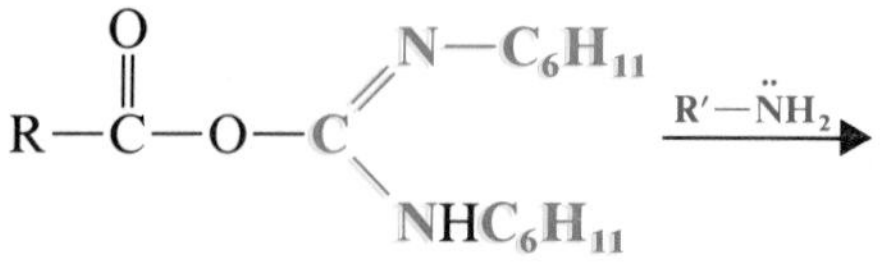

$$R-\overset{\displaystyle O}{\overset{\|}{C}}-O-C \overset{\displaystyle N-C_6H_{11}}{\underset{\displaystyle NHC_6H_{11}}{\big\langle}} \xrightarrow{R'-\ddot{N}H_2}$$

Reactive
intermediate

$$\text{R}-\underset{\underset{\text{R}'}{\overset{\displaystyle |}{\underset{\text{NH}_2^+}{|}}}}{\overset{\overset{\displaystyle \text{O}^-}{|}}{\text{C}}}-\text{O}-\underset{\text{NHC}_6\text{H}_{11}}{\overset{\text{N}-\text{C}_6\text{H}_{11}}{\text{C}}} \quad \xrightarrow[\text{(H}^+ \text{ shift)}]{} \quad \text{R}-\underset{\text{NHR}'}{\overset{\displaystyle \text{O}}{\text{C}}} + \text{O}=\underset{\text{NHC}_6\text{H}_{11}}{\overset{\text{NHC}_6\text{H}_{11}}{\text{C}}}$$

An amide   N,N-Dicyclohexylurea

The intermediate in this synthesis does not need to be isolated, and both steps take place at room temperature. Amides are produced in very high yield. In Chapter 23 we shall see how dicyclohexylcarbodiimide can be used in an automated synthesis of proteins.

18.8F Hydrolysis of Amides

Amides undergo hydrolysis when they are heated with aqueous acid or aqueous base.

Acidic Hydrolysis

$$\text{R}-\underset{\ddot{\text{N}}\text{H}_2}{\overset{\displaystyle \text{O}}{\text{C}}} + \text{H}_3\text{O}^+ \xrightarrow[\text{heat}]{\text{H}_2\text{O}} \text{R}-\underset{\text{OH}}{\overset{\displaystyle \text{O}}{\text{C}}} + \text{NH}_4^+$$

Basic Hydrolysis

$$\text{R}-\underset{\ddot{\text{N}}\text{H}_2}{\overset{\displaystyle \text{O}}{\text{C}}} + \text{Na}^+\text{OH}^- \xrightarrow[\text{heat}]{\text{H}_2\text{O}} \text{R}-\underset{\text{O}^-\text{Na}^+}{\overset{\displaystyle \text{O}}{\text{C}}} + \ddot{\text{N}}\text{H}_3$$

N-Substituted amides and *N,N*-disubstituted amides also undergo hydrolysis in aqueous acid or base. Amide hydrolysis by either method takes place more slowly than the corresponding hydrolysis of an ester. Thus, amide hydrolyses generally require more forcing conditions.

The mechanism for acid hydrolysis of an amide is similar to that given in Section 18.7A for the acid hydrolysis of an ester. Water acts as a nucleophile and attacks the protonated amide. The leaving group in the acidic hydrolysis of an amide is ammonia (or an amine).

$$\text{R}-\underset{\ddot{\text{N}}\text{H}_2}{\overset{\displaystyle \ddot{\text{O}}\text{:}}{\text{C}}} + \text{H}^+ \rightleftharpoons \text{R}-\underset{\ddot{\text{N}}\text{H}_2}{\overset{\displaystyle \overset{+}{\text{O}}-\text{H}}{\text{C}}} \xleftarrow{\text{:ÖH}_2} \text{R}-\underset{\text{:NH}_2}{\overset{\overset{\displaystyle \text{:Ö}-\text{H}}{|}}{\underset{|}{\text{C}}}}-\overset{+}{\ddot{\text{O}}}\text{H}_2 \rightleftharpoons \text{R}-\underset{\overset{+}{\text{NH}_3}}{\overset{\overset{\displaystyle \text{:Ö}-\text{H}}{|}}{\underset{|}{\text{C}}}}-\text{O}-\text{H}$$

$$\Updownarrow$$

$$\text{R}-\underset{\underset{\ddot{\text{O}}-\text{H}}{}}{\overset{\displaystyle \overset{+}{\ddot{\text{O}}}-\text{H}}{\text{C}}} + \ddot{\text{N}}\text{H}_3$$

$$\downarrow$$

$$\text{R}-\underset{\text{OH}}{\overset{\displaystyle \ddot{\text{O}}\text{:}}{\text{C}}} + \overset{+}{\text{N}}\text{H}_4$$

There is evidence that in basic hydrolyses of amides, hydroxide ions act both as nucleophiles and as bases. In the first step (in the following reaction) a hydroxide ion attacks the acyl carbon of the amide. In the second step, a hydroxide ion removes a proton to give a dianion. In the final step, the dianion loses a molecule of ammonia (or an amine); this step is synchronized with a proton transfer from water.

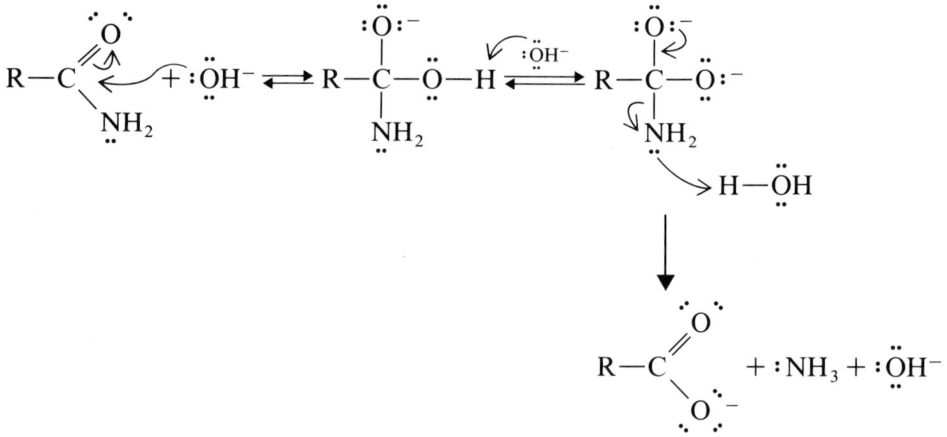

PROBLEM 18.15

What products would you obtain from acidic and basic hydrolysis of each of the following amides?

(a) *N,N*-Diethylbenzamide

(b)

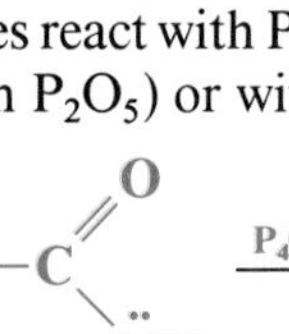

(c) 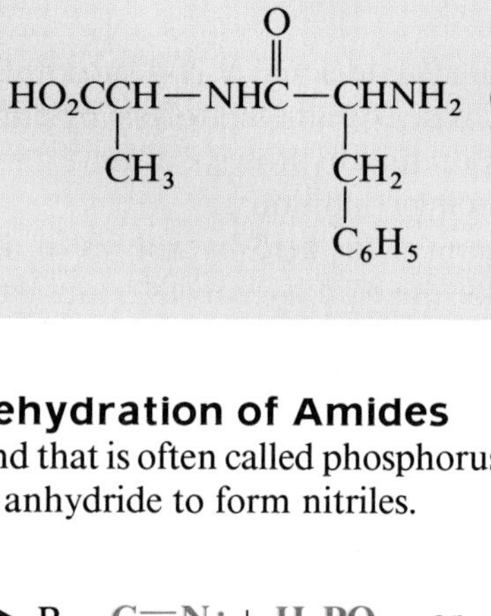 $HO_2CCH-NHC-CHNH_2$ (a dipeptide)

with CH_3 and CH_2 / C_6H_5 substituents and a $C=O$ on the central carbon.

18.8G Nitriles from the Dehydration of Amides

Amides react with P_4O_{10} (a compound that is often called phosphorus pentoxide and written P_2O_5) or with boiling acetic anhydride to form nitriles.

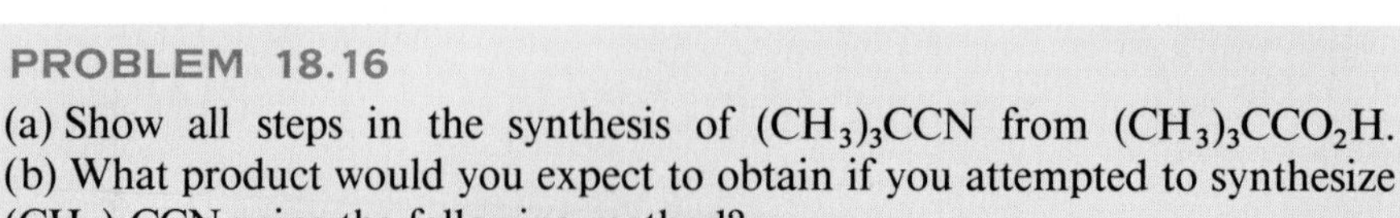

$$R-C(=O)NH_2 \xrightarrow[\text{heat} \ (-H_2O)]{P_4O_{10} \text{ or } (CH_3CO)_2O} R-C\equiv N: + H_3PO_4 \text{ or } CH_3CO_2H$$

A nitrile

This is a useful synthetic method for preparing nitriles that are not available by nucleophilic substitution reactions between alkyl halides and cyanide ion.

PROBLEM 18.16

(a) Show all steps in the synthesis of $(CH_3)_3CCN$ from $(CH_3)_3CCO_2H$.
(b) What product would you expect to obtain if you attempted to synthesize $(CH_3)_3CCN$ using the following method?

$$(CH_3)_3C-Br + CN^- \longrightarrow$$

18.8H Hydrolysis of Nitriles

Although nitriles do not contain a carbonyl group they are usually considered to be derivatives of carboxylic acids because complete hydrolysis of a nitrile produces a carboxylic acid or a carboxylate ion (Sections 16.9A and 18.3).

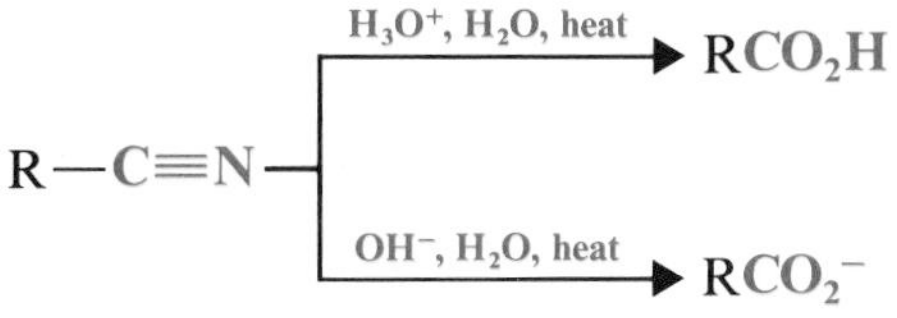

$$R-C\equiv N \longrightarrow \begin{cases} \xrightarrow{H_3O^+, H_2O, \text{ heat}} RCO_2H \\ \xrightarrow{OH^-, H_2O, \text{ heat}} RCO_2^- \end{cases}$$

The mechanisms for these hydrolyses are related to those for the acidic and basic hydrolyses of amides. In **acidic hydrolysis** of a nitrile the first step is protonation of the nitrogen atom. This protonation (in the following sequence) polarizes the nitrile group and makes the carbon atom more susceptible to nucleophilic attack by the weak nucleophile, water. The loss of a proton from the oxygen atom then produces a tautomeric form of an amide. Gain of a proton at the nitrogen atom gives a **protonated amide** and from this point on the steps are the same as those given for the acidic hydrolysis of an amide in Section 18.8F.

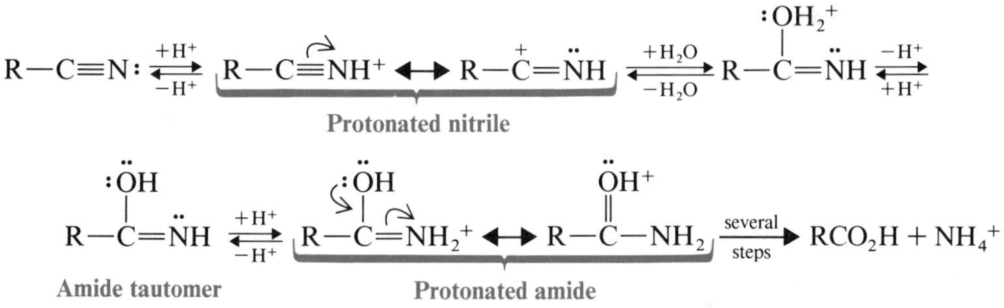

In **basic hydrolysis,** a hydroxide ion attacks the nitrile carbon atom and subsequent protonation leads to the amide tautomer. Further attack by hydroxide ion leads to hydrolysis in a manner analogous to that for the basic hydrolysis of an amide (Section 18.8F). (Under the appropriate conditions, amides can be isolated when nitriles are hydrolyzed.)

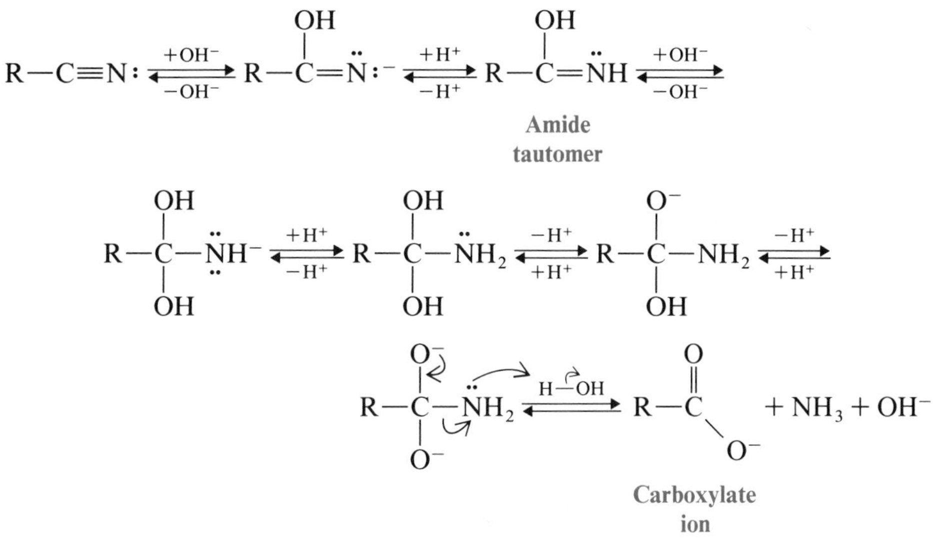

18.8I Lactams

Cyclic amides are called lactams. The size of the lactam ring is designated by Greek letters in a way that is analogous to lactone nomenclature.

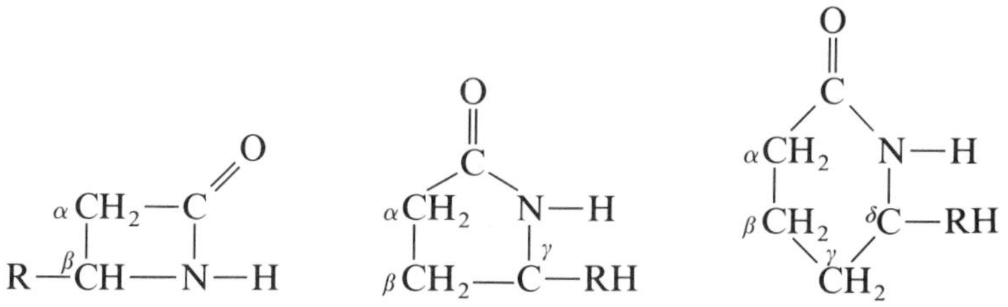

$$
\begin{array}{ccc}
\text{A } \beta\text{-lactam} & \text{A } \gamma\text{-lactam} & \text{A } \delta\text{-lactam}
\end{array}
$$

γ-Lactams and δ-lactams often form spontaneously from γ- and δ-amino acids. β-Lactams, however, are highly reactive; their strained four-membered rings open easily in the presence of nucleophilic reagents. The penicillin antibiotics (see following structures) contain a β-lactam ring.

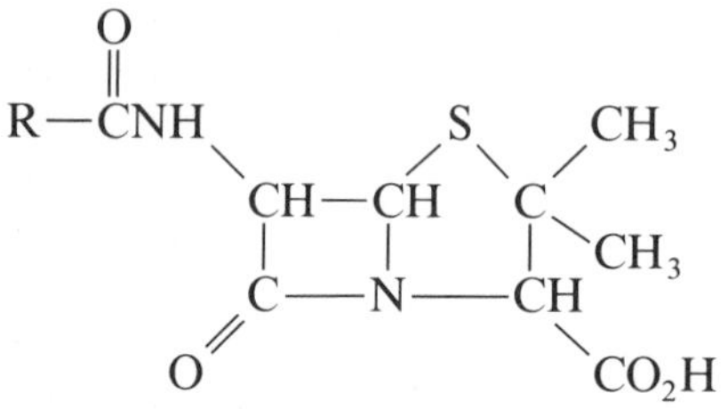

$$
\begin{array}{ll}
R = C_6H_5CH_2- & \text{(penicillin G)} \\
R = C_6H_5CH- & \text{(ampicillin)} \\
\quad\quad\;\; NH_2 & \\
R = C_6H_5OCH_2- & \text{(penicillin V)}
\end{array}
$$

The penicillins apparently act by interfering with the synthesis of the bacterial cell walls. It is thought that they do this by reacting with an amino group of an essential enzyme of the cell wall biosynthetic pathway. This reaction, which involves ring opening of the β-lactam and acylation of the amino group, inactivates the enzyme.

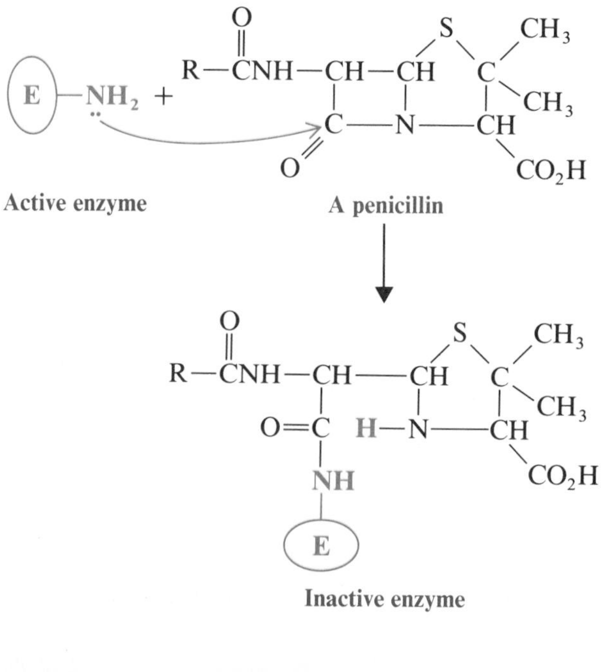

Active enzyme   A penicillin

Inactive enzyme

18.9 α-HALO ACIDS: THE HELL– VOLHARD–ZELINSKI REACTION

Aliphatic carboxylic acids react with bromine or chlorine in the presence of phosphorus (or a phosphorus halide) to give α-halo acids through a reaction known as the Hell–Volhard–Zelinski reaction.

General Reaction

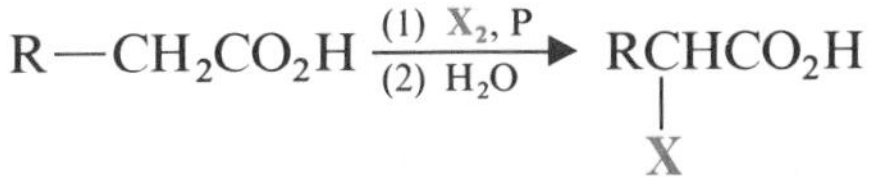

$$R—CH_2CO_2H \xrightarrow[\text{(2) H}_2\text{O}]{\text{(1) X}_2,\ \text{P}} RCHCO_2H$$
$$\underset{X}{|}$$

α-Halo acid

Specific Example

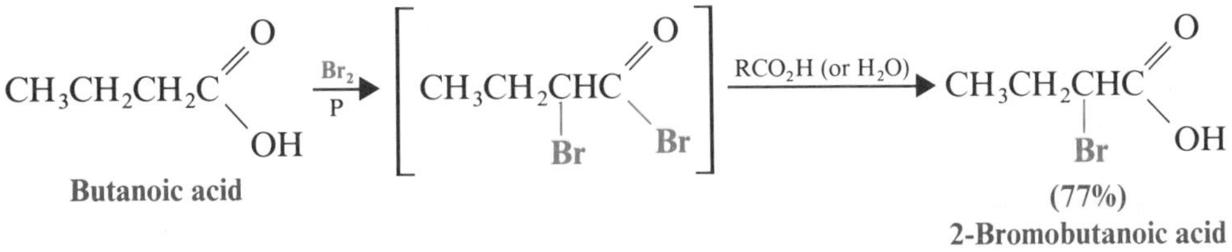

$$CH_3CH_2CH_2C\overset{O}{\underset{OH}{\diagdown}} \xrightarrow[\text{P}]{\text{Br}_2} \left[CH_3CH_2CHC\overset{O}{\underset{\underset{Br}{|}}{\diagdown}}_{Br} \right] \xrightarrow{\text{RCO}_2\text{H (or H}_2\text{O)}} CH_3CH_2CHC\overset{O}{\underset{\underset{Br}{|}}{\diagdown}}_{OH}$$

Butanoic acid   (77%)
2-Bromobutanoic acid

Halogenation occurs specifically at the α carbon. If more that one molar equivalent of bromine or chlorine is used in the reaction, the products obtained are α,α-dihalo acids or α,α,α-trihalo acids.

The mechanism for the Hell–Volhard–Zelinski reaction is outlined here. The key step involves the formation of an enol from an acyl halide. (Carboxylic acids do not form enols readily.) Enol formation accounts for specific halogenation at the α position.

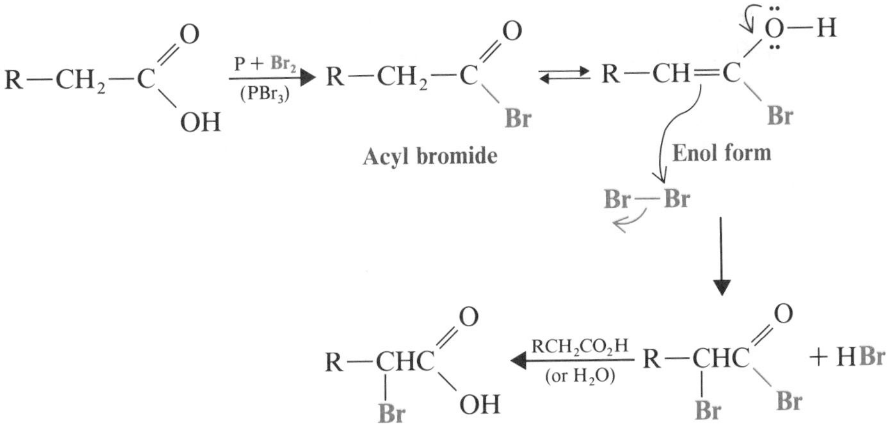

$$R—CH_2—C\overset{O}{\underset{OH}{\diagdown}} \xrightarrow[\text{(PBr}_3)]{\text{P + Br}_2} R—CH_2—C\overset{O}{\underset{Br}{\diagdown}} \rightleftharpoons R—CH=C\overset{\ddot{O}—H}{\underset{Br}{\diagdown}}$$

Acyl bromide   Enol form

$$Br—Br$$

$$R—CHC\overset{O}{\underset{Br}{\overset{}{\diagdown}}_{OH}} \xleftarrow[\text{(or H}_2\text{O)}]{\text{RCH}_2\text{CO}_2\text{H}} R—CHC\overset{O}{\underset{\underset{Br}{|}}{\diagdown}}_{Br} + HBr$$

α-Halo acids are important synthetic intermediates because they are capable of reacting with a variety of nucleophiles:

Conversion to α-Hydroxy Acids

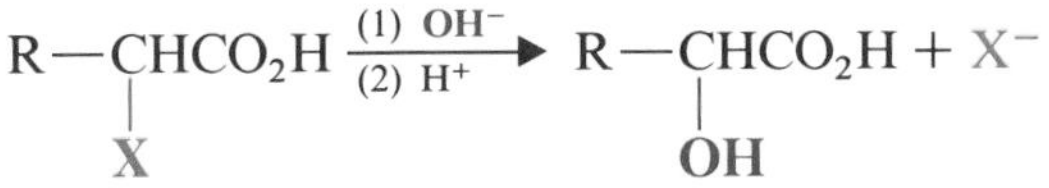

$$R—CHCO_2H \xrightarrow[\text{(2) H}^+]{\text{(1) OH}^-} R—CHCO_2H + X^-$$
$$\underset{X}{|} \qquad\qquad\qquad \underset{OH}{|}$$

α-Halo acid   α-Hydroxy acid

EXAMPLE

$$\text{CH}_3\text{CH}_2\underset{\overset{|}{\text{Br}}}{\text{CH}}\text{CO}_2\text{H} \xrightarrow[\substack{(2)\ \text{H}^+}]{\substack{(1)\ \text{K}_2\text{CO}_3,\ \text{H}_2\text{O} \\ 100\ °\text{C}}} \text{CH}_3\text{CH}_2\underset{\overset{|}{\text{OH}}}{\text{CH}}\text{CO}_2\text{H}$$

(69%)

2-Hydroxybutanoic acid

Conversion to α-Amino Acids

$$\underset{\overset{|}{\text{X}}}{\text{R}-\text{CH}}\text{CO}_2\text{H} + 2\text{NH}_3 \longrightarrow \underset{\overset{|}{\text{NH}_3^+}}{\text{RCH}}\text{CO}_2^- + \text{NH}_4\text{X} \ ■$$

α-Halo   α-Amino
acid   acid

EXAMPLE

$$\underset{\overset{|}{\text{Br}}}{\text{CH}_2}\text{CO}_2\text{H} + \text{NH}_3(\text{excess}) \longrightarrow \underset{\overset{|}{\text{NH}_3^+}}{\text{CH}_2}\text{CO}_2^- + \text{NH}_4\text{Br} \ ■$$

(60–64%)

Aminoacetic acid
(glycine)

18.10 DERIVATIVES OF CARBONIC ACID

Carbonic acid, $\text{HO}\overset{\overset{\text{O}}{\|}}{\text{C}}\text{OH}$, is an unstable compound that decomposes spontaneously (to produce carbon dioxide and water) and, therefore, cannot be isolated. However, many acyl chlorides, esters, and amides that are derived from carbonic acid (on paper, not in the laboratory) are stable compounds that have important applications.

Carbonyl dichloride (ClCOCl), a compound that is also called *phosgene,* can be thought of as the diacyl chloride of carbonic acid. Carbonyl dichloride reacts with two molar equivalents of an alcohol to yield a **dialkyl carbonate.**

$$\text{Cl}-\overset{\overset{\text{O}}{\|}}{\text{C}}-\text{Cl} + 2\text{CH}_3\text{CH}_2\text{OH} \longrightarrow \text{CH}_3\text{CH}_2\text{O}\overset{\overset{\text{O}}{\|}}{\text{C}}\text{OCH}_2\text{CH}_3 + 2\text{HCl}$$

Carbonyl   Diethyl carbonate
dichloride

A tertiary amine is usually added to the reaction to neutralize the hydrogen chloride that is produced.

Carbonyl dichloride reacts with ammonia to yield **urea** (Section 1.2A).

$$\text{Cl}-\overset{\overset{\text{O}}{\|}}{\text{C}}-\text{Cl} + 4\text{NH}_3 \longrightarrow \text{H}_2\text{N}\overset{\overset{\text{O}}{\|}}{\text{C}}\text{NH}_2 + 2\text{NH}_4\text{Cl}$$

Urea

Urea is the end product of the metabolism of nitrogen-containing compounds in most mammals and is excreted in the urine.

18.10A Alkyl Chloroformates and Carbamates (Urethanes)

Treating carbonyl dichloride with one molar equivalent of an alcohol leads to the formation of an alkyl chloroformate:

$$\text{ROH} + \text{Cl}-\overset{\overset{\textstyle O}{\|}}{\text{C}}-\text{Cl} \longrightarrow \underset{\substack{\text{Alkyl} \\ \text{chloroformate}}}{\text{RO}-\overset{\overset{\textstyle O}{\|}}{\text{C}}-\text{Cl}} + \text{HCl}$$

Specific Example

$$\text{C}_6\text{H}_5\text{CH}_2\text{OH} + \text{Cl}-\overset{\overset{\textstyle O}{\|}}{\text{C}}-\text{Cl} \longrightarrow \underset{\substack{\text{Benzyl} \\ \text{chloroformate}}}{\text{C}_6\text{H}_5\text{CH}_2\text{O}-\overset{\overset{\textstyle O}{\|}}{\text{C}}-\text{Cl}} + \text{HCl}$$

Alkyl chloroformates react with ammonia or amines to yield compounds called *carbamates* or *urethanes:*

$$\text{RO}-\overset{\overset{\textstyle O}{\|}}{\text{C}}-\text{Cl} + \text{R}'\text{NH}_2 \xrightarrow{\text{OH}^-} \underset{\substack{\text{A carbamate} \\ \text{(or urethane)}}}{\text{RO}-\overset{\overset{\textstyle O}{\|}}{\text{C}}-\text{NHR}'}$$

Benzyl chloroformate is used to install a protecting group on an amino group called the benzyloxycarbonyl group. We shall see in Section 23.7 how use is made of this protecting group in the synthesis of peptides and proteins. One advantage of the benzyloxycarbonyl group is that it can be removed under mild conditions. Treating the benzyloxycarbonyl derivative with hydrogen and a catalyst or with cold HBr in acetic acid removes the protecting group.

$$\text{R}-\text{NH}_2 + \text{C}_6\text{H}_5\text{CH}_2\text{O}\overset{\overset{\textstyle O}{\|}}{\text{C}}\text{Cl} \xrightarrow{\text{OH}^-} \left. \text{R}-\text{NH}-\overset{\overset{\textstyle O}{\|}}{\text{C}}\text{OCH}_2\text{C}_6\text{H}_5 \right\} \begin{array}{l}\text{Protected} \\ \text{amine}\end{array}$$

$$\text{R}-\text{NH}-\overset{\overset{\textstyle O}{\|}}{\text{C}}\text{OCH}_2\text{C}_6\text{H}_5 \begin{cases} \xrightarrow{\text{H}_2,\ \text{Pd}} \text{R}-\text{NH}_2 + \text{CO}_2 + \text{C}_6\text{H}_5\text{CH}_3 \\ \\ \xrightarrow{\text{HBr, CH}_3\text{CO}_2\text{H}} \text{R}-\text{NH}_3^+ + \text{CO}_2 + \text{C}_6\text{H}_5\text{CH}_2\text{Br} \end{cases}$$

Carbamates can also be synthesized by allowing an alcohol to react with an isocyanate, R—N=C=O. The reaction is an example of nucleophilic addition to the acyl carbon.

$$ROH + C_6H_5-N=C=O \longrightarrow RO\overset{\displaystyle O}{\overset{\|}{C}}-NH-C_6H_5$$

Phenyl
isocyanate

Carbamates tend to be nicely crystalline solids and are useful derivatives for identifying alcohols. The formation of a carbamate is also central to the synthesis of polymers called polyurethanes (see Special Topic J).

PROBLEM 18.17

Write structures for the products of the following reactions:

(a) $C_6H_5CH_2OH + C_6H_5N=C=O \longrightarrow$

(b) $ClCOCl + \text{excess } CH_3NH_2 \longrightarrow$

(c) Glycine $(H_3\overset{+}{N}CH_2CO_2^-) + C_6H_5CH_2OCOCl \xrightarrow{OH^-}$

(d) Product of (c) $+ H_2$, Pd $\longrightarrow$

(e) Product of (c) $+$ cold HBr, $CH_3CO_2H \longrightarrow$

(f) Urea $+ OH^-$, H_2O, heat

Although alkyl chloroformates ($RO\overset{\displaystyle O}{\overset{\|}{C}}Cl$), dialkyl carbonates ($RO\overset{\displaystyle O}{\overset{\|}{C}}OR$), and carbamates ($RO\overset{\displaystyle O}{\overset{\|}{C}}NH_2$, $RO\overset{\displaystyle O}{\overset{\|}{C}}NHR$, etc.) are stable, chloroformic acid ($HO\overset{\displaystyle O}{\overset{\|}{C}}Cl$), alkyl hydrogen carbonates ($RO\overset{\displaystyle O}{\overset{\|}{C}}OH$), and carbamic acid ($HO\overset{\displaystyle O}{\overset{\|}{C}}NH_2$) are not. These latter compounds decompose spontaneously to liberate carbon dioxide.

$$HO-\overset{\displaystyle O}{\overset{\|}{C}}-Cl \longrightarrow HCl + CO_2$$
Unstable

$$RO-\overset{\displaystyle O}{\overset{\|}{C}}-OH \longrightarrow ROH + CO_2$$
Unstable

$$HO-\overset{\displaystyle O}{\overset{\|}{C}}-NH_2 \longrightarrow NH_3 + CO_2$$
Unstable

This instability is a characteristic that these compounds share with their functional parent, carbonic acid.

$$HO-\overset{\overset{\displaystyle O}{\|}}{C}-OH \longrightarrow H_2O + CO_2$$
Unstable

18.11 DECARBOXYLATION OF CARBOXYLIC ACIDS

The reaction whereby a carboxylic acid loses CO_2 is called a *decarboxylation.*

$$R-\overset{\overset{\displaystyle O}{\|}}{C}-OH \xrightarrow{\text{decarboxylation}} R-H + CO_2$$

Although the unusual stability of carbon dioxide means that decarboxylation of most acids is exothermic, in practice the reaction is not always easy to carry out. Special groups usually have to be present in the molecule for decarboxylation to be synthetically useful.

Acids whose molecules have a carbonyl group one carbon removed from the carboxylic acid group, **called β-keto acids,** decarboxylate readily when they are heated to 100–150 °C. (Some β-keto acids even decarboxylate slowly at room temperature.)

$$R\overset{\overset{\displaystyle O}{\|}}{C}CH_2\overset{\overset{\displaystyle O}{\|}}{C}OH \xrightarrow{100-150\ °C} R\overset{\overset{\displaystyle O}{\|}}{C}CH_3 + CO_2$$
A β-keto acid

There are two reasons for this ease of decarboxylation:

1. When the carboxylate ion decarboxylates, it forms a resonance-stabilized anion:

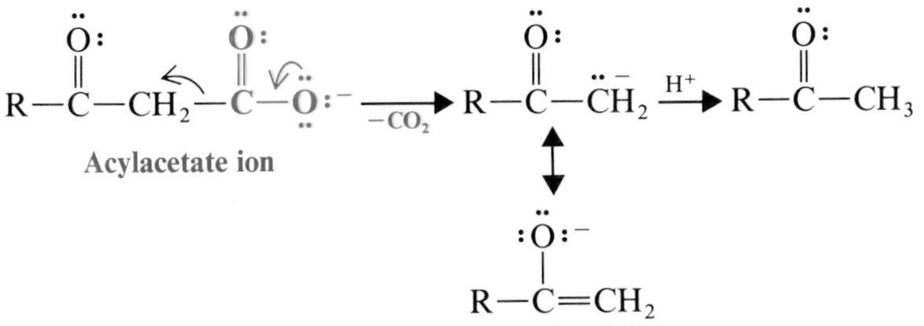

$$R-\overset{\overset{\displaystyle \ddot{O}:}{\|}}{C}-CH_2-\overset{\overset{\displaystyle \ddot{O}:}{\|}}{C}-\ddot{O}:^- \xrightarrow{-CO_2} R-\overset{\overset{\displaystyle \ddot{O}:}{\|}}{C}-\ddot{C}H_2^- \xrightarrow{H^+} R-\overset{\overset{\displaystyle \ddot{O}:}{\|}}{C}-CH_3$$

Acylacetate ion

$$\updownarrow$$

$$R-\overset{\overset{\displaystyle :\ddot{O}:^-}{|}}{C}=CH_2$$
Resonance-stabilized anion

This anion is much more stable than the anion $RCH_2:^-$ that would be produced by decarboxylation of an ordinary carboxylic acid anion.

2. When the acid itself decarboxylates, it can do so through a six-membered cyclic transition state:

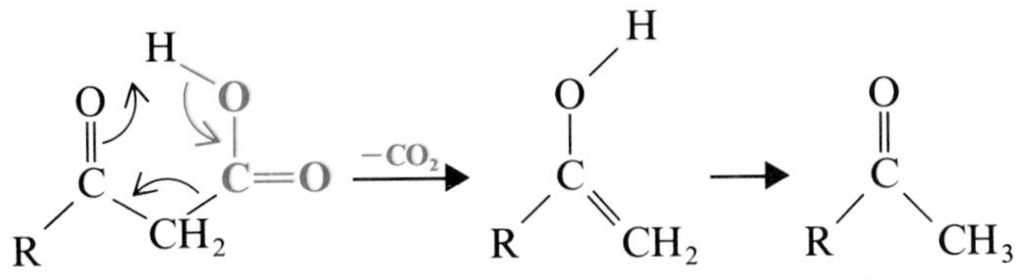

β-Keto acid   **Enol**   **Ketone**

This reaction produces an enol directly and avoids an anionic intermediate. The enol then tautomerizes to a methyl ketone.

Malonic acids also decarboxylate readily and for similar reasons.

$$\underset{\text{A malonic acid}}{\text{HOC}-\underset{R}{\overset{R}{\underset{|}{\overset{|}{C}}}}-\text{COH}} \xrightarrow{100-150\ °C} \text{H}-\underset{R}{\overset{R}{\underset{|}{\overset{|}{C}}}}-\text{COH} + CO_2$$

Notice that malonic acids undergo decarboxylation so readily that they do not form cyclic anhydrides (Section 18.6).

We shall see in Chapter 20 how decarboxylations of β-keto acids and malonic acids are synthetically useful.

Aromatic carboxylic acids decarboxylate when their salts are heated with copper and quinoline (the structure of quinoline is given in Section 19.1B):

$$\text{ArCO}_2^- + OH^- \xrightarrow[\text{heat}]{\text{Cu-quinoline}} \text{ArH} + CO_3^{2-}$$

These are extremely forcing conditions. The mechanism of the reaction is not known.

18.11A Decarboxylation of Carboxyl Radicals

Although the carboxylate ions (RCO_2^-) of simple aliphatic acids do not decarboxylate readily, carboxyl radicals ($\text{RCO}_2\cdot$) do. They decarboxylate by losing CO_2 and producing alkyl radicals:

$$\text{RCO}_2\cdot \longrightarrow \text{R}\cdot + CO_2$$

Carboxyl radicals can be generated by electrolysis, in a reaction known as the *Kolbe electrolysis,* or they can be generated chemically in a reaction known as the *Hunsdiecker reaction.*

In the **Kolbe electrolysis** an aqueous solution of the sodium or potassium salt of a carboxylic acid is subjected to electrolysis. At the anode the carboxylate ion loses an electron to become a carboxyl radical.

Step 1 $\quad \text{R}-\overset{\overset{\cdot\cdot}{\text{O}}}{\underset{}{\overset{\|}{C}}}-\overset{\cdot\cdot}{\underset{\cdot\cdot}{\text{O}}}:^- \xrightarrow[(-e^-)]{\text{anode}} \text{R}-\overset{\overset{\cdot\cdot}{\text{O}}}{\underset{}{\overset{\|}{C}}}-\overset{\cdot\cdot}{\underset{\cdot\cdot}{\text{O}}}\cdot$

Then the carboxyl radical decarboxylates and the alkyl radicals that are produced combine to form an alkane.

Step 2 $\quad \text{R}-\overset{\overset{\cdot\cdot}{\text{O}}}{\underset{}{\overset{\|}{C}}}-\overset{\cdot\cdot}{\underset{\cdot\cdot}{\text{O}}}\cdot \longrightarrow \text{R}\cdot + CO_2$

Step 3 $\quad 2\text{R}\cdot \longrightarrow \text{R}-\text{R}$

In the **Hunsdiecker reaction** the silver salt of a carboxylic acid is heated with bromine in CCl_4. A carboxyl radical is produced in a two-step process as follows:

$$\text{Step 1} \qquad R-\overset{\overset{\displaystyle O}{\|}}{C}-OAg + Br_2 \xrightarrow{CCl_4} R\overset{\overset{\displaystyle O}{\|}}{C}-OBr + AgBr$$

$$\text{Step 2} \qquad R-\overset{\overset{\displaystyle O}{\|}}{C}-OBr \longrightarrow R-\overset{\overset{\displaystyle O}{\|}}{C}-O\cdot + Br\cdot$$

Then the carboxyl radical decarboxylates. The resulting alkyl radical abstracts a bromine atom from $R\overset{\overset{\displaystyle O}{\|}}{C}OBr$ to produce an alkyl bromide and regenerate a carboxyl radical.

$$\text{Step 3} \qquad R-\overset{\overset{\displaystyle O}{\|}}{C}-O\cdot \longrightarrow R\cdot + CO_2$$

$$\text{Step 4} \qquad R\cdot + R-\overset{\overset{\displaystyle O}{\|}}{C}-OBr \longrightarrow R-Br + R-\overset{\overset{\displaystyle O}{\|}}{C}-O\cdot$$

Then steps 3 and 4 are repeated, and so on.

Overall the Hunsdiecker reaction amounts to the following:

$$R-CO_2Ag + Br_2 \xrightarrow[\text{heat}]{CCl_4} RBr + CO_2 + AgBr$$

PROBLEM 18.18

Using decarboxylation reactions, outline a synthesis of each of the following from appropriate starting materials.

(a) Decane   (d) Benzyl bromide   (g) Cyclohexanone

(b) 2-Hexanone   (e) 2-Butanone   (h) Pentanoic acid

(c) 2-Methylbutanoic acid (f) Cyclohexane

PROBLEM 18.19

Diacyl peroxides, $R\overset{\overset{\displaystyle O}{\|}}{C}-O-O-\overset{\overset{\displaystyle O}{\|}}{C}R$, decompose readily when heated. (a) What factor accounts for this instability? (b) The decomposition of a diacyl peroxide produces CO_2. How is it formed? (c) Diacyl peroxides are often used to initiate free radical reactions, for example, the polymerization of an alkene:

$$nCH_2{=}CH_2 \xrightarrow[\text{(−CO}_2)]{R\overset{\overset{\displaystyle O}{\|}}{C}-O-O-\overset{\overset{\displaystyle O}{\|}}{C}R} R-(CH_2CH_2)_n-H$$

Show the steps involved.

18.12 CHEMICAL TESTS FOR ACYL COMPOUNDS

Carboxylic acids are weak acids and their acidity helps us to detect them. Aqueous solutions of water-soluble carboxylic acids give an acid test with blue litmus paper. Water-insoluble carboxylic acids dissolve in aqueous sodium hydroxide and aqueous sodium bicarbonate (cf. 18.2C). The latter reagent helps us distinguish carboxylic acids from most phenols. Except for the di- and trinitrophenols, phenols do not dissolve in aqueous sodium bicarbonate. Carboxylic acids not only dissolve in aqueous sodium bicarbonate, they also cause the evolution of carbon dioxide.

In identifying a carboxylic acid, it is often helpful to determine its equivalent weight by titrating a measured quantity of the acid with a standard solution of sodium hydroxide. For a monocarboxylic acid, the equivalent weight equals the molecular weight; for a dicarboxylic acid, the equivalent weight is one half the molecular weight, and so on.

All acid derivatives can be hydrolyzed to carboxylic acids. The conditions required to bring about hydrolysis vary greatly, with acyl chlorides being the easiest to hydrolyze and amides being the most difficult.

Acyl chlorides hydrolyze in water and thus give a precipitate when treated with aqueous silver nitrate. Acid anhydrides dissolve when heated briefly with aqueous sodium hydroxide.

Esters and amides hydrolyze slowly when they are refluxed with sodium hydroxide: An ester produces a carboxylate ion and an alcohol; an amide produces a carboxylate ion and an amine or ammonia. The hydrolysis products, the acid and the alcohol or amine, can be isolated and identified. Since base-promoted hydrolysis of an unsubstituted amide produces ammonia, this ammonia can often be detected by holding moist red litmus in the vapors above the reaction mixture.

Base-promoted hydrolysis of an ester consumes one molar equivalent of hydroxide ion for each molar equivalent of the ester. It is often convenient, therefore, to carry out the hydrolysis quantitatively.

$$\underset{\substack{\text{1 molar} \\ \text{equivalent}}}{RCO_2R'} + \underset{\substack{\text{1 molar} \\ \text{equivalent}}}{OH^-} \longrightarrow RCO_2^- + R'OH$$

This reaction allows us to determine the *equivalent weight* of the ester. We can make this determination by hydrolyzing a known weight of the ester with an excess of a standard solution of sodium hydroxide. After the hydrolysis is complete, we can titrate the excess sodium hydroxide with a standard acid. For an ester containing one —CO_2R group the equivalent weight will equal the molecular weight.

Amides can be distinguished from amines with dilute HCl. Most amines dissolve in dilute HCl, whereas most amides do not (cf. Problem 18.38).

18.13 SUMMARY OF THE REACTIONS OF CARBOXYLIC ACIDS AND THEIR DERIVATIVES

The reactions of carboxylic acids and their derivatives are summarized here.

18.13A Reactions of Carboxylic Acids

1. As acids (discussed in Section 18.2A).

$$RCO_2H + NaOH \longrightarrow RCO_2^-Na^+ + H_2O$$

$$RCO_2H + NaHCO_3 \longrightarrow RCO_2^-Na^+ + H_2O + CO_2$$

2. Reduction (discussed in Section 15.2).

$$RCO_2H + LiAlH_4 \xrightarrow[(2)\ H_2O]{(1)\ ether} RCH_2OH$$

3. Conversion to acyl chlorides (discussed in Section 18.5).

$$RCO_2H + SOCl_2 \longrightarrow RCOCl + SO_2 + HCl$$

$$3RCO_2H + PCl_3 \longrightarrow 3RCOCl + H_3PO_3$$

$$RCO_2H + PCl_5 \longrightarrow RCOCl + POCl_3 + HCl$$

4. Conversion to acid anhydrides (discussed in Section 18.6).

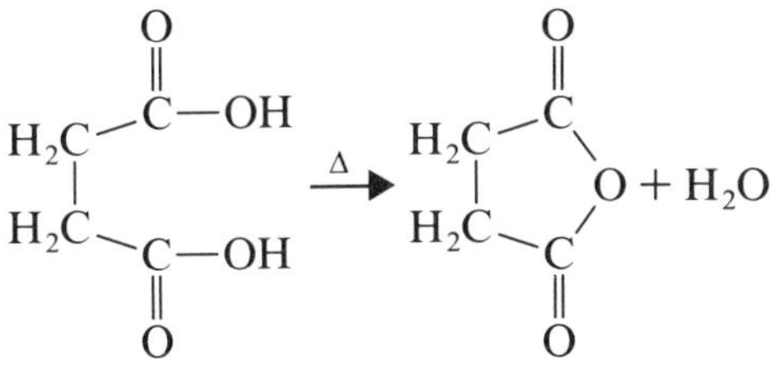

5. Conversion to esters (discussed in Section 18.7).

$$R{-}\overset{\overset{\displaystyle O}{\|}}{C}OH + R'{-}OH \underset{}{\overset{H^+}{\rightleftharpoons}} R{-}\overset{\overset{\displaystyle O}{\|}}{C}{-}OR' + H_2O$$

6. Conversion to lactones (discussed in Section 18.7).

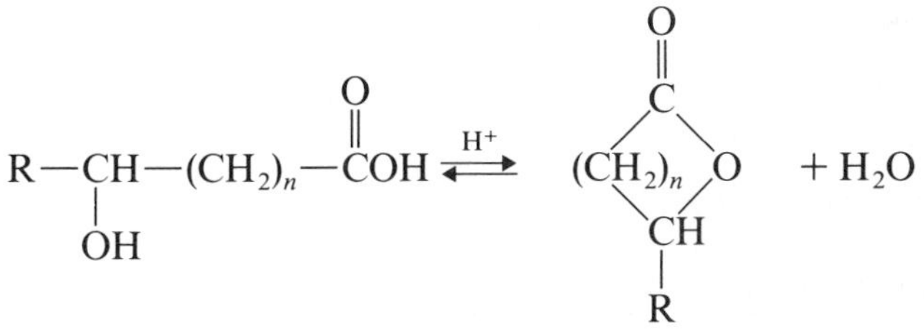

$n = 2$, a γ-lactone
$n = 3$, a δ-lactone

7. Conversion to amides and imides (discussed in Section 18.8).

$$R{-}\overset{\overset{\displaystyle O}{\|}}{C}{-}OH + NH_3 \rightleftharpoons R{-}\overset{\overset{\displaystyle O}{\|}}{C}{-}O^-NH_4^+ \xrightarrow{heat} R{-}\overset{\overset{\displaystyle O}{\|}}{C}{-}NH_2 + H_2O$$

An amide

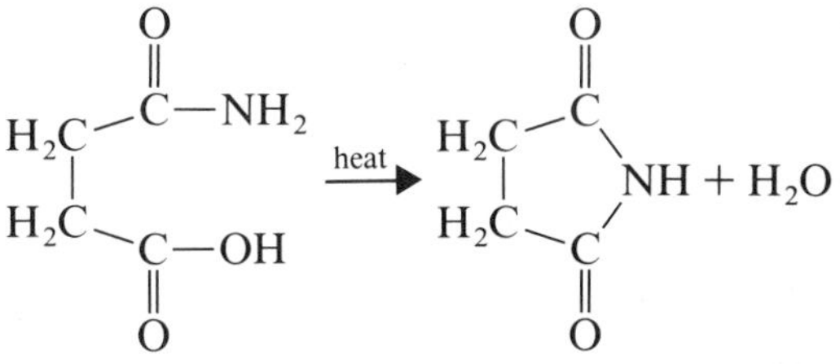

A cyclic imide

8. Conversion to lactams (discussed in Section 18.8).

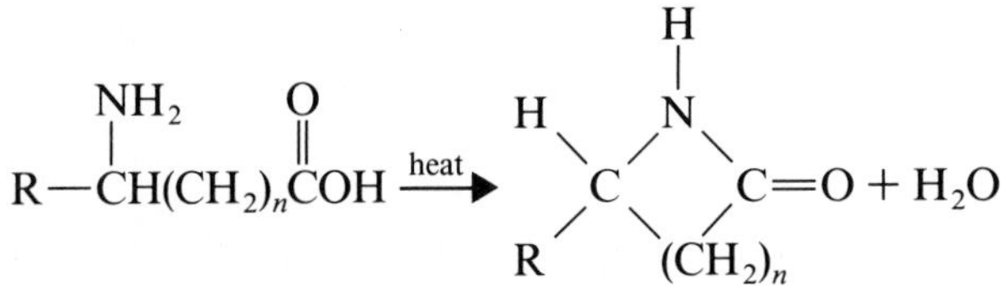

$$R-\underset{\underset{NH_2}{|}}{C}H(CH_2)_n\overset{O}{\underset{||}{C}}OH \xrightarrow{heat} \underset{R}{\overset{H}{\diagdown}}\underset{(CH_2)_n}{C}\underset{N}{\overset{H}{|}}C=O + H_2O$$

n = 2, a γ-lactam
n = 3, a δ-lactam

9. α-Halogenation (discussed in Section 18.9).

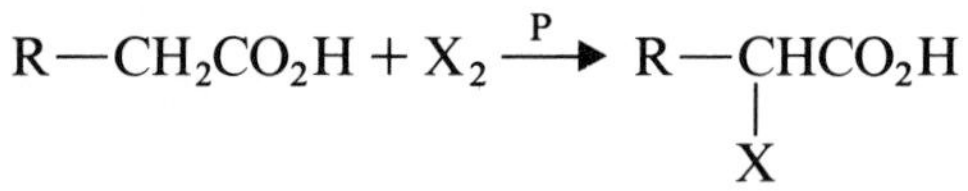

$$R-CH_2CO_2H + X_2 \xrightarrow{P} R-\underset{\underset{X}{|}}{C}HCO_2H$$

$X_2 = Cl_2$　or　Br_2

10. Decarboxylation (discussed in Section 18.11).

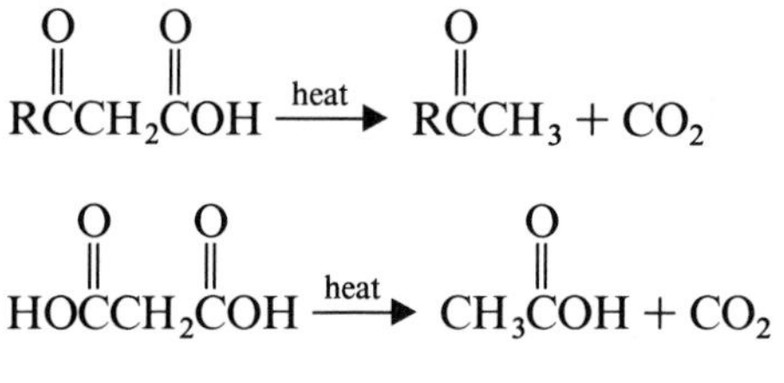

$$R\overset{O}{\overset{||}{C}}CH_2\overset{O}{\overset{||}{C}}OH \xrightarrow{heat} R\overset{O}{\overset{||}{C}}CH_3 + CO_2$$

$$HO\overset{O}{\overset{||}{C}}CH_2\overset{O}{\overset{||}{C}}OH \xrightarrow{heat} CH_3\overset{O}{\overset{||}{C}}OH + CO_2$$

18.13B Reactions of Acyl Chlorides

1. Conversion to acids (discussed in Section 18.4).

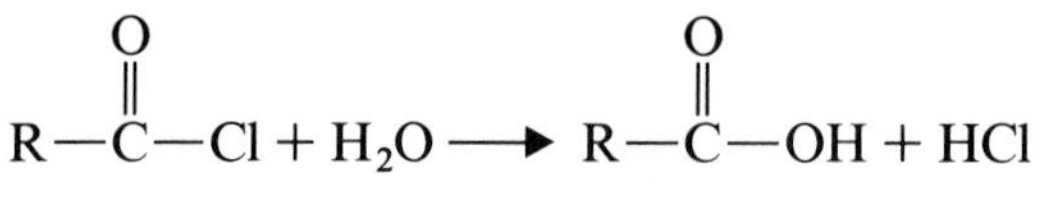

$$R-\overset{O}{\overset{||}{C}}-Cl + H_2O \longrightarrow R-\overset{O}{\overset{||}{C}}-OH + HCl$$

2. Conversion to anhydrides (discussed in Section 18.6).

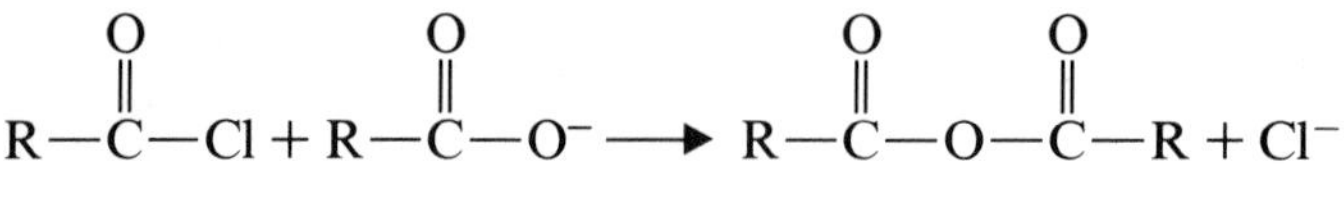

$$R-\overset{O}{\overset{||}{C}}-Cl + R-\overset{O}{\overset{||}{C}}-O^- \longrightarrow R-\overset{O}{\overset{||}{C}}-O-\overset{O}{\overset{||}{C}}-R + Cl^-$$

3. Conversion to esters (discussed in Section 18.7).

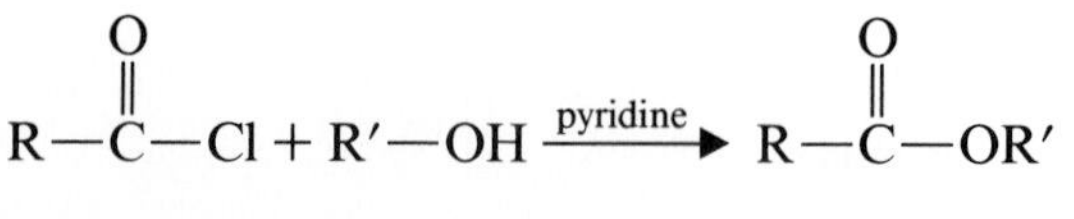

$$R-\overset{O}{\overset{||}{C}}-Cl + R'-OH \xrightarrow{pyridine} R-\overset{O}{\overset{||}{C}}-OR'$$

4. Conversion to amides (discussed in Section 18.8).

$$R-\overset{\overset{\displaystyle O}{\|}}{C}-Cl + NH_3(excess) \longrightarrow R-\overset{\overset{\displaystyle O}{\|}}{C}-NH_2 + NH_4Cl$$

$$R-\overset{\overset{\displaystyle O}{\|}}{C}-Cl + NH_2R'(excess) \longrightarrow R-\overset{\overset{\displaystyle O}{\|}}{C}-NHR' + R'NH_3Cl$$

$$R-\overset{\overset{\displaystyle O}{\|}}{C}-Cl + NHR_2'(excess) \longrightarrow R-\overset{\overset{\displaystyle O}{\|}}{C}-NR_2' + R_2'NH_2Cl$$

5. Conversion to ketones

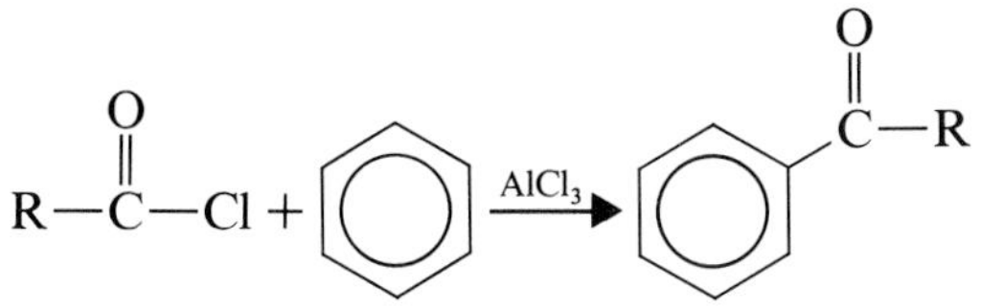

(discussed in Section 12.7)

$$R-\overset{\overset{\displaystyle O}{\|}}{C}-Cl + R_2'CuLi \longrightarrow R-\overset{\overset{\displaystyle O}{\|}}{C}-R' \qquad \text{(discussed in Section 16.5)}$$

6. Conversion to aldehydes (discussed in Section 16.4).

$$R-\overset{\overset{\displaystyle O}{\|}}{C}-Cl + LiAlH[OC(CH_3)_3]_3 \xrightarrow[\text{(2) } H_2O]{\text{(1) ether}} R-\overset{\overset{\displaystyle O}{\|}}{C}-H$$

18.13C Reactions of Acid Anhydrides

1. Conversion to acids (cf. Section 18.12).

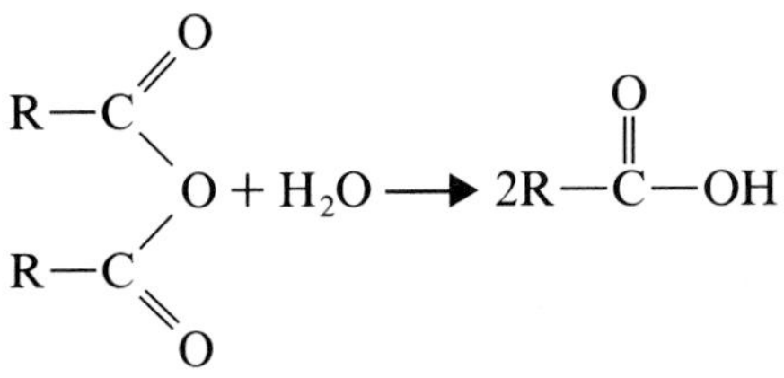

2. Conversion to esters (discussed in Sections 18.4 and 18.7).

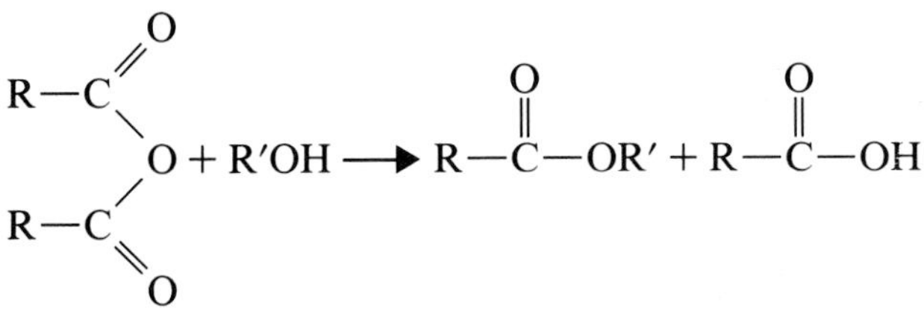

3. Conversion to amides and imides (discussed in Section 18.8).

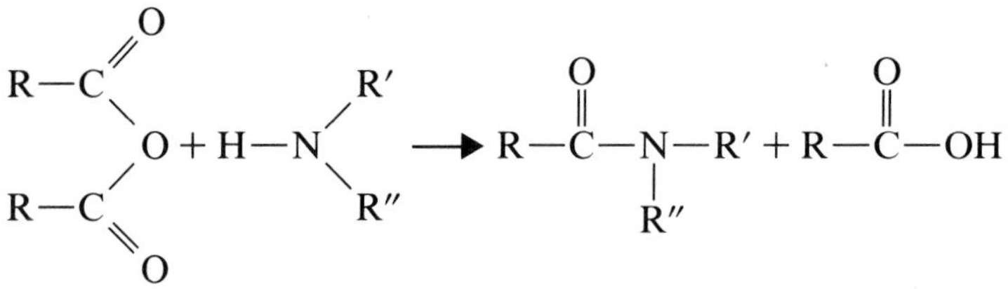

R′ and/or R″ may be H

or

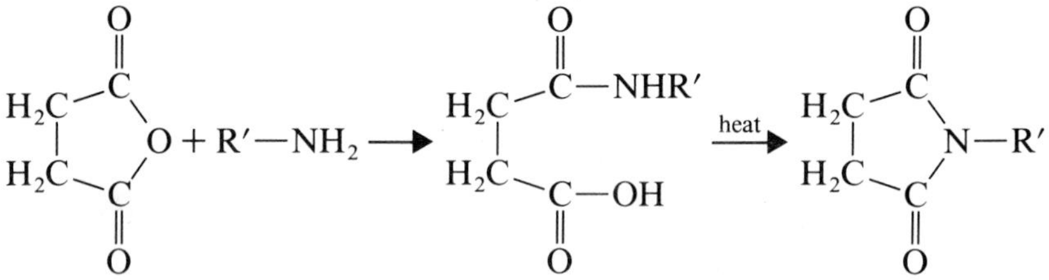

R′ may be H

4. Conversion to ketones (discussed in Section 12.7).

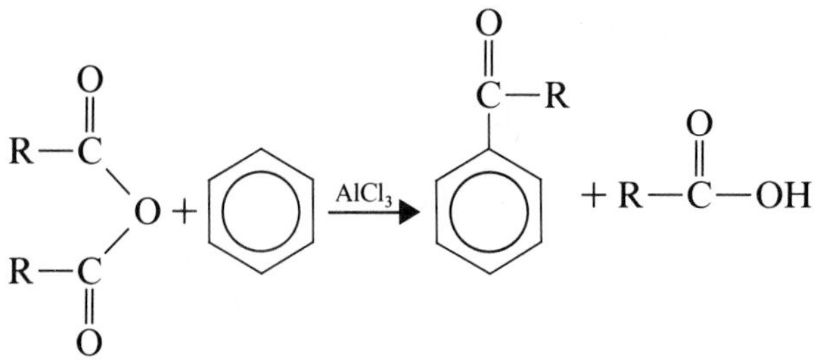

18.13D Reactions of Esters

1. Hydrolysis (discussed in Section 18.7).

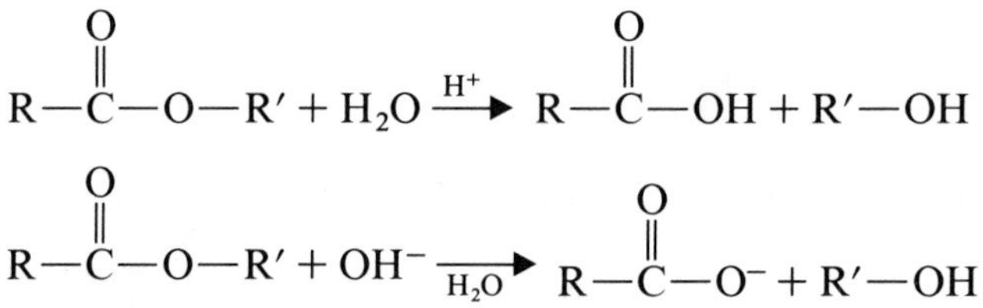

2. Conversion to other esters: transesterification (discussed in Problem 18.12).

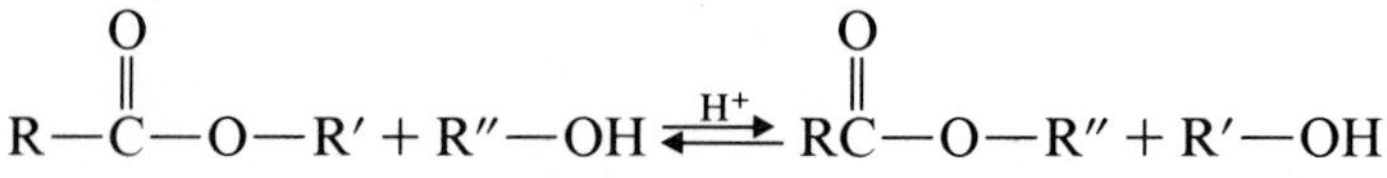

3. Conversion to amides (discussed in Section 18.8).

$$R-\overset{\overset{\displaystyle O}{\|}}{C}-OR' + HN\overset{\displaystyle R''}{\underset{\displaystyle R'''}{\diagdown}} \longrightarrow R-\overset{\overset{\displaystyle O}{\|}}{C}-N\overset{\displaystyle R''}{\underset{\displaystyle R'''}{\diagdown}} + R'-OH$$

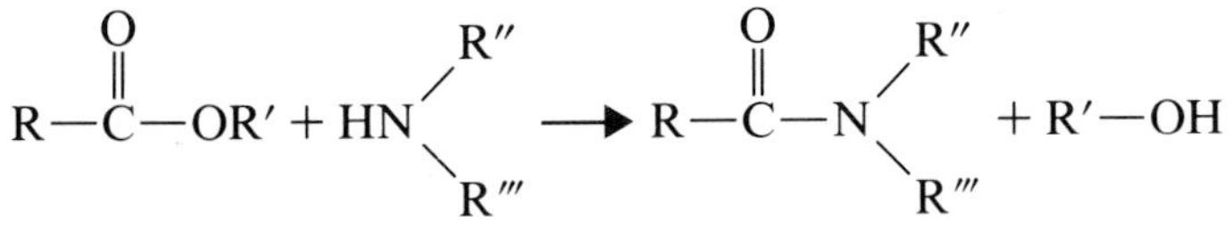

R, R'' and/or R''' may be H

4. Reaction with Grignard reagents (discussed in Section 15.7).

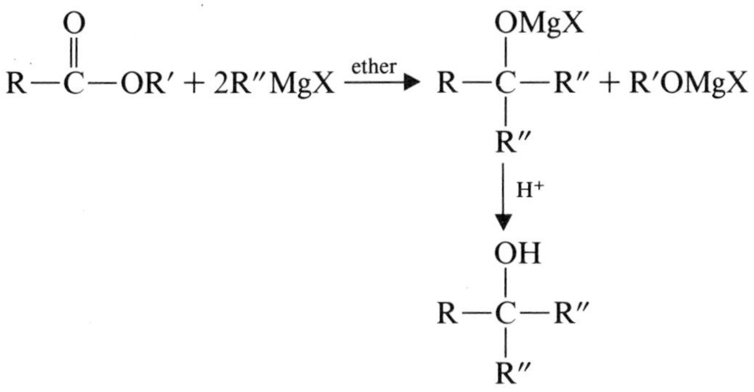

$$R-\overset{\overset{\displaystyle O}{\|}}{C}-OR' + 2R''MgX \xrightarrow{\text{ether}} R-\overset{\overset{\displaystyle OMgX}{|}}{\underset{\overset{|}{R''}}{C}}-R'' + R'OMgX$$

$$\downarrow \text{H}^+$$

$$R-\overset{\overset{\displaystyle OH}{|}}{\underset{\overset{|}{R''}}{C}}-R''$$

5. Reduction (discussed in Section 15.2).

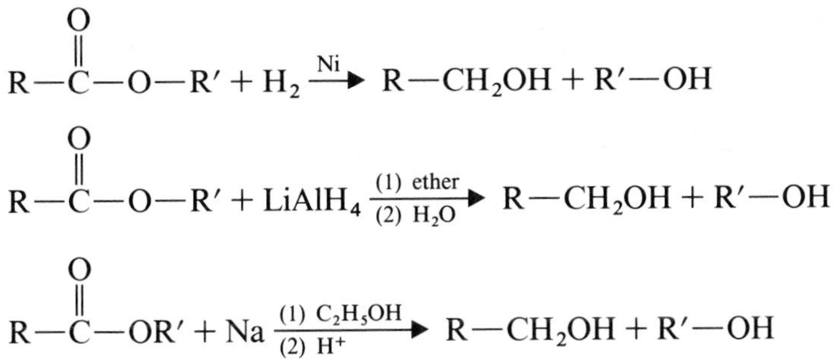

$$R-\overset{\overset{\displaystyle O}{\|}}{C}-O-R' + H_2 \xrightarrow{\text{Ni}} R-CH_2OH + R'-OH$$

$$R-\overset{\overset{\displaystyle O}{\|}}{C}-O-R' + LiAlH_4 \xrightarrow[(2)\ H_2O]{(1)\ \text{ether}} R-CH_2OH + R'-OH$$

$$R-\overset{\overset{\displaystyle O}{\|}}{C}-OR' + Na \xrightarrow[(2)\ H^+]{(1)\ C_2H_5OH} R-CH_2OH + R'-OH$$

18.13E Reactions of Amides

1. Hydrolysis (discussed in Section 18.8).

$$R-\overset{\overset{\displaystyle O}{\|}}{\underset{\overset{|}{R''}}{C}}-NR' + H_3O^+ \xrightarrow{H_2O} R-\overset{\overset{\displaystyle O}{\|}}{C}-OH + R'-\overset{|}{\underset{\overset{|}{R''}}{N}}H_2{}^+$$

$$R-\overset{\overset{\displaystyle O}{\|}}{\underset{\overset{|}{R''}}{C}}-NR' + OH^- \xrightarrow{H_2O} R\overset{\overset{\displaystyle O}{\|}}{C}-O^- + R'-\overset{|}{\underset{\overset{|}{R''}}{N}}H$$

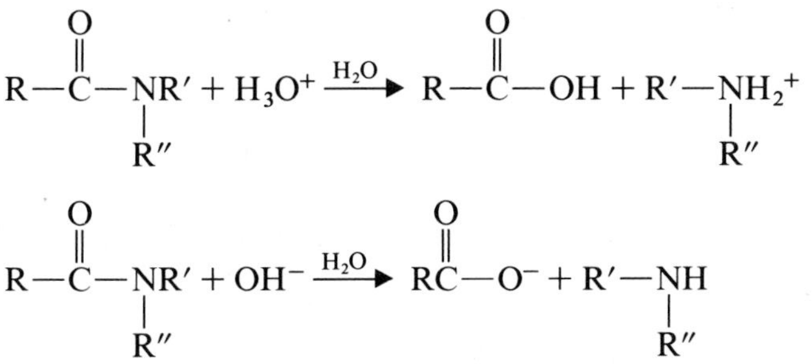

R, R', and/or R'' may be H

2. Conversion to nitriles: dehydration (discussed in Section 18.8).

$$R-\overset{\overset{\displaystyle O}{\|}}{C}NH_2 \xrightarrow[\substack{\text{heat} \\ (-H_2O)}]{P_4O_{10}} R-C\equiv N$$

3. Conversion to imides (discussed in Section 18.8).

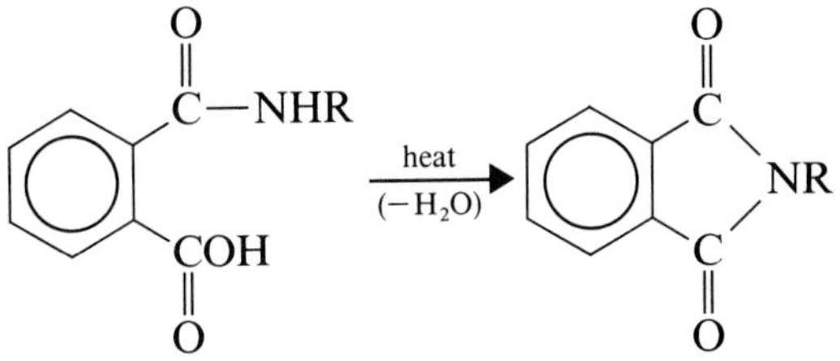

ADDITIONAL PROBLEMS

18.20
Write a structural formula for each of the following compounds:

(a) Hexanoic acid

(b) Hexanamide

(c) N-Ethylhexanamide

(d) N,N-Diethylhexanamide

(e) 3-Hexenoic acid

(f) 2-Methyl-4-hexenoic acid

(g) Hexanedioic acid

(h) Phthalic acid

(i) Isophthalic acid

(j) Terephthalic acid

(k) Diethyl oxalate

(l) Diethyl adipate

(m) Isobutyl propanoate

(n) 2-Naphthoic acid

(o) Maleic acid

(p) 2-Hydroxybutanedioic acid (malic acid)

(q) Fumaric acid

(r) Succinic acid

(s) Succinimide

(t) Malonic acid

(u) Diethyl malonate

18.21
Give an IUPAC systematic or common name for each of the following compounds:

(a) $C_6H_5CO_2H$

(b) C_6H_5COCl

(c) $C_6H_5CONH_2$

(d) $(C_6H_5CO)_2O$

(e) $C_6H_5CO_2CH_2C_6H_5$

(f) $C_6H_5CO_2C_6H_5$

(g) $CH_3CO_2CH(CH_3)_2$

(h) $CH_3CON(CH_3)_2$

(i) CH_3CN

(j)

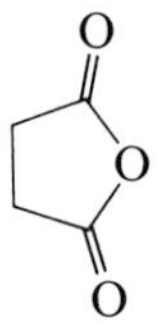

(k)

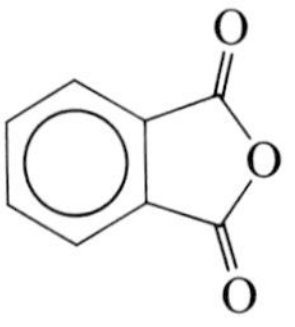

(l)

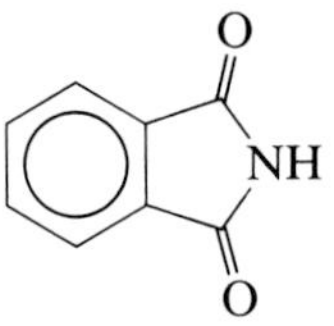

(m)

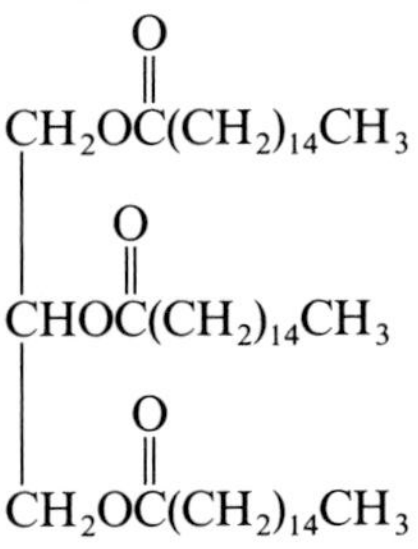

(n) $HO_2CCH_2CCO_2H$ (with C=O)

(o) (structure: benzene ring with OH and $COCH_3$ group, ketone)

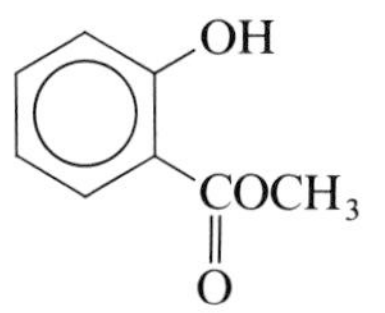

18.22
Show how benzoic acid can be synthesized from each of the following:

(a) Bromobenzene   (d) Acetophenone   (g) Benzyl alcohol

(b) Toluene   (e) Benzaldehyde

(c) Benzonitrile, C_6H_5CN (f) Styrene

18.23
Show how phenylacetic acid can be prepared from each of the following:

(a) Phenylacetaldehyde   (b) Benzyl bromide (two ways)

18.24
Show how pentanoic acid can be prepared from each of the following:

(a) 1-Pentanol   (c) 5-Decene

(b) 1-Bromobutane (two ways)   (d) Pentanal

18.25
What major organic product would you expect to obtain when acetyl chloride reacts with each of the following?

(a) H_2O (i) CH_3NH_2 (excess)

(b) $AgNO_3/H_2O$ (j) $C_6H_5NH_2$ (excess)

(c) $CH_3(CH_2)_2CH_2OH$ and pyridine   (k) $(CH_3)_2NH$ (excess)

(d) NH_3 (excess)   (l) CH_3CH_2OH and pyridine

(e) $C_6H_5CH_3$ and $AlCl_3$ (m) $CH_3CO_2^-Na^+$

(f) $LiAlH[OC(CH_3)_3]_3$ (n) CH_3CO_2H and pyridine

(g) $(CH_3)_2CuLi$ (o) Phenol and pyridine

(h) $NaOH/H_2O$

18.26
What major organic product would you expect to obtain when acetic anhydride reacts with each of the following?

(a) NH_3 (excess)   (c) $CH_3CH_2CH_2OH$ (e) $CH_3CH_2NH_2$ (excess)

(b) H_2O (d) $C_6H_6 + AlCl_3$ (f) $(CH_3CH_2)_2NH$ (excess)

18.27
What major organic product would you expect to obtain when succinic anhydride reacts with each of the reagents given in Problem 18.26?

18.28
Show how you might carry out the following transformations:

(a) Succinic anhydride ⟶ (structure: bicyclic — benzene ring fused to cyclohexanone ring, with O)

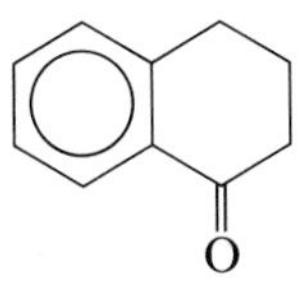

(b)

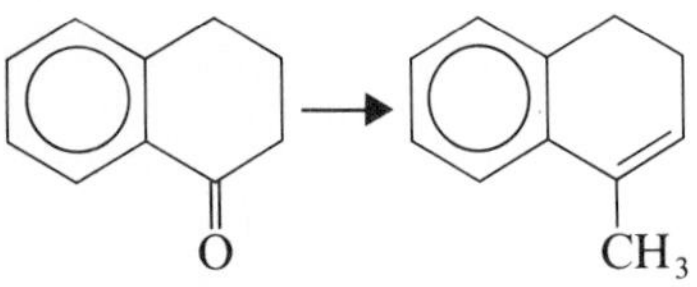

(c)

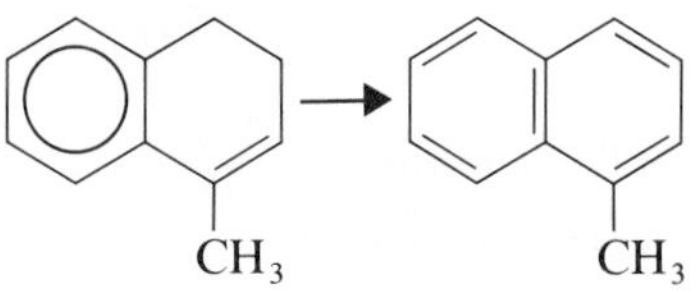

(d) Phthalic anhydride ⟶

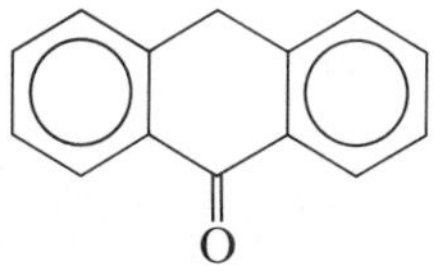

(e) Phthalic anhydride ⟶ *N*-methylphthalimide

(f) Maleic anhydride ⟶

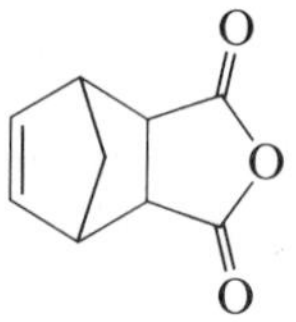

(g) Maleic anhydride ⟶

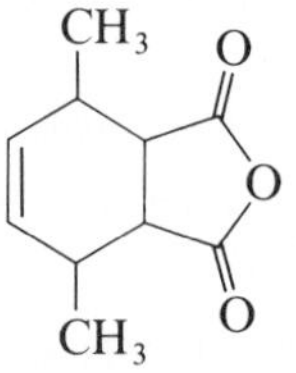

18.29

What products would you expect to obtain when ethyl propanoate reacts with each of the following?

(a) H_3O^+, H_2O (c) 1-Octanol, HCl   (e) $LiAlH_4$, then H_2O

(b) OH^-, H_2O (d) CH_3NH_2 (f) C_6H_5MgBr, then H_2O

18.30

What products would you expect to obtain when propanamide reacts with each of the following?

(a) H_3O^+, H_2O (b) OH^-, H_2O (c) P_4O_{10} and heat

18.31

Outline a simple chemical test that would serve to distinguish between

(a) Benzoic acid and methyl benzoate   (e) Ethyl benzoate and benzamide

(b) Benzoic acid and benzoyl chloride   (f) Benzoic acid and cinnamic acid

(c) Benzoic acid and benzamide   (g) Ethyl benzoate and benzoyl chloride

(d) Benzoic acid and *p*-cresol   (h) 2-Chlorobutanoic acid and butanoic acid

18.32

What products would you expect to obtain when each of the following compounds is heated?

(a) 4-Hydroxybutanoic acid   (c) 2-Hydroxybutanoic acid

(b) 3-Hydroxybutanoic acid   (d) Glutaric acid

(e) $CH_3CHCH_2CH_2CH_2CO^-$
      $\underset{NH_3{}^+}{|}$ with $\overset{O}{\overset{\|}{}}$ on the carbonyl

(f) benzene ring with $\overset{O}{\overset{\|}{C}}-O^-$ and $CH_2NH_3{}^+$

18.33

Give stereochemical formulas for compounds **A** to **Q**.

(a) (R)-$(-)$-2-Butanol $\xrightarrow[\text{pyridine}]{\text{Ts Cl}}$ **A** $\xrightarrow{CN^-}$ **B**(C_5H_9N) $\xrightarrow[H_2O]{H_2SO_4}$

   $(+)$-**C**$(C_5H_{10}O_2)$ $\xrightarrow[(2)\ H_2O]{(1)\ LiAlH_4}$ $(-)$-**D**$(C_5H_{12}O)$

(b) (R)-$(-)$-2-Butanol $\xrightarrow[\text{pyridine}]{PBr_3}$ **E**(C_4H_9Br) $\xrightarrow{CN^-}$ **F**(C_5H_9N) $\xrightarrow[H_2O]{H_2SO_4}$

   $(-)$-**C**$(C_5H_{10}O_2)$ $\xrightarrow[(2)\ H_2O]{(1)\ LiAlH_4}$ $(+)$-**D**$(C_5H_{12}O)$

(c) **A** $\xrightarrow{CH_3CO_2{}^-}$ **G**$(C_6H_{12}O_2)$ $\xrightarrow{OH^-}$ $(+)$-**H**$(C_4H_{10}O)$ + $CH_3CO_2{}^-$

(d) $(-)$-**D** $\xrightarrow{PBr_3}$ **J**$(C_5H_{11}Br)$ $\xrightarrow[\text{ether}]{Mg}$ **K**$(C_5H_{11}MgBr)$ $\xrightarrow[(2)\ H^+]{(1)\ CO_2}$ **L**$(C_6H_{12}O_2)$

(e) (R)-$(+)$-Glyceraldehyde $\xrightarrow{HCN}$ **M**$(C_4H_7NO_3)$ + **N**$(C_4H_7NO_3)$

   Diastereomers, separated
   by fractional crystallization

(f) **M** $\xrightarrow[H_2O]{H_2SO_4}$ **P**$(C_4H_8O_5)$ $\xrightarrow[HNO_3]{(O)}$ *meso*-tartaric acid

(g) **N** $\xrightarrow[H_2O]{H_2SO_4}$ **Q**$(C_4H_8O_5)$ $\xrightarrow[HNO_3]{(O)}$ $(-)$-tartaric acid

18.34

(a) $(\pm)$-Pantetheine and $(\pm)$-pantothenic acid, important intermediates in the synthesis of coenzyme A, were prepared by the following route. Give structures for compounds **A** to **D**.

$$\underset{\overset{|}{CH_3}}{CH_3CHCHO} + H\overset{O}{\overset{\|}{C}}H \xrightarrow[H_2O]{K_2CO_3} (\pm)\text{-}\mathbf{A}(C_5H_{10}O_2) \xrightarrow{HCN}$$

$$(\pm)\text{-}\mathbf{B}(C_6H_{11}NO_2) \xrightarrow{H_3O^+} [(\pm)\text{-}\mathbf{C}(C_6H_{12}O_4)] \xrightarrow{-H_2O}$$

$$(\pm)\text{-}\mathbf{D}(C_6H_{10}O_3) \xrightarrow{H_3{}^+NCH_2CH_2CO^-} (CH_3)_2\overset{\overset{CH_2OH}{|}}{C}\underset{\underset{OH}{|}}{-CHC}\overset{O}{\overset{\|}{}}-NHCH_2CH_2\overset{O}{\overset{\|}{C}}OH$$

A γ-lactone $(\pm)$-Pantothenic acid

$$\downarrow H_2NCH_2CH_2\overset{O}{\overset{\|}{C}}NHCH_2CH_2SH$$

$$(CH_3)_2\overset{\overset{CH_2OH}{|}}{C}\underset{\underset{OH}{|}}{-CH}-\overset{O}{\overset{\|}{C}}-NHCH_2CH_2\overset{O}{\overset{\|}{C}}NHCH_2CH_2SH$$

$(\pm)$-**Pantetheine**

(b) The γ-lactone, ($\pm$) **D**, can be resolved. If the ($-$)-γ-lactone is used in the last step, the pantotheine that is obtained is identical with that obtained naturally. The ($-$)-γ-lactone has the (R) configuration. What is the stereochemistry of naturally occurring pantetheine? (c) What products would you expect to obtain when ($\pm$)-pantetheine is heated with aqueous sodium hydroxide?

18.35

The infrared and proton nmr spectra of phenacetin ($C_{10}H_{13}NO_2$) are given in Fig. 18.2. Phenacetin is an analgesic and antipyretic compound, and is the P of A-P-C tablets (**A**spirin – **P**henacetin – **C**affeine). When phenacetin is heated with aqueous sodium hydroxide, it yields phenetidine ($C_8H_{11}NO$) and sodium acetate. Propose structures for phenacetin and phenetidine.

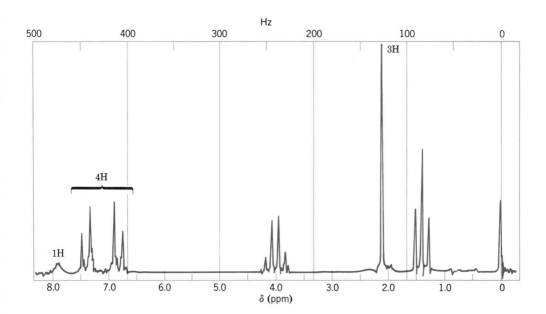

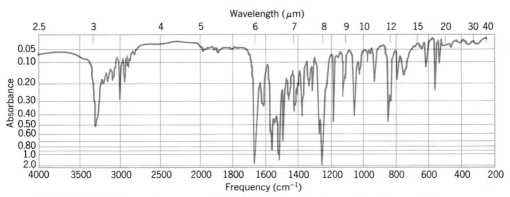

FIGURE 18.2 The proton nmr and infrared spectra of phenacetin. (The proton nmr spectrum, courtesy Varian Associates, Palo Alto, CA. Infrared spectrum, courtesy Sadtler Research Laboratories, Philadelphia.)

18.36

Given here are the proton nmr spectra and carbonyl absorption peaks of five acyl compounds. Propose structures for each.

(a) $C_8H_{14}O_4$

proton nmr spectrum		IR spectrum
Triplet	$\delta 1.2$ (6H)	1740 cm^{-1}
Singlet	$\delta 2.5$ (4H)	
Quartet	$\delta 4.1$ (4H)	

(b) $C_{11}H_{14}O_2$

proton nmr spectrum		IR spectrum
Doublet	$\delta 1.0$ (6H)	1720 cm^{-1}
Multiplet	$\delta 2.1$ (1H)	
Doublet	$\delta 4.1$ (2H)	
Multiplet	$\delta 7.8$ (5H)	

(c) $C_{10}H_{12}O_2$

proton nmr spectrum		IR spectrum
Triplet	$\delta 1.2$ (3H)	1740 cm^{-1}
Singlet	$\delta 3.5$ (2H)	
Quartet	$\delta 4.1$ (2H)	
Multiplet	$\delta 7.3$ (5H)	

(d) $C_2H_2Cl_2O_2$

proton nmr spectrum		IR spectrum
Singlet	$\delta 6.0$	Broad peak 2500–2700 cm^{-1}
Singlet	$\delta 11.70$	1705 cm^{-1}

(e) $C_4H_7ClO_2$

proton nmr spectrum		IR spectrum
Triplet	$\delta 1.3$	1745 cm^{-1}
Singlet	$\delta 4.0$	
Quartet	$\delta 4.2$	

18.37

The active ingredient of the insect repellent "Off" is N,N-diethyl-m-toluamide, m-CH$_3$C$_6$H$_4$CON(CH$_2$CH$_3$)$_2$. Outline a synthesis of this compound starting with m-toluic acid.

18.38

Amides are weaker bases than corresponding amines. For example, most water-insoluble amines (RNH$_2$) will dissolve in dilute aqueous acids (e.g., aqueous HCl, H$_2$SO$_4$, etc.) by forming water-soluble alkylammonium salts (RNH$_3$$^+X^-$). Corresponding amides (RCONH$_2$) *do not dissolve in dilute aqueous acids,* however. Propose an explanation for the much lower basicity of amides when compared to amines.

18.39

While amides are much less basic than amines, they are much stronger acids. Amides have acidity constants (K_a's) in the range 10^{-14}–10^{-16}, while for amines, $K_a = 10^{-33}$–10^{-35}. (a) What factor accounts for the much greater acidity of amides? (b) *Imides,* that is, com-

pounds with the structure $(RC)_2NH$, with the carbonyl O above, are even stronger acids than amides. For imides $K_a = 10^{-9}$–10^{-10}, and, as a consequence, water-insoluble imides dissolve in aqueous NaOH by forming soluble sodium salts. What extra factor accounts for the greater acidity of imides?

18.40

Compound X($C_7H_{12}O_4$) is insoluble in aqueous sodium bicarbonate. The infrared spectrum of X has a strong absorption peak near 1740 cm^{-1}, and its proton-decoupled ^{13}C spectrum is given in Fig. 18.3. Propose a structure for X.

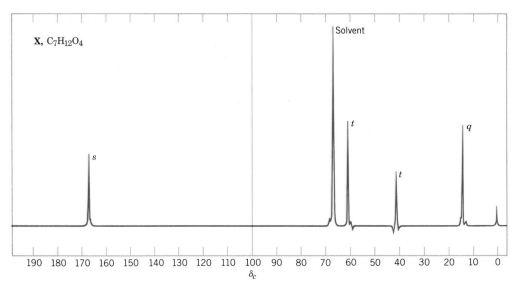

FIGURE 18.3 The proton-decoupled ^{13}C nmr spectrum of compound **X**, Problem 18.40. The letters *s*, *t*, and *q* refer to the signal splitting (singlet, triplet, and quartet) in the proton off-resonance decoupled spectrum. (Adapted from L. F. Johnson and W. C. Jankowski, *Carbon-13 NMR Spectra: A Collection of Assigned, Coded, and Indexed Spectra*, Wiley–Interscience, New York, 1972.)

*18.41

Alkyl thiolacetates, $CH_3\overset{\overset{O}{\|}}{C}SCH_2CH_2R$, can be prepared by a peroxide-initiated reaction between thiolacetic acid, $CH_3\overset{\overset{O}{\|}}{C}SH$, and an alkene, $CH_2{=}CHR$.   (a) Outline a reasonable mechanism for this reaction.   (b) Show how you might use this reaction in a synthesis of 3-methyl-2-butanethiol from 2-methyl-2-butene.

*18.42

On heating, *cis*-4-hydroxycyclohexanecarboxylic acid forms a lactone but *trans*-4-hydroxy-cyclohexanecarboxylic acid does not. Explain.

*18.43

(R)-(+)-Glyceraldehyde can be transformed into (+)-malic acid by the following synthetic route. Give stereochemical structures for the products of each step.

(R)-(+)-Glyceraldehyde $\xrightarrow[\text{oxidation}]{\text{Br}_2,\ \text{H}_2\text{O}}$ (−)-glyceric acid $\xrightarrow{\text{PBr}_3}$

$\qquad$ (−)-3-bromo-2-hydroxypropanoic acid $\xrightarrow{\text{NaCN}}$ $C_4H_5NO_3$ $\xrightarrow[\text{heat}]{\text{H}_3\text{O}^+}$ (+)-malic acid

*18.44

(R)-(+)-Glyceraldehyde can also be transformed into (−)-malic acid. This synthesis begins with the conversion of (R)-(+)-glyceraldehyde into (−)-tartaric acid as shown in Problem 18.33 parts (e) and (g). Then (−)-tartaric acid is allowed to react with phosphorus tribromide in order to replace one alcoholic —OH group with —Br. This step takes place with inversion of configuration at the carbon that undergoes attack. Treating the product of this reaction with zinc and acid produces (−)-malic acid.   (a) Outline all steps in this synthesis by writing stereochemical structures for each intermediate. (b) The step in which (−)-tartaric acid is treated with phosphorus tribromide produces only one stereoisomer even though there are two replaceable —OH groups. How is this possible? (c) Suppose that the step in which (−)-tartaric acid is treated with phosphorus tribromide had taken place with "mixed" stereochemistry— with both inversion and retention at the carbon under attack. How many stereoisomers would have been produced? (d) What difference would this have made to the overall outcome of the synthesis?

*18.45

Cantharidin is a powerful vesicant that can be isolated from dried beetles (*Cantharis vesicatoria* or "Spanish fly"). Outlined here is the stereospecific synthesis of cantharidin reported by Gilbert Stork of Columbia University in 1953. Supply the missing reagents (a) to (n).

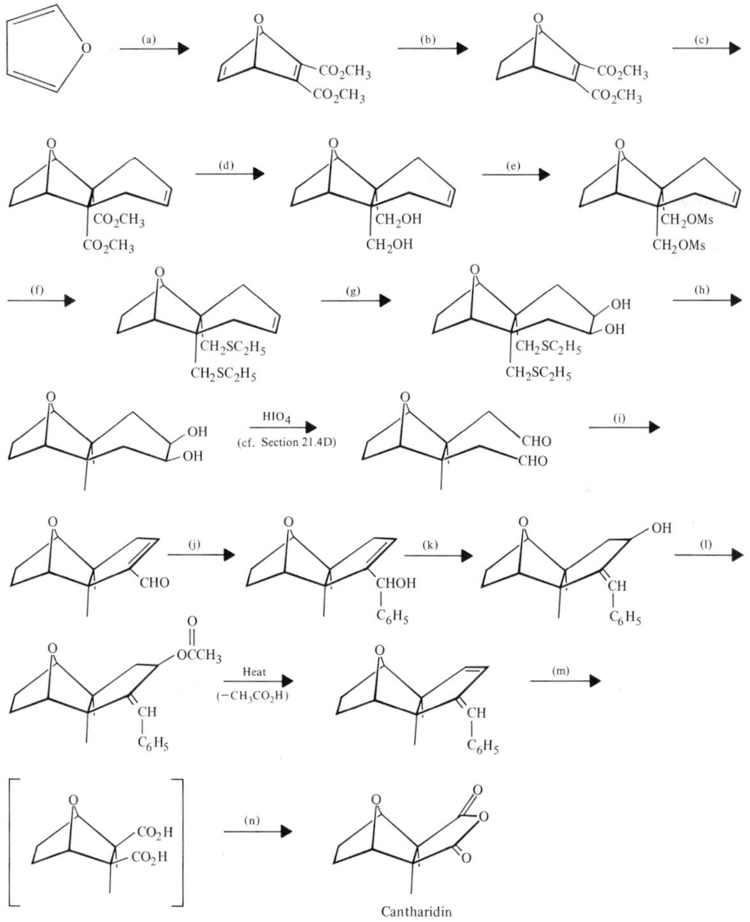

Cantharidin

*18.46

Examine the structure of cantharidin (Problem 18.45) carefully and (a) suggest a possible two-step synthesis of cantharidin starting with furan (Section 11.10). (b) F. von Bruchhausen and H. W. Bersch at the University of Münster attempted this two-step synthesis in 1928 only a few months after Diels and Alder published their first paper describing their new diene addition and found that the expected addition failed to take place. Von Bruchhausen and Bersch also found that although cantharidin is stable at relatively high temperatures, heating cantharidin with a palladium catalyst causes cantharidin to decompose. They identified furan and

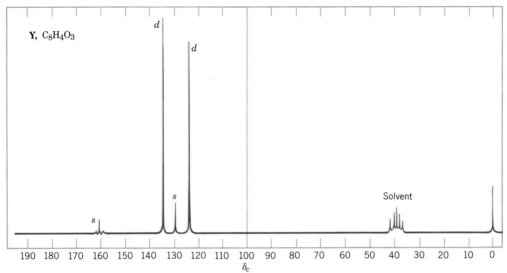

FIGURE 18.4 The proton-decoupled ^{13}C nmr spectrum of compound **Y**, Problem 18.47. The letters *s* and *d* refer to the signal splitting (singlet and doublet) in the proton off-resonance decoupled spectrum. (Adapted from L. F. Johnson and W. C. Jankowski, *Carbon-13 NMR Spectra: A Collection of Assigned, Coded and Indexed Spectra*, Wiley – Interscience, New York, 1972.)

dimethylmaleic anhydride among the decomposition products. What has happened in the decomposition and what does this suggest about why the first step of their attempted synthesis failed?

18.47

Compound **Y** ($C_8H_4O_3$) dissolves slowly when warmed with aqueous sodium bicarbonate. The infrared spectrum of **Y** has strong peaks at 1779 and at 1854 cm^{-1}. The proton-decoupled ^{13}C spectrum of **Y** is given in Fig. 18.4. Propose a structure for **Y**.

CONDENSATION POLYMERS

We saw, in Special Topic D, that large molecules with many repeating subunits—called *polymers*—can be prepared by addition reactions of alkenes. These polymers, we noted, are called *addition polymers.*

Another broad group of polymers has been called *condensation polymers.* These polymers, as their name suggests, are prepared by condensation reactions—reactions in which monomeric subunits are joined through intermolecular eliminations of small molecules such as water or alcohols. Among the most important condensation polymers are *polyamides, polyesters, polyurethanes,* and *formaldehyde resins.*

J.1 Polyamides

Silk and wool are two naturally occurring polymers that man has used for centuries to fabricate articles of clothing. They are examples of a family of compounds that are called *proteins*—a group of compounds that we shall discuss in detail in Chapter 23. At this point we need only to notice (below) that the repeating subunits of proteins are derived from α-amino acids and that these subunits are joined by amide linkages. Proteins, therefore, are polyamides.

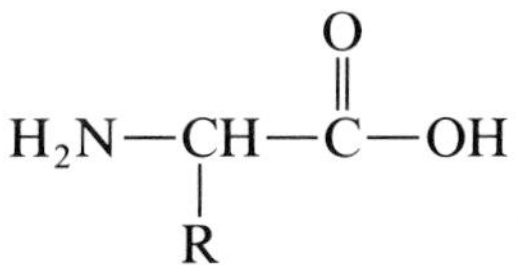

$$H_2N-\underset{\underset{R}{|}}{CH}-\overset{\overset{O}{||}}{C}-OH$$

An α-amino acid

Amide linkages

$$-NH-\underset{\underset{R}{|}}{CH}-\overset{\overset{O}{||}}{C}-NH-\underset{\underset{R}{|}}{CH}-\overset{\overset{O}{||}}{C}-NH-\underset{\underset{R}{|}}{CH}-\overset{\overset{O}{||}}{C}-NH-\underset{\underset{R}{|}}{CH}-$$

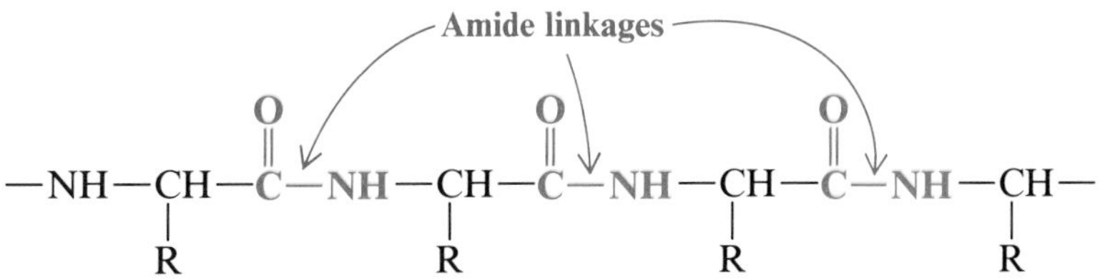

A portion of a polyamide chain as
it might occur in a protein

The search for a synthetic material with properties similar to those of silk led to the discovery of a family of synthetic polyamides called nylons.

One of the most important nylons, called *nylon 6,6,* can be prepared from the six-carbon dicarboxylic acid, adipic acid, and the six-carbon diamine, hexamethylenediamine. In the commercial process these two compounds are allowed to react in equimolar proportions in order to produce a 1 : 1 salt,

$$n\text{HOC} \overset{\displaystyle O}{\overset{\|}{\text{C}}} \overset{}{+}\text{CH}_2 \overset{}{\rightarrowtail_4} \overset{\displaystyle O}{\overset{\|}{\text{C}}}\text{OH} \; + \; n\text{H}_2\text{N} \overset{}{+}\text{CH}_2 \overset{}{\rightarrowtail_6}\text{NH}_2 \longrightarrow$$

Adipic acid   Hexamethylenediamine

$$n\left[\overset{\displaystyle O}{\overset{\|}{^-\text{OC}}} \overset{}{+}\text{CH}_2 \overset{}{\rightarrowtail_4} \overset{\displaystyle O}{\overset{\|}{\text{C}}}\text{—O}^- \quad \text{H}_3\overset{+}{\text{N}} \overset{}{+}\text{CH}_2 \overset{}{\rightarrowtail_6}\overset{+}{\text{N}}\text{H}_3 \right] \xrightarrow[\text{(polymerization)}]{\text{heat}}$$

1 : 1 salt (nylon salt)

$$\overset{\displaystyle O}{\overset{\|}{^-\text{OC}}} \overset{}{+}\text{CH}_2 \overset{}{\rightarrowtail_4} \overset{\displaystyle O}{\overset{\|}{\text{C}}}\left[\text{—NH—}(\text{CH}_2\overset{}{\rightarrowtail_6}\text{NH—}\overset{\displaystyle O}{\overset{\|}{\text{C}}} \overset{}{+}\text{CH}_2 \overset{}{\rightarrowtail_4}\overset{\displaystyle O}{\overset{\|}{\text{C}}} \right]_{n-1}\text{NH—}(\text{CH}_2)_6\text{—NH}_3{}^+$$

Nylon 6,6
(a polyamide)

$$+ (2n - 1)\text{H}_2\text{O}$$

Then, heating the 1 : 1 salt (nylon salt) to a temperature of 270 °C at a pressure of 250 psi (pounds per square inch) causes a polymerization to take place. Water molecules are lost as condensation reactions occur between $\overset{\displaystyle O}{\overset{\|}{\text{—C—O}^-}}$ and $\text{—NH}_3{}^+$ groups of the salt to give the polyamide.

The nylon 6,6 produced in this way has a molecular weight of $\sim 10{,}000$, has a melting point of ~ 250 °C, and when molten can be spun into fibers from a melt. The fibers are then stretched to about four times their original length. This orients the linear polyamide molecules so that they are parallel to the fiber axis and allows hydrogen bonds to form between —NH— and C=O groups on adjacent chains. Called "cold drawing," stretching greatly increases the fibers' strength.

Another type of nylon, nylon 6, can be prepared by a ring-opening polymerization of ε-caprolactam:

$$\overset{\displaystyle O}{\overset{\|}{\Big\langle}}\text{NH} \xrightarrow{\text{H}_2\text{O}} \overset{\displaystyle O}{\overset{\|}{^-\text{OC}}}\text{—}(\text{CH}_2)_5\text{—}\overset{+}{\text{N}}\text{H}_3 \; + \; \overset{\displaystyle O}{\overset{\|}{\Big\langle}}\text{NH} \xrightarrow[250\,°\text{C}]{(-\text{H}_2\text{O})}$$

ε-Caprolactam
(a cyclic amide)

$$\text{—NH}\left[\overset{\displaystyle O}{\overset{\|}{\text{C}}}(\text{CH}_2)_5\text{—NH—}\overset{\displaystyle O}{\overset{\|}{\text{C}}}\text{—}(\text{CH}_2)_5\text{—NH} \right]_n\overset{\displaystyle O}{\overset{\|}{\text{C}}}\text{—}$$

Nylon 6

In this process ε-caprolactam is allowed to react with water, converting some of it to ε-aminocaproic acid. Then heating this mixture at 250 °C drives off water as ε-caprolactam and ε-aminocaproic acid react to produce the polyamide. Nylon 6 can also be converted into fibers by melt spinning.

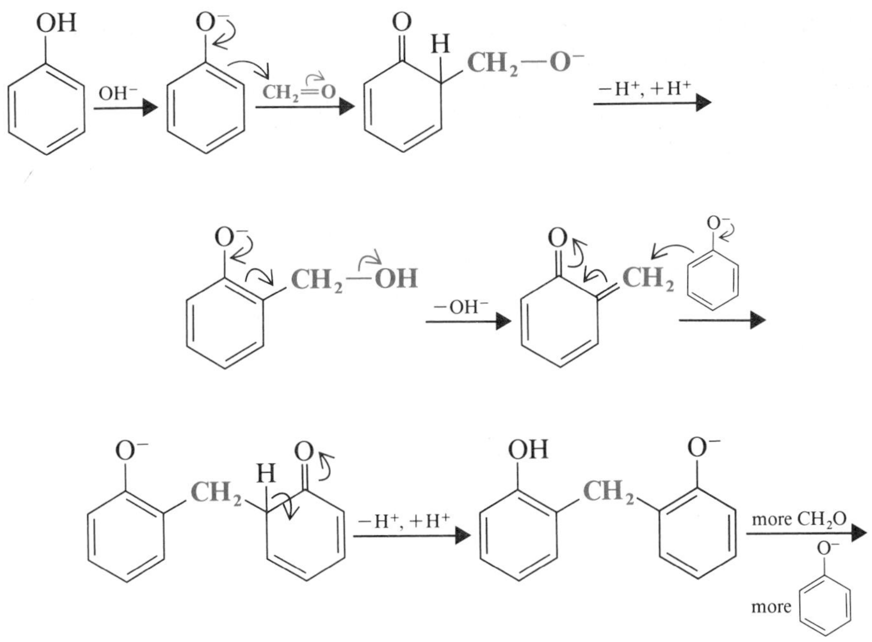

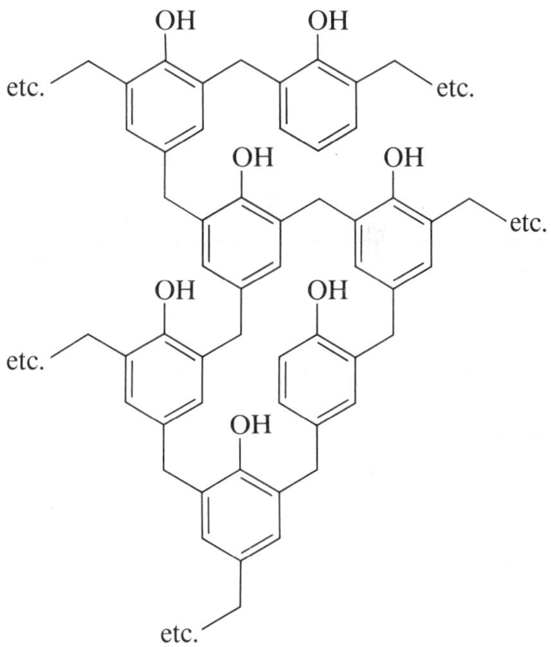

Bakelite

Generally, the polymerization is carried out in two stages. The first polymerization produces a low-molecular-weight fusible (meltable) polymer called a *resole*. The resole can be molded to the desired shape, and then further polymerization produces a very high-molecular-weight polymer, which, because it is highly cross linked, is infusible.

PROBLEM J.8

Using a para-substituted phenol such as *p*-cresol yields a phenol–formaldehyde polymer that is *thermoplastic* rather than *thermosetting*. That is, the polymer remains fusible; it does *not* become impossible to melt. What accounts for this?

PROBLEM J.9

Outline a general mechanism for acid-catalyzed polymerization of phenol and formaldehyde.

CHAPTER NINETEEN

2,4,6-Tribromoaniline and *p*-nitrophenylacetic acid crystals.
2,4,6-Tribromoaniline can be prepared by treating aniline with
excess bromine in water (Section 12.8A).

Amines

19.1 NOMENCLATURE

In common nomenclature most primary amines are named as *alkylamines*. In systematic nomenclature (in parentheses below) they are named by adding the suffix *-amine* to the name of the chain or ring system to which the NH_2 group is attached with elision of the final *e*.

Primary Amines

$$CH_3\overset{..}{N}H_2$$
Methylamine
(methanamine)

$$CH_3CH_2\overset{..}{N}H_2$$
Ethylamine
(ethanamine)

$$CH_3CHCH_2\overset{..}{N}H_2$$
$$|$$
$$CH_3$$
Isobutylamine
(2-methyl-1-propanamine)

—$\overset{..}{N}H_2$
Cyclohexylamine
(cyclohexanamine)

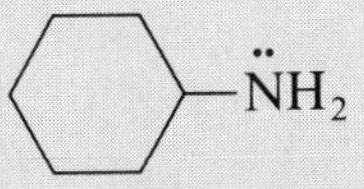

Most secondary and tertiary amines are named in the same general way. In common nomenclature we either designate the organic groups individually if they are different, or use the prefixes *di-* or *tri-* if they are the same. In systematic nomenclature we use the locant *N* to designate substituents attached to a nitrogen atom.

Secondary Amines

$$CH_3\overset{..}{N}HCH_2CH_3$$
Ethylmethylamine
(*N*-methylethanamine)

$$(CH_3CH_2)_2\overset{..}{N}H$$
Diethylamine
(*N*-ethylethanamine)

Tertiary Amines

$$(CH_3CH_2)_3\overset{..}{N}$$
Triethylamine
(*N,N*-diethylethanamine)

$$CH_2CH_3$$
$$|$$
$$CH_3\overset{..}{N}CH_2CH_2CH_3$$
Ethylmethylpropylamine
(*N*-ethyl-*N*-methyl-1-propanamine)

In the IUPAC system, the substituent —NH_2 is called the *amino* group. We often use this system for naming amines containing an OH group or a CO_2H group.

$$H_2\overset{..}{N}CH_2CH_2OH$$
2-Aminoethanol

$$H_2\overset{..}{N}CH_2CH_2\overset{O}{\overset{||}{C}}OH$$
3-Aminopropanoic acid

19.1A Arylamines

Three common arylamines have the following names:

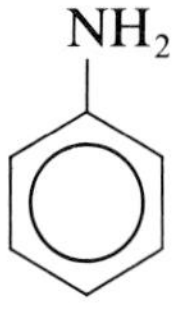

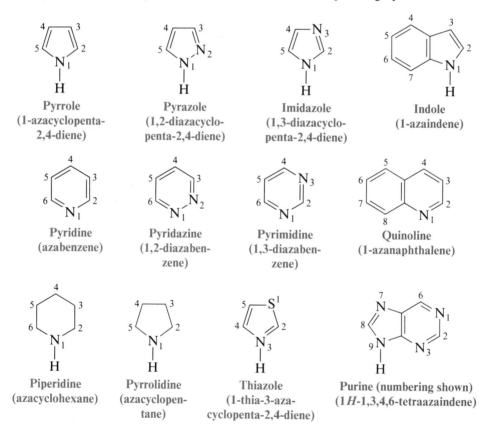

| Aniline | N-Methylaniline | p-Toluidine |
| (benzenamine) | (N-methylben-
zenamine) | (4-methylben-
zenamine) |

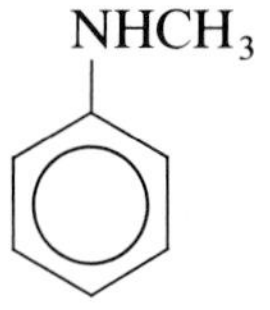

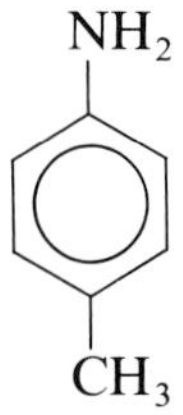

19.1B Heterocyclic Amines

The important *heterocyclic* amines all have common names. In systematic replacement nomenclature the prefixes *aza-, diaza-,* and *triaza-* are used to indicate that nitrogen atoms have replaced carbon atoms in the corresponding hydrocarbon.

Pyrrole
(1-azacyclopenta-
2,4-diene)

Pyrazole
(1,2-diazacyclo-
penta-2,4-diene)

Imidazole
(1,3-diazacyclo-
penta-2,4-diene)

Indole
(1-azaindene)

Pyridine
(azabenzene)

Pyridazine
(1,2-diazaben-
zene)

Pyrimidine
(1,3-diazaben-
zene)

Quinoline
(1-azanaphthalene)

Piperidine
(azacyclohexane)

Pyrrolidine
(azacyclopen-
tane)

Thiazole
(1-thia-3-aza-
cyclopenta-2,4-diene)

Purine (numbering shown)
(1H-1,3,4,6-tetraazaindene)

19.2 PHYSICAL PROPERTIES OF AMINES

Amines are moderately polar substances; they have boiling points that are higher than those of alkanes but generally lower than those of alcohols of comparable molecular weight. Molecules of primary and secondary amines can form strong hydrogen bonds to each other and to water. Molecules of tertiary amines cannot form hydrogen bonds to each other, but they can form hydrogen bonds to molecules of water. As a result, tertiary amines generally boil at lower temperatures than primary and secondary amines of comparable molecular weight, but all low-molecular-weight amines are very water soluble.

Table 19.1 lists the physical properties of some common amines.

19.3 BASICITY OF AMINES: AMINE SALTS

Amines are relatively weak bases. They are stronger bases than water but are far weaker bases than hydroxide ions, alkoxide ions, and carbanions.

A convenient expression for relating basicities is a quantity called the *basicity constant*, K_b, or its negative logarithm, pK_b. When an amine dissolves in water, the following equilibrium is established:

$$R\ddot{N}H_2 + H_2O \rightleftarrows RNH_3^+ + OH^-$$

K_b and pK_b are given by the expressions:

$$K_b = \frac{[RNH_3^+][OH^-]}{[RNH_2]} \qquad pK_b = -\log K_b$$

The larger the value of K_b (or the *smaller* the value of pK_b), the greater is the tendency of the amine to accept a proton from water and, thus, the greater will be the concentrations of RNH_3^+ and OH^- in the solution. Larger values of K_b, therefore, are associated with those amines that are stronger bases, and smaller values of K_b are associated with those amines that are weaker bases. (Just the opposite is true for values of pK_b.)

The basicity constant of ammonia at 25 °C is 1.8×10^{-5}.

$$\ddot{N}H_3 + H_2O \rightleftarrows \overset{+}{N}H_4 + OH^-$$

$$K_b = 1.8 \times 10^{-5} = \frac{[NH_4^+][OH^-]}{[NH_3]} \qquad pK_b = -\log(1.8 \times 10^{-5}) = 4.74$$

Another way to compare the base strength of amines is to compare the acidity constants (or pK_a's) of their conjugate acids, the alkylaminium ions. The expression for this acidity constant is as follows:

$$RNH_3^+ + H_2O \rightleftarrows RNH_2 + H_3O^+$$

$$K_a = \frac{[RNH_2][H_3O^+]}{[RNH_3^+]} \qquad pK_a = -\log K_a$$

Multiplying the expression for the K_b of an amine times the expression for the K_a of its conjugate acid, the alkylaminium ion, yields the expression for the ion product constant of water, which is equal to $1.0 \times 10^{-14} \ M^2$.

$$K_a K_b = \frac{[RNH_2][H_3O^+]}{[RNH_3^+]} \times \frac{[RNH_3^+][OH^-]}{[RNH_2]} = [H_3O^+][OH^-] = 10^{-14} \ M^2$$

therefore, $\qquad\qquad\qquad K_a K_b = 1.0 \times 10^{-14} \ M^2$

and $\qquad\qquad\qquad\qquad pK_a + pK_b = 14$

TABLE 19.1 Physical properties of amines

NAME	STRUCTURE	mp (°C)	bp (°C)	WATER SOLUBILITY (25 °C) (g/100 mL)	K_b (25 °C)
Primary Amines					
Methylamine	CH_3NH_2	−94	−6	Very soluble	4.4×10^{-4}
Ethylamine	$CH_3CH_2NH_2$	−81	17	Very soluble	5.6×10^{-4}
Propylamine	$CH_3CH_2CH_2NH_2$	−83	49	Very soluble	4.7×10^{-4}
Isopropylamine	$(CH_3)_2CHNH_2$	−101	33	Very soluble	5.3×10^{-4}
Butylamine	$CH_3(CH_2)_2CH_2NH_2$	−51	78	Very soluble	4.1×10^{-4}
Isobutylamine	$(CH_3)_2CHCH_2NH_2$	−86	68	Very soluble	3.1×10^{-4}
sec-Butylamine	$CH_3CH_2CH(CH_3)NH_2$	−104	63	Very soluble	3.6×10^{-4}
tert-Butylamine	$(CH_3)_3CNH_2$	−68	45	Very soluble	2.8×10^{-4}
Cyclohexylamine	$Cyclo\text{-}C_6H_{11}NH_2$	−18	134	Sl. soluble	4.4×10^{-4}
Benzylamine	$C_6H_5CH_2NH_2$	10	185	Sl. soluble	2.0×10^{-5}
Aniline	$C_6H_5NH_2$	−6	184	3.7	3.8×10^{-10}
p-Toluidine	$p\text{-}CH_3C_6H_4NH_2$	44	200	Sl. soluble	1.2×10^{-9}
p-Anisidine	$p\text{-}CH_3OC_6H_4NH_2$	57	244	V. sl. soluble	2.0×10^{-9}
p-Chloroaniline	$p\text{-}ClC_6H_4NH_2$	73	232	Insoluble	1.0×10^{-10}
p-Nitroaniline	$p\text{-}NO_2C_6H_4NH_2$	148	332	Insoluble	1.0×10^{-13}
Secondary Amines					
Dimethylamine	$(CH_3)_2NH$	−92	7	Very soluble	5.2×10^{-4}
Diethylamine	$(CH_3CH_2)_2NH$	−48	56	Very soluble	9.6×10^{-4}
Dipropylamine	$(CH_3CH_2CH_2)_2NH$	−40	110	Very soluble	9.5×10^{-4}
N-Methylaniline	$C_6H_5NHCH_3$	−57	196	Sl. soluble	5.0×10^{-10}
Diphenylamine	$(C_6H_5)_2NH$	53	302	Insoluble	6.0×10^{-14}
Tertiary Amines					
Trimethylamine	$(CH_3)_3N$	−117	2.9	Very soluble	5.0×10^{-5}
Triethylamine	$(CH_3CH_2)_3N$	−115	90	14	5.7×10^{-4}
Tripropylamine	$(CH_3CH_2CH_2)_3N$	−93	156	Sl. soluble	4.4×10^{-4}
N,N-Dimethylaniline	$C_6H_5N(CH_3)_2$	3	194	Sl. soluble	11.5×10^{-10}

When we examine the basicity constants of the amines given in Table 19.1, we see that most primary aliphatic amines (e.g., methylamine, ethylamine) are somewhat stronger bases than ammonia:

	$\ddot{N}H_3$	$CH_3\ddot{N}H_2$	$CH_3CH_2\ddot{N}H_2$	$CH_3CH_2CH_2\ddot{N}H_2$
K_b	1.8×10^{-5}	4.4×10^{-4}	5.6×10^{-4}	4.7×10^{-4}
pK_b	4.74	3.36	3.25	3.33

We can account for this on the basis of the electron-releasing ability of an alkyl group. An alkyl group releases electrons, and it *stabilizes* the alkylaminium ion that results from the acid–base reaction *by dispersing its positive charge.* It stabilizes the alkylaminium ion to a greater extent than it stabilizes the amine.

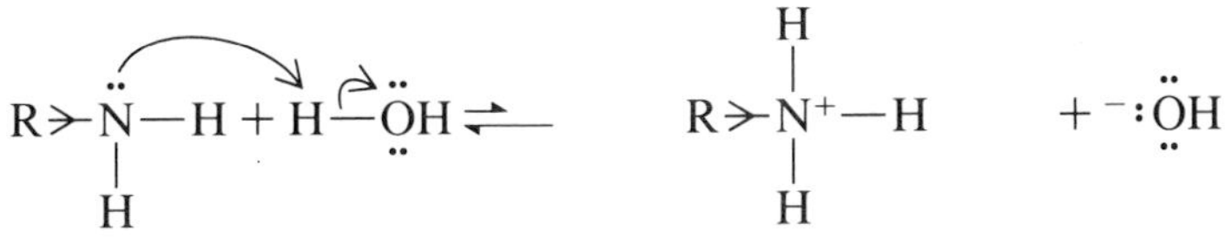

By releasing electrons, R⊱
stabilizes the alkylaminium ion
through dispersal of charge

This explanation is supported by measurements showing that in the *gas phase* the basicities of the following amines increase with increasing methyl substitution:

$$(CH_3)_3N > (CH_3)_2NH > CH_3NH_2 > NH_3$$

However, this is not the order of basicity of these amines in aqueous solution (cf. Table 19.1). In aqueous solution the aminium ions formed from primary and secondary amines are stabilized by solvation much more effectively than are the ions formed from tertiary amines. Therefore, in aqueous solution tertiary amines are less basic than corresponding secondary amines.

19.3A Basicity of Arylamines

When we examine the basicity constants of the aromatic amines (e.g., aniline, *p*-toluidine) in Table 19.1, we see that they are much weaker bases than the corresponding nonaromatic amine, cyclohexylamine.

	Cyclo-$C_6H_{11}NH_2$	$C_6H_5NH_2$	p-$CH_3C_6H_4NH_2$
K_b	4.4×10^{-4}	3.8×10^{-10}	1.2×10^{-9}
pK_b	3.36	9.42	8.92

We can account for this effect on the basis of resonance contributions to the overall hybrid of an arylamine amine. For aniline, the following contributors are important.

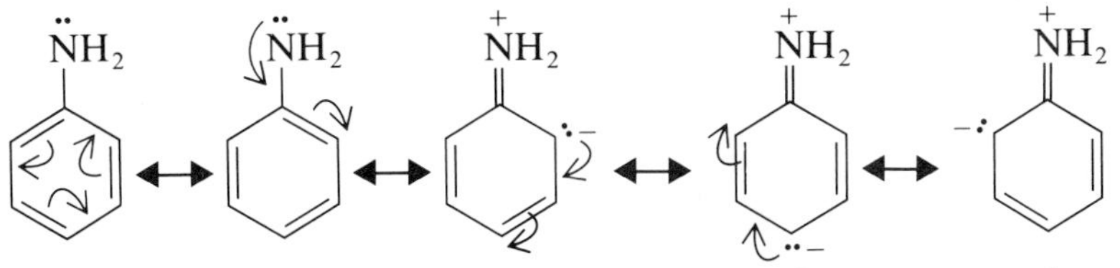

Structures **1** and **2** are the Kekulé structures that contribute to any benzene derivative. Structures **3–5**, however, *delocalize* the unshared electron pair of the nitrogen
over the ortho and para positions of the ring. This delocalization of the electron pairs
makes it less available to a proton but, more importantly, *delocalization of the
electron pair stabilizes aniline.*
    When aniline accepts a proton it becomes an anilinium ion.

$$C_6H_5\ddot{N}H_2 + H_2O \rightleftharpoons C_6H_5\overset{+}{N}H_3 + \overset{-}{O}H$$

**Anilinium
ion**

Since the electron pair of the nitrogen atom accepts the proton, we are able to write
only *two* resonance structures for the anilinium ion — the two Kekulé structures:

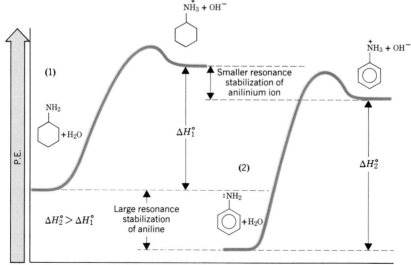

Structures corresponding to **3–5** are not possible for the anilinium ion and, consequently, resonance does not stabilize the anilinium ion to as great an extent as it does
aniline itself. This greater stabilization of the reactant (aniline) when compared to
that of the product (anilinium ion) means that ΔH° for the reaction,

$$\text{Aniline} + H_2O \longrightarrow \text{anilinium ion} + OH^- \qquad \text{(Fig. 19.1)}$$

will be a larger positive quantity than that for the reaction,

$$\text{Cyclohexylamine} + H_2O \longrightarrow \text{cyclohexylaminium ion} + OH^-$$

Aniline, as a result, is the weaker base.

FIGURE 19.1 Potential energy diagram for the reaction of cyclohexylamine
with H_2O (1), and for the reaction of an aniline with H_2O (2). (The curves are
aligned for comparison only.)

19.3B Amines versus Amides

Although amides are superficially similar to amines, they are far less basic (even less basic than arylamines). The K_b of a typical amide is about 10^{-14} ($pK_b = 14$).

This lower basicity of amides when compared to amines can also be understood in terms of resonance. An amide is stabilized by resonance involving the nonbonding pair of electrons on the nitrogen atom. However, an amide protonated on its nitrogen atom lacks this type of resonance stabilization. (Both species are stabilized by resonance structures involving the carbonyl group.)

Amide

$$\overset{\ddots}{:}\overset{\textstyle O}{\underset{\textstyle \parallel}{}} \quad\quad :\overset{\ddots}{O}:^- \quad\quad :\overset{\ddots}{O}:^-$$

$$R-\overset{\parallel}{C}-\overset{\ddots}{N}H_2 \;\longleftrightarrow\; R-\underset{+}{C}-\overset{\ddots}{N}H_2 \;\longleftrightarrow\; R-C{=}NH_2{}^+ \quad\quad \begin{array}{l}\text{Larger}\\\text{resonance}\\\text{stabilization}\end{array}$$

N-Protonated Amide

$$:\overset{\ddots}{O} \quad\quad\quad :\overset{\ddots}{O}:^-$$

$$R-\overset{\parallel}{C}-NH_3{}^+ \;\longleftrightarrow\; R-\underset{+}{C}-NH_3{}^+ \quad\quad \begin{array}{l}\text{Smaller}\\\text{resonance}\\\text{stabilization}\end{array}$$

Consequently, an amide is stabilized by resonance much more than its conjugate acid. This greater stabilization of the reactant (the amide) when compared to that of the product (the N-protonated amide) means that ΔH for the reaction,

$$R-\overset{\overset{\textstyle O}{\parallel}}{C}-NH_2 + H_2O \longrightarrow R-\overset{\overset{\textstyle O}{\parallel}}{C}-NH_3{}^+ + OH^-$$

will be a larger positive quantity than that for the reaction,

$$R-\overset{\ddots}{N}H_2 + H_2O \longrightarrow R-NH_3{}^+ + OH^-$$

and explains why the amide is the weaker base.

> The nitrogen atoms of amides are so weakly basic that when an amide accepts a proton, it does so on its oxygen atom instead. Protonation on the oxygen atom occurs even though oxygen atoms (because of their greater electronegativity) are typically less basic than nitrogen atoms. Notice, however, that if an amide accepts a proton on its oxygen atom, resonance stabilization involving the nonbonding electron pair of the nitrogen atom is possible.

$$:\overset{\textstyle OH^+}{\underset{\parallel}{}} \quad\quad :\overset{\ddots}{O}H \quad\quad :\overset{\ddots}{O}H$$

$$R-\overset{\parallel}{C}-\overset{\ddots}{N}H_2 \;\longleftrightarrow\; R-\underset{+}{C}-\overset{\ddots}{N}H_2 \;\longleftrightarrow\; R-C{=}NH_2{}^+$$

19.3C Aminium Salts and Quaternary Ammonium Salts

When primary, secondary, and tertiary amines act as bases and react with acids, they form compounds called **aminium salts.** In an aminium salt the positively charged nitrogen atom is attached to at least one hydrogen atom.

$$CH_3CH_2\overset{\cdot\cdot}{N}H_2 + HCl \xrightarrow{H_2O} CH_3CH_2\overset{+}{N}H_3 \quad Cl^-$$

Ethylaminium chloride
(an amine salt)

$$(CH_3CH_2)_2\overset{\cdot\cdot}{N}H + HBr \xrightarrow{H_2O} (CH_3CH_2)_2\overset{+}{N}H_2 \quad Br^-$$

Diethylaminium bromide

$$(CH_3CH_2)_3\overset{\cdot\cdot}{N} + HI \xrightarrow{H_2O} (CH_3CH_2)_3\overset{+}{N}H \quad I^-$$

Triethylaminium iodide

When the central nitrogen atom of a compound is positively charged *but is not attached to a hydrogen atom* the compound is called a **quaternary ammonium salt.** For example,

$$\begin{array}{c} CH_2CH_3 \\ | \\ CH_3CH_2-\overset{+}{N}-CH_2CH_3 \quad Br^- \\ | \\ CH_2CH_3 \end{array}$$

Tetraethylammonium bromide
(a quaternary ammonium salt)

Quaternary ammonium halides — because they do not have an unshared electron pair on the nitrogen atom — cannot act as bases.

$$(CH_3CH_2)_4\overset{+}{N} \quad Br^-$$

Tetraethylammonium bromide
(does not undergo reaction with acid)

Quaternary ammonium *hydroxides,* however, are strong bases. As solids, or in solution, they consist *entirely* of quaternary ammonium cations (R_4N^+) and hydroxide ions (OH^-); they are, therefore, strong bases — as strong as sodium or potassium hydroxide. Quaternary ammonium hydroxides react with acids to form quaternary ammonium salts:

$$(CH_3)_4\overset{+}{N} \quad OH^- + HCl \longrightarrow (CH_3)_4\overset{+}{N} \quad Cl^- + H_2O$$

Almost all alkylaminium chlorides, bromides, iodides, and sulfates are soluble in water. Thus, primary, secondary, or tertiary amines that are not soluble in water will dissolve in dilute aqueous HCl, HBr, HI, or H_2SO_4. Solubility in dilute acid provides a convenient chemical method for distinguishing amines from nonbasic compounds that are insoluble in water. Solubility in dilute acid also gives us a useful method for separating amines from nonbasic compounds that are insoluble in water.

$$\overset{\diagdown}{\underset{\diagup}{}}N: \quad + \quad HX \quad \longrightarrow \quad \overset{\diagdown}{\underset{\diagup}{}}\overset{+}{N}-H \quad X^-$$
$$\text{(or } H_2SO_4) \qquad\qquad \text{(or } HSO_4^+)$$

Water-insoluble   **Water-soluble**
amine   **aminium salt**

Because amides are far less basic than amines, water-insoluble amides *do not dissolve* in dilute aqueous HCl, HBr, HI, or H_2SO_4.

$$\begin{array}{c} O \\ \| \\ R-C-NH_2 \end{array}$$

Water-insoluble amide
(not soluble in aqueous acids)

PROBLEM 19.1

Outline a procedure for separating hexylamine from cyclohexane using dilute HCl, aqueous NaOH, and ether.

PROBLEM 19.2

Outline a procedure for separating a mixture of benzoic acid, *p*-cresol, aniline, and benzene using acids, bases, and organic solvents.

19.3D Optically Active Amines as Resolving Agents

Optically active amines are often used to resolve racemic forms of acidic compounds. We can illustrate the principles involved in this procedure by showing how a racemic form of an organic acid might be resolved (separated) into its enantiomers with the single enantiomer of an amine (Fig. 19.2) used as a resolving agent.

In this procedure the single enantiomer of an amine (*R*)-1-phenylethylamine is added to the racemic form of a water-insoluble acid. The salts that form are not enantiomers: They are diastereomers. (The stereocenters of the acid portion of the salts are enantiomerically related to each other, but the stereocenters of the amine portion are not.) The diastereomers have different solubilities and can be separated by careful crystallization. The separated salts are then acidified with hydrochloric

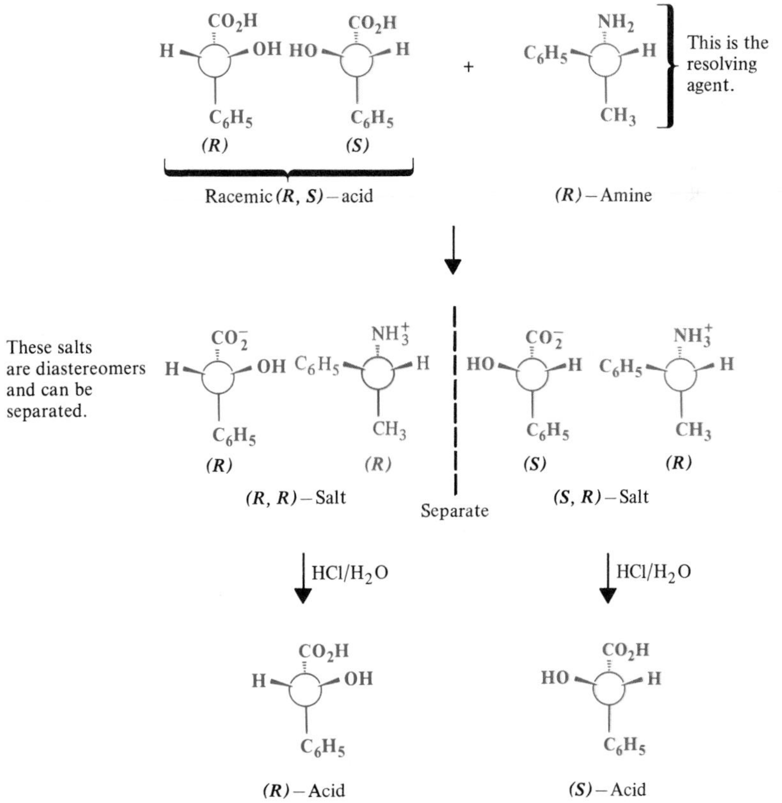

FIGURE 19.2 The resolution of the racemic form of an organic acid by the use of an optically active amine. Acidification of the separated diastereomeric salts causes the enantiomeric acids to precipitate (assuming they are insoluble in water) and leaves the resolving agent in solution as its conjugate acid.

acid and the enantiomeric acids precipitate from the separate solutions. The amine remains in solutions as its hydrochloride salt.

　　Single enantiomers that are employed as resolving agents are often readily available from natural sources. Because most of the chiral organic molecules that occur in living organisms are synthesized by enzymatically catalyzed reactions, most of them occur as single enantiomers. Naturally occurring optically active amines such as (−)-quinine, (−)-strychnine, and (−)-brucine are often employed as resolvng agents for racemic acids. Acids such as (+)- or (−)-tartaric acid are often used for resolving racemic bases.

19.4 SOME BIOLOGICALLY IMPORTANT AMINES

A large number of medically and biologically important compounds are amines. Listed here are some important examples.

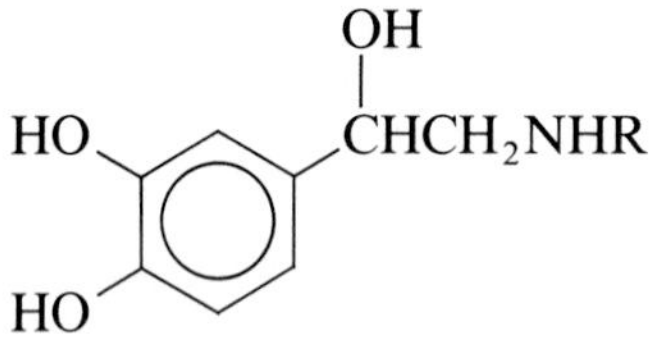

R = CH₃, Adrenaline (epinephrine)
R = H, Noradrenaline (norepinephrine)

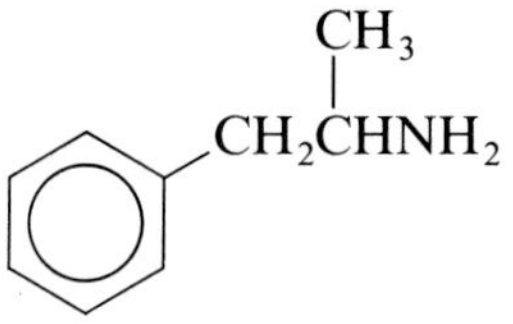

Amphetamine
(benzedrine)

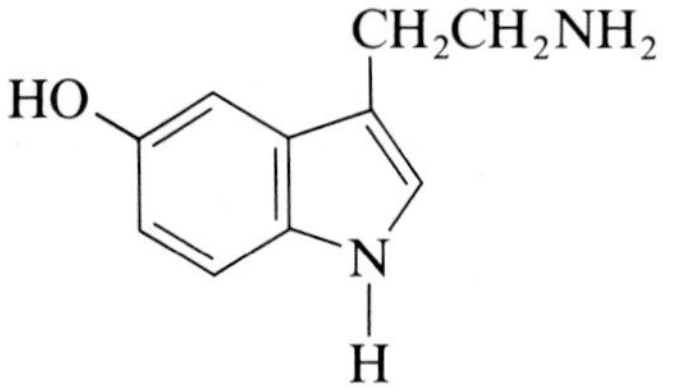

Serotonin

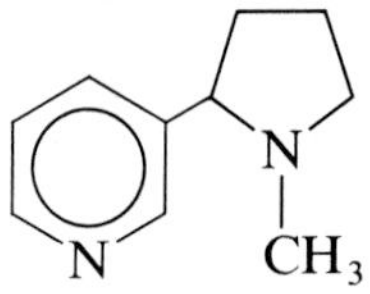

Nicotine

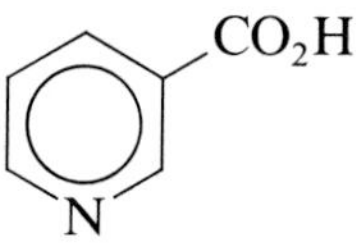

Nicotinic acid
(niacin)

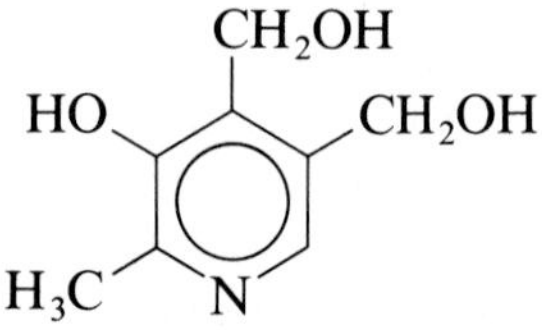

Pyridoxine
(vitamin B₆)

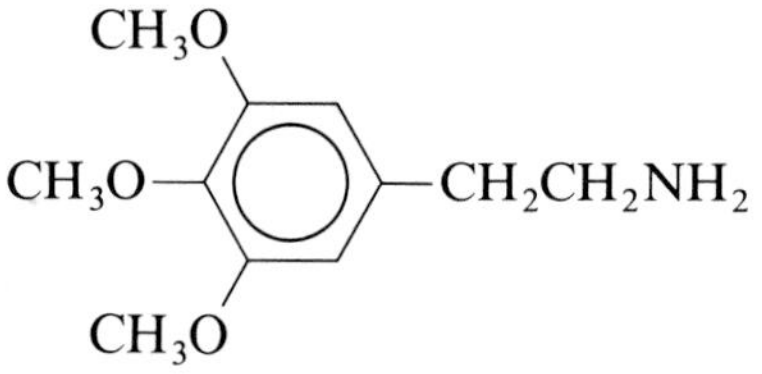

Mescaline

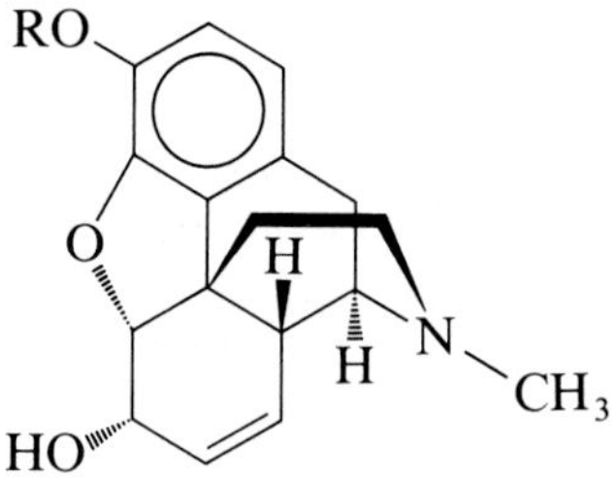

Morphine (R = H)
Codeine (R = CH₃)

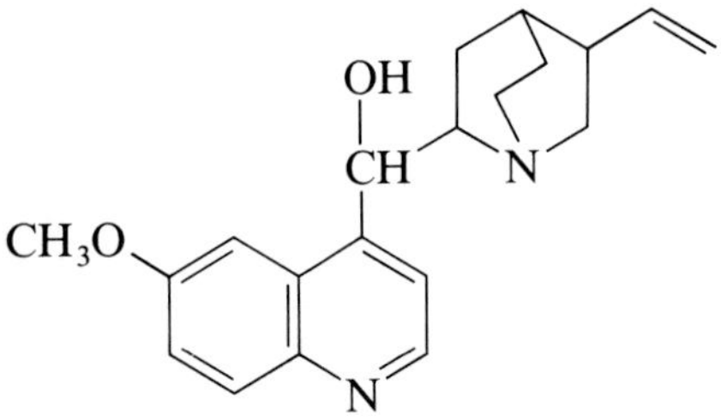

Quinine

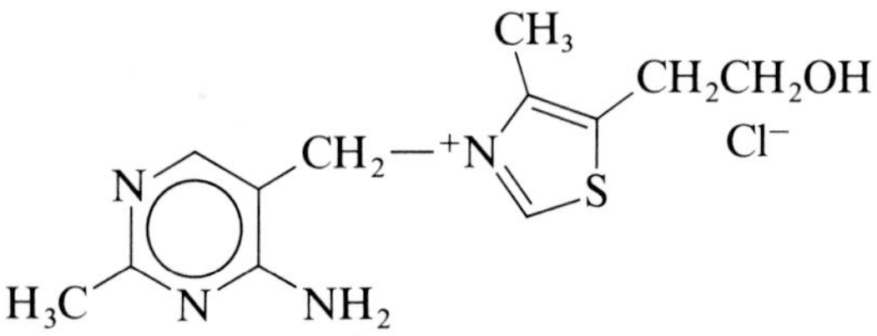

Thiamine chloride
(vitamin B₁)

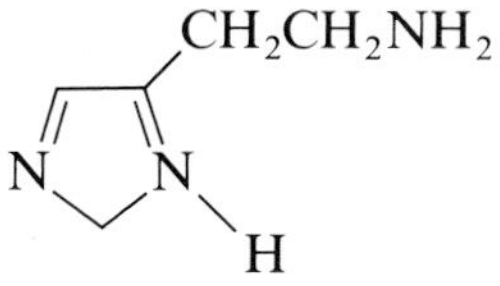

$$CH_2CH_2NH_2$$

Histamine

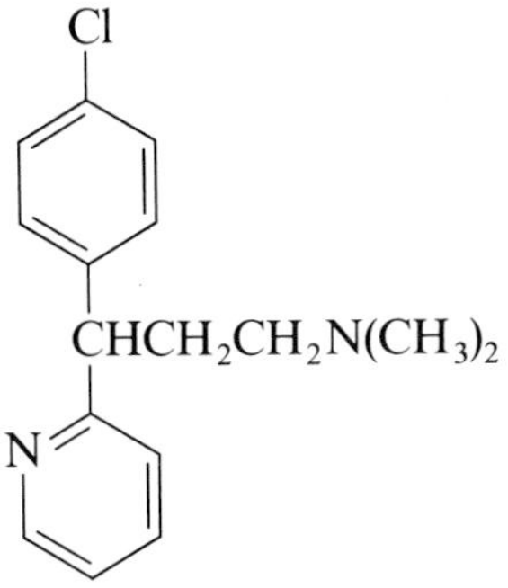

Chlorpheniramine

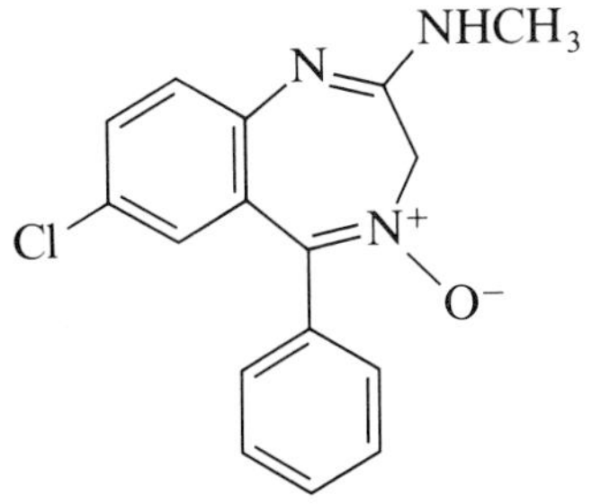

Chlorodiazepoxide
(Librium)

Many of these compounds have powerful physiological and psychological effects. Adrenaline and noradrenaline are two hormones secreted in the medulla of the adrenal gland. Released into the bloodstream when an animal senses danger, adrenaline causes an increase in blood pressure, a strengthening of the heart rate, and a widening of the passages of the lungs. All of these effects prepare the animal to fight or flee. Noradrenaline also causes an increase in blood pressure, and it is involved in the transmission of impulses from the end of one nerve fiber to the next. Serotonin is a compound of particular interest because it appears to be important in maintaining stable mental processes. It has been suggested that the mental disorder schizophrenia may be connected with abnormalities in the metabolism of serotonin.

Amphetamine (a powerful stimulant) and mescaline (a hallucinogen) have structures similar to those of serotonin, adrenaline, and noradrenaline. They are all derivatives of 2-phenylethylamine (see following diagram). (In serotonin the nitrogen is connected to the benzene ring to create a five-membered ring.) The structural similarities of these compounds must be related to their physiological and psychological effects because many other compounds with similar properties are also derivatives of 2-phenylethylamine. Examples (not shown) are *N*-methylamphetamine and LSD. Even morphine and codeine, two powerful analgesics, have a 2-phenylethylamine system as a part of their structures. (Morphine and codeine are examples of compounds called alkaloids, which are discussed in more detail in Special Topic L. Try to locate the 2-phenylethylamine system in their structures now, however.)

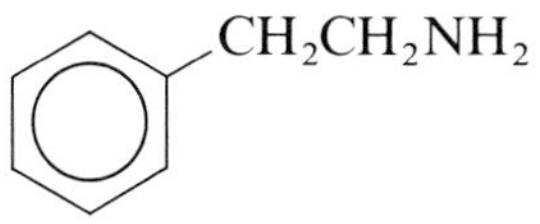

$$CH_2CH_2NH_2$$

2-Phenylethylamine

A number of amines are vitamins. These include nicotinic acid and nicotinamide (the antipellagra factors), pyridoxine (vitamin B_6), and thiamine chloride (vitamin B$_1$). Nicotine is a toxic alkaloid found in tobacco that makes smoking habit forming. Histamine, another toxic amine, is found bound to proteins in nearly all tissues of the body. Release of free histamine causes the symptoms associated with allergic reactions and the common cold. Chlorpheniramine, an "antihistamine," is an ingredient of many over-the-counter cold remedies.

Chlorodiazepoxide, an interesting compound with a seven-membered ring, is one of the most widely prescribed tranquilizers. (Chlorodiazepoxide also contains a positively charged nitrogen, present as an N-oxide.)

Acetylcholine and choline (see following reaction) contain a quaternary ammonium group. Being small and ionic, both compounds are highly soluble in water. Acetylcholine is vital in the process by which impulses are transmitted across junctions between nerves in muscles. After acetylcholine is released by the nerve and moves to a receptor site, contraction of the muscle is stimulated. For the muscle to contract again, the acetylcholine must be removed. This is done by an enzyme, cholinesterase, which hydrolyzes acetylcholine to choline and acetic acid (or acetate ion).

$$(CH_3)_3\overset{+}{N}CH_2CH_2O\overset{\overset{\displaystyle O}{\parallel}}{C}CH_3 + H_2O \xrightarrow{\text{cholinesterase}} (CH_3)_3\overset{+}{N}CH_2CH_2OH + CH_3CO_2H$$

$$\text{Acetylcholine} \qquad\qquad\qquad \text{Choline}$$

The group that binds acetylcholine to the enzyme is the quaternary ammonium group $(CH_3)_3\overset{+}{N}CH_2$—. Other compounds that have this grouping can inhibit cholinesterase. Included among them are compounds used in surgery as muscle relaxants:

$$(CH_3)_3\overset{+}{N}CH_2CH_2CH_2CH_2CH_2CH_2CH_2CH_2CH_2CH_2\overset{+}{N}(CH_3)_3 \quad 2Br^-$$
*Decamethonium bromide**

$$(CH_3)_3\overset{+}{N}CH_2CH_2O\overset{\overset{\displaystyle O}{\parallel}}{C}CH_2CH_2\overset{\overset{\displaystyle O}{\parallel}}{C}OCH_2CH_2\overset{+}{N}(CH_3)_3 \quad 2Br^-$$
*Succinylcholine bromide**

Decamethonium bromide has a relatively long-lasting effect. Succinylcholine bromide, because it is an ester and can be hydrolyzed, has a muscle-relaxing effect of much shorter duration.

19.5 PREPARATION OF AMINES

19.5A Through Nucleophilic Substitution Reactions

Salts of primary amines can be prepared from ammonia and alkyl halides by nucleophilic substitution reactions. Subsequent treatment of the resulting amnium salts with base gives primary amines.

---

*These names are nonsystematic names.

$$\ddot{N}H_3 + R \overset{\frown}{-} X \longrightarrow R - NH_3^+ X^- \xrightarrow{OH^-} RNH_2$$

This method is of very limited synthetic application because multiple alkylations occur. When ethyl bromide reacts with ammonia, for example, the ethylammonium bromide that is produced initially can react with ammonia to liberate ethylamine. Ethylamine can then compete with ammonia and react with ethyl bromide to give diethylammonium bromide. Repetitions of acid–base and alkylation reactions ultimately produce some tertiary amines and even some quaternary ammonium salts if the alkyl halide is present in excess.

$$\ddot{N}H_3 + CH_3CH_2 \overset{\frown}{-} Br \longrightarrow CH_3CH_2 - NH_3^+ + Br^-$$

$$CH_3CH_2 - \overset{\overset{H}{|}}{\underset{\underset{H}{|}}{N}} - H^+ + :NH_3 \longrightarrow CH_3CH_2\ddot{N}H_2 + NH_4^+$$

$$CH_3CH_2\ddot{N}H_2 + CH_3CH_2 \overset{\frown}{-} Br \longrightarrow (CH_3CH_2)_2\overset{+}{N}H_2 + Br^- \text{ etc.}$$

Multiple alkylations can be minimized by using a large excess of ammonia. (Why?) An example of this technique can be seen in the synthesis of alanine from 2-bromopropanoic acid:

$$\underset{\substack{| \\ Br \\ (1\ mole)}}{CH_3CHCO_2H} + \underset{(70\ moles)}{NH_3} \longrightarrow \underset{\substack{| \\ NH_2 \\ (65-70\%) \\ \textbf{Alanine}}}{CH_3CHCO_2^-NH_4^+}$$

A much better method for preparing a primary amine from an alkyl halide is first to convert the alkyl halide to an alkyl azide ($R - N_3$) by a nucleophilic substitution reaction:

$$R - X + \underset{\substack{\textbf{Azide} \\ \textbf{ion}}}{N_3^-} \xrightarrow[(-X^-)]{S_N 2} \underset{\substack{\textbf{Alkyl} \\ \textbf{azide}}}{R - N \overset{+}{=} N \overset{-}{=} N} \xrightarrow[\substack{or \\ LiAlH_4}]{Na/alcohol} RNH_2$$

Then the alkyl azide can be reduced to a primary amine with sodium and alcohol or with lithium aluminum hydride. A word of *caution:* Alkyl azides are explosive and low-molecular-weight alkyl azides should not be isolated but should be kept in solution.

Potassium phthalimide (see following reaction) can also be used to prepare primary amines by a method known as the *Gabriel synthesis.* This synthesis also avoids the complications of multiple alkylations that occur when alkyl halides are treated with ammonia:

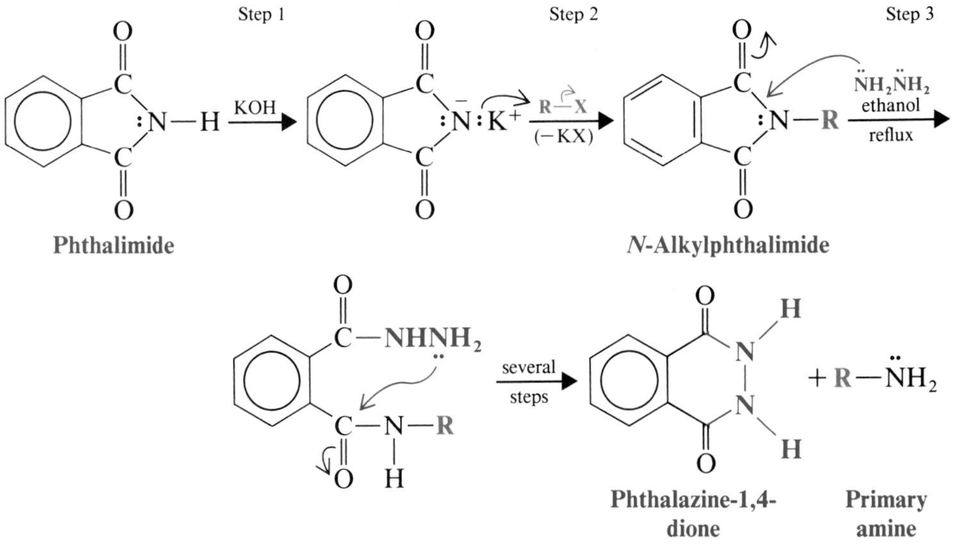

Phthalimide   **N-Alkylphthalimide**

Phthalimide is quite acidic ($K_a \approx 10^{-9}$); it can be converted to potassium phthalimide by potassium hydroxide (step 1). The phthalimide anion is a strong nucleophile and (in step 2) it reacts with an alkyl halide to give an N-alkylphthalimide. At this point, the N-alkylphthalimide can be hydrolyzed with aqueous acid or base, but the hydrolysis is often difficult. It is often more convenient to treat the N-alkylphthalimide with hydrazine (NH_2NH_2) in refluxing ethanol (step 3) to give a primary amine and phthalazine-1,4-dione.

Syntheses of amines using the Gabriel synthesis are, as we might expect, re-stricted to the use of methyl, primary, and secondary alkyl halides. The use of tertiary halides leads almost exclusively to eliminations.

PROBLEM 19.3

(a) Write resonance structures for the phthalimide anion that will account for the acidity of phthalimide.   (b) Would you expect phthalimide to be more or less acidic than benzamide? Why?

PROBLEM 19.4

Outline a preparation of benzylamine using the Gabriel synthesis.

Multiple alkylations are not a problem when tertiary amines are alkylated with methyl or primary halides. Reactions such as the following take place in good yield.

$$R_3N : + RCH_2 - Br \longrightarrow R_3N - CH_2R^+ \; Br^-$$

19.5B Preparation of Amines Through Reduction of Nitro Compounds

The most widely used method for preparing aromatic amines involves nitration of the ring and subsequent reduction of the nitro group to an amino group.

$$Ar - H \xrightarrow[H_2SO_4]{HNO_3} Ar - NO_2 \xrightarrow{(H)} Ar - NH_2$$

We studied ring nitration in Chapter 12 and saw there that it is applicable to a wide variety of aromatic compounds. Reduction of the nitro group can also be carried out in a number of ways. The most frequently used methods employ catalytic hydrogenation, or treatment of the nitro compound with acid and iron. (Zinc, or tin, or a metal salt such as $SnCl_2$ can also be used.)

$$Ar-NO_2 \xrightarrow[\text{or \ (1) Fe, HCl \ (2) OH}^-]{\text{H}_2,\text{ cat.}} Ar-NH_2$$

Specific Example

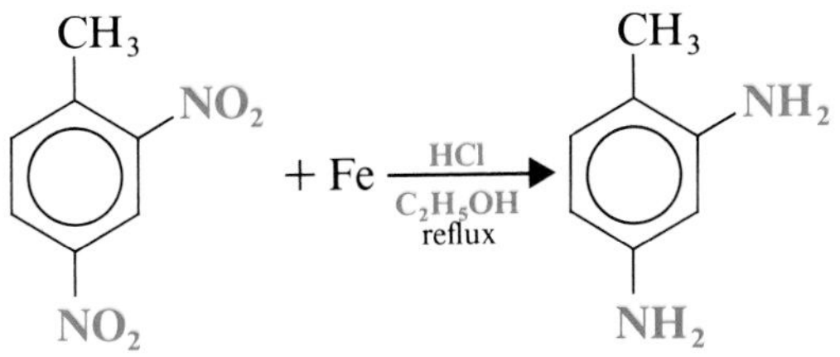

(74%)

Selective reduction of one nitro group of a dinitro compound can often be achieved through the use of hydrogen sulfide in aqueous (or alcoholic) ammonia:

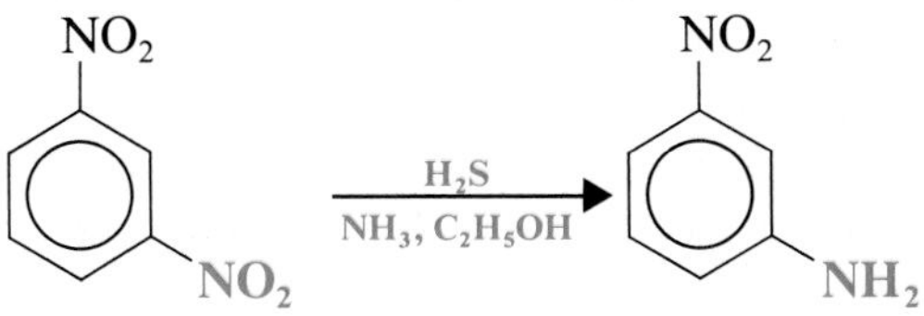

m-Dinitrobenzene

(70–80%)
m-Nitroaniline

When this method is used, the amount of the hydrogen sulfide must be carefully measured because the use of an excess may result in the reduction of more than one nitro group.

It is not always possible to predict just which nitro group will be reduced, however. Treating 2,4-dinitrotoluene with hydrogen sulfide and ammonia results in reduction of the 4-nitro group:

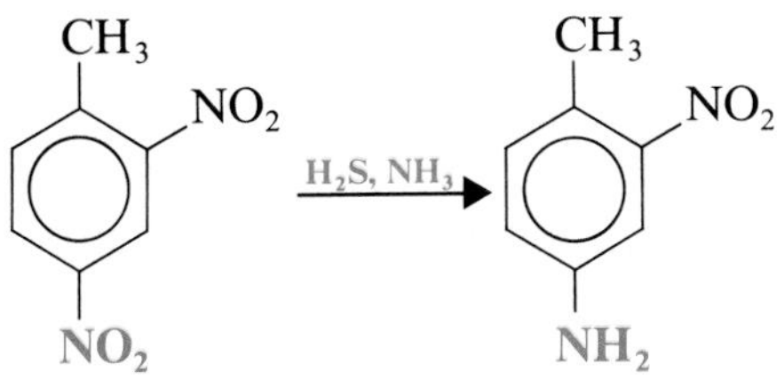

On the other hand, monoreduction of 2,4-dinitroaniline causes reduction of the 2-nitro group:

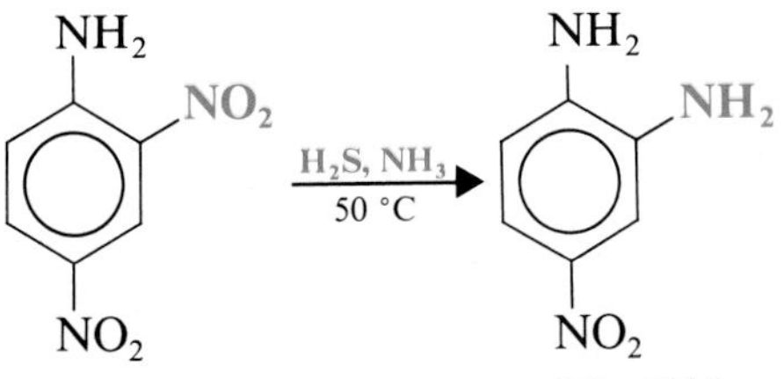

(52–58%)

19.5C Preparation of Amines Through Reductive Amination

Aldehydes and ketones can be converted to primary amines through catalytic or chemical reduction in the presence of ammonia. This process, called *reductive amination*, proceeds through the formation of an imine. The imine is not isolated, however, and in practice only one laboratory step is required.

General Reaction

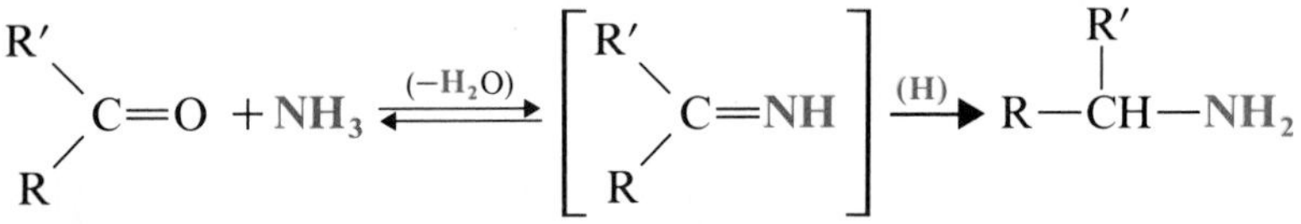

$$\begin{matrix} R' \\ \diagdown \\ C{=}O \\ \diagup \\ R \end{matrix} + NH_3 \underset{}{\overset{(-H_2O)}{\rightleftharpoons}} \left[\begin{matrix} R' \\ \diagdown \\ C{=}NH \\ \diagup \\ R \end{matrix} \right] \overset{(H)}{\longrightarrow} \begin{matrix} R' \\ | \\ R{-}CH{-}NH_2 \end{matrix}$$

$\qquad$ Aldehyde $\qquad\qquad\qquad$ Imine $\qquad\qquad$ A 1° amine
$\qquad$ or ketone

Specific Example

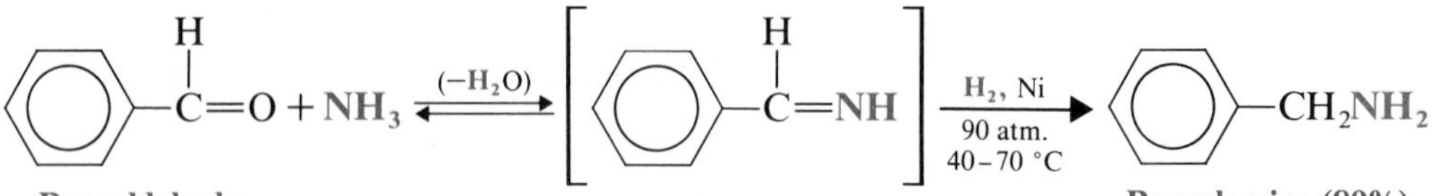

Benzaldehyde $\qquad\qquad\qquad\qquad\qquad\qquad\qquad\qquad$ Benzylamine (89%)

A secondary amine can be prepared through reductive amination of an aldehyde or ketone in the presence of a primary amine.

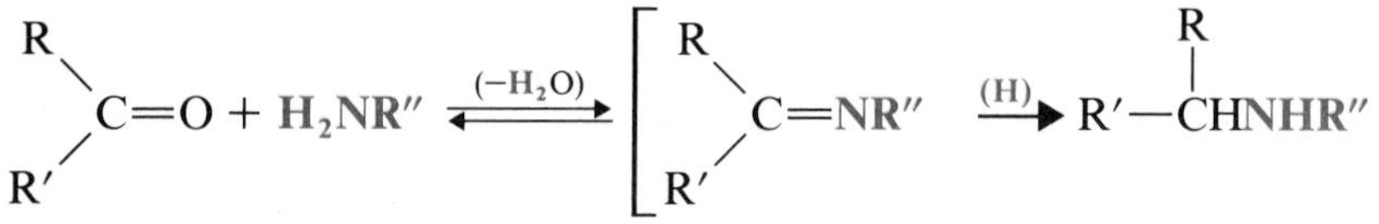

$$\begin{matrix} R \\ \diagdown \\ C{=}O \\ \diagup \\ R' \end{matrix} + H_2NR'' \underset{}{\overset{(-H_2O)}{\rightleftharpoons}} \left[\begin{matrix} R \\ \diagdown \\ C{=}NR'' \\ \diagup \\ R' \end{matrix} \right] \overset{(H)}{\longrightarrow} \begin{matrix} R \\ | \\ R'{-}CHNHR'' \end{matrix}$$

$\quad$ Aldehyde $\quad$ 1° Amine $\qquad\qquad$ An imine $\qquad\qquad$ 2° Amine
$\quad$ or ketone

An example is the synthesis of benzylethylamine from benzaldehyde and ethylamine:

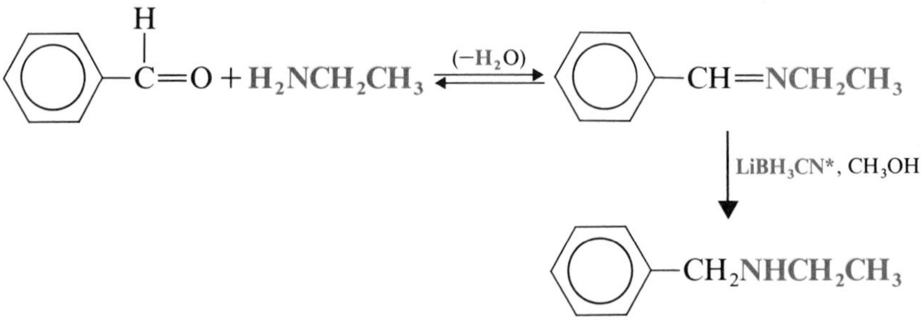

(72%)

*A reducing agent similar to $NaBH_4$. $LiBH_3CN$ and $NaBH_3CN$ reduce imine groups more rapidly than they reduce carbonyl groups.

PROBLEM 19.5

Show how you might prepare each of the following amines through reductive amination:

(a) $CH_3(CH_2)_3CH_2NH_2$ (c) $CH_3(CH_2)_4CH_2NHC_6H_5$

(b) $C_6H_5\underset{\underset{NH_2}{|}}{C}HCH_3$

PROBLEM 19.6

Reductive amination of a ketone is almost always a better method for the

preparation of amines of the type $R\underset{\underset{H}{|}}{\overset{\overset{R'}{|}}{C}}NH_2$, than treatment of an alkyl halide with ammonia. Why would this be true?

19.5D Preparation of Amines Through Reduction of Amides, Oximes, and Nitriles

Amides, oximes, and nitriles can be reduced to amines. Reduction of a nitrile or an oxime yields a primary amine; reduction of an amide can yield a primary, secondary, or tertiary amine.

$$\underset{\text{Nitrile}}{R-C\equiv N} \xrightarrow{(H)} \underset{1° \text{ Amine}}{RCH_2NH_2}$$

{ Nitriles can be prepared from alkyl halides and CN^- (Section 18.3) or from aldehydes and ketones as cyanohydrins (Section 16.9A) }

$$\underset{\text{Oxime}}{RCH=NOH} \xrightarrow{(H)} \underset{1° \text{ Amine}}{RCH_2NH_2}$$

{ Oximes can be prepared from aldehydes and ketones (Section 16.8A) }

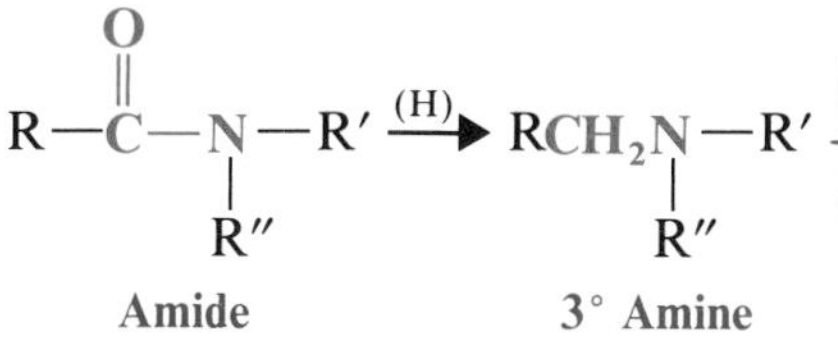

$$\underset{\text{Amide}}{R-\overset{\overset{O}{\|}}{C}-\underset{\underset{R''}{|}}{N}-R'} \xrightarrow{(H)} \underset{3° \text{ Amine}}{RCH_2\underset{\underset{R''}{|}}{N}-R'}$$

{ Amides can be prepared from acid chlorides, acid anhydrides, and esters (Section 18.8) }

(If R′ and R″ = H, the product is a 1° amine, if R′ = H, the product is a 2° amine.)

All of these reductions can be carried out with hydrogen and a catalyst or with LiAlH$_4$. Oximes are also conveniently reduced with sodium in alcohol—a safer method than the use of LiAlH$_4$.

Specific examples follow:

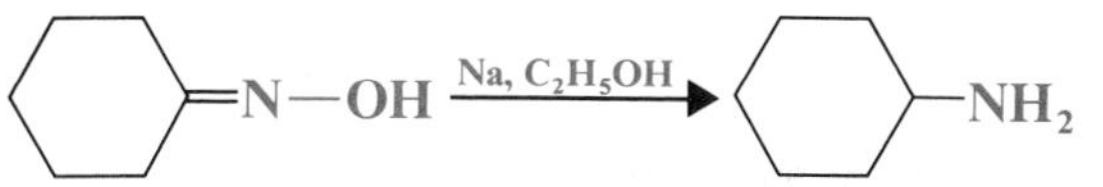

(50–60%)

Benzyl cyanide

$CH_2C{\equiv}N + 2H_2 \xrightarrow[140\ °C]{Raney\ Ni}$ → $CH_2CH_2NH_2$

(71%)
2-Phenylethylamine

N-Methylacetanilide

$\xrightarrow[(2)\ H_2O]{ether}$

N-Ethyl-N-methylaniline

Reduction of an amide is the last step in a useful procedure for **monoalkylation of an amine.** The process begins with *acylation* of the amine using an acyl chloride or acid anhydride; then the amide is reduced with lithium aluminum hydride. For example,

$$C_6H_5CH_2NH_2 \xrightarrow[base]{CH_3COCl} C_6H_5CH_2NHCCH_3 \xrightarrow[(2)\ H_2O]{(1)\ LiAlH_4,\ ether} C_6H_5CH_2NHCH_2CH_3$$

Benzylamine   Benzylethylamine

PROBLEM 19.7

Show how you might utilize the reduction of an amide, oxime, or nitrile to carry out each of the following transformations:
(a) Benzoic acid ⟶ benzylethylamine
(b) 1-Bromopentane ⟶ hexylamine
(c) Propanoic acid ⟶ tripropylamine
(d) 2-Butanone ⟶ *sec*-butylamine

19.5E Preparation of Amines Through the Hofmann Rearrangement

Amides with no substituent on the nitrogen react with solutions of bromine or chlorine in sodium hydroxide to yield amines through a reaction known as the *Hofmann rearrangement* or *Hofmann degradation:*

$$R{-}\overset{O}{\overset{\|}{C}}NH_2 + Br_2 + 4NaOH \xrightarrow{H_2O} RNH_2 + 2NaBr + Na_2CO_3 + 2H_2O$$

From this equation we can see that the carbonyl carbon atom of the amide is lost (as CO_3^{2-}) and that the R group of the amide becomes attached to the nitrogen of the amine. Primary amines made this way are not contaminated by 2° or 3° amines.

The mechanism for this interesting reaction involves the following steps: In step 1, the amide undergoes a base-promoted *N-bromination.* In step 2, the *N*-bromo amide reacts with base to yield an unstable intermediate called an *acylnitrene.* (Nitrenes are analogous to carbenes; see Special Topic C.) In step 3 the acylnitrene undergoes rearrangement to give an *isocyanate.* Finally, in step 4 the isocyanate undergoes hydrolysis to yield an amine and CO_3^{2-} ion.

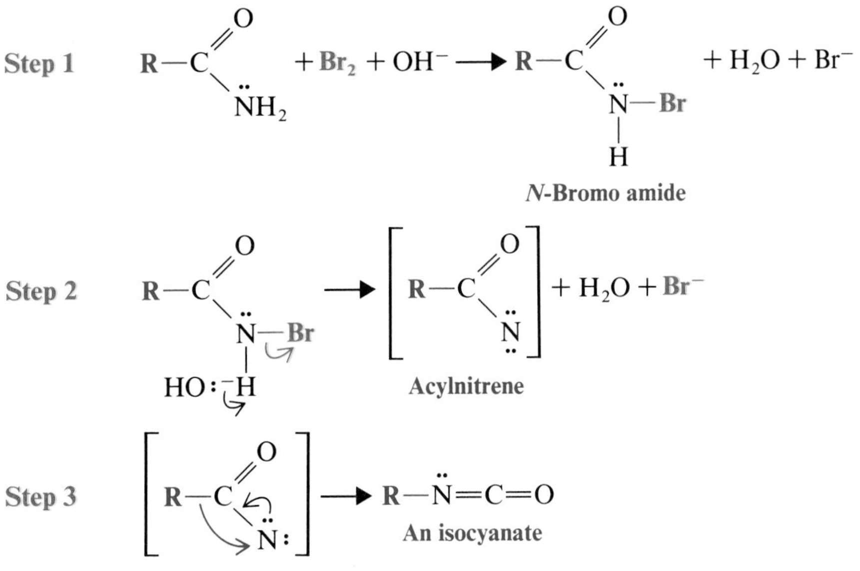

Step 1

$$R-C\overset{O}{\underset{\ddot{N}H_2}{\diagup}} + Br_2 + OH^- \longrightarrow R-C\overset{O}{\underset{\underset{H}{N-Br}}{\diagup}} + H_2O + Br^-$$

N-Bromo amide

Step 2

$$R-C\overset{O}{\underset{\underset{HO:^-H}{\ddot{N}-Br}}{\diagup}} \longrightarrow \left[R-C\overset{O}{\underset{\ddot{N}}{\diagup}} \right] + H_2O + Br^-$$

Acylnitrene

Step 3

$$\left[R-C\overset{O}{\underset{\ddot{N}:}{\diagup}} \right] \longrightarrow R-\ddot{N}=C=O$$

An isocyanate

Step 4

$$R-\ddot{N}=C=O + 2\,OH^- \longrightarrow R-\ddot{N}H_2 + CO_3^{2-}$$

An examination of steps 1 and 2 of this mechanism shows that, initially, two hydrogen atoms must be present on the nitrogen of the amide for the reaction to occur. Consequently, the Hofmann rearrangement is limited to amides of the type $RCONH_2$.

> Studies of Hofmann rearrangement of optically active amides in which the stereocenter is directly attached to the carbonyl group have shown that these reactions occur with *retention of configuration.* Thus, the R group migrates to nitrogen with its electrons, *but without inversion.*

PROBLEM 19.8

Using a different method for each part, but taking care in each case to select a *good* method, show how each of the following transformations might be accomplished:

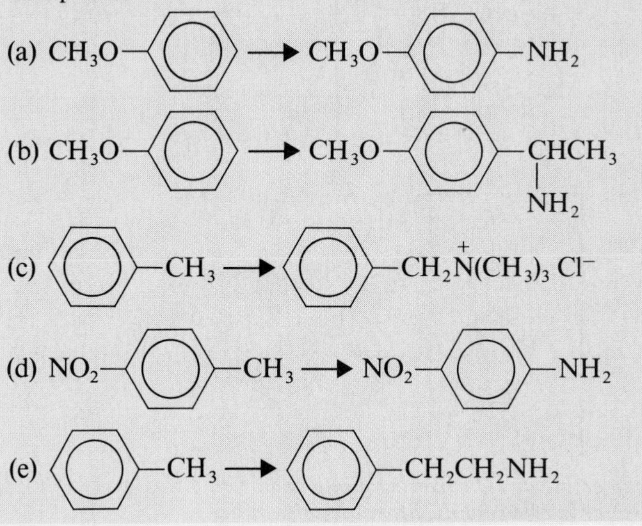

(a) CH_3O-⬡ $\longrightarrow$ CH_3O-⬡$-NH_2$

(b) CH_3O-⬡ $\longrightarrow$ CH_3O-⬡$-\underset{NH_2}{CHCH_3}$

(c) ⬡$-CH_3 \longrightarrow$ ⬡$-CH_2\overset{+}{N}(CH_3)_3\ Cl^-$

(d) NO_2-⬡$-CH_3 \longrightarrow NO_2-$⬡$-NH_2$

(e) ⬡$-CH_3 \longrightarrow$ ⬡$-CH_2CH_2NH_2$

19.6 REACTIONS OF AMINES

We have encountered a number of important reactions of amines in earlier sections of this book. In Section 19.3 we saw reactions in which primary, secondary, and tertiary amines act *as bases.* In Section 19.5 we saw their reactions as *nucleophiles* in *alkylation reactions,* and in Chapter 18 as *nucleophiles* in *acylation reactions.* In Chapter 12 we saw that an amino group on an aromatic ring acts as a powerful *activating group* and as an *ortho–para director.*

The structural feature of amines that underlies all of these reactions and that forms a basis for our understanding of most of the chemistry of amines is the ability of nitrogen to share an electron pair:

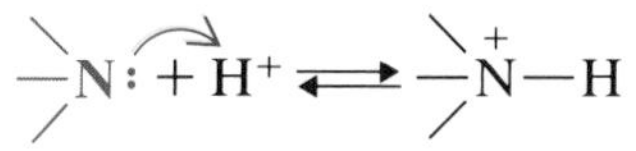

$$-\overset{|}{\underset{|}{N}}: + H^+ \rightleftharpoons -\overset{|}{\underset{|}{\overset{+}{N}}}-H$$

An amine acting as a base

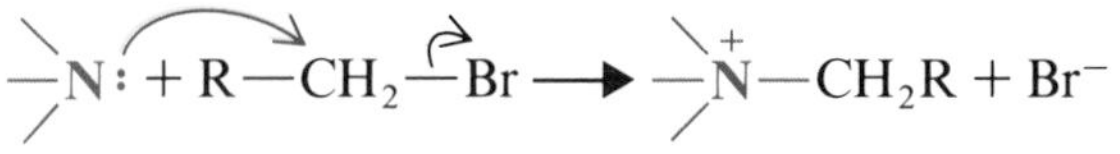

$$-\overset{|}{\underset{|}{N}}: + R-CH_2-Br \longrightarrow -\overset{|}{\underset{|}{\overset{+}{N}}}-CH_2R + Br^-$$

An amine acting as a nucleophile in an alkylation reaction

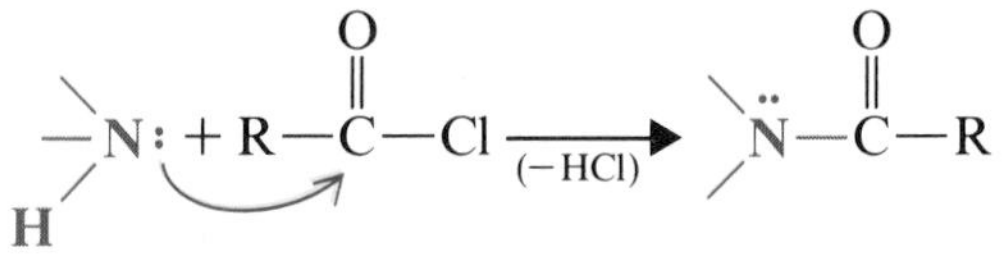

$$-\overset{|}{\underset{H}{N}}: + R-\overset{\overset{O}{\|}}{C}-Cl \xrightarrow[(-HCl)]{} \overset{\cdot\cdot}{\underset{|}{N}}-\overset{\overset{O}{\|}}{C}-R$$

An amine acting as a nucleophile in an acylation reaction

In the preceding examples the amine acts as a nucleophile by donating its electron pair to an electrophilic reagent. In the following example resonance contributions involving the nitrogen electron pair make *carbon* atoms nucleophilic.

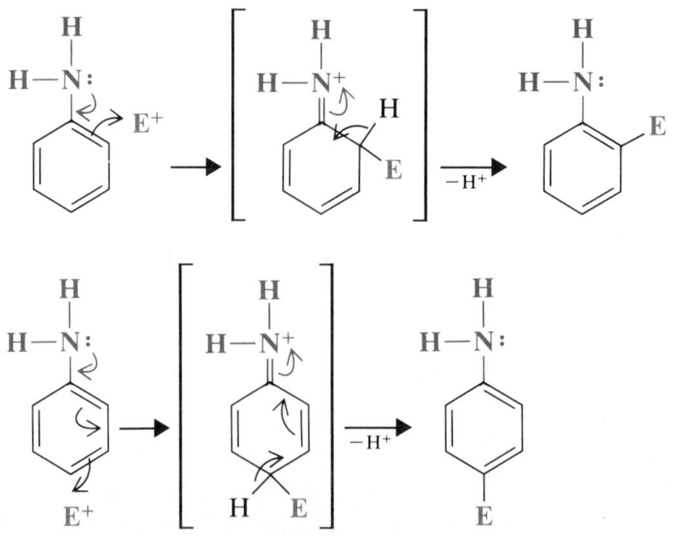

The amino group acting as an activating group and as an ortho–para director in electrophilic aromatic substitution

19.6A Oxidation of Amines

Primary and secondary aliphatic amines are subject to oxidation, although in most instances useful products are not obtained. Complicated side reactions often occur, causing the formation of complex mixtures.

Tertiary amines can be oxidized cleanly to tertiary amine oxides. This transformation can be brought about by using hydrogen peroxide or a peroxy acid.

$$R_3N: \xrightarrow[\text{}]{H_2O_2 \quad \text{or} \quad \overset{\overset{\textstyle O}{\|}}{R C O O H}} R_3\overset{+}{N}-\overset{..}{\underset{..}{O}}:^-$$

A tertiary amine
oxide

Tertiary amine oxides undergo a useful elimination reaction as discussed in Section 19.13B.

Arylamines are very easily oxidized by a variety of reagents including the oxygen in air. Oxidation is not confined to the amino group but also occurs in the ring. (The amino group through its electron-donating ability makes the ring electron rich and hence especially susceptible to oxidation.) The oxidation of other functional groups on an aromatic ring cannot usually be accomplished when an amino group is present on the ring, because oxidation of the ring takes place first.

19.7 REACTIONS OF AMINES WITH NITROUS ACID

Nitrous acid ($HONO$) is a weak, unstable acid. It is always prepared *in situ*, usually treating sodium nitrite ($NaNO_2$) with an aqueous solution of a strong acid:

$$HCl_{(aq)} + NaNO_{2(aq)} \longrightarrow HONO_{(aq)} + NaCl_{(aq)}$$
$$H_2SO_4 + 2NaNO_{2(aq)} \longrightarrow 2HONO_{(aq)} + Na_2SO_{4(aq)}$$

Nitrous acid reacts with all classes of amines. The products that we obtain from these reactions depend on whether the amine is primary, secondary, or tertiary and whether the amine is aliphatic or aromatic.

19.7A Reactions of Primary Aliphatic Amines with Nitrous Acid

Primary aliphatic amines react with nitrous acid through a reaction called *diazotization* to yield highly unstable aliphatic *diazonium salts.* Even at low temperatures, *aliphatic* diazonium salts decompose spontaneously by losing nitrogen to form carbocations. The carbocations go on to produce mixtures of alkenes, alcohols, and alkyl halides by elimination of H^+, reaction with H_2O, and reaction with X^-.

General Reaction

$$R-NH_2 + NaNO_2 + HX \xrightarrow[\text{H}_2\text{O}]{\text{(HONO)}} \left[R-\overset{+}{N}\equiv N\colon \ X^- \right]$$

1° Aliphatic   Aliphatic diazonium salt
amine   *(highly unstable)*

$$\downarrow -N_2 \text{ (i.e., :}N\equiv N\text{:)}$$

$$R^+ + X^-$$

$$\downarrow$$

alkenes, alcohols, alkyl halides

Diazotizations of primary aliphatic amines are of little synthetic importance because they yield such a complex mixture of products. Diazotizations of primary aliphatic amines are used in some analytical procedures, however, because the evolution of nitrogen is quantitative. They can also be used to generate and thus study the behavior of carbocations in water, acetic acid, and other solvents.

19.7B Reactions of Primary Arylamines with Nitrous Acid

Primary arylamines react with nitrous acid to give arenediazonium salts. While arenediazonium salts are unstable, they are far more stable than aliphatic diazonium salts; they do not decompose at an appreciable rate when the temperature of the reaction mixture is kept below 5 °C.

$$Ar-NH_2 + 2NaNO_2 + 2HX \longrightarrow Ar-\overset{+}{N}\equiv N\colon \ X^- + NaX + 2H_2O$$

Primary arylamine   Arenediazonium
salt
(stable if kept below
5° C)

Diazotization of a primary amine takes place through a series of steps. In the presence of strong acid, nitrous acid dissociates to produce ^+NO ions. These ions then react with the nitrogen of the amine to form an unstable *N*-nitrosoammonium ion as an intermediate. This intermediate then loses a proton to form an *N*-nitrosoamine, which, in turn, tautomerizes by a proton shift to a diazohydroxide in a reaction that is similar to keto–enol tautomerization. Then, in the presence of acid, the diazohydroxide loses water to form the diazonium ion.

$$HO\ddot{N}O + H_3O^+ \rightleftharpoons H_2O^+-\ddot{N}O + H_2O \rightleftharpoons 2H_2O + {}^+\ddot{N}=O$$

$$\underset{\substack{\text{1° Aryl- or}\\\text{alkyl-}\\\text{amine}}}{\overset{\displaystyle H}{\underset{\displaystyle H}{Ar-\overset{|}{\underset{|}{N}}\colon}}} + {}^+N=O \longrightarrow \underset{\substack{N\text{-Nitroso-}\\\text{ammonium}\\\text{ion}}}{\overset{\displaystyle H}{\underset{\displaystyle H}{Ar-\overset{|}{\underset{|}{N^+}}-\ddot{N}=O}}} \xrightarrow{-H^+} \underset{N\text{-Nitrosoamine}}{\overset{\displaystyle H}{\underset{\displaystyle H}{Ar-\overset{|}{\underset{|}{\ddot{N}}}-\ddot{N}=O}}} \xrightarrow[\text{shift}]{\text{proton}}$$

$$Ar-\ddot{N}=\ddot{N}-OH \underset{}{\overset{+H^+}{\rightleftharpoons}} Ar-\ddot{N}=\ddot{N}-\overset{\frown}{O}H_2{}^+ \rightleftharpoons$$

Diazohydroxide

$$\boxed{Ar-\overset{+}{N}\equiv N\colon \longleftrightarrow Ar-\ddot{N}=\overset{\cdot\cdot}{N^+} + H_2O}$$

Diazonium ion

Diazotization reactions of primary arylamines are of considerable synthetic importance because the diazonium group, $-\overset{+}{N}\equiv N\colon$, can be replaced by a variety of other functional groups. We shall examine these reactions in Section 19.8.

19.7C Reactions of Secondary Amines with Nitrous Acid

Secondary amines—both aryl and alkyl—react with nitrous acid to yield *N*-nitrosoamines. *N*-Nitrosoamines usually separate from the reaction mixture as oily yellow liquids.

Specific Examples

$$(CH_3)_2\overset{..}{N}H \;+\; HCl \;+\; NaNO_2 \;\xrightarrow[\;H_2O\;]{(HONO)}\; (CH_3)_2\overset{..}{N}-\overset{..}{N}=O$$

Dimethylamine　　　　　　　　　　　　　　　　　　*N*-Nitrosodimethylamine
　　　　　　　　　　　　　　　　　　　　　　　　　　(a yellow oil)

N-Methylaniline $+ HCl + NaNO_2 \xrightarrow[\;H_2O\;]{(HONO)}$

(87–93%, a yellow oil)
N-Nitroso-*N*-methyl-aniline

N-Nitrosoamines are very powerful carcinogens which many scientists fear may be present in many foods, especially in cooked meats that have been cured with sodium nitrite. Sodium nitrite is added to many meats (e.g., bacon, ham, frankfurters, sausages, and corned beef) to inhibit the growth of *Clostridium botulinum* (the bacterium that produces botulinus toxin) and to keep red meats from turning brown. (Food poisoning by botulinus toxin is often fatal.) In the presence of acid or under the influence of heat, sodium nitrite reacts with amines always present in the meat to produce *N*-nitrosoamines. Cooked bacon, for example, has been shown to contain *N*-nitrosodimethylamine and *N*-nitrosopyrrolidine. There is also concern that nitrites from food may produce nitrosoamines when they react with amines in the presence of the acid found in the stomach. In 1976, the FDA reduced the permissible amount of nitrite allowed in cured meats from 200 parts per million (ppm) to 50–125 ppm. Nitrites (and nitrates that can be converted to nitrites by bacteria) also occur naturally in many foods. Cigarette smoke is known to contain *N*-nitrosodimethylamine. Someone smoking a pack of cigarettes a day inhales about 0.8 μg of *N*-nitrosodimethylamine and even more has been shown to be present in the side-stream smoke.

19.7D Reactions of Tertiary Amines with Nitrous Acid

When an tertiary aliphatic amine is mixed with nitrous acid, an equilibrium is established between the tertiary amine, its salt, and an *N*-nitrosoammonium compound.

$$2R_3N\colon \;+\; HX \;+\; NaNO_2 \;\rightleftharpoons\; R_3\overset{+}{N}H \;\; X^- \;+\; R_3\overset{+}{N}-N=O \;\; X^-$$

Tertiary aliphatic　　　　　　　　　　Amine salt　　　*N*-Nitrosoammonium
amine　　　　　　　　　　　　　　　　　　　　　　　　compound

While *N*-nitrosoammonium compounds are stable at low temperatures, at higher temperatures and in aqueous acid they decompose to produce aldehydes or ketones. These reactions are of little synthetic importance, however.

Tertiary arylamines react with nitrous acid to form *C*-nitroso aromatic compounds. Nitrosation takes place almost exclusively at the para position if it is open; and, if not, at the ortho position.

Specific Example

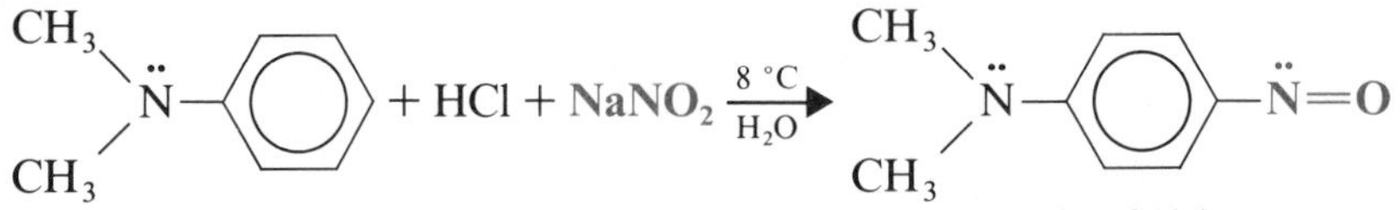

$$CH_3\!-\!\overset{..}{N}\!-\!\!\!\bigcirc\!\!\!\! + HCl + NaNO_2 \xrightarrow[\text{H}_2\text{O}]{8\,°\text{C}} CH_3\!-\!\overset{..}{N}\!-\!\!\!\bigcirc\!\!\!\!-\!\overset{..}{N}\!=\!O$$

(80–90%)
p-Nitroso-*N,N*-dimethylaniline

> **PROBLEM 19.10**
>
> Para nitrosation of *N,N*-dimethylaniline (*C*-nitrosation) is believed to take place through an electrophilic attack of NO^+ ions. (a) Show how NO^+ ions might be formed in an aqueous solution of $NaNO_2$ and HCl. (b) Write a mechanism for *p*-nitrosation of *N,N*-dimethylaniline. (c) Tertiary aromatic amines and phenols undergo *C*-nitrosation reaction, whereas most other benzene derivatives do not. How can you account for this difference?

19.8 REPLACEMENT REACTIONS OF ARENEDIAZONIUM SALTS

Diazonium salts are highly useful intermediates in the synthesis of aromatic compounds, because the diazonium group can be replaced by any one of a number of other atoms or groups, including —F, —Cl, —Br, —I, —CN, —OH, and —H.

Diazonium salts are almost always prepared by diazotizing primary aromatic amines. Primary arylamines can be synthesized through reduction of nitro compounds that are readily available through direct nitration reactions.

19.8A Syntheses Using Diazonium Salts

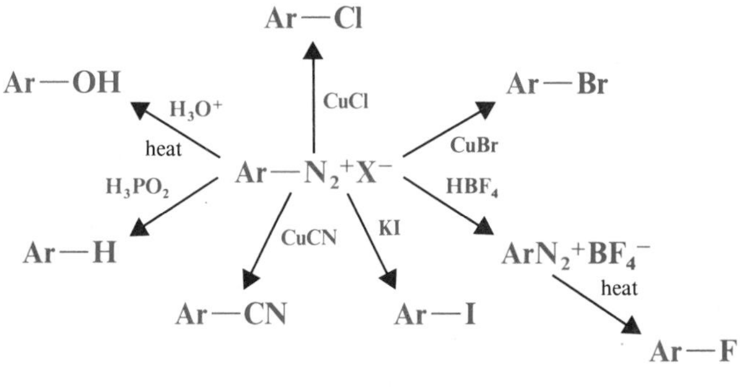

Most arenediazonium salts are unstable at temperatures above 5–10 °C, and many explode when dry. Fortunately, however, most of the replacement reactions of diazonium salts do not require their isolation. We simply add another reagent (CuCl, CuBr, KI, and so on) to the mixture, gently warm the solution, and the replacement (accompanied by the evolution of nitrogen) takes place.

Only in the replacement of the diazonium group by —F need we isolate a diazonium salt. We do this by adding HBF_4 to the mixture, causing the sparingly soluble and reasonably stable arenediazonium fluoroborate, $ArN_2^+\ BF_4^-$, to precipitate.

19.8B The Sandmeyer Reaction: Replacement of the Diazonium Group by —Cl, —Br, or —CN

Arenediazonium salts react with cuprous chloride, cuprous bromide, and cuprous cyanide to give products in which the diazonium group has been replaced by —Cl, —Br, and —CN, respectively. These reactions are known generally as *Sandmeyer reactions.* Several specific examples follow:

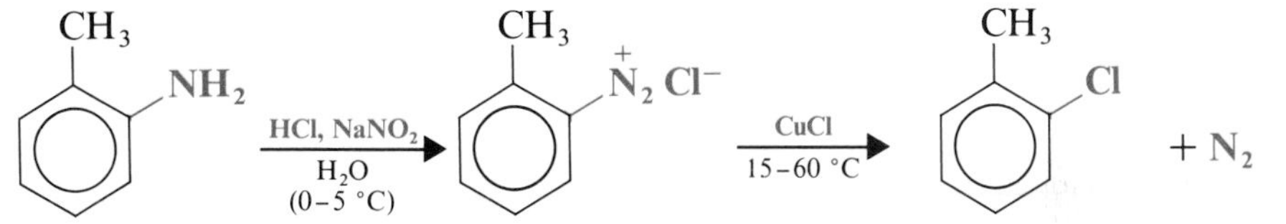

o-Toluidine

(74–79% overall)
o-Chlorotoluene

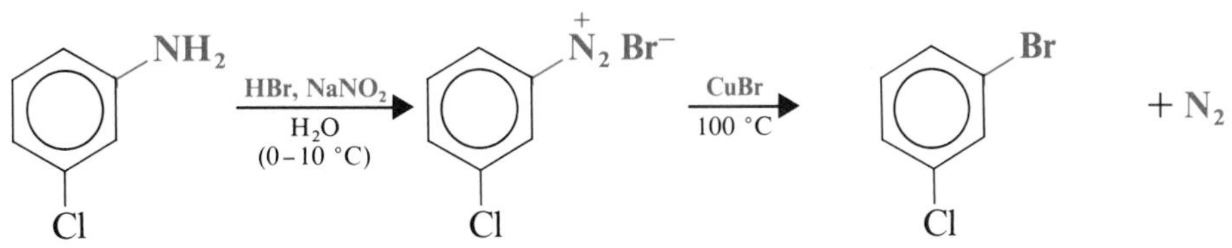

m-Chloroaniline

(70% overall)
m-Bromochlorobenzene

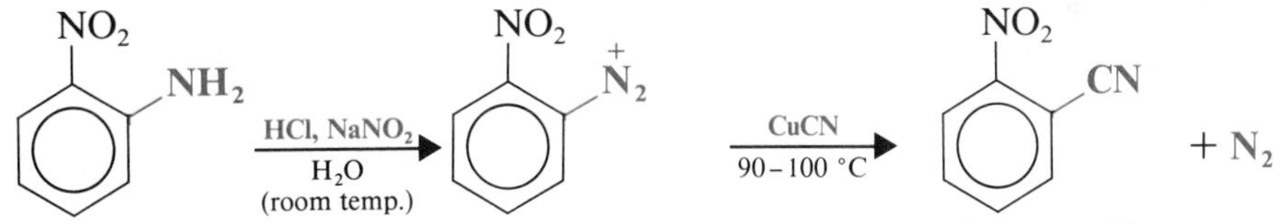

o-Nitroaniline

(65% overall)
o-Nitrobenzonitrile

19.8C Replacement by —I

Arenediazonium salts react with potassium iodide to give products in which the diazonium group has been replaced by —I. An example is the synthesis of *p*-iodonitrobenzene:

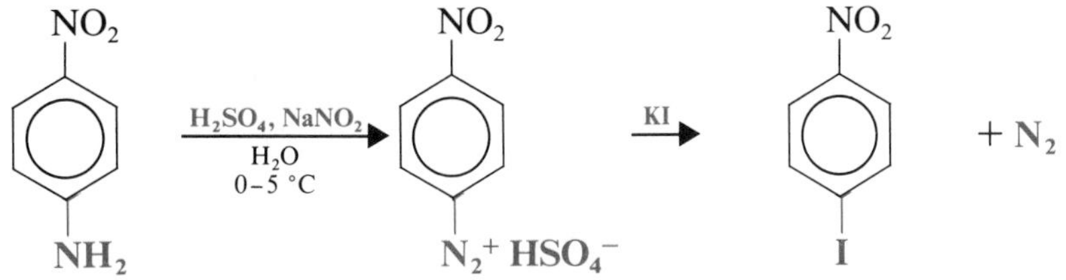

p-Nitroaniline

(81% overall)
p-Iodonitrobenzene

19.8D Replacement by —F

The diazonium group can be replaced by fluorine by treating the diazonium salt with fluoroboric acid (HBF_4). The diazonium fluoroborate that precipitates is isolated, dried, and heated until decomposition occurs. An aryl fluoride is produced.

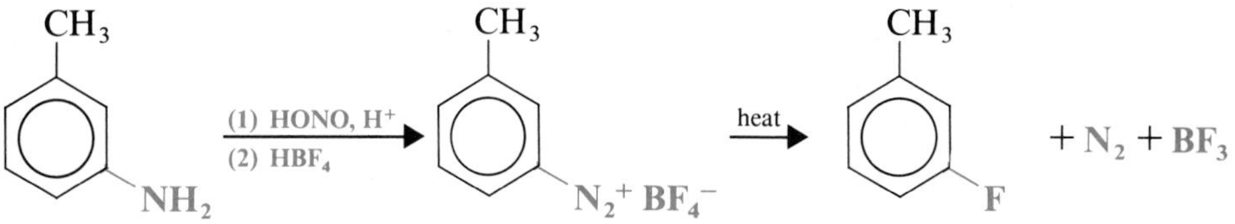

CH₃ structure diagram:

m-Toluidine → (1) HONO, H⁺ (2) HBF₄ → m-Toluenediazonium fluoroborate (79%) → heat → m-Fluorotoluene (69%) + N_2 + BF_3

19.8E Replacement by —OH

The diazonium group can be replaced by a hydroxyl group simply by acidifying the aqueous mixture strongly and heating it:

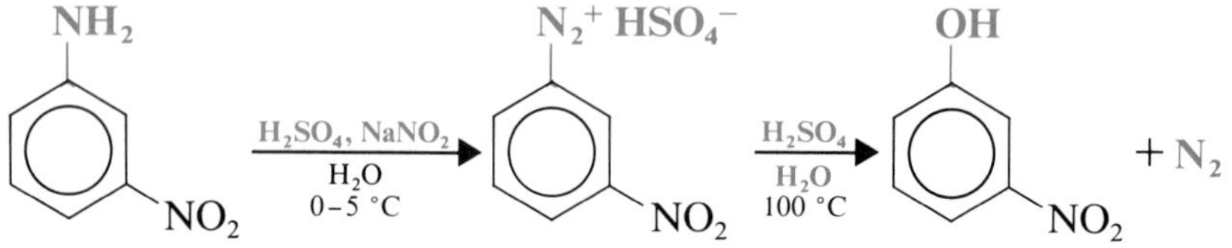

m-Nitroaniline → H_2SO_4, $NaNO_2$, H_2O, 0–5 °C → (N_2^+ HSO_4^-) → H_2SO_4, H_2O, 100 °C → m-Nitrophenol (74–79%) + N_2

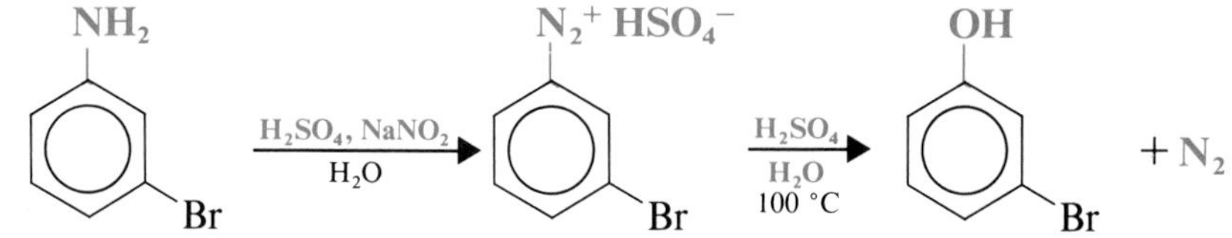

m-Bromoaniline → H_2SO_4, $NaNO_2$, H_2O → (N_2^+ HSO_4^-) → H_2SO_4, H_2O, 100 °C → m-Bromophenol (78% overall) + N_2

Sulfuric acid is used for the diazotization, because HSO_4^- competes poorly with water in the second step of the sequence.

PROBLEM 19.11

In the preceding examples of diazonium reactions, we have illustrated syntheses beginning with the compounds (a)–(e) here. Show how you might prepare each of the following compounds from benzene.

(a) m-Nitroaniline   (c) m-Bromoaniline   (e) p-Nitroaniline

(b) m-Chloroaniline   (d) o-Nitroaniline

19.8F Replacement by —H.
Deamination by Diazotization

Arenediazonium salts react with hypophosphorous acid (H_3PO_2) to yield products in which the diazonium group has been replaced by —H.

Since we usually begin a synthesis using diazonium salts by nitrating an aromatic compound, that is, replacing —H by —NO₂ and then by —NH₂, it may seem strange that we would ever want to replace a diazonium group by —H. However, replacement of the diazonium group by —H can be a useful reaction. We can introduce an amino group into an aromatic ring to influence the orientation of a subsequent reaction. Later we can remove the amino group (i.e., carry out a *deamination*) by diazotizing it and treating the diazonium salt with H₃PO₂.

We can see an example of the usefulness of a deamination reaction in the following synthesis of *m*-bromotoluene. We cannot prepare *m*-bromotoluene by

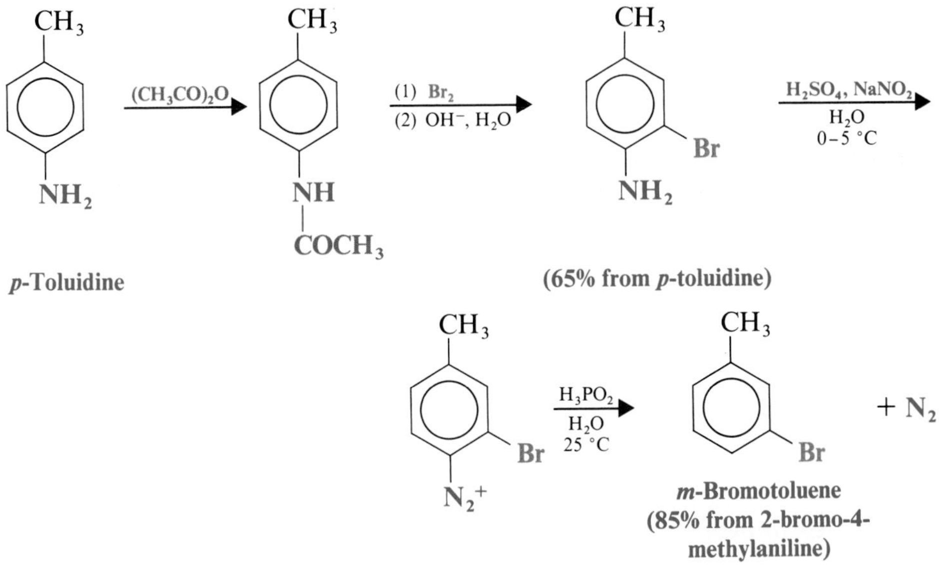

p-Toluidine

(65% from *p*-toluidine)

m-Bromotoluene
(85% from 2-bromo-4-
methylaniline)

direct bromination of toluene or by a Friedel–Crafts alkylation of bromobenzene because both reactions give *o*- and *p*-bromotoluene. (Both CH₃— and Br— are ortho–para directors.) However, if we begin with *p*-toluidine (prepared by nitrating toluene, separating the para isomer, and reducing the nitro group), we can carry out the sequence of reactions shown and obtain *m*-bromotoluene in good yield. The first step, synthesis of the *N*-acetyl derivative of *p*-toluidine, is done to reduce the activating effect of the amino group. (Otherwise both ortho positions would be brominated.) Later, the acetyl group is removed by hydrolysis.

PROBLEM 19.12

Suggest how you might modify the preceding synthesis in order to prepare 3,5-dibromotoluene.

PROBLEM 19.13

(a) In Section 19.8D we showed a synthesis of *m*-fluorotoluene starting with *m*-toluidine. How would you prepare *m*-toluidine from toluene? (b) How would you prepare *m*-chlorotoluene? (c) *m*-Bromotoluene? (d) *m*-Iodotoluene? (e) *m*-Tolunitrile (*m*-CH₃C₆H₄CN)? (f) *m*-Toluic acid?

19.9 COUPLING REACTIONS OF DIAZONIUM SALTS

Diazonium ions are weak electrophiles; they react with highly reactive aromatic compounds—with phenols and tertiary arylamines—to yield *azo* compounds. This electrophilic aromatic substitution is often called a *diazo coupling reaction.*

General Reaction

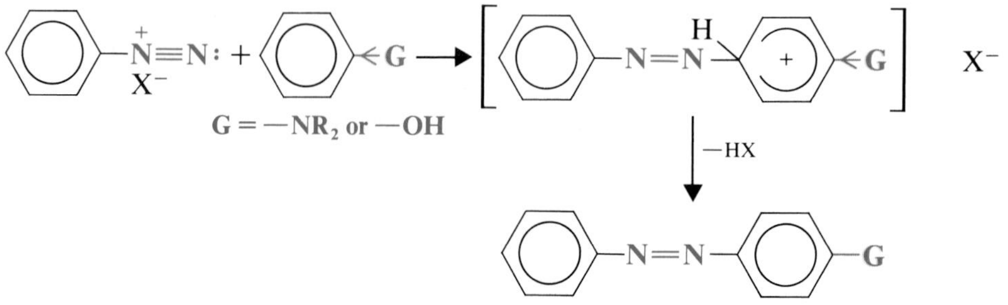

$G = -NR_2$ or $-OH$

An azo compound

Specific Examples

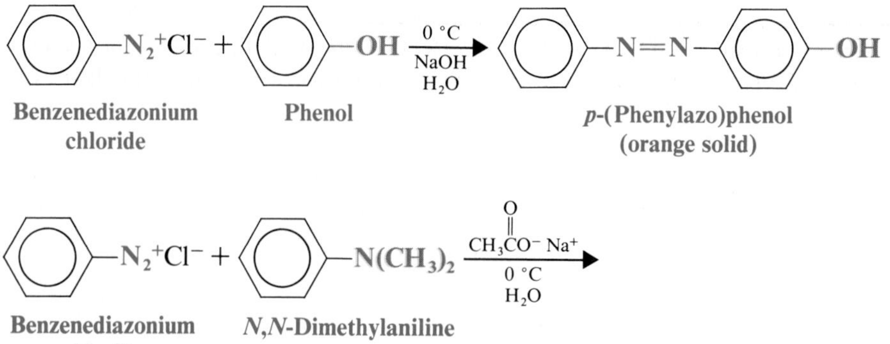

Benzenediazonium   Phenol   *p*-(Phenylazo)phenol
chloride   (orange solid)

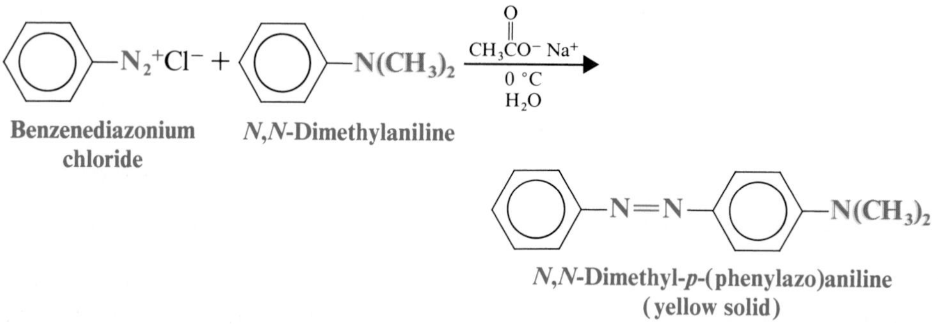

Benzenediazonium   *N,N*-Dimethylaniline
chloride

N,N-Dimethyl-*p*-(phenylazo)aniline
(yellow solid)

Couplings between diazonium cations and phenols take place most rapidly in *slightly* alkaline solution. Under these conditions an appreciable amount of the phenol is present as a phenoxide ion, ArO^-, and phenoxide ions are even more reactive toward electrophilic substitution than are phenols themselves. (Why?) If the solution is too alkaline, however (pH > 10), the diazonium salt itself reacts with hydroxide ion to form a relatively unreactive diazohydroxide or diazotate ion:

:ÖH   :Ö:⁻

$$\text{Phenol} \underset{H^+}{\overset{OH^-}{\rightleftharpoons}} \text{Phenoxide ion}$$

Phenol
(couples slowly)

Phenoxide ion
(couples rapidly)

$$\overset{+}{Ar-N}\equiv N: \underset{H^+}{\overset{OH^-}{\rightleftharpoons}} Ar-\ddot{N}=\ddot{N}-OH \underset{H^+}{\overset{OH^-}{\rightleftharpoons}} Ar-\ddot{N}=\ddot{N}-\ddot{O}:^-$$

Diazonium   **Diazohydroxide**   **Diazoate ion**
ion   **(does not couple)**   **(does not couple)**
(couples)

Couplings between diazonium cations and amines take place most rapidly in slightly acidic solutions (pH 5–7). Under these conditions the concentration of the diazonium salt is at a maximum; at the same time an excessive amount of the amine has not been converted to an unreactive amminium salt:

:NR$_2$ NR$_2$H$^+$

$$\underset{OH^-}{\overset{H^+}{\rightleftharpoons}}$$

Amine   **Aminium salt**
(couples)   **(does not couple)**

If the pH of the solution is lower than 5, the rate of amine coupling is low.

With phenols and aniline derivatives, coupling takes place almost exclusively at the para position if it is open. If it is not, coupling takes place at the ortho position.

$$\overset{+}{N_2} Cl^- + \underset{CH_3}{\overset{OH}{\bigcirc}} \xrightarrow[H_2O]{NaOH} \bigcirc-N=N-\underset{CH_3}{\overset{OH}{\bigcirc}}$$

(*p*-cresol)   4-Methyl-2-(phenylazo)phenol

Azo compounds are usually intensely colored because the azo (diazenediyl) linkage —N=N— brings the two aromatic rings into conjugation. This gives an extended system of delocalized π electrons and allows absorption of light in the visible region. Azo compounds, because of their intense colors, and because they can be synthesized from relatively inexpensive compounds, are used extensively as *dyes*.

Azo dyes almost always contain one or more —SO$_3^-$ Na$^+$ groups to confer water solubility on the dye and assist in binding the dye to the surfaces of polar fibers (wool, cotton, or nylon). Many dyes are made by coupling reactions of naphthyl-amines and naphthols.

Orange II, a dye introduced in 1876, is made from β-naphthol.

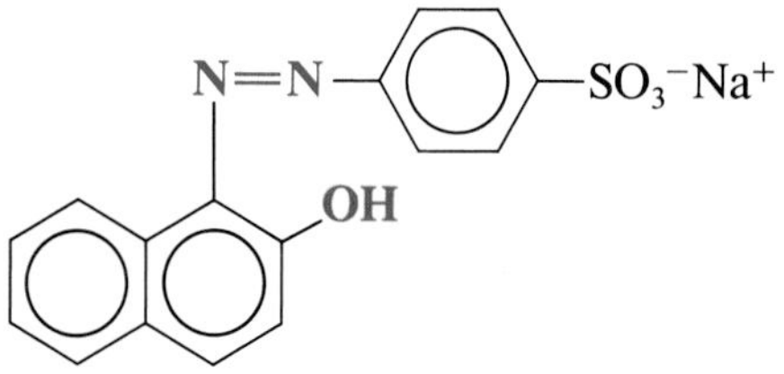

Orange II

PROBLEM 19.15

Outline a synthesis of Orange II from β-naphthol and p-aminobenzenesulfonic acid.

PROBLEM 19.16

Butter Yellow is a dye once used to color margarine. It has since been shown to be carcinogenic, and its use in food is no longer permitted. Outline a synthesis of Butter Yellow from benzene and N,N-dimethylaniline.

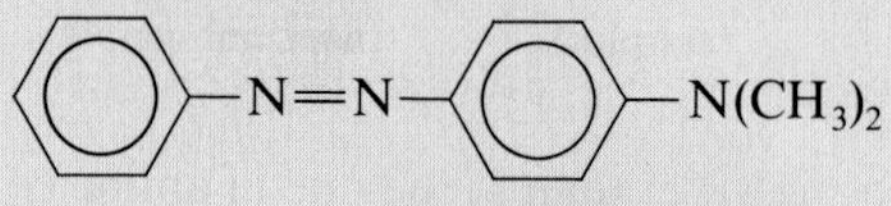

Butter Yellow

PROBLEM 19.17

Azo compounds can be reduced to amines by a variety of reagents including stannous chloride ($SnCl_2$).

$$Ar-N=N-Ar' \xrightarrow{SnCl_2} ArNH_2 + Ar'NH_2$$

This reduction can be useful in synthesis as the following example shows:

4-Ethoxyaniline $\xrightarrow[\text{(2) phenol, OH}^-]{\text{(1) HONO, H}_3\text{O}^+}$ **A** ($C_{14}H_{14}N_2O_2$) $\xrightarrow{\text{NaOH, CH}_3\text{CH}_2\text{Br}}$

B ($C_{16}H_{18}N_2O_2$) $\xrightarrow{SnCl_2}$ two molar equivalents of

C ($C_8H_{11}NO$) $\xrightarrow{\text{acetic anhydride}}$ phenacetin ($C_{10}H_{13}NO_2$)

Give a structure for phenacetin and for the intermediates **A, B, C**. (Phenacetin, an analgesic, is also the subject of Problem 18.35.)

19.10 REACTIONS OF AMINES WITH SULFONYL CHLORIDES

Primary and secondary amines react with sulfonyl chlorides to form *sulfonamides.*

$$
\underset{\substack{1° \text{ Amine}}}{R-\overset{\overset{\displaystyle H}{|}}{\underset{\displaystyle ..}{N}}-H} + \underset{\substack{\text{Sulfonyl} \\ \text{chloride}}}{Cl-\overset{\overset{\displaystyle O}{\|}}{\underset{\displaystyle \|}{\underset{O}{S}}}-Ar} \xrightarrow[(-\text{HCl})]{} \underset{\substack{N\text{-Substituted} \\ \text{sulfonamide}}}{R-\overset{\overset{\displaystyle H}{|}}{\underset{\displaystyle ..}{N}}-\overset{\overset{\displaystyle O}{\|}}{\underset{\displaystyle \underset{O}{\|}}{S}}-Ar}
$$

$$
\underset{\substack{2° \text{ Amine}}}{R-\overset{\overset{\displaystyle R}{|}}{\underset{\displaystyle ..}{N}}-H} + Cl-\overset{\overset{\displaystyle O}{\|}}{\underset{\displaystyle \underset{O}{\|}}{S}}-Ar \xrightarrow[(-\text{HCl})]{} \underset{\substack{N,N\text{-Disubstituted} \\ \text{sulfonamide}}}{R-\overset{\overset{\displaystyle R}{|}}{\underset{\displaystyle ..}{N}}-\overset{\overset{\displaystyle O}{\|}}{\underset{\displaystyle \underset{O}{\|}}{S}}-Ar}
$$

When heated with aqueous acid, sulfonamides are hydrolyzed to amines:

$$
R-\overset{\overset{\displaystyle R}{|}}{\underset{\displaystyle ..}{N}}-\overset{\overset{\displaystyle O}{\|}}{\underset{\displaystyle \underset{O}{\|}}{S}}-Ar \xrightarrow[(2)\ OH^-]{(1)\ H_3O^+,\text{heat}} R-\overset{\overset{\displaystyle R}{|}}{\underset{\displaystyle ..}{N}}-H + {}^-O-\overset{\overset{\displaystyle O}{\|}}{\underset{\displaystyle \underset{O}{\|}}{S}}-Ar
$$

This hydrolysis is much slower, however, than hydrolysis of carboxamides.

19.10A The Hinsberg Test

Sulfonamide formation is the basis for a chemical test, called the Hinsberg test, that can be used to demonstrate whether an amine is primary, secondary, or tertiary. A Hinsberg test involves two steps. First, a mixture containing a small amount of the amine and benzenesulfonyl chloride is shaken with *excess* potassium hydroxide. Next, after allowing time for a reaction to take place, the mixture is acidified. Each type of amine—primary, secondary, or tertiary—gives a different set of *visible* results after each of these two stages of the test.

Primary amines react with benzenesulfonyl chloride to form *N*-substituted benzenesulfonamides. These, in turn, undergo acid–base reactions with the excess potassium hydroxide to form water-soluble potassium salts. (These reactions take place because the hydrogen attached to nitrogen is made acidic by the strongly electron-withdrawing —SO_2— group). At this stage our test tube will contain a clear solution. Acidification of this solution will, in the next stage, cause the water-insoluble *N*-substituted sulfonamide to precipitate.

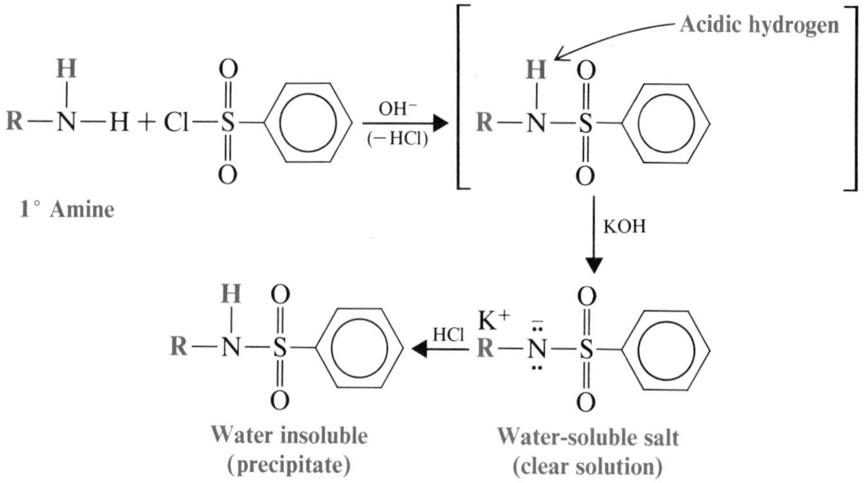

1° Amine

Water insoluble
(precipitate)

Water-soluble salt
(clear solution)

Secondary amines react with benzenesulfonyl chloride in aqueous potassium hydroxide to form insoluble N,N-disubstituted sulfonamides that precipitate after the first stage. N,N-Disubstituted sulfonamides do not dissolve in aqueous potassium hydroxide because they do not have an acidic hydrogen. Acidification of the mixture obtained from a secondary amine produces no visible result—the nonbasic N,N-di-substituted sulfonamide remains as a precipitate.

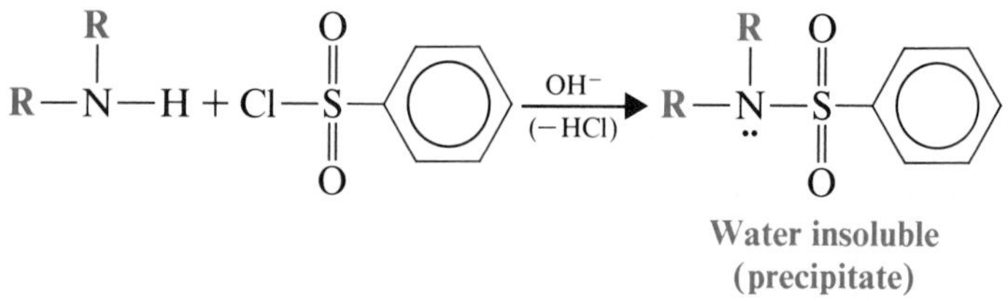

Water insoluble
(precipitate)

If the amine is a tertiary amine and if it is water insoluble, no apparent change will take place in the mixture as we shake it with benzenesulfonyl chloride and aqueous KOH. When we acidify the mixture, the tertiary amine will dissolve because it will form a water-soluble salt.

PROBLEM 19.18

An amine **A** has the molecular formula C_7H_9N. **A** reacts with benzenesulfonyl chloride in aqueous potassium hydroxide to give a clear solution; acidification of the solution gives a precipitate. When **A** is treated with $NaNO_2$ and HCl at $0-5\ °C$, and then with 2-naphthol, an intensely colored compound is formed. **A** gives a single strong absorption peak in the $680-840\text{-cm}^{-1}$ region at $815\ \text{cm}^{-1}$. What is the structure of **A**?

PROBLEM 19.19

Sulfonamides of primary amines are often used to synthesize *pure* secondary amines. Suggest how this synthesis is carried out.

19.11 THE SULFA DRUGS: SULFANILAMIDE

19.11A Chemotherapy

Chemotherapy is defined as the use of chemical agents selectively to destroy infectious organisms without simultaneously destroying the host. Although it may be difficult to believe in this age of "wonder drugs," chemotherapy is a relatively modern phenomenon. Prior to 1900 only three specific chemical remedies were known: mercury (for syphilis — but often with disastrous results), cinchona bark (for malaria), and ipecacuanha (for dysentery).

Modern chemotherapy began with the work of Paul Ehrlich early in this century — particularly with his discovery in 1907 of the curative properties of a dye called Trypan Red I when used against experimental trypanosomiasis and with his discovery in 1909 of Salvarsan as a remedy for syphilis (Special Topic H). Ehrlich invented the term "chemotherapy," and in his research sought what he called "magic bullets," that is, chemicals that would be toxic to infectious microorganisms but harmless to humans.*

As a medical student, Ehrlich had been impressed with the ability of certain dyes to stain tissues selectively. Working on the idea that "staining" was a result of a chemical reaction between the tissue and the dye, Ehrlich sought dyes with selective affinities for microorganisms. He hoped that in this way he might find a dye that could be modified so as to render it specifically lethal to microorganisms.

19.11B Sulfa Drugs

Between 1909 and 1935, tens of thousands of chemicals, including many dyes, were tested by Ehrlich and others in a search for such "magic bullets." Very few compounds, however, were found to have any promising effect. Then, in 1935, an amazing event happened. The daughter of Gerhard Domagk, a doctor employed by a German dye manufacturer, contracted a streptococcal infection from a pin prick. As his daughter neared death, Domagk decided to give her an oral dose of a dye called Prontosil. Prontosil had been developed at Domagk's firm (I. G. Farbenindustrie) and tests with mice had shown that Prontosil inhibited the growth of streptococci. Within a short time the little girl recovered. Domagk's gamble not only saved his daughter's life, but it also initiated a new and spectacularly productive phase in modern chemotherapy.†

A year later, in 1936, Ernest Fourneau of the Pasteur Institute in Paris demonstrated that Prontosil breaks down in the human body to produce sulfanilamide, and that sulfanilamide is the actual active agent against streptococci.

Prontosil   Sulfanilamide

---

*P. Ehrlich was awarded the Nobel Prize for medicine in 1908.
†G. Domagk was awarded the Nobel Prize for medicine in 1939 but was unable to accept it until 1947.

Fourneau's announcement of this result set in motion a search for other chemicals (related to sulfanilamide) that might have even better chemotherapeutic effects. Literally thousands of chemical variations were played on the sulfanilamide theme; the structure of sulfanilamide was varied in almost every imaginable way. The best therapeutic results were obtained from compounds in which one hydrogen of the $-SO_2NH_2$ group was replaced by some other group, usually a heterocyclic amine. Among the most successful variations were the following compounds. Sulfanilamide itself is too toxic for general use.

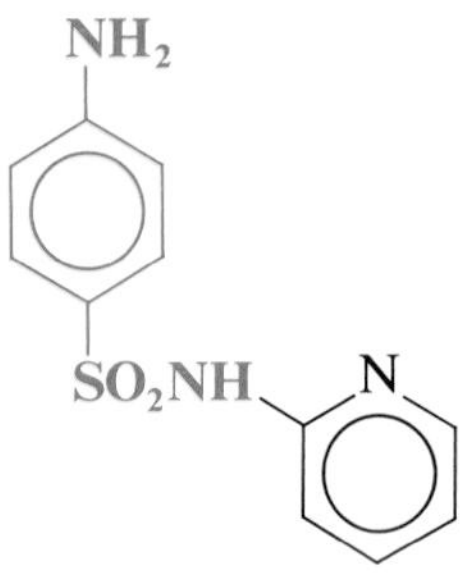

Sulfapyridine 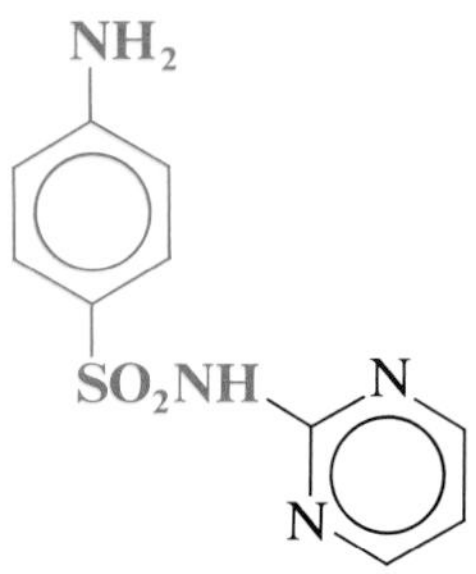   Sulfadiazine

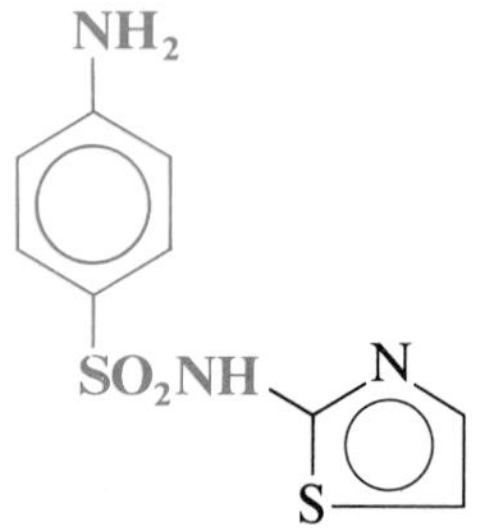

Sulfathiazole 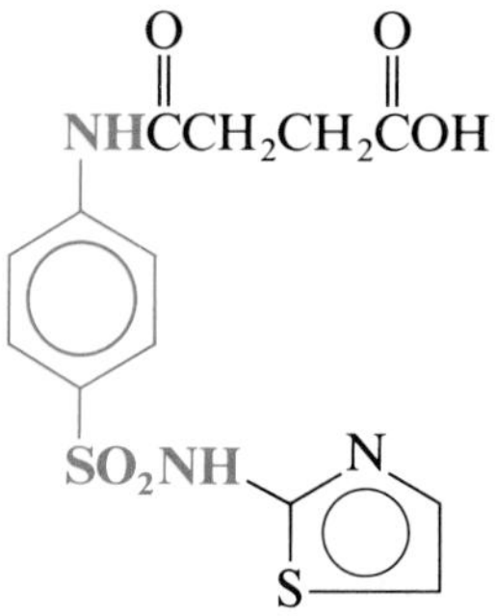   Succinylsulfathiazole 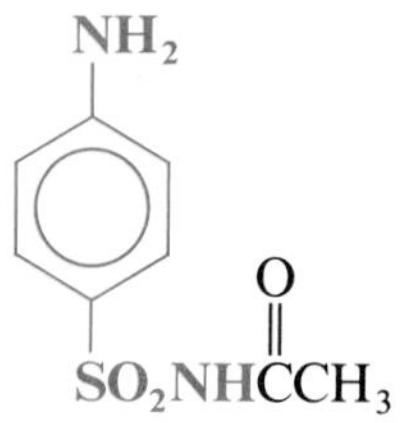   Sulfacetamide

Sulfapyridine was shown to be effective against pneumonia in 1938. (Prior to that time pneumonia epidemics had brought death to tens of thousands.) Sulfacetamide was used successfully in treating urinary tract infections in 1941. Succinoylsulfathiazole and the related compound phthalylsulfathiazole were used as chemotherapeutic agents against infections of the gastrointestinal tract beginning in 1942. (Both compounds are slowly hydrolyzed internally to sulfathiazole.) Sulfathiazole saved the lives of countless wounded soldiers during World War II.

In 1940 a discovery by D. D. Woods laid the groundwork for our understanding of how the sulfa drugs work. Woods observed that the inhibition of growth of certain microorganisms by sulfanilamide is competitively overcome by *p*-aminobenzoic acid. Woods noticed the structural similarity between the two compounds (Fig. 19.3) and reasoned that the two compounds compete with each other in some essential metabolic process.

19.11C Essential Nutrients and Antimetabolites

All higher animals and many microorganisms lack the biochemical ability to synthesize certain essential organic compounds. These essential nutrients include vitamins, certain amino acids, unsaturated carboxylic acids, purines, and pyrimidines. The

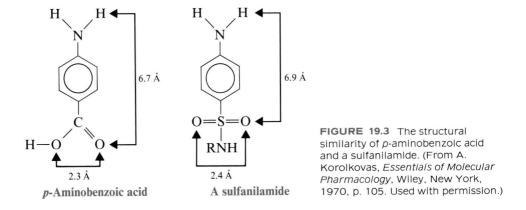

FIGURE 19.3 The structural similarity of *p*-aminobenzoic acid and a sulfanilamide. (From A. Korolkovas, *Essentials of Molecular Pharmacology*, Wiley, New York, 1970, p. 105. Used with permission.)

aromatic amine *p*-aminobenzoic acid is an essential nutrient for those bacteria that are sensitive to sulfanilamide therapy. Enzymes within these bacteria use *p*-aminobenzoic acid to synthesize another essential compound called *folic acid.*

Folic acid

Chemicals that inhibit the growth of microbes are called *antimetabolites.* The sulfanilamides are antimetabolites for those bacteria that require *p*-aminobenzoic acid. The sulfanilamides apparently inhibit those enzymatic steps of the bacteria that are involved in the synthesis of folic acid. The bacterial enzymes are apparently unable to distinguish between a molecule of a sulfanilamide and a molecule of *p*-aminobenzoic acid; thus, sulfanilamide "inhibits" the bacterial enzyme. Because the microorganism is unable to synthesize enough folic acid when sulfanilamide is present, it dies. Humans are unaffected by sulfanilamide therapy because we derive our folic acid from dietary sources (folic acid is a vitamin) and do not synthesize it from *p*-aminobenzoic acid.

The discovery of the mode of action of the sulfanilamides has led to the discovery of many new and effective antimetabolites. One example is *methotrexate,* a derivative of folic acid that has been used successfully in treating certain carcinomas:

Methotrexate

Methotrexate, by virtue of its resemblance to folic acid, can enter into some of the same reactions as folic acid, but it cannot serve the same function, particularly in important reactions involved in cell division. Although methotrexate is toxic to all dividing cells, those cells that divide most rapidly — *cancer cells* — are most vulnerable to its effect.

19.11D Synthesis of Sulfa Drugs

Sulfanilamides can be synthesized from aniline through the following sequence of reactions.

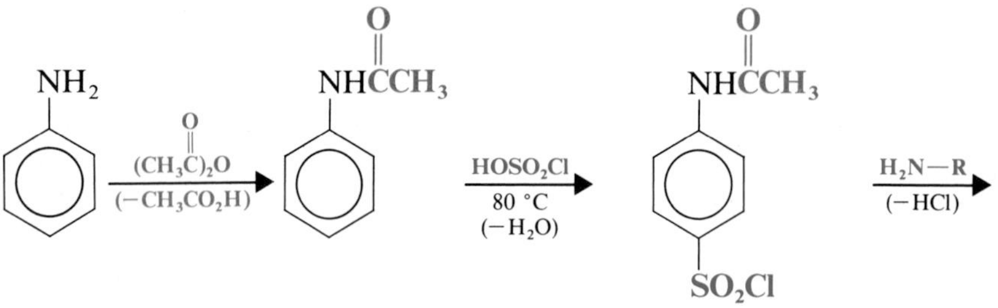

Aniline
1

Acetanilide
2

p-Acetamidobenzene-
sulfonyl chloride
3

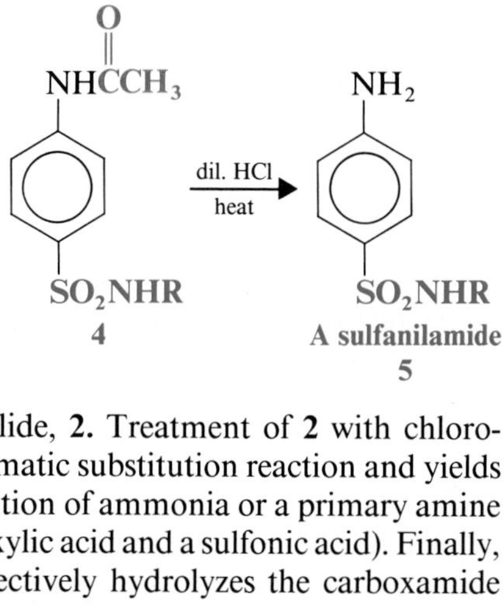

4

A sulfanilamide
5

Acetylation of aniline produces acetanilide, **2**. Treatment of **2** with chlorosulfonic acid brings about an electrophilic aromatic substitution reaction and yields *p*-acetamidobenzenesulfonyl chloride, **3**. Addition of ammonia or a primary amine gives the diamide, **4** (an amide of both a carboxylic acid and a sulfonic acid). Finally, refluxing **4** with dilute hydrochloric acid selectively hydrolyzes the carboxamide linkage and produces sulfanilamide. (Hydrolysis of carboxamides is much more rapid than that of sulfonamides.)

PROBLEM 19.20

(a) Starting with aniline and assuming that you have 2-aminothiazole available, show how you would synthesize sulfathiazole.   (b) How would you convert sulfathiazole to succinylsulfathiazole?

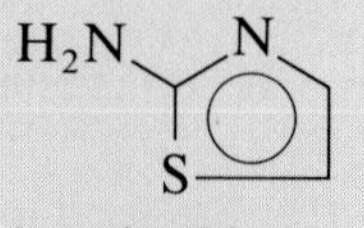

2-Aminothiazole

19.12 ANALYSIS OF AMINES

19.12A Chemical Analysis

Amines are characterized by their basicity, and thus, by their ability to dissolve in dilute aqueous acid (Section 19.3A). Primary, secondary, and tertiary amines can be distinguished from each other on the basis of the Hinsberg test (Section 19.10A). Primary aromatic amines are often detected through diazonium salt formation and subsequent coupling with β-naphthol to form a brightly colored azo dye (Section 19.9).

19.12B Spectroscopic Analysis

The *infrared spectra* of primary and secondary amines are characterized by absorption bands in the 3300–3555-cm^{-1} region that arise from N—H stretching vibration. Primary amines give two bands in this region; secondary amines generally give only one. Absorption bands arising from C—N stretching vibrations of aliphatic amines occur in the 1020–1220-cm^{-1} region but are usually weak and difficult to identify. Aromatic amines generally give a strong C—N stretching band in the 1250–1360-cm^{-1} region.

The *proton nmr spectra* of primary and secondary amines show N—H proton absorptions in the region 1–5 δ. These peaks are sometimes difficult to identify and are best detected by proton counting.

19.13 ELIMINATIONS INVOLVING AMMONIUM COMPOUNDS

19.13A The Hofmann Elimination

All of the eliminations that we have described so far have involved electrically neutral substrates. However, eliminations are known in which the substrate bears a positive charge. One of the most important of these is the elimination that takes place when a quaternary ammonium hydroxide is heated. The products are an alkene, water, and tertiary amine.

$$\overset{\cdot\cdot}{HO}:^{-}\ H$$

$$-\overset{|}{\underset{|}{C}}-\overset{|}{\underset{|}{C}}-\overset{+}{N}R_3 \longrightarrow \ \ \overset{\diagdown}{\diagup}C=C\overset{\diagup}{\diagdown}\ +\ HOH\ +\ :NR_3$$

A quaternary $\longrightarrow$ an alkene + water + a tertiary
ammonium hydroxide amine

This reaction was discovered in 1851 by August W. von Hofmann and has since come to bear his name.

Quaternary ammonium hydroxides can be prepared from quaternary ammonium halides in aqueous solution through the use of silver oxide or an ion exchange resin.

$$2RCH_2CH_2\overset{+}{N}(CH_3)_3X^- + Ag_2O + H_2O \longrightarrow 2RCH_2CH_2\overset{+}{N}(CH_3)_3\ OH^- + 2AgX\downarrow$$

A quaternary ammonium Quaternary ammonium
halide hydroxide

Silver halide precipitates from the solution and can be removed by filtration. The quaternary ammonium hydroxide can then be obtained by evaporation of the water.

While most eliminations involving neutral substrates tend to follow the *Zaitsev rule* (Section 6.12A), eliminations with charged substrates tend to follow what is called the *Hofmann rule* and *yield mainly the least substituted alkene.* We can see an example of this behavior if we compare the following reactions.

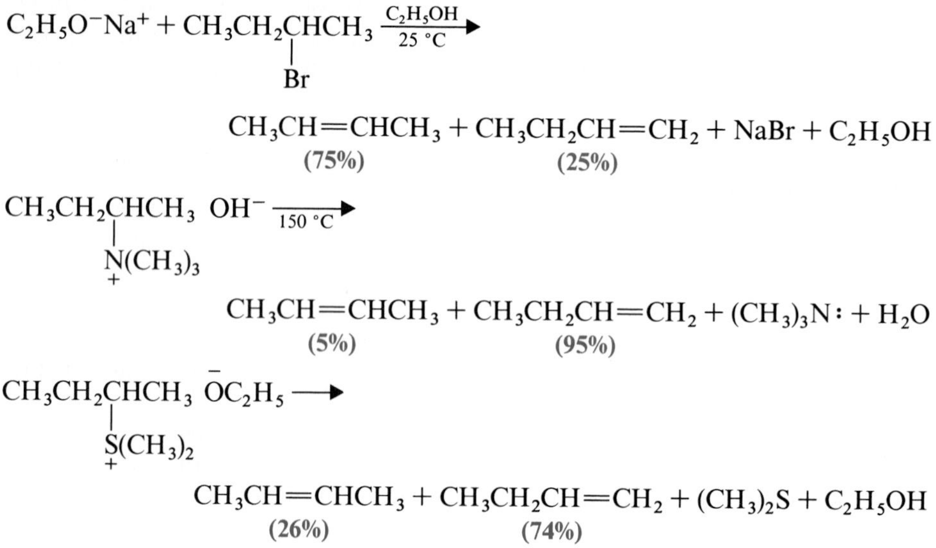

$$C_2H_5O^-Na^+ + CH_3CH_2CHCH_3 \xrightarrow[\text{25 °C}]{C_2H_5OH}$$
$$\underset{\displaystyle|}{} Br$$

$$CH_3CH=CHCH_3 + CH_3CH_2CH=CH_2 + NaBr + C_2H_5OH$$
$$\quad\quad (75\%) \quad\quad\quad\quad\quad (25\%)$$

$$CH_3CH_2CHCH_3 \;\; OH^- \xrightarrow[150\,°C]{}$$
$$\underset{\overset{|}{\underset{+}{N(CH_3)_3}}}{}$$

$$CH_3CH=CHCH_3 + CH_3CH_2CH=CH_2 + (CH_3)_3N\!:\! + H_2O$$
$$\quad\quad (5\%) \quad\quad\quad\quad\quad (95\%)$$

$$CH_3CH_2CHCH_3 \;\; \overset{-}{O}C_2H_5 \longrightarrow$$
$$\underset{\overset{|}{\underset{+}{S(CH_3)_2}}}{}$$

$$CH_3CH=CHCH_3 + CH_3CH_2CH=CH_2 + (CH_3)_2S + C_2H_5OH$$
$$\quad\quad (26\%) \quad\quad\quad\quad\quad (74\%)$$

The precise mechanistic reasons for these differences are complex and are not yet fully understood. One possible explanation is that the transition states of elimination reactions with charged substrates have considerable carbanion character. Therefore, these transition states show little resemblance to the final alkene product and, thus, are not stabilized appreciably by a developing double bond.

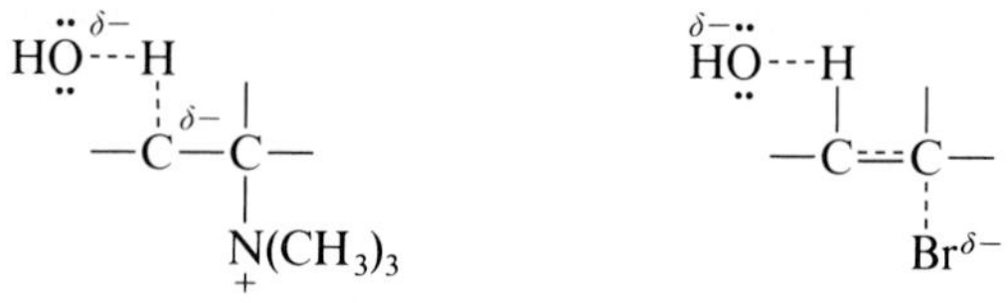

Carbanionlike transition state
(gives Hofmann orientation)

Alkenelike transition state
(gives Zaitsev orientation)

With a charged substrate, the base attacks the most acidic hydrogen instead. A primary hydrogen atom is more acidic because its carbon atom bears only one electron-releasing group.

19.13B The Cope Elimination
Tertiary amine oxides undergo the elimination of a dialkylhydroxylamine when they are heated. This reaction is called the Cope elimination.

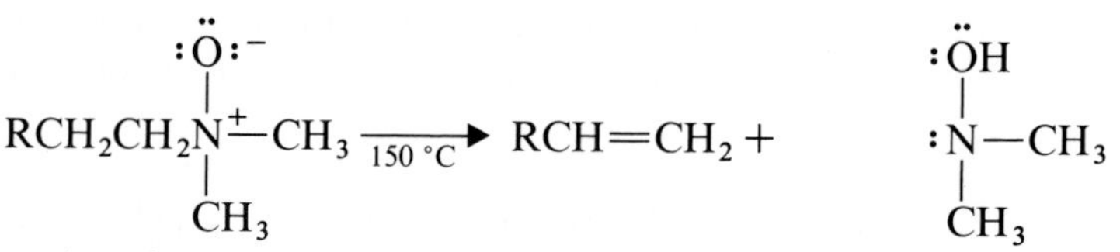

$$RCH_2CH_2\overset{+}{N}\!-CH_3 \xrightarrow[150\,°C]{} RCH=CH_2 + \quad :N\!-CH_3$$

A tertiary amine
oxide

An alkene

N,N-Dimethylhydroxylamine

The Cope elimination is a syn elimination and proceeds through a cyclic transition state:

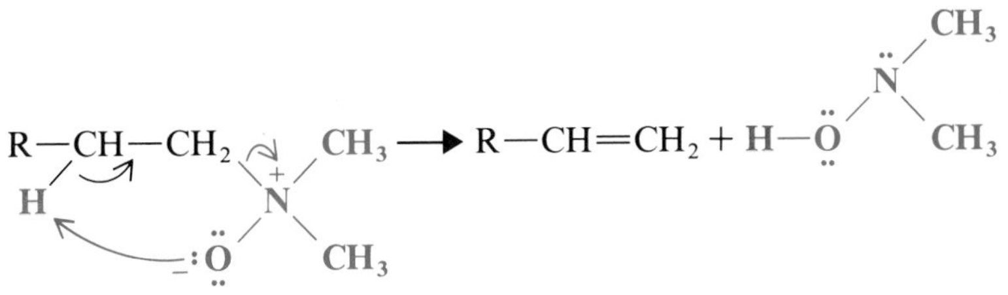

Tertiary amine oxides are easily prepared by treating tertiary amines with hydrogen peroxide (Section 19.6A).

ADDITIONAL PROBLEMS

19.21
Write structural formulas for each of the following compounds:

(a) Benzylmethylamine

(b) Triisopropylamine

(c) N-Ethyl-N-methylaniline

(d) m-Toluidine

(e) 2-Methylpyrrole

(f) N-Ethylpiperidine

(g) N-Ethylpyridinium bromide

(h) 3-Pyridinecarboxylic acid

(i) Indole

(j) Acetanilide

(k) Dimethylamminium chloride

(l) 2-Methylimidazole

(m) 3-Amino-1-propanol

(n) Tetrapropylammonium chloride

(o) Pyrrolidine

(p) N,N-Dimethyl-p-toluidine

(q) 4-Methoxyaniline

(r) Tetramethylammonium hydroxide

(s) p-Aminobenzoic acid

(t) N-Methylaniline

19.22
Give common or systematic names for each of the following compounds:

(a) $CH_3CH_2CH_2NH_2$

(b) $C_6H_5NHCH_3$

(c) $(CH_3)_2CH\overset{+}{N}(CH_3)_3\ I^-$

(d) o-$CH_3C_6H_4NH_2$

(e) o-$CH_3OC_6H_4NH_2$

(i) $C_6H_5N(CH_2CH_2CH_3)_2$

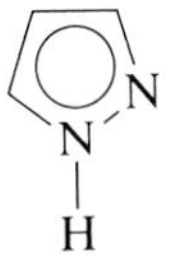

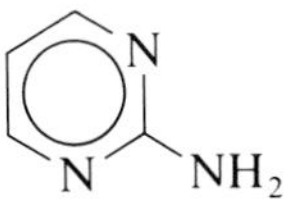

(j) $C_6H_5SO_2NH_2$

(k) $CH_3NH_3{}^+CH_3CO_2{}^-$

(l) $HOCH_2CH_2CH_2NH_2$

(f)

(g)

(m)

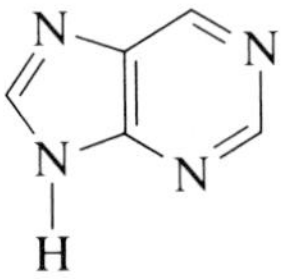

(n)

(h) $C_6H_5CH_2NH_3{}^+\ Cl^-$

19.23

Show how you might prepare benzylamine from each of the following compounds:

(a) Benzonitrile   (d) Benzyl tosylate   (g) Phenylacetamide

(b) Benzamide   (e) Benzaldehyde

(c) Benzyl bromide (two ways)   (f) Phenylnitromethane

19.24

Show how you might prepare aniline from each of the following compounds:

(a) Benzene   (b) Bromobenzene   (c) Benzamide

19.25

Show how you might synthesize each of the following compounds from butyl alcohol:

(a) Butylamine (free of 2° and 3° amines)   (c) Propylamine

(b) Pentylamine   (d) Butylmethylamine

19.26

Show how you might convert aniline into each of the following compounds. (You need not repeat steps carried out in earlier parts of this problem.)

(a) Acetanilide   (i) Iodobenzene

(b) N-Phenylphthalimide   (j) Benzonitrile

(c) p-Nitroaniline   (k) Benzoic acid

(d) Sulfanilamide   (l) Phenol

(e) N,N-Dimethylaniline   (m) Benzene

(f) Fluorobenzene   (n) p-(Phenylazo)phenol

(g) Chlorobenzene   (o) p-N,N-Dimethyl-p-(phenylazo)aniline

(h) Bromobenzene

19.27

What products would you expect to be formed when each of the following amines reacts with aqueous sodium nitrite and hydrochloric acid?

(a) Propylamine   (c) N-Propylaniline   (e) p-Propylaniline

(b) Dipropylamine   (d) N,N-Dipropylaniline

19.28

(a) What products would you expect to be formed when each of the amines in the preceding problem reacts with benzenesulfonyl chloride and excess aqueous potassium hydroxide? (b) What would you observe in each reaction? (c) What would you observe when the resulting solution or mixture is acidified?

19.29

(a) What product would you expect to obtain from the reaction of piperidine with aqueous sodium nitrite and hydrochloric acid? (b) From the reaction of piperidine and benzenesulfonyl chloride in excess aqueous potassium hydroxide?

19.30

Give structures for the products of each of the following reactions:

(a) Ethylamine + benzoyl chloride ⟶

(b) Methylamine + acetic anhydride ⟶

(c) Methylamine + succinic anhydride $\longrightarrow$

(d) Product of (c) $\xrightarrow{\text{heat}}$

(e) Pyrrolidine + phthalic anhydride $\longrightarrow$

(f) Pyrrole + acetic anhydride $\longrightarrow$

(g) Aniline + propanoyl chloride $\longrightarrow$

(h) Tetraethylammonium hydroxide $\xrightarrow{\text{heat}}$

(i) m-Dinitrobenzene + H_2S $\xrightarrow[C_2H_5OH]{NH_3}$

(j) p-Toluidine + Br_2(excess) $\xrightarrow{H_2O}$

19.31

Starting with benzene or toluene, outline a synthesis of each of the following compounds using diazonium salts as intermediates. (You need not repeat syntheses carried out in earlier parts of this problem.)

(a) o-Cresol

(b) m-Cresol

(c) p-Cresol

(d) m-Dichlorobenzene

(e) m-$C_6H_4(CN)_2$

(f) m-Iodophenol

(g) m-Bromobenzonitrile

(h) 1,3-Dibromo-5-nitrobenzene

(i) 3,5-Dibromoaniline

(j) 3,4,5-Tribromophenol

(k) 3,4,5-Tribromobenzonitrile

(l) 2,6-Dibromobenzoic acid

(m) 1,3-Dibromo-2-iodobenzene

(n) 4-Bromo-2-nitrotoluene

(o) 4-Methyl-3-nitrophenol

(p)

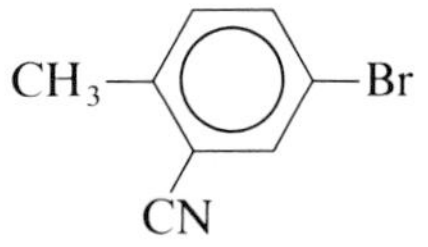

(q)

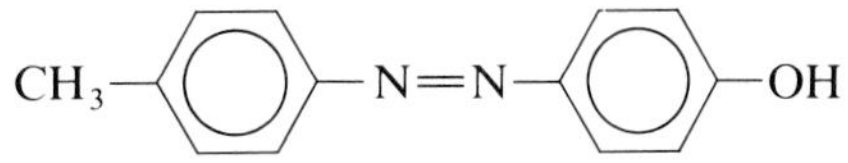

(r)

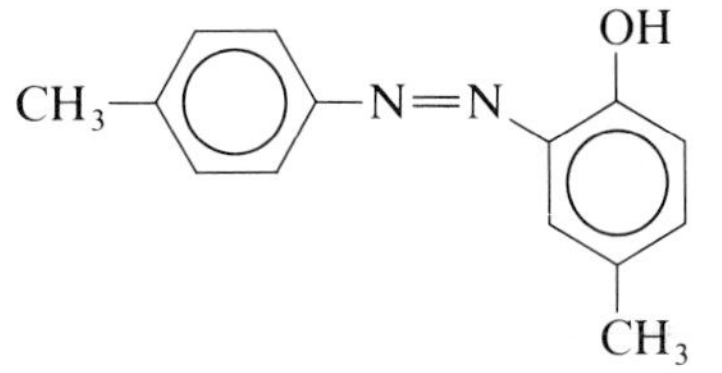

19.32

Write equations for simple chemical tests that would distinguish between

(a) Benzylamine and benzamide

(b) Allylamine and propylamine

(c) p-Toluidine and N-methylaniline

(d) Cyclohexylamine and piperidine

(e) Pyridine and benzene

(f) Cyclohexylamine and aniline

(g) Triethylamine and diethylamine

(h) Tripropylaminium chloride and tetrapropylammonium chloride

(i) Tetrapropylammonium chloride and tetrapropylammonium hydroxide

19.33

Describe with equations how you might separate a mixture of aniline, p-cresol, benzoic acid, and toluene using ordinary laboratory reagents.

19.34

Show how you might synthesize β-aminopropionic acid ($H_3\overset{+}{N}CH_2CH_2CO_2^-$) from succinic anhydride. (β-Aminopropionic acid is used in the synthesis of pantothenic acid; cf. Problem 18.34.)

19.35

Show how you might synthesize each of the following from the compounds indicated and any other needed reagents.

(a) Decamethonium bromide (Section 19.4) from 1,10-decanediol

(b) Succinylcholine bromide from succinic acid, 2-bromoethanol, and trimethylamine

(c) Acetylcholine chloride from ethylene oxide

19.36

A commercial synthesis of folic acid consists of heating the following three compounds with aqueous sodium bicarbonate. Propose reasonable mechanisms for the reactions that lead to folic acid.

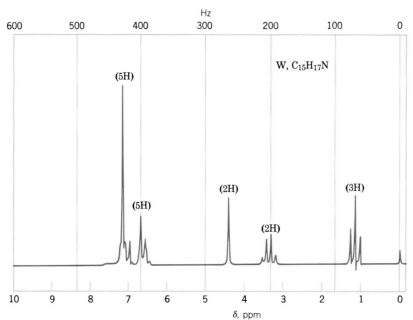

19.37

When compound W ($C_{15}H_{17}N$) is treated with benzenesulfonyl chloride and aqueous potassium hydroxide, no apparent change occurs. Acidification of this mixture gives a clear solution. The proton nmr spectrum of W is shown in Fig. 19.4. Propose a structure for W.

FIGURE 19.4 The proton nmr spectrum of W, Problem 19.37. (Courtesy Aldrich Chemical Company Inc., Milwaukee, WI.)

19.38
Propose structures for compounds **X**, **Y**, and **Z**.

$$X(C_7H_7Br) \xrightarrow{NaCN} Y(C_8H_7N) \xrightarrow{LiAlH_4} Z(C_8H_{11}N)$$

The proton nmr spectrum of **X** gives two signals, a multiplet at $\delta 7.3$ (5H) and a singlet at $\delta 4.25$ (2H); the $680-840\text{-cm}^{-1}$ region of the infrared spectrum of **X** shows peaks at 690 and 770 cm^{-1}. The proton nmr spectrum of **Y** is similar to that of **X**: multiplet $\delta 7.3$ (5H), singlet $\delta 3.7$ (2H). The proton nmr and infrared spectra of **Z** are shown in Fig. 19.5.

*19.39
Using reactions that we have studied in this chapter, propose a mechanism that accounts for the following reaction:

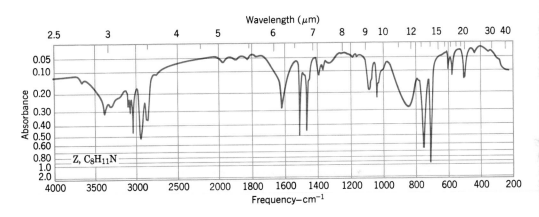

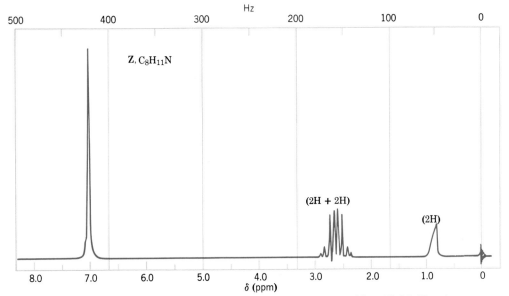

FIGURE 19.5 Infrared and proton nmr spectra for compound **Z**, Problem 19.38. (Courtesy Sadtler Research Laboratories, Inc., Philadelphia.)

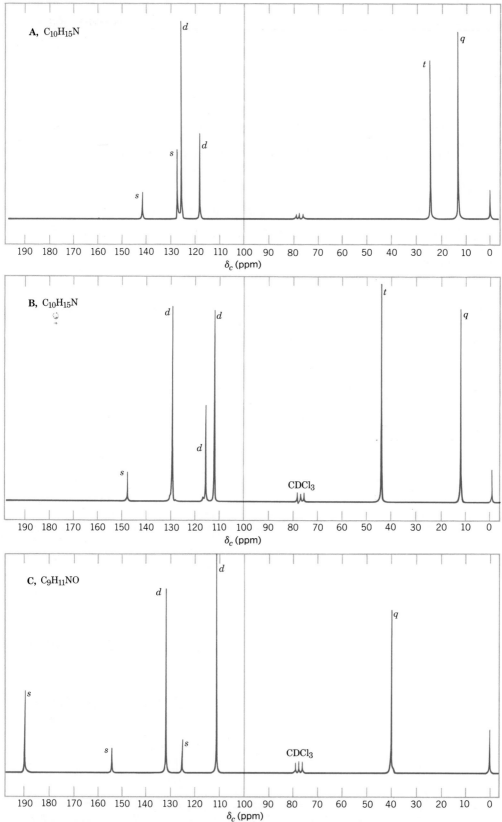

FIGURE 19.6 The proton-decoupled ^{13}C nmr spectra of compounds **A**, **B**, and **C**, Problems 19.41, 19.42, and 19.43. The letters s, d, t, and q refer to the signal splitting in the proton off-resonance decoupled spectra. (Adapted from L. F. Johnson and W. C. Jankowski, *Carbon-13 NMR Spectra: A Collection of Assigned, Coded, and Indexed Spectra*, Wiley– Interscience, New York, 1972.)

19.40
Give structures for compounds $R-W$:

$$N\text{-Methylpiperidine} + CH_3I \longrightarrow R\ (C_7H_{16}NI) \xrightarrow[\text{H}_2\text{O}]{\text{Ag}_2\text{O}}$$

$$S\ (C_7H_{17}NO) \xrightarrow[(-H_2O)]{\text{heat}} T\ (C_7H_{15}N) \xrightarrow{CH_3I} U\ (C_8H_{18}NI) \xrightarrow[\text{H}_2\text{O}]{\text{Ag}_2\text{O}}$$

$$V\ (C_8H_{19}NO) \xrightarrow{\text{heat}} W\ (C_5H_8) + H_2O + (CH_3)_3N$$

19.41
Compound A ($C_{10}H_{15}N$) is soluble in dilute HCl. The infrared absorption spectrum shows two bands in the $3300-3500\text{-cm}^{-1}$ region. The proton-decoupled ^{13}C spectrum of A is given in Fig. 19.6. Propose a structure for A.

19.42
Compound B, an isomer of A (Problem 19.41), is also soluble in dilute HCl. The infrared spectrum of B shows no bands in the $3300-3500\text{-cm}^{-1}$ region. The proton-decoupled ^{13}C spectrum of B is given in Fig. 19.6. Propose a structure for B.

19.43
Compound C ($C_9H_{11}NO$) gives a positive Tollens' test and is soluble in dilute HCl. The infrared spectrum of C shows a strong band near 1695 cm^{-1} but shows no bands in the $3300-3500\text{-cm}^{-1}$ region. The proton-decoupled ^{13}C nmr spectrum of C is shown in Fig. 19.6. Propose a structure for C.

SPECIAL TOPIC K

REACTIONS AND SYNTHESIS
OF HETEROCYCLIC AMINES

Heterocyclic amines undergo many reactions that are similar to those of the amines that we have studied in earlier chapters.

K.1 Heterocyclic Amines as Bases

Nonaromatic heterocyclic amines have basicity constants that are approximately the same as those of acyclic amines.

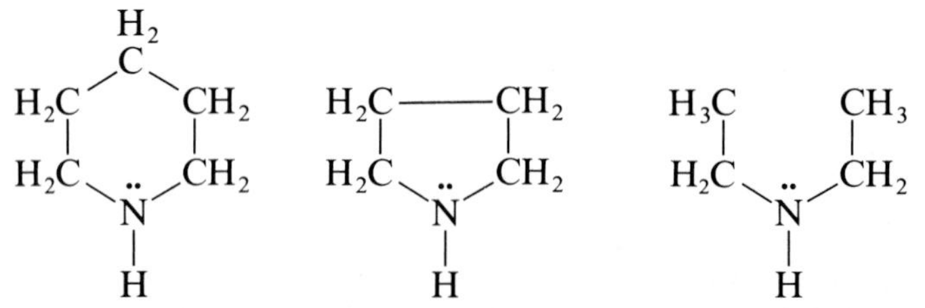

Piperidine
$K_b = 1.6 \times 10^{-3}$

Pyrrolidine
$K_b = 1.3 \times 10^{-3}$

Diethylamine
$K_b = 9.6 \times 10^{-4}$

In aqueous solution, aromatic heterocyclic amines such as pyridine, pyrimidine, and pyrrole are much weaker bases than nonaromatic amines or ammonia ($K_b = 1.8 \times 10^{-5}$). (In the gas phase, however, pyridine and pyrrole are more basic than ammonia, indicating that solvation has a very important effect on their relative basicities, cf. Section 19.3.)

Pyridine
$K_b = 1.7 \times 10^{-9}$

Pyrimidine
$K_b = 5 \times 10^{-12}$

Pyrrole
$K_b = 2.5 \times 10^{-14}$

K.2 Heterocyclic Amines as Nucleophiles
in Alkylation and Acylation Reactions

Most heterocyclic amines undergo alkylation and acylation reactions in much the same way as acyclic amines.

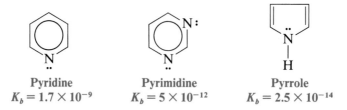

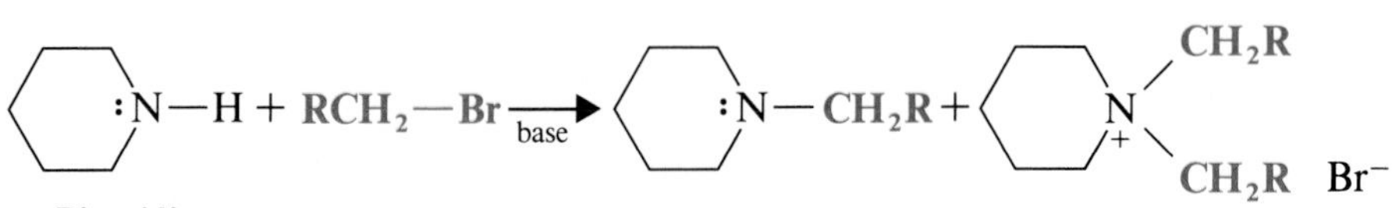

Piperidine

N-Alkylpiperidine

N,N-Dialkylpiper-
idinium bromide

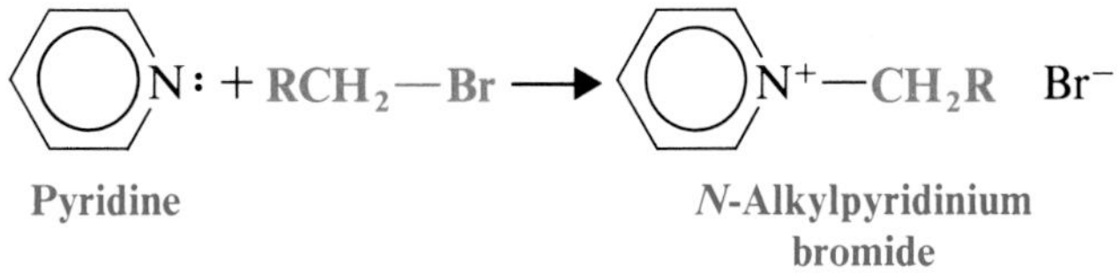

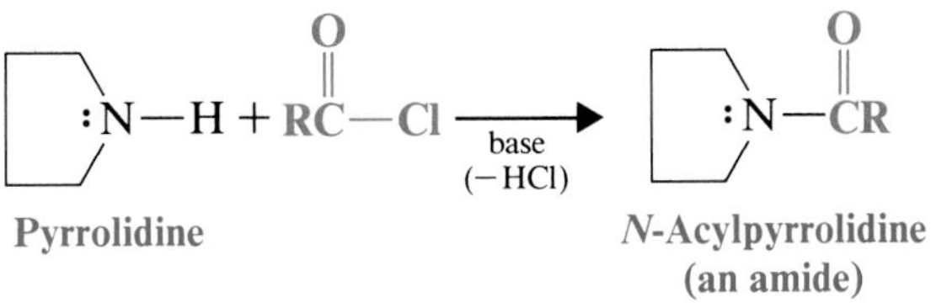

Pyridine   N-Alkylpyridinium
                                    bromide

Pyrrolidine   N-Acylpyrrolidine
                              (an amide)

PROBLEM K.1

What products would you expect to obtain from the following reactions?

(a) Piperidine + acetic anhydride ⟶

(b) Pyridine + methyl iodide ⟶

(c) Pyrrolidine + phthalic anhydride ⟶

(d) Pyrrolidine + (excess) methyl iodide $\xrightarrow[\text{(base)}]{}$

(e) Product of (d) + Ag_2O, H_2O, then heat ⟶

K.3 Electrophilic Substitution Reactions of Aromatic Heterocyclic Amines

Pyrrole is highly reactive toward electrophilic substitution and substitution takes place primarily at position 2.

General Reaction

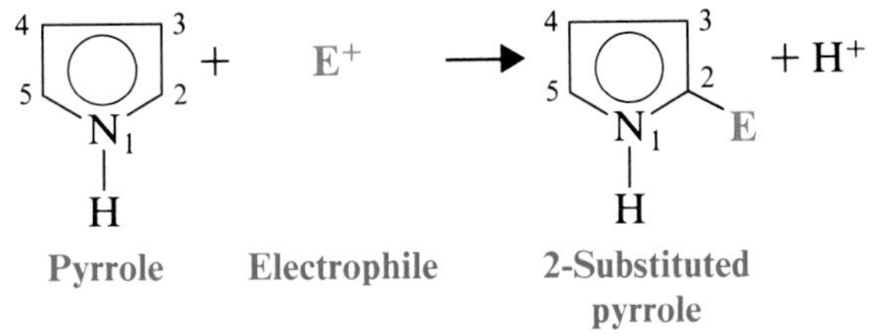

Pyrrole   Electrophile   2-Substituted
                                 pyrrole

Specific Example

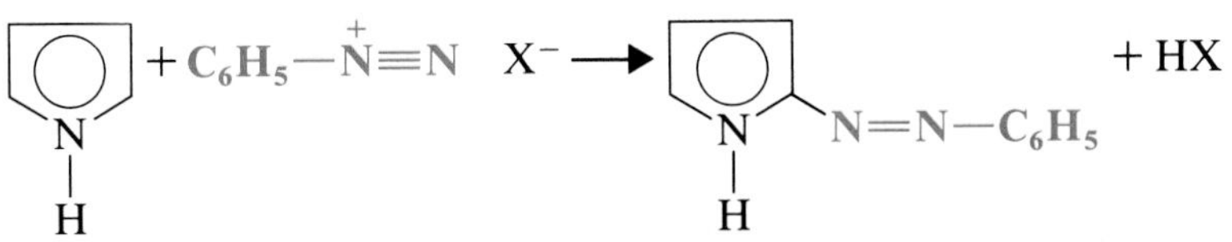

We can understand why electrophilic substitution at the 2 position is preferred if we examine the resonance structures that follow:

Substitution at the 2 position of pyrrole

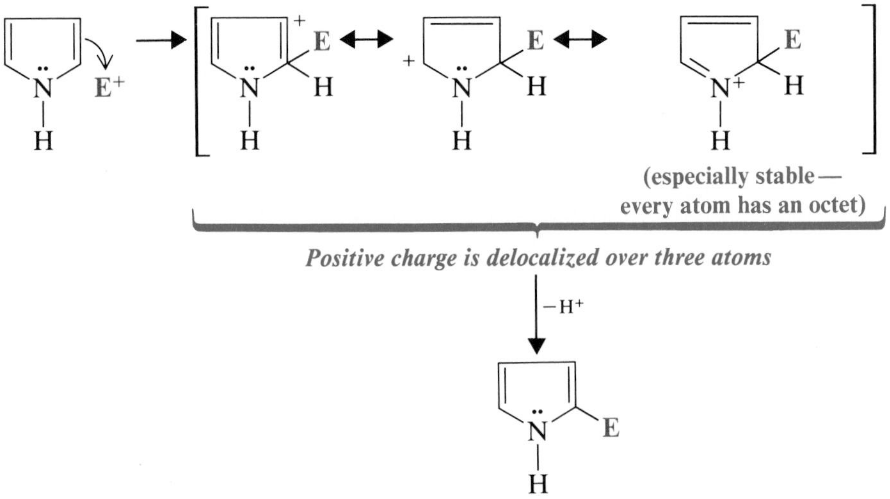

(especially stable —
every atom has an octet)

Positive charge is delocalized over three atoms

$-H^+$

Substitution at the 3 position of pyrrole

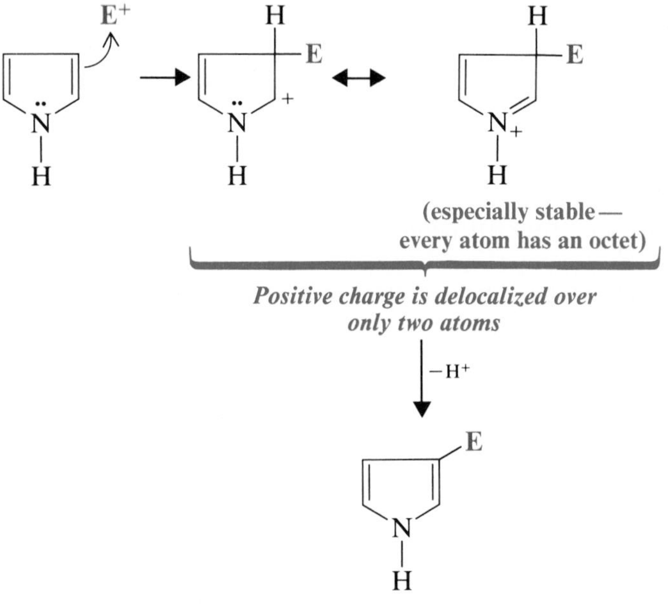

(especially stable —
every atom has an octet)

*Positive charge is delocalized over
only two atoms*

$-H^+$

We see that while a relatively stable structure contributes to the hybrid for both intermediates, the intermediate arising from attack at the 2 position is stabilized by one additional resonance structure, and the positive charge is delocalized over three atoms rather than two. This means that this intermediate is more stable, and that attack at the 2 position has a lower activation energy.

Pyridine is much less reactive toward electrophilic substitution than benzene. Pyridine does not undergo Friedel–Crafts acylation or alkylation; it does not couple with diazonium compounds. Bromination of pyridine can be accomplished but only in the vapor phase at 200 °C where a free radical mechanism may operate. Nitration and sulfonation also require forcing conditions. Electrophilic substitution, when it occurs, nearly always takes place at the 3 position.

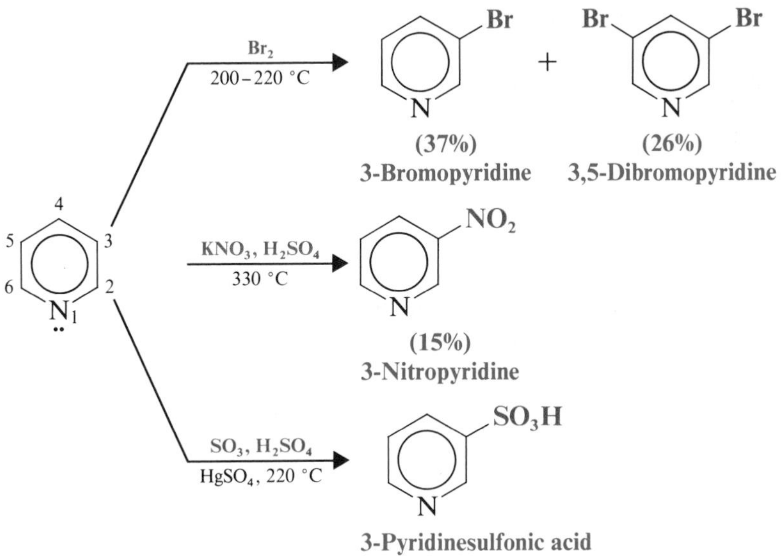

Br_2
200–220 °C

Br

N

(37%)
3-Bromopyridine

+

Br   Br

N

(26%)
3,5-Dibromopyridine

KNO_3, H_2SO_4
330 °C

NO_2

N

(15%)
3-Nitropyridine

SO_3, H_2SO_4
$HgSO_4$, 220 °C

SO_3H

N

3-Pyridinesulfonic acid

We can, in part, attribute the lower reactivity of pyridine (when compared to benzene) to the greater electronegativity of nitrogen (when compared to carbon). Nitrogen, being more electronegative, is less able to accommodate the electron deficiency that characterizes the transition state leading to the positively charged ion (similar to an arenium ion) in electrophilic substitution.

$+ E^+ \longrightarrow$

Pyridine

*Transition state
is of higher energy
because of greater
electronegativity
of nitrogen*

*Similar to
an arenium
ion*

$+ E^+ \longrightarrow$

Benzene

*Transition state
is of lower energy
because of lower
electronegativity
of carbon*

Arenium ion

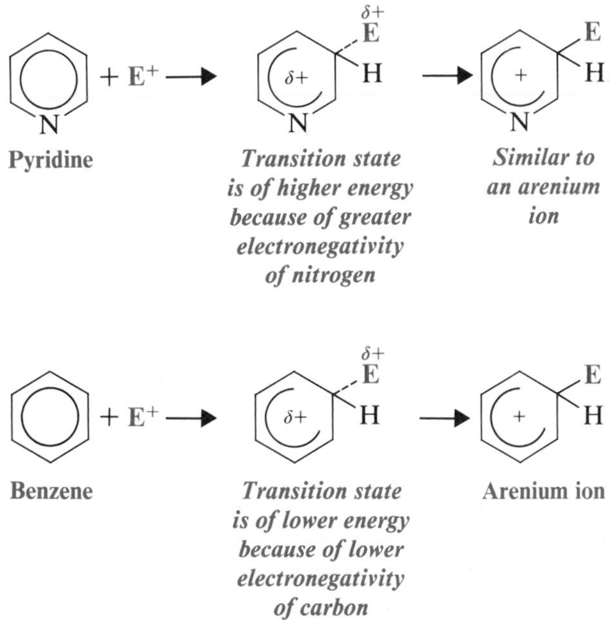

The low reactivity of pyridine toward electrophilic substitution may arise mainly from the fact that pyridine is converted initially to a pyridinium ion by a proton or other electrophile.

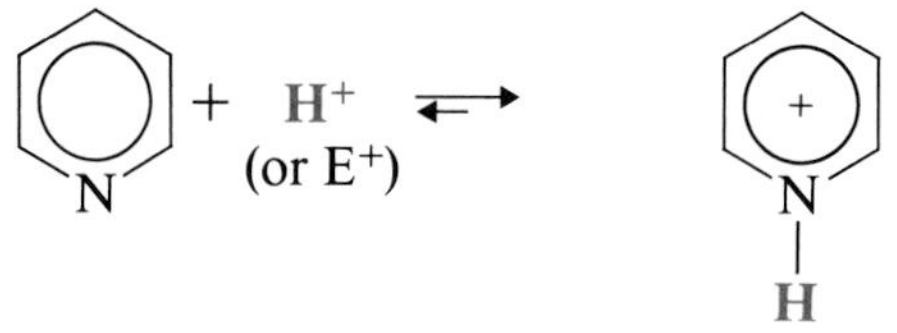

Pyridinium ion
*(highly unreactive because
of positive charge)*

Electrophilic attack at the 4 position (or the 2 position) is unfavorable because an especially unstable resonance structure contributes to the intermediate hybrid.

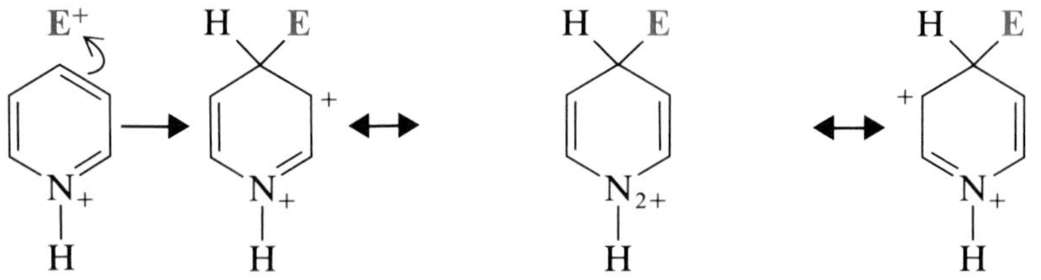

*Especially unstable because
nitrogen has a sextet
and two positive charges*

Similar resonance structures can be written for attack at the 2 position.

No especially unstable *or stable* structure contributes to the hybrid arising from attack at the 3 position; as a result, attack at the 3 position is preferred but occurs slowly.

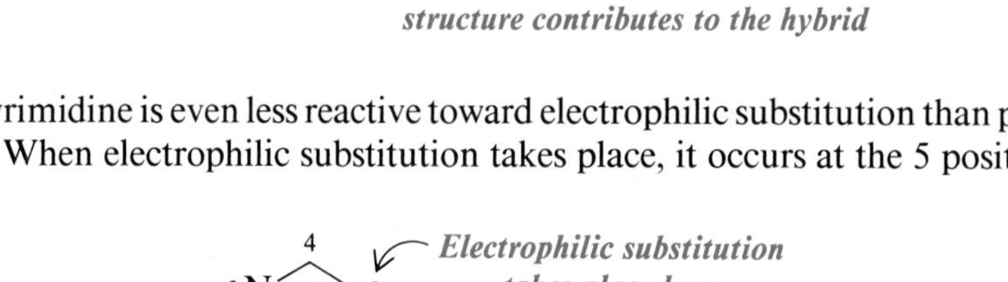

*No especially unstable or stable
structure contributes to the hybrid*

Pyrimidine is even less reactive toward electrophilic substitution than pyridine. (Why?) When electrophilic substitution takes place, it occurs at the 5 position.

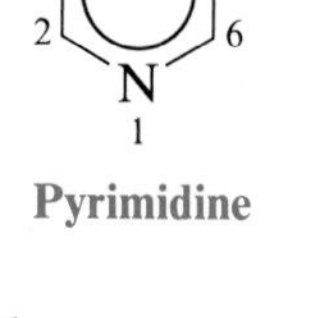

*Electrophilic substitution
takes place here*

Pyrimidine

Imidazole is much more susceptible to electrophilic substitution than pyridine or pyrimidine, but is less reactive than pyrrole. Imidazoles with 1 substituents undergo electrophilic substitution at the 4 position.

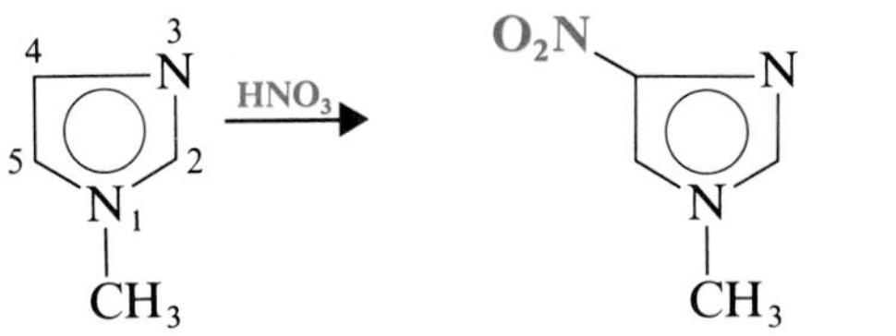

1-Methyl-4-nitroimidazole

Imidazole, itself, undergoes electrophilic substitution in a similar fashion. Tautomerism, however, makes the 4 and 5 positions equivalent.

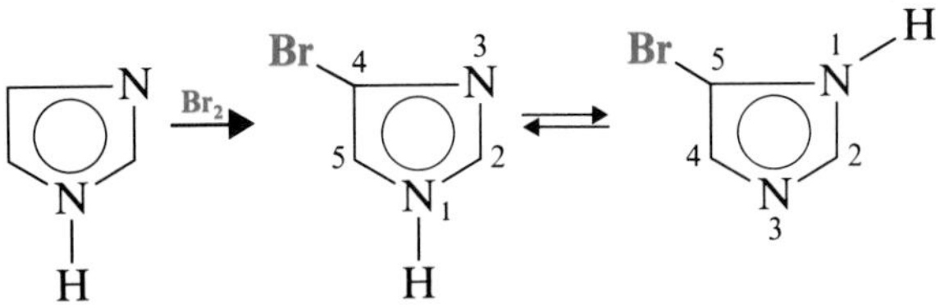

4-(5)-Bromoimidazole

PROBLEM K.2

Both pyrrole and imidazole are weak acids; they react with strong bases to form anions:

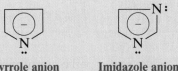

Pyrrole anion   Imidazole anion

(a) These anions resemble a carbocyclic anion that we have studied before. What is it?   (b) Write resonance structures that account for the stabilities of pyrrole and imidazole anions.

K.4 Nucleophilic Substitutions of Pyridine

In its reactions, the pyridine ring resembles a benzene ring with a strong electron-withdrawing group; pyridine is relatively unreactive toward electrophilic substitution but appreciably reactive toward nucleophilic substitution.

In the previous section we compared the reactivity of pyridine and benzene toward electrophilic substitution and there we attributed pyridine's lower reactivity to the greater electronegativity of its ring nitrogen. Because nitrogen is more electronegative than carbon, it is less able to accommodate the electron deficiency in the transition state of the rate-determining step in electrophilic aromatic substitution. On the other hand, nitrogen's greater electronegativity makes it *more* able to accommodate the excess *negative* charge that an aromatic ring must accept in *nucleophilic substitution.*

Pyridine reacts with sodium amide, for example, to form 2-aminopyridine: In this remarkable reaction (called the Chichibabin reaction) amide ion (NH_2^-) displaces a hydride ion (H^-).

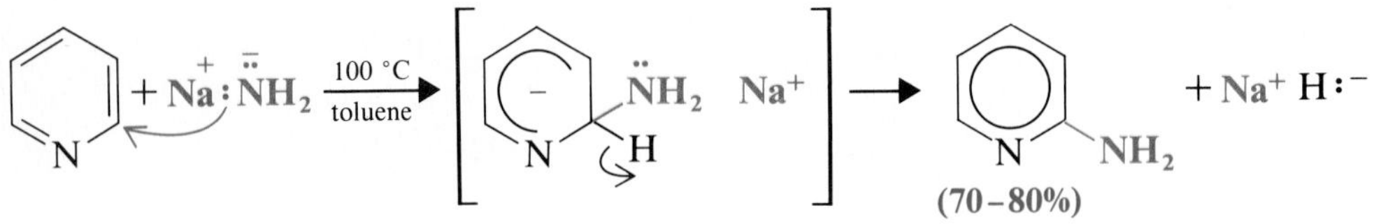

If we examine the resonance structures that contribute to the intermediate in this reaction we shall be able to see how the ring nitrogen atom accommodates the negative charge:

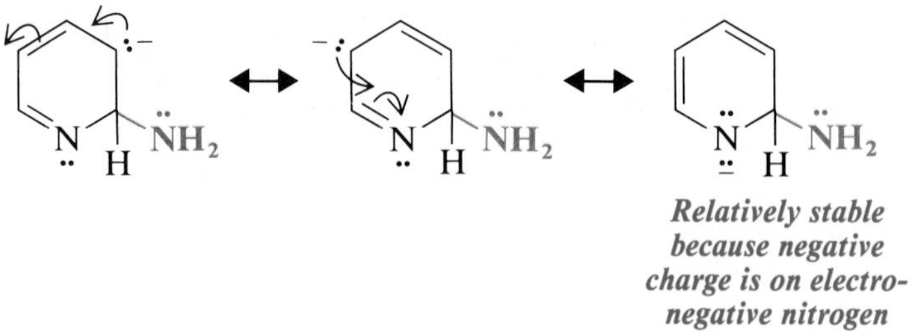

Relatively stable because negative charge is on electro- negative nitrogen

In the next step the intermediate loses a hydride ion and becomes 2-amino-pyridine.*

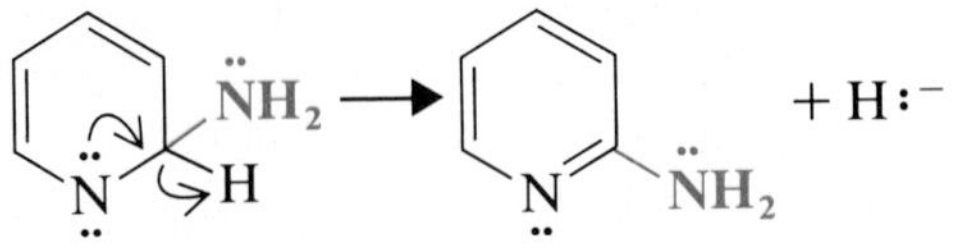

Pyridine undergoes similar nucleophilic substitution reactions with phenyllith-ium, butyllithium, and potassium hydroxide.

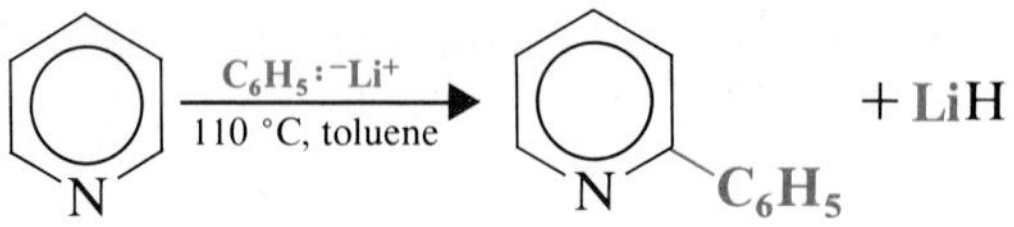

2-Phenylpyridine

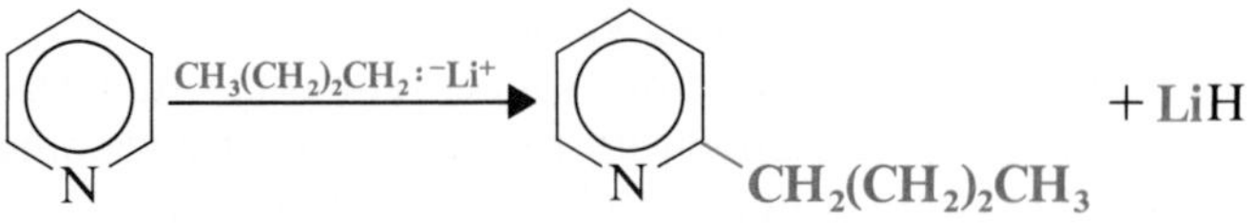

2-Butylpyridine

* In practice, a subsequent reaction occurs; 2-aminopyridine reacts with the sodium hydride to produce a sodio derivative:

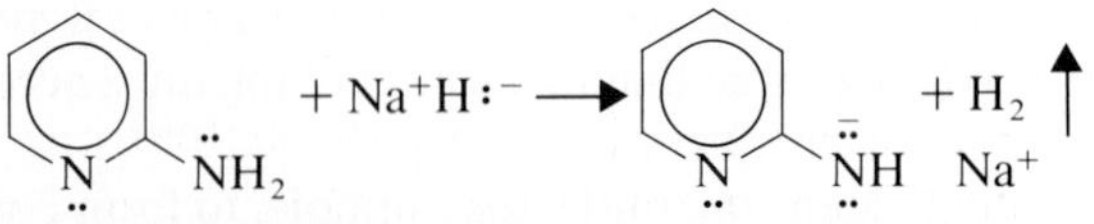

When the reaction is over, the addition of cold water to the reaction mixture converts the sodio derivative to 2-aminopyridine.

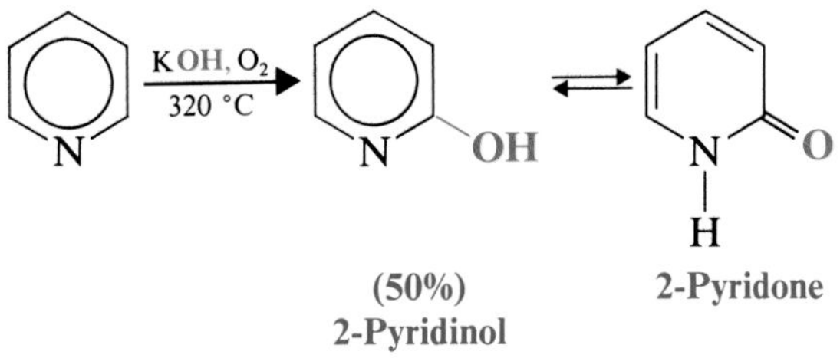

(50%)   2-Pyridone
2-Pyridinol

2-Chloropyridine reacts with sodium methoxide to yield 2-methoxypyridine:

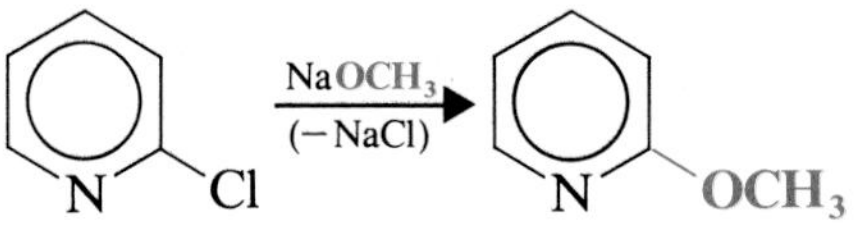

PROBLEM K.3

An alternative mechanism to the one given for the amination of pyridine in Section K.4, involves a "pyridyne" intermediate, that is,

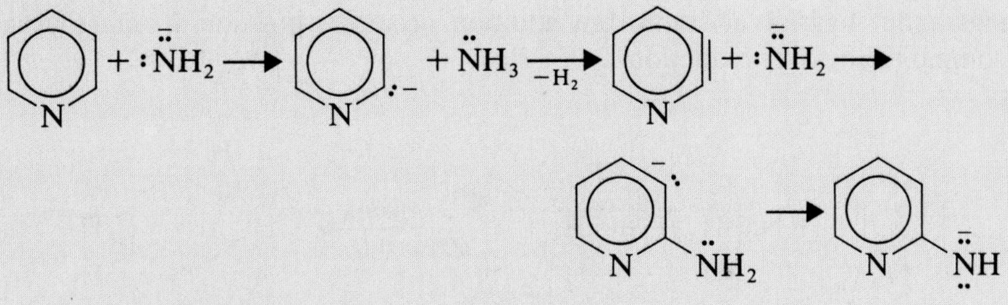

This mechanism was disallowed on the basis of an experiment in which 3-deuteriopyridine was allowed to react with sodium amide. Consider the fate of deuterium in both mechanisms and explain.

PROBLEM K.4

2-Halopyridines undergo nucleophilic substitution much more readily than pyridine itself. What factor accounts for this?

K.5 Nucleophilic Additions to Pyridinium Ions

Pyridinium ions are especially susceptible to nucleophilic attack at the 2 or 4 position because of the contributions of the resonance forms shown here.

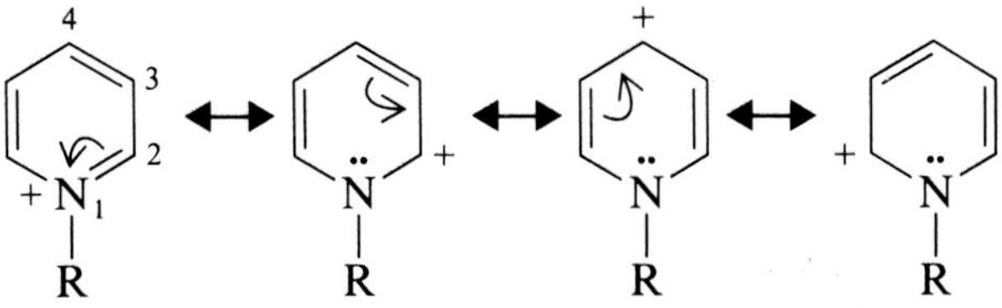

N-Alkylpyridinium halides, for example, react with hydroxide ions primarily at position 2; this causes the formation of an addition product called a *pseudo base.*

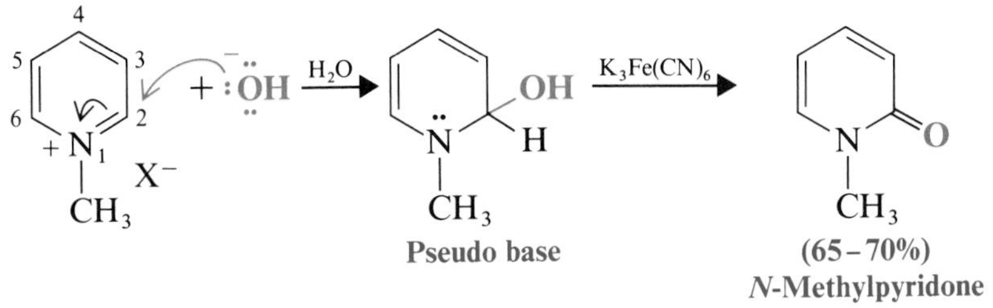

Pseudo base

(65–70%)
N-Methylpyridone

Oxidation of the pseudo base with potassium ferricyanide (see previous reaction) produces an *N*-alkylpyridone.

Nucleophilic additions to pyridinium ions, especially the addition of *hydride ions,* have been of considerable interest to chemists because these reactions resemble the biological reduction of the important coenzyme nicotinamide adenine dinucleotide (NAD^+) (Section 11.11).

A number of model reactions have been carried out in connection with these studies. Treating an *N*-alkylpyridinium ion with sodium borohydride, for example, brings about hydride addition, but addition occurs at position 2 and is usually accompanied by over reduction:

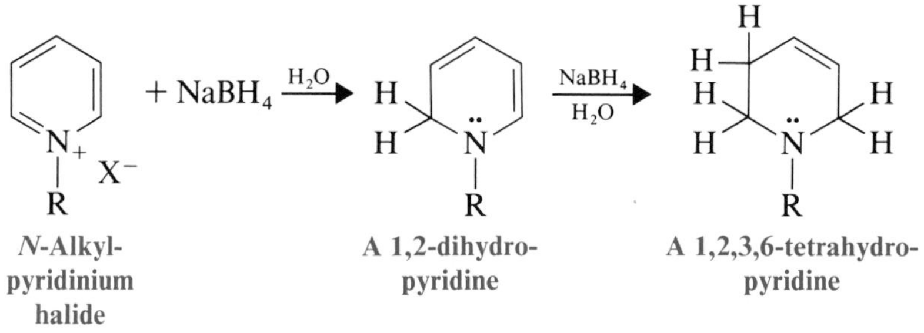

N-Alkyl-
pyridinium
halide

A 1,2-dihydro-
pyridine

A 1,2,3,6-tetrahydro-
pyridine

Treating a pyridinium ion with basic sodium dithionite ($Na_2S_2O_4$), however, brings about specific addition to position 4:

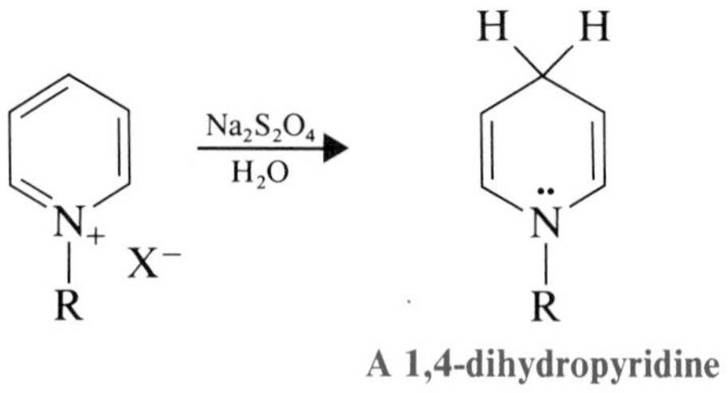

A 1,4-dihydropyridine

Sodium dithionite in aqueous base also reduces NAD^+ to NADH. The NADH formed by dithionite reduction has been shown to be biologically active and can be oxidized to NAD^+ with potassium ferricyanide.

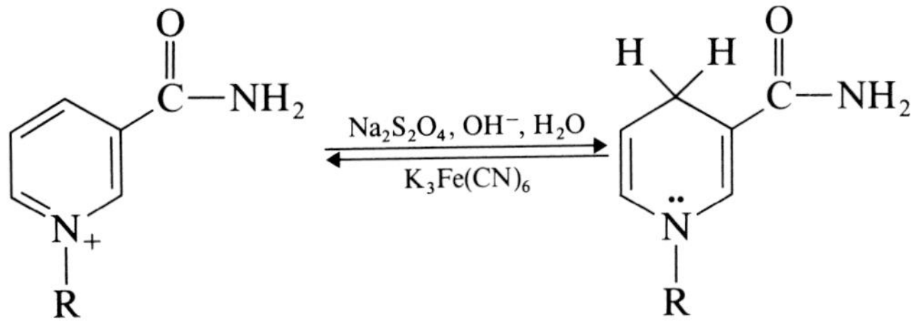

NAD$^+$
(see Section 11.10 for
the structure of R)

NADH

K.6 Synthesis of Heterocyclic Amines

The most general and widely used method for synthesizing pyrroles is to condense an α-amino ketone or α-amino-β-keto ester with a ketone or keto ester. This reaction, called the Knorr synthesis, is catalyzed by acids or bases. Two examples are shown here.

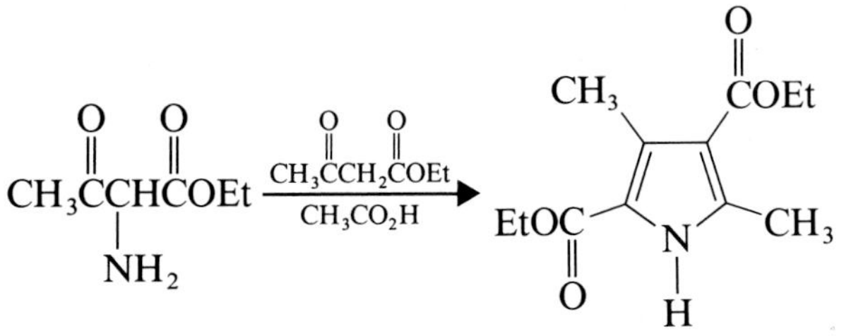

(57–64%)

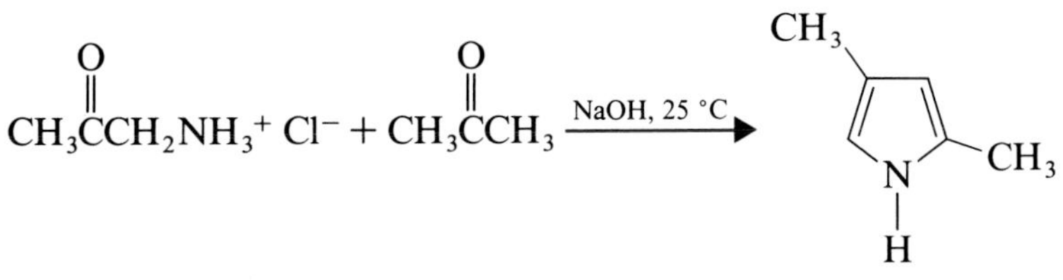

PROBLEM K.5

Propose reasonable mechanisms for the two syntheses of substituted pyrroles just given.

Pyridine and many of its derivatives can be isolated from coal tar. Many pyridine derivatives are prepared from these coal-tar derivatives through substitution reactions. The most general overall pyridine synthesis is one called the Hantzsch synthesis. In this method a β-keto ester, an aldehyde, and ammonia are allowed to condense to produce a dihydropyridine; oxidation of the dihydropyridine yields the substituted pyridine. An example is the following:

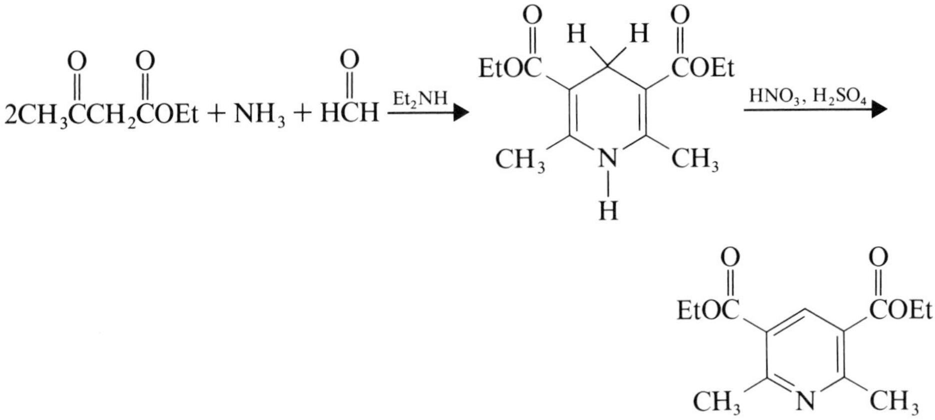

The most general quinoline synthesis is the Skraup synthesis. In this method, aniline is heated with glycerol in the presence of sulfuric acid and an oxidizing agent. Various oxidizing agents have been used including nitrobenzene and air.

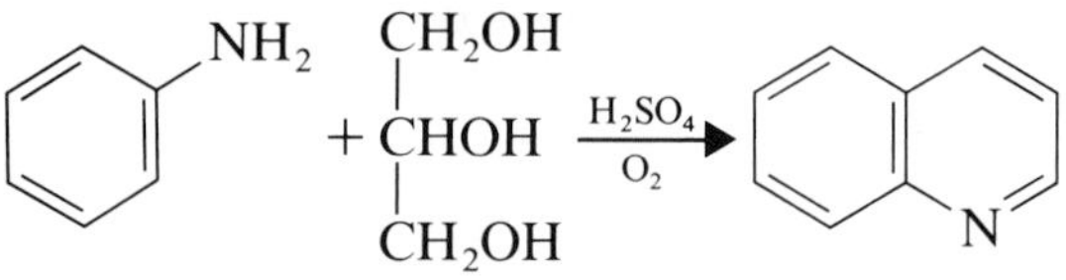

The mechanism for this reaction consists of the following steps:

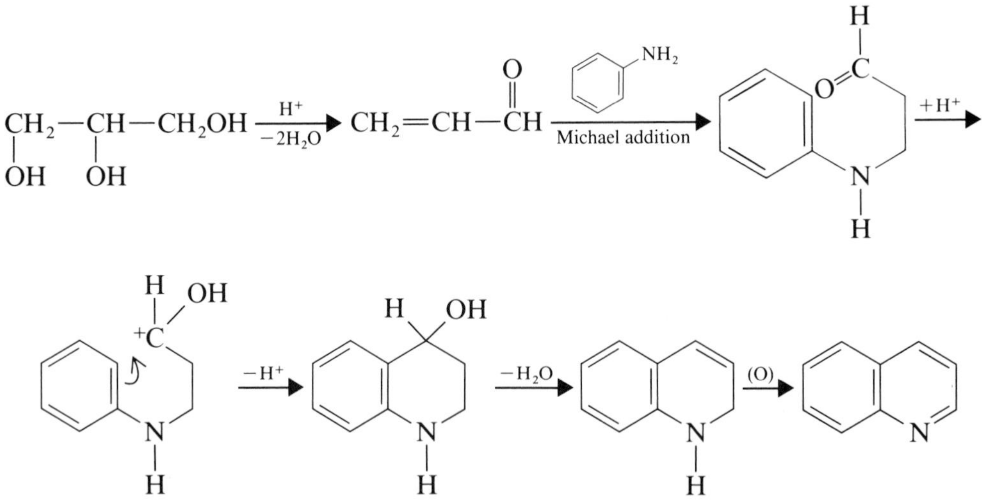

In the first step glycerol dehydrates in the presence of the acid to produce propenal (acrolein). Then a Michael addition of aniline to the propenal is followed by an acid-catalyzed cyclization to yield dihydroquinoline. Finally, oxidation of the dihydroquinoline produces quinoline.

PROBLEM K.6

Give structures for compounds **A**–**H**.

(a) 2,5-Hexanedione + $(NH_4)_2CO_3$ $\xrightarrow{100\ °C}$ **A** (C_6H_9N)
  A pyrrole

(b) $CH_3\overset{\overset{\displaystyle O}{\|}}{C}CH_2NH_2$ + acetone $\xrightarrow{base}$ **B** (C_6H_9N)
  An isomer of A

(c) CH_3NHNH_2 + $(CH_3O)_2CHCH_2CH(OCH_3)_2$ $\xrightarrow[H_2O]{H^+}$ **C** $(C_4H_6N_2)$
  A pyrazole

(d) 2,5-Hexanedione + hydrazine $\xrightarrow{heat}$ **D** $(C_6H_{10}N_2)$ $\xrightarrow{O_2}$ **E** $(C_6H_8N_2)$
  A dihydropyridazine   **A pyridazine**

(e) Aniline + $CH_2{=}CH\overset{\overset{\displaystyle O}{\|}}{C}CH_3$ $\xrightarrow[FeCl_3]{ZnCl_2}$ **F** $(C_{10}H_9N)$
  A quinoline

(f) 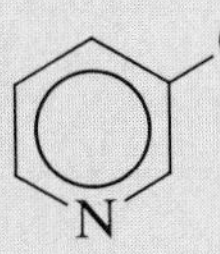 $\underset{N}{\underset{\displaystyle\bigodot}{}}\overset{\displaystyle CH(CH_2)_3NHCH_3}{\underset{\displaystyle I}{|}}$ $\xrightarrow{heat}$ $\xrightarrow{OH^-}$ **G** $(C_{10}H_{14}N_2)$, $\xrightarrow[(2)\ H^+]{(1)\ KMnO_4,\ OH^-}$
  Nicotine

H $(C_6H_5NO_2)$
Nicotinic acid

CHAPTER TWENTY

Polarized crystals of *sym*-diisopropylurea. *sym*-Diisopropyl-
urea is $(CH_3)_2CHNHCONHOH(CH_3)_2$.

Synthesis and Reactions of β-Dicarbonyl Compounds: More Chemistry of Enolate Ions

20.1 INTRODUCTION

Compounds having two carbonyl groups separated by an intervening carbon

$$
\underset{\beta}{-}\overset{\overset{\displaystyle O}{\|}}{C}-\underset{|\atop\alpha}{C}-\overset{\overset{\displaystyle O}{\|}}{C}-
$$

The β-dicarbonyl system

$$
\underset{\beta\ \alpha}{R\overset{\overset{\displaystyle O}{\|}}{C}CH_2\overset{\overset{\displaystyle O}{\|}}{C}OR'}
$$

A β-keto ester
(Section 20.2)

$$
\underset{\beta\ \alpha}{RO\overset{\overset{\displaystyle O}{\|}}{C}CH_2\overset{\overset{\displaystyle O}{\|}}{C}OR}
$$

A malonic ester
(Section 20.4)

are called β-dicarbonyl compounds, and these compounds are highly versatile reagents for organic synthesis. In this chapter we shall explore some of the methods for preparing β-dicarbonyl compounds and some of their important reactions.

Central to the chemistry of β-dicarbonyl compounds is the acidity of protons located on the carbon between two carbonyl groups. The K_a for such a proton is in the range 10^{-10}–10^{-14}.

$$
-\overset{\overset{\displaystyle O}{\|}}{C}-\underset{\underset{\displaystyle H}{|}}{C}-\overset{\overset{\displaystyle O}{\|}}{C}- \quad K_a = 10^{-10} - 10^{-14}
$$

Early in this chapter we shall see how the acidity of these protons allows the synthesis of β-dicarbonyl compounds through reactions that are called *Claisen syntheses* (Section 20.2). Later in the chapter we shall study the *acetoacetic ester synthesis* (Section 20.3) and the *malonic ester synthesis* (Section 20.4), in which the acidity of these hydrogen atoms forms the basis for the synthesis of substituted acetones and substituted acetic acids. The acidity of the hydrogen atoms of a carbon located between two carbonyl groups allows easy conversion of the compound to an enolate ion, and these enolate ions can be alkylated and acylated. Similar chemistry underlies syntheses using a variety of other useful reactions (Section 20.5) including the Knoevenagel condensation (Section 20.7).

One other feature that will appear again and again in the syntheses that we study here is the decarboxylation of a β-keto acid:

941

$$-\overset{\displaystyle O}{\overset{\displaystyle \|}{C}}-\overset{\displaystyle}{\underset{\displaystyle}{C}}-\overset{\displaystyle O}{\overset{\displaystyle \|}{C}}OH \xrightarrow{\text{heat}} -\overset{\displaystyle O}{\overset{\displaystyle \|}{C}}-\overset{\displaystyle}{\underset{\displaystyle}{C}}-H + CO_2$$

We learned in Section 18.11 that these decarboxylations occur at relatively low temperatures, and it is this ease of decarboxylation that makes many of the syntheses in this chapter such useful ones.

20.2 THE CLAISEN CONDENSATION: THE SYNTHESIS OF β-KETO ESTERS

When ethyl acetate reacts with sodium ethoxide, it undergoes *a condensation reaction*. After acidification, the product is a β-keto ester, ethyl acetoacetate (commonly called *acetoacetic ester*).

$$2CH_3\overset{\displaystyle O}{\overset{\displaystyle \|}{C}}OC_2H_5 \xrightarrow{\text{NaOC}_2\text{H}_5} \left[CH_3\overset{\displaystyle O}{\overset{\displaystyle \|}{C}}\overset{\displaystyle O}{\underset{\displaystyle \ddot{}}{C}}H\overset{\displaystyle O}{\overset{\displaystyle \|}{C}}OC_2H_5 \right] + C_2H_5OH$$

Na⁺

Sodioacetoacetic   (removed by
ester   distillation)

↓ HCl

$$CH_3\overset{\displaystyle O}{\overset{\displaystyle \|}{C}}CH_2\overset{\displaystyle O}{\overset{\displaystyle \|}{C}}OC_2H_5 \quad (75–76\%)$$

Ethyl acetoacetate
(acetoacetic ester)

Condensations of this type occur with many other esters and are known generally as *Claisen condensations*. Like the aldol condensation (Section 17.5), Claisen condensations involve the α carbon of one molecule and the carbonyl group of another. Ethyl pentanoate, for example, reacts with sodium ethoxide to give the β-keto ester that follows:

$$2CH_3CH_2CH_2CH_2\overset{\displaystyle O}{\overset{\displaystyle \|}{C}}OC_2H_5 \xrightarrow{\text{NaOCH}_2\text{CH}_2} \left[CH_3CH_2CH_2CH_2\overset{\displaystyle O}{\overset{\displaystyle \|}{C}}-\overset{\displaystyle \text{Na}^+}{\underset{\displaystyle}{\underset{\displaystyle \ddot{C}}{}}}-\overset{\displaystyle O}{\overset{\displaystyle \|}{C}}OC_2H_5 \right] + C_2H_5OH$$

Ethyl pentanoate

CH₂
CH₂
CH₃

↓ CH₃CO₂H

$$CH_3CH_2CH_2CH_2\overset{\displaystyle O}{\overset{\displaystyle \|}{C}}-CH-\overset{\displaystyle O}{\overset{\displaystyle \|}{C}}OC_2H_5$$

CH₂
CH₂
CH₃

(77%)

If we look closely at these examples, we can see that, overall, both reactions involve a condensation in which one ester loses an α hydrogen and the other loses an ethoxide ion; that is,

$$R-CH_2\overset{\overset{\displaystyle O}{\|}}{C}\!-\!OC_2H_5 + H\!-\!\overset{\overset{\displaystyle O}{\|}}{\underset{\underset{\displaystyle R}{|}}{C}}HC\!-\!OC_2H_5 \xrightarrow[\text{(2) H}^+]{\text{(1) NaOC}_2\text{H}_5}$$

(R may also be H)

$$R-CH_2\overset{\overset{\displaystyle O}{\|}}{C}\!-\!\overset{\overset{\displaystyle O}{\|}}{\underset{\underset{\displaystyle R}{|}}{C}}HCOC_2H_5 + C_2H_5OH$$

A β-keto ester

We can understand how this happens if we examine the reaction mechanism.

The first step of a Claisen condensation resembles that of an aldol addition. Ethoxide ion abstracts an α proton from the ester. Although the α protons of an ester are not as acidic as those of aldehydes and ketones, the enolate anion that forms is stabilized by resonance in a similar way.

Step 1

$$R\overset{\alpha}{C}H-\overset{\overset{\displaystyle O}{\|}}{C}OC_2H_5 + :\!\overset{..}{\underset{..}{O}}C_2H_5 \rightleftharpoons RCH\!-\!\overset{\overset{\displaystyle \overset{..}{O}:}{\|}}{C}OC_2H_5 + C_2H_5OH$$

$$\overset{..}{:}\!O\!:^-$$
$$RCH\!=\!COC_2H_5$$

In the second step the enolate anion attacks the carbonyl carbon atom of a second molecule of the ester. It is at this point that the Claisen condensation and the aldol addition *differ,* and they differ in an understandable way: In the aldol reaction nucleophilic attack leads to *addition;* in the Claisen condensation it leads to *substitution.*

Step 2

$$RCH_2\overset{\overset{\displaystyle \overset{..}{O}}{\diagup\!\diagup}}{C}\underset{\displaystyle OC_2H_5}{\diagdown} + {}^-\!:CH\!-\!\overset{\overset{\displaystyle \overset{..}{O}:}{\|}}{\underset{\underset{\displaystyle R}{|}}{C}}OC_2H_5 \rightleftharpoons RCH_2\overset{\overset{\displaystyle :\overset{..}{O}:^-}{|}}{\underset{\underset{\displaystyle C_2H_5\overset{..}{O}:}{|}}{C}}\!-\!\overset{}{\underset{\underset{\displaystyle R}{|}}{C}}H\!-\!\overset{\overset{\displaystyle \overset{..}{O}:}{\|}}{C}OC_2H_5$$

$$RCH_2\overset{\overset{\displaystyle :\overset{..}{O}}{\|}}{C}\!-\!\overset{}{\underset{\underset{\displaystyle R}{|}}{C}}H\!-\!\overset{\overset{\displaystyle \overset{..}{O}:}{\|}}{C}OC_2H_5$$

$$+ {}^-\!:\overset{..}{\underset{..}{O}}C_2H_5$$

Although the products of this second step are a β-keto ester and ethoxide ion, all of the equilibria up to this point have been unfavorable. Very little product would be formed if this were the last step in the reaction.

The final step of a Claisen condensation is an acid–base reaction that takes place between ethoxide ion and the β-keto ester. *The position of equilibrium for this step is favorable,* and we can make it even more favorable by distilling ethanol from the reaction mixture as it forms.

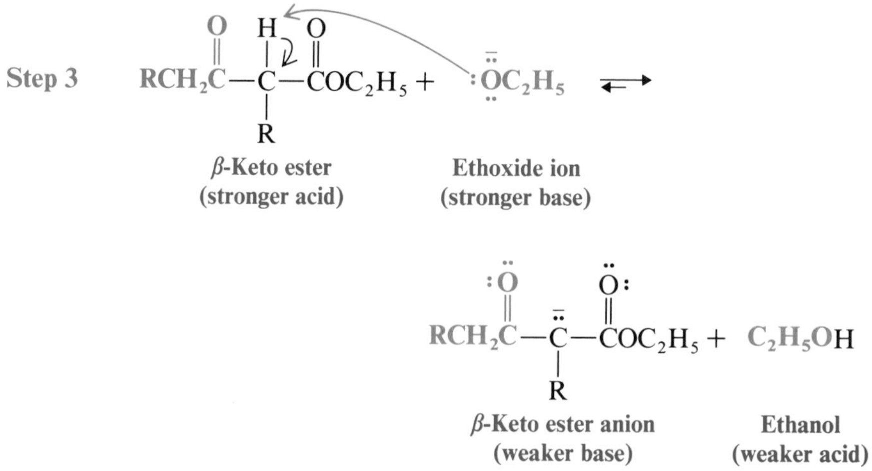

Step 3
$$RCH_2C-C-COC_2H_5 + :OC_2H_5 \rightleftharpoons$$

β-Keto ester
(stronger acid)

Ethoxide ion
(stronger base)

$$RCH_2C-C-COC_2H_5 + C_2H_5OH$$

β-Keto ester anion
(weaker base)

Ethanol
(weaker acid)

β-Keto esters are stronger acids than ethanol. They react with ethoxide ion almost quantitatively to produce ethanol and anions of β-keto esters. (It is this reaction that pulls the equilibrium to the right.) β-Keto esters are much more acidic than ordinary esters, because their enolate anions are more stabilized by resonance: Their negative charge is delocalized into two carbonyl groups:

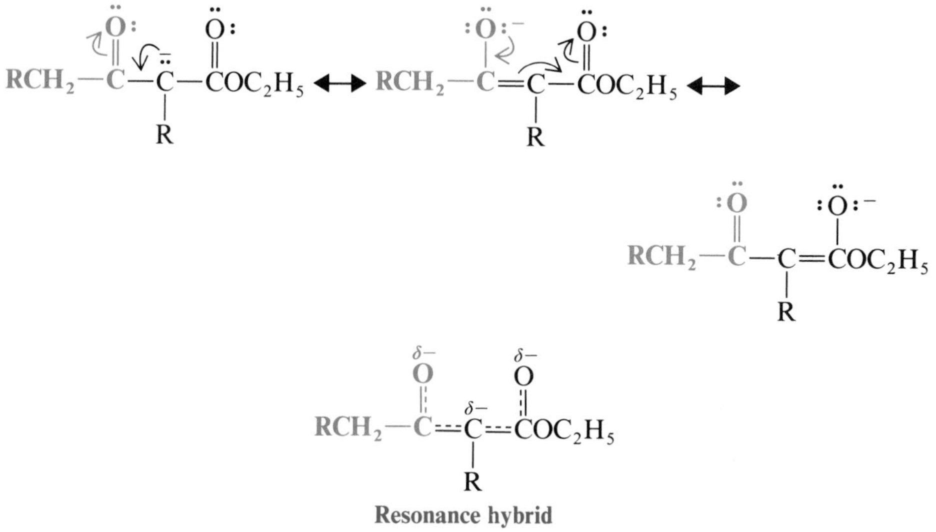

$$RCH_2-C-C-COC_2H_5 \longleftrightarrow RCH_2-C=C-COC_2H_5 \longleftrightarrow$$

$$RCH_2-C-C=COC_2H_5$$

$$RCH_2-C=C=COC_2H_5$$

Resonance hybrid

After steps 1–3 of a Claisen condensation have taken place, we add an acid to the reaction mixture. This addition brings about a rapid protonation of the anion and produces the β-keto ester as an equilibrium mixture of its keto and enol forms.

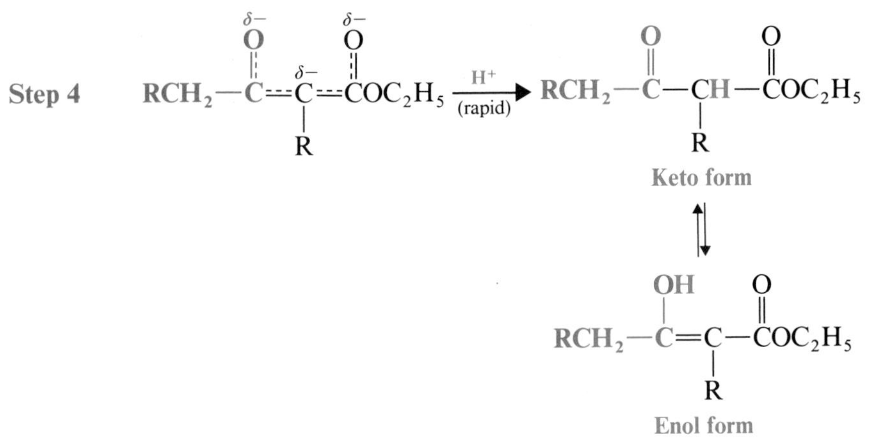

Step 4

$$RCH_2-\overset{\overset{\delta-}{\overset{O}{\|}}}{C}\overset{\delta-}{=\!=\!=}\overset{\underset{R}{|}}{\overset{\delta-}{C}}\overset{\overset{O}{\|}}{=\!=\!=COC_2H_5} \xrightarrow[\text{(rapid)}]{H^+} RCH_2-\overset{\overset{O}{\|}}{C}-\overset{\underset{R}{|}}{CH}-\overset{\overset{O}{\|}}{C}OC_2H_5$$

Keto form

$$RCH_2-\overset{\overset{OH}{|}}{C}=\overset{\underset{R}{|}}{C}-\overset{\overset{O}{\|}}{C}OC_2H_5$$

Enol form

Esters that have only one α hydrogen do not undergo the usual Claisen condensation. An example of an ester that does not react in a normal Claisen condensation is ethyl 2-methylpropanoate.

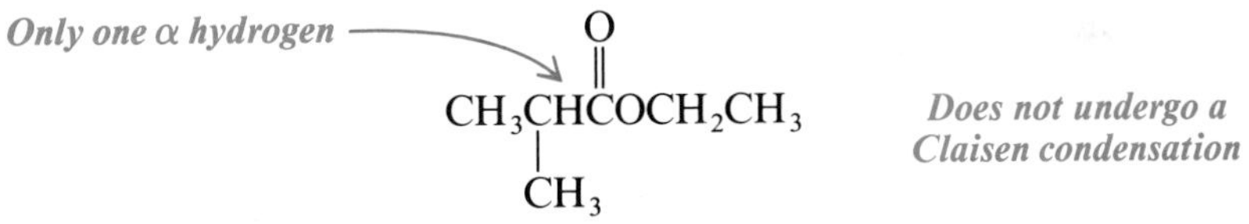

Only one α hydrogen

$$CH_3\overset{\underset{CH_3}{|}}{CH}\overset{\overset{O}{\|}}{C}OCH_2CH_3$$

Does not undergo a Claisen condensation

Ethyl 2-methylpropanoate

Inspection of the mechanism just given will make clear why this is so. An ester with only one α hydrogen will not have an acidic hydrogen when step 3 is reached, and step 3 provides the favorable equilibrium that ensures the success of the reaction. (In Section 20.2A we shall see how esters with only one α hydrogen can be converted to a β-keto ester through the use of very strong bases.)

PROBLEM 20.1

(a) Write a mechanism for all steps of the Claisen condensation that takes place when ethyl propanoate reacts with ethoxide ion.   (b) What products form when the reaction mixture is acidified?

When diethyl hexanedioate is heated with sodium ethoxide, subsequent acidification of the reaction mixture gives ethyl 2-oxocyclopentanecarboxylate.

$$C_2H_5O\overset{\overset{O}{\|}}{C}(CH_2)_4\overset{\overset{O}{\|}}{C}OC_2H_5 \xrightarrow[\text{(2) }H^+]{\text{(1) NaOC}_2H_5}$$

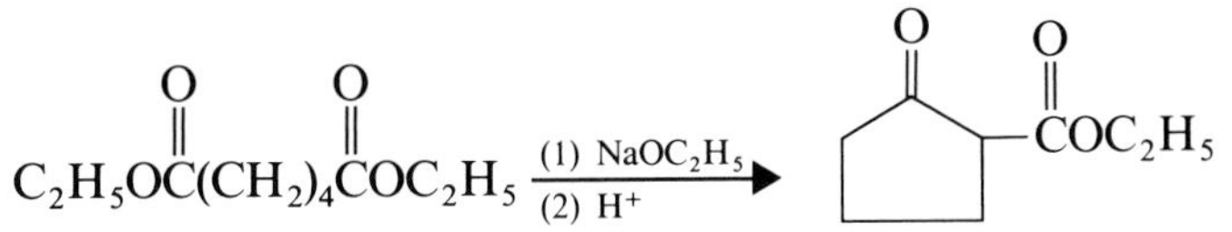

Diethyl hexanedioate
(diethyl adipate)

(74–81%)
Ethyl 2-oxocyclopentane-
carboxylate

This reaction, called the *Dieckmann condensation,* is an intramolecular Claisen condensation. The α carbon atom and the ester group for the condensation come from the same molecule.

PROBLEM 20.2

(a) Show all steps in the mechanism for the Dieckmann condensation. (b) What product would you expect from a Dieckmann condensation of diethyl heptanedioate (diethyl pimelate)? (c) Can you account for the fact that diethyl pentanedioate (diethyl glutarate) does not undergo a Dieckmann condensation?

20.2A Crossed Claisen Condensations

Crossed Claisen condensations (like crossed aldol condensations) are possible **when one ester component has no α hydrogens** and is, therefore, unable to undergo self-condensation. Ethyl benzoate, for example, condenses with ethyl acetate to give ethyl benzoylacetate.

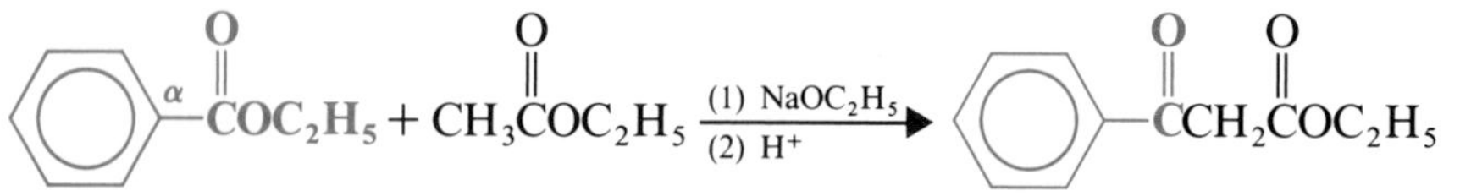

Ethyl benzoate   (60%)
(no α hydrogen)   Ethyl benzoylacetate

Ethyl phenylacetate condenses with diethyl carbonate to give diethyl phenylmalonate.

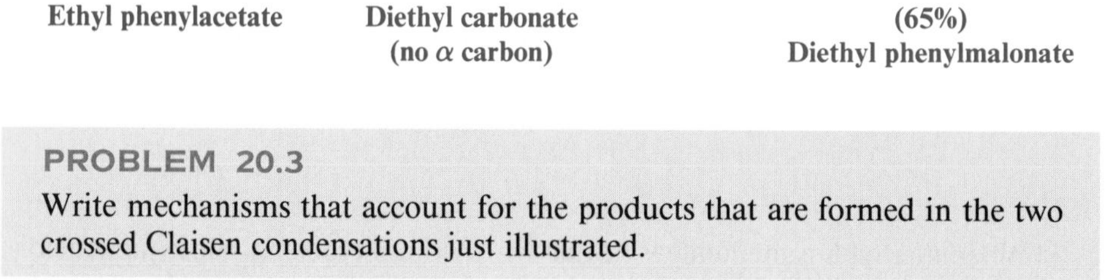

Ethyl phenylacetate   Diethyl carbonate   (65%)
        (no α carbon)   Diethyl phenylmalonate

PROBLEM 20.3

Write mechanisms that account for the products that are formed in the two crossed Claisen condensations just illustrated.

PROBLEM 20.4

What products would you expect to obtain from each of the following crossed Claisen condensations?

(a) Ethyl propanoate + diethyl oxalate $\xrightarrow[\text{(2) H}^+]{\text{(1) NaOCH}_2\text{CH}_3}$

(b) Ethyl acetate + ethyl formate $\xrightarrow[\text{(2) H}^+]{\text{(1) NaOCH}_2\text{CH}_3}$

As we learned earlier in this section, esters that have only one α hydrogen cannot be converted to β-keto esters by sodium ethoxide. However, they can be converted to β-keto esters by reactions that use very strong bases. The strong base converts the ester to its enolate anion in nearly quantitative yield. This allows us to *acylate* the enolate anion by treating it with an acyl chloride or an ester. An example of this technique that makes use of the very powerful base sodium triphenylmethanide is shown next.

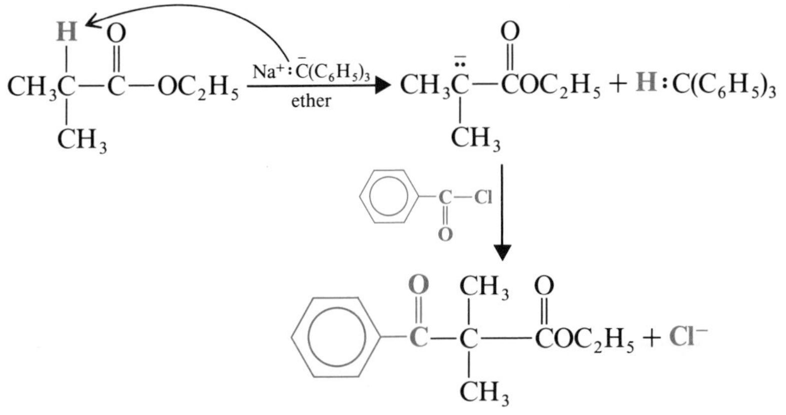

Ethyl 2,2-dimethyl-3-oxo-3-phenylpropanoate

20.2B Acylation of Other Carbanions

Enolate anions derived from ketones also react with esters in nucleophilic substitution reactions that resemble Claisen condensations. In the following first example, although two anions are possible from the reaction of the ketone with sodium amide, the major product is derived from the primary carbanion. The primary α hydrogens are more acidic than the secondary α hydrogens.

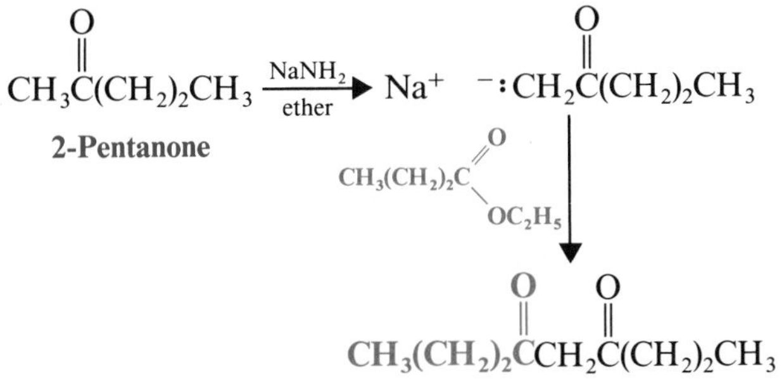

2-Pentanone

(76%)

4,6-Nonanedione

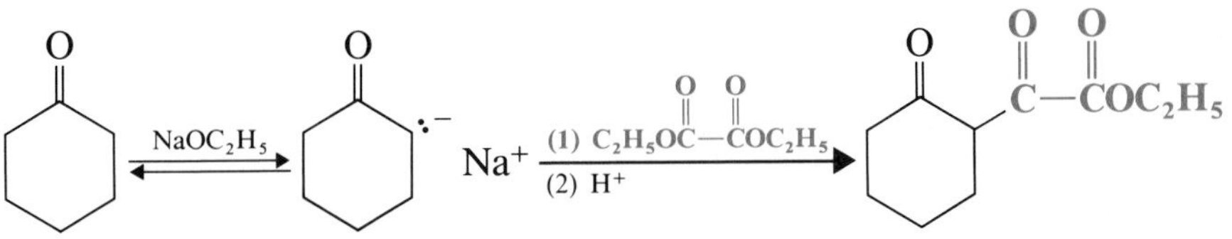

(67%)

PROBLEM 20.5

Show how you might synthesize each of the following compounds using, as your starting materials, esters, ketones, acyl halides, and so on.

(a) 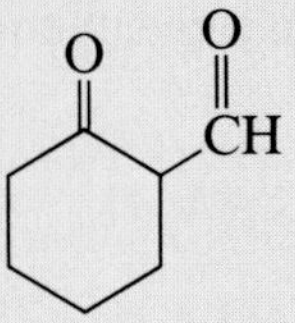   (b) 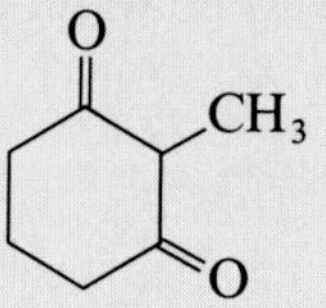   (c)

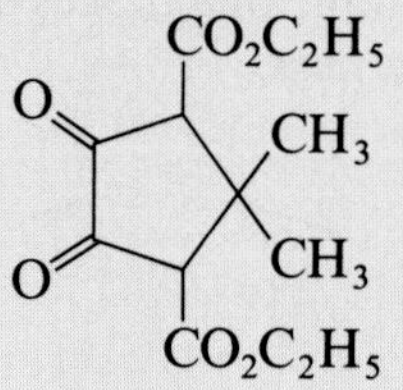

PROBLEM 20.6

Keto esters are capable of undergoing cyclization reactions similar to the Dieckmann condensation. Write a mechanism that accounts for the product formed in the following reaction:

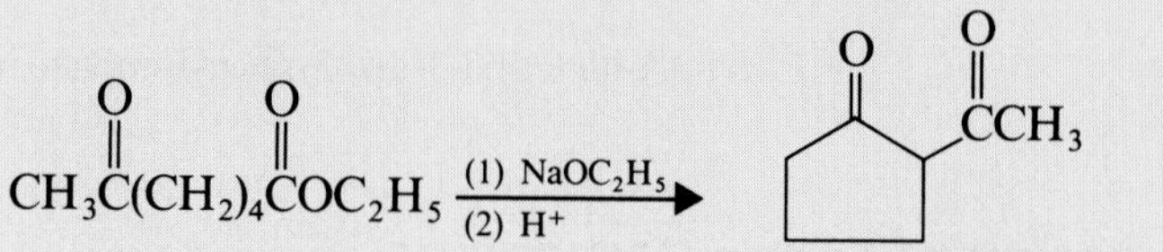

$$CH_3\overset{O}{\overset{\|}{C}}(CH_2)_4\overset{O}{\overset{\|}{C}}OC_2H_5 \xrightarrow[(2)\ H^+]{(1)\ NaOC_2H_5}$$

2-Acetylcyclopentanone

20.3 THE ACETOACETIC ESTER SYNTHESIS. SYNTHESIS OF SUBSTITUTED ACETONES

Acetoacetic esters are useful reagents for the preparation of methyl ketones of the types shown here:

$$CH_3-\overset{O}{\overset{\|}{C}}-CH_2-R \quad or \quad CH_3-\overset{O}{\overset{\|}{C}}-\underset{R}{\overset{|}{C}}H-R$$

Monosubstituted   Disubstituted
    acetone   acetone

Two factors make such syntheses practical: (1) The methylene protons of β-keto esters are appreciably acidic and (2) β-keto acids decarboxylate readily (cf. Section 18.11).

As we have seen (Section 20.2) the methylene protons of acetoacetic ester are more acidic than the —OH proton of ethanol because they are located between two carbonyl groups and yield a highly stabilized enolate anion. This acidity means that we can convert acetoacetic ester to an enolate anion using sodium ethoxide as a base. We can then carry out an alkylation reaction by treating the enolate anion with an alkyl halide.

Since the alkylation (see following reaction) is an S_N2 reaction, best yields are obtained from the use of primary alkyl halides (including primary allylic and benzylic halides) or methyl halides. Secondary halides give lower yields, and tertiary halides give only elimination.

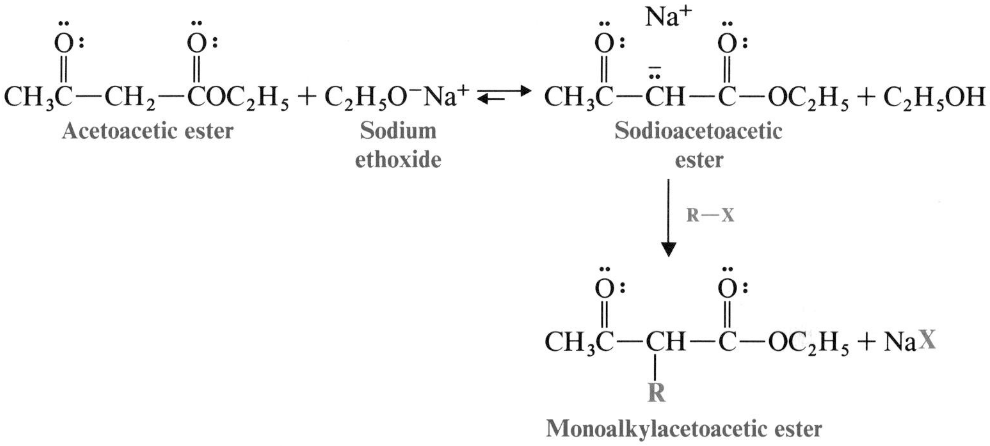

$$CH_3\overset{\overset{\ddot{O}:}{\|}}{C}-CH_2-\overset{\overset{\ddot{O}:}{\|}}{C}OC_2H_5 + C_2H_5O^-Na^+ \rightleftharpoons CH_3\overset{\overset{\ddot{O}:}{\|}}{C}-\overset{\overset{\ddot{}}{\ddot{C}}H}{}-\overset{\overset{\ddot{O}:}{\|}}{C}-OC_2H_5 + C_2H_5OH$$

Acetoacetic ester   Sodium ethoxide   Sodioacetic ester (Na⁺)

$$\downarrow R-X$$

$$CH_3\overset{\overset{\ddot{O}:}{\|}}{C}-\underset{\underset{R}{|}}{CH}-\overset{\overset{\ddot{O}:}{\|}}{C}-OC_2H_5 + NaX$$

Monoalkylacetoacetic ester

The monoalkylacetoacetic ester still has one appreciably acidic hydrogen and, if we desire, we can carry out a second alkylation. Because the monoalkylacetoacetic ester is somewhat less acidic than acetoacetic ester itself (why?), it is usually helpful to use a base stronger than ethoxide ion.

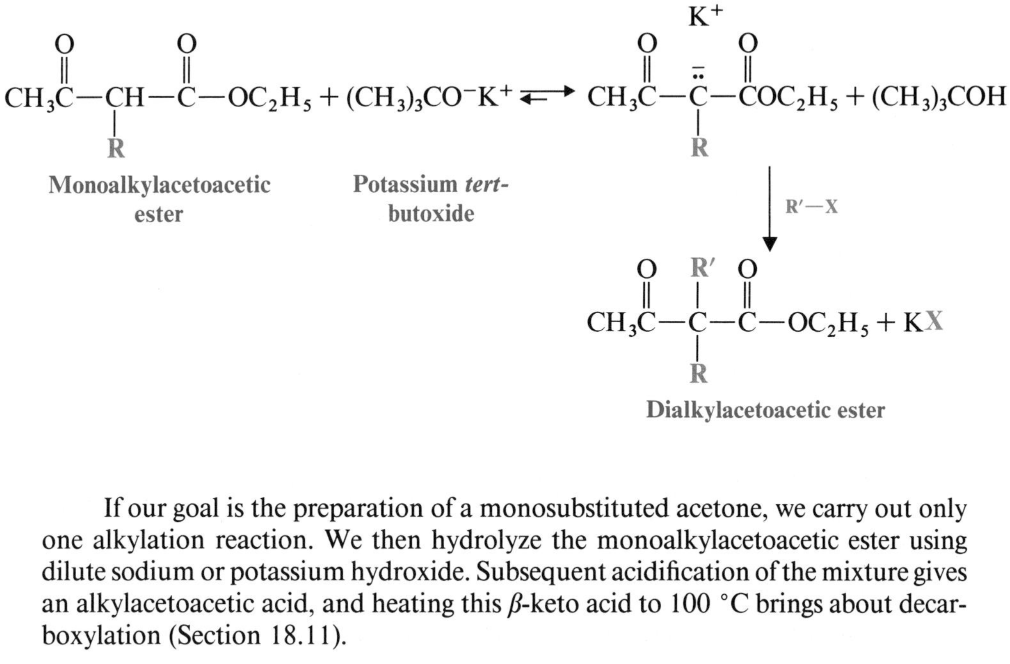

$$CH_3\overset{\overset{O}{\|}}{C}-\underset{\underset{R}{|}}{CH}-\overset{\overset{O}{\|}}{C}-OC_2H_5 + (CH_3)_3CO^-K^+ \rightleftharpoons CH_3\overset{\overset{O}{\|}}{C}-\underset{\underset{R}{|}}{\overset{..}{C}}-\overset{\overset{O}{\|}}{C}OC_2H_5 (K^+) + (CH_3)_3COH$$

Monoalkylacetoacetic ester   Potassium *tert*-butoxide

$$\downarrow R'-X$$

$$CH_3\overset{\overset{O}{\|}}{C}-\underset{\underset{R}{|}}{\overset{\overset{R'}{|}}{C}}-\overset{\overset{O}{\|}}{C}-OC_2H_5 + KX$$

Dialkylacetoacetic ester

If our goal is the preparation of a monosubstituted acetone, we carry out only one alkylation reaction. We then hydrolyze the monoalkylacetoacetic ester using dilute sodium or potassium hydroxide. Subsequent acidification of the mixture gives an alkylacetoacetic acid, and heating this β-keto acid to 100 °C brings about decarboxylation (Section 18.11).

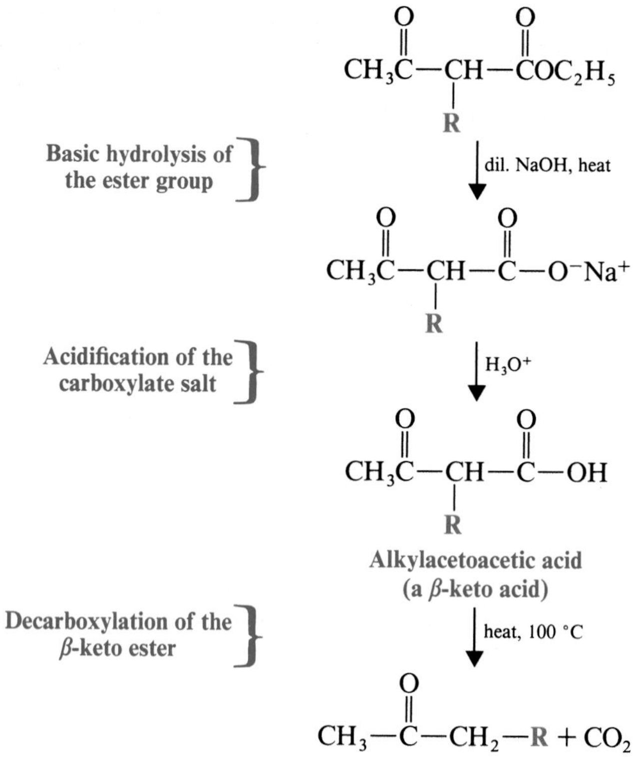

$$\underset{\substack{\text{R}}}{\overset{\displaystyle O \qquad\qquad O}{CH_3C-CH-COC_2H_5}}$$

Basic hydrolysis of }   ↓ dil. NaOH, heat
the ester group

$$\underset{\substack{\text{R}}}{\overset{\displaystyle O \qquad\qquad O}{CH_3C-CH-C-O^-Na^+}}$$

Acidification of the }   ↓ H_3O^+
carboxylate salt

$$\underset{\substack{\text{R}}}{\overset{\displaystyle O \qquad\qquad O}{CH_3C-CH-C-OH}}$$

Alkylacetoacetic acid
(a β-keto acid)

Decarboxylation of the }   ↓ heat, 100 °C
β-keto ester

$$\overset{\displaystyle O}{CH_3-C-CH_2-R} + CO_2$$

A specific example is the following synthesis of 2-heptanone:

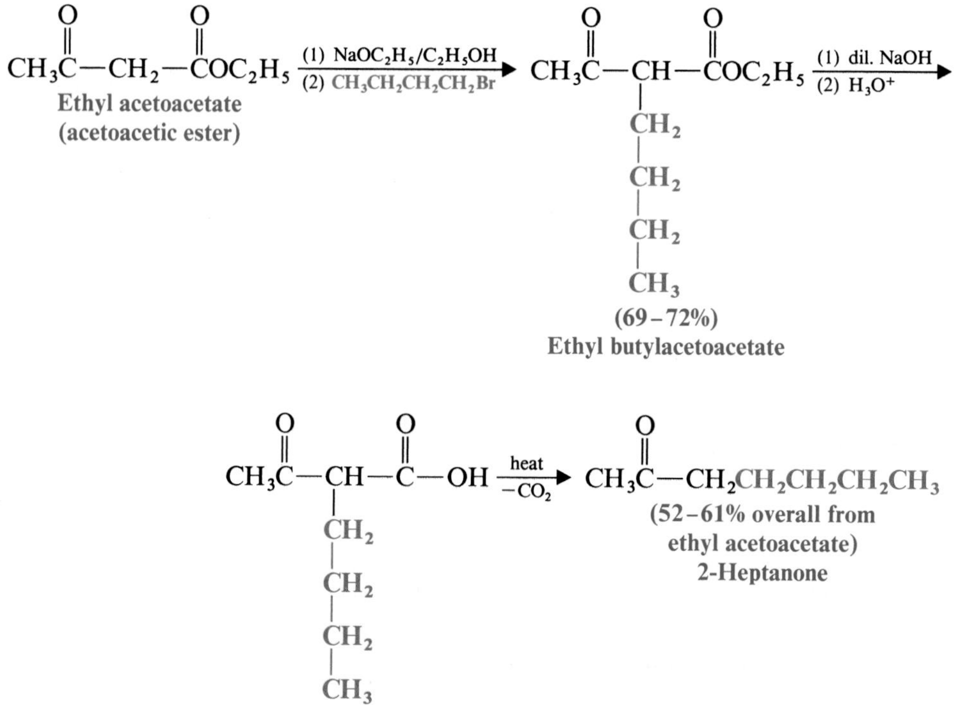

$$\overset{\displaystyle O \qquad\qquad O}{CH_3C-CH_2-COC_2H_5} \xrightarrow[\text{(2) } CH_3CH_2CH_2CH_2Br]{\text{(1) } NaOC_2H_5/C_2H_5OH} \underset{\substack{CH_2\\|\\CH_2\\|\\CH_2\\|\\CH_3}}{\overset{\displaystyle O \qquad\qquad O}{CH_3C-CH-COC_2H_5}} \xrightarrow[\text{(2) } H_3O^+]{\text{(1) dil. NaOH}}$$

Ethyl acetoacetate
(acetoacetic ester)

(69–72%)
Ethyl butylacetoacetate

$$\underset{\substack{CH_2\\|\\CH_2\\|\\CH_2\\|\\CH_3}}{\overset{\displaystyle O \qquad\qquad O}{CH_3C-CH-C-OH}} \xrightarrow[-CO_2]{\text{heat}} \overset{\displaystyle O}{CH_3C-CH_2CH_2CH_2CH_2CH_3}$$

(52–61% overall from
ethyl acetoacetate)
2-Heptanone

If our goal is the preparation of a disubstituted acetone, we carry out two successive alkylations, we hydrolyze the dialkylacetoacetic ester that is produced, and then we decarboxylate the dialkylacetoacetic acid. An example of this procedure is the synthesis of 3-butyl-2-heptanone.

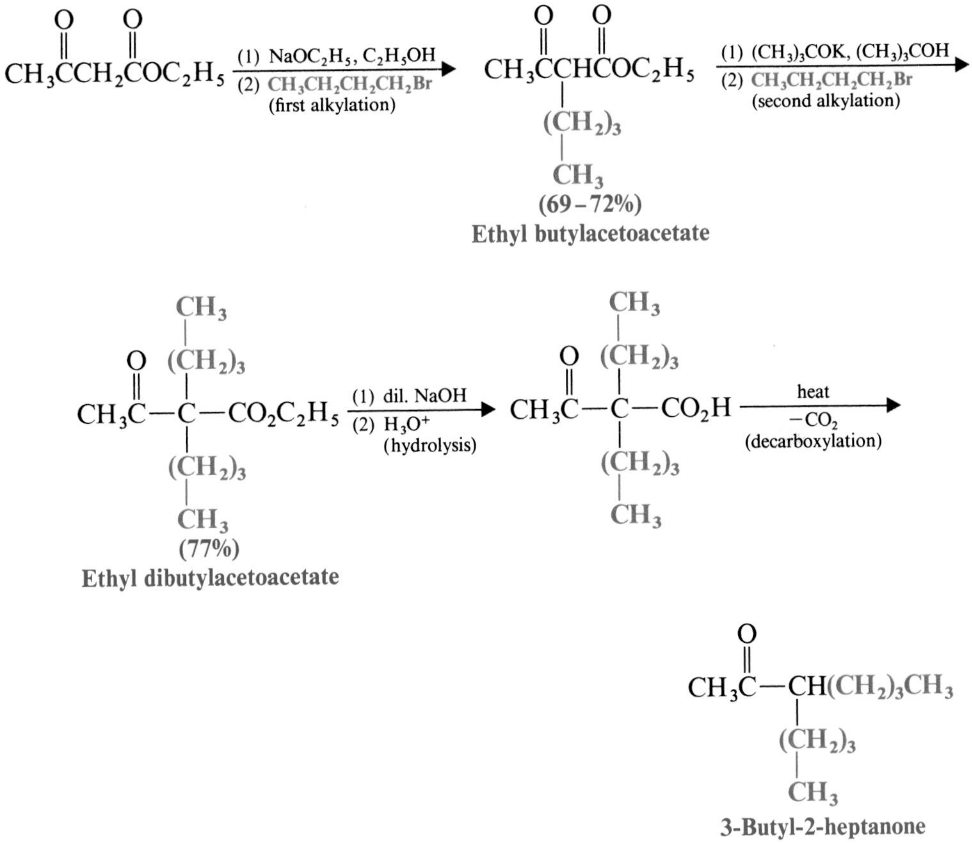

$$
\underset{\text{CH}_3\overset{\text{O}}{\overset{\|}{\text{C}}}\text{CH}_2\overset{\text{O}}{\overset{\|}{\text{C}}}\text{OC}_2\text{H}_5}{} \xrightarrow[\substack{\text{(2) CH}_3\text{CH}_2\text{CH}_2\text{CH}_2\text{Br} \\ \text{(first alkylation)}}]{\text{(1) NaOC}_2\text{H}_5,\ \text{C}_2\text{H}_5\text{OH}} \underset{\substack{(\text{CH}_2)_3 \\ | \\ \text{CH}_3 \\ (69-72\%)}}{\text{CH}_3\overset{\text{O}}{\overset{\|}{\text{C}}}\overset{|}{\text{C}}\text{H}\overset{\text{O}}{\overset{\|}{\text{C}}}\text{OC}_2\text{H}_5} \xrightarrow[\substack{\text{(2) CH}_3\text{CH}_2\text{CH}_2\text{CH}_2\text{Br} \\ \text{(second alkylation)}}]{\text{(1) (CH}_3)_3\text{COK, (CH}_3)_3\text{COH}}
$$

Ethyl butylacetoacetate

$$
\underset{\substack{(\text{CH}_2)_3 \\ | \\ \text{CH}_3 \\ (77\%)}}{\overset{\substack{\text{CH}_3 \\ | \\ (\text{CH}_2)_3}}{\text{CH}_3\overset{\text{O}}{\overset{\|}{\text{C}}}-\overset{|}{\underset{|}{\text{C}}}-\text{CO}_2\text{C}_2\text{H}_5}} \xrightarrow[\substack{\text{(2) H}_3\text{O}^+ \\ \text{(hydrolysis)}}]{\text{(1) dil. NaOH}} \underset{\substack{(\text{CH}_2)_3 \\ | \\ \text{CH}_3}}{\overset{\substack{\text{CH}_3 \\ | \\ (\text{CH}_2)_3}}{\text{CH}_3\overset{\text{O}}{\overset{\|}{\text{C}}}-\overset{|}{\underset{|}{\text{C}}}-\text{CO}_2\text{H}}} \xrightarrow[\substack{-\text{CO}_2 \\ \text{(decarboxylation)}}]{\text{heat}}
$$

Ethyl dibutylacetoacetate

$$
\underset{\substack{(\text{CH}_2)_3 \\ | \\ \text{CH}_3}}{\text{CH}_3\overset{\text{O}}{\overset{\|}{\text{C}}}-\text{CH}(\text{CH}_2)_3\text{CH}_3}
$$

3-Butyl-2-heptanone

Although both alkylations in the example just given were carried out with the same alkyl halide, we could have used different alkyl halides if our synthesis had required it.

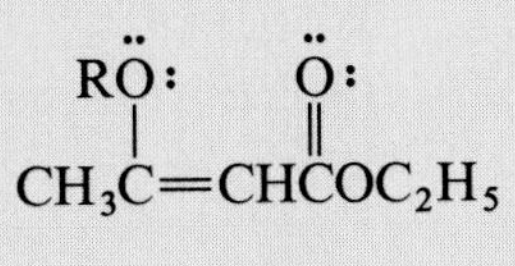

PROBLEM 20.8

Show how you would use the acetoacetic ester synthesis to prepare each of the following:
(a) 2-pentanone,   (b) 3-propyl-2-hexanone, and   (c) 4-phenyl-2-butanone.

PROBLEM 20.9

The acetoacetic ester synthesis generally gives best yields when primary halides are used in the alkylation step. Secondary halides give low yields and tertiary halides give practically no alkylation product at all.   (a) Explain.   (b) What products would you expect from the reaction of sodioacetoacetic ester and *tert*-butyl bromide?   (c) Bromobenzene cannot be used as an arylating agent in an acetoacetic ester synthesis in the manner we have just described. Why not?

PROBLEM 20.10

Since the products obtained from Claisen condensations are *β*-keto esters, subsequent hydrolysis and decarboxylation of these products gives a general method for the synthesis of ketones. Show how you would employ this technique in a synthesis of 4-heptanone.

   The acetoacetic ester synthesis can also be carried out using halo esters and halo ketones. The use of an *α*-halo ester provides a convenient synthesis of *γ*-keto acids:

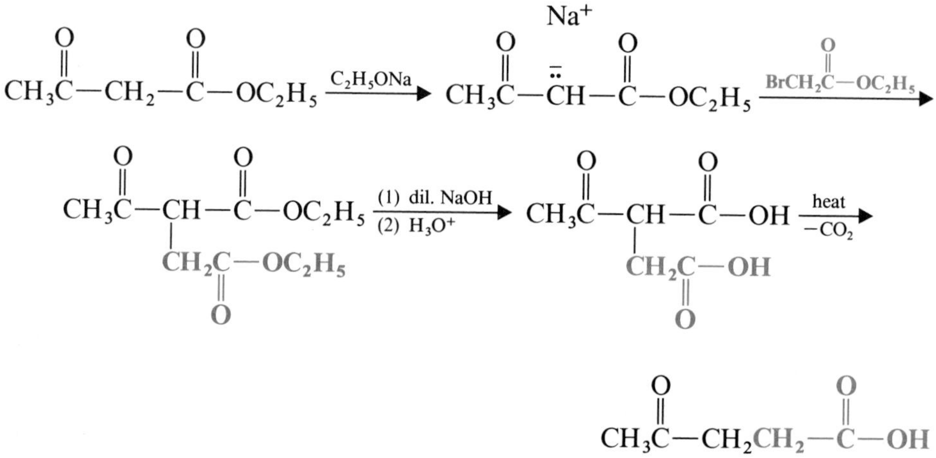

4-Oxopentanoic acid

The use of an α-halo ketone in an acetoacetic ester synthesis provides a general method for preparing γ-diketones:

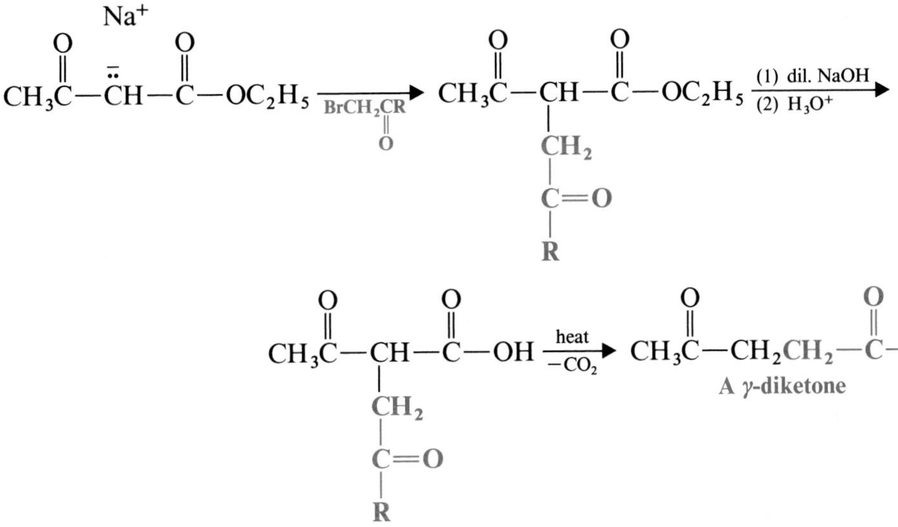

$$
\underset{\substack{\text{Na}^+}}{CH_3\overset{O}{\overset{||}{C}}-\overset{\cdot\cdot}{CH}-\overset{O}{\overset{||}{C}}-OC_2H_5} \xrightarrow{\underset{BrCH_2CR}{}}
$$

The use of an α-halo ketone reaction scheme showing:

CH₃C(=O)—CH—C(=O)—OC₂H₅ with CH₂—C(=O)—R substituent, then (1) dil. NaOH, (2) H₃O⁺

CH₃C(=O)—CH—C(=O)—OH with CH₂—C(=O)—R substituent, then heat, −CO₂ gives CH₃C(=O)—CH₂CH₂—C(=O)—R A γ-diketone

PROBLEM 20.11

In the synthesis of the keto acid just given, the dicarboxylic acid decarboxylates in a specific way; it gives

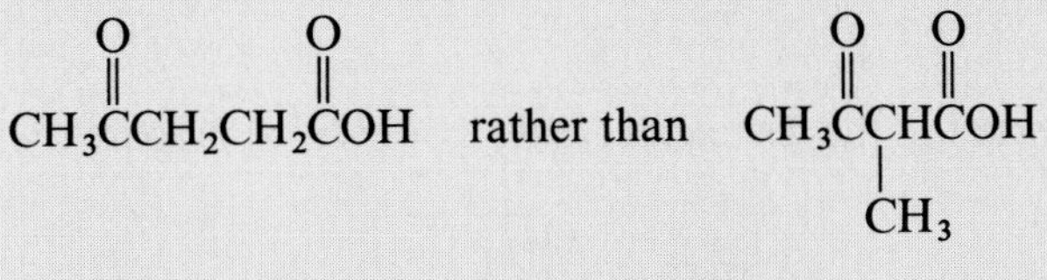

$$CH_3\overset{O}{\overset{||}{C}}CH_2CH_2\overset{O}{\overset{||}{C}}OH \quad \text{rather than} \quad CH_3\overset{O}{\overset{||}{C}}\underset{\substack{|\\CH_3}}{CH}\overset{O}{\overset{||}{C}}OH$$

Explain.

PROBLEM 20.12

How would you use the acetoacetic ester synthesis to prepare the following?

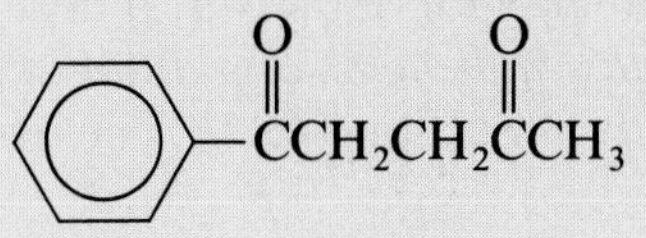

$$\text{[benzene ring]}-\overset{O}{\overset{||}{C}}CH_2CH_2\overset{O}{\overset{||}{C}}CH_3$$

Anions obtained from acetoacetic esters undergo acylation when they are treated with acyl chlorides or acid anhydrides. Because both of these acylating agents react with alcohols, acylation reactions cannot be carried out in ethanol and must be carried out in aprotic solvents such as DMF, DMSO, or HMPT (Section 5.13C). (If the reaction were to be carried out in ethanol, using sodium ethoxide, for example, then the acyl chloride would be rapidly converted to an ethyl ester and the ethoxide ion would be neutralized.) Sodium hydride can be used to generate the enolate anion in an aprotic solvent.

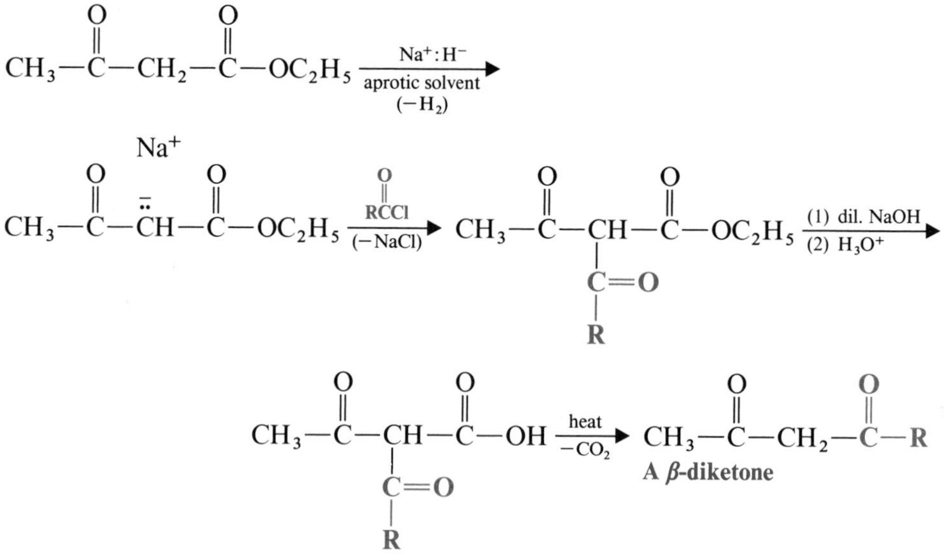

$$
CH_3-\overset{\overset{O}{\|}}{C}-CH_2-\overset{\overset{O}{\|}}{C}-OC_2H_5 \xrightarrow[\substack{\text{aprotic solvent} \\ (-H_2)}]{Na^+:H^-}
$$

$$
CH_3-\overset{\overset{O}{\|}}{C}-\overset{\overset{Na^+}{\ddot{}}}{C}H-\overset{\overset{O}{\|}}{C}-OC_2H_5 \xrightarrow[(-NaCl)]{R\overset{O}{\overset{\|}{C}}Cl} CH_3-\overset{\overset{O}{\|}}{C}-\underset{\underset{R}{\underset{\|}{C=O}}}{CH}-\overset{\overset{O}{\|}}{C}-OC_2H_5 \xrightarrow[(2)\ H_3O^+]{(1)\ \text{dil. NaOH}}
$$

$$
CH_3-\overset{\overset{O}{\|}}{C}-\underset{\underset{R}{\underset{\|}{C=O}}}{CH}-\overset{\overset{O}{\|}}{C}-OH \xrightarrow[-CO_2]{\text{heat}} \underset{\text{A β-diketone}}{CH_3-\overset{\overset{O}{\|}}{C}-CH_2-\overset{\overset{O}{\|}}{C}-R}
$$

Acylations of acetoacetic esters followed by hydrolysis and decarboxylation give us a method for preparing β-diketones.

PROBLEM 20.13

How would you use the acetoacetic ester synthesis to prepare the following?

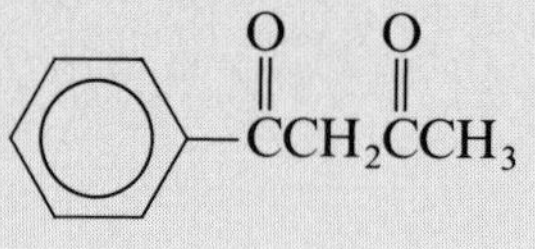

Acetoacetic ester cannot be phenylated in a manner analogous to the alkylation reactions we have studied because bromobenzene is not susceptible to S_N2 reactions [Section 5.14A and Problem 20.9(c)]. However, if acetoacetic ester is treated with bromobenzene and *two molar equivalents of sodium amide,* then phenylation does occur *by a benzyne mechanism* (Section 14.11). The overall reaction is as follows:

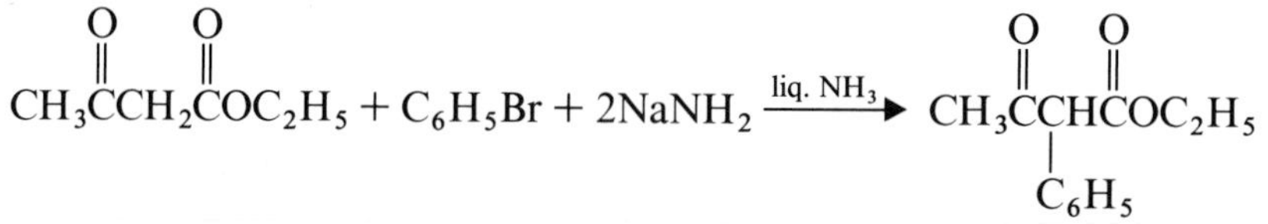

$$
CH_3\overset{\overset{O}{\|}}{C}CH_2\overset{\overset{O}{\|}}{C}OC_2H_5 + C_6H_5Br + 2NaNH_2 \xrightarrow{\text{liq. NH}_3} CH_3\overset{\overset{O}{\|}}{C}\underset{\underset{C_6H_5}{|}}{C}H\overset{\overset{O}{\|}}{C}OC_2H_5
$$

Malonic esters, $RO\overset{\overset{O}{\|}}{C}CH_2\overset{\overset{O}{\|}}{C}OR$, can be phenylated in an analogous way.

PROBLEM 20.14

(a) Outline a step-by-step mechanism for the phenylation of acetoacetic ester by bromobenzene and two molar equivalents of sodium amide. (Why are two molar equivalents of $NaNH_2$ necessary?) (b) What product would be otained by hydrolysis and decarboxylation of the phenylated acetoacetic ester? (c) How would you prepare phenylacetic acid from malonic ester?

One further variation of the acetoacetic ester synthesis involves the conversion of an acetoacetic ester to a resonance-stabilized *dianion* by using a very strong base such as potassium amide in liquid ammonia.

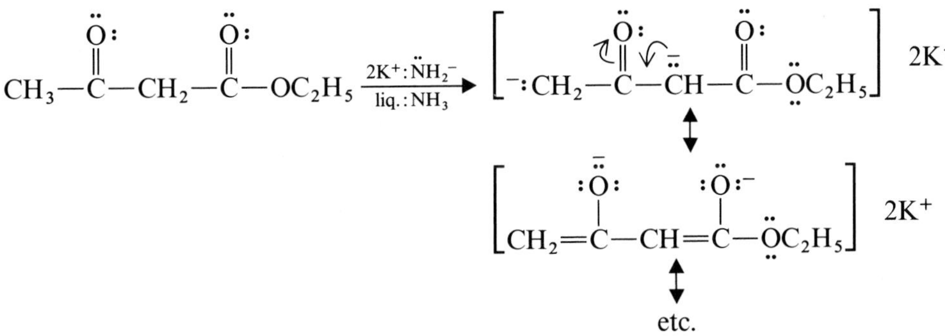

etc.

When this dianion is treated with 1 mole of a primary (or methyl) halide, it undergoes alkylation at its terminal carbon rather than at its interior one. This orientation of the alkylation reaction apparently results from the greater basicity (and thus nucleophilicity) of the terminal carbanion. This carbanion is more basic because it is stabilized by only one adjacent carbonyl group. After monoalkylation has taken place, the anion that remains can be protonated by adding ammonium chloride.

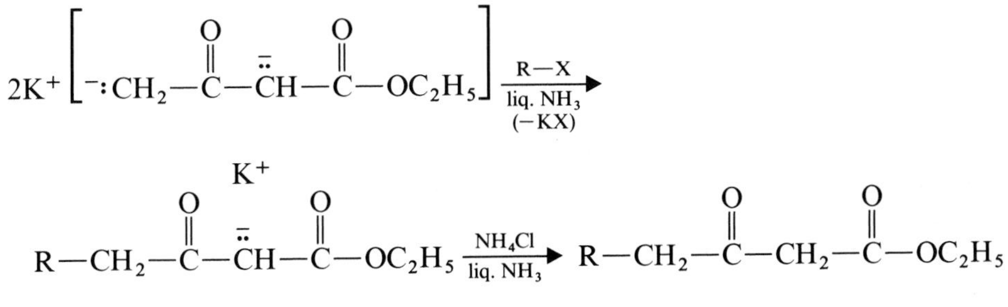

PROBLEM 20.15

Show how you could use ethyl acetoacetate in a synthesis of

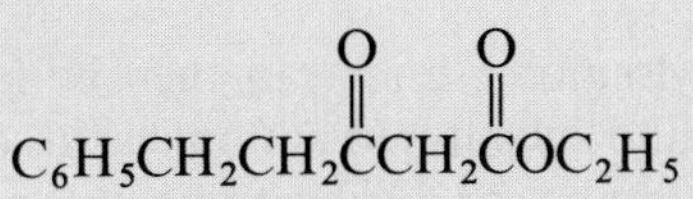

$$C_6H_5CH_2CH_2CCH_2COC_2H_5$$

20.4 THE MALONIC ESTER SYNTHESIS.
SYNTHESIS OF SUBSTITUTED ACETIC ACIDS

A useful counterpart of the acetoacetic ester synthesis — one that allows the synthesis of *mono- and disubstituted acetic acids* — is called the *malonic ester synthesis.*

The malonic ester synthesis resembles the acetoacetic ester synthesis in several respects.

1. Diethyl malonate, the starting compound, forms a relatively stable enolate ion:

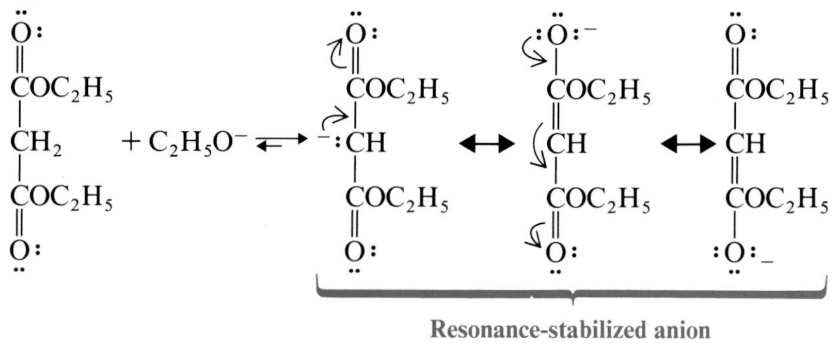

Resonance-stabilized anion

2. This enolate ion can be alkylated,

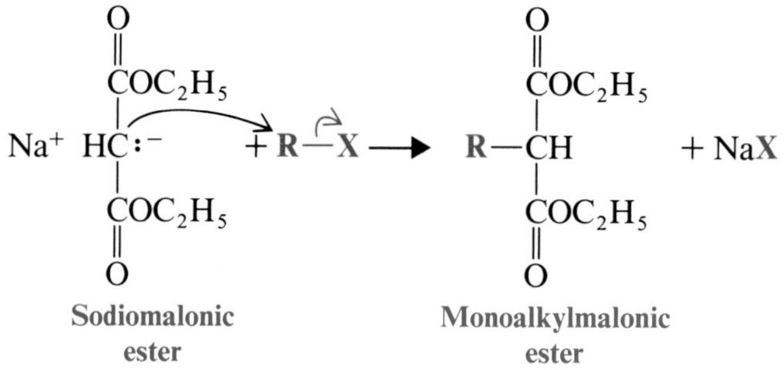

| Sodiomalonic ester | Monoalkylmalonic ester |

and the product can be alkylated again if our synthesis requires it:

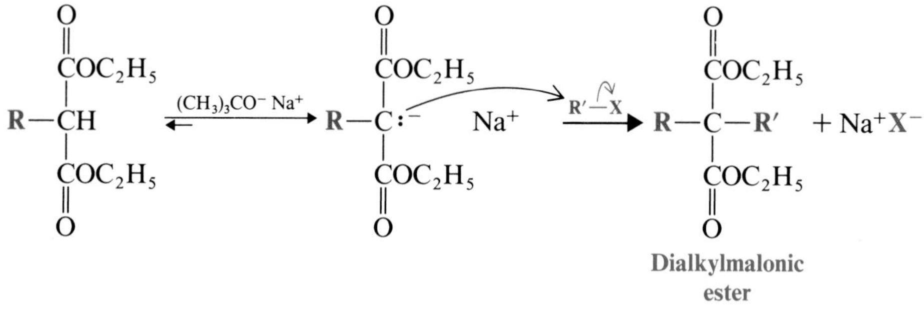

Dialkylmalonic ester

3. The mono- or dialkylmalonic ester can then be hydrolyzed to a mono- or dialkylmalonic acid, and substituted malonic acids decarboxylate readily. Decarboxylation gives a mono- or disubstituted acetic acid.

$$
\underset{\substack{\text{Monoalkylmalonic}\\\text{ester}}}{\begin{array}{c}\text{O}\\\parallel\\\text{C}-\text{OC}_2\text{H}_5\\|\\\text{R}\overset{|}{\text{C}}\text{H}\\|\\\text{C}-\text{OC}_2\text{H}_5\\\parallel\\\text{O}\end{array}}
\xrightarrow[\text{(2) H}_3\text{O}^+]{\text{(1) OH}^-,\text{H}_2\text{O}}
\begin{array}{c}\text{O}\\\parallel\\\text{C}-\text{OH}\\|\\\text{R}\overset{|}{\text{C}}\text{H}\\|\\\text{C}-\text{OH}\\\parallel\\\text{O}\end{array}
\xrightarrow[-\text{CO}_2]{\text{heat}}
\underset{\substack{\text{Monoalkyl-}\\\text{acetic acid}}}{\text{RCH}_2\text{CO}_2\text{H}}
$$

$$
\underset{\substack{\text{Dialkylmalonic}\\\text{ester}}}{\begin{array}{c}\text{O}\\\parallel\\\text{COC}_2\text{H}_5\\|\\\text{R}-\overset{|}{\text{C}}-\text{R}'\\|\\\text{COC}_2\text{H}_5\\\parallel\\\text{O}\end{array}}
\xrightarrow[\text{(2) H}_3\text{O}^+]{\text{(1) OH}^-,\text{H}_2\text{O}}
\begin{array}{c}\text{O}\\\parallel\\\text{C}-\text{OH}\\|\\\text{R}-\overset{|}{\text{C}}-\text{R}'\\|\\\text{C}-\text{OH}\\\parallel\\\text{O}\end{array}
\xrightarrow[-\text{CO}_2]{\text{heat}}
\underset{\substack{\text{Dialkylacetic}\\\text{acid}}}{\text{R}-\underset{\underset{\text{R}'}{|}}{\text{C}}\text{HCO}_2\text{H}}
$$

Two specific examples of the malonic ester synthesis are the syntheses of hexanoic acid and 2-ethylpentanoic acid given here.

$$
\begin{array}{c}\text{O}\\\parallel\\\text{COC}_2\text{H}_5\\|\\\text{CH}_2\\|\\\text{COC}_2\text{H}_5\\\parallel\\\text{O}\end{array}
\;+\;
\xrightarrow[\text{(2) CH}_3\text{CH}_2\text{CH}_2\text{CH}_2\text{Br}]{\text{(1) NaOC}_2\text{H}_5}
$$

$$
\underset{\substack{(80-90\%)\\\textbf{Diethyl butylmalonate}}}{\text{CH}_3\text{CH}_2\text{CH}_2\text{CH}_2\overset{\overset{\textstyle\text{O}\atop\parallel}{\text{COC}_2\text{H}_5}}{\underset{\underset{\parallel\atop\textstyle\text{O}}{\text{COC}_2\text{H}_5}}{\text{CH}}}}
\xrightarrow[\substack{\text{(2) dil. H}_2\text{SO}_4,\text{ reflux}\\(-\text{CO}_2)}]{\text{(1) 50\% KOH, reflux}}
\underset{\substack{(75\%)\\\textbf{Hexanoic acid}}}{\text{CH}_3\text{CH}_2\text{CH}_2\text{CH}_2\text{CH}_2\text{CO}_2\text{H}}
$$

$$
\begin{array}{c}\text{O}\\\parallel\\\text{COC}_2\text{H}_5\\|\\\text{CH}_2\\|\\\text{COC}_2\text{H}_5\\\parallel\\\text{O}\end{array}
\xrightarrow[\text{(2) CH}_3\text{CH}_2\text{I}]{\text{(1) NaOC}_2\text{H}_5}
\underset{\textbf{Diethyl ethylmalonate}}{\text{CH}_3\text{CH}_2\overset{\overset{\textstyle\text{O}\atop\parallel}{\text{COC}_2\text{H}_5}}{\underset{\underset{\parallel\atop\textstyle\text{O}}{\text{COC}_2\text{H}_5}}{\text{CH}}}}
\xrightarrow[\text{(2) CH}_3\text{CH}_2\text{CH}_2\text{Br}]{\text{(1) NaOC(CH}_3)_3}
$$

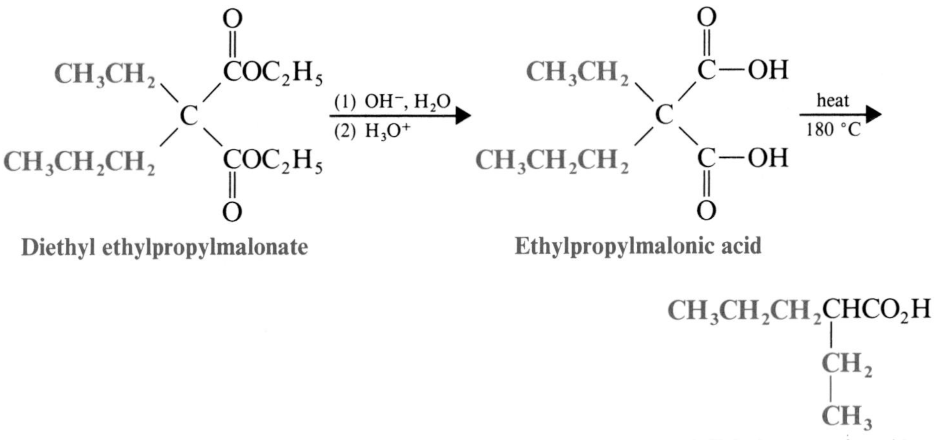

Diethyl ethylpropylmalonate → Ethylpropylmalonic acid

$$CH_3CH_2CH_2\underset{\underset{\underset{CH_3}{|}}{\overset{|}{CH_2}}}{CH}CO_2H$$

2-Ethylpentanoic acid

PROBLEM 20.16

Outline all steps in a malonic ester synthesis of each of the following:
(a) pentanoic acid, (b) 2-methylpentanoic acid, and (c) 4-methylpentanoic acid.

Two variations of the malonic ester synthesis make use of dihaloalkanes. In the first of these, two molar equivalents of sodiomalonic ester are allowed to react with a dihaloalkane. Two consecutive alkylations occur giving a tetraester; hydrolysis and decarboxylation of the tetraester yield a dicarboxylic acid. An example is the synthesis of glutaric acid:

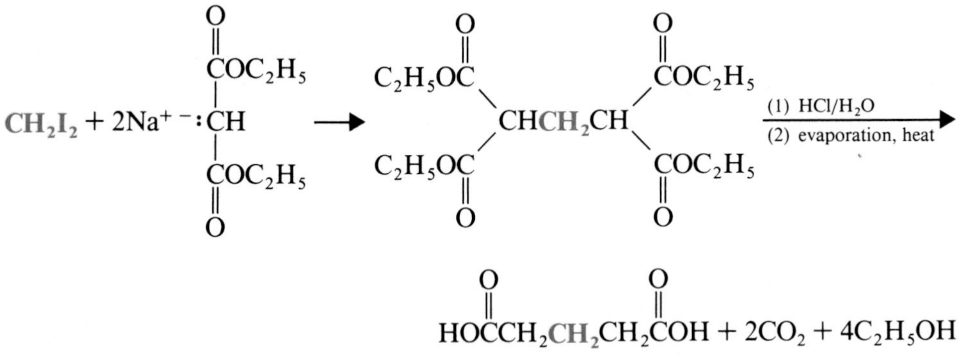

$$\underset{\overset{\overset{O}{\|}}{HOCCH_2CH_2CH_2COH}}{} + 2CO_2 + 4C_2H_5OH$$
(80% from tetraester)
Glutaric acid

In a second variation, one molar equivalent of sodiomalonic ester is allowed to react with one molar equivalent of a dihaloalkane. This reaction gives a haloalkyl-malonic ester, which when treated with sodium ethoxide, undergoes an internal alkylation reaction. This method has been used to prepare three-, four-, five-, and six-membered rings. An example is the synthesis of cyclobutanecarboxylic acid.

Cyclobutanecarboxylic acid

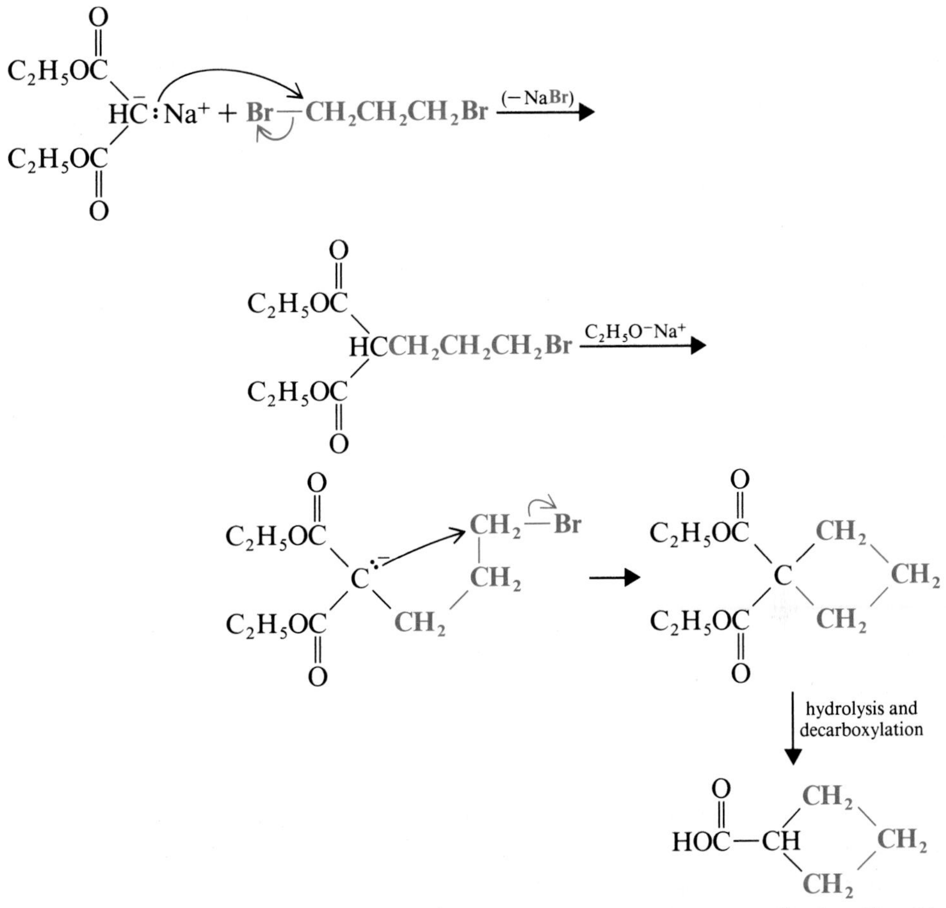

20.5 FURTHER REACTIONS OF ACTIVE HYDROGEN COMPOUNDS

Because of the acidity of their methylene hydrogens, malonic esters, acetoacetic esters, and similar compounds are often called *active hydrogen compounds* or *active methylene compounds.* Generally speaking, active hydrogen compounds have two electron-withdrawing groups attached to the same carbon atom:

$$Z{-}CH_2{-}Z'$$

Active hydrogen compound
(Z and Z′ are electron-withdrawing groups)

The electron-withdrawing groups can be a variety of substituents including:

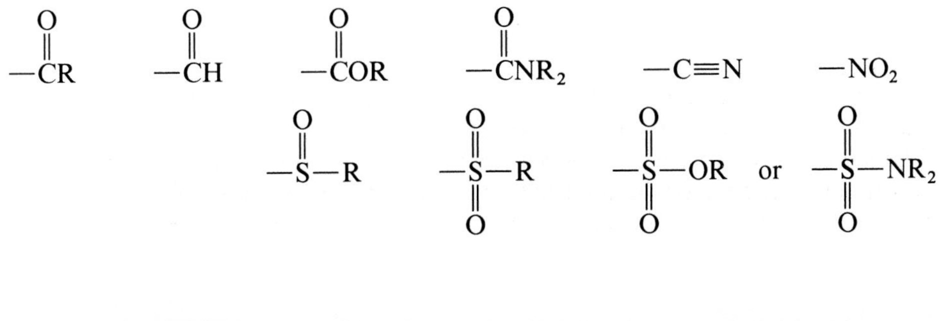

For example, ethyl cyanoacetate reacts with base to yield a resonance-stabilized anion:

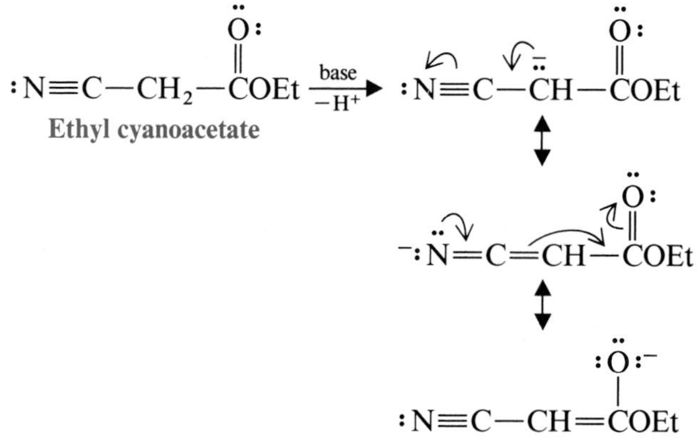

$$
:N\equiv C-CH_2-\overset{\overset{\displaystyle \ddot O:}{\|}}{C}OEt \xrightarrow[-H^+]{base} :N\equiv C-\overset{\overset{\displaystyle \ddot O:}{\|}}{C}H-COEt
$$

Ethyl cyanoacetate

Ethyl cyanoacetate anions also undergo alkylations. They can be dialkylated with isopropyl iodide, for example.

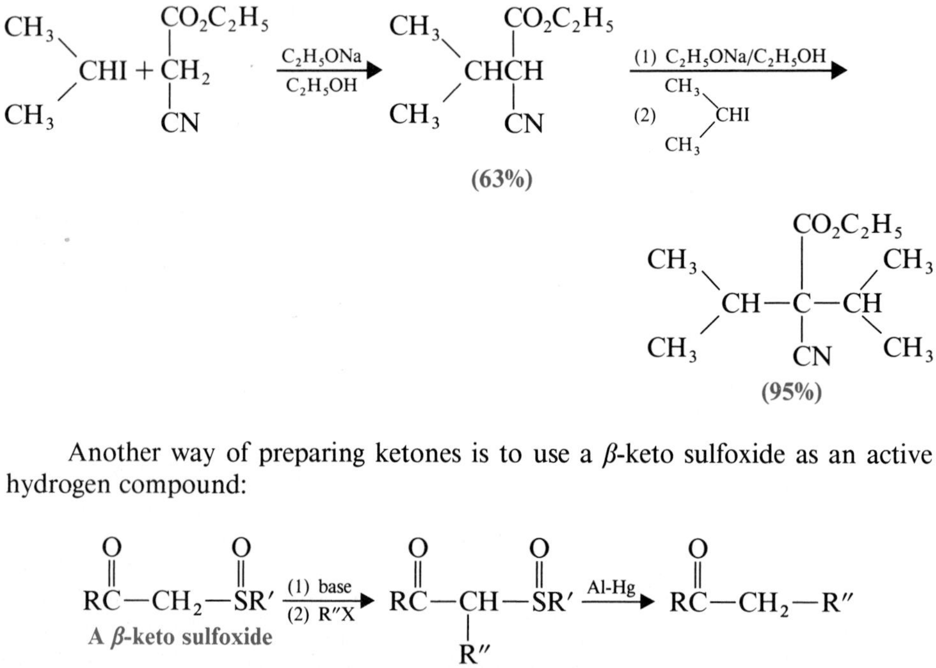

(63%)

(95%)

Another way of preparing ketones is to use a β-keto sulfoxide as an active hydrogen compound:

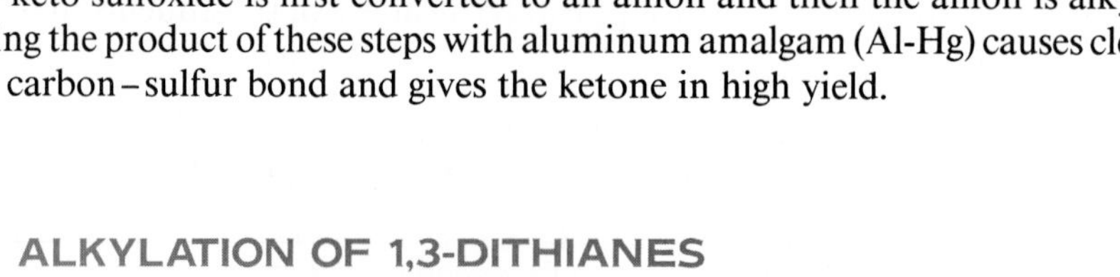

$$
\overset{\overset{\displaystyle O}{\|}}{R}C-CH_2-\overset{\overset{\displaystyle O}{\|}}{S}R' \xrightarrow[(2)\ R''X]{(1)\ base} \overset{\overset{\displaystyle O}{\|}}{R}C-\underset{\underset{\displaystyle R''}{|}}{C}H-\overset{\overset{\displaystyle O}{\|}}{S}R' \xrightarrow{Al\text{-}Hg} \overset{\overset{\displaystyle O}{\|}}{R}C-CH_2-R''
$$

A β-keto sulfoxide

The β-keto sulfoxide is first converted to an anion and then the anion is alkylated. Treating the product of these steps with aluminum amalgam (Al-Hg) causes cleavage at the carbon–sulfur bond and gives the ketone in high yield.

20.6 ALKYLATION OF 1,3-DITHIANES

Two sulfur atoms attached to the same carbon of 1,3-dithiane cause the hydrogen atoms of that carbon to be more acidic ($K_a \simeq 10^{-32}$) than those of most alkyl carbon atoms.

$$\underset{\substack{\text{1,3-Dithiane}\\(K_a \simeq 10^{-32})}}{\overset{\displaystyle \underset{\text{H}\quad\text{H}}{\underset{|}{\overset{|}{\underset{\diagdown}{\overset{\diagup}{\text{C}}}}}}{\overset{\text{S}\diagdown\quad\diagup\text{S}}{}}}$$

1,3-Dithiane
($K_a \simeq 10^{-32}$)

Sulfur atoms, because they are easily polarized, can aid in stabilizing the negative charge of the anion (cf. Special Topic F). Strong bases such as butyllithium are usually used to convert a dithiane to its anion.

$$\underset{\text{H}\quad\text{H}}{\text{S}\diagdown\quad\diagup\text{S}} + \overset{\delta-\ \delta+}{\text{C}_4\text{H}_9\text{Li}} \longrightarrow \underset{\text{H}\quad\text{Li}^+}{\text{S}\diagdown\quad\diagup\text{S}^{\cdot-}} + \text{C}_4\text{H}_{10}$$

1,3-Dithianes are thioacetals (cf. Section 16.7E); they can be prepared by treating an aldehyde with 1,3-propanedithiol in the presence of a trace of acid.

$$\underset{\text{RCH}}{\overset{\text{O}}{\overset{\|}{}}} + \text{HSCH}_2\text{CH}_2\text{CH}_2\text{SH} \xrightarrow{\text{H}^+} \underset{\text{R}\quad\text{H}}{\text{S}\diagdown\quad\diagup\text{S}}$$

A 1,3-dithiane

Alkylating the 1,3-dithiane and then hydrolyzing the product (a thioketal) is a method for converting an aldehyde to a ketone. Hydrolysis is usually carried out by using HgCl_2 either in methanol or in aqueous CH_3CN.

$$\underset{\text{R}\quad\text{H}}{\text{S}\diagdown\quad\diagup\text{S}} \xrightarrow[\text{(2) R'X}(-\text{LiX})]{\text{(1) C}_4\text{H}_9\text{Li}(-\text{C}_4\text{H}_{10})} \underset{\text{R}\quad\text{R}'}{\text{S}\diagdown\quad\diagup\text{S}} \xrightarrow[(-\text{HSCH}_2\text{CH}_2\text{CH}_2\text{SH})]{\text{HgCl}_2,\text{CH}_3\text{OH},\text{H}_2\text{O}} \underset{}{\overset{\text{O}}{\overset{\|}{\text{R}-\text{C}-\text{R}'}}}$$

Thioketal   Ketone

Notice that in these 1,3-dithiane syntheses the usual mode of reaction of an aldehyde is reversed. Normally the carbonyl carbon atom of an aldehyde is partially positive; it is electrophilic and, consequently, it reacts with nucleophiles. When the aldehyde is converted to a 1,3-dithiane and treated with butyllithium, this same carbon atom becomes negatively charged and reacts with electrophiles. This reversal of polarity of the carbonyl carbon atom is called **umpolung** (German for **polarity reversal**).

$$\underset{\text{R}\quad\text{H}}{\overset{\text{O}^{\delta-}}{\overset{\|}{\overset{\text{C}^{\delta+}}{}}}} \xrightarrow[\text{(2) C}_4\text{H}_9\text{Li}]{\text{(1) HSCH}_2\text{CH}_2\text{CH}_2\text{SH, H}^+} \underset{\text{R}}{\text{S}\diagdown\ \underset{\diagup}{\text{C}\colon^-}\diagup\text{S}}$$

Aldehyde

Umpolung

The synthetic use of 1,3-dithianes was developed by E. J. Corey (Section 15.9) and D. Seebach and is often called the *Corey–Seebach* method.

PROBLEM 20.17

(a) Which aldehyde would you use to prepare 1,3-dithiane itself?   (b) How would you synthesize $C_6H_5CH_2CHO$ using a 1,3-dithiane as an intermediate? (c) How would you convert benzaldehyde to acetophenone?

PROBLEM 20.18

The Corey–Seebach method can also be used to synthesize molecules with the structure RCH_2R'. How might this be done?

PROBLEM 20.19

(a) The Corey–Seebach method has been used to prepare the following highly strained molecule called a metaparacyclophane. What are the structures of the intermediates **A – D**?

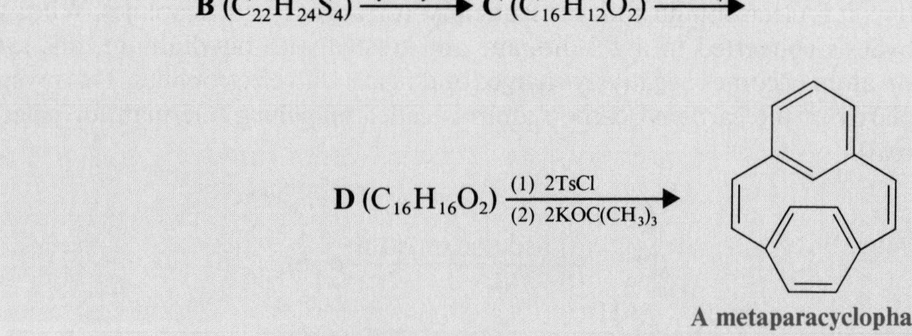

A metaparacyclophane

(b) What compound would be obtained by treating **B** with excess Raney Ni?

20.7 THE KNOEVENAGEL CONDENSATION

Active hydrogen compounds condense with aldehydes and ketones. Known as Knoevenagel condensations, these aldol-like condensations are catalyzed by weak bases. An example is the following:

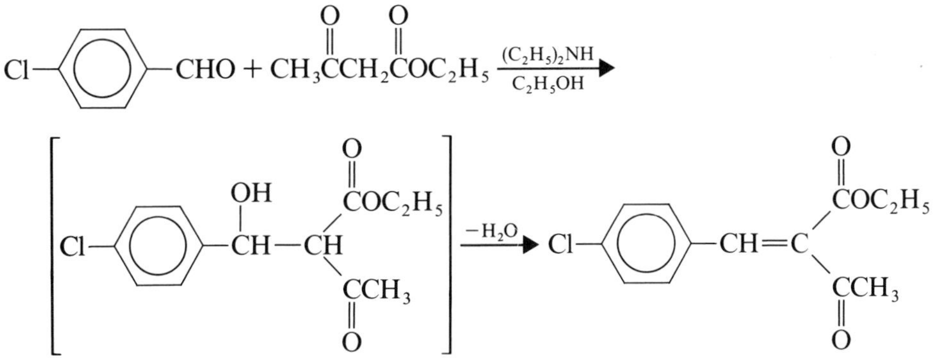

(86%)

20.8 MICHAEL ADDITIONS

Active hydrogen compounds also undergo conjugate additions to α,β-unsaturated carbonyl compounds. These reactions are known as Michael additions, a reaction that we studied in Section 17.9B. An example of the Michael addition of an active hydrogen compound is the following:

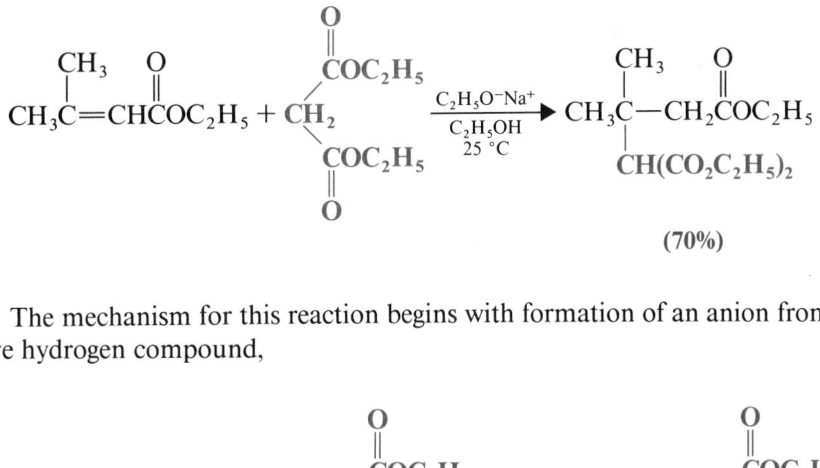

(70%)

The mechanism for this reaction begins with formation of an anion from the active hydrogen compound,

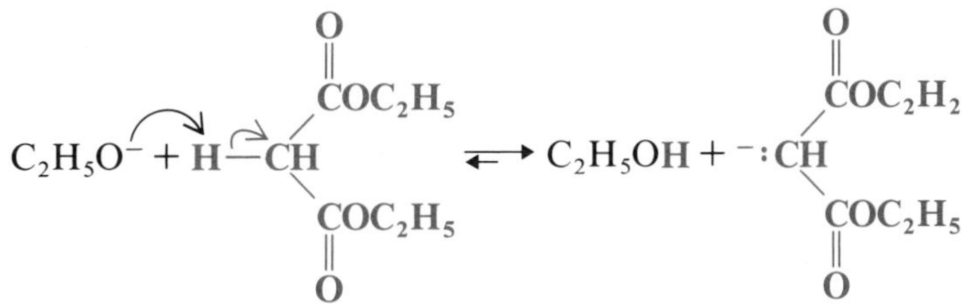

then conjugate addition of the anion to the α,β-unsaturated ester (step 2) is followed by the acceptance of a proton (step 3).

Step 2

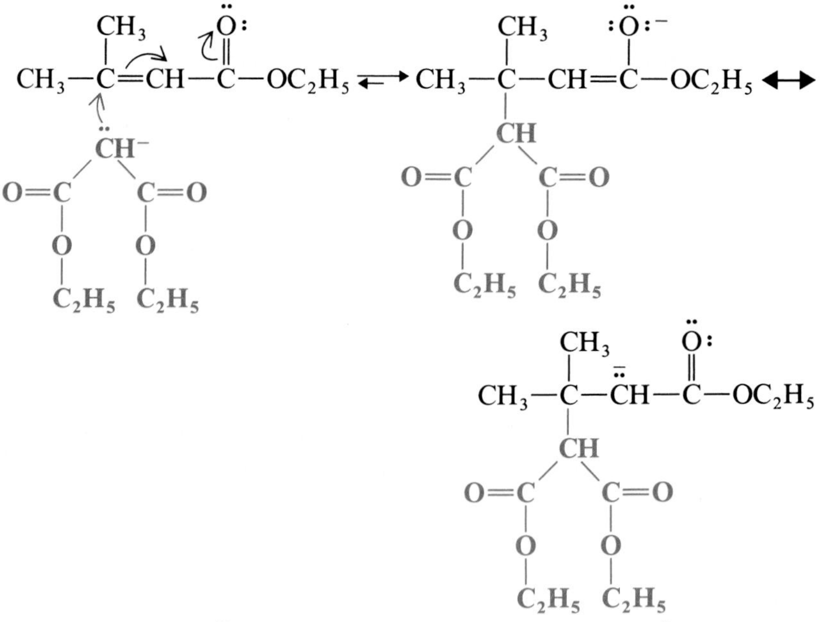

$$CH_3-\overset{\overset{\textstyle CH_3}{|}}{C}=CH-\overset{\overset{\textstyle \ddot{O}:}{||}}{C}-OC_2H_5 \;\rightleftharpoons\; CH_3-\overset{\overset{\textstyle CH_3}{|}}{\underset{\underset{\textstyle CH}{|}}{C}}-CH=\overset{\overset{\textstyle :\ddot{O}:^-}{|}}{C}-OC_2H_5 \;\longleftrightarrow$$

$$\overset{\overset{\textstyle \ddot{C}H^-}{|}}{\underset{}{}} $$

$$O=C \qquad C=O$$

$$O \qquad O$$

$$C_2H_5 \quad C_2H_5$$

$$CH_3-\overset{\overset{\textstyle CH_3}{|}}{\underset{\underset{\textstyle CH}{|}}{C}}-\overset{\overset{\textstyle \ddot{..}}{\underline{\ddot{C}}H}}{}-\overset{\overset{\textstyle \ddot{O}:}{||}}{C}-OC_2H_5$$

$$O=C \qquad C=O$$

$$O \qquad O$$

$$C_2H_5 \quad C_2H_5$$

Step 3

$$CH_3-\overset{\overset{\textstyle CH_3}{|}}{\underset{\underset{\textstyle CH}{|}}{C}}-\underline{\ddot{C}}H-\overset{\overset{\textstyle \ddot{O}:}{||}}{C}-OC_2H_5 \xrightarrow{+H^+} CH_3-\overset{\overset{\textstyle CH_3}{|}}{\underset{\underset{\textstyle CH}{|}}{C}}-CH_2-\overset{\overset{\textstyle \ddot{O}:}{||}}{C}-OC_2H_5$$

$$O=C \qquad C=O \qquad\qquad O=C \qquad C=O$$

$$O \qquad O \qquad\qquad\qquad O \qquad O$$

$$C_2H_5 \quad C_2H_5 \qquad\qquad\quad C_2H_5 \quad C_2H_5$$

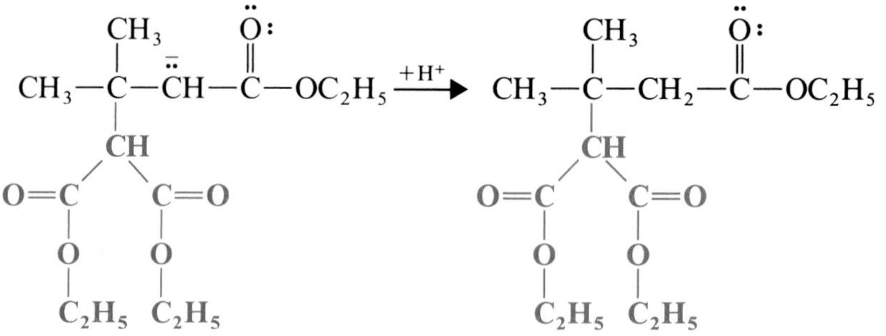

PROBLEM 20.20

How would you prepare $\overset{\overset{\textstyle O}{||}}{HOC}CH_2\overset{\overset{\textstyle CH_3}{|}}{\underset{\underset{\textstyle CH_3}{|}}{C}}CH_2\overset{\overset{\textstyle O}{||}}{COH}$ from the product of the Michael addition given previously?

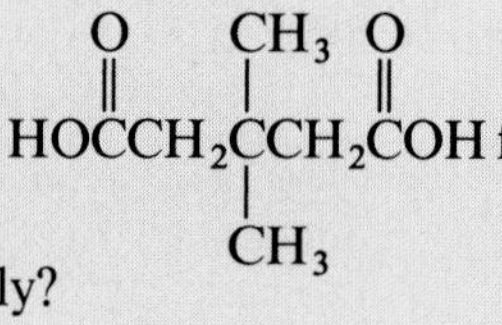

Michael additions take place with a variety of other reagents; these include acetylenic esters and α,β-unsaturated nitriles:

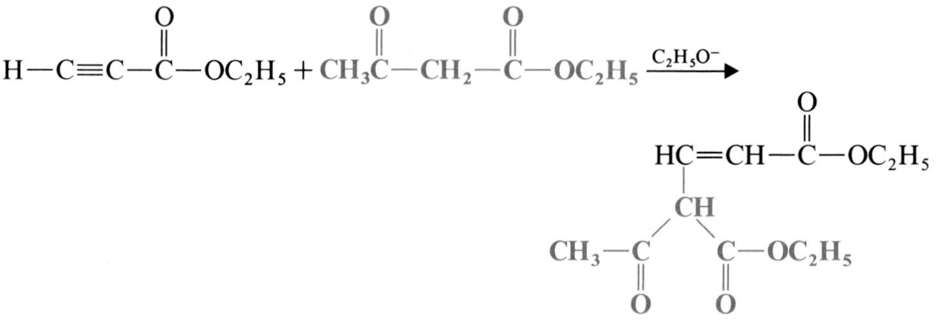

$$H-C\equiv C-\overset{\overset{\textstyle O}{||}}{C}-OC_2H_5 + CH_3\overset{\overset{\textstyle O}{||}}{C}-CH_2-\overset{\overset{\textstyle O}{||}}{C}-OC_2H_5 \xrightarrow{C_2H_5O^-}$$

$$HC=CH-\overset{\overset{\textstyle O}{||}}{C}-OC_2H_5$$

$$\underset{\underset{\textstyle CH}{|}}{}$$

$$CH_3-\underset{\underset{\textstyle O}{||}}{C} \qquad \underset{\underset{\textstyle O}{||}}{C}-OC_2H_5$$

Cyclohexanone, for example, reacts with pyrrolidine in the following way:

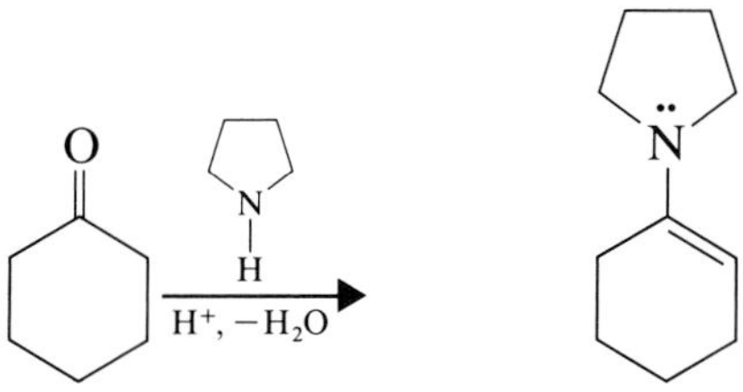

N-(1-Cyclohexenyl)pyrrolidine
(an enamine)

Enamines are good nucleophiles, and an examination of the resonance structures that follow will show us that we should expect enamines to have both a nucleophilic nitrogen and a *nucleophilic carbon.*

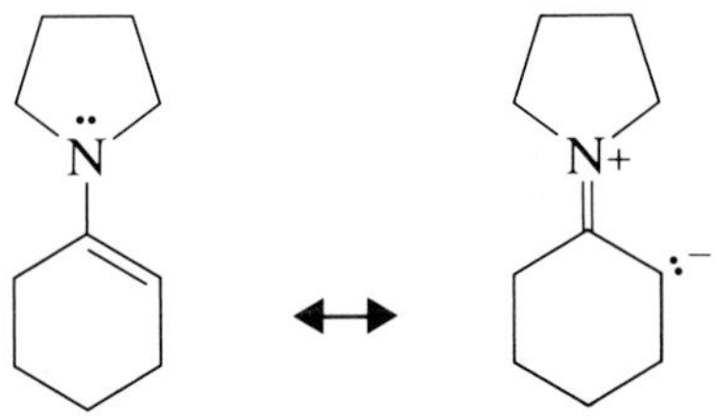

*Contribution to the
hybrid made by this
structure confers
nucleophilicity on
nitrogen*

*Contribution to the
hybrid made by this
structure confers
nucleophilicity on
carbon and decreases
nucleophilicity of
nitrogen*

The nucleophilicity of the carbon of enamines makes them particularly useful reagents in organic synthesis because they can be **acylated, alkylated,** and used in **Michael additions.** Development of these techniques originated with the work of Gilbert Stork of Columbia University and in his honor they have come to be known as **Stork enamine reactions.**

When an enamine reacts with an acyl halide or an acid anhydride, the product is the *C*-acylated compound. The iminium ion that forms hydrolyzes when water is added and the overall reaction provides a synthesis of β-diketones.

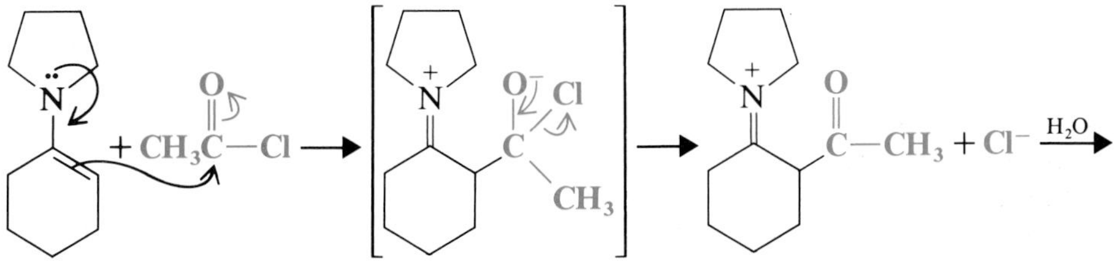

Iminium salt

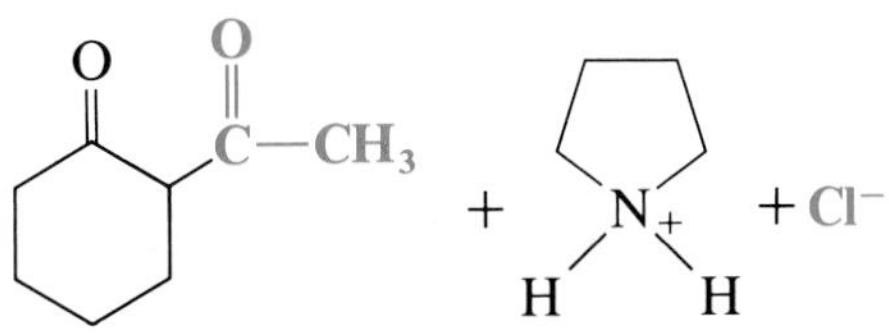

2-Acetylcyclohexanone
(a β-diketone)

Although *N*-acylation may occur in this synthesis, the *N*-acyl product is unstable and can act as an acylating agent itself.

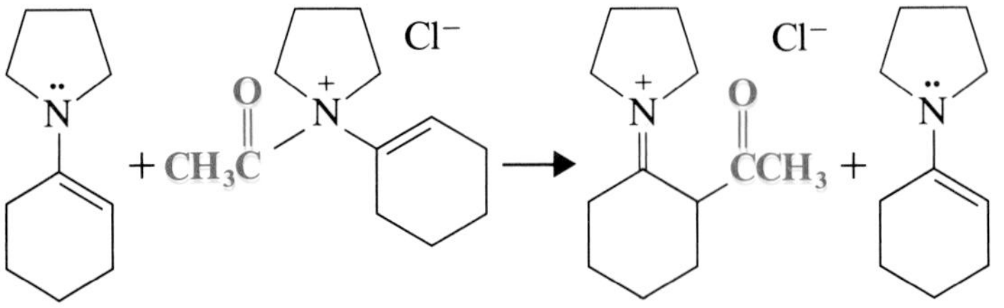

| Enamine | *N*-Acylated enamine | *C*-Acylated iminium salt | Enamine |

As a consequence, the yields of *C*-acylated products are generally high.

Enamines can be alkylated as well as acylated. While alkylation may lead to the formation of a considerable amount of *N*-alkylated product, heating the *N*-alkylated product often converts it to a *C*-alkyl compound. This rearrangement is particularly favored when the alkyl halide is an allylic halide, benzylic halide, or *α*-haloacetic ester.

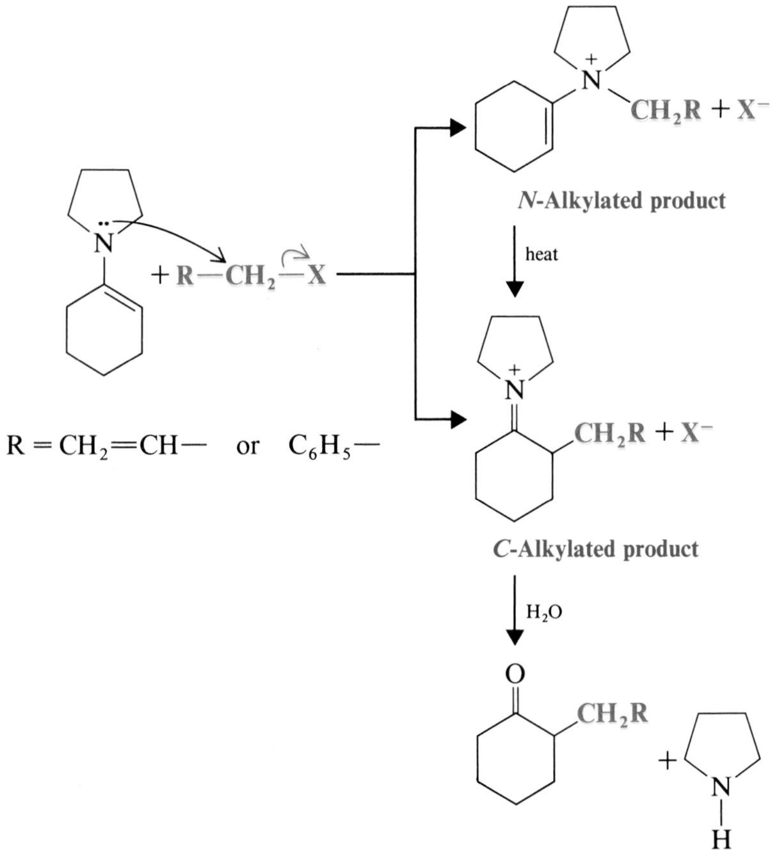

N-Alkylated product

heat

C-Alkylated product

$R = CH_2{=}CH-$ or C_6H_5-

H_2O

Enamine alkylations are S_N2 reactions; thus, when we choose our alkylating agents, we are usually restricted to the use of methyl, primary, allylic, and benzylic halides. *α*-Halo esters can also be used as the alkylating agents, and this reaction provides a convenient synthesis of *γ*-keto esters:

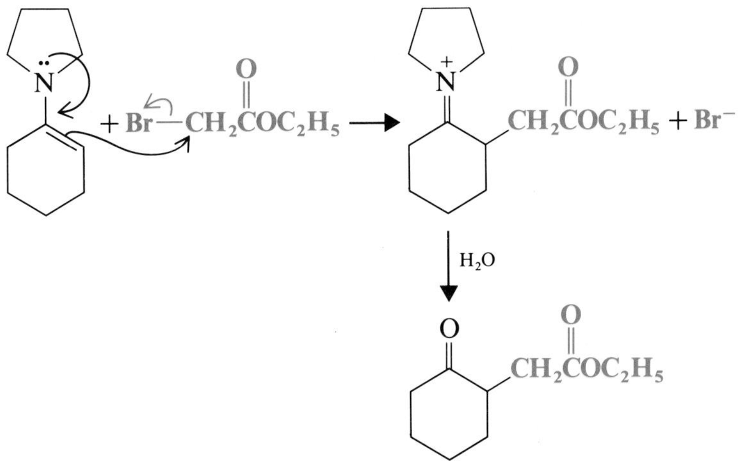
N
+ Br—CH₂COC₂H₅
O
N
CH₂COC₂H₅ + Br⁻
H₂O
O
CH₂COC₂H₅

(75%)
(a γ-keto ester)

PROBLEM 20.22

Show how you could employ enamines in syntheses of the following compounds:

(a) 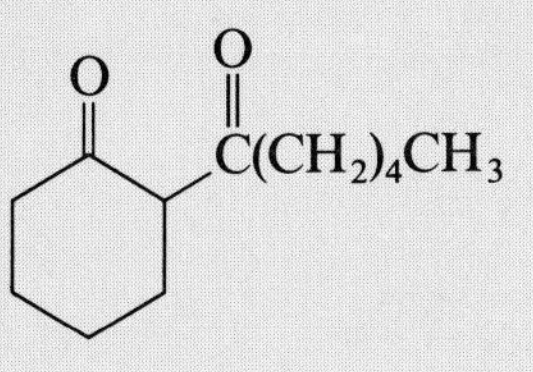
O
O
C(CH₂)₄CH₃

(c) 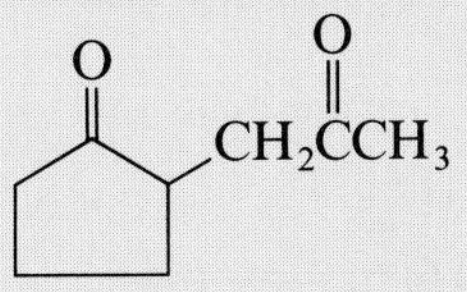
O
O
CH₂CCH₃

(b) 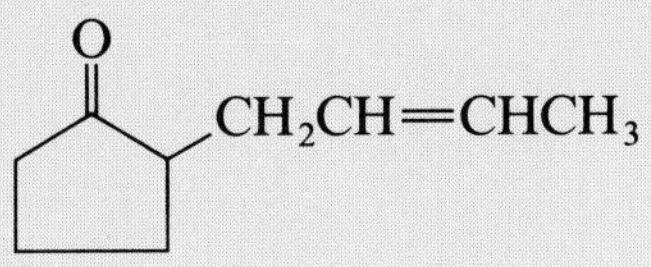
O
CH₂CH=CHCH₃

(d) 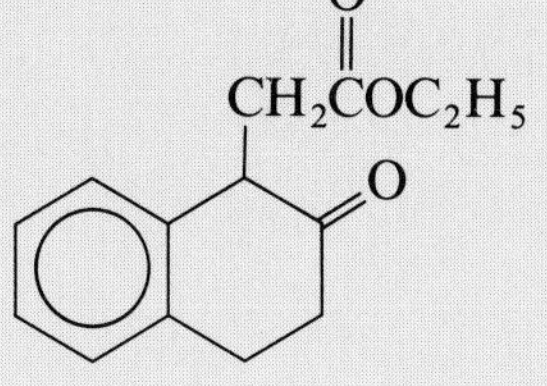
O
CH₂COC₂H₅
O

An especially interesting set of enamine alkylations is shown in the following reactions (developed by J. K. Whitesell of the University of Texas at Austin). The enamine (prepared from a single enantiomer of the secondary amine) is chiral. Alkylation from the bottom of the enamine is severely hindered by the methyl group. (Notice that this hindrance will exist even if rotation of the groups takes place about the bond connecting the two rings.) Consequently, alkylation takes place much more rapidly from the top side. This reaction yields (after hydrolysis) 2-substituted cyclohexanones consisting almost entirely of a single enantiomer.

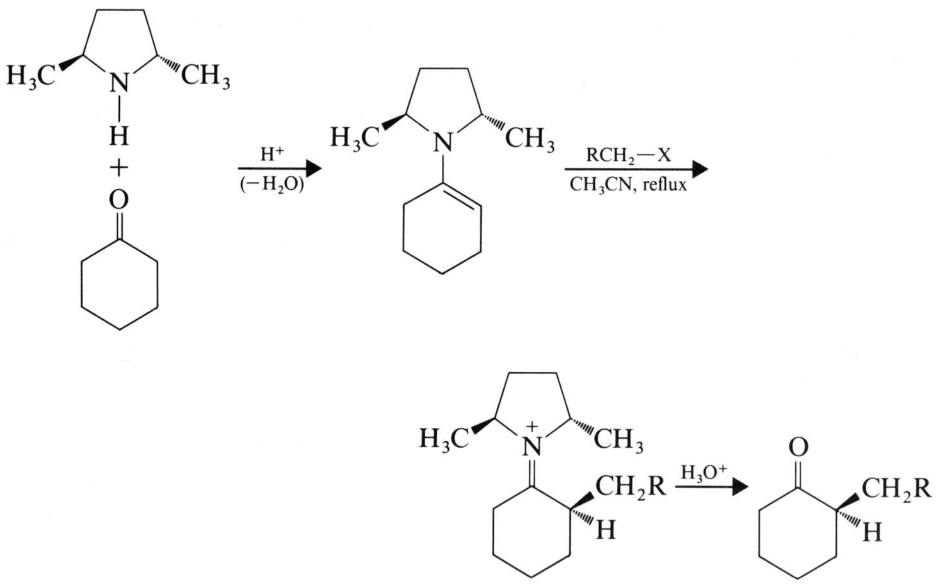

R GROUP	CHEMICAL YIELD (%)	OPTICAL PURITY (%)
H—	50	83
CH₃CH₂—	57	93
CH₂=CH—	80	82

Enamines can also be used in Michael additions. An example is the following:

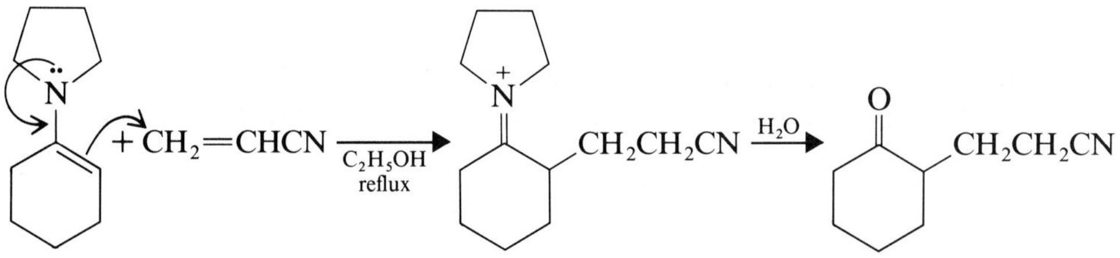

20.11 BARBITURATES

In the presence of sodium ethoxide, diethyl malonate reacts with urea to yield a compound called barbituric acid.

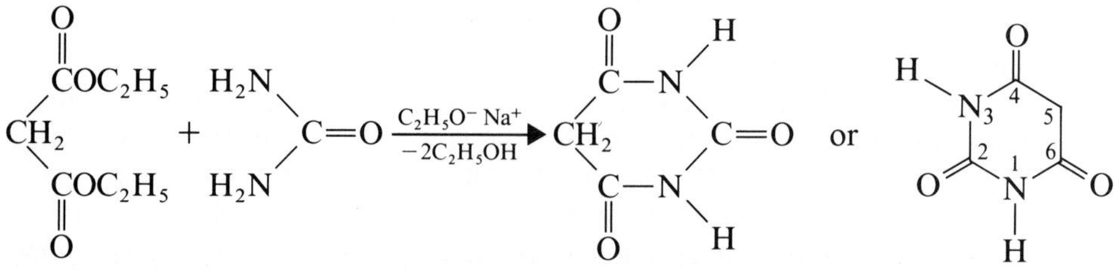

Barbituric acid

Barbituric acid is a pyrimidine (cf. Section 19.1), and it exists in several tautomeric forms including one with an aromatic ring.

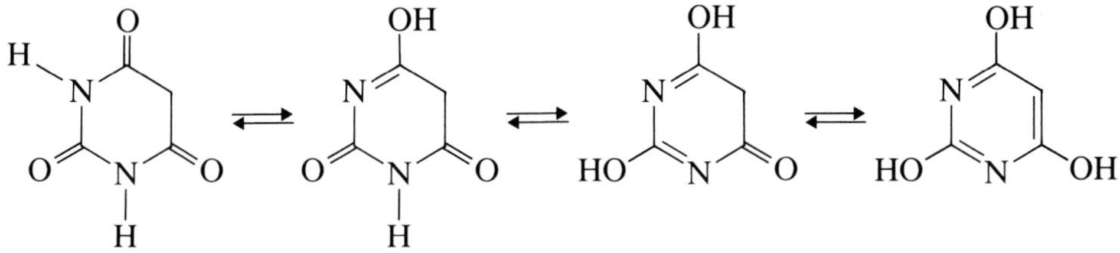

As its name suggests barbituric acid is a moderately strong acid, stronger even than acetic acid.* Its anion is highly resonance stabilized.

Derivatives of barbituric acid are *barbiturates.* Barbiturates have been used in medicine as soporifics (sleep producers) since 1903. One of the earliest barbiturates introduced into medical use is the compound veronal (5,5-diethylbarbituric acid). Veronal is usually used as its sodium salt. Other barbiturates are seconal and phenobarbital.

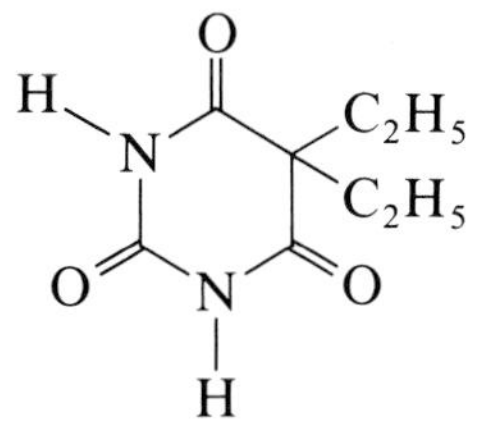

Veronal	Seconal	Phenobarbital
(5,5-diethylbarbituric acid)	[5-allyl-5-(1-methylbutyl) barbituric acid]	(5-ethyl-5-phenylbarbituric acid)

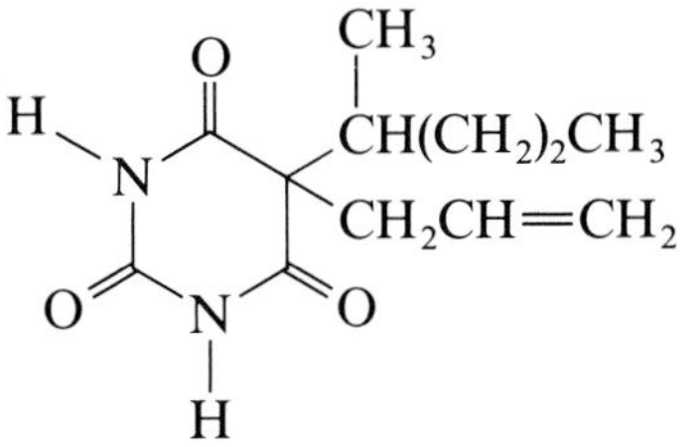

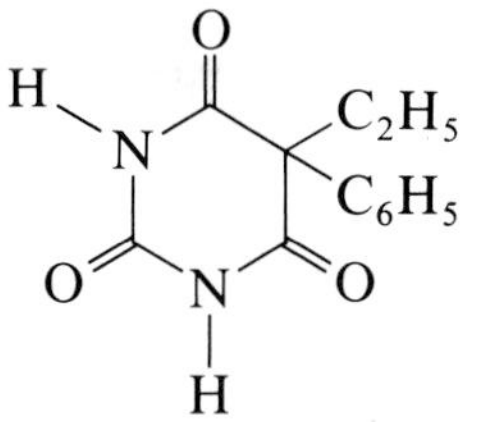

Although barbiturates are very effective soporifics, their use is also hazardous. They are addictive, and overdosage, often with fatal results, is common.

PROBLEM 20.23

Outlined here is a synthesis of phenobarbital.
(a) What are compounds **A – F**? (b) Propose an alternative synthesis of **E** from diethyl malonate.

$$C_6H_5{-}CH_3 \xrightarrow[CCl_4]{NBS} A\ (C_7H_7Br) \xrightarrow[\text{(2) } CO_2,\ \text{then } H^+]{\text{(1) Mg, Et}_2O} B\ (C_8H_8O_2) \xrightarrow{SOCl_2}$$

$$C\ (C_8H_7ClO) \xrightarrow{EtOH} D\ (C_{10}H_{12}O_2) \xrightarrow[\text{NaOEt}]{EtO\overset{O}{\overset{\|}{C}}OEt} E\ (C_{13}H_{16}O_4) \xrightarrow[CH_3CH_2Br]{KOC(CH_3)_3}$$

$$F\ (C_{15}H_{20}O_4) \xrightarrow[\text{H}_2N\overset{O}{\overset{\|}{C}}NH_2,\ \text{NaOEt}]{} \text{phenobarbital}$$

*Barbituric acid was given its name by Adolf von Baeyer in 1864. The barbituric part of the name is thought to have been motivated by Baeyer's gallantry toward a friend named Barbara.

PROBLEM 20.24

Starting with diethyl malonate, urea, and any other required reagents, outline a synthesis of veronal and seconal.

ADDITIONAL PROBLEMS
20.25

Show all steps in the following syntheses. You may use any other needed reagents but should begin with the compound given. You need not repeat steps carried out in earlier parts of this exercise.

(a) 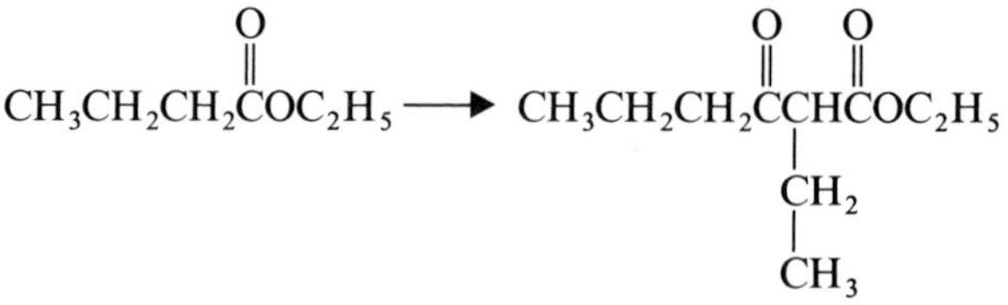 $CH_3CH_2CH_2\overset{O}{\overset{\|}{C}}OC_2H_5 \longrightarrow CH_3CH_2CH_2\overset{O}{\overset{\|}{C}}\overset{O}{\overset{\|}{C}}HCOC_2H_5$

with substituent $\underset{CH_3}{\overset{CH_2}{|}}$

(b) $CH_3CH_2CH_2\overset{O}{\overset{\|}{C}}OC_2H_5 \longrightarrow CH_3CH_2CH_2\overset{O}{\overset{\|}{C}}CH_2CH_2CH_3$

(c) $C_6H_5CH_2\overset{O}{\overset{\|}{C}}OC_2H_5 \longrightarrow C_6H_5\overset{CH_3}{\overset{|}{C}}HCO_2H$

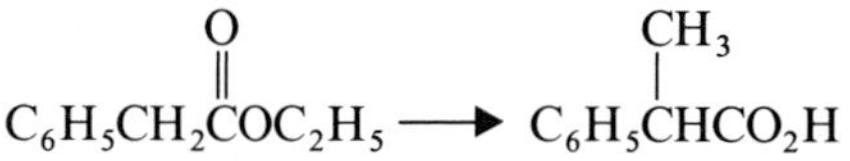

(d) 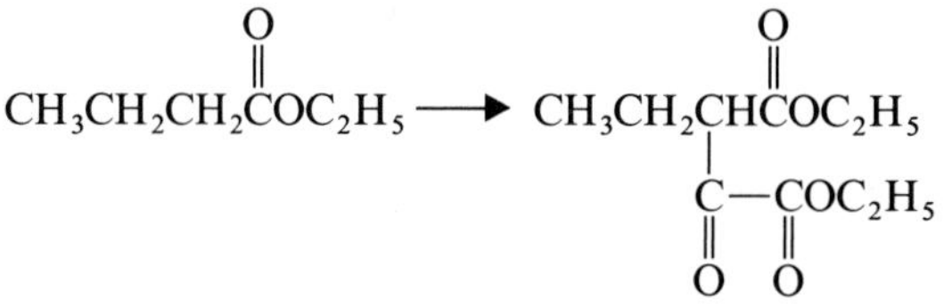 $CH_3CH_2CH_2\overset{O}{\overset{\|}{C}}OC_2H_5 \longrightarrow CH_3CH_2\overset{O}{\overset{\|}{C}}HCOC_2H_5$

with substituent $\overset{}{\underset{\overset{\|}{O}}{C}}-\overset{}{\underset{\overset{\|}{O}}{C}}OC_2H_5$

(e) 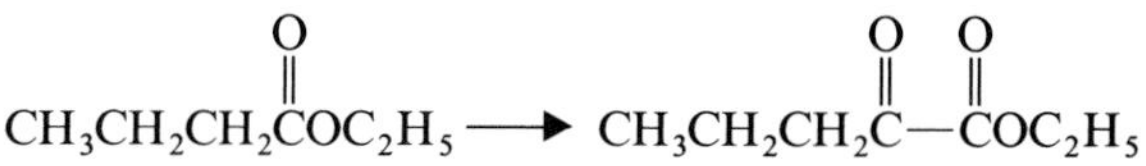 $CH_3CH_2CH_2\overset{O}{\overset{\|}{C}}OC_2H_5 \longrightarrow CH_3CH_2CH_2\overset{O}{\overset{\|}{C}}-\overset{O}{\overset{\|}{C}}OC_2H_5$

(f) 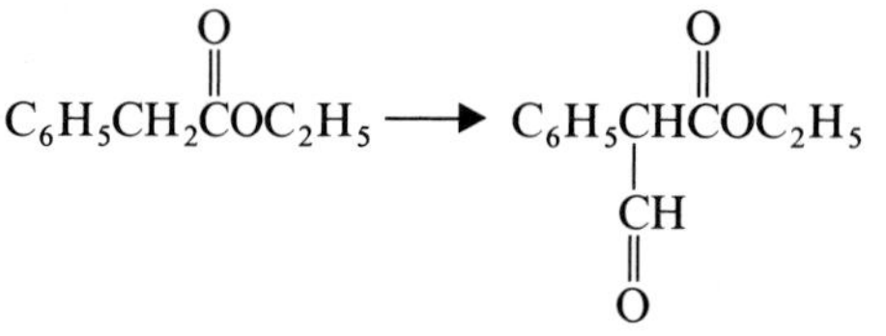 $C_6H_5CH_2\overset{O}{\overset{\|}{C}}OC_2H_5 \longrightarrow C_6H_5\overset{O}{\overset{\|}{C}}HCOC_2H_5$

with substituent $\overset{CH}{\underset{\overset{\|}{O}}{}}$

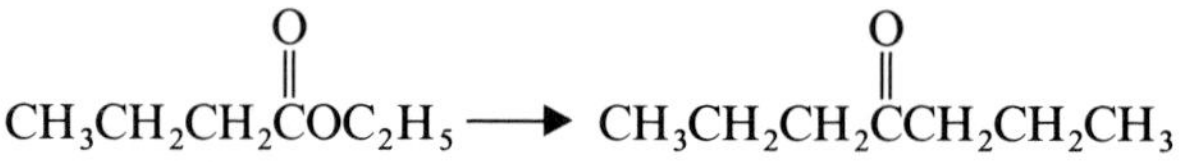

(g)

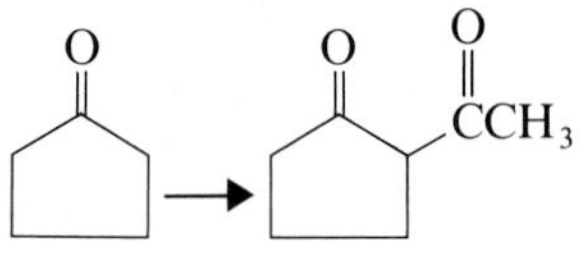

(h)

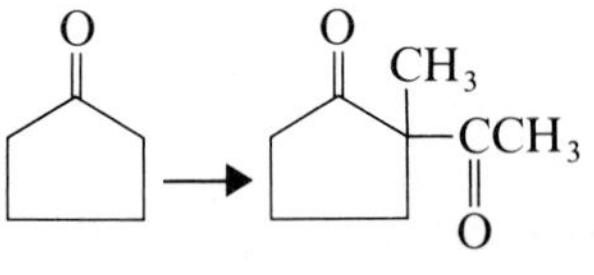

(i)

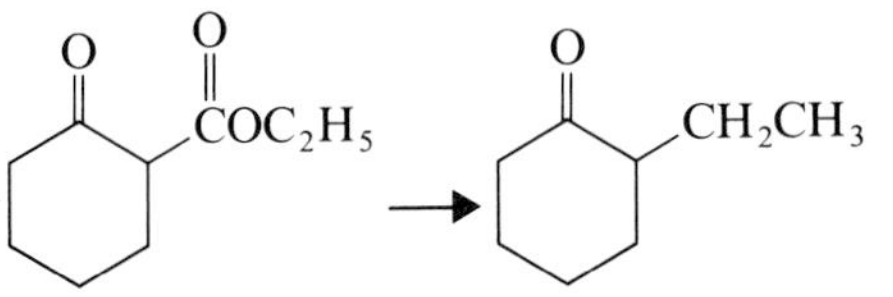

20.26
Outline syntheses of each of the following from acetoacetic ester and any other required reagents.

(a) Methyl *tert*-butyl ketone   (c) 2,5-Hexanedione   (e) 2-Ethyl-1,3-butanediol

(b) 2-Hexanone   (d) 4-Hydroxypentanoic acid (f) 1-Phenyl-1,3-butanediol

20.27
Outline syntheses of each of the following from diethyl malonate and any other required reagents.

(a) 2-Methylbutanoic acid   (d) $HOCH_2CH_2CH_2CH_2OH$

(b) 4-Methyl-1-pentanol

(c) $CH_3CH_2CHCH_2OH$
       |
       CH_2OH

20.28
The synthesis of cyclobutanecarboxylic acid given on p. 959 was first carried out by William Perkin, Jr., in 1883, and it represented one of the first syntheses of an organic compound with a ring smaller than six carbon atoms. (There was a general feeling at the time that such compounds would be too unstable to exist.) Earlier in 1883, Perkin reported what he mistakenly believed to be a cyclobutane derivative obtained from the reaction of acetoacetic ester and 1,3-dibromopropane. The reaction that Perkin had expected to take place was the following:

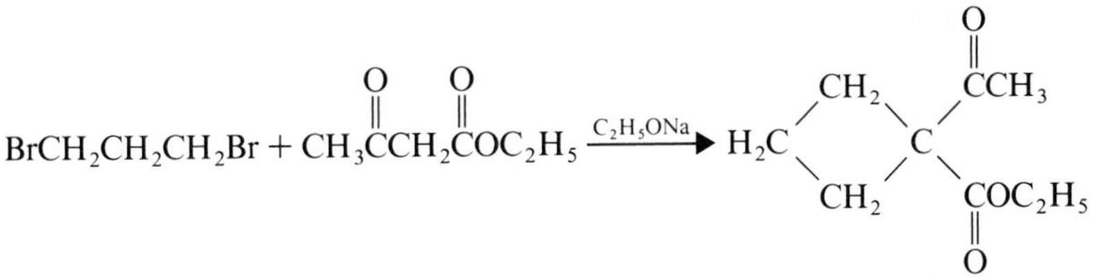

The molecular formula for his product agreed with the formulation given in the preceding reaction, and alkaline hydrolysis and acidification gave a nicely crystalline acid (also having the expected molecular formula). The acid, however, was quite stable to heat and resisted decarboxylation. Perkin later found that both the ester and the acid contained six-membered rings (five carbon atoms and one oxygen atom). Recall the charge distribution in the enolate ion obtained from acetoacetic ester and propose structures for Perkin's ester and acid.

20.29
(a) In 1884 Perkin achieved a successful synthesis of cyclopropanecarboxylic acid from sodiomalonic ester and 1,2-dibromoethane. Outline the reactions involved in this synthesis.   (b) In 1885 Perkin synthesized five-membered carbocyclic compounds **D** and **E** in the following way:

$$2Na^+ : CH(CO_2C_2H_5)_2 + BrCH_2CH_2CH_2Br \longrightarrow A\ (C_{17}H_{28}O_8) \xrightarrow{2C_2H_5O^-Na^+} \xrightarrow{Br_2}$$

$$B\ (C_{17}H_{26}O_8) \xrightarrow[(2)\ H_3O^+]{(1)\ OH^-/H_2O} C\ (C_9H_{10}O_8) \xrightarrow{heat} D\ (C_7H_{10}O_4) + E\ (C_7H_{10}O_4)$$

D and **E** are diastereomers; **D** can be resolved into enantiomeric forms while **E** cannot. What are the structures of **A, B, C, D,** and **E**?   (c) Ten years later Perkin was able to synthesize 1,4-dibromobutane; he later used this compound and diethyl malonate to prepare cyclopentanecarboxylic acid. Show the reactions involved.

20.30
Write mechanisms that account for the products of the following reactions:

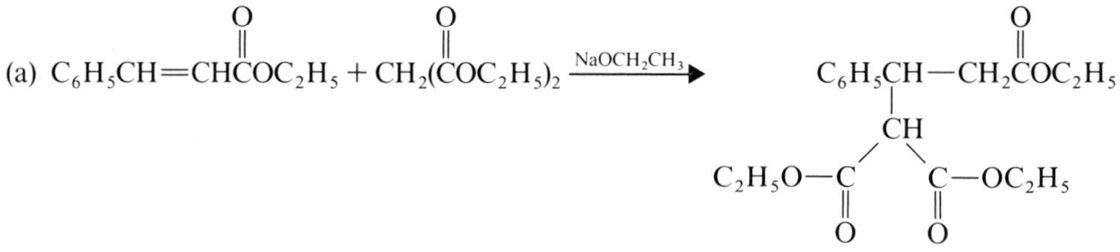

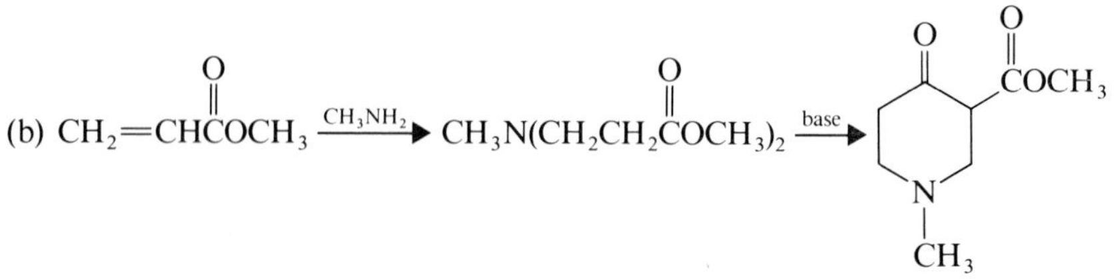

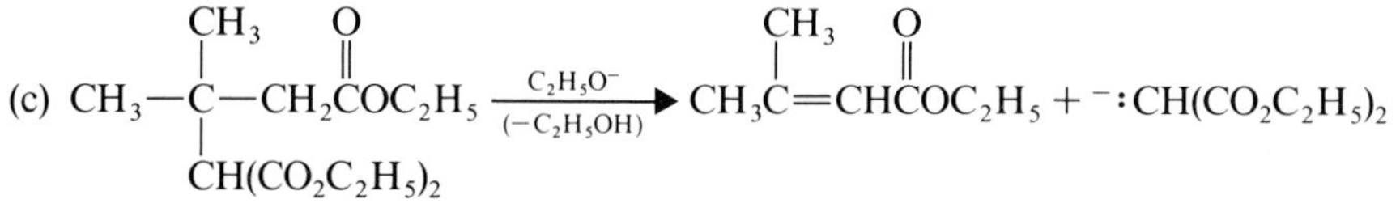

(a) $C_6H_5CH{=}CHCOC_2H_5 + CH_2(COC_2H_5)_2 \xrightarrow{\ \text{NaOCH}_2\text{CH}_3\ }$

$$C_6H_5CH{-}CH_2COC_2H_5$$
$$\underset{|}{\phantom{C_6H_5CH{-}CH_2COC_2H_5}}$$
$$\text{CH}$$
$$C_2H_5O{-}\underset{\underset{O}{\|}}{C}\qquad \underset{\underset{O}{\|}}{C}{-}OC_2H_5$$

(b) $CH_2{=}CHCOCH_3 \xrightarrow{\ \text{CH}_3\text{NH}_2\ } CH_3N(CH_2CH_2COCH_3)_2 \xrightarrow{\ \text{base}\ }$

(with product: N-methyl piperidine ring bearing $=O$ and $COCH_3$ substituent, N–CH$_3$)

(c) $\underset{\underset{CH(CO_2C_2H_5)_2}{|}}{\overset{\overset{CH_3}{|}}{CH_3}}{-}\overset{O}{\underset{\|}{C}}{-}CH_2\overset{O}{\underset{\|}{C}}OC_2H_5 \xrightarrow[(-C_2H_5OH)]{\ C_2H_5O^-\ } CH_3C{=}CHCOC_2H_5 + {}^-{:}CH(CO_2C_2H_5)_2$

20.31
Knoevenagel condensations in which the active hydrogen compound is a β-keto ester or a β-diketone often yield products that result from one molecule of aldehyde or ketone and two molecules of the active methylene component. For example,

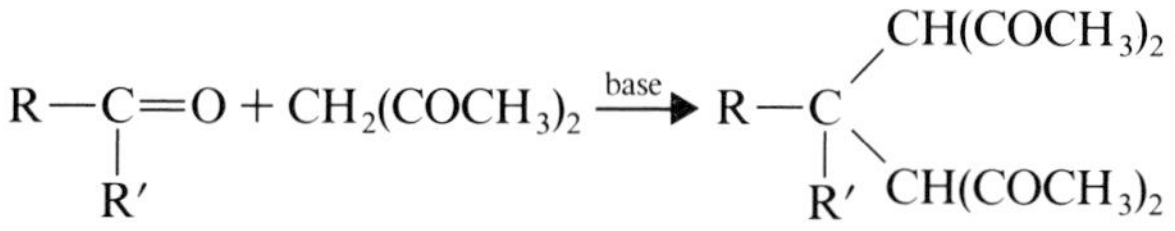

$$R{-}C{=}O + CH_2(COCH_3)_2 \xrightarrow{\ \text{base}\ } R{-}\underset{\underset{R'}{|}}{C}\overset{\diagup CH(COCH_3)_2}{\underset{\diagdown CH(COCH_3)_2}{}}$$

Suggest a reasonable mechanism that will account for the formation of these products.

20.32
Thymine is one of the heterocyclic bases found in DNA. Starting with ethyl propanoate and using any other needed reagents, show how you might synthesize thymine.

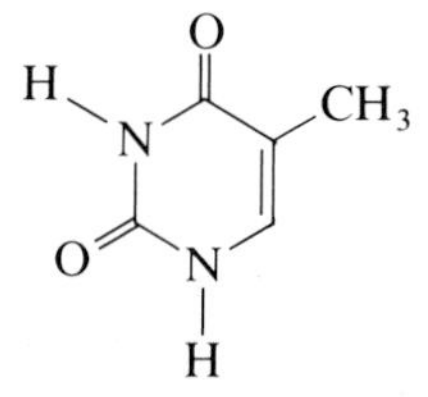

(structure of Thymine)

Thymine

20.33
The mandibular glands of queen bees secrete a fluid that contains a remarkable compound known as "queen substance." When even an exceedingly small amount of the queen substance is transferred to worker bees, it inhibits the development of their ovaries and prevents the workers from bearing new queens. Queen substance, a monocarboxylic acid with the molecular formula $C_{10}H_{16}O_3$, has been synthesized by the following route:

Cycloheptanone $\xrightarrow[\text{(2) } H_3O^+]{\text{(1) } CH_3MgI}$ **A** $(C_8H_{16}O)$ $\xrightarrow{H^+, \text{ heat}}$ **B** (C_8H_{14}) $\xrightarrow[\text{(2) } Zn, H_2O]{\text{(1) } O_3}$

C $(C_8H_{14}O_2)$ $\xrightarrow[\text{pyridine}]{CH_2(CO_2H)_2}$ queen substance $(C_{10}H_{16}O_3)$

On catalytic hydrogenation queen substance yields compound **D**, which, on treatment with iodine in sodium hydroxide and subsequent acidification, yields a dicarboxylic acid **E**; that is,

Queen substance $\xrightarrow[\text{Pd}]{H_2}$ **D** $(C_{10}H_{18}O_3)$ $\xrightarrow[\text{(2) } H_3O^+]{\text{(1) } I_2/NaOH}$ **E** $(C_9H_{16}O_4)$

Provide structures for the queen substance and compounds **A**–**E**.

20.34
Linalool, a fragrant compound that can be isolated from a variety of plants, is 3,7-dimethyl-1,6-octadien-3-ol. Linalool is used in making perfumes and it can be synthesized in the following way:

$CH_2{=}C{-}CH{=}CH_2 + HBr \longrightarrow$ **F** (C_5H_9Br) $\xrightarrow[\text{ester}]{\text{sodioacetoacetic}}$
  $\phantom{CH_2{=}C{-}}|$
  $\phantom{CH_2{=}C{-}}CH_3$

G $(C_{11}H_{18}O_3)$ $\xrightarrow[\text{(2) } H_3O^+, \text{ (3) heat}]{\text{(1) dil. NaOH}}$ **H** $(C_8H_{14}O)$ $\xrightarrow[\text{(2) } H_3O^+]{\text{(1) } LiC{\equiv}CH}$

I $(C_{10}H_{16}O)$ $\xrightarrow[\substack{\text{Lindlar's} \\ \text{catalyst}}]{H_2}$ linalool

Outline the reactions involved. (*Hint:* Compound **F** is the more stable isomer capable of being produced in the first step.)

*****20.35**
Compound **J**, a compound with two four-membered rings, has been synthesized by the following route. Outline the steps that are involved.

$NaCH(CO_2C_2H_5)_2 + BrCH_2CH_2CH_2Br \longrightarrow (C_{10}H_{17}BrO_4) \xrightarrow{NaOC_2H_5}$

$C_{10}H_{16}O_4 \xrightarrow[\text{(2) } H_2O]{\text{(1) } LiAlH_4} C_6H_{12}O_2 \xrightarrow{HBr} C_6H_{10}Br_2 \xrightarrow[2NaOC_2H_5]{CH_2(CO_2C_2H_5)_2}$

$C_{13}H_{20}O_4 \xrightarrow[\text{(2) } H_3O^+]{\text{(1) } OH^-, H_2O} C_9H_{12}O_4 \xrightarrow{\text{heat}} \textbf{J} (C_8H_{12}O_2) + CO_2$

20.36
When an aldehyde or a ketone is condensed with ethyl α-chloroacetate in the presence of sodium ethoxide, the product is an α,β-epoxy ester called a *glycidic ester*. The synthesis is called the Darzens condensation.

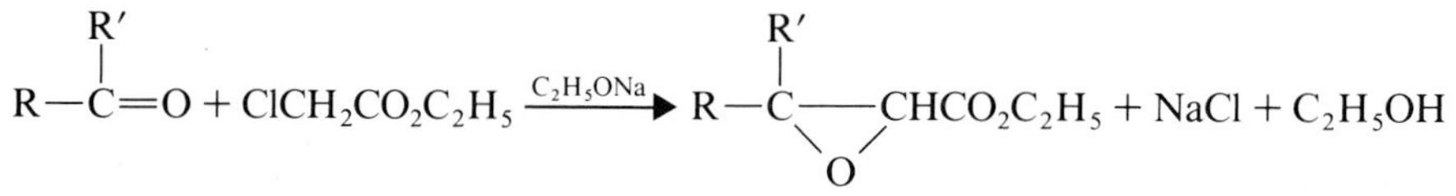

$$R-\overset{\underset{\displaystyle |}{R'}}{C}=O + ClCH_2CO_2C_2H_5 \xrightarrow{C_2H_5ONa} R-\overset{\underset{\displaystyle \diagdown O \diagup}{\overset{\displaystyle |}{C}}}{}-CHCO_2C_2H_5 + NaCl + C_2H_5OH$$

A glycidic ester

(a) Outline a reasonable mechanism for the Darzens condensation. (b) Hydrolysis of the epoxy ester leads to an epoxy acid that, on heating with pyridine, furnishes an aldehyde.

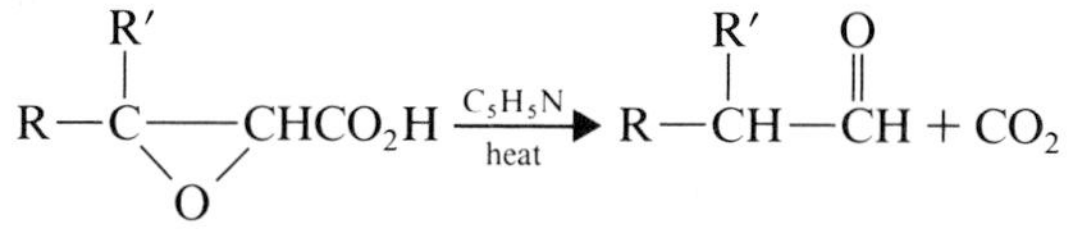

$$R-\overset{\underset{\displaystyle \diagdown O \diagup}{\overset{\displaystyle |}{C}}}{}-CHCO_2H \xrightarrow[\text{heat}]{C_5H_5N} R-\overset{\underset{\displaystyle |}{R'}}{C}H-\overset{\displaystyle O}{\overset{\displaystyle \|}{C}}H + CO_2$$

What is happening here? (c) Starting with β-ionone (Problem 17.13), show how you might synthesize the following aldehyde. (This aldehyde is an intermediate in an industrial synthesis of vitamin A.)

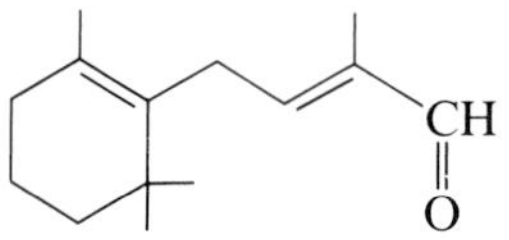

*20.37

The *Perkin condensation* is an aldol-type condensation in which an aromatic aldehyde (ArCHO) reacts with a carboxylic acid anhydride $(RCH_2CO)_2O$, to give an α,β-unsaturated acid $(ArCH{=}CRCO_2H)$. The catalyst that is usually employed is the potassium salt of the carboxylic acid (RCH_2CO_2K). (a) Outline the Perkin condensation that takes place when benzaldehyde reacts with propanoic anhydride in the presence of potassium propanoate. (b) How would you use a Perkin condensation to prepare p-chlorocinnamic acid, $p\text{-}ClC_6H_4CH{=}CHCO_2H$?

*20.38

(+)-Fenchone is a terpenoid that can be isolated from fennel oil. (±)-Fenchone has been synthesized through the following route. Supply the missing intermediates and reagents.

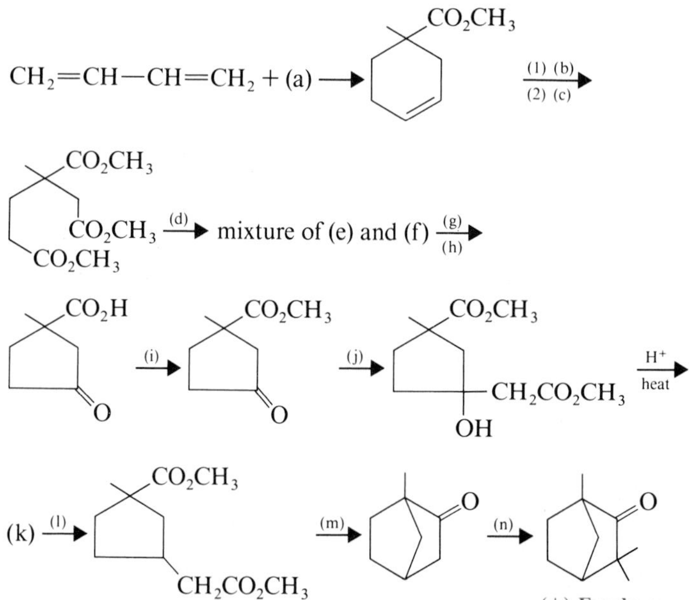

SECOND REVIEW PROBLEM SET

1. Arrange the compounds of each of the following series in order of increasing acidity.

(a) CH_3CH_2OH, $CH_3\overset{O}{\overset{\|}{C}}OH$, $CH_3O\overset{O}{\overset{\|}{C}}CH_2\overset{O}{\overset{\|}{C}}OCH_3$, $CH_3\overset{O}{\overset{\|}{C}}CH_3$

(b) 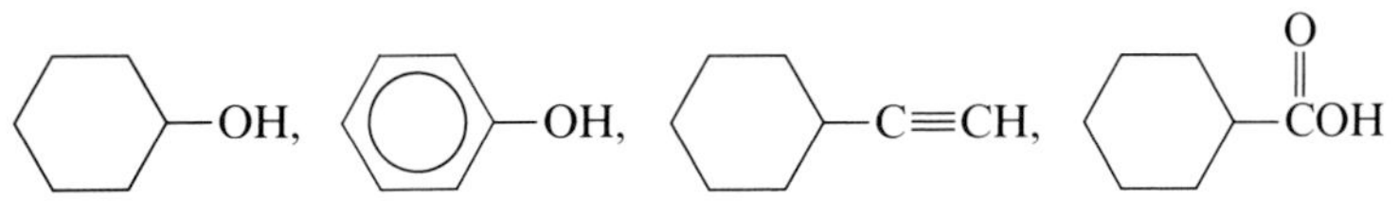 ⬡—OH,   ◯—OH,   ◯—C≡CH,   ◯—$\overset{O}{\overset{\|}{C}}OH$

(c) $(CH_3)_3\overset{+}{N}$—◯—$\overset{O}{\overset{\|}{C}}OH$, $(CH_3)_3C$—◯—$\overset{O}{\overset{\|}{C}}OH$,   ◯—$\overset{O}{\overset{\|}{C}}OH$

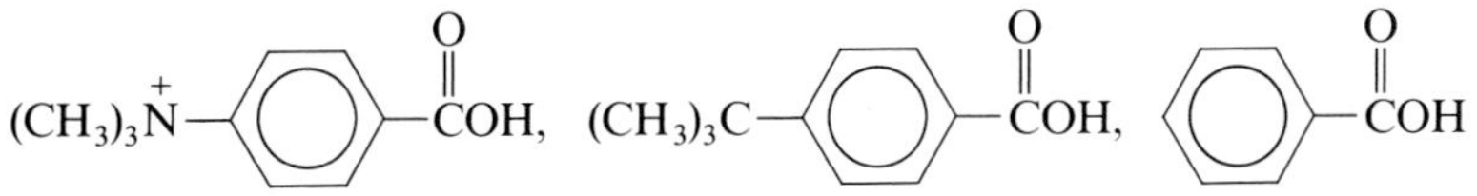

(d) $CH_3CCl_2\overset{O}{\overset{\|}{C}}OH$, $CH_3CH_2\overset{O}{\overset{\|}{C}}OH$, $CH_3CHCl\overset{O}{\overset{\|}{C}}OH$

(e) 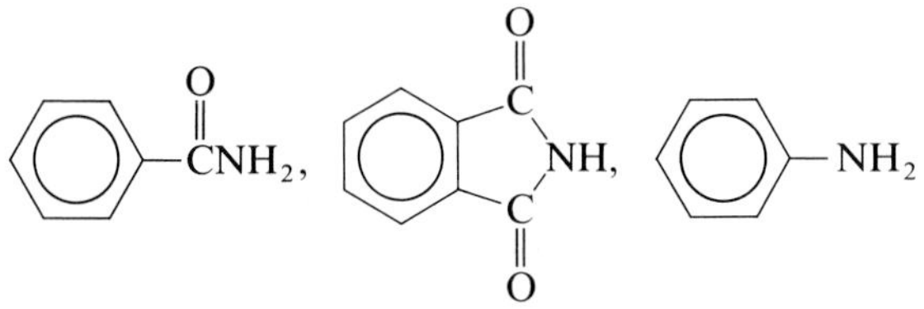 ◯—$\overset{O}{\overset{\|}{C}}NH_2$, ◯$\overset{\overset{O}{\overset{\|}{C}}}{\underset{\underset{O}{\overset{\|}{C}}}{}}NH$, ◯—$NH_2$

2. Arrange the compounds of each of the following series in order of increasing basicity.

(a) $CH_3\overset{O}{\overset{\|}{C}}NH_2$, $CH_3CH_2NH_2$, NH_3

(b) 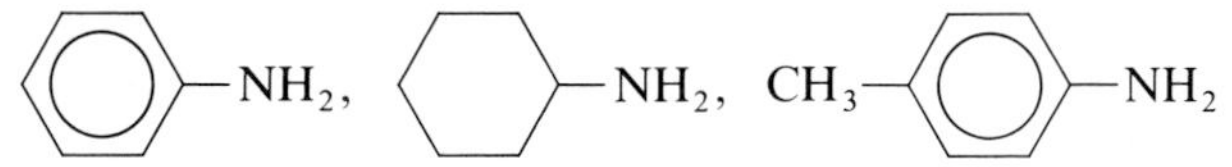 ◯—NH_2,   ⬡—NH_2, CH_3—◯—NH_2

(c) 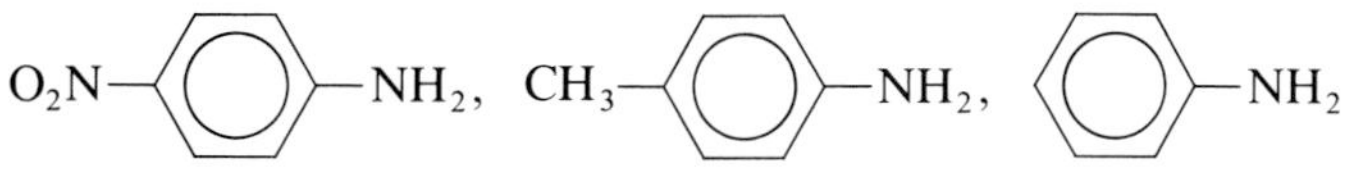 O_2N—◯—NH_2, CH_3—◯—NH_2, ◯—NH_2

(d) $CH_3CH_2CH_3$, CH_3NHCH_3, CH_3OCH_3

3. Starting with 1-butanol and using any other required reagents, outline a synthesis of each of the following compounds. You need not repeat steps carried out in earlier parts of this problem.

(a) Butyl bromide   (e) Pentanoic acid   (i) Propylamine
(b) Butylamine   (f) Butanoyl chloride   (j) Butylbenzene
(c) Pentylamine   (g) Butanamide   (k) Butanoic anhydride
(d) Butanoic acid   (h) Butyl butanoate   (l) Hexanoic acid

4. Starting with benzene, toluene, or aniline and any other required reagents, outline a synthesis of each of the following.

(a) 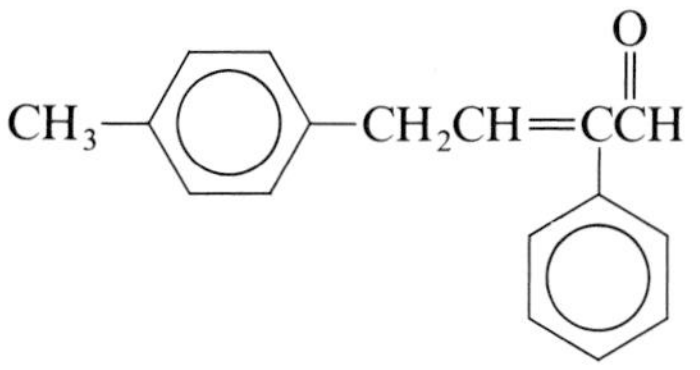 CH_3—◯—$CH_2CH=\overset{O}{\overset{\|}{C}}CH$ (b) 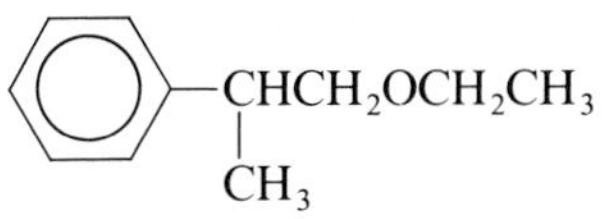 ◯—$\underset{\underset{CH_3}{|}}{C}HCH_2OCH_2CH_3$

(with phenyl ring attached below in (a))

(c)

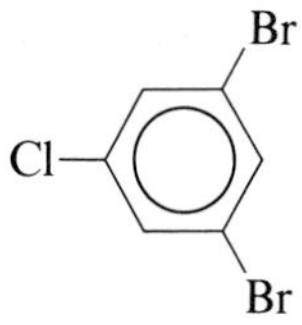

(d) O$_2$N—⟨benzene⟩—CH=CHC(=O)—⟨benzene⟩

(e) C$_6$H$_5$CH=CCH$_3$ with CO$_2$H

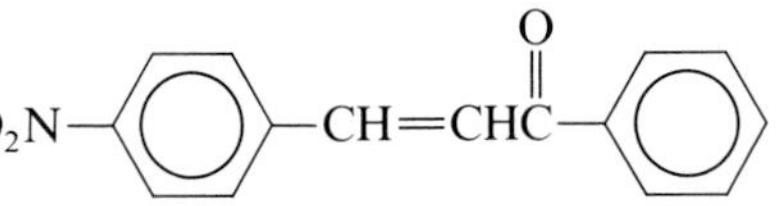

5. Give stereochemical structures for compounds **A–D**.

2-Methyl-1,3-butadiene + diethyl fumarate ⟶ **A** (C$_{13}$H$_{20}$O$_4$) $\xrightarrow{\text{(1) LiAlH}_4\text{, (2) H}_2\text{O}}$

B (C$_9$H$_{16}$O$_2$) $\xrightarrow{\text{PBr}_3}$ **C** (C$_9$H$_{14}$Br$_2$) $\xrightarrow{\text{Zn, H}^+}$ **D** (C$_9$H$_{16}$)

6. A Grignard reagent that is a key intermediate in an industrial synthesis of vitamin A (Section 17.6B) can be prepared in the following way:

HC≡CLi + CH$_2$=CHCCH$_3$ (with =O) $\xrightarrow[\text{(2) NH}_4^+]{\text{(1) liq. NH}_3}$ **A** (C$_6$H$_8$O) $\xrightarrow{\text{H}^+}$

B HOCH$_2$CH=C—C≡CH (with CH$_3$) $\xrightarrow{\text{2C}_2\text{H}_5\text{MgBr}}$ **C** (C$_6$H$_6$Mg$_2$Br$_2$O)

(a) What are the structures of compounds **A** and **C**?
(b) The acid-catalyzed rearrangement of **A** to **B** takes place very readily. What two factors account for this?

7. The remaining steps in the industrial synthesis of vitamin A (as an acetate) are as follows: The Grignard reagent **C** from Problem 6 is allowed to react with the aldehyde shown here.

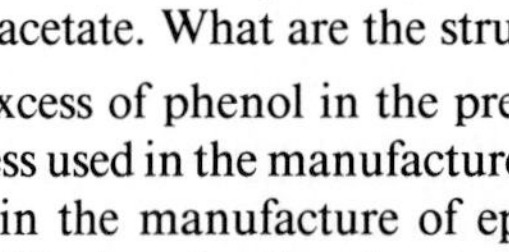

After acidification, the product obtained from this step is a diol **D**. Selective hydrogenation of the triple bond of **D** using Ni$_2$B (P-2) catalyst yields **E** (C$_{20}$H$_{32}$O$_2$). Treating **E** with one molar equivalent of acetic anhydride yields a monoacetate (**F**) and dehydration of **F** yields vitamin A acetate. What are the structures of **D** to **F**?

8. Heating acetone with an excess of phenol in the presence of hydrogen chloride is the basis for an industrial process used in the manufacture of a compound called "bisphenol A." (Bisphenol A is used in the manufacture of epoxy resins and a polymer called "Lexan," cf. Problem J.5.) Bisphenol A has the molecular formula C$_{15}$H$_{16}$O$_2$ and the reactions involved in its formation are similar to those involved in the synthesis of DDT (Problem H.1). Write out these reactions and give the structure of bisphenol A.

9. Outlined here is a synthesis of the local anesthetic *procaine*. Provide structures for procaine and the intermediates **A–C**.

p-Nitrotoluene $\xrightarrow[\text{(2) H}_3\text{O}^+]{\text{(1) KMnO}_4\text{, OH}^-\text{, heat}}$ **A** (C$_7$H$_5$NO$_4$) $\xrightarrow{\text{SOCl}_2}$

B (C$_7$H$_4$ClNO$_3$) $\xrightarrow{\text{HOCH}_2\text{CH}_2\text{N(C}_2\text{H}_5)_2}$ **C** (C$_{13}$H$_{18}$N$_2$O$_4$) $\xrightarrow{\text{H}_2\text{, cat.}}$ procaine (C$_{13}$H$_{20}$N$_2$O$_2$)

10. The sedative-hypnotic *ethinamate* can be synthesized by the following route. Provide structures for ethinamate and the intermediates **A** and **B**.

Cyclohexanone $\xrightarrow{\text{(1) HC}\equiv\text{CNa. (2) H}^+}$ **A** ($C_8H_{12}O$) $\xrightarrow{\text{ClCOCl}}$

B ($C_9H_{11}ClO_2$) $\xrightarrow{\text{NH}_3}$ ethinamate ($C_9H_{13}NO_2$)

11. The prototype of the antihistamines, *diphenhydramine* (also called Benadryl), can be synthesized by the following sequence of reactions. (a) Give structures for diphenhydramine and for the intermediates **A** and **B**. (b) Comment on a possible mechanism for the last step of the synthesis.

Benzaldehyde $\xrightarrow{\text{(1) C}_6\text{H}_5\text{MgBr, (2) H}_3\text{O}^+}$ **A** ($C_{13}H_{12}O$) $\xrightarrow{\text{PBr}_3}$

B ($C_{13}H_{11}Br$) $\xrightarrow{\text{(CH}_3)_2\text{NCH}_2\text{CH}_2\text{OH}}$ diphenhydramine ($C_{17}H_{21}NO$)

12. Show how you would modify the synthesis given in the previous problem to synthesize the following drugs.

(a) Br—⟨O⟩—CHOCH$_2$CH$_2$N(CH$_3$)$_2$ *Bromodiphenhydramine* (an antihistamine)
             |
             C$_6$H$_5$

(b)   CH$_3$
  ⟨O⟩—CHOCH$_2$CH$_2$N(CH$_3$)$_2$ *Orphenadrine* (an antispasmodic, used in controlling Parkinson's disease)
             |
             C$_6$H$_5$

13. Outlined here is a synthesis of 3-oxo-2-methylcyclopentanecarboxylic acid. Give the structure of each intermediate.

CH$_3$CHCO$_2$C$_2$H$_5$ $\xrightarrow{\text{CH}_2(\text{CO}_2\text{C}_2\text{H}_5)_2,\ \text{EtO}^-}$ **A** ($C_{12}H_{20}O_6$) $\xrightarrow{\text{CH}_2=\text{CHCN, EtO}^-}$
  |
  Br

B ($C_{15}H_{23}NO_6$) $\xrightarrow{\text{EtOH, H}^+}$ **C** ($C_{17}H_{28}O_8$) $\xrightarrow{\text{EtO}^-}$ **D** ($C_{15}H_{22}O_7$) $\xrightarrow[\text{(2) H}_3\text{O}^+,\ \text{(3) heat}]{\text{(1) OH}^-,\ \text{H}_2\text{O, heat}}$

14. Give structures for compounds **A–D**. Compound **D** decolorizes bromine in carbon tetrachloride and gives a strong infrared absorption band near 1720 cm^{-1}.

CH$_3$CCH$_3$ $\xrightarrow{\text{HCl}}$ **A** ($C_6H_{10}O$) $\xrightarrow[\text{base}]{\text{CH}_3\text{CCH}_2\text{COC}_2\text{H}_5}$ [**B** ($C_{12}H_{20}O_4$)] $\xrightarrow{\text{base}}$

C ($C_{12}H_{18}O_3$) $\xrightarrow{\text{H}^+,\ \text{H}_2\text{O, heat}}$ **D** ($C_9H_{14}O$)

15. A synthesis of the broad-spectrum antibiotic *chloramphenicol* is shown here. In the last step basic hydrolysis selectively hydrolyzes ester linkages in the presence of an amide group. What are the intermediates **A–E**?

$$\text{Benzaldehyde} + HOCH_2CH_2NO_2 \xrightarrow{\text{EtO}^-} \textbf{A} (C_9H_{11}NO_4) \xrightarrow{\text{H}_2, \text{cat.}}$$

$$\textbf{B} (C_9H_{13}NO_2) \xrightarrow{\text{Cl}_2\text{CHCOCl}} \textbf{C} (C_{11}H_{13}Cl_2NO_3) \xrightarrow{\text{excess (CH}_3\text{CO)}_2\text{O}}$$

$$\textbf{D} (C_{15}H_{17}Cl_2NO_5) \xrightarrow{\text{HNO}_3, \text{H}_2\text{SO}_4} \textbf{E} (C_{15}H_{16}Cl_2N_2O_7) \xrightarrow{\text{OH}^-, \text{H}_2\text{O}}$$

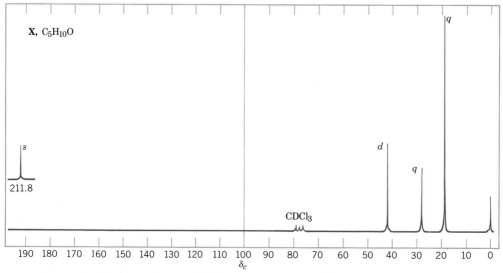

Chloramphenicol

16. The tranquilizing drug *meprobamate* (Equanil or Miltown) can be synthesized from 2-methylpentanal as follows. Give structures for meprobamate and for the intermediates **A–C**.

$$CH_3CH_2CH_2\underset{\underset{CH_3}{|}}{CH}CH \overset{O}{\overset{\|}{}} \xrightarrow{\text{HCHO, OH}^-} [\textbf{A} (C_7H_{14}O_2)] \xrightarrow{\text{HCHO}} \textbf{B} (C_7H_{16}O_2) \xrightarrow{\text{ClCOCl}}$$

$$\textbf{C} (C_9H_{14}Cl_2O_4) \xrightarrow{\text{NH}_3} \text{meprobamate} (C_9H_{18}N_2O_4)$$

17. What are compounds **A–C**? Compound **C** is useful as an insect repellent.

$$\text{Succinic anhydride} \xrightarrow{\text{CH}_3\text{CH}_2\text{CH}_2\text{OH}} \textbf{A} (C_7H_{12}O_4) \xrightarrow{\text{SOCl}_2}$$

$$\textbf{B} (C_7H_{11}ClO_3) \xrightarrow{(\text{CH}_3\text{CH}_2)_2\text{NH}} \textbf{C} (C_{11}H_{21}NO_3)$$

X, C₅H₁₀O

s

211.8

CDCl₃

d

q

q

190 180 170 160 150 140 130 120 110 100 90 80 70 60 50 40 30 20 10 0
δ_C

FIGURE 1 The proton-decoupled ¹³C nmr spectra of compound **X** (Problem 21). The letters *s, d, t*, and *q*, stand for the multiplicity of the peaks (singlet, doublet, triplet, and quartet) in the proton off-resonance decoupled spectrum. The signal marked *x* arises from an impurity and should be ignored. Spectra adapted from L. F. Johnson and W. C. Jankowski, *Carbon-13 NMR Spectra; A Collection of Assigned, Coded, and Indexed Spectra*, Wiley–Interscience, New York, 1972.

18. Outlined here is the synthesis of a central nervous system stimulant called *fencamfamine*. Provide structural formulas for each intermediate and for fencamfamine itself.

1,3-Cyclopentadiene + (*E*)-$C_6H_5CH{=}CHNO_2$ $\longrightarrow$ **A** ($C_{13}H_{13}NO_2$) $\xrightarrow{H_2, Pt}$

B ($C_{13}H_{17}N$) $\xrightarrow{CH_3CHO}$ [**C** ($C_{15}H_{19}N$)] $\xrightarrow{H_2, Ni}$ fencamfamine ($C_{15}H_{21}N$)

19. What are compounds **A** and **B**? Compound **B** has a strong infrared absorption band in the 1650–1730-cm^{-1} region and a broad strong band in the 3200–3550-cm^{-1} region.

1-Methylcyclohexene $\xrightarrow[\text{(2) NaHSO}_3]{\text{(1) OsO}_4}$ **A** ($C_7H_{14}O_2$) $\xrightarrow[\text{CH}_3\text{CO}_2\text{H}]{\text{CrO}_3}$ **B** ($C_7H_{12}O_2$)

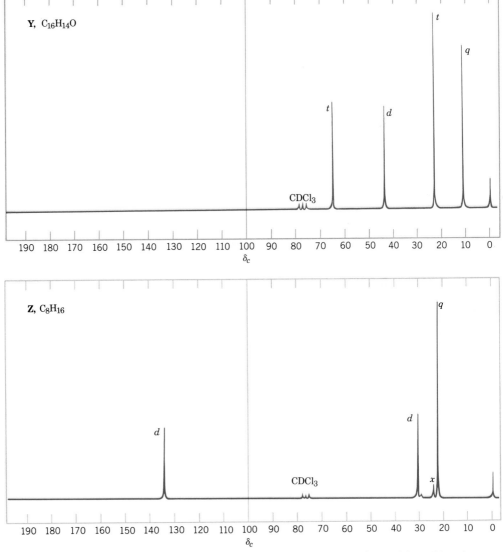

FIGURE 2 The proton-decoupled ^{13}C nmr spectra of compounds **Y** and **Z** (Problems 22 and 23). The letters *s*, *d*, *t*, and *q*, stand for the multiplicity of the peaks (singlet, doublet, triplet, and quartet) in the proton off-resonance decoupled spectrum. The signal marked *x* arises from an impurity and should be ignored. Spectra adapted from L. F. Johnson and W. C. Jankowski, *Carbon-13 NMR Spectra: A Collection of Assigned, Coded, and Indexed Spectra*, Wiley–Interscience, New York, 1972.

20. Starting with phenol, outline a stereoselective synthesis of methyl *trans*-4-isopropyl-cyclohexanecarboxylate, that is,

21. Compound **X** ($C_5H_{10}O$) gives a positive iodoform test and shows a strong infrared absorption band near 1710 cm^{-1}. The proton-decoupled ^{13}C nmr spectrum of **X** is shown in Fig. 1. Propose a structure for **X**.

22. Compound **Y** ($C_6H_{14}O$) gives a green opaque solution when treated with CrO_3 in aqueous H_2SO_4. Compound **Y** gives a negative iodoform test. The proton-decoupled ^{13}C nmr spectrum of **Y** is given in Fig. 2. Propose a structure for **Y**.

23. Compound **Z** (C_8H_{16}) decolorizes bromine in carbon tetrachloride and is the more stable of a pair of stereoisomers. Ozonolysis of **Z** gives a single product. The proton-decoupled ^{13}C nmr spectrum of **Z** is given in Fig. 2. Propose a structure for **Z**.

ALKALOIDS

Extracting the bark, roots, leaves, berries, and fruits of plants often yields nitrogen-containing bases called *alkaloids*. The name alkaloid comes from the fact that these substances are "alkalilike"; that is, since alkaloids are amines they often react with acids to yield soluble salts. The nitrogen atoms of most alkaloids are present in heterocyclic rings. In a few instances, however, nitrogen may be present as a primary amine or as a quaternary ammonium group.

When administered to animals most alkaloids produce striking physiological effects, which effects *vary greatly* from alkaloid to alkaloid. Some alkaloids stimulate the central nervous system, others cause paralysis; some alkaloids elevate blood pressure, others lower it. Certain alkaloids act as pain relievers; others act as tranquilizers; still others act against infectious microorganisms. Most alkaloids are toxic when their dosage is large enough, and with some this dosage is very small. In spite of this, many alkaloids find use in medicine.

Systematic names are seldom used for alkaloids, and their common names have a variety of origins. In many instances the common name reflects the botanical source of the compound. The alkaloid strychnine, for example, comes from the seeds of the *Strychnos* plants. In other instances the names are more whimsical: the name of the opium alkaloid morphine comes from Morpheus, the ancient Greek god of dreams; the name of the tobacco alkaloid nicotine comes from Nicot, an early French ambassador who sent tobacco seeds to France. The one characteristic that alkaloid names have in common is the ending *-ine*, reflecting the fact that they are all amines.

Alkaloids have been of interest to chemists for nearly two centuries, and in that time thousands of alkaloids have been isolated. Most of these have had their structures determined through the application of chemical and physical methods, and in many instances these structures have been confirmed by independent synthesis. A complete account of the chemistry of the alkaloids would (and does) occupy volumes; here we have space to consider only a few representative examples.

L.1 Alkaloids Containing a Pyridine or Reduced Pyridine Ring

The predominant alkaloid of the tobacco plant is nicotine:

Nicotine   Nicotinic acid

In very small doses nicotine acts as a stimulant, but in larger doses it causes depression, nausea, and vomiting. In still larger doses it is a violent poison. Nicotine salts are used as insecticides.

Oxidation of nicotine by concentrated nitric acid produces pyridine-3-carboxylic acid—a compound that is called *nicotinic acid.* While the consumption of nicotine is of no benefit to humans, nicotinic acid is a vitamin; it is incorporated into the important coenzyme, nicotinamide adenine dinucleotide.

PROBLEM L.1

Nicotine has been synthesized by the following route. All of the steps involve reactions that we have seen before. Suggest reagents that could be used for each.

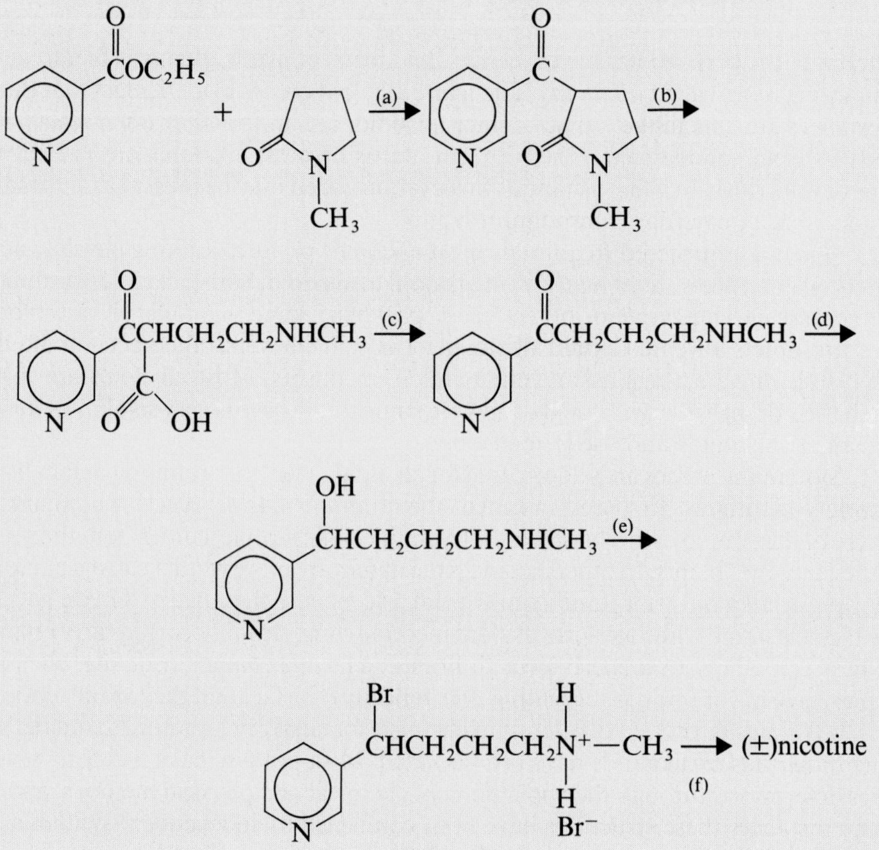

A number of alkaloids contain a piperidine ring. These include coniine (from the poison hemlock, *Conium maculatum,* a member of the carrot family, Umbelliferae), atropine (from *Atropa belladonna* and other genera of the plant family Solanaceae), and cocaine (from *Erythroxylon coca*).

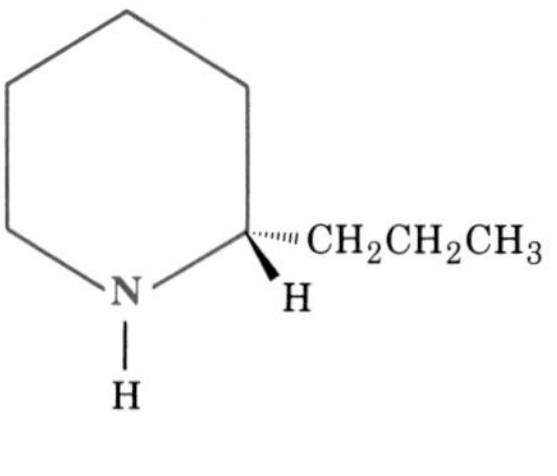

Coniine
[(+)-2-propylpiperidine]

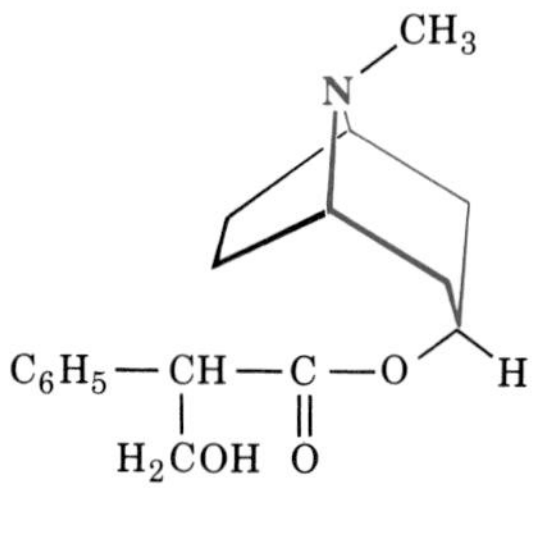

Atropine

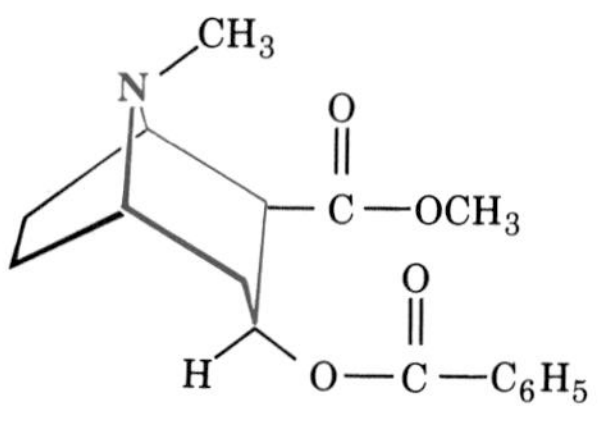

Cocaine

Coniine is toxic; its ingestion may cause weakness, drowsiness, nausea, labored respiration, paralysis, and death. Coniine was one toxic substance of the "hemlock" used in the execution of Socrates (other poisons may have been included as well).

In small doses cocaine decreases fatigue, increases mental activity, and gives a general feeling of well being. Prolonged use of cocaine, however, leads to physical addiction and to periods of deep depression. Cocaine is also a local anesthetic and, for a time, it was used medically in that capacity. When its tendency to cause addiction was recognized, efforts were made to develop other local anesthetics. This led, in 1905, to the synthesis of novocaine, a compound that has some of the same structural features as cocaine (i.e., its benzoic ester and tertiary amine groups).

$$CH_3CH_2 \diagdown \atop CH_3CH_2 \diagup N - CH_2CH_2 - O - \overset{\overset{O}{\|}}{C} - \langle\bigcirc\rangle - NH_2$$

Novocaine

Atropine is an intense poison. In dilute solutions (0.5 – 1.0%) it is used to dilate the pupil of the eye in ophthalmic examinations. Compounds related to atropine are contained in the 12-h continuous-release capsules used to relieve symptoms of the common cold.

PROBLEM L.2

The principal alkaloid of *Atropa belladonna* is the optically active alkaloid *hyoscyamine*. During its isolation hyoscyamine is often racemized by bases to optically inactive atropine. (a) What stereocenter is likely to be involved in the racemization? (b) In hyoscyamine this stereocenter has the (*S*) configuration. Write a three-dimensional structure for hyoscyamine.

PROBLEM L.3

Hydrolysis of atropine gives tropine and (±)-tropic acid. (a) What are their structures? (b) Even though tropine has a stereocenter, it is optically inactive. Explain. (c) An isomeric form of tropine called ψ-tropine has also been prepared by heating tropine with base. ψ-Tropine is also optically inactive. What is its structure?

PROBLEM L.4

In 1891 G. Merling transformed tropine (cf. Problem L.3) into 1,3,5-cycloheptatriene (tropylidene) through the following sequence of reactions.

$$\text{Tropine } (C_8H_{15}NO) \xrightarrow{-H_2O} C_8H_{13}N \xrightarrow{CH_3I} C_9H_{16}NI \xrightarrow[\text{(2) heat}]{\text{(1) Ag}_2\text{O/H}_2\text{O}}$$

$$C_9H_{15}N \xrightarrow{CH_3I} C_{10}H_{18}NI \xrightarrow[\text{(2) heat}]{\text{(1) Ag}_2\text{O/H}_2\text{O}}$$

$$\text{1,3,5-cycloheptatriene} + (CH_3)_3N + H_2O$$

Write out all of the reactions that take place.

PROBLEM L.5

Many alkaloids appear to be synthesized in plants by reactions that resemble the Mannich reaction (Section 20.9). Recognition of this (by R. Robinson in 1917) led to a synthesis of tropinone that takes place under "physiological conditions," that is, at room temperature and at pH values near neutrality. This synthesis is shown here. Propose reasonable mechanisms that account for the overall course of the reaction.

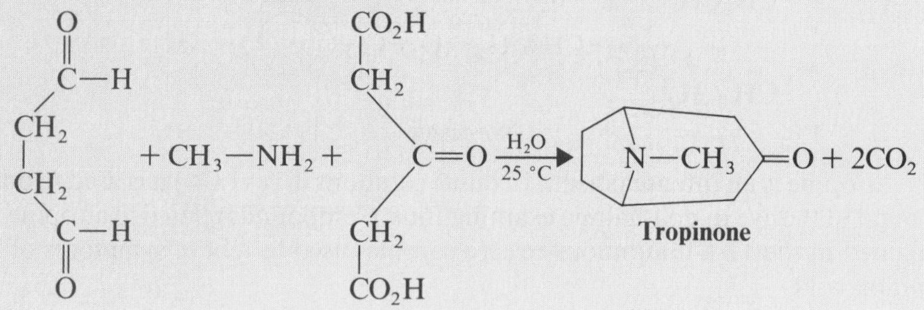

L.2 Alkaloids Containing an Isoquinoline
or Reduced Isoquinoline Ring

Papaverine, morphine, and codeine are all alkaloids obtained from the opium poppy, *Papaver somniferum.*

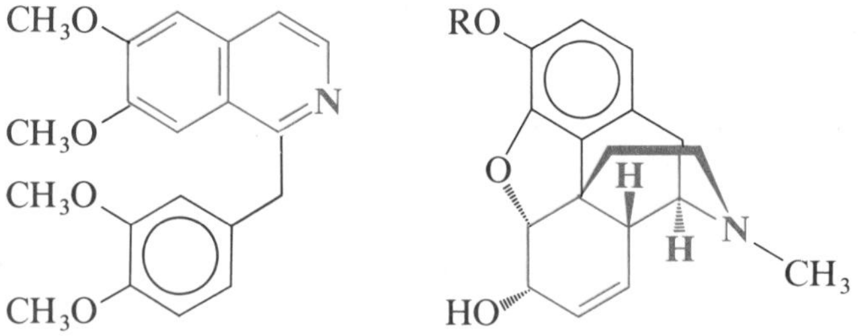

Papaverine

Morphine (R = H)
Codeine (R = CH₃)

Papaverine has an isoquinoline ring; in morphine and codeine the isoquinoline ring is partially hydrogenated (reduced).

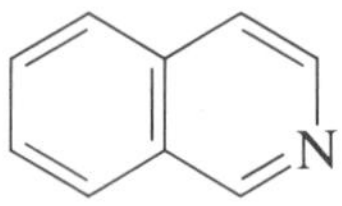

Isoquinoline

Opium has been used since earliest recorded history. Morphine was first isolated from opium in 1803, and its isolation represented one of the first instances of the purification of the active principle of a drug. One hundred and twenty years were to pass, however, before the complicated structure of morphine was deduced, and its final confirmation through independent synthesis (by Marshall Gates of the University of Rochester) did not take place until 1952.

Morphine is one of the most potent analgesics known, and it is still used extensively in medicine to relieve pain, especially "deep" pain. Its greatest drawbacks, however, are its tendencies to lead to addiction and to depress respiration. These disadvantages have brought about a search for morphinelike compounds that do not have these disadvantages. One of the newest candidates is the compound pentazocine. Pentazocine is a highly effective analgesic and is nonaddictive; unfortunately however, like morphine, it depresses respiration.

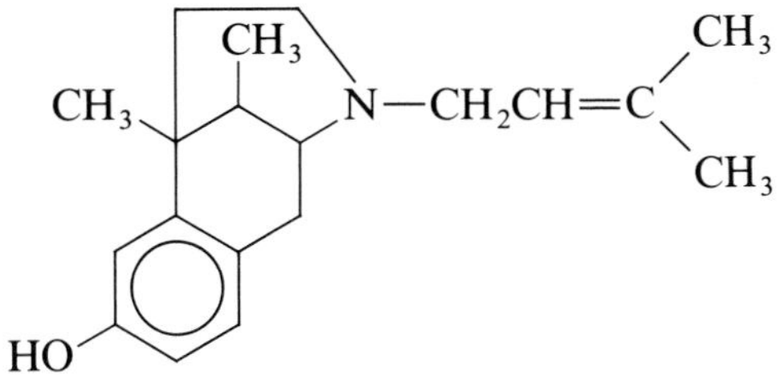

Pentazocine

PROBLEM L.6

Papaverine has been synthesized by the following route:

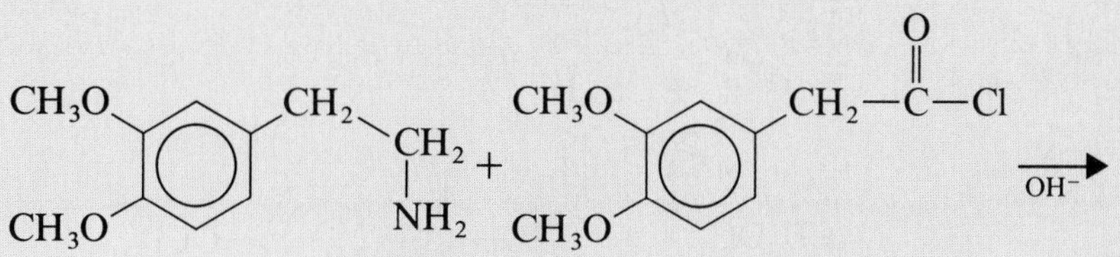

$$C_{20}H_{25}NO_5 \xrightarrow[\substack{heat \\ (-H_2O)}]{P_4O_{10}} \text{dihydropapaverine} \xrightarrow[\substack{heat \\ (-H_2)}]{Pd} \text{papaverine}$$

Outline the reactions involved.

PROBLEM L.7

One of the important steps in the synthesis of morphine involved the following transformation:

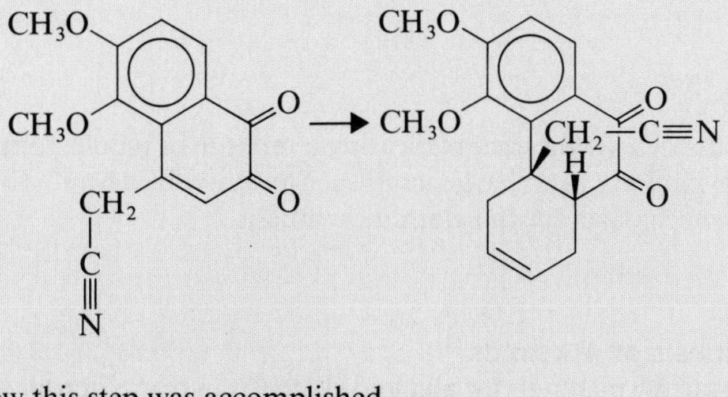

Suggest how this step was accomplished.

PROBLEM L.8

When morphine reacts with two moles of acetic anhydride, it is transformed into the highly addictive narcotic, heroin. What is the structure of heroin?

L.3 Alkaloids Containing Indole or Reduced Indole Rings

A large number of alkaloids are derivatives of an indole ring system. These range from the relatively simple *gramine* to the highly complicated structures of *strychnine* and *reserpine*.

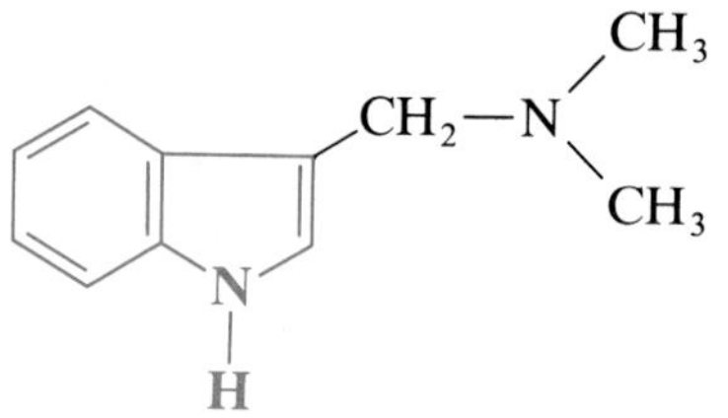

Gramine

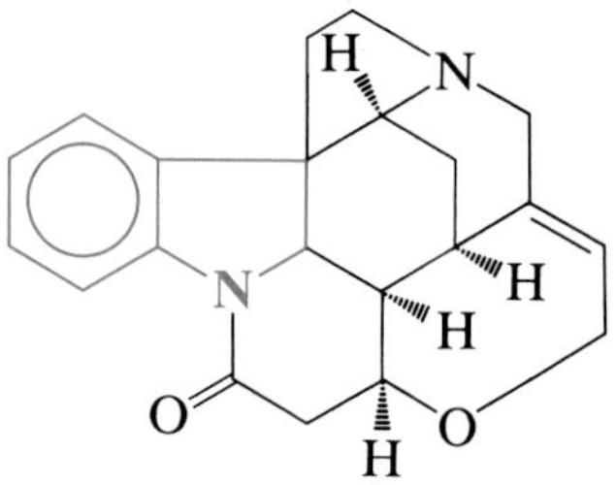

Strychnine

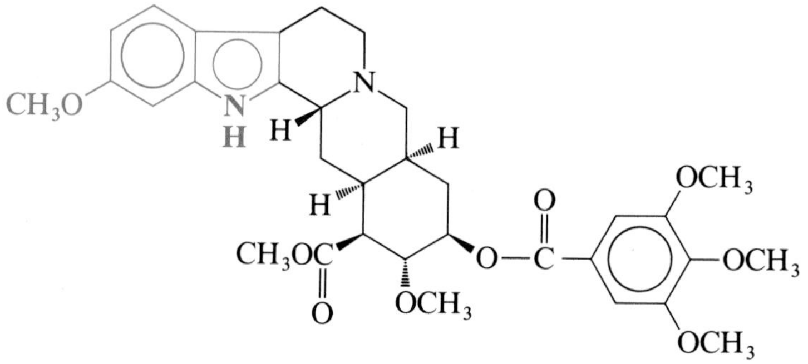

Reserpine

Gramine can be obtained from chlorophyll-deficient mutants of barley. Strychnine, a very bitter and highly poisonous compound, comes from the seeds of *Strychnos nux-vomica.* Strychnine is a central nervous system stimulant and has been used medically (in low dosage) to counteract poisoning by central nervous system depressants. Reserpine can be obtained from the Indian snakeroot *Rauwolfia serpentina,* a plant that has been used in native medicine for centuries. Reserpine is used in modern medicine as a tranquilizer and as an agent to lower blood pressure.

PROBLEM L.9

Gramine has been synthesized by heating a mixture of indole, formaldehyde, and dimethylamine. (a) What general reaction is involved here? (b) Outline a reasonable mechanism for the gramine synthesis.

L.4 Biosynthesis of Alkaloids

The primary starting materials for alkaloid synthesis in plants appear to be α-amino acids. More than 20 different α-amino acids occur naturally; they are the main building blocks for proteins (Chapter 23). Two amino acids, in particular, are important in alkaloid biosynthesis. These are tyrosine and tryptophan:

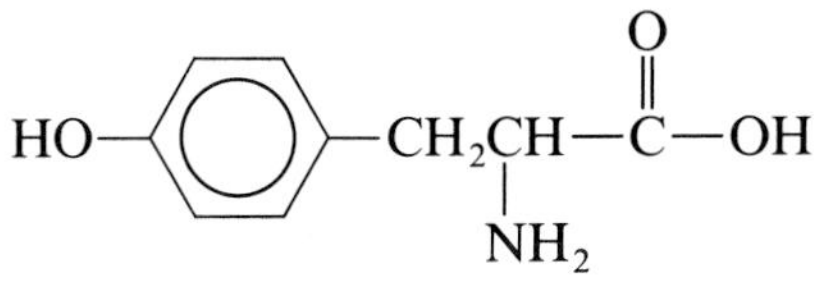

Tyrosine 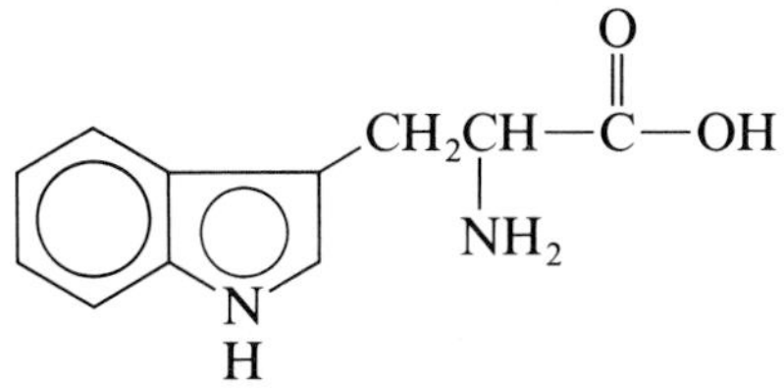   Tryptophan

Two general reactions appear to be of central importance in alkaloid biosynthesis—the Mannich reaction and the oxidative coupling of phenols. We studied the Mannich reaction in Section 20.9 (cf. also Problem L.5). The oxidative coupling of phenols is a free radical process that is catalyzed by enzymes in plants and that can also be carried out (usually less successfully) in the laboratory. A simple formulation of an oxidative phenol coupling is outlined using phenol itself as the starting compound. Loss of an electron and a proton from phenol leads to a resonance-stabilized free radical. Two free radicals can then undergo coupling in a variety of ways:

Oxidation

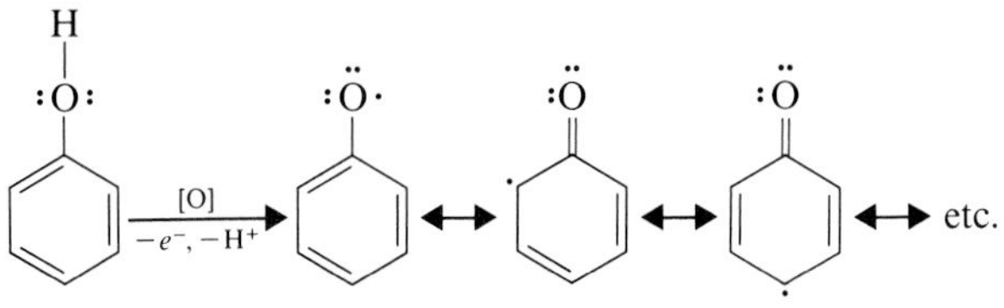

ortho–ortho Coupling

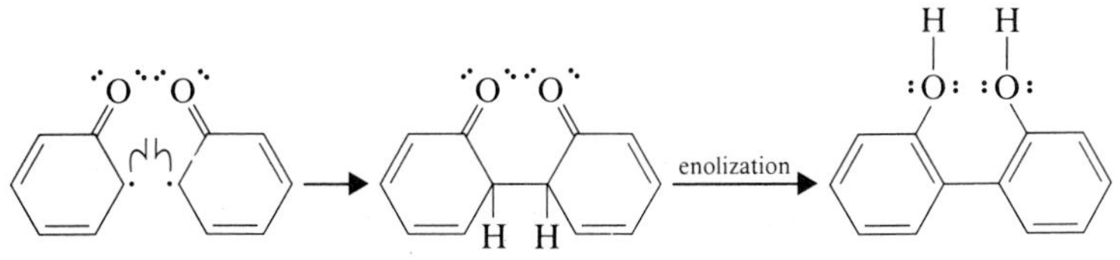

ortho–para Coupling

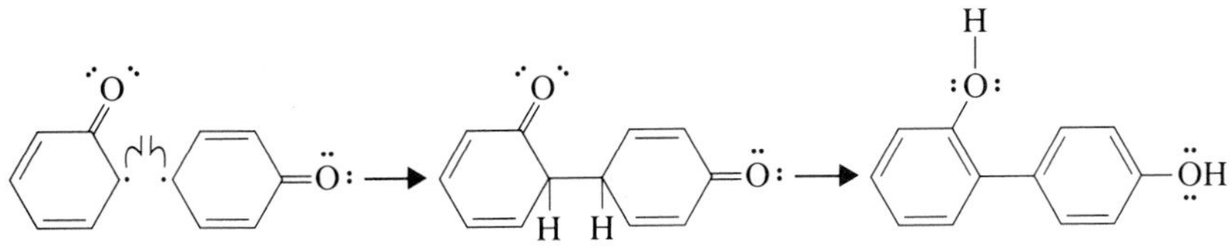

para–para Coupling

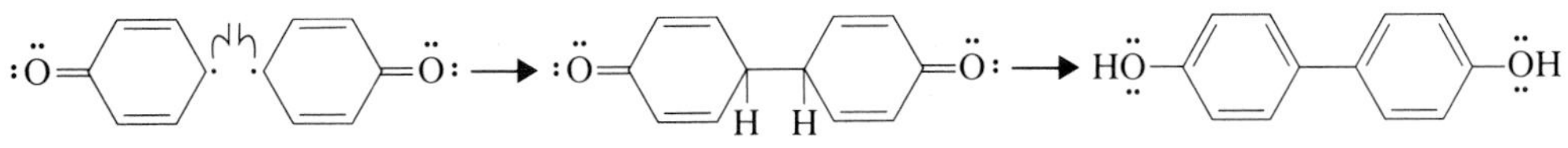

In most cases, oxidative coupling occurs intramolecularly.

The biosynthetic route that the opium poppy uses to synthesize morphine is now known. Most of the morphine molecule, it turns out, is constructed from two molecules of tyrosine.

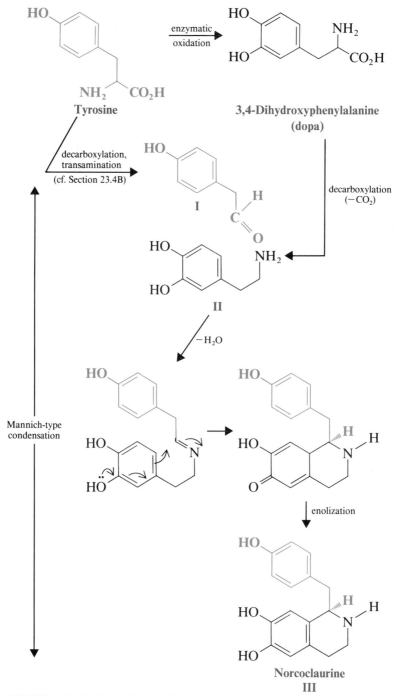

FIGURE L.1 The biosynthesis of norcoclaurine **(III)** from 2 moles of 3,4-dihydroxyphenylalanine.

The synthesis begins with the oxidation of tyrosine to 3,4-dihydroxyphenylalanine (dopa):

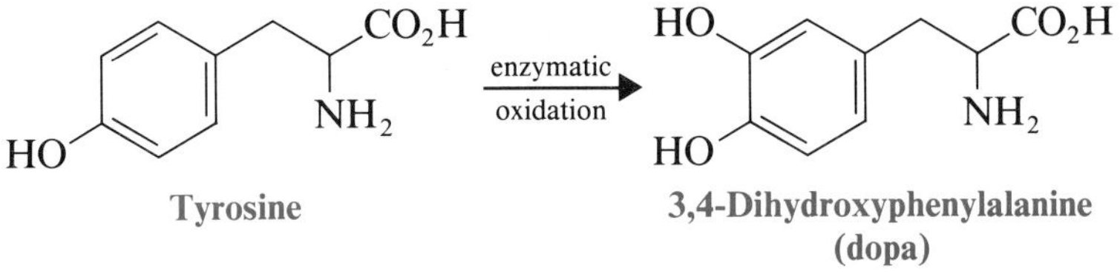

Further enzyme-catalyzed reactions transform 3,4-dihydroxyphenylalanine into 3,4-dihydroxyphenylethylamine [dopamine (**II**); see Figure L.1]. The other part of the synthesis involves the conversion of tyrosine into *p*-hydroxyphenylacetaldehyde (**I**). These two molecules then react in a Mannich-type condensation to yield norcoclaurine (**III**).

Hydroxylation of the benzene ring derived from *p*-hydroxyphenylacetaldehyde and methylation of norcoclaurine at two of its —OH groups and at its —N—H group yield reticuline (**IVa**) (Fig. L.2). A reticuline molecule can be twisted into

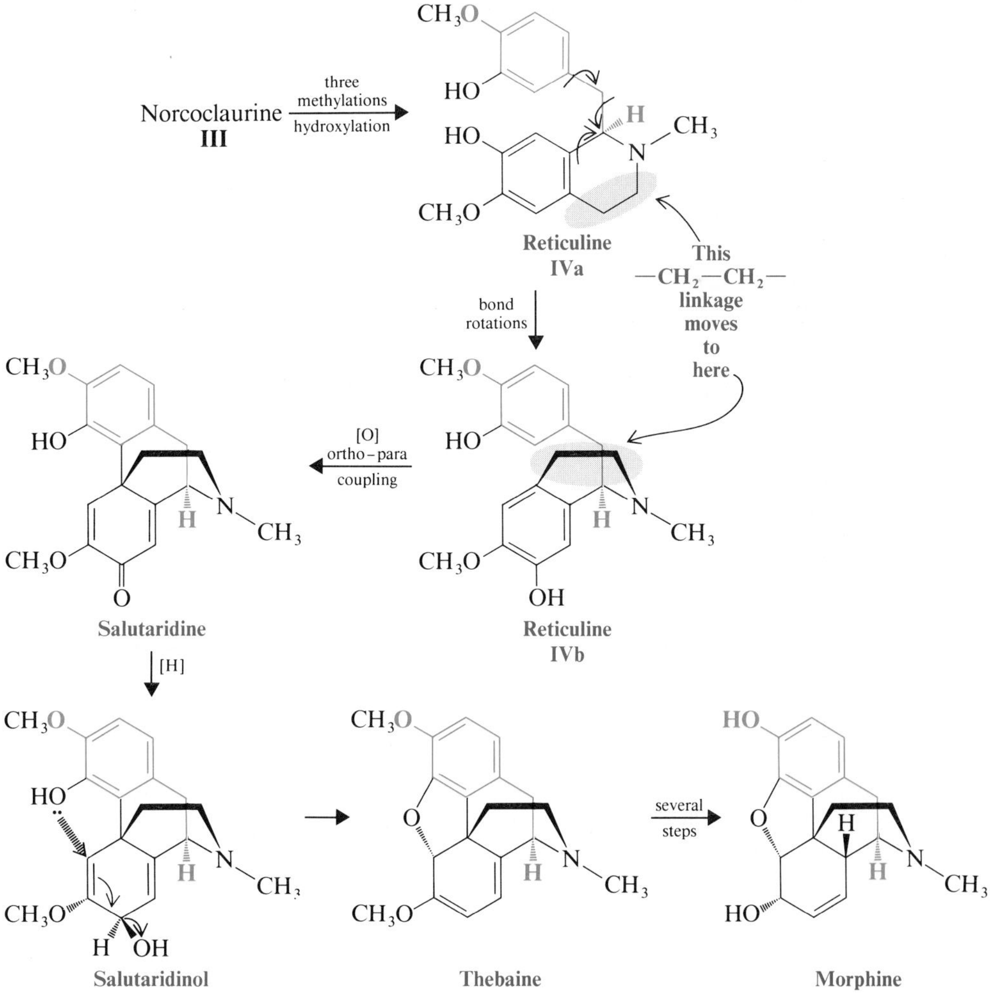

FIGURE L.2 The biosynthesis of morphine from norcoclaurine (**III**).

conformation **(IVb),** one that allows an ortho–para oxidative phenolic coupling to take place yielding salutaridine. Reduction of salutaridine produces salutaridinol. Then salutaridinol is transformed into thebaine. (In this highly unusual step the oxygen bridge is installed through a reaction that is accompanied by *the displacement of a hydroxide ion.*) Finally, several additional enzymatic reactions transform thebaine into morphine.

PROBLEM L.10

There is considerable evidence that oxidative phenol couplings are important in the biosynthesis of bulbocapnine and glaucine (see following structures). (These two alkaloids have what is called an *apomorphine* ring system.) Both compounds appear to arise from reticuline. Show the type of oxidative phenol coupling that is involved in each biosynthesis. (Assume that methylation of —OH groups in both alkaloids and synthesis of the —O—CH$_2$—O— bridge in bulbocapnine occur after the oxidative phenol couplings.)

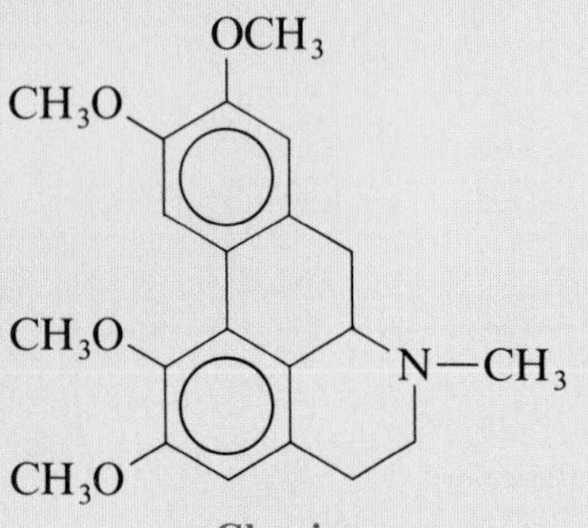

Glaucine 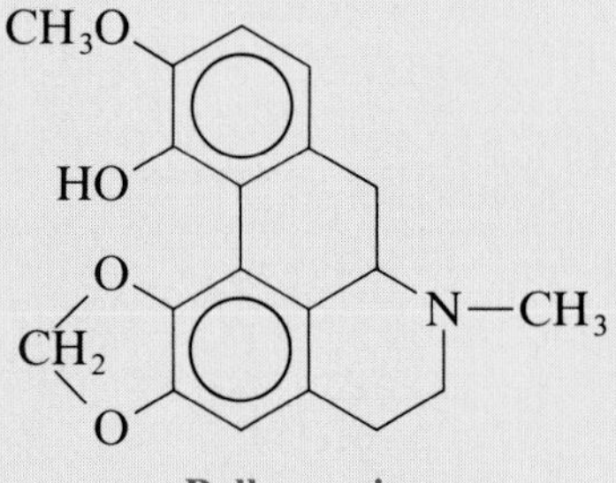   Bulbocapnine

---

PROBLEM L.11

Harmine is an alkaloid isolated from *Peganum harmala* L. When tryptophan and pyruvic acid labeled in the positions shown were fed to the plant, the harmine produced had the labeling pattern indicated.

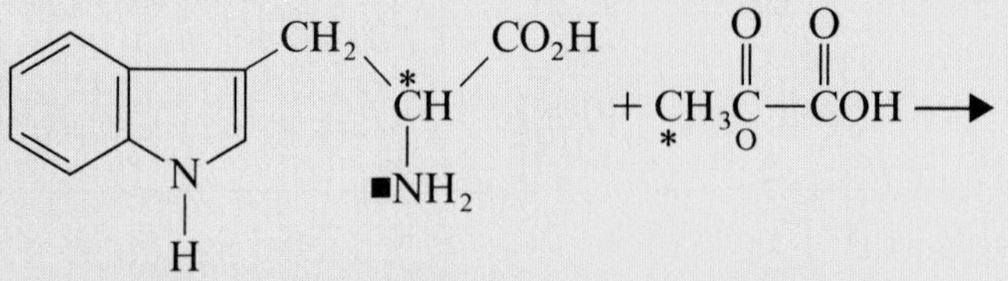

Tryptophan
* and ○ are ^{14}C labels
■ is an ^{15}N label

Pyruvic acid

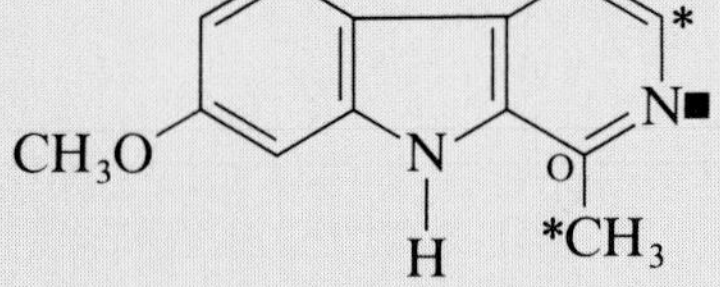

Harmine

Show how these results are consistent with the following pathway: (1) decarbox-ylation of tryptophan (to tryptamine), (2) a Mannich-type condensation of tryptamine and pyruvic acid, then (3) dehydrogenation, (4) hydroxylation, and (5) methylation.

CHAPTER TWENTY ONE

Sucrose and pyrogallic acid crystals. Sucrose is ordinary table sugar. We shall describe its chemistry in detail in this chapter. Pyrogallic acid is 1,2,3-trihydroxybenzene.

Carbohydrates

21.1 INTRODUCTION

21.1A Classification of Carbohydrates

The group of compounds known as carbohydrates received their general name because of early observations that they often have the formula $C_x(H_2O)_y$—that is, they appear to be "hydrates of carbon." Simple carbohydrates are also known as sugars or saccharides (Latin *saccharum,* sugar) and the ending of the names of most sugars is *-ose.* Thus we have such names as *sucrose* for ordinary table sugar, *glucose* for the principal sugar in blood, and *maltose* for malt sugar.

Carbohydrates are usually defined as *polyhydroxy aldehydes and ketones or substances that hydrolyze to yield polyhydroxy aldehydes and ketones.* Although this definition draws attention to the important functional groups of carbohydrates, it is not entirely satisfactory. We shall later find that because carbohydrates contain

$\diagdown$ C=O groups and —OH groups, they exist, primarily, as *hemiacetals* and *acetals* $\diagup$

or as *hemiketals* and *ketals* (Section 16.7).

The simplest carbohydrates, those that cannot be hydrolyzed into smaller simpler carbohydrates, are called *monosaccharides.* On a molecular basis, carbohydrates that undergo hydrolysis to produce only two molecules of a monosaccharide are called *disaccharides;* those that yield three molecules of a monosaccharide are called *trisaccharides;* and so on. (Carbohydrates that hydrolyze to yield 2 to 10 molecules of a monosaccharide are sometimes called *oligosaccharides.*) Carbohydrates that yield a large number of molecules of a monosaccharide (> 10) are known as *polysaccharides.*

Maltose and sucrose are examples of disaccharides. On hydrolysis, a mole of maltose yields 2 moles of the monosaccharide glucose; sucrose undergoes hydrolysis to yield 1 mole of glucose and 1 mole of the monosaccharide fructose. Starch and cellulose are examples of polysaccharides; both are glucose polymers. Hydrolysis of either yields a large number of glucose units.

$$1 \text{ mole of maltose} \xrightarrow[\text{H}^+]{\text{H}_2\text{O}} 2 \text{ moles of glucose}$$

(a disaccharide)   **(a monosaccharide)**

$$\text{1 mole of sucrose} \xrightarrow[\text{H}^+]{\text{H}_2\text{O}} \text{1 mole of glucose} + \text{1 mole of fructose}$$

(a disaccharide)　　　　　　　　　　　　(monosaccharides)

$$\begin{matrix}\text{1 mole of starch}\\ \text{or}\\ \text{1 mole of cellulose}\end{matrix} \xrightarrow[\text{H}^+]{\text{H}_2\text{O}} \text{many moles of glucose}$$

(polysaccharides)　　　　　　　　　(monosaccharide)

Carbohydrates are the most abundant organic constituents of plants. They not only serve as an important source of chemical energy for living organisms (sugars and starches are important in this respect), but also in plants and in some animals they serve as important constituents of supporting tissues (this is the primary function of the cellulose found in wood, cotton, and flax, for example).

We encounter carbohydrates at almost every turn of our daily lives. The paper on which this book is printed is largely cellulose; so too is the cotton of our clothes and the wood of our houses. The flour from which we make bread is mainly starch, and starch is also a major constituent of many other foodstuffs, such as potatoes, rice, beans, corn, and peas.

21.1B Photosynthesis and Carbohydrate Metabolism

Carbohydrates are synthesized in green plants by *photosynthesis*—a process that uses solar energy to reduce, or "fix," carbon dioxide. The overall equation for photosynthesis can be written as follows:

$$x\text{CO}_2 + y\text{H}_2\text{O} + \text{solar energy} \longrightarrow \text{C}_x(\text{H}_2\text{O})_y + x\text{O}_2$$

Carbohydrate

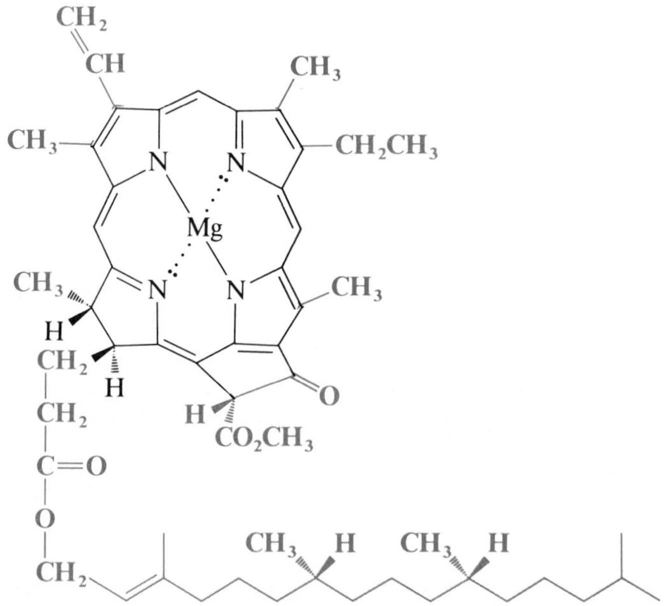

FIGURE 21.1　Chlorophyll-*a*. [The structure of chlorophyll-*a* was established largely through the work of H. Fischer (Munich), R. Willstätter (Munich), and J. B. Conant (Harvard). A synthesis of chlorophyll-*a* from simple organic compounds was achieved by R. B. Woodward (Harvard) in 1960, who won the Nobel Prize in 1965 for his outstanding contributions to synthetic organic chemistry.]

Many individual enzyme-catalyzed reactions take place in the general photosynthetic process and not all are fully understood. We know, however, that photosynthesis begins with the absorption of light by the important green pigment of plants, chlorophyll (Fig. 21.1). The green color of chlorophyll and, therefore, its ability to absorb sunlight in the visible region are due primarily to its extended conjugated system. As photons of sunlight are trapped by chlorophyll, energy becomes available to the plant in a chemical form that can be used to carry out the reactions that reduce carbon dioxide to carbohydrates and oxidize water to oxygen.

Carbohydrates act as a major chemical repository for solar energy. Their energy is released when animals or plants metabolize carbohydrates to carbon dioxide and water.

$$C_x(H_2O)_y + xO_2 \longrightarrow xCO_2 + yH_2O + \text{energy}$$

The metabolism of carbohydrates also takes place through a series of enzyme-catalyzed reactions in which each energy-yielding step is an oxidation (or the consequence of an oxidation).

Although some of the energy released in the oxidation of carbohydrates is inevitably converted to heat, much of it is conserved in a new chemical form through reactions that are coupled to the synthesis of adenosine triphosphate (ATP) from adenosine diphosphate (ADP) and inorganic phosphate (P_i) (Fig. 21.2). The phos-

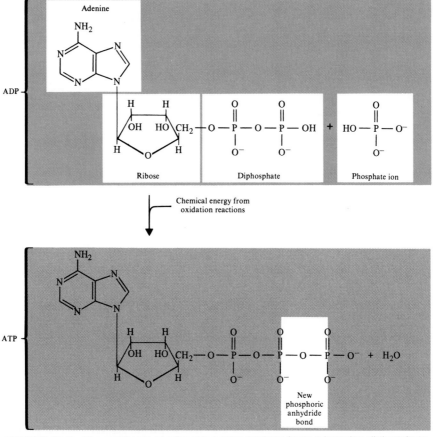

FIGURE 21.2 The synthesis of adenosine triphosphate from adenosine diphosphate and hydrogen phosphate ion. This reaction takes place in all living organisms, and adenosine triphosphate is the major compound into which the chemical energy released by biological oxidations is transformed.

phoric anhydride bond that forms between the terminal phosphate group of ADP and the phosphate ion becomes another repository of chemical energy. Plants and animals can use the conserved energy of ATP (or very similar substances) to carry out all of their energy-requiring processes: the contraction of a muscle, the synthesis of a macromolecule, and so on. When the energy in ATP is used, a coupled reaction takes place in which ATP is hydrolyzed:

$$\text{ATP} + \text{H}_2\text{O} \xrightarrow{\text{(− energy)}} \text{ADP} + \text{P}_i$$

or a new anhydride linkage is created:

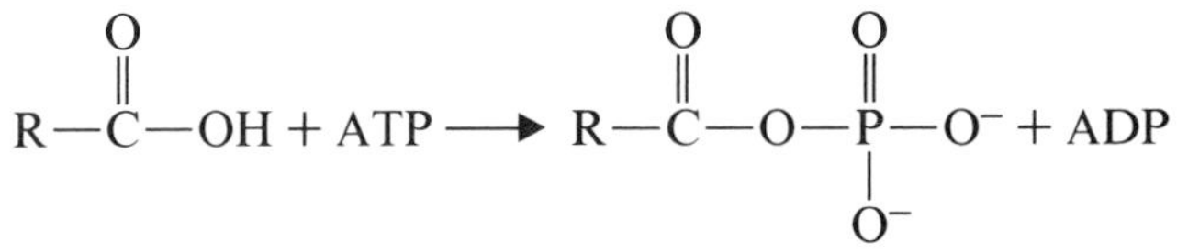

$$\underset{\text{Acyl phosphate}}{R-\overset{\displaystyle O}{\overset{\|}{C}}-OH + ATP \longrightarrow R-\overset{\displaystyle O}{\overset{\|}{C}}-O-\overset{\displaystyle O}{\underset{\underset{\displaystyle O^-}{|}}{\overset{\|}{P}}}-O^- + ADP}$$

21.2 MONOSACCHARIDES

21.2A Classification of Monosaccharides

Monosaccharides are classified according to (1) the number of carbon atoms present in the molecule and (2) whether they contain an aldehyde or keto group. Thus, a monosaccharide containing three carbon atoms is called a *triose;* one containing four carbon atoms is called a *tetrose;* one containing five carbon atoms is a *pentose;* and one containing six carbon atoms is a *hexose.* A monosaccharide containing an aldehyde group is called an *aldose;* and one containing a keto group is called a *ketose.* These two classifications are frequently combined: A four-carbon aldose, for example, is called an *aldotetrose,* a five-carbon ketose is called a *ketopentose.*

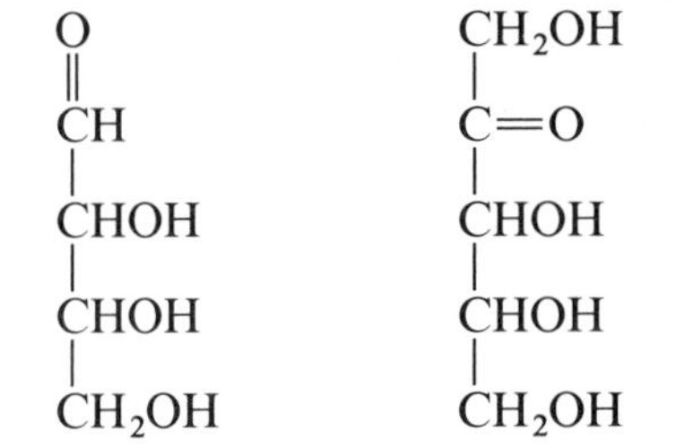

$$\begin{array}{cc}
\overset{\displaystyle O}{\overset{\|}{CH}} & CH_2OH \\
| & | \\
CHOH & C=O \\
| & | \\
CHOH & CHOH \\
| & | \\
CH_2OH & CHOH \\
 & | \\
 & CH_2OH \\
\text{An aldotetrose} & \text{A ketopentose}
\end{array}$$

PROBLEM 21.1

How many stereocenters are contained by the (a) aldotetrose and (b) keto-pentose given previously? (c) How many stereoisomers would you expect from each general structure?

21.2B D and L Designations of Monosaccharides

The simplest monosaccharides are the compounds glyceraldehyde and dihydroxy-acetone (see following structures). Of these two compounds, only glyceraldehyde contains a stereocenter.

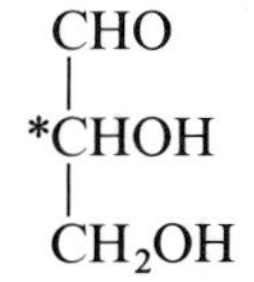

CHO
|
*CHOH
|
CH₂OH
Glyceraldehyde
(an aldotriose)

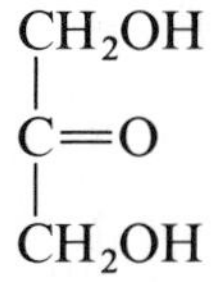

CH₂OH
|
C=O
|
CH₂OH
Dihydroxyacetone
(a ketotriose)

Glyceraldehyde exists, therefore, in two enantiomeric forms which are known to have the absolute configurations shown here.

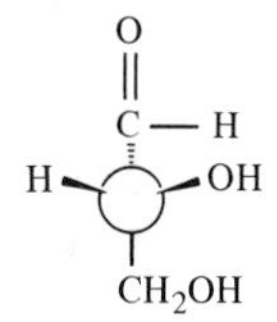

O
‖
C — H
H — OH
CH₂OH

(+)-Glyceraldehyde

and

O
‖
C — H
HO — H
CH₂OH

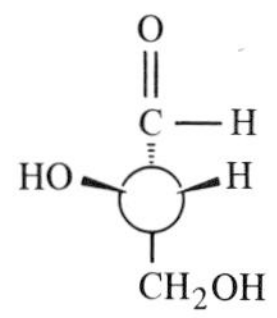

(−)-Glyceraldehyde

We saw in Section 4.5 that according to the Cahn–Ingold–Prelog convention, (+)-glyceraldehyde should be designated (*R*)-(+)-glyceraldehyde and (−)-glyceraldehyde should be designated (*S*)-(−)-glyceraldehyde.

Early in this century, before the absolute configurations of any organic compounds were known, another system of stereochemical designations was introduced. According to this system (first suggested by M. A. Rosanoff of New York University in 1906), (+)-glyceraldehyde is designated D-(+)-glyceraldehyde and (−)-glyceraldehyde is designated L-(−)-glyceraldehyde. These two compounds, moreover, serve as configurational standards for all monosaccharides: A monosaccharide *whose highest-numbered stereocenter* has the same configuration as D-(+)-glyceraldehyde is designated as a D sugar; one whose highest-numbered stereocenter has the same configuration as L-glyceraldehyde is designated as an L sugar.

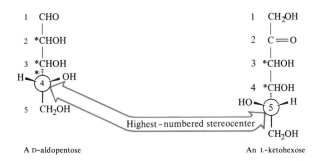

1 CHO
|
2 *CHOH
|
3 *CHOH
H — 4 — OH
5 CH₂OH

A D-aldopentose

Highest-numbered stereocenter

1 CH₂OH
|
2 C=O
|
3 *CHOH
|
4 *CHOH
HO — 5 — H
|
CH₂OH

An L-ketohexose

D and L designations are like (*R*) and (*S*) designations in that they are not necessarily related to the optical rotations of the sugars to which they are applied. Thus, one may encounter other sugars that are D-(+)- or D-(−)- and that are L-(+)- or L-(−)-.

The D–L system of stereochemical designations is thoroughly entrenched in the literature of carbohydrate chemistry, and even though it has the disadvantage of specifying the configuration of only one stereocenter — that of the highest-numbered stereocenter — we shall employ the D–L system in our designations of carbohydrates.

PROBLEM 21.2

Write three-dimensional formulas for each aldotetrose and ketopentose isomer in Problem 21.1 and designate each as a D or L sugar.

21.2C Structural Formulas for Monosaccharides

Later in this chapter we shall see how the great carbohydrate chemist Emil Fischer* was able to establish the stereochemical configuration of the aldohexose D-(+)-glucose, the most abundant monosaccharide. In the meantime, however, we can use D-(+)-glucose as an example illustrating the various ways of representing the structures of monosaccharides.

Fischer represented the structure of D-(+)-glucose with the cross formulation (1) in Fig. 21.3. This type of formulation is now called a Fischer projection formula (Section 4.11) and is still useful for carbohydrates. In Fischer projection formulas, by convention, *horizontal lines project out towards the reader and vertical lines project behind the plane of the page. When we use Fischer projection formulas, however, we must not* (in our mind's eye) *remove them from the plane of the page in order to test their superposability.* In terms of more familiar formulations, the Fischer projection formula translates into formulas **2** and **3**.†

> In IUPAC nomenclature and with the Cahn–Ingold–Prelog system of stereochemical designations, the open-chain form of D-(+)-glucose is (2R, 3S, 4R, 5R)-2,3,4,5,6-pentahydroxyhexanal.

Although many of the properties of D-(+)-glucose can be explained in terms of an open-chain structure (**1, 2,** or **3**), a considerable body of evidence indicates that the open-chain structure exists, primarily, in equilibrium with two cyclic forms. These can be represented by structures **4** and **5** or **6** and **7**. The cyclic forms of D-(+)-glucose are *hemiacetals* formed by an intramolecular reaction of the —OH group at C-5 with the aldehyde group (Fig. 21.4). Cyclization creates a new stereocenter at C-1 and this stereocenter explains how two cyclic forms are possible. These two cyclic forms are *diastereomers* that differ only in the configuration of C-1. In carbohydrate chemistry diastereomers of this type are called *anomers,* and the hemiacetal carbon atom is called the *anomeric carbon atom.*

---

*Emil Fischer (1852–1919) was professor of organic chemistry at the University of Berlin. In addition to monumental work in the field of carbohydrate chemistry, where Fischer and his co-workers established the configuration of most of the monosaccharides, Fischer also made important contributions to studies of amino acids, proteins, purines, indoles, and stereochemistry generally. As a graduate student Fischer discovered phenylhydrazine, a reagent that was highly important in his later work with carbohydrates. Fischer was the second recipient (in 1902) of the Nobel Prize for Chemistry.

†The meaning of formulas **1, 2,** and **3** can be seen best through the use of molecular models: We first construct a chain of six carbon atoms with the —CHO group at the top and a —CH₂OH group at the bottom. We then bring the CH₂OH group up behind the chain until it almost touches the —CHO group. Holding this model so that the —CHO and —CH₂OH groups are directed generally away from us, we then begin placing —H and —OH groups on each of the four remaining carbon atoms. The —OH group of carbon-2 is placed on the right; that of carbon-3 on the left; and those of carbons-4 and -5 on the right.

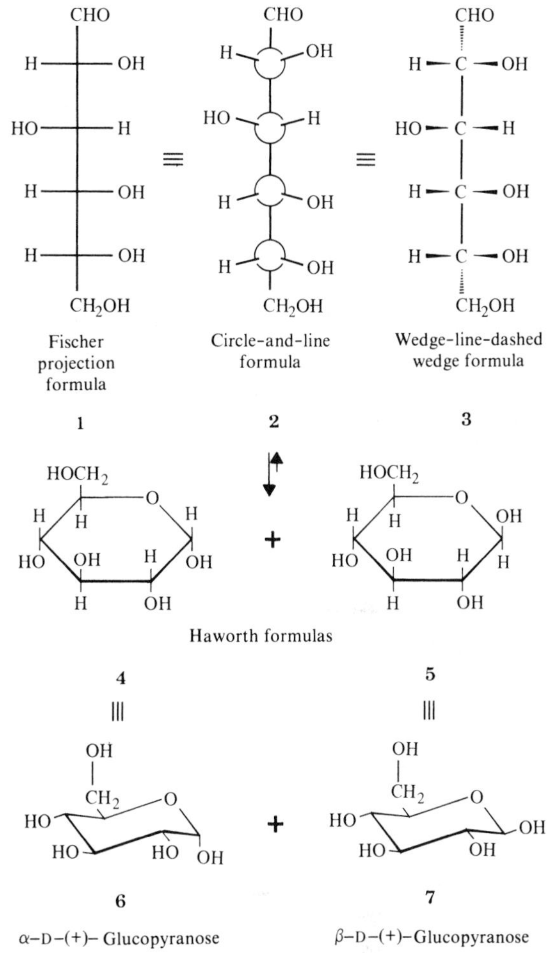

FIGURE 21.3 (*a*) **1–3** are formulas used for the open-chain structure of D-(+)-glucose. (*b*) **4–7** are formulas used for the two cyclic hemiacetal forms of D-(+)-glucose.

Structures **4** and **5** for the glucose anomers are called Haworth formulas* and although they do not give an accurate picture of the shape of the six-membered ring, they have many practical uses. Figure 21.4 demonstrates how the representation of each stereocenter of the open-chain form can be correlated with its representation in the Haworth formula.

Each glucose anomer is designated as an α *anomer* or a β *anomer* depending on the location of the —OH group of C-1. When we draw the cyclic forms of a D sugar in the orientation shown in Figs. 21.3 or 21.4, the α anomer has the —OH *down* and the β anomer has the —OH *up*.

*After the English chemist W. N. Haworth (University of Birmingham) who, in 1926, along with E. L. Hirst, demonstrated that the cyclic form of glucose acetals consists of a six-membered ring. Haworth received the Nobel Prize for his work in carbohydrate chemistry in 1937. For an excellent discussion of Haworth formulas and their relation to open-chain forms see the following article: D. M. S. Wheeler, M. M. Wheeler, and T. S. Wheeler, "The Conversion of Open Chain Structures of Monosaccharides into the Corresponding Haworth Formulas," *J. Chem. Educ.,* **59,** 969 (1982).

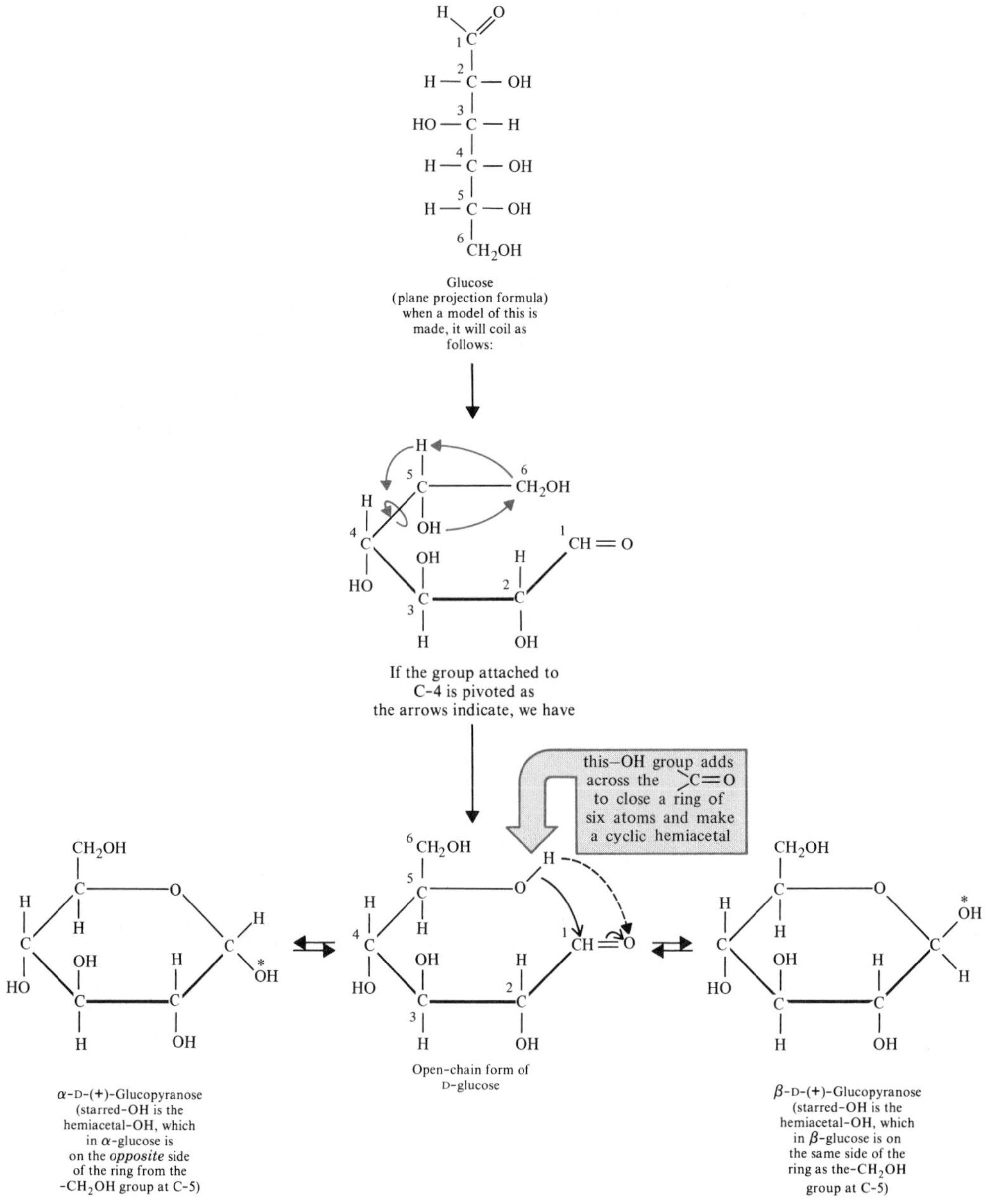

FIGURE 21.4 The Haworth formulas for the cyclic hemiacetal forms of D-(+)-glucose and their relation to the open-chain polyhydroxy aldehyde structure. [From John R. Holum, *Organic Chemistry: A Brief Course*, Wiley, New York, (1975), p. 332. Used by permission.]

Studies of the structures of the cyclic hemiacetal forms of D-(+)-glucose using X-ray analysis have demonstrated that the actual conformations of the rings are the chair forms represented by conformational formulas **6** and **7** in Fig. 21.3. This shape is exactly what we would expect from our studies of the conformations of cyclohexane (Chapter 3), and it is especially interesting to notice that in the β anomer of D-glucose all of the large substituents, —OH or —CH$_2$OH, are equatorial. In the α anomer the only bulky axial substituent is the —OH at C-1.

It is convenient at times to represent the cyclic structures of a monosaccharide without specifying whether the configuration of the anomeric carbon atom is α or β.

When we do this we shall use formulas such as the following:

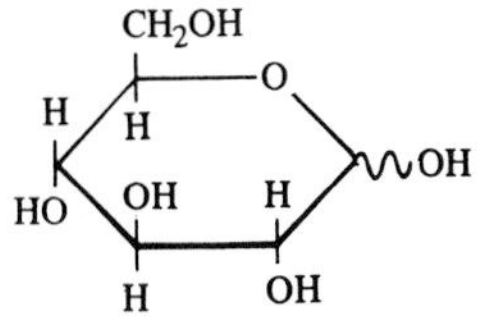

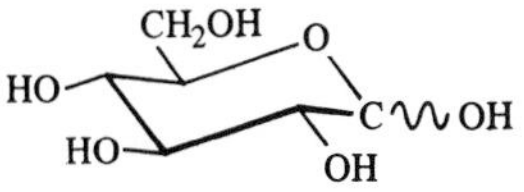

Not all carbohydrates exist in equilibrium with six-membered hemiacetal rings; in several instances the ring is five membered. (Even glucose exists, to a small extent, in equilibrium with five-membered hemiacetal rings.) Because of this variation, a system of nomenclature has been introduced to allow designation of the ring size. If the monosaccharide ring is six membered, the compound is called a *pyranose;* if the ring is five membered the compound is designated as a *furanose.** Thus, the full name of compound **4** (or **6**) is α-D-(+)-glucopyranose, while that of **5** (or **7**) is β-D-(+)-glucopyranose.

21.3 MUTAROTATION AND GLYCOSIDE FORMATION

Part of the evidence for the cyclic hemiacetal structure for D-(+)-glucose comes from experiments in which both α and β forms have been isolated. Ordinary D-(+)-glucose has a melting point of 146 °C. However, when D-(+)-glucose is crystallized by evaporating an aqueous solution kept above 98 °C, a second form of D-(+)-glucose with a melting point of 150 °C can be obtained. When the optical rotations of these two forms are measured, they are found to be significantly different, but when an aqueous solution of either form is allowed to stand, its rotation changes — the specific rotation of one form decreases and the rotation of the other increases — *until both solutions show the same value.* A solution of ordinary D-(+)-glucose (mp, 146 °C) has an initial specific rotation of $+112°$ but ultimately the specific rotation of this solution falls to $+52.7°$. A solution of the second form of D-(+)-glucose (mp, 150 °C) has an initial specific rotation of $+19°$; but slowly, the specific rotation of this solution rises to $+52.7°$. This change in rotation towards an equilibrium value is called *mutarotation.*

The explanation for this mutarotation lies in the existence of an equilibrium between the open-chain form of D-(+)-glucose and the α and β forms of the cyclic hemiacetals.

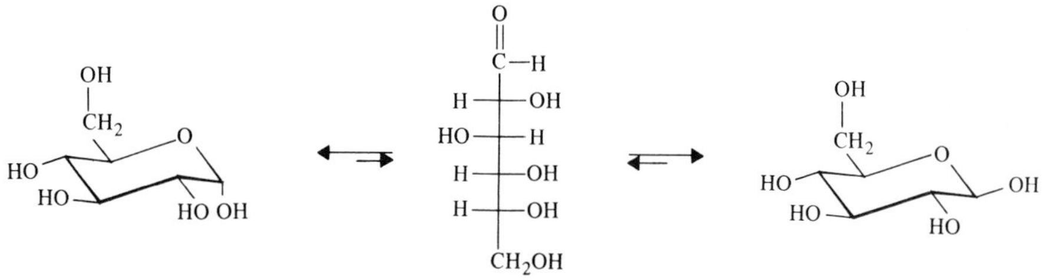

α-D-(+)-Glucopyranose
(mp, 146°C $[\alpha]_D^{25} = +112°$)

Open–chain
form of
D-(+)-glucose

β-D-(+)-Glucopyranose
(mp, 150°C $[\alpha]_D^{25} = +18.7°$)

*These names come from the names of the oxygen heterocycles *pyran* and *furan* + *ose.*

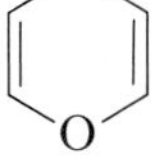

A pyran 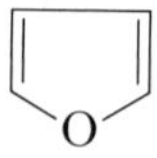   Furan

X-ray analysis has confirmed that ordinary D-(+)-glucose has the α configuration at the anomeric carbon atom and that the higher-melting form has the β configuration.

The concentration of open-chain D-(+)-glucose in solution at equilibrium is very small. Solutions of D-(+)-glucose give no observable ultraviolet or infrared absorption band for a carbonyl group, and solutions of D-(+)-glucose give a negative test with Schiff's reagent—a special reagent that requires a relatively high concentration of a free aldehyde group (rather than a hemiacetal) in order to give a positive test.

Assuming that the concentration of the open-chain form is negligible, one can, by use of the specific rotations in the preceding figures, calculate the percentages of the α and β anomers present at equilibrium. These percentages, 36% α anomer and 64% β anomer, are in accord with a greater stability for β-D-(+)-glucopyranose. This preference is what we might expect on the basis of its having only equatorial groups.

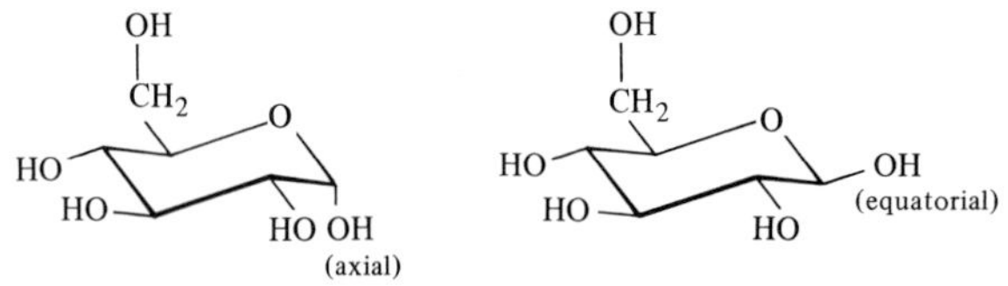

α-D-(+)-Glucopyranose
(36% at equilibrium)

β-D-(+)-Glucopyranose
(64% at equilibrium)

The β anomer of a pyranose is not always the more stable, however. With D-mannose the equilibrium favors the α anomer and this result—termed an *anomeric effect*—is at present not fully understood.

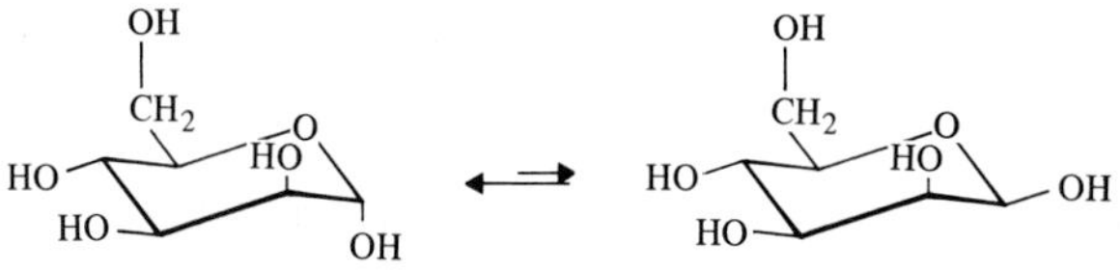

α-D-Mannopyranose
(69% at equilibrium)

β-D-Mannopyranose
(31% at equilibrium)

When a small amount of gaseous hydrogen chloride is passed into a solution of D-(+)-glucose in methanol, a reaction takes place that results in the formation of anomeric methyl *acetals*.

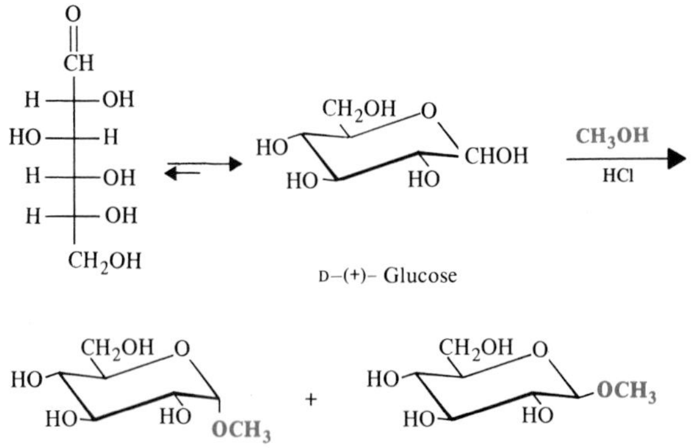

D-(+)- Glucose

Methyl α-D-glucopyranoside
(mp, 165°C $[α]_D^{25} = +158°$)

Methyl β-D-glucopyranoside
(mp, 107°C $[α]_D^{25} = -33°$)

Carbohydrate acetals, generally, are called *glycosides* and an acetal of glucose is called a *glucoside.* (Acetals of mannose are *mannosides,* ketals of fructose are *fructosides,* and so on.) The methyl D-glucosides have been shown to have six-membered rings (Section 21.10) so they are properly named methyl α-D-glucopyranoside and methyl β-D-glucopyranoside.

The mechanism for the formation of the methyl glucosides (starting arbitrarily with β-D-glucopyranose) is as follows:

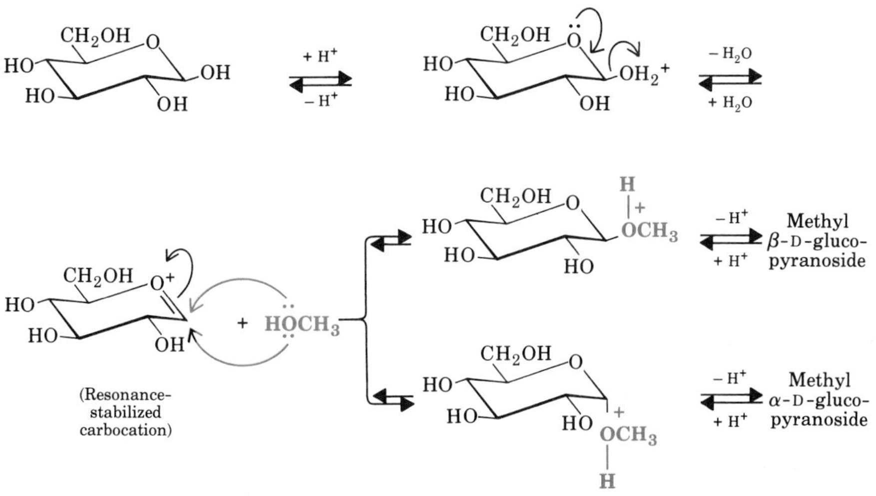

You should review the mechanism for acetal formation given in Section 16.7 and compare it with the steps given here. Notice, again, the important role played by the electron pair of the adjacent oxygen atom in stabilizing the carbocation that forms in the second step.

Since glycosides are acetals, they exist in equilibrium with an open-chain form only in acidic media. Under neutral or basic conditions glycosides do not equilibrate with the open-chain form, and do not show mutarotation.

PROBLEM 21.3

Solutions of glycosides *do* exhibit mutarotation in aqueous acid. Explain.

PROBLEM 21.4

Write conformational formulas for   (a) methyl α-D-glucopyranoside and (b) methyl β-D-glucopyranoside.

PROBLEM 21.5

Write a conformational formula for methyl α-D-mannopyranoside.

21.4 OXIDATION REACTIONS OF MONOSACCHARIDES

A number of oxidizing agents are used to identify functional groups of carbohydrates, in elucidating their structures, and for syntheses. The most important are (1) Benedict's or Tollens' reagents, (2) bromine water, (3) nitric acid, and (4) periodic acid. Each of these reagents produces a different and usually specific effect when it is allowed to react with a monosaccharide. We should now examine what these effects are.

21.4A Benedict's or Tollens' Reagents: Reducing Sugars

Benedict's reagent (an alkaline solution containing a cupric citrate complex ion) and Tollens' solution [$Ag(NH_3)_2{}^+OH^-$] oxidize and thus give positive tests with alde-hydes *and with α-hydroxy ketones.* Both reagents, therefore, give positive tests with aldoses *and ketoses.* This is true even though aldoses and ketoses exist primarily as cyclic hemiacetals.

We studied the use of Tollens' silver mirror test in Section 16.13. Benedict's solution (and the related Fehling's solution that contains a cupric tartrate complex ion) give brick-red precipitates of Cu_2O when they oxidize an aldehyde. (In alkaline solution α-hydroxy ketones are converted to aldehydes, which are then oxidized by the cupric complexes.) Since the solutions of cupric tartrates and citrates are blue, the appearance of a brick-red precipitate is a vivid and unmistakable indication of a positive test.

$$\underset{\substack{\text{Benedict's} \\ \text{solution} \\ \text{(blue)}}}{Cu^{2+} \text{ (complex)}} + \underset{\substack{\text{(aldehyde} \\ \text{or aldose)}}}{\overset{\overset{\displaystyle O}{\overset{\|}{\text{—}}}}{R—CH}} \text{ or } \underset{\substack{\text{(α-hydroxy ketone} \\ \text{or ketose)}}}{\underset{\underset{\displaystyle O}{\|}}{R—CCH_2OH}} \longrightarrow \underset{\substack{\text{(brick-red} \\ \text{reduction} \\ \text{product)}}}{Cu_2O} + \text{oxidation products}$$

Sugars that give positive tests with Tollens' or Benedict's solutions are known as *reducing sugars,* and all carbohydrates that contain a *hemiacetal group* or a *hemi-ketal group* give positive tests. In aqueous solution these hemiacetals or hemiketals exist in equilibrium with relatively small, but not insignificant, concentrations of noncyclic aldehydes or α-hydroxy ketones. It is the latter that undergo the oxidation, perturbing the equilibrium to produce more aldehyde or α-hydroxy ketone, which undergoes oxidation and so forth, until one reactant is exhausted.

Carbohydrates that contain only acetal or ketal groups do not give positive tests with Benedict's or Tollens' solution, and they are called *nonreducing sugars.* Acetals or ketals do not exist in equilibrium with aldehydes or α-hydroxy ketones in the basic aqueous media of the test reagents.

Reducing Sugar   **Nonreducing Sugar**

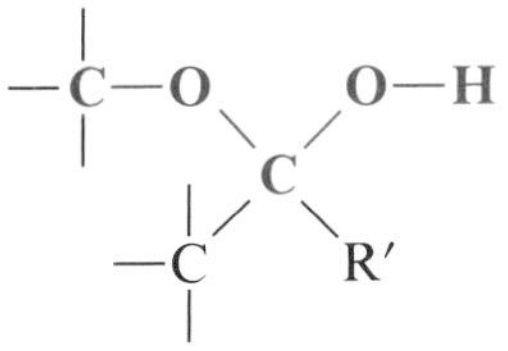

Hemiacetal (R′ = H;
or hemiketal, R′ ≠ H)
(gives positive Tollens'
or Benedict's test)

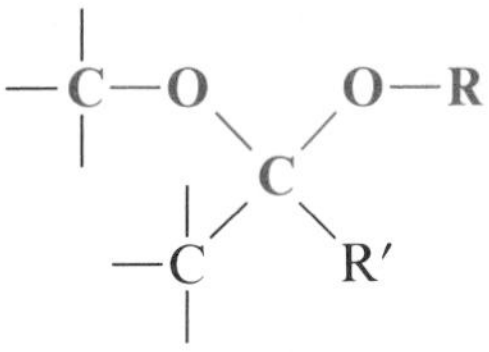

Acetal (R′ = H;
or ketal, R′ ≠ H)
(does not give a
positive Tollens' or
Benedict's test)

PROBLEM 21.6

How might you distinguish between α-D-glucopyranose (i.e., D-glucose) and methyl α-D-glucopyranoside?

Although Benedict's and Tollens' reagents have some use as diagnostic tools (Benedict's solution can be used in quantitative determinations of glucose in blood or urine), neither of these reagents is useful as a preparative reagent in carbohydrate oxidations. Oxidations with both reagents take place in alkaline solution, *and in alkaline solutions sugars undergo a complex series of reactions that lead to isomerizations and fragmentations.*

PROBLEM 21.7

If D-glucose is treated with aqueous calcium hydroxide and the solution is allowed to stand for several days, a mixture of products results, including D-mannose, D-fructose, glycolic aldehyde, and D-erythrose.

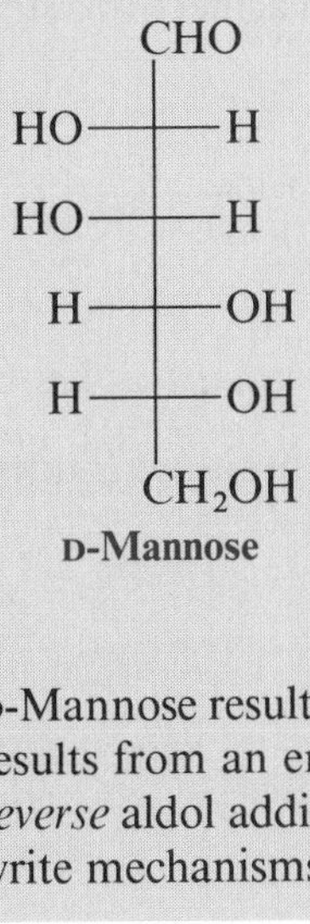

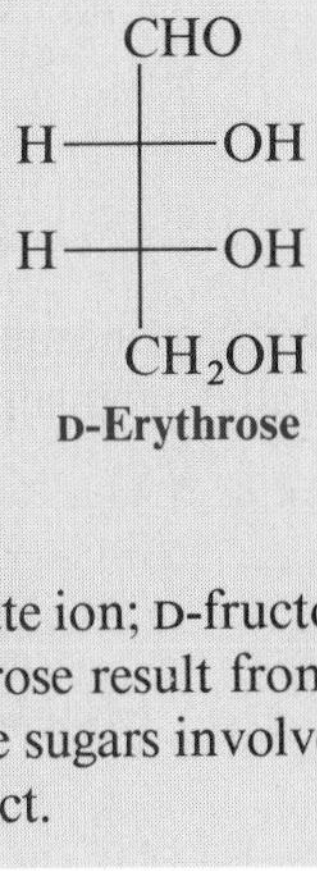

D-Mannose 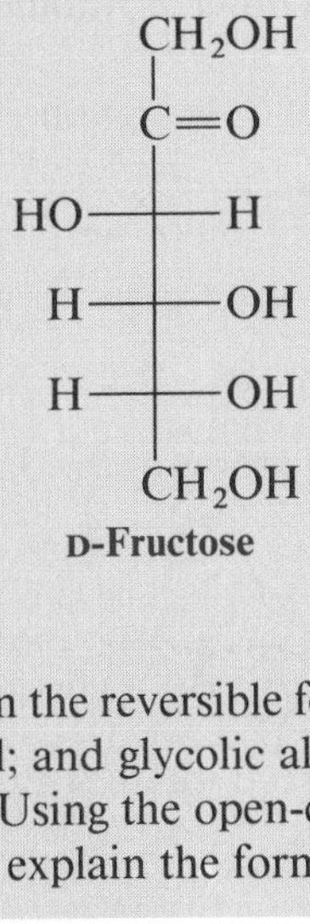   D-Fructose 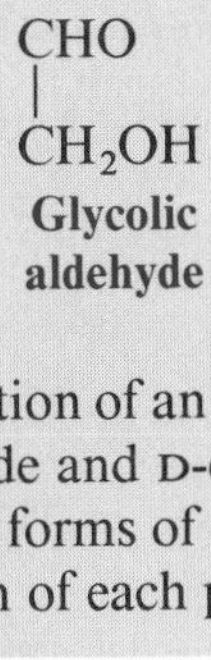   Glycolic   D-Erythrose
                                    aldehyde

D-Mannose results from the reversible formation of an enolate ion; D-fructose results from an enediol; and glycolic aldehyde and D-erythrose result from a *reverse* aldol addition. Using the open-chain forms of all the sugars involved, write mechanisms that explain the formation of each product.

21.4B Bromine Water:
The Synthesis of Aldonic Acids

Monosaccharides do not undergo isomerization and fragmentation reactions in mildly acidic solution. Thus, a useful oxidizing reagent for preparative purposes is bromine in water (pH 6.0). Bromine water is a general reagent that selectively oxidizes the —CHO group to a —CO_2H group. It converts an aldose to an *aldonic acid:*

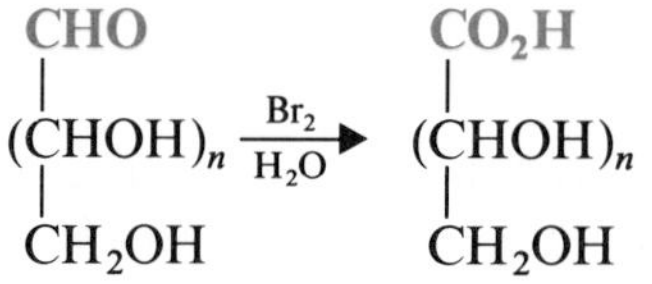

$$\text{CHO} \atop \text{(CHOH)}_n \atop \text{CH}_2\text{OH} \quad \xrightarrow[\text{H}_2\text{O}]{\text{Br}_2} \quad \text{CO}_2\text{H} \atop \text{(CHOH)}_n \atop \text{CH}_2\text{OH}$$

Aldose   Aldonic acid

Experiments with aldopyranoses have shown that the actual course of the reaction is somewhat more complex than we have indicated above. Bromine water specifically oxidizes the β anomer, and the initial product that forms is a *δ-aldonolactone.* This compound may then hydrolyze to an aldonic acid, and the aldonic acid may undergo a subsequent ring closure to form a *γ-aldonolactone.*

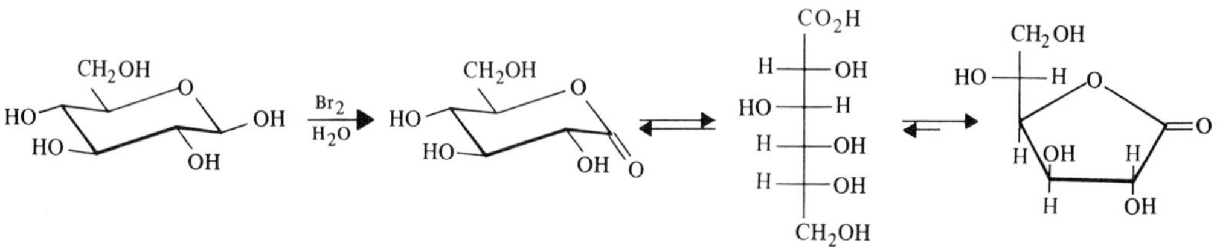

β-D-Glucopyranose   D-Glucono-δ-lactone   D-Gluconic acid   D-Glucono-γ-lactone

Although α-aldopyranoses also undergo this oxidation, they react much more slowly. The oxidation of β-D-glucopyranose, for example, is 250 times as fast as that of α-D-glucopyranose. The very low rate of oxidation of α-D-glucopyranose reflects its conversion to the β anomer (followed by oxidation) rather than direct oxidation of the α anomer.

A mechanism that accounts for this behavior requires attack by Br^+ at the anomeric oxygen followed by loss of HBr from an intermediate in which the C—H and O—Br bonds are *antiparallel* to each other. β Anomers can achieve this stereochemistry easily.

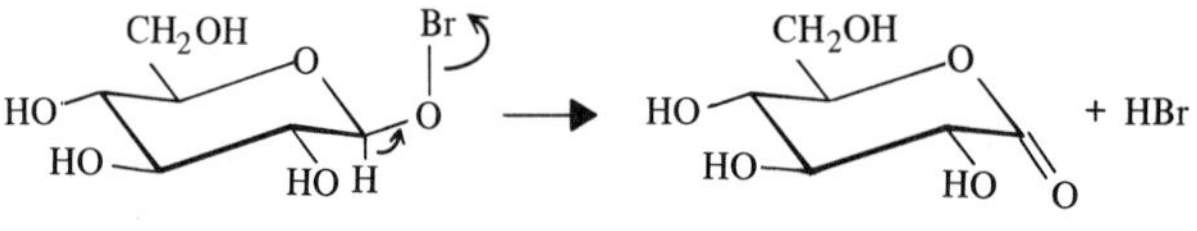

(from β anomer)
Antiparallel arrangement

With α anomers the axial hydrogen atoms at the 3 and 5 positions hinder an antiparallel arrangement:

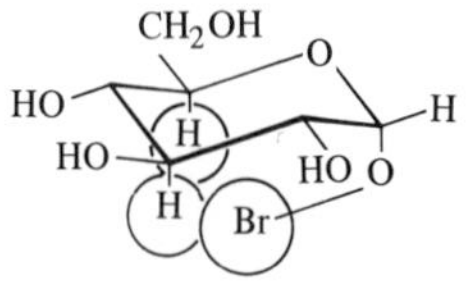

(from α anomer)
Antiparallel arrangement is hindered

PROBLEM 21.8

Write a sequence of equations to illustrate the oxidation of β-D-mannopyranose by bromine water (cf. Section 21.3). The name of the aldonic acid is D-mannonic acid and the names of the aldonolactones are δ- and γ-D-mannonolactone.

21.4C Nitric Acid Oxidation: Aldaric Acids

Dilute nitric acid—a stronger oxidizing agent than bromine water—oxidizes both the —CHO group and the terminal $—CH_2OH$ group of an aldose to $—CO_2H$ groups. These dicarboxylic acids are known as *aldaric acids.*

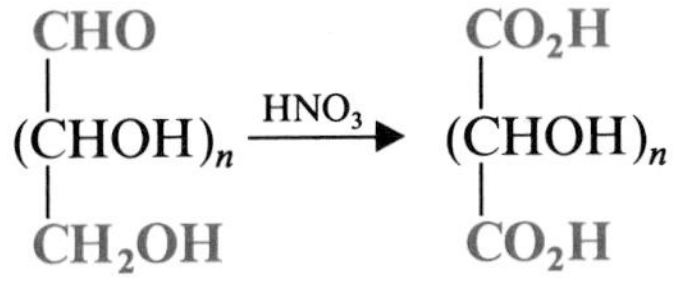
CHO
(CHOH)n
CH2OH
HNO3
CO2H
(CHOH)n
CO2H
Aldose   Aldaric acid

It is not known whether a lactone is an intermediate in the oxidation of an aldose to an aldaric acid; however, aldaric acids form γ- and δ-lactones readily.

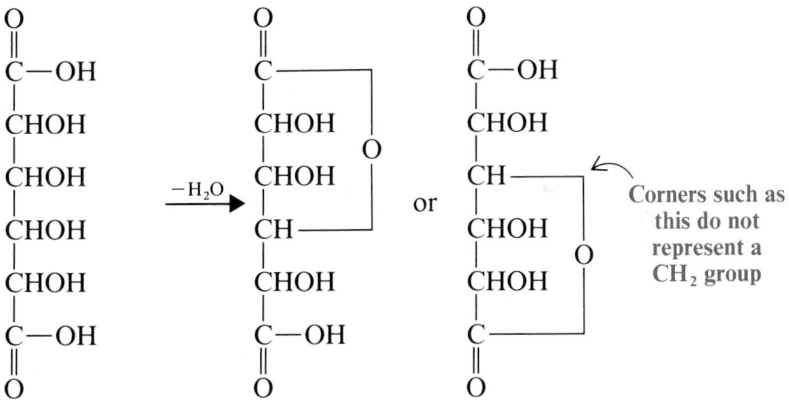
−H2O
or
Corners such as this do not represent a CH2 group
Aldaric acid (from an aldohexose)   γ-Lactones of an aldaric acid

The aldaric acid obtained from D-glucose is called D-glucaric acid.*

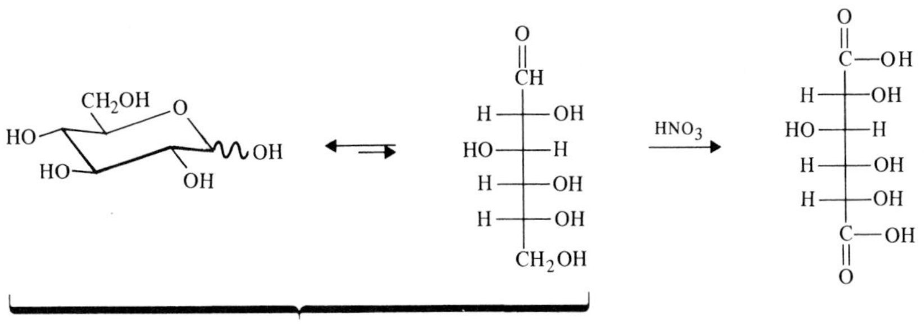
HNO3
D-Glucose   D-Glucaric acid

*Older terms for an aldaric acid are a *glycaric* acid or a *saccharic* acid.

PROBLEM 21.9

(a) Would you expect D-glucaric acid to be optically active? (b) Write the open-chain structure for the aldaric acid (mannaric acid) that would be obtained by nitric acid oxidation of D-mannose. (c) Would you expect it to be optically active? (d) What aldaric acid would you expect to obtain from D-erythrose (cf. Problem 21.7)? (e) Would it show optical activity? (f) D-Threose, a diastereomer of D-erythrose, yields an optically active aldaric acid when it is subjected to nitric acid oxidation. Write Fischer projection formulas for D-threose and its nitric acid oxidation product. (g) What are the names of the aldaric acids obtained from D-erythrose and D-threose? (See Section 4.13A.)

PROBLEM 21.10

D-Glucaric acid undergoes lactonization to yield two different γ-lactones. What are their structures?

21.4D Periodate Oxidations: Oxidative Cleavage of Polyhydroxy Compounds

Compounds that have hydroxyl groups on adjacent atoms undergo oxidative cleavage when they are treated with aqueous periodic acid (HIO_4). The reaction breaks carbon–carbon bonds and produces carbonyl compounds (aldehydes, ketones, or acids). The stoichiometry of the reaction is

$$
\begin{array}{c}
-\overset{|}{\underset{}{C}}-OH \\
\text{-----}|\text{-------} \quad + HIO_4 \longrightarrow 2-\overset{|}{\underset{}{C}}=O + HIO_3 + H_2O \\
-\overset{|}{\underset{}{C}}-OH
\end{array}
$$

Since the reaction usually takes place in quantitative yield, valuable information can often be gained by measuring the number of molar equivalents of periodic acid that are consumed in the reaction as well as by identifying the carbonyl products.*

Periodate oxidations are thought to take place through a cyclic intermediate:

$$
\begin{array}{c}
-\overset{|}{\underset{}{C}}-OH \\
\\
-\overset{|}{\underset{}{C}}-OH
\end{array}
+ IO_4^- \xrightarrow{(-H_2O)}
\quad\cdots\quad
\longrightarrow
\begin{array}{c}
-\overset{|}{\underset{}{C}}=O \\
\\
-\overset{|}{\underset{}{C}}=O
\end{array}
+ IO_3^-
$$

Before we discuss the use of periodic acid in carbohydrate chemistry, we should illustrate the course of the reaction with several simple examples. Notice in these

*The reagent lead tetraacetate, $Pb(O_2CCH_3)_4$, brings about cleavage reactions similar to those of periodic acid. The two reagents are complementary; periodic acid works well in aqueous solutions and lead tetraacetate gives good results in organic solvents.

periodate oxidations that *for every C—C bond broken, a C—O bond is formed at each carbon.*

1. When 1,2-propanediol is subjected to this oxidative cleavage, the products are formaldehyde and acetaldehyde.

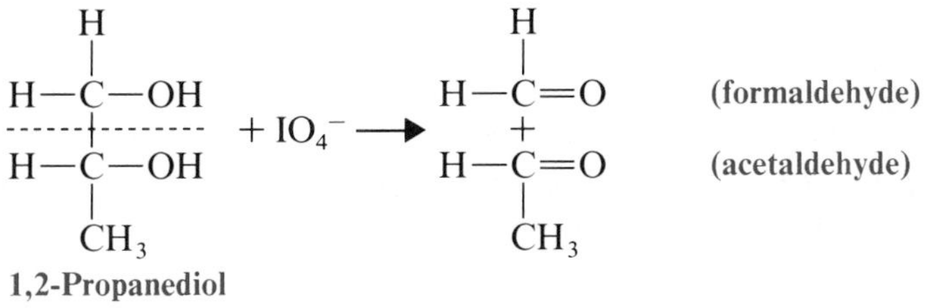

$$
\begin{array}{ccc}
\underset{\text{1,2-Propanediol}}{\overset{\displaystyle \text{H}\atop\displaystyle |}{\underset{\displaystyle |}{\text{H—C—OH}}}} & & \\
\overline{\vdots} + \text{IO}_4^- \longrightarrow & \\
\overset{\displaystyle |}{\underset{\displaystyle \text{CH}_3}{\text{H—C—OH}}} & &
\end{array}
$$

H—C=O   (formaldehyde)
+
H—C=O   (acetaldehyde)
|
CH₃

2. When three or more —CHOH groups are contiguous, the internal ones are obtained as *formic acid.* Periodate oxidation of glycerol, for example, gives two molar equivalents of formaldehyde and one molar equivalent of formic acid.

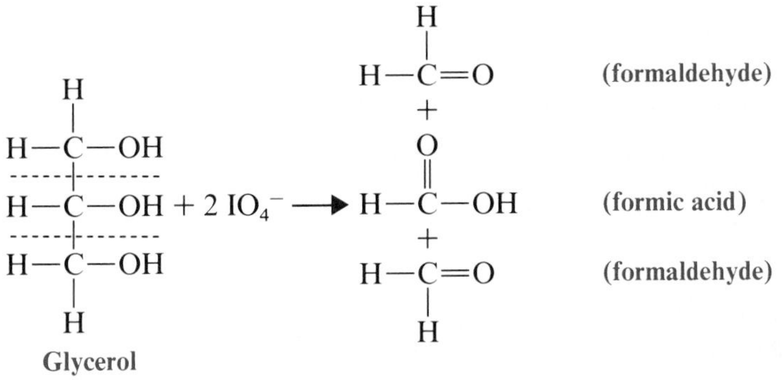

Glycerol + 2 IO₄⁻ ⟶

H—C=O   (formaldehyde)
+
O
‖
H—C—OH   (formic acid)
+
H—C=O   (formaldehyde)
|
H

3. Oxidative cleavage also takes place when an —OH group is adjacent to the carbonyl group of an aldehyde or ketone (but not that of an acid or an ester). Glyceraldehyde yields two molar equivalents of formic acid and one molar

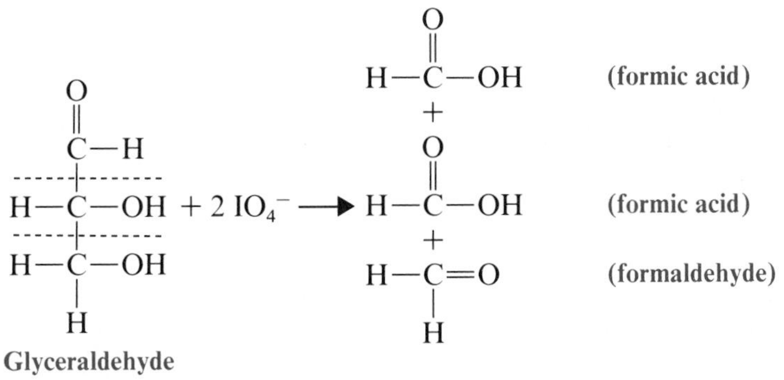

Glyceraldehyde + 2 IO₄⁻ ⟶

O
‖
H—C—OH   (formic acid)
+
O
‖
H—C—OH   (formic acid)
+
H—C=O   (formaldehyde)
|
H

equivalent of formaldehyde, while dihydroxyacetone gives two molar equivalents of formaldehyde and one molar equivalent of carbon dioxide.

$$
\begin{array}{ccccc}
& \text{H} & & \text{H} & \\
& | & & | & \\
& \text{H}-\text{C}-\text{OH} & & \text{H}-\text{C}=\text{O} & \text{(formaldehyde)} \\
& \text{----+-------} & & + & \\
& \text{C}=\text{O} & +\ 2\ \text{IO}_4^- \longrightarrow & \text{O}=\text{C}=\text{O} & \text{(carbon dioxide)} \\
& \text{----+-------} & & + & \\
& \text{H}-\text{C}-\text{OH} & & \text{H}-\text{C}=\text{O} & \text{(formaldehyde)} \\
& | & & | & \\
& \text{H} & & \text{H} & \\
\end{array}
$$

Dihydroxyacetone

4. Periodic acid does not cleave compounds in which the hydroxyl groups are separated by an intervening $-\text{CH}_2-$ group, nor those in which a hydroxyl group is adjacent to an ether or acetal function.

$$
\begin{array}{llll}
\text{CH}_2\text{OH} & & \text{CH}_2\text{OCH}_3 & \\
| & & | & \\
\text{CH}_2 & + \text{IO}_4^- \longrightarrow \text{no cleavage} & \text{CHOH} & + \text{IO}_4^- \longrightarrow \text{no cleavage} \\
| & & | & \\
\text{CH}_2\text{OH} & & \text{CH}_2\text{R} & \\
\end{array}
$$

PROBLEM 21.11

What products would you expect to be formed when each of the following compounds is treated with an appropriate amount of periodic acid? How many molar equivalents of HIO_4 would be consumed in each case?

(a) 2,3-Butanediol

(b) 1,2,3-Butanetriol

(c) $\text{CH}_2\text{OHCHOHCH(OCH}_3)_2$

(d) $\text{CH}_2\text{OHCHOHCOCH}_3$

(e) $\text{CH}_3\text{COCHOHCOCH}_3$

(f) *cis*-1,2-Cyclopentanediol

$$
\begin{array}{l}
\qquad\qquad \text{CH}_3 \\
\qquad\qquad | \\
\text{(g)}\ \text{CH}_3\text{C}-\text{CH}_2 \\
\qquad\quad |\quad\ | \\
\qquad\ \ \text{OH OH}
\end{array}
$$

(h) D-Erythrose

PROBLEM 21.12

Show how periodic acid could be used to distinguish between an aldohexose and a ketohexose. What products would you obtain from each, and how many molar equivalents of HIO_4 would be consumed?

In 1936 E. L. Jackson and C. S. Hudson (U.S. Public Health Service) used periodate oxidations in an elegant procedure for determining the ring size of glycosides. Their method also permits us to demonstrate that different glycosides of the α-D-hexose series have the same configurations at C-1 and at C-5. This method is outlined in Fig. 21.5.

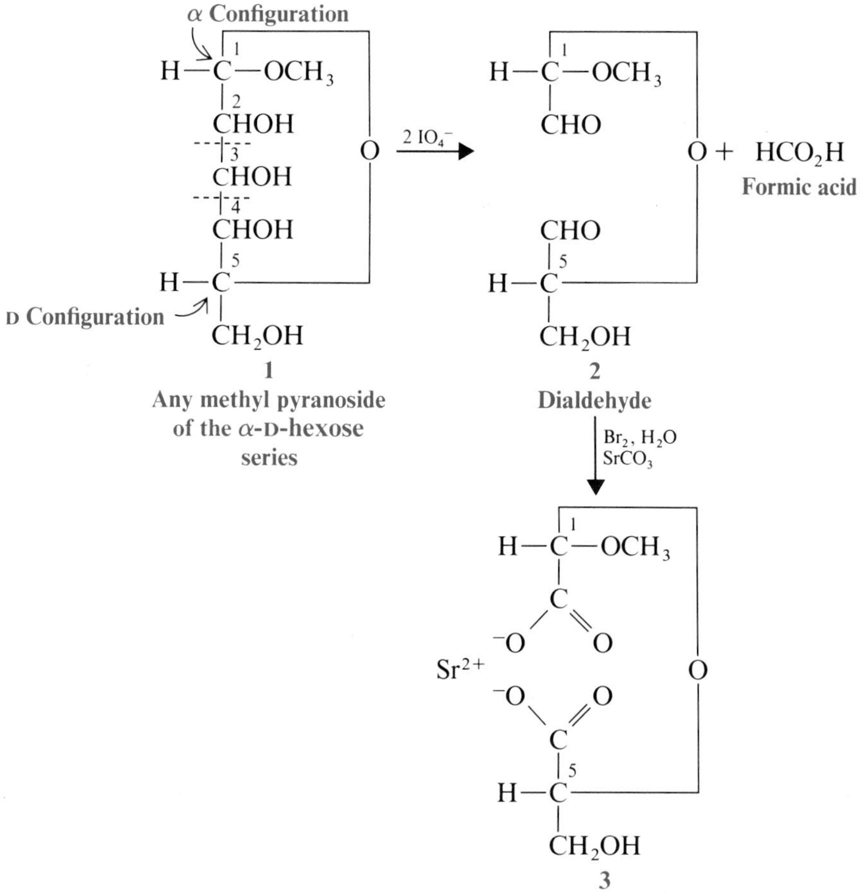

FIGURE 21.5 Jackson and Hudson's method. Placing the —OCH₃ group on the right in structure **1** indicates that the glycoside is an α anomer. The configuration of C-2, C-3, and C-4 is not specified; thus, **1** is any α-methyl-D-hexopyranoside.

The fact that the methyl glycoside **(1)** consumes two molar equivalents of periodate ion and produces one molar equivalent of formic acid demonstrates that the glycoside contains a six-membered ring. The oxidation also eliminates three stereocenters in **1**. (C-2 and C-4 are oxidized to aldehyde groups and C-3 is eliminated as formic acid.) The fact that all methyl pyranosides of the α-D-hexose series yield the same dialdehyde **(2)**, as can be shown by conversion to the same strontium salt **(3)**, demonstrates that they all have the same configuration at C-1 and C-5.

PROBLEM 21.13

When methyl furanosides of the α-D-pentose series are subjected to Jackson and Hudson's procedure they, too, yield the same dialdehyde **(2)** and strontium salt **(3)**. (a) Outline the reactions involved. (b) In what respect would you obtain a different result from applying Jackson and Hudson's method to a methyl α-D-pentofuranoside and a methyl α-D-hexopyranoside?

21.5 REDUCTION OF MONOSACCHARIDES: ALDITOLS

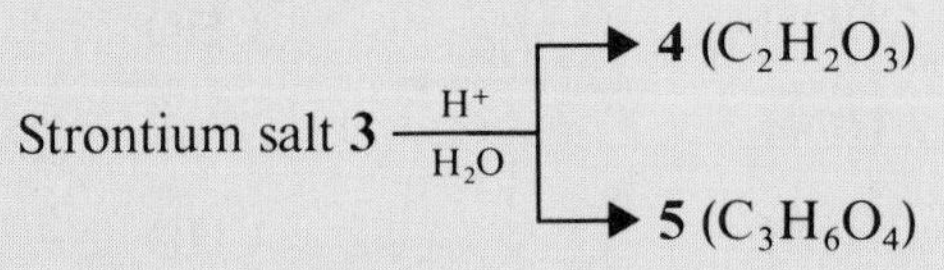

Aldoses (and ketoses) can be reduced with sodium borohydride to compounds called *alditols*.

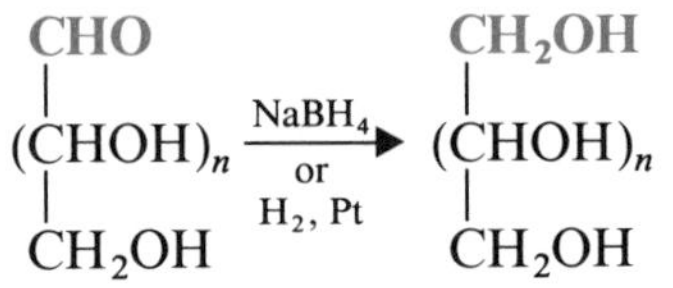

$$\begin{array}{c} \text{CHO} \\ | \\ (\text{CHOH})_n \\ | \\ \text{CH}_2\text{OH} \\ \text{Aldose} \end{array} \xrightarrow[\substack{\text{or} \\ \text{H}_2,\ \text{Pt}}]{\text{NaBH}_4} \begin{array}{c} \text{CH}_2\text{OH} \\ | \\ (\text{CHOH})_n \\ | \\ \text{CH}_2\text{OH} \\ \text{Alditol} \end{array}$$

Reduction of D-glucose, for example, yields D-glucitol.

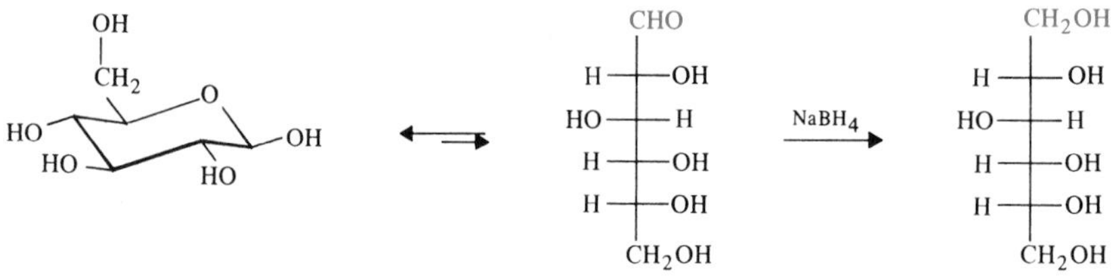

$$\begin{array}{c} \text{CHO} \\ \text{H}\!-\!\!-\!\text{OH} \\ \text{HO}\!-\!\!-\!\text{H} \\ \text{H}\!-\!\!-\!\text{OH} \\ \text{H}\!-\!\!-\!\text{OH} \\ \text{CH}_2\text{OH} \end{array} \xrightarrow{\text{NaBH}_4} \begin{array}{c} \text{CH}_2\text{OH} \\ \text{H}\!-\!\!-\!\text{OH} \\ \text{HO}\!-\!\!-\!\text{H} \\ \text{H}\!-\!\!-\!\text{OH} \\ \text{H}\!-\!\!-\!\text{OH} \\ \text{CH}_2\text{OH} \end{array}$$

D-Glucitol

21.6 REACTIONS OF MONOSACCHARIDES WITH PHENYLHYDRAZINE: OSAZONES

The aldehyde group of an aldose reacts with such carbonyl reagents as hydroxylamine and phenylhydrazine (Section 16.8). With hydroxylamine, the product is the expected oxime. With enough phenylhydrazine, however, three molar equivalents of phenylhydrazine are consumed and a second phenylhydrazone group is introduced at C-2. The product is called a *phenylosazone.*

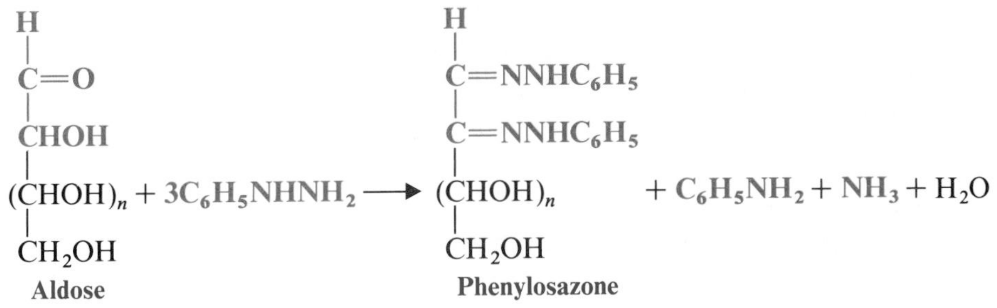

$$
\begin{array}{l}
\text{H} \\
| \\
\text{C}=\text{O} \\
| \\
\text{CHOH} \\
| \\
(\text{CHOH})_n \\
| \\
\text{CH}_2\text{OH}
\end{array}
+ 3\text{C}_6\text{H}_5\text{NHNH}_2 \longrightarrow
\begin{array}{l}
\text{H} \\
| \\
\text{C}=\text{NNHC}_6\text{H}_5 \\
| \\
\text{C}=\text{NNHC}_6\text{H}_5 \\
| \\
(\text{CHOH})_n \\
| \\
\text{CH}_2\text{OH}
\end{array}
+ \text{C}_6\text{H}_5\text{NH}_2 + \text{NH}_3 + \text{H}_2\text{O}
$$

      Aldose   **Phenylosazone**

Although the mechanism for osazone formation is not known with certainty, it probably depends on a series of reactions in which $\text{C}=\text{N}-$ behaves very much like $\text{C}=\text{O}$ in giving a nitrogen version of an enol.

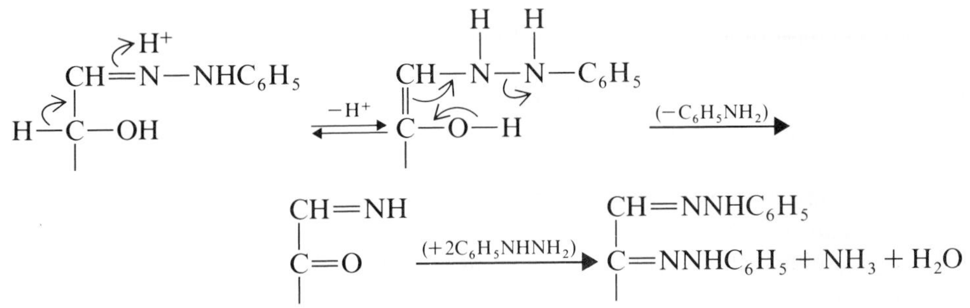

Osazone formation results in a loss of the stereocenter at C-2 but does not affect other stereocenters; D-glucose and D-mannose, for example, yield the same phenylosazone:

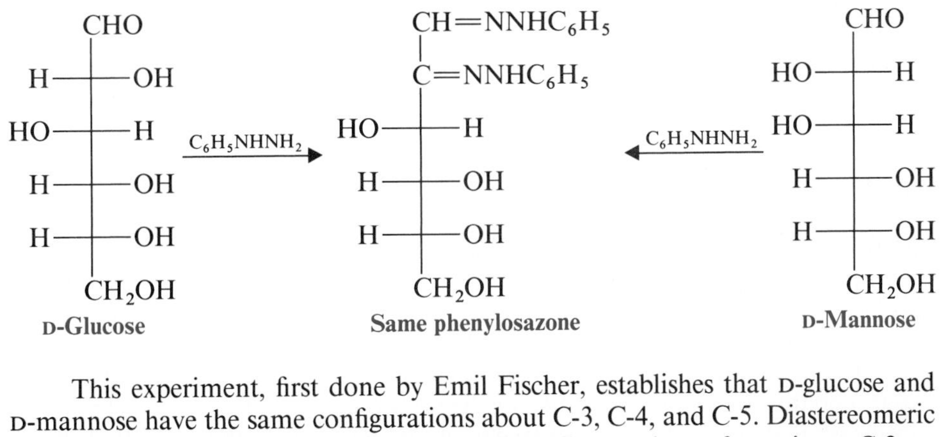

    D-Glucose   **Same phenylosazone**   **D-Mannose**

This experiment, first done by Emil Fischer, establishes that D-glucose and D-mannose have the same configurations about C-3, C-4, and C-5. Diastereomeric aldoses (such as D-glucose and D-mannose) that differ only in configuration at C-2 are called *epimers.**

---

*The term *epimer* has taken on a broader meaning and is now often applied to any pair of diastereomers that differ only in the configuration at a single atom.

PROBLEM 21.16

Although D-fructose is not an epimer of D-glucose or D-mannose (D-fructose is a ketohexose), all three yield the same phenylosazone. (a) Using Fischer projection formulas, write an equation for the reaction of fructose with phenylhydrazine. (b) What information about the stereochemistry of D-fructose does this experiment yield?

21.7 SYNTHESIS AND DEGRADATION OF MONOSACCHARIDES

21.7A Kiliani–Fischer Synthesis

In 1885 Heinrich Kiliani (Freiburg, Germany) discovered that an aldose can be converted to the epimeric aldonic acids having one additional carbon through the addition of hydrogen cyanide and subsequent hydrolysis of the epimeric cyanohydrins. Fischer later extended this method by showing that aldonolactones obtained from the aldonic acids can be reduced to aldoses. Today this method for lengthening the carbon chain of an aldose is called the Kiliani–Fischer synthesis.

We can illustrate the Kiliani–Fischer synthesis with the synthesis of D-threose and D-erythrose (aldotetroses) from D-glyceraldehyde (an aldotriose) in Fig. 21.6.

Addition of hydrogen cyanide to glyceraldehyde produces two epimeric cyanohydrins because the reaction creates a new stereocenter. The cyanohydrins can be separated easily (since they are diastereomers) and each can be converted to an aldose through hydrolysis, acidification, lactonization, and reduction. One cyanohydrin ultimately yields D-(−)-erythrose and the other yields D-(−)-threose.

We can be sure that the aldotetroses that we obtain from this Kiliani–Fischer synthesis are both D sugars because the starting compound is D-glyceraldehyde and its stereocenter is unaffected by the synthesis. On the basis of the Kiliani–Fischer synthesis we cannot know just which aldotetrose has both —OH groups on the right and which has the top —OH on the left in the Fischer projection formulas. However, if we oxidize both aldotetroses to aldaric acids, one [D-(−)-erythrose] will yield an *optically inactive* product while the other [D-(−)-threose] will yield a product that is *optically active* (cf. Problem 21.9).

PROBLEM 21.17

(a) What are the structures of L-(+)-threose and L-(+)-erythrose? (b) What aldotriose would you use to prepare them in a Kiliani–Fischer synthesis?

PROBLEM 21.18

(a) Outline a Kiliani–Fischer synthesis of epimeric aldopentoses starting with D-(−)-erythrose (use Fischer projection formulas). (b) The two epimeric aldopentoses that one obtains are D-(−)-arabinose and D-(−)-ribose. Nitric acid oxidation of D-(−)-ribose yields an optically inactive aldaric acid, while similar oxidation of D-(−)-arabinose yields an optically active product. On the basis of this information alone, which Fischer projection formula represents D-(−)-arabinose and which represents D-(−)-ribose?

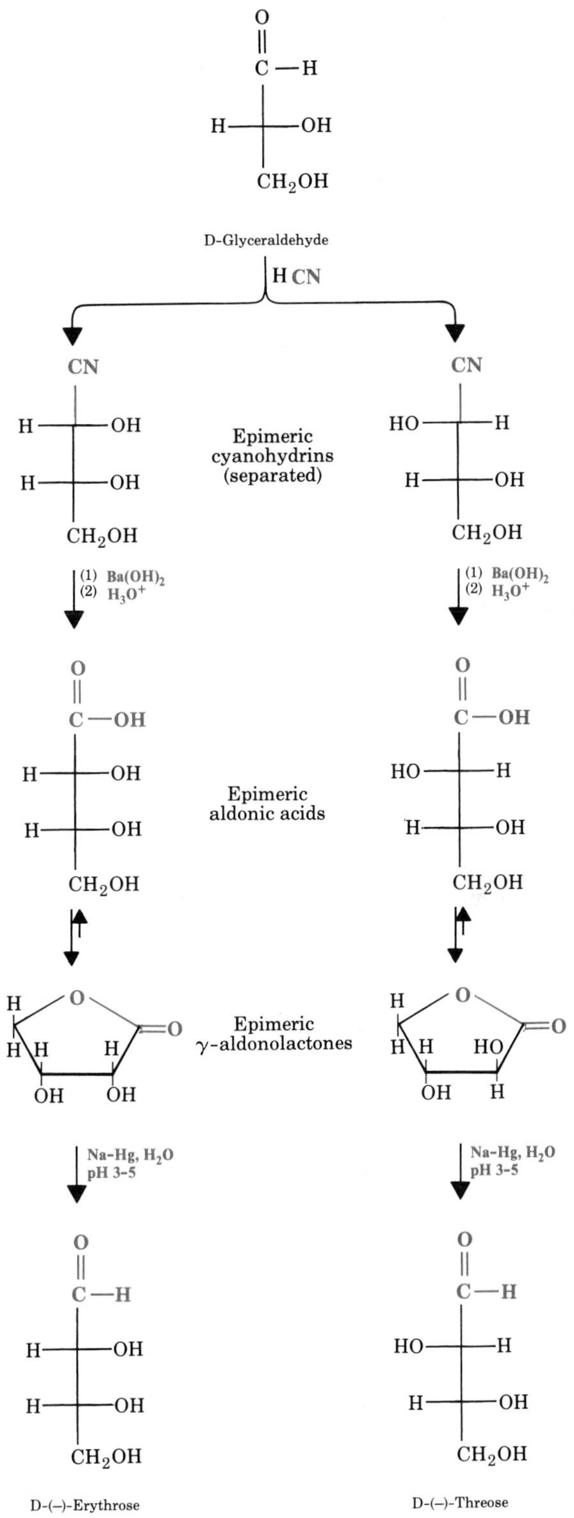

FIGURE 21.6 A Kiliani–Fischer synthesis of D-(−)-erythrose and D-(−)-threose from D-glyceraldehyde.

21.7B The Ruff Degradation

Just as the Kiliani–Fischer synthesis can be used to lengthen the chain of an aldose by one carbon atom, the Ruff degradation* can be used to shorten the chain by a similar unit. The Ruff degradation involves (1) oxidation of the aldose to an aldonic acid using bromine water and (2) oxidative decarboxylation of the aldonic acid to the next lower aldose using hydrogen peroxide and ferric sulfate. D-(−)-Ribose, for example, can be degraded to D-(−)-erythrose:

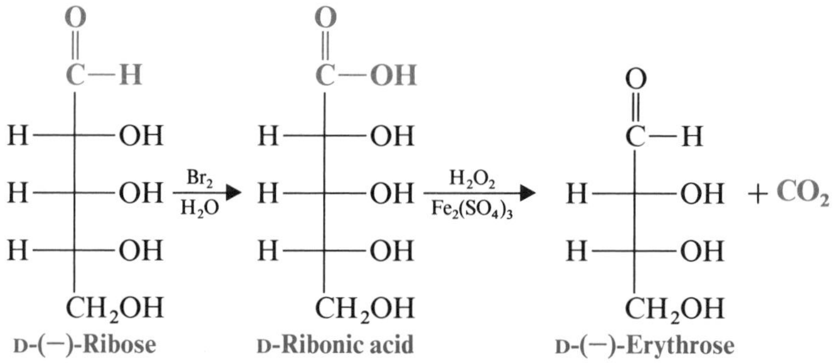

$$
\begin{array}{ccccc}
\underset{\text{D-(−)-Ribose}}{\begin{array}{c} O \\ \parallel \\ C-H \\ H\!-\!\!-\!OH \\ H\!-\!\!-\!OH \\ H\!-\!\!-\!OH \\ CH_2OH \end{array}}
& \xrightarrow[\text{H}_2\text{O}]{\text{Br}_2}
& \underset{\text{D-Ribonic acid}}{\begin{array}{c} O \\ \parallel \\ C-OH \\ H\!-\!\!-\!OH \\ H\!-\!\!-\!OH \\ H\!-\!\!-\!OH \\ CH_2OH \end{array}}
& \xrightarrow[\text{Fe}_2(\text{SO}_4)_3]{\text{H}_2\text{O}_2}
& \underset{\text{D-(−)-Erythrose}}{\begin{array}{c} O \\ \parallel \\ C-H \\ H\!-\!\!-\!OH \\ H\!-\!\!-\!OH \\ CH_2OH \end{array}} \ + \ CO_2
\end{array}
$$

21.8 THE D-FAMILY OF ALDOSES

The Ruff degradation and the Kiliani–Fischer synthesis allow us to place all of the aldoses into families or "family trees" based on their relation to D- or L-glyceraldehyde. Such a tree is constructed in Fig. 21.7 and includes the structures of the D-aldohexoses, **1–8**.

*Developed by Otto Ruff, a German chemist, 1871–1939.

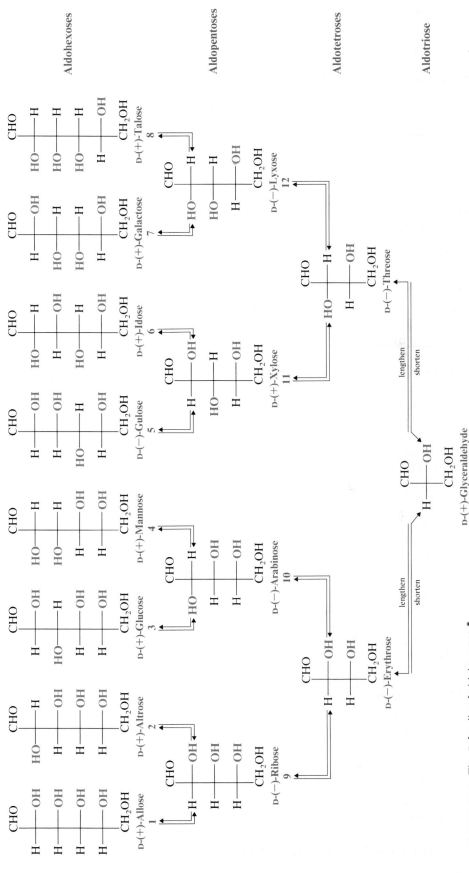

FIGURE 21.7 The D family of aldohexoses.*

*A useful mnemonic for the D-aldohexoses: All altruists gladly make gum in gallon tanks. Write the names in a line and above each write CH_2OH. Then, for C-5 write OH to the right all the way across. For C-4 write OH to the right four times, then four to the left; for C-3, write OH twice to the right, twice to the left, and repeat; C-2, alternate OH and H to the right. (From L. F. Fieser and Mary Fieser, *Organic Chemistry*, Reinhold, New York, 1956, p. 359.)

Most, but not all, of the naturally occurring aldoses belong to the D family with D-(+)-glucose being by far the most common. D-(+)-Galactose can be obtained from milk sugar (lactose); but L-(−)-galactose occurs in a polysaccharide obtained from the vineyard snail, *Helix pomatia.* L-(+)-Arabinose is found widely, but D-(−)-arabinose is scarce, being found only in certain bacteria and sponges. Threose, lyxose, gulose, and allose do not occur naturally, but one or both forms (D or L) of each have been synthesized.

21.9 FISCHER'S PROOF OF THE CONFIGURATION OF D-(+)-GLUCOSE

Emil Fischer began his work on the stereochemistry of (+)-glucose in 1888, only 12 years after van't Hoff and Le Bel had made their proposal concerning the tetrahedral structure of carbon. Only a small body of data was available to Fischer at the beginning: Only a few monosaccharides were known, including (+)-glucose, (+)-arabinose, and (+)-mannose. [(+)-Mannose had just been synthesized by Fischer.] The sugars (+)-glucose and (+)-mannose were known to be aldohexoses; (+)-arabinose was known to be an aldopentose.

Since an aldohexose has four stereocenters, 2^4 (or 16) stereoisomers are possible — *one of which is* (+)-*glucose.* Fischer arbitrarily decided to limit his attention to the eight structures with the D configuration given in Fig. 21.7 (structures **1 – 8**). Fischer realized that he would be unable to differentiate between enantiomeric configurations because methods for determining the absolute configuration of organic compounds had not been developed. It was not until 1951, when Bijvoet (Section 4.13A) determined the absolute configuration of L-(+)-tartaric acid [and, hence, D-(+)-glyceraldehyde] that Fischer's arbitrary assignment of (+)-glucose to the family we call the D family was known to be correct.

Fischer's assignment of structure **3** to (+)-glucose was based on the following reasoning:

1. Nitric acid oxidation of (+)-glucose gives an optically active aldaric acid. This eliminates structures **1** and **7** from consideration because both compounds would yield *meso*-aldaric acids.

2. *Degradation of* (+)-*glucose gives* (−)-*arabinose, and nitric acid oxidation of* (−)-*arabinose gives an optically active aldaric acid.* This means that (−)-arabinose cannot have configurations **9** or **11** and must have either structure **10** or **12**. It also establishes that (+)-glucose cannot have configuration **2, 5,** or **6**. This leaves structures **3, 4,** and **8** as possibilities for (+)-glucose.

3. A Kiliani–Fischer synthesis beginning with (−)-arabinose gives (+)-glucose and (+)-mannose; nitric acid oxidation of (+)-mannose gives an optically active aldaric acid. This together with the fact that (+)-glucose yields a different but also optically active aldaric acid establishes structure **10** as the structure of (−)-arabinose and eliminates structure **8** as a possible structure for (+)-glucose. Had (−)-arabinose been represented by structure **12**, a Kiliani–Fischer synthesis would have given the two aldohexoses, **7** and **8**, one of which (**7**) would yield an optically inactive aldaric acid on nitric acid oxidation.

4. Two structures now remain, **3** and **4**; one structure represents (+)-glucose and one represents (+)-mannose. Fischer realized that (+)-glucose and (+)-mannose were epimeric (at C-2), but a decision as to which compound was represented by which structure was most difficult.

5. Fischer had already developed a method for effectively *interchanging the two end groups* (CHO and CH_2OH) *of an aldose chain.* And, with brilliant logic, Fischer realized that if (+)-glucose has structure **4**, an interchange of end groups *will yield the same aldohexose:*

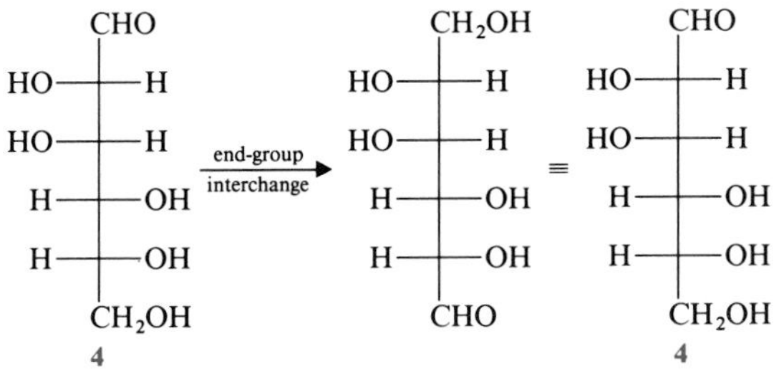

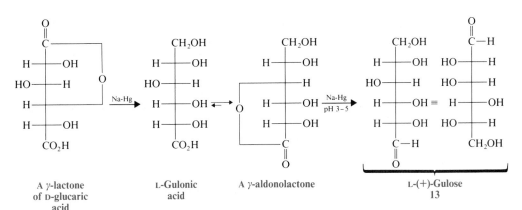

On the other hand, if (+)-glucose has structure **3**, *an end-group interchange will yield a different aldohexose,* **13**:

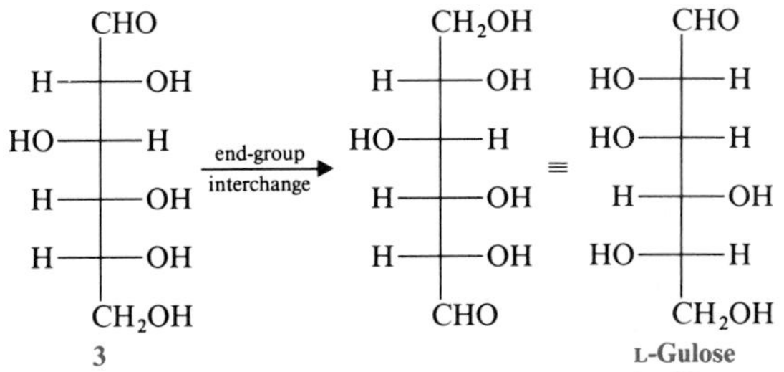

This new aldohexose, if it were formed, would be an L-sugar and it would be the mirror reflection of D-gulose. Thus its name would be L-gulose.

Fischer carried out the end-group interchange starting with (+)-glucose and *the product was the new aldohexose* **13**. This outcome proved that (+)-glucose has structure **3**. It also established **4** as the structure for (+)-mannose, and it proved the structure of L-(+)-glucose as **13**.

The procedure Fischer used for interchanging the ends of the (+)-glucose chain began with one of the γ-lactones of D-glucaric acid (cf. Problem 21.10) and was carried out as follows:

A γ-lactone of D-glucaric acid

L-Gulonic acid

A γ-aldonolactone

L-(+)-Gulose

13

Notice in this synthesis that the second reduction with Na-Hg is carried out at pH 3–5. Under these conditions reduction of the lactone yields an aldehyde and not a primary alcohol.

PROBLEM 21.22

Fischer actually had to subject both γ-lactones of D-glucaric acid (Problem 21.10) to the procedure just outlined. What product does the other γ-lactone yield?

21.10 METHYLATION OF MONOSACCHARIDES

Since methyl glycosides are acetals, they do not undergo isomerization or fragmentation in basic media. Thus, a methyl glucoside can be converted to a pentamethyl derivative by treating it with excess dimethyl sulfate in aqueous sodium hydroxide. Subsequent acid-catalyzed hydrolysis of the glycosidic methyl produces a tetramethyl-D-glucose.

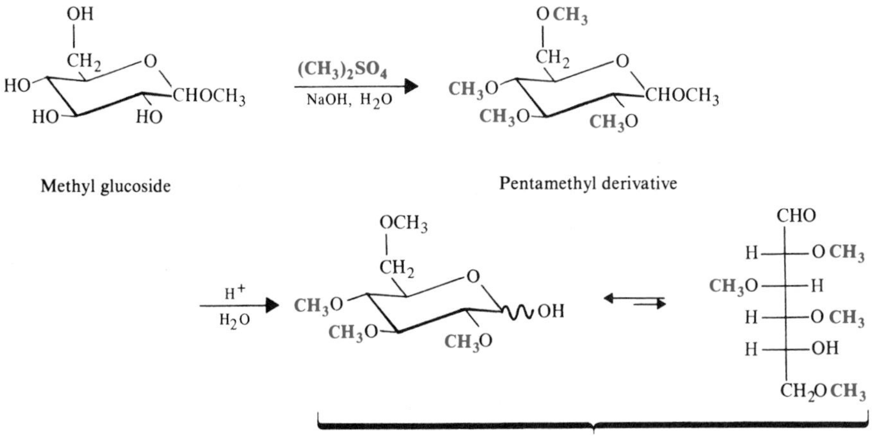

Methyl glucoside   Pentamethyl derivative

(2, 3, 4, 6-tetra-*O*-Methyl-D-glucose)*

This method (developed by Haworth) allowed Hirst (in 1926) to demonstrate that the methyl glucosides have six-membered rings. Hirst oxidized the tetramethyl ether given by the reaction above and obtained a trimethoxyglutaric acid **A** and a dimethoxysuccinic acid **B**. These two products must come from cleavage at one of the two carbon–carbon bonds of the carbon bearing the hydroxyl group (in the open chain) and demonstrate that the 5-OH was involved in the oxide ring. (When the bond between C-5 and C-6 breaks, C-5 becomes a CO_2H group and the product is **A**. When the bond between C-4 and C-5 breaks, C-4 becomes a CO_2H group and the product is **B**.)

---

*The designation ". . . -*O*-methyl . . ." in this name means that the methyl groups are attached to oxygen atoms.

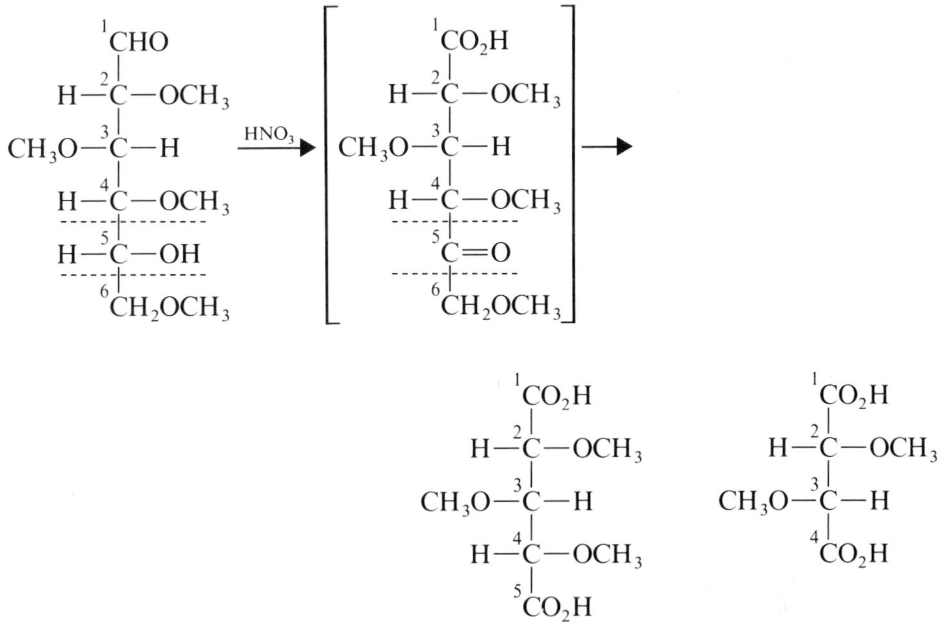

$$
\begin{array}{c}
\overset{1}{C}HO \\
H-\overset{2}{C}-OCH_3 \\
CH_3O-\overset{3}{C}-H \\
H-\overset{4}{C}-OCH_3 \\
H-\overset{5}{C}-OH \\
\overset{6}{C}H_2OCH_3
\end{array}
\xrightarrow{HNO_3}
\left[
\begin{array}{c}
\overset{1}{C}O_2H \\
H-\overset{2}{C}-OCH_3 \\
CH_3O-\overset{3}{C}-H \\
H-\overset{4}{C}-OCH_3 \\
\overset{5}{C}=O \\
\overset{6}{C}H_2OCH_3
\end{array}
\right]
\longrightarrow
$$

$$
\begin{array}{c}
\overset{1}{C}O_2H \\
H-\overset{2}{C}-OCH_3 \\
CH_3O-\overset{3}{C}-H \\
H-\overset{4}{C}-OCH_3 \\
\overset{5}{C}O_2H
\end{array}
\qquad\qquad
\begin{array}{c}
\overset{1}{C}O_2H \\
H-\overset{2}{C}-OCH_3 \\
CH_3O-\overset{3}{C}-H \\
\overset{4}{C}O_2H
\end{array}
$$

A trimethoxy-
glutaric acid
A

A dimethoxy-
succinic acid
B

PROBLEM 21.23

What products would have been obtained from the Haworth–Hirst procedure if the methyl glucoside had been a furanoside?

Methylation procedures, as we shall see in the next section, are especially important in deciphering the structures of disaccharides.

21.11 DISACCHARIDES

21.11A Sucrose

Ordinary table sugar is a disaccharide called *sucrose*. Sucrose, the most widely occurring disaccharide, is found in all photosynthetic plants and is obtained commercially from sugar cane or sugar beets. Sucrose has the structure shown in Fig. 21.8.

The structure of sucrose is based on the following evidence:

1. Sucrose has the molecular formula $C_{12}H_{22}O_{11}$.

2. Acid-catalyzed hydrolysis of 1 mole of sucrose yields 1 mole of D-glucose and 1 mole of D-fructose.

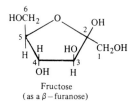

FIGURE 21.8 Two representations of the formula for (+)-sucrose (α-D-glucopyranosyl β-D-fructofuranoside).

Fructose
(as a β – furanose)

3. Sucrose is a nonreducing sugar; it gives negative tests with Benedict's and Tollens' solutions. Sucrose does not form an osazone and does not undergo mutarotation. These facts mean that neither the glucose nor the fructose portions of sucrose has a hemiacetal or hemiketal group. Thus, the two hexoses must have a glycoside linkage that involves C-1 of glucose and C-2 of fructose, for only in this way will both carbonyl groups be present as full acetals or ketals (i.e., as glycosides).

4. The stereochemistry of the glycoside linkages can be inferred from experiments done with enzymes. Sucrose is hydrolyzed by an *α-glucosidase* obtained from yeast but not by *β*-glucosidases. This hydrolysis indicates *an α configuration at the glucoside portion.* Sucrose is also hydrolyzed by *sucrase,* an enzyme known to hydrolyze *β*-fructofuranosides but not *α*-fructofuranosides. This hydrolysis indicates *a β configuration at the fructoside portion.*

5. Methylation of sucrose gives an octamethyl derivative that, on hydrolysis, gives 2,3,4,6-tetra-*O*-methyl-D-glucose and 1,3,4,6-tetra-*O*-methyl-D-fructose. The identities of these two products demonstrate that the glucose portion is a *pyranoside* and that the fructose portion is a *furanoside.*

The structure of sucrose has been confirmed by X-ray analysis and by an unambiguous synthesis.

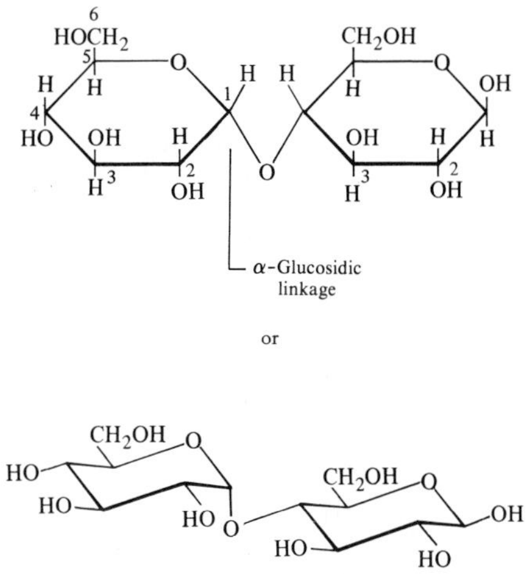

FIGURE 21.9 Two representations of the structure of the β anomer of (+)-maltose, 4-O-(α-D-glucopyranosyl)-β-D-glucopyranose.

21.11B Maltose

When starch (Section 21.12A) is hydrolyzed by the enzyme *diastase,* one product is a disaccharide known as *maltose* (Fig. 21.9).

1. When 1 mole of maltose is subjected to acid-catalyzed hydrolysis it yields 2 moles of D-(+)-glucose.

2. Unlike sucrose, *maltose is a reducing sugar;* it gives positive tests with Fehling's, Benedict's, and Tollens' solutions. Maltose also reacts with phenylhydrazine to form a monophenylosazone (i.e., it incorporates two molecules of phenylhydrazine).

3. Maltose exists in two anomeric forms; α-maltose, $[\alpha]_D^{25} = 168°$, and β-maltose, $[\alpha]_D^{25} = +112°$. The maltose anomers undergo mutarotation to yield an equilibrium mixture, $[\alpha]_D^{25} = +136°$.

Facts **2** and **3** demonstrate that one of the glucose residues of maltose is present in a hemiacetal form; the other, therefore, must be present as a glucoside. The configuration of this glucosidic linkage can be inferred as α, because maltose is hydrolyzed by α-glucosidases and not by β-glucosidases.

4. Maltose reacts with bromine water to form a monocarboxylic acid, maltonic acid (Fig. 21.10*a*). This fact, too, is consistent with the presence of only one hemiacetal group.

5. Methylation of maltonic acid followed by hydrolysis gives 2,3,4,6-tetra-*O*-methyl-D-glucose and 2,3,5,6-tetra-*O*-methyl-D-gluconic acid. That the first product has a free —OH at C-5 indicates that the nonreducing glucose portion is present as a pyranoside; that the second product, 2,3,5,6-tetra-*O*-methyl-D-gluconic acid, has a free —OH at C-4 indicates that this position was involved in a glycosidic linkage with the nonreducing glucose.

Only the size of the reducing glucose ring needs to be determined.

6. Methylation of maltose itself, followed by hydrolysis (Fig. 21.10*b*), gives 2,3,4,6-tetra-*O*-methyl-D-glucose and 2,3,6-tri-*O*-methyl-D-glucose. The free —OH at C-5 in the latter product indicates that it must have been involved in the oxide ring and that the reducing glucose is present as a *pyranose.*

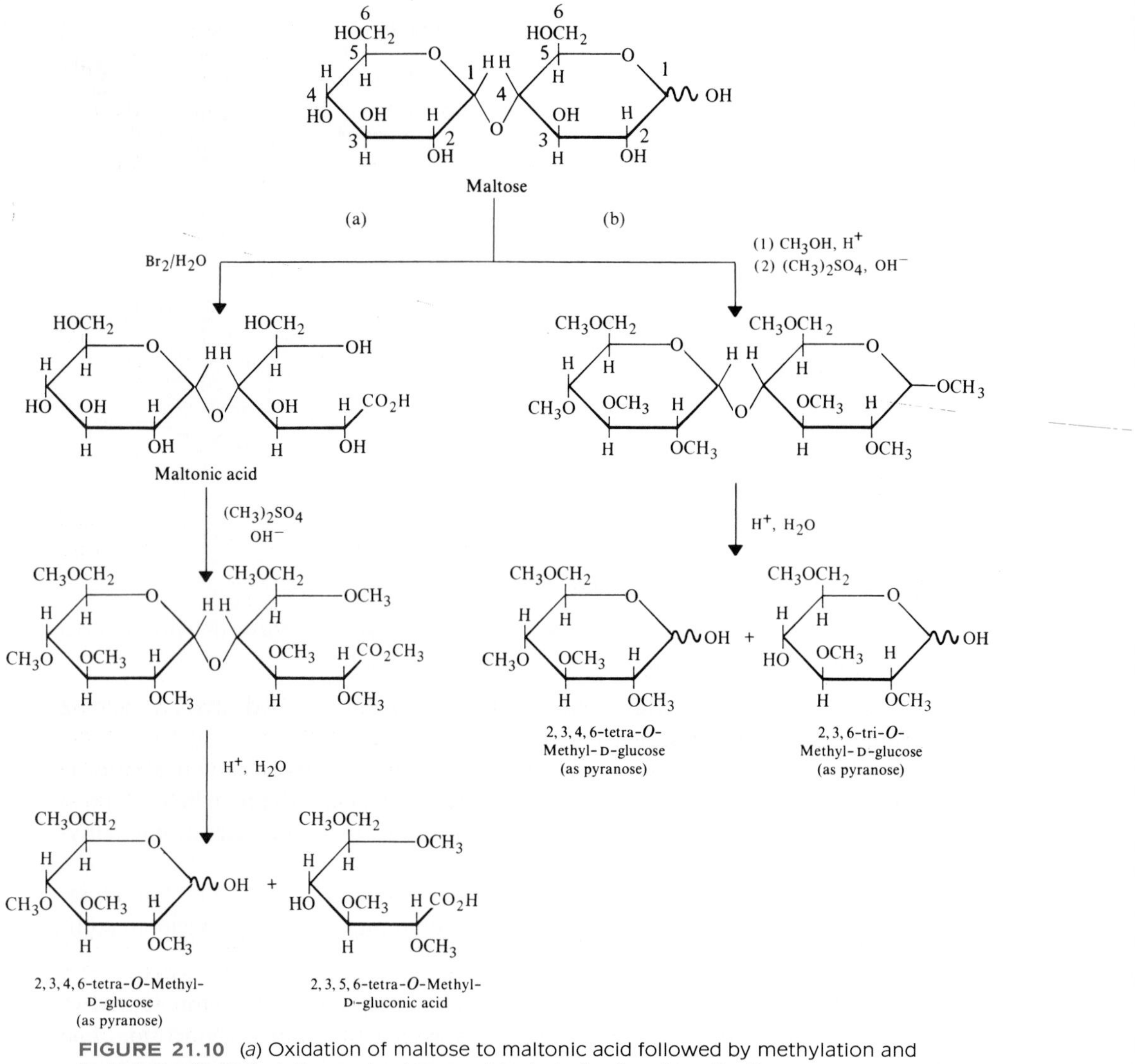

FIGURE 21.10 (a) Oxidation of maltose to maltonic acid followed by methylation and hydrolysis. (b) Methylation and subsequent hydrolysis of maltose itself.

21.11C Cellobiose

Partial hydrolysis of cellulose (Section 21.12C) gives the disaccharide cellobiose ($C_{12}H_{22}O_{11}$) (Fig. 21.11). Cellobiose resembles maltose in every respect except one: The configuration of its glycosidic linkage.

Cellobiose, like maltose, is a reducing sugar that, on acid-catalyzed hydrolysis, yields two molar equivalents of D-glucose. Cellobiose also undergoes mutarotation and forms a phenylosazone. Methylation studies show that C-1 of one glucose unit is connected in glycosidic linkage with C-4 of the other and that both rings are six membered. Unlike maltose, however, cellobiose is hydrolyzed by β-*glucosidases* and not by α-glucosidases: This indicates that the glycosidic linkage in cellobiose is β (Fig. 21.11).

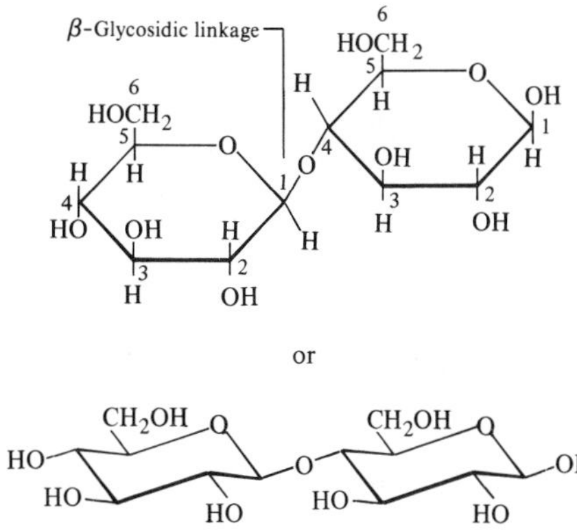

FIGURE 21.11 Two representations of the β anomer of cellobiose, 4-O-(β-D-glucopyrano-syl)-β-D-glucopyranose.

21.11D Lactose

Lactose (Fig. 21.12) is a disaccharide present in the milk of humans, cows, and almost all other mammals. Lactose is a reducing sugar that hydrolyzes to yield D-glucose and D-galactose; the glycosidic linkage is β.

21.12 POLYSACCHARIDES

Three important polysaccharides, all of which are polymers of D-glucose, are starch, glycogen, and cellulose. Starch is the principal food reserve of plants; glycogen functions as a carbohydrate reserve for animals; and cellulose serves as structural material in plants. As we examine the structures of these three polysaccharides we shall be able to see how each is especially suited for its function.

21.12A Starch

Starch occurs as microscopic granules in the roots, tubers, and seeds of plants. Corn, potatoes, wheat, and rice are important commercial sources. Heating starch with water causes the granules to swell and produce a colloidal suspension from which two major components can be isolated. One fraction is called *amylose* and the other *amylopectin*. Most starches yield 10–20% amylose and 80–90% amylopectin.

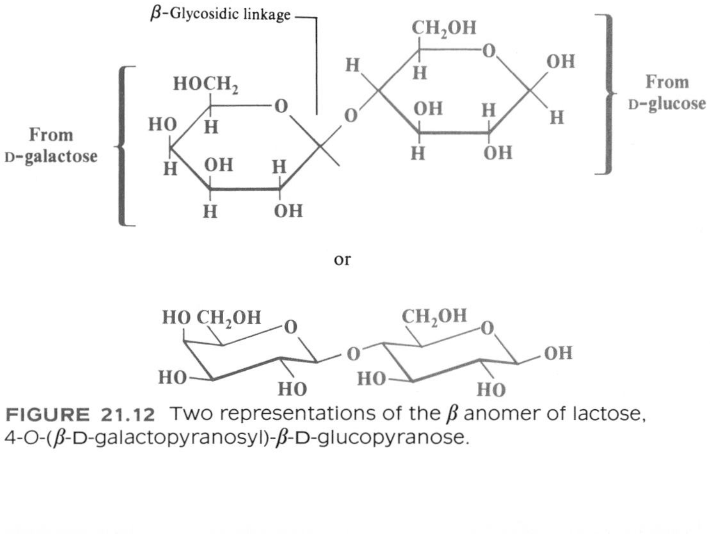

FIGURE 21.12 Two representations of the β anomer of lactose, 4-O-(β-D-galactopyranosyl)-β-D-glucopyranose.

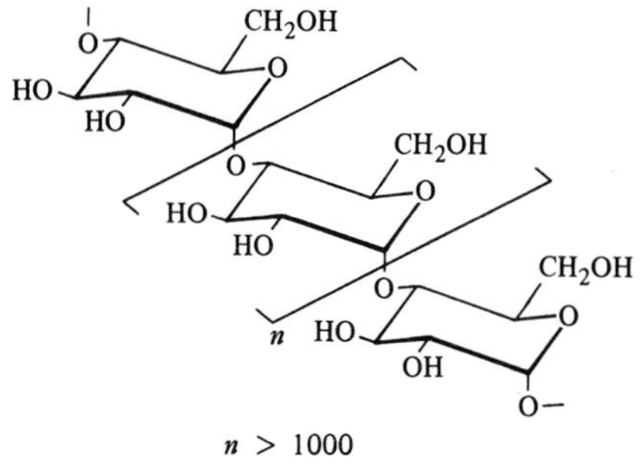

FIGURE 21.13 Partial structure of amylose, an unbranched polymer of D-glucose connected in $\alpha, 1:4$-glycosidic linkages.

Physical measurements show that amylose typically consists of more than 1000 D-glucopyranoside units *connected in α linkages* between C-1 of one unit and C-4 of the next (Fig. 21.13). Thus, in the ring size of its glucose units and in the configuration of the glycosidic linkages between them, amylose resembles maltose.

Chains of D-glucose units with α-glycosidic linkages such as those of amylose tend to assume a helical arrangement. This arrangement results in a compact shape for the amylose molecule even though its molecular weight is quite large (150,000– 600,000).

Amylopectin has a structure similar to that of amylose (i.e., $\alpha, 1:4$ links), with the exception that in amylopectin the chains are branched. Branching takes place between C-6 of one glucose unit and C-1 of another and occurs at intervals of 20–25 glucose units (Fig. 21.14). Physical measurements indicate that amylopectin has a molecular weight of one to six million; thus amylopectin consists of hundreds of interconnecting chains of 20–25 glucose units each.

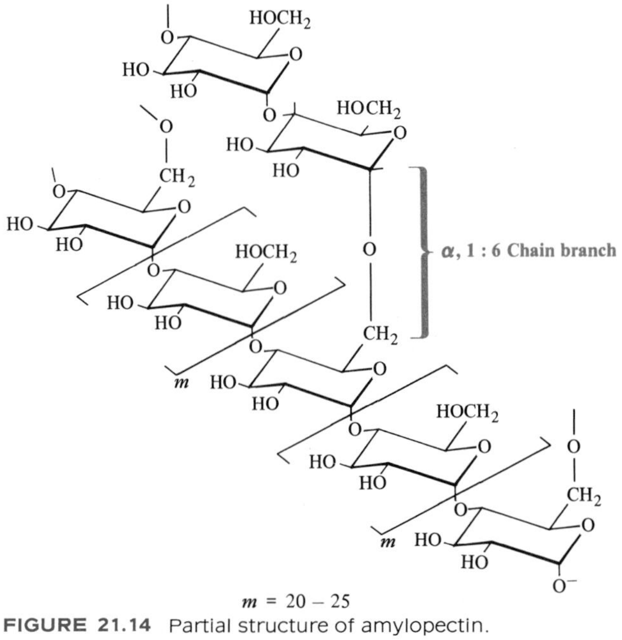

$m = 20 - 25$
FIGURE 21.14 Partial structure of amylopectin.

21.12B Glycogen

Glycogen has a structure very much like that of amylopectin; however, in glycogen the chains are much more highly branched. Methylation and hydrolysis of glycogen indicates that there is one end group for every 10 to 12 glucose units; branches may occur as often as every 6. Glycogen has a very high molecular weight. Studies of glycogens isolated under conditions that minimize the likelihood of hydrolysis indicate molecular weights as high as 100 million.

The size and structure of glycogen beautifully suit its function as reserve carbohydrate for animals. First, its size makes it too large to diffuse across cell membranes; thus, glycogen remains inside the cell where it is needed as an energy source. Second, because glycogen incorporates tens of thousands of glucose units in a single molecule, it solves an important osmotic problem for the cell. Were so many glucose units present in the cell as individual molecules, the osmotic pressure within the cell would be enormous—so large that the cell membrane would almost certainly break.* Finally, the localization of glucose units within a large, highly branched structure simplifies one of the cell's logistical problems: That of having a ready source of glucose when cellular glucose concentrations are low and of being able to store glucose rapidly when cellular glucose concentrations are high. There are enzymes within the cell that catalyze the reactions by which glucose units are detached from (or attached to) glycogen. These enzymes operate at end groups by hydrolyzing (or forming) $\alpha,1:4$ glycosidic linkages. Because glycogen is so highly branched, a very large number of end groups are available at which these enzymes can operate. At the same time the overall concentration of glycogen (in moles per liter) is quite low because of its enormous molecular weight.

Amylopectin presumably serves a similar function in plants. The fact that amylopectin is less highly branched than glycogen is, however, not a serious disadvantage. Plants have a much lower metabolic rate than animals—and plants, of course, do not require sudden bursts of energy.

Animals store energy as fats (glyceryl trialkanoates) as well as in glycogen. Fats, because they are more highly reduced, are capable of furnishing much more energy. The metabolism of a typical fatty acid, for example, liberates more than twice as much energy per carbon as glucose or glycogen. Why then, we might ask, has Nature developed two different repositories? Glucose (from glycogen) is readily available and is highly water soluble.† Glucose, as a result, diffuses rapidly through the aqueous medium of the cell and serves as an ideal source of "ready energy." Long-chain fatty acids, by contrast, are almost insoluble in water and their concentration inside the cell could never be very high. They would be a poor source of energy if the cell were in an energy pinch. On the other hand, fatty acids (as glyceryl trialkanoates) because of their caloric richness are an excellent energy repository for long-term energy storage.

21.12C Cellulose

When we examine the structure of cellulose we find another example of a polysaccharide in which nature has arranged monomeric glucose units in a manner that suits its

---

*The phenomenon of osmotic pressure occurs whenever two solutions of different concentrations are separated by a membrane that will allow penetration (by osmosis) of the solvent but not of the solute. The osmotic pressure, π, on one side of the membrane is related to the number of moles of solute particles, n, the volume of the solution, V, and RT (the gas constant times the absolute temperature): $\pi V = nRT$.

†Glucose is actually liberated as glucose-6-phosphate, which is also water soluble.

function. Cellulose contains D-glucopyranoside units linked in 1:4 fashion in very long unbranched chains. Unlike starch and glycogen, however, the linkages in cellulose are *β-glycosidic linkages* (Fig. 21.15). This configuration of the anomeric carbon atoms of cellulose makes cellulose chains essentially linear; they do not tend to coil into helical structures as do glucose polymers when linked in an α,1:4 manner.

The linear arrangement of β-linked glucose units in cellulose presents a uniform distribution of —OH groups on the outside of each chain. When two or more cellulose chains make contact, the hydroxyl groups are ideally situated to "zip" the chains together by forming hydrogen bonds. Zipping many cellulose chains together in this way gives a highly insoluble, rigid, and fibrous polymer that is ideal as cell-wall material for plants.

This special property of cellulose chains, we should emphasize, is not just a result of β,1:4 glycosidic linkages; it is also a consequence of the precise stereochemistry of D-glucose at each stereocenter. Were D-galactose or D-allose units linked in a similar fashion, they almost certainly would not give rise to a polymer with properties like cellulose. Thus, we get another glimpse of why D-glucose occupies such a special position in the chemistry of plants and animals. Not only is it the most stable aldohexose (because it can exist in a chair conformation that allows all of its bulky groups to occupy equatorial positions), but its special stereochemistry also allows it to form helical structures when α linked as in starches, and rigid linear structures when β linked as in cellulose.

> Another interesting and important fact about cellulose: The digestive enzymes of humans cannot attack its β,1:4 linkages. Hence, cellulose cannot serve as a food source for humans, as can starch. Cows and termites, however, can use cellulose (of grass and wood) as a food source because symbiotic bacteria in their digestive systems furnish β-glucosidases.

Perhaps we should ask ourselves one other question: Why has Nature "chosen" D-(+)-glucose for its special role rather than L-(−)-glucose, its mirror image? Here an answer cannot be given with any certainty. The selection of D-(+)-glucose may simply have been a random event early in the course of the evolution of enzyme catalysts. Once this selection was made, however, the stereogenicity of the active sites of the enzymes involved would retain a bias toward D-(+)-glucose and against L-(−)-glucose (because of the improper fit of the latter). Once introduced, this bias would be perpetuated and extended to other catalysts.

Finally, when we speak of Nature selecting or choosing a particular molecule for a given function, we do not mean to imply that evolution operates on a molecular level. Evolution, of course, takes place at the level of organism populations, and molecules are selected only in the sense that their use gives the organism an increased likelihood of surviving and procreating.

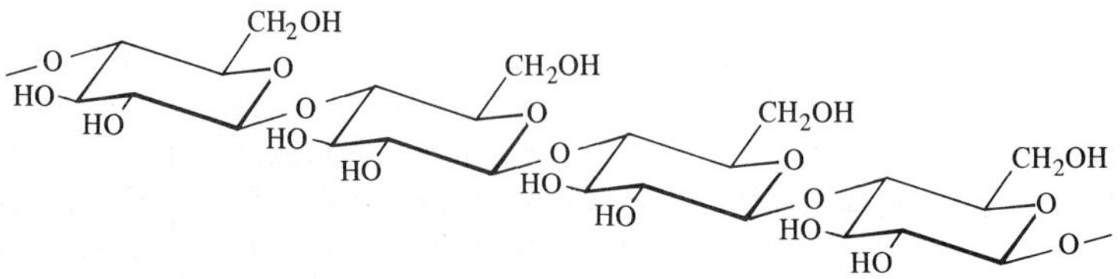

FIGURE 21.15 A portion of a cellulose chain.

21.12D Cellulose Derivatives

A number of derivatives of cellulose are used commercially. Most of these are compounds in which two or three of the free hydroxyl groups of each glucose unit have been converted to an ester or an ether. This conversion substantially alters the physical properties of the material, making it more soluble in organic solvents and allowing it to be made into fibers and films. Treating cellulose with acetic anhydride produces the triacetate known as "Arnel" or "acetate," used widely in the textile industry. Cellulose trinitrate, also called "gun cotton" or nitrocellulose, is used in explosives.

Rayon is made by treating cellulose (from cotton or wood pulp) with carbon disulfide in a basic solution. This reaction converts cellulose to a soluble xanthate:

$$\text{Cellulose}-\text{OH} + \text{CS}_2 \xrightarrow{\text{NaOH}} \text{cellulose}-\text{O}-\overset{\overset{\displaystyle S}{\|}}{\text{C}}-\text{S}^-\text{Na}^+$$

Cellulose xanthate

The solution of cellulose xanthate is then passed through a small orifice or slit into an acidic solution. This operation regenerates the —OH groups of cellulose causing it to precipitate as a fiber or a sheet.

$$\text{Cellulose}-\text{O}-\overset{\overset{\displaystyle S}{\|}}{\text{C}}-\text{S}^-\text{Na}^+ \xrightarrow{\text{H}^+} \text{cellulose}-\text{OH}$$

Rayon or cellophane

The fibers are *rayon;* the sheets, after softening with glycerol, are *cellophane.*

21.13 SUGARS THAT CONTAIN NITROGEN

21.13A Glycosylamines

A sugar in which an amino group replaces the anomeric —OH is called a glycosylamine. Examples are β-D-glucopyranosylamine and adenosine (see following figures).

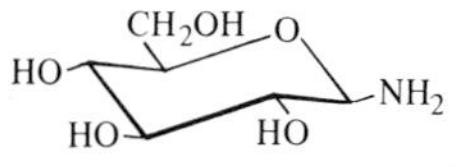

β−D− Glucopyranosyl amine 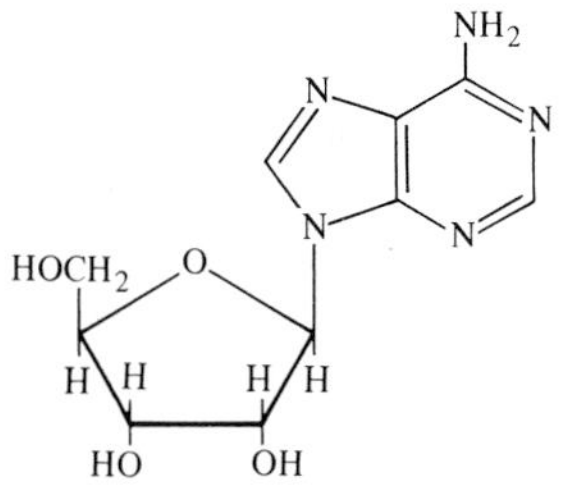   Adenosine

Adenosine is an example of a glycosylamine that is also called a **nucleoside.** Nucleosides are glycosylamines in which the amino component is a pyrimidine or a purine (Section 19.1B) and in which the sugar component is either D-ribose or 2-deoxy-D-ribose (i.e., D-ribose minus the oxygen at the 2 position). Nucleosides are the important components of RNA (ribonucleic acid) and DNA (deoxyribonucleic acid). We shall describe their properties in detail in Section 23.10.

21.13B Amino Sugars

A sugar in which an amino group replaces a nonanomeric —OH group is called amino sugar. Amino sugars occur widely in nature. Two frequently encountered are D-glucosamine and D-galactosamine. D-Glucosamine can be obtained by hydrolysis of chitin, a polysaccharide found in the shells of lobsters and crabs and the external skeletons of insects. D-Galactosamine can be prepared by hydrolysis of chondroitin, a polysaccharide found in cartilage and nasal mucus.

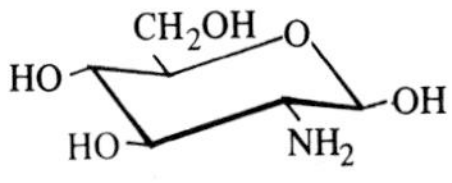

β–D–Glucosamine
(β–2–amino-2-deoxy-
D–glucopyranose)

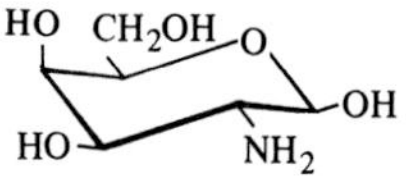

β–D–Galactosamine
(β–2–amino-2-deoxy-
D–galactopyranose)

21.13C Carbohydrate Antibiotics

One of the important discoveries in carbohydrate chemistry was the isolation (in 1944) of the carbohydrate antibiotic called *streptomycin*. Streptomycin is made up of the following three subunits:

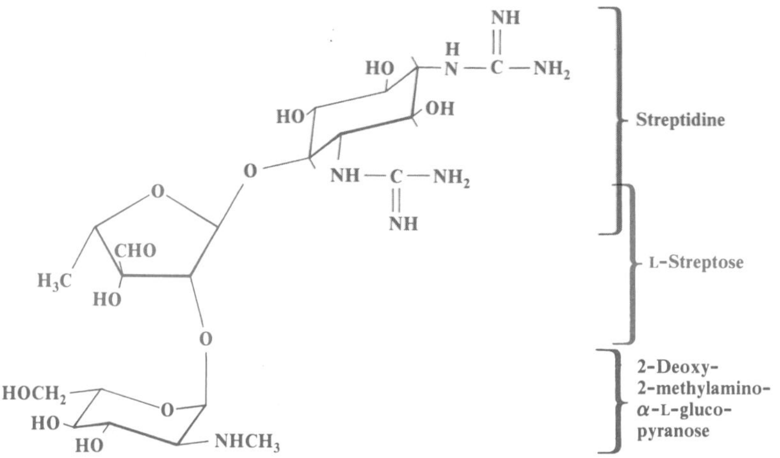

All three components are unusual: The amino sugar is based on L-glucose; streptose is a branched-chain monosaccharide; and streptidine is not a sugar at all, but is a cyclohexane derivative called an amino cyclitol.

Other members of this family are antibiotics called kanamycins, neomycins, and gentamicins (not shown). All are based on an amino cyclitol linked to one or more amino sugars. The glycosidic linkage is nearly always α. These antibiotics are especially useful against bacteria that are resistant to penicillins.

ADDITIONAL PROBLEMS
21.24

Give appropriate structural formulas to illustrate each of the following:

(a) An aldopentose (c) An L-monosaccharide (e) An aldonic acid

(b) A ketohexose (d) A glycoside (f) An aldaric acid

(g) An aldonolactone	(k) A pyranoside	(o) A phenylosazone
(h) A pyranose	(l) A furanoside	(p) A disaccharide
(i) A furanose	(m) Epimers	(q) A polysaccharide
(j) A reducing sugar	(n) Anomers	(r) A nonreducing sugar

21.25
Draw conformational formulas for each of the following: (a) α-D-allopyranose, (b) methyl β-D-allopyranoside, and (c) methyl 2,3,4,6-tetra-O-methyl-β-D-allopyranoside.

21.26
Draw structures for furanose and pyranose forms of D-ribose. Show how you could use periodate oxidation to distinguish between a methyl ribofuranoside and a methyl ribopyranoside.

21.27
One reference book lists D-mannose as being dextrorotatory; another lists it as being levorotatory. Both references are correct. Explain.

21.28
The starting material for a commercial synthesis of vitamin C is L-sorbose (see following reaction); it can be synthesized from D-glucose through the following reaction sequence:

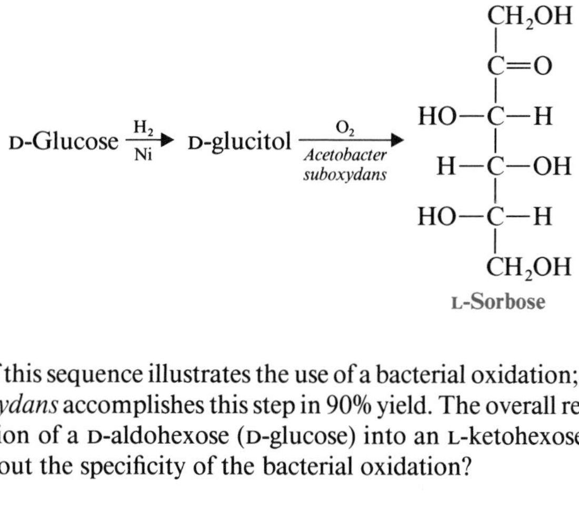

$$\text{D-Glucose} \xrightarrow[\text{Ni}]{H_2} \text{D-glucitol} \xrightarrow[\substack{Acetobacter \\ suboxydans}]{O_2}$$

$$\begin{array}{c} CH_2OH \\ | \\ C{=}O \\ | \\ HO-C-H \\ | \\ H-C-OH \\ | \\ HO-C-H \\ | \\ CH_2OH \end{array}$$

L-Sorbose

The second step of this sequence illustrates the use of a bacterial oxidation; the microorganism *Acetobacter suboxydans* accomplishes this step in 90% yield. The overall result of the synthesis is the transformation of a D-aldohexose (D-glucose) into an L-ketohexose (L-sorbose). What does this mean about the specificity of the bacterial oxidation?

21.29
What two aldoses would yield the same phenylosazone as L-sorbose (Problem 21.28)?

21.30
In addition to fructose (Problem 21.16) and sorbose (Problem 21.28) there are two other 2-ketohexoses, *psicose* and *tagatose*. D-Psicose yields the same phenylosazone as D-allose (or D-altrose); D-tagatose yields the same osazone as D-galactose (or D-talose). What are the structures of D-psicose and D-tagatose?

21.31
A, **B**, and **C** are three aldohexoses. **A** and **B** yield the same optically active alditol when they are reduced with hydrogen and a catalyst; **A** and **B** yield different phenylosazones when treated with phenylhydrazine; **B** and **C** give the same phenylosazone but different alditols. Assuming that all are D-sugars, give names and structures for **A**, **B**, and **C**.

21.32

Although monosaccharides undergo complex isomerizations in base (cf. Problem 21.7), aldonic acids are epimerized specifically at C-2 when they are heated with pyridine. Show how you could make use of this reaction in a synthesis of D-mannose from D-glucose.

21.33

(a) The most stable conformation of most aldopyranoses is one in which the largest group — the —CH$_2$OH group — is equatorial. However, D-idopyranose exists primarily in a conformation with an axial —CH$_2$OH group. Write formulas for the two chair conformations of α-D-idopyranose (one with the —CH$_2$OH group axial and one with the —CH$_2$OH group equatorial) and provide an explanation.

21.34

(a) Heating D-altrose with dilute acid produces a nonreducing *anhydro sugar* (C$_6$H$_{10}$O$_5$). Methylation of the anhydro sugar followed by acid hydrolysis yields 2,3,4-tri-*O*-methyl-D-altrose. The formation of the anhydro sugar takes place through a chair conformation of β-D-altropyranose in which the —CH$_2$OH group is axial. What is the structure of the anhydro sugar and how is it formed? (b) D-Glucose also forms an anhydro sugar but the conditions required are much more drastic than for the corresponding reaction of D-altrose. Explain.

21.35

Show how the following experimental evidence can be used to deduce the structure of lactose (Section 21.11D).

1. Acid hydrolysis of lactose (C$_{12}$H$_{22}$O$_{11}$) gives equimolar quantities of D-glucose and D-galactose. Lactose undergoes a similar hydrolysis in the presence of a β-*galactosidase.*

2. Lactose is a reducing sugar and forms a phenylosazone; it also undergoes mutarotation.

3. Oxidation of lactose with bromine water followed by hydrolysis with dilute acid gives D-galactose and D-gluconic acid.

4. Bromine water oxidation of lactose followed by methylation and hydrolysis gives 2,3,6-tri-*O*-methylgluconolactone and 2,3,4,6-tetra-*O*-methyl-D-galactose.

5. Methylation and hydrolysis of lactose gives 2,3,6-tri-*O*-methyl-D-glucose and 2,3,4,6-tetra-*O*-methyl-D-galactose.

21.36

Deduce the structure of the disaccharide *melibiose* from the following data:

1. Melibiose is a reducing sugar that undergoes mutarotation and forms a phenylosazone.

2. Hydrolysis of melibiose with acid or with an α-*galactosidase* gives D-galactose and D-glucose.

3. Bromine water oxidation of melibiose gives *melibionic acid.* Hydrolysis of melibionic acid gives D-galactose and D-gluconic acid. Methylation of melibionic acid followed by hydrolysis gives 2,3,4,6-tetra-*O*-methyl-D-galactose and 2,3,4,5-tetra-*O*-methyl-D-gluconic acid.

4. Methylation and hydrolysis of melibiose gives 2,3,4,6-tetra-*O*-methyl-D-galactose and 2,3,4-tri-*O*-methyl-D-glucose.

21.37

Trehalose is a disaccharide that can be obtained from yeasts, fungi, sea urchins, algae, and insects. Deduce the structure of trehalose from the following information:

1. Acid hydrolysis of trehalose yields only D-glucose.

2. Trehalose is hydrolyzed by α-glucosidases but not by β-glucosidases.

3. Trehalose is a nonreducing sugar; it does not mutarotate, form a phenylosazone, or react with bromine water.

4. Methylation of trehalose followed by hydrolysis yields two molar equivalents of 2,3,4,6-tetra-O-methyl-D-glucose.

21.38
Outline chemical tests that will distinguish between each of the following:
(a) D-Glucose and D-glucitol
(b) D-Glucitol and D-glucaric acid
(c) D-Glucose and D-fructose
(d) D-Glucose and D-galactose
(e) Sucrose and maltose
(f) Maltose and maltonic acid
(g) Methyl β-D-glucopyranoside and 2,3,4,6-tetra-O-methyl-β-D-glucopyranose
(h) Methyl α-D-ribofuranoside (**I**) and methyl 2-deoxy-α-D-ribofuranoside (**II**)

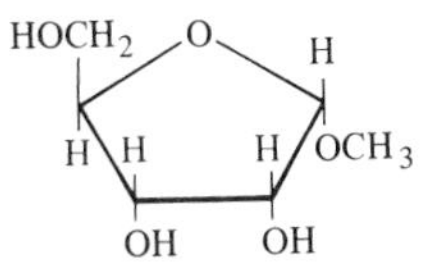

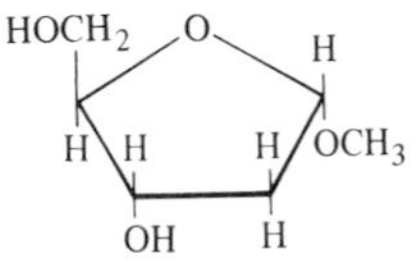

*21.39
A group of oligosaccharides called *Schardinger dextrins* can be isolated from *Bacillus macerans* when the bacillus is grown on a medium rich in amylose. These oligosaccharides are all *nonreducing*. A typical Schardinger dextrin undergoes hydrolysis when treated with an acid or an α-glucosidase to yield six, seven, or eight molecules of D-glucose. Complete methylation of a Schardinger dextrin followed by acid hydrolysis yields only 2,3,6-tri-O-methyl-D-glucose. Propose a general structure for a Schardinger dextrin.

*21.40
Isomaltose is a disaccharide that can be obtained by enzymatic hydrolysis of amylopectin. Deduce the structure of isomaltose from the following data:

1. Hydrolysis of 1 mole of isomaltose by acid or by an α-glucosidase gives 2 moles of D-glucose.

2. Isomaltose is a reducing sugar.

3. Isomaltose is oxidized by bromine water to isomaltonic acid. Methylation of isomaltonic acid and subsequent hydrolysis yields 2,3,4,6-tetra-O-methyl-D-glucose and 2,3,4,5-tetra-O-methyl-D-gluconic acid.

4. Methylation of isomaltose itself followed by hydrolysis gives 2,3,4,6-tetra-O-methyl-D-glucose and 2,3,4-tri-O-methyl-D-glucose.

21.41
Stachyose occurs in the roots of several species of plants. Deduce the structure of stachyose from the following data:

1. Acidic hydrolysis of 1 mole of stachyose yields 2 moles of D-galactose, 1 mole of D-glucose, and 1 mole of D-fructose.

2. Stachyose is a nonreducing sugar.

3. Treating stachyose with an α-galactosidase produces a mixture containing D-galactose, sucrose, and a nonreducing trisaccharide called *raffinose*.

4. Acidic hydrolysis of raffinose gives D-glucose, D-fructose, and D-galactose. Treating raffinose with an α-galactosidase yields D-galactose and sucrose. Treating raffinose with invertase (an enzyme that hydrolyzes sucrose) yields fructose and *melibiose* (cf. Problem 21.36).

5. Methylation of stachyose followed by hydrolysis yields 2,3,4,6-tetra-*O*-methyl-D-galactose, 2,3,4-tri-*O*-methyl-D-galactose, 2,3,4-tri-*O*-methyl-D-glucose, and 1,3,4,6-tetra-*O*-methyl-D-fructose.

*21.42

Arbutin, a compound that can be isolated from the leaves of bearberry, cranberry, and pear trees, has the molecular formula $C_{12}H_{16}O_7$. When arbutin is treated with aqueous acid or with a β-glucosidase, the reaction produces D-glucose and a compound **X** with the molecular formula $C_6H_6O_2$. The proton nmr spectrum of compound **X** consists of two singlets, one at $\delta 6.8$ (4H) and one at $\delta 7.9$ (2H). Methylation of arbutin followed by acidic hydrolysis yields 2,3,4,6-tetra-*O*-methyl-D-glucose and a compound **Y** ($C_7H_8O_2$). Compound **Y** is soluble in dilute aqueous NaOH but is insoluble in aqueous $NaHCO_3$. The proton nmr spectrum of **Y** shows a singlet at $\delta 3.9$ (3H), a singlet at $\delta 4.8$ (1H), and a multiplet (that resembles a singlet) at $\delta 6.8$ (4H). Treating compound **Y** with aqueous NaOH and $(CH_3)_2SO_4$ produces compound **Z** ($C_8H_{10}O_2$). The proton nmr spectrum of **Z** consists of two singlets, one at $\delta 3.75$ (6H) and one at $\delta 6.8$ (4H). Propose structures for arbutin and for compounds **X**, **Y**, and **Z**.

*21.43

D-Glucose reacts with acetone in the presence of sulfuric acid to yield a compound with the molecular formula $C_{12}H_{20}O_6$, which has been given the common name "diacetone glucose." D-Galactose undergoes a similar reaction to yield an isomeric product. "Diacetone glucose" has three five-membered rings; the corresponding compound obtained from D-galactose has two five-membered rings and a six-membered ring. (a) Write structures for these two compounds and (b) explain their formation. (*Hint:* Use models.)

CHAPTER TWENTY TWO

Estrogen crystals. Estrogen, the female sex hormone, is discussed in Section 22.4C.

Lipids

22.1 INTRODUCTION

When plant or animal tissues are extracted with a nonpolar solvent (e.g., ether, chloroform, benzene, or an alkane) a portion of the material usually dissolves. The components of this soluble fraction are called **lipids.***

Lipids include a wide variety of structural types including the following:

Carboxylic acids (or "fatty" acids)

Glyceryl trialkanoates (or neutral fats)

Phospholipids

Glycolipids

Waxes

Terpenes

Steroids

Prostaglandins

We shall now examine the properties of the members of each lipid group.

22.2 FATTY ACIDS AND GLYCERYL TRIALKANOATES

Only a small portion of the total lipid fraction consists of free carboxylic acids. Most of the carboxylic acids in the lipid fraction are found as *esters of glycerol,* that is, as **glyceryl trialkanoates.**† The most common are *glyceryl trialkanoates of long-chain carboxylic acids* (Fig. 22.1)

Glyceryl trialkanoates are the oils and fats of plant or animal origin. They include such common substrates as peanut oil, olive oil, soybean oil, corn oil, linseed oil, butter, lard, and tallow. Those glyceryl trialkanoates that are liquids at room temperature are generally known as *oils;* those that are solids are usually called *fats.*

*Lipids are defined in terms of the physical operation that we use to isolate them. In this respect, the definition of a lipid differs from that of a protein (Chapter 23) or a carbohydrate (Chapter 21). The latter are defined on the basis of their structures.

†In older literature glyceryl trialkanoates were referred to as triacylglycerols, triglycerides, or, simply as glycerides.

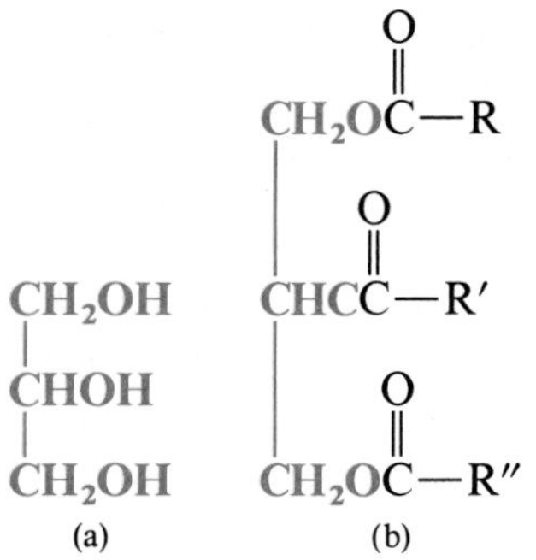

FIGURE 22.1 (*a*) Glycerol. (*b*) A glyceryl trialkanoate. R, R′, and R″ are usually long-chain alkyl groups. R, R′, and R″ may also contain one or more carbon–carbon double bonds. In a glyceryl trialkanoate R, R′, and R″ may all be different.

TABLE 22.1 Common fatty acids

	mp (°C)
Saturated Carboxylic Acids	
$CH_3(CH_2)_{12}CO_2H$ Myristic acid (tetradecanoic acid)	54
$CH_3(CH_2)_{14}CO_2H$ Palmitic acid (hexadecanoic acid)	63
$CH_3(CH_2)_{16}CO_2H$ Stearic acid (octadecanoic acid)	70
Unsaturated Carboxylic Acids	

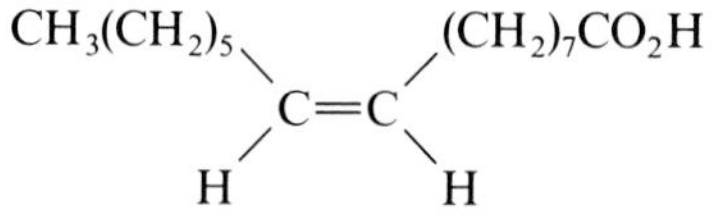

Palmitoleic acid
(*cis*-9-hexadecenoic acid)   32

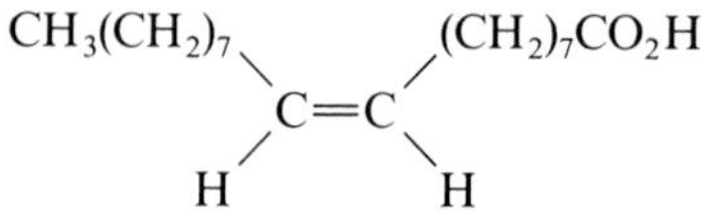

Oleic acid
(*cis*-9-octadecenoic acid)   4

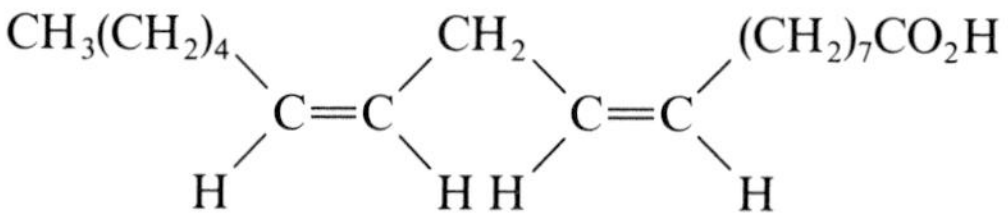

Linoleic acid
(*cis,cis*-9,12-octadecadienoic acid)   −5

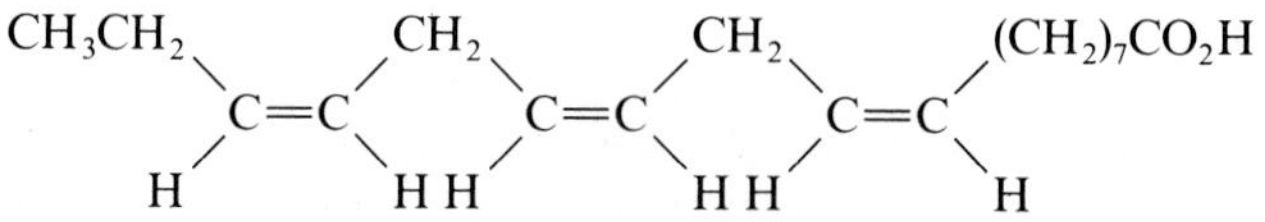

Linolenic acid
(*cis,cis,cis*-9,12,15-octadecatrienoic acid)   −11

The carboxylic acids that are obtained by hydrolysis of naturally occurring fats and oils usually have unbranched chains with an even number of carbon atoms. (We shall see, later, that these facts give us important clues as to how they are synthesized in plants and animals.) The most common carboxylic acids obtained from fats and oils are the C_{14}-, C_{16}-, and C_{18}-acids shown in Table 22.1.

In addition to most of the common fatty acids given in Table 22.1, the hydrolysis of butter gives small amounts of saturated even-numbered carboxylic acids in the C_4–C_{12} range. These are butyric (butanoic), caproic (hexanoic), caprylic (octanoic), capric (decanoic), and lauric (dodecanoic) acids. The hydrolysis of coconut oil also gives short-chain carboxylic acids and a large amount of lauric acid.

Other less common fatty acids and their sources are given in Table 22.2. Table 22.3 gives the fatty acid composition of a number of common fats and oils.

22.2A Hydrogenation of Glyceryl Trialkanoates

Most oils are made up of esters formed largely from unsaturated fatty acids. The fact that oils have lower melting points than fats is related to this factor; hydrogenation of an oil produces a solid fat. The cis double bonds of unsaturated fatty acids cause their carbon chains to assume conformations that do not fit easily into an orderly crystal structure of a solid. The saturated chains produced by hydrogenating an oil fit much better.

$$
\begin{array}{l}
\overset{O}{\overset{\|}{\text{CH}_2\text{OC}}}(\text{CH}_2)_7\text{CH}=\text{CH}(\text{CH}_2)_7\text{CH}_3 \quad \text{(from oleic acid)} \\[2mm]
\overset{O}{\overset{\|}{\text{CHOC}}}(\text{CH}_2)_7\text{CH}=\text{CHCH}_2\text{CH}=\text{CH}(\text{CH}_2)_4\text{CH}_3 \\[2mm]
\overset{O}{\overset{\|}{\text{CH}_2\text{OC}}}(\text{CH}_2)_7\text{CH}=\text{CHCH}_2\text{CH}=\text{CH}(\text{CH}_2)_4\text{CH}_3
\end{array}
\left.\begin{array}{l} \\[2mm] \\[2mm] \\ \end{array}\right\} \text{(from linoleic acid)}
$$

A glyceryl trialkanoate of a typical vegetable oil
(a liquid)

$$\Big\downarrow \; \text{H}_2, \text{Ni}$$

$$
\begin{array}{l}
\overset{O}{\overset{\|}{\text{CH}_2\text{OC}}}(\text{CH}_2)_{16}\text{CH}_3 \\[2mm]
\overset{O}{\overset{\|}{\text{CHOC}}}(\text{CH}_2)_{16}\text{CH}_3 \\[2mm]
\overset{O}{\overset{\|}{\text{CH}_2\text{OC}}}(\text{CH}_2)_{16}\text{CH}_3
\end{array}
$$

Glyceryl tristearate
(a solid fat)

Commercial cooking fats such as Crisco and Spry, for example, are manufactured in just this way. Vegetable oils are hydrogenated until a semisolid of an appealing consistency is obtained. Complete hydrogenation is avoided because a completely saturated glyceryl trialkanoate is very hard and brittle.

TABLE 22.2 Some less common fatty acids

NUMBER OF CARBON ATOMS	NAME	STRUCTURE	SOURCE
18	Eleostearic acid	$CH_3(CH_2)_3$ — C=C / C=C / C=C — $(CH_2)_7CO_2H$ (with H substituents)	The main (80%) fatty acid obtained by hydrolysis of tung oil
18	Ricinoleic acid	$CH_3(CH_2)_5CHCH_2$ (OH) — C=C — $(CH_2)_7CO_2H$ (with H substituents)	The main (80%) fatty acid obtained by hydrolysis of castor oil
18	Chaulmoogric acid	cyclopentene ring —$(CH_2)_{12}CO_2H$	From oil of *Chaulmoogra* (used in the treatment of leprosy)
19	Sterculic acid	$CH_3(CH_2)_7$ — C=C (CH$_2$) — $(CH_2)_7CO_2H$	From kernel oil of *Sterculia foetida*

19	Lactobacillic acid	$CH_3(CH_2)_5$—$\overset{\displaystyle HC—CH}{\underset{\displaystyle CH_2}{\diagdown\diagup}}$—$(CH_2)_9CO_2H$	From phospholipids of bacteria
19	Tuberculostearic acid	$CH_3(CH_2)_7\underset{\displaystyle CH_3}{CH}(CH_2)_8CO_2H$	From tubercle bacilli
20	Arachidonic acid	$CH_3(CH_2)_4(CH=CHCH_2)_4(CH_2)_2CO_2H$	From human fat (0.3 – 1.0%), liver, lecithins
22	Cetoleic acid	$CH_3(CH_2)_9CH=CH(CH_2)_9CO_2H$	From fish oils
22	Erucic acid	$CH_3(CH_2)_7CH=CH(CH_2)_{11}CO_2H$ (cis)	From seed oils of rape, wallflower, nasturtium, and mustard
24	Nervonic acid	$CH_3(CH_2)_7CH=CH(CH_2)_{13}CO_2H$ (cis)	From fish oils and brain tissues
27	Mycolipenic acid	$CH_3(CH_2)_{17}\underset{\displaystyle CH_3}{CH}CH_2\underset{\displaystyle CH_3}{CH}CH=\underset{\displaystyle CH_3}{C}—CO_2H$	From tubercle bacilli

1043

TABLE 22.3 Fatty acid composition obtained by hydrolysis of common fats and oils[a]

AVERAGE COMPOSITION OF FATTY ACIDS (mole %)

FAT OR OIL	SATURATED								UNSATURATED			
	C_4 BUTYRIC ACID	C_6 CAPROIC ACID	C_8 CAPRYLIC ACID	C_{10} CAPRIC ACID	C_{12} LAURIC ACID	C_{14} MYRISTIC ACID	C_{16} PALMITIC ACID	C_{18} STEARIC ACID	C_{16} PALMITOLEIC ACID	C_{18} OLEIC ACID	C_{18} LINOLEIC ACID	C_{18} LINOLENIC ACID
Animal Fats												
Butter	3–4	1–2	0–1	2–3	2–5	8–15	25–29	9–12	4–6	18–33	2–4	
Lard						1–2	25–30	12–18	4–6	48–60	6–12	0–1
Beef tallow						2–5	24–34	15–30		35–45	1–3	0–1
Vegetable Oils												
Olive						0–1	5–15	1–4		67–84	8–12	
Peanut							7–12	2–6		30–60	20–38	
Corn						1–2	7–11	3–4	1–2	25–35	50–60	
Cottonseed						1–2	18–25	1–2	1–3	17–38	45–55	
Soybean						1–2	6–10	2–4		20–30	50–58	5–10
Linseed							4–7	2–4		14–30	14–25	45–60
Coconut		0–1	5–7	7–9	40–50	15–20	9–12	2–4	0–1	6–9	0–1	
Marine Oils												
Cod liver						5–7	8–10	0–1	18–22	27–33	27–32	

[a] Data adapted from John R. Holum, *Organic and Biological Chemistry*, Wiley, New York, 1978, p. 220, and from *Biology Data Book*, Philip L. Altman and Dorothy S. Ditmer, Eds., Federation of American Societies for Experimental Biology, Washington, DC, 1964.

22.2B Saponification of Glyceryl Trialkanoates

Alkaline hydrolysis (i.e., saponification) of glyceryl trialkanoates produces glycerol and a mixture of salts of long-chain carboxylic acids:

$$
\begin{array}{c}
\underset{\displaystyle\text{CH}_2\text{OCR}}{\overset{\displaystyle\text{O}\atop\displaystyle\|}{}} \\[1em]
\underset{\displaystyle\text{CHOCR}'}{\overset{\displaystyle\text{O}\atop\displaystyle\|}{}} + 3\text{NaOH} \\[1em]
\underset{\displaystyle\text{CH}_2\text{OCR}''}{\overset{\displaystyle\text{O}\atop\displaystyle\|}{}}
\end{array}
\longrightarrow
\begin{array}{c}
\text{CH}_2\text{OH} \\[1em]
\text{CHOH} \\[1em]
\text{CH}_2\text{OH} \\[0.3em]
\text{Glycerol}
\end{array}
\quad
\begin{array}{c}
+\ \ \underset{\displaystyle}{\overset{\displaystyle\text{O}\atop\displaystyle\|}{\text{RCO}^-}}\ \ \text{Na}^+ \\[1em]
\underset{\displaystyle}{\overset{\displaystyle\text{O}\atop\displaystyle\|}{\text{R}'\text{CO}^-}}\ \ \text{Na}^+ \\[1em]
\underset{\displaystyle}{\overset{\displaystyle\text{O}\atop\displaystyle\|}{\text{R}''\text{CO}^-}}\ \ \text{Na}^+ \\[0.3em]
\text{Sodium carboxylates} \\
\text{"soap"}
\end{array}
$$

These salts of long-chain carboxylic acids are **soaps,** and this saponification reaction is the way most soaps are manufactured. Fats and oils are boiled in aqueous sodium hydroxide until hydrolysis is complete. Adding sodium chloride to the mixture then causes the soap to precipitate. (After the soap has been separated, glycerol can be isolated from the aqueous phase by distillation.) Crude soaps are usually purified by several reprecipitations. Perfumes can be added if a toilet soap is the desired product. Sand, sodium carbonate, and other fillers can be added to make a scouring soap, and air can be blown into the molten soap if the manufacturer wants to market a soap that floats.

The sodium salts of long-chain carboxylic acids (soaps) are almost completely miscible with water. However, they do not dissolve as we might expect, that is, as individual ions. Except in very dilute solutions, soaps exist as **micelles** (Fig. 22.2). Soap micelles are usually spherical clusters of carboxylate ions that are dispersed throughout the aqueous phase. The carboxylate ions are packed together with their

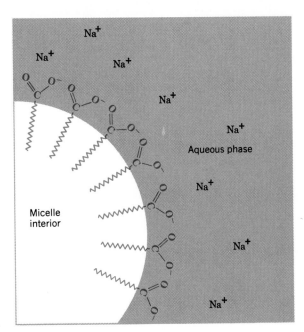

FIGURE 22.2 A portion of a soap micelle showing its interface with the polar dispersing medium.

negatively charged (and thus, *polar*) carboxylate groups at the surface and with their nonpolar hydrocarbon chains on the interior. The sodium ions are scattered throughout the aqueous phase as individual solvated ions.

Micelle formation accounts for the fact that soaps dissolve in water. The nonpolar (and thus, **hydrophobic**) alkyl chains of the soap remain in a nonpolar environment — in the interior of the micelle. The polar (and therefore, **hydrophilic**) carboxylate groups are exposed to a polar environment — that of the aqueous phase. Because the surfaces of the micelles are negatively charged, individual micelles repel each other and remain dispersed throughout the aqueous phase.

Soaps serve their function as "dirt removers" in a similar way. Most dirt particles (on the skin, for example) become surrounded by a layer of an oil or fat. Water molecules alone are unable to disperse these greasy globules because they are unable to penetrate the oily layer and separate the individual particles from each other or from the surface to which they are stuck. Soap solutions, however, *are* able to separate the individual particles because their hydrocarbon chains can "dissolve" in the oily layer (Fig. 22.3). As this happens each individual particle develops an outer layer of carboxylate ions and presents the aqueous phase with a much more compatible exterior — a polar surface. The individual globules now repel each other and thus become dispersed throughout the aqueous phase. Shortly thereafter they make their way down the drain.

Synthetic detergents (Fig. 22.4) function in the same way as soaps; they have long nonpolar alkane chains with polar groups at the end. The polar groups of most synthetic detergents are sodium sulfonates or sodium sulfates.

Synthetic detergents offer an advantage over soaps; they function well in "hard" water, that is, water containing Ca^{2+}, Fe^{2+}, Fe^{3+}, and Mg^{2+} ions. Calcium, iron, and magnesium salts of alkanesulfonates and alkyl hydrogen sulfates are largely water soluble and, thus, synthetic detergents remain in solution. Soaps, by contrast, form precipitates — the ring around the bathtub — when they are used in hard water.

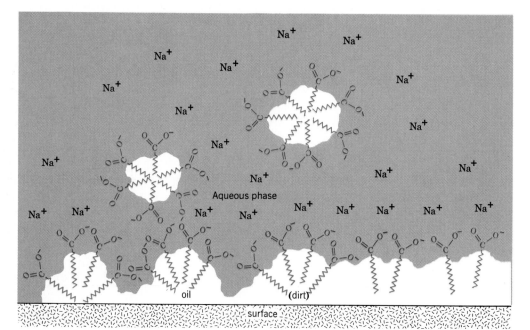

FIGURE 22.3 Dispersal of oil-coated dirt particles by a soap.

$$CH_3(CH_2)_nCH_2SO_2O^- \; Na^+$$
Sodium alkanesulfonate

$$CH_3(CH_2)_nCH_2OSO_2O^- \; Na^+$$
Sodium alkyl sulfate

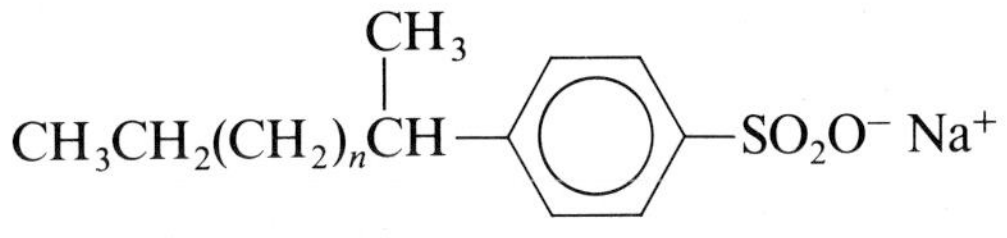

$$CH_3CH_2(CH_2)_n\overset{\overset{\displaystyle CH_3}{|}}{CH}-\!\!\bigcirc\!\!-SO_2O^- \; Na^+$$

Sodium alkylbenzenesulfonate

FIGURE 22.4 Typical synthetic detergents (n = 10).

22.2C Reactions of the Carboxyl Group of Fatty Acids

Fatty acids, as we might expect, undergo reactions typical of carboxylic acids. They react with LiAlH$_4$ to form alcohols, with alcohols and mineral acid to form esters, and with thionyl chloride to form acyl chlorides:

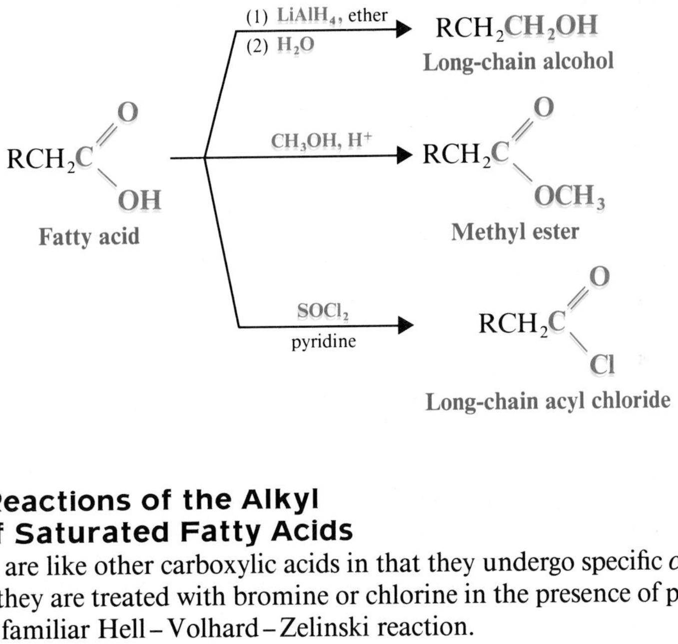

22.2D Reactions of the Alkyl Chain of Saturated Fatty Acids

Fatty acids are like other carboxylic acids in that they undergo specific α-halogenation when they are treated with bromine or chlorine in the presence of phosphorus. This is the familiar Hell–Volhard–Zelinski reaction.

$$\underset{\text{Fatty acid}}{RCH_2\overset{\overset{\displaystyle O}{||}}{C}OH} + \xrightarrow[\text{(2) } H_2O]{\text{(1) } X_2,\, P_4} RCH\overset{\overset{\displaystyle O}{||}}{C}OH + HX$$
$$\underset{X}{|}$$

22.2E Reactions of the Alkenyl Chain of Unsaturated Fatty Acids

The double bonds of the carbon chains of fatty acids undergo characteristic alkene addition reactions:

$$CH_3(CH_2)_nCH=CH(CH_2)_mCO_2H \longrightarrow$$

$\xrightarrow[\text{Ni}]{H_2,} CH_3(CH_2)_nCH_2-CH_2(CH_2)_mCO_2H$

$\xrightarrow[\text{CCl}_4]{Br_2,} CH_3(CH_2)_nCHBr\,CHBr(CH_2)_mCO_2H$

$\xrightarrow[\text{(2) NaHSO}_3]{\text{(1) OsO}_4} CH_3(CH_2)_nCH\ CH(CH_2)_mCO_2H$
$\quad\quad\quad\quad\quad\quad\quad\quad\quad\quad | \quad |$
$\quad\quad\quad\quad\quad\quad\quad\quad\quad\ OH\ OH$

$\xrightarrow{HBr} CH_3(CH_2)_nCH_2CHBr(CH_2)_mCO_2H$
$\quad\quad\quad\quad\quad\quad + $
$\quad\quad CH_3(CH_2)_nCHBrCH_2(CH_2)_mCO_2H$

PROBLEM 22.1

(a) How many stereoisomers are possible for 9,10-dibromohexadecanoic acid? (b) The addition of bromine to palmitoleic acid yields primarily one set of enantiomers, (±)-*threo*-9,10-dibromohexadecanoic acid. The addition of bromine is an *anti* addition to the double bond (i.e., it apparently takes place through a bromonium ion intermediate). Taking into account the cis stereochemistry of the double bond of palmitoleic acid and the stereochemistry of the bromine addition, write three-dimensional structures for the (±)-*threo*-9,10-dibromohexadecanoic acids.

22.2F Biological Function of Glyceryl Trialkanoates

The primary function of the glyceryl trialkanoates of mammals is as a source of chemical energy. When glyceryl trialkanoates are converted to carbon dioxide and water by biochemical reactions (i.e., when glyceryl trialkanoates are *metabolized*), they yield more than twice as many kilocalories per gram as do carbohydrates or proteins. This is largely because of the high proportion of carbon–hydrogen bonds per molecule.

Glyceryl trialkanoates are distributed throughout nearly all types of body cells but they are stored primarily as *body fat* in certain depots of specialized connective tissue known as *adipose tissue.*

The saturated glyceryl trialkanoates of the body can be synthesized from all three major foodstuffs: proteins, carbohydrates, and fats or oils. Certain polyunsaturated fatty acids, however, are essential in the diets of higher animals.

22.3 TERPENES AND TERPENOIDS

People have isolated organic compounds from plants since antiquity. By gently heating or by steam distilling certain plant materials, one can obtain mixtures of odoriferous compounds known as *essential oils.* These compounds have had a variety of uses, particularly in early medicine and in the making of perfumes.

As the science of organic chemistry developed, chemists separated the various components of these mixtures and determined their molecular formulas and then later their structural formulas. Even today these natural products offer challenging problems for chemists interested in structure determination and synthesis. Research in this area has also given us important information about the ways the plants, themselves, synthesize these compounds.

Hydrocarbons known generally as **terpenes** and oxygen-containing compounds called **terpenoids** are the most important constituents of essential oils. Most terpenes have skeletons of 10, 15, 20, or 30 carbon atoms and are classified in the following way:

NUMBER OF CARBON ATOMS	CLASS
10	Monoterpenes
15	Sesquiterpenes
20	Diterpenes
30	Triterpenes

One can view terpenes as being built up from two or more five-carbon units known as *isoprene units*. Isoprene is 2-methyl-1,3-butadiene. Isoprene and the isoprene unit can be represented in various ways.

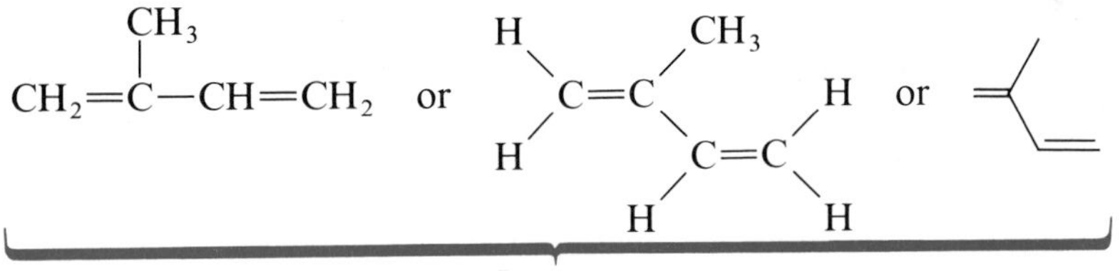

Isoprene

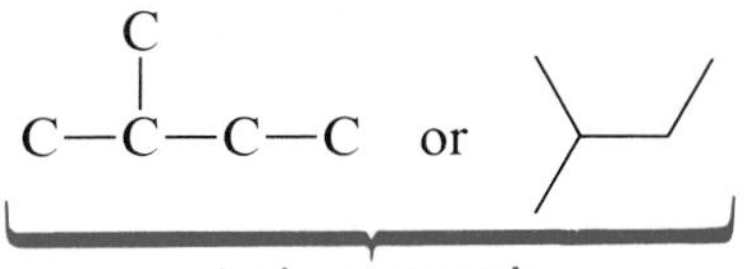

An isoprene unit

We now know that plants do not synthesize terpenes from isoprene (Special Topic M.3). However, recognition of the isoprene unit as a component of the structure of terpenes has been a great aid in elucidating their structures. We can see how, if we examine the following structures.

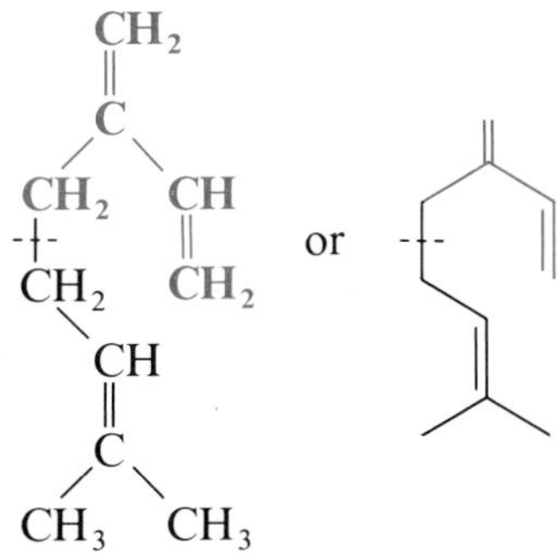

Myrcene
(isolated from bay oil)

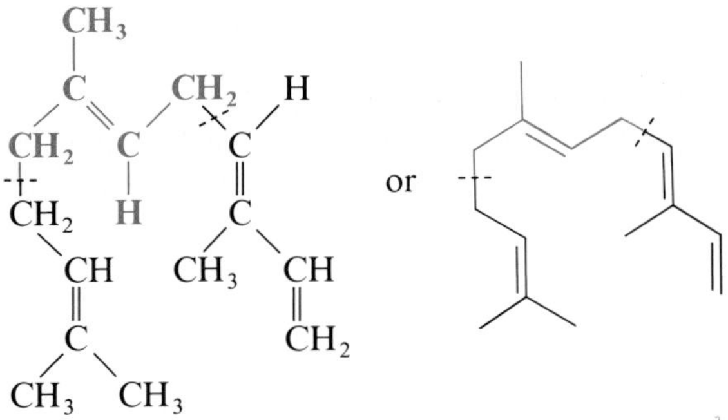

α-Farnesene
(from natural coating of apples)

Using dashed lines to separate isoprene units we can see that the monoterpene (myrcene) has two isoprene units, and that the sesquiterpene (α-farnesene) has three. In both compounds the isoprene units are linked head to tail.

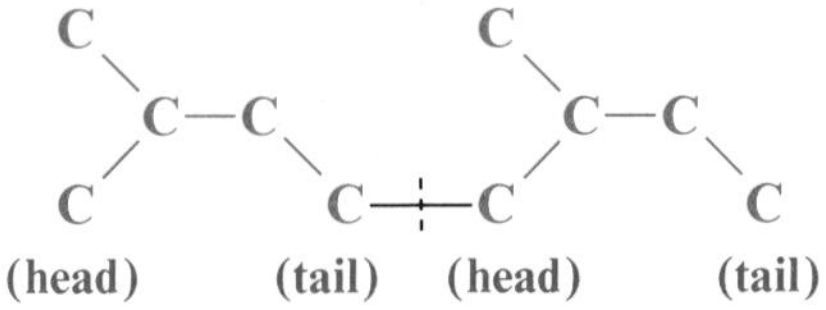

(head)   (tail)   (head)   (tail)

Many terpenes also have isoprene units linked in rings, and others (terpenoids) contain oxygen.

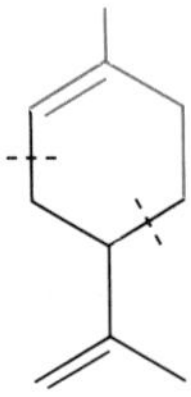

Limonene
(from oil of lemon or orange)

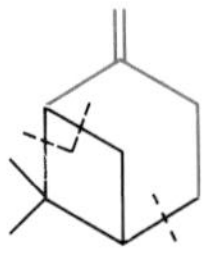

β-Pinene
(from oil of turpentine)

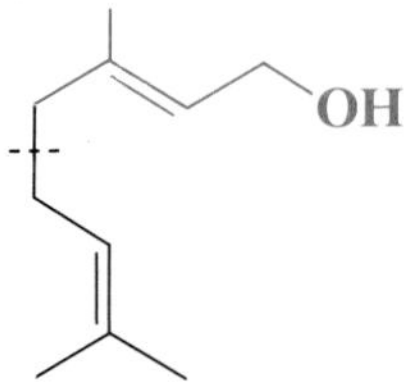

Geraniol
(from roses and other flowers)

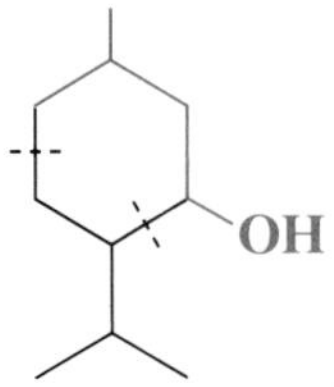

Menthol
(from peppermint)

PROBLEM 22.2

(a) Show the isoprene units in each of the following terpenes. (b) Classify each as a monoterpene, sesquiterpene, diterpene, and so on.

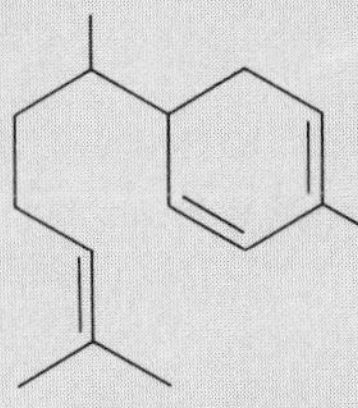

Zingiberene
(from oil of ginger)

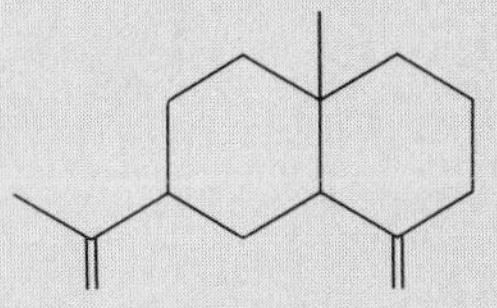

β-Selinene
(from oil of celery)

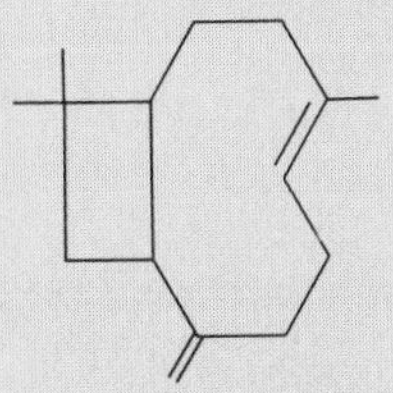

Caryophyllene
(from oil of cloves)

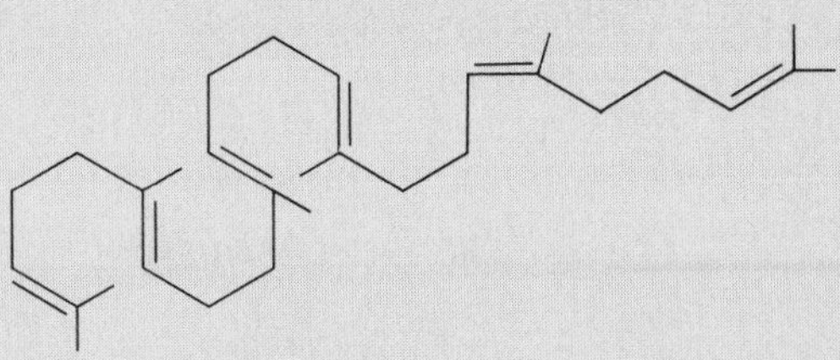

Squalene
(from shark liver oil)

PROBLEM 22.3

What products would you expect to obtain if each of the following terpenes were subjected to ozonization and subsequent treatment with zinc and water?

(a) Myrcene   (c) α-Farnesene   (e) Squalene

(b) Limonene   (d) Geraniol

PROBLEM 22.4

Give structural formulas for the products that you would expect from the following reactions:

(a) β-Pinene + hot KMnO$_4$ $\longrightarrow$

(b) Zingiberene + H$_2$ $\xrightarrow{\text{Pt}}$

(c) Caryophyllene + HCl $\longrightarrow$

(d) β-Selinene + 2 THF:BH$_3$ $\xrightarrow{\text{(2) H}_2\text{O}_2,\ \text{OH}^-}$

PROBLEM 22.5

What simple chemical test could you use to distinguish between geraniol and menthol?

The carotenes are tetraterpenes. They can be thought of as two diterpenes linked in tail-to-tail fashion.

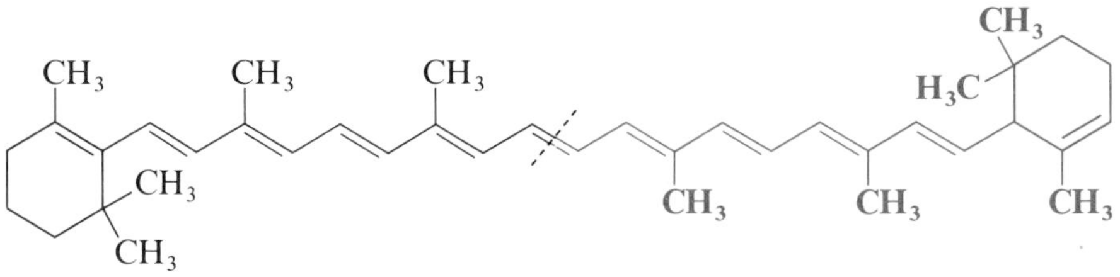

α-Carotene

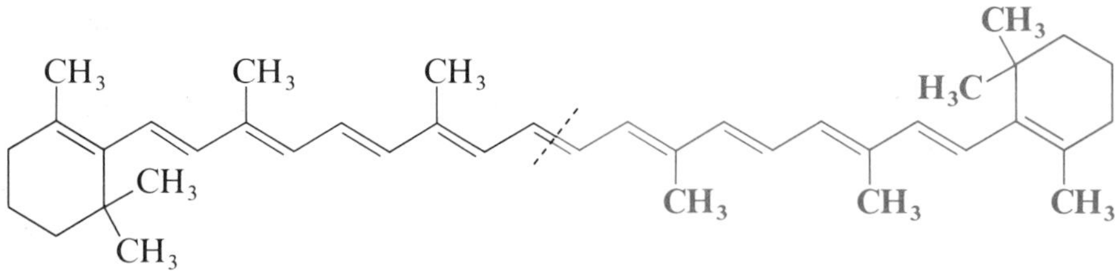

β-Carotene

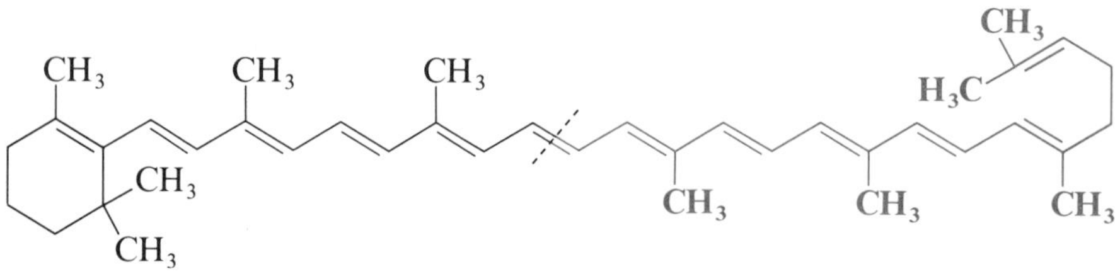

γ-Carotene

The carotenes are present in almost all green plants. All three carotenes serve as precursors for vitamin A, for they all can be converted to vitamin A by enzymes in the liver.

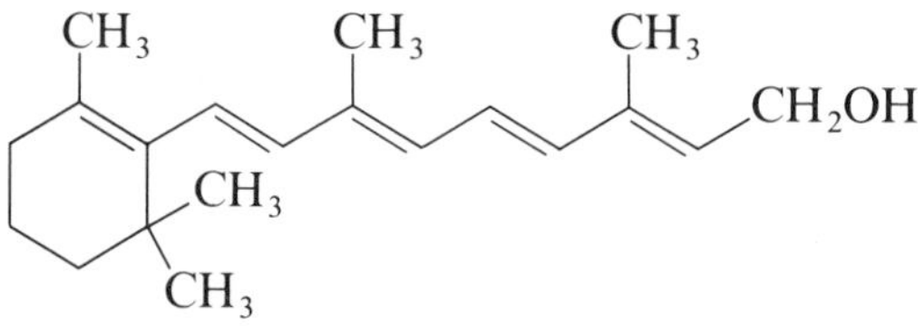

Vitamin A

In this conversion, one molecule of β-carotene yields two of vitamin A: α- and γ-carotene give only one. Because β-carotene became available before vitamin A, β-carotene was adopted as the standard for vitamin A content in foods. One international unit (IU) of vitamin A is equivalent to the vitamin A activity of 0.6 μg of crystalline β-carotene. Vitamin A is important not only in vision but in many other ways as well. For example, young animals whose diets are deficient in vitamin A fail to grow.

22.3A Natural Rubber

Natural rubber can be viewed as a 1,4-addition polymer of isoprene. In fact, pyrolysis degrades natural rubber to isoprene. Pyrolysis (Greek: *pyros,* a fire + *lysis*) is heating something in the absence of air until it decomposes. The isoprene units of natural rubber are all linked in a head-to-tail fashion and all of the double bonds are cis.

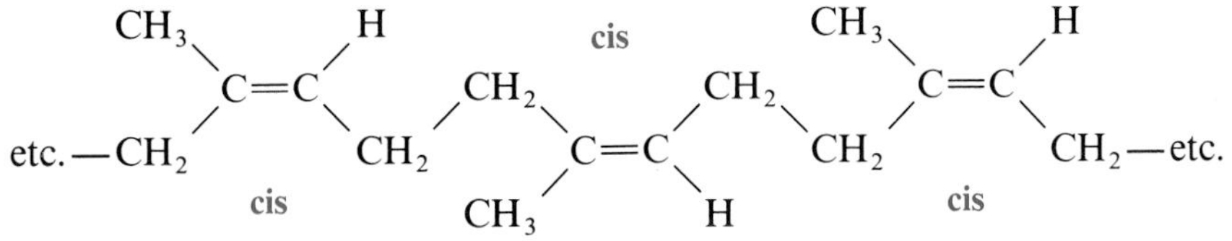

Natural rubber
(*cis*-1,4-polyisoprene)

Ziegler–Natta catalysts (Special Topic D) make it possible to polymerize isoprene and obtain a synthetic product that is identical with the rubber obtained from natural sources.

Pure natural rubber is soft and tacky. To be useful, natural rubber has to be *vulcanized.* In vulcanization, natural rubber is heated with sulfur. A reaction takes place that produces cross links between the *cis*-polyisoprene chains and makes the rubber much harder. Sulfur reacts both at the double bonds and at allylic hydrogen atoms.

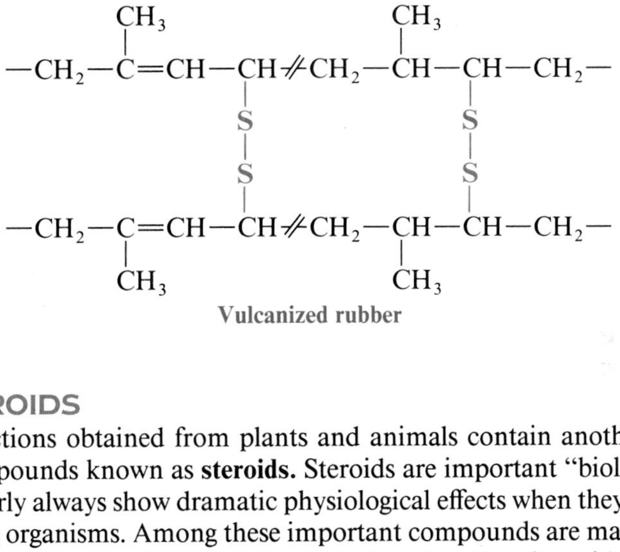

Vulcanized rubber

22.4 STEROIDS

The lipid fractions obtained from plants and animals contain another important group of compounds known as **steroids.** Steroids are important "biological regulators" that nearly always show dramatic physiological effects when they are administered to living organisms. Among these important compounds are male and female sex hormones, adrenocortical hormones, D vitamins, the bile acids, and certain cardiac poisons.

22.4A Structure and Systematic Nomenclature of Steroids

Steroids are derivatives of the following perhydrocyclopentanophenanthrene ring system.

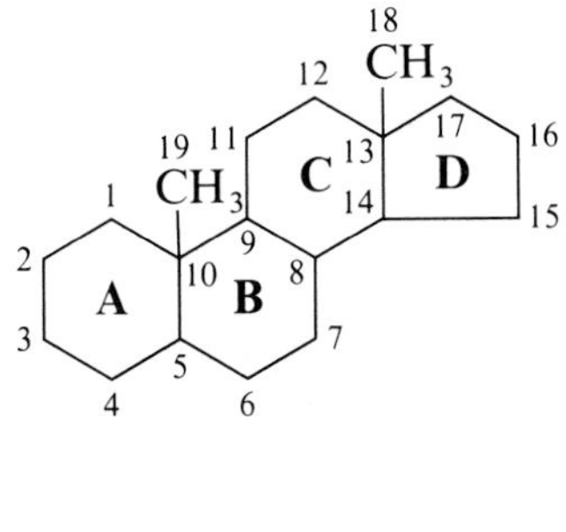

The carbon atoms of this ring system are numbered as shown. The four rings are designated with letters.

In most steroids the **B,C** and **C,D** ring junctions are trans. The **A,B** ring junction, however, may be either cis or trans and this possibility gives rise to two general groups of steroids having the three-dimensional structures shown in Fig. 22.5.

PROBLEM 22.6

Draw the two basic ring systems given in Fig. 22.5 for the 5α and 5β series showing all hydrogen atoms of the cyclohexane rings. Label each hydrogen atom as to whether it is axial or equatorial.

The methyl groups that are attached at points of ring junction (i.e., those numbered 18 and 19) are called **angular methyl groups** and they serve as important reference points for stereochemical designations. The angular methyl groups protrude above the general plane of the ring system when it is written in the manner shown in Fig. 22.5. By convention, other groups that lie on the same general side of the molecule as the angular methyl groups (i.e., on the top side) are designated as β **substituents** (these are written with a solid wedge). Groups that lie generally on the bottom (i.e., are trans to the angular methyl groups) are designated as α **substituents** (these are written with a dashed wedge). When α and β designations are applied to the hydrogen atom at position 5, the ring system in which the **A,B** ring junction is trans becomes the 5α series; and the ring system in which the **A,B** ring junction is cis becomes the 5β series.

In systematic nomenclature the nature of the R group at position 17 determines (primarily) the base name of an individual steroid. These names are derived from the steroid hydrocarbon names given in Table 22.4.

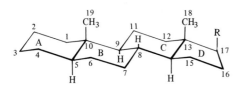

5α Series of steroids
(all ring junctions are trans)

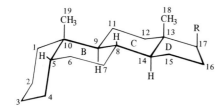

5β Series of steroids
(A, B ring junction is cis)

FIGURE 22.5 The basic ring systems of the 5α and 5β series of steroids.

TABLE 22.4 Names of steroid hydrocarbons

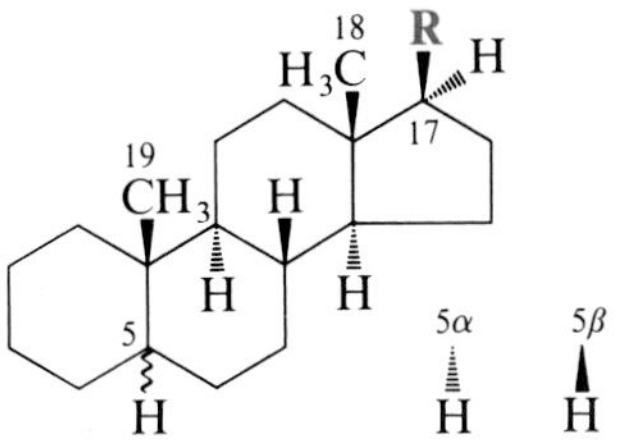

R	NAME
—H	Androstane
—H (with —H also replacing $\overset{19}{-CH_3}$)	Estrane
$\overset{20}{-CH_2}\overset{21}{CH_3}$	Pregnane
$-\overset{20}{C}H\overset{22}{CH_2}\overset{23}{CH_2}\overset{24}{CH_3}$ $\underset{21}{CH_3}$	Cholane
$-\overset{20}{C}H\overset{22}{CH_2}\overset{23}{CH_2}\overset{24}{CH_2}\overset{25}{C}H\overset{26}{CH_3}$ $\underset{21}{CH_3}$ $\underset{27}{CH_3}$	Cholestane

The following two examples illustrate the way these base names are used.

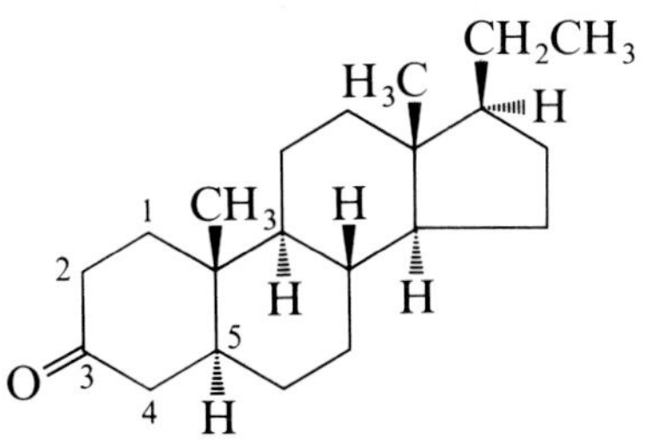

5α-Pregnan-3-one

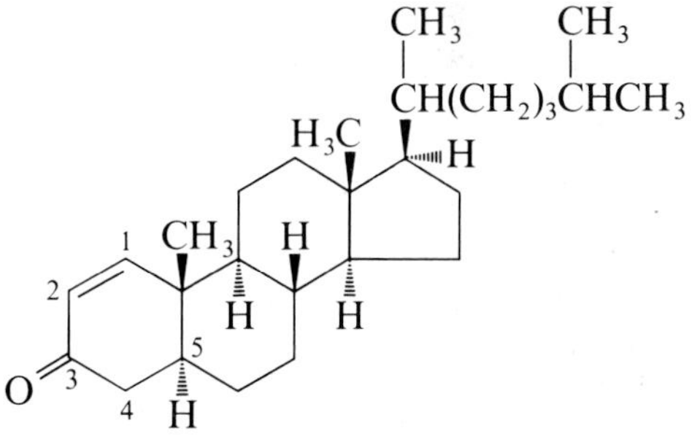

5α-Cholest-1-en-3-one

We shall see that many steroids also have common names and that the names of the steroid hydrocarbons given in Table 22.4 are derived from these common names.

PROBLEM 22.7

(a) Androsterone, a secondary male sex hormone, has the systematic name 3α-hydroxy-5α-androstan-17-one. Give a three-dimensional formula for androsterone. (b) Norethynodrel, a synthetic steroid that has been widely used in oral contraceptives, has the systematic name 17α-ethynyl-17β-hydroxy-5(10)-estren-3-one. Give a three-dimensional formula for norethynodrel.

22.4B Cholesterol

Cholesterol, one of the most widely occurring steroids, can be isolated by extraction of nearly all animal tissues. Human gallstones are a particularly rich source.

Cholesterol was first isolated in 1770. In the 1920s two German chemists, Adolf Windaus (University of Göttingen) and Heinrich Wieland (University of Munich), were responsible for outlining a structure for cholesterol; they received Nobel Prizes for their work in 1927 and 1928.*

Part of the difficulty in assigning an absolute structure to cholesterol is that cholesterol contains *eight* tetrahedral stereocenters. This feature means that 2^8 or 256 possible stereoisomeric forms of the basic structure are possible, *only one of which is cholesterol.*

Absolute configuration of cholesterol
(5-cholesten-3β-ol)

PROBLEM 22.8

Designate with an asterisk the eight stereocenters of cholesterol.

Cholesterol occurs widely in the human body, but not all of the biological functions of cholesterol are yet known. Cholesterol is known to serve as an intermediate in the biosynthesis of all of the steroids of the body. Cholesterol, therefore, is essential to life. We do not need to have cholesterol in our diet, however, because our body can synthesize all we need. When we ingest cholesterol, our body synthesizes less than if we ate none at all, but the total cholesterol is more than if we ate none at all. Far more cholesterol is present in the body than is necessary for steroid biosynthesis. High levels of blood cholesterol have been implicated in the development of arteriosclerosis (hardening of the arteries) and in heart attacks that occur when cholesterol-containing plaques block arteries of the heart. Considerable research is being carried out in the area of cholesterol metabolism with the hope of finding ways of minimizing cholesterol levels through the use of dietary adjustments or drugs.

---

*The original cholesterol structure proposed by Windaus and Wieland was incorrect. This became evident in 1932 as a result of X-ray diffraction studies done by the British physicist J. D. Bernal. By the end of 1932, however, English scientists, and Wieland himself, using Bernal's results, were able to outline the correct structure of cholesterol.

22.4C Sex Hormones

The sex hormones can be classified into three major groups: (1) the female sex hormones, or **estrogens,** (2) the male sex hormones, or **androgens,** and (3) the pregnancy hormones, or **progestins.**

The first sex hormone to be isolated was an estrogen, *estrone.* Working independently, Adolf Butenandt (in Germany at the University of Göttingen) and Edward Doisy (in the United States at St. Louis University) isolated estrone from the urine of pregnant women. They published their discoveries in 1929. Later, Doisy was able to isolate the much more potent estrogen, *estradiol.* In this research Doisy had to extract *4 tons* of sow ovaries in order to obtain just 12 mg of estradiol. Estradiol, it turns out, is the true female sex hormone, and estrone is a metabolized form of estradiol that is excreted.

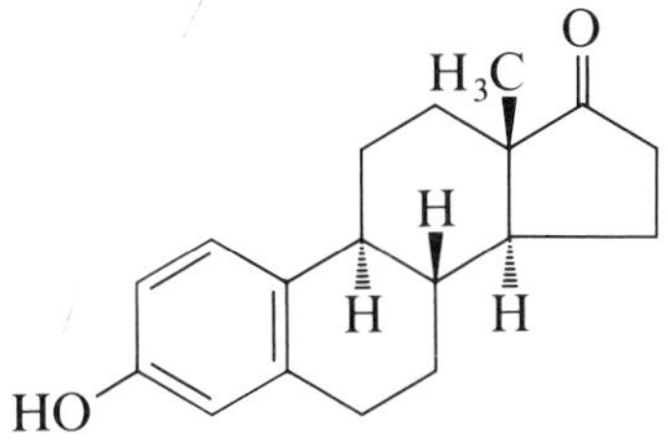

Estrone
[3-hydroxy-1,3,5(10)-
estratrien-17-one]

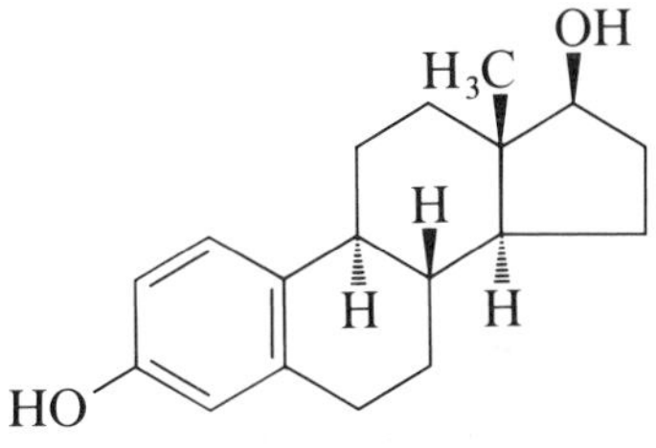

Estradiol
[1,3,5(10)-estra-
triene-3,17β-diol]

Estradiol is secreted by the ovaries and promotes the development of the secondary female characteristics that appear at the onset of puberty. Estrogens also stimulate the development of the mammary glands during pregnancy and induce estrus (heat) in animals.

In 1931 Butenandt and Kurt Tscherning isolated the first androgen, *androsterone.* They were able to obtain 15 mg of this hormone by extracting approximately 15,000 L of male urine. Soon afterwards (in 1935), Ernest Laqueur (in Holland) isolated another male sex hormone, *testosterone,* from bull testes. It soon became clear that testosterone is the true male sex hormone and that androsterone is a metabolized form of testosterone that is excreted in the urine.

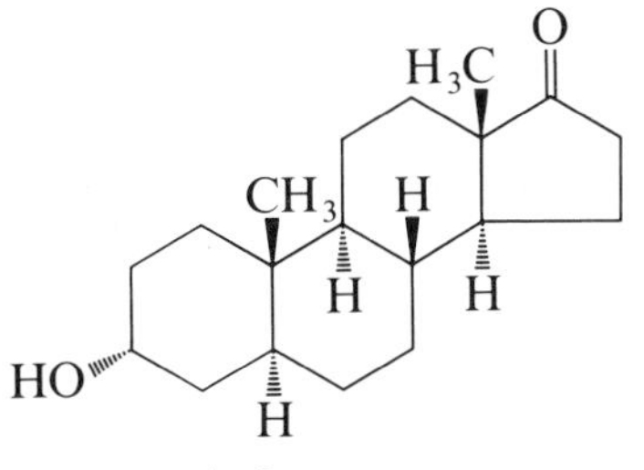

Androsterone
(3α-hydroxy-5α-androstan-17-one)

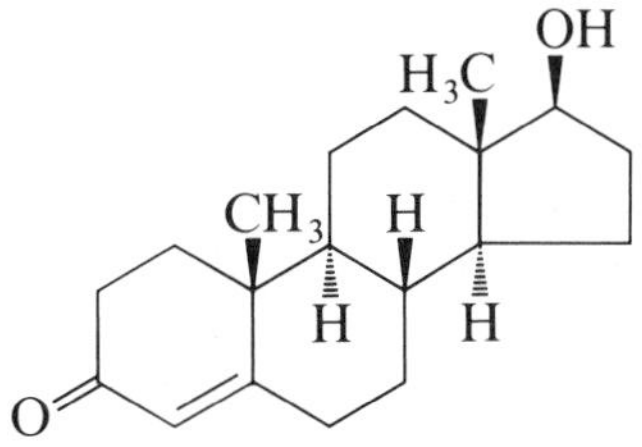

Testosterone
(17β-hydroxy-4-androsten-3-one)

Testosterone, secreted by the testes, is the hormone that promotes the development of secondary male characteristics: the growth of facial and body hair; the deepening of the voice; muscular development; and the maturation of the male sex organs.

Testosterone and estradiol, then, are the chemical compounds from which "maleness" and "femaleness" are derived. It is especially interesting to examine their structural formulas and see how very slightly these two compounds differ. Testosterone has an angular methyl group at the **A,B** ring junction that is missing in estradiol. Ring **A** of estradiol is a benzene ring and, as a result, estradiol is a phenol. Ring **A** of testosterone contains an α,β-unsaturated keto group.

PROBLEM 22.9

The estrogens (estrone and estradiol) are easily separated from the androgens (androsterone and testosterone) on the basis of one of their chemical properties. What is the property and how could such a separation be accomplished?

Progesterone
(4-pregnene-3,20-dione)

Progesterone is the most important *progestin* (pregnancy hormone). After ovulation occurs, the remnant of the ruptured ovarian follicle (called the *corpus luteum*) begins to secrete progesterone. This hormone prepares the lining of the uterus for implantation of the fertilized ovum, and continued progesterone secretion is necessary for the completion of pregnancy. (Progesterone is secreted by the placenta after secretion by the corpus luteum declines.)

Progesterone *also suppresses ovulation,* and it is the chemical agent that apparently accounts for the fact that pregnant women do not conceive again while pregnant. It was this observation that led to the search for synthetic progestins that could be used as oral contraceptives. (Progesterone, itself, requires very large doses to be effective in suppressing ovulation when taken orally because it is degraded in the intestinal tract.) A number of such compounds have been developed and are now widely used. In addition to norethynodrel (cf. Problem 22.7), another widely used synthetic progestin is its double-bond isomer, *norethindrone.*

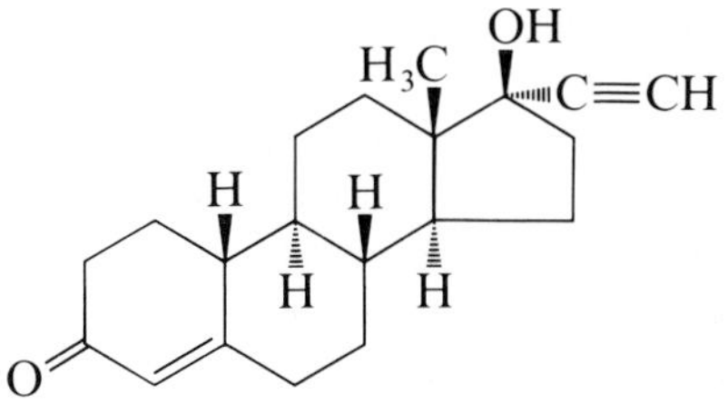

Norethindrone
(17α-ethynyl-17-β-hydroxy-4-estren-3-one)

Synthetic estrogens have also been developed and these are often used in oral contraceptives in combination with synthetic progestins. A very potent synthetic estrogen is the compound called *ethynylestradiol* or *novestrol.*

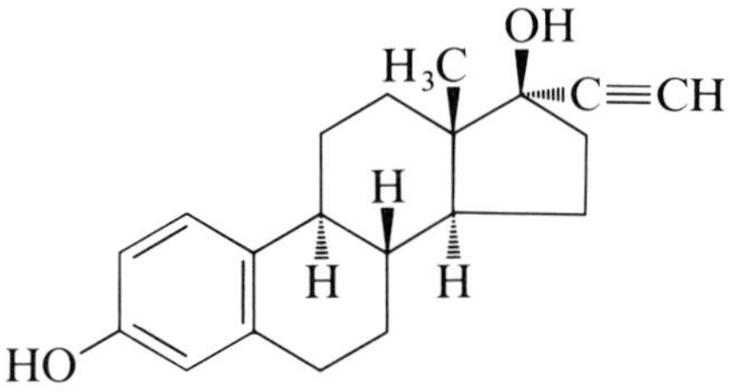

Ethynylestradiol
[17α-ethynyl-1,3,5(10)-estratriene-3,17β-diol]

22.4D Adrenocortical Hormones

At least 28 different hormones have been isolated from the adrenal cortex. Included in this group are the following two steroids.

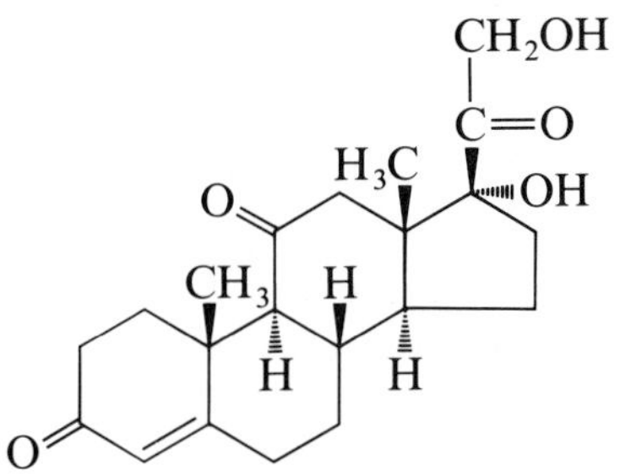

Cortisone
(17α,21-dihydroxy-4-pregnene-
3,11,20-trione)

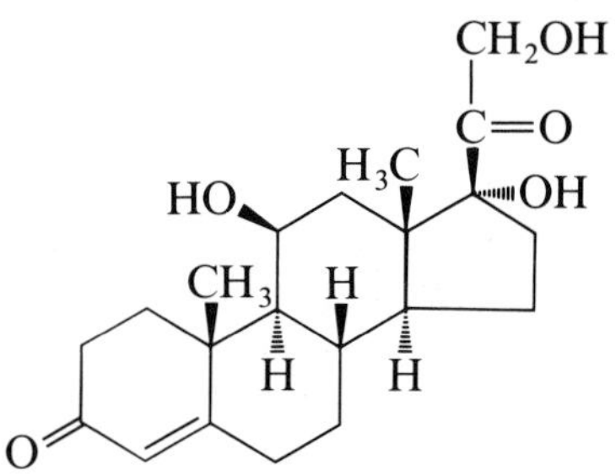

Cortisol
(11β,17α,21-trihydroxy-4-pregnene-
3,20-dione)

Most of the adrenocortical steroids have an oxygen function at position 11 (a keto group in cortisone, for example, and a β-hydroxyl in cortisol). Cortisol is the major hormone synthesized by the human adrenal cortex.

The adrenocortical steroids are apparently involved in the regulation of a large number of biological activities including carbohydrate, protein, and lipid metabolism, water and electrolyte balance, and reactions to allergic and inflammatory phenomena. Recognition of the antiinflammatory effect of cortisone and its usefulness in the treatment of rheumatoid arthritis, in 1949, has led to extensive research in this area. Many 11-oxygenated steroids are now used in the treatment of a variety of disorders ranging from Addison's disease, to asthma, and to skin inflammations.

22.4E D Vitamins

The demonstration, in 1919, that sunlight helped cure rickets — a childhood disease characterized by poor bone growth — began a long search for a chemical explanation. Soon it was discovered that irradiation of certain foodstuffs increased their antirachitic properties and, in 1930, the search led to a steroid that can be isolated from yeast, called *ergosterol.* Irradiation of ergosterol was found to produce a highly active material. In 1932 Windaus (Section 22.4B) and his co-workers in Germany demonstrated that this highly active substance was vitamin D_2. The photochemical reaction that takes place is one in which the dienoid ring **B** of ergosterol opens to produce a conjugated triene:

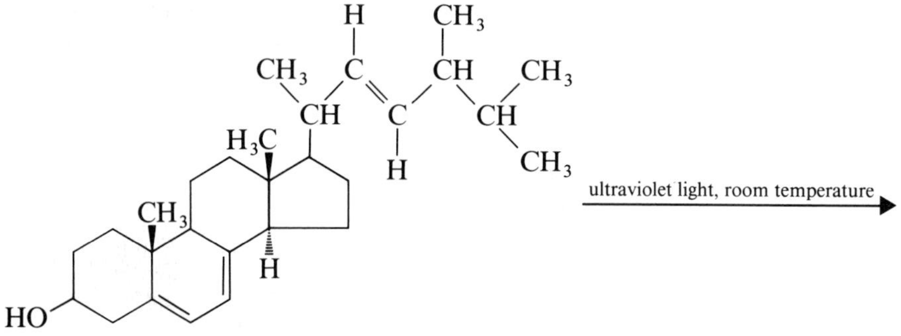

Ergosterol

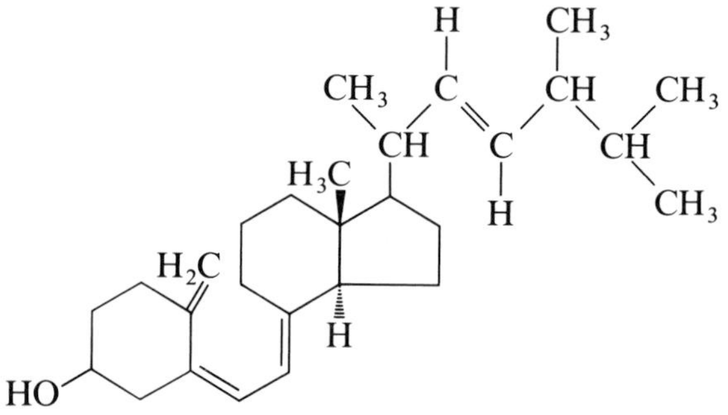

ultraviolet light, room temperature

Vitamin D_2

22.4F Other Steroids

The structures, sources, and physiological properties of a number of other important steroids are given in Table 22.5.

22.4G Reactions of Steroids

Steroids undergo all of the reactions that we might expect of molecules containing double bonds, hydroxyl groups, keto groups, and so on. While the stereochemistry of

TABLE 22.5 Other important steroids

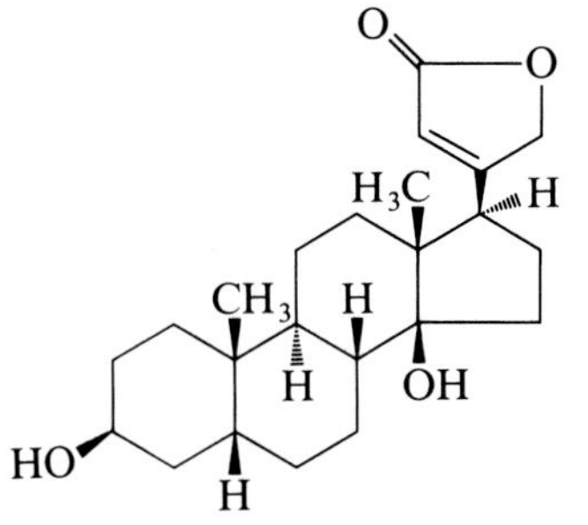

Digitoxigenin

Digitoxigenin is a cardiac aglycone that can be isolated by hydrolysis of digitalis, a pharmaceutical that has been used in treating heart disease since 1785. In digitalis, sugar molecules are joined in acetal linkages to the 3-OH of the steroid. In small doses digitalis strengthens the heart muscle; in larger doses it is a powerful heart poison.

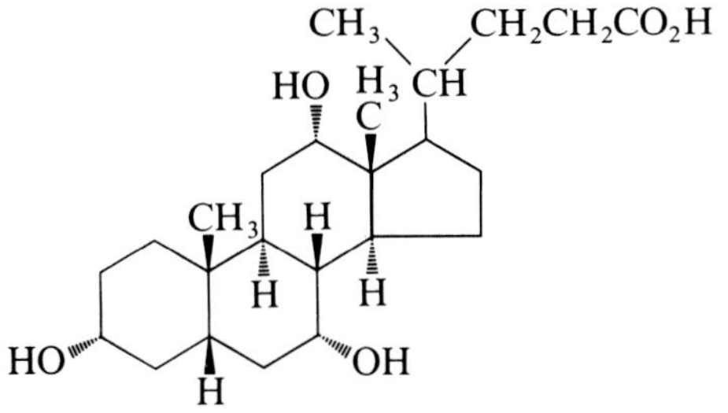

Cholic acid

Cholic acid is the most abundant acid obtained from the hydrolysis of human or ox bile. Bile is produced by the liver and stored in the gall bladder. When secreted into the small intestine, bile emulsifies lipids by acting as a soap. This action aids in the digestive process.

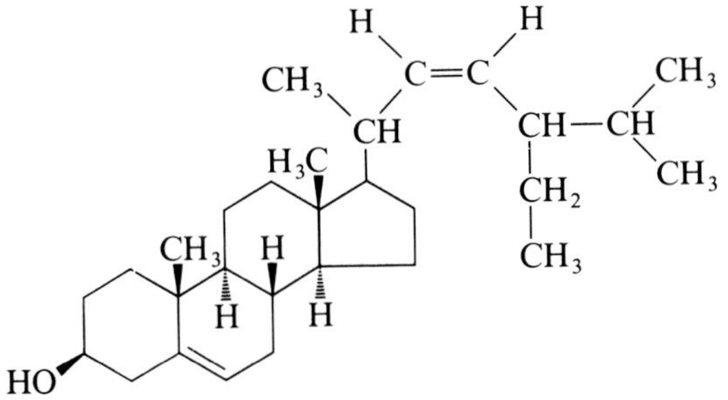

Stigmasterol

Stigmasterol is a widely occurring plant steroid that is obtained commercially from soybean oil.

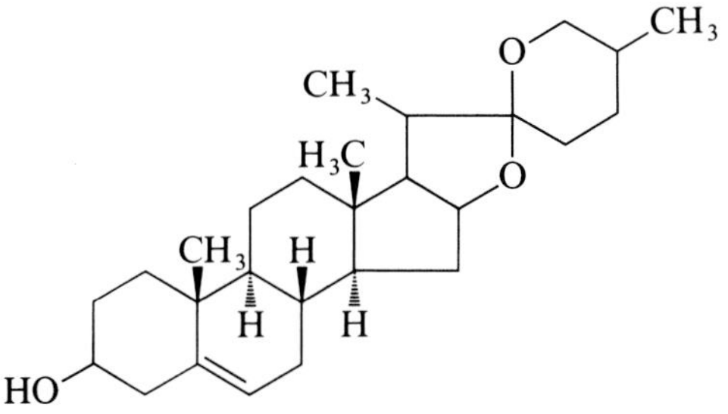

Diosgenin

Diosgenin, obtained from a particular species of yams, is used as the starting material for a commercial synthesis of cortisone and sex hormones.

steroid reactions is often quite complex, it is many times strongly influenced by the steric hindrance presented at the β face of the molecule by the angular methyl groups. Many reagents react preferentially at the relatively unhindered α face especially when the reaction takes place at a functional group very near an angular methyl group and when the attacking reagent is bulky. Examples that illustrate this tendency are shown in the following reaction.

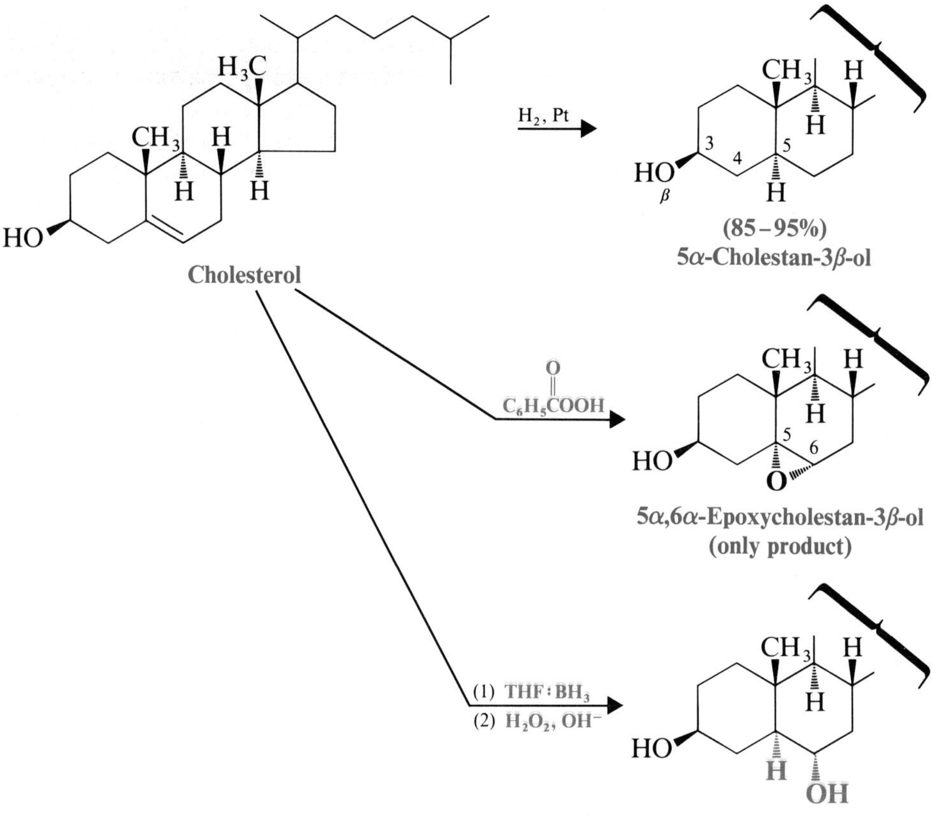

Cholesterol

H₂, Pt →

(85–95%)
5α-Cholestan-3β-ol

C_6H_5COOH →

5α,6α-Epoxycholestan-3β-ol
(only product)

(1) THF : BH₃
(2) H₂O₂, OH⁻ →

(78%)
5α-Cholestane-3β,6α-diol

When the epoxide ring of 5α,6α-epoxycholestan-3β-ol (see following reaction) is opened, attack by chloride ion must occur from the β face, but it takes place at the more open 6 position. Notice that the 5- and 6- substituents in the product are *diaxial* (Section 7.7).

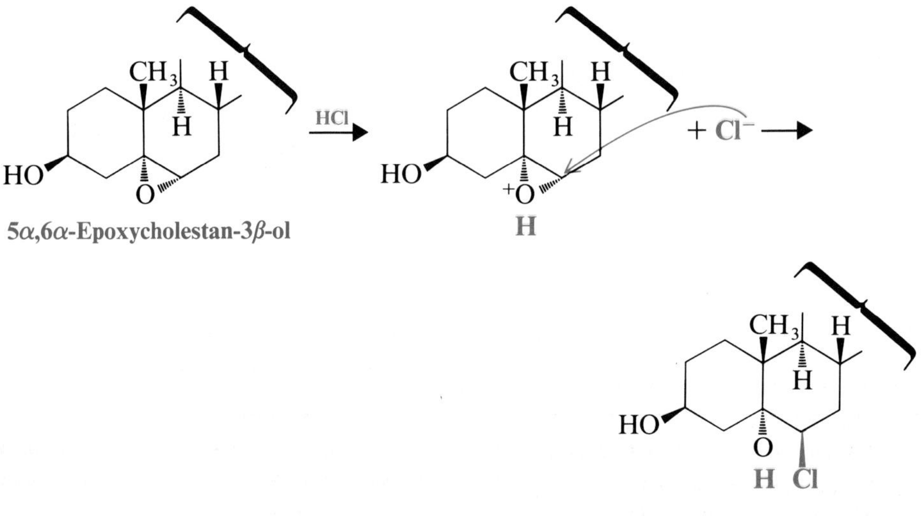

5α,6α-Epoxycholestan-3β-ol

HCl →

+ Cl⁻ →

PROBLEM 22.10

Show how you might convert cholesterol into each of the following compounds:

(a) 5α,6β-Dibromocholestan-3β-ol   (d) 6α-Deuterio-5α-cholestan-3β-ol

(b) Cholestane-3β,5α,6β-triol   (e) 6β-Bromocholestane-3β,5α-diol

(c) 5α-Cholestan-3-one

The relative openness of equatorial groups (when compared to axial groups) also influences the stereochemical course of steroid reactions. When 5α-cholestane-3β,7α-diol (see following reaction) is treated with excess ethyl chloroformate (C₂H₅OCOCl), only the equatorial 3β-hydroxyl becomes esterified. The axial 7α-hydroxyl is unaffected by the reaction.

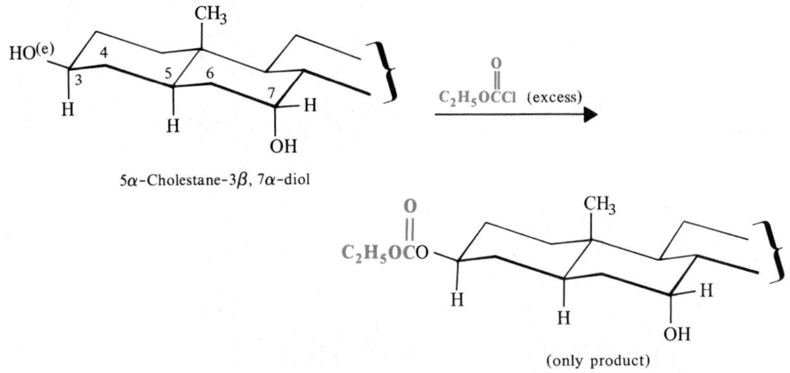

5α-Cholestane-3β, 7α-diol

(only product)

By contrast, treating 5α-cholestane-3β,7β-diol with excess ethyl chloroformate esterifies both hydroxyl groups. In this instance both groups are equatorial.

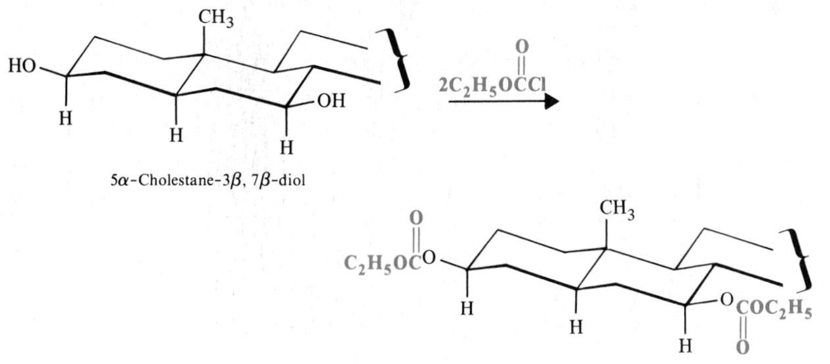

5α-Cholestane-3β, 7β-diol

22.5 PROSTAGLANDINS

One very active area of current research is concerned with a group of lipids called prostaglandins.* Prostaglandins are C₂₀-carboxylic acids that contain a five-mem-

---

*The 1982 Nobel Prize in physiology or medicine was awarded to S. K. Bergström and B. I. Samuelson (of the Karolinska Institute, Stockholm, Sweden) and to J. R. Vane (of the Wellcome Foundation, Bechenham, England) for their work on prostaglandins.

bered ring, at least one double bond, and several oxygen-containing functional groups. Two of the most biologically active prostaglandins are prostaglandin E_2 and prostaglandin $F_{1\alpha}$.

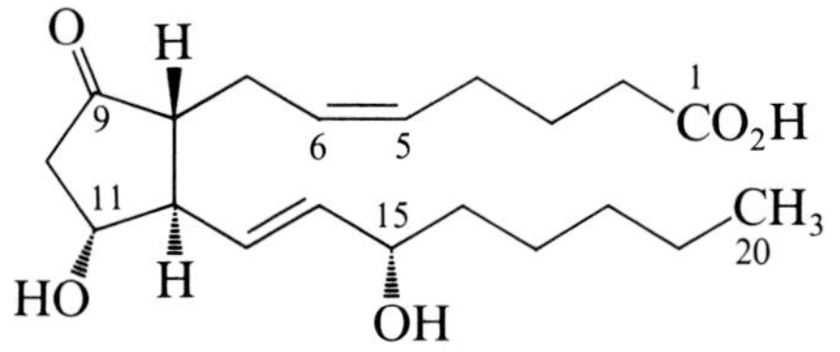

Prostaglandin E_2*
(PGE$_2$*)

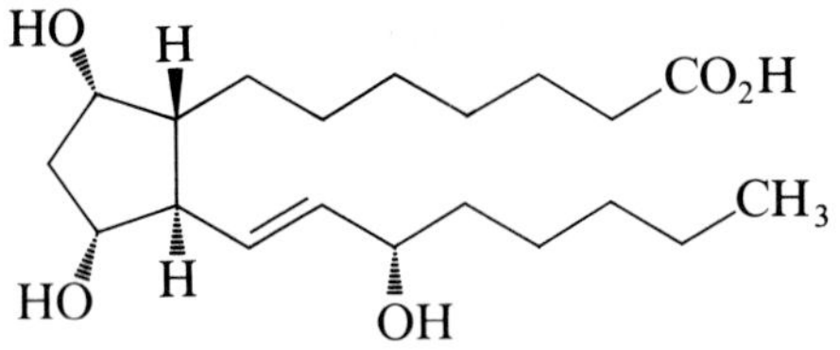

Prostaglandin $F_{1\alpha}$
(PGF$_{1\alpha}$)

Prostaglandins of the E type have a carbonyl group at C-9 and a hydroxyl group at C-11; those of the F type have hydroxyl groups at both positions. Prostaglandins of the 2 series have a double bond between C-5 and C-6; in the 1 series this bond is a single bond.

First isolated from seminal fluid, prostaglandins have since been found in almost all animal tissues. The amounts vary from tissue to tissue but are almost always very small. Most prostaglandins have powerful physiological activity, however, and this activity covers a broad spectrum of effects. Prostaglandins are known to affect heart rate, blood pressure, blood clotting, conception, fertility, and allergic responses.

The recent finding that prostaglandins can prevent formation of blood clots has great clinical significance, because heart attacks and strokes often result from the formation of abnormal clots in blood vessels. An understanding of how prostaglandins affect the formation of clots may lead to the development of drugs to prevent heart attacks and strokes.

The biosynthesis of prostaglandins of the 2 series begins with a 20-carbon polyenoic acid, arachidonic acid. (Synthesis of prostaglandins of the 1 series begins with a fatty acid with one fewer double bond.) The first step requires two molecules of oxygen and is catalyzed by an enzyme called *cyclooxygenase*.

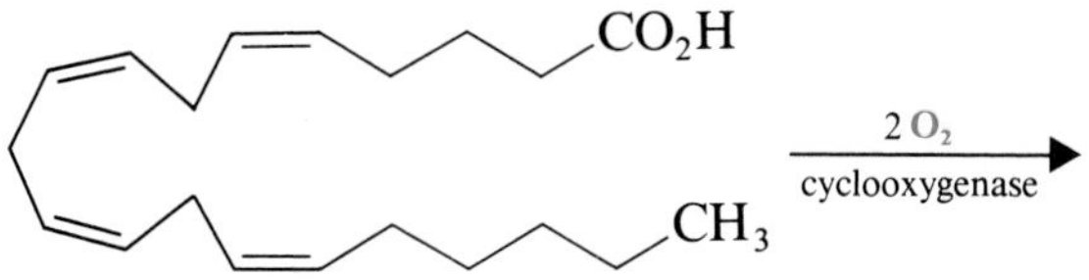

Arachidonic acid

*These names are code designations used by workers in the field; systematic names are seldom used for prostaglandins.

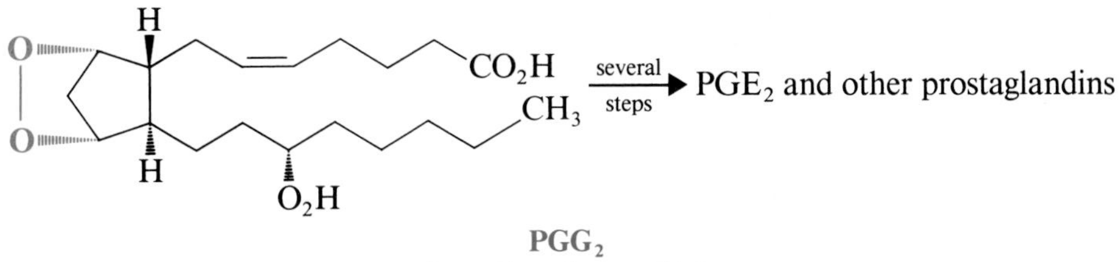

PGG$_2$
(a cyclic endoperoxide)

The involvement of prostaglandins in allergic and inflammation responses has also been of special interest. Some prostaglandins induce inflammation; others relieve it. The most widely used antiinflammatory drug is ordinary aspirin (cf. Section 14.8). Aspirin blocks the synthesis of prostaglandins from arachidonic acid, apparently by acetylating the enzyme cyclooxygenase, thus rendering it inactive. This reaction may represent the origin of aspirin's antiinflammatory properties. Another prostaglandin (PGE$_1$) is a potent fever-inducing agent (pyrogen), and aspirin's ability to reduce fever may also arise from its inhibition of prostaglandin synthesis.

22.6 PHOSPHOLIPIDS

Another large class of lipids are those called *phospholipids*. Most phospholipids are structurally derived from a glycerol derivative known as a *phosphatidic acid*. In a phosphatidic acid, two hydroxyl groups of glycerol are joined in ester linkages to fatty acids and one terminal hydroxyl group is joined in an ester linkage to *phosphoric acid*.

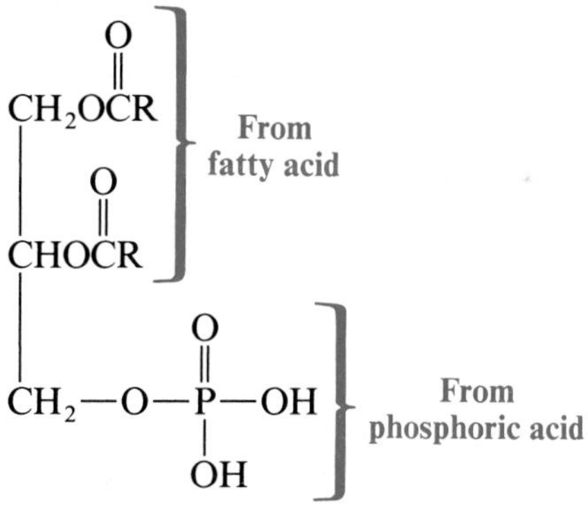

A phosphatidic acid
(a diacylglyceryl phosphate)

22.6A Phosphatides

In *phosphatides*, the phosphate group of a phosphatidic acid is bound through another phosphate ester linkage to one of the following nitrogen-containing compounds.

$$HOCH_2CH_2\overset{+}{N}(CH_3)_3 \quad OH^- \qquad HOCH_2CH_2NH_2 \qquad HOCH_2\underset{\underset{CO_2^-}{|}}{C}HNH_3^+$$

Choline   2-Aminoethanol   L-Serine
                 (ethanolamine)

The most important phosphatides are the **lecithins,** the **cephalins, phosphatidyl serines,** and the **plasmalogens** (a phosphatidyl derivative). Their general structures are shown in Table 22.6.

TABLE 22.6 Phosphatides

Lecithins

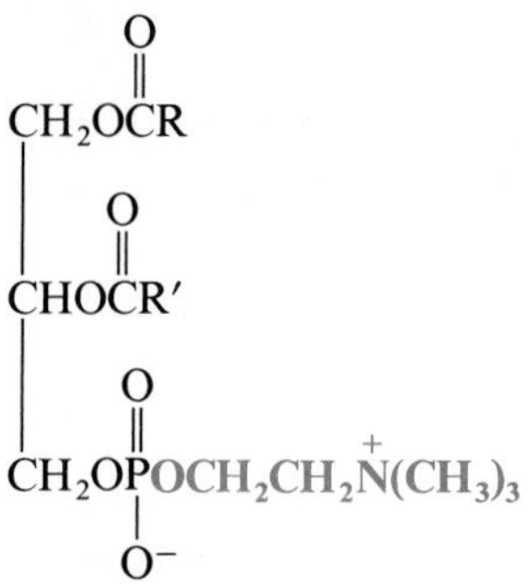

$$CH_2OCR$$ (with O double-bonded above)

$$CHOCR'$$ (with O double-bonded above)

$$CH_2OPOCH_2CH_2\overset{+}{N}(CH_3)_3 \quad \text{(from choline)}$$
with O double-bonded above P and O^- below P

R is saturated and R′ is unsaturated

Cephalins

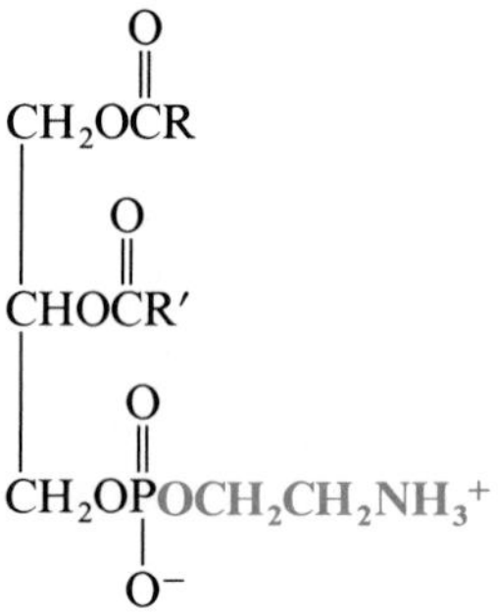

$$CH_2OCR$$ (with O double-bonded above)

$$CHOCR'$$ (with O double-bonded above)

$$CH_2OPOCH_2CH_2NH_3^+ \quad \text{(from 2-aminoethanol)}$$
with O double-bonded above P and O^- below P

Phosphatidyl Serines

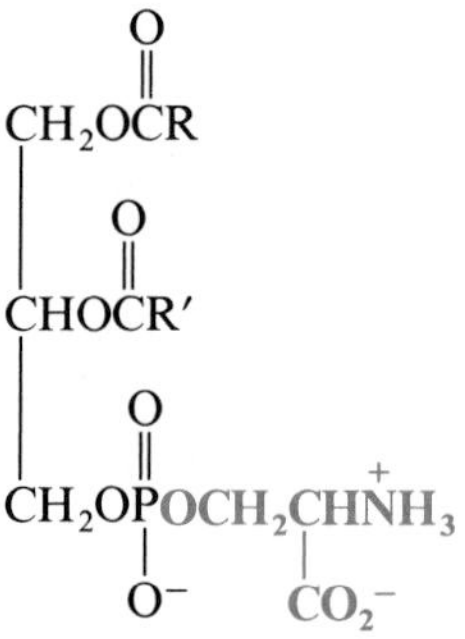

$$CH_2OCR$$ (with O double-bonded above)

$$CHOCR'$$ (with O double-bonded above)

$$CH_2OPOCH_2\overset{+}{C}HNH_3 \quad \text{(from L-serine)}$$
with O double-bonded above P, O^- below P, and CO_2^- below the CH

R is saturated and R′ is unsaturated

Plasmalogens

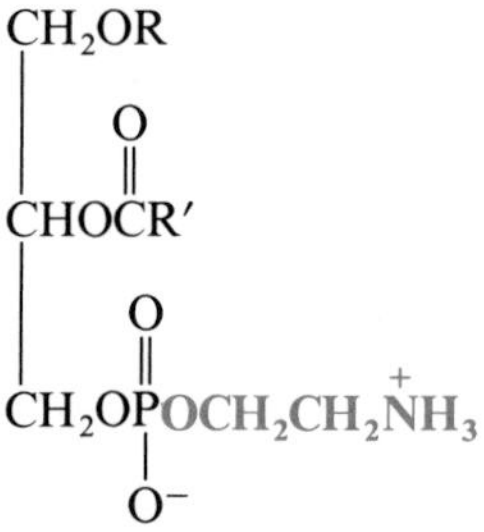

$$CH_2OR$$

$$CHOCR'$$ (with O double-bonded above)

$$CH_2OPOCH_2CH_2\overset{+}{N}H_3 \quad \text{(from 2-aminoethanol)} \quad \text{or} \quad OCH_2CH_2\overset{+}{N}(CH_3)_3 \quad \text{(from choline)}$$
with O double-bonded above P and O^- below P

R is $-CH=CH(CH_3)_nCH_3$. (This linkage is that of an α,β-unsaturated ether.)

R′ is that of an unsaturated fatty acid.

Phosphatides resemble soaps and detergents in that they are molecules having both polar and nonpolar groups (Fig. 22.6a). Like soaps and detergents, too, phosphatides "dissolve" in aqueous media by forming micelles. There is evidence that in biological systems the preferred micelles consist of three-dimensional arrays of "stacked" bimolecular micelles (Fig. 22.6).

The hydrophilic and hydrophobic portions of phosphatides make them perfectly suited for one of their most important biological functions: They form a portion of a structural unit that creates an interface between an organic and an aqueous environment. This structure is found in cell walls and membranes where phosphatides are often found associated with proteins. Phosphatides also appear to be an essential factor in the formation of blood clots.

PROBLEM 22.11

Under suitable conditions all of the ester (and ether) linkages of a phosphatide can be hydrolyzed. What organic compounds would you expect to obtain from the complete hydrolysis of (a) a lecithin, (b) cephalin, (c) a choline-based plasmalogen? [*Note:* Pay particular attention to the fate of the α,β-unsaturated ether in part (c).]

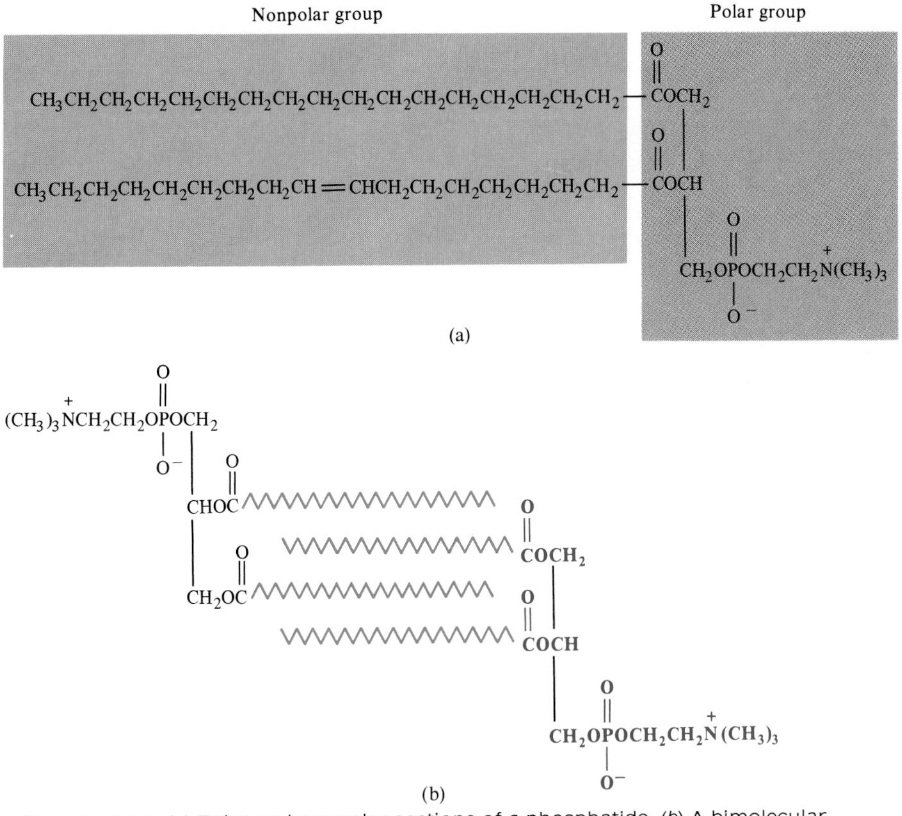

FIGURE 22.6 (a) Polar and nonpolar sections of a phosphatide. (b) A bimolecular phosphatide micelle.

22.6B Derivatives of Sphingosine

Another important group of lipids is derived from **sphingosine,** the derivatives are called **sphingolipids.** Two sphingolipids, a typical *sphingomyelin* and a typical *cerebroside,* are shown in Fig. 22.7.

On hydrolysis sphingomyelins yield sphingosine, choline, phosphoric acid, and a 24-carbon fatty acid called lignoceric acid. In a sphingomyelin this last component is bound to the $-NH_2$ group of sphingosine. The sphingolipids do not yield glycerol when they are hydrolyzed.

The cerebroside shown in Fig. 22.7 is an example of a **glycolipid.** Glycolipids have a polar group that is contributed by a *carbohydrate.* They do not yield phosphoric acid or choline when they are hydrolyzed.

The sphingolipids, together with proteins and polysaccharides, make up **myelin,** the protective coating that encloses nerve fibers or **axons.** The axons of nerve cells carry electrical nerve impulses. Myelin has been described as having a function relative to the axon similar to that of the insulation on an ordinary electric wire.

22.7 WAXES

Most waxes are esters of long-chain fatty acids and long-chain alcohols. Waxes are found as protective coatings on the skin, fur, or feathers of animals, and on the leaves

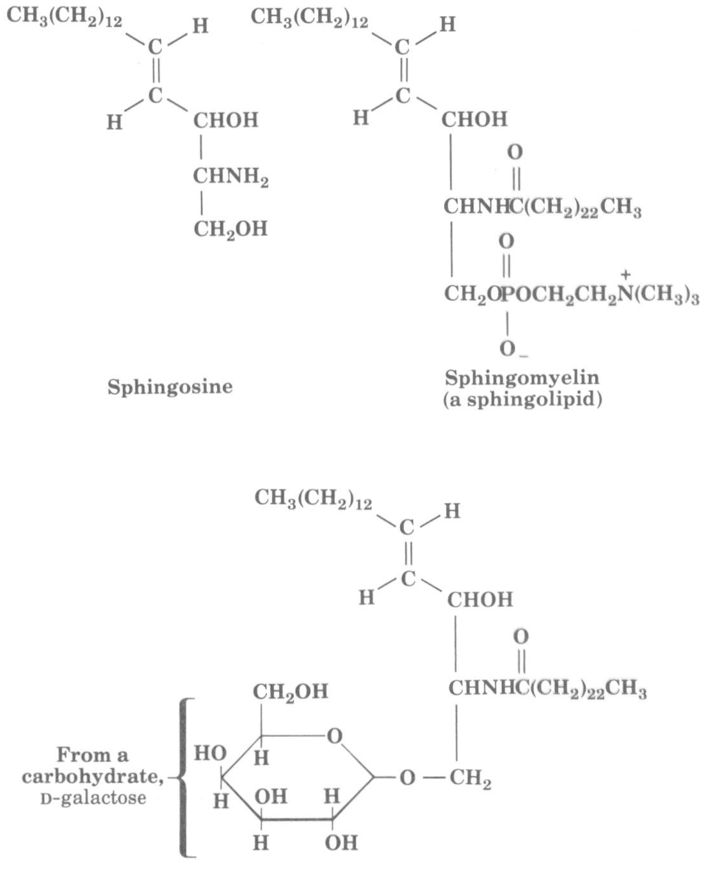

FIGURE 22.7 A sphingosine and two sphingolipids.

and fruits of plants. Several esters isolated from waxes are the following:

$$CH_3(CH_2)_{14}\overset{\overset{\displaystyle O}{\|}}{C}OCH_2(CH_2)_{14}CH_3$$

Cetyl palmitate
(from spermaceti)

$$CH_3(CH_2)_n\overset{\overset{\displaystyle O}{\|}}{C}OCH_2(CH_2)_mCH_3$$

$n = 24$ and 26; $m = 28$ and 30
(from beeswax)

$$HOCH_2(CH_2)_n\overset{\overset{\displaystyle O}{\|}}{C}-OCH_2(CH_2)_mCH_3$$

$n = 16 - 28$; $m = 30$ and 32
(from carnauba wax)

ADDITIONAL PROBLEMS

22.12

How would you convert stearic acid, $CH_3(CH_2)_{16}CO_2H$, into each of the following?

(a) Ethyl stearate, $CH_3(CH_2)_{16}CO_2C_2H_5$ (two ways)

(b) *tert*-Butyl stearate, $CH_3(CH_2)_{16}CO_2C(CH_3)_3$

(c) Stearamide, $CH_3(CH_2)_{16}CONH_2$

(d) *N,N*-Dimethylstearamide, $CH_3(CH_2)_{16}CON(CH_3)_2$

(e) Octadecylamine, $CH_3(CH_2)_{16}CH_2NH_2$

(f) Heptadecylamine, $CH_3(CH_2)_{15}CH_2NH_2$

(g) Octadecanal, $CH_3(CH_2)_{16}CHO$

(h) Octadecyl stearate, $CH_3(CH_2)_{16}\overset{\overset{\displaystyle O}{\|}}{C}OCH_2(CH_2)_{16}CH_3$

(i) 1-Octadecanol, $CH_3(CH_2)_{16}CH_2OH$ (two ways)

(j) 2-Nonadecanone, $CH_3(CH_2)_{16}\overset{\overset{\displaystyle O}{\|}}{C}CH_3$

(k) 1-Bromooctadecane, $CH_3(CH_2)_{16}CH_2Br$

(l) Nonadecanoic acid, $CH_3(CH_2)_{16}CH_2CO_2H$

22.13

How would you transform myristic acid into each of the following?

(a) $CH_3(CH_2)_{11}\underset{\underset{\displaystyle Br}{|}}{C}HCO_2H$

(c) $CH_3(CH_2)_{11}\underset{\underset{\displaystyle CN}{|}}{C}HCO_2H$

(b) $CH_3(CH_2)_{11}\underset{\underset{\displaystyle OH}{|}}{C}HCO_2H$

(d) $CH_3(CH_2)_{11}\underset{\underset{\displaystyle NH_3^+}{|}}{C}HCO_2^-$

22.14

Using palmitoleic acid as an example and neglecting stereochemistry, illustrate each of the following reactions of the double bond.

(a) Addition of iodine

(c) Hydroxylation

(b) Addition of hydrogen

(d) Addition of HCl

22.15

When oleic acid is heated to 180–200 °C (in the presence of a small amount of selenium), an equilibrium is established between oleic acid (33%) and an isomeric compound called elaidic acid (67%). Suggest a possible structure for elaidic acid.

22.16

Gadoleic acid $(C_{20}H_{38}O_2)$, a fatty acid that can be isolated from cod-liver oil, can be cleaved by hydroxylation and subsequent treatment with periodic acid to $CH_3(CH_2)_9CHO$ and $OHC(CH_2)_7CO_2H$. (a) What two stereoisomeric structures are possible for gadoleic acid? (b) What spectroscopic technique would make possible a decision as to the actual structure of gadoleic acid? (c) What peaks would you look for?

22.17

When limonene (p. 1050) is heated strongly, it yields 2 moles of isoprene. What kind of reaction is involved here?

22.18

α-Phellandrene and β-phellandrene are isomeric compounds that are minor constituents of spearmint oil; they have the molecular formula $C_{10}H_{16}$. Each compound has an ultraviolet absorption maximum in the 230–270-nm range. On catalytic hydrogenation each compound yields 1-isopropyl-4-methylcyclohexane. On vigorous oxidation with potassium permanga-

nate, α-phellandrene yields $CH_3\overset{\overset{\displaystyle O}{\|}}{C}CO_2H$ and $CH_3\overset{|}{C}HCH(CO_2H)CH_2CO_2H$. A similar oxida-
$\underset{\displaystyle CH_3}{}$

tion of β-phellandrene yields $CH_3\overset{|}{C}HCH(CO_2H)CH_2CH_2\overset{\overset{\displaystyle O}{\|}}{C}CO_2H$ as the only isolable prod-
$\underset{\displaystyle CH_3}{}$

uct. Propose structures for α- and β-phellandrene.

22.19

Vaccenic acid, a constitutional isomer of oleic acid, has been synthesized through the following reaction sequence:

1-Octyne + NaNH$_2$ $\xrightarrow{\text{NH}_3}$ **A** $(C_8H_{13}Na)$ $\xrightarrow{\text{ICH}_2(\text{CH}_2)_7\text{CH}_2\text{Cl}}$

 **B** $(C_{17}H_{31}Cl)$ $\xrightarrow{\text{NaCN}}$ **C** $(C_{18}H_{31}N)$ $\xrightarrow{\text{KOH,H}_2\text{O}}$ **D** $(C_{18}H_{31}O_2K)$ $\xrightarrow{\text{H}_3\text{O}^+}$

 **E** $(C_{18}H_{32}O_2)$ $\xrightarrow[\text{BaSO}_4]{\text{H}_2,\text{Pd}}$ vaccenic acid $(C_{18}H_{34}O_2)$

Propose a structure for vaccenic acid and for the intermediates **A–E**.

22.20

ω-Fluorooleic acid can be isolated from a shrub, *Dechapetalum toxicarium,* that grows in Sierra Leone. The compound is highly toxic to warm-blooded animals; it has found use as an arrow poison in tribal warfare, in poisoning enemy water supplies, and by witch doctors "for terrorizing the native population." Powdered fruit of the plant has been used as a rat poison, hence ω-fluorooleic acid has the common name "ratsbane." A synthesis of ω-fluorooleic acid is outlined here. Give structures for compounds **F–I**.

1-Bromo-8-fluorooctane + sodium acetylide $\longrightarrow$ **F** ($C_{10}H_{17}F$) $\xrightarrow[(2)\ I(CH_2)_7Cl]{(1)\ NaNH_2}$

G ($C_{17}H_{30}FCl$) $\xrightarrow{NaCN}$ **H** ($C_{18}H_{30}NF$) $\xrightarrow[(2)\ H^+]{(1)\ KOH}$ **I** ($C_{18}H_{31}O_2F$) $\xrightarrow{H_2}{Ni_2B\ (P-2)}$

$$F-(CH_2)_8 \underset{H}{\overset{}{\diagdown}} C=C \underset{H}{\overset{(CH_2)_7COH}{\diagup}}$$

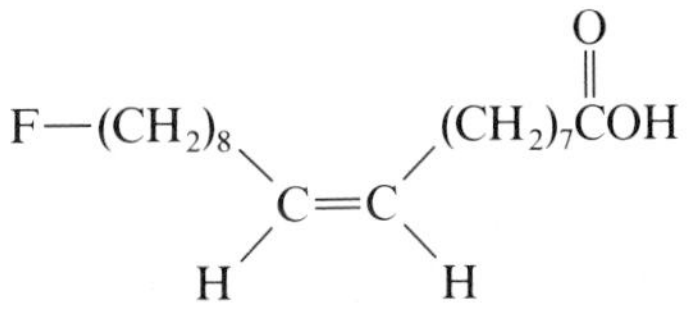

ω-Fluorooleic acid
(46% yield, overall)

22.21

Give formulas and names for compounds **A** and **B**.

$$5\alpha\text{-Cholest-2-ene} \xrightarrow{C_6H_5\overset{O}{\overset{\|}{C}}OOH} \textbf{A (an epoxide)} \xrightarrow{HBr} \textbf{B}$$

(*Hint:* **B** is not the most stable stereoisomer.)

*22.22

One of the first laboratory syntheses of cholesterol was achieved by R. B. Woodward and his students at Harvard University in 1951. Many of the steps of this synthesis are outlined here. Supply the missing reagents.

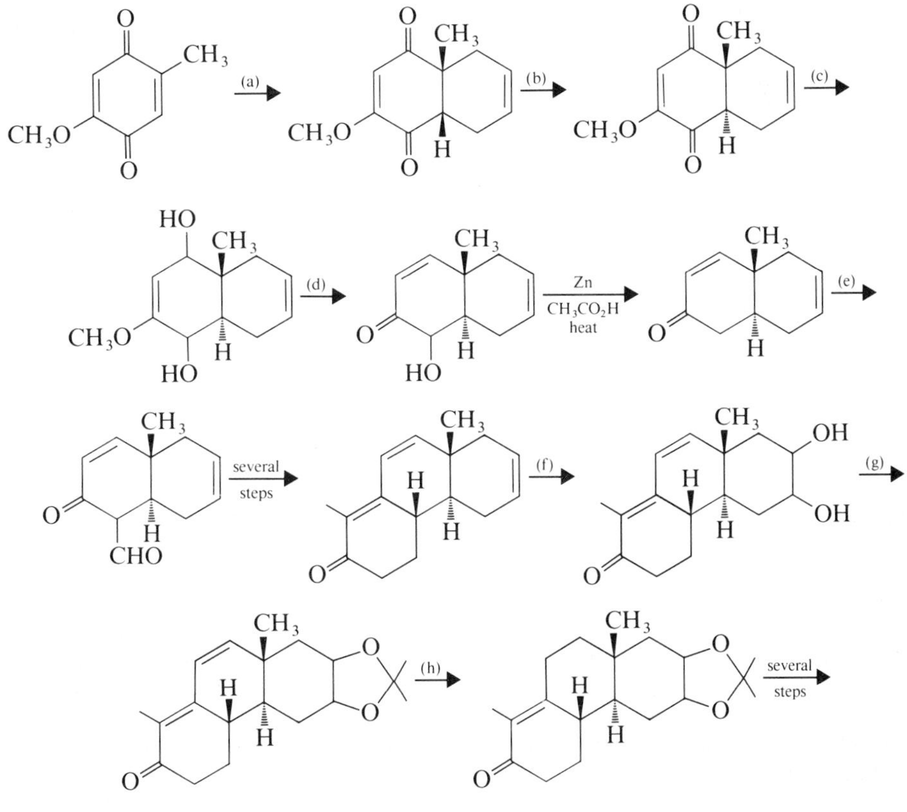

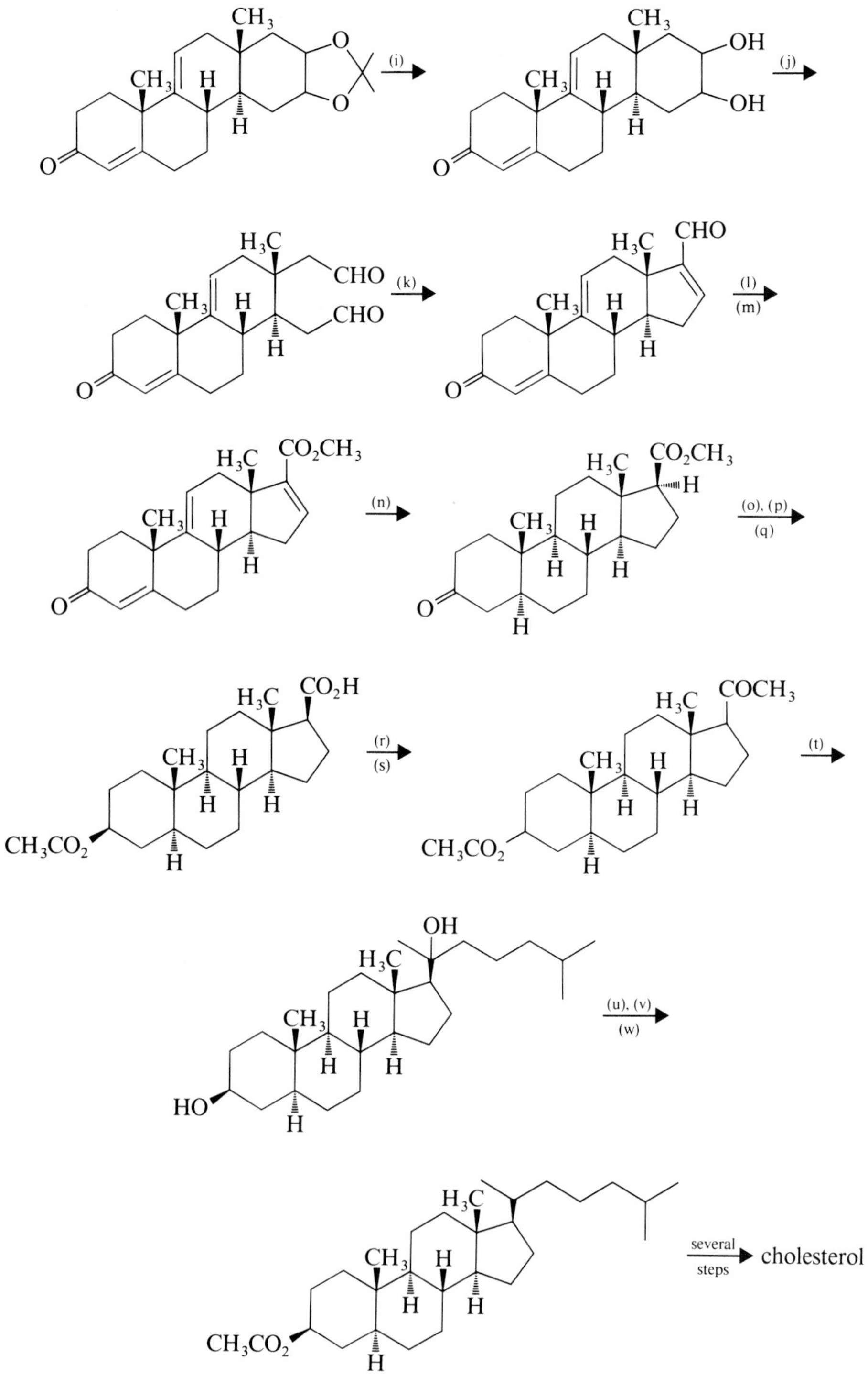

22.23
The initial steps of a laboratory synthesis of several prostaglandins reported by E. J. Corey (Section 15.9) and his co-workers in 1968 are outlined here. Supply each of the missing reagents.

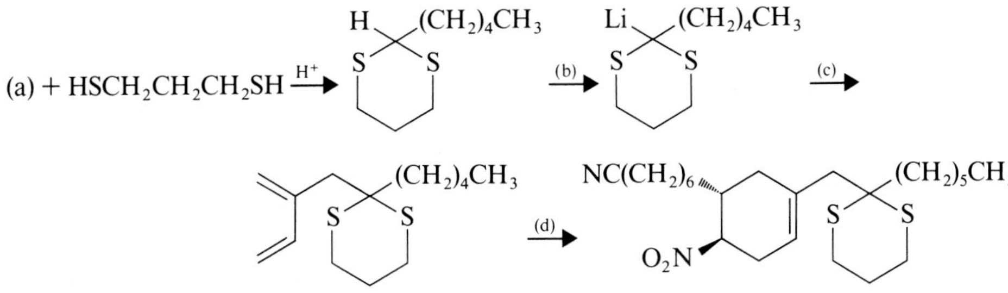

(a) + $HSCH_2CH_2CH_2SH$ $\xrightarrow{H^+}$

(b) $\rightarrow$

(c) $\rightarrow$

(d) $\rightarrow$

(e) The initial step in another prostaglandin synthesis is shown in the following reaction. What kind of reaction — and catalyst — is needed here?

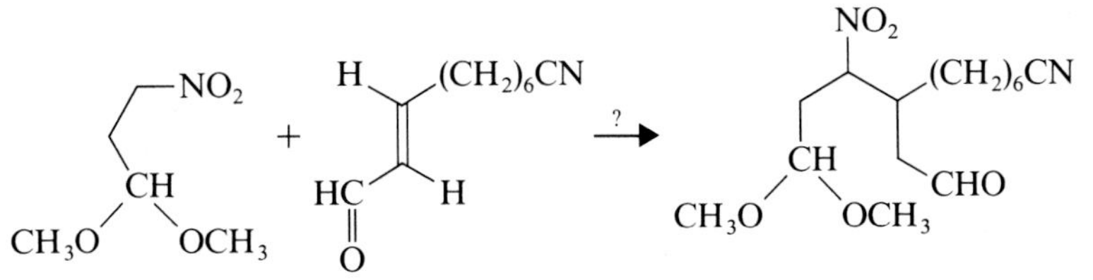

$\xrightarrow{?}$

22.24

A useful synthesis of sesquiterpene ketones, called *cyperones,* was accomplished through a modification of the following Robinson annulation procedure (Section 17.9B)

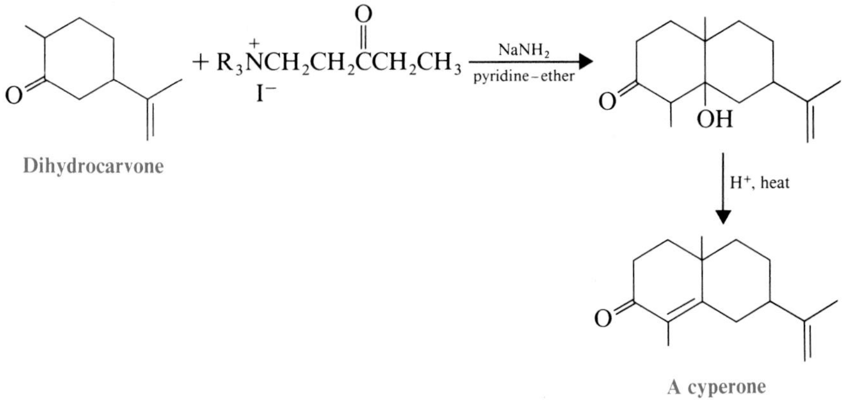

$+ R_3\overset{+}{N}CH_2CH_2\overset{O}{\overset{\|}{C}}CH_2CH_3$ I^- $\xrightarrow[\text{pyridine–ether}]{\text{NaNH}_2}$

Dihydrocarvone

$\downarrow$ H$^+$, heat

A cyperone

Write a mechanism that accounts for each step of this synthesis.

SPECIAL TOPIC M

THIOL ESTERS AND LIPID BIOSYNTHESIS

M.1 Thiol Esters

Thiol esters can be prepared by reactions of a thiol with an acyl chloride.

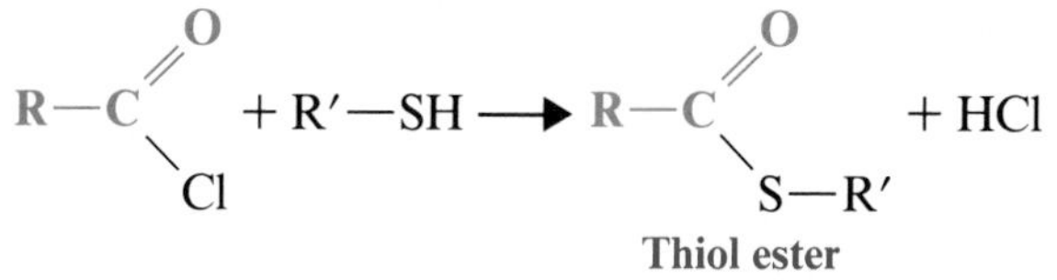
R—C(=O)Cl + R′—SH → R—C(=O)S—R′ + HCl
Thiol ester

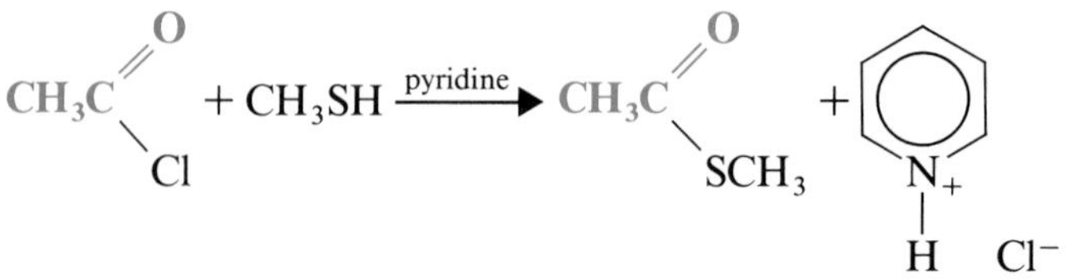
CH3C(=O)Cl + CH3SH
pyridine
CH3C(=O)SCH3 +
N+
H Cl−

Although thiol esters are not often used in laboratory syntheses, they are of great importance in syntheses that occur within living cells. One of the important thiol esters in biochemistry is “acetyl coenzyme A.”

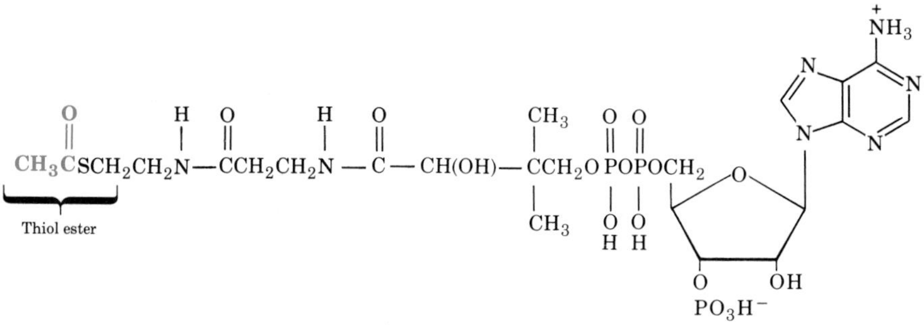
Thiol ester
CH3
CH3
NH3
N
N
N
N
O
OH
PO3H−

Acetyl coenzyme A

The important part of this rather complicated structure is the thiol ester at the beginning of the chain; because of this, acetyl coenzyme A is usually abbreviated as follows:

$$CoA{-}S\overset{\overset{\displaystyle O}{\|}}{C}CH_3$$

and coenzyme A, itself, is abbreviated:

$$CoA{-}SH$$

In certain biochemical reactions, an *acyl* coenzyme A operates as an *acylating agent;* it transfers an acyl group to another nucleophile in a reaction that involves a nucleophilic attack at the acyl carbon of the thiol ester. For example:

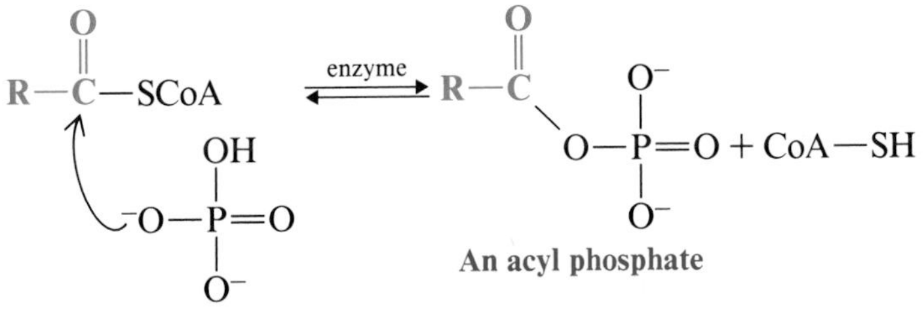

An acyl phosphate

This reaction is catalyzed by the enzyme *phosphotransacetylase.*

The α hydrogens of the acetyl group of acetyl coenzyme A are appreciably acidic. Acetyl coenzyme A, as a result, also functions as an *alkylating agent.* Acetyl coenzyme A, for example, reacts with oxaloacetate ion to form citrate ion in a reaction that resembles an aldol addition.

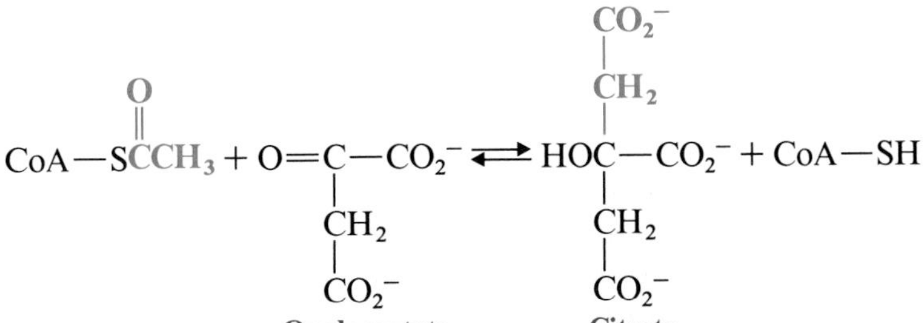

| | Oxaloacetate ion | Citrate ion |

One might well ask, "Why has nature made such prominent use of thiol esters?" Or, "In contrast to ordinary esters, what advantages do thiol esters offer the cell?" In answering these questions we can consider three factors:

1. Resonance contributions of type (*b*) in the following reaction stabilize an ordinary ester and make the carbonyl group less susceptible to nucleophilic attack.

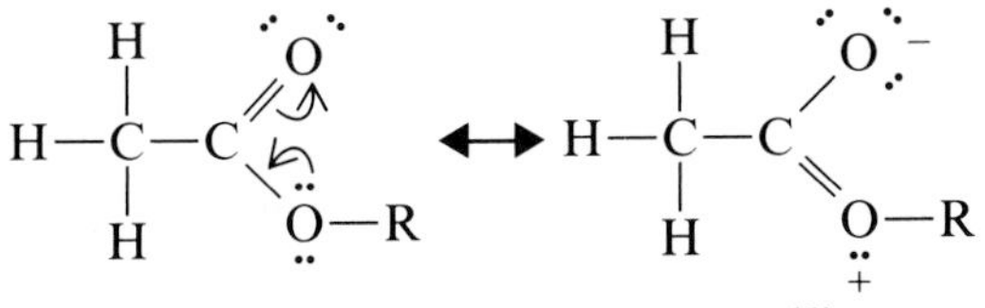

(*a*)   (*b*)
*This structure makes
an important contribution*

By contrast, thiol esters are not as effectively stabilized by a similar resonance contribution because structure (*d*) among the following ones requires overlap between the $3p$ orbital of sulfur and a $2p$ orbital of carbon. Since this overlap is not large, resonance stabilization by (*d*) is not as effective. Structure (*e*) does, however, make an important contribution—one that makes the carbonyl group more susceptible to nucleophilic attack.

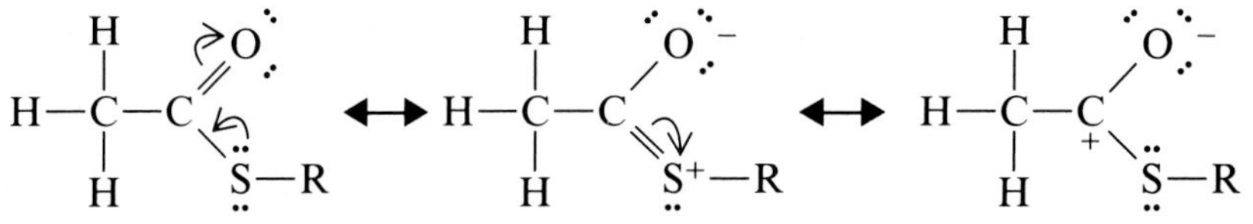

(c)

(d)
*This structure is
not an important
contributor*

(e)
*This structure
makes the carbonyl
carbon atom susceptible
to nucleophilic attack*

2. A resonance contribution from the similar structure (*g*) makes the α hydrogens of thiol esters more acidic than those of ordinary esters.

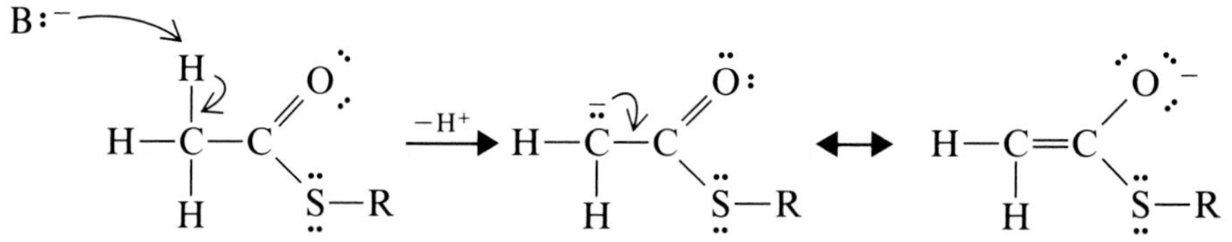

(f)

(g)
*This structure's
contribution stabilizes
the anion of a thiol ester*

3. The carbon–sulfur bond of a thiol ester is weaker than the carbon–oxygen bond of an ordinary ester; $^-\!:S$—R is a better leaving group than $^-\!:OR$.

Factors **1** and **3** make thiol esters effective *acylating agents;* factor **2** makes them effective *alkylating* agents. Thus, we should not be surprised when we encounter reactions similar to the following one:

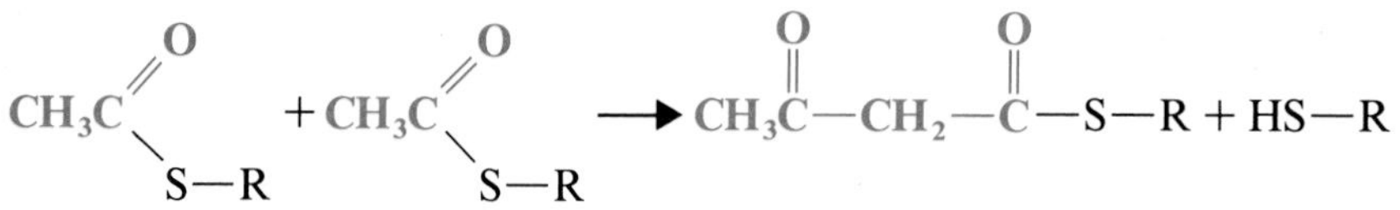

In this reaction 1 mole of a thiol ester acts as an acylating agent and the other acts as an alkylating agent (cf. Section M.2).

M.2 Biosynthesis of Fatty Acids

The fact that most naturally occurring fatty acids are made up of an even number of carbon atoms suggests that they are assembled from two-carbon units. The idea that these might be acetate ($CH_3CO_2^-$) units was put forth as early as 1893. Many years later, when radioactively labeled compounds became available, it became possible to test and confirm this hypothesis.

When an animal is fed acetic acid labeled with carbon-14 at the carboxyl group, the fatty acids that the animal synthesizes contain the label at alternate carbon atoms beginning with the carboxyl carbon:

$$\overset{*}{C}H_3CO_2H \qquad CH_3CH_2\overset{*}{C}H_2CH_2\overset{*}{C}H_2CH_2\overset{*}{C}H_2CH_2\overset{*}{C}H_2CH_2\overset{*}{C}H_2CH_2\overset{*}{C}H_2CH_2\overset{*}{C}O_2H$$

Feeding
carboxyl-
labeled
acetic acid
($C^* = {}^{14}C$) · · ·   yields palmitic acid labeled at these positions

Conversely, feeding acetic acid labeled at the methyl carbon yields a fatty acid labeled at the other set of alternate carbon atoms:

$$\overset{*}{C}H_3CO_2H \qquad \overset{*}{C}H_3CH_2\overset{*}{C}H_2CH_2\overset{*}{C}H_2CH_2\overset{*}{C}H_2CH_2\overset{*}{C}H_2CH_2\overset{*}{C}H_2CH_2\overset{*}{C}H_2CH_2\overset{*}{C}H_2CO_2H$$

Feeding
methyl-
labeled
acetic acid · · ·   **yields palmitic acid labeled at these positions**

The biosynthesis of fatty acids is now known to begin with acetyl coenzyme A:

$$\underset{}{CH_3\overset{\overset{\textstyle O}{\|}}{C}-S-CoA}$$

The acetyl portion of acetyl coenzyme A can be synthesized in the cell from acetic acid; it can also be synthesized from carbohydrates, proteins, and other fats.

$$CH_3\overset{\overset{\textstyle O}{\|}}{C}OH$$

Carbohydrates

Proteins

Fats

$$\xrightarrow{CoA-SH} \quad CH_3\overset{\overset{\textstyle O}{\|}}{C}S-CoA$$

Acetyl coenzyme A

Although the methyl group of acetyl coenzyme A is already activated toward condensation reactions by virtue of its being a part of a thiol ester (Section M.1), nature activates it again by converting it to *malonyl coenzyme A.*

$$CH_3\overset{\overset{\textstyle O}{\|}}{C}S-CoA + CO_2 \underset{}{\overset{\text{acetyl CoA carboxylase*}}{\rightleftharpoons}} HO\overset{\overset{\textstyle O}{\|}}{C}CH_2\overset{\overset{\textstyle O}{\|}}{C}S-CoA$$

Acetyl CoA   **Malonyl CoA**

The next steps in fatty acid synthesis involve the transfer of acyl groups of malonyl CoA and acetyl coenzyme A to the thiol group of a coenzyme called *acyl carrier protein* or ACP—SH.

$$HO\overset{\overset{\textstyle O}{\|}}{C}CH_2\overset{\overset{\textstyle O}{\|}}{C}S-CoA + ACP-SH \rightleftharpoons HO\overset{\overset{\textstyle O}{\|}}{C}CH_2\overset{\overset{\textstyle O}{\|}}{C}S-ACP + CoA-SH$$

Malonyl CoA   **Malonyl—S—ACP**

$$CH_3\overset{\overset{\textstyle O}{\|}}{C}S-CoA + ACP-SH \rightleftharpoons CH_3\overset{\overset{\textstyle O}{\|}}{C}S-ACP + CoA-SH$$

Acetyl CoA   **Acetyl—S—ACP**

*This step also requires 1 mole of adenosine triphosphate (Section 21.1B) and an enzyme that transfers the carbon dioxide.

Acetyl-S-ACP and malonyl-S-ACP then condense with each other to form acetoacetyl-S-ACP:

$$\underset{\text{Acetyl-S-ACP}}{CH_3\overset{O}{\overset{\|}{C}}S-ACP} + \underset{\text{Malonyl-S-ACP}}{HO\overset{O}{\overset{\|}{C}}CH_2\overset{O}{\overset{\|}{C}}S-ACP} \rightleftharpoons \underset{\text{Acetoacetyl-S-ACP}}{CH_3\overset{O}{\overset{\|}{C}}CH_2\overset{O}{\overset{\|}{C}}S-ACP} + CO_2 + ACP-SH$$

The molecule of CO_2 that is lost in this reaction is the same molecule that was incorporated into malonyl CoA in the acetyl CoA carboxylase reaction.

This remarkable reaction bears a strong resemblance to the malonic ester syntheses that we saw earlier (Section 20.4) and it deserves special comment. One can imagine, for example, a more economical synthesis of acetoacetyl-S-ACP, that is, a simple condensation between 2 moles of acetyl-S-ACP.

$$CH_3\overset{O}{\overset{\|}{C}}S-ACP + CH_3\overset{O}{\overset{\|}{C}}S-ACP \rightleftharpoons CH_3\overset{O}{\overset{\|}{C}}CH_2\overset{O}{\overset{\|}{C}}S-ACP + ACP-SH$$

Studies of this last reaction, however, have revealed that it is highly *endothermic* and that the position of equilibrium lies very far to the left. By contrast, the condensation of acetyl-S-ACP and malonyl-S-ACP is highly exothermic, and the position of equilibrium lies far to the right. The favorable thermodynamics of the condensation utilizing malonyl-S-ACP comes about because *the reaction also produces a highly stable substance: Carbon dioxide.* Thus, decarboxylation of the malonyl group provides the condensation with thermodynamic assistance.

The next three steps in fatty acid synthesis transform the acetoacetyl group of acetoacetyl-S-ACP into a butyryl (butanoyl) group. These steps involve (1) reduction of the keto group (utilizing NADPH* as the reducing agent), (2) dehydration of an alcohol, and (3) reduction of a double bond (again utilizing NADPH).

Reduction of the Keto Group

$$\underset{\text{Acetoacetyl-S-ACP}}{CH_3\overset{O}{\overset{\|}{C}}CH_2\overset{O}{\overset{\|}{C}}S-ACP} + NADPH + H^+ \rightleftharpoons \underset{\beta\text{-Hydroxybutyryl-S-ACP}}{CH_3\overset{OH}{\overset{|}{C}}HCH_2\overset{O}{\overset{\|}{C}}S-ACP} + NADP^+$$

Dehydration of the Alcohol

$$\underset{\beta\text{-Hydroxybutyryl-S-ACP}}{CH_3\overset{OH}{\overset{|}{C}}HCH_2\overset{O}{\overset{\|}{C}}S-ACP} \rightleftharpoons \underset{\text{Crotonyl-S-ACP}}{CH_3CH=CH\overset{O}{\overset{\|}{C}}S-ACP} + H_2O$$

Reduction of the Double Bond

$$\underset{\text{Crotonyl-S-ACP}}{CH_3CH=CH\overset{O}{\overset{\|}{C}}S-ACP} + NADPH + H^+ \rightleftharpoons \underset{\text{Butyryl-S-ACP}}{CH_3CH_2CH_2\overset{O}{\overset{\|}{C}}S-ACP} + NADP^+$$

*NADPH is *nicotinamide adenine dinucleotide phosphate (reduced form),* a coenzyme that is very similar in structure and function to NADH, Section 11.11.

These steps complete one cycle of the overall fatty acid synthesis. Their net result is the conversion of two acetate units into the four-carbon butyrate unit of butyryl-S-ACP. (This conversion requires, of course, the crucial intervention of a molecule of carbon dioxide.) At this point, another cycle begins and the chain is lengthened by two more carbon atoms:

Condensation

$$CH_3CH_2CH_2\overset{\overset{\displaystyle O}{\|}}{C}S-ACP + HO\overset{\overset{\displaystyle O}{\|}}{C}CH_2\overset{\overset{\displaystyle O}{\|}}{C}S-ACP \longrightarrow$$

$$CH_3CH_2CH_2\overset{\overset{\displaystyle O}{\|}}{C}CH_2\overset{\overset{\displaystyle O}{\|}}{C}S-ACP + CO_2 + ACP-SH$$

(four carbon atoms)

Reduction

$$CH_3CH_2CH_2\overset{\overset{\displaystyle O}{\|}}{C}CH_2\overset{\overset{\displaystyle O}{\|}}{C}S-ACP \xrightarrow[\text{NADPH NADP}^+]{\text{+H}^+} CH_3CH_2CH_2\overset{\overset{\displaystyle OH}{|}}{C}HCH_2\overset{\overset{\displaystyle O}{\|}}{C}S-ACP$$

Dehydration

$$CH_3CH_2CH_2\overset{\overset{\displaystyle OH}{|}}{C}HCH_2\overset{\overset{\displaystyle O}{\|}}{C}S-ACP \xrightarrow{(-H_2O)} CH_3CH_2CH_2CH=CH\overset{\overset{\displaystyle O}{\|}}{C}S-ACP$$

Reduction

$$CH_3CH_2CH_2CH=CH\overset{\overset{\displaystyle O}{\|}}{C}S-ACP \xrightarrow[\text{NADPH NADP}^+]{\text{+H}^+} CH_3CH_2CH_2CH_2CH_2\overset{\overset{\displaystyle O}{\|}}{C}S-ACP$$

(six carbon atoms)

Subsequent turns of the cycle continue to lengthen the chain by two-carbon units until a long-chain fatty acid is produced. The overall equation for the synthesis of palmitic acid, for example, can be written as follows:

$$CH_3\overset{\overset{\displaystyle O}{\|}}{C}S-CoA + 7HO\overset{\overset{\displaystyle O}{\|}}{C}CH_2\overset{\overset{\displaystyle O}{\|}}{C}S-CoA + 14NADH + 14H^+ \longrightarrow$$

$$CH_3(CH_2)_{14}CO_2H + 7CO_2 + 8CoA-SH + 14NAD^+ + 6H_2O$$

One of the most remarkable aspects of fatty acid synthesis is that the entire cycle appears to be carried out by a complex of enzymes that are clustered into a single unit. The molecular weight of this cluster of proteins, called *fatty acid synthetase,* has been estimated as 2,300,000.* The synthesis begins with a single molecule of acetyl-S-ACP serving as a primer. Then, in the synthesis of palmitic acid, for example, successive

---

*As isolated from yeast cells. Fatty acid synthetases from different sources have different molecular weights; that from pigeon liver, for example, has a molecular weight of 450,000.

condensations of seven molecules of malonyl-S-ACP occur with each condensation followed by reduction, dehydration, and reduction. All of these steps, which result in the synthesis of a 16-carbon chain, take place before the fatty acid is released from the enzyme cluster.

The acyl carrier protein has been isolated and purified; its molecular weight is approximately 10,000. The protein contains a chain of groups called a *phospho-pantetheine group* that is identical to that of coenzyme A (Section M.1). In ACP this chain is attached to a protein (rather than to an adenosine phosphate as it is in coenzyme A):

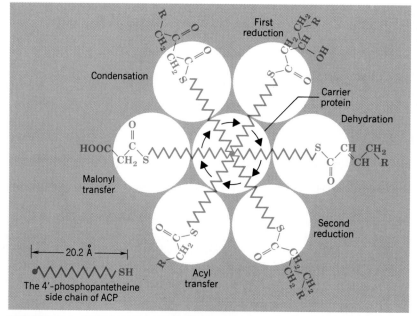

The length of the phosphopantetheine group is 20.2 Å, and it has been postulated that it acts as a "swinging arm" in transferring the growing acyl chain from one enzyme of the cluster to the next (Fig. M.1).

M.3 Biosynthesis of Isoprenoids

The basic building block for the synthesis of terpenes and terpenoids (Section 22.3) is 3-methyl-3-butenyl pyrophosphate. The five carbon atoms of this compound are the source of the "isoprene units" of all "isoprenoids." The pyrophosphate group is a group that nature relies upon for a vast number of chemical processes. In the reac-

FIGURE M.1 The phosphopantetheine group as a swinging arm in the fatty acid synthetase complex. (Adapted from A. L. Lehninger, *Biochemistry*, Worth, New York, 1970, p. 519. Used with permission.)

tions that we shall soon see the pyrophosphate group functions as a natural "leaving group" in enzymatic reactions that resemble the alkylation reactions of alkenes that we studied in Section 7.11.

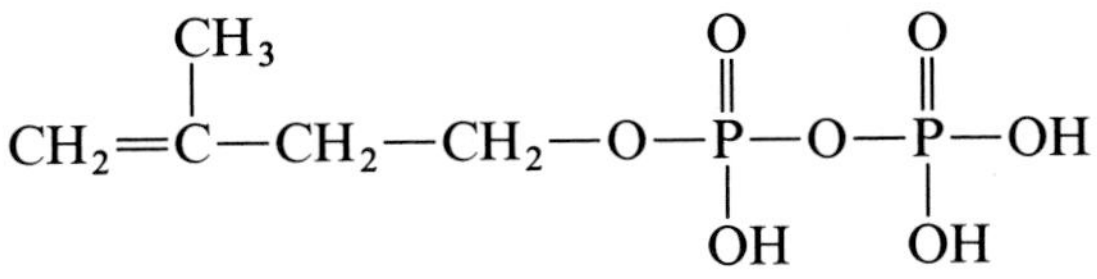

$$CH_2 = \overset{\overset{\displaystyle CH_3}{|}}{C} - CH_2 - CH_2 - O - \overset{\overset{\displaystyle O}{\|}}{\underset{\underset{\displaystyle OH}{|}}{P}} - O - \overset{\overset{\displaystyle O}{\|}}{\underset{\underset{\displaystyle OH}{|}}{P}} - OH$$

3-Methyl-3-butenyl pyrophosphate

3-Methyl-3-butenyl pyrophosphate is isomerized by an enzyme to 3-methyl-2-butenyl pyrophosphate. The isomerization establishes an equilibrium that makes both compounds available to the cell.

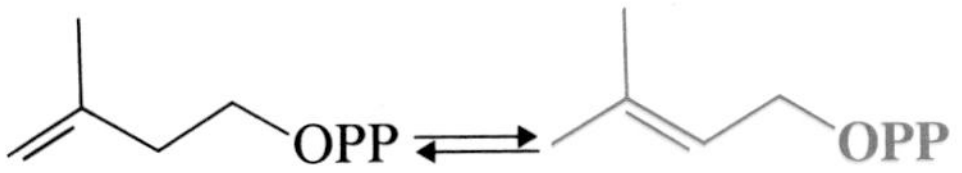

| **3-Methyl-3-butenyl**
pyrophosphate | **3-Methyl-2-butenyl**
pyrophosphate | OPP = Pyrophosphate |

These two 5-carbon compounds condense with each other in another enzymatic reaction to yield the 10-carbon compound, geranyl pyrophosphate. The first step involves the formation of an allylic cation.

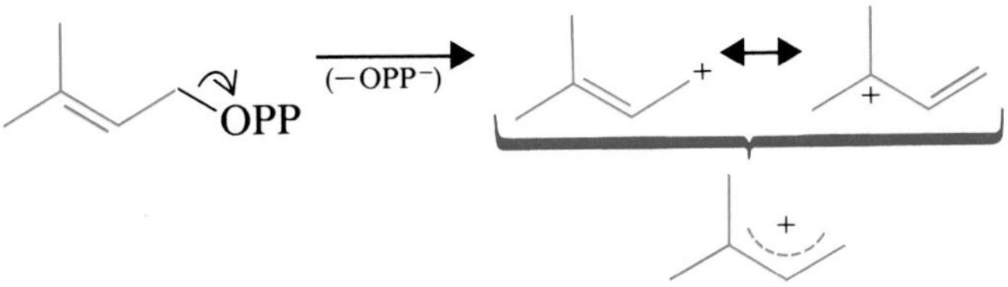

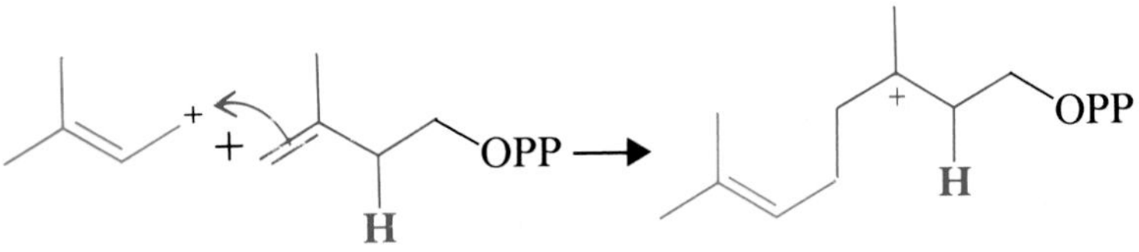

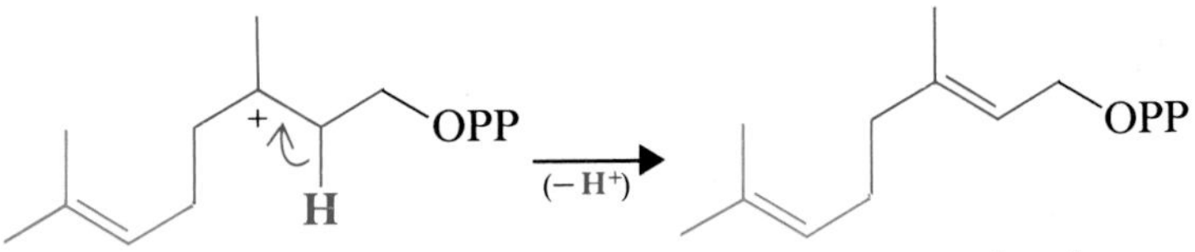

Geranyl pyrophosphate

Geranyl pyrophosphate is the precursor of the monoterpenes; hydrolysis of geranyl pyrophosphate, for example, yields geraniol.

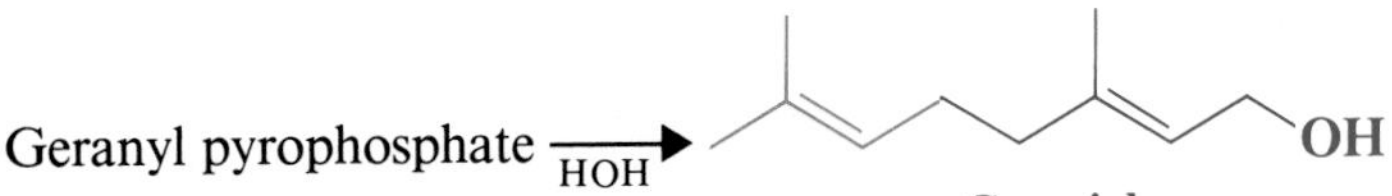

Geraniol

Geranyl pyrophosphate can also condense with 3-methyl-3-butenyl pyrophosphate to form the 15-carbon precursor for sesquiterpenes, farnesyl pyrophosphate.

Geranyl pyrophosphate

$-OPP^-$

Farnesyl pyrophosphate

Farnesol

other sesquiterpenes

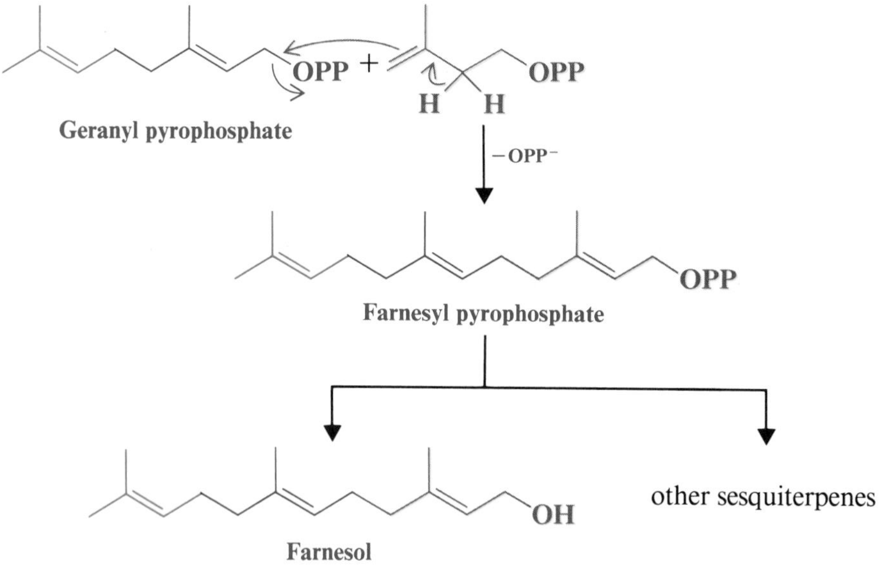

Farnesol has been isolated from ambrette oil. It has the odor of lily of the valley. Farnesol also functions as a hormone in certain insects and initiates the change from caterpillar to pupa to moth.

Similar condensation reactions yield the precursors for all of the other terpenes (Fig. M.2). In addition, a tail-to-tail reductive coupling of two molecules of farnesyl pyrophosphate produces squalene, the precursor for the important group of isoprenoids known as *steroids* (cf. Sections 22.4 and M.4).

PROBLEM M.1

When farnesol is treated with sulfuric acid, it is converted to bisabolene. Outline a possible mechanism for this reaction.

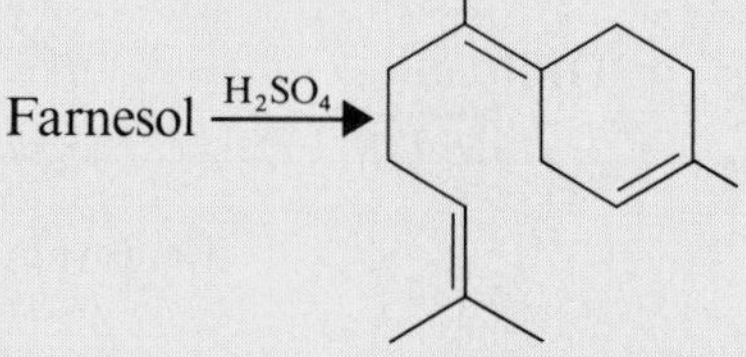

Farnesol $\xrightarrow{\text{H}_2\text{SO}_4}$

Bisabolene

M.4 Biosynthesis of Steroids

We saw in the previous section that the five-carbon compound, 3-methyl-3-butenyl pyrophosphate, is the actual "isoprene unit" that nature uses in constructing terpen-

Monoterpenes ◄── Geranyl pyrophosphate
(C_{10})   (C_{10}-pyrophosphate)

3-methyl-3-butenyl
pyrophosphate

Sesquiterpenes ◄── Farnesyl pyrophosphate ──► Squalene
(C_{15})   (C_{15}-pyrophosphate)   (C_{30})

3-methyl-3-butenyl
pyrophosphate

Diterpenes ◄── C_{20}-Pyrophosphate   Lanosterol
(C_{20})   (cf. p. 1086)

Tetraterpenes   Cholesterol
(C_{40})   (a steroid)

FIGURE M.2 The biosynthetic paths for terpenes and steroids.

oids and carotenoids. We can now extend that biosynthetic pathway in two directions. We can show how 3-methyl-3-butenyl pyrophosphate (like the fatty acids) is ultimately derived from acetate units, and how cholesterol, the precursor of most of the important steroids, is synthesized from 3-methyl-3-butenyl pyrophosphate.

In the 1940s Konrad Bloch of Harvard University used labeling experiments to demonstrate that all of the carbon atoms of cholesterol can be derived from acetic acid. Using *methyl-labeled* acetic acid, for example, Bloch found the following label distribution in the cholesterol that was synthesized.

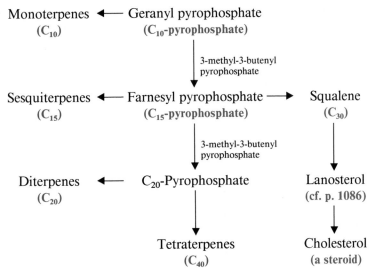

Bloch also found that feeding *carboxyl-labeled* acetic acid led to incorporation of the label into all of the other carbon atoms of cholesterol (the unstarred carbon atoms of the previous formula).

Subsequent research by a number of investigators has shown that 3-methyl-3-butenyl pyrophosphate is synthesized from acetate units through the following sequence of reactions:

$$\underset{\substack{(C_2)\\ \text{Acetyl CoA}}}{CH_3\overset{\overset{O}{\|}}{C}S{-}CoA} + \underset{\substack{(C_4)\\ \text{Acetoacetyl CoA}}}{CH_3\overset{\overset{O}{\|}}{C}CH_2\overset{\overset{O}{\|}}{C}S{-}CoA} \xrightarrow{\;\;\overset{\displaystyle CoA{-}SH}{}\;\;} \underset{\substack{(C_6)\\ \beta\text{-Hydroxy-}\beta\text{-methylglutaryl CoA}}}{HO\overset{\overset{O}{\|}}{C}CH_2\underset{\underset{\displaystyle CH_3\;\;\;OH}{|}}{\overset{|}{C}}CH_2\overset{\overset{O}{\|}}{C}S{-}CoA}$$

$$\Big\downarrow\; \begin{array}{l}2NADPH + 2H^+ \\ \searrow\; 2NADP^+ + CoA{-}SH\end{array}$$

$$HO\overset{\overset{O}{\|}}{C}CH_2\underset{\underset{\displaystyle CH_3\;\;\;OH}{|}}{\overset{|}{C}}CH_2CH_2OH$$

(C₆)
Mevalonic acid

$$\Big\downarrow\; \begin{array}{l}\curvearrowright 3ATP \\ \searrow 3ADP\end{array}\Big\} \text{Three successive steps}$$

$$H{-}O{-}\overset{\overset{O}{\|}}{C}{-}CH_2\underset{\underset{\displaystyle CH_3\;\;\;OPO_3H^-}{|}}{\overset{|}{C}}CH_2CH_2O{-}\overset{\overset{O}{\|}}{\underset{\underset{O^-}{|}}{P}}{-}O{-}\overset{\overset{O}{\|}}{\underset{\underset{O^-}{|}}{P}}{-}O^-$$

(C₆)
3-Phospho-5-pyrophosphomevalonic acid

$$\Big\downarrow\; \blacktriangleright CO_2 + H_2PO_3^-$$

$$\underset{CH_3}{\overset{CH_2}{\diagdown\!\!\diagup}}C{-}CH_2CH_2{-}O{-}\overset{\overset{O}{\|}}{\underset{\underset{O^-}{|}}{P}}{-}O{-}\overset{\overset{O}{\|}}{\underset{\underset{O^-}{|}}{P}}{-}O^-$$

(C₅)
3-Methyl-3-butenyl pyrophosphate

The first step of this synthetic pathway is straightforward. Acetyl CoA (from 1 mole of acetate) and acetoacetyl CoA (from 2 moles of acetate) condense to form the six-carbon compound, β-hydroxy-β-methylglutaryl CoA. This step is followed by an enzymatic reduction of the thiol ester group of β-hydroxy-β-methylglutaryl CoA to the primary alcohol of mevalonic acid. The key to finding this pathway was the discovery that mevalonic acid was an intermediate and that this six-carbon compound could be transformed into the five-carbon 3-methyl-3-butenyl pyrophosphate by successive phosphorylations and decarboxylation.

Then 3-methyl-3-butenyl pyrophosphate isomerizes to produce an equilibrium mixture that contains 3-methyl-2-butenyl pyrophosphate, and these two compounds condense to form geranyl pyrophosphate, a 10-carbon compound. Geranyl pyrophosphate subsequently condenses with another mole of 3-methyl-3-butenyl pyrophosphate to form farnesyl pyrophosphate, a 15-carbon compound. (Geranyl pyrophosphate and farnesyl pyrophosphate are the precursors of the mono- and sequiterpenes, cf. Section M.3.)

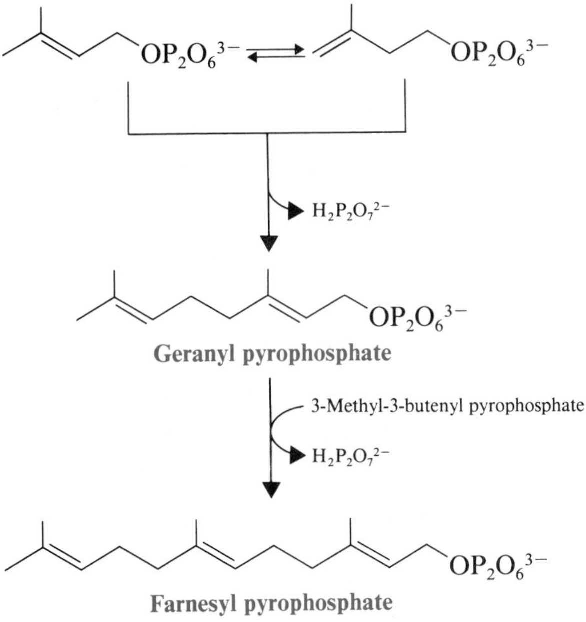

Geranyl pyrophosphate

Farnesyl pyrophosphate

Two molecules of farnesyl pyrophosphate then undergo a reductive condensation to produce squalene.

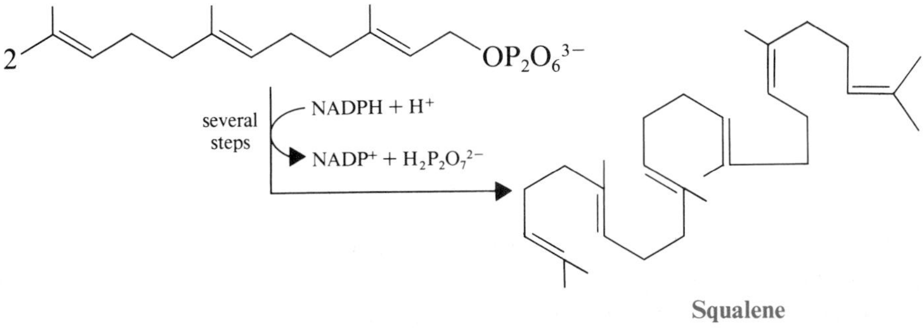

Squalene

Squalene is the direct precursor of cholesterol. Oxidation of squalene yields squalene 2,3-epoxide, which undergoes a remarkable series of ring closures accompanied by concerted methyl and hydride migrations to yield lanosterol. Lanosterol is then converted to cholesterol through a series of enzyme-catalyzed reactions.

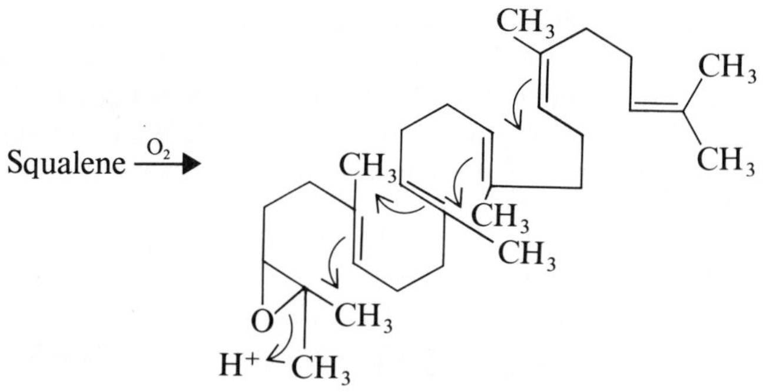

Squalene $\xrightarrow{\text{O}_2}$

Squalene 2,3-epoxide

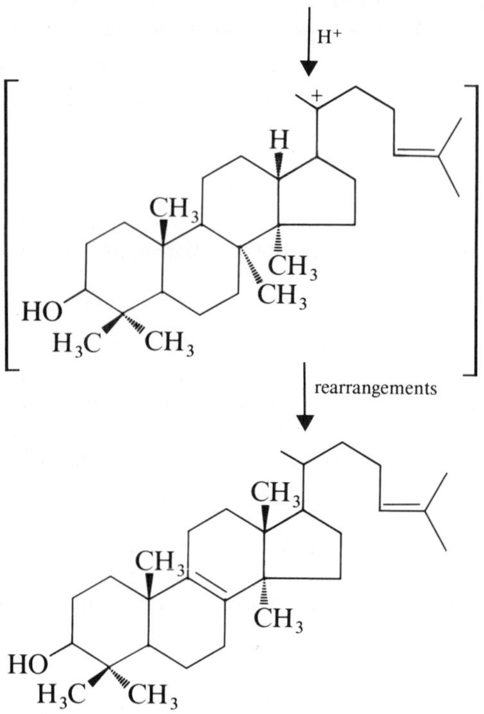

$\downarrow$ H$^+$

$\downarrow$ rearrangements

Lanosterol

$\downarrow$ several steps

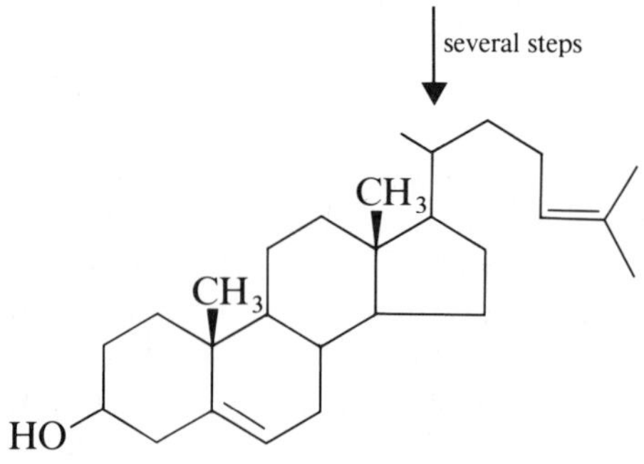

Cholesterol

PHOTOCHEMISTRY OF VISION

The chemical changes that occur when light impinges on the retina of the eye involve several of the phenomena that we have studied in earlier chapters. Central to an understanding of the visual process at the molecular level are two phenomena in particular: the absorption of light by conjugated polyenes and the interconversion of cis–trans isomers.

The retina of the human eye contains two types of receptor cells. Because of their shapes, these cells have been named *rods* and *cones.* Rods are located primarily at the periphery of the retina and are responsible for vision in dim light. Rods, however, are color-blind and "see" only in shades of gray. Cones are found mainly in the center of the retina and are responsible for vision in bright light. Cones also possess the pigments that are responsible for color vision.

Some animals do not possess both rods and cones. The retinas of pigeons contain only cones. Thus, while pigeons have color vision, they see only in the bright light of day. The retinas of owls, on the other hand, have only rods; owls see very well in dim light, but are color blind.

> A recent discovery about the vision of cats is of interest here. Considerable confusion had existed as to whether or not cats are color-blind (as one might expect for a nocturnal predator). Experiments have shown, however, that cats have enough cones to distinguish colors easily if the object is large enough (about the size of a credit card). The report of these experiments concluded with the statement that, as far as cats are concerned, "apples are red, but cherries are gray."

The chemical changes that occur in rods are much better understood than those in cones. For this reason we shall concern ourselves here with rod vision alone.

When light strikes rod cells, it is absorbed by a compound called rhodopsin. This initiates a series of chemical events that ultimately results in the transmission of a nerve impulse to the brain.

Our understanding of the chemical nature of rhodopsin and the conformational changes that occur when rhodopsin absorbs light has resulted largely from the research of George Wald and co-workers at Harvard University. Wald's research began in 1933 when he was a graduate student in Berlin; work with rhodopsin, however, began much earlier in other laboratories.

Rhodopsin was discovered in 1877 by the German physiologist Franz Boll. Boll noticed that the initial red-purple color of a pigment in the retina of frogs was "bleached" by the action of light. The bleaching process led first to a yellow retina and then to a colorless one. A year later another German scientist, Willy Kuhne, isolated the red-purple pigment and named it, because of its color, *Sehpurpur* or "visual purple." The name visual purple is still commonly used for rhodopsin.

In 1952 Wald and one of his students, Ruth Hubbard, showed that the chromophore (light-absorbing group) of rhodopsin is the polyunsaturated aldehyde, 11-*cis*-retinal and a protein called opsin (Fig. N.1). The reaction is between the aldehyde group of 11-*cis*-retinal and an amino group on the chain of the protein and involves the loss of a molecule of water. Other secondary interactions involving —SH groups of the protein probably also hold the *cis*-retinal in place. The site on the chain of the protein is one on which *cis*-retinal fits precisely.

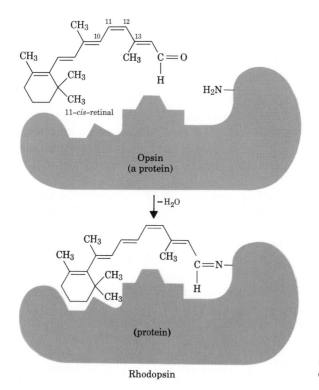

FIGURE N.1 The formation of rhodopsin from 11-*cis*-retinal and opsin.

The conjugated polyunsaturated chain of 11-*cis*-retinal gives rhodopsin the ability to absorb light over a broad region of the visible spectrum. Figure N.2 shows the absorption curve of rhodopsin in the visible region and compares it with the sensitivity curve for human rod vision. The fact that these two curves coincide provides strong evidence that rhodopsin is the light-sensitive material in rod vision.

When rhodopsin absorbs a photon of light, the visual process begins. Two very important phenomena accompany the absorption of the photon: A nerve impulse is generated, and 11-*cis*-retinal of rhodopsin is ultimately isomerized to the all-trans form of metarhodopsin II. The all-trans configuration of retinal does not fit the site on the chain of the protein and, because it does not, hydrolysis of the $-CH=N-$ group occurs. Hydrolysis produces opsin and all-*trans*-retinal. These steps are illustrated in Fig. N.3.

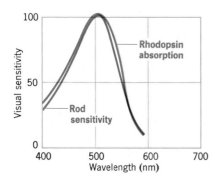

FIGURE N.2 A comparison of the visible absorption spectrum of rhodopsin and the sensitivity curve for rod vision. [Adapted from S. Hecht, S. Shlaer, and M. H. Pirenne, *J. Gen. Chem. Physiol.*, **25**, 819 (1942).]

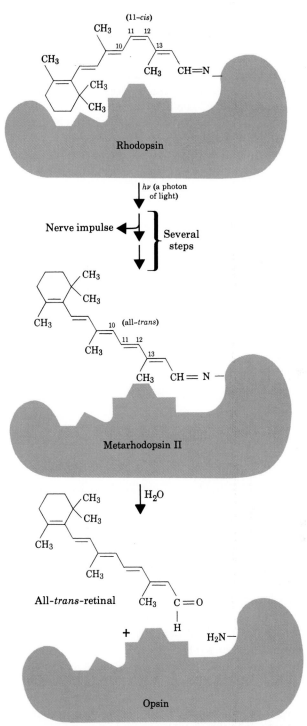

FIGURE N.3 The important chemical steps of the visual process. Absorption of a photon of light by the 11-*cis*-retinal portion of rhodopsin generates a nerve impulse and sets off an isomerization that leads, through a series of steps, to metarhodopsin II. Then hydrolysis of metarhodopsin II produces all-*trans*-retinal and opsin. This illustration greatly oversimplifies the shape of rhodopsin; the retinal portion is actually embedded in the center of a very complex protein structure. For a much more detailed representation of the structure of rhodopsin, and for a description of how a cascade of reactions results in a nerve signal, see L. Stryer, "The Molecules of Visual Excitation," *Scientific American*, **257**, 32 (1987).

Rhodopsin has an absorption maximum at 498 nm. This gives rhodopsin its red-purple color. Together, all-*trans*-retinal and opsin have an absorbance maximum at 387 nm and, thus, are yellow. The light-initiated transformation of rhodopsin to all-*trans*-retinal and opsin corresponds to the initial bleaching that Boll observed in the retinas of frogs. Further bleaching to a colorless form occurs when all-*trans*-retinal is reduced enzymatically to all-*trans*-vitamin A. This reduction converts the aldehyde group of retinal to the primary alcohol of vitamin A.

all-*trans*-Retinal

[H] | enzyme

all-*trans*-Vitamin A

N.1 Regeneration of Rhodopsin

If the retina of a live animal is subjected to constant irradiation, a steady-state concentration of rhodopsin in the retina develops: Rhodopsin is created at the same rate that it is destroyed. This is important because rhodopsin is essential to vision.

If an animal is placed in the dark for about 25 min, a process of "dark adaptation" takes place and retinal rhodopsin reaches its maximum value.

Rhodopsin regeneration occurs in two important ways: one (in light) occurs in the retina itself; the other (in the dark) involves enzymes of the liver. Let us begin by describing how rhodopsin regeneration occurs in light.

N.2 Photoregeneration of Rhodopsin

Several different chemical intermediates that occur between rhodopsin and meta-rhodopsin II have been identified on the basis of their absorption spectra. These are shown in Table N.1.

Isomerization of the 11-cis double bond of the rhodopsin occurs in the first step, rhodopsin ⟶ prelumirhodopsin. The changes from prelumirhodopsin to meta-rhodopsin II probably involve conformational changes of the protein, particularly at the site of its attachment to the all-*trans*-retinal. Prelumirhodopsin and lumirhodopsin have lifetimes so short that they cannot be detected in solution at room temperature.

Metarhodopsin I absorbs light in the same general region as rhodopsin itself, and there is evidence that when the retina is exposed to strong irradiation (i.e., when a large number of photons fall on the rod cells), metarhodopsin I is converted back to rhodopsin. This regeneration of rhodopsin from metarhodopsin I is a photochemical process, too. Metarhodopsin I absorbs a photon of light and its all-trans double bond

TABLE N.1

COMPOUND	λ_{max}
Rhodopsin	498 nm
Prelumirhodopsin	534 nm
Lumirhodopsin	500 nm
Metarhodopsin I	478 nm
Metarhodopsin II	380 nm

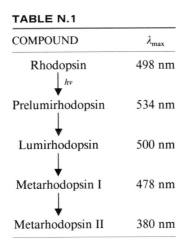

structure is reconverted to the 11-cis form of rhodopsin. We can summarize the reactions that occur in the retina in the following way (Figure N.4).

The bleaching sequence begins when rhodopsin absorbs a photon of light and is converted to metarhodopsin I. If, however, metarhodopsin I absorbs a second photon of equal energy, it can be converted back to rhodopsin before it is converted to metarhodopsin II. This is not likely to occur in dim light when the number of photons striking the retina is low. However, in bright light photoreversal becomes important because the number of photons striking the retina is large, and the probability that the same molecule will absorb a second photon before it is further transformed is also large. Under these conditions the amount of rhodopsin being bleached reaches a limiting value.

The all-*trans*-retinal that is formed when metarhodopsin II is hydrolyzed can be isomerized to 11-*cis*-retinal by an enzyme in the retina. This enzymatic reaction also requires light but it requires light of shorter wavelength than that for the photoregeneration of rhodopsin from metarhodopsin I. Rhodopsin is resynthesized in the retina when the 11-*cis*-retinal recombines with opsin.

N.3 Regeneration of Rhodopsin in the Dark

The all-*trans*-retinal that is not isomerized to 11-*cis*-retinal in the retina is reduced to all-*trans*-vitamin A by an enzyme, retinal reductase. The all-*trans*-vitamin A is then transported to the liver and the next step in the dark synthesis of rhodopsin occurs there.

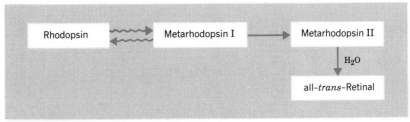

FIGURE N.4 A summary of the reactions occurring in the retina. Wavy arrows, ⤳ are used to represent reactions involving light; and straight arrows, → are used to represent reactions that occur in the dark.

Enzymes in the liver convert all-*trans*-vitamin A into 11-*cis*-vitamin A.

all-*trans*-Vitamin A

11-*cis*-Vitamin A

Then, 11-*cis*-vitamin A is returned to the eye where it is reoxidized to 11-*cis*-retinal and used to synthesize rhodopsin.

The entire visual cycle is summarized in Fig. N.5.

N.4 Photochemical Isomerization of Alkenes

The photochemical isomerizations of cis and trans alkenes that play such an important part in the visual process require extra comment. The barrier to rotation of groups joined by a carbon–carbon double bond is quite large. The energy of activation for the interconversion of 11-*cis*-retinal and all-*trans*-retinal, for example, has been estimated to be approximately 25 kcal/mole. Yet, when rhodopsin absorbs light of the proper wavelength, the reaction occurs very rapidly.

We can understand how light absorption brings about the rapid interconversion of cis–trans isomers if we examine the molecular orbitals given in Fig. N.6.

In the ground state of an alkene, both electrons are in the bonding molecular orbital, π. In this bonding orbital the electrons are located in a region of space above and below but generally between the two carbon atoms. In the excited state, however,

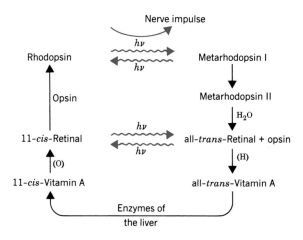

FIGURE N.5 The visual cycle. (All of the photoreactions require light of a wavelength that can be absorbed by reacting molecules.)

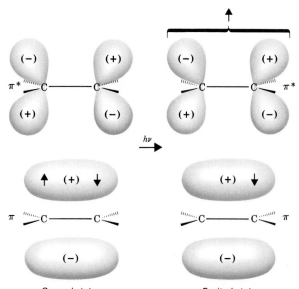

FIGURE N.6 The ground state and an excited state of an alkene.

one electron is promoted to the antibonding orbital, π^*. The antibonding orbital has four lobes that are directed generally away from the area between the two carbon atoms. Thus in the excited state the carbon–carbon bond is much more like that of a single bond and has a very low barrier to rotation, as a result.

   This does not mean, however, that in photoisomerization we have avoided the energy barrier. The energy for the isomerization is provided by the photon, and it is supplied just where it is needed: At the carbon–carbon double bond.

N.5 The Sensitivity of the Eye

The human eye is truly an amazing instrument. Although each rod contains at least 10 billion molecules of rhodopsin, it has been shown that the absorption of as few as *five* photons of light by five molecules of rhodopsin is detectable if the rhodopsin molecules are in different rods. Each rod apparently possesses an extraordinary mechanism for amplification of the nerve impulse generated by absorption of the photon.

CHAPTER TWENTY THREE

Computer-generated graphics showing the domains of a synthetic molecular cleft (see Section 23.10), prepared at the University of Pittsburgh in the laboratories of Professor Julius Rebek, Jr. Regions of high polarity (red and blue) in the lining of the cleft combined with hydrophobic methyl groups (yellow) and an aromatic acridine nucleus (green). These domains provide surfaces for binding smaller molecules of complementary size, shape, and functionality in studies of molecular recognition.

Amino Acids and Proteins

23.1 INTRODUCTION

The three groups of biological polymers are polysaccharides, proteins, and nucleic acids. We studied polysaccharides in Chapter 21 and saw that they function primarily as energy reserves and, in plants, as structural materials. When we study nucleic acids in Chapter 24 we shall find that they serve two major purposes: storage and transmission of information. Of the three groups of biopolymers, proteins have the most diverse functions. As enzymes and hormones, proteins catalyze and regulate the reactions that occur in the body; as muscles and tendons they provide the body with the means for movement; as skin and hair they give it an outer covering; as hemoglobins they transfer all-important oxygen to its most remote corners; as antibodies they provide it with a means of protection against disease; and in combination with other substances in bone they provide it with structural support.

Given such diversity of functions, we should not be surprised to find that proteins come in all sizes and shapes. By the standard of most of the molecules we have studied, even small proteins have very high molecular weights. Lysozyme, an enzyme, is a relatively small protein and yet its molecular weight is 14,600. The molecular weights of most proteins are much larger. Their shapes cover a range from the globular proteins such as lysozyme and hemoglobin to the helical coils of α-keratin (hair, nails, wool) and the pleated sheets of silk fibroin.

And yet, in spite of such diversity of size, shape, and function, all proteins have common features that allow us to deduce their structures and understand their properties. Later in this chapter we shall see how this is done.

Proteins are **polyamides** and their monomeric units are about 20 different α-amino acids:

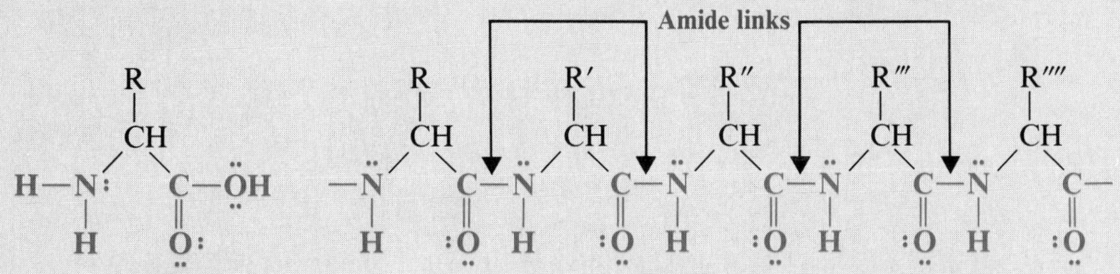

An α-amino acid   A portion of a protein molecule

Cells use the different α-amino acids to synthesize proteins. The exact sequence of the different α-amino acids along the protein chain is called the **primary structure** of the protein. This primary structure, as its name suggests, is of fundamental importance. For the protein to carry out its particular function, the primary structure must be correct. We shall see later that when the primary structure is correct, the polyamide chain folds in certain particular ways to give it the shape it needs for its particular task. This folding of the polyamide chain gives rise to higher levels of complexity called the **secondary** and **tertiary structure** of the protein.

Hydrolysis of proteins with acid or base yields a mixture of the different amino acids. Although hydrolysis of naturally occurring proteins may yield as many as 22 different amino acids, the amino acids have an important structural feature in common: With the exception of glycine (whose molecules are achiral), almost all naturally occurring amino acids have the L configuration at the α carbon.* That is, they have the same relative configuration as L-glyceraldehyde:

$$
\begin{array}{ccc}
\text{CO}_2\text{H} & \text{CHO} & \\
\text{H}_2\text{N}\!-\!\text{C}\!-\!\text{H} & \text{HO}\!-\!\text{C}\!-\!\text{H} & \text{CH}_2\text{CO}_2\text{H} \\
\text{R} & \text{CH}_2\text{OH} & \text{NH}_2 \\
\text{An L-}\alpha\text{-amino acid} & \text{L-Glyceraldehyde} & \text{Glycine} \\
\text{[usually an }(S)\text{-}\alpha\text{-amino acid]} & \text{[}(S)\text{-glyceraldehyde]} &
\end{array}
$$

23.2 AMINO ACIDS

23.2A Structures and Names

The 22 α-amino acids that can be obtained from proteins can be subdivided into five different groups on the basis of the structures of their side chains, R. These are given in Table 23.1.

Only 20 of the 22 α-amino acids in Table 23.1 are actually used by cells when they synthesize proteins. Two amino acids are synthesized after the polyamide chain is intact. Hydroxyproline (present mainly in collagen) is synthesized from proline, and cystine (present in most proteins) is synthesized from cysteine.

This conversion of cysteine to cystine requires additional comment. The —SH group of cysteine makes cysteine a *thiol* (Special Topic F). One property of thiols is that they can be converted to disulfides by mild oxidizing agents. This conversion, moreover, can be reversed by mild reducing agents.

$$
2\text{R}\!-\!\text{S}\!-\!\text{H} \underset{[\text{H}]}{\overset{[\text{O}]}{\rightleftarrows}} \text{R}\!-\!\text{S}\!-\!\text{S}\!-\!\text{R}
$$

$$
\text{Thiol} \qquad\qquad \text{Disulfide}
$$

Disulfide linkage

$$
2\text{HO}_2\text{CCHCH}_2\text{SH} \underset{[\text{H}]}{\overset{[\text{O}]}{\rightleftarrows}} \text{HO}_2\text{CCHCH}_2\text{S}\!-\!\text{SCH}_2\text{CHCO}_2\text{H}
$$

$$
\begin{array}{ccc}
\text{NH}_2 & \text{NH}_2 & \text{NH}_2 \\
\text{Cysteine} & & \text{Cystine}
\end{array}
$$

*Some D-amino acids have been obtained from the material comprising the cell walls of bacteria, and by hydrolysis of certain antibiotics.

We shall see later how the disulfide linkage between cysteine units in a protein chain contributes to the overall structure and shape of the protein.

23.2B Essential Amino Acids

Amino acids can be synthesized by all living organisms, plants and animals. Many higher animals, however, are deficient in their ability to synthesize all of the amino acids they need for their proteins. Thus these higher animals require certain amino acids as a part of their diet. For adult humans there are eight essential amino acids; these are designated with the superscript *e* in Table 23.1.

23.2C Amino Acids as Dipolar Ions

Since amino acids contain both a basic group ($-NH_2$) and an acidic group ($-CO_2H$), they are amphoteric. In the dry solid state, amino acids exist as **dipolar ions,** a form in which the carboxyl group is present as a carboxylate ion, $-CO_2^-$, and the amino group is present as an aminium group, $-NH_3^+$. (Dipolar ions are also called **zwitterions.**) In aqueous solution, an equilibrium exists between the dipolar ion and the anionic and cationic forms of an amino acid.

$$\overset{+}{\text{H}_3\text{NCHCO}_2\text{H}} \underset{+\text{H}^+}{\overset{-\text{H}^+}{\rightleftharpoons}} \overset{+}{\text{H}_3\text{NCHCO}_2^-} \underset{+\text{H}^+}{\overset{-\text{H}^+}{\rightleftharpoons}} \text{H}_2\text{NCHCO}_2^-$$
$$\quad\quad\text{R} \quad\quad\quad\quad\quad \text{R} \quad\quad\quad\quad\quad \text{R}$$

Cationic form	Dipolar ion	Anionic form
(predominant in strongly acidic solutions, e.g., at pH 0)		(predominant in strongly basic solutions, e.g., at pH 14)

The predominant form of the amino acid present in a solution depends on the pH of the solution and on the nature of the amino acid. In strongly acidic solutions all amino acids are present primarily as cations; in strongly basic solutions they are present as anions. At some intermediate pH, called the *isoelectric point, pI,* the concentration of the dipolar ion is at its maximum and the concentrations of the anions and cations are equal. Each amino acid has a particular isoelectric point. These are given in Table 23.1.

Let us consider first an amino acid with a side chain that contains neither acidic nor basic groups — an amino acid, for example, such as alanine.

If alanine is present in a strongly acidic solution (e.g., at pH 0), it is present mainly in the following cationic form. The acidity constant, K_a, for the carboxyl group of cationic form is 5×10^{-3}. This is considerably higher than the K_a of a corresponding carboxylic acid (e.g., propanoic acid) and indicates that the cationic form of alanine is the stronger acid. But we should expect it to be. After all it is a positively charged species and therefore should lose a proton more readily.

$$\text{CH}_3\text{CHCO}_2\text{H} \quad\quad\quad\quad \text{CH}_3\text{CH}_2\text{CO}_2\text{H}$$
$$\quad\; | $$
$$\quad \text{NH}_3^+$$

Cationic form of alanine
$K_{a_1} = 5 \times 10^{-3}$
$pK_{a_1} = 2.3$

Propanoic acid
$K_a = 1.3 \times 10^{-5}$
$pK_a = 4.89$

TABLE 23.1 L-Amino acids found in proteins

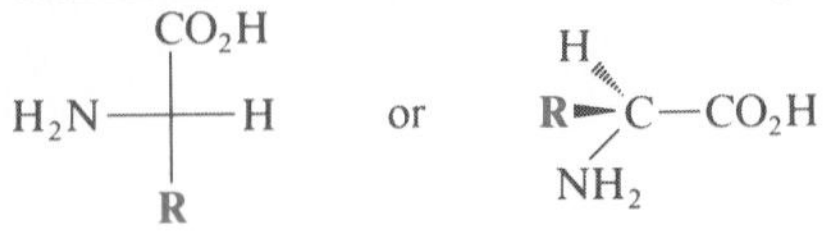

STRUCTURE OF R	NAME	ABBRE-VIATION	pK_{a_1} α-CO$_2$H	pK_{a_2} α-NH$_3^+$	pK_{a_3} R GROUP	pI
Neutral Amino Acids						
—H	Glycine	Gly	2.3	9.6		6.0
—CH$_3$	Alanine	Ala	2.3	9.7		6.0
—CH(CH$_3$)$_2$	Valine^e	Val	2.3	9.6		6.0
—CH$_2$CH(CH$_3$)$_2$	Leucine^e	Leu	2.4	9.6		6.0
—CHCH$_2$CH$_3$ CH$_3$	Isoleucine^e	Ile	2.4	9.7		6.1
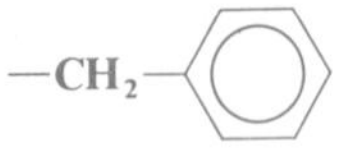 —CH$_2$—⬡	Phenylalanine^e	Phe	1.8	9.1		5.5
—CH$_2$CONH$_2$	Asparagine	Asn	2.0	8.8		5.4
—CH$_2$CH$_2$CONH$_2$	Glutamine	Gln	2.2	9.1		5.7
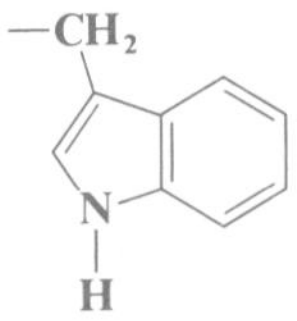 —CH$_2$ (indole)	Tryptophan^e	Trp	2.4	9.4		5.9
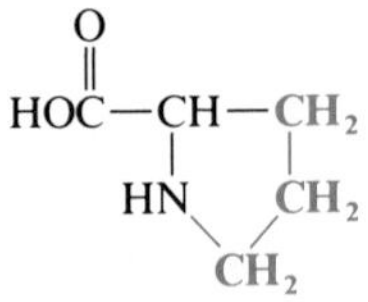 HOC—CH—CH$_2$ (Proline complete structure)	Proline	Pro	2.0	10.6		6.3
—CH$_2$OH	Serine	Ser	2.2	9.2		5.7
—CHOH CH$_3$	Threonine^e	Thr	2.6	10.4		6.5
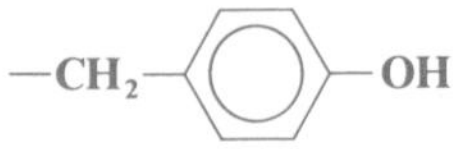 —CH$_2$—⬡—OH	Tyrosine	Tyr	2.2	9.1	10.1	5.7
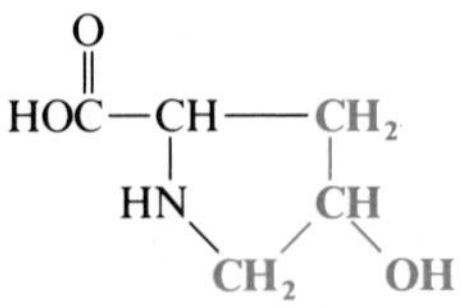 HOC—CH—CH$_2$ (Hydroxyproline complete structure)	Hydroxyproline	Hyp	1.9	9.7		6.3
—CH$_2$SH	Cysteine	Cys	1.7	10.8	8.3	5.0

TABLE 23.1 (continued)

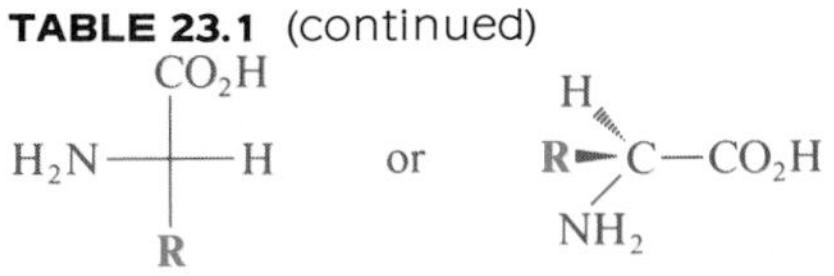

$$H_2N \overset{\displaystyle CO_2H}{\underset{\displaystyle R}{\rule[0.5em]{0pt}{1em}\vert\!\!\vert}} H \quad \text{or} \quad R \overset{H}{\underset{NH_2}{\triangleright}} C - CO_2H$$

STRUCTURE OF R	NAME	ABBRE-VIATION	pK_{a_1} α-CO_2H	pK_{a_2} α-NH_3^+	pK_{a_3} R GROUP	pI
$-CH_2-S$ $\quad\vert$ $-CH_2-S$	Cystine	Cys-Cys	$\begin{cases}1.6\\2.3\end{cases}$	$\begin{cases}7.9\\9.9\end{cases}$		5.1
$-CH_2CH_2SCH_3$	Methionine^e	Met	2.3	9.2		5.8
R Contains a Carboxyl Group						
$-CH_2CO_2H$	Aspartic acid	Asp	2.1	9.8	3.9	3.0
$-CH_2CH_2CO_2H$	Glutamic acid	Glu	2.2	9.7	4.3	3.2
R Contains a Basic Amino Group						
$-CH_2CH_2CH_2CH_2NH_2$	Lysine^e	Lys	2.2	9.0	10.5*	9.8
$-CH_2CH_2CH_2NH-\overset{\displaystyle NH}{\overset{\displaystyle \|}{C}}-NH_2$	Arginine	Arg	2.2	9.0	12.5*	10.8
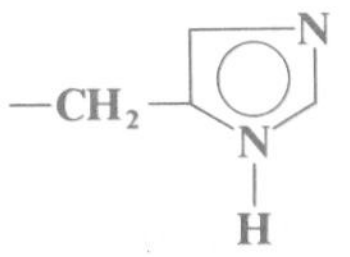 $-CH_2-\!\!\underset{\displaystyle H}{\overset{N}{\text{(imidazole)}}}$	Histidine	His	1.8	9.2	6.0*	7.6

e = essential amino acids

* pK_a is of protonated amine of R group.

The pK_a of an acid is the negative logarithm of the K_a:

$$pK_a = -\log K_a = \log \frac{1}{K_a}$$

For the cationic form of alanine, $pK_{a_1} = 2.3$; for propanoic acid, $pK_a = 4.8$.
The dipolar ion form of an amino acid is also a potential acid because the $-NH_3^+$ group can donate a proton. The pK_a of the dipolar ion form of alanine is 9.7 ($K_a = 2 \times 10^{-10}$).

$$CH_3\overset{\displaystyle CHCO_2^-}{\underset{\displaystyle NH_3^+}{\vert}}$$
$$K_{a_2} = 2 \times 10^{-10}$$
$$pK_{a_2} = 9.7$$

The isoelectric point, pI, of an amino acid such as alanine is the average of pK_{a_1} and pK_{a_2}.

$$pI = \frac{2.3 + 9.7}{2} = 6.0 \qquad \text{(isoelectric point of alanine)}$$

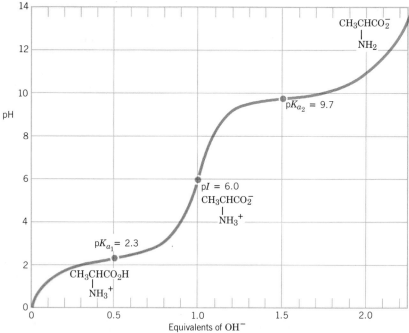

FIGURE 23.1 A titration curve for $\underset{\underset{NH_3^+}{|}}{CH_3CHCO_2H}$.

What does this mean about the behavior of alanine as the pH of a strongly acidic solution containing it is gradually raised by adding a base (i.e., OH^-)? At first (pH 0) (Fig. 23.1), the predominant form will be the cationic form. But then as the acidity reaches pH 2.3 (the pK_a of the cationic form, pK_{a_1}), half of the cationic form will be converted to the dipolar ion.* As the pH increases further—from pH 2.3 to pH 9.7—the predominant form will be the dipolar ion. At pH 6.0, the pH equals pI and the concentration of the dipolar ion is at its maximum.

$$CH_3\underset{\underset{NH_3^+}{|}}{CH}CO_2H \underset{H^+}{\overset{OH^-}{\rightleftarrows}} CH_3\underset{\underset{NH_3^+}{|}}{CH}CO_2^- \underset{H^+}{\overset{OH^-}{\rightleftarrows}} CH_3\underset{\underset{NH_2}{|}}{CH}CO_2^-$$

Cationic form   Dipolar ion   Anionic form
($pK_{a_1} = 2.3$)   ($pK_{a_2} = 9.7$)

When the pH rises to pH 9.7 (the pK_a of the dipolar ion), the dipolar ion will be half-converted to the anionic form. Then, as the pH approaches pH 14, the anionic form becomes the predominant form present in the solution.

*It is easy to show that for an acid:

$$pK_a = pH + \log \frac{[acid]}{[conjugate\ base]}$$

When the acid is half-neutralized, [acid] = [conjugate base] and $\log \dfrac{[acid]}{[conjugate\ base]} = 0$; thus $pH = pK_a$.

If the side chain of an amino acid contains an extra acidic or basic group, then the equilibria are more complex. Consider lysine, for example, an amino acid that has an extra $-NH_2$ group on its ε carbon. In strongly acidic solution, lysine will be present as a dication because both amino groups will be protonated. The first proton to be lost as the pH is raised is a proton of the carboxyl group ($pK_a = 2.2$), the next is from the α-aminium group ($pK_a = 9.0$), and the last is from the ε-aminium group.

$$\overset{+}{H_3N}(CH_2)_4CHCO_2H \underset{H^+}{\overset{OH^-}{\rightleftarrows}} \overset{+}{H_3N}(CH_2)_4CHCO_2^- \underset{H^+}{\overset{OH^-}{\rightleftarrows}}$$

$$\underset{NH_3^+}{|} \qquad\qquad \underset{NH_3^+}{|}$$

Dicationic form of   Monocationic
lysine   form
($pK_{a_1} = 2.2$)   ($pK_{a_2} = 9.0$)

$$\overset{+}{H_3N}(CH_2)_4CHCO_2^- \underset{H^+}{\overset{OH^-}{\rightleftarrows}} H_2N(CH_2)_4CHCO_2^-$$

$$\underset{NH_2}{|} \qquad\qquad\qquad \underset{NH_2}{|}$$

Dipolar ion   Anionic form
($pK_{a_3} = 10.5$)

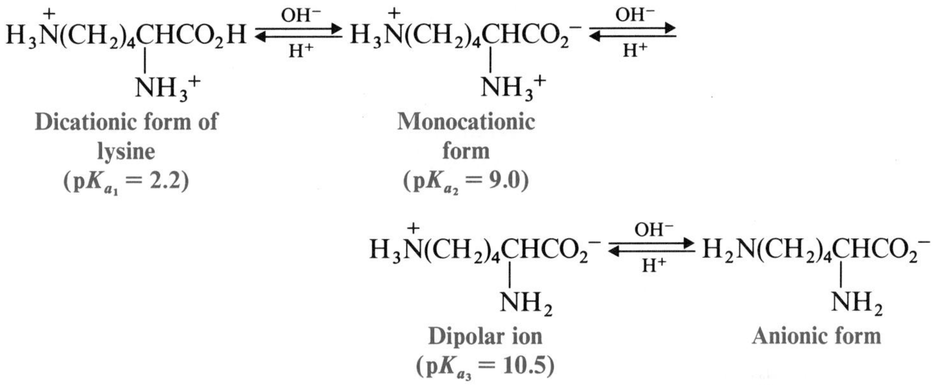

The isoelectric point of lysine is the average of pK_{a_2} (the monocation) and pK_{a_3} (the dipolar ion).

$$pI = \frac{9.0 + 10.5}{2} = 9.8 \qquad \text{(isoelectric point of lysine)}$$

PROBLEM 23.1

What form of glutamic acid would you expect to predominate in   (a) strongly acid solution?   (b) Strongly basic solution?   (c) At its isoelectric point (pI 3.2)? (d) The isoelectric point of glutamine (pI 5.7) is considerably higher than that of glutamic acid. Explain.

PROBLEM 23.2

$$\underset{\overset{\|}{NH}}{}$$

The guanidino group, $-NH-C-NH_2$, of arginine is one of the most strongly basic of all organic groups. Explain.

23.3 LABORATORY SYNTHESIS OF α-AMINO ACIDS

A variety of methods have been developed for the laboratory synthesis of α-amino acids. We shall describe here three general methods, all of which are based on reactions we have seen before.

23.3A Direct Ammonolysis of an α-Halo Acid

$$R\!-\!CH_2CO_2H \xrightarrow[\text{(2) }H_2O]{\text{(1) }X_2,\,P_4} \underset{\underset{X}{|}}{RCHCO_2H} \xrightarrow{NH_3\text{ (excess)}} \underset{\underset{NH_3{}^+}{|}}{R\!-\!CHCO_2{}^-}$$

This method is probably used least often because yields tend to be poor. We saw an example of this method in Section 18.9.

23.3B From Potassium Phthalimide

This method is a modification of the Gabriel synthesis of amines (Section 19.5A). The yields are usually high and the products are easily purified.

Potassium   Ethyl chloroacetate
phthalimide

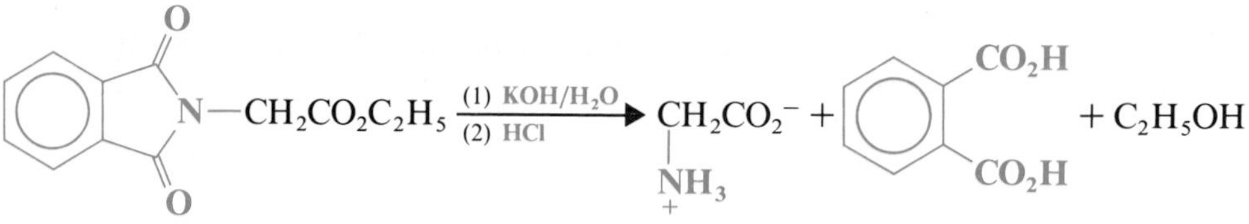

(97%)   (85%)   Phthalic
                                         Glycine   acid

A variation of this procedure uses potassium phthalimide and diethyl α-bromomalonate to prepare an *imido* malonic ester. This method is illustrated with a synthesis of methionine.

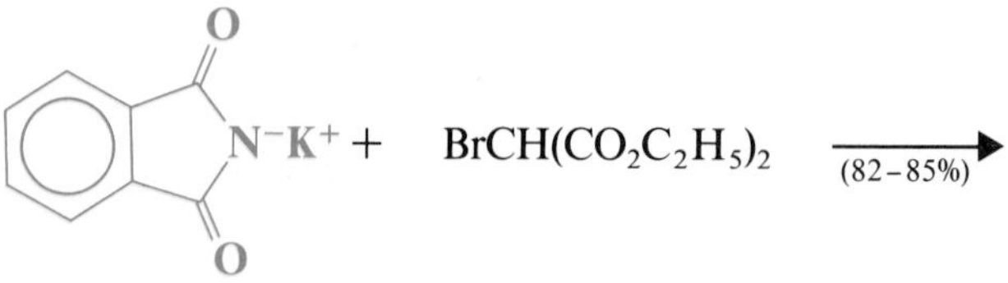

Diethyl α-bromomalonate

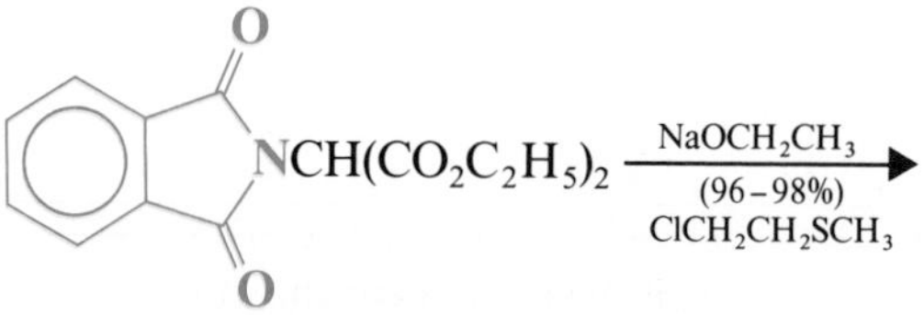

Phthalimidomalonic
ester

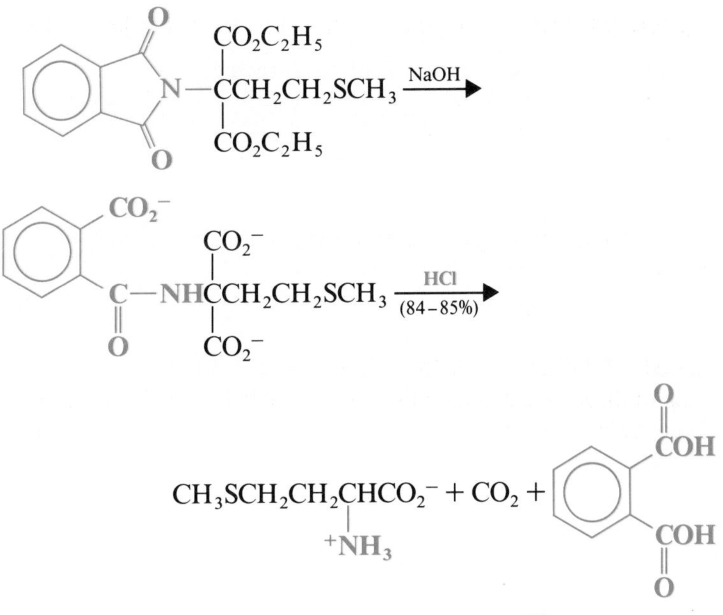

$$CH_3SCH_2CH_2CHCO_2^- + CO_2 + [phthalic\ acid]$$

with $^+NH_3$ below the $CHCO_2^-$

DL-Methionine

PROBLEM 23.3

Starting with diethyl α-bromomalonate and potassium phthalimide and using any other necessary reagents show how you might synthesize: (a) DL-leucine, (b) DL-alanine, and (c) DL-phenylalanine.

23.3C The Strecker Synthesis

Treating an aldehyde with ammonia and hydrogen cyanide produces an α-amino nitrile. Hydrolysis of the nitrile group (Section 18.3) of the α-amino nitrile converts the latter to an α-amino acid. This synthesis is called the Strecker synthesis.

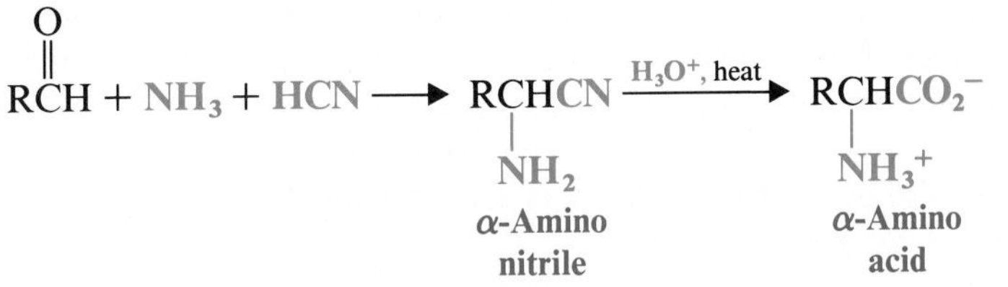

The first step of this synthesis probably involves the initial formation of an imine from the aldehyde and ammonia, followed by the addition of hydrogen cyanide.

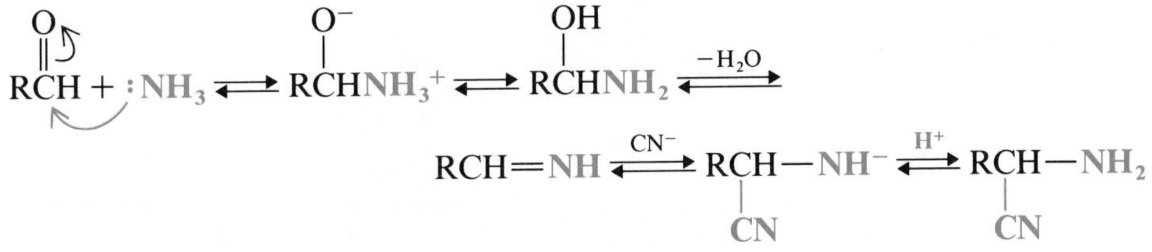

SAMPLE PROBLEM

Outline a Strecker synthesis of DL-tyrosine.

Answer:

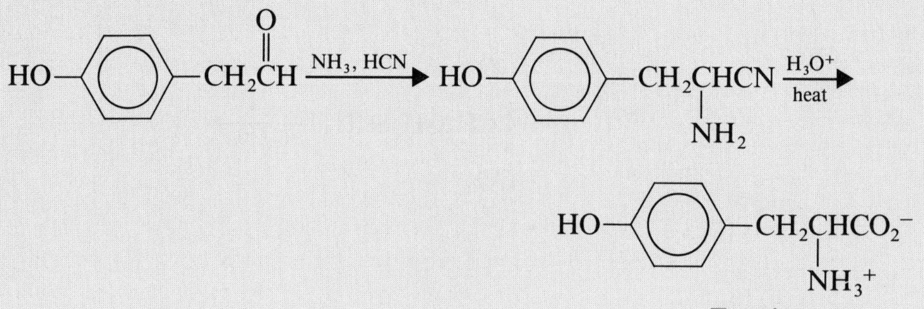

DL-Tyrosine

PROBLEM 23.4

(a) Outline a Strecker synthesis of DL-phenylalanine. (b) DL-Methionine can also be synthesized by a Strecker synthesis. The required starting aldehyde can be prepared from acrolein ($CH_2{=}CHCHO$) and methanethiol (CH_3SH). Outline all steps in this synthesis of DL-methionine.

23.3D Resolution of DL-Amino Acids

With the exception of glycine, which has no stereocenter, the amino acids that are produced by the methods we have outlined are all produced as racemic forms. In order to obtain the naturally occurring L-amino acid we must, of course, resolve the racemic form. This can be done in a variety of ways including the methods outlined in Section 4.14.

One especially interesting method for resolving amino acids is based on the use of enzymes called *deacylases.* These enzymes catalyze the hydrolysis of *N-acylamino acids* in living organisms. Since the active site of the enzyme is chiral, it hydrolyzes only *N*-acylamino acids of the L configuration. When it is exposed to a racemic modification of *N*-acylamino acids, only the L-amino acid is affected and the products, as a result, are separated easily.

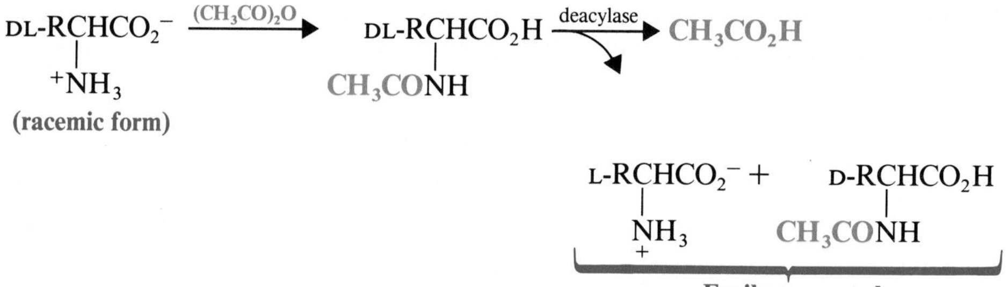

(racemic form)

Easily separated

23.3E Stereoselective Syntheses of Amino Acids

The ideal synthesis of an amino acid, of course, would be one that would produce only the naturally occurring L-amino acid. This ideal has now been realized through the use of chiral hydrogenation catalysts derived from transition metals (Special Topic G). A variety of catalysts has been used. One developed by B. Bosnich (of the University of Toronto) is based on a rhodium complex with (R)-1,2-bis(diphenyl-phosphino)propane, a compound that is called "(R)-prophos." When a rhodium

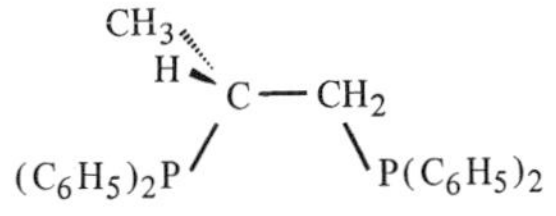

(R)—Prophos

complex of norbornadiene (NBD) is treated with (R)-prophos, the (R)-prophos replaces one of the molecules of norbornadiene surrounding the rhodium atom to produce a *chiral* rhodium complex.

$$[Rh(NBD)_2]ClO_4 + (R)\text{-prophos} \longrightarrow [Rh((R)\text{-prophos})(NBD)]ClO_4 + NBD$$
Chiral rhodium complex

Treating this rhodium complex with hydrogen in a solvent such as ethanol yields a solution containing an active *chiral* hydrogenation catalyst, which probably has the composition $[Rh((R)\text{-prophos})(H)_2(EtOH)_2]^+$.

When 2-acetylaminopropenoic acid is added to this solution and hydrogenation is carried out, the product of the reaction is the N-acetyl derivative of L-alanine in 90% enantiomeric excess. Hydrolysis of the N-acetyl group yields L-alanine. Because the hydrogenation catalyst is chiral, it transfers its hydrogen atoms in a stereoselective way. This type of reaction is often called an **asymmetric synthesis** or **enantioselective synthesis.**

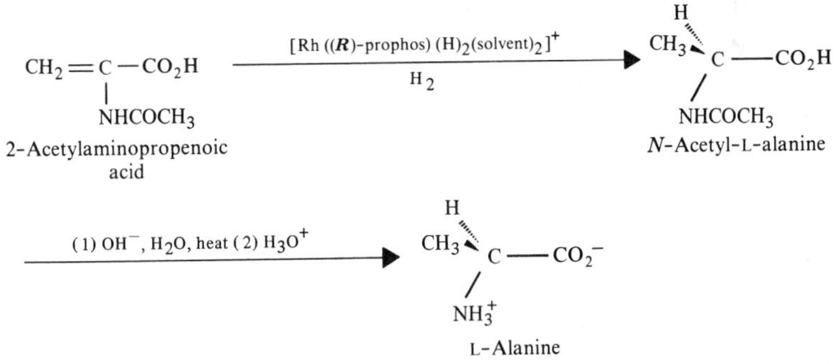

This same procedure has been used to synthesize several other L-amino acids from 2-acetylaminopropenoic acids that have substituents at the 3 position. Use of the (R)-prophos catalyst in hydrogenation of the (Z)-isomer yields the L-amino acid with an enantiomeric excess of 87–93%.

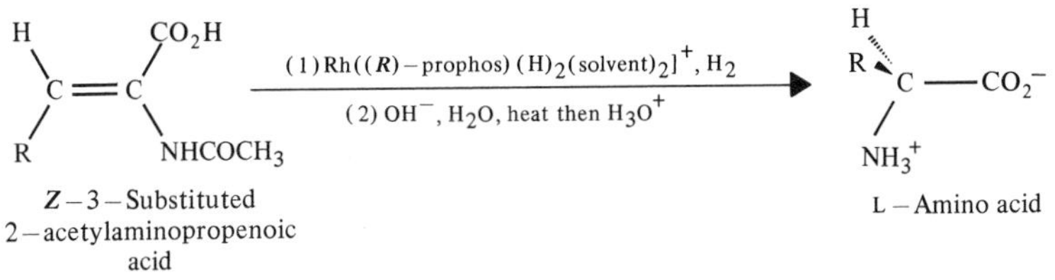

23.4 BIOSYNTHESIS OF AMINO ACIDS

Two of the most important methods used by living cells for the synthesis of amino acids are described in the following sections.

23.4A Reductive Amination

This biosynthesis bears a remarkable resemblance to one laboratory method that we have seen for the synthesis of amines (Section 19.5C). In enzymatic reductive amination, α-ketoglutaric acid combines with ammonia in the presence of the reducing agent NADH. The product is L-glutamic acid.

$$HO_2CCH_2CH_2\overset{\overset{\displaystyle O}{\|}}{C}CO_2H + H^+ + NH_3 \underset{NAD^+}{\overset{NADH}{\rightleftharpoons}} HO_2CCH_2CH_2\overset{\underset{\displaystyle +NH_3}{|}}{C}HCO_2^- + H_2O$$

α-Ketoglutaric acid   L-Glutamic acid

   The cell has α-ketoglutaric acid readily available because it is an intermediate in the metabolism of carbohydrates.

23.4B Transamination

The α-amino group of L-glutamic acid can be transferred to other α-keto acids whose carbon chains correspond in structure to those of other naturally occurring amino acids. These reactions are catalyzed by enzymes called *transaminases.* An example is the biosynthesis of L-aspartic acid from L-glutamic acid and oxaloacetic acid.

$$HO_2CCH_2CH_2\overset{\underset{\displaystyle +NH_3}{|}}{C}HCO_2^- + HO_2CCH_2\overset{\overset{\displaystyle O}{\|}}{C}CO_2H \overset{transaminase}{\rightleftharpoons}$$

L-Glutamic acid   Oxaloacetic acid

$$HO_2CCH_2CH_2\overset{\overset{\displaystyle O}{\|}}{C}CO_2H + HO_2CCH_2\overset{\underset{\displaystyle +NH_3}{|}}{C}HCO_2^-$$

α-Ketoglutaric acid   L-Aspartic acid

Oxaloacetic acid is also an intermediate in the metabolism of carbohydrates.

23.5 ANALYSIS OF AMINO ACID MIXTURES

The amide linkages that join α-amino acids in proteins are commonly called **peptide linkages,** and α-amino acid polymers with molecular weights less than 10,000 are usually called **polypeptides.** Those with molecular weights greater than 10,000 are called **proteins.** This division is quite arbitrary and the two terms are often interchanged. In actuality, both proteins and polypeptides are **polyamides.**

   We can represent the structures of polypeptides using the symbols for the amino acids. The dipeptide glycylvaline, for example, is represented by Gly · Val, and the dipeptide valylglycine as Val · Gly. In each case the amino acid whose carboxyl group is involved in an amide linkage is placed first.

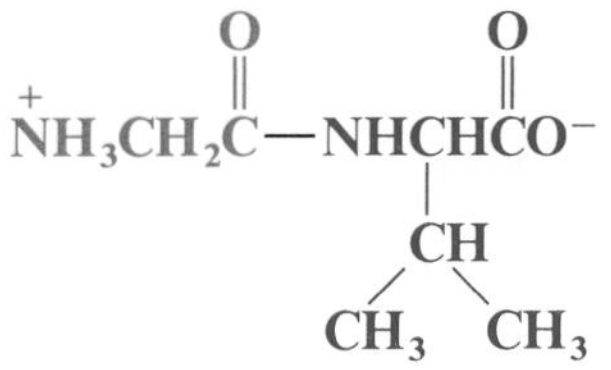

$$\overset{+}{N}H_3CH_2\overset{\overset{O}{\|}}{C}-NHCHC\overset{O^-}{\overset{\|}{}}$$

$$\overset{CH}{\overset{|}{}}$$

$$CH_3\quad CH_3$$

Glycylvaline
(Gly · Val)

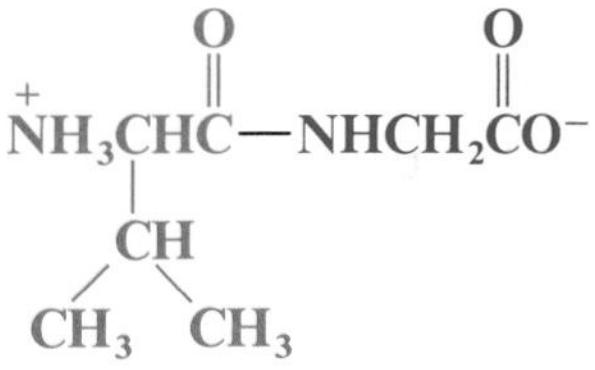

$$\overset{+}{N}H_3CH\overset{\overset{O}{\|}}{C}-NHCH_2\overset{O}{\overset{\|}{C}}O^-$$

$$\overset{CH}{\overset{|}{}}$$

$$CH_3\quad CH_3$$

Valylglycine
(Val · Gly)

The tripeptide glycylvalylphenylalanine can be represented in the following way:

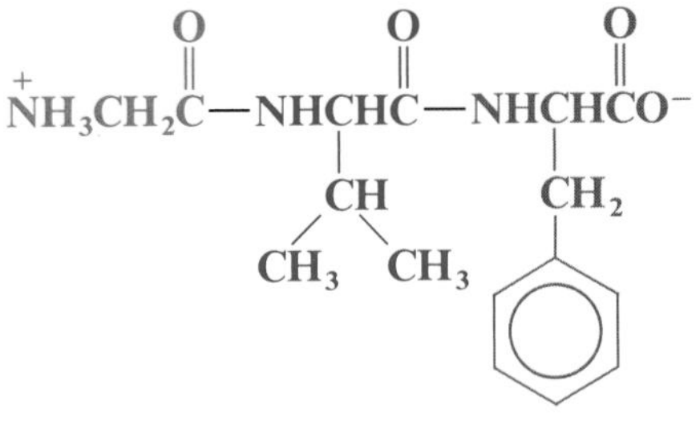

$$\overset{+}{N}H_3CH_2\overset{\overset{O}{\|}}{C}-NHCH\overset{\overset{O}{\|}}{C}-NHCH\overset{O}{\overset{\|}{C}}O^-$$

$$\overset{CH}{\overset{|}{}}\qquad\qquad\overset{CH_2}{\overset{|}{}}$$

$$CH_3\quad CH_3$$

Glycylvalylphenylalanine
(Gly · Val · Phe)

When a protein or polypeptide is refluxed with 6 M hydrochloric acid for 24 h, hydrolysis of all of the amide linkages usually takes place, and this produces a mixture of amino acids. One of the first tasks that we face when we attempt to determine the structure of a polypeptide or protein is the separation and identification of the individual amino acids in such a mixture. Since as many as 22 different amino acids may be present, this could be a formidable task if we are restricted to conventional methods.

Fortunately, techniques have been developed, based on the principle of elution chromatography, that simplify this problem immensely and even allow its solution to be automated. Automatic amino acid analyzers were developed at the Rockefeller Institute in 1950 and have since become commercially available. They are based on the use of insoluble polymers containing sulfonate groups, called *cation-exchange resins* (Fig. 23.2).

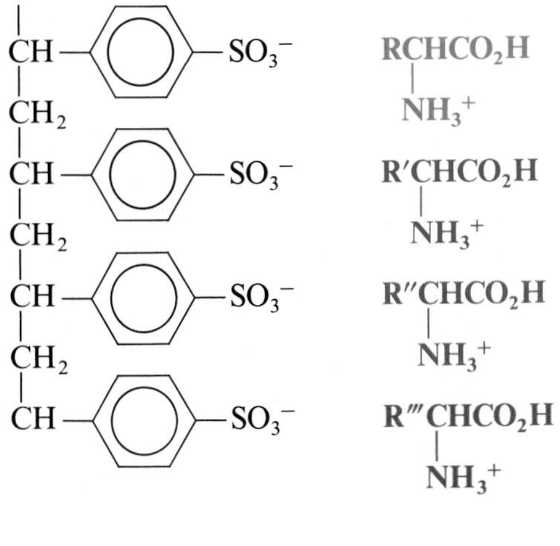

FIGURE 23.2 A section of a cation-exchange resin with adsorbed amino acids.

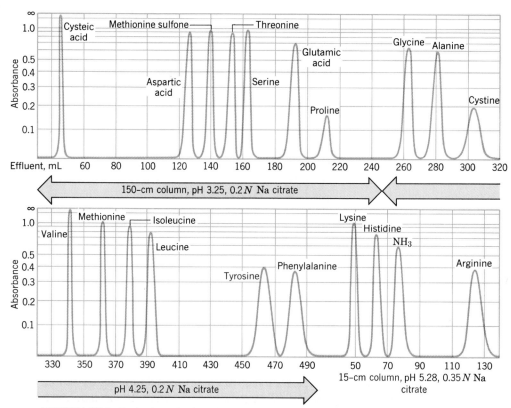

FIGURE 23.3 Typical result given by an automatic amino acid analyzer. [Adapted with permission from D. H. Spackman, W. H. Stein, and S. Moore, *Anal. Chem.*, **30,** 1190 (1958). Copyright by the American Chemical Society.]

If an acidic solution containing a mixture of amino acids is passed through a column packed with a cation-exchange resin, the amino acids will be adsorbed by the resin because of attractive forces between the negatively charged sulfonate groups and the positively charged amino acids. The strength of the adsorption will vary with the basicity of the individual amino acids; those that are most basic will be held most strongly. If the column is then washed with a buffered solution at a given pH, the individual amino acids will move down the column at different rates and ultimately become separated. At the end of the column the eluate is allowed to mix with **ninhydrin,** a reagent that reacts with most amino acids to give a derivative with an intense purple color (λ_{max} 570 nm). The amino acid analyzer is designed so that it can measure the absorbance of the eluate (at 570 nm) continuously and record this absorbance as a function of the volume of the effluent.

A typical graph obtained from an automatic amino acid analyzer is shown in Fig. 23.3. When the procedure is standardized, the positions of the peaks are characteristic of the individual amino acids and the areas under the peaks correspond to their relative amounts.

Ninhydrin is the hydrate of indane-1,2,3-trione. With the exception of proline and hydroxyproline, all of the α-amino acids found in proteins react with ninhydrin to give the same intensely colored purple anion (λ_{max} 570 nm). We shall not go into the mechanism here, but notice that the only portion of the anion that is derived from the α-amino acid is the nitrogen.

Indane-1,2,3-trione   Ninhydrin

Purple anion

Proline and hydroxyproline do not react with ninhydrin in the same way because their α-amino groups are part of a five-membered ring.

23.6 AMINO ACID SEQUENCE OF PROTEINS AND POLYPEPTIDES

Once we have determined the amino acid composition of a protein or a polypeptide, we should then determine its molecular weight. Various methods are available for doing this, including chemical methods, ultracentrifugation, light scattering, osmotic pressure, and X-ray diffraction. With the molecular weight and amino acid composition we shall now be able to calculate the *molecular formula* of the protein; that is, we shall know how many of each type of amino acid are present as **amino acid residues** $\left(\text{i.e., } \underset{-\text{NH}}{\text{RCHCO}}\right)$ in each protein molecule. Unfortunately, however, we have only begun our task of determining its structure. The next step is a formidable one, indeed. We must determine the order in which the amino acids are connected; that is, we must determine the ***covalent structure of the polyamide.***

A simple tripeptide composed of 3 different amino acids can have 6 different amino acid sequences; a tetrapeptide composed of 4 different amino acids can have as many as 24. For a protein with a molecular weight of 10,000 or more, composed of 20 different amino acids, the number of possibilities is extremely large.

In spite of this, a number of methods have been developed that allow the amino acid sequences to be determined and these, as we shall see, have been applied with amazing success. In our discussion here we shall limit our attention to two methods that illustrate how sequence determinations can be done: **terminal residue analysis** and **partial hydrolysis.**

23.6A Terminal Residue Analysis

One of the ends of a polypeptide chain terminates in an amino acid residue that has a free $-\text{NH}_2$ group; the other terminates in an amino acid residue with a free $-\text{CO}_2\text{H}$ group. These two amino acids are called the ***N*-terminal** residue and the ***C*-terminal** residue, respectively.

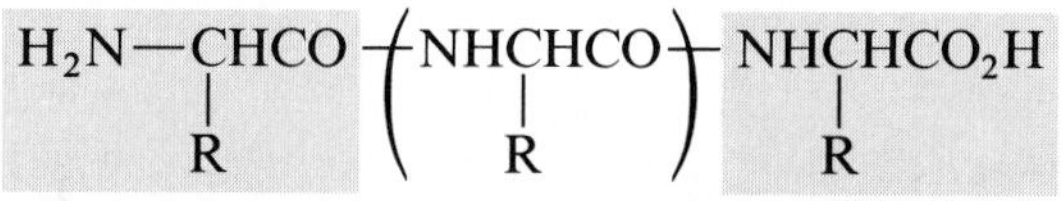

$$H_2N-\underset{\underset{R}{|}}{CH}CO \left(\underset{\underset{R}{|}}{NHCH}CO\right) \underset{\underset{R}{|}}{NHCH}CO_2H$$

N-Terminal   C-Terminal
residue   residue

One very useful method for determining the N-terminal amino acid residue, called the **Sanger method,** is based on the use of 2,4-dinitrofluorobenzene (DNFB).* When a polypeptide is treated with DNFB in mildly basic solution, a nucleophilic aromatic substitution reaction takes place involving the free amino group of the N-terminal residue. Subsequent hydrolysis of the polypeptide gives a mixture of amino acids in which the N-terminal amino acid bears a label, *the 2,4-dinitrophenyl group.* As a result, after separating this amino acid from the mixture, we can identify it.

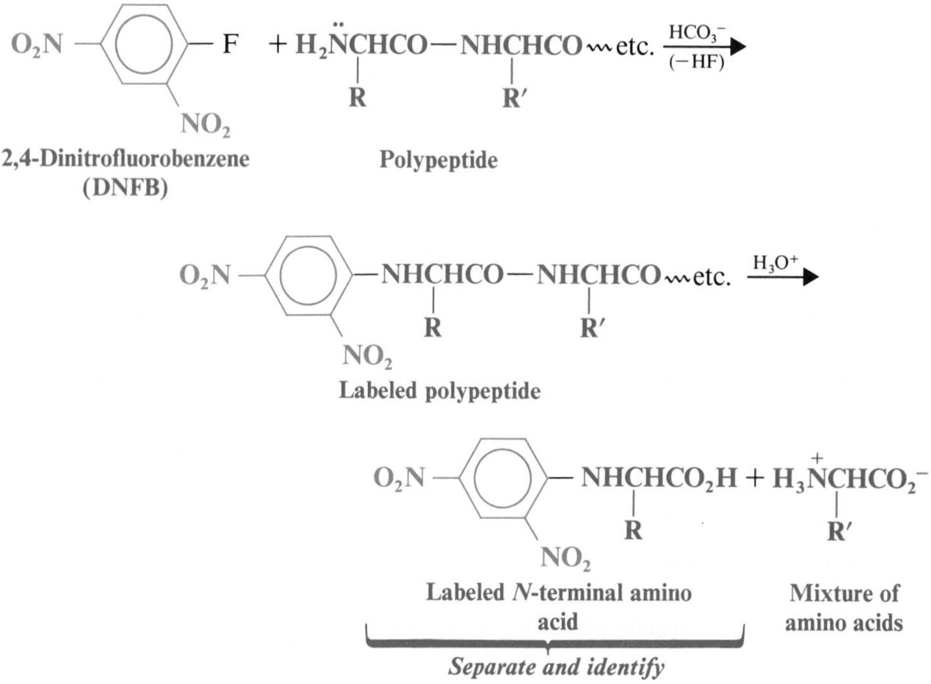

$$O_2N-\text{⟨⟩}-F \quad + H_2\overset{..}{N}\underset{\underset{R}{|}}{CH}CO-\underset{\underset{R'}{|}}{NHCH}CO\rightsquigarrow etc. \xrightarrow[(-HF)]{HCO_3^-}$$

NO₂

2,4-Dinitrofluorobenzene   **Polypeptide**
(DNFB)

$$O_2N-\text{⟨⟩}-\underset{\underset{R}{|}}{NHCH}CO-\underset{\underset{R'}{|}}{NHCH}CO\rightsquigarrow etc. \xrightarrow{H_3O^+}$$

NO₂
Labeled polypeptide

$$O_2N-\text{⟨⟩}-\underset{\underset{R}{|}}{NHCH}CO_2H + H_3\overset{+}{N}\underset{\underset{R'}{|}}{CH}CO_2^-$$

NO₂

Labeled N-terminal amino   **Mixture of**
acid   **amino acids**

Separate and identify

PROBLEM 23.5

The electron-withdrawing property of the 2,4-dinitrophenyl group makes separation of the labeled amino acid very easy. Suggest how this is done.

Of course, 2,4-dinitrofluorobenzene will react with any free amino group present in a polypeptide including the ε-amino group of lysine. But only the N-terminal amino acid residue will bear the label at the α-amino group.

*This method was introduced by Frederick Sanger of Cambridge University in 1945. Sanger made extensive use of this procedure in his determination of the amino acid sequence of insulin and won a Nobel Prize for the work in 1958.

A second method of *N*-terminal analysis is the *Edman degradation* (developed by Pehr Edman of the University of Lund, Sweden). This method offers an advantage over the Sanger method in that it removes the *N*-terminal residue and leaves the remainder of the peptide chain intact. The Edman degradation is based on a labeling reaction between the *N*-terminal amino group and phenyl isothiocyanate, $C_6H_5N{=}C{=}S$ (see the following reactions). When the labeled polypeptide is treated with acid, the *N*-terminal amino acid residue splits off as an unstable intermediate that undergoes rearrangement to a phenylthiohydantoin. This last product can be identified by comparison with phenylthiohydantoins prepared from standard amino acids.

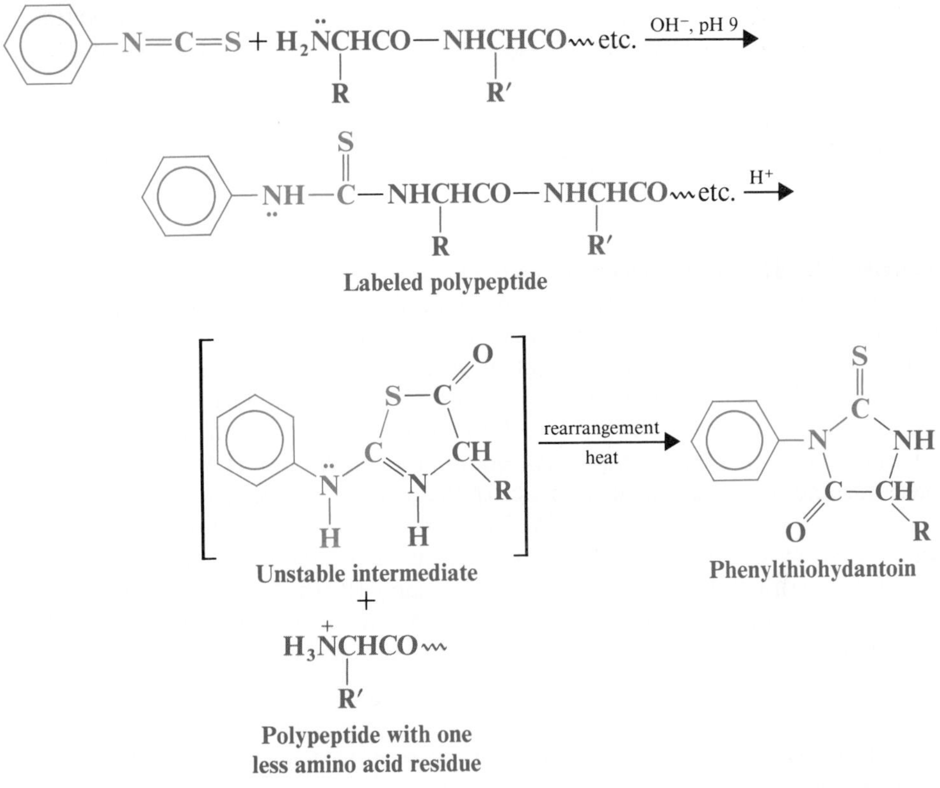

$$\text{C}_6\text{H}_5\text{—N}{=}\text{C}{=}\text{S} + \text{H}_2\ddot{\text{N}}\text{CHCO—NHCHCO}\text{\sim\sim etc.} \xrightarrow{\text{OH}^-,\ \text{pH 9}}$$

with R below the first CHCO and R′ below the second CHCO

$$\text{C}_6\text{H}_5\text{—NH—}\overset{\displaystyle S}{\overset{\|}{\text{C}}}\text{—NHCHCO—NHCHCO}\text{\sim\sim etc.} \xrightarrow{\text{H}^+}$$

with R below the first CHCO and R′ below the second CHCO

Labeled polypeptide

$$\left[\ \text{Unstable intermediate structure: phenyl ring attached to N(H)—C, five-membered ring containing S—C(=O), CH—R, N(H)}\ \right] \xrightarrow[\text{heat}]{\text{rearrangement}} \text{Phenylthiohydantoin}$$

Unstable intermediate

+

$$\overset{+}{\text{H}_3\text{N}}\text{CHCO}\text{\sim\sim}$$

with R′ below

Polypeptide with one less amino acid residue

Phenylthiohydantoin

The polypeptide that remains after the first Edman degradation can be submitted to another degradation to identify the next amino acid in the sequence, and this process has even been automated. Unfortunately, Edman degradations cannot be repeated indefinitely. As residues are successively removed, amino acids formed by hydrolysis during the acid treatment accumulate in the reaction mixture and interfere with the procedure. The Edman degradation, however, has been automated into what is called a **sequenator.** Each amino acid is automatically detected as it is removed. This technique has been successfully applied to polypeptides with as many as 60 amino acid residues.

C-Terminal residues can be identified through the use of digestive enzymes called *carboxypeptidases.* These enzymes specifically catalyze the hydrolysis of the amide bond of the amino acid residue containing a free $—CO_2H$ group, liberating it as a free amino acid. A carboxypeptidase, however, will continue to attack the polypeptide chain that remains, successively lopping off *C*-terminal residues. As a consequence, it is necessary to follow the amino acids released as a function of time.

The procedure can be applied to only a limited amino acid sequence, for at best, after a time the situation becomes too confused to sort out.

PROBLEM 23.6

(a) Write a reaction showing how 2,4-dinitrofluorobenzene could be used to identify the N-terminal amino acid of Val·Ala·Gly. (b) What products would you expect (after hydrolysis) when Val·Lys·Gly is treated with 2,4-dinitrofluorobenzene?

PROBLEM 23.7

Write the reactions involved in a sequential Edman degradation of Met·Ile·Arg.

23.6B Partial Hydrolysis

Sequential analysis using the Edman degradation or carboxypeptidase becomes impractical with proteins or polypeptides of appreciable size. Fortunately, however, we can resort to another technique, that of **partial hydrolysis.** Using dilute acids or enzymes, we attempt to break the polypeptide chain into small fragments, ones that we can identify using DNFB or the Edman degradation. Then we examine the structures of these smaller fragments looking for points of overlap and attempt to piece together the amino acid sequence of the original polypeptide.

Consider a simple example: We are given a pentapeptide known to contain valine (two residues), leucine (one residue), histidine (one residue), and phenylalanine (one residue). With this information we can write the "molecular formula" of the protein in the following way, using commas to indicate that the sequence is unknown.

$$\text{Val}_2, \text{Leu}, \text{His}, \text{Phe}$$

Then let us assume that by using 2,4-dinitrofluorobenzene and carboxypeptidase, we discover that valine and leucine are the N-terminal and C-terminal residues, respectively. So far we know the following:

$$\text{Val (Val, His, Phe) Leu}$$

But the sequence of the three nonterminal amino acids is still unknown.

We then subject the pentapeptide to partial acid hydrolysis and obtain the following dipeptides. (We also get individual amino acids and larger pieces, i.e., tripeptides and tetrapeptides.)

$$\text{Val·His + His·Val + Val·Phe + Phe·Leu}$$

The points of overlap of the dipeptides (i.e., His, Val, and Phe) tell us that the original pentapeptide must have been the following:

$$\text{Val·His·Val·Phe·Leu}$$

Two enzymes are also frequently used to cleave certain bonds in a large protein. *Trypsin* preferentially catalyzes the hydrolysis of peptide bonds in which the carboxyl group is a part of a lysine or arginine residue. *Chymotrypsin* preferentially catalyzes the hydrolysis of peptide bonds at the carboxyl groups of phenylalanine, tyrosine, and tryptophan. It will also attack the peptide bonds at the carboxyl groups of leucine, methionine, asparagine, and glutamine. Treating a large protein with trypsin or chymotrypsin will break it into smaller pieces. Then each smaller piece can be subjected to an Edman degradation or to labeling followed by partial hydrolysis.

PROBLEM 23.8

Glutathione is a tripeptide found in most living cells. Partial acid-catalyzed hydrolysis of glutathione yields two dipeptides, Cys·Gly and one composed of Glu and Cys. When this second dipeptide was treated with DNFB, acid hydrolysis gave *N*-labeled Glu. (a) Based on this information alone, what structures are possible for glutathione? (b) Synthetic experiments have shown that the second dipeptide has the following structure:

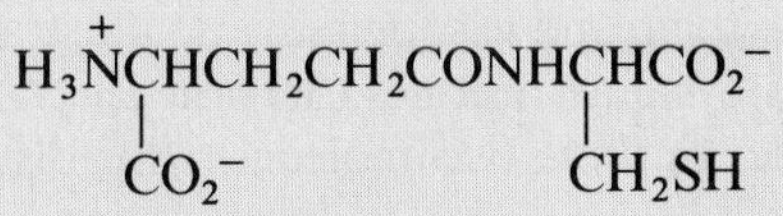

$$\overset{+}{\text{H}_3}\text{NCHCH}_2\text{CH}_2\text{CONHCHCO}_2^-$$
$$\underset{\text{CO}_2^-}{|} \qquad\qquad \underset{\text{CH}_2\text{SH}}{|}$$

What is the structure of glutathione?

PROBLEM 23.9

Give the amino acid sequence of the following polypeptides using only the data given by partial acidic hydrolysis.

(a) Ser, Hyp, Pro, Thr $\xrightarrow[\text{H}_2\text{O}]{\text{H}^+}$ Ser·Thr + Thr·Hyp + Pro·Ser

(b) Ala, Arg, Cys, Val, Leu $\xrightarrow[\text{H}_2\text{O}]{\text{H}^+}$ Ala·Cys· + Cys·Arg + Arg·Val + Leu·Ala

---

23.7 PRIMARY STRUCTURES OF POLYPEPTIDES AND PROTEINS

The covalent structure of a protein or polypeptide is called its *primary structure* (Fig. 23.4). By using the techniques we described in the previous sections, chemists have had remarkable success in determining the primary structures of polypeptides and proteins. The compounds described in the following pages are important examples.

23.7A Oxytocin and Vasopressin

Oxytocin and vasopressin (Fig. 23.5) are two rather small polypeptides with strikingly similar structures (where oxytocin has leucine, vasopressin has arginine and where oxytocin has isoleucine, vasopressin has phenylalanine). In spite of the similarity of their amino acid sequences, these two polypeptides have quite different physio-

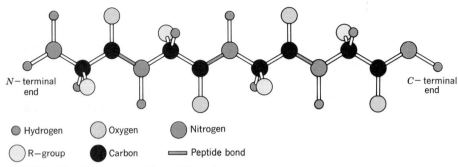

N—terminal end

C—terminal end

● Hydrogen   ○ Oxygen   ● Nitrogen

○ R—group   ● Carbon   ▱▱▱ Peptide bond

FIGURE 23.4 A representation of the primary structure of a tetrapeptide.

logical effects. Oxytocin occurs only in the female of a species and stimulates uterine contractions during childbirth. Vasopressin occurs in males and females; it causes contraction of peripheral blood vessels and an increase in blood pressure. Its major function, however, is as an *antidiuretic;* physiologists often refer to vasopressin as an *antidiuretic hormone.*

The structures of oxytocin and vasopressin also illustrate the importance of the disulfide linkage between cysteine residues (Section 23.2A) in the overall primary structure of a polypeptide. In these two molecules this disulfide linkage leads to a cyclic structure.*

PROBLEM 23.10

Treating oxytocin with certain reducing agents (e.g., sodium in liquid ammonia) brings about a single chemical change that can be reversed by air oxidation. What chemical changes are involved?

23.7B Insulin

Insulin, a hormone secreted by the pancreas, regulates glucose metabolism. Insulin deficiency in humans is the major problem in diabetes mellitus.

The amino acid sequence of bovine insulin (Fig. 23.6) was determined by Sanger in 1953. Bovine insulin has a total of 51 amino acid residues in two polypeptide chains, called the A and B chains. These chains are joined by two disulfide linkages. The A chain contains an additional disulfide linkage between cysteine residues at positions 6 and 11.

Human insulin differs from bovine insulin at only three amino acid residues: Threonine replaces alanine once in the A chain (residue 8) and once in the B chain (residue 30), and isoleucine replaces valine once in the A chain (residue 10). Insulin from most mammals is similar.

---

*Vincent du Vigneaud of Cornell Medical College synthesized oxytocin and vasopressin in 1953; he received the Nobel Prize in 1955.

Oxytocin

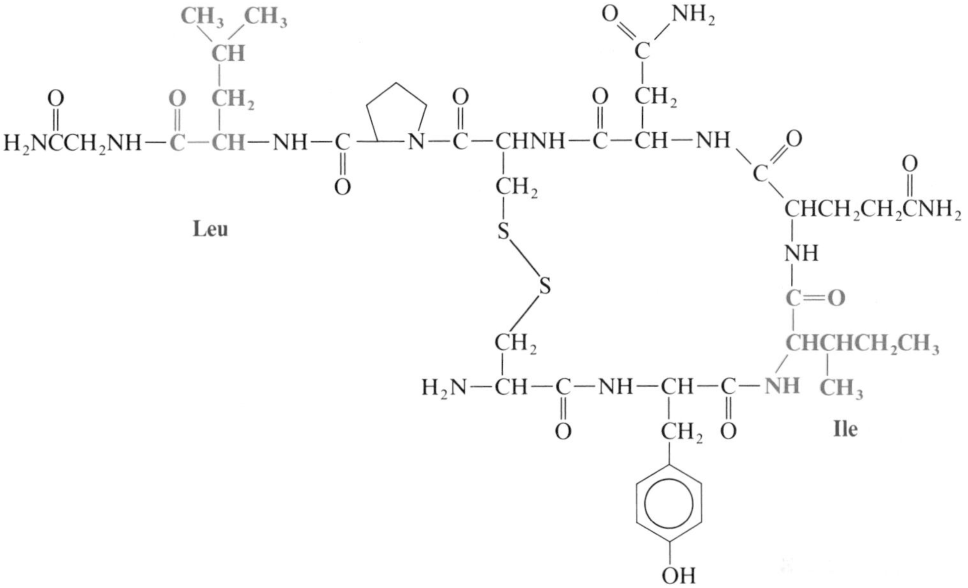

Vasopressin

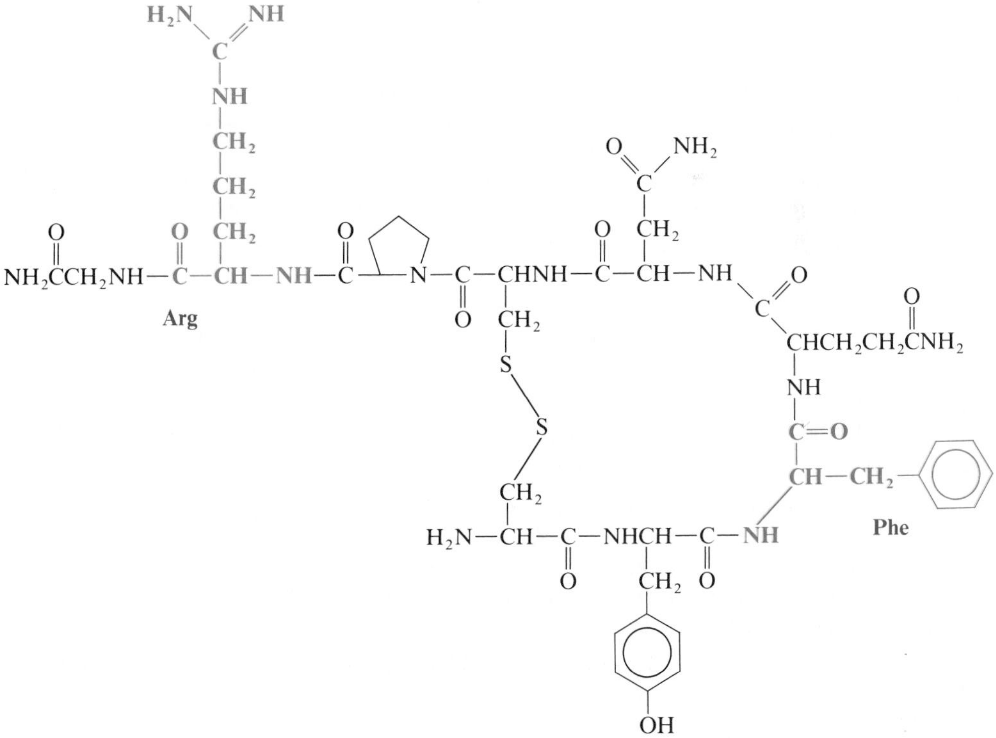

FIGURE 23.5 The structures of oxytocin and vasopressin.

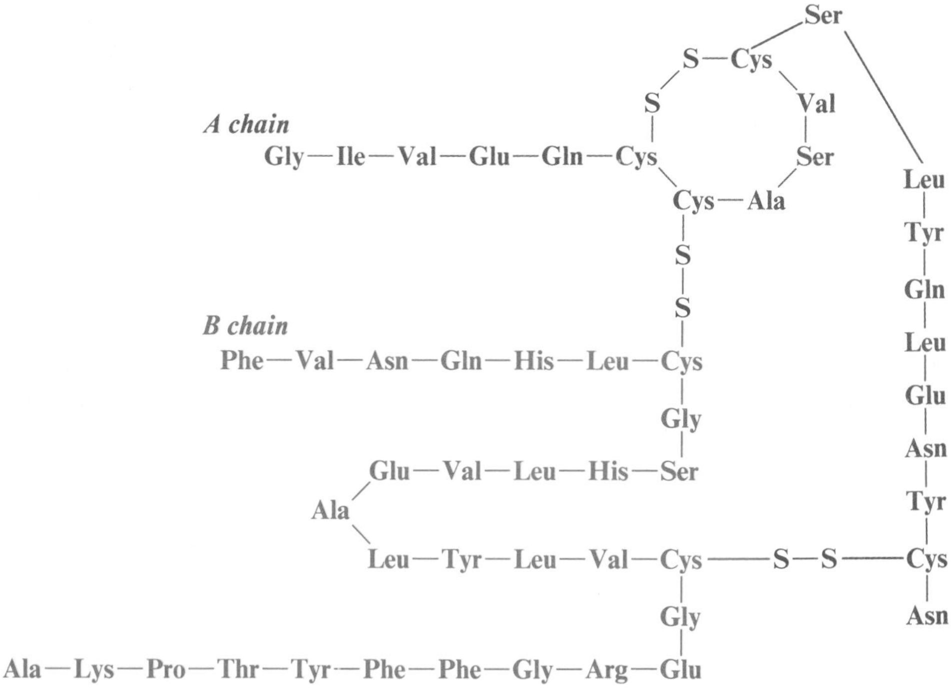

FIGURE 23.6 The amino acid sequence of bovine insulin.

23.7C Other Polypeptides and Proteins

Successful sequential analyses have now been achieved with hundreds of other polypeptides and proteins including the following:

1. **Bovine ribonuclease.** This enzyme, which catalyzes the hydrolysis of ribonucleic acid (Chapter 24), has a single chain of 124 amino acid residues and four intrachain disulfide linkages.

2. **Human hemoglobin.** There are four peptide chains in this important oxygen-carrying protein. Two identical α chains have 141 residues each, and two identical β chains have 146 residues. The genetically based disease sickle cell anemia results from a single amino acid error in the β chain. In normal hemoglobin, position 6 has a glutamic acid residue, while in sickle cell hemoglobin position 6 is occupied by valine.

Red blood cells (erythrocytes) containing hemoglobin with this amino acid residue error tend to become crescent shaped ("sickle") when the partial pressure of oxygen is low, as it is in venous blood. These distorted cells are more difficult for the heart to pump through small capillaries. They may even block capillaries by clumping together; at other times the red cells may even split open.

Children who inherit this genetic trait from both parents suffer from a severe form of the disease and usually do not live past the age of two. Children who inherit the disease from only one parent generally have a much milder form.

Sickle cell anemia arose among the populations of central and western Africa where, ironically, it may have had a beneficial effect. People with a mild form of the disease are far less susceptible to malaria than those with normal hemoglobin. Malaria, a disease caused by an infectious microorganism, is especially prevalent in central and western Africa.

Mutational changes such as those that give rise to sickle cell anemia are very common. Approximately 150 different types of mutant hemoglobin have been detected in humans; fortunately, most are harmless.

3. **Bovine trypsinogen and chymotrypsinogen.** These two enzyme precursors have single chains of 229 and 245 residues, respectively.
4. **Gamma globulin.** This immunoprotein has a total of 1320 amino acid residues in four chains. Two chains have 214 residues each; the other two have 446.

23.8 PROTEIN AND POLYPEPTIDE SYNTHESIS

We saw in Chapter 18 that the synthesis of an amide linkage is a relatively simple one. We must first "activate" the carboxyl group of an acid by converting it to an anhydride or acid chloride and then allow it to react with an amine:

$$
\underset{\text{Anhydride}}{R-\overset{\overset{\displaystyle O}{\|}}{C}-O-\overset{\overset{\displaystyle O}{\|}}{C}-R} + \underset{\text{Amine}}{R'-NH_2} \longrightarrow \underset{\text{Amide}}{R-\overset{\overset{\displaystyle O}{\|}}{C}-NHR'} + R-CO_2H
$$

The problem becomes somewhat more complicated, however, when both the acid group and the amino group are present in the same molecule, as they are in an amino acid and, especially, when our goal is the synthesis of a naturally occurring polyamide where the sequence of different amino acids is all-important. Let us consider, as an example, the synthesis of the simple dipeptide alanylglycine, Ala·Gly. We might first activate the carboxyl group of alanine by converting it to an acid chloride, and then we might allow it to react with glycine. Unfortunately, however, we cannot prevent alanyl chloride from reacting with itself. So our reaction would yield not only Ala·Gly but also Ala·Ala. It could also lead to Ala·Ala·Ala and Ala·Ala·Gly, and so on. The yield of our desired product would be low, and we would also have a difficult problem separating the dipeptides, tripeptides, and so on.

$$
\underset{\substack{\overset{\displaystyle |}{NH_3}\\ + \\ \text{Ala}}}{CH_3CHCO^-} \xrightarrow[\text{(2) } H_3NCH_2CO_2^-]{\text{(1) } SOCl_2} \underset{\substack{\overset{\displaystyle |}{NH_3}\\ + \\ \text{Ala·Gly}}}{CH_3\overset{\overset{\displaystyle O}{\|}}{C}HCNHCH_2CO_2^-} + \underset{\substack{\overset{\displaystyle |}{NH_3}\quad\overset{\displaystyle |}{CH_3}\\ + \\ \text{Ala·Ala}}}{CH_3CHCNHCHCO_2^-}
$$

(1) SOCl₂, (2) H₃NCH₂CO₂⁻, Gly

$$
+ \underset{\substack{\overset{\displaystyle |}{NH_3}\quad\overset{\displaystyle |}{CH_3}\quad\overset{\displaystyle |}{CH_3}\\ + \\ \text{Ala·Ala·Ala}}}{CH_3CHCNHCHCNHCHCO_2^-} + \underset{\substack{\overset{\displaystyle |}{NH_3}\quad\overset{\displaystyle |}{CH_3}\\ + \\ \text{Ala·Ala·Gly}}}{CH_3CHCNHCHCNHCH_2CO_2^-}
$$

23.8A Protecting Groups

The solution of this problem is to "protect" the amino group of the first amino acid before we activate it and allow it to react with the second. By protecting the amino group we mean that we must convert it to some other group of low nucleophilicity — *one that will not react with a reactive acyl derivative*. The protecting group must be carefully chosen because after we have synthesized the amide linkage between the first amino acid and the second we will want to be able to remove the protecting group without disturbing the new amide bond.

A number of reagents have been developed to meet these requirements. Two that are often used are *benzyl chloroformate* and di-*tert*-butyl carbonate.

$$C_6H_5CH_2-O-\overset{\overset{\displaystyle O}{\|}}{C}-Cl \qquad (CH_3)_3C-O-\overset{\overset{\displaystyle O}{\|}}{C}-O-C(CH_3)_3$$

Benzyl chloroformate Di-*tert*-butyl carbonate

Both reagents react with the following amino groups to form derivatives that are unreactive toward further acylation. Both derivatives, however, are of a type that allow removal of the protecting group under conditions that do not affect peptide bonds. The benzyloxycarbonyl group (abbreviated Z-) can be removed by catalytic hydrogenation or by treating the derivative with cold HBr in acetic acid. The *tert*-butyloxycarbonyl group (abbreviated Boc-) can be removed through treatment with HCl or CF_3CO_2H in acetic acid.

Benzyloxycarbonyl Group

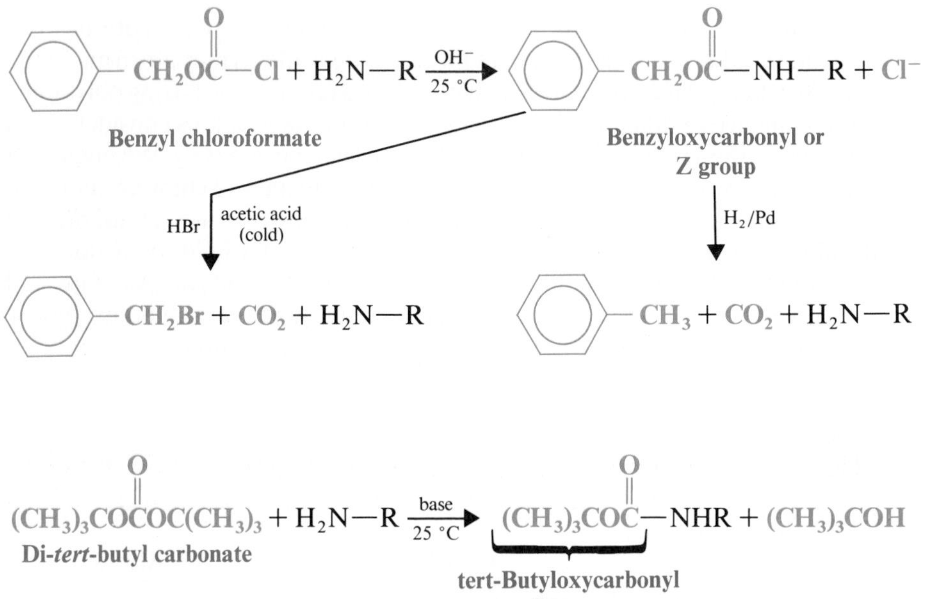

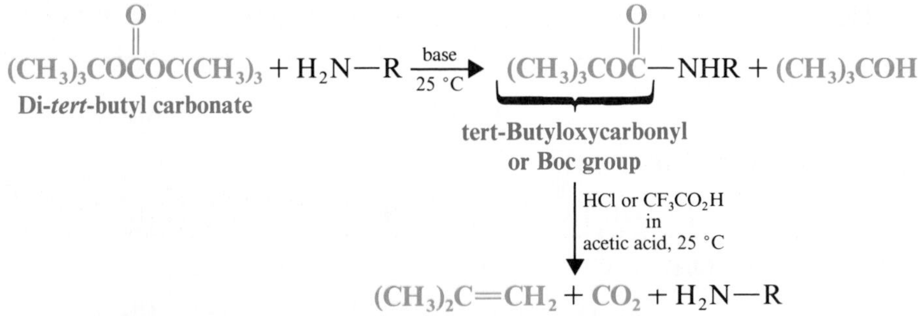

The easy removal of both groups (Z- and Boc-) in acidic media results from the exceptional stability of the carbocations that are formed initially. The benzyloxy-carbonyl group gives a *benzyl cation;* the *tert*-butyloxycarbonyl group yields, initially, a *tert*-butyl cation.

Removal of the benzyloxycarbonyl group with hydrogen and a catalyst depends on the fact that benzyl–oxygen bonds are weak and are subject to hydrogenolysis at low temperatures.

$$C_6H_5CH_2-O\overset{\overset{\displaystyle O}{\|}}{C}R \xrightarrow[25\ °C]{H_2,\ Pd} C_6H_5CH_3 + HO\overset{\overset{\displaystyle O}{\|}}{C}R$$

A benzyl ester

23.8B Activation of the Carboxyl Group

Perhaps the most obvious way to activate a carboxyl group is to convert it to an acyl chloride. This method was used in early peptide syntheses, but acyl chlorides are actually more reactive than necessary. As a result, their use leads to complicating side reactions. A much better method is to convert the carboxyl group of the "protected" amino acid to a mixed anhydride using ethyl chloroformate, $Cl-\overset{\overset{\displaystyle O}{\|}}{C}-OC_2H_5$

$$Z-NHCH\overset{\overset{\displaystyle O}{\|}}{C}-OH \underset{R}{\quad} \xrightarrow[\text{(2) ClCO}_2\text{C}_2\text{H}_5]{\text{(1) (C}_2\text{H}_5)_3\text{N}} Z-NHCH\underset{R}{-}\overset{\overset{\displaystyle O}{\|}}{C}-O-\overset{\overset{\displaystyle O}{\|}}{C}-OC_2H_5$$

"Mixed anhydride"

The mixed anhydride can then be used to acylate another amino acid and form a peptide linkage.

$$Z-NHCH\underset{R}{\overset{\overset{\displaystyle O}{\|}}{C}}-O-\overset{\overset{\displaystyle O}{\|}}{C}OC_2H_5 \xrightarrow{\overset{+}{H_3}N-\underset{R'}{CHCO_2^-}}$$

$$Z-NHCH\underset{R}{\overset{\overset{\displaystyle O}{\|}}{C}}-NH\underset{R'}{CHCO_2H} + CO_2 + C_2H_5OH$$

Dicyclohexylcarbodiimide (Section 18.8E) can also be used to activate the carboxyl group of an amino acid. In Section 23.8D we shall see how it is used in an automated peptide synthesis.

23.8C Peptide Synthesis

Let us examine now how we might use these reagents in the preparation of the simple dipeptide, Ala·Leu. The principles involved here can, of course, be extended to the synthesis of much longer polypeptide chains.

$$CH_3CHCO_2^- \underset{\overset{+}{N}H_3}{} + C_6H_5CH_2O\overset{\overset{\displaystyle O}{\|}}{C}-Cl \xrightarrow[25\ °C]{OH^-} CH_3CH-CO_2H \xrightarrow[\text{(2) ClCO}_2\text{C}_2\text{H}_5]{\text{(1) (C}_2\text{H}_5)_3\text{N}}$$

$$\begin{array}{c} NH \\ | \\ C=O \\ | \\ C_6H_5CH_2O \end{array}$$

| Ala | Benzyl chloroformate | Z-Ala |

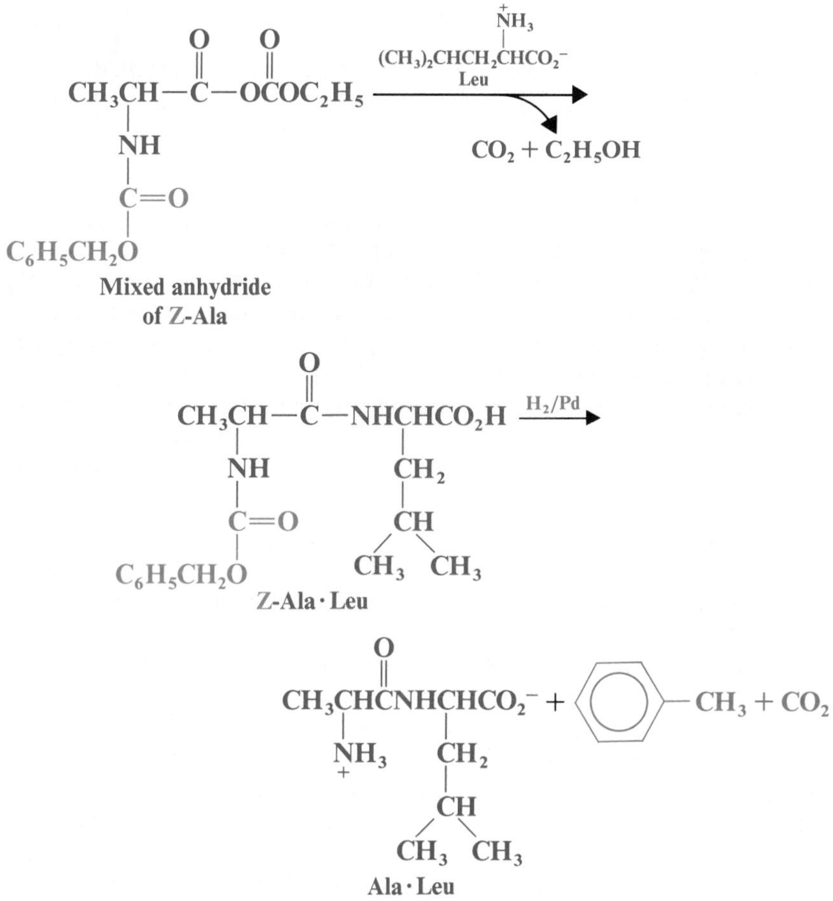

$$CH_3CH-\overset{\overset{\displaystyle O}{\|}}{C}-O\overset{\overset{\displaystyle O}{\|}}{C}OC_2H_5 \xrightarrow[\text{Leu}]{\overset{\overset{+}{NH_3}}{|} (CH_3)_2CHCH_2CHCO_2^-}$$

with

$$\begin{array}{c} NH \\ | \\ C=O \\ | \\ C_6H_5CH_2O \end{array}$$

Mixed anhydride
of Z-Ala

$$CO_2 + C_2H_5OH$$

$$CH_3CH-\overset{\overset{\displaystyle O}{\|}}{C}-NHCHCO_2H \xrightarrow{H_2/Pd}$$

with side groups

$$\begin{array}{cc} NH & CH_2 \\ | & | \\ C=O & CH \\ | & \diagup\diagdown \\ C_6H_5CH_2O & CH_3 \quad CH_3 \end{array}$$

Z-Ala·Leu

$$CH_3CH\overset{\overset{\displaystyle O}{\|}}{C}NHCHCO_2^- + \left\langle\bigcirc\right\rangle-CH_3 + CO_2$$

with side groups

$$\begin{array}{cc} \overset{+}{N}H_3 & CH_2 \\ & | \\ & CH \\ & \diagup\diagdown \\ & CH_3 \quad CH_3 \end{array}$$

Ala·Leu

PROBLEM 23.11

Show all steps in the synthesis of Gly · Val · Ala using the *tert*-butyloxycarbonyl (Boc-) group as a protecting group.

PROBLEM 23.12

The synthesis of a polypeptide containing lysine requires the protection of both amino groups. (a) Show how you might do this in a synthesis of Lys · Ile using the benzyloxycarbonyl group as a protecting group. (b) The benzyloxycarbonyl group can also be used to protect the guanidino group, $-NH\overset{\overset{\displaystyle NH}{\|}}{C}-NH_2$, of arginine. Show a synthesis of Arg · Ala.

PROBLEM 23.13
The terminal carboxyl groups of glutamic acid and aspartic acid are often protected through their conversion to benzyl esters. What mild method could be used for removal of this protecting group?

23.8D Automated Peptide Synthesis

Although the methods that we have described thus far have been used to synthesize a number of polypeptides including ones as large as insulin, they are extremely time consuming. One must isolate and purify the product at almost every stage. Thus, a real advance in peptide synthesis came with the development by R. B. Merrifield (at Rockefeller University) of a procedure for automating peptide synthesis. Merrifield received the Nobel Prize in chemistry in 1984 for this work.

The Merrifield method is based on the use of a polystyrene resin similar to the one we saw in Fig. 23.2, *but one that contains* $-CH_2Cl$ *groups* instead of sulfonic acid groups. This resin is used in the form of small beads and is insoluble in most solvents.

The first step in automated peptide synthesis (Fig. 23.7) involves a reaction that attaches the first protected amino acid residue to the resin beads. After this step is complete, the protecting group is removed and the next amino acid (also protected) is condensed with the first using dicyclohexylcarbodiimide (Section 18.8E) to activate its carboxyl group. Then removal of the protecting group of the second residue readies the resin-dipeptide for the next step.

The great advantage of this procedure is that purification of the resin with its attached polypeptide can be carried out at each stage by simply washing the resin with an appropriate solvent. Impurities, because they are not attached to the insoluble resin, are simply carried away by the solvent. In the automated procedure each cycle of the "protein-making machine" requires only four hours and attaches one new amino acid residue.*

The Merrifield technique has been applied successfully to the synthesis of ribonuclease, a protein with 124 amino acid residues. The synthesis involved 369 chemical reactions and 11,931 automated steps — all were carried out without isolating an intermediate. The synthetic ribonuclease not only had the same physical characteristics as the natural enzyme; it possessed the biological activity as well. The overall yield was 17%, which means that the average yield of each individual step was greater than 99%.

PROBLEM 23.14
The resin for the Merrifield procedure is prepared by treating polystyrene, $-\left(CH_2CH\right)_n-$ with C_6H_5, CH_3OCH_2Cl and a Lewis acid catalyst. (a) What reaction is involved? (b) After purification, the complete polypeptide or protein can be detached from the resin by treating it with HBr in trifluoroacetic acid under conditions mild enough not to affect the amide linkages. What structural feature of the resin makes this possible?

---

*Protein synthesis in the body catalyzed by enzymes and directed by DNA/RNA takes only 1 min to add 150 amino acids in a specific sequence (cf. Section 24.5).

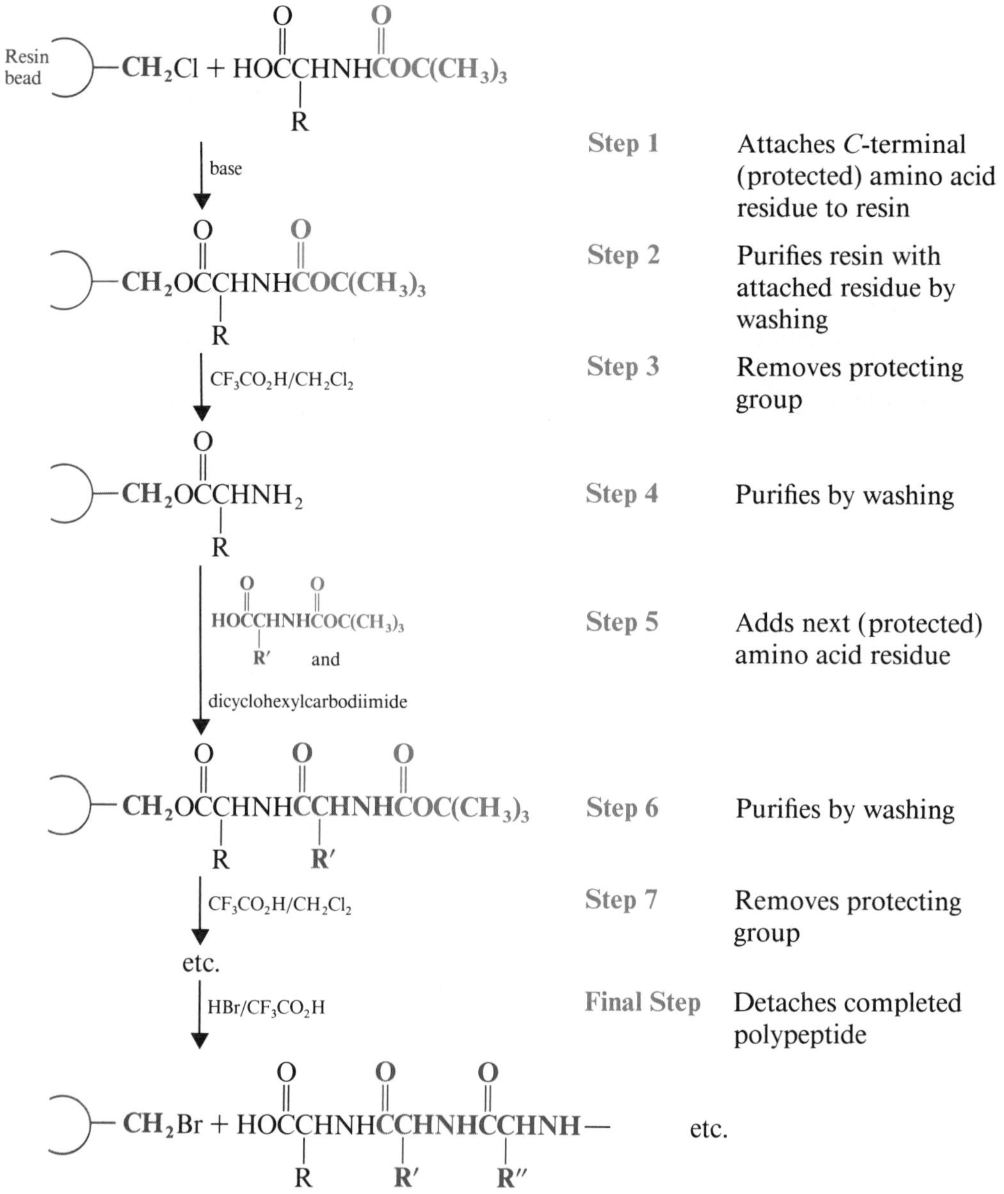

$$\text{Resin bead} \rangle\!-\!CH_2Cl + HO\overset{O}{\overset{\|}{C}}CHNH\overset{O}{\overset{\|}{C}}OC(CH_3)_3$$
$$\underset{R}{|}$$

Step 1   Attaches *C*-terminal (protected) amino acid residue to resin

↓ base

$$\rangle\!-\!CH_2O\overset{O}{\overset{\|}{C}}CHNH\overset{O}{\overset{\|}{C}}OC(CH_3)_3$$
$$\underset{R}{|}$$

Step 2   Purifies resin with attached residue by washing

↓ CF₃CO₂H/CH₂Cl₂

Step 3   Removes protecting group

$$\rangle\!-\!CH_2O\overset{O}{\overset{\|}{C}}CHNH_2$$
$$\underset{R}{|}$$

Step 4   Purifies by washing

$$HO\overset{O}{\overset{\|}{C}}CHNH\overset{O}{\overset{\|}{C}}OC(CH_3)_3$$
$$\underset{R'}{|} \quad and$$

dicyclohexylcarbodiimide

Step 5   Adds next (protected) amino acid residue

$$\rangle\!-\!CH_2O\overset{O}{\overset{\|}{C}}CHNH\overset{O}{\overset{\|}{C}}CHNH\overset{O}{\overset{\|}{C}}OC(CH_3)_3$$
$$\underset{R}{|} \quad \underset{R'}{|}$$

Step 6   Purifies by washing

↓ CF₃CO₂H/CH₂Cl₂

Step 7   Removes protecting group

etc.

↓ HBr/CF₃CO₂H

Final Step   Detaches completed polypeptide

$$\rangle\!-\!CH_2Br + HO\overset{O}{\overset{\|}{C}}CHNH\overset{O}{\overset{\|}{C}}CHNH\overset{O}{\overset{\|}{C}}CHNH\!-\! \quad etc.$$
$$\underset{R}{|} \quad \underset{R'}{|} \quad \underset{R''}{|}$$

FIGURE 23.7 The Merrifield method for automated protein synthesis.

PROBLEM 23.15

Outline the steps in the synthesis of Lys·Phe·Ala using the Merrifield procedure.

23.9 SECONDARY AND TERTIARY STRUCTURES OF PROTEINS

We have seen how amide and disulfide linkages constitute the covalent or *primary structure* of proteins. Of equal importance in understanding how proteins function is knowledge of the way in which the peptide chains are arranged in three dimensions. Involved here are the secondary and tertiary structures of proteins.

23.9A Secondary Structure

The major experimental technique that has been used in elucidating the secondary structures of proteins is X-ray analysis.

When X-rays pass through a crystalline substance they produce diffraction pa ns. Analysis of these patterns indicates a regular repetition of particular structural units with certain specific distances between them, called **repeat distances.** The complete X-ray analysis of a molecule as complex as a protein often takes years of painstaking work. Nonetheless, many X-ray analyses have been done and they have revealed that the polypeptide chain of a natural protein can interact with itself in two major ways: through formation of a **pleated sheet** and an **α helix.***

To understand how these interactions occur let us look first at what X-ray analysis has revealed about the geometry at the peptide bond itself. Peptide bonds tend to assume a geometry such that six atoms of the amide linkage are coplanar (Fig. 23.8). The carbon–nitrogen bond of the amine linkage is unusually short, indicating that resonance contributions of the type shown here are important.

$$\ddot{\underset{/}{N}}-C\overset{O}{\diagup} \quad \longleftrightarrow \quad \overset{+}{\underset{/}{N}}=C\overset{O^-}{\diagup}$$

The carbon–nitrogen bond, consequently, has considerable double-bond character (~40%) and rotations of groups about this bond are severely hindered.

Rotations of groups attached to the amide nitrogen and the carbonyl carbon are relatively free, however, and these rotations allow peptide chains to form different conformations.

The transoid arrangement of groups around the relatively rigid amide bond would cause the R groups to alternate from side to side of a single fully extended peptide chain:

$$\longleftarrow \text{7.2 Å} \longrightarrow$$

*Pioneers in the X-ray analysis of proteins were two American scientists, Linus Pauling (Section 8.16) and Robert B. Corey. Beginning in 1939, Pauling and Corey initiated a long series of studies of the conformations of peptide chains. At first they used crystals of single amino acids, then dipeptides and tripeptides, and so on. Moving on to larger and larger molecules and using the precisely constructed molecular models, they were able to understand the secondary structures of proteins for the first time.

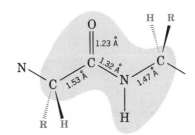

FIGURE 23.8 The geometry and bond lengths of the peptide linkage. The six enclosed atoms tend to be coplanar and assume a "transoid" arrangement.

Calculations show that such a polypeptide chain would have a repeat distance (i.e., distance between alternating units) of 7.2 Å.

Fully extended polypeptide chains could conceivably form a flat-sheet structure with each alternating amino acid in each chain forming two hydrogen bonds with an amino acid in the adjacent chain:

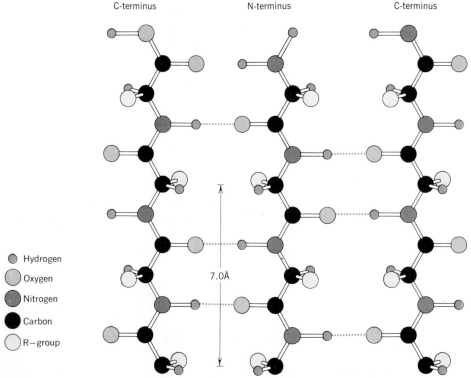

Hypothetical flat—sheet structure

This structure does not exist in naturally occurring proteins because of the crowding that would exist between R groups. If such a structure did exist, it would have the same repeat distance as the fully extended peptide chain, that is, 7.2 Å.

Slight rotations of bonds, however, can transform a flat-sheet structure into what is called the **pleated-sheet** or **β configuration** (Fig. 23.9). The pleated-sheet

C-terminus   N-terminus   C-terminus

○ Hydrogen
○ Oxygen
○ Nitrogen
● Carbon
○ R—group

7.0Å

FIGURE 23.9 The pleated sheet or β configuration of a protein.

structure gives small- and medium-sized R groups room enough to avoid van der Waals repulsions and is the predominant structure of silk fibroin (48% glycine and 38% serine and alanine residues). The pleated-sheet structure has a slightly shorter repeat distance, 7.0 Å, than the flat sheet.

Of far more importance in naturally occurring proteins is the secondary structure called the α helix (Fig. 23.10). This structure is a right-handed helix with 3.6 amino acid residues per turn. Each amide group in the chain has a hydrogen bond to an amide group at a distance of three amino acid residues in either direction, and the R groups all extend away from the axis of the helix. The repeat distance of the α helix is 1.5 Å.

The α-helical structure is found in many proteins; it is the predominant structure of the polypeptide chains of fibrous proteins such as *myosin,* the protein of muscle, and of *α-keratin,* the protein of hair, unstretched wool, and nails.

Not all peptide chains can exist in an α-helical form. Certain peptide chains assume what is called a **random coil arrangement,** a structure that is flexible, chang-

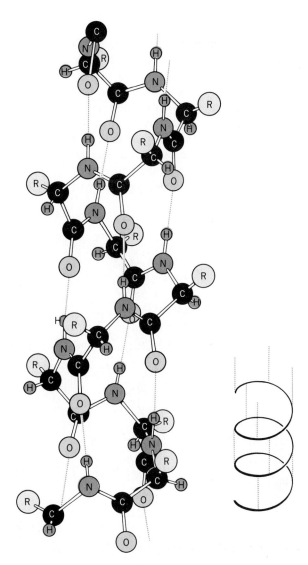

FIGURE 23.10 A representation of the α-helical structure of a polypeptide. Hydrogen bonds are denoted by dotted lines.

ing, and statistically random. Synthetic polylysine, for example, exists as a random coil and does not normally form an α helix. At pH 7 the ε-amino groups of the lysine residues are positively charged and, as a result, repulsive forces between them are so large that they overcome any stabilization that would be gained through hydrogen bond formation of an α helix. At pH 12, however, the ε-amino groups are uncharged and polylysine spontaneously forms an α helix.

The presence of proline or hydroxyproline residues in polypeptide chains produces another striking effect: Because the nitrogen atoms of these amino acids are part of five-membered rings, the groups attached by the nitrogen — α-carbon bond cannot rotate enough to allow an α-helical structure. Wherever proline or hydroxyproline occur in a peptide chain, their presence causes a kink or bend and interrupts the α helix.

The polypeptide chains of globular proteins such as hemoglobin, ribonuclease, α-chymotrypsin, and lysozyme contain segments of α helix and segments of random coil. Proline and hydroxyproline are often found in those regions of the structure where the conformation changes.

23.9B Tertiary Structure

The tertiary structure of a protein is its three-dimensional shape that arises from further foldings of its polypeptide chains, foldings superimposed on the coils of the α helixes. These foldings do not occur randomly: Under the proper environmental conditions they occur in one particular way — a way that is characteristic of a particular protein and one that is often highly important to its function.

Various forces are involved in stabilizing tertiary structures including the disulfide bonds of the primary structure. One characteristic of most proteins is that the folding takes place in such a way as to expose the maximum number of polar (hydrophilic) groups to the aqueous environment and enclose a maximum number of nonpolar (hydrophobic) groups within its interior.

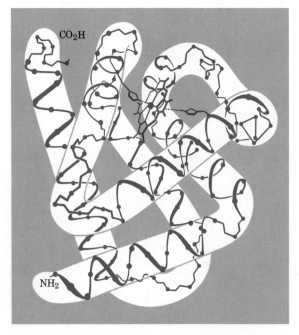

FIGURE 23.11 The three-dimensional structure of myoglobin. (Adapted from R. E. Dickerson, *The Proteins II*, H. Neurath, Ed., Academic Press, New York, 1964, p. 634. Used with permission.)

The soluble globular proteins tend to be much more highly folded than fibrous proteins. However, fibrous proteins also have a tertiary structure; the α-helical strands of α-keratin, for example, are wound together into a "super helix." This super helix has a repeat distance of 5.1-Å units indicating that the super helix makes one complete turn for each 35 turns of the α helix. The tertiary structure does not end here, however. Even the super helixes can be wound together to give a ropelike structure of seven strands.

Myoglobin (Fig. 23.11) and hemoglobin (Section 23.11) were the first proteins (in 1957 and 1959) to be subjected to a completely successful X-ray analysis. This work was accomplished by J. C. Kendrew and Max Perutz at Cambridge University in England. (They received the Nobel Prize in 1962.) Since then a number of other proteins including lysozyme, ribonuclease, and α-chymotrypsin have yielded to complete structural analysis.

23.10 LYSOZYME: MODE OF ACTION OF AN ENZYME

Lysozyme is made up of 129 amino acid residues (Fig. 23.12). Three short segments of the chain between residues 5–15, 24–34, and 88–96 have the structure of an α helix; the residues between 41–45 and 50–54 form pleated sheets, and a hairpin turn occurs at residues 46–49. The remaining polypeptide segments of lysozyme have a random coil arrangement.

The discovery of lysozyme is an interesting story in itself:

> One day in 1922 Alexander Fleming was suffering from a cold. This is not unusual in London, but Fleming was a most unusual man and he took advantage of the cold in a characteristic way. He allowed a few drops of his nasal mucus to fall on a culture of bacteria he was working with and then put the plate to one side to see what would happen. Imagine his excitement when he discovered some time later that the bacteria near the mucus had dissolved away. For a while he thought his ambition of finding a universal antibiotic had been realized. In a burst of activity he quickly established that the antibacterial action of the mucus was due to the presence of an enzyme; he called this substance lysozyme because of its capacity to lyse, or dissolve, the bacterial cells. Lysozyme was soon discovered in many tissues and secretions of the human body, in plants and most plentifully of all in the white of an egg. Unfortunately Fleming found that it is not effective against the most harmful bacteria. He had to wait seven years before a strangely similar experiment revealed the existence of a genuinely effective antibiotic: penicillin.

This story was related by Professor David C. Phillips of Oxford University who many years later used X-ray analysis to discover the three-dimensional structure of lysozyme.*

Phillips' X-ray diffraction studies of lysozyme are especially interesting because they have also revealed important information about how this enzyme acts on its substrate. Lysozyme's substrate is a polysaccharide of amino sugars that makes up part of the bacterial cell wall. An oligosaccharide that has the same general structure as the cell wall polysaccharide is shown in Fig. 23.13.

By using oligosaccharides (made up of *N*-acetylglucosamine units only) on which lysozyme acts very slowly, Phillips and his co-workers were able to discover how the substrate fits into the enzyme's active site. This site is a deep cleft in the

---

*Quotation from David C. Phillips, *The Three-Dimensional Structure of an Enzyme Molecule.*
Copyright © 1966 by Scientific American, Inc. All rights reserved.

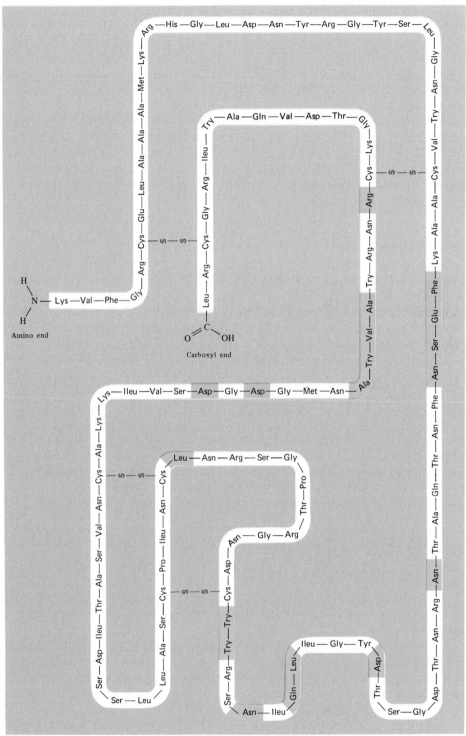

FIGURE 23.12 The covalent structure of lysozyme. The amino acids that line the active site of lysozyme are shown in color.

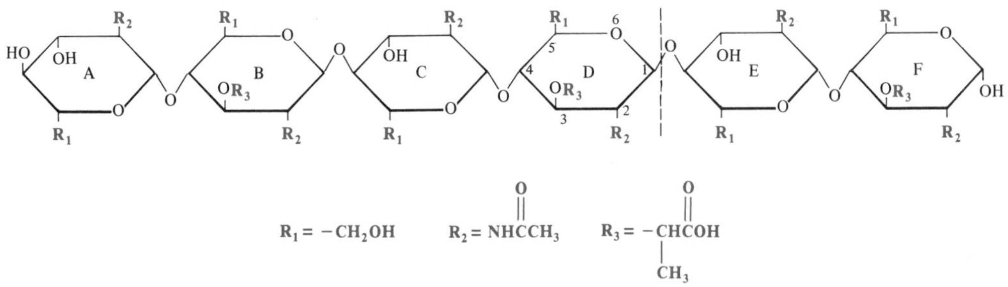

$$R_1 = -CH_2OH \qquad R_2 = NH\overset{\overset{\displaystyle O}{\|}}{C}CH_3 \qquad R_3 = -\underset{\underset{\displaystyle CH_3}{|}}{\overset{\overset{\displaystyle O}{\|}}{CH}}COH$$

FIGURE 23.13 A hexasaccharide that has the same general structure as the cell wall polysaccharide on which lysozyme acts. Two different amino sugars are present: rings **A**, **C**, and **E** are derived from a monosaccharide called *N*-acetylglucosamine; rings **B**, **D**, and **F** are derived from a monosaccharide called *N*-acetylmuramic acid. When lysozyme acts on this oligosaccharide, hydrolysis takes place and results in cleavage at the glycosidic linkage between rings **D** and **E**.

lysozyme structure (Fig. 23.14*a*). The oligosaccharide is held in this cleft by hydrogen bonds, and as the enzyme binds the substrate two important changes take place: The cleft in the enzyme closes slightly and ring **D** of the oligosaccharide is "flattened" out of its stable chair conformation. This flattening causes atoms 1, 2, 5, and 6 of ring **D** to become coplanar; it also distorts ring **D** in such a way as to make the glycosidic linkage between it and ring **E** more susceptible to hydrolysis.*

Hydrolysis of the glycosidic linkage probably takes the course illustrated in Fig. 23.14*b*. The carboxyl group of glutamic acid (residue number 35) donates a proton to the oxygen between rings **D** and **E**. Protonation leads to cleavage at the glycosidic link and to the formation of a carbocation at C-1 of ring **D**. This carbocation is stabilized by the negatively charged carboxylate group of aspartic acid (residue number 52), which lies in close proximity. A water molecule diffuses in and supplies an OH⁻ ion to the carbocation and a proton to replace that lost by glutamic acid.

When the polysaccharide is a part of a bacterial cell wall, lysozyme probably first attaches itself to the cell wall by hydrogen bonds. After hydrolysis has taken place, lysozyme falls away leaving behind a bacterium with a punctured cell wall.

23.11 HEMOGLOBIN: A CONJUGATED PROTEIN

Some proteins contain as a part of their structure a nonprotein group called a **prosthetic group.** An example is the oxygen-carrying protein, hemoglobin. Each of the four polypeptide chains of hemoglobin is bound to a prosthetic group called *heme* (Fig. 23.15). The four polypeptide chains of hemoglobin are wound in such a way as to give hemoglobin a roughly spherical shape (Fig. 23.16). Moreover, each heme group lies in a crevice with the hydrophobic vinyl groups of its porphyrin structure surrounded by side chains of hydrophobic amino residues. The two propanoate side chains of heme lie near positively charged amino groups of lysine and arginine residues.

The iron of the heme group is in the 2 + (ferrous) oxidation state and it forms a coordinate bond to a nitrogen of the imidazole group of histidine of the polypeptide

*R. H. Lemieux and G. Huber while with the National Research Council of Canada showed that when an aldohexose is converted to a carbocation the ring of the carbocation assumes just this flattened conformation.

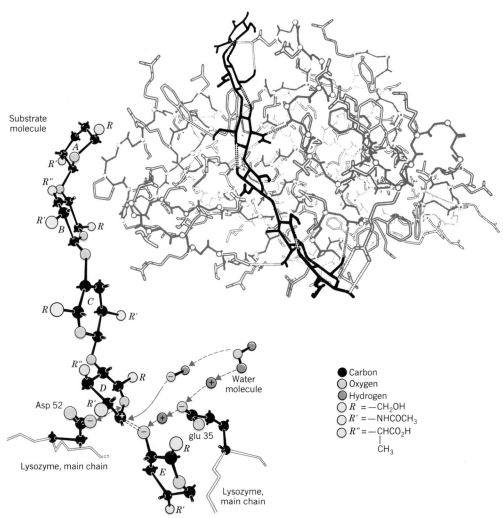

Substrate molecule

Lysozyme, main chain

Asp 52

Water molecule

glu 35

Lysozyme, main chain

● Carbon
◐ Oxygen
● Hydrogen
◯ R = —CH₂OH
◯ R' = —NHCOCH₃
◯ R" = —CHCO₂H
      |
      CH₃

FIGURE 23.14 (*a*) This drawing shows the backbones of the lysozyme-substrate complex. The substrate (in this drawing a hexasaccharide) fits into a cleft in the lysozyme structure and is held in place by hydrogen bonds. As lysozyme binds the oligosaccharide, the cleft in its structure closes slightly. (Adapted with permission from *Atlas of Protein Sequence and Structure*, 1969, Margaret O. Dayoff, Ed. National Biomedical Research Foundation, Washington, DC, 1969. The drawing was made by Irving Geis, based on his perspective painting of the molecule, which appeared in *Scientific American*, November 1966. The painting was made of an actual model assembled at the Royal Institution, London, by D. C. Phillips and his colleagues, based on their X-ray crystallography results.)) (*b*) A possible mechanism for lysozyme action. This drawing shows an expanded portion of part (*a*) and illustrates how hydrolysis of the acetal linkage between rings **D** and **E** of the substrate may occur. Glutamic acid (residue 35) donates a proton to the intervening oxygen atom. This causes the formation of a carbocation that is stabilized by the carboxylate ion aspartic acid (residue 52). A water molecule supplies an OH⁻ to the carbonium ion and H⁺ to glutamic acid. (Adapted with permission from *The Three-Dimensional Structures of an Enzyme Molecule*, by David C. Phillips, Copyright © Nov. 1966 by Scientific American, Inc., All rights reserved.)

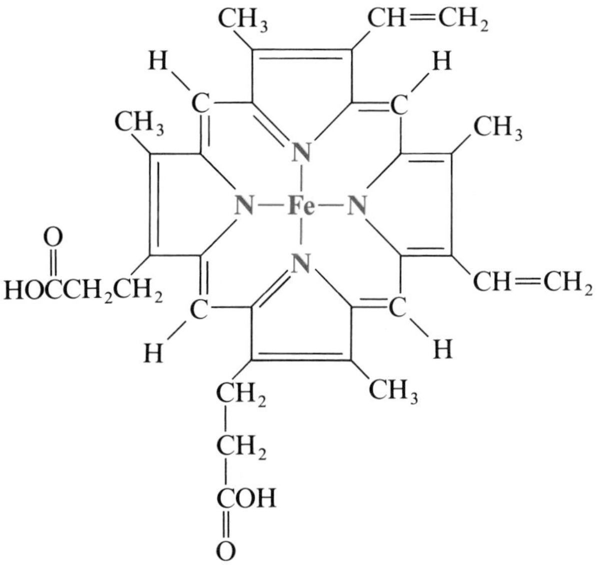

FIGURE 23.15 The structure of heme, the prosthetic group of hemoglobin. Heme has a structure similar to that of chlorophyll (Fig. 21.1) in that each is derived from the heterocyclic ring, porphyrin. The iron of heme is in the ferrous (2+) oxidation state.

chain. This leaves one valence of the ferrous ion free to combine with oxygen as follows:

A portion of oxygenated hemoglobin

The fact that the ferrous ion of the heme group combines with oxygen is not particularly remarkable; many similar compounds do the same thing. What is re-

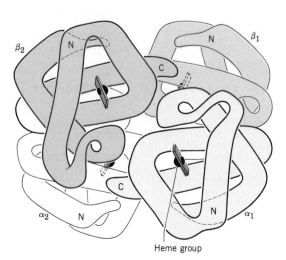

FIGURE 23.16 The hemoglobin molecule.

Heme group

markable about hemoglobin is that when the heme combines with oxygen the ferrous ion does not become readily oxidized to the ferric state. Studies with model heme compounds in water, for example, show that they undergo a rapid combination with oxygen but they also undergo a rapid oxidation of the iron from Fe^{2+} to Fe^{3+}. When these same compounds are embedded in the hydrophobic environment of a polystyrene resin, however, the iron is easily oxygenated and deoxygenated and this occurs *with no change in oxidation state of iron.* In this respect, it is especially interesting to note that X-ray studies of hemoglobin have revealed that the polypeptide chains provide each heme group with a similar hydrophobic environment.

ADDITIONAL PROBLEMS

23.16
(a) Which amino acids in Table 23.1 have more than one stereocenter? (b) Write Fischer projection formulas for the isomers of each of these amino acids that would have the L configuration at the α carbon. (c) What kind of isomers have you drawn in each case?

23.17
(a) Which amino acid in Table 23.1 could react with nitrous acid (i.e., a solution of $NaNO_2$ and HCl) to yield lactic acid? (b) All of the amino acids in Table 23.1 liberate nitrogen when they are treated with nitrous acid except two; which are these? (c) What product would you expect to obtain from treating tyrosine with excess bromine water? (d) What product would you expect to be formed in the reaction of phenylalanine with ethanol in the presence of hydrogen chloride? (e) What product would you expect from the reaction of alanine and benzoyl chloride in aqueous base?

23.18
(a) On the basis of the following sequence of reactions Emil Fischer was able to show that (−)-serine and L-(+)-alanine have the same configuration. Write Fischer projection formulas for the intermediates **A–C.**

$$(-)\text{-Serine} \xrightarrow[\text{CH}_3\text{OH}]{\text{HCl}} \textbf{A } (C_4H_{10}ClNO_3) \xrightarrow{\text{PCl}_5} \textbf{B } (C_4H_9Cl_2NO_2) \xrightarrow[\text{(2) OH}^-]{\text{(1) H}_3\text{O}^+, \text{H}_2\text{O, heat}}$$

$$\textbf{C } (C_3H_6ClNO_2) \xrightarrow[\text{dil. H}_3\text{O}^+]{\text{Na-Hg}} \text{L-(+)-alanine}$$

(b) The configuration of L-(−)-cysteine can be related to that of L-(−)-serine through the following reactions. Write Fischer projection formulas for **D** and **E.**

$$\textbf{B } [\text{from part (a)}] \xrightarrow{\text{OH}^-} \textbf{D } (C_4H_8ClNO_2) \xrightarrow{\text{NaSH}}$$

$$\textbf{E } (C_4H_9NO_2S) \xrightarrow[\text{(2) OH}^-]{\text{(1) H}_3\text{O}^+, \text{H}_2\text{O, heat}} \text{L-(+)-cysteine}$$

(c) The configuration of L-(−)-asparagine can be related to that of L-(−)-serine in the following way. What is the structure of **F**?

$$\text{L-(−)-Asparagine} \xrightarrow[\text{Hofmann rearrangement}]{\text{NaOBr/OH}^-} \textbf{F } (C_3H_7N_2O_2)$$

$$\uparrow \text{NH}_3$$

$$\textbf{C } [\text{from part (a)}]$$

23.19

(a) DL-Glutamic acid has been synthesized from diethyl acetamidomalonate in the following way: Outline the reactions involved.

$$CH_3\overset{\overset{\displaystyle O}{\|}}{C}NHCH(CO_2C_2H_5)_2 + CH_2{=}CH{-}C{\equiv}N \xrightarrow[\substack{C_2H_5OH \\ (95\% \text{ yield})}]{NaOC_2H_5}$$

Diethyl acetamido-
malonate

$$G\ (C_{12}H_{18}N_2O_5) \xrightarrow[\substack{\text{reflux 6 h} \\ (66\% \text{ yield})}]{\text{conc. HCl}} DL\text{-glutamic acid}$$

(b) Compound **G** has also been used to prepare the amino acid DL-ornithine through the following route. Outline the reaction involved here.

$$G\ (C_{12}H_{18}N_2O_5) \xrightarrow[\substack{68\ ^\circ C,\ 1000\ \text{psi} \\ (90\% \text{ yield})}]{H_2,\ Ni} H\ (C_{10}H_{16}N_2O_4,\ \text{a }\delta\text{-lactam}) \xrightarrow[\substack{\text{reflux 4 h} \\ (97\% \text{ yield})}]{\text{conc. HCl}}$$

$$DL\text{-ornithine hydrochloride } (C_5H_{13}ClN_2O_2)$$

(L-Ornithine is a naturally occurring amino acid but does not occur in proteins. In one metabolic pathway L-ornithine serves as a precursor for L-arginine.)

23.20

Bradykinin is a nonapeptide released by blood plasma globulins in response to a wasp sting. It is a very potent pain-causing agent. Its molecular formula is Arg_2, Gly, Phe_2, Pro_3, Ser. The use of 2,4-dinitrofluorobenzene and carboxypeptidase show that both terminal residues are arginine. Partial acid hydrolysis of bradykinin gives the following di- and tripeptides:

$$Phe \cdot Ser + Pro \cdot Gly \cdot Phe + Pro \cdot Pro + Ser \cdot Pro \cdot Phe + Phe \cdot Arg + Arg \cdot Pro$$

What is the amino acid sequence of bradykinin?

23.21

Complete hydrolysis of a heptapeptide showed that it had the following molecular formula:

$$Ala_2,\ Glu,\ Leu,\ Lys,\ Phe,\ Val$$

Deduce the amino acid sequence of this heptapeptide from the following data.

1. Treatment of the heptapeptide with 2,4-dinitrofluorobenzene followed by incomplete hydrolysis gave, among other products: Val labeled at the α-amino group, lysine labeled at the ε-amino group, and a dipeptide, DNP—Val·Leu (DNP = 2,4-dinitrophenyl-).
2. Hydrolysis of the heptapeptide with carboxypeptidase gives an initial high concentration of alanine, followed by a rising concentration of glutamic acid.
3. Partial enzymatic hydrolysis of the heptapeptide gave a dipeptide (**A**) and a tripeptide (**B**).
   a. Treatment of **A** with 2,4-dinitrofluorobenzene followed by hydrolysis gave DNP-labeled leucine and lysine labeled only at the ε-amino group.
   b. Complete hydrolysis of **B** gave phenylalanine, glutamic acid, and alanine. When **B** was allowed to react with carboxypeptidase, the solution showed an initial high concentration of glutamic acid. Treatment of **B** with 2,4-dinitrofluorobenzene followed by hydrolysis gave labeled phenylalanine.

23.22

Synthetic polyglutamic acid exists as an α helix in solution at pH 2 – 3. When the pH of such a solution is gradually raised through the addition of base, a dramatic change in optical rotation takes place at pH 5. This change has been associated with the unfolding of the α helix and the formation of a random coil. What structural feature of polyglutamic acid, and what chemical change, can you suggest as an explanation of this transformation?

*23.23

Part of the evidence for restricted rotation about the carbon – nitrogen bond in a peptide linkage (see Section 23.9A) comes from proton nmr studies done with simple amides. For example, at room temperature and with the instrument operating at 60 MHz, the proton nmr spectrum of N,N-dimethylformamide $(CH_3)_2NCHO$, shows a doublet at $\delta 2.80$ (3H), a doublet at $\delta 2.95$ (3H), and a multiplet at $\delta 8.05$ (1H). When the spectrum is determined at lower magnetic field strength (i.e., with the instrument operating at 30 MHz), the doublets are found to have shifted so that the distance (in hertz) that separates one doublet from the other is smaller. When the temperature at which the spectrum is determined is raised, the doublets persist until a temperature of 111 °C is reached; then the doublets coalesce to become a single signal. Explain in detail how these observations are consistent with the existence of a relatively large barrier to rotation about the carbon – nitrogen bond of N,N-dimethylformamide.

CHAPTER TWENTY FOUR

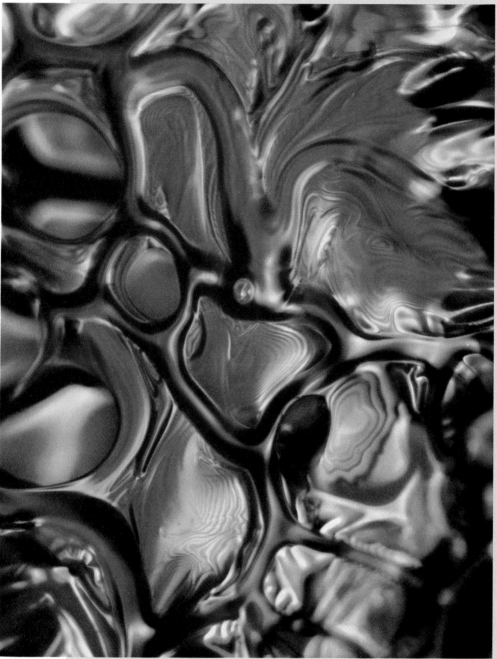

Neomycin sulfate crystals. Neomycin is an antibiotic produced by *Streptomyces fradiae*.

Nucleic Acids and Protein Synthesis

> . . . I cannot help wondering whether some day an enthusiastic scientist will christen his newborn twins Adenine and Thymine.
>
> F. H. C. Crick*

24.1 INTRODUCTION

The molecules that preserve hereditary information and that transcribe and translate that information in a way that allows the synthesis of all the varied enzymes of the cell are the nucleic acids, deoxyribonucleic acid (DNA) and ribonucleic acid (RNA). These biological polymers are sometimes found associated with proteins and in this form they are known as **nucleoproteins.**

It has been from studies of nucleic acids themselves that has come much of our knowledge of how genetic information is preserved, how it is passed on to succeeding generations of the organism, and how it is transformed into the working parts of the cell. For these reasons we shall focus our attention on the structures and properties of nucleic acids and of their components, **nucleotides** and **nucleosides.**

24.2 NUCLEOTIDES AND NUCLEOSIDES

Mild degradations of nucleic acids yield their monomeric units, compounds that are called **nucleotides.** A general formula for a nucleotide and the specific structure of one, called adenylic acid, are shown in Fig. 24.1.

Complete hydrolysis of a nucleotide furnishes:

1. A heterocyclic base, either a purine or pyrimidine.
2. A five-carbon monosaccharide, either D-ribose or 2-deoxy-D-ribose.
3. A phosphate ion.

The central portion of the nucleotide is the monosaccharide and it is always present as a five-membered ring, that is, as a furanoside. The heterocyclic base of a nucleotide is attached through an N-glycosidic linkage to C-1′ of the ribose or deoxy-ribose unit and this linkage is always β. The phosphate group of a nucleotide is present as a phosphate ester and it may be attached at C-5′ or C-3′. (In nucleotides the carbons of the monosaccharide portion are designated with primed numbers, i.e., 1′, 2′, 3′, and so on.)

---

*Who along with J. D. Watson and Maurice Wilkins shared the Nobel Prize in 1962 for their proposal of (and evidence for) the double helix structure of DNA. (Taken from F. H. C. Crick, "The Structure of the Hereditary Material," *Sci. Am.,* October 1954.)

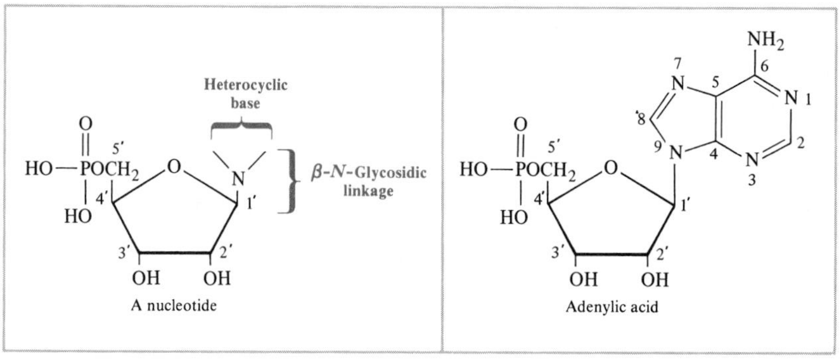

(a) (b)

FIGURE 24.1 (*a*) General structure of a nucleotide obtained from RNA. The heterocyclic base is a purine or pyrimidine. In nucleotides obtained from DNA, the sugar component is 2′-deoxyribose, that is, the —OH at position 2′ is replaced by —H. The phosphate group of the nucleotide is shown attached to the 5′-carbon; it may instead be attached to the 3′-carbon atom. The heterocyclic base is always attached through a β-glycosidic linkage at C-1′. (*b*) Adenylic acid, a typical nucleotide.

Removal of the phosphate group of a nucleotide converts it to a compound known as a *nucleoside* (Section 21.13A). The nucleosides that can be obtained from DNA all contain 2-deoxy-D-ribose as their sugar component and one of four heterocyclic bases, either adenine, guanine, cytosine, or thymine:

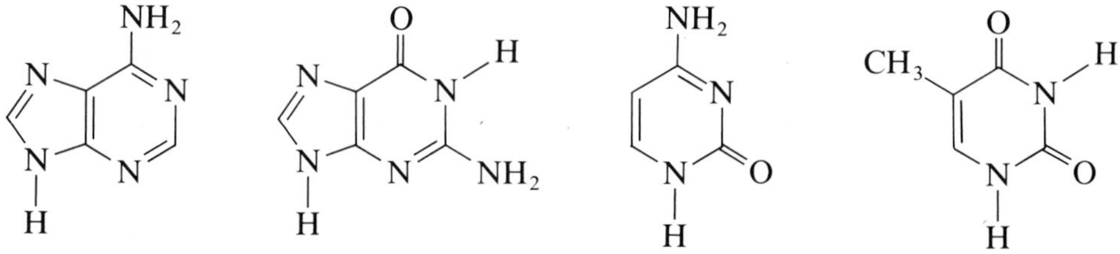

| **Adenine** | **Guanine** | **Cytosine** | **Thymine** |
| (A) | (G) | (C) | (T) |

◄──────── Purines ────────► ◄──────── Pyrimidines ────────►

The nucleosides obtained from RNA contain D-ribose as their sugar component and either adenine, guanine, cytosine, or uracil as their heterocyclic base.*

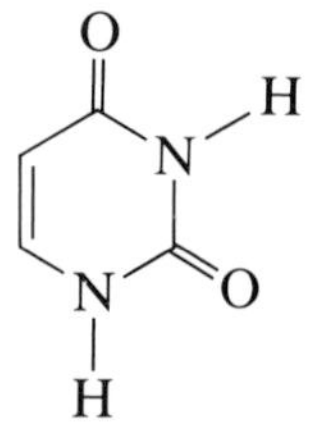

Uracil
(a pyrimidine)

The heterocyclic bases obtained from nucleosides are capable of existing in more than one tautomeric form. The forms that we have shown are the predominant forms that the bases assume when they are present in nucleic acids.

*Notice that in an RNA nucleoside (or nucleotide) uracil replaces thymine. (Some nucleosides obtained from specialized forms of RNA may also contain other, but similar, purines and pyrimidines.)

PROBLEM 24.1

Write the structures of other tautomeric forms of adenine, guanine, cytosine, thymine, and uracil.

The names and structures of the nucleosides found in DNA are shown in Fig. 24.2; those found in RNA are given in Fig. 24.3.

PROBLEM 24.2

The nucleosides shown in Figs. 24.2 and 24.3 are stable in dilute base. In dilute acid, however, they undergo rapid hydrolysis yielding a sugar (deoxyribose or ribose) and a heterocyclic base. (a) What structural feature of the nucleoside accounts for this behavior? (b) Propose a reasonable mechanism for the hydrolysis.

Nucleotides are named in several ways. Adenylic acid (Fig. 24.1), for example, is sometimes called 5'-adenylic acid in order to designate the position of the phosphate group; it is also called adenosine 5'-phosphate, or simply adenosine monophosphate (AMP). Uridylic acid is called 5'-uridylic acid, uridine 5'-phosphate, or uridine monophosphate (UMP), and so on.

Nucleosides and nucleotides are found in places other than as part of the structure of DNA and RNA. We have seen, for example, that adenosine units are part of the structures of two important coenzymes, NADH and coenzyme A (Special

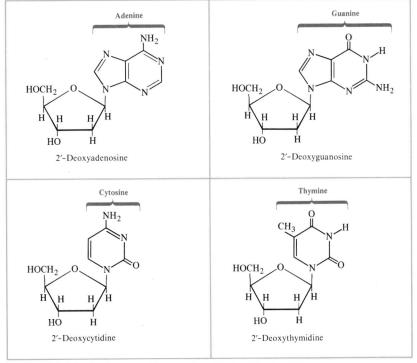

FIGURE 24.2 Nucleosides that can be obtained from DNA.

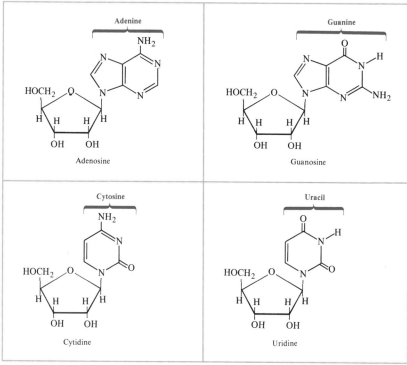

FIGURE 24.3 Nucleosides that can be obtained from RNA.

Topic M). The 5'-triphosphate of adenosine is, of course, the important energy source, ATP (Section 21.1B). The compound called 3',5'-cyclic adenylic acid (or cyclic AMP) (Fig. 24.4) is an important regulator of hormone activity. Cells synthesize this compound from ATP through the action of an enzyme, *adenyl cyclase*. In the laboratory, 3',5'-cyclic adenylic acid can be prepared through dehydration of 5'-adenylic acid with dicyclohexylcarbodimide.

PROBLEM 24.3

When 3',5'-cyclic adenylic acid is treated with aqueous sodium hydroxide, the major product that is obtained is 3'-adenylic (adenosine 3'-phosphate) rather than 5'-adenylic acid. Suggest an explanation that accounts for the course of this reaction.

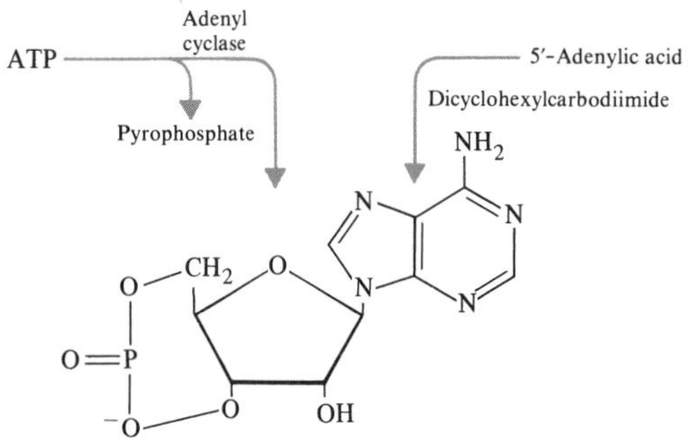

ATP ——— Adenyl cyclase ——— 5'-Adenylic acid

Pyrophosphate

Dicyclohexylcarbodiimide

FIGURE 24.4 3',5'-Cyclic adenylic acid and its biosynthesis and laboratory synthesis.

24.3 LABORATORY SYNTHESIS OF NUCLEOSIDES AND NUCLEOTIDES

A variety of methods have been developed for the synthesis of nucleosides. One technique uses reactions that assemble the nucleoside from suitably activated and protected ribose derivatives and heterocyclic bases. An example is the following synthesis of adenosine:

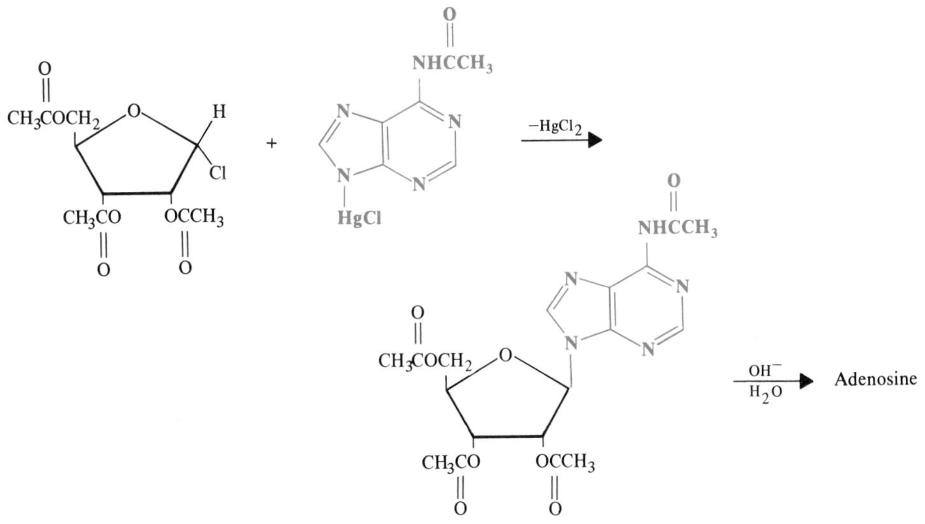

Another technique involves formation of the heterocyclic base on a protected ribosylamine derivative:

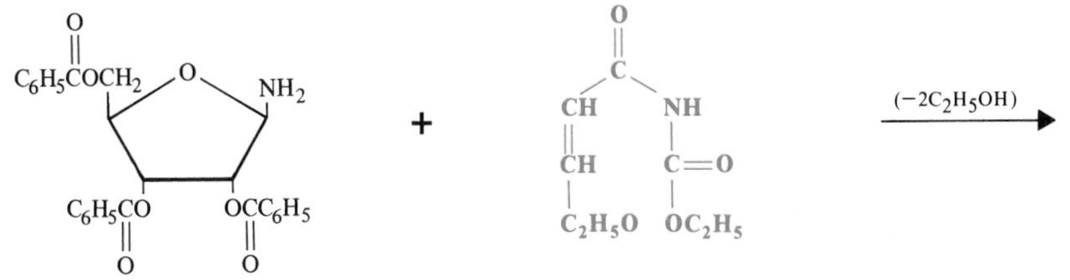

2, 3, 5-tri-*O*-Benzoyl-
β-D-ribofuranosylamine

β-Ethoxy-*N*-ethoxy-
carbonyllacrylamide

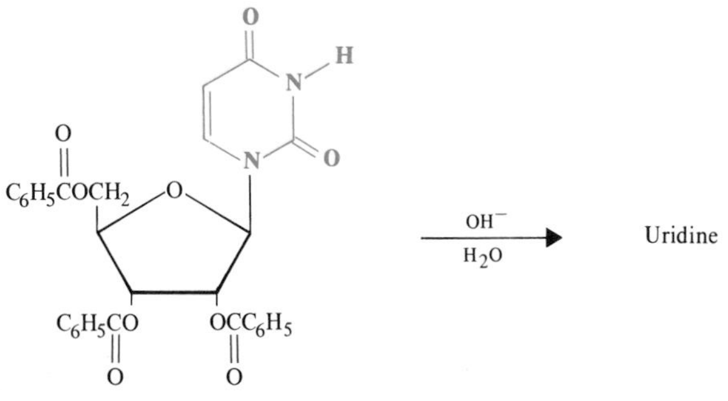

Uridine

Still a third technique involves the synthesis of a nucleoside with a substituent in the heterocyclic ring that can be replaced with other groups. This method has been used extensively to synthesize unusual nucleosides that do not necessarily occur naturally. The following example makes use of a 6-chloropurine derivative obtained from the appropriate ribofuranosyl chloride and chloromercuripurine.

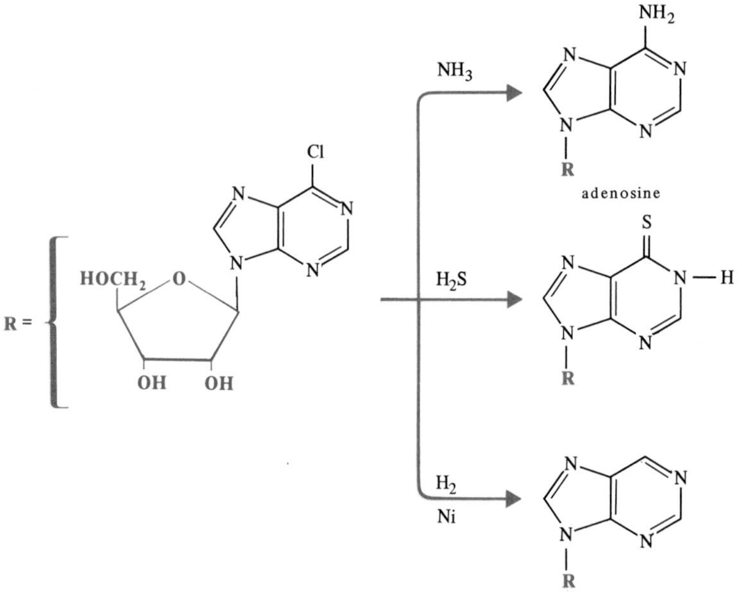

Numerous phosphorylating agents have been used to convert nucleosides to nucleotides. One of the most useful is dibenzyl phosphochloridate.

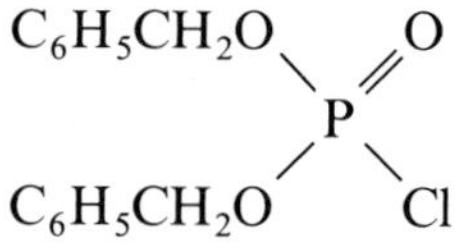

Dibenzyl phosphochloridate

Specific phosphorylation of the 5′-OH can be achieved if the 2′- and 3′-OH groups of the nucleoside are protected by an isopropylidine group (see following figure).

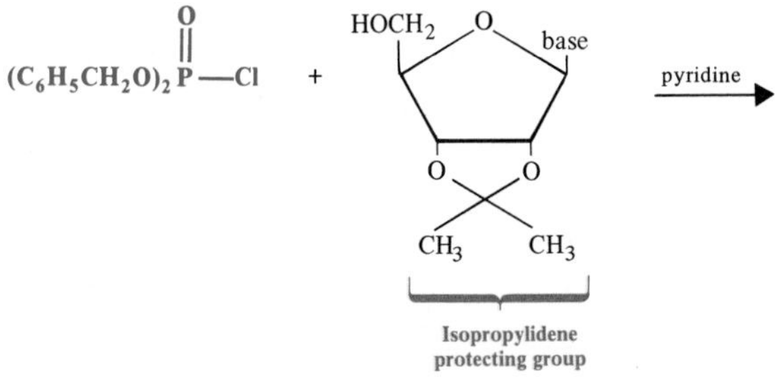

Isopropylidene
protecting group

Mild acid-catalyzed hydrolysis removes the isopropylidene group, and hydrogen-olysis cleaves the benzyl phosphate bonds.

PROBLEM 24.5

(a) What kind of linkage is involved in the isopropylidene-protected nucleoside and why is it susceptible to mild acid-catalyzed hydrolysis? (b) How might such a protecting group be installed?

24.4 DEOXYRIBONUCLEIC ACID: DNA

24.4A Primary Structure

Nucleotides bear the same relation to a nucleic acid that amino acids do to a protein; they are its monomeric units. The connecting links in proteins are amide groups; in nucleic acids they are phosphate ester linkages. Phosphate esters link the 3′-hydroxyl of one ribose (or deoxyribose) with the 5′-hydroxyl of another. This makes the nucleic acid a long unbranched chain with a "backbone" of sugar and phosphate units with heterocyclic bases protruding from the chain at regular intervals (Fig. 24.5).

It is, as we shall see, the **base sequence** along the chain of DNA that contains the encoded genetic information. The sequence of bases can be determined through techniques based on selective enzymatic hydrolyses. The actual base sequences have been worked out for a number of smaller nucleic acids.

24.4B Secondary Structure

It was the now-classic proposal of Watson and Crick (made in 1953 and verified shortly thereafter by the X-ray analysis of Wilkins) that gave a model for the secondary structure of DNA. The secondary structure of DNA is especially important because it enables us to understand how the genetic information is preserved, how it can be passed on during the process of cell division, and how it can be transcribed to provide a template for protein synthesis.

Of prime importance to Watson and Crick's proposal was an earlier observation (late 1940s) by E. Chargaff that certain regularities can be seen in the percentages of heterocyclic bases obtained from the DNA of a variety of species. Table 24.1 gives results that are typical of those that can be obtained.

Chargaff pointed out that for all species examined:

1. The total mole percentage of purines is approximately equal to that of the pyrimidines, that is, $(\%G + \%A)/(\%C + \%T) \simeq 1$.

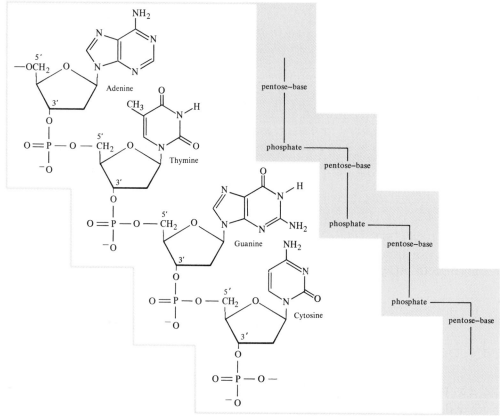

FIGURE 24.5 Hypothetical segment of a single DNA chain showing how phosphate ester groups link the 3′- and 5′-hydroxyl groups of deoxyribose units. RNA has a similar structure with two exceptions: A hydroxyl replaces a hydrogen atom at the 2′-position of each ribose unit and uracil replaces thymine.

2. The mole percentage of adenine is nearly equal to that of thymine (i.e., %A/%T ≃ 1) and that the mole percentage of guanine is nearly equal to that of cytosine (i.e., %G/%C ≃ 1).

Chargaff also noted that the ratio that varies from species to species is the ratio (%A + %T)/(%G + %C). He noted, moreover, that while this ratio is characteristic of the DNA of a given species, it is the same for DNA obtained from different tissues of the same animal, and does not vary appreciably with the age or conditions of growth of individual organisms within the same species.

Watson and Crick also had X-ray data that gave them the bond lengths and angles of the purines and pyrimidines of model compounds. In addition they had data from Wilkins that indicated an unusually long repeat distance, 34 Å, in natural DNA.

Reasoning from these data, Watson and Crick proposed a double helix as a model for the secondary structure of DNA. According to this model, two nucleic acid chains are held together by hydrogen bonds between base pairs on opposite strands. This double chain is wound into a helix with both chains sharing the same axis. The base pairs are on the inside of the helix and the sugar–phosphate backbone on the outside (Fig. 24.6). The pitch of the helix is such that 10 successive nucleotide pairs

TABLE 24.1 DNA composition of various species

	BASE PROPORTIONS (mole %)							
SPECIES	G	A	C	T	$\dfrac{G+A}{C+T}$	$\dfrac{A+T}{G+C}$	$\dfrac{A}{T}$	$\dfrac{G}{C}$
Sarcina lutea	37.1	13.4	37.1	12.4	1.02	0.35	1.08	1.00
Escherichia coli K12	24.9	26.0	25.2	23.9	1.08	1.00	1.09	0.99
Wheat germ	22.7	27.3	22.8[a]	27.1	1.00	1.19	1.01	1.00
Bovine thymus	21.5	28.2	22.5[a]	27.8	0.96	1.27	1.01	0.96
Staphylococcus aureus	21.0	30.8	19.0	29.2	1.11	1.50	1.05	1.11
Human thymus	19.9	30.9	19.8	29.4	1.01	1.52	1.05	1.01
Human liver	19.5	30.3	19.9	30.3	0.98	1.54	1.00	0.98

[a] Cytosine + methylcytosine.
From *Principles of Biochemistry* by A. White, P. Handler, and E. L. Smith. Copyright © 1964 by McGraw-Hill, Inc. Used with permission of McGraw-Hill Book Company, New York.

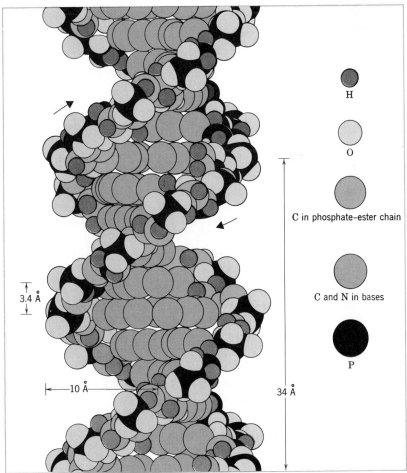

3.4 Å

10 Å

34 Å

H

O

C in phosphate–ester chain

C and N in bases

P

FIGURE 24.6 A molecular model of a portion of the DNA double helix. (Adapted from *Chemistry and Biochemistry: A Comprehensive Introduction* by A. L. Neal. Copyright © 1971 by McGraw-Hill Inc. Used with permission of McGraw-Hill Book Company, New York.)

give rise to one complete turn in 34 Å (the repeat distance). The exterior width of the spiral is ⁓20 Å and the internal distance between 1'-positions of ribose units on opposite chains is ⁓11 Å.

Using molecular scale models Watson and Crick observed that the internal distance of the double helix is such that it allows only a purine–pyrimidine type of hydrogen bonding between base pairs. Purine–purine base pairs do not occur because they would be too large to fit, and pyrimidine–pyrimidine base pairs do not occur because they would be too far apart to form effective hydrogen bonds.

Watson and Crick went one crucial step further in their proposal. Assuming that the oxygen-containing heterocyclic bases existed in keto forms, they argued that base pairing through hydrogen bonds can occur in only a specific way:

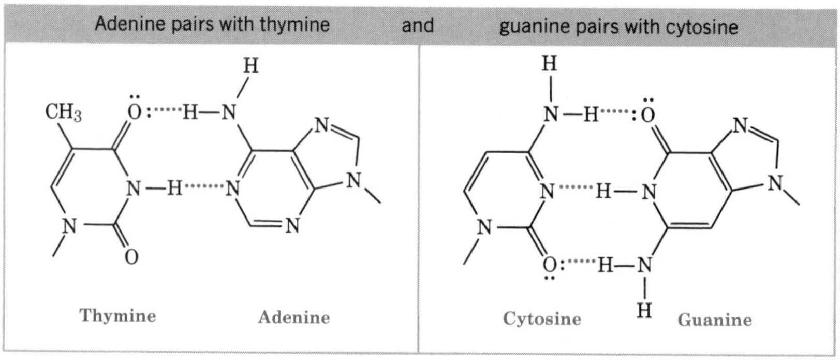

Adenine pairs with thymine   and   guanine pairs with cytosine

Thymine   Adenine   Cytosine   Guanine

The bond lengths of these base pairs are shown in Fig. 24.7.

Specific base pairing of this kind is consistent with Chargaff's finding that %A/%T ≃ 1 and that %G/%C ≃ 1.

Specific base pairing also means that the two chains of DNA are complementary. Wherever adenine appears in one chain, thymine must appear opposite it in the other; wherever cytosine appears in one chain, guanine must appear in the other (Fig. 24.8).

Notice that while the sugar–phosphate backbone of DNA is completely regular, the sequence of heterocyclic base pairs along the backbone can assume many different permutations. This is important because it is the precise sequence of base pairs that carries the genetic information. Notice, too, that one chain of the double strand is the complement of the other. By knowing the sequence of bases along one chain, one could write down the sequence along the other, because A always pairs with T and G always pairs with C. It is this complementarity of the two strands that explains how a DNA molecule replicates itself at the time of cell division and thereby passes on the genetic information to each of the two daughter cells.

24.4C Replication of DNA

Just prior to cell division the double strand of DNA begins to unwind at one end. Complementary strands are formed along each chain (Fig. 24.9). Each chain acts, in effect, as a template for the formation of its complement. When unwinding and duplication is complete there are two identical DNA molecules where only one had existed before. These two molecules can then be passed on, one to each daughter cell.

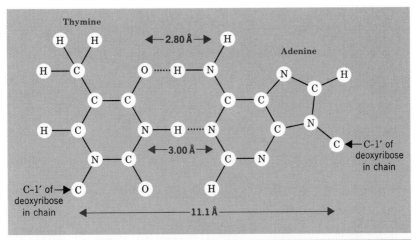

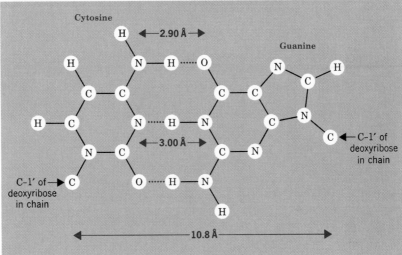

FIGURE 24.7 Dimensions of thymine–adenine and cytosine–guanine base pairs. The dimensions are such that they allow the formation of strong hydrogen bonds and also allow the base pairs to fit inside the two phosphate–ribose chains of the double helix. [Adapted from L. Pauling and R. B. Corey, *Arch. Biochem. Biophys.*, **65**, 164 (1956).]

PROBLEM 24.6

(a) There are approximately 6 billion base pairs in the DNA of a single human cell. Assuming that this DNA exists as a double helix, calculate the length of all the DNA contained in a human cell.   (b) The weight of DNA in a single human cell is 6×10^{-12} g. Assuming that the earth's population is about 3 billion, we can conclude that all of the genetic information that gave rise to all human beings now alive was once contained in the DNA of a corresponding number of fertilized ova. What is the total weight of this DNA? (The volume that this DNA would occupy is approximately that of a raindrop, yet if the individual molecules were laid end to end they would stretch to the moon and back almost eight times.)

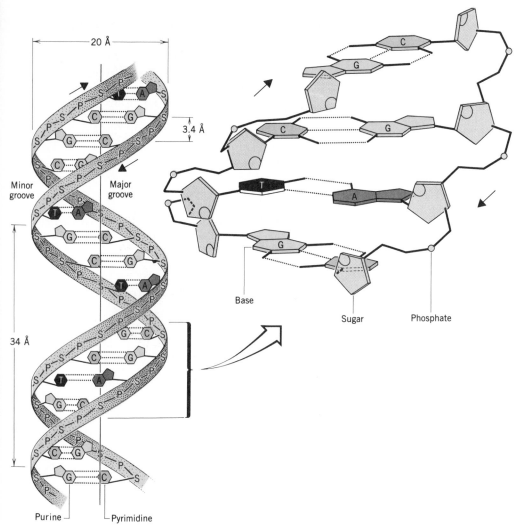

FIGURE 24.8 Diagram of the DNA double helix showing complementary base pairing.

PROBLEM 24.7

(a) The most stable tautomeric form of guanine is the lactam form. This is the form normally present in DNA and, as we have seen, it pairs specifically with cytosine. If guanine tautomerizes to the abnormal lactim form, it pairs with thymine instead. Write structural formulas showing the hydrogen bonds in this abnormal base pair.

(b) Improper base pairings that result from tautomerizations occurring during the process of DNA replication have been suggested as a source of spontaneous mutations. We saw in part (a) that if a tautomerization of guanine occurred at the proper moment it could lead to the introduction of thymine (instead of cytosine) into its complementary DNA chain. What error would this new DNA chain introduce into *its* complementary strand during the next replication even if no further tautomerizations take place?

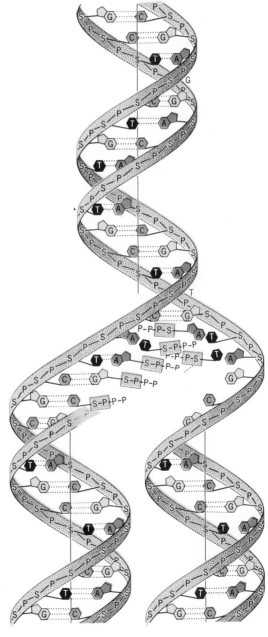

FIGURE 24.9 Replication of DNA. The double strand unwinds from one end and complementary strands are formed along each chain.

PROBLEM 24.8

Mutations can also be caused chemically, and nitrous acid is one of the most potent chemical **mutagens.** One explanation that has been suggested for the mutagenic effect of nitrous acid is the deamination reactions that it produces with purines and pyrimidines bearing amino groups. When, for example, an adenine-containing nucleotide is treated with nitrous acid, it is converted to a hypoxanthine derivative:

| Adenine | Hypoxanthine |
| nucleotide | nucleotide |

(a) Basing your answer on reactions you have seen before, what are likely intermediates in the adenine ⟶ hypoxanthine interconversion? (b) Adenine normally pairs with thymine in DNA, but hypoxanthine pairs with cytosine. Show the hydrogen bonds of a hypoxanthine–cytosine base pair. (c) Show what errors an adenine ⟶ hypoxanthine interconversion would generate in DNA through two replications.

24.5 RNA AND PROTEIN SYNTHESIS

Soon after the Watson–Crick hypothesis was published, scientists began to extend it to yield what Crick has called "the central dogma of molecular genetics." This dogma states that genetic information flows from:

$$\text{DNA} \longrightarrow \text{RNA} \longrightarrow \text{proteins*}$$

The synthesis of protein is, of course, all important to a cell's function because proteins (as enzymes) catalyze all its reactions. Even the very primitive cells of bacteria require as many as 3000 different enzymes. This means that the DNA molecules of these cells must contain a corresponding number of genes to direct the synthesis of these proteins. A gene is that segment of the DNA molecule that contains the information necessary to direct the synthesis of one protein (or one polypeptide).

DNA is found primarily in the nucleus of the cell. Protein synthesis takes place primarily in that part of the cell called the *cytoplasm.* Protein synthesis requires that two major processes take place; the first takes place in the cell nucleus, the second in the cytoplasm. The first is **transcription,** a process in which the genetic message is transcribed on to a form of RNA called messenger RNA (mRNA). The second process involves two other forms of RNA, called ribosomal RNA (rRNA) and transfer RNA (tRNA).

---

*Recent evidence has shown that in certain tumor viruses, information also flows back from RNA to DNA.

24.5A Messenger RNA Synthesis — Transcription

Protein synthesis begins in the cell nucleus with the synthesis of messenger RNA. Part of the DNA double helix unwinds sufficiently to expose on a single chain a portion corresponding to at least one gene. Ribonucleotides, present in the cell nucleus, assemble along the exposed DNA chain pairing with the bases of DNA. The pairing patterns are the same as those in DNA with the exception that in RNA uracil replaces thymine. The ribonucleotide units of messenger RNA are joined into a chain by an enzyme called *RNA polymerase.* This process is illustrated in Fig. 24.10.

> **PROBLEM 24.9**
>
> Write structural formulas showing how the keto form of uracil (Section 24.2) in messenger RNA can pair with adenine in DNA through hydrogen bond formation.

After messenger RNA has been synthesized in the cell nucleus, it migrates into the cytoplasm where, as we shall see, it acts as a template for protein synthesis.

24.5B Ribosomes — rRNA

Scattered throughout the cytoplasm of most cells are small bodies called ribosomes. Ribosomes of *Escherichia coli (E. coli),* for example, are about 180 Å in diameter and are composed of approximately 60% RNA (ribosomal RNA) and 40% protein. They apparently exist as two associated subunits called the 50S and 30S subunits (Fig. 24.11); together they form a 70S ribosome.* Although the ribosomes are at the site of

*S stands for svedberg unit; it is used in describing the behavior of proteins in an ultracentrifuge.

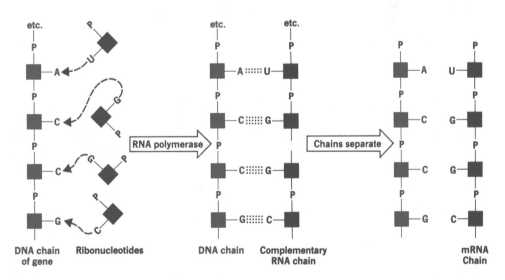

P = Phosphate ester linkage   A = Adenine   U = Uracil
■ = Deoxyribose   C = Cytosine
■ = Ribose   G = Guanine

FIGURE 24.10 Transcription of the genetic code from DNA to messenger RNA.

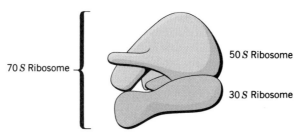

FIGURE 24.11 A 70S ribosome showing the two subunits.

protein synthesis, ribosomal RNA itself does not direct protein synthesis. Instead, a number of ribosomes become attached to a chain of mRNA and form what is called a **polysome**. It is along the polysome—with mRNA acting as the template—that protein synthesis takes place. One of the functions of ribosomal RNA is to bind the ribosome to the messenger RNA chain.

24.5C Transfer RNA

Transfer RNA has a very low molecular weight when compared to that of messenger RNA or ribosomal RNA. Transfer RNA, consequently, is much more soluble than mRNA or rRNA and is sometimes referred to as soluble RNA. The function of tRNA is to transport amino acids to specific areas of the mRNA of the polysome. There are, therefore, at least 20 different forms of tRNA, one for each of the 20 amino acids that are incorporated into proteins.*

The structures of most tRNAs have been determined. They are composed of a relatively small number of nucleotide units (70–90 units) folded into several loops or arms through base pairing along the chain (Fig. 24.12). One arm always terminates in the sequence cytosine–cytosine–adenine. It is to this arm that a specific amino acid becomes attached *through an ester* linkage to the 3′-OH of the terminal adenosine. This attachment reaction is catalyzed by an enzyme that is specific for the tRNA and for the amino acid. The specificity may grow out of the enzyme's ability to recognize base sequences along other arms of the transfer RNA.

At the loop of still another arm is a specific sequence of bases, called the **anticodon**. The anticodon is highly important because it allows the tRNA to bind with a specific site—called the **codon**—of mRNA. The order in which amino acids are brought by their tRNA units to the mRNA strand is therefore determined by the sequence of codons. This sequence, therefore, constitutes a genetic message. Individual units of that message (the individual words, each corresponding to an amino acid) are triplets of nucleotides.

24.5D The Genetic Code

Which triplet on mRNA corresponds to which amino acid is called the genetic code (see Table 24.2). The code must be in the form of three bases, not one or two because there are 20 different amino acids used in protein synthesis but there are only four different bases in mRNA. If only two bases were used, there would be only 4^2 or 16 possible combinations, a number too small to accommodate all of the possible amino

*Although proteins are composed of 22 different amino acids, protein synthesis requires only 20. Proline is converted to hydroxyproline and cysteine is converted to cystine after synthesis of the polypeptide chain has taken place.

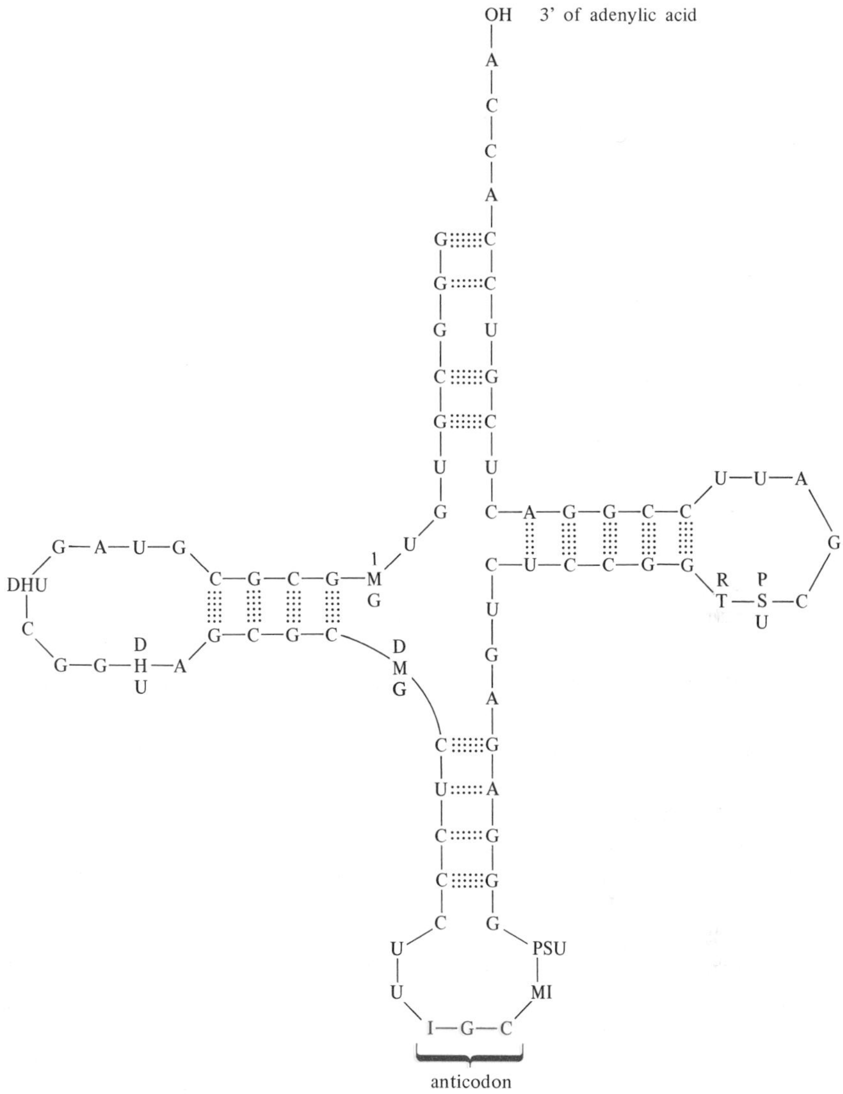

FIGURE 24.12 Structure of a transfer RNA isolated from yeast that has the specific function of transferring alanine residues. Transfer RNAs often contain unusual nucleosides. PSU = pseudouridine, RT = ribothymidine, MI = 1-methylinosine, I = inosine, DMG = N^2-methylguanosine, DHU = 4,5-dihydrouridine, 1 MG = 1-methylguanosine.

acids. However, with a three-base code, 4^3 or 64 different sequences are possible. This is far more than are needed, and it allows for multiple ways of specifying an amino acid. It also allows for sequences that punctuate protein synthesis, sequences that say, in effect, "start here" and "end here."

Both methionine (Met) and N-formylmethionine (Met$_{formyl}$) have the same messenger RNA code (AUG); however, N-formylmethionine is carried by a different transfer RNA from that which carries methionine. N-Formylmethionine appears to be the first amino acid incorporated into the chain of proteins in bacteria, and the transfer RNA that carries Met$_{formyl}$ appears to be the punctuation mark that says "start here." Before the polypeptide synthesis is complete, N-formylmethionine is removed from the protein chain by an enzymatic hydrolysis.

TABLE 24.2 The messenger RNA genetic code

AMINO ACID	BASE SEQUENCE	AMINO ACID	BASE SEQUENCE	AMINO ACID	BASE SEQUENCE
Ala	GCA	His	CAC	Ser	AGC
	GCC		CAU		AGU
	GCG				UCA
	GCU	Ile	AUA		UCG
			AUC		UCC
Arg	AGA		AUU		UCU
	AGG				
	CGA	Leu	CUA	Thr	ACA
	CGC		CUC		ACC
	CGG		CUG		ACG
	CGU		CUU		ACU
			UUA	Trp	UGG
Asn	AAC		UUG	Tyr	UGG
	AAU				UAC
		Lys	AAA		UAU
Asp	GAC		AAG		
	GAU			Val	GUA
					GUG
Cys	UGC	Met	AUG		GUC
	UGU				GUU
		Phe	UUU		
			UUC		
Gln	CAA			Chain initiation	
	CAG	Pro	CCA		
			CCC	Met$_{formyl}$	AUG
Glu	GAA		CCG		
	GAG		CCU		
Gly	GGA				
	GGC			Chain termination	UAA
	GGG				UAG
	GGU				

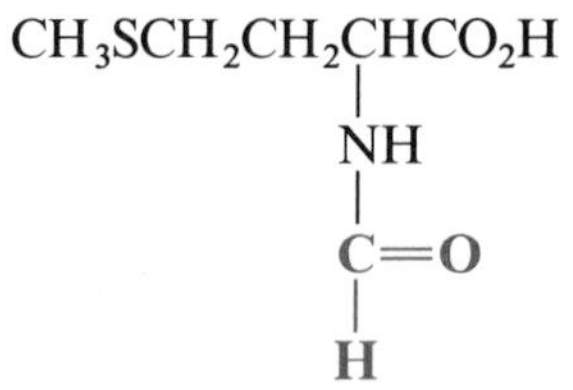

$$CH_3SCH_2CH_2CHCO_2H$$
$$|$$
$$NH$$
$$|$$
$$C{=}O$$
$$|$$
$$H$$

N-Formylmethionine

We are now in a position to see how the synthesis of a hypothetical polypeptide might take place. Let us imagine that a long strand of messenger RNA is in the cytoplasm of a cell and that it is in contact with ribosomes. Also in the cytoplasm are the 20 different amino acids, each acylated to its own specific transfer RNA.

As shown in Fig. 24.13, a transfer RNA bearing Met$_{formyl}$ uses its anticodon to associate with the proper codon (AUG) on that portion of messenger RNA that is in contact with a ribosome. The next triplet of bases on this particular messenger RNA

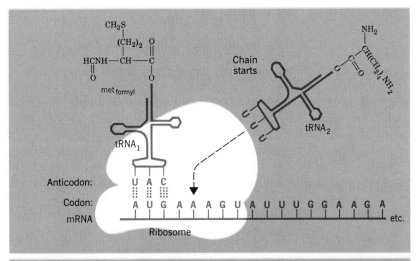

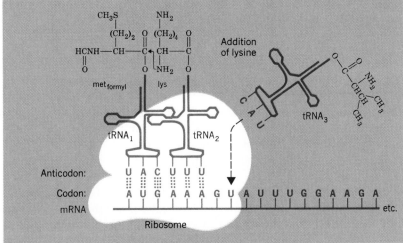

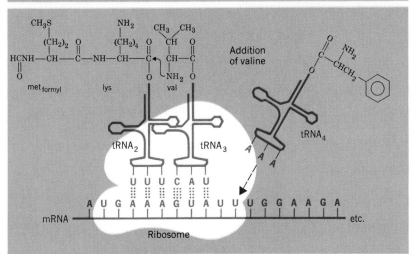

FIGURE 24.13 Step-by-step growth of a polypeptide chain with messenger RNA acting as a template. Transfer RNAs carry amino acid residues to the site of mRNA that is in contact with a ribosome. Codon–anticodon pairing occurs between mRNA and RNA at the ribosomal surface. An enzymatic reaction joins the amino acid residues through an amide linkage. After the first amide bond is formed the ribosome moves to the next codon on mRNA. A new tRNA arrives, pairs, and transfers its amino acid residue to the growing peptide chain, and so on.

chain is AAA; this is the codon that specifies lysine. A lysyl-transfer RNA with the matching anticodon UUU attaches itself to this site. The two amino acids, Met$_{formyl}$ and Lys, are now in the proper position for an enzyme to join them in peptide linkage. After this happens, the ribosome moves down the chain so that it is in contact with the next codon. This one, GUA, specifies valine. A transfer RNA bearing valine (and with the proper anticodon) binds itself to this site. Another enzymatic reaction takes place attaching valine to the polypeptide chain. Then the whole process repeats itself again and again. The ribosome moves along the messenger RNA chain, other transfer RNAs move up with their amino acids, new peptide bonds are formed, and the polypeptide chain grows. At some point an enzymatic reaction removes Met$_{formyl}$ from the beginning of the chain. Finally, when the chain is the proper length the ribosome reaches a punctuation mark, UAA, saying "stop here." The ribosome separates from the messenger RNA chain and so, too, does the protein.

Even before the polypeptide chain is fully grown, it begins to form its own specific secondary and tertiary structure (Fig. 24.14). This happens because its primary structure is correct—its amino acids are ordered in just the right way. Hydrogen bonds form, giving rise to specific segments of α helix, pleated sheet, and random coil. Then the whole thing folds and bends; enzymes install disulfide linkages, so that when the chain is fully grown, the whole protein has just the shape it needs to do its job.

If this protein happens to be lysozyme it has a deep cleft, or jaw, where a specific polysaccharide fits. And if it is lysozyme, and a certain bacterium wanders by, that jaw begins to work; it bites its first polysaccharide in half.

In the meantime other ribosomes nearer the beginning of the messenger RNA chain are already moving along, each one synthesizing another molecule of the polypeptide. The time required to synthesize a protein depends, of course, on the number of amino residues it contains, but indications are that each ribosome can cause 150 peptide bonds to be formed each minute. Thus, a protein, such as lysozyme, with 129 amino acid residues requires less than a minute for its synthesis. However, if four ribosomes are working their way along a single messenger RNA chain, the polysome can produce a lysozyme molecule every 13 s.

But why, we might ask, is all this protein synthesis necessary—particularly in a fully grown organism? The answer is that proteins are not permanent; they are not synthesized once and then left intact in the cell for the lifetime of the organism. They are synthesized when and where they are needed. Then they are taken apart, back to amino acids; enzymes disassemble enzymes. Some amino acids are metabolized for energy; others—new ones—come in from the food that is eaten and the whole process begins again.

PROBLEM 24.10

A segment of DNA has the following sequence of bases:

$$...\ A\ C\ C\ C\ C\ A\ A\ A\ A\ T\ G\ T\ C\ G\ ...$$

(a) What sequence of bases would apear in messenger RNA transcribed from this segment? (b) Assume that the first base in this mRNA is the beginning of a codon. What order of amino acids would be translated into a polypeptide synthesized along this segment? (c) Give anticodons for each transfer RNA associated with the translation in part (b).

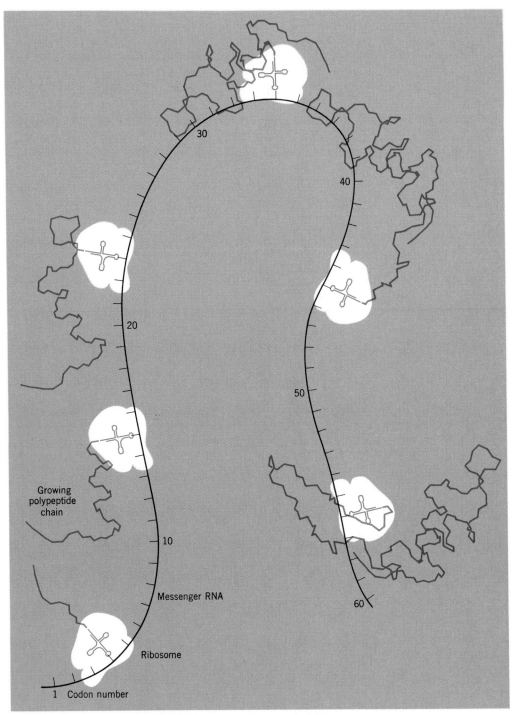

30

40

20

50

Growing
polypeptide
chain

10

Messenger RNA

60

Ribosome

1 Codon number

FIGURE 24.14 The folding of a protein molecule as it is synthesized. [Adapted with permission from D. C. Phillips, "The Three-Dimensional Structure of an Enzyme Molecule." Copyright © 1966 Scientific American, Inc. All rights reserved.]

PROBLEM 24.11

(a) Using the first codon given for each amino acid in Table 24.2, write the base sequence of messenger RNA that would translate the synthesis of the following pentapeptide:

Arg · Ile · Cys · Tyr · Val

(b) What base sequence in DNA would transcribe a synthesis of the messenger RNA? (c) What anticodons would appear in the transfer RNAs involved in the pentapeptide synthesis?

PROBLEM 24.12

Explain how an error of a single base in each strand of DNA could bring about the amino acid residue error that causes sickle cell anemia (Section 23.7C).

REACTIONS CONTROLLED BY ORBITAL SYMMETRY

O.1 Introduction

In recent years chemists have found that there are many reactions where certain symmetry characteristics of molecular orbitals control the overall course of the reaction. These reactions are often called *pericyclic reactions* because they take place through cyclic transition states. Now that we have a background knowledge of molecular orbital theory — especially as it applies to conjugated polyenes (dienes, trienes, etc.) — we are in a position to examine some of the intriguing aspects of these reactions. We shall look in detail at two basic types: *electrocyclic reactions* and *cycloaddition reactions.*

O.2 Electrocyclic Reactions

A number of reactions, like the one shown here, transform a conjugated polyene into a cyclic compound.

1,3-Butadiene   Cyclobutene

In many other reactions the ring of a cyclic compound opens and a conjugated polyene forms.

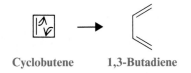

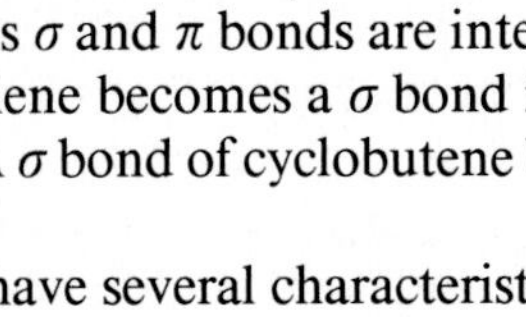

Cyclobutene   1,3-Butadiene

Reactions of either type are called *electrocyclic reactions.*

In electrocyclic reactions σ and π bonds are interconverted. In our first example, one π bond of 1,3-butadiene becomes a σ bond in cyclobutene. In our second example, the reverse is true: A σ bond of cyclobutene becomes a π bond in 1,3-butadiene.

Electrocyclic reactions have several characteristic features:

1. They require only heat or light for initiation.
2. Their mechanisms do not involve free radical or ionic intermediates.
3. Bonds are made and broken in *a single concerted step involving a cyclic transition state.*
4. The reactions are *highly stereospecific.*

The examples that follow demonstrate this last characteristic of electrocyclic reactions.

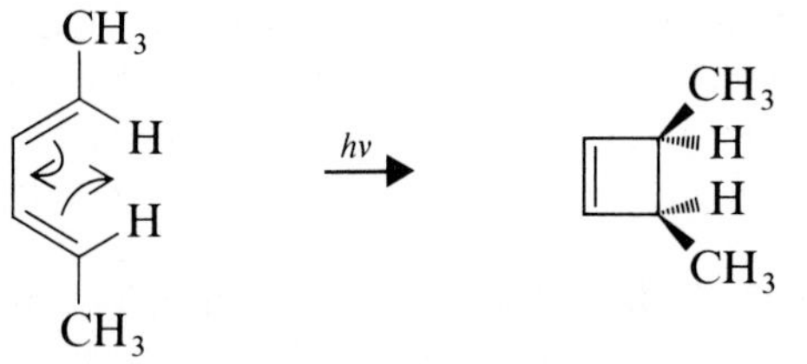

trans,trans-2,4-Hexadiene   *cis*-3,4-Dimethylcyclobutene

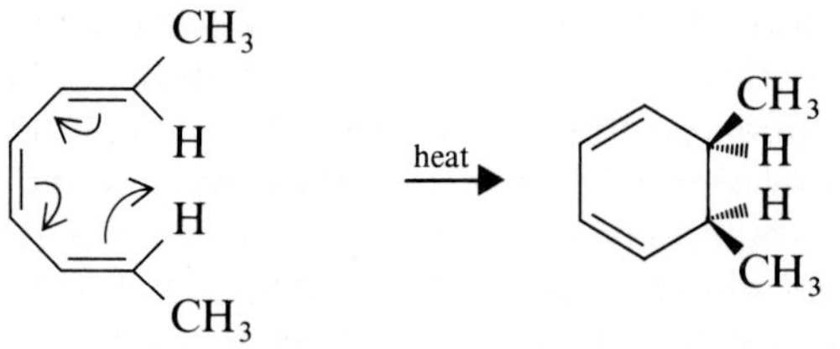

trans,cis,trans-2,4,6-Octatriene   *cis*-5,6-Dimethyl-1,3-
cyclohexadiene

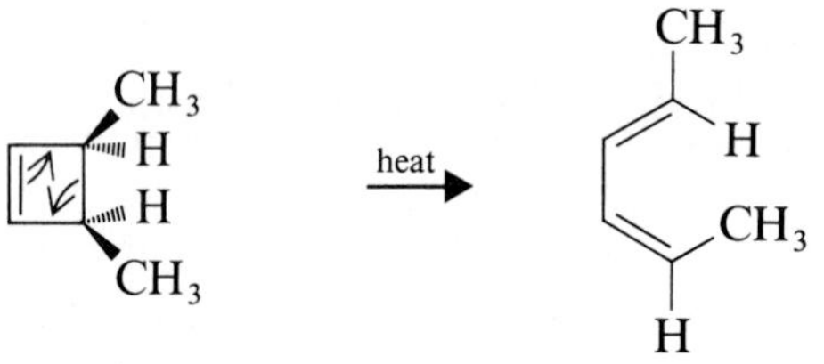

cis-3,4-Dimethylcyclobutene   *cis,trans*-2,4-Hexadiene

In each of these three examples a single stereoisomeric form of the reactant yields a single stereoisomeric form of the product. The concerted photochemical cyclization of *trans,trans*-2,4-hexadiene, for example, yields only *cis*-3,4-dimethylcyclobutene; it does not yield *trans*-3,4-dimethylcyclobutene.

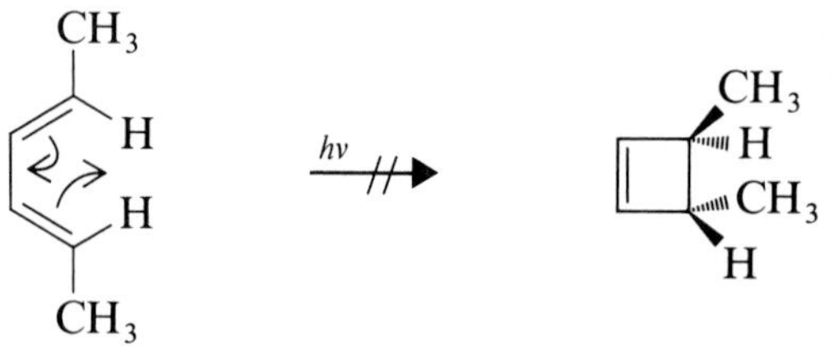

(not formed)

trans,trans-2,4-Hexadiene   *trans*-3,4-Dimethylcyclobutene

The other two concerted reactions are characterized by the same stereospecificity.

The electrocyclic reactions that we shall study here and the concerted cyclo-addition reactions that we shall study in the next section were poorly understood by chemists before 1960. In the years that followed, several scientists, most notably K. Fukui in Japan, H. C. Longuet-Higgins in England, and R. B. Woodward and R. Hoffmann in the United States provided us with a basis for understanding how these reactions occur and why they take place with such remarkable stereospecificity.

All of these men worked from molecular orbital theory. In 1965, Woodward and Hoffmann formulated their theoretical insights into a set of rules that not only enabled chemists to understand reactions that were already known but that correctly predicted the outcome of many reactions that had not been attempted.

The Woodward–Hoffmann rules are formulated for concerted reactions only. Concerted reactions are reactions in which bonds are broken and formed simultaneously and thus no intermediates occur. The Woodward–Hoffmann rules are based on this hypothesis: *In concerted reactions molecular orbitals of the reactant are continuously converted into molecular orbitals of the product.* This conversion of molecular orbitals is not a random one, however. Molecular orbitals have symmetry characteristics. Because they do, restrictions exist on which molecular orbitals of the reactant may be transformed into particular molecular orbitals of the product.

According to Woodward and Hoffmann, certain reaction paths are said to be *symmetry allowed* while others are said to be *symmetry forbidden.* To say that a particular path is symmetry forbidden does not necessarily mean, however, that the reaction will not occur. It simply means that if the reaction were to occur through a symmetry-forbidden path, the concerted reaction would have a much higher energy of activation. The reaction may occur, but it will probably do so in a different way: Through another path that is symmetry allowed or through a nonconcerted path.

A complete analysis of electrocyclic reactions using the Woodward–Hoffmann rules requires a correlation of symmetry characteristics of *all* of the molecular orbitals of the reactants and product. Such analyses are beyond the scope of our discussion here. We shall find, however, that a simplified approach can be undertaken, one that will be easy to visualize and, at the same time, will be accurate in most instances. In this simplified approach to electrocyclic reactions we focus our attention only on the *highest occupied molecular orbital (HOMO) of the conjugated polyene.* This approach is based on a method developed by Fukui called the *frontier orbital method.*

O.2A   Electrocyclic Reactions of 4n π-Electron Systems
Let us begin with an analysis of the thermal interconversion of *cis*-3,4-dimethylcyclobutene and *cis,trans*-2,4-hexadiene shown here.

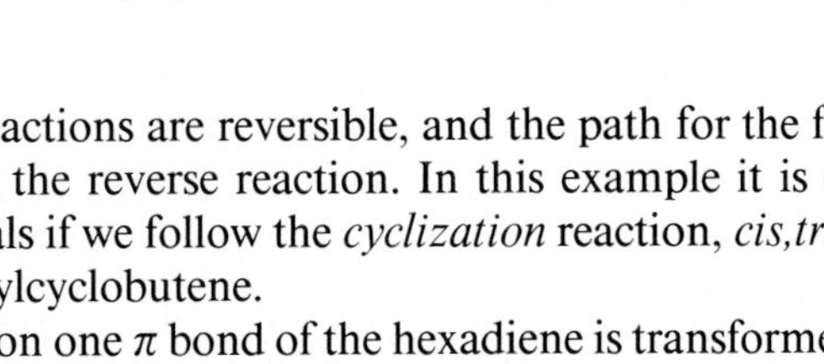

cis-3,4-Dimethylcyclobutene   *cis,trans*-2,4-Hexadiene

Electrocyclic reactions are reversible, and the path for the forward reaction is the same as that for the reverse reaction. In this example it is easier to see what happens to the orbitals if we follow the *cyclization* reaction, *cis,trans*-2,4-hexadiene ⟶ *cis*-3,4-dimethylcyclobutene.

In this cyclization one π bond of the hexadiene is transformed into a σ bond of the cyclobutene. But which π bond? And, how does the conversion occur?

Let us begin by examining the π molecular orbitals of 2,4-hexadiene and, in particular, let us look at *the highest occupied molecular orbital of the ground state* (Fig. O.1).

The cyclization that we are concerned with now, *cis,trans*-2,4-hexadiene ⇌ *cis*-3,4-dimethylcyclobutene, requires heat alone. We conclude, therefore, that excited states of the hexadiene are not involved, for these would require the absorption of light. If we focus our attention on Ψ_2 — the highest occupied molecular

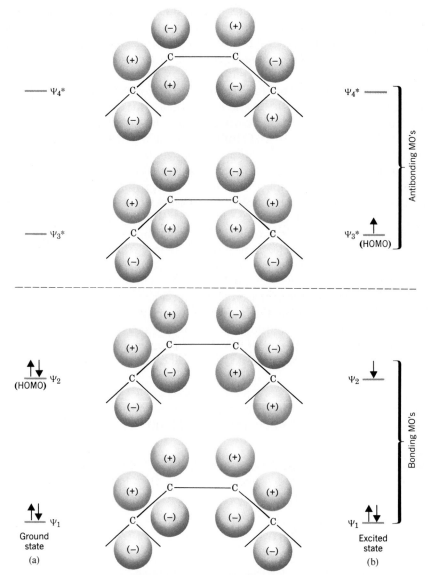

FIGURE O.1 The π molecular orbitals of a 2,4-hexadiene. (a) The electron distribution of the ground state. (b) The electronic distribution of the first excited state. (The first excited state is formed when the molecule absorbs a photon of light of the proper wavelength.) Notice that the orbitals of a 2,4-hexadiene are like those of 1,3-butadiene shown in Fig. 10.5.

orbital of the ground state — we can see how the p orbitals at carbon-2 and carbon-5 can be transformed into a σ bond in the cyclobutene.

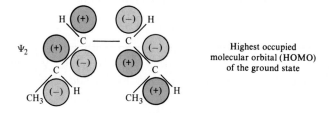

A bonding σ-molecular orbital between C-2 and C-5 is formed when the *p* orbitals *rotate in the same direction* (both clockwise, as shown, or both counterclockwise, which leads to an equivalent result). The term *conrotatory* is used to describe this type of motion of the two *p* orbitals relative to each other.

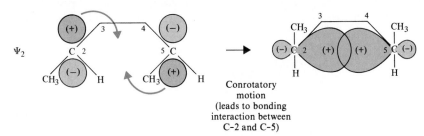

Conrotatory
motion
(leads to bonding
interaction between
C-2 and C-5)

Conrotatory motion allows *p*-orbital lobes of the *same phase sign* to overlap. It also places the two methyl groups on the same side of the molecule in the product, that is, in the cis configuration.*

The pathway with conrotatory motion of the methyl groups is consistent with what we know from experiments to be true: The *thermal reaction* results in the interconversion of *cis*-3,4-dimethylcyclobutene and *cis,trans*-2,4-hexadiene.

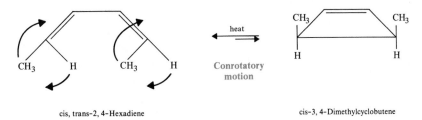

heat

Conrotatory
motion

cis, trans-2, 4-Hexadiene

cis-3, 4-Dimethylcyclobutene

We can now examine another 2,4-hexadiene ⇌ 3,4-dimethylcyclobutene interconversion: one that takes place under the influence of light. This reaction is shown here.

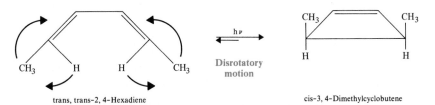

hν

Disrotatory
motion

trans, trans-2, 4-Hexadiene

cis-3, 4-Dimethylcyclobutene

In the photochemical reaction *cis*-3,4-dimethylcyclobutene and *trans,trans*-2,4-hexadiene are interconverted. The photochemical interconversion occurs with the

---

*Notice that if conrotatory motion occurs in the opposite (counterclockwise) direction, lobes of the same phase sign still overlap, and the methyl groups are still cis.

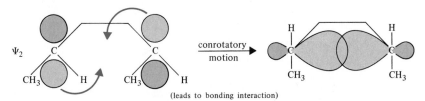

conrotatory
motion

(leads to bonding interaction)

methyl groups rotating in *opposite directions,* that is, with the methyl groups undergoing *disrotatory motion.*

The photochemical reaction can also be understood by considering orbitals of the 2,4-hexadiene. In this reaction, however—since the absorption of light is involved—we want to look at the first *excited state* of the hexadiene. We want to examine Ψ_3^*, because in the first excited state Ψ_3^* *is the highest occupied molecular orbital.*

Highest occupied
molecular orbital of
the first excited state

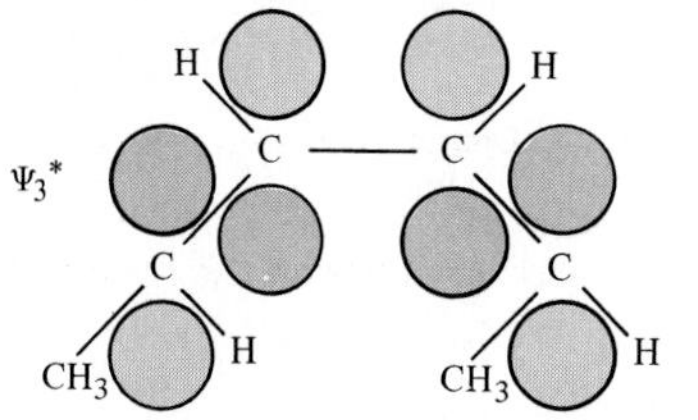

We find that disrotatory motion of the orbitals at carbons-2 and -5 of Ψ_3^* allows lobes of the same sign to overlap and form a bonding sigma molecular orbital between them. Disrotatory motion of the orbitals, of course, also requires disrotatory motion of the methyl groups and, once again, this is consistent with what we find experimentally. The *photochemical reaction* results in the interconversion of *cis*-3,4-dimethyl-cyclobutene and *trans,trans*-2,4-hexadiene.

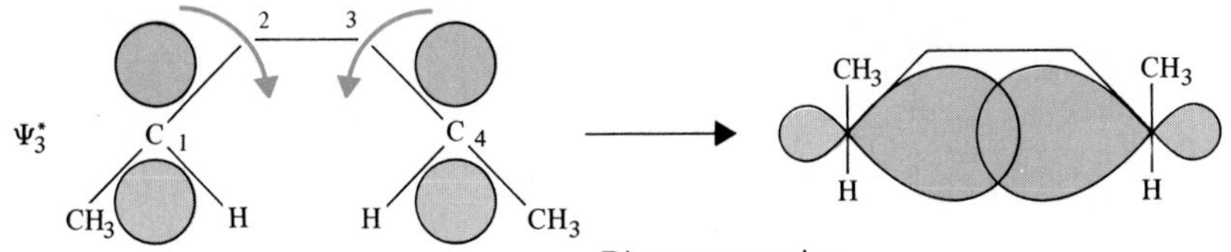

Disrotatory motion
(leads to bonding
interaction between
C-2 and C-5)

Since both of the interconversions that we have presented so far involve *cis*-3,4-dimethylcyclobutene, we can summarize them in the following way:

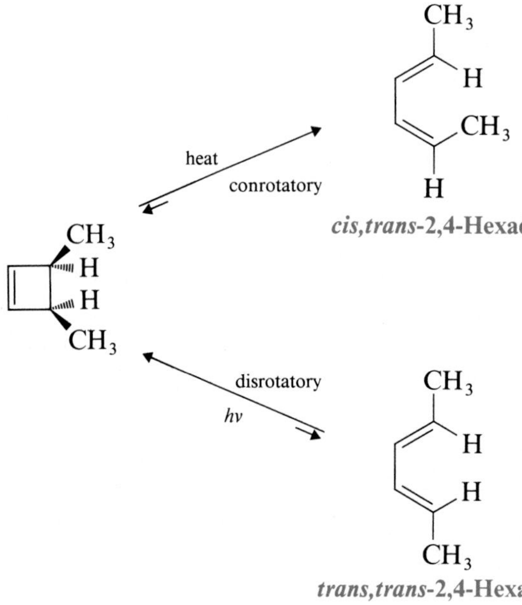

heat
conrotatory

cis,trans-**2,4-Hexadiene**

disrotatory
hv

trans,trans-**2,4-Hexadiene**

We see that these two interconversions occur with precisely opposite stereochemistry. We also see that the stereochemistry of the interconversions depends on whether or not the reaction is brought about by the application of heat or light.

The first Woodward–Hoffmann rule can be stated as follows:

1. **A thermal electrocyclic reaction involving $4n$ π electrons (where $n = 1, 2, 3,$ . . .) proceeds with conrotatory motion; the photochemical reaction proceeds with disrotatory motion.**

Both of the interconversions that we have studied involve systems of 4 π electrons and both follow this rule. Many other $4n$ π-electron systems have been studied since Woodward and Hoffmann stated their rule. Virtually all have been found to follow it.

Before we leave the subject of $4n$ π-electron systems let us illustrate an application of the rule with one other example.

When *trans*-3,4-dimethylcyclobutene is heated, ring opening occurs and *trans,trans*-2,4-hexadiene is formed.

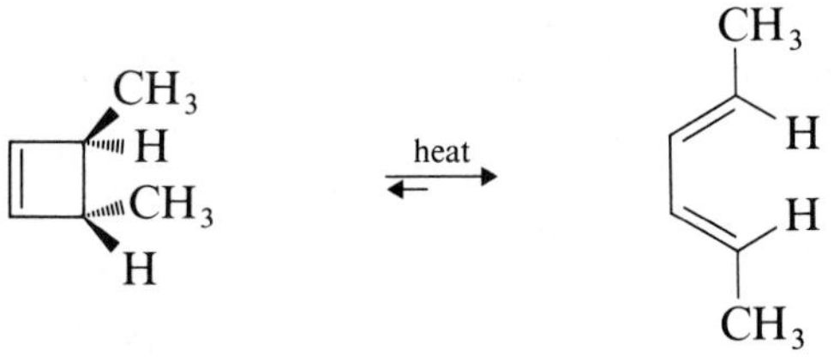

trans-3,4-Dimethylcyclobutene   *trans,trans*-2,4-Hexadiene

According to the Woodward–Hoffmann rule, this thermal reaction of a 4 π-electron system should occur with *conrotatory motion; and this is precisely what happens.* *trans*-3,4-Dimethylcyclobutene is transformed into *trans,trans*-2,4-hexadiene.

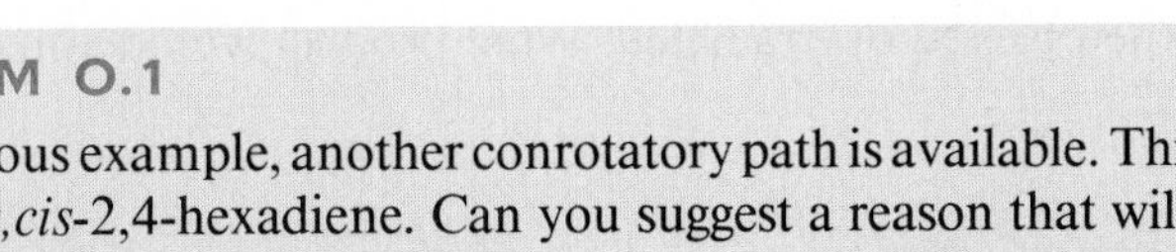

trans,trans-2,4-Hexadiene

PROBLEM O.1

In the previous example, another conrotatory path is available. This path would produce *cis,cis*-2,4-hexadiene. Can you suggest a reason that will account for the fact that this path is not followed to any appreciable extent?

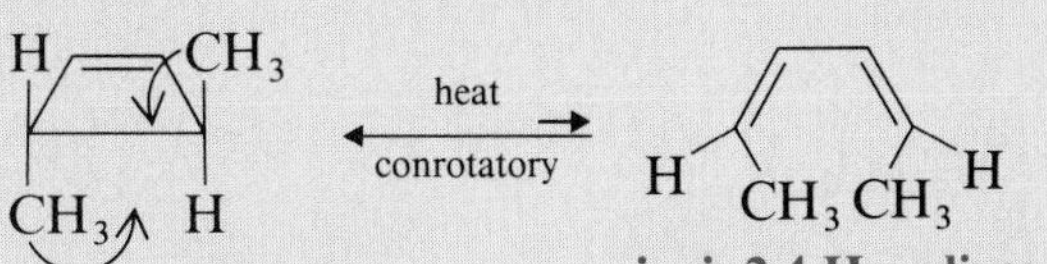

cis,cis-2,4-Hexadiene

PROBLEM O.2

What product would you expect from a concerted photochemical cyclization of *cis,trans*-2,4-hexadiene?

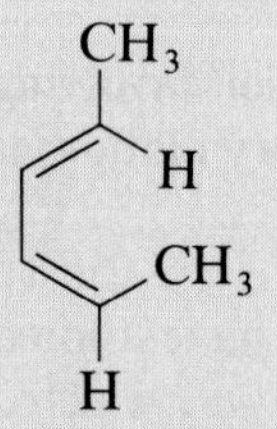

CH3
H
CH3
H

***cis,trans*-2,4-Hexadiene**

PROBLEM O.3

(a) Show the orbitals involved in the following thermal electrocyclic reaction.

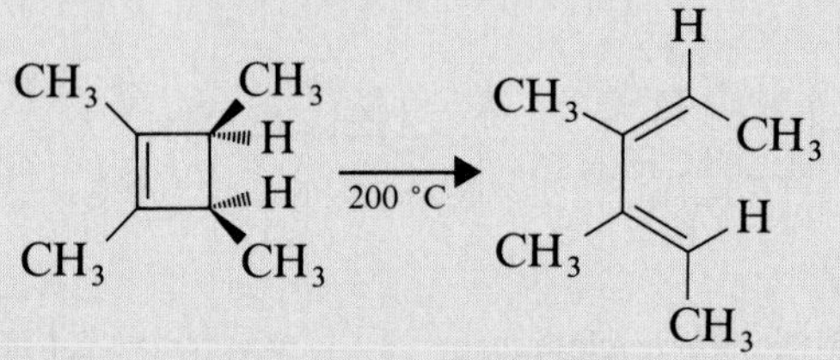

CH3
CH3
H
H
CH3
CH3
200 °C
H
CH3
CH3
H
CH3
CH3

(b) Do the groups rotate in a conrotatory or disrotatory manner?

PROBLEM O.4

Can you suggest a method for carrying out a stereospecific conversion of *trans,trans*-2,4-hexadiene into *cis,trans*-2,4-hexadiene?

PROBLEM O.5

The following 2,4,6,8-decatetraenes undergo ring closure to dimethylcyclooctatrienes when heated or irradiated. What product would you expect from each reaction?

(a) 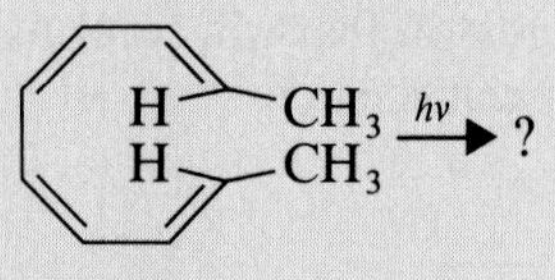

H
CH3
H
CH3
hv
?

(b) 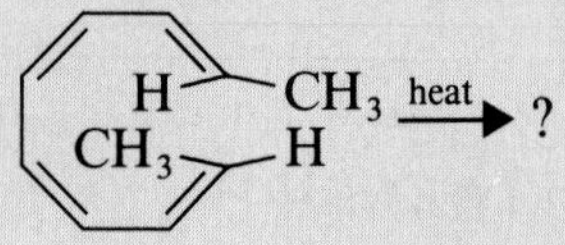

H
CH3
CH3
H
heat
?

PROBLEM O.6

(a) For each of the following reactions, state whether conrotatory or disrotatory motion of the groups is involved and   (b) state whether you would expect the reaction to occur under the influence of heat or light.

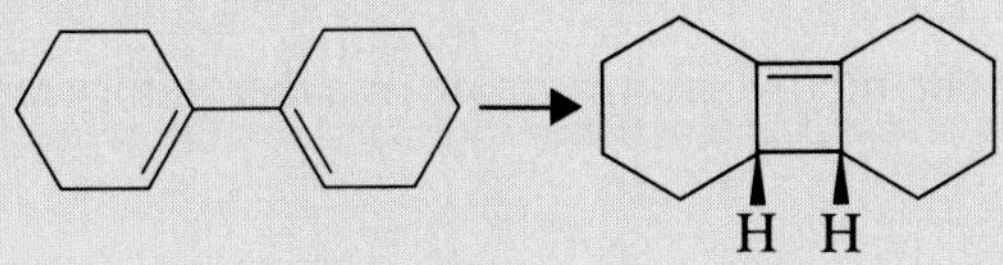

O.2B Electrocyclic Reactions of (4n + 2) π-Electron Systems

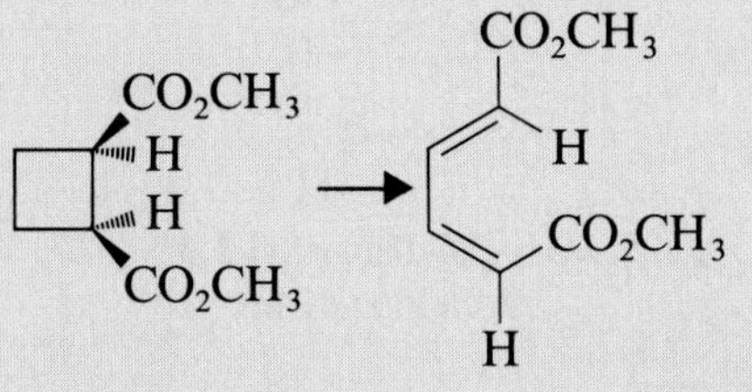

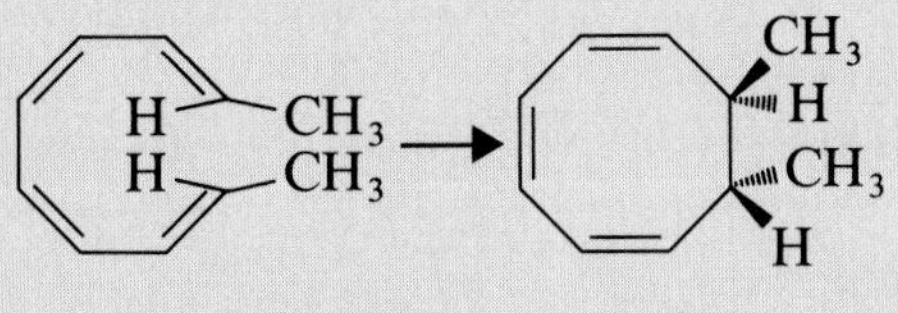

The second Woodward–Hoffmann rule for electrocyclic reactions is stated as follows:

> 2. **A thermal electrocyclic reaction involving (4n + 2) π electrons (where n = 0, 1, 2, . . .) proceeds with disrotatory motion; the photochemical reaction proceeds with conrotatory motion.**

According to this rule the direction of rotation of the thermal and photochemical reactions of $(4n + 2)$ π-electron systems is the opposite of that for corresponding $4n$ systems. Thus, we can summarize both systems in the way shown in Table O.1.

TABLE O.1 Woodward–Hoffmann rules for electrocyclic reactions

NUMBER OF ELECTRONS	MOTION	RULE
$4n$	Conrotatory	Thermally allowed, photochemically forbidden
$4n$	Disrotatory	Photochemically allowed, thermally forbidden
$4n + 2$	Disrotatory	Thermally allowed, photochemically forbidden
$4n + 2$	Conrotatory	Photochemically allowed, thermally forbidden

The interconversions of *trans*-5,6-dimethyl-1,3-cyclohexadiene and the two different 2,4,6-octatrienes that follow illustrate thermal and photochemical interconversions of 6 π-electron systems ($4n + 2$ where $n = 1$).

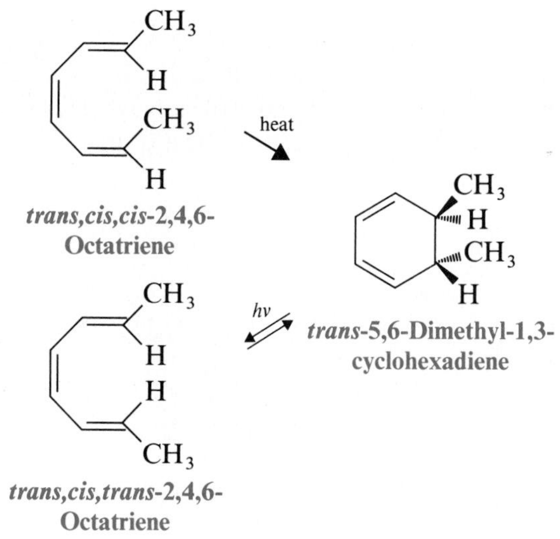

trans,cis,cis-2,4,6-
Octatriene

trans-5,6-Dimethyl-1,3-
cyclohexadiene

trans,cis,trans-2,4,6-
Octatriene

In the following thermal reaction the methyl groups rotate in a disrotatory fashion.

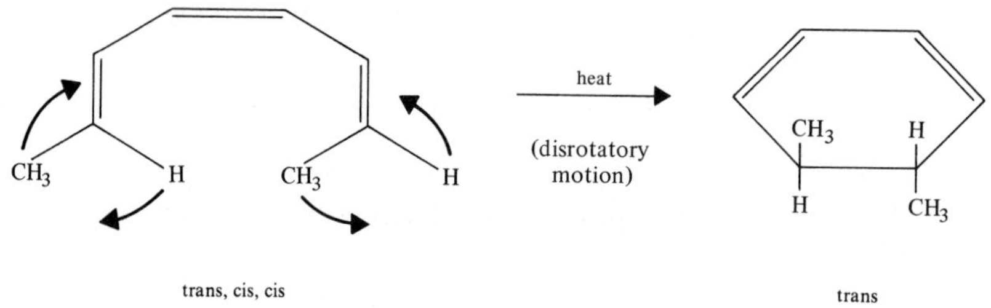

trans, cis, cis

(disrotatory motion)

trans

In the photochemical reaction the groups rotate in a conrotatory way.

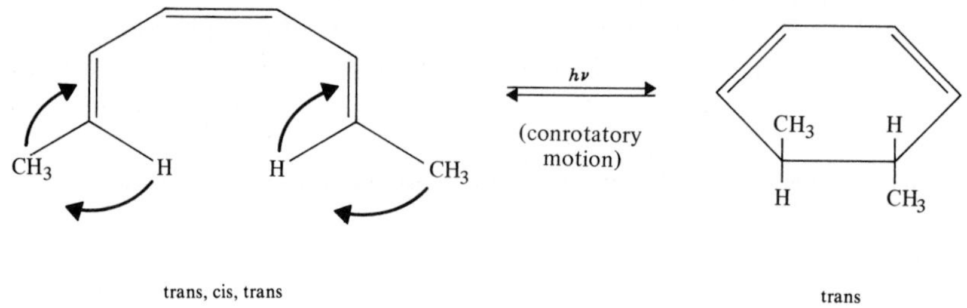

trans, cis, trans

(conrotatory motion)

trans

We can understand how these reactions occur if we examine the π molecular orbitals shown in Fig. O.2. Once again we want to pay attention to the highest occupied molecular orbitals. For the thermal reaction of a 2,4,6-octatriene, the highest occupied orbital is Ψ_3 because the molecule reacts in its ground state.

We see in the following figure that disrotatory motion of orbitals at carbons-2 and -7 of Ψ_3 allows the formation of a bonding sigma molecular orbital between them. Disrotatory motion of the orbitals, of course, also requires disrotatory motion

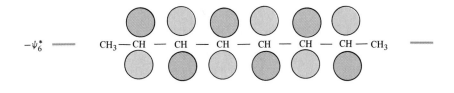

FIGURE O.2 The π molecular orbitals of a 2,4,6-octatriene. The first excited state is formed when the molecule absorbs light of the proper wavelength. (These molecular orbitals are obtained from calculations that are beyond the scope of our discussions.)

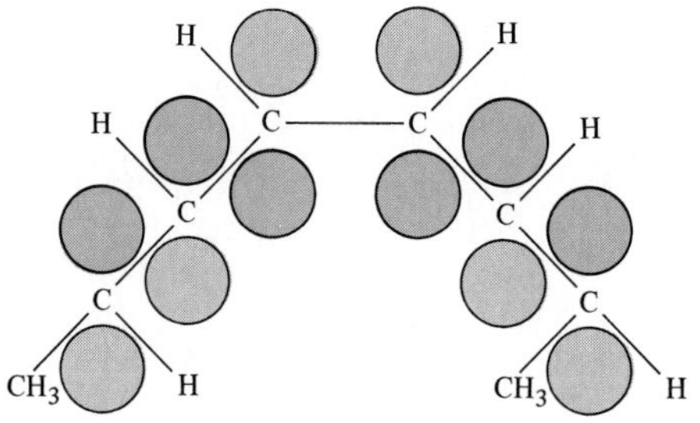

Ψ_3 of *trans, cis, cis,*$-2, 4, 6-$octatriene

of the groups attached to carbons-2 and -7. And disrotatory motion of the groups is what we observe in the thermal reaction: *trans,cis,cis*-2,4,6-octatriene ⟶ *trans*-5,6-dimethyl-1,3-cyclohexadiene.

HOMO of ground state

Ψ_3

trans, cis, cis

heat

Disrotatory motion leads to bonding interaction

trans

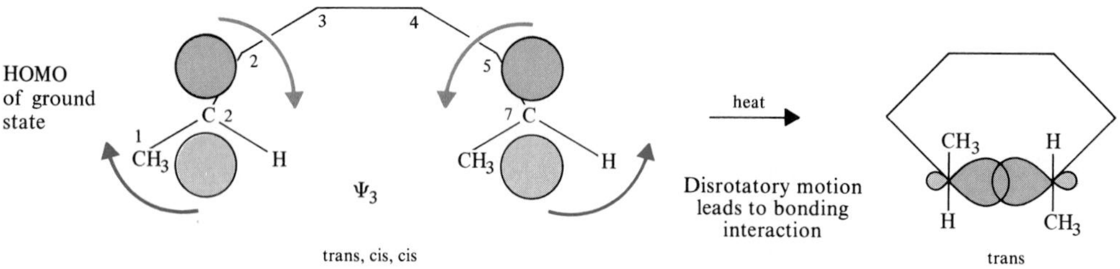

When we consider the photochemical reaction, *trans,cis,trans*-2,4,6-octa-triene ⇌ *trans*-5,6-dimethyl-1,3-cyclohexadiene, we want to focus our attention on Ψ_4^*. In the photochemical reaction, light causes the promotion of an electron from Ψ_3 to Ψ_4^*, and thus Ψ_4^* becomes the highest occupied molecular orbital. We also want to look at the symmetry of the orbitals at C-2 and C-7 of Ψ_4^*, for these are the orbitals that form a σ bond. In the interconversion shown here, conrotatory motion of the orbitals allows lobes of the same sign to overlap. Thus, we can understand why conrotatory motion of the groups is what we observe in the photochemical reaction.

HOMO of first excited state

Ψ_4 of trans, cis, trans-2, 4, 6-Octatriene

$h\nu$

Conrotatory motion leads to bonding interaction

trans

PROBLEM O.7

Give the stereochemistry of the product that you would expect from each of the following electrocyclic reactions.

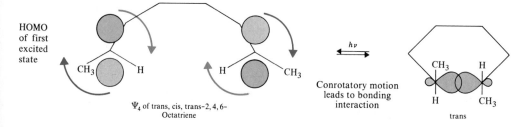

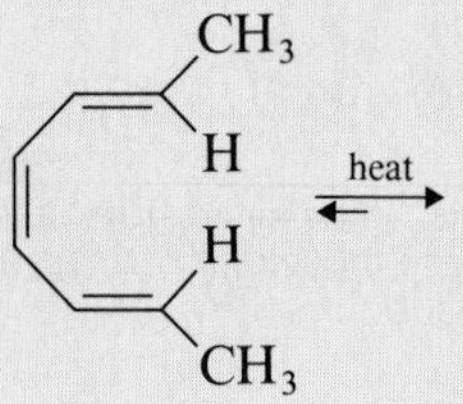

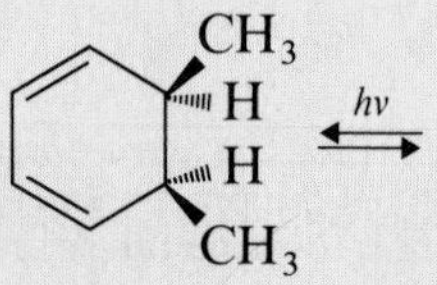

PROBLEM O.8

Can you suggest a stereospecific method for converting *trans*-5,6-dimethyl-1,3-cyclohexadiene into *cis*-5,6-dimethyl-1,3-cyclohexadiene?

PROBLEM O.9

When compound **A** is heated, compound **B** can be isolated from the reaction mixture. A sequence of two electrocyclic reactions occurs; the first involves a 4 π-electron system, the second involves a 6 π-electron system. Outline both electrocyclic reactions and give the structure of the intermediate that intervenes.

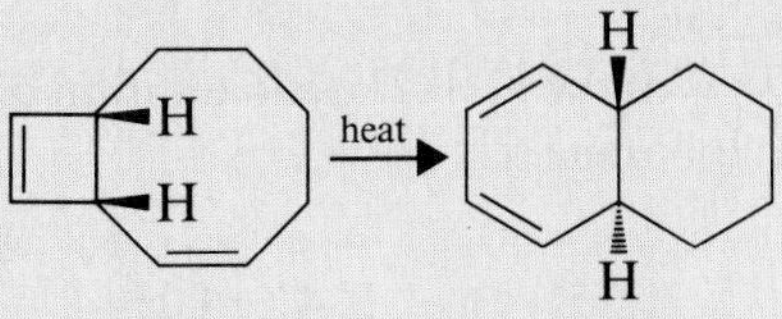

A   B

O.3 Cycloaddition Reactions

There are a number of reactions of alkenes and polyenes in which two molecules react to form a cyclic product. These reactions, called *cycloaddition* reactions, are shown next.

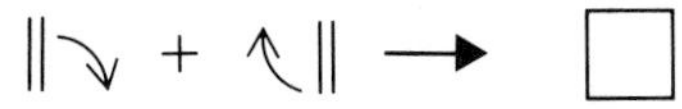

Alkene   Alkene   Cyclobutane   A [2 + 2] cycloaddition

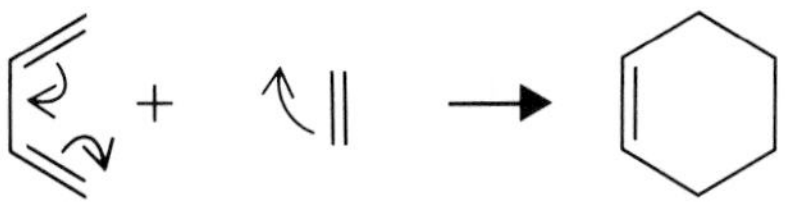

Diene   Alkene   Cyclohexene   A [4 + 2] cycloaddition
           (dienophile)   (adduct)

Chemists classify cycloaddition reactions on the basis of the number of π electrons involved in each component. The reaction of two alkenes to form a cyclobutane is a [2 + 2] cycloaddition; the reaction of a diene and an alkene to form a cyclohexene is called a [4 + 2] cycloaddition. We are already familiar with the [4 + 2] cycloaddition, because it is the Diels–Alder reaction that we studied in Section 10.12.

Cycloaddition reactions resemble electrocyclic reactions in the following important ways:

1. Sigma and pi bonds are interconverted.
2. Cycloaddition reactions require only heat or light for initiation.
3. Free radicals and ionic intermediates are not involved in the mechanisms for cycloadditions.
4. Bonds are made and broken in a single concerted step involving a cyclic transition state.
5. Cycloaddition reactions are highly stereospecific.

As we might expect, concerted cycloaddition reactions resemble electrocyclic reactions in still another important way: The symmetry elements of the interacting molecular orbitals allow us to account for their stereochemistry. The symmetry elements of the interacting molecular orbitals also allow us to account for two other observations that have been made about cycloaddition reactions:

1. *Photochemical [2 + 2] cycloaddition reactions occur readily while thermal [2 + 2] cycloadditions take place only under extreme conditions.* When thermal [2 + 2] cycloadditions do take place, they occur through free radical (or ionic) mechanisms, not through a concerted process.
2. *Thermal [4 + 2] cycloaddition reactions occur readily and photochemical [4 + 2] cycloadditions are difficult.*

O.3A [2 + 2] Cycloadditions

Let us begin with an analysis of the [2 + 2] cycloaddition of two ethylene molecules to form a molecule of cyclobutane.

$$2 \; \begin{matrix} CH_2 \\ \| \\ CH_2 \end{matrix} \longrightarrow \begin{matrix} H_2C - CH_2 \\ | \quad\quad | \\ H_2C - CH_2 \end{matrix}$$

In this reaction we see that two π bonds are converted into two σ bonds. But how does this conversion take place? One way of answering this question is by examining the frontier orbitals of the reactants. The frontier orbitals are the highest occupied molecular orbital (HOMO) of one reactant and the lowest unoccupied molecular orbital (LUMO) of the other.

We can see how frontier orbital interactions come into play if we examine the possibility of a *concerted thermal* conversion of two ethene molecules into cyclobutane.

Thermal reactions involve molecules reacting in their ground states. The following is the orbital diagram for ethene in its ground state.

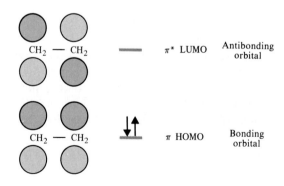

The ground state of ethene

The highest occupied molecular orbital (HOMO) of ethene in its ground state is the π orbital. Since this orbital contains two electrons, it interacts with an *unoccupied* molecular orbital of another ethene molecule. The lowest unoccupied molecular orbital (LUMO) of the ground state of ethene is, of course, π^*.

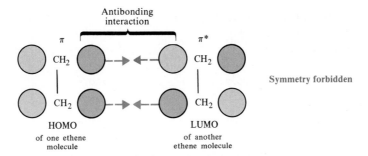

We see from the previous diagram, however, that overlapping the π orbital of one ethene molecule with the π^* orbital of another does not lead to bonding between both sets of carbon atoms because orbitals of opposite signs overlap between the top pair of carbon atoms. This reaction is said to be *symmetry forbidden*. What does this mean? It means that a thermal (or ground state) cycloaddition of ethene would be unlikely to occur in a concerted process. This is exactly what we find experimentally; thermal cycloadditions of ethene, when they occur, take place through nonconcerted, free radical mechanisms.

What, then, can we decide about the other possibility—a photochemical [2 + 2] cycloaddition? If an ethene molecule absorbs a photon of light of the proper wavelength, an electron is promoted from π to π^*. In this excited state the highest occupied molecular orbital of an ethene molecule is π^*. The following diagram shows how the highest occupied molecular orbital of an excited state ethene molecule interacts with the lowest unoccupied molecular orbital of a ground state ethene molecule.

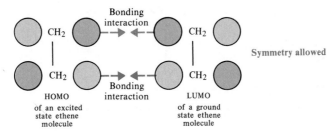

Here we find that bonding interactions occur between both CH_2 groups, that is, lobes of the same sign overlap between both sets of carbons. Complete correlation diagrams also show that the photochemical reaction is *symmetry allowed* and should occur readily through a concerted process. This, moreover, is what we observe experimentally: Ethene reacts readily in a *photochemical* cycloaddition.

The analysis that we have given for the [2 + 2] ethene cycloaddition can be made for any alkene [2 + 2] cycloaddition because the symmetry elements of the π and π^* orbitals of all alkenes are the same.

PROBLEM O.10

What products would you expect from the following concerted cycloaddition reactions? (Give stereochemical formulas.)

(a) *cis*-2-Butene $\xrightarrow{\ hv\ }$

(b) *trans*-2-Butene $\xrightarrow{\ hv\ }$

PROBLEM O.11

Show what happens in the following reaction:

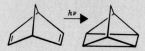

O.3B [4 + 2] Cycloadditions

Concerted [4 + 2] cycloadditions—Diels–Alder reactions—are *thermal reactions.* Considerations of orbital interactions allow us to account for this fact as well. To see how, let us consider the diagrams shown in Figure O.3.

Both modes of orbital overlap shown in Fig. O.3 lead to bonding interactions and both involve *ground states* of the reactants. The ground state of a diene has two electrons in Ψ_2 (its highest occupied molecular orbital). The overlap shown in part (*a*) allows these two electrons to flow into the lowest unoccupied molecular orbital, π^*, of the dienophile. The overlap shown in part (*b*) allows two electrons to flow from the highest occupied molecular orbital of the dienophile, π, into the lowest unoccupied molecular orbital of the diene, Ψ_3^*. This thermal reaction is said to be symmetry allowed.

In Section 10.12 we saw that the Diels–Alder reaction proceeds with retention of configuration of the dienophile. Because the Diels–Alder reaction is usually concerted, it also proceeds with retention of configuration of the diene.

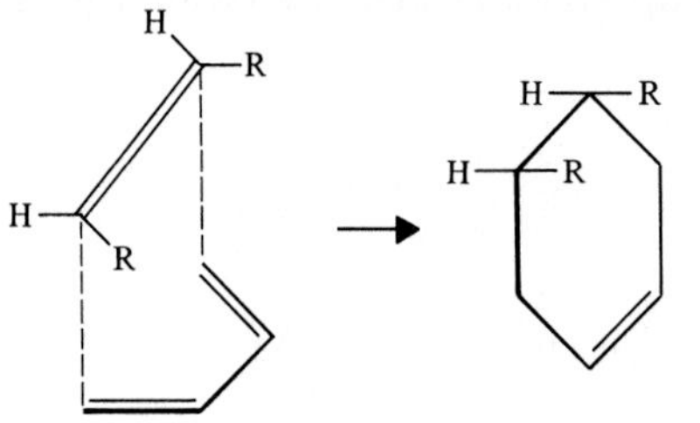

Retention of configuration of the dienophile

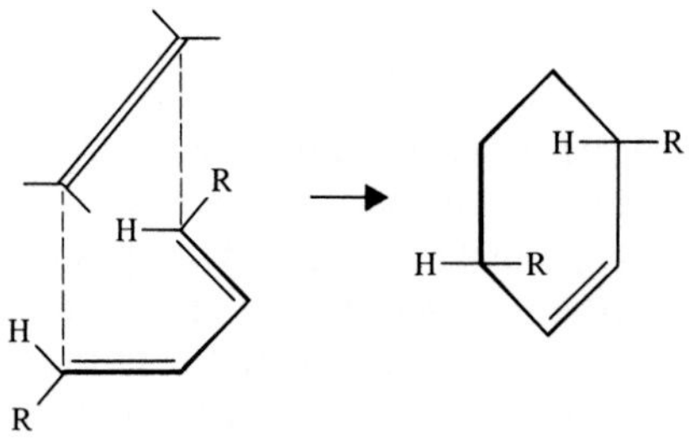

Retention of configuration of the diene

PROBLEM O.12

What products would you expect from the following reaction?

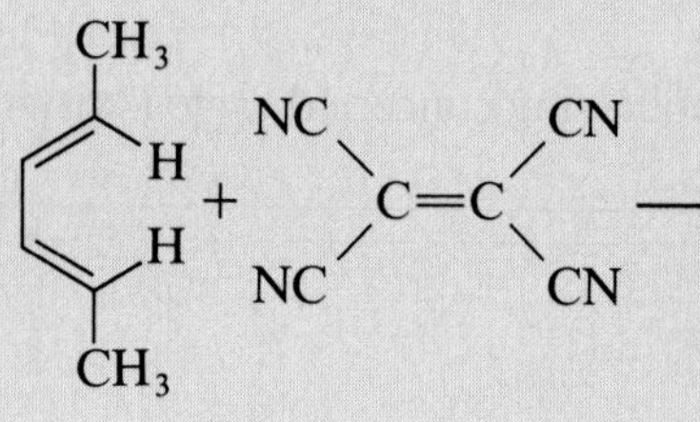

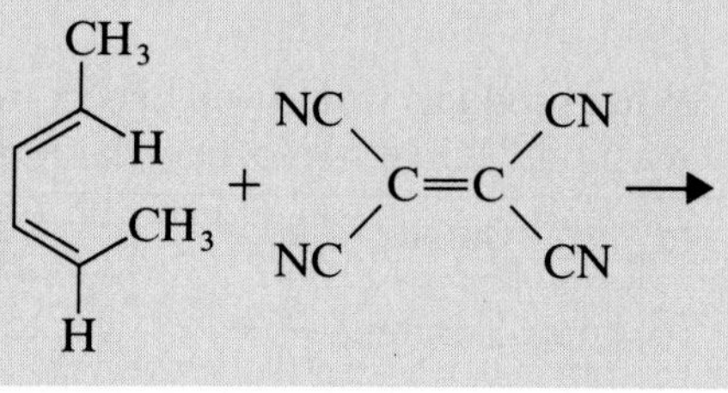

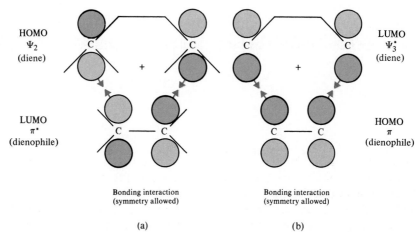

HOMO
Ψ_2
(diene)

LUMO
Ψ_3^*
(diene)

LUMO
π^*
(dienophile)

HOMO
π
(dienophile)

Bonding interaction
(symmetry allowed)

Bonding interaction
(symmetry allowed)

(a)

(b)

FIGURE O.3 Two symmetry-allowed interactions for a thermal [4 + 2] cycloaddition. (*a*) Bonding interaction between the highest occupied molecular orbital of a diene and the lowest unoccupied molecular obital of a dienophile. (*b*) Bonding interaction between the lowest unoccupied molecular orbital of the diene and the highest occupied molecular orbital of the dienophile.

PROBLEM O.13

What are compounds **3**, **4**, and **5** in the following reaction sequence?

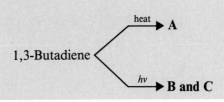

---

*PROBLEM O.14

Propose structures for compounds **A**, **B**, and **C**.

1,3-Butadiene

heat → **A**

$h\nu$ → **B and C**

*PROBLEM O.15

What are the intermediates **A** and **B** in the following synthesis of basketene?

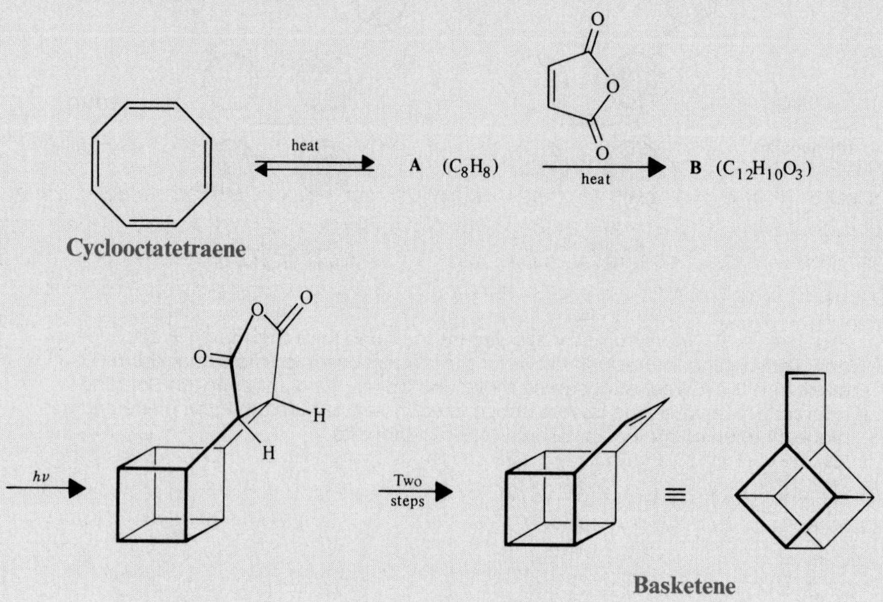

heat
A (C_8H_8)
O
O
O
heat
B ($C_{12}H_{10}O_3$)
Cyclooctatetraene
O
O
O
H
H
$h\nu$
Two steps
≡
Basketene

SPECIAL TOPIC P

NUCLEOPHILIC SUBSTITUTION REACTIONS—ANOTHER LOOK

P.1 S$_N$2 Reactions: The Role of Ion Pairs

For many years the Ingold mechanism (Section 5.6) was widely accepted as the only mechanism for bimolecular nucleophilic substitution reactions. In recent years, however, evidence has been advanced to support another possible mechanism: one involving *ion pairs.* The ion pair mechanism seems to be particularly important in those nucleophilic substitutions called *solvolyses;* however, it may operate in other S$_N$2 reactions. *A solvolysis is a reaction in which the nucleophile is a molecule of the solvent* (solvent + *lysis:* Cleavage by solvent).

> Solvolytic reactions are often described as being pseudofirst order since, by convention, the concentration of the solvent is not included in the rate expression. Even if it were included, we might not be able to detect a variation in rate with a change in concentration of the solvent since the solvent concentration is usually very large and is, therefore, essentially constant. As we shall see, however, some solvolyses are actually bimolecular nucleophilic substitution reactions and occur with complete inversion of configuration.

> The first clear demonstration of ion pair involvement in solvolytic reactions was given by S. Winstein (of the University of California, Los Angeles).

Two examples of solvolytic reactions for which ion pair mechanisms have been proposed are the hydrolysis and acetolysis of methylheptyl sulfonates.

Hydrolysis of a 1-Methylheptyl Sulfonate

Acetolysis of a 1-Methylheptyl Sulfonate

Both of these reactions have been shown to take place with *complete inversion of configuration.*

On the basis of kinetic evidence (beyond the scope of our treatment here) it has been proposed that these reactions take place with the rapid but reversible formation of an "*intimate*" (or tight) ion pair followed by a slow reaction with a solvent molecule. An intimate ion pair is an ionic intermediate in which the cation and anion are in close proximity and are not separated by solvent molecules.

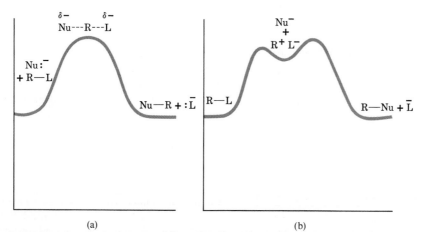

The intimate ion pair retains the configuration of the original sulfonate. However, when it reacts with the solvent, the displacement takes place from the back side and produces an inversion of configuration. In all likelihood, inversion occurs because the intimate ion pair still has partial bonding between the carbon and the leaving group, and attack of the nucleophile must occur from the back side in the same manner as for an S_N2 reaction of a neutral compound.

The difference between the Ingold mechanism and the ion pair mechanism is most apparent in their potential energy diagrams (Fig. P.1). In the one-step displacement mechanism there is a single transition state; in the ion pair mechanism there are two transition states and the ion pair occupies an energy minimum lying between them.

At the time of this writing not enough evidence is available to distinguish between these two mechanistic possibilities for most S_N2 reactions. However, with either mechanism the stereochemistry of S_N2 reactions is clear: S_N2 reactions, whether they take place in one step or through the formation of ion pairs, *occur with inversion of configuration.*

P.2 Ion Pairs and S_N1 Reactions

When 1-phenylethyl chloride reacts with water in aqueous acetone, that is,

FIGURE P.1 Potential energy diagrams for the one-step mechanism (the Ingold mechanism) for an S_N2 reaction (*a*), and for the intimate ion pair mechanism (*b*).

the reaction rate is first order; it depends only on the concentration of 1-phenylethyl chloride and is essentially independent of the concentration of water.

$$\text{Rate} = k \left[\underset{\underset{\displaystyle C_6H_5\overset{\textstyle |}{C}HCH_3}{}}{\overset{\textstyle Cl}{}} \right]$$

The stereochemistry of the reaction is shown here.

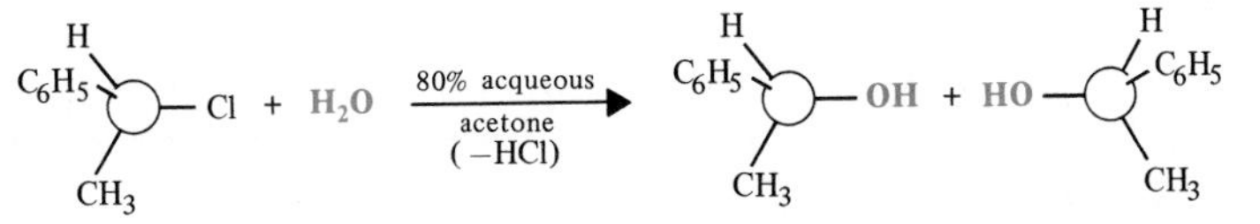

(S)-1-Phenylethyl chloride (optically pure)

(S)-1-Phenylethanol   (R)-1-Phenylethanol

98% Racemization
2% Net inversion

We see from this equation that the reaction of 1-phenylethyl chloride of the (S) configuration gives a product, 1-phenylethanol, of which 51% has the opposite (R) configuration and 49% has the same (S) configuration. In other words, 51% of the 1-phenylethyl chloride molecules have had their configuration inverted by the reaction, while the remainder (49%) have retained their original configuration. We describe this situation by saying that the reaction has taken place with 98% *racemization* and 2% *net* inversion.

We can account for the fact that this reaction is first order if we assume that the rate-determining step (or slow step) for the reaction involves the organic halide alone. A general mechanism is the following:

Step 1 $R-X \xrightarrow{\text{slow}} R^+ + X^-$

Step 2 $R^+ + H_2O \xrightarrow{\text{fast}} R\overset{+}{O}H_2$

Step 3 $R\overset{+}{O}H_2 \xrightarrow[-H^+]{\text{fast}} ROH$

Step 1, the formation of a carbocation, is the slow step. Step 2 is a rapid reaction of the carbocation with water and step 3 is the rapid loss of a proton.

Since step 1 involves the organic halide alone (we are, for the moment, neglecting the involvement of solvent molecules) the overall rate of the reaction must correspond to the rate of this step,

$$\text{Rate} = k[RX]$$

and the reaction as a whole must show first-order kinetics.

A more detailed mechanism (shown in the following figure) illustrates one way in which we can account for the overall stereochemistry of the hydrolysis of 1-phenylethyl chloride.

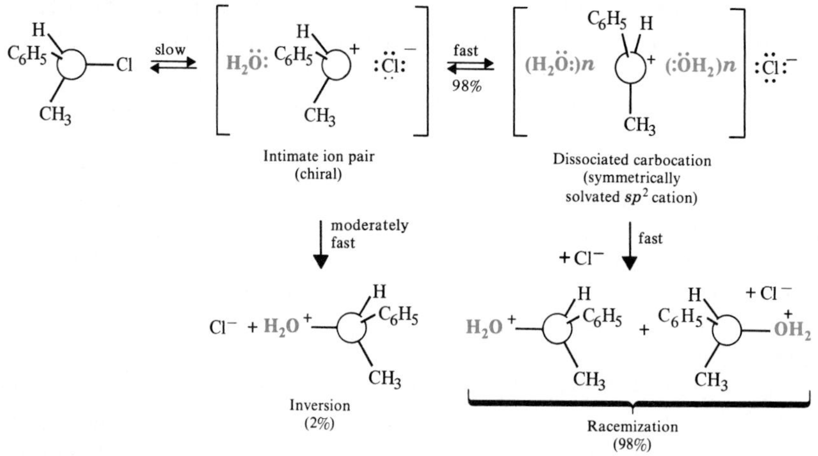

Here we see the important part played by solvent molecules and we also see the formation of two different cationic intermediates: an intimate ion pair and a more dissociated carbocation. The intimate ion pairs are formed first and their carbocations are solvated only on their back sides. A relatively small number of the intimate ion pairs collapse to give an inverted product. However, since the developing cation in this reaction is a relatively stable *benzylic* carbocation, most of the intimate ion pairs survive long enough to become more dissociated. The positive carbon atoms of the dissociated carbocations are sp^2 hybridized and they are solvated on both the front and back sides. They react equally rapidly with water molecules at either face to give a racemic modification of the protonated alcohol.

Some evidence, however, suggests that a third type of cationic intermediate called a "solvent-separated" ion pair may intervene between the intimate ion pair and the dissociated carbocation in reactions of this type. A solvent-separated ion pair, as its name suggests, is one in which a molecule of solvent is situated between the carbocation and the anion. There is also evidence that solvent-separated ion pairs

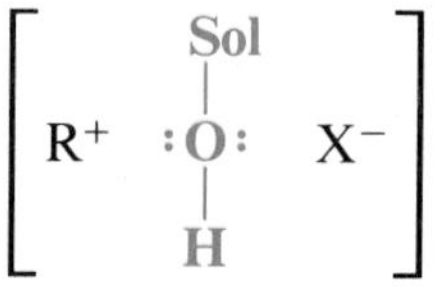

A solvent-separated ion pair

may preferentially react with the intervening solvent molecule by a mechanism that operates with *retention of configuration:*

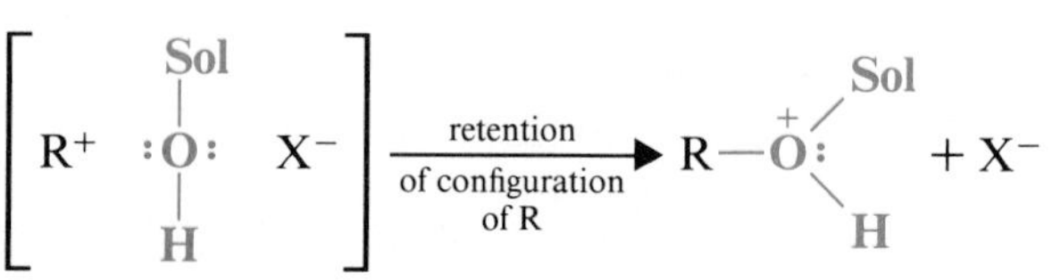

Therefore, it is possible that not all of the racemic product comes from dissociated carbocations and that more than 2% of the reaction may take place through intimate ion pairs. Part of the racemic product may result from a balanced collapse of intimate ion pairs (with inversion) and of solvent-separated ion pairs (with retention).

P.3 Summary of S_N Reaction Mechanisms at a Saturated Carbon

Nucleophilic substitution reactions may very well take place through a spectrum of mechanisms ranging from a one-step displacement mechanism at one end and a mechanism involving fully dissociated carbocations at the other. Intervening between these limits may be mechanisms involving at least two kinds of ion pairs, intimate ion pairs and solvent-separated ion pairs. We can represent this spectrum as shown in Fig. P.2.

Whether a reaction gives a first-order or second-order rate equation will depend on just which of these mechanisms operates. We shall obtain a second-order rate equation for those reactions that take place by a one-step displacement mechanism (the Ingold mechanism) or through an intimate ion pair since, in these instances, the rate-determining step is bimolecular. We describe these reactions as being bimolecular or S_N2 reactions.

> The mechanism involving an intimate ion pair is a true bimolecular reaction since the transition state of the rate-determining step,
>
> $$R^+ L^- + Nu\!: \longrightarrow NuR + L^-$$
>
> or $\qquad R^+ L^- + \text{Sol-OH} \longrightarrow \text{Sol} - \overset{+}{\underset{\overset{\displaystyle |}{H}}{O}} - R + L^-$
>
> involves two species: the intimate ion pair and the nucleophile or solvent.

We shall obtain a first-order rate equation for those reactions that involve dissociated carbocations or solvent-separated ion pairs, for in these reactions the rate-determining step is unimolecular. We describe these reactions as being unimolecular or S_N1.

> The only exception to these generalizations is a solvolysis. A solvolysis can be *bimolecular* because the transition state of its rate-determining step can involve two species: the substrate and the solvent. Such a solvolysis, however, will show *pseudofirst-order kinetics* because the solvent concentration is very large and is, consequently, essentially constant.

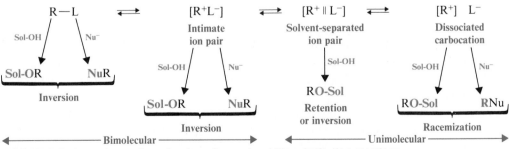

FIGURE P.2 A spectrum of mechanisms for nucleophilic substitution reactions.

TABLE P.1 The stereochemistry of nucleophilic substitution reactions

MOLECULARITY	SUBSTRATE	REPRESENTATION	STEREOCHEMISTRY
S_N2	 R—L Alkyl halide, tosylate, etc.	R—L	Nucleophilic attack by the solvent or the nucleophile from the back side gives an inverted product by a one-step displacement mechanism
S_N2	 R—L Intimate ion pair	$[R^+L^-]$	Nucleophilic attack by the solvent or the nucleophile from the back side gives an inverted product
S_N1	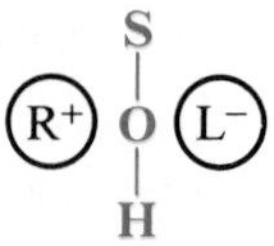 R^+ O L^- with S—O—H Solvent-separated ion pair	$[R^+\|L^-]$	Nucleophilic attack by the solvent from the front side occurs with retention of configuration. (Attack by another nucleophile may occur with inversion)
S_N1	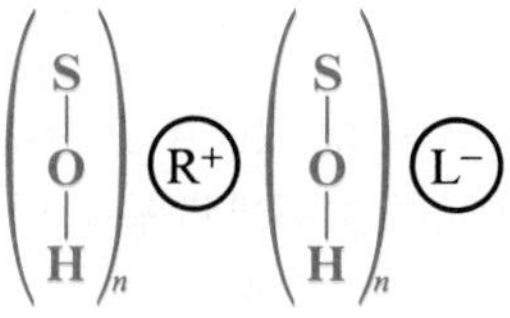 $(S-O-H)_n$ R^+ $(S-O-H)_n$ L^- Dissociated carbocation	$[R^+]\,L^-$	Nucleophilic attack by the solvent or the nucleophile from the front or back side gives a racemic product

The stereochemical possibilities for nucleophilic substitution reactions are summarized in Table P.1.

P.4 Neighboring Group Participation in Nucleophilic Substitution Reactions

Not all nucleophilic substitutions (Sections P.1–P.3) take place with racemization or with inversion of configuration. Some take place with overall *retention of configuration.*

One factor that can lead to retention of configuration in a nucleophilic substitution is a phenomenon known as *neighboring group participation.* Let us see how this operates by examining the stereochemistry of two reactions in which 2-bromopropanoic acid is converted to lactic acid.

$$CH_3CHCO_2H \longrightarrow CH_3CHCO_2H$$

Br	OH
2-Bromopropanoic acid	Lactic acid

When (S)-2-bromopropanoic acid is treated with concentrated sodium hydroxide, the reaction is *bimolecular* and it takes place with *inversion of configuration*. This, of course, is the normal stereochemical result for an S_N2 reaction.

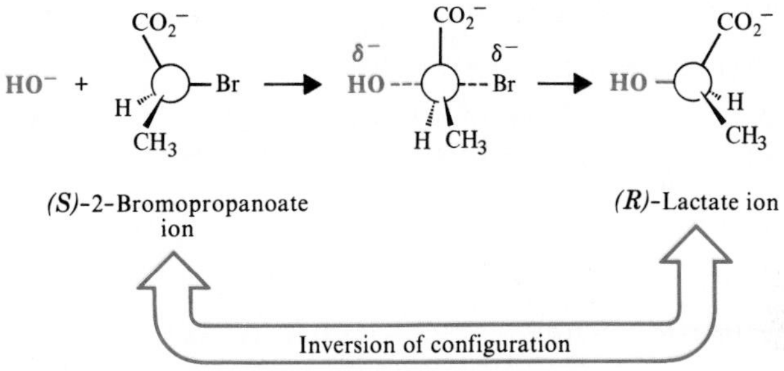

(S)-2-Bromopropanoate ion   *(R)*-Lactate ion

Inversion of configuration

However, when the same reaction is carried out with a low concentration of hydroxide ion in the presence of Ag_2O, it takes place with an overall *retention of configuration*. In this case, the mechanism for the reaction involves the participation of the carboxylate group. In step 1 (see the following reaction) an oxygen of the carboxylate group attacks the stereocenter from the back side and displaces bromide ion. (Silver ion aids in this process in much the same way that protonation assists the ionization of an alcohol.) The configuration of the stereocenter inverts in step 1, and a cyclic ester called an α-lactone forms.

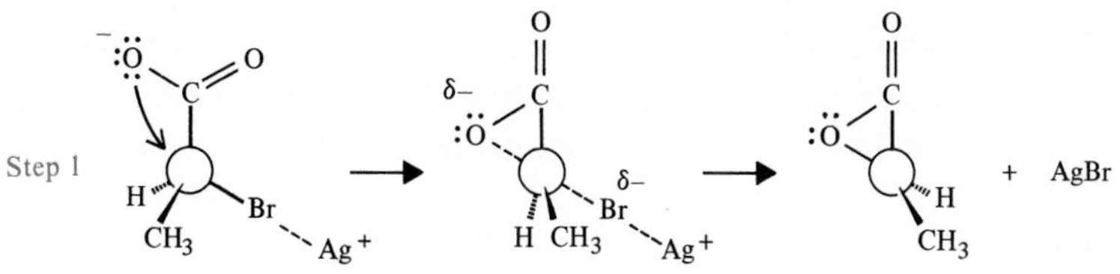

Step 1   + AgBr

An α-lactone

The highly strained three-membered ring of the α-lactone opens when it is attacked by a water molecule in step 2. *This step also takes place with an inversion of configuration.*

Step 2

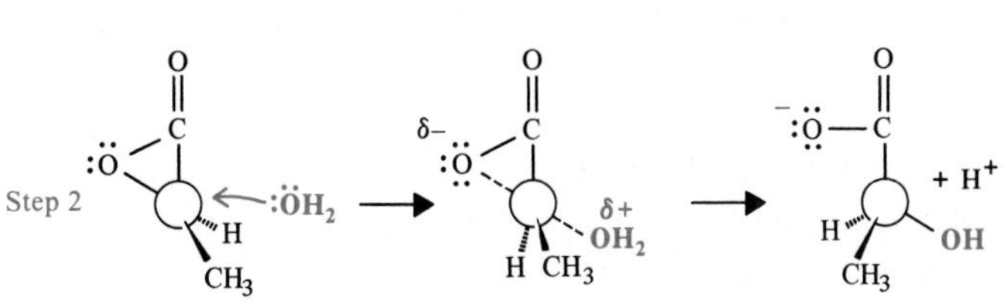

The net result of two inversions (in steps 1 and 2) is an overall *retention of configuration.*

PROBLEM P.1

The phenomenon of configuration inversion in a chemical reaction was discovered in 1896 by Paul von Walden. (Configuration inversions are still called Walden inversions in his honor.) Walden's proof of configuration inversion was based on the following cycle:

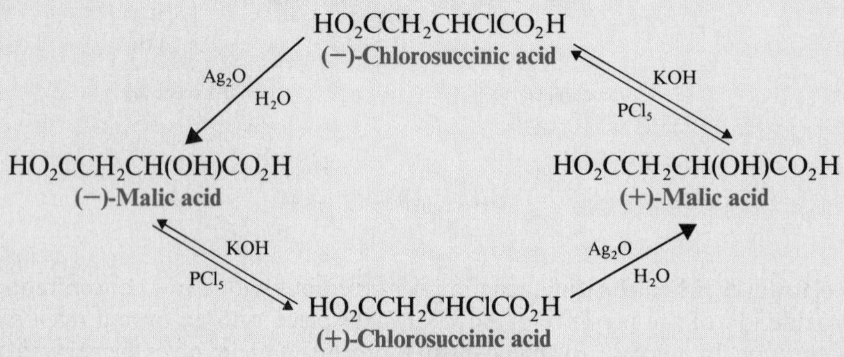

$$HO_2CCH_2CHClCO_2H$$
(−)-Chlorosuccinic acid

Ag$_2$O
H$_2$O

KOH
PCl$_5$

$$HO_2CCH_2CH(OH)CO_2H$$
(−)-Malic acid

$$HO_2CCH_2CH(OH)CO_2H$$
(+)-Malic acid

KOH
PCl$_5$

Ag$_2$O
H$_2$O

$$HO_2CCH_2CHClCO_2H$$
(+)-Chlorosuccinic acid

The Walden cycle

(a) Basing your answer on the preceding discussion, which reactions of the Walden cycle are likely to take place with overall inversion of configuration and which are likely to occur with overall retention of configuration? (b) Malic acid with a negative optical rotation is now known to have the (*S*) configuration. What are the configurations of the other compounds in the Walden cycle? (c) Walden also found that when (+)-malic acid is treated with thionyl chloride (rather than PCl$_5$) the product of the reaction is (+)-chlorosuccinic acid. How can you explain this result? (d) Assuming that the reaction of (−)-malic acid and thionyl chloride has the same stereochemistry, outline a Walden cycle based on the use of thionyl chloride instead of PCl$_5$.

Neighboring group participation can also lead to *cyclization reactions.* Epoxides, for example, can be prepared from 2-bromo alcohols by treating them with sodium hydroxide. This reaction involves the following steps:

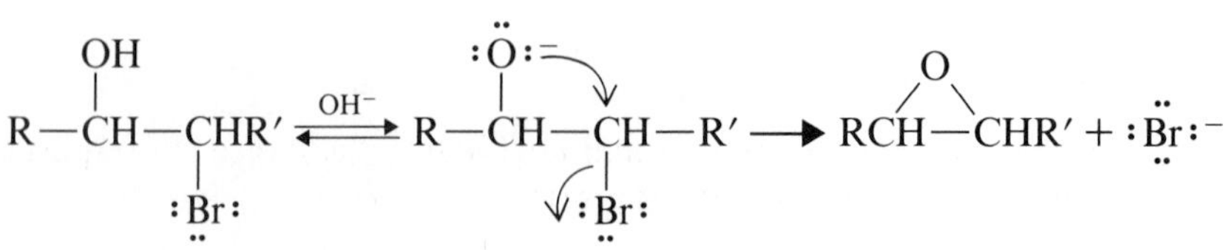

OH
|
R—CH—CHR' $\xrightarrow{OH^-}$ R—CH—CH—R' $\longrightarrow$ RCH—CHR' + :Br:$^-$
| |
:Br: :Br:

When neighboring group participation occurs during the rate-determining step for a reaction, the rate is often markedly increased. This effect, called *anchimeric assistance* (Greek *anchi* + *meros,* meaning neighboring parts), can be seen in the relative rates of S_N1 solvolysis reactions of isobutyl chloride and 2-phenyl-1-chloro-

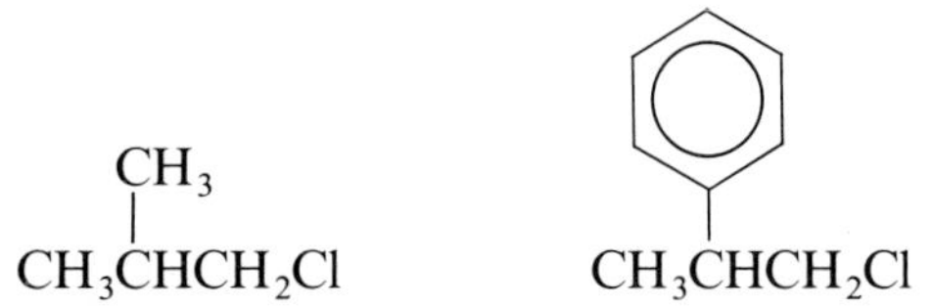

$$CH_3$$
$$|$$
$$CH_3CHCH_2Cl$$
Isobutyl chloride

$$CH_3CHCH_2Cl$$
2-Phenyl-1-chloropropane

propane. When 2-phenyl-1-chloropropane undergoes S_N1 solvolysis, it reacts more rapidly than isobutyl chloride. The phenyl group is thought to assist in the ionization step by stabilizing the transition state leading to the phenonium ion intermediate. The methyl group of isobutyl chloride is apparently unable to provide a similar kind of assistance when it undergoes solvolysis.

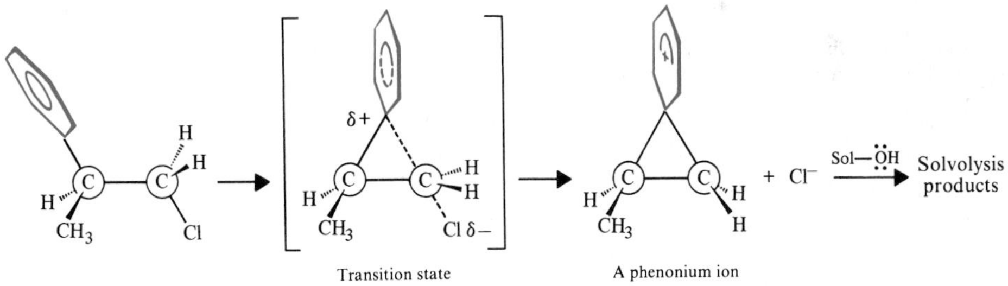

Transition state   A phenonium ion

Neighboring group participation and anchimeric assistance are important in many reactions that are catalyzed by enzymes.

PROBLEM P.4

In 1949, D. J. Cram published the first of a series of papers on the solvolysis of 1-methyl-2-phenylpropyl tosylates, **A** and **C**. These reactions displayed a remarkable stereospecificity: When the optically active tosylate **A** was heated in acetic acid, the reaction yielded almost exclusively the optically active acetate **B**. On the other hand, heating the optically active tosylate **C** in acetic acid gave the racemic acetate **D** and **E**. Provide an explanation for these results.

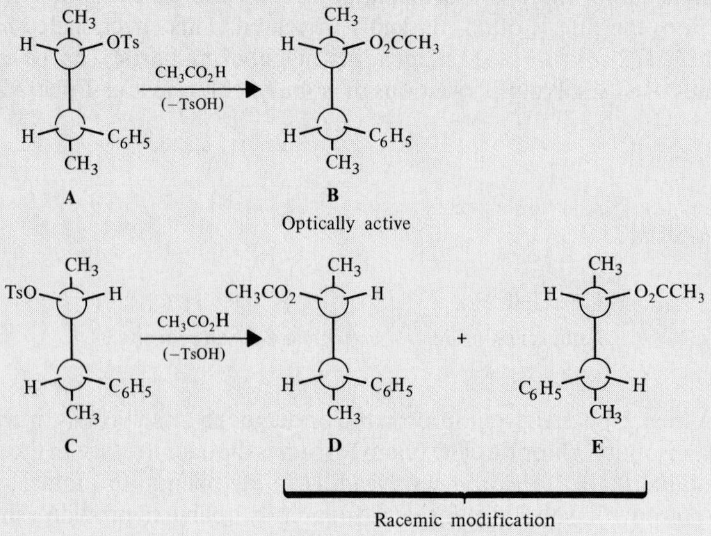

Optically active

Racemic modification

CHAPTER 1

1.5 (a) $\xrightarrow{\text{H}-\text{Br;}}$; (b) $\xrightarrow{\text{I}-\text{Cl;}}$;
(c) $H_2, \mu = 0$; (d) $Cl_2, \mu = 0$.

1.10 (a), (b), (d), (e), (g) tetrahedral;
(c), (i) trigonal planar; (f), (h) linear.

1.12 The bond moments cancel.

1.15 Trigonal planar structure causes
bond moments to cancel.

1.18 (a) and (d), (e) and (f).

1.32 (a) An sp^3 orbital; (b) an sp^3 orbital.

1.38 The carbon atom of the methyl
cation is sp^2 hybridized and uses sp^2 orbitals to form bonds to each hydrogen. The
carbon also has a vacant p orbital.

CHAPTER 2

2.6 (a) RCH_2OH; (b) R_2CHOH;
(c) R_3COH

2.16 Molecules of trimethylamine cannot
form hydrogen bonds to each other,
whereas molecules of propylamine can.

2.18 (a) Alkyne; (d) aldehyde.

2.23 (c) Tertiary; (e) secondary.

2.24 (a) Secondary; (c) tertiary.

2.25 (b) $CH_3CH_2CH_2OH$
(e) $CH_3CH_2CH_2CH_2X$
(l) $CH_3N(CH_3)CH_2CH_3$

2.28 (b) Ethylene glycol; (f) propionic acid

2.34 Ester

CHAPTER 3

3.6 (d) 7-Methylbicyclo[4.2.1]nonane

3.17 (a) $CH_3CHClCHClCH_2CH_3$
(k) $CH_3CH(CH_3)CH_2CH_2CH_2Cl$
(m) $CH_3C(CH_3)_2CH_2Cl$
(n) $CH_3CH(CH_3)CH_2CH_3$

3.18 (a) 3,4-Dimethylhexane;
(f) cyclopentylcyclopentane.

3.19 (a) Neopentane (or 2,2-dimethylpro-
pane); (d) cyclopentane.

3.20 $CH_2{=}\overset{\overset{\textstyle CH_3}{|}}{C}CH_2CH_3$, $CH_3\overset{\overset{\textstyle CH_3}{|}}{C}{=}CHCH_3$,
and $CH_3\overset{\overset{\textstyle CH_3}{|}}{C}HCH{=}CH_2$

3.28 (c)

3.31 (d) Chloroethane, (e) ethyl alcohol.

3.34 (c) *trans*-1,4-Dimethylcyclohexane.

CHAPTER 4

4.1 Chiral: (a), (e)–(h); achiral: (b)–(d).

4.6 (b)–(d).

4.13 75% (S)-(+)-2-Butanol and 25%
(R)-(−)-2-butanol.

4.15 (a) Diastereomers; (b) diastereomers; (c) diastereomers; (e) yes; (f) no.

4.16 (a) **A** alone would be optically active.

4.21 (a) No; (b) yes; (c) no; (d) no;
(e), (f) diastereomers.

4.27 (a) Enantiomers; (d) diastereomers;
(g) two molecules of the same compound;
(j) enantiomers; (n) constitutional
isomers; (p) diastereomers;
(q) enantiomers.

CHAPTER 5

5.1 (a) $CH_3CH_2\overset{\cdot\cdot}{O}H$; (c) $:NH_3$;
(e) $^-:C{\equiv}N:$

5.2 *cis*-3-Methylcyclopentanol.

5.5 (a) $CH_3CH_2OC(CH_3)_3$

5.8 (a) NH_2^-; (b) RS^-; (c) PH_3

5.13 (a) $CH_3CH_2CH_2CH_2Br$ because it is a primary halide; (c) $CH_3CH_2CH_2Br$ because bromide ion is a better leaving group.

5.15 (b) $(CH_3)_3CBr + H_2O \longrightarrow$
$$(CH_3)_3COH + HBr$$
because water is a more polar medium than CH_3OH.

5.19 Reaction (2) because the substrate for the S_N2 reaction is a methyl halide.

5.24 (a) CH_3NH^- because it is the stronger base.

CHAPTER 6

6.1 (a) 2-Methyl-2-butene; (d) 4-methylcyclohexene.

6.5 (a) one; (b) one; (c) no; (d) no; (e) two.

6.10 (a) 2-Butene, the more highly substituted alkene. (b) *trans*-2-butene.

6.31 (c), (e), and (i) can exist as cis-trans isomers.

6.34 (a) cis-trans Isomerization. This happens because at 300 °C the molecules have enough energy to surmount the rotational barrier of the carbon-carbon double bond. (b) *trans*-2-Butene because it is more stable.

6.39 (a) *cis*-1,2-Dimethylcyclopentane; (b) *cis*-1,2-dimethylcyclohexane.

6.42 (a) IHD = 4; 2 double bonds; (b) 2 rings.

CHAPTER 7

7.1 2-Chloro-1-iodopropane.

7.13 (a) *meso*-2,3-Butanediol

7.20 (c) Cyclopentanol (i) cyclopentene.

7.29 2-Methylpropene > propene > ethene.

7.32 4-Methylcyclohexene.

7.36

$$\overset{\overset{\displaystyle CH_3}{|}}{CH_3C}=CHCH_2CH_2\overset{\overset{\displaystyle CH_3}{|}}{C}=CHCH_2CH_2\overset{\overset{\displaystyle CH_2}{\|}}{C}CH=CH_2$$

CHAPTER 8

8.2 Ethylene glycol is more highly associated, because having two —OH groups it can form more hydrogen bonds.

8.17 *trans*-2-Pentene, because it is more stable.

8.29 The reaction is an S_N2 reaction and, therefore, nucleophilic attack takes place more rapidly at the primary carbon atom.

8.33 (a) 3,3-Dimethyl-1-butanol; (e) 1-methyl-2-cyclopenten-1-ol.

8.41 (a) $CH_3Br + CH_3CH_2Br$ (c) $BrCH_2CH_2CH_2CH_2Br$

CHAPTER 9

9.1 (a) $\Delta H° = -25$ kcal/mole; (c) $\Delta H° = +9.5$ kcal/mole.

9.14 Good yields can be obtained when all of the hydrogen atoms of the compound are equivalent.

9.17 (b) Diastereomers. (c) no, the (R,S)-isomer is a meso form; (e) yes, because diastereomers have different physical properties.

9.18 (b) All fractions are optically inactive.

CHAPTER 10

10.1 (a) $^{14}CH_2$=CH—CH_2—X + X—$^{14}CH_2$—CH=CH_2; (c) in equal amounts.

10.10 (b) 1,4-Cyclohexadiene is an isolated diene.

10.17 (a) 1,4-Dibromobutane + $(CH_3)_3COK$, and heat; (g) HC≡CCH=CH_2 + H_2, Ne_2B (P-2).

10.20 (a) 1-Butene + *N*-bromosuccinimide, then $(CH_3)_3COK$ and heat; (e) cyclopentane + Br_2, *hv*, then $(CH_3)_3COK$ and heat, then *N*-bromosuccinimide.

10.22 (a) $Ag(NH_3)_2OH$; (c) H_2SO_4; (e) $AgNO_3$ in alcohol.

10.29 This is another example of rate versus equilibrium control of a reaction.

The *endo* adduct, **G**, is formed faster, and at the lower temperature it is the major product. The *exo* adduct, **H**, is more stable, and at the higher temperature it is the major product.

CHAPTER 11

11.1 (a) None; (b) none.

11.5 Tropylium bromide is a largely ionic compound consisting of the cycloheptatrienyl (tropylium) cation and a bromide anion.

11.14 The nitrogen atoms at positions 1-, 3-, and 7- are of the pyridine type. The nitrogen at position 9- is of the pyrrole type.

11.20 Compound **V** is the cyclooctatetraenyl dianion, a 10 π electron aromatic system.

11.21 (a)–(d) Would not be aromatic; (e)–(h) would be aromatic.

CHAPTER 12

12.2 $HO{-}NO_2 + HO{-}NO_2 \rightleftarrows$
$H_2O^+{-}NO_2 + {^-}O{-}NO_2$
$H_2O^+{-}NO_2^+ \rightleftarrows H_2O + NO_2^+$

12.17 Introduce the chlorine into the benzene ring first, otherwise the double bond will undergo addition of chlorine when ring chlorination is attempted.

12.21 (a) benzene +

$$CH_3CH_2CH_2CH_2CH_2\overset{\overset{\displaystyle O}{\|}}{C}Cl \xrightarrow{AlCl_3}$$

$$C_6H_5\overset{\overset{\displaystyle O}{\|}}{C}CH_2CH_2CH_2CH_2CH_3 \xrightarrow[HCl]{Zn(Hg)}$$

$$C_6H_5CH_2CH_2CH_2CH_2CH_2CH_3$$

12.23 (a) 5-Acetyl-2-methylbenzenesulfonic acid; (c) 2,4-dimethoxynitrobenzene; (e) 4-hydroxy-3-nitrobenzenesulfonic acid.

12.27 (a) Toluene, KMnO$_4$, OH$^-$, heat, then H$_3$O$^+$, then Cl$_2$, FeCl$_3$.
(b) toluene, CH$_3$COCl, AlCl$_3$, then isolate *para* isomer.
(c) toluene, HNO$_3$, H$_2$SO$_4$, then isolate *para* isomer, then Br$_2$, FeBr$_3$.

12.30 p-NO$_2$C$_6$H$_4$O$-\overset{\overset{\displaystyle O}{\|}}{C}C_6H_5$ and

o-NO$_2$C$_6$H$_4$O$-\overset{\overset{\displaystyle O}{\|}}{C}C_6H_5$

12.33

CHAPTER 13

13.4 (a) One; (b) two; (c) one; (d) three; (e) two; (f) three.

13.9 A doublet (3H) downfield; a quartet (1H) upfield.

13.10 (a) CH$_3$CHICH$_3$; (b) CH$_3$CHBr$_2$; (c) CH$_2$ClCH$_2$CH$_2$Cl

13.14 (a) C$_6$H$_5$CH(CH$_3$)$_2$; (b) C$_6$H$_5$CH(NH$_2$)CH$_3$;
(c)

13.18 **A**, *o*-Bromotoluene; **B**, *p*-bromotoluene; **C**, *m*-bromotoluene; **D**, benzyl bromide.

13.20 Phenylacetylene.

13.22

13.23 **G**, CH$_3$CH$_2$CHBrCH$_3$
H, CH$_2$=CBrCH$_2$Br

13.33 **R** is bicyclo[2.2.1]heptane.

13.36 **X** is *m*-xylene.

CHAPTER 14

14.3 (d), (e), and (f) are all stronger acids than H$_2$CO$_3$ and would be converted to soluble sodium salts by aqueous NaHCO$_3$.

14.4 (a) The *para*-sulfonated phenol.

14.7   (a)

14.9   (a)   OCH$_3$ **(b)**   NO$_2$

NO$_2$

(c)   NHC$_6$H$_5$

NO$_2$

14.22   Z is 3-methyl-2-buten-1-ol.

CHAPTER 15

15.1   (c) A change from -2 to 0;
(d) an oxidation;
(e) a reduction from $+6$ to $+3$.

15.5   (a) PCC/CH$_2$Cl$_2$;
(b) KMnO$_4$, OH$^-$, H$_2$O, heat;
(c) H$_2$CrO$_4$/acetone;
(d) (1) O$_3$ (2) Zn, H$_2$O.

15.22   (a) Br$_2$/CCl$_4$ or KMnO$_4$/H$_2$O
(c) alcoholic AgNO$_3$
(e) alcoholic AgNO$_3$

15.23   (a) C$_6$H$_5$CH=CH$_2$, H$_2$O, H$^+$, and
heat, or C$_6$H$_5$CH=CH$_2$, Hg(O$_2$CCH$_3$)$_2$,
H$_2$O, then NaBH$_4$, OH$^-$;
(e) C$_6$H$_5$CH$_2$CO$_2$H, LiAlH$_4$, ether;
(h) C$_6$H$_6$, Br$_2$, FeBr$_3$, then Mg, Et$_2$O, then
ethylene oxide, then H$_2$O.

CHAPTER 16

16.2   (a) 1-Pentanol;   (c) pentanal;
(e) benzyl alcohol.

16.8   A hydride ion.

16.18   (b) CH$_3$CH$_2$Br + (C$_6$H$_5$)$_3$P, then
strong base, then C$_6$H$_5$COCH$_3$;
(d) CH$_3$I + (C$_6$H$_5$)$_3$P, then strong base,
then cyclopentanone;

(f) CH$_2$=CHCH$_2$Br + (C$_6$H$_5$)$_3$P, then
strong base, then C$_6$H$_5$CHO.

16.27   (a) CH$_3$CH$_2$CH$_2$OH
(c) CH$_3$CH$_2$CH$_2$OH
(h) CH$_3$CH$_2$CH=CHCH$_3$
(j) CH$_3$CH$_2$CO$_2^-$NH$_4^+$ + Ag↓
(l) CH$_3$CH$_2$CH=NNHCONH$_2$
(n) CH$_3$CH$_2$CO$_2$H

16.40   (a) Tollens' reagent;   (e) Br$_2$/CCl$_4$;
(f) Tollens' reagent;   (h) Tollens' reagent.

16.46

X   is

16.47   Y is 1-phenyl-2-butanone;   **Z** is
4-phenyl-2-butanone.

CHAPTER 17

17.1   The enol form is phenol. It is
especially stable because it is aromatic.

17.4   Base is consumed as the reaction
takes place. A catalyst, by definition is not
consumed.

17.6   (a), (b), (d), (f), (h), (i).

17.10   (b) CH$_3$CCH=C(CH$_3$)$_2$ (with C=O above)

17.11   C$_6$H$_5$CHO + OH$^- \xrightarrow[\text{heat}]{\text{CH}_3\text{CHO}}$

C$_6$H$_5$CH=CHCHO

17.14   (b) CH$_3$NO$_2$ + HCH $\xrightarrow{\text{OH}^-}$

HOCH$_2$CH$_2$NO$_2$

17.23   (a) CH$_3$CH$_2$CH(OH)CHCHO
                                                        |
                                                      CH$_3$

(b) C$_6$H$_5$CH=CCHO
                              |
                            CH$_3$

(k) CH$_3$CH$_2$CH(OH)C$_6$H$_5$

(l) CH$_3$CH$_2$CH(OH)C≡CH

17.28   B is CH$_3$C—C—CH$_3$ (with O, CH$_3$, OH substituents)

CHAPTER 18

18.4 (a) CH_2FCO_2H; (c) CH_2ClCO_2H; (e) $CH_3CH_2CHClCO_2H$;

(g) CF_3—⟨○⟩—CO_2H

18.8 (a) $C_6H_5CH_2Br + Mg +$ ether, then CO_2, then H_3O^+; (c) $CH_2{=}CHCH_2Br + Mg +$ ether, then CO_2, then H_3O^+.

18.9 (a), (c), and (e).

18.11 In the carboxyl group of benzoic acid.

18.16 (a) $(CH_3)_3CCO_2H + SOCl_2$, then NH_3, then P_4O_{10}, heat; (b) $CH_2{=}\overset{\displaystyle |}{C}{-}CH_3$
$\underset{CH_3}{}$

18.25 (a) CH_3CO_2H
(c) $CH_3CO_2CH_2(CH_2)_2CH_3$
(e) $p\text{-}CH_3COC_6H_4CH_3 +$
$o\text{-}CH_3COC_6H_4CH_3$
(g) CH_3COCH_3;
(i) $CH_3CONHCH_3$
(k) $CH_3CON(CH_3)_2$
(m) $(CH_3CO)_2O$
(o) $CH_3CO_2C_6H_5$

18.31 (a) $NaHCO_3/H_2O$; (c) $NaHCO_3/H_2O$; (e) OH^-/H_2O, heat, detect NH_3 with litmus paper; (g) $AgNO_3/$alcohol.

18.33 (a) Diethyl succinate; (c) ethyl phenylacetate; (e) ethyl chloroacetate.

18.40 **X** is diethyl malonate.

CHAPTER 19

19.5 (a) $CH_3(CH_2)_3CHO + NH_3 \xrightarrow{H_2,\,Ni}$
$CH_3(CH_2)_3CH_2NH_2$
(c) $CH_3(CH_2)_4CHO + C_6H_5NH_2 \xrightarrow{H_2,\,Ni}$
$CH_3(CH_2)_4CH_2NHC_6H_5$

19.6 The reaction of a secondary halide with ammonia is almost always accompanied by some elimination.

19.8 (a) Methoxybenzene + HNO_3 + H_2SO_4, then Fe + HCl; (b) Methoxybenzene + CH_3COCl + $AlCl_3$, then $NH_3 + H_2 + Ni$; (c) toluene + Cl_2 and light, then $(CH_3)_3N$; (d) p-nitrotoluene + $KMnO_4 + OH^-$, then H_3O^+, then $SOCl_2$

followed by NH_3, then NaOBr (Br_2 in NaOH); (e) toluene + N-bromosuccinimide in CCl_4, then KCN, then $LiAlH_4$.

19.14 p-Nitroaniline + Br_2 + Fe, followed by $H_2SO_4/NaNO_2$ followed by CuBr, then Fe/HCl, then $H_2SO_4/NaNO_2$ followed by H_3PO_2.

19.37 **W** is N-benzyl-N-ethylaniline.

CHAPTER 20

20.4 (a) $CH_3\overset{\displaystyle |}{C}HCOCO_2C_2H_5$
$\underset{CO_2C_2H_5}{}$

(b) $H\overset{\displaystyle O}{\overset{\displaystyle \|}{C}}CH_2CO_2C_2H_5$

20.7 O-alkylation that results from the oxygen of the enolate ion acting as a nucleophile.

20.9 (a) Reactivity is the same as with any S_N2 reaction. With primary halides substitution is highly favored, with secondary halides elimination competes with substitution, and with tertiary halides elimination is the exclusive course of the reaction. (b) Acetoacetic ester and 2-methylpropene. (c) Bromobenzene is unreactive toward nucleophilic substitution.

20.27 **D** is racemic *trans*-1,2-cyclopentanedicarboxylic acid, **E** is *cis*-1,2-cyclopentanedicarboxylic acid a, meso compound.

20.38 (a) $CH_2{=}C(CH_3)CO_2CH_3$
(b) $KMnO_4$, OH^-, H_3O^+
(c) CH_3OH, H^+
(d) CH_3ONa, then H^+
(e) and (f)

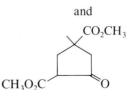

and

(g) OH^-, H_2O, then H_3O^+
(h) heat $(-CO_2)$
(i) CH_3OH, H^+
(j) Zn, $BrCH_2CO_2CH_3$, ether, then H_3O^+

(k)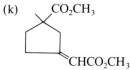

(l) H$_2$, Pt
(m) CH$_3$ONa, then H$^+$
(n) 2NaNH$_2$ + 2CH$_3$I

CHAPTER 21

21.1 (a) Two; (b) two; (c) four.

21.3 Acid catalyzes hydrolysis of the glycosidic (acetal) group.

21.11 (a) 2CH$_3$CHO, one molar equivalent HIO$_4$
(b) HCHO + HCO$_2$H + CH$_3$CHO, two molar equivalents HIO$_4$
(c) HCHO + OHCCH(OCH$_3$)$_2$, one molar equivalent HIO$_4$
(d) HCHO + HCO$_2$H + CH$_3$CO$_2$H, two molar equivalents HIO$_4$
(e) 2CH$_3$CO$_2$H + HCO$_2$H, two molar equivalents, HIO$_4$

21.22 D-(+)-Glucose.

21.27 One anomeric form of D-mannose is dextrorotatory ($[\alpha]_D = +29.3°$), the other is levorotatory ($[\alpha]_D = -17.0°$).

21.28 The microorganism selectively oxidizes the —CHOH group of D-glucitol that corresponds to C-5 of D-glucose.

21.31 **A** is D-altrose; **B** is D-talose, **C** is D-galactose.

CHAPTER 22

22.5 Br$_2$ in CCl$_4$ would be decolorized by geraniol but not by menthol.

21.12 (a) C$_2$H$_5$OH, H$^+$, heat or SOCl$_2$, then C$_2$H$_5$OH
(d) SOCl$_2$, then (CH$_3$)$_2$NH
(g) SOCl$_2$, then LiAlH[OCC(CH$_3$)$_3$]
(j) SOCl$_2$, then (CH$_3$)$_2$CuLi

22.15 Elaidic acid is *trans*-9-octadecenoic acid.

22.19 **A** is CH$_3$(CH$_2$)$_5$C≡CNa
B is CH$_3$(CH$_2$)$_5$C≡CCH$_2$(CH$_2$)$_7$CH$_2$Cl

C is CH$_3$(CH$_2$)$_5$C≡CCH$_2$(CH$_2$)$_7$CH$_2$CN
E is CH$_3$(CH$_2$)$_5$C≡CCH$_2$(CH$_2$)$_7$CH$_2$CO$_2$H
Vaccenic acid is

$$CH_3(CH_2)_5 \diagdown \qquad \diagup (CH_2)_9CO_2H$$
$$C=C$$
$$H \diagup \qquad \diagdown H$$

22.20 **F** is FCH$_2$(CH$_2$)$_6$CH$_2$C≡CH
G is FCH$_2$(CH$_2$)$_6$CH$_2$C≡C(CH$_2$)$_7$Cl
H is FCH$_2$(CH$_2$)$_6$CH$_2$C≡C(CH$_2$)$_7$CH$_2$CN
I is FCH$_2$(CH$_2$)$_7$C≡C(CH$_2$)$_7$CO$_2$H

CHAPTER 23

23.5 The labeled amino acid no longer has a basic —NH$_2$ group; it is, therefore, insoluble in aqueous acid.

23.8 Glutathione is

$$\overset{+}{H_3N}CHCH_2CH_2CONHCHCONHCH_2CO_2H$$
$$\quad | \qquad\qquad\qquad\qquad |$$
$$\quad CO_2^- \qquad\qquad\qquad CH_2SH$$

23.20 Arg·Pro·Pro·Gly·Phe·Ser·Pro·Phe·Arg

23.21 Val·Leu·Lys·Phe·Ala·Glu·Ala

CHAPTER 24

24.2 (a) The nucleosides have an *N*-glycosidic linkage that (like an *O*-glycosidic linkage) is rapidly hydrolyzed by aqueous acid, but one that is stable in aqueous base.

24.3 The reaction appears to take place through an S$_N$2 mechanism. Attack occurs preferentially at the primary 5′-carbon rather than at the secondary 3′-carbon.

24.5 (a) The isopropylidene group is a cyclic ketal
(b) By treating the nucleoside with acetone and a trace of acid.

24.7 (b) Thymine would pair with adenine, and, therefore, adenine would be introduced into the complementary strand where guanine should occur.

24.12 A change from C-T to C-A-T, or a change from C-T-C to C-A-C.

Bibliography of Suggested Readings

CHAPTER ONE

L. SALEM, "A Faithful Couple: The Electron Pair," *J. Chem. Educ.,* **55,** 344 (1978).

M. B. HALL, "Valence Shell Electron Pair Repulsion and the Pauli Exclusion Principle," *J. Am. Chem. Soc.,* **100,** 6333 (1978).

D. KOLB, "The Chemical Formula, Part I: Development," *J. Chem. Educ.,* **55,** 44 (1978).

J. E. FERNANDEZ and ROBERT D. WHITAKER, *An Introduction to Chemical Principles.* Macmillan, New York, 1975.

J. E. BRADY and G. E. HUMISTON, *General Chemistry: Principles and Structure,* 3rd ed., Wiley, New York, 1981.

O. T. BENFEY, *From Vital Force to Structural Formulas,* Houghton Mifflin, Boston, 1964.

R. J. GILLESPIE, "The Electron-Pair Repulsion Model for Molecular Geometry," *J. Chem. Educ.,* **47,** 18 (1970).

P. E. VERKADE, "August Kekulé," *Proc. Chem. Soc.,* 205 (1958).

L. PAULING, *The Nature of the Chemical Bond,* 3rd ed., Cornell University Press, Ithaca, NY, 1960.

G. W. WHELAND, *Resonance in Organic Chemistry,* Wiley, New York, 1955.

R. H. MAYBURY, "The Language of Quantum Mechanics," *J. Chem. Educ.,* **39,** 367 (1962).

L. C. PAULING, *The Chemical Bond; A Brief Introduction to Modern Structural Chemistry,* Cornell University Press, Ithaca, NY, 1967.

J. A. CARROLL, "Drawing Lewis Structures without Anticipating Octets," *J. Chem. Educ.,* **63,** 28 (1986).

CHAPTER TWO

D. KOLB, "Acids and Bases," *J. Chem. Educ.,* **55,** 459 (1978).

O. T. BENFEY, *The Names and Structures of Organic Compounds,* Wiley, New York, 1966.

J. D. ROBERTS, R. STEWART, and M. C. CASERIO, *Organic Chemistry: Methane to Macromolecules,* Benjamin, New York, 1971, Chapter 2.

CHAPTER THREE

D. KOLB, "The Chemical Formula, Part II: Determination," *J. Chem. Educ.,* **55,** 109 (1978).

J. H. FLETCHER, O. C. DERMER, and R. B. FOX, *Nomenclature of Organic Compounds,* American Chemical Society, Washington, DC, 1973.

E. L. ELIEL, *Conformational Analyses,* McGraw-Hill, New York, 1965.

G. W. WHELAND, *Advanced Organic Chemistry,* 3rd ed., Wiley, New York, 1960.

LLOYD N. FERGUSON, "Ring Strain and Reactivity of Alicycles," *J. Chem. Educ.,* **47,** 46 (1970).

C. A. COULSON, *Valence,* Oxford University Press, New York, 1952, Chapter VIII.

J. MARCH, *Advanced Organic Chemistry,* 2nd ed., McGraw-Hill, New York, 1977, pp. 24–133.

G. H. POSNER, "Substitution Reactions Using Organocopper Reagents," *Organic Reactions,* Vol. 22, Wiley, New York, 1975.

A. MOYANO, F. SERRATOSE, P. CAMPS, and J. M. DRUDIS, "The IUPAC Names of the Regular Polyhedranes," *J. Chem. Educ.,* **55,** 126. (1982).

R. S. CAHN and O. C. DERMER, *Introduction to Chemical Nomenclature,* 5th ed., Butterworths, Boston, 1979.

CHAPTER FOUR

M. GIELEN, "From the Concept of Relative Configuration to the Definition of *Erythro* and *Threo,*" *J. Chem. Educ.,* **54,** 673 (1977).

J. MARCH, *Advanced Organic Chemistry,* 2nd ed., McGraw-Hill, New York, 1977, Chapter 4.

E. L. ELIEL, *Stereochemistry of Carbon Compounds,* McGraw-Hill, New York, 1962.

G. B. KAUFMANN, "Resolvability and the Tetrahedral Configuration of Carbon," *J. Chem. Educ.,* **60,** 402 (1983).

J. P. IDOUX, "A Simple Method for Specifying the *R/S* Configuration about a Chiral Center," *J. Chem. Educ.,* **59,** 553 (1982).

O. B. RAMSAY, *Stereochemistry,* Heyden, Philadelphia, 1981.

E. L. ELIEL, "Recent Advances in Stereochemical Nomenclature," *J. Chem. Educ.,* **48,** 163 (1971).

"IUPAC Tentative Rules for the Nomenclature of Organic Chemistry, Section E. Fundamental Stereochemistry," *J. Org. Chem.,* **35,** 2849 (1970).

E. L. ELIEL, *Elements of Stereochemistry,* Wiley, New York, 1969.

K. MISLOW, *Introduction to Stereochemistry,* Benjamin, New York, 1965.

D. F. MOWERY, JR., "Criteria for Optical Activity in Organic Molecules," *J. Chem. Educ.,* **46,** 269 (1969).

D. WHITTAKER, *Stereochemistry and Mechanism,* Clarendon Press, Oxford, 1973, Chapters 1, 2, and 5.

J. E. HUHEEY, "A Novel Method for Assigning *R, S* Labels to Enantiomers," *J. Chem. Educ.,* **63,** 598 (1986).

M. P. AALUND and J. A. PINCOCK, "A Simple Method for Cahn-Ingold-Prélog Assignment of *R* and *S* Configuration to Chiral Carbons," *J. Chem. Educ.,* **63,** 600 (1986).

J. H. BREWSTER, "Stereochemistry and the Origins of Life," *J. Chem. Educ.,* **63,** 667 (1986).

CHAPTER FIVE

W. H. SAUNDERS, JR., *Ionic Aliphatic Reactions,* Prentice-Hall, Englewood Cliffs, NJ, 1965, Chapters 3–5.

R. K. Boyd, "Some Common Oversimplifications in Teaching Chemical Kinetics," *J. Chem. Educ.*, **55**, 84 (1978).

C. K. Ingold, *Structure and Mechanism in Organic Chemistry*, 2nd ed., Cornell University Press, Ithaca, New York, 1969, Chapters 7 and 9.

J. G. Traynam, "Carbonium Ion: Waxing and Waning of a Name," *J. Chem. Educ.*, **63**, 931 (1986).

CHAPTER SIX

S. I. Miller, "Dissociation Energies of Pi Bonds," *J. Chem. Educ.*, **55**, 778 (1978).

J. J. Lagowski, "The Chemistry of Liquid Ammonia," *J. Chem. Educ.*, **55**, 752 (1978).

T. F. Rutledge, *Acetylenic Compounds: Preparation and Substitution Reactions*, Reinhold, New York, 1968.

G. A. Olah, *Carbocation and Electrophilic Reactions*, Wiley, New York 1973.

G. A. Olah and P. V. R. Schleyer, Eds., *Carbonium Ions*, Wiley, New York, 1968.

M. Orchin and H. H. Jaffé, *The Importance of Antibonding Orbitals*, Houghton Mifflin, Boston, 1967.

CHAPTER SEVEN

F. C. Whitmore and J. M. Church, *J. Am. Chem. Soc.*, **54**, 3710 (1932).

O. T. Benfey, *Introduction to Organic Reaction Mechanisms*, McGraw-Hill, New York, 1970, Chapter 5.

W. H. Saunders, *Ionic Aliphatic Reactions*, Prentice-Hall, Englewood Cliffs, NJ, 1965, Chapter 2.

T. F. Rutledge, *Acetylenes and Allenes: Addition Cyclization and Polymerization Reactions*, Reinhold, New York, 1969.

R. L. Shriner, R. C. Fuson, D. Y. Curtin, *and* T. C. Morrill, *Systematic Identification of Organic Compounds*, 6th ed., Wiley, New York, 1980.

T. L. Jacobs, "The Synthesis of Acetylenes," *Organic Reactions*, Vol. 5, Wiley, New York, 1949.

W. S. Johnson, "Non-enzymic Biogenetic-like Olefinic Cyclizations," *Acc. Chem. Res.*, **1**, 1 (1968).

J. G. MacConnell and Robert M. Silverstein, "Recent Results in Insect Pheromone Chemistry," *Angew. Chem. Int. Ed. Engl.*, **12**, 644 (1973).

SPECIAL TOPIC C

M. Jones, Jr., "Carbenes," *Scientific American*, **276**, No. 2 (1976), p. 101.

J. Hine, *Divalent Carbon*, Ronald Press, New York, 1964

G. L. Closs, "Structures of Carbenes and the Stereochemistry of Carbene Additions to Olefins," *Topics in Stereochemistry*, Vol. 3, Wiley, New York, 1968.

W. E. Parham and E. E. Schweizer, "Halocyclopropanes from Halocarbenes," *Organic Reactions*, Vol. 13, Wiley, New York, 1963.

H. E. Simmons, T. L. Cairns, S. A. Vladuchick, and C. M. Hoiness, "Cyclopropanes from Unsaturated Compounds, Methylene Iodide, and Zinc-Copper Couple," *Organic Reactions*, Vol. 20, Wiley, New York, 1973.

CHAPTER EIGHT

H. C. Brown, *Organic Syntheses via Boranes*, Wiley, New York, 1975.

H. C. Brown, *Hydroboration*, Benjamin, New York, 1962.

H. C. Brown and P. J. Geoghegan, Jr., "Solvomercuration–Demercuration. I.," *J. Org. Chem.*, **35**, 1844 (1970).

G. Zweifel and H. C. Brown, "Hydration of Olefins, Dienes, and Acetylenes, via Hydroboration," *Org. React.*, Vol. 13, Wiley, NY, 1963.

N. Isenberg and M. Grdinic, "A Modern Look at Markovnikov's Rule and the Peroxide Effect," *J. Chem. Educ.*, **46**, 601 (1969).

S. Patai, Ed., *Chemistry of the Hydroxyl Group*, Wiley, New York, 1971.

S. Patai, Ed., *Chemistry of the Ether Linkage*, Wiley, New York, 1967.

L. B. Clapp, *The Chemistry of the OH Group*, Prentice-Hall, Englewood Cliffs, NJ, 1967.

W. P. Weber and G. W. Gokel, "Phase Transfer Catalysis," *J. Chem. Educ.*, **55**, 350 (Part I) and 429 (Part II), 1978.

R. West and T. J. Barton, "Organosilicon Chemistry," *J. Chem. Educ.*, **57**, 334 (1980).

H. C. Brown, "The Borane Adventure—Past, Present and Future," *Isr. J. Chem.*, **25**, 84 (1985).

G. W. Kabalka, "Incorporation of Stable and Radioactive Isotopes via Organoborane Chemistry," *Acc. Chem. Res.*, **17**, 215 (1984).

CHAPTER NINE

J. March, *Advanced Organic Chemistry*, 2nd ed., McGraw-Hill, New York, 1977, Chapters 6 and 14.

S. W. Benson, "Bond Energies," *J. Chem. Educ.*, **42**, 502 (1965).

W. A. Pryor, *Free Radicals*, McGraw-Hill, New York, 1965.

E. S. Huyser, *Free-Radical Chain Reaction*, Wiley, New York, 1970.

C. Walling, *Free Radicals in Solution*, Wiley, New York, 1957.

W. A. Pryor, *Introduction to Free Radical Chemistry*, Prentice-Hall, Englewood Cliffs, New Jersey, 1965.

E. Baer, "Advanced Polymers," *Scientific American*, **255**, No. 4, 178 (1986).

C. Walling, "The Development of Free Radical Chemistry," *J. Chem. Educ.*, **63** 99 (1986).

R. A. Kerr, "Has Stratospheric Ozone Started to Disappear?" *Science*, **237**, 131 (1987).

SPECIAL TOPIC D

R. P. QUIRK, "Stereochemistry and Macromolecules," *J. Chem. Educ.,* **58,** 540 (1981).

F. W. BILLMEYER, *Textbook of Polymer Science,* 2nd ed., Wiley, New York, 1973.

L. R. G. TRELOAR, *Introduction to Polymer Science,* Springer-Verlag, New York, 1970.

J. Chem. Educ., **58** Nov. 1981. (An entire issue devoted to polymer chemistry.)

CHAPTER TEN

J. MARCH, *Advanced Organic Chemistry,* 2nd ed., McGraw-Hill, New York, 1977, pp. 29–41.

H. H. JAFFÉ and M. ORCHIN, *Theory and Applications of Ultraviolet Spectroscopy,* Wiley, New York, 1962.

A. LIBERLES, *Introduction to Molecular Orbital Theory,* Holt, Rinehart, and Winston, New York, 1966.

M. ORCHIN and H. H. JAFFÉ, *The Importance of Antibonding Orbitals,* Houghton Mifflin, 1967.

J. SAUER, "Diels–Alder Reactions, Part I," *Angew. Chem. Int. Ed. Engl.,* **5,** 211 (1966); "Part II," *Ibid.,* **6,** 16 (1967).

CHAPTER ELEVEN

J. MARCH, *Advanced Organic Chemistry,* 2nd ed., McGraw-Hill, New York, 1977, pp. 41–69.

G. M. BADGER, *Aromatic Character and Aromaticity,* Cambridge University Press, 1969.

R. BRESLOW, "Antiaromaticity," *Acc. Chem. Res.,* **6,** 393 (1973).

F. SONDHEIMER, "The Annulenes," *Acc. Chem. Res.,* **5,** 81 (1972).

L. J. SCHAAD and B. A. HESS, "Hückel Theory and Aromaticity," *J. Chem. Educ.,* **51,** 640 (1974).

J., AIHARA, "A New Definition of Dewar-Type Resonance Energies," *J. Am. Chem. Soc.,* **98,** 2750 (1976).

R. G. HARVEY, "Activated Metabolites of Carcinogenic Hydrocarbons," *Acc. Chem. Res.,* **14,** 218 (1981).

C. GLIDEWELL and D. LLOYD, "The Arithmetic of Aromaticity," *J. Chem. Educ.,* **63,** 306 (1986).

CHAPTER TWELVE

J. MARCH, *Advanced Organic Chemistry,* 2nd ed., McGraw-Hill, New York, 1977, Chapter 11.

G. A. OLAH, *Friedel–Crafts and Related Reactions,* Vol. I, Wiley, New York, 1963.

W. R. DOLBIER, JR., "Electrophilic Additions to Alkenes," *J. Chem. Educ.,* **46,** 342 (1969).

E. C. TAYLOR and A. MCKILLOP, "Thallium in Organic Synthesis," *Acc. Chem. Res.,* **3,** 338 (1970).

CHAPTER THIRTEEN

P. L. FUCHS and C. A. BUNNELL, *Carbon-13 NMR Based Organic Spectral Problems,* Wiley, New York, 1979.

G. C. LEVY and G. L. NELSON, *Carbon-13 Nuclear Magnetic Resonance for Organic Chemists,* Wiley, New York, 1972.

L. J. BELLAMY, *The Infrared Spectra of Complex Molecules,* 3rd ed., Wiley, New York, 1975.

J. D. ROBERTS, *An Introduction to Spin-Spin Splitting in High Resolution Nuclear Magnetic Resonance Spectra,* Benjamin, Menlo Park, CA, 1961.

F. A. BOVEY, *Nuclear Magnetic Resonance Spectroscopy,* Academic Press, New York, 1969.

J. D. ROBERTS and M. C. CASERIO, *Basic Principles of Organic Chemistry,* 2nd ed., Benjamin, Menlo Park, CA, 1977, Chapters 9 and 27.

R. M. SILVERSTEIN and G. C. BASSLER, *Spectrometric Identification of Organic Compounds,* Wiley, New York, 1967.

J. R. DYER, *Applications of Absorption Spectroscopy of Organic Compounds,* Prentice-Hall, Englewood Cliffs, NJ, 1965.

J. D. ROBERTS, *Nuclear Magnetic Resonance,* McGraw-Hill, New York, 1959.

E. R. ANDREW, "NMR Imaging," *Acc. Chem. Res.,* **16,** 114 (1983).

SPECIAL TOPIC E

W. F. MACLAFFERTY, *Interpretation of Mass Spectroscopy,* 2nd ed., Benjamin, Reading, Mass., 1973.

CHAPTER FOURTEEN

J. F. BUNNETT, "The Remarkable Reactivity of Aryl Halides with Nucleophiles," *J. Chem. Educ.,* **51,** 312 (1974).

S. PATAI, Ed., *Chemistry of the Hydroxyl Group,* Wiley, New York, 1971.

CHAPTER FIFTEEN

H. C. BROWN, "Hydride Reductions: A 40-Year Revolution in Organic Chemistry," *Chem. Eng. News,* March 5, 1979, p. 24.

SPECIAL TOPIC G

J. SCHWARTZ and J. A. LABINGER, "Patterns in Organometallic Chemistry with Application in Organic Synthesis," *J. Chem. Educ.,* **57,** 170 (1980).

M. ORCHIN, "HCo(CO)₄, The Quintessential Catalyst," *Acc. Chem. Res.,* **14,** 259 (1981).

J. E. ELLIS, "The Teaching of Organometallic Chemistry to Undergraduates," *J. Chem. Educ.,* **53,** 2(1976).

J. P. COLLMAN, "Patterns of Organometallic Reactions Related to Homogeneous Catalysis," *Acc. Chem. Res.,* **1,** 136 (1968).

J. P. COLLMAN, "Disodium Tetracarbonylferrate —A Transition-Metal Analog of a Grignard Reagent," *Acc. Chem. Res.,* **8,** 342 (1975).

SPECIAL TOPIC H

D. L. RABENSTEIN, "The Chemistry of Methylmercury Toxicology," *J. Chem. Educ.,* **55,** 292 (1978).

J. R. HOLUM, *Topics and Terms in Environmental Problems,* Wiley, New York, 1977.

Chem. Eng. News, June 6, 1983. (An entire issue devoted to Dioxin.)

R. E. BEYLER and V. K. MEYERS "What Every Chemist Should know about Teratogens–Chemicals that Cause Birth Defects," *J. Chem. Educ.* **59,** 759 (1982).

F. H. TSCHIRLEY, "Dioxin," *Scientific American,* **254,** No. 2, 29 (1986).

CHAPTER SIXTEEN

C. A. BUEHLER and D. E. PEARSON, *Survey of Organic Synthesis,* Wiley, New York, 1970.

H. O. HOUSE, *Modern Synthetic Reactions,* 2nd ed., Benjamin, New York, 1972.

S. PATAI, Ed., *The Chemistry of the Carbonyl Group,* Vol. 1, Wiley, New York, 1966.

S. PATAI AND J. ZABICKY, Eds., *The Chemistry of the Carbonyl Group,* Vol. 2, Wiley, New York, 1970.

E. VEDEJS, "Clemmensen Reduction of Ketones in Anhydrous Organic Solvents," *Organic Reactions,* Vol. 22, Wiley, New York, 1975.

M. W. RATHKE, "The Reformatsky Reaction," *Organic Reactions,* Vol. 22, Wiley, New York, 1975.

C. H. HASSALL, "The Baeyer–Villiger Oxidation of Aldehydes and Ketones," *Organic Reactions,* Vol. 9, Wiley, New York, 1957.

CHAPTER SEVENTEEN

A. J. NIELSON and W. J. HOULIHAN, "The Aldol Condensation," *Organic Reactions,* Vol. 16, Wiley, New York, 1968.

G. H. POSNER, "Conjugate Addition Reactions of Organocopper Reagents," *Organic Reactions,* Vol. 19, Wiley, New York, 1972.

SPECIAL TOPIC I

T. MUKAIYAMA, "The Directed Aldol Reaction," *Organic Reactions,* Vol. 28, 203, Wiley, New York, 1982.

I. KUWAJIMA and E. NAKAMURA, "Reactive Enolates from Enol Silyl Ethers," *Acc. Chem. Res.,* **18,** 181 (1985).

G. STORK and P. F. HUDRLIK, "Isolation of Ketone Enolates as Trialkylsilyl Ethers, *J. Am. Chem. Soc.* **90,** 4462 (1968).

H. J. REICH, "Functional Group Manipulation Using Organoselenium Reagents," *Acc. Chem. Res.,* **12,** 22 (1979).

D. L. J. CLIVE, "Selenium Reagents for Organic Synthesis," *Aldrichimica Acta,* **11,** 43 (1978).

D. LIOTTA, "New Organoselenium Methodology," *Acc. Chem. Res.,* **17,** 28 (1984).

CHAPTER EIGHTEEN

S. PATAI, Ed., *The Chemistry of Carboxylic Acids and Esters,* Wiley, New York, 1969.

L. F. FIESER and M. FIESER, *Advanced Organic Chemistry,* Reinhold, New York, 1961, Chapters 11, 23, and 24.

S. PATAI, Ed., *The Chemistry of Amides,* Wiley, New York, 1969.

C. D. GUTSCHE, *The Chemistry of Carbonyl Compounds,* Prentice-Hall, Englewood Cliffs, NJ, 1967.

SPECIAL TOPIC J

J. K. STILLE, *Industrial Organic Chemistry,* Prentice-Hall, Englewood Cliffs, NJ, 1968.

CHAPTER NINETEEN

G. B. KAUFFMAN, "Isoniazid-Destroyer of the White Plague," *J. Chem. Educ.,* **55,** 448–449 (1978).

S. PATAI, Ed., *The Chemistry of the Amino Group,* Wiley, New York, 1968.

L. F. FIESER and M. FIESER, *Advanced Organic Chemistry,* Reinhold, New York, 1961, Chapters 14 and 21.

H. K. PORTER, "The Zinin Reduction of Nitroarenes," *Organic Reactions,* Vol. 20, Wiley, New York, 1973.

H. ZOLLINGER, *Diazo and Azo Chemistry,* Wiley, New York, 1961.

L. A. PAGUETTE, *Principles of Modern Heterocyclic Chemistry Q,* Benjamin, New York, 1968.

SPECIAL TOPIC K

L. A. PAGUETTE, *Principles of Modern Heterocyclic Chemistry Q,* Benjamin, New York, 1968.

CHAPTER TWENTY

C. R. HAUSER and B. E. HUDSON, "The Acetoacetic Ester Condensation and Certain Related Reactions," *Organic Reactions,* Vol. 1, Wiley, New York, 1942.

H. O. HOUSE, *Modern Synthetic Reactions,* Benjamin, New York, 1965, Chapters 7 and 9.

W. McCRAE, *Basic Organic Reactions,* Heyden and Son, Ltd., London, 1973, Chapters 3 and 4.

J. P. SCHAEFER and J. J. BLOOMFIELD, "The Dieckmann Condensation," *Organic Reactions,* Vol. 15, Wiley, New York, 1967.

G. JONES, "The Knoevenagel Condensation," *Organic Reactions,* Vol. 15, Wiley, New York, 1967.

T. M. HARRIS and C. M. HARRIS, "The γ-Alkylation and γ-Arylation of Dianions of β-Dicarbonyl Compounds," *Organic Reactions,* Vol. 17, Wiley, New York, 1969.

A. G. COOK, *Enamines: Synthesis, Structure, and Reactions,* Dekker, New York, 1969.

V. BOEKELHEIDE, "[2n] Cyclophanes: Paracyclophane to Superphane," *Acc. Chem. Res.,* **13,** 67 (1980).

J. K. WHITESELL, "New Perspectives in Asymmetric Induction," *Acc. of Chem. Res.,* **18,** 280 (1985).

SPECIAL TOPIC L

G. A. SWAN, *An Introduction to Alkaloids,* Wiley, New York, 1967.

T. A. GEISSMAN and D. H. G. CROUT, *Organic Chemistry of Secondary Plant Metabolism*, Freeman, Cooper and Co., San Francisco, 1969, Chapters 16 to 19.

H. HART and J. L. REILLY, "Oxidative Coupling of Phenols," *J. Chem. Educ.*, **55**, 120 (1978).

CHAPTER TWENTY-ONE

R. J. BERGERON, "Cycloamyloses," *J. Chem. Educ.*, **54**, 204 (1977).

L. N. FERGUSON et al., "Sweet Organic Chemistry," *J. Chem. Educ.*, **55**, 281 (1978).

G. B. KAUFFMAN and R. P. CIULA, "Emil Fischer's Discovery of Phenylhydrazine," *J. Chem. Educ.*, **54**, 295 (1977).

C. R. NOLLER, *Chemistry of Organic Compounds*, 3rd ed., Saunders, New York, 1965, Chapter 18.

D. E. GREEN and R. F. GOLDBERGER, *Molecular Insights into the Living Process*, Academic Press, 1967, Chapters 2 and 3.

C. S. HUDSON, "Emil Fischer's Discovery of the Configuration of Glucose," *J. Chem. Educ.*, **18**, 353 (1941).

R. BARKER, *Organic Chemistry of Biological Compounds*, Prentice-Hall, Englewood Cliffs, NJ, 1971, Chapter 5.

I. TABUSHI, "Cyclodextrin Catalysis as a Model for Enzyme Action," *Acc. Chem. Res.*, **15**, 66 (1982).

A. CERAMI, H. VLASSARA, and M. BROWNLEE, "Glucose and Aging," *Scientific American*, **256**, No. 5, 90 (1987).

R. BENTLEY and J. L. POPP, "Configurations of Glucose and Other Aldoses," *J. Chem. Educ.*, **64**, 15 (1987).

CHAPTER TWENTY-TWO

D. KOLB, "A Pill for Birth Control," *J. Chem. Educ.*, **55**, 591 (1978).

L. F. FIESER, "Steroids," *Bio-organic Chemistry: Readings from Scientific American*, M. Calvin and M. Jorgenson, Eds., Freeman, San Francisco, 1968.

E. E. CONN and P. K. STUMPF, *Outlines of Biochemistry*, 3rd ed., Wiley, New York, 1972, Chapters 3 and 12.

J. R. HANSON, *Introduction to Steroid Chemistry*, Pergamon Press, New York, 1968.

F. M. MENGER, "On the Structure of Micelles," *Acc. Chem. Res.*, **12**, 111 (1979).

R. BRESLOW, "Biomimetic Control of Chemical Selectivity," *Acc. Chem. Res.*, **13**, 170 (1980).

S. HAKOMORI, "Glycosphingolipids," *Scientific American*, **254**, No. 5, 44 (1986).

SPECIAL TOPIC M

W. S. JOHNSON, "Nonenzymic Biogenetic-like Olefin Cyclizations," *Acc. Chem. Res.*, **1**, 1 (1968).

C. D. POULTER and H. C. RILLING, "The Prenyl Transfer Reaction. Enzymatic and Mechanistic Studies of 1'–4 Coupling Reaction in Terpene Biosynthetic Pathway," *Acc. Chem. Res.*, **11**, 307 (1978).

J. W. CORNFORTH, "Terpenoid Biosynthesis," *Chem. Br.*, **4**, 102 (1968).

J. B. HENDRICKSON, *The Molecules of Nature*, W. A. Benjamin, Menlo Park, CA, 1965.

SPECIAL TOPIC N

R. HUBBARD and A. KROPF, "Molecular Isomers in Vision," *Bio-organic Chemistry: Readings from Scientific American*, M. Calvin and M. Jorgenson, Eds., Freeman, San Francisco, 1968.

R. H. JOHNSON and T. P. WILLIAMS, "Action of Light upon the Visual Pigment Rhodopsin," *J. Chem. Educ.*, **47**, 736 (1970).

E. L. MENGER, Ed., "Special Issue on the Chemistry of Vision," *Acc. Chem. Res.*, **8**, (3), 81–112, (1975).

L. STRYER, "The Molecules of Visual Excitation," *Scientific American*, **257**, No. 1, 42 (1987).

CHAPTER TWENTY-THREE

N. M. SENOZAN and R. L. HUNT, "Hemoglobin: Its Occurrence, Structure, and Adaptation," *J. Chem. Educ.*, **59**, 173 (1982).

R. BRESLOW, "Artificial Enzymes," *Science*, **218**, 532 (1982).

J. R. HOLUM, *Organic Chemistry: A Brief Course*, Wiley, New York, 1975, Chapter 15.

The following articles from *Bio-organic Chemistry: Readings from Scientific American*, M. Calvin and M. Jorgenson, Eds., Freeman, San Francisco, 1968:

P. Doty, "Proteins," p. 15.

W. H. STEIN and S. MOORE, "The Chemical Structure of Proteins," p. 23.

E. O. P. THOMPSON, "The Insulin Molecule," p. 34.

M. F. PERUTZ, "The Hemoglobin Molecule," p. 41.

E. ZUCKERKANDL, "The Evolution of Hemoglobin," p. 53.

D. C. PHILLIPS, "The Three-Dimensional Structure of an Enzyme Molecule," p. 67.

H. D. LAW, *The Organic Chemistry of Peptides*, Wiley, New York, 1970.

D. E. GREEN and R. F. GOLDBERGER, *Molecular Insights into the Living Process*, Academic Press, 1967, Chapters 4 and 5.

E. E. CONN and P. K. STUMPF, *Outlines of Biochemistry*, 3rd ed., Wiley, New York, 1972, Chapter 4.

R. E. DICKERSON and I. GEIS,, *The Structure and Action of Proteins*, Harper and Row, New York, 1969.

M. D. FRYZUK and B. BOSNICH, "Asymmetric Synthesis. Production of Optically Active Amino Acids by Catalytic Hydrogenation," *J. Am. Chem. Soc.*, **99**, 6262 (1977).

W. S. KNOWLES, "Asymmetric Hydrogenation," *Acc. Chem. Res.*, **16**, 106 (1983).

B. MERRIFIELD, "Solid Phase Synthesis," *Science*, **232**, 341 (1986).

J. REBEK, "Model Studies in Molecular Recognition," *Science,* **235,** 1478 (1987).

R. F. DOOLITTLE, "Proteins," *Scientific American,* **253,** No. 4, 88 (1985).

V. T. D'SOUZA and M. L. BENDER, "Miniature Organic Models of Enzymes," *Acc. Chem. Res,* **20,** 146 (1987).

CHAPTER TWENTY-FOUR

J. D. WATSON, *Molecular Biology of the Gene,* 2nd ed., Benjamin, New York, 1970.

The following articles in *Bio-organic Chemistry: Readings from Scientific American,* M. Calvin and M. J. Jorgenson, Eds., Freeman, San Francisco, 1968:

F. H. C. CRICK, "The Structure of the Hereditary Material," p. 75.

R. W. Holley, "The Nucleotide Sequence of a Nucleic Acid," p. 82.

R. A. WEINBERG, "The Molecules of Life," *Scientific American,* **253,** No. 4, 48 (1985).

G. FELSENFELD, "DNA," *Scientific American,* **253,** No. 4, 58 (1985).

J. E. DARNELL, "RNA," *Scientific American,* **253,** No. 4, 68 (1985).

SPECIAL TOPIC O

K. N. HOUK, "The Frontier Molecular Orbital Theory of Cycloaddition Reactions," *Acc. Chem. Res.,* **8,** 361 (1975).

R. W. WOODWARD and R. HOFFMAN, *The Conservation of Orbital Symmetry,* Academic Press, New York, 1970.

SPECIAL TOPIC P

W. H. SAUNDERS, JR. and A. F. COCKERILL, *Mechanisms of Elimination Reactions,* Wiley, New York, 1973.

W. H. SAUNDERS, JR., "Distinguishing between Concerted and Nonconcerted Eliminations," *Acc. Chem. Res.,* **9,** 19 (1976).

D. J. RABER and J. M. HARRIS, "Nucleophilic Substitution Reactions at Secondary Carbon Atoms," *J. Chem. Educ.,* **49,** 60 (1972).

R. A. SNEEN, "Organic Ion Pairs as Intermediates in Nucleophilic Substitution and Elimination Reactions," *Acc. Chem. Res.,* **6,** 46 (1973).

F. G. BORDWELL, "How Common are Base Initiated, Concerted 1,2-Eliminations?" *Acc. Chem. Res.,* **5,** 374 (1972).

Index

TABLE 13.2 Approximate proton chemical shifts

TYPE OF PROTON	CHEMICAL SHIFT (δ, ppm)
1° Alkyl, RCH_3	0.8–1.0
2° Alkyl, RCH_2R	1.2–1.4
3° Alkyl, R_3CH	1.4–1.7
Allylic, $R_2C=C-CH_3$ $\quad\quad\quad\quad\mid$ $\quad\quad\quad\quad R$	1.6–1.9
Benzylic, $ArCH_3$	2.2–2.5
Alkyl chloride, RCH_2Cl	3.6–3.8
Alkyl bromide, RCH_2Br	3.4–3.6
Alkyl iodide, RCH_2I	3.1–3.3
Ether, $ROCH_2R$	3.3–3.9
Alcohol, $HOCH_2R$	3.3–4.0
Ketone, $RCCH_3$ $\quad\quad\quad\overset{\displaystyle O}{\|\|}$	2.1–2.6
Aldehyde, RCH $\quad\quad\quad\quad\overset{\displaystyle O}{\|\|}$	9.5–9.6
Vinylic, $R_2C=CH_2$	4.6–5.0
Vinylic, $R_2C=CH$ $\quad\quad\quad\quad\quad\mid$ $\quad\quad\quad\quad\quad R$	5.2–5.7
Aromatic, ArH	6.0–9.5
Acetylenic, $RC\equiv CH$	2.5–3.1
Alcohol hydroxyl, ROH	0.5–6.0[a]
Carboxylic, $RCOH$ $\quad\quad\quad\quad\overset{\displaystyle O}{\|\|}$	10–13[a]
Phenolic, $ArOH$	4.5–7.7[a]
Amino, $R-NH_2$	1.0–5.0[a]

[a]The chemical shifts of these protons vary in different solvents and with temperature and concentration.

TABLE 13.3 Approximate carbon-13 chemical shifts

TYPE OF CARBON ATOM	CHEMICAL SHIFT (δ, ppm)
1° Alkyl, RCH_3	0–40
2° Alkyl, RCH_2R	10–50
3° Alkyl, $RCHR_2$	15–50
Alkyl halide or amine, $-\underset{\mid}{\overset{\mid}{C}}-X$ ($X = Cl$, Br, or $-\underset{\mid}{N}-$)	10–65
Alcohol or ether, $-\underset{\mid}{\overset{\mid}{C}}-O$	50–90
Alkyne, $-C\equiv$	60–90
Alkene, $\overset{\displaystyle /}{\underset{\displaystyle \diagdown}{C}}=$	100–170
Aryl, 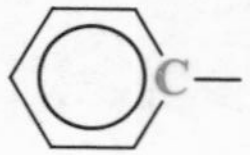	100–170
Nitriles, $-C\equiv N$	120–130
Amides, $-\overset{\displaystyle O}{\overset{\|\|}{C}}-\underset{\mid}{N}-$	150–180
Carboxylic acids, esters, $-\overset{\displaystyle O}{\overset{\|\|}{C}}-O$	160–185
Aldehydes, ketones, $-\overset{\displaystyle O}{\overset{\|\|}{C}}-$	182–215